Loss Prevention in the Process Industries

Volume 1

This book is dedicated to

Herbert Douglas Lees (1860–1944), gas engineer;
Frank Priestman Lees (1890–1916), gas engineer;
Herbert Douglas Lees (1897–1955), gas engineer;
David John Lees (1936–), agricultural engineer;
Frank Lyman MacCallum (1893–1955), mining engineer and missionary;
Vivien Clare Lees (1960–), plastic and hand surgeon
Harry Douglas Lees (1962–), restaurateur
and their families

'They do not preach that their God will rouse them a little before
the nuts work loose.
They do not teach that His Pity allows them to drop their job when
they dam'-well choose.
As in the thronged and the lighted ways, so in the dark and the
desert they stand,
Wary and watchful all their days that their brethren's days may be
long in the land.'

Rudyard Kipling (The Sons of Martha, 1907)

Wo einer kommt and saget an,
Er hat es allen recht getan,
So bitten wir diesen lieben Herrn,
Er wöll uns solche Kunste auch lehrn

(Whoever is able to say to us
'I have done everything right',
We beg that honest gentleman
To show us how it is done)

Inscription over the 'Zwischenbau' adjoining the Rathaus in Brandenburg-on-the-Haven
(quoted by Prince B.H.M. von Bulow in *Memoirs*, 1932)

If the honeye that the bees gather out of so manye floure of herbes... that are growing in other mennis medowes... may justly be called the bees' honeye... so maye I call it that I have... gathered of manye good autores... my booke.
William Turner (quoted by A. Scott-James in *The Language of the Garden: A Personal Anthology*)

By the same author:

A.W. Cox, F.P. Lees and M.L. Ang (1990): *Classification of Hazardous Locations* (Rugby: Institution of Chemical Engineers)

Elwyn Edwards and Frank P. Lees (1973): *Man and Computer in Process Control* (London: Institution of Chemical Engineers)

Elwyn Edwards and Frank P. Lees (eds) (1974): *The Human Operator in Process Control* (London: Taylor & Francis)

Frank P. Lees and M.L. Ang (1989): *Safety Cases* (London: Butterworths)

Loss Prevention in the Process Industries

Hazard Identification, Assessment and Control

Volume 1

Second edition

Frank P. Lees

Emeritus Professor of Chemical Engineering, Department of Chemical Engineering, Loughborough University, United Kingdom

Butterworth-Heinemann
Linacre House, Jordan Hill, Oxford OX2 8DP
A division of Reed Educational and Professional Publishing

A member of the Reed Elsevier plc group

OXFORD BOSTON JOHANNESBURG
MELBOURNE NEW DELHI SINGAPORE

First published 1980
Second edition 1996

British Library Cataloguing in Publication Data
Lees, Frank P.
Loss Prevention in the Process
Industries: Hazard Identification,
Assessment and Control. – 2Rev.ed
I. Title
658.47

ISBN 0 7506 1547 8

Library of Congress Cataloguing in Publication Data
Lees, Frank P.
Loss prevention in the process industries : hazard
identification, assessment, and control / Frank P. Lees. – 2nd ed.
p. cm.
Rev. ed. of : Loss prevention in the process industries. 1980.
Includes bibliographical references and index.
ISBN 0 7506 1547 8
1. Petroleum chemicals industry–Great Britain–Safety measures.
I. Lees, Frank P. Loss prevention in the process industries.
II. title.
TP690.6.L43 1995
600'.2804–dc20

Printed in Great Britain

Preface to Second Edition

The first edition of this book appeared in 1980, at the end of a decade of rapid growth and development in loss prevention. After another decade and a half the subject is more mature, although development continues apace. In preparing this second edition it has been even more difficult than before to decide what to put in and what to leave out.

The importance of loss prevention has been underlined by a number of disasters. Those at San Carlos, Mexico City, Bhopal and Pasadena are perhaps the best known, but there have been several others with death tolls exceeding 100. There have also been major incidents in related areas, such as those on the Piper Alpha oil platform and at the nuclear power stations at Three Mile Island and Chernobyl.

Apart from the human tragedy, it has become clear that a major accident can seriously damage even a large international company and may even threaten its existence, rendering it liable to severe damages and vulnerable to takeover.

Accidents in the process industries have given impetus to the creation of regulatory controls. In the UK the Advisory Committee on Major Hazards made its third and final report in 1983. At the same time the European Community was developing its own controls which appeared as the EC Directive on Major Accident Hazards. The resulting UK legislation is the NIHHS Regulations 1982 and the CIMAH Regulations 1984. Other members of the EC have brought in their own legislation to implement the Directive. There have been corresponding developments in planning controls...

An important tool for decision-making on hazards is hazard assessment. The application of quantitative methods has played a crucial role in the development of loss prevention, but there has been lively debate on the proper application of such assessment, and particularly on the estimation and evaluation of the risk to the public.

Hazard assessment involves the assessment both of the frequency and of the consequences of hazardous events. In frequency estimation progress has been made in the collection of data and creation of data banks and in fault tree synthesis and analysis, including computer aids. In consequence assessment there has been a high level of activity in developing physical models for emission, vaporization and gas dispersion, particularly dense gas dispersion; for pool fires, fireballs, jet flames and engulfing fires; for vapour cloud explosions; and for boiling liquid expanding vapour explosions (BLEVEs). Work has also been done on injury models for thermal radiation, explosion overpressure and toxic concentration, on models of the density and other characteristics of the exposed population, and on shelter and escape.

Some of these topics require experimental work on a large scale and involving international cooperation. Large scale tests have been carried out at several sites on dense gas dispersion and on vapour cloud fires and explosions. Another major cooperative research programme has been that of DIERS on venting of chemical reactors.

The basic approach developed for fixed installations on shore has also been increasingly applied in other fields. For transport in the UK the *Transport Hazards Report* of the Advisory Committee on Dangerous Substances represents an important landmark. Another application is in the offshore oil and gas industry, for which the report on the Piper Alpha disaster, the *Cullen Report*, constitutes a watershed.

As elsewhere in engineering, computers are in widespread use in the design of process plants, where computer aided design (CAD) covers physical properties, flowsheeting, piping and instrument diagrams, unit operations and plant layout. There is increasing use of computers for failure data retrieval and analysis, reliability and availability studies, fault tree synthesis and analysis and consequence modelling, while more elusive safety expertise is being captured by computer-based expert systems.

The subject of this book is the process industries, but the process aspects of related industries, notably nuclear power and oil and gas platforms are briefly touched on. The process industries themselves are continually changing. In the last decade one of the main changes has been increased emphasis on products such as pharmaceuticals and agrochemicals made by batch processes, which have their own particular hazards.

All this knowledge is of little use unless it reaches the right people. The institutions which educate the engineers who will be responsible for the design and operation of plants handling hazardous materials have a duty to make their students aware of the hazards and at least to make a start in gaining competence in handling them.

I would like again to thank for their encouragement the heads of the Department of Chemical Engineering at Loughborough, Professors D.C. Freshwater, B.W. Brooks and M. Streat; our Industrial Professors T.A. Kletz and H.A. Duxbury and Visiting Professor S.M. Richardson; my colleagues, past and present, in the Plant Engineering Group, Mr R.J. Aird, Dr P.K. Andow, Dr M.L. Ang, Dr P.W.H. Chung, Dr D.W. Edwards, Dr P. Rice and Dr A.G. Rushton – I owe a particular debt to the latter; the members of the ACMH, chaired by Professor B.H. Harvey; the sometime directors of Technica Ltd, Dr D.H. Slater, Mr P. Charsley, Dr P.J. Comer, Dr R.A. Cox, Mr T. Gjerstad, Dr M.A.F. Pyman, Mr C.G. Ramsay, Mr M.A. Seaman and Dr R. Whitehouse; the members of the IChemE Loss Prevention Panel; the IChemE's former Loss Prevention Officer, Mr B.M. Hancock; the members of the IChemE Loss Prevention Study Group and of the Register of Safety Professionals; the editorial staff of the IChemE, in particular Mr B. Brammer; numerous members of the Health and Safety Executive, especially Dr A.C. Barrell, Mr J. Barton, Dr D.A. Carter, Mr K. Cassidy, Mr P.J.

Crossthwaite, Dr N.W. Hurst, Dr S.F. Jagger, Dr J. McQuaid, Dr K. Moodie, Dr C. Nussey, Dr R.P. Pape, Dr A.F. Roberts and Dr N.F. Scilly; workers at the Safety and Reliability Directorate, particularly Dr A.T.D. Butland, Mr I. Hymes, Dr D.W. Phillips and Dr D.M. Webber; staff at Shell Thornton Research Centre, including Dr D.C. Bull and Dr A.C. Chamberlain; staff at British Gas, including Dr J.D. Andrews, Dr M.J. Harris, Mr H. Hopkins, Dr J.M. Morgan and Dr D.J. Smith; staff at the Ministry of Defence, Explosives Storage and Transport Committee, including Mr M.A. Gould, Mr J. Henderson and Mr P. Stone; and colleagues who have taught on post-experience courses at Loughborough, in particular Dr C.D. Jones, Dr D.J. Lewis and Mr J. Madden; BP International and Mr R. Malpas for allowing me to spend a period of study leave with the company in 1985–86 and Mr F.D.H. Moysen, Mr G. Hately, Mr M. Hough, Mr R. Fearon and others in the Central Safety Group and in Engineering Department; the Honourable Lord Cullen, my fellow Technical Assessors on the Piper Alpha Inquiry, Mr B. Appleton and Mr G.M. Ford and the Cremer and Warner team at the inquiry, in particular Mr G. Kenney and Mr R. Sylvester-Evans; other professional colleagues Dr L.J. Bellamy, Professor B.A. Buffham, Dr D.A. Crowl, Mr T.J. Gilbert, Mr D.O. Hagon, Dr D.J. Hall, Mr K.M. Hill, Professor T.M. Husband, Mr M. Kneale, Dr V.C. Marshall, Dr M.L. Preston, Dr J. Rasmussen, Dr J.R. Roach, Dr J.R. Taylor, Dr V.M. Trbojevic, Mr H.M. Tweeddale, Dr G.L. Wells and Dr A.J. Wilday; my research colleagues Dr C.P. Murphy, Mrs J.I. Petts, Dr D.J. Sherwin, Mr R.M.J. Withers and Dr H. Zerkani; my research students Mr M. Aldersey, Mr D.C. Arulanantham, Dr A. Bunn, Dr M.A. Cox, Dr P.A. Davies, Dr S.M. Gilbert, Mr P. Heino, Dr A. Hunt, Dr B.E. Kelly, Dr G.P.S. Marrs, Dr J.S. Mullhi, Dr J.C. Parmar, Mr B. Poblete, Dr A. Shafaghi and Dr A.J. Trenchard as well as colleagues' research students Mr E.J. Broomfield, Mr R. Goodwin, Mr M.J. Jefferson, Dr F.D. Larkin, Mr S.A. McCoy, Dr K. Plamping, Mr J. Soutter, Dr P. Thorpe and Mr S.J. Wakeman; the office staff of the Department, Mrs E.M. Barradell, Mr D.M. Blake, Miss H.J. Bryers and Miss Y. Kosar; the staff of the University Library, in particular Miss S.F. Pilkington; and my wife Elizabeth, whose contribution has been many-faceted and in scale with this book.

FRANK P. LEES
Loughborough,
1994

Preface to First Edition

Within the past ten or fifteen years the chemical and petroleum industries have undergone considerable changes. Process conditions such as pressure and temperature have become more severe. The concentration of stored energy has increased. Plants have grown in size and are often single-stream. Storage has been reduced and interlinking with other plants has increased. The response of the process is often faster. The plant contains very large items of equipment. The scale of possible fire, explosion or toxic release has grown and so has the area which might be affected by such events, especially outside the works boundary.

These factors have greatly increased the potential for loss both in human and in economic terms. This is clear both from the increasing concern of the industry and its insurers and from the historical loss statistics.

The industry has always paid much attention to safety and has a relatively good record. But with the growing scale and complexity involved in modern plants the danger of serious large-scale incidents has been a source of increasing concern and the adequacy of existing procedures has been subjected to an increasingly critical examination.

Developments in other related areas have also had an influence. During the period considered there has been growing public concern about the various forms of pollution, including gaseous and liquid effluents and solid wastes and noise.

It is against this background that the loss prevention approach has developed. It is characteristic of this approach that it is primarily concerned with the problems caused by the depth of technology involved in modern processes and that it adopts essentially an engineering approach to them. As far as possible both the hazards and the protection are evaluated quantitatively.

The clear recognition by senior management of the importance of the loss prevention problem has been crucial to these developments. Progress has been made because management has been prepared to assign to this work many senior and capable personnel and to allocate the other resources necessary.

The management system is fundamental to loss prevention. This involves a clear management structure with well defined line and advisory responsibilities staffed by competent people. It requires the use of appropriate procedures, codes of practice and standards in the design and operation of plant. It provides for the identification, evaluation and reduction of hazards through all stages of a project from research to operation. It includes planning for emergencies.

The development of loss prevention can be clearly traced through the literature. In 1960 the Institution of Chemical Engineers held the first of a periodic series of symposia on *Chemical Process Hazards with Special Reference to Plant Design*. The Dow Chemical Company published its *Process Safety Manual* in 1964. The American Institute of Chemical Engineers started in 1967 an annual series of symposia on Loss Prevention. The European Federation of Chemical Engineers' symposium on *Major Loss Prevention in the Process Industries* at Newcastle in 1971 and the Federation's symposium on *Loss Prevention and Safety Promotion in the Process Industries* (Buschmann, 1974) at Delft are further milestones.

Another indicator is the creation in 1973 by the Institution of Chemical Engineers Engineering Practice Committee of a Loss Prevention Panel under the chairmanship of Mr T.A. Kantyka.

In the United Kingdom the Health and Safety at Work etc. Act 1974 has given further impetus to loss prevention. The philosophy of the *Robens Report* (1972), which is embodied in the Act, is that of self-regulation by industry. It is the responsibility of industry to take all reasonable measures to assure safety. This philosophy is particularly appropriate to complex technological systems and the Act provides a flexible framework for the development of the loss prevention approach.

The disaster at Flixborough in 1974 has proved a turning point. This event has led to a much more widespread and intense concern with the loss prevention problem. It has also caused the government to set up in 1975 an Advisory Committee on Major Hazards. This committee has made far-reaching recommendations for the identification and control of major hazard installations.

It will be apparent that loss prevention differs somewhat from safety as traditionally conceived in the process industries. The essential difference is the much greater engineering content in loss prevention.

This is illustrated by the relative effectiveness of inspection in different processes. In fairly simple plants much can be done to improve safety by visual inspection. This approach is not adequate, however, for the more technological aspects of complex processes.

For the reasons given above loss prevention is currently a somewhat fashionable subject. It is as well to emphasize, therefore, that much of it is not new, but has been developed over many years by engineers whose patient work in an often apparently unrewarding but vital field is the mark of true professionalism.

It is appropriate to emphasize, moreover, that accidents arising from relatively mundane situations and activities are still responsible for many more deaths and injuries than those due to advanced technology.

Nevertheless, loss prevention has developed in response to the growth of a new problem, the hazard of high technology processes, and it does have a distinctive approach and some novel techniques. Particularly characteristic are the emphasis on matching the management system to the depth of technology in the installation, the techniques developed for identifying hazards, the principle and methods of quantifying hazards, the application of reliability assessment, the practice of planning for emergencies and the critique of traditional practices or existing codes, standards or regulations where these are outdated by technological change.

There is an enormous, indeed intimidating, literature on safety and loss prevention. In addition to the symposia already referred to, mention may be made of the *Handbook of Safety and Accident Prevention in Chemical Operations* by Fawcett and Wood (1965); the *Handbook of Industrial Loss Prevention* by the Factory Mutual Engineering Corporation (1967); and the *Industrial Safety Handbook* by Handley (1969, 1977). These publications, which are by multiple authors, are invaluable source material.

There is a need, however, in the author's view for a balanced and integrated textbook on loss prevention in the process industries which presents the basic elements of the subject, which covers the recent period of intense development and which gives a reasonably comprehensive bibliography. The present book is an attempt to meet this need.

The book is based on lectures given to undergraduate and postgraduate students at Loughborough over a period of years and the author gladly acknowledges their contribution.

Loss prevention is a wide and rapidly developing field and is therefore not an easy subject for a book. Nevertheless, it is precisely for these reasons that the engineer needs the assistance of a textbook and that the attempt has been considered justified.

The structure of the book is as follows. Chapter 1 deals with the background to the historical development of loss prevention, the problem of large, single-stream plants, and the differences between loss prevention and conventional safety, and between loss prevention and total loss control; Chapter 2 with hazard, accident and loss, including historical statistics; Chapter 3 with the legislation and legal background; Chapter 4 with the control of major hazards; Chapter 5 with economic and insurance aspects; Chapter 6 with management systems, including management structure, competent persons, systems and procedures, standards and codes of practice, documentation and auditing arrangements; Chapter 7 with reliability engineering, including its application in the process industires; Chapter 8 with the spectrum of techniques for identifying hazards from research through to operation; Chapter 9 with the assessment of hazards, including the question of acceptable risk; Chapter 10 with the siting and layout of plant; Chapter 11 with process design, including application of principles such as limitation of inventory, consideration of known hazards associated with chemical reactors, unit processes, unit operations and equipments, operating conditions, utilities, particular chemicals and particular processes and plants, and checking of operational deviations; Chapter 12 with pressure system design, including properties of materials, design of pressure vessels and pipework, pressure vessel standards and codes, equipment such as heat exchangers, fired heaters and rotating machinery, pressure relief and blowdown arrangements, and failure in pressure systems; Chapter 13 with design of instrumentation and control systems, including regular instrumentation, process computers and protective systems; Chapter 14 with human factors in process control, process operators, computer aids and human error; Chapter 15 with loss of containment and dispersion of material; Chapter 16 with fire, flammability characteristics, ignition sources, flames and particular types of process fire, effects of fire and fire prevention, protection and control; Chapter 17 with explosion, explosives, explosion energy, particular types of process explosion such as confined explosions, unconfined vapour cloud explosions and dust explosions, effects of explosion and explosion prevention, protection and relief; Chapter 18 with toxicity of chemicals, toxic release and effects of toxic release; Chapter 19 with commissioning and inspection of plant; Chapter 20 with plant operation; Chapter 21 with plant maintenance and modification; Chapter 22 with storage; Chapter 23 with transport, particularly by road, rail and pipeline; Chapter 24 with emergency planning both for works and transport emergencies; Chapter 25 with various aspects of personal safety such as occupational health and industrial hygiene, dust and radiation hazards, machinery and electrical hazards, protective clothing and equipment, and rescue and first aid; Chapter 26 with accident research; Chapter 27 with feedback of information and learning from accidents; Chapter 28 with safety systems, including the roles of safety managers and safety committees and representatives. There are appendices on Flixborough, Seveso, case histories, standards and codes, institutional publications, information sources, laboratories and pilot plants, pollution and noise, failure and event data, Canvey, model licence conditions for certain hazardous plants, and units and unit coversions.

Many of the matters dealt with, such as pressure vessels or process control, are major subject areas in their own right. It is stressed, therefore, that the treatment given is strictly limited to loss prevention aspects. The emphasis is on deviations and faults which may give rise to loss.

In engineering in general and in loss prevention in particular there is a conflict between the demand for a statement of basic principles and that for detailed instructions. In general, the first of these approaches has been adopted, but the latter is extremely important in safety, and a considerable amount of detailed material is given and references are provided to further material.

The book is intended as a contribution to the academic education of professional chemical and other engineers. Both educational and professional institutions have long recognized the importance of education in safety. But until recently the rather qualitative, and indeed often exhortatory, nature of the subject frequently seemed to present difficulties in teaching at degree level. The recent quantitative development of the subject goes far towards removing these objections and to integrating it more closely with other topics such as engineering design.

In other words, loss prevention is capable of development as a subject presenting intellectual challenge. This is all to the good, but a note of caution is appropriate. It remains true that safety and loss prevention depend primarily on the hard and usually unglamorous work of engineers with a strong sense of responsibility, and it is important that this central fact should not be obscured.

For this reason the book does not attempt to select particular topics merely because a quantitative treatment is possible or to give such a treatment as an academic exercise. The subject is too important for such an approach. Rather the aim has been to give a balanced treatment of the different aspects and a lead in to further reading.

It is also hoped that the book will be useful to practising engineers in providing an orientation and entry to unfamiliar areas. It is emphasized, however, that in this subject above all others, the specialized texts should be consulted for detailed design work.

Certain topics which are often associated with loss prevention, for example included in loss prevention symposia, have not been treated in detail. These include, for example, pollution and noise. The book does not attempt to deal in detail with total loss control, but a brief account of this is given.

The treatment of loss prevention given is based mainly on the chemical, petrochemical and petroleum industries, but much of it is relevant to other process industries, such as electrical power generation (conventional and nuclear), iron and steel, gas, cement, glass, paper and food.

The book is written from the viewpoint of the United Kingdom and, where differences exist within the UK, of England. This point is relevant mainly to legislation.

Reference is made to a large number of procedures and techniques. These do not all have the same status. Some are well established and perhaps incorporated in standards or codes of practice. Others are more tentative. As far as possible the attempt has been made to give some indication of the extent to which particular items are generally accepted.

There are probably also some instances where there is a degree of contradiction between two approaches given. In particular, this may occur where one is based on engineering principles and the other on relatively arbitrary rules-of-thumb.

The book does not attempt to follow standards and codes of practice in drawing a distinction between the words *should*, *shall* and *must* in recommending particular practices and generally uses only the former. The distinction is important, however, in standards and codes of practice and it is described in Appendix 4[a].

An explanation of some of the terms used is in order at this point. Unfortunately there is at present no accepted terminology in this field. In general, the problems considered are those of loss, either of life or property. The term ***hazard*** is used to describe the object or situation which constitutes the threat of such loss. The consequences which might occur if the threat is realized are the ***hazard potential***. Associated with the hazard there is a risk, which is the probability of the loss occurring. Such a risk is expressed as a ***probability*** or as a ***frequency***. Probability is expressed as a number in the range 0 to 1 and is dimensionless; frequency is expressed in terms of events per unit time, or sometimes in other units such as events per cycle or per occasion. Rate is also used as an alternative to frequency and has the same units.

The analysis of hazards involves qualitative ***hazard identification*** and quantitative ***hazard assessment***. The latter term is used to describe both the assessment of hazard potential and of risk. The assessment of risk only is described as ***risk assessment***.

In accident statistics the term ***Fatal Accident Frequency Rate*** (FAFR) has some currency. The last two terms are tautologous and the quantity is here referred to as ***Fatal Accident Rate*** (FAR).

Further treatments of terminology in this field are given by BS 4200: 1967, by Green and Bourne (1962), by the Council for Science and Society (1977) and by Harvey (1979b).

Notation is defined for the particular chapter at the point where the symbols first occur. In general, a consistent notation is used, but well established equations from standards, codes and elsewhere are usually given in the original notation. A consolidated list of the notation is given at the end of chapters in which a large number of symbols is used.

The units used are in principle SI, but the exceptions are fairly numerous. These exceptions are dimensional equations, equations in standards and codes, and other equations and data given by other workers where conversion has seemed undesirable for some reason. In cases of conversion from a round number it is often not clear what degree of rounding off is appropriate. In cases of description of particular situations it appears pedantic to make the conversion where a writer has referred, for example, to a 1 inch pipe.

Notes on some of the units used are given in Appendix 12[a]. For convenience a unit conversion table is included in this appendix. Numerical values given by other authors are generally quoted without change and numerical values arising from conversion of the units of data given by other authors are sometimes quoted with an additional significant figure in order to avoid excessive rounding of values.

Some cost data are quoted in the book. These are given in pounds or US dollars for the year quoted.

A particular feature of the book is a fairly extensive bibliography of some 5000 references. These references are consolidated at the end of the book rather than at the end of chapters, because many items are referred to in a number of chapters. Lists of selected references on particular topics are given in table form in the relevant chapters.

Certain institutions, however, have a rather large number of publications which it is more convenient to treat in a different manner. These are tabulated in Appendices 4[a] and 5[a], which contain some 2000 references. There is a cross-reference to the institution in the main reference list.

In many cases institutions and other organizations are referred to by their initials. In all cases the first reference in the book gives the full title of the organization. The initials may also be looked up in the Author Index, which gives the full title.

A reference is normally given by quoting the author and, in brackets, the date, e.g. Kletz (1971). Publications by the same author in the same year are denoted by letters of the alphabet a, b, c, etc., e.g. Allen (1977a), while publications by authors of the same surname and in the same year are indicated for convenience by an asterisk against the year in the list of references. In addition, the author's initials are given in the main text in cases where there may still be ambiguity. Where a date has not been determined this is indicated as n.d.

[a]Appendices 4, 5 and 12 in the first edition correspond to Appendices 27, 28 and 30, respectively, in this second edition.

In the case of institutional publications listed in Appendices 4[a] and 5[a] the reference is given by quoting the insitution and, in brackets, the date, the publication series, e.g. HSE (1965 HSW Bklt 34) or the item number, e.g. IChemE (1971 Item 7). For institutional publications with a named author the reference is generally given by quoting the author and, in brackets, the initials of the institution, the date and the publication series or item number, e.g. Eames (UKAEA 1965 Item 4).

The field of loss prevention is currently subject to very rapid change. In particular, there is a continuous evolution of standards and codes of practice and legislation. It is important, therefore, that the reader should make any necessary checks on changes which may have occurred.

I would like to thank for their encouragement in this project Professor D.C. Freshwater and the publishers, and to acknowledge the work of many authors which I haved used directly or indirectly, particularly that of Dr J.H. Burgoyne and of Professor T.A. Kletz. I have learned much from my colleagues on the Loss Prevention Panel of the Institution of Chemical Engineers, in particular Mr T.A. Kantyka and Mr F. Hearfield, and on the Advisory Committee on Major Hazards, especially the chairman Professor B.H. Harvey, the secretary Mr H.E. Lewis, my fellow group chairmen Professor F.R. Farmer and Professor J.L.M. Morrison and the members of Group 2, Mr K. Briscoe, Dr J.H. Burgoyne, Mr E.J. Challis, Mr S. Hope, Mr M.A. McTaggart, Professor J.F. Richardson, Mr J.R.H. Schenkel, Mr R. Sheath and Mr M.J. Turner, and also from my university colleagues Dr P.K. Andow, Mr R.J. Aird and Dr D.J. Sherwin and students Dr S.N. Anyakora, Dr B. Bellingham, Mr C.A. Marpegan and Dr G.A. Martin-Solis. I am much indebted to Professor T.A. Kletz for his criticisms and suggestions on the text. My thanks are due also to the Institution of Plant Engineers, which has supported plant engineering activities at Loughborough, to the Leverhulme Trust which awarded a Research Fellowship to study *Loss Prevention in the Process Industries* and to the Science Research Council, which has supported some of my own work in this area. I have received invaluable help with the references from Mrs C.M. Lincoln, Mrs W. Davison, Mrs P. Graham, Mr R. Rhodes and Mrs M.A. Rowlatt, with the typing from Mrs E.M. Barradell, Mrs P. Jackson and, in particular, Mrs J. Astley, and with the production from Mr R.L. Pearson and Mr T. Mould. As always in these matters the responsibility for the final text is mine alone.

FRANK P. LEES
Loughborough,
1979

Acknowledgements

For permission to reproduce material in this book the author would like to acknowledge in particular:

Academic Press; Adam Hilger; Addison-Wesley Publishers; AGEMA Infrared Systems; the Air Pollution Control Association; the American Chemical Society; the American Gas Association; the American Institute of Chemical Engineers; the American Petroleum Institute; the American Society of Heating, Refrigerating and Air Conditioning Engineers; the American Society of Mechanical Engineers; Anderson Greenwood Company; Anderson Studios; *Archivum Combustionis*; Associated Press; the Association of the British Pharmaceutical Industry; *Atmospheric Environment*; the Badische Anilin und Soda Fabrik; the Bettman Archive; the Boots Company; Borowski Image Communications; BP Chemicals Ltd; BP Petroleum Company Ltd; BP Petroleum Development Ltd; the British Ceramics Research Association; the British Chemical Industry Safety Council; British Gas plc; *British Medical Journal*; the British Standards Institution; the Brookhaven National Laboratory; Brooks/Cole Publishers; the Building Research Establishment; the Bureau of Mines; Business Books; Business India; Butterworth-Heinemann; Cambridge University Press; Castle House Publications; the Center for Chemical Process Safety; Champaign Fire Department; Chapman and Hall; the Chartered Institute of Building Services Engineers; *Chemical Engineering*; *Chemical Engineering Science*; the Chemical Industries Association; the Chemical Industry Safety and Health Council; Chemical Processing; the Chicago Bridge and Iron Company; the Christian Michelsen Institute; Cigna Insurance; Clarendon Press; *Combustion and Flame*; the Combustion Institute; *Combustion Science and Technology*; Compagnie General d'Edition et de Presse; Crosby Valve and Engineering Ltd; DECHEMA; Marcel Dekker; the Dow Chemical Company; Dupont Safer Emergency Systems; the Electrical Power Research Institute; the Electrochemical Society; Elsevier Publishing Company, Elsevier Sequoia SA and Elsevier Science Publishers; the Engineering Equipment Manufacturers and Users Association; the Enterprise Publishing Company, Blair, Nebraska; the Environmental Protection Agency; Expert Verlag; The Faraday Society; Ferranti Ltd; the Federal Power Commission; *Filtration and Separation*; *Fire Prevention Science and Technology*; the Fire Protection Association; *Fire Safety Journal*; *Fire Technology*; the Foxboro Company; Gastech; Gordon and Breach Science Publishers; Gower Press; de Gruyter Verlag; Gulf Publishing Corporation; the *Guiness Book of Records*; the Health and Safety Executive; Hemisphere Publishing Company; the High Pressure Technology Association; the Controller of HM Stationery Office; *Hydrocarbon Processing*; Imperial Chemical Industries plc; Industrial Risk Insurers; the Institute of Electrical and Electronic Engineers; the Institute of Measurement and Control; the Institute of Petroleum; the Institute of Physics; the Institution of Chemical Engineers; the Institution of Electrical Engineers; the Institution of Gas Engineers; the Institution of Marine Engineers; the Institution of Mechanical Engineers; the Instrument Society of America; International Atomic Energy Agency; *International Journal of Air and Water Pollution*; ITAR-TASS; *Journal of Applied Physics*; *Journal of Colloid and Interface Science*; *Journal of Electrostatics*; *Journal of Occupational Psychology*; *Journal of Ship Research*; William Kaufman Inc.; Kent Instruments Ltd; Kluwer Academic Publishers; the McGraw-Hill Book Company; Marston Excelsior Ltd; the Meteorological Office; Mr C.A. Miller; the Ministry of Social Affairs, the Netherlands; Ministry of Supply; the National Academy of Sciences; the National Fire Protection Association; the National Radiological Protection Board; the National Transportation Safety Board; *Nature*; New York Academy of Sciences; North Holland Publishing Company; the Norwegian Society of Chartered Engineers; the Nuclear Regulatory Commission; *Nuclear Safety*; the Occupational Safety and Health Administration; the *Oil Gas Journal*; Oxford University Press; Pergamon Press; *Philosophical Magazine*; Plenum Press; Prentice-Hall; *Process Engineering*; *Progress in Energy and Combustion Science*; PSC Freysinnet; Reidel Publishing Company; Research Study Press; Riso National Laboratory; Mr A. Ritchie; the Rijnmond Public Authority; the Royal Meteorological Society; the Royal Society; the Safety in Mines Research Establishment; *Science*; the Scientific Instrument Research Association; Scottish Technical Developments Ltd and the University of Strathclyde; Shell International Petroleum Company Ltd; Shell Research Ltd; Signs and Labels Ltd; Skandia International and the State of Mexico; the Society of Chemical Industry; the Society of Gas Tanker and Terminal Operators; *Sound and Vibration*; Frank Spooner Pictures; Springer Verlag; Taylor and Francis Ltd; TNO; Technica Ltd; *Trade and Industry*; the UK Atomic Energy Authority, Safety and Reliability Directorate; the Union of Concerned Scientists; Union Carbide; the United Nations; United Press International; the University of Nottingham; the US Atomic Energy Commission; the US Coast Guard; Mr T. Vanus; Warren Spring Laboratory; the Watt Committee; Whessoe Ltd; John Wilay and Sons Inc.

Extensive use is acknowledged of the reports of AEA Technology (mostly referenced as UKAEA Safety and Reliability Directorate). Much of this research has been supported by the Health and Safety Executive as part of a long-term programme of work on safety, particularly on major hazards.

The author would also like to acknowledge his use of material from *Refining Process Safety Booklets* of the Amoco Oil Company (formerly the American Oil

Company, Chicago), in particular Booklet No. 4, *Safe Ups and Downs of Refinery Units*, Copyright 1960 and 1963 The American Oil Company and Booklet No. 9, *Safe Operation of Air, Ammonia and Ammonium Nitrate Plants*, Copyright 1964 The American Oil Company. Quoted material is used with the permission of the copyright owner.

Professor H.A. Duxbury and Dr A.J. Wilday have been good enough comment on Chapter 17, Sections 17.16–17.21. Professor Duxbury has also contributed Appendix 13 on safety factors in simple relief systems.

The responsibility for the text is mine alone.

Terminology

Attention is drawn to the availability in the literature of a number of glossaries and other aids to terminology. Some British Standard glossaries are given in Appendix 27 and other glossaries are listed in Table 1.1.

Notation

In each chapter a given symbol is defined at the point where it is first introduced. The definition may be repeated if there has been a significant gap since it was last used. The definitions are summarized in the notation given at the end of the chapter. The notation is global to the chapter unless redefined for a section. Similarly, it is global to a section unless redefined for a subsection and global to a subsection unless redefined for a set of equations or a single equation. Where appropriate, the units are given, otherwise a consistent system of units should be used, SI being the preferred system. Generally the units of constants are not given; where this is the case it should not be assumed that a constant is dimcnsionless.

Use of References

The main list of references is given in the section entitled References, towards the end of the book. There are three other locations where references are to be found. These are Appendix 27 on standards and codes; Appendix 28 on institutional publications; and in the section entitled Loss Prevention Bulletin which follows the References.

The basic method of referencing an author is by surname and date, e.g. Beranek (1960). Where there would otherwise be ambiguity, or where there are numerous references to the same surname, e.g. Jones, the first author's initials are included, e.g. A. Jones (1984). Further guidance on names is given at the head of the section References.

References in Appendices 27 and 28 are by institution or author. Some items in these appendices have a code number assigned by the institution itself, e.g. API (1990 Publ. 421), but where such a code number is lacking, use is generally made of an item number separated from the date by a slash, e.g. IChemE (1971/13). Thus typical entries are

API Std 2000: 1992	a standard, found in Appendix 27 under American Petroleum Institute
API (1990 Publ. 421)	an institutional publication, found in Appendix 28 under American Petroleum Institute
HSE (1990 HS(G) 51)	an institutional publication, found in Appendix 28 under Health and Safety Executive, Guidance Booklets, HS(G) series
Coward and Jones (1952 BM Bull. 503)	an institutional publication, found in Appendix 28 under Bureau of Mines, Bulletins

Institutional acronyms are given in the section Acronyms which precedes the Author Index.

There are several points of detail which require mention concerning Appendix 28. (1) The first part of the appendix contains publications of a number of institutions and the second part those of the Nuclear Regulatory Commission. (2) The Fire Protection Association publications include a number of series which are collected in the *Compendium of Fire Safety Data* (CFSD). A typical reference to this is FPA (1989 CFSD FS 6011). (3) The entries for the Health and Safety Executive are quite extensive and care may be needed in locating the relevant series. (4) The publications of the Safety and Reliability Directorate appear under the UK Atomic Energy Authority, Safety and Reliability Directorate. A typical reference is Ramskill and Hunt (1987 SRD R354). These publications are immediately preceded by the publications of other bodies related to the UKAEA, such as the Health and Safety Branch, the Systems Reliability Service and the National Centre for Systems Reliability.

References to authors in the IChemE *Loss Prevention Bulletin* are in the style Eddershaw (1989 LPB 88), which refers to issue 88 of the bulletin.

Contents of Volume 1

Contents of Volume 2

Contents of Volume 3

1 Introduction

Contents

Over the last three decades there has developed in the process industries a distinctive approach to hazards and failures that cause loss of life and property. This approach is commonly called 'loss prevention'. It involves putting much greater emphasis on technological measures to control hazards and on trying to get things right first time. An understanding of loss prevention requires some appreciation of its historical development against a background of heightened public awareness of safety, and environmental problems, of its relation to traditional safety and also to a number of other developments. Selected references on safety and loss prevention are given in Table 1.1.

Table 1.1 *Selected references on safety and loss prevention*

General safety

ABCM (n.d./1, 1964/3); AIChE (see Appendix 28); Creber (n.d.); IChemE (see Appendix 28); IOSHIC (Information Sheet 15); NSC (n.d./2, 4, 6, 7, 1992/11); RoSPA (IS/72, IS/106); Ramazzini (1713); Blake (1943); Rust and Ebert (1947); Plumbe (1953); Gugger *et al.* (1954); Guelich (1956); Harvey and Murray (1958); Armistead (1959); Coates (1960); Thackara *et al.* (1960); ILO (1961, 1972); Meyer and Church (1961); AIA (1962); Devauchelle and Ney (1962); Ducommun (1962); Shearon (1962); Kirk and Othmer (1963–, 1978–, 1991–); Simonds and Grimaldi (1963); Gilbert (1964); Vervalin (1964a, 1973a, 1976c, 1981a,b, 1983); G.T. Austin (1965a); Christian (1965); H.H. Fawcett (1965a, 1981, 1982a, 1985); H.H. Fawcett and Wood (1965, 1982); Gagliardi (1965); Gilmore (1965); Gimbel (1965a); Gordon (1965); Kac and Strizak (1965); Voigtlander (1965); Emerick (1967); FMEC (1967); McPherson (1967); Sands and Bulkley (1967); Tarrants (1967); Badger (1968); J.E. Browning (1968); CBI (1968); Fowler and Spiegelman (1968); Leeah (1968–); Packman (1968); Berry (1969, 1977); Everett (1969); Handley (1969, 1977); Klaassen (1969, 1971, 1979); Maas (1969); Northcott (1969); Preddy (1969); Davidson (1970, 1974); Hearfield (1970, 1974, 1976); D.L. Katz (1970); MCA (1970–/18); Queener (1970); N.T. Freeman and Pickbourne (1971); ILO (1971–72); Kletz (1971, 1975a,c,e, 1976a,c,e,f, 1977f, 1978b, 1979c,e,k, 1980k, 1981c, 1983c,f, 1984b,g,j, 1984 LPB 59, 1985a, 1986g, 1987i, 1988j, 1990d, 1991k, 1994 LPB 120); Rodgers (1971); Tye and Ullyet (1971); Hammer (1972); R.Y. Levine (1972b); SCI (1972); Burns (1973); CAPITB (1973/1, 1975 Information Paper 16); Holder (1973); Kinnersly (1973); Kirven and Handke (1973); Ludwig (1973a,b); Orloff (1973); Society of the Plastics Industry (1973); Widner (1973); C.A.J. Young (1973); Buschmann (1974); Critchfield (1974); Kantyka (1974a,b); Lees (1974a,b, 1980); Malasky (1974); D. Turner (1974); Anon. (1975l); Barber (1975); Boyes (1975); D. Farmer (1975); Gardner and Taylor (1975); HSE 1975 (HSW Booklet 35); Institute of Fuel (1975); D. Petersen (1975, 1982a, 1984, 1988a,b, 1989); TUC (1975, 1978, 1986); Bean (1976a,b); D.B. Brown (1976); Koetsier (1976); Marti (1976); V.C. Marshall (1976d, 1990d); Singleton (1976b); Arscott and Armstrong (1977); Atherley (1977b,c, 1978); Barbieri *et al.* (1977); Blohm (1977); IP (1977); J.Jones (1977); Leuchter (1977); Lugenheim (1977); McCrindle (1977, 1981); Nicolaescu (1977); Rogojina (1977); Sisman and Gheorgiu (1977); Wakabayashi (1977); Webster (1977); Allianz Versicherung (1978); M.E. Green (1978, 1979); Napier (1978); Anon. (1979d); BASF (1979); Birkhahn and Wallis (1979); Hagenkötter (1979); Kerr (LPB 25 1979); R. King and Magid (1979); Menzies and Strong (1979); Peine (1979); Schaeffer (1979); Schierwater (1979); H. Clarke (1980); Heinrich *et al.* (1980); Krishman and Ganesh (1980); Kumar *et al.* (1980); Napier (1980); Spiegelman (1980); Srinivasan *et al.* (1980); Wells (1980); Chowdhury (1981); McCrindle (1981); Teja (1981); AGA (1982/7); Laitinen (1982); Ormsby (1982, 1990); Carter (LPB 50 1983); W.B. Howard (1983, 1984, 1985, 1989); Parmeggiani (1983); Preece (1983); Ridley (1983–); Sinnott (1983); Tailby (LPB 50 1983); Warner (LPB 50 1983); Carson and Jones (1984); AD Little (1984); Ross (1984); Sellers (1984); Zanetti (1984a); Burgoyne (1985b, 1986b); EFCE (LPB 66 1985); Gagliardi (1985); Hildebrand (1985); McKechnie (1985); Munson (1985); Nordic Liaison Committee (1985 NKA/LIT(85)3); Packer (1985); Pilarski (1985); APCA (1986); T.O. Gibson (1986); Grollier-Baron (1986); Joschek (1986, 1987); Lihou (1986); DnV (1987 RP C201); Romer (1987); Scheid (1987); Hoyos and Zimolong (1988); Kharbanda and Stallworthy (1988); Carson and Mumford (1989); Crawley (1990 LPB 91); Hastings (1990); Jochum (1990); Koh (1990); Krishnaiah *et al.* (1990); H.C.D. Phillips (1990); Renshaw (1990); Dupont, Theodore and Reynolds (1991); Eades (1991); L. Hunt (1991); McQuaid (1991); Rasmussen (1991); Whiston (1991); Pasman *et al.* (1992); Andrews (1993); Fisk and Howes (1993); Cullen (1994); Donald and Canter (1994)

Total loss control

BSC (n.d./6, 7); Heinrich (1959); Bird (1966, 1974); Bird and Germain (1966, 1985); Gilmore (1970); Goforth (1970); J. Tye (1970); J.A. Fletcher and Douglas (1971); Webster (1974, 1976); Anon. (1975 LPB 4, p. 1); Hearfield (1975); D.G. King (1975); F.E. Davis (1976); Ling (1976, 1979); D.H. Farmer (1978); Planer (1979); Heinrich, Petersen and Roos (1980); Dave (1987)

Company policies

Aalbersberg (1991); Auger (1993); Whiston (1993)

Responsible care

CIA (RC51, RC52, 1990 RC23, 1992 RC53); Belanger (1990); Kavasmaneck (1990); Whiston (1991); Jacob (1992)

Product stewardship

Rausch (1990)

Organizational initiatives

IChemE: Hancock (1983); Street, Evans and Hancock (1984)

SRD: Clifton (1983 LPB 52)

CEFIC: Jourdan (1990)

CCPS: CCPS (1985); Carmody (1988, 1989, 1990a,b); R.A. Freeman (1990); K.A. Friedman (1990); Schreiber (1990)

EPSC: Anon. (1992 LPB 103); Anon. (1993 LPB 111, p. 0); EPSC (1993); Anon. (1994 LPB 115); Hancock (1994 LPB 120)

Organization guides

NSC (1974); ILO (1989, 1991); OECD (1992)

Contractors
Kletz (1991m)

Terminology
IP (Oil Data Sht 3); Harvey (1979b); Burgoyne (1980); Berthold and Loeffler (1981); V.C. Marshall (1981a,b, 1990c); A.E. Green (1982); Kletz (1983c, 1984c); IChemE (1985/78, 1992/98); ACDS (1991) BS (Appendix 27 Glossaries)

Bibliographies
Commonwealth Department of Productivity (1979); Lees (1980); Vervalin (1981a,b); Lees and Ang (1989a)

Critiques
Wallick (1972); Commoner (1973); L.N. Davis (1979, 1984); Howlett (1982)

Other related fields
Pugsley (1966); Ingles (1980) (civil engineering); Thurston (1980) (aviation); Garrick and Caplan (1982); V.M. Thomas (1982); Grigoriu (1984) (structural engineering)

1.1 Management Leadership

By the mid-1960s it was becoming increasingly clear that there were considerable differences in the performance of companies in terms of occupational safety. These disparities could be attributed only to differences in management. There appeared at this time a number of reports on safety in chemical plants arising from studies by the British chemical industry of the safety performance in the US industry, where certain US companies appeared to have achieved an impressive record. These reports included *Safety and Management* by the Association of British Chemical Manufacturers (ABCM) (1964, 3), *Safe and Sound* and *Safety Audits* by the British Chemical Industry Safety Council (BCISC) (1969, 9; 1973 12). The companies concerned attributed their success entirely to good management and this theme was reflected in the reports.

1.2 Industrial Safety and Loss Trends

About 1970 it became increasingly recognized that there was a world-wide trend for losses due to accidents to rise more rapidly than gross national product (GNP).

This may be illustrated by the situation in the UK. The first half of this century saw a falling trend in personal accidents in British factories, but about 1960 this fall bottomed out. Over the next decade very little progress was made; in fact there was some regression. Figure 1.1 shows the number of fatal accidents and the total number of accidents in factories over the period 1961–74. The Robens Committee on Health and Safety at Work, commenting on these trends in 1972, suggested that part of the reason was perhaps the increasingly complex technology employed by industry (Robens, 1972).

Another important index is that of fire loss. The estimated fire damage loss in factories and elsewhere in the UK for the period 1964–74 is shown in Figure 1.2.

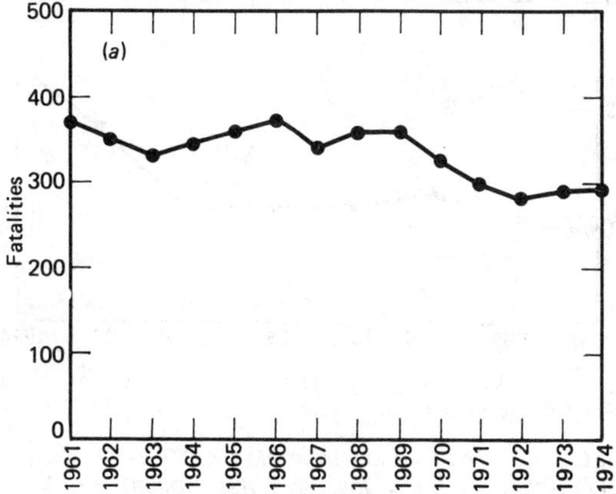

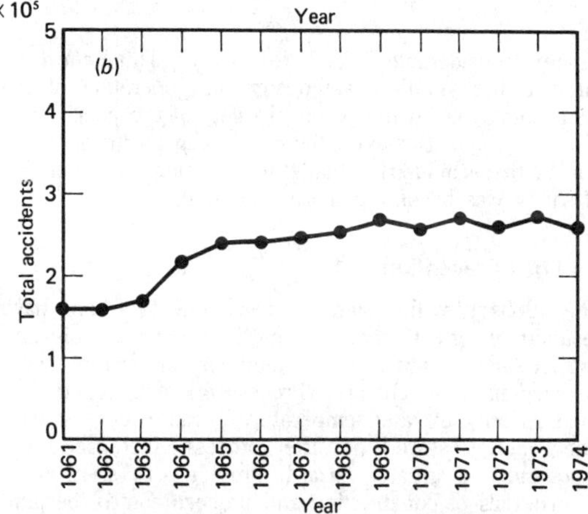

Sources: Robens (1972); HM Chief Inspector of Factories (1974)

Figure 1.1 *Fatalities (a) and total accidents (b) in factories in the UK, 1961–74 (Courtesy of the Health and Safety Executive)*

1.3 Safety and Environment Concerns

There was also at this time growing public awareness and concern regarding the threat to people and to the environment from industrial activities, particularly those in which the process industries are engaged. Taking the UK as an illustration, the massive vapour cloud explosion at Flixborough in 1974 highlighted the problem of major hazards. This led to the setting up of the Advisory Committee on Major Hazards (ACMH) which sat from 1975 to 1983, and to the introduction of legislation to control major hazard installations. Likewise, there was a continuous flow of legislation to tighten up both on emissions from industrial installations and on exposure of workers to noxious substances at those installations.

It is against this background, therefore, that the particular problems of the process industries should be viewed. The chemical, oil and petrochemical industries handle hazardous substances and have always had to

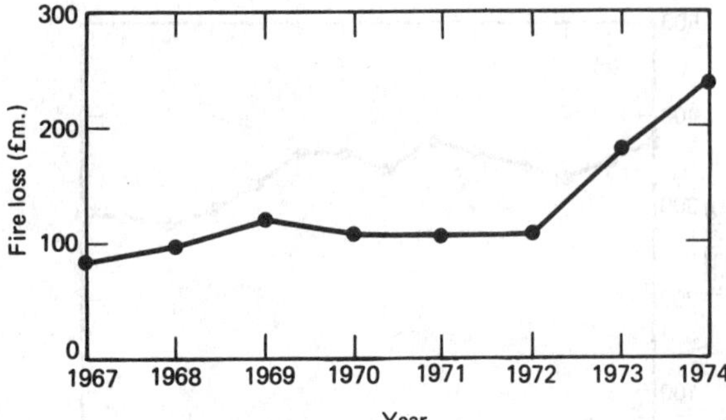

Figure 1.2 *Total fire losses in the UK, 1967–74 (Courtesy of the British Insurance Association)*

devote considerable effort to safety. This effort is directed both to the safe design and operation of the installations and to the personal safety of the people who work on them. However, there was a growing appreciation in these industries that the technological dimension of safety was becoming more important.

1.4 Loss Prevention – 1

The 1960s saw the start of developments which have resulted in great changes in the chemical, oil and petrochemical industries. A number of factors were involved in these changes. Process operating conditions such as pressure and temperature became more severe. The energy stored in the process increased and represented a greater hazard. Problems in areas such as materials of construction and process control became more taxing. At the same time plants grew in size, typically by a factor of about 10, and were often single stream. As a result they contained huge items of equipment, such as compressors and distillation columns. Storage, both of raw materials and products and of intermediates, was drastically reduced. There was a high degree of interlinking with other plants through the exchange of by-products.

The operation of such plants is relatively difficult. Whereas previously chemical plants were small and could be started up and shut down with comparative ease, the start-up and shut-down of a large, single-stream plant on an integrated site is a much more complex and expensive matter. These factors resulted in an increased potential for loss – both human and economic. Such loss may occur in various ways. The most obvious is the major accident, usually arising from loss of containment and taking the form of a serious fire, explosion or toxic release. But loss due to such situations as delays in commissioning and downtime in operating is also important.

The chemical and oil industries have always paid much attention to safety and have a relatively good record in this respect. In the UK, for example, the fatal accident rate for the chemical industry has been about equal to that for industry generally, which in view of the nature of the industry may be regarded as reasonable. These are high technology industries and there has always been a strong technological element in their approach to safety. However, the increasing scale and technology of modern plants caused the chemical industry to re-examine its approach to the problem of safety and loss. If the historical development of this concern in the UK is considered, there are several problem areas which can be seen, in retrospect, to have given particular impetus to the development of loss prevention.

One of these is the problem of operating a process under extreme conditions and close to the limits of safety. This is usually possible only through the provision of relatively sophisticated instrumentation. About the mid-1960s, several such systems were developed. One of the most sophisticated, influential and well documented was the high integrity protective system developed by R. M. Stewart (1971) for the ethylene oxide process. About the same time many difficulties were being experienced in the commissioning and operation of large, single-stream plants, such as ethylene and ammonia plants, involving quite severe financial loss. On the design side, too, there was a major problem in getting value for money in expenditure aimed at improving safety and reducing loss. It was increasingly apparent that a more cost-effective approach was needed.

These developments did not take place in isolation. The social context was changing also and other themes, notably pollution, including effluent and waste disposal and noise, were becoming of increasing concern to the public and the Government. In consequence, the industry was obliged to examine the effects of its operations on the public outside the factory fence and, in particular, to analyse more carefully the possible hazards and to reduce emissions and noise. Another matter of concern was the increasing quantities of chemicals transported around the country by road, rail and pipeline. The industry had to take steps to show that these operations were conducted with due regard to safety. In sum, by the 1970s these problems became a major preoccupation of senior management. Management's recognition of the problems and its willingness to assign to their solution many senior and capable people as well as other resources has been fundamental in the development of loss prevention.

The existence of expertise in related areas has been of great value. In the UK, the UK Atomic Energy Authority (UKAEA), initially through its Health and Safety Branch and then through its Safety and Reliability Directorate (SRD), was able to advise on reliability assessment. Industry adopted UKAEA techniques in the assessment of major hazards and of protective instrumentation and data on failure rates. Many firms in the industry now have their own reliability engineers.

The historical development of loss prevention is illustrated by some of the milestones listed in Table 1.2. The impact of events has been different in different countries and the table gives a UK perspective. Within the industry loss prevention emerged as a theme of technical meetings which indicated an increasingly sophisticated technological approach. The Institution of Chemical Engineers (IChemE) established a Loss Prevention Panel which operates an information-exchange scheme and publishes the *Loss Prevention Bulletin*. This growing industrial activity was matched in the regulatory sphere. The Robens Committee (1972)

Plate 1 Installation of an ethylene fractionation column (Courtesy of PSC Freysinnet Ltd)

Plate 2 Flange protection for control of emissions (Courtesy of Badische Anilin und Soda Fabrik)

Plate 3 Control room of Magnus platform (Courtesy of British Petroleum Company plc)

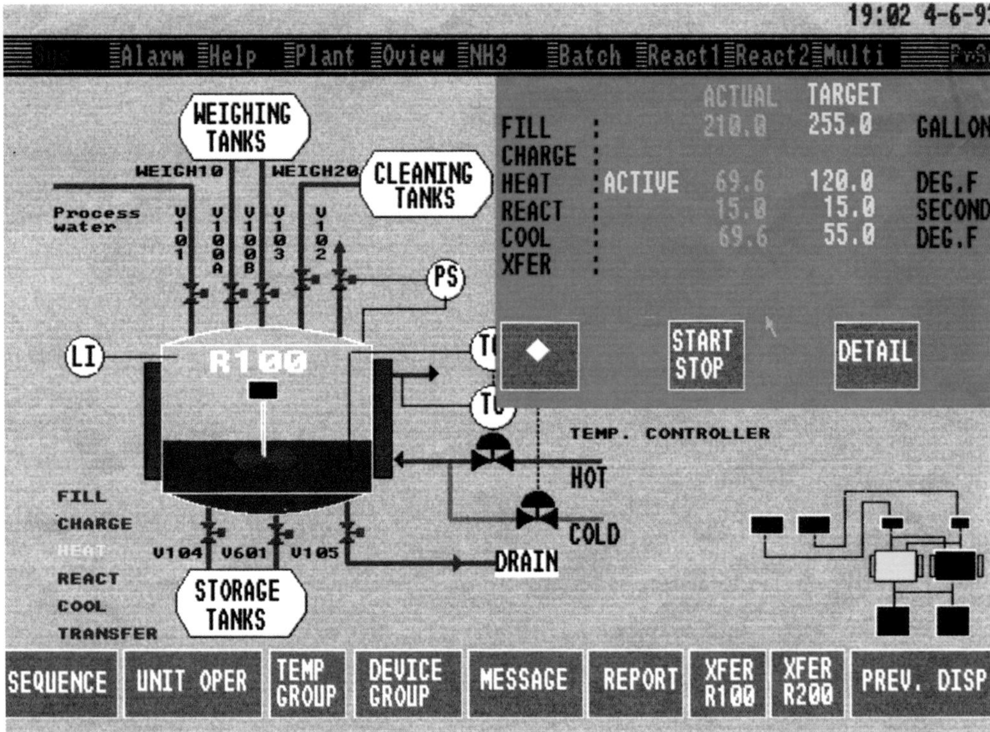

Plate 4 VDU display on a computer controlled batch process (Courtesy of Boots Company plc)

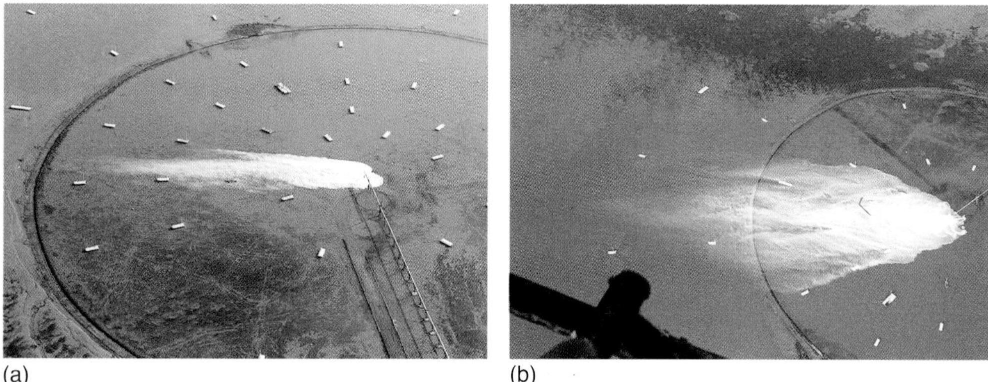

(a) (b)

Plate 5 Maplin Sands trials of vaporization and dispersion of refrigerated liquefied gases: continuous releases of refrigerated liquid propane (Puttock, Colenbrander and Blackmore, 1983): (a) Spill 46; and (b) Spill 54. Spill 46: release of 2.8 m^3/min for 8 minutes with wind speed 8.1 m/s. Spill 54: release of 2.3 m^3/min with wind speed 3.8 m/s (Reproduced by permission of Shell Research Ltd)

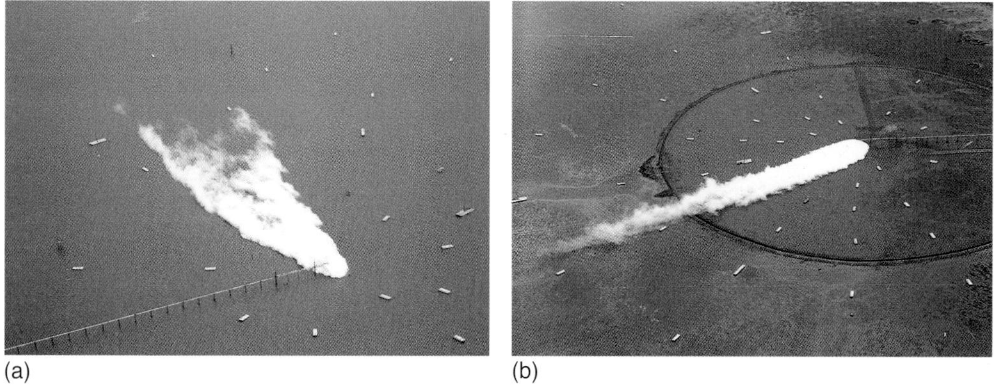

(a) (b)

Plate 6 Maplin Sands trials of vaporization and dispersion of refrigerated liquefied gases: continuous releases of LNG (Puttock, Blackmore and Colenbrander, 1982; Puttock, Colenbrander and Blackmore, 1982): (a) Spill 29; and (b) Spill 56. Spill 29: release of 4.1 m^3/min for 225 seconds with wind speed 7.4 m/s. Spill 56: release of 2.5 m^3/min with wind speed 4.8 m/s (Reproduced by permission of Shell Research Ltd)

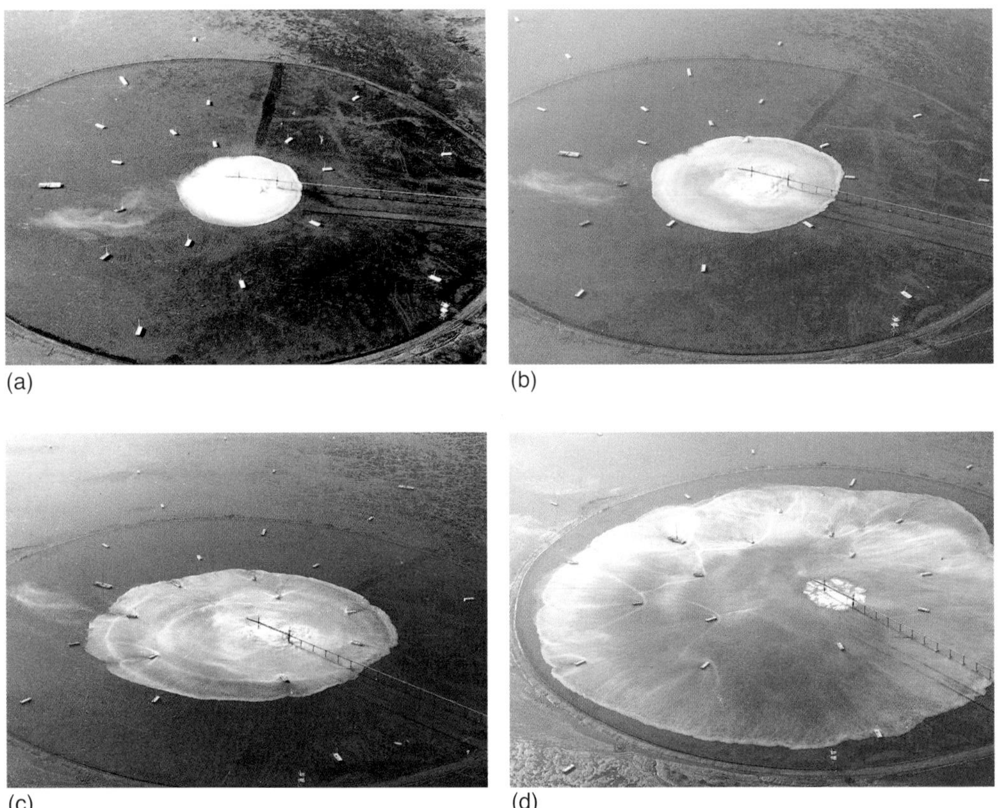

(a) (b) (c) (d)

Plate 7 Maplin Sands trials of vaporization and dispersion of refrigerated liquefied gases: instantaneous release of liquid propane (Puttock, Blackmore and Colenbrander, 1982): (a)-(d) successive times (Reproduced by permission of Shell Research Ltd)

(a) (b) (c) (d)

Plate 8 Thorney Island trials of dispersion of heavy gases: (a)-(c) successive times, elevation views; (d) plan view (Courtesy of the Health and Safety Executive)

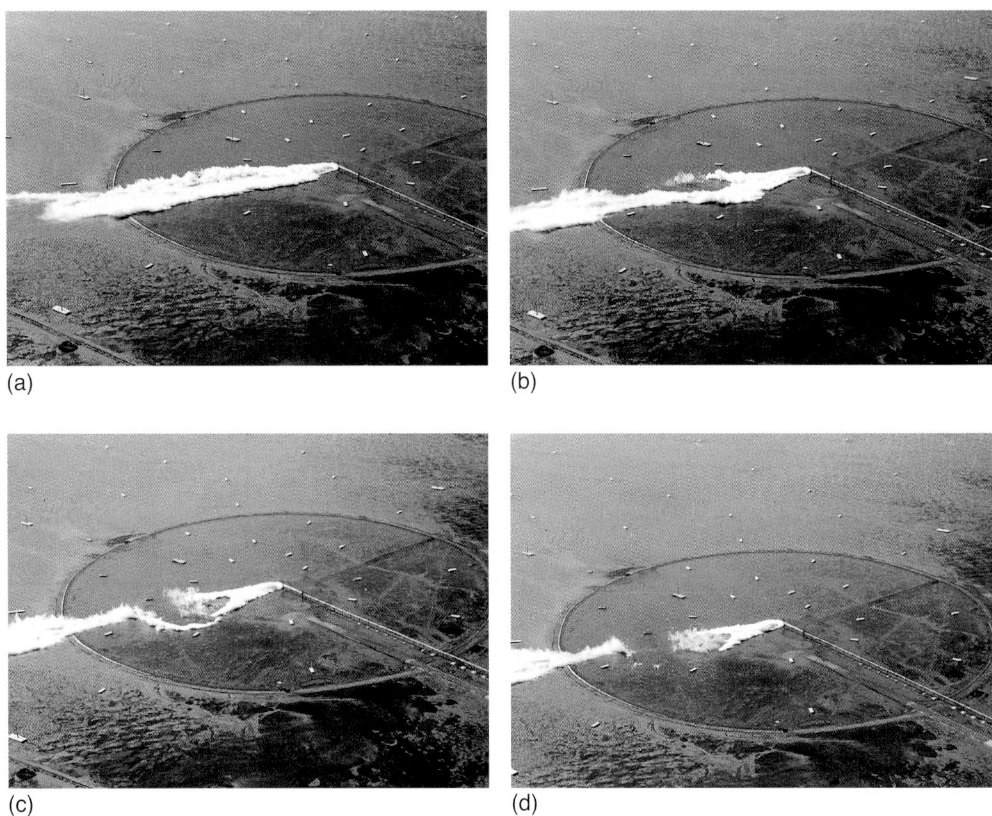

(a) (b) (c) (d)

Plate 9 Maplin Sands trials of combustion of vapour clouds (Blackmore, Eyre and Summers, 1982; Hirst and Eyre, 1983): Trial 27: times from ignition: (a) 0 seconds; (b) 9 seconds; (c) 14 seconds; and (d) 16 seconds. Release of 32 m^3/min with wind speed 6 m/s (Reproduced by permission of Shell Research Ltd)

Plate 10 Kerosene pool fire (Mizner and Eyre, 1982) (Reproduced by permission of Shell Research Ltd)

Plate 11 LNG pool fire (Mizner and Eyre, 1982) (Reproduced by permission of Shell Research Ltd)

Plate 12 Jet flame from a release of liquid propane (Tam and Cowley, 1989) (Reproduced by permission of Shell Research Ltd)

Plate 13 Fire-engulfed vessel with jet flame issuing from the pressure relief valve (Courtesy of the Health and Safety Executive)

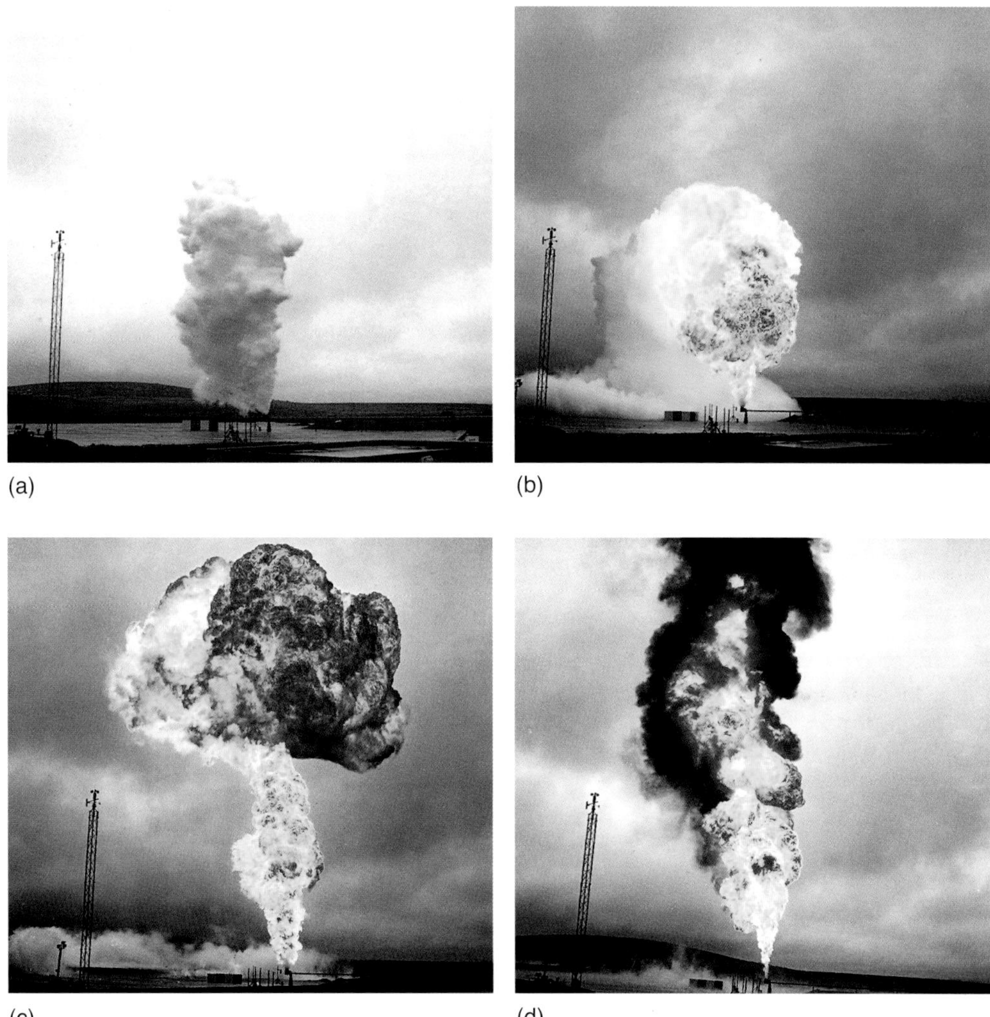

(a) (b) (c) (d)

Plate 14 Leak, ignition of leak and fireball on a pipeline from a release of liquid propane in Test 23 (Hirst, 1984, 1986): time relative to ignition: (a) –10 seconds; (b) +1 second; (c) +3 seconds; (d) +8 seconds. Release of 41 kg/s with wind speed <0.5 m/s (Reproduced by permission of Shell Research Ltd)

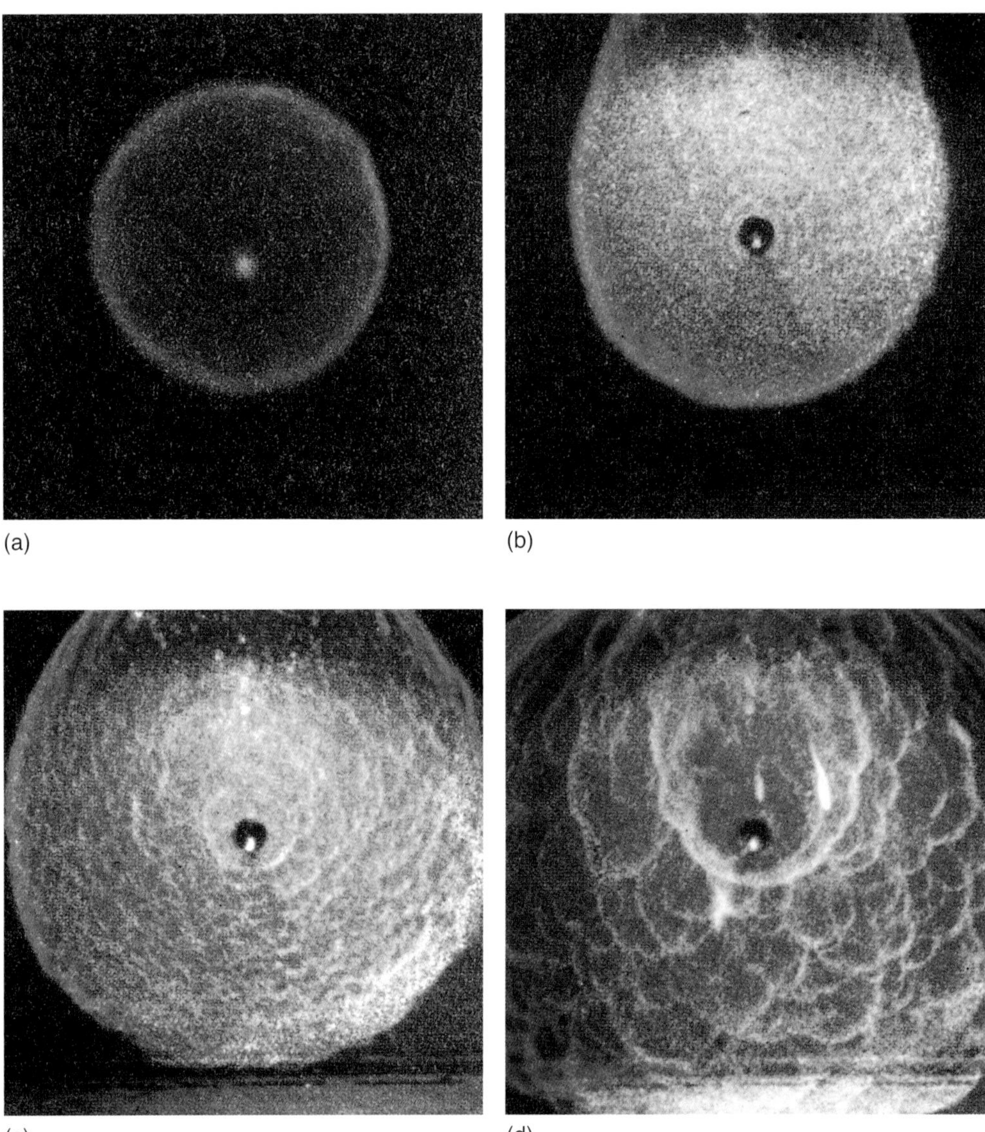

Plate 15 Stages in flame development during a vented explosion in a vessel (R.J.Harris, 1983): times after ignition: (a) 95 ms; (b) 130 ms, flow through vent begins; (c) 210 ms, flame almost reaching maximum flame area; (d) 320 ms, flame fills enclosure. Mixture: natural gas-air. Vent located at top. (Courtesy of British Gas plc)

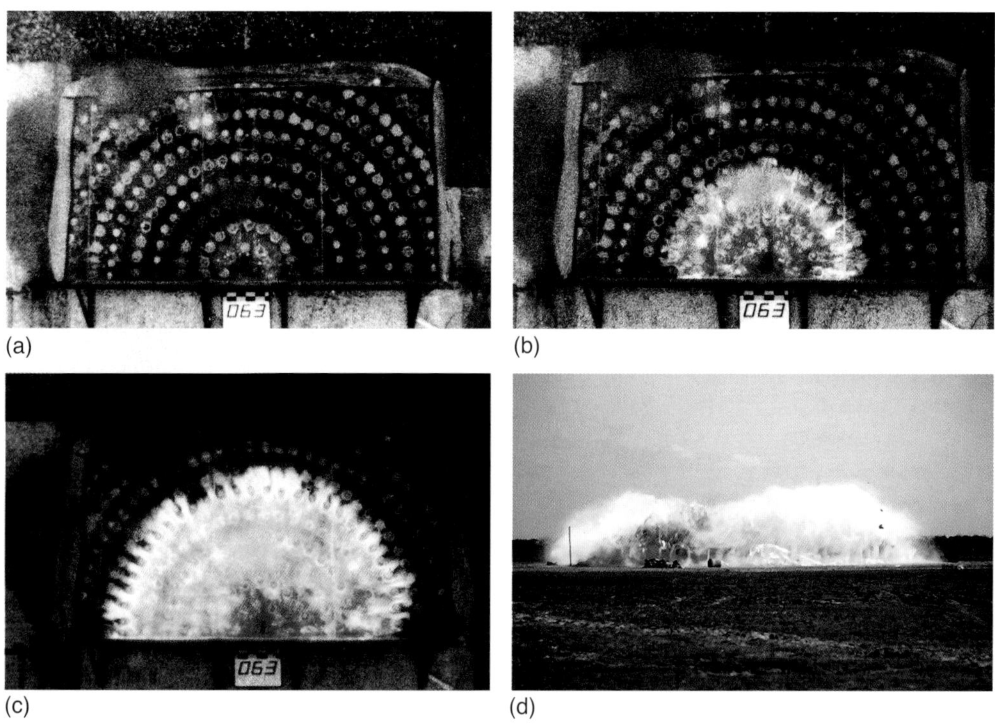

Plate 16 Vapour cloud explosion trials (Zeeuwen, van Wingerden and Dauwe, 1983; van Wingerden, 1989): (a)-(c) flame development, plan view; (d) elevation view (Courtesy of TNO)

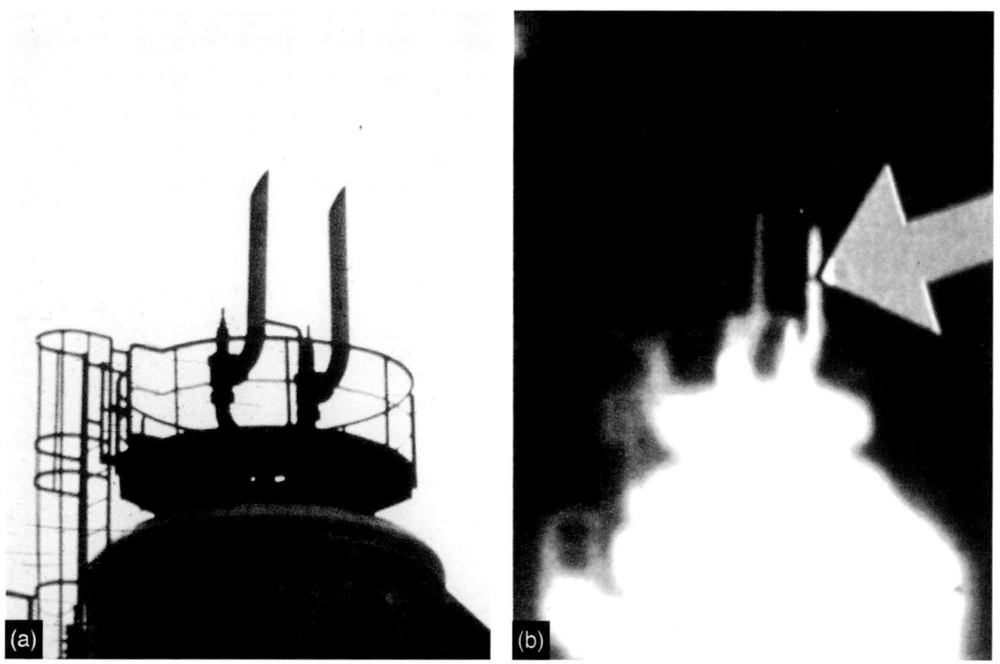

Plate 17 Thermographic pictures of pressure relief valves: (a) normal photgraph: (b) thermographic photograph, showing a leaking valve (Courtesy of AGEMA Infrared Systems)

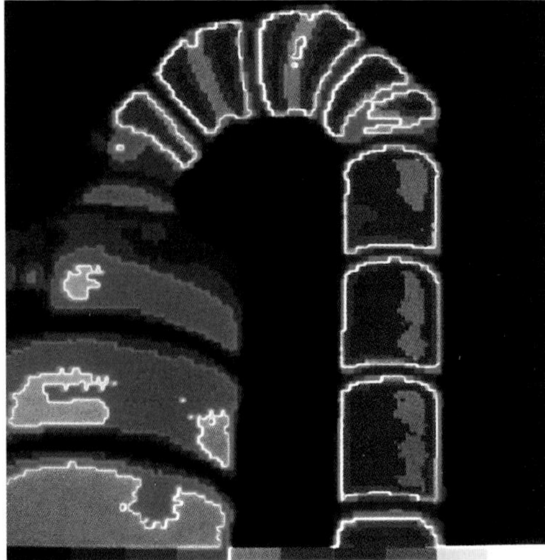

Plate 18 Thermographic picture of a distillation column (Courtesy of AGEMA Infrared Systems)

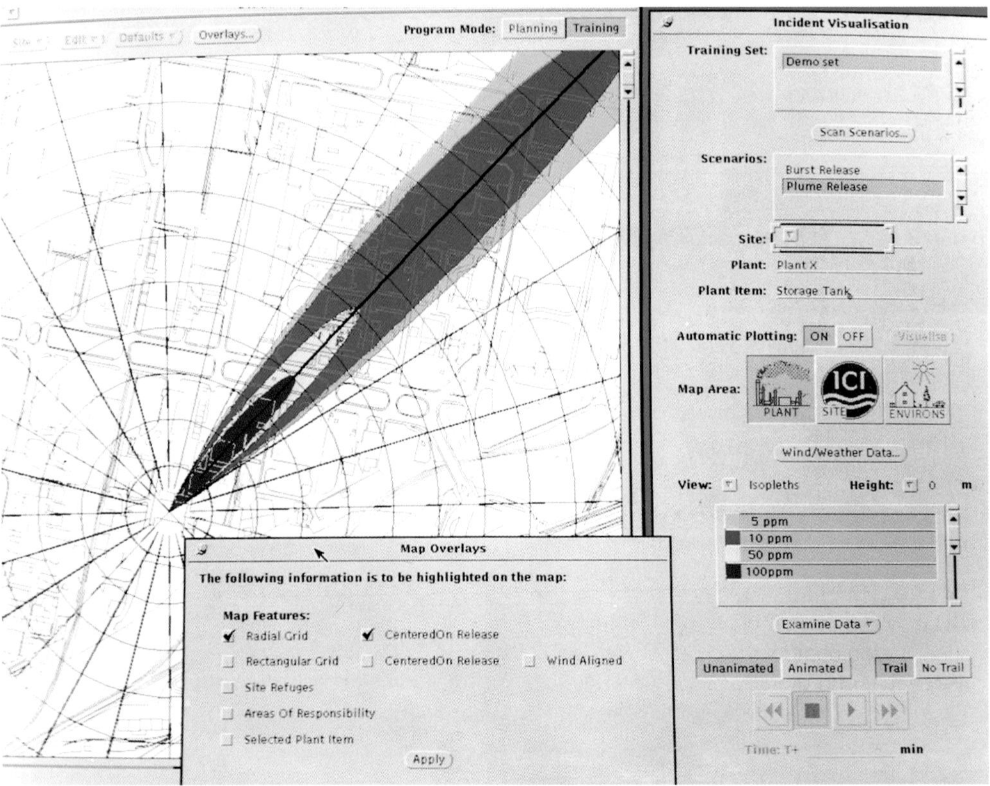

Plate 19 Computer displays for handling an emergency: ICI DISCOVER system (Preston, 1993): real-time VDU display of gas cloud concentration contours superimposed on site plan (Courtesy of ICI plc)

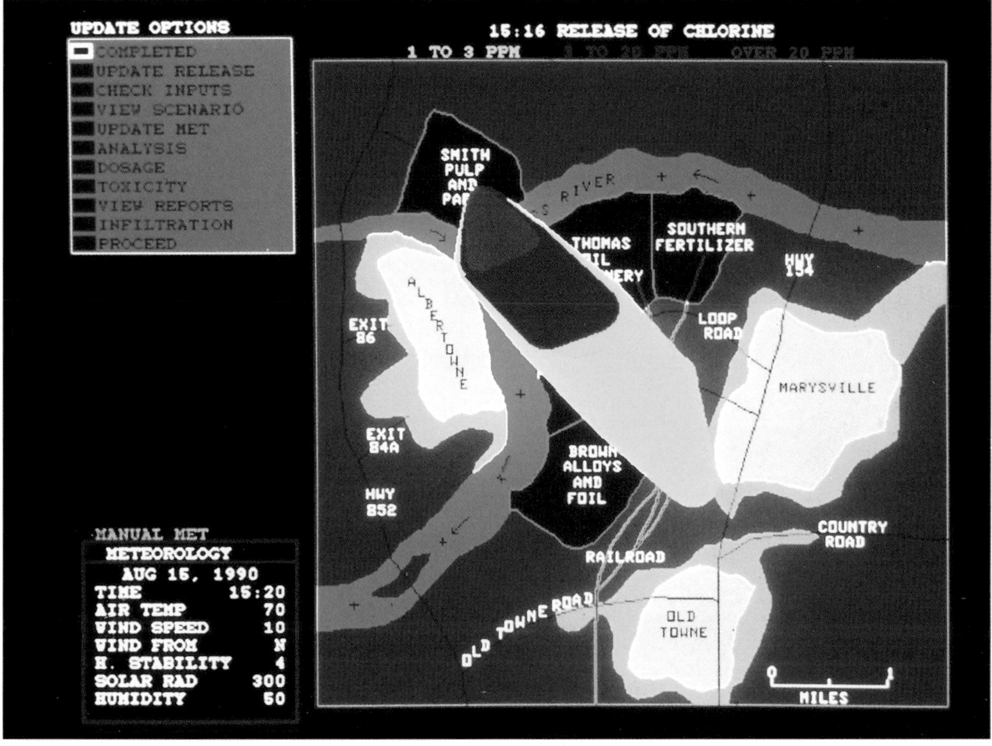

(a)

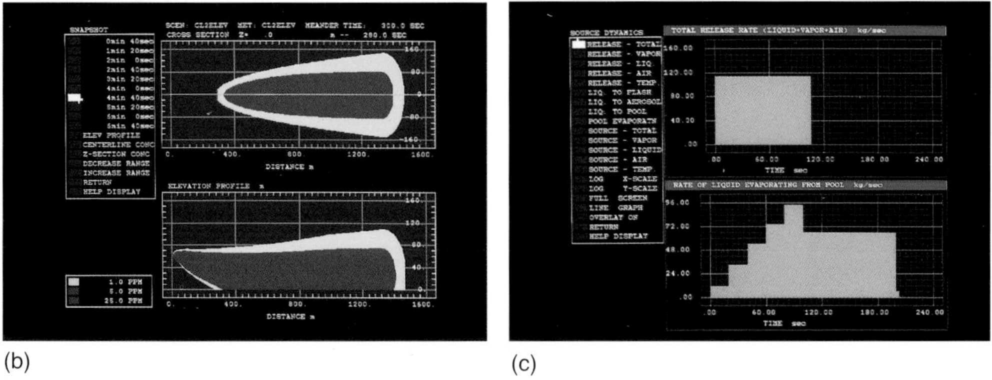

(b) (c)

Plate 20 Computer displays for handling an emergency: DuPont SAFER system: real-time VDU displays of: (a) chlorine gas concentration profile superimposed on site plan; (b) gas concentration contours from an elevated release, in plan and elevation; (c) time profiles of total release rate and of vaporization rate (Courtesy of Dupont Safer Emergency Systems Inc.)

KNOW YOUR HAZARD WARNING SIGNS

Flammable gas. A substance which has a critical temperature below 50°C or which at 50°C has a vapour pressure of more than 3 bars absolute and is flammable.

Flammable liquid. A liquid with a flash point of 55°C or below except a liquid which - (a) has a flash point equal to or more than 21°C and less than or equal to 55°C; and (b) when tested at 55°C in the manner described in Schedule 2 to the Highly Flammable Liquids and liquefied Petroleum Gases Regulations 1972 (a) does not support combustion.

Flammable solid. A solid which is readily combustible under conditions encountered in conveyance by road or which may cause or contribute to fire through friction.

Spontaneously combustible substance. A substance which is liable to spontaneous heating under conditions encountered in conveyance by road or to heating in contact with air being then liable to catch fire.

Substance which in contact with water emits flammable gas. A substance which in contact with water is liable to become spontaneously combustible or to give off a flammable gas.

Oxidizing substance. A substance other than an organic peroxide which, although not itself necessarily combustible, may by yielding oxygen or by a similar process cause or contribute to the combustion of other material.

Organic peroxide. A substance which is - (a) an organic peroxide; and (b) an unstable substance which may undergo exothermic self - accelerating decomposition.

KNOW YOUR HAZARD WARNING SIGNS

Non-flammable compressed gas. A substance which - (a) has a critical temperature below 50°C or which at 50°C has a vapour pressure of more than 3 bars absolute: and (b) is conveyed by road at a pressure of more than 500 millibars above atmospheric pressure or in liquefied form, other than a toxic gas or a flammable gas.

Toxic gas. A substance which has a critical temperature below 50°C or which at 50°C has a vapour pressure of more than 3 bars absolute and which is toxic.

Toxic substance. A substance known to be so toxic to man as to afford a hazard to health during conveyance or which, in the absence of adequate data on humam toxicity, is presumed to be toxic to man.

Harmful substance. A substance known to be toxic to man or, in the absence of adequate data on human toxicity, is presumed to be toxic to man but which is unlikely to afford a serious acute hazard to health during conveyance.

Corrosive substance. A substance which by chemical action will - (a) cause severe damage when in contact with living tissue: (b) materially damage other freight or equipment if leakage occurs.

Other dangerous substance. A substance which is listed in Part 1A of the approved list and which may create a risk to the health or safety of persons in the conditions encountered in conveyance by road, whether or not it has any of the characteristic properties set out above.

Mixed hazards. Packages containing two or more dangerous substances which have different characteristic properties.

Plate 21 Hazard warning signs (Courtesy of British Standards Institution and of Signs and Labels Ltd)

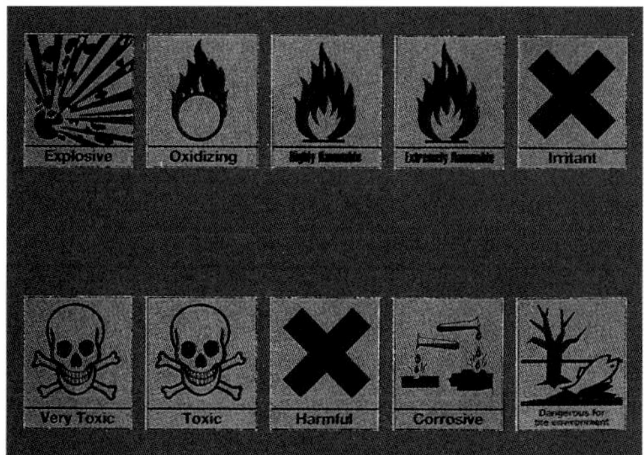

Plate 22 Hazard signs (Courtesy of British Standards Institution and of Signs and Labels Ltd)

Plate 24 Hazchem sign (Courtesy of Chemical Industries Association and of Signs and Labels Ltd)

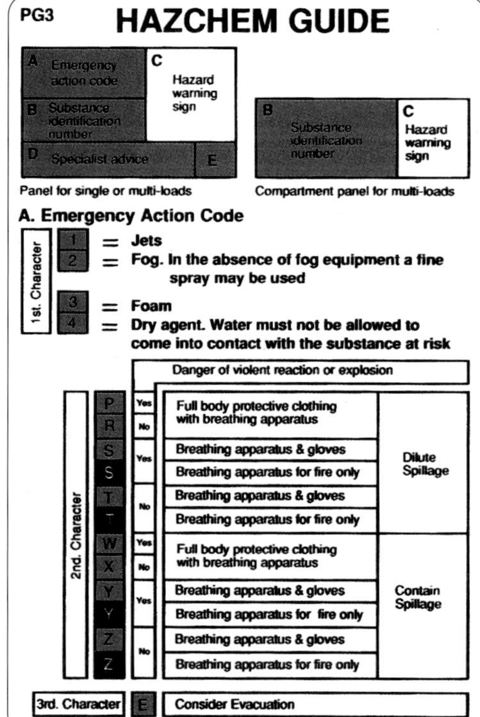

Plate 23 Hazchem guide (Courtesy of Chemical Industries Association and of Signs and Labels Ltd)

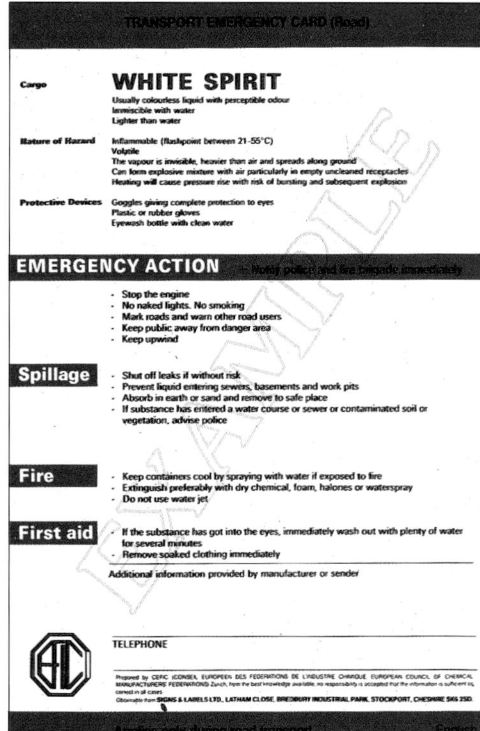

Plate 25 Tremcard (Courtesy of Chemical Industries Association and of Signs and Labels Ltd)

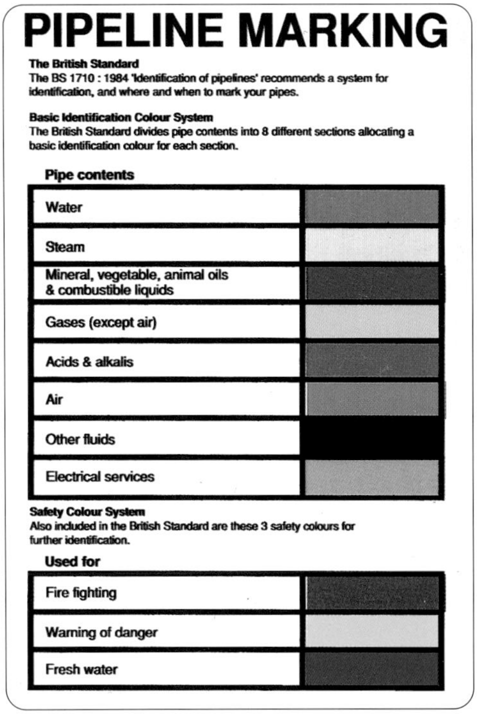

PIPELINE MARKING

The British Standard
The BS 1710 : 1984 'Identification of pipelines' recommends a system for identification, and where and when to mark your pipes.

Basic Identification Colour System
The British Standard divides pipe contents into 8 different sections allocating a basic identification colour for each section.

Pipe contents	
Water	
Steam	
Mineral, vegetable, animal oils & combustible liquids	
Gases (except air)	
Acids & alkalis	
Air	
Other fluids	
Electrical services	

Safety Colour System
Also included in the British Standard are these 3 safety colours for further identification.

Used for	
Fire fighting	
Warning of danger	
Fresh water	

Plate 26 Pipeline markings (Courtesy of British Standards Institution and of Signs and Labels Ltd)

Plate 27 Computer simulation of combustion in an offshore module, showing computer display of plant layout and development of the flame (Courtesy of the Christian Michelsen Institute)

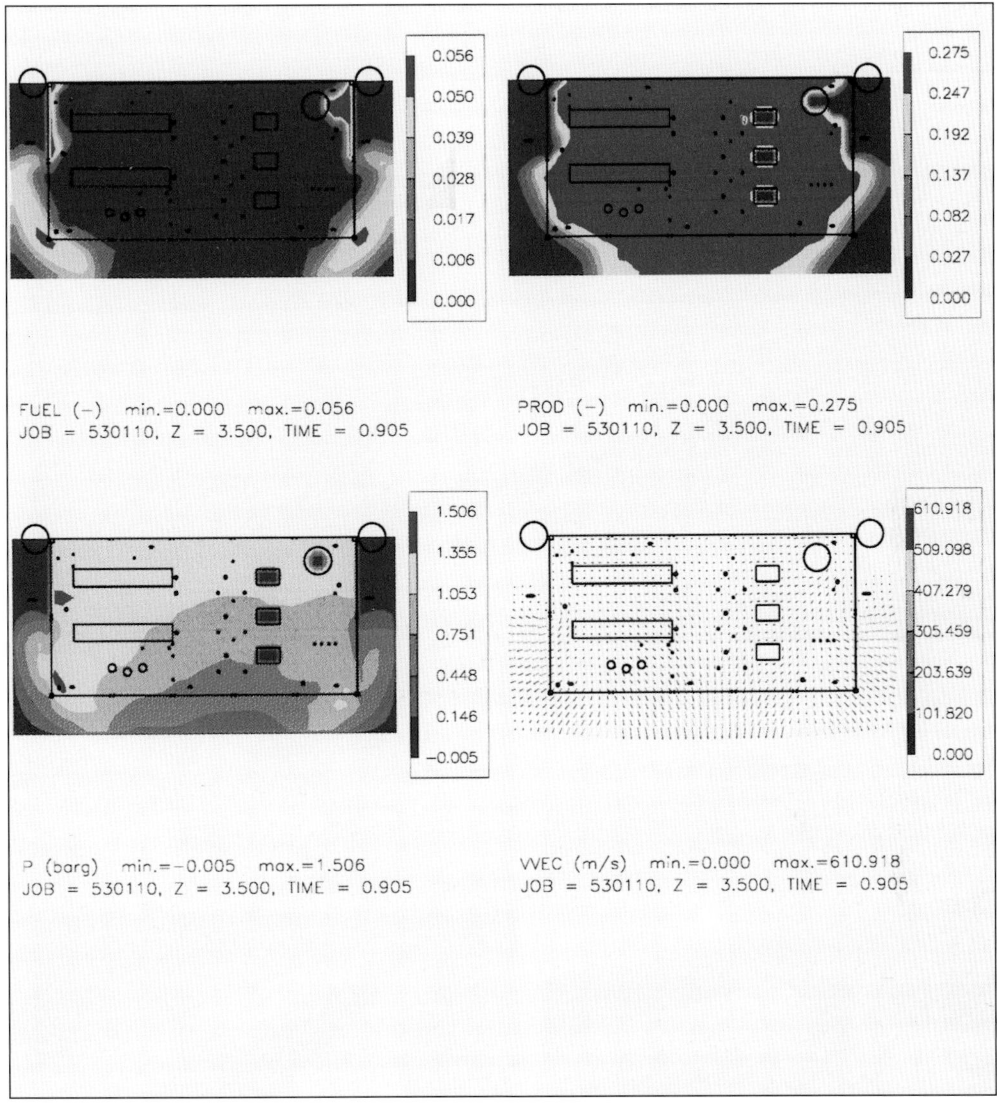

Plate 28 Computer simulation of an explosion in an offshore module: (a) concentration contours of fuel; (b) concentration contours of combustion products; (c) contours of overpressure; (d) velocity vectors. Height 3.5 m. Time after ignition 0.905 seconds (Courtesy of the Christian Michelsen Institute)

Plate 29 An LNG tanker (Reproduced by permission of Shell International Petroleum Company Ltd)

Plate 30 A chlorine road tanker (Courtesy of ICI plc)

Plate 31 LPG rail tank cars (Courtesy of British Petroleum Company plc)

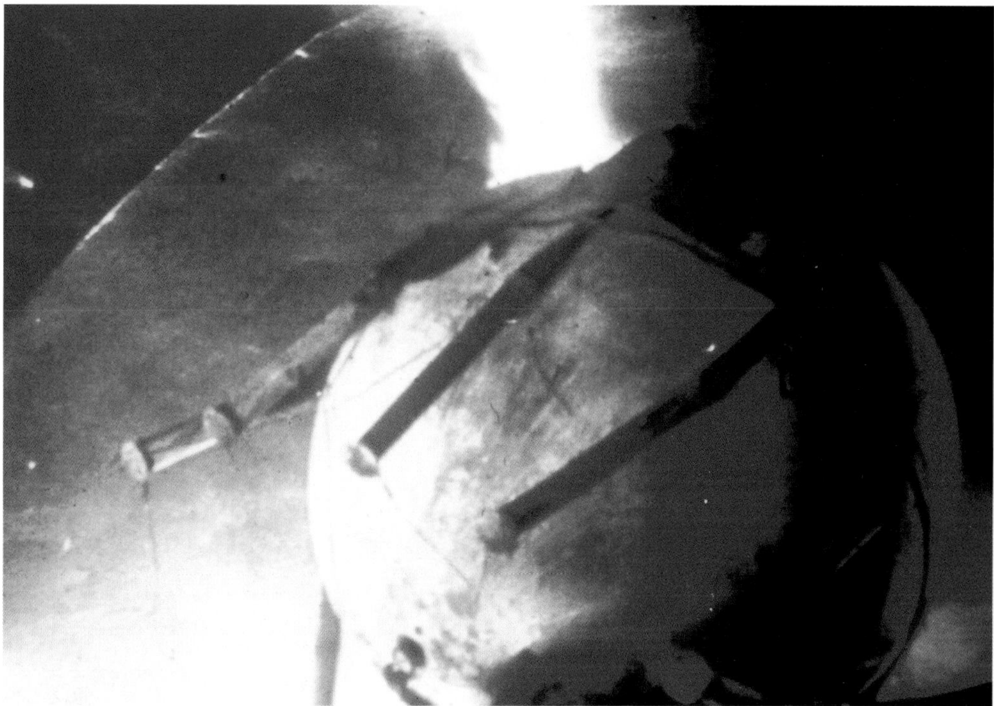

Plate 32 Feyzin, France, 4 January 1966: fire-engulfed LPG storage sphere following explosion

Plate 33 Crescent City, 12 June 1970: LPG fireball (Courtesy of Anderson Studios: photograph by Richard E. Anderson)

Plate 34 Flixborough, UK, 1 June 1974: wreckage of reactor train following vapour cloud explosion (Courtesy of the Health and Safety Executive)

Plate 35 Houston, Texas, 11 May 1976: vegetation damage following release of ammonia from road tanker (Fryer and Kaiser 1979 SRD R152) (Courtesy of the Safety and Reliability Directorate)

(a) (b)

Plate 36 Mexico City, 19 November 1984: (a)-(c) wreckage of terminal following fire and BLEVEs (Courtesy of Skandia Int.)

Plate 36 (c)

Plate 37 Ufa, USSR, 3 June 1989: train wreckage and vegetation damage following vapour cloud explosion (Frank Spooner Pictures)

(a)

(b)

Plate 38 Pasadena, Texas, 23 October 1989: (a) fire following vapour cloud explosion; (b) wreckage of plant (Associated Press)

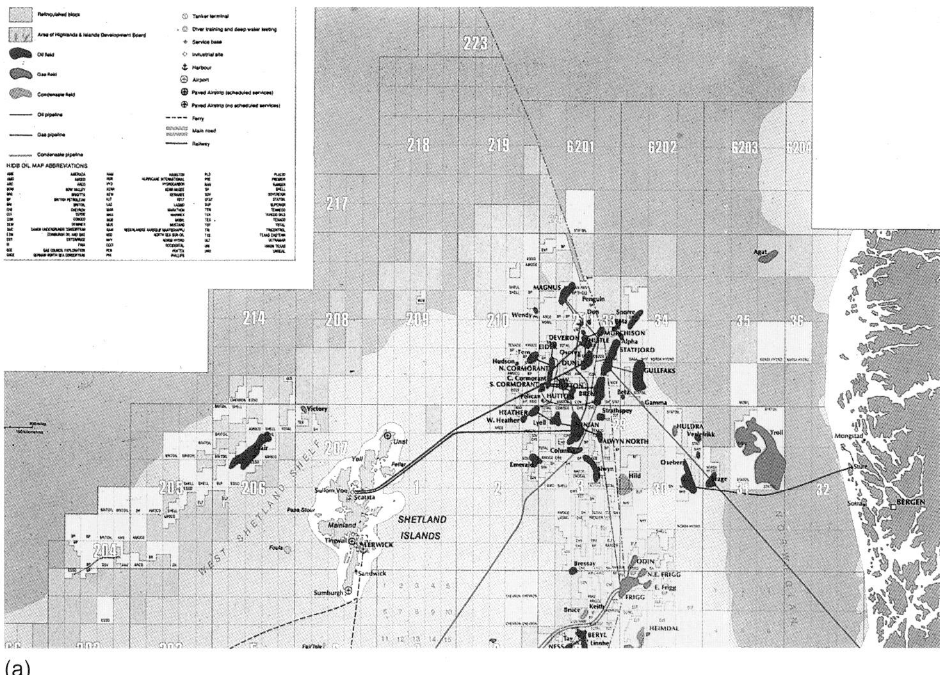

(a)

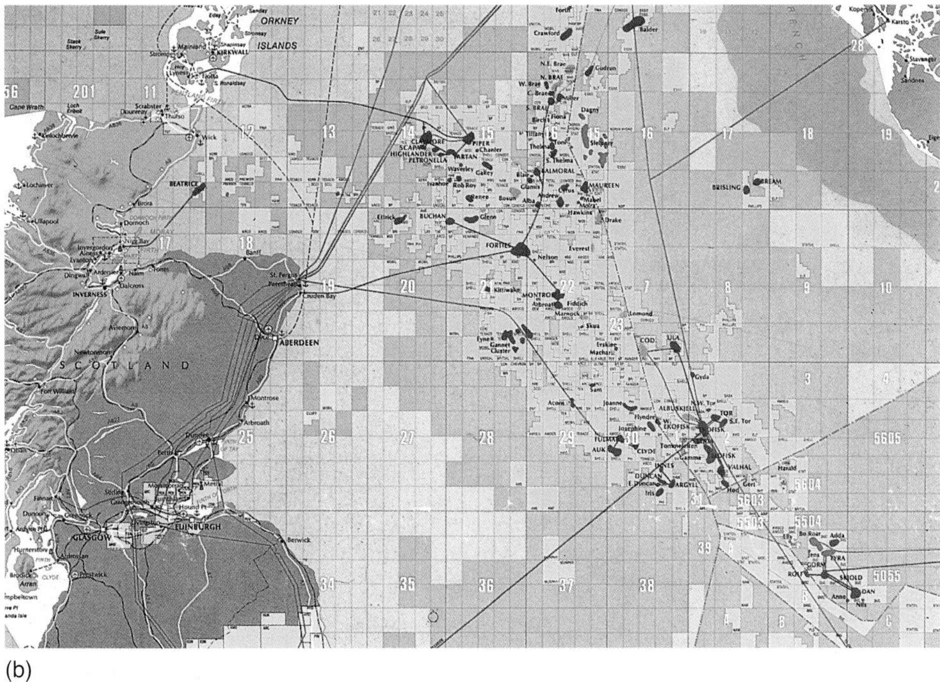

(b)

Plate 39 Map showing oil and gas fields in the Northern North Sea (Cullen, 1990) (Courtesy of HM Stationery Office)

Plate 40 The Magnus platform jacket under tow (Courtesy of British Petroleum Company plc)

Plate 41 The Heerema crane barge *Thor* during module lifting operations in the Forties field (Courtesy of British Petroleum Company plc)

Plate 42 The pipelaying barge *Viking Piper* laying the submarine pipeline from the Ninian field (Courtesy of British Petroleum Company plc)

Plate 43 The emergency support vessel *Lolair* at the Forties Charlie platform (Courtesy of British Petroleum Company plc)

Plate 44 The Buchan Alpha floating production platform (Courtesy of British Petroleum Company plc)

Plate 45 A bridge-linked platform in the Southern North Sea (Courtesy of British Petroleum Company plc)

Plate 46 The Brent B platform under tow after top-decking in Norway (Courtesy of British Petroleum Company plc)

Plate 47 The unmanned platform Beatrix C (Courtesy of British Petroleum Company plc)

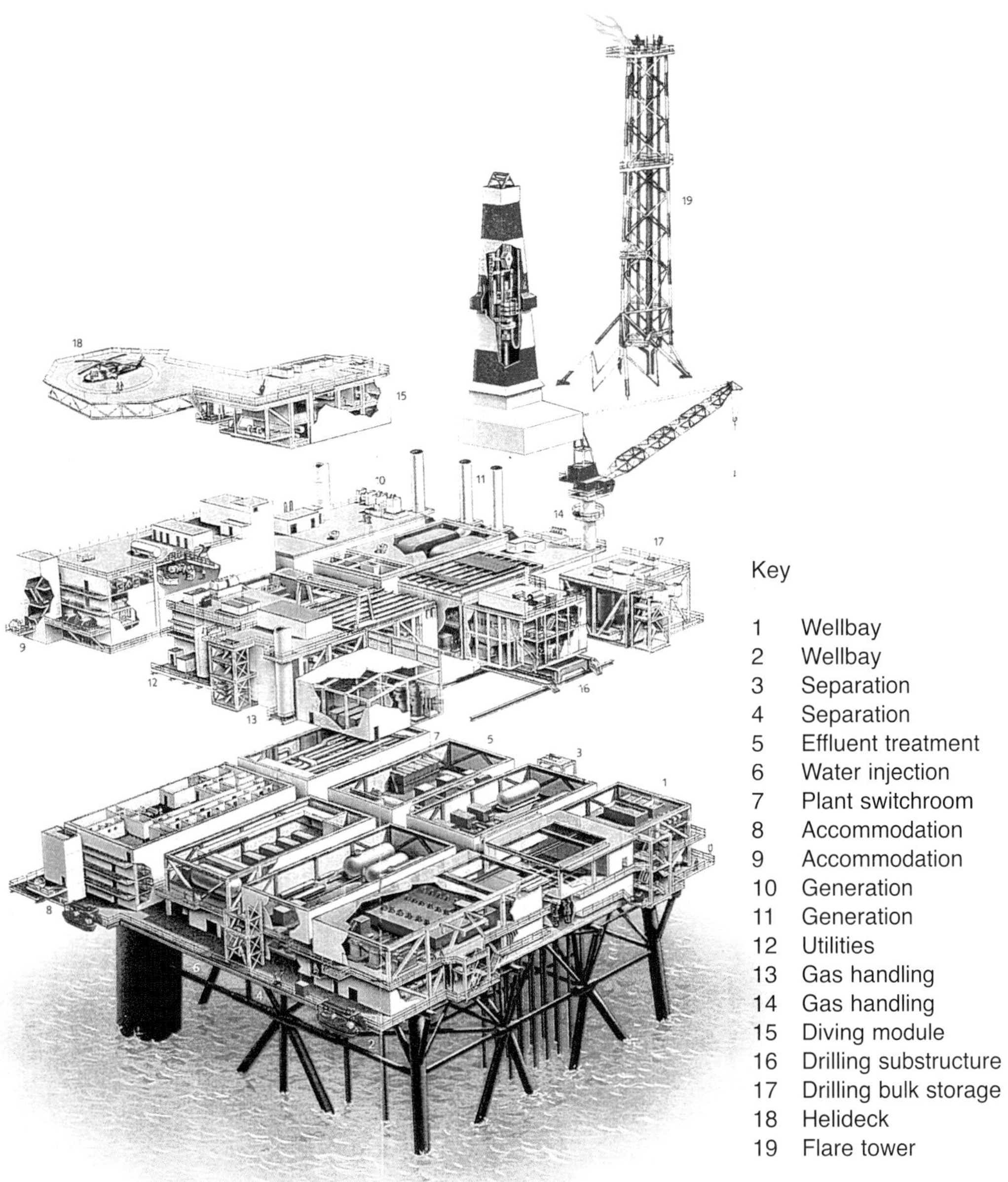

Plate 48 The Magnus platform: exploded view (Courtesy of British Petroleum Company plc)

Plate 49 Wellhead module on the Clyde platform (Courtesy of British Petroleum Company plc)

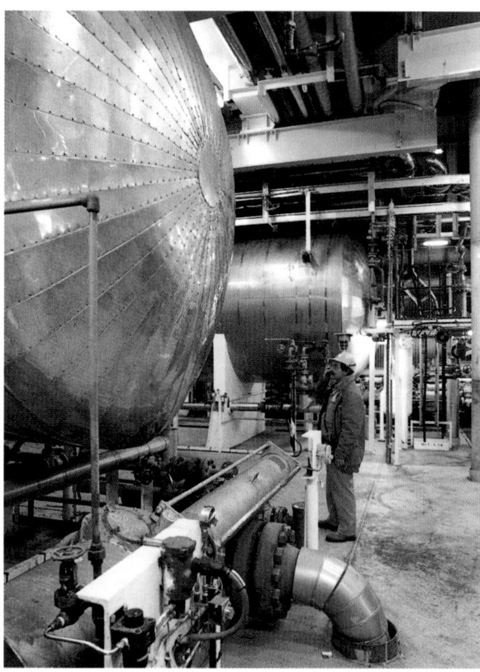

Plate 50 Separation module on the Magnus platform(Courtesy of British Petroleum Company plc)

Plate 51 Gas compression module on the Buchan Alpha platform (Courtesy of British Petroleum Company plc)

Plate 52 Main oil line pumps on the Clyde platform (Courtesy of British Petroleum Company plc)

Plate 53 Methanol injection facilities on a Southern North Sea platform (Courtesy of British Petroleum Company plc)

Plate 54 One of the three diesel fire pumps on the Magnus platform (Courtesy of British Petroleum Company plc)

Plate 55 Kaldair flares on a platform in the Forties field (Courtesy of British Petroleum Company plc)

Plate 56 A blowout preventer on the rig *Sea Conquest* (Courtesy of British Petroleum Company plc)

Plate 57 Piper Alpha, 6 July 1988 (Cullen, 1990): the Piper Alpha platform

Plate 58 Piper Alpha, 6 July 1988 (Cullen, 1990): model of C Module looking east (Courtesy of Borowski Image Communications Ltd)

Plate 59 Piper Alpha, 6 July 1988 (Cullen, 1990): fireball issuing from west side, taken some 15 seconds after the initial explosion (Courtesy of Mr C. A. Miller)

Plate 60 Piper Alpha, 6 July 1988 (Cullen, 1990): fires on the platform before riser rupture seen from east side (taken by Mr T. Vanus)

Plate 61 Piper Alpha, 6 July 1988 (Cullen, 1990): rupture of the Tartan riser (Courtesy of Mr A. Ritchie)

Plate 62 Piper Alpha, 6 July 1988 (Cullen, 1990): a fast rescue craft at the platform some time after the first riser rupture (Courtesy of Mr C. A. Miller)

Table 1.2 Some milestones in the development of loss prevention

1960	First Institution of Chemical Engineers symposium on Chemical Process Hazards with Special Reference to Plant Design
1966	Dow Chemical Company's Process Safety Manual
1967	First American Institute of Chemical Engineers symposium on Loss Prevention
1968	First ICI Safety Newsletter; American Insurance Association Hazard Survey of the Chemical and Allied Industries
1971	European Federation of Chemical Engineering symposium on Major Loss Prevention in the Process Industries
1972	Report of Robens Committee on Safety and Health at Work
1973	Institution of Chemical Engineers Loss Prevention Panel, information exchange scheme and, in 1975, Loss Prevention Bulletin
1974	Health and Safety at Work etc. Act 1974; First International Symposium on Loss Prevention and Safety Promotion in the Process Industries; Rasmussen Report; Flixborough disaster
1975	Flixborough Report
1976	First Report of Advisory Committee on Major Hazards; Seveso accident
1978	First Canvey Report; San Carlos disaster
1979	Second Report of Advisory Committee on Major Hazards; Three Mile Island accident
1981	Norwegian Guidelines for Safety Evaluation of Platform Conceptual Design
1982	EC Directive on Control of Industrial Major Accident Hazards
1984	Third Report of Advisory Committee on Major Hazards; Control of Industrial Accident Hazards Regulations 1984; Bhopal disaster; Mexico City disaster
1985	American Institute of Chemical Engineers Center for Chemical Process Safety
1986	Chernobyl disaster
1988	Piper Alpha disaster
1990	Piper Alpha Report
1992	Offshore Safety Act 1992; Offshore Installations (Safety Cases) Regulations 1992

emphasized the need for an approach to industrial safety more adapted to modern technology, and recommended self-regulation by industry as opposed to regulation from outside. This philosophy is embodied in the Health and Safety at Work etc. Act 1974 (HSWA), which provides the framework for such an approach. The Act does more, however, than this. It lays a definite statutory requirement on industry to assess its hazards and demonstrate the effectiveness of its safety systems. It is enforced by the Health and Safety Executive (HSE).

The disastrous explosion at Flixborough in 1974 has proved a watershed. Taken in conjunction with the Act, it has greatly raised the level of concern for safety and loss prevention in the industries affected. It also led, as mentioned, to the setting up of the ACMH. The accident at Seveso in 1976 has been equally influential. It had a profound impact in Continental Europe and was the stimulus for the development of the EC Directive on Control of Industrial Major Accident Hazards in 1982.

Further disasters such as those at San Carlos, Bhopal and Mexico City have reinforced these developments. The toxic release at Bhopal in India, which was by far the worst disaster ever to occur at a chemical plant and which involved an American company, has been a major contributor to a fundamental change of approach in the USA itself.

1.5 Large Single-Stream Plants

For some decades up to about 1980, there was a strong trend for the size of plants to increase. The problems associated with large, single-stream plants are a major reason for the development of loss prevention. These problems are now considered in more detail in order to illustrate some of the factors underlying its growth. Selected references on large, single-stream plants are given in Table 1.3.

The increase in the size of plant in the period in question for two principal chemicals is shown in Figure 1.3, which gives the size of the largest ethylene and ammonia plants built by a major contractor and the year in which they came on stream. Thus, whereas in 1962 an ethylene plant with a capacity in excess of 100 000 ton/year was exceptional, by 1969 this had become the minimum size ordered, whilst the newest UK plant, the No. 5 Unit of Imperial Chemical Industries (ICI), had a capacity of 450 000 ton/year. Although briefly the largest single-stream naphtha cracker in the world, it was soon overtaken by other plants of 500 000 ton/year capacity or more. Similar trends occurred in ammonia plants.

This increase in plant size took place in a rapidly expanding market. Thus the growth in world demand for

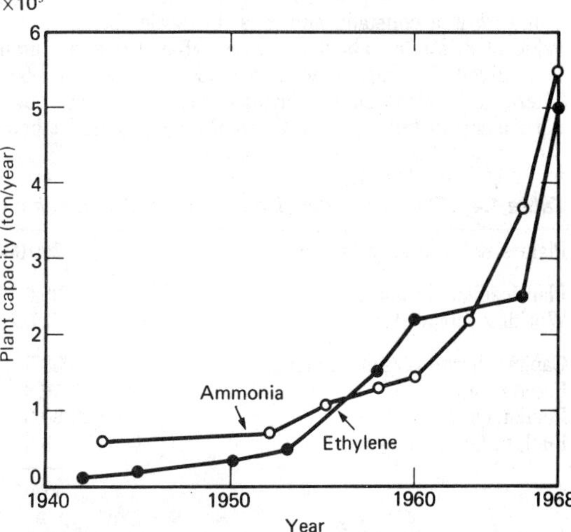

Figure 1.3 Capacity of ethylene and ammonia plants, 1940–68 (after Axelrod, Daze and Wickham, 1968) (Courtesy of the American Institute of Chemical Engineers)

Table 1.3 *Selected references on large, single-stream plants*

Frank and Lambrix (1966); Quigley (1966); Anon. (1967d); Davidson (1967); Holroyd (1967); R.L. Miller (1967); H.S. Robinson (1967a); Axelrod *et al.* (1968); Deschner *et al.* (1968); Mapstone (1968); Mayo (1968); S.P. Rose (1968); Lofthouse (1969); Chase (1970); Dailey (1970); W.E. Tucker and Cline (1970, 1971); Coulter and Morello (1971, 1972); Platten (1971); Walley and Robinson (1972); Knight (1973); Ball and Steward (1974); Huettner (1974); Woodhouse *et al.* (1974); Anon. (1975j); Baba and Kennedy (1976); Ball and Pearson (1976); Pilz and van Herck (1976); Ennis and Lesur (1977); Holland and Watson (1977); J.C. Davis (1978); Froment (1979); Hammock (1979); Anon. (1984nn); Remer and Chai (1990); Brennan (1992); Garnett and Patience (1993)

ethylene was given as follows (Walley and Robinson, 1972):

1965	8.5 megatonnes/year
1970	20 megatonnes/year
1975 (estimated)	39 megatonnes/year

The basic cause of the increase in size was concern for capital cost. Up to the early 1950s the chemical industry did not give minimization of capital cost particular priority. But with the move to naphtha feedstocks and the growth of petrochemicals, capital came to represent a much more significant cost than before compared with other costs such as raw materials and labour.

The relationship between size and capital cost is given by the well-known equation

$$C = kP^n \qquad [1.5.1]$$

where C is the capital cost of the plant, P its design output, k is a constant, and n is the scale-up index. The value of n is often about 0.6–0.7 so that if design output is doubled the capital cost increases by only 50–60%. There are also some savings on operating costs, particularly in terms of the thermal economy and labour.

Most calculations of the savings obtainable by building big showed the savings to be large; therefore, it was difficult in a competitive industry to avoid doing so, despite the recognized risks. The world price tended to be set by the larger plants, so that smaller plants became obsolete. Although calculations showed savings from large plants, there was considerable variation in the size of these savings, depending on the assumptions made. One figure given was a reduction in the cost of ethylene to about 50% in going from a 150 000 to a 450 000 ton/year plant (Holroyd, 1967); another calculation gave a reduction to about 80% in going from a 200 000 to a 400 000 ton/year plant (Lofthouse, 1969).

The calculation depends on many factors and is complex. This is particularly so for ethylene, which is only one of a number of products from an ethylene plant. The values of the propylene and butadiene produced were each about as great as that of the ethylene. Much therefore depended on whether these and other by-products could be sold at full or only at fuel value. A typical calculation of the effect of plant size and co-product credit on the economics of an ethylene plant at that time is shown in Table 1.4.

Another important factor for an ethylene plant is the acquisition of load. The best situation is where the ethylene is take away by gas pipeline. The alternative of sea transport, which involves liquefaction, is more expensive and may well cancel the cost advantage of a large plant. However, ethylene pipelines are difficult to justify except where there is a long-term contract, and it is not easy to create an ethylene grid into which other users can tap.

The economics of a large plant are also badly affected if it fails to reach full output immediately, either on account of commissioning difficulties or of lack of demand. A typical calculation of the effect of plant size, lateness and underloading on ammonia plant economics is shown in Table 1.5.

Despite these problems, the general assessment around, say, 1966 was that the building of large plants was justified. However, there continued to be a lively debate as to whether the expected economies of scale were realizable. Several difficulties became apparent. The economies of scale depend very largely on the retention of the single-stream philosophy. This gives economies

Table 1.4 *Effects of plant size and co-product credit on ethylene plant economics (after Walley and Robinson, 1972)*

Plant size (ton/year)	200 000	300 000	400 000	500 000
Plant capital (£m.)	25.0	33.0	40.0	47.0
Working capital (£m.)	2.0	2.9	3.8	4.6
Capital charge (£/ton ethylene)	33.5	29.8	27.5	25.9
Fixed costs	9.3	7.7	6.8	6.2
Feedstock	30.5	30.5	30.5	30.5
Fuel, utilities, catalysts	8.6	8.6	8.6	8.6
	81.9	76.6	73.4	71.2
Co-product credit (£/ton ethylene)	34.2	34.2	34.2	34.2
Ethylene cost (£/ton)	47.7	42.4	39.2	37.0
Co-product credit (£/ton ethylene)	34.2	29.1	26.6	25.1
Ethylene cost (£/ton)	47.7	47.5	46.8	46.1

Table 1.5 *Effects of plant size, lateness and underloading on ammonia plant economics (after Holroyd, 1967)*

Plant	Output pattern (% design capacity)	DCF return: No delay (%)	6 month delay (%)	12 month delay (%)
3 × 333 t/day suitably phased	Year 1 onwards 100%	4	—	—
	Year 1 60% 2 80% 3 onwards 100%	$\frac{1}{2}$	—	—
1 × 1000 t/day	Year 1 onwards 100%	26	23	21
	Year 1 60% 2 80% 3 onwards 100%	16	14	13
	Year 1 30% 2 80% 3 90% 4 onwards 100%	12	10	9
	Year 1 30% 2 70% 3 onwards 90%	7	6	5

not only in the cost of the equipment itself but of the associated pipework, heat exchangers, instrumentation, civil engineering, etc. These are largely lost if there is resort to duplication. However, in some cases the equipment appeared to be nearing the limits of size. This had long been so for the furnaces on an ethylene plant, which typically had multiple furnaces with outputs of about 30 000 ton/year. The main compressor was very large – that on ICI's No. 5 plant being 35 000 h.p. The main distillation column was also very big and had to be fabricated on site.

At large sizes the value of the scale-up index n is subject to some modification. A common value of n for ethylene plants was 0.65 (Lofthouse, 1969). However, as already mentioned, the cracking section of the plant, which accounts for about 30% of the capital cost, offers little scope for scale-up economies. Other items such as compressors, distillation columns and heat exchangers, may be near their limits and may have a higher index. Some facilities such as storage and effluent control may even have an index greater than 1.0. Thus it was suggested (Walley and Robinson, 1972) that a more realistic actual value of n is 0.9.

Storage is a serious problem with large plants. For the size of plant described, storage, whether of raw materials or products, is extremely expensive and has to be kept to a minimum.

The reliability of large plants was often unsatisfactory, there being a number of reasons for this. Compact layout made maintenance more difficult and increased the vulnerability to fire. Arrangements aimed at thermal economy increased the degree of interdependence in the plant. Economies were made in capital cost in areas such as materials of construction and duplication of equipment. As Holroyd (1967) comments:

> Faulty welding has resulted in leaky high pressure piping systems and there has been little improvement as regards jackets, dinners and footballs left in equipment despite the much more serious effect this sort of thing has in high cost, high capacity single plant units. There have been many examples of faulty equipment – faulty fabrication of interchangers, improper assembly of compressors and high pressure reactors which have failed under test. Complete shutdown of the plant due to failure of a simple piece of equipment, such as a pump of established design, and of negligible cost in itself, has been a frequent experience. Faulty supervision and human error in operation has shown up in not following the proper sequence of actions in emergencies, inadvertent tripping out of machines, failure to isolate equipment under maintenance and carelessness in checking instruments and with regard to such matters as water treatment.

As a result of such factors there were many cases of difficulties and delays in commissioning and operation of such plants throughout the world. A breakdown of the causes of such problems in ammonia plants is given in Table 1.6.

Large plants also have some undesirable features from the safety point of view. Particularly significant is the scale of the inventory. The amount of material in the main distillation column of a large ethylene plant exceeds

Table 1.6 *Causes of commissioning and operating problems in ammonia plants (after Holroyd, 1967)*

Faults	M.W. Kellogg experience (%)	ICI experience (%)
Design	10	10
Erection	20	16
Equipment	40	61
Operating	30	13

that formerly contained in the storage vessels of smaller plants. Plate 1 shows a distillation column under erection in the early 1990s and illustrates the scale of such items in modern plant.

Problems of pollution and noise, including flares and pressure relief, also appear to increase in severity rather rapidly for large plant sizes. Some more specific faults on ammonia plants have been quoted (Holroyd, 1967). One of these, migration of silica under the more severe operating conditions, is a matter of advanced technology. Most of the others are more mundane.

It is not surprising, therefore, that the trend to large scale plants was criticized and the economies of scale questioned. The debate was particularly lively over the period 1967–72. Even at the start of the debate the large, single-stream plant had its defenders (Holroyd, 1967; Lofthouse, 1969). The cost sensitivity and other problems of large plants were admitted. It was also conceded that the economies of scale were not as great as initial estimates suggested. But these factors were not considered sufficient to negate the economies of large plants or to make the return to small ones appear attractive.

It was agreed that part of the problem was that the increase in size of plants had been accompanied by a drive to reduce capital costs which may have been taken too far:

> This drive took many other forms besides increase of scale. It tended to lead to elimination of duplicates of even minor items of plant, to economies which sometimes proved to be false economies in the provision of services such as steam and power, to extreme sophistication in energy recovery which sometimes added so much to the complexity of the plants as to make them difficult to run, and to the concentration of the plant within very much smaller areas which increased their vulnerability to fire and also complicated maintenance. (Lofthouse, 1969)

Moreover, it was reasonable to claim that the industry had already learnt much from the troubles of the first generation of large plants and that it would be able to avoid many of these in the future.

In effect, the argument was that failures in large plants were not primarily due to the size of items of equipment, the single-stream arrangement or the use of high technology. It was conceded, however, and indeed emphasized, that the penalties of failures on large, single-stream plants are very great and that it is essential for their success to put maximum effort into ensuring high reliability and good operation. Particular emphasis was laid on the effectiveness of the operation and maintenance in large plants. Here the main factor is the quality of personnel at all levels: management, process operators and maintenance crews. Close supervision and rigorous inspection have an important part to play, but are no substitute for well-educated and well-trained people.

The large, single-stream plant offers substantial rewards for success, but this does not come easily. It can be achieved only by first class management and engineering in design and operation. The large plant places a premium on these factors and is thus a means by which a firm which possesses them gains a competitive advantage.

The growth in the size of plants has now slowed appreciably. The typical size of a new ethylene plant in 1978 was about 500 000 ton/year. This is still typical for a new plant world-wide. A plant for 700 000 ton/year still ranks among the largest plants (Mahoney, 1990). There is no marked trend to yet larger plants, but equally there is little sign of reversion to smaller ones.

1.6 Loss Prevention – 2

The area of concern and the type of approach which goes by the name of 'loss prevention' is a development of safety work. But it is a response to a changing situation and need, and it has certain particular characteristics and emphases. The essential problem which loss prevention addresses is the scale, depth and pace of technology. The effect of these features is to increase the size of the hazards, to make their control more difficult and to reduce the appropriateness of learning by trial and error. In fact, control of such hazards is possible only through effective management. The prime emphasis in loss prevention is, therefore, on the management system. This has always been true, of course, with regard to safety. But high technology systems are particularly demanding in terms of formal management organization, competent persons, systems and procedures, and standards and codes of practice.

The method of approach is also somewhat different. When systems are small scale and relatively unhazardous and change slowly over the years, they are able to evolve by trial and error. This is often simply not possible with modern systems, where the pace of change is too fast and the penalties of failure are too great. It becomes necessary to apply forethought to try to ensure that the system is right first time. In a related area the widespread application of human factors to large man–machine systems has been motivated by very similar considerations. A major hazard in a modern process plant usually materializes due to loss of containment. The three big major hazards are fire, explosion and toxic release.

Thus loss prevention is characterized by

(1) an emphasis on management and management systems, particularly for technology;
(2) a concern with hazards arising from technology;
(3) a concern with major hazards;
(4) a concern for integrity of containment;
(5) a systems rather than a trial-and-error approach.

Some other features which are characteristic of loss prevention are the use of

(6) techniques for identification of hazards;
(7) a quantitative approach to hazards;
(8) quantitative assessment of hazards and their evaluation against risk criteria;
(9) techniques of reliability engineering;
(10) the principle of independence in critical assessments and inspections;
(11) planning for emergencies;
(12) incident investigation;

together with

(13) a critique of traditional practices or existing regulations, standards or codes where these appear outdated by technological change.

The identification of hazards is obviously important, since the battle is often half won if the hazard is recognized. A number of new and effective techniques have been developed for identifying hazards at different stages of a project. These include hazard indices, chemicals screening, hazard and operability studies, and plant safety audits.

Basic to loss prevention is a quantitative approach, which seeks to make a quantitative assessment, however elementary. This has many parallels with the early development of operational research. This quantitative approach is embodied in the use of quantitative risk assessment (QRA). The assessment produces numerical values of the risk involved. These risks are then evaluated against risk criteria. However, the production of numerical risk values is not the only, or even the most important, aspect. A QRA necessarily involves a thorough examination of the design and operation of the system. It lays bare the underlying assumptions and the conditions which must be met for success, and usually reveals possible alternative approaches. It is therefore an aid to decision-making on risk, the value of which goes far beyond the risk numbers obtained.

Reliability engineering is now a well-developed discipline. Loss prevention makes extensive use of the techniques of reliability engineering. It also uses other types of probabilistic calculation which are not usually included in conventional treatments of reliability, such as probabilities of weather conditions or effectiveness of evacuation. Certain aspects of a system may be particularly critical and may require an independent check. Examples are independent assessment of the reliability of protective systems, independent audit of plant safety, and independent inspection of pressure vessels.

Planning for emergencies is a prominent feature of loss prevention work. This includes both works and transport emergencies.

Investigation of incidents plays an important part in loss prevention. Frequently there is some aspect of technology involved. But the recurring theme is the responsibility of management.

The loss prevention approach takes a critical view of existing regulations, standards, rules or traditional practices where these appear to be outdated by changing technology. Illustrations are criticisms of incident reporting requirements and of requirements for protection of pressure vessels. These developments taken as a whole do constitute a new approach and it is this which characterizes loss prevention.

It might perhaps be inferred from the foregoing that the problems which have received special emphasis in loss prevention are regarded somehow as more important than the aspects, particularly personal accidents, with which traditionally safety work in the process industries has been largely concerned. Nothing could be further from the truth. It cannot be too strongly emphasized that mundane accidents are responsible for many more injuries than and as many deaths as those arising from high technology.

1.7 Total Loss Control

It is now necessary to consider some other developments which have contributed to the modern approach to safety and loss prevention. The first of these is total loss control. The basic concept underlying total loss control is that loss due to personal accident and injury is only the tip of the iceberg of the full loss arising from incidents. It follows that attention should be paid, and controls applied, to all losses. Early work in this area is described in *Industrial Accident Prevention* (Heinrich, 1959) and by Bird (1966). Accounts are given in *Total Loss Control* (J.A. Fletcher and Douglas, 1971) and *Practical Loss Control Leadership* (Bird and Germain, 1985).

Total loss control, like loss prevention, is concerned with losses associated with hazards and accidents or other incidents. It is not concerned with losses which do not have a hazard element. It may be seen, therefore, as an extension of the activity of the safety manager. The attraction of such a change of emphasis is that the safety manager becomes involved in a major cost area which is of concern to other managers and thus increases his influence, with consequent benefit to safety.

The ratio between different types of accident is a key concept in loss control. Early work on this was done by Heinrich (1959), who gave the following ratio for the different types of accident:

Major or lost time injury/Minor injury/No injury = 1:29:300

These ratios are frequently shown in the form of the accident pyramid illustrated in Figure 1.4.

The numbers of accidents on which the above ratios were based were evidently fairly limited. Later studies have been done involving much larger numbers. Bird and Germain (1985) report a study of some 1.75 million accidents from 297 co-operating organizations in which the ratio was:

Disabling injury/Minor injury/Property damage/No injury or damage = 1:10:30:600

Tye and Pearson (HSE, 1991b) give the results of an investigation of almost 1 million accidents:

Disabling injury /minor injury/First-aid injury/Property damage/No injury or damage = 1:3:50:80:400

The accident definitions and ratios tend to vary between different studies, but the relationship between the

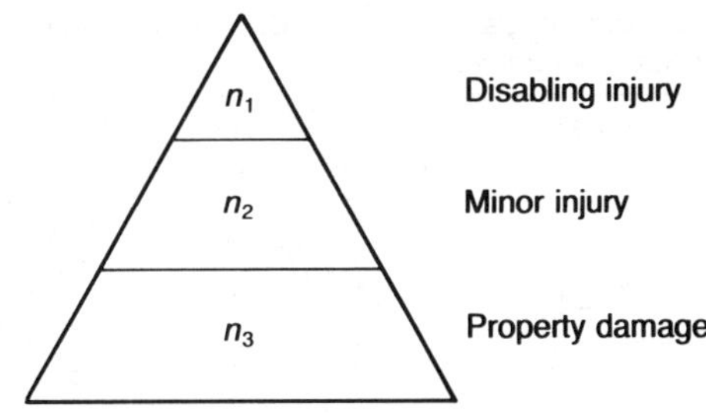

Figure 1.4 *The accident pyramid (after Bird and Germain, 1966)*

different kinds of event is usually consistent in a given study.

In another study, *The Costs of Accidents at Work*, (HSE,1993a), an investigation was made of accidents at a creamery, a construction site, a transport company, a hospital and an oil platform. For the oil platform, the most relevant here, the ratios obtained were as follows (the values given in parentheses being those for all the studies except the construction site):

Over 3-day injury/Minor injury/Non-injury accident = 1:4(7):126(189)

Another ratio quoted by Heinrich is the ratio of uninsured to insured costs of accidents, which he gives as about 4:1. However, J.A. Fletcher and Douglas (1971) report studies in which this ratio ranges from 2.3 to 101 and suggest that it is necessary to investigate this in any particular work places.

An illustration, given by Fletcher and Douglas, of a typical situation in a medium sized factory is shown in Table 1.7. The actual numbers of disabling and minor injury accidents were as indicated, while the number of property-damage accidents was computed from the number of disabling accidents using Bird's ratio of 500. In this case the insured costs were the workmen's compensation costs of US$208 300, while, as shown, the uninsured costs were US$1 273 518, of which the major part was property-damage costs. The ratio of uninsured to insured costs was about 6.1.

The relative importance of the damage costs is enough to justify paying attention to damage accidents. But there is the further point that assessment of such costs is straightforward compared with the assessment of injury costs, which frequently involves some rather dubious assumptions. The policy suggested, therefore, is one of monitoring not only injury accidents but also damage accidents. For this it is necessary to have damage control centres which give general information on damage accidents in much the same way as first-aid points and medical centres generate information on injury accidents. The existing maintenance system can normally be adapted for this purpose.

As an illustration of the application of such a policy J.A. Fletcher and Douglas (1971) describe a study on the reporting of damage accidents by crane drivers. In the 3 years prior to the institution of the policy, the number of damage accidents reported was rather less than 25 per year. During the 10 years after its institution some 1643 accidents per year were reported. Reporting was enforced with an average of nine warnings and three work suspensions per year. Although many accidents were of a minor nature, many others revealed the need for prompt action to avert a more serious accident.

Total loss control areas are listed as: (1) business interruption; (2) injury; (3) property damage; (4) fire; (5) security; (6) health hygiene; (7) pollution; and (8) product liability. Each of these areas is treated as follows:

(1) identification – of possible loss producing situations;
(2) measurement – of such losses;
(3) selection – of methods to minimize loss;
(4) implementation – of methods within the capability of the organization.

The application of total loss control in the chemical industry has been described by Webster (1974), Hearfield (1975) and Ling (1976, 1979).

In the company described by Hearfield, incidents reported in 1 year on a particular plant cost about £25 000, including consequential losses. It was estimated that if all the unreported incidents had been included, this figure might have been double. A total loss control programme was instituted, which revealed that the estimated loss in the works was about £350 000. One source of loss on the plant was associated with steam condensate systems. Following an incident involving condensate, a full plant investigation of the use of condensate and demineralized water was undertaken. A 20% deficiency in the demineralized water balance was reported. It was found that the orifice plate was completely missing from one of the water flowmeters. The investigation was extended to the works. Expenditure of £40 000 was incurred on pipework and instrumentation modification. In the first year of operation savings of £130 000 were made due to reduced consumption of demineralized water, and capital expenditure of £200 000 on new demineralization plant was deferred.

Table 1.7 *Losses in a medium sized factory (after J.A. Fletcher and Douglas, 1971) (Courtesy of Associated Business Programmes Ltd)*

Type of accident	*No. of accidents per year*	*Cost per accident (US$)*	*Total uninsured cost (US$)*
Disabling injury	71	52	3 692
Minor injury	10 122	3.86	39 032
Property damage (est.)	35 500	34.67	1 230 794
Total			1 273 518

1.8 Quality Assurance

Another development which shapes the modern approach to safety and loss prevention is quality assurance and total quality management. Methods of ensuring that a product meets the required quality standards are the province of quality control and quality assurance. The first of these terms in particular tends to be used with different meanings.

Quality control of products has long been standard practice in a range of industries, from those manufacturing cars to those making ice cream. In many cases the emphasis in such quality control has traditionally been on inspection of the product.

The creation of systems to ensure that the product meets the required standards is the role of *quality assurance*. These systems are applied not only to the intermediate and final products made within the company but also to the inputs, raw materials and products, purchased from outside, since unless these meet their specifications, it may be impossible for the company to meet its own quality standards.

The 1980s have seen a strong movement in industry world-wide to adopt quality assurance systems and to seek accreditation to a recognized standard. The international standard is ISO 9000 and the corresponding British Standard is BS 5750: 1987: *Quality Systems.* As just indicated, a move by one firm to seek accreditation creates a chain reaction which obliges its suppliers to do likewise. Adoption of quality assurance involves the creation, and documentation, of a set of systems designed to ensure quality outputs from all the activities of the company.

1.9 Total Quality Management

A related, but not identical, development is total quality management (TQM). The concept underlying TQM is that the problem of failures and their effects has an influence on company performance which is far greater than is generally appreciated, defining failure in a broad way. TQM has its origins in quality control on production lines. It has spread to industry generally, including the process industries.

Most accounts of TQM highlight the role played by a number of individuals who have been influential in its development. Some principal texts are *Quality is Free* (Crosby, 1979), *Quality Control Handbook* (Juran, 1979), *Quality Planning and Analysis* (Juran and Gryna, 1980), *Quality, Productivity and Competitive Position* (Deming, 1982) and *Out of the Crisis* (Deming, 1986). An overview, including an account of the different schools, is given in *Total Quality Management* (Oakland, 1989).

The TQM approach has been pioneered in Japan, particularly in the motor industry. Accounts of the Japanese approach are given by Ishikawa Kaoru (1976, 1985), Taguchi (1979, 1981) and Singo Shigeo (1986).

In many countries industry is well advanced with the adoption of TQM concepts. World-wide this typically involves implementation of the requirements of ISO 9000. Guidance on this process is given in *ISO 9000* (P.L. Johnson, 1993). In the UK, the relevant standard to which organizations are accredited is BS 5750:1987. Guidance is given in *Implementing BS 5750* (Holmes, 1991). A guide to the associated audit is *Quality Management System Audit* (C.A. Moore, 1992).

As stated earlier, the basic concept of TQM is that failures of various kinds have effects which are much more widespread, damaging and costly than has been generally appreciated. Failures have effects both internal and external to the company. The internal effects experienced by the company include loss of production, waste of materials, damage to equipment, and inefficient use of manpower. The external effects experienced by the customer include the same features. Failures undermine the competitiveness of both parties.

The starting point in tackling this problem is to review the product which is required and the system which is to produce it. The requirement for the product is 'fitness for purpose'. In determining fitness for purpose it is necessary to consider both the effectiveness of the product whilst it is operating and its reliability of operation. In order to achieve fitness for purpose it is necessary to ensure that the design of the product is suitable and that the product made conforms to that design. These are two separate aspects and success depends on getting them both right. It is of little use to make a product which conforms well with the design if the design is defective, or to make a product which has a good design but which is produced in such a way that it does not conform to that design.

Given a product which is well designed, it is necessary to consider the system which will deliver conformability to the design. Traditionally, industry has placed much emphasis on an approach to quality control based on inspection. The approach taken in TQM is radically different. The attempt to inspect quality into an inherently defective production process is regarded as ineffective. Instead the emphasis is on prevention. The basic question asked therefore, is whether the production process is actually capable of producing to the quality required. In many cases it has been concluded that it was not. Attempts were being made to deal with the situation by intensifying the inspection effort, attempts which were largely futile.

The spirit of the TQM approach has been summarized in *Right First Time* (F. Price, 1985) as:

(1) Can we make it OK?
(2) Are we making it OK?
(3) Have we made it OK?
(4) Could we make it better?

The TQM approach seeks to root out failures in all aspects of the company's operation, not just failures of equipment but also in all aspects of the company's operation, including systems, documentation, communications, purchasing and maintenance. TQM is therefore concerned with both products and activities. In assessing the quality of a product or activity it seeks to identify the 'customers' and to make sure that their requirements are properly defined and met. An important technique for reducing failures is the involvement of the workforce, who are encouraged to not allow failures to persist but to report them and make proposals for their elimination.

The prime responsibility for TQM lies with management, and management leadership is essential to its success. It is management which is responsible for the necessary features such as organization, personnel, systems, design, planning and training. All employees are involved, however, in dealing with failure. An important part of TQM, therefore, is the involvement and motivation of the workforce. The quality circles (QCs) developed in Japan are an example of this. Another aspect of TQM is the 'just-in-time' (JIT) approach developed in Japan, notably in the motor industry. Each section of the production line operates with little or no inventory of input components, but takes them as required from the upstream supplier. This system requires an intolerance of failure, with rapid detection and rectification.

One way of expressing this intolerance is to adopt 'zero defects' as the performance standard, as advocated by Crosby (1979). In the variant of TQM given by Crosby, the overall approach adopted is summarized as:

Definition	Conformance to requirements
System	Prevention
Performance standard	Zero defects
Measurement	Price of non-conformance

It will be apparent that, insofar as the process industries are concerned, this emphasis on, and intolerance of, failures constitutes an approach very close to that of loss prevention.

Accounts of the relationship between quality management and management of safety, health and environment have been given by Berkey, Dowd and Jones (1993), Rooney, Smith and Arendt (1993) and Olsen (1994).

1.10 Risk Management

Another related development is that of risk management. Any industrial project involves risk. But some of the developments described above, such as the increase in the technological dimension and the growth of public concern over safety and pollution, have tended to introduce further dimensions of unpredictability. Risk management addresses these latter risks and provides a means of assessing and managing them along with the normal commercial risks.

Accounts of risk management are given in *Managing Risks* (Grose, 1987), *Risk Management* (Jardine Insurance Brokers Ltd, 1988), *Managing Risk* (Bannister, 1989), *Risk Assessment and Risk Management for the Chemical Process Industry* (Greenberg and Cramer, 1991) and *Reliability, Safety and Risk Management* (S.J. Cox and Tait, 1991), and by Turney (1990a,b) and Frohlich and Rosen (1992b).

The variety of risks and the vulnerability of a business to these risks is now such that they need to be managed explicitly. Individual risks are recognized by line managers, but are frequently not fully addressed by them. There is a tendency to defer consideration and to take too optimistic a view. In any case, the individual line manager sees only part of the picture. It is for such reasons that risk management has emerged as a discipline in its own right. A systematic review of hazards is now normal practice in the management of projects in the process industries. A complementary system for the review of the totality of risks (commercial, legal and technological) is increasingly common.

1.11 Safety-Critical Systems

Another concept which is gaining increasing prominence is that of safety-critical systems. These are the systems critical to the safe operation of some larger system, whether this be a nuclear power station or a vehicle. In a modern aircraft, particularly of the fly-by-wire type, the computer system is safety critical. An account of safety critical systems is given by P.A. Bennett (1991a,b).

1.12 Environment

Another major concern of the process industries is the protection of the environment. Developments in environmental protection (EP) have run in parallel with those in safety and loss prevention (SLP). These two aspects of process plant design and operation have much in common. In recent years there has been a trend to assign to the same person responsibilities for both. There are also some situations where there is a potential conflict between the two. The environment, and pollution of the environment, are considered in Appendix 11.

1.13 Professional Institutions and Bodies

The professional engineering institutions have responded to the problems described above with a number of initiatives.

As stated earlier, in 1973 the IChemE created a Loss Prevention Panel. The Institution itself publishes a range of monographs and books on safety and loss prevention and the panel publishes the *Loss Prevention Bulletin* and a range of aids for teaching and training.

In 1985 the American Institute of Chemical Engineers (AIChE) formed the Center for Chemical Process Safety (CCPS). The Center publishes a series of guidelines on safety and loss prevention issues.

At the European level, the European Process Safety Centre (EPSC) was set up in 1992 to disseminate information on safety matters, including legislation, research and development, and education and training (EPSC, 1993).

The work of these and other bodies is described in Chapter 27. Much of the material referred to in this book derives from these sources.

1.14 Responsible Care

In a number of countries, the chemical industry has responded to safety, health and environmental concerns with the Responsible Care initiative, which was developed in the early 1980s by the Canadian Chemical Producers Association and was then taken up in 1988 in the USA by the Chemical Manufacturers Association (CMA), in 1989 in the UK by the Chemical Industries Association (CIA), and elsewhere.

Companies participating in Responsible Care commit themselves to achieving certain standards in terms of safety, health and environment.

Guidance is given in *Responsible Care* (CIA, 1992 RC53) and *Responsible Care Management Systems* (CIA, 1992 RC51). An account of the development of Responsible Care is given by Jacob (1992). The Canadian perspective is described by Buzzelli (1990) and Creedy (1990).

1.15 Overview

The modern approach to the avoidance of injury and loss in the process industries is the outcome of the various developments just described. Central to this approach is leadership by management, starting with senior management. Such leadership is an indispensable condition for success. It is not, however, a sufficient condition. Management must also identify the right objectives. The contribution of total loss control, quality assurance and total quality management is to identify as key management objectives the elimination of failures of all kinds and the conduct of activities so that they are satisfactory to those affected by them. These disciplines provide in quality circles a tool for meeting these objectives.

As far as the process industries are concerned, it is the contribution of loss prevention to handle the technological dimension and to provide methods by which failure is eliminated. In the modern approach to safety and loss prevention these themes come together. The ends are the safety of personnel and the avoidance of loss. The means to achieve both these aims is leadership by management, informed by an understanding of the technology and directed to elimination of failures of all kinds.

2 Hazard, Accident and Loss

A rational approach to loss prevention must be based on an understanding of the nature of accidents and of the types of loss which actually occur. Therefore, in this chapter, first the nature of the accident process is considered and then the accident and loss statistics are reviewed to give an indication of the problem. Selected references on accident and loss experience are given in Table 2.1. In addition, many other tables of data are given in other chapters. Cross-references to some of these tables are given in Table 2.2.

Table 2.1 *Selected references on accident and loss experience*

Natural and man-made hazards, disasters
Thygerson (1972, 1977); Walker (1973); G.F. White (1974); Bignell *et al.* (1977); Münchener Rück (1978); B.A. Turner (1978); ASCE (1979/9); Ferrara (1979); Whittow (1980); Perry (1981); Rossi *et al.* (1983); Simkim and Fiske (1983); Perrow (1984); Wijkman and Timberlake (1984); McWhirter (1985); Cairns (1986); Sir R. Jackson (1986); E.A. Bryant (1991); Guinness Publishing Co. (1991); K. Smith (1992); R. Smith (1992)

Process hazards, accidents
Matheson (1960); Vervalin (1964a, 1973a); BCISC (1968/7); Fowler and Spiegelman (1968); W.H. Doyle (1969); Spiegelman (1969, 1980); CIA (1970/3); Cornett and Jones (1970); Rasbash (1970b); Houston (1971); H.D. Taylor and Redpath (1971, 1972); R.L. Browning (1973); Walker (1973); FPA (1974b, 1976); AEC (1975); N. McWhirter (1976); J.R. Nash (1976); McIntire (1977); AIA (1979); Harvey (1979); Carson and Mumford (1979); Ferrara (1979); R. King and Magid (1979); Kletz and Turner (1979); Lees (1980); Pastorini *et al.* (1980); Mance (1984); Manuele (1984 LPB 58); Hawkins (1985); D. McWhirter (1985); APCA (1986); Kletz (1986b); V.C. Marshall (1986a, 1988c); Garrison (1988a,b); Instone (1989); Mahoney (1990); Anon. (1991 LPB 99, p. 1); O'Donovan (1991 LPB 99); K.N. Palmer (1991 LPB 99); Guinness Publishing Co. (1991); Marsh and McLennan (1992); Pastorini *et al.* (1992); Bisio (1993); Crooks (1994 LPB 115)

Accident models
Surry (1969b); Houston (1971); Macdonald (1972); Haddon (1973a, b); W.G. Johnson (1973a,b, 1980); de Jong (1980); Rasmussen (1982b, 1983); Haastrup (1983); Benner (1984); A.R. Hale and Glendon (1987); Wells *et al.* (1991); Bond (1994 LPB 120)

Accident ranking
Keller *et al.* (1990); Keller and Wilson (1991)

Annual or periodic reports, statistical summaries
AGA (Appendix 28 *Pipeline Incident Reports*); API (Appendix *28 Annual Summaries*); BIA (annual report); BRE (annual statistics, Appendix *28 UK Fire and Loss Statistics*); FPA (annual report); HM Chief Inspector of Explosives (annual report); HM Chief Inspector of Factories (annual report, annual analysis of accidents); HM Senior Electrical Inspector of Factories (annual report); M&M Protection Consultants (periodic); NTSB (Appendix 28); NSC (n.d./1); ABCM (1930–64/2); MCA (1962–/1–4, 1971–/20, 1975–/23); BCISC (1965–/4); CISHC (1975–/5); CONCAWE (1977 9/77, 1992 4/92); HSC (1977); HSE (1977d, 1986c, 1992b); ILO (1992)

Fire
BRE (annual statistics); FPA (annual report, 1974, 1976, 1991); W.H. Doyle (1969); Spiegelman (1969, 1980); H.D. Taylor and Redpath (1971, 1972); FRS (1972 Fire Research Note 920); P. Nash (1972b) Vervalin (1963a, 1972c, 1973a, 1974a, 1975c, 1976b, 1977, 1978a,b, 1986b); Duff (1975); Redpath (1976); Rutstein (1979a, b); Rutstein and Clarke (1979); Banks and Rardin (1982); Norstrom (1982a); Gebhardt (1984); Uehara and Hasegawa (1986); Mahoney (1990); Home Office (1992)

Explosion
Eggleston (1967); Doyle (1969); Spiegelman (1969, 1980); Duff (1975); Davenport (1977b, 1981b); Norstrom (1982b); Uehara and Hasegawa (1986); Vervalin (1986b); Mahoney (1990); Lenoir and Davenport (1993)

Refineries
Anon. (1970a); McFatter (1972); W.L. Nelson (1974); McIntire (1977); Mahoney (1990)

Ammonia plants
Holroyd (1967); Axelrod *et al.* (1968); Sawyer *et al.* (1972); G.P. Williams and Sawyer (1974); G.P. Williams (1978); G.P. Williams and Hoehing (1983); G.P. Williams *et al.* (1987)

Educational institutions
Bowes (1985)

Table 2.2 *Cross-references to other accident and loss data*

Major fires	Section 16.38
Major condensed phase explosions	Table 17.24; Appendix 1
Major vapour cloud explosions (VCEs)	Section 17.28; Table 17.30
Major boiling liquid expanding vapour explosions (BLEVEs)	Section 17.29 Table 17.37
Major missile incidents	Section 17.34 Table 17.49
Major dust explosions	Section 17.43; Table 17.63
Major toxic releases	Section 18.27; Tables 18.30, 18.31
Case histories	Appendix 1
Failure data	Appendix 14

2.1 The Accident Process

There are certain themes which recur in the investigation of accidents and which reveal much about the accident process. First, although in some reporting schemes the investigator is required to determine the cause of the accident, it frequently appears meaningless to assign a single cause as the accident has arisen from a particular combination of circumstances. Second, it is often found that the accident has been preceded by other incidents which have been 'near misses'. These are cases where most but not all of the conditions for the accident were met. A third characteristic of accidents is that when the critical event has occurred there are wide variations in the consequences. In one case there may be no injury or damage, whilst in another case which is similar in most respects there is some key circumstance which results in severe loss of life or property. These and other features of accidents are discussed in *Man-Made Disasters* (B.A. Turner, 1978). It is in the nature of disasters that they tend to occur only as the result of the combination of a number of events and to have a long incubation period before such a conjunction occurs.

It is helpful to model the accident process in order to understand more clearly the factors which contribute to accidents and the steps which can be taken to avoid them. One type of model, discussed by Houston (1971), is the classical one developed by lawyers and insurers which focuses attention on the 'proximate cause'. It is recognized that many factors contribute to an accident, but for practical, and particularly for legal, purposes a principal cause is identified. This approach has a number of defects: there is no objective criterion for distinguishing the principal cause; the relationships between causes are not explained; and there is no way of knowing if the cause list is complete.

There is need for accident models which bring out with greater clarity the common pattern in accidents. Some models of the accident process which may be helpful in accident investigation and prevention are given below, with emphasis on the management and engineering aspects. Further accident models are discussed in Chapters 26–28.

2.1.1 The Houston model

The model given by Houston (1971, 1977) is shown schematically in Figure 2.1. Three input factors are necessary for the accident to occur: (1) target, (2) driving force, and (3) trigger. Principal driving forces are energy and toxins. The target has a threshold intensity θ below which the driving force has no effect. The trigger also has a threshold level θ' below which it does not operate.

The development of the accident is determined by a number of parameters. The contact probability p is the probability that all the necessary input factors are present. The contact efficiency ϵ defines the fraction of the driving force which actually reaches the target, and the contact effectiveness η is the ratio of damage done to the target under the actual conditions to that done under standard conditions. The contact time t is the duration of the process.

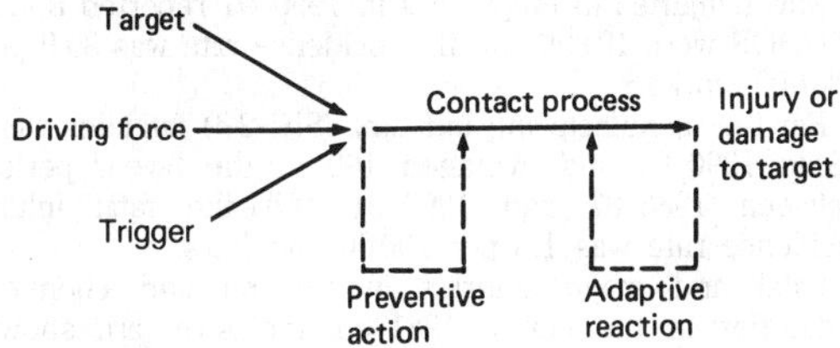

Figure 2.1 *Houston model of the accident process (after Houston, 1977)*

Table 2.3 *Some energy control strategies (after Haddon, 1973a)*

1. To prevent the initial marshalling of the form of energy
2. To reduce the amount of energy marshalled
3. To prevent the release of energy
4. To modify the rate or spatial distribution of release of energy from its source
5. To separate in space or time the energy being released from the susceptible structure
6. To separate the energy being released from the susceptible structure by interposition of a material barrier
7. To modify the contact surface, subsurface, or basic structure which can be impacted
8. To strengthen the living or non-living structure which might be damaged by energy transfer
9. To move rapidly in detection and evaluation of damage and to counter its continuation and extension
10. All those measures which fall between the emergency period following the damaging energy exchange and the final stabilization of the process (including intermediate and long-term reparative rehabilitative measures)

The model indicates a number of ways in which the probability or severity of the accident may be reduced. One of the input factors (target, driving force or trigger) may be removed. The contact probability may be minimized by preventive action. The contact efficiency and contact effectiveness may be reduced by adaptive reaction.

Work by Haddon (1973a,b) emphasizes prevention of accidents by control of the energy. His list of energy control strategies is given in Table 2.3. Failure of one or more of these modes of control is a normal feature of an accident and hence of accident models.

2.1.2 The fault tree model

A simple fault tree model of an accident is given in Figure 2.2. An initiating event occurs which constitutes a potential accident, but often only if some enabling event occurs, or has already occurred. This part of the tree is the 'demand' tree, since it puts a demand on the protective features. The potential accident is realized only if prevention by protective equipment and human action fails. An accident occurs which develops into a more severe accident only if mitigation fails.

A somewhat similar model has been proposed by Wells *et al.* (1992).

2.1.3 The MORT model

A more complex fault tree model is that used in the management oversight and risk tree (MORT) developed by W.G. Johnson (1980) and shown in Figure 2.3. This tree is the basis of a complete safety system, which is described further in Chapter 28.

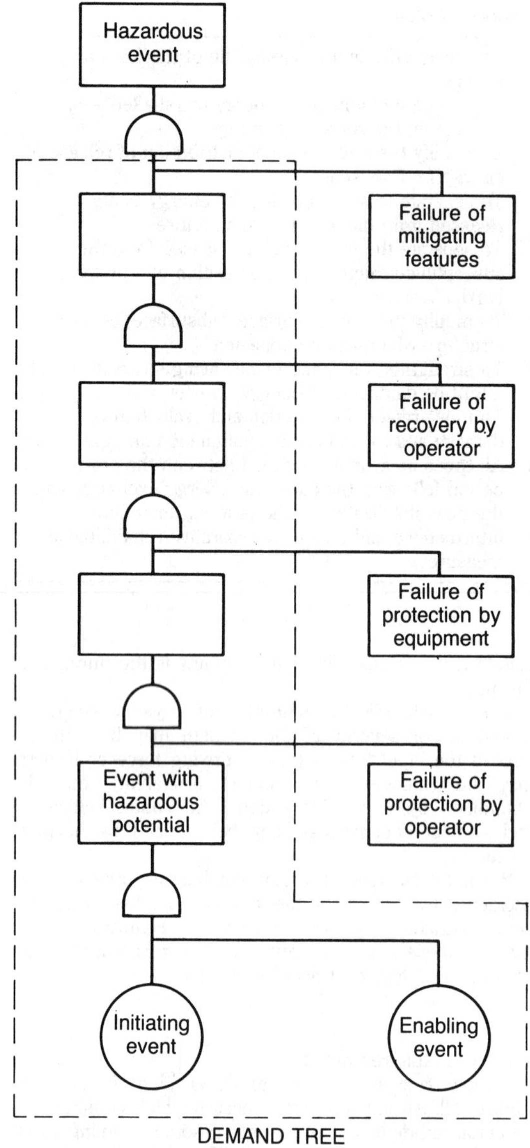

Figure 2.2 *Fault tree model of the accident process*

2.1.4 The Rasmussen model

Accident models which show the role of human error have been developed by Rasmussen (1982a,b). Figure 2.4 shows such a model. The role of human error in causing accidents is considered in more detail in Chapter 14.

2.1.5 The ACSNI model

Figure 2.5 shows a model proposed by the Advisory Committee on the Safety of Nuclear Installations (ACSNI, 1991). The model provides a general framework which can be used to identify latent failures that are likely to lead to critical errors.

2.1.6 The Bellamy and Geyer model

A model which emphasizes the broader, socio-technical background to accidents has been developed by Geyer and Bellamy (1991) as shown in Figure 2.6. Figure 2.6(a) gives the generic model and Figure 2.6(b) shows the application of the model to a refinery incident.

2.1.7 The Kletz model

Another approach is that taken by Kletz (1988h), who has developed a model oriented to accident investigation. The model is based essentially on the sequence of decisions and actions which lead up to an accident, and shows against each step the recommendations arising from the investigation. An example is shown in Figure 2.7, which refers to an incident involving a small fire on a pump.

2.2 Standard Industrial Classification

Statistics of injuries and damage in the UK are generally classified according to the Standard Industrial Classification 1980 (SIC 80). In this classification the classes relevant here are: Class 1 Energy and water industries; Class 2 Extraction of minerals and ores other than fuels, manufacture of metals, mineral products and chemicals; Class 3 Metal goods, engineering and vehicles industries; and Class 4 Other manufacturing industries. The mineral oil processing industry falls into Class 1, Subclass 14, and the chemical industry in Class 2, Subclass 25.

2.3 Injury Statistics

Accident statistics are available in the *Annual Report* of HM Chief Inspector of Factories (HMCIF), or its current equivalent, and the annual *Health and Safety Statistics.* The former also gives occasional detailed studies for particular industrial sectors.

The definition of a major injury changed with the introduction of the Reporting of Injuries, Diseases and Dangerous Occurrences Regulations 1985 (RIDDOR). *The Health and Safety Statistics 1990-91* (HSE, 1992b) show that in 1990–91 there were 572 fatalities reported under RIDDOR, of which 346 were to employees, 87 to the self-employed and 139 to members of the public. The fatal injury incidence rate for employees was 1.6 per 100 000 workers.

Major injuries to employees in 1990–91 reported under RIDDOR were 19 896 and the incidence rate was 89.9 per 100 000 workers.

For the manufacturing industry (SIC 2-4) fatalities were 88 in 1990–91 and averaged 100 in the 5-year period between 1986–87 and 1990–91 and the fatal injury incidence rate was 1.8 per 100 000 workers.

Fatal and major injuries in the oil and chemical industries in the period 1981–85, inclusive, are shown in Table 2.4. For 1990–91 the *Health and Safety Statistics 1990–91* show that in the oil and chemical industries the accidents to employees were as shown in Table 2.5.

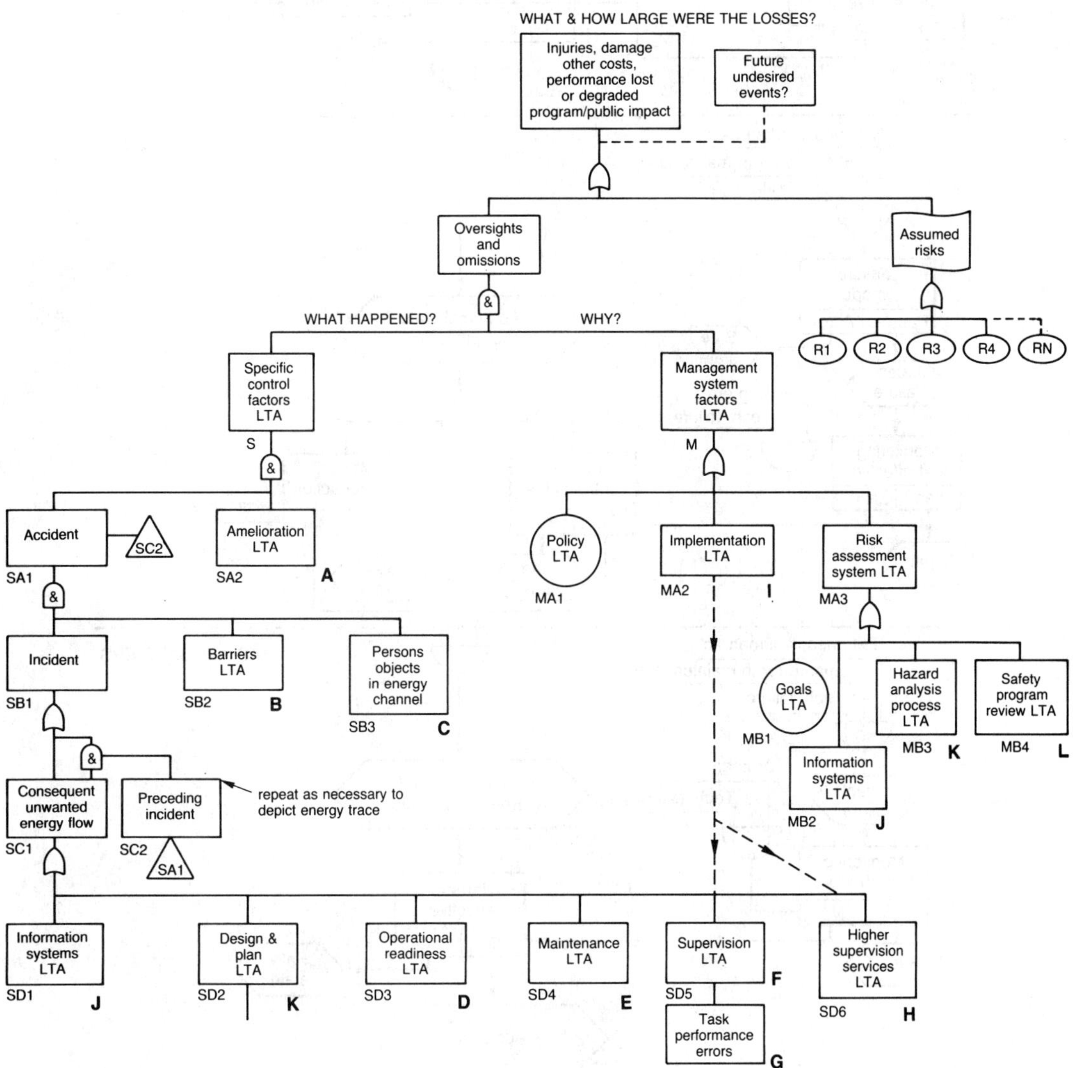

Figure 2.3 *MORT model of the accident process (W.G. Johnson, 1980). The letters A–H refer to further subtrees. LTA, less than adequate (Courtesy of Marcel Dekker)*

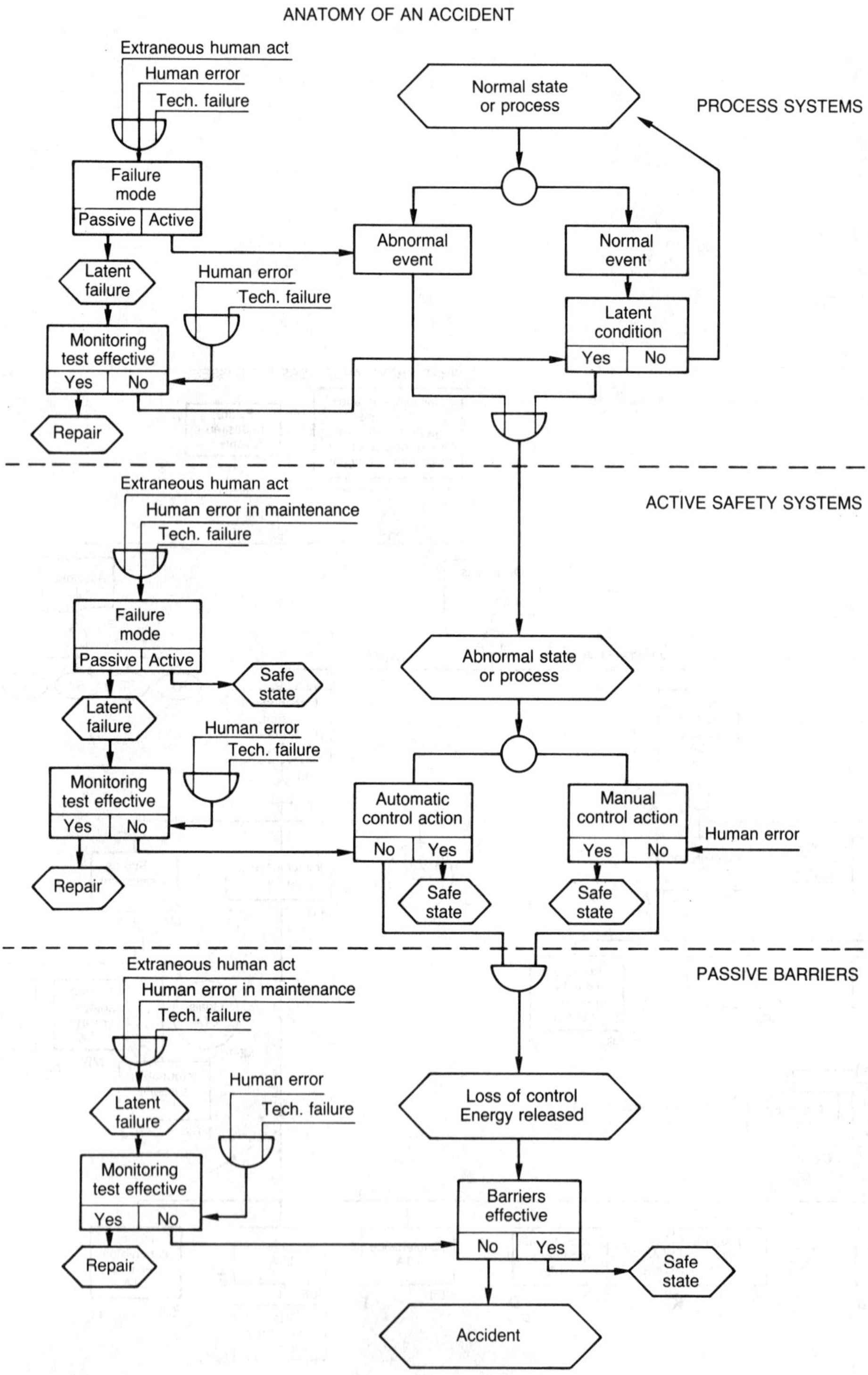

Figure 2.4 *Rasmussen model of the accident process (Rasmussen, 1982b) (Reproduced by permission from High Risk Safety Technology by A.E. Green, copyright John Wiley)*

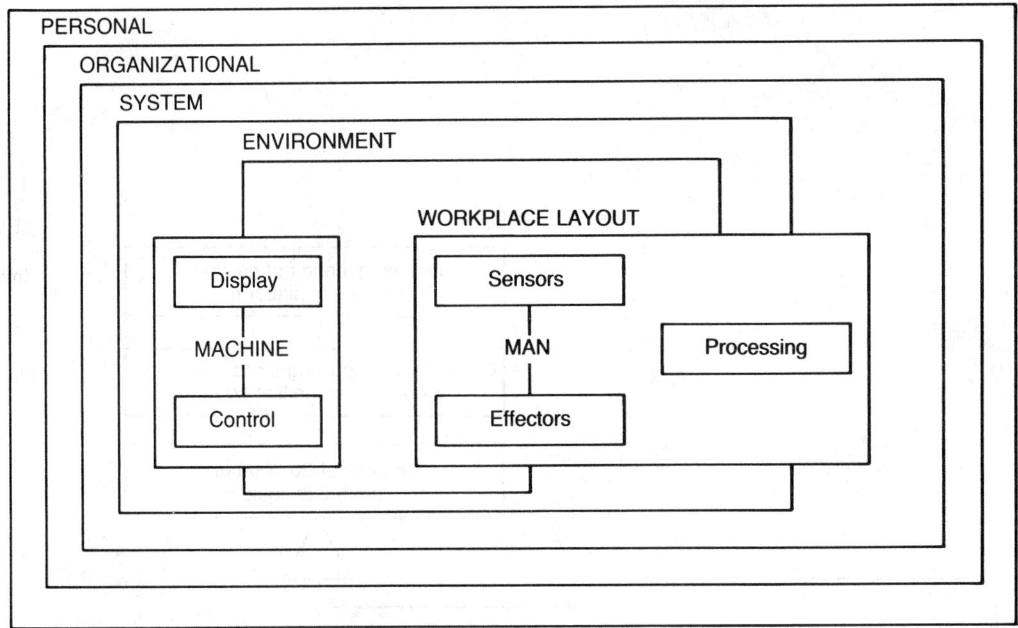

Figure 2.5 *ACSNI model of the accident process (ACSNI, 1991) (Courtesy of HM Stationery Office)*

Table 2.4 *Fatal and major accidents in chemical and petroleum factories in the UK 1981–85 (Cox, Lees and Ang, 1990) (Courtesy of Institution of Chemical Engineers)*

A Period 1981–85: number of fatal (F) and major (M) injuries

	1981		*1982*		*1983*		*1984*		*1985*		*Total*	
	F	*M*	*F*	*M*	*F*	*M*	*F*	*M*	*F*	*M*	*F*	*M*
Chemicals	8	321	6	344	10	374	5	370	5	390	34	1799
Mineral oil processing	3	28	3	36	1	29	1	24	1	24	9	141
Total	11	349	9	380	11	403	6	394	6	414	43	1940

B Period 1981–85: incidence rates of fatal and major injuries

	Incidence per 10^5 employees				
	1981	*1982*	*1983*	*1984*	*1985*
Chemicals	89.4	100.3	115.2	112.7	117.2
Mineral oil processing	108.4	154.2	136.4	130.2	139.7

C 1984

Industry	*No. employees*	*Fatalities*	*Major injuries*	*Fatal and major injuries per 10^5 persons*
Chemicals	360 000	5	349	98.4
Other chemical processes	38 200	1	38	102.1
Mineral oil processing	18 200	1	14	82.4

Sources: Health and Safety Executive (1986c); HM Chief Inspector of Factories (1986a).

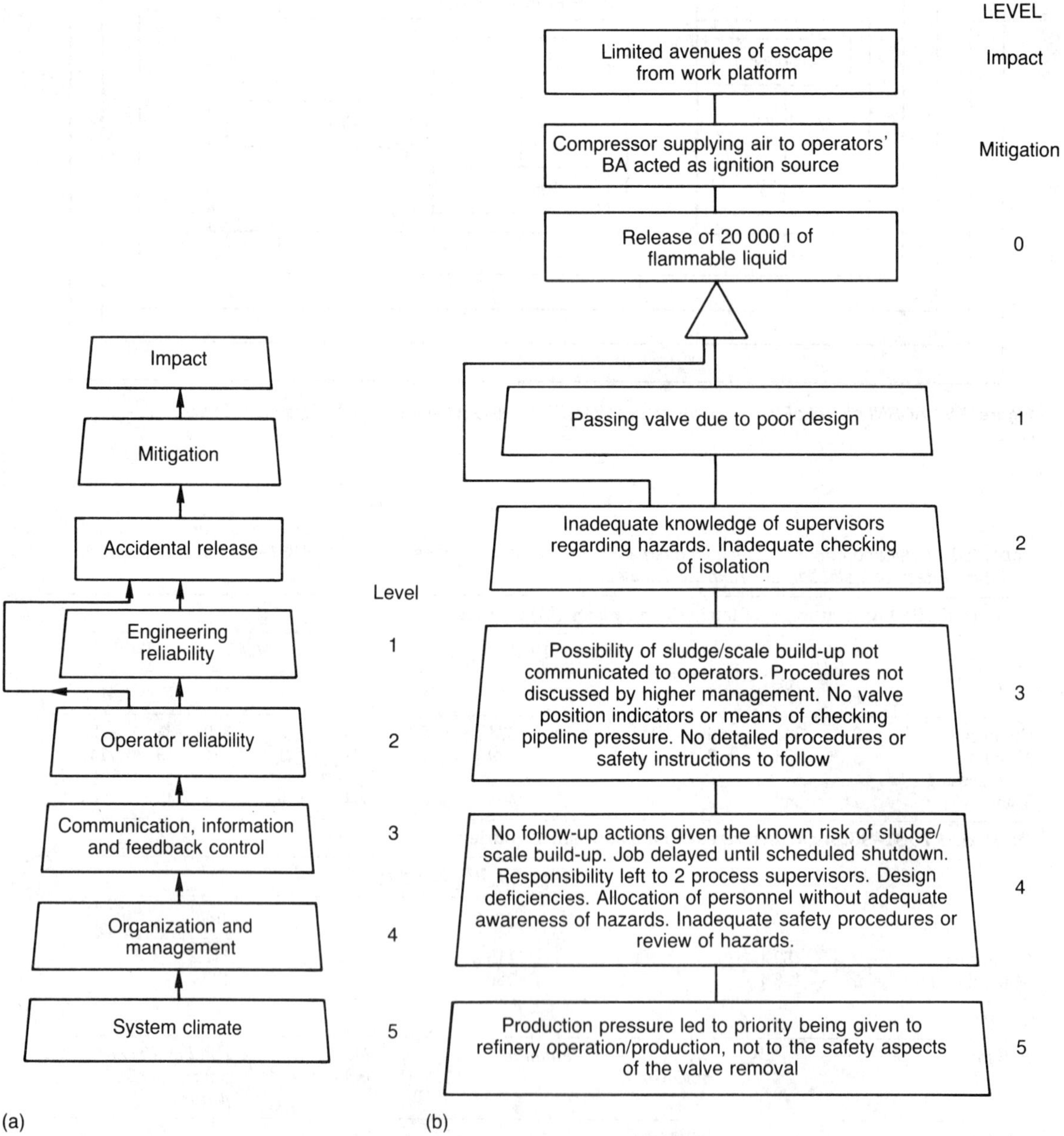

Figure 2.6 *Bellamy and Geyer model of the accident process; (a) generic model; (b) model applied to a refinery incident (Geyer and Bellamy, 1991) (Courtesy of the Health and Safety Executive)*

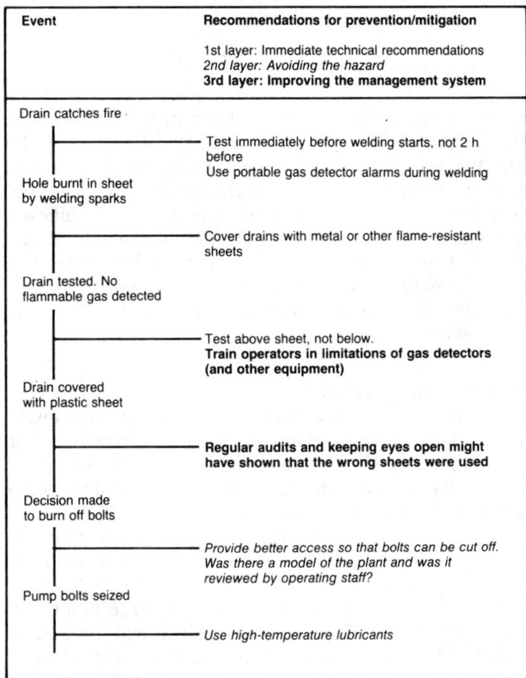

Event
Recommendations for prevention/mitigation
1st layer: Immediate technical recommendations
2nd layer: Avoiding the hazard
3rd layer: Improving the management system
Drain catches fire
Test immediately before welding starts, not 2 h before
Use portable gas detector alarms during welding
Hole burnt in sheet by welding sparks
Cover drains with metal or other flame-resistant sheets
Drain tested. No flammable gas detected
Test above sheet, not below.
Train operators in limitations of gas detectors (and other equipment)
Drain covered with plastic sheet
Regular audits and keeping eyes open might have shown that the wrong sheets were used
Decision made to burn off bolts
Provide better access so that bolts can be cut off.
Was there a model of the plant and was it reviewed by operating staff?
Pump bolts seized
Use high-temperature lubricants

Figure 2.7 *Kletz model of the accident process (Kletz, 1988h) (Courtesy of Butterworths)*

Table 2.5 *Fatal and major injuries in the oil and chemical industries in the UK 1981–85 (after HSE, 1992b)*

Industry	*Type of accident*			
	Fatal	*Non-fatal major*	*Over 3 days*	*All reportable*
Mineral oil processing	0	33	147	180
Chemicals	5	503	3427	3935

The comparative incidence of fatalities in some principal industries and jobs in the UK is given in Table 2.6. The table shows that there is a wide variation between industries. It also shows a downward trend. The fatal accident rate of the chemical industry is approximately the same as that for all manufacturing industry. The injury statistics can be dramatically changed, however, by a single major disaster. In the process industries the worst disaster since 1945 was the vapour cloud explosion at Flixborough in 1974, which killed 28 people. Offshore in the British sector of the North Sea, the Piper Alpha disaster resulted in 167 deaths. The comparative incidence of fatalities in some leading industrial countries, mainly in 1983, is given in Table 2.7.

2.4 Major Disasters

It is appropriate at this point briefly to consider major disasters. A list of the worst disasters in certain principal categories, both for the world as a whole and for the UK, is given in Table 2.8.

Those which are of primary concern in the present context are fire, explosion and toxic release. Both of the worst fires listed occurred in theatres. The explosion at Halifax which killed 1963 people was that of a ship carrying explosives. The Chilwell explosion, in which 134 people died, was in an explosives factory. The toxic gas release at Bhopal, where the death toll was some 2500, was an escape of methylisocyanate from a storage tank.

There are available a number of accounts of disasters, both natural and man-made, and these are summarized in Table 2.9. *Disasters* (Walker, 1973), *Darkest Hours* (J.R. Nash, 1976), *Man-Made Disasters* (B.A. Turner, 1978), *The Disaster File: The 1970s* (Ferrara, 1979), *Disasters* (Whittow, 1980) and *Catastrophes and Disasters* (R. Smith, 1992) all contain large numbers of disaster case histories, including those from the process industries. Rail accidents in the British Isles are described in *Red for Danger* (Rolt, 1982) and accidents in the process industries are

Table 2.6 *Annual risk and fatal accident rate (FAR) in different industries and jobs in the UK*

Industry or activity	*1974–78*		*1987–90*	
	Annual risk [a,b]	*FAR* [c,d]	*Annual risk* [a]	*FAR* [b]
Deep sea fishing	280	140	84	42
Offshore oil and gas	165	82	125	62
Coal mining	21	10.5	14.5	7.3
Railways	18	9	9.6	4.8
Construction	15	7.5	10	5
Agriculture	11	5.5	7.4	3.7
Chemical and allied industries	8.5	4.3	2.4	1.2
Premises covered by Factories Act	–	–	8[(e)]	4
All manufacturing industry	–	–	2.3[(f)]	1.2
Vehicle manufacture	1.5	0.75	1.2	0.6
Clothing manufacture	0.5	0.25	0.09	0.05

[a] Annual risk is given as probability of death in 10^5 years.
[b] Health and Safety Executive, quoted in the Royal Society (1992).
[c] Fatal accident rate is defined as probability of death in 10^8 hours of exposure.
[d] Some values from Kletz (1992b), evidently obtained from annual risk; remainder obtained in like manner by author.
[e] British Medical Association (1987).
[f] HSE (1988c).

***Table 2.7** Fatal accidents in manufacturing industry in different countries[a]*

	Fatality rate	
	Deaths per 1000 man-years[b,c]	*Deaths per 100 000 workers per year*[d]
Argentina	0.020	
Austria	0.142	
Belgium	0.140 (1979)	
Canada	0.080	14 (1971–74)
Czechoslovakia	0.061	
Eire		9 (1971–75)
France	0.068 (1982)	11 (1971–74)
Germany (FRG)	0.120 (1982)	17 (1971–75)
Germany (GDR)	0.030	
Italy		8 (1971–73)
Japan	0.010	5 (1971–75)
Netherlands	0.009	4 (1971–73)
Norway	0.050	
Poland	0.066 (1984)	
Spain	0.109	
Switzerland	0.080	
UK	0.020	4 (1971–75)
USA	0.022	7 (1971–74)

[a] The basis of the calculation differs somewhat between countries and the original references should be consulted for further details.
[b] International Labour Office (1985b).
[c] For 1983 unless otherwise stated.
[d] HSE (1977d).

described in *Chemical Industry Hazards* (V.C. Marshall, 1987).

2.5 Major Process Hazards

The major hazards with which the chemical industry is concerned are fire, explosion and toxic release. Of these three, fire is the most common but, as shown later, explosion is particularly significant in terms of fatalities and loss. As already mentioned, in the UK the explosion at Flixborough killed 28 people, while offshore 167 men died in the explosion and fire on the Piper Alpha oil platform. Toxic release has perhaps the greatest potential to kill a large number of people. Large toxic releases are very rare but, as Bhopal indicates, the death toll can be very high. There have been no major toxic release disasters in the UK.

The problem of avoiding major hazards is essentially that of avoiding loss of containment. This includes not only preventing an escape of materials from leaks, etc., but also avoidance of an explosion inside the plant vessels and pipework. Some factors which determine the scale of the hazard are:

(1) the inventory;
(2) the energy factor;
(3) the time factor;
(4) the intensity–distance relations;
(5) the exposure factor; and
(6) the intensity–damage and intensity–injury relationships.

These factors are described below.

2.5.1 The inventory

The most fundamental factor which determines the scale of the hazard is the inventory of the hazardous material. The larger the inventory of material, the greater the potential loss. As plants have grown in size and output, so inventory in process and in storage has grown. In the early days of this growth, there was perhaps insufficient appreciation of the increase in the magnitude of the hazard. There is now, however, much wider recognition of the importance of inventory. At the same time, it is important to emphasize that inventory is not the only factor which determines the scale of the hazard.

2.5.2 The energy factor

For an inventory of hazardous material to explode inside the plant or to disperse in the form of a flammable or toxic vapour cloud there must be energy. In most cases this energy is stored in the material itself as the energy either of chemical reaction or of material state.

In particular, a material which is held as a liquid above its normal boiling point at high pressure and temperature, in other words superheated, contains large quantities of physical energy, which cause a large proportion of it to vaporize by instantaneous flash-off and to disperse if there is loss of containment. On the other hand, a material which is held as a refrigerated liquid at atmospheric pressure contains much less physical energy and does not vaporize to anything like the same extent if containment is lost. In this case the energy necessary for vaporization has to be supplied by the ground and the air, which is a relatively slow process. Similarly, the hazard presented by an ultratoxic material depends very largely on whether there is energy available for its dispersion. There is a hazard, for example, if an ultratoxic substance is produced as a by-product in a chemical reactor in which a runaway exothermic reaction may occur. But if there is no such source of energy, the hazard is much less.

The energy requirement is thus another fundamental feature. Unless it is taken into account in the calculation, the scenarios considered may be not merely unlikely, but literally physically impossible.

2.5.3 The time factor

Another fundamental factor is the development of the hazard in time. The time factor affects both the rate of release and the warning time.

The nature and scale of the hazard is often determined by the rate of release rather than by the inventory. Thus it is the rate of release which determines the size of a flammable gas cloud formed from a jet of flashing hydrocarbon liquid, such as occurred at Flixborough. Similarly, the hazard presented by an escape of toxic gas depends on the rate of release. There is a considerable difference in the concentrations attained between an instantaneous and a continuous release of toxic gas.

The warning time available to take emergency countermeasures and reduce the number of people exposed is also very important. An explosion gives a warning time which is usually measured only in seconds and may be zero, whereas a toxic release gives a warning which is often measured in minutes.

Table 2.8 *Some of the worst non-industrial and industrial disasters world-wide and in the UK (Material from Guinness Book of Records, copyright © reproduced by permission of the publishers)*

A Worst world disasters			
Earthquake	Near East and East Mediterranean	1201	1 100 000
Volcanic eruption	Tambora Sumbawa, Indonesia	1815	92 000
Landslide	Kansu Province, China	1920	180 000
Avalanche	Yungay, Juascaran, Peru	1970	≈18 000
Circular storm	Ganges Delta Islands, Bangladesh	1970	1,000,000
Tornado	Shaturia, Bangladesh	1989	≈1 300
Flood	Hwang-ho river, China	1887	900 000
Lightning	Hut in Chinamasa Krael nr Umtali, Zimbabwe (single bolt)	1975	21
Smog	London fog, UK (excess deaths)	1951	2 850
Panic	Chungking (Zhong qing), air raid shelter, China	1941	≈4 000
Dam burst	Manchu River Dam, Morvi, Gujarat, India	1979	≈5 000
Fire (single building)	The Theatre, Canton, China	1845	1 670
Explosion	Halifax, Nova Scotia, Canada	1917	1 963
Mining	Hankeiko Colliery, China (coal dust explosion)	1942	1 572
Industrial	Union Carbide methylisocyanate plant, Bhopal, India	1984	≈2 500
Offshore platform	Piper Alpha, North Sea	1988	167
Nuclear reactor	Chernobyl Reactor No. 4	1986	31[a]
Aircraft	KLM-Pan Am, Boeing 747 crash, Tenerife	1977	583
Marine (single ship)	*Wilhelm Gustloff*, German liner torpedoed off Danzig by Soviet submarine S-13	1945	≈7 700
Rail	Bagmati River, Bihar, India	1981	>800
Road	Petrol tanker explosion inside Salang Tunnel, Afghanistan	1982	≈1 100
Atomic bomb	Hiroshima, Japan	1945	141 000
Conventional bombing	Tokyo, Japan	1945	≈140 000
B UK			
Earthquake	London earthquake, Christ's Hospital, Newgate	1580	2
Landslide	Pantglas coal tip No. 7, Aberfan, Mid-Glamorgan	1966	144
Avalanche	Lewes, East Sussex	1836	8
Circular storm	'The Channel Storm'	1703	≈8 000
Tornado	Tay Bridge collapsed under impact of two tornadic vortices	1879	75
Flood	Severn Estuary	1606	≈2 000
Smog	London fog (excess deaths)	1951	2 850
Panic	Victoria Hall, Sunderland	1883	183
Dam burst	Bradfield Reservoir, Dale Dyke, near Sheffield (embankment burst)	1864	250
Fire (single building)	Theatre Royal, Exeter	1887	188
Explosion	Chilwell, Nottinghamshire (explosives factory)	1918	134
Mining	Universal Colliery, Senghenydd, Mid-Glamorgan	1913	439
Offshore platform	Piper Alpha, North Sea	1988	167
Nuclear reactor	Windscale (now Sellafield), Cumbria (cancer deaths)	1957	–[b]
Marine (single ship)	HMS *Royal George*, off Spithead	1782	≈800
Rail	Triple collision, Quintinshill, Dumfries	1915	227
Road	Coach crash, River Dibb, near Grassington, North Yorkshire	1975	33
Conventional bombing	London, 10–1 May	1941	1 436

[a]The *Guinness Book of Records* states: 'Thirty one was the official Soviet total of immediate deaths. On 25 April 1991 Vladimir Shovkoshitny stated in the Ukrainian Parliament that 7000 "clean-up" workers had already died from radiation. The estimate for the eventual death toll has been put as high as 75 000 by Dr Robert Gale, a US bone transplant specialist.'

[b]The *Guinness Book of Records* states: 'There were no deaths as a direct result of the fire, but the number of cancer deaths which might be attributed to it was estimated by the National Radiological Protection Board in 1989 to be 100.'

Table 2.9 *Coverage of some books on disasters*

	Bignell, Peters and Pym (1977)	Ferrara (1979)	R. Smith (1993)	Thygerson (1977)	Walker (1973)	Whittow (1980)
Natural hazards						
Subterranean stress:						
Earthquakes		x	x	x	x	x
Volcanoes		x	x	x		x
Tsunamis			x	x		x
Surface instability:						
Landslides and avalanches	x		x	x	x	x
Ground surface collapse						x
Weather:						
Wind, storm		x	x			x
Tornadoes		x	x			x
Hurricanes			x	x		x
Floods (river, sea)		x	x	x	x	x
Fires (forest, grass)		x	x	x		
Man-made hazards						
Structures:						
Buildings	x		x		x	
Dams			x		x	
Bridges	x		x		x	
Building fires	x	x		x		x
Gas explosions		x	x		x	
Industrial:						
Mines		x	x		x	
Fire		x	x		x	
Explosion		x	x		x	
Transport:						
Air	x	x	x		x	
Rail	x	x	x		x	
Road		x	x	x	x	
Sea	x	x	x		x	

2.5.4 The intensity–distance relationship

An important characteristic of the hazard is the distance over which it may cause injury and/or damage. In general, fire has the shortest potential range, then explosion and then toxic release, but this statement needs considerable qualification. The range of a fireball is appreciable and the range of a fire or explosion from a vapour cloud is much extended if the cloud drifts away from its source.

It is possible to derive from the simpler physical models for different hazards analytical expressions which give the variation of the intensity of the physical effect (thermal radiation, overpressure, toxic concentration) with distance. For some models the variation follows approximately the inverse square law. This aspect is discussed in Chapter 9.

With regard to the exposure of the public to process hazards, it is of interest to know the distance at which there might be a significant number of fatalities or injuries and the maximum distance at which any fatality or injury might occur. Estimates of the distance necessary to reduce the risk of fire and explosion to members of the public to a level which is assumed to be not unacceptable, based on criteria such as thermal radiation from fire and overpressure from explosion, are generally of the order of 250–500 m for a major plant handling hydrocarbons, but may be less or more. Estimates of the distance necessary to reduce the risk from toxic release tend to be somewhat greater.

The maximum distance at which there might conceivably be fatalities or injuries cannot be determined with any great accuracy. The explosion effect which can occur at the greatest distance is the shattering of glass – this has happened at distances of up to 20 miles from a very large explosion. But in such cases the energy of the glass fragments is low and very rarely causes injury. Similarly, cases of injury from toxic gas at large distances, say over 10 miles, are rare but are reported to have occurred.

The effects of fire, explosion and toxic release are discussed further in Chapters 16–18. Although a potential effect of a hazard is often expressed as a function of distance, it is the area covered by the effect which determines the number of people at risk.

2.5.5 The exposure factor

A factor which can greatly mitigate the potential effects of an accident is the reduction of exposure of the people who are in the affected area. This reduction of exposure may be due to features which apply before the hazard

Table 2.10 Large fires in the chemical and petroleum industries[a] in Great Britain, 1963–75: number and cost (Fire Protection Association, 1974; Redpath, 1976)

Year	No. of fires	Cost (£m.)
1963	44	2.3
1964	43	2.9
1965	40	3.0
1966	43	2.5
1967	51	4.4
1968	53	3.1
1969	44	3.2
1970	67	6.2
1971	65	6.2
1972	66	3.5
1973	80	12
1974	45	43
1975	46	6.6

[a] The chemical and petroleum industries are taken as Standard Industrial Classification Order 4 (Coal and petroleum products) and 5 (Chemical and allied industries)

develops, or to emergency measures which are taken after the hazard is recognized.

The principal mitigating features are shelter and escape. Escape may be by personal initiative or by preplanned evacuation. It should not be assumed that emergency measures are synonymous with evacuation. For a toxic release, for example, the emergency instructions may be to evacuate the area but are more likely to be to stay indoors and seal the house. Emergency measures may be of great value in reducing the toll of casualties from a major accident.

For an explosion which gives no advance warning there is no time for emergency measures such as evacuation. This does not mean, however, that evacuation has no role to play as far as fire and explosion are concerned. On the contrary, although the initial event may be sudden, there are frequently further fire and explosion hazards. Evacuation may then be applicable.

Measures which can be taken to mitigate exposure are discussed in Chapter 24.

2.5.6 The intensity–damage and intensity–injury relationships

The range of the hazard depends also on the relationships between the intensity of the physical effect and the proportion of people who suffer injury at that level of the effect. The annular zone within which injury occurs is determined by the spread of the injury distribution. If the spread is small, the injury zone will be relatively narrow, while if it is large the zone may extend much further out. Similar considerations apply to damage. This aspect is discussed further in Chapter 9.

2.6 Fire Loss Statistics

The loss statistics of interest are primarily those for fire loss. In the UK, principal sources of statistics on such losses are the Home Office, the Fire Protection Association, the Loss Prevention Council and the insurance companies. These organizations produce annual statistics for fire losses. Loss due to explosions is generally included in that for fire loss. There are no regular loss statistics on toxic release, since this is a rare event and usually causes minimal damage to property.

Fire loss data are given in *Fire Statistics United Kingdom 1990* (Home Office, 1992), in *FPA Large Fire Analysis for 1989* (Fire Protection Association (FPA), 1991) and in *Insurance Statistics 1987–91* (Association of British Insurers (ABI), 1992).

The Home Office data show that in 1990 fire brigades were called to some 467 000 fires, of which 108 000 were in buildings. The FPA defines a large loss fire as one involving a loss of £50 000 or more. In 1990 there were 739 such fires, with a total cost of £282 million and an average cost of £0.382 million. In the chemical and allied industries in 1990 there were eight large loss fires, with a total cost of £5.24 million and an average cost of £0.656 million. There were no large loss fires in the coal and petroleum industries class.

A more detailed analysis of fires in the chemical and petroleum industries given by the FPA in 1974 is shown in Tables 2.10 and 2.11, the former giving analysis by number and cost and the latter by number and occupancy, by place of origin, by ignition source, by material first ignited and by time of day. There are a number of significant points in the tables. The chemical industry had the largest number of fires, but the oil

Table 2.11 Large fires in the chemical and petroleum industries in Great Britain, 1971–73 (Fire Protection Association, 1974b)

A Number and cost by occupancy

Occupancy	No. of fires £10 000–£39 000	£40 000–£99 000	£100 000–£249 000	>£250 000	Total No.	Total cost (£m.)	Average cost (£m.)
Chemicals	1	12	13	7	33	7.215	0.219
Oil and tar	0	1	2	6	9	6.225	0.692
Paint and varnish	1	2	2	1	6	0.738	0.123
Fertilizer	1	1	2	1	5	0.625	0.125
Agricultural products	0	0	3	0	3	0.525	0.175
Plastics	1	1	0	0	2	0.051	0.025
Others	2	7	1	1	11	0.823	0.075
Total	6	24	23	16	69	16.202	0.235

continued

B Place of origin

	No. of fires
Storage:	
Warehouse or open site	21
Tank	12
Leakage:	
from fractured pipe	15
from leaking coupling, flange or seal	6
from electrical equipment	6
unspecified	2
Reactor or mixer	4
Steam drier	2
Spray booth	1
Cooling tower	1
Unreported	9
Total	*79*

C Ignition source

	No. of fires
Hot surfaces	8
Burner flames	8
Electrical equipment	6
Spontaneous ignition	6
Friction heat and sparks	6
Flame cutting	4
Children with matches	3
Malicious ignition	2
Static electricity	1
Unknown	35
Total	*79*

D Material first ignited

	No. of fires
Classification by phase	
Gas	10
Vapour	16
Liquid	20
Solid	23
Unknown	10
Total	*79*
Classification by material	
Hydrocarbons:	
Gas	3
Liquid/vapour	18
Solid	2
Other organics, etc.:	
Liquid/vapour	16
Solid	7
Cellulosic solids (timber, paper, cardboard, fireboard)	6
Hydrogen	7
Steel	2
Sulphur	1
Unknown	17
Total	*79*

E Time of day

Time of day	*No. of fires*	*Time of day*	*No. of fires*
24.00–1.00 h (midnight)	4	12.00–13.00 h	1
1.00–2.00	0	13.00–14.00	3
2.00–3.00	4	14.00–15.00	2
3.00–4.00	2	15.00–16.00	5
4.00–5.00	3	16.00–17.00	3
5.00–6.00	2	17.00–18.00	6
6.00–7.00	1	18.00–19.00	5
7.00–8.00	4	19.00–20.00	2
8.00–9.00	3	20.00–21.00	5
9.00–10.00	3	21.00–22.00	3
10.00–11.00	3	22.00–23.00	0
1.00–12.00	3	23.00–24.00 (midnight)	2

industry the most expensive. The origin of the fires is predominantly in storage and in leakages. The sources of ignition are fairly evenly spread. There is only one fire attributed to static electricity and two to arson. But in 35 cases, i.e. 44%, the ignition source was unknown. The material phase first ignited also shows a balanced spread, with the solid phase actually being predominant. There is no marked trend in the time of day of the fires, although the number is somewhat higher during the day shift.

Further analyses of fire statistics have been given by H.D. Taylor and Redpath (1971), the FPA (1976) and Redpath (1976). In particular, these sources present additional data of the type given in Tables 2.10 and 2.11. These data are a useful general pointer. They constitute, however, a rather small sample. Moreover, they are not necessarily representative of the type of fire or explosion which constitutes a major disaster.

Work on the occurrence of fires in industrial buildings has been described by Rutstein and Clarke (1979) and Rutstein (1979b). The probability of fire was found to increase with the size of building according to the equation

$$P = aB^c \qquad [2.6.1]$$

where B is the floorspace (m^2) and P is the probability of fire per year, and a and c are constants. For production buildings in manufacturing industry generally (SIC Classes 3–19) the values of the constants a and c were 0.0017 and 0.53 and in the chemical and allied industries (SIC 4) they were 0.0069 and 0.60, respectively. (These SIC numbers refer to the Standard Industrial Classification at the time). The probability of fire in a 1500 m^2 production building was thus 0.083 for manufacturing industry and 0.21 for the chemical industry.

An analysis of some 2000 large loss claims at Cigna Insurance has been given by Instone (1989) and is summarised in Table 2.12.

A study of the contribution of human factors to failures of pipework and in-line equipment by Bellamy, Geyer and Astley (1989) contains a large amount of information characterising releases, as shown in Table 2.13.

As with fatal injuries so with fire losses, a single accident may dominate the process industries loss for a

***Table 2.12** Large losses in the process industries insured by Cigna Insurance (after Instone, 1989) (Courtesy of Cigna Insurance)*

	Proportion (%)
A Operations type	
Refinery complex	48
Petrochemicals	20
General properties	10
Storage terminals	8
Onshore production plants	7
Jetty installations	3
Offshore production plant	2
Coal mines	2
Total	*100*
B Plant type	
Storage tanks and pipelines	23
Refineries	22
Oil/gas production	13
Offsites	10
Monomers and polymers	6
Petrochemicals	6
Fertilizers	5
Coal products	3
Gas processing	3
Refinery feedstock and products	3
Acids, glycols, etc.	3
Aromatics	2
Alcohols, ketones, etc.	1
Paraffins	0
Total	*100*
C Process unit type	
Furnace or heater	10
Pipework	9
Storage tank – unspecified	7
Drilling rig	5
Pump	5
Compressor	5
Boiler	5
Heat exchanger	4
Cone roof tank	4
Distillation column	3
Warehouse	3
Process vessel	3
Floating roof tank	3
Electrical substation	3
Reactor	3
Reformer	2
Conveyor belt	2
Riser pipe	2
Jetty or buoy	2
Gas turbine generator	<2
Control room	<2
Hopper	<2
Transformer	<2
Flare; furnace stack; relief system (2); air cooler; centrifuge; filter; extruder; drier; refrigeration circuit; cooling tower; incinerator; electric motor; meters; instrument analyser; valve; crane; API (American Petroleum Institute) separator (2); hydrogen; loading arm – vessels; loading arm – road vehicles; road tanker; hose; laboratory; office; computer	<1
D Plant status	
Normal operations	50
Plant start-up	15
Plant under maintenance	10
Filling tanks or vessels	6
Well drilling	5
Plant in shut-down state	3
Process upset	2
Emptying tanks or vessels	2
Construction	<2
Well workover	<2
Plant shutting down	<1
Ship berthing/sailing	<1
Plant commissioning	<1
Blending operation	<1
Well logging	<1
E Loss type	
Fire	33
Explosion	12
Explosion and fire	10
Wind, storm and flood	10
Well blowout	5
Mechanical	5
Ship impact	5
Contamination	4
Machinery breakdown	4
Product loss	3
Electric cable fire	3
Construction defect; collapse; subsidence/collapse; earthquake; impact; implosion; floating roof sunk; pool fire; vapour fire; vapour cloud explosion; theft	<1
F Cause of loss	
Operator error	17
Wind, storm and flood	16
Pipe/weld failure	7
Tube failure	7
Machinery breakdown	7
Electrical short-circuit	4
Valve leak	4
Pipe flange leak	4
Instrumentation failure	4
Vehicle/digger impact	3
Corrosion	3
Seal failure	3
Lightning	<2
Anchor/hull damage to jetty	<2
Power failure	<2
Storage tank overflow	<2
Tank vent blocked	<2
Pressure increase	<2
Sabotage	<2
Burst vessel	<2
Compressor leak	<2
Runaway reaction	<2
Lost well circulation fluid	<2
Modification/design error; erosion; feedstock; process vessel overflow; flare carryover; water hammer; insufficient air purge; lubrication failure; instrument air failure; pyrophoric iron sulphide; gauge broken open; corroded electrical contacts; flame impingement; fire water leak; malicious damage	<1

G Material ignited	
Refinery feedstock and products	41
Petrochemicals	31
Gas processing	13
Acids, glycols, additives	6
Monomers and polymers	3
Alcohols, ketones, ethers; aromatics; paraffins; fertilisers; coal products	<1
H Source of ignition	
Material not ignited	35
Furnace	12
Hot surface	12
Autoignition	7
Electrical arcing	3
Operator error	3
Other electrical apparatus	3
Welding	3
Friction	<3
Static electricity	<3
Pyrophoric matter; spontaneous combustion; matches; spark; electric motor; boiler; flare; explosive device; vehicle; lightning; spread of fire; hot embers	<2

***Table 2.13** Characteristics of releases in study of failure of pipework and in-line equipment (Bellamy, Geyer and Antley, 1989) (Courtesy of the Health and Safety Executive)*

	No. of incidents
A Location type	
Chemical plant	278
Refinery	96
Factory	187
Storage depot	47
Tank yard	28
Fuel station	15
Other	38
Unknown	232
Total	*921*
B Site status	
Normal operations	343
Storage	103
Loading/unloading	33
Maintenance	146
Modification	8
Contractor work	18
Testing	5
Unknown	128
Other	40
Start-up	42
Shut-down	18
Total	*884*
C Materials released	
Ammonia	54
Hydrocarbons (unspecified)	54
Chlorine	50
Hydrogen	37
Benzene	33
Crude oil	28
Steam	25
Natural gas	24
Propane	20
Butane	18
Fuel oil	18
Hydrochloric acid	16
Sulphuric acid	16
Ethylene	16
Hydrogen sulphide	14
Water	13
Nitrogen	13
Oxygen	13
Vinyl chloride	12
LPG	12
Styrene	11
Naphtha petroleum	10
Total	*507*
D Material phase	
Liquid	393
Gas	260
Vapour	13
Solid	9
Liquid + gas/vapour	120
Solid + gas/vapour	3
Total	*798*
E Unignited material dispersion	
Flammable	127
Toxic	123
Flammable/toxic	47
Corrosive	97
Irritant	1
Unignited gas	96
Vapour cloud	180
Liquid	212
Spill	186
Jet/spurt	8
Spray	10
Total	*1087*
F Fire or explosion event	
Fire	145
Flash fire	11
Pool fire	4
Jet fire	1
Fireball	7
BLEVE	4
Explosion	63
Explosion followed by fire	77
Explosion followed by flash fire	2
Total	*314*

Note: BLEVE, boiling liquid expanding vapour explosion.

particular year. In the UK, the Flixborough disaster constituted a significant proportion of the fire loss in 1974, whilst the Piper Alpha disaster dominated that for 1988. Fire losses are considered further in Chapter 5.

Table 2.14 *Large losses in the chemical industry insured by the Factory Insurance Association: fire, explosion and other loss (after W.H. Doyle, 1969) (Courtesy of the American Institute of Chemical Engineers)*

Year	*Fires*		*Explosions*		*Other*		*Total*	
	No.	*Loss (%)*	*No.*	*Loss (%)*	*No.*	*Loss (%)*	*No.*	*Loss (%)*
1964	3	1.6	6	13.4	1	0.4	10	15.4
1965	4	1.9	8	8.7	0	0	12	10.6
1966	7	9.2	6	9.9	1	0.6	14	19.7
1967	8	5.9	12	22.4	2	1.1	22	29.4
1968	13	11.6	12	13.3	0	0	25	24.9
Total	*35*	*30.2*	*44*	*67.7*	*4*	*2.1*	*83*	*100.0*

Table 2.15 *Large losses in the chemical industry insured by the Factory Insurance Association: types of explosion (W.H. Doyle, 1969) (Courtesy of the American Institute of Chemical Engineers)*

	No.	*Loss (%)*
Combustion:		
In equipment	13	10.5
Outside equipment, in building	8	24.4
In open	1	3.3
Subtotal	22	38.2
Reaction:		
Explosive liquid or solid	12	16.8
Runaway reaction	4	6.5
Subtotal	16	23.3
Metal failure:		
Corrosion	1	1.4
Overheating	3	4.1
Accidental overpressure	2	1.0
Subtotal	6	6.5
Total	*44*	*68.0*

Table 2.16 *Large fires in the chemical and allied industries insured by Industrial Risk Insurers (Norstrom, 1982a) (Courtesy of the American Institute of Chemical Engineers)*

	Proportion (%)
A Cause	
Flammable liquid or gas (release, overflow)	17.8
Overheating, hot surfaces, etc.	15.6
Pipe or fitting failure	11.1
Electrical breakdown	11.1
Cutting and welding	11.1
Arson	4.4
Others	28.7
B Location	
Enclosed process or manufacturing buildings	42.2
Outdoor structures	33.3
Warehouses	6.7
Others	17.8
C Occupancy	
Mixing/blending	8.9
Storage	8.9
Distillation	6.7
Control/computer rooms	6.7
Chemical reaction, batch	4.4
Chemical reaction, continuous	4.4
Heating	4.4
Drying	4.4
Others	51.2
D Contributing factors	
Sprinkler or water spray lacking	35.6
Human element	15.6
Presence of flammable liquids	11.1
Rupture of vessel or equipment	8.9
Excessive residue	8.9
Production bottleneck	6.7
Sprinkler or water spray inadequate or impaired	6.7
E Area protection	
Protected building or structure	26.7
Unprotected building or structure	62.2
Outside building or structure	11.1

2.7 Fire and Explosion

So far no distinction has been made between fire and explosion losses. The latter are normally included in the overall fire statistics. In fact it is explosions which cause the most serious losses. This is illustrated by Table 2.14, which shows losses in the chemical industry insured by the Factory Insurance Association (FIA) of the USA (W.H. Doyle, 1969). Some two-thirds of the loss is attributable to explosions. The nature of these explosions is shown in Table 2.15. Over three-quarters of the explosions involve combustion or explosive materials.

***Table 2.17** Large explosions in the chemical and allied industries insured by Industrial Risk Insurers (Norstrom, 1982a) (Courtesy of the American Institute of Chemical Engineers)*

	Proportion (%)
A Cause	
Chemical reaction uncontrolled	20.0
Chemical reaction accidental	15.0
Combustion explosion in equipment	13.3
Unconfined vapour cloud	10.0
Overpressure	8.3
Decomposition	5.0
Combustion sparks	5.0
Pressure vessel failure	3.3
Improper operation	3.3
Others	16.8
B Location	
Enclosed process or manufacturing buildings	46.7
Outdoor structures	31.7
Yard	6.7
Tank farm	3.3
Boiler house	3.3
Others	8.3
C Occupancy	
Chemical reaction process, batch	26.7
Storage tank	10.0
Boiler	8.3
Chemical reaction process, continuous	6.7
Compressor	5.0
Evaporation	3.3
Recovery	3.3
Transfer	3.3
Liquefaction	3.3
Others	25.1
D Contributing factors	
Rupture of equipment	26.7
Human element	18.3
Improper procedures	18.3
Faulty design	11.7
Vapour-laden atmosphere	11.7
Congestion	11.7
Flammable liquids	8.3
Long replacement time	6.7
Inadequate venting	6.7
Inadequate combustion controls	5.0
Inadequate explosion relief	5.0
E Area protection	
Protected building or structure	43.3
Unprotected building or structure	36.7
Outside building or structure	20.0

Further analyses of fires and explosions, treated separately, have been made by Norstrom (1982a), as shown in Tables 2.16 and 2.17.

The problem of large fires or explosions in vapour clouds is considered in Chapter 17. The data given in Table 17.33 by V.C. Marshall (1976a), in which the incidents are ranked in terms of the amount of vapour released, suggest that the large releases often result in explosions rather than fires.

2.8 Causes of Loss

There are almost as many analyses of the causes of loss as there are investigators. Unfortunately, there is no accepted taxonomy, so that it is often difficult to reconcile different analyses. Two typical breakdowns of the causes of loss are given in Table 2.18 (W.H. Doyle, 1969) and in Table 2.19 (American Insurance Association (AIA), 1979), the latter being an up-date of an earlier table (Spiegelman, 1969). Doyle emphasizes the importance of poor maintenance, followed by poor design and layout of equipment and inadequate knowledge of the properties of chemicals. Spiegelman (1969) gives, in addition to the table referred to, a fairly detailed breakdown of the factors considered under each heading. His category of equipment failures evidently has a large element of poor maintenance.

***Table 2.18** Large losses in the chemical industry insured by the Factory Insurance Association: causes of loss (W.H. Doyle, 1969) (Courtesy of the American Institute of Chemical Engineers)*

	No.	*Loss (%)*
Incomplete knowledge of the properties of a specific chemical	6	11.2
Incomplete knowledge of the chemical system or process	6	3.5
Poor design or layout of equipment	13	20.5
Maintenance failure	14	31.0
Operator error	5	6.9
Total	*44*	*73.1*

***Table 2.19** Hazard factors for 465 fires and explosions in the chemical industry 1960–77 (American Insurance Association, 1979)*

	No. of times assigned	*Proportion (%)*
Equipment failure	223	29.2
Operational failure	160	20.9
Inadequate material evaluation	120	15.7
Chemical process problems	83	10.9
Material movement problems	69	9.0
Ineffective loss prevention programme	47	6.2
Plant site problems	27	3.5
Inadequate plant layout	18	2.4
Structures not in conformity with use requirements	17	2.2
Total	*764*	*100.0*

Further information on causes of loss is provided by the data of Norstrom (1982a) given in Tables 2.16 and 2.17.

2.9 Downtime Losses

The losses considered so far are insured losses arising from fire and explosion. Another important, but often uninsured, loss arises from plant shut-down and downtime. Once again there are many different analyses of shut-down and downtime, and its causes on different types of plant. Three sets of statistics on shut-down and downtime are given in Tables 2.20–2.22 for a refinery (McFatter, 1972), in Tables 2.23 and 2.24 for another refinery (Anon, 1970b), and in Tables 2.25–2.28 (G.P. Williams, Hoehing and Byington 1987). In addition, data on down-time causes in ammonia plants were given earlier in Table 1.5.

Table 2.20 *Shut-downs in a refinery*[a] *(after McFatter, 1972) (Courtesy of the American Institute of Chemical Engineers)*

Year	*Scheduled*	*Unscheduled*
1961	1	1
1962	4	1
1963	3	1
1964	3	1
1965	4	1
1966	1	0
1967	3	10
1968	2	3
1969	4	1
1970	2	5

[a] There are seven units in the refinery.

Table 2.21 *Causes of non-scheduled shut-downs in a refinery (after McFatter, 1972) (Courtesy of the American Institute of Chemical Engineers)*

	%
Line leaks and fires	28
Line leaks	28
Process	12
Utilities	20
Miscellaneous	12
Total	*100*

Table 2.22 *Causes of significant non-shut-down failures in a refinery (after McFatter, 1972) (Courtesy of the American Institute of Chemical Engineers)*

	%
Compressors	30
Furnaces	18
Exchangers	17
Towers	5
Process	18
Process integration	7
Miscellaneous	5
Total	*100*

Table 2.23 *Causes of downtime in a refinery (after Anon., 1970b)*

	%
Scheduled maintenance	71.6
Process problems	15.6
Problems in other linked plants	7.6
Utility problems	5.2
Total	*100*

Table 2.24 *Significant equipment failures in a refinery (after Anon., 1970b)*

	All failures	
	(No.)	(%)
Pumps and compressors	35	33.9
Furnaces	14	13.6
Piping	11	10.7
Towers and reactors	9	8.8
Exchangers	7	6.8
Utilities	23	22.3
Other	4	3.9
Total	*103*	*100*

McFatter's data in Tables 2.20–2.22 are for a single refinery consisting of seven units over a 10-year period. During this time the overall availability of all the units was 96.4%. In other words the units were down for scheduled or unscheduled shut-down 3.6% of the time. But in addition there were some 122 failures which were significant in that they resulted in 1313 equivalent days of lost production, although they did not cause shut-down. The large number of unscheduled shut-downs in 1967 was due to a major hydrocarbon line leak and fire, which resulted in complete shut-down of the refinery.

The importance of compressors, furnaces and heat exchangers is clear from Table 2.22. Other data given by McFatter show a rising trend of equivalent days lost due to failures. He comments that this is due to large increases in throughput, reduction in sparage of critical items, e.g. compressors and pumps, and lack of opportunity to take items, e.g. furnaces, off for maintenance. The data for another refinery given in Tables 2.23 and 2.24 show an essentially similar picture; as do the data given by McIntire (1977a,b).

Surveys of ammonia plant shut-downs have been published by Sawyer, Williams and Clegg (1972), G.P. Williams and Sawyer (1974), G.P. Williams (1978), G.P. Williams and Hoehing (1983) and G.P. Williams, Hoehing and Byington (1987). The 1987 survey covers 136 plants, divided into three groups: (1) large-tonnage, single-stream, centrifugal compressor type plants, world-wide in North America, Europe and the rest of the

***Table 2.25** Shut-down and downtime in ammonia plants (after G.P. Williams, Hoehing and Byington, 1987) (Courtesy of the American Institute of Chemical Engineers)*

	Large tonnage plants					
	North America	*Europe*	*Rest of world*	*Total*	*Reciprocating plants*	*Partial oxidation plants*
A Downtime and availability						
Total downtime (days/plant/year)	71.1	48.6	84.5	70.1	60.2	71.3
Unavoidable downtime (days/plant/year):						
Feedstock curtailment	0.8	0.6	20.0	7.6	11.2	0.5
Inventory control	45.4	8.7	15.2	25.2	20.8	9.7
Other	0.9	2.0	4.2	2.3	3.4	19.1
Total	47.1	11.2	39.4	35.1	35.4	29.3
Net avoidable (days/plant/year)	24.1	37.4	45.2	35.0	24.7	41.9
Service factor (%)	92.4	89.4	87.7	89.4	92.5	87.5
B Turnarounds						
Time between turnarounds (days)	21.7	33.2	33.7	30	30.8	21.2
Actual frequency (months)	24	20.5	13.5	18	17	16
Desired frequency (months)	23.5	25	14.4	20.6	16	12
C Classification of shut-downs						
Avoidable shut-downs (shut-downs/plant/year):						
Instrument failure	1.2	1.4	1.9	1.5	0.8	1.9
Equipment failure	3.2	3.5	5.5	4.1	10.7	8.0
Turnarounds	0.5	0.4	0.7	0.5	0.5	0.7
Other	0.7	0.4	1.0	0.8	1.1	1.8
Total	5.6	5.8	9.1	6.9	13.1	12.3
Unavoidable shut-downs (shut-downs/plant/year):						
Feedstock curtailment	0.14	0.08	0.74	0.34	0.31	0.41
Inventory control	0.34	0.07	0.23	0.23	0.29	0.48
Electrical failure	0.36	0.17	1.37	0.67	2.31	2.45
Other	0.27	0.11	0.19	0.20	0.46	3.14
Total	1.11	0.43	2.53	1.44	3.37	6.48
Total shut-downs (shut-downs/plant/year)	6.7	6.3	11.6	8.3	16.5	18.8
D Classification of downtime						
Avoidable shut-downs (days/plant/year):						
Instrument failure	1.1	1.4	1.1	1.2	0.4	2.0
Equipment failure	11.4	20.8	20.3	16.7	9.1	20.0
Turnarounds	10.5	14.6	20.8	15.2	14.0	16.0
Other	1.0	0.7	2.9	1.6	1.4	4.0
Total	24.1	37.4	45.2	35.0	24.7	41.9
Unavoidable shut-downs (days/plant/year):						
Feedstock curtailment	0.8	0.6	20.0	7.6	11.2	0.5
Inventory control	45.4	8.7	15.2	25.2	20.8	9.7
Electrical failure	0.2	0.5	1.7	0.8	2.1	3.1
Other	0.7	1.4	2.5	1.5	1.4	16.0
Total	47.1	11.2	39.4	35.1	35.4	29.3
Total downtime (days/plant/year)	71.1	48.6	84.5	70.1	60.2	71.3

world; (2) small plants, usually with multiple parallel streams of reciprocating compressors; and (3) partial oxidation plants. The general nature of the shut-downs and downtime is shown in Table 2.25. Part of the downtime is due to non-plant problems (gas curtailment and market problems) and part is due to plant problems. The service factor is the availability net of non-plant problems. The contributions to shut-downs and downtime are given in Table 2.26 in terms of plant area and type of equipment and in Table 2.27 in terms of individual items of equipment. Table 2.28 gives data on fires in the plants.

Table 2.26 *Major equipment failures causing shut-down and downtime in ammonia plants[a] (after G.P. Williams, Hoehing and Byington, 1987) (Courtesy of the American Institute of Chemical Engineers)*

	Large tonnage plants				*Reciprocating plants*	*Partial oxidation plants*
	North America	*Europe*	*Rest of world*	*Total*		
A Equipment failures causing shut-downs by plant area						
Primary reforming	3.0 (0.70)	3.0 (0.52)	4.8 (0.68)	3.7 (0.65)	1.1 (0.50)	–
Secondary reforming (including waste heat boilers)	1.3 (0.14)	5.4 (0.24)	1.3 (0.13)	2.3 (0.16)	0.8 (0.10)	–
Purification	1.1 (0.44)	1.9 (0.39)	1.9 (0.68)	1.6 (0.51)	1.6 (0.59)	4.3 (1.74)
Synloop	1.3 (0.48)	2.2 (0.34)	1.1 (0.36)	1.5 (0.40)	0.8 (0.52)	3.7 (0.60)
Compression	3.5 (1.20)	3.1 (0.91)	6.1 (2.32)	4.3 (1.52)	1.7 (7.71)	3.3 (1.45)
Miscellaneous	1.3 (0.24)	5.2 (1.10)	5.2 (1.36)	3.7 (0.86)	3.1 (1.23)	8.7 (4.21)
Total	11.4 (3.21)	20.8 (3.51)	20.3 (5.53)	17.0 (4.10)	9.1 (10.65)	19.9 (8.00)
B Equipment failures by type of equipment						
Reformer piping	2.3 (0.48)	1.5 (0.27)	4.1 (0.45)	2.8 (0.42)	0.8 (0.32)	–
Other piping	0.6 (0.22)	2.0 (0.48)	1.1 (0.43)	1.1 (0.36)	0.8 (0.49)	3.3 (0.81)
Valves	0.4 (0.22)	0.2 (0.33)	0.7 (0.53)	0.4 (0.36)	0.4 (0.35)	0.9 (0.57)
Compression:						
Centrifugal	2.6 (1.02)	2.8 (0.60)	5.1 (1.20)	3.5 (0.98)	0.1 (0.17)	1.1 (0.38)
Reciprocating	–	–	–	–	1.7 (7.65)	1.0 (0.02)
Pumps and drives	0.1 (0.10)	0.4 (0.10)	0.3 (0.19)	0.3 (0.13)	0.1 (0.10)	0.3 (0.33)
Exchangers	3.3 (0.66)	5.8 (0.91)	6.1 (1.17)	4.9 (0.91)	2.9 (1.13)	4.0 (1.19)
Vessels	0.7 (0.10)	6.1 (0.24)	1.3 (0.13)	2.3 (0.15)	1.7 (0.23)	5.2 (2.93)
Miscellaneous	1.5 (0.31)	1.8 (0.58)	1.8 (1.40)	1.7 (0.79)	0.6 (0.21)	4.2 (1.8)
Total	11.4 (3.21)	20.8 (3.51)	20.3 (5.50)	17.0 (4.10)	9.1 (10.65)	19.9 (8.00)

[a] Shut-down in shut-downs/plant/year; downtime in days/plant/year, in parentheses.

Table 2.27 *Equipment and other failures in ammonia plants[a] (after G.P. Williams, Hoehing and Byington, 1987) (Courtesy of the American Institute of Chemical Engineers)*

	North America		*Europe*		*Rest of world*		*Total*	
	DT	*SD*	*DT*	*SD*	*DT*	*SD*	*DT*	*SD*
A Equipment failures								
Primary reforming:								
Tube–riser–pig	1.99	0.44	1.25	0.24	2.82	0.35	2.10	0.36
Transfer header	0.32	0.04	0.45	0.04	1.25	0.10	0.68	0.06
Convection section	0.40	0.07	0.38	0.05	0.12	0.03	0.30	0.05
ID–FD fan	0.18	0.12	0.75	0.15	0.45	0.14	0.42	0.13
Miscellaneous	0.12	0.03	0.21	0.04	0.20	0.06	0.17	0.04
Total	3.01	0.70	3.04	0.52	4.84	0.68	3.67	0.65
Secondary reforming:								
Primary waste heat boiler	0.74	0.06	1.76	0.12	0.17	0.01	0.80	0.06
Secondary waste heat boiler	0.35	0.06	0.41	0.05	0.16	0.05	0.30	0.05
Secondary reformer	0.20	0.01	3.05	0.06	0.88	0.13	1.17	0.07
Miscellaneous	0.00	0.01	0.14	0.01	0.04	0.08	0.05	0.03
Total	1.29	0.14	5.36	0.24	1.25	0.27	2.32	0.21
Purification:								
Exchangers	0.41	0.18	0.28	0.07	1.04	0.26	0.60	0.18
Pumps and drives	0.06	0.07	0.25	0.05	0.05	0.05	0.11	0.06
Piping and valves	0.27	0.14	0.08	0.12	0.59	0.29	0.34	0.19
Vessels	0.36	0.04	1.24	0.13	0.18	0.05	0.52	0.07
Miscellaneous	0.01	0.01	0.05	0.02	0.04	0.03	0.03	0.02
Total	1.11	0.44	1.90	0.39	1.90	0.68	1.59	0.51

continued

Feed compressor:								
Driver	0.03	0.02	0.00	0.00	0.00	0.00	0.01	0.01
Compressor	0.03	0.01	0.01	0.01	0.02	0.03	0.02	0.02
Miscellaneous	0.00	0.01	0.16	0.02	0.02	0.01	0.05	0.01
Total	0.06	0.04	0.17	0.03	0.04	0.04	0.08	0.04
Air compressor:								
Driver	0.25	0.09	0.14	0.05	0.61	0.02	0.35	0.05
Compressor	0.44	0.15	0.14	0.07	0.69	0.13	0.45	0.12
Miscellaneous	0.11	0.07	0.04	0.09	0.37	0.24	0.18	0.14
Total	0.80	0.31	0.32	0.21	1.67	0.39	0.99	0.31
Syngas compressor:								
Driver	0.50	0.21	0.33	0.07	0.76	0.16	0.55	0.16
Compressor	0.73	0.25	1.38	0.22	2.01	0.58	1.35	0.36
Miscellaneous	0.26	0.19	0.62	0.28	0.93	0.83	0.59	0.44
Total	1.49	0.65	2.33	0.57	3.70	1.57	2.49	0.96
Syngas circulator:								
Driver	0.01	0.01	0.00	0.00	0.00	0.00	0.00	0.00
Compressor	0.00	0.01	0.00	0.00	0.01	0.02	0.00	0.01
Miscellaneous	0.00	0.00	0.00	0.00	0.01	0.01	0.00	0.00
Total	0.01	0.02	0.00	0.00	0.02	0.03	0.01	0.02
Refrigeration compressor:								
Driver	0.14	0.05	0.01	0.02	0.37	0.09	0.19	0.06
Compressor	0.33	0.09	0.07	0.01	0.17	0.04	0.21	0.05
Miscellaneous	0.66	0.04	0.17	0.07	0.09	0.16	0.33	0.09
Total	1.13	0.18	0.25	0.10	0.63	0.29	0.73	0.20
Synloop and refrigeration:								
Exchangers	0.72	0.24	0.26	0.09	0.81	0.15	0.63	0.17
Piping and valves	0.39	0.20	1.51	0.24	0.30	0.19	0.65	0.21
Vessels	0.07	0.03	0.40	0.01	0.01	0.02	0.13	0.02
Miscellaneous	0.10	0.01	0.00	0.00	0.00	0.00	0.04	0.00
Total	1.28	0.48	2.17	0.34	1.12	0.36	1.45	0.40
Miscellaneous equipment:								
Auxiliary boiler	0.48	0.02	0.31	0.03	0.45	0.07	0.43	0.04
Piping and valves	0.27	0.10	0.63	0.44	0.90	0.49	0.59	0.33
Exchangers	0.17	0.02	2.59	0.51	3.31	0.59	1.91	0.35
Vessels	0.06	0.01	1.40	0.04	0.19	0.04	0.45	0.03
Pumps and drives	0.08	0.04	0.17	0.05	0.23	0.14	0.16	0.08
Miscellaneous	0.20	0.05	0.12	0.03	0.08	0.03	0.14	0.04
Total	1.26	0.24	5.22	1.10	5.16	1.36	3.66	0.86
Grand total	11.44	3.20	20.76	3.50	20.33	5.67	16.99	4.16
B Electrical, instrument and other shut-down causes								
Electrical:								
External	0.18	0.36	0.53	0.17	1.70	1.37	0.81	0.67
Internal	0.49	0.22	0.19	0.20	1.70	0.35	0.84	0.26
Total	0.67	0.58	0.72	0.37	3.40	1.72	1.65	0.93
Instruments:								
Feed compressor	0.03	0.04	0.00	0.00	0.02	0.02	0.02	0.02
Air compressor	0.14	0.19	0.08	0.15	0.06	0.20	0.10	0.18
Syngas compressor	0.17	0.33	0.76	0.47	0.37	0.68	0.39	0.49
Refrigeration compressor	0.07	0.09	0.10	0.10	0.06	0.18	0.07	0.12
Miscellaneous	0.64	0.51	0.47	0.70	0.60	0.79	0.58	0.66
Total	1.05	1.16	1.41	1.42	1.11	1.87	1.16	1.48

continued

Other:								
Strikes	0.00	0.00	1.35	0.05	0.10	0.01	0.38	0.02
Weather	0.66	0.18	0.04	0.03	1.19	0.05	0.69	0.10
Operator error	0.01	0.18	0.12	0.21	0.08	0.13	0.10	0.17
Catalyst	0.12	0.12	0.23	0.02	0.42	0.08	0.25	0.08
Miscellaneous avoidable	0.24	0.03	0.16	0.09	0.40	0.38	0.28	0.17
Miscellaneous unavoidable	0.05	0.34	0.03	0.01	1.24	0.13	0.47	0.18
Total	1.17	0.85	1.93	0.41	3.43	0.78	2.17	0.71
Feedstock/market curtailment:								
Feedstock curtailment	0.80	0.14	0.55	0.08	19.97	0.74	7.56	0.34
Inventory control	45.37	0.34	8.69	0.07	15.16	0.23	25.21	0.23
Total	46.17	0.48	9.24	0.15	35.13	0.97	32.77	0.57
Grand total	49.06	3.07	13.30	2.35	43.07	5.34	37.75	3.69

[a] DT, downtime (days/plant/year); ID–FD, induced draft-forced draft; SD, shut-downs (shut-downs/plant/year).

Table 2.28 *Fires in ammonia plants (after G.P. Williams, Hoehing and Byington, 1987) (Courtesy of the American Institute of Chemical Engineers)*

	1973–76	*1977–81*	*1981–85*
A No. of plant fires			
No. of plants having no fires	2	22	41
No. of plants having fires	27	74	95
No. of fires	125	257	520
Frequency of fires (fires/month)	11.1	14.6	12.2
B Classification of fires (%)			
Flange	36	32	31
Valve packing	8	4	8
Oil leaks	20	29	19
Transfer header	7	–	1
Piping	10	9	11
Electrical	2	3	3
Miscellaneous	17	23	27

Table 2.29 *Trend in the number of major accidents in UK, Europe and world-wide (after Keller and Wilson, 1991)*

	1970–79		*1980–87*		*1970–87*		
No. of fatalities	*Europe*	*World-wide*	*Europe*	*World-wide*	*UK*	*Europe*	*World-wide*
A Number of accidents							
5-10	12	53	6	40	3	18	93
10-100	10	45	5	30	2	15	75
100-10^2	2	3	0	2	0	2	5
10^2-10^3	0	0	0	1	0	0	1
B Frequency of accidents (accidents/year)							
5-10	1.2	5.3	0.75	5.0			
10-100	1.0	4.5	0.63	3.8			
100-10^2	0.2	0.3	0	0.25			
10^2-10^3	0	0	0	0.13			

2.10 Trend of injuries

The long-term trend of injury rates in the process industry is downwards. The trend in the UK may be seen in Table 2.6, which shows that between 1970–74 and 1987–90 the fatal accident rate for the chemical and allied industries fell from 4.3 to 1.2. However, as Table 2.4 shows, the actual number of fatalities in this industry is small. Consequently, one large accident would have a significant effect on the figures, even if its effect were absorbed over quite a long period such as 10 years.

Evidence is available that the efforts devoted to safety and loss prevention by some companies have borne fruit. Figure 2.8 Hawksley (1984), shows the trend of the fatal accident rate in ICI over the period 1960–82. The data represent the 5-year moving average. They show that particular success was achieved in reducing the fatal accidents associated with the process risks, the reduction factor being about 15. The success achieved by certain US companies in reducing the lost time injury rate (LTIR) is illustrated in Figure 2.9 (Brian, 1988). The LTIR is the percentage of workers in the organization who, in a given year, suffer an injury so severe that it causes the worker to lose some work days. In one case the reduction factor is 280.

The trend in multiple fatality accidents has been studied by Keller and Wilson (1991), who obtained the results given in Table 2.29. The table shows a broadly stable situation.

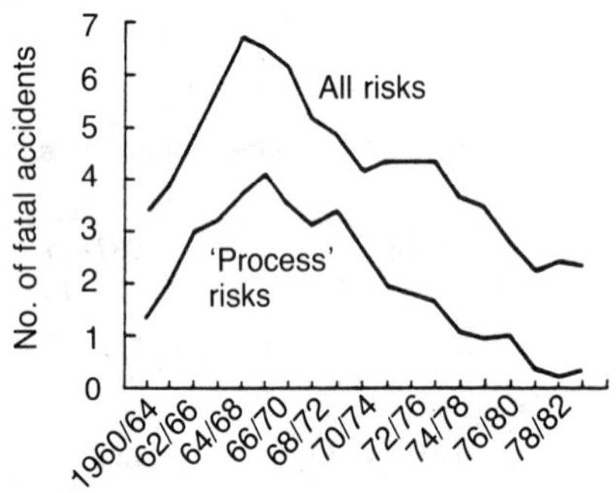

Figure 2.8 Trend of the fatal accident rate in ICI, 1960–82 (Hawksley, 1984). The number of fatal accidents is given as the number in 10^8 working hours or the number in 1000 men in a working lifetime expressed as a 5-year moving average

2.11 Trend of Losses

There have been a number of studies addressing the question of whether the number of major accidents in the chemical and oil industries is increasing. Early work on these lines was done by V.C. Marshall (1975c) and Kletz and Turner (1979). The insurers Marsh and McLennan (M&M) publish periodically a list of the 100 largest losses in the chemical and oil industries worldwide over a running 30-year period, (e.g. Garrison, 1988b; Mahoney, 1990; Marsh and McLennan, 1992). The 1987 edition contains the analysis shown in Table 2.30. This indicates a trend which is rising up to 1986.

A further review of loss trends is that given by Crook (1994, LPB 115). He comments that the M&M data suggests that until 1989 losses were doubling each decade. He gives further data from a study drawing on the *Lloyds Weekly Casualty Reports* (LWCRs). The frequency for the incidents which he considers, has risen from some 28/year in 1971 to some 103/year in 1991.

A particular type of incident which has received individual attention is the vapour cloud explosion

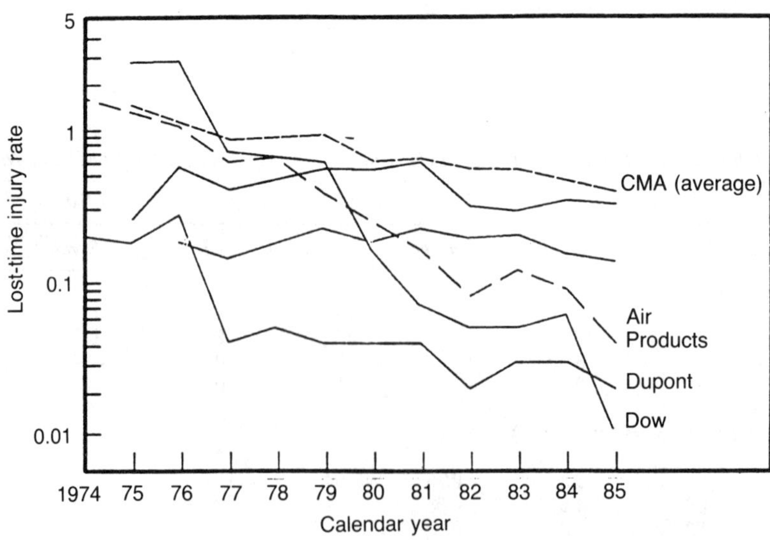

Figure 2.9 Trend of lost-time injury rate in some US companies, Chemical Manufacturers Association 1974–85 (Brian, 1988) (Courtesy of the Institution of Chemical Engineers)

Table 2.30 *The 100 largest losses in the chemical and oil industries world-wide (after Marsh and MacLennan, 1987)*

Period	*No. of losses*	*Average loss (US$m.)*
1957–66	15	28.5
1967–76	29	38.2
1977–86	56	36.6

Table 2.31 *Trend in number of vapour cloud explosions world-wide (after Lenoir and Davenport, 1992)*

Period	*No. of VCEs*
1930–34	0
1935–39	1
1940–44	1
1945–49	2
1950–54	5
1955–59	8
1960–64	11
1965–69	12
1970–74	20
1975–79	22
1980–84	8
1985–89	15
1990–July 1991	7

(VCE), formerly generally referred to as unconfined vapour cloud explosion (UVCE). In the UK the VCE at Flixborough resulted in a heightened awareness. A study at that time by V.C. Marshall (1975c) showed that the number of VCEs world-wide was indeed on an apparently increasing trend. Similarly, in his book on VCEs, Gugan (1979) stated: 'The trends in frequency, proportion of incidents producing blast and fatalities in UVCEs are all upwards.' This trend appears now to have been arrested. Industrial Risk Insurers (IRI) publish a periodic survey of VCE incidents. The number of VCEs obtained from this survey is given in Table 2.31.

2.12 Case Histories

The generalized statistics may be supplemented by individual case histories. These are treated in Appendix 1, which describes the various sources and gives specific case histories. The sources include: the accident reports of the Health and Safety Executive (HSE) and the National Transportation Safety Board (NTSB); the collections of the Manufacturing Chemists Association (MCA) and the American Petroleum Institute (API); the periodic reviews of insurers, such as the *100 Large Losses* (Marsh and McLennan); the *NFPA Quarterly* of the National Fire Protection Association (NFPA), the *Chemical Safety Summary* of the Chemical Industry Safety and Health Council (CISHC) and its successor, the Chemical Industries Association (CIA), and the *Loss Prevention Bulletin* of the Institution of Chemical Engineers (IChemE), all of which publish case histories.

Accounts of major accidents are given in: Appendices 2-6 on Flixborough, Seveso, Mexico City, Bhopal and Pasadena, respectively; Appendix 16 on San Carlos; Appendix 19 on Piper Alpha; and Appendices 21 and 22 on Three Mile Island and Chernobyl.

3 Legislation and Law

It is the object of this chapter to give a brief account of the principal legislation and of the legal background relevant to safety and loss prevention. The account is largely restricted to the situation in the UK. Within the UK, the law of Scotland is distinct from that of England and Wales and where differences exist the account given is based on the latter. There is some reference to legislation within the European Community and in the USA. Selected references on legislation and law relevant to safety and loss prevention are given in Table 3.1.

Table 3.1 *Selected references on legislation and law*

A Britain

Industrial law, employer's liability
Legge (1934); Allsop (1962); Fife and Machin (1963, 1972, 1982, 1990); Whincup (1963, 1968, 1976); Wedderburn (1965); Ison (1967); F.A. Robinson and Amies (1967); Samuels (1967, 1969); Winn (1968); Gold (1969); Atiyah (1970); Law Commission (1970a,b); Society of Labour Lawyers (1970); Broadhurst (1971); Munkman (1971); Robens (1972); Kinnersly (1973); D. Farmer (1977); Locke (1977); W.E. Cox and Walker (1978); J. Jackson (1979); Anon. (1980 LPB 35, p. 13); Staunton (1982); Farquhar (1986); Anon. (1987d): A.L. Jones (1994)

Historical background
M.W. Thomas (1948); BASF (1965); Hutchins and Harrison (1966); Anon. (1972b); Robens (1972); V.C. Marshall (1975b); RoSPA (1976 IS/108); HSE (1983d)

Chemical industry
H. Watts (1956); J.R. Hughes (1970); Elphick (1972); R. King and Magid (1979); Cullis and Firth (1981); Burgoyne (1985); Jarratt (1986); Laing (1986); C. Martin (1986); D. Stevens (1986); Viney (1990); Pantony (1991 LPB 100); Dykes (1992)

Factories Act 1961
HM Chief Inspector of Factories (annual report); HMSO (periodic); Ministry of Labour (1964); DoEm (1972/2); Fife and Machin (1972, 1988); McKown (1974); HSE (1992/31)

Factory Inspectorate
HMSO (periodic); HSE (Appendix 28, inc. *HSE Leaflets*, 1972/4, 1976/7, 1980 G 1, 1983c, 1984, 1985 OP 11, 1987/23, 1988 HSE 22); Legge (1934); Djang (1942); Symons (1969); Robens (1972); Anon. (1974d); Blakey (1974); HM Chief Inspector of Factories (1974, 1986a,b, 1988a); Critchley (1981); Church (1986).
Medical Inspectorate: HSE (1979/9, 1985 HSE 5)

Robens Report
Robens (1972); Beeby (1974a,b); Chicken (1975)

Self-regulation
Dawson *et al.* (1988)

Health and Safety at Work etc. Act 1974, including implementation
HSC (HSC 8–11, 1975 HSC 1-7, 1977, 1985, 1986/5); RoSPA (IS/103, 1976 IS/108); Powell-Smith (1974); CAPITB (1975 Information Paper 16); Cartwright (1975); Cushworth (1975); D. Farmer (1975); J. Jackson (1975); W.T. Jones (1975); V.C. Marshall (1975a,b); E. Mitchell (1975); Riston (1976); HSE (1977d, 1980b, 1983 HS(R) 6, 1986 HSE 16, 1989/27, 1991 IDD(G) 117(L), 118(L), 1992/32); IMechE (1977/36); Warwick (1977); Buchanan (1979): Kletz (1979d); J.Q. Wilson (1980); Anon. (1981i); Anon. (1982l); Anon. (1984u,mm); Dewis (1985b); Baker-Counsell (1986b); FPA (1989 CFSD OR 3); Goldsmith (1992a); IBC (1993/99)

Regulations
HSE (Appendix 28 Regulations Booklets)

Safety policies
BSC (n.d./8); CIA and CBI (CIA 1974/7); CBI (1974); Egan (1975, 1979); HSC (1975 HSC 6, 1991 HSC 6); HSE (1980/10, 1986/19)

Safety committees, safety representatives
HSC (HSC 8, 1977 HSC 3); HSE (1990 CRR 201, 1991 IND(G) 119(L))

Approved Codes of Practice
HSC (1975 HSC 7); Kletz (1985m); Sewell (1987); Dykes (1992)

National standards
Perry (1985)

Alkali Act, Alkali Inspectorate
HM Chief Alkali Inspector (annual); Ireland (n.d.); Ministry of Housing and Local Government (1959); Garner and Crow (1969); Napier (1974a); Garner (1975a,b); Social Audit (1974); Tunnicliffe (1975); Anon. (1979a)

Major hazards
Chicken (1975); Harvey (1976, 1979b, 1984); Beveridge (1979); IBC (1982/33); HSE (1983 HS(R) 16, 1985 HS(R) 21); Barrell (1984, 1985, 1990, 1992); Bennett (1988); Cassidy (1988c); Deieso *et al.* (1988)

Particular hazards, activities and systems
Equipment: HSE (1977 GS 8, 1982 IND(G) 1(L))
Building Regulations: Anon. (1992d); CONDAM
Regulations: Ottewell (1993)
Construction regulations: RoSPA (IS/13)
Pressure vessels, pressure systems: Warwick (1969); J.R. Hughes (1970); Pilborough (1971, 1989); Toogood (1972); HSC (1977/1); Heathershaw (1978); Anon. (1990j); HSE (1990 HS(R) 30)
Electricity: HSE (1984 HS(R) 18, 1989 HS(R) 25)
Flammables: Bryce *et al.* (1982); Winston (1984)
Explosives: Ministry of Transport (n.d.); J.H. Thomson (1941); H. Watts (1954); ICI (1961); HSE (1991 IND(G) 115(L))
Toxics: HSC (1977); Langley (1978); Cavallo (1979); George, Phillips and Silk (1980); C.R. Pearson (1982); Pitblado (1989)
Carcinogens: J.L. Fox (1980b)
Asbestos: IBC (1985/61)
Notification of New Substances Regulations: van den Heuvel (1982); HSE (1982 HS(R) 14)

Fire, fire certificates: Langdon-Thomas (1967a,b); Everton (1972); HSE (1976/6); BRE (1986 BR85); FPA (1989 CFSD FPDG 1, OR 1); Everton *et al.* (1990)
Storage: H. Watts (1951); J.R. Hughes (1970); Anon. (1985cc); M.R. Wright (1985)
Classification, labelling and packaging, CLP Regulations: HSE (1978 HS(R) 1, 1983 HS(R) 17, 1985 HS(R) 22); IBC (1983/44)
Transport: British Rail (n.d.); HSE (1979 HS(R) 3, 1987 HS(R) 24); Ministry of Transport (n.d.); British Transport Commn, British Waterways (1962); Home Office (1967/9); Gross (1968); J.R. Hughes (1970); A.W. Clarke (1971a,b); A.B. Kelly (1971); Walmsley (1973); PITB (1975/3); HSE (1985 PML 6); IMechE (1975/23, 1976/30); Savage (1979); K. Warner (1981); Bailey (1984); Sewell (1984a, b); Anon. (1985aa); Williamson (1985); M.R. Wright (1985); ACDS (1991); B. Davies (1993); Crooks (1994)
Ports: HSE (1988 HS(R) 27)
Occupational health, including COSHH Regulations 1988: IBC (1985/60); Blain (1987); Carson and Mumford (1987 LPB 76); Waterman (1987); Boniface (1990b); Wrightson (1990 LPB 90); Anon. (1991j); Simpson and Simpson (1991)
Ionizing radiations: Dewis (1985c)
Product safety, liability, including Consumer Protection Act 1987: CAPITB (1979); Kainer (1984); Anon. (1986aa); Appleton (1986, 1988); Dewis (1986); CIA (1992 RC46)
Accident reporting: RIDDOR: HSE (1986 HS(R) 23, HSE 11, 17, 1988 HSE 21)
Laboratories: Marshall (1980 LPB 32)

Pollution
HM Chief Alkali Inspector (annual report); Nonhebel (1958); McLoughlin (1972, 1973, 1976, 1980); Bingham (1973); Tunnicliffe (1973, 1975); McKnight *et al.* (1974); Napier (1974a); Social Audit (1974); G.G. Jones (1977); Pocock and Docherty (1978); Anon. (1979a); Hawkins (1979); Tetlow (1979); D.D. Young (1979); Ireland (1981, 1984, 1987); M.A. Smith (1985); R.E., Smith (1985); C. Tayler (1985f); Harper (1986); Anon. (1991h); Bristows Environmental Law Group (1991); Goldsmith (1992a); Gardner (1994); Petts and Eduljee (1994)

Noise
Kerse (1975)

Offshore
Burgoyne (1980); Petrie (1983); DoEn (1984); Barrett, Howells and Hindley (1987); Aarstad (1990); Cullen (1990); Kenney (1991); Salter (1994)

Safety professionals
V.C. Marshall (1983a); A.L. Jones (1984)

Engineer's liability
Anon (1980f); J.S. Parkinson (1980); Dauerman (1982); IChemE (1987a); K. Taylor (1988); Kletz (1991b); D. Davis (1993)

Other related fields
A. Bryan (1975) (mines); Street and Frame (1966) (nuclear energy)

B USA

Anon. (1976f); D.E. Miller (1976); Marion (1978a,b); Moon (1978); Bevirt (1979); Gilbert (1979); J. Harris (1980); W.G. Johnson (1980); Whittaker (1981); Breyer (1982); Cahan *et al.* (1982); Okrent and Wilson (1982); R. Nash (1984); Oreffice (1984); Breyer and Stewart (1985); R.K. Johnson (1985); Okrent *et al.* (1985); Moskowitz (1986); Halley (1977); Olsen (1987); Bisio (1989, 1991b); Horner (1989); Waldo (1989); Burk (1990); Majumdar (1990); OSHA (1990b); A.S. West (1991); Milburn and Cameron (1992)

Occupational Safety and Health Act 1970
OSHA (OSHA 2001, 2098, 2253, 2254, 1988/4); American Society for Personnel Administration (1970); Anon. (1971c,g); Hodgson (1971); S. Ross (1971, 1972b); Vervalin (1971–, 1972b,d); Nilsen (1972a); L.J. White (1972, 1973b, 1974); Anon. (1973d,e); Ludwig (1973a,b); Anon (1974h); Peck (1974); Stender (1974); Hopf (1975); D. Petersen (1975); Corn (1976); Anon (1977m); Demery (1977); Bureau of National Affairs (1978); Moran (1979); Ladou (1981); ACGIH (1990/50); A. Foster and Bund (1993)

Occupational Safety and Health Administration
J. Shaw (1978, 1979); R.D. Morgan (1979); Cahan (1982b); H. Bradford (1986b); Spiegelman (1987); AGA (1990/14); Burk (1990); Seymour (1992)

Environmental Protection Agency
J. Shaw (1978); Porter (1988); K.A. Friedman (1989a,b); Stephan and Atcheson (1989)

National standards
van Atta (1982); W.J. Bradford (1985); M.F. Henry (1985); R.K. Johnson (1985); Short (1985)

Major hazards
Shortreed and Stewart (1988); Brooks *et al.* (1988)

Environmental impact statements
Willis and Henebry (1976, 1982–, 1983)

Special Emphasis Program
Hanson (1986)

Process Safety Management
OSHA (1990b); Anon. (1992e); Chowdhury and Parkinson (1992); Kuryla and Yohay (1992); Bisio (1993); Lopinto (1993)

Coastal Zone Management Act
Knecht (1978); R.F. Nelson (1978)

Toxics
Anon. (1973f); Lepkowski (1973); Nilsen (1974); Vervalin (1975b); Ricci (1976b, c); Dominguez (1977, 1978a,b); C.W. Smith (1977); D. Brown (1978); Doniger (1978); A.S. West (1979, 1982, 1986, 1993b); Kuhn (1980a); Vandergrift (1980); Corn and Corn (1984); Rodricks and Tardiff (1984); Anon. (1987j); Colen (1987a, 1988); Lipsett (1987); Rosenberg (1987); Dombrowski (1988)
Carcinogens: Cahan (1982a); Trewhitt and Cahan (1983); Deisler (1984a); Rodricks and Tardiff (1984)

Storage
Anon. (1987w)

Toxic Substances Control Act 1976
Dominguez (1977, 1978); Marion (1978a,b); J. Smith (1978b); EPA (1979, 1992); Recio *et al.* (1979); A.S. West (1979, 1982, 1986, 1993b); W.D. Muir (1980); Trewhitt (1981); Tabershaw (1982); US Congress, OTA (1983); Burch (1986); Hansch (1991); Kuryla (1993)

Emergency Planning and Community Right-to-Know Act 1986
Makris (1988); Bowman (1989); Bisio (1992b)

New Jersey Toxic Catastrophe Prevention Act 1986
Florio (1987); Somerville (1990)

Pollution
Chementator (*see* Table A3.3); Yocom (1962); Kirk (1967); L.W. Ross (1967, 1972b); Anon. (1968a); Remirez (1968c); Sheldrick (1968); Coulter (1972); S.S. Ross (1972b); Vervalin (1972d); Anon. (1973g); Weisberg (1973); L.J. White (1973a, 1974); Zarytkiewicz (1975); Rich (1976); Steymann (1976); Kohn (1977a); E.B. Harrison (1978a,b); Henebry and Brown (1978); Passow (1978); J.I. Lewis and Tarsey (1978); Booth (1979); Dobrzynski (1979); Weismantel and Parkinson (1980); Fabian (1982b); Bacow and Wheeler (1984); Ihnen (1984); Trewhitt (1985); H. Bradford (1986a); Caruana (1986); Anon. (1987o); Molton (1988); G. Parkinson (1988); Stevens (1988); Neff (1990); Anon. (1991e); Ruzzo (1991); Shelby *et al.* (1991); Bisio (1992); Davenport (1992a–c)
Fugitive emissions: Colyer and Meyer (1991); Davenport (1992b)
Hazardous waste: J.W. Lynch (1980); J.S. Shaw (1980a); J.L. Fox (1981); H.H. Fawcett (1982b); Kahane *et al.* (1985); Colen (1987b); Caruana (1988); Melamed (1989); Davenport (1992a)
Hazardous spills: G.F. Bennett and Wilder (1981); Henrichs and Hallenbeck (1986)

Clean Air Act
E.B. Harrison (1978b); Siegel *et al.* (1979); G.A. Brown *et al.* (1988); Zahodakian (1990); Matthiessen (1992); Kaiser (1993)

Resource Conservation and Recovery Act
Glaubinger (1979); Sobel (1979); Basta (1981c); H.H. Fawcett (1982b); Hoppe (1982)

Superfund
J.S. Shaw (1980b); Resen (1984); Casler (1985); Sidley and Austin (1987); Habicht (1988); Nott (1988); Bowman (1989); Hirschorn and Oldenburg (1989); Melamed (1989); Rhein (1989); Bisio (1992a)

SARA
K.A. Friedman (1989a,b); Anon (1990k); ACGIH (1990/50); Curran and Kizior (1992); Fillo and Keyworth (1992); Heinold (1992); Keyworth *et al.* (1992)

Nuclear
T.W. Evans (1979)

Consumer protection, product liability
Kolb and Ross (1980)

Other legislation
Chowdhury (1982); Sabatier and Mazmanian (1983)

Engineer's liability
T.D. Kent (1978)

C European Community

CEC (EUP 13699 EN); J. Smith (1978b,c); European Parliament (1979); Trowbridge (1979); Vinck (1982); Blokker (1986, 1987); HSE (1986/8, 1991/29); Kremer (1987); Pilz (1987, 1989); CONCAWE (1989 89/55, 1991 91/55, 1992 92/55); Anon. (1991j); IBC (1992/90, 92, 1994/114); Milburn and Cameron (1992); Agra Europe (1994a,b)

'Dangerous Substances' Directive
Chipperfield (1979); Cussell (1979); Malle (1979); Otter (1979); Furlong (1991)

'Packaging and Labelling' Directive
C. Mattin and Ranby (1982); Williamson (1985)

'Major Accident Hazards' Directive
IBC (1984/52); Dewis (1985a); Vermeulen and Hands (1993)

Toxics
Chipperfield (1979); Cussell (1979); Malle (1979); Otter (1979)

Pollution
CONCAWE (1977 2/77); J. Smith (1978); Provinciale Waterstaat Groningen (1979); Tetlow (1979); McLoughlin (1980); Salamitou and Robin (1980); Szelinski (1980); B.D. Clark *et al.* (1984); D. Williams (1985); Fairclough (1986); Häberle (1986); Anon. (1988e); O'Riordan (1989); Anon. (1993b)

Other countries
Skole and Pilarski (1978); Anon. (1980w)

D General

Engineering ethics, engineer's responsibility
Kornhauser (1962); Siekevitz (1972); Tamplin (1973); G.D. Friedlander (1974, 1975); Burck (1975); Hughson and Kohn (1980); Mingle and Reagan (1980); Thring (1980); F. Warner (1981a); Flores (1983); Lindauer and Hagerty (1983); Tucker (1983); Wilcox (1983); Kohn and Hughson (1985); Gunn and Vesilind (1986); Matley and Greene (1987); Matley *et al.* (1987); M.W. Martin and Schintzing (1989); Mingle (1989); C. Butcher *et al.* (1991); Kletz (1991b); Mascone *et al.* (1991a, b); Rosenzweig and Butcher (1992a,b); Wells (1992); D. Davis (1993)

Engineering Council codes
Engineering Council (1991, 1992, 1993)

Corporate responsibility, corporate manslaughter
C. Wells (1992); V.C. Marshall (1994 LPB 120)

Whistleblowing
Anon. (1985w)

Engineer as expert witness
de Nevers (1988)

3.1 Factory legislation

The current regulatory regime and legislation for health and safety in the UK is best understood as a continuation of developments which have taken place over the best part of 200 years. Accounts of the development of factories legislation are given in *Safety and Health at Work* (Robens, 1972) (the *Robens Report*), the *HM Chief Inspector of Factories Annual Report 1974* (HMCIF, 1974), *Her Majesty's Inspectors of Factories 1833–1983* (HSE, 1983d) and *The Work of a Factory Inspector* (HSE, 1987b). There are further accounts by M.W. Thomas (1948), Hutchins and Harrison (1966), V.C. Marshall (1975b) and the Royal Society for the Prevention of Accidents (RoSPA) (1976, IS/108).

Factories legislation has its roots in the Industrial Revolution and in particular in the cotton mills, and in the conditions and reforming movements to which these gave rise. The spirit of the time favoured a *laissez faire* approach and there was little regulation of industry by the state. Nevertheless, the older paternalist tradition was alive also. The appalling working conditions of children in some of the cotton mills led in 1802 to the first Factories Act, the Health and Morals of Apprentices Act. This limited the working hours of apprentices in such mills and laid down certain standards of heating, lighting, ventilation, etc. It relied for enforcement on visitors appointed by local Justices of the Peace. The Act was generally ineffective, its provisions were often misunderstood or simply flouted, and control by magistrates, who were often closely associated with mill owners, was weak.

During the early part of the century water power in the mills was replaced by steam power and factories were built in towns rather than in the country. Further statutes in 1819, 1825 and 1831 tried to strengthen or extend control of the 1802 Act to meet this changing situation. In 1830 there began the Short Hours Movement. This was the start of a campaign for the statutory limitation of the working hours of women and young persons which was to last some 20 years and was to have a profound influence on the development of factory legislation. The 1831 Factories Act introduced a 12-h day for young persons in cotton mills, but did not abate the agitation for reform. A Ten Hour Bill was brought in by Lord Ashley and the government countered this by appointing a Royal Commission on child labour in factories, which resulted in 1833 in Lord Althorpe's Factories Act.

This Act was a watershed in factory legislation through its provisions for enforcement and its creation of a Factory Inspectorate. It referred specifically to the failure of enforcement of the provisions of the 1802 Act, and it gave the Inspectors the right to enter any factory and to make such regulations, rules and orders as were necessary for enforcement. Nevertheless, much criticism of the Act came from reformers, who feared too close a relationship between the new Inspectors and the mill owners. In the period following this Act the Inspectorate developed its activities, seeking to improve conditions by persuasion and education rather than by confrontation, but gradually becoming more rigorous and plugging more loopholes.

So far the legislation had dealt with working hours and environmental conditions, but not with safety. The Inspectors, who saw in their daily work the appalling accidents which occurred in factories, pressed strongly for safety to be included. The Factories Act 1844 introduced the first safety legislation with its provisions for the guarding of machinery such as flywheels and transmission shafts. The guarding of transmission shafts became a matter of some controversy. The 1844 Act required the guarding of all such shafts, but owners argued that shafts positioned high out of reach were safe. The Inspectorate accepted this and did not insist on strict compliance, but within a few years their experience indicated that 'safety by position' could not be relied on, and in 1854 a circular called for stricter enforcement.

A further five Acts were passed between 1844 and 1856, mainly as a result of difficulties in enforcement or of campaigns to end some particular hazard. During this period opposition by some owners had built up, centring on the National Association of Factory Occupiers – dubbed by Dickens 'The Association for the Mangling of Operatives'. The Factories Act 1856 weakened safety measures by accepting safety by position and by encouraging appeals to arbitration. It needs to be said, however, that many owners did not oppose factory legislation and some were far in advance of their time in trying to provide humane conditions both inside and outside the factories.

The extension of protection to outside the textile industry did not start until the Factories Act in 1864, which brought in a further six trades. In 1867, some heavy metallurgical processes were encompassed by a Factories Act, as were factories employing more than 50 people. Also in 1867, the Workshop Regulation Act extended a degree of protection to workers in smaller establishments.

A recurring theme in factory legislation was the distinction between accidents caused by machinery and other accidents. The 1844 Factories Act required reporting of all accidents, whether associated with machinery or not. The reporting of non-machinery accidents, however, appeared to many Inspectors of doubtful value. As one of them, Horner, put it, such 'accidents might have happened in any dwelling house and the investigation of them could have no tendency whatever to diminish their recurrence.' In consequence, the Factories Act 1871 distinguished between machinery and non-machinery accidents and laid emphasis on reporting and investigation of the former. A distinction developed between premises with and those without machinery. The phrase 'steam, water or other mechanical power' occurs repeatedly in the legislation.

The Factory and Workshop Act 1878 consolidated the existing legislation and gave coverage to virtually all factories. It was followed in 1880 by the Employers Liability Act, which further strengthened the legislation. The Factories Act 1891 gave power to make special regulations for dangerous trades. Industrial activities outside factories proper were generally covered by the extension of legislation to 'notional factories' – docks in 1904, shipyards in 1914, some building sites in 1926 and all building sites in 1948.

Further factory legislation was essentially designed to deal with particular problems, with periodic consolidation. There was also a change in the nature of the Acts insofar as the attempt to use Acts of Parliament to regulate detail was largely abandoned and was entrusted instead to regulations made by a Minister under an Act. The next consolidating Act was the Factories Act 1901, which encompassed five statutes passed since 1878. It was followed by a large number of regulations, some of which have lasted most of the century.

The Factories Act 1937 was an important measure in that it was the first major factories legislation for nearly 40 years and it incorporated several changes of emphasis. It abolished the outmoded distinction between textile mills and other factories. The distinction between premises depending on whether or not they used mechanical power was also ended. Provisions related to non-machinery accidents were emphasized, such as means of access, maintenance of floors and stairs, etc. Additional requirements on other technical hazards such as pressure vessels and flammable and toxic gases and dusts were introduced. Machinery hazards were not neglected, however, and there were various new provisions on machinery, including cranes. The Factories Acts of 1948 and 1959 made relatively few changes, except that the latter did considerably strengthen fire precautions.

Until 1940 the administration of factory legislation lay with the Home Office, but in that year it was transferred to the more appropriate Ministry of Labour, now the Department of Employment.

The most recent Factories Act, in 1961, was a further consolidating measure. It is described in more detail in Section 3.2.

It is clear from this account that British factory legislation developed in response to the particular problems of the Industrial Revolution and has continued to be influenced by this fact. The implications of this in relation to safety and loss prevention are well described by V.C. Marshall (1975b):

> What were the essential features of a factory in the nineteenth century? It employed large numbers of people in enormous rooms. Each operative tended a machine or group of machines which might be there to prepare raw cotton, to spin it or to weave it. The machines were driven by a line shaft, which in its turn was driven by a steam engine with its flywheel. The engine took its steam from a boiler, probably of the well known Lancashire type.
>
> Such factories were untidy, ill ventilated, ill lit, verminous, insanitary, and the moving machinery presented serious dangers of injury or death. They were devoid of mess rooms and wash rooms. There were serious fire hazards, the engine flywheel might burst and the boiler presented the hazard of explosion. The factory employed the cheapest labour it could hire and after the introduction of gas lighting the owner would try to run it day and night particularly when trade was brisk.
>
> An examination of the Factories Acts and related legislation will bear out the view that the model given above is the one which was the basis for such legislation. It would be wrong to pretend that factories similar to this model may not exist today here and there, but how does such a model match up to the realities of the present day process industries? Scarcely at all.
>
> Many of the hazards associated with the nineteenth century factory have never existed in modern factories or they have been swept away by technological change. But they have been replaced by other hazards perhaps of a more serious kind. There is another basic difference, the significance of which is sometimes underestimated, between the nineteenth century and today. And that is the difference between the source of demand for control and legislation in the past and in the present.
>
> In the nineteenth century the source of demand lay in philanthropy. Factory reform, like prison reform or poor law reform, was a branch of philanthropy. It was essentially paternalistic, whether it originated in the work of an enlightened capitalist like Sir Titus Salt or an aristocrat like Lord Shaftesbury. It was paternalistic because *self help* had no place in it. Self help implied combination by work people and this with its connotations of 'restraint of trade' was anathema to the well-to-do reformer.
>
> Finally we should note that the nineteenth century brought also onto the agenda the impact of industry on the public.
>
> Firstly there was the question of boiler explosions which were endemic in the nineteenth century. Under the conditions then existing factory premises were themselves cramped and they were hemmed in by the humble houses of the workpeople. Boiler explosions not only killed workpeople but also their families and other people. Legislation could and did eventually enforce 'best practice' in the realm of boiler management with a significant factor being the influence of those who insured the boilers. But legislation could not go beyond 'best practice' neither then nor can it today.
>
> Secondly there was the question of emissions to the atmosphere. The textile industry not only brought into being the factory but also the large scale chemical industry because of its demands for detergents, bleaches and dyestuffs. One of the first large-scale industries was the Leblanc process with its concomitant production of 'saltcake'. In its original form the saltcake process discharged to atmosphere *one mol of HCl per mol of saltcake manufactured* and it would not, I think, be too much to claim that atmospheric pollution in South Lancashire in those days was one thousand times worse than anything we put up with today. This crying scandal gave birth to the Alkali Works Act to which all later legislation on atmospheric pollution owes its origin. Thus at the beginning of this century and indeed until the beginning of this year legislation to control hazards to the lives and health of work people and to the general public arising from industry were essentially resting on nineteenth century foundations.
>
> *(Reproduced with permission of the author)*

***Figure 3.1** The main lathe shop of Badische Anilin und Soda Fabrik works, Ludwigshafen, about 1921 (BASF, 1965)*

The kind of factory conditions to which much of the factory legislation referred, in the UK and abroad, are illustrated in Figure 3.1, which shows the main lathe shop of the Badische Anilin und Soda Fabrik (BASF) in Ludwigshafen about 1921.

Selected Acts of Parliament and Statutory Instruments, essentially Regulations and Orders, relevant to safety and loss prevention are given in Tables 3.2 and 3.3, respectively.

3.2 Factories Act 1961

The Factories Act 1961 is the last of the long line of Factories Acts. The Act is supplemented by numerous regulations, etc., made under it. Together these form a large body of factory legislation. The Act is described in the *Guide to the Factories Act* (HSE, 1992a). Although the Act is in the process of being superseded by legislation under the Health and Safety at Work etc. Act 1974 (HSWA), its statutory provisions remain in force until they are replaced. The HSWA does, however, change the means of enforcing these provisions. The provisions of the Factories Act 1961 may be considered under the following heads: (1) health, (2) safety, (3) welfare, (4) other matters, and (5) enforcement.

The health provisions are concerned with environmental aspects such as cleanliness; overcrowding, temperature, ventilation, lighting, drainage and underground rooms; sanitary and meal arrangements; processes involving lead and other noxious chemicals; and medical examination.

The safety provisions deal with: fencing and other safeguards on machinery and the cleaning of machines; lifting equipment such as ropes, chains, hoists and cranes; floors, stairs and passages, and means of access and escape; work in confined spaces; fire and explosion; steam boilers and receivers and air receivers; eye protection; and the training of young persons.

The welfare provisions require facilities such as those for drinking water, washing, storing clothes, sitting down and first aid.

Further provisions place certain restrictions on the work which can be done and on the hours which can be worked by women and young persons, while others deal with matters such as outwork, piecework, and deductions from wages.

The enforcement provisions require the keeping of a General Register and certain other records. They give the Factory Inspectorate power to inspect the factory and its records, to take samples, to direct investigation of

Table 3.2 *Selected Acts of Parliament relevant to safety and loss prevention*[a]

Year	Act
1802	Factories Act – Health and Morals of Apprentices Act
1819	Factories Act (also 1825, 1831)
1833	Factories Act
1844–56	Factories Acts (seven)
1860	Factories Act (also 1864, 1867, 1871)
1863	Alkali Act
1875	Explosives Act
1878	Factories Act
1880	Employers Liability Act
1882	Boiler Explosions Act
	Electricity Supply Act
1883	Explosive Substances Act
1890	Boiler Explosions Act
1891	Factories Act (also 1895)
1897	Workmens Compensation Act
1901	Factories Act
1906	Alkali etc. Works Regulation Act
1923	Explosive Substances Act
1926	Lead Paint (Protection against Poisoning) Act
	Public Health (Smoke Abatement) Act
1928	Petroleum (Consolidation) Act
1933	Pharmacy and Poisons Act
1934	Petroleum (Production) Act
1936	Petroleum (Transfer of Licences) Act
	Public Health Act
1937	Factories Act
	Public Health (Drainage of Trade Premises) Act
1945	Water Act
1946	Atomic Energy Act
	National Insurance (Industrial Injuries) Act
	Trunk Roads Act
1947	Fire Services Act
1948	Civil Defence Act
	Factories Act
	Radioactive Substances Act
	River Boards Act
1949	Coast Protection Act
	Special Roads Act
1951	Alkali etc. Works Regulation (Scotland) Act
	Rivers (Prevention of Pollution) Act
1954	Atomic Energy Authority Act
1954–61	Mines and Quarries Act
1955	Food and Drugs Act
	Oil in Navigable Waters Act
1956	Clean Air Act
	Traffic Act
	Nurses Act
1957	Thermal Insulation (Industrial Buildings) Act
1959	Factories Act
	Fire Services Act
	Highways Act
	Nuclear Installations (Licensing and Insurance) Act
	Occupiers Liability Act
1960	Clean Rivers (Estuaries and Tidal Waters) Act
	Radioactive Substances Act
	Road Traffic Act
1961	Factories Act
	Public Health Act
	Rivers (Prevention of Pollution) Act
1962	Pipelines Act
	Poisons Act
1963	Offices, Shops and Railway Premises Act
	Water Resources Act
1964	Continental Shelf Act
	Dangerous Drugs Act
1965	Gas Act
	Merchant Shipping Act
	National Insurance (Industrial Injuries) Act
	Nuclear Installations Act
1967	Companies Act
	Road Safety Act
1968	Clean Air Act
	Medicines Act
	Transport Act
1969	Employers' Liability (Compulsory Insurance) Act
	Mines and Quarries (Tips) Act
	Nuclear Installations Act
	Nuclear Installations (Licensing and Insurance) Act
1970	Merchant Shipping Act
	Rivers (Prevention of Pollution) Act
1971	Civil Aviation Act
	Fire Precautions Act
	Mineral Workings (Offshore Installations) Act
	Mines Management Act
	Prevention of Oil Pollution Act
	Town and Country Planning Act
1972	Deposit of Poisonous Wastes Act
	Employment Medical Advisory Services Act
	Local Government Act
	National Insurance Act
	Road Traffic Act
	Town and Country Planning (Scotland) Act
1973	Water Act
1974	Control of Pollution Act
	Dumping at Sea Act
	Health and Safety at Work etc. Act
	Merchant Shipping Act
1975	Airports Authority (Consolidation) Act
	Employment Protection Act
	Petroleum and Submarine Pipe-lines Act
1976	Atomic Energy Authority (Special Constables) Act
1977	Merchant Shipping (Safety Convention) Act
	Town and Country Planning (Scotland) Act
1978	Civil Aviation Act
	Transport Act
1979	Merchant Shipping Act
1980	Civil Aviation Act
	Transport Act
1981	Merchant Shipping Act
	Transport Act
1982	Oil and Gas (Enterprise) Act
	Transport Act
1983	Merchant Shipping Act
	Building Act
1985	Local Government Access to Information Act
	Oil and Pipeline Act
	Town and Country Planning (Amendment) Act
	Transport Act
1986	Civil Protection in Peacetime Act
	Gas Act
	Housing and Planning Act
	Prevention of Oil Pollution Act

	Safety at Sea Act
1987	Consumer Protection Act
	Petroleum Act
1988	Merchant Shipping Act
1988	Road Traffic Act
1989	Atomic Energy Act
	Water Act
1990	Planning (Hazardous Substances) Act
	Town and Country Planning Act
1991	Planning and Compensation Act
	Radioactive Material (Road Traffic) Act
	Road Traffic Act
1992	Offshore Safety Act
	Offshore Safety (Protection against Victimisation) Act
1993	Clean Air Act

Sources: Public General Acts and Measures (annual) and *Statutory Instruments* (annual) (both London: HM Stationery Office).
The table gives the Acts in chronological order and includes some which have been superseded.

accidents or diseases, and to extend notification to dangerous occurrences.

Further reference to particular provisions of the Act is made in the following sections. A commentary on the interpretation of the Act is given in *Redgrave's Factories Acts* (Fife and Machin, 1982). The Act makes frequent reference to the phrases 'practicable' and 'reasonably practicable'. *Redgrave* gives the following comments on these terms:

> *Practicable.* In the Factories Act, 1961 and subordinate legislation an obligation is frequently qualified by the phrase 'so far as reasonably practicable', or by the phrase 'so far as practicable'. Each of these phrases affects in a different manner the obligation which it qualifies. 'Reasonably practicable' is a narrower term than 'physically possible', and implies that a computation must be made in which the quantum of risk is placed in one scale and the sacrifice involved in the measures necessary for averting the risk (whether in money, time or trouble) is placed in the other, and that, if it be shown that there is a gross disproportion between them – the risk being insignificant in relation to the sacrifice – the defendants discharge the onus upon them. Moreover, this computation falls to be made by the owner at the point of time anterior to the accident . . .
>
> Where the statutory obligation is qualified solely by the word 'practicable' a stricter standard is imposed. Measures may be 'practicable' which are not 'reasonably practicable'. . ., but, none the less, 'practicable' means something other than 'physically possible'. The measures must be possible in the light of current knowledge and invention . . .
>
> *Reproduced with permission from I. Fife and E.A. Machin,* Redgraves Health and Safety in Factories, 2nd edn., *1982, Butterworths, Oxford.*

***Table 3.3** Selected Statutory Regulations and Orders and Statutory Instruments (SIs) relevant to safety and loss prevention*[a]

Date	*SI No.*	
1908	1312	Electricity (Factories Act) Special Regulations 1989/635
1911	752	Lead Smelting and Manufacturing Regulations
1921	1825	Celluloid Manufacture etc. Regulations
1922	731	Chemical Works Regulations 1988/1657
1928	26	Alkali etc. Works Order 1966/1143
1929	992	Petroleum (Carbide of Calcium) Order
	993	Petroleum (Mixtures) Order
	952	Petroleum Spirit (Motor Vehicles) Regulations
1930	34	Petroleum (Compressed Gases) Order
1931	1140	Asbestos Industry Regulations 1969/690
	679	Gas Cylinder (Conveyance) Regulations 1989/2169
1934	279	Docks Regulations 1988/1655
1937	54	Acetylene, restriction – Order in Council
1938	599	Chains, Ropes and Lifting Tackle (Registration) Order
	598	Gasholders (Record of Examination) Order
	641	Operations at Unfenced Machinery Regulations
1944	739	Electricity (Factories Act) Special Regulations 1989/635
1947	805	Compressed Acetylene Order
	31	Dangerous Occurrences (Notification) Regulations 1980/809
	1594	Gas Cylinder (Conveyance) Regulations 1989/2169
	1442	Petroleum (Carbide of Calcium) Order
	1443	Petroleum (Inflammable Liquids and other Dangerous Substances) Order 1986/1951
1951	1163	Stores for Explosives Order
1954	921	Dangerous Machines (Training of Young Persons) Order
1957	191	Petroleum Spirit (Conveyance by Road) Regulations 1986/1951
	859	Petroleum (Liquid Methane) Order
1958	257	Petroleum (Carbon Disulphide) Order 1986/1951
	962	Petroleum Spirit (Conveyance by Road) Regulations 1986/1951
	61	Work in Compressed Air Special Regulations
1959	1919	Gas Cylinder (Conveyance) Regulations 1989/2169
1960	1932	Shipbuilding and Ship Repairing Regulations
1961	1345	Breathing Apparatus etc. (Report on Examination) Order
	1580	Construction (General Provisions) Regulations
	1581	Construction (Lifting Operations) Regulations
1962	2527	Carbon Disulphide (Conveyance by Road) Regulations 1986/1951
	224	Construction (General Provisions) Reports Order

1963	2003	Hoist and Lifts (Report of Examinations) Order 1992/195
	1382	Lifting Machines (Particulars of Examination) Order 1992/195
1964	781	Examination of Steam Boilers Regulations 1989/2169
	1070	Examination of Steam Boilers Reports (No.1) Order 1989/2169
1965	1373	Building Regulations 1972/317
	1824	Nuclear Installations (Dangerous Occurrences) Regulations
	1441	Power Presses Regulations
1966	1143	Alkali etc. Works Order 1983/943
	95	Construction (Health and Welfare) Regulations
	94	Construction (Working Places) Regulations
1967	879	Carcinogenic Substances Regulations 1988/1657
	1675	Carcinogenic Substances (Prohibition of Importation) Regulations 1988/1657
1968	928	Inflammable Substances (Conveyance by Road) (Labelling) Regulations 1971/1062
	780	Ionizing Radiations (Unsealed Radioactive Substances) Regulations 1985/1333
	571	Petroleum (Carbon Disulphide) Order 1986/1951
	570	Petroleum (Inflammable Liquids) Order 1971/1040
1969	690	Asbestos Regulations 1987/2115
	808	Ionizing Radiations (Sealed Substances) Regulations 1985/1333
1970	1826	Radioactive Substances (Carriage by Road) (Great Britain) Regulations 1974/1735
	1945	Petroleum (Corrosive Substances) Order 1986/1951
1971	960	Clean Air Alkali etc. Works Order 1983/943
	618	Corrosive Substances (Conveyance by Road) Regulations 1986/1951
	1061	Inflammable Liquids (Conveyance by Road) Regulations 1986/1951
	1062	Inflammable Substances (Conveyance by Road) (Labelling) Regulations 1986/1951
	1040	Petroleum (Inflammable Liquids) Order 1986/1951
1972	317	Building Regulations 1976/1676
	1017	Deposit of Poisonous Waste (Notification of Removal or Deposit) Regulations
	1178	Gas Safety Regulations
	917	Highly Flammable Liquids and Liquefied Petroleum Gases Regulations
	703	Offshore Installations (Managers) Regulations
	1938	Poisons List Order
	1385	Town and Country Planning (Use Classes) Order 1987/764
1973	8	Factories Act General Registration Order
	6	Notice of Industrial Diseases Order 1985/2023
	1842	Offshore Installations (Inspections and Casualties) Regulations 1986/1951
	2221	Organic Peroxides (Conveyance by Road) Regulations 1986/1951
	1897	Petroleum (Organic Peroxides) Order
	5	Work in Compressed Gases (Health Register) Order
1974	289	Offshore Installations (Construction and Survey) Regulations
	1681	Protection of the Eyes Regulations 1992/2966
	1735	Radioactive Substances (Carriage by Road) (Great Britain) Regulations
	903	Woodworking Machinery Regulations
1975	2111	International Carriage of Dangerous Goods (Rear Marking of Motor Vehicles) Regulations
	303	Protection of the Eyes (Amendment) Regulations
1976	1676	Building Regulations 1985/1065
	2003	Fire Certificate (Special Premises) Regulations
	1542	Offshore Installations (Emergency Procedures) Regulations
	1019	Offshore Installations (Operational Safety, Health and Welfare) Regulations
	923	Submarine pipelines (Diving Operations) Regulations
1977	890	Conveyance in Harbours of Military Explosives Regulations 1987/37
	889	Conveyance by Rail of Military Explosives Regulations
	888	Conveyance by Road of Military Explosives Regulations 1986/615
	746	Health and Safety (Enforcing Authority) Regulations 1989/1903
	486	Offshore Installations (Life-saving Appliances) Regulations
	500	Safety Representatives and Safety Committees Regulations
	835	Submarine Pipe-lines (Inspectors etc.) Regulations
	289	Town and Country Planning General Development Order 1988/1813
1978	1126	Factories (Standard of Lighting) (Revn) Regulations
	1702	Hazardous Substances (Labelling of Road Tankers) Regulations 1981/1059
	611	Offshore Installations (Fire Fighting Equipment) Regulations
	209	Packaging and Labelling of Dangerous Substances Regulations 1984/1244
1979	427	Petroleum (Consolidation) Act 1928 (Enforcement) Regulations
1980	1248	Control of Lead at Work Regulations
	1709	Control of Pollution (Special Waste) Regulations
	804	Notification of Accidents and Dangerous Occurrences Regulations 1985/2023
	1759	Offshore Installations (Well Control) Regulations
	1471	Safety Signs Regulations
1981	1709	Control of Pollution (Special Wastes) Regulations
	1059	Dangerous Substances (Conveyance by Road in Road Tankers and Tank Containers) Regulations 1992/743
	399	Diving Operations at Work Regulations
	917	Health and Safety (First-aid) Regulations
	1077	Merchant Shipping – The Merchant Shipping (Tankers) (EEC Requirements)

Year	No.	Title
		Regulations
1982	876	Merchant Shipping (Safety Officials and Reporting of Accidents and Dangerous Occurrences) Regulations
	1357	Notification of Installations Handling Hazardous Substances Regulations
	1496	Notification of New Substances Regulations 1993/3050
	630	Petroleum Spirit (Plastic Containers) Regulations
	218	The Poisons Rules
	1513	Submarine Pipelines Safety Regulations
	555	Town and Country Planning (Structure and Local Plans) Regulations
1983	1649	Asbestos (Licensing) Regulations
	1140	Classification and Labelling of Explosives (CLER) Regulations
	943	Health and Safety (Emissions into the Atmosphere) Regulations
	1615	Town and Country Planning (Amendment) Order 1988/1813
	1614	Town and Country Planning Use Classes (Amendment) Order 1987/764
1984	1244	Classification, Packaging and Labelling of Dangerous Substances Regulations (CPLR) 1993/1746
	1902	Control of Industrial Major Accident Hazards Regulations (CIMAH)
	1890	Freight Containers (Safety Convention) Regulations
	1358	Gas Safety (Installation and Use) Regulations
	1217	Merchant Shipping (Cargo Ship Construction and Survey) Regulations
	1218	Merchant Shipping (Fire Protection) Regulations
	237	Town and Country Planning (General Development) (Amendment) Order
1985	1939	Air Navigation (Dangerous Goods) Regulations
	910	Asbestos (Prohibitions) Regulations
	1065	Building Regulations 1991/2764
	1333	Ionizing Radiation Regulations 1992/743
	1218	Merchant Shipping (Fire Protection) (Ships Built before 25th May 1980) Regulations
	1884	Waste Regulation and Disposal (Authorities) Regulations
	2023	Reporting of Injuries, Diseases and Dangerous Occurrences Regulations
1986	1068	Merchant Shipping (Chemical Tankers) Regulations 1987/549
	1073	Merchant Shipping (Gas Carrier) Regulations
	1951	Road Traffic (Carriage of Dangerous Substances in Packages, etc.) Regulations 1992/742
1987	2115	Control of Asbestos at Work Regulations (CAWR)
	37	Dangerous Substances in Harbour Areas Regulations
	549	Merchant Shipping (BCH Code) Regulations
	550	Merchant Shipping (IBC Code) Regulations
	1331	Offshore Installations (Safety Zones) Regulations
	764	Town and Country Planning (Use Classes) Order
1988	819	Collection and Disposal of Waste Regulations
	1462	Control of Industrial Major Accident Hazards (Amendment) Regulations
	1657	Control of Substances Hazardous to Health Regulations (COSHH)
	1655	Docks Regulations
	896	Pressure Vessels (Verification) Regulations
	1199	Town and Country Planning (Assessment of Environmental Effects) Regulations
	1813	Town and Country Planning General Development Order
1989	2004	Air Navigation Order
	317	Air Quality Standards Regulations
	1119	Building Regulations (Amendment) Regulations
	318	Control of Industrial Air Pollution (Registration of Works) Regulations
	635	Electricity at Work Regulations
	76	Fire Precautions (Factories, Offices, Shops and Railway Premises) Regulations
	1903	Health and Safety (Enforcing Authority) Regulations
	682	Health and Safety Information for Employees Regulations
	840	Health and Safety at Work etc. Act 1974 (Application outside Great Britain) Regulations
	1790	Noise at Work Regulations
	1029	Offshore Installations (Emergency Pipeline Valves) Regulations
	971	Offshore Installations (Safety Representatives and Safety Committees) Regulations
	2169	Pressure Systems and Transportable Gas Containers Regulations (PSR)
	105	Road Traffic (Carriage of Dangerous Goods in Packages etc.) (Amendment) Regulations 1992/742
	615	Road Traffic (Carriage of Explosives) Regulations
1990	1255	Classification, Packaging and Labelling of Dangerous Substances (Amendment) Regulations
	556	Control of Asbestos in the Air Regulations
	2325	Control of Industrial Major Accident Hazards (Amendment) Regulations
	2026	Control of Substances Hazardous to Health (Amendment) Regulations
	13	Electrical Equipment for Explosive Atmospheres (Certification) Regulations
	304	Dangerous Substances (Notification and Marking of Sites) Regulations
	2605	Merchant Shipping (Dangerous Goods and Marine Pollutants) Regulations
1991	2768	Building Regulations
	1531	Control of Explosives Regulations
	2431	Control of Substances Hazardous to Health (Amendment) Regulations
	2570	Electrical Equipment for Explosive Atmospheres (Certification) (Amendment) Regulations

	472	Environmental Protection (Prescribed Processes and Substances) Regulations
	2097	Packaging of Explosives for Carriage Regulations
	2749	Simple Pressure Vessels (Safety) Regulations
1992	3067	Asbestos (Prohibitions) Regulations
	2415	Export of Dangerous Chemicals Regulations
	195	Lifting Plant and Equipment (Records of Test and Examination etc.) Regulations
	1914	Notification of New Substances (Amendment) Regulations
	2051	Management of Health and Safety at Work Regulations
	2793	Manual Handling Regulations
	2885	Offshore Installations (Safety Case) Regulations
	2966	Personal Protective Equipment at Work Regulations
	656	Planning (Hazardous Substances) Regulations
	3139	Protective Personal Equipment (EC Directive) Regulations
	2932	Provision and Use of Work Equipment Regulations
	1213	Road Traffic (Carriage of Dangerous Goods and Substances) (Amendment)Regulations
	742	Road Traffic (Carriage of Dangerous Substances in Packages etc.) Regulations (PGR)
	743	Road Traffic (Carriage of Dangerous Substances in Road Tankers and Tank Containers) Regulations (RTR)
	744	Road Traffic (Training of Drivers Carrying Dangerous Goods) Regulations
	3073	Supply of Machinery (Safety) Regulations
	1492	Town and Country Planning General Regulations
	3004	Workplace (Health, Safety and Welfare) Regulations
1993	1746	Chemicals (Hazard Information and Packaging) Regulations (CHIP)
	1812	Civil Defence (General Local Authority Functions) Regulations
	2379	Ionizing Radiations (Outside Workers) Regulations
	3050	Notification of New Substances Regulations
1994	670	Carriage of Dangerous Goods by Rail Regulations (CGD Rail)
	237	Railways (Safety Case) Regulations
	299	Railways (Safety Critical Work) Regulations (RSCW)

Sources: Statutory Instruments (annual) and *Table of Government Orders* (annual) (both London: HM Stationery Office).

[a] The table gives the regulations in chronological order and includes some which have been superseded. In some cases the instrument revoking the regulations is indicated after the title by its date/SI number

3.3 Factory Inspectorate

The original inspectorate is the Factory Inspectorate set up under the 1833 Act. This is the largest Inspectorate and that which is of principal interest in the present context. References to accounts of the work of a Factory Inspector were given in Section 3.1. The development of the Factory Inspectorate and its current work are described in the *HM Chief Inspector of Factories Annual Report 1974* (HMCIF, 1974), *Her Majesty's Inspectors of Factories 1833–1983* (HSE, 1983d) and *The Work of a Factory Inspector* (HSE, 1987d), and by Djang (1942) and Symons (1969).

The function of the Factory Inspectorate has been to enforce and to give advice on factory legislation, notably the Factories Act 1961 and the Offices, Shops and Railway Premises Act 1963 (OS&RP Act). This work covers (1) accidents, (2) occupational diseases, (3) environmental conditions, and (4) hours of work – in other words safety, health and welfare.

The Inspectorate's approach to accident prevention has been strongly influenced by the fact that the overwhelming proportion of serious accidents are due to identifiable and often preventable causes. Thus Symonds (1969) states that some 90% of fatalities (although only some 20% of reported accidents) are due to (1) power-driven machinery, (2) technical incidents and (3) long falls. The power-driven machinery includes works internal transport. The technical incidents include those due to fires, explosions, toxic or inert gases, and electricity. 'Long falls' are falls of greater than about 2 m of persons or articles. In general, these are the types of accident which are more amenable to reduction by legislation and enforcement and it is on such hazards that the Inspectorate has very largely concentrated.

Fire hazards are an important concern. There are two main kinds of fire hazard. One is that from very flammable materials, which may give rise quickly to a dangerous fire or explosion. The other is that from fires in buildings, which may trap people by the spread of fire or smoke.

Hazards from chemicals with noxious long-term effects are another of the Inspectorate's main preoccupations. There are now a large number of chemicals in use in general industry. Many of these are hazardous if incorrectly handled, although safe if used properly. The firms using them may be small and lacking in specialist knowledge, and may need the advice of the Inspectorate.

The Factory Inspectorate has to operate within the limits of its resources; its staff has never been particularly numerous. The 1833 Act allowed for four Inspectors, although they were empowered to appoint assistants and did in fact appoint eight. In 1993 the Inspectorate numbered about 650 general Inspectors. These Inspectors were responsible for some 400 000 premises. Inevitably, therefore, the resources of the Inspectorate are fairly thinly spread.

The growth of specialist functions within the Inspectorate is indicated by the creation of a Medical Inspector in 1898, an Engineering Adviser in 1899, an Electrical Inspector in 1902, and an Inspector of Dangerous Trades in 1903. In 1972 the Employment Medical Advisory Service (EMAS) was created to provide a much more comprehensive coverage of health at work.

The Inspectorate has developed further specialized branches, such as the Chemical Branch, which deals with the more technical aspects of certain hazards. It also has specialized units, such as the Industrial Hygiene Division, which is specially equipped to conduct atmospheric analyses and other tests.

The contribution of individual Inspectors is illustrated by the classic works *Redgrave's Factories Acts* (Fife and Machin, 1972), by Alexander Redgrave, the first Chief Inspector of Factories appointed in 1878, and *Industrial Maladies* by Thomas Legge (1934), the first Medical Inspector appointed in 1898. Legge believed that 'All workmen should be told something of the danger of the materials with which they come in contact and not be left to find out for themselves . . . sometimes at the cost of their lives.' This elementary right is now embodied in legislation.

The Factory Inspectorate is responsible for the enforcement of factory law. In discharging this duty its approach has, in general, been to persuade and to educate and it has undertaken prosecution only as a last resort.

3.4 *Robens Report*

By the late 1960s the need for an overhaul of British legislation on health and safety at work was apparent. In part the need was for revision and consolidation. But it was recognized that the question was deeper than this, that the law was lagging behind changes in industrial practice and that it was not dealing effectively with problems of scale and technology. There was also excessive fragmentation in the law and its administration. Some nine separate sets of legislation were administered by five Government Departments with seven separate Inspectorates, and local authorities were also involved.

Accordingly, in 1970 a committee was set up under the chairmanship of Lord Robens, a former trade union official and Chairman of the National Coal Board (NCB), to consider the matter. The terms of reference embraced the whole provision made for health and safety of people at work (except for transport work). The committee issued its report *Safety and Health at Work* (the *Robens Report*) in 1972 (Robens, 1972). It made fundamental criticisms of existing arrangements and recommendations for change.

The committee's starting point was the fact that, on the evidence of the decade to 1970, there was no discernible trend in industrial accident statistics and little to suggest progress in reducing accidents. This is shown in the data for the number of fatal accidents and of total accidents in factories already given in Figure 1.1. The committee posed the question of whether this was due simply to the law of diminishing returns or whether the means used were perhaps increasingly inappropriate to the real problems. The report's critique of the existing arrangements, its emphasis on self-regulation and some of its main specific points are illustrated in the following passage (Robens, 1972, paragraphs 457–459):

> There is a lack of balance between the regulatory and voluntary elements of the overall 'system' of provision for safety and health at work. The primary responsibility for doing something about present levels of occupational accidents and diseases lies with those who create the risks and those who work with them. The statutory arrangements should be reformed with this in mind. The present approach tends to encourage people to think and behave as if safety and health at work were primarily a matter of detailed regulation by external agencies.
>
> Present regulatory provisions follow a style and pattern developed in an earlier and different social and technological context. Their piecemeal development has led to a haphazard mass of law which is intricate in detail, unprogressive, often difficult to comprehend and difficult to amend and keep up to date. It pays insufficient regard to human and organizational factors in accident prevention, does not cover all workpeople, and does not deal comprehensively and effectively with some sources of serious hazard. These defects are compounded and perpetuated by excessively fragmented administrative arrangements.
>
> A more effectively self-regulating system is needed. Reform should be aimed at two fundamental and closely related objectives. First, the statutory arrangements should be revised and reorganized to increase the efficiency of the State's contribution to safety and health at work. Secondly, the new statutory arrangements should be designed to provide a framework for better self-regulation.

The basic philosophy of the report is that control which emphasizes detailed regulation is not appropriate to modern technology and that self-regulation by industry itself, exercising a more open-ended duty of care, is likely to be more satisfactory. Thus the report supported the development and use by industry of voluntary standards and codes of practice.

The report emphasized the need for a comprehensive approach to safety and the importance of management attitude and organizational systems in achieving this. It advocated more explicit and formal safety systems, including statements of safety policy, analyses of hazards, etc. It also reflected the social context in calling for more disclosure of information on hazards and for involvement of workpeople in monitoring hazards and safety arrangements. Similarly, it stressed the need to consider the interests not only of employees but of the public also.

The report recognized the problem of 'major hazards' associated with high technology and recommended comprehensive provisions to deal with explosive and flammable substances and with toxic substances. It recommended also the creation of a major hazards branch within the Inspectorate and the use of standing advisory committees on these hazards.

The report recommended the creation of a single unified authority to deal with health and safety at work, with a comprehensive range of functions and powers. It envisaged that the Inspectorate could be reorganized so as to strengthen its ability to give expert and impartial advice, and to exercise tighter control of serious problems. A critical view was taken of the efficiency of existing means of enforcement. In particular, the use of prosecutions in the courts was seen as relatively cumbersome and ineffective.

As the following section indicates, the spirit and principal recommendations of the *Robens Report* are embodied in the Health and Safety at Work etc. Act 1974. An account of the work of the Robens Committee is given by Beeby (1974a).

3.5 Health and Safety at Work etc. Act 1974

The Health and Safety at Work etc. Act 1974 (HSWA) established a Health and Safety Commission (HSC), which is responsible for the development of policy under the Act. The Commission consists of representatives of management and workers in industry and of local authorities. The enforcement of the HSWA is the responsibility of the Health and Safety Executive (HSE), which is part of the Department of Employment (DoEm). The Act is described in *A Guide to the Health and Safety at Work etc. Act 1974* (HSE, 1990a).

The Act is a quite different kind of Act from its predecessor, the Factories Act 1961. It is essentially an enabling measure superimposed on existing health and safety legislation. Its purpose is to provide the legislative framework for a comprehensive system of law to deal with the health and safety of virtually all persons at work.

The enabling provisions in the Act were superimposed on some 31 Acts. The new Act provided for the gradual replacement of existing health and safety legislation by a system of regulations and approved codes of practice so as to create an integrated system of law on the subject. In the interim the main provisions of these Acts have continued to apply. The HSWA thus lays down broad principles. It is relatively free of technical details; these are dealt with in the Factories Act 1961 and in the other Acts and regulations.

The provisions of the Act may be considered under the following headings: (1) scope, (2) regulations, (3) codes of practice, (4) general duties, (5) enforcement, and (6) the HSC.

The Act covers all persons at work, including employers, employees and the self-employed, but excluding domestic servants. It extends the protection of health and safety legislation to some 5 million people in the UK who were not previously covered, such as those employed in education, medicine and leisure industries and some parts of the transport industry. The protection of the Act extends not only to people at work, but also to the general public insofar as they are affected by work activities. This feature distinguishes the HSWA from the Factories Act 1961 and from factory legislation in many other countries.

The Act covers the keeping and use of dangerous substances and makes possible an integrated system of control of such substances. It also covers the airborne emission of noxious substances from prescribed premises, even those which are not a danger to health but which would cause a nuisance or damage to the environment. Although the Act is an enabling one and in general avoids detailed regulatory provisions of the type contained in the Factories Act 1961, it does contain in Schedule 3 comprehensive powers to allow detailed regulations to be made. Regulations under the Act are made by the appropriate Minister, usually on the basis of proposals by the HSC formulated after consultation with interested parties. While most regulations are likely to originate from the Department of Employment, there are several other departments concerned.

Approved codes of practice are used to supplement regulations and have a special legal status. They are not statutory instruments, but they may be used in criminal proceedings in evidence that their requirements have been disregarded. Approval of a code of practice for this purpose is the subject of consultation between the Commission and interested parties.

General duties are imposed by the Act on the following: (1) employers, (2) employees, (3) the self-employed, and (4) manufacturers and suppliers.

The Act lays on the employer the statutory duty of safeguarding as far as reasonably practicable the health, safety and welfare of employees. This is similar to, but seeks to strengthen, the general duty of care in common law. In particular, there is a requirement to provide and maintain safe plant and safe systems of work. The employer is also required to make provisions for the following aspects of safety: (1) safety policies, organization and arrangements; and (2) safety information and training. Thus the employer has a duty to provide a statement of safety policy, to update it and to bring it to employees' notice. He is also required to provide safe systems of work. These include systems for the safe control of activities, such as permit-to-work systems used to control maintenance work. The employer is required to furnish employees with information on safety matters. This includes technical and legal information and information on hazards. The employee has a 'right to know'. In addition, it is the duty of the employer to train employees in safe practices. The employer also has a duty to others who may not be employees but who may be at risk, such as self-employed workers or contractors' employees and the general public.

The employees have a duty to take reasonable care to avoid injury to themselves or others by their work activities and to co-operate with the employer or others in meeting statutory requirements. The employee is also required not to interfere with or misuse any protection provided in compliance with the Act.

The self-employed have a duty similar to that laid on employers to avoid risk to the health and safety of themselves and others.

The manufacturer or supplier, which includes the designer or importer, of articles or substances for use at work, must ensure as far as reasonably practicable that these are safe when properly used. This involves testing and providing instructions for use, where applicable.

The responsibility for enforcement of the Act does not in all cases lie with the HSE. For many activities, particularly those associated with shops and offices, restaurants, hotels and consumer services, it is the Local Authority which is responsible. The division of responsibility is defined in the Health and Safety (Enforcing Authority) Regulations 1989.

The HSE is the enforcing authority for factories. Under the Act the section of the HSE which deals with factories is the Factory Inspectorate, as before. The Act provides the Factory Inspector with much more positive means of enforcement. The means available to him under the Act are (1) seizure, (2) improvement notice, (3) prohibition notice, and (4) prosecution.

If the Inspector considers that an article or substance gives rise to a risk of serious personal injury, the article can be seized, neutralized or destroyed.

If the inspector discovers a contravention of the relevant statutory provisions, an improvement notice can be issued. This is served on the person who is deemed to be in contravention or any person who has responsibility, whether employer, employee or supplier. Such a notice requires that the fault be remedied within a specified time.

However, if the contravention is an activity which involves risk of serious personal injury, the Inspector can issue a prohibition notice. This is served on the person carrying out the activity or on the person in control of it at the time of serving the notice. Such a notice requires that the activity cease immediately and that it be not resumed until the fault is remedied.

Instead of, or in addition to, serving a notice, the Inspector may bring a prosecution for contravention of a relevant statutory provision. Contravention of some requirements may lead to summary prosecution in a Magistrates Court and of others to summary prosecution or to prosecution on indictment in the Crown Court. Failure to comply with an improvement or prohibition notice may result in prosecution and in the latter case in imprisonment. A person on whom a notice is served may appeal against it to an industrial tribunal.

The Act contains a number of other provisions, notably those concerned with the EMAS and with building trades.

3.6 Health and Safety Commission

As described above, the HSC consists of representatives of management and workers in industry and of local authorities. It keeps under review the whole legislative regime and is advised by a number of Industry Advisory Committees and Special Advisory Committees.

The work of the HSC and the operation of the HSWA 1974 during the initial period are described in the *Report 1974–76* by the HSC (1977) and in the *Health and Safety Statistics 1975* HSE (1977d). In the present context, features of interest in this period are: the formulation of codes of practice; the creation of the safety representatives and safety committees system; the development of the Advisory Committees to advise the Commission in specialized areas such as major hazards, dangerous substances, toxic substances and asbestos; the work on the problems of major hazard plants, asbestos and vinyl chloride; and the enforcement procedures.

The HSC issues each year an *Annual Report* and a *Plan of Work.* There are also annual *Health and Safety Statistics.* HM Chief Inspector of Factories has also issued an annual report, but the last such report was that for the year 1986–87.

Legislation on dangerous substances is kept under continuous review by the Advisory Committee on Dangerous Substances (ACDS), and that on toxic chemicals is reviewed by the Advisory Committee on Toxic Substances (ACTS).

3.7 Health and Safety Executive

The HSE forms part of the Department of Employment. It is responsible for enforcing the relevant legislation and for providing advice. The HSWA 1974 provides a unifying framework for a number of different existing Acts. The HSE brings together a number of inspectorates and other functions which previously operated separately under these Acts. The structure of the Commission and Executive is shown in Figure 3.2.

There has been a tendency for other inspectorates to be consolidated within the HSE. Thus the inspectorates for the railways and for offshore are now part of that body. The Executive therefore comprises not only the Factory Inspectorate but also the Inspectorate of Mines and Quarries, the Nuclear Installations Inspectorate (NII) and the Offshore Safety Division (OSD). The inspectorate dealing with pollution, the Alkali Inspectorate, became part of the HSE on the formation of the latter, but in 1987 became the separate HM Inspectorate of Pollution (HMIP).

The HSE has access to a variety of Government, or in some cases former Government, research and consultancy establishments. A selected list of these is shown in Table 3.4.

3.8 Regulatory Regime

The duty of care owed by an employer to an employee under common law is made a statutory duty by the HSWA 1974. In fulfilment of this duty the employer is required to take all measures which are reasonably practicable. His duty of care is not limited to adherence to regulations but is open-ended.

The system created by the Act is one of self-regulation. The employer is required to identify, evaluate and eliminate or control hazards. In this system the role of the regulator is to create a framework and instruments which promote good practice and to ensure that effective self-regulation is taking place. *The Health and Safety System in Great Britain* (HSE, 1992c) gives an overview of the system operated by the enforcing authority. The detailed features of the HSE's approach in applying the Act are described by Dykes (1992).

Reliance on the duty of care alone is not satisfactory. The regulatory body is liable to have difficulty in making a prosecution for failure to fulfil the duty of care stick. Industry for its part needs to know more clearly what is required of it. Without further guidance the duty is liable

Table 3.4 *Some Government or former Government establishments with services available to the Health and Safety Executive*

AEA Technology, Harwell
AEA Technology, Safety and Reliability Directorate, Culcheth
British Hydromechanics Research Association
Building Research Establishment, Borehamwood
Electrical Equipment Certification Service, Buxton
Explosion and Flame Laboratory, Buxton
Fire Research Station, Borehamwood
Health and Hygiene Laboratories, Sheffield
HSE Research Laboratories, Sheffield
Meteorological Office, Bracknell
National Engineering Laboratory, East Kilbride
National Physical Laboratory, Teddington
Warren Spring Laboratory

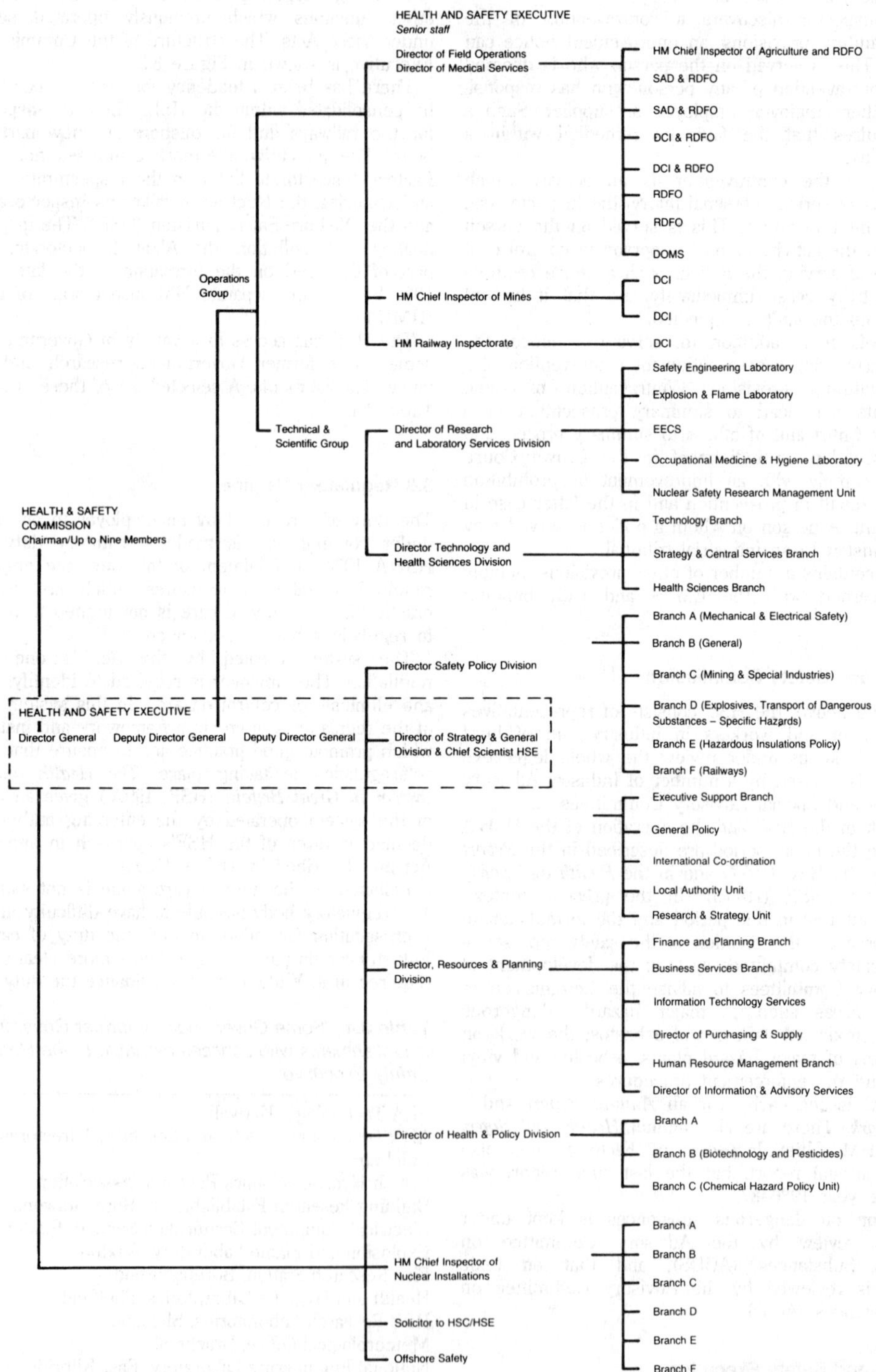
HEALTH AND SAFETY EXECUTIVE
Senior staff
Director of Field Operations
Director of Medical Services
HM Chief Inspector of Agriculture and RDFO
SAD & RDFO
SAD & RDFO
DCI & RDFO
DCI & RDFO
RDFO
RDFO
DOMS
DCI
DCI
DCI
Operations Group
HM Chief Inspector of Mines
HM Railway Inspectorate
Safety Engineering Laboratory
Explosion & Flame Laboratory
Technical & Scientific Group
Director of Research and Laboratory Services Division
EECS
Occupational Medicine & Hygiene Laboratory
Nuclear Safety Research Management Unit
Technology Branch
HEALTH & SAFETY COMMISSION
Chairman/Up to Nine Members
Director Technology and Health Sciences Division
Strategy & Central Services Branch
Health Sciences Branch
Branch A (Mechanical & Electrical Safety)
Branch B (General)
Director Safety Policy Division
Branch C (Mining & Special Industries)
Branch D (Explosives, Transport of Dangerous Substances – Specific Hazards)
HEALTH AND SAFETY EXECUTIVE
Director General
Deputy Director General
Deputy Director General
Director of Strategy & General Division & Chief Scientist HSE
Branch E (Hazardous Insulations Policy)
Branch F (Railways)
Executive Support Branch
General Policy
International Co-ordination
Local Authority Unit
Research & Strategy Unit
Finance and Planning Branch
Director, Resources & Planning Division
Business Services Branch
Information Technology Services
Director of Purchasing & Supply
Human Resource Management Branch
Director of Information & Advisory Services
Branch A
Director of Health & Policy Division
Branch B (Biotechnology and Pesticides)
Branch C (Chemical Hazard Policy Unit)
Branch A
Branch B
HM Chief Inspector of Nuclear Installations
Branch C
Branch D
Solicitor to HSC/HSE
Branch E
Offshore Safety
Branch F

***Figure 3.2** A chart giving the outline of the main organization of the Health and Safety Executive.*

to be subject to different interpretations by different inspectors. The HSE has therefore developed a hierarchy of instruments to put flesh on the bones of the Act. These are (1) the HSWA 1974 itself; (2) regulations; (3) Approved Codes of Practice; and (4) guidance.

On the second tier are regulations. These govern a specific hazard or group of hazards. They are more specific than the Act. But increasingly they set goals rather than prescribe details.

The third tier is the Approved Codes of Practice (ACOPs) issued in support of the HSWA itself or of a set of regulations. Failure to adhere to an ACOP is not an offence as such, but the onus is then on the party concerned to demonstrate compliance with the substantive requirement. Use of an ACOP is unsuitable where it is desirable to retain flexibility of options. The fourth tier, therefore, consists of guidance. This may take various forms. A large proportion of regulations are accompanied by guidance. Other guidance comes in the form of codes and standards prepared by the HSE, professional bodies or trade associations. Such guidance is not statutory and courts are free not to admit it in legal proceedings.

It has always been the philosophy of the Factory Inspectorate to seek improvement in safety by persuasion rather than by confrontation, and to undertake legal proceedings only as a last resort. The HSE has adhered to the same philosophy. This philosophy finds application at several levels. In making regulations the HSE consults interested parties and Advisory Committees beforehand. Frequently it encourages contributions through conferences and other forms of participation. The draft regulations are then subject to statutory consultation. Similarly, in technical matters, such as the physical models to be used in hazard assessment, the HSE is active in promoting interchange and the development of consensus. At the level of the individual firm, the HSE seeks first to obtain compliance by advice and persuasion. This approach is generally effective and resort to improvement or prohibition notices or to prosecution is necessary in only a small proportion of cases.

The activities of the HSE are of three broad types: (1) policy-making; (2) technological, scientific and medical matters; and (3) field operations.

The formulation of policy at the highest level is done by the HSC and its Advisory Committees, as already described. These committees need to be serviced and policy translated into regulatory instruments, codes and guidance.

The HSE is involved in a wide range of technical, scientific and medical activities, both in its own research establishments and through research sponsorship, and in the dissemination of the results of this work, and in many of these areas it is an international leader.

The Field Operations Division (FOD) has the oversight of the premises which come under the HSWA 1974 and enforces compliance by audit and inspection. This division also maintains Field Consultancy Groups of specialist inspectors and National Interest Groups (NIGs) for specific industries. One of these is the Chemical Manufacturing (CM) NIG. Another is the Hazardous Installations and Transport (HIT) NIG. The NIGs are one of the means by which the guidance described earlier is generated.

The HSC's *Plan of Work 1985/86* (HSC, 1985) proposes a change of emphasis in enforcement policy. It distinguishes between more and less hazardous types of work establishment and suggests that the HSE should concentrate more on the former, whilst the latter should be covered by 'self-assurance' schemes. It argues in effect that where practice is good, inspection is of less value. It recognizes, however, the problem of major hazards and no relaxation is proposed for these.

Some topics which have required the particular attention of the HSE at some time in the last two decades include:

(1) toxic substances in the workplace,
(2) asbestos,
(3) vinyl chloride,
(4) major hazard installations,
(5) transport of hazardous substances,
(6) warehouse storage of hazardous substances,

and, recently,

(7) offshore installations.

The approach taken by the HSE has for some time laid increasing emphasis on management and on human factors. The growing emphasis on these aspects has received additional impetus from the general movement towards total quality management and from the specific recommendation of the report of the inquiry into the Piper Alpha disaster (Cullen, 1990) that the safety case for an offshore installation should include a safety management system.

The offshore regime provides further illustration of the principles and problems of regulation, particularly for major hazards. The regulatory arrangements offshore are described in more detail in Section 3.24.

3.9 Offshore Regime

By the early 1970s, oil and gas activities on the UK continental shelf were well under way. The regulatory regime in the North Sea was based on the Mineral Workings (Offshore Installations) Act 1971 (MWA) administered by the Department of Energy (DoEn) which both supervised the commercial developments and regulated their safety. These offshore activities came within the scope of the HSWA 1974 and their safety became the concern of the HSC, but under an agency agreement with the latter the DoEn continued to act as the regulatory body, making regulations mainly under the MWA 1971.

In 1980 there was published the report *Offshore Safety* by a committee chaired by Dr J.H. Burgoyne (1980) (the *Burgoyne Report*). The report took cognizance of the special conditions and novel technology in, and of the expertise within, the DoEn on North Sea activities and recommended *inter alia* that the supervision of safety in the North Sea should continue to be done by the DoEn.

This regime continued until the Piper Alpha disaster in 1988. This was the subject of a public inquiry chaired by Lord Cullen. The report *The Public Inquiry into the Piper Alpha Disaster* by Lord Cullen (1990) (the *Piper Alpha* or *Cullen Report*) found that the regulatory regime in the North Sea had a number of defects. Supervision by the DoEn placed too much emphasis on inspection of installations and too little on audit of management. The number of staff dealing with safety in the DoEn was relatively small and reliance was placed on checking of designs by certifying authorities. The regulations made tended to be prescriptive. The guidance which accompanied the regulations tended not to achieve its aim of introducing flexibility but to be treated as further prescriptive requirements. There was no regulatory requirement for the formal safety assessment of major hydrocarbon hazards and practice was patchy.

The *Cullen Report* recommended that responsibility for the regulation of offshore safety be taken from the DoEn and given to a separate division within the HSE. This was accepted by government and an Offshore Safety Division (OSD) was created within the HSE which took over from the DoEn in 1991. This was followed by the Offshore Safety Act 1992. The report also made a number of other recommendations. It recommended that the new regime should be based on goal-setting rather than prescriptive regulations, that formal safety assessment should be undertaken and that it should be a means of demonstrating compliance, that there should be a safety case for each installation and that this safety case should demonstrate that the company possessed a suitable safety management system.

The offshore regime is a good illustration of a continuing problem in regulation, particularly of the process industries. This is the question of goal-setting versus prescriptive regulations. This has been touched on in this and in the previous section; it is discussed further in Section 3.25 and Appendix 18.

3.10 Specific Legislation

Some topics on which there is specific legislation that is relevant in the present context include

(1) general topics;
(2) pressure systems;
(3) new chemicals;
(4) toxic chemicals;
(5) flammables;
(6) explosives;
(7) fire;
(8) storage;
(9) major hazards;
(10) planning;
(11) environment;
(12) ionizing radiations;
(13) buildings and construction;
(14) electricity;
(15) personal safety;
(16) transport;
(17) offshore.

It has been the policy of the HSE to consolidate the subordinate legislation under the HSWA into a much smaller number of regulations and the last decade has seen considerable progress towards this goal. On the other hand it is necessary to make regulations to comply with Directives of the European Community (EC). In some cases there is a one-to-one correspondence between a Directive and a set of regulations, whereas in others the requirements of a Directive are met through several sets of regulations or one set of regulations is used to implement several Directives or parts thereof.

For many of the regulations, the HSE issues a *Guidance on Regulations* (HS(R) or L series) and/or an *Approved Code of Practice* (COP or L series). Some of these two series of publications are given in Table 3.5.

Current information on the legislation is given in *The Health and Safety Factbook* (A.L. Jones, 1994), which is updated twice a year.

The topics listed above are now considered in turn.

Table 3.5 *Selected guidance on regulations and Approved Codes of Practice*

A Guidance on Regulations		
HS(R) 4 (rev.)	1989	A guide to the OSRP Act 1963
HS(R) 13	1981	A guide to the Dangerous Substances (Conveyance by Road in Road Tankers and Tank Containers) Regulations 1981
HS(R) 14 (rev.)	1988	A guide to the Notification of New Substances Regulations
HS(R) 15 (rev.)	1987	Administrative guidance on the European Community 'Explosive Atmospheres' Directive (76/117/EEC and 79/196/EEC) and related Directives
HS(R) 16	1983	A guide to the Notification of Installations Handling Hazardous Substances Regulations 1982
HS(R) 17	1983	A guide to the Classification, Packaging and Labelling of Explosives Regulations 1984
HS(R) 19 (rev.)	1989	A guide to the Asbestos (Licensing) Regulations 1983
HS(R) 21 (rev.)	1990	A guide to the Control of Major Accident Hazards Regulations 1984
HS(R) 22	1985	A guide to the Classification, Packaging and Labelling of Dangerous Substances Regulations 1984
HS(R) 23	1986	A guide to the Reporting of Injuries, Diseases and Dangerous Occurrences Regulations 1985
HS(R) 24	1987	A guide to the Road Traffic (Carriage of Dangerous Substances in Packages etc.) Regulations 1986
HS(R) 25	1989	Memorandum of guidance on the Electricity at Work Regulations 1989
HS(R) 27	1988	A guide to the Dangerous Substances in Harbour Areas Regulations 1987
HS(R) 29	1990	Notification and Marking of Sites. The Dangerous Substances (Notification and Marking of Sites) Regulations
HS(R) 30	1990	A guide to the Pressure Systems and Transportable Gas Containers Regulations 1989

B Approved Codes of Practice

COP 1	1988	Safety representatives and safety committees
COP 2	1988	Control of lead at work
COP 3	1988	Work with asbestos insulation, asbestos coating and asbestos insulating board. Revised approved code of practice
COP 7	1982	Principles of good laboratory practice. Notification of New Substances Regulations 1982
COP 11	1983	Operational provisions of the Dangerous Substances (Conveyance by Road in Road Tankers and Tank Containers) Regulations 1981
COP 14	1985	Road tanker testing: examination, testing and certification of the carrying tanks of road tankers and of tank containers used for the conveyance of dangerous substances by road (in support of SI 1981/1059)
COP 16	1985	Protection of persons against ionizing radiations from any work activity. The Ionizing Radiations Regulations 1985
COP 17	1987	Notice of approval of the operational provisions of the Road Traffic (Carriage of Dangerous Substances in Packages etc.) Regulations 1986 by the Health and Safety Commission
COP 18	1987	Dangerous substances in harbour areas. The Dangerous Substances in Harbour Areas Regulations 1987
COP 19	1990	Classification and labelling of dangerous substances for conveyance by road in tankers, tank containers and packages (revision 1): – Dangerous Substances (Conveyance by Road in Road Tankers and Tank Containers) Regulations 1981 – Classification, Packaging and Labelling of Dangerous Substances Regulations 1984 – Road Traffic (Carriage of Dangerous Substances in Packages etc.) Regulations 1986
COP 21	1988	Control of asbestos at work. The Control of Asbestos at Work Regulations 1987
COP 31	1988	Control of vinyl chloride at work
COP 33	1989	Transport of compressed gases in tube trailers and tube containers. Dangerous Substances (Conveyance by Road in Road Tankers and Tanker Containers) Regulations 1981
COP 36	1989	Carriage of explosives by road. Road Traffic (Carriage of Explosives by Road) Regulations 1989
COP 37	1990	Safety of pressure systems. Pressure Systems and Transportable Gas Containers Regulations 1989
COP 38	1990	Safety of transportable gas containers. Pressure Systems and Transportable Gas Containers Regulations 1989
COP 40	1990	Packaging and labelling of dangerous substances for conveyance by road (revision 2). Classification, Packaging and Labelling of Dangerous Substances Regulations 1984
COP 42	1990	First aid at work. Health and Safety (First-aid) Regulations 1981

C Legal Series (gradually superseding both COP and HS(R) series)

L1	1990	A guide to the Health and Safety at Work etc. Act 1974, 4th edition
L5	1993	Control of substances hazardous to health and Control of carcinogenic substances. Control of Substances Hazardous to Health Regulations 1988, 4th ed. ACOPs
L6	1991	Diving operations at work. The Diving Operations at Work Regulations 1981 as amended by the Diving Operations at Work (Amendment) Regulations 1990
L7	1991	Dose limitation – restriction of exposure. Additional guidance on Regulation 6 of the Ionizing Radiation Regulations 1985. Approved Code of Practice – Part 4
L10	1991	A guide to the Control of Explosives Regulations 1991
L11	1991	A guide to the Asbestos (Licensing) Regulations 1983
L13	1991	A guide to the Packaging of Explosives for Carriage Regulations 1991
L14	1993	Classification and labelling of dangerous substances for carriage by road in tankers, tank containers and packages. ACOP
L16	1993	Design and construction of vented non-pressure road tankers used for the carriage of flammable liquids. ACOP
L17	1993	Design and construction of road tankers used for the carriage of carbon disulphide. ACOP
L18	1993	Design and construction of vacuum insulated road tankers used for the carriage of non-toxic

		deeply refrigerated gases. ACOP
L19	1993	Design and construction of vacuum operated road tankers used for the carriage of hazardous wastes. ACOP
L21	1992	Management of health and safety at work. Management of Health and Safety at Work Regulations 1992. ACOP
L22	1992	Work equipment. Provision and Use of Work Equipment Regulations 1992. Guidance on regulations
L23	1992	Manual handling. Manual Handling Operations Regulations 1992. Guidance on regulations
L24	1992	Workplace health, safety and welfare. Workplace (Health, Safety and Welfare) Regulations 1992. Approved code of practice and guidance.
L25	1992	Personal protective equipment at work. Personal Protective Equipment at Work Regulations 1992. Guidance on regulations
L27	1993	The control of asbestos at work. Control of Asbestos at Work Regulations 1987, 2nd edition ACOP
L30	1992	A guide to the Offshore Installations (Safety Case) Regulations 1992
L37	1993	Safety data sheets for substances and preparations dangerous for supply
L38	1993	Approved guide to the classification and labelling of substances and preparations dangerous for supply
L50	1994	Railway safety critical work. Railways (Safety Critical Work) Regulations 1994. Guidance on regulations
L51	1994	Carriage of dangerous goods by rail. Carriage of Dangerous Goods by Rail Regulations 1994. Guidance on regulations
L52	1994	Railway safety cases. Railways (Safety Cases) Regulations 1994. Guidance on regulations

3.11 Workplace Legislation

3.11.1 Safety representatives and safety committees

The Safety Representatives and Safety Committees Regulations 1977 require the appointment of safety representatives and, on request from the latter, of safety committees. A safety representative is appointed by a recognized trade union. The representative has a right to obtain information and to inspect the workplace in relation to safety matters.

As described above, the HSWA 1974 creates a special category of codes of practice – the Approved Codes of Practice. The first such code was the ACOP on Safety Representatives.

Safety representatives and safety committees are discussed in more detail in Chapter 28.

3.11.2 Information for employees

The provision of information to employees about hazards and precautions is a requirement of the HSWA 1974. This duty is strengthened by the Health and Safety Information for Employees Regulations 1989.

3.11.3 Workplace

The Workplace (Health, Safety and Welfare) Regulations 1992 tidy up some 38 pieces of legislation concerned with the workplace, including parts of the Factories Act 1961 and the Offices, Shops and Railway Premises Act 1963. They deal with such matters as the working environment, floors and stairs, facilities and housekeeping.

3.11.4 Work equipment

The Provision and Use of Work Equipment Regulations 1992 draw together legislation on work equipment, concerning such matters as the suitability of equipment, its proper application, its maintenance and training in its use.

3.11.5 Management

The Management of Health and Safety at Work Regulations 1992 deal with risks to employees. They are accompanied by the ACOP *Management of Health and Safety at Work* (HSE, 1992 L21). The regulations require an employer to assess the risks to the health and safety of employees and to take appropriate precautions. Matters covered include provision of information to employees, appointment of competent people, health surveillance, training and contractors.

3.11.6 Incident reporting

Requirements for the reporting of incidents of various kinds were contained in many of the principal Acts including the Factories Act 1961, the Boiler• Explosions Acts 1882 and 1890, the Petroleum (Consolidation) Act 1928, and the Pipelines Act 1962, while the Dangerous Occurrences (Notification) Regulations 1947 dealt specifically with notification of incidents.

These latter regulations and the reporting requirements in the Acts mentioned have been superseded by the Notification of Accidents and Dangerous Occurrences Regulations 1980 (NADOR). These regulations in turn were superseded by the Reporting of Injuries, Diseases and Dangerous Occurrences Regulations 1985 (RIDDOR).

RIDDOR requires the reporting of injuries resulting from accidents which cause incapacity for more than 3 days, of certain diseases and of certain dangerous occurrences which may be summarized as follows:

(1) Collapse, overturning or failure of any load bearing part of a lift, hoist, crane, excavator or piledriver, etc.

(2) (Incidents at fun fairs).
(3) Explosion, collapse or bursting of a vessel, including a boiler, liable to cause injury or involving more than 24 h stoppage of work.
(4) Electrical short circuit or overload attended by fire or explosion liable to cause injury or involving more than 24 h stoppage of work.
(5) Explosion or fire involving more than 24 h stoppage of work.
(6) Sudden, uncontrolled release of 1 tonne or more of flammable gas or certain flammable liquids (highly flammable or flashing liquid).
(7) Collapse or partial collapse of certain scaffolding.
(8) Unintended collapse or partial collapse of certain structures.
(9) Uncontrolled or accidental release of hazardous materials liable to cause injury or harm health.
(10) Unintended ignition or explosion of explosives.
(11) Failure of certain freight containers.
(12) Following incidents defined in Pipelines Act 1972:
 (a) bursting, explosion or collapse of a pipeline;
 (b) ignition of anything which is or has just been in a pipeline.
(13) Overturning of, damage to, release from or fire on certain road tankers or tank containers containing hazardous materials.
(14) Release from or fire on certain other vehicles carrying hazardous materials by road.
(15) Malfunction of certain breathing apparatus.
(16) Unintentional contact with, or discharge between, plant or equipment and certain uninsulated overhead high voltage electric lines.
(17) Accidental collision between locomotive or train and a vehicle liable to cause injury.

There are a few reporting requirements in other pieces of legislation, such as those covering merchant shipping and ionizing radiations. These are listed in Schedule 6 of RIDDOR.

3.12 Pressure Systems Legislation

3.12.1 The Chemical Works Regulations 1922

The Chemical Works Regulations 1922 (CWR) represented the application to the chemical industry of the general approach taken in the Factories Acts. These regulations have now been superseded, but they had a long life and reference may well be found to them, particularly in relation to overpressure protection, entry into vessels and permits to work. The processes covered by the CWR are defined in a schedule. They are for the most part the more traditional processes for the manufacture of basic chemicals.

The regulations deal with such matters as: plant spacing and layout; ventilation of noxious processes; control of dust; lighting and sources of ignition; pressure relief valves; breathing apparatus; entry into vessels; first aid equipment and training, and ambulance rooms; amenities; and dangerous practices.

The requirement for a pressure relief valve or other equally efficient means to relieve pressure on any vessel in which pressure is liable to rise to a dangerous degree is important in the present context.

The precautions to be taken on entry into vessels are specified in the regulations, which thus give this particular operation something of a special status.

There are also requirements associated with specific processes. These are mainly the older type processes making the traditional products of the industry, e.g. sulphuric acid, caustic soda and chlorine.

3.12.2 Pressure vessels and systems legislation

The frequency with which boiler explosions occurred during the nineteenth century led to the Boiler Explosions Acts 1882 and 1890. These Acts created the duty of notifying boiler explosions. Notification might be followed by an inquiry and the establishment of negligence. The Acts required the fitting to boilers of safety valves for the relief of overpressure, but did not create any requirement for regular inspection.

The definition of a boiler under these Acts is of interest: 'Any closed vessel used for generating steam or for heating other liquids or into which steam is admitted for heating, steaming, boiling or other purposes.'

A legal requirement for periodic examination was not introduced until the Factories and Workshops Act 1901.

On its introduction, the Factories Act 1961 became the main legislation on pressure vessels. This dealt principally with (1) steam boilers (Sections 32–34), (2) steam receivers (Section 35), air receivers (Section 36) and (4) gasholders (Section 39). The Act lays down requirements for the construction and maintenance of these vessels. The vessel has to be soundly constructed and properly maintained. The fittings required include: on a steam boiler, a water level gauge, a pressure gauge, a stop valve and a safety valve; on a steam receiver which cannot withstand the boiler pressure, a pressure gauge, a stop valve, a reducing valve and a safety valve; and on an air receiver, a pressure gauge, a safety valve and, if unable to withstand the compressor pressure, a reducing valve.

The requirements for periodic examination of steam boilers under the Act were given in the Examination of Steam Boilers Regulations 1964, together with the Examination of Steam Boilers Reports (No. 1) Order 1964. Under these regulations, the normal maximum interval between thorough examinations is 14 months. But a longer interval of 26 months is permitted for large boilers with an evaporation rate of more than 50 000 lb/h, since such boilers are expensive plant where maintenance may be expected to be good. The maximum examination interval for steam receivers and air receivers is given in the Act as 26 months, except for solid drawn air receivers, for which it is 14 months. For gasholders the Act gives a maximum examination interval of 2 years. In principle, the system had provision for exemptions to be granted by the Factory Inspectorate to allow increase in the inspection interval in certain cases, but in practice such exemption has not been readily obtained.

The legislation just described was limited in scope and was not well adapted to the needs of the process industries. It is oriented to steam boilers and receivers and air receivers. It concentrates on steam and air, on pressure vessels and on aspects such as design, construction and inspection. It does not deal with the other more hazardous fluids handled, with the other components of the pressure system, which tend to be more vulnerable than the pressure vessels themselves, or with aspects such as operation, maintenance or modification. Moreover, the legislation was fragmented, so that pressure systems were dealt with under various Acts and their associated regulations. There were also statutory requirements on the construction and maintenance of gas cylinders and gas containers. They included the Gas Cylinders (Conveyance) Regulations 1931, 1947 and 1959. Gas cylinders were also the subject of the *Report of the Home Office Gas Cylinders and Containers Committee* (1968).

The explosion at Flixborough in 1974, which was caused by a massive leak of cyclohexane from a large pipe which had been the subject of a modification, highlighted the weaknesses of the existing legislative controls, which covered neither the hazardous fluid, nor the pressure system component, nor the operation of the pressure system.

Industry deals with all these aspects as a matter of good industrial practice. And insofar as good practice is a requirement of the HSWA 1974, they may be held to have been covered. Nevertheless, it was considered necessary to introduce more comprehensive legislation. Therefore, in 1977, the HSC introduced its *Proposals for New Legislation for Pressurised Systems* (HSC, 1977f). This document outlined an entirely different approach to legislation for pressurized systems, based on the philosophy of self-regulation. It proposed that there should be general regulations for pressurized systems, which it defined as covering all fluids and virtually all pressures and components in the pressurized system. It included transportable vessels and systems. Also in 1977, the EC brought in the framework Directive 77/767/EEC on pressure vessels and their inspection.

The HSE proposals were far-reaching and they did not see the light of day until 1989, in the form of the Pressure Systems and Transportable Gas Containers Regulations 1989 (the Pressure Systems Regulations 1989). The regulations are supplemented by two ACOPs, *Safety of Pressure Systems* (HSE, 1990 COP 37) and *Safety of Transportable Gas Containers* (HSE, 1990 COP 38). These regulations create a comprehensive system of controls for pressure systems and consolidate existing legislation in this area. They revoke a large number of existing regulations and other legislation, including: the Factories Act 1961, Sections 32, 33, 35 and 36; the Examination of Steam Boilers Regulations 1964 and the Examination of Steam Boilers Reports (No. 1) Order 1964; and the Gas Cylinders (Conveyance) Regulations 1931, 1947 and 1959. The principal contents of the Pressure Systems Regulations 1989 are given in Chapter 12.

Two other regulations dealing with pressure systems are the Pressure Vessels (Verification) Regulations 1988 and the Simple Pressure Vessels (Safety) Regulations 1991. The former implement provisions of EC Directive 77/767/EEC on pressure vessels and of separate EC Directives (84/525/EEC, 84/526/EEC and 84/527/EEC) on gas cylinders, insofar as they relate to the verification and inspection of such vessels. The Simple Pressure Vessels (Safety) Regulations 1991 implement EC Directive 87/404/EEC (amended by 90/488/EEC) on simple pressure vessels. They give requirements for simple pressure vessels manufactured in series.

3.13 New Chemicals and Toxic Substances Legislation

A large number of new chemicals are introduced each year by the chemical industry and it is necessary that they be properly controlled. The control of new chemicals is the subject of EC Directive 79/831/EEC (the Sixth Amendment to the Dangerous Substances Directive), which was implemented in Britain by the Notification of New Substances Regulations 1982. Many of these chemicals are to some degree toxic, and it is with this aspect that these regulations are primarily concerned. An account of the background to the regulations has been given by van den Heuvel (1982). These regulations are superseded by the Notification of New Substances Regulations 1993.

There are EC Directives on the exposure of personnel to toxic chemicals. Directive 89/677/EEC deals with toxic chemicals and Directive 90/394/EEC with carcinogens. The control of exposure of personnel to toxic chemicals in the workplace is effected through the Control of Substances Hazardous to Health Regulations 1988 (COSHH), as amended 1990. An account of the background to the regulations has been given by P. Lewis (1985b).

Poisons are governed by the Poisons List Order 1972 and there are poisons rules which are periodically updated, the current rules being the Poisons Rules 1982. There are several substances, or classes, which are governed by specific legislative controls. The Lead Smelting and Manufacturing Regulations 1911 deal with the processing of lead. A more recent set of controls is the Control of Lead at Work Regulations 1980.

Asbestos is another substance which early on attracted controls. These include the Asbestos Industry Regulations 1931 and the Asbestos Regulations 1969, now both superseded by: the Control of Asbestos at Work Regulations 1987; the Asbestos (Licensing) Regulations 1983; the Asbestos (Prohibition of Importation) Regulations 1985; the Control of Asbestos in the Air Regulations 1990; and the Asbestos (Prohibitions) Regulations 1992.

In the mid-1960s concern over cancer-inducing chemicals led to the Carcinogenic Substances Regulations 1967 and the Carcinogenic Substances (Prohibition of Importation) Regulations 1967. These are both now superseded by the COSHH Regulations 1988.

Also relevant are the Export of Dangerous Chemicals Regulations 1992.

There are also EC Directives on hazards information, namely 88/379/EEC, the 'Preparations' Directive and 91/155/EEC, the 'Safety Data Sheets' Directive. The materials safety data sheet (MSDS) requirements are implemented by Regulation 6 of the Chemicals (Hazards Information and Packaging) Regulations 1993.

3.14 Flammables, Explosives and Fire Legislation

3.14.1 The Petroleum (Consolidation) Act 1928

The Petroleum (Consolidation) Act 1928 (P(C)A) was brought in to deal with the storage and transport of 'petroleum spirit'. Petroleum spirit is defined as petroleum with a flash point equal to or less than 22.8°C (73°F). Although old, the Act is still in force.

The P(C)A deals mainly with the licensing of premises for the storage of petroleum spirit. By its extension to other flammable substances it represented for a long time the principal means of control of such chemicals. The application for a licence is to the local authority and is a planning matter. If the applicant is refused a licence or is offered one with restrictive conditions, it may appeal to the Minister of Employment.

The Act also deals with the transport of petroleum spirit, as described in Section 3.22. An early application to motor vehicles was through the Petroleum Spirit (Motor Vehicles) Regulations 1929.

The instrument for the extension of the Act to other substances is Section 19. The Petroleum (Mixtures) Order 1929 applied the Act to mixtures of petroleum with other substances where the mixture has a flash point of less than 22.8°C (73°F). The Petroleum (Carbide of Calcium) Orders 1929 and 1947, the Petroleum (Compressed Gases) Order 1930 and the Petroleum (Liquid Methane) Order 1957 illustrate further applications.

The Act was applied through a number of further regulations and orders, many of them dealing with transport. A large proportion of these instruments have been revoked by the subsequent legislation, notably the Highly Flammable Liquids and Liquefied Petroleum Gases Regulations 1972 and the Road Traffic (Carriage of Dangerous Substances) Regulations 1986.

3.14.2 Flammables

There is a large body of additional legislation on flammable substances. It may conveniently be considered under the following headings: (1) piped gas; (2) flammable gases (other than piped gas); and (3) flammable liquids.

Gas pipelines come under the Pipelines Act 1962, which is considered in Section 3.22. Essentially, the enforcement of this Act now lies with the HSE.

The Gas Act 1986 extends the responsibility of the HSE to protection of the public from injury from transmission and distribution of gas through pipes and from use of gas supplied by pipes.

Control of gas distribution and use is effected by the Gas Safety Regulations 1972 (GSR) and the Gas Safety (Installation and Use) Regulations 1984 (GS(I&U)R). The requirements of the latter regulations are unnecessarily restrictive for industry, and mines and factories have exemption from them.

Other principal flammable gases are (1) liquefied petroleum gas (LPG) and (2) acetylene. LPG is covered by the Highly Flammable Liquids and Liquefied Petroleum Gases Regulations 1972 (HFLR), which deal with the storage of flammable liquids, including LPG. The Gas Safety (Installation and Use) Regulations 1984 apply to LPG which is stored centrally and distributed to a number of premises.

Acetylene has long been subject to controls based on the traditional view that it is explosive when subject to a pressure exceeding 0.62 barg (9 psig). Restrictions were set by Order in Council 30 under the Explosives Act 1875 on the handling of acetylene in the pressure range 0.62–1.5 barg (9–22 psig). Acetylene has also been governed by the Compressed Acetylene Order 1947.

As just stated, flammable liquids are covered by the HFLR 1972 which deal with the storage of such liquids. Early guidance on flammable liquid storage was given in the *Model Code of Practice* (Home Office, 1968) and there is more recent guidance in HSE *Guidance Notes*. The controls on the transport of flammable liquids are described in Section 3.22.

3.14.3 Explosives

The manufacture, handling, storage and transport of explosives have been governed by the Explosives Acts 1875 and 1923 and the associated regulations. These include the Stores for Explosives Order 1951.

The legislation is mainly concerned with: the construction of buildings for the manufacture of and magazines for the storage of explosives; the spacing of magazines to avoid sympathetic detonation; the limitation of the numbers of people exposed in manufacture; and the design of vehicles for and precautions to be taken in the conveyancing of explosives. The requirements for plant layout and for limitation of exposure of personnel are especially relevant in the present context.

These controls are updated by the Control of Explosives Regulations 1991 (COER). The regulations deal particularly with controls on people who keep and handle explosives and on the security of explosives. The controls on the transport of explosives are described in Section 3.22.

3.14.4 Fire

Legislation dealing with fire in general includes the Fire Services Acts 1947 and 1959 and the Fire Precautions Act 1971. These Acts are concerned with the fire services, minimum standards of construction, control of fire outbreaks and safeguarding of life. As already described, there have also been provisions on fire hazard in the Chemical Works Regulations 1922 and the Petroleum (Consolidation) Act 1928.

Under the Fire Precautions Act 1971 premises may be designated as requiring a fire certificate. Such a certificate is normally issued by the local Fire Authority. The Fire Certificates (Special Premises) Regulations 1976, however, require that for certain special premises which store or process flammable, explosive and/or toxic materials the fire certificate should be issued by the HSE. The regulations give a schedule of the premises concerned.

The main legislation covering fire in the chemical industry, however, is the Factories Act 1961 and its associated regulations. The Factories Act 1961 includes provisions on the following:

(1) precautions for plant using flammable gas, vapour and dusts (Section 31);
(2) provision and maintenance of means of escape (Sections 40–47);
(3) safety in case of fire (Section 48);
(4) instruction on escape (Section (49);
(5) provision for the making of special regulations for fire prevention (Section 50);
(6) provision, maintenance and testing of fire fighting equipment (Sections 51–52);
(7) dangerous conditions and practices (Section 54).

Also relevant are the Fire Precautions (Factories, Offices, Shops and Railway Premises) Regulations 1989.

3.15 Storage Legislation

The principal legislation relevant to the storage of flammable substances has already been outlined, namely the Petroleum (Consolidation) Act 1928 and the Highly Flammable Liquids and Liquefied Petroleum Gases Regulations 1972. There is relatively little in the way of specific legislation for the storage of toxic substances. Storage of both flammable and toxic substances is well covered, however, by HSE *Guidance Notes* and industry codes.

Notification of the storage of hazardous substances and marking of the premises is required by the Dangerous Substances (Notification and Marking of Sites) Regulations 1990 (NMS). These regulations cover not only factories but also warehouses.

3.16 Major Hazards Legislation

The problem of major hazards was reviewed by the Advisory Committee on Major Hazards (ACMH), which sat from 1975 to 1983, and advised the HSC on legislative initiatives in this area. There are special arrangements for the control of major hazards. Notification of installations with more than a certain inventory of hazardous materials is required under the Notification of Installations Handling Hazardous Substances Regulations 1982 (NIHHS).

The control of major hazards is the subject of EC Directive 82/501/EEC (the 'Major Accident Hazards' Directive), amended by Directives 87/216/EEC and 88/610/EEC. The first of these Directives is implemented in Britain by the Control of Industrial Major Accident Hazards Regulations 1984 (CIMAH), and the other two are implemented by amendments in 1988 and 1990. The CIMAH regulations include requirements for notification, a safety case, emergency planning and provision of information to the public. Legislation on major hazards is discussed further in Chapter 4.

3.17 Planning Legislation

It is convenient at this point to consider the legislation governing land use planning. Whereas most of the other legislation considered here is made under the HSWA 1974 and is enforced by the HSE, the planning legislation is separate and is enforced by the Department of the Environment (DoE).

Accounts of planning legislation with special reference to major hazards are given by Petts (1988b, 1989) and to hazardous waste by Petts and Eduljee (1994). A system of planning control over new major hazards and of development in the vicinity has existed in the UK since 1972.

Planning control of major hazards is governed in England and Wales by the Town and Country Planning Act 1990 (TCPA), the Planning (Hazardous Substances) Act 1990 and the Planning and Compensation Act 1991. These three acts have revised and extended previous controls under the TCPA 1971. There is separate legislation in Scotland.

Planning control primarily relates to forward planning through the production of development plans and to the control of new development through the granting of planning permission. The Town and Country Planning (Structure and Local Plans) Regulations 1982 require authorities to create and update such plans.

As far as major hazards are concerned, the ACMH identified as crucial the problems associated with the definition of 'development' in the TCPA 1971 and the accompanying legislation, the Town and Country Planning Act (Use Classes) Order 1972 and the Town and Country Planning Act General Development Order 1977. The TCPA 1971 and the associated orders allowed a hazardous use to commence on land and not necessarily constitute development so that no planning permission was required.

Amending regulations which went some way to improve the control of major hazards were introduced in the Town and Country Planning Use Classes (Amendment) Order 1983 and the Town and Country Planning General Development (Amendment) Order 1983. These orders were in turn superseded by the Town and Country Planning (Use Classes) Order 1987 and the Town and Country Planning General Development Order 1988, although the major hazard arrangements were essentially unaffected.

More effective planning controls on major hazards are now in place. The Planning and Compensation Act 1991 has strengthened the role of the development plan in controlling the siting of new major hazards and of development in the vicinity by making plans a specific 'material consideration' in the granting of planning permission and giving statutory status to the adoption of local district plans, as well as to structure plans (at county council or equivalent level).

The Planning (Hazardous Substances) Act 1990 (PHSA) and the Planning (Hazardous Substances) Regulations 1992 close the loophole associated with the concept of development. Whilst the definition of development has not been amended under the TCPA 1990, the PSHA 1990 lays down the basic requirement that the presence of a hazardous substance requires the 'consent' of the local Hazardous Substance Authority, the planning authority, in addition to any planning permission required for the use of the land.

The HSE is a statutory consultee for all planning applications for new major hazards and for development in the vicinity of existing installations under the TCPA 1990 and for all hazardous substances consents granted under the PHSA 1990.

The Town and Country Planning (Assessment of Environmental Effects) Regulations 1988 require that a planning application submitted for an 'integrated chemical plant' include an environmental statement. For other types of plant which may be classified as a major hazard, one may be requested by the local authority.

Legislation on planning is discussed further in Chapter 4.

3.18 Environmental Legislation

The three main forms of pollution are: (1) discharge of gaseous effluents to the atmosphere; (2) discharge of liquid effluents to drains, thence to rivers and canals, and ultimately to the sea; and (3) dumping of wastes.

Environmental problems in the UK have been the subject of a series of *Reports of the Royal Commission on Environmental Pollution,* the initial chairmen being Sir Eric Ashby and Sir Brian Flowers. There have now been some 18 such reports, the titles of which are given in Appendix 28.

Accounts of pollution legislation include those in *The Health and Safety Factbook* (A.L. Jones, 1994) and by McLoughlin (1972, 1973, 1976, 1980), Bingham (1973), McKnight, Marstrand and Sinclair (1974), Pocock and Docherty (1978), Ireland (1981, 1984, 1987), Harper (1986) and Petts and Eduljee (1994). EC legislation is given in *European Community Environment Legislation for Industry* (Agra Europe, 1994b).

The control of emission of toxic gases was for a long time based on the Alkali Acts. The first Alkali Act was in 1863 and was prompted by the problem of pollution by hydrogen chloride. The last such Act was the Alkali etc. Works Regulation Act 1906. Under this Act were made the Alkali etc. Works Orders 1928 and 1966 and the Clean Air Alkali etc. Works Orders 1971. The enforcing authority for the Act was the Alkali Inspectorate. The regulation of emissions by the Inspectorate was based on a small number of statutory limits for gases such as hydrogen chloride and on a larger number of presumptive limits.

Petrochemical plants were brought within the scope of the Alkali Act 1906 by the Clean Air Alkali etc. Works Orders 1971. The Clean Air Acts of 1956 and 1968 dealt with the emission of smoke, dust and grit from boilers and industrial plant, the administration of the Acts being by the Alkali and Clean Air Inspectorates and by the local authorities.

In addition to this main body of legislation the local authorities were empowered under the Public Health Act 1936 to deal with airborne emissions which cause a nuisance or are the subject of complaint. The approach taken by the Alkali Inspectorate was the use of 'best practicable means' (BPM). 'Practicable' is defined in the Clean Air Act as: '"Practicable" means reasonably practicable having regard amongst other things to local conditions and circumstances, to the financial implications and to the current state of technical knowledge.' It was the practice of the Alkali Inspectorate to give guidance on best practicable means for certain processes. Thus the *Annual Report on Alkali etc. Works 1974* gives notes on the BPM for synthetic nitric acid plants, crude oil refineries, lead works and electricity works.

The control of liquid effluents discharged into rivers was regulated by a series of Acts passed between 1937 and 1970, including the Rivers (Prevention of Pollution) Act 1970. These Acts are administered by the water authorities. Other relevant legislation included the Clean Rivers, Salmon Rivers and Estuarial Waters Acts.

The control of wastes was exercised by the Department of the Environment (DoE) and by the local authorities through the Deposit of Poisonous Wastes Act 1972 and the Deposit of Poisonous Waste (Notification of Removal or Deposit) Regulations 1972.

A new framework for pollution control was introduced with the Control of Pollution Act 1974 (COPA). The Act has the following principal parts: Part I, Waste on Land; Part II, Pollution of Water; Part III, Noise; and Part IV, Pollution of the Atmosphere.

In addition, it may be noted that whereas the Factories Act 1961 was concerned exclusively with the health and safety of people at work, the HSWA 1974 also covers the effect of industrial activities on the health and safety of the general public. With the introduction of the HSWA 1974 and the establishment of the HSE, the Alkali Inspectorate became part of the HSE.

More recent subordinate legislation includes: for air pollution, the Health and Safety (Emissions into the Atmosphere) Regulations 1983; the Control of Industrial Air Pollution (Registration of Works) Regulations 1989; and the Air Quality Standards Regulations 1989.

In 1987 the inspectorate dealing with pollution again became a separate one – HM Inspectorate of Pollution (HMIP).

The Water Act 1989 created the National Rivers Authority (NRA) and transferred to it, from the water authorities, the responsibility for water standards.

The Environmental Protection Act 1990 (EPA) is a further framework Act and introduces a new regime for the control of pollution. The parts of this Act which are relevant here are: Part I, Integrated Pollution Control and Air Pollution Control by Local Authorities; Part II, Waste on Land, and Part III, Statutory Nuisances and Clean Air.

The EPA 1990 establishes a system of integrated pollution control (IPC). For materials discharged to the environment it requires the selection of the best practical environmental option (BPEO). It introduces the concept of the use of the best available technique not involving excessive cost (BATNEEC).

The Clean Air Act 1993 extends the controls on smoke, dust, grit and fumes to processes not otherwise covered.

Hazardous waste is controlled by the COPA 1974 and the EPA 1990 as well as by the 1972 legislation just mentioned and by the Control of Pollution (Special Wastes) Regulations 1980 and 1981 and the Collection and Disposal of Waste Regulations 1988.

A particular form of pollution is noise. The problem of noise was reviewed in the *Report of the Committee on the Problem of Noise* chaired by Sir Alan Wilson (1963) (the *Wilson Report*). The control of noise at work has a basis in the general requirement to provide a healthy and safe working environment. In addition, noise at work is now the subject of specific controls in the form of the Noise at Work Regulations 1989, which implement the requirements of the EC Directive 86/188/EEC.

The control of noise affecting the public has been dealt with by the local authorities using Noise Abatement Orders. Noise is one of the forms of pollution covered by the COPA 1974 and the EPA 1990.

Further accounts of legislation on environmental protection and on noise are given in Appendices 11 and 12, respectively.

3.19 Ionizing Radiations Legislation

Up to 1985 legislation on ionizing radiations relevant to factories included the Radioactive Substances Act 1960, the Ionizing Radiations (Unsealed Radioactive Substances) Regulations 1968, the Ionizing Radiations (Sealed Sources) Regulations 1969 and the Radioactive Substances (Road Transport Workers) (Great Britain) Regulations 1970. The last three sets of regulations were superseded by the Ionizing Radiations Regulations 1985. An account of ionizing radiations legislation is given by Dewis (1985c).

The regulations require that measures be taken to restrict the exposure of workers to ionizing radiations.

These measures include: engineering controls to minimize exposure, such as containment of radioactive substances, prevention of contamination and shielding, and warning devices and interlocks; designation of controlled areas; safe systems of work, including systems to minimize exposure and to ensure that workers have protective equipment; monitoring of exposure and monitoring of biological effects by medical examinations; classification of employees liable to higher doses; appointment of radiation protection advisers where higher doses may occur; keeping of records of quantity and location of radioactive substances.

The Radioactive Substances Act 1960 is mainly concerned with the storage and disposal of radioactive waste. It includes the requirement that, with certain specified exceptions, all holders of radioactive material and mobile radioactive apparatus should be registered.

Transport of radioactive substances has in general been governed at national and international level primarily by the regulations and conditions of acceptance of the carriers, but also sometimes by government legislation. The International Atomic Energy Agency (IAEA) issues regulations which are adopted by most countries, including the UK.

Also relevant are the Ionizing Radiations (Outside Workers) Regulations 1993.

3.20 Building, Construction and Electricity Legislation

3.20.1 Building and construction

Building is governed by a series of Building Regulations – 1965, 1972, 1976, 1985 and now 1991.

Regulations on construction include the Construction (General Provisions) Regulations 1961, the Construction (Lifting Operations) Regulations 1961, the Construction (General Provisions) Order 1962, the Construction (Health and Welfare) Regulations 1966, and the Construction (Working Places) Regulations 1966.

Regulations on lifting and related equipment, particularly on examinations of such equipment, include the Chains, Ropes and Lifting Tackle (Registration) Order 1938, the Hoists and Lifts (Report of Examination) Order 1963, and the Lifting Machines (Particulars of Examination) Order 1963.

There are currently in draft the Construction (Design and Management) Regulations (CONDAM) which *inter alia* clarify the responsibilities of the various parties in a construction project.

3.20.2 Electricity

The Electricity (Factories Act) Special Regulations 1908 and 1944 were for a long time the basic legislation on electricity in industry. They have now been superseded by the Electricity at Work Regulations 1989.

Lighting is covered by the Factories (Standard of Lighting) (Revn) Regulations 1978.

The certification of electrical equipment for use in hazardous areas is controlled by the Electrical Equipment for Explosive Atmospheres (Certification) Regulations 1990, as amended 1991.

3.21 Personal Safety Legislation

There is a large amount of legislation concerned with various aspects of personal safety, including legislation on certain types of hazardous work (e.g. construction), and activity (e.g. lifting operations); on particular sources of hazard (e.g. electrical equipment, compressed air, ionizing radiations); and on personal safety equipment (e.g. protective clothing, eye protection, breathing apparatus).

Some principal regulations are: the Operations at Unfenced Machinery Regulations 1938; the Dangerous Machines (Training of Young Persons) Order 1954; the work in Compressed Air Special Regulations 1958; the Breathing Apparatus (Report on Examination) Order 1961; the Power Presses Regulations 1965; the Work in Compressed Gases (Health Register) Order 1973; the Woodworking Machinery Regulations 1974; the Safety Signs Regulations 1980; the Health and Safety (First Aid) Regulations 1983; the Manual Handling Operations Regulations 1992; the Personal Protective Equipment at Work Regulations 1992; and the Protective Personal Equipment (EC Directive) Regulations 1992. Further reference is made to some of this legislation in the following chapters.

3.22 Transport Legislation

Transport legislation relates partly to aspects which are common to more than one mode of transport, such as classification, packaging and labelling (CPL) and the various types of container, and partly to specific modes of transport.

Reviews of the legislation governing the different modes of transport are given in *Major Hazard Aspects of the Transport of Dangerous Substances* (ACDS, 1991). Further accounts include those by J.R. Hughes (1970), A.W. Clarke (1971a,b), Walmsley (1973), Sewell (1984b), Williamson (1985), M.R. Wright (1985) and Crooks (1994).

Much transport is international and hence international requirements such as those given in the codes of the United Nations (UN) or the International Maritime Organization (IMO) have a strong influence on national legislation.

The major hazards aspects of transport are kept under review by a subcommittee of the ACDS.

Containers for the various modes of transport are covered by the Freight Containers (Safety Convention) Regulations 1984.

3.22.1 Road

The carriage of dangerous chemicals by road was originally regulated by the Home Office, mainly through the Petroleum (Consolidation) Act 1928 and associated regulations. These regulatory functions now reside with the HSE. There are three main sets of regulations, introduced in 1992–93. For the first set of regulations, those applicable until 1993 were the Classification, Packaging and Labelling of Dangerous Substances Regulations 1984 (CPLR), which cover these aspects as they apply to road transport. An account of the background to the regulations has been given by Williamson (1985). The CPLR were superseded by the Chemical (Hazard Information and Packaging) Regulations 1993 (CHIP), which fulfil an essentially similar function.

Until 1992, the two other sets of regulations were as follows. The second set of regulations was the Dangerous Substances (Conveyance by Road in Road Tanker and Tank Containers) Regulations 1981 (the Road Tanker Regulations). The regulations lay duties on the operators of road tankers, of tank containers and of vehicles conveying a tank container, and on drivers. These include: construction, maintenance and testing of tankers and tank containers; marking of vehicles with hazard warning panels; checking that substances may be lawfully conveyed; precautions against fire and explosion; and instructions to and training of drivers. Associated with these regulations are three approved codes of practice on classification, operational provision and road tanker testing.

The third set of regulations applicable until 1992 was the Road Traffic (Carriage of Dangerous Substances in Packages, etc.) Regulations 1986. These regulations revoke some 15 previous regulations, including: some concerned with petroleum spirit, the Petroleum Spirit (Inflammable Liquids and Other Dangerous Substances) Order 1947, the Petroleum Spirit (Conveyance by Road) Regulations 1957, 1958 and 1966; some concerned with other flammable liquids, the Petroleum (Inflammable Liquids) Order 1971, the Inflammable Liquids (Conveyance by Road) Order 1971, the Petroleum (Corrosive Substances) Order 1970 and the Corrosive Substances (Conveyance by Road) Regulations 1971; some with carbon disulphide, the Petroleum (Carbon Disulphide) Order 1958, the Carbon Disulphide (Conveyance by Road) Regulations 1958, the Petroleum (Carbon Disulphide) Order 1968; and some with organic peroxides, the Petroleum (Organic Peroxides) Order 1973 and the Organic Peroxides (Conveyance by Road) Regulations 1973. Guidance on the regulations is given in HS(R) 24 (HSE, 1987). The regulations place limitations on the carriage of certain substances and contain three schedules. Schedule 1 gives specifications of types of dangerous substance such as flammable liquids, flammable solids and toxic substances; Schedules 2 and 3 give maximum temperatures for the conveyance of organic peroxides and flammable solids, respectively. Matters dealt with in the regulations, which cover much of the same ground as the Road Tanker Regulations, include: the design, construction and maintenance of vehicles; the marking of vehicles; loading, stowage and unloading; precautions against fire and explosion; information for and training of the driver; and information for the police.

In 1992 these last two sets of regulations were superseded by two new ones, the Road Traffic (Carriage of Dangerous Substances in Road Tanker and Tank Containers) Regulations 1992 and the Road Traffic (Carriage of Dangerous Substances in Packages, etc.) Regulations 1992. The matters covered in these regulations are broadly similar to those in the regulations which they replace.

The road transport of compressed gases has long been regulated by the Petroleum (Compressed Gases) Order 1930 and the Gas Cylinder (Conveyance) Regulations 1931, 1947 and 1959.

The international transport of dangerous chemicals by road in the countries of the EC is carried out under the terms of the *European Agreement Concerning the International Carriage of Dangerous Goods by Road* (ADR).

3.22.2 Rail

The carriage of dangerous chemicals by rail is decided by British Rail (BR). A general account of its requirements is given in *Dangerous Goods by Freight and Passenger Train*, but bulk carriage is a matter for commercial negotiation.

Regulations on rail transport include: the Carriage of Dangerous Goods by Rail Regulations 1994; the Railways (Safety Case) Regulations 1994; and the Railways (Safety Critical Work) Regulations 1994.

The international transport of dangerous chemicals by rail in the EC countries is carried out according to the *International Regulations Concerning the Carriage of Dangerous Goods by Rail*, Annexe I (RID).

3.22.3 Inland waterways

British Transport Commission (BTC) bye-laws apply to the transport of dangerous chemicals on its canals.

The international carriage of dangerous chemicals by inland waterway in EC countries is carried out under the terms of the *European Agreement for the International Transport of Dangerous Goods* (ADN) and the *European Agreement for the International Transport of Dangerous Goods (River Rhine)* (ADNR).

3.22.4 Marine

Traditionally, the carriage of dangerous chemicals by sea has been regulated by the Department of Trade through the Merchant Shipping Act 1965 and the Merchant Shipping (Dangerous Goods) Rules 1965, with the practices accepted as constituting compliance with this legislation given in the *Report of the Advisory Committee on the Carriage of Dangerous Goods by Sea* (the 'Blue Book').

The *International Maritime Dangerous Goods Code* (IMDG) of the International Maritime Organization (IMO) (1992 IMO-200) covers the international movement of dangerous chemicals by sea. Other IMO codes include the *Code for the Construction and Equipment of Ships Carrying Dangerous Chemicals in Bulk* (IMO 1990 IMO-772) and the *International Code for the Construction and Equipment of Ships Carrying Liquefied Gases in Bulk (IGC)* (IMO 1993 IMO-104).

Ships are regulated by: the Merchant Shipping (Safety Officials and Reporting of Accidents and Dangerous Occurrences) Regulations 1982; the Merchant Shipping (Cargo Ship Construction and Survey) Regulations 1984; the Merchant Shipping (Fire Protection) Regulations 1984; the Merchant Shipping (Fire Protection) (Ships Built before 25th May 1980) Regulations 1985; and the Merchant Shipping (Dangerous Goods and Marine Pollutants) Regulations 1990.

Tankers are governed by: the Merchant Shipping (Tankers) (EEC Requirements) Regulations 1981; the Merchant Shipping (Gas Carrier) Regulations 1986; the Merchant Shipping (IBC Code) Regulations 1987; and the Merchant Shipping (BHC Code) Regulations 1987. There were also the Merchant Shipping (Chemical Tankers) Regulations 1986, but these were superseded the following year by the Merchant Shipping (BHC Code) Regulations 1987.

Docks are controlled by the Docks Regulations 1934, and now 1988. The Petroleum (Consolidation) Act 1928 contains requirements for harbour masters to make byelaws for the handling of petroleum spirit at harbours, these being confirmed by the Department of Transport. More comprehensive controls on dangerous substances at ports are contained in the Dangerous Substances in Harbour Areas Regulations 1987.

The Shipbuilding and Ship Repairing Regulations 1960 deal with measures to prevent fire/explosion during ship repair.

The controls on the transport of hazardous goods are considered further in Chapter 23.

3.22.5 Pipelines

The movement of dangerous chemicals by pipeline is covered, as stated earlier, by the Pipelines Act 1962. The Act distinguishes between local pipelines (less than 10 miles long) and cross-country pipelines (greater than 10 miles long). The *Petroleum Pipeline Safety Code* of the Institute of Petroleum (IP) (1982 MCSP Pt 6) is available for guidance. Until its demise, the Department of Energy was the enforcing authority for this Act. In 1991 its functions were split, with the planning functions passing to the Department of Trade and Industry (DTI) and enforcement functions to the HSE. The Act is enforced by the Pipelines Inspectorate.

British Gas pipelines are exempt from the Act, although the Notification of Installations Handling Hazardous Substances Regulations 1982 require notification to the HSE of the British Gas high pressure transmission system.

3.22.6 Air

Air transport is governed by the series of Civil Aviation Acts – 1971, 1978 and 1980. The relevant subordinate legislation is the Air Navigation Order 1989 and the Air Navigation (Dangerous Goods) Order 1985.

3.22.7 Commercial explosives

Controls on the classification, packaging and labelling of explosives are the Classification and Labelling of Explosives Regulations 1983 and the Packaging of Explosives for Carriage Regulations 1991. The road transport of explosives is governed by the Road Traffic (Carriage of Explosives) Regulations 1989. Control at docks is effected by the Dangerous Substances in Harbour Area Regulations 1987.

3.22.8 Military explosives

The transport of military explosives was originally covered by separate arrangements from those governing commercial explosives, but there is now a unified regime. The transport of military explosives by rail is still covered by separate regulations, the Conveyance by Rail of Military Explosives Regulations 1977. But the transport by road of such explosives, formerly controlled by the Conveyance by Road of Military Explosives Regulations 1977, is now governed by the Road Traffic (Carriage of Explosives) Regulations 1989. Similarly, the control of military explosives in docks, formerly effected by the Conveyance in Harbours of Military Explosives Regulations 1977, is now governed by the Dangerous Substances in Harbour Areas Regulations 1987.

3.23 Enforcement Practice

Information on the enforcement activities of the HSE is given in the annual reports of the HSC and in the annual *Health and Safety Statistics.*

The *Health and Safety Statistics 1990–91* (HSE, 1992b) gives data for enforcement by all enforcement agencies. The number of notices issued in 1990–91 was as follows: improvement notices 19 079; deferred prohibition notices 467; immediate prohibition notices 6222; and total notices 25 764.

Data more specific to industry are given for the Field Operations Division (FOD) of the HSE, which comprises the Factory and Agricultural Inspectorates and, from 1990–91, the Quarries Inspectorate. The number of offences for which the FOD issued notices in 1990–91 included the following: general safety, 12 425; dusts, toxic substances, ionizing radiations and noise, 2434; safety organization 2200; dangerous materials, 829; fire, 578; health, 400; and protective equipment, 209. There were in total 19 478 offences covered by 12 653 notices.

With regard to the particular legislation under which action was taken, in 1990–91 the number of requirements for action issued by the Factory and Agricultural Inspectorates included the following: Control of Substances Hazardous to Health Regulations 1988 2654; Construction (Working Places) Regulations 1966, 2318; Electricity at Work Regulations 1989, 822; Noise at Work Regulations 1989, 543; and Highly Flammable Liquids and Liquefied Petroleum Gases Regulations 1972, 421.

In respect of prosecutions the Factory and Agricultural Inspectorates laid 876 informations and obtained 806 convictions.

3.24 Offshore Legislation

Until 1991 the regulator of oil and gas activities in the British sector of the North Sea was the Department of Energy (DoEn), operating for the HSC under an agency agreement.

The HSWA 1974 is applicable offshore in the British sector of the North Sea, its applicability being governed by the Health and Safety at Work etc. Act 1974 (Application outside Great Britain) Regulations 1989, replacing similar regulations of 1977.

However, until 1988 most regulations were made under the Mineral Workings (Offshore Installations) Act 1971 (MWA). Principal regulations are the Offshore Installations (Managers) Regulations 1972, the Offshore Installations (Inspections and Casualties) Regulations 1973, the Offshore Installations (Construction and Use) Regulations 1974, the Offshore Installations (Operational Safety, Health and Welfare) Regulations 1976, the Offshore Installations (Emergency Procedures) Regulations 1976, the Offshore Installations (Life-Saving Appliances) Regulations 1977, the Offshore Installations (Fire Fighting Equipment) Regulations 1978, the Offshore Installations (Well Control) Regulations 1980, and the Offshore Installations (Safety Zones) Regulations 1987.

Pipelines are governed by the Petroleum and Submarine Pipelines Act 1975, the Submarine Pipeline (Inspectors etc.) Regulations 1977 and the Submarine Pipelines Safety Regulations 1982.

Diving activities are governed by the Submarine Pipelines (Diving operations) Regulations 1976 and the Diving Operations at Work Regulations 1981.

Following the Piper Alpha disaster in 1988, two additional sets of regulations were introduced, the Offshore Installations (Emergency Pipeline Valves) Regulations 1989 and the Offshore Installations (Safety Representatives and Safety Committees) Regulations 1989.

The *Cullen Report* on Piper Alpha (Cullen, 1990) called for a complete change in the regulatory regime. It recommended that the regulations should be goal-setting rather than prescriptive, that the means of demonstrating compliance should include formal safety assessment (FSA), that an operator should be required to submit a safety case for both existing and new installations, that this regime should be overseen by the HSE rather than the Department of Energy, and that an offshore safety division should be created within the HSE for this purpose.

The Offshore Safety Act 1992 initiates this new regime. The Offshore Installations (Safety Case) Regulations 1992 require the submission of a safety case which describes the Safety Management System (SMS) and gives a quantitative risk assessment. The Offshore Safety (Protection against Victimization) Act 1992 seeks to provide a degree of protection for persons who raise safety issues.

3.25 Goal-Setting Regulations

As already described, the philosophy underlying the HSWA 1974 is that the responsibility for the identification and control of hazards lies with the organization which creates them. In accordance with this philosophy, it is in general the policy of the HSE to move towards a system of regulations which are goal-setting rather than prescriptive. The *Cullen Report* on the Piper Alpha disaster was critical of what it regarded as the prescriptive approach taken by the Department of Energy in the North Sea and it recommended instead the use of goal-setting regulations.

Some factors bearing on the appropriate style of regulations have been discussed by Lees (1992a). This author makes the point that another relevant distinction is between regulations which are concerned with hardware and those dealing with software. It is in respect of hardware that the arguments against prescriptive regulations are strongest. These arguments do not usually have the same force in respect of software.

Since the *Robens Report* in 1972, the UK has tended to move away from regulations prescriptive of hardware. The reasoning behind this is that prescription of hardware, and hardware solutions, tends to lead to a loss of flexibility which is inimical to good design and to technological innovation. The last 15 years have in fact seen the introduction of a very large number of regulations in the UK, but they are for the most part prescriptive of software rather than of hardware. A good example is the Pressure Systems and Transportable Gas Containers Regulations 1989. These regulations include requirements for design, construction, operation, maintenance and modification. In particular, they are relatively detailed in respect of the written scheme of examination, of examination in accordance with the written scheme and of the keeping of records.

Limiting consideration to hardware, Lees attempts to identify some factors which bear on the most appropriate style of regulations. Regulatory requirements will generally fall into one of the following five categories. They may state

(1) a general goal;
(2) default means with alternative options;
(3) prescribed means with flexibility on parameters;
(4) prescribed means with default parameters;
(5) prescribed means with fixed parameters.

These categories give rise to five methods of formulating a set of regulations (Methods 1–5). Factors listed by Lees as relevant to the style of regulatory requirements are

(1) level of knowledge;
(2) demand (posed by the hazard);
(3) options (available to control the hazard);
(4) interactions and side-effects (of options);
(5) practicality (of options)
  (a) mature technology,
  (b) widespread use;
(6) dependability (of options)
  (a) general reliability,
  (b) passive measure;
(7) design skill (required);
(8) cost;
(9) cost-effectiveness;
(10) longevity.

The level of knowledge about the problem and its potential solutions must be the first factor in the list. It needs to be considered in relation to each of the other items.

A particular hazard places a certain demand on the features which may be provided to protect against it. In order to design to meet the demand it is necessary to have some qualitative and quantitative understanding of the demand itself and of the factors which can give rise to it. The options available to control the hazard, or meet the demand, may be few or many. Again some qualitative and quantitative understanding is needed of the extent to which different options can handle the demand. If there is more than one option, it may be appropriate to select

Table 3.6 *Factors affecting the prescriptive approach to regulation of a hazard (after Lees, 1992a)*

Feature	*Unsuitable for prescriptive approach*	*Not unsuitable for prescriptive approach*
Level of knowledge	Low	High
Demand	Complex	Simple
Options	Multiple	Single
Interactions/side-effects	Significant	Negligible
Practicability	Low	High
Dependability	Low	High
Design skill	High	Low
Cost	High	Low
Cost-effectiveness	Low	High
Longevity	Low	High

one or to use several in combination. On the other hand, there may be only one practicable option.

The problem considered may be relatively simple or may have a rather deeper structure. A particular solution may have side-effects. Before it can be made the subject of a regulatory requirement, an option needs to be practicable. One indication of this is that the option involves mature technology. Another is that it is in widespread use. These two factors are often the same, but not always.

The option selected must be dependable, the term 'dependability' encompassing both capability and reliability.

The distinction between passive and active protection is important in relation to dependability. In general, passive protection will tend to be more reliable, though the case may not always be clear-cut.

The degree of design skill required to implement the option is also important. Designs which involve assessment of a number of options before a choice is made necessarily tend to be more complex. Individual options differ in the extent to which they test the skill of the designer.

Cost is always a relevant factor. It includes not only the cost of the option itself, but of any interactions and side-effects. In general, the higher the cost, the less appropriate the prescriptive approach.

A related, but separate, factor is cost-effectiveness.

The option selected should have an appropriate life-span in the sense that it does not become significantly inferior to some other option until some long time period has elapsed. This time period relates to the frequency with which it is reasonable to revise the requirements.

The regulator must consider all these factors, plus the question of monitoring and enforcement. Lees lists the characteristics of problems which lie at the two extreme ends of the spectrum described. The characteristics which point to Method 1 are

Method 1

(1) Level of knowledge of problem poor.
(2) Demand and its causes difficult to define; demand difficult to handle.
(3) Several options available; use of options in combination attractive.
(4) Options tend to have strong interactions and/or side-effects.
(5) Practicability of options not straightforward, conditional.
(6) Dependability of options not straightforward, conditional.
(7) Design skill of high order required.
(8) Options tend to involve major hardware items.
(9) Cost implications of options appreciable.
(10) Any prescribed solution likely to be short-lived.

The characteristics which point to Method 5 are:

Method 5

(1) Level of knowledge of problem good.
(2) Demand and its causes readily defined; demand readily handled.
(3) Prescribed measure accepted as preferred solution.
(4) Prescribed measure has negligible interaction or side effects.
(5) Practicability of prescribed measure high.
(6) Dependability of prescribed measure high.
(7) Design skills of low order.
(8) Prescribed measure does not involve major hardware.
(9) Prescribed measure relatively cheap.
(10) Prescribed measure likely to be long-lasting.

A particular hazard might be characterized by rating it on a scale such as that shown in Table 3.6. A problem suitable for treatment by Method 5 would thus be one where there is a simple demand, for which there is a clear preferred solution with negligible interactions or side-effects, which is practical and dependable, which requires relatively little design skill, and which is cheap and cost-effective and will be long-lived.

Generally, regulations are accompanied by guidance. Even where the regulations themselves are in a goal-setting style, the guidance tends to be more prescriptive. There is an in-built tendency for the system of regulations plus guidance to become prescriptive. If it is intended that it should not be so, specific steps need to be taken to avoid such an outcome.

A critique of goal-setting regulations is given by Kenney (1991). The style of regulations is considered further in Appendix 18.

3.26 Self-Regulation

As already discussed, the philosophy underlying British safety legislation, derived from the *Robens Report*, is that the organization which creates the hazards is responsible for identifying and controlling them. In this sense it is one of self-regulation.

A critical study of self-regulation is given in *Safety at Work: the Limits of Self-Regulation* by S. Dawson *et al.* (1988). The authors identify the core elements of local self-regulation in the Robens conception as effective management, specialist advice and workforce involvement. They describe three case studies of the effects of a regime of self-regulation, in the chemical industry, construction and retailing.

Dawson *et al.* draw four general conclusions which they summarize as follows:

> The first is that self-regulation of safety at local level can be effective, but only if adequately resourced, if related to nationally established standards and if supported by the knowledge that failures of self-regulation will lead to enforcement. The second is that effective local self-regulation of health and safety will not be developed and maintained 'naturally' out of the operation of deregulated market forces; elements of government regulation are an essential prerequisite for systems of self-regulation at national, industrial and local levels. The third general conclusion is that the overall performance of the system of local self-regulation has deteriorated since 1981 as indicated by published statistics. Fourthly, we have suggested that the deterioration is marked in specific sectors characterised by small firms, subcontracting, low pay, weak trade unionism and productivity improvements. (S. Dawson *et al.*, 1988, p.268)

The comments on pay and productivity evidently relate to an analysis by the authors of the fatal and major accidents rate in 1981–84, which showed the rate to be higher in industries which had undergone higher than average productivity increases and in industries where low pay was prevalent. Since this period covered a recession, it is pertinent to note that the authors found no simple effect of recession.

3.27 Standards and Codes of Practice

Standards and codes of practice normally represent a consensus of good industrial practice. In effect, therefore, they are aids to the discharge of the legal duty of reasonable professional care and criteria against which this can be judged.

The standing of a code, legally or otherwise, is, of course, a function of the body which has produced it. Certain approved codes have the further legal status of being referred to in Acts or in statutory instruments. The HSWA 1974 recognizes Approved Codes of Practice (ACOPs) and accords them a special legal status. Failure to adhere to an ACOP places the onus on the party concerned to demonstrate compliance with the substantive requirement of the legislation. The declared practice is to discuss such codes with the interested parties before giving them the status of Approved Codes under the Act. For some time there was only one Approved Code of Practice, that on safety representatives and safety committees, but there are now a number of them, as shown in Table 3.5.

The use of codes of practice for legal purposes involves a potential difficulty. Such codes are normally developed by industry for its own use and are adopted voluntarily. This approach is favourable to the development of codes which go beyond the lowest common denominator. It is desirable that the creation of effective codes is not inhibited by emphasis on the legal dimension and that it proceed with the greatest possible measure of agreement by industry. The initiative for the creation of codes may be taken not only by industry but also by the HSE, which has in fact been responsible for encouraging the writing of a number of important codes.

Standards and codes of practice are treated more fully in Appendix 27.

3.28 Industrial Law

Industrial law is partly common law and partly statute law. Statute law consists of Acts of Parliament together with the statutory instruments made under these Acts. Common law prescribes certain obligations, such as a general duty of care by an employer for an employee. Such general duties are then interpreted in the light of case law. Contravention of statue law may lead to criminal proceedings. It may also result in a civil action for compensation. Liability under common law may give rise to a civil action.

There is a good deal of uncertainty surrounding industrial law. Commenting on the law on strict liability the Law Commission (1970a) referred to the 'complexity, uncertainty and inconsistency of the law'. Only a brief discussion is given here. Fuller accounts relevant to industrial safety and loss prevention are given by Gold (1969) and Broadhurst (1971). Both these accounts quote a number of legal judgements on finer points of law.

Common law is one of the main streams of industrial law. Two parts of the common law which are particularly important in the present context are the laws of negligence and of strict liability. The relationship between an employer and an employee is a voluntary contract. By entering into this the employer assumes a duty of care. The part of the common law which deals with this is the law of negligence. This is far-reaching and open-ended: 'The categories of negligence are never closed.'

Negligence is essentially the omission of something which a reasonable person would do or the commission of something which he would not. The concept of the 'reasonable person' is an important one in this context. In particular, where a person is exercising a profession or skill the law demands the standard of care which is to be expected of a reasonable person in that situation; it expects, for example, a technical competence which it would not require from the average person.

In judging what is reasonable, account is taken of cost. An employer is not required to spend enormous sums of money to reduce the probability of the realization of an already remote and minor hazard. Liability also depends on the foreseeability of the hazard. Failure to take precautions against a foreseeable hazard may result in liability for the direct consequences if this hazard materializes. It is not necessary that the precise way in which the hazard is realized be foreseeable.

The duty of care may be broken down into a number of more specific duties. These include the duties to provide (1) safe premises, (2) safe systems of work, (3) safe machinery, (4) safe materials, and (5) competent personnel.

The provision of safe systems of work covers, among other things, organization of the work itself with respect to such features as plant layout and sequence of operations, provision of safe working conditions and protective equipment and clothing, instructions and documentation appropriate to the job, training in the operations of the job and in safety procedures, and the provision of warnings and notices. If equipment is obtained from a reputable maker but proves faulty, it is normally the manufacturer who is held liable. However, this may not be the case if the equipment has been in use for some time and has not been properly maintained. The requirement to have competent personnel is largely a matter of selection and training. But in addition the employer has a duty to dismiss anyone whom the employer has reason to believe may through negligence injure his fellow employees.

There is in general no liability without fault. Thus the plaintiff must prove that the defendant was negligent. In order to do this the plaintiff must prove that the defendant owed him a duty of care, that he failed to discharge it and that as a result he suffered damage. There are many factors which may determine whether an accusation of negligence is upheld. If a particular job is generally recognized as dangerous, this does not relieve the employer of taking appropriate precautions. On the contrary, he is obliged to take additional safety measures. Some allowance appears to be made where an employer faces a dilemma: this includes shutting down production. Thus in *Latimer v. AEC Ltd,* management kept a factory going after flooding had made the floors slippery, because shut-down was considered unduly drastic. An employee who slipped and was injured did not succeed in establishing negligence. In other cases where there has been no element of dilemma, similar actions have succeeded.

It is not necessarily sufficient to follow the general practice of the trade, even if this is quite well established. One defence against an action for negligence is the principle of ***volenti non fit injuria***, which is that the employee has voluntarily accepted the risk in the job. Another is contributory negligence by the employee.

Common law also contains, in addition to the law of negligence, another major branch, which is the law relating to strict liability. In essence, if the occupier of land has something dangerous on it, he is *prima facie* responsible for the damage which is the natural consequence of its escape. This applies whether or not there has been negligence on his part, and is thus strict liability. This principle was established in a famous case, ***Rylands v. Fletcher***. It should be emphasized that the law of strict liability just described is part of civil rather than criminal law. Both the law of negligence and that of strict liability are highly relevant to the problems which arise in the process industries.

Statute law is the other main stream of industrial law. There is a considerable body of legislation relevant to safety and loss prevention in the form of Acts of Parliament and of detailed statutory instruments made under these Acts. The objectives of such Acts, which may be protection of life or property, are important, because they influence the interpretation of the law.

Acts in the field tend to be of two principal types, depending on whether they deal with places or materials. The Factories Act 1961 and the Petroleum (Consolidation) Act 1928 are examples of the former and latter, respectively. There is inevitably a degree of overlap in industrial legislation. The differing bases for the Acts just described are a prime cause, but are by no means the only one.

The HSWA 1974 is part of statute law. This Act creates specific statutory duties of care for employees and of avoidance of injury to the public. It therefore strengthens statute law and hence criminal law in these areas.

The law needs constantly to be updated. Common law develops through a continuous stream of judgements. In the case of statute law adaptation is achieved by wording the principal Acts in fairly general terms and supplementing these with statutory instruments, and, when necessary, by passing new Acts.

3.29 EC Directives

The European Community (EC) issues to its member states Directives which require the governments of those states to enact legislation in accordance with these Directives. The object of the Directives is to create a degree of conformity between the legislative provisions of the member states. EC legislation is given in ***Eurochem Monitor*** (Agra Europe, 1994a) and ***European Community Environment Legislation for Industry*** (Agra Europe, 1994b).

An EC Directive is addressed to governments. It does not itself have the force of law within the member states. Its provisions become effective through their embodiment in the legislation of those states. Selected EC Directives relevant to safety and loss prevention are

***Table 3.7** Selected EC Directives relevant to safety and loss prevention*[a]

Directive	*Description*
67/548/EEC 27 June	On the approximation of the laws of member states on the classification, labelling and packaging of dangerous substances. OJ L196, 16.8.67. (The 'Dangerous Substances 'Directive)
75/442/EEC 15 July	On waste. OJ L194/39, 25.7.75
76/117/EEC 18 December	On the approximation of the laws of the member states concerning electrical equipment for use in potentially explosive atmospheres. OJ L24/45, 30.1.76. (The 'Explosive Atmospheres' Directive)
76/464/EEC 4 May	On pollution caused by certain dangerous substances discharged into the aquatic environment of the Community. OJ L129/23, 18.5.76
76/769/EEC 27 July	On the approximation of the laws, regulations and administrative provisions of the member states relating to the marketing and use of certain dangerous substances and provisions. OJ L262/201, 27.9.76
77/767/EEC 27 July	On the approximation of the laws of the member states relating to common provisions for pressure vessels and methods of inspecting them. OJ L262, 27.9.76
78/319/EEC 20 March	On toxic and dangerous waste. OJ L84/43, 31.3.78
78/610/EEC 29 June	On the approximation of the laws, regulations and administrative provisions of the member states on the protection of the health of workers exposed to vinyl chloride monomer. OJ L197/12, 22.7.78
79/196/EEC 6 February	On the approximation of the laws of the member states concerning electrical equipment for use in explosive atmospheres employing certain types of protection. OJ L43/20, 20.2.79
79/831/EEC 18 September	Amending for the sixth time the Directive 67/548/EEC on the approximation of the laws, regulations and administrative provisions relating to the classification, packaging and labelling of dangerous substances. OJ L259/10, 15.10.79. (The 'Sixth Amendment', to the 'Dangerous Substances' Directive)
80/68/EEC 17 December 1979	On the protection of ground water against pollution caused by certain dangerous substances. OJ L20/43, 26.1.80
80/836/Euratom 15 July	Amending the Directives laying down the basic safety standards for the health protection of the general public

	and workers against the dangers of ionizing radiations. OJ L246/1, 17.9.80
80/876/EEC 15 July	On the approximation of the laws of the member states relating to straight ammonium nitrate fertilizers of high nitrogen content. OJ L250/23, 23.9.80
80/1107/EEC 27 November	On the protection of workers from the risk related to exposure to chemical, physical and biological agents at work. OJ L327/8, 3.12.80
82/501/EEC 24 June	On the major-accident hazards of certain industrial activities. OJ L230/1, 5.8.82; also OJ L289/35, 13.10.82 (The 'Major Accident Hazards' Directive).
83/189/EEC 28 March	Laying down procedures for the provision of information in the field of technical standards and regulations. OJ L109/8, 26.4.83
84/360/EEC 28 June	On the combating of air pollution from industrial plants. OJ L188/20, 16.7.84
84/449/EEC 25 April	Adapting to technical progress for the sixth time Council Directive 67/548/EEC on the approximation of the laws, regulations and administrative provisions of the member states relating to the classification, packaging and labelling of dangerous substances. OJ L251/1, 19.9.84
84/467/EEC 3 September	Amending Directive 80/836/Euratom as regards the basic safety standards for the health protection of the general public and workers against the dangers of ionizing radiations. OJ L265/4, 5.10.84
84/631/EEC 6 December	On the supervision and control within the European Community of transfrontier shipment of hazardous waste. OJ L326/31, 13.12.84
85/337/EEC June 27	On the assessment of the effects of certain public and private projects on the environment. OJ L175/40, 5.7.85
86/188/EEC 12 May	On the protection of workers from risks related to exposure to noise at work. OJ L137/28, 24.5.86
86/280/EEC 12 June	On limit values and quality objectives for discharges of certain dangerous substances included in List I of the Annex to Directive 76/464/EEC. OJ L181/16, 4.7.86
87/216/EEC 19 March	Amending Directive 82/501/EEC on the major-accident hazards of certain industrial activities. OJ L85/30, 28.3.87 (first amendment)
87/217/EEC 19 March	On the prevention and reduction of environmental pollution by asbestos. OJ L85/40, 28.3.87
87/404/EEC 25 June	On the harmonization of the laws of the member states relating to simple pressure vessels. OJ L220/48, 8.8.87
88/379/EEC 7 June	On the approximation of the laws, regulations and administrative provisions of the member states relating to the classification, packaging and labelling of dangerous preparations. OJ L187/14, 16.7.88. (the 'Preparations' Directive)
88/609/EEC 24 November	On the limitation of emissions of certain pollutants into the air from large combustion plants. OJ L336/1, 7.12.88
88/610/EEC 24 November	Amending Directive 82/501/EEC on the major-accident hazards of certain industrial activities. OJ L336, 7.12.88 (second amendment)
89/178/EEC 22 February	Adapting to technical progress Council Directive 88/379/EEC on the approximation of the laws, regulations and administrative provisions of the member states relating to the classification, packaging and labelling of dangerous preparations. OJ L64/16, 8.3.89
89/391/EEC 12 June	On the introduction of measures to encourage improvements in the safety and health of workers at work. OJ L183/1, 29.6.89. (The 'Framework' Directive)
89/392/EEC 14 June	On the approximation of the laws of the member states relating to machinery. OJ L183/13, 30.6.89 (the 'Machinery' Directive)
89/677/EEC 21 December	Amending for the eighth time the Directive 76/769/EEC on the approximation of the laws, regulations and administrative provisions of the member states relating to the marketing and use of certain dangerous substances and preparations. OJ L398/19, 30.12.89
90/394/EEC 28 June	On the protection of workers from the risks related to exposure to carcinogens at work. OJ L196/1, 26.7.90
90/488/EEC 17 September	Amending Directive 87/404/EEC on the harmonization of the laws of the member states relating to simple pressure vessels. OJ L270/25, 2.10.90
90/492/EEC 5 September	Adapting to technical progress Council Directive 88/379/EEC on the approximation of the laws, regulations and administrative provisions of the member states relating to the classification, packaging and labelling of dangerous preparations. OJ L275/35, 5.10.90
91/155/EEC 5 March	Defining and laying down the detailed arrangements for the system of specific information relating to dangerous preparations in implementation of Article 10 of Directive 88/379/EEC. OJ L76/35, 22.3.91. (the 'Safety Data Sheets' Directive)
91/383/EEC 25 June	Supplementing the measures to encourage improvements in the safety and health at work of workers with a fixed-duration employment relationship or a temporary employment

	relationship. OJ L206/19, 29.7.91. (The 'Temporary Workers' Directive).
91/689/EEC 12 December	On hazardous waste. OJ L377/20, 31.12.91

[a] There are numerous directives concerned with limitation of pollution. They include: 76/403/EEC (PCB/PCTs); 80/779/EEC (sulphur dioxide and particulates); 82/176/EEC (mercury); 82/884/EEC (lead); 83/513/EEC (cadmium); 84/156/EEC (mercury); 85/203/EEC (nitrogen oxides); and 92/72/EEC (ozone).
Sources: Official Journal of the European Community and Official Journal of the European Community. Annual Alphabetical and Methodological Table (Luxembourg).

given in Table 3.7. There are a number of further directives on pollution, some of which are briefly listed at the foot of the table.

Legislation in the EC is described in *The European Community Legislative Process* by CONCAWE (1991 91/55). An account of EC Directives and of the burden of responding to them is given by Trowbridge (1979). In the period January 1976 to April 1979, on matters of specific interest to the chemical industry and in areas of health, safety, use of chemicals, transport and distribution, energy and environment, there were 137 actual Directives and Decisions or amendments to them, 95 actual Regulations or amendments to them, and 76 proposals for Directives, Regulations and amendments. Trowbridge discusses, in particular, the problems posed by the 'Sixth Amendment' to the 'Dangerous Substances' Directive, described below.

3.29.1 'Dangerous Substances' Directive

The 'Dangerous Substances' Directive 67/548/EEC deals with the classification, packaging and labelling of dangerous substances. A uniform approach to these features facilitates international trade. The principal British legislation implementing the Directive was the Classification, Packaging and Labelling of Dangerous Substances Regulations 1984, now superseded by the Chemicals (Hazard Information and Packaging) Regulations 1993.

3.29.2 'Safety Data Sheets' Directive

Requirements for materials safety data sheets, contained in Directive 91/155/EEC, are in the Chemicals (Hazards Information and Packaging) Regulations 1993.

3.29.3 'Explosive Atmospheres' Directive

The 'Explosive Atmospheres' Directive 76/117/EEC, deals with electrical equipment to be used in hazardous areas.

3.29.4 Pressure vessels Directives

Directive 77/767/EEC is a framework Directive for pressure vessels and their inspection. Another Directive on pressure vessels is Directive 87/404/EEC on simple pressure vessels, essentially those manufactured in series.

3.29.5 'Dangerous Substances' Directive: Sixth Amendment

The Sixth Amendment to the 'Dangerous Substances' Directive 79/831/EEC includes amendments relating to classification, packaging and labelling, but also has other far-reaching provisions.

One set of requirements created by the Directive is for the notification of new substances. This aspect was implemented in British legislation in the Notification of New Substances Regulations 1982, now superseded by the Notification of New Substances Regulations 1993.

Another set of requirements created by the Directive is for the reduction of pollution of the aquatic environment. It gives two categories of substance, List I and List II. For List I the method of control is to set specific limits on the concentration of the pollutants, while for List II it is to set limits to their discharge.

The Directive makes use of Environmental Quality Objectives (EQOs). These provide a basis for setting discharge limits and may also be used to set concentration limits. Accounts of this Directive are given by Chipperfield (1979), Cussell (1979), Malle (1979), Otter (1979) and Trowbridge (1979).

3.29.6 'Major Accident Hazards' Directive

The Major Accident Hazards Directive 82/501/EEC arose from concern at events such as Flixborough and, in particular, Seveso, and was initially often referred to as the 'Seveso Directive'. Accounts of this directive have been given by Dewis (1985a) and Vermeulen and Hands (1993).

The British legislation implementing the Directive is the Control of Industrial Major Accidents Hazards Regulations 1984 (CIMAH). There have been two amendments to the Directive, by Directives 87/216/EEC and 88/610/EEC, resulting in amendments to the CIMAH Regulations in 1988 and 1990.

3.29.7 'Noise at Work' Directive

Directive 86/188/EEC, the 'Noise at Work' Directive, is implemented in the UK by the Noise at Work Regulations 1989.

3.29.8 Environmental impact Directive

Directive 85/337/EEC introduces a requirement for an environmental impact statement. This requirement is incorporated in British legislation in the Town and Country Planning (Assessment of Environmental Effects) Regulations 1988.

3.30 US Legislation

It is appropriate to give a brief account of US legislation in the area of safety and loss prevention, both because this allows a comparison to be made with legislation in the UK and the EC, and because it provides a background to US literature in this field.

In the 1970s, there occurred in the USA a very rapid development of public awareness of problems of health and safety, and of environmental problems, and a corresponding growth in the creation of controls in this area. In the USA both State and Federal Governments are sources of legislation. In recent years the involvement of the Federal Government in health and safety legislation has greatly increased. Some areas of legisla-

Table 3.8 *Selected US legislation relevant to safety and loss prevention*

1899	Rivers and Harbors Act
1908	Explosives Transportation Act
1909	Explosives Transportation Act
1921	Explosives Transportation Act
1936	Tank Vessels Act
1938	Federal Food, Drug and Cosmetics Act
	Natural Gas Act
1947	Federal Insecticide, Fungicide and Rodenticide Act
1952	Dangerous Cargo Act
1956	Federal Administrative Procedures Act
	Federal Water Pollution Control Act
1958	Federal Aviation Act
1962	Hazardous Substances Labeling Act
1963	Clean Air Act
1965	Solid Waste Disposal Act
	Water Quality Act
1967	Federal Hazardous Substances Act
1968	Natural Gas Pipeline Safety Act
1969	National Environmental Policy Act (NEPA)
1970	Bulk Flammable Combustible Liquids Act
	Clean Air Act (CAA)[a]
	Consumer Product Safety Act (CPSA)
	Dangerous Cargo Act
	Federal Environmental Pollution Control Act
	Hazardous Materials Transportation Act (HMTA)
	Occupational Safety and Health Act (OSHA)
	Resource Recovery Act
1972	Coastal Zone Management Act
	Federal Water Pollution Control Act (FWPCA)
	Noise Control Act
	Ports and Waterways Safety Act
1974	Safe Drinking Water Act (SDWA)
1975	Hazardous Materials Transportation Act (HMTA)
1976	Federal Technology Transfer Act
	Resource Conservation and Recovery Act (RCRA)
	Toxic Substances Control Act (TSCA)
1980	Comprehensive Environmental Response, Compensation and Liability Act (CERCLA, or Superfund)
1986	Emergency Planning and Community Right-to-Know Act (EPCRA)
	Superfund Amendments and Reauthorization Act (SARA)

[a] Major amendments were made to the Act; reference is made both to the Clean Air Act 1970 and to the Clean Air Act Amendments 1970.

tion which are relevant here include (1) occupational health and safety, (2) environment, (3) toxic substances, (4) accidental releases and (5) transport.

The development of legislation on health and safety in the USA may be followed through the references given in Table A1.4 in Appendix 1. In particular, attention is drawn to: the problems of toxic chemicals in the workplace, of pollution and of transport; to the cases of abandonment of actual or planned production and of litigation; and to the resulting legislation.

In general, it is probably fair to say that legislation has been tougher in the UK in some areas and in the USA in others, and also that the rate at which controls have been tightened has often been different in the two countries. It is worth emphasizing also that in these areas it is not sufficient merely to pass legislation – it is necessary to enforce it. There has been, in both countries, a considerable tightening of methods of enforcement.

Both State and Federal legislation make use of national standards and codes of practice. In particular, standards such as those of the American National Standards Institute (ANSI) and the *National Fire Codes* of the National Fire Protection Association (NFPA) are widely quoted in legislation.

Selected US legislation relevant to safety and loss prevention is given in Table 3.8.

3.31 Occupational Safety and Health Act 1970

The Occupational Safety and Health Act 1970 (OSHA) is the framework Act. It provides that the Secretary of Labor should promulgate safety standards, inspect workplaces and assess penalties. There is a separate commission which deals with violations of the Act.

The Act has some similarity with the HSWA 1974 in the UK, in that it is essentially an enabling measure which is superimposed on existing health and safety legislation, lays duties on employers and employees to comply with standards and regulations, and has provisions for inspection and enforcement.

The Act is administered by the Secretary of Labor and the Secretary of Health, Education and Welfare. The former is primarily responsible for enforcement aspects and the latter for developing standards. The Act is enforced by the Occupational Safety and Health Administration (OSHA also).

Standards are issued under the Act by the Secretary of Labor, who may issue any established Federal standard or any national consensus standard. The initial standards promulgated under the Act were published in the *Federal Register* 29 May, 1971.

The Act established the National Institute for Occupational Safety and Health (NIOSH) to develop and establish recommended occupational safety and health standards.

The Act places the onus for compliance with the law primarily on the employer. This applies even in cases where employees are not working according to instructions which have been issued to them to provide for their safety. An employer has the right to seek a variance from a standard, and may file a petition challenging the validity of a standard.

Accounts of the Act are given by Anon. (1971c), Hodgson (1971), Nilsen (1972a), L.J. White (1973b) and Corn (1976).

3.32 US Regulatory Agencies

The Occupational Safety and Health Administration (OSHA) was created by the OSHA 1970 in that year and has responsibility for enforcing that Act. OSHA comes under the Secretary of Labor and is headed by an

Assistant Secretary of Labor. As described above, its activities include the adoption of standards and making of rules, the inspection of workplaces and the investigation of accidents.

The Environmental Protection Agency (EPA), created by the National Environmental Policy Act 1969, is responsible for environmental legislation, including that on air pollution, water pollution and hazardous wastes.

The areas of responsibility between the OSHA and the EPA are not clear-cut. Although the OSHA is concerned with the workplace and EPA with the environment, there are some areas where both agencies are involved. One of these is accidental releases and another is toxic substances, as described below. The relation between the two bodies is discussed by Spiegelman (1987) and Burk (1990).

3.33 US Environmental Legislation

Of the issues which have assumed an increasingly high profile in the last two decades, pollution attained an early prominence and has given rise to a large volume of legislation. Legislation of particular importance in this field includes: the Water Quality Act 1965, the National Environmental Policy Act 1969, the Clean Air Act 1970, the Federal Environmental Pollution Control Act 1970, the Federal Water Pollution Control Act 1972, the Safe Drinking Water Act, 1974 and the Noise Control Act 1972.

The National Environmental Policy Act 1969 (NEPA) created the EPA.

The Clean Air Act 1970 (CAA) gives the EPA power to adopt and enforce air pollution standards. The EPA has promulgated air quality standards which set the maximum concentrations for various gaseous pollutants. There have been amendments to the Clean Air Act in 1977, 1981 and 1990. Accounts of air pollution legislation are given by Vervalin (1972d), Zarytkiewicz (1975), Halley (1977), Lipton and Lynch (1987), G.A. Brown, Cramer and Samela (1988), Veselind, Peirce and Weiner (1990) and Asante-Duah (1993).

The Federal Water Pollution Control Act 1972 (FWPCA) gives the EPA power to adopt and enforce water pollution standards. The EPA has promulgated effluent quality standards and guidelines which set the maximum concentrations for all liquid pollutants. Accounts of water pollution legislation are given by Weisberg (1973), L.J. White (1974), Veselind, Peirce and Weiner (1990) and Asante-Duah (1993).

Hazardous wastes are regulated by the Resource Conservation and Recovery Act 1976 (RCRA). There are rules governing the storage of such wastes in storage tanks, including underground storage tanks. The Act is administered by the EPA.

The Comprehensive Environmental Response, Compensation and Liability Act 1980 (CERCLA, or Superfund) strengthens the regulation of hazardous wastes by the EPA. It creates controls on hazardous waste sites and enforces clean-up of existing and abandoned sites. It seeks to identify potentially responsible persons (PRPs) and, where such parties are not found, it supervises a clean-up paid for from the Superfund. The Act has to be renewed periodically. The Superfund Amendments and Reauthorization Act 1986 (SARA) renews the Superfund. The part of the Act known as SARA Title III extends its application to accidental releases, as described below. Accounts of hazardous waste legislation and of Superfund are given by Caruana (1986), Edelman (1987), Melamed (1989), Veselind, Peirce and Weiner (1990) and Asante-Duah (1993).

The development of legislation on pollution may be followed through the references given in Table A1.4. These references show the evolution of the air and water quality standards and of their application. The NEPA 1969 imposes a duty on Federal agencies to obtain from a developer an environmental impact statement (EIS) before they take any action which affects the environment, such as granting a permit. A number of other Acts also include a requirement for an EIS. A regulation of the President's Commission on Environmental Quality (CEQ) made in 1978 defines requirements for an EIS. Permits may be required from a number of agencies at federal, state and local level.

Accounts of environmental impact statements are given by Willis and Henebry (1976, 1982–). Environmental legislation in the USA is described further in Appendix 11.

3.34 US Toxic Substances Legislation

The Toxic Substances Control Act 1976 (TSCA) is a framework Act which creates a comprehensive system of controls over toxic substances. It empowers the EPA to regulate the manufacture, processing, distribution, use and disposal of existing and new chemicals in order to avoid unreasonable risk to health or the environment and to delay or ban manufacture or marketing. It provides for the creation of an inventory of existing chemicals and for notification by a manufacturer of a new chemical or of new uses for a chemical. There are requirements for the toxicity testing of chemicals and for Premanufacture Notices (PMNs). The EPA has an Office of Toxic Substances (OTS) dealing with the Act. Accounts of the Act are given by Ricci (1976c), C.W. Smith (1977) and A.S. West (1979, 1982, 1986, 1993b).

Section 313 of SARA Title III, described below, also deals with toxic substances, but essentially in relation to high level emissions. It requires the EPA to compile a toxic chemical inventory data base, the Toxic Release Inventory (TRI). Section 313 is discussed by Bowman (1989). The control of toxic substances in the workplace and in the atmosphere is also covered by the OSHA 1970 and by pollution legislation.

The development of legislation on toxic substances in the workplace and in the atmosphere may be followed through the references given in Table A1.4. The case of vinyl chloride is of particular interest.

3.35 US Accidental Release Legislation

The accidents at Flixborough, at Seveso and, above all, at Bhopal, and the development of major hazard controls in Europe led in the second half of the 1980s to US legislation on accidental releases. Accidental releases are covered by the Emergency Planning and Community Right to Know Act 1986 (ECPRA). This is Title III of the Superfund Amendments and Reauthorization Act 1986 (SARA). It is commonly known as SARA Title III and was signed into law in 1988. It contains four main parts:

(1) emergency planning, (2) emergency notification, (3) community right-to-know, and (4) toxic chemicals inventory. The Act is enforced by the EPA. An account of the EPA's chemical accident release prevention program is given by Matthiesen (1994).

Controls on accidental releases have also been introduced by the OSHA in the form of a rule for the Process Safety Management of Highly Hazardous Chemicals 1990. This introduces a requirement for a process safety management system to identify, evaluate and control such hazards. An account of OSHA's process safety management rule is given by Donnelly (1994).

Individual states have also brought in their own legislation. The New Jersey Toxic Catastrophe Prevention Act 1985 (TCPA) creates a requirement for a risk management programme for facilities handling certain toxic chemicals above a given inventory. The California Hazardous Materials Planning Program is based on a codification of four laws within the California Health and Safety Program.

Accounts of legislation on hazardous chemicals with potential impact off site are given by Brooks *et al.* (1988) and Horner (1989).

3.36 US Transport Legislation

Another major concern in the USA in recent years has been the problem of the transport of hazardous materials, and this has given rise to a large volume of legislation.

Some principal items of legislation in the field include: the early Rivers and Harbors Act 1899, the Explosives Transportation Acts 1908, 1909 and 1921, the Tank Vessels Act 1936, the Natural Gas Act 1938, the Federal Aviation Act 1958, the Natural Gas Pipeline Safety Act 1968, the Dangerous Cargo Act 1970, the Hazardous Materials Transportation Act 1970, the Coastal Zone Management Act 1972, the Ports and Waterways Safety Act 1972, and the Hazardous Materials Transportation Act 1975. This legislation is considered further in Chapter 23.

3.37 Engineering Ethics

It is appropriate to end this chapter with a brief consideration of professional ethics, particularly as applied to safety and loss prevention. The general topic is discussed in *Ethics in Engineering* (M.W. Martin and Schinzinger, 1989) and by Mingle and Reagan (1980), Flores (1983) and Mingle (1989). Environmental ethics are considered in *Environmental Ethics for Engineers* (Gunn and Vesilind, 1986). The teaching of engineering ethics is discussed by Lindauer and Hagerty (1983), Tucker (1983) and Wilcox (1983).

Accounts of engineering ethics tend to treat the topic under the following broad headings: (1) professional competence, (2) observance, (3) contribution and communication, (4) risk issues, and (5) ethical dilemmas.

Clearly the engineer's first duty is professional competence, which includes education and training to enhance it. It is also the engineer's duty to observe the law and to adhere to standards. But beyond this engineers should assume their wider responsibilities by making their contribution within their organization, communicating with colleagues and, where appropriate, with the general public. Engineers should understand the issues which arise in relation to the hazards to which people, both employees and the public, may be exposed as a result of engineering activities, should aim for openness in the declaration and treatment of these issues, and should seek to arrive at an informed consent.

3.37.1 Engineering Council Codes

In the UK, the Engineering Council (1992) has published a *Code of Professional Practice on Risk Issues.* Table 3.9 lists the 10 points of the code. A *Code of Professional Practice on Engineers and the Environment* has also been issued by the Council (Engineering Council, 1993).

3.37.2 Risk issues

A major area of engineering ethics is that of the risks to employees and to the public from the hazards with which the engineer deals. In large part this issue turns on the question of what risks are acceptable, or at least tolerable. This topic is dealt with later, principally in Chapters 4 and 9.

3.37.3 Ethical dilemmas

An investigation of ethical dilemmas such as arise in the process industries has been described by Matley and coworkers (Matley and Greene, 1987; Matley, Greene and McCauley, 1987). They devised a set of nine such dilemmas with multiple-choice answers and conducted a survey of the responses of US professional engineers to these dilemmas. The stance of the majority of respondents, given an unavoidable choice, was: that health, safety and the environment should come before profits; that loyalty to the company should not override professionalism; that persuasion was far preferable to whistle-blowing; and that the engineer may have to take some career risk to uphold standards.

3.37.4 Company support

Although most discussions of engineering ethics tend to centre around the dilemmas which face the individual, there is a company dimension also. The extent to which

Table 3.9 *Outline of Engineering Council* Code of Professional Practice on Risk Issues *(Engineering Council, 1992)*

(1) Professional responsibility: exercise reasonable professional skill and care
(2) Law: know about and comply with the law
(3) Conduct: act in accordance with the codes of conduct
(4) Approach: take a systematic approach to risk issues
(5) Judgement: use professional judgement and experience
(6) Communication: communicate within your organization
(7) Management: contribute effectively to corporate risk management
(8) Evaluation: assess the risk implications of alternatives
(9) Professional development: keep up to date by seeking education and training
(10) Public awareness: encourage public understanding of risk issues

the individual is faced by such dilemmas, and their intensity, is in large part a function of the company culture. A culture that encourages good practice shields its employees from dilemmas which arise in one which does not. The employee needs to feel at ease in making the ethical choice and to know that he will be backed up.

There are also specific steps which the company can take to assist. One is the establishment of a procedure for dealing with violations of the law. It is in the interest of the company to have and retain engineers with a high degree of professionalism.

3.37.5 Regulatory support

There is also a regulatory dimension. Legislation which is based on good industrial practice and is developed by consultation is likely to gain greater respect and consent that that which is imposed. Actions by individuals who have little respect for some particular piece of legislation are a common source of ethical dilemmas for others.

The professionalism of the regulators is another important aspect. A prompt, authoritative and constructive response may often avert the adoption of poor practice or a short cut. The regulatory body can contribute further by responding positively when a company is open with it about a violation or other misdemeanour which has occurred.

4 Major Hazard Control

Some of the installations operated by the chemical industry constitute major hazards. In the early days of the industry there were some major disasters, particularly ammonium nitrate explosions, involving hundreds of fatalities. The period 1950–80, however, was relatively free of such disasters, but since then there have been several accidents with death tolls of over 100, of which Bhopal was by far the worst.

Process operations are one of a number of activities which are of concern to communities and governments. In particular, high technology, which sometimes brings unforeseen hazards, causes some unease. Inevitably, therefore, the evolution of hazard control policy for the chemical industry must be influenced by developments in public policy for the control of hazards generally.

One area of activity which attracts hazard controls is transport. The evolution of hazard controls on the railways is very instructive and has many lessons for the chemical industry, while air transport illustrates the controls developed for an industry in which high technology is particularly important. Another area of activity closer to the process industries where there has long been a well-developed hazard control policy is the nuclear industry. This has often been suggested, therefore, as a model for the control of hazards in the process industries. Other models include the hazard control systems used in mines, in the explosives industry and in the pharmaceuticals industry.

Selected references on hazard control are given in Table 4.1.

4.1 Superstar Technologies

Within the last two decades there has been a growing awareness of the problems associated with industrial growth and technological development. This has manifested itself in debate and conflict on such topics as

(1) environment (amenities, countryside, pollution, waste disposal, noise);
(2) resources (oil, metals);
(3) transport (road building, road accidents, aircraft crashes, tanker shipwrecks);
(4) buildings and structures (high rise flats, bridges);
(5) oil and chemical industry (pollution, noxious release, noise);
(6) nuclear power (major accidents, low level release, waste disposal);
(7) toxic substances (drug and food additive side-effects, materials at work).

It is also increasingly recognized that these problems have certain common features and that there is a generic problem in the assessment and control of technological developments. The question is considered in *Superstar Technologies* (Council for Science and Society, 1976) and the following discussion draws on this work.

The existence of a problem is due to the combination of a number of factors. Human artefacts are now often large and complex. Society also is complex, interdependent and vulnerable. Technological developments are often rapid and of large scale, and economic pressures militate against extensive testing. Devices are scaled up by extrapolation into untested areas and there is no

Table 4.1 *Selected references on hazard control*

Major projects
Sage (1977); NAE (1986); P.W.G. Morris and Hough (1987)

Catastrophic failures
Bignell, *et al.* (1977); B.A. Turner (1978); Michaelis (1986).

Disaster technology
Manning (1976); Cuny (1983); Comfort (1988)

Hazard control policy
Bock (1965); N. Kaplan (1965); NAE (1970); Greenberg (1972); Goldsmith (1973); Bettman (1974); Brodeur (1974); S.S. Epstein and Grudy (1974); Schillmoeller (1974); Chicken (1975); Council for Science and Society (1976, 1977); F. Warner (1976); Hodgkins (1977); IBC (1978/2,); Wearne (1979); Collingridge (1980); Gusman *et al.* (1980); Lagadec (1982); Nilsson (1983a); Anon. (1984b); Beveridge and Waite (1985); D. Williams (1985); Kaufman *et al.* (1986); Kharbanda and Stallworthy (1986b); Ostrom (1986); Solomon (1987); Majone (1989); Carnino *et al.* (1990); Funtowicz and Ravetz (1992); The Royal Society (1992)
Safety cases: Mellit (1995) (London Underground)

Interest groups
Stewart (1958); Moodie and Studdert-Kennedy (1970); G.K. Roberts (1970); Wootton (1970); Olson (1971); Wraith and Lamb (1971)

Particular hazards
Aircraft: Wheatcroft (1964); Lowell (1967); Pardoe (1968); W. Tye (1970); Warren (1977)
Drugs: Sjorstrom (1972)
Explosives: L. Allen (1977a)
Motor cars: Nader (1965); Plowden (1971)
Railways: Rolt (1982)
Ships: Cabinet Office (1967)

Hazard assessment
Fussell *et al.* (1974); AEC (1975); Council for Science and Society (1977)
Critiques: Brooks (1972); Weinberg (1972a,b); Mazur (1973); Hafele (1974); The Royal Society (1992)

Warning of disasters
Lees (1982b); The Fellowship of Engineering (1991a,b); J.B. Cook (1991); Cullen (1991); Derrington (1991); Hambly (1991); Sampson (1991); Street (1991)

Planning
Chalk (1973); Dundee Department of Town and Regional Planning (1979); B.D. Clark *et al.* (1981); IBC (1981/17); Petts and Eduljee (1994)
Planning inquiries: Sieghart (1979); Barker (1984); Kemp *et al.* (1984); R.W. King (1984)

Environment
Nicholson (1970); WHO (1972); NAS/NRC (1975); Burton *et al.* (1978); Kates (1978); Conway (1982); Petts and Eduljee (1994)

Cost–benefit, risk–benefit analysis
NAE (1972); Putnam (1980); The Royal Society (1992)

Science and society
von Helmholtz (1895); Steffens (1931); Huff (1954); Cardwell (1957); D. de S. Price (1963); Weinberg (1963, 1967); Hagstrom (1965); N. Kaplan (1965); Commoner (1966); Greenberg (1967); Nicholson (1967, 1970); Hersch (1968); Crowe (1969); Arendt (1970); Ravetz (1971, 1974a); Rawls (1971); Schrader-Frechette (1984a,b); The Royal Society (1992)
Engineer's, scientist's responsibility (*see also* Table 3.1): Gilpin and Wright (1964); Hagstrom (1965); D.K. Price (1965); Ferkiss (1969); Mayo (1970); Nader (1971); Ravetz (1971); Lederberg (1972); Feinberg (1974); Primack and von Hippel (1974); Edsall (1975); Fishbein and Ajzen (1975)

opportunity for gradual evolution and learning by trial and error.

At the same time, some of the former moderating influences have grown weaker. The complexity of projects involving numerous teams dilutes individual responsibility. There are many new specialisms which tend not to have the same ethos as the older professions. In some of the more important areas the project is so specialized that the organization concerned has a virtual monopoly of the experts available. The combination of the variety and depth of technology increases the difficulty of monitoring by government inspectors.

The degree of acquiescence by the public in technological development has changed considerably in recent years. The increased militancy of interest groups from trade unions to neighbourhood associations shows that people are not prepared to put up with the hazards and nuisances which were tolerated in the past.

Part of the problem is that high technology projects tend to be vulnerable to failures of all kinds. There is only one way in which they can go right, but many in which they can go wrong. In reliability terms they are series systems in which all the elements must work if the system is to be a success. There is a particular problem in identifying all the ways in which the system may fail. Another aspect of the problem is the question of unforeseen dangerous side-effects, of which there have been numerous examples. These include the side-effects of drugs and of industrial materials such as asbestos and vinyl chloride.

Occasionally a disaster occurs which is sufficiently dramatic to result in action. Some major technological disasters in the UK in recent years have been the crashes of the *Comet* aircraft, the release of radioactive material from Windscale, the side-effects of the drug *Thalidomide,* the defects in the Ronan Point high rise flats, the oil pollution from the *Torrey Canyon* tanker, the failures of box girder bridges, the collapse of structures built in high alumina cement, the vapour cloud explosion at Flixborough, the sinking of the *Herald of Free Enterprise,* the fire at the Kings Cross Underground station, the rail crash at Clapham Junction, and the explosion and fire on the Piper Alpha oil platform. Some of these have led to improvements of safety procedures (e.g. Windscale), and some to the abandonment of the technology (e.g. Ronan Point). All parties, including the public, the professions and the government tend to react more strongly to such disasters than they do to smaller scale failures, even where these are very frequent.

It is possible to identify a number of factors which allow high technology projects to develop to the point where a disaster can result. At the intellectual level there is the difficulty of trying to foresee the potential problems and of judging where knowledge is inadequate or experiment is required. It is necessary, but not easy, to keep abreast of the technical literature and to maintain contact with experts in other fields. In the moral sphere there are the difficulties of maintaining a conscientious approach to the frequently tedious requirements of safety and of speaking out when doubts arise on safety features. The problem has an organizational aspect in that it tends to require a strong personality to run a major project, but such a person is frequently intolerant of criticism. As a consequence, colleagues are discouraged from dissenting.

Studies of major failures in man-made systems have been described in *Catastrophic Failures* by Bignell, Peters and Pym, (1977), and *Man-Made Disasters* (B.A. Turner, 1978), and by Wearne (1979).

4.2 Hazard Monitoring

The Council for Science and Society argues for the creation of mechanisms by which the community can monitor and control hazards. Monitoring is likely to be effective only if it is generally accepted as a necessary and constructive activity which not only protects society but also contributes to the project. There is an analogy here with the role of 'Her Majesty's Loyal Opposition' in Parliament.

The project needs to be monitored at all stages: conception, decision and operation. At the conceptual stage the need is to discover the problems, the hazards and the disbenefits. There should be testing of the basic assumptions, speculative reviews of hazards and study of various scenarios. The debate should involve the public as well as the experts. Argument continues in the decision stage, but now centres on detailed proposals and designs of the project. Here the discussion is mainly between the project team and the other agencies involved. At the operational stage the emphasis changes to enforcement of operational requirements and monitoring of operating hazards.

Monitoring should be carried out both within the organization itself and through external agencies and fora. Internal monitoring may be conducted by internal but independent functions, such as a pressure vessel inspection section, and by consultants. External monitoring takes place through the activity of interest groups and government agencies. An interest group requires a forum in which to operate; typical fora are courts of law, public inquiries and the news media.

Monitoring by government agencies has a crucial role, but again there are certain difficulties. Despite the resources of government, it is not easy for a government agency to deploy the full range of up-to-date expertise which ideally it should possess. Like any other activity, monitoring may become routine, and complacency may set in. The agency must expect to be subject to the internal politics of the civil service machine. Government itself is not infrequently involved in some way in the projects that require to be monitored.

There are certain relatively detached institutions, such as universities and the professional bodies, which contain expertise and which have a role to play in the monitoring process. It does not follow, however, that they should necessarily themselves be monitors. Their function is rather to provide a source of independent expertise.

An explicit attempt needs to be made to enhance the effectiveness of the individual as a monitor. One important measure is to increase the backing which professional standards can give to the individual. Another is to afford protection to the individual who 'blows the whistle' on a hazard or an abuse.

The process of communal decision-making also places an obligation on those who claim to speak on behalf of the community to be well informed and responsible in their arguments.

4.3 Risk Issues

High technology projects involve risks. These risks raise a number of issues including:

(1) risk perception;
(2) risk criteria;
(3) risk management;
(4) risk estimation.

There is a growing literature on risk issues, which includes *How Safe is Safe?* (Koshland, 1974), *Of Acceptable Risk* (Lowrance, 1976), *The Acceptability of Risks* (Council for Science and Society, 1977), *An Anatomy of Risk* (Rowe, 1977), *Society, Technology and Risk Assessment* (Conrad, 1980), *Societal Risk Assessment* (Schwing and Albers, 1980), *Science, Technology and the Human Prospect* (Starr and Ritterbush, 1980), *Technological Risk* (Dierkes, Edwards and Coppock, 1980), *Acceptable Risk* (Fischhof *et al.*, 1981), *The Assessment and Perception of Risk* (F. Warner and Slater, 1981), *Dealing with Risk* (R.F. Griffiths, 1981), *Risk in the Technological Society* (Hohenemser and Kasperson, 1982), *Major Technological Risk* (Lagadec, 1982), *Technological Risk* (Lind, 1982) and *Risk in Society* (Jouhar, 1984).

The HSE has published *The Tolerability of Risks from Nuclear Power Stations* (HSE, 1988c), *Risk Criteria for Land Use Planning in the Vicinity of Major Hazard Installations* (HSE, 1989e) and *Quantified Risk Aassessment: Its Input to Decision Making* (HSE, 1989c).

A review of risk issues is given in *Risk: Analysis, Perception and Management* (Royal Society, 1992) (the *RSSG Report*) which provides the framework for the following discussion.

The treatment of this topic here is necessarily brief and is limited to an account of some of the principal themes. A further discussion of risk and in particular of risk criteria for hazard control is given in Chapter 9.

4.4 Risk Perception

Work on risk perception shows that people tend not to think in terms of an abstract concept of risk, but rather to evaluate the characteristics of a hazard and to perceive risk in a multi-dimensional way. The idea that risk is an inherent property of the hazard has been criticized by Watson (1981) as the 'phlogiston theory of risk'. Various authors have attempted to define risk. One such attempt is that of Vlek and Keren (1991), who have given 10 different definitions.

4.4.1 Acceptable risk

This question of risk is often treated in the same way as that of 'acceptable risk', although the latter term is itself contentious, raising as it does the question: Acceptable to whom? Nevertheless, the term 'acceptable risk' has been widely used to describe this generic problem. A discussion of the problem has been given in *The Acceptability of Risks* by the Council for Science and Society, (1977). Some of the considerations affecting judgements about risks given by Lowrance (1976) are shown in Table 4.2.

The Council for Science and Society take up the argument advanced by earlier workers that the distinction between risks which are assumed voluntarily and those which are borne involuntarily is a crucial one. In general, people are prepared to tolerate higher levels of risk for hazards to which they expose themselves voluntarily. The risk to which a member of the public is exposed from an industrial activity is an involuntary one. It is a common view that the risk to which an employee is exposed from this activity is in some degree voluntarily assumed. Thus the Council's report states

> Some hazards are accepted voluntarily, even when the risk is high. At one extreme we may say that the risk is 'embraced' when it is an integral part of the challenge in a hazardous sport, such as pot-holing or motor-racing . . . .
>
> 'The personal attitude is different in emergencies, where the risk may be said to be 'defied' in the course of a response to a call for help. Rescue operations are the prime example of such hazards. In them, the balance of costs and benefits may be very different from that which prevails in ordinary life, as we can see from the lengths to which kind-hearted people will go in retrieving lost or trapped animals . . . .
>
> When a serious hazard is encountered involuntarily, acceptance may extend only to a much lower level of risk than otherwise. When, in addition, the sufferer feels impotent in the face of danger, tolerance is further reduced. Accidents in trains seem peculiarly

Table 4.2 *Some considerations affecting judgements on safety (after Lowrance, 1976) (Reproduced with permission from* Of Acceptable Risk, *by W.W. Lowrance. Copyright © 1976, William Kaufman Inc., Los Altos, CA 94022. All rights reserved)*

Risk assumed voluntarily/Risk borne involuntarily
Effect immediate/Effect delayed
No alternative available/Many alternatives available
Risk known with certainty/Risk not known
Exposure is an essential/Exposure is a luxury
Encountered occupationally/Encountered non-occupationally
Common hazard/'Dread' hazard
Affects average people/Affects especially sensitive people
Will be used as intended/Likely to be misused
Consequences reversible/Consequences irreversible

unacceptable, perhaps more so than accidents in aeroplanes, where rightly or wrongly the passengers are generally considered to have taken the risk on themselves for the sake of the extra benefit of the time saved. In underground tube-trains only absolute safety seems to be good enough; perhaps the enclosed environment exerts a strong psychological influence. The stark terror of impotence in the face of impending destruction is an important part of the evaluation of such risks. For a strong contrast, we notice how the illusion of control by a driver in a private motor car who is under the influence of alcohol makes very high risk levels acceptable to him.

Perhaps the most difficult class of hazards for judgements of acceptability are those called 'major hazards'. These are defined by a low probability of realization, combined with the likelihood of very great harm if the hazard is realized. In this case, the intuitions derived from ordinary experience provide little help in conceiving the hazard; and the experts themselves may well disagree even over the probabilities. When harm is liable to be inflicted on people who have neither any conceivable power to avert it, nor any responsibility for its occurrence, the judgement of the 'acceptability' of the risk is at its most difficult.

This variety of perceptions may well be a cause of irritation to the expert on risk assessment. Policy decisions would be so much easier if a purely quantitative analysis were enough or nearly so. The subjective perceptions of risk can have enormous political importance, possibly to the extent of distorting priorities in programmes for coping with the real risks that society encounters. (This problem is most noticeable in medicine.) There is occasionally a temptation on the part of the expert in technological risks to throw in a couple of extra orders of magnitude of restriction of unacceptable hazards and then perhaps to be annoyed when even this does not render them acceptable to critics.

In the last analysis, the Council continues, there is only one sort of risk 'that is truly "acceptable" in the ethical sense: the risk that is judged *worthwhile* (in some estimation of costs and benefits), and is incurred by a deliberate choice made *by its potential victims* in preference to feasible alternatives.' The important question, therefore, is *'under what conditions, if any, is someone in society entitled to impose a risk on someone else on behalf of a supposed benefit to yet others?'*

The Council concludes that it is not practical to give the individual an absolute veto over any development which may be a hazard to him or her. The public interest must also be considered. This does not necessarily mean, however, that it is sufficient to leave the judgement on the acceptability of risk to the organization concerned or to government. The Council cites a number of cases in which it considers that control by government inspection agencies has not achieved the proper balance of risk. It suggests:

> Rather than continuing to demand an answer to the question whether a risk is fair *in itself*, we should redirect our attention to the ways that risks come about and are controlled. That is, we should focus on the *procedures* by which the decisions are taken on the creation or persistence of risks, and ask whether these procedures are fair.

The Council advocates that decisions on projects involving hazards should be openly arrived at and that those affected should be able to participate in the decision through the medium of public inquiries and other similar fora.

The report contains a number of appendices which deal with risk in aircraft (D.V. Warren, 1977), in nuclear radiation (Burhop, 1977), with explosives (R.L. Allen, 1977a), at Flixborough (R.L. Allen, 1977b), with asbestos (Woolf, 1977) and with the use of mathematics in risk assessment (Ravetz, 1977b). Many of these are discussed further in later chapters.

4.4.2 Acceptable versus tolerable risk

The account of the acceptability of risk in the *RSSG Report* describes some of the approaches taken to the problem of acceptable risk and the distinction between acceptable and tolerable risk.

Fischhoff *et al.* (1981) suggest that there are three main methods used to resolve risk questions. These are professional judgements by experts, formal cost–benefit analyses and methods based on comparisons with the risks accepted from existing hazards. The authors suggest that the relevant factors are whether a method is: comprehensive; logically consistent; practical; open to evaluation; politically acceptable; compatible with existing institutions; conducive to society's learning about risk; and whether it improves decision-making in the long run. Fischhoff *et al.* use these seven criteria to evaluate the approaches described.

The concept of acceptable risk came in for criticism from Sir Frank Layfield, chairman of the Sizewell B Inquiry. He argued that the concept understates the seriousness of the problem and that the term 'tolerable risk' is preferable. He expressed the view that the opinions of the public should underlie the evaluation of risk, although he saw no practical way in which they could reliably be so used. *The Tolerability of Risk from Nuclear Power Stations,* produced following this inquiry, (HSE, 1988b) gives the following definition:

> 'Tolerability' does not mean 'acceptability'. It refers to the willingness to live with a risk to secure certain benefits and in the confidence that it is being properly controlled. To tolerate a risk means that we do not regard it as negligible or something we might ignore, but rather as something we need to keep under review and reduce still further if and as we can.

4.4.3 Actual versus perceived risk

One aspect of the risk debate concerns the relationship between objective, or statistical, risk and subjective, or perceived, risk. One line of inquiry draws a distinction between actual and perceived risks, and seeks to investigate the extent to which perceived risks differ from actual ones. The opposing school of thought holds that such a distinction is misconceived.

The attempt to assess the risks arising from a hazard involves a number of areas of considerable difficulty. The *RSSG Report* highlights two of these: (1) consequences and (2) uncertainty.

Scenarios for the realization of a hazard generally involve a whole range of consequences. There may be injuries, which may be fatal or non-fatal, prompt or delayed. There may be environmental damage where the effects may appear promptly or only after a delay. In other words, the consequences are multi-dimensional.

The assessment of the risk is subject to uncertainty. Traditionally, uncertainty has been discussed in terms of probability, but this does not exhaust the matter. In the taxonomy of ignorance given by Smithson (1989) probability is one aspect along with others such as incompleteness and distortion. Even the concept of probability has different meanings, four conventional interpretations being the classical, frequency, logical and Bayesian.

One type of uncertainty which arises in risk assessments concerns the values to be assigned to the parameters. A more fundamental uncertainty arises over the completeness and correctness of the structure of the analysis. There are both parametric and systemic uncertainties – or as C. H. Green, Tunstall and Fordham (1991) put it 'What you know you don't know' and 'What you don't know you don't know'.

Expert risk assessments do not provide an unambiguous and reproducible measure of risk. The validity of the assessment depends very largely on its pedigree, but this can be mixed, with highly expert and less expert elements, such that the latter may well be the weak link. A study by Lathrop and Linnerooth (1983) found widely differing assumptions underlying three separate studies of the same proposed facility in the USA. Considerations of this kind undermine the assumption that it is possible to determine an objective measure of 'risk' which can then be used as a basis for judging perceptions of risk.

4.4.4 Psychological issues

One of the main fields of study in psychology is sense perception and the methodology for this is one starting point for investigating risk perception. In such work an investigation is made of the extent to which the subject's perception of objective stimuli corresponds with reality. The difficulty in applying these methods to risk perception is that risk is not an objective property.

There are, however, several specific topics bearing on risk where progress has been made using the psychological approach. The *RSSG Report* describes three such areas: (1) estimation of fatality, (2) characterization of hazards, and (3) differences between individual and group perceptions.

Lichtenstein *et al.* (1978) investigated the estimation by laymen of the annual number of fatalities in the USA from 40 different hazards. The estimates made by the laymen showed two biases. One was that they overestimated the number of deaths due to infrequent causes such as tornadoes, but underestimated those due to frequent causes such as diabetes. The other was that the infrequent events for which overestimates were highest tended to be those of which laymen have a vivid picture and which are salient in their memory. The authors term this the 'availability heuristic'. The correspondence between the lay estimates and the statistical data were better in respect of rank ordering than in respect of the absolute number of deaths. The conclusion drawn in the *Report* from this and other work is that individuals are quite capable of making well-founded judgements of the relative magnitude of expected fatalities.

A study of the perception of the qualitative characteristics of hazards was made by Slovic, Fischhoff and Lichtenstein (1980). In this study respondents were asked to rate 90 hazards in respect of 18 qualitative characteristics. The latter were then reduced to three dominant factors of which the first two were 'dread risk' and 'unknown risk'. Dread has to do with features such as whether the hazard is uncontrollable and exposure to it involuntary, unknowability has to do with features such as whether the hazard is a familiar one and whether its effects are immediate or delayed. The results obtained are shown in Figure 4.1. The third factor was the number of people exposed to the hazard. The findings of this and subsequent studies have also shown that respondents tend not to be satisfied with existing trade-offs between risks and benefits and to desire stricter regulation of hazards.

An individual's perception of risk is likely to be affected by his personality but also by his membership of a group. A number of studies have been done of the effect of membership of groups differentiated by culture, gender or age. Work on groups also includes investigation of the attitudes of individuals within groups with different stances on a particular issue such as pro- and anti-nuclear groups. One finding of this work is that individuals in the two groups had rather similar attitudes to particular undesirable characteristics such as increased centralization of power or imposition of involuntary risk, but differed rather in the extent to which they perceived these characteristics as applying to nuclear power.

Another finding of work in this area is that an attitude which favours a technology is often accompanied by a desire that it be more closely regulated and, therefore, that a call for stricter regulation is not necessarily evidence of hostility.

4.4.5 Social science issues

A starting point for the social sciences is that there is every reason to expect that a person's attitude to hazards is related to his overall system of values and beliefs and that it is affected by broad social factors. Three insights of social science which bear on risk issues described in the *RSSG Report* are (1) cultural theory, (2) social amplification and (3) social framing.

Cultural theory, developed by Douglas (1966, 1982, 1985, 1990, 1992), holds that attitudes such as that to risks vary systematically with cultural bias. The cultural bias of an individual depends partly on the extent to which he is part of a bounded group ('group') and partly on social interactions which are conducted according to rules rather than negotiated ('grid'). Four principal biases are recognized: hierarchists (high grid/high group), egalitarians (low grid/high group), fatalists (high grid/low group) and individualists (low grid/low group). The *Report* describes these different types as follows:

> The argument goes that individualists see risk and opportunity as going hand-in-hand; fatalists do not knowingly take risks but accept what is in store for them; hierarchists are willing to set acceptable risks at high levels so long as decisions are made by experts or in other socially approved ways; but egalitarians accentuate the risks of technological development and economic growth so as to defend their own way of life and attribute blame to those who hold to other cosmologies.

It continues:

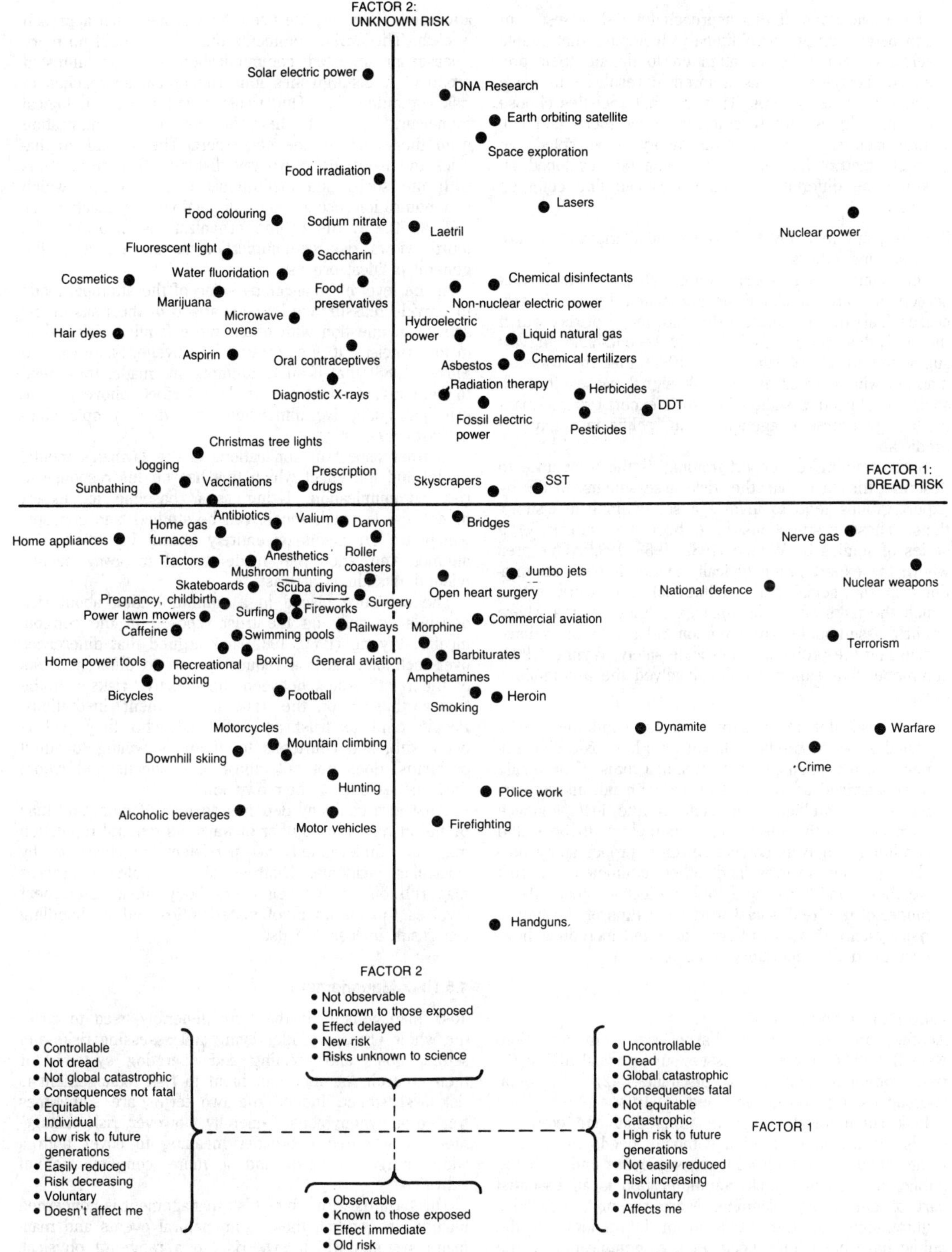

Figure 4.1 *Perception of different hazards: 'dread risk' versus 'unknown risk' (Slovic, Fischhoff and Lichtenstein, 1980): locations of 90 hazards on factor 1 (dread risk) and factor 2 (unknown risk) of the three-dimensional space derived from the interrelationships among 18 risk characteristics. Factor 3 (not shown) reflects the number of people exposed to the hazard and the degree of one's personal exposure (Courtesy of Plenum Press)*

The implications of this approach for risk assessment and perception are revolutionary. It implies that people select certain risks for attention to defend their preferred lifestyles and as a forensic resource to place blame on other groups. That is, what societies choose to call risky is largely determined by social and cultural factors. If the cultural theory is accepted, no 'single metric' for risk analysis can be developed on which the different cultural biases can find common ground.

Studies point to a divide between egalitarians, who are anti-risk, and others.

The concept of social amplification is that risk perceptions are amplified or attenuated by social processes. This occurs because the knowledge of risk which the individual possesses is largely second-hand. It was suggested by Kasperson *et al.* (1988) that the transformations which occur as the risk signal passes through social 'amplification stations' such as expert groups, mass media, government agencies and politicians may be predictable.

'Social framing' or 'expert framing' is the term used to describe the fact that the risk assessments made by expert groups tend to involve a set of implicit assumptions. These assumptions have been the subject of a series of studies by Wynne (1982, 1989, 1992). One area where the expert group typically makes implicit assumptions is the social and institutional framework within which the risks are to be managed. Another area where implicit assumptions are common relates to operations. Discussing the problem of pesticide safety, Wynne (1992) comments that expert framing involved the assumptions that

> . . . pesticides manufacture process conditions never varied so as to produce dioxin or other toxic contaminants of the main product stream; drums of herbicide always arrived at the point of use with full instructions intact and intelligible; in spite of the inconvenience farmers and other users would comply with the stated conditions, such as correct solvents, proper spray nozzles, pressure valves and other equipment, correct weather conditions, and full protective gear. As a model of the 'real' social world and thus of the typical risk system, this was utterly naive and incredible however good the laboratory science.

4.4.6 Risk communication

Another area of study is that of risk communication. Here the *RSSG Report* addresses four principal topics: (1) risk communication as a discipline, (2) conceptual approaches, (3) applications, and (4) trust.

Risk communication occurs at national and company levels. At the latter level a company needs to communicate about risks both with its employees and with the public. In particular, risk communication is an essential part of emergency planning. More recently, legislative requirements for the provision of information to the public have been introduced. Risk communication at the national level is seen most clearly when opposition arises to developments such as hazardous installations or waste disposal sites.

Risk communication in such a context is now recognized as a discipline, but its study is not well advanced. However, the need for a professional approach is clear. The *Report* comments that 'One should no more release an untested communication than an untested product.' It distinguishes four conceptual approaches to risk communication. One treats it within a technological framework in which there is one-way communication from the expert to the non-expert. The second emphasizes the need for a two-way dialogue. The third deals with the social and institutional context within which communication occurs and the actions, or inaction, as well as the words of the organizations involved. The fourth views risk communication as an aspect of the general political process.

At the level of applications some of the attempts made to provide reassurance by comparison of the risks of the activity in question with other, more familiar risks come in for criticism. It is suggested by Covello, Sandman and Slovic (1988) that if such attempts are made, they need to take into account the work described above on the principal qualitative dimensions by which people characterize hazards.

Another aspect of applications is the layman's mental model and the effect which this has on his response to risk communication. Using as a baseline an expert model, the layman's model can be studied, and concepts which are erroneous or entirely absent identified. The attempt can then be made to clear away factual misunderstandings at least.

The credence placed in a communication about risk depends crucially on the trust reposed in the communicator. Wynne (1980, 1982) has argued that differences over technological risk reduce in part to different views of the relationships between the effective risks and the trustworthiness of the risk management institutions. People tend to trust an individual who they feel is open with, and courteous to, them, is willing to admit problems, does not talk above their heads and whom they see as one of their own kind.

Trust in a company depends on many factors and may be put at risk in a number of ways. Its general reputation may be undermined by persistent practice or by particular incidents. Distrust of the political process may rub off on the regulatory body. Risk assessment involves the treatment of uncertainties and mishandling can result in loss of trust.

4.5 Risk Management

'Risk management' is the term generally used to cover the whole process of identifying and assessing risks and setting goals and creating and operating systems for their control. An important input to risk management is risk assessment. Indeed the two terms are sometimes treated as synonymous. Generally, however, risk management is accorded a broader meaning to cover both a wider range of risks and a more complete set of activities.

The hazards with which risk management is concerned include, in general, those from natural events and man-made systems which give rise to a range of physical, financial, legal and social risks. The main concern here is with the risks from technological systems.

Risk management is practised at government as well as company level and, indeed, many of the important issues in risk management have to do with the design of the institutions required for public risk management and with the associated risk evaluation.

4.5.1 Public risk management

Ultimately it is up to government to decide which activities can be allowed and which must be banned and how any disbenefits from the former are handled. Thus government is involved in risk management, but the extent to, and manner in, which it should be is a matter of debate.

Many activities involve external 'costs' which are imposed on other parties. These externalities include danger and environmental pollution. A market solution is to require the party responsible to pay these costs. In the case of pollution, this is the principle of 'polluter pays'. Another market approach is the use of insurance to pool risks. However, there are a number of difficulties in these approaches. The party responsible may not be readily identifiable and may have limited resources. This is illustrated by the difficulties in enforcing clean-up of hazardous waste sites.

Insurance also has it problems. Insurers may be reluctant to provide cover for perils which are poorly defined such as pollution. Even where insurance is available, there are recognized problems. Adverse selection occurs where parties know their own risks better than the insurers and insure only the high risks. Moral hazard occurs where the possession of insurance lessens the incentive to reduce the risk.

Public risk management is subject to two types of error. One is the prohibition of an activity or product which may be harmless; the other is failure to prohibit an activity or product which is harmful. By analogy with the terms applied to hypotheses, these two errors are termed Type I and Type II errors, respectively. A Type II error can lead to public outcry, whereas a Type I error, which may involve foregoing great economic benefit, is unlikely to do so.

Risk management is practised not only at national level but at the supranational and local government levels also. The *RSSG Report* identifies nine types of institutional 'player', as shown in Table 4.3. It describes the implications as follows:

> As soon as more than one organization enters the risk management scene, different priorities and preoccupations must be resolved through bargaining. Problems of deadlock, jurisdictional tensions, ordering of preferences and varying levels of preference intensity become more acute. The policy 'packaging' processes involved in trading preferences, particularly in multi-level bargaining situations such as those involved in EC decision-making, may result in outcomes that do not match the preferences of any of the negotiating parties.

Many of the public bodies such as advisory committees concerned with risk management have an advisory rather than an executive role. In this they are analogous to a staff function in a company.

The *RSSG Report* identifies the following four modes of operation available in public risk management: (1) information, (2) resources, (3) legislation, and (4) direct action. Information is of particular importance in relation to emergency warnings of impending natural hazards, although the problem of liability for false alarms is a severe one. Where resources are required and unavailable elsewhere the state may provide them, as in the clean-up of hazardous waste sites. Direct action, by police or other agencies, is largely limited to emergencies.

The approach taken may rely on prevention or on mitigation. For a major hazard at a fixed installation this typically means either preventing an incident or providing an adequate distance between the plant and the public offsite. However, the applications of the two approaches are much wider. Mitigation, for example, may include financial compensation.

4.5.2 Risk management rules

Some of the dimensions of risk management systems have been studied by Ostrom (1986), as described in the

Table 4.3 *Some types of player in institutional risk management (The Royal Society, 1992) (Courtesy of The Royal Society)*

	Institutional type		
Territorial level	*Core executive bodies*	*Independent public bodies*	*Private or independent bodies*
Supranational	Example: EC Commission	Example: EC Court of Justice	Example: Greenpeace
National	Example: national Parliaments and ministerial departments	Example: national courts and independent regulatory bodies, like the NRA[a]	Example: National Associations of Insurers
Subnational	Example: state or local governments	Example: independent regional/local statutory bodies	Example: local firms and activists

[a] National Rivers Authority.

RSSG Report. Table 4.4 shows six such dimensions and their characteristic trends.

The boundary rules determine the parties allowed to participate in the risk management process. The decision process within the regulatory body typically involves the internal experts, consulting with approved expert groups; it is not readily accessible to other parties and becomes less so at supranational level. The boundaries are a matter of debate. They are challenged, for example, by interest groups seeking to participate.

The scope rules determine the type of decision which the risk management body can take. Typical choices are fact-finding versus blame attribution, national versus international measures in sea and air transport, integrated pollution control versus control of particular vectors.

The position rules determine the hierarchy and relationships of the decision points in the risk management structure, including the power of appointment and dismissal. Thus in the UK many regulatory bodies are independent with direct accountability to Parliament, and Ministers are readily able to dismiss them.

The information rules determine access to information. This has many aspects. The risk management system needs to be able to obtain information, such as the location of major hazard installations or sources of pollution, which it can obtain only by specific legislation. The public needs information on the hazards to which it is exposed and on what to do in an emergency. In this case, the information may be made available on a 'right-to-know' or a 'need-to-know' basis. Again, legislation is required to ensure this information is made available.

The authority and procedural rules determine the mode of decision-making, its timing and the evidence admissible. Typical choices, or emphases, are adversarial versus inquisitorial procedures, anticipatory action versus incident investigation, and qualitative versus quantitative assessments.

The preference-merging rules determine the process by which the individual inputs contribute to the collective decision. Two principal rules are unanimity and majority voting. Both are exemplified in the decision-making processes of the European Community.

4.5.3 Risk management issues

The *RSSG Report* identifies, but does not attempt to resolve, seven areas of debate in risk management. These issues are: (1) anticipation, (2) liability and blame, (3) quantitative risk assessment, (4) institutional design, (5) risk reduction costs, (6) participation, and (7) regulatory goals. They are summarized in Table 4.5. For each issue the *Report* presents both sides of the argument, but with *caveats* that the issues are not usually so starkly posed, that opposing views are not necessarily irreconcilable and that one of the two viewpoints may be much more accepted than the other.

The anticipation issue concerns the relative priorities to be given to prevention and to mitigation. The term 'anticipation' has a rather wide meaning and covers different aspects, depending on the context. For accidents, anticipation has to do with the management systems which assure effective hazard control. For pollution, it has to do with caution in not permitting activities or products the effects of which are not fully

Table 4.4 *Some dimensions of, and putative trends in, risk management (Royal Society, 1992; after Ostrom, 1986) (Courtesy of de Gruyter Verlag and The Royal Society)*

Rule type	*Explication*	*Range of key types*	*Characteristic trends*
Boundary	Who is counted as a player?	Technocratic/ participative	More participation
Scope	What is managed and what can be decided?	Broad/narrow	Extension of scope
Position	What is the hierarchy of players?	Single organization/ multi-organization	More multi-organizations
Information	Who is entitled to know what from whom?	Open/closed	More open
Authority and procedure	Under what conditions must decisions be made?	Formal/informal	More formal
Preference merging	How are individual preferences aggregated into collective decisions?	Consensus (integration)/ conflict (aggregation)	More conflict

Table 4.5 *Some issues in risk management (Royal Society, 1992) (Courtesy of The Royal Society)*

Doctrine	*Justificatory argument*	*Counter-doctrine*	*Justificatory argument*
Anticipationism	Apply causal knowledge of system failure to *ex ante* actions for better risk management	Resilienism	Complex system failures not predictable in advance and anticipation makes things worse
Absolutionism	A 'no-fault' approach to blame avoids distortion of information and helps learning	Blamism	Targeted blame gives strong incentives for taking care on the part of key decision-makers
Quantificationism	Quantification promotes understanding and rationality, exposes special pleading	Qualitativism	Proper weight needs to be given to the inherently unquantifiable factors in risk management
Designism	Apply the accumulated knowledge available for institutional design	Design agnosticism	There is no secure knowledge base or real market for institutional design
Complementarism	Safety and other goals go hand in hand under good management	Trade-offism	Safety must be explicitly traded off against other goals
Narrow participationism	Discussion is most effective when confined to expert participants	Broad participationism	Broader discussion, better tests assumptions and avoids errors
Outcome specificationism	The regulatory process should concentrate on specifying structures or products	Process specificationism	The regulatory process should concentrate on specifying institutional processes

known. Two complementary arguments are advanced against anticipation. One is that it is not totally effective, so that the alternative (resilience) is necessary anyway; the other is that emphasis on anticipation reduces the motivation and resources needed to promote resilience.

The liability and blame issue concerns the desirability or otherwise of attributing blame where investigation of an incident indicates that someone has made an error. One school of thought argues that the prevention of incidents requires strict discipline and the attribution of blame for error, whilst the other holds that incidents arising from human error are conditioned by the work situation and that the rectification of defects in that situation requires the free flow of information which is inhibited by a blame culture.

Studies are quoted to the effect that such information flow is essential to the efficient functioning of complex organizations (Laporte, 1982; K.H. Roberts, 1989; K.H. Roberts and Gargano, 1989). Mention is made of the explicit adoption of a no-blame policy by Shell International (1988).

The argument concerning blame also extends to the corporate level. In 1990, a suit was brought against P&O Ferries for corporate manslaughter in respect of the sinking of the *Herald of Free Enterprise*. This case collapsed, but it may well indicate a trend.

The quantitative risk assessment (QRA) issue concerns the contribution of QRA to decision-making about risks. Supporters of QRA advance it as an aid to, though not the sole determinant of, rational decision-making and argue that it is an effective means of understanding complex technological systems, and also natural hazards, and that there is no real alternative method of ordering priorities. The opponents' critique develops two main themes. One is that QRA contains value judgements which are not made explicit. The other is that QRA is beset by uncertainty, which extends not just to values of the parameters but to the whole structure and completeness of the assessment. The more radical argue that QRA is liable to mislead and is positively harmful.

The institutional design issue concerns the feasibility of designing institutions for risk management and the extent to which good practice in this area can be defined. Changes in risk management systems are typically among the principal recommendations of accident inquiries. Those who believe good practice can be articulated refer to the safety management systems and safety culture in major companies such as those in the oil and chemical industries. Sceptics point to wide differences of practice, or even contradictory practices, which they perceive to yield similar outcomes, such as the blame and no-blame cultures mentioned earlier.

The risk reduction cost issue concerns the cost of risk reduction and trade-offs of risk against other goals. One view is that reduction of risks incurs additional costs and that there has to be a trade-off between risk reduction and other goals. The types of risk cover a wide spectrum, including both safety and environmental risks. The opposing viewpoint is essentially that good practice in respect of risks is generally profitable also, but the arguments which support this appear to differ, depending on the type of risk considered. In safety, the argument is that the good management which ensures high safety also delivers high profits. In terms of the environment, the argument relies rather on examples where the move to more environmentally friendly products appears not to have disadvantaged the company.

The participation issue concerns the extent of participation appropriate in risk management. The general proposal for wider participation embraces a number of different types of participant and does so for different reasons. Thus it is suggested that human factors experts should participate more in engineering projects to avert the problems which otherwise are liable to arise. Participation by lay groups is held to force consideration of wider issues and to improve the quality of the decision-making in large technological projects. Another argument for the involvement of the latter is to improve public accountability and to broaden the base of those committed to the project. Others are more sceptical and fear that the process is liable to result in less rational decisions and to dispersion of responsibility.

The regulatory goals issue concerns the emphasis to be placed on outputs and on process in respect of risks. The engineering, and regulatory, culture tends to concentrate on outputs in the form of physical products or systems. These are typically designed with safety margins which are implicitly based on doing what is 'practical' and 'reasonable'. Critics argue that the risks of these outputs are subject to such uncertainty that this approach is inadequate, and call for more emphasis on process by the application of quality management concepts.

As stated earlier, whilst it is convenient for purposes of exposition to present these viewpoints as thesis and antithesis, there is on all these issues a spectrum of views, and in many cases a synthesis can be envisioned.

4.5.4 Cost–benefit analysis

The *RSSG Report* contains an appendix on cost-benefit analysis in relation to risks by Marin (1992), with particular emphasis on risks to people. The application of cost–benefit analysis to such risks is based on the principle of reducing the risks until the marginal cost equals the marginal benefit. In order to do this costs and benefits have to be expressed in the same units, which means in money terms.

The first requirement is therefore to devise methods of estimating the 'value of a statistical life' or for short, the 'value of a life'. Marin describes both the wide variety of approaches which have been used and the perceived ethical problems. The process of putting a price on life is apt to appear cold-blooded or even unethical. In practice, the allocation of scarce resources demands ordering of priorities. Administrative decisions are constantly being made in which a value is placed on life, if not explicitly then implicitly. Thus, at present, society mainly leaves this problem in the hands of special groups.

There are certain approaches to the determination of the value of a life which have not proved fruitful. One is to base the assessment on the individual's earning power, the 'human capital' approach. This implies that those who cannot work, for whatever reason, are without value and is unethical. It is also incompatible with the normal approach in cost–benefit analysis, which is to ascertain what the course of action proposed is worth to the person(s) affected. On the other hand, it is also not helpful to ask an individual what sum he would be prepared to pay in exchange for his life. This sum may be regarded as infinity.

A more fruitful approach is to consider small increments of risk and then determine what people will pay to eliminate these. Two main methods are available to make this determination. One is to ask people, the other is to observe their behaviour. Typical studies make comparisons of choices such as those between modes of transport or between jobs of different degrees of safety.

Marin quotes two UK studies which both put the value of a life at £3 million at 1990 values, one based on fatal accident rates in different occupations (Marin and Psacharopoulos, 1982) and the other on willingness to pay for improved safety (Jones-Lee, Hammerton and Philips, 1985). For comparison, the estimate given based on the human capital approach is about £235 000.

It is also possible to consider large increments of risk, in particular those which are at the threshold where they are not acceptable to the individual, whatever the financial compensation. This aspect is not well developed, but it is relevant to situations such as those which occur in nuclear accidents where people may be asked to do work in conditions where the levels of radiation are high.

Modes of fatality, non-fatal injury and age pose further problems in determining the value of a life.

4.6 Hazard Control Policy

It is appropriate at this point to consider some of the elements which go to make up a hazard control policy. An account of this has been given in *Hazard Control Policy in Britain* (Chicken, 1975). The legislation referred to is that current at the time. Chicken considers five fields which illustrate hazard control policy. These are (1) road transport, (2) air transport, (3) factories, (4) nuclear power, and (5) air pollution. These fields represent very different hazard control situations.

His argument runs as follows. In road transport, hazard control is mainly by legislation aimed at general improvement. Research is conducted by the Road Research Laboratory (RRL), manufacturers are encouraged to improve design standards, road construction takes account of safety knowledge and compulsory testing of vehicles is required.

Air transport, by contrast, is controlled not only by legislation (Civil Aviation Act 1971) but by a strict system of licensing operated by the Civil Aviation Authority (CAA). Aircraft are technologically much more complex and give rise to hazards which provoke much greater public outcry. The hazards are subjected to quantitative assessment and are reduced to a minimal level by sophisticated technology.

Control of factories is carried out through legislation (Factories Act 1961) which is enforced by a rather thinly spread inspectorate. The level of technology, the degree of hazard and the sophistication in control vary widely.

Nuclear power is again controlled not only by legislation (Nuclear Installations Act 1965) but by a strict licensing system operated by a well-staffed Nuclear Installations Inspectorate. Again, nuclear reactors are technologically complex and constitute major hazards, and full hazard assessment and control is carried out.

Control of air pollution by legislation (Alkali etc. Works Regulation Act 1906, Clean Air Acts 1956 and 1968) is enforced again by a somewhat thinly spread Alkali Inspectorate.

Thus arrangements for hazard control for factories and for air pollution are rather similar, but they are very different from those for air transport and nuclear power, which are also similar. The situation in road transport does not constitute a hazard control arrangement in the same sense.

It is apparent that there are wide differences in the voluntary component in these policies. In general, this is greatest where a large number of units are involved, where the controls were first applied to an established activity, where other regulatory bodies such as local authorities are involved and where there are strong interest groups.

Thus the controls are most stringent in air transport and nuclear power, where the number of units or operators is small, the activity is a relatively new technology, there is strong central government involvement and interest groups opposing regulation are absent. It is relevant to note also that these are the two high technology activities considered.

The elements of hazard control policy are (1) identification, (2) legislation and control arrangements, (3) research, (4) consultative arrangements, and (5) priorities in resource allocation.

Hazard control policy appears to develop in response to (1) demands for improved control; (2) developments in hazard reduction; (3) defects in existing policies; and (4) developments abroad. Study of the way in which hazard control policy evolves indicates that the initiating role of political parties and Parliament is relatively small and that it is most directly influenced by the Civil Service in the departments primarily involved, with the various inspectorates playing a leading role. When major developments are being considered, extensive use is made of independent specialist committees, such as the Edwards Committee on Air Transport or the Robens Committee on Health and Safety at Work. It is largely through these committees that the interest groups are able to influence policy.

Interest groups have an important role to play in the development of hazard control policy. The study of interest groups is a well-established aspect of social science (e.g. Moodie and Studdert-Kennedy, 1970; G.K. Roberts, 1970; Wootton, 1970). Wootton classifies interest groups as economic (e.g. Confederation of British Industry (CBI)), integrative (e.g. Institution of Chemical Engineers) or cultural (e.g. National Trust), and as first, second or third order, depending on whether they operate at local, regional or national level. This role is well illustrated by the proceedings of the Robens Committee, whose report contains much material from such organizations.

It is suggested by Chicken (1975) that the most influential interest groups are the economic (e.g. industrial) and integrative (e.g. professional) groups which operate at national level. While this is no doubt true in terms of detailed positive recommendations, the effect of agitation by local groups and of news media campaigns by groups at all levels on the general spirit in which policy-makers approach their task should not be underestimated.

The broad conclusions of this study remain valid.

4.7 Nuclear Hazard Control

Hazard control is well established in nuclear energy, which is an obvious model for control of chemical industry hazards. It is helpful, therefore, to consider briefly the arrangements in the nuclear industry. A short account of nuclear energy is given in Appendix 20 and the treatment here is confined to hazard control policy for the industry.

One of the principal features of the control of nuclear hazards is the role played by hazard assessment. Many of the hazard assessment techniques used in the process industries were developed in the nuclear industry. Nuclear hazard assessment is one of the topics discussed in Appendix 20.

Selected references on nuclear hazard control are given in Table 4.6.

4.7.1 Nuclear plant licensing

The early work on nuclear energy in the UK was for military purposes and was controlled by the Ministry of Supply, but in 1954 the UK Atomic Energy Authority (UKAEA) was established and began to operate nuclear power reactors at Calder Hall. Although these were again primarily for defence, they confirmed the feasibility of nuclear power. In 1958, the Central Electricity Generating

***Table 4.6** Selected references on nuclear hazard control (see also Table A20.1)*

Nuclear hazard assessment and control
NRC (Appendix 28 Regulation); F.R. Farmer (1967a,b, 1969a,b, 1970, 1971); Dale and Harrison (1971); F.R. Farmer and Gilby (1971); Kirk and Taylor (1971); Weinberg (1972b); Gronow and Gausden (1973); EPA (1974); Karam and Morgan (1974); AEC (1975); Chicken (1975); Freudenthal (1976); Rust and Weaver (1976); Fussell and Burdick (1977); HM Chief Inspector of Nuclear Installations (1977); HSE (1977c, 1992d); NRC (1977a–c); Fremlin (1978, 1983); Higson (1978); R.J. Parker (1978); E.E. Lewis (1979); I.K.G. Williams (1979); Windscale Local Liaison Committee (1979); Bradbury (1980); Fells (1980, 1981); Joksimovic and Vesely (1980); Strong and Menzies (1980); Charlesworth *et al.* (1981); D. Pearce (1981); Sagan (1981); Schrader-Frechette (1981); Chicken (1982); Openshaw (1982); Amendola (1983a,b); R.F. Griffiths (1984b); C. Tayler (1985f); UKAEA and BNFL (1985); Lester (1986); Andrews (1988); Solomon and Kastenberg (1988); Ballard (1989); Carnino *et al.* (1990); Wu and Apostolakis (1992)

Debate, critics, public relations
Weil (n.d.); Hedgpath (1965); Bryerton (1970); Novick (1970); R. Lewis (1972); Metzger (1972); Dunster (1973); Gofman and Tamplin (1973); Cochran (1974); Ebbin and Kasper (1974); Kendall and Moglewer (1974); Kendall (1975); Lovins (1975, 1977); Primack (1975); J. Hill (1976, 1979b); Patterson (1976, 1988); Brookles (1977); Hayes (1977); Icerman (1977); Puiseaux (1977); Breach (1978); Foley and van Buren (1978); Grossman (1980); Nickel (1980); Sweet (1980); Dunster (1981); Howlett (1982); B.L. Cohen (1983, 1989); Cannell and Chudleigh (1984); T. Hall (1986); McGill (1987); V.C. Marshall (1988b); Aubrey (1991); Slovic *et al.* (1991)

Board (CEGB) was created and became the main operator of nuclear power reactors. These two operators were later joined by two others, the South of Scotland Electricity Board (SSEB) and British Nuclear Fuels Ltd (BNFL).

The responsibility for the control of nuclear power reactors was shared between the UKAEA and the Nuclear Installations Inspectorate (NII). The Atomic Energy Act 1954, which established the UKAEA, gave the latter responsibility for the safety of its own reactors. Reasons for this rather unusual arrangement include the facts that the initial purpose of the UKAEA was to produce nuclear weapons and that the UKAEA had virtually all the experts. The coincidence of the occurrence in 1957 of the Windscale incident and of the creation in 1958 of the CEGB, which was to operate nuclear power reactors for civil use, led the Fleck Committee, which investigated the former, to recommend a separate nuclear inspectorate. The Nuclear Installations (Licensing and Insurance) Act 1959 created the NII and also laid down that reactors other than those of the UKAEA must be insured to cover all claims which might arise from the release of radioactivity. The situation that developed was one in which the UKAEA was responsible for the safety of its own reactors, with the NII controlling other nuclear reactors through a licensing system.

The nuclear licence which the operator of a nuclear power reactor is required to obtain represents a very stringent system of control. The licensee must furnish design and operating documents which effectively cover all aspects of the design and operation of the reactor. A Nuclear Safety Committee must be created to advise the licensee and the names and qualifications of the persons nominated submitted to the Minister for approval. Individuals must also be nominated as Competent Persons to execute particular tasks.

Licensing is not a once-for-all affair but an on-going procedure. The first stage in the granting of a licence is simply the permission, in principle, to use a particular piece of land for a nuclear power reactor. Subsequent stages result in the granting of permission to build and to operate. Inevitably the licensee will wish to make changes in design features, operating procedures and nominated personnel, and these must then be agreed with the licensing authority.

This licensing system has some characteristic features which are important in relation to hazard control. It is flexible and adaptable to change and avoids the disadvantages of general and often irrelevant regulations. But is highly detailed and involves considerable effort in making changes. And it makes heavy demands on manpower both for the licensee and the controlling authority. It also inevitably tends to shift some of the responsibility from the operator to the inspectorate. It is a highly effective but resource intensive system.

The particular approach taken to the control of nuclear hazards reflects the uncertainties in the early days of nuclear energy, say the mid-1950s, about the hazards of this very high technology industry. It is debatable how applicable this approach is to other hazards.

4.7.2 Nuclear plant siting

In the UK the sites which might be used in a power network vary by a factor of 10 at most in population density. Hazard assessments for nuclear plants are accurate only within a similar factor. Thus reactors cannot be matched to sites with any precision. In effect, the design must be to a standard which, it may be argued, then permits complete freedom of siting.

Hazard assessments have been presented by F.R. Farmer (1967a,b, 1971) to support the relaxation of the historical restrictions requiring remote siting and to permit siting nearer urban areas. However, it is not government policy to allow this degree of freedom in siting nuclear plants. The siting of nuclear reactors in or near large centres of population is not permitted. A major determinant of this policy is the difficulty of devising effective emergency countermeasures to the hazards of a radioactive release.

An account of UK siting policy is given by Gronow and Gausden (1973).

4.7.3 Nuclear safety case

An important feature of the nuclear licensing system is the safety case. The nuclear safety case is broadly similar, *mutatis mutandis*, to the process safety case described below.

The safety issues to be addressed are outlined in *Nuclear safety. Safety Assessment Principles for Nuclear Reactors* (HSE, 1979d) and, for pressurized water reactors, in *PWR* (HSE, 1979e).

Quantification of the risks has long been a feature of the nuclear safety case. The evaluation of these risks is addressed in *The Tolerability of Risks from Nuclear Power Stations* (HSE, 1988c).

4.8 Process Hazard Control: Background

The growth of major industrial hazards, especially those associated with the process industries, and the means of controlling them, was one of the main concerns of the Robens Committee. The subsequent Health and Safety at Work etc. Act 1974 (HSWA) provides a much improved framework for control. Even before this, a Department of the Environment (DoE) circular DoE 72/1 had drawn attention to the need to take account of major hazards in planning.

The disaster at Flixborough in 1974 caused a great surge of public concern over such hazards and represents a watershed as far as process industry hazards are concerned. Following Flixborough the Health and Safety Commission took certain steps to improve control of major hazard installations. A Major Hazards Unit was set up within the Health and Safety Executive (HSE) and an Advisory Committee on Major Hazards was appointed. Within the Major Hazards Unit, a Risk Appraisal Group was formed to give advice to local planning authorities on planning applications for major hazard installations. The supervision of these installations by the Factory Inspectorate was intensified. These arrangements were interim measures taken pending the recommendations of the committee.

Selected references on major hazard control in the process industries are given in Table 4.7.

4.9 Process Hazard Control: Advisory Committee on Major Hazards

The Advisory Committee on Major Hazards (ACMH) under the chairmanship of Professor B.H. Harvey, a former Deputy Director-General of the HSE, was set up soon after Flixborough in 1974. The terms of reference of the committee were:

> To identify types of installations (excluding nuclear installations) which have the potential to present major hazards to employees or to the public or the environment, and to advise on measures of control, appropriate to the nature and degree of hazard, over the establishment, siting, layout, design, operation, maintenance and development of such installations, as well as over all development, both industrial and non-industrial, in the vicinity of such installations.

The work of the Committee, which ended in 1983, is described in the three *Reports of the Advisory Committee on Major Hazards* (Harvey 1976, 1979b, 1984). Some of the main themes of these reports are given in Table 4.8.

The first task of the Committee was to define, or rather identify, major hazards. It concluded that, with few exceptions, a major hazard arises only if there is a release from containment. The hazard which then arises is a release which is flammable, toxic or both. A

Table 4.7 *Selected references on major hazard control in the chemical industry*

Harvey (1976, 1979b, 1984); Solomon (1976); Fairley and Mosteller (1977); Lees (1977b, 1980d, 1982c); Locke (1977); Napier (1977b); Schenkel (1977); Wakabayashi (1977); AERA (1978 Harwell Environmental Seminar 1, 1979 Harwell Environmental Seminar 2); Cremer and Warner (1978); Yaroch (1978); Schoch (1979); Beveridge (LPB 31 1980); Levitt (1980); Kunreuther *et al.* (1981); Barrell and Scott (1982); F.R. Farmer (1982); Helsby, *et al.* (1982); Kunreuther and Ley (1982); Labour Inspectorate (1982); Lagadec (1982); V.C. Marshall (1982b, 1982 LPB 46, 1985, 1987, 1988c, 1989c, 1990d); Pantony and Smith (1982); Ramsay *et al.* (1982); Pape (LPB 51 1983); Anon. (1984b,m,w); Crossthwaite (1984, 1986); Hawksley (1984); Kunreuther (1984); D.J. Lewis (1984f); Anon. (1985h); Barrell (1985, 1990, 1992); Beveridge and White (1985); K.R. Davies (1985); Dewis (1985a); ILO (1985a, 1989, 1991); R. King (1985); C. Martin (1985); Merriman (1985); Moysen (1985); Otway and Peltu (1985); Wang (1985); Anon. (1986z); Air Pollution Control Association (1986); J.C. Consultancy Ltd (1986); Raman, Stephens and Haddad (1986); Slater *et al.* (1986); Paté-Cornell (1987); Ripple (1987); J.G. Collier (1988); HMCIF (1988b); Withers (1988); Challis (1989); Machida (1990); van Mynen (1990); Renshaw (1990); Shortreed (1990); Zimmerman (1990); Keller and Wilson (1991); Crawley *et al.* (1992); A.J. Williams (1992); Donegani and Jones (1993); Rimington (1993); Rosenthal (1993); CEC (1994)

Natural gas, liquefied natural gas

US Congress, OTA (1977); Cofield (1978); DoEn (1978); UN (1978); Sutcliffe (1980); Kunreuther *et al.* (1981)

Community involvement, citizen groups, public relations (*see also* Table 9.1)

Chopey (1967b); F.C. Price (1974b); Sutcliffe (1980); Rodgers (1987); Buzelli (1990); Delbridge (1990); Poje (1990); Grollier-Baron (1992c); Marsili, Volloni and Zaponi (1992); Renn (1992); The Royal Society (1992)

Planning

DoE (1972 Circular 1/72, 1980 Circular 22/80, 1983, 1984a–c, 1984 Circular 22/84); Anon. (1975h); Harvey (1976, 1979b, 1984); Salter and Thomas (1981); Anon. (1984w); Crossthwaite (1984, 1986); IBC (1984/57); Pape (1984); Petts (1984a,c, 1985a–b, 1987, 1988a,b, 1989, 1992); Milburn and Cameron (1992); C. Miller and Fricker (1993)

Planning inquiries: Sieghart (1979); Barker (1984); Wang and Parker (1984); Petts (1985a,b); Petts *et al.*, (1986)

Mossmorran: Barrell (1988b); Sellers (1988)

Safety cases

Cassidy (1987, 1988a–c, 1989, 1990); R. Clark (1987); Dyson (1987); Eberlein (1987); Hawksley (1987); Orrell and Cryan (1987); CIA, Chlorine Users Group (1989); R. Clark *et al.* (1989); FMA (1989); Fullam (1989); Lees and Ang (1989b); Pape (1989); Petts (1989); Singleton (1989); IBC (1995/115); Fullam (1995); Myers (1995)

Safety cases and environment: Cassidy (1990)

Offshore (*see* Table A18.1)

USA
Brooks *et al.* (1988); Keffer (1991)

Germany
Kirsch (1979); Nagel *et al.* (1979); Anon. (1980w); Kremer (1981); Land Nordrhein-Westfalen (1981); Schleifenbaum (1981); Wohlleben and Vahrenholt (1981); Braubach (1982); Pilz (1982, 1987, 1989)

The Netherlands
Meppelder (1977); van de Putte and Meppelder (1980); van de Putte (1981, 1983); Versteeg (1988); Husmann and Ens (1989); Oh and Husmann (1988); Oh and Albers (1989); Ale (1991, 1992); Bottelsburgh (1995)

---

flammable release may give rise to a jet or pool fire or a fireball, or the vapour cloud may burn as a flash fire, or it may give a vapour cloud explosion.

Although there is little in the three reports on numerical risk criteria, the discussion on identification of major hazards in the *First Report* does contain the following statement:

> Since we cannot achieve or expect to achieve no risk of failure in any of the areas discussed, we feel bound to put forward some quantitative objective. If, for instance, such tentative conclusions indicated with reasonable confidence that in a particular plant a serious accident was unlikely to occur more often than once in 10 000 years (or – to put it another way – a 1 in 10 000 chance in any one year), this might perhaps be regarded as just on the borderline of acceptability, bearing in mind the known background of risks faced every day by the general public.

The first step in the control of major hazards was clearly that they should be notified, and an outline notification scheme was presented in the *First Report.* The principle adopted was notification of hazardous substances above a certain level of inventory. It was recognized that inventory is by no means the only factor which determines the hazard, but it was nevertheless adopted as the basis because it is simple to administer. Other characteristics of the materials were to some extent taken into account by the pragmatic approach of specifying different inventories for the principal hazardous materials.

At this basic level of inventory only notification was proposed, but it was envisaged that for a level of inventory 10 times the notification level a hazard survey would be required and that the HSE would also have the further power to call for a detailed assessment if it judged this necessary. The Committee held strongly to the view that it should set priorities and should confine its proposals to major hazards. It was appreciated that the various cut-offs were to some degree arbitrary.

The Committee strongly endorsed the principle of self-regulation by industry which it considered particularly appropriate to major hazard plants with their high level of technology. Self-regulation is a constant theme in the reports. The arrangements proposed for the control of major hazards are therefore intended to be part of a continuum of controls for all types of hazard under the HSWA and to avoid introducing a discontinuity specifically for major hazards. On the other hand it was clear that the community expects not only that such hazards should be under control but also that they should be seen to be so. Consequently, the view was taken that some form of monitoring by the HSE was necessary.

The *First Report* also began consideration of the measures necessary to prevent major accidents, against the background of Flixborough. Here the factor on which the Committee placed most emphasis was the management. Unless the management are competent to operate a major hazard plant, other measures are likely to be rendered ineffective.

The Flixborough explosion occurred because the plant held a very large inventory of flammable material which on release would undergo partial flashing to form a vapour cloud. The report takes up the theme of the need to limit such hazardous inventories. Given that the plant does contain an inventory of such material it must be the aim of management to avoid loss of containment.

Table 4.8 *Some principal themes in the reports of the Advisory Committee on Major Hazards*

First Report 1976
Identification of major hazards
Notification of major hazard installations
Review of types of regulatory control
Control of major hazards by self-regulation
Monitoring of control by the HSE
Management of major hazards
Prevention of loss of containment
Design and operation of pressure systems
Limitation of hazardous inventory
Limitation of exposure
Planning controls for major hazards
Second Report 1979
Notification of major hazard installations
Comparison of theoretical potential and historical experience
Bounds on consequences of major releases
Unconfined vapour cloud explosions
Limitation of exposure of workforce
Design and location of control rooms
Planning controls for major hazards
Research on major hazards
Model licence conditions
Third Report 1984
Discussion of risk
Role of quantitative assessment
Inherently safer design
High reliability plant
Monitoring of warning events
Separation of major hazards and public
Information to the public
Emergency planning
Planning controls for major hazards
Education of engineers and scientists
Research on major hazards

The design and operation of pressure systems is therefore of particular importance. Here the report places emphasis on pressure systems, including pipes, valves, pumps, etc., as well as pressure vessels and on operation as well as design.

The Committee held strongly to the view that planning powers even for developments involving major hazards should continue to reside with the local planning authority (LPA). There are many factors besides safety which affect the planning decision and which need to be taken into account in arriving at a balanced decision and only the LPA is able to do this.

The Committee also identified the problems in planning control over major hazards. These largely centre around the definition of 'development' in the Town and Country Planning Act 1971. The preferred solution was to amend the Act, but the implications of a change in definition were very far-reaching, and alternative though less satisfactory proposals were also made. A major difficulty in planning identified in the report is that of the compensation due if planning permission already granted is revoked. The sums involved can be very large.

The *Second Report* gave revised proposals for notification. This revision took into account the problem of ultratoxic materials, which had been highlighted by the Seveso disaster. Various models of regulatory control were considered, including those for the nuclear industry, mines and pharmaceuticals. The possibility of a licensing scheme similar to that applied to nuclear installations was given particular consideration but, in the event, the Committee decided against licensing.

Much of the work described in the report is concerned with comparison of theoretical estimates of major hazard scenarios with historical experience and with putting some bounds to the effects to be expected from major releases. The problem is that for some hazards taking the most pessimistic scenarios and models yields theoretical estimates of casualties which are very high and barely credible in the light of the historical record. Work on this aspect included collection of incident data and development of a mortality index relating the number of fatalities to the quantity released historically. Work was also done to set some bounds on the range of certain specific hazards which it seemed might have lethal effects at rather far distances, such as vapour cloud explosions and glass breakage. The Flixborough disaster was a vapour cloud explosion, and the report goes into some detail on the technical aspects of such explosions and on methods of estimating their effects.

The influence of Flixborough is also seen in the proposals in the report on the design and location of control rooms for major hazard plants. It became apparent in considering control rooms that this problem is part of the wider question of the limitation of exposure of the workforce. The report gives detailed recommendations for limitation of exposure both by location of the work base and by control of access to the high hazard zone.

The vapour cloud explosion problem highlighted the fact that many major hazard phenomena were imperfectly understood. Another important problem area was that of heavy gas dispersion. The Committee therefore gave encouragement to research on these topics.

Background studies to the reports are described by V.C. Marshall (1976a,c, 1977b, 1978, 1980d, 1982d, 1987).

The *Second Report* gives as an appendix a set of model licence conditions for a possible licensing scheme for major hazard plants. Although the Committee did not in fact opt for licensing, it retained the model licence conditions and proposed that these be used as a kind of code of practice for such plants. These model licence conditions are given in Appendix 24.

The *Third Report* opens with a discussion of risk. It does not give any specific numerical criteria, but states certain principles which may assist in deriving such criteria. The report brings together the overall system of hazard control proposed by the Committee. The essential elements are identification, avoidance and mitigation of the hazard and planning for it. In the context of hazard identification, it gives a discussion of the historical development of quantitative hazard assessment in the process industries and of its role in major hazard control.

The report gives support to the avoidance of hazard by inherently safer design, a generalization of the earlier theme of limitation of inventory.

It also discusses the measures necessary to ensure high plant reliability. Another aspect of avoidance is the monitoring of and learning from warnings, specifically the application of hazard warning concepts.

Measures to mitigate consequences which are discussed in the report are separation distances between the hazard and the public and emergency planning. The committee broadly endorsed methods then being devised by the HSE for the control over development at major hazard sites aimed at stabilizing and, where practical, reducing the population at risk.

The Committee took the view that, although a major hazard plant should be designed to high standards, it is nevertheless prudent to seek to have some separation between the plant and the public as a further line of defence. This may be regarded as the extension to the public of the principle of limitation of exposure applied earlier to the workforce.

Both the *Second* and *Third Reports* re-echo the basic proposals on planning outlined in the *First Report*. The *Third Report* calls again for research on major hazards and also for education on major hazards for the engineers and scientists who are likely to have responsibility for major hazard plants.

4.10 Process Hazard Control: Major Hazards Arrangements

The first legislative initiative in response to the ACMH's recommendations was proposals for the Hazardous Installations (Notification and Survey) Regulations 1978. These draft regulations contained requirements for the notification of installations holding a specified inventory of listed hazardous materials and for a hazard survey for installations containing 10 times the notifiable level.

These legislative proposals were overtaken by the EC Directive on the Major-Accident Hazards of Certain Industrial Activities 1982 (82/501/EEC) (the Major Accident Hazards Directive) and were never implemented. They were replaced by two sets of regulations, the Notification of Installations Handling Hazardous Substances Regulations 1982 (NIHHS) and the Control of Industrial Major Accident Hazards Regulations 1984 (CIMAH) .

4.10.1 NIHHS Regulations 1982

The NIHHS Regulations implement a notification scheme based on the ACMH proposals and similar to that of the original Hazardous Installations Regulations proposals. They are confined essentially to notification and do not contain requirements for a hazard survey, which is now covered by the CIMAH Regulations. The NIHHS Regulations provide for the notification of the installations which the authorities in Britain wish to see notified and provide the basis for planning controls over these installations. The schedule of hazardous installations is given in Table 4.9. A critique of inventory levels for substances covered by the NIHHS Regulations is given by D.C. Wilson (1980).

4.10.2 CIMAH Regulations 1984

The CIMAH Regulations are confined to those requirements necessary to implement the EC Directive. The industrial activities covered by the regulations are defined in terms of processes and of storage involving specified hazardous materials.

The EC Major Accident Hazards Directive 1982 has been amended twice, in 1987 (87/216/EEC) and in 1988 (88/610/EEC). The first amendment, prompted by the disastrous toxic release at Bhopal, made a revision of some of the threshold inventories. The second, following the pollution of the Rhine by chemicals from the Sandoz warehouse fire, modified the controls on storage. These amendments of the Directive were implemented by amendments to the CIMAH Regulations in 1988 and 1990.

The regulations apply to defined industrial activities (Regulation 2). There are certain activities to which the regulations do not apply, notably defence, explosives and nuclear installations (Regulation 3). Two levels of activity are defined. For the first level the requirements are to take the precautions necessary to prevent a major accident and to limit the consequences and generally to demonstrate safe operation (Regulation 4) and to report any major accident which does occur (Regulation 5). For the higher level of activity (defined in Regulation 6) more extensive requirements apply (Regulations 7–12). These include the requirements to submit a safety report (Regulation 7), to update the safety report (Regulation 8), to provide on request additional information (Regulation 9), to prepare an on-site emergency plan (Regulation 10) and to provide information to the public (Regulation 12). The local authority is required to prepare an off-site emergency plan (Regulation 11).

The initial directive made a distinction between storage associated with a process and isolated storage. In accordance with the directive the CIMAH Regulations 1984 gave separate lists of inventories for the application of the regulations for isolated storage (Schedule 2) and other storage (Schedule 3). The 1990 revision again gives lists of inventories as Schedules 2 and 3, but the type of storage to which Schedule 2 applies has been broadened to include certain process-associated storage. Schedule 2 gives two sets of inventories. Inventories at either level attract Regulation 4, but only an inventory in the higher level attracts Regulations 7–12.

An outline of the regulations is given in Table 4.10 and details of Regulations 4, 6, 7 and 8 are shown in Table 4.11. Tables 4.12–4.17 give details of Schedules 1–4, 6 and 8, respectively. Schedule 1 gives the indicative

Table 4.9 *Notification of Installations Handling Hazardous Substances Regulations 1982: Schedule 1: notifiable inventories*

PART I Named substances

Substance	*Notifiable quantity (tonne)*
Liquefied petroleum gas, such as commercial propane and commercial butane, and any mixtures thereof held at a pressure greater than 1.4 bar absolute	25
Liquefied petroleum gas, such as commercial propane and commercial butane, and any mixture thereof held under refrigeration at a pressure of 1.4 bar absolute or less	50
Phosgene	2
Chlorine	10
Hydrogen fluoride	10
Sulphur trioxide	15
Acrylonitrile	20
Hydrogen cyanide	20
Carbon disulphide	20
Sulphur dioxide	20
Bromine	40
Ammonia (anhydrous or as solution containing more than 50% by weight of ammonia)	100
Hydrogen	2
Ethylene oxide	5
Propylene oxide	5
tert-Butyl peroxyacetate	5
tert-Butyl peroxyisobutyrate	5
tert-Butyl peroxymaleate	5
tert-Butyl peroxisopropyl carbonate	5
Dibenzylperoxydicarbonate	5
2,2-bis(*tert*-Butylperoxy)butane	5
1,1-bis(*tert*-Butylperoxy)cyclohexane	5
Di-*sec*-butylperoxydicarbonate	5
2,2-Dihydroperoxypropane	5
Di-*n*-propylperoxydicarbonate	5
Methylethyl ketone peroxide	5
Sodium chlorate	25
Cellulose nitrate other than: (a) cellulose nitrate to which the Explosives Act 1875 applies or (b) solutions of cellulose nitrate where the nitrogen content of the cellulose nitrate does not exceed 12.3% by weight and the solution contains not more than 55 parts of	

cellulose nitrate per 100 parts by weight of solution	50
Ammonium nitrate and mixtures of ammonium nitrate where the nitrogen content derived from the ammonium nitrate exceeds 28% of the mixture by weight other than: (a) mixtures to which the Explosives Act 1875 applies, or (b) ammonium nitrate based products manufactured chemically for use as fertilizer which comply with Council Directive 80/876/EEC	500
Aqueous solutions containing more than 90 parts by weight of ammonium nitrate per 100 parts by weight of solution	500
Liquid oxygen	500

PART II Classes of substance not specifically named in Part I

Class of substance	*Notifiable quantity (tonne)*
1. Gas or any mixture of gases which is flammable in air and is held in the installation as a gas	15
2. A substance or any mixture of substances which is flammable in air and is normally held in the installation above its boiling point (measured at 1 bar absolute) as a liquid or as a mixture of liquid and gas at a pressure of more than 1.4 bar absolute	25 being the total quantity of substances above the boiling points whether held singly or in mixtures
3. A liquefied gas or any mixture of liquefied gases, which is flammable in air, has a boiling point of less than 0°C (measured at 1 bar absolute) and is normally held in the installation under refrigeration or cooling at a pressure of 1.4 bar absolute or less	50 being the total quantity of substances having boiling points below 0°C whether held singly or in mixtures
4. A liquid or any mixture of liquids not included in items 1 to 3 above, which has a flashpoint of less than 21°C	10 000

criteria for the application of Regulations 2 and 4. Schedule 3 gives the list of substances, with associated inventories, for the application of Regulations 7–12. As just described, Schedule 2 gives two lists of inventories, both attracting Regulation 4, but only the higher level attracting Regulations 7–12. Schedule 4 lists the types of industrial installation other than isolated storage which come within Regulation 2. Schedule 6 gives the information to be included in the safety report (Regulation 7)

Table 4.10 *Control of Industrial Major Accident Hazards Regulations 1984: outline of regulations*[a]

Regulation 2	Interpretation
Regulation 3	Application of these Regulations
Regulation 4	Demonstration of safe operation
Regulation 5	Notification of major accidents
Regulation 6	Industrial activities to which Regulations 7 to 12 apply
Regulation 7	Reports on industrial activities
Regulation 8	Updating of reports under Regulation 7
Regulation 9	Requirement of further information to be sent to the Executive
Regulation 10	Preparation of on-site emergency plan by the manufacturer
Regulation 11	Preparation of off-site emergency plan by the local authority
Regulation 12	Information to the public
Regulation 13	Disclosure of information notified under these Regulations
Regulation 14	Enforcement
Regulation 15	Charge by the local authority for off-site emergency plan
Schedule 1	Indicative criteria
Schedule 2	Storage other than of substances listed in Schedule 3 associated with an installation referred to in Schedule 4
Schedule 3	List of substances for the application of Regulations 7 to 12
Schedule 4	Industrial installations within the meaning of Regulation 2(1)
Schedule 5	Information to be supplied to the Commission of the European Communities by the Member States pursuant to Regulation 5(2)
Schedule 6	Information to be included in a report under Regulation 7(1)
Schedule 7	Preliminary information to be sent to the Executive under Regulation 7(3)
Schedule 8	Items of information to be communicated to the public in the application of Regulation 12

[a] This table and the following seven tables incorporate the 1990 amendments to the regulations.

and Schedule 8 that to be given to the public (Regulation 12).

Guidance on the CIMAH Regulations is given in *A Guide to the Control of Industrial Major Accident Hazards Regulations 1984* (HSE 1990 HS(R) 21 rev.). A decision tree for the application of the regulations is shown in Figure 4.2.

The requirements for the safety report are discussed below, those for emergency planning, both on site and off site, in Chapter 24 and those for provision of information to the public in Section 4.11.

An account of the CIMAH arrangements has been given by Cassidy and Pantony (1988). The sites covered are classified as large inventory top tier sites (LITTSs), holding large inventories of flammable and/or toxic materials, and small inventory top tier sites (SITTSs), holding smaller quantities of more toxic substances.

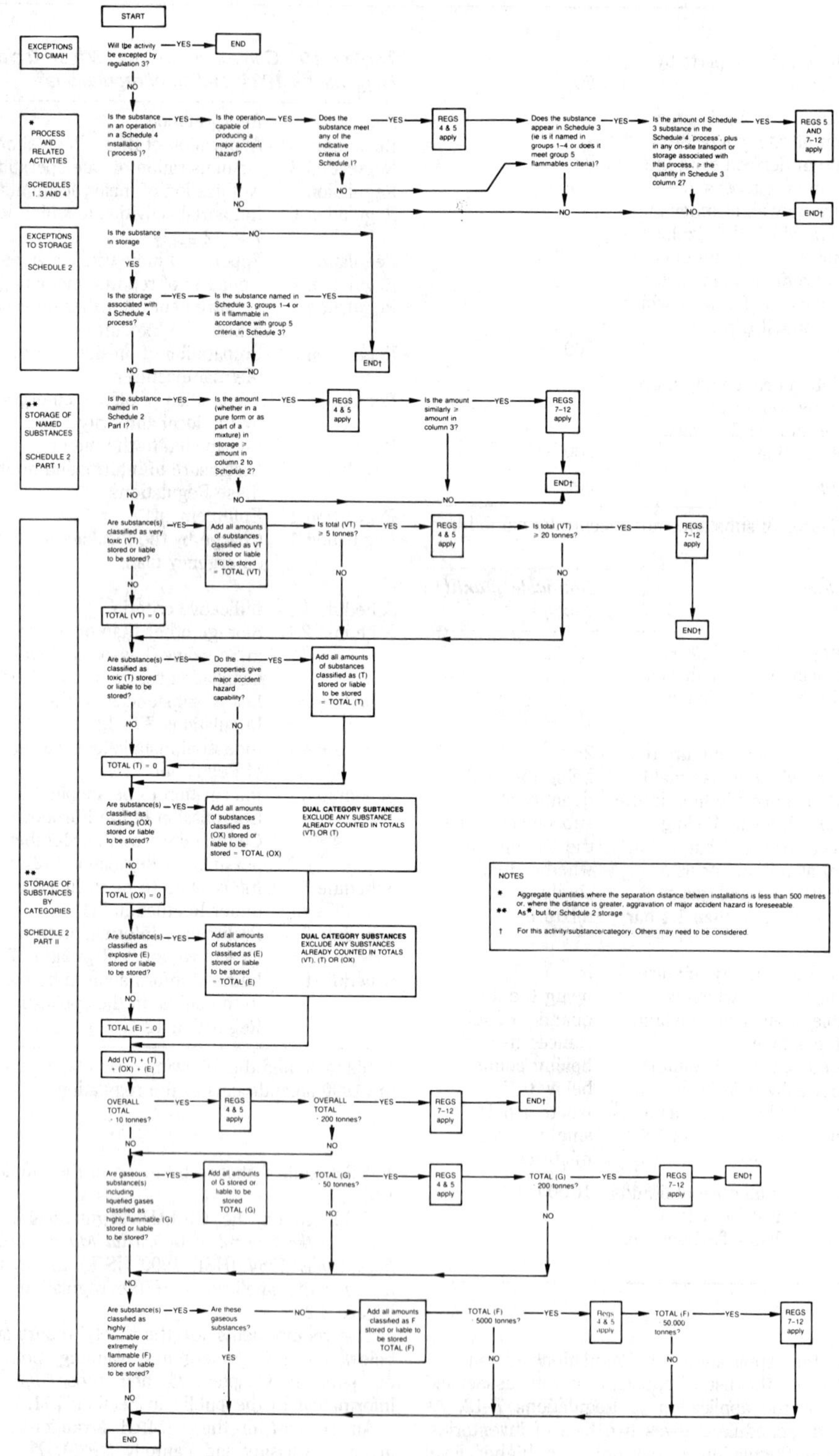

***Figure 4.2** Control of Industrial Major Accident Hazards Regulations 1990: decision tree for application of the regulations (Courtesy of HM Stationery Office)*

Table 4.11 Control of Industrial Major Accident Hazards Regulations 1984: Regulations 4, 6, 7 and 8

Regulation 4 Demonstration of safe operation

(1) This Regulation shall apply to:
  (a) An industrial activity to which subparagraph (a) of the definition of industrial activity in Regulation 2(1) applies and in which a substance which satisfies any of the criteria laid down in Schedule 1 is involved or is liable to be involved; and
  (b) An industrial activity to which subparagraph (b) of that definition applies and in which there is involved, or liable to be involved –
    (i) For a substance specified in column 1 of Part I of Schedule 2, a quantity of that substance which is equal to or more than the quantity specified in the entry for that substance in column 2 of that Part.

(2) A manufacturer who has control of an industrial activity to which this Regulation applies shall at any time provide evidence including documents to show that he has:
  (a) Identified the major accident hazards; and
  (b) Taken adequate steps to:
    (i) Prevent such major accidents and to limit their consequences to persons and the environment, and
    (ii) Provide persons working on the site with the information, training and equipment necessary to ensure their safety.

Regulation 6 Industrial activities to which Regulations 7 to 12 apply

(1) Regulations 7 to 12 shall apply to:
  (a) An industrial activity to which subparagraph (a) of the definition of industrial activity in Regulation 2(1) applies and in which there is involved, or liable to be involved, a substance listed in column 1 of Schedule 3 in a quantity which is equal to or more than the quantity specified in the entry for that substance in column 2 of that Schedule; and
  (b) An industrial activity to which subparagraph (b) of that definition applies and in which there is involved, or liable to be involved —
    (i) For a substance specified in column 1 of Part I of Schedule 2, a quantity of that substance which is equal to or more than the quantity specified in the entry for that substance in column 3 of that Part;
    (ii) For substances and preparations falling within a category or categories specified in an entry in column 1 of Part II of Schedule 2, a total quantity of such substances and preparations in the category or categories in that entry which is equal to or more than the quantity for that entry specified in column 3 of that Part.

(2) For the purposes of Regulations 7 to 11:
  (a) A 'new industrial activity' means an industrial activity which —
    (i) Was commended after the date of the coming into operation of this Regulation, or
    (ii) If commenced before that date, is an industrial activity in which there has been since that date a modification which would be likely to have important implications for major accident hazards, and that activity shall be deemed to have been commenced on the date on which the change was made;
  (b) an 'existing industrial activity' means an industrial activity which is not a new industrial activity.

Regulation 7 Report on industrial activities

Regulation 7(1) gives the requirement for the submission of a written safety report, 7(2) and 7(3) deal with new activities and existing activities, respectively, and 7(4) with exemptions granted by the HSE.

Regulation 8 Updating of reports under Regulation 7

(1) Where a manufacturer has made a report in accordance with Regulation 7(1), he shall not make any modification to the industrial activity to which that report relates which could materially affect the particulars in that report, unless he has made a further report to take account of those changes and has sent a copy of that report to the Executive at least 3 months before making those changes or before such shorter time as the Executive may agree in writing.

(2) Where a manufacturer has made a report in accordance with Regulation 7(1), paragraph (1) of this Regulation or this paragraph, and that industrial activity is continuing, the manufacturer shall within three years of the date of the last such report, make a further report which shall have regard in particular to new technical knowledge which materially affects the particulars in the previous report relating to safety and developments in the knowledge of hazard assessment, and shall within one month, or in such longer time as the Executive may agree, send a copy of the report to the Executive.

(3) A certificate of exemption issued under Regulation 7(4), shall apply to reports or declarations made under this Regulation as it applies to reports made under Regulation 7(1).

They state that in the UK there are over 1700 NIHHS sites and 'several hundred' CIMAH sites.

4.10.3 CIMAH safety case

A central feature of the regulations is the safety report, commonly called the 'safety case'. Two aspects of particular interest are the requirements concerning the management system and the quantification of the hazards. With regard to the former, Schedule 6 of the original CIMAH Regulations 1984 requires that the report provide information on the management system and specifically on the staffing arrangements, including: for certain responsibilities, the names of the persons assigned; the arrangements for safe operation; and the arrangements for training. The wording of the 1990 Regulations is identical. However, the *Guide* to the latter

Table 4.12 Control of Industrial Major Accident Hazards Regulations 1984: Schedule 1: indicative criteria

Regulations 2(1) and 4(1)

(a) Very toxic substances:
- Substances which correspond to the first line of the table below,
- Substances which correspond to the second line of the table below and which, owing to their physical and chemical properties, are capable of producing major accident hazards similar to those caused by the substance mentioned in the first line.

	LD_{50} (oral)[(1)] (mg/kg body weight)	LD_{50} (cutaneous)[(2)] (mg/kg body weight)	LC_{50}[(3)] (mg/l (inhalation)
1	$LD_{50} \leq 5$	$LD_{50} \leq 10$	$LC_{50} \leq 0.1$
2	$5 < LD_{50} \leq 25$	$10 < LD_{50} \leq 50$	$0.1 < LC_{50} \leq 0.5$

(1) LD_{50} oral in rats.
(2) LD_{50} cutaneous in rats or rabbits.
(3) LC_{50} by inhalation (4 h) in rats.

(b) Other toxic substances:
The substances showing the following values of acute toxicity and having physical and chemical properties capable of producing major accident hazards.

LD_{50} (oral)[(1)] (mg/kg body weight)	LD_{50} (cutaneous)[(2)] (mg/kg body weight)	LC_{50}[(3)] (mg/l (inhalation)
$25 < LD_{50} \leq 200$	$50 < LD_{50} \leq 400$	$0.5 < LC_{50} \leq 2$

(1) LD_{50} oral in rats.
(2) LD_{50} cutaneous in rats or rabbits.
(3) LC_{50} by inhalation (4 h) in rats.

(c) *Flammable substances:*
(i) *Flammable gases:*
Substances which in the gaseous state at normal pressure and mixed with air become flammable and the boiling point of which at normal pressure is 20°C or below;
(ii) *Highly flammable liquids:*
Substances which have a flash point lower than 21°C and the boiling point of which at normal pressure is above 20°C;
(iii) *Flammable liquids:*
Substances which have a flash point lower than 55°C and which remain liquid under pressure, where particular processing conditions, such as high pressure and high temperature, may create major accident hazards.

(d) *Explosive substances:*
Substances which may explode under the effect of flame or which are more sensitive to shocks or friction than dinitrobenzene.

(e) *Oxidizing substances:*
Substances which give rise to highly exothermic reaction when in contact with other substances, particularly flammable substances.

Table 4.13 Control of Industrial Major Accident Hazards Regulations 1984: Schedule 2: storage other than of substances listed in Schedule 3 associated with an installation referred to in Schedule 4

Regulations 2(1), 4(1) and 6(1)

This Schedule applies to storage of dangerous substances and/or preparations at any place, installation, premises, building, or area of land, isolated or within an establishment, being a site used for the purpose of storage, except where that storage is associated with an installation covered by Schedule 4 and where the substances in question appear in Schedule 3.
The quantities set out below in Parts I and II relate to each store or group of stores belonging to the same manufacturer where the distance between the stores is not sufficient to avoid, in foreseeable circumstances, any aggravation of major accident hazards. These quantities apply in any case to each group of stores belonging to the same manufacturer where the distance between the stores is less than 500 m.
The quantities to be considered are the maximum quantities which are, or are liable to be, in storage at any one time.

Part I Named substances
Where a substance (or a group of substances) listed in Part I also falls within a category of Part II the quantities set out in Part I shall be used.

Substances or groups of substances (Column 1)	Quantities (tonnes) *For application of Regulation 4* (Column 2)	*For application of Regulations 7–12* (Column 3)
Acetylene	5	50
Acrolein (2-propenal)	20	200
Acrylonitrile	20	200
Ammonia	50	500
Ammonium nitrate[a]	350	2 500
Ammonium nitrate in the form of fertilizers[b]	1 250	10 000
Bromine	50	500
Carbon disulphide	20	200
Chlorine	10	75
Diphenylmethane di-isocyanate (MDi)	20	200
Ethylene dibromide (1,2-dibromoethane)	5	50
Ethylene oxide	5	50
Formaldehyde (concentration ≥90%)	5	50
Hydrogen	5	50
Hydrogen chloride (liquefied gas)	25	250

Substances or groups of substances (Column 1)	Quantities (tonnes) For application of Regulation 4 (Column 2)	For application of Regulations 7 to 12 (Column 3)
Hydrogen cyanide	5	20
Hydrogen fluoride	5	50
Hydrogen sulphide	5	50
Methyl bromide (Bromomethane)	20	200
Methylisocyanate	0.15 (150 kg)	0.15 (150 kg)
Oxygen	200	2 000
Phosgene (carbonyl chloride)	0.75 (750 kg)	0.75 (750 kg)
Propylene oxide	5	50
Sodium chlorate	25	250
Sulphur dioxide	25	250
Sulphur trioxide	15	100
Tetraethyl lead or tetramethyl lead	5	50
Toluene di-isocyanate (TDI)	10	100

[a] This applies to ammonium nitrate and mixtures of ammonium nitrate where the nitrogen content derived from the ammonium nitrate is >28% by weight and to aqueous solutions of ammonium nitrate where the concentration of ammonium nitrate is >90% by weight.
[b] This applies to straight ammonium nitrate fertilizers which comply with Council Directive 8/876/EEC (OJ No. L250, 23.9.80, p.7) 'on the approximation of laws of the Member States relating to straight ammonium nitrate fertilizers of high nitrogen content' and to compound fertilizers where the nitrogen content derived from the ammonium nitrate is >28% by weight (a compound fertilizer contains ammonium nitrate together with phosphate and/or potash).

Part II Categories of substances and preparations not specifically named in Part I

The quantities of different substances and preparations of the same category are cumulative. Where there is more than one category specified in the same entry, the quantities of all substances and preparations of the specified categories in that entry shall be summed up.

Categories of substances and preparations (Column 1)	Quantities (tonne) For application of Regulation 4 (Column 2)	For application of Regulations 7–12 (Column 3)
1 Substances and preparations that are classified as 'very toxic'	5	20
2 Substances and preparations that are classified as 'very toxic', 'toxic',[a] 'oxidizing' or 'explosive'	10	200
3 Gaseous substances and preparations including those in liquefied form, which are gaseous at normal pressure and which are classified as 'highly flammable'[b]	50	20
4 Substances and preparations (excluding gaseous substances and preparations covered under item 3 above) which are classified as 'highly flammable' or 'extremely flammable'[c]	5 000	50 000

[a] Where the substances and preparations are in a state which gives them properties capable of producing a major accident hazard.
[b] This includes flammable gases as defined in paragraph c(i) of Schedule 1.
[c] This includes highly flammable liquids as defined in paragraph (c)(ii) of Schedule 1.

Substances and preparations shall be assigned categories in accordance with the classification provided for by Regulation 5 of the Classification, Packaging and Labelling of Dangerous Substances Regulations 1984 (SI 1984/1244, amended by SI 1986/1922, SI 1988/766, SI 1989/2208 and SI 1990/1255) whether or not the substance or preparation is required to be classified for the purposes of those Regulations, or, in the case of a pesticide approved under the Food and Environment Protection Act 1985 (c.48), in accordance with the classification assigned to it by that approval.

(HSE 1990 HS(R) 21 rev.) is much more explicit on the matters to be covered and is in effect a description of good practice in respect of a safety management system.

When the regulations were brought in, it was a matter of some concern to know the extent to which the HSE would require quantitative risk assessment. The guidance to the original regulations (HSE 1985 HS(R) 21) stated:

> Whilst it may be possible for manufacturers to write a safety case in qualitative terms, HSE may well find it easier to accept conclusions which are supported by quantitative arguments. A quantitative assessment is also a convenient way of limiting the scope of the safety case by demonstrating either that an adverse event has a very remote probability of occurring or

Table 4.14 *Control of Industrial Major Accident Hazards Regulations 1984: Schedule 3: list of substances for the application of Regulations 7 to 12*

The quantities set out below relate to each installation or group of installations belonging to the same manufacturer where the distance between the installations is not sufficient to avoid, in foreseeable circumstances, any aggravation of major-accident hazards. These quantities apply in any case to each group of installations belonging to the same manufacturer where the distance between the installations is less then 500 metres

Substance *(Column 1)*	*Quantity (for application of Regulations 7 to 12)* *(Column 2)*
Group 1 Toxic substances (quantity ≤1 tonne)[a]	
	(kg)
Arsine	10
Methyl isocyanate	150[c]
Parathion	100
Phosgene	750[d]
2, 3, 7, 8-Tetrachlorodibenzo-p-dioxin (TCDD)	1
Group 2 Toxic substances (quantity >1 tonne)[b]	
	(tonne)
Acetone cyanohydrin	200
Acrolein	200
Acrylonitrile	200
Allyl alcohol	200
Allylamine	200
Ammonia	500
Bromine	500
Carbon disulphide	200
Chlorine	25[e]
Ethylene dibromide	50
Ethyleneimine	50
Formaldehyde (concentration ≥ 90%)	50
Hydrogen chloride (liquefied gas)	250
Hydrogen cyanide	20
Hydrogen fluoride	50
Hydrogen sulphide	50
Methyl bromide	200
Nitrogen oxides	50
Propyleneimine	50
Sulphur dioxide	250[f]
Tetraethyl lead	50
Tetramethyl lead	50
Group 3 Highly reactive substances[a]	
	(tonne)
Acetylene	50
Ammonium nitrate[g]	2500
Ammonium nitrate in the form of fertilizers[h]	5000
Ethylene oxide	50
Hydrogen	50
Oxygen (liquid)	2000
Propylene oxide	50
Sodium chlorate	250
Group 4 Explosive substances[a]	
	(tonne)
Cellulose nitrate (containing > 12.6% nitrogen)	100
Lead azide	50
Mercury fulminate	10
Nitroglycerine	10
Pentaerythritol tetranitrate	50
Picric acid	50
2, 4, 6-Trinitrotoluene	50
Group 5 Flammable substances[b]	
	(tonne)
Flammable substances as defined in Schedule 1, paragraph c(i)	200
Flammable substances as defined in Schedule 1, paragraph c(ii)	50 000
Flammable substances as defined in Schedule 1, paragraph c(iii)	200

[a]Selection from complete list for this group.
[b]Complete list for this group
[c]Reduced from inventory of 1 tonne in original CIMAH Regulations 1984
[d]Reduced from inventory of 20 tonne in original CIMAH Regulations 1984
[e]Reduced from inventory of 50 tonne in original CIMAH Regulations 1984
[f]Reduced from inventory of 1000 tonne in original CIMAH Regulations 1984
[g]This applies to ammonium nitrate and mixtures of ammonium nitrate where the nitrogen content derived from the ammonium nitrate is >28% by weight and aqueous solutions of ammonium nitrate where the concentration of ammonium nitrate is > 90%
[h]This applies to ammonium nitrate fertilizers which comply with Council Directive 80/876/EEC and to compound fertilizers where the nitrogen content derived from the ammonium nitrate is > 28% by weight (a compound fertilizer contains ammonium nitrate together with phosphate and/or potash)

> that a particular consequence is relatively minor. (p. 34)

The 1990 *Guide* states that

> The nature and extent of the consequence assessment needed depends upon the magnitude of the hazard, its likelihood of occurrence and, most importantly, the need to describe the adequacy of the prevention, control and mitigatory measures.

It later continues

> In many cases explicit quantification of the consequences of a major accident combined with relatively broad, but justifiable, qualitative predictions about the likelihood of occurrence will be sufficient to enable both the manufacturer and HSE to judge whether

Table 4.15 *Control of Industrial Major Accident Hazards Regulations 1984: Schedule 4: industrial installations within the meaning of Regulation 2(1)*

Regulation 2(1)

1 Installations for the production, processing or treatment of organic or inorganic chemicals using for this purpose, amongst others:
- Alkylation
- Amination by ammonolysis
- Carbonylation
- Condensation
- Dehydrogenation
- Esterification
- Halogenation and manufacture of halogens
- Hydrogenation
- Hydrolysis
- Oxidation
- Polymerization
- Sulphonation
- Desulphurization, manufacture and transformation of sulphur-containing compounds
- Nitration and manufacture of nitrogen-containing compounds
- Manufacture of phosphorus-containing compounds
- Formulation of pesticides and of pharmaceutical products
- Distillation
- Extraction
- Solvation
- Mixing.

2 Installations for distillation, refining or other processing of petroleum or petroleum products.

3 Installations for the total or partial disposal of solid or liquid substances by incineration or chemical decomposition.

4 Installations for the production, processing or treatment of energy gases, for example, LPG, LNG, SNG

5 Installations for the dry distillation of coal or lignite.

6 Installations for the production of metals or non-metals by a wet process or by means of electrical energy.

Table 4.16 *Control of Industrial Major Accident Hazards Regulations 1984: Schedule 6: information to be included in a report under Regulation 7(1)*

Regulation 7(1)

1 The report required under Regulation 7(1) shall contain the following information.

2 Information relating to every dangerous substance involved in the activity in a relevant quantity as listed in Schedule 2 column 3 or Schedule 3, namely:
(a) the name of the dangerous substance as given in Schedule 2 or 3 or, for a dangerous substance included in either of those Schedules under a general designation, the name corresponding to the chemical formula of the dangerous substance;
(b) A general description of the analytical methods available to the manufacturer for determining the presence of the dangerous substance, or references to such methods in the scientific literature;
(c) A brief description of the hazards which may be created by the dangerous substance;
(d) The degree of purity of the dangerous substance, and the names of the main impurities and their percentages.

3 Information relating to the installation, namely:
(a) A map of the site and its surrounding area to a scale large enough to show any features that may be significant in the assessment of the hazard or risk associated with the site;
(b) A scale plan of the size showing the locations and quantities of all significant inventories of the dangerous substance;
(c) A description of the processes or storage involving the dangerous substance and an indication of the conditions under which it is normally held;
(d) The maximum number of persons likely to be present on site;
(e) Information about the nature of the land use and the size and distribution of the population in the vicinity of the industrial activity to which the report relates.

4 Information relating to the management system for controlling the industrial activity, namely:
(a) The staffing arrangements for controlling the industrial activity with the name of the person responsible for safety on the site and the names of those who are authorized to set emergency procedures in motion and to inform outside authorities;
(b) The arrangements made to ensure that the means provided for the safe operation of the industrial activity are properly designed, constructed, tested, operated, inspected and maintained;
(c) The arrangements for training of persons working on the site.

5 Information relating to the potential major accidents; namely:
(a) A description of the potential sources of a major accident and the conditions or events which could be significant in bringing one about;
(b) A diagram of any plant in which the industrial activity is carried on, sufficient to show the features which are significant as regards the potential for a major accident or its prevention or control;
(c) A description of the measures taken to prevent, control or minimize the consequences of any major accident;
(d) Information about the emergency procedures laid down for dealing with a major accident occurring at the site;
(e) Information about prevailing meteorological conditions in the vicinity of the site;

(f) An estimate of the number of people on site who may be exposed to the hazards considered in the report.

6 In the case of the storage of substances and preparations to which Part II of Schedule 2 applies, paragraphs 2(a), (b) and (d) and 5(b) of this Schedule shall apply so far as is appropriate.

Table 4.17 *Control of Industrial Major Accident Hazards Regulations 1984: Schedule 8: Items of information to be communicated to the public in the application of Regulation 12*

Regulations 2(1) and 12(1)

(a) Name of manufacturer and address of site.
(b) Identification, by position held, of person giving the information.
(c) Confirmation that the site is subject to these Regulations and that the report referred to in Regulation 7(1) or at least the information required by Regulation 7(3) has been submitted to the Executive.
(d) An explanation in simple terms of the activity undertaken on the site.
(e) The common names, or in the case of storage covered by Part II of Schedule 2 the generic names or the general danger classification, of the substances and preparations involved on site which could give rise to a major accident, with an indication of their principal dangerous characteristics.
(f) General information relating to the nature of the major accident hazards, including their potential effects on the population and the environment.
(g) Adequate information on how the population concerned will be warned and kept informed in the event of an accident.
(h) Adequate information on the actions the population concerned should take and on the behaviour they should adopt in the event of an accident.
(i) Confirmation that the manufacturer is required to make adequate arrangements on site, including liaison with the emergency services, to deal with accidents and to minimise their effects.
(j) A reference to the off-site emergency plan drawn up to cope with any off-site effects from an accident. This shall include advice to co-operate with any instructions or requests from the emergency services at the time of an accident.
(k) Details of where further relevant information can be obtained, subject to the requirements of confidentiality laid down in national legislation.

the precautionary measures are adequate. However, there will be some activities involving highly hazardous and difficult to control processes, e.g. the oxidation of ethylene to produce ethylene oxide, where it may be necessary for the manufacturer to be more precise about the frequency of the various events which could lead to a major accident . . . (p.65)

4.11 Process Hazard Control: Planning

Some appreciation of the planning system is necessary for the understanding of overall arrangements for the control of major hazards. The treatment given here is limited to providing this essential background. Fuller accounts are given by Petts (1984b,c, 1985a,b, 1988a,b, 1989, 1992). Legislation relevant to planning and major hazards is discussed in Chapter 3.

4.11.1 Planning system

The essential function of planning is the control of land use. This control may be used to prevent incompatible uses of adjacent pieces of land, but such control is much easier to exercise before development has taken place than after it. The control of land use is effected through the structure plans of the counties, which set the overall framework, and through the local plans of the local planning authorities (LPAs), which deal in specific developments. New developments are required to relate to these plans. Structure and local plans are essentially instruments for forward planning. Difficulties can arise when the problem is one not of a greenfield development but of an existing installation.

An example of positive forward planning to accommodate hazardous activities is the *National Planning Guidelines* issued in 1977 by the Scottish Development Department. The guidelines give detailed advice on the incorporation of such activities in local plans.

The main responsibility for planning lies with the LPA. In arriving at a planning decision on an industrial activity the LPA has to take into account a large number of factors, including employment, safety and amenity. The LPA is publicly accountable both through the elected members on the local planning committee and through the procedures allowing for individual representations which it is required to follow.

Planning control has been exercised through the powers embodied in the Town and Country Planning Act 1971 (TCPA). In general, this allows new development to be controlled by withholding planning permission or by attaching planning conditions. There is provision for a voluntary agreement between the developer and the LPA. There are limited powers to revoke or discontinue planning permission which has been granted.

The TCPA 1971 gives an LPA certain powers to control 'development' but these are circumscribed by the definition of 'development' in the Act and on the subsequent interpretation of this in the courts. The term development is defined as 'the carrying out of building, engineering, mining or other operations, in, on, over or under land or the making of any material change in the use of land'. However, while planning powers exist for the control of the general type of use, they do not extend to the control of particular uses within the general type.

Since the definition of 'development' is so wide, it has been necessary in order not to overload the planning system to grant certain general permissions. Under the Town and Country Planning General Development Order 1977 (GDO) a number of changes of land use are classed as 'permitted' developments, which do not require planning permission. Further, under the Town and Country Planning (Use Classes) Order 1972 (UCO), changes within a particular Use Class are permitted development. The definition of a Use Class, however, is very broad, e.g. light industry. A change of use of a piece of land or a building is not classed as development unless it constitutes both a 'material change' and a change from one Use Class to another.

For large developments it is usual to seek planning permission in two stages: 'outline' permission and then 'detail' permission. This allows the developer to ascertain that it has permission before incurring costs of development. Outline planning permission is full planning permission. Permission is granted subject to 'reserved matters', which cover such aspects as building design, plant design, landscaping and access. These are the only matters which the LPA is subsequently entitled to consider.

In granting outline planning permission the LPA may stipulate planning conditions. This is an important power, but the conditions must be reasonable and the use of such conditions to deal with matters which are properly the concern of the HSE is discouraged.

A developer which has been refused planning permission or had permission granted subject to planning conditions may appeal to the Secretary of State for the Environment. Alternatively, where the land has become incapable of reasonably beneficial use, it may serve on the LPA a purchase notice.

Planning permission may be revoked before it has been implemented or discontinued for an existing installation or it may be modified but in such cases the LPA may be liable to pay compensation. The sums involved in such compensation can be very large and are a strong deterrent to revocation.

The Secretary of State has powers to approve all structure and certain local plans, to consider appeals against planning decisions, to call in a planning application for his own consideration and to confirm or reject revocations, modifications and discontinuances of planning permission.

This remains the general framework under the TCPA (1990).

4.11.2 Planning and major hazards

The problem of developments involving major hazards was recognised well before the Flixborough disaster. In 1972 the DoE issued the first of a number of circulars to LPAs on hazardous installations. This circular, DoE 72/1, gave a list of hazardous inventories for which it was recommended that the LPA should seek advice. Some of the DoE circulars are listed in Table 4.18.

Following Flixborough, the HSE experienced an increased number of inquiries from LPAs concerning developments involving hazardous installations and set up a risk appraisal group (RAG) to give advice in response to these queries. In 1974, the ACMH was set up with a membership which included planners. Planning was one of the principal topics considered by the Committee which made a number of recommendations in this area.

Decisions on planning for major hazard installations as for other developments are taken by the LPA. The alternative that the decision might be made by the HSE was considered at some length by the ACMH but rejected. The case for leaving the decision with the LPA was made in the committee's *First Report* in the following terms:

> It might be argued that the siting of potentially hazardous installations should be controlled by the Health and Safety Executive as part of their application of comprehensive safety controls in general, but we hold very firmly to the view that siting of all industrial developments should remain a matter for planning authorities to determine since the safety implications, however important, cannot be divorced from other planning considerations. (paragraph 72)

In its *Second* and *Third Reports* the Committee recognized that this opinion was not universally shared, but reaffirmed its view:

> When a planning application is being considered a balanced view should be taken of all aspects including social and economic factors and not just health and safety. Our view is that the HSE ought to provide a clear assessment of the risks associated with the development and to ensure that plant standards are appropriate for those risks. The decision on whether or not to grant planning permission for an installation which meets health and safety criteria should however rest with the local authority who, on behalf of the local community, attempts to come to a balanced decision, having taken all the factors into consideration. In some cases where an installation is of wider regional or national significance, the final decision may be taken by central government. (*Third Report*, paragraph 111)

On siting policy the committee stated:

> The overall objective should always be to reduce the number of people at risk, and in the case of people who unavoidably remain at risk, to reduce the likelihood and the extent of harm if loss of control or containment occurs. (*Third Report*, paragraph 109)

The Committee distinguished between several different development situations involving major hazards:

(1) initial introduction of hazards to 'greenfield' sites;
(2) initial introduction of hazards to existing installations;
(3) intensification of hazards at existing installations;
(4) proposed development in the vicinity of hazardous installations;
(5) existing development in the vicinity of hazardous installations.

In the Committee's view, the existing planning system was well able to handle the initial introduction of hazards to a greenfield site. They recommended that the LPA should impose a standard planning condition prohibiting without specific consent the introduction onto a site of notifiable hazards.

The problems of the initial introduction of a hazard or of intensification of an existing hazard were recognized as more difficult. The ideal solution would have been to amend the definition of 'development' in the TCPA, but this was a fundamental change which in the short term at least the DoE was unable to accept. An alternative approach to the initial introduction of hazard which the Committee recommended in default of amendment of the TCPA was to amend the GDO and UCO, the former to exclude from permitted development modifications which have the effect of creating a notifiable installation, and the latter to omit notifiable activities altogether from the

***Table 4.18** Department of the Environment Circulars relevant to planning and major hazards*

Circular	
DoE 1/72 (WO 3/72)	Development involving the use or storage in bulk of hazardous materials
DoE 71/73	Publicity for planning applications, appeals and other proposals for development
DoE 4/79	Memorandum on structure and local plans
DoE 22/80 (WO 40/80)	Development control – policy and practice
DoE 26/82	Hazardous substances
DoE 9/84 (WO 17/84)	Planning controls over hazardous development
DoE 22/84	Memorandum on structure and local plans
DoE 1/85	The use of conditions in planning permissions
DoE 2/85	Planning control over oil and gas operations
DoE 15/88	Environmental assessment

Use Classes. These proposals did not, however, solve the problem of the intensification of an existing hazard.

With regard to development in the vicinity of a hazardous installation, the Committee endorsed the aim of first stabilizing and then reducing the number of people exposed to the hazard. It recommended that for proposed development the LPA should consult the HSE. It also drew attention to the power of the LPA to enter into a voluntary agreement with the owners of land to restrict the use to which the land is put.

The Committee recognized that, in a limited number of cases, the LPA might consider it necessary to revoke or discontinue planning permission and recommended that government review its discretionary powers to make payments to local authorities to meet compensation liabilities.

With regard to existing development the Committee stated:

> The HSE is also frequently asked to comment on proposals to develop or redevelop land in the neighbourhood of an existing hazardous undertaking where there may be other land users who are closer and possibly incompatible. In these cases the HSE tell us that it takes the view, which we fully endorse, that the existence of intervening development should not in any way affect the advice that it gives about the possible effects of that activity on proposed developments which may appear to be less at risk than the existing ones. In other words the existing situation should never be regarded as providing grounds for failing to draw attention to the implications for development at a greater distance. (*Second Report*, paragraph 108)

The ACMH emphasized the importance of LPAs having available adequate information on which to reach its decision and on the need for the authority to consult the HSE on developments involving hazardous installations and on the necessity for the HSE's advice to be in a form appropriate to a planning authority which is not expert in major hazards.

Following the work of the ACMH, legislative changes have been made to the planning arrangements, as described below. In parallel with these legislative developments have gone administrative changes. The HSE has upgraded its capabilities to provide advice to LPAs. The ability of the HSE to provide advice has been much enhanced by the creation of the Major Hazards Assessment Unit (MHAU), while the local Factory Inspectorate, who have always had the prime responsibility for the control of hazardous installations, has been reorganized so that offices dealing with the process industries have inspectors who specialize in those industries. The problem of major hazards has proved a troublesome one for LPAs. Major hazards involve both high technology and difficult issues of risk. Most local planners have had little training on major hazards matters.

4.11.3 Planning reforms

A number of changes have been made to the planning system to deal with the problem of major hazards. The first step is to ensure that the LPA knows of the existence of hazardous installations in its area. This is effected by the NIHHS Regulations 1982 and the CIMAH Regulations, originally 1984 and now 1990. Consultation by the LPA with the HSE is now mandatory for any new hazardous development and recommended for other development.

Under the Town and Country Planning General Development (Amendment) Order 1983 development of new hazardous installations and development at an existing notifiable site involving a three-fold increase in inventory are excluded from the general permissions of the GDO, while under the Town and Country Planning Use Classes (Amendment) Order 1983 uses involving notifiable inventories are excluded from the provisions of the UCO. The notification referred to here is that applicable under the NIHHS Regulations 1982.

The Housing and Planning Act 1986 creates a requirement for a written consent from the LPA for a new notifiable inventory. This consent is separate from planning permission. The effect of this written consent procedure is to obviate for such inventories the difficulties of redefining development under planning legislation, to freeze existing hazards at their notified level and to remove liability for compensation.

The Town and Country Planning (Assessment of Environmental Effects) Regulations 1988 require that an environmental impact assessment be submitted to the LPA along with the planning application.

The loophole associated with the concept of development is closed by the provisions of the Planning (Hazardous Substances) Act 1990 and the Planning (Hazardous Substances) Regulations 1992. The Act requires that the presence of a hazardous substance needs the consent of the Hazardous Substance Authority, or LPA.

The Planning and Compensation Act 1991 strengthens the function of the local plan as a means of planning for major hazard installations.

4.11.4 HSE consultation and advice

As stated above, the policy of the HSE is to stabilize and then reduce the number of people at risk from a hazardous installation. The HSE implements this policy primarily through its advice to LPAs. An early account of the HSE's approach was given as an appendix in the *Third Report* of the ACMH. It is reproduced here in Appendix 25. A later description has been given by Pape (1984).

The vast majority of planning decisions are concerned not with development on industrial sites but with development in the vicinity of such sites. The HSE issues to LPAs guidelines on the size of 'consultation zones' around these sites and asks to be consulted on developments within these zones. Consultation with the HSE is not appropriate for minor planning applications. Guidance on the type of application for which consultation is required is given in DoE Circular 9/84.

Initially, the HSE set a general consultation distance of 2 km around a hazardous site. This proved to be excessive and the HSE has since refined its assessment methods and now specifies consultation distances which are related to the particular hazard. In particular, it has developed model approaches for two base cases, liquefied petroleum gas and chlorine which have been described by Crossthwaite (1984) and Pape and Nussey (1985), respectively. These model cases are considered in more detail in Chapter 9. Here consideration is limited to the consultation distances and advice which result from them.

The HSE distinguishes three types of development: A, B and S (special). Type A developments are those where control may well be appropriate. They are situations where people would be present most of the time or where large numbers might be present quite often and the people are not highly protected. Examples are housing, markets, large shops, transport termini and sports stadia. Type B developments are those where the risk would not normally be great enough to warrant control. This includes situations where individuals may be present regularly but not full time even though the total number of people is not necessarily small, and situations where people might be readily incorporated into an emergency plan and where they are not particularly vulnerable to the hazard. Examples are relatively low density factories or warehouses, small office buildings, small shops, sport fields and roads. Type S is intended to cover developments which are significantly more vulnerable than type A. Examples are hospitals, old people's homes and schools.

The replies from the HSE following consultation by the LPA are generally in terms of the risk categories given in DoE 9/84: 'negligible', 'marginal' and 'substantial'. Negligible risk means that the HSE assessment of the consequences of realization of the hazard is that it is unlikely, if not inconceivable, that people in the development would be killed and that if someone was killed this would be a 'freak' effect. Marginal risk means that the probability of people being affected by a major release is remote, that in such a case they might be seriously affected but that death is unlikely. Safety reasons would not in themselves justify a refusal of planning permission, but might justifiably be among those contributing to it. Substantial risk means that people might be seriously injured by a major release. The risk could in itself justify refusal of planning permission and safety should be a major factor in the decision. If there were other factors strongly favouring granting permission, the HSE would suggest the LPA should hear a more detailed explanation of its assessment. In quantitative terms, negligible risk means an individual risk of less than the 'trivial' level of 10^{-6}/year, marginal risk an individual risk of about the

Table 4.19 *HSE guidelines on consultation and related zones and risk categories for LPG storage (after Crossthwaite, 1984)*

A Distances to zone boundaries

Zone	Tank size (tonne)						
	6–10	11–15	16–25	26–40	41–80	81–120	121–300
Consultation zone	150	175	250	300	400	500	600
Middle zone	125	150	200	250	300	400	500
Inner zone	50	50	50	75	100	125	150

B Risk categories in zones

Development type	Zone		
	Inner	*Middle*	*Consultation*
A	Substantial	Assessment[a]	Negligible
B	Negligible	Negligible	Negligible
S(i)	Assessment	Assessment	Negligible
S(ii)	Substantial	Assessment	Assessment

[a] Individual assessment necessary.

trivial level and low societal risk, and substantial risk is an individual risk above the trivial level and more substantial societal risk.

Table 4.19 shows the HSE guidelines on consultation zones for LPG storage installations. The table gives guidance not only on the consultation zone but also on zones within this. Section A of Table 4.19 gives the distances to the zone boundaries for different tank sizes and Section B the risk categories for these zones.

The effectiveness of advice from the HSE to the LPAs has been studied by Petts (1987). An account of the use of hazard assessment by the HSE in relation to its advice to planning authorities is described by Pantony and Smith (1982).

4.11.5 Emergency planning

There is a statutory requirement for counties and equivalent authorities to undertake emergency planning for civil defence, and the tendency has been to extend these arrangements to cover process site and transport emergencies. The extent of emergency planning undertaken has been variable. Some authorities whose areas contain major chemical complexes have long had comprehensive emergency plans, while others have not.

The CIMAH Regulations 1984 require such authorities to have emergency plans for installations notifiable under CIMAH (though not under NIHHS). Off-site emergency plans are considered in Chapter 24.

4.11.6 Information to public

The CIMAH Regulations 1984 contain a requirement that the public be given information about the hazards to which they are exposed and the action which they should take in an emergency. The choice of body to be responsible for informing the public proved a difficult one. The arrangements decided on are as follows. It is intended that the body which informs the public is the local authority. The manufacturer is required to try to reach with the local council an agreement on the information to be provided and then to furnish this to the council, which then issues it to the public. If, however, agreement cannot be reached, the manufacturer is responsible for informing the public directly.

The information provided must include a statement that the activity is notifiable under the CIMAH Regulations and has been notified to the HSE and information on the nature of the hazard (e.g. fire, explosion or toxic release) and on the safety measures and actions to take in the event of an accident.

4.11.7 Public inquiries

A planning application may become the subject of a public inquiry. This may occur if the developer is refused planning permission and appeals or if the application is called in by the Secretary of State. In some cases this call-in may be at the behest of the HSE.

It is only relatively rarely that a planning application for a process plant becomes the subject of a formal planning or public inquiry. Nevertheless, there have been a number of public inquiries on such developments. They include

(1) the Mossmorran Inquiry (1977), dealing with proposals by Shell and Esso to build a natural gas liquids plant and ethylene cracker (McGill, 1982);
(2) the Canvey Inquiries (1980 and 1982) into United Refineries' permission for a new refinery and British Gas operation of a methane terminal (Petts, 1985a);
(3) the Pheasant Wood Inquiry (1980) into housing development near an ICI chlorine and phosgene plant (Petts, 1985b).

These inquiries have shown a trend towards the increasing use of hazard assessment. They have also brought out some of the problems in the use of an adversarial forum for the discussion of technical matters.

A discussion of the problems posed by public inquiries with special reference to the use of hazard assessment and to the presentation of expert evidence has been given by Petts, Withers and Lees (1986).

Experience at inquiries indicates that the use of hazard assessments has been fraught with difficulties, but that many of these relate to obscurity, incompleteness and inconsistency in the assessments. Another difficult area is the public perception of risk.

The adversarial approach used in inquiries poses another problem. Many engineers feel it to be a poor way of arriving at the truth in scientific and engineering matters as well as being unfamiliar and intimidating. The Flixborough Inquiry gave rise to some debate on this question (e.g. Mecklenburgh, 1977a; R. King, 1990). It is considered further in relation to accident inquiries in Chapter 27.

The study by Petts, Withers and Lees (1986) suggests that many of the problems of inquiries into hazardous installations can be greatly mitigated if certain practical steps are taken. The adversarial approach is supported as the least imperfect means of arriving at the truth on technical matters and of allowing public participation, but the use is encouraged of informal meetings between parties before and during the inquiry.

The problems experienced over hazard assessment are regarded as capable of mitigation, provided that the deficiencies of hazard assessments in past inquiries are remedied, and provided that the necessary information is available to all the parties and on a suitable schedule. The means proposed is that for any installation to which the NIHHS or CIMAH Regulations apply and which goes to a planning inquiry a form of open safety case should be made available by industry.

Although public perception of risk at inquiries remains a continuing problem, there is evidence that much of it is related to deficiencies in the hazard assessments and also to the existence of risks which assessment shows to be rather high. A fuller account of public inquiries is given in Appendix 26.

4.11.8 Planner's viewpoint

The viewpoint of the local planning authority is given in *Major Hazard Installations: Planning and Assessment* (Petts, 1984b) and *Major Hazard Installations: Planning Implications and Problems* (Petts, 1984c).

Certain LPAs have been conscious from an early date of the need to provide for major hazard installations in their area. Accounts of the measures taken following Flixborough by Halton Borough Council have been given by Brough (1981) and Payne (1981).

4.12 Process Hazard Control: European Community

The arrangements for the control of major hazards in the EC and in certain other European countries are now briefly reviewed. Comparative accounts are given in *Risk Assessment for Hazardous Installations* (J.C. Consultancy Ltd, 1986), and by Beveridge and Waite (1985) and Milburn and Cameron (1992).

4.12.1 European Community

Controls over major hazards in the EC are established by Directive 82/501/EEC, the Major Accident Hazards Directive, amended by 87/216/EEC and 88/610/EEC, as described in Section 4.10. The Directive contains requirements for notification of installations, for a safety report and for emergency planning.

A critique of the inventory levels set for particular hazardous substances is given by V.C. Marshall (1982b).

A fundamental revision of the major hazard arrangements is in prospect, as outlined in *Proposal for a Council Directive on the Control of Major-Accident Hazards Involving Dangerous Substances (COMAH)* (Commission of the European Communities, 1994). Features emphasized in the document are land use planning, management and human factors, and the safety report. It proposes requirements on: land use planning; the safety management system; a major accident prevention plan (MAPP); and inspection. It is proposed that the safety report should include management systems, should be harmonized in style and should be available to the public, subject to certain safeguards.

4.12.2 Germany

In Germany, basic safety requirements are embodied in a number of measures which include the Industrial Code 1869–1978, the Imperial Insurance Code 1911, the rules for Prevention of Accidents to Man and the Technical Means of Work Act 1968. Selected legislation relevant to major hazards in Germany is given in Table 4.20.

The Federal Emission Control Act 1974 (FECA) (Bundesimmissionsschützgesetz) is the basic legislation on environmental protection and on major hazards. It creates a system of licensing for installations liable to cause 'harmful effects' to the environment or 'considerable disadvantages' to the public.

These terms are vague and gave rise to difficulties of interpretation. More specific requirements are given in the Ordinance on Installations Subject to Licensing 1975 (ISLO), which lists 58 types of installation with a hazard potential sufficient to attract licensing.

Control of major hazards is effected, still under the FECA, through the Hazardous Incident Ordinance 1980 (HIO) (Störfallverordnung). The ordinance gives lists of processes and chemicals. The licensing requirements apply to installations which carry out these processes and use these chemicals.

Exemption from the HIO is allowed if the inventory of the hazardous substance is very small. The levels of inventory below which exemption is given are specified in the First General Administrative Regulation 1981 to the ordinance.

The HIO requires the preparation of a safety analysis. The requirements for this are specified in greater detail in the Second General Administrative Regulation 1982. The information supplied must be such as to allow the regulatory authority to assess adequately the compliance by the operator with its safety obligations. Where this involves calculations, the documentation must show that these have been done. There is no stated requirement for quantitative risk assessment.

Table 4.20 *Selected legislation relevant to major hazards in Germany*

Substances at Work Act 1939
Technical Means of Work Act 1968
Explosives Act 1976
Chemicals Act 1980

Federal Emission Control Act
(Bundesimmissionsschützgesetz) 1974
Fourth Ordinance for the Implementation of the Federal Emission Control Law: Ordinance on Installations Subject to Licensing 1975 Ordinance on Principles of Licensing Procedure 1977
Twelfth Ordinance for the Implementation of the Federal Emission Control Law: Hazardous Incident Ordinance 1908
First General Administrative Regulation on the Hazardous Incident Ordinance 1981
Second General Administrative Regulation on the Hazardous Incident Ordinance 1982

An account of the legislation is given by Stahl (1988) and an industrial viewpoint is given by Pilz (1982, 1987, 1989).

4.12.3 France

In France, safety legislation is based on the Imperial Decree of 1810, which has been subject to amending Laws in 1917, 1932 and 1961. This legislation was extensively modified in the Law of 19 July 1976 on Registered Works for Environment Protection (Installations Classes) and the associated Decree of 10 September 1977. Under this legislation a distinction is made between authorized and declared installations. Installations subject to authorization, which is effectively licensing, are defined by inventory of hazardous chemicals.

A safety study is required for an authorized installation. The study must include a justification of the measures taken to reduce the probability of realization of the hazard and its effects. No requirement is stated for a probabilistic risk assessment. The coverage of these arrangements is wide. There are some 50 000 authorized installations and the number of declared installations is presumably greater still.

4.12.4 The Netherlands

In the Netherlands there are separate systems for the assurance of safety of employees at the workplace and of the public. The first is the responsibility of the Ministry of Social Affairs, and specifically the Labour Inspectorate, and the second that of the Ministry of the Environment. This division is reflected in the requirements for two safety reports, an occupational safety report and an external safety report.

Safety legislation in the workplace is based on the employee safety law of 1934. The Statute of 23 November

1977 creates a requirement for hazardous installations to submit a safety report. This occupational safety report statute is the basic legislation on major hazards at the workplace. Accounts of the safety report have been given by Meppelder (1977), van de Putte and Meppelder (1980), van de Putte (1981, 1983), Husmann and Ens (1989) and Oh and Albers (1989).

The installations for which a safety report is required are defined not by inventory but on a 'threshold value', which is defined as the quantity of material which would present a serious danger to human life at a distance of over 100 m from the point of release. There is a specified method of calculating this value.

The report must contain an assessment of the hazards in the form of a specified fire and explosion index and a toxicity index, derived from the Dow Index method. With regard to quantitative risk assessment, the account given by van de Putte (1981) indicates a view that quantification of probabilities and consequences is desirable, but so is a recognition of the difficulties in the current state of the art. Husmann and Ens (1989) state that a quantitative study involving risk assessment should be performed where novel technology is used.

The external safety report is the concern of the Ministry of the Environment. It deals with the effects of major hazards outside the factory. Accounts are given by Versteeg (1988) and Ale (1991, 1992).

The Ministry of the Environment is deeply involved in quantitative risk assessment. It was one of the sponsors of the *Rijnmond Report* on hazardous installations in the Rijnmond. It has sponsored the development of the computer based SAFETI package by Technica (see Ale and Whitehouse, 1986).

4.13 Process Hazard Control: USA

The arrangements for the control of major hazards in the USA are now briefly described. Overviews are given by Brooks *et al.* (1988) and Horner (1989). US legislation on safety has been outlined in Chapter 3. The account given here is confined to accidental releases, which include major hazards.

4.13.1 Plant siting

The development of these controls needs to be seen against the background of the growing problems experienced in the siting of new hazardous installations since the late 1970s.

A case in point at this time was the siting of liquefied natural gas (LNG) facilities, as described in the *Transportation of Liquefied Natural Gas* by the Office of Technology Assessment (OTA) of the US Congress (1977). The Federal Power Commission (FPC) was the lead agency in determining whether an individual LNG project should be allowed. It was the practice of the FPC to make decisions on siting on a case-by-case basis. The OTA identified several problems in this approach. The FPC was a regulatory rather than a policy-making body and there was no national siting policy to which it could refer. The criteria on which the FPC made its decisions on siting were not known to the industry, since no guidelines were issued. There had been pressure from the state legal authorities and from industry for the issue of uniform siting criteria. In some cases the FPA had made its approval contingent on the receipt of state and local approval, so that the criteria which the industry had to satisfy had become even more obscure.

There were several areas of overlapping jurisdiction. The Office of Pipeline Safety Operations (OPSO) was involved through the Natural Gas Pipeline Safety Act 1968, although there appeared to be a statutory provision against OPSO standards prescribing the location of LNG facilities. In the past, the two agencies had clashed directly on the standards required. The FPC required a temporary shut-down on an LNG facility which OPSO had inspected and approved. The US Coast Guard (USCG) was also involved through the Coastal Zone Management Act 1972. This Act required an applicant for a Federal licence or permit for an activity in a coastal zone of a state to certify the project to be consistent with the state's programme.

Siting was also affected by other legislation. The National Environmental Policy Act 1969 required an environmental impact statement.

4.13.2 Accidental releases

The US legislation which covers major hazards is framed in terms of accidental releases. It has been decisively influenced by releases at home and abroad, and particularly by the toxic gas disaster at Bhopal.

The principal Federal legislation is the Emergency Planning and Community Right-to-Know Act 1986 (EPCRA), which is Title III of the Superfund Amendments and Reauthorization Act 1986 (SARA), generally known as SARA Title III, and is enforced by the Environmental Protection Agency (EPA). Other relevant federal legislation is the Occupational Health and Safety Administration (OSHA) rule for Process Safety Management of Highly Hazardous Chemicals 1990. There are also the state laws in New Jersey and California. One focus of concern is the control of high toxic hazard materials (HTHMs).

4.13.3 SARA Title III

SARA Title III is the Federal legislation which creates controls on accidental releases of hazardous substances. The legislation followed soon after Bhopal. Accounts are given by Brooks *et al.* (1988), Bowman (1989), Horner (1989), Burk (1990) and Fillo and Keyworth (1992). SARA Title III has four main parts dealing with emergency planning, emergency notification, community right-to-know, and toxic chemicals inventory.

For each state the governor appoints an Emergency Response Commission (ERC) which in turn designates local emergency planning committees charged with the preparation of an emergency plan. The facilities which attract such an emergency plan are those containing certain chemicals above specified inventory levels. These chemicals and inventories are given in the *List of Extremely Hazardous Substances and Their Threshold Planning Quantities* (TPQ) developed by the EPA. Management is required to provide any information needed by the local emergency planning committee to implement the emergency plan. The wording has deliberately been left broad enough for the local emergency planning committee to request a quantitative risk assessment (Florio, 1987).

The emergency notification arrangements require the management of the facility to report immediately to the local planning committee any accidental release and the area likely to be affected.

With regard to the right-to-know requirements, for any hazardous chemical which attracts the requirement for a Material Safety Data Sheet (MSDS) under the OSHA 1970, the facility has to submit either the data sheets or a list of the chemicals to the local emergency planning committee, the state Emergency Response Commission and the responsible fire department. The facility must also submit to the same three bodies information on the hazardous inventory.

Section 313 contains the requirement to submit information on the 'toxic chemical release emission inventory', or Toxic Release Inventory (TRI), on the Toxic Chemical Release Form. The data required in this form are concerned primarily with passage of toxic chemicals into the environment. For each chemical a mass balance and the quantity emitted annually is required.

The subjects of the principal sections of SARA Title III may be summarized as follows: Section 301, state Emergency Response Commission; Section 302, notification requirements; Section 304, accidental release reporting; Section 305, study of safety capabilities; Sections 311 and 312, information for the public; and Section 313, toxic emissions.

4.13.4 Special Emphasis Program

In 1986, OSHA initiated a Special Emphasis Program for installations with large inventories of hazardous chemicals. The purpose of the programme was to ensure that such facilities were subject to a systems safety inspection (Hanson, 1986). The systems safety inspection emphasizes four main areas: risk management, hazard identification and assessment, process design and control, and emergency planning.

4.13.5 Process Safety Management Rule

The OSHA has further extended its controls to prevent accidental releases with the rule for Process Safety Management of Highly Hazardous Chemicals (the Process Safety Management rule). The governing document (Federal Register 1990 July 17, 29 CFR Part 1901) refers to major accidents such as Flixborough, Seveso and Bhopal, and to EC controls on major hazard installations in the 'Seveso' Directive and to the need to strengthen such controls in the USA, which otherwise largely rely on the OSHA 1970 and on standards. As its name implies, the Process Safety Management rule places particular emphasis on management. The rule covers hazard identification, evaluation and control.

4.13.6 New Jersey Toxic Catastrophe Prevention Act

The New Jersey Toxic Catastrophe Prevention Act 1985 (TCPA) is also addressed to the identification, evaluation and control of hazards, for facilities handling extraordinarily hazardous substances (EHSs), there being 11 such chemicals on the initial list. It creates a requirement for a risk management plan (RMP). The RMP covers risk management, safety review, risk assessment and emergency planning.

4.13.7 California Hazardous Materials Planning Program

The California Hazardous Materials Planning Program draws together four complementary laws which are codified in Chapter 6.95 of the state Health and Safety Code. It requires a facility handling more than a specified minimum inventory of a hazardous material to establish a 'business plan for emergency response', giving details of the inventory and the emergency plan. The program requires a business which handles a defined 'acutely hazardous material' to register with the administrative authority. If a business is notified by the authority that a Risk Management and Prevention Program (RMPP) is required, it must then prepare one.

4.13.8 Regulatory agencies

As just described, both the EPA and the OSHA are involved in the regulation of accidental releases, primarily in the case of the former through SARA Title III and of the latter through the OSHA 1970 and, more recently, the Process Safety Management rule. The division of responsibility between the two agencies is not clear-cut.

The Jeffreys Amendment of 1990 defined roles for OSHA and EPA in preventing accidental releases. OSHA was directed to bring in the regulation on Process Safety Management of Highly Hazardous Chemicals and EPA to bring in a regulation on accidental releases which would complement the OSHA requirements and would focus on community issues.

4.13.9 Voluntary initiatives

Mention should also be made of some of the voluntary initiatives which have been taken in the USA in this area. Since 1985 the Chemical Manufacturers Association (CMA) has recommended to its members the Community Awareness and Emergency Response program (CAER). This voluntary programme deals particularly with emergency planning. The creation in 1985 of the Center for Chemical Process Safety (CCPS) of the AIChE, described in Chapter 1, is another such initiative.

5 Economics and Insurance

5.1 Economics of Loss Prevention

Loss prevention is concerned with the avoidance both of personal injury and of economic loss. In both spheres there is an economic balance to be struck. But there are also quite difficult problems in making the economic assessments. There is no doubt, however, about the economic importance of loss prevention. Some costs arise through failure to take proper loss prevention measures; others are incurred through uninformed and unnecessarily expensive measures. Both types of cost are numerous and can involve major items. The financial viability of a project is often determined by loss prevention factors.

Selected references on process economics and on economics of loss prevention are given in Table 5.1.

***Table 5.1** Selected references on process economics and economics of loss prevention*

Peters (1948); Perry (1954); Aries and Newton (1955); Schweyer (1955); Happel (1958); Vilbrandt and Dryden (1959); Baumann (1964); Foord (1967–68); ICI (1968); Peters and Timmerhaus (1968); Raiffa (1968); Rudd and Watson (1968); IChemE (1969/48); IProdE and ICWA (1971); D.H. Allen (1968, 1972 IChemE/52, 1975, 1988, 1990 IChemE/85); H. Gibson and Jackson (1972); Coward (1973 FRS Fire Research Note, 1982); Holland *et al.* (1974); L.M. Rose (1976); Windebank (1976); R. King and Magid (1979); de la Mare (1982); Kurtz (1984); C. Tayler (1984b); T.E. Powell (1985); S.E. Dale (1987); Drangeid (1989); Garrett (1989)

Costs
NRC (Appendix 28 *Cost Estimation*); *Chemical Engineering* (1958–, 1963, 1978c, 1979a); Anon (1963); Weinberger (1963); Dow Chemical Co. (1964, 1966a,b, 1976, 1980, 1987, 1994b); Anon. (1965b); Roth (1965); Dybdal (1966); J.T. Gallagher (1967); Guthrie (1968, 1969, 1970, 1971, 1974); Mendel (1968); Chase (1970); Liptak (1970); Alonso (1971); Frost (1971); S.P. Marshall and Brandt (1971, 1974); Popper (1971); Pitkin (1972); W.L. Nelson (1976); Pikulik and Diaz (1977); Hasselbarth (1977); Baltzell (1978); R. Kern (1978b); Kohn (1978a); Bridgewater (1979); C.A. Miller (1979); Vatavuk and Neveril (1980–); Barrett (1981); Cran (1981); Desai (1981); L.D. Epstein (1981); Mulet *et al.* (1981); Humphreys and Katell (1981); Corripio *et al.* (1982a,b); S.R. Hall *et al.* (1982); Lonsdale and Mundy (1982); Mcintyre (1982); de la Mare (1982); Matley (1982); Purohit (1983); Vogel and Martin (1983–); K.E. Lunde (1984); T.J. Ward (1984); Anon. (1985j); AIChE (1985/82); F.D. Clark and Lorenzoni (1985); Klumpar and Slavsky (1985a–c); A. Rose (1985); H.W. Russell (1985); Anon. (1987f,y); IChemE (1988/82); Kletz (1988g); March (1989); Samid (1990); Kerridge (1992); Lindley and Floyd (1993); Rodriguez and Coronel (1992); Ulrich (1992); Garnett and Patience (1993); Plavsic (1993); Gerrard (1994)

Loss statistics
BIA (periodic); BRE (periodic, see Appendix 28); FRS (periodic, see Appendix 28); HM Chief Inspector of Factories (annual); Ministry of Technology (1969); Munich Re (1991)

Costs of accidents
Badger (1968); Beddoe (1969); Calabresi (1970); Mealey (1970); Robens (1972); Sinclair (1972); Munich Re (1991); HSE (1993 HS(G) 96); Anon. (1994 LPB 116); N.V. Davies and Teasdale (1994)

Costs of plant unreliability, downtime
Ministry of Technology (1969); Jenkins *et al.* (1971); Tucker and Cline (1971); C.F. King and Rudd (1972); D.H. Allen (1973); J.A. Richardson and Templeton (1973); de la Mare (1975, 1976); Reynolds (1976); Sancaktar (1983)

Costs of prevention
FRS (1973 Fire Research Note 982); Pratt (1974); Amson (1976); Windebank (1976); C.W. Smith (1977); Amson and Goodier (1978); Ricci (1978c); Anon. (1982b)

Level of loss prevention expenditure
Redington (1965); Sinclair (1972); Melinek (1974 BRE CP 88/74); Kramer (1974); S.B. Gibson (1976d); Kletz (1976d, 1980a, 1980–, 1981f, 1988g); Anon. (1981u); Witter (1982); Berenblut and Whitehouse (1988); Wakeling (1988); Hirano (1990a)
Decision aids: G.H. Mitchell (1972); R.E. Hanna (1979)
Expenditure decisions: Oakland (1982)

Inherently safer plants
D.W. Edwards and Lawrence (1993); Rushton *et al.* (1994)

5.2 Cost of Losses

The result of not taking adequate loss prevention measures is to give rise to losses and costs such as those of (1) accidents; (2) damage; (3) plant design delays; (4) plant commissioning delays; (5) plant downtime, restricted output; (6) equipment repairs; (7) loss of markets; (8) public reaction; and (9) insurance. A rational approach to loss prevention requires some understanding of the economic importance of these factors and of the means of estimating them.

Unless otherwise stated, the cost figures given below are those for the year referred to. The cost in real terms may be obtained by correcting the values given using the appropriate inflation index. There may be no single index which is totally applicable, but in some cases the retail price index (RPI) and in others a plant cost index may be the most suitable. There is also the further complication of exchange rate conversion.

5.2.1 National level

It is convenient to consider first the cost of these losses at national level in the UK. Estimates may be based on information on the following: (1) accident statistics; (2) fire loss statistics; (3) insurance statistics; (4) maintenance and downtime statistics; and (5) major accidents.

Accident statistics are fairly readily available from the *Annual Report* of HM Chief Inspector of Factories and the annual *Health and Safety Statistics*. Then, making

Table 5.2 *Estimated national resource cost of accidents (after Robens, 1972)*

	Cost (£ million)
Fatalities	8.3
Industrial accidents[a]	51.5
Under-reporting	18.6
Non-reportable accidents	14.5
Total	92.9

[a]Damage, lost output, hospital.

certain assumptions, the national resource cost, or cost to the nation, of these accidents may be estimated.

The *Robens Report* (Robens, 1972) went into this problem in some detail. It quotes estimates of the national resource costs of accidents as between £200 million and £900 million. It gives an estimate of national resource costs for industrial accidents and prescribed industrial diseases based on statistics of the Department of Health and Social Security as £209 million in 1969. It also gives another estimate of the national resource costs of accidents based on HM Factory Inspectorate data. Taking the mid-range value of the estimated damage cost per accident of £40, the national resource cost is then obtained as £93 million, built up as shown in Table 5.2.

The total number of fatal accidents given in the report for 1969 is 1070. Of this total, there were 649 fatalities in Factories Act premises, and of these 357 were in factories proper.

A more recent analysis is that given in *The costs to the British Economy of Work Accidents and Work-Related Ill Health* (N.V. Davies and Teasdale, 1994). A summary of their results is shown in Table 5.3. The table gives a breakdown into the costs of accidents and those of ill health. Damage costs are given in three categories: costs of injury accidents, costs of non-injury accidents and costs of low frequency major accidents, the latter being estimated from insured losses. The medical costs are divided into short-term and long-term costs. Loss of output costs are classified as those due to temporary absence and those due to withdrawal from the workforce. Business interruption is evidently covered only via the insurance costs mentioned. The figures given do not include the additional subjective costs, borne by individuals, which are estimated at some £4300 million. The table shows the estimated national costs of accidents, as opposed to ill health, to be in the range £4500 to £9600 million.

Table 5.3 *Estimated costs to the British economy of accidents and work-related ill health (after N.V. Davies and Teasdale, 1994)*

	Estimated annual cost (£ million)		
	Accidents	*Ill health*	*Total*
Damage:			
injury accidents	15–140	–	15–140
non-injury accidents	2152–6499	–	2152–6499
major accidents	430[a]	–	430
Medical:			
short-term	55–241	46–207	101–448
long-term	–[b]	15[b]	15
Loss of output:			
absence	944	603	1547
withdrawal from workforce	421	1305	1726
Administration	514–930	163–296	677–1226
Total	4531–9605	2132–2426	6663–12031

[a] Costs of low frequency major accidents obtained from insured losses.
[b] Long-term medical costs allocated entirely to ill health (Lees).

Fires and explosions account for a dominant proportion of the property damage and business interruption losses. Statistics for fires are published by the Home Office in the annual *Fire Statistics United Kingdom* and by the Fire Protection Association (FPA). The FPA defines a large fire as one costing £50 000 or more and the Home Office uses this definition. The *Fire Statistics United Kingdom 1990* (Home Office, 1992) gives the annual loss due to large fires over the 4-year period 1986–1989 as some £290 million.

The *FPA Large Fire Analysis 1989* (FPA, 1991) gives the number of large loss fires as 739, with an average loss of £0.382 million, making a total loss of £282 million. In the chemical and allied industries there were eight such fires, with an average loss of £0.656 million, making a total loss of £5.2 million, which was 1.9% of the total.

Another angle from which the problem may be approached is the expenditure on insurance. Many of the fire loss data already given derive from the insurance world. But, in addition, a first-order estimate may be obtained as the multiple of the replacement cost of the national inventory of plant and the premium rate, making suitable allowance for the insurers' costs, and their profits and losses. However, the computation of the national replacement cost is not straightforward. From the national viewpoint, insurance costs are in general not additions to, but reflections of, accident costs.

The national costs of machine repair and maintenance, both in terms of direct maintenance costs and of lost production, have been the subject of a major survey by the Ministry of Technology (1969b). The survey concluded that these two costs were approximately £1100 million and £200–300 million, respectively. The estimate of direct maintenance costs of the chemical and allied industries was £215 million.

An estimate of the direct cost of maintenance in British industry made for the successor ministry, the Department of Trade and Industry (DTI), in 1988 put these costs at about £8 billion per annum, or 3.7% of turnover (Anon. 1987f; March, 1989).

The cost of downtime in these industries at national level is less well documented, although the cost of particular accidents is sometimes quoted. Thus the *Annual Report on Alkali etc. Works 1974* states that in 1974 a major stoppage at one refinery cost the company £10 million. Some further examples are given below.

There is little information available from which the other costs, such as those of design and commissioning delays or loss of markets can be established.

Table 5.4 *Material damage and business interruption losses worldwide, 1985–89 (after Munich Re, 1991)*

	No. of losses	*Loss (US $million)*	
		Unindexed	*Indexed*
Material Damage Losses			
1985	198	457	571
1986	149	184	204
1987	135	877	947
1988	121	671	698
1989	115	1336	1350
Business Interruption Losses			
1985	53	146	178
1986	22	71	85
1987	31	304	349
1988	31	284	309
1989	56	1862	1937

A study by Dupont (Chementator, 27 April 1987) showed that on the basis of data from the National Safety Council (NSC) in 1985 and a performance review of client companies the cost of a disabling injury in the USA was about $18 650. There were some 2 million such injuries nationally and the cost to the nation was thus computed as $37.3 billion.

Data on world-wide fire and explosion losses have been given in *Losses in the Oil, Petrochemical and Chemical Industries* by Munich Re (1991). The material damage and business interruption losses world-wide in the Munich Re data base for the years 1985–89 are given in Table 5.4.

It has been estimated that the UK capacity in chemical and petrochemical plants and oil refineries is some 5.8% of that of continental Western Europe and the USA. Assuming that the two latter are 70% of world-wide capacity, the UK capacity is 4% of that world-wide. This figure may be used to obtain another estimate of the national loss costs.

Even at the international level, large losses such as those of the Piper Alpha oil platform in 1988 and of Pasadena in 1989 have a major impact on the cost of losses in a given year. This is even more true at national level. This needs to be taken into account if annual average figures for periods when no such loss occurs are not to be misleading.

Public reaction to incidents involves the industry in additional costs. A major disaster such as Flixborough may result in legislation or other measures which affect costs throughout the industry. This may occur even if the disaster is in another country. Seveso had a marked influence in Europe and Bhopal world-wide, not least in the USA, the base of the parent company. In other cases the reaction is at local level. Intermediate cases occur when public reaction causes changes in the practice of a particular branch of the industry. The costs due to public reaction have been discussed in general terms by S.B. Gibson (1976d), but the estimation of such costs is very much a matter of judgement.

Further information relevant to the cost of losses is given in Appendix 1. This appendix lists, in Table A1.4, numerous cases of fires and explosions which resulted in property damage and or business interruption.

Also listed in Table A1.4 are numerous instances where, for reasons of hazard and/or pollution, a company has either ceased production at an existing plant or has not proceeded, or has not been allowed to proceed, with a planned development of production. Such cases often involve, of course, heavy financial losses.

5.2.2 Company and works level

If now the methods of determining these costs are considered at the level of the individual company or works, estimates may again be based on the same types of information as before: (1) accident statistics; (2) fire loss statistics; (3) insurance statistics; (4) maintenance and downtime statistics; and (5) major accidents. There are some important differences, however, between company and national level. The two principal points are that the company generally has access to higher quality information than is available at national level. The other is that the company is much more vulnerable to a single major incident.

An illustration of the cost of accidents at company level has been given in the Dupont study already mentioned. This study was referred to by McKee of Conoco at the Piper Alpha Inquiry as evidence that safety is good business (1990). World-wide, in 1985 Conoco had 39 lost work-day cases. On the basis of the industry average, the expected value was 372, so that 333 cases, and their associated costs, were avoided. This was equivalent to a saving of some $6.2 million.

The Costs of Accidents at Work by the HSE (1993a) describes a set of five case studies of accident ratios, on the lines described in Chapter 1, and gives corresponding cost estimates.

Company level fire loss statistics may be used, but the effect of a possible single large loss fire needs to be borne in mind.

The insurance costs incurred by the company are another measure of the cost of losses. Insurance costs need to be considered both in terms of the rates which are currently quoted and of those which might be obtained by the adoption of further loss prevention measures. Actual insurance rates may be obtained from the literature (e.g. Neil, 1971; Drewitt, 1975; Alexander, 1990a), from company data and from insurer's quotations. The extent to which reduced rates may be obtained which give credit for loss prevention measures is considered below.

Costs which the company is especially well placed to assess as those of commissioning delays, downtime and maintenance in its own plants. Delays in design or commissioning related to plant safety tend to arise mainly from failure to recognize serious hazards sufficiently early. The influence of plant lateness on profitability is a well-developed topic in process economics and methods are available to estimate it. Estimates of the effect of plant lateness on ammonia plant economics were given in Table 1.5.

Plant unavailability, or downtime, may or may not incur serious cost penalties, depending on whether there is a ready market for the products. Similar considerations apply to partial unavailability or limited output. Moreover, the penalties of unavailability are not necessarily linear. It is quite possible to live permanently with a degree of undesirable downtime, but a single protracted stoppage may be catastrophic in terms of loss of profits and markets. Methods of estimating both the mean and spread of plant availabilities are described in Chapter 7.

The effect of downtime on plant profitability is another standard topic in process economics (e.g. D.H. Allen, 1973). A theoretical study, which considers the effect of downtime, and in particular the point in the plant life when this occurs, has been made by J.A. Richardson and Templeton (1973). A case study of the effect of downtime on the profitability of an actual petrochemical plant has been described by Jenkins, Ottley and Packer, (1971).

Data on equipment failures are available as generic failure data given in the literature or as data obtained from the works; similarly for equipment repair times and costs. Often the works accounting system can yield much valuable information in this area (R.P. Reynolds, 1976). It is then a relatively straightforward matter to estimate the direct costs of repairing failures.

If an incident occurs at a works, public reaction must be expected and is likely to give rise to expenditure.

The costs outlined are not all mutually exclusive categories. For example, property damage costs as determined by a total loss control programme also appear as equipment repair costs, but the costs have been listed in this way to illustrate the different approaches which may be made to their assessment.

5.3 Cost of Prevention

Some of the areas in which costs tend to be incurred to prevent loss are: (1) management effort; (2) research effort; (3) design effort; (4) process route; (5) operational constraints; (6) plant siting; (7) plant layout; (8) plant equipment (safety margins, materials, duplication); (9) process instrumentation (trip systems); (10) fire protection; (11) inspection effort; and (12) emergency planning.

Since all aspects of a project have safety implications, any such list is somewhat arbitrary. Others are given in the literature (e.g. Pratt, 1974). The areas listed are those in which additional costs attributable to loss prevention are particularly likely to occur.

Loss prevention requires considerable additional effort in management generally (see Chapters 1, 6, 20 and 21), hazard identification (Chapter 8), process and pressure system design (Chapters 11 and 12), plant inspection (Chapter 19), emergency planning (Chapter 24) and research (Chapter 27).

Considerations of safety may well determine the process route, i.e. the basic chemical reactions, and define the operating limits for the process parameters, e.g. pressure, temperature, concentration (see Chapter 11). Obviously these factors are of fundamental importance to the economics of the process. Process economics are also greatly affected by plant siting and layout (see Chapter 10). The site which is otherwise most attractive may be ruled out on safety grounds. A layout which requires large separation distances may be expensive in land and pipework.

The various safety factors which are incorporated in the plant design greatly increase costs. These include design with thicker vessel walls, use of more costly materials of construction, selection of more expensive equipments and duplication of items (see Chapter 12). Additional instrumentation such as trip systems (see Chapter 13) and additional fire protection and fire fighting equipment (Chapters 10 and 16) constitute further costs. While expenditure in all these areas is unavoidable, it is the aim of loss prevention to get value for money in this expenditure.

There are available some global estimates for expenditure attributable to safety and loss prevention, and also to environmental protection. Kletz (1988g) quotes three sets of estimates. One estimate for safety and environmental costs for plant in the UK given by a Chemical Industries Association (CIA) source is 12–15% of total investment. A second estimate derives from his own survey of 10 ICI plants, which showed that on average 10–15% of the capital cost was spent on safety measures over and above the irreducible minimum necessary for a workable plant; items such as relief valves are regarded as basic equipment and are not included in this figure. Another 5% was spent on pollution control. The proportions spent varied widely between plants; in one case the

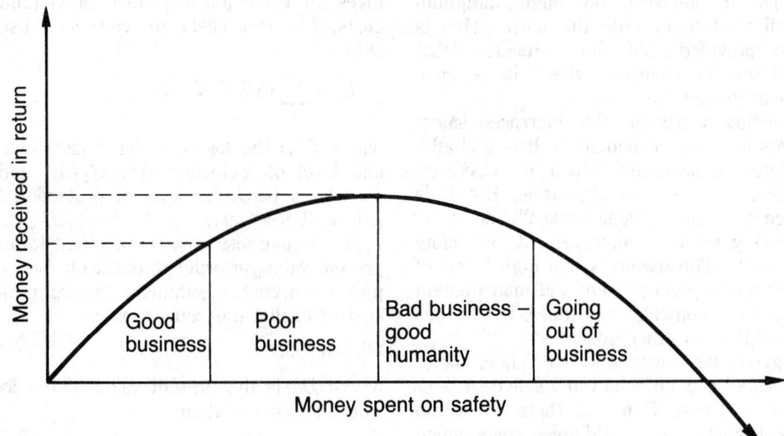

Figure 5.1 *Effects of increasing expenditure on safety and loss prevention (Kletz, 1986d) (Courtesy of the Institution of Chemical Engineers)*

expenditure on safety was 50% of capital cost. The third estimate is that of a US source to the effect that in the hydrocarbon processing industry (HPI) some 18% of capital is spent on fire protection and 12% on pollution prevention. A sum equal to 9% of capital is spent on health and safety, though a large part of this is on revenue items.

5.4 Level of Loss Prevention Expenditure

5.4.1 General considerations

The effect of different levels of expenditure on safety is shown in general terms in Figure 5.1 (Kletz, 1986d). The expenditure shown refers to that required over and above that necessary for a workable plant. At the left-hand end of the graph, expenditure on safety is also good business, but as the expenditure is increased the returns diminish and lead eventually to the company going out of business.

The loss prevention approach has altered considerably the terms of the long-standing debate on the optimum level of safety expenditure. An essential element in this approach is the quantitative assessment of hazard. This in turn makes it necessary to take a view of the levels of risk which are unacceptable and which must be removed by expenditure. Moreover, quantification also provides the information on which can be based rational decisions on the cost-effectiveness of additional expenditure in reducing the risk to human life.

The question of safety expenditure then largely revolves around the decisions on these primary and secondary criteria, the levels of unacceptable risk and the expenditure to save a life. Both criteria can be compared with those used explicitly or implicitly in other industries and activities. The question of acceptable risk and risk criteria is considered in more detail in Chapters 4 and 9.

In fact, the area of choice in safety expenditure in the chemical industry is not as wide as sometimes appears from general discussions of optimum levels of expenditure. The standards of safety which are already obligatory are high and have tended to rise. Since in the chemical industry safety is both relatively good and relatively expensive to improve, the high minimum standard may well tend to become the norm. This is not unreasonable provided, but only provided, that additional expenditure is incurred where it is cost-effective in terms of human life.

It can be misleading to assume that increased safety necessarily involves increased expenditure. It is undoubtedly true that safety expenditure which is based on reaction to incidents can be very expensive. But it is general experience that good practice usually costs no more than bad and gives both improvements in safety and reductions in costs. The reason is that high levels of safety require high quality management and management systems, and that if the company has these, it also has the capability to achieve its other goals.

Thus, given good loss prevention practice, large improvements in safety can be achieved relative to poor practice at no additional cost. That said, there are within the context of good practice some additional costs which are attributable only to safety and which must be borne.

Another aspect of safety expenditure is the estimation of the effect which additional expenditure already incurred has had in improving safety. This is considered in Chapter 27.

5.4.2 Decision aids

The various techniques of economic analysis such as net present value (NPV) and discounted Cash Flow (DCF) may also be used in support of assessment of the economic level of loss prevention expenditure (D.H. Allen, 1973). The difficulties lie principally in assessing the savings.

Some decisions on the level of loss prevention expenditure are amenable to treatment by other formal decision-making techniques, in particular decision analysis. Accounts of decision analysis are given by G. H. Mitchell (1972) and Hastings and Mellow (1978). In a typical decision analysis there is a set of possible outcomes, or states of nature, and a set of possible options for action, or decisions. The aim is to choose the decision(s) which maximize the gain, or utility, or minimize the loss, or regret. A common form of the problem is that there is a possibility of obtaining better information, which will improve the decision but will itself involve a cost.

The application of this technique to loss prevention has been illustrated by R. C. Hanna (1979). The problem considered is the appropriate level of fire protection for a process unit handling liquefied petroleum gas (LPG). Three states of nature are considered – S_1, no fire; S_2, a small fire and S_3 a major fire incident – in a given year. There are also three decisions: d_1, no fire protection; d_2, fireproofing and sprinklers; and d_3, fireproofing, sprinklers and surveillance and control instrumentation. There is the option to do a hazard survey which should provide improved information on which to make the decision. The three outcomes of the survey, or indicators, I_1–I_3 correspond to recommendations to implement decisions d_1–d_3, respectively.

The data required for the analysis are shown in Table 5.5. Sections A and B give, respectively, the estimates made of the costs of preventive measures and of probabilities $P(S_j)$ of no fire, a small fire and a major fire, the latter being based on historical data. Section C gives the estimated loss costs, prevention costs and total costs. The loss costs are calculated using the relationship:

$$D(d_i) = \sum_j P(S_j)C(d_i, S_j) \qquad [5.4.1]$$

where C is the loss cost for a particular state of nature and level of protection (US\$k/year) and D is the loss cost for a particular level of protection for all states of nature (US\$k/year).

The annual loss cost is converted to a capital sum, the present value, in order to put it on the same basis as the prevention cost. Assuming a 20-year project life and 15% cost of capital the conversion is

$$D^* = 6.259D \qquad [5.4.2]$$

where D^* is the present value of the loss cost (US\$k). The total cost is then

$$T^* = M^* + D^* \qquad [5.4.3]$$

where M^* is the prevention cost (US\$k) and T^* is the total cost (US\$k).

Table 5.5 *Decision analysis for level of fire prevention expenditure (after R.C. Hanna, 1979)*

A Prevention costs

	Cost M^* (US$)
d_1	0
d_2	25 000
d_3	100 000

B State of nature probabilities

	States of nature		
	S_1	S_2	S_3
$P(S_j)$	0.93	0.05	0.02

C Loss costs, prevention costs and total costs (without survey)

	Loss cost						
	Annual						
	C (US$k/year) S_1	S_2	S_3	D (US$k/year) Total	Present value D^* (US$k)	Prevention cost, M^* (US$k)	Total cost, T^* (US$k)
d_1	0	100	1,000	25	156.5	0	156.5
d_2	0	50	200	6.5	40.7	25	65.7
d_3	0	10	50	1.5	9.4	100	109.4

D Probabilities of survey correctness

	State of nature		
	S_1	S_2	S_3
$P(I_1\|S_j)$	0.80	0.05	0.03
$P(I_2\|S_j)$	0.15	0.85	0.07
$P(I_3\|S_j)$	0.05	0.10	0.90
Total	1.00	1.00	1.00

E Branch probabilities

Survey branches

$P(I_1)$	0.747
$P(I_2)$	0.183
$P(I_3)$	0.070

State of nature branches

	States of nature			
	S_1	S_2	S_3	Total
$P(S_j\|I_1)$	0.996	0.003	0.001	1.000
$P(S_j\|I_2)$	0.760	0.232	0.008	1.000
$P(S_j\|I_3)$	0.669	0.072	0.259	1.000

F Loss costs, prevention costs and total costs (with survey)

		Loss cost			
		Annual, D (US$k/year)	Present value, D^* (US$k)	Prevention cost, M^* (US$k)	Total cost, T^* (US$k)
I_1	d_1	1.3	8.14	0	8.14
	d_2	0.35	2.19	25	27.19
	d_3	0.08	0.5	100	100.5
I_2	d_1	31.2	195.28	0	195.28
	d_2	13.2	82.62	25	107.62
	d_3	2.72	17.02	100	117.02
I_3	d_1	266.2	1666.15	0	1666.15
	d_2	55.4	346.75	25	371.75
	d_3	13.67	85.56	100	185.56

It is possible to make a decision on the basis of the information in Section C. The minimum value in the total cost, or expected cost of the decision, is that which corresponds to decision d_2 and is US$65.7k.

Alternatively, the decision may be deferred until more information has been obtained by conducting the hazard survey. In this case it is necessary to estimate also the probability that the survey will correspond with reality. Section D of Table 5.5 gives the estimates made for the conditional probabilities $P(I_k|S_j)$ of the survey recommendations given the states of nature. The probabilities $P(I_k)$ of the survey recommendations and the conditional probabilities $P(S_j|I_k)$ of the states of nature given the survey recommendations must also be calculated as shown in Section E. The relationships used are:

$$P(I_k) = \sum_j P(I_k|S_j)P(S_j) \quad [5.4]$$

$$P(S_j|I_k) = \sum_k P(I_k \mid S_j)P(S_j)/P(I_k) \quad [5.5]$$

Equation 5.5[3] is a form of Bayes' theorem, which is described in Chapter 7.

Then, using the probabilities given in Section E of Table 5.5, the loss cost may be recalculated using the relationship:

$$D(I_k, d_i) = C(d_i, S_j)P(S_j|I_k) \quad [5.6]$$

The present value of the loss cost and the total cost may then be recalculated using Equations 5.4.2 and 5.4.3. The results are shown in Section F of Table 5.5. The analysis is also illustrated in the decision tree diagram shown in Figure 5.2.

The expected cost of the decision is then given by the relationship:

$$E^* = \sum_k P(I_k)(T_k^*)_{min} \quad [5.7]$$

Where E^* is the loss cost (with survey) (US$k). The value of E^* is US$38.76. This contrasts with the earlier value of US$65.7k. The difference of US$26.94k is a measure of the value of the survey.

5.4.3 Major hazards

While it is generally true in the process industries that good practice costs no more than bad, in the case of

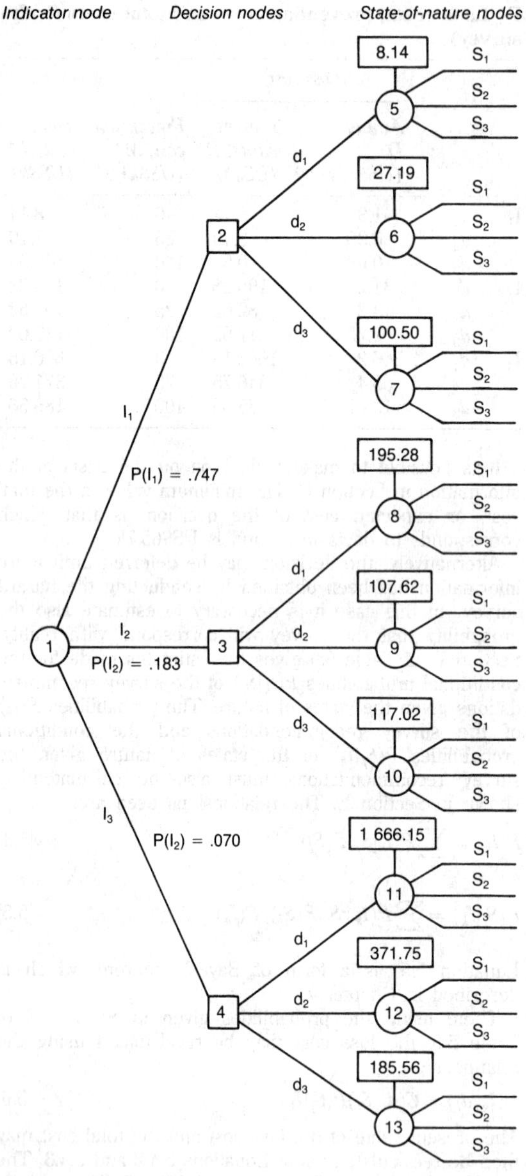

Figure 5.2 *Decision tree for level of fire prevention expenditure (after R.C. Hanna, 1979) (Courtesy of Hydrocarbon Processing)*

major hazards it is often necessary to incur expenditure on safety which is not self-financing. This is illustrated by the following simple calculation given by Lowe (1980). Consider an installation on which an incident causing US$10 million damage is estimated to occur once in 10^4 years so that the average annual loss is US$1000. Expenditure to reduce the frequency of the event to once in 10^5 years would yield a saving of US$900/year. This would justify only about US$2000 capital plus US$200/year operating costs. It is highly unlikely that a 10-fold reduction in incident frequency could be obtained for this expenditure. For major hazards, therefore, some decisions on the level of expenditure are influenced rather by risk criteria for such hazards.

5.5 Insurance of Process Plant

There are a number of ways in which questions of insurance impinge on safety and loss prevention. It is the business of the insurer to carry out assessments of risk in order to set premium rates. The insurer is also increasingly concerned with the management of risk and with the effectiveness of loss prevention measures. The economics and the design of the process operated by the insured are affected by insurance factors.

The company operating a hazardous plant must either itself bear the whole financial risk or seek outside insurance for this risk. The decision is an important but difficult one and it needs to be related to the overall approach taken by the company to loss prevention. Insurance practices vary, of course, from country to country and it is essentially the practice in the UK which is described here.

The insurance scene is constantly changing. There may be some restriction on insurance capabilities where there are novel risks or bad loss experience, but the market is usually able to adapt. The insurance which is of principal interest here is essentially fire insurance, which effectively covers fire and explosion risks.

Accounts of insurance of process plants are given by Spiegelman (1969, 1987), Neil (1971), Drewitt (1975), Cloughton (1981), Norstrom (1982b, 1986), Redmond (1984, 1990), Davenport (1987), di Gesso (1989), Instone (1989, 1990), Alexander (1990a) and Hallam (1990).

Selected references on insurance of process plants are given in Table 5.6.

5.5.1 The insurance process

The basic principle of insurance is to spread the risk so that losses incurred by the few are borne by the many.

Table 5.6 *Selected references on insurance of process plants*

FPA (CFSD FPDG 8); Collier (1964); Dow Chemical Co. (1964, 1966a,b, 1976, 1980, 1987, 1994b); Eckert (1965); Houghton (1965); Williamson (1966); FMEC (1967); H.S. Robinson (1967a); Warwick (1967); Badger (1968); W.H. Doyle (1969); Spiegelman (1969, 1981, 1987); Ashe (1971); Neil (1971); Weldon (1971); Chilver (1973); Ackroyd (1974); Bowman and Palmer (1974); BIA (1975); Drewitt (1975); Earle (1975); H.D. Taylor (1976); R.W. Nelson (1977); Temple (1977a); R. King and Magid (1979); W.D.J. Price (1979/80); Munday *et al.* (1980); Wessels (1980); Cloughton (1981); Norstrom (1982b, 1986); Colt (1984); Redmond (1984, 1990); Anon. (1985ii); Alpert (1985); IOI (1985); J. Siddle (1986a); Davenport (1987); Kleindorfer and Kunreuther (1987); Canaway (1989, 1991, 1992); CIA (1989 RC19); di Gesso (1989, 1992 LPB 103); Instone (1989, 1990); Alexander (1990a); Hallam (1990); O'Donovan (1991 LPB 99); ABI (1992)

For industrial fire insurance this is done by a combination of co-insurance and re-insurance. In co-insurance the insurers take direct responsibility for providing a specified proportion of the cover. The prime responsibility for assessing the risk is taken by a 'leading' company. In re-insurance the co-insurers 'lay off' their risk with other insurers. For a large risk the re-insurers may themselves reinsure further. The re-insurance business is an international one.

The extent of an individual insurer's participation is usually determined not by the full sum insured but by a loss measure such as the probable maximum loss (PML) or estimated maximum loss (EML), as described below. The underwriter determines his share of the insurance as a proportion of the PML; he usually does this using his company's table of PML limits. He may exercise his discretion to reduce this share, if he has doubts about either the magnitude or the probability of the PML. On this basis the underwriter writes his 'net line'. This net line may then be augmented if re-insurance facilities are brought into play, so that the gross acceptance notified to the buyer may be several times the net line.

The procedure adopted by the insurer in response to an insurance inquiry is to do a breakdown of insurance values by site and by installation, and to obtain a report from the insurance surveyor. Inspection is a normal preliminary to insurance. In general, the industry operates on the principle of 'no insurance without inspection'.

5.5.2 Insurance policies

An insurance policy covers a named peril for a specified period. The perils may be standard or special perils. Some of the types of insurance available in a multi-line policy in the USA are given by Spiegelman (1987): (1) public liability, (2) product liability, (3) workers' compensation, (4) motor vehicle, (5) business interruption, (6) professional liability, (7) boiler and machinery, and (8) property damage.

The principal peril relevant here is fire. Industrial fire insurance almost invariably includes fire, explosion and lightning, and may include, depending on the geographical location, windstorm, flood and earthquake. The effects of fire include both property damage and business interruption. Insurance may be limited to the first or may cover both.

For damage insurance, the sum assured may cover either the current value or the replacement cost.

5.5.3 Loss measures

The insurance policy may provide cover for the full sum at risk, but, as just described, insurers generally find it convenient to work in terms of a more limited sum. There are various measures which are used, notably normal maximum loss (NML), PML and EML, but the definitions of the terms tend to vary. These terms are discussed further below.

5.5.4 Insurance surveyors

It is the function of the insurance surveyor to provide the insurer with a report on the risk to be insured. Insurance surveyors are employees of the insurance companies and, in the UK, are typically experienced people, but are not necessarily qualified engineers. Insurance surveyors have wide experience in risk assessment and there is considerable consensus on the critical features and on good and bad practice in loss prevention.

5.5.5 Insurance surveyor's report

The insurance surveyor's report includes an assessment of the loss measure used such as the PML and serves as a guide to the underwriter in setting the premium. The report involves assessing the fire load and fire risk on a plant and then making allowance for protective features.

5.5.6 Tariff and non-tariff systems

The traditional approach to setting premium rates for industrial fire damage risks has been classification by trade or occupation. Classifications run to as many as 300 trades. For many trades there is a detailed tariff which consists of a basic rate together with increments or decrements for factors which increase or decrease the risk, respectively. However, for some trades there is no tariff. In particular, there has never been a tariff for the chemical industry. Instead, in this industry the premium rates are set according to a scale of relative risk as assessed by the insurance surveyor's report.

5.5.7 Fire insurance

Throughout the 1960s there was a history of growing fire loss in many trades in the UK. Many of these fires were in industries where fire losses have always been high. Insurers dealt with this by applying to each trade appropriate premium weighting factors. This approach was presumably adopted in the belief that the problem was a temporary one. In the event this proved not to be the case so that by the end of the decade the weighing factors in use ranged as high as 7.5. This suggested that the rating structure was no longer an appropriate one.

5.5.8 Business interruption insurance

Until about the 1960s much fire insurance was virtually synonymous with property damage (PD) insurance. There then came into prominence another type of insurance – consequential loss or business interruption (BI) insurance. The need for this is due mainly to the very heavy losses which can occur if a large, single-stream plant is out of action for more than a short period. The costs of losses due to, and of insurance for, business interruption now tend to exceed several times those of damage.

5.5.9 Large, single-stream plants

As described in Chapter 1, the 1960s saw the growth of the large, single-stream plant. This has presented many problems to insurers. In considering the problems posed by such plants, it is necessary to distinguish between average and maximum losses and between damage and business interruption losses. There does not appear to be much evidence that for damage the average losses are any worse than for other plants, but for business interruption the situation is probably different, although it is difficult to assess, because business interruption insurance is largely limited to such plants. The PML is much greater, of course, on such plants for both damage and business interruption insurance.

Large plants pose several problems to insurers. One is the magnitude of the PML for both damage and business interruption. Another is the high premium rates required to provide cover for business interruption. Another is the difficulty of assessing the adequacy of the management and of the technology.

Insurers have been virtually unanimous in expressing their dislike of large plant risks and in suggesting to the industry that it may have gone too far in putting its eggs in one basket. On the other hand, whilst the early generations of large plants gave rise to many problems, it is arguable that the management and technology of such plants are now better understood, not least due to the development of loss prevention.

Some perspective on this is also provided by the fact that many of the worst losses in the 1960s were on medium-sized, medium-technology plants. Neil (1971) suggests that this may have been due partly to the diversion of management resources to the large plants.

5.5.10 Insurance market

The insurance market is an international one and the insurance of a large plant will normally be spread with insurance companies around the world. Just as business is subject to the business cycle, so the insurance industry is subject to the insurance cycle. This is described by Norstrom (1982b). Expansion of capacity in the insurance industry creates competitive forces which drive premium rates down and reduce profitability, thus causing a contraction of capacity. At some times the business and insurance cycles come together. This was the case, for example, during the period 1977–81 in the USA.

When interest rates are high and loss experience is good, the insurance industry is more profitable and capacity expands, whilst when interest rates fall and losses are heavy, profitability and capacity fall. The insurance market is in a continuous state of change and adaptation to new risks. The development of seepage and pollution insurance is an example of the growth of insurance capacity in a new area. The industry has also been able to create capacity to cover very large loss potentials, as described below.

5.5.11 Loss experience

As just indicated, the loss experience of the insurance industry is variable and this has an impact both on capacity and on premium rates.

In the UK, the late 1960s and the 1970s saw loss on an increasingly serious scale in the chemical industry, which up to this time had had a good record. The experience of the insurance companies was therefore not a happy one. In general, in fire insurance, premium income did not cover losses sustained and administrative costs, so that underwriting was in deficit, although insurance companies benefited from investment income. The insurance market is very competitive and in too many cases business was taken at rates which later experience showed to have been unrealistic.

Likewise in the USA the insurance industry experienced large losses around 1967 with the vapour cloud explosion at Lake Charles in 1967 and other major incidents. The period 1974-76 also saw heavy losses.

5.5.12 Insurance capacity

The capacity of the insurance market is usually given in terms of the extent of cover which can be obtained for a single potential loss. As just described, this capacity fluctuates. Although generally the market is able to provide insurance even for very large loss potentials, there is some degree of restriction.

In 1971, the capacity of the world-wide insurance market was described by Neil (1971) as quite limited. He estimated that the theoretical capacity was about £30 million per loss, but also stated that it was difficult to obtain insurance for a PML exceeding £6 million for damage or £10 million for damage and business interruption combined. This was certainly not sufficient for complete insurance cover on a large ethylene plant. This limitation of the market was evidently a consequence of the adverse experience in industrial fire insurance in the 1960–70s.

The vapour cloud explosion at Flixborough had an influence on the insurance market. Some insurers recognized the possibility of such an explosion, either by allowing for it in their assessment of PML or by buying specific re-insurance. A loss of US$100 million from damage and business interruption due to a major vapour cloud explosion was considered within the realm of possibility.

A later estimate of insurance market capacity by Neil (1978) put the capacity for PML at about £40 million for companies with a good loss record and good reputation for management and design. After 1975 the bulk of oil and petrochemical risks owned by oil companies and to be insured on the London market was directed to Industrial Oil Insurers (IOI), a body set up by the leading British insurers, including Lloyds, to underwrite such business. The capacity of IOI was estimated by Neil (1978) as about US$15 million on a PML basis.

Redmond (1990) describes the capacity available through the Oil Industry Insurance (OII) mutual together with the London Master Energy Line Slip, which is a facility whereby many insurers commit their capacity automatically. He gives the capacity of this market as some US$1.4 billion. However, the disasters of Piper Alpha, Pasadena and the *Exxon Valdez* have tended to cause some reduction of capacity. The estimate given by Redmond for capacity subsequent to these events is some US$1.0 billion.

5.5.13 Insurance restrictions

In periods when the capacity of the insurance market is restricted, there may be pressure on the insured to bear part of the risk. There are various ways in which this may be effected. The restrictions may take the form of exclusions, limited coverage for particular activities, or the application of deductibles. If deductibles, or excesses, are applied, the insured company itself is required to meet a fixed degree of loss. Typically, for property damage this is a cash sum, whilst for business interruption it is a number of days of downtime. Alternatively, the operating company may insure itself. Either the company with the risk carries that risk itself, or it is insured by a subsidiary insurance company.

5.5.14 Self-insurance

A company may often choose not to seek full insurance cover but to bear some or all of the losses itself. Indeed, the demands placed on the world insurance market by the chemical industry have tended to leave companies with little choice but to carry some of the risk themselves. Policy on the purchase of insurance is a complex matter and depends very much on the circumstances of the individual firm, so that only a few general points are made here. Perhaps the most obvious and important is that the scale of potential loss is such that it would be severely damaging to even the largest firm, so that some degree of outside insurance is advisable.

Insurers usually emphasize that it is not reasonable that industry should seek to place only selected risks while withholding from the insurers the 'bread-and-butter' business; the former alone cannot sustain a healthy insurance industry. On the other hand, the industry does expect to receive credit for high standards of management and technology and for loss prevention measures, and its decision to purchase insurance may be influenced by its confidence in the assessment made of these by the insurers.

5.5.15 Vapour cloud explosions

A significant proportion of the loss potential on process plants is attributable to vapour cloud explosions. Alexander (1990a) quotes his own estimate made in about 1980 that this proportion was in the range 5–10%.

A method for the assessment of the EML due to a vapour cloud explosion has been developed by the IOI (1985). A discussion of insurance with special reference to vapour cloud explosions is given by Davenport (1987).

5.5.16 Major disasters

Disasters such as Piper Alpha in 1988 and Pasadena in 1989 have a significant impact on the insurance world.

The losses incurred in these two accidents have been discussed by Redmond (1990). Piper Alpha was not simply a single oil platform, but the centre of an oilfield, with pipelines coming onto it from other platforms and from satellite fields. The interim estimates of loss given are: for physical damage US\$680 million; business interruption, US\$275 million; fatalities and injuries, US\$160 million; removal of wreckage, US\$100 million and other items US\$155 million. These items total US\$1370 million. The interim estimates for the loss at Pasadena quoted by Redmond are very similar, about US\$1400 million, divided almost equally between property damage and business interruption. The HSE (1993a) has estimated the loss due to Piper Alpha as amounting to some £2 billion, including about £750 million in direct insured losses.

5.6 Damage Insurance

There are a variety of factors which are taken into account by insurance surveyors to assess property damage risk for the purpose of premium rating. Some of the features which are frequently mentioned by insurers are: (1) size of plant; (2) novelty of technology; (3) process materials (flammable liquids at high pressures and temperatures); (4) process features; (5) plant layout; (6) building design; (7) fire protection of structures; (8) fire fighting arrangements; and (9) housekeeping.

5.6.1 Loss measures

As already mentioned, a number of measures of loss are utilized, but different meanings are attached to the terms used. Alexander (1990a) utilizes the following terms: average loss; normal maximum loss (NML), or probable maximum loss (PML); estimated maximum loss (EML), or possible maximum loss; and maximum credible loss (MCL), or maximum foreseeable loss. The average loss is self-explanatory. NML is the maximum loss which would be expected if all protective functions work, EML is the maximum loss if one critical item of protection does not work, and MCL is the maximum loss in circumstances where a number of critical items of protection do not work or where a catastrophic event occurs which is just credible. He gives the following approximate frequencies:

NML $\geq 10^{-3}$/year
EML 10^{-4}–10^{-3}/year
MCL 10^{-5}–10^{-4}/year

5.6.2 Risk profiles

The estimates just given for the frequency of the various levels of loss are general ones. It is also possible, and instructive, to construct the loss profile for a particular plant. The construction and use of such profiles is discussed by Alexander (1990a). The method which he describes is the use of what may be termed a frequency–loss (FL) plot. This plot has similarities with the frequency–number (FN) plot for major accidents described in Chapter 9 but, since the loss levels plotted on it are simply the single values of NML, EML and MCL, the frequency is an absolute rather than a cumulative one.

Figure 5.3 illustrates profiles for two hypothetical plants. For plant (A), a well-protected plant, the frequency of the EML is much less than that of the NML and the frequency for the MCL much less again, but for plant (B), an unprotected plant, the EML has the same frequency as the NML.

It is also possible to construct FL curves for the industry as whole. In this case the use of a cumulative frequency is more appropriate. Such curves might be constructed from the data published periodically by insurers (e.g. Norstrom, 1982b; Munich Re, 1991; Marsh and MacClellan, 1992).

5.6.3 Risk assessment methods

Insurance assessors use a variety of methods to assess the risk and to arrive at a premium rating. These

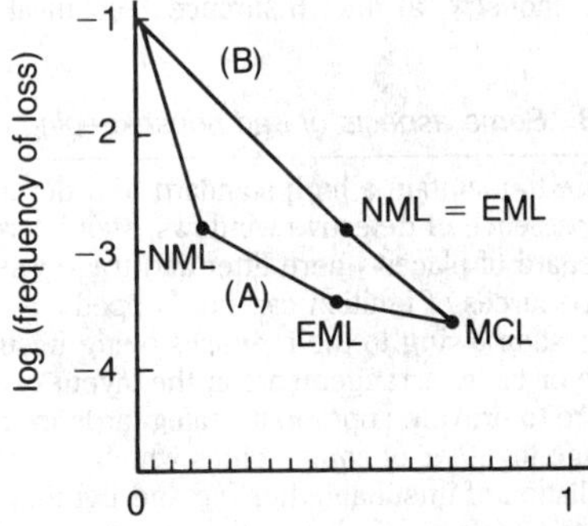

Figure 5.3 Frequency–loss profiles of two hypothetical process plants (Alexander, 1990a) (Courtesy of the Institution of Chemical Engineers)

Table 5.7 Headings of a checklist for insurance assessment of a chemical plant (after Norstrom, 1982b; Spiegelman, 1987)

Plant siting
Plant layout
Buildings and structures
Process materials
Chemical process
Materials movement
Operations
Equipment (design, testing and maintenance)
Loss prevention programme

methods include the use of checklists, of various types of hazard index, and of formal premium rating plans. The last two are closely related. Increasingly, such hazard indices and premium rating plans emphasize management aspects. A number of specific management audit methods have been developed.

5.6.4 Checklists

The insurance surveyor tends particularly to use checklists of features to be assessed. A series of such checklists is given by Spiegelman (1969). Later versions of such checklists are given by Norstrom (1982b) and Spiegelman (1987), the latter in the context of gas plants; the two checklists are similar and may be summarized as shown in Table 5.7. The original checklists are much more detailed. Further checklists are given by Ashe (1971) and Ackroyd (1974). A checklist of undesirable features given by Ashe is shown in Table 5.8.

5.6.5 Hazard indices

From such checklists there have evolved a number of hazard indices and premium rating plans. The best known hazard index is probably the Dow Index, given in the so-called *Dow Guide*, which has gone through a number of versions. The most recent is the *Dow Fire and Explosion Index Hazard Classification Guide* (AIChE, 1994). Another hazard index is the Mond Index, which is an adaptation of the Dow Index (D.J. Lewis, 1979). A third index is the instantaneous fractional annual loss (IFAL). This index was specifically developed for the insurance industry at the Insurance Technical Bureau (Munday *et al.*, 1980). These three indices are described further in Chapter 8.

5.6.6 Premium rating plans

There are a number of formal premium rating plans which are closely related to, and in some cases are based on, hazard indices. The premium rating scheme described here is that of Imperial Chemicals Insurance, which takes the lead in the insurance of property of Imperial Chemical Industries around the world. This scheme, which is outlined by Drewitt (1975), is more than usually oriented to the loss prevention approach. It applies to plants with an estimated NML of £1 million or more at 1975 values.

First an assessment is made of the NML. This is expressed as a percentage of the total insured value and as an absolute sum of money. The NML is an estimate of the probable loss from a major fire or explosion and is determined primarily by the segregation of value at risk either by distance or by physical barriers.

Then evaluations are carried out of the three basic factors shown in Table 5.9: NML, or L factor; the inherent hazard classification, or C factor; and the HSF factor, which covers hardware, software and fire fighting facilities.

The L factor is a number in the range 1–5 and is a function of the NML, as shown in the table. Intermediate values of the L factor, e.g. 1.5, 2.5, etc., are also used.

The C factor is assessed as a number in the range 1–12. This rating is determined by assigning values in the range 0–3 (or N/A (not applicable)) to the C factor parameters shown in the table, summing these and dividing the result by 2. Since there are 8 parameters, each with a maximum value of 3 points, the highest value of the C factor is 12. Fractional values are normally rounded upwards. In addition, an adjustment may be made to the C factor to give credit in this factor for above-average preventive and protective measures.

The HSF factor is assigned a value in the range 0–45. This number is determined by assessing the HSF parameters on a scale 0–3 in ascending order of merit, summing these and halving the result. There are 30 parameters altogether, each with a maximum value of 3 points, so that after division by 2, the highest value of the HSF factor is 45.

The quantities so calculated are then applied to the calculation of the basic premium rate and the discounted

Table 5.8 Some aspects of bad housekeeping (Ashe, 1971) (Courtesy of the Institution of Chemical Engineers)

1. Failure to maintain a high standard of order and cleanliness, both inside and outside the premises
2. The presence of defective windows, stall boards, pavement lights, trap doors, etc.
3. Disregard of places where litter and trade waste can accumulate and into which cigarette ends, lighted matches or other sources of ignition can be dropped
4. Congestion owing to the premises being unsuited to the purposes for which they are used, or to insufficient floor space or faulty arrangements in the layout of stocks, plant, machinery, fixtures and fittings
5. Failure to provide appropriate safeguards in the storage and use of hazardous materials
6. Storage together of commodities which may produce spontaneous heating
7. Installation of unsuitable heating and lighting appliances
8. Failure to safeguard all supplies of power, lighting and heating when premises are left unattended
9. Failure to provide fire protection equipment, the provision of an inadequate number of appliances or the wrong type, or faulty distribution and maintenance
10. Buildings, boundary walls, fences and gates being allowed to fall into a poor state of repair, which may render the premises more accessible to trespassers

Table 5.9 *Outline of a premium rating plan for damage insurance on large capital value plants (Drewitt, 1975)*

Rating factor
L factor Normal maximum loss
C factor Inherent hazard classification
HSF factor Hardware (H), software (S), Fire fighting facilities (F)

L factor

Normal maximum loss (%)	(£m.)	*L factor*
$\not>20$	$\not>2$	1
$33\frac{1}{3}$	4	2
50	6	3
$66\frac{2}{3}$	8	4
$>66\frac{2}{3}$	>8	5

If the % and £m. NMLs give different L values an average may be taken.

C factor

	C factor
Low risk plants	1–3
Medium risk plants	4–6
Relatively high risk plants	7–9
High risk plants	10–12

C factor parameters:

1. Nature of raw materials	0 Non-flammables 1 Flammable solids, heavy flammable liquids or vapours, dusts at ambient conditions. 2 Light flammable liquids, hydrogen, combustible gases, naphtha, petrol 3 LPG-type materials or other flashing liquids (any substance which is flashing above 5%) materials in use above their ignition point
2. Nature of products	As (1) above
3. Process type	Rather difficult to define but must range from the innocuous mixing of solids and liquids under ambient conditions to oxidation reactions, reactions carried out at high temperatures and pressures, reactions involving the use of acetylene above 20 psi, reactions involving the use of explosive materials, special hazards ancillary to the main process, e.g. catalyst preparation, catalyst changing or equipment cleaning
4. Heat content	In considering any plant, cognizance must be taken of the inherent heat content. This involves consideration of heats of combustion of the raw materials and products (hydrocarbons have heats of combustion about 10 000 tcal/t, while most other organic materials are below this value), heat of reaction or heat of polymerization. Associated with these considerations should also be the quantity effect. If we are dealing with large stocks and inventories, then the total 'heat content potential' in the event of an incident must affect the plant hazard rating. The highest of these should be used as the assessment factor. The extent to which inventories of flammable materials within process areas are in excess of actual practical requirements must be taken into account in assessing the hazard in relationship to the minimum unavoidable risk
5. Reaction temperature	0 Atmospheric temperature 1 Up to 100°C or above flash point of materials involved 2 In range 100–250°C or above boiling point of materials involved 3 Above 250°C or above autoignition point of materials involved
6. Reaction pressure	0 Atmospheric pressure 1 Vacuum or up to 5 at pressure 2 Up to 100 at pressure 3 Above 100 at pressure

C factor – continued

7. Corrosion and erosion hazards	On a well-designed plant due consideration will have been given to the corrosive and erosive nature of the materials being handled. However, it is accepted that corrosive fluids are difficult to contain and that leakages due to corrosion, particularly at moving parts, do occur. Similarly, the erosive nature of moving fluids and particles is well known
8. Domino effect	This factor is included to take account of the hazard that can occur if leakage of a hazardous material can lead to the escalation of the incident, e.g. a small leak which fires and damages by flame impingement at larger pipe or vessel with subsequent spillage of a large inventory or hazardous material

HSF factor

H factor parameters:

1. Location	Is the plant well sited in relation to dwelling houses, other industries, office blocks, warehouses, other plants, etc.? Is the site well drained? Are soil characteristics suitable for plant requirements?
2. Construction	Is the plant well designed and constructed with adequate safety and contingency factors? Are the materials of construction suitable for the duties involved? Are girders protected with fire resisting cement? Are vessels, particularly up to 20 ft from ground level, protected with fire resisting cement?
3. Separation	Is adequate spacing available between sections of the plant (minimum 20 m) and between items of equipment within the sections of the plant? Are control rooms well sited so that adequate protection is afforded in the event of an incident? Are storage tanks sited away from operating plant? Are storage tanks adequately bunded? Are plants suitably provided with dikes to retain any inadvertent spillage of hazardous material?
4. Ventilation	Is the plant well ventilated? Are machines operating inside buildings or out in the open? Are compressors housed in 'Dutch barn' type structures? Can an inadvertent leak of gas or liquid disperse easily? Are buildings equipped with adequate fire break walls and these fitted with adequate fire doors? Are buildings fitted, where necessary, with explosion relief panels to deal with explosions that may arise from leakage of combustible gases, liquids or dusts?
5. Instrumentation	Is the plant adequately instrumented? Are the instruments kept in A1 condition? Are there sufficient trips and alarms to deal with all the eventualities on the plant (see also S factor parameters, item (4), and F factor parameters, item (10))
6. Process valves and lines	Are there sufficient valves and lines to eliminate the possibility of contamination through common-pipe usage? Are there adequate facilities for washing out pipes and valves? Is it possible to divert inventories of hazardous materials to other vessels or to portable containers in the event of an incident? (See also F factor parameters, item (9).) Are there dumping and blowdown facilities available so that inventories of hazardous materials can be safely dispersed in the event of an incident?
7. Cooling	Is there sufficient reserve in the cooling system to deal with happenings outside the normal run of events? Are there adequate facilities to deal with a total cooling water failure?
8. Electrics	Are the electrics well designed and installed to conform with present-day thoughts of area classification? Are there adequate facilities to deal with a complete failure of electricity supply? Are the electric cables adequately protected against fire damage? Are alternative cable runs installed?

HSF factor/H factor – continued	
9. Emergency supplies	Are there sufficient supplies of nitrogen and inert gases to deal with all blanketing and purging problems? Is there an adequate supply of compressed air and instrument air? Is there an emergency power supply?
10. Flammable waste disposal	Are full facilities available for dealing with the safe disposal of all flammable wastes?
S factor parameters:	
1. Management attitude	What is the management attitude to safety? Are expenditure proposals sympathetically received and treated? Does management take an active interest in the organization and running of safety committees? Does management make a regular inspection of plants to ensure that safety standards are being maintained?
2. Housekeeping	Is the plant clean and truly free from debris, pieces of pipe, paper, plastic bags, wood, etc.?
3. Maintenance	Is the plant well maintained? Is it free from leakage of steam, water and hazardous fluids? Is maintenance conducted in a clean and orderly manner?
4. Plant operating instructions	Are plant operating instructions written in a clear and concise manner? Are they available to all operators? Are they kept up to date and revised at regular intervals? Are they obeyed? Are automatic safety devices tested on a regular schedule and with programmed monitoring on display? Have emergency instructions been written to deal with plant abnormalities?
5. Use of permit-to-work system	Is the permit-to-work system clearly defined and rigidly followed? (Consider competence of issuing authority)
6. Reporting of accidents	Are all accidents (and near misses) reported accurately and in detail? Are these reports circulated to those in authority? Are they effective in reducing accidents in the future? Are the lessons learned from accidents brought to the attention of new starters on the plants? (Accidents should be considered as abnormal occurrences and it should not be assumed that injury always results)
7. Labour	What is the quality of plant labour? What is the attitude of plant labour to work, discipline and safety? Is there a high turnover of labour?
8. Personnel training in plant operation	Are operators well trained in plant operation and tested before given a position of responsibility? Do they have frequent refresher courses?
9. Personnel training in safety	Is there adequate training in plant safety? Are regular Works Safety Meetings called?
10. Frequency of technical audits	Are technical audits of plant safety conducted at frequent intervals by an outside body?
F factor parameters:	
1. Works facilities	Is there a works fire brigade? Is there a full-time Chief Fire Officer? Are adequate fire fighting machines available on site? Are adequate stocks of foam-forming compounds retained on site?
2. Municipal fire brigade	Is there good liaison with the local municipal fire brigade? How long does it take the municipal brigade to reach the site? Do the works and municipal brigades carry out combined exercises on site?
3. Water supplies	Is there adequate water supply on site for fire fighting purposes? Is it a separate system or can supplies be tapped off for process use?

F factor – continued	
4. First aid extinguishers	Is the plant well provided with first aid extinguishers? Are plant personnel familiar with the use of the extinguishers?
5. Fixed protection	Is the plant provided with fixed fire protection equipment appropriate to the work, e.g. a sprinkler system, a steam fire curtain, a water curtain, fixed monitors?
6. Automatic fire alarms	Are adequate fire alarms installed throughout the plant? Are adequate gas leak detectors installed throughout the plant?
7. Communications	Are communications good between the plant and the shift manager and between the plant and the fire services so that incidents can be reported with the minimum of delay?
8. Training of first aid fire fighters	Are there adequate first aid fire fighters on site? Are they well trained in dealing with the type of incident that might be expected on their plant? Are they familiar with the protective devices installed? Do they have frequent refresher courses?
9. Remote isolation valves	Is the plant well provided with remote isolation valves so that large inventories of hazardous materials can be isolated in the event of inadvertent leakage, say due to a fractured pipe?
10. Testing of protective devices	Are the protective devices installed on the plant tested at regular intervals according to an agreed schedule?

premium rate as follows. The basic premium rate R is related to the L and C factors by a simple equation. The increase of R with increase in C is the sum of linear and exponential terms. The premium rates are quoted as ‰ or value per £1000.

The discounted premium rate DR is then determined from R and the HSF factor. An HSF factor of 30 is regarded as mediocre and a value below this attracts a penalty. For higher values each mark between 30 and 40 attracts a 3% discount and each mark between 40 and 45 a 4% discount, so that a maximum HSF discount of 50% is possible. DR is calculated by subtracting the HSF discount from the basic premium rate. The discounted annual premium is then determined by multiplying the insured value by DR. This is the basic rating plan. Two illustrative examples are given in Table 5.10. The plan is modified in practice by the application of further, commercial discounts.

The premium rating scheme has been described in some detail, because it gives a particularly clear illustration of the effect of loss prevention measures on insurance assessments.

Another premium rating method with a somewhat similar structure is described by Redmond (1984). The factors considered are the process factor P, the engineering factor E and the management factor M. The hazard index R is the product of these

$$R = P \times E \times M \qquad [5.6.1]$$

Another scheme which is relevant to premium rating for damage insurance is the Dow Index (AIChE, 1994). The scheme involves the determination of the Fire and Explosion Index (F&EI) and then of the Maximum Probable Property Damage (MPPD). The Dow F&EI, including determination of the MPPD, is considered in detail in Chapter 8.

5.6.7 Insurance Technical Bureau

In the early 1970s, the British insurance industry recognized the need for technological back-up and supported the formation of the Insurance Technical Bureau (ITB) (Chilver, 1973). The IFAL method, mentioned above and described in Chapter 8, was developed by the ITB.

5.6.8 Management audits

The increasing emphasis on management is reflected in the use of audit methods which concentrate particularly

Table 5.10 *Illustrative examples for premium rating plant for damage insurance for large capital value plants (Drewitt, 1975)*

	Example 1	*Example 2*
Total insured value	£15 m.	£30 m.
NML	20% and £3 m.	$66\frac{2}{3}$% and £20 m.
L factor	1.5	4.5
C factor	7	8
Basic premium rate R	3.38‰[a]	6.21‰
HSF factor	36	27
HSF discount	18%	−9% (loading)
Discounted premium rate	2.78‰	6.77‰
Discounted annual premium	£41 700 p.a.	£203 100 p.a.

[a] ‰ = rate per £1000 insured.

on management. Redmond (1984) gives a checklist by which the management may be assessed. A more detailed discussion of the assessment of management for insurance purposes is given by Instone (1990).

5.6.9 Estimation of EML

Some of the problems of assessing the EML are discussed by Redmond (1984). In general, this assessment is relatively easy for indoor plant with subdivisions each of which is provided with fire protection. It is more difficult to determine for outdoor plant, where there are no such physical barriers. Modes of escalation which must be taken into account include, in particular, running liquid fires and vapour cloud explosions.

5.6.10 Risk assessment approaches

There are variations of practice in the methods used to determine the premium rating. As indicated, some insurers have developed formal quantitative methods. Others consider that such quantitative methods offer little advantage over qualitative approaches. This is the view argued by Hallam (1990).

5.7 Business Interruption Insurance

The need for business interruption insurance did not really arise until the mid-1960s, when the first large, single-stream plants began to make their appearance. It then became apparent that consequential losses could be as costly as, or more costly than, property damage losses. The problem of business interruption insurance is a difficult one. The viability of such insurance depends on the charging of realistic premiums, on the evolution of effective assessment methods and on the progress made in the reduction of loss. Accounts of business interruption insurance are given by Neil (1971), Cloughton (1981) and di Gesso (1989).

5.7.1 Large, single-stream plants

As just indicated, the advent of the large, single-stream plant changed the relationship between property damage and business interruption insurance. The consequential losses which may occur with such plants apply not only to the plant itself, but also to the suppliers of the plant's inputs and the customers for its outputs. The nature and vulnerabilities of such plants have been discussed in Chapter 1.

5.7.2 Minor incidents

Serious business interruption can be caused by incidents in which the property damage is relatively minor. Typical incidents which may involve only minor damage but which can cause business interruption for a number of weeks include damage to control rooms, instrument cables, or individual items of equipment. Examples cited are a flash fire resulting from an escape of flashing hydrocarbon liquid and causing damage to instrument cables, and damage to a large gas compressor causing a large section of plant to be shut down.

5.7.3 Business interruption insurance

The initial approach to business interruption loss was to assess it as a proportion of damage loss. There is in fact only a weak connection between the two, the business interruption loss depending rather on the extent to which the loss interrupts a complex business chain. Moreover, the relative magnitude of the two losses has changed dramatically. Whereas business interruption loss was at first evaluated as a fraction of damage loss, it is now commonly estimated at several times damage loss.

Thus business interruption insurance merits separate treatment. Basically the insurance provides cover for a fixed period and against an insured peril, and relates to the margin between expected sales income and variable costs on loss of sales suffered due to the interruption at the insured's plant and, if extended, interruption at suppliers' and customers' plants. It also covers increased working costs incurred in attempts to minimize loss of margin.

Business interruption insurance is for a specified maximum indemnity period. Typical periods are 12 or 18 months. There is a tendency for insurers to require the insured to bear, as an excess, the losses sustained over some initial period such as 5, 10 or 20 days. Insurance may be extended to cover also the contingent liability for losses due to incidents at supplier and customer plants. The cover generally applies provided the losses are due to damage from a peril for which the insured's plant is covered.

Di Gesso (1989) quotes the case of a vapour cloud explosion at a natural gas liquids (NGL) plant which supplied feedstock to a petrochemical complex in 1987. A massive leak of propane led to a vapour cloud explosion which put the NGL plant out of commission for several months.

***Table 5.11** Some aspects of an interruption survey (after Neil, 1971) (Courtesy of the Institution of Chemical Engineers)*

1. Whether the process is single- or multi-stream, batch or continuous, innovatory or proven
2. The protection of essential services, routing of cables, etc.
3. Dependency on a key unit and the extent to which spares or standby facilities are available, and the time-scale involved in achieving replacement or repair
4. The size of buffer stocks, over and above the minimum necessary to cover shutdown for planned maintenance, and the possibility of alternative sources of supply to keep downstream plants in operation
5. If the plant is a replacement, whether existing plants and production will remain available as insurance against teething troubles and whether there will be an excess of capacity over demand in the short and long term
6. Whether the use of a computer is to control or maximize efficiency of the plant

5.7.4 Business interruption survey

It is normal to carry out a damage survey as a basis for damage insurance. It is appropriate to conduct an interruption survey for the purposes of business interruption insurance. This type of survey is, however, rather less well developed.

An interruption survey may extend the assessment of a plant to include features additional to those covered in the damage survey. The factors listed by Neil (1971) are given in Table 5.11. The main object of the interruption survey is normally the assessment of PML on the business interruption. Survey for business interruption should not be a one-off exercise, but should be repeated at intervals to ensure that the information is up to date.

Further aspects of business interruption surveys have been described by Cloughton (1981). Essentially the problem is that of recovery from the event causing the interruption. People are one of the most significant resources in enabling the recovery to take place. A resourceful purchasing manager is particularly important. Another vital factor is preservation of information on all aspects of the business, from design drawings to customer lists.

In assessing the vulnerability of a single-stream operation, Cloughton somewhat discounts in the UK context problems of availability of services and of fuels, but emphasizes weak points in single-stream operation. In particular, he highlights vulnerability to loss of particular pieces of equipment. Here he draws a distinction between a sole supplier and a unique source. A company may choose to buy from a single supplier, but there may nevertheless be other potential suppliers in the marketplace. In some cases, however, there is a unique source which for some reason may be unable to supply promptly.

Vulnerability to interruption depends not only on features of the production process but also potential loss of stocks in the same incident. Another relevant aspect is the expected profile of the recovery over time. This may be a short total interruption followed by a gradual recovery. Or it may be a partial interruption with production and sales partially maintained but at increased cost.

If a particular weakness is revealed, such as high exposure to loss of a particular piece of equipment, the solution is not necessarily more insurance. It may make more sense to remedy the exposure.

5.7.5 Estimation of EML

It will be apparent that the assessment of the EML for business interruption is not straightforward. It involves first determining the downtime of the plant, or sections of it, and then converting this loss of production to loss of profit. This last step may be particularly problematic. The assessment of EML for business interruption is discussed by di Gesso (1989), who gives an illustrative example.

5.7.6 Property damage and business interruption

The world-wide data from Munich Re (1991) quoted above show that over the 5-year period 1985–89 for material damage the annual average number of incidents was 144 and the average annual loss US$754 million. The corresponding figures for business interruption are annual number of incidents 39 and annual loss US$572 million.

Table 5.12 *Examples of claims for property damage and business interruption (after di Gesso, 1989)*

Case[a]	*Property damage (US$m.)*	*Downtime (months)*	*Business interruption (US$m.)*
1	–	2	22
2	4	4	32
3	9	10	200

[a]See text

At the level of the individual plant, some examples of claims for property damage and business interruption have been given by di Gesso (1989). He quotes the following three cases. Case 1, in 1981, involved a tower which failed during a hydrostatic test with water at a temperature of 10°C rather than at 20°C as recommended. In Case 2, also in 1981, a compressor trip led to a furnace tube burst and damage to a common stack. In Case 3, in 1984, a fire in a cable trench damaged a control room. The losses sustained are given in Table 5.12.

5.7.7 Business interruption insurance capacity

At the start of the 1960s it was possible to effect business interruption insurance. It rapidly became apparent, however, that the premiums asked for this class of business had been set at an unrealistically low level. There were some very large claims and some insurers sustained heavy losses. The nature of the problem can be seen from the fact that Neil (1971) estimated the total business interruption insurance premium income at £35 million. Such a sum could easily be swallowed up by one or two disasters. And in fact, as already mentioned, his estimate of the combined damage and business interruption insurance obtainable in practice was £10 million per loss.

As already described, since then the market has adapted and has been able, for example, to cover the large business interruption risks of offshore oil platforms. For Piper Alpha, one estimate of the business interruption loss, given above, is US$275 million.

Nevertheless, business interruption insurance poses a problem for the insurer. The report by Munich Re (1991) just quoted details the problem.

5.8 Other Insurance Aspects

5.8.1 Insurance credit

The extent to which insurers should encourage and give credit for good practice in general and for loss prevention measures is a long-standing question. Insurers have always been very much involved in measures against fire. The first organized fire brigades in the UK were run by the insurance companies. In the nineteenth century, the Fire Offices Committee was formed and set standards for fire construction, alarms, sprinklers, etc., which have been widely applied. In 1946, the Joint Fire Research Organization (JFRO) was set up jointly by the Fire Offices Committee and the government and this was followed in 1947 by the formation of the Fire Protection Association.

Insurers usually emphasize that they prefer to keep both risks and premiums down. Buyers of insurance are equally keen to hold premiums down and seek credit for loss prevention measures. This is well illustrated by the growth of the Factory Mutual Insurance Corporation, which is one of the largest fire insurance groups in the USA and which originated from a group of manufacturers who were dissatisfied with the credit which existing insurers were prepared to allow for such measures and who therefore formed their own insurance co-operative. The loss prevention measures which constitute the Factory Mutual System are described in the *Handbook of Industrial Loss Prevention* (Factory Mutual Engineering Corporation (FMEC), 1967). There is one particular protective measure on which insurers have placed considerable emphasis and for which they have given generous credit in premium ratings. This is the use of sprinklers. The effectiveness of sprinklers is well proven across a wide range of fire risks, including hotels, schools, offices, factories and warehouses. The premium recognition given to this particular protective measure appears, however, somewhat exaggerated relative to other measures as far as concerns the process industries. These industries have developed a whole range of loss prevention measures and an expectation that they should receive credit for them appears reasonable.

Clearly a precondition for a system of premium rating which allows such credits is the possession by the insurer of a method of risk assessment which it is confident gives a reliable discrimination between risks. In fact, for a period in the 1970s, the trend was rather towards a reduction of differentials. The range of risks in chemical plants is wide. The insurance industry was reluctant, however, to reduce the minimum premium for low risk plants. In consequence, the chemical industry, whose loss experience up to the 1960s had been relatively good, put pressure on insurers to reduce premium rates on the more hazardous plants. The result was an excessive narrowing of the rating scale between the lowest and highest risks. The extent to which the chemical industry does now receive credit for loss prevention measures in the form of reductions in premium rating appears variable. The success of ICI in obtaining reduced insurance premiums for damage cover is illustrated by Figure 5.4 (Hawksley, 1984).

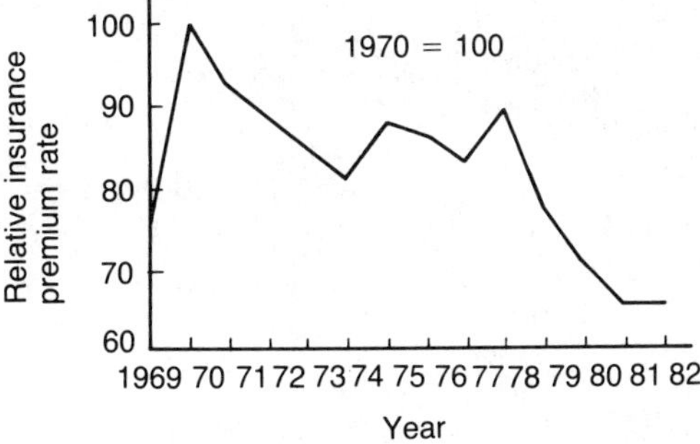

Figure 5.4 *Relative insurance premium rate (1970 = 100) for material damage cover for ICI UK (Hawksley, 1984)*

5.8.2 Insurance in design

The degree to which reduction of insurance costs is a design objective is discussed by Alexander (1990a), who suggests that it is not a major aim. The factors governing the safety and loss prevention features of the design, he argues, are, in order: (1) the attitude of the planning and licensing authorities; (2) the policy of the company; (3) the business interruption assessment; (4) the acceptability as an insurance risk; and (5) the cost of insurance.

He considers the example of the layout of a large petrochemical plant. One design option might be a more generous layout. For this an increase in capital cost of 10% would be a reasonable estimate. Assuming the need for a real return of 15% on this additional capital, the annual cost would be 1.5% of the total replacement cost. Yet the premium rate for the insurance might well be no more than 0.5% of this latter cost. This example suggests that it is not economic to seek to reduce the insurance costs by additional capital expenditure.

5.8.3 Insurers' advice

The experience of the insurer is available to the operating company. There are several forms which advice from the insurer may take. Spiegelman (1987) emphasizes that the primary and only real function of the insurance surveyor is to make a report to the underwriter. Although it is often assumed that he is also there to offer a free consulting service, this is a misconception. However, as described by Redmond (1984), an insurance survey report generally contains recommendations for improvements. Hallam (1990) enumerates some of the recommendations typically made in such reports. In addition an insurer may also offer a separate consulting service. Thus a large proportion of the major insurers offer inspections in respect of OSHA compliance.

It is constantly emphasized by insurance surveyors that the surveyor should be called in at the design stage, when modifications can be done relatively easily.

5.8.4 Loss adjusters

When an incident involving loss has occurred, it is the job of the loss adjuster to assess the financial value of that loss. The loss adjuster is usually employed by a firm specializing in this business and is independent of the insurer and insured. The report of the loss adjuster is primarily a financial rather than a technical document.

5.8.5 Loss data and analysis

The insurance industry is uniquely placed to provide information on loss and is in fact a prime source of loss data. This is illustrated by the large proportion of the loss data given in Chapter 2 which derives from insurance sources. Data sources include the annual report of bodies such as the Association of British

Insurers and the Loss Prevention Council in the UK, the corresponding bodies in other countries, the periodic reports *Large Property Damage Losses in the Hydrocarbon-Chemical Industries: A Thirty Year Review* (Marsh and McClennan, 1992), earlier versions being those by Garrison (1988b) and Mahoney (1990), as well as occasional publications such as *Losses in the Oil, Petrochemical and Chemical Industries* (Munich Re, 1991).

Insurers are also well placed to assess the significant sources of loss in the process industries. This has already been discussed in general terms in Chapter 2, but brief reference may be made here to the features particularly mentioned by insurers. Explosions are identified by W.H. Doyle (1969) as the main type of loss in the chemical industry. Many of his data have already been given in Chapter 2. Weldon (1971) singles out as the predominant initiating factors deviation from standard operating procedures and failure to perform proper maintenance. Almost invariably, insurers emphasize that most losses are caused by common hazards rather than by high technology.

Notation

C	loss cost for a particular state of nature and level of protection (US$k/year)
d	decision
D	loss cost for a particular level of production for all states of nature (US$k/year)
D*	loss cost (present value) (without survey) (US$k)
E*	loss cost (with survey) (US$k)
i	counter for decisions
I	outcome, or indicator
j	counter for of states of nature
k	counter for outcomes
M*	prevention cost (US$k)
P	probability
S	state of nature
T*	total cost (US$k)

Subscript

min	minimum

6 Management and Management Systems

The preceding chapters have indicated the nature of and the background to the problem of safety and loss prevention (SLP). The starting point for its solution is the management and the management system.

The importance of the management system has been stressed in a number of reports on safety in the chemical industry, including *Safety and Management* by the Association of British Chemical Manufacturers (ABCMs) (1964/3), in *Safe and Sound* and *Safety Audits* by the British Chemical Industry Safety Council (BCISC) (1969/9, 1973/12). It was the main lesson drawn from the Flixborough disaster and is the principal theme of the three reports of the Advisory Committee on Major Hazards (ACMH) (Harvey, 1976, 1976b, 1984). The Cullen Report on the Piper Alpha disaster (Cullen, 1990) has a similar emphasis. Accounts of process safety management include *Safety Management* (D. Petersen, 1988b) and *Techniques of Safety Management* (D. Petersen, 1989).

The importance of management has long been recognized by the Health and Safety Executive (HSE), which has produced a series of publications on safety management and on its regulation, including *Success and Failure in Accident Prevention* (HSE, 1976d), *Managing Safety* (HSE, 1981d), *Monitoring Safety* (HSE, 1985c), *Effective Policies for Health and Safety* (HSE, 1986a), *Successful Health and Safety Management* (HSE, 1991b) and *The Costs of Accidents at Work* (HSE, 1993a).

Other relevant publications include *The Management of Health and Safety* (Industrial Society, 1988) and *Developing a Safety Culture* by the Confederation of British Industry (CBI, 1990).

Management is also well represented in the publications of the Center for Chemical Process Safety (CCPS), which are described in Section 6.28.

The management system is in part specific to the particular process and should be matched to it. This is a point emphasized in the *Code of Practice for Particular Chemicals with Major Hazards: Chlorine* issued by the BCISC (1975/1).

Many treatments of safety and management touch on virtually the whole range of subjects dealt with in this book. It is convenient here, however, to limit consideration of the management system to the specific management topics outlined below. The present chapter is structured in the following way. Sections 6.1–6.3 start by laying emphasis on:

(1) management attitude;
(2) management leadership;
(3) management organization.

Sections 6.4–6.11 describe elements of the management system for major hazard installations and in large part reflect the work of the ACMH. These are:

(4) competent people;
(5) systems and procedures;
(6) project safety reviews;
(7) management of change;
(8) standards and codes of practice;
(9) pressure systems;
(10) documentation;
(11) audit systems;
(12) independent checks;
(13) major hazards.

The next part of the chapter describes the management system in more detail and reflects the impact of developments in management generally on the management of safety and loss prevention. It is introduced by Sections 6.14 and 6.15 on:

(14) quality management;
(15) safety management.

Sections 6.16–6.20 draw on the accounts of management systems in works on safety management such as *Safety Management* (D. Petersen, 1988b) and *Successful Health and Safety Management* (HSE, 1991b). The elements of the management system described are:

(16) policy;
(17) organization;
(18) planning;
(19) measurement;
(20) control;
(21) audit.

Sections 6.22–6.25 deal with four additional topics:

(22) process knowledge;
(23) safety strategies;
(24) human factors;
(25) contractors.

Sections 6.26–6.28 treat:

(26) safety management systems;
(27) process safety management;
(28) CCPS Management Guidelines.

Section 6.29 covers:

(29) regulatory control;

and Section 6.30 covers the HSE management audit system:

(30) STATAS.

Treatments of the management system typically cover a number of topics which in this book are dealt with elsewhere. These include hazard identification and assessment (Chapters 8 and 9), process design (Chapter 11), human factors and training (Chapter 14) and the safety system (Chapter 28).

Selected references on management and organization of SLP are given in Table 6.1.

6.1 Management attitude

Safety and loss prevention in an organization stand or fall by the attitude of senior management. This fact is simply stated, but it is difficult to overemphasize and it has far-reaching implications.

6.1.1 Safety culture

It is the duty of senior management to ensure that this attitude to safety and loss prevention is realized throughout the company by the creation of a safety culture in which the company's way of doing things is also the safe way of doing things. It is not easy to create a proper attitude to safety. The most effective approach

Table 6.1 *Selected references on management and organization of safety and loss prevention*

ABCM (n.d./1, 1964/3); NRC (Appendix 28 Management and Management Systems); Guelich (1956); Argyris (1957); Black (1958); Christian (1960); Simmonds and Grimaldi (1963); Tver (1964); H.H. Fawcett and Wood (1965, 1982); Pope and Cresswell (1965); BIM (1966, 1990); Herzberg (1966); Likert (1967); CBI (1968, 1974, 1990); Leeah (1968); Porter and Lawler (1968); BCISC (1969/9, 10, 1973/12, 13); Handley (1969, 1977); Gilmore (1970); N.T. Freeman and Pickbourne (1971); Hoffman (1971); Lloyd and Roberts (1971); R.L. Miller and Howard (1971); D. Petersen (1971, 1980, 1984, 1988a,b, 1989); Sachere (1971); D. Williams (1971); D.J. Smith (1972, 1981, 1985, 1991); Gardner (1973); W.G. Johnson (1973a, 1977, 1980); Kramers and Meijnen (1974); D. Farmer (1975); Harvey (1976, 1979b, 1984); Hope (1976); Arscott and Armstrong (1977); Briscoe and Nertney (1977); Luck (1977); Hawthorne (1978); Nertney (1978); Challis (1979); Kletz (1979i, 1982d, 1984g, 1988c,e); Anon. (1980e); Findlay and Kuhlman (1980); Chissick and Derricott (1981); HSE (1981d, 1981 OP 3, 1991 HS(G) 65), 1985c, 1986a); Laporte (1982); London (1982); Ormsby (1982, 1990); BG (1984 Communication 1233, 1986 Communication 1304); Anthony (1985); Livingston-Booth *et al.* (1985); Tweeddale (1985, 1990); C. Mill (1986); Kharbanda and Stallworthy (1986b); Winkler (1986); Boyen *et al.* (1987); Kemp (1987); Sulzer-Azaroff (1987); ACGIH (1988/17, 1990/45, 47, 1992/81); Brian (1988); Coke (1988); Hawksley (1988); W.B. Howard (1988); ILO (1988, 1991); Industrial Society (1988); Knowlton (1988, 1990); Luck and Howe (1988); Rasmussen (1988); Shell Int. (1988); Tombs (1988); Watson and Oakes (1988); Weise (1988); Whalley and Lihou (1988); Bond (1989a); CONCAWE (1989 4/89); Institute of Materials (1989 PR 1001, 1992 PR 1002); Ognedal (1989); Oh and Albers (1989); K.H. Roberts (1989); K.H. Roberts and Gargano (1989); Schreiber and Sweeny (1989); Verhagen (1989); Belanger (1990); Burk and Smith (1990); P.C. Campbell (1990); Emerson (1990); Frohlich (1990); van Hemel *et al.* (1990); Krause *et al.* (1990); Lihou (1990a, 1992 LPB 103); McKee (1990); Perry (1990); Rowe (1990); Toth (1990); Turney (1990a,b); Wade (1990); Anon. (1991a); Anon. (1991 LPB 98); Hurst (1991); IAEA (1991); B.A. Turner (1991); Argent *et al.* (1992); Bird (1992 LPB 103); Bleeze (1992); Burge and Scott (1992); Engineering Council (1992, 1993); HSE, APAU (1992); J. King (1992); McKeever and Lawrenson (1992); NSC (1992/11); A.J. Smith *et al.* (1992); Ball and Proctor (1993 LPB 111); Hoskins and Worm (1993); IMechE (1993/160); Nawar and Samsudin (1993); Pahlow and Dendy (1993); Ramsey *et al.* (1993); Sanders (1993b); Walker *et al.* (1993); Westerberg (1993)

CCPS Guidelines
Schreiber (1991); Sweeney (1992)

IChemE Guide
Cloke (1988)

Total quality management
Ishikawa Kaoru (1976, 1985); Crosby (1979, 1984, 1986); Juran (1979); Taguchi (1979, 1981); Juran and Gryna (1980); Deming (1982, 1986); Laporte (1982); Feigenbaum (1983); F. Price (1985); DTI (1986); Singo Shigeo (1986); Shell Int. (1988); Baguley (1989, 1990); Bond (1989b, 1990a); Hodge and Whiston (1989); Oakland (1989, 1993); K.H. Roberts (1989); K.H. Roberts and Gargano (1989); Whiston and Eddershaw (1989); Cairns and Garrett (1990); Institute of Materials (1990 PR 1004); Knox (1990); Choppin (1991, 1993); Klein (1991a,b); D.J. Smith and Edge (1991); Turney (1991); Zairi (1991); C.A. Moore (1992); Schreurs *et al.* (1992); Abbott (1993); ASME (1993/205); Woodruff (1993)
BS (Appendix 27 *Quality Assurance, Total Quality Management*)

Quality assurance, ISO 9000, BS 5750
ASME (Appendix 27); ISO (Appendix 27, 1987 ISO 9000); IMechE (1982/65, 66, 1990/122); ASCE (1985/22); IChemE (1986/124, 1988/133); Allen and Nixon (1987 SRD R455); ICI (1987); Atkinson (1988a); Weismantel (1990); CIA (1991 CE6); Holmes (1991); Fouhy *et al.* (1992); Graham (1992); Hockman and Erdman (1993); P.L. Johnson (1993); Love (1993); Owen and Maidment (1993); Weightman (1993); Mancine (1994)
BS (Appendix 27 Quality Assurance), BS Handbook 22: 1990, BS PD series

Safety management systems
Cullen (1990); Stricoff (1990); D.J. Griffiths (1993); EPSC (1994)

Process safety management
API (1990 RP 750); Early (1991); Hawks and Merian (1991); Mallett (1992); Barrish (1993); Bisio (1993); Drake and Thurston (1993); Migihon *et al.* (1993); Lohry (1994)

Management of change
Sankaran (1993); Anon. (1994 LPB 119, p. 17)

Risk management
R.N. Hugh (1976); J.Q. Wilson (1980); Wynne (1980, 1982, 1989, 1992); Rica *et al.* (1984); Wildavsky (1985); Ostrom (1986); Pitblado (1986); Caputo (1987); Grose (1987); Kleindorfer and Kunreuther (1987); Jardine Insurance Brokers (1988); Bannister (1989); Hubert and Pages (1989); Weick (1989); O'Riordan (1990); Page (1990); Rowe (1990); Roy (1990); Shortreed (1990); Turney (1990a, b); S.J. Cox and Tait (1991); Greenberg and Cramer (1991); C. Wells (1991); Bergman *et al.* (1992); Frohlich and Rosen (1992b); D.J. Parker and Handmer (1992); Clagett *et al.* (1993); P.L. Johnson (1993); Purdy and Waselewski (1994)

CCPA Code of Manufacturing Practice
Creedy (1990)

Contractors
Sachere (1971); Cullen (1990); Bisio (1991b); Kletz (1991m); Webb (1993); Whitaker (1993)

Research models
J. Powell and Canter (1984); Dunford (1990 LPB 93); HSE (1992 CRR 33, 34, 38); Hurst *et al.* (1992); J.C. Williams and Hurst (1992)

STATAS
Ratcliffe (1993 LPB 112, LPB 113, LPB 114)

appears to be to treat safety as a matter of professionalism. This fits particularly well with the development of loss prevention with its more technical and quantitative approach. The discipline required to create and maintain a safety culture brings benefits not only to safety but also to the other aspects of the operation of the company. At this point the culture becomes self-reinforcing.

6.1.2 Will to safety

There is a danger that the emphasis in loss prevention on technological aspects of hazards and their control may obscure a very simple truth, that in many cases an accident occurs because the will to avoid it was lacking. This theme has been developed by W.B. Howard (1983, 1984), who has given a number of telling examples of accidents where the essential cause was a conscious decision to take a particular course of action. Among the accidents cited are instances where an operator: was put in charge of equipment for which he had no training; omitted to check trip systems thus allowing them to degrade; chose to bypass trips and alarms to obtain increased production; and failed to provide a necessary trip on a temporary filling system. He also instances the installation of the temporary bypass pipe at Flixborough.

Direction of attention to the role of conscious decisions in accidents is a necessary corrective. The specific techniques associated with loss prevention are necessary, but they are not sufficiently alone. It is crucial that there also exists the will to safe design and operation.

6.2 Management Leadership

The creation and maintenance of a safety culture requires strong leadership by senior management. This means that the attitude of senior management must be demonstrated in practical ways so that all concerned are convinced of its commitment. An account of some of the ways in which this is done was given by McKee (1990) of Conoco, part of Dupont, in evidence to the Piper Alpha Inquiry. They include giving safety a high profile, giving managers safety objectives, backing managers who give priority to safety in their decisions, operating an active audit system and responding to deficiencies and incidents.

Safety is given a high profile in a number of ways. This is achieved by the measures just mentioned and also by putting it as the first item on the agenda of the appropriate meetings and making sure that personnel are aware of management's actions both in initiation of, and in response to, safety matters. McKee states:

> I keep the safety priority in front of the organisation on a daily basis, mainly through my close interest in safety reporting and my continuous 'drumbeat' about the safety priority.

Managers are given specific safety objectives and are assessed on their performance in achieving them. These objectives cover both personal safety and major hazards. But whereas the former may be monitored in terms of personal accident statistics, the latter are realized so rarely that event statistics are not the appropriate measure and adherence to systems and procedures is a better one. In McKee's words:[6] 'He is accountable not only for what is achieved but also for how it is attempted.'

One of the situations most revealing of senior management's real attitude to safety is its response to a manager's decision in a specific case not to act in a way which he considers unsafe even though there is an immediate financial penalty. As McKee puts it:

> I have to do my part in helping people instinctively feel comfortable about the boundaries within which they make everyday decisions. In Conoco we all try to let safety be the number one influence on every decision or action.

With regard to audits, McKee states:

> I receive safety audit reports and react to them if required by raising the issues involved directly with the departmental director involved. I conduct my own informal audits on my frequent visits to our installations. I participate in the London safety action committee and I specifically discuss the audit system in general with managers and employees alike.

Another principle is the prompt correction of deficiencies. A deficiency should not be allowed to persist, it should be corrected. But in addition the correction should be prompt and any delay explicitly justified.

Closely related to this is the investigation of, and response to, incidents. An incident should be investigated and the lessons drawn. But, in addition, the involvement of senior management should be explicitly demonstrated. McKee states that he receives a daily report on safety from his safety manager, who is the only manager to report daily to him. If an incident occurs, the manager informs him immediately: 'He interrupts whatever I am doing to do so, and that would apply whether or not I happened to be with the Minister for Energy or the Dupont chairman at the time.' In sum, in McKee's words: 'The fastest way to fail in our company is to do something unsafe, illegal or environmentally unsound.'

The attitude and leadership of senior management, then, are vital, but they are not in themselves sufficient. Appropriate organization, competent people and effective systems are equally necessary.

6.3 Management Organization

The discharge of senior management's duty to exercise due care for the safety of its employees, of other people on the site and of the public requires that it create a rather comprehensive and formal system and that it be active in creating, operating, maintaining, auditing and adapting the system. It is necessary to define clearly the management structure. The distinction between the line, or executive, functions and the staff, or advisory, functions is a crucial one and is a well-established aspect of good management practice, but it is particularly important in safety and loss prevention. Responsibility needs to be assigned unambiguously. There should be a job description for each of the jobs shown in the management structure. Once a job is defined, it is possible to select a competent person to fill it.

These are essential steps, but they are not sufficient alone. There are other aspects to which it is necessary to pay attention if the organization is to be fully supportive of the efforts of individuals.

One important matter is arrangements to ensure that there is cover available for key jobs. The importance of this point was underlined by the Flixborough disaster. Such cover involves both short-term stand-by measures and long-term career planning.

A more subtle point concerns discrepancies between function and seniority. This has frequently arisen, for example, in relation to the safety officer, who in many cases has lacked the authority to carry out his functions with full effectiveness. It is necessary to consider the possibility of such problems and to take appropriate steps.

6.4 Competent People

The design and operation of hazardous processes requires competent people. Here it is necessary to try to define carefully what is required. The question is particularly important in respect of the technical executive management which has prime responsibility for the process. The statement of the management structure and the job description is a necessary first step, but does not in itself take the matter very far.

It is generally common ground that academic qualifications, practical training, recent relevant experience and personal qualities are all important. The educational requirement raises questions both of level and content. As far as the former is concerned, the greatest need is for breadth of technological understanding. The person concerned should have his own area of expertise, but should also be able to appreciate what he does not know and where he needs help from others. In the UK, a first degree or equivalent in engineering or science is commonly held to be some assurance of such a broad technological education, although it may not be the only route.

With regard to the most suitable content, this may depend on the process. Much emphasis was placed following the Flixborough disaster on maintaining the integrity of the plant, which favours chemical or mechanical engineering. In some cases, on the other hand, it may be as important to maintain the integrity of the process, which points to chemistry. Each case should be treated on its merits.

As far as training and experience are concerned, distinctions may be made between design and operation and between different processes. Experience in design does not necessarily fit someone for operation nor does experience in operation of a chlorine cellroom necessarily make him suitable to operate an ethylene cracker. It is necessary that the experience be not only relevant but recent. This is particularly important in processes with a large content of complex and rapidly changing technology. On the other hand, these requirements should not be interpreted so restrictively that career development is hindered.

Personal qualities are equally important, particularly in relation to operation of major hazard plants. Effectiveness in this job is in large part a question of such qualities as temperament. Particular emphasis is placed on the operation of plants, especially major hazard plants. Here recent relevant experience and personal qualities are particularly important.

6.5 Systems and Procedures

It is fundamental that responsibility for safety and loss prevention should be shared by all concerned in the project. It cannot be delegated to a separate safety function. This does not mean, however, that reliance should be placed simply on individual competence and conscientiousness. It is essential to support the competent people with appropriate systems of work. Experience indicates that effective systems require quite a high degree of formality.

The purpose of these systems of work is to ensure a personal and collective discipline, to exploit the experience gained by the organization, and to provide checks to minimize problems and errors. The framework of such systems is typically a set of standing orders or instructions which lay down requirements for the conduct of particular activities at the different stages of the project from research and development through to operation.

Some key systems are those which are concerned with (1) identification of hazards, (2) assessment of hazards, (3) operation of plant (normal, emergency), (4) control of access to plant, (5) control of plant maintenance (permits-to-work), (6) control of plant modification (referral procedures), (7) inspection of plant equipment, (8) emergency planning and (9) incident reporting. All these systems are considered in detail in other chapters.

It is important that the various activities be properly phased and matched to the project stages. Thus, for example, hazard identification should be undertaken at a stage early enough to identify any serious hazards which might threaten the viability of the whole project, but such an investigation can be on a relatively coarse scale. As design progresses, other more detailed hazard identification procedures are required.

There is, in principle, no limit to the number of systems which may be devised, but the aim is to have the optimum number which leaves no serious gaps and yet avoids confusion due to overlapping. Inevitably formal systems sometimes appear tedious or unnecessary. They need to be well thought out so as to minimize this reaction. But experience shows that a large proportion of incidents which occur are due to lack of proper procedures or to non-observance of them.

6.6 Project Safety Reviews

The management system should include a formal system of project safety reviews for the identification, assessment and evaluation of hazards. The system of reviews should be comprehensive in that it covers all aspects of the project and does so over the whole life cycle.

A typical set of project safety reviews is (1) pre-project review, (2) design proposal review, (3) detailed design review, (4) construction review, (5) pre-commissioning review and (6) post-commissioning review. Project safety reviews are considered in Chapter 8.

6.7 Management of Change

A large project is subject throughout its life cycle to numerous changes, or modifications. It is necessary to have systems which will manage these changes satisfactorily. In particular, it is necessary keep control of modifications, which may occur in design or operation.

These modifications may be to the process or the plant. Either way, modification control systems are required to detect intent to modify, to refer the modification to the appropriate function for checking, to record a modification authorized, to inform others of the modification and to follow up any implications such as a need for training.

Modifications may be proposed during the design of a new plant or for existing plant. The latter also embraces major plant extensions. Different modification control systems are required for the design and operation stages. Management of change and control of modifications is discussed in Chapters 11 and 21.

6.8 Standards and Codes of Practice

An important aspect of the procedures is the use of standards and codes of practice, both external and internal. These are referred to extensively throughout this book and it is difficult to overemphasize their importance. They represent a distillation of industry's experience and are not to be disregarded lightly.

Although the majority of standards and codes relate to design, there are also many which are concerned with operation.

6.9 Pressure Systems

Central to SLP is the problem of loss of containment. The management system for the design and operation of pressure systems is therefore of crucial importance. Major failure of a properly designed, fabricated, constructed, tested, inspected and operated pressure vessel is very rare. But failures do occur in pressure systems. They tend to be failures of other pressure system components such as pipework and fittings, pumps and heat exchangers or failures due to maloperation of the system. The management systems for the control and monitoring of a pressure system should be in two parts, covering the two broad areas of design and operation and administered by two separate authorities.

The design authority should be responsible for systems for control of design, fabrication, testing and inspection, and the operating authority for those for control of commissioning, operation, maintenance and modification. The design authority should define the parameters within which the system is to operate, should specify the design codes, should execute the actual design, should identify and assess hazards, should specify standards for fabrication, construction and testing of the systems and should prescribe the documentation required on all these aspects. It should also ensure the inspection of the plant during fabrication and construction. This inspection may be done by the design authority itself or by an external body such as an insurance company or an engineering inspection agency. The inspection should be done to the specified standards and should be recorded in the prescribed way.

The operating authority should prepare written operating instructions based on the method of operation envisaged in the design, the design parameters and the hazard studies, and covering both normal and abnormal conditions. It should also create a system for the control of modifications in order to ensure the integrity both of the plant and of the process. The system should distinguish between proposed modifications which might affect the integrity of the plant and routine maintenance, and should require that the former be referred back to the design authority for checking. Similarly, the system should identify proposed modifications of operating procedures which go beyond established practice and might affect the integrity of the process, and should require that these also be referred back for checking by the design authority.

The operating authority should also provide a code for the regular inspection of the pressure system. There should be documentation on each component giving its unique identification, location, engineering description, operating conditions, inspection interval and maintenance record. Inspection intervals for each class of equipment should be specified and rules stated for the alteration of inspection intervals. The code should ensure that the inspection authority is independent of the operating authority.

6.10 Documentation

A project in the chemical industry involves a large amount of documentation. A list of some principal items is given in Table 6.2.

Some of the material is general documentation on company systems and on standards and codes, but most is specific to the particular process and plant. As the list given indicates, a large project requires a considerable amount of documentation. The table shows essentially the basic elements of the documentation. Selected items may be put together to form manuals for various purposes. Typical manuals are the Design Manual, the Plant Manual and the Operating Manual.

6.11 Audit System

It is essential that there is a mechanism to monitor the system as a whole and to make sure that it is working correctly and is not falling into decay; in other words, an audit system. Audit differs from control. In control the outputs of a system are measured and corrections made as necessary. Audit examines the control system itself to check that it is still fit for purpose. It takes place over a much longer time-scale.

The need for audit may vary. It may be fairly clear without the need for an audit whether certain systems or procedures are working properly. But for others audit is required. Thus, for example, an audit of a permit-to-work system is generally necessary to prevent it from degenerating. The audit system may include a specific instruction that the plant manager should each week examine a sample of the permits issued. The essential point is that the approach adopted is not simply to exhort the plant manager to monitor the permit system but to build the instructions to do an audit into the system.

The audit function is considered further in Section 6.21. The essential point to make at this stage is the need for an audit system.

6.12 Independent Checks

The principle of independent checks is extremely important in ensuring reliability. It is already widely practised in relation to pressure vessel inspection. Other

***Table 6.2** Some principal items of documentation in a large chemical plant project*

Subject area	*Documentation*
Systems	Documents on company systems (e.g. see Section 6.5)
Standards and codes, legal requirements	National standards and codes applicable to the design. In-house standards and codes applicable to the design, including guidance on situations where national standards and codes do not apply. Legal requirements, statutory approvals
Organization	Organization chart of personnel Job descriptions and duties of personnel, including: (a) Process operators (b) Maintenance personnel (c) Supervisory staff
Process design	Description and history of process Design basis for plant including economics, output, yield, availability, storage, siting, pollution, loss prevention aspects Design data for process, including: (a) Process reactions and reaction kinetics, including possible reactions under abnormal conditions (b) Physical and chemical properties of materials including flammable, explosive and toxic features (c) Specifications for raw materials, products, by-products and effluents (d) Data relevant to selection of materials of construction (e) Data from pilot plant Process flowsheet giving main items of equipment, quantities of materials and services, inventories and operating conditions Process flow diagram giving all items of equipment, flow rates and other operating conditions Process design data sheets for items of equipment giving design basis, operating conditions, characterizing parameters, safety factors, equipment dimensions Inventories of hazardous materials in process and in storage Sources of information (people, literature) Some of this documentation is usually collected together in a Design Manual
Plant layout	Site and plant layout diagrams Site and plant pipework drawings Equipment and pipework identification schemes Documents on hazardous area classification, means of escape, site vehicle routing
Mechanical design	Mechanical job specifications, design data sheets and general arrangement, detail and layout drawings, parts lists for items of equipment Piping job specifications, piping and instrument diagrams, piping plan, elevation and isometric drawings, piping stress and piping support documents, piping parts lists Documentation on pressure relief, blowdown and flare systems
Services design	Job specifications, including arrangements for loss of services covering electricity, steam, cooling water, process water, instrument air, process air, nitrogen and any other service required
Electrical, civil, structural design	Job specifications, general arrangement, detail and layout drawings, and materials or parts lists as appropriate
Plant buildings	Design basis for control room and other buildings Documents on siting, layout, construction and explosion resistance, ventilation, fire limitation, explosion relief as appropriate
Control and instrumentation	Design basis for process control Instrument job specifications, piping and instrument diagrams, instrument design data sheets Documents on alarm, trip and interlock systems
Effluents, waste disposal, noise	Effluent, waste disposal and noise specifications and requirements of regulatory agencies Pollution, waste disposal and noise surveys of plant

***Table 6.2** – continued*

Fire protection	Design basis for fire protection system Documents on fire protection system, including fire water supply
Plant operation	Works and plant safety rules Chemical data sheets Operating instructions, including instructions on: (a) Normal operation, including all sequential operations (b) Normal start-up and shut-down, including variations dependent on length of shut-down (c) Start-up on new plant (d) Shut-down under abnormal or emergency conditions (e) Trip systems Sampling instructions, including location and identification of sample points, sampling frequency, sampling method, safety precautions Instructions on dealing with leaks and spillages Instructions for reporting incidents Operating documents, including process operator's log, plant manager's log, storage logs, standard calculations of efficiency, costs, etc. Some of this documentation and some documentation from other sections, e.g. process design, is usually collected together in a Plant Manual and/or Operating Manual
Training	Documents for process operator training (e.g. see Table 14.10) Documents for safety training (e.g. see Table 28.5)
Safety equipment	Documents on safety equipment covering for installed equipment: (a) Equipment location guide (b) Equipment inspection schedules (c) Equipment use manual (d) Equipment maintenance manual and for equipment in stock: (e) Equipment stock control (f) Instructions on equipment issue
Hazard identification and assessment	Records from pilot operation, of work on screening of chemicals or reactions, of hazard and operability studies, of hazard assessments, of safety audits Checklists of all kinds (e.g. see Table 8.4)
Security	Lists of permanent personnel and of temporary personnel, e.g. construction personnel Security passes
Plant maintenance	Documents on plant maintenance, including: (a) Code for maintenance and modifications with supporting documents (permits to work, clearance certificates, modifications forms, etc.) (b) Equipment identification and location guide (c) Equipment inspection and lubrication schedules (d) Equipment maintenance manuals and instructions (e) Instructions on lubricants, gaskets, valve packings, pump seals (f) Maintenace stock requirements (g) Maintenance stock control (h) Standard repair times (i) Equipment turnround schedules
Plant inspection	Code for inspection of pressure systems Records of equipment, including identification, location, engineering description, operating conditions, inspection intervals and maintenance history Records of non-destructive or condition monitoring tests
Emergency planning	Documents on emergency planning (e.g. see Chapter 24)
Environmental control	Environmental standards (e.g. OESs) Sampling schedules
Medical	Schedules for special medicals for personnel

areas where use is already made of independent checks are audits of management systems, hazard assessment of the plant as a whole and reliability assessment of protective instrumentation.

It is not essential in order to ensure independence to go to an outside organization. An independent function can exist within the organization – but it must be genuinely independent. In fact many organizations do rely on outside bodies, such as insurers for the inspection of pressure vessels or consultants for hazard assessments of plants.

Another independent check is furnished by the Factory Inspectorate. The most important function which the Inspectorate performs is to provide a further independent audit of the management and management system. It is emphasized, however, in accordance with the principle of self-regulation, that it is the responsibility of management to audit itself and that it should not rely on the Inspectorate to do this.

6.13 Major Hazards

The principles just outlined apply with special force to the management of major hazard installations. But there are also some particular considerations which need to be borne in mind in dealing with major hazards. These have been discussed by Challis (1979). What Challis particularly emphasizes is the combination on major hazard plants of high technology and people. If such plant is to be operated safely, it is necessary that there be strong leadership by management. The level of management which is particularly crucial is the first level of executive technical management.

The people who operate and maintain the plant cannot always be expected to have a full appreciation of all aspects of the complex technology involved. It is the responsibility of the manager on such a plant to address this problem. The workforce needs to have an understanding of the process and the plant, of the hazards involved and of the actions required. It should be well trained. It should be provided with clear instructions, both for normal operation and for emergencies. The manager on such an installation needs to be 'out and about' rather than in the office. There should be a strong executive atmosphere and discipline on the plant.

6.14 Quality Management

Increasingly, SLP is subject to the influence of quality assurance and total quality management. An account of these was given in Chapter 1.

6.14.1 Quality assurance

As described in Chapter 1, there is a strong trend for companies to adopt quality assurance (QA) throughout their operations and to seek accreditation. In the UK this means accreditation to BS 5750:1987. Since QA covers inputs to the company's activities, this requires that subcontractors and suppliers also address QA.

QA affects all aspects of a company's activities and it will not generally be introduced solely as a means of improving SLP. But given that it is introduced, it will have an impact on the latter. Some SLP activities to which QA has been applied include, in particular, hazop (hazard and operability) studies and quantitative risk assessment and the work of consultants.

6.14.2 Total quality management

Essentially similar considerations apply to total quality management (TQM). This also is unlikely to be introduced for the sole purpose of SLP, but if adopted it will have an impact. It is of interest that the Piper Alpha Inquiry was invited to recommend that TQM be adopted by offshore operators as a means of improving safety, but that the Cullen Report (Cullen, 1990) did not do so, considering that although this might be so, the implications of a requirement for TQM went far beyond safety.

The concept underlying loss prevention is that the problem of failures and their effects has an influence on company performance which goes far beyond safety of personnel. Essentially the same concept underlies the TQM approach. Thus TQM and SLP have much in common.

There is some tendency for those involved with total quality management to treat SLP as a subset of TQM. Insofar as TQM covers the whole range of a company's activities whereas SLP deals with one aspect, this view might be held to have some force. Against this, SLP has developed within its own field the basic principle of TQM and has evolved a highly successful approach of some depth and sophistication. This evolution may be expected to continue within the context of TQM.

The degree of sophistication of SLP in the process industries is well illustrated by developments in the nuclear power industry. A glance through the publications of the Nuclear Regulatory Commission (NRC) given in Appendix 28 shows clearly the efforts made to foresee and forestall failures. Accounts of the application of TQM in the chemical industry have been given by Whiston and Eddershaw (1989) and by Whiston (1991). A description of the application of TQM to hazop studies has been given by Turney (1991).

6.15 Safety Management

An account is now given of some principles of a process safety management system. Figure 6.1 illustrates the key elements and structure of a safety management system as given in *Successful Health and Safety Management* (HSE, 1991b).

The elements described here are similar, except that measurement and control are considered as separate elements, whilst review is treated as an aspect of control. The structure considered here is therefore:

(1) policy;
(2) organization;
(3) planning;
(4) measurement;
(5) control;
(6) audit.

Some of the insets and appendices given in the HSE publication in support of the above topics are listed in Table 6.3.

6.16 Policy

Policy on safety should aim to set appropriate goals and objectives, to organize and plan to achieve these objectives in a cost-effective way, and to ensure by

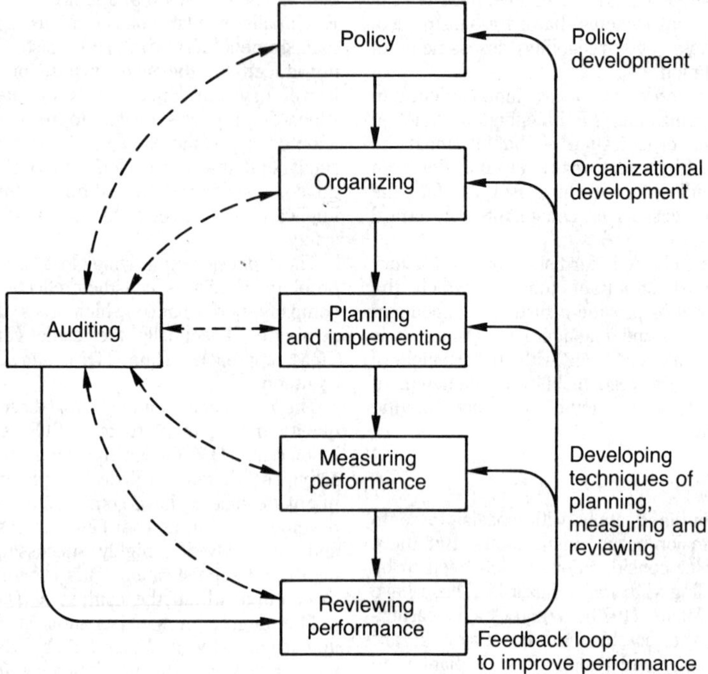

***Figure 6.1** Key elements of health and safety management (HSE, 1991b) (Courtesy of HM Stationery Office)*

***Table 6.3** Insets and appendices given in Successful Health and Safety Management (HSE, 1991b)*

Inset
1. Accident ratio studies
2. Human factors in industrial safety
3. The impact of effective health and safety policies on business thinking
4. Performance standards
5. Supervision
6. Examples of statements of health and safety philosophy
7. An outline for statements of health and safety policy
8. Training for health and safety
9. Role and functions of health and safety advisers
10. A framework for setting performance standards
11. Controlling health risks
12. Assessing the relative importance of health and safety risks
13. 'So far as is reasonably practicable', 'So far as is practicable' and 'Best practicable means'
14. Inspection
15. Key data to be covered in accident, ill health and incident reports
16. Effective health and safety audits

Appendix
1. Terminology
2. Organizing for health and safety
3. Minimum objectives for performance standards
4. Accident incidence and frequency rates

systems of measurement and control and of audit that the plan is implemented. The safety policy should be integrated with the other policies of the business, with the aim of achieving high standards of performance across the whole range of the activities undertaken. It should be the goal of safety policy to ensure that activities take place in a controlled manner and to eliminate deficiencies and failures which result in undesired events, some of which escalate to cause damage and/or injury. Safety policy should take full recognition of the importance of human resources and human factors, in other words of the people, the job they do and the organization within which they work, as shown in Figure 6.2.

6.17 Organization

Organization for safety is considered by the HSE under the following headings:

(1) control;
(2) co-operation;
(3) communication;
(4) competence.

6.17.1 Control

The safety goal is to ensure that activities take place in a controlled, and therefore safe, manner. This means putting in place systems of control. Such a control system is analogous to the typical feedback control system used in process control. The variable to be controlled is identified, the variable is measured, the

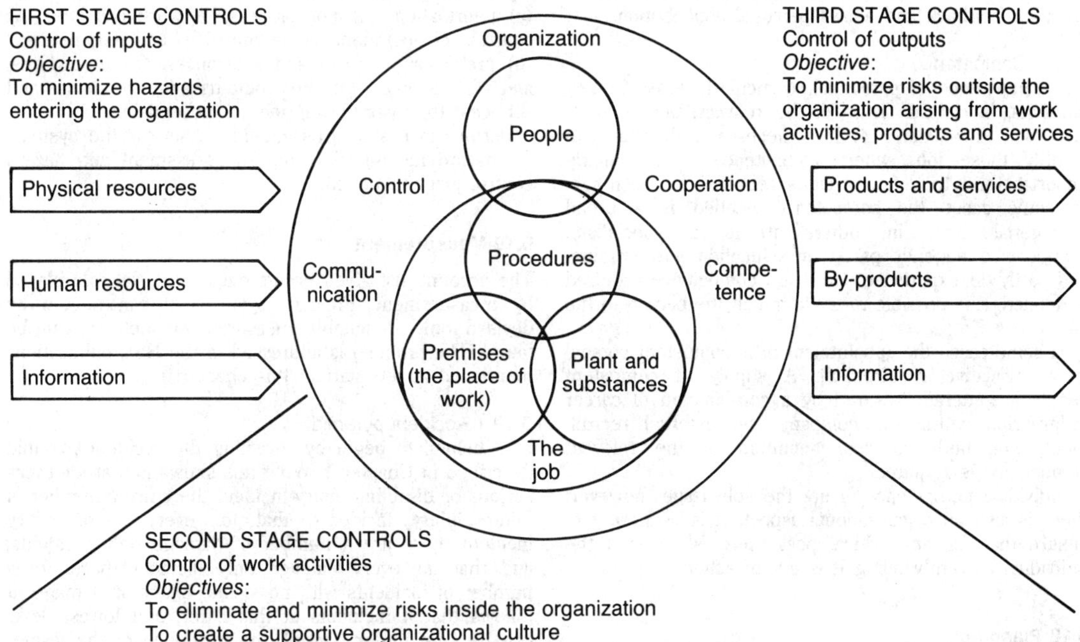

Figure 6.2 *Human factors in industrial safety (HSE, 1991b) (Courtesy of HM Stationery Office)*

measured value is compared with a desired value, and if necessary control action is taken to effect correction. Furthermore, there may be a system of nested control loops, with the higher loops controlling higher level functions and operating on longer time-scales.

Measurement is an essential function in any control system. It is necessary to decide first what specific variables need to be regulated in order to bring the system as a whole under a satisfactory degree of control and then to decide whether for a given variable it is practical to measure it directly or whether it is necessary to infer it from some other measurement. Given that the variable can be measured, it is necessary to decide what value is acceptable, by establishing a desired value or performance standard.

In the control systems considered here the control function is exercised by people. It is therefore essential to define responsibilities. There is normally a hierarchy of responsibilities down through the successive levels of line management. It is then necessary to establish accountability for the discharge of responsibility at each level. Accountability may be reinforced by making it part of the job description, using it as an indicator of individual performance, and identifying lapses.

Motivation is crucial to good performance. The most effective motivation is the existence of a safety culture which provides strong reinforcement. Training which enhances understanding and motivation also has an important contribution.

Supervision provides guidance, example and discipline, and furnishes the external element which complements the internal motivation.

6.17.2 Co-operation

A safety culture necessarily involves the co-operation of the whole workforce. Specific measures have to be taken to obtain this. A system of safety committees and safety representatives, such as is required by UK legislation, is one means whereby such involvement can take place. This system alone, however, does not generally yield the degree of co-operation obtained by a TQM culture in which the people at all levels are encouraged to identify and correct deficiencies and failures, by participation in safety circles (analogous to quality circles) and like methods.

6.17.3 Communication

Effective operation of the control systems described involves large flows of information. The forms of communication distinguished by the HSE are: (1) information inputs; (2) internal information flows; (3) visible behaviour; (4) written communications; (5) face-to-face discussion; and (6) information outputs. The operation of a safety management system actually requires a considerable input of information into the organization.

Good communication within the organization is critical. Effective communication uses three complementary modes: visible behaviour, written communications, and face-to-face discussion. Management needs to demonstrate leadership which is visible to all. Written communications are the principal means of transmitting information. The information needs of the recipients differ, however, and such communication is most effective if tailored to those needs. Face-to-face contact provides feedback and allows personal contribution.

6.17.4 Competence

The competence of personnel, particularly those in key positions, is crucial. Some of the relevant factors were discussed in Section 6.4. It is necessary, therefore, to identify those jobs where competence is particularly important and to define what is required for each job. In some cases the competence needed is a broad managerial one; in others it is a specialism. Competence depends partly on education and training, and partly on experience. Once the competence required is defined, it is possible to try to match the person to the job.

In some cases the appointment of a 'competent person' is a legislative requirement. A supply of competent people is generally assured by a combination of career progression within the company and external recruitment. For both sources, definition of the relevant competence is required.

Individual competence is not the sole issue, however; there is also an organizational aspect. It is necessary to ensure that for any critical post there is cover if the individual currently filling it is out of action.

6.18 Planning

Planning of the safety management system involves:

(1) setting goals and objectives;
(2) defining control systems;
(3) setting performance standards.

6.18.1 Goals and objectives

Broad safety goals need to be translated into specific objectives. The objectives which it is practical to set are essentially determined by the variables which can be measured.

6.18.2 Control systems

In management theory a distinction is commonly made between the following three stages of an activity to which controls may be applied: (1) input, (2) process and (3) output. Again, on the analogy of process control, the first two may be likened to open loop control in that the required output is obtained if the process gain remains constant and the input is maintained at an appropriate value. Alternatively, the output can be measured, and closed loop (or feedback) control applied. The application of controls to all three stages therefore involves a degree of redundancy. As redundancy enhances reliability, this is a desirable feature.

The systems required for hazards include systems for (1) hazard identification, (2) hazard assessment and (3) hazard control.

6.18.3 Performance standards

Once objectives have been defined, suitable measurements, or metrics, have been selected and control systems based on those devised, the performance standards which these control systems should meet can be set. The performance standards should cover inputs, process and outputs. Some of the performance standards required are identified by the HSE as follows: (1) For inputs: (a) physical resources; (b) human resources, and (c) information. (2) For process, or work activities: (a) control, co-operation, communication and competence; (b) premises; (c) plant and substances; (d) procedures; and (e) people. (3) For outputs: (a) products and services; (b) by-products; and (c) information.

Performance standards should be set for the systems for hazard identification, hazard assessment and hazard control just mentioned.

6.19 Measurement

The account just given brings out the crucial role played by measurement. Effective systems of control can be devised only if suitable measures, or metrics, can be found. The problem is addressed in the HSE publications mentioned at the start of this chapter.

6.19.1 Accident pyramid

It is helpful to begin by revisiting the accident pyramid described in Chapter 1. What this shows is that for every serious or disabling injury incident there are a number of minor injury incidents; that for every minor injury incident there are a number of damage only incidents; and that for every damage only incident there are a number of incidents with no visible injury or damage. It follows that if incidents at this latter, and lowest, level can be kept under control, the incidence of the former will be much reduced. This is true, although there can be some practical problems. Since such incidents have no visible effect, it is not easy either to define them or to persuade people that they should be reported. The accident pyramid is the centrepiece of the study in *The Costs of Accidents at Work* (HSE, 1993a), which gives five case studies, including one of an offshore oil platform.

6.19.2 Frequent and rare events

There are various definitions of accidents. One of the best known is the lost time accident (LTA). Statistics of such accidents can be compiled and monitored. If the accidents are occurring with a sufficiently high frequency, the accident statistics provide a suitable measure of performance. Some definitions of accidents and the monitoring of accident statistics is discussed in Chapters 26 and 27.

The goal of the safety management system, however, is to eliminate accidents. The greater the success achieved, the less useful become the statistics of the more serious types of accident. This is seen most clearly in respect of major hazards, since the realization of a major hazard is a rare event in any company. The fact that there has been no such accident is a poor indicator of the company's performance in the control of major hazards. In dealing with major hazards the aim is to reduce the already low probability of a realization. Thus the performance of the system in this regard has to be judged rather in terms of its effectiveness in operating and maintaining safety systems. This points to the need to measure inputs.

It is pertinent to ask, however, whether performance in respect of accidents for which meaningful statistics are available correlates with performance in respect of major hazards. The answer appears to be that it does. Although realizations of major hazards occur too infrequently for their statistics to be a practical measure of the performance of an individual manager, except perhaps at the highest level, over a period statistics of major accidents and near-misses may be compiled for a company and are then a measure of its performance. This point has been discussed by Brian (1988). A correlation does exist between the level of minor accidents and that of major accidents. The common factor is people, their attitude and discipline.

The upshot of this is that for control systems for frequent events there is a need to measure both inputs and outputs, whereas for rare events the only option is to measure inputs.

6.19.3 Proactive monitoring

Turning then to specific measures, some items which may be made the subject of proactive monitoring include (1) achievement of objectives, (2) adherence to systems and procedures, (3) conduct of auditing, (4) state of plant and equipment and (5) state of documentation.

A manager, for example, will be concerned with personal safety and may have responsibility for one or more major hazards. He will have objectives which have been set. There will be various systems and procedures to which he is required to adhere. He needs to audit the performance of his subordinates in achieving their objectives and adhering to systems and procedures.

The state of plant and equipment is monitored by inspection. This is a feature on which monitoring has traditionally concentrated. The state of the associated documentation can also be very revealing.

If there are noxious substances in the plant environment, reduction of the levels of these should be an objective. In this case there is something physical to measure and monitor.

6.19.4 Reactive measures

Some items which may be made the subject of reactive monitoring include (1) injuries, (2) damage, (3) other losses, (4) incidents, and (5) workplace deteriorations.

6.20 Control

The purpose of the measurement activities just described is to provide the basis for control actions to correct deviations and for review of the performance of the system.

6.20.1 Control and action

If a deviation is detected, control action is required to correct it. In some cases the corrective action needed is obvious and it can then be taken. In other cases it is necessary to carry out an investigation. The purpose of the investigation is to discover not a single 'cause' but the total situation which has given rise to the deviation and the corrective action, or actions, required to prevent it.

6.20.2 Review

The other aspect of the control system is the review of performance to ensure that the performance standards are being met. Review is continuous and is performed by line management. These features make it a control rather than an auditing activity.

6.21 Audit

The whole management system just described needs to be examined periodically to ensure that its goals and objectives remain appropriate and that the control systems to achieve these objectives are working. This is the function of the audit system. Continuing the analogy of process control, the audit system constitutes a set of outer, higher level loops around the control systems.

The purpose of an audit is to detect degradations and defects in a management system. This may be the management system of the company as a whole, a subsidiary company, a works or a plant. An audit is performed by competent people who are independent of the organization being audited. It may be carried out by an individual or a team. It is a task requiring its own particular skills.

A full audit of the management system will address the complete set of activities just described: policy, organization, planning, measurement and control. It is more usual to conduct audits with a more limited scope. Two approaches may be taken. One is to examine a specific function such as maintenance or fire protection; this is sometimes termed the vertical approach. The other is to examine a specific level of activity such as planning; this is the horizontal approach.

At plant level, there will typically be a post-commissioning audit followed by audits at a fixed interval. Where the installation is owned by a partnership but operated by a single company, as is typical of many offshore oil platforms, the other partners will usually require audits.

It is not easy to perform an effective audit. That this is so was brought out at the Piper Alpha Inquiry. One of the crucial systems of work on the platform was the permit-to-work system. The installation had been subject to a number of audits, but critical defects in the permit system had not been picked up. Yet, in less than 2h of evidence from the witness who was describing the system, these defects were laid bare.

Audits are discussed further in Chapters 8 and 28.

6.21.1 Proprietary systems

There exist proprietary audit systems which may be used within a company to promote safety and loss prevention. One of the best known is the International Safety Rating System (ISRS) (Bird and Germain, 1985). This system is described in Chapter 28.

A critique of packaged audit systems is given by D. Petersen (1989). Essentially the criticism is that in such systems the items on which the rating is based tend to be arbitrary and do not accord well with items identified in other studies as being crucial to safety performance.

6.22 Process Knowledge

Knowledge of its processes and plants is one of the prime assets of a company, but the management of this asset often appears to be relatively neglected.

6.22.1 Organizational memory

In safety, the problem of knowledge management manifests itself most obviously in the repetition of similar incidents, which suggests defects in the collective memory. It is a matter of experience that incidents continue to occur despite the fact that similar, even virtually identical, incidents have occurred before, were fully investigated and reported, and should in principle be known to the profession. The collective memory of the profession is deficient. Likewise, at company level incidents are repeated even though in the past there has been full knowledge of the problem in the company.

This phenomenon has been discussed and illustrated by Kletz (1980g). One particularly telling case which he quotes is where two similar accidents occurred within a couple of hundred metres, one in 1948 and one in 1968, though in this case in areas which were under separate management.

6.22.2 Awareness

The first requirement in process safety knowledge is simply awareness of hazards. Familiarity with a large number of case histories is one of the most effective ways of acquiring this awareness. Consideration should be given to any additional measures required to promote awareness in personnel entering into new responsibilities or undertaking new tasks.

6.22.3 Understanding

Knowledge is held within the organization in a variety of forms, including systems and procedures, in-house standards and codes, design manuals and other forms of codifications of practice. In sum, these generally represent an enormous store of knowledge. Such traditional forms are now supplemented by computer programs and data bases, and in some cases by expert systems.

In almost all cases these forms constitute abstractions from the original raw data. As necessarily happens when this is performed, some important features of these data disappear. The immediacy of the individual accident case history is lost. The reason why a code contains a particular recommendation is no longer clear. Information is lost in the process of abstraction. The promotion of understanding of the reasons for particular features in codes enhances their effectiveness.

6.22.4 Issues

In the course of a design there arise a large number of issues, some of which are of considerable importance. Issues may also arise at other stages of the life cycle of the plant. These issues need to be managed. There are two problems. One is to ensure that when the issue first arises it is satisfactorily handled. An important aspect of this is ensuring that the right people participate in its resolution. The other problem is to be able to retrieve information about the issue at a later date, so that the reasons underlying the decisions taken can be understood. It is not uncommon for a plant to have design features the purposes or limitations of which are not fully understood.

6.22.5 Operating procedures

Design of a plant proceeds on assumptions about the way in which it is to be operated, which may or may not be made fully explicit. Most incidents which occur are related in some way to plant operation or maintenance. Any mismatch between the design and operation of the plant is to be avoided.

6.22.6 Knowledge base

The safe design and operation of process plants requires an effective knowledge base within the company. This knowledge base has some similarities to a living organism. Much attention is devoted to data bases. The knowledge base is at least as important. Among other things, it determines the effectiveness of the utilization of the data bases.

The knowledge base takes various forms. These include in-house standards, codes and manuals, as already described. But above all it consists of the expertise of individuals.

The maintenance of the knowledge base depends on a certain critical level of activity within the company, and within its individual functions. If this is not sustained, the knowledge base can decay quite rapidly. This is a particular danger if the company goes through a bad patch.

Certain forms of knowledge can be documented, as described. Expertise derived from experience is more difficult to capture, although there is increasing interest in methods of doing so.

6.22.7 Knowledge utilization

There are factors which favour the maintenance, and enhancement, of the knowledge base. The most effective way to maintain it is to utilize it. In effect, it is a case of 'use it or lose it'.

A crucial factor, therefore, is a safety culture in which people are motivated and have occasion to draw on the knowledge base.

A safety culture as such is not enough, however. Specific steps need to be taken to make the necessary knowledge available by an inflow of information into the organization from publications and personal interchange, by converting information to transparent and utilizable forms, by facilitating access and by dissemination. Such arrangements will permit a company of moderate size to maintain a knowledge base which would otherwise be available only to a much larger organisation.

6.22.8 Incident investigation

A important aspect of this activity is learning. It is usual in accounts of safety management to assign a prominent place to incident investigation. The point has already been made that incidents tend to repeat themselves. In many cases investigation of the incident reveals little that is genuinely new. Nevertheless, the emphasis on incident investigation is not misplaced. It has a significant role in educating and motivating people and in keeping alive the knowledge base.

6.23 Safety Strategies

For a given hazard, it is necessary to devise a suitable safety strategy. The strategy should fit the particular hazard, but there are certain principles which can be stated. At the most general level, the basic elements of a hazard control strategy are:

(1) elimination;
(2) control
  (a) reduction of frequency,
  (b) reduction of consequence;
(3) mitigation.

If practical, the hazard should be completely eliminated. Otherwise, it should be controlled by reducing both the frequency and consequence of realization of the immediate event. Other measures to mitigate the effect of this event should be taken as practical. In process plants the event of particular concern tends to be loss of containment. The strategy will then be built in large part around measures to reduce the frequency of such emission and to reduce the quantity released.

There is frequently a choice of measures which may be taken to ensure safe operation. It is good practice, therefore, to define a 'basis of safety' and then to ensure that the necessary measures are taken. The basis of safety approach has found particular application in relation to the safety of chemical reactors. Another concept is that of the 'design basis accident'. In some cases it is not practical to have a design which will withstand all eventualities. Instead a level of accident is defined which the plant is designed to withstand. The safety strategy is particularly important for major hazards.

6.24 Human Factors

The increasing emphasis on management is accompanied by a growing recognition of the importance of human factors. The human element in incidents is illustrated in the numerous case histories described by Kletz in *Learning from Accidents in Industry* (Kletz, 1988h), *What Went Wrong?* (Kletz, 1988n), *Improving Chemical Engineering Practices* (Kletz, 1990d), *An Engineer's View of Human Error* (Kletz, 1991e), and numerous other publications. The essential message is that attribution of such incidents to human error is not an appropriate response and that it is more constructive to treat them as arising from the work situation, which is the responsibility of management. Human factors is the subject of a number of HSE publications, including *Deadly Maintenance* (HSE, 1985b), *Dangerous Maintenance* (HSE, 1987a) and *Human Factors in Industrial Safety* (HSE, 1989b). Human factors are treated in Chapter 14.

6.25 Contractors

A particular problem is posed by contractors. In some sections of the process industries contractors constitute a large proportion of the workforce. They carry out not only construction work, but specialist engineering and, in some cases, serve as process operators or maintenance personnel. These contractors have to conform to the systems and procedures of the company. They must therefore be trained in these and then subject to a discipline which ensures that they comply with them.

Contractors constitute a high proportion, typically some 70%, of the workforce in the UK sector of the North Sea. The problems associated with this surfaced at the inquiry into the Piper Alpha disaster, where the gas release which led to the explosion occurred from the site of a pressure safety valve. The valve had been removed and it was held that the flange assembly had not been made leak-tight. The man leading the two-man specialist team overhauling the valve was on his first tour as a supervisor and was not fully familiar with the permit-to-work system operated on that particular platform.

Good practice requires that a company apply to contractors the same philosophy as it applies to its own personnel. This should cover all aspects, including safety culture, training, discipline and audits. There are certain steps which a company can take to ensure that the work of contractors is satisfactory. One is to specify the standards to be met and to enforce adherence to them as part of a QA system. Another is to be prepared to pay a price which is realistic and not to drive it down below the point at which the contractor is able to deliver the quality required; in some cases this means not accepting the lowest bid.

6.26 Safety Management Systems

The system which delivers the approach to process SLP described above is in effect a safety management system. A requirement for a formal safety management system has been introduced in UK legislation following the Piper Alpha disaster. The inquiry into the disaster was urged to strengthen the management of safety by requiring offshore operators to adopt TQM, but such a recommendation would have had implications which go far beyond safety. The *Cullen Report* (Cullen, 1990) recommended instead that for offshore installations the safety case should include a formal safety management system (SMS) and this requirement is embodied in the Offshore Installations (Safety Case) Regulations 1992. The corresponding regulations for onshore major hazard plants, the Control of Industrial Major Accident Hazards (CIMAH) Regulations 1984, contained no such requirement. The Cullen Report states that *inter alia* the SMS should cover the elements listed in Table 6.4.

An account of industrial practice in safety management is given in *Safety Management Systems* by the European Process Safety Centre (EPSC) (1994).

6.27 Process Safety Management

6.27.1 OSHA PSM Rule

In the USA certain hazardous chemicals attract the statutory requirement for a process safety management (PSM) system, introduced by the Occupational Health

Table 6.4 *Elements of the safety management system listed in the Cullen Report (Cullen, 1990)*

Organizational structure
Management personnel
Training for operations and emergencies
Safety assessment
Design procedures
Procedures for operations, maintenance, modifications and emergencies
Management of safety by contractors in respect of their work
The involvement of the workforce (operators and contractors) in safety
Accident and incident reporting, investigation and follow-up
Monitoring and auditing of the operation of the system
Systematic re-appraisal of the system in the light of experience of the operator and industry

and Safety Administration (OSHA) rule Process Safety Management of Highly Hazardous Chemicals (Federal Register 29 CFR 1910.119). An account is given by Ozog and Stickles (1992).

The PSM system rule specifies national performance standards in 14 elements: (1) employee participation, (2) process safety information, (3) process hazard analysis, (4) operating procedures, (5) training, (6) contractors, (7) pre-start-up safety review, (8) mechanical integrity, (9) hot work permit, (10) management of change, (11) incident investigation, (12) emergency planning and response, (13) compliance audits and (14) trade secrets.

6.27.2 EPA Risk Management Program

In addition to the PSM system given in the OSHA Rule, there are several other American systems. One is the EPA Risk Management Program, the components of which are as follows: (1) hazard assessment, (2) prevention programme, (3) emergency response programme and (4) risk management plan. The prevention programme covers: (1) management system, (2) process hazard analysis, (3) process safety information, (4) standard operating procedures, (5) training, (6) maintenance of mechanical integrity, (7) pre-start-up review, (8) management of change, (9) safety audits and (10) accident investigation.

6.27.3 API RP 750

Another PSM system is that given in API RP 750: 1990 *Management of Process Hazards*. The elements of the American Petroleum Institute (API) system are (1) process safety information, (2) process hazards analysis, (3) management of change, (4) operating procedures, (5) safe work practices, (6) training, (7) assurance of the quality and mechanical integrity of critical equipment, (8) pre-start-up safety review, (9) emergency response and control, (10) investigation of process-related incidents and (11) audit of process hazards management system.

6.27.4 CMA system

The PSM system developed by the Chemical Manufacturers Association (CMA) has four main parts: (1) management leadership in process safety, (2) process safety management of technology, (3) process safety management of facilities and (4) managing personnel for safety. Detailed elements are given under each of these headings.

6.27.5 CCPS system

Finally, there is the PSM system of the CCPS, which is now described.

6.28 CCPS Management Guidelines

The CCPS has published several sets of guidance on management. The first of these was *A Challenge to Commitment* (CCPS, 1985) addressed to senior management in the industry. There have since followed a number of guidelines on management systems, as described below.

6.28.1 *Guidelines for Technical Management of Chemical Process Safety*

The *Guidelines for Technical Management of Chemical Process Safety* (the *Technical Management Guidelines*) by the CCPS (1989/7) give a comprehensive treatment of the management of safety and loss prevention. An account of these *Guidelines* is given by Schreiber (1991). The main headings of the *Guidelines* may be summarized as follows:

(1) overview;
(2) management and management systems;
(3) goals and objectives;
(4) process knowledge and documentation;
(5) process safety reviews;
(6) process risk management;
(7) management of change;
(8) process and equipment integrity;
(9) human factors;
(10) training and performance;
(11) incident investigation;
(12) company standards, codes and regulations;
(13) audits and corrective action;
(14) enhancement of process safety knowledge.

The *Guidelines* are concerned with process safety management. Process safety is defined as 'the operation of facilities that handle, use, process, or store hazardous materials in a manner free from episodic or catastrophic incidents' and process safety management (PSM) as 'the application of management systems to the identification, understanding, and control of process hazards to prevent process-related injuries and accidents'. A distinction is drawn between process and personal safety, the scope of the work being confined to the former. The starting point is the recognition that 'major accidents could not be prevented by technology-oriented solutions alone'.

The *Guidelines* identify 12 key elements of process safety, each of which is broken down into a number of components. These are shown in Table 6.5, together with the location in this book where they are treated. Management leadership is emphasized. Management systems and management of process safety are treated in terms of the functions of (1) planning, (2) organization, (3) implementation, and (4) control. Each of these functions operates are three levels: (1) strategic, (2) managerial and (3) task. The format of planning, organization, implementation and control (POIC) is applied to the other elements of the system.

Accountability needs to be: planned so that the general goals are converted into specific objectives owned, or sponsored, by individuals; organized so that roles are defined, lines of authority are established and formal systems specified; implemented by demonstration of compliance to senior management; and controlled by performance measurement and audit.

Process knowledge and documentation provides the basis for much of the safety programme. Examples are given of topics on which documentation is required such as chemical data and Material Safety Data Sheets (MSDSs), process definition, process and equipment design, and protective systems.

Process safety review procedures are required for capital projects, whether for new or existing plants, extensions or acquisitions. The review procedure described is a five-stage one: (1) conceptual engineering, (2) basic engineering, (3) detail design, (4) equipment procurement and construction and (5) commissioning. A comparison is given of hazard identi-

Table 6.5 *Elements and components of CCPS Technical Management Guidelines (CCPS, 1989/7)*

	Chapter[a]
1. *Accountability: Objectives and Goals*	6
Continuity of operations	
Continuity of systems (resources and funding)	
Continuity of organizations	
Company expectations (vision or master plan)	
Quality process	
Control of exceptions	
Alternative methods (performance vs specification)	
Management accessibility	
Communications	
2. *Process Knowledge and Documentation*	6, 11
Process definition and design criteria	
Process and equipment design	
Company memory	
Documentation of risk management decisions	
Protective systems	13
Normal and upset conditions	11, 20
Chemical and occupational health hazards	18, 25
3. *Capital Project Review and Design Procedures*[b]	6, 8, 11
Appropriation request procedures	
Risk assessment for investment purposes	
Hazards review (including worst credible cases)	
Siting (relative to risk management)	10
Plot plan	10
Process design and review procedures	
Project management procedures	
4. *Process Risk Management*	5, 6, 8, 9
Hazard identification	
Risk assessment of existing operations	
Reduction of risk	
Residual risk management (in-plant emergency response and mitigation)	20, 24
Process management during emergencies	20, 40
Encouraging client and supplier companies to adopt similar risk management practices	
Selection of businesses with acceptable risks	
5. *Management of Change*	6, 11, 21
Change of technology	
Change of facility	
Organizational changes that may have an impact on process safety	
Variance procedures	
Temporary changes	
Permanent changes	
6. *Process and Equipment Integrity*	11, 12
Reliability engineering	7
Materials of construction	12
Fabrication and inspection procedures	12
Installation procedures	19
Preventive maintenance	7, 21
Process, hardware and systems inspections and testing (pre-startup safety review)	19
7. *Human Factors*	14
Human error assessment	
Operator–process and operator–equipment interface	
Administrative controls vs hardware	
8. *Training and Performance*	14, 20, 21, 28
Definition of skills and knowledge	
Training programs (e.g. new employees, contractors, technical employees)	
Design of operating and maintenance procedures	
Initial qualification assessment	
Ongoing performance and refresher training	
Instructor programme	
Records management	
9. *Incident Investigation*	27
Major incidents	
Near-miss reporting	
Follow-up and resolution	
Communication	
Incident reporting	
Third-party participation, as needed	
10. *Standards, Codes, and Laws*	11, 12, App. 27
Internal standards, guidelines, and practices (past history, flexible performance standards, amendments and upgrades)	
External standards, guidelines and practices	
11. *Audits and Corrective Actions*	6, 8
Process safety audits and compliance reviews	
Resolutions and close-out procedures	
12. *Enhancement of Process Safety Knowledge*	27
Internal and external research	
Improved predictive systems	
Process safety reference library	

[a] Chapter in this book in which this topic is principally addressed.
[b] For new or existing plants, expansions and acquisitions.

fication, or review, procedures such as hazard and operability studies, failure modes and effects analysis and fault trees, and of their role in the review process. Examples are given of topics covered in different types of process safety review such as the design stage, review or commissioning review.

Process risk management involves a process of hazard identification, risk analysis, risk reduction, residual risk management and emergency planning. An account is given of the ranking of risks and of risk acceptability. The point is made that the analysis may reveal that the risks of the project may be simply too great to be acceptable.

Management of change requires that mechanisms be in place to recognize and handle changes in the process, the plant or the organization. The features which prompt changes to the process or the plant and the effect of personnel changes on the organization are described.

Process and equipment integrity is assured by measures which include: reliability engineering; materials selection; fabrication and inspection procedures; installation procedures; preventive maintenance; inspection and testing, of process, hardware and software; maintenance procedures; alarm and instrument management; and demolition procedures.

Human factors need to be taken into account in: the operator–process and operator–equipment interfaces; the choice between administrative control (procedures) and hardware controls (interlocks); and the assessment of human error.

Training is necessary to provide both knowledge and motivation. Some principal topics requiring training are listed, including operating and maintenance procedures. The approach to training is described, including definition of the knowledge and skills required, assessment before and after training, and keeping of training records.

Incident investigation should be approached in terms of failure of the management system rather than of human error. An account is given of the incident investigation process, including: preparation; team selection; recording, reporting and analysis; follow-up and resolution; and dissemination of results.

Company standards, codes and regulations provide a framework of requirements and guidance. Use is made of both external and in-house codes. The principal US codes relevant to the process industries are outlined. A plan is given for the repositories of external codes and regulations within a company.

Audits are required to show the company whether its systems and procedures and its practices are adequate and are being adhered to, so that corrective action can be taken if they are not. The reasons for establishing an audit programme are described. They include the raising of safety awareness, acceleration of the development of safety control systems, improvement of safety performance, and optimization of safety resources. A particular type of audit is the compliance review, which is carried out to ensure that the company is meeting the legal requirements.

Enhancement of process safety knowledge is an ongoing activity and involves the utilization of a range of resources, some of which are enumerated. The direction in which this work may progress is outlined by Sweeney (1992). The emphasis is on the development of measures of performance.

6.28.2 *Plant Guidelines for Technical Management of Chemical Process Safety*

The *Plant Guidelines for Technical Management of Chemical Process Safety* (the *Plant Technical Management Guidelines*) by the CCPS (1992/11) have the same basic structure as the *Technical Guidelines*, but emphasize concrete examples.

These are principally given in the appendices, the titles of which are listed in Table 6.6.

6.28.3 *Guidelines for Implementing Process Safety Management Systems*

The implementation of a PSM system is covered in *Guidelines for Implementing Process Safety Management Systems* (CCPS 1994/13). The elements of the system are those given in the *Technical Management Guidelines* and shown in Table 6.5. The treatment of implementation is under the headings: (1) management commitment, (2) definition of goals, (3) evaluation of the present status, (4) development of a plan, (5) development of specific PSM systems, (6) implementation of the system, (7) measurement and monitoring of system installation and (8) expansion beyond the original scope. The guidelines describe a case study.

6.28.4 *Guidelines for Auditing Process Safety Management Systems*

The auditing of a PSM system is dealt with in *Guidelines for Auditing Process Safety Management Systems* (CCPS 1993/12). These guidelines cover topics closely aligned to, but not identical with, the 12 elements of the CCPS PSM system, the headings being: (1) management of PSM systems audits, (2) audit techniques, (3) accountability and responsiblity, (4) process safety knowledge, (5) project safety reviews, (6) management of change, (7) process equipment integrity, (8) process risk management, (9) incident investigation, (10) human factors, (11) training and performance and (12) emergency response planning.

6.29 Regulatory Control

The crucial role of the management and management system in SLP means that it is this which is, or should be, of prime concern to the regulatory body. The organization and activities of that body should reflect this.

6.29.1 Evolution of policy

The HSE has for a long time placed emphasis on the crucial role of management, but its approach has naturally undergone a process of evolution. The process has been described by Bleeze (1992). This evolution is documented in the series of publications on management already referred to: *Managing Safety* (HSE, 1981d), *Monitoring Safety* (HSE, 1985c), *Effective Policies for Health and Safety* (HSE, 1986a) and *Successful Health and Safety Management* (HSE, 1991b).

6.29.2 Inspection versus audit

In the UK, as elsewhere, the traditional approach to enforcement has been by inspection. This is reflected in the title of the arm of the HSE which operates at works level, the Factory Inspectorate.

Table 6.6 *Appendices of CCPS Plant Management Guidelines (CCPS, 1992/11)*

2A	Characteristics of a management system
3A	Example of the management of process hazards
4A	Example of typical Material Safety Data Sheet
4B	Lead questions (hazardous conditions)
4C	Example of process definition/design criteria contents
4D	Example of typical protective systems/equipment data
4E	Example of components included in process knowledge and documentation
5A	Plant example of organization of process hazard review for appropriation requests
5B	Plant example of plant procedures: 'Description of hazard review program for appropriation requests'
5C	Example of plant procedures for appropriation request information and approvals
5D	Plant example of 'Standard practice for process reviews'
6A	Plant example of hazard identification
6B	Plant 'X' risk analysis of operations
6C	Plant 'Y' risk analysis of operations
6D	Example of process management in emergencies
7A	Management of change policy
7B	Guidelines for review of plant changes or modifications
7C	Control of change – safety management practices
7D	Control of changes
7E	Example of change of process technology
7G	Example of safety assessment form for plant modification work
7H	Safety guidelines variance request
7I	Permanent change considerations
7J	Examples of jumpers/bypass logging
8A	Example of plant management system for materials of construction (MOC)
8B	Example of test and field inspection equipment and procedures
8C	Example of field inspection and testing of process safety systems
8D	Example of a hot work permit
8E	Example of criteria for test and inspection of safety relief devices
8F	Example of management system for critical and unique safety features
8G	Table of contents from Chemical Manufacturers Association Fixed Equipment Inspection Guide
10A	Example of operator process safety training programme
10B	Example of maintenance training program for process safety
10C	Example of operations technical staff training programme
10D	Example of maintenance training implementation
10E	Example of instructional standards
10F	Example of refresher training course frequencies
11A	Example of extraordinary event notification, investigation and reporting
11B	Example of plant follow-up procedures for accident or incident investigation recommendations
11C	Plant example of unplanned incidents causal factors analysis
11D	Example of plant accident or incident reporting procedures
12A	Example of operational safety standards guidelines
12B	Example of critical operating parameters: interpretation guidelines
12C	Example of manufacturing standards
12D	Example of hazardous systems (existing system)
13A	Protocol for estimating progress in implementation of CCPS *Guidelines for Technical Management of Chemical Process Safety*
13B	'Key questions' for the elements of process hazards management
13C	Example of process safety audit
13D	Example of safety and property protection procedures
14A	Professional industry organizations offering process safety enhancement
14B	Center for Chemical Process Safety Guidelines and conferences
14C	Examples of subjects covered in process safety libraries

This approach is increasingly being supplanted by the auditing of the management and management systems. The features which are of interest to the regulator are now those such as the arrangements for the identification and assessment of hazards, for the proof testing of protective systems or for the investigation of incidents.

The deficiencies of an approach to regulation based on inspection were highlighted by the Cullen Report on the Piper Alpha disaster and the report recommended a change in the regulatory regime away from one based on inspection and towards one based on audit of the safety management system (Cullen, 1990).

6.29.3 Accident Prevention Advisory Unit

In order to provide assistance to inspectors in the assessment of management and management systems, the HSE has created the Accident Prevention Advisory Unit (APAU). Much of the HSE work described above derives from this unit.

6.29.4 Management assessment

As described above, the HSE has given guidance on good practice in the management of process safety. It may be assumed to apply much the same criteria in the assessments which it undertakes. It is also active in the evaluation of research models of management effectiveness, including as proprietary systems. An account of one such study has been given by J.C. Williams and Hurst (1992). Against this background, the HSE has been developing a method for auditing a safety management system (SMS). This is now described.

6.30 STATAS

The HSE is well advanced with the development of a methodology for the audit of safety management systems. The method goes by the name of Structured Audit Technique for the Assessment of Safety Management Systems (STATAS) and is described by Ratcliffe (1993, LPB 112, 113, 114) and Hurst and

Ratcliffe (1994). The starting point is a field study of failures in vessels and pipework (Bellamy, Geyer and Astley, 1989; Geyer *et al.*, 1990; Geyer and Bellamy, 1991; Hurst *et al.*, 1991). An account of the study and of the pipework failure data is given in Chapter 12, whilst the human error aspects are considered in Chapter 14.

From this study a loss of containment model has been derived. The pipework failures are classified on three dimensions:

(1) origin of failure (basic or underlying cause);
(2) direct (immediate) cause;
(3) recovery (preventive) mechanism.

The origins of failure are divided into nine categories:

(1) design;
(2) manufacture/assembly;
(3) construction/installation;
(4) normal operation;
(5) maintenance;
(6) natural causes;
(7) domino;
(8) sabotage;
(9) unknown.

There are 12 categories of direct cause:

(1) corrosion;
(2) erosion;
(3) external load;
(4) impact;
(5) overpressure;
(6) vibration;
(7) temperature (high, low);
(8) wrong equipment;
(9) defective equipment;
(10) human error;
(11) other;
(12) unknown.

There are four preventive mechanism categories. Two relate to equipment:

(1) hazard identification and assessment (HAZ);
(2) routine checking and testing (ROUT).

and two to people:

(3) human factors review (HF);
(4) task checking (TCHECK).

The influencing factors which bear on the failures are represented by a socio-technical model. The structure of this model is shown in the socio-technical pyramid given in Figure 2.6(a). The model has six levels as follows:

Level	
5	System climate
4	Organization and management
3	Communication and feedback
2	Operator reliability
1	Engineering reliability
0	Loss of containment and mitigation

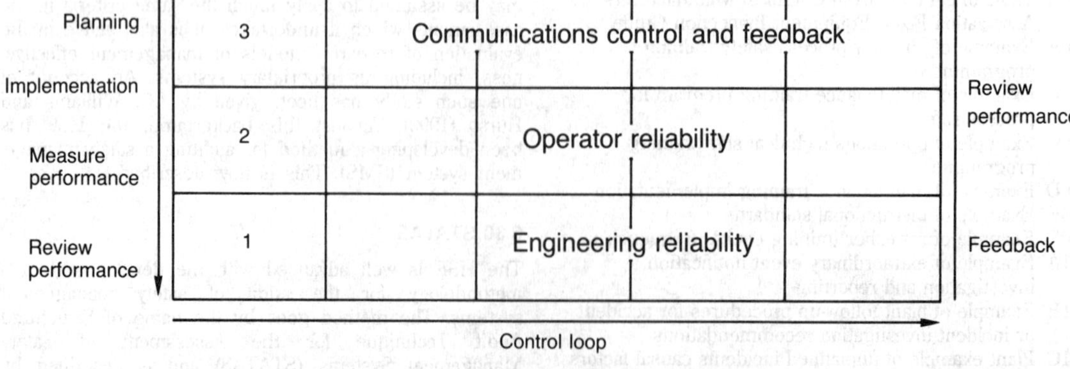

Figure 6.3 *Arrangement of question cell sets for audit of topic MAINT/ROUT in STATAS (Ratcliffe, 1993 LPB 114)*

Some influencing factors at Level 5 are background, legislation, regulation and resources; at Level 4, policies, management structure, formal systems, assigned responsibilities, performance criteria and emergency response organization; at Level 3, job descriptions, written procedures, performance evaluation, safety audits, incident reporting and follow-up, and meetings; at Level 2, design, procedures, competence, manning and shifts; at Level 1 plant layout and equipment design.

In STATAS these models are combined with an audit scheme. The five phases of the plant life cycle are:

(1) design (DES);
(2) manufacture (MANF);
(3) construction (CON);
(4) operation (OP);
(5) maintenance (MAINT).

Since there are four recovery, or preventive, categories, there are in principle 20 combinations of plant phase and preventive category. The field study showed, however, that the great majority of failures are associated with just eight combinations. These eight cover 83% of vessel failures, the proportion recoverable being assessed as 87%; the corresponding figures for pipework are 84% and 92%. The eight combinations are

(1) DES/HAZ;
(2) CON/TCHECK;
(3) MAINT/ROUT;
(4) MAINT/HF;
(5) MAINT/TCHECK;
(6) OP/HAZ;
(7) OP/HF;
(8) OP/TCHECK.

Then, for each combination, an audit question set is devised. The themes of the questions are

(1) structures, systems and procedures,
(2) standards and criteria,
(3) mitigation of pressures,
(4) availability and use of resources.

Each question set contains some 40–80 questions. The questions are grouped in cells containing between 2 and 10 questions which explore a particular topic.

The conduct of the audit aims to take samples 'horizontally' and 'vertically' through the system, concentrating mainly on Levels 1–4. The method is illustrated in Figure 6.3, which shows the arrangement of question cell sets for the topic MAINT/ROUT. Work on the question sets links with that of Wells, Hurst and coworkers described in Chapter 8.

7 Reliability Engineering

Loss prevention is in large part the application of probabilistic methods to the problems of failure in the process industries. The discipline which is concerned with the probabilistic treatment of failure in systems in general is reliability engineering. This chapter gives an account of reliability engineering and of some reliability techniques.

Selected references on reliability engineering are given in Table 7.1. There are numerous books on the subject. These include *Reliability Theory and Practice* (Bazovsky, 1961), *Reliability Principles and Practices* (Calabro, 1962), *Reliability: Management, Methods and Mathematics* (Lloyd and Lipow, 1962), *System Reliability Engineering* (Sandler, 1963), *Reliability Engineering* (von Alven, 1964), *Reliability Engineering for Electronic Systems* (Myers *et al.*, 1964), *Mathematical Theory of Reliability* (Barlow and Proschan, 1965), *Probabilistic Reliability: An Engineering Approach* (Shooman, 1968a), *Probabilistic Systems Analysis* (Breipohl, 1970), *Mechanical Reliability* (A.D.S. Carter, 1972; 2nd edn, 1986), *Reliability*

Table 7.1 *Selected references on reliability engineering*

IEC (Std 1078); NRC (Appendix 28 *Reliability Engineering*); UKAEA (Appendix 28); US Armed Forces (Appendix 27 *MIL Standards, Handbooks*); Fry (1928); Feller (1951); Siegel (1956); Bazovsky (1960, 1961); Chorofas (1960); Hosford (1960); Parzen (1960); Reza (1961); Blackett (1962); Calabro (1962); D.R. Cox (1962); Lloyd and Lipow (1962); Machol and Grey (1962); Savage (1962a,b, 1971); Pieruschka (1963); Sandler (1963); Zelen (1963); von Alven (1964); Myers *et al.* (1964); N.H. Roberts (1964); Barlow and Proschan (1965, 1975); P.L. Meyer (1965); I. Miller and Freund (1965a); A.D.S. Carter (1972, 1973, 1976, 1979, 1980, 1983, 1986); A.E. Green and Bourne (1966 UKAEA AHSB(S) R117, 1972); Nathan (1966); Vance (1966); Hahn and Shapiro (1967); Haugen (1968); Hofmann (1968); Ireson (1968); Polovko (1968); Shooman (1968a); Gnedenko *et al.* (1969); A.E. Green (1969–70, 1971 SRS/GR/2, 1972, 1973, 1974a,b, 1976, 1982b, 1983); IEE (1969 Conference Publication 60); C.S. Smith (1969); Thomason (1969); Tribus (1969); Breipohl (1970); Bourne (1970, 1971 UKAEA SRS/GR/4, 1972, 1975 NCSR R7, 1982); Kozlov and Ushakov (1970); R. LeWis (1970); Rau (1970); Truscott (1970); Buffham *et al.* (1971); Eames (1971 SRD R1); R.A. Howard (1971); Jenkins and Youle (1971); Amstadter (1972); Caplen (1972); D.R. Cox and Miller (1972); Cunningham and Cox (1972); Ross (1972); D.J. Smith (1972, 1981, 1985a, 1991); Bompas-Smith (1973); Henley and Williams (1973); Kletz (1973b, 1979l); Locks (1973); Reinschke (1973); Schneeweiss (1973); D.J. Smith and Babb (1973); Apostolakis (1974); de Finetti (1974); J.R. Taylor (1974b, 1975b, 1976b, 1979); Mann *et al.* (1974); AEC (1975); H.M. Wagner (1975); Henley and Lynn (1976); Moss (1976 NCSR R12); C.O. Smith (1976); Lievens (1976); D.J. Bennett (1977); Dunster (1977); F.R. Farmer (1977a); Kapur and Lamberson (1977); Keller (1977); Mosteller (1977); C. Singh and Billinton (1977); Sorenson and Besuner (1977); Halpern (1978); Hastings and Mellow (1978); Rodin (1978); Richards (1980); Sinha and Kale (1980); Court (1981); Dhillon and Singh (1981); J.W. Foster (1981); Nieuwhof (1981, 1984, 1985a,c, 1986); O'Connor (1981, 1984); Sherif (1981); Durr (1982); S.T. Parkinson (1982); Amendola and Melis (1983); Billinton and Hossain (1983); Hutchinson (1983); MoD (1983); Sayles (1983); Frankel (1984); K.A.P. Brown (1985); Serra and Barlow (1986); Ballard (1987); M.G. Singh et al. (1987); Veevers (1989/90); Misra (1992, 1994); F.R. Nash (1993); Andrews and Moss (1993); Sherwin and Bossche (1993); BS 5760: 1981–

Terminology
US Armed Forces (MIL-STD-721B, MIL-STD-781B); IEC (1969); BS 5760: Part 0: 1986

Background mathematics, statistical distributions
Fry (1928); Clopper and Pearson (1934); Kendall (1948, 1970); Arkin and Colton (1950); Hald (1952a,b); Burr (1953); Meehl (1954); K. Pearson and Hartley (1954); Moroney (1956); K. Pearson and Hartley (1956); Siegel (1956); Beers (1957); Kendall and Stewart (1958–); Bowker and Lieberman (1959); Pantony (1961); Raiffa and Schlaifer (1961); Reichman (1961); D.B. Owen (1962); Savage (1962a,b, 1964); Kron (1963); Abromowitz and Stegun (1964); E.L. Grant (1964); N.L. Johnson and Leone (1964); Kyburg and Smokler (1964); Lindley and Miller (1964); Lindley (1965); P.L. Meyer (1965); Papoulis (1965); Conway *et al.* (1967); Yamane (1967); Beyer (1968, 1978, 1984); Winkler (1968, 1981); Aitcheson and Brown (1969); N.L. Johnson and Katz (1969); O'Brien (1969); Schmitt (1969); Woodcock and Eames (1970 AHSB(S) R179); J.R. King (1971); Barks (1972); FRS (1972 Fire Research Note 909); Bolz and Tuve (1973); Lomnicki (1973); Hastings and Peacock (1974); Lehman (1975); Linstone and Turoff (1975); Patel *et al.* (1976); Worledge (1976 SRD R68); Beck and Arnold (1977); Parry *et al.* (1977 SRD R80, 1979 SRD R129); BRE (1978 CP8/78); Paradine and Rivett (1980); Govil and Agarwala (1982, 1983a,b); Nelson and Rasmuson (1982); Patel and Read (1982); Sherif (1982a); Ichikawa (1983, 1984a,b, 1986); Moran (1984); A.T. White (1984); Kececioglu and Dingjun (1985).
Error function approximation: Karlsson and Bjerle (1980); Ernst (1992)

Log–normal distribution, error factor
Chambers *et al.* (n.d.); Finney (1941, 1971); Gaddum (1945); Brownlee (1949); Day (1949); Aitcheson and Brown (1969); AEC (1975); Kline (1984); Siu and Apostolakis (1985); Murty and Verma (1986); Savoie (1988)

Weibull distribution
Weibull (1951); L.S. Nelson (1967); R.A. Mitchell (1967); Hastings (1967–68); Shooman (1968a); Truscott (1970); Steiger (1971); Hinds *et al.* (1977); Kapur and Lamberson (1977); O'Connor (1977); Sherwin and Lees (1980); Guida (1985); Kekecioglu and Jacks (1985); P.W. Hale (1987); Lihou and Spence (1988)

Extreme value theory, distribution
R.A. Fisher and Tippett (1928); Cramer (1946); Gumbel (1958); B. Epstein (1960); Lloyd and Lipow (1962); Wiesner (1964); Ramachandran (FRS 1972 Fire Research Notes 910, 929, 943; 1973 Fire Research Note 991); Singpurwalla (1972); Bompas-Smith (1973); Mann *et al.* (1974); C.W. Anderson (1976); E.M. Roberts (1979);

Perry (1981); Schueller (1982); R.L. Smith (1986); Surman *et al.* (1987)

Bayesian methods
Savage (1962b); Lindley (1965); Schmitt (1968); Weir (1968); Brand (1980); Colombo and Saracco (1983); Kaplan (1983); Puccini (1983); Yuan (1987); Hauptmans and Hömke (1989)

Early failure, burn-in
R. Ward (1972); de la Mare and Ball (1981); Jensen (1982); Jensen and Petersen (1982)

Wearout failure
Newby (1986)

Repair times
Sandler (1963); AEC (1975); Kline (1984)

Dependent failure
NRC (Appendix 28 *Common Cause Failure*); Apostolakis (1975, 1976); J.R. Taylor (1976a); Heising and Luciani (1977); Edwards and Watson (1979 SRD R146); P. Martin (1980); A.M. Smith and Watson (1980); Watson (1980); Heising (1983); M.G.K. Evans *et al.* (1984); Platz (1984); Games *et al.* (1985); Crellin *et al.* (1986); Teichman (1986); Yun and Bai (1986); G.T. Edwards (1987a,b); Johnston (1987a,b); Yuan (1987); Ballard (1988); Andrews and Moss (1993)

Markov methods
Kemeny and Snell (1960); Sandler (1963); Shooman (1968); R.A. Howard (1971); Platz (1984); J.N.P. Gray (1985); Andrews and Moss (1993)

Monte Carlo methods
Metropolis and Ulam (1949); NBS (1951); Kahn (1957); Hammersley and Handscomb (1964); Shreider (1964); Tayyabkhan and Richardson (1965); Muth (1967); Rudd and Watson (1968); Shooman (1968a); Schmitt (1969); Ang and Tang (1974–75); Sobol (1974); AEC (1975); Lanore and Kalli (1977); EPRI (1981b); Rubinstein (1981); E.E. Lewis and Tu Zhuguo (1986); Soon Chang *et al.* (1986)

Physics of failure
Kao (1965); Shooman (1968a); Birnbaum and Saunders (1969); Yost and Hall (1976)

Catastrophe theory
Zeeman (1972, 1977); Hilton (1976); Poston and Stewart (1978); Saunders (1980); Gilmore (1981)

Rare events
Selvidge (1972)

Reliability specification, apportionment
US Armed Forces (MIL-S-38130); Petkar (1980); Sledge (1982); MoD (1983); EEMUA (1986 Publication 148)

Equipment testing
US Armed Forces (MIL-STD-781A, MIL-STD-810); Kececioglu and Jacks (1985); P.W. Hale (1986); Irwing (1986); Lydersen and Rausand (1987); BS 5760: Part 10: 1993, BS DD 57: 1978

Reliability growth modelling
US Armed Forces (MIL-HDBK-189, MIL-STD-1635); Madansky and Peisakoff (1960); Duane (1962); Kamins and Gross (1967); Codier (1968); R.C.F. Hill (1977); Sheppard (1983, 1985); Catchpole *et al.* (1984); Halliday and Devereux (1984); BS 5760: Part 6: 1991

Proportional hazards modelling
D.R. Cox (1972); Kalbfleisch and Prentice (1980); Prentice *et al.* (1981); Andersen (1982); Lawless (1982); Bendell (1985); C.J. Dale (1985); Wightman and Bendell (1986)

Expert opinion
NRC (Appendix 28 *Expert Judgement*); Thurstone (1927, 1931); Wherry (1938); Kendall (1948, 1955, 1970); Guilford (1954); Khan (1957); C. Peterson and Miller (1964); B. Brown (1964); Pontecorvo (1965); Schmid (1966); C.R. Peterson and Beach (1967); Torgerson (1967); Winkler (1968, 1981, 1986); Alpert and Raiffa (1969); B. Brown *et al.* (1969); Dalkey (1969); Keats (1971); Pill (1971); Klee (1972); L.L. Philipson (1974a); Tversky and Kahneman (1974); Linstone and Turoff (1975); Spetzler and Stahl von Holstein (1975); Nachmias (1976, 1992); Lichtenstein *et al.* (1977); Saaty (1977, 1980, 1982); Winkler and Murphy (1978); Apostolakis *et al.* (1980); Hunns (1980, 1982); Apostolakis (1982, 1985a,b, 1988, 1990); Mosleh and Apostolakis (1982, 1985, 1986); Uppuluri (1983); Martz (1984); Mosleh *et al.* (1987, 1988); Shields *et al.* (1987); Clarotti and Lindley (1988); Goossens *et al.* (1989); NRC (1989); van Steen *et al.* (1989); Svenson (1989); Wheeler et al. (1989); Meyer and Booker (1990); van Steen (1992).
Interviewing: Khan (1957); Oppenheim (1966); Gorden (1969)
Applications: Minarick and Kukielka (1982); Pickard *et al.* (1983); Swain and Guttman (1983); Cottrell *et al.* (1984); Embrey *et al.* (1984); IEEE (1984 Standard 500); Veneziano *et al.* (1984); Minarik *et al.* (1985); NRC, Steam Explosion Review Group (1985); Benjamin *et al.* (1986); EPRI (1986); Hannaman *et al.* (1986); Siu and Apostolakis (1982, 1985, 1986); Mosley, Bier and Apostolakis (1987); Wheeler *et al.* (1989)

Complex systems, large systems
Drenick (1960); Kron (1963); Htun (1965); Shooman (1968a); Nelson *et al.* (1970); Batts (1971); Colombo (1973); C. Singh (1974); NCSR (1975 NCSR R6); Blin *et al.* (1977); Hunns (1977); M.J. Harris and Rowe (1980); Kontoleon (1982); Laviron *et al.* (1982); Windebank (1982); Pickup (1983); Laviron (1985, 1986); Laviron and Heising (1985)

Special topics
J.A. Baker (1963); Buzacott *et al.* (1967); D.B. Brown (1971); Koen and Carnino (1974); Kontoleon and Kontoleon (1974); Bourne (1975 NCSR R7); Buzacott (1976); N.D. Cox (1976); Astolfi and Elbaz (1977); G.O. Davies (1977); Mosteller (1977); Parry and Worledge (1977 SRD R95); Colombo *et al.* (1978); Parry (1979 SRD R143); Kontoleon (1980, 1981); Platz (1980); Hohenbichler and Rackwitz (1981); Moieni *et al.* (1981); Nicolescu and Weber (1981); J.H. Powell (1981); Sharma (1981); Gopalan and Natesan (1982); Govil and Agarwal (1982, 1983a,b); Husseiny *et al.* (1982); Misra and Gadani

(1982); Brooks (1983); Hirschman *et al.* (1983); Ichikawa (1983); J.W.H. Price (1983); Sayles (1983); Walker (1983); Bendell and Ansell (1984); Connors (1984); Aven (1985, 1986); Jain and Gopal (1985); Reiser (1985a,b); Witt (1985); Gopalan and Venkateswarlu (1986a,b); Gupta *et al.* (1986); Pickles (1986); Dörre (1987); Knezevich (1987); Limnios (1987); McCormick (1987); D.B. Parkinson (1987); Rushdi (1987); Walls and Bendell (1987)

Standby systems
Signoret *et al.* (1983); Vaurio (1985)

Quality assurance, control (see also Table 6.1)
NRC (Appendix 28 *Quality Assurance*); US Air Force (n.d.); US Armed Forces (Appendix 27); Calabro (1962); C.S. Smith (1969); Thomason (1969); F. Nixon (1971); Schmitt and Wellein (1980); Kolff and Mertens (1984); A. Smith (1991)

Mechanical reliability
ASME (Appendix 28, 1975/129, 1977/17, 1981/18, 1982/161, 1983/9, 1990 PVP 193, 1993 DE 55); NRC (Appendix 28 *Failures*); Haviland (1964); IMechE (1970/3, 1974/10, 1975/16, 1984/79, 1988/105, 1994/172); A.D.S. Carter (1972, 1973, 1976, 1979, 1980, 1983, 1986); API (1973–, Refinery Inspection Guide); Bompas-Smith (1973); Eames (1973 SRSD/GR/12); Eames and Fothergill (1973 (SRS/GR/13); Hensley (1973 SRS/GR/1); Pronikov (1973); D. Scott and Smith (1975); P. Martin (1976a,b); C.O. Smith (1976); Yost and Hall (1976); Kapur and Lamberson (1977); Venton (1977, 1980); Moss (1980); Sherwin and Lees (1980); Martin *et al.* (1983); ASCE (1985/23); Parkhouse (1987); Davidson (1988)

Failure data analysis
NRC (Appendix 28 *Failure Data Analysis*); Shaw (n.d.); Weibull (1951); Carhart (1953); Karassik (1959); L.G. Johnson (1964); Hastings (1967–68); R.A. Mitchell (1967); L.S. Nelson (1967); Kivenson (1971); Fercho and Ringer (1972); Bompas-Smith (1973); Jardine and Kirkham (1973); Ryerson (1973); Whitaker (1973a,b); J.C. Moore (1974); Nino (1974); de la Mare (1976); Moss (1976 NCSR R12, 1977 NCSR 11); Aird (1977a, 1978); Berg (1977); Blanks (1977); Hinds *et al.* (1977); Kapur and Lamberson (1977); O'Connor (1977); Sherwin (1978, 1983); Hastings and Jardine (1979); Sherwin and Lees (1980); Stokoe *et al.* (1981); Crellin and Smith (1982); Colombo and Jaarsma (1983); Harries *et al.* (1983); Puccini (1983); Bendell and Walls (1985); Nieuwhof (1985b); Lamerse and Bosman (1985); Walls and Bendell (1985); Coit *et al.* (1986); Crellin *et al.* (1986); Schiffman (1986); Vaurio (1986); Bendell (1987); Andrews and Moss (1993)
Bath tub curve: Carhart (1953); A.D.S. Carter (1973); Talbot (1977); Aird (1978); Veevers (1989/90)

Availability
H. Smith and Grace (1961); Gibbons (1962); Sandler (1963); R.E. Jackson *et al.* (1965); H.L. Gray and Lewis (1967); W.N. Smith (1968); Jenkins (1969); Sherry (1969–70); Buzacott (1970a, b); Dailey (1970); Henley (1971); Jenkins *et al.* (1971); Konoki (1971); Kuist and Fife (1971); Roth and Fiedler (1971); Cason (1972); McFatter (1972); G.H. Mitchell (1972); Ufford (1972); Gaddy and Culbertson (1973); Kardos and Vondran (1973); Locks (1973); Yaro (1973); Inone *et al.* (1974); Kafarov *et al.* (1974); Kardos *et al.* (1974–); Rosen and Henley (1974); Sawyer and Williams (1974); Walsham (1974); Cowan (1975); G.D.M. Pearson (1975, 1977); Vondran and Kardos (1975); NCSR (1976 NCSR R8); Apostolakis and Bansal (1976, 1977); D.H. Allen and Pearson (1976); Caceres and Henley (1976); Cowan *et al.* (1976); Holmes (1976); Platz (1976, 1977); Blin *et al.* (1977); Cherry *et al.* (1977); D.H. Allen and Coker (1977); G.O. Davies (1977); Henley and Hoshino (1977); Siddons (1977); Coker (1978); Parry and Worledge (1978 SRD R113); Ong and Henley (1979); Apostolakis and Chu (1980); Leblanc *et al.* (1980); J.W. Foster (1981); M.J. Phillips (1981); Heising (1983); Nieuwhof (1983b); Piccinnini and Anatra (1983); Sherwin (1984); Brouwers (1986); Brouwers *et al.* (1987); Evans (1987 NCSR/GR/68); Watanabe (1987); Fairclough (1988); Odi and Karimi (1988); E.S. Lee and Reklaitis (1989); Dougan and Reilly (1993); Sherwin and Bossche (1993)

Maintenance
NRC (Appendix 28 *Maintenance, Maintenance Personnel Reliability Model*); US Armed Forces (MIL-HDBK-472, MIL-STD-470, MIL-STD-471); J.C. Moore (1960, 1966, 1974); Bartholemew (1963); Dean (1963); J.E. Miller and Blood (1963); Goldman and Slattery (1964); Nathan (1966); Newborough (1967); Woodman (1967); Blanchard and Lowery (1969); IMechE (1969/2, 1973/6, 1975/16,17, 1994/174); Armitage (1970); Geraerds (1970); Kelly and Harris (1971); D.J. Smith (1972, 1981, 1985a, 1991); Gradon (1973); Hastings (1973); Jardine (1970a, 1973a, 1976); D.J. Smith and Babb (1973); Trotter (1973); Clifton (1974); Priel (1974); Reynolds (1974); Corder (1976); Husband (1976); André (1977); Johns and Sadlowski (1977); Kapur and Lamberson (1977); de la Mare (1979 NCSR 21); Sherwin and Lees (1980); J.W. Foster (1981); Barry and Hudson (1983); G.T. Edwards (1983); MoD (1983); Sherwin (1983); van Aken *et al.* (1984); Backert and Rippin (1985); Kelly (1986); L.C. Thomas (1986); Henry (1990, 1993a,b); Factory Mutual Int. (1991a); Rao (1992); Andrews and Moss (1993); Whetton (1993).
Life cycle costing: Sherif (1982b); Lees (1983); Jambulingam and Jardine (1986)
Reliability-centred maintenance: US Armed Forces (MIL-HDBK-472); Nail and Nair (1965); Nowlan and Heap (1978); Jambulingam and Jardine (1986); Anderson and Neri (1990); Moubray (1991); Sandtory (1991)
Spares holdings: Messinger and Shooman (1976); G.H. Mitchell (1972)

Computer system, software reliability
(*see also* Table 13.3)
NRC (Appendix 28 *Computer Software Reliability*); Aiken (1958); Anon. (1960b); Naur (1966); Floyd (1967); London (1968); Manna and Pnueli (1969); Bauer (1975a,b); Goos (1975); Poole (1975); Tsichritzis (1975a,b); Fagan (1976); Meyers (1976); Shooman (1976, 1983); J.L. Peterson (1977); R.B. Anderson (1979); Daniels (1979 NCSR 17, 1983, 1986, 1987); Boulton and Kittler (1979); Daniels and Hughes (1979 NCSR 16); Glass (1979); Kopetz (1979); Cho (1980); Longbottom (1980); Dempster *et al.* (1981); Kersken and Ehrenberger (1981); J. Peterson (1981); Dunn and Ullman (1982);

McGettrick (1982); Becker (1983); Gubitz (1983); Leveson and Harvey (1983); Leveson and Stolzy (1983); Lord (1983); Bennett (1984, 1991a,b, 1993); Dunn (1984); Goldsack (1985); IEE (1985, 1989); Quirk (1985); Backhouse (1986); CEC (1986); Helps (1986); C.B. Jones (1986); Leveson (1986, 1987); O. Anderson *et al.* (1987); T. Anderson (1987); Baber (1987); Barry (1987); C. Dale (1987); Dale *et al.* (1987 NCSR/GR/65); Ehrenburger (1987); Hennell (1987); Humphreys (1987); IEC (1987/1, 1991 SC65A WG9); Littlewood (1987a,b); Macro and Buxton (1987); MoD (1987); Musa *et al.* (1987); National Computer Centre (1987, 1989); Nordic Council of Ministers (1987); Pyle (1987, 1991, 1993); Schagen (1987); Schagen and Sallih (1987); Schulmeyer and MacManus (1987); Voges (1987); Heilbrunner (1988); IChemE (1988/132); E. Johnson (1988); Redmill (1988, 1989); Humphrey (1989); Lasher (1989); Littlewood (1989); McDermid (1989a,b, 1991, 1993); Sennett (1989); Somerville (1989); D.J. Smith and Wood (1989); R. Clarke (1989–90); Bishop (1990); DTI (1990); Frederickson and Beckman (1990); C.B. Jones and Shaw (1990); C. Morgan (1990); Rook (1990, 1991); Schulmeyer (1990); P. Bennett (1991); Facey (1991); Fergus *et al.* (1991); P.A.V. Hall (1991); Hunns and Wainwright (1991); IEE (1991, 1991 Coll. Dig. 91/3, 1992); Ince (1991a,b); L. Jones (1991); Littlewood and Miller (1991); Paula and Roberts (1991); Rook (1991); Shrivastava (1991); A. Smith (1991); J.T. Webb (1991); Woodward (1991); Chudleigh and Catmar (1992); Wichmann (1992); Bologna (1993); Cluley (1993); Danielsen (1993); Drake and Thurston (1993); Ehrenberger (1993); Fink *et al.* (1993); Johnston (1993); Malcolm (1993); Rata (1993); Redmill and Anderson (1993); Rowland (1993);Walton (1993); P. Woods (1993)
BS (Appendix 27 *Computers*)

Nuclear systems (*see also* Tables 9.1 and A20.1)
NRC (Appendix 28); UKAEA (Appendix 28); Siddall (1959); Eames (1966); F.R. Farmer (1967a,b, 1969a,b, 1971, 1977a); Hensley (1968); A.E. Green (1969–70, 1972, 1973, 1974a,b, 1976); Bourne (1970); A.E. Green and Bourne (1972); Ablitt (1975); Carnino (1976); Eames *et al.* (1976); Rudolph (1977); Schmitt and Wellein (1980); Welsh and Lundberg (1980); Bowers *et al.* (1981); Ballard (1989)
Analysis of faults and abnormal occurrences: J.C. Moore (1960, 1966, 1974); R.L. Scott (1971); J.R. Taylor (1974a, 1975c); AEC (1975); NEA (1977)

Electrical power systems
Sherry (1969–70); Billinton (1970); Billinton *et al.* (1973); Wakeman and Laughton (1976); Siddons (1977); C. Singh and Billinton (1977); Sherif (1982c); Snaith (1982); Billinton and Allan (1983); Yip *et al.* (1984); Allan *et al.* (1986); Systems Engineering Comm. (1986)

Process industry systems
Rudd (1962); Rudd and Watson (1968); Weisman (1968); Browning (1969a–c, 1970, 1973); Buffham and Freshwater (1969); Freshwater and Buffham (1969); Bently and Reid (1970); Cornett and Jones (1970); Lenz (1970); Loftus (1970); Pan (1970); H.L. Williams and Russell (1970); Buffham, *et al.* (1971); Patterson and Clark (1971); C.F. King and Rudd (1972); Low and Noltingk (1972); D.H. Allen (1973); N.D. Cox (1973); Eames (1973 UKAEA SRS/GR/12); Eames and Fothergill (1973 SRS/GR/13); Hensley (1973 UKAEA SRS/GR/1); Whitaker (1973a,b); S.B. Gibson (1974, 1976c); D.R. Wood *et al.* (1974); Henley and Gandhi (1975); Anon. (1976 LPB 11, p. 18; LPB 12, p. 17); Birbara (1976); Campbell and Gaddy (1976); Lees (1976b, 1977a); Sayers (1976); Bennett (1977); Berg (1977); J.H. Bowen (1977); Cannon (1977); Cerda and Napier (1977); Craker and Mobbs (1977); Doering and Gaddy (1977); McIntire (1977); R.W. Nelson (1977); Coltharp *et al.* (1978); Senior (1978); Doig and Reinten (1979); Isaszegi and Timar (1979); Kumamoto and Henley (1979); Mundo (1979); Aird (1980, 1981, 1984); M.J. Harris and Rowe (1980); Leblanc *et al.* (1980); Burton (1980); Sherwin and Lees (1980); Temple (1980); Butterfield (1981); Craker (1981); Dransfield and Lowe (1981); Keller and Stipho (1981); Stokoe *et al.* (1981); Sturrock (1981); Bradbury (1982); Corran and Witt (1982); Piccinnini and Anatra (1983); Sherwin (1983); Bosman (1985); Jebens (1986); Churchley (1987); Keey (1987); Fairclough (1988); Khadke *et al.* (1989); Beckman (1992b)

Other industries and systems
Burgess (1974); F.H. Thomas (1977); Blockley (1980); Kinkead (1982); Tregelles and Worthington (1982); Yao (1985); Harr (1987)

Technology (A.E. Green and Bourne, 1972), *Reliability, Maintainability and Availability Assessment* (Locks, 1973), *Reliability Engineering in Design* (Kapur and Lamberson, 1977), *Engineering Reliability* (Dhillon and Singh, 1981), *Practical Reliability Engineering* (O'Connor, 1981; 2nd edn, 1984), *Reliability and Maintainability in Perspective* (D.J. Smith, 1988) *Reliability, Maintainability and Risk* (D.J. Smith, 1991), *Reliability Analysis and Prediction* (Misra, 1992), *Reliability and Risk* (Andrews and Moss, 1993), *The Reliability Availability and Productiveness of Systems* (Sherwin and Bossche, 1993) and *New Trends in System Reliability Evaluation* (Misra, 1994). It is intended that the material given in this chapter be supplemented by this literature.

7.1 Development of Reliability Engineering

Some of the earliest developments in reliability engineering occurred during the Second World War. The Germans had problems with the reliability of the V-1 missile (Bazovsky, 1961). The project team leader, Lusser, has described how the first approach taken to the problem was based on the argument that a chain is no stronger than its weakest link. This concentrated attention on the small number of low reliability components. But this approach was not successful. It was then pointed out by a mathematician, Pieruschka, that the probability of success p in a system in which all the components must work if the system is to work is the product of the individual probabilities of success p_i:

$$p = \prod_{i=1}^{n} p_i \qquad [7.1.1]$$

This drew attention to the need to improve the reliability of the many medium reliability components. This approach was much more successful in improving missile reliability. Equation 7.1.1 is known as *Lusser's product law of reliabilities.*

On the other side of the Channel, Blackett (1962) was drawing attention to the significance of Equation 7.1.1 to military operations in general:

> In the simplest case of air attack on a ship, the four main probabilities are (1) the chance of a sighting, (2) the chance the aircraft gets in an attack, (3) the chance of a hit on a ship, and (4) the chance that the hit causes the ship to sink.

This work was the beginning of operational research.

The US armed forces also had serious reliability problems, particularly with vacuum tubes used in electronic equipment. Studies of electronic equipment reliability at the end of the War showed some startling situations (Shooman, 1968a). In the Navy the number of vacuum tubes in a destroyer had risen from 60 in 1937 to 3200 in 1952. A study conducted during manoeuvres revealed that equipment was operational only 30% of the time. An Army study showed that equipment was broken down between two-thirds and three-quarters of the time. The Air Force found that over a 5-year period maintenance and repair costs of equipment exceeded the initial cost by a factor of 10. It was also discovered that for every vacuum tube in use there was one held as spare and seven in transit, and that one electronics technician was needed for every 250 vacuum tubes. These studies illustrate well the typical problems in reliability engineering, which is concerned not only with reliability but also with availability, maintenance and so on.

From these early beginnings the study of reliability has become a fully developed discipline. It has received particular impetus from the reliability requirements in the fields of defence and aerospace, and electronics and computers. Perhaps the most spectacular example is the moonshots, which depended crucially on reliability technology.

One of the main fields of application of reliability engineering has been in electronic equipment. Such equipment typically has a large number of components. Initially the reliability of electronic equipment was much less than that of mechanical equipment. But the application of reliability engineering to electronic equipment has now made it generally as reliable.

Another area in which reliability engineering has been widely use is nuclear energy. Methods have had to be developed to assess the hazards of nuclear reactors and to design instrument trip systems to shut them down safely.

In the UK, work on reliability of nuclear reactors has been done by the UK Atomic Energy Authority (UKAEA), originally through its Health and Safety Branch and subsequently through the Safety and Reliability Directorate (SRD), which runs the National Centre for Systems Reliability (NCSR) and the Systems Reliability Service (SRS). The latter operates a consultancy service on industrial reliability problems and a failure data bank.

The development of reliability engineering and of loss prevention in the chemical and petroleum industries has been described in Chapter 1.

7.2 Reliability Engineering in the Process Industries

7.2.1 Applicability of reliability techniques

It is entirely right that the process industries should seek to apply the techniques and obtain the benefits of reliability engineering. But it is important to recognize that reliability engineering as a discipline has grown up outside these industries. It is to be expected, therefore, that the techniques will need to be adapted to and developed for the problems of the process industries. A similar adaptation was necessary in order to apply control engineering to process problems. The process industries are particularly concerned with mechanical equipment reliability. This is proving a rather more intractable problem.

7.2.2 Reliability assessment and improvement

Reliability engineering involves an iterative process of reliability assessment and improvement, and the relationship between these two aspects is important. Work on the reliability of a system necessarily involves assessment of that reliability. In some cases the assessment shows that the system is sufficiently reliable. In other cases the reliability is found to be inadequate, but the assessment work reveals ways in which the reliability can be improved. It is generally agreed that the value of reliability assessment lies not in the figure obtained for system reliability, but in the discovery of the ways in which reliability can be improved.

The reliability engineer, however, cannot wait until his fellow engineers have solved all their reliability problems. It is his job to identify the areas where improvements are essential for success. But for the rest, he is obliged to accept, as given, the levels of reliability currently being achieved. This is not the case, however, with other engineers. It is they who are in a position to reduce the number of failures. They too should use reliability techniques. But it would be disastrous if they were to take the existing level of reliability as unalterable.

7.2.3 Reliability and other probabilistic methods

Loss prevention makes use of a wide range of probabilistic methods. A large proportion of these are reliability techniques as conventionally defined, but in addition use is made of other probabilities. In particular, there is considerable emphasis in loss prevention on the assessment of the consequences of failures, taking into account factors such as numbers of people exposed, weather conditions, etc.

7.2.4 Reliability and quality control

There is a close link between reliability and quality control of equipment. But the two are not identical and the distinction between them is important. An equipment is likely to be unreliable unless there is good quality control over its manufacture. In general, quality control is a necessary condition for reliability. It is not, however, a sufficient condition. Deficiencies in specification, design or application are also causes of unreliability. A badly designed equipment may be manufactured with good quality control, but it will remain unreliable. This aspect is discussed in more detail by F. Nixon (1971).

7.2.5 Reliability standards

The principal British Standards dealing with reliability are BS 4778:1979 *Quality Vocabulary*, for terminology, and BS 5760: 1981 *Reliability of Systems, Equipment and Components*. BS 4200: 1967 *Guide on the Reliability of Electronic Equipment and Parts Used Therein*, which dealt both with terminology and other matters, is now withdrawn.

The constituent parts of BS 5760 are Part 0: 1986 *Introductory Guide to Reliability*, Part 1: 1985 *Guide to Reliability and Maintainability Programme Management*, Part 2: 1981 *Guide to the Assessment of Reliability*, Part 3: 1982 *Guide to Reliability Practices: Examples*, Part 4: 1986 *Guide to Specification Clauses Relating to the Achievement and Development of Reliability in New and Existing Items*, Part 5: 1991 *Guide to Failure Modes, Effects and Criticality Analysis (FMEA and FMECA)*, Part 6: 1991 *Guide to Programmes for Reliability Growth*, Part 7: 1991 *Guide to Fault Tree Analysis*, Part 9: 1991 *Guide to the Block Diagram Technique*, and Part 10: 1993 *Guide to Reliability Testing*.

Another widely quoted set of standards are the US Armed Services MIL standards, and handbooks. Terminology is given in MIL-STD-721B and MIL-STD-781B. Standards dealing with reliability include MIL-STD-756 *Reliability Prediction*, MIL-STD-781A *Reliability Tests: Exponential Distribution*, MIL-STD-781C *Reliability Design Qualification and Production Acceptance Tests*, MIL-STD-785 *Requirements for Reliability Program*, and MIL-STD-1635 *Reliability Growth Theory*, and handbooks MIL-HDBK-189 *Reliability Growth Management*, and MIL-HDBK-217 *Reliability Prediction of Electronic Equipment*. Maintainability is covered in MIL-STD-470 *Maintainability Program Requirements*, MIL-STD-471 *Maintainability Demonstration* and MIL-HDBK-472 *Maintainability Prediction*.

7.3 Definition of Reliability

Definitions of reliability are given in various British Standards dealing with terminology on quality and reliability. BS 4778: Part 1: 1987 defines reliability as

> The ability of an item to perform a required function under stated conditions for a stated period of time.

This supersedes the definition given in BS 4200: Part 2: 1974 which has now been withdrawn. Further, essentially similar, definitions are given in BS 4778: Section 3.1: 1991 and BS 4778: Section 3.2: 1991. BS 4778: Section 3.1: 1991 states that in definitions relating to reliability, the word 'time' may be replaced by 'distance', 'cycles', or other quantities or units, as appropriate.

An alternative definition of reliability is: 'The probability that an item will perform a required function under stated conditions for a stated period of time'. This definition brings out several important points about reliability: (1) it is a probability, (2) it is a function of time, (3) it is a function of defined conditions, and (4) it is a function of the definition of failure.

Some definitions of failure are: (1) failure in operation, (2) failure to operate on demand, (3) operation before demand, and (4) operation after demand to cease. The first of these definitions is applicable to an equipment which operates continuously, while the other definitions are applicable to one which operates on demand.

7.4 Meanings of Probability

It may appear intuitively obvious what is meant by probability, but the word in fact has several meanings. Moreover, the distinctions are of some practical importance. They are relevant, for example, to the question of the relative weight which should be attached to field data and to other information available to individuals.

7.4.1 Equal likelihood

One definition of probability derives from the principle of equal likelihood. If a situation has n equally likely and mutually exclusive outcomes, and if n_A of these outcomes are event A, then the probability $P(A)$ of event A is:

$$P(A) = \frac{n_A}{n} \qquad [7.4.1]$$

This probability can be calculated *a priori* and without doing experiments.

The example usually given is the throw of an unbiased die, which has six equally likely outcomes: the probability of throwing a one is 1/6. Another example is the withdrawal of a ball from a bag containing four white balls and two red ones: the probability of withdrawing a red one is 1/3. The principle of equal likelihood applies to the second case also, because, although the likelihood of withdrawing a red ball and a white one are unequal, the likelihood of withdrawing any individual ball is equal.

This definition of probability is often of limited usefulness in engineering because of the difficulty of defining situations with equally likely and mutually exclusive outcomes.

7.4.2 Relative frequency

The second definition of probability is based on the concept of relative frequency. If an experiment is performed n times and if the event A occurs on n_A of these occasions, then the probability $P(A)$ of event A is:

$$P(A) = \lim_{n \to \infty} \frac{n_A}{n} \qquad [7.4.2]$$

This probability can only be determined by experiment.

This definition of reliability is the one which is most widely used in engineering. In particular, it is this definition which is implied in the estimation of probability from field failure data.

7.4.3 Personal probability

A third definition of probability is degree of belief. It is the numerical measure of the belief which a person has that the event will occur. Often this corresponds to the relative frequency of the event. But this is not always so, for several reasons. One is that the relative frequency data available to the individual may be limited or non-existent. Another is that even if he has such data, he may have other information which causes him to think that the data are not the whole truth. There are many possible reasons for this. The individual may doubt the applicability of the data to the case under consideration, or he may have information which suggests that the situation has changed since these data were collected.

It is entirely legitimate to take into account such personal probabilities:

> Personal probability was cast into disrespect during the nineteenth century when science was believed to be absolute truth, because, with this definition the results will depend upon the person solving the problem. However, this objection to subjectivity has recently been countered quite effectively by Savage, Janes and other physical scientists. (Breipohl, 1970, p. 6)

There are several branches of probability theory which attempt to accommodate personal probability. These include ranking techniques (e.g. Siegel, 1956), which give the numerical encoding of judgements on the probability of ranking of items, and Bayesian methods (e.g. Breipohl, 1970), which allow probabilities to be modified in the light of additional information.

Further discussions of personal probability are given by Savage (1962) and by Tribus (1969).

7.5 Some Probability Relationships

It is appropriate to give, at this point, a brief treatment of some basic probability relationships. These are important in the present context not only because they are the basis of reliability expressions, but also because they are needed for work in areas such as fault trees.

7.5.1 Sets and Boolean algebra

Some set theory definitions, operations and laws are given in Table 7.2 and some of these are illustrated by the Venn diagrams given in Figure 7.1.

The union and intersection of sets may be written, respectively, in the notation

$$C = A \cup B \qquad [7.5.1]$$

and

$$C = A \cap B \qquad [7.5.2]$$

or in the notation

$$C = A + B \qquad [7.5.3]$$

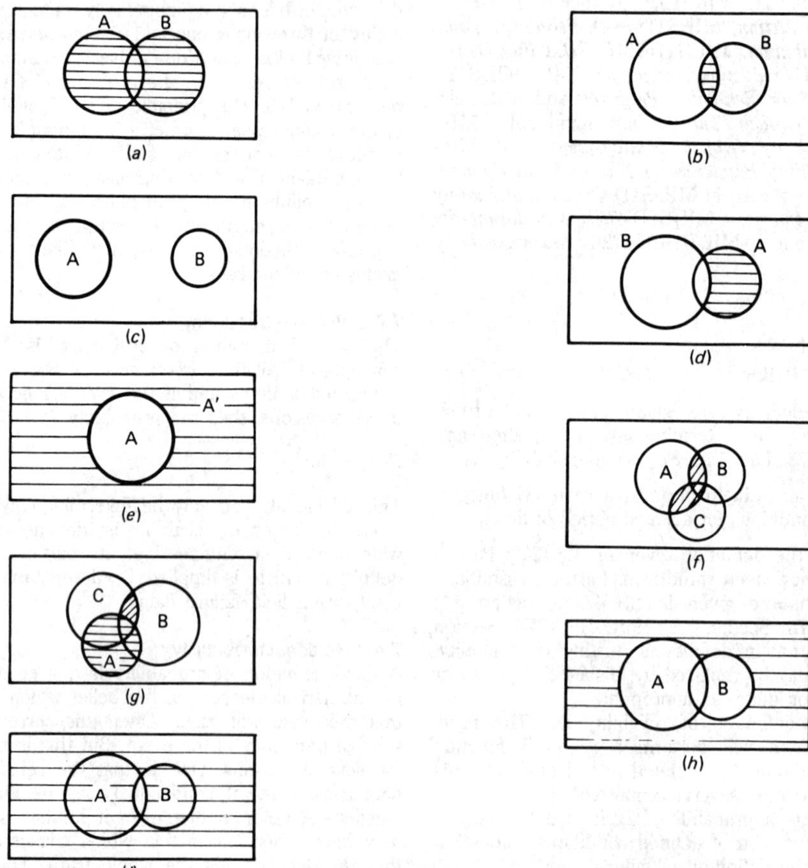

Figure 7.1 Some set theory definitions, operations and laws: (a) union of sets, A + B; (b) intersection of sets, AB; (c) disjoint sets; (d) difference of sets, A − B; (e) complement of a set; (f) distributive law, A(B + C) = A + BC; (g) distributive law, (A + B)(A + C) = A + BC; (h) dualization law (A + B)′ = A′B′; (i) dualization law (AB)′ = A′ + B′

Table 7.2 Some set theory definitions, operations and laws

Definitions:	
Null set $\varnothing$	$\varnothing$ is a set with no elements
Sample space set S	S is the set containing all the elements in the sample space
Elements	
$a \in A$	a is an element of A
Subsets	
$A \subset B$	A is a subset of B
$B \supset A$	B contains A
$A \subseteq B$	A is a subset of B or is equal to B
Equality	
$A = B$	A has the same elements as B
$A = \varnothing$	A has no elements
Operations:	
Union of sets	
$C = A \cup B$	C contains all the elements of A and B
also written	
$C = A + B$	
Intersection of sets	
$C = A \cap B$	C contains only the elements common to A and B
also written	
$C = AB$ or $A . B$	
$C = A$ and B	
Disjoint sets (mutually exclusive sets)	
$C = A \cap B = \varnothing$	
Difference of sets	
$C = A - B$	C contains all the elements of A which are not elements of B
Complement of a set[a]	
$A' = S - A$	A' contains all the elements of S which are not elements of A
Laws:	
Commutative laws	
$A + B = B + A$	
$AB = BA$	
Associative laws	
$(A + B) + C = A + (B + C) = A + B + C$	
$(AB)C = A(BC) = ABC$	
Distributive laws	
$A(B + C) = AB + AC$	
$A + BC = (A + B)(A + C)$	
Absorption laws	
$A + A = A$	
$AA = A$	
Dualization (de Morgan's) laws	
$(A + B)' = A'B'$	
$(AB)' = A' + B'$	

[a] Use may also be made of the notation $\bar{A}$ to signify 'not A'.

and

$$C = A \cdot B = AB \qquad [7.5.4]$$

It is mainly the latter notation which is used here.

Attention is drawn in particular to the distributive and absorption laws, which differ somewhat from their apparent analogues in normal algebra. Set theory and Boolean algebra are described in standard texts (e.g. Reza, 1961; Shooman, 1968a; Breipohl, 1970). Probability and set theory are closely related, because the outcomes of a situation constitute, in effect, a set.

7.5.2 Probability of unions

The probability of an event X which occurs if any of the events A_i occur and is thus the union of those events is:

$$P(X) = P\left(\bigcup_{i=1}^{n} A_i\right) \qquad [7.5.5]$$

If there are two events:

$$P(X) = P(A_1) + P(A_2) - P(A_1A_2) \qquad [7.5.6]$$

If there are three events:

$$P(X) = P(A_1) + P(A_2) + P(A_3) - P(A_1A_2) - P(A_1A_3) - P(A_2A_3) + P(A_1A_2A_3) \quad [7.5.7]$$

If there are n events:

$$P(X) = P(A_1) + (PA_2) + \cdots + P(A_n) - P(A_1A_2) - P(A_1A_3) - \cdots - P(A_{n-1}A_n) + P(A_1A_2A_3) + P(A_1A_2A_4) + \cdots + P(A_{n-2}A_{n-1}A_n) \cdots (-1)^{n-1}P(A_1A_2\cdots A_n) \quad [7.5.8]$$

If the events are mutually exclusive, Equation 7.5.5 simplifies to:

$$P(X) = \sum_{i=1}^{n} P(A_i) \quad [7.5.9]$$

In general:

$$P\left(\bigcup_{i=1}^{n} A_i\right) \leq P\sum_{i=1}^{n} P(A_i) \quad [7.5.10]$$

For events not mutually exclusive but of low probability the error in using Equation 7.5.9 instead of Equation 7.5.5 is small. Equation 7.5.9 is sometimes called the low probability, or rare event, approximation. The estimate of probability given by Equation 7.5.9 errs on the high side and hence is conservative in calculating failure probabilities, but is not conservative in calculating success probabilities or reliabilities.

7.5.3 Joint and marginal probability

So far the events considered are the outcomes of a single experiment. Consideration is now given to events which are the outcome of several subexperiments.

The probability of an event X which occurs only if all the n events A_i occur and is thus the intersection of these events is:

$$P(X) = P(A_1 \ldots A_n) = P\left(\bigcap_{i=1}^{n} A_i\right) \quad (7.5.11)$$

$P(A_1 \ldots A_n)$ is the joint probability of the event.

The probability of an event X which occurs if any of the mutually exclusive and exhaustive n events A_i in one subexperiment occurs and the event B in a second subexperiment occurs is:

$$P(X) = P\left(\bigcup_{i=1}^{n} A_i B\right) \quad [7.5.12]$$

$$= \sum_{i=1}^{n} P(A_i)P(B) \quad [7.5.13]$$

$$= P(B) \quad [7.5.14]$$

P(B) is the marginal probability of the event.

7.5.4 Conditional probability

The probability of an event X which occurs if the event A occurs in one subexperiment and the event B occurs in a second subexperiment where the event A depends on the event B is:

$$P(X) = P(AB) = P(A|B)P(B) \quad [7.5.15]$$

$P(A|B)$ is the conditional probability of A, given B.

The probability obtained in Equation 7.5.15 is a joint probability. Marginal probabilities may also be obtained from conditional probabilities:

$$P(X) = P(B) = \sum_{i=1}^{n} P(B|A_i)P(A_i) \quad [7.5.16]$$

7.5.5 Independence and conditional independence

If events A and B are independent:

$$P(AB) = P(A|B)P(B) = P(A)P(B) \quad [7.5.17]$$

which is equivalent to

$$P(A|B) = P(A) \quad [7.5.18]$$

The probability of an event X which occurs only if all n events A_i occur is given by Equation 7.5.11. If all n events are independent:

$$P(X) = P(A_1 \ldots A_n) = \Pi_{i=1}^{n} P(A_i) \quad [7.5.19]$$

Two events A and B are conditionally dependent if their relationship with a third event C is:

$$P(AB|C) = P(A|C)P(B|C) \quad [7.5.20]$$

Conditional independence does not imply independence.

7.5.6 Bayes' theorem

The relationship given in Equation 7.5.15 is a form of Bayes' theorem. This theorem is extremely important in probability work. It appears in various forms, some of which are:

$$P(AB) = P(A|B)P(B) = P(B|A)P(A) \quad [7.5.21]$$

$$P(A|B) = \frac{P(AB)}{P(B)} = \frac{P(B|A)P(A)}{P(B)} \quad [7.5.22]$$

Or, rewriting the denominator as a marginal probability, and A as A_k

$$P(A_k|B) = \frac{P(B|A_k)P(A_k)}{\sum_{i=1}^{n} P(B|A_i)P(A_i)} \quad [7.5.23]$$

where $P(A_k|B)$ is the posterior probability, $P(B|A_k)$ is the likelihood, and $P(A_k)$ is the prior probability.

Further treatment of Bayes' theorem is given in Section 7.14.

7.6 Some Reliability Relationships

7.6.1 Reliability function and hazard rate

If n equipments operate without replacement, then after time t the numbers which have survived and failed are $n_s(t)$ and $n_f(t)$, respectively, and the probability of survival, or reliability, $R(t)$ is:

$$R(t) = 1 - \frac{n_f(t)}{n} \quad [7.6.1]$$

The instantaneous failure rate, or failure rate expressed as a function of the number of equipments surviving, $z(t)$ is:

$$z(t) = \frac{1}{n - n_f}\frac{dn_f(t)}{dt} = -\frac{1}{R(t)}\frac{dr(t)}{dt} = -\frac{d[ln\ R(t)]}{dt} \quad [7.6.2]$$

$z(t)$ is also called the 'hazard rate', or just the 'failure rate'.

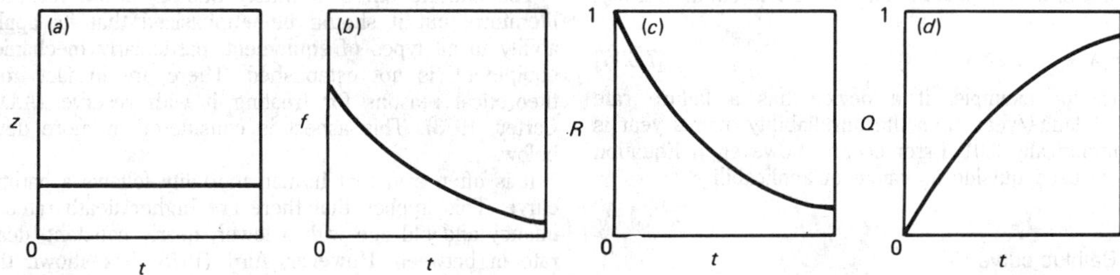

Figure 7.2 The exponential distribution: (a) hazard-rate; (b) failure density; (c) reliability; (d) unreliability

The cumulative hazard function $H(t)$ is:

$$H(t) = \int_0^t z(t)\,\mathrm{d}t \qquad [7.6.3]$$

Then, by integration of Equation 7.6.2 the reliability $R(t)$ is

$$R(t) = \exp\left(-\int_0^t z(t)\,\mathrm{d}t\right) = \exp[-H(t)] \qquad (7.6.4)$$

$R(t)$ is also called the 'reliability function'.

7.6.2 Failure density and failure distribution functions

The overall failure rate, or failure rate expressed as a function of the original number of equipments, $f(t)$ is:

$$f(t) = \frac{1}{n}\frac{\mathrm{d}n_f(t)}{\mathrm{d}t} = -\frac{\mathrm{d}R(t)}{\mathrm{d}t} \qquad (7.6.5)$$

$f(t)$ is also called the 'failure density function'.

The complement of the reliability, or the unreliability, $Q(t)$ is:

$$Q(t) = 1 - R(t) \qquad [7.6.6]$$

$Q(t)$ is also called the 'failure distribution function' and is then commonly written as $F(t)$.

The failure density function and failure distribution function are often referred to, respectively, as the 'failure density' or the 'density function' and the 'failure distribution', the 'distribution function' or the 'cumulative distribution function'.

7.6.3 Relationships between basic functions

The following relationships can readily be derived from Equations 7.6.1–7.6.6 and are particularly useful:

$$z(t) = \frac{f(t)}{R(t)} \qquad [7.6.7a]$$

$$= \frac{f(t)}{1 - Q(t)} = \frac{f(t)}{1 - F(t)} \qquad [7.6.7b]$$

$$R(t) = \int_t^\infty f(t)\,\mathrm{d}t \qquad [7.6.8]$$

$$Q(t) = F(t) = \int_0^t f(t)\,\mathrm{d}t \qquad [7.6.9]$$

7.6.4 Exponential distribution

An important special case is that in which the hazard rate $z(t)$ is constant:

$$z(t) = \lambda \qquad [7.6.10]$$

Then,

$$R(t) = \exp(-\lambda t) \qquad [7.6.11]$$

$$f(t) = \lambda \exp(-\lambda t) \qquad [7.6.12]$$

$$Q(t) = 1 - \exp(-\lambda t) \qquad [7.6.13]$$

These four quantities are shown in Figure 7.2.

In Figure 7.2, the vertical axes for the reliability $R(t)$ and the unreliability $Q(t)$ have the range 0 to 1, but those for the hazard rate $z(t)$ and the failure density function $f(t)$ are proportional to λ. At low values of λt:

$$R(t) = 1 - \lambda t \qquad \lambda t \ll 1 \qquad [7.6.14]$$

$$Q(t) = \lambda t \qquad \lambda t \ll 1 \qquad [7.6.15]$$

Equation 7.6.14 is useful in obtaining accuracy in numerical computation of high values of the reliability $R(t)$. Equation 7.6.15 is useful in making simple computations of the unreliability $Q(t)$. This latter point is further discussed below.

The assumption of constant hazard rate is that normally made in the absence of other information. This is therefore a special case of particular importance.

The failure distribution with constant hazard rate is called the 'exponential distribution' or, more accurately but less commonly, the 'negative exponential distribution'. It is also referred to as the 'random failure distribution', although in most cases it is a moot point whether the failures are appropriately called random.

7.6.5 Probability and event rate

The relationship between unreliability and hazard rate represented by Equation 7.6.13, or more generally the relationship between the probability $P(t)$ and the event rate ζ

$$P(t) = 1 - \exp(-\zeta t) \qquad [7.6.16]$$

is an important one. It is frequently necessary in reliability work to calculate a probability from a rate, or vice versa, and it is this equation which then applies.

7.6.6 Unreliability and failure rate

There is sometimes confusion between unreliability Q(t) and failure rate λ. This confusion arises because, given certain assumptions, the two are numerically identical.

This occurs if in Equation 7.6.15 time t is equal to unity. Then,

$$Q(t) = \lambda \qquad \lambda \ll 1 \qquad [7.6.17]$$

Thus, for example, if a device has a failure rate $\lambda = 0.01$ faults/year, then the unreliability over a year is also numerically 0.01. Error occurs, however, if Equation 7.6.15 is used outside its range of applicability.

7.6.7 Bathtub curve

In general, the failure behaviour of an equipment exhibits three stages: initially during commissioning the rate is high, then it declines during normal operation, and finally it rises again as deterioration sets in. For many equipments, particularly electronic equipments, the rate has been found to form a bathtub curve, as shown in Figure 7.3(a) (e.g. Carhart, 1953). This curve has three regimes: (1) early failure, (2) constant failure, and (3) wear-out failure.

Early failure, or infant mortality, is usually due to such factors as defective equipment, incorrect installation, etc. It also tends to reflect the learning curve of the equipment user. Constant failure, or so-called 'random failure', is often caused by random fluctuations of load which exceed the design strength of the equipment. A constant failure characteristic is also shown by an equipment which has a number of components that individually exhibit different failure distributions. Wear-out failure is self-explanatory. The corresponding curve for the failure density function is shown in Figure 7.3(b).

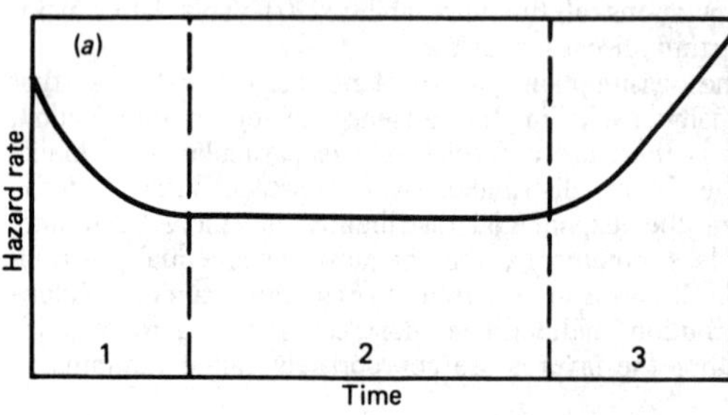

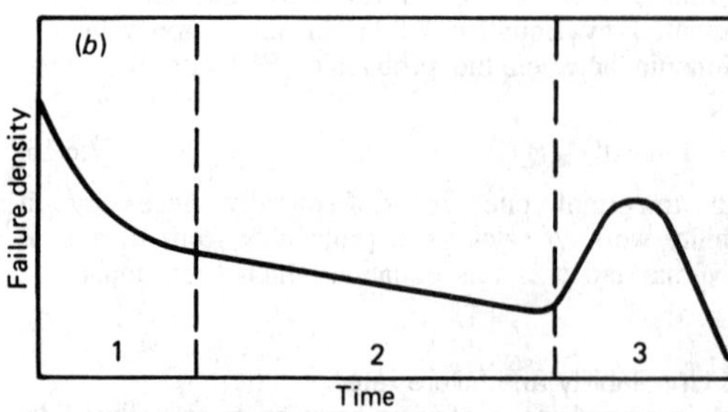

Figure 7.3 The bathtub curve: (a) hazard rate; (b) failure density

The bathtub curve is widely quoted in the reliability literature, but it should be emphasized that its applicability to all types of equipment, particularly mechanical equipment, is not established. There are in fact good theoretical reasons for treating it with reserve (A.D.S. Carter, 1973). This aspect is considered in more detail below.

It is often said that human mortality follows a bathtub curve. This implies that there are higher death rates in infancy and old age with a lower, nearly constant, death rate in between. However, Aird (1978) has shown that for humans the infant mortality period is followed immediately by the onset of the wear-out period, so that there is effectively no constant failure period.

7.6.8 Mean life

The mean life m is defined as the first moment of the failure density function

$$m = \int_0^\infty tf \, dt \qquad [7.6.18]$$

Alternatively, from Equations 7.6.5 and 7.6.18

$$m = -\int_0^\infty t\frac{dR}{dt} \, dt \qquad [7.6.19a]$$

$$= -[tR]_0^\infty + \int_0^\infty R \, dt \qquad [7.6.19b]$$

Since

$$[tR]_0^\infty = 0$$

provided that

$$\lim_{t\to\infty} (tR) = 0$$

then

$$m = \int_0^\infty R \, dt \qquad [7.6.20]$$

Alternatively, use may be made of the relationships between the moments μ_i and the Laplace transform $\bar{g}(s)$ of a function $g(t)$:

$$\mathcal{L}[g(t)] = \bar{g}(s) = \int_0^\infty g(t) \exp(-st) \, dt \qquad [7.6.21]$$

Hence:

$$\frac{d^i\bar{g}(s)}{ds^i} = (-1)^i \int_0^\infty t^i g(t) \exp(-st) \, dt \qquad [7.6.22]$$

and

$$\left(\frac{d^i\bar{g}(s)}{ds^i}\right)_{s=0} = (-1)^i \int_0^\infty t^i g(t) \, dt \qquad [7.6.23a]$$

$$= (-1)^i \mu_i \qquad [7.6.23b]$$

Applying these relationships to the present case by substituting dR/dt for $g(t)$ in Equation 7.6.23a,

$$m = \left(\frac{d[sR - R(0)]}{ds}\right)_{s=0} = (R)_{s=0} \qquad [7.6.24]$$

For the exponential distribution

$$m = \int_0^{\infty} t\lambda \exp(-\lambda t)\, dt \qquad [7.6.25a]$$

$$= \frac{1}{\lambda} \qquad [7.6.25b]$$

Other terms used in addition to 'mean life' are the mean time between failures (MTBF), the mean time to failure (MTTF) and the mean time to first failure (MTTFF). These times are sometimes used interchangeably, but they are not identical.

The most widely used is probably MTBF. MTBF has meaning only when applied to a population of components, equipments or systems in which there is repair. It is the total operating time of the items divided by the total number of failures. It is also the mean of the failure distribution, regardless of its form.

MTTF is applied to items without repair and is the mean of the distribution of times to failure. MTTFF is applied to items with repair and is the mean of the distribution of times to first failure.

MTTF and MTTFF are applied particularly to systems. For an n-component parallel system with an exponential failure distribution of the individual components and without repair,

$$\mathrm{MTTF} = \sum_{i=1}^{n} \frac{1}{i\lambda} \qquad [7.6.26]$$

and for an n-component parallel system with repair

$$\mathrm{MTTFF} = \frac{1}{\lambda} \cdot \sum_{i=0}^{n-1} \frac{(1+\mu/\lambda)^i}{i+1} \qquad [7.6.27]$$

where λ is the failure rate of equipment, and μ is the repair rate of equipment. Repair rates are discussed in Section 7.10. If the repair rate μ is zero, Equation 7.6.27 reduces to Equation 7.6.26.

Formal definitions of MTBF and MTTF are given in BS 4200: Part 2: 1967. For repairable systems used is also made of mean time to repair (MTTR). A more detailed discussion of these quantities is given by Myers *et al.*, (1964).

7.6.9 Expected value

Use is frequently made in reliability engineering of the concept of the expected value. The expected value of a distribution is its mean value. The use of the expected value concept allows convenient manipulation, as the following treatment demonstrates. (A treatment of expected value is given by Breipohl (1970).)

For a variable t, the expected value, or mean value, is:

$$m = E[t] \qquad [7.6.28]$$

The variance is:

$$\sigma^2 = E[(t-m)^2] \qquad [7.6.29]$$

The variance may then be expressed in the following alternative forms:

Table 7.3 *Some properties of failure distributions*

Distribution	*Failure density function* $f(t)$	*Mean* $m = \int_{-\infty}^{\infty} t f(t)\, dt$	*Median* $= t$ where $\int_{-\infty}^{\tau} f(\tau)\, d\tau = 0.5$	*Mode* $= t$ where $\frac{df(t)}{dt} = 0$	*Variance* $\sigma^2 = \int_{-\infty}^{\infty} (t-m)^2 f(t)\, dt$
Exponential	$\lambda \exp(-\lambda t)$	$\frac{1}{\lambda}$	$\frac{1}{\lambda} \ln 2$	0	$\frac{1}{\lambda}^2$
Normal	$\frac{1}{\sigma(2\pi)^{1/2}} \exp\left[-\frac{(t-m)^2}{2\sigma^2}\right]$	m	m	m	σ^2
Log–normal	$\frac{1}{\sigma t(2\pi)^{1/2}} \exp\left[-\frac{[\ln(t)-m^*]^2}{2\sigma^2}\right]$	$\exp\left(m^* + \frac{\sigma^2}{2}\right)$	$\exp(m^*)$	$\exp(m^* - \sigma^2)$	$\exp(2m^* + \sigma^2)[\exp(\sigma^2) - 1]$
Weibull (two-parameter)	$\frac{\beta}{\eta}\left(\frac{t}{\eta}\right)^{\beta-1} \exp\left[-\left(\frac{t}{\eta}\right)^{\beta}\right]$	$\eta\Gamma\left(1+\frac{1}{\beta}\right)$	$\eta(\ln 2)^{1/\beta}$	$\eta\left(1-\frac{1}{\beta}\right)^{1/\beta}$, $\beta > 1$ 0 $\beta \leq 1$	$\eta^2\left\{\Gamma\left(1+\frac{2}{\beta}\right) - \left[\Gamma\left(1+\frac{1}{\beta}\right)\right]^2\right\}$
Rectangular	$\frac{1}{b}$	$a + \frac{b}{2}$	$a + \frac{b}{2}$	—	$\frac{b^2}{12}$
Gamma	$\frac{1}{b\Gamma(a)}\left(\frac{t}{b}\right)^{a-1} \exp\left(-\frac{t}{b}\right)$	ba	—[a]	$b(a-1)$, $a \geq 1$	$b^2 a$
Pareto	$at^{-(a+1)}$	$\frac{a}{a-1}$, $a > 1$	$2^{1/a}$	0	$\frac{a}{a-2} - \left(\frac{a}{a-1}\right)^2$, $a > 2$
Extreme value	$\frac{1}{b} \exp\left(\frac{t-a}{b}\right) \exp\left[-\exp\left(\frac{t-a}{b}\right)\right]$	$a + b\,\Gamma'(1)$[b]	$a + b \ln \ln 2$	a	$\frac{b^2\pi^2}{6}$

[a] No simple expression is available.
[b] $\Gamma'(1) = -0.577\,21$ is the first derivative of the gamma function.

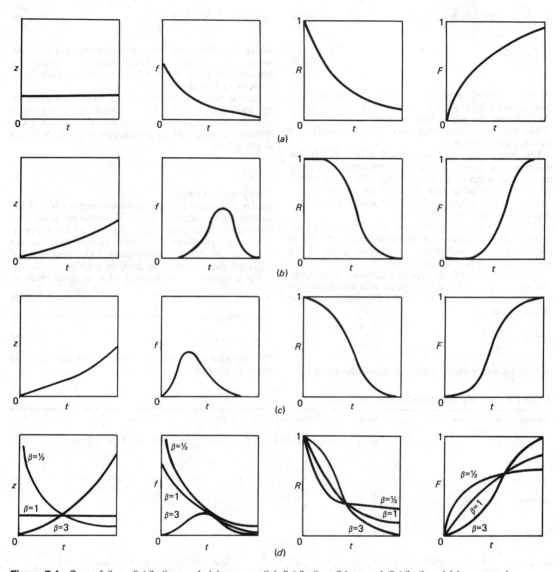

Figure 7.4 *Some failure distributions – 1: (a) exponential distribution; (b) normal distribution; (c) log–normal distribution; (d) Weibull distribution*

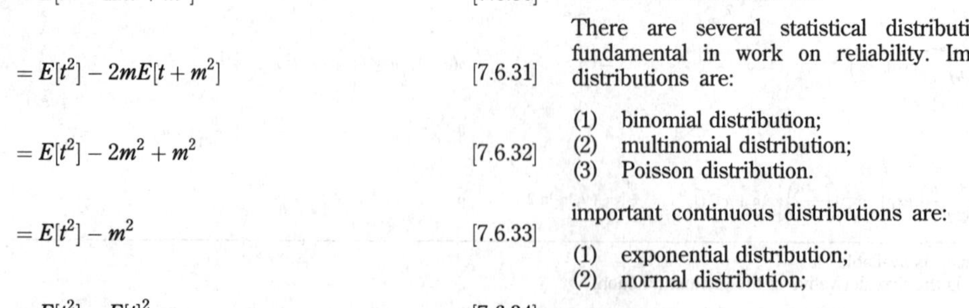

$$\sigma^2 = E[t^2 - 2mt + m^2] \qquad [7.6.30]$$

$$= E[t^2] - 2mE[t + m^2] \qquad [7.6.31]$$

$$= E[t^2] - 2m^2 + m^2 \qquad [7.6.32]$$

$$= E[t^2] - m^2 \qquad [7.6.33]$$

$$= E[t^2] - E[t]^2 \qquad [7.6.34]$$

7.7 Failure Distributions

There are several statistical distributions which are fundamental in work on reliability. Important discrete distributions are:

(1) binomial distribution;
(2) multinomial distribution;
(3) Poisson distribution.

important continuous distributions are:

(1) exponential distribution;
(2) normal distribution;

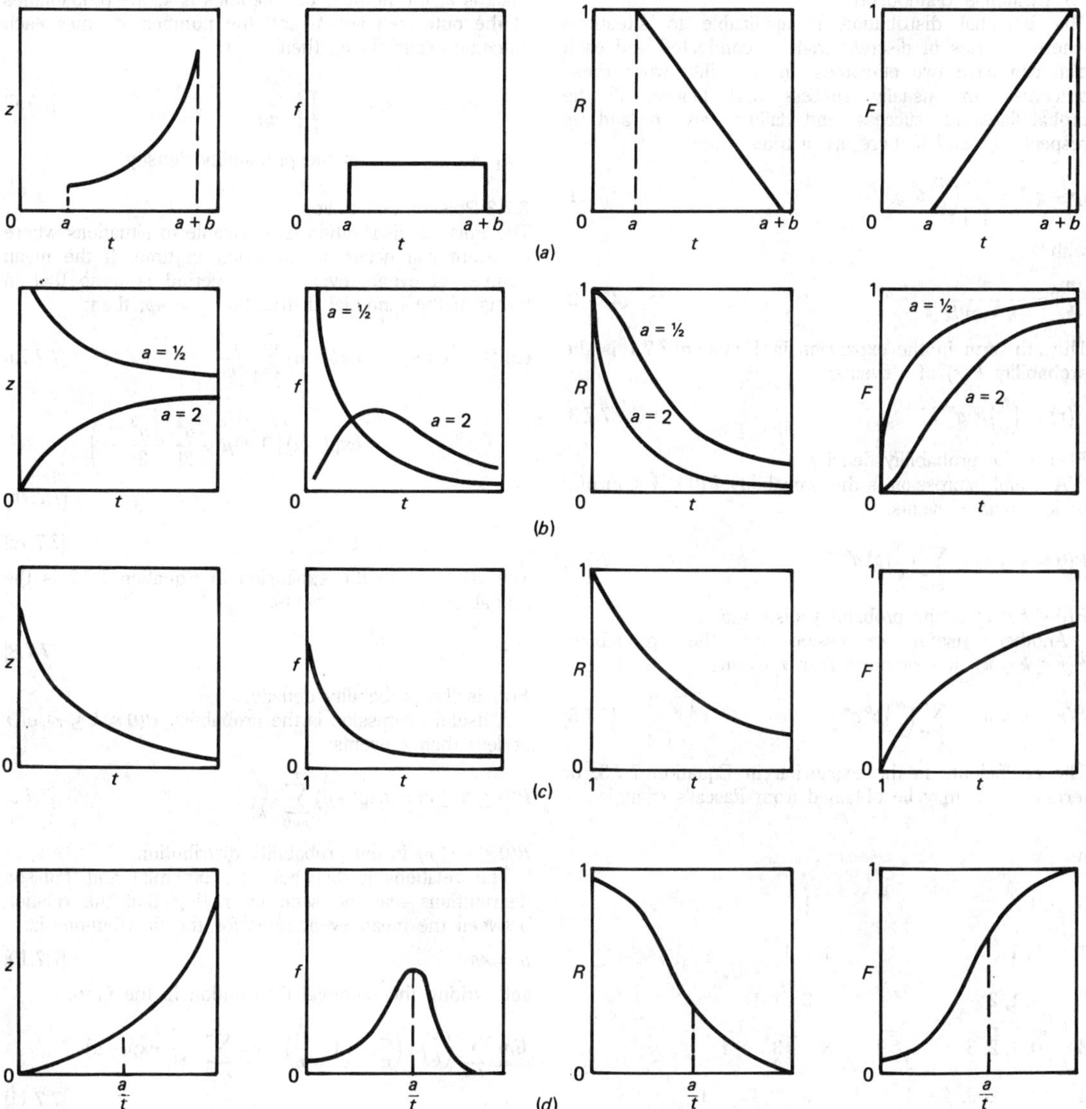

Figure 7.5 *Some failure distributions – 2: (a) rectangular distribution; (b) gamma distribution; (c) Pareto distribution; (d) extreme value distribution*

(3) log–normal distribution;
(4) Weibull distribution;
(5) rectangular distribution;
(6) gamma distribution;
(7) Pareto distribution;
(8) extreme value distribution.

Some properties of these distributions are given in Table 7.3 and in Figures 7.4 and 7.5.

The distributions are best regarded as statistical distributions and the independent variable t as a generalized variable, although in the present context the primary interest in many of these distributions is as failure distributions, with time as the independent variable. The distribution function is given as F rather than Q, but the two are identical.

Accounts of the properties of statistical distributions are given in most texts on reliability engineering and in the Center for Chemical Process Safety (CCPS) *QRA Guidelines* (1989/5). Additional treatments for particular distributions are mentioned below. A rather comprehensive summary of the principal distributions is given by Hastings and Peacock (1974).

7.7.1 Binomial distribution

The binomial distribution is applicable to situations where a series of discrete trials is conducted and each trial can have two outcomes. In reliability work these outcomes are usually success and failure. If the probabilities of success and failure are p and q, respectively, and if there are n trials, then:

$$(p+q)^n = \sum_{k=0}^{n} \binom{n}{k} p^k q^{n-k} = 1 \qquad [7.7.1]$$

with

$$\binom{n}{k} = \frac{n!}{(n-k)!k!} \qquad [7.7.2]$$

The rth term in the expansion in Equation 7.7.1 is the probability $P(r)$ of r events:

$$P(r) = \binom{n}{r} p^r q^{n-r} \qquad [7.7.3]$$

$P(r)$ is the probability density.

A useful expression is the probability $P(0 \leq k \leq r)$ of r or less than r events:

$$P(0 \leq k \leq r) = \sum_{k=0}^{r} \binom{n}{k} p^k q^{n-k} \qquad [7.7.4]$$

$P(0 \leq k \leq r)$ is the probability distribution.

Another useful expression is the probability $P(r \leq k \leq n)$ of r or more than r events:

$$P(r \leq k \leq n) = \sum_{k=r}^{n} \binom{n}{k} p^k q^{n-k} \qquad [7.7.5]$$

The coefficients in the expansion in Equation 7.7.3, or terms of $\binom{n}{r}$, may be obtained from Pascal's triangle

n	r									
0	0					1				
1	0, 1				1		1			
2	0, 1, 2			1		2		1		
3	0, 1, 2, 3		1		3		3		1	
4	0, 1, 2, 3, 4	1		4		6		4		1

A number in the triangle is obtained by summing the two numbers directly above it to the left and the right.

The mean of the binomial distribution is np and the variance is npq. The binomial distribution approximates to the normal distribution for large values of n and does so most rapidly for $p = q = 0.5$. The approximation holds for $p \leq 0.5$ and $np \geq 5$ or for $p > 0.5$ and $nq > 5$. In terms of n and p the mean and variance of the normal distribution are np and $np(1-p)$, respectively. The binomial distribution approximates to the Poisson distribution for $p \leq 0.05$ and $n \geq 20$. In terms of n and p, the mean of the Poisson distribution is np.

7.7.2 Multinomial distribution

The multinomial distribution is applicable to situations where a series of trials is conducted and each trial can have more than two outcomes. If the total number of trials is n, the number of outcomes is k, the probabilities of the outcomes are p_i and the number of times each outcome occurs is n_i, then:

$$f(n_1, n_2, \ldots, n_k) = n! \prod_{i=1}^{k} \frac{p_i^{n_i}}{n_i!} \qquad [7.7.6]$$

$f(n_1, n_2, \ldots, n_k)$ is the probability density.

7.7.3 Poisson distribution

The Poisson distribution is applicable to situations where an event can occur at any point in time. If the mean number of events over a time period is μ so that in terms of the binomial distribution $\mu = np$, then:

$$\exp(-\mu)\exp(\mu) = \exp(-\mu)\sum_{k=0}^{\infty} \frac{\mu^k}{k!} \qquad [7.7.7a]$$

$$= \exp(-\mu)\left(1 + \mu + \frac{\mu^2}{2!} + \frac{\mu^3}{3!} \cdots\right) \qquad [7.7.7b]$$

$$= 1 \qquad [7.7.7c]$$

The rth term in the expansion in Equation 7.7.7 is the probability $P(r)$ of r events:

$$P(r) = \exp(-\mu)\frac{\mu^r}{r!} \qquad [7.7.8]$$

$P(r)$ is the probability density.

A useful expression is the probability $P(0 \leq k \leq r)$ of r or less than r events:

$$P(0 \leq k \leq r) = \exp(-\mu)\sum_{k=0}^{r} \frac{\mu^k}{k!} \qquad [7.7.9]$$

$P(0 \leq k \leq r)$ is the probability distribution.

The relationship between the binomial and Poisson distributions may be seen by noting that the relation between the mean event rates for the distributions is:

$$\mu = np \qquad [7.7.10]$$

and writing the binomial distribution in the form:

$$\lim_{n\to\infty} \sum_{k=0}^{n} \binom{n}{k} \cdot \left(\frac{\mu}{n}\right)^k \cdot \left(1 - \frac{\mu}{n}\right)^{n-k} = \sum_{k=0}^{\infty} \frac{\mu^k}{k!} \exp(-\mu) \qquad [7.7.11]$$

A derivation of the Poisson distribution from Markov models is given below.

7.7.4 Exponential distribution

For the exponential distribution, the characteristics hazard rate z, failure density f, reliability R and failure distribution F have been derived above, and are:

$$z = \lambda \qquad [7.7.12]$$

$$f = \lambda \exp(-\lambda t) \qquad [7.7.13]$$

$$R = \exp(-\lambda t) \qquad [7.7.14]$$

$$F = 1 - \exp(-\lambda t) \qquad [7.7.15]$$

for the range $0 \leq t \leq \infty$. These quantities are shown in Figure 7.4(a). The distribution is characterized by a single parameter, the hazard rate λ.

The exponential distribution, which has a constant hazard rate, is the distribution usually applied to data in the absence of other information and is the most widely used in reliability work.

7.7.5 Normal distribution

For the normal distribution the characteristics are:

$$z = \frac{\exp\left[-\dfrac{(t-m)^2}{2\sigma^2}\right]}{\displaystyle\int_t^{\infty} \exp\left[-\dfrac{(t-m)^2}{2\sigma^2}\right] dt} \qquad [7.7.16]$$

$$f = \frac{1}{\sigma(2\pi)^{\frac{1}{2}}} \cdot \exp\left[-\frac{(t-m)^2}{2\sigma^2}\right] \qquad [7.7.17]$$

$$R = \frac{1}{\sigma(2\pi)^{\frac{1}{2}}} \cdot \int_t^{\infty} \exp\left[-\frac{(t-m)^2}{2\sigma^2}\right] dt \qquad [7.7.18]$$

for the range $-\infty \leq t \leq \infty$. These quantities are shown in Figure 7.4(b).

The failure density f is the quantity most commonly used to define the normal distribution. It may be noted that Equations 7.7.16 and 7.7.18 describing the other quantities are readily derived from Equations 7.6.7a and 7.6.8, respectively.

The distribution has two parameters, the mean m and the standard deviation σ. It is the natural distribution which is used to characterize data that lie about a mean value and deviate from it by absolute amounts. The normal distribution is widely used in reliability engineering, particularly to fit certain types of failure such as those due to wear out and to fit repair times.

7.7.6 Log-normal distribution

$$z = \frac{\dfrac{1}{t}\exp\left\{-\dfrac{[\ln(t)-m^*]^2}{2\sigma^2}\right\}}{\displaystyle\int_t^{\infty} \frac{1}{t}\exp\left\{-\dfrac{[\ln(t)-m^*]^2}{2\sigma^2}\right\} dt} \qquad [7.7.19]$$

$$f = \frac{1}{\sigma t(2\pi)^{\frac{1}{2}}} \exp\left\{-\frac{[\ln(t)-m^*]^2}{2\sigma^2}\right\} \qquad [7.7.20]$$

$$R = \frac{1}{\sigma(2\pi)^{\frac{1}{2}}} \int_t^{\infty} \exp\left\{-\frac{[\ln(t)-m^*]^2}{2\sigma^2}\right\} dt \qquad [7.7.21]$$

for the range $0 \leq t \leq \infty$. These quantities are shown in Figure 7.4(c).

The distribution has two parameters, m^* and σ. It is the natural distribution to use when deviations from the model value are by factors, proportions or percentages rather than by absolute values as in the normal distribution.

The log–normal distribution has a number of uses in reliability work. It is used to fit certain types of failure such as fatigue failures and to fit repair times. It is also used to describe the range of possible hazard rates of equipment where there is uncertainty about these. In this case the independent variable is the hazard rate z ($= \lambda$ for the exponential distribution) rather than time:

$$f(z) = \frac{1}{\sigma_z z(2\pi)^{\frac{1}{2}}} \exp\left\{-\frac{[\ln(z)-m_z^*]^2}{2\sigma_z^2}\right\} \qquad [7.7.22]$$

The properties of the log–normal distribution are described in more detail by Aitcheson and Brown (1969).

7.7.7 Weibull distribution

For the Weibull distribution the characteristics are:

$$z = \frac{\beta}{\eta}\left(\frac{t-\gamma}{\eta}\right)^{\beta-1} \qquad 7.7.23]$$

$$f = \frac{\beta}{\eta}\left(\frac{t-\gamma}{\eta}\right)^{\beta-1} \exp\left[-\left(\frac{t-\gamma}{\eta}\right)^{\beta}\right] \qquad [7.7.24]$$

$$R = \exp\left[-\left(\frac{t-\gamma}{\eta}\right)^{\beta}\right] \qquad [7.7.25]$$

for the range $\gamma \leq t \leq \infty$. These quantities are shown in Figure 7.4(d) for values of the parameter $\beta = 0.5$, 1.0 and 3.0.

The distribution given is the three-parameter one with the characteristic life η, the shape factor β and the location parameter γ. If the latter is set to zero, the two-parameter distribution is obtained.

The significance of the shape factor β is:

$\beta < 1$ hazard rate decreasing;

$\beta = 1$ hazard rate constant;

$\beta > 1$ hazard rate increasing

By suitable choice of the shape factor β the Weibull distribution may be made to equal or approximate other distributions:

$\beta = 1$ exponential distribution;

$\beta = 2$ Rayleigh distribution.

A value of $\beta \approx 3.4$ corresponds approximately to a normal distribution in which the standard deviation is one-third of the mean.

There are several other versions of the two-parameter Weibull distribution. An alternative form (Truscott, 1970) has the characteristics:

$$z = \frac{\beta}{\alpha} t^{\beta-1} \qquad [7.7.26]$$

$$f = \frac{\beta}{\alpha} t^{\beta-1} \exp\left(-\frac{t^{\beta}}{\alpha}\right) \qquad [7.7.27]$$

$$R = \exp\left(-\frac{t^{\beta}}{\alpha}\right) \qquad [7.7.28]$$

This is related to the previous form by:

$$\alpha = \eta^{\beta} \qquad [7.7.29]$$

A third form (Shooman, 1968a) has the characteristics:

$$z = kt^m \qquad [7.7.30]$$

$$f = kt^m \exp\left(-\frac{kt^{m+1}}{m+1}\right) \quad [7.7.31]$$

$$R = \exp\left(-\frac{kt^{m+1}}{m+1}\right) \quad [7.7.32]$$

This is related to the first form given by:

$$k = \frac{\beta}{\eta} \quad [7.7.33]$$

$$m = \beta - 1 \quad [7.7.34]$$

The Weibull distribution is frequently used in reliability work to fit failure data, because it is flexible enough to handle decreasing, constant and increasing failure rates. This use of the Weibull distribution is considered in more detail below.

The properties of the Weibull distribution are described in the original paper by Weibull (1951) and in standard texts (e.g. Bompas-Smith, 1973; Kapur and Lamberson, 1977).

7.7.8 Rectangular distribution

For the rectangular distribution the characteristics are:

$$z = \frac{1}{a + b - 1} \quad [7.7.35]$$

$$f = \frac{1}{b} \quad [7.7.36]$$

$$R = \frac{a + b - t}{b} \quad [7.7.37]$$

for the range $a \leq t \leq a + b$. These quantities are shown in Figure 7.5(a).

The distribution is also called the 'uniform distribution' and has two parameters, the location parameter a and the scaling factor b.

The rectangular distribution is used in reliability work mainly to give random variables a uniform distribution across a specified interval. It is used, for example, in Monte Carlo simulation work. Computers generate pseudo-random numbers with a rectangular distribution. Random numbers with other distributions can be obtained from these, using the inverse cumulative distribution, where available, and other techniques.

7.7.9 Gamma distribution

For the gamma distribution the characteristics are:

$$z = \frac{t^{a-1} \exp(-t/b)}{\int_t^\infty t^{a-1} \exp(-t/b)\, dt} \quad [7.7.38]$$

$$f = \frac{1}{b^a \Gamma(a)} t^{a-1} \exp(-t/b) \quad [7.7.39]$$

$$R = \frac{1}{b^a \Gamma(a)} \int_t^\infty t^{a-1} \exp(-t/b)\, dt \quad [7.7.40]$$

for the range $0 \leq t \leq \infty$, with

$$\Gamma(a) = \int_0^\infty t^{a-1} \exp(-t)\, dt \quad [7.7.41]$$

or, if a is an integer,

$$\Gamma(a) = (a - 1)! \quad [7.7.42]$$

The quantities given in Equations 7.7.38–7.7.40 are shown in Figure 7.5(b) for values of the parameter $a = 0.5$ and 2.0.

The distribution has two parameters, the shape factor a and the scaling factor b.

The gamma distribution is another distribution used in reliability work to fit failure data, because it is sufficiently flexible to deal with decreasing, constant and increasing failure rates, but the Weibull distribution is more generally used.

7.7.10 Beta distribution

Use is also sometimes made in reliability engineering of the beta distribution. The properties of the beta distribution are given by Hastings and Peacock (1974). Its characteristics are:

$$f = \frac{x^{v-1}(1 - x)^{w-1}}{B(v, w)} \quad [7.7.43]$$

$$F = \int_0^x \frac{x^{v-1}(1 - x)^{w-1}}{B(v, w)}\, dx \quad [7.7.44]$$

with

$$B(v, w) = \int_0^1 u^{v-1}(1 - u)^{w-1}\, du \quad [7.7.45]$$

for the range $0 \geq x \geq 1$. The distribution has two parameters, v (> 0) and w (> 0).

The beta function is related to the gamma function:

$$\Gamma(v)\Gamma(w) = B(v, w)\Gamma(v + w) \quad [7.7.46]$$

for positive values of v and w.

7.7.11 Pareto distribution

For the Pareto distribution the characteristics are:

$$z = \frac{a}{t} \quad [7.7.47]$$

$$f = at^{-(a+1)} \quad [7.7.48]$$

$$R = t^{-a} \quad [7.7.49]$$

for the range $1 \leq t \leq \infty$. These quantities are shown in Figure 7.5(c). The distribution has a single parameter, the shape factor a.

The Pareto distribution is commonly used in reliability work in discrete form to describe the distribution of the numbers of failures in different modes. In this case the independent variable is the failure mode.

7.7.12 Extreme value distribution

For the extreme value distribution the characteristics are:

$$z = \frac{1}{b} \exp\left(\frac{t - a}{b}\right) \quad [7.7.50]$$

$$f = \frac{1}{b} \exp\left(\frac{t - a}{b}\right) \exp\left[-\exp\left(\frac{t - a}{b}\right)\right] \quad [7.7.51]$$

$$R = \exp\left[-\exp\left(\frac{t - a}{b}\right)\right] \quad [7.7.52]$$

for the range $-\infty \leq t \leq \infty$. These quantities are shown in Figure 7.5(d).

Equations 7.7.50–7.7.52 are for the distribution of the smallest extreme. For the distribution of the largest extreme the sign of t should be reversed.

The distribution has two parameters, the location parameter a and the scaling factor b.

The extreme value distribution is used in reliability work to investigate extreme, and therefore very rare, values of phenomena.

7.7.13 Hyperexponential distribution

Another distribution which is sometimes used in reliability engineering is the hyperexponential distribution. This is described by Jardine (1970). Its characteristics are

$$z = \frac{2\lambda\{\sigma^2 + (1-\sigma^2)\exp[-2(1-2\sigma)\lambda t]\}}{\sigma + (1-\sigma)\exp[-2(1-2\sigma)\lambda t]} \qquad [7.7.53]$$

$$f = 2\sigma^2\lambda\exp[-2\sigma\lambda t] + 2\lambda(1-\sigma^2)\exp[-2(1-\sigma)\lambda t] \qquad [7.7.54]$$

$$F = 1 - \sigma\exp[-2\sigma\lambda t] + (1-\sigma)\exp[-2(1-\sigma)\lambda t] \qquad [7.7.55]$$

for $t \geq 0$ and $0 < \sigma \leq 0.5$. The distribution has two parameters, λ and σ, the former being the mean arrival rate of failures.

The hyperexponential distribution is suitable for the case where the times to failure are either very short or very long.

7.7.14 Error function

A mathematical function which occurs both in some of the distributions described and also in other topics treated here is the error function. Accounts of the error function are given in standard mathematical texts (e.g. Jensen and Jeffreys, 1963; Paradine and Rivett, 1980).

The error function erf is defined as:

$$\mathrm{erf}(x) = \frac{2}{\pi^{\frac{1}{2}}}\int_0^x \exp(-u^2)\,du \qquad [7.7.56]$$

The term $(2/\pi^{\frac{1}{2}})$ is a normalizing factor such as to yield

$$\mathrm{erf}(\infty) = 1 \qquad [7.7.57]$$

The error function complement erfc is defined as

$$\mathrm{erfc}(x) = 1 - \mathrm{erf}(x) \qquad [7.7.58]$$

$$\mathrm{erfc}(x) = \frac{2}{\pi^{\frac{1}{2}}}\int_x^\infty \exp(-u^2)\,du \qquad [7.7.59]$$

The integral which constitutes the error function is similar in form to those which occur in the distribution function of the normal distribution and in the log–normal distribution, as comparison of Equation 7.7.56 with Equations 7.7.18 and 7.7.21 shows. A similar integral also occurs in the definition of the probit Y described in Chapter 9.

Furthermore, in passive gas dispersion, the gas concentrations follow a normal, or Gaussian, distribution so that the models used for such dispersion are often referred to as 'Gaussian models'. These models are described in Chapter 15.

7.7.15 Error function approximations

The error function may be looked up in the tables given in standard texts on statistics, but it is convenient for computation to have it in analytical function form. A review of approximations of the error function has been given by Karlsson and Bjerle (1980). Two desirable properties of an expression for the error function are that it should represent the function with acceptable accuracy and should be capable of being inverted.

Karlsson and Bjerle give the following approximate expression of their own for the error function:

$$\mathrm{erf}(x) = \frac{2}{\pi^{\frac{1}{2}}}\sin[\sin(x)] \qquad 0 \leq x \leq 1 \qquad [7.7.60a]$$

$$= \sin(x^{\frac{2}{3}}) \qquad 1 \leq x \leq 2 \qquad [7.7.60b]$$

$$= 1 \qquad 2 \leq x \leq \infty \qquad [7.7.60c]$$

They also give the following approximation for the error function inverse

$$\mathrm{erf}^{-1}(y) = \sin^{-1}\left[\sin^{-1}\left(\frac{y\pi^{\frac{1}{2}}}{2}\right)\right] \qquad 0 \leq y \leq 0.84 \qquad [7.7.61a]$$

$$= [\sin^{-1}(y)]^{\frac{3}{2}} \qquad 0.84 \leq y \leq 0.995 \qquad [7.7.61b]$$

7.8 Reliability of Some Standard Systems

The reliability of some standard systems is now considered. It is assumed that the exponential failure distribution applies, unless otherwise stated.

7.8.1 Series systems

A series system is one which operates only if all its components operate. It is not implied that the components are necessarily laid out physically in series configuration. For a series system the reliability R is the product of the reliabilities R_i of the components:

$$R = \prod_{i=1}^{n} R_i \qquad [7.8.1]$$

This follows directly from Equation 7.5.19.

For the exponential distribution, if the failure rates of the components are constants λ_i,

$$R_i = \exp(-\lambda_i) \qquad [7.8.2]$$

and hence

$$R = \prod_{i=1}^{n}\exp(-\lambda_i t) = \exp\left(-\sum_{i=1}^{n}\lambda_i t\right) \qquad [7.8.3]$$

If the overall failure rate of the system is a constant λ,

$$\mathrm{R} = \exp(-\lambda t) \qquad [7.8.4]$$

and hence

$$\lambda = \sum_{i=1}^{n}\lambda_i \qquad [7.8.5]$$

7.8.2 Parallel systems

A parallel system is one which fails to operate only if all its components fail to operate. Again it is not implied that the components are necessarily laid out physically in a parallel configuration. For a parallel system, the unreliability Q is the product of the unreliabilities Q_i of the components:

$$Q = \prod_{i=1}^{n} Q_i \quad [7.8.6]$$

Again this follows directly from Equation 7.5.19. The reliability R of the system is:

$$R = 1 - \prod_{i=1}^{n}(1 - R_i) \quad [7.8.7]$$

Since parallel configurations incorporate redundancy, they are also referred to as 'parallel redundant systems'.

For the exponential distribution, Equation 7.8.2 applies, and hence:

$$R = 1 - \prod_{i=1}^{n}[1 - \exp(-\lambda_i t)] \quad [7.8.8]$$

If the overall failure rate of the system is a constant λ, Equation 7.8.4 is applicable. In this case there is no simple general relationship between the system and component failure rates. But, if the component failure rates are all the same,

$$\frac{1}{\lambda} = \sum_{i=1}^{n} \frac{1}{i\lambda i} \quad [7.8.9]$$

If the failure rates are different, then for two components

$$\frac{1}{\lambda} = \frac{1}{\lambda_1} + \frac{1}{\lambda_2} - \frac{1}{\lambda_1 + \lambda_2} \quad [7.8.10]$$

for three components

$$\frac{1}{\lambda} = \frac{1}{\lambda_1} + \frac{1}{\lambda_2} - \frac{1}{\lambda_1 + \lambda_2} - \frac{1}{\lambda_1 + \lambda_3} - \frac{1}{\lambda_2 + \lambda_3} + \frac{1}{\lambda_1 + \lambda_2 + \lambda_3} \quad [7.8.11]$$

and for n components

$$\frac{1}{\lambda} = \left(\frac{1}{\lambda_1} + \frac{1}{\lambda_2} + \ldots + \frac{1}{\lambda_n}\right) - \left(\frac{1}{\lambda_1 + \lambda_2} + \frac{1}{\lambda_1 + \lambda_3} + \ldots + \frac{1}{\lambda_{n-1} + \lambda_n}\right)$$

$$+ \left(\frac{1}{\lambda_1 + \lambda_2 + \lambda_3} + \frac{1}{\lambda_1 + \lambda_2 + \lambda_4} + \ldots + \frac{1}{\lambda_{n-2} + \lambda_{n-1} + \lambda_n}\right)$$

$$- \ldots + (-1)^{n-1} \frac{1}{\sum_{i=1}^{n} \lambda_i} \quad [7.8.12]$$

7.8.3 r-out-of-n parallel systems

For a system which has n components in parallel and operates as long as r components survive, the binomial distribution is applicable. Thus for an r-out-of-n system the reliability R is obtained from the component reliability R_x:

$$R(r \le k \le n) = \sum_{k=r}^{n} \binom{n}{k} R_x^k Q_x^{n-k} \quad [7.8.13]$$

As an illustration, consider the reliability of a 2-out-of-4 system. Expanding

$$(R_x + Q_x)^n = 1 \quad [7.8.14]$$

for the terms between n and r to get the individual terms of Equation 7.8.13 gives

Probability of 0 failures $= R_x^4$

Probability of 1 failure $= 4R_x^3 Q_x$

Probability of 2 failures $= 6R_x^2 Q_x^2$

The coefficients of the terms may be obtained from Pascal's triangle. The system survives provided no more than two failures occur, and hence

$$R = R_x^4 + 4R_x^3 Q_x + 6R_x^2 Q_x^2 \quad [7.8.15]$$

An r-out-of-n system reduces in the limiting cases to a series or a parallel system. If all the components must operate for the system to survive, it becomes a series system

$$R = R_x^n \quad [7.8.16]$$

whilst if it is sufficient for one component to operate for the system to survive, it becomes a parallel system

$$Q = Q_x^n \quad [7.8.17]$$

7.8.4 Stand-by systems

The simplest stand-by system is one in which there is one component operating and one or more on stand-by, all components have the same failure rate in the operational mode, the stand-by components have zero failure rate in the stand-by mode, and there is perfect switchover. For this case, the Poisson distribution can be used. Thus, for such a stand-by system, the failure rate is λ, the number of failures which the system can withstand is r, and the system reliability R is

$$R(0 \le k \le r) = \exp(-\lambda t) \sum_{k=0}^{r} \frac{(\lambda t)^k}{k!} \quad [7.8.18a]$$

$$= \exp(-\lambda t)\left[1 + \lambda t + \frac{(\lambda t)^2}{2!} + \ldots + \frac{(\lambda t)^r}{r!}\right] \quad [7.8.18b]$$

As an illustration, consider the reliability of a stand-by system with one component operating and one on stand-by, so that the system can survive one failure. Then

$$R = \exp(-\lambda t)(1 + \lambda t) \quad [7.8.19]$$

It is relatively easy in this case to take into account imperfect switchover. If the reliability of switchover is R_{SW}, the system reliability R is:

$$R = \exp(-\lambda t)(1 + R_{SW}\lambda t) \quad [7.8.20]$$

Stand-by systems may be more complex than this simple case in a number of ways. The stand-by component may be different from the component normally operating and may have a different failure rate in the operational mode. It may also have a finite failure rate in the stand-by mode, and there may be imperfect switchover.

7.8.5 Systems with repair

If it is possible to carry out repair on a system, a much higher reliability can be achieved. The repair time which is used here is the total time from initial failure to final repair, and therefore includes any time required for detection of the failure and organization of the repair. Obviously, the determination of the reliability of systems

***Table 7.4** Mean life for some standard systems assuming exponential failure distribution*

System	Failure rates	Mean life
Series	Same	$\dfrac{1}{n\lambda}$
	Different	$\dfrac{1}{\sum_{i=1}^{n}\lambda_i}$
Parallel	Same	$\dfrac{1}{\lambda\sum_{i=1}^{n}\frac{1}{i}}$
	Different	$\left(\frac{1}{\lambda_1}+\frac{1}{\lambda_2}+\cdots+\frac{1}{\lambda_n}\right)-\left(\frac{1}{\lambda_1+\lambda_2}+\frac{1}{\lambda_1+\lambda_3}+\cdots+\frac{1}{\lambda_{n-1}+\lambda_n}\right)$ $+\left(\frac{1}{\lambda_1+\lambda_2+\lambda_3}+\frac{1}{\lambda_1+\lambda_2+\lambda_4}+\cdots+\frac{1}{\lambda_{n-2}+\lambda_{n-1}+\lambda_n}\right)-\cdots+(-1)^{n-1}\frac{1}{\sum_{i=1}^{n}\lambda_i}$
Stand-by	Same	$\dfrac{n}{\lambda}$
	Different	$\sum_{i=1}^{n}\dfrac{1}{\lambda_i}$
r-out-of-n	Same	—[a]
Parallel with repair (n repairmen)	Same	$\dfrac{1}{\lambda}\sum_{i=0}^{n-1}\dfrac{(1+\mu/\lambda)^i}{i+1}$
Stand-by with repair (n repairmen)	Same	$\dfrac{1}{\lambda}\sum_{i=0}^{n-1}\dfrac{n!}{(i+1)(n-(i+1))!}\left(\dfrac{\mu}{\lambda}\right)^i$

[a] The mean life for this case may be obtained by applying the binomial expansion and using Equation 7.6.20

with repair requires data on repair times as well as failure rates.

Repair times, like failure times, may fit various distributions. It is sometimes assumed that the repair time is a constant τ, but a more common assumption is that the repair times t_r have an exponential distribution with constant repair rate μ:

$$f_r(t_r) = \mu \exp(-\mu t_r) \quad [7.8.21]$$

For the exponential distribution the mean repair time m_r is

$$m_r = \frac{1}{\mu} \quad [7.8.22]$$

Equations 7.8.21 and 7.8.22 are analogous to Equations 7.6.12 and 7.6.25b for failure times and mean life. A more detailed discussion of repair times is given below.

The reliability of some standard systems with repair is now considered. It is assumed that the exponential repair time distribution applies, unless otherwise stated. In particular, this assumption is made in the Markov models described below.

7.8.6 Parallel systems with repair

For a system which has n components in parallel, and which survives provided that at least one component operates, and for which the failure and repair rates are λ and μ, respectively, expressions for the reliability R can be derived using Markov models, as described below. For a parallel system with two components it can be shown that the reliability R is:

$$R = \frac{1}{r_1 - r_2}[(3\lambda+\mu+r_1)\exp(r_1 t) - (3\lambda+\mu+r_2)\exp(r_2 t)] \quad [7.8.23]$$

with

$$r_1, r_2 = \frac{-(3\lambda+\mu) \pm [(3\lambda+\mu)^2 - 8\lambda^2]^{\frac{1}{2}}}{2} \quad [7.8.24]$$

7.8.7 Stand-by systems with repair

For a stand-by system with two components in which there is one component operating and one is on stand-by, both components have the same failure rate in the operational mode, the stand-by component has zero failure rate in the standby mode, the switchover is perfect, and there is repair, it can be shown using Markov models that the reliability R is:

$$R = \frac{1}{r_1 - r_2}[(2\lambda + \mu + r_1)\exp(r_1 t) - (2\lambda + \mu + r_2)\exp(r_2 t)] \quad [7.8.25]$$

with

$$r_1, r_2 = \frac{-(2\lambda + \mu) \pm [(2\lambda + \mu)^2 - 4\lambda^2]^{\frac{1}{2}}}{2} \quad [7.8.26]$$

7.8.8 Constant repair time

As just described, the usual assumption for repair times is a constant repair rate μ, which implies a distribution of repair times. An alternative assumption is a single constant repair time τ.

For a parallel system of n components with repair, the probability of the system being in the 'as new' condition and then suffering an initial failure a given number of times, is given by the Poisson distribution:

$$\exp(-n\lambda t)\left[1 + n\lambda t + \frac{(n\lambda t)^2}{2!} + \frac{(n\lambda t)^3}{3!}\ldots\right] = 1 \quad [7.8.27]$$

The probability of the system failing completely after an initial failure and before repair to the as-new condition is:

$$Q(\tau) = [1 - \exp(-\lambda t)]^{n-1} \quad \lambda t \ll 1 \quad [7.8.28]$$

Then the probability of the system suffering one initial failure followed by system failure is:

$$P(t) = \exp(-n\lambda t)n\lambda t\, Q(\tau) \quad [7.8.29]$$

The probabilities of the system suffering high numbers of initial failures followed by system failure may be obtained in a similar manner. Hence it can be shown that the reliability of the system is:

$$R = \exp(-\lambda' t) \quad [7.8.30]$$

with

$$\lambda' = n\lambda Q(\tau) \quad \lambda t \ll 1 \quad [7.8.31]$$

7.8.9 Mean life

Some methods of determining the mean life have been outlined in Section 7.6. The Markov methods described below are a further powerful method for obtaining the mean life and also the variance. The expressions for the mean life of some standard systems, assuming the exponential distribution, are shown in Table 7.4. For systems without repair the expressions allow for different failure rates for each equipment, but for systems with repair they assume the same failure rate. Some expressions for the mean life assuming the Weibull distribution are given by Shooman (1968a).

7.9 Reliability of complex systems

7.9.1 System reliability analysis

The reliability analysis of systems is carried out first to obtain an assessment of system reliability and to identify critical features, and then to achieve necessary improvements. A full analysis is likely to require considerable effort. It is important, therefore, to identify those subsystems which particularly affect the overall system reliability. In other words, the direction of effort is assisted by a sensitivity analysis. If the reliability of a particular subsystem does not greatly affect that of the system, it is not necessary to analyse it in great detail.

Reliability calculations require data not only on failure rates but also on repair times. The sensitivity of the system reliability determines the accuracy required in the data in a given case. In some instances it is necessary to obtain a rather accurate estimate from field data. In others an engineering estimate made without any field data may be quite sufficient. The matching of the data to the application is an important element in the art of the reliability engineer.

There are a number of methods which can be used to analyse more complex systems and to decompose them into their subsystems. These techniques may be illustrated by reference to Figure 7.6. Figure 7.6(a) illustrates a system consisting of two intermediate storage tanks and three pumps. Figure 7.6(b) shows the three pumps all in parallel, with complete interchangeability between them. Figure 7.6(c) shows two separate streams, with one tank and one pump in each, and with the third pump available to either stream. The system is

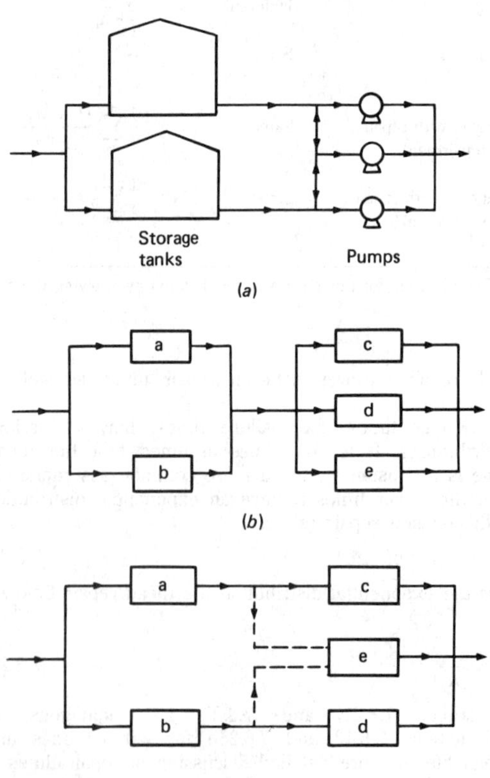

Figure 7.6 A system to illustrate methods of analysing complex systems: (a) storage tank pump system; (b) system with three pumps in parallel; (c) system with separate streams, but with a common spare pump

defined as successful provided flow is maintained by pumping with at least one pump from one tank.

7.9.2 Reliability graphs

The structure of a system may be represented in terms of simple reliability graphs. Each branch of the graph represents a component. Some examples are given in Figure 7.7. Figure 7.7(a) shows the reliability graph for three components in series, and Figure 7.7(b) shows that for three components in parallel. Figure 7.7(c) shows the reliability graph for two components in parallel, in series with three components in parallel, which corresponds to the storage tank pump system shown in Figure 7.6(a), assuming the configuration given in Figure 7.6(b). The system reliability may then be determined by computing the reliability of the subsystems and then that of the system itself. Reliability graphs are discussed in more detail by Shooman (1968a).

7.9.3 Logic flow diagrams

The most widely used method of graphical representation is the logic flow diagram, also called the 'logic sequence diagram' or simply the 'logic diagram'. Figure 7.8 shows a logic flow diagram for the storage tank pump system with the configuration shown in Figure 7.6(b). In this case the top event considered is a success, but in other cases it may be a failure. Some typical applications of logic flow diagrams are given in Table 7.5. Logic flow diagrams can be used for systems of considerable complexity. Further examples of logic flow diagrams are given by A.E. Green and Bourne (1972).

One of the most widely used types of logic flow diagram is the fault tree. Fault trees are discussed in Chapter 9. An important development in the area of fault trees is kinetic tree theory, which has been developed by Vesely and co-workers (e.g. Vesely, 1969, 1970b; Vesely and Narum, 1970) and which permits the determination of system characteristics such as reliability and availability in time-varying systems.

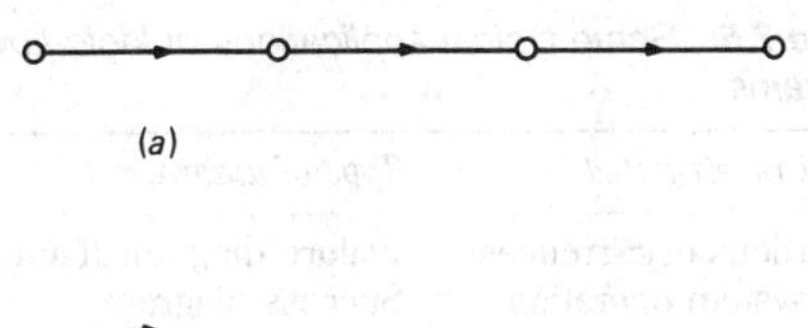

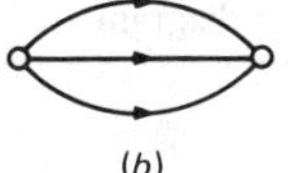

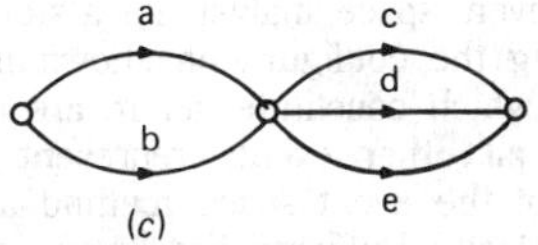

Figure 7.7 *Reliability graphs: (a) for a three-component series system; (b) for a three-component parallel system; (c) for a storage tank pump system (configuration as in Figure 7.6(b))*

7.9.4 Event space method

The event space method involves making a comprehensive list of all the possible states of the system and determining which of these events correspond to success. The events E_i so listed are mutually exclusive and system reliability R is:

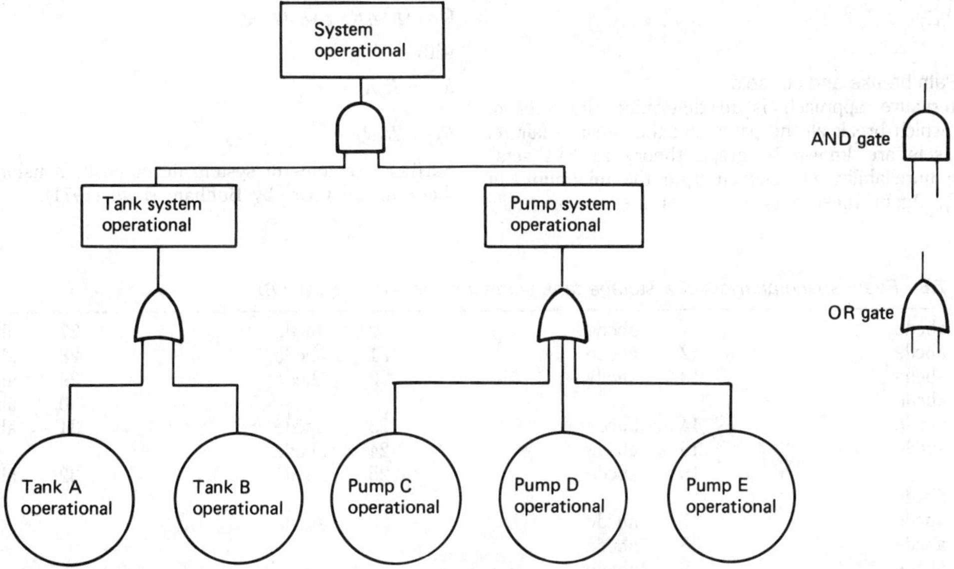

Figure 7.8 *Logic flow diagram for storage tank pump system (configuration as in Figure 7.6(b))*

Table 7.5 *Some typical applications of logic flow diagrams*

Event investigated	*Type of diagram*
Hazardous occurrences	Failure diagram (fault tree)
Trip system operation	Success diagram
Plant availability	Success diagram

$$R = \sum_{i=1}^{n} P(\mathrm{E_i}) \qquad [7.9.1]$$

Table 7.6 gives an event space analysis for a storage tank pump system having the configuration shown in Figure 7.6(b). The events which constitute failure are events 7, 17–19 and 26–32; all other events represent success. Further examples of the event space method are given by Shooman (1968a) and Buffham, Freshwater and Lees (1971).

7.9.5 Tree diagrams

Another graphical representation is the tree diagram. Figure 7.9 shows such a diagram for a storage tank pump system having the configuration given in Figure 7.6(c). The diagrams can be used for simple cases, but rapidly become rather cumbersome as systems become more complex. Further examples of tree diagrams are given by von Alven (1964) and Breipohl (1970).

7.9.6 Path tracing and tie sets

The path tracing method consists of tracing through the system the paths which constitute success. These paths are known in graph theory as 'tie sets'. System reliability is obtained from the minimum tie sets T_i. These sets of events are not mutually exclusive and the system reliability is given by:

$$R = P\left(\bigcup_{i=1}^{n} T_i\right) \qquad [7.9.2]$$

7.9.7 Path breaks and cut sets

An alternative approach is to determine the sets of events which break all the paths and thus ensure failure. These sets are known in graph theory as 'cut sets'. System unreliability is obtained from the minimum cut sets C_i. Again these sets of events are not mutually exclusive and the system reliability is given by:

$$R = 1 - \left(\bigcup_{i=1}^{n} C_i\right) \qquad [7.9.3]$$

If the event space method is applied to a storage tank pump system having the configuration shown in Figure 7.6(c) rather than that given in Figure 7.6(b), so that paths *ad* and *bc* are not permissible, then the complete cut sets correspond to events 7, 17–19, 22, 24 and 26–32, and are:

$\bar{a}\bar{b}, \bar{a}\bar{b}\bar{c}, \bar{a}\bar{b}\bar{d}, \bar{a}\bar{b}\bar{e}, \bar{a}\bar{d}\bar{e}, \bar{b}\bar{c}\bar{e}, \bar{c}\bar{d}\bar{e},$

$\bar{b}\bar{c}\bar{d}\bar{e}, \bar{a}\bar{c}\bar{d}\bar{e}, \bar{a}\bar{b}\bar{d}\bar{e}, \bar{a}\bar{b}\bar{c}\bar{e}, \bar{a}\bar{b}\bar{c}\bar{d}, \bar{a}\bar{b}\bar{c}\bar{d}\bar{e}$

However, a cut set which contains within itself a smaller cut set is not a minimum. The minimum cut sets are:

$\bar{a}\bar{b}, a\bar{d}\bar{e}, b\bar{c}\bar{e}, c\bar{d}\bar{e}$

Minimum cut sets are important in fault tree work and are considered further in Chapter 9.

7.9.8 System decomposition

It is often possible to simplify the analysis of a system by selecting a key component which makes it possible to decompose the system into subsystems. In such a case, one approach is the use of Bayes' theorem. This theorem is considered in more detail in Section 7.14. Here it is sufficient to note that the theorem may be written in the form:

$$P(\mathrm{A}) = P(\mathrm{A}|\mathrm{B})P(\mathrm{B}) + \mathrm{P}(\mathrm{A}|\bar{\mathrm{B}})P(\bar{\mathrm{B}}) \qquad [7.9.4]$$

Thus the reliability $R(\mathrm{S})$ of the system is related to the reliability $R(\mathrm{X})$ of the key component as follows:

$$R(\mathrm{S}) = R(\mathrm{S} \mid \mathrm{X})R(\mathrm{X}) + R(\mathrm{S} \mid \bar{\mathrm{X}})Q(\bar{\mathrm{X}}) \qquad [7.9.5]$$

For a storage tank pump system having the configuration shown in Figure 7.6(c) the system unreliability Q is:

$$Q = Q_a Q_b R_e + Q_{ac} Q_{bd} Q_e \qquad [7.9.6]$$

with

$$R_{ac} = R_a R_c \qquad [7.9.7a]$$

$$R_{bd} = R_b R_d \qquad [7.9.7b]$$

Further examples of system decomposition using Bayes' theorem are given by Buffham *et al.* (1971).

Table 7.6 *Event space analysis of a storage tank pump system (see Figure 7.6)*

1	abcde	11	$a\bar{b}\bar{c}de$	20	$\bar{a}b\bar{c}\bar{d}e$	27	$a\bar{b}\bar{c}\bar{d}\bar{e}$
2	$\bar{a}bcde$	12	$a\bar{b}c\bar{d}e$	21	$\bar{a}b\bar{c}d\bar{e}$	28	$\bar{a}b\bar{c}\bar{d}\bar{e}$
3	$a\bar{b}cde$	13	$a\bar{b}cd\bar{e}$	22	$\bar{a}bc\bar{d}\bar{e}$	29	$\bar{a}\bar{b}c\bar{d}\bar{e}$
4	$ab\bar{c}de$					30	$\bar{a}\bar{b}\bar{c}d\bar{e}$
5	$abc\bar{d}e$	14	$ab\bar{c}\bar{d}e$	23	$a\bar{b}\bar{c}\bar{d}e$	31	$\bar{a}\bar{b}\bar{c}\bar{d}e$
6	$abcd\bar{e}$	15	$ab\bar{c}d\bar{e}$	24	$a\bar{b}\bar{c}d\bar{e}$		
		16	$abc\bar{d}\bar{e}$	25	$a\bar{b}c\bar{d}\bar{e}$	32	$\bar{a}\bar{b}\bar{c}\bar{d}\bar{e}$
7	$\bar{a}\bar{b}cde$						
8	$\bar{a}b\bar{c}de$	17	$\bar{a}\bar{b}\bar{c}de$	26	$ab\bar{c}\bar{d}\bar{e}$		
9	$\bar{a}bc\bar{d}e$	18	$\bar{a}\bar{b}c\bar{d}e$				
10	$\bar{a}bcd\bar{e}$	19	$\bar{a}\bar{b}cd\bar{e}$				

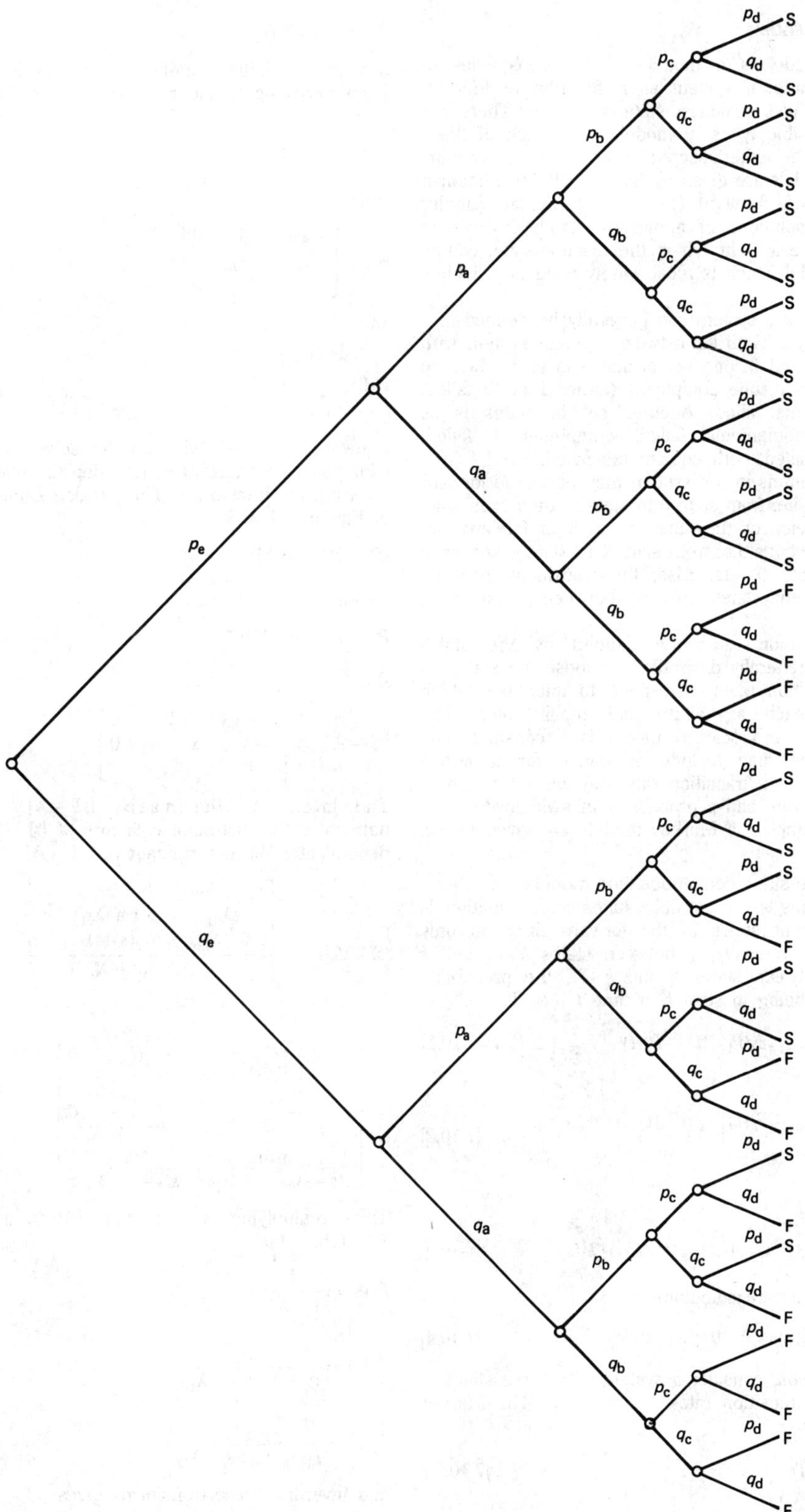

Figure 7.9 *Tree diagram for storage tank pump system (configuration as in Figure 7.6(c)): S, system success; F, system failure*

7.10 Markov Models

A Markov model is a model of the probabilities of different states of a system as a function of time. It therefore has two variables, state and time. There are thus four possible types of model, since each of these variables may be either discrete or continuous. Accounts of Markov models are given by Sandler (1963), Shooman (1968a) and R.A. Howard (1971). In particular, Sandler (1963) gives solutions of numerous standard cases in reliability and availability. It is the discrete-state, continuous-time model which is most widely used in reliability work.

The state of a system can generally be defined in a number of ways. Thus for a two-equipment system with equipments 1 and 2, one set of states is as follows: no equipments failed, one equipment (either 1 or 2) failed, both equipments failed. Another set of states is as follows: no equipments failed, equipment 1 failed, equipment 2 failed, both equipments failed.

The transitions in the system may be in the forward direction only, i.e. from state 0 to state 1, then from state 1 to state 2, etc., or they may be in both forward and backward directions, i.e. from state 0 to state 1 and from state 1 to state 0, etc. Also, the transitions may be between adjacent states only or between non-adjacent states.

The transition rates are defined by the states chosen. It is generally desirable to choose the states so that the transition rates correspond to quantities which are known, such as failure and repair rates. The transition rate of a Markov model is a constant. The transition rates may include allowance for imperfect switchover. Thus a transition rate may be a function of both a failure rate and a probability of switchover.

Some examples of Markov models are given below.

7.10.1 Discrete-state, continuous-time models

If a system has $n+1$ possible states and transition is between adjacent states in the forward direction only with transition rates $\lambda_{k-1,k}$ between states $k-1$ and k and $\lambda_{k,k+1}$ between states k and $k+1$, the probability $P_k(t+\Delta t)$ of being in state k at time $t+\Delta t$ is:

$$P_k(t+\Delta t) = \lambda_{k-1,k}\Delta t P_{k-1}(t) + P_k(t) - \lambda_{k,k+1}\Delta t P_k(t) \quad [7.10.1]$$

But

$$\lim_{\Delta t \to 0}\left[\frac{P_k(t+\Delta t) - P_k(t)}{\Delta t}\right] = \frac{\mathrm{d}P_k(t)}{\mathrm{d}t} \quad [7.10.2]$$

Hence

$$\frac{\mathrm{d}P_k(t)}{\mathrm{d}t} = \dot{P}_k(t) = \lambda_{k-1,k}P_{k-1}(t) - \lambda_{k,k+1}P_k(t) \quad [7.10.3]$$

Initial conditions are commonly:

$$P_0(0) = 1; \quad P_k(0) = 0 \quad 0 < k \le n \quad [7.10.4]$$

As an illustration, consider a system with three states 0, 1 and 2 and transition rates λ_{01} and λ_{12}. The Markov equations are:

$$\dot{P}_0(t) = \lambda_{01}P_0(t) \quad [7.10.5a]$$

$$\dot{P}_1(t) = \lambda_{01}P_0(t) - \lambda_{12}P_1(t) \quad [7.10.5b]$$

$$\dot{P}_2(t) = \lambda_{12}P_1(t) \quad [7.10.5c]$$

Assume the initial conditions given in Equation 7.10.4. Then, rewriting Equations 7.10.5a–c and 7.10.4 in matrix form:

$$\mathbf{P}(t) = \mathbf{AP}(t) \quad [7.10.6]$$

where

$$\mathbf{A} = \begin{bmatrix} -\lambda_{01} & 0 & 0 \\ \lambda_{01} & -\lambda_{12} & 0 \\ 0 & \lambda_{12} & 0 \end{bmatrix} \quad [7.10.7]$$

and

$$\mathbf{P}(0) = \begin{bmatrix} 1 \\ 0 \\ 0 \end{bmatrix} \quad [7.10.8]$$

Equations 7.10.6–7.10.8 may be solved analytically by taking Laplace transforms, inverting the matrix, and then inverting the transforms. Thus, taking Laplace transforms of Equation 7.10.6:

$$s\overline{\mathbf{P}} - \mathbf{P}(0) = \mathbf{A}\overline{\mathbf{P}} \quad [7.10.9]$$

Hence

$$\overline{\mathbf{P}} = [s\mathbf{I} - \mathbf{A}]^{-1}\mathbf{P}(0) \quad [7.10.10]$$

But

$$[s\mathbf{I} - \mathbf{A}] = \begin{bmatrix} s+\lambda_{01} & 0 & 0 \\ -\lambda_{01} & s+\lambda_{12} & 0 \\ 0 & -\lambda_{12} & s \end{bmatrix} \quad [7.10.11]$$

The inverse of the matrix, $[s\mathbf{I} - \mathbf{A}]^{-1}$, has as its numerator the matrix of cofactors of $[s\mathbf{I} - \mathbf{A}]$ and as its denominator the determinant of $[s\mathbf{I} - \mathbf{A}]$. Thus:

$$[s\mathbf{I} - \mathbf{A}]^{-1} = \frac{\begin{bmatrix} s(s+\lambda_{12}) & 0 & 0 \\ s\lambda_{01} & s(s+\lambda_{01}) & 0 \\ \lambda_{01}\lambda_{12} & \lambda_{12}(s+\lambda_{01}) & (s+\lambda_{01})(s+\lambda_{12}) \end{bmatrix}}{s(s+\lambda_{01})(s+\lambda_{12})} \quad [7.10.12a]$$

$$= \begin{bmatrix} \dfrac{1}{s+\lambda_{01}} & 0 & 0 \\ \dfrac{\lambda_{01}}{(s+\lambda_{01})(s+\lambda_{12})} & \dfrac{1}{s+\lambda_{12}} & 0 \\ \dfrac{\lambda_{01}\lambda_{12}}{s(s+\lambda_{01})(s+\lambda_{12})} & \dfrac{\lambda_{12}}{s(s+\lambda_{12})} & \dfrac{1}{s} \end{bmatrix} \quad [7.10.12b]$$

Then, multiplying out Equation 7.10.10 using Equation 7.10.12b:

$$\bar{P}_0(s) = \frac{1}{s+\lambda_{01}} \quad [7.10.13a]$$

$$\bar{P}_1(s) = \frac{\lambda_{01}}{(s+\lambda_{01})(s+\lambda_{12})} \quad [7.10.13b]$$

$$\bar{P}_2(s) = \frac{\lambda_{01}\lambda_{12}}{s(s+\lambda_{01})(s+\lambda_{12})} \quad [7.10.13c]$$

and inverting these transforms gives

$$P_0(t) = \exp(-\lambda_{01}t) \quad [7.10.14a]$$

$$P_1(t) = \frac{\lambda_{01}}{\lambda_{12} - \lambda_{01}}[\exp(-\lambda_{01}t) - \exp(-\lambda_{12}t)] \quad [7.10.14b]$$

$$P_2(t) = 1 - \frac{1}{\lambda_{12} - \lambda_{01}}[\lambda_{12}\exp(-\lambda_{01}t) - \lambda_{01}\exp(-\lambda_{12}t)] \quad [7.10.14c]$$

This example illustrates a number of practical points about Markov models:

(1) The sum of the terms on the right-hand side of all the original model equations is zero.
(2) The sum of the probabilities $P(t)$ is unity and the sum of their transforms $\bar{P}(s)$ is $1/s$.
(3) The initial conditions are frequently $P_0(0) = 1$ and $P_k(0) = 0$ $(0 < k \leq n)$, and in this case only the first column of the matrix $[s\mathbf{I} - \mathbf{A}]^{-1}$ is of interest.
(4) If there are no transitions back from the last state, the equation for this may be omitted and the probability calculated at the end from the other probabilities; this is particularly useful in third- and fourth-order models.

7.10.2 Poisson process

If in Equation 7.10.3 the transition rate from one state to the next is a constant λ, the Markov equations are:

$$\dot{P}_0(t) = -\lambda P_0(t) \quad [7.10.15a]$$

$$\dot{P}_1(t) = \lambda P_0(t) - \lambda P_1(t) \quad [7.10.15b]$$

$$\vdots$$

$$\dot{P}_n(t) = \lambda P_{n-1}(t) - \lambda P_n(t) \quad [7.8.15c]$$

Assume the initial conditions given in Equation 7.10.4. Then the solution is:

$$P_0(t) = \exp(-\lambda t) \quad [7.10.16a]$$

$$P_1(t) = \exp(-\lambda t)\lambda t \quad [7.10.16b]$$

$$\vdots$$

$$P_n(t) = \exp(-\lambda t)\frac{(\lambda t)^n}{n!} \quad [7.10.16c]$$

This is a Poisson process.

7.10.3 Two-equipment systems

If a system consists of two components then, whatever its configuration, one formulation of the states in which it can be is as follows: no equipments failed, equipment 1 failed, equipment 2 failed, and both equipments failed. Let $P_0(t)$, $P_1(t)$, $P_2(t)$ and $P_3(t)$ be, respectively, the probabilities of: no failures, failure of component 1, failure of component 2, and failure of both components. In addition, let λ_1, λ_2, λ_3 and λ_4 be, respectively, the transition rates from: states 0 to 1, 0 to 2, 1 to 3 and 2 to 3. Then the Markov equations are:

$$\dot{P}_0(t) = -(\lambda_1 + \lambda_2)P_0(t) \quad [7.10.17a]$$

$$\dot{P}_1(t) = \lambda_1 P_0(t) - \lambda_3 P_1(t) \quad [7.10.17b]$$

$$\dot{P}_2(t) = \lambda_2 P_0(t) - \lambda_4 P_2(t) \quad [7.10.17c]$$

$$\dot{P}_3(t) = \lambda_3 P_1(t) + \lambda_4 P_2(t) \quad [7.10.17d]$$

Assume the initial conditions given in Equation 7.10.4. Then the solution is:

$$P_0(t) = \exp[-(\lambda_1 + \lambda_2)t] \quad [7.10.18a]$$

$$P_1(t) = \frac{\lambda_1}{\lambda_1 + \lambda_2 - \lambda_3}\{\exp(-\lambda_3 t) - \exp[-(\lambda_1 + \lambda_2)t]\} \quad [7.10.18b]$$

$$P_2(t) = \frac{\lambda_2}{\lambda_1 + \lambda_2 - \lambda_4}\{\exp(-\lambda_4 t) - \exp[-(\lambda_1 + \lambda_2)t]\} \quad [7.10.18c]$$

$$P_3(t) = 1 - [P_0(t) + P_1(t) + P_2(t)] \quad [7.10.18d]$$

Thus far no particular configuration has been assumed. The following cases can be treated to obtain the system reliability $R(t)$:

(1) Series system, failure rates of two equipments identical:

$$\lambda_1 = \lambda_2 = \lambda$$

$$R(t) = P_0(t) = \exp(-2\lambda t) \quad [7.10.19]$$

(2) Series system, failure rates of two equipments different:

$$R(t) = P_0(t) = \exp[-(\lambda_1 + \lambda_2)t] \quad [7.10.20]$$

(3) Parallel system, failure rates of two equipments identical:

$$\lambda_1 = \lambda_2 = \lambda_3 = \lambda_4 = \lambda$$

$$R(t) = P_0(t) + P_1(t) + P_2(t)$$

$$= 2\exp(-\lambda t) - \exp(-2\lambda t) \quad [7.10.21]$$

(4) Parallel system, failure rates of two equipments different

$$\lambda_1 = \lambda_4 = \lambda\,; \quad \lambda_2 = \lambda_3 = \lambda'$$

$$R(t) = P_0(t) + P_1(t) + P_2(t)$$

$$= \exp(-\lambda t) + \exp(-\lambda' t) - \exp[-(\lambda + \lambda')t] \quad [7.10.22]$$

(5) Parallel system, failure rates of two equipments operating together identical, but failure rate of one equipment surviving different:

$$\lambda_1 = \lambda_2 = \lambda\,; \quad \lambda_3 = \lambda_4 = \lambda'$$

$$R(t) = P_0(t) + P_1(t) + P_2(t)$$

$$= \frac{1}{2\lambda - \lambda'}[2\lambda\exp(-\lambda' t) - \lambda' \exp(-2\lambda t)] \quad [7.10.23]$$

(6) Stand-by system, failure rates of two equipments identical in operating mode, no failure in stand-by mode and no switchover failure:

$$\lambda_1 = \lambda_3 = \lambda\,; \quad \lambda_2 = \lambda_4 = 0$$

$$R(t) = P_0(t) + P_1(t)$$

In this case the expression for $P_1(t)$ is indeterminate since the denominator is zero, but applying de l'Hopital's rule of differentiating both numerator and denominator by λ_1 gives

$$R(t) = \exp(-\lambda t)(1 + \lambda t) \quad [7.10.24]$$

This example illustrates the versatility of the method.

7.10.4 Single-equipment systems with repair

In the system considered so far the transitions have all been unidirectional towards some failed state. In fact in most systems there is repair. The transition rate in the reverse direction is the repair rate.

It is normally assumed in Markov models that there is a constant repair rate μ, and hence an exponential distribution of repair times t_r, as given in Equation 7.8.21. If a system comprises a single equipment with repair, there are two possible states it can take. Let $P_0(t)$ and $P_1(t)$ be the probabilities of being operational and failed, respectively, and let λ and μ be the failure and repair rates, respectively. The Markov equations are:

$$\dot{P}_0(t) = -\lambda P_0(t) + \mu P_1(t) \quad [7.10.25a]$$

$$\dot{P}_0(t) = \lambda P_0(t) - \mu P_1(t) \quad [7.10.25b]$$

Assume the initial conditions given in Equation 7.10.4. Then the solution is:

$$P_0(t) = \frac{\mu}{\lambda+\mu} + \frac{\lambda}{\lambda+\mu}\exp[-(\lambda+\mu)t] \quad [7.10.26a]$$

$$P_1(t) = \frac{\lambda}{\lambda+\mu} - \frac{\lambda}{\lambda+\mu}\exp[-(\lambda+\mu)t] \quad [7.10.26b]$$

Since there is now transition in both directions, $P_0(t)$ represents the system availability $A(t)$

$$A(t) = \frac{\mu}{\lambda+\mu} + \frac{\lambda}{\lambda+\mu}\exp[-(\lambda+\mu)t] \quad [7.10.27]$$

Thus the availability is, in general, a function of time. But is it usually the steady state availability $A(\infty)$ which is of most interest:

$$A(\infty) = \frac{\mu}{\lambda+\mu} \quad [7.10.28]$$

7.10.5 Reliability and availability formulations

The availability $A(t)$ introduced in the previous section is the probability that the system will be in the operational state. This is not the same as the reliability $R(t)$, which is the probability that the system will not leave the operational state. The difference lies in the fact that, in determining availability, transition back by repair from the non-operational state is permitted. The availability for a given system is always greater than or equal to the reliability:

$$A(t) \geq R(t) \quad [7.10.29]$$

Different Markov models are required for reliability and availability. For reliability, repair transition from the states constituting failure is not permissible, although repair transition from other states is. For availability, repair transition from all states is permitted.

For an *r*-out-of-*n* system, in which the states up to and including the $(n-r)$th state give survival, reliability is obtained from the Markov model as:

$$R(t) = \sum_{i=0}^{n-r} P_i(t) \quad [7.10.30]$$

Availability is obtained as

$$A(t) = \sum_{i=0}^{n-r} P_i(t) \quad [7.10.31]$$

when the states 0 to $n-r$ again constitute survival. But the formulation of the equations for the various states is different in the two types of model. For the reliability formulation, the repair transition rates in equations $n-r+1$ to n are zero.

A state which it is possible to enter but not to leave is an absorbing state. If there is a single absorbing state, the steady state probability of being in that state is unity. The reliability formulations of Markov models have an absorbing state, but availability formulations do not. If the last state of a Markov model is an absorbing state, so that there is no interaction of this state and the preceding states, the last state probability may be omitted in the formulation and determined at the end of the calculation from the other state probabilities. This is particularly useful in the analytical solution or manual calculation of third- and fourth-order models.

7.10.6 Repair rates

The repair rate used in Markov models for systems with more than one equipment needs to be carefully specified. There are two aspects to consider. One is the repair transition rate in systems with n equipments from states in which there is more than one failure. This transition rate will normally lie somewhere in the range μ to $n\mu$. The other aspect is the increase in repair transition rate for a single equipment from μ to $\xi\mu$ which may be obtained by allocating additional effort.

These repair transition rates are often considered in terms of the number of repair men, it being assumed that the normal case is one repair man for one equipment. Thus for systems with n equipments and with states with multiple failures the repair transition rate in the fully failed state may be $n\mu$. This is often called the '*n* repairmen case'. But the number of repairmen may also affect the repair transition rate for a single equipment $\xi\mu$. Thus Sandler (1963) frequently uses $\xi = 1.5$ where two repair men are available to work on one equipment.

As an illustration, consider a system which consists of two equipments with identical failure rates. Then there are three possible states: no equipments failed, one equipment failed and both equipments failed. Let $P_0(t)$, $P_1(t)$ and $P_2(t)$ be the probabilities of no, one and two failures, respectively, and let λ be the failure rate and μ the basic repair rate. If repair resources allow the repair of only one equipment at a time, i.e. the one repair man case, then:

$$\dot{P}_0(t) = -2\lambda P_0(t) + \mu P_1(t) \quad [7.10.32a]$$

$$\dot{P}_1(t) = 2\lambda P_0(t) - (\lambda+\mu)P_1(t) + \mu P_2(t) \quad [7.10.32a]$$

$$\dot{P}_2(t) = \lambda P_1(t) - \mu P_2(t) \quad [7.10.32c]$$

Assume the initial conditions given in Equation 7.10.4. Then the system availability, defined as having both equipments operating, can be shown to be:

$$A(\infty) = P_0(\infty) = \frac{\mu^2}{2\lambda^2 + 2\lambda\mu + \mu^2} \quad [7.10.33]$$

If, on the other hand, the repair resources allow the repair of two equipments simultaneously, i.e. the two repair men case, then:

$$\dot{P}_0(t) = -2\lambda P_0(t) + \mu P_1(t) \quad [7.10.34a]$$

$$\dot{P}_1(t) = 2\lambda P_0(t) - (\lambda + \mu)P_1(t) + 2\mu P_2(t) \quad [7.10.34b]$$

$$\dot{P}_2(t) = \lambda P_1(t) - 2\mu P_2(t) \quad [7.10.34c]$$

and, assuming the same initial conditions and definition of availability,

$$A(\infty) = \frac{\mu^2}{(\lambda + \mu)^2} = \frac{\mu^2}{\lambda^2 + 2\lambda\mu + \mu^2} \quad [7.10.35]$$

It may be noted that there is a difference between the number of repair men that can usefully be used, depending on whether it is reliability or availability which is being calculated. In a system with n equipments which fails when $n - r + 1$ equipments have failed, it is possible to use n repair men to maximize availability, but only $n - r$ repair men to maximize reliability.

7.10.7 Mean life

Markov models offer a powerful method of obtaining the mean life and the variance of the distribution of complex systems. This is discussed in more detail by Sandler (1963).

7.10.8 Discrete-state, discrete-time models

The discrete-state, discrete-time Markov model is also useful in some applications. If a system has a number of possible states and transition occurs between these states over a given time interval, then the vectors of state probabilities before and after the transition $\boldsymbol{P}(0)$ and $\boldsymbol{P}(1)$ are related by the equation:

$$\boldsymbol{P}(1) = \boldsymbol{AP}(0) \quad [7.10.36]$$

where $\mathbf{A}$ is the matrix of transition probabilities. Then it can be shown that, as the number of transitions tends to infinity:

$$\boldsymbol{P}(\infty)\mathbf{A} = \boldsymbol{P}(\infty) \quad [7.10.37]$$

The final state probabilities are independent of the initial state probabilities. In addition,

$$\Sigma P_i = 1 \quad [7.10.38]$$

The final states of the system may be obtained by solution of Equations 7.10.37 and 7.10.38.

An account of discrete-state, discrete-time Markov models is given by Sandler (1963). An application involving a discrete-state, discrete-time Markov model is given in Section 7.13.

7.11 Joint Density Functions

An alternative to Markov models is the use of the joint density function approach. The joint density function method is convenient where a failure distribution other than the exponential distribution has to be handled. It is also useful where the failure rate changes depending on the failures which have already occurred. The joint density function method is discussed by Lloyd and Lipow (1962) and Shooman (1968a).

If a system consists of two subsystems and if the second subsystem takes over the system function when the first fails at time τ, then the joint density function of the system $\phi(\tau, t)$ is

$$\phi(\tau, t) = f_1(\tau)f_2(t|\tau) \quad [7.11.1]$$

where $f_1(\tau)$ is a density function, and $f_2(t|\tau)$ is a conditional density function. The associated marginal density function $f(t)$ is:

$$f(t) = \int_0^\infty \phi(\tau, t)\ \mathrm{d}\tau \quad [7.11.2]$$

The application of the joint density function is now considered.

7.11.1 Parallel systems with variable failure rates

If a system consists of two equipments in parallel and both equipments have the same failure rate λ_1 when both are operating but have a different failure rate λ_2 after the first equipment has failed, then the density functions are:

$$f_1(t) = 2\lambda_1 \exp(-2\lambda_1 t) \quad [7.11.3]$$

$$f_2(t|\tau) = \lambda_2 \exp[-\lambda_2(t - \tau)] \quad 0 < \tau < t \quad [7.11.4a]$$

$$= 0 \quad \tau > t \quad [7.11.4b]$$

The joint density function is:

$$\phi(\tau, t) = f_1(\tau)f_2(t|\tau) \quad [7.11.5]$$

and the marginal density function is:

$$f(t) = \int_0^t 2\lambda_1 \exp(-2\lambda_1\tau)\lambda_2 \exp[-\lambda_2(t - \tau)]\ \mathrm{d}\tau \quad [7.11.6a]$$

$$= \frac{2\lambda_1\lambda_2}{2\lambda_1 - \lambda_2}[\exp(-\lambda_2 t) - \exp(-2\lambda_1 t)] \quad [7.11.6b]$$

The reliability is

$$R(t) = 1 - \int_0^t f(t)\ \mathrm{d}t \quad [7.11.7a]$$

$$= \frac{1}{2\lambda_1 - \lambda_2}[2\lambda_1 \exp(-\lambda_2 t) - \lambda_2 \exp(-2\lambda_1 t)] \quad [7.11.7b]$$

7.11.2 Stand-by systems with variable failure rates

If a system consists of two equipments, one operating and one on stand-by, with perfect switchover, and if the failure rates of the operating and the stand-by equipments are λ_1 and λ_2, respectively, then the density functions are:

$$f_1(\tau) = \lambda_1 \exp(-\lambda_1\tau) \quad [7.11.8]$$

$$f_2(t|\tau) = \lambda_2 \exp[-\lambda_2(t - \tau)] \quad 0 < \tau < t \quad [7.11.9a]$$

$$= 0 \quad \tau > t \quad [7.11.9b]$$

and hence

$$f(t) = \frac{\lambda_1 \lambda_2}{\lambda_1 - \lambda_2}[\exp(-\lambda_2 t) - \exp(-\lambda_1 t)] \qquad [7.11.10]$$

$$R(t) = \frac{1}{\lambda_1 - \lambda_2}[\lambda_1 \exp(-\lambda_2 t) - \lambda_2 \exp(-\lambda_1 t)] \qquad [7.11.11]$$

7.12 Monte Carlo Simulation

Many practical problems cannot be solved by any of the analytical methods described and are only soluble by simulation. The principal method used is Monte Carlo simulation. Accounts of Monte Carlo simulation include that given in *Method of Statistical Testing. Monte Carlo Method* (Shreider, 1964) and that by Shooman (1968a).

7.12.1 Simulation method

The application of Monte Carlo simulation to the determination of the reliability and/or availability of complex systems proceeds broadly as follows.

The system configuration, the failure and repair characteristics of the components, and the time period of interest are specified. The time period is divided into small increments. The first trial is then carried out covering the time period specified. At the first time increment the probability of failure of the first equipment is calculated. This probability is compared with a random number in the range 0 to 1 generated from a uniform distribution: if the random number is less than or equal to the failure probability the equipment fails, otherwise it survives. The state of all the equipments is computed in like manner. The situation is then reviewed to determine the overall state of the system. If failures have occurred, there may be various changes in the system, such as initiation of repair of failed equipments, alteration in failure rates of more highly loaded surviving equipments, etc. The system behaviour at the second time increment is then computed and the process is continued until the specified time period is complete. This whole calculation constitutes only a single trial.

The result of this trial is that the system either has or has not met its reliability/availability objective. It is then necessary to carry out further trials to obtain the full simulation. The ratio of successful trials to total trials asymptotically approaches a constant value and this is the required system reliability/availability result. The equipment characteristics such as failure and repair rates obtained in the series of simulation trials naturally also approach asymptotically the values originally specified.

7.12.2 Illustrative example

As an illustration, consider the determination of the reliability of a single equipment over a specified period of time, assuming the exponential failure distribution. This simple case is chosen for illustration because there is also an analytical solution which is given by Equation 7.6.11. Let the time period to be considered be 10 months and let the failure rate λ be 0.5 faults/year. Then, in Equation 7.6.11,

$\lambda = 0.5$ faults/year

$t = 10$ months $= 0.83$ year

Hence

$R = 0.66$

Now consider the alternative solution using Monte Carlo simulation. The time period is divided into suitable increments, which in this case are shown as 1 month. The probability of failure q in one time increment δt is then calculated as:

$$q = \lambda \delta t$$

$$= 0.5 \times 0.0833$$

$$= 0.0417$$

The Monte Carlo simulation is then carried out using a set of random numbers. A suitable series of random numbers, which in this case have been generated by a computer and are therefore strictly pseudo-random numbers, is given in Table 7.7. The first trial is carried out as follows. For the first month the probability of failure q $(= 0.0417)$ is compared with the random number ν. The occurrence of survival or failure is determined as follows:

$\nu > q$ survival;

$\nu \leq q$ failure.

Thus, taking the random numbers from the first line of the table, in the first month, since $\nu = 0.0417$, the

Table 7.7 *A series of random numbers*

				Random numbers					
0.7710	0.7818	0.7517	0.4740	0.0782	0.2032	0.5159	0.2664	0.9556	0.3355
0.4129	0.4574	0.0284	0.0538	0.0677	0.9217	0.9214	0.2330	0.1053	0.5345
0.2598	0.7481	0.1507	0.1707	0.6685	0.4742	0.8289	0.7054	0.7725	0.2862
0.7649	0.0135	0.1968	0.0592	0.5838	0.9701	0.5664	0.6681	0.9104	0.4500
0.5064	0.9879	0.3699	0.3285	0.6416	0.8936	0.5872	0.4803	0.5972	0.2603
0.1873	0.7806	0.9981	0.9630	0.7952	0.1043	0.4693	0.8766	0.0359	0.3265
0.6358	0.8761	0.5345	0.3221	0.1226	0.8365	0.9158	0.9662	0.5547	0.6329
0.8050	0.1336	0.5572	0.1402	0.8269	0.7000	0.7549	0.2335	0.6067	0.5389
0.7730	0.7877	0.7693	0.5267	0.2364	0.6778	0.9397	0.5379	0.7701	0.7789
0.7430	0.4476	0.9991	0.9659	0.8040	0.1307	0.5484	0.1139	0.7478	0.4622

Table 7.8 *Monte Carlo simulation example: survival or failure of equipment in trials 1–10*

Trial	*Survival/failure*
1	1
2	0
3	1
4	0
5	1
6	0
7	1
8	1
9	1
10	1

equipment survives. The same process is repeated for each of the remaining 9 months. Since in all cases $\nu > q$, the equipment survives the whole 10-month period. This completes the first trial, in which the equipment has survived.

Further trials are then carried out on the same lines. The second, fourth and sixth lines of the random number table contain values of the random number ν which are less than q; these are shown underlined in Table 7.7. In the second, fourth and sixth trials, therefore, the equipment fails, but in all the other trials it survives. The trials which constitute survival and failure may then be tabulated as shown in Table 7.8, assigning 1 to a survival and 0 to a failure.

The reliability of the equipment is then determined as the ratio of the number of trials in which the equipment survives to the total number of trials. In this example the ratio is 0.7. Since only 10 trials have been done, the reliability can be calculated only to one place of decimals. This value of the reliability compares with the value of 0.66 calculated earlier from the analytical relationship.

7.12.3 Applications of the method

As an illustration of the sort of problem to which Monte Carlo methods are applicable consider the olefins plant system shown in Figure 7.10. The system consists of four cracking furnaces in parallel followed by a compressor and then by a separation train. The furnaces have the characteristics shown in Table 7.9. The failure rate of the compressor is 0.3 faults/year and that of the separation train is 3 faults/year. The repair characteristics of the equipment are given in Table 7.10. The furnace repair team is able to work on only one furnace at a time, but there is no other restriction on the repair work. The information required from the simulation might be, say: (1) the probability of achieving full availability, (2) the mean availability achieved, (3) the range of availabilities and the corresponding probabilities, over a period of 90 days.

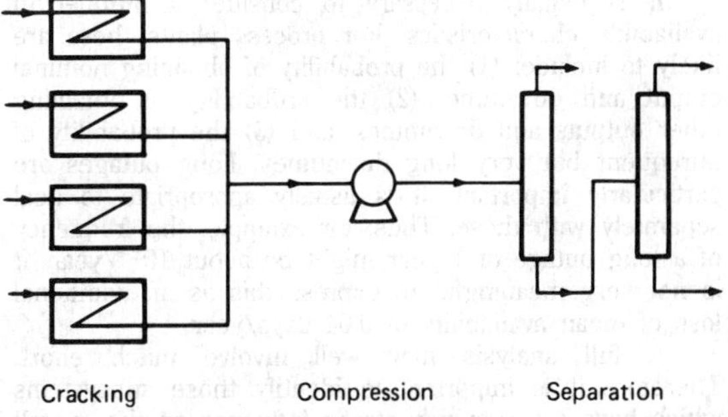

Figure 7.10 *Olefins plant system*

Table 7.9 *Monte Carlo simulation example: characteristics of the furnaces*

Time since start-up (days)	*Proportion of furnaces failing*
0–50	0.08
50–100	0.05
100–200	0.05
200–300	0.10
300–400	0.10

Table 7.10 *Monte Carlo simulation example: repair characteristics of the equipment*

	Repair time (days)	*Proportion of occasions*
Furnace	20	0.7
	30	0.2
	40	0.1
Compressor	10	0.8
	30	0.2
Separation train	10	0.7
	20	0.3

7.12.4 Features of the method

It will be apparent from this example that it is necessary to resort to Monte Carlo simulation on account of the complexity not so much of the system configuration, which is in fact rather simple, but rather of that of the failure and repair data and other rules and constraints. A single simulation is capable of giving a number of results. Thus in the example quoted the results required were, in effect, the system reliability, its mean availability and its availability density function.

The failure data are frequently provided in the form of a failure histogram. The quantity which is required in the simulation for comparison with the random number is the probability of failure q. If the fraction failing over one time increment is δF, then

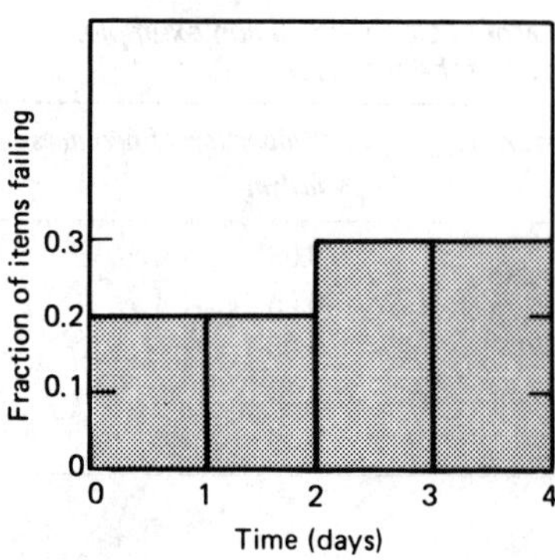

Figure 7.11 Failure data histogram

$$q = \frac{\delta F}{1 - F} \quad [7.12.1]$$

Thus consider the failure histogram shown in Figure 7.11. The probabilities of failure for the 4 days obtained from Equation 7.12.1 are 0.2, 0.25, 0.5 and 1.0.

It is often convenient to perform one set of trials, examine the results and then perform further sets. If this is done, however, it is necessary that the subsequent sets should not start with the same random number as the first set, otherwise the effect is simply to repeat that set continuously.

7.12.5 Number of trials

The Monte Carlo method is extremely powerful, but it does have one serious drawback with regard to the number of trials required to obtain an accurate result, especially where the probability investigated is high. The number of trials required to achieve a given confidence level on the probability cannot, in general, be calculated. An approximate estimate of the number of trials can be obtained, however, if certain simplifying assumptions are made. The following method of estimation has been described by Shooman (1968a).

In determining the value of a single probability, such as a reliability, the simulation may be regarded as a series of n trials which have a binomial distribution with a true probability of success p and of failure q. The binomial distribution approaches the normal distribution for $p > 0.5$ and $nq > 5$. The point estimate of the probability of failure q is given by n_f/n, where n_f is the number of failures recorded in n trials. Then the fractional error e between the estimated and true probabilities of failure is:

$$e = \frac{n_f/n - q}{q} \quad [7.12.2]$$

If μ and σ are the expected number of failures and the expected standard deviation, then for a $(1 - \alpha)$ confidence level:

Table 7.11 Monte Carlo simulation: number of trials required in example to give an error of ≯5% at a 95% confidence level

Probability of success	*No. of trials required*
0.9	14 400
0.99	158 400
0.999	1 598 400

$$P(-\gamma\sigma < n_f - \mu < \gamma\sigma) = 1 - \alpha \quad [7.12.3]$$

where γ is the number of standard deviations corresponding to that level. In terms of the parameters n, p and q of the binomial distribution:

$$\mu = nq \quad [7.12.4]$$

$$\sigma = (npq)^{\frac{1}{2}} \quad [7.12.5]$$

Then

$$P[-\gamma(npq)^{\frac{1}{2}} < n_f - n_q < \gamma(npq)^{\frac{1}{2}}] = 1 - \alpha \quad [7.12.6]$$

$$P[-\gamma(p/nq)^{\frac{1}{2}} < e < \gamma(p/nq)^{\frac{1}{2}}] = 1 - \alpha \quad [7.12.7]$$

Thus at the 95% confidence level, for which $\gamma = 2$:

$$e > 200(p/nq)^{\frac{1}{2}}\% \quad [7.12.8]$$

The very large number of trials required for a system with very high reliability can be seen by considering the number required to give an error of ≯ 5% at a 95% confidence level (see Table 7.11). Thus, despite the undoubted power of the Monte Carlo method, it remains attractive to use and develop analytical techniques wherever possible. A further discussion of the number of trials required in Monte Carlo simulation is given by Shreider (1964).

Monte Carlo simulation also has other applications in reliability work. Its use to generate hazard rates from distributions of hazard rates is described in Chapter 9 in connection with fault tree work.

7.13 Availability

7.13.1 System availability analysis

The availability analysis of systems, like reliability analysis, has as its objectives: (1) to assess system availability and to identify critical aspects; and (2) to effect the required improvements.

It is usually necessary to consider a number of availability characteristics. For process plants these are likely to include: (1) the probability of obtaining nominal output and downtime, (2) the probability of obtaining other outputs and downtimes, and (3) the probability of infrequent but very long downtimes. Long outages are particularly important. It is usually appropriate to deal separately with these. Thus, for example, the frequency of a long outage of 1 year might be about 10^{-4}/year. It is not very meaningful to express this as an additional loss of mean availability of 0.04 days/year.

A full analysis may well involve much effort. Therefore it is important to identify those subsystems which have a particularly strong influence on the overall system availability. Those subsystems which do not greatly affect system availability can be treated in less detail.

There are two main types of availability data which may be used. Sometimes data on the availability of whole subsystems can be obtained. In other cases it is necessary to calculate system availability from failure rate and repair time data. In either case it is important that the accuracy of the data be matched to the sensitivity of the system availability to errors in the data.

As far as possible data on the availability of a process unit should describe the performance of that unit independent of other units. Downtime of the unit due to unavailability of other units or to other extraneous causes should not be included.

Particular attention should be paid to the availability of process raw materials and services. It is also often necessary to consider the availability of other plants which receive from the plant under consideration outputs such as by-products, electricity or steam.

There are a number of methods available for the analysis of more complex systems, and some of these are described below.

7.13.2 The availability function

In general, availability *A(t)* is a function of time. It may be expressed in terms of the uptime *u(t)* and downtime *d(t)*:

$$A(t) = \frac{u(t)}{u(t) + d(t)} \qquad [7.13.1]$$

The unavailability *V(t)* is

$$V(t) = 1 - A(t) \qquad [7.13.2]$$

$$= \frac{d(t)}{u(t) + d(t)} \qquad [7.13.3]$$

The point availability *A(t)* can be integrated over the time period T to give the mean availability over that period:

$$A(T) = \frac{1}{T}\int_0^T A(t)\ \mathrm{d}t \qquad [7.13.4]$$

Often, however, it is the long-term, or steady-state, availability $A(\infty)$ which is of most interest:

$$A(\infty) = \frac{u(\infty)}{u(\infty) + d(\infty)} \qquad [7.13.5]$$

The long-term availability may also be expressed in terms of the mean uptime m_u and the mean downtime m_d:

$$A(\infty) = \frac{m_u}{m_u + m_d} \qquad [7.13.6]$$

For a single equipment or system with exponential failure and repair distributions, failure rate λ and repair rate μ:

$$m_u = \frac{1}{\lambda} \qquad [7.13.7]$$

$$m_d = \frac{1}{\mu} \qquad [7.13.8]$$

$$= m_r \qquad [7.13.9]$$

Then, from Equations 7.13.6–7.13.9:

$$A(\infty) = \frac{\mu}{\lambda + \mu} \qquad [7.13.10a]$$

$$= \frac{1}{1 + \lambda m_r} \qquad [7.13.10b]$$

$$= 1 - \lambda m_r \quad \lambda m_r \ll 1 \qquad [7.13.10c]$$

This result was given earlier in Equation 7.10.28, and the long-term unavailability is:

$$V(\infty) = \frac{\lambda}{\lambda + \mu} \qquad [7.13.11a]$$

$$= \frac{\lambda m_r}{1 + \lambda m_r} \qquad [7.13.11b]$$

$$= \lambda m_r \quad \lambda m_r \ll 1 \qquad [7.13.11c]$$

7.13.3 The repair time density function

It is appropriate at this point to consider the distribution of repair times. The mean repair time m_r is defined as the first moment of the repair time density function $f_r(t_r)$:

$$m_r = \int_0^\infty t_r f_r(t_r)\ \mathrm{d}t_r \qquad [7.13.12]$$

So far, the repair time distribution which has been considered is the exponential. For this distribution the repair time density function $f_r(t_r)$ is

$$f_r(t_r) = \mu\ \exp(-\mu t_r) \qquad [7.13.13]$$

The repair time distribution function $F_r(t_r)$ is:

$$F_r(t_r) = \int_0^{t_r} f_r(t_r)\ \mathrm{d}t_r \qquad [7.13.14a]$$

$$= 1 - \exp(-\mu t_r) \qquad [7.13.14b]$$

The mean repair time is:

$$m_r = \int_0^\infty t_r \mu\ \exp(-\mu t_r)\ \mathrm{d}t_r \qquad [7.13.15a]$$

$$= \frac{1}{\mu} \qquad [7.13.15b]$$

For the normal distribution:

$$f_r(t_r) = \frac{1}{\sigma_r (2\pi)^{\frac{1}{2}}}\ \exp\left(-\frac{(t_r - m_r)^2}{2\sigma_r^2}\right) \qquad [7.13.16]$$

where the mean repair time is m_r.

For the log–normal distribution:

$$f_r(t_r) = \frac{1}{\sigma_r t_r (2\pi)^{\frac{1}{2}}}\ \exp\left(-\frac{[\ln(t_r) - m_r^*]^2}{2\sigma_r^2}\right) \qquad [7.13.17]$$

Repair time data often fit a log–normal distribution.

7.13.4 The downtime density function

For a single equipment the repair time and the downtime are the same. For a system it is more usual to speak of the downtime.

So far the downtime used in the expressions for availability has been the mean downtime m_d. But, just as

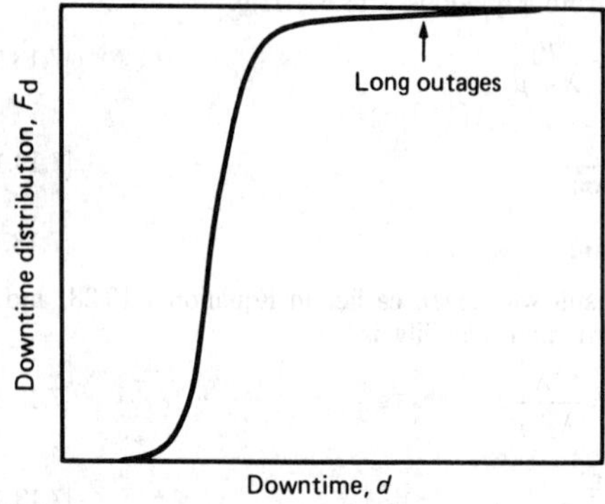

Figure 7.12 Downtime distribution showing a tail corresponding to infrequent, but long, outages

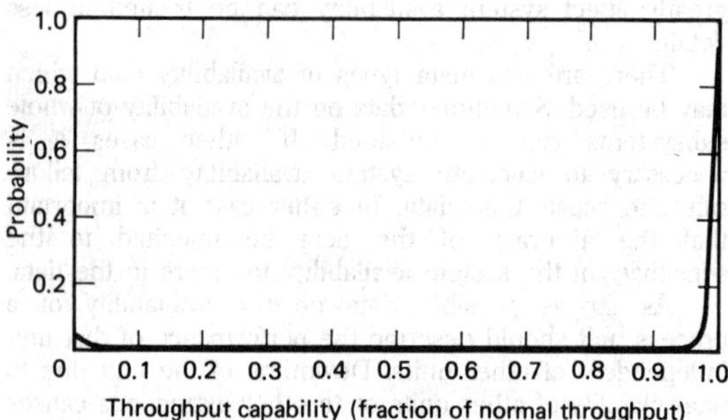

Figure 7.13 Throughput density for a process unit (G.D.M. Pearson, 1975) (Reproduced by permission of the University of Nottingham)

there is a repair time distribution, so there is a downtime distribution. It is possible to define a downtime density function $f_d(d)$ and distribution function $F_d(d)$.

Particularly important are long downtimes. In the downtime distribution shown in Figure 7.12, the tail in the distribution is due to infrequent, but long, outages.

7.13.5 The throughput density function

So far it has been assumed that either an item is available or it is not. In fact, however, in the process industries many items may exhibit a range of throughputs. This partial availability can be expressed by a throughput density function $f_w(w)$.

Typical throughput data from a process unit given by G.D.M. Pearson (1975) are shown in Table 7.12 and plotted in Figure 7.13. These data constitute a throughput density function. Similarly, the throughput of a whole plant can be expressed as a throughput density function.

7.13.6 Markov models

The derivation of expressions for availability from Markov models has been described briefly above. This method is now considered in more detail.

It is normally assumed in such models that the repair time distribution is exponential with constant repair rate μ. Usually it is the long-term availability which is of interest. In this case the Markov equations become steady-state equations. Thus, for example, Equation 7.10.25 becomes

$$0 = -\lambda P_0 + \mu P_1 \quad [7.13.18a]$$

but Equation 7.10.25b becomes

$$1 = P_0 + P_1 \quad [7.13.18b]$$

Equations 7.13.18a,b then yield equation [7.10.28] directly.

The definition of availability used depends on the configuration of the system. Considering systems with n equipments, for a series system only the first state P_0 constitutes survival:

$$A(\infty) = P_0 \quad [7.13.19]$$

For a parallel system only the last state represents failure:

$$A(\infty) = 1 - P_n \quad [7.13.20]$$

For an r-out-of-n system the states up to and including the $(n - r)$th state give survival:

$$A(\infty) = \sum_{i=0}^{n-r} P_i \quad [7.13.21]$$

Long-term availabilities for some standard systems assuming exponential failure and repair time distributions are given in Table 7.13. The availabilities are defined by Equations 7.13.19–7.13.21. The repair man assumption is that each repair man repairs one equipment, so that the repair transition rates between states are limited by the number of repair men.

Another important and general case is that in which there are n equipments but only m repair men. If there are k equipments down then, at steady state:

$$0 = (n - k + 1)\lambda P_{k-1} - [(n - k)\lambda + k\mu]P_k + (k + 1)\mu P_{k+1} \qquad k < m \quad [7.13.22a]$$

$$0 = (n - k + 1)\lambda P_{k-1} - [(n - k)\lambda + m\mu]P_k + m\mu P_{k+1} \qquad k \geq m \quad [7.13.22b]$$

$$1 = \sum_{i=0}^{n} P_i \quad [7.13.22c]$$

Table 7.12 Throughput capabilities of a process unit (G.D.M. Pearson, 1975) (Courtesy of the University of Nottingham)

Throughput capability (fraction of normal throughput)	*Probability*
0	0.035
0.025	0.015
0.050	0.005
0.800	0.001
0.850	0.002
0.900	0.005
0.925	0.007
0.950	0.010
0.975	0.020
1.000	0.900

Table 7.13 *Long-term availabilities for some standard systems assuming exponential failure and repair time distributions*

System	*No. of components*	*No. of repairmen*	*Availability*
—	1	1	$\frac{\mu}{\lambda+\mu}$
Series	2	1	$\frac{\mu^2}{2\lambda^2+2\lambda\mu+\mu^2}$
Series	2	2	$\frac{\mu^2}{(\lambda+\mu)^2}$
Series	n	1	$\frac{\left(\frac{\mu}{\lambda}\right)^n}{n!\sum_{i=0}^{n}\frac{1}{i!}\left(\frac{\mu}{\lambda}\right)^i}$
Series	n	n	$\frac{\mu^n}{(\lambda+\mu)^n}$
Parallel	2	1	$\frac{\mu^2+2\lambda\mu}{2\lambda^2+2\lambda\mu+\mu^2}$
Parallel	2	2	$\frac{\mu^2+2\lambda\mu}{\lambda^2+2\lambda\mu+\mu^2}$
Parallel	n	1	$1-\frac{1}{\sum_{i=0}^{n}\frac{1}{i!}\left(\frac{\mu}{\lambda}\right)^i}$
r-out-of-n	n	n	$\sum_{i=r}^{n}\binom{n}{i}\left(\frac{\mu}{\lambda+\mu}\right)^i\left(\frac{\lambda}{\lambda+\mu}\right)^{n-i}$
Stand-by	2	1	$\frac{\mu^2+\lambda\mu}{\lambda^2+\lambda\mu+\mu^2}$
Stand-by	2	2	$\frac{2\mu^2+2\lambda\mu}{\lambda^2+2\lambda\mu+2\mu^2}$
Stand-by	n	1	$1-\frac{1}{\sum_{i=0}^{n}\left(\frac{\mu}{\lambda}\right)^i}$
Stand-by	n	n	$1-\frac{1}{\sum_{i=0}^{n}\frac{n}{i!}\left(\frac{\mu}{\lambda}\right)^{n-i}}$

Equations 7.13.22a–c have the solution:

$$P_0 = \left[\sum_{k=0}^{m-1}\frac{n!}{(n-k)!k!}\rho^k + \sum_{k=m}^{n}\frac{n!}{(n-k)!m!}\rho^m\left(\frac{\rho}{m}\right)^{k-1}\right]^{-1} \qquad [7.13.23a]$$

$$P_k = \frac{n!}{(n-k)!k!}\rho^k P_0 \quad k < m \qquad [7.13.23b]$$

$$= \frac{n!}{(n-k)!m!}\rho^m\left(\frac{\rho}{m}\right)^{k-m}P_0 \quad k \geq m \qquad [7.13.23c]$$

with

$$\rho = \frac{\lambda}{\mu} \qquad [7.13.24]$$

The long-term availability of an r-out-of-n system with m repair men is given by:

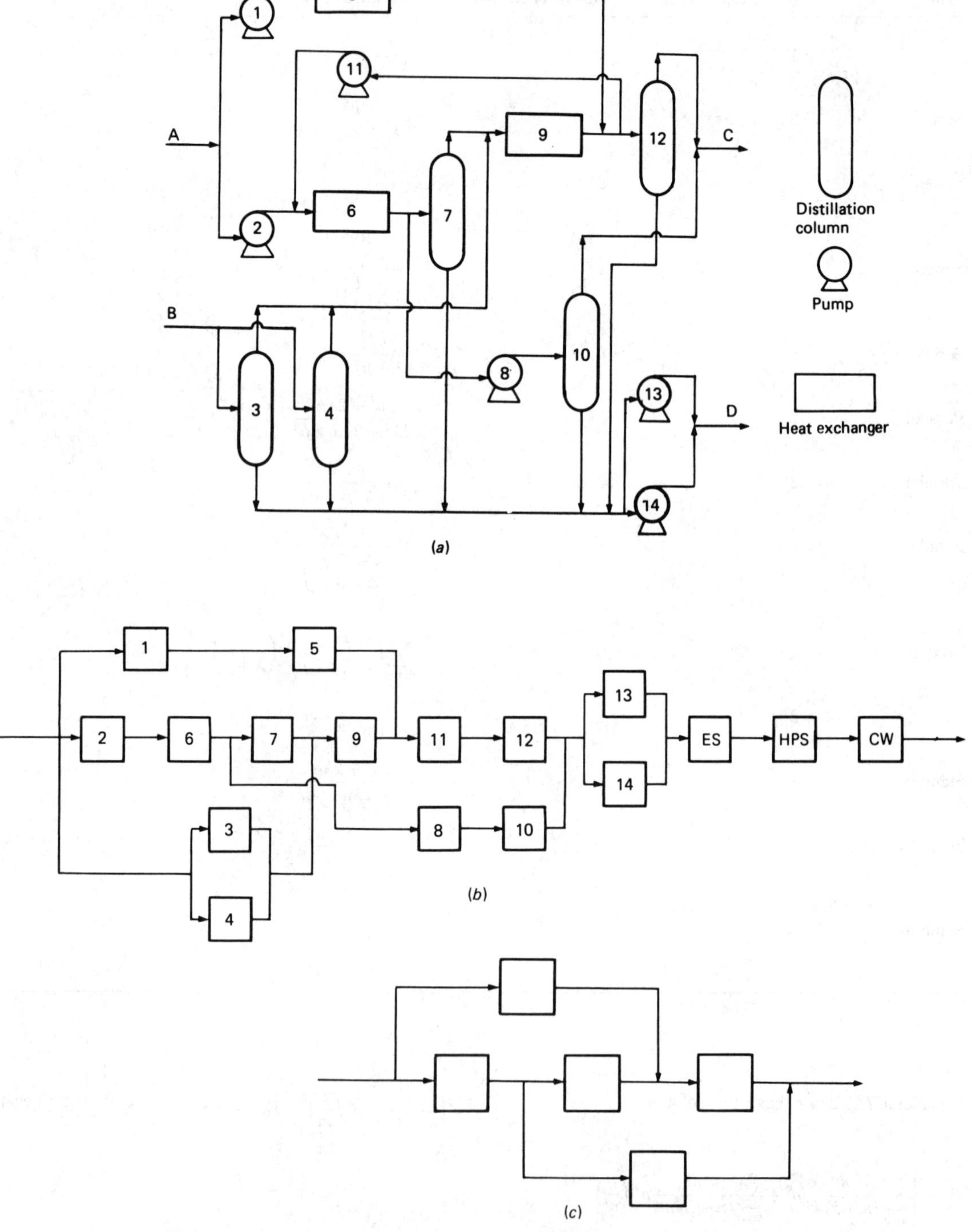

Figure 7.14 Distillation plant system (G.D.M. Pearson, 1975): (a) flow diagram of system; (b) availability block diagram of system; (c) equivalent block diagram of units 1–12. ES, electricity supply; HPS, high pressure steam; CW, cooling water (Reproduced by permission of the University of Nottingham)

$$A(\infty) = \sum_{i=0}^{n-r} P_i \qquad [7.13.25]$$

using Equations 7.13.23a–c.

These and numerous other cases are treated in more detail by Sandler (1963).

7.13.7 Logic flow diagrams

Logic flow diagrams are often used to represent system availability. Thus the logic flow diagram shown in Figure 7.8 for the storage tank pump system with the configuration given in Figure 7.6(b) is in fact an availability diagram. In order to use the logic flow diagram method it is necessary to represent the plant flow diagram in logical form. It is necessary, therefore, to convert the plant flow diagram into a network of relatively simple blocks such as series or parallel units.

The plant availability determined using a logic flow diagram may be the mean availability or the availability density function. Usually the calculation is limited to the former. There are two main methods of determining the mean availability of the plant. In many cases the availability data are the mean availabilities of the units which constitute the plant. Typical units might be a tank–pump system, a complete distillation column or a set of centrifuges. The mean unavailability of a unit is effectively a fractional dead time or probability of being unavailable. Then the mean availability of the plant may be calculated from the mean unavailabilities of these units.

In other cases the availability data obtainable are the failure rates and repair times of the equipments which constitute the units. Typical equipments are a pump, a heat exchanger or a drier. The mean unavailability, or fractional dead time, of an equipment may be calculated from the failure rate and repair time data using Equation (7.13.10). The fractional dead time of the unit may be calculated from those of the constituent equipments. Then the mean unavailability of the plant may again be calculated from the unavailabilities of the units. The mean repair time of the plant may be determined by calculating the failure rate and the mean unavailability.

It is also possible, however, to express the unit availabilities as throughput density functions, and in this case the plant availability is also a throughput density function. This may be obtained from the logic flow diagram by Monte Carlo simulation in a manner similar to that used in fault tree evaluation, as described in Chapter 9.

7.13.8 Throughput capability method

There is an alternative method, which allows the system availabilities to be expressed as throughput densities and still avoid the use of simulation (G.D.M. Pearson, 1975). The method involves converting the system into a series system and determining successively the limiting throughputs.

As an illustration, consider the distillation plant system flow diagram given by Pearson and shown in Figure 7.14(a). The corresponding availability block diagram is given in Figure 7.14(b). The path (11) affects the operation of paths (2, 6, 7, 9) and (1, 5) and therefore takes the position shown, while side paths (13) and (14) affect the operation of the main forward path and are therefore shown there. The main forward path is

Table 7.14 *Throughput capabilities of a process system (G.D.M. Pearson, 1975) (Courtesy of the University of Nottingham)*

Throughput capability (fraction of normal throughput)	*Probability*
0	0.300
0.025	0.102
0.050	0.031
0.800	0.006
0.850	0.012
0.900	0.029
0.925	0.038
0.950	0.050
0.975	0.086
1.000	0.349

also affected by the single supplies of electricity, high pressure steam and cooling water. The units (1–2) can be reduced to an equivalent but simplified block diagram, as shown in Figure 7.14(c). This is a complex subsystem, which cannot be resolved into simple series or parallel systems.

The throughput capability of the system is evaluated by series and parallel state enumeration techniques. The series enumeration technique is as follows. All the stages of the system, i.e. the units or the subsystems, are treated together as a series system. The throughput levels of the system are considered in turn, starting at zero throughput and rising to full throughput. At each level there is a stage which has the lowest throughput capability and limits the system throughput capability; this is the 'bottleneck' stage. The probability that the system has this throughput is determined by the probability that this bottleneck stage is in this state. The throughput capability of the bottleneck stage is then increased to its next discrete level. A new bottleneck stage is then identified and the procedure is repeated. Table 7.14 shows the throughput capabilities for a system of 10 identical process units with throughput capabilities as given in Table 7.13. The corresponding plot of system throughput capabilities is illustrated in Figure 7.15. This is the throughput density function of the system. The mean throughput capability of a single unit is 0.943 and that of the system is 0.557.

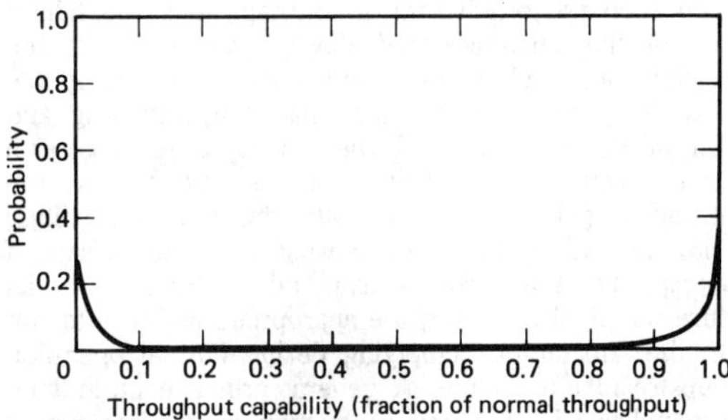

Figure 7.15 *Throughput density for a process system (G.D.M. Pearson, 1975) (Reproduced by permission of the University of Nottingham)*

The parallel path enumeration technique is as follows. Only two stages of the system are considered at a time. The throughput density function of this subsystem is obtained by summing the discrete throughput capabilities of the two stages and multiplying the corresponding probabilities. If in the original system there are more than two stages in parallel, this subsystem and the next parallel stage are then considered and the procedure repeated.

Further details of the throughput capability method, including the application of the state enumeration technique to complex subsystems, are given by G.D.M. Pearson (1975).

7.13.9 Flowsheeting and simulation methods

Flowsheeting methods and, in particular, computer flowsheeting programs, may also be used to determine plant availability. The use of this approach has been described by Holmes (1976).

7.13.10 Storage

Plant availability is affected by storage, which introduces flexibility, and may allow units upstream or downstream of a failed unit to continue in operation for a time. Methods of determining the effect of storage on plant availability have been discussed by Holmes (1976) and by Henley and Hoshino (1977).

There are two main classes of storage tank: (1) raw material and product storage tanks, and (2) intermediate storage tanks. Storage tanks are provided for various reasons, some of which have nothing to do with plant availability. Raw material and product storage tanks are used to smooth deliveries and collections and to blend materials. Intermediate storage tanks are also used to smooth fluctuations and to blend intermediates.

As an illustration, consider the simplest system with intermediate storage capacity of an upstream unit, an intermediate storage tank and a downstream unit. This system and its operating policies have been investigated by Henley and Hoshino (1977).

Some basic operating policies are: (1) to keep the tank filled if possible, (2) to keep the tank empty if possible, (3) to keep the tank half full if possible, and (4) to allow the tank level to fluctuate. The appropriate operating policy depends on the objectives to be achieved by the use of the storage. It is convenient to consider initially only the improvement of plant availability.

The policy necessary to maximize plant availability depends on the relative maximum capacities and failure rates of the upstream and downstream units. If the capacities and failure rates are equal, the appropriate policy is to run with the tank half full, emptying and filling during unit failures. If the capacity of the upstream unit is higher but its failure rate is also higher, the appropriate policy is to run with the tank full, filling during normal operation or downstream unit failure. If the capacity of the downstream unit is higher, but its failure rate is also higher, the appropriate policy is to run with the tank empty, emptying during normal operation or upstream unit failure. A general principle underlying these policies is that unit production is not interrupted merely in order to fill or empty the tank.

It may be necessary, however, to adopt other policies which have objectives other than improvement of plant availability. Thus if the shut-down of the downstream unit is relatively hazardous or difficult or if the output of the upstream unit needs to be blended, the policy adopted may be to run with the tank full. On the other hand, if the shut-down of the upstream unit is relatively hazardous or difficult, the policy adopted may be to run with the tank empty.

For certain defined cases, a relatively simple treatment is possible. One such case, described by Henley and Hoshino (1977), is that of an upstream unit with excess capacity. It is assumed that the time required to fill the tank using this excess capacity is much less than the mean time to failure of the upstream unit and that the operating policy is to keep the tank full. Then it can be shown that the following relationships apply:

$$V_s = \frac{\lambda_u m_{ru}}{1 + \lambda_u m_{ru}} \quad [7.13.26]$$

$$V_{st} = kV_s \quad [7.13.27]$$

$$k = \exp(-T_e/m_{ru}) \quad [7.13.28]$$

$$\lambda_{st} = \frac{k[\lambda_u/(1 + \lambda_u m_{ru})]}{1 - k[\lambda_u m_{ru}/(1 + \lambda_u m_{ru})]} \quad [7.13.29]$$

where m_{ru} is the mean repair time of the upstream unit, T_e is the time required to empty the tank, V_s is the unavailability of the upstream unit, V_{st} is the unavailability of the upstream unit + the tank, λ_{st} is the failure rate of the upstream unit + the tank, and λ_u is the failure rate of the upstream unit.

A fuller treatment of this and other cases is given by Henley and Hoshino (1977).

The problem of the effect of storage has also been treated by Coker (1978), who has extended the throughput capability method of Pearson to take account of storage. The system considered is an upstream unit, an intermediate storage tank and a downstream unit. The states of the storage are defined as discrete levels of storage, or hold-ups. These states are then determined using a discrete-state, discrete-time steady-state Markov model, as described in Section 7.10.

As an illustration of the method, consider the system given by Coker in which the throughput capabilities of the upstream and downstream units are as shown in Table 7.15 and in which the tank capacity is 15 m^3. Then the probability of a net inflow of 100 m^3 into the tank is $0.95 \times 0.04 = 0.038$. The probabilities of the other net

Table 7.15 *Plant availability example: unit throughput capabilities*

	Throughput capability (m^3/h)	*Probability*
Upstream unit	0	0.02
	95	0.03
	100	0.95
Downstream unit	0	0.04
	90	0.06
	100	0.90

inflows may be calculated in a similar manner. It should be noted that a zero net inflow is obtained both for a flow of 0 m^3/h in both units and for one of 100 m^3/h in both units. The probability distribution of net inflow is then as shown in Table 7.16.

Now assume that the states of the tank are defined as hold-ups of 0, 5, 10 and 15 m^3. The transition probabilities, which are the elements of the matrix **A** in Equation 7.10.37, are determined by considering the possible changes in tank hold-up and the net flow differences which can produce such changes, and then summing the probabilities of these net flows. Thus for a hold-up 'change' of 0 to 0 m^3 the net flows which can cause this change are 0, -5, -90 and -100 m^3/h. Then the transition probability between an initial hold-up state of 0 m^3 and a final state of 0 m^3 is:

$$0.8558 + 0.027 + 0.0012 + 0.018 = 0.902$$

The other transition probabilities may be calculated in a similar manner. The transition matrix is then:

$$\mathbf{A} = \begin{bmatrix} 0.9020 & 0.0018 & 0.0570 & 0.0392 \\ 0.0462 & 0.8558 & 0.0018 & 0.0962 \\ 0.0192 & 0.0270 & 0.8558 & 0.0980 \\ 0.0192 & 0 & 0.0270 & 0.9538 \end{bmatrix}$$

Then application of Equation 7.10.37 gives the probabilities of tank hold-up listed in Table 7.17.

The throughput capability of the system is now calculated. A throughput capability of 10 m^3/h occurs when the throughput capabilities of the upstream and downstream units are 0 m^3 and 90 or 100 m^3, respectively. The probability of a system throughput capability of 10 m^3/h is:

$$0.02 \times 0.1826 \times (0.06 + 0.90) = 0.0035$$

Table 7.16 *Plant availability example: probability of net inflow*

Net inflow into tank (m^3/h)	*Probability*
100	0.0380
95	0.0012
10	0.0570
5	0.0018
0	0.8558
−5	0.0270
−90	0.0012
−100	0.0180

Table 7.17 *Plant availability example: probabilities of tank holdup*

Tank holdup (m^3)	*Probability*
0	0.1722
5	0.0363
10	0.1826
15	0.6089

Table 7.18 *Plant availability example: system throughput capabilities calculated in the example*

Throughput capability of system (m^3/h)	*Probability*
0	0.0433
5	0.0007
10	0.0035
15	0.0167
90	0.0588
95	0.0046
100	0.8774

The other throughput capabilities may be calculated in a similar manner. The result obtained is given in Table 7.18.

7.14 Bayes' Theorem

It frequently occurs in reliability engineering that there is a need to test a hypothesis or to revise an estimate in the light of additional information. A powerful tool for handling such problems is Bayes' theorem. Accounts of Bayes' theorem include those given in *Introduction to Probability and Statistics from a Bayesian Viewpoint* (Lindley, 1965), and those by Savage (1962a), Brand (1980) and Kaplan (1983). Its use in expert systems is discussed by Michie (1979).

7.14.1 Basic formulation

A basic formulation of Bayes' theorem is as follows. Consider the probability of the joint event AB. If the event A has two outcomes A and $\bar{A}$:

$$P(\mathrm{AB}) = P(\mathrm{A|B})P(\mathrm{B}) \qquad [7.14.1]$$

$$= P(\mathrm{B|A})P(\mathrm{A}) \qquad [7.14.2]$$

Then:

$$P(\mathrm{A|B}) = \frac{P(\mathrm{B|A})P(\mathrm{A})}{P(\mathrm{B|A})P(\mathrm{A}) + P(\mathrm{B}|\bar{\mathrm{A}})P(\bar{\mathrm{A}})} \qquad [7.14.3]$$

In this equation the term $P(\mathrm{A})$ is known as the prior probability, $P(\mathrm{A|B})$ as the posterior probability and $P(\mathrm{B|A})$ as the likelihood.

Equation 7.14.3 may also be reformulated to accommodate multiple outcomes. If the event A has n outcomes

$$P(\mathrm{A}_j|\mathrm{B}) = \frac{P(\mathrm{B|A}_j)P(\mathrm{A}_j)}{\sum_{i=1}^{n} P(\mathrm{B|A}_i)P(\mathrm{A}_i)} \qquad [7.14.4]$$

7.14.2 Hypothesis formulation

Equation 7.14.3 may also be written in terms of a hypothesis H and an event E, which serves to confirm or deny the hypothesis:

$$P(\mathrm{H|E}) = \frac{P(\mathrm{E|H})P(\mathrm{H})}{P(\mathrm{E|H})P(\mathrm{H}) + P(\mathrm{E}|\bar{\mathrm{H}})P(\bar{\mathrm{H}})} \qquad [7.14.5]$$

7.14.3 Continuous formulation

There is also a continuous formulation of Bayes' theorem in terms of distributions. If X is a random variable, its density function may be written as $f_X(x)$. Then a formulation of Bayes' theorem in terms of such distributions is:

$$f_{Y|X}(y|x) = \frac{f_{X|Y}(x|y)f_Y(y)}{f_X(x)} \qquad [7.14.6]$$

But

$$f_X(x) = \int_{-\infty}^{\infty} f_{X|Y}(x|y)f_Y(y)\ \mathrm{d}y \qquad [7.14.7]$$

Hence

$$f_{Y|X}(y|x) = \frac{f_{X|Y}(x|y)f_Y(y)}{\int_{-\infty}^{\infty} f_{X|Y}(x|y)f_Y(y)\ \mathrm{d}y} \qquad [7.14.8]$$

7.14.4 Failure rate estimation

A common problem in reliability work is the estimation of a failure rate and the revision of this estimate as more information becomes available. A typical situation might be that the initial estimate is one obtained from the literature which it is desired to revise in the light of works data.

If the failure rate is considered as a random variable, its density function may be written as $f_\Lambda(\lambda)$. The field data are assumed to be available as times to failure $t_1, \ldots, t_n$. Then from Equation 7.14.8, for the first time to failure t_1:

$$f_{\Lambda|T1}(\lambda|t_1) = \frac{f_{T1|\Lambda}(t_1|\lambda)f_\Lambda(\lambda)}{\int_{-\infty}^{\infty} f_{T1|\Lambda}(t_1|\lambda)f_\Lambda(\lambda)\ \mathrm{d}\lambda} \qquad [7.14.9]$$

At this stage the formulation is a general one. Its application may be illustrated by considering the common case of the revision of an initial estimate of the failure rate in the exponential distribution. In the terminology used here the exponential distribution itself may be written in terms of the random variable T:

$$f_T(t) = \Lambda\ \exp(-\Lambda t) \qquad t \geq 0 \qquad [7.14.10a]$$

$$= 0 \qquad t < 0 \qquad [7.14.10b]$$

If Λ is now considered to be a random variable, to be estimated on the basis of observed values of T, the exponential distribution just given may be rewritten as

$$f_{T|\Lambda}(t|\lambda) = \lambda\ \exp(-\lambda t) \qquad t \geq 0 \qquad [7.14.11a]$$

$$= 0 \qquad t < 0 \qquad [7.14.11b]$$

Since the failure rate is to be treated as a random variable, it is necessary to specify the form of the distribution which it follows. The distribution assumed is again the exponential distribution, so that

$$f_\Lambda(\lambda) = \mu\ \exp(-\mu\lambda) \qquad [7.14.12]$$

where μ is the parameter of this distribution. Its dimensions are time, or lifetime, being the inverse of those of λ.

Then, from Equations 7.14.9, 7.14.11 and 7.14.12

$$f_{\Lambda|T1}(\lambda|t_1) = \frac{\lambda\ \exp(-\lambda t_1)\mu\ \exp(-\mu\lambda)}{\int_0^{\infty} \lambda\ \exp(-\lambda t_1)\mu\ \exp(-\mu\lambda)\ \mathrm{d}\lambda} \qquad [7.14.13]$$

which yields

$$f_{\Lambda|T1}(\lambda|t_1) = \lambda(\mu + t_1)^2\ \exp[-(\mu + t_1)\lambda]\lambda \geq 0 \qquad [7.14.14a]$$

$$= 0 \qquad \lambda < 0 \qquad [7.14.14b]$$

By extension to further times to failure $t_2, \ldots, t_n$, it can be shown that

$$f_{\Lambda|T1,\ldots,Tn}(\lambda|t_1, \ldots, t_n) = \frac{\lambda^n}{n!}\left(\mu + \sum_{i=1}^{n} t_i\right)^{n+1} \exp\left[-\left(\mu + \sum_{i=1}^{n} t_i\right)\lambda\right] \qquad [7.14.15]$$

From this density function the mean value is determined in the usual way, the quantity obtained being the estimate $\hat{\Lambda}$ of the mean:

$$\hat{\Lambda} = \int_0^{\infty} f_{\Lambda|T1,\ldots,Tn}(\lambda|t_1, \ldots, t_n)\ \mathrm{d}\lambda \qquad [7.14.16]$$

The required result is thus:

$$\hat{\Lambda} = \frac{n+1}{\mu + \sum_{i=1}^{n} t_i} \qquad [7.14.17]$$

The estimate $\hat{t}$ of the mean life, or MTBF, is obtained as the inverse of that of the failure rate:

$$\hat{t} = \frac{\mu + \sum_{i=1}^{n} t_i}{n+1} \qquad [7.14.18]$$

This equation shows that for the case of the exponential distribution the estimate of the mean life is the average of the prior estimate and of the observed values times to failure.

As an illustration, consider the case where there is for an equipment a prior estimate μ of the time to failure of 8 months which is to be revised in the light of four further measurements ($n = 4$) of the times to failure of 17, 13, 18 and 12 (t_1 to t_4). Then from Equation 7.14.18 the estimated values are:

$$\hat{t} = \frac{8 + (17 + 13 + 18 + 12)}{4 + 1}$$

$$= 13.6 \text{ months}$$

$$\hat{\Lambda} = 0.074 \text{ failures/month}$$

7.15 Renewal Theory

Another technique for the analysis of repairable systems is renewal theory. Accounts of renewal theory are given in *Renewal Theory* (D.R. Cox, 1962) and by Shooman (1968a) and G.H. Mitchell (1972).

Renewal theory is illustrated here by considering the case of a single component system with instantaneous replacement of the component when it fails by another component. The theory is developed initially for the case where each replacement may be by a component with different failure characteristics.

Let system failures occur at time intervals $t_1, \ldots, t_2, \ldots, t_n$. Then the system operating time τ_n to n renewals is:

$$\tau_n = t_1 + t_2 \ldots t_n \qquad [7.15.1]$$

The density function $f_{\tau n}(t)$ of τ_n is obtained by convolution of the individual density functions $f_1(t), f_2(t), \ldots, f_n(t)$. This is handled most easily in the Laplace domain. Then taking Laplace transforms, the convolution becomes:

$$f_{\tau n}(s) = f_1(s) \cdot f_2(s) \ldots f_n(s) \quad [7.15.2]$$

For simplicity, consideration is now limited to the case where replacement is by identical components. Then Equation 7.15.2 becomes:

$$f_{\tau n}(s) = [f(s)]^n \quad [7.15.3]$$

The probability that the nth renewal occurs before or at time t is given by the distribution function:

$$F_{\tau n}(t) = \int_0^t f_{\tau n}(t)\mathrm{d}t \quad [7.15.4]$$

The mean, or expected, value of the time for n renewals is:

$$E[\tau_n] = \int_0^\infty t f_{\tau n}(t)\mathrm{d}t \quad [7.15.5]$$

The probability of the number N of renewals occurring exactly at time t is:

$$P(N = n) = F_{\tau n}(t) - F_{\tau n+1}(t) \quad [7.15.6]$$

The mean, or expected, value of N renewals in time t is:

$$E[N] = \sum_{n=0}^\infty nP(N = n) \quad [7.15.7]$$

From Equations 7.15.6 and 7.15.7

$$E[N] = \sum_{n=0}^\infty n[F_{\tau n}(t) - F_{\tau n+1}(t)] \quad [7.15.8]$$

Expanding the terms in Equation 7.15.8, it can be shown that:

$$E[N] = \sum_{n=1}^\infty F_{\tau n(t)} \quad [7.15.9]$$

The application of these relationships may be illustrated by considering the case where the components have an exponential failure distribution with constant failure rate λ, so that the density function is:

$$f(t) = \lambda\exp(-\lambda t) \quad [7.15.10]$$

Taking Laplace transforms

$$f(s) = \frac{\lambda}{s + \lambda} \quad [7.15.11]$$

and then substituting Equation 7.15.11 into Equation 7.15.3 yields

$$f_{\tau n}(s) = \left(\frac{\lambda}{s + \lambda}\right)^n \quad [7.15.12]$$

Inverting back into the time domain gives:

$$f_{\tau n}(t) = \lambda\frac{(\lambda t)^{n-1}}{(n-1)!} \exp(-\lambda t) \quad [7.15.13]$$

This equation is the Erlangian operating time density function.

The distribution function of Equation 7.15.4 may be integrated using the expression for the density function given in Equation 7.15.13. For example, for the first renewal ($n = 1$)

$$f_{\tau 1}(t) = \lambda \exp(-\lambda t) \quad [7.15.14]$$

Hence the distribution function is:

$$F_{\tau 1}(t) = \int_0^t \lambda \exp[-(\lambda t)] \quad [7.15.15]$$

$$= 1 - \exp(-\lambda t) \quad [7.15.16]$$

and from Equation 7.15.5 the mean value

$$E[\tau_1] = \int_0^\infty t\lambda \exp(-\lambda t)\mathrm{d}t \quad [7.15.17]$$

$$= 1/\lambda \quad [7.15.18]$$

Furthermore, the probability given in Equation 7.15.6 may be written, by using Equation 7.15.4, as:

$$P(N = n) = \int_0^t f_{\tau n}(t)\mathrm{d}t - \int_0^t f_{\tau n+1}(t)\mathrm{d}t \quad [7.15.19]$$

Substituting Equation 7.15.13 in this equation yields

$$P(N = n) = \int_0^t \lambda\frac{(\lambda t)^{n-1}}{(n-1)!} \exp(-\lambda t)\mathrm{d}t - \int_0^t \lambda\frac{(\lambda t)^n}{n!} \exp(-\lambda t) \quad [7.15.20]$$

which can be shown to reduce to

$$P(N = n) = \lambda\frac{(\lambda t)^n}{n!} \exp(-\lambda t) \quad [7.15.21]$$

Again, for the case of the first renewal ($n = 1$)

$$P(N = 1) = \lambda t \exp(-\lambda t) \quad [7.15.22]$$

and

$$E[N] = \sum_{n=1}^\infty F_{\tau n}(t) \quad [7.15.23]$$

But, using Equation 7.15.4 and then taking Laplace transforms:

$$F_{\tau n}(s) = \frac{f_{\tau n}(s)}{s} \quad [7.15.24]$$

From Equation 7.15.3

$$F_{\tau n}(s) = \frac{[f(s)]^n}{s} \quad [7.15.25]$$

Then

$$E[N(s)] = \frac{1}{s}\sum_{n=1}^\infty [f(s)]^n \quad [7.15.26]$$

$$= \frac{1}{s}[f(s) + f(s)^2 \ldots] \quad [7.15.27]$$

$$= \frac{1}{s}\frac{f(s)}{1 - f(s)} \quad [7.15.28]$$

Substituting from Equation 7.15.11:

$$E[N(s)] = \frac{\frac{1}{s}\frac{\lambda}{s+\lambda}}{\frac{\lambda}{s+\lambda}} \quad [7.15.29]$$

$$= \lambda/s^2 \quad [7.15.30]$$

Inverting into the time domain:

$$E[N] = \lambda t \quad [7.15.31]$$

7.16 Replacement Models

One of the principal decisions in the maintenance of plant is whether to replace an equipment. Support for such decisions is provided by replacement theory. Treatments of replacement are given by Dean (1963), Hastings (1970), Jardine (1970a,b) and Mitchell (1972).

Some reasons for replacing an equipment are that: (1) it has failed, (2) it is about to fail, (3) it has deteriorated, and (4) an improved version has become available.

In most replacement models the quantity sought is the overall economic optimum. Some models treat the problem solely in terms of the equipment, whilst others take into account the effects on the process either of outright failure or of deterioration of the equipment. Another distinction is between models which allow for the time to effect the replacement and those which do not.

A somewhat different model is required where the case considered is that exemplified by the lightbulb replacement problem. Here the cost of the actual replacement actions is high and has a significant influence on the result. This type is known as a 'group replacement model'. Treatments are given by Bartholemew (1963) and Naik and Nair (1965).

Other cases to be considered include replacement as opportunity arises, possibly during a major shut-down, and replacement of an equipment which cannot be repaired to the 'as new' condition.

In general, for simple items which are used in large numbers, what is required is straightforward rules, while for more complex individual equipment a more detailed analysis may be justified.

If the time-scale over which replacement occurs extends over a number of years, it may be necessary to allow for the change in the value of money using a technique such as net present value.

A common type of replacement model is that which considers only the equipment and determines the optimum trade-off between capital cost and operating cost. It is necessarily assumed that the latter increases with time, since if it does not, the best policy is to run the equipment until it fails. A replacement model of this type has been given by G.H. Mitchell (1972) and is as follows:

$$C = \frac{C_c - S}{n} + \frac{1}{n}\int_0^n f(t)\mathrm{d}t \quad [7.16.1]$$

where C is the average annual cost of operation, C_c is the capital cost of the equipment, $f(t)$ is the rate of expenditure on maintenance, t is time, and S is the scrap value of the equipment. As just stated, $f(t)$ is taken to increase with time t. Then it can be shown by differentiating Equation 7.16.1 that the average annual cost is a minimum when

$$f(n) = \frac{C_c - S}{n} + \frac{1}{n}\int_0^n f(t)\mathrm{d}t = C \quad [7.16.2]$$

Thus the equipment should be replaced when the current maintenance cost becomes equal to the average annual cost to date.

Further replacement models are given by Jardine (1970a) who treats the cases of failure replacement + preventive replacement, where the latter is (1) fixed-term and (2) based on equipment age.

7.17 Models of Failure: Strength–Load Interaction

7.17.1 Strength and load

Failure occurs when the load on a component exceeds the strength or, more generally, when the duty exceeds the capability. There are various factors which may cause the load to be higher, or the strength lower, than expected. Factors which increase stress include: internal residual stresses and stress raisers; reductions of cross-sectional area due to size variations in manufacture, wear and corrosion; and applications of the component different from that for which it was designed. Factors which reduce strength include quality variations in manufacture, defects, fatigue and creep.

7.17.2 Safety factor and safety margin

The traditional method of allowing for factors which increase stress or reduce strength is the use of a safety factor. The approach has been to define the safety factor as the ratio of the mean of the strength to that of the load, both mean values being assumed constant. This is a deterministic approach, and increasingly it is being superseded by a probabilistic approach which allows for the variability of both strength and load. This does not mean, however, that concepts of the safety factor, and the associated safety margin, are abandoned, but merely that they are redefined, as described below.

7.17.3 Deterministic approach: mean of strength and load

Traditional safety factors are shown in Figure 7.16. The design safety margin (the difference between the design strength and the design stress) and the design safety factor (the ratio of the design strength and the design stress) are normally substantial. However, by the time account is taken of the margin of ignorance, which is the sum of the increases in stress and the decreases in strength, the actual safety margin and safety factor are greatly reduced.

This reduction in the design safety factor is important. The actual safety factor is often much less than might be supposed and the potential causes of reduction in the safety factor need to be given close attention. As stated earlier, this approach is based essentially on the means of the strength and of the load.

7.17.4 Probabilistic approach: variability of strength and load

The weakness of the approach just described is that it does not take into account the variability of the strength and of the load. A more realistic representation of failure is shown in Figure 7.17. Both the strength and the load

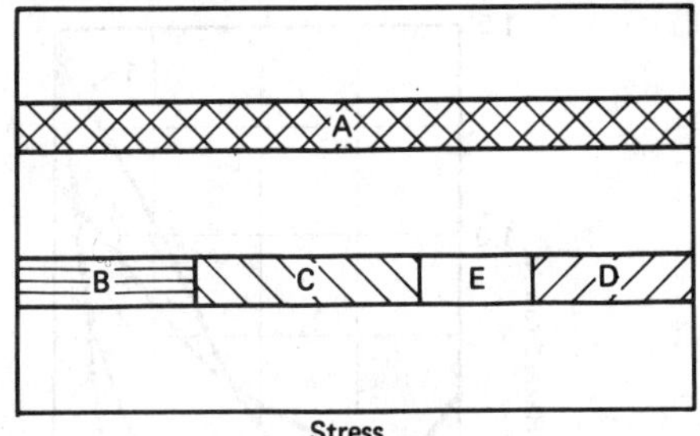

A Assumed static strength
B Design working stress
C Factors increasing stress: internal residual stresses, stress raisers, size variations in manufacture, wear, corrosion
D Factors reducing strength: quality variations in manufacture, defects, fatigue, creep

Design safety margin = A − B
Margin of ignorance = C + D
Actual safety margin = A − (B + C + D) = E
Design safety factor = A/B
Actual safety factor = A/(B + C + D)

Figure 7.16 *Design and actual safety factors and margins*

can be expressed as density functions characterized by a mean and a spread. Failure occurs when the tails of the distributions overlap and create an area where the load exceeds the strength.

7.17.5 Interference theory

The effect of this interaction between load and strength can be quantified using interference theory. Accounts of interference theory are given by Lloyd and Lipow (1962) and von Alven (1964). In this subsection an account is given of the basic theory. A failure model based on the interference approach is given in Section 7.17.

The basic approach may be illustrated by considering the case where the variabilities of the load and the strength can each be represented by a normal distribution. Then the joint density function $g(z)$ for the load to exceed the strength is:

$$g(z) = \int_b^a f(x)f(x-z)\mathrm{d}x\mathrm{d}z \qquad [7.17.1]$$

with

$$z = x - y \qquad [7.17.2]$$

where $f(x)$ is the density function of the load x, $f(y)$ is the density function of the strength y $(= x - z)$, and z is the difference between x and y.

A safety parameter may be defined as:

$$\phi = S/L \qquad [7.17.3]$$

where L is the load, S is the strength, and ϕ is the safety parameter. The variability of the load and the strength may be written as:

$$L = m_x + n_x\sigma_x \qquad [7.17.4]$$

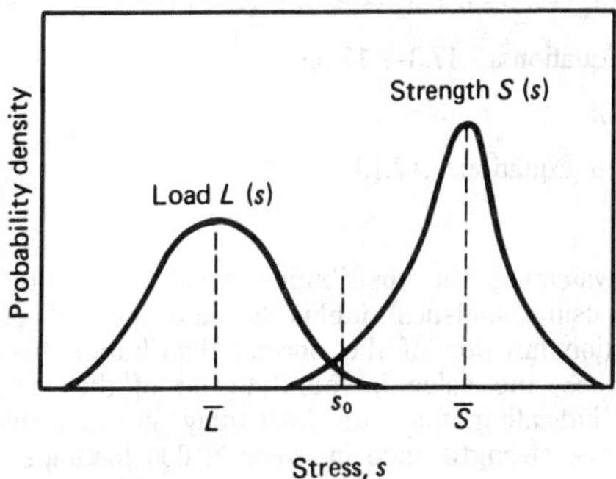

Figure 7.17 *Effect of variability of strength and load on failure (A.D.S. Carter, 1973) (Reproduced by permission of the Institution of Mechanical Engineers)*

$$S = m_y - n_y\sigma_y \qquad [7.17.5]$$

Since the load has a normal distribution, its density function is:

$$f(x) = \frac{1}{\sigma_x(2\pi)^{\frac{1}{2}}} \exp\left[-\frac{(x-m_x)^2}{2\sigma_x^2}\right] \qquad [7.17.6]$$

Similarly, the density function of the strength is:

$$f(y) = \frac{1}{\sigma_y(2\pi)^{\frac{1}{2}}} \exp\left[-\frac{(y-m_y)^2}{2\sigma_y^2}\right] \qquad [7.17.7]$$

From Equations 7.17.1, 7.17.6 and 7.17.7, it can be shown that:

$$g(z) = \frac{1}{[2\pi(\sigma_x^2+\sigma_y^2)]^{\frac{1}{2}}} \exp\left[-\frac{(z+m_y-m_x)^2}{2(\sigma_x^2+\sigma_y^2)}\right] \qquad [7.17.8]$$

which itself is the density function of a normal distribution with a mean value

$$m_z = m_x - m_y \qquad [7.17.9]$$

and standard deviation

$$\sigma_z = (\sigma_x^2 + \sigma_y^2)^{\frac{1}{2}} \qquad [7.17.10]$$

Then the probability that the load exceeds the strength is given by the distribution function

$$G(z) = \int_0^\infty g(z)\mathrm{d}z \qquad [7.17.11]$$

which may be integrated to give

$$G(z) = \frac{1}{(2\pi)^{\frac{1}{2}}}\int_u^\infty \exp(-v^2/2)\mathrm{d}v \qquad [7.17.12]$$

with

$$u = \frac{m_y - m_x}{(\sigma_x^2+\sigma_y^2)^{\frac{1}{2}}} \qquad [7.17.13]$$

$$v = \frac{z + m_y - m_x}{(\sigma_x^2+\sigma_y^2)^{\frac{1}{2}}} \qquad [7.17.14]$$

As an illustration, consider the example given by von Alven (1964):

$m_x = 1; \sigma_x = \sigma_y = 0.4; n_x = n_y = 2; \phi = 1.25$

From Equations 7.17.3–7.17.5,

$m_y = 3.05$

and from Equation 7.17.13

$u = 3.6$

Then evaluating the distribution function in Equation 7.17.12 using statistical tables for the integral of the distribution function of the normal distribution between 3.6 and ∞, the value is found to be of the order of 0.0001, indicating that the load may be expected to exceed the strength once in every 10 000 loadings.

7.17.6 Carter's rough loading model

A theory of failure based on the overlap of the strength and load distributions has been developed by A.D.S. Carter (1973). The approach is illustrated in Figure 7.17 and is broadly as follows.

Initially, a component has a strength s_0, but this gradually deteriorates so that by the ith application of load the strength is reduced to $s_0 - \Delta s_i$. The probability (P_1) of obtaining a component with initial strength lying between s_0 and $s_0 + \mathrm{d}s$ is:

$$P_1 = S(s_0)\mathrm{d}s \qquad [7.17.15]$$

The probability (P_2) of encountering a load which is less than $s_0 - \Delta s_i$ is:

$$P_2 = \int_0^{s_0 - \Delta s_i} L(s)\mathrm{d}s \qquad [7.17.16]$$

Then the probability (P_3) of encountering a load which is less than the current strength of the component the initial strength of which is s_0 in any of the n applications of load is:

$$P_3 = \prod_{i=1}^{n} \int_0^{s_0 - \Delta s_i} L(s)\mathrm{d}s \qquad [7.17.17]$$

And the probability (P_4) of obtaining a component of initial strength s_0 and of encountering a load which over n applications is always less than its strength:

$$P_4 = S(s_0)\mathrm{d}s \prod_{i=1}^{n} \int_0^{s_0 - \Delta s_i} L(s)\mathrm{d}s \qquad [7.17.18]$$

Then the reliability of the component is:

$$R(n) = \int_0^{\infty} S(s) \prod_{i=1}^{n} \int_0^{s_0 - \Delta s_i} [L(s)\ \mathrm{d}s]\mathrm{d}s \qquad [7.17.19]$$

The decrements of strength Δs_i may be expressed in terms of a constant scaling factor σ and a shaping factor $f(i)$:

$$\Delta s_i = \sigma f(i) \qquad [7.17.20]$$

with

$$f(i) = (i - 1)p \qquad [7.17.21]$$

where $p < 1$ gives a fatigue-type curve; $p = 1$ gives a corrosion deterioration-type curve; and $p > 1$ gives an erosion/wear-type curve.

The hazard rate z is then obtained from the relationship:

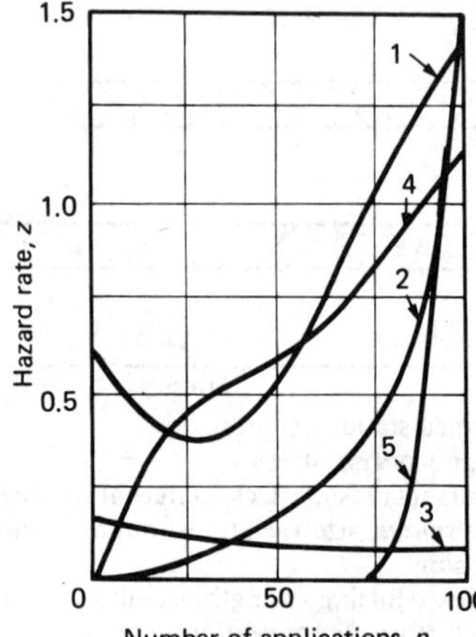

curve 1 $\bar{S} - \bar{L} = 1.0$ $p = 0.2$ medium loading
curve 2 $\bar{S} - \bar{L} = 1.0$ $p = 1.0$ medium loading
curve 3 $\bar{S} - \bar{L} = 1.333$ $p = 0.2$ smooth loading
curve 4 $\bar{S} - \bar{L} = 1.666$ $p = 0.2$ rough loading
curve 5 $\bar{S} - \bar{L} = 1.666$ $p = 3.0$ rough loading

Figure 7.18 *Effect of mean and spread of strength and load on hazard rate (A.D.S Carter, 1973) (Reproduced by permission of the Institution of Mechanical Engineers)*

$$z(n) = -\frac{1}{R(n)}\frac{\mathrm{d}R(n)}{\mathrm{d}t} \qquad [7.17.22]$$

If preferred, n may be replaced by kt to give the reliability as a function of time.

The quantities in Equation 7.17.19 are not normally known, but the theory can be used to explore the implications of making various assumptions. In Carter's work this was done by expressing $S(s)$ and $L(s)$ as Weibull distributions with the shape factor β given a value of 3.44 to approximate normal distributions. Various values of the means $\bar{S}$ and $\bar{L}$ and standard deviations σ_s and σ_L of the strength and load distributions were investigated. Three regimes were distinguished

$\sigma_L \ll \Delta_s$ smooth loading

$\sigma_L = \sigma_s$ medium loading

$\sigma_L \gg \sigma_s$ rough loading

The following quantities were defined:

$\bar{S}/\bar{L}$ safety factor

σ_L/σ_s roughness of loading

$\dfrac{\bar{S} - \bar{L}}{\sqrt{(\sigma_s^2 + \sigma_L^2)^{\frac{1}{2}}}}$ safety margin

This definition of safety margin is different from that conventionally used, as described above.

Some examples of hazard rate curves obtained by Carter are shown in Figure 7.18. The values of σ_s and σ_L were chosen to give a safety margin of six standard deviations and the scaling factor σ was chosen to give $\Delta s_{100} = 2.333$. These results throw considerable doubt on the applicability of the bathtub curve to mechanical components.

7.18 Models of Failure: Some Other Models

7.18.1 Bathtub curve revisited

As already indicated, the original bathtub curve concept has come in for a good deal of criticism. Frequently it appears that it is honoured more in the breach than the observance. The rough loading theory of Carter gives one indication why this should be so.

A modern critique of the concept is that by Moubray (1991), who gives instead the failure patterns shown in Figure 7.19. Moubray states that studies on civil aircraft have shown the following distribution of patterns of failure in terms of the curves in Figure 7.19:

A	4%	*D*	7%
B	2%	*E*	14%
C	5%	*F*	68%

Discussion of these features is taken up again in Section 7.19.

7.18.2 Lightbulb curve

Systems with large numbers of components often tend to exhibit a constant hazard rate. This is readily understood intuitively in terms of a smoothing out of particular hazard rate characteristics of individual components.

The phenomenon is well illustrated by the so-called 'lightbulb curve' shown in Figure 7.20. If a building is provided with new lightbulbs throughout, then assuming that these fail by wearout with a normal distribution and that the failed bulbs are replaced each day, the number of failures recorded per day will be as shown in the figure. The dotted curves (a)–(e) show the numbers actually failing, and the horizontal full line shows the constant number of failures per day to which the system is tending. Each successive curve has a smaller modal value and higher spread than the previous one. Eventually the number of failures recorded per day will become constant. If the lighting system of the building taken as a whole is considered as a series system, then when the number of bulb failures has become constant, the system hazard rate is also constant, so that the system itself appears to have an exponential failure distribution, although its components have a normal failure distribution.

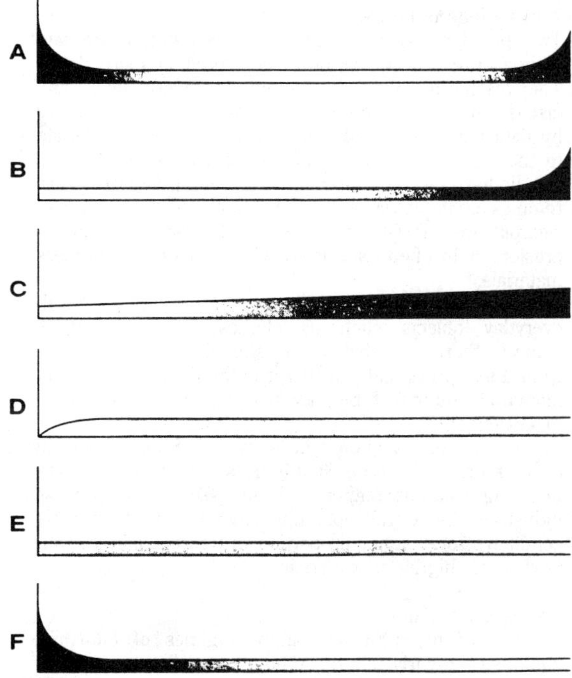

Figure 7.19 Some patterns of equipment hazard rate, including and additional to the bathtub curve (Moubray, 1991) (Courtesy of Butterworth-Heinemann)

7.18.3 Drenick's theorem

A theoretical demonstration that systems with many components tend to have a constant hazard rate is given by Drenick's theorem (Drenick, 1960). A modified version of this theorem has been given by Shooman (1968a). For an n-component series system, from Equation (7.6.4):

$$\ln R(t) = -\sum_{i=1}^{n} H_i(t) = -\sum_{i=1}^{n} \int_0^t z_i(t)\,\mathrm{d}t \qquad [7.18.1]$$

Assume that the hazard rates for small values of t can be expressed as

$$z_i = a_i + b_i t^m \qquad [7.18.2]$$

the value of m being the highest found for the n components. Then

$$\ln R(\tau) = -\tau - k\frac{\tau^{m+1}}{\bar{\lambda}^m} \qquad [7.18.3]$$

with

$$\bar{\lambda} = \sum_{i=1}^{n} a_i \qquad [7.18.4]$$

$$k = \frac{1}{(m+1)\bar{\lambda}}\sum_{i=1}^{n} b_i \qquad [7.18.5]$$

$$\tau = \bar{\lambda}t \qquad [7.18.6]$$

Assuming that as $n \to \infty$, $\bar{\lambda} \to \infty$, and that k is bounded, then as $n \to \infty$

$$\ln R(\tau) = -\tau \qquad [7.18.7]$$

and hence

$$R(t) = \exp(-\bar{\lambda}t) \qquad [7.18.8]$$

The theorem assumes that the system has a series structure. If the system contains parallel paths, the treatment is pessimistic. In practice, there are many systems which are well fitted by the theorem, but also many which are not.

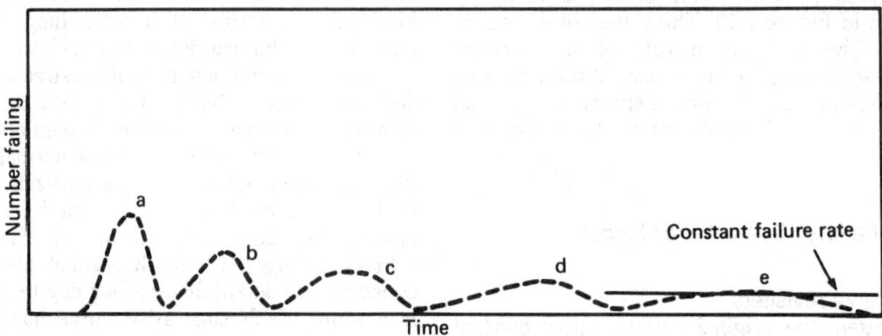

Figure 7.20 *The lightbulb curve*

The equation

$$k = \sum_{i=1}^{n} b_i/(m+1)\bar{\lambda}$$

is bounded, provided at least one $z_i(0) \neq 0$. Its rate of convergence depends, however, on the number of components for which $z_i(0) = 0$. If these are numerous, the convergence is slow.

The implications of this theorem are considered in more detail by Shooman (1968a).

7.19 Failure Behaviour and Regimes

7.19.1 Failure regimes

The measures which can be taken to reduce failure depend very largely on the failure regime of the equipment. The three regimes are: (1) decreasing hazard rate, (2) constant hazard rate, and (3) increasing hazard rate. The bathtub curve illustrates these three regimes but, as indicated, this model is honoured as much in the breach as in the observance. However, this does not alter the fact that the failure regime must be one or other of those just listed.

The failure regime of an equipment may be identified using the methods described in Section 7.20. In particular, the failure data may be fitted to the Weibull distribution. The relationship between the failure regime and the Weibull shape factor β is:

$\beta < 1$ decreasing hazard rate (early failure);

$\beta = 1$ constant hazard rate (constant failure);

$\beta > 1$ increasing hazard rate (wearout failure).

7.19.2 Catastrophic failure vs tolerance failure

Some failures occur suddenly, whilst others develop much more gradually. In early work the two types were often called 'catastrophic failure' and 'tolerance failure'. These terms were not well chosen. There is no implication that a catastrophic failure has any consequences wider than the failure itself. Nor is there any implication that catastrophic failures occur early in life or tolerance failures late in life.

Tolerance failures are associated with the interaction between the component and the system. Catastrophic failures have a greater tendency to occur independently of the system.

7.19.3 Bimodal failure distributions

Analysis of failure data not infrequently yields a failure density function with two distinct peaks, indicating a bimodal distribution. Such a density function may be written as

$$f(t) = pf_1(t) + (1-p)f_2(t) \quad [7.19.1]$$

where p is a weighting factor. As described below, early failure can give rise to a bimodal failure distribution.

7.19.4 Wearout failure

Two principal types of failure are wearout failure and early failure. These are now considered in more detail. The discussion here is limited to failure itself and discussion of the determination of the type of failure by data analysis and of maintenance policies appropriate to the types of failure is deferred to later sections.

Failure by wearout is one of the principal failure regimes of industrial equipment. Some of the causes of wearout are: (1) fatigue, (2) wear, (3) corrosion and (4) erosion. It is often associated with contact with process materials.

Wearout failure is a common experience with everyday objects such as clothes and tools. It is natural, therefore, that there should be carried over from this a preconception that it is the dominant mode in industrial equipment but, as just indicated, this can be misleading.

In the results quoted in Section 7.18 for failures in civil aircraft, only some 6% are classified as in some way involving wearout (patterns A and B). In the process industries the overall operating environment is probably conducive to a higher proportion of wearout failures, but how much higher is uncertain.

7.19.5 Early failure

Another of the principal failure regimes of industrial equipment is early failure. Although initially perceived as failure due essentially to premature failure of weak components, there is now better understanding that there is more to early failure than this, and that it can be the dominant mode. In particular, it is now appreciated that early failure in the sense of a decreasing hazard rate is a phenomenon which is not confined to new equipment but is often exhibited by old equipment. The proportion of early failure in the results for civil aircraft quoted in Section 7.18 is 68% (pattern F).

Causes of early failure may be grouped under three headings: (1) those which result in failure of newly commissioned equipment, (2) those which result in failure of equipment which has been operating for some time, and (3) those associated with start-up and shut-down stresses, whether on new or old equipment. Among the former causes, in design and commissioning, are: (1) incorrect design specification, (2) incorrect design, (3) incorrect user specification and selection, (4) incorrect manufacture, (5) incorrect installation, and (6) incorrect commissioning and initial operation.

Causes of early failure in equipment with a longer operating history centre on the quality of the maintenance and include: (1) incorrect fault identification, (2) incorrect repair technique, (3) incorrect replacement parts, (4) incorrect reassembly and alignment, (5) dirty working conditions, and (6) disturbance to other parts of the equipment. The effect of such failings is that the equipment is not fully restored to the 'as new' condition after repair or overhaul. Underlying these causes of early failure due to maintenance are two more fundamental causes. One is the conduct of unnecessary preventive maintenance. The other is inadequate training and/or discipline of maintenance personnel.

The third group of causes of early failure is the mechanical and thermal stresses associated with plant start-up and shut-down.

Evidence of early failure in process equipment was found in a study by Berg (1977) who investigated some 600 centrifugal and air vacuum pumps and some 200 bottom run-off valves and found values of the shape parameter β ranging from 0.735 to 1.072. Aird (1977b) obtained values of β as low as 0.5 for some mechanical equipment in process plants.

In a further study Sherwin and Lees (1980) found early failure to be prevalent in both process plant equipment and hospital autoclaves. The process plant investigation was done in two stages. In the first stage failure data were collected and analysed and the maintenance tasks were observed. Recommendations were then made for improvement to the conduct of maintenance. In the second stage, 2 years later, further data collection and analysis was carried out. The items of plant equipment studied were: (1) acid pumps, (2) water pumps, (3) vacuum pumps, (4) agitators, (5) screw conveyers, (6) filters, (7) fans, (8) heat exchangers, (9) evaporator flash vessels, (10) other tanks and vessels, (11) pipes and ducts, and (12) other items. The failure modes were classified as: (1) blockages, (2) leaks, (3) drives, (4) electrical, (5) holes and breaks, (6) instruments, (7) valves, and (8) other faults. The authors found that with few exceptions the failures could be characterized as early failures with β values less than unity. In the first stage, values of β obtained on six types of equipment ranged from 0.61 to 1.07, whilst β values for five failure modes of the acid pumps ranged from 0.48 to 0.76 with the overall value for the eight failure modes being 0.63. As stated, the maintenance work was also observed. The research revealed many of the problems of maintenance work listed above, such as incorrect use of consumables, use of incorrect parts, failure to realign correctly and failure to adjust clearances correctly, as well as working in dirty conditions. Above all, it pointed to a lack of training. Recommendations were made to rectify these shortcomings, in particular proposals on training, and these were acted on. Analysis of the failure data from the second stage found that of the 12 types of equipment, 8 showed improvements in MTBF and 3 deteriorations in MTBF, with one type classed as N/A (not applicable). The MTBFs of the acid, water and vacuum pumps, for example, were increased from 47.5 to 108, from 75.8 to 382 and from 57.3 to 117 days, respectively, and that of the agitators from 910 days to 4641 days.

The hospital study was on autoclaves used for sterilization, which were breaking down at an unacceptable frequency. The study yielded results remarkably similar to those on the process plant. Again early failure ($\beta < 1$) was the dominant regime. Recommendations were made and acted on, particularly with respect to training, and the overall failure rate was reduced from 0.065 to 0.014 failures/cycle.

These studies indicate that as a result of early failure the overall failure rate of the equipment is higher than it need be and often never settles down to the lower constant failure rate which might be regarded as the more 'normal' condition. The concept of 'maintenance induced' failure has long been familiar to plant engineers. The investigations described give it statistical support.

7.19.6 Burn-in

In the foregoing discussion wearout and early failure have been viewed as failure regimes. It is also of interest to develop models of failure in these regimes. An account of such modelling is given in *Burn-In* (Jensen and Petersen, 1982).

The starting point is the observation that analysis of failure data for components often gives a bimodal distribution, such that the failure density function exhibits two distinct peaks: the first for 'freak' failures and the second for the main failures. An example is the work of Herr, Baker and Fox (1968) on the failure of electronic components.

Jensen and Petersen have developed a model for the freak failures which is based on two assumptions: (1) the strength of all components deteriorates with time, and (2) the strength of weak components deteriorates faster. The more rapid deterioration of the weaker components results in a sharper separation between the peaks of the two failure distributions.

The foregoing refers to components in isolation. Another concept developed by Jensen and Petersen is that once incorporated in a system, the susceptibility to failure of some of the components may be drastically increased. They term this phenomenon 'infant mortality' and treat it as distinct from freak failures. In their terminology, the early failures are made up of the two separate sets of failures (freak and infant mortality). It follows that there is potential for a failure distribution which is not just bimodal but trimodal.

The implications of such concepts of early failure for burn-in policies are discussed in Section 7.25.

7.20 Failure data analysis

Guidance on current performance in equipment reliability and on measures which may be taken to improve it can be obtained from the analysis of failure data.

7.20.1 Need for failure data

The application of reliability techniques creates a demand for data on equipment failure and repair time, on other failure-related events and on human error. These data may be obtained from the literature, from data banks or within the works. Usually it is possible to obtain approximate data fairly readily, but the determination of accurate data tends to involve much more effort. It is wasteful, therefore, to seek for greater accuracy in the data than the problem warrants.

The accuracy required in the data varies considerably between different types of reliability calculation and even between different parts of the same calculation. In general, where the reliability problem has a structure, as in a comparison of a simple equipment with a parallel or stand-by system or in a fault tree analysis, less accurate data may often be used, at least in some parts of the calculation. Thus in a fault tree study, for example, some branches of the tree may be sensitive to the failure rates used, whilst others may not be. On the other hand, where the problem is a straight comparison, as between the failure rates of two instruments, the accuracy required is clearly greater.

Often, for the solution of the reliability problem to be clear it is sufficient to know that the failure rate lies between certain broad limits. In such cases it may be sufficient to rely on expert judgement or on relatively crude failure data rather than on accurate data.

7.20.2 Types of failure data

The failure information required for reliability work includes not only data on (1) overall failure rates, but also data on (2) failure rates in individual modes, (3) variation of failure rates with time and (4) repair times.

Failure modes can be classified in several ways. Some important categories of failure mode are (1) condition, (2) performance, (3) safety and (4) detection. In the failure classification based on condition, a failure mode is exemplified by a faulty seal on a pump or by a defective gearbox on an agitator. It is the failure classification which is normally used in maintenance. In the classification by performance, illustrations of failure are inability of a heat exchanger to achieve the required heat transfer or of a pump to give the specified head. The safety classification divides failure into fail-safe and fail-to-danger in the context of the process. The detection classification makes a distinction between revealed and unrevealed failure. These different categories of failure mode are particularly important in instruments and are discussed further in Chapter 13.

Generally, information on the variation of failure rate with time is not available and therefore is not used much in ordinary reliability work, but it is important in relation to maintenance problems. It is also necessary to have information on human error rates. This aspect is discussed in Chapter 14.

Other information required relates to other events which have a bearing on failure. Typical examples are the probability of a worker being nearby when an explosion occurs in an equipment, the distance travelled by a vapour cloud before it finds a source of ignition and the time taken for the fire services to reach the scene of a tanker crash.

For availability studies the type of data used varies. In some cases the availability is determined from the failure rates and repair times of the basic equipment. In other cases data on the availability of blocks of plant are used.

Failure rates may be expressed in terms of several parameters. The parameter most commonly used is time, but it may be appropriate to distinguish between operating time and non-operating time, and other measures such as the number of cycles or the number of demands are appropriate in some cases. In this connection it should be emphasized that it is often not known which is the important measure in a particular case. Thus misleading results may be obtained if a particular failure is expressed as failures/year when failures/cycle would be more appropriate.

7.20.3 Failure data sources, data collection and data banks

The sources of failure data are essentially external sources such as the literature and data banks. Alternatively, data may be collected within the works. Often it is the works which is the most dependable source of applicable data, and in this case it is necessary to have some form of data collection system. In other cases, particularly for rare events, it will be necessary to make use of the external sources.

The acquisition of data from these various sources is not straightforward. The data are of use only if they are of the right type, originate from a dependable source and apply to the case in hand. These and other issues related to failure data are discussed in Appendix 14.

7.20.4 Fundamentals of failure data analysis

Analysis of failure data should begin with consideration of the fundamental qualitative factors. One of these factors is the appropriateness of lumping together, as a homogenous set, equipments which may differ in design, manufacture, system function and process environment. Another factor is the parameter by which the failure rate is to be measured. As mentioned, this is usually time, but in some cases another measure such as the number of cycles is more suitable.

The qualitative analysis which can be done depends on the information available. Some items which can be calculated are:

(1) overall failure rates;
(2) failure rates in individual modes;
(3) confidence limits and bounds on failure rates;
(4) failure distributions;
(5) repair time distributions.

Frequently, the only information available is the total number of failures and the total operating time of the equipments, and the analysis is restricted to calculation of the overall failure rate and corresponding mean life.

7.20.5 Failure rates in individual modes

If the failure modes and their relative frequency are available, they are often analysed in the form of a

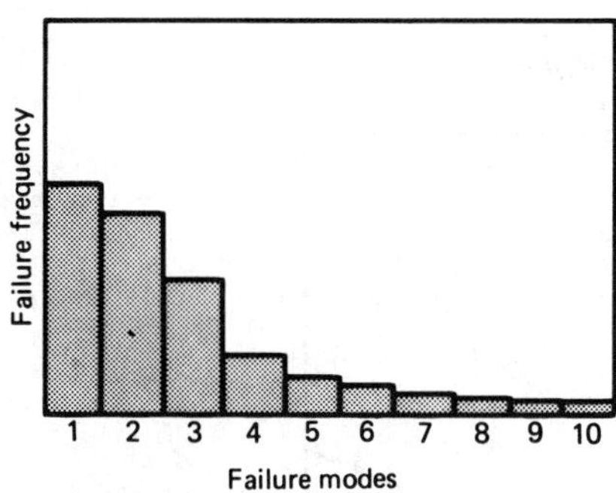

Figure 7.21 *Failure modes of an equipment*

histogram, as shown in Figure 7.21. In many cases there are a few failure modes which have a high relative frequency followed by a larger number of low frequency modes. This is generally referred to as a 'Pareto-type distribution' of failure modes.

It is the failure rates in the individual modes which are the data primarily required for fault tree analysis and for improvement of equipment performance by design and by maintenance.

7.20.6 Confidence limits on failure frequency and mean life

The confidence which can be placed in the estimate of the mean life depends on the number of failures recorded. If this number is small, it may be useful to determine the confidence limits. For the estimate of the mean life in the exponential distribution, the confidence limit may be determined using the χ^2 distribution.

It is necessary to distinguish between the case where the end of the test period T coincides with the final failure n and that where it does not. In the latter case it is conservative to assume that a further failure was just due to occur when the test was terminated.

For the failure-terminated test with replacement, the estimate of the mean life is

$$\hat{m}_1 = \frac{T}{n} \qquad [7.20.1]$$

and the degrees of freedom f are

$$f_1 = 2n \qquad [7.20.2]$$

whilst for the time-terminated test

$$\hat{m}_2 = \frac{T}{n+1} \qquad [7.20.3]$$

and

$$f_2 = 2(n+1) \qquad [7.20.4]$$

Then for the failure-terminated test it can be shown that the ratio $f_1\hat{m}_1/m$ has a χ^2 distribution. For a two-sided confidence interval at a confidence level $(1-\alpha)$

$$P\left(\chi^2_{1-(\alpha/2):f1} \leq \frac{f_1\hat{m}_1}{m} \leq \chi^2_{\alpha/2:f1}\right) = 1-\alpha \qquad [7.20.5]$$

Then the lower limit L and the upper limit U of the ratio m/m are:

$$L\left(\frac{m}{\hat{m}}\right) = \frac{f_1}{\chi^2_{\alpha/2:f1}} \qquad [7.20.6]$$

and

$$U\left(\frac{m}{\hat{m}}\right) = \frac{f_1}{\chi^2_{1-\alpha/2:f1}} \qquad [7.20.7]$$

with

$$\hat{m} = \hat{m}_1 = \frac{T}{n} \qquad [7.20.8]$$

For the time-terminated test it can be shown that:

$$L\left(\frac{m}{\hat{m}}\right) = \frac{f_1}{\chi^2_{\alpha/2:f2}} \qquad [7.20.9]$$

and

$$U\left(\frac{m}{\hat{m}}\right) = \frac{f_1}{\chi^2_{1-(\alpha/2):f1}} \qquad [7.20.10]$$

again with $\hat{m}$ defined by Equation 7.20.8.

Values of the percentage points of the χ^2 distribution may be obtained from standard tables (e.g. Yamane, 1967, p. 879). It should be noted that some tables give the α percentage points and others the $(1-\alpha)$ percentage points. A graph showing the confidence limits on $m/\hat{m}$ according to Equations 7.20.9 and 7.20.10 is given in Figure 7.22.

As an illustration, consider the situation in which five failures have been recorded on 10 pumps over a period of half a year, with the final failure occurring before the end of the observation period, and in which it is required to estimate the confidence limits within which the ratio $m/\hat{m}$ lies at a 90% confidence level. Then

$n = 5$ failures

From Equation 7.20.8

$$\hat{m} = \frac{0.5 \times 10}{5} = 1 \text{ year}$$

From Equations 7.20.2 and 7.20.4

$f_1 = 10$

$f_2 = 12$

For a 90% confidence level, $\alpha = 0.1$. From tables of percentage points of the χ^2 distribution

$\chi^2_{\alpha/2:f2} = \chi^2_{0.05:\,12} = 21.0$

$\chi^2_{1-(\alpha/2):f1} = \chi^2_{0.95:\,10} = 3.94$

Then from Equations 7.20.9 and 7.20.10, the lower and upper confidence limits on the ratio $m/\hat{m}$ are:

$$L\left(\frac{m}{\hat{m}}\right) = \frac{10}{21.0} = 0.476$$

and

$$U\left(\frac{m}{\hat{m}}\right) = \frac{10}{3.94} = 2.54$$

Alternatively, the confidence limits may be obtained from Figure 7.22. Then

$L(m) = 0.476$ year

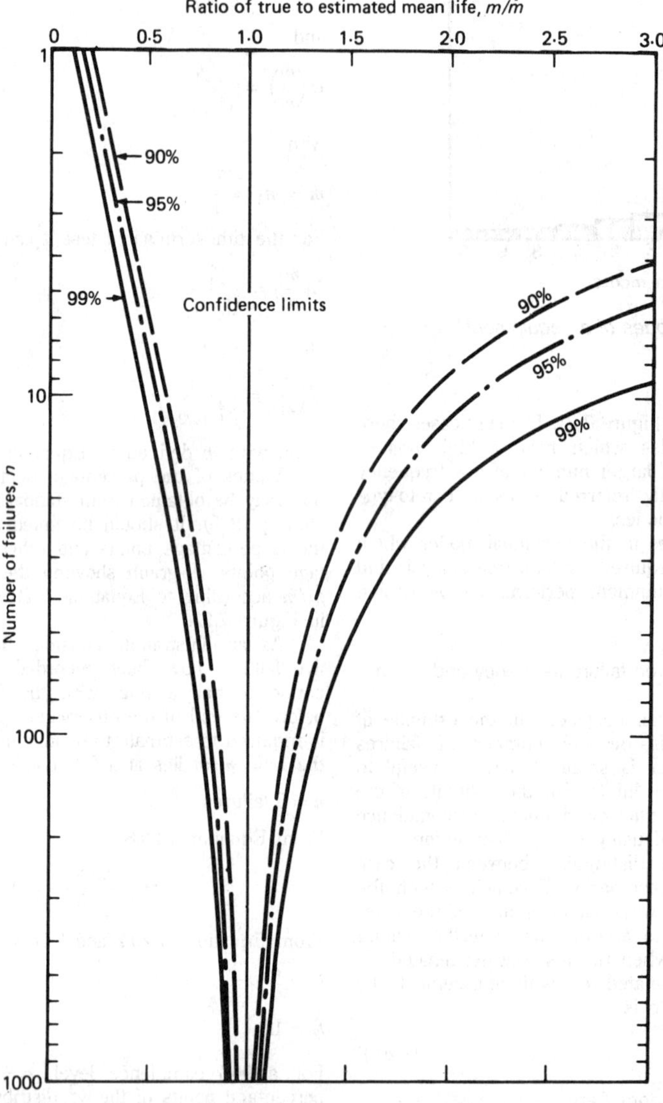

Figure 7.22 Two-sided confidence limits for the ratio of true to estimated mean life $m/\hat{m}$ vs number of failures n in a time-terminated test with an exponential failure distribution (Buffham, Freshwater and Lees, 1971) (Reproduced by permission of the Institution of Chemical Engineers)

$U(m) = 2.54$ year

In some cases no failures are observed. Nevertheless, it is still possible to make an estimate of the lower limit of the ratio $m/\hat{m}$. For a one-sided confidence limit at a confidence level $(1 - \alpha)$:

$$L\left(\frac{m}{\hat{m}}\right) = \frac{f_2}{\chi^2_{\alpha:f2}} \quad [7.20.11]$$

with

$$\hat{m} = \hat{m}_2 = \frac{T}{n+1} \quad [7.20.12]$$

$$f_2 = 2 \quad [7.20.13]$$

$$\chi^2_{\alpha;\,2} = 4.61$$

Then from Equation 7.20.11 the lower confidence limit on the ratio $m/\hat{m}$ is

$$L\left(\frac{m}{\hat{m}}\right) = \frac{2}{4.61} = 0.434$$

and that on the mean life is

$L(m) = 4.34$ years

Further accounts of confidence limits on failure data are given by Bazovsky (1961) and A.E. Green and Bourne (1972).

There is often a quite wide spread in failure rates reported by various sources. The variability of failure rates may be described by a log-normal distribution. Since this distribution has two parameters, it is determined by the end-points of a suitably defined range. This method of describing failure data was extensively used in the *Rasmussen Report* (AEC, 1975). The use of these log-normal distributions in fault trees in this study is discussed in Chapter 9.

7.20.7 Confidence limits on failure probability

The treatment just given refers to failure frequencies. It may also be necessary to determine the confidence limits on failure probabilities. A treatment of this problem is given by von Alven (1964). The approach taken in this case is the use of the binomial theorem. Use may be made of the binomial theorem itself or of charts constructed from it.

Such a plot is given in Figure 7.23, which shows, for the case where there are S successes in N trials, the 95% confidence limits as a function of the ratio S/N, or success probability. As an illustration, consider the case where there are 16 successes in 20 trials. Then the estimated value of the probability of success is 0.8 and the upper and lower confidence limit values are 0.98 and 0.6. The failure probabilities are the complements of these values.

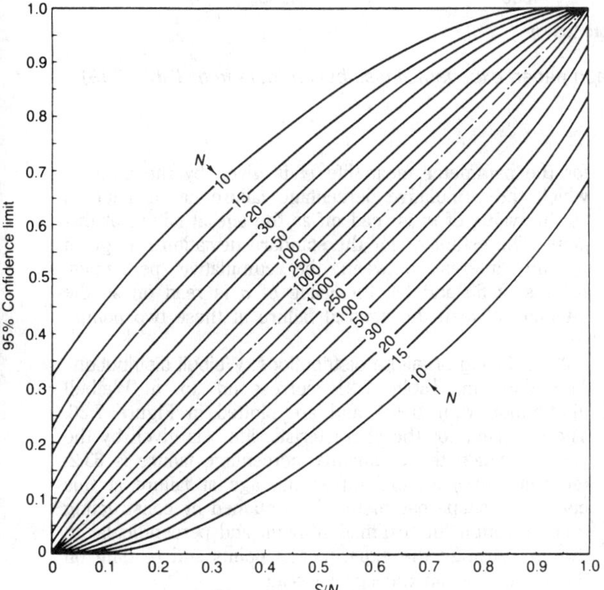

Figure 7.23 *Two-sided 95% confidence limits for the success probability S/N vs number of trials* N *(Clopper and Pearson, 1934)*

7.20.8 Fitting of failure distributions: graphical methods

Information on the variation of a failure rate with time, i.e. the failure distribution, can lead to its reduction through improvement in design and maintenance. The data required for the determination of the failure distribution are the individual times to failure of the equipment.

The procedure is to convert the data to a set of values of the failure distribution $F(t)$ vs times to failure t, and to plot the latter against that function of $F(t)$ on a scale which corresponds to the distribution to be fitted.

For the exponential distribution:

$$F(t) = 1 - \exp(-\lambda t) \qquad [7.20.14]$$

and hence

$$t = \frac{1}{\lambda} \ln\left[\frac{1}{1 - F(t)}\right] \qquad [7.20.15]$$

Thus a plot of $1/[1 - F(t)]$ on a log scale vs t on a linear scale gives a straight line.

For the Weibull distribution:

$$\ln(t - \gamma) = \frac{1}{\beta} \ln\left\{\ln\left[\frac{1}{1 - F(t)}\right]\right\} + \ln \eta \qquad [7.20.16]$$

For these common distributions special graph papers are available which allow $F(t)$ vs t to be plotted directly. Figures 7.24–7.27 illustrate such graphs for the exponential, the normal and the Weibull distributions (two cases), respectively.

The determination of the failure distribution $F(t)$ from the experimental data is not entirely straightforward, however, and requires further discussion.

7.20.9 Fitting of failure distributions: ranking

If a test is considered in which n equipments operate without replacement and all fail during the test and if $n_f(t)$ equipments have failed at time t, then intuitively the failure distribution is simply the proportion which have failed:

$$F(t) = \frac{n_f(t)}{n} \qquad [7.20.17]$$

But the matter is not quite so simple. While Equation 7.20.17 holds for the n items tested, it is not the best estimate if these items tested are a sample from a large population. In this case a better estimate is the expected value:

$$F(t) = \frac{n_f(t)}{n + 1} \qquad [7.20.18]$$

An alternative approach is the use of ranking. This involves calculating the order number m_j of the jth failure. The following ranking expressions are used:

$$F_{\text{mean}} = \frac{m_j}{n + 1} \qquad \text{mean rank} \qquad [7.20.19]$$

$$F_{\text{med}} = \frac{m_j - 0.3}{n + 0.4} \qquad \text{median rank} \qquad [7.20.20]$$

It is usual to calculate the failure distribution $F(t)$ as either the mean rank F_{mean} or the median rank F_{med}. The use of ranking is essential when the number of items n under test is small. The mean is commonly

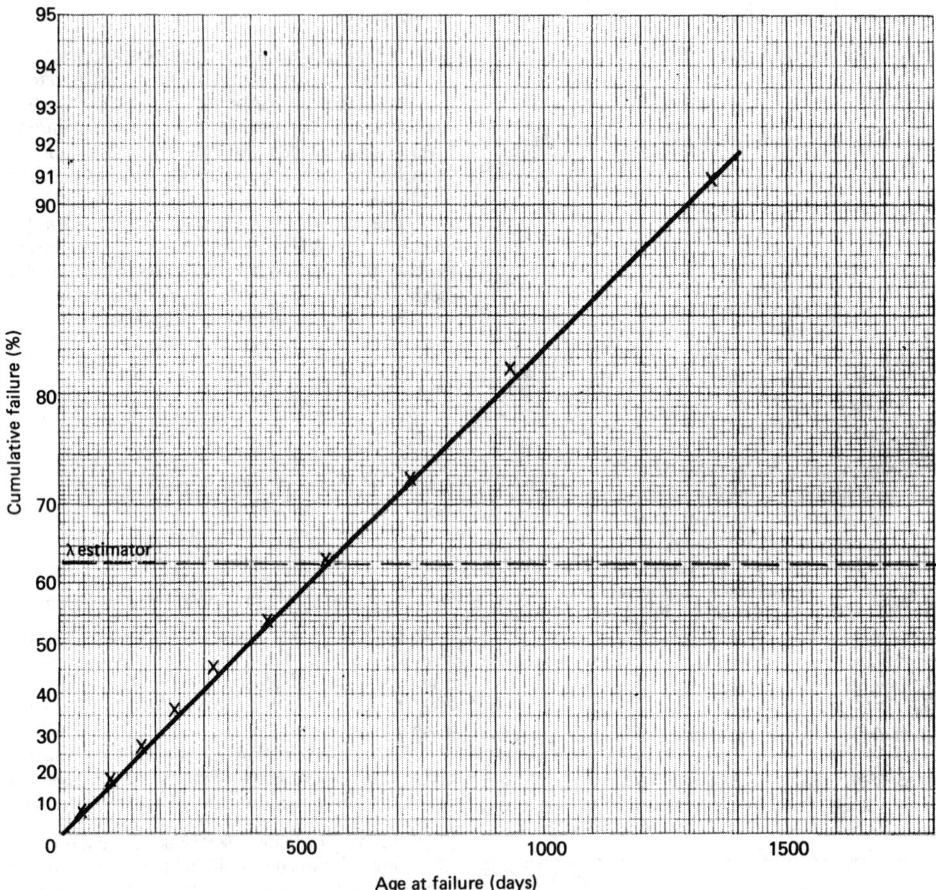

Figure 7.24 *Fitting of failure distributions using special graph paper: exponential distribution (data from Table 7.19)*

taken as a representative descriptor of a sample from a distribution. In the case of a highly skewed distribution, however, a better description may be the median. A discussion of the difference between mean rank and median rank and of the choice between them is given by Kapur and Lamberson (1977). Illustrative examples involving the use of the mean rank and median rank of a set of failure data are given in Tables 7.19–7.22.

7.20.10 Fitting of failure distributions: exponential distribution

The data in Table 7.19 correspond to an exponential failure distribution and are plotted in Figure 7.24. The estimator for the parameter failure rate λ in this distribution is given by the point at which the cumulative percentage failure has the value 63.2; the value of λ is then read off as the age at failure at this point, which in this case is 550 days.

7.20.11 Fitting of failure distributions: normal distribution

The data in Table 7.20 correspond to a normal failure distribution and are plotted in Figure 7.25. The estimator for the parameter mean life m is given by the point at which the cumulative percentage failure has a value of 50; the value of m is read off as the age at failure at this point. The estimator for the standard deviation σ is given by the interval between the cumulative percentage failures of 50 and 84; the value of σ is read off as the interval between the ages at failure at these two points.

7.20.12 Fitting of failure distributions: Weibull distribution

The data in Table 7.21 correspond to a Weibull distribution with $\beta = 1$ and are plotted in Figure 7.26. The estimator for the characteristic life η is given by the point at which the cumulative percentage failure is 63.2; the value of η is read off as the age at failure at this point. The shape parameter β is obtained by constructing a line through the 'estimation' point and perpendicular to the line through the experimental points and reading off the value of β on the special scale.

The mean of the distribution may be obtained as follows. The value of the cumulative probability P_μ is read off from the construction line just described. Then the mean of the distribution is obtained by entering the

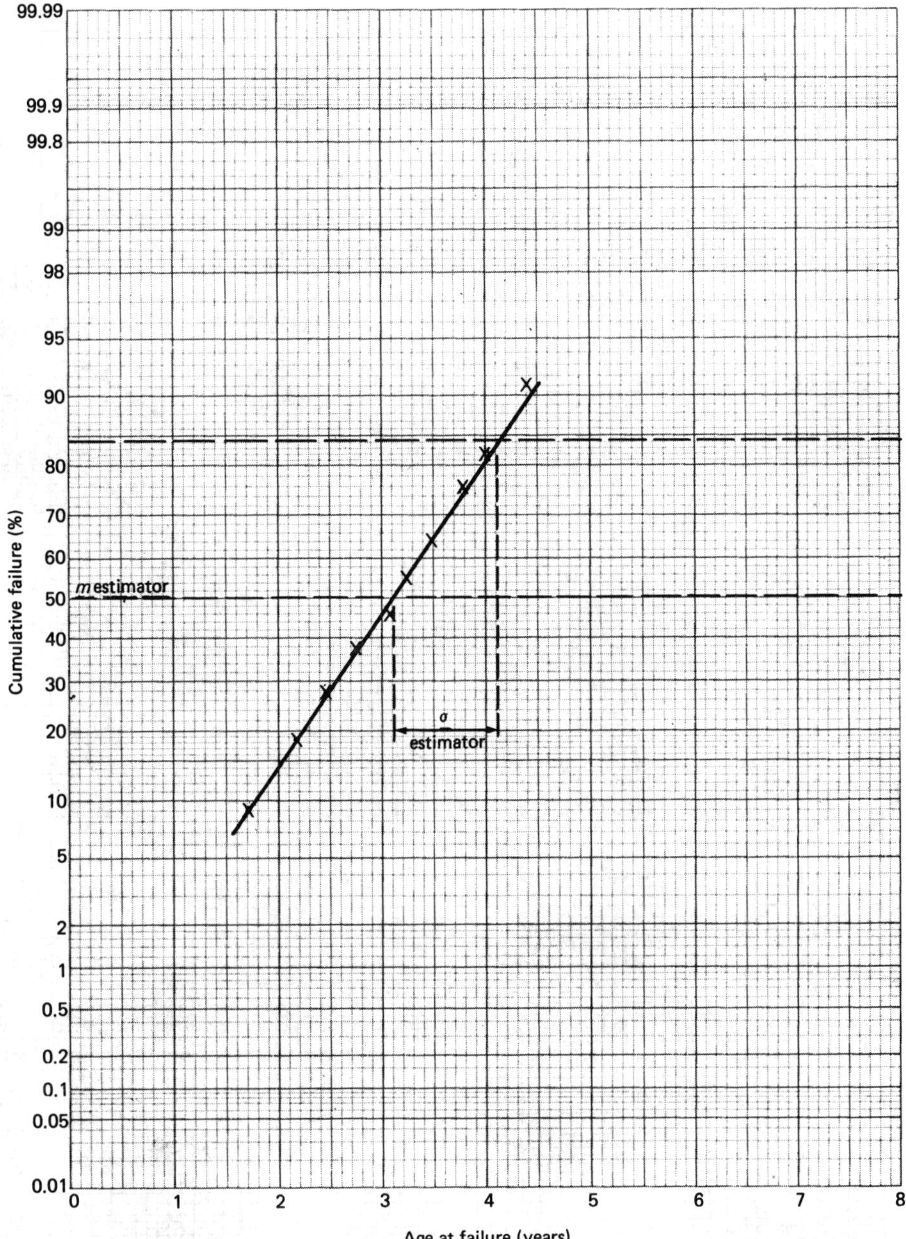

Figure 7.25 *Fitting of failure distributions using special graph paper: normal distribution (data from Table 7.20)*

vertical scale at this value of P_μ, going to the data line and reading off on the horizontal scale.

Also shown in Figure 7.26 are the 90% confidence limits, which are obtained in the following manner. Use is made of the appropriate percentage ranks for the particular sample size. A table of such ranks is given in Appendix VIII of Kapur and Lamberson (1977). The values of the median rank, the 5% rank and the 95% rank for a sample size of $n = 5$ are given in Table 7.22. These values have been used to plot the confidence limits shown.

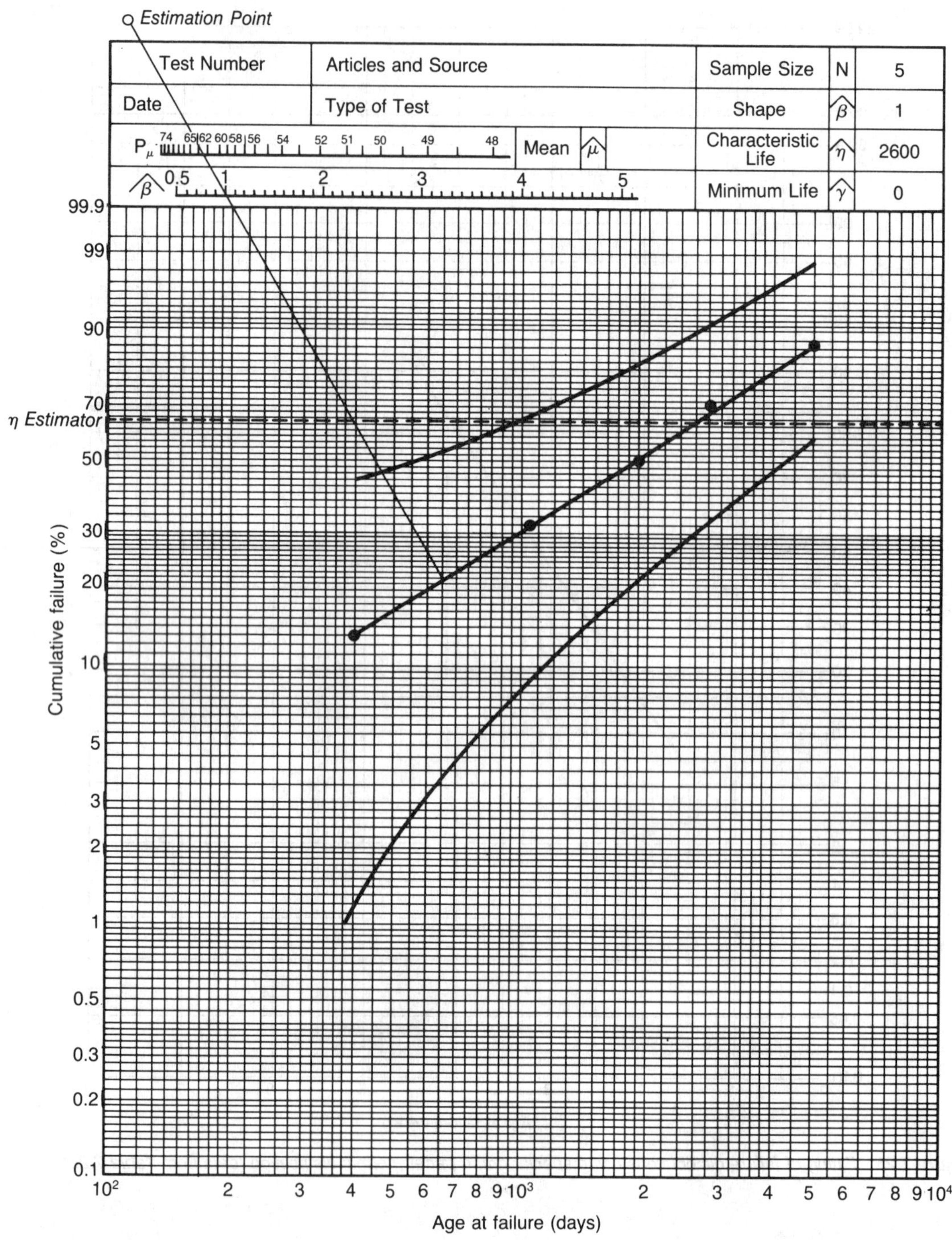

Figure 7.26 *Fitting of failure distributions using special graph paper: Weibull distribution (data from Table 7.21)*

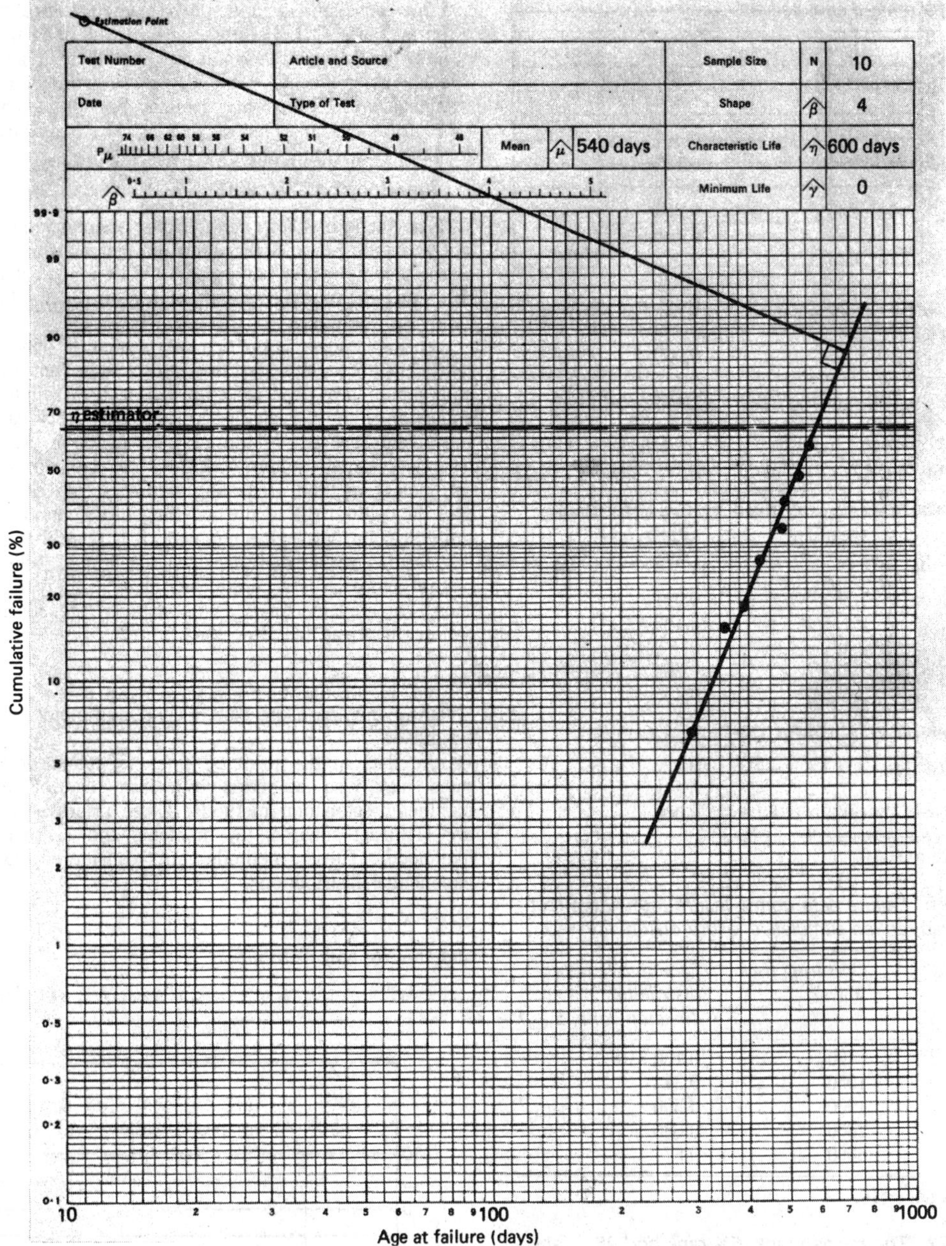

Figure 7.27 Fitting of failure distributions using special graph paper: Weibull distribution for censored data (data from Table 7.23)

7.20.13 Fitting of failure distributions: censored data

It frequently happens that items are withdrawn from test before they have failed. The test is then a censored one. Such a test has both failures and survivals. The mean order number m_j is calculated as

$$m_j = m_{j-1} + \Delta m \qquad [7.20.21]$$

$m_0 = 0$ and $m_1 = 1$ if the sequence begins with a failure. The mean order number increment Δm, which is only recalculated for a failure following a survival, is

$$\Delta m = \frac{n + 1 - m_{j-1}}{n + 1 - x} \qquad [7.20.22]$$

where x is the number of items before the jth failure.

Table 7.19 *Failure data from a failure-terminated test and derived ranked cumulative failure distribution – 1*

Failure No.[a]	*Operating time* (days)	*Cumulative failure distribution (mean rank, F_{mean})*
1	40	0.091
2	98	0.182
3	165	0.273
4	235	0.364
5	312	0.455
6	428	0.545
7	547	0.636
8	720	0.727
9	925	0.818
10	1340	0.909

[a] Original no. = 10.

Table 7.20 *Failure data from a failure-terminated test and derived ranked cumulative failure distribution – 2*

Failure No.[a]	*Operating time* (years)	*Cumulative failure distribution (mean rank, F_{mean})*
1	1.69	0.091
2	2.11	0.182
3	2.42	0.273
4	2.73	0.364
5	3.05	0.455
6	3.20	0.545
7	3.44	0.636
8	3.76	0.727
9	4.02	0.818
10	4.37	0.909

[a] Original no. = 10.

Table 7.21 *Failure data from a failure-terminated test and derived ranked cumulative failure distribution – 3*

Failure No.[a]	*Operating time* (days)	*Cumulative failure distribution (median rank, F_{med})*
1	400	0.130
2	1050	0.314
3	1900	0.500
4	2800	0.686
5	5000	0.871

[a] Original no. = 10.

Table 7.22 *The median rank, 5% rank and 95% rank for sample size n = 5 used to plot the confidence limits in Figure 7.26*

Failure No.	*Median rank*	*5% rank*	*95% rank*
1	12.95	1.02	45.07
2	31.38	7.64	65.74
3	50.00	18.93	81.08
4	68.62	34.26	92.36
5	87.06	54.93	98.98

An illustrative example of the determination of mean and median ranks of a set of censored failure data is given in Table 7.23. The data in Table 7.23 can be fitted by a Weibull distribution and are plotted in Figure 7.27. The characteristic life η and the shape parameter β are obtained in the same way as in the previous example.

Further discussions of ranking are given by L.G. Johnson (1964) and Bompas-Smith (1973).

7.20.14 Fitting of failure distributions: observation window

These methods are appropriate for test situations, such as the testing of new equipments by the manufacturer. In this case, the equipments are known to be in the new condition at the start of the test.

In the process industries the position is often rather different. Frequently the test situation is that working equipments are observed through a test 'window', as shown in Figure 7.28 (Aird, 1977b). In this case the equipments are not in the 'as new' condition when the test is started. There appears to be no entirely satisfactory method of dealing with this case at present. The difficulty arises in the handling of the period before the first failure of such equipment. It is possible either to neglect this period or to treat it as a time to failure, but neither approach is accurate. The error in the parameters thus estimated increases as the ratio of the observation window to mean life decreases.

7.20.15 Fitting of failure distributions: parameter estimation

Failure data in the form of times to failure or hazard rates may also be fitted to failure distributions by parameter estimation methods. Three principal methods are (1) least squares, (2) moments and (3) maximum likelihood. These methods are considered for the exponential distribution

$$z(t) = \lambda \qquad [7.20.23]$$

and the Weibull distribution

$$z(t) = kt^m \qquad [7.20.24]$$

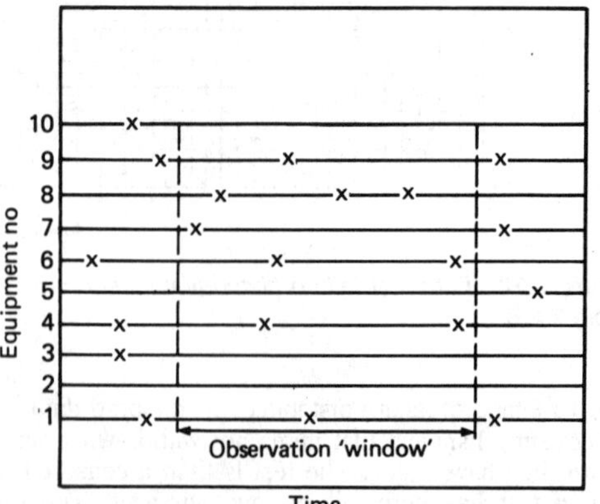

Figure 7.28 *Observation of failures in an operational system (after Aird, 1977b): ×, failure*

Table 7.23 *Failure data from a time-terminated censored test and derived ranked cumulative failure distribution*

Event No.[a]	*Operating time* (days)	*No. left*	*Increment*	*Mean order No.*	*Cumulative failure distribution (mean rank, F_{mean})*	*Cumulative failure distribution (median rank, F_{med})*
1F	300	14	1	1	0.063	0.045
2F	360	13	1	2	0.125	0.110
3F	400	12	1	3	0.188	0.175
4W	420	11	—	—	—	—
5F	440	10	1.083	4.083	0.255	0.246
6F	480	9	1.083	5.166	0.323	0.316
7W	490	8	—	—	—	—
8F	500	7	1.204	6.370	0.398	0.394
9F	540	6	1.204	7.574	0.473	0.472
10F	570	5	1.204	8.778	0.549	0.551

[a] F, failure; W, withdrawal; original no. 15.

The least squares method may be applied in various ways. It is often used to fit data in the form of hazard rates. The hazard rate *z(t)* and the failure density *f(t)* over each time interval Δt between failures are:

$$z(t) = \frac{1}{n_s \Delta t} \quad [7.20.25]$$

$$f(t) = \frac{1}{n \Delta t} \quad [7.20.26]$$

where n and $n_s(t)$ are, respectively, the initial number of equipments and the number of equipments surviving. An illustrative example of the determination of hazard rate and failure density of a set of failure data is given in Table 7.24.

In the least squares method the error e is defined as:

$$e = \Sigma(z - z_c)^2 \quad [7.20.27]$$

where z and z_c are the observed and calculated hazard rates. Then if

$$z = f(a_1 \ldots a_p) \quad [7.20.28]$$

the parameters a_i are given by the equations:

$$\frac{\partial e}{\partial a_1} = 0 \quad [7.20.29a]$$

$$\vdots$$

$$\frac{\partial e}{\partial a_p} = 0 \quad [7.20.29b]$$

A simple relationship for the hazard rate is:

$$z = a + b(t - \bar{t}) \quad [7.20.30]$$

where $\bar{t}$ is the mean time to failure. From Equation 7.20.27

$$e = \sum_{i=1}^{n} \{z_i - [a + b(t_i - \bar{t})]\}^2 \quad [7.20.31]$$

Then from Equation 7.20.28

$$a = \frac{1}{n}\sum_{i=1}^{n} z_i \quad [7.20.32]$$

$$b = \left[\sum_{i=1}^{n}(t_i - \bar{t})z_i\right] \Big/ \left[\sum_{i=1}^{n}(t_i - \bar{t})^2\right] \quad [7.20.33]$$

For the exponential distribution the parameters in Equation 7.20.30 are

$$a = \lambda$$

$$b = 0$$

For the Weibull distribution, from Equation 7.20.27

$$e = \sum_{i=1}^{n}(z_i - kt_i^m)^2 \quad [7.20.34]$$

$$k = \left(\sum_{i=1}^{n} z_i t_i^m\right) \Big/ \left(\sum_{i=1}^{n} t_i^{2m}\right) \quad [7.20.35]$$

m is obtained from the implicit equation:

$$\sum_{i=1}^{n} z_i t_i^m \ln t_i - \sum_{i=1}^{n} k t_i^{2m} \ln t_i = 0 \quad [7.20.36]$$

The moments method is used to fit data in the form of times to failure. The moments of the data m_k are:

$$m_k = \frac{1}{n}\sum_{i=1}^{n} t_i^k \quad [7.20.37]$$

The moments of the distribution μ_k are:

Table 7.24 Failure data from a failure-terminated test and derived hazard rates and failure densities

Failure No.[a]	*Operating time* (days)	*Time interval,* Δt (days)	*Hazard rate,* z ($\times 10^2$)	*Failure density,* f ($\times 10^2$)
1	40	40	0.250	0.25
2	98	58	0.192	0.172
3	165	67	0.187	0.149
4	235	70	0.204	0.143
5	312	77	0.216	0.130
6	428	116	0.172	0.086
7	547	119	0.210	0.084
8	720	173	0.193	0.058
9	925	205	0.244	0.049
10	1340	415	0.241	0.024

[a] Original No. = 10.

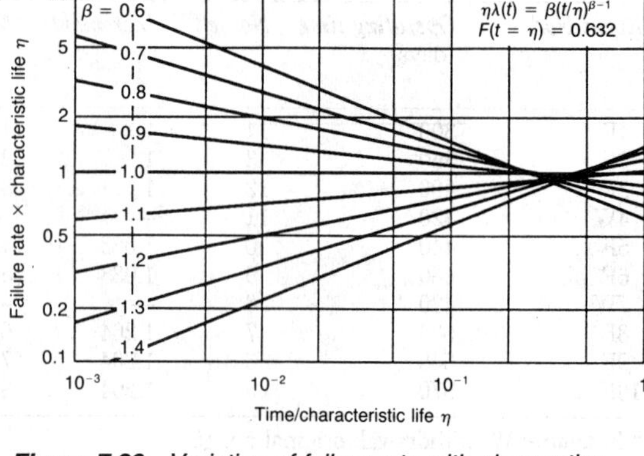

Figure 7.29 Variation of failure rate with observation time, in terms of the Weibull distribution (Blanks, 1977)

$$\mu_k = \int_0^\infty t_k f(t)\mathrm{d}t \quad [7.20.38]$$

The method is to equate the moments of the data and those of the distribution:

$$\mu_k = m_k \quad [7.20.39]$$

For the exponential distribution:

$$\mu_1 = \int_0^\infty t\lambda\exp(-\lambda t)\mathrm{d}t = \frac{1}{\lambda} = m_1 \quad [7.20.40]$$

For the Weibull distribution:

$$\mu_k = \int_0^\infty t^k kt^m \exp\left(-\frac{kt^{m+1}}{m+1}\right)\mathrm{d}t \quad [7.20.41a]$$

$$= \left(\frac{k}{m+1}\right)^{-k/m+1} \Gamma\left(\frac{k+m+1}{m+1}\right) \quad [7.20.41b]$$

m and k are obtained from the implicit equations

$$\mu_1 = \left(\frac{k}{m+1}\right)^{-1/(m+1)} \Gamma\left(\frac{m+2}{m+1}\right) = m_1 \quad [7.20.42a]$$

$$\mu_2 = \left(\frac{k}{m+1}\right)^{-2/(m+1)} \Gamma\left(\frac{m+3}{m+1}\right) = m_2 \quad [7.20.42b]$$

The maximum likelihood method is also used to fit data in the form of times to failure. The likelihood function L is defined as:

$$L(t_1 \ldots t_n;\ a_1 \ldots a_p)$$

$$= f(t_1;\ a_1 \ldots a_p) \cdot f(t_2 \ldots;\ a_1 \ldots a_p) \ldots f(t_n;\ a_1 \ldots a_p) \quad [7.20.43]$$

where f is the density function. Then the maximum likelihood estimate is obtained by maximizing L or $\mathcal{L} = \ln L$ with respect to the parameters a_i

$$\frac{\partial L}{\partial a_i} = 0 \quad [7.20.44a]$$

or

$$\frac{\partial \mathcal{L}}{\partial a_1} = 0 \quad [7.20.44b]$$

$$\vdots$$

$$\frac{\partial L}{\partial a_p} = 0 \quad [7.20.44c]$$

or

$$\frac{\partial \mathcal{L}}{\partial a_p} = 0 \quad [7.20.44d]$$

For the exponential distribution:

$$L = \prod_{i=1}^{n} \lambda\ \exp(-\lambda t_i) \quad [7.20.45]$$

or

$$\mathcal{L} = \ln L = n \ln \lambda - \lambda \prod_{i=1}^{n} t_i \quad [7.20.46]$$

Then from Equations 7.20.44 and 7.20.46:

$$\frac{\partial \mathcal{L}}{\partial \lambda} = \frac{n}{\lambda} - \sum_{i=1}^{n} t_i = 0 \quad [7.20.47]$$

$$\lambda = \frac{1}{\frac{1}{n}\sum_{i=1}^{n} t_i} \quad [7.20.48]$$

For the Weibull distribution:

$$L = \prod_{i=1}^{n} kt_i^m\ \exp\left(-\frac{kt_i^{m+1}}{m+1}\right) \quad [7.20.49]$$

or

$$L = n \ln k + m \sum_{i=1}^{n} \ln t_i - \frac{k}{m+1} \sum_{i=1}^{n} t_i^{m+1} \quad [7.20.50]$$

Then from Equations 7.20.44 and 7.20.50:

$$k = n(m+1)/\left(\sum_{i=1}^{n} t_i^{m+1}\right) \quad [7.20.51a]$$

$$\left[\left(\sum_{i=1}^{n} t_i^{m+1} \ln t_i\right)/\left(\sum_{i=1}^{n} t_i^{m+1}\right)\right] - \left(\frac{1}{m+1}\right)$$
$$-\left[\left(\sum_{i=1}^{n} \ln t_i\right)/n\right] = 0 \quad [7.20.51b]$$

Once the parameters of the distribution have been estimated by one of the methods described, an assessment can be made of the goodness of fit. A common approach is to use the χ^2 distribution.

7.20.16 Generation of failure distributions

It is sometimes necessary for purposes such as the testing of failure analysis schemes to generate sets of times to failure obtained from particular distributions. This is normally done by rearranging the equation of the failure distribution $F(t)$ to give the time to failure t as a function of the probability, and then generating the set of times to failure by replacing the probability by a set of random numbers ν in the range 0 to 1 from the uniform distribution.

For the exponential distribution, from Equation 7.7.14

$$t = -\frac{1}{\lambda} \ln(1-\nu) \quad [7.20.52a]$$

or, since ν and $(1-\nu)$ are both uniform distributions with the same range,

$$t = -\frac{1}{\lambda} \ln \nu \quad [7.20.52b]$$

For the normal distribution, with zero mean and unit variance from Equation 7.7.18, the following approximation may be used:

$$t = \lim_{n\to\infty} \left[\left(\frac{12}{n}\right)^{\frac{1}{2}} \cdot \left(\sum_{i=1}^{n} \nu_i\right) - \frac{n}{2}\right] \quad [7.20.53]$$

It is normally sufficient to take $n = 12$, in which case

$$t = \sum_{i=1}^{12} \nu_i - 6 \quad [7.20.54]$$

For the Weibull distribution, from Equation 7.7.25

$$t = \eta(-\ln \nu)^{1/\beta} + \gamma \quad [7.20.55]$$

The generation of failure distributions is described in more detail by Hastings and Peacock (1974).

7.20.17 Proportional hazards modelling

Another technique useful in failure data analysis is proportional hazards modelling. This is a method of determining the relative effects of influencing factors on the failure rate. Proportional hazards modelling was proposed by D.R. Cox (1972), who suggested that it might have application to data in the fields of medicine and reliability engineering. Its initial development was in the latter. Accounts are given in *The Statistical Analysis of Failure Time Data* (Kalbfleisch and Prentice, 1980) and *Statistical Models and Methods for Lifetime Data* (Lawless, 1982). The application of the method to reliability engineering is described by Bendell (1985) and Dale (1985).

The basic relationship in proportional hazards modelling is:

$$z(t) = z_0(t)\exp\left(\sum_{i=1}^{n} \beta_i z_i\right) \quad [7.20.56]$$

where $z(t)$ is the hazard rate, z_i the ith influencing variable, or explanatory factor, $z_0(t)$ the baseline hazard function, and β_i the coefficient of z_i.

Thus the action of the explanatory factors on the hazard rate is, in this model, assumed to be multiplicative, in contrast to the more common correlation models, in which it is taken to be additive. A multiplicative effect appears more realistic. The baseline hazard rate $z_0(t)$ is, in principle, distribution free.

Examples of the application of proportional hazards modelling are given by Dale (1985).

7.20.18 Time-varying failure data

The foregoing treatment is based on the assumption that the failure distribution and its parameters do not vary with time. Some guidance on the effect of variation of the failure rate with time is available from work by Blanks (1977). He shows that significant error occurs where failure rates are estimated from exposure periods which are very much less than the representative life, which for the log–normal distribution is the median life and for the Weibull distribution the characteristic life.

This is illustrated in Figure 7.29, which shows that for the Weibull distribution the product (Failure rate × Characteristic life) diverges significantly from unity for low values of the time/characteristic life ratio.

7.20.19 Combination of failure data

A common situation is that some failure data are available, typically from the literature, whilst further data have been obtained within the company. A method is needed for combining these two sets of data to produce a single value for the failure rate. In this case use may be made of Bayes' theorem, as described in Section 7.14. The problem is also discussed by the Center for Chemical Process Safety (CCPS, 1989/1).

7.20.20 Role of failure data analysis

Failure data analysis can be a powerful tool for improvement in reliability and in maintenance. It is most effective if used as one prong of an integrated approach. It also has certain limitations. For work on reliability, as opposed to maintenance, the acquisition of data is clearly essential.

For maintenance, at the most elementary level collection and analysis of failure data can highlight which equipments have high failure rates and the most frequent failure modes of such equipment. At a more sophisticated level, determination of failure distributions can identify the failure regime and can point to particular types of remedial action or maintenance policy.

Such an analysis is more effective if it is combined with other methods. In the study by Sherwin and Lees (1980) described above, the authors emphasize the value of combining failure data analysis and observation of the maintenance task. They state that this combination is more powerful than either method on its own.

Failure data analysis has it limitations, however. A works commonly contains a large number of equipments, each with a number, often a large number, of failure modes. The collection and analysis of failure data for these is a serious undertaking. Furthermore, failure data analysis is effectively a form of epidemiology, and is thus subject to criticisms which are directed at the latter, namely that it is a method which comes into play only after failures, and possibly harm, have occurred.

An alternative approach which seeks to get things right first time is that of reliability-centred maintenance. This is described below. Here it is sufficient to note its critique of failure data analysis. Failure data analysis appears to be of most value in application to significant items of equipment with a few dominant failure modes and perhaps also to lesser items present in large numbers.

7.20.21 Repair times

Information on repair times is also required in reliability work. Repair time data are sometimes difficult to obtain accurately as there is a tendency for nominal repair times to be recorded, particularly when tasks are of short duration. It is frequently found that repair times fit a log–normal distribution.

7.21 Reliability in Design

An outline is now given of the design and development of an equipment which is to have high reliability. It draws largely on accounts of the practice of suppliers and users in high technology industries such as defence and aerospace where the users have stringent reliability requirements.

7.21.1 Design process

An overview of the design process for mechanical systems with specific reference to reliability has been given by A.D.S. Carter (1986). He distinguishes three broad stages: (1) the formulation of objectives, (2) the conventional design process, and (3) the development process.

7.21.2 Manufacturer–user relationship

Before considering the three stages of the design process, it is relevant to recognize that the relationship between the manufacturer and the user can have a strong influence on the reliability requirements. In some cases, such as defence, there may be a single manufacturer and single user who work closely together to define the requirements. At the other extreme there is a market in which there are several manufacturers and numerous users. The users may be imprecise about their requirements so that the manufacturers must take their own view about what is needed.

In principle, a user should be seeking to minimize the life cycle costs. In practice, he may well not do this, but may concentrate on some less global objective such as minimization of the combined capital and operating costs of the equipment, or even of the capital cost alone.

The interests of the manufacturer and user do not necessarily coincide. However, in principle, both parties have an interest in the formulation of the reliability goals and specification.

7.21.3 Reliability targets

The reliability requirement is formulated in the first instance as a set of broad targets. In addition to the normal design considerations targets for reliability, these will include availability and maintainability. The priority accorded to reliability proper, the minimization of outright failure, will depend on the consequences of failure, and in particular on the extent to which the item is safety critical. The targets should ideally include the minimization of life cycle cost, but the extent to which the manufacturer does this will depend on the information about use available to him and on commercial judgement.

7.21.4 Reliability prediction

System design and development need to be supported by a continuous process of reliability prediction. This can be significant activity in its own right, and consideration of this is deferred to the next section.

7.21.5 Reliability feasibility study

Before embarking on detailed specification and design, it is usual to carry out a feasibility study. The purpose of this is to confirm that the proposed design is capable of meeting the reliability goals.

7.21.6 Reliability specification

Once established, the reliability targets need to be converted into reliability specifications. There will be a reliability specification for the system as a whole and further specifications for its component parts. The specification process involves sharper definition of features, such as the precise performance requirements for the system, the limits of the conditions and loads which it is required to withstand and the maintenance requirements and the criteria for success in meeting these various requirements. An account of the practicalities of reliability specification is given by Calabro (1962).

7.21.7 Conventional design

Within his overall account of reliability in design, Carter (1986) gives a detailed description of the conventional design process. The potential system failures should be identified and evaluated. For safety-critical features, techniques such as fail-safe behaviour or redundancy may be appropriate.

In most cases it is possible to make use of an existing design, either unchanged or with relatively minor modification. In this case information should be sought on its reliability in the field under the conditions of interest.

If a new design is indicated, care should be taken to avoid known generic causes of unreliability. These include features which constitute stress raisers, promote corrosion, result in poor connections, allow rupture of pipes, hoses and wires, and so on. The design should be suitable for the environment in which it is to be used. The components should be compatible with each other. A.D.S. Carter gives checklists covering this aspect. Efforts should be made to obtain information on the field reliability of similar designs.

The design should adhere to some basic principles. It should be kept simple. And it should be adequately conservative.

Care should be taken that 'improvements' to an existing or proposed design, whether arising from activities such as value engineering or otherwise, do not in fact result in an unwarranted decrease in reliability.

The maintainability of the system should be considered as well as its reliability.

The components and subsystems used in the design should be fully specified, and there should be inspection and test procedures for them.

Prior to entry into service the system may be subject to handling, packing, transport and storage, which may cause damage or deterioration, and attention should be paid to this aspect.

A quantitative evaluation should be made of the reliability of the system, as much for the sake of stimulating a critical review as for the determination of a numerical value.

7.21.8 Mechanical design

Traditional mechanical design of equipment is based on deterministic methods. It is also possible, however, to take a probabilistic approach using concepts such as those of interference theory. A.D.S. Carter (1986) gives illustrations of the application of this latter approach to problems of stress-rupture phenomena and of fatigue.

7.21.9 Design reviews

The design function should have a formal procedure for the review of the reliability of the systems and equipment under design. This procedure should cover not just the design stage proper but also the testing and development stages, and follow-up at each stage.

7.21.10 Problem identification

The design reviews should make use of the various techniques for problem identification, such as critical examination, checklists, failure modes and effects analysis, fault tree analysis, and so on. These methods are described in Chapter 8 and are therefore not discussed further here.

7.22 Reliability Prediction

As just indicated, design for reliability requires the supporting activity of reliability prediction. Reliability prediction is not simply the estimation of numerical values of reliability using suitable models. The term is used to cover the various and much wider range of activities running through the design and development stages, and possibly beyond. Reliability prediction is carried out through this life cycle as a guide to action. Accounts of reliability prediction include those given by von Alven (1964) and D.J. Smith (1985a).

7.22.1 Stages of reliability prediction

Reliability prediction passes through three broad stages, generally characterized as (1) pre-design, (2) interim and (3) final reliability prediction.

7.22.2 Pre-design reliability prediction

The pre-design reliability prediction has to be performed with limited information and involves in particular the feasibility evaluation and the reliability apportionment.

7.22.3 Reliability feasibility study

One of the first reliability prediction activities is the feasibility study. A pre-design estimate is made of the system reliability and is compared with the specification. This indicates whether or not the proposed design is at least within hailing distance of meeting the specification. If it is not, possible responses are modification to the design, or to the specification, or to both. Even if the reliability prediction indicates that the design will meet the specification, it is good practice to explore alternatives. The next step is to set reliability specifications for the subsystems.

7.22.4 Reliability apportionment

The fundamental reliability specification is that for the system as a whole. However, in order to achieve this it is necessary to decompose the system into subsystems and to determine the reliability specifications for the constituent parts. This is known as 'reliability apportionment', or 'reliability allocation'.

There are a number of models which may be used to apportion the reliability targets among the parts, as described by Anderson and Neri (1990). A model for the case where the parts, or subsystems, are similar is:

$$R_x = R^{1/n} \qquad [7.22.1]$$

where n is the number of subsystems, R the reliability of the system, and R_x the reliability of a single subsystem.

For the case where the subsystems are different one model is:

$$R_x = R^{W_i} \qquad [7.22.2]$$

with

$$W_i = C_i / \left(\sum_{j=1}^{n} C_j \right) \qquad [7.22.3]$$

where C_i is a parameter representing the complexity of the ith subsystem and W_i is the weighting factor for that subsystem.

7.22.5 Interim reliability prediction

Once the design is under way, reliability prediction is used in support. As is usual in design, the process is iterative and involves various trade-offs.

7.22.6 Final reliability prediction

The final reliability prediction is made as part of the overall specification of the product. This prediction also serves as a measure of the success of the design process. There may be complications in producing a single final reliability prediction. The product may come in several versions and, over time, may be subject to modification.

7.23 Reliability Growth, Testing and Demonstration

In the achievement of reliability, design for reliability and reliability prediction need to be complemented by a programme of reliability testing and demonstration. In large part, reliability prediction and testing are directed to the growth in the reliability of the equipment. Reliability growth is therefore conveniently considered along with testing.

7.23.1 Reliability testing

In reliability testing a distinction is made between (1) reliability development and demonstration testing, (2) qualification and acceptance testing and (3) operational testing.

Reliability development and demonstration testing is performed to determine (1) whether the design needs improvement, (2) to identify modifications which might lead to improvement and (3) to verify that the modifications made have led to improvement.

In high technology activities it is generally required that a design be accepted as qualified for the intended application. An example is the seismic qualification of equipment for nuclear plants. Similarly, there may be a requirement that the design be qualified with respect to reliability. This is the purpose of qualification testing.

Acceptance testing is concerned with the quality of manufacture of the equipment.

Operational testing provides information on the operation of the equipment, as verification of the design and feedback of information on performance for future use by the designers. This type of testing is less likely to be subject to the continuous process of modification that is characteristic of development testing.

Reliability testing is governed by a formal programme involving subcontractors as well as the lead manufacturer. The programme covers the stages of (1) design concept, (2) production prototype and (3) production testing. It is generally subject to frequent modification.

Further treatment of acceptance testing is deferred to later in the text in order to consider reliability growth.

7.23.2 Reliability growth

If the reliability development programme is effective, it should result in an increase in the reliability of the equipment. Experience with different products shows that the process of reliability growth has certain characteristic features. Accounts of reliability growth are given by Lloyd and Lipow (1962) and D.J. Smith (1985a). A relevant standard is BS 5760: Part 6: 1991 *Guide to Programs for Reliability Growth.*

The situation considered is the development phase of an equipment in which the manufacturer conducts field tests involving a continuous process of feedback of information on reliability performance and modification of the equipment to improve this reliability by reducing design-related failures.

Various models have been developed for reliability growth. Such models may be used to extrapolate in time to the point when the failure rate of an equipment under development should reach the specified value. They may also be used to make estimates of development times.

7.23.3 Duane method

A widely quoted method for the analysis of reliability growth during field testing has been given by Duane (1962). Further discussions of the method are given by Codier (1968), R.C.F. Hill (1977) and D.J. Smith (1985a).

Duane found that a log-log plot of cumulative failure rate vs cumulative testing time tended to give a straight line. Codier (1968) modified this plot, replacing cumulative failure rate by cumulative MTBF, as shown in Figure 7.30. A straight line on this plot may be expressed as:

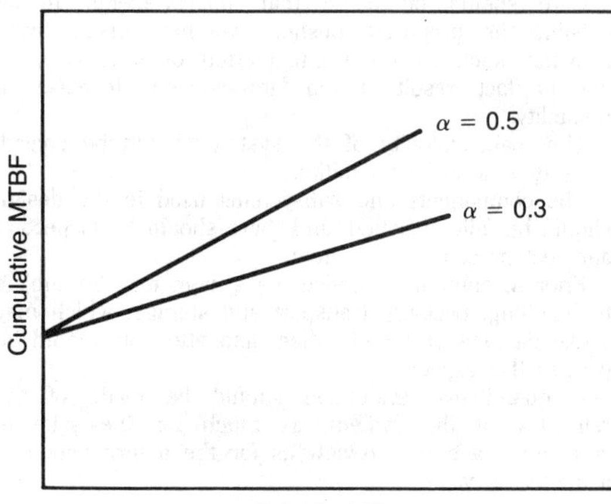

Figure 7.30 *Reliability growth plots: after Duane*

$$MTBF \propto T^{\alpha} \qquad [7.23.1]$$

The current MTBF is given by:

$$MTBF_{cur} = \frac{MTBF_{cum}}{1-\alpha} \qquad [7.23.2]$$

where the subscripts cum and cur indicate the cumulative and current values. Values of α tend to lie in the range 0.1–0.65.

The use of cumulative time in the Duane plot means that at long testing times changes in the reliability growth tend to be smoothed out and are difficult to discern.

7.23.4 Reliability acceptance testing

The decision of whether to accept or reject items on reliability grounds needs to be governed by a formal system of acceptance testing. Accounts of acceptance testing are given in texts on quality control (e.g. Burr, 1953; Moroney, 1956; E.L. Grant, 1964). Treatments in the context of reliability engineering are given by von Alven (1964), Breipohl (1970) and D.J. Smith (1985a).

Acceptance testing is generally conducted by reference to an operating characteristic (OC) curve. For the case where such testing is performed by taking a sample of items from a production batch, the operating curve may be constructed from the binomial distribution:

$$P = \sum_{k=0}^{c} \binom{n}{k} q^k p^{n-k} \qquad [7.23.3]$$

where c is the allowable number of defective items in a sample of size n, p and q are the proportions in the batch of good and defective items, respectively ($p + q = 1$), and P is the probability of acceptance. The OC curve is a plot of P vs q for given values of n and c. A typical OC curve of this kind is shown in Figure 7.31(a).

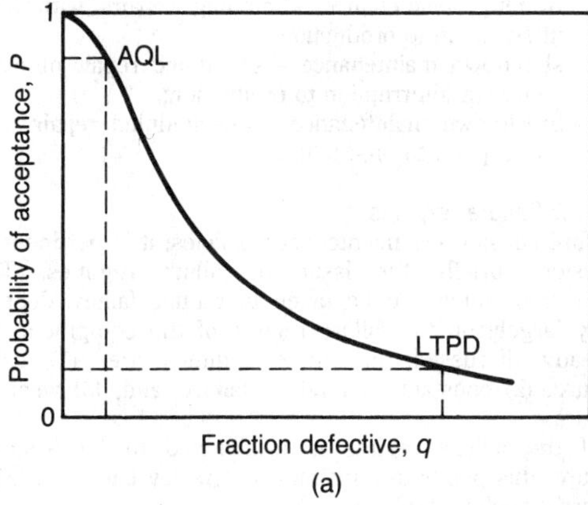

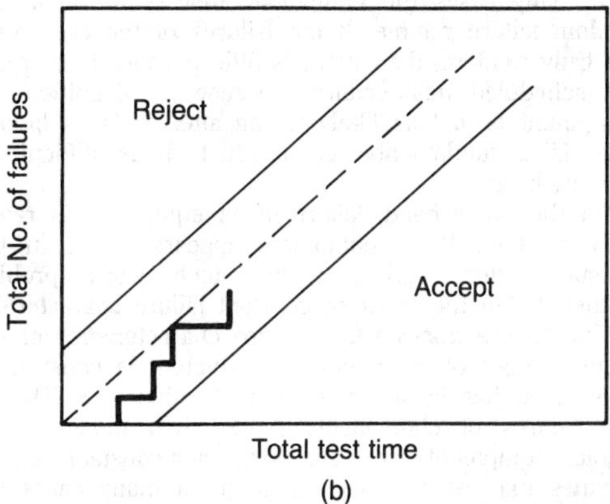

Figure 7.31 *Acceptance testing schemes: (a) operating characteristic curve for single sample plan; (b) operating limits for sequential sampling plan*

In using a sample as the basis of acceptance, both producer and consumer accept a certain risk. The producer's risk is the probability that the batch is acceptable but fails the test, whilst the consumer's risk is the probability that the batch is unacceptable but passes the test.

The usual procedure is to set values for the acceptable quality level (AQL) and the lot tolerance per-cent defective (LTPD) which are entered on the OC curve as shown in Figure 7.31(a). The AQL is the batch, or lot, quality (q value) for which the probability of acceptance has some (high) specified value. The complement of this probability is the producer's risk (α). A typical value of the AQL is 95%, giving a producer's risk of 5% ($\alpha = 0.05$). Likewise, the LTPD is the lot quality (q value) for which the probability of acceptance has some (low) specified value. This probability is the consumer's risk (β). A typical value of LTPD is 10%, giving a consumer's risk of 10% ($\beta = 0.1$).

As just described, this scheme refers to the sampling of a single batch and provides lot quality protection (LQP). If the batch is then to be mixed with other batches, this degree of protection may not be necessary. An alternative scheme may be used which offers average quality protection (AQP).

An approach to acceptance testing for the case where the acceptance criterion is the failure rate of the equipment, which is based on the Poisson distribution, is described by Smith.

The binomial scheme just described involves single sampling. Another approach is the use of sequential sampling in which a series of tests is conducted, the decision after each test being (1) accept, (2) continue testing or (2) reject. This approach is illustrated in Figure 7.31(b).

An important issue in the design of an acceptance testing scheme is the degree of discrimination which it provides. This is discussed by Smith.

7.24 Maintainability

Accounts of reliability engineering generally include as a third theme, in addition to those of reliability and availability, that of maintainability. Maintainability is treated in the main reliability engineering texts cited here and in particular in *Maintainability* (Goldman and Slattery, 1964) and *Reliability, Maintainability and Availability Assessment* (Locks, 1973), and by von Alven (1964) and D.J. Smith (1985a).

The cost of maintaining equipment is a recurring theme in reliability work and was one of the original motivations for it. Although estimates vary, they are generally in agreement that the cost is high. Accounts of maintainability generally treat it as a property of the equipment. 'Maintainability' is therefore not synonymous with maintenance effectiveness, which includes other aspects such as organization and training.

7.24.1 Maintainability measures

The principal measures of maintainability of an equipment are its failure rate and repair time. Attention tends to centre mainly on the repair time, but the overall cost is also strongly influenced by the frequency at which repair is necessary.

The overall repair time is usually decomposed into several parts, a typical division being (1) active repair time, (2) logistics time and (3) administrative time. Essentially, logistics time is that spent waiting for spare parts, and administrative time is the balance of time spent on the repair which is not required for either active repair or logistics.

7.24.2 Maintainability principles

Some basic principles of maintainability are: (1) provision of good access; (2) minimization of the complexity of the task, tools and test equipment; (3) provision of good maintenance manuals; (4) clear criteria for recognition of faults or marginal performance; and (5) optimization of the diagnosis and repair task.

The latter typically involves (1) preparation, (2) malfunction verification, (3) fault location, (4) fault isolation, (5) disassembly, (6) part procurement, (7) part interchange, (8) reassembly, (9) alignment and (10) checkout.

In many cases an approach based on simple checks to identify the malfunctioning module and replacement of that module is an appropriate strategy.

7.24.3 Maintainability prediction

Methods have been developed for the prediction of maintainability as measured by the repair time. The basic approach is to synthesize estimates of the repair time from a data bank of times for element of the repair task. Von Alven (1964) describes a number of data collections developed for this purpose. One is the ASB-4 system. Another system is that developed by the American Institute for Research (AIR).

7.24.4 Maintainability testing and demonstration

Major clients such as the armed services require the supplier to demonstrate maintainability as well as reliability. MIL-STD-471 *Maintainability Demonstration* states requirements for the US Armed Services, and is widely quoted.

Maintainability is generally demonstrated by conducting a series of tests. In the typical test a defined maintenance task is undertaken and measurements are made of the times taken to perform it. A plot is made of the probability of passing the test vs the MTTR (mean time to repair) which is then evaluated against a predefined criterion.

Comparability between tests depends crucially on holding constant the various factors which are known to exert a strong influence. These are listed by D.J. Smith (1985a) as (1) method of selecting the demonstration task, (2) tools and test equipment available, (3) preventive maintenance given to test system, (4) maintenance documentation, (5) environment during test and (6) skill level and training of test subject.

7.25 Maintenance Activities and Policies

The maintenance analysis of a system has as its objective the identification of failure situations and the formulation of appropriate maintenance policies.

The theory of maintenance is a well-developed aspect of reliability engineering. Accounts are given in *Maintainability* (Goldman and Slattery, 1964), *Maintainability, Principles and Practice* (Blanchard and Lowery, 1969), *Maintenance, Replacement and Reliability* (Jardine, 1973) and *Maintainability Engineering* (D.J. Smith and Babb, 1973).

The management of maintenance is described in *Modern Maintenance* (J.E. Miller and Blood, 1963), *Effective Maintenance Organisation* (Newborough, 1967), *Maintenance Engineering and Organisation* (Gradon, 1973), *Principles of Planned Maintenance* (Clifton, 1974), *Systematic Maintenance Organisation* (Priel, 1974), *Maintenance Management Techniques* (Corder, 1976), and *Maintenance Management and Terotechnology* (Husband, 1976).

The maintenance of plant to reduce failures and stoppages is discussed below. Some wider aspects of the maintenance function and of the information which it generates are described in Chapter 21.

7.25.1 Maintenance activities

Some maintenance activities which are carried out on process plants have traditionally been:

(1) production assistance – adjusting machine settings;
(2) servicing – replacement of consumables, including lubrication;
(3) running maintenance – running repairs with little interruption to production;
(4) shut-down maintenance – scheduled repair or overhaul with interruption to production;
(5) breakdown maintenance – unscheduled repair with interruption to production.

7.25.2 Failure regimes

Before considering maintenance policies, it is pertinent to consider briefly the issue of failure regimes. The measures which can be taken to reduce failure depend very largely on the failure regime of the equipment. As already discussed, the three regimes are: (1) early failure; (2) constant, or random, failure, and; (3) wearout failure.

If the failure regime is established to be wearout failure, this points to a maintenance policy based on fixed interval replacement.

In many cases the equipment appears to be in the random failure regime. If the failures of the equipment are truly random, then there is little purpose in carrying out scheduled maintenance, because by definition the equipment is just as likely to fail afterwards as before. This is a fundamental point which it is difficult to overemphasize.

On the other hand, failure of an equipment is rarely truly random. If an equipment appears to be in the constant failure regime, it is usually worth probing further. Often the apparent constant failure characteristic is due to the non-constant failure characteristics of the large number of components of which it is constituted. This effect has been discussed in Section 7.18. Thus if the failures of components rather than those of the whole equipment are analysed, non-constant failure regimes may be revealed, although in many cases the number of failures of each component may be so small that analysis is difficult.

If the failure regime is established to be early failure, the choice of maintenance policy is again complex. In formulating a policy, account should be taken of the various factors which contribute to early failure, as described in Section 7.19. One of these may be failure induced by unnecessary scheduled maintenance.

7.25.3 Repair, reconditioning and replacement

Another preliminary matter is the issue of repair, reconditioning and replacement. If an equipment is found to have deteriorated to a point where its condition or performance is unsatisfactory, it is sometimes replaced, but is more often repaired or reconditioned with the aim of restoring it to the 'as new' condition.

Often this is not achieved. Failure to make full restoration can be serious. A policy where this tends to be the outcome can result in a higher overall level of faults on the plant. It is essential, therefore, that for equipments with fine tolerances the requirements for repair or reconditioning be properly specified. There should be a written specification for the task. The specification not only should quote the original design information such as a drawing number but also should give the permissible wear tolerances.

It should be borne in mind that maintenance personnel deal with a wide range of equipment. On some of this equipment tolerances may not be particularly critical. It is important that they do not transfer the rough-and-ready methods which may be adequate there to equipment which requires more careful treatment.

Certain equipment designs allow for the rectification of a fault by simply removing a defective module from the equipment and replacing it with a new one. This is then generally the best action.

7.25.4 Maintenance policies

Some policies for maintenance are:

(1) scheduled maintenance;
  (a) inspection;
  (b) minor adjustment;
  (c) replacement;
(2) on-condition maintenance;
(3) opportunity maintenance;
(4) breakdown maintenance.

Maintenance policies can be classed as either preventive maintenance (PM) or breakdown maintenance. All the policies except the last may be regarded as forms of preventive maintenance.

7.25.5 Planned maintenance

In general, it is desirable that maintenance tasks be planned rather than undertaken *ad hoc*. The term 'planned maintenance' is widely used and refers to any maintenance which is planned. It includes some breakdown maintenance where the latter can be planned in advance.

7.25.6 Scheduled maintenance

A subset of preventive maintenance policies is that comprising the variations on scheduled maintenance. There are several forms of scheduled maintenance, the distinction between them being the actions prescribed in advance. These may be (1) inspection, (2) adjustment or (3) replacement, the term 'adjustment' covering virtually any action beyond inspection but short of replacement.

If the policy is scheduled inspection and no problem is found, no further action is taken. However, if the inspection indicates that further action is required, it is taken, whether it be adjustment or replacement. Likewise, scheduled adjustment generally includes inspection and may involve replacement if the inspection so indicates.

7.25.7 Scheduled replacement

If it is established that an item exhibits wearout failure, it is possible in principle to determine the failure distribution and to adopt a policy of scheduled replacement based on this. If the parameters of the distribution are known, the probability of failure prior to replacement which is tolerable can be specified and the time interval for replacement read off from the failure distribution function, as shown in Figure 7.32.

Alternatively, the failure distribution may be assumed. In this case a common assumption is that the failure distribution is normal. This assumption together with knowledge of the value of the mean life and a further assumption about the variance allows an estimate to be made of the replacement interval. If the mean life to wearout is m and the corresponding standard deviation is

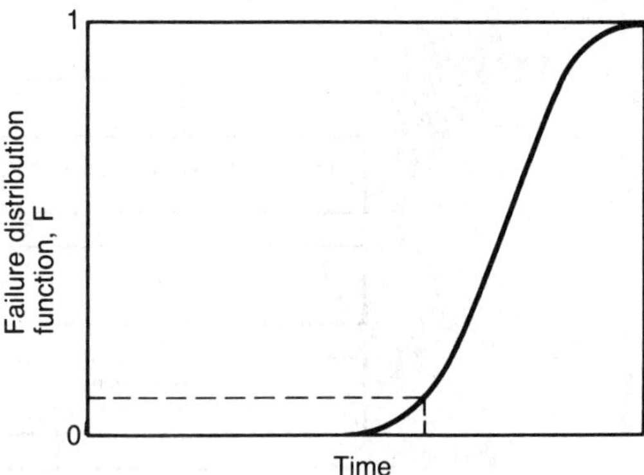

Figure 7.32 *Determination of the scheduled replacement interval from the failure distribution*

σ, then the maintenance policy is to replace at a time m_0 where

$$m_0 = m - k\sigma \qquad [7.25.1]$$

As a rule-of-thumb the value of σ is often taken as

$$\sigma = 0.1m \qquad [7.25.2]$$

A typical value of k is 6, so that Equation 7.25.1 becomes

$$m_0 = 0.4m \qquad [7.25.3]$$

Thus, for example, bearings are commonly replaced at a proportion of their mean life, the so-called 'B10 life'.

7.25.8 On-condition maintenance

An alternative to scheduled maintenance is on-condition or condition-based maintenance. On-condition maintenance involves periodic inspection, and to this extent resembles scheduled inspection. Insofar as a difference can be drawn, the distinguishing features of on-condition maintenance (OCM) is that it tends to draw on the many powerful techniques now available for condition monitoring and, when deterioration is detected, to involve prediction of the expected time to failure. On-condition maintenance has the advantage of averting failures by early action if inspection detects deterioration and of avoiding unnecessary maintenance if inspection reveals no deterioration.

An account of on-condition maintenance is given by Moubray (1991). The essential conditions for its practice are there is a well-defined potential failure (PF) condition and that it is feasible to construct a dependable PF curve. In a PF diagram the fractional margin of acceptable performance (from 1 to 0) is plotted against time. This PF curve can then be used to predict the time of occurrence of total loss of margin, or failure. The time between first detection of PF and actual failure is the PF interval. For on-condition maintenance to be practical, the PF curve must be reasonably reproducible. An inspection interval can then be chosen which is less than the PF interval. The time available for action to be taken is the

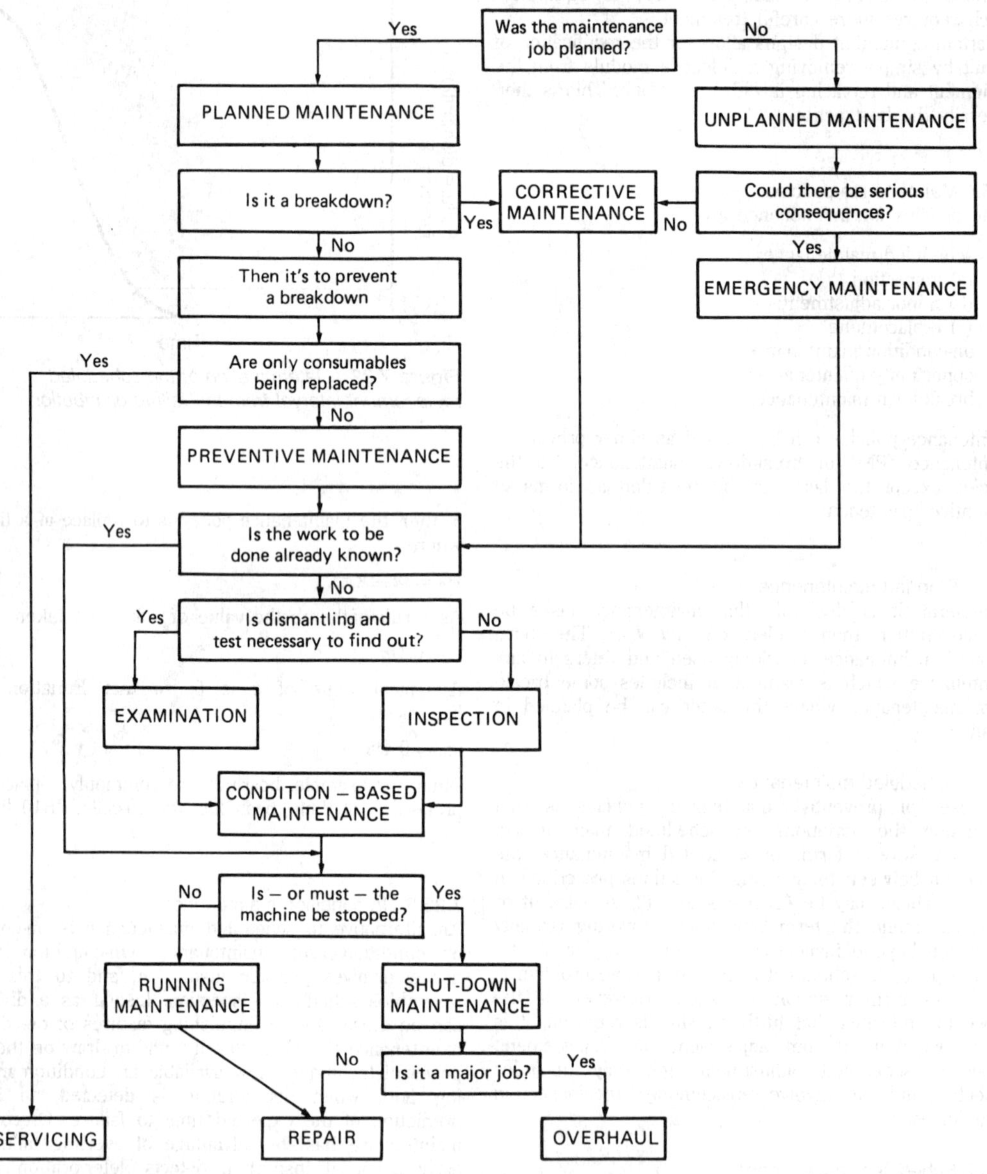

Figure 7.33 *Decision tree illustrating the relationship between the various forms of maintenance (Department of Industry, 1975/4). Terminology used is from BS 3811: 1974*

difference between the PF and inspection intervals, or the net PF interval.

Moubray (1991) also gives an appendix containing brief summaries of the various condition monitoring techniques and their applications. A further account of these techniques is given in Chapter 19.

On-condition maintenance is a feasible policy only if an appropriate inspection or monitoring technique is available, but the development of new techniques is continually widening the scope of items which can be inspected for deterioration. However, since inspection incurs cost, there is an optimum degree of inspection. Optimal inspection policies are dealt with in standard texts on maintenance.

7.25.9 Opportunity maintenance

For equipments which are normally inaccessible, breakdowns or stoppages provide the occasion to practice 'opportunity maintenance'. This consists of doing preventive maintenance work as and when the item becomes available.

Thus, for example, it is often normal practice that when a pump is opened up a check is made on alignment, bearings, seals and cooling water flow, but this check may be extended to other features such as the casing and the small bore connections.

7.25.10 Burn-in

As described above, one of the causes of early failure in a system with a large number of new components can be the presence of a proportion of weak components. An account of methods for dealing with this problem is given in *Burn-In* (Jensen and Petersen, 1982). These methods might be regarded as another type of maintenance policy. Jensen and Petersen describe a methodology for dealing with this kind of early failure problem by burn-in. Interest centres in large part on the prediction of the burn-in period required.

7.25.11 Selection of maintenance policies

There are considerable differences in the maintenance policies which may be followed on process plant. At one extreme the policy may be one of breakdown maintenance only, whilst at the other the policy may be one of extensive preventive maintenance.

The terminology used for maintenance policies is defined in BS 3811: 1984 *Glossary of Maintenance Terms used in Terotechnology*. An account of maintenance policies is given in *Maintenance Aspects of Terotechnology. 1, Planned Maintenance* (Department of Industry (DoI), 1975/4). A decision tree illustrating the relationships between the various forms of maintenance is given in Figure 7.33.

Over the years, maintenance policies have been subject to change, even fashions. Moubray (1991) distinguishes three stages. The first stage is that of breakdown maintenance, characterized by plant which was relatively simple, which was designed with large safety margins and which was usually not safety-critical. The second stage is that of preventive maintenance, in response to pressures for longer equipment life, high availability and lower costs. The third stage is that of reliability-centred maintenance, in response to intensification of the same pressure, as well as those of safety, health and environment.

At one time preventive maintenance was very much in vogue. It was often taken to uneconomic lengths and, in consequence, the fashion faded. Most present policies in the process industries are a mix of preventive and breakdown maintenance.

If maintenance policies are considered from the safety and loss prevention viewpoint, it is apparent that it is desirable to forestall breakdown as far as possible and that it is prudent to identify critical equipments and to carry out appropriate inspection and preventive maintenance on them.

Process plants tend to run for long periods between scheduled shutdowns, or turnarounds. There are many equipments, however, which are not accessible for maintenance except at shut-down. Thus the shut-down is extremely important from the maintenance viewpoint. In particular, the shut-down allows an opportunity for the maintenance of plant utilities.

7.25.12 Maintenance information

The formulation of maintenance policies and the monitoring of maintenance activities require that there be a suitable maintenance information system. The maintenance information system has as its main purpose the provision of data on which to base maintenance decision-making, but it also has the further important function of feeding back information on failure to the design department and information on failure and repair to the reliability engineering section.

Some information which is relevant to maintenance includes identification and quantification of (1) failures responsible for hazardous conditions, (2) failures responsible for plant downtime and (3) failures responsible for direct repair costs.

7.25.13 Maintenance effectiveness

Measurement and monitoring of the effectiveness of the maintenance system is necessary, but is often not straightforward. Maintenance costs break down into direct costs and indirect costs. The direct costs are (1) materials and (2) labour. Some indirect costs are those due to (1) plant unavailability, (2) loss of plant throughput, (3) loss of plant efficiency and (4) spares holdings. The overall effect of poor maintenance may result in a degree of disruption, the cost of which can exceeds the apparent sum of these factors.

These considerations are the basis of approaches to maintenance which concentrate on the three 'top tens', these being the top ten items which contribute most in each of the three categories: (1) direct cost, (2) indirect cost and (3) troublesomeness.

7.25.14 Maintenance indices

It is common practice to evaluate maintenance performance in terms of ratios, or indices. Such indices typically measure (1) relative maintenance cost, (2) proportion of preventive maintenance, (3) plant availability, (4) breakdown rate, (5) breakdown repair time and (6) breakdown cost. Normalizing factors used include (1) plant capital cost, (2) plant operating cost, (3) maintenance man-hours and (4) number of breakdowns.

$$Index\ 1 = \frac{Total\ cost\ of\ maintenance}{Total\ capital\ cost\ of\ plant}$$

$$Index\ 2 = \frac{Total\ cost\ of\ maintenance}{Total\ cost\ of\ plant\ operation}$$

$$Index\ 3 = \frac{Total\ man\text{-}hours\ of\ preventive\ maintenance}{Total\ man\text{-}hours\ on\ all\ maintenance}$$

$$Index\ 4 = \frac{Plant\ uptime}{Plant\ uptime + Plant\ downtime}$$

$$Index\ 5 = \frac{Plant\ uptime}{Number\ of\ breakdowns} = MTBF$$

$$Index\ 6 = \frac{Plant\ downtime}{Number\ of\ breakdowns} = MTTR$$

$$Index\ 7 = \frac{Cost\ of\ breakdowns}{Number\ of\ breakdowns}$$

Such indices can be useful tools, but they may also imply unwarranted assumptions. Index 3, for example, appears to assume that preventive maintenance is always a good thing, which is not necessarily the case.

A maintenance index may be evaluated by reference either to some absolute criterion value or to the historical trend. Both approaches have their uses.

7.26 Reliability-Centred Maintenance

An approach to maintenance which is finding increasing acceptance is reliability-centred maintenance (RCM). RCM is not a specific maintenance policy such as scheduled replacement or on-condition maintenance, but an integrated method by which to select such policies. Accounts of this approach to maintenance are given in *Reliability-Centred Maintenance* (Anderson and Neri, 1990) and *Reliability-Centred Maintenance* (Moubray, 1991) and by Nowlan and Heap (1978), Sandtory and Rausand (1991) and Pradhan (1993).

RCM developed within the civil aircraft industry. In 1974, the Department of Defense commissioned a report on maintenance practice in that industry, published as *Reliability-centred Maintenance* (Nowlan and Heap, 1978).

The elements of RCM may be considered under the following headings: (1) functions, (2) function failures, (3) failures modes, (4) failure effects, (5) failure consequences, (6) preventive tasks and (7) default actions. This description suggests that RCM may be regarded as being based on a form of failure modes and effects analysis (FMEA). This is broadly true, but it takes an approach to FMEA which places particular emphasis on function, function failure and prevention.

As described in the next chapter, FMEA is one of the prime tools of hazard identification used in safety and loss prevention work. Here it is sufficient to note that its growing use in maintenance is a major influence for the integration of SLP and maintenance.

7.26.1 Functions, functional failures and performance standards

Taking these elements in turn, RCM starts by establishing the functions of the equipment. The equipment will have some primary function which is its *raison d'etre*, but in addition it will generally have secondary functions. Thus, for example, a pump has the primary function of raising the pressure of the liquid, but it also has the secondary function of containing it. Satisfactory performance with respect to safety, health and environment may be treated as other secondary functions.

Increasingly, equipment comes with various kinds of protective device which serve to protect one or other of these functions. Protective devices are used to maintain the primary function by means such as bringing in stand-by equipment when the main equipment fails or to provide protection in relation to secondary functions such as safety or environment. For each function and each protective device there is a performance standard.

7.26.2 Failure modes and causes

It is necessary to identify the failure modes of the equipment. In general, failure modes may be described in terms of failure of performance or failure of condition. In RCM the former is the failure of function just described and it is the latter which is the failure mode. For example, if a pump becomes incapable of pumping the design flow of liquid, this might be expressed as a failure of pumping capability or a failure due to impeller wear. It is the failure of condition, the impeller wear, which is treated as the failure mode.

In the analysis of the failure modes it is important to uncover the root causes. Typical root causes include incorrect lubrication or maloperation. In the case of the pump impeller wear, the root cause might be cavitation due to partial blockages in the suction line.

7.26.3 Failure effects and consequences

The failure effects describe what happens when the failure occurs. They include the symptoms of failure and the effects on the process, on other equipment, safety and the environment. These failure effects are evaluated in order to assess the failure consequences. Here four types of consequence are distinguished: (1) hidden failure consequences, (2) safety and environmental consequences, (3) operational consequences and (4) non-operational consequences.

The term 'hidden failure' is that generally used in RCM to denote the unrevealed failure of protective devices. Most protective devices tend to fail without signalling the fact. In such cases the device then remains in the failed state until the failure is revealed either because the device is checked or because a demand occurs against which the device no longer provides protection. In other words, protective devices with this characteristic are not fail-safe. The treatment of unrevealed failure of protective devices is discussed in Chapter 13.

A hidden failure means that there is a risk of a demand occurring against which there is no protection, and hence of what in RCM is called 'multiple failure'.

The safety and environmental, operational and non-operational consequences are largely self-explanatory. The latter covers any damage to the equipment

7.26.4 Preventive tasks and default actions

The next step is to determine for each failure mode in turn whether a preventive maintenance task exists which can eliminate or ameliorate the failure. For main equipment the approach followed is essentially that already described of establishing the failure characteristic (e.g. early failure, wearout failure) and identifying a corresponding appropriate preventive task. Consideration is also given to the appropriateness of on-condition maintenance. For protective equipment, where the aim is to maintain an adequate availability, the main preventive action is to reduce the failure rate. The alternative measure is to alter the inspection interval, which counts as a default action.

In the RCM context, whether or not a preventive task is appropriate depends first on its technical feasibility and then on its worthwhileness. A task is undertaken only if it is feasible and worthwhile.

It is not appropriate to try to deal with every failure mode by a preventive maintenance task. Where such a solution is not available, a default action is taken. Principal default actions are (1) redesign, (2) periodic inspection and (3) no action.

The typical account of RCM describes a rather formalized system with decision flow diagrams, FMEA worksheets, and so on. Commonly used abbreviations are: F, failure; FF, function failure; FM, failure mode; H, hidden failure; S, safety; E, environment; O, operation.

7.26.5 Illustrative example

The application of RCM to the process industries is illustrated by an example given by Pradhan (1993). The author deals with the use of RCM on a set of main reactor charge pumps handling vinyl chloride monomer (VCM) on a polyvinyl chloride (PVC) plant. Aims included the improvement of the pump system reliability and avoidance of the need for breakdown maintenance, and the prevention of VCM leaks and fires. He gives sample functions and failure analysis tables akin to those produced in FMEA, discusses in particular the hidden failure of protective devices, and describes the outcomes of the study in terms of the preventive tasks and default actions.

7.27 Life Cycle Costing

The concept of life cycle costing is that ownership of an asset involves not only capital but also other costs, and that the aim should be to minimize these costs over the life cycle of the asset. An account of life cycle costing (LCC) in the context of process plants is given by Lees (1983d).

7.27.1 Management of life cycle costing

The practice of LCC is likely to be effective only if fully supported by management, whose attitude is crucial. Frequently the issue which will reveal management's attitude will be the trade-off between higher initial cost and overall LCC. The role of management is to establish the policy of LCC and to set up formal systems to ensure that the policy is followed.

There are also other ways in which management can promote LCC. One is the rotation of personnel between the plant and the design function, so that plant experience is fed back into design. Failure to achieve such feedback, by this or other means, can be very costly.

There are a number of other activities to which LCC is related. They include (1) reliability and maintainability engineering, (2) value engineering and (3) process economics. It is unlikely that the practice of LCC will flourish unless the organization has a sufficient level of activity in such areas.

A practical policy for LCC needs to define clearly the items to which it is to be applied and needs to avoid excessive complication.

7.27.2 Applicability of life cycle costing

LCC is most readily applied to an asset which is 'free-standing', that is, one whose failure does not cause significant disruption. In this case the optimization of the life cycle cost is essentially a trade-off between the higher capital cost of an equipment with superior performance, including reliability, and the direct costs of inferior performance, including unreliability. Whatever the case with other equipment, and a truly free-standing equipment is perhaps not as common as might appear at first sight, that used in process plant does not fall into this category.

LCC is subject to the law of diminishing returns. A policy for LCC therefore needs to include some sort of stopping rule. The returns are likely to be greatest for two types of equipment: (1) major items and (2) minor items used in large numbers.

7.27.3 Decision process for life cycle costing

It is not uncommon in the acquisition of an asset that the decision is made by the wrong person at the wrong time on the basis of inadequate information, or even that it is made essentially by default. For LCC the decision process covers the stages of (1) specification, (2) design and (3) procurement.

Often the key decisions are those taken early in the project. Thereafter, many of the options which have the strongest influence on the LCC are foreclosed. That said, the importance of the procurement stage should not be underestimated.

7.27.4 Factors contributing to life cycle cost

Some of the factors which contribute to LCC are: for the equipment, (1) capital cost, (2) maintenance cost and (3) equipment life; and, for the plant, (1) plant availability, (2) plant instantaneous throughput, (3) plant efficiency and (4) plant flexibility. Safety, health, environment and other factors act as constraints which may have a strong influence on the LCC.

Often the main elements from this list are (1) capital cost, (2) maintenance cost and (3) plant availability.

7.27.5 Information for life cycle costing

The practice of LCC requires information on *inter alia* (1) failure rate, (2) repair time and (3) plant unavailability.

It is possible to construct a long list of information which in principle bears on LCC. However, not all this information is readily observable, at least without excessive cost. If the acquisition of information is considered in terms of the two attributes of importance and observability, the policy adopted might well be the following:

		Observability	
		High	Low
Importance:	High	Y	?
	Low	?	N

where Y indicates that the information is sought, N that it is not sought and ? that the case is considered on its merits.

7.27.6 Essential and optional equipment functions

The functions of equipment divide into the essential and the desirable. Examples of essential functions include containment by materials of construction, material transfer by pumps and compressors, control by instruments and computers, protection by protective devices and thermal insulation by lagging. Other functions may be less essential, even optional. Examples are equipment for heat recovery or improved process yield.

The computation of LCC in these two cases is different. Where the function is essential, the decision is concerned with the alternative ways of providing the function, but where it is optional the decision is rather whether to provide it at all. The latter decision may well be more self-contained and easier to make.

7.27.7 Equipment functions involving availability

A further distinction to be made is that between those decisions which involve plant availability and those which do not. The effect of an equipment on plant availability is often quite difficult to assess. Intermediate storage is provided to improve plant availability, but its effectiveness is a function of the storage operating policy adopted. Stand-by equipment is provided to cut in when the main equipment fails, but its reliability in responding to this demand depends largely on management factors.

A further complication arises when the plant is not operating at full capacity and its availability is at a discount. It is difficult to generalize about this situation, but one point is worth making. It can be of value to demonstrate that the plant can operate dependably at high availability, even if not currently required to do so, since this means that it is known that when demand picks up it can be met from the existing plant, thus allowing expenditure on a further plant to be deferred.

7.27.8 Equipment quality vs system configuration

It often occurs that a particular function can be provided by a single equipment or by several equipments in some suitable configuration. For a given level of reliability, the single equipment will need to be of higher quality. There is therefore a choice to be made between a single, high quality item and several items which could be of lower quality.

In making this decision, use may be made of simple models of system reliability and availability such as those given above, particularly in Sections 7.8 and 7.13. If the case is considered of a single equipment vs two parallel redundant equipments, the reliability of the single item needs to be very high to match that of a parallel system of two items of moderate reliability.

7.27.9 Assessment of equipment quality

The ability to assess the quality of equipment is clearly crucial to the practice of LCC. Some factors relevant to the assessment are (1) user experience, (2) vendor reputation, (3) design status (new, established), (4) capital cost, (5) material of construction and (6) engineering principles. Engineering principles include (1) the number of subsystems and components, (2) electronic vs mechanical subsystems, (3) moving parts and (4) stressing of parts. Also relevant are any quality assurance arrangements which exist between the vendor and the user.

7.28 Notation

Sections 7.1–7.13

a	constant; parameter in rectangular, gamma, Pareto and extreme value distributions
A	event A
$A(t)$	availability
A	matrices defined by Equations 7.10.6 and 7.10.36
b	constant; parameter in rectangular, gamma and extreme value distributions
B	event B
C	event C; minimum cut set
$d(t)$	downtime
e	error
E	event E
$f(t)$	failure density function
$f_d(d)$	downtime density function
$f_r(t_r)$	repair time density function
$f_w(w)$	throughput density function
$f_1(t), f_2(t)$	failure density function of subsystem
$F(t)$	failure distribution function
$F_d(d)$	downtime distribution function
$F_r(t_r)$	repair time distribution function
$\bar{g}(s)$	Laplace transform of $g(t)$
$H(t)$	cumulative hazard function
I	identity matrix
k	counter; state counter; parameter in Weibull distribution (alternative form); parameters defined by Equations 7.13.28 and 7.18.5
m	mean life; parameter in Weibull distribution (alternative form); mean of normal distribution; number of repair-men
$m*$	location parameter in log–normal distribution
m_d	mean downtime
m_i	ith moment (experimental)
m_r	mean repair time; mean of normal distribution of repair times
m_r*	location parameter in log-normal distribution of repair times
m_{ru}	mean repair time of upstream unit
m_u	mean uptime
m_z*	parameter in log–normal distribution of hazard rates
n	counter; number of items; number of states; number or trials; number of repairmen; number of applications of load
$n_f(t)$	number of items failed
n_i	number of times outcome i occurs
$n_s(t)$	number of items surviving
p	probability; probability of success
P	probability; probability of success
$P(\|)$	conditional probability
P_k	probability of being in state k
P	vector of probabilities
q	probability of failure
$Q(t)$	probability of failure, unreliability, failure distribution function
r	counter
$r_1 r_2$	equation roots defined by Equations 7.8.24 and 7.8.26
$R(t)$	probability of success or survival, reliability, reliability function
R_{sw}	probability of switchover
R_x	reliability of component
s	Laplace operator
t	time; failure time
t_r	repair time
T	time period; minimum tie set
T_e	time to empty tank
$u(t)$	uptime
$V(t)$	unavailability
V_s	unavailability of upstream unit
V_{st}	unavailability of upstream unit + tank
w	throughput
$z(t)$	hazard rate
$\mathcal{L}(\ )$	Laplace transform
α	parameter in Weibull distribution (alternative form); confidence level parameter

β shape factor in Weibull distribution
γ location parameter in Weibull distribution; number of standard deviations
Γ gamma function
ζ event rate
η characteristic life in Weibull distribution
λ failure rate; transition rate
λ' failure rate
λ_{st} failure rate of upstream unit + tank
λ_u failure rate of upstream unit
μ moment; mean number of events; repair rate
μ_i *i*th moment (theoretical)
ξ repair rate enhancement factor
ρ parameter defined by Equation 7.13.24
σ standard deviation of normal distribution; shape factor in log–normal distribution
σ_L standard deviation of loading
σ_r standard deviation of repair times; shape parameter in log–normal distribution of repair times
σ_s standard deviation of strength
σ_z parameter in log–normal distribution of hazard rates
τ time; repair time; parameter defined by equation 7.18.6
$\phi(\tau, t)$ joint density function

Superscripts

\- Laplace transform
^ estimated value

Section 7.14

E event
$f_{T1}(t_1)$ density function of t_1
$f_{T1}|_{\Lambda}(t_1|\lambda)$ density function of $t_1|\lambda$
$f_X(x)$ density function of x
$f_{X|Y}(x|y)$ density function of x|y
$f_Y(y)$ density function of y
$f_{Y|X}(y|x)$ density function of y|x
$f_{\Lambda(\lambda)}$ density function of λ
$f_{\Lambda}|_{T1}(\lambda|t_1)$ density function of $\lambda|t_1$
H hypothesis
n number of failures
t time
t_1 specific value of T_1 (time to failure)
T_1 random variable (time to failure)
x specific value of X
X random variable
y specific value of Y
Y random variable
λ specific value of Λ (failure rate)
Λ random variable (failure rate)
μ parameter in distribution of λ

Superscript

^ estimated value

Section 7.15

E( ) expected value
f_i density function of component i
$f_{\tau n}$ density function of τ_n
$F_{\tau n}$ distribution of function of τ_n
n counter for renewals
N number of renewals
P probability
s Laplace operator
t time
t_i time to failure of component i

Superscript

– Laplace transform

Section 7.16

C average annualized cost of operation
C_c capital cost of equipment
$f(n)$ rate of expenditure on maintenance
$f(t)$ rate of expenditure on maintenance
n number of years equipment is to be in use
S scrap value of equipment
t time equipment is to bc in use

Section 7.17

Subsection 7.17.5

$f(x)$ density function of load x
$f(y)$ density function of strength y
g(z) density function of z
G(z) distribution function of z
L load
m_x mean value of load x
m_y mean value of strength y
m_z mean value of z
n_x constant in Equation 7.17.4
n_y constant in Equation 7.17.5
S strength
u parameter defined by Equation 7.17.13
v parameter defined by Equation 7.17.14
x load
y strength
z difference between load and strength
σ_x standard deviation of load x
σ_y standard deviation of strength y
σ_z standard deviation of z
ϕ safety parameter

Subsection 7.17.6

f(i) shaping factor
L load
n number of applications of load
p parameter defining shaping factor in Equation 7.17.21
R(n) probability of survival after n applications of load
s stress (or strength)
s_0 initial strength
Δs_i decrement of strength
S strength
z(t) hazard rate
σ scaling factor

Section 7.18

a,b constants
H(t) cumulative hazard function
k parameter in Weibull distribution (alternative form)
m parameter in Weibull distribution (alternative form)
R(t) probability of success or survival, reliability
$\bar{\lambda}$ parameter defined by Equation 7.18.4
τ parameter defined by Equation 7.18.6

Section 7.19

f(t)	density function
$f_1(t), f_2(t)$	constituent density functions
p	weighting factor

Section 7.20

Subsection 7.20.6

f	degrees of freedom
f_1, f_2	degrees of freedom
L( )	lower limit
m	mean life
n	number of trials
N	number of trials
S	number of successes
T	period of test
U( )	upper limit
α	confidence parameter
χ^2	chi square distribution

Superscript

^	estimated value

Subsection 7.20.8

as Sections 7.1–7.13 plus

F_{mean}	mean rank
F_{med}	median rank
m_j	mean order number
m_o	replacement time
Δm	mean order number increment
n	number of items
$n_f(t)$	number of items failed
x	number of items before jth failure

Subsection 7.20.11

as Sections 7.1–7.13 plus

a	parameter
a_i	parameters
b	parameter
L	likelihood function
t	time to failure
z_c	calculated hazard rate
μ_k	kth moment of distribution
$\mathcal{L}$	logarithm of likelihood function (=ln L)

Subsection 7.20.16

as Sections 7.1–7.13 plus

ν	random number

Subsection 7.20.17

z(t)	hazard rate
z_i	ith influencinbg variable, or explanatory factor
$z_o(t)$	baseline hazard function
β_i	coefficient of z_i

Section 7.22

as Sections 7.1–7.13 plus

C_i	parameter representing complexity of ith subsystem
W_i	weighting factor for ith subsystem

Section 7.23

Subsection 7.23.3

T	period of test
α	index

Subscripts

cum	cumulative
cur	current

Subsection 7.2.3.4

c	allowable number of defects
n	number of items in sample
p	proportion of good items
P	probability of acceptance
q	proportion of defective items

Section 7.25

k	constant
m	mean life of component
m_o	replacement time of component
σ	standard deviation of life of component

8 Hazard Identification

Contents

The identification of areas of vulnerability and of specific hazards is of fundamental importance in loss prevention. Once these have been identified, the battle is more than half won. Such identification is not a simple matter, however. In many ways it has become more difficult as the depth of technology has increased. Loss prevention tends increasingly to depend on the management system and it is not always easy to discover the weaknesses in this. The physical hazards also no longer lie on the surface, accessible to simple visual inspection. On the other hand, there is now available a whole battery of hazard identification methods which may be used to solve these problems. Selected references on hazard identification are given in Table 8.1.

Different methods are required at different stages of a project. Table 8.2 lists some of these stages and the corresponding hazard identification techniques. The list is illustrative and, in particular, a technique quoted for one stage may be applicable also to another stage. There

Table 8.1 *Selected references on hazard identification*

Albisser and Silver (1960, 1964); W.H. Richardson (1962, 1963); Carpenter (1964); Coulter (1965); Dow Chemical Co. (1966a,b, 1976, 1994); Fowler and Spegelman (1968); Leeah (1968–); BCISC (1969/9, 10, 1973/12, 13); N.T. Freeman and Pickbourne (1971); Voigtsberger (1973); DOT, CG (1974a–d); D.W. Jones (1975); Harvey (1976, 1979); Heron (1976); Gelburd (1977); W.G. Johnson (1980); L. Kaplan (1981b); Gorbell (1982); Husman and van de Putte (1982); Kletz and Lawley (1982); Engineering Council (1983); W.B. Howard (1983, 1984); Kletz (1983d, 1984i, 1984 LPB 59, 1985e, 1986d, 1992b); Solomon (1983 LPB 52, 1984 LPB 54); Anon. (1984 LPB 60, p. 40); Embling (1986); CCPS (1985); Hoffmann (1985); Hoffmann and Maser (1985); van Horn (1985); Ducsherer and Molzala (1986); Flothmann and Mjaavatten (1986); Parry (1986 SRD R379); Pikaar *et al.* (1986); Pitblado (1986); Vervalin (1986b); Ozog and Bendixen (1987); Waite *et al.* (1988); Andreasen and Rasmussen (1990); Galagher and Tweeddale (1990, 1992); Tweeddale (1990a–c); Ahmed and Khan (1992); Argent *et al.* (1992); Kavianian *et al.* (1992); Pontiggia (1992); Sharkey et al. (1992); Toola (1992); Sankaran (1983); Knowlton (1993); Taylor (1993 LPB 112); Vincoli (1993)

CCPS Hazard Evaluation Guidelines
Witter (1992)

Physical and chemical properties (*see also* Tables 11.1, 11.18, 16.1, 16.3, 17.3, 17.62, 18.1)
NIOSH (Crit. Doc. series, 1990/18); Nuckolls (1929); Mellan (1950, 1957, 1977); MCA (1952-SD series, 1972/21); NSC (1952–, Safety Data Sheets); DECHEMA (1953–); Sax (1957–, 1981, 1986); Kirk and Othmer (1963–, 1978–, 1991–); Marsden (1963); Shabica (1963); D.T. Smith (1965a); Anon. (1966–67); CGA (1966/1); G.D. Muir (1971–); Bahme (1972); FPA (1972–, CFSD, Vol. 4, H series); Meidl (1972); Bretherick (1974, 1975, 1976, 1987a,b, 1990); DOT, CG (1974a,b); Int. Tech. Inf. Inst. (Japan) (1975); DoE and Chemical Society (1976); Shieler and Pauze (1976); Redeker and Schebsdat (1977); Sittig (1979); Weiss (1981); Anon. (1985p); Sax and Lewis (1986); Reale and Young (1987); A. Allen (1988); Walsh (1988); NFPA (1991/27); Carson and Mumford (1994)

Safety evaluation of chemicals
Cumberland and Hebden (1975); Brannegan (1985); A.S. West (1986, 1993a,b); Kohlbrand (1990)

Reactivity, instability, explosibility
ASTM (STP 394, 1975); Lothrop and Handrick (1949); Boynton *et al.* (1959); Steele and Duggan (1959); van Dolah (1961, 1965, 1969a); Mackenzie (1962, 1974); van Dolah *et al.* (1963); Shabica (1963); Wilcox and Bromley (1963); Cruise (1964); J.R. Marshall (1964); G.T. Austin (1965b); Garn (1965); Snyder (1965, 1982); Platt and Ford (1966); Sykes (1966); Prugh (1967); Silver (1967); Wankel (1967); Adams (1968); Lindeijer (1968); Roburn (1968); Settles (1968); Woodworth (1968); Brinkley (1969); Lindeijer and Pasman (1969); Spiegelman (1969); Flynn and Rossow (1970a); Stull (1970, 1973, 1976, 1977); Bersier *et al.* (1971); Gordon and McBride (1971, 1976); V.J. Clancey (1972a, 1974b); Coffee (1972, 1973, 1982a); Rollet and Bouzis (1972); Blazek (1973); Daniels (1973); E.J. Davis and Ake (1973); Dehn (1973); Sheldon and Kochi (1973); Treweek *et al.* (1973); Bowes (1974b); Carver *et al.* (1974); Groothuizen *et al.* (1974); A.J. Owen and von Zahn (1974); Porter (1974); Wendlandt (1974); Berthold *et al.* (1975); Burleson *et al.* (1975); Denn (1975); Keattch and Dollimore (1975); O'Driscoll (1975b); O'Neill (1975); Berthold and Löffler (1976); Brasie (1976a); Janin (1976); Lemke (1976); Riethmann *et al.* (1976); Grewer (1977); Pesata *et al.* (1977); Townsend (1977); Noronha and Juba (1980); G.F.P. Harris *et al.* (1981); Verhoeff (1981, 1983a,b); Hoffmann (1985); Hoffmann and Maser (1985); Bretherick (1987a, 1990); Hofelich and Thomas (1989); Kohlbrand (1990); Merrifield and Roberts (1991); NFPA (1991 NFPA 49, 325M, 491M); T.A. Roberts and Royle (1991); Frurip (1992); H.E. Fisher and Goetz (1993); Siwek and Cesana (1993); Fierz (1994); Wang, van Kiang and Merkl (1994)
Combustible solids, thermal explosion: (see also Table 16.28): Groothuizen *et al.* (1977); Beever (1981); Beever and Thorne (1981): Bishop (1981); Boddington *et al.* (1981); Bowes (1981); E.J. Davis *et al.* (1981); N. Gibson and Harper (1981b); Napier and Vlatis (1981); D.W. Smith (1982); Grewer (1987a,b); Cardillo (1988); de Faveri, Zonato *et al.* (1989); Fierz and Zwahlen (1989); Kotoyori (1989a); Tharmalingam (1989b); Liang and Tanaka (1990)
Test methods: V.J. Clancey (n.d.); A.K. Gupta (1949); Rapean *et al.* (1959); Koenen *et al.* (1961); Snyder (1965, 1982); Coffee (1969, 1972, 1973, 1976, 1982); MoD (1972); Connor (1974); Seaton *et al.* (1974); Berthold *et al.* (1975); Berthold and Löffler (1976); Eigenmann (1976); Janin (1976); Lemke (1976); Treweek *et al.* (1976); May (1978); Anon. (LPB 30 1979, p. 159); Dahn (1980); Myers (1980); Seyler (1980); D.W. Smith *et al.* (1980); Storey (1980, 1981); ABPI (1981, 1989); de Groot and Hupkens van der Elst (1981); Short *et al.* (1981); Townsend (1981); Mohan *et al.* (1982); Duch *et al.* (1982); Daugherty (1983); Pickard (1983); Fenlon (1984, 1987); C.A. Davies *et al.* (1985); Kohlbrand (1985, 1987a, b, 1989, 1990); Yoshida et al. (1985); Cutler (1986); UN (1986, 1988, 1990); Craven (1987); Cronin and Nolan (1987a,b); Grewer (1987a,b); Mellor *et al.* (1987); Snee

(1987); Ducros and Sanner (1988); HSE (1988 CRR 5); Mellor *et al.* (1988); Thomson and Zahn-Ullmann (1988); Boddington *et al.* (1989); Frurip *et al.* (1989); Hasegawa *et al.* (1989); Seaton and Harrison (1990); Uehara and Kitamura (1990); Ando *et al.* (1991); Cardillo and Cattaneo (1991); Cardillo and Nebuloni (1991); H.G. Fisher and Goetz (1991); Townsend and Valder (1993); Whitmore and Wilberforce (1993); Kossoy *et al.* (1994); Mores *et al.* (1994); Wan *et al.* (1994); Whitmore (1994)
Classification of materials: ACDS (1991); Merrifield and Roberts (1991); T.A. Roberts and Royle (1991); UN (1991); Hofelich, Power and Frurip (1994)

Process chemical reactions
Schierwater (1971); Grewer (1974, 1975, 1976, 1991); Berthold and Löffler (1976); Eigenmann (1976, 1977); Janin (1976); Regenass (1976); Schleich (1976); Schofield (1976); Schildknecht (1977); Stull (1977); Husain and Hamilcar (1978); Dehaven (1979); Hub *et al.* (1979); Holzer *et al.* (1980); Noronha *et al.* (1980); D.W. Smith (1982, 1984); Hoffmann and Maser (1985); Zeller (1985); Grewer and Klais (1988); B. Rasmussen (1988); Verhoeff (1988); Gygax (1989, 1990); Palazzi and Ferraiolo (1989); Barton and Rogers (1993); Gustin (1993); Landau and Cutro (1993); S. Schwartz (1993); Stoessel (1993); Cardillo (1994)
Test methods: Snyder (1965, 1982); Hub (1975, 1976, 1977a–c, 1980, 1981); Berthold and Löffler (1976); Eigenmann (1976); Regenass (1976); Noronha *et al.* (1979); Grewer and Klais (1980, 1988); Hakl (1980, 1981); Noronha and Juba (1980); Townsend and Tou (1980); Verhoeff and Janswoude (1980); Brogli *et al.* (1981); Dollimore (1981); Grewer (1981, 1987a,b); Wilberforce (1982, 1984); Fierz *et al.* (1983, 1984); Giger *et al.* (1983); Grewer and Duch (1983); de Haven (1983); Klais and Grewer (1983); Coates (1984a,b); N. Gibson (1984); Verhoeff and Heemskerk (1984); Hub and Jones (1986); Ottaway (1986); Spence et al. (1986); Wendlandt (1986); T.K. Wright and Rogers (1986); Kirch *et al.* (1987); T.K. Wright and Butterworth (1987); Spence and Noronha (1988); Chaineaux and Dunnin (1989); Dixon-Jackson (1989); Fauske *et al.* (1989); Hofelich (1989); Hoppe and Grob (1989); Kohlbrand (1989); Kotoyori (1989b); Mix (1989); Rogers (1989a); Steele and Nolan (1989); Duswalt (1991); Vilchez and Casal (1991); Hoppe (1992); Lambert *et al.* (1992); Landau and Williams (1992); J. Singh and Boey (1992); Snee and Hare (1992, 1994); Suter *et al.* (1992); Cardillo *et al.* (1993); Gustin (1993); Snee *et al.* (1993); Steele *et al.* (1993)
CCPS program: A.S. West (1993a)

Laboratories, pilot plants (*see also* Table A7.2)
Fleming (1958); Gernand (1965); Gorman *et al.* (1965); Hudson (1967); Perciful and Edwards (1967); Katzen (1968); Anon. (1969a); Conn (1971); Katell (1973); Dollimore (1981); Hoffmann (1985); van Horn (1985); Kohlbrand (1985); Brummnel (1989)

Hazard indices
Pratt (1965); Stull (1970, 1973, 1977); DOT, CG (1974a,b); D.E. Miller (1976); C.J. Jones (1978); Pitt (1982); Embling (1986); Kletz (1988f); de Graaf (1989); Lapp (1990); NFPA (1991 NFPA 49); Kavianian *et al.* (1992)
Dow Index, Dow Guide: Dow Chemical Co. (1964, 1966a,b, 1976, 1980, 1987, 1994); van Gaalen (1974); Scheffler (1994)
Mond Index: D.J. Lewis (1979, 1980b, 1989b); Doran and Greig (1984 LPB 55); ICI (1985); Tyler (1985)
IFAL Index: J. Singh and Munday (1979); Munday *et al.* (1980); ITB (1981); Menashe and Berenblut (1981); Berenblut and Menashe (1982); Whitehouse (1985); Berenblut and Whitehouse (1988)
FIRST: Mudan (1989c)

Insurance assessments
Dow Chemical Co. (1964, 1966a,b, 1976, 1980, 1987, 1994); Fowler and Spiegelman (1968); Spiegelman (1969); Neil (1971); Drewitt (1975)

Ranking methods, rapid ranking
Mumford and Lihou (LPB 59 1984); Gillett (1985); Keey (1991); G. Stevens and Stickles (1992)

Process safety reviews, hazard studies
R.W. Henry (1979); Wells (1981); Anthony (1985); IMechE (1985/88); Oh and Husmann (1986); Vogler (1986); van Mynen (1990); Turney (1990a,b, 1991); Desjardin *et al.* (1991); Grossman and Fromm (1991); Wells *et al.* (1991); Burk (1992); Kolodji (1993); Heetveld (1993); R. James and Wells (1994); Wells *et al.* (1994)

Hazard and operability (hazop) studies
Binsted (1960); Birchall (1960); Elliott and Owen (1968); Kletz (1972a, 1983d, 1984i, 1985e,i, 1986d, 1988b, 1992b); S.B. Gibson (1974, 1976a,b); ICI (1974); Lawley (1974a,b); Cowie (1976); Henderson and Kletz (1976); Knowlton (1976, 1979, 1981, 1989, 1992); Lock (1976); CIA (1988 RC18); CISHC (1977/6); Rushford (1977); Himmelblau (1978); Sachs (1978); Ministry of Social Affairs (1979a); A.F. Johnston, McQuaid and Games (1980); Lihou (1980a,b, 1990b, 1985 LPB 66); Wells (1980); Merriman (1982 LPB 48); Scott (LPB 48 1982); Anon. (LPB 49 1983, p. 1); IBC (1983/42, 1984/53); Illidge (1983 LPB 53); Lowe and Solomon (1983); Sinnott (1983); Solomon (1983 LPB 52, 1984 LPB 54); Bendixen and O'Neill (1984); Piccinnini and Levy (1984); Ozog (1985); Shafaghi and Gibson (1985); Flothmann and Mjaavatten (1986); Suokas (1986); Qureshi (1988); ILO (1989); Anon. (1991 LPB 98, p. 1); R.A. Freeman (1991); W.J. Kelly (1991); Black and Ponton (1992); Deshotel and Goyal (1992); Hendershot (1992); Isalski *et al.* (1992); Kavianian *et al.* (1992); G.C. Stevens and Humphreys (1992); D. Jones (1992); McCluer and Whittle (1992); Mitchell (1992 LPB 105); D. Scott and Crawley (1992); G.C. Stevens *et al.* (1992); Wells and Phang (1992); Willis (1992 LPB 108); Charsley and Brown (1993); Pully (1993); Sweeney (1993); Taylor (1993 LPB 112)
CCPS Guidelines: Witter (1992)

Safety audits, including management audits, plant audits
MCA (SG-20); Stapleton (1963); J.U. Parker (1967, 1973); Whitehorn and Brown (1967, 1973); BCISC (1969/9, 10, 1973/12, 13); D. Williams (1971); Harvey (1976, 1979b); Kletz *et al.* (1977); R. King and Magid (1979); Kletz (1977h, 1981l); Anon. (LPB 60 1984); Conrad (1984); Scheid (1987); Anon. (1988 LPB 79, p. 25); Atallah and

Guzman (1988); Luck and Howe (1988); Madhavan and Kirsten (1988a,b); Madhavan and Landry (1988); Monk (1988); Camp (1989); Stallworthy (1989); Kase and Wiese (1990); Scheimann (1990); Tweeddale (1990a, 1993d, 1994, 1994 LPB 117); I.G. Wallace (1990); NSW Government (1991); L.W. Ross (1991); Galagher and Tweeddale (1992); Ozog and Stickles (1992); D. Scott and Crawley (1992); Lockwood (1993); Ozog (1993); Sankaran (1993); Hurst and Ratcliffe (1994); Hurst *et al.* (1994)

Failure modes and effects analysis *(also failure modes, effects and criticality analysis)*
IEC (Std 812); US Armed Forces (MIL-STD-1629); Recht (1966b); Ostrander (1971); C.F. King and Rudd (1972); Jordan (1972); Taylor (1973, 1974c, 1975a); Himmelblau (1978); Lambert (1978a); Aldwinkle and Slater (1983); Flothmann and Mjaavatten (1986); Rooney *et al.* (1988); Arendt and Lorenzo (1991); Kavianian, Rao and Brown (1992); D. Scott and Crawley (1992); Andrews and Moss (1993); Goyal (1993); Vincoli (1993); BS 5760: Part 5: 1991

'What if?' method
Burk (1992); Zoller and Esping (1993)

Fault trees, event trees and cause-consequence diagrams (*see* Table 9.1)

Sneak path analysis
E.J. Hill and Bose (1975); Dore (1991); Hahn *et al.* (1991); Whetton (1993)

HAZCHECK
Wells and Phang (1992 LPB 105)

Hazard analysis
Kletz (1971, 1972a, 1973a, 1976d, 1983d, 1984b, 1986d, 1992b); S.B. Gibson (1976a, 1977a); Illidge (1983 LPB 53)

Computer hazop
Technica Ltd (1988); Andow (1991); Malagoli *et al.*. (1992); Willis (1992 LPB 102); Burns and Pitblado (1993); Broomfield and Chung (1994); Chung and Broomfield (1994)

Operator task analysis
Annett *et al.* (1971); Duncan (1974); Duncan and Gray (1975a); Duncan and Shepherd (1975a,b)

Emergency planning
Duff and Husband (1974)

Contractor hazard analysis
Whitaker (1993)

Pre-acquisition audits
Lloyds Register of Shipping (1992)

Quality assurance, completeness of identification
J.R. Taylor (1979, 1992); VTT (1985/1, 1988 Research Report 516); Rouhianen (1990)

Selection of method
CCPS (1989/7); Pontiggia (1992)

is no single ideal system of hazard identification procedures. The most appropriate system varies to some extent with the type of industry and process. Thus, for example, a firm involved in the batch manufacture of a large number of organic chemicals is likely to be much more interested in techniques of screening and testing chemicals and reactions than one operating ethylene plants.

The choice of hazard identification technique also depends on the purpose for which the study is done. For the identification of hazards and operating problems on a plant a hazard and operability study is suitable, while for the identification of sources of release for a hazard assessment it is necessary to carry out a specific review of such sources. It is only sensible in hazard identification to make use of past experience. The use of standards and codes, of course, helps to avoid hazards of which people may not even be aware. As far as hazard identification is concerned, however, the principal means of transmitting this experience in readily usable form is the checklist.

Many of the hazard identification techniques used, however, are concerned with situations which have some element of novelty. Methods for screening and testing are used to detect hazardous runaways and decompositions in new chemicals and reactions. Hazard and operability (hazop) studies are used to identify how often familiar hazards may be specifically realized in a new plant.

8.1 Safety Audits

One of the first systematic methods of hazard identification used in the chemical industry was the safety audit. Audits of various types are a normal management tool and are of considerable importance in safety.

An early account of safety audits in the UK was the British Chemical Industry Safety Council (BCISC) report *Safe and Sound* (1969/9), which drew on the experience of the US chemical industry, where the safety audit was established as a prime means of ensuring safety. A full description of such an audit is given in the BCISC's *Safety Audits* (1973/12). A safety audit subjects every area of the organization's activity to a systematic critical examination. It aims, like other audits, to reveal the strengths and weaknesses and the areas of vulnerability. It is carried out by professionals and results in a formal report and action plan.

Safety Audits draws a distinction between a *safety audit* and a *safety survey*, which is a detailed examination of a narrow field such as specific procedures or a particular plant; *safety inspection*, which is a scheduled inspection of a unit carried out by the unit's own personnel; *safety tour*, which is an unscheduled tour of a unit carried out by an outsider such as the works manager or a safety representative; and *safety sampling*, which is a specific application of a safety inspection/tour designed to measure accident potential.

A safety audit examines and assesses in detail the standards of all facets of a particular activity. It extends from complex technical operations and emergency procedures to clearance certificates, job descriptions, housekeeping and attitudes. It involves labour relations since people are asked about their training, their under-

***Table 8.2** Hazard identification techniques appropriate to different project stages*

Project stage	*Hazard identification technique*
All stages	Management and safety system audits
	Checklists
	Feedback from workforce
Research and development	Screening and testing for
	Chemicals (toxicity, instability, explosibility)
	Reactions (explosibility)
	Impurities
	Pilot plant
Pre-design	Hazard indices
	Insurance assessments
	Hazard studies (coarse scale)
Design	Process design checks
	Unit processes
	Unit operations
	Plant equipments
	Pressure systems
	Instrument systems
	Hazard and operability studies (fine scale)
	Failure modes and effects analysis
	Fault trees and event trees
	Hazard analysis
	Reliability assessments
	Operator task analysis and operating instructions
Commissioning	Checks against design, inspection, examination, testing
	Non-destructive testing, condition monitoring
	Plant safety audits
	Emergency planning
Operation	Inspection, testing
	Non-destructive testing, condition monitoring, corrosion monitoring, malfunction detection, plant degradation audits
	Plant safety audits

standing of works policy and whether they think they are making their contribution in the right way.

An audit might cover a company-wide problem or a total works situation (say, its emergency procedures or effluent systems) or simply a single plant activity.

At any level an audit is carried out thoroughly by a team made up of people immediately concerned, assisted by experts in various fields and experienced people not connected with the area of audit, to ensure that a fresh and unbiased look is injected into the inspection process.

Safety audits thus cover the following main areas:

(1) site level
  (a) management system;
  (b) specific technical features;
(2) plant level
  (a) management system.

The audit has five main elements;

(1) *identification* of possible hazards;
(2) *assessment* of potential consequences of realization of these hazards:
(3) *selection* of measures to minimize frequency and consequences of realization;
(4) *implementation* of these measures within the organization;
(5) *monitoring* of the changes.

The audit is conducted in a fairly formal way, using a checklist. Table 8.3, taken from *Safety Audits*, gives a list of activities and activity standards. These standards can be used as the basis of a numerical rating scheme. If a safety audit is conducted, it is essential that management act on it. This does not necessarily mean that every recommendation must be implemented, but there should be a reasoned response.

Safety audits are an important part of the activity of the safety function. In addition to identifying and rectifying hazards, they play a major role in educating personnel at all levels. These aspects are discussed in more detail in Chapter 28. Safety audits of the organization and management system at site and at plant level are considered below.

Another form of audit is the insurance survey. Insurance surveys have been described in Chapter 5. Although the insurer is not able to go down to the lower levels of detail, his contribution is valuable in that it represents an independent assessment.

Table 8.3 *Activity standards for safety audits (British Chemical Industry Safety Council, 1973/12)*

Activity	Poor	Fair	Good	Excellent
A Organization and administration				
1 Statement of policy, responsibilities assigned	No statement of loss control policy. Responsibility and accountability not assigned	A general understanding of loss control, responsibilities and accountability, but not written	Loss control policy and responsibilities written and distributed to supervisors	In addition to 'good', loss control policy is reviewed annually and is posted. Responsibility and accountability is emphasized in supervisory performance evaluations
2 Safe operating procedures (SOPs)	No written SOPs	Written SOPs for some, but not all, hazardous operations	Written SOPs for all hazardous operations	All hazardous operations covered by a procedure, posted at the job location, with an annual documented review to determined adequacy
3 Employee selection and placement	Only pre-employment physical examination given	In addition, an aptitude test is administered to new employees	In addition to 'fair', new employees' past safety record is considered in their employment	In addition to 'good', when employees are considered for promotion, their safety attitude and record is considered
4 Emergency and disaster control plans	No plan or procedures	Verbal understanding on emergency procedures	Written plan outlining the minimum requirements	All types of emergency covered with written procedures. Responsibilities are defined with back-up personnel provisions
5 Direct management involvement	No measurable activity	Follow-up on accident problems	In addition to 'fair', management reviews all injury and property damage reports and holds supervision accountable for verifying firm corrective measures	In addition to 'good', reviews all investigation reports. Loss control problems are treated as other operational problems in staff meeting
6 Plant safety rules	No written rule	Plant safety rules have been developed and posted	Plant safety rules are incorporated in the plant work rules	In addition, plant work rules are firmly enforced and updated at least annually
B Industrial hazard control				
1 Housekeeping storage of materials, etc.	Housekeeping is generally poor. Raw materials, items being processed and finished materials are poorly stored	Housekeeping is fair. Some attempts to adequately store materials are being made	Housekeeping and storage of materials are orderly. Heavy and bulky objects well stored out of aisles, etc.	Housekeeping and storage of materials are ideally controlled

2	Machine guarding	Little attempt is made to control hazardous points on machinery	Partial but inadequate or ineffective attempts at control are in evidence	There is evidence of control which meets applicable Federal and State requirements, but improvement may still be made	Machine hazards are effectively controlled to the extent that injury is unlikely. Safety of operator is given prime consideration at time of process design
3	General area guarding	Little attempt is made to control such hazards as: unprotected floor openings, slippery or defective floors, stairway surfaces, inadequate illumination, etc.	Partial but inadequate attempts to control these hazards are evidenced	There is evidence of control which meets applicable Federal and State requirements – but further improvement may still be made	These hazards are effectively controlled to the extent that injury is unlikely
4	Maintenance of equipment, guards, handtools, etc.	No systematic programme of maintaining guards, handtools, controls and other safety features of equipment, etc.	Partial but inadequate or ineffective maintenance	Maintenance programme for equipment and safety features is adequate. Electrical handtools are tested and inspected before issuance, and on a routine basis	In addition to 'good', a preventive maintenance system is programmed for hazardous equipment and devices. Safety reports filed and safety department consulted when abnormal conditions are found
5	Material handling – hand and mechanized	Little attempt is made to minimize possibility of injury from the handling of materials	Partial but inadequate or ineffective attempts at control are in evidence	Loads are limited as to size and shape for handling by hand, and mechanization is provided for heavy or bulky loads	In addition to controls for both hand and mechanized handling, adequate measures prevail to prevent conflict between other workers and material being moved
6	Personal protective equipment – adequacy and use	Proper equipment not provided or is not adequate for specific hazards	Partial but inadequate or ineffective provision, distribution and use of personal protective equipment	Proper equipment is provided. Equipment identified for special hazards, distribution of equipment is controlled by supervisor. Employee is required to use protective equipment	Equipment provided complies with standards. Close control maintained by supervision. Use of safety equipment recognized as an employment requirement. Injury record bears this out
C Fire control and industrial hygiene					
1	Chemical hazard control references	No knowledge or use of reference data	Data available and used by foremen when needed	In addition to 'fair', additional standards have been requested when necessary	Data posted and followed where needed. Additional standards have been promulgated, reviewed with employees involved and posted

Table 8.3 *Continued*

	Activity	*Poor*	*Fair*	*Good*	*Excellent*
C Fire control and industrial hygiene – continued					
2	Flammable and explosive materials control	Storage facilities do not meet fire regulations. Containers do not carry name of contents. Approved dispensing equipment not used. Excessive quantities permitted in manufacturing areas	Some storage facilities meet minimum fire regulations. Most containers carry name of contents. Some approved dispensing equipment in use	Storage facilities meet minimum fire regulations. Most containers carry name of contents. Approved equipment generally is used. Supply at work area is limited to one day requirement. Containers are kept in approved storage cabinets	In addition to 'good', storage facilities exceed the minimum fire regulations and containers are always labelled. A strong policy is in evidence relative to the control of the handling, storage and use of flammable materials
3	Ventilation – fumes, smoke and dust control	Ventilation rates are below industrial hygiene standards in areas where there is an industrial hygiene exposure	Ventilation rates in exposure areas meet minimum standards	In addition to 'fair', ventilation rates are periodically measured, recorded and maintained at approved levels	In addition to 'good', equipment is properly selected and maintained close to maximum efficiency
4	Skin contamination control	Little attempt at control or elimination of skin irritation exposures	Partial but incomplete programme for protecting workers. First-aid reports on skin problems are followed up on an individual basis for determination of cause	The majority of workmen instructed concerning skin-irritating materials. Workmen provided with approved personal protective equipment or devices. Use of this equipment is enforced	All workmen informed about skin-irritating materials. Workmen in all cases provided with approved personal protective equipment or devices. Use of proper equipment enforced and facilities available for maintenance. Workers are encouraged to wash skin frequently. Injury record indicates good control
5	Fire control measures	Do not meet minimum insurance or municipal requirements	Meet minimum requirements	In addition to 'fair', additional fire hoses and/or extinguishers are provided. Welding permits issued. Extinguishers on all welding carts	In addition to 'good', a fire crew is organized and trained in emergency procedures and in the use of fire fighting equipment
6	Waste – trash collection and disposal, air/water pollution	Control measures are inadequate	Some controls exist for disposal of harmful wastes or trash. Controls exist but are ineffective in methods or procedures of collection and disposal. Further study is necessary	Most waste disposal problems have been identified and control programmes instituted. There is room for further improvement	Waste disposal hazards are effectively controlled. Air/water pollution potential is minimal

D Supervisory participation, motivation and training					
1	Line supervisor safety training	All supervisors have not received basic safety training	All shop supervisors have received some safety training	All supervisors participate in division safety training session a minimum of twice a year	In addition, specialized sessions conducted on specific problems
2	Indoctrination of new employees	No programme covering the health and safety job requirements	Verbal only	A written handout to assist in indoctrination	A formal indoctrination programme to orientate new employees is in effect
3	Job hazard analysis	No written programme	Job hazard analysis (JHA) programme being implemented on some jobs	JHA conducted on majority of operations	In addition, job hazard analyses performed on a regular basis and safety procedures written and posted for all operations
4	Training for specialized operations (fork trucks, grinding, press brakes, punch presses, solvent handling, etc.)	Inadequate training given for specialized operations	An occasional training programme given for specialized operations	Safety training is given for all specialized operations on a regular basis and retraining given periodically to review correct procedures	In addition to 'good', an evaluation is performed annually to determine training needs
5	Internal self-inspection	No written programme to identify and evaluate hazardous practices and/or conditions	Plant relies on outside sources, i.e. Insurance Safety Engineer, and assumes each supervisor inspects his area	A written programme outlining inspection guidelines, responsibilities, frequency and follow-up is in effect	Inspection programme is measured by results, i.e. reduction in accidents and costs. Inspection results are followed up by top management
6	Safety promotion and publicity	Bulletin boards and posters are considered the primary means for safety promotion	Additional safety displays, demonstrations, films, are used infrequently	Safety displays and demonstrations are used on a regular basis	Special display cabinets, windows, etc., are provided. Displays are used regularly and are keyed to special themes
7	Employee-Supervisor safety contact and communication	Little or no attempt made by supervisor to discuss safety with employees	Infrequent safety discussions between supervisor and employees	supervisors regularly cover safety when reviewing work practices with individual employees	In addition to items covered under 'good', supervisors make good use of the shop safety plan and regularly review job safety requirements with each worker. They contact at least one employee daily to discuss safe job performance

Table 8.3 *Continued*

	Activity	*Poor*	*Fair*	*Good*	*Excellent*
E Accident investigation, statistics and reporting procedures					
1	Accident investigation by line personnel	No accident investigation made by line supervision	Line supervision makes investigations of only medical injuries	Line supervision trained and makes complete and effective investigations of all accidents; the cause is determined; corrective measures initiated immediately with a completion date firmly established	In addition to items covered under 'good', investigation is made of every accident within 24 h of occurrence. Reports are reviewed by the department manager and plant manager
2	Accident cause and injury location analysis and statistics	No analysis of disabling and medical cases to identify prevalent causes of accidents and location where they occur	Effective analysis by both cause and location maintained on medical and first-aid cases	In addition to effective accident analysis, results are used to pinpoint accident causes so accident prevention objectives can be established	Accident causes and injuries are graphically illustrated to develop the trends and evaluate performance. Management is kept informed on status
3	Investigation of property damage	No programme	Verbal requirement or general practice to inquire about property damage accidents	Written requirement that all property damage accidents of US $50 and more will be investigated	In addition, management requires a vigorous investigation effort on all property damage accidents
4	Proper reporting of accidents and contact with carrier	Accident reporting procedures are inadequate	Accidents are correctly reported on a timely basis	In addition to 'fair', accident records are maintained for analysis purposes	In addition to 'good', there is a close liaison with the insurance carrier

8.2 Management System Audits

The management system is crucial to loss prevention and it is essential that this system itself be monitored. The management system, and its auditing, have been described in Chapter 6. Management audit is so important, however, that a brief, summary account is in order at this point.

The management system may be audited in several ways. These include (1) self-checking procedures, (2) internal audit and (3) external audit. An illustration of a specific check built into the system is a formal requirement that a plant manager audit a proportion of the permits-to-work each week.

The safety audits described above include an internal audit of the overall system. In particular, there are checks on:

(1) overall management attitude, policies, systems and procedures, personnel selection;
(2) plant level management, attitude, systems, training and feedback;
(3) incident reporting, investigation and statistics.

External audit may be carried out by an outside body such as a firm of consultants.

There are now available a number of audit systems, including proprietary systems. Some of these yield quantitative factors applicable in quantitative risk assessment. Accounts of audit systems are given in Chapters 6, 9 and 28.

The Health and Safety Executive (HSE) is also concerned with auditing the management system. Indeed it is fair to say that this is its most important function. When visiting a factory an Inspector is very much concerned with the way the system works in practice. Audit by the HSE is another aspect discussed in Chapter 6.

8.3 Checklists

One of the most useful tools of hazard identification is the checklist. Like a standard or a code of practice, a checklist is a means of passing on hard-won experience. It is impossible to envisage high standards in hazard control unless this experience is effectively utilized. The checklist is one of the main tools available to assist in this.

Checklists are applicable to management systems in general and to a project throughout all its stages. Obviously the checklist must be appropriate to the stage of the project, starting with checklists of basic materials properties and process features, continuing on to checklists for detailed design and terminating with operations audit checklists.

There are a large number of checklists given in the literature – indeed a paper on practical engineering is quite likely to include a checklist. Selected references on checklists are given in Table 8.4, but these are only a tiny fraction of those available. A number of checklists are given in the present volume. These are collated in the index.

The proper use of checklists is discussed by R.L. Miller and Howard (1971). A checklist should be used for just one purpose only – as a final check that nothing has been neglected. Also, it is more effective if the

Table 8.4 *Selected references on checklists*

Dow Chemical Co. (1964, 1966a,b, 1976, 1980, 1987, 1994); Hettig (1966); Yelland (1966); J.U. Parker (1967, 1973); Fowler and Spiegelman (1968); Preddy (1969); MCA (1970/18); Klaassen (1971); R.L. Miller and Howard (1971); Whitehorn and Brown (1973); Balemans *et al.* (1974); Eberlein (1974); Kline *et al.* (1974); Henderson and Kletz (1976); Wells *et al.* (1976, 1977); Webb (1977); J.C. Rose *et al.* (1978); Wells (1980); Gorbell (1982); Flothmann and Mjaavatten (1986); Vervalin (1986b); Gordon and Moscone (1988); Wells *et al.* (1990); Kongso (1992); D. Scott and Crawley (1992); Oliver (1993); Anon. (1994 LPB 120, p. 13)

Insurance: Fowler and Spiegelman (1968); Spiegelman (1969); Ashe (1971); Neil (1971); Drewitt (1975)

Management systems: BCISC (1973/12, 13)

Physical and chemical properties: Dow Chemical Co. (1964, 1966a,b, 1976, 1980, 1987, 1994); Spiegelman (1969); Wells (1980)

Audits: Eley (1992)

HAZCHECK: Wells (1980); Wells *et al.* (1991)

Plant siting: Dow Chemical Co. (1964, 1966a,b, 1976, 1980, 1987, 1994); Spiegelman (1969); Balemans *et al.* (1974)

Plant layout: McGarry (1958); Dow Chemical Co. (1964, 1966a,b, 1976, 1980, 1987, 1994); Fowler and Spiegelman (1968); Spiegelman (1969); Balemans *et al.* (1974); Drewitt (1975); Wells (1980); Gorbell (1982); D. Scott and Crawley (1992)

Ventilation systems: BOHS (1987 TG7); HSE (1990 HS(G) 54)

Process design: Dow Chemical Co. (1964, 1966a,b, 1976, 1980, 1987, 1994); Spiegelman (1969); Wells *et al.* (1976); W.G. Johnson (1980); Wells (1980); Gorbell (1982); D. Scott and Crawley (1992)

Operating parameters and deviations: Wells *et al.* (1976); Wells (1980); Gorbell (1982); Trask (1990)

Documentation: Wells *et al.* (1976); J.C. Rose *et al.* (1978); Wells (1980)

Statutory approvals: J.C. Rose *et al.* (1978)

Chemical reactors: Gorbell (1982)

Dust-handling plant: Anon. (1990 LPB 95, p. 7)

Utilities: Gorbell (1982)

Pressure systems design: Spiegelman (1969); Gorbell (1982); D. Scott and Crawley (1992)

Pressure relief: Gorbell (1982);

Instrumentation: Gorbell (1982)

Human factors, human error: SRD, Human Factors in Reliability Group (1985 SRD R347); HSE (1989 HS(G) 48); Brazendale (1990 SRD 510)

Training: FPA (1974/23);

Fire and fire protection: Dow Chemical Co. (1964, 1966a,b, 1976, 1980, 1987, 1994); FPA (1965/3, 1974/23); Wells (1980); Gorbell (1982); D. Scott and Crawley (1992)

Plant commissioning: Matley (1969); Buyers (1972); J.D. Baker (1974); Mackey (1974); Turnbull *et al.* (1974); Unwin *et al.* (1974); L. Pearson (1977a,b); J.C. Rose *et al.* (1978); Wells (1980); D. Scott and Crawley (1992)

Plant start-up and shut-down: Wells (1980); D. Scott and Crawley (1992)

Plant operation: D. Scott and Crawley (1992)

Plant maintenance: D. Scott and Crawley (1992)

Demolition: Stuart (1974)

Plant modifications: Henderson and Kletz (1976); Wells (1980); Anon. (1994 LPB 120, p. 13)
Storage: Dow Chemical Co. (1964, 1966a,b, 1976, 1980, 1987, 1994); Spiegelman (1969); Balemans *et al.* (1974); Drewitt (1975); Jager and Haferkamp (1992)
Transport: Spiegelman (1969)
Personal safety: Gorbell (1982)
Pilot plants: Carson and Mumford (1989 LPB 89)
Incident investigation: M.E. Lynch (1967, 1973)

questions cannot be answered by a simple 'yes' or 'no' but require some thought in formulating an answer.

Checklists are effective only if they are used. There is often a tendency for them to be left to gather dust on the shelf. This is perhaps part of the reason for the development of other techniques such as hazard and operability studies, as described below.

8.4 Materials Properties

8.4.1 Physical and chemical properties

It is obviously necessary to have comprehensive information on all the chemicals in the process: raw materials, intermediates and final products. A list of some important physical and chemical properties of any chemical is given in Table 8.5. The significance of most of the items in the list is self-evident.

Compilations of properties of hazardous substances include those by the National Fire Protection Association (NFPA) (1994 NFPA 49), the Manufacturing Chemists Association (MCA) (now the Chemical Manufacturers Association (CMA)) (Appendix 28) and the National Institute for Occupational Safety and Health (NIOSH) (Appendix 28), and those by Sax (1957–), Marsden (1963), Muir (1971–), Bahme (1972), Meidl (1972), and Bretherick (1975–).

Corrosivity, flammability, toxicity and radioactivity are considered in more detail in Chapters 12, 16, 18 and 25, respectively. Reactivity, instability and explosibility and

Table 8.5 *Some important physical and chemical properties of a chemical*

General properties	Molecular structure Freezing point Vapour pressure, boiling point Critical pressure, temperature, volume Vapour density, specific heat, viscosity, thermal conductivity Liquid density, specific heat, viscosity, thermal conductivity Latent heats of vaporization and fusion Dielectric constant, electrical conductivity
Flammability	Flammability limits Flash point Autoignition temperature Minimum ignition energy Maximum experimental safe gap Self-heating
Corrosion	Corrosiveness to materials of construction Incompatibility with particular materials
Polymerization, decomposition	Polymerization characteristics Decomposition, hydrolysis characteristics
Impurities	Impurities in: raw material plant material Mutual solubilities with water
Reaction, explosion	Heats of formation, combustion, decomposition Energy hazard potential Thermal stability Impact sensitivity
Toxicity	Threshold limit values, emergency exposure limits Lethal concentration LC_{50}, lethal dose LD_{50} Exposure effects (inhalation, ingestion, skin and eye contact) Long-term low exposure effects Warning levels (smell)
Radioactivity	Radiation survey α–particle, β–, γ–ray exposures

ANILINE

Common Synonyms	Oily liquid	Colourless to yellowish brown	Aniline odour
Aminobenzene Aniline oil Phenylalanine Blue oil	Sinks slowly in water		

AVOID CONTACT WITH LIQUID AND VAPOUR. KEEP PEOPLE AWAY.
Wear chemical protective suit with self-contained breathing apparatus.
Stop discharge if possible.
Stay upwind and use water spray to 'knock down' vapour.
Call fire department.
Isolate and remove discharged material.
Notify local health and pollution control agencies.

Fire	Combustible. POISONOUS GAS IS PRODUCED WHEN HEATED. Vapour may explode if ignited in an enclosed area. Wear chemical protective suit with self-contained breathing apparatus. Extinguish with water, dry chemical, foam, or carbon dioxide. Cool exposed containers with water.
Exposure	CALL FOR MEDICAL AID. LIQUID POISONOUS IF SWALLOWED OR IF SKIN IS EXPOSED. Irritating to eyes. Remove contaminated clothing and shoes. Flush affected areas with plenty of water. IF IN EYES, hold eyelids open and flush with plenty of water. IF SWALLOWED and victim is CONSCIOUS, have victim drink water or milk.
Water Pollution	Dangerous to aquatic life in high concentrations. May be dangerous if it enters water intakes. Notify local health and wildlife officials. Notify operators of nearby water intakes.

1. RESPONSE TO DISCHARGE

(See *Response Methods Handbook*)

Issue warning–poison, water contaminant
Restrict access
Should be removed
Chemical and physical treatment

2. LABEL

2.1 Category: Poison
2.2 Class: 6

3. CHEMICAL DESIGNATIONS

3.1 CG Compatibility Class: Aromatic aniline
3.2 Formula: $C_2N_2NH_2$
3.3 IMO/UN Designation: 6.1/1547
3.4 DOT ID No: 1547
3.5 CAS Registry No: 62.53–3

4. OBSERVABLE CHARACTERISTICS

4.1 Physical State (as shipped): Liquid
4.2 Colour: Colourless to pale brown
4.3 Odour: Aromatic aniline-like; characteristic, peculiar: strongly aniline-like

5. HEALTH HAZARDS

5.1 Personal Protective Equipment: Respirator for organic vapours, splashproof goggles, rubber gloves, boots.
5.2 Symptoms following exposure: ACUTE EXPOSURE: Blue discolouration of finger-tips, cheeks, lips and nose: nausea, vomiting, headache and drowsiness followed by delirium, coma and shock. CHRONIC EXPOSURE: Loss of appetite, loss of weight, headaches, visual disturbances; skin lesions
5.3 Treatment of Exposure: Remove victim to fresh air and call a physician at once. SKIN, EYE CONTACT: immediately flush skin or eyes with plenty of water for at least 15 min. If cyanosis is present, shower with soap and warm water, with special attention to scalp and fingernails. Administer oxygen until physician arrives.
5.4 Threshold Limit Value: 2 ppm
5.5 Short Term Inhalation Limits: 50 ppm for 30 min: 5 ppm for 8 h.
5.6 Toxicity by Ingestion: Grade 3; LD_{30} = 50 to 500 mg/kg.
5.7 Late Toxicity: None recognized
5.8 Vapour (Gas) Irritant Characteristics: Vapours cause a slight smarting of the eyes or respiratory system if present in high concentrations. The effect is temporary.
5.9 Liquid or Solid Irritant Characteristics: if spilled on clothing or allowed to remain, may cause smarting and reddening of the skin.
5.10 Odour thresholds: 0.5 ppm
5.11 IOLH Value: 100 ppm

6. FIRE HAZARDS

6.1 Flush P.: 168°F O.C. 158°F C.C.
6.2 Flammable Limits in Air: 1.3%–11%
6.3 Fire Extinguishing Agents: Water, foam, dry chemical, or carbon dioxide
6.4 Fire Extinguishing Agents Not to be Used: Not pertinent
6.5 Special Hazards of Combustion Products: Toxic vapours are generated when heated.
6.6 Behaviour in Fire: Not pertinent
6.7 Ignition Temperature: 1418°F
6.8 Electrical Hazard: Not pertinent
6.9 Burning Rate: 3.0 mm/min.
6.10 Adiabatic Flame Temperature: Data not available
6.11 Stoichiometric Air to Fuel Ratio: Data not available
6.12 Flame Temperature: DAta not available

7. CHEMICAL REACTIVITY

7.1 Reactivity with Water: No reaction
7.2 Reactivity with Common Materials: No reaction
7.3 Stability During Transport: Stable
7.4 Neutralizing Agents for Acids and Caustics: Flush with water and rinse with dilute acetic acid
7.5 Polymortication: Not pertinent
7.6 Inhibitor of Polymortication: Not pertinent
7.7 Motor Ratio (Reactant to Product): Data no available
7.8 Reactivity Group: 9

8. WATER POLLUTION

8.1 Aquatic Toxicity:
1020 ppm/1 h/sunfish/killed/fresh water
10 ppm/96 h/scenedeemus/TL_2/fresh water
8.2 Waterfowl Toxicity: Data not available
8.3 Biological Oxygen Demand (BOD): 150%, 5 days
8.4 Food Chain Concentration Potential: None

9. SHIPPING INFORMATION

9.1 Grades of Purity: Commercial: 99.5%
9.2 Storage Temperature: Ambient
9.3 Inert Atmosphere: No requirement
9.4 Venting: Pressure-vacuum

10. HAZARD ASSESSMENT CODE

(See Hazard Assessment Handbook)
A-P-Q-T-U-X-Y

11. HAZARD CLASSIFICATIONS

11.1 Code of Federal Regulations: Poison, B
11.2 NAS Hazard Rating for Bulk Water Transportation:

Category	Rating
Fire	1
Health	
Vapour irritant	1
Liquid or Solid Irritant	1
Poisons	3
Water Pollution	
Human Toxicity	2
Aquatic Toxicity	3
Aesthetic Effect	2
Reactivity	
Other Chemicals	4
Water	3
Self Reaction	0

11.3 NFPA Hazard Classification:

Category	Classification
Health Hazard (Blue)	3
Flammability (Red)	2
Reactivity (Yellow)	0

12. PHYSICAL AND CHEMICAL PROPERTIES

12.1 Physical State at 15°C and 1 atm: Liquid
12.2 Molecular Weight: 93.13
12.3 Boiling Point at 1 atm: 363 6°F = 184.2°C = 457.4°K
12.4 Freezing Point: 21°F = −6.1°C = 267.1°K
12.5 Critical Temperature: 798.1°F = 425.6°C = 698.8°K
12.6 Critical Pressure: 770 ppm = 52.4 stm = 5.31 Mn/m^3
12.7 Specific Gravity: 1.022 at 20°C (liquid)
12.8 Liquid Surface Tension: 45.5 dynes/cm = .0455 N/m at 20°C
12.9 Liquid Water Interfacial Tension: 5.8 dynes/cm = 0.0058 N/m at 20°C
12.10 Vapour (Gas) Specific Gravity: Not pertinent
12.11 Ratio of Specific Heats of Vapour (Gas): 1.1
12.12 Latent Heat of Vapourization: 198 Btu/lb = 110 cal/g = 4.61 × 10^4 J/kg
12.13 Heat of Combustion: −14.980 Btu/lb = −8320 cal/g = −348.3 × 10^4 J/kg
12.14 Heat of Decomposition: Not pertinent
12.15 Heat of Solution: Not pertinent
12.16 Heat of Polymerization: Not pertinent
12.25 Heat of Fusion: Data not available
12.26 Limiting Value: Data not available
12.27 Reid Vapour Pressure: 0.02 psu

NOTES

Figure 8.1 *A safety data sheet (de Renzo, 1986) (Courtesy of Noyes Publishers)*

process chemical reactions are dealt with in the following sections.

8.4.2 Materials safety data sheets

Information on the physical and chemical properties of chemicals and on the associated hazards and the appropriate precautions cast in standard format is available in the form of materials safety data sheets (MSDSs).

Safety data sheet compilations include those of: Keith and Walters (1985–); de Renzo (1986); A. Allen (1988) and Walsh (1988); and Kluwer Publishers (1992). NFPA 49: 1994 consists largely of a compilation of Hazardous Chemical Data Sheets. It also contains a guide to their preparation.

For petroleum products guidance has been available in *Health and Safety Data Sheets for Petroleum Products* (CONCAWE, 1983 3/83).

A typical material safety data sheet is shown in Figure 8.1. Another exemplar is that given in the Center for Chemical Process Safety (CCPS) *Technical Management Guidelines* (1992/11).

There is now an EC Directive (91/155/EEC) which lays down the contents of a safety data sheet. In the UK this is implemented by Regulation 6 of the Chemical (Hazard Information and Packaging) (CHIP) Regulations 1993. HSE guidance is given in L37: *Safety Data Sheets for Substances and Preparations Dangerous for Supply* (1993) and further guidance is available in *The EC Safety Data Sheet Directive* by CONCAWE (1992 92/55).

The Directive requires that the contents of a safety data sheet cover

(1) identification of the substance/preparation and company;
(2) composition/information on ingredients;
(3) hazards identification;
(4) first-aid measures;
(5) first-fighting measures;
(6) accidental release measures;
(7) handling and storage;
(8) exposure controls/personal protection;
(9) physical and chemical properties;
(10) stability and reactivity;
(11) toxicological information;
(12) ecological information;
(13) disposal considerations;
(14) transport information;
(15) regulatory information;
(16) other information.

Both the guides mentioned give detailed guidance on the material to be included under each of these heads. It is the responsibility of the supplier of a chemical to supply the safety data sheet. L37 details this responsibility. Whilst data sheet compilations published in the open literature give some information under most or all of these headings, they do not necessarily conform with the guidance in L37.

8.4.3 Impurities

Impurities, foreseen or unforeseen, can have serious adverse effects on many aspects of process operation, including safety and loss prevention. These effects are discussed in more detail in Chapter 11. Here consideration is confined to the identification of possible problems with impurities.

There is no established methodology of identifying impurities which may arise in the process, that is comparable with that for checking the reactivity of a chemical. Reliance usually has to be placed on the literature and on experience. As discussed below, however, indication of possible impurities and of their adverse effects is one of the particularly valuable features of a pilot plant.

***Table 8.6** Degrees of reactivity (instability) hazard as defined in NFPA: 1994 49 (Reproduced with permission from NFPA 49 Hazardous Chemicals Data, Copyright, 1994, National Fire Protection Association, Batterymarch Park, Quincy, MA 02269, USA)*

Degree	*Description*
4	Materials that in themselves are readily capable of detonation or explosive decomposition or explosive reaction at normal temperatures and pressures. This degree usually includes materials that are sensitive to localized thermal or mechanical shock at normal temperatures and pressures
3	Materials that in themselves are capable of detonation or explosive decomposition or explosive reaction, but that require a strong initiating source or that must be heated under confinement before initiation. This degree usually includes: Materials that are sensitive to thermal or mechanical shock at elevated temperatures and pressures; Materials that react explosively with water without requiring heat or confinement
2	Materials that readily undergo violent chemical change at elevated temperatures and pressures. This degree usually includes: Materials that exhibit an exotherm at temperatures less than or equal to 150°C when tested by differential scanning calorimetry; Materials that may react violently with water or form potentially explosive mixtures with water
1	Materials that in themselves are normally stable, but that can become unstable at elevated temperatures and pressure. This degree usually includes: Materials that change or decompose on exposure to air, light, or moisture; Materials that exhibit an exotherm at temperatures greater than 150°C but less than or equal to 300°C, when tested by differential scanning calorimetry
0	Materials that in themselves are normally stable, even under fire conditions. This degree usually includes: Materials that do not react with water; Materials that exhibit an exotherm at temperatures greater than 300°C but $\leq$ 500°C, when tested by differential scanning calorimetry; Materials that do not exhibit an exotherm at temperatures $\leq$ 500°C when tested by differential scanning calorimetry

8.5 Instability of Materials

It is important to know whether a chemical can be safely handled both in process and in storage and transport. The problem has two main aspects. One is the hazard of instability of a single component such as a reactant, intermediate or product when undergoing physical processing on the plant and when stored and transported. The other aspect is the hazard of instability of the reaction mass in a chemical reactor. These two aspects of instability are dealt with here and in Chapter 11, respectively. This section deals with thermal instability and Section 8.6 with explosibility.

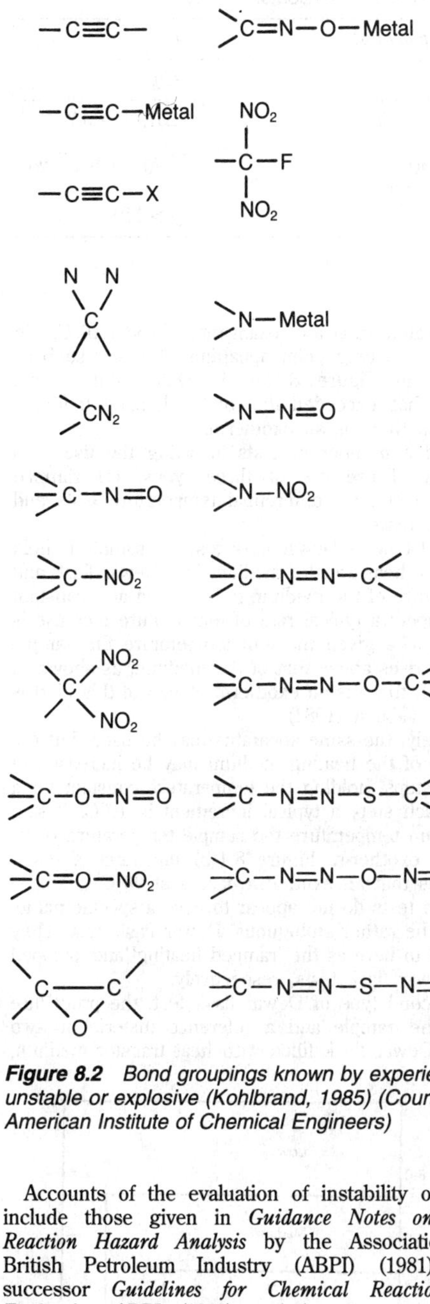

—C≡C—   >C=N—O—Metal

—C≡C—Metal NO_2—C(F)—NO_2

—C≡C—X

>C(N)N   >N—Metal

>CN_2 >N—N=O

≥C—N=O   >N—NO_2

≥C—NO_2 ≥C—N=N—C≤

>C(NO_2)(NO_2)   ≥C—N=N—O—C≤

≥C—O—N=O   ≥C—N=N—S—C≤

≥C=O—NO_2 ≥C—N=N—O—N=N—C≤

>C—C< (epoxide, O)   ≥C—N=N—S—N=N—C≤

Figure 8.2 *Bond groupings known by experience to be unstable or explosive (Kohlbrand, 1985) (Courtesy of the American Institute of Chemical Engineers)*

Accounts of the evaluation of instability of materials include those given in *Guidance Notes on Chemical Reaction Hazard Analysis* by the Association of the British Petroleum Industry (ABPI) (1981) and its successor *Guidelines for Chemical Reaction Hazard Evaluation* ABPI (1989) and by Coates and Riddell (1981a), N. Gibson, Harper and Rogers (1985) and Cronin, Nolan and Barton (1987a).

8.5.1 Thermal stability

Identification of instability of a single component should be based on a coherent strategy. This will normally involve 'desk' screening, preliminary tests and substantial tests. If an exotherm is identified, tests may be done to characterize it. Other aspects may also need to be investigated such as the extent of gas evolution and the behaviour of the substance under conditions of processing such as drying.

8.5.2 Desk screening

Initial desk screening will provide information on the physical and chemical properties and the reactivity of the chemical. Where specific data are lacking, features such as group structure, thermodynamic data, oxygen balance, and reactivity of analogous substances provide pointers. Discussions of this aspect are given by Snyder (1965) and by the ABPI (1981, 1989).

The starting point should always be a thorough search to determine the data already available on the properties of the chemical. Material frequently mentioned in this connection includes: Sax (1957–), Kirk-Othmer (1963–, 1978–, 1991–) and Bretherick (1975–, 1987a); NFPA 49 and 491M; FM data sheets 7-19 and 7-23; and CMA (MCA) case histories.

A general method of estimating reactivity and instability is given in NFPA: 1994 49 *Hazardous Chemicals Data*, which gives materials a Chemical Reactivity Rating as shown in Table 8.6. The code lists the rating for a large number of chemicals. There are certain groups which are known by experience to be unstable or explosive, as shown in Figure 8.2 (Kohlbrand, 1985).

A substance may release energy by combustion in air or by decomposition; the main hazard is decomposition. Rapid energy release occurs if the constitution of a substance is such that its carbon and hydrogen are able to react with its own oxygen without needing to obtain oxygen from the surrounding air. If there is just enough oxygen to give a stoichiometric reaction of all the carbon to carbon dioxide and hydrogen to water, there is said to be a zero oxygen balance. The more reactive substances, such as explosives, typically contain enough oxygen to give such decomposition.

The oxygen balance (OB) is an important indicator of stability. For an organic compound of formula $C_xH_yO_z$ it has been defined by Lothrop and Handrick (1949) as:

$$OB = -1600(2x + y/2 - z)/M \qquad [8.5.1]$$

where M is the molecular weight. Thus an unstable, or explosive, material having perfect balance to yield carbon dioxide and water has a zero OB value, one lacking sufficient oxygen has a negative OB value, and one containing excess oxygen has a positive OB value.

In principle, therefore, the ideal OB for a substance suitable as an explosive would appear to be zero. Most explosives actually have a negative OB, but Lothrop and Handrick did find that the power of explosives increases as the OB increases and approaches a value of zero. Values of the oxygen balance for some materials of recognized instability are nitrobenzene (– 163), dinitrotoluene (– 114) and glycerol trinitrate (+ 3.5).

An attempt has been made by Stull (1970) to find some theoretical basis for reactivity. With unstable materials there is relatively little difference between the heat of decomposition ΔH_d and the heat of combustion ΔH_o under stoichiometric conditions, or between the adiabatic decomposition and combustion temperatures. Thus Stull gives an empirical correlation between the difference between these two heats $\Delta H_o - \Delta H_d$, or between the temperatures and the NFPA Chemical Reactivity Rating.

Table 8.7 *Energy hazard potential criteria in CHETAH (after Treweek, Claydon and Seaton, 1973)*

Criterion	*Energy hazard potential*		
	Low	*Medium*	*High*
1	$\Delta H_d > -0.3$	$-0.7 < \Delta H_d < -0.3$	$\Delta H_d < -0.7$
2	See Figure 8.3		
3	$OB > 240$	$120 < OB < 240$	$-80 < OB < 120$
	$OB < -160$	$-160 < OB < -80$	
4	$y < 30$	$30 < y < 110$	$y > 110$

In a refinement of this work, Stull (1973) has introduced kinetic as well as equilibrium considerations and has derived an empirical Reaction Hazard Index (RHI):

$$RHI = \frac{10T_d}{T_d + 30E_a} \qquad [8.5.2]$$

where E_a is the activation energy (kcal/mol), and T_d is the adiabatic decomposition temperature (K). The RHI correlates broadly with the Chemical Reactivity Index.

There are a number of thermodynamic schemes, and corresponding computer programs, which have been developed to calculate the equilibrium products and heats of combustion and/or decomposition reactions. These include those of Cruise (1964), of the National Aeronautics and Space Administration (NASA) (Gordon and McBride, 1971, 1976) and of the American Society for Testing of Materials (ASTM) (E.J. Davis and Ake, 1973; Treweek, Claydon and Seaton, 1973). These programs can then be used to determine the energy hazard potential of materials.

The ASTM program is named CHETAH (CHEmical Thermodynamics And energy Hazard evaluation). The program calculates four quantities which may be taken as criteria of energy hazard potential (Treweek, Claydon and Seaton, 1973):

(1) heat of decomposition ΔH_o;
(2) heat of combustion $\Delta H_o -$ heat of decomposition ΔH_d;
(3) oxygen balance;
(4) y criterion.

The y criterion is given by:

$$y = 10\Delta H_d^2 M \qquad [8.5.3]$$

The energy hazard potentials are related to criteria (1)–(4) as shown in Table 8.7 (see Figure 8.3). These criteria correlate reasonably well with data on shock sensitivity.

An account of the updating of the CHETAH program has been given by Frurip, Freedman and Hertel, (1989). CHETAH is now available as an interactive PC program. It includes an interpreter subroutine which uses pattern recognition techniques to combine the four hazard criteria into an overall hazard rating and a data bank.

8.5.3 Preliminary tests

Considering first the preliminary tests, the starting point is usually to test for the occurrence of an exothermic reaction when the substance is heated. A simple if crude test is the capillary melting point tube test in which the sample is heated up at a constant rate of about 10°C/min in a standard melting point apparatus. A result such as that shown in Figure 8.4(a), in which the sample temperature has exceeded that of the heating medium, indicates that there is an exotherm.

There are a number of tests involving the use of a Dewar flask. These are of three types: (1) ramped temperature tests, (2) differential temperature tests and (3) adiabatic tests.

In the first type of Dewar flask test the sample is held in a tube in a heat transfer medium in a Dewar flask and the temperature of the medium is increased at a constant rate, or ramped; a typical rate of temperature increase is 2°C/min. If at a given medium temperature the sample temperature rises above that of the medium, as shown in Figure 8.4(a), there is an exotherm. A plot of this kind is given by N. Gibson (1984).

Alternatively, the same apparatus may be used, but the temperature of the heating medium may be increased in a series of steps, holding the temperature constant for a period at each step; a typical increment is 10°C. If at a given medium temperature the sample temperature rises, there is an exotherm. Figure 8.4(b) illustrates a result which shows that an exotherm occurs at 90°C.

These two tests do not appear to have a specific name, other than the rather ambiguous 'Dewar flask' test. They are referred to here as the 'ramped heating' and 'stepped heating' Dewar flask tests, respectively.

In the second type of Dewar flask test, the procedure is to put the sample and a reference material in two tubes in a Dewar flask filled with heat transfer medium,

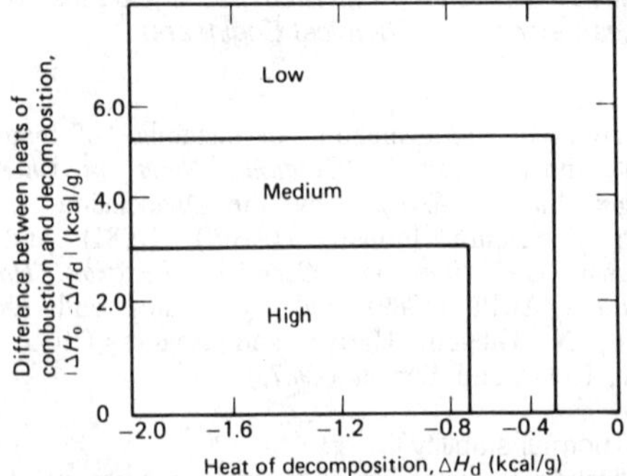

Figure 8.3 *Energy hazard potential as a function of the heats of combustion and decomposition (after Treweek, Claydon and Seaton, 1973) (Courtesy of the American Institute of Chemical Engineers)*

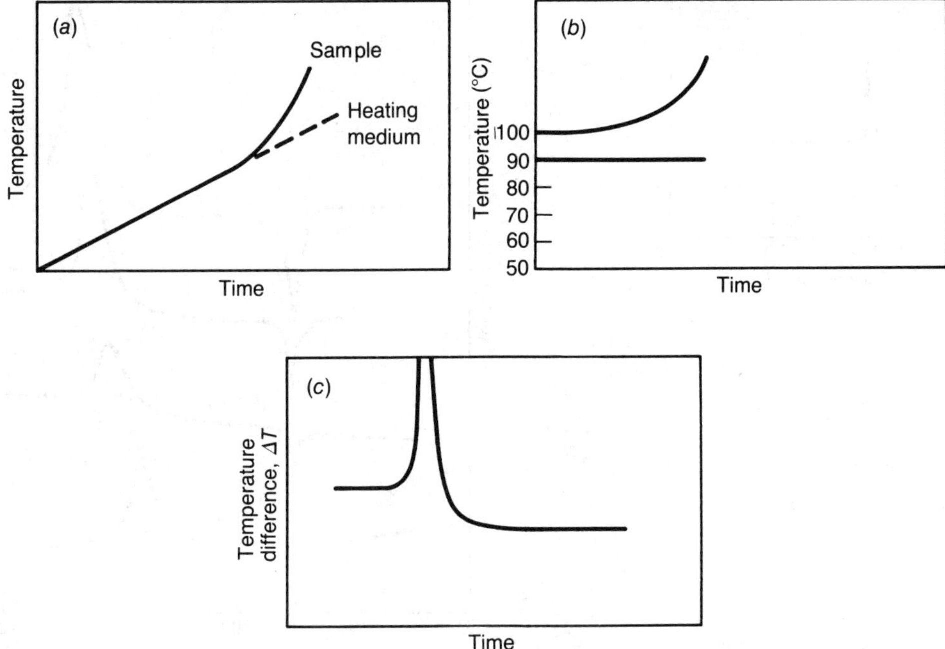

Figure 8.4 *Exotherm profiles in tests for thermal stability of individual substances: (a) sample and heating medium temperatures in melting point tube test or Dewar flask ramped temperature test; (b) sample temperature in Dewar flask stepped temperature test; (c) temperature difference between sample and reference material in Dewar flask dynamic heating test*

to heat the medium up at a constant rate and to measure the temperature difference between the sample and the reference material. A plot is made of the temperature difference ΔT between the sample and the reference material vs the temperature T of the heat transfer medium. A typical result for a material is illustrated in Figure 8.4(c). This test is referred to by the ABPI (1981), which gives details and plots, as the dynamic heating test, and by Coates and Riddell (1981a) as the 'simple exotherm test' and the 'dynamic heating test'.

The third type of Dewar flask test is the adiabatic test. The sample is held in a Dewar flask in an oven, the temperature of the oven is controlled to follow closely that of the sample, and any exotherm is observed. The test is described by T.K. Wright and Rogers (1986).

An alternative to the Dewar flask for some tests is an open or a closed tube. A Carius tube is often used. If gas evolution occurs, a closed tube is more suitable. It is desirable to establish the extent of any gas evolution in the initial screening. This may be done by a pressure–time test using a closed Carius tube.

A closed tube test may be used to detect onset of an exotherm and gas evolution. The sample is held in the tube in an oven and the temperature of the oven is ramped up, a typical rate being 2°C/min. The test is described by N. Gibson, Rogers and Wright (1987).

A ramped heating test using a tube held in an oven is described by N. Gibson (1984). He refers to it as the Imperial Chemical Industries (ICI) programmed screening test.

Another test is the delayed onset detection test. The sample, held in a lagged Carius tube placed in an isothermal oven, is maintained about 10°C above the oven temperature using a small electric heater. An exotherm is detected by a fall in the electrical power requirement. The test is described by the ABPI (1981). The tests just outlined can be performed using in-house apparatus and avoid the use of specialist instrumentation, which can be expensive.

If there is reason to suspect that the substance has the potential to explode, it may be subjected to tests for explosibility. These include an impact sensitivity test such as the drop weight test and a detonatability test. Explosibility tests are described below.

A detailed account of initial screening tests, of the precautions to be taken and of the interpretation of the results is given by Snyder (1965).

8.5.4 Hot storage tests

It may also be necessary to test the behaviour of the material when held at a high temperature for a long period under adiabatic conditions. In the hot storage test the sample is held in a container in a Dewar flask under quasi-adiabatic conditions and any exotherm is recorded. The test period is usually at least 7 days. The test is also referred to as the heat accumulation test. The hot storage test is carried out if the material is to be held on the plant at a temperature relatively close to the exotherm onset temperature T_e, say at $T_e - 50$°C.

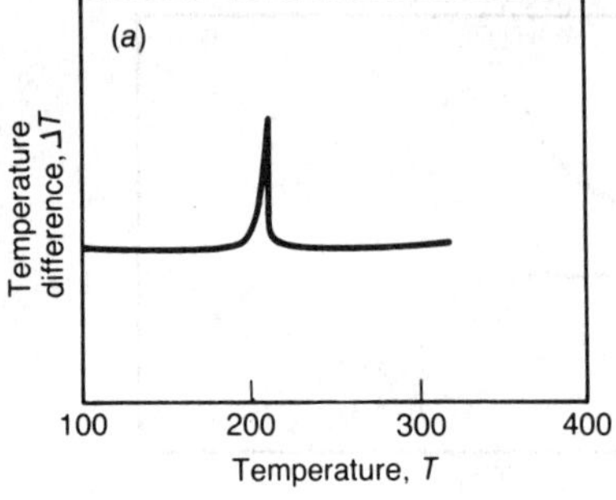

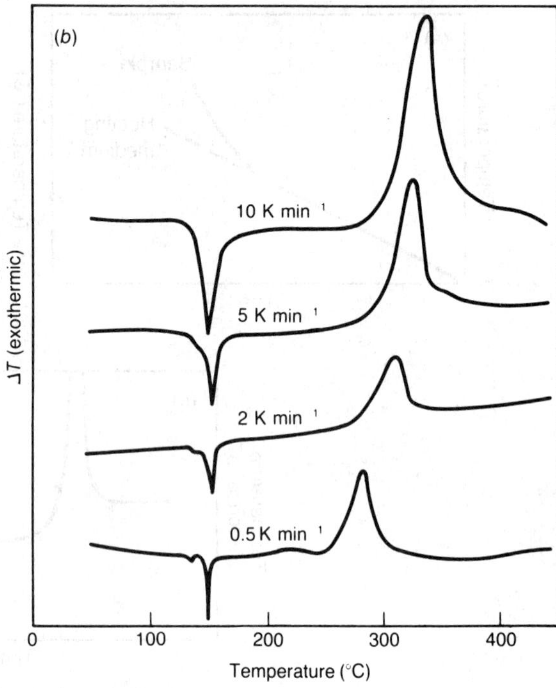

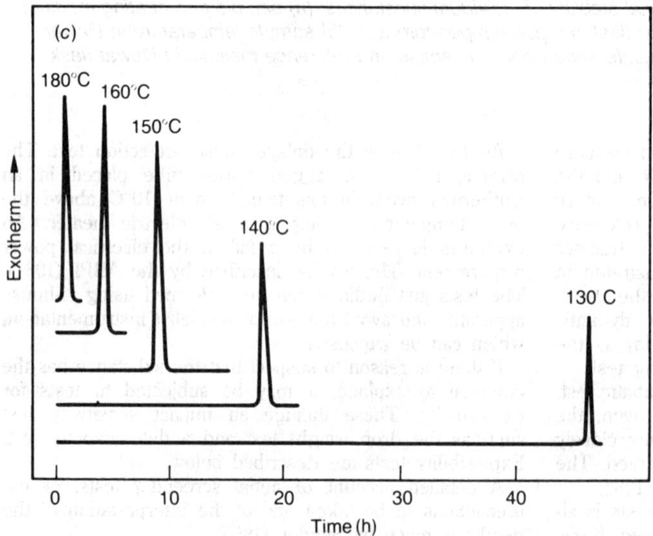

***Figure 8.5** Exotherm profiles in tests for thermal stability of individual substances using differential thermal analysis (DTA): (a) conventional ramped temperature test; (b) ramped temperature tests, showing effect of rate of temperature increase (Grewer et al., 1989) (c) isothermal tests, showing effect of time-dependent material stability (Duval, 1985) (Courtesy of the American Chemical Society)*

8.5.5 Tests using special instruments and calorimeters

There are a number of rather more elaborate techniques which may be employed to detect an exotherm and for which special instrumentation has been developed. The principal techniques are differential thermal analysis (DTA), differential scanning calorimetry (DSC) and accelerating rate calorimetry (ARC).

In DTA the sample is held in a tube in a vessel surrounded by a heat transfer medium and is heated at a constant rate. A typical heating rate is 2°C/min. A plot is made of the temperature difference ΔT between the sample and the heat transfer medium vs the temperature T of the latter. A typical result for a material giving an exotherm at 200°C is illustrated in Figure 8.5(a). The use

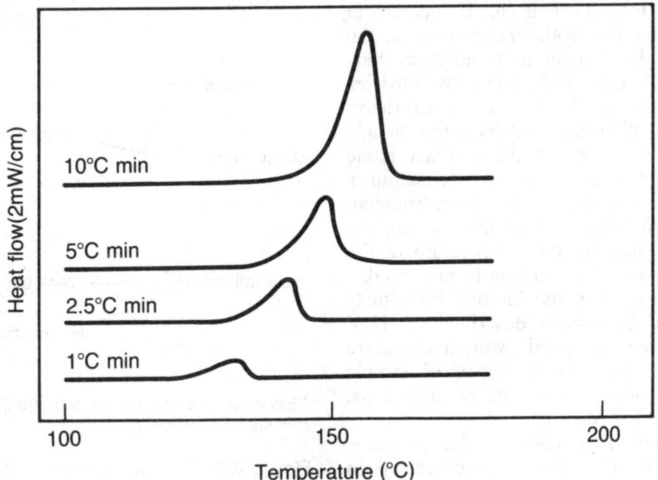

Figure 8.6 *Exotherm profile in test for thermal stability of individual substances using differential scanning calorimetry (DSC) (Duval, 1985) (Courtesy of the American Chemical Society)*

of DTA to detect thermal instability is described by Silver (1967) and Coffee (1969).

DTA may be carried out with open or closed tubes, the latter being more appropriate where gas evolution occurs.

In conventional DTA, the temperature is programmed to rise at a constant rate. Figure 8.5(b) (Grewer *et al.*, 1989) illustrates a set of traces obtained for different heating rates.

Where the thermal stability of a substance may be a function of time, DTA may be performed isothermally. Figure 8.5(c) (Duval, 1985) shows a set of traces for DTA of a substance exhibiting time-dependent thermal instability. The use of isothermal DTA is described by Schulz, Pilz and Schacke (1983) and by Duval (1985).

DTA may also be performed using in-house apparatus. Cronin, Nolan and Barton (1987) have described an 'insulated exotherm test' which is based on DTA.

In DSC the sample and a reference material are held in tubes in a vessel and are heated at a constant rate. The heating rate is typically 5–10°C/min. The heating is carried out by a control system which maintains the sample and the reference material at the same temperature. The variation in the heat which has to be supplied to the sample to keep it at the same temperature as the reference material gives a quantitative measure of any exotherm in the sample. A plot is made of the rate of change of heat input to the sample with time dH/dt vs the temperature T of the reference sample. Again, open or closed tube methods may be used. An account of DSC is given by Wendlandt (1986). Figure 8.6 (Duval, 1985) shows a typical DSC plot.

In ARC the sample is held in a bomb calorimeter under adiabatic conditions. ARC may be done in different ways, but one of the most common is the 'heat, wait and search' method. The temperature is increased in steps, a typical increment being 10°C. The sample is then held for a period at that temperature. The rate of any temperature increase is observed to see if it exceeds a set value, typically about 0.02°C/min. If it does not, the procedure is repeated. If the temperature rises at a high rate, there is an exotherm. Accounts of ARC are given by Coates (1984a,b) and Ottaway (1986). A typical result, given by N. Gibson *et al.* (1987), for a material giving an exotherm at about 80°C is shown in Figure 8.7. A widely used ARC instrument is the CSI-ARC instrument.

Another instrumental technique is differential thermogravimetry (DTG), but this is less widely used.

There are also available several special calorimeters developed for the study of thermal stability and reaction exotherms. One is the SIKAREX described by Hub (1976, 1977a,b). The calorimeter consists of a tube held

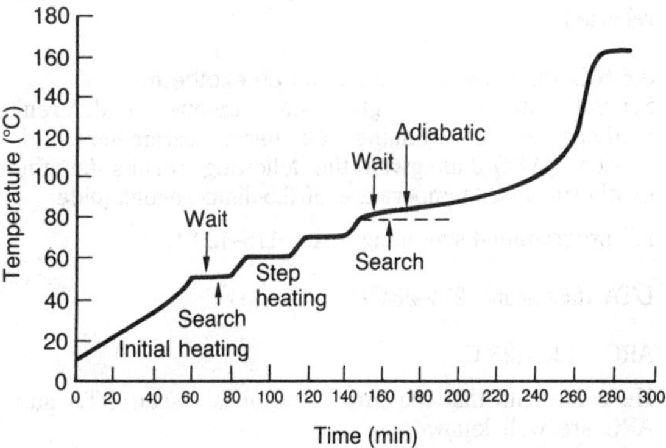

Figure 8.7 *Exotherm profile in test for thermal stability of individual substances using accelerating rate calorimetry (ARC) (N. Gibson, Rogers and Wright, 1987) (Courtesy of the Institution of Chemical Engineers)*

in a temperature controlled environment and with a heating element on the tube itself. It can be operatedin quasi-isothermal, adiabatic and isothermal modes. In the quasi-isothermal mode the sample is brought to temperature equilibrium with the environment, the environment temperature is increased and, if an exotherm occurs, the temperature difference between the sample and the environment is measured. In the adiabatic mode the environment is maintained at the same temperature as the sample. In the isothermal mode the environment is held at a temperature below that of the sample, the tube heater is used to maintain the temperature of the sample, and the heat required to do this is measured.

Another special calorimeter is the SEDEX (Sensitivity Detector of Exothermic Processes) described by Hakl (1981). This is an oven equipped with temperature control and adapted to take various types of sample container with inlet and outlet tubes, stirrers, and so on. The system may be used in various modes.

Some substances need to be tested under pressure. One reason for this is that a volatile substance can reach higher temperatures if held under pressure and thus the temperature range over which an exotherm may occur is extended. Another reason is that pressure may enhance an exotherm, especially in air. The conduct of DTA and DSC with the sample under pressure in a pressure cell and the use of a test scheme based on atmospheric DTA(DSC), pressure DTA(DSC) and a confinement test have been described by Seyler (1980). The confinement test is designed to measure the temperature at which a material under confinement will generate heat and pressure while subjected to a constant temperature increase of 1–2°C/min. In the scheme described by Seyler (illustrated in Figure 8.8), if atmospheric DTA(DSC) shows a moderate to severe exotherm, a thermal hazard is established; pressure DTA(DSC) may be bypassed but a confinement test is required. If atmospheric DTA(DSC) shows a weak exotherm or none, pressure DTA(DSC) is carried out and if a strong exotherm is detected, a confinement test is performed.

The information obtained in the various types of test varies. Some tests merely indicate that an exotherm exists, whilst others permit quantitative evaluation of the exotherm parameters such as the heat evolved. This information gives a quantitative measure of the heat released.

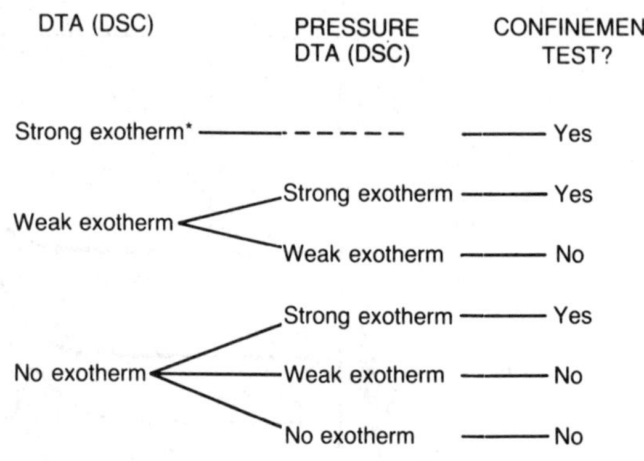

Figure 8.8 *Test scheme using pressure DTA/DSC to identify need for confinement test (Seyler, 1980) (Courtesy of Elsevier Science Publishers)*

8.5.6 Comparison of methods for an exotherm

Several authors have given comparisons of different methods of determining exotherm parameters. N. Gibson (1984) has given the following results for the exotherm onset temperature of 3,5-dinitro-*o*-toluamide:

ICI programmed screening test 115–120°C

DTA (fast scan) 274–284°C

ARC 120–125°C

He comments that the limitations of fast scan DTA and ARC are well known.

A comparison of exotherm onset temperatures for a large number of substances as determined by ARC and by other methods has been given by Fenlon (1984).

Hakl (1981) has given a comparison of exotherm onset temperatures for five substances determined by DSC and by SEDEX. The values determined by the former are consistently higher, the difference ranging from 25°C (80°C against 55°C) for *p*-xylylchloride with 0.02% Fe and 85°C (380°C against 295°C) for 1-nitroathrachinone.

8.5.7 Characterization of an exotherm

Some parameters which characterize an exotherm are the exotherm onset temperature, the heat of reaction, the reaction velocity constant and activation energy, the adiabatic temperature rise, the adiabatic self-heat rate and the adiabatic induction time. These parameters may be obtained from an adiabatic Dewar flask test or ARC test. The theoretical analysis is essentially similar to that for a process chemical reaction exotherm, as given in Chapter 11.

8.5.8 Other tests

Tests to detect and characterize an exotherm may need to be supplemented by other tests. Mention has already been made of explosibility/detonability and of deflagration tests.

A combustion test for preliminary screening is the train firing test which indicates the ability of a flame to propagate through the material in powder form. A train of finely ground material is laid on a heat insulating surface in a line and ignited. In the version described by ABPI (1981), ignition is by heated wire, in that described by N. Gibson, Harper and Rogers (1985) it is by a gas flame. The extent of fire spread is observed and categorized as shown in Table 8.8, the degree of hazard increasing with the rating number.

8.5.9 Tests for chemical reactions

The tests so far described are those applicable to individual substances. A somewhat similar set of tests is required for the evaluation of the hazard of conducting a chemical reaction in a reactor. In this case it is still necessary to determine the behaviour of the individual components – reactant, intermediates and products – but in addition it is necessary to examine the reaction as a

Table 8.8 *Interpretation of train firing test (after N. Gibson, Harper and Rogers 1986)*

Fire spread	*Result*	*Rating*
No	No ignition	1
No	Brief ignition, rapid extinction	2
No	Local combustion, at most smouldering, slight spread	3
Yes	Hot red glow or slow decomposition without flame	4
Yes	Burning like a firework or slow steady burning with flame	5
Yes	Very rapid combustion with flame or rapid decomposition without flame	6

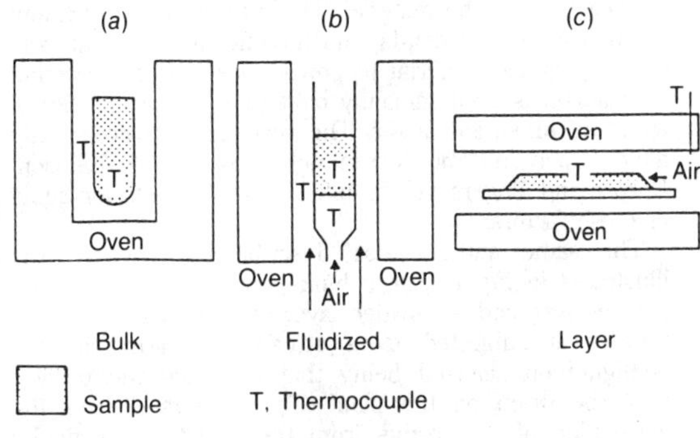

Figure 8.9 *Tests for thermal stability in driers: (a) bulk form; (b) fluidized form; and (c) layer form (after N. Gibson, 1984) (Courtesy of the Society of Chemical Industry)*

whole, including not only the main, desired reaction, but also secondary reactions.

For the conduct of a chemical reaction, however, it is desirable not only to obtain a qualitative understanding of the behaviour of the reaction, but also to obtain quantitative data on various parameters which are needed for the specification of the process. Discussion of chemical reaction hazard evaluation is therefore deferred until Chapter 11.

8.5.10 Tests for drying of powders

The main group of special tests which are performed are drying tests on powders.

The ABPI (1981) describes two tests, a 'through air' drying test in which air is passed through a bed, and an 'over air' test in which air is passed over a layer.

A more detailed account of the problem is given by N. Gibson, Harper and Rogers (1985, 1986). Some factors which affect the behaviour of a powder in drying include

(1) chemical composition;
(2) impurities;
(3) effective thermal conductivity;
(4) effect of air;
(5) time-dependent effects.

The material to be dried may be complex chemically, perhaps incorporating special effect agents. It may

Table 8.9 *Some test schemes for thermal stability of a single substance*[a]

Test	*Snyder (1965)*	*ABPI (1981)*	*Coates and Riddell (1981a)*	*Duval (1985)*	*N. Gibson (1984)*	*Cronin, and (1987)*	*Grewer et al. (1989)*	*ABPI (1989)*
Melting point tube	x[b]	x[b]						
Dewar flask:								
Hot holding	x[b]							
Dynamic heating		x[b]	x[b]					
Hot storage		x[b]	x[b]				x	x[b]
Pressure tube	x[b]				x	x	x	x[b]
Delayed onset								x[b]
DTA				x		x	x	
DSC				x				x[b]
ARC								x[b]
Thermal stability bomb	x[b]							
Explosion/ detonation	x[b]	x	x		x			x
Deflagration		x[b]	x				x	x
Drying		x[b]	x[b]		x			x

[a] The table gives the tests principally discussed by the authors; it should not be assumed that no other tests are done.
[b] These references give an account of the test apparatus and method.

contain impurities. Both of these factors may influence the reactivity of the material. The rate of heat loss if any reaction occurs depends on the effective thermal conductivity of the material in powder form. If the reaction mechanism is predominantly oxidation, the availability of air is an important factor. The reaction may occur only after a certain time, or induction period. N. Gibson, Harper and Rogers (1985, 1986) describe tests for some of these factors.

The same authors also describe the set of tests illustrated in Figure 8.9: a bulk powder test, an aerated powder test and a powder layer test. In each test the powder is subjected to a stream of warm air, the configuration selected being that in which the powder is to be dried on the plant. The authors discuss the application of the results from these tests to particular drying processes. This aspect is considered further in Chapter 11.

8.5.11. Test schemes

As already stated, test schemes for thermal instability of materials have been described by a number of authors. Table 8.9 summarizes some of the principal schemes. A test scheme given by Kohlbrand (1985) is shown in Figure 8.10(a) and another given by Cronin, Nolan and Burton (1987) is shown in Figure 8.10(b). Test schemes for powders have been given by N. Gibson, Harper and Rogers (1985, 1986), as shown in Figure 8.11. Test schemes for chemical reaction hazard evaluation are given in Chapter 11.

8.5.12 ABPI test scheme

The *Guidelines for Chemical Reaction Hazard Evaluation* (ABPI, 1989), although concerned with chemical reaction hazards, include as part of the overall scheme for evaluating such hazards a subscheme for the screening of the substances involved in the reaction: raw materials, intermediates and products. The overall scheme is illustrated in Figure 8.12. Its application to reaction hazard evaluation is described in Chapter 11. This scheme for individual substances covers (1) explosive properties, (2) unexpected decomposition and (3) maximum safe temperature for storage.

For explosive decomposition the scheme involves assessment of explosive properties from (1) chemical structure, (2) oxygen balance and (3) energy of decomposition, (4) preliminary small scale tests (e.g. DSC), and, if necessary, larger scale Koenen tube or Trauzl lead block tests.

For unexpected decomposition five tests are given:

(1) small scale testing using ramped temperatures –
  (a) DSC,
  (b) closed tube test;
(2) isothermal testing for delayed onset –

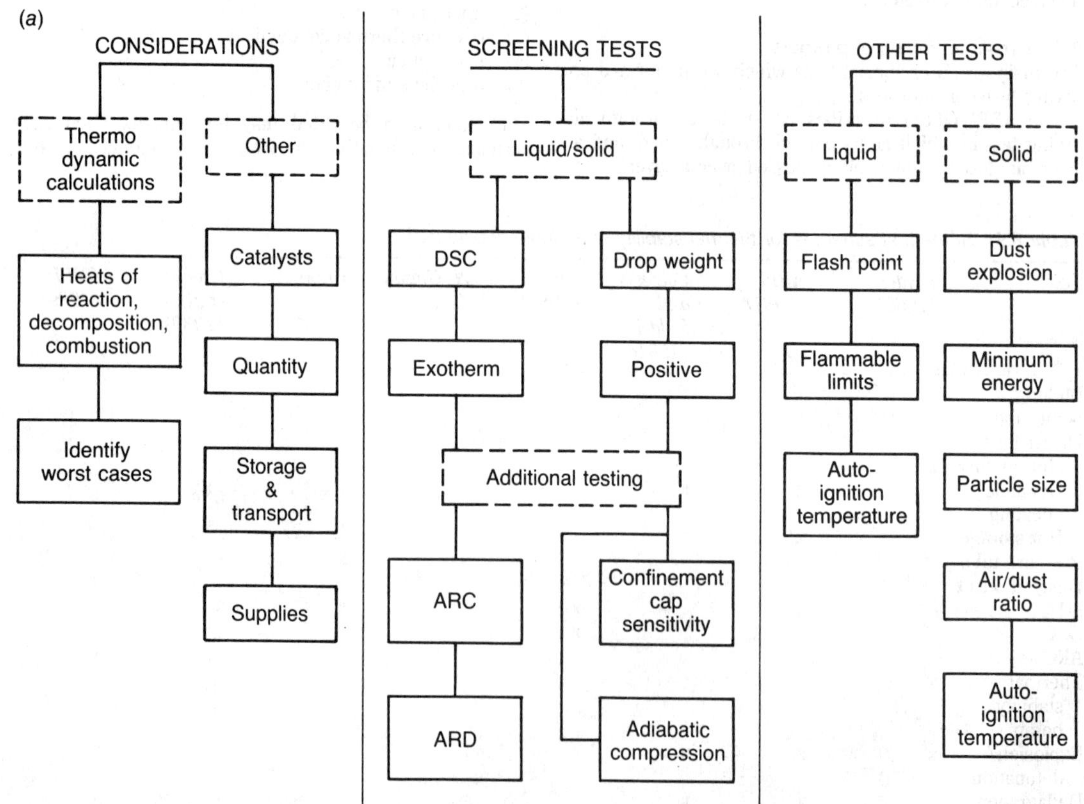

Figure 8.10 *Test schemes for thermal stability of a material: (a) test scheme of Kohlbrand (1985) (Courtesy of the American Institute of Chemical Engineers); and (b) test scheme of Cronin, Nolan and Burton (1987) (Courtesy of the Institution of Chemical Engineers)*

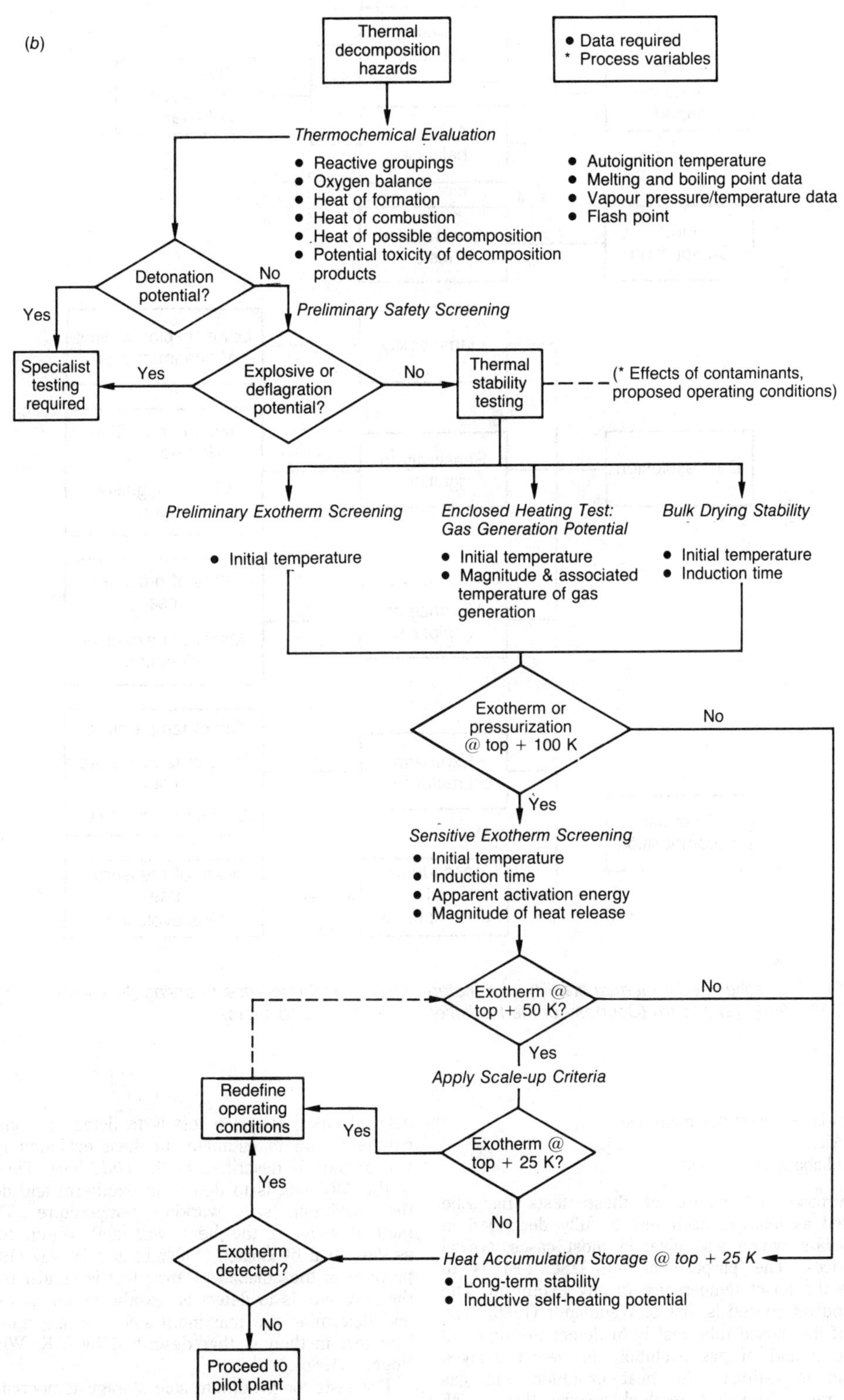

Figure 8.10 *continued*

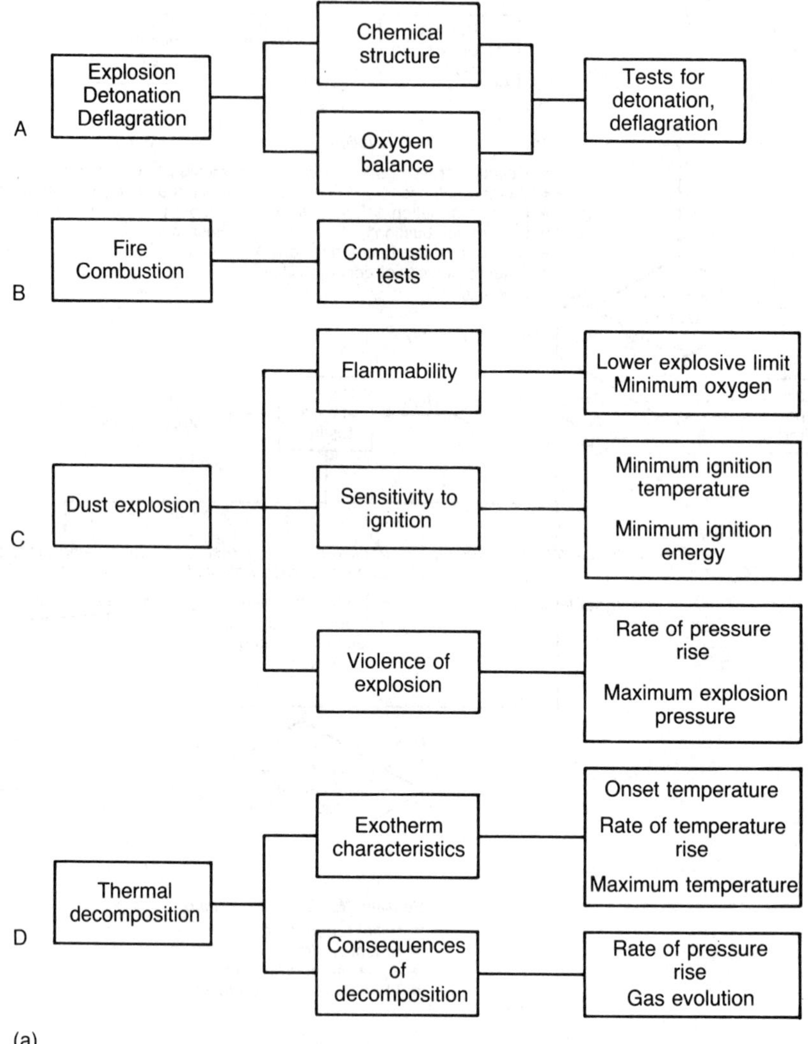

Figure 8.11 *Test schemes for thermal stability of a material with special reference to drying (N. Gibson, Harper and Rogers (1985, 1986): (a) and (b) (Courtesy of the Institution of Chemical Engineers)*

(a) delayed onset detection test,
(b) ARC,
(c) adiabatic Dewar test.

The purpose and nature of these tests may be summarized as follows. Each test is fully described in the *Guidelines*, which also give, in most cases, typical output traces. The purpose of the DSC test is to determine the onset temperature of any exotherm. The test method referenced is that of Wendlandt (1986). The purpose of the closed tube test is to detect the onset of an exotherm and of gas evolution, to detect delayed onset, and to estimate the heat evolution and gas evolution rates. The test method follows that of N. Gibson, Rogers and Wright (1987). The purpose of the delayed onset detection test is to detect the onset of an exotherm and to estimate the heat evolution rate. The test method is described in the *Guidelines*. The purpose of the ARC test is to detect an exotherm and determine the maximum safe working temperature. The ARC method given is the heat, wait and search technique, as described by Coates (1984a,b) and Ottway (1986). The purpose of the adiabatic Dewar test is similar to that for the ARC and is to detect an exotherm and gas evolution and determine the maximum safe working temperature. The test method is that described by T.K. Wright and Rogers (1986).

The tests for maximum safe storage temperature apply only to liquids. For solids the temperature may not be uniform and it is then necessary to treat the problem as

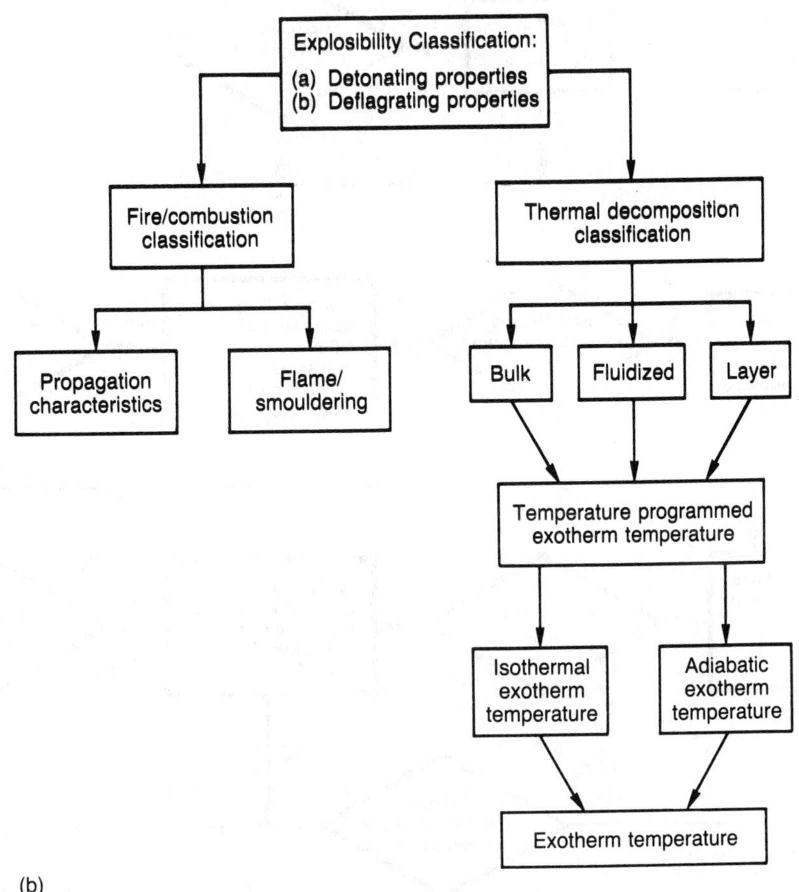

Figure 8.11 *continued*

one of self-heating. An account of self-heating is given in Chapter 16.

For powders, the *Guidelines* refer to the methods of N. Gibson, Harper and Rogers (1985).

8.6 Explosibility of Materials

There are a large number of tests for explosibility which, apart from their use to characterize explosives and to provide preliminary screening as described above, are used mainly, but not exclusively, to assure the safe transport of materials.

Explosibility testing facilities exist in many countries and each has tended to have its own mix of tests. However, the growing international trade in explosive substances is imposing on this scene an increasing degree of uniformity.

In the UK, the Ministry of Defence (MoD) has developed a large number of tests. The Waltham Abbey establishment has issued a *Manual of Tests* for explosive hazard assessment. The Royal Armament Research and Development Establishment (RARDE), EM2, also has a series of tests for explosives. The RARDE tests which are applied to industrial explosives have been described by V.J. Clancey (1974b). Tests intended to investigate the possible explosive characteristics of substances which are not regular explosives have been described by Connor (1974). The tests are used for liquids and solids, but not for gases. The Explosion and Flame Laboratory of the HSE has also

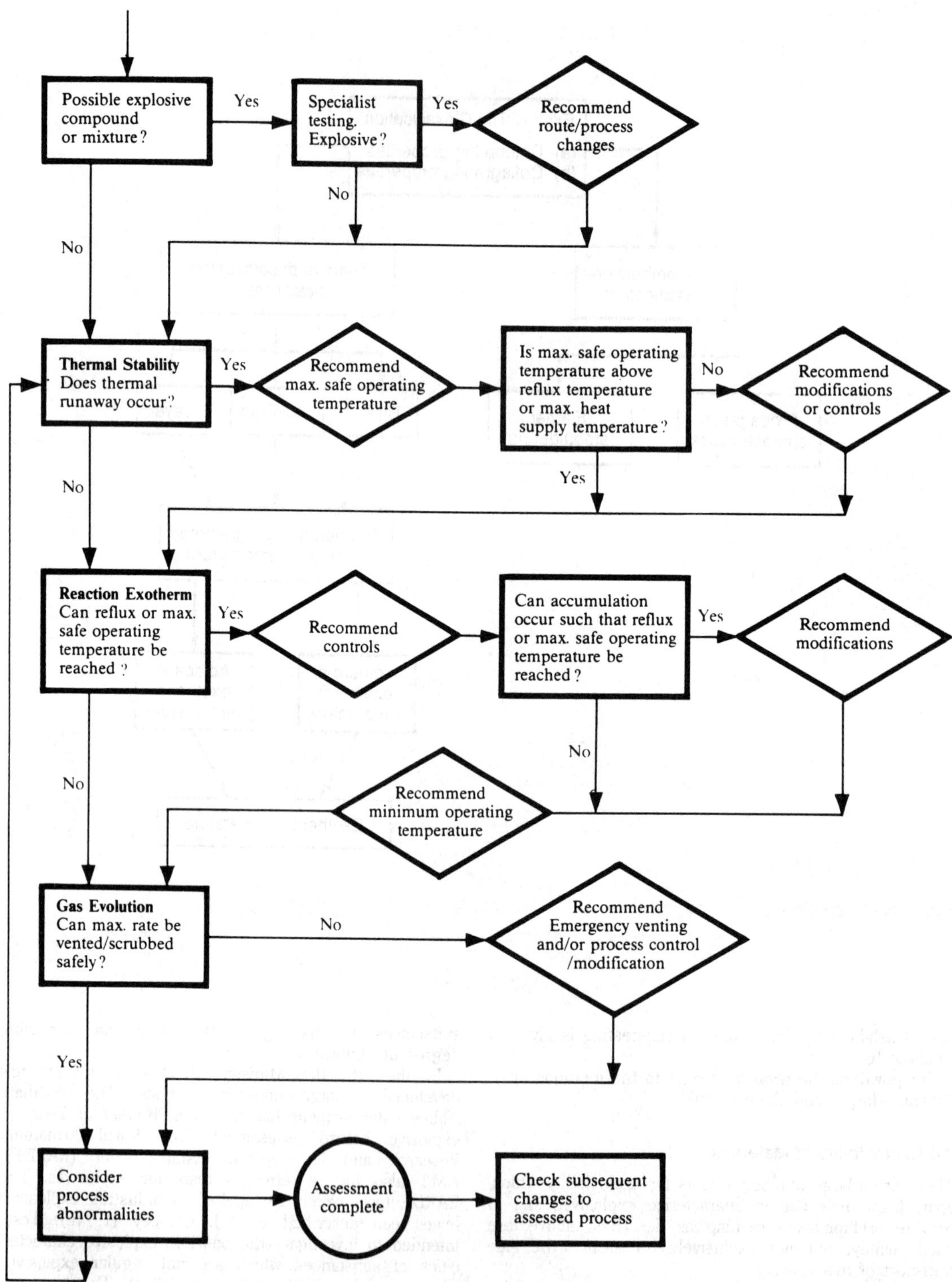

Figure 8.12 *Test scheme for assessment of thermal stability of reactant materials (ABPI, 1989).*

developed its own set of tests, which is described by Cutler (1986). Another set of tests which are widely used are those developed by the Bundesanstalt für Materialprüfung (BAM) in Germany.

Internationally, the development of tests has been stimulated by the requirements for uniform standards for the transport of hazardous materials as expressed in the United Nations (UN) classification codes and the International Maritime Organization (IMO) requirements at the international level and in European Community (EC) legislation and the RID and ADR requirements at the European level, as well as in the work of the OECD Group of Experts on the Explosion Risks of Unstable Substances (OECD-IGUS).

Testing for explosibility is not a simple matter. The tests are empirical ones. The mechanisms of explosion are sometimes unclear. This makes it difficult to extrapolate the results of small-scale tests to the full scale. The state of the sample can have a marked influence on the test result obtained. Water tends to have a marked effect, as may impurities.

In carrying out tests the objective is generally to obtain a positive result, i.e. an explosion, whenever possible. If necessary, fairly extreme conditions should be used. It may then be possible to arrange for the material to be handled safely under less extreme conditions. The tests seek to investigate the various aspects of explosive decomposition: initiation, nature, propagation and influencing factors.

Reviews of tests used in the UK and their interpretation have been given by V.J. Clancey (1974b), Connor (1974), Cutler (1987) and T.A. Roberts and Royle (1991).

It is convenient to begin by outlining the UN and EC classification systems and tests, and then to describe some of the individual tests.

8.6.1 UN transport classification tests: main test series

The UN system of tests is in support of the transport of potentially explosive materials. The governing code is the *Recommendations on the Transport of Dangerous Goods* (the UN *Transport Code*) of the UN (1988a, 1991). The tests in the main series and tests for self-accelerating decomposition temperature were given in a supplement to the earlier editions of the code *Recommendations on the Transport of Dangerous Goods: Tests and Criteria* (the UN *Tests and Criteria*) (UN, 1986) and tests for organic peroxides in a second supplement *Recommendations on the Transport of Dangerous Goods: Tests and Criteria, Addendum 1* (UN, 1988b). All three sets of tests are now given as Parts I–III, respectively, of the 1990 edition of *Tests and Criteria* (UN, 1990), which also contains in Part IV tests for fertilizers. Details of the UN *Transport Code* classification are given in Chapter 23. Here consideration is limited to explosives and substances with explosion potential.

The main series of tests is given in *Tests and Criteria*, Part I, and is summarized in Table 8.10. Sections A and B of the table give the objectives and general nature of the tests, and Section C gives the specific tests.

In the UN system, the category embracing explosives is Class 1. Substances and articles in this class are subdivided into the following Hazard Divisions:

1.1 substances and articles which have a mass explosion hazard;

Table 8.10 *UN transport classification tests: main test series and tests for self-accelerating decomposition temperature (UN, 1990)*

A Main test series: objectives of tests

Test Series 1 Is the substance explosive?
Test Series 2 Is the substance too insensitive for acceptance into Class 1?
Test Series 3 Is the substance too hazardous for transport (in the form in which it was tested)?
Test Series 4 Is the article, packaged article or packaged substance too hazardous for transport?
Test Series 5 Is the product a very insensitive explosive substance (with a mass explosion hazard)?
Test Series 6 Which hazard division should the product be assigned to? Should the product be excluded from Class 1?

B Main test series: general nature of tests

Test Series 1
(a) Shock tests with a defined booster and confinement to determine the ability of the substance to propagate a detonation
(b) Combustion or thermal tests to determine the thermal response

Test Series 2
(a) Shock tests with a defined booster and confinement to determine the sensitivity to shock
(b) Combustion or thermal tests to determine the thermal sensitivity

Test Series 3
(a) Tests to determine the sensitiveness to impact
(b) Tests to determine the sensitiveness to friction, including impacted friction
(c) Tests to determine the thermal stability
(d) Tests to determine the response of the substance to flame, i.e. ease of deflagration to detonation transition in small quantities when unconfined

Test Series 4
(a) Tests to determine the thermal stability of the packaged substance, packaged article(s) or unpackaged article
(b) Tests to determine the effect of dropping the explosive from a height of a few metres

Test Series 5
(a) Shock tests to determine the sensitivity to detonation by a standard detonator
(b) Tests to determine the tendency of transition from deflagration to detonation
(c) Tests to determine if the substance, when in large quantities, explodes when subjected to a large fire
(d) Tests to determine if the substance ignites easily when subject to an incendiary spark

Test Series 6
(a) Test on a single package to determine:
- whether initiation or ignition in the package causes burning or explosion and whether burning or explosion is propagated in the package; and

– in what way the surroundings could be endangered by these effects
(b) Test on a stack of packages of an explosive product or a stack of non-packaged explosive articles to determine:
– whether burning or explosion in the stack is propagated from one package to another or from a non-packaged article to another; and
– in what way the surroundings could be endangered by this event
(c) Test on a stack of packages of an explosive product or a stack of non-packaged explosive articles to determine:
– how the packages or non-packaged articles in the stack behave when involved in an external fire simulating a realistic accident; and
- whether and in what way the surroundings are endangered by blast waves, heat radiation and/or fragment projection

C Main test series: specific tests

Test No.		*Country of origin*[a]
1(a)	(i) BAM 50/60 Steel Tube Test	D
	(ii) TNO 50/70 Steel Tube Test	NL
	(iii) Gap test for solids and liquids	USA
	(iv) Gap test for solids and liquids	F
1(b)	(i) Koenen Test[b]	D
	(ii) Internal Ignition Test	USA
	(iii) SCB[c] Test	USA
2(a)	(i) BAM 50/60 Steel Tube Test	D
	(ii) TNO 50/70 Steel Tube Test	NL
	(iii) Gap test for solids and liquids	USA
	(iv) Gap test for solids and liquids	F
2(b)	(i) Koenen Test	D
	(ii) Internal Ignition Test	USA
	(iii) Time/Pressure Test	GB
	(iv) SCB Test	USA
3(a)	(i) Bureau of Explosives Machine	USA
	(ii) BAM Fallhammer Test	D
	(iii) Rotter Test	UK
	(iv) 30 kg Fallhammer Test	F
	(v) Modified Type 12 Impact Tool	C
3(b)	(i) BAM Friction Apparatus	D
	(ii) Rotary Friction Test	GB
	(iii) ABL Friction Test	USA
3(c)	Thermal stability test at 75°C	USA
3(d)	(i) Small-Scale Burning Test	USA
	(ii) Small-Scale Burning Test	F
4(a)	Thermal Stability Test for Articles and Packaged Articles	USA
4(b)	(i) Steel Tube Drop Test for liquids	F
	(ii) 12-m Drop Test for Articles and Solid Substances	USA
5(a)	Cap Sensitivity Test	D/USA
5(b)	(i) DDT[d] Test	F
	(ii) DDT Test	USA
5(c)	External Fire Test for Hazard Division 1.5	–
5 (d)	Princess Incendiary Spark Test	GB
6(a)	Single Package Test	–
6(b)	Stack Test	–
6(c)	External Fire (Bonfire) Test	–

D Tests for self-accelerating decomposition temperature (SADT)

1. US SADT Test
2. Adiabatic Storage Test (ADT)
3. Isothermal Storage Test
4. Heat Accumulation Storage Test

[a] C, Canada; D, Germany; F, France; NL, The Netherlands.
[b] This test has often been referred to as the Koenen Steel Tube Test, but is now generally known simply as the Koenen Test.
[c] Small Scale Cook-Off Bomb.
[d] Deflagration to detonation transition.

1.2 substances and articles which have a projection hazard but not a mass explosion hazard;
1.3 substances and articles which have a fire hazard and either a minor blast hazard or a minor projection hazard, but not a mass explosion hazard;
1.4 substances and articles which present no significant hazard;
1.5 very insensitive substances which have mass explosion hazard;
1.6 extremely insensitive articles which do not have a mass explosion hazard.

Also relevant in relation to substances with explosive potential are the categories

4.1 flammable solids;
5.2 organic peroxides;
9 miscellaneous dangerous substances.

The main series of tests is concerned with classification of the substance or article in relation to Class 1. This proceeds in two stages: acceptance into Class 1 and assignment within Class 1.

Figure 8.13 gives an outline flow chart for the classification of a substance or article. It includes the procedure for accepting the item into Class 1. At the acceptance stage the item may be accepted into Class 1, rejected as too hazardous for transport, or rejected because it does not constitute a Class 1 hazard. If the item is accepted, it is then tested to assign it to a Hazard Division. It is also assigned to a Compatibility Group.

The procedure for acceptance into Class 1 is shown in more detail in the flow chart in Figure 8.14. At this point the acceptance is provisional. The item may still be rejected for Class 1 if the further information given by the assignment testing leads to this conclusion. The Test Series used for this procedure are 1–4. Test Series 1 determines whether the item is explosive, Test Series 2 whether it is too insensitive and should be rejected from Class 1, Test Series 3 whether it is too hazardous for transport and should be rejected from Class 1, and Test

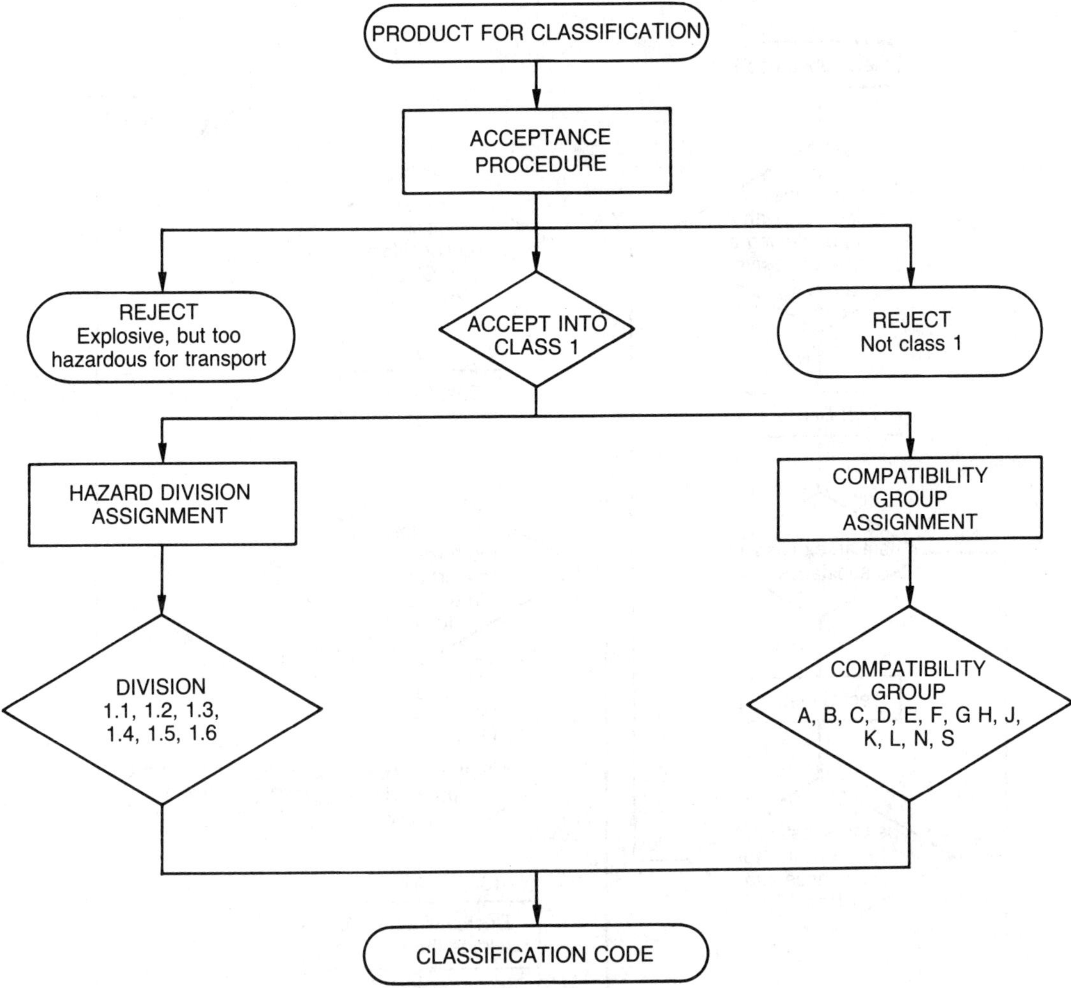

Figure 8.13 *UN transport classification tests: flow chart for classification of a substance or article as Class 1 (UN, 1991)*

Series 4 whether the item in packaged form is too hazardous for transport and should be rejected from Class 1.

Figure 8.15 shows the flow chart for assignment within Class 1. The Test Series used in this procedure are 5–7. Test Series 5 determines whether the assignment is to HD1.5, Test Series 6 whether the assignment is within HD1.1-HD1.4 or rejection from Class 1, and Test Series 7 the assignment to HD1.6.

The UN *Transport Code* also includes a flow chart for assigning self-reactive substances to Division 4.1.

8.6.2 UN transport classification tests: self-accelerating decomposition tests

The tests for determination of self-accelerating decomposition temperature (SADT) of organic peroxides and other thermally unstable substances are given in *Tests and Criteria*, Part II. They are summarized in Table 8.10, Section D.

8.6.3 UN transport classification tests: tests for organic peroxides

The tests for classification of organic peroxides (Class 5.2) are given in *Tests and Criteria*, Part III, and are summarized in Table 8.11. The flow chart for classification of organic peroxides is shown in Figure 8.16.

8.6.4 UN transport classification tests: tests for fertilizers

The tests for determination of self-sustaining exothermic decomposition of fertilizers containing nitrates are given in *Tests and Criteria*, Part IV.

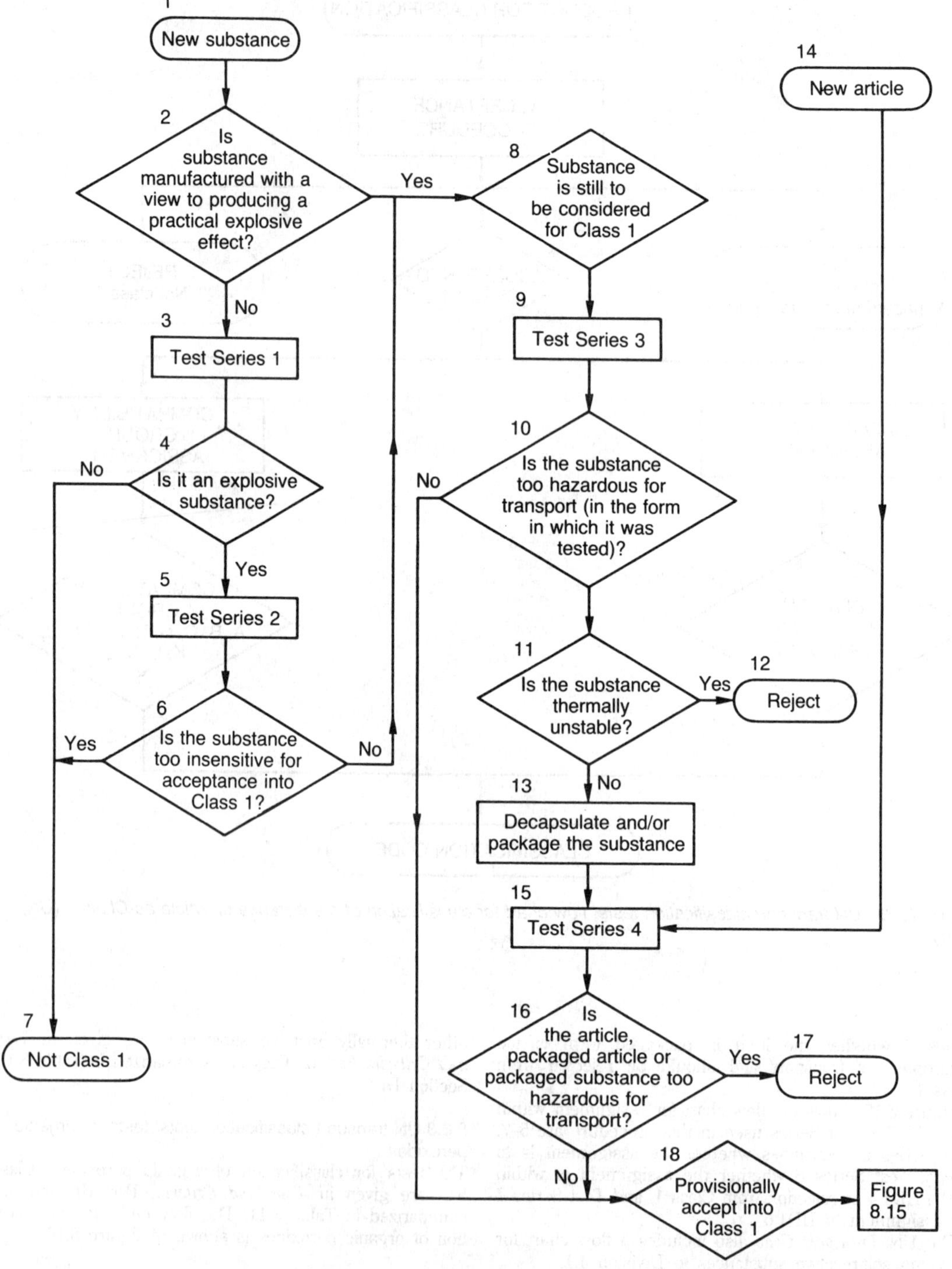

Figure 8.14 *UN transport classification tests: flow chart for acceptance of a substance or article into Class 1 (UN, 1990)*

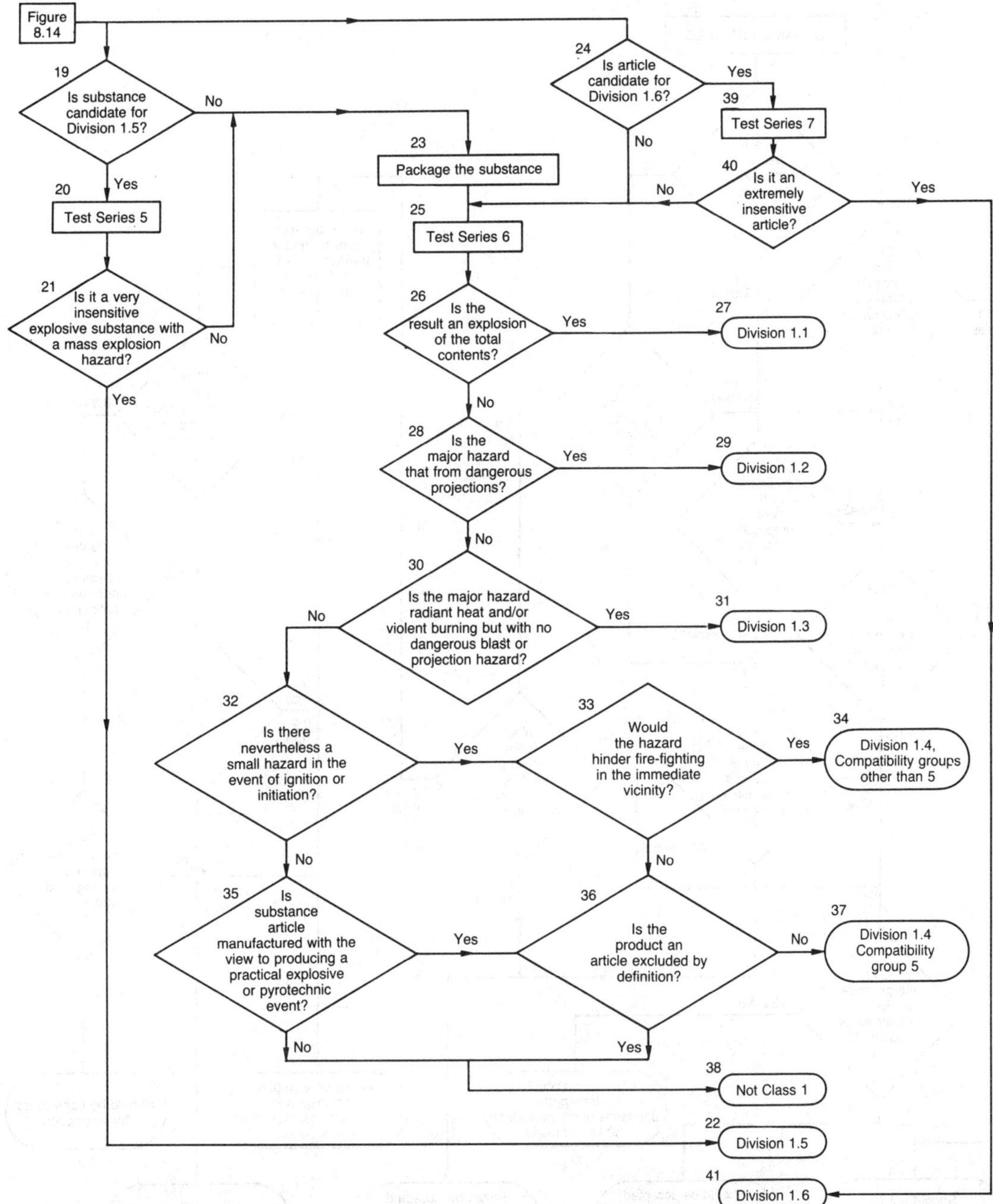

Figure 8.15 *UN transport classification tests: decision tree for assignment of a substance or article within Class 1 (UN, 1990)*

8.6.5 UN transport classification tests: details of tests

Tests and Criteria gives for each test the following details: an introduction, apparatus (including diagrams) and materials, procedure, criteria and method of assessment, and examples of results. The test results are typically of the following types:

(1) observation of state of containment;
(2) simple positive/negative result;
(3) positive/negative result at some specified value of a parameter;
(4) value of parameter at which specified proportion (typically 50%) of results are positive;

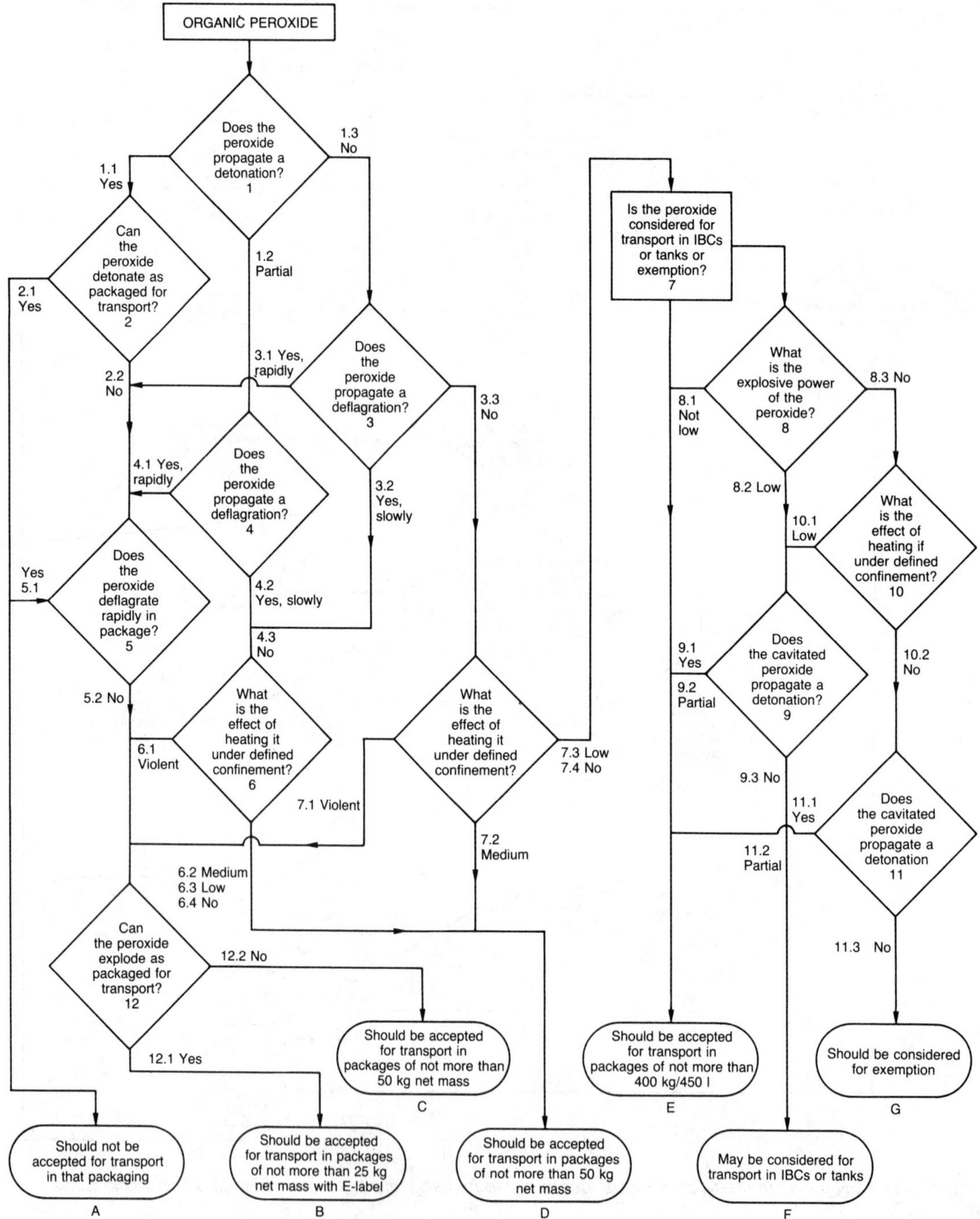

Figure 8.16 *UN transport classification tests: decision tree for classification of an organic peroxide (UN, 1991). IBC, intermediate bulk container*

Table 8.11 *UN transport classification tests: tests for organic peroxides (UN, 1990)*

	Country of origin[a]
Test Series A Detonation	
A.1 BAM 50/60 Steel Tube Test	D
A.2 TNO 50/70 Steel Tube Test	NL
A.3 Gap Test for Organic Peroxides	USA
A.4 Gap Test	F
Test Series B Detonation in package	
B.1 Detonation Test in Package	NL
Test Series C Deflagration	
C.1 Time/Pressure Test	GB
C.2 Deflagration Test	NL
Test Series D Deflagration in package	
D.1 Deflagration Test in Package	NL
Test series E Heating under confinement	
E.1 Koenen Test	D
E.2 Dutch Pressure Vessel Test	NL
E.3 US Pressure Vessel Test	USA
E.4 Thermal Explosion Vessel Test	NL
Test Series F Explosive power	
F.1 Ballistic Mortar Mk IIId Test	GB
F.2 Ballistic Mortar Test	F
F.3 BAM Trauzl Lead Block Test	D
F.4 Modified Trauzl Test for Organic Peroxides	USA
F.5 High Pressure Autoclave	NL
Test Series G Explosion in package	
G.1 Thermal Explosion Test in Package	NL
G.2 Organic Peroxide Package Test	USA

[a] D, Germany; F, France; NL, The Netherlands.

(5) quantitative measurement of value of one or more parameters.

Use may be made of the 'Bruceton staircase' technique. The value of the governing parameter is set well above the value at which a positive result is obtained. The value is then successively reduced until the event is just obtained. In some cases a single test is carried out, in others some other specified number. Where more than one test is required, the rules for determining how many tests constitute a positive result vary.

The following account gives a bare outline of the tests. In most cases the test conditions are specified in some detail. Details of the instrumentation are omitted, so are details of the safety precautions.

The form in which the sample is tested is relevant; essentially, a test applies only to that form. Some tests are more sensitive to this than others.

Part I

Test Series 1

In the BAM 50/60 Steel Tube Test (Test 1(a)(i)) the sample is placed in a steel tube and initiated using a detonator and a booster. The detonability of the substance is assessed from the fragmentation of the tube.

The TNO 50/70 Steel Tube Test (Test 1(a)(ii)) also involves a sample placed in a steel tube and initiated by a detonator and a booster. The detonability of the substance is assessed primarily from the fragmentation of the tube, but if even fragmentation of the tube is incomplete, the substance is still held to have detonated, provided that certain conditions on the velocity of propagation are satisfied. In this test, it is also possible to measure the relationship between the mass of booster and the velocity of propagation, though this is not part of the UN criteria.

In the US Gap Test for Solids and Liquids (Test 1(a)(iii)) the sample is placed in a vertical tube. A witness plate is put on the top of the cylinder and separated from it by thin spacers. A booster is put in the tube below the sample and separated from it by a small gap. The sample is initiated by a detonator and the booster. The detonability of the substance is confirmed if two of the following three events occur: (1) a stable velocity of propagation is attained exceeding the speed of sound in the substance; (2) a hole is punched through the witness plate; and (3) the sample tube is fragmented along its entire length.

In the Koenen Test (Test 1(b)(i)) the sample is placed in a steel tube with an orifice plate at one end and subjected to external heating. A set of orifice plates of different diameters is used. Eight modes of fragmentation are defined. No explosion is deemed to occur if the result is one of the first five, milder modes, and an explosion is deemed to occur if it is one of the last three, severer modes. The largest diameter at which an explosion occurs, the 'limiting diameter', is determined. A limiting diameter of 1–1.5 mm indicates that the substance has some thermal explosive properties and one of ≥ 2 mm that it is thermally sensitive. This test has a crucial role in the assignment scheme.

The Internal Ignition Test (Test 1(b)(ii)) is designed to determine the response of the substance to rapidly rising temperature. The sample is placed in a vertical steel tube and ignited by an igniter consisting of a small quantity of black powder located at the centre of the tube. The result is deemed positive if either the tube or at least one of the end caps is fragmented into two or more pieces.

The Small-scale Cook-off Bomb (SCB) Test (Test 1(b)(iii)) is designed to determine the response of the substance to slowly rising temperature. The sample is placed in a steel vessel with a witness plate below and heated electrically. The result is deemed positive if the bomb is ruptured or fragmented, or if the witness plate is deformed or punctured.

Test Series 2

Versions of each of the foregoing tests are used in Test Series 2. The BAM Steel Test Tube (2(a)(i)), the TNO Steel Tube Test (2(a)(ii)) and the US Gap Test for Liquids and Solids (2(a)(iii)) are similar to their Series 1 versions, except that in the first the booster is omitted, in the second a different booster is used and in the third the gap is different. In all three cases the test criteria are the same as in the Series 1 versions.

The Koenen Test (2(b)(i)) in Test Series 2 is similar to its Series 1 version. If the limiting diameter is < 2 mm the substance is deemed too thermally insensitive to be accepted into Class 1.

Test Series 2 contains a test additional to those in Test Series 1, the Time/Pressure Test (2(b)(iii)). In this test the sample is placed in a steel pressure vessel fitted with a bursting disc and is ignited by an internal pyrotechnic ignition system. The profile of pressure vs time is measured. The time for the gauge pressure to rise from 690 kPa (100 psig) to 2070 kPa (300 psig) is determined. The substance is deemed to undergo explosion by deflagration if the time interval between these two pressures is < 30 ms.

Test Series 3

In the BAM Fallhammer Test (3(a)(ii)) a drop weight is dropped onto an impact device containing the sample placed on an anvil. An event is deemed to occur if there is an explosion, flashes or flames. A series of trials is done with different impact energies and the lowest impact energy at which an event is obtained is recorded. If this energy is ≤ 2 J, the result is positive and the substance is deemed too sensitive for transport.

In the Rotter Test (3(a)(iii)) the sample is placed in a chamber and the quantity of gas evolved in any explosion is measured in a gas burette. The results obtained are expressed as a figure of insensitiveness (FoI) which is compared with the FoI of 80 of the reference substance RDX. This test is relevant particularly where a substance may be nipped between two hard surfaces.

In the BAM Friction Apparatus (3(b)(i)) the sample is spread on a porcelain plate which is then dragged under a fixed porcelain peg. An event is deemed to occur if there is an explosion, flashes or flames. The force on the peg is measured and a series of trials is done with different forces on the peg. If this force is < 80 N, the result is positive and the substance is deemed too sensitive for transport.

In the Rotary Friction Test (3(b)(ii)) the sample is held between a flat bar and a rotating wheel. An event is deemed to occur if there is an explosion, flames or smoke. The force on the wheel is kept constant and the angular velocity of the wheel is measured and a series of trials done with different angular velocities. The results obtained are expressed as a figure of friction (FoF) which is compared with the FoF of 3.0 of the reference substance RDX. If this FoF is ≤ 3.0, the result is positive and the substance is deemed too sensitive for transport.

In the Thermal Stability Test at 75°C (3(c)) the sample and an inert reference substance are placed in separate vessels in an oven held at 75°C for 48 h. The temperature difference is measured. The substance is deemed thermally unstable if ignition or explosion of the sample occurs or if the temperature difference is ≥ 3°C.

Test Series 5

In the Princess Incendiary Spark Test (5(d)) the sample is placed in vertical glass tube and subjected to sparks from a safety fuse. The result is deemed positive if the sample ignites, explodes or emits coloured fumes or odour.

Test Series 6

In the External Fire (Bonfire) Test (6(c)) a stack of packages of an explosive product or a stack of non-packaged articles is placed on a support and subjected to a firewood or liquid fuel fire beneath it. Depending on its behaviour in this test, the object is assigned to one of the Hazard Divisions (HD1.1–HD1.4), using the detailed criteria given in *Tests and Criteria.* The assignment to a Hazard Division using tests in Series 6 is illustrated in Figure 8.15.

Part II

Turning to tests given in *Tests and Criteria*, Part II, for self-accelerating decomposition temperature (SADT), this is defined as the lowest temperature at which self-accelerating decomposition occurs.

In the US SADT Test, the package is held in a test chamber and maintained at constant temperature for 7 days. The temperature in the oven is measured. A series of trials is done at different temperatures.

In the Adiabatic Storage Test (ADT) the sample is held in a Dewar flask in an oven at a constant temperature. The temperature of the sample is measured. Any temperature rise of the sample is measured and the corresponding heat generated is calculated. A profile of heat generated vs time is obtained.

In the Isothermal Storage Test (IST) the sample is held in an aluminium block, which is maintained at a constant temperature. The heat flow from the sample is measured. A profile of heat generated vs time is obtained.

In the Heat Accumulation Storage Test the sample is held in a test chamber and maintained at constant temperature for 7 days. The temperature in the oven is measured. A series of trials is done at different temperatures.

Part III

Moving on to tests for organic peroxides given in Part III, in the Dutch Pressure Vessel Test (E.2) the sample is placed in a vessel fitted with a bursting disc and with an orifice plate. The vessel is heated by a butane burner. The results are given in terms of the largest diameter orifice with which the bursting disc is ruptured. A series of trials is conducted using 10 g and 50 g of sample and orifice plates of different diameters. The largest diameter at which failure of the disc occurs, the 'limiting diameter', is determined. The results are classed as 'violent', 'medium or low explosions' or 'no'.

In the Thermal Explosion Vessel Test (E.4) a sample is placed in a sealed steel vessel and heated on a hot plate. The temperature and pressure in the vessel are measured. The result is expressed in terms of the product of the maximum pressure (MPa) and the maximum rate of pressure rise (MPa/s). The results are classed as 'violent', 'medium or low explosions' or 'no'.

In the Ballistic Mortar Mk IIId Test (F.1) the sample is placed heavily confined in a tube suspended on a rope and ignited by a detonator. The recoil of the mortar is measured and compared with that for the reference substance picric acid. The explosive power is classed as 'not low', 'low' and 'no' for explosive powers > 7, 1–7% and ≤ 1% of picric acid, respectively.

In the BAM Trauzl Lead Block Test (F.3) the sample is placed in a lead block and ignited by a detonator. The expansion of the lead block is measured. The explosive power is classed as 'not low', 'low' or 'no' for expansions

Table 8.12 *UN transport classification tests: illustrative example (after T.A. Roberts and Royle, 1991) (Courtesy of the Institution of Chemical Engineers)*

Test	*Substance*[a]						
	AN	*AP*	*AC*	*AZDN*	*DNB*	*NM*	*77% TNP*
BAM Fallhammer	>50	5[b]	>50	3[c]	20[c]	40[b]	>50
BAM Friction Test	>363	>363	>363	363[c]	>363	ND[d]	>363
Ballistic Mortar Test	Not low	Not low	No	Low	Not low	Not low	No
HP Autoclave	Low	PR[e]	Low	Low	VR[f]	Not low	VR
BAM 50/60 Tube Test - 1[g]	Partial	Detn[h]	No	Defn[i]	Detn	Detn	Detn
BAM 50/60 Tube Test - 2[j]	ND	No	ND	ND	Detn	Partial?	No
Time/Pressure Test	No	Yes, rapidly	Yes, slowly	Yes, rapidly	No	Yes, slowly	No
Deflagration Test	No	No	Yes, slowly	Yes, slowly	No	No	No
Koenen Test	Low	Violent	Medium	Medium	No	No	Violent
DPV Test	Low	Medium	Low	Violent	Low	No	Medium
Thermal Explosion Vessel Test	Low	Violent	Medium	Medium	No[k]	No	Violent
Assessment:							
Thermally stable?	No[l]	No[l]	No[l]	Yes	No[l]	No[l]	No[l]
Too hazardous?	No	No	No	Yes[m]	No	No	No
Too insensitive?	Yes	No	Yes	No	No	Yes	No
Explosive substance[n]?	Yes	Yes	Yes	Yes	Yes	Yes	Yes

[a] AN, ammonium nitrate; AP, ammonium perchlorate; AC, azodicarbonamide; AZDN, 2,2'-azodi(isobutyronitrile); DNB, 1,3-dinitrobenzene; NM, nitromethane; TNP, 2,4,6-trinitrophenol.
[b] Report.
[c] Smoke.
[d] Not done.
[e] Partial reaction.
[f] Violent reaction.
[g] Test with detonator and booster.
[h] Detonation.
[i] Deflagration.
[j] Test with detonator only.
[k] No effect at 320°C.
[l] Not tested – result estimated from DSC results and literature values.
[m] As a Class 1 substance, but not as a self-reactive substance.
[n] Some explosive properties, but not necessarily an explosive.

of the lead block with a 10 g sample mass of ≥ 25 cm^3, 10–25 cm^3 and < 10 cm^3, respectively.

In the High Pressure Autoclave Test (F.5) the sample is placed in a steel pressure vessel and is heated by an electrical resistance wire inside the vessel. The profile of pressure vs time is measured. The specific energy is determined from the maximum pressure rise. The explosive power is classed as 'not low', 'low' or 'no' for values of the specific energy of > 100 J/g, 5–100 J/g and < 5 J/g, respectively.

Part IV

Finally, for the determination of the self-sustaining exothermic decomposition of fertilizers containing nitrates there is the Trough Test. In this test localized decomposition is initiated in a sample held in a horizontal trough and the propagation of the decomposition is observed. The substance is considered capable of self-sustained decomposition if, but only if, the propagation of decomposition continues throughout the entire trough.

8.6.6 EC transport classification tests

EC Directive 84/449/EEC specifies tests in support of Directive 67/548/EEC on classification, packaging and labelling of dangerous substances. The tests relevant here are the test series given in the Annex, Part A, Section A.14 Explosive Properties.

Three tests are specified: (a) thermal sensitivity, (b) mechanical sensitivity with respect to shock and (c) mechanical sensitivity with respect to friction. The thermal sensitivity test is a steel tube test, the shock sensitivity test is a falling mass test, and the friction sensitivity test is a rod and plate test.

8.6.7 Some tests used by the HSE: overview

A number of other tests used by the HSE have been described by Connor (1974), Cutler (1986) and T.A. Roberts and Royle (1991). Connor mentions a cartridge case test for detonation, a drop weight test for impact sensitiveness and a pendulum test for friction sensitiveness, but these tests are evidently obsolete.

For detonability use is made of the 50/60 and 50/70 Steel Tube Tests. Other steel tube tests may also be used, sometimes on an *ad hoc* basis to investigate particular problems.

For explosive power there is a family of ballistic mortar tests of which the Mk IIId is the UN test. The Trauzl Lead Block Test may also be used, but tends to be expensive, although there is a smaller and more economic US version. In any event, there is relatively good correlation between results from the ballistic mortar and lead block tests, and it is generally more convenient to use the former.

For thermal initiation by direct ignition sources use is made of an electric spark test and of the Princess Incendiary Spark Test.

The ease of ignition of the material by sparks, hot surfaces and flames is tested. One test is the Princess Incendiary Spark Test. Another test is an electric spark test with energy levels of 4.5, 0.45 and 0.045 J. The results obtained are a relative rather than an absolute guide. If ignition occurs at the lowest of these energy levels, further testing for the hazard of static electricity is indicated.

Use is made of train tests in which material is laid out in a shaped train and ignited at one end. In an EC version of the test the material is deemed flammable if the flame propagates. In the UN version one end is wetted, and if the flame crosses into the wetted zone further tests are indicated. Another MoD train test is no longer used.

For the flammability of liquids the basic test is the flash point. There is also a trough test, developed originally by the MoD.

For thermal initiation by indirect heating, tests used include the Koenen Test and the Time/Pressure Test. Other tests which have been used are the RARDE time/temperature/pressure test and the steel box test. In the RARDE time/temperature/pressure test a sample is placed inside a vessel which is heated externally. The temperature rise when ignition occurs and the pressure rise thereafter are measured. In the steel box test the sample is placed in a steel cubical box with a crimped on lid and heated externally. It has been used in cases where a scale-up problem with the Koenen test was suspected, but is not currently used.

For impact sensitiveness the BAM Fallhammer and Rotter Tests are used, and for friction sensitiveness the BAM Friction Apparatus and the Rotary Friction Test are used. Use is also made of a mallet test for impacted friction. The mallet test involves striking the sample on an anvil with a mallet. Results are recorded in terms of whether firing occurs with different types of anvil, mallet and blow.

For organic peroxides, tests used include tests in the UN main test series (the Adiabatic Storage Test and the Heat Accumulation Storage Test) and the Dutch Pressure Vessel Test, the Thermal Explosion Vessel Test and the High Pressure Autoclave Test.

8.6.8 Some tests used by the HSE: illustrative example

Extensive test results for a set of some seven substances have been given by T.A. Roberts and Royle (1991). Table 8.12 summarizes some test results quoted by these authors together with their conclusions on acceptance of the substances into Class 1.

8.7 Pilot Plants

Scale-up from laboratory to full-size plant has always been a difficult problem in the chemical industry. Indeed it is this problem which in large part is responsible for the profession of chemical engineering. An important tool for tackling the problem is the pilot plant. In the words of L.H. Baekeland the philosophy of the pilot plant is: 'Make your mistakes on a small scale and your profits on a large scale.'

There are many reasons beside safety for using a pilot plant. They include the determination of the most economic design and operating parameters, the production of product for evaluation and trial marketing. Nevertheless, safety considerations constitute a substantial part of the case for the pilot plant.

A pilot plant is used principally to assist in the scale-up of the process design rather than the mechanical design, e.g. reaction conditions rather than pump specifications. A number of writers have discussed the general reasons for using pilot plants (e.g. Conn, 1971). Progress in chemical engineering has led to some questioning of the need for the pilot plant stage (e.g. Gorman, Morrow and Anhorn, 1965; Katzen, 1968; Anon., 1969a). But the usefulness of the pilot plant in providing information relevant to safety has been strongly argued (e.g. Hudson, 1965; Perciful and Edwards, 1967; Conn, 1971).

Some types of information relevant to safety which a pilot plant may yield include:

(1) operating conditions (e.g. pressure, temperature);
(2) design parameters (e.g. reaction rates, heat transfer coefficients);
(3) reactor problems;
(4) unit operations problems;
(5) materials handling problems (e.g. foaming, solids handling, catalyst handling);
(6) decomposition;
(7) impurities;
(8) corrosion;
(9) fouling;
(10) analytical problems;
(11) operating problems;
(12) working environment problems (e.g. toxics);
(13) effluent and waste disposal problems.

Impurities can cause many problems, and a pilot plant is one of the most effective means of identifying these. There may be impurities in the feedstock which give unexpected effects. Impurities may arise through side reactions, decomposition or polymerization. Leaks into the system may bring in other impurities such as pump lubricant and seal fluids or heat transfer media, including water. Air and water from other sources are common impurities or sources of impurity. Impurities which in a one-pass process might not be significant can build up if there is recycle. The product may be contaminated with any of these impurities. Impurities may cause fouling, blockages, etc. Some impurities can explode or catalyse explosive reactions. Problems may arise in effluent disposal due to impurities. Operation of a pilot plant offers a high probability of detecting impurities problems.

Similarly, a pilot plant often reveals corrosion problems. These are frequently associated not with the main materials and bulk chemicals, but with minor

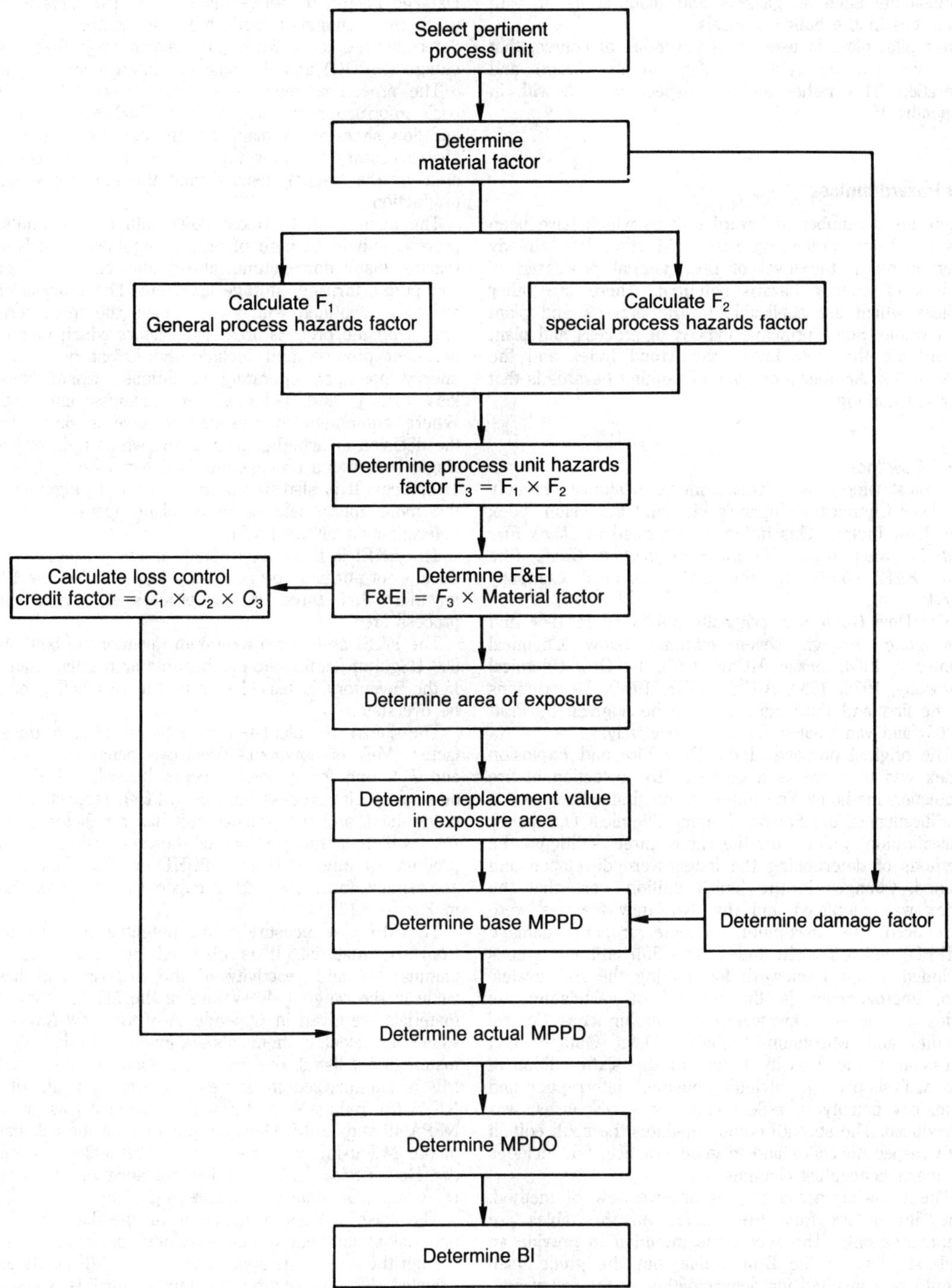

***Figure 8.17** Dow Fire and Explosion Index: procedure for calculating the Fire and Explosion Index and other quantities (Dow Chemical Company, 1994) (Courtesy of the Dow Chemical Company)*

components such as gaskets and diaphragms or with impurities in the bulk chemicals.

If a pilot plant is used, it is essential, of course, that due attention be paid to safety in its design and operation. This rather different aspect is dealt with in Appendix 10.

8.8 Hazard Indices

There are a number of hazard indices which have been developed for various purposes. Mention has already been made at the level of the material processed of indices of energy hazard potential. There are other indices which are applicable to the process and plant as a whole. Some principal indices of process and plant hazard are the Dow Index, the Mond Index and the IFAL Index. Another technique of ranking hazards is that of rapid ranking.

8.8.1 Dow Index

The most widely used hazard index is almost certainly the Dow Chemical Company's Fire and Explosion Index (the Dow Index). This index is described in *Dow's Fire and Explosion Index. Hazard Classification Guide* (the Dow *F&EI Guide*) by the Dow Chemical Company (1994).

The Dow *Guide* was originally published in 1964 and has gone through seven editions (Dow Chemical Company, 1964, 1966a; AIChE, 1973, 40; Dow Chemical Company, 1976, 1980; AIChE, 1987, 1994). Descriptions of the first and third editions have been given by Pratt (1965) and van Gaalen (1974), respectively.

The original purpose of the Dow Fire and Explosion Index was to serve as a guide to the selection of fire protection methods. The index in the first edition was a modification of the Factory Mutual Chemical Occupancy Classification guide. In the first three editions the methods of determining the index were developed and refined. Changes in the fourth edition were that the index was simplified and that two new features were introduced, the maximum probable property damage (MPPD) and a Toxicity Index. The fifth edition changes included a new framework for making the risk evaluation, improvements in the method of calculating the index and several new features, including Loss Control Credits and Maximum Probable Days Outage, and omission of the Toxicity Index. In the sixth edition, a risk analysis package, including business interruption and a toxicity penalty to reflect emergency responses, was introduced. The seventh edition up-dates the sixth edition with respect to codes and to good practice, but includes no major conceptual changes.

The following account gives an overview of method. The figures are from the *Guide*, but the tables are summaries only. The account is intended to provide an understanding of the Dow *Guide*, but the guide itself should be consulted for determination of the index and, in particular, for fuller explanations, qualifications and illustrative examples.

The overall structure of the method is shown in Figure 8.17. The procedure is to calculate the Fire and Explosion Index (F&EI) and to use this to determine fire protection measures and, in combination with a Damage Factor, to derive the base MPPD. This is then used, in combination with the loss control credits, to determine the actual MPPD, the maximum probable days outage (MPDO) and the business interruption (BI) loss.

The procedure may be summarized as follows. The basic information required for the FE&I is the plot plan and flow sheet of the plant. For the economic losses it is also necessary to know the value of the plant per unit area, or the 'capital density', and the monthly value of production.

The plant is first broken down into process units. A process unit is an item of process equipment such as a reactor, distillation column, absorption column, compressor, pump, furnace, storage tank, etc. The process units are then identified which could have the most serious impact on the process area. The factors which identify a pertinent process unit include the inventory, chemical energy potential, operating conditions, capital density, loss history and potential for business interruption. Where equipment is arranged to give a single train, the decision on whether to treat the whole train or just a single vessel as a process unit is determined by the type of process. It is also necessary to make a judgement on the most appropriate stage of plant operation for the determination of the F&EI.

The F&EI is then determined for the process units. It should not normally be necessary to determine the F&EI for more than three or four process units in a single process area.

The F&EI assumes a minimum inventory of 5000 lb or 600 USgal of flammable, combustible or reactive material. If the inventory is less than this, the risk will generally be overstated.

The F&EI is calculated as follows. First a material factor (MF) is obtained. Then two penalty factors (F_1 and F_2), one for general process hazards (GPHs) and one for special process hazards (SPHs), respectively, are determined, and the process unit hazards factor (PUHF) (F_3), which is the product of these, is calculated. The product of the MF and PUHF is the F&EI. The assessment form used to calculate the index is shown in Figure 8.18.

The MF is a measure of the potential energy release from the material. It is obtained by considering the flammability and reactivity of the material and has a value in the range 1–40. Values of the MF for some 216 materials are given in Appendix A of the Dow *Guide* and values for selected chemicals are given in Table 8.13. For materials not listed, the procedure used to determine the MF is summarized in Table 8.14. Use is made of the NFPA fire rating N_F and reactivity rating N_R as given in NFPA 49 and 325M. Detailed guidance on the calculation of the MF using the procedure in the table is given in the Dow *Guide*. This includes guidance on the use of DTA and DSC data to estimate N_R.

The base MF is a measure of the hazard of the material at ambient temperature and pressure. If these are not the conditions applying, the base MF needs to be adjusted. The correction for temperature is shown in Table 8.15. Pressure is taken into account through the SPH. The Dow *Guide* also gives a method for the determination of the MF for a mixture.

The penalty factors applicable for GPHs and for SPHs are given in Tables 8.16 and 8.17 and Figure 8.19, respectively. The GPH factor (F_1) is the sum of all the

FIRE & EXPLOSION INDEX

AREA / COUNTRY	DIVISION	LOCATION	DATE
SITE	MANUFACTURING UNIT	PROCESS UNIT	
PREPARED BY:	APPROVED BY: (Superintendent)	BUILDING	
REVIEWED BY: (Management)	REVIEWED BY: (Technology Center)	REVIEWED BY: (Safety & Loss Prevention)	
MATERIALS IN PROCESS UNIT			
STATE OF OPERATION __ DESIGN __ START UP __ NORMAL OPERATION __ SHUTDOWN	BASIC MATERIAL(S) FOR MATERIAL FACTOR		

MATERIAL FACTOR (See Table 1 or Appendices A or B) Note requirements when unit temperature over 140 °F (60 °C)		
1. General Process Hazards	**Penalty Factor Range**	**Penalty Factor Used(1)**
Base Factor	1.00	1.00
A. Exothermic Chemical Reactions	0.30 to 1.25	
B. Endothermic Processes	0.20 to 0.40	
C. Material Handling and Transfer	0.25 to 1.05	
D. Enclosed or Indoor Process Units	0.25 to 0.90	
E. Access	0.20 to 0.35	
F. Drainage and Spill Control ________ gal or cu.m.	0.25 to 0.50	
General Process Hazards Factor (F_1)		
2. Special Process Hazards		
Base Factor	1.00	1.00
A. Toxic Material(s)	0.20 to 0.80	
B. Sub-Atmospheric Pressure (< 500 mm Hg)	0.50	
C. Operation In or Near Flammable Range __ Inerted __ Not Inerted		
1. Tank Farms Storage Flammable Liquids	0.50	
2. Process Upset or Purge Failure	0.30	
3. Always in Flammable Range	0.80	
D. Dust Explosion (See Table 3)	0.25 to 2.00	
E. Pressure (See Figure 2) Operating Pressure ______ psig or kPa gauge Relief Setting ______ psig or kPa gauge		
F. Low Temperature	0.20 to 0.30	
G. Quantity of Flammable/Unstable Material: Quantity ____ lb or kg H_C = ____ BTU/lb or kcal/kg		
1. Liquids or Gases in Process (See Figure 3)		
2. Liquids or Gases in Storage (See Figure 4)		
3. Combustible Solids in Storage, Dust in Process (See Figure 5)		
H. Corrosion and Erosion	0.10 to 0.75	
I. Leakage – Joints and Packing	0.10 to 1.50	
J. Use of Fired Equipment (See Figure 6)		
K. Hot Oil Heat Exchange System (See Table 5)	0.15 to 1.15	
L. Rotating Equipment	0.50	
Special Process Hazards Factor (F_2)		
Process Unit Hazards Factor (F_1 x F_2) = F_3		
Fire and Explosion Index (F_3 x MF = F&EI)		

(1) For no penalty use 0.00.

Figure 8.18 *Dow Fire and Explosion Index: assessment form for Fire and Explosion Index and other quantities (Dow Chemical Company, 1994) (Courtesy of the Dow Chemical Company). (Figure and table references are to the Dow Guide)*

Table 8.13 *Dow Fire and Explosion Index: material factors for selected chemicals (Dow Chemical Company, 1994) (Courtesy of the Dow Chemical Company)*

	Heat of combustion[a] H_c (BTU/lb)	*NFPA Classification* *Health,* N_H	*Fire,* N_F	*Reactivity,* N_R	*Material Factor,* *MF*
Acetone	12 300	1	3	0	16
Acetylene	20 700	0	4	3	29
Acrylonitrile	13 700	4	3	2	24
Ammonia	8 000	3	1	0	4
Ammonium nitrate	12 400[b]	0	0	3	29
Benzene	17 300	2	3	0	16
Benzylperoxide	12 000	1	3	4	40
1, 3-Butadiene	19 200	2	4	2	24
Butane	19 700	1	4	0	21
t-Butylhydroperoxide	11 900	1	4	4	40
Carbon disulphide	6 100	3	4	0	21
Carbon monoxide	4 300	3	4	0	21
Chlorine	0	4	0	0	1
Cyclohexane	18 700	1	3	0	16
Diesel fuel	18 700	0	2	0	10
Diethyl ether	14 500	2	4	1	21
Ethane	20 400	1	4	0	21
Ethylene	20 800	1	4	2	24
Ethylene dichloride	4 600	2	3	0	16
Ethylene oxide	11 700	3	4	3	29
Hydrogen	51 600	0	4	0	21
Kerosene	18 700	0	2	0	10
Methane	21 500	1	4	0	21
Methyl ethyl ketone	13 500	1	3	0	16
Naphtha (VM&P, regular)	18 000	1	3	0	16
Propane	19 900	1	4	0	21
Propylene	19 700	1	4	1	21
Styrene	17 400	2	3	2	24
Toluene	17 400	2	3	0	16
Vinyl acetate	9 700	2	3	2	24
Vinyl chloride	8 000	2	4	2	24

[a] This is the net heat of combustion, which is the value obtained when the water formed is considered to be in the vapour state.
[b] Heat of combustion H_c is equivalent to six times the heat of decomposition H_d.

GPH penalties, and the SPH factor (F_1) is the sum of all the SPH penalties. The PUHF factor (F_3) is the product of the GPH and SPH factors:

$$F_3 = F_1 \times F_2 \qquad [8.8.1]$$

The F&EI is the product of the MF and the PUHF:

$$FEI = F_3 \times MF \qquad [8.8.2]$$

The F&EI is of importance in its own right as a guide to the fire and explosion hazard.

The degree of hazard implied by the FE&I is shown by the relationships given in Table 8.18. The further use of the F&EI to undertake a process risk analysis, covering property damage and business interruption, is then as follows. The FE&I is used by itself to obtain the exposure radius (ER) and hence the area of exposure. The value of the area of exposure (VAE) is obtained from the area of exposure and the capital density, as described below. The FE&I and the PUHF are used in combination to obtain the damage factor (DF). The base MPPD is then obtained as the product of the VAE and the damage factor.

The exposure radius is a linear function of the F&EI, being zero and 168 at values of the F&EI of zero and 200, respectively. In other words:

$$ER = 0.84 \times FEI \qquad [8.8.3]$$

The value of the area of exposure is obtained as follows. The original value (OV) is obtained as the product of the area of exposure and the original capital density. The replacement value (RV) is then a function of the OV and the escalation factor (EF):

$$RV = 0.82 \times OV \times EF \qquad [8.8.4]$$

where 0.82 is an allowance for the cost of items which will not require replacement. The Dow *Guide* gives detailed guidance on the appropriate costs to use.

Table 8.14 *Dow Fire and Explosion Index: material factor determination guide (Dow Chemical Company, 1994) (Courtesy of the Dow Chemical Company)*

Liquids & Gases Flammability or Combustibility[a]	*NFPA 325M or 49*	Reactivity or instability: $N_R = 0$	$N_R = 1$	$N_R = 2$	$N_R = 3$	$N_R = 4$
Non-combustible[b]	$N_F = 0$	1	14	24	29	40
FP > 200°F(> 93.3°C)	$N_F = 1$	4	14	24	29	40
FP > 100°F(> 37.8°C) ≤ 200°F(≤ 93.3°C)	$N_F = 2$	10	14	24	29	40
FP ≥ 73°F(≥ 22.8°C) < 100°F(< 37.8°C) or FP < 73°F(< 22.8°C) & BP ≥ 100°F(≥ 37.8°C)	$N_F = 3$	16	16	24	29	40
FP < 73°F(< 22.8°C) & BP < 100°F < 37.8°C)	$N_F = 4$	21	21	24	29	40
Combustible dust or mist[c]						
St-1 ($K_{St} \leq 200$ bar m/s)		16	16	24	29	40
St-2 ($K_{St} = 201$–300 bar m/s)		21	21	24	29	40
St-3 ($K_{St} > 300$ bar m/s)		24	24	24	29	40
Combustible solids						
Dense > 40 mm thick[d]	$N_F = 1$	4	14	24	29	40
Open < 40 mm thick[e]	$N_F = 2$	10	14	24	29	40
Foam, fibre, powder, etc[f]	$N_F = 3$	16	16	24	29	40

[a] Includes volatile solids. FP, Flash point, closed cup; BP, boiling point at standard temperatures and pressure (STP).
[b] Will not burn in air when exposed to a temperature of 1500°F (816°C) for a period of 5 min.
[c] K_{St} values are for a 16 l or larger closed test vessel with strong ignition source. See NFPA 68: *Guide for Venting of Deflagrations.*
[d] Includes wood – 2 in. nominal thickness, magnesium ingots, tight stacks of solids and tight rolls of paper or plastic film. Example: SARAN WRAP®.
[e] Includes coarse granular material such as plastic pellets, rack storage, wood pallets and non-dusting ground material such as polystyrene.
[f] Includes rubber goods such as tyres and boots. Styrofoam® brand plastic foam and fine material such as Methocel® cellulose ethers in dust/leak-free packages.

Table 8.15 *Dow Fire and Explosion Index: material factor temperature adjustment*[a] *(Dow Chemical Company, 1994) (Courtesy of the Dow Chemical Company)*

	N_F	*St*	N_R
(a) Enter N_F (*St* for dusts) and N_R.			
(b) If temperature less than 140°F(60°C), go to (e)			
(c) If temperature above flash point or if temperature greater than 140°F (60°C), enter 1 under N_F			
(d) If temperature above exotherm start or autoignition, enter 1 under N_R.			
(e) Add each column but enter 4 where total is 5.			

(f) Using (e) and Table 8.14 determine material factor (MF) and enter on F&EI Form and Manufacturing Unit Risk Analysis Summary

[a] 140°F(60°C) can be reached in storage due to layering and solar heat.

Table 8.16 *Dow Fire and Explosion Index: penalties for General Process Hazards (after Dow Chemical Company, 1994) (Courtesy of the Dow Chemical Company)*

A Exothermic chemical reactions:

For reactions taking place in a reactor:

(1) Mild exotherms, e.g. hydrogenation, hydrolysis, isomerization, sulphonation and neutralization, penalty = 0.30
(2) Moderate exotherms, e.g. alkylation, esterification, oxidation,[a] polymerization, condensation, addition reactions,[b] penalty = 0.50
(3) Critical-to-control exotherms, e.g. halogenation, penalty = 1.00
(4) Particularly sensitive exotherms, e.g. nitration, penalty = 1.25

B Endothermic processes:

(1) Endothermic process taking place in a reactor, penalty = 0.2
(2) Endothermic process taking place in a reactor where energy source is provided by combustion of solid, liquid or gaseous fuel, e.g. calcination, direct-fired pyrolysis, penalty = 0.4

C Material handling and transfer

For hazard of fire involving process unit during handling, transfer and warehousing of materials:

(1) Loading and unloading of Class I flammable liquids or liquefied petroleum gas (LPG)-type materials where transfer lines are connected and disconnected, penalty = 0.50
(2) Processes where introduction of air during manual addition may create a flammable mixture or reactivity hazard, e.g. centrifuges, batch reactions or batch mixing, penalty = 0.50
(3) Warehousing and yard storage:
(a) Flammable liquids or gases N_F = 3 or 4,[c] penalty = 0.85
(b) Combustible solids N_F = 3, penalty = 0.65
(c) Combustible solids N_F = 2, penalty = 0.40
(d) Combustible liquids (closed cup flash point>100°F (37.8°C) and <140°F (60°C)), penalty = 0.25

If any of (a)–(d) are stored on racks without in-rack sprinklers, add 0.20 to penalty

D Enclosed or indoor process units

For enclosed area, defined as any roofed area with three or more sides or an area enclosed by roofless structure with walls on all sides:

(1) Dust filters or collectors in an enclosed area, penalty = 0.50
(2) Process in which flammable liquids are handled above flash point in an enclosed area, penalty = 0.30; for quantities in excess of 10 M lb, penalty = 0.45
(3) Process in which LPG or flammable liquids are handled above boiling point in an enclosed area, penalty = 0.60; for quantities in excess of 10 000 lb (≈ 1000 USgal), penalty = 0.90

If properly designed mechanical ventilation is installed, reduce penalties in (1)–(3) above by 50%

E Access

For inadequate access[d]

(1) Process area >10 000 ft^2 (925 m^2) with inadequate access, penalty = 0.35
(2) Warehouse > 25 000 ft^2 (2312 m^2) with inadequate access, penalty = 0.35

F Drainage and spill control

For a flammable material in a process unit with a flash point <140°F or being processed above its flash point:

(1) Diking which exposes all equipment within dike to potential fire, penalty = 0.50
(2) Flat area around process unit which allows spills to spread, exposing large areas to potential fire, penalty = 0.5
(3) Diking around process unit which surrounds three sides of the area and directs spills to an impounding basin or non-exposing drainage trench, no penalty provided:
(i) Slope to basin or trench is at least 2% for earthen surfaces or 1% for hard surfaces
(ii) Distance to equipment from nearest edge of trench or basin is at least 50 ft (15 m). This distance can be reduced if a fire wall is installed
(iii) The impounding basin has a capacity at least equal to the combined volume of flammable liquid and fire water[e]
(4) Basin or trench which exposes utility lines or does not meet distance requirements, penalty = 0.5

[a] In oxidation reactions involving combustion processes or vigorous oxidizing agents such as chlorates, nitric acid, hypochlorous acids and salts, etc., increase penalty to 1.00.
[b] When the acid is a strong reacting material, increase penalty to 0.75.
[c] This includes drums, cylinders and aerosol cans.
[d] For access to be adequate there should be access from at least two sides. At least one of these should be a roadway. A monitor nozzle which would remain easily accessible and operational during a fire could be considered a second access. Strong consideration should be given to this penalty for process units in an enclosed area.
[e] This volume is the sum of volume of the flammable liquid, which for process and storage facilities is 100% of the capacity of the unit's largest tank + 10% of the capacity of the next largest tank, and of the volume of a 30 min flow of fire water.

Table 8.17 *Dow Fire and Explosion Index: penalties for Special Process Hazards (after Dow Chemical Company, 1994) (Courtesy of the Dow Chemical Company)*

A Toxic materials

Penalty of $0.2N_h$ is applied for a process involving toxic materials[a]

B Subatmospheric pressure[b]

Penalty of 0.5 is applied for processes in which air inleak could create a hazard and which operate at an absolute pressure <500 mmHg[c]

C Operation in or near flammable range

For processes where air might enter the system and form a flammable mixture:

(1) Tank storage of $N_F = 3$ or 4 flammable liquids where air can be breathed into the tank during pump-out or sudden cooling of the tank, penalty = 0.50
For open vent or non-inert gas padded operating pressure-vacuum relief system, penalty = 0.50
Storage of combustible liquids above closed cup flash point, penalty = 0.50
No penalty is applied if an inerted, closed vapour recovery system is used and its air-tightness is assured
(2) Process equipment or process storage tanks which could be in or near the flammable range only in the event of instrument or equipment failure,[d] penalty = 0.30
(3) Processes or operations which are by nature always in or near the flammable range either because purging is not practical or because it was elected not to purge, penalty = 0.8

D Dust explosion[e]

For any process unit involving dust handling operations such as transferring, blending, grinding, bagging, etc.

Particle size[f] (mm)	*Tyler mesh size*	*Penalty[g]*
175+	60–80	0.25
150-175	80–100	0.50
100–150	100–150	0.75
75–100	150–200	1.25
<75	>200	2.00

E Relief pressure

For a process operating at a pressure above atmospheric, a penalty is applied which is a function of the operating pressure. Two pressure ranges are considered. For operating pressures in the range 0–1000 psig the penalty is obtained using Figure 8.19(a)[1] in the manner described below. For pressures above 1000 psig the penalty is

Pressure (psig)	*Penalty*
1000	0.86
1500	0.92
2000	0.96
2500	0.98
3000–10 000	1.0
>10 000	1.5

For the lower pressure range the curve shown in Figure 8.19(a)[1] gives an initial penalty which is applicable for flammable and combustible liquids with flash point <140°F (60°C). For other materials the initial penalty given by the curve is adjusted by a material adjustment factor as follows:

(1) Highly viscous materials[h], multiply penalty by 0.7
(2) Compressed gases used alone or flammable liquids pressurized with any gas above 15 psig, multiply penalty by 1.2
(3) Liquefied flammable gases[h], multiply penalty by 1.3

The actual penalty is obtained as follows. First determine from Figure 8.19(a)[1] the penalties associated with the operating pressure and with the set pressure of the relief device. Divide the operating pressure penalty by the set pressure penalty to obtain the final pressure penalty adjustment factor. Now multiply the operating pressure by the material adjustment factor given in (1)–(3) above and then by the final pressure penalty adjustment factor to obtain the final pressure penalty.

Example:
Vessel design pressure = 150 psig
Normal operating pressure = 100 psig; initial penalty = 0.31 (from Figure 8.19(a)[1])
Rupture disc set pressure = 125 psig; initial penalty = 0.34 (from Figure 8.19(a)[1])
Process with viscous material; penalty adjustment = 0.70
Adjusted initial penalty = 0.7 x 0.31 = 0.22
Actual penalty = 0.22 x (0.31/0.34) = 0.20

F Low temperature

For a process for which it has not been shown by that there is no possibility of temperatures below the transition temperature due to normal or abnormal operating conditions:

(1) Process utilizing carbon steel and operated at or below the ductile/brittle transition temperature,[j] penalty = 0.30
(2) Process utilizing materials other than carbon steel where the operating temperature is at or below the transition temperature, penalty = 0.20

G Quantity of flammable and unstable material

A process involving flammable or unstable materials is classed as either

(1) Liquids or gases in process
(2) Liquids or gases in storage
(3) Combustible solids in storage or dust encountered in process

and a single penalty is applied based on the material selected for the MF as follows:

The penalty is a function of the quantity of material which it is estimated might be released within 10 min.

For case (3) above, this quantity is used directly, whilst for cases 1 and 2 above it is used in the form of the potential heat release.

Liquids or gases in process

This category applies to:

(a) Flammable liquids and those combustible liquids with flash point <140°F (60°C)
(b) Flammable gases
(c) Liquefied flammable gases
(d) Combustible liquids with closed cup flash points >140°F (60°C)
(e) Reactive materials, regardless of the flammability

The penalty is a function of the potential heat release, which is the product of the quantity of material which it is estimated might be released within 10 min[k] and of the heat quantity H_c (BTU/lb),[l] as given in Figure 8.19(b)[1].

Liquids or gases in storage

This category applies to liquids or gases in storage and thus outside the process area. The penalty is a function of the potential heat release, which is the product of the quantity of material in storage[m] and of H_c[n] as given in Figure 8.19(c)[1].[o]

Combustible solids in storage or dust encountered in process

This category applies to combustible solids in storage and to dust in a process unit. The penalty is a function of the quantity of material[p] as given in Figure 8.19(d)[1].

H Corrosion and erosion[q]

For a corrosion rate, defined as the sum of the external and internal corrosion rates:

(1) Corrosion rate <0.5 mil/year (0.127 mm/year) with risk of pitting or local corrosion, penalty = 0.10
(2) Corrosion rate >0.5 mil/year (0.127 mm/year) and <1 mm/year, penalty = 0.20
(3.) Corrosion rate >1 mil/year (0.254 mm/year), penalty = 0.50
(4) Risk of stress corrosion cracking developing, penalty = 0.75
(5) Lining required to prevent corrosion,[r] penalty = 0.20

I Leakage – joints and packing

(1) Pump and gland seals likely to give some leakages of a minor nature, penalty = 0.10
(2) Process known to give regular leakage problems at pumps, compressors and flange joints, penalty = 0.30
(3) Process in which thermal and pressure cycling occurs, penalty = 0.3
(4) Process material penetrating in nature or an abrasive slurry which cause intermittent problems of sealing and process utilizing a rotating shaft seal or packing, penalty = 0.40
(5) Process with sight glasses, bellows assemblies or expansion joints, penalty = 1.50

J Use of fired equipment

For a process unit which is in the vicinity of a fired heater or which is itself a fired heater, a penalty is applied which is a function of the distance from the air intake of the fired heater to a probable leak point as given in Figure 8.19(e)[1]. The two curves A-1 and A-2 are applied as follows, where the material referred to is the material of the MF:

Curve A-1 is used for

(a) A process unit in which the material could be released above its flash point
(b) A process unit in which the material is a combustible dust

Curve A-2 is used for:

(a) A process unit in which the material could be released above its boiling point

Note: If the material is below its flash point, the only situation which attracts a penalty is where the fired heater itself is the process unit being evaluated. In this case the distance from the possible leak source is zero and even if the material is below its flash point, a penalty of 1.0 is applied.

K Hot oil exchange system

The penalty in this section is not applied:

(1) If the hot oil exchange system is the process unit being evaluated.
(2) If the hot oil is non-combustible or, if combustible, is always used below its flash point

Otherwise, for a process unit utilizing a hot oil exchange system in which the heat exchange fluid is combustible:

Quantity[s]		*Penalty*	
(USgal)	(m^3)	*Oil above flash point*	*Oil at or above boiling point*
<5000	<18.9	0.15	0.25
5000–10 000	18.9–37.9	0.30	0.45
10 000–25 000	37.9–94.6	0.50	0.75
>25 000	>94.6	0.75	1.15

Note: (1) It is recommended that the F&EI also be determined for the hot oil system itself, including the process tank, pumps and distribution/return piping, but not the storage tank. (2) If a fired hot oil heat exchange system is actually located in the area of the process unit, the penalty in Section J applies.

L Rotating equipment[t]

Penalty of 0.5 is applied for a process unit which utilizes, or is

(1) A compressor in excess of 600 hp
(2) A pump in excess of 75 hp
(3) Agitators and circulating pumps in which failure could produce a process exotherm[u]
(4) Other large, high speed rotating equipment with a significant loss history, e.g. centrifuge

---

[a] For mixtures the highest N_h value of the individual components should be used.
[b] If the penalty is applied, the penalty specified in Section C or E should not be duplicated or repeated
[c] The main process units in this category are most stripping operations, some compressor operations and a few distillation operations
[d] For a process unit which relies on inert purge to keep it out of the flammable range, penalty = 0.30. This penalty also applies to padded barges or tank cars. (Note: no penalty is applied here if the penalty in Section B has already been taken).
[e] Penalties should be applied unless it has been shown by testing that no dust explosion hazard exists.
[f] The particle size to be used is the 10% particle size, i.e. that at which 90% of the dust is coarser and 10% finer.
[g] If the dust is handled in an inert gas the penalty should be halved.
[h] This includes materials such as tars, bitumen, heavy lubricating oils and asphalts.
[i] This includes all other flammable materials stored above their boiling point.
[j] If no data are available, a 50°F (10°C) transition temperature should be assumed.
[k] This figure should be based on the user's own judgement but, as a guide, experience shows that a reasonable estimate is the larger of the quantity of material (1) in the process unit or (2) in the largest connected unit, disregarding any connected unit which can be isolated by an emergency isolation valve;
[l] for most materials H_c is the heat of combustion, but for unstable materials ($N_r \geq 2$) it is six times the heat of decomposition or the heat of combustion, whichever is larger.
[m] In the case of portable drums, the relevant quantity is the total quantity in all the drums and in the case where two or more vessels are in a common dike which would not drain adequately into an impounding basin it is the quantity in all the vessels.
[n] See footnote *l*.
[o] If there is more than one class of material, the total heat release should be utilized in conjunction with the highest curve applicable to any of the individual materials.
[p] For unstable materials ($N_R \geq 2$) the quantity which should be used is the actual mass multiplied by six and the curve to be used is curve A.
[q] The *Guide* gives guidance on situations where corrosion or erosion is likely.
[r] This penalty does not apply if the lining is simply to prevent discoloration of the product.
[s] This quantity is defined as follows. It is the lesser of (1) a 15-minute spill caused by a break in the lines servicing the process unit, or (2) the oil inventory within the active circulating hot oil system, where the portion of the exchange system classed as storage is not included unless it is connected much of the time to the process unit.
[t] Formulae have not been developed for all types and sizes of rotating equipment, but there is evidence that equipments above a certain size are liable to contribute to an incident.
[u] Where the exotherm is due to lack of cooling from interrupted mixing or circulation of coolant or to interrupted and resumed mixing.

Equations of curves
The equations given in the *Guide* for the curves in Figures 8.19(a)–(e)[1] are as follows.

Figure 8.19(a)[1]:

$$Y = 0.16109 + 1.61503\,(X/1000) - 1.42879\,(X/1000)^2 + 0.5172\,(X/1000)^3 \qquad 0 < X < 1000$$

Figure 8.19(b)[1]:

$$\log_{10}Y = 0.17179 + 0.42988\,\log_{10}X - 0.37244(\log_{10}X)^2 + 0.17712(\log_{10}X)^3 - 0.029984(\log_{10}X)^4$$

Figure 8.19(c)[1]:
Curve A

$$\log_{10}Y = -0.289069 + 0.472171\,\log_{10}X - 0.074585(\log_{10}X)^2 - 0.018641(\log_{10}X)^3$$

Curve B

$$\log_{10}Y = -0.403115 + 0.378703\,\log_{10}X - 0.046402(\log_{10}X)^2 - 0.015379(\log_{10}X)^3$$

Curve C

$$\log_{10}Y = -0.558394 + 0.363321\,\log_{10}X - 0.057296(\log_{10}X)^2 - 0.010759(\log_{10}X)^3$$

Figure 8.19(d)[1]:
Curve A

$$\log_{10}Y = 0.280423 + 0.464559\,\log_{10}X - 0.28291(\log_{10}X)^2 + 0.066218(\log_{10}X)^3$$

Curve B

$$\log_{10}Y = -0.358311 + 0.459926\,\log_{10}X - 0.141022(\log_{10}X)^2 - 0.02276(\log_{10}X)^3$$

Figure 8.19(e)[1]:

Curve A-1
$$\log_{10}Y = -3.3243(X/210) + 3.75127(X/210)^2 - 1.42523(X/210)^3$$

Curve A-2
$$\log_{10}Y = -0.3745(X/210) - 2.70212(X/210)^2 + 2.09171(X/210)^3$$

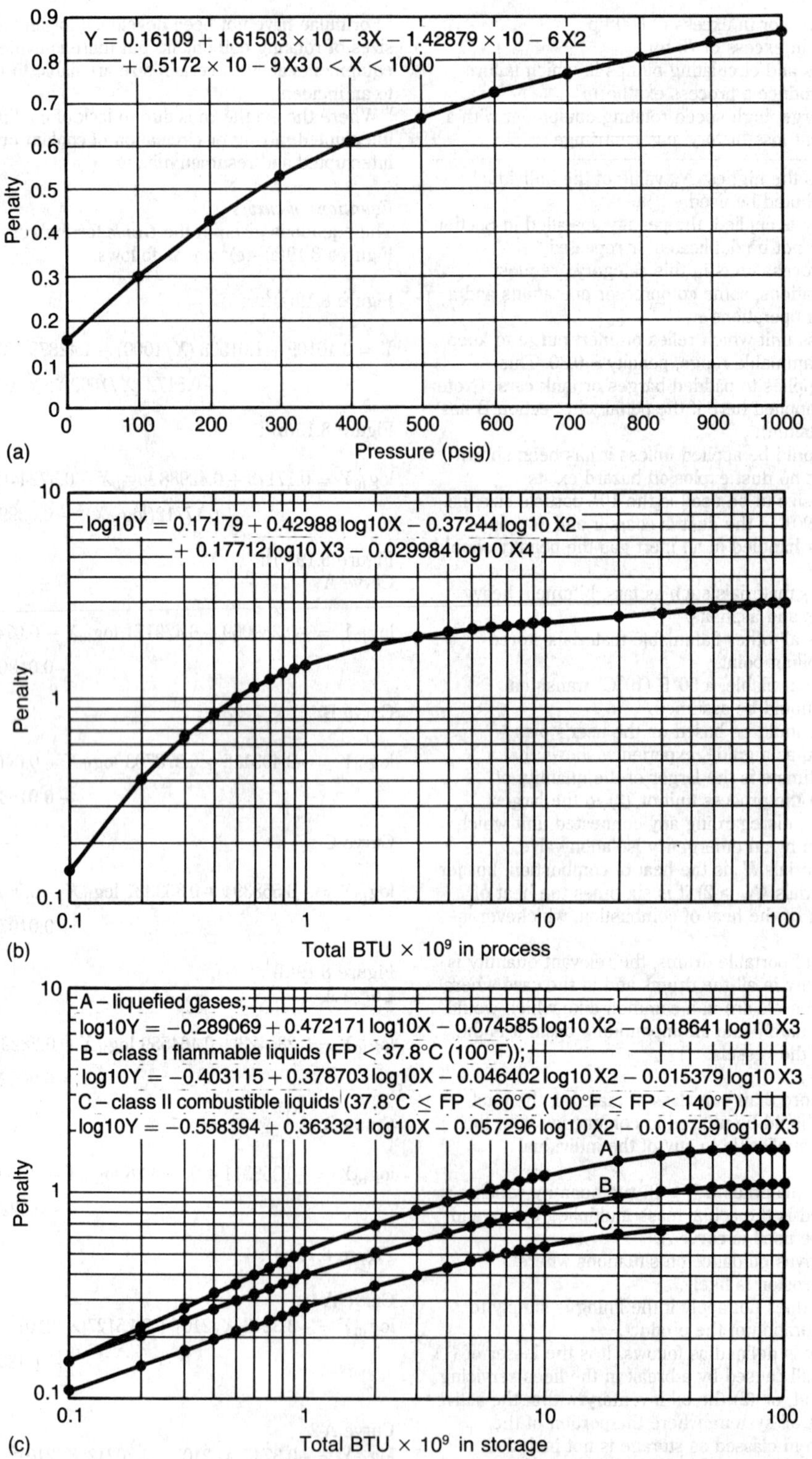

***Figure 8.19[1]** Dow Fire and Explosion Index: penalties for special process hazards (Dow Chemical Company, 1994): (a) penalty for pressure of flammable and combustible liquids; (b) penalty for potential heat release for liquids or gases in process; (c) penalty for potential heat release for liquids or gases in storage; (d) penalty for quantity of combustible solids in storage/dust in process; (e) penalty for location of fired equipment (Courtesy of the Dow Chemical Company)*

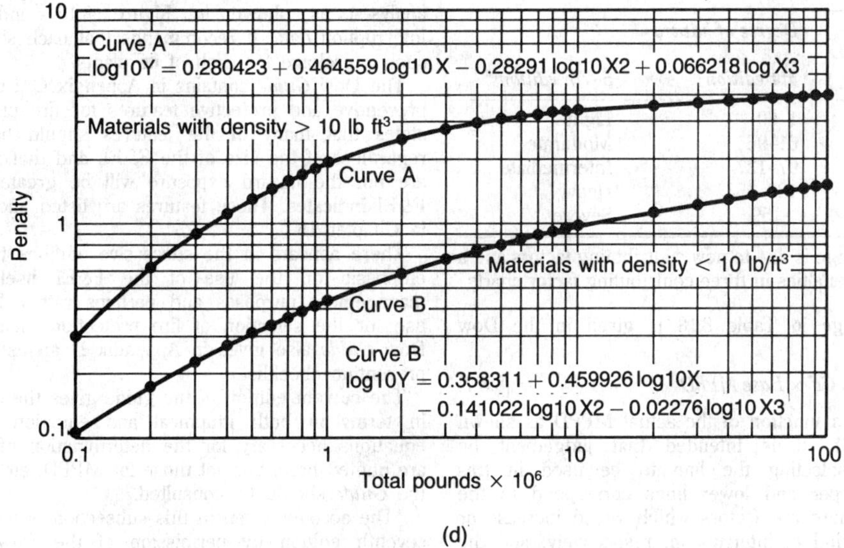

(d)

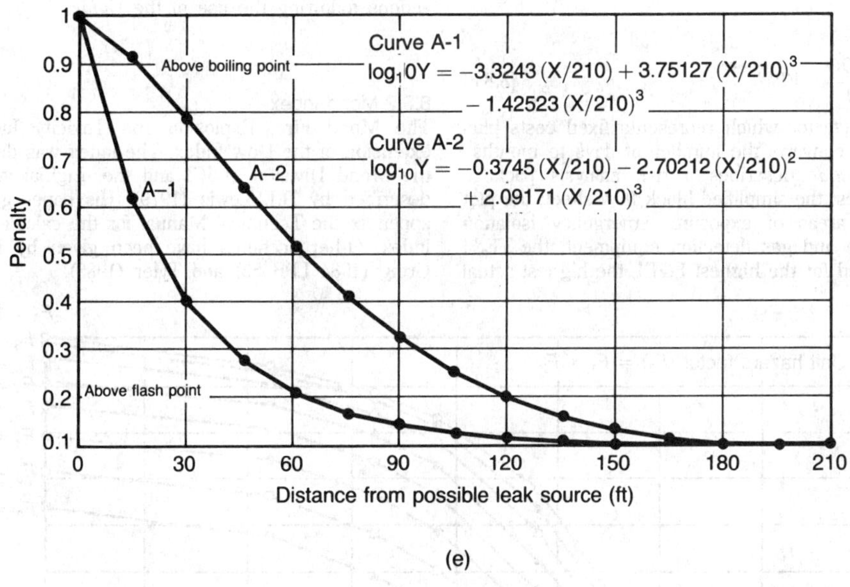

(e)

***Figure 8.19** continued*

The damage factor is a function of the FE&I and the PUHF, as shown in Figure 8.20. Then

$$Base\ MPPD = VAE \times DF \qquad [8.8.5]$$

A credit factor (CF) is then obtained as the product of three individual loss control credit factors ($C_1 - C_3$). The product of the base MPPD and the credit factor is the actual MPPD. The MPDO is a function of the actual MPPD. The BI is then obtained from the MPDO and the value of production per month (VPM). The credit factor is:

$$CF = C_1 \times C_2 \times C_3 \qquad [8.8.6]$$

The three loss control credit factors $C_1 - C_3$ are each the product of a set of individual loss control credits. The ranges of these credit items are given in Table 8.19. Detailed guidance on the precise value of each credit

Table 8.18 Dow Fire and Explosion Index: degree of hazard

F&E Index range	Degree of hazard	
	4th edition	5–7th editions[a]
1–50	1–60	Light
51–81	61–96	Moderate
82–107	97–127	Intermediate
108–133	128–158	Heavy
≥134	≥159	Severe

[a] Increase in range is due to new contributing factors and a correction for penalties in three contributing factor charts

within the range in Table 8.18 is given in the Dow *Guide*. Then

$$Actual\ MPPD = CF \times Base\ MPPD \qquad [8.8.7]$$

The MPDO is a function of the actual MPPD as shown in Figure 8.21. It is intended that judgement be exercised in selecting the line to be used in this figure. The upper and lower lines correspond to the cases where there are factors which would increase or reduce the period of interruption, respectively, and the central line corresponds to the case where neither of these applies.

The BI is:

$$BI = 0.7 \times \frac{MPDO}{30} \times VPM \qquad [8.8.8]$$

where 0.7 is a factor which represents fixed costs plus profits, and 30 converts the number of days to months.

The Dow *Guide* describes a risk analysis package which comprises: the simplified block flow sheet; the plot plan showing areas of exposure, emergency isolation valves, and fire and gas detection equipment; the FE&I forms completed for the highest F&EI, the highest actual MPPD and the highest MPDO and BI; a risk analysis summary for the plant, based on the process units analyses as shown in Figure 8.18; and business interruption data. It recommends that each site maintain such a package for each of its plants.

The Dow *Guide* contains in Appendix C a set of basic preventive and protective features for fire protection. It states that many of the features should be provided regardless of the size of the F&EI, and that where they are not the hazard exposure will be greater than the F&EI indicates. These features are listed and discussed in Chapter 16.

There appears in the successive editions to be more emphasis on the use of the F&EI itself for risk management purposes and perhaps rather less on its use for the selection of fire protection measures. The Dow *Guide* also gives in Appendix D an extensive loss prevention checklist.

The current edition of the guide gives the correlations in terms in both graphical and equation form. The equations necessary for the determination of the F&EI are quoted here, but not those for MPPD, etc., for which the *Guide* should be consulted.

*The account given in this subsection is based on the seventh edition by permission of the Dow Chemical Company. The company accepts no liability for results of actions following the use of the *Guide*.

8.8.2 Mond Index

The Mond Fire, Explosion and Toxicity Index is an extension of the Dow Index. The index was developed at the Mond Division of ICI and the original version was described by D.J. Lewis (1979). His paper gives as an appendix the Technical Manual for the calculation of the index. Other accounts have been given by Doran and Greig (1984 LPB 55) and Tyler (1985).

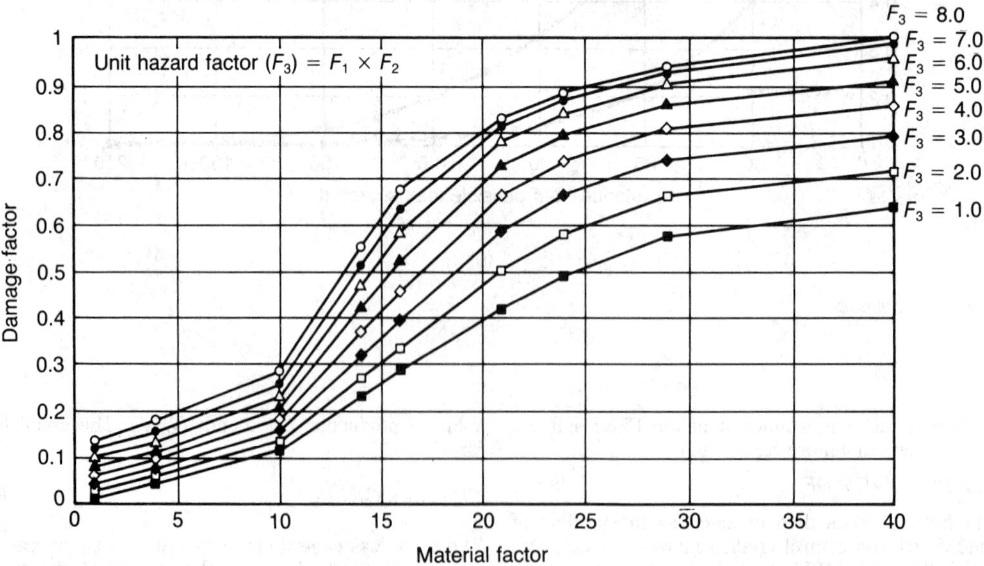

Figure 8.20 Dow Fire and Explosion Index: damage factor (Dow Chemical Company, 1994) (Courtesy of the Dow Chemical Company)

Table 8.19 *Dow Fire and Explosion Index: loss control credit factor (Dow Chemical Company, 1994) (Courtesy of the Dow Chemical Company)*

A Process control (C_1)			
(a) Emergency power	0.98	(f) Inert gas	0.94–0.96
(b) Cooling	0.97–0.99	(g) Operating procedures	0.91–0.99
(c) Explosion control	0.84–0.98	(h) Reactive chemical review	0.91–0.98
(d) Emergency shutdown	0.96–0.99	(i) Other process hazard analysis	0.91–0.98
(e) Computer control	0.93–0.99		
B Material isolation (C_2)			
(a) Remote control valve	0.96–0.98	(c) Drainage	0.91–0.97
(b) Dump/blowdown	0.96–0.98	(d) Interlock	0.98
C Fire protection (C_3)			
(a) Leak detection	0.94–0.98	(f) Water curtains	0.97–0.98
(b) Structural steel	0.95–0.98	(g) Foam	0.92–0.97
(c) Water supply	0.94–0.97	(h) Hand extinguishers/monitors	0.93–0.98
(d) Special systems	0.91	(i) Cable protection	0.94–0.98
(e) Sprinkler systems	0.74–0.97		

The principal modifications to the Dow method made in the Mond Index are:

(1) to enable a wider range of processes and storage installations to be studied;
(2) to cover the processing of chemicals which are recognized as having explosive properties;
(3) to offset difficulties raised by high heat of combustion per unit mass of hydrogen and to enable distinctions to be made between processes where a given fuel is reacted with different reactants;
(4) to include a number of additional special process type of hazard considerations that have been shown by a study of incidents to affect the level of hazard significantly;
(5) to allow aspects of toxicity to be included in the assessment;

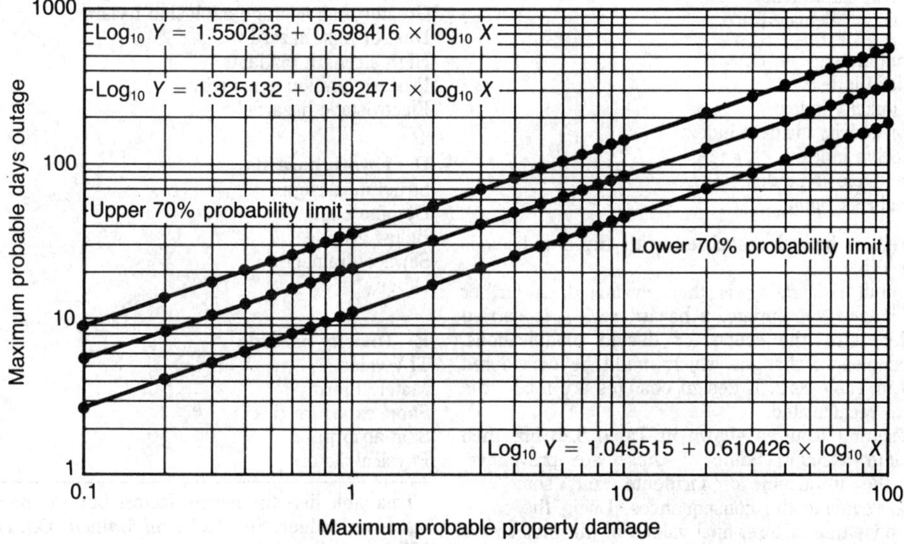

Figure 8.21 *Dow Fire and Explosion Index: maximum probable days outage. The MPPD is given as the actual MPPD in 10^{12} US\$, 1986 basis. To update to 1993 basis, multiply by 359.9/318.4 = 1.130 (based on Chemical Engineering Plant Cost Index) (Dow Chemical Company, 1994) (Courtesy of the Dow Chemical Company)*

(6) to include a range of offsetting factors for good design of plant and control/safety instrumentation systems to enable realistic hazard levels to be assessed for plant units under varying levels of safety features;
(7) to indicate how the results of using the method can be applied logically in the design of plants having a greater degree of 'inherent safety'.

The Mond Index was developed from the 1973 version of the Dow Index and the comparisons between the two given here relate to that version. Since then, there have been a number of developments in the Dow Index itself, in particular the introduction of the loss control credit factors.

The Mond method involves making an initial assessment of hazard in a manner similar to that used in the Dow Index, but taking into account additional hazard considerations. The potential hazard is expressed in terms of the initial value of a set of indices for fire, explosion and toxicity. A hazard factor review is then carried out to see if there is scope to reduce the hazard by making design changes, and intermediate values of the indices are determined. Offsetting factors for preventive and protective features are applied and the final values of the indices, or offset indices, are calculated.

In outline, the method of determining the Mond Index is as follows. The plant is divided into units, the demarcation between units being based on the feasibility of locating a separating barrier (open space, wall or floor) between the unit and its neighbours.

The material factor is determined as in the Dow method, but in addition special material hazards factors are introduced. Again as in the Dow method, use is made of factors for GPHs and SPHs although the particular factors are different. A quantity factor, based on the inventory of material, and layout hazard factors are also introduced. There are also factors for toxicity hazard. The features taken into account in these factors are shown in Table 8.20.

The indices calculated are:

(1) overall index;
(2) fire load index;
(3) unit toxicity index;
(4) major toxicity incident index;
(5) explosion index;
(6) aerial explosion index;
(7) overall risk rating.

The degree of hazard associated with these indices is shown in Table 8.21.

The overall risk rating is then evaluated. If further action is judged appropriate, a hazard review is carried out to determine the scope for design modifications. Some examples of changes which might be considered are given in Table 8.22. If design changes are made, the indices are recalculated.

The offsetting features shown in Table 8.23 are then brought into account. Some of these are preventive, reducing the frequency of incidents, and some are mitigating, reducing the consequences. Using the corresponding offsetting factors, final values of the indices are calculated.

The features to be taken into account in calculating the various factors are specified in detail in the *Technical*

Table 8.20 *Mond Index: hazard factors (after D.J. Lewis, 1979)*[a]

A Special material hazards
Oxidizing materials
Reacts with water to produce a combustible gas
Mixing and dispersion characteristics
Subject to spontaneous heating
Subject to rapid spontaneous polymerization
Ignition sensitivity
Subject to explosive decomposition
Condensed phase properties
Other

B General process hazards
Handling and physical changes only
Single continuous reactions
Single batch reactions
Multiplicity of reactions or different process operations carried out in same equipment
Material transfer
Transportable containers

C Special process hazards
Low pressure (below 15 psia)
High pressure
Low temperature:
  1: Carbon steel +10°C to -10°C
  2: Carbon steel below -10°C
  3: Other materials
High temperature:
  1: Flammability
  2: Construction materials
Corrosion and erosion
Joint and packing leakages
Vibration, load cycling, etc.
Processes or reactions difficult to control
Operation in or near flammable range
Greater than average explosion hazard
Dust or mist hazard
High strength oxidants
Process ignition sensitivity
Electrostatic hazards

D Layout hazards
Structure design
Domino effect
Below ground
Surface drainage
Other

E Toxicity hazards
TLV value
Material form
Short exposure risk
Skin absorption
Physical factors

[a] This table lists the hazard factors but not the associated numerical values. See *Technical Manual* (D.J. Lewis, 1979) for full details.

Table 8.21 *Mond Index: evaluation of index (after D.J. Lewis, 1979)*

A Overall Index D

Range	*Overall degree of hazard*
0–20	Mild
20–40	Light
40–60	Moderate
60–75	Moderately heavy
75–90	Heavy
90–115	Extreme
115–150	Very extreme
150–200	Potentially catastrophic
>200	Highly catastrophic

B Fire Load Index F

Fire load (10^3 BTU/ft^2)	*Expected fire duration* (h)	*Category*	*Comments*
0–50	0.25–05	Light	
50–100	0.5–1	Low	Dwellings
100–200	1–2	Moderate	Factories
200–400	2–4	High	Factories
400–1000	4–10	Very high	Maximum for occupied buildings
1000–2000	10–20	Intensive	Rubber warehouses
2000–5000	20–50	Extreme	
5000–10 000	50–100	Very extreme	

C Explosion Indices

Internal Unit Explosion Index, E range	*Aerial Explosion Index,* A range	*Category*
0–1	0–10	Light
1–2.5	10–30	Low
2.5–4	30–100	Moderate
4–6	100–500	High
>6	>500	Very high

D Toxicity Indices

Unit Toxicity Index, U range	*Major Toxicity Incident Index,* C range	*Category*
0–1	0–20	Light
1–3	20–50	Low
3–6	50–200	Moderate
6-10	200–500	High
>10	>500	Very high

E Overall Risk Factor R

Range	*Category*
0–20	Mild
20–100	Low
100–500	Moderate
500–1100	High (Group 1)
1100–2500	High (Group 2)
2500–12 500	Very high
12 500–65 000	Extreme
>65 000	Very extreme

Table 8.22 *Mond Index: some examples of potentially beneficial design changes (after D.J. Lewis, 1979)*

1. Changes to process whereby the key material is diluted by an inert at all times so reducing the material factor
2. Use of effective stabilizer in all parts of the unit where polymerization hazards can exist
3. Alteration in process conditions (i.e. pressure, temperature or mixture composition) to avoid Special Material Hazards such as explosive decomposition, gaseous detonation (e.g. acetylene), condensed phase behaviour, etc.
4. Separation of unit into two or more units of reduced capacity or into units carrying out separate stages when such smaller units can be effectively separated from each other
5. Elimination of heated two phase storages
6. Elimination of multiple reactions within the same equipment
7. Replacement of removable connection systems by fixed fully closed pipework
8. Avoidance of use of open or semi-open equipment
9. Operation under less arduous pressure conditions (either vacuum or high pressure)
10. Selection of less arduous temperature conditions (avoidance of low temperature and high temperature operational hazards)
11. Use of materials of construction having reduced corrosion potential for vessels, pipework and fittings
12. Reduction in numbers of joint and packing leakage points and use of superior joint and packing designs and materials
13. Changes in design to reduce vibration or thermal cycling effects (e.g. elimination of bellows)
14. Adoption of more effective and safe control systems for the process
15. Change in operation to take conditions further away from flammable range
16. Use of some inert dilutents where high strength oxidants are involved (i.e. oxygen enriched air in place of oxygen)
17. Selection of equipment requiring a smaller inventory of the key material in the unit
18. Changes to eliminate as much high level or below ground storage of flammable materials within the unit plan area as possible
19. Separation of as much storage capacity from process operations as possible
20. Placement of specific items behind blast or fire resistant walls
21. Addition of effective second containment walls to storage units
22. Changes to unit ventilation requirements

Manual, which contains a wealth of practical information, particularly in respect of the offsetting features.

As already mentioned, the plant is divided into individual units on the basis of the feasibility of creating separating barriers and one of the factors taken into account in the index is plant layout. The use of the method to assist in layout design is an important

Table 8.23 Mond Index: offsetting factors (after D.J. Lewis, 1979)[a]

A Containment system
Pressure vessels
Non-pressure vertical storage tanks
Transfer pipelines
Additional containment vessels, sleeves and bund walls
Leakage detection systems and response
Disposal of relief, vented or dumped material

B Process control
Alarm systems
Emergency power supplies
Process cooling systems
Inert gas systems
Hazard study activities
Safety shut-down systems
Computer control
Explosion and incorrect reaction protection
Operating instructions
Plant supervision

C Safety attitude
Management involvement in safety
Safety training
Maintenance and safety procedures

D Fire protection
Structural fire protection
Fire walls, barriers and equivalent devices
Equipment fire protection

E Material isolation
Valve systems
Ventilation

F Fire fighting
Fire alarms
Hand fire extinguishers
Water supply
Installed sprinkler, water spray or monitor systems
Foam and inerting installations
Fire brigade attendance
Site co-operation in fire fighting
Smoke ventilators

[a] This table lists the offsetting factors but not the associated numerical values. See *Technical Manual* (D.J. Lewis, 1979) for full details.

application. It has been described by D.J. Lewis (1980, 1984) and is considered in more detail in Chapter 10.

8.8.3 IFAL Index

The instantaneous fractional annual loss (IFAL) index is a separate index developed by the Insurance Technical Bureau primarily for insurance assessment purposes. Accounts have been given by J. Singh and Munday (1979), Munday *et al.* (1980) and H.B. Whitehouse (1985). Calculation of the index is described in the *IFAL p Factor Workbook* (Insurance Technical Bureau, 1981).

The index was developed in order to provide a means of assessment that was more scientifically based and satisfactory than the historical loss record, which tends to be subject to chance fluctuations. An outline flow chart of the method is shown in Figure 8.22.

The method involves dividing the plant into blocks and examining each major item of process equipment in turn to assess its contribution to the index. The main hazards which contribute to the index are

(1) pool fires;
(2) vapour fires;
(3) unconfined vapour cloud explosions;
(4) confined vapour cloud explosions;
(5) internal explosions.

For hazards (2)–(5) emission frequency and hole size distributions are used, and for each case emission, ignition, fire and explosion are modelled and the damage effects are estimated.

The IFAL Index is the product of the process factor p and of two modifying factors, the engineering factor e and the management factor m. For a process engineered and managed to 'standard good practice' the two latter factors are unity and the IFAL Index equals the p factor.

In contrast to the Dow and Mond Index methods, the IFAL Index method is too complex for manual calculation and is carried out using a computer program.

Table 8.24 shows some results obtained using the IFAL method. Section A gives some relative p factors for 100 000 te/year petrochemical plants, and Section B gives the relative contribution of the five hazards to losses on an ethylbenzene plant.

8.8.4 Dow Chemical Exposure Index

As stated earlier, the main Dow Index contained at one time a Toxicity Index. Dow now uses a separate Chemical Exposure Index (CEI). An account of this index is given in *Dow's Chemical Exposure Index Guide* (AIChE 1994/39) (the Dow *CEI Guide*).

The CEI is a measure of the relative acute toxicity risk. It is used by Dow for initial process hazard analysis (PHA), in the calculation of its Distribution Ranking Index (DRI) and in emergency response planning.

The information needed for the calculation of the CEI is: the physical and chemical properties of the material; a simplified process flow sheet, showing vessels and major pipework with inventories; and an accurate plot plan of the plant and the surrounding area.

It is also necessary to have toxicity limits in the form of the American Industrial Hygiene Association (AIHA) Emergency Response Planning Guidelines (ERPGs) for the material or, where these do not exist, Dow's own Emergency Exposure Planning Guidelines (EEPGs). There are three levels of ERPG (ERPG-1, ERPG-2 and ERPG-3), which are essentially the maximum airborne concentrations below which it is believed that nearly all individuals could be exposed without experiencing some defined effect, the effects being: for level 1, experience of something more than a mild transient health effect or odour; for level 2, serious or irreversible health effects; and for level 3, life-threatening effects. ERPGs are considered further in Chapter 18.

The procedure for the calculation of the CEI is to determine (1) the release scenarios, (2) the ERPG-2, (3)

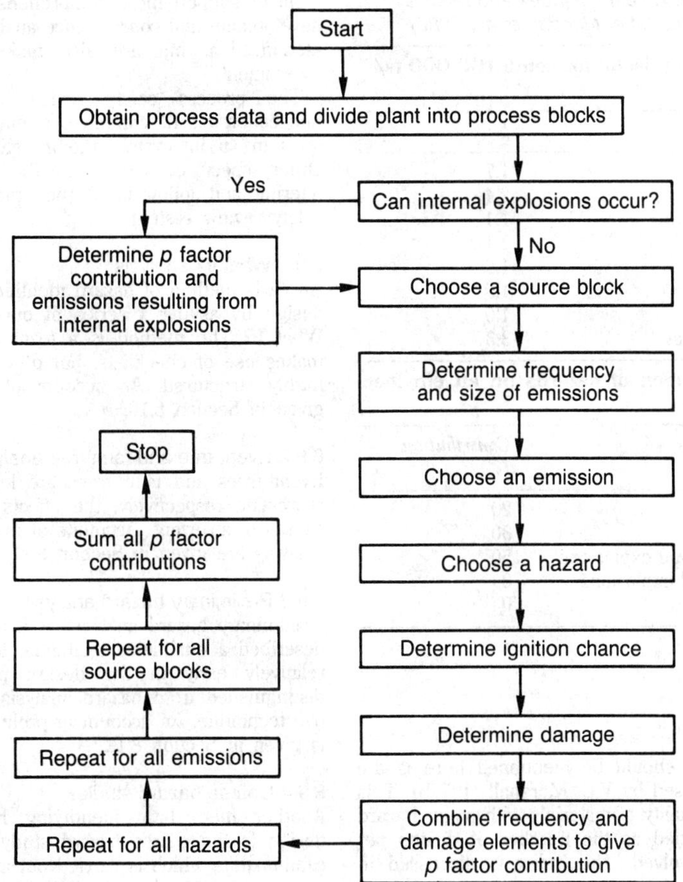

***Figure 8.22** IFAL Index: procedure for calculating* p *factor contributions (Munday* et al.*, 1980)*

the airborne quantities, (4) the CEI and (5) the hazard distance. The CEI is based on the scenario giving the largest airborne quantity (AQ).

The Dow *CEI Guide* gives a set of rules for defining release scenarios and of equations for estimating the AQ based on gas or liquid flow from vessels and pipes, liquid flash-off, and pool formation and evaporation. These, in effect, constitute a release model, description of which is deferred to Chapter 15.

The CEI is then calculated from the relationship:

$$\mathrm{CEI} = 655.1\left(\frac{\mathrm{AQ}}{\mathrm{ERPG\text{-}2}}\right)^{\frac{1}{2}} \qquad [8.8.9]$$

where AQ is the airborne quantity (kg/s) and ERPG-2 is the relevant toxic limit (mg/m^3). If the calculated value of the CEI is greater than 1000, the CEI is set equal to 1000.

The hazard distance is then determined as:

$$\mathrm{HD} = 6551\left(\frac{\mathrm{AG}}{\mathrm{ERPG}}\right)^{\frac{1}{2}} \qquad [8.8.10]$$

where HD is the hazard distance (m) and ERPG is any one of three ERPG toxic limits (mg/m^3). If the calculated value of HD is greater than 10 000 m, HD is set equal to 10 000 m. There is thus a set of three HD values corresponding to the three levels of ERPG.

The Dow *CEI Guide* contains tabulations of relevant physical properties and ERPG values. It also gives CEI values for releases of different chemicals from a 2 in. hole together with worked examples. It provides a containment and mitigation checklist.

There have been earlier accounts of the Dow CEI in the literature (e.g. R.A. Smith and Miller, 1988). The Dow *CEI Guide* states that the index has been revised and that values calculated by the method described therein are not comparable with those obtained from earlier versions.

Table 8.24 IFAL Index: value of index and relative contributions of hazards (after Munday et al., 1980)

A Relative values of p factor for some 100 000 te/year plants

Plant	*p*
Acetic acid	1.7
Ethanol	2.4
Ethyl benzene	1.1
Ethylene oxide	2.1
Isopropanol	1.6
Styrene (1)	1.9
Styrene (2)	1.0
Vinyl chloride monomer	3.3

B Relative contribution of hazards on an ethylbenzene plant

Event	*Contribution (%)*
Pool fires	20
Vapour fires	30
Unconfined vapour cloud explosions	50
Confined vapour cloud explosions	0
Internal explosions	0

8.8.5 Mortality index

Another index which should be mentioned here is the Mortality Index proposed by V.C. Marshall (1977b). This is an index of the lethality of materials with major hazard potential, and is defined as the number of deaths per tonne of material involved. The index is discussed in Chapters 16–18.

8.9 Hazard Studies

There are now a number of methods that have been developed for the identification of hazards. A brief overview of these is given in this section, and a more detailed account of each method is then presented in the succeeding sections.

Overviews of hazard identification methods are given in the Chemical Industries Association (CIA) *Guide to Hazard and Operability Studies* (CIHSC, 1977/3) (the CIA *Hazop Study Guide*), *Guidelines for Hazard Evaluation Procedures* (CCPS, 1985/1, 1992/9) (the CCPS *Hazard Evaluation Guidelines*) and *A Manual of Hazard and Operability Studies* (Knowlton, 1992) (the Chemetics *Hazard Study Manual*), and by Kavianian, Rao and Brown, *et al.* (1992).

Some of the methods described are complete methods. Methods in this category are What If? analysis, preliminary hazard analysis, coarse hazard studies, hazard and operability (hazop) studies and failure modes, effects and criticality analysis. Of these, preliminary hazard analysis and coarse hazard studies are designed for use at an early stage in the design.

Other methods, such as event tree and fault tree analysis, sneak analysis, computer hazop and human error analysis, are specialist techniques used to complement or support more comprehensive methods. Scenario development and consequence analysis are part of hazard identification, but are also major features in hazard assessment.

The process safety review system described in Subsection 8.9.12 combines a number of these techniques in an integrated, though still developing, system. Other aspects considered are the choice of method, the filtering and follow-up of the study results and formal safety review systems.

8.9.1 What if? analysis

An early method of hazard identification is to review the design by asking a series of questions beginning with What If? The method is a team exercise and typically makes use of checklists, but otherwise tends not to be highly structured. An account of What If? analysis is given in Section 8.10.

8.9.2 Event tree and fault tree analysis

Event trees and fault trees are logic diagrams used to represent, respectively, the effects of an event and the causes of an event. Accounts of event tree and fault tree analysis are given in Section 8.11 and Chapter 9.

8.9.3 Preliminary hazard analysis

'Preliminary hazard analysis' is a term normally used to described a qualitative technique for identifying hazards relatively early in the design process. It is to be distinguished from hazard analysis, which is a quantitative technique. An account of preliminary hazard analysis is given in Section 8.12.

8.9.4 Coarse hazard studies

Another method of identifying hazards early in the design is the coarse hazard study or creative checklist examination, which is carried out as a team exercise. An account of coarse hazard studies is given in Section 8.13.

8.9.5 Hazard and operability studies

A method now widely used for the identification of hazards at, or close to, the engineering line diagram stage is the hazard and operability (hazop)study. This technique is a development from critical examination. It is a team exercise which involves examining the design intent in the light of guide words. The technique has itself been subject to numerous variations.

For the process industries, the hazop study might fairly be described as the jewel in the crown of hazard identification methods. An account of hazard and operability studies is given in Section 8.14.

8.9.6 Failure modes, effects and criticality analysis

Another method of hazard identification which is often used as an alternative to a hazop study is failure modes, effects and criticality analysis. It involves the analysis of the failure modes of an entity, their causes and effects. An account of failure modes, effects and criticality analysis is given in Section 8.15.

8.9.7 Sneak analysis

There are a number of ways in which a hazard can 'sneak' into a system. Sneak analysis is essentially a set of techniques for dealing with the various types of sneak. It is a complementary rather than a comprehensive

Table 8.25 *What if? method: checklist for simplified process hazard analysis (Burk, 1992) (Courtesy of the American Institute of Chemical Engineers)*

STORAGE OF RAW MATERIALS, PRODUCTS, INTERMEDIATES	
Storage Tanks	Design Separation, Inerting, Materials of Construction.
Dikes	Capacity, Drainage
Emergency Valves	Remote Control-Hazardous Materials
Inspections	Flash Arresters, Relief Devices
Procedures	Contamination Prevention, Analysis
Specifications	Chemical, Physical, Quality, Stability
Limitations	Temperature, Time, Quantity
MATERIALS HANDLING	
Pumps	Relief, Reverse Rotation, Identification of Materials of Construction
Ducts	Explosion Relief, Fire Protection, Support
Conveyors, Mills	Stop Devices, Coasting, Guards
Procedures	Spills, Leaks, Decontamination
Piping	Rating, Codes, Cross-connections, Materials of Construction
PROCESS EQUIPMENT, FACILITIES AND PROCEDURES	
Procedures	Start-up, Normal, Shut-down, Emergency
Conformance	Job Audits, Shortcuts, Suggestions
Loss of Utilities	Electricity, Heating, Coolant Air, Inerts, Agitators
Vessels	Design, Materials, Codes, Access, Materials of Construction
Identification	Vessels, Piping, Switches, Valves
Relief Devices	Reactors, Exchangers, Glassware
Review of Incidents	Plant, Company, Industry
Inspections, Tests	Vessels, Relief Devices, Corrosion
Hazards	Hang-fires, Runaways
Electrical	Area Classification, Conformance, Purging
Process	Description, Test Authorizations
Operating Ranges	Temperature, Pressure, Flows, Ratios, Concentrations, Densities, Levels, Time, Sequence
Ignition Sources	Peroxides, Acetylides, Friction, Fouling, Compressors, Static Electricity, Valves, Heaters
Compatibility	Heating Media, Lubricants, Flushes, Packings
Safety Margins	Cooling, Contamination
PERSONNEL PROTECTION	
Protection	Barricades, Personal, Shower, Escape Aids,
Ventilation	General, Local, Air intakes, Rate
Exposures	Other Processes, Public Environment
Utilities	Isolation: Air, Water, Inerts, Steam
Hazards Manual	Toxicity, Flammability, Reactivity, Corrosion, Symptoms, First Aid
Environment	Sampling, Vapors, Dusts, Noise, Radiation
CONTROLS AND EMERGENCY DEVICES	
Controls	Ranges, Redundancy, Fail-Safe
Calibration, Inspection	Frequency, Adequacy
Alarms	Adequacy, Limits, Fire, Fumes
Interlocks	Tests, Bypass Procedures
Relief Devices	Adequacy, Vent Size, Discharge, Drain, Support
Emergencies	Dump, Drown, Inhibit, Dilute
Process Isolation	Block Valves, Fire-Safe Valves, Purging
Instruments	Air Quality, Time Lag, Reset Windup, Materials of Construction
WASTE DISPOSAL	
Hatches	Flame Traps, Reactions, Exposures, Solids
Vents	Discharge, Dispersion, Radiation, Mists
Characteristics	Sludges, Residues, Fouling Materials
SAMPLING FACILITIES	
Sampling points	Accessibility, Ventilation, Valving

Procedures	Pluggage, Purging
Samples	Containers, Storage, Disposal
Analysis	Procedures, Records, Feedback
MAINTENANCE	
Decontamination	Solutions, Equipment, Procedures
Vessel Openings	Size, Obstructions, Access
Procedures	Vessel Entry, Welding, Lockout
FIRE PROTECTION	
Fixed Protection	Fire Areas, Water Demands, Distribution System, Sprinklers, Deluge, Monitors, Inspection, Testing, Procedures, Adequacy
Extinguishers	Type, Location, Training
Fire Walls	Adequacy, Condition, Doors, Ducts
Drainage	Slope, Drain Rate
Emergency Response	Fire Brigades, Staffing, Training Equipment

Table 8.26 *What if? method: results for a high pressure/low density polyethylene plant (Kavianian, Rao and Brown, 1992)*

What if	*Consequence/hazard*	*Recommendations*
Coolant pump to reactor fails	Runaway condition in reactor explosion/fatality	• Provide accurate temperature monitoring in reactor • Employ backup pump/high temperature alarm • Relieve reactor pressure in reactor through automatic control to stop reactions • Provide automatic shut off of ethylene flow
Coolant temperature to jacket is high	Eventual runaway condition in reactor	• Provide adequate temperature control on coolant line • Use heat exchanger flow control to adjust inlet temperature
Runaway condition in reactor	Explosion; fire/fatality	• Provide adequate temperature control on coolant line • Use heat exchanger flow control to adjust inlet temperature • Install rupture disk/relief valve to relieve pressure to stop reactions • Emergency shut-down procedure
Recycle gas compressor 1 or 2 fails	None likely	• Provide spare compressor or shut-down procedure
Melt pump fails	High level in reactor causing more polymerization: runaway reaction eventually exceeds design pressure	• Provide level and flow control schemes to activate spare pump or shut the flow of monomer • Shut down procedure if no spare pump
Leak at suction or discharge of compressors	Fire; explosion	• Use monitoring devices to ensure no flammable gas is released
Ethylene leaks out of process lines	Fire; explosion	• Provide adequate flammable gas monitoring devices
Monomer/initiator ratio out of control	Eventual runaway reaction causing fire and explosion	• Provide flow control on the initiator and monomer lines

method. An account of sneak analysis is given in Section 8.16.

8.9.8 Computer hazop

The conventional hazop study was developed for plants in which the control system was based on analogue controllers. It is not itself oriented to failures associated with computer control systems. Computer hazop has been developed to augment regular hazop in this respect. An account of computer hazop is given in Section 8.17.

8.9.9 Human error analysis

It is well established that human error plays a large role in causing accidents. Human error analysis is used to take this aspect into account. As a hazard identification technique human error analysis is qualitative, although a similar term is also sometimes used to describe a quantitative method. An account of human error analysis is given in Section 8.18.

8.9.10 Scenario development

Examination of different scenarios of plant disturbance and loss of containment is a central activity in hazard identification. The scenarios may relate to the events before release or to events involved in escalation after release. Hazard identification, therefore, includes methods of developing and structuring scenarios. Accounts of scenario development are given in Sections 8.19 and 8.21.

8.9.11 Consequence modelling

Hazard identification involves a degree of filtering out of those events which are identified but assessed as having negligible consequences. Such assessment requires a degree of consequence modelling, however minimal. An account of consequence modelling is given in Section 8.20.

8.9.12 Process safety review system

A set of methods of hazard identification and hazard assessment is the family of techniques developed by Wells and co-workers, described here as a process safety review system. An account of this process safety review system is given in Section 8.21.

8.9.13 Choice of method

The number of techniques available for hazard identification is large and is still growing. Each has its own field of application. One relevant distinction is between methods which are comprehensive and those which are not but which supplement the former. Another is between methods for use early and late in the design. A third is in the type of problem to which a method is best suited. Guidance on choice of method is given in Section 8.22.

8.9.14 Filtering and follow-up

Further aspects of hazard identification are the filtering of the hazard identified to determine those on which action is required, and the recording and follow-up of these actions. An account of filtering and follow-up is given in Section 8.23.

8.9.15 Safety review systems

The arrangements for hazard identification need to be embodied in a formal system which ensures that the necessary studies are done and which covers the life cycle of the plant. An account of safety review systems is given in Section 8.24.

8.10 What If? Analysis

The What If? method involves asking a series of questions beginning with this phrase as a means of identifying hazards. Accounts of What If? analysis are given in the CCPS *Hazard Evaluation Guidelines* and by Burk (1992) and Kavianian, Rao and Brown (1992). Apart from checklists, What If? analysis is possibly the oldest method of hazard identification. The method is to ask questions such as

What if the pumps stops?
What if the temperature sensor fails?

The questions posed need not necessarily all start with What If?; other phrases may be used.

The method involves review of the whole design by a team using questions of this type, often using a checklist. The few accounts in the literature give little information beyond this. A checklist for use in What If? analysis is given by Burk (1992) and is shown in Table 8.25.

Table 8.26 gives an example of the results from a What If? study reported by Kavianian, Rao and Brown (1992).

8.11 Event Tree and Fault Tree Analysis

The event tree and fault tree methods may be used either qualitatively or quantitatively. The concept of the fault tree was introduced in Chapter 7 and a fuller treatment of both event trees and fault trees, including their application to quantitative assessment, is given in Chapter 9. Here it is sufficient to draw attention to the value of tree methods, particularly fault trees, in hazard identification.

An event tree involves the development of the consequences of an event. The overall approach is similar to that adopted in failure modes and effects analysis, which is described below.

A fault tree involves the development of the causes of an undesirable event, often a hazard. The possibility of this event must be foreseen before the fault tree can be constructed. What the fault tree helps to reveal are the possible causes of the hazard, some of which may not have been foreseen.

Fault trees are used extensively in hazard assessment, but they are also of great value in hazard identification. In many cases it is sufficient to be able to identify the fault paths and the base events which can give rise to the top event, it being unnecessary to quantify the frequency of occurrence of these events.

In a failure modes and effects analysis or an event tree the approach is 'bottom up', while in a fault tree it is 'top down'. The hazop method involves both approaches, starting with the deviations and tracing down to the causes and up to the consequences.

8.12 Preliminary Hazard Analysis

Preliminary hazard analysis (PHA) is a method for the identification of hazards at an early stage in the design process. Accounts of PHA are given in the CCPS *Hazard Evaluation Guidelines* and by Kavianian, Rao and Brown (1992). PHA is a requirement of the MIL-STD-882 *System Safety Program.*

Since the early identification of hazards is of prime importance, many companies have some sort of technique for this purpose, a portion of which are called 'preliminary hazard analysis'; the use of the term tends to be fairly loose.

The CCPS *Guidelines* state that PHA is intended for use only in the preliminary stage of plant development in cases where past experience provides little insight into the potential hazards, as with a new process. The

***Table 8.27** Preliminary hazard analysis (Kavianian, Rao and Brown 1992)*

Hazard	*Cause*	*Major effects*	*Corrective/preventive measures*
Damage to feed reactor tubes	Feed compressor failure (no endothermic reactions in reactor)	Capital loss, downtime	• Provide spare compressor with automatic switch-off control
		Damage to the furnace coils due to high temperature	• Develop emergency response system
Explosion, fire	Pressure build-up in the reactor due to plug in transfer lines	Fatalities, injuries	• Provide pressure relief valve on the reactor tubes • Provide warning system for pressure fluctuations (high-pressure alarm) • Provide auxiliary lines with automatic switch off
	Violent reaction of H_2 to acetylene converter with air in presence of ignition source	Potential for injuries and fatalities due to fire or explosion	• Provide warning system (hydrogen analyser) • Eliminate all sources of ignition near hydrogen gas storage area • Develop emergency fire response • Automatically shut off the H_2 feed • Provide fire fighting equipment
Flammable gas release	Ethane storage tank ruptures	Potential for injuries and fatalities due to fire or explosion	• Provide warning control system (pressure control) • Minimize on-site storage • Develop procedure for tank inspection • Develop emergency response system • Provide gas monitoring system
Flammable gas release	CH_4 storage tank (line) leak/ rupture (fuel for the furnace)	Potential for injuries and fatalities due to fire or explosion	• Provide warning system • Minimize on-site storage • Develop procedure for tank inspection • Develop emergency response system • Provide gas monitoring system
Flammable gas release	Radiant tube rupture in the furnace	Potential for injuries and fatalities due to fire	• Improve reactor materials of construction • Monitor design vs operating reactor temperature • Provide temperature control instrument
Employee exposure to benzene (carcinogen)	Leak in knock-out pots or during handling benzene	Chronic health hazard	• Install warning signs in the area • Provide appropriate PPE • Develop safety procedures for handling and cleanup • Monitor concentration of benzene in area to meet TLV requirements
Fire/explosion in acetylene converter	Runaway reaction (exothermic)	Fatality, injury, or loss of capital	• Install temperature control on converter • Install pressure relief on reactor responding to temperature control
Flammable atmosphere	Leak in transfer lines	Fire/explosion	• Install combustible gas meter in sensitive areas • Provide adequate fire fighting equipment • Provide for emergency shut-down • Educate and train personnel on emergency procedures

information required for the study is the design criteria, the material and equipment specification, and so on. The *Guidelines* list the entities examined for hazards as: (1) raw materials, intermediates and final products; (2) plant equipment; (3) facilities; (4) safety equipment; (5) interfaces between system components; (6) operating environment; and (7) operations (maintenance, testing, etc.).

The results from a PHA are illustrated in Table 8.27 (Kavianian, Rao and Brown, 1992).

8.13 Coarse Hazard Studies

One of the principal methods of hazard identification is the hazard and operability (hazop) study, which is a technique for examining the design at the stage of the engineering line diagram. Before describing this, however, it is appropriate to consider first the method developed for use in conjunction with, but earlier than, hazop.

The method was first described in the CIA *Hazop Study Guide*, and a much expanded account is given in the Chemetics *Hazard Study Manual*. This method is the coarse hazard study (not hazop study) or, as described in the Chemetics *Manual*, the creative checklist examination. It may be regarded as a form of, or as an alternative to, PHA.

This coarse hazard study is done at the stage in the design where there is a block layout of plant items. The object is to determine whether there are problems in areas such as data on the chemicals, information about the hazards, basic features of the process design, or layout and siting of the plant.

There are a number of advantages to be gained from conducting a coarse hazard study prior to the hazop study itself. It assists in the identification of those hazards and other problems which are quite basic and which are therefore capable, in principle, of being identified at this earlier stage. It reveals deficiencies in the design information. And it exposes hazards due to interactions between the plant and other plants or the environment, at which the hazop study is less effective. These features contribute to removing potential delays on the critical path of the project.

The conduct of a coarse hazard study involves compilation of two lists: (1) a database of hazards and nuisance properties of the material and (2) a list of potential hazards, nuisances and other matters of concern. The study therefore covers, albeit in a different way from a hazop study, the two essential features of the design intent: (1) materials and (2) equipment and activities.

A coarse hazard study is a team exercise and is led by a study leader. Working documents for a coarse hazard study are a block layout of the equipment, a site plan and, where necessary, a map of the neighbourhood.

Use is made of a list of suitable guide words. Table 8.28 illustrates a minimal list. Other guide words may be generated by considering:

(1) energy;
(2) interactions;
(3) environment.

Each of these headings may be broken down further. A suitable energy set is a list of the main forms of energy. For interactions the *Manual* gives:

(1) people;
(2) materials;
(3) equipment.

For environment it gives

(1) climate;
(2) geothermal/geotechnical;
(3) biological.

The geothermal/geotechnical category is broken down into mainly natural hazards such as earthquakes and tidal waves, and the biological one into living creatures such as birds and insects.

The study results in a set of work assignments to follow up the queries raised. As an illustration, the block layout might show a storage tank containing flammable liquids. The application of the guide word 'fire' to this tank might raise a number of queries. Amongst those listed by the *Manual* are (1) reduction of frequency of fire, (2) detection of fire, (3) access for firefighting, (4) method of firefighting, (5) source and disposal of fire water, (6) use of alternative firefighting agents, (7) nature of combustion products, (8) effects of radiant heat, and, (9) need for a disaster plan.

Table 8.28 *Coarse hazard study: guide words*[a] *(after Knowlton, 1992)*

FIRE
EXPLOSION
TOXICITY
CORROSION
SMELL
EFFLUENT

[a] The original list in the CIA *Hazop Study Guide* also included detonation, radiation, noise, vibration, noxious material, electrocution, asphyxia and mechanical failure.

8.14 Hazard and Operability Studies

The hazard and operability (hazop) study is carried out when the engineering line diagram of the plant becomes available. Accounts of hazop studies are given in the CIA *Guide to Hazard and Operability Studies* (the CIA *Hazop Study Guide*) (CISHC, 1977/3), in *Hazop and Hazan* (Kletz, 1983d, 1986d, 1992b), *Guidelines for Hazard Evaluation Procedures* (CCPS, 1985/1, 1992/9) (the CCPS *Hazard Evaluation Guidelines*) and *A Manual of Hazard and Operability Studies* (Knowlton, 1992) (the Chemetics Hazard Study Manual), and by Kletz (1972a), S.B. Gibson (1974, 1976b,e), Lawley (1974a,b) and Kavianian, Rao and Brown (1992).

The following description is based on those given in the CIA *Hazop Study Guide*, the *Chemetics Hazard Study Manual* and by Lawley (1974b). The study is carried out by a multidisciplinary team, who review the process to discover potential hazards and operability problems using a guide word approach. It is essentially an application of the technique of critical examination.

The basis of such a study may, in principle, be a word model, a process flowsheet, a plant layout or a flow diagram, or other information which reveals the design intent. The basis of a hazop study is generally the engineering line diagram supplemented, as appropriate, by other information such as operating instructions.

The hazop study technique is not a substitute for good design. There is something fundamentally wrong if the application of the method consistently reveals too many basic design faults.

8.14.1 Origins of hazop studies

The origins of hazop studies were in ICI in the 1960s. As with many techniques, there is more than one source which can lay claim to have been influential in its development. Accounts of the development of hazop have been given by Kletz (1986d) and in the Chemetics *Manual* (Knowlton, 1992).

The techniques which eventually grew into hazop developed at a time when the applications of method study, including critical examination, were being explored. Accounts of work in the Heavy Organic Chemicals Division of ICI (Binsted, 1960) and in the Mond Division (Elliott and Owen, 1968) describe the development of hazop in terms of critical examination.

8.14.2 Principle of hazop studies

The basic concept of the hazop study is to take a full description of the process and to question every part of it to discover what deviations from the intention of the design can occur and what the causes and consequences of these deviations may be. This is done systematically by applying suitable guide words. Thus important features of the study are:

(1) design intent;
(2) deviations from intent;
(3) causes of deviations;
(4) consequences
  (a) hazards;
  (b) operating difficulties.

8.14.3 Design intent and entities examined

The design intent is examined in respect of the following entities:

(1) material;
(2) activity;
(3) equipment;
(4) source;
(5) destination.

In some applications other relevant entities are

(6) time;
(7) space.

8.14.4 Guide words

To each of these above entities there is applied a basic set of guide words. These guide words and their meanings are:

NO or NOT	Negation of intention
MORE	Quantitative increase
LESS	Quantitative decrease
AS WELL AS	Qualitative increase
PART OF	Qualitative decrease
REVERSE	Logical opposite of intention
OTHER THAN	Complete substitution

The application of these guide words is illustrated by some of the examples discussed below. It is usually possible to apply all the guide words intelligibly to activities and, with the possible exception of REVERSE, to substances.

In applying the guide words to time, the following aspects may be relevant: duration, frequency, absolute time, and sequence. The guide words MORE and LESS are applicable for duration or frequency, while the guide words SOONER and LATER may be more applicable than OTHER THAN for absolute time or sequence.

Similarly, in applying the guide words to space, or place, the following aspects may be relevant: position, source, and destination. The guide words HIGHER or LOWER may be more applicable than MORE or LESS for elevation, while the guide word WHERE ELSE may be more applicable than OTHER THAN for position, source or destination.

Some additional guide words include

STARTING
STOPPING
CONTROLLING
ISOLATION
INGRESS
ESCAPE
DECONTAMINATION

8.14.5 Illustrative example: reactor transfer system

The *Hazop Study Guide* gives the following example, which serves to illustrate both the basic examination principle and the detailed use of the guide words. The flow sheet of a reactor system shows that raw material streams A and B are transferred by pump to the reactor, where they react to give product C. The flow of B should not exceed that of A otherwise an explosion may occur. The flow sheet thus shows an intention to

TRANSFER A

to the reactor at the design flow. Applying the first guide word NO, NOT or DON'T to this intention gives:

DON'T TRANSFER A

This is a deviation from intent. Some causes might be that the supply tank is empty, that the pipe is fractured or that an isolation valve is closed. A consequence might be an explosion due to an excess of B. Thus the examination has discovered a hazard. The study is continued using further guide words such as MORE, LESS, etc.

The application of the guide words may be illustrated by the example already described: TRANSFER A. The guide words may be applicable to either the word TRANSFER or to the word A. Thus for this case the meanings might include the following:

NO or NOT	No flow of A
MORE	Flow of A greater than design flow
LESS	Flow of A less than design

	flow
AS WELL AS	Transfer of some component other than A
	Occurrence of some operation/event additional to TRANSFER
PART OF	Failure to transfer all components of A
	Failure to achieve all that is implied by TRANSFER
REVERSE	Flow of A in direction opposite to design direction
OTHER THAN	Transfer of some material other than A
	Occurrence of some operation/event other than TRANSFER

A detailed commentary on this example is given in the Chemetics *Manual.*

8.14.6 Organization and conduct of hazop studies

The stages in the conduct of a hazop study are:

(1) definition of objectives;
(2) selection of team;
(3) preparation;
(4) conduct;
(5) follow-up;
(6) recording.

In general, the objectives of a hazop study as originally defined are to check the design and the operating procedures in order to identify hazards and operability problems, to which, increasingly, are added environmental problems.

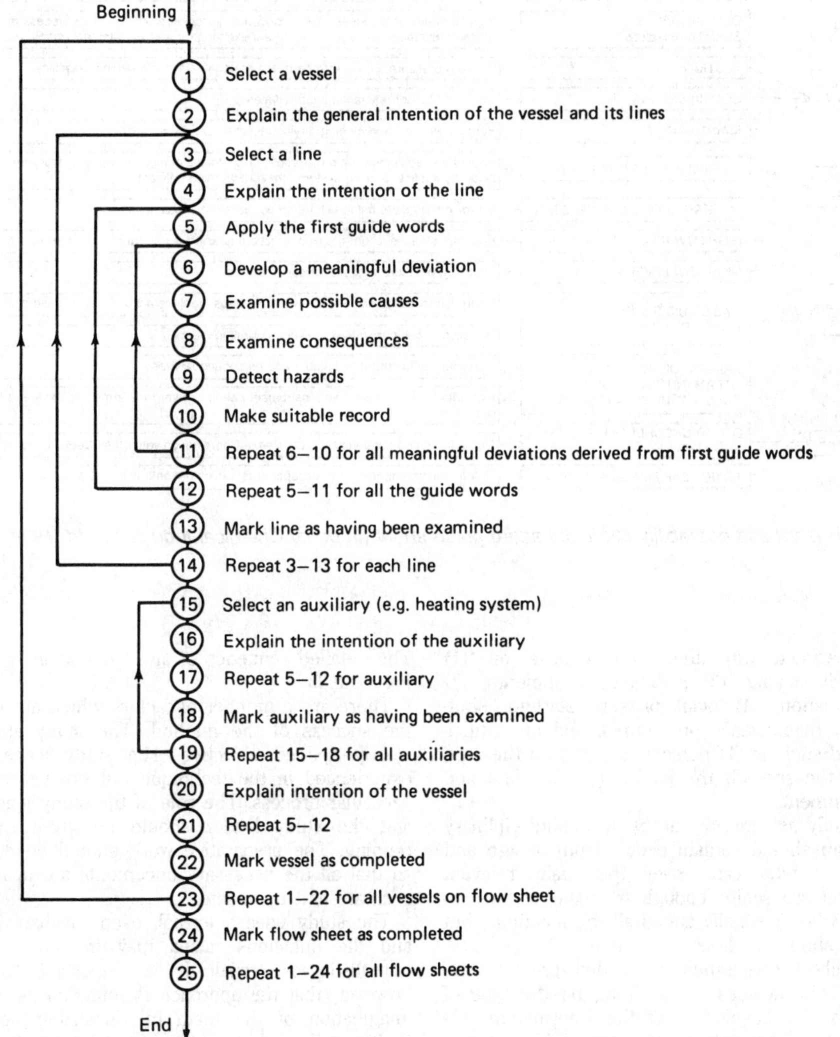
Beginning
1 Select a vessel
2 Explain the general intention of the vessel and its lines
3 Select a line
4 Explain the intention of the line
5 Apply the first guide words
6 Develop a meaningful deviation
7 Examine possible causes
8 Examine consequences
9 Detect hazards
10 Make suitable record
11 Repeat 6–10 for all meaningful deviations derived from first guide words
12 Repeat 5–11 for all the guide words
13 Mark line as having been examined
14 Repeat 3–13 for each line
15 Select an auxiliary (e.g. heating system)
16 Explain the intention of the auxiliary
17 Repeat 5–12 for auxiliary
18 Mark auxiliary as having been examined
19 Repeat 15–18 for all auxiliaries
20 Explain intention of the vessel
21 Repeat 5–12
22 Mark vessel as completed
23 Repeat 1–22 for all vessels on flow sheet
24 Mark flow sheet as completed
25 Repeat 1–24 for all flow sheets
End

Figure 8.23 *Hazard and operability studies: detailed sequence of examination (Chemical Industry Safety and Health Council, 1977/3)*

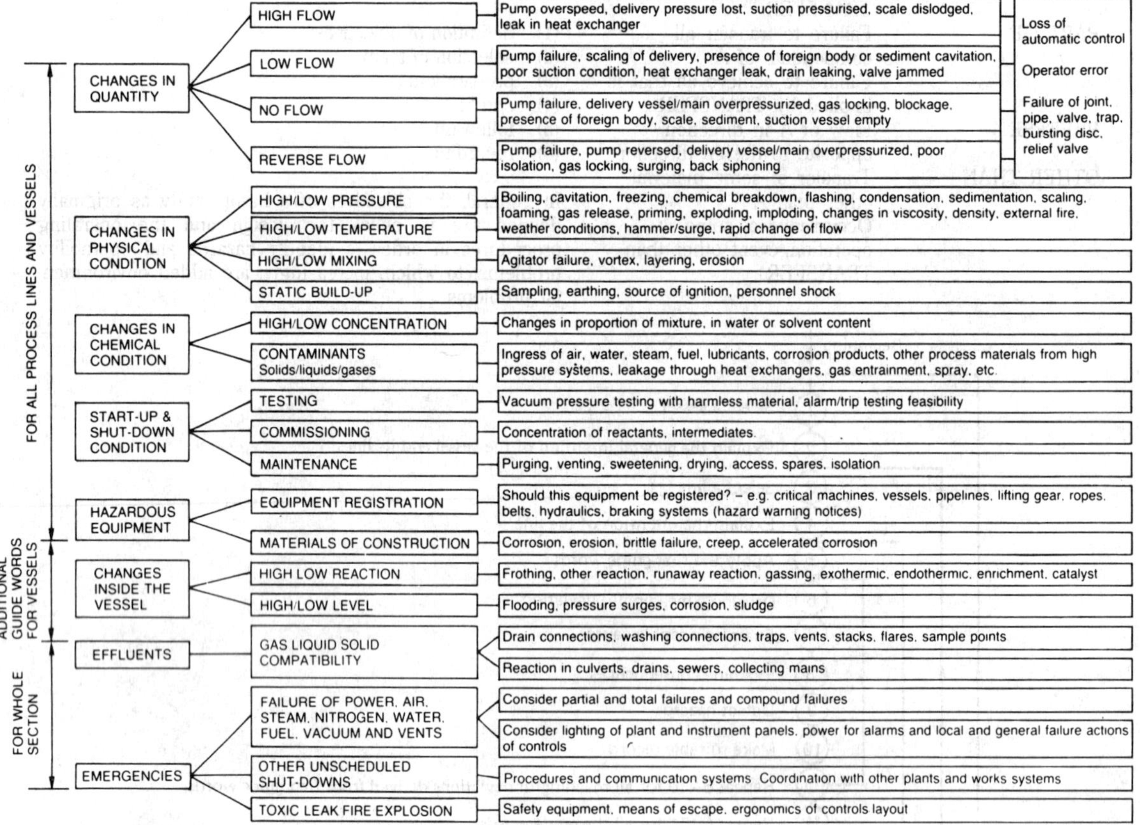

Figure 8.24 *Hazard and operability studies: hazard guide sheet (Imperial Chemical Industries Ltd, 1994)*

Closely associated with these are checks on (1) information still lacking, (2) particular equipments, (3) supplier information, (4) plant phases (start-up, shut-down) and (5) maintenance procedures, and on entities to be protected such as (1) persons working on the unit, (2) others on the site, (3) the public, (4) the plant and (5) the environment.

A hazop study is carried out by a multidisciplinary team. The team should contain people from design and from operations who can cover the main relevant disciplines, who are senior enough to make on-the-spot decisions and who personally attend all the meetings, but the team size should be kept fairly small.

A good deal of preparation is needed prior to the hazop study. This includes (1) deciding on the type of study required, (2) acquisition of the information, (3) validation of the information, (4) conversion of the information to suitable form, (5) planning of the study sequence and (6) arrangement the schedule of meetings.

The detailed sequence of an examination is illustrated in Figure 8.23.

There are a number of factors which are important for the success of the method. The study should have a clearly defined objective. The study leader should be experienced in the technique but not necessarily in the particular process. The role of the study leader is crucial and the study leader should be given the necessary training. The preparative work should be done carefully so that all the necessary documents are to hand and are accurate and up-to-date.

The study uses a formal, even mechanistic, approach and the questions raised may in some cases appear unrealistic or trivial. It is important to emphasize, however, that the approach is intended as an aid to the imagination of the team in visualizing deviations and their causes and consequences. The effectiveness of the technique depends very much on the spirit in which it is done.

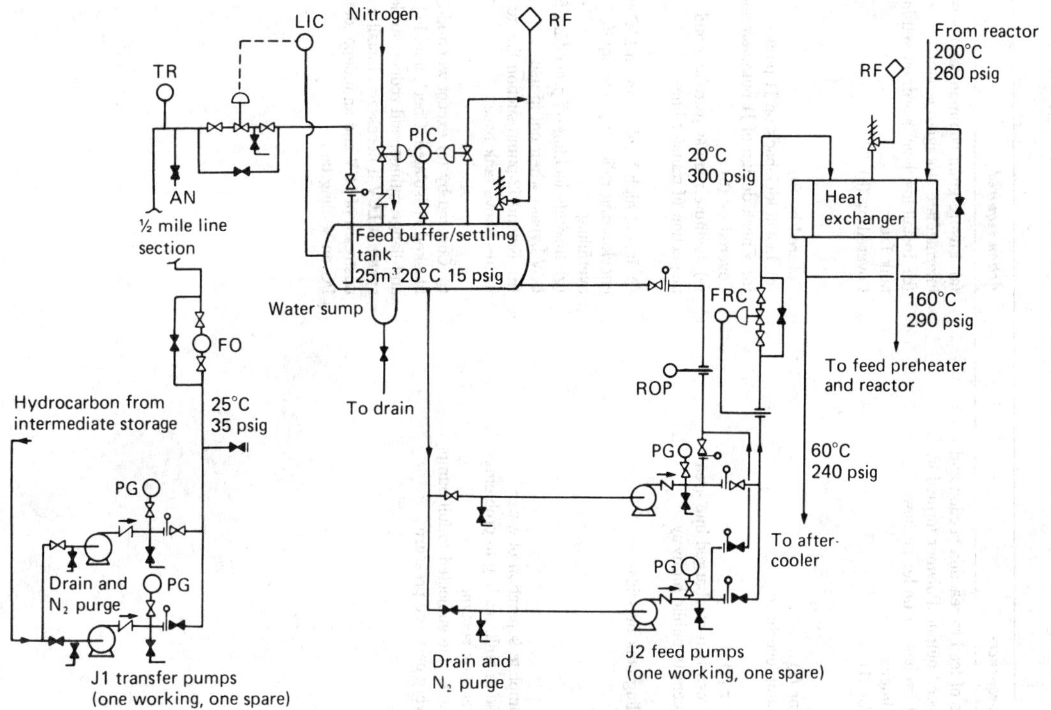

***Figure 8.25** Hazard and operability studies: feed section of proposed alkene dimerization plant (Lawley, 1974b) (Courtesy of the American Institute of Chemical Engineers)*

Fuller discussions of the organizational aspects of hazop studies are given in the CIA *Guide* and the Chemetics *Manual.*

8.14.7 Parametric method

As so far described, the focus of a hazop study is on the design intent. The Chemetics *Manual* distinguishes between this classic form and what it calls the 'parametric method', which concentrates on the deviations from design conditions.

As described in the *Manual,* this approach was originally developed in ICI Mond Division in response to the need to carry out a large number of hazop studies on existing plants. The method uses a hazard guide sheet. An early version of this guide sheet was reproduced in the first edition of this book, a version is given in the *Manual* and a more recent version from ICI is shown in Figure 8.24.

8.14.8 Illustrative example: continuous plant

As an illustration of a hazop study of a continuous process, consider the example given by Lawley (1974b) and reproduced in the *Hazop Study Guide.* The proposed design is shown in Figure 8.25.

The process description is:

> An alkene/alkane fraction containing small amounts of suspended water is continuously pumped from bulk intermediate storage via a half-mile pipeline section into a buffer/settling tank. Residual water is settled out prior to passing via a feed/product heat exchanger and preheater to the reactor section. The water, which has an adverse effect on the dimerization reaction, is run off manually from the settling tank at intervals. Residence time in the reaction section must be held within closely defined limits to ensure adequate conversion of the alkene and to avoid excessive formation of polymer.

A summary of the results of a study for the first line section from the intermediate storage up to the buffer/settling tank is given in Table 8.29. A further table showing the results for the second line section is given in Lawley's paper.

For a continuous plant, the study should include the operability of the plant during commissioning and during regular start-up and shut-down. There are generally a number of operations which are carried out only during these periods.

8.14.9 Time in hazop studies: sequences and batch processes

The hazop technique is applicable both to batch processes and to sequential operations on continuous

***Table 8.29** Hazard and operability studies: results for feed section of proposed alkene dimerization plant from intermediate storage to buffer/setting tank (Lawley, 1974b) (Courtesy of the American Institute of Chemical Engineers)*

Guide word	*Deviation*	*Possible causes*	*Consequences*	*Action required*
NONE	No flow	(1) No hydrocarbon available at intermediate storage	Loss of feed to reaction section and reduced output. Polymer formed in heat exchanger under no flow conditions	(a) Ensure good communications with intermediate storage operator. (b) Install low level alarm on settling tank LIC
		(2) J1 pump fails (motor fault, loss of drive, impeller corroded away, etc.)	As for (1)	Covered by (b)
		(3) Line blockage, isolation closed in error, or LCV fails shut	As for (1) J1 pump overheats	Covered by (b) (c) Install kick-back on J1 pumps (d) Check design of J1 pump strainers
		(4) Line fracture	As for (1) Hydrocarbon discharged into area adjacent to public highway	Covered by (b) (e) Institute regular patrolling and inspection of transfer line
MORE OF	More flow	(5) LCV fails open or LCV bypass open in error	Settling tank overfills	(f) Install high level alarm on LIC and check sizing relief opposite liquid overfilling (g) Institute locking off procedure for LCV bypass when not in use
			Incomplete separation of water phase in tank, leading to problems on reaction section	(h) Extend J2 pump suction line to 12 inches above tank base
	More pressure	(6) Isolation valve closed in error or LCV closes, with J1 pump running	Transfer line subjected to full pump delivery or surge pressure	(j) Covered by (c) except when kick-back blocked or isolated. Check line, FQ and flange ratings, and reduce stroking speed of LCV if necessary. Install a PG upstream of LCV and an independent PG on settling tank

		(7) Thermal expansion in an isolated valved section due to fire or strong sunlight	Line fracture or flange leak	(k) Install thermal expansion relief on valved section (relief discharge route to be decided later in study)
	More temperature	(8) High intermediate storage temperature	Higher pressure in transfer line and settling tank	(l) Check whether there is adequate warning of high temperature at intermediate storage. If not, install
LESS OF	Less flow	(9) Leaking flange or valved stub not blanked and leaking	Material loss adjacent to public highway	Covered by (e) and the checks in (j)
	Less temperature	10) Winter conditions	Water sump and drain line freeze up	(m) Lag water sump down to drain valve, and steam trace drain valve and drain line downstream
PART OF	High water concentration in stream	(11) High water level in intermediate storage tank	Water sump fills up more quickly. Increased chance of water phase passing to reaction section	(n) Arrange for frequent draining off of water from intermediate storage tank. Install high interface level alarm on sump
	High concentration of lower alkanes or alkenes in stream	(12) Disturbance on distillation columns upstream of intermediate storage	Higher system pressure	(p) Check that design of settling tank and associated pipework, including relief valve sizing, will cope with sudden ingress of more volatile hydrocarbons
MORE THAN	Organic acids present	(13) As for (12)	Increased rate of corrosion of tank base, sump and drain line	(q) Check suitability of materials of construction
OTHER	Maintenance	(14) Equipment failure, flange leak, etc.	Line cannot be completely drained or purged	(r) Install low point drain and N_2 purge point downstream of LCV. Also N_2 vent on settling tank

plants such as start-up and shut-down. This introduces time as an additional factor to be taken into account. An adaptation of the set of guide words applicable to time has been given in Section 8.14.4.

For batch and sequential operations the *Manual* gives the following classification of activities:

(1) make ready;
(2) key;
(3) put away.

'Key' is the activity which progresses the process, 'make ready' is the activity needed to prepare for the key activity, and 'put away' is the activity needed on completion of the key activity.

8.14.10 Illustrative example: batch plant

For a batch plant, the working documents comprise not only the flow diagram but also the operating procedures. Several documents may be needed to cover the latter, such as tables giving the operating sequences, bar charts showing the states of equipment during the cycle and flow charts indicating the operator's movements.

As an illustration of a hazop study of a batch plant, consider the example given in the *Hazop Study Guide.* The design is shown in a simplified diagram in Figure 8.26. The plant consists of two measure vessels, four reaction vessels, a condenser, an absorption tower with its circulation system and a Nutsche filter with its filtrate receiver.

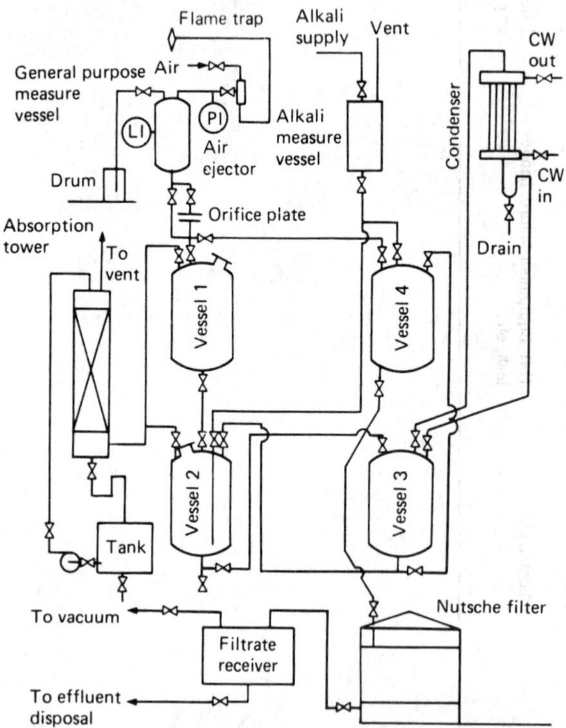

Figure 8.26 *Hazard and operability studies: batch plant (simplified diagram) (Chemical Industry Safety and Health Council, 1977/3)*

The development of the study depends on the kind of sequence to be followed. In this case it is possible to follow a sequence derived from the flow diagram or one derived from the operating instructions. Assuming that the latter is used, an instruction part way through the sequence might be:

Instruction 23	Charge 100 l of material C from drum to the general purpose vessel using the air ejector.

Such an instruction is quite adequate for operational purposes, but it is too complex for the generation of deviations in the study, and needs to be broken down into the part dealing with the air ejector and that dealing with the liquid transfer. Thus the leader may ask a member of the team to describe the purpose of the air ejector which might be stated as: 'Remove some air from measure vessel'. The guide words may then be applied to the statement.

A summary of the results of a study of this batch plant for the operation 'Remove some air from measure vessel' is given in Table 8.30. Further tables for the operations 'Charge 100 l of material C to measure vessel' and 'Transfer 100 l of material C from the general purpose measure vessel to vessel 1 at a controlled rate' are given in the *Hazop Study Guide.*

8.14.11 Illustrative example: proprietary equipment

As an illustration of a study of proprietary equipment, consider the example given in the *Hazop Study Guide.* The equipment is a sterilization autoclave and the design is shown in Figure 8.27.

The process description is:

> Sensitization of stillage-loaded materials is achieved by treatment with steam humidified sterilizing gas in a jacketed autoclave chamber under specified conditions. Two entries are provided to the chamber – from the sterile and from the non-sterile working areas of the facility.
>
> Steam is admitted to the chamber via a let-down system and sterilizing gas via a vaporizer. The chamber may be evacuated via a cooler either directly to drain or via a luted sealed catchpot to a vent stack. Filtered atmospheric air may be admitted via a non-return valve. A relief valve is fitted to the chamber which exhausts to the vent stack and may be by-passed by opening a vent valve if it is required to dump the contents to stack. Water is circulated through the jacket and heated indirectly by means of steam.
>
> Once the autoclave is charged and the doors closed, automatic sequence control takes over and programmes the process as shown in Figure 8.28. The machine itself checks the progress of the process cycle, monitoring the status of the chamber and auxiliaries. Certain checks [see Figure 8.28: Autochecks] control progress in association with timers.

For an equipment of this kind it is also possible to produce a flow process chart for the different operations

Table 8.30 *Hazard and operability studies: results for batch plant for operation 'Remove some air from measure vessel' (Chemical Industry Safety and Health Council, 1977/3)*

Deviations	*Causes*	*Consequences*
DON'T REMOVE AIR	No air supply Faulty ejector Valve shut	Process inconvenience but no hazard
REMOVE MORE AIR	Completely evacuate measure vessel	Can vessel stand full vacuum?
REMOVE LESS AIR	Insufficient suck to transfer contents of drum	Process inconvenience but no hazard
AS WELL AS REMOVE AIR	Pull droplets of material C or other materials from drums or vessels 1 or 4 along exhaust line	Fire hazard? Static hazard? Corrosion hazard? Blocked flame trap? Will material be a hazard after leaving the flame trap? Where does it go?
REMOVE PART OF AIR	Remove oxygen or nitrogen only: not possible	
REVERSE REMOVAL OF AIR	If line from air ejector is blocked compressed air will flow into measure vessel	Overpressure vessel? Blow air into drums and spray out contents? Put air into vessels 1 or 4?
OTHER THAN REMOVE AIR	Put air ejector on when measure vessel full	Spray contents along line and out through flame trap. Similar hazards to AS WELL AS

of the cycle. A chart for the operation 'Charging autoclave' is shown in Figure 8.29.

The development of this study again depends on the choice of sequence to be followed. In this case there are a number of possible sequences that might be followed. Three of these are the flow diagram, the sequence diagram (dot chart) and the flow process chart shown in Figures 8.27–8.29, respectively.

A summary of the results of a study of this equipment for the operation 'Humidify autoclave chamber' derived from the flow diagram is given in Table 8.31. A further table for the operation 'Extends forks to charge stillage into chamber' derived from the flow process chart is given in the *Hazop Study Guide*.

8.14.12 Space and interactions in hazop studies: hybrid studies

In their classic form, hazop studies are essentially concerned with conditions inside the plant, including loss of containment from it. They do not address, however, any interaction in space between the plant and other plants nearby or the environment.

A technique for dealing with such problems is outlined in the Chemetics *Manual*, which outlines a method that it describes as a hybrid approach, utilizing both creative checklists and guide words. The method involves nominating a 'top event', which is then examined to identify its causes, using both a checklist of potential causes and guide words. The method is illustrated in the *Manual* using as a top event a flammable leak from the fuel supply to a boiler and gas turbine system.

8.14.13 Critical examination and hazop studies

As stated at the outset, hazop studies developed from the technique of critical examination. In some applications it may be useful to revert to the latter. An account of this approach is given in the Chemetics *Manual*. A critical examination may be carried out by examining the design intent in respect of material, activities, equipment, source and destination.

The *Manual* gives the classic guide words of critical examination:

WHAT	The design intent
HOW	Method and resources: materials, activities, equipments
WHEN	Time aspects: sequence, frequency, duration, absolute time
WHERE	Space aspects: relative location, sources and destinations, dimensions, absolute location
WHO	Persons or controls: person, skills, organization, control system

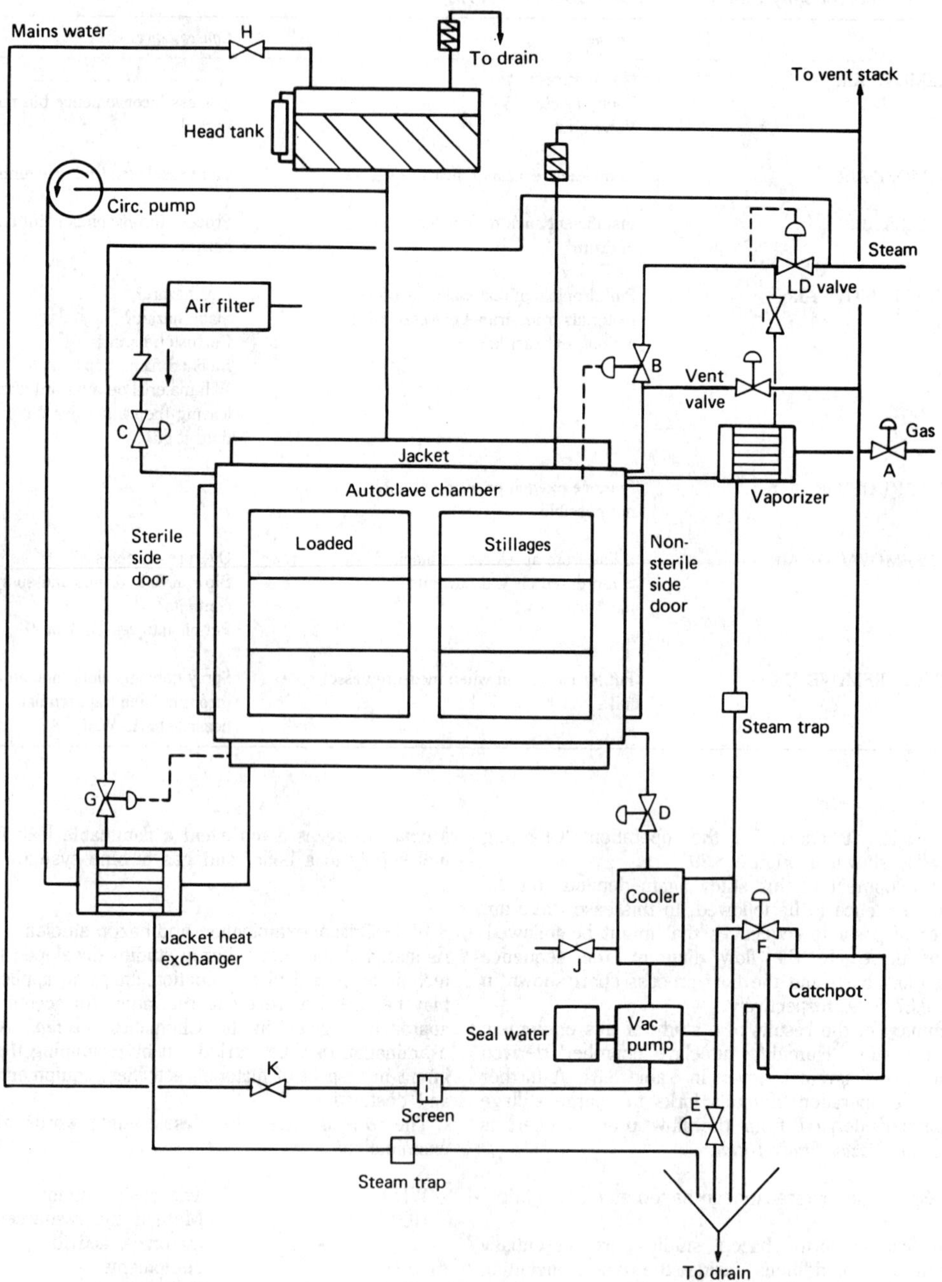

Figure 8.27 *Hazard and operability studies: autoclave arrangement (Chemical Industry Safety and Health Council, 1977/3)*

8.14.14 Control systems in hazop studies

The hazop study method was developed originally for plant, continuous or batch, operating under the control of analogue controllers. The method can be adapted to the examination of a plant which is taken through a sequence, whether under manual or automatic control, using the dot chart method described in Section 8.14.11. The Manual discusses other approaches to various types

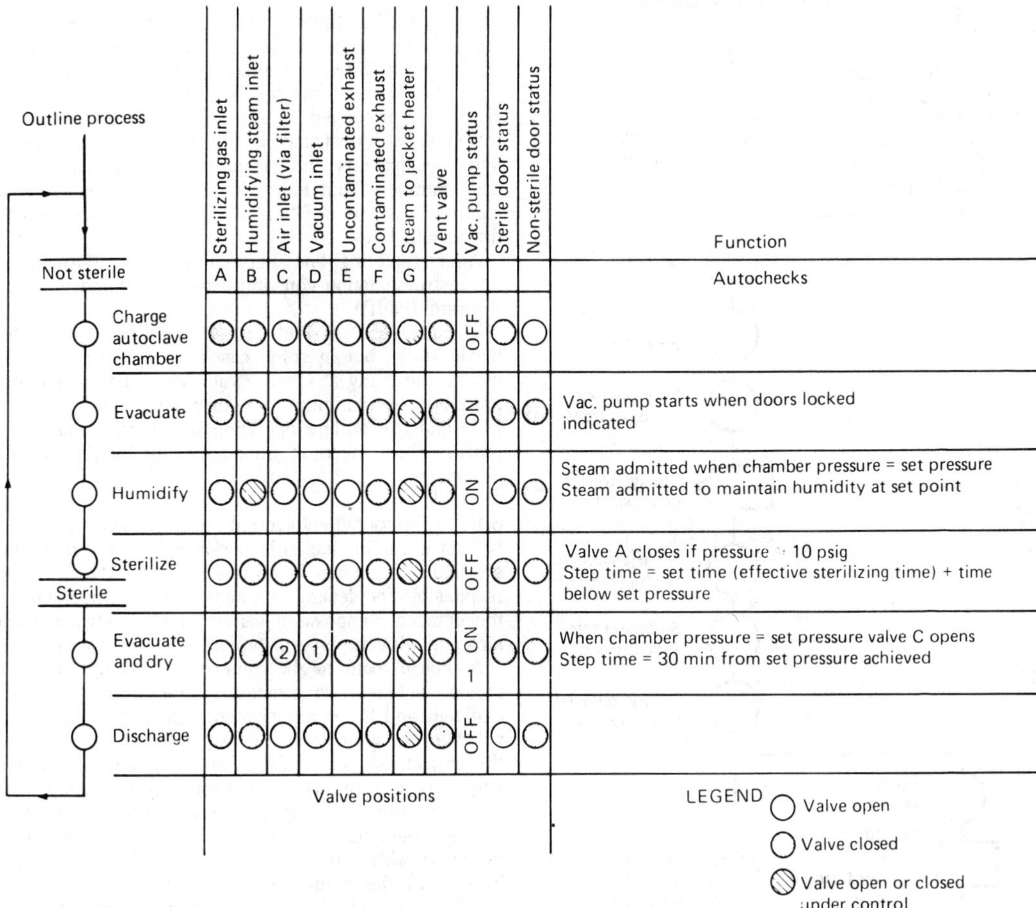

Figure 8.28 *Hazard and operability studies: sequence diagram (dot chart) for autoclave arrangement (Chemical Industry Safety and Health Council, 1977/3)*

of problem associated with control systems, including computer control. The latter is also considered in Section 8.17.

8.14.15 Electrical systems in hazop studies

The treatment of electronic control and interlock systems and electrical distribution systems in a hazop has been described by Mitchell (1992 LPB 105).

The headings used in his account for electronic control and interlock systems are: current, voltage, temperature, contamination, high current protection, instruments, sampling, corrosion, services failure, maintenance, abnormal operations, spares, static electricity and safety. A short checklist is given under each heading. Where appropriate, as with current and voltage, guide words such as NO, REDUCED, INCREASED, etc., are used. A basically similar list is used for electrical distribution systems.

8.14.16 Additional parameters and guide words

Wider application of the hazop approach, to systems such as materials handling systems, has led to the creation of further lists of parameters and guide words such as the one shown in Table 8.32, derived from work by Tweeddale (1993e).

8.14.17 Timing of hazop studies

The timing of a hazop study is an important issue. The disadvantages of an early study are that some of the information necessary to its effective conduct is not available and that there is a greater risk of design changes being made to the detriment of operability and/or safety. The advantages are that it is easier to incorporate in the design any changes arising out of the study and, in particular, to achieve an inherently safer design.

Conversely, the disadvantages of a late study are that the design response options will have been rendered more difficult and expensive to implement, or sometimes

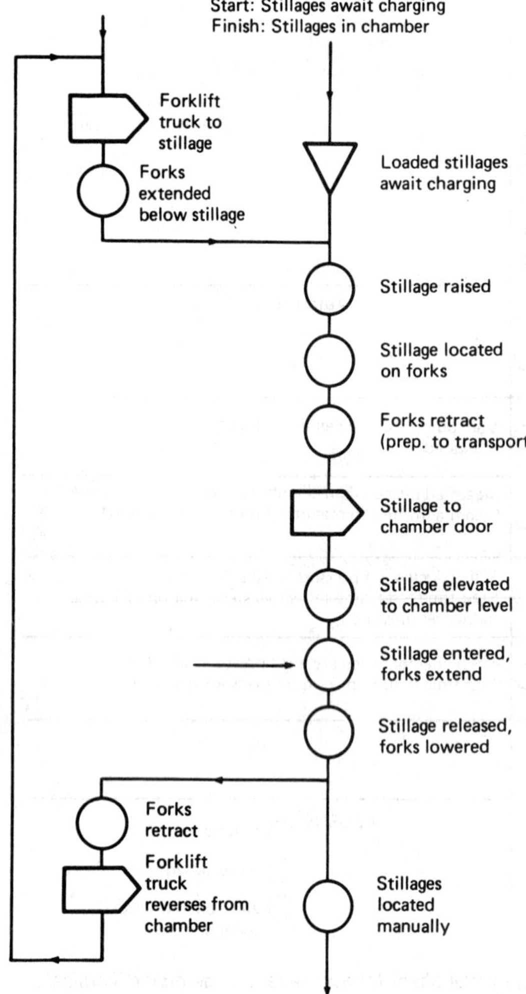

Figure 8.29 *Hazard and operability studies: flow process chart for autoclave arrangement for operation 'Charging autoclave' (Chemical Industry Safety and Health Council, 1977/3)*

foreclosed altogether, and consequently there is a stronger tendency to solve safety problems by 'add-on' measures.

8.14.18 Documentation for hazard studies

As the design progresses, the documentation produced, including the process and plant diagrams, becomes firmer and more detailed.

The most primitive document is a block diagram of the whole process, in which each block represents some operation, possibly a whole plant. Addition of flow quantities and utility usages to this diagram turns it into a quantities flowsheet.

Next come, for each of the blocks in the block diagram, the various versions of the process flow diagram (PFD), these typically being concept, preliminary and final versions. The PFD shows the main items of equipment, utilities and effluents, and control loops.

As the design firms up, the engineering line diagram (ELD) is produced, showing the in-line items, the materials of construction, the equipment sizings, the line specifications and sizings, the equipment and line code numbers, and more details of instrumentation and control loops, as well as of pipework features such as drains and lutes.

An engineering line diagram with full information on the instrumentation constitutes the piping and instrument diagram (P&ID).

The firm ELD or P&ID is the preferred diagram for a hazop study, but in some cases it may be necessary to initiate the study at some earlier stage. For a continuous process the minimum documentation required is the ELD and the outline operating procedures. For a batch process these need to be augmented by the operating sequence.

8.14.19 Personnel involved in hazop studies

As stated earlier, hazop is performed by a multidisciplinary team. The team is normally composed of personnel responsible for design, commissioning and operation, but the detailed composition varies both as between companies and with the type of project within a given company.

A typical team might comprise the study leader, the project engineer, a process engineer, an instrument engineer and the commissioning manager. Other personnel who are often included, depending on the nature of the project, are a chemist, a civil engineer, an electrical engineer, a materials technologist, an operations supervisor, an equipment supplier's representative, and so on.

The Chemetics *Manual* discusses the roles of the following within the hazop study itself: (1) the study leader, (2) the study secretary, (3) the technical team members, and (4) the follow-up co-ordinator. It also deals with the roles of various parties outside the study itself, such as the client for the particular study and the personnel who have responsibilities for hazop within the company.

The technical team members provide the technical input in response to the guide words. They are also able to amplify the information about the plant design given in the plant diagrams, operating instructions, etc. They may receive work assignments arising from the meetings and, as indicated below, one of their number may be appointed to act as the follow-up co-ordinator.

8.14.20 Leadership of hazop studies

The role of the study leader is to act as a facilitator to bring to bear the expert knowledge of the technical team members in a structured interaction. It is not his role to identify hazards and operability problems, but rather to ensure that such identification takes place.

The study leader should be someone not directly involved in the design, but with skills as a hazop leader. The effectiveness of a hazop study is highly dependent on the skill of the study leader. Trained and experienced leaders are crucial.

The study leader is responsible for the definition of the project hazop; for the preparation of the meetings to

***Table 8.31** Hazard and operability studies: results for autoclave arrangement for operation 'Humidify autoclave chamber' (Chemical Industry Safety and Health Council, 1977/3)*

Deviations	*Causes*	*Consequences*
DON'T HUMIDIFY	Valve B not open LD valve closed Steam line choked Vent valve open All steam to vaporizer Line fractured	Hazard to product – sterilizing gas not effective dry
MORE HUMIDIFY	Too much steam – LD valve failed open	Could chamber be overpressured? Is relief valve sized for full bore ingress of steam to chamber?
	Too high steam pressure/temperature	Is product temperature/pressure sensitive? Effect of high temperature and pressure on sealing components of autoclaves, e.g. door seals?
LESS HUMIDIFY	Too little steam, too low steam pressure/temperature	Is condensation on product deleterious?
AS WELL AS HUMIDIFY	Contaminants in steam, e.g. CO_2, condensation, air, rust, etc.	Effect on product?
	Sterilizing gas (valve A passing)	Waste gas by evacuation. Gas into plant atmosphere – is second isolation valve required?
	Air (valve C passing)	Reduces effectiveness of humidification
	Contaminants from vent stack (vent valve passing)	Might induce gases from other autoclaves
	Plant atmosphere (door seals leaking)	No hazard at this stage, but could be leak at sterilizing stage
	Vacuum pump fails	Contaminants from drain enter autoclave
PART OF HUMIDIFY	Steam in but loss of vacuum pump (or valve D closes)	Chamber will be partly pressurized. Load very wet, temperature too high. Chamber partly filled with water. Cycle will continue and sterilizing gas may not enter due to back pressure
	Vacuum, but no steam in	Cycle will continue; if lack of humidity is not detected and action taken, load will *not* be sterile
REVERSE HUMIDIFY	Vacuum drying	No steam in, as above
OTHER THAN HUMIDIFY	Sterilize by omission of humidifying step	Cycle will continue – load will not be sterile (as above)

be held; for arranging the schedules of meetings, and hence their timing and pacing; for assembling an appropriate team and ensuring that they understand their role, and receive training if necessary; for the provision to each study meeting of the necessary documents and other information; for the conduct and recording of the meetings; and for follow-up of matters raised during the meetings.

At the definition stage the study leader should ensure that there is a satisfactory liaison with the client such that the latter will follow up the results emerging from the study.

In the preparation stage the study leader should review the extent to which the plant under consideration is similar to one already studied and how this should affect the study to be conducted. In many cases the design is not a completely new one but constitutes a modification of an existing design. The study leader will then define the features which are novel – raw materials, plant equipment, materials of construction, environmental

Table 8.32 Hazard and operability studies: some additional parameters and guide words (after Tweedale, 1993e)

Process variables	Flow	High; low; reverse; two-phase; leak
	Level	High; low
	Temperature	High; low
	Pressure	High; low; vacuum
	Load	High, low
	Viscosity	High; low
	Quality	Concentration/proportion; impurities; cross contamination, side reactions; particle size; viscosity; water content; inspection and testing
Plant states	Commissioning and startup	Statutory approvals; compliance checking; sequence of steps; supervision, training
	Shutdown	Isolation; purging; cleaning
	Breakdown	Fail-safe response; loss of utilities; emergency procedures
Production	Throughput	Sources of unreliability/unavailability; bottlenecks
	Efficiency	
Materials of construction		Corrosion; erosion; wear; chilling; compatibility; sparking
Plant layout	Access	Operation; maintenance; escape; emergency response
	Space	(For housekeeping, work in progress, escape) Cramped; wasted
	Electrical safety	Hazardous area classification; electrostatic discharge and earthing; lightning protection
Utilities		Air; nitrogen; steam; electrical power; water: process, hot/cold, demineralized, drinking; drainage
Machinery	Machinery	Overload; malfunction; foreign body; rotation: fast, slow; jamming/seizing; frictional overheating; mechanical failure/fracture; impact; valve blockage: mechanical, low temperature; interlocks
	Assembly	Component missing, extra, or wrong; assembly sequence wrong; screwed wrong: (too tight, too loose); crossed/contaminated thread; shielding
	Incompatibility	Tools and equipment material; foreign objects
Materials handling	Speed	Fast; slow; unbalanced
	Packaging	Filling (over, under); damage (external, internal); poor sealing; legal requirements; labelling
	Physical damage	Impact; dropping; vibration
	Stoppages	Breakage; blockage; jamming; loss of feed; loss of packaging; advance warning; rectification
	Direction	Upwards; downwards; to one side; reverse
	Spillage	Spillage: into product, into other materials. Spillage: outside equipment, outside plant
	Location	Wrong: vertical, horizontal; orientation
Stockholding	Stockholding	Failed stock rotation; poor storage (water, vermin); storage of consumables (spares, tools, paints/solvents); storage of raw materials
Hazards	Fire and explosion	Prevention; detection; separation; protection; control
	Toxicity	Acute; long-term; ventilation; control
	Ignition	Friction; impact; static electricity; degraded electricals; failed earthing; mechanical sparking; misalignment; high temperature; dropped objects
	Environment	Housekeeping; dust, spillage, scrap/residues; humidity high, low; ventilation failure
	Environmental control	Effluents: gaseous, liquid, solid; noise; monitoring
Targets	Severity	Quantity exposed; protection and adequacy; venting; personnel escape; unplanned exposure: dust, maintenance, protection malfunction, propagation via duct
	Operator injury	Heavy lifting; repetitive motion; exposure: dust, fume, heat; falling, slipping, tripping
Reactions	Reaction rate	Fast; slow
Control and protection	Control	Sensor and display location; response speed; interlocks
	Protection	Response speed; element common with control loop: sensor, valve, operator; remote actuation; venting; testing
	Safety equipment	Personal equipment; showers
Sequences	Timing	Duration/dwell; rate of approach; sequence; start too early, late; stop too early, late
Testing	Testing	Raw materials; products; equipment; protective instrumentation: alarms, interlocks, trips; protective equipment

conditions, etc. – and decide whether examination of certain parts can be omitted as unproductive. He should also ensure as part of the preparation that the necessary information is available and is correct.

The most elusive skill of the study leader lies in the conduct of the hazop meeting itself. One essential requirement is to ensure that the examination neither becomes too superficial nor gets bogged down in detail so that it identifies all the hazards but within a reasonable time-scale. Another is to manage the personal interactions between the team members, to obtain balanced contributions and to minimize the effect on individuals when the design is subject to criticism. These requirements are easily stated but constitute a significant skill.

A detailed account of the role of the study leader is given in the Chemetics *Manual*.

8.14.21 Follow-up of hazard studies

The output of a hazop study is a set of queries concerning the design. There needs to be a formal arrangement to ensure that these are followed up. One method of doing this is to arrange that at each meeting every question raised is assigned to a specific individual to follow up and, in addition, to appoint a follow-up co-ordinator, preferably a member of the study team, to ensure that this is done. A common practice is to assign to this role a person with line responsibility such as the project manager.

Follow-up is one of the prime functions of computer packages developed to assist hazop.

8.14.22 Computer aids for hazop studies

There are a number of computer programs available to assist in the housekeeping aspects of hazop studies. These are covered in Chapter 29. These programs are distinct from codes which perform hazard identification by emulating hazop studies. Developments in this area are described in Chapter 30.

8.14.23 Experience and further development of hazop studies

The hazop study has been found to be an effective technique for discovering potential hazards and operating difficulties at the design stage. Reductions of at least an order of magnitude in the number of hazards and problems encountered in operation due to such studies have been claimed.

The technique has become firmly established in the process industries, in small as well as large companies, as a prime, or the prime, method of hazard identification. In those companies where the method is used, hazop studies have tended to become an appreciable proportion of the design effort.

The account of the technique given above has followed the original emphasis on the deviations of process variables. Application of the technique has led to its extension particularly to the various activities of operation and maintenance which are the source of many other deviations.

Not surprisingly, the widespread use of the technique has also led to the development of local variants and to adaptations to the needs of the particular companies using it. The hazop study technique has also been extended to cover other concerns such as the environment. In some cases the same study is used to cover, on the one hand, safety and operability and, on the other, environment. Some hazop studies given in the literature are listed in Table 8.33.

Table 8.33 *Selected hazard and operability studies*

Author(s)	*Subject of study*
Lawley (1974b)	Feed section of olefine dimerization plant
Lawley (1976)	Ethylene oxide feed system for batch reactors
CIHSC (1977/3)	Reactor transfer system; feed section of olefine dimerization plant; batch plant; autoclave
Rushford (1977)	Section of cracker unit
Austin and Jeffreys (1979)	Reactor section
Sinnott (1983)	Reactor section of nitric acid plant
Piccinini and Levy (1984)	Ethylene oxide reactor system
Kletz (1985e)	Liquid propane cross-country pipeline
Ozog (1985)	Flammable liquid storage tank
Flothmann and Mjaavatten (1986)	Refrigerated liquid ammonia storage
Kavianen *et al.* (1992)	High pressure/low density polyethylene plant; metal organic chemical vapour deposition process
Knowlton (1992)	Reactor transfer system; batch plant; autoclave; interstage cooler; emergency shutdown and blowdown system; gas supply and oil storage system (hybrid study)

8.14.24 Activities in hazop studies

It is of interest to know the actual activities that are undertaken in a hazop study, particularly for the development of the hazop technique, including computer aided activities. A study of these activities has been made by Roach and Lees (1981), who recorded and analysed the verbal protocol from an industrial hazop.

Figure 8.30 gives a flow chart for a typical study. A large proportion of the protocol consisted, in addition to selection of the item of plant and the parameter deviation for study, of the following activities:

(1) generation of possible causes and consequences of initial deviation;
(2) explanation of features of process material or plant item;
(3) estimation of quantitative aspects of plant dynamics, plant reliability and availability, hazard and risk;
(4) checks on features of design and operation and of potential hazard and operability problems, including detectability of deviations/faults and final consequence hazards and methods for starting, controlling and stopping the operation;
(5) specification of any action to be taken.

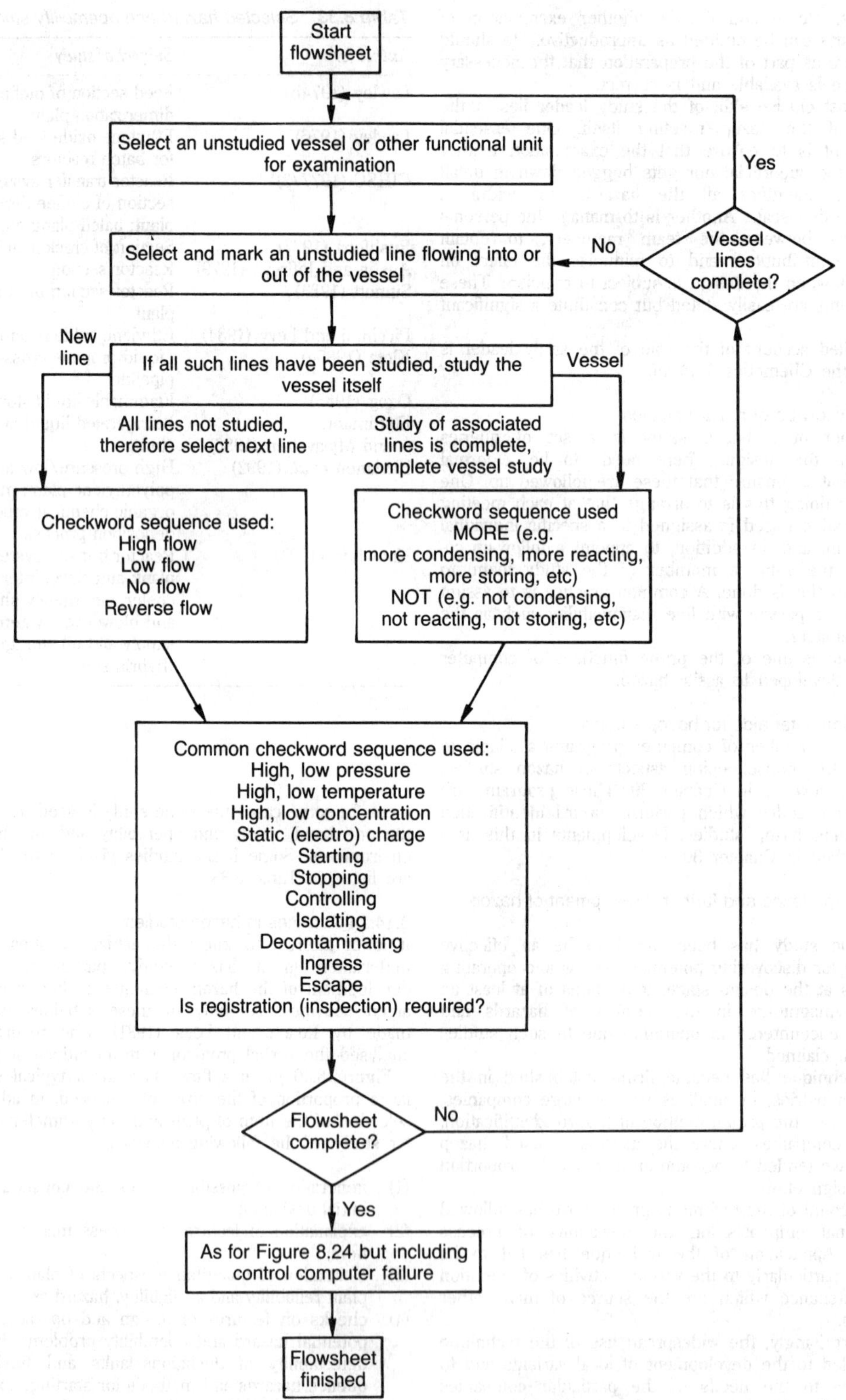

***Figure 8.30** Hazard and operability studies: sequence of activities in a study (Roach and Lees, 1981) (Courtesy of the Institution of Chemical Engineers)*

Table 8.34 *Hazard and operability studies: activities in a study*[a] *(Roach and Lees, 1981)*

A. Features of design and operation

- Plant and pipework configuration:
  - Potential alternative pipework configuration
  - Pipework configuration relative to intended flow
- Equipment capacity, underdesign:
  - Condenser
  - Distillation column trays
  - Pipe (6)
  - Pressure relief valve
  - Vent
- Equipment overdesign:
  - Pump
  - Valve (2)
- Plant turndown (2)
- Materials of construction:
  - Corrosion (2)
  - High temperature
  - Low temperature effect on lined pipe
  - Low temperature embrittlement
  - Product discoloration
- Lagging:
  - Frost protection (5)
  - Heat conservation
  - Personnel protection (4)
- Equipment features and facilities:
  - Access
  - Analysis point (5)
  - Break tank
  - Bund
  - Disengagement space
  - Drain point (2)
  - Droplet knockout facilities
  - Drying facilities
  - Gas balancing
  - Isolation (4)
  - Lute (4)
  - NPSH
  - Purging facilities (2)
  - Self-draining:
    - Condenser fall
    - Pipe fall (3)
    - Restrictor orifice, effect of
  - Siphon breaker (2)
  - Slip plate facilities
  - Spool piece (3)
  - Venting facilities (3)
  - Warming facilities
- Instrumentation and control:
  - Alarms on pumps
  - Alarm dynamic response
  - Alarm setting
  - Flow measurement – full bore flow in pipe
  - Interlocks (3)
  - Potential correction
  - Pressure deviation – effect on distillation column controllers
  - Temperature measurement – full immersion of probe
  - Trip dynamic response
  - Trip setting
- Pressure equipment:
  - Pressure relief
  - Pressure vessel design assumptions
  - Pump delivery pipework rating for high pump suction pressure
  - Rating of vessel as a pressure vessel
- Pumps:
  - Automatic *vs* manual changeover
  - Configuration and layout
  - Kickback
  - Maintenance error–pump wrongly piped up
  - Operating instructions
  - Overdesign
  - Pressure relief valve (positive displacement pump)
  - Replacement error – pump of twice design capacity installed
- Valves:
  - Action on air failure
  - Valve and bypass:
    - Interlock
    - Operating instructions
  - Valve lock open
  - Valve selection – isolation valve *vs* instrument valve
  - Valve trim size
- Power supply
- Earthing, electrical continuity
- Plant layout:
  - Access (3)
  - Floor levels, relative to operations
  - Goggles area
  - Manholes
  - Overflows
- Operating instructions:
  - Additives
  - Distillation column high level
  - Effluent system operation
  - Lute maintenance
  - Pipe function and flow
  - Pump maximum flow
  - Pump operation
  - Reverse flow hazard
  - Valve and bypass
- Sequential operations:
  - Charging of acid and water
- Additives:
  - Operating instructions
  - Pipework
  - Pump interlocks
- Pressure system register:
  - Hazardous pipelines (3)
  - Pipe acting as vent
  - Vent
- Maintenance, maintainability:
  - Column repacking
  - Equipment removal
  - Pressurised equipment (2)

B. Hazards and operability problems

- Air ingress (3)
- Blockage (6)
  - Freezing (2)
  - Icing
  - Polymerization
- Chemical reaction
- Decomposition
- Effluents
  - Impurities
  - Vented materials, fate of
- Erosion
- Explosion
- Explosive impurities
- External fire
- Flammable atmosphere
- Fluid flow phenomena:
  - Boiling (2)
  - Entrainment
  - Erosion
  - Flashing
  - Flooding:
    - Condenser
    - Distillation column (2)
    - Gas absorption column
  - Frothing
  - Gas breakthrough
  - Gas evolution from liquid (2)
  - Gas, vapour lock (2)
  - Hammerblow
  - Inert gas blanketing of heat transfer surface
  - Layering
  - Liquid slugs, surges (2)
  - Pressure surges (2)
  - Reaction forces
  - Siphon (3)
  - Suction due to high downflow
  - Suckback
  - Vortexing
- Heat effects:
  - Heat of mixing (3)
- Impurities:
  - Effluents
  - Process water
  - Product
- Leaks:
  - to Environment
  - within Heat exchanger (4)
  - through Valve (3)
- Lutes
- Maloperation:
  - Pumping of fluid from one pump back through another
- Material deposits
- Sample points:
  - Access (2)
  - Air ingress
  - Liquid circulation
  - Pressurised system
  - Wrong location
- Sabotage
- Static electricity (2)
- Trip action effects
- Utilities failure:
  - Cooling water (3)
  - Electrical power
  - Instrument air
  - Nitrogen
- Vacuum:
  - Absorption of soluble vapour
  - Collapse of pipe lining
- Valves:
  - Heating up of fluid trapped between two closed valves
  - Installed or replaced wrong way round
  - Leaks
- Water ingress (2)

[a] The categories listed are not all mutually exclusive, and a small proportion of checks are listed under more than one entry.

The activity of generating causes and consequences triggered by the application of the guide word to the parameter was relatively unsystematic. The team explored fault paths as they occurred to individuals.

Explanation of features of process materials or plant items emerged as a significant activity. Generally, in a design there are (1) features that are ambiguous, (2) features that have to be modified and (3) features that have characteristics which have not been considered. These features may require further explanation for several reasons: (1) the information may be documented but it is more convenient to retrieve it by verbal explanation, e.g. philosophy of the design calculation; (2) the information is, or should be, part of the design, but it is not yet documented, e.g. intended operating procedure; (3) the information is generated by the hazop team during the study. Some topics on which explanations were given are shown in Table 8.34, Section C.

Another significant activity was estimation of quantitative or semi-quantitative aspects of plant dynamics, plant

***Table 8.35** Hazard and operability studies: checks made in a study[a] (Roach and Lees, 1981) (Courtesy of the Institution of Chemical Engineers)*

A Check words applicable to deviation of process variables

NO, NOT	Negation of intention
MORE	Quantitative increase
LESS	Quantitative decrease
AS WELL AS	Qualitative increase
PART OF	Qualitative decrease
REVERSE	Logical opposite of intention
OTHER THAN	Complete substitution

B Other check words

STARTING
STOPPING
CONTROLLING
ISOLATION
ESCAPE
DECONTAMINATION
INGRESS

C Some topics on which explanations are given

Flowsheet discrepancies
Function of a unit:
  Drain/flushing point
  Valve
Modes of operation of unit
Design capacity/overdesign of unit
Materials of construction
Lagging
Physical characteristics of process materials:
  Miscibility
  Solubility
Chemical characteristics/reactions of process materials:
  Decomposition
Additives, functions of
Impurities, fate of
Pressures and pressure drops:
  Gravity *vs* pumped flows (2)
  Units at atmospheric pressure
System arrangements:
  Bypasses (2)
  Gas balancing
  Inserting/purging
  Isolation
  Lutes
  Purging (2)
  Sampling (2)
  Venting
Instrumentation, control and protection
  Conventional control system configuration, logic and behaviour (3)
  Computer control configuration logic and behaviour (3)
  Manual control loops
  Trip *vs* alarm setting
  Trip or pressure relief setting
Quality standards:
  Products
  Effluents (2)
Potential deviation:
  Pressure
Failure effects:
  Flow meter failure
  Pump corrosion
  Pump start failure
  Pump valve configuration errors
Failure data:
  Failure rate
  Repair time
Company practices:
  Lagging
  Overpressure protection
Statutory requirements:
  Break tank

D Some features on which estimates are made

Fraction of time spent by unit in different operational modes
Magnitude of deviations (4)
Steady-state gain relating consequence and cause deviations
Time for deviation/fault to give rise to consequences (3)
Time for deviation/fault to be detected (2)
Performance of plant unit in face of deviation
Failure rate:
  Utilities (2)
Event rate
Repair time
Fractional dead time (3)
Degree of hazard/inoperability (4)
Frequency/probability of hazard/inoperability (8)

E Checks on detectability of deviations/faults and of correct operation

Deviations/faults checked for detectability

Process variable deviations:
  High flow (2)
  Low flow
  Reverse flow (2)
  High level (4)
  Low level (3)
  High pressure
  Low pressure (2)
  High temperature
  Low temperature
Contamination (3)
Additive deviations
Blockages:
  Blockages (3)
  Freezing (3)
Leaks
Flooding
Utility failures:
  Cooling water (2)
  Refrigeration set
  Steam
Instrument failure
Pump failure (2)
Valve fails open
Valve misdirected shut (2)
Valve installation errors on pump
Lute problems

Flow at pumps and valves checked for detectability

Pumps
Valves

F Principal final and intermediate consequences of deviations/faults

Overpressure (5)
Underpressure, vacuum collapse (2)
Overtemperature (3)
Undertemperature (2)
Corrosion (5)
Explosive mixture, confined
Explosion, confined
Equipment damage
Overflow (liquids) (11)
Boilover
Escape (gases)
Escape (liquids)
Product contamination (4)
Effluent (flow normally exists):
  Contamination (3)
  High flow
  High temperature
  pH deviation (2)
Effluent (occasional venting flow occurs) (3):
  High flow (2)
Forced shut-down:
  Manual
  Automatic trip (2)
  Equipment failure
Personnel injury (6)

G Decisions on simultaneous deviations/faults

One of two pump motors down for maintenance AND electrical supply phased – no action
Process upset AND cooling water failure – no action
Steam flow high AND cooling water failure – no action
Pump failure AND low flow alarm failure – further investigation
Deviation in pH AND deviation in chemical composition with consequent possibility of SO_2 generation – no action

H Uncommunicated features, ambiguities and variations in the design

Uncommunicated design features
Uncommunicated design changes
Ambiguity about vessel design pressure
Equipment designed for throughputs other than flow sheet one
Potential variations in method of plant operation not communicated

[a] Numbers in parentheses indicate the number of times an item occurs in the protocol. If no number is given, an item occurs once. Entries followed by a colon are headings, not specific items.

reliability and availability, and hazard and risk. Some features on which estimates were made are given in Table 8.34, Section D.

The central activity was checks on features of design and operation and on hazards and operability problems. Some of these checks are listed in Table 8.34. The checks given are limited to those arising directly or indirectly from use of the guide words which were not instantly dismissed. An important class of check was those on the detectability of deviations/faults. A large proportion of operability problems are problems of deviation/fault detection. Some of the checks of this type are listed in Table 8.34, Section E.

Most of the hazards listed in Table 8.34 are intermediate events rather than final consequences. Table 8.34, Section F, gives some of the principal final consequences of ultimate interest in the study.

Some of the hazards considered would be identified only if two deviations/faults were to occur simultaneously. Some of the decisions made on such simultaneous deviations/faults are given in Table 8.34, Section G. In four out of the five cases shown in the table the probability of simultaneous occurrence was assessed as negligible. In the fifth case one of the faults is failure of a low flow alarm system. This type of failure could well occur before the low flow deviation but remain unrevealed so that the probability of 'simultaneous' deviations/faults could be much higher in this case unless the instrument was proof tested. The problem of unrevealed failure was evidently recognized by the team in assessing this risk as not negligible.

The analysis revealed several problems of design communication. Some of these are shown in Table 8.34, Section H.

8.14.25 Filtering in hazop studies

The account given so far of hazop studies has emphasized primarily the generation of deviations. But hazop is not in fact a pure identification technique. A significant part of the activity in a hazop is concerned with the filtering of the deviations identified. It is not the function of the hazop team to carry out 'instant design'. But it is a proper function to carry out a coarse filtering process on the deviations generated in order to prevent the design systems being swamped with insignificant items. The contribution of experienced people is probably at least as much in filtering as in generation.

8.14.26 Quality assurance

A survey of hazop studies conducted as part of the safety case for oil platforms in the North Sea has been described by Rushton *et al.* (1994). The survey covered the context, policy, practical aspects and quality assurance of hazop.

A more detailed account of possible approaches to the quality assurance of hazop is given in a further report by Rushton (1995).

Quality assurance is discussed further in Sections 8.29–8.31.

8.14.27 Limitations of hazop studies

The hazop technique has been widely adopted and is the centrepiece of the hazard identification system in many companies. It does, however, have some limitations.

These limitations are of two kinds. The first type arises from the assumptions underlying the method and is an intended limitation of scope. In its original form, the method assumes that the design has been carried out in accordance with the appropriate codes. Thus, for example, it is assumed that the design caters for the pressures at normal operating conditions and intended relief conditions. It is then the function of hazop to identify pressure deviations which may not have been foreseen.

The other type of limitation is that which is not intended, or desirable, but is simply inherent in the method. Hazop is not, for example, particularly well suited to deal with spatial features associated with plant layout and their resultant effects.

8.14.28 Variants of hazop studies

There are a large number of variants on the original hazop study. Many, perhaps most, companies have adapted the technique to their own needs. Many of these adaptations extend the method in respect of non-normal operation or of other activities such as maintenance, or to take account of other concerns such as environmental effects. Another type of variant is the application of the basic technique of hazop to systems other than process vessels and pipework. Some of these are described in Section 8.14.29. In this regard it should be remembered that hazop itself was an application of method study, and it may often be better to step back to this origin and then forward again from there rather than to try to apply hazop *per se*.

8.14.29 Other applications of hazop methodology

The methodology of hazop, or its parent critical examination, has been applied to a number of other situations, many of which are of interest in the process industries. These include (1) plant modification, (2) plant commissioning, (3) plant maintenance, (4) emergency shut-down and emergency systems, (5) mechanical handling, (6) tanker loading and unloading, (7) works traffic, (8) construction and demolition and (9) buildings and building services.

8.15 Failure Modes, Effects and Criticality Analysis

Another major method of hazard identification is failure modes and effects analysis and its extension failure modes, effects and criticality analysis.

Accounts of failure modes and effects analysis (FMEA) and of failure modes, effects and criticality analysis (FMECA) are given in the CCPS *Hazard Evaluation Guidelines* (1985/1, 1992/9) and by Recht (1966b), J.R. Taylor (1973, 1974c, 1975a), Himmelblau (1978), Lambert (1978a), A.E. Green (1983), Flothmann and Mjaavatten (1986), Moubray (1991), Kavianian, Rao and Brown (1992), D. Scott and Crawley (1992) and Goyal (1993). A relevant code is BS 5760, which is described below.

Failure modes and effects analysis involves reviewing systems to discover the mode of failure which may occur and the effects of each failure mode. The technique is oriented towards equipment rather than process parameters. Table 8.36 shows a set of typical results from an FMEA study on a process plant given by Recht (1966b).

8.15.1 BS 5760

Guidance on failure modes and effects analysis is given in BS 5760 *Reliability of Systems, Equipment and Components*, Part 5:1991 *Guide to Failure Modes, Effects and Criticality Analysis (FMEA and FMECA)*. BS 5760: Part 5: 1991 deals with the purposes, principles, procedure and applications of FMEA, with its limitations and its relationship to other methods of hazard identification, and gives examples.

***Table 8.36** Failure modes and effects analysis: result for a process plant (after Recht, 1966b; Himmelblau, 1978)*

Component	*Failure or error mode*	*Effects on other components*	*Effects on whole system*	*Hazard class*[a] *1*	*2*	*3*	*4*	*Failure frequency*	*Detection methods*	*Compensating provisions and remarks*
Pressure relief valve	Jammed open	Increased operation of temperature sensing controller, and gas flow, due to hot water loss	Loss of hot water; greater cold water input, and greater gas consumption	1				Reasonably probable	Observe at pressure-relief valve	Shut off water supply, reseat or replace relief valve
	Jammed closed	None	None	1				Probable	Manual testing	Unless combined with other component failure, this failure has no consequence
Gas valve	Jammed open	Burner continues to operate. Pressure-relief valve opens	Water temperature and pressure increase. Water → steam			3		Reasonably probable	Water at faucet too hot. Pressure-relief valve open (observation)	Open hot water faucet to relieve pressure. Shut off gas supply. Pressure-relief valve compensates
	Jammed closed	Burner ceases to operate	System fails to produce hot water	1				Remote	Observe at output (water temperature too low)	
Temperature measuring and comparing device	Fails to react to temperature rise above preset level	Controller gas valve, burner continue to function 'on'. Pressure-relief valve opens	Water temperature too high. Water → steam			3		Remote	Observe at output (faucet)	Pressure-relief valve compensates. Open hot water faucet to relieve pressure. Shut off gas supply
Temperature measuring and comparing device	Fails to react to temperature drop below preset level	Controller, gas valve, burner continue to function off	Water temperature too low	1				Remote	Observe at output faucets	
Flue	Blocked	Incomplete combustion at burner	Inefficiency. Production of toxic gasses				4	Remote	Possibly smell products of incomplete combustion	No compensation built in. Shut down system
Pressure-relief valve and gas valve	Jammed closed Jammed open	Burner continues to operate, pressure increases	Increased pressure cannot bleed at relief valve. Water → steam. If pressure cannot back up cold water inlet, system may rupture violently				4	Probable Reasonably probable Reasonably probable	Manual testing of relief valve Observe water output Temperature too high	Open hot water faucet. Shut off gas supply. Pressure might be able to back up into cold water supply, providing pressure in supply is not greater than failure pressure of system

[a] 1, Negligible effects; 2, marginal effects; 3, critical effects; 4, catastrophic effects.

FMECA is an enhancement of FMEA in which a criticality analysis is performed. Criticality is a function of the severity of the effect and the frequency with which it is expected to occur. The criticality analysis involves assigning to each failure mode a frequency and to each failure effect a severity.

The purpose of a FMEA is to identify the failures which have undesired effects on system operation. Its objectives include: (1) identification of each failure mode, of the sequence of events associated with it and of its causes and effects; (2) a classification of each failure mode by relevant characteristics, including detectability, diagnosability, testability, item replaceability, compensating and operating provisions; and, for FMECA, (3) an assessment of the criticality of each failure mode.

The standard lists the information required for an FMEA, under the headings: (1) system structure; (2) system initiation, operation, control and maintenance; (3)

system environment; (4) system modelling; (5) system software; (6) system boundary; (7) system functional structure; (8) system functional structure representation; (9) block diagrams; and (10) failure significance and compensating provisions.

Core information on the items studied is the (1) name, (2) function, (3) identification, (4) failure modes, (5) failure causes, (6) failure effects on system, (7) failure detection methods, (8) compensating provisions, (9) severity of effects and (10) comments.

The main documentation used in an FMEA is the functional diagram. Use may also be made of reliability block diagrams.

The identification of the failure modes, causes and effects is assisted by the preparation of a list of the expected failure modes in the light of (1) the use of the system, (2) the element involved, (3) the mode of operation, (4) the operation specification, (5) the time constraints and (6) the environment.

The failure modes may be described at two levels: generic failure modes and specific failure modes. The standard gives as an example of a set of generic failure modes: (1) failure during operation, (2) failure to operate at a prescribed time, (3) failure to cease operation at a prescribed time and (4) premature operation. As examples of specific failure modes the standard gives: (1) cracked/fractured, (2) distorted, (3) undersized, and so on.

The failure causes associated with each mode should be identified. BS 5760 gives a checklist of potential failure causes under the headings (1) specification, (2) design, (3) manufacture, (4) installation, (5) operation, (6) maintenance, (7) environment and (8) uncontrollable forces.

The failure effects involve changes in the operation, function or status of the system and these should be identified by the analysis. Failure effects can be classified as local or as end effects. Local effects refer to the consequences at the level of the element under consideration and end effects to those at the highest level of the system.

Where FMEA is to be applied within a hierarchical structure, it is preferable to restrict it to two levels only and to perform separate analyses at the different levels. The failure effects identified at one level may be used as the failure modes of the next level up, and so on.

BS 5760: Part 5: 1991 contains in Appendix B a number of tables illustrating the results of FMECAs conducted on various types of system. They include analysis of a subsystem of a motor generator set and of the fire protection system of an electric locomotive. Taking the latter as being most representative of process applications, the table is headed by a declaration that the system considered is the fire protection system and that its function is to detect and extinguish a fire within the locomotive. The columns of the table are headed as follows: (1) item, (2) item failure mode, (3) failure cause, (4) block function description, (5) functional failure mode, (6) effect on subsystem outputs, (7) effect on system reliability, (8) effect on system safety, (9) preventive action (design), (10) preventive action (quality assurance), (11) comment, (12) severity and (13) failure rate. Item failure mode is exemplified by 'failed to open circuit conditions', and functional failure mode by 'detection of false fire'.

FMEA is an efficient method of analysing elements which can cause failure of the whole, or of a large part, of a system. It works best where the failure logic is essentially a series one. It is much less suitable where complex logic is required to described system failure.

FMEA is an inductive method. A complementary deductive method is provided by fault tree analysis, which is the more suitable where analysis of complex failure logic is required.

BS 5760: Part 5: 1991 states that FMEA can be a laborious and inefficient process unless judiciously applied. The uses to which the results are to be put should be defined. FMEA should not be included in specifications indiscriminately.

8.15.2 Application of FMEA

The range of applications of FMEA is very wide. At one end of the spectrum, A.E. Green (1983) has applied it to a pneumatic differential pressure transmitter, whilst at the other Aldwinkle and Slater (1983) have described its application to a liquefied natural gas terminal. Table 8.37 lists some of the FMEA studies given in the literature.

8.16 Sneak Analysis

In contrast to general methods such as hazop and FMECA, there are also niche methods. One of these is sneak analysis. Accounts of sneak analysis are given by E.J. Hill and Bose (1975), J.R. Taylor (1979, 1992), Rankin (1984), Dore (1991), Hahn *et al.* (1991), Hokstad, Aro and Taylor (1991) and Whetton (1993b).

Sneak analysis originated in sneak circuit analysis (SCA), a method of identifying design errors in electronic circuits (E.J. Hill and Bose, 1975). The technique was

Table 8.37 *Failure modes and effects analysis: selected studies*

Author(s)	*Subject of study*
Recht (1966b)	Hot water system
C.F. King and Rudd (1972)	Heavy water recovery plant
Eames (1973 UKAEA SRS/GR/12)	Nuclear instrumentation
Lambert (1978a)	
Aldwinkle and Slater (1983)	Liquified natural gas terminal
A.E. Green (1983)	Pneumatic differential pressure transmitter
Flothmann and Mjaavatten (1986)	Refrigerated liquid ammonia storage
BS 5760: Part 5:1991	Fire protection system of an electric locomotive; subsystem of a motor generator set
Kavianian, *et al.* (1992)	Metal organic chemical vapour deposition process

developed within a particular company and was slow to spread. An early use in the process industry was in the work of J.R. Taylor (1979) on sneak path analysis. A lead in encouraging the wider use of sneak analysis has been taken by the European Space Agency (ESA), as described by Dore (1991).

A review of sneak analysis is given by Whetton (1993b), who gives as a rough definition: 'A sneak is an undesired condition which occurs as a consequence of a design error, sometimes, but not necessarily, in conjunction with a failure.'

In sneak analysis generally, four recognized categories of sneak have emerged: (1) path, (2) indication, (3) label and (4) timing. Whetton refers to the following categories, which have developed from process work: (1) flow, (2) indication, (3) energy, (4) procedure, (5) reaction and (6) label.

8.16.1 Types of sneak

The different types of sneak which can occur are discussed by Whetton, with examples.

Sneak flow
Sneak flow is unintended flow from a source to a target. An example is the case where two vessels containing liquids, one at higher pressure than the other, are connected to a common drain header. If the bottom outlet valves on both vessels are open at the same time, liquid can flow from the higher to the lower pressure vessel.

Sneak indication
A sneak indication is one which is incorrect or ambiguous. An example is the power-operated relief valve (PORV) at Three Mile Island, where there was displayed an indication of 'valve position' which was in fact the signal to, rather than the actual position of, the valve.

Sneak label
Likewise, a sneak label is one which is incorrect or ambiguous.

Sneak energy
Sneak energy is the unintended presence or absence of energy. An example is the energy release attendant on the restart of agitation in a reactor where the agitator has been temporarily switched off.

Sneak reaction
Sneak reaction is an unintended reaction. Examples are a side reaction in a reactor and an undesired reaction between a fluid and a material of construction.

Sneak procedure or sequence
A sneak procedure or sequence is the occurrence of events in an unintended or conflicting order.

8.16.2 Sneak analysis methods

Whetton distinguishes three basic methods of sneak analysis: (1) topological, (2) path and (3) clue.

The original sneak analysis method was based on decomposing electrical circuits into standard subnetworks. According to Whetton, the method has proved difficult to adapt to process plants in that successful application depends on the ability of the analyst to cast process phenomena in the form of electrical analogues.

Sneak path analysis, developed by J.R. Taylor (1979, 1992), is the investigation of unintended flow between a source and a target. It is performed in a systematic manner by decomposing the piping and instrument diagram into functionally independent sections and identifying sources and targets, source–target pairs and paths between sources and targets. One method of doing this is the use of coloured markers to trace the various paths, sometimes known as the 'rainbow' technique.

The clue method involves the use of a structured checklist. It is applicable to all types of sneak and examines each type by means of a question set.

8.16.3 Sneak-augmented hazop

Sneak analysis may be used as an enhancement of hazop. A method of effecting this is described by Whetton. The techniques used are the path and clue methods. Whetton gives a detailed procedure for the conduct of a sneak-augmented hazop. All drawings used should be as-built. Information on labels should cover the as-built condition. Essentially, the sneak analysis aspect of the procedure involves an examination of the various types of sneak against the appropriate clue list. Among the other points made are the following:

Sneak flow
Examine all paths including utilities and drains, sewers and vents
Examine, in turn, all plant states such as normal operation, start-up, shut-down, etc.
Examine for each valve in turn effect of unintended deviation in valve state

Sneak indications
Examine each instrument
Examine control room, including humble lamps

Sneak labels
Examine consistency of labels against as-built drawings
Examine consistency of labels on instruments

Sneak procedures
Examine all procedures, including start-up, shut-down, etc., and maintenance

Sneak energy
Examine, in turn, for all plant states, including start-up, shut-down, etc., and maintenance

8.17 Computer Hazop

The hazop study method just described was developed for plants in which predominantly the control system was based on analogue controllers. The advent of computer control has created the need for some method which addresses the specific problems of this form of control. There are now the beginnings of a methodology for the hazop-like study of computer controlled process systems, or computer hazop (CHAZOP). This is quite distinct from computer-aiding of hazop.

Accounts of approaches to computer hazop include those by Andow (1991), Burns and Pitblado (1993), Fink *et al.* (1993), Lear (1993), Nimmo *et al.* (1993 LPB 111),

Table 8.38 *Computer hazop: task considerations and attributes for task 'intervention' (Chung and Broomfield, 1994)*

Task	*Task consideration*	*Task attributes*	*Questions*	*Incidents*
Intervention	Specification	Definition	...	...
		Objective	...	...
		Options	...	...
		Input/outps	...	...
		Timing/control	...	...
		...	...	...
	Association	Tasks	...	...
		Devices	...	...
	Implementation	Selection	...	...
		Installation	...	...
		Testing	...	...
		Maintenance	...	...
		Environment	...	...
		Utilities	...	...
	Fail safe/protection	Failure detection	...	...
		Interlocks	...	...
		Trips	...	...
		Recovery procedure	...	...
		...	...	...
	Failure modes	Not initialized	...	...
		Incorrectly executed	...	...
		Not terminated	...	...
		...	...	...

Broomfield and Chung (1994) and Chung and Broomfield (1994).

An early expression of concern was that by Kletz (1982g), who highlights the human problem associated with computer control. The HSE has identified this as a topic requiring attention. One outcome has been its work on programmable electronic systems, leading to the HSE *PES Guide*, which deals with the reliability of the computer system itself, but the concern is wider than this, as described by P.G. Jones (1989, 1991).

8.17.1 Checklist and guide word methods

A method for computer hazop which has the same general approach as conventional hazop has been described by Andow (1991). Like hazop it is intended to be carried out by a multidisciplinary team following a systematic methodology and using standard guidewords and questions.

Andow describes a two-part study format, comprising a preliminary study and then the main study. The purpose of the preliminary study is to identify critical features early in the design. The study covers (1) system architecture, (2) safety related features, (3) system failure and (4) utilities failure. In the full study attention is directed to (1) the computer-system environment, (2) input/output signals and (3) complex control schemes.

Study of the computer system environment involves examination of the machine itself and of cabinet, crates, etc.; control, input/output cards; communication links; operator consoles; power supplies; watchdog timers; and other utilities. It uses questions to establish design intent and then a question set to investigate failures:

(1) Does it matter?
(2) Will the operator know?
(3) What should he do?

The input/output signals are studied following rather more closely the hazop format, with examination of items such as the signal using guide words such as LOW, HIGH, INVARIANT, DRIFTING and BAD followed up with the failure question set just quoted. Another item examined is the actuator.

Complex control schemes are examined using a list of potential weak points and corresponding problems:

Control tuning	Initialization and windup
Points of operator access	Set-points, cascades that may be made or broken
Interaction with other schemes	Start-up, normal operation, shut-down; synchronization and timing issues; expected or required operator actions

The approach described by Burns and Pitblado (1993) also follows that of hazop; it applies to features, such as signal or information, guide words such as NO, MORE, LESS and WRONG.

8.17.2 Task analysis method

A quite different approach is that described by Broomfield and Chung (Broomfield and Chung, 1994; Chung and Broomfield, 1994). The methodology developed by these authors is based on an analysis of some 300 incidents in two organizations – one a process industry and one an avionics company. As an example of a process industry incident, they quote the premature start of a control sequence due to inadequate specification of the preconditions for initiation.

The underlying premise of the method is that most incidents are the result of deficiencies in the software–hardware interface. The authors state that attempts to conduct separate analyses of the hardware and the software in order to simplify the task, actually complicates it.

This interface is represented by four concentric functional levels. Each level is associated with certain system components. Each component has a defined task. Starting at the periphery and working in to the centre, these are as follows:

Functional level	*Component*	*Task*
Intervention	Utility	Intervention
	Operator	Intervention
Input/output	Sensor	Input
	Human input device	Input
	Actuator	Output
	Display	Output
Communication	Communication link	Communication
Control and processing	Computer	Control and processing

There is for each task a guide word and question set. Those for the intervention task are shown in Table 8.38.

8.18 Human Error Analysis

An important source of hazards and losses is maloperation of the plant, and other forms of operator error. There are a number of methods for addressing this problem. One of these is the hazop study which, as just described, utilizes aids such as the sequence diagram to discover potential for maloperation. Other methods which may be mentioned here are task analysis and action error analysis. The problem of human error is, however, very complex. It is considered in more detail in Chapter 14.

8.18.1 Task analysis

Task analysis is a technique which was originally developed as a training tool. As applied to the process operator, it involves breaking down the task of running the plant into separate operations carried out according to a plan. The object is to discover potential difficulties, and hence errors, in executing the individual operations or the overall plan. A detailed account is given in Chapter 14.

Clearly such task analysis ranks also as a hazard identification technique. Task analysis has been applied to process plants with good results, but is not in widespread use. It may also be noted in passing that the writing of operating instructions constitutes a less formalized type of task analysis, and may likewise serve to identify hazards.

Table 8.39 *Action error analysis: errors handled (J.R. Taylor, 1979)*

Cessation of a procedure
Excessive delay in carrying out an action or omission of an action
Premature execution of an action – timing error
Premature execution of an action – preconditions not fulfilled
Execution on wrong object of action
Single extraneous action
In making a decision explicitly included in a procedure, taking the wrong alternative
In making an adjustment or an instrument reading, making an error outside tolerance limits

8.18.2 Action error analysis

Another technique of identifying operating errors is action error analysis, developed by J.R. Taylor (1979). This is a method of analysing the operating procedures to discover possible errors in carrying them out.

The actions to be carried out on the process interface are listed in turn, each action being followed by its effects on the plant, so that a sequence is obtained of the form:

Action – Effect on plant – Action – Effect on plant . . .

The actions are interventions on the plant such as pushing buttons, moving valves, etc.

The errors handled in the method are shown in Table 8.39. The effects of possible errors are then examined using guide words for action. The main guide words are:

TOO EARLY
TOO LATE
TOO MUCH
TOO LITTLE
TOO LONG
TOO SHORT
WRONG DIRECTION
ON WRONG OBJECT
WRONG ACTION

In analysing the effects of an error, an important consideration is whether the effect can be observed and corrected.

It is necessary to limit the scope of the analysis. Consideration of actions on wrong objects may be confined to those which are physically or psychologically close to the correct object. The number of wrong actions to be considered has to be kept small.

Multiple errors are considered only to a limited extent. Thus in a large operating procedure it would normally be practical to take into account only single initial errors, but in a few cases it may be possible to use heuristic rules to identify double errors, a factor which it is worthwhile to explore. For example, one error may result in material being left in a vessel, while a second error may result in an accident arising from this.

The technique of action error analysis appears to have been quite widely used in Nordic countries, but much less so elsewhere.

8.19 Scenario Development

8.19.1 Release scenarios

The methods just described are concerned with hazard and operability problems which occur on most plants. A rather different question is the identification of potential sources of major release of hazardous materials which may pose a risk to the workforce and/or the plant, and in unfavourable circumstances may even affect the public.

Information on release sources is required for hazard assessment and for emergency planning. In order to identify such sources it is necessary to carry out a review. In principle, virtually all elements of the pressure system (vessels, pumps, pipework) are points at which a release may occur. Table 8.40 gives a list of some of these release sources. A more detailed checklist of release sources is given in the CCPS *QRA Guidelines* (1989/5). The identification of release sources in a hazard assessment is illustrated by the studies given in the *Rijnmond Report* described in Appendix 8.

The role of the hazop study in identifying such sources merits mention. A hazop study starts from the assumption that the plant is basically well designed mechanically and concentrates primarily on process parameter deviations. The type of release which a hazop study might identify would be overfilling of a tank as a result of maloperation, or brittle fracture of a pipe due to contact with cold liquefied gas, but the study is not normally concerned with mechanical failures which occur while the process parameters are within their design range. The points on the plant at which a release may occur can generally be identified by the release source review. What the hazop study does is to identify specific credible realizations of release from these points.

For a potential release, it is necessary not only to identify a source but also to decide on the nature of the release which could occur. The tendency is to assume that the fluid released is at its design conditions. A hazop study may reveal that this is not so. For example, materials may be released from a storage tank as a result of a chemical reaction in the tank which causes the liquid to heat up considerably above its storage temperature and hence when released to give a much larger fraction of vapour.

Table 8.40 *Some sources of release on a plant*

Pipework:
- Pipe rupture
- Pipe flanges
- Valves
- Hoses

Compressors
Pumps
Agitators
Vessels and tanks:
- Vessel/tank rupture
- Vessel/tank overfilling

Reactors
Drain points
Sample points
Drains
Relief devices:
- Pressure relief valves
- Bursting discs

Vents
Flares
Operational release
Maintenance release

8.19.2 Escalation scenarios

The identification process should not stop at the point where a release occurs, but in principle should be continued to embrace the consequences of the release and the failures and other events which may allow these consequences to escalate. In practice, this aspect is usually treated as part of the hazard assessment. Whilst this is a reasonable approach, there is a danger that, unless the identification of the escalation modes is treated with a thoroughness matching that applied to the identification of the release modes, features which permit escalation and against which measures might be taken will be missed. The identification of the modes of escalation appears to be a rather neglected topic.

8.20 Consequence Modelling

For major hazard plants it is common practice following the development of the scenarios just described to carry out some degree of quantitative modelling of the consequences. Such consequence modelling is described in more detail in the next chapter. Here it is sufficient to note its effect in giving a much better understanding of the likely development of the release and its potential for the identification both of hazards and of mitigatory measures.

8.21 Process Safety Review System

It is convenient at this point to consider under the heading of process safety reviews the family of techniques for hazard identification, hazard assessment, audit and accident investigation developed by Wells and co-workers.

8.21.1 HAZCHECK

The first element in the system is HAZCHECK, which might be described as an enhanced checklist system. Accounts of HAZCHECK are given by Reeves, Wells and Linkens (1989), Hurst and Reeves (1990), Wells, Phang and Reeves (1991) and Wells (1992 LPB 105).

HAZCHECK is essentially a structured checklist, or set of checklists. The main headings are given in the master list and are expanded in subordinate lists. The master list has been given in a number of versions. It is convenient here to give that shown in Table 8.41, which is in a form structured to give a representation of a general incident scenario. HAZCHECK is implemented as a computer program.

Table 8.41 *Process safety review system: HAZCHECK – master checklist (Allum and Wells, 1993) (Courtesy of the Institution of Chemical Engineers)*

Damage and harm Consequences from appreciable to catastrophic Minor consequences or near miss	
Further escalation Post-incident damage Further dispersion on ground Further dispersion in air Damage by chemicals Damage by missiles or impact Damage by fire or explosion	*Failure to prevent further escalation* Inadequate post-incident response Failure of public response Failure of off-site emergency response Failure of on-site emergency response
Escalation of events Damage and harm on escalation Escalation by fire or explosion Ignition of flammable mixture Dispersion of chemicals	*Failure to mitigate or prevent escalation* Failure of emergency response to prevent escalation Failure of emergency response to mitigate effects
Significant release of material Release of material causes damage/harm Release creates hazard or hazardous condition	*Failure to recover situation after release* Release fails to disperse safely Accumulation after release Release fails to attenuate Immediate emergency response inadequate Inadequate protection, passive protection
Release of material Rupture of plant with release Discharge of process material	*Failure to recover situation before release* Operator action fails Control systems fail to recover situation
Dangerous disturbance of plant Disturbance ultimately exceeding critical defect or deterioration in construction Flow through abnormal opening to atmosphere Change in planned discharge or vent	*Inadequate emergency control or action* Emergency control system fails to correct
Hazardous disturbance of plant Hazardous trend in operation conditions Construction defective or deteriorated in service Abnormal opening in equipment Change in planned discharge or vent	*Inadequate emergency control or action* Normal control systems fail to correct the situation Operators fail to correct the situation Maintenance fail to correct the situation
Immediate causes of failure or disturbance Action by plant personnel inadequate Defects directly cause loss of plant integrity Plant or equipment inadequate or inoperable Control system or emergency control inadequate Change from design intent Environmental and external causes of disturbance	*Root causes of failure disturbance* Site and plant facilities Operator performance Information systems and procedures Management performance Resource provision Organization and management systems System climate External systems

The application of HAZCHECK to a refinery flare system incident is described by Wells (1992 LPB 105).

8.21.2 General incident scenario

The next element is a method of defining incident scenarios, which is a necessary part of both hazard assessment and accident investigation. An account of this is given by Wells *et al.* (1992). The general incident scenario (GIS) described by these authors is a structure intended to be applicable to any incident. As just stated, it has essentially the same entities as the HAZCHECK master list, but lays them out in a more structured way. The logic of an incident is also represented by the general incident structure shown in Figure 8.31. The GIS may be used for hazard identification, preliminary hazard analysis or incident investigation.

8.21.3 Preliminary safety analysis

The general incident scenario provides the basis for a set of techniques for preliminary safety analysis. Accounts of preliminary safety analysis (PSA) are given by Wells, Wardman and Whetton (1993) and Wells, Phang and Wardman (1994). The PSA methods described are:

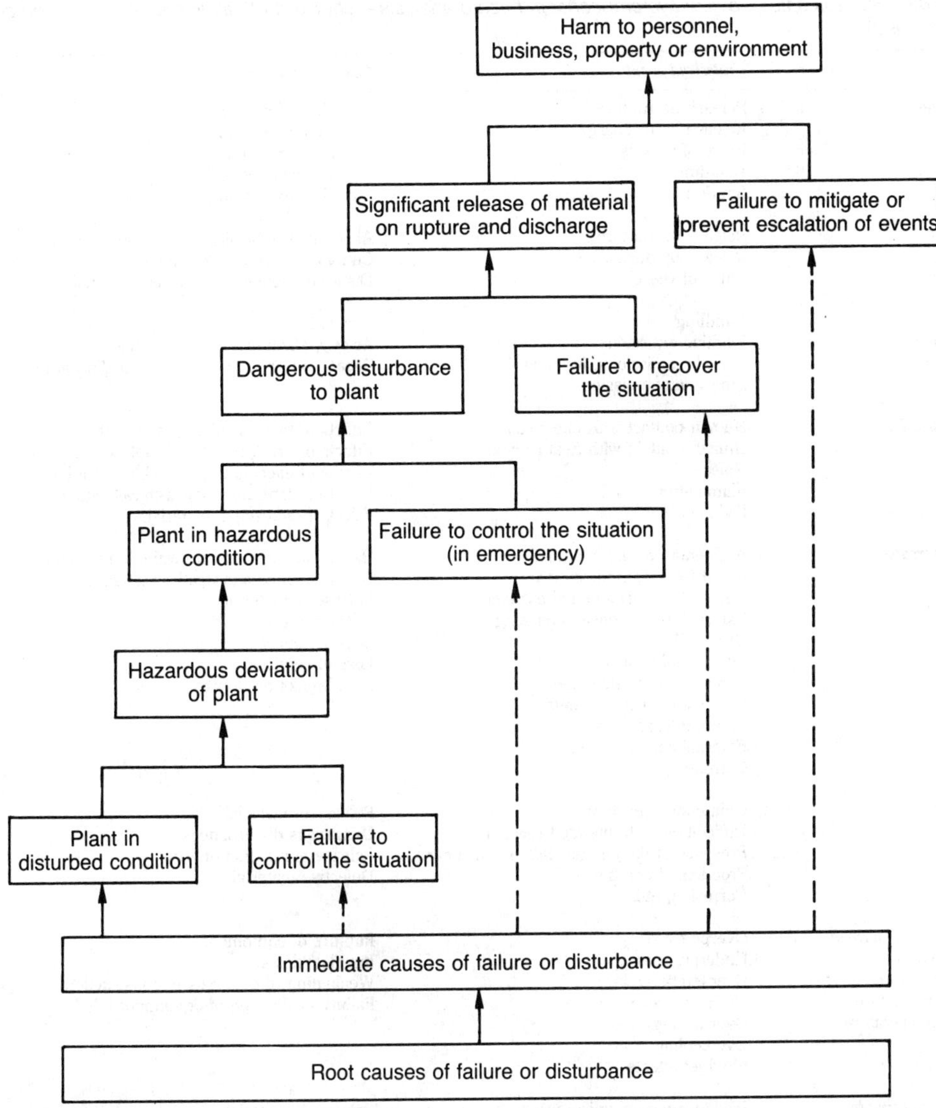

Figure 8.31 *Process safety review system: general incident structure (Wells et al., 1992) (Courtesy of the Institution of Chemical Engineers)*

(1) concept safety review (CSR);
(2) concept hazard analysis (CHA);
(3) critical examination (CE);
(4) preliminary consequence analysis (PCA);
(5) preliminary hazard analysis (PHA).

These methods are now considered in turn, based on the account by Wells, Wardman and Whetton (1993).

8.21.4 Concept safety review

The concept safety review (CSR) is carried out as early as practical, possibly during the process development stage. Features of the CSR are review of the process and the alternatives, the safety, health and environment (SHE) information requirements, any previous incidents within the company and elsewhere, movement of raw materials and products, and organizational aspects.

8.21.5 Concept hazard analysis

The concept hazard analysis (CHA) is an examination of the hazards of the process and is commenced as soon as the preliminary process flow diagram becomes available. A checklist of key words for a CHA is given in Table 8.42. Wells, Wardman and Whetton (1993) distinguish between a hazard and a hazardous condition. The former is purely qualitative, but the latter has a quantitative element. For example, a hazardous material constitutes a hazard.

Table 8.42 *Process safety review system: concept hazard analysis – checklist of keywords (Wells, Wardman and Whetton, 1993)*

Keywords	*Undesired event*	*Consequences/problems*
Flammables Ignition Fire Explosion	Release on rupture Release by discharge Entry of vessels Handling Ignition	Fire: flash, torch, pool Chemical explosion Physical explosion Vapour cloud explosion Electrical explosion
Chemicals Toxicity Corrosion	Release on rupture Release by discharge Entry of vessels	Absorption, inhalation, ingestion Contamination of environment Disposal, incineration, storage, landfill
Pollutants Emissions Effluents Waste	Handling Fugitive emissions Periodic emissions, washings Emergency emissions	Asphyxia Acidity, alkalinity, exposure effects Separation or accumulation after discharge
Health hazards	Human contact with chemicals Human contact with heat or cold Noise Illumination Radiation	Effects of toxicity, biological activity Effects of fire, contact with hot bodies, cold surfaces Effects of exceeding acceptable noise levels Effect of glare, mist, fog, contrast, smoke Effect of radioactive materials
External threats	Accidental impact, vibration Act of God, natural causes Abnormal environmental extreme External interference, loosening Drop, fail Theft, hooliganism Force majeure, sabotage External energetic events External toxic events External contamination Corrosion, erosion	Harm, damage and removal of equipment Harm, damage and death of personnel Release of material Adverse discharge Loss of supply Loss of services Item breaks on impact
Reactions	Unintended reactions Difficulties with intended reactions Presence of dangerous (toxic) substances Products of combustion Corrosion, etc.	Release of material Dangerous disturbances Release of reaction energy Defective materials
Thermodynamic hazards Overpressure Underpressure Over-temperature Under-temperature	Overpressure Underpressure Over-temperature Under-temperature Overheating Overcooling Fluid jet effects	Rupture of equipment Impulse blows Weakening of materials of construction Failure or damage of equipment
Abnormal opening to atmosphere	Inadequate mechanical integrity Corrosion, erosion Wrong status of equipment, valves, emergency relief etc.	Release of material Change in planned or emergency discharge
Mechanical hazards Structural hazards Collapse, drop	Overload/stress/tension Mechanical energy/inertia Mechanical weakness Loss of structural integrity	Rupture of equipment, change in material properties Failure of equipment or structure, transient effects, forces Impulse blows, fragility, vibration Failure of structure, collapse, object dropped
Electrical hazards	Charge, current, magnetism High voltage	Explosion, spark, shock, heat transfer, ionization Shock to personnel
Equipment problems	Dangerous disturbances or incident initiators	Release of material Off-specification material
Mode of operation Start-up Shutdown Maintenance Abnormal Emergency	Any notable disturbances or incident initiators Loss of supply Loss of services	Release of material Common cause failures Off-specification material

***Table 8.43** Process safety review system: critical examination – checklist of keywords (Wells, Wardman and Whetton, 1993) (Courtesy of Butterworth-Heinemann)*

Keyword	*Examples of use*
Eliminate	Eliminate by a completely different method or part of a method Eliminate certain chemicals, change the route, use a lean technology Eliminate additives, solvents, heat exchange mediums, additives Change the equipment or processing method Eliminate leakage points; use a weld not a bolted fitting, etc. Eliminate a prime mover or heat exchange or agitator Eliminate a separation stage or step Eliminate intermediate storage Eliminate an installed spare Eliminate manual handling Eliminate sneak paths, openings to atmosphere Eliminate waste Eliminate entry into vessels or disconnection Eliminate products that are harmful in use Eliminate an ignition source, particularly permanent flame
Avoid	Avoid extremes of operating conditions Avoid operating in a flammable atmosphere Avoid possible layering of materials, inadequate mixing Avoid flashing liquids, particularly in extensive heat exchanger networks Avoid production of large quantities of dangerous intermediates Avoid unwanted reactions in and outside reactors Avoid operating near extremes of materials of construction Avoid operating conditions leading to rapid deterioration of plant Avoid maintenance on demand and in short time periods Avoid items of plant readily toppled by explosions Avoid stage, step or activity by doing something as well as or instead of
Modify	Modify any topics above Modify batch operation to continuous operation, or vice versa
Alter	Alter the composition of waste, emission and effluents Alter the sequence, method of working Alter the time or duration of an activity (faster/slower, earlier, later?) Alter the frequency of an activity (more/less, why then?) Alter quality, quantity, rate, ratio, speed of any part of an operation or activity Alter who does an activity (why them? more/less people)
Prevent	Prevent emissions and exposure by totally enclosed processes and handling systems Prevent exposure by use of remote control
Increase	Increase heat transfer and separation efficiency or capacity Increase conversion in reactions
Reduce	Reduce inventory; less storage, hold-up, smaller size of equipment, less piping Reduce amount of energy in system Reduce pressure and temperature above ambient Reduce emissions and exposure by improved containment, piped vapour return, covers, condensation of return, use of reactive liquids, wetting dust Reduce frequency of opening, improve ventilation, change dilution or mixing Reduce size of possible openings to atmosphere
Segregate	Segregate by distance, barriers, duration and time of day Segregate plant items to avoid certain common-mode failures Segregate fragile items from roads, etc.
Isolate	Isolate plant by shut-down systems, emergency isolation valves
Improve	Improve plant integrity, reliability and availability Improve control or computer control; use user-friendly controls Improve response Improve quality of engineering, construction, manufacture and assembly

The hazardous condition is determined by the quantity of the material. The CHA is akin to a coarse hazard study.

8.21.6 Critical examination

The critical examination (CE) of system safety is in large part a critical examination of the inherent safety of the system, and is therefore different from the critical examination involved in conventional hazop. The CE examines how the proposal is to be achieved and, in particular, (1) the materials, (2) the method and (3) the equipment. It also examines any dangerous condition and its cause. A checklist of keywords for the CE is shown in Table 8.43.

8.21.7 Preliminary consequence analysis

The preliminary consequence analysis (PCA), which may be started when the preliminary process flow diagram becomes available, is a hazard assessment kept at a fairly simple level. The main hazardous events considered are: (1) fire; (2) explosion, including missiles; and (3) toxic release. The analysis covers the effects on the environment as well as on humans.

The analysis seeks to identify worst-case scenarios, to establish whether these could have impact outside the site and to define the types of emergency which might

***Table 8.44** Process safety review system: results from a preliminary consequence analysis (Wells, Wardman and Whetton, 1993)*

What if undesired event?	*FR/PR*	*GA*	*Failure to mitigate*	*PR*	*L*	*S*	*P*	*Consequences*
Significant release of material: catastrophic failure of methanator circuit	10^{-3}	And	Countermeasures for a release fail: insufficient time for response	1	3	1		Release causes hazardous condition: cloud of flammable material
Ignition of flammable mixture: ignition and torch fire	0.5	And	Countermeasures fail to control fire: fire too great to be put out immediately	1	4	3	B	Escalation by torch fire
Escalation by fire: further spread of fire to pipe rack	0.2	And	Countermeasures fail to control fire: fire brigade fails to put out fire (no barrier)	0.5	5			Escalation by further release of material
Ignition of flammable mixture: ignition and pool fire	1.0	And	Countermeasures fail to control fire: fire brigade fails to put out fire	0.2	6	3	C	Escalation by pool fire, generating possible explosion with missiles
Escalation by explosion: missiles land on C plant	0.2	And	Countermeasures for a release fail	1	6	4	C	Escalation by further release of flammable material. Aromatics washed into sewer

Note: GA, gate; FR, frequency; L, likelihood; P, priority; PR, probability; S, severity.

arise. The results from a PCA are illustrated in Table 8.44.

8.21.8 Preliminary hazard analysis

The preliminary hazard analysis (PHA) is an analysis which centres on the dangerous disturbance entry in the GIS. It involves estimating the frequency/probability of certain events and thus, in contrast to some other PHAs described in Section 8.12, it is a simplified form of conventional hazard analysis.

The results of a PHA are presented in tabular form, the core of which is the events:

(1) significant event to be prevented;
(2) failure to recover the situation;
(3) dangerous disturbance;
(4) inadequate emergency control or action;
(5) hazardous disturbances;

together with expansions at both ends, into causes of the set of hazardous disturbances and into escalation of the significant event. For each event an estimate is given of the frequency/probability.

8.21.9 Short-cut risk assessment

The short-cut risk assessment method (SCRAM) is essentially an extension of PHA. It is described by Allum and Wells (1993). The SCRA generates a set of significant events to be prevented, or top events, each of which is described in a risk evaluation sheet, which is essentially the tabulation of the PHA as just outlined. The risk assessment is developed using:

(1) target risk;
(2) severity categories;
(3) priority categories.

The target risk TR is defined as:

$$\mathrm{TR} = L + S \qquad [8.21.1]$$

where L is the exponent of likelihood as measured by frequency, and S is the severity category (described below).

The likelihood L is thus a negative number. If the frequency of the event is 10^2, the likelihood has the

***Table 8.45** Process safety review system: short-cut risk assessment – adjustments to frequency estimates to allow for non-average conditions (Allum and Wells, 1993) (Courtesy of the Institution of Chemical Engineers)*

• Determine whether the *plant conditions* represent:
(a) Average duty or appropriate baseline failure rate data
(b) Excellent case conditions in which the internal duty is clean and maintenance performance is good on a well-established plant
(c) Worst-case conditions in which the duty is severe or maintenance performance poor or the plant is of novel technology or various combinations of similar effects

• Adjust the *likelihood* of failure in average duties as follows:
(a) In the worst case multiply by 10
(b) In an average case multiply by 1
(c) In an excellent case multiply by 0.5

• Modify the *ineffectiveness* probabilities of control and mitigation:
(a) If a protective system is in the failed state, $P = 1$.
(b) In the worst case *increase* values as follows: $0.01 \rightarrow 0.1$; $0.1 \rightarrow 0.5$; $0.5 \rightarrow 0.9$
(c) In an average case do not adjust probabilities
(d) In an excellent case *reduce* values as follows: 0.001 (no change); 0.01 to 0.005; 0.1 to 0.05; 0.5 to 0.1

Table 8.46 *Process safety review system: short-cut risk assessment – severity categories (Allum and Wells, 1993) (Courtesy of the Institution of Chemical Engineers)*

Catastrophic consequences: Severity 5
Catastrophic damage and severe clean-up costs
On-site: loss of normal occupancy > 3 months
Off-site: loss of normal occupancy > 1 month
Severe national pressure to shut down
Three or more fatalities of plant personnel
Fatality of member of public or at least five injuries
Damage to Site of Special Scientific Interest or historic building
Severe permanent or long-term environmental damage in a significant area of land

Acceptable frequency 0.00001 per year

Severe consequences: Severity 4
Severe damage and major clean-up
Major effect on business with loss of occupancy up to 3 months
Possible damage to public property
Single fatality or injuries to more than five plant personnel
A 1 in 10 chance of a public fatality
Short-term environmental damage over a significant area of land
Severe media reaction

Acceptable frequency 0.0001 per year

Major consequences: Severity 3
Major damage and minor clean-up
Minor effect on business but no loss of building occupancy
Injuries to a maximum of five plant personnel with a 1 in 10 chance of fatality
Some hospitalization of public
Short-term environmental damage to water, land, flora or fauna
Considerable media reaction

Acceptable frequency 0.001 per year

Appreciable consequences: Severity 2
Appreciable damage to plant
No effect on business
Reportable near miss incident under CIMAH [a]
Injury to plant personnel
Minor annoyance to public

Acceptable frequency 0.01 per year

Minor consequences/near miss: Severity 1
Near-miss incident with significant quantity released
Minor damage to plant
No effect on business
Possible injury to plant personnel
No effect on public, possible smell

Acceptable frequency 0.1 per year

[a] Control of Industrial Major Accident Hazards Regulations 1984.

value −2. The severity is the severity category number, which lies in the range 1–5. The authors give guidance on the estimation of event frequencies. This includes the adjustments to be made where the case under consideration is judged to depart from the average condition, as shown in Table 8.45. The target risk is taken as acceptable only if it is zero or less.

The severity categories are shown in Table 8.46. The severity category assigned to a given incident scenario is based on the highest level indicated by the entries in the category. The categories 'catastrophic' and 'severe' relate to very rare events, with a frequency of less than, say, 10^{-5}/year. The severity categories are not to be interpreted as implying that an event in a lower category is less serious than one in a higher category. The categorization is for use solely as a practical design tool.

The priority categories are a function of the likelihood values and the severity categories. They are:

None No action
A Immediate attention needed
B Further study probably required
C Further study may be necessary

The prioritization of risk is

Severity Category	*Value of risk*			
	-2	-1	0	1
1	None	None	None	C
2	None	None	C	B
3	None	C/B	A/B	
4	C	B/C	B	A
5	B	B/C	A	A

The SCRA provides guidance on priorities to be addressed in the design. It also provides an indication as to whether more detailed quantitative risk assessment is appropriate.

8.21.10 Goal-oriented failure analysis

The goal oriented failure analysis (GOFA) technique is another based on the HAZCHECK and GIS checklists. An account is given by Reeves, Wells and Linken (1989). The method is described as a combination of fault tree analysis and checklists, with the latter available in the HAZCHECK computer program. The example given by the authors is loss of containment, the causes of which are traced through the checklist structure.

8.21.11 Socio-technical systems analysis

A socio-technical systems analysis is used to examine the preconditions for failure. An account is given by Wells, Phang and Wardman (1994). These authors describe the procedure and give the primary checklist. The main headings of this list are:

(1) system climate;
(2) communication and information systems;
(3) working environment;
(4) organization and management;
(5) management control;
(6) operator performance;

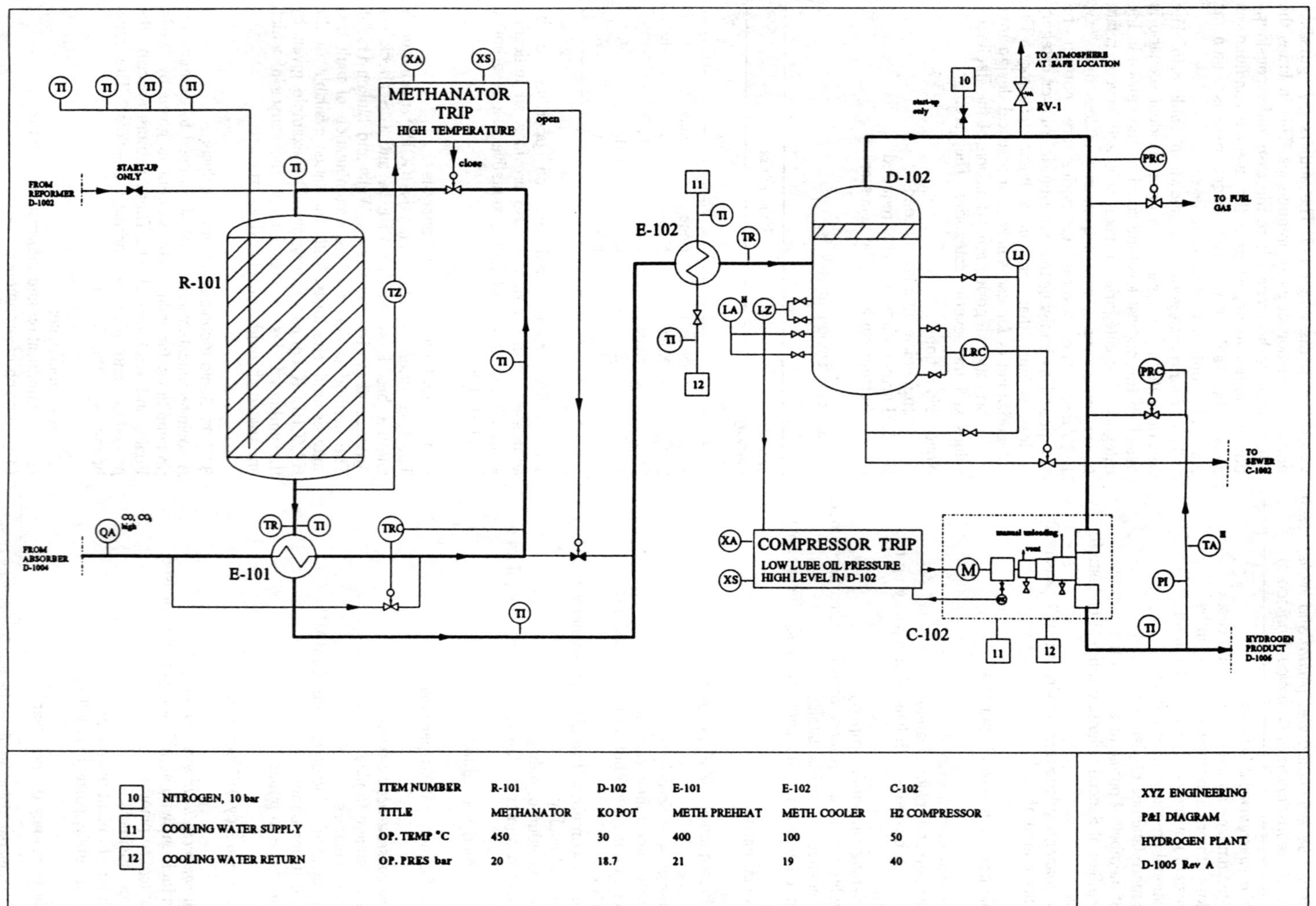

Figure 8.32 *Process safety review system: illustrative example – simplified piping and instrument diagram of a methanator system (Wells et al., 1992) (Courtesy of the Institution of Chemical Engineers)*

(7) external systems;
(8) procedures and practice;
(9) site and plant facilities;
(10) equipment integrity.

They also give the subsidiary checklist for external systems.

8.21.12 Illustrative example: methanator system

Some of the techniques which constitute the process safety review system just described have been illustrated by Wells and co-workers using the methanator system shown in Figure 8.32. Treatments which include examples involving the methanator system are given by Wells *et al.* (1992), Wells, Wardman and Whetton and Allum and Wells (1993).

The methanator is a reactor, the feed to which contains H_2 97%, oxides of carbon 2% and methane 1% and in which the oxides of carbon are converted to methane in a reaction which is exothermic. On occasion, the oxides of carbon in the feed can rise to 10–15%. In the examples quoted, the two dangerous disturbances explored for the reactor are overtemperature and overpressure.

The results of a concept hazard analysis for the system are shown in Table 8.47, those of a critical examination in Table 8.48 and those of a preliminary hazard analysis in Table 8.49, as given by Wells, Wardman and Whetton (1993). The latter leads in to the risk evaluation sheet, given by Wells *et al.* (1992), shown in Table 8.50.

8.21.13 Audit applications

The application of the process safety review system so far described is to plant design. Other applications are to audit and incident investigation. The HAZCHECK and GIS checklists have been developed with these applications in mind. They are not intended to be exhaustive. Rather they are pitched at that level of detail which an auditor can handle within an interview.

8.21.14 Incident investigation applications

Likewise, the checklists are adapted to use during a meeting concerned with investigation of an incident.

8.22 Choice of Method

A variety of methods of hazard identification applicable at the different stages of the project have been described. An overview of these methods is given in Table 8.2. In many cases the aspect of hazard identification covered by a particular method is self-evident, but in others it is not. Some methods are alternatives or are complementary. Table 8.51 lists some of the principal methods as a guide to selection. The table gives the prime purpose(s) of each method. With regard to checklists, the applications shown in the table are limited, but some of the other methods such as hazop may also make use of a checklist specific to that method.

8.23 Filtering and Follow-up

Although it is convenient for purposes of exposition to treat hazard identification and assessment separately, the two are often interwoven. A process of filtering of the hazards takes place. Two forms of filtering, which occur with many of the identification methods described, may be mentioned. The first involves discarding a hazard which is considered to be unrealizable.

In the second form of filtering a hazard is discarded if it is considered not to be sufficiently important. A measure of the need for action on an identified hazard which is widely used, albeit often not explicitly, is the product of the magnitude and frequency of the event. It is this criterion which underlies the method of ranking described in Section 8.21.9. It is also used in task analysis as a stopping rule to decide how far to carry the process of subdivision of the task.

Filtering is an important feature of a hazop study. In such a study a filtering process occurs whereby the hazards identified are assessed on the spot and only a proportion are passed on for further assessment.

Once a hazard has been identified and accepted as worth further consideration, it is necessary to decide what to do about it. Figure 8.33 indicates four principal outcomes of the identification process. In many cases, once the problem has been identified the solution is obvious or is known from experience. In many others it is taken care of by codes of practice. In a small proportion of cases it may be appropriate to carry out a hazard analysis.

8.24 Safety Review Systems

Formal systems for the conduct of safety reviews appropriate to the stages of the project are now widespread in the process industries. These systems go by different names.

8.24.1 Hazard study systems

The system used in Mond Division of ICI has been described by several authors (S.B. Gibson, 1976a; N.C. Harris, 1975; Hawksley, 1984) and is shown in Figure 8.34 and Table 8.52. The system involves six hazard studies.

Hazard Study I is the first formal study, carried out at the project exploration stage. Its purpose is to ensure that there is full understanding of the materials involved in the process and their potential interactions, and also the constraints on the project due to the particular site, not only in terms of safety but also in terms of environmental factors, including pollution and noise. This type of study was referred to above as a 'coarse hazard study'.

Hazard Study II is performed using the process flow diagrams. Each section such as reaction, distillation, etc., is studied in turn and potential hazardous events such as fire, explosion, etc., are identified. At this stage it is often possible to eliminate such a hazard by means of design changes or to protect against it by the use of protective systems. In some instances it may be necessary to study an event further. In this case use may be made of fault tree analysis, with the hazardous event the top event in the tree.

Hazard Study III is the hazard and operability (hazop) study, which has already been described.

Hazard Study IV involves a review of the actions generated in Hazard Studies I–III by the plant or commissioning manager, who also ensures that operating procedures, safe systems of work and emergency procedures are available.

Hazard Study V is the plant inspection carried out to ensure that the plant complies with the legal require-

Table 8.47 *Process safety review system: methanator system – results of a concept hazard analysis (extract) (Wells, Wardman and Whetton, 1992)*

Ref. no	*Keyword*	*Dangerous disturbance*	*Consequences*	*Suggested safeguards*	*Comments/action*
1	Flammables	Release on rupture	Release may self-ignite torch fire. Escalation to pipe rack likely. Missile could affect C plant	Segregation by distance. Depressure or steam purge	Study best way of reducing damage
2	Flammables	Release on emergency discharge	Release at safe height: possible ignition and fire	Segregation by distance	Check possible radiation levels
3	Flammables	Normal discharge to sewer	Chemical explosion in sewer	Vent sewer with standpipes	Check other plants for incidents
4	Reaction	Exothermic runaway reaction in methanator	High level of oxides of carbon cause runaway with rupture and possible physical explosion	More robust design of absorber. Trip methanator on high temperature. Alarms on temperature and CO_2 high	Check action if trip fails
5	Reaction	Air in combustion vessels	Combustion in vessels. Causes chemical explosion	Purge plant before start-up. Ensure catalyst covered by N_2 as replaced	Get more information on catalyst
6	Reaction	Inadequate reaction Catalyst failure Low temperature feed Methanator bypassed	Off-specification H_2 to downstream plant. This can cause runaway reaction with chemical explosion	Alarm on high CO_2 outlet. Vent if off-specification and shut-down compressor. Ensure methanator warm enough to start up. Connect methanator trip to compressor trip	Design heat exchanger circuit to preheat methanator
7	Pollutants	Effluent to sewer	Water with high sodium salts	Sewer to effluent treatment	Check effect on current treatment
8	Pollutants	Effluent caused by firewater	Fire-water will flood. River receives minor contamination		Check other sewers in area for contamination
9	Pollutants	Noise	Noise in compressor area	Building would cause explosion hazard	Operators to wear protection in danger area
10	Overpressure	Overpressure in hydrogen plant	High pressure caused by inadequate release of excess gas to fuel gas or blockage or incorrect valve status causes explosion	Two relief valves in circuit. High pressure alarm	Flare may be needed on fuel gas if demand low
11	Over-temperature	Over-temperature in methanator	Runaway reaction (see above)		
12	Over-temperature	Over-temperature in compressor	Excess recycle of hydrogen around compressor can result in physical explosion	High temperature alarm on loop	Evaluate as no safeguard provided
13	Mechanical hazard: overload	Overload of compressor	Stress in compressor caused by two phase feed due to liquid blowby from KO Pot can result in physical explosion	Trip on high level in KO Pot. Level alarm in KO Pot	Explosion unlikely but note compressor may be damaged
14	Abnormal opening	Vibration at compressor	Loosening of flange gives release. Possible torch fire	Vibration probe	
15	Abnormal opening	Spurious relief	Loss of material to safe point. Could ignite as minor torch fire		Consider need for lock open valve after RVs or bursting disc before RVs.

Table 8.48 *Process safety review system: methanator system – results of a critical examination (Wells, Wardman and Whetton, 1992)*

DESIGN INTENT: A fixed-bed catalytic reactor, operating at 20 bar, 400°C inlet, 450°C outlet, converts the small amounts of oxides of carbon (maximum 2%) in a stream of hydrogen into a hydrogen product stream containing 10 ppm maximum of oxides of carbon

Query proposal	*Response*	*Generate alternatives*	*Comments*	*Recommendations*
Why remove oxides of carbon?	Oxides of carbon affect downstream catalyst on aromatics plant	Eliminate methanator here and install on aromatics plant only	No real saving on risk overall	Reject or change downstream catalyst
Why this process?	No addition of further materials	Alter by using pressure-swing adsorption system upstream	Lower yield of hydrogen but cheaper plant	Review for plant after next
Why at 400°C?	Optimized design for this catalyst	Alter the catalyst or use a larger size of bed	No safety advantage	Reject
Dangerous condition	*Cause*	*Modification/control*	*Comments*	*Recommendations*
Reactor runaway due to excess oxides of carbon in feed leading to reactor	Failure of absorption system	Increase capacity of absorption unit using an absorption train	Expensive but robust solution	Evaluate using quantitative risk analysis
		Isolate the methanator by shut-down system	Requires the diversion of upstream flow from methanator by shut-down system	Install bypass and vent off-specification material
Catastrophic failure of methanator circuit	Overtemperature due to reactor runaway	Improve metallurgy of reactor to withstand maximum temperature during upset condition	Long-term effort required	Review later
		Increasing cooling of reactor by external quench	Weakness in circuit may well not be the reactor	Consider under preliminary hazard analysis

ments on such matters as access and escape, guarding, emergency equipment, etc.

Hazard Study VI is undertaken during the commissioning and early operation of the plant and involves a review of changes made to the plant and its operating procedures, and of any differences between design intent and actual operation.

Once in operation the plant is subject to further audits, but these fall outside the system as described.

8.24.2 Project safety review system

A broadly similar system is the project safety review system operated by BP. The reviews conducted in this system are:

(1) pre-project review;
(2) design proposal review;
(3) detailed design review;
(4) construction review;
(5) pre-commissioning review;
(6) post-commissioning review.

The two sets of safety review systems described above typify those in use.

8.25 Hazard Ranking Methods

There are available a number of hazard ranking methods for assigning priorities among hazards. Accounts include those of Gillett (1985), Ashmore and Sharma (1988), Rooney, Turner and Arendt (1988), Casada, Kirkman and Paula (1990), Fraser, Bussey and Johnston (1990).

8.25.1 Rapid ranking

Typical of these methods is that of rapid ranking described by Gillett (1985). The basis of the method is a trade-off between the consequence and the frequency of the event considered. The more serious the consequence, the lower the frequency which is tolerable for the event. In principle, therefore, the criterion for action may be taken, as a first approximation, as some threshold level of the product of consequence and frequency.

The rapid ranking method actually utilizes a mixture of quantitative and qualitative measures. The cost of damage to the plant and the frequency of occurrence are expressed quantitatively and the other consequences qualitatively.

In the method, five hazard categories are defined. Table 8.53 defines the hazard categories in terms of the consequences. In determining the features of a particular

Table 8.49 *Process safety review system: methanator system – preliminary hazard analysis (Wells, Wardman and Whetton, 1992)*

PROJECT: TOMHID PLANT: HYDROGEN UNIT: METHANATOR SECTION	REFERENCE: GLW LOCATION: SHEFFIELD EQUIPMENT: METHANATOR/PREHEAT	Priority			
		C			
FUNCTION: Fixed bed reactor converting oxides of carbon to hydrogen		*S*	*L*	*S*	*L*
Consequences of escalation	Fire escalates to pipe rack and C plant	4	E–6[a]	4	7
Failure to prevent further escalation	Failure to avoid domino due to lack of time and ineffective fire-fighting		0.01		0.01
Consequences of significant event	Torch fire on section of plant	3	E–4	3	E–5
Failure to mitigate or avoid escalation	Failure to avoid ignition: self-ignites as hot and release not attenuated in 15 min	3			1

Significant event to be prevented	Release through overtemperature	E–4	Release through overpressure	E–5
Failure to recover the situation	Operator fails to stop all plant flows	* 0.1	Operator action to depressure fails	0.1
Plant in danger: dangerous disturbance	Overtemperature in reactor	E–3	Overpressure in reactor	E–4
Inadequate emergency control or action	Failure of operator to stop flow to methanator	0.1	Pressure relief system fails	0.01
	Failure of shut-down system	0.2		
Hazardous disturbance	High temperature in reactor	0.1	High pressure in section	E–2
Inadequate control or action	Operator fails to reduce trend on TAH	1		
Immediate causes	High inlet temperature (slow propagation)	E–1	Downstream blockage (clean duty)	E–4
Hazardous disturbance				
Inadequate control or action	Operator fails to reduce trend on CO_2 alarm or TAH	0.8	PRC closed Fuel gas overpressure	0.01 0.01
Immediate causes	High CO_2 in stream from absorber	E–1	Lack of demand for product	E–0
Hazardous disturbance				
Inadequate control or action	Operator fails to reduce trend on CO_2 alarm or TAH or PAH	0.1	PRC closed Fuel gas overpressure	0.01 0.01
Immediate causes	Impurities in feed: sneak path down start-up line	E–2	Off-specification product	E–0

Recommendations/comments/actions:
1. Public not affected by domino escalation.
2. Business damage would be extensive if spread to complex.

*The operator can increase the probability of a release by wrong action and special supervision is required on any methanator problem.

1. Do not depressurize on high temperature unless sure of no flow through methanator.
2. Operator alert by several alarms. New TAH in and out.
3. Check if start-up line needed if heat exchange circuit modified.
4. Alter outlet location of start-up line. Add PAH and TR. Double block and bleed.
5. Check catalyst activation.
6. Improve absorber design to enhance reliability.

1. Operator also alerted by PAH.
2. Two relief valves in system and hydrogen is exceptionally free-flowing.
3. Add RV-1 and depressuring valve: locate before methanator.

[a] $E - n = 10^{-n}$

Table 8.50 *Process safety review system: methanator system – risk evaluation sheet (Wells et al., 1992) (Courtesy of the Institution of Chemical Engineers)*

PSA: RISK EVALUATION SHEET	DATE: 11.7.92 PAGE: D1 of 2	
PROJECT: TOMHID PLANT: HYDROGEN UNIT: METHANATOR SECTION	REFERENCE: GLW LOCATION: SHEFFIELD EQUIPMENT: METHANATOR/PREHEAT	PRIORITY
FUNCTION: Fixed bed reactor converting oxides of carbon to hydrogen		

		C			
		S	*L*	*S*	*L*
CONSEQUENCES OF ESCALATION	Fire escalates to pipe rack and C Plant	4	E–6[a]	4	7
FAILURE TO PREVENT FURTHER ESCALATION	Failure to avoid domino due to lack of time and ineffective fire-fighting		0.01		0.01
CONSEQUENCES OF SIGNIFICANT EVENT	Torch fire on section of plant	3	E–4	3	E–5
FAILURE TO MITIGATE OR AVOID ESCALATION	Failure to avoid ignition: self-ignites as hot AND release not attenuated in 15 min	3			1

SIGNIFICANT EVENT TO BE PREVENTED	Release through overtemperature	E–4	Release through overpressure	E–5
FAILURE TO RECOVER THE SITUATION	Operator fails to stop all plant flows	* 0.1	Operator action to depressure fails	0.1
PLANT IN DANGER: DANGEROUS DISTURBANCE	Overtemperature in reactor	E–3	Overpressure in reactor	E–4
INADEQUATE EMERGENCY CONTROL OR ACTION	Failure of operator to stop flow to Methanator Failure of shutdown system	0.1 0.2	Pressure relief system fails	0.01
HAZARDOUS DISTURBANCE	High temperature in reactor	0.1	High pressure in section	E–2
INADEQUATE CONTROL OR ACTION	Operator fails to reduce trend on TAH	1		
IMMEDIATE CAUSES	High inlet temperature (slow propagation)	E–1	Downstream blockage (clean duty)	E–4
HAZARDOUS DISTURBANCE				
INADEQUATE CONTROL OR ACTION	Operator fails to reduce trend on CO_2 alarm or TAH	0.8	PRC closed Fuel gas overpressure	0.01 0.01
IMMEDIATE CAUSES	High CO_2 in stream from absorber	E–1	Lack of demand for product	E–0
HAZARDOUS DISTURBANCE				
INADEQUATE CONTROL OR ACTION	Operator fails to reduce trend on CO_2 alarm or TAH or PAH	0.1	PRC closed Fuel gas overpressure	0.01 0.01
IMMEDIATE CAUSES	Impurities in feed: sneak path down start-up line	E–2	Offspec product	E–0

RECOMMENDATIONS/COMMENTS/ ACTIONS:
1. Public not affected by domino escalation.
2. Business damage would be extensive if spread to complex.

*The operator can increase the probability of a release by wrong action and special supervision is required on any Methanator problem.

1. Do not depressurize on high temperature unless sure of no flow through Methanator.
2. Operator alert by several alarms. *New TAH in and out*
3. Check if start-up line needed if heat exchange circuit modified.
4. *Alter outlet location of start-up line. Add PAH and TR. Double block and bleed.*
5. Check catalyst activation.
6. Improve absorber design to enhance reliability.

1. Operator also alerted by PAH.
2. Two relief valves in system and hydrogen is exceptionally free-flowing.
3. *Add RV-1 and depressuring valve:* locate before *methanator.*

E–2 $= 10^{-2}$

[a] $E - n = 10^{-n}$

Table 8.51 *Application of some techniques for hazard identification*[a]

Purpose	Checklist	Safety review	Scenario development[b]	Hazard indices	Hazard ranking methods	Coarse hazard study[c]	Hazop	Failure modes, effects and criticality analysis[d]	Human error analysis[e]	Fault tree analysis	Event tree analysis	Consequence modelling	Hazard warning analysis
Identification of:													
Deviations from good practice	Y	Y											
Hazards	Y	Y	Y	Y		Y	Y						
Hazards liable to threaten project viability			Y			Y							
Hazards with potential for large property damage loss				Y									
Hazards requiring priority					Y								
Worst-case accidents			Y	Y	Y								
Initiating events							Y	Y	Y	Y			
Pre-release accident path							Y	Y	Y	Y			
Measures to reduce probability of enabling condition, frequency of initiating event							Y	Y	Y	Y			
Post-release escalation paths and outcomes											Y		
Measures to reduce frequency of consequences											Y	Y	
Measures to mitigate effects of consequences												Y	
Precursor 'warning' events													Y

[a] The table is oriented towards hazard identification techniques for loss of containment events. It does not include hazard identification techniques the purpose which is obvious, such as reaction screening or commissioning inspections.
[b] Also techniques such as 'What if?' method.
[c] Also techniques such as preliminary problems analysis (PPA).
[d] Also techniques such as action error analysis (AEA).
[e] Also techniques such as common mode failure analysis and cause–consequence analysis (CCA).

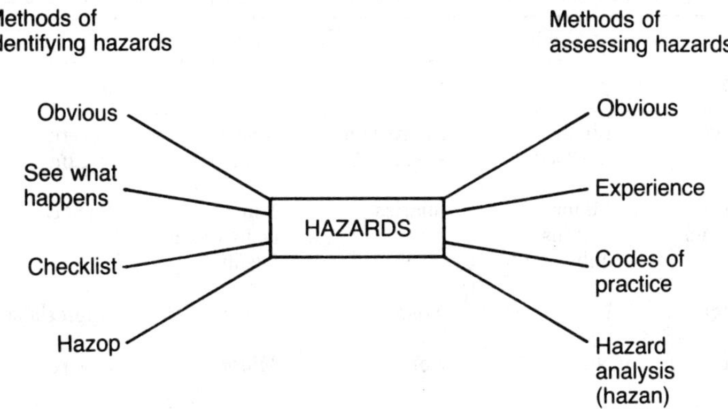

Figure 8.33 *Process of hazard identification and assessment (Kletz, 1983d) (Courtesy of the Institution of Chemical Engineers)*

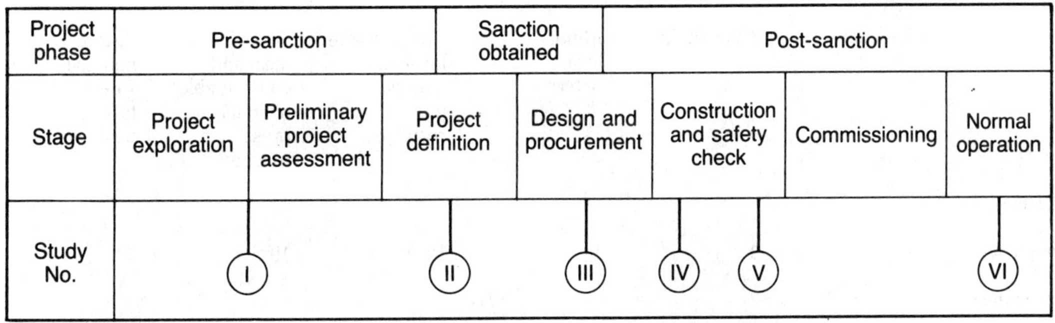

Figure 8.34 *System of hazard studies (after S.B. Gibson, 1976; Illidge and Wolstenholme, 1979) (Reproduced by permission of the Institution of Chemical Engineers)*

hazard for comparison with the entries in the table it is intended that the damage costs be estimated and that the other entries be obtained from expert judgement. The table also gives both relative guide frequencies and typical absolute frequencies for each hazard category.

For a particular hazard a hazard category is obtained on the basis of the consequences. If different consequences yield different hazard categories, judgement must be exercised. The estimated frequencies derived from historical data are then compared with the absolute guide frequencies given in Table 8.53 and a hazard ranking obtained using Table 8.54.

Examples of the technique are given by Mumford and Lihou (1984 LPB 59). As an illustrative example, consider the hazard of fire in a pipe trench. In a particular application it is estimated that a pipe trench fire will cause on average £10 000, damage and the following estimates are made of the frequency of the two hazards – (1) a fire without operator fatality, and (2) a fire with a chance of an operator fatality:

Frequency of pipe trench fires without operator fatality = 0.5/year

Frequency of pipe trench fires with 1 in 10 chance of operator fatality = 1.5×10^{-2}/year

Hazard 1, with an estimated damage cost of £10 000, is given by Table 8.53 as hazard category 2 and therefore has guide frequency of 10^{-1}/year. The expected frequency is 5×10^{-1}/year and thus exceeds the guide value by a factor of 5. From Table 8.54 the hazard ranking is A. Hazard 2, with the same damage cost but a 1 in 10 chance of fatality, is hazard category 3 and therefore has a guide frequency of 10^{-2}/year. The estimated frequency is 1.5×10^{-1}/year and thus exceeds the guide value by a factor of 1.5. The hazard ranking is again A, but only just. Action is indicated in both cases.

8.26 Hazard Warning Analysis

In some cases analyses are carried out to identify the lesser events which may serve as precursors to, or warnings of, more serious events. Such analyses clearly rank as hazard identification techniques. One such method is the technique of hazard warning analysis described by Lees (1982b). Like fault tree analysis, on

Table 8.52 *Rapid ranking of hazards: hazard categories (Gillett, 1985) (Courtesy of Process Engineering)*

Area at risk	*Description of risk*	*Hazard category*				
		1	*2*	*3*	*4*	*5*
Plant	Damage	Minor < £2000	Appreciable < £20 000	Major < £200 000	Severe < £2m	Total destruction > £2m
	Effect on personnel	Minor injuries only	Injuries	1 in 10 chance of a fatality	Fatality	Multiple fatalities
Works	Damage	None	None	Minor	Appreciable	Severe
Business	Business loss	None	None	Minor	Severe	Total loss of business
Public	Damage	None	Very minor	Minor	Appreciable	Severe
	Effects on people	None (smells)	Minor	Some hospitalization	1 in 10 change of public fatality	Fatality
	Reaction	None/mild	Minor local outcry	Considerable local and national press reaction	Severe local and considerable national press reaction	Severe national (pressure to stop business)
Relative guide frequency of occurrence	–	1	10^{-1}	10^{-2}	10^{-3}	10^{-4}
Typical[a] judgmental values for a plant/ small works (years)	–	1	1/10	1/100	1/1000	$1/10^4$

[a] These typical comparative figures are given for illustration and should not be taken as applicable to all situations or taken to indicate absolute levels of acceptability.

which it is based, this technique may be used to obtain quantitative results. It is therefore described in the next chapter.

8.27 Plant Safety Audits

The conduct of some form of plant safety review is a normal part of plant commissioning. Traditionally this is a joint inspection by the plant manager and the plant safety officer and is concerned mainly with checking that the company complies with the legal and company safety requirements. Attention is directed to features such as: access and means of escape; walkways, stairs and floors; and fire fighting and protective equipment.

There has developed from this, however, the more comprehensive plant safety audit. There is no uniform system for such audits. The BCISC publications *Safe and Sound* (1969/9) and *Safety Audits* (1973/12) give information on several systems. D. Williams (1971) gives a detailed description of one system, including a comparison of one of the audits from *Safe and Sound* with a later audit by the same company which illustrates the evolution of the method.

The general approach is on the following lines. The plant safety audit is conducted by a small, interdisciplinary team who are not connected with the plant. The members may come from the same works or from other works. The team may be led by a quite senior manager. The audit is carried out first during the plant commissioning, but is also repeated later at intervals. Typical intervals are a year after initial start-up and every 5 years thereafter.

***Table 8.53** Rapid ranking of hazards: hazard ranking (Gillett, 1985) (Courtesy of Process Engineering)*

Hazard category (from Table 8.52)	*Expected frequency compared with guide frequency*			
	Smaller *(−)*	*Same* *(=)*	*Greater* *(+)*	*Uncertain* *(U)*
1	D	D	D/C at team's discretion	D/C at team's discretion
2	D	Normally C, but if upper end of frequency/ potential raised to B at team's discretion	Equally damaging hazard as those below, A, but if lower end of frequency/ potential could be lowered to B at team's discretion	B Frequency estimate should not be difficult at this category; may be a lack of fundamental knowledge which requires research
3	C	B	A Major hazard	A/B at team's discretion. Such potential should be better understood
4 & 5	B/C at team's discretion	B, but can be raised to A at team's discretion	A Major hazard	A Such potential should be better understood

The topics covered by the audit are illustrated by the checklists given in Tables 8.55 and 8.56. The former gives topics listed in *Safety Audits*. The latter is a checklist for both audits and surveys given by Williams. As already emphasized, it is essential that management act on any audit made.

8.28 Other Methods

8.28.1 Feedback from workforce

It is the responsibility of management to create systems and attitudes which lead to the effective identification of hazards, but it is the responsibility of everyone involved in the design and operation of the process to play his role in identifying and reporting any hazard of which he becomes aware.

This aspect has already been considered in the context of total loss control in Chapter 1 and of the safety system in Chapter 6, and is discussed again in Chapters 11 and 28.

8.28.2 Process design checks

There is an enormous amount of experience on the hazards associated with unit processes, unit operations, plant equipment, pressure systems, instrumentation, etc., and it is essential that this be utilized. It is not enough, therefore, to rely only on generalized methods of hazard identification. Full use needs to be made of what is already known. This may be done by methods such as the use of checklists for such processes, operations, etc. This aspect is discussed in more detail in Chapter 11.

8.28.3 Plant equipment checks

There are numerous types of check which are carried out on the plant equipment at various stages in the project. There are checks done before and during plant commissioning, including testing in manufacturers' works, checking the correct installation of the equipment, testing *in situ*, etc. There is a whole range of inspection and examination techniques, including non-destructive testing and condition monitoring, which are applied both during commissioning and in operation. Many items are checked or tested on a scheduled basis. While the results of some tests are unambiguous, others require careful interpretation and the application of sophisticated inspection standards.

8.28.4 Emergency planning

If a works contains hazardous plant, it is essential to plan for possible emergency situations. The first step in emergency planning is identification of those hazards which are sufficiently serious to warrant such planning. Such a study is not necessarily an exercise separate from the other hazard identification procedures, but it does emphasize particular aspects, especially the development and scale of the emergency, the effects inside and outside the works boundary, the parties involved in dealing with the situation, etc. Emergency planning is treated in Chapter 24.

8.29 Quality Assurance

The quality of the hazard identification system should be assured by a suitable system of quality assurance. Normally this will be part of an accredited system covering the quality assurance of all aspects of company operation. Principles of quality assurance are discussed

Table 8.54 *Checklist for a plant safety audit (after British Chemical Industry Safety Council, 1973/12)*

(1) Statutory requirements in the area under consideration
(2) Methods of process operation. Hazards of the materials involved and of the technology in use
(3) Material handling
(4) Tools, machinery, maintenance equipment
(5) Permit-to-work system, schedules for regular inspection of emergency equipment
(6) Personal protective equipment, its condition, care and suitability
(7) Plant tidiness, condition of floors, stairs, walkways, cleanliness of toilets and washing facilities, environmental factors, waste disposal
(8) Fire prevention and protection, alarm systems, emergency exits, flammable material storage
(9) Unsafe practices
(10) Arrangements for treating injuries, condition of safety showers, eye- and hand-wash cabinets, resuscitation equipment
(11) Involvement of employees in safety activities, their knowledge of the safety organization, attitudes, condition of displays and notices

Also

Electrostatic hazards
Alarms and trip systems and testing
Pressure vessels and relief valve inspections
Cable protection
Radiation hazards
Operation and maintenance of internal and external transport
Occupational health and hygiene standards
Electrical equipment maintenance
Major emergency arrangements
New projects and processes
Adherence to codes of practice (of internal or external origin)
Equipment faults
Labelling of products
Adherence to approved work and operational procedures
Training procedures/programmes and their effectiveness
Adequacy of safety procedures in office, administrative and ancillary buildings
Methods of communication and their success
Receipt, storage and transportation of chemicals and other materials
Extensions/alterations to existing production and storage facilities
Pollution/environmental control methods
Liaison with outside contractors
Safety standards in research, development and laboratory work
Consideration of safety and loss prevention at the design stage
Security arrangements covering the storage and issue of harmful substances
Safety advisory service to customers, covering products, their handling and use

in Chapters 1 and 6 are therefore not repeated here, but it is appropriate to consider some aspects which apply to hazard identification.

There should be a formal system for hazard identification. For each stage of a project a set of hazard identification techniques should be selected. The system should then ensure that these techniques are used.

Quality assurance may be based on inputs and process and/or outputs. It is generally prudent to utilize both. Inputs and process for hazard identification include: the qualifications, training, experience and plant knowledge of the team performing the task; the methods of hazard identification used and their suitability; the project documentation available, its comprehensiveness and accuracy; and the time available for the task.

Outputs include the whole range of information generated by the team. They are therefore not confined just to items such as release scenarios or deviations with associated causes and consequences, but extend also to the statements of scope, the report on the analysis and the follow-up measures.

Quality assurance of hazard identification requires documentation both of the hazard identification system and of the hazard identification activities conducted in a particular project. It should leave an audit trail which can be followed later if necessary.

8.30 Quality Assurance: Completeness

The quality of hazard identification virtually parallels the completeness of identification. It is of interest, therefore, to consider attempts made to assess the completeness of hazard identification. Some identified hazards are discarded as unrealizable. In discarding such hazards discrimination is exercised. It is desirable to have some measure of this also.

A study of completeness and discrimination in hazard identification has been made by J.R. Taylor (1979; 1981

***Table 8.55** Checklist for a plant safety audit or survey (D. Williams, 1971) (Courtesy of the Institution of Chemical Engineers)*

Buildings and surroundings:
(1) Are emergency exits adequate and accessible?
(2) Is lighting adequate for operations, walking, material handling?
(3) Is fire fighting equipment adequate and accessible?
(4) Are aisles large enough? Marked? Clear?
(5) Is floor free of tripping or slipping hazards?
(6) Are floor loads within safe working loads? Are safe loading limits posted where necessary?
(7) Is general ventilation sufficient?
(8) Are stairs and platforms free of tripping hazards? Equipped with necessary guard rails?
(9) Are there any openings in fire walls? Are fire doors operable?
(10) Are hazardous areas provided with adequate explosion venting?
(11) Does building electrical service appear adequate? Does visual inspection reveal any maintenance problems?
(12) Is building trim (window sash, doors, rain gutters, fire escapes, etc.) in safe condition?
(13) Are lifts regularly inspected and adequately maintained between inspections? Are safe lifts loads posted? Adhered to?
(14) Are building approaches (outside steps, platforms, paths, etc.) in a safe condition?
(15) Are safety signs adequate and in good condition?

Equipment:
(1) Are danger points adequately guarded? Rotating parts? Pinch or nip points? Belts? Cutting edges? Hot surfaces? Open flames?
(2) Are operating controls positioned for safe use?
(3) Is there room for the operator to work safely?
(4) Are valves, switches, instruments, and other operating controls clearly identified? Are pilot lights working?
(5) Are safety alarms properly identified and tested regularly?
(6) Do automatic controls fail safe? (Can a hazardous condition be created by the failure of automatic controls?)
(7) Do wiring and piping appear to be in good condition and adequate for intended use?
(8) Is equipment rated and inspected for the service in which it is used? (Area engineers to provide list of pressure equipment rating.)
(9) If process presents corrosion problems, is equipment inspected regularly to anticipate and prevent failure in service?
(10) Are pressure relief devices adequately sized and accessible for inspection?
(11) Are overflow lines and breather vents adequate and clear? Equipped with flame arresters, liquid seals where needed? Do overflows and vents discharge at a safe location?
(12) Does equipment conform to grounding policy? Are grounds tested regularly? (Area supervisor to specify frequency of test.)
(13) Are drop lights and extension cords in good condition? Meet electrical standards?
(14) Are hand tools, portable tools and ladders inspected regularly? Are they proper for the intended use? In good condition?
(15) Can equipment be made safe for routine maintenance work? Electric lockouts? Lines blanked? Sufficient safe working area for mechanics? Adequate room and facilities for rescue?

Process:
(1) Is process safe under normal conditions?
(a) Are toxic vapours normally present? Are safeguards adequate?
(b) Are flammable vapours normally present? Are vapours confined? Are sources of ignition controlled?
(c) Are operators or other employees normally exposed to toxic, corrosive or hot liquids? Is protective equipment properly stored, maintained and used? Are special physical examinations of employees conducted as recommended by the medical department?
(d) Are sampling procedures safe?
(e) Are safe storage conditions provided for hazardous raw materials and intermediates?
(2) Are abnormal conditions provided for?
(a) Loss of utilities: Electricity – agitation, circulation, controls and instruments, light. Steam – heating, vacuum, pumps, steam snuffers. Air – instruments, pumps. Water – cooling, quenching. Gas – inert and illuminating.
(b) Mechanical failures.
(c) Operator errors – what can happen?
(d) Are emergency venting facilities adequate? Inspected?
(e) What sources of ignition may develop in this process? Are they guarded against?
(f) Can a fire be controlled in this process?
(g) Are operators or other employees unnecessarily exposed to serious injury in event of probable fire, explosion, large spills or mechanical failures?
(h) Are operators protected from probable leaks of steam, toxic or corrosive chemicals?

Table 8.55 *(continued)*

(i) Is quantity of exposed flammable or toxic liquids in populated areas kept to an absolute minimum consistent with production requirements?
(j) Is process subject to abnormal reactions?

ARE THERE ANY HAZARDS WHICH WE HAVE NOT GUARDED AGAINST AND FROM WHICH OUR ONLY PROTECTION IS THE LAW OF CHANCE?

Operating instructions:
(1) Are operating instructions available in a written, legible, easily identified form?
(2) Are they up to date?
(3) Is there a procedure for keeping them up to date?
(4) Are they reviewed regularly with each employee?
(5) Are they accessible to operators for reference?
(6) Do the operating instructions spell out hazards of job and how to avoid them?
(7) Do operating instructions provide definite procedures for emergency situations where required?
(8) Is operator's performance checked regularly by supervision?
(9) Is there an effective system for notifying operators of non-standard conditions?
(10) Are operators permitted to take unauthorized short cuts in operations without reprimand?
(11) Are operators working in isolated locations checked regularly by supervision or the guard force?
(12) Do department supervisors know immediate first aid treatment required for exposure to toxic chemicals used in their departments?

Table 8.56 *System of hazard studies (after N.C. Harris, 1975; S.B. Gibson, 1976a; and Hawksley, 1984)*

Study	*Project phase*	
I	Exploratory	Carried out at inception of project. Review of hazards of materials and process, of effluent and environmental problems and of site
II	Definition, preliminary design	Completed before sanction is sought. Examination of process flowsheets and assessment of ability of design to meet specifications
III	Detailed design	Takes place about time plant is sanctioned. Critical examination of hazards and operability of plant using Piping and instrument diagram, operating instructions, maintenance procedures, etc.
IV	Construction	Takes place during later stages of construction and early stages of commissioning. Review to ensure provisions of previous studies have been implemented
V	Commissioning	Carried out before initial startup. Inspection to ensure safety systems are installed and operate as intended.
VI	Operation	Carried out within a year of handover to operating team. Review of plant modifications, operating instructions, documentation

LPB 46). It is possible to specify a list of hazards which is complete. According to Taylor, process plant hazards are covered by the following list: fires, explosions, toxic releases, asphyxiations, drownings, and mechanical impact and cutting accidents. However, the price paid for completeness is a list so general that it is of little practical use. As soon as the list is developed into more specific categories, the degree of completeness is liable to decrease.

8.30.1 Criteria of completeness

The following criteria of completeness are defined:

Criterion	*Definition*
1	N_{id}/N_{tot}
2	$(N_{id})_m/N_{chl}$
3	$(\Sigma f_i c_i)_{id}/(S f_i c_i)_{chl}$
4	N_{rl}/N_{id}

where c is a measure of consequence, f is the frequency of occurrence, and N is the number of hazards and where subscripts refer to the following:

id	identified
chl	identified in a case history list
m	identified by method m
rl	realisable
tot	total existing

Criterion 1 is the simple measure of completeness. Criterion 2 is a more practical criterion which measures completeness obtained by a particular method in relation to hazards which are ascertainable as having occurred historically. This requires the use of a large list of case histories.

Some hazards are expressly excluded from consideration in order to reduce the work involved. This is the case where a cut-off is applied to exclude combinations of hazards of order higher than 1 or 2. Criterion 3 is intended for use in such situations.

Some hazard identification methods possess internal completeness. They treat a specific class of hazard which is logically complete. This is the case with a failure modes and effects analysis, where the set of failure modes is complete. Sources of incompleteness in such a method are errors in using the method or hazards which lie outside the scope of the method.

Criterion 4 is a measure of discrimination. A hazard which has been identified may actually not be realizable, as may be demonstrated by further analysis.

There is a degree of conflict in the attempt to achieve both completeness and discrimination, particularly in industrial work where the time available for analysis is limited.

8.30.2 Studies of completeness

Taylor has extended this work to encompass a comparative study of the completeness of hazard identification on particular plants. The investigation covered four full-scale risk analysis projects plus a research study on a batch distillation plant.

In the batch plant study, various hazard identification techniques were applied successively. Starting with the hazop method, using a supporting checklist, the number of hazards identified was 30. A further two were identified by making some use of fault trees to explore causes and of consequence diagrams to explore consequences. Application of action error analysis then identified a further 82 hazards. Inspection of the actual plant using checklists identified a further 10 hazards. This gives a total of 124 hazards identified. A further three hazards were identified from disturbances during commissioning, making a total of 127.

The completeness of hazard identification was checked against a generic fault tree based on case histories. A further two hazards were identified using the tree. On this basis, therefore, the completeness of hazard identification obtained using the hazop, action error analysis and inspection was 97%.

Taylor also describes evaluation of completeness using manually and automatically constructed fault trees. For the former he quotes a completeness of 80% as typical for analysis of a piping diagram, but states that the degree of completeness is much less for sequential operations and operating problems. For the latter, work was still in progress, but he quotes a completeness at the time of writing of some 94%.

The hazop method used in the study was based on vessels rather than lines and appears not to have been a version adapted for sequential operations, so that, as the author states, the results do not do justice to the hazop method.

8.31 Quality Assurance: QUASA

A formal method for quality assessment assurance of safety analyses (QUASA) has been developed by Rouhiainen (1990). The method covers both qualitative and quantitative aspects, but with emphasis on hazard identification, and it is therefore appropriate to consider it at this point.

8.31.1 Safety analysis

The content of safety analysis is considered in terms of four main elements: (1) hazard identification, (2) quantitative risk assessment, (3) remedial measures and (4) resources.

The quality assessment examines fitness for purpose. The purposes of safety analysis considered include: assessment of risks in relation to plant siting; identification, elimination or reduction, and control of hazards in design of a new plant; and the same for an existing plant.

Safety analysis has potential deficiencies which it is the function of quality assessment to uncover. Some of these deficiencies are listed by Rouhiainen, as shown in Table 8.57.

8.31.2 QUASA

QUASA is based on quality theory. In accordance with this theory methods for the assessment of the quality of a safety analysis may be based essentially on inputs and process and/or on outputs. The features assessed in QUASA are given in Table 8.58. As stated earlier, they include not just the immediate results of the hazard identification and risk assessment but other features such

Table 8.57 *Some potential deficiencies in a safety analysis (Rouhiainen, 1990)*

Phase of analysis	*Examples of possible deficiencies*
System definition	Correlation between the plant and its descriptions not checked Important parts of the production system not included in the analysis Important production situations not taken into account
Hazard identification	Important hazard types omitted Methods used do not cover all hazards
Accident modelling	Important accident chains not modelled Important contributing factors omitted
Estimation of accident frequencies	The data used inaccurate Incorrect subjective judgments used Accuracy of the results not evaluated
Estimation of accident consequences	Simplifications made incorrect Model used not suitable to the situation studied
Risk estimation	More reliable toxicity data might be available Emergency provisions overestimated or underestimated
Documentation	Description of the object inadequate Initial assumptions not presented
Measures after the analysis	Measures presented not implemented No plan for updating the analysis

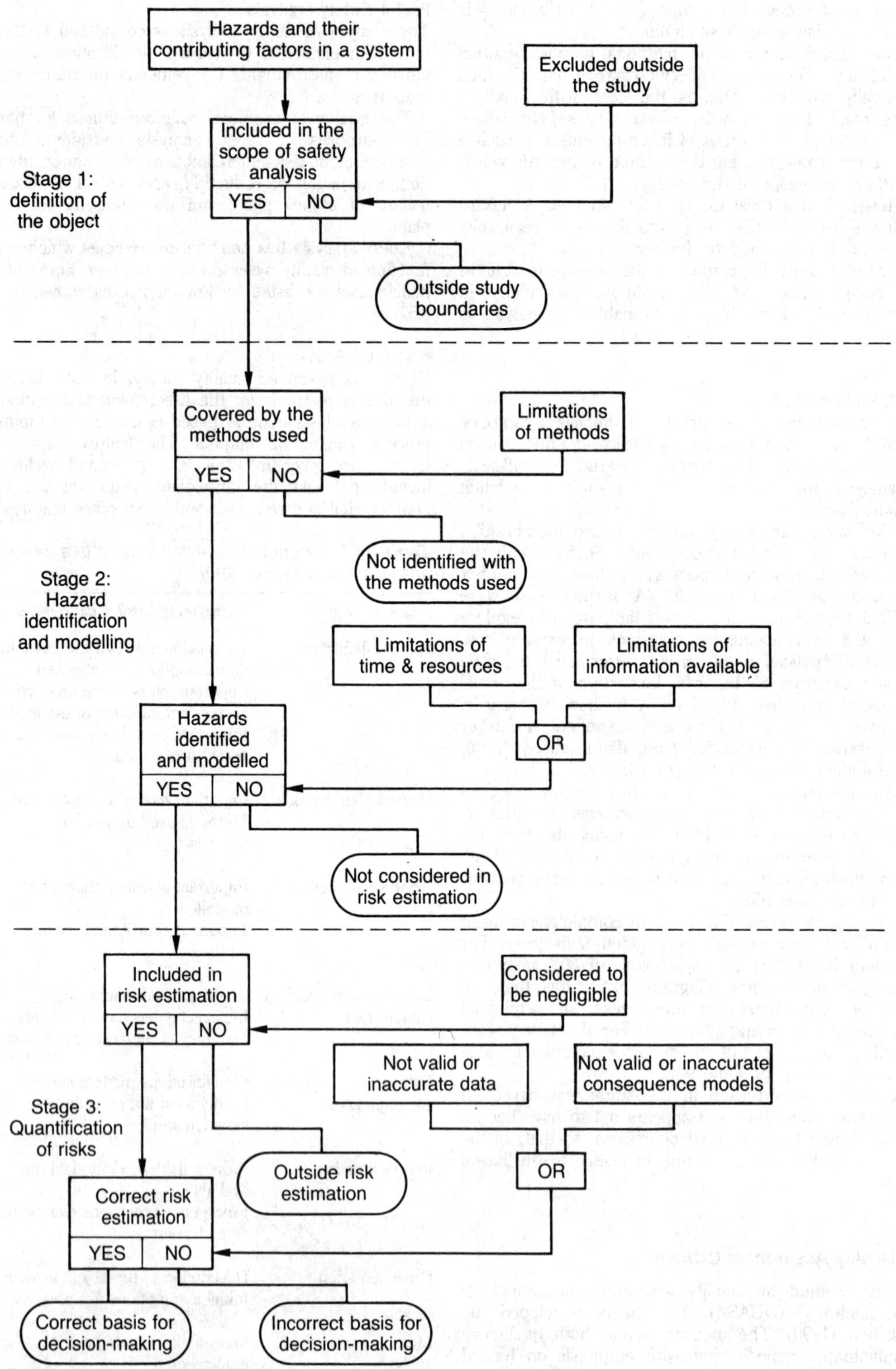

Figure 8.35 *Some potential limitations and deficiencies in a safety analysis (Rouhiainen, 1990)*

Table 8.58 *Checklist for quality assessment of a safety analysis (Rouhiainen, 1990)*

(1) Preparation of the analysis
  (a) Selection of the object
  (b) Restriction and definition of the object
  (c) Definition of the goals
  (d) Organization of the safety analysis
(2) Initiating the safety analysis
(3) Selection of methods and performance of the analysis
  (a) Identification of hazards
    (i) General aspects
    (ii) Equipment
    (iii) Processes and materials
    (iv) Organization of the object
    (v) Human activities
  (b) Modelling of accidents
  (c) Estimation of accident frequencies
  (d) Estimation of consequences
    (i) Calculation of consequences
    (ii) Emergency preparedness
  (e) Estimation of risk
  (f) Planning of remedial measures
(4) Report of the safety analysis
  (a) Description of the object
  (b) Description of the analysis
  (c) Results of the analysis
  (d) Description of the proposed measures
(5) Measures after the analysis
  (a) Measures to be taken in the object under analysis
  (b) Information about analysis results
  (c) Plans for reviewing and updating the analysis
  (d) Follow-up

as the statements of scope, the report on the analysis and the follow-up measures. The emphasis in the method is thus on outputs, broadly defined.

A core feature of the method is the assessment of the quality of the hazard identification and the quantitative risk assessment (QRA). Here consideration is confined to hazard identification, the treatment of the QRA in the method being deferred to the next chapter. The quality assessment in QUASA may be made by a single assessor or by a group of assessors.

8.31.3 Assessment models

Rouhiainen gives the following model of the accident process:

Hazard → Exposure → Consequences

with contributory factors influencing each stage. Two types of contributory factor are distinguished: determining factors and deviations. Essentially, determining factors are those governing the design of the system and deviations are excursions in its operation.

For the assessment of a safety analysis use is made of the further model shown in Figure 8.35, which gives an overview of limitations and deficiencies of the safety analysis process, cast in the form of a cause–consequence diagram.

8.31.4 Assessment checklist

The QUASA method is based on the checklist shown in Table 8.58. The checklist covers the various stages of the safety analysis, its preparation and initiation, conduct, reporting and follow-up. It draws on the method given in ISO 9002: 1987 for the assessment of the quality assurance system of a subcontractor.

8.31.5 Assessment measures

Assessment of quality involves a comparison of actual performance against some performance standard. Performance standards for some parts of the safety analysis are difficult to define. This is true particularly for hazard identification, since the complete list of hazards is not known. The approach taken in QUASA is to use a number of complementary methods of generating data alternative to that yielded by the original safety analysis with which the latter can then be compared. Some methods of generating such information for assessment of the hazard identification outputs include

(1) parallel analysis
  (a) complete,
  (b) partial;
(2) comparison with operating experience;
(3) comparison with incident case histories.

Parallel analysis involves repeating the analysis using another analyst, either completely or in part. Full repeat analyses are time-consuming and are rarely undertaken except in benchmark exercises, but sampling of the problem and partial analysis are more common.

A comparison may be made between the hazards identified in the safety analysis and incidents drawn from the operating experience of the plant. Since these cannot be expected to give the full range of potential incidents, particularly for rare events, a further comparison may be made by drawing on incident case histories for plants with similar features.

The safety analysis identifies deficiencies which in a proportion of cases will lead to incidents, but does not predict incidents as such. The same applies to the quality assessment of the safety analysis. The quality assessment, therefore, has to proceed by making a judgement as to which deficiencies have caused a particular incident. This then establishes the link between deficiencies and incidents.

8.31.6 Safety analysis performance

Thus, in general, there will be a number of deficiencies identified in the safety analysis, which may be compared with two other sets of identified deficiencies: those found by the quality assessor(s) and those inferred from incidents. There is no guarantee that these three sets comprise the total universe of deficiencies, but nevertheless the comparison gives a useful indicator of quality.

8.31.7 Assessor performance

Two indices of assessor performance are used which apply when there is more than one assessor. If a group of assessors is used, it is possible to compare the performance of the individual assessors. An interassessor reliability for a given assessor is defined as the ratio of the number of deficiencies identified by that assessor to

***Table 8.59** Identification of factors contributing to incidents by safety analysts and by quality assurance assessors: study of 117 incidents (after Rouhiainen, 1990)*

		Not identified in safety analysis		*Total*
	Identified in safety analysis	*Identified by at least one assessor*	*Not identified by any assessor*	
Physical subsystem	49	63	15	127
Environmental subsystem	7	4	2	13
Human subsystem	4	5	0	9
Managerial subsystem	2	10	1	13
Total	62	82	18	162
N	44	59	14	117

the total number identified by all the assessors. The other index used is that of validity. This is defined as the ratio of the number of incidents illustrating deficiencies in the analysis identified by a given assessor to the total number identified by all the assessors. A high score by an assessor in interassessor reliability and a low score in validity means that a large proportion of the deficiencies identified are not significant causes of incidents.

8.31.8 Assessment studies

Four quality assessment studies were conducted and used to improve the checklist. The studies covered: (1) a pulp manufacturing process; (2) a chlorine liquefaction plant; (3) an liquefied petroleum gas storage and distribution system; and (4) the First *Canvey Report*. All the studies involved quality assessment of the safety analysis as a whole, but in the first three the emphasis was on the quality of hazard identification, whilst in the fourth it was on the quality of the quantitative risk analysis.

In the original safety analysis of the pulp manufacturing process, the hazard identification methods used had been potential problem analysis (PPA), hazop and action error analysis (AEA). The PPA identified 30 hazards. The hazop identified over 150 hazards and gave 120 proposals for improvement. The AEA identified 25 opportunities for human error and gave 22 improvements.

In the quality assessment of this safety analysis, assessments were made by three assessors. The total number of deficiencies identified by the three assessors was as follows: system definition, 11; hazard identification, 24; accident modelling, 1; estimation of accident frequencies, 1; estimation of accident consequences, 3; risk estimation, 3; documentation, 9; and follow-up measures, 5. Assessor A found 36 deficiencies, assessor B found 17 of these same deficiencies and 13 new ones, assessor C found 20 of the deficiencies found by A and B and 8 new ones.

As far as concerns the checklist, out of the 138 questions, 70 were useful in revealing deficiencies in the safety analysis. Of these, 45 led to identification of deficiencies by one assessor, 28 to identification by two assessors and only 2 to identification by all three assessors.

The performance both of the safety analysis and of the quality assessors was checked against data obtained on operating experience on the plant. From these data the assessors identified some 280 trivial incidents, which they classified under five headings, and 112 non-trivial incidents, giving a total number of 117 different incidents in all. Of these 117 incidents, 73 had not been identified in the safety analysis.

Rouhiainen (1990) gives evaluations of the interassessor reliability and of the validity. The interassessor reliabilities were 58%, 53% and 49%. With regard to validity, of the 117 incidents, 59 illustrated deficiencies identified by at least one assessor and 30 deficiencies identified by all three assessors.

The number of factors considered to have contributed to these 117 incidents is given in Table 8.59.

The physical subsystem covered failures and malfunctions, structural defects, leaks within and from the system, pipe blockage and conveyor jamming, and process disturbances. The environmental subsystem covered environmental factors, weather and impurities.

Further aspects of the incidents identified were also analysed. Of the total 177 incidents the effects on personnel safety were found to be as follows: leak of hazardous substance, 12; another immediate accident hazard, 5; hazard with a parallel failure, 19; increase in probability of hazard, 25; and no direct effect on safety, 56. The effects on plant operation were as follows: shut-down of whole plant, 8; shut-down of part of plant, 53; operating disturbance, 40; decrease in quality of pulp, 4; extra work only, 8; and no effect on production, 4. The account given by Rouhiainen (1990) of the other two plant studies follows this broad pattern.

After the second study the checklist was revised. The aim of the revisions was to repair omissions, overlaps and obscurities and variability in level of detail.

The study of the *Canvey Report* was concerned with the quality of the quantitative risk assessment rather than with hazard identification. The checklist was applied by a single assessor and the deficiencies identified, which are shown in Table 8.60, were compared with those given in other critiques of the report, including the Second *Canvey Report*.

Table 8.60 *Some principal deficiencies in the First Canvey Report identified in two quality assessments (Rouhiainen, 1990)*

Author[a]	*(Analysis . . . 1980)*
Definition of the object	
– Definition of the area not clear	– Geographical limits should be applied only to fixed installations
– No definition of the residual regions	– Minor events ignored
– Employees and minor accidents excluded	– Office and factory workers excluded
Definition of the goals	
– No definition of the concept 'major interaction'	
– No definition of the types of hazards excluded	
– No criteria for choosing the accidents modelled	
– Cut-off criteria for risks which can be excluded not presented	
	– The analysis could have gone further
Organization of the risk analysis and methods used	
– Background of the team members not presented	
– Knowledge and experience of the team not presented	
– Incident statistics of the plants not used	– Comparison with accident statistics should have been made
Identification of hazards	
– Hazard identification not systematic	– Reference presents two hazards overlooked
– Not all factors affecting event-chains considered	– Smaller initiating events not considered
– Hazards caused outside the area not studied	– The same deficiency noted
– Not all materials taken into account	– Not all toxic chemicals included in the report
Estimation of accident frequencies	
– Not all assumptions and simplifications presented	– Assumptions and calculations criticized
– Reference for the data used not presented	– Noted that this prevents assessment of the data
– Statistical data used may in some cases be incorrect	– Reference presents examples of incorrect failure frequencies
– Effect of speed limit overestimated	– Noted as a source of inaccuracy
	– Basis for subjective judgments incorrect
	– Frequencies overestimated
Estimation of consequences	
– Not all principles and limitations of the models used presented	– Calculation methods not presented in enough detail
– Estimation based only on average amounts of materials	
– The effect of the changes in the source term not estimated	
	– Reference presents many mistakes in the assumptions, simplifications and methods
Estimation of risks	
– Not all assumptions, simplifications and calculations presented	– Many mistakes presented
	– Some serious arithmetic errors noted
– Cryogenic commission of LNG[b] tankers excluded	– The same deficiency noted
– Risk estimated only for some regions	
– Not all aspects concerning emergency planning studied	– Avoiding action of people not taken into account
Report of the risk analysis	
– Some characteristics of the analysis object not described	– The same deficiency noted
	– Results not presented in enough detail
– Accuracy of the results not presented	– No logical basis for selecting the remedies
	– No guidance for the future
	– The total impact of the hazards not estimated

[a] Author, single assessor; (Analysis, 1980), other critiques
[b] Liquefied natural gas.

8.32 Notation

Section 8.5

E_a	activation energy (kcal/mol)
H	heat input
ΔH_d	heat of decomposition
ΔH_o	heat of combustion
M	molecular weight
T	absolute temperature
T_d	adiabatic decomposition temperature (K)
T_e	exotherm onset temperature
x	number of carbon atoms in molecule
y	number of hydrogen atoms in molecule (Equation 8.5.1); y criterion (Equation 8.5.3)
z	number of oxygen atoms in molecule

Section 8.8

AQ	airborne quantity (kg/s)
BI	Business Interruption Loss
C	Loss Control Credit Factor
CEI	Chemical Exposure Index
CF	Credit Factor
DF	Damage Factor
EF	Escalation Factor
EPRG	toxic limit (mg/m^3)
ER	Exposure Radius
F_1	GPH factor
F_2	SPH factor
F_3	PUHF factor
F&EI	Fire and Explosion Index
GPH	General Process Hazard
HD	Hazard Distance (m)
MF	Material Factor
MPDO	Maximum Probable Days Outage
MPPD	Maximum Probable Property Damage
OV	Original Value
PUHF	Process Unit Hazards Factor
RV	Replacement Value
SPH	Special Process Hazard
VAE	Value of the Area Exposed
VPM	Value of Production per Month

Section 8.21

L	Exponent of likelihood (where latter is expressed as a frequency)
S	severity category
TR	target risk

Section 8.30

c	measure of consequence
f	frequency of occurrence
N	number of hazards

Subscripts

chd	identified in a large case history list
id	identified
m	identified by method m
rl	realisable
tot	total existing

9 Hazard Assessment

Contents

In the previous chapter an account was given of the process of identifying hazards, of the methods available and of the follow-up. For the great majority of hazards the action to be taken is clear once the hazard is known. There are, however, some hazards which require further assessment. Originally, such hazard assessment was undertaken in order to assist in making engineering decisions in the 'grey' areas where further investigation is needed in order to decide on the most cost-effective measures. This is still the prime purpose of hazard assessment, but increasingly in recent years this activity has been enlarged to deal with major hazards which are realized only very rarely but which present a threat to the public.

The aims and nature of these two activities are somewhat different. The terminology used to describe them also differs. In the UK the practitioners of the first type of hazard assessment have tended to call it 'hazard analysis' (HAZAN). The second type is often called 'probabilistic risk analysis' (PRA), a term which came into use in the US nuclear industry, or simply 'risk analysis'.

Hazard assessment is introduced here, therefore, with an account of how it has developed historically. The use of hazard analysis as an aid to engineering decision-making is then described. As this account shows, the methods used in hazard analysis are often relatively simple.

Risk assessment is a more complex undertaking, as the account given indicates. The starting point is the identification of the hazards and definition of the hazard scenarios to be considered. For each hazard the frequency and consequences are then estimated. The estimation of frequency may require the use of fault trees, event trees and/or cause–consequence diagrams. The estimation of the consequences involves the study of a sequence of events. Usually this is an emission of hazardous material which gives rise to certain physical effects. The estimation of these effects involves the use of a wide range of hazard models.

Some hazard assessments terminate at this point. If a full risk assessment giving risks to the public is required, however, it is necessary also to have models that define the characteristics of the population at risk, in order to allow for factors such as shelter and escape, which modify exposure to the physical effects, and then to estimate the proportion of people who suffer injury.

There are a number of forms in which the results of a risk assessment may be presented. It is desirable that information be presented on the degree of confidence in the results of the risk assessment. This is important both because there are inherent uncertainties in risk assessment and because the methodology is still developing. Although the numerical estimates produced by a quantitative hazard assessment may not be the most important outputs, it is necessary to have risk criteria by which to evaluate them.

Model hazard assessments for major hazards have been carried out by the Health and Safety Executive (HSE) to assist in land use planning and by the Chemical Industries Association (CIA) to assist in emergency planning and in preparation of a safety case. The impact of a hazard on the surrounding area may be estimated using a hazard impact model.

A full risk assessment is a major undertaking. Various simplified techniques have been developed, involving a greater or lesser degree of approximation. These range from short-cut versions of the normal method to very simple and approximate methods. A major accident is usually preceded by a number of lesser events which may serve as warnings. The hazard warning structure may be analysed and used to enhance confidence that the hazard is under control.

A number of major risk assessments have been carried out, notably the two *Canvey Reports* and the *Rijnmond Report*. These are described in Appendices 7 and 8, respectively. Selected references on hazard assessment are given in Table 9.1.

Table 9.1 *Selected references on hazard assessment*

NRC (Appendix 28 *Probabilistic Risk Assessment*); Green and Bourne (1965 UKAEA AHSB(S) R91,1966 UKAEA AHSB(S) R117,1972); Recht (1965); F.R. Farmer (1967b, 1969a, 1971, 1975, 1981); Hensley (1967 UKAEA AHSB(S) R136, 1968); Buchanan and Hutton (1968); Fowler and Spiegelman (1968); A.E. Green (1968, 1969 UKAEA AHSB(S) R172, 1969-70, 1970, 1972, 1973, 1974a,b, 1976, 1982, 1983); R.L. Browning (1969a–c, 1970, 1973, 1980); O'Sell and Bird (1969); Benjamin and Cornell (1970); MCA (1970/18); Fine (1971, 1973); Houston (1971); Kletz (1971, 1972a, 1973a, 1974c, 1976d, 1977a,b,e,f, 1978a, 1979l,m, 1981d, 1983e,h,i, 1984b,h,i); Malloy (1971); R.M. Stewart (1971); J.R. Taylor (1973, 1974b,d, 1975a,b, 1976b,c); Apostolakis (1974, 1975, 1978, 1981, 1986, 1991); G.D. Bell and Beattie (1974); Bulloch (1974); DOT, CG (1974a–d); S.B. Gibson (1974, 1977b); Hoffman (1974); Krasner (1974); Lawley (1974a,b, 1976, 1980); AEC (1975); Burgoyne (1975); Eisenberg, Lynch and Breeding (1975); Erdmann *et al.* (1975, 1977); D.O. Cooper and Davidson (1976); Critchley (1976); Kastenburg, McKone and Okrent (1976); M.J. Katz (1976); Sather (1976); Erdmann *et al.* (1977); Gangadharan and Brown (1977); Ravetz (1977b); Saeks and Liberty (1977); US Army Matériel and Development and Readiness Command (1977); US Congress, OTA (1977); Anon. (1978g); Cremer and Warner (1978); HSE (1978b, 1981a); Kumamoto and Henley (1978); Albaugh *et al.* (1979); Hanna (1979); R. King and Magid (1979); V.C. Marshall (1979a, 1980b, 1982a,b); Mumford and Lihou (1979a–c); Riso National Laboratory (1979/1); Anon. (1980m); Apostolakis, Garribba and Volta (1980); Bjordal (1980); Burke and Weiss (1980); Comer, Cox and Sylvester-Evans (1980); Garribba and Ovi (1980); Jager (1980, 1983); Lawley (1980); Lees (1980b,c, 1981b, 1982c); Opschoor and Hoftijzer (1980); Abramson (1981a,b); Apostolakis and Kaplan (1981); EPA (1981, 1987a); R.F. Griffiths (1981a,b, 1989, 1991a); Henley and Kumamoto (1981, 1985, 1992); Kaplan and Garrick (1981); Napier (1981); D.W. Pearce (1981); Slater, Ramsay and Cox (1981); J.M.T. Thompson (1981); W.A. Thompson (1981); F. Warner (1981b); Chatwin (1982a,b); CONCAWE (1982 10/82, 1988 88/56); Covello, Menkes and Nehnevajsa (1982); L.A. Cox (1982); R.A. Cox (1982a,b, 1986, 1987, 1989a, 1990); R.A. Cox and Comer (1982); N.C. Harris (1982); Helmers and Schaller (1982); IBC (1982/38, 1989/77, 1991/85, 86, 1992/95, 1993/103);

Kaiser (1982c, 1993); S. Kaplan (1982b); NRC (1982a); Rijnmond Public Authority (1982); Sancaktar (1982, 1983); R.F. White (1982); Barrell, Daniels and Hagon (1983); Borse (1983a-c); Cardinale, Grillo and Messina (1983); A.P. Cox and Pasman (1983); Coxon and Gilbert (1983); Daniels and Holden (1983); Eddershaw (1983 LPB 52); R.A. Freeman (1983, 1985, 1989); IMechE (1983); Joschek (1983); Lans and Bjordal (1983); Lathrop and Linerooth (1983); S. Levine, Joksimovic and Stetson (1983); Mancini and Volta (1983); Opschoor and Schecker (1983); The Royal Society (1983, 1992); Considine (1984 SRD R310); Frankel (1984); Hawksley (1984, 1989); Lowe (1984); T. Redmond (1984); Waller and Covello (1984); Cullingford, Shah and Gittus (1985); Holden (1985); Nieuwhof (1985c); Olkkola *et al.* (1986); Rasbash (1986a); J.R. Taylor *et al.* (1986); Apostolakis and Moieni (1987); Boykin and Kazarians (1987); Guymer, Kaiser and McKelvey (1987); Kanury (1987); Ozog and Bendixen (1987); Heising (1987); Paté-Cornell (1987); J.R. Thomson (1987); Andrews (1988); Burns (1988); CPD (1988, 1992a,b); Emerson, Pitblado and Sharifi (1988); Nelms (1988); Suokas (1988); Waite, Shillito and Sylvester-Evans (1988); Wakeling (1988); Withers (1988); J.L. Woodward and Silvestro (1988); Andow (1989); Bourdeau and Green (1989); FEMA (1989); Geary (1989); R.F. Griffiths (1989); Holden (1989 SRD R504); HSC (1989); ILO (1989); van Loo and Opschoor (1989); Prugh (1989); L.E.J. Roberts (1989); Rouhianen and Suokas (1989); Suokas and Kakko (1989); B. Rasmussen and Smith-Hansen (1989); Suokas and Rouhianen (1989); Anon (1990h); Arendt (1990); Castleman (1990); Edmondson (1990); Fraser, Bussey and Johnstone (1990); French, Olsen and Peloquin (1990a,b); Hirano (1990b); Pitblado, Williams and Slater (1990); Schaller (1990); Smith (1990 LPB 93); Turney (1990a,b); Bowonder, Arvind and Myake (1991); A. Fisher (1991); Greenberg and Cramer (1991); Hurst (1991); S. Brown (1991); Basta (1992); CIA (1992 SRD RC54, RC55, RC56); S.J. Cox (1992); Fryer and Mackenzie (1992); Goodner (1992); Heinold (1992); IAEA (1992); Papazoglou *et al.* (1992); Peacock and White (1992); J.R. Taylor (1992); Tweeddale (1992, 1993a,c); Twisdale *et al.* (1992); Ziegler (1992); Claggett *et al.* (1993); Frank, Giffin and Hendershot (1993); Goldsmith and Schubach (1993); Melchers and Stewart (1993); Rimington (1993); M.W. Wright, Bellamy and Cox (1993); Christen, Bohnenblust and Seitz (1994); Eastwood (1994); Melville (1994); Pitblado (1994b); Rosenthal and Lewis (1994)

Terminology: Higson (1981); F. Warner (1981b); V.C. Marshall (1981a,b, 1982c); A.E. Green (1982a); R. Mitchell (1982); Rasbash (1982b); D. Jones (1992); Hotson (1994 LPB 120)

ISGRA: ISGRA (1982, 1985); Holden, Lowe and Opschoor (1985); Clerinx (1986)

CONCAWE: Hope (1983)

CCPS: Goldthwaite *et al.* (1987); R.A. Freeman (1990)

Incident scenarios

Wells *et al.* (1992)

Hazard assessment priorities, preliminary hazard assessment

Harrington *et al.* (1986); Holloway (1989); Keey (1991); Bergmann (1993); Allum and Wells (1993); Ganger and Bearrow (1993); Wells, Wardman and Whetton (1993)

Event data (*see* Table A14.1)

Fault trees

IEC (Std 1205); NRC (Appendix 28 *Fault Tree Analysis*); Esary and Proschan (1963); Boeing Company (1965); Feutz and Waldeck (1965); Haasl (1965); Mearns (1965); Michels (1965); Recht (1966c); W.Q. Smith and Lien (1968); Crosetti and Bruce (1970); Mieze (1970); Vesely (1970, 1977b, 1983); Crosetti (1971); Houston (1971); NTSB (1971); R.M. Stewart (1971, 1974a,b); Bennetts (1972, 1975); Fussell and Vesely (1972); Andow (1973, 1976, 1980, 1981, 1989); Barlow and Chatterjee (1973); Fussell (1973a,b, 1975, 1976, 1978b); Henley and Williams (1973); Fussell, Henry and Marshall (1971); R.A. Evans (1974, 1975); Fussell, Powers and Bennetts (1974); Lawley (1974a,b, 1980); Malasky (1974); Phibbs and Kumamoto (1974); Powers and Tompkins (1974a,b, 1976); AEC (1975); Andow and Lees (1975); Barlow, Fussell and Singpurwalla (1975); Esary and Ziehms (1975); Lambert (1975, 1976, 1977, 1978b); R.L. Browning (1975, 1976, 1979, 1980); Nieuwhof (1975); Caceres and Henley (1976); Carnino (1976); Lambert and Yadigaroglu (1976); Y.T. Lee and Apostolakis (1976); Powers and Lapp (1976); Salem, Apostolakis and Okrent (1975a,b, 1977); Apostolakis and Lee (1977); Brock (1977 NCSR R14); Burdick *et al.* (1977); Doering and Gaddy (1977); Gangadharan, Rao and Sundarajan (1977); Garriba *et al.* (1977); Kolodner (1977b); Lambert and Gilman (1977a,b); Lapp and Powers (1977a, 1979); Martin-Solis, Andow and Lees (1977); Mingle, Chawla and Person (1977); NEA (1977); G.D.M. Pearson (1977); Vaillant (1977); Astolfi, Contini and van den Muyzenberg (1978); Astolfi *et al.* (1978); Chamow (1978); Himmelblau (1978); J.R. Taylor (1976c, 1978b, 1979, 1982); D.M. Brown and Ball (1980); Dasarathy and Yang (1980); Hauptmanns (1980, 1986); Piccinini *et al.* (1980); Pilz (1980a,b); D.J. Allen (1981); Prugh (1981, 1982, 1992a); Welsh and Lundberg (1980); Arendt and Fussel (1981); Henley and Kumamoto (1981, 1985, 1992); Kumamoto, Inoue and Henley (1981); Vesely *et al.* (1981); Kletz and Lawley (1982); Schreiber (1982); Alesso (1983, 1985); Hauptmanns and Yllera (1983); Johnston and Matthews (1983 SRD R245); Bendixen and O'Neill (1984); Doelp *et al.* (1984); J.M. Morgan and Andrews (1984); Shafaghi, Andow and Lees (1984a,b); Slater (1984); Zipf (1984); Alesso, Prasinos and Smith (1985); Andrews and Morgan (1985, 1986); Bendixen, Dale and O'Neill (1985); BG (1985 Communication 1242); Camarinopoulos and Yllera (1985); Jiang Mingxiang (1985); Keey and Smith (1984); Laviron and Heising (1985); W.S. Lee *et al.* (1985); J.M. Morgan (1985); Yampolsky, Adam and Karahalios (1985); Flothmann and Mjaavatten (1986); Harron (1986); Page and Perry (1986); Ruan Keqiang (1986); Vervalin (1986b); R.P. Hughes (1987a); Kwang Sub Jeong, Soon Heung Chang and Tae Woon Kim (1987); Limnios (1987b); Kohda and Henley (1988); Kohda, Henley and Inoue (1989); Kumar, Chidambaram and Gopalan (1989); Singer (1990); Aldersey, Lees and Rushton (1991); H. James, Harris and Hall (1992); Kavianian, Rao and Brown (1992); Minton and Johnson (1992); Andrews and Moss

(1993); Bueker *et al.* (1993); D.A. Lee and Browne (1993); Vincoli (1993); ANSI Y32.14-1973; BS 5760: Part 7: 1991
Graph theory, graphical methods: Karnopp and Rosenberg (1968); Harary, Norman and Cartwright (1975); Biggs, Lloyd and Wilson (1976); Carre (1979); Chachra, Ghare and Moore (1979); Umeda *et al.* (1979, 1980); Boffey (1981); Temperley (1981); Khoda and Henley (1988)
Kinetic tree theory: Vesely (1969, 1970b); Vesely and Narum (1970); Vesely and Goldberg (1977a,b)
GO methodology: R.L. Williams (1977)

Fault tree synthesis (*see also* Table 30.1)
Lapp–Powers method: Henley and Kumamoto (1977); Lapp and Powers (1977a, 1979); Locks (1979, 1980); Powers (1977); Shaeiwitz, Lapp and Powers (1977); Lambert (1979); Yellman (1979); Cummings, Lapp and Powers (1983); Ulerich and Powers (1988)

Event trees
NRC (Appendix 28 *Event Trees*); von Alven (1964); N.C. Rasmussen (1974); AEC (1975); Andow (1976); Vesely (1977b); HSE (1978b); Worledge (1979 SRD R128); Welsh and Lundberg (1980); Kaplan (1982a); Schreiber (1982); Slater (1984); Limnios and Jeannette (1987); Melo, Lima and Oliviera (1987); Aldersey, Lees and Rushton (1991); H. James, Harris and Hall (1992); Piccinini, Scarrone and Ciarambino (1994)

Cause–consequence diagrams
D.S. Nielsen (1971, 1974, 1975); R.A. Evans (1974); J.R. Taylor (1974c,d, 1975a, 1976c, 1978a); D.S. Nielsen, Platz and Runge (1975); J.R. Taylor and Hollo (1977a,b); Himmelblau (1978)

Dependent failures
Epler (1969); Fleming (1974); AEC (1975); R.A. Evans (1975); Apostolakis (1975); Fleming and Hannaman (1976); J.R. Taylor (1976a); NEA (1977); Vesely (1977a); D.P. Wagner, Cate and Fussell (1977); D.P. Wagner and Fussell (1977); Fussell (1978a); Edwards and Watson (1979 SRD R146); Welsh and Lundberg (1980); Bourne *et al.* (1981 SRD R196); Johnston and Crackett (1985 SRD R383); Crackett (1986 SRD R411); B.D. Johnston (1987a,b); Humphreys (1987); R.P. Hughes (1987b); Humphreys and Johnson (1987 SRD R418); B.R. Martin and Wright (1987); G.T. Edwards (1988); Mosleh *et al.* (1988); Attwood (1991); Andrews and Moss (1993)

Escalation, domino effects
Labath and Amendola (1989); Bagster and Pitblado (1991); Four Elements Ltd (1991); Purdy, Pitblado and Bagster (1992); Latha, Gautam and Raghavan (1992); Scilly and Crowther (1992); Pettit, Schumacher and Seeley (1993)

Expert judgement (*see* Table 7.1)

Rare events
Pressure vessel failure (*see* Table 12.1)

External threats
NRC (Appendix 28 *External Hazards*)
Aircraft crashes: Chelapati, Kennedy and Wall (1972); Marriott (1987); Phillips (1987 SRD R435)
Weather: Eagleman, Muirhead and Williams (1975); Eagleman (1983); Page and Lebens (1986); Peters and Hansel (1992)
Earthquakes (see Table A15.1)
Floods: McCullough (1968)
Hurricanes: L.R. Russell and Schueller (1974)
Landslides: Zaruba and Mencl (1982)
Tornadoes: Twisdale *et al.* (1978); Peters and Hansel (1992); Rutch *et al.* (1992)

Management systems
Barrell and Thomas (1982); R.A. Cox and Comer (1982); Powell and Canter (1984); Pitblado, Williams and Slater (1990); Burge and Scott (1992); J.C. Williams and Hurst (1992); A.J. Smith (1992); Bellamy, Wright and Hurst (1993)

Human factors (*see also* Table 14.11)
Howland (1980); Ingles (1980); Welsh and Lundberg (1980); Spangler (1982); Rasmussen and Pedersen (1983); Melchers (1984); Ramsden (1985); Bellamy, Kirwan and Cox (1986); Bellamy and Geyer (1988); Delboy, Dubnansky and Lapp (1991); Banks and Wells (1992); Bridges, Kirkman and Lorenzo (1992); Yukimachi, Nagasaka and Sasou (1992); Zimolong (1992); Cameron (1993); Nawar and Samsudin (1993)

Hazard models (*see also* tables in Chapters 15–18)
Longinow *et al.* (1973); Dunn (1974); AD Little Inc. (1974a); Benedict (1978); Cremer and Warner (1978); HSE (1978b, 1981a); R.A. Cox *et al.* (1980); Rijnmond Public Authority (1982); Bello and Romano (1983a); Crocker and Napier (1988a,b); Chhiba, Apostolakis and Okrent (1991); Pietersen (1990); Bagster (1993); Geeta, Tripathi and Narasimhan (1993); Mallet (1993); K.E. Petersen (1994)

Hazard model systems
Harding, Parnarouskis and Potts (1978); Jäger, Diedershagen and Kühnreich (1989); FEMA (1989); Pitblado and Nalpanis (1989); Raj (1991); Papazoglou, Nivolianitou *et al.* (1992)
CPD, TNO: Opschoor (1979); CPD (1992a,b);
DYLAM: Nivolianitou, Amendola and Reina (1986); Labath and Amendola (1989)
RISKAT: Pape and Nussey (1985, 1989); Hurst, Nussey and Pape (1989); Nussey, Mercer and Clay (1990); Nussey, Pantony and Smallwood (1992, 1993)
SAFETI: Pitblado and Nalpanis (1989); Pitblado, Shaw and Stevens (1990)
Vulnerability model: DOT, CG (1974a–d); Raj and Kalelkar (1974); Eisenberg, Lynch and Breeding (1975); Enviro Control Inc. (1976, 1977); Rausch, Tsao and Rowley (1977); Rowley and Rausch (1977); Rausch, Tsao and Rowley (1977); Rausch, Eisenberg and Lynch (1977); Tsao and Perry (1979); USCG (1979); Parnarouskis, Perry and Articola (1980); Perry and Articola (1980); Dodge *et al.* (1983)
WHAZAN: Technica (1985); World Bank, Office of Environmental and Science Affairs (1985)
ORA Toolkit: Technica (1991)

Population characteristics, exposure
NRC (Appendix 28 *Population Distribution*); J.P. Robinson and Converse (1966); F.R. Farmer (1967b); Athey, Tell and Janes (1973); AEC (1975); Eisenberg, Lynch and Breeding (1975); Hewitt (1976); Chartered Institute of Building Services (1977b); Fitzpatrick and Goddard (1977); HSE (1978b); Hushon and Ghovanlou (1980); OPCS (1980, 1981a–c); Rhind (1983); CSO (1985); Schewe and Carvitti (1986); Petts, Withers and Lees (1987); van Loo and Opschoor (1989); ACDS (1991)

Mitigation of exposure
NRC (Appendix 28 *Mitigation Systems*)
Evacuation: NRC (Appendix 28 *Evacuation*); Hans and Sell (1974); Westbrook (1974); Solomon, Rubin and Okrent (1976); HSE (1978b); Urbanik *et al.* (1980); A.F.C. Wallace (1980); Sorensen (1987); Duclos, Binder and Rieter (1989); G.O. Rogers and Sorensen (1989); P.J. Harrison and Bellamy (1989/90)
Shelter (see Table 15.10)
Other measures: Tsuchiya *et al.* (1990)

Injury relations
Lees (1980b); Bourdeau and Green (1989); CPD (1992b); CCPS (1994/15)
Probit methods: R.A. Fisher and Yates (1957); Finney (1971); Eisenberg, Lynch and Breeding (1975); HSE (1978b, 1981a); Lees (1980b); Paradine and Rivett (1980); CPD (1988, 1992a,b); Emerson, Pitblado and Sharifi (1988); MacFarlane and Ewing (1990); V.C. Marshall (1989b); J. Singh and McBride (1990); R.F. Griffiths (1991b); de Weger, Pietersen and Reuzel (1991); Opschoor, van Loo and Pasman (1992)

Presentation of results
Rothschild (1993)
Frequency–number (FN) tables, curves: F.R. Farmer (1967b); Provinciale Waterstaat Groningen (1979); R.F. Griffiths (1981e); Okrent (1981); Rijnmond Public Authority (1982); Hagon (1983, 1984, 1986); Vervalin (1986b); Ormsby and Le (1988); J.L. Woodward and Silvestro (1988); Saccomano, Shortreed and Mehta (1990); ACDS (1991); Pietersen and van het Veld (1992); Prugh (1992c)

Uncertainty, confidence and sensitivity
AEC (1975); Jacobsen (1980); Parry and Winter (1980 SRD R190); Amendola (1983b); Nussey (1983); Olivi (1983); Unwin (1984 SRD R301); Siu and Apostolakis (1985); Crick, Morrey and Hill (1986); Garlick (1987 SRD R443); Baybutt (1989); I. Cook and Unwin (1989); Goosens, Cooke and van Steen (1989); S.R. Hanna, Chang and Strimaitis (1990); ACDS (1991); Chhiba, Apostolakis and Okrent (1991); S.R. Hanna, Strimaitis and Chang (1991b); Amendola, Contini and Ziomas (1992a,b); Kortner (1992); van Wees and Mercx (1992); Shevenell and Hoffman (1993); Quelch and Cameron (1994)
Error propagation: Barry (1978); Karlsson and Bjerle (1980); Park and Himmelblau (1980); Soon Chang, Joo Park and Myung Kim (1985); Keey and Smith (1984); Asbjornsen (1986); Dohnal *et al.* (1992); Melchers (1993a); Quelch and Cameron (1993)

Evaluation of results (*see also* Table 9.36)
Harvey (1979b, 1984); Kastenberg and Solomon (1985); R.F. Griffiths (1989); HSE (1989c); Casada, Kirkman and Paula (1990); Lapp (1990)

Quality assurance of hazard assessment
J.R. Taylor (1979); Welsh and Debenham (1986); Rouhiainen (1991d); Pitblado (1994a)

Validation of hazard assessments
A. Taylor (1979, 1981); Jenssen (1993); R.F. Evans (1994); Nussey (1994)

Follow-up, outcome of hazard assessments
HSE (1978b, 1981a); Rijnmond Public Authority (1982); Baybutt (1983, 1986); J.R. Taylor (1984); Arendt (1986); Vestergaard and Rasmussen (1988); Desaedeleer *et al.* (1989); Cullen (1990); Schaller (1990); Kakko, Virtanen and Lautkaski (1992); Tweedale (1993b)

Hazard assessment in decision-making
HSE (1983e, 1989c); Kelly and Hemming (1983); Harbison and Kelly (1985); Blokker (1986)

Hazard assessment and emergency planning
Burns (1988); J. Singh and McBride (1990); Tavel, Maraven and Taylor (1989); Essery (1991)

Hazard impact model
Poblete, Lees and Simpson (1984); Lees, Poblete and Simpson (1986); Lees (1987)

Hazard warning
Warning events analysis: J.H. Bowen (1978); Page (1979); W.G. Johnson (1980); Slater and Cox (1984); van Hemel, Connelly and Haas (1990)
Hazard warning structure: Lees (1982b, 1983b, 1985); Keey (1986a,b); Lake (1986); Pitblado and Lake (1986, 1987); E. Smith and Harris (1990)

Planning (*see also* Table 4.1)
Batstone and Tomi (1980); Brough (1981); Pantony and Smith (1982)

HSE
Harvey (1976, 1979b, 1984); HSE (1978b, 1981a, 1989c,e); Barrell (1980); Barrell and Thomas (1982); Pantony and Smith (1982); ACDS (1991)

Offshore
Borse (1979); Pyman and Gjerstad (1983); Vinnem (1983); Cullen (1990)

Civil and structural engineering
Blockley (1980); Ingles (1980); Melchers (1984)

Hazard assessment applications and case histories
McGillivray (1963a,b); Kletz (1971, 1972a); Ybarrando, Solbrig and Isbin (1972 AIChE/119); Siccama (1973); Lawley (1974a,b, 1976, 1980); N.C. Rasmussen (1974); AEC (1975); Dicken (1975); J.R. Campbell and Gaddy (1976); Moser, Moel and Heckard (1976); Sellers (1976, 1988); Shell UK Exploration and Production (1976); Lundquist and Laufke (1977); D.S. Nielsen (1977); D.S. Nielsen, Platz and Kongso (1977); Okrent (1977);

Rasbash (1977b); HSE (1978b,d, 1981a); Blokker *et al.* (1980); Joschek *et al.* (1980); de Ruiter and van Lookeren Campagne (1980); van der Schaaf and Opschoor (1980); Sutcliffe (1980); van Vliet *et al.* (1980); Helmeste and Phillips (1981); Considine, Grint and Holden (1982); Huberich (1982); Piccinini *et al.* (1982); Solberg and Skramstad (1982); Bello and Romano (1983b); Bergmann and Riegel (1983); Considine (1983, 1986); Jager (1983); D.A. Jones (1983); D.S. Nielsen and Platz (1983); White (1983 SRD R273); Arendt *et al.* (1984); O'Mara and Burns (1984); Prijatel (1984); L.B. Grant (1985); Arendt (1986); Arendt, Casada and Rooney (1986); Blything (1986); Gebhart and Caldwell (1986); D.A. Jones and Fearnehough (1986); Labath *et al.* (1986); Rochina (1986); Shea and Jelinek (1986); Al-Abdullally, Al-Shuwaib and Gupta (1987); Page (1987); R.A. Cox (1988, 1989b); Crossthwaite, Fitzpatrick and Hurst (1988); Ormsby and Le (1988); Rooney, Turner and Arendt (1988); Sellers and Picciolo (1988); J.L. Woodward and Silvestro (1988); Anon. (1989 LPB 96, p. 19); Cassidy (1989, 1990); CIA, Chlorine Sector Group (1989); M.M. Grant (1989); Klug (1989); Kumar, Chidambaram and Gopalan (1989); Grint and Purdy (1990); Schaller (1990); Shei and Conradi (1990); M.P. Singh (1990); Tweeddale and Woods (1990); ACDS (1991); S. Brown (1991); Duong (1991); R.J. Clarke and Nicholson (1992); Myers, Mudan and Hachmuth (1992); Sorenson, Carnes and Rogers (1992); Goyal (1993, 1993 LPB 1123); Mant (1993)

ISPRA Benchmark Studies: CEC (EUR 13386 EN, EUR 13597 EN); Amendola (1983a,b); Amendola, Contini and Ziomas (1992a,b)

Safety cases: Lees and Ang (1989b); Wiedeman (1992); R.F. Evans (1994)

Guide assessments, including safety case assessments

BCGA (1984); Crossthwaite (1984, 1986); Pape and Nussey (1985); CIA Chlorine Sector Gp (1986, 1989); Clay *et al.* (1988); LPGITA (1988); FMA (1989); Lees and Ang (1989b)

9.1 Background

An understanding of the way in which hazard assessment has developed historically is important in appreciating its present role. Hazard analysis was developed in the 1960s as part of the response to the problems which stimulated the development of loss prevention in general, as described in Chapter 1.

One problem area was the availability of large, single stream chemical plants. Application of the techniques of reliability engineering was one of the measures taken to correct this. Closely related was the problem of high hazard plants. Several major protective systems were implemented. These were designed using fault trees and instrument failure data, and risk criteria were devised to evaluate them. Another matter of concern was the increasing quantity of chemicals being transported around the country by road, rail and pipeline. Assessments were made of these operations.

The industry was initially assisted in the use of reliability engineering by the UK Atomic Energy Authority (UKAEA), but soon began to develop its own approach. One innovation introduced was the use of the hazard and operability study as a means of identifying hazards. This technique was developed by the chemical industry. The industry also developed a rather distinctive approach to hazard assessment. It was clear that in order to improve plant reliability and safety it was necessary to modify the plant designs, by using more reliable equipment and/or configurations and by incorporating additional protective devices. There was a severe problem, however, in selecting from the large number of measures which might be taken, those which were most cost effective.

In many cases once a hazard had been identified, it was clear what action should be taken, but there were sometimes 'grey' areas where the decision was not clear-cut. This is the problem to which hazard analysis addresses. Typically, the use of hazard analysis involves estimating the frequency of realization of the hazard without and with some protective measure, and then evaluating the results using an appropriate criterion. The frequency of the hazard is usually estimated directly from data for that event, but it sometimes has to be synthesized from a fault tree using data for events lower down the tree.

The spirit in which the early hazard analysis was developed was to use quantitative methods but to keep it simple. Fault trees were used, but they were usually nothing like as complex as those developed for nuclear plants. Often the calculations were almost of a 'back of the envelope' type. In this, the approach had much in common with that of early operational research. Some examples of such hazard analysis are given in the next section. They are estimates made in support of engineering decision-making.

9.2 Hazard Analysis

Hazard analysis is now a well established method for aiding engineering decision-making. An account is given in *Hazop and Hazan: Notes on the Identification and Assessment of Hazards* (Kletz, 1983d, 1986d, 1992b). The activity of hazard analysis may best be described in terms of illustrative examples. A collection of such examples has been given by Kletz (1971). Before giving these examples, however, it is necessary to describe the risk criterion used in them.

9.2.1 Fatal Accident Rate criterion

A risk criterion is required for the evaluation of the estimates made in hazard analysis. The criterion which was developed in the original work and which is still in almost exclusive use is the Fatal Accident Rate (FAR). The FAR is defined as the number of fatalities per 10^8 exposed hours. The meaning of the FAR may perhaps be best understood by saying that if the actual FAR is, say, 4, and if 1000 men go into a chemical works at the age of 20 years, then at the age of 60 years 996 men will leave the works alive. At the time when it was developed as a criterion, the actual FAR for the UK chemical industry was about 3.5.

The way in which the FAR is applied is then as follows. The total FAR is divided into 10 parts. Five of these are allocated for everyday hazards such as falling down stairs or having a spanner dropped on the head. The other five parts are allocated to technological hazards specific to the plant on which the employee works. It is assumed that there are no more than five of these on a given plant. This then gives one part available for each technological hazard on that plant. Thus if, for example, the actual overall FAR of 3.5 is also taken as the overall target value which is not to be exceeded, the target FAR for a single, specific technological hazard will be 0.35.

The FAR was originally called the Fatal Accident Frequency Rate (FAFR), but since this is tautologous, the term Fatal Accident Rate (FAR) is preferred. The FAR criterion is discussed further in Section 9.21.

9.2.2 Pipeline fracture

A hazard identified on one particular plant was contact of refrigerated liquefied gas with a mild steel pipeline, leading to brittle fracture. One method of avoiding this hazard was to use stainless steel, which would be very expensive. A much cheaper method was to use a trip system to prevent the cold gas from contacting the line.

The use of hazard analysis to tackle this problem is described by Kletz (1971) as follows:

> Using data on the reliability of the various components of the trip system and assuming the operator will ignore the initial alarm system on one occasion out of four, the 'fractional dead time' of the whole system (that is, the fraction of time it is not operating) was calculated. This depends on the frequency with which the trips are tested. The 'demand rate' on the system, the number of times per year it is called on to operate, was estimated from previous experience.
>
> The failure rate of the whole system was then estimated as once in 10 000 years or once in 2500 years for the whole plant which contained four similar systems.
>
> It was assumed that one tenth of the occasions on which the cold gas reached mild steel would result in a leak, an explosion and a fatality – almost certainly an overestimate. A fatality will then occur once in 25 000 years giving an FAFR of 0.45. It was therefore agreed that the control system was satisfactory and it was not necessary to replace the mild steel by stainless steel.
>
> It might be argued that the latter course is preferable as it is 100% safe, the FAFR is zero. Had we done this, the cost of reducing the hazard rate to zero would have been equivalent to £150 000 000 per life saved.

This example illustrates many of the characteristic features of hazard analysis. The aim of the analysis is to select a measure to deal with a hazard. An estimate is made of the frequency of realization of the hazard, based on a simple calculation using field data. The estimate made is somewhat conservative. The result is evaluated using the FAFR criterion, this criterion not being applied slavishly but with regard to the uncertainty in the estimate. A further check is done on the value of a life implied by the decision alternatives.

9.2.3 Crankcase explosions

Another example is the problem of possible crankcase explosions in a building housing 25 compressors. Data obtained from a survey showed that there was one explosion per 500 machine-years. In order to achieve a FAR of 0.35 the proportion of explosions causing fatalities would not have had to exceed 1 in 1650. As there were normally several men in the building, the risk was clearly too high. The risk had to be reduced, the method actually used being to fit a suitable relief valve.

9.2.4 Hazardous area classification

A third example relates to hazardous area classification. Flameproof or intrinsically safe electrical equipment is required in Division 1 areas in which flammable gas or vapour is likely to be present under normal operating conditions. Totally enclosed 'non-sparking' equipment may be used in Division 2 areas, in which flammable gas or vapour is likely to be present under abnormal conditions. It is more economic to use this latter type of equipment and it is therefore desirable to apply the Division 1 classification only to those areas where it is really necessary. Records indicate that Division 2 motors develop a fault which causes them to spark or overheat at a rate of once per 100 years. Observation indicates that the probability of a man being within 10 ft (3 m) of a given motor in a plant area is 1 in 20. If it is assumed that anyone within this distance is killed when an explosion occurs, that on half the occasions when a flammable atmosphere occurs there is no explosion because diffusion into the equipment is slow and short-lived concentrations of gas/vapour can be ignored, and that there are 100 motors on the plant, then an area may be classed as Division 2 if it has a flammable mixture for less than 10 h/year. This is the figure which corresponds to a FAR of 0.35.

9.3 Risk Assessment

The hazards considered in hazard analysis as just described are typically accidents with the potential to cause one or two fatalities. By contrast, quantitative risk assessment (QRA), also known as probabilistic risk assessment (PRA) or probabilistic safety assessment (PSA), usually deals with major hazards which could cause a high death toll.

A full risk assessment involves the estimation of the frequency and consequences of a range of hazard scenarios and of individual and societal risk. This is a major undertaking. An outline of the classical procedure of risk assessment is given in Figure 9.1. The diagram shows the principal elements and the broad structure, but there are variations and there is a degree of iterations, which is not shown. A risk assessment should be undertaken only if the purpose of the study has been well defined. If this is not done, it is unlikely that full value will be derived from the results obtained.

Identification of the hazards is the essential first step. The quality of the assessment depends crucially on the comprehensiveness of the hazard identification. Techniques of hazard identification have been discussed in Chapter 8. The method most relevant to risk assessment is the review of release sources. It is also necessary to identify the vulnerable targets, both persons and property.

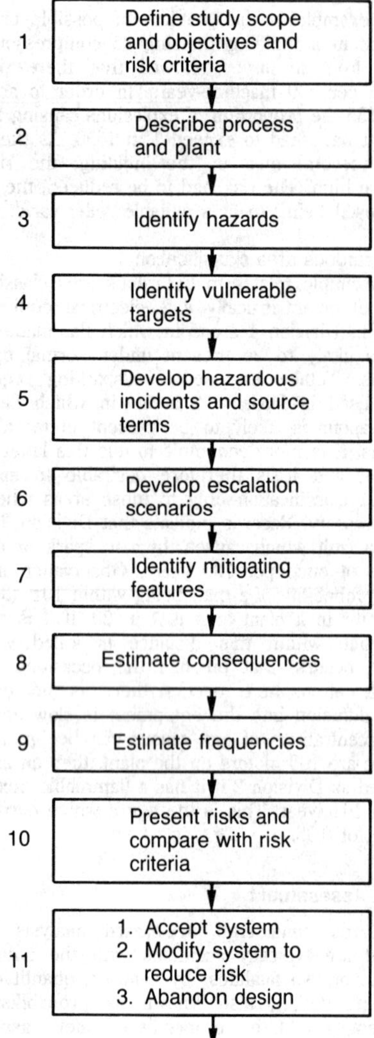
1
Define study scope and objectives and risk criteria
2
Describe process and plant
3
Identify hazards
4
Identify vulnerable targets
5
Develop hazardous incidents and source terms
6
Develop escalation scenarios
7
Identify mitigating features
8
Estimate consequences
9
Estimate frequencies
10
Present risks and compare with risk criteria
11
1. Accept system
2. Modify system to reduce risk
3. Abandon design

Figure 9.1 *The risk assessment process*

A scenario is then developed for each potential release. The scenario defines the nature of the release and thus determines the subsequent chain of consequences. Since there is an infinite number of potential releases, it is necessary to arrange them in a limited number of groups of broadly similar nature. Some factors affecting the scenarios include the hole size and the duration, which may depend on any emergency shut-off. This procedure gives the source terms.

For each release scenario a set of consequence chains is developed showing the escalation of the event. The different chains arise from the influence of conditions that affect the development of the event, such as the wind direction and velocity and stability conditions and, for flammables, the ignition sources. A typical consequence chain is shown in Figure 9.2. The dotted line in the figure represents the widening of the confidence bounds as the successive events become more remote from the initial release.

The set of consequence chains is conveniently represented in the form of an event tree. Each of the branches of the event tree represents a single consequence chain. The outcomes are generally defined as events such as flash fire, vapour cloud explosion or toxic cloud. It is then necessary to identify any features which may mitigate the effects of these events on people or property. Factors that modify the exposure of the people include shelter, escape and evacuation.

In order to estimate injury or damage it is necessary to model the sequence of events leading up to each outcome. Two typical chains of events are shown in Table 9.2, one for a flammable release and the other for a toxic release. Physical models are required which describe each stage in the sequence. Using these hazard models for the successive stages of the chain the intensity of the physical effect is estimated.

The effect of a given outcome event on persons or property is estimated from the intensity of the harmful physical effect. These effects include:

(1) thermal radiation;
(2) explosion
 (a) overpressure,
 (b) impulse;
(3) toxic concentration
 (a) dosage,
 (b) concentration–time function.

The risk of injury is determined using injury relations which give the probability of a defined degree of injury as a function of the intensity of the physical effect. Similar procedures apply to property damage. Some hazard assessments are terminated at this point. The consequences are evaluated.

In other cases an estimate is made of the frequencies of some or all of the outcomes. This involves first the estimation of the frequency of the release scenario. It may be possible to do this directly from failure data or it may be necessary to synthesize a value using fault tree methods. Then, using the event tree framework, an estimate is made of the probability of the various events constituting the branch points in the event tree and hence the frequencies of the outcomes. Other hazard

Table 9.2 *Typical sequences of events in a risk assessment*

A Flammable release
Emission
Vaporization (if liquid)
Air entrainment
Gas dispersion
Ignition
Flash fire or vapour cloud explosion

B Toxic release
Emission
Vaporization
Air entrainment
Gas dispersion

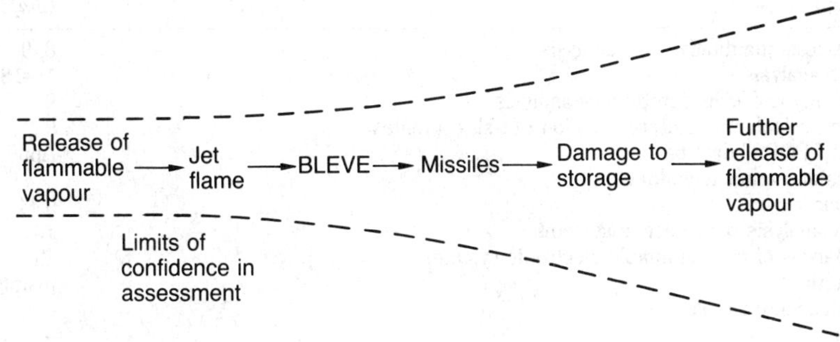

Figure 9.2 *A consequence chain*

assessments are terminated at this stage. The frequencies of the various consequences are evaluated.

Alternatively, the assessment is continued to provide a full risk assessment yielding individual and societal risks. Societal risk is expressed as a set of points representing increasingly serious events and relating the numbers of persons affected to the frequencies of the events. Estimation of societal risk involves determining the population at risk. The results of the assessment may be presented in a number of different forms such as risk contours on a map of the site or tables or graphs of individual and societal risks. Each stage of the assessment is attended by uncertainty. It is necessary, therefore, to provide measures of this uncertainty. Thus the frequency of an outcome may be expressed as a frequency distribution or range of values rather than as a single figure. Finally the results so obtained are evaluated using risk criteria.

The foregoing account is a simplified one. In particular, the information from a particular stage may lead to modification of the design and an iteration in which the results from that stage are recalculated. As indicated, the hazard assessment may be terminated short of a full risk assessment following assessment of the consequences. The consequences may be judged tolerably low for any foreseeable frequency.

In the scheme outlined, the estimation of the frequencies is deferred until a relatively late stage. An alternative approach is to perform this stage rather earlier. The hazard assessment may also be terminated following assessment of the frequencies. The frequencies may be judged tolerably low for any foreseeable consequences.

The individual aspects of quantitative risk assessment are discussed in more detail in the following sections.

There are differences of view on the value of such risk assessments. Some criticisms are directed to the basic philosophy underlying the methodology, others to the evaluation of the risk estimates generated. An account of the debate on risk assessment is given Section 9.28.

9.3.1 NRC *PRA Procedures Guide*

Initially, guidance on the conduct of risk assessments had to be sought mainly in published studies such as the two *Canvey Reports*, the *Rijnmond Report* and the *Rasmussen Report*. More formal guidance appeared in the *PRA Procedures Guide* of the Nuclear Regulatory Commission (NRC, 1982). The coverage of the *PRA Procedures Guide* is indicated by the list of contents given in Table 9.3.

9.3.2 IAEA *PSA Procedures Guide*

A more recent nuclear industry guide is *Procedures for Conducting Probabilistic Safety Assessment of Nuclear Power Plant* by the International Atomic Energy Agency (IAEA, 1992). This guide is particularly strong on the development of accident scenarios.

9.3.3 CONCAWE *Guide*

Outline guidance for the process industries was published in *Methodologies for Hazard Analysis and Risk Assessment in the Petroleum Refining and Storage Industry* by CONCAWE (1982 10/82).

Table 9.3 *Principal contents of NRC PRA Procedures Guide (Nuclear Regulatory Commission, 1982a)*

1. Introduction
2. PRA organization
3. Accident-sequence definition and system modelling
4. Human reliability analysis
5. Data base development
6. Accident sequence quantification
7. Physical processes of core melt accidents
8. Radionuclide release and transport
9. Environmental transport and consequence analysis
10. Analysis of external events
11. Seismic, fire and flood risk analyses
12. Uncertainty and sensitivity analysis
13. Development and interpretation of results

Appendices, including:
C Sources indexes for availability and risk data
D Live issues in dispersion and deposition calculations
E Evacuation and sheltering
F Liquid pathway consequence analysis

Table 9.4 Principal contents of CCPS QRA Guidelines (Centre for Chemical Process Safety 1989/5)

	Chapter/section[a]
1. Chemical process quantitative risk analysis	8, 9
2. Consequence analysis	15–18
3. Event probability and failure frequency analysis	9
4. Measurement, calculation and presentation of risk estimates	9
5. Creation of a CPQRA[b] data base	Appendix 14
6. Special topics and other techniques	
Domino effects	9.14
Unavailability analysis of protective systems	13
Reliability analysis of programmable electronic systems	13
Other techniques	mainly 7–9
7. CPQRA application examples	–
8. Case studies	–
9. Future developments	passim
Appendices	
A Loss-of-containment causes in the chemical industry	15
B Training programs	14, 28
C Sample outline CPQRA results	9.19
D Minimal cut set analysis	7, 9.5
E Approximation methods for quantifying fault trees	9.5
F Probability distributions, parameters, and terminology	7
G Statistical distributions available for use as failure rate models	7
H Errors from assuming that time-related equipment failure rates are constant	7
I Data reduction techniques: distribution identification and testing methods	7
J Procedures for combining available generic data and plant-specific data	7

[a] Chapter or section of this book in which the topic is principally addressed.
[b] Chemical process quantitative risk analysis.

9.3.4 CCPS *QRA Guidelines*

A much more detailed QRA guide is given in *Guidelines for Chemical Process Quantitative Risk Analysis* by the Center for Chemical Process Safety (CCPS, 1989/5) (the CCPS *QRA Guidelines*). The principal contents of the CCPS *Guidelines* are shown in Table 9.4 together with the location in this book where they are treated.

The *Guidelines* are oriented to the QRA practitioner. Discussion of the various consequence models and frequency estimation techniques typically has the following structure: (1) background (purpose, philosophy/technology, application), (2) description (description of technique, theoretical foundation, input requirements and availability, outputs, simplified approaches), (3) sample problem and (4) discussion (strengths and weaknesses, identification and treatment of errors, utility, resources needed, available computer codes).

9.3.5 QRA in safety cases

In the UK, one of the principal applications of QRA in the process industries is in the safety case. An account of safety cases and, in particular, of the extent to which QRA is required is given in Chapter 4.

9.3.6 QRA and decision-making

A review by the HSE of the role of QRA in decision-making is given in *Quantified Risk Assessment: Its Input to Decision-making* (HSE, 1989c). The document is part of the HSE's response to the *Layfield Report* (Layfield, 1987) on Sizewell B, which called for studies both of the tolerability of risk from nuclear power stations and of the relationship of safety standards in the nuclear industry to those in other industries. It deals with the second of these topics, the first being treated in *The Tolerability of Risks from Nuclear Power Stations* (HSE, 1988c).

The study considers some 16 cases of the use of QRA. They include: in the process industries, the *Canvey Reports* (HSE, 1978b, 1981a), the St Fergus to Moss Morran pipeline (HSE, 1978a), the Moss Morran/Braefoot Bay development (HSE, 1983e), housing development near an ammonium nitrate plant in the Goole–Hook plan, major retail development near a chemical plant at Ellesmere Port, a harbour with a liquefied petroleum gas (LPG) shipping risk, an explosives wharf, a sulphonation reactor and a plastics injection moulding machine; in the nuclear industry, the Heysham/Torness AGR and Sizewell B; in civil engineering, Ronan Point and the River Thames Flood Defence scheme; in mining, Markham Colliery and, in the USA, Sunshine Mine; and in rail transport, automatic level crossings. In all cases, except apparently Sunshine Mine, the QRA was used as an input to decision-making. Details of these cases are given in Appendix 2 of the study.

There is, in general, considerable variation between one QRA and another. One cause of variation is the nature of the problem addressed. This may be, for example, a single machine, a whole plant or even a complex of plants. Another cause is the degree of conservatism in the estimates made. Furthermore, QRA is subject to quite large margins of error. The study makes the point that a QRA does not necessarily have to be a complex exercise. In some cases a relatively simple QRA suffices.

In the cases considered, the main forms in which the risks were presented are individual fatal risk and societal fatal risk, the latter in the form of frequency–number (FN) curves. The study emphasizes the need to distinguish between the different bases on which FN curves are constructed. An FN curve may be based on (1) historical data, (2) design requirements or (3) predicted values. These three bases are quite different, and this needs to be borne in mind when using the curves derived.

The two latter types are illustrated by considering two FN curves produced for Sizewell B. One curve, by Harbison and Kelly (1985), is based on design requirements implied in the safety assessment principles of the Nuclear Installations Inspectorate (NII). The other curve, produced by the Central Electricity Generating Board (CEGB) (G.N. Kelly and Hemming, 1983) represents the predictions arising from the actual design. Another difference in the basis of FN curves is the fatalities on which the curve is based, which may be confined to immediate deaths or may also include delayed deaths.

The study gives a review of risk criteria, dealing in particular with the HSE criterion for a major civil nuclear accident and the Advisory Committee on Major Hazards (ACMH) criterion for a major accident. The HSE nuclear criterion may be summarized as the requirement that an uncontrolled release from any of a family of reactors nationwide capable of giving doses of 100 mSv at 3 km, which pessimistically might cause eventual deaths of about 100 people, should be no more frequent than 10^{-4}/year and the ACMH major hazard criterion that a major accident such as Flixborough, involving say $\geqslant 10$ deaths, should occur nationwide with a frequency no greater than 10^{-4}/year.

In Appendix 1, the study lists some 42 factors that seem important in judging the tolerability of risk, under the four headings of: (1) the hazard, the consequential risk and the consequential benefits; (2) the nature of the assessment; (3) the factors of importance to those generating the risk, to government or to regulators; and (4) public attitudes.

Although the results of a QRA are typically expressed in terms of deaths or of casualties, appropriately defined, there are generally other consequences which need to be considered. The study lists (1) the write-off of the plant, (2) the impact on the surrounding area, (3) the anxiety factor, (4) consequential detriments and (5) the 'What if?' factor. The write-off of a plant such as occurred at Three Mile Island or Chernobyl is costly in itself and harms the whole industry. The impact on the surrounding area can be severe, especially where land is rendered radioactive, and there may be consequent detrimental outcomes such as the increased number of miscarriages which took place after Seveso. The occurrence of a major accident may lead to concern arising from asking 'What if an even larger scale one were to occur?'

For the cases described, the study reviews the nature of the decision to be made, the risk estimates yielded by the QRAs and the decisions taken. Many of the decisions concerned the granting of planning permission. Others related to the safety of particular equipment or civil engineering designs. Two, Canvey and Sizewell, were much more complex than might appear. Canvey started from a planning application for a new plant but grew into an examination of the risks from an existing complex. Sizewell inevitably became a generic inquiry into the case for nuclear power in general and the pressurized water reactor (PWR) design in particular.

The study tabulates the QRA estimates of individual risk and gives for the societal risk estimates a FN curve in which the curves differ widely. For example, the frequency of $\geqslant 10$ deaths varies from 10^{-1}/year for the historical data for flats of the Ronan Point type to somewhat over 10^{-7}/year for the implied design requirements of Sizewell B, with most cases clustering between about 10^{-3} and 10^{-4}/year.

The decision outcomes, as a decision proper or advice on a decision, include: granting of planning permission to a new development, without or with improvements (e.g. St Fergus to Moss Morran pipeline; Moss Morran Braefoot Bay development, harbour with LPG shipping, explosives wharf); requirement to make improvements to existing situation (e.g. Canvey); acceptance of a design (sulphonation reactor); and rejection of a design (e.g. Ronan Point).

Principal conclusions of the study are as follows. First,

> QRA is an element that cannot be ignored in decision making about risk since it is the only discipline capable, however imperfectly, of enabling a number to be applied and comparisons of a sort to be made, other than of a purely qualitative kind. This said, the numerical element must be viewed with great caution and treated as only one parameter in an essentially judgmental exercise. Moreover, since any judgement upon risk is distributional, risk being caused to some as an outcome of the activity of others, it is therefore essentially political in the widest sense of the word.

Second, QRA illuminates some important components of a safety assessment but needs to be supplemented by other approaches particularly in the areas of management and human factors. Third, the fact that other factors need to be taken into account does not detract from the value of QRA as a decision input. Fourth, it is not legitimate to 'read across' QRA-derived risk figures from one hazardous situation to another. It follows that it is not possible to specify a universal upper limit for all societal risks. Fifth, it needs to be borne in mind that, although QRA sometimes gives predictions of very severe accidents, such accidents do occur, as instanced by events such as Mexico City, Bhopal and Piper Alpha.

9.4 Event Data

The methods which are used to estimate the frequency of an event depend somewhat on the nature of the event. The aim is usually to base the estimate on historical experience, but the precise way in which this is done varies.

In the simplest case there may be available frequency data which apply directly to the event in question. This is likely to be the case if the event of interest is one which does sometimes occur, e.g. the failure of a single item of equipment such as a pipeline fracture or pump leak.

In other cases it may not be possible to obtain such direct data. This tends to be the case for complex systems and/or systems with multiple layers of protection. Thus the frequency of failure of a nuclear power reactor is a case in point. The presence of the protective systems has the effect both of rendering each system more or less unique and of reducing the frequency of system failure to a very low value. For both these reasons historical data on failure frequency are lacking. The frequency therefore has to be synthesized using a fault tree. Again historical data are used, but in this case the data are for the base events in the tree and so are used indirectly.

The choice of method for the estimation of frequency depends on the hazard under consideration. This is illustrated very clearly in the different methodologies used in the *Canvey Reports* and the *Rasmussen Report*. In the former, the typical events of interest were emissions from vessels, pipelines and pumps. For these events historical frequency data exist and were used. In the latter study, however, the events of interest were accidents on nuclear power plants. Historical data for the frequency of these events did not exist and estimates had to be synthesized. Thus, whereas the *Canvey Reports* contain very few fault trees, the *Rasmussen Report* has an appendix several centimetres thick and full of fault trees.

A particularly difficult problem is posed by the estimation of the frequency of a rare event where this is not amenable to synthetic methods. A typical example is the outright failure of a high quality pressure vessel. In many cases the estimates required are probabilities rather than frequencies. In the following most of the comments made about frequency apply also to probability. Similarly, it is convenient to refer to failure frequency, but the account given applies in most instances to the frequency of other events.

There are a number of sources of historical data which can be used to obtain a frequency estimate. They include the literature, the works and data banks. The data which are often most easily obtained are the total number of events over a given time period. These do not in themselves give an event rate. To determine the event rate it is necessary to know the total number of items to which the data apply. It can be quite difficult to determine this inventory.

Given that these data are available, frequency is typically estimated by dividing the total number of events by the total number of equipment-years. The assumption underlying this procedure is that a constant event rate, or the exponential distribution, applies.

Where failure data are concerned, it is often necessary to have information not only on the overall frequency of failure but also on the composition of this frequency in terms of the individual failure modes. In hazards work the fail-to-danger mode is usually required.

The estimation of the frequency from historical data is in some cases straightforward, but in others not. Areas of difficulty include

(1) inapplicability of data;
(2) sparseness of data;
(3) status of data.

The first point which has to be ascertained is whether the data available are applicable to the case in hand. There are various reasons why they may not be. The data may apply to equipment and/or conditions which are out of date. There may be significant differences between the equipment and/or its situation.

The design and environment in which equipment is used have a major influence on its failure rate. High failure rates can occur if equipment is highly rated or used in the wrong application. Operating practices affect the failure rate. Above all, failure rates can vary widely, depending on the maintenance regime.

In some instances it is possible to identify specific ways in which the situation to which the data apply differs from that under consideration. A not uncommon case is where a particular failure mode is known to be much less frequent or even inapplicable. For example, pipeline failures due to corrosion have been much reduced by the use of cathodic protection. Similarly, changes to the design of rail tank cars for hazardous materials in the USA have reduced the probability of impact damage. It is, in principle, legitimate to adjust an estimate based on historical failure data to allow for changed conditions, although this should be done with care.

Another common problem is that the data available are sparse. As described in Chapter 7, methods exist which allow estimates to be made of the confidence bounds on such data, but the bounds may be wide, and in any case uncertainty in the basic data complicates the assessment.

Usually the frequency data used come from a number of different sources and have varying statuses. Data statuses include

(1) historical data
    (a) numerous,
    (b) sparse;
(2) synthesized values;
(3) expert judgements
    (a) amenable to improvement,
    (b) unamenable to improvement.

If it is not practical to obtain direct data on the frequency of an event, it is often possible to obtain a value indirectly by synthesis using tree methods. The most commonly used method is the fault tree, but the event tree can also be used.

A fault tree is typically used to estimate the frequency where the failure is such that the system or situation which gives rise to it is sufficiently complex as to render it relatively unique. In such a case historical failure data will not be available. One important feature which tends to render a system unique is the use of protective systems. Fault tree methods are described in Section 9.5.

Events which present a threat to the public are also relatively rare and unique, and again it may be necessary to resort to synthesis to obtain a value for the frequency. Such an event is the outcome of a chain of events following a release. The frequency may be obtained using an event tree. Event tree methods are described in Section 9.6.

In some cases historical data are not obtainable. In this case an approach which is increasingly used is that of expert judgement. Expert judgement methods are described in Section 9.9.

9.5 Fault Trees

A fault tree is used to develop the causes of an event. It starts with the event of interest, the top event, such as a hazardous event or equipment failure, and is developed from the top down.

Accounts of fault trees are given in *Reliability and Fault Tree Analysis* (Barlow, Fussell and Singpurwalla, 1975), *Fault Tree Handbook* (Vesely *et al.*, 1981), *Engineering Reliability* (Dhillon and Singh, 1981), *Reliability Engineering and Risk Assessment* (Henley and Kumamoto, 1981), *Designing for Reliability and Safety Control* (Henley and Kumamoto, 1985) and *Probabilistic Risk Assessment, Reliability Engineering, Design and Analysis* (Henley and Kumamoto, 1992), and by Vesely (1969, 1970a,b), Vesely and Narum (1970), Fussell (1973a, 1975, 1976, 1978b), Lawley (1974b, 1980), Lapp and Powers (1977a, 1979), Vesely and Goldberg (1977b) and Kletz and Lawley (1982).

The fault tree is both a qualitative and a quantitative technique. Qualitatively it is used to identify the individual paths which led to the top event, while quantitatively it is used to estimate the frequency or probability of that event.

The identification of hazards is usually carried out using a method such as a hazard and operability (hazop) study. This may then throw up cases, generally small in number, where a more detailed study is required, and fault tree analysis is one of the methods which may then be used.

Fault tree analysis is also used for large systems where high reliability is required and where the design is to incorporate many layers of protection, such as in nuclear reactor systems.

With regard to the estimation of the frequency of events, the first choice is generally to base an estimate on historical data, and to turn to fault tree analysis only where data are lacking and an estimate has to be obtained synthetically.

9.5.1 Fault tree analysis

The original concept of fault tree analysis was developed at the Bell Telephone Laboratories in work on the safety evaluation of the Minuteman Launch Control System in the early 1960s, and wider interest in the technique is usually dated from a symposium in 1965 in which workers from that company (e.g. Mearns) and from the Boeing Company (e.g. Haasl, Feutz, Waldeck) described their work on fault trees (Boeing Company, 1965).

Developments in the methodology have been in the synthesis of the tree, the analysis of the tree to produce minimum cut sets for the top event, and in the evaluation of the frequency or probability of the top event. There have also been developments related to trees with special features, including repair, secondary failures, time features, etc.

A general account of fault tree methods has been given by Fussell (1976). He sees fault tree analysis as being of major value in

(1) directing the analyst to ferret out failures deductively;
(2) pointing out the aspects of the system important in respect of the failure of interest;
(3) providing a graphical aid giving visibility to those in system management who are removed from system design changes;
(4) providing options for qualitative or quantitative system reliability analysis;
(5) allowing the analyst to concentrate on one particular system failure at a time;
(6) providing the analyst with genuine insight into system behaviour.

He also draws attention to some of the difficulties in fault tree work. Fault tree analysis is a sophisticated form of reliability assessment and it requires considerable time and effort by skilled analysts. Although it is the best tool available for a comprehensive analysis, it is not foolproof and, in particular, it does not of itself assure detection of all failures, especially common cause failures.

9.5.2 Basic fault tree concepts

A logic tree for system behaviour may be oriented to success or failure. A fault tree is of the latter type, being a tree in which an undesired or fault event is considered and its causes are developed. A distinction is made between a failure of and a fault in a component. A fault is an incorrect state which may be due to a failure of that component or may be induced by some outside influence. Thus fault is a wider concept than failure. All failures are faults, but not all faults are failures.

A component of a fault tree has one of two binary states: essentially it is either in the correct state or in a fault state. In other words, the continuous spectrum of states from total integrity to total failure is reduced to just two states. The component state which constitutes a fault is essentially that state which induces the fault that is being developed.

As a logic tree, a fault tree is a representation of the sets of states of the system which are consistent with the top event at a particular point in time. In practice, a fault tree is generally used to represent a system state which has developed over a finite period of time, however short. This point is relevant to the application of Boolean algebra. Strictly, the implication of the use of Boolean algebra is that the states of the system are contemporaneous.

Faults may be classed as primary faults, secondary faults or command faults. A primary fault is one which occurs when the component is experiencing conditions for which it is designed, or qualified. A secondary fault is one which occurs when the component is experiencing conditions for which it is unqualified. A command fault involves the proper operation of the component at the wrong time or in the wrong place.

A distinction is made between failure mechanism, failure mode and failure effect. The failure mechanism is the cause of the failure in a particular mode and the failure effect is the effect of such failure. For example, failure of a light switch may occur as follows:

Failure mode – high contact resistance
Failure mechanism – corrosion
Failure effect – switch fails to make contact

Some components are passive and others active. Items such as vessels and pipes are passive, whilst those such as valves and pumps are active. A passive component is a transmitter or recipient in the fault propagation process, an active one can be an initiator. In broad terms, the failure rate of a passive component is commonly two or three orders of magnitude less than that of an active component.

There is a distinction to be made between the occurrence of a fault and the existence of a fault. Interest may centre on the frequency with which or probability that a fault occurs, i.e. on the unreliability, or on the probability that at any given moment the system is in a fault state, i.e. on the unavailability. The distinction between reliability and availability was discussed in detail in Chapter 7.

The simplest case is the determination of the reliability of a non-repairable system. This is sometimes known as the 'mission problem': the system is sent on a mission in which components that fail are not repaired. The obvious example is space missions, but there are cases in the process industries which may approximate to this, such as remote pumping stations or offshore subsea modules. The availability of a non-repairable system may also be determined, but the long-term availability, which is usually the quantity of interest, tends to zero.

Generally, however, process systems are repairable systems, and for these both reliability and availability may be of interest. If concern centres on the frequency of realization of a hazard, it is the reliability which is relevant. If, on the other hand, the concern is with the fractional downtime of some equipment, it is the availability which is required.

A fault tree may be analysed to obtain the minimum cut sets, or minimum sets of events which can cause the top event to occur. Discussion of minimum cut sets is deferred to Section 9.5.8, but it is necessary to mention them at this point since some reference to them in relation to fault tree construction is unavoidable.

9.5.3 Fault tree elements and symbols

The basic elements of a fault tree may be classed as (1) the top event, (2) primary events, (3) intermediate events and (4) logic gates.

The symbols most widely used in process industry fault trees are shown in Table 9.5. The British Standard symbols are given in BS 5760 *Reliability of Systems, Equipment and Components*, Part 7: 1991 *Guide to Fault Tree Analysis*. For the most part the symbols shown in Table 9.5 correspond to those in the standard, but in several cases the symbols in the table are the Standard's alternative rather than preferred symbols.

The top event is normally some undesired event. Typical top events are flammable or toxic releases, fires, explosion and failures of various kinds.

Primary events are events which are not developed further. One type of primary event is a basic event, which is an event that requires no further development. Another is an undeveloped event, which is an event that could be developed, but has not been. One common reason for not developing an event is that its causes lie outside the system boundary. The symbol for such an undeveloped event is a diamond and this type is therefore often called a 'diamond event'. A third type of primary event is a conditioning event, which specifies conditions applicable to a logic gate. A fourth type of event is an external event, which is an event that is normally expected to occur.

Intermediate events are the events in the tree between the top event and the primary events at the bottom of the tree.

Logic gates define the logic relating the inputs to the outputs. The two principal gates are the AND gate and the OR gate. The output of an AND gate exists only if all the inputs exist. The output of an OR gate exists provided at least one of the inputs exists. The probability relations associated with these two gates are shown in Table 9.6, Section A. Other gates are the EXCLUSIVE OR gate, the PRIORITY AND gate and the INHIBIT gate. The output of an EXCLUSIVE OR gate exists if one, and only one, input exists. The output of a PRIORITY AND gate exists if the inputs occur in the sequence specified by the associated conditioning event. The output of an INHIBIT gate exists if the (single) input exists in the presence of the associated conditioning event. There are also symbols for TRANSFER IN and TRANSFER OUT, which allow a large fault tree to be drawn as a set of smaller trees.

9.5.4 AND gates

One of the two principal logic gates in a fault tree is the AND gate. AND gates are used to represent a number of different situations and therefore require further explanation. The following typical situations can be distinguished:

(1) output exists given an input and fault on a protective action;
(2) output exists given an input and fault on a protective device;
(3) output exists given faults on two devices operating in parallel;
(4) output exists given faults on two devices, one operating and one on stand-by.

In constructing the fault tree the differences between these systems present no problem, but difficulties arise at the evaluation stage.

As already described, the probability p_0 that the output of a two-input AND gate exists, given that the probabilities of the inputs are p_1 and p_2, is

$$p_0 = p_1 p_2 \quad [9.5.1]$$

The occurrence of events may be expressed quantitatively in terms of frequency or of probability. Failure of equipment is normally expressed as a frequency and failure of a protective action or device as a probability.

A protective device is normally subject to unrevealed failure and needs therefore to be given a periodic proof test. Data for the failure of such a device may be available either as probability of failure on demand, or as frequency of failure. As described in Chapter 13, it can be shown that, subject to certain assumptions, the relationship between the two is

$$p = \lambda\tau_p/2 \quad [9.5.2]$$

where p is the probability of failure, λ is the failure rate, and τ_p is the proof test interval.

Then for a Type 1 situation the frequency λ_0 of a fault is

$$\lambda_0 = \lambda p \quad [9.5.3]$$

***Table 9.5** Fault tree event and logic symbols*

A Events

Symbol	
	Primary, or base, event – basic fault event requiring no further development
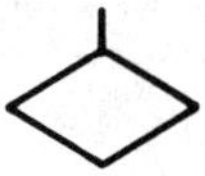	Undeveloped, or diamond, event – fault event which has not been further developed
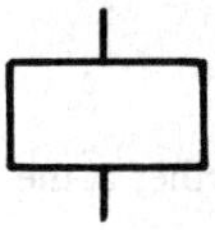	Intermediate event – fault event which occurs due to antecedent causes acting through a logic gate
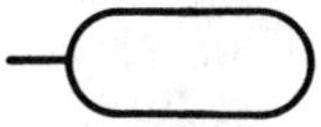	Conditioning event – specific condition which applies to a logic gate (used mainly with PRIORITY AND and INHIBIT gates)
	External, or house, event – event which is normally expected to occur[a]

B Logic gates, etc.

Symbol	*Alternative symbol*	
		AND gate – output exists only if all inputs exist
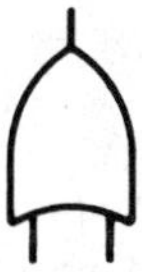		OR gate – output exists if one or more inputs exist
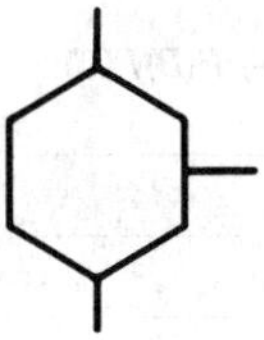		INHIBIT gate – output exists if input occurs in presence of the specific enabling condition (specified by conditioning event to right of gate)
	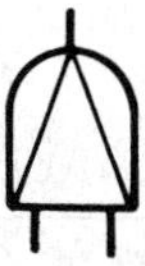	PRIORITY AND gate – output exists if all inputs occur in a specific sequence (specified by conditioning event to right of gate)

Table 9.5 Continued

	EXCLUSIVE OR gate – output exists if one, and only one, input exists
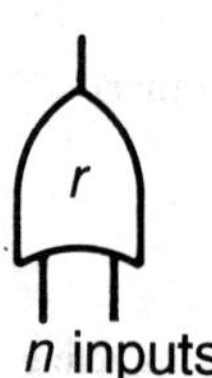	VOTING gate – output exists if there exist *r*-out-of-*n* inputs[b]
	TRANSFER IN – symbol indicating that the tree is developed further at the corresponding TRANSFER OUT symbol
	TRANSFER OUT – symbol indicating that the portion of the tree below the symbol is to be attached to the main tree at the corresponding TRANSFER IN symbol

[a] This is the definition given by Vesely *et al.* (1981). Other authors such as Henley and Kumamoto (1981) use this symbol for an event which is expected to occur or not to occur.
[b] See Chapter 13.

Table 9.6 Probability and frequency relations for fault tree logic gates (output A; inputs B and C)

A Basic probability relations[a]

Logic symbol	*Reliability graph*	*Boolean algebra relation*	*Probability relations*
		A = BC	$P(A) = P(B)P(C)$
		A = B + C	$P(A) = P(B) + P(C) - P(B)P(C)$

B Relations involving frequencies and/or probabilities[a]

Gate	*Inputs*	*Output*
OR	P_B OR P_C	$P_A = P_B + P_C - P_B P_C \approx P_B + P_C$
	F_B OR F_C	$F_A = F_B + F_C$
	F_B OR P_C	Not permitted
AND	P_B AND P_C	$P_A = P_B P_C$
	F_B AND F_C	Not permitted; reformulate
	F_B AND P_C	$F_A = F_B P_C$

[a] F, frequency; P, probability.

where p is the probability of failure of the protective action, λ is the frequency of the input event, and λ_0 is the frequency of the output event.

For a Type 2 situation, Equation 9.5.3 is again applicable, with the probability p of failure of protective action in this case being obtained from Equation 9.5.2.

The evaluation of a Type 3 situation is less straightforward. For this, use may be made of the appropriate parallel system model derived from either the Markov or joint density function methods, described in Chapter 7. These give the probability of the output, event given the frequency of the input events. Where applicable, the rare event approximation may be used to convert from probability to frequency:

$$\lambda = p/t \qquad [9.5.4]$$

Similarly, for a Type 4 situation use may be made of the appropriate stand-by system model.

9.5.5 Fault tree construction

The construction of a fault tree appears a relatively simple exercise, but it is not always as straightforward as it seems and there are a number of pitfalls. Guidance on good practice in fault tree construction is given in the *Fault Tree Handbook*. Other accounts include that in the CCPS *QRA Guidelines*, and those by Lawley (1974b, 1980), Fussell (1976) and Doelp *et al.* (1984).

An essential preliminary to construction of the fault tree is definition and understanding of the system. Both the system itself and its bounds need to be clearly defined. Information on the system is generally available in the form of functional diagrams such as piping and instrument diagrams and more detailed instrumentation and electrical diagrams. There will also be other information required on the equipment and its operation, and on the environment. The quality of the final tree depends crucially on a good understanding of the system, and time spent on this stage is well repaid.

It is emphasized by Fussell (1976) that the system boundary conditions should not be confused with the physical bounds of the system. The system boundary conditions define the situation for which the fault tree is to be constructed. An important system boundary condition is the top event. The initial system configuration constitutes additional boundary conditions. This configuration should represent the system in the unfailed state. Where a component has more than one operational state, an initial condition needs to be specified for that component. Furthermore, there may be fault events declared to exist and other fault events not to be considered, these being termed by Fussell the 'existing system boundary conditions' and the 'not-allowed system boundary conditions', respectively.

Fault trees for process plants fall into two main groups, distinguished by the top event considered. The first group comprises those trees where the top event is a fault within the plant, including faults which can result in a release or an internal explosion. In the second group the top event is a hazardous event outside the plant, essentially fires and explosions.

If the top event of the fault tree is an equipment failure, it is necessary to decide whether it is the reliability, availability, or both, which is of interest. Closely related to this is the extent to which the components in the system are to be treated as non-repairable or repairable.

As already described, the principal elements in fault trees are the top event, primary events and intermediate events, and the AND and OR gates. The *Handbook* gives five basic rules for fault tree construction:

Ground Rule 1	Write the statements that are entered in the event boxes as faults; state precisely what the fault is and when it occurs.
Ground Rule 2	If the answer to the question, 'Can this fault consist of a component failure?' is 'Yes', classify the event as a 'state-of-component fault'. If the answer is 'No', classify the event as a 'state-of-system fault'.
No Miracles Rule	If the normal functioning of a component propagates a fault sequence, then it is assumed that the component functions normally.
Complete-the-Gate Rule	All inputs to a particular gate should be completely defined before further analysis of any one of them is undertaken.
No Gate-to-Gate Rule	Gate inputs should be properly defined fault events, and gates should not be directly connected to other gates.

Each event in the tree, whether a top, intermediate or primary event, should be carefully defined. Failure to observe a proper discipline in the definition of events can lead to confusion and an incorrect tree.

The identifiers assigned to events are also important. If a single event is given two identifiers, the fault tree itself may be correct, if slightly confusing, but in the minimum cut sets the event will appear as two separate events, which is incorrect.

For a process system, the top event will normally be a failure mode of an equipment. The immediate causes will be the failure mechanisms for that particular failure. These in turn constitute the failure modes of the contributing subsystems, and so on.

The procedure followed in constructing the fault tree needs to ensure that the tree is consistent. Two types of consistency may be distinguished: series consistency within one branch and parallel consistency between two or more branches. Account needs also to be taken of events which are certain to occur and those which are impossible.

The development of a fault tree is a creative process. It involves identification of failure effects, modes and mechanisms. Although it is often regarded primarily as a means of quantifying hazardous events, which it is, the fault tree is of equal importance as a means of hazard identification. It follows also that fault trees created by different analysts will tend to differ. The differences may be due to style, judgement and/or omissions and errors.

It is generally desirable that a fault tree have a well defined structure. In many cases such a structure arises naturally. It is common to create a 'demand tree', which shows the propagation of the faults in the absence of protective systems, and then to add branches, representing protection by instrumentation and by the process operator, which are connected by AND gates at points in the demand tree. An example of a fault tree constructed in this way has been given in Figure 2.2. Essentially the same fault tree may be drawn in several different ways, depending particularly on the location of certain events which appear under AND gates.

9.5.6 Dependence

A fundamental assumption in work on reliability generally, and on fault trees in particular, is that the events considered are independent, unless stated otherwise. Formally, the events are assumed to be statistically independent, or 's-independent'. In practice, there are many types of situation where events are not completely independent. In fault tree work this problem was originally known as 'common mode failure', then as 'common cause failure', and now more usually as 'dependent failure'.

The problem is particularly acute in systems, such as nuclear reactor systems, where a very high degree of reliability is sought. The method of achieving this is through the use of protective systems incorporating a high degree of redundancy. On paper, the assessed reliabilities of such systems are very high. But there has been a nagging worry that this protection may be defeated by the phenomenon of dependent failure, which may take many and subtle forms. Concern with dependent failure is therefore high in work on fault trees for nuclear reactors.

Dependent failure takes various forms. In most cases it requires that there be a common susceptibility in the component concerned. Some situations which can cause dependent failure include: (1) a common utility; (2) a common defect in manufacture; (3) a common defect in application; common exposure to (4) a degrading factor, (5) an external influence, or (6) a hazardous event; (7) inappropriate operation; and (8) inappropriate maintenance.

Perhaps the most obvious dependency is supply from a common utility such as electric power or instrument air. Equipment may suffer common defects either due to manufacture or to specification and application. Common degrading factors are vibration, corrosion, dust, humidity, and extremes of weather and temperature. External influences include such events as vehicle impacts or earthquakes. An event such as a fire or explosion may disable a number of equipments. Equipment may suffer abuse from operators or may be maintained incorrectly. It will be clear that in such cases redundancy may be an inadequate defence.

Generally, a common location is a factor in dependent failure, interpreting this fairly broadly. But it is by no means essential. In particular, incorrect actions by a maintenance fitter can disable similar equipments even though the separation between the items is appreciable.

A type of dependent failure that is important in the present context is that resulting from a process accident. A large proportion of equipments, including protective and fire fighting systems, may be susceptible to a major fire or explosion, just at the time when they are required.

There is some evidence that dependent failure is associated particularly with components where the fault is unrevealed. Thus a study of nuclear reactor accident reports by J.R. Taylor (1978b) showed that of the dependent failures considered only one was not associated with a stand-by or intermittently operated system.

Not all dependent failure involves redundant equipment. Another significant type of dependent failure is the overload which can occur when one equipment fails and throws a higher load on another operating equipment. Failures caused by domino effects, and escalation faults generally, may also be regarded as dependent failures.

Dependent failure, then, is a crucial problem in high reliability systems. A more detailed account is therefore given in Section 9.8. Here further discussion is confined to fault tree aspects.

Dependent failure can be taken into account in a fault tree only if the potential for it is first recognized. Given that this potential has been identified, there are two ways of representing it in the tree. One is to continue to enter each fault separately as it occurs in the tree, but ensuring that each such entry is assigned the same identifier, so that the minimum cut sets are determined correctly. The other approach is to enter the effect as a single fault under an AND gate higher up the tree. A further measure which may be taken to identify dependent failure is to examine the minimum cut sets for common susceptibilities or common locations.

9.5.7 Illustrative example: instrument air receiver system

As an illustration of fault tree analysis, consider the system shown in Figure 9.3(a). The vessel is an air receiver for an instrument air supply system. Air is let down from the receiver to the supply through a pressure reducing valve. The pressure in the receiver is controlled by a pneumatic control loop which starts up an air compressor when the receiver pressure falls below a certain value. The instrument air supply to the control loop is taken from the instrument air supply described, and if the pressure in the supply system falls below a certain value this too causes the control loop to start up the compressor. There is a pressure relief valve on the receiver. There is also a pressure relief valve (not shown) on the instrument air supply system. The design intent is that the pressure relief valve on the air receiver is sized to discharge the full throughput of the compressor and is set to open at a pressure below the danger level and that the pressure reducing valve is sized to pass the full throughput of the compressor if the instrument air pressure downstream falls to a very low value. One of the main causes of failure in the system is likely to be dirt.

The top event considered is the explosion of the air receiver due to overpressure. A fault tree for the top event of 'Receiver explosion' is shown in Figure 9.3(b).

One fault event occurs in two places – 'Pressure reducing valve partially or completely seized shut or blocked'. This is drawn as a subtree. One primary failure event appears at several points in the tree – 'Dirt'. As shown, this is treated in the tree as separate primary failures for the pressure reducing valve and the pressure relief valve.

Two of the events in the tree are mutually exclusive. These are 'Instrument air system pressure abnormally high' and 'Instrument air pressure abnormally low'. These events are denoted by B and B*, respectively.

The analysis of this fault tree to obtain the minimum cut sets and the probability of occurrence of the top event is described below.

9.5.8 Minimum cut sets

A fault tree may be analysed to obtain the minimum cut sets. A cut set is a set of primary events, that is of basic or undeveloped faults, which can give rise to the top event. A minimum cut set is one which does not contain within itself another cut set. The complete set of minimum cut sets is the set of principal fault modes for the top event.

The minimum cut sets may be determined by the application of Boolean algebra. The procedure may be illustrated by reference to the fault tree shown in Figure 9.3(b). This may be represented in Boolean form as:

$$T = (A+B+C+D)\ (B^*+F)\ (G+H+I) \qquad [9.5.5]$$

Then substituting

$$B^* = C+D+E$$

and noting that:

$$BB^* = 0$$

$$CC = C;\ DD - D$$

$$AC,\ CD,\ CE,\ CF \subset C$$

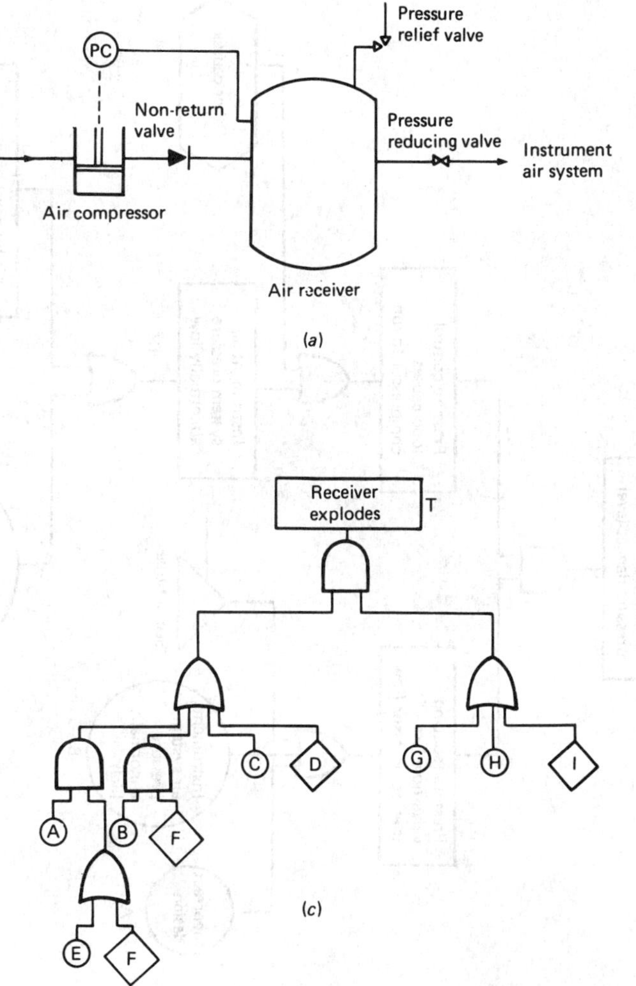

Figure 9.3 Instrument air receiver system: flow diagram and fault trees for the explosion of an air receiver: (a) instrument air receiver system; (b) fault tree for top event 'Receiver explodes'; (c) equivalent but simplified fault tree for top event 'Receiver explodes'

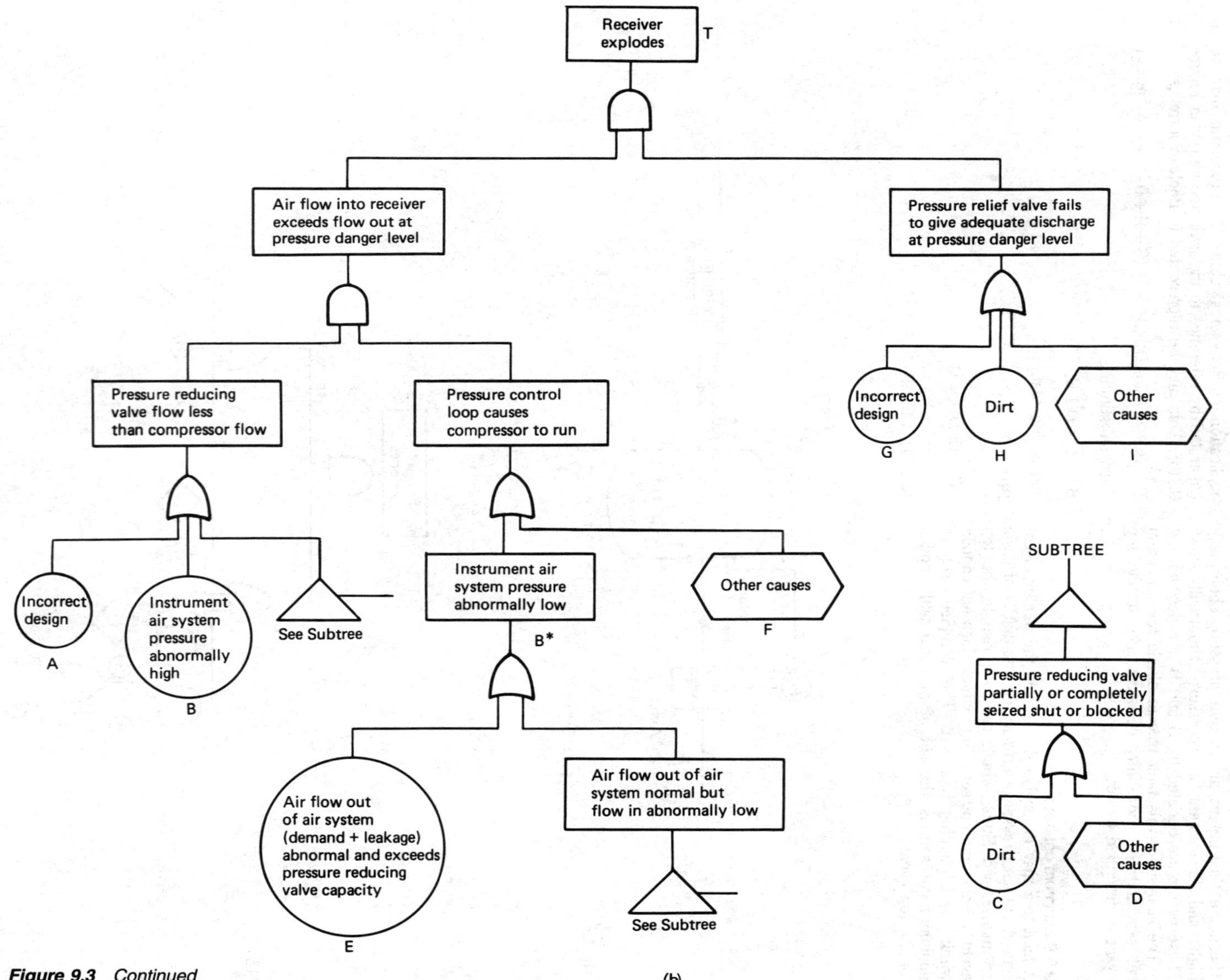

Figure 9.3 *Continued*

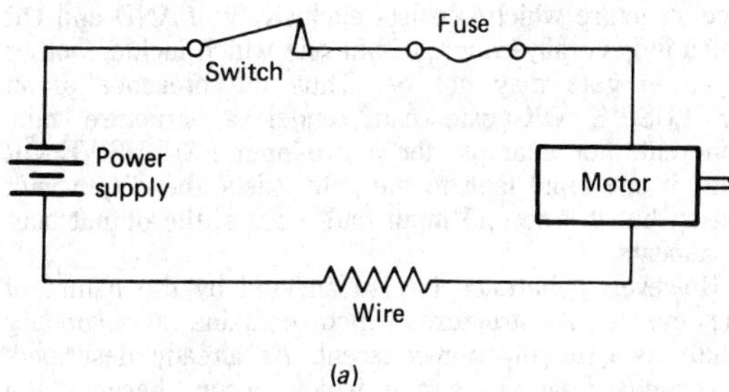

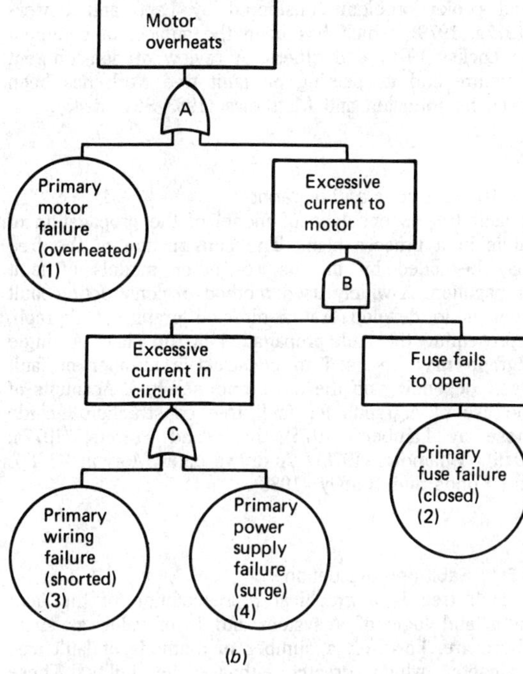

Figure 9.4 *Motor system: system diagram and fault tree for overheating of the motor (Fussell, 1976): (a) motor system; and (b) fault tree for top event 'Motor overheats' (Courtesy of the Sijthoff and Noordhoff International Publishing Company)*

AD, DC, DE, DF $\subset$ D

gives

$$T = (AE + AF + BF + C + D) \cdot (G + H + I) \quad [9.5.6a]$$

$$= [A \cdot (E + F) + BF + C + D] \cdot (G + H + I) \quad [9.5.6b]$$

and thus the minimum cut sets are:

AEG	AEH	AEI
AFG	AFH	AFI
BFG	BFH	BFI
CG	CH	CI
DG	DH	DI

A simplified fault tree which corresponds to Equation 9.5.6b is shown in Figure 9.3(c).

Since fault trees for industrial systems are often large, it is necessary to have systematic methods of determining the minimum cut sets. Such a method is that described by Fussell (1976). As an illustration of this method, consider the motor system which is described by this author and which is shown in Figure 9.4(a). The top event considered is the overheating of the motor. The fault tree for this event is shown in Figure 9.4(b). The structure of the tree is:

Gate	*Gate type*	*No. of inputs*	*Input code No.*	
A	OR	2	1	B
B	AND	2	C	2
C	OR	2	4	3

The procedure is based on successive elimination of the gates. The analysis starts with a matrix containing the first gate, gate A, in the top left-hand corner:

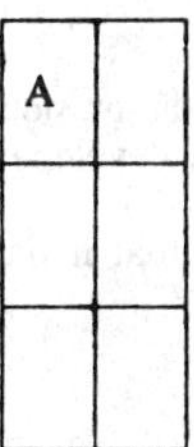

A is an OR gate and is replaced by its inputs listed vertically:

1	
B	

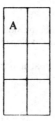

B is an AND gate and is replaced by its inputs listed horizontally:

1	
C	2

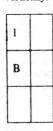

C is an OR gate and is replaced by its inputs listed vertically:

1	
4	2
3	2

It should be noted that when C is replaced by 4 and 3, the event 2, which is linked to C by an AND gate, is listed with both events 4 and 3. The minimum cut sets are then:

(1); (4, 2); (3, 2)

There are now a large number of methods available for the determination of the minimum cut sets of a fault tree. Methods include those described by Vesely (1969, 1970b), Gangadharan, Rao and Sundararajan (1977), Zipf (1984) and Camarinopoulos and Yllera (1985).

There are also a number of computer codes for minimum cut set determination. One of the most commonly used is the code set PREP and KITT. Another widely used minimum cut set code is FTAP.

9.5.9 Coherence of tree

The structure of a fault tree is either coherent or non-coherent. The property of coherence has been developed by Esary and Proschan (1963) as follows.

They define a structure function $\phi(\mathbf{x})$ such that it has three properties.

Property 1:

$$\phi(\mathbf{x}) \geqslant \phi(\mathbf{y}) \qquad \mathbf{x} \geqslant \mathbf{y} \qquad [9.5.7]$$

Property 2:

$$\phi(\mathbf{x}) = 1 \qquad \mathbf{x} = 1 \qquad [9.5.8]$$

Property 3:

$$\phi(\mathbf{x}) = 0 \qquad \mathbf{x} = 0 \qquad [9.5.9]$$

where $\mathbf{x}$ is the vector $[x_1, x_2, \ldots, x_n]$ and $\mathbf{y}$ the vector $[y_1, y_2, \ldots, y_n]$. For each component i, x_i has the value 1 if the component is functioning and 0 if it is failed; $\mathbf{x} \geqslant \mathbf{y}$ signifies $x_i \geqslant y_i$ for all i. A structure which has such a structure function is coherent or monotonic.

In plain language, this means that:

(1) *Property 1* If sufficient components are functioning for the system to be functioning, the functioning of an additional component will not cause the system to fail; and if sufficient components are failed for the system to be failed, the failure of an additional component will not cause the system to function.
(2) *Property 2* If all the components are functioning, the system is functioning.
(3) *Property 3* If all the components are failed, the system is failed.

Coherence is of interest to reliability engineers because, given the property of coherence, it is possible to derive a number of useful relationships for bounds on the behaviour of large systems.

For fault trees the property of coherence depends partly on the logic gates and partly on the events. A fault tree structure which consists exclusively of AND and OR gates is generally coherent, but one which includes other types of gate may not be. Thus the presence of an EXCLUSIVE OR gate can render a structure non-coherent. For example for a two-input EXCLUSIVE OR gate, if one input fault to the gate exists, the output fault exists, but if a second input fault occurs, the output fault disappears.

However, coherence is also affected by the nature of the events. A structure which contains a secondary failure is generally non-coherent. As already described, a secondary failure is one which occurs because the component is overstressed. This component is not necessarily restored by restoration of the other components.

An EXCLUSIVE OR gate was used in the classic nitric acid cooler problem considered by Lapp and Powers (1977a, 1979), which has been the subject of comment by Locks (1979) and others. A review of non-coherent structure and its bearing on fault tree work has been given by Johnston and Matthews (1983 SRD R245).

9.5.10 Fault trees and digraphs

A fault tree is one type of model of the propagation of faults in a process plant. The construction of the tree may be aided by the use of other models of fault propagation. A widely used method of constructing fault trees is to develop first a directed graph, or digraph, representing the fault propagation in the plant. A single digraph may be used to construct a number of fault trees, depending on the top events selected. Accounts of the use of digraphs for fault tree construction include those by Lambert (1975), Lapp and Powers (1977a, 1979), Chamow (1978), Andrews and Morgan (1986) and Kohda and Henley (1988).

9.5.11 Fault tree evaluation

A fault tree is a graphical representation of the fault paths and logic of a system and is of value as such. There are, however, a number of methods of fault tree evaluation which greatly enhance its utility. These methods are both qualitative and quantitative.

Qualitative evaluation is largely based on the determination of the minimum cut sets. The minimum cut sets and methods for their determination have been described in Section 9.5.8. One form of qualitative evaluation which is particularly useful is the determination of the importance of the primary events. This is described in Section 9.5.12.

Quantitative evaluation of a fault tree requires that numbers be put to the frequency or probability of the primary events. Given these quantitative data, there are several options for the evaluation of the frequency or probability of the top event.

Three methods of evaluation are considered here. These are the use of (1) the minimum cut sets, (2) the gate-by-gate method and (3) Monte Carlo simulation.

In the first of these methods, the probability of the top event may be evaluated from the probabilities of the minimum cut sets C_i

$$P(\mathrm{T}) = P\left(\bigcup_{i=1}^{n} \mathrm{C}_i\right) \qquad [9.5.10]$$

The events are not mutually exclusive and, therefore, strictly

$$P(T) = P(C_1) + P(C_2) + \ldots + P(C_n) - P(C_1)P(C_2)$$

$$- P(C_1)P(C_3) - \ldots - P(C_{n-1})P(C_n)$$

$$+ P(C_1)P(C_2)P(C_3) + P(C_1)P(C_2)P(C_4)$$

$$+ \ldots + P(C_{n-2})P(C_{n-1})P(C_n)$$

$$+ \ldots + (-1)^{n-1} \prod_{i=1}^{n} P(C_i) \qquad [9.5.11]$$

But usually it is sufficient to use the low probability, or rare event, approximation

$$P(T) = \sum_{i=1}^{n} P(C_i) \qquad [9.5.12]$$

Equation 9.5.12 always gives a higher probability than Equation 9.5.11 and thus for failure oriented logic it is conservative.

As an illustration, consider the evaluation of the probability of the top event in the fault tree for the instrument air receiver system as given in Figure 9.3(c). The minimum cut sets have been listed above. Assume that

$P_A = 10^{-5}$ $\quad P_D = 2 \times 10^{-3}$ $\quad P_G = 10^{-3}$
$P_B = 3 \times 10^{-5}$ $\quad P_E = 10^{-3}$ $\quad P_H = 8 \times 10^{-3}$
$P_C = 2 \times 10^{-3}$ $\quad P_F = 5 \times 10^{-4}$ $\quad P_I = 2 \times 10^{-3}$

Then, for the first minimum cut set AEG,

$$P(C_1) = P_A P_E P_G$$

$$= 10^{-5} \times 10^{-3} \times 10^{-3}$$

$$= 10^{-11}$$

and so on for the other 14 cut sets. Alternatively, from Equation 9.5.12,

$$P(T) = \sum_{i=1}^{n} P(C_i) = [P_A(P_E + P_F) + P_B P_F + P_C + P_D]$$

$$\times (P_G + P_H + P_I)$$

$$= [10^{-5}(10^{-3} + 5 \times 10^{-4}) + (3 \times 10^{-5} \times 5 \times 10^{-4})$$

$$+ (2 \times 10^{-3} + 2 \times 10^{-3})] \times (10^{-3} + 8 \times 10^{-3} + 2 \times 10^{-3})$$

$$= 4.4 \times 10^{-5}$$

This form of the minimum cut set method is applicable where the occurrence of the primary events can be characterized by simple probabilities. More commonly the occurrence of these events is characterized by a mix of frequencies and probabilities. Furthermore, it may be necessary to take into account repair of components. An account of an approximate method of handling systems in which event rates are expressed as frequencies and which involve repair is given in Section 9.5.15.

The second method is to work up the tree gate by gate from the bottom calculating the frequency or probability of the output event of each gate from those of the input events. The procedure is straightforward except where there occur at the gate some input faults which are expressed in terms of frequency rather than probability.

The output–input relations which are permitted and those which are not are given for a two-input gate in Table 9.6, Section B. The main problem arises where the two inputs to an AND gate both have the dimensions of frequency. The output from the gate must also have the dimension of frequency. It is not permissible to multiply the two frequencies together.

The handling of an AND gate where the two inputs both have the dimension of frequency is discussed in Section 9.5.4. Two main situations occur: one is where both the components whose faults are the inputs to the gate are operating, and the other is where one component is operating and the other is on stand-by. As indicated in the Section 9.5.4, both situations can be handled by the use of suitable models derived from Markov modelling.

Use may also be made of the alternative model given in Equations 7.8.28–7.8.31, which in effect utilizes a repair time to convert one of the frequencies to a fractional downtime, or probability.

Accounts of the gate-by-gate method are given in the CCPS *QRA Guidelines* and by Doelp *et al.* (1984). The method is illustrated in the fault trees given by Lawley (1974b, 1980), one of which is described in Section 9.5.14.

There are advantages and disadvantages to both the gate-by-gate and minimum cut set methods. The former is the method traditionally used in manual evaluation of trees. It gives an evaluation of all the intermediate events as well as of the top event, which can be valuable. On the other hand, it tends to be error prone and becomes tedious for large trees.

The third method is the use of Monte Carlo simulation. The basic principle of this method has been described in Chapter 7. Its application to fault tree evaluation involves a series of trials. In a given trial each primary event either does or does not occur, the occurrence being determined by sampling, where the values returned by the sampling are, on average, in accordance with the frequency or probability data supplied. The outcome of each trial is the occurrence or non-occurrence of the top event. Provided a sufficient number of trials is used, the frequency or probability of the top event is then given by the proportion of trials in which it occurs. An account of the use of this method for fault tree evaluation has been given by Hauptmanns and Yllera (1983).

The use of Monte Carlo simulation is virtually unavoidable if the primary events are characterized by a range of values of probability, or frequency. The probability of a failure may be given not by a point probability value but by a probability density function. Obviously if the probabilities of the other events are expressed as density functions, then the probability of the top event must be expressed as a density function also.

A given Monte Carlo trial generates a set of probabilities for the primary failure events. The probability of the top event may then be evaluated by an analytical method, such as the minimum cut set method. The result of the series of trials is a probability density function for the top event. The principle of the method is

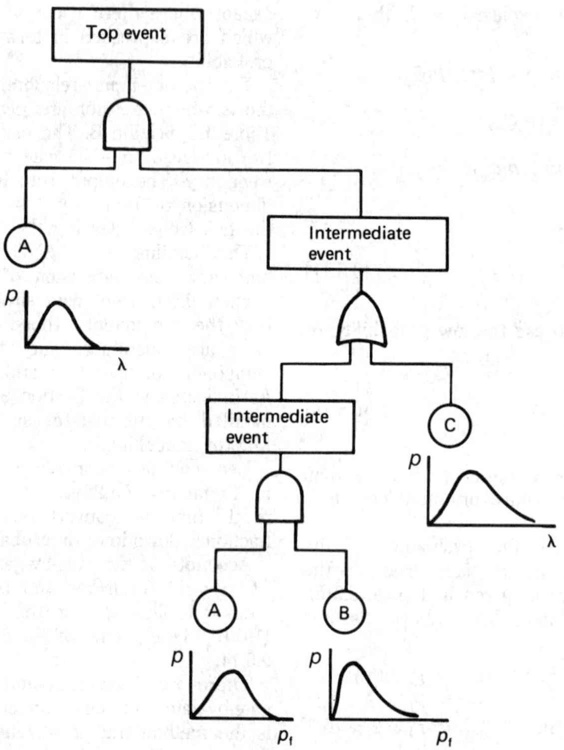

Figure 9.5 Fault tree illustrating the principle of probabilities expressed in the form of density functions and of evaluation by Monte Carlo simulation. λ, Failure frequency; p_f, failure probability; p, probability of given value of λ or p_f

illustrated in Figure 9.5. Monte Carlo simulation was used extensively for fault tree evaluation in the *Rasmussen Report.*

9.5.12 Importance of events

Since the number of primary events which, either serially or in combination, can cause the top event to occur can be large, it is very desirable to have means of assessing their relative significance. Primary events have the property of 'importance'. There are a number of measures of importance. Some of these rely solely on structural considerations, while others are based on probabilities. In other words, there are both qualitative and quantitative measures.

A simple approach to structural importance may be illustrated by the following minimum cut sets of events given in Section 9.5.8: (1); (4, 2); (3, 2).

The order of importance is then: events in one-event minimum cuts, events common to two-event minimum cut sets, events in two-event minimum cut sets, and so on. Thus, in the example given, the order of importance is:

1

2

3, 4

There are a number of quantitative measures of importance which may be calculated if frequency or probability data are available. These include the Birnbaum criterion, which is the incremental reduction in the probability of the top event where the probability of the primary event is reduced incrementally, and the Vesely–Fussell criterion, which is the probability that a cut set containing the primary event has occurred given that the top event has occurred.

Formal methods of determining importance have been given by Lambert and Gilman (1977a,b), who have developed the computer code IMPORTANCE for such analysis.

9.5.13 Illustrative example: pressure tank system

As another illustration, consider the fault tree for the pressure tank system described by Vesely *et al.* (1981). This system is shown in Figure 9.6. The authors describe the system as follows:

> The function of the control system is to regulate the operation of the pump. The latter pumps fluid from an infinitely large reservoir into the tank. We shall assume that it takes 60 seconds to pressurize the tank. The pressure switch has contacts which are closed when the tank is empty. When the threshold pressure has been reached, the pressure switch contacts open, removing

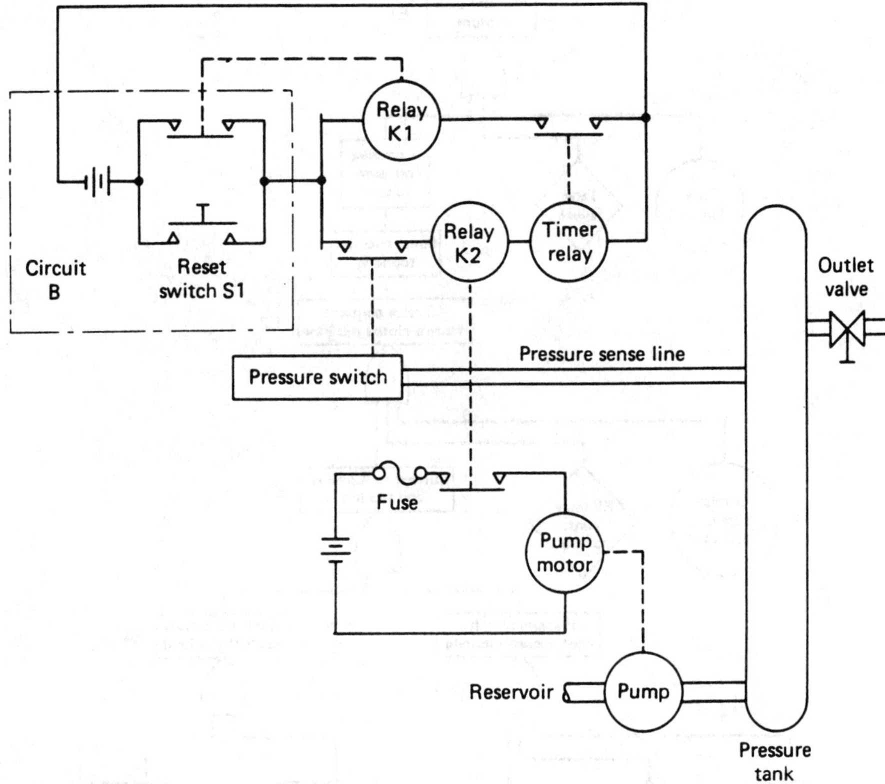

Figure 9.6 *Pressure tank system: system diagram (Vesely et al., 1981)*

power from the pump, causing the pump motor to cease operation. The tank is fitted with an outlet valve that drains the entire tank in a negligible time; the outlet valve, however, is not a pressure relief valve. When the tank is empty, the pressure switch contacts close, and the cycle is repeated.

Initially the system is considered to be in its dormant mode: switch S1 contact open, relay K1 contacts open, and relay K2 contacts open: i.e., the control system is de-energized. In this de-energized state the contacts of the timer relay are closed. We will also assume that the tank is empty and the pressure switch contacts are therefore closed.

System operation is started by momentarily depressing switch S1. This applies power to the coil of relay K1, thus closing relay K1 contacts. Relay K1 is now electrically self-latched. The closure of relay K1 contacts allows power to be applied to the coil of relay K2, whose contacts close to start up the pump motor.

The timer relay has been provided to allow emergency shutdown in the event that the pressure switch fails closed. Initially the timer relay contacts are closed and the timer relay coil is de-energized. Power is applied to the timer coil as soon as relay K1 contacts are closed. This starts a clock in the timer. If the clock registers 60 seconds of *continuous* power application to the timer relay coil, the timer relay contacts open (and latch in that position), breaking the circuit to the K1 relay coil (previously latch closed) and thus producing system shutdown. In normal operation, when the pressure switch contacts open (and consequently relay K2 contacts open), the timer resets to 0 seconds.

The top event considered is 'Rupture of pressure tank after the start of pumping'. The fault tree developed by Vesely *et al.* for this event is shown in Figure 9.7. The detailed development of this fault tree is described by Vesely *et al.* and need not be given here, but the following points may be noted.

If the undeveloped events are disregarded, this fault tree reduces to that indicated by the events E1–E5, K1, K2, R, S, S1 and T shown in Figure 9.7. The top event E1 and the intermediate events E2–E5 in this tree are:

E1 pressure tank rupture
E2 excessive pressure to tank
E3 current to K2 relay coil for too long
E4 circuit B carries current for too long
E5 current to K1 relay coil for too long

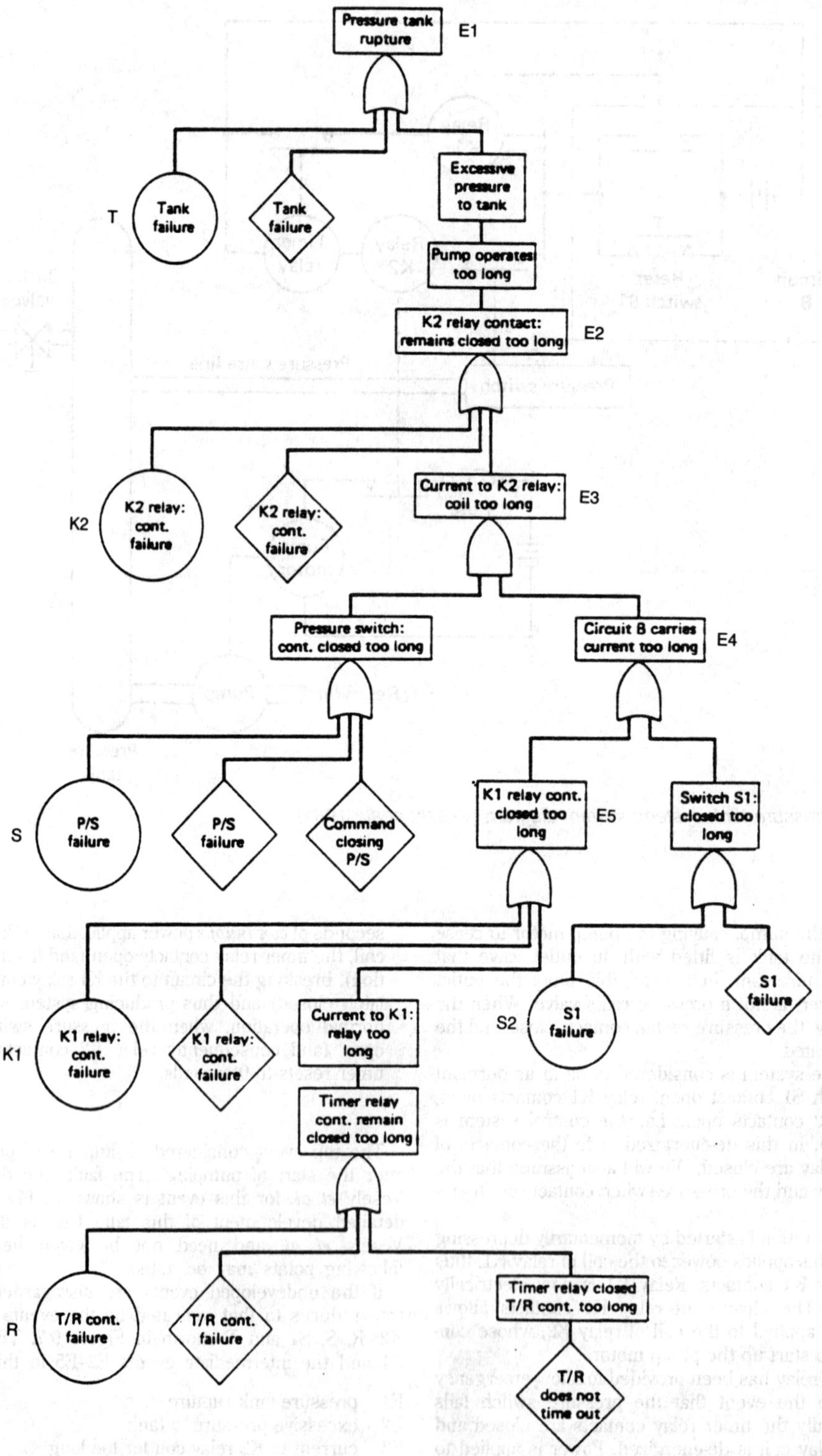

Figure 9.7 *Pressure tank system: fault tree for rupture of pressure tank (after Vesely et al., 1981; Henley and Kumamoto, 1981)*

E2 is equivalent to K2 relay contacts closed for too long, and E5 to timer relay contacts failing to open when pressure switch contacts have been closed for too long.

The cut sets for the top event of this fault tree may be obtained by applying Boolean algebra. Starting with the top event E1:

$$E1 = T + E2$$

$$= T + (K2 + E3)$$

$$= T + K2 + (S \cdot E4)$$

$$= T + K2 + S \cdot (S1 + E5)$$

$$= T + K2 + (S \cdot S1) + (S \cdot E5)$$

$$= T + K2 + (S \cdot S1) + S \cdot (K1 + R)$$

$$= T + K2 + (S \cdot S1) + (S \cdot K1) + (S \cdot R)$$

There are therefore five minimal cut sets: T, K2, (S·S1), (S·K1) and (S·R).

Analysis of the tree at this stage gives qualitative information about the design of the system. The pressure tank system has in effect a single protective device, the timer relay, which protects against failure of the pressure switch. This is reflected in the fault tree which has a single AND gate.

With regard to the qualitative importance of the events, K2 and T both occur in single event cut sets, but K2 is an active failure and T a passive one, and hence K2 ranks higher than T. Events S, S1, K1 and R each occur in two-event cut sets, but of these S occurs in all three cut sets, whereas the other three faults each occur only in one cut set and hence S ranks higher than the other three, which all rank equally. The order of importance is thus: K2; T; S; and S1, K1 and R.

The failure probabilities given by Vesely *et al.* are:

Component	*Symbol*	*Failure probability*
Pressure tank	T	5×10^{-6}
Relay K2	K2	3×10^{-5}
Pressure switch	S	1×10^{-4}
Relay K1	K1	3×10^{-5}
Timer relay	R	1×10^{-4}
Switch S1	S1	3×10^{-5}

Then the probabilities of the minimum cut sets are:

$$P(T) = 5 \times 10^{-6}$$

$$P(K2) = 3 \times 10^{-5}$$

$$P(S \cdot K1) = 1 \times 10^{-4} \times 3 \times 10^{-5} = 3 \times 10^{-9}$$

$$P(S \cdot R) = 1 \times 10^{-4} \times 1 \times 10^{-4} = 1 \times 10^{-8}$$

$$P(S \cdot S1) = 1 \times 10^{-4} \times 3 \times 10^{-5} = 3 \times 10^{-9}$$

and the probability of the top event is the sum of these

$$P(E1) = 3.5 \times 10^{-5}$$

The contribution of these minimum cut sets to the top event is a measure of their quantitative importance, and is:

Cut set	*Importance* (%)
T	14
K2	86
Other	<0.1

9.5.14 Illustrative example: crystallizer system

As a third illustrative example, consider the proposed crystallization plant system described by Lawley (1974b). This system is shown in Figure 9.8. The problem is to estimate the expected frequency of emission of the toxic and corrosive crystallizer slurry via the stack to the atmosphere. There are two ways in which this can occur – by discharge at about atmospheric pressure or by discharge at high pressure – but only the former is considered here. If the stack base and luted drain are properly designed for the flow from the three reactors, a discharge of slurry at only a few pounds per square inch pressure would normally pass via the pressure control valve (PCV) header or the relief header to the stack and hence to the drain. If a blockage occurs in the stack or the drain, the discharge from the crystallizer will result in a build-up of liquid in the stack and eventually to entrainment of the slurry to the atmosphere.

The top event considered is 'Low pressure discharge of slurry to the atmosphere'. The fault tree developed by the author for this event is shown in Figure 9.9. The tree is constructed by working from the top down, but it is convenient here to explain it from the bottom up. The level in No. 1 crystallizer can rise only if the outflow to No. 2 crystallizer is restricted. This can occur in three ways:

(1) Discharge line choked with slurry; frequency estimated from plant experience of similar materials, 5 occasions/year.
(2) Crystallizer level control valve (LCV) on discharge line shut or misdirected shut; frequency estimated from literature data, 0.5 occasions/year.
(3) Manual isolation valve on discharge line shut in error; frequency estimated from general plant experience, 0.1 occasions/year.

Thus the estimated frequency of restricted outflow is 5.6 occasions/year. The level will continue to rise only if:

(1) The operator fails to take action opposite a rising level signal; frequency estimated 4% of occasions.
(2) The level meter fails dangerous; failure rate estimated 2 failures/year, downtime 4 h/failure, dead time $2 \times 4 = 8$ h/year, fractional dead time $8/8760 \approx 0.1\%$ of time.

The slurry will only overflow into the PCV header and thence to the stack if

(1) The operator fails to take action opposite a high level alarm signal; frequency estimated 3% of occasions.
(2) The high level alarm fails dangerous; fractional dead time estimated 1.25% of time.

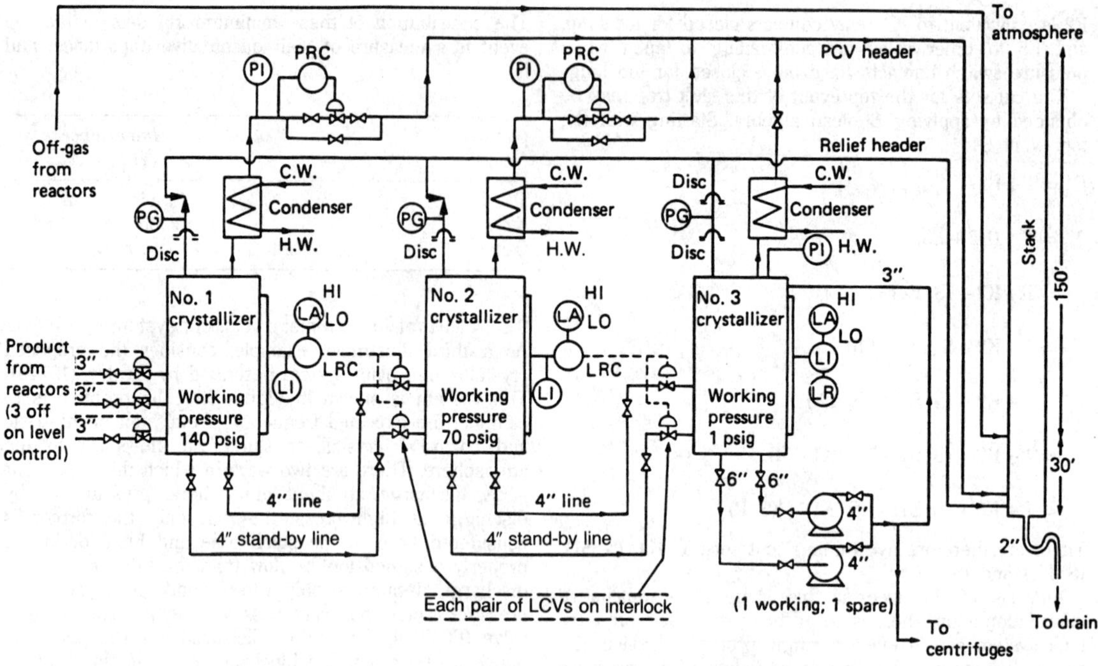

Figure 9.8 Proposed crystallization plant: flow diagram (Lawley, 1974b) CW, cold water; HI, high; HW, hot water; LA, level alarm; LD, low; LRC, level recorder controller; PCV, pressure control valve; PG, pressure gauge; PI, pressure indicator; PRC, pressure recorder controller (Courtesy of the American Institute of Chemical Engineers)

This fractional dead time calculation is as follows. The failure rate of the high level alarm (excluding the meter) is estimated as 0.2 failures/year and the proof test interval is 6 weeks. Thus the fractional dead time is:

$$\frac{1}{2} \times 0.2 \times \frac{6}{52} = 1.15 \times 10^{-2}$$

$$= 1.15\%$$

The fractional dead time of the meter has already been given as 0.1%. Thus the fractional dead time of the level meter and the high level alarm together is 1.25%.

If the slurry overflows, it is assumed that it continues to pass to the stack for up to 2 h maximum and thus for 1 h/occasion, on average.

Similarly, the level in No. 2 crystallizer can rise only if the outflow to No. 3 crystallizer is restricted. The treatment is identical to that for No. 1 crystallizer.

The level in No. 3 crystallizer can rise only if the outflow is restricted. This can occur in four ways:

(1) Discharge line or centrifuge feed line blocked; frequency estimated from plant experience of similar materials, 0.2 occasions/year.
(2) Failure of on-line circulation pump followed by failure to commission stand-by circulation pump in time; frequency estimated, 0.1 occasion/year.
(3) The operator fails to reduce the feed to the reaction section when there are problems on the centrifuges; frequency estimated 0.2 occasions/year.
(4) Manual isolation valve on discharge line shut in error; frequency estimated 0.1 occasions/year.

Thus the estimated frequency of restricted outflow is 0.6 occasions/year.

Since No. 1 crystallizer has restricted outflow on 5.6 occasions/year and overflow on 9.7×10^{-3} occasions/year, No. 3 crystallizer, which has restricted outflow on 0.6 occasions/year, will have overflow on:

$$\frac{0.6}{5.6} \times 9.7 \times 10^{-3} = 1 \times 10^{-3} \text{ occasions/year}$$

Thus the low pressure discharge of slurry from one of the crystallizers reaches the stack via the PCV header on

$$2 \times 9.7 \times 10^{-3} + 1 \times 10^{-3} = 20.4 \times 10^{-3} \text{ occasions/year}$$

Noxious discharge from the top of the stack occurs only if the stack base or drain is blocked. It is estimated that blockage will occur on 2 occasions/year. The stack drain is checked for blockage once per day so that, on average, a blockage would be unrevealed for 12 h. Allowing 6 h for clearing the blockage gives a mean downtime for the drain of 18 h/occasion. Then the low pressure discharge of slurry to atmosphere is

$$2 \times 20.4 \times 10^{-3} \left(\frac{1+18}{8760}\right) = 8.9 \times 10^{-5} \text{ occasions/year}$$

This is a fairly typical illustration of an industrial fault tree analysis. Some additional examples related to protective systems are given in Chapter 13.

9.5.15 Repairable systems

The treatment given so far has not taken into account the repair of failed components. For lack of exact solutions, approximate treatments have been developed which give the upper bounds on reliability and avail-

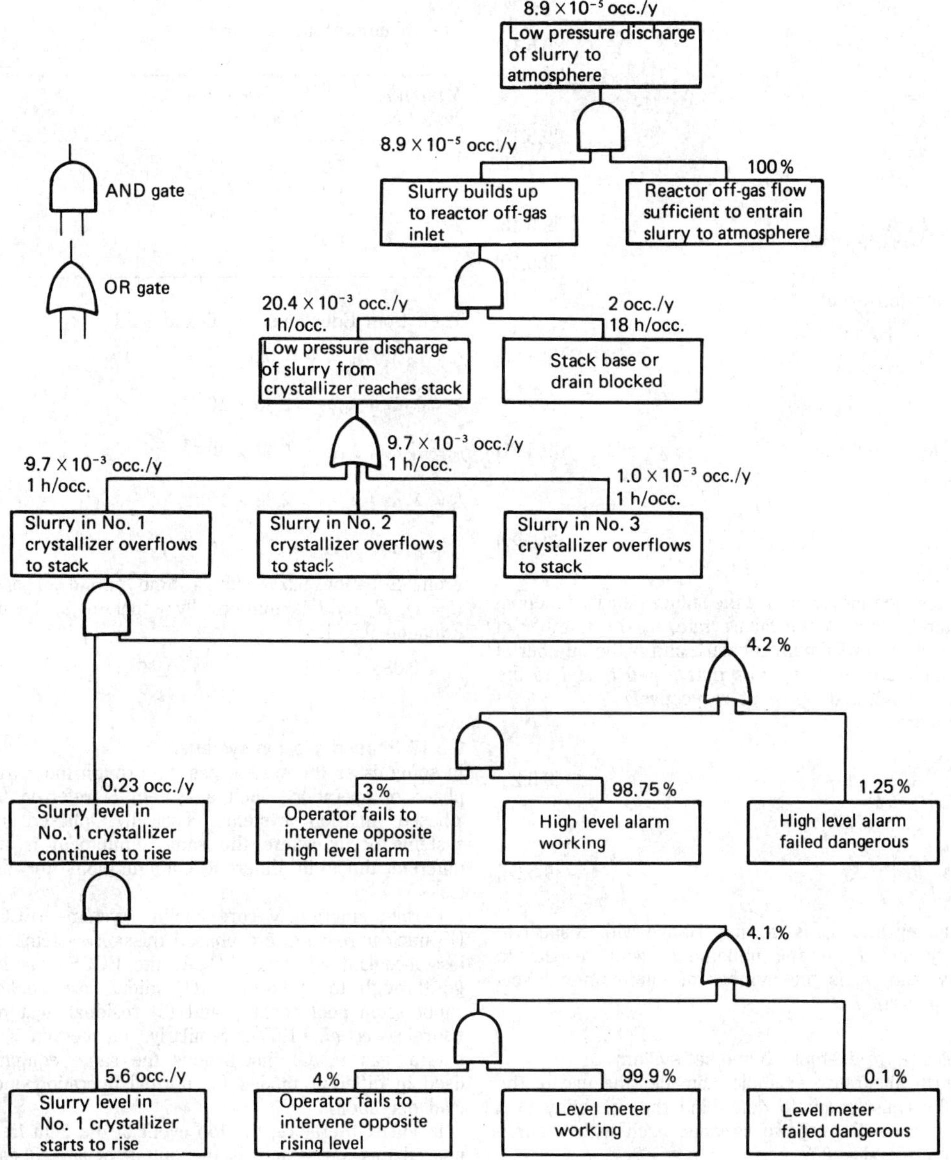

Figure 9.9 *Proposed crystallization plant: fault tree for low pressure discharge of crystallizer slurry to stack and entrainment to atmosphere (Lawley, 1974b) (Courtesy of the American Institute of Chemical Engineers)*

ability. A method of determining the reliability and availability for the top event which is based on this approach and which is suitable for manual calculation has been described by Fussell (1975).

The method is based on the crucial minimum cut sets, or collection of minimum cut sets which can be taken as adequately representing for quantitative evaluations the top event fault logic. It utilizes the rare event approximation, the principal condition being $0.1 < \lambda_i t$, where λ_i is the failure rate of component i. The account given here is confined to the presentation of the results obtained by Fussell; full derivations are given by that author.

For the primary events:

$$\bar{r}_i \leqslant \lambda_i t \quad \text{non-repairable and repairable components} \qquad [9.5.13]$$

$$\bar{a}_i \leqslant \lambda_i t \quad \text{non-repairable components} \qquad [9.5.14a]$$

$$= r_i \qquad [9.5.14b]$$

$$\bar{a}_i \leqslant \lambda_i \tau_i \quad \text{repairable components} \qquad [9.5.15]$$

where, for component i, a_i and $\bar{a}_i$ are the availability and unavailability, r_i and $\bar{r}_i$ the reliability and unreliability, and λ_i is the failure rate and τ is the restoration time.

For the minimum cut sets:

$$\bar{A}_k = \sum_{i=1}^{n_k} \bar{a}_i \quad [9.5.16]$$

$$f_k \leq \bar{A} \sum_{i=1}^{n_k} \frac{\lambda_i}{\bar{a}_i} \quad [9.5.17]$$

$$\bar{R}_k = \int_0^t f_k(t')\, dt' \quad [9.5.18]$$

$$\Lambda_k = f_k/\bar{R} \quad [9.5.19]$$

and for the top event

$$\bar{A}_t \leq \sum_{k=1}^{N} \bar{A}_k \quad [9.5.20]$$

$$\bar{R}_t \leq \sum_{k=1}^{N} \bar{R}_k \quad [9.5.21]$$

$$\Lambda_t \approx \sum_{k=1}^{N} \bar{\Lambda}_k \quad [9.5.22]$$

where, $\bar{A}$ is the unavailability, f the failure density function, $\bar{R}$ the unreliability, Λ the failure rate, n_k the number of primary events in minimum cut set k, and N the number of crucial minimum cut sets; subscripts k and t refer to the minimum cut set and top event, respectively.

For importance:

$$I_{ia} \approx \left(\sum_{k=1}^{m_i} \bar{A}_k\right) \Big/ \bar{A}_t \quad [9.5.23]$$

$$I_{ir} \approx \left(\sum_{k=1}^{m_i} \bar{R}_k\right) \Big/ \bar{R}_t \quad [9.5.24]$$

where, for event i, I_{ia} is the importance with respect to availability and I_{ir} is the importance with respect to reliability, and m_i is the number of minimum cut sets containing event i.

9.5.16 Illustrative example: pump set system

As a fourth illustrative example, consider the use of the relationships just given to determine the reliability of a pump set, consisting of two streams, each with a pump and a control valve (CV).

The basic components, their failure rates and restoration times, and hence their unavailabilities, obtained from Equation 9.5.15, are:

Component	*Component No.*	*Failure rate, λ_i (year^{-1})*	*Restoration time, τ_i*		*Unavailability, $\bar{a}_i$*
			hours	*years*	
Power supply	1	0.08	1	1.1×10^{-4}	8.8×10^{-6}
Pump 1	2	2	6	6.8×10^{-4}	1.36×10^{-3}
Pump 2	3	2	6	6.8×10^{-4}	1.36×10^{-3}
Control valve 1	4	0.02	3	3.4×10^{-4}	6.8×10^{-6}
Control valve 2	5	0.02	3	3.4×10^{-4}	6.8×10^{-6}

The minimum cut sets are:

Minimum cut set No.	*Component failure No.*
1	1
2	2, 5
3	2, 3
4	3, 4
5	4, 5

Then from Equations 9.5.16 and 9.5.17:

$f_1 = \bar{a}_1 \lambda_1/\bar{a}_1 = \lambda_1 = 0.08$

$f_2 = \lambda_2 \bar{a}_5 + \lambda_5 \bar{a}_2 = 2.86 \times 10^{-5}$

$f_3 = \lambda_2 \bar{a}_3 + \lambda_3 \bar{a}_2 = 5.44 \times 10^{-3}$

$f_4 = \lambda_3 \bar{a}_4 + \lambda_4 \bar{a}_3 = 2.86 \times 10^{-5}$

$f_5 = \lambda_4 \bar{a}_5 + \lambda_5 \bar{a}_4 = 2.72 \times 10^{-7}$

From Equation 9.5.18 for a time period of one year ($t = 1$), $\bar{R}_k = f_k t$ is numerically equal to f_k. Then, from Equation 9.5.21,

$\bar{R}_t \leqslant 0.085$

9.5.17 Phased mission systems

In some cases the system passes through more than one phase of operation. Such a system is referred to as a 'phased mission system'. Typically, phased mission systems occur where the same equipment is used at different times in different configurations for different tasks.

Certain emergency core cooling systems (ECCSs) in US nuclear reactors are phased mission systems. After a loss-of-coolant accident (LOCA) the ECCS may have to go through the phases of (1) initial core cooling, (2) suppression pool cooling, and (3) residual heat removal (Burdick *et al.*, 1977). Similarly, on certain liquefied natural gas (LNG) installations the same equipment is used in different modes for normal operation and peak load operation.

In effect, therefore, the top event of the fault tree for a phased mission system is the output of an OR gate, the inputs to which are the top events of the several phases of the mission. This particular OR gate differs, however, from the conventional type in that its inputs are separated in time.

Although a phased mission system can be treated by constructing fault trees for the separate phases, there are certain problems which arise at the boundaries. Such a treatment is not conservative.

For non-repairable systems only, Esary and Ziehms (1975) have developed a transformation which converts the fault tree under the special type of OR gate just mentioned to a conventional fault tree. This transformation is not applicable, however, to repairable systems.

9.5.18 Protective systems

A large proportion of fault tree applications are for protective systems. The determination of the characteristics of systems of this type is therefore of particular interest. A treatment of this problem has been given by Kumamoto, Inoue and Henley (1981). The method given treats the system as a whole rather than the individual protective devices.

The states of the system are: 0, operational; 1, under a normal trip; 2, under a spurious trip; and 3, subject to destructive hazard. The system may then be described by the following equations:

$$w_{01}(t) = f_{01}(t, 0) + \int_0^t f_{01}(t, u)[v_{10}(u) + v_{20}(u)]du \quad [9.5.25]$$

$$w_{02}(t) = f_{02}(t, 0) + \int_0^t f_{02}(t, u)[v_{10}(u) + v_{20}(u)]du \quad [9.5.26]$$

$$w_{03}(t) = f_{03}(t, 0) + \int_0^t f_{03}(t, u)[v_{10}(u) + v_{20}(u)]du \quad [9.5.27]$$

with

$$v_{10}(u) = w_{01}(u - \tau_n) \quad [9.5.28]$$

$$v_{20}(u) = w_{02}(u - \tau_s) \quad [9.5.29]$$

where: $F_{0j}(t,u)$ is the probability at time t that the system is in state j given that it entered state 0 at time u; $f_{0j}(t,u)$ is the time differential of $F_{0j}(t,u)$; $v_{j0}(t)$ is the transition rate, or intensity, at time t from state j to state 0; $w_{0j}(t)$ is the transition rate, or intensity, at time t from state 0 to state j; τ_n is the restoration time after a normal trip; and τ_s is the restoration time after a spurious trip.

Then the expected numbers of the transition events are as follows:

$$N_n = \int_0^t w_{01}(u)du \quad [9.5.30]$$

$$N_s = \int_0^t w_{02}(u)du \quad [9.5.31]$$

$$N_h = \int_0^t w_{03}(u)du \quad [9.5.32]$$

where N_h is the number of realized destructive hazards, N_n is the number of normal trips and N_s is the number of spurious trips.

Equations 9.5.25–9.5.27 are solved by numerical integration. The authors have written a code, PROTECT, for this purpose. An application of this method, using PROTECT, has been described by Kumar, Chidambaram and Gopalan (1989).

Table 9.7 Some fault trees for process industry systems

System	Reference
Ethylene oxide plant trip system	R.M. Stewart (1971)
Pressure tank system	Fussell (1973a,b); Vesely *et al.* (1981)
Crystallization plant	Lawley (1974b)
Reactor system	Powers and Tompkins (1974a)
Nitric acid cooler	Lapp and Powers (1977a, 1979); Shaewitz, Lapp and Powers (1977)
Sulphur trioxide reactor	Lambert (1977)
Liquid propane pipeline system	Lawley (1980)
Motor system	Vesely *et al.* (1981)
Distillation column system	Kletz and Lawley (1982)
Power distribution network	Cummings, Lapp and Powers (1983)
Storage tank system	Ozog (1985)
Butane vaporizer system	Andrews and Morgan (1986)

9.5.19 Fault tree applications

Reviews of work on fault trees have been given by Arendt and Fussell (1981) and W.S. Lee *et al.* (1985). Accounts of industrial application of fault trees include those of Caceres and Henley (1976), Lihou (1980a,b), Pilz (1980a), Prugh (1981, 1982, 1992a) and Schreiber (1982). Some fault tree applications described in the literature are listed in Table 9.7.

9.6 Event Trees

An event tree is used to develop the consequences of an event. It starts with a particular initial event such as a power failure or pipe rupture and is developed from the bottom up. The event tree is both a qualitative and a quantitative technique. Qualitatively it is used to identify the individual outcomes of the initial event, while quantitatively it is used to estimate the frequency or probability of each outcome.

An event tree is constructed by defining an initial event and the possible consequences which flow from this. The initial event is usually placed on the left and branches are drawn to the right, each branch representing a different sequence of events and terminating in an outcome. The main elements of the tree are event definitions and branch points, or logic vertices. The initial event is usually expressed as a frequency (events/year) and the subsequent splits as probabilities (events/demand), so that the final outcomes are expressed also as frequencies (events/year).

Each branch of the event tree represents a particular scenario. The tree is a means of estimating the frequency of the outcome for that scenario. It is used in conjunction with a set of hazard models for the

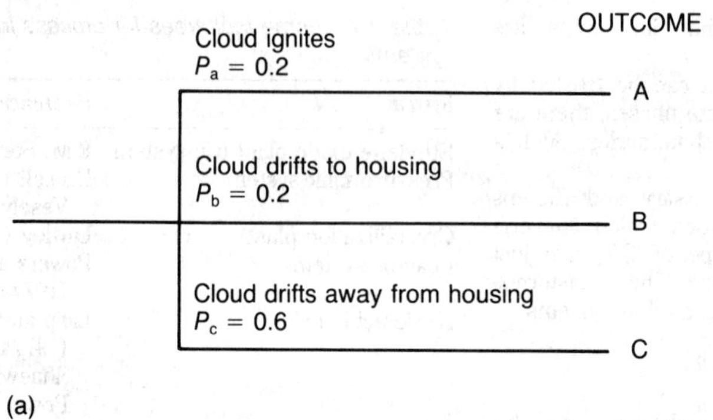

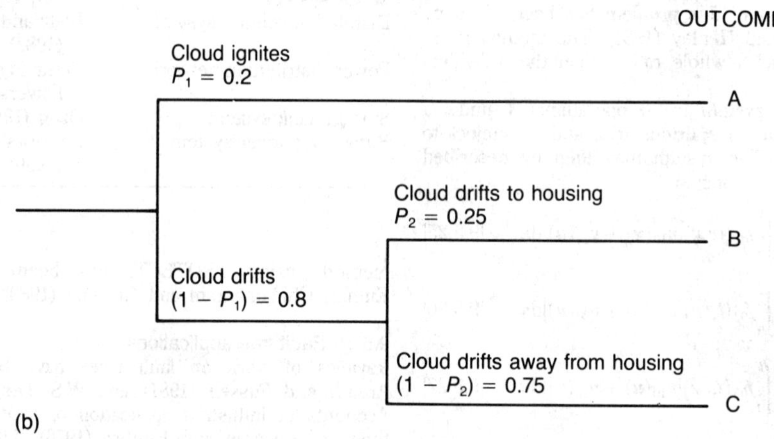

Figure 9.10 *Construction of an event tree: (a) multiple splits; (b) binary splits*

determination of the physical consequences of that scenario. For example, for a flammable release a typical series of models might be those for emission, gas dispersion, ignition, explosion and explosion injury.

In constructing an event tree the normal convention is to use binary rather than multiple splits. For example, consider the representation of a release for which the immediate outcomes are that (a) the vapour cloud is ignited at the point of release, (b) it drifts towards housing and (c) it drifts away from housing, where the probabilities p_a, p_b and p_c of these three outcomes are 0.2, 0.2 and 0.6, respectively. Figure 9.10 shows two ways of representing this situation, but it is Figure 9.10(b) which is the preferred form. It should be noted that in this case the probability p_2 is obtained from

$$p_2 = p_b/(1 - p_a) \qquad [9.6.1]$$

In certain cases one event tree can be reduced to another, equivalent event tree. Examples of such reduction have been given by Schreiber (1982).

Figures 9.11–9.13 show event trees for the response of an operator to a nuclear reactor transient, a loss of grid power supply and a release of LPG, respectively. In the plant power supply system analysed in the event tree in Figure 9.12 power is supplied to a plant from the National Grid, but in the event of grid loss a stand-by generator is started up to supply power. The problem is to estimate the frequency of loss of power to the plant. In the event of diesel failure, a backup power supply is brought in. It is assumed in the example that the frequency of grid loss is 1×10^{-1} events/year, that the probability of diesel failure is 2.1×10^{-2} failures/demand, made up of 2×10^{-2} failures/demand for start-up and 1×10^{-3} failures/demand for running for the required period, and that the backup power supply failure probability is 1×10^{-2} failures/demand. The event tree shows that the frequency of works power loss is 2.1×10^{-5} events/year.

Figure 9.13 shows an event tree analysis of a release of LPG near a detergent alkylate plant (DAP). The possibilities for the development of the release are then as follows. The release may ignite and, if it does, it will either explode or give a fireball. If it does not ignite, it may be borne by the wind towards the DAP and may then ignite and explode at the DAP. Alternatively, the

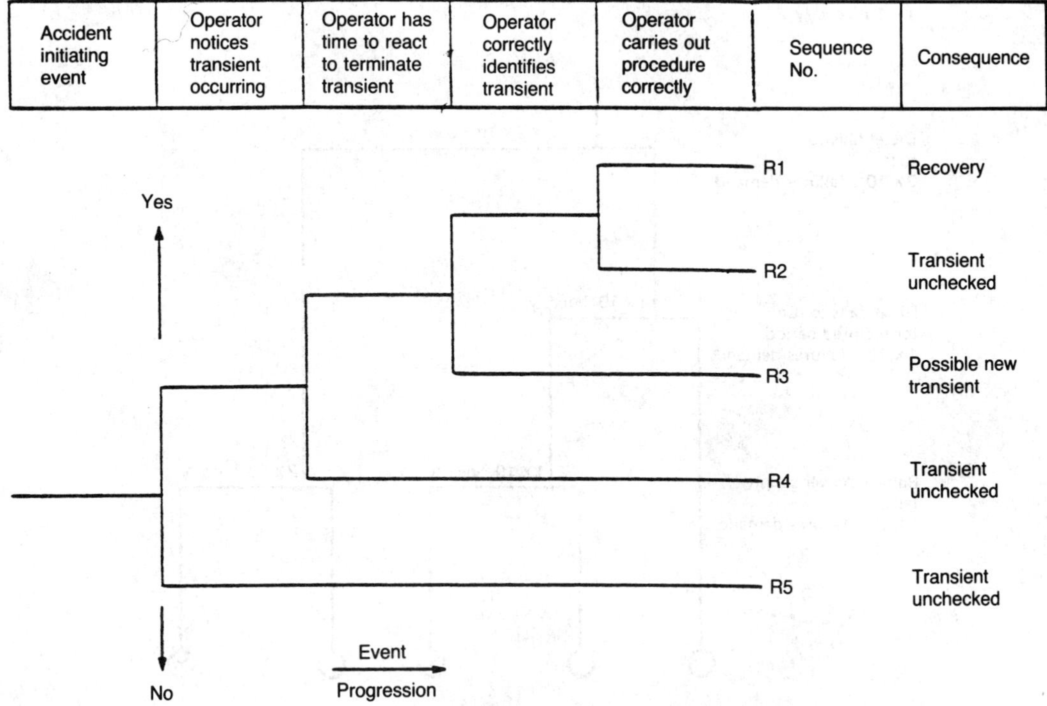

Figure 9.11 *Event tree for response of operator to nuclear reactor transient (Welsh and Lundberg, 1980) (Courtesy of Elsevier Science Publishers)*

cloud may drift in some other direction. The estimates are:

Frequency of large LPG release 1 = 68×10^{-6}/year

Probability of ignition $p_1 = 0.9$

Probability of explosion given ignition $p_2 = 0.1$

Probability of wind to DAP $p_3 = 0.4$

Probability of ignition and explosion at DAP $p_4 = 0.9$

Then the probabilities of the outcomes are:

Probability of hydrogen fluoride (HF) release (independent of wind direction) $p_5 = p_1 p_2 = 0.9 \times 0.1 = 0.09$

Probability of fireball $p_6 = p_1 \times (1 - p_2) = 0.9 \times 0.9 = 0.81$

Probability of HF release drifting north-east
$p_7 = (1 - p_1) \times p_3 \times p_4 = 0.1 \times 0.4 \times 0.9 = 0.036$

Probability of drifting cloud $p_{10} = p_8 + p_9 = (1 - p_1) \times [p_3 \times (1 - p_4) + (1 - p_3)] = 0.1 \times (0.4 \times 0.1 + 0.6) = 0.064$

The outcome frequencies are then obtained by multiplying the initial event frequency by these outcome frequencies. Thus

Frequency of HF release (independent of wind direction)

$\lambda_5 = 0.09 \times 68 \times 10^{-6} = 6.1 \times 10^{-6}$/year

and so on.

The outcome probabilities should sum to unity and the outcome frequencies to that of the initial event.

This last example also illustrates the need for care in the definition and interpretation of events. Outcome C is an HF release drifting north-east, but this is not the only outcome where this occurs. It occurs implicitly also in outcome A, so that the frequency of such a release is that of outcome C plus a contribution from that of outcome A.

Event trees are used particularly, but not exclusively, to analyse failures of utilities and outcomes of releases from plants. Other event trees are given in the literature (e.g. von Alven, 1964; N.C. Rasmussen, 1974; AEC, 1975; HSE, 1978b, 1981a; Cross, 1982; Rijnmond Public Authority, 1982).

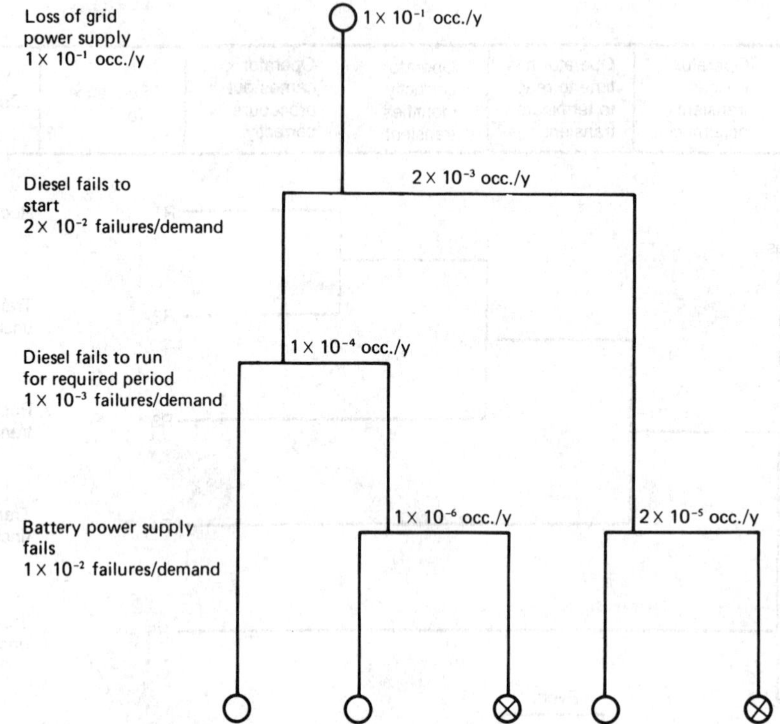

Figure 9.12 *Event tree for loss of grid power supply (after Andow, 1976)*

9.7 Cause–Consequence Diagrams

A third technique which incorporates features both of the fault tree and of the event tree is the cause–consequence diagram. The cause–consequence diagram has been developed by D.S. Nielsen (1971, 1974, 1975) and by J.R. Taylor (1974d, 1978a).

A cause–consequence diagram is constructed by defining a critical event and then developing the causes and consequences of this event. The forward development has the features of an event or a decision tree and the backward development those of a fault tree. The main elements of the diagram are therefore event and conditions definitions and logic gates and vertices.

Some cause–consequence diagram symbols are given in Table 9.8. The logic symbols include both gates, which describe the relations between cause events, and vertices, which describe the relations between consequences. The event and condition symbols describe the type of event or condition. The symbols given are those which have been used by Nielsen and by Taylor.

The main logic gates are the AND gate and the OR gate. There are corresponding logic vertices in the form of the AND vertex and the OR vertex.

The EITHER/OR vertex, or decision box, is also very useful. It is utilized in particular to determine the effect of an event or condition on the paths which the system takes. If, as is often the case, the 'No' output from the decision box is the result of an abnormal condition, then the fault tree for this condition is developed. Thus fault trees occur on the diagram not only for the critical event but also for abnormal conditions throughout the diagram.

Some important features of the cause–consequence diagram are its ability to handle alternative consequence paths and time delay and time order.

Examples of the use of cause–consequence diagrams in hazard assessment have been given by Nielsen and by Taylor. As an illustration consider the example of the surge tank system shown in Figure 9.14(a), which is adapted from a similar example by J.R. Taylor (1978a). The problem is to develop the diagram for the critical event of 'Flow controller fails'. The cause–consequence diagram is shown in Figure 9.14(b).

The cause–consequence diagram may also be used for quantitative assessment.

9.8 Dependent Failures

In the design of systems for reliability, use is made of redundancy and diversity of subsystems and components. In principle, the use of redundancy allows very high reliability to be obtained. The condition for this, however, is that the failures which occur are independent. If this is not so, then the design intent is defeated and the

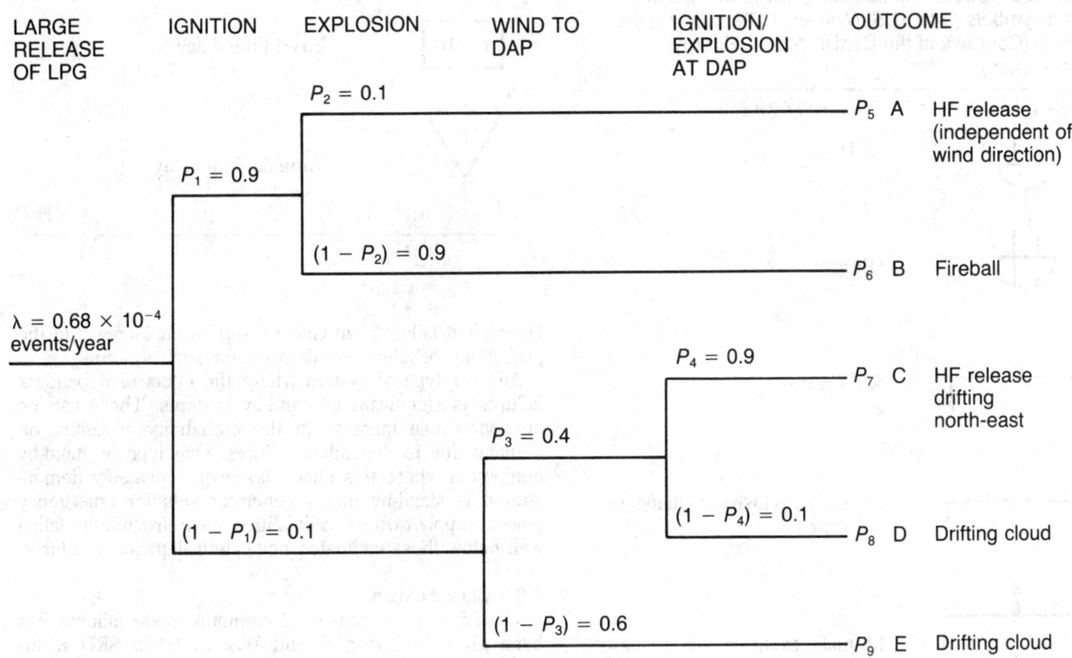

Figure 9.13 *Event tree for release of LPG (after HSE, 1981a)*

reliability may be less, sometimes dramatically less. In other words, the phenomenon of dependent failure tends to set a limit on the reliability which can be achieved. It is therefore a major concern for designers of high reliability systems.

Work on dependent failures has been described in *A Study of Common Mode Failures* (Edwards and Watson, 1979 SRD R146) and on defences against it in *Defences Against Common Mode Failures in Redundancy Systems* (Bourne *et al.*, 1981 SRD R196) and *SRD Dependent Failure Procedures Guide* (Humphreys and Johnston, 1987 SRD R418).

The terminology used in the field has changed somewhat over the years. Initially, the term generally used was 'common mode failure' (CMF). The term 'common cause failure' (CCF) then gained currency. Some debate arose as to the difference between the two. The term 'dependent failure' (DF) is now preferred. A set of definitions is given by Humphreys and Johnston. They define dependent failure as 'the failure of a set of events, the probability of which cannot be expressed as the simple product of the unconditional failure probabilities of the individual events'. They treat common mode failure as a subset of common cause failure in which the failures are in the same mode and due to the same cause. They also refer to and include a fourth category, cascade failures, or propagating failures.

9.8.1 Occurrence

Dependent failures are liable to occur in all sorts of ways. One of the most obvious is loss of a utility, such as failure of the power supply and hence simultaneous loss of power to all the instruments in a particular system. Another readily envisaged type of dependent failure is that due to some influence from the external environment. This may be freezing due to cold weather, or shock caused by an earthquake.

There are other causes of dependent failure which are less readily appreciated but which are just as important. One of these is human error, specifically systematic error. A systematic error in maintenance or testing may disable a complete system of redundant devices.

In general, a particular component will be potentially subject to a number of types of dependent failure. Typical dependencies are manufacture, maintenance and external events. Another form of dependent failure is damage caused by the accident event itself.

9.8.2 Significance

The significance of dependent failures is readily appreciated. There is a limit to the degree of reliability which can be achieved in a system by the use of single items of high reliability. For high system reliability it is necessary to exploit redundancy and diversity. In the absence of dependent failures, very high system reliabilities can be achieved by these means. The effect of dependent failures, however, is to set a limit to the increase in reliability which is obtainable in practice.

One type of system where this is important is protective, or trip, systems. High reliability protective systems tend to utilize more than one channel, typical configurations being 1/2 systems and 2/3 voting systems.

Table 9.8 *Cause–consequence diagram logic and event symbols (after D.S. Nielsen, 1974; J.R. Taylor, 1974d)* (Courtesy of the Danish Atomic Energy Commission)

Logic symbol	*Meaning of symbols*
	AND gate
	OR gate
	AND vertex
	Mutually exclusive, exhaustive OR vertex
	Mutually exclusive OR vertex (used after time delays)
No / Yes	EITHER/OR vertex, decision box
$t > 10$ No / Yes	Condition vertex
Event and condition symbols	
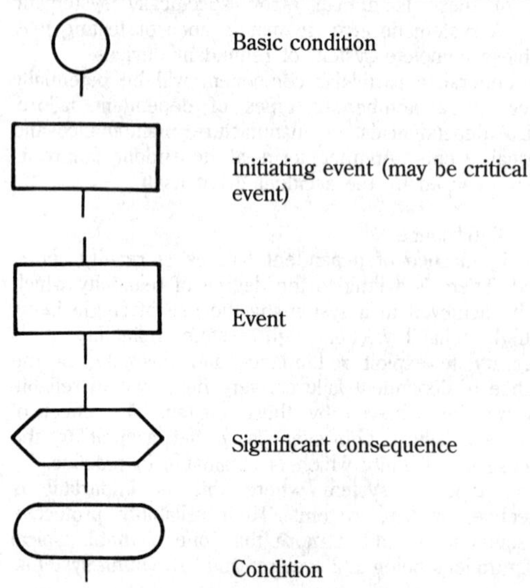	Basic condition
	Initiating event (may be critical event)
	Event
	Significant consequence
	Condition
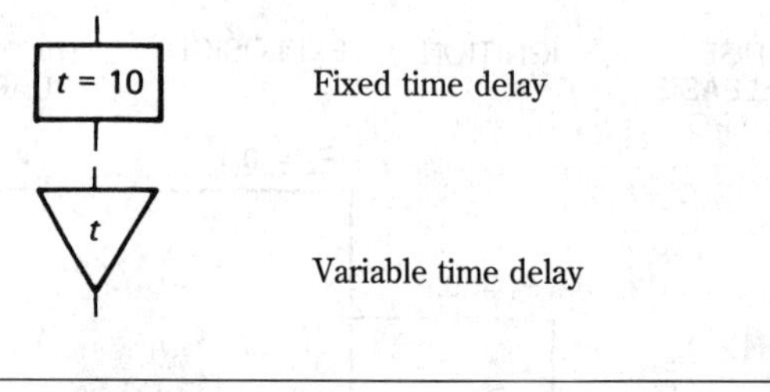 $t = 10$	Fixed time delay
t	Variable time delay

Dependent failures can cause a significant increase in the probability of failure on demand for such systems.

Another type of system where the effect of dependent failures is significant is stand-by systems. There can be an appreciable increase in the probability of failure on demand due to dependent failures. One type of stand-by equipment where this effect has been repeatedly demonstrated is stand-by diesel generator sets for emergency power supply, where reliabilities have frequently fallen well below those estimated neglecting dependent failures.

9.8.3 Classification

A classification of causes of common mode failures has been given by Edwards and Watson (1979 SRD R146) and is shown in Figure 9.15. Engineering failures are classed as those of design (ED), which is subdivided into functional deficiencies (EDF) and realization (EDR), and those of construction (EC), subdivided into manufacturing faults (ECM) and installation and commissioning faults (ECI). Operations failures are classed as procedural (OP), subdivided into maintenance and test (OPM) and operation (OPO), and environmental (OE), subdivided into normal extremes (OEN) and energetic events (OEE). Each of these classes is broken down into further detailed categories.

9.8.4 Failure data

Edwards and Watson review the data available on common mode failures. Most of these data relate to aircraft systems and to nuclear reactor systems in the USA, France, Germany and the UK.

Data on abnormal occurrences in US nuclear power plants are available in the form of Licensee Event Reports (LERs). Two reactor systems for which high reliability is required are the automatic protective system (APS) and the emergency core cooling system (ECCS). The authors analyse the reports to identify CMFs.

CMF also figures prominently in the *Rasmussen Report* (AEC, 1975). This report contains (in Appendix III) CMFs identified in various data sources on reactor incidents and operation. Analysis of the data in the *Rasmussen Report* shows that of a total of 303 faults some 32 were CMFs. The causes of these CMFs include EDR (12 faults), OPM (6 faults) and OEN (5 faults).

Further discussions of data on dependent failures are given by Humphreys (1987) and G.T. Edwards (1988).

9.8.5 Beta factor

A quantitative measure of the prevalence of dependent failure is the ratio β of the dependent failures to the total failures, introduced by Fleming (1974). It is defined as:

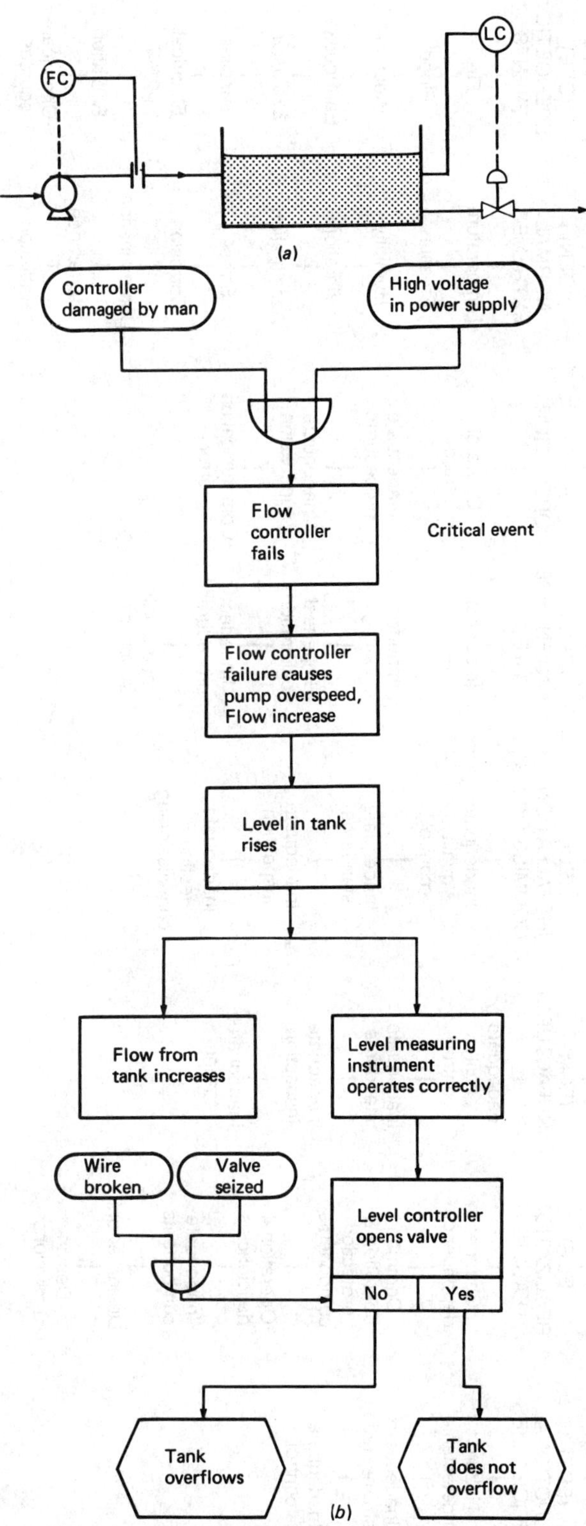

Figure 9.14 *Surge tank system and cause–consequence diagram for overfilling of a surge tank (after J.R. Taylor, 1978a): (a) surge tank system; (b) cause–consequence diagram of the surge tank system for the critical event 'Flow controller fails' (Courtesy of the Danish Atomic Energy Commission)*

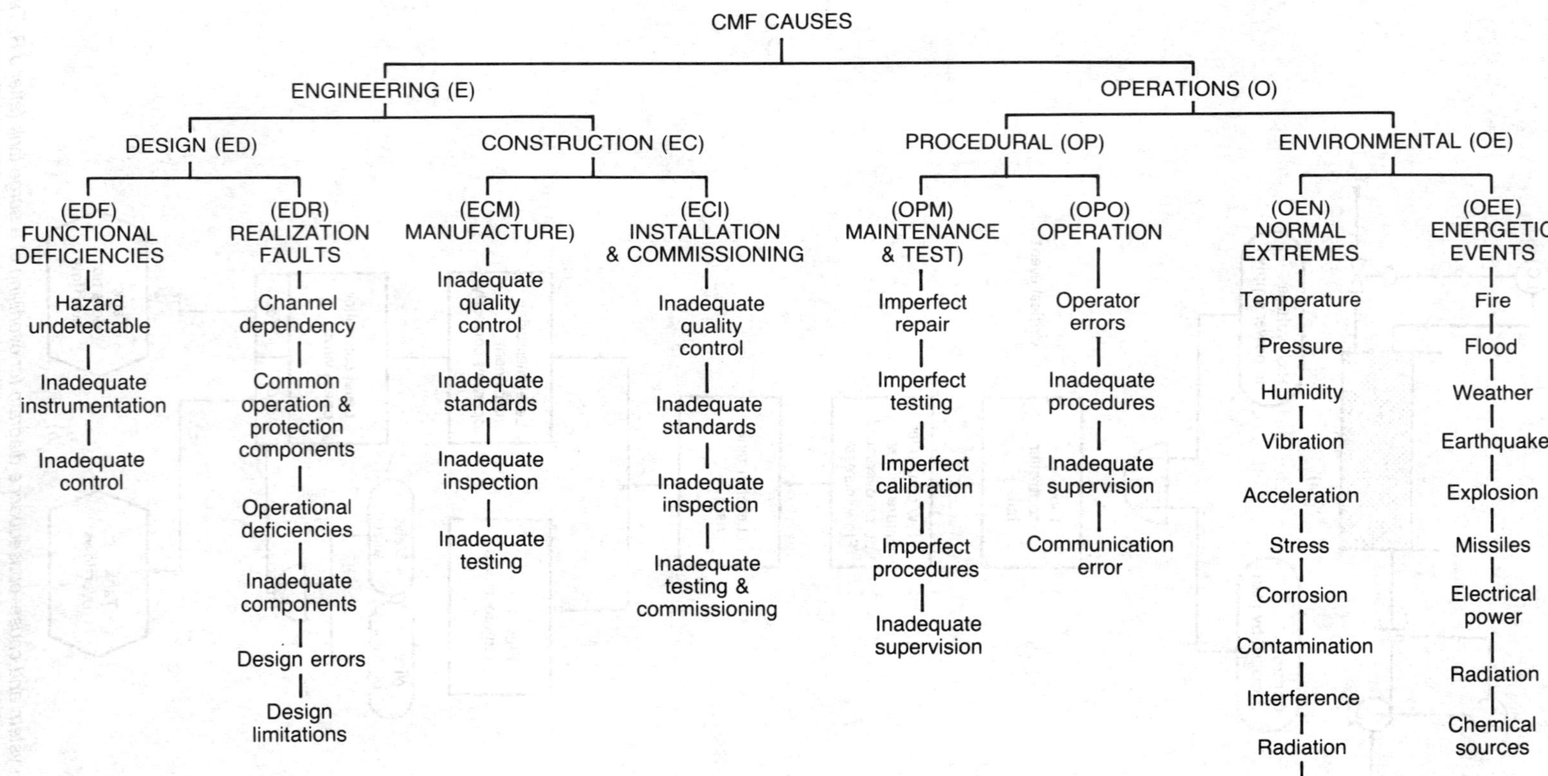

Figure 9.15 *Classification of common mode failures (Courtesy of the UKAEA Safety and Reliability Directorate)*

$$\beta = \frac{\lambda_d}{\lambda_i + \lambda_d} + \frac{\lambda_d}{\lambda} \quad [9.8.1]$$

with

$$\lambda = \lambda_i + \lambda_d \quad [9.8.2]$$

where λ is the overall failure rate, λ_d is the rate of dependent failure, and λ_i is the rate of independent failure.

The beta factor may be used to quantify the contribution of dependent failures to overall system failure. Edwards and Watson have analysed the data and derived values of the beta factor for a number of subsystems and components. Particularly important are protective systems with redundancy. For such redundant systems typical values of the beta factor are 0.05 to 0.25.

Another important item is diesel generator sets for emergency power supply. In their study the authors identified 24 nuclear power plants which yielded usable data. For the diesel generators in these plants there were 67 failures recorded over a period of 238 operating years. There were 6 dependent failures over a period of 118 operating years. Each diesel generator was broken down into 8 subsystems. The overall subsystem failure rate was therefore 0.035 failures/year ($67/(8 \times 238)$) and the dependent failure rate 0.0064 ($6/(8 \times 118)$). The resultant value of β is 0.18 (0.0064/0.035).

A further study of beta factors has been described by Humphreys (1987).

9.8.6 Protective systems

The type of system for which dependent failures are of particular concern is protective systems with a high degree of redundancy. The characteristics of such systems have been discussed by Bourne *et al.* (1981 SRD R196). The authors recognize five basic types of configuration for such systems: (1) a single channel, (2) a redundant system, (3) a partly diverse system, (4) a fully diverse system and (5) two diverse systems.

Thus, considering the measurement channels for a trip system, the authors give as an example of a redundant system a system with n identical redundant channels and m/n majority voting. As examples of partly diverse and fully diverse systems they give systems with two parallel channels, in which the channels are partly diverse and fully diverse, respectively. Two diverse systems are exemplified by a system with two complete subsystems in parallel, each of which is a half-identical redundant channel subsystem.

Bourne *et al.* give the following approximate ranges for the probability of failure of these configurations, based on a typical proof test interval of 1–3 months:

Single channel	>0.5 to 5×10^{-3}
Redundant system	5×10^{-2} to 5×10^{-4}
Partly diverse system	10^{-2} to 5×10^{-5}
Fully diverse system	10^{-3} to 5×10^{-6}
Two diverse systems	5×10^{-5} to $<10^{-6}$

The probabilities of failure of the systems therefore show considerable overlap.

9.8.7 Modelling for redundancy

A number of methods have been developed for the modelling of dependent failures in redundant systems. Accounts of such modelling are given by Edwards and Watson (1979 SRD R146), G.T. Edwards (1987a,b) and Humphreys and Johnston (1987 SRD R418). The methods include the beta factor method, the multiple Greek letter method, the binomial failure rate method and the Marshall-Olkin method. An account of these and other methods is given by Humphreys and Johnston.

One of the most widely used is the beta factor method. This has a single parameter β. Some of the other methods have several parameters. The use of a single parameter has the merit that only one parameter needs to be estimated from the data, but the corresponding limitation that it cannot, in principle, model the data as accurately as a multiple parameter method.

9.8.8 Beta factor method

As just stated, one of the most widely used models of dependent failure is the beta factor method of Fleming (1974). The beta factor was defined in Section 9.8.5. For a single component the overall failure rate may be written in terms of the beta factor as:

$$\lambda = \lambda_i + \lambda_d \quad [9.8.3]$$

Then, writing

$$\lambda_i = (1 - \beta)\lambda \quad [9.8.4]$$

$$\lambda_d = \beta\lambda \quad [9.8.5]$$

gives

$$\lambda = (1 - \beta)\lambda + \beta\lambda \quad [9.8.6]$$

The simplest system to which common cause failure is applicable is the two-component parallel redundant system shown in Figure 9.16(a). The failure rate for a single component is λ. Then, applying the beta method and separating out the contributions of independent and common cause failures, gives the system shown in Figure 9.16(b). The system now consists of a two-component parallel subsystem in series with a single component subsystem. The failure rate for the system λ_s is then given by the relation:

$$\lambda_s = \frac{1}{\frac{1}{(1-\beta)\lambda} + \frac{1}{2(1-\beta)\lambda}} + \beta\lambda \quad [9.8.7]$$

The fault trees for the two cases are given in Figures 9.16(c) and (d), respectively.

Models such as the beta model are applied particularly to protective systems. For such a system consisting of multiple redundant items in parallel, which is normally not operational but which is required to operate when a demand is placed on it, the probability of failure is equal to its unavailability, or fractional dead time. Since the probability of failure increases with time and since for such a system the failure will normally be unrevealed, it is necessary to perform a proof test at some fixed interval.

The application of the beta model may be illustrated as follows. It is shown in Chapter 13 that, subject to certain conditions, the fractional dead time for a single channel system is $\phi = \lambda\tau_p/2$ and that for a system of two parallel redundant channels it is $\phi = (\lambda\tau_p)^{2/3}/3$, where ϕ is the

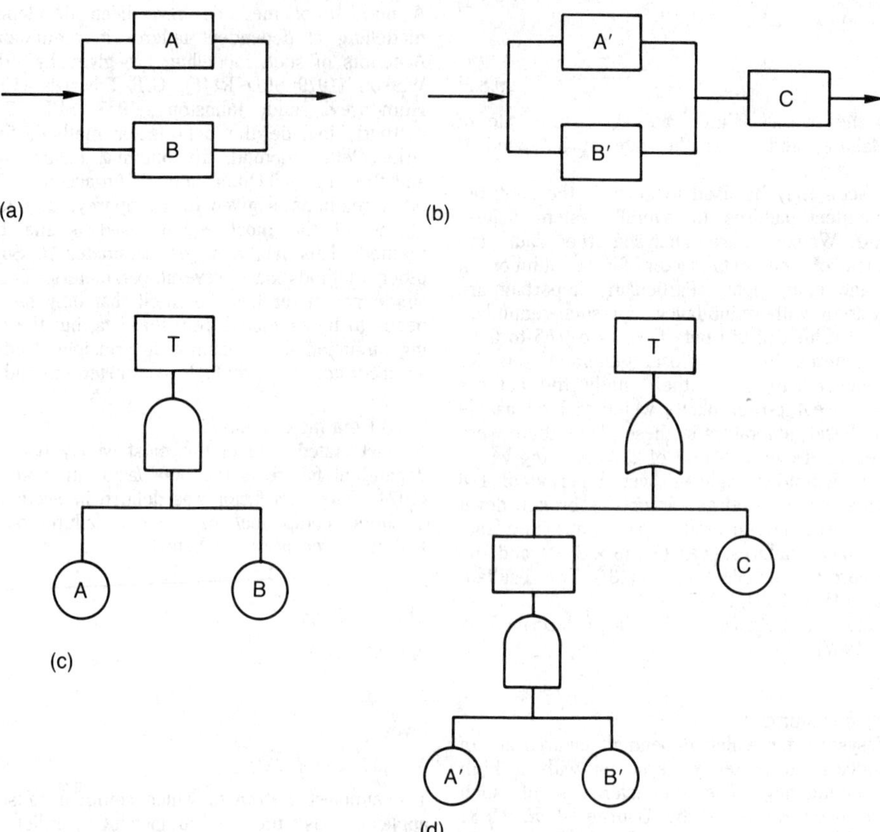

Figure 9.16 Representation of dependent failure: (a) a two-component parallel system; (b) the two-component parallel system distinguishing between independent and common cause failures; (c) fault tree for system (a); (d) fault tree for system (b)

fractional dead time and τ_p the proof test interval. The corresponding expression for the latter case in terms of the beta factor is:

$$\phi = \frac{[(1-\beta)\lambda\tau_p]^2}{3} + \frac{\beta\lambda\tau_p}{2} \qquad [9.8.8]$$

The use of the beta factor method requires that a suitable value of β be known. As stated earlier, typical values for redundant systems are 0.05 to 0.25.

9.8.9 Geometric method

Another method of handling common cause failure is the geometric method. This utilizes the upper and lower bounds of failure of the system. The lower bound λ_1 of the system failure rate is that assuming that there are no failures due to a common cause failure, and the upper bound λ_u is that assuming that all failures are due to the common cause. The system failure rate λ_s is

$$\lambda_s = (\lambda_1^n \times \lambda_u)^{1/(n+1)} \qquad [9.8.9]$$

where n is an index. The value of n is a function of the diversity of the components in the system and its value usually lies in the range 1–4.

9.8.10 Partial beta factor method

The partial beta factor method is not actually a separate method but rather a method of obtaining a more accurate value of the beta factor. The method is described by Humphreys and Johnston (1987 SRD R418). The authors suggest that a practical limit of the beta factor in a redundant subsystem is about 2×10^{-2} and that in a diverse subsystem is about 10^{-3}. They take the latter as the lower limit for a subsystem for which very high reliability is required and in which all defences are exploited.

The beta factor is decomposed into a set of partial beta factors, each of which corresponds to a particular defence. The beta factor is computed as the product of the partial beta factors. The procedure is as follows. For each partial beta factor a minimum value β_{pm} is assigned. The product of the minimum values is 10^{-3}. Then for each partial beta factor the user assigns an actual value β_{pa}. The maximum in each case is unity, so that if this maximum is assigned in every instance, the product of these values is unity. Thus, in general, the beta value computed from the partial beta values will lie between 10^{-3} and 1.0.

The authors give a proforma for the determination of the beta factor from the partial beta factors, in which account is taken both of the causes of, and defences against, dependent failures. A further discussion of the partial beta factor method is given by B.D. Johnston (1987a,b).

9.8.11 Multiple Greek letter method

The beta factor method does not distinguish between different levels of redundancy in a subsystem. This deficiency is overcome in the multiple Greek letter method, in which a separate Greek letter is introduced for each component after the first. Thus for a four-component system one would use the letters β, γ and δ, where β is the conditional probability that a cause of component failure will be shared by one or more components, γ is the conditional probability that the cause of component failure which is shared by one or more components will be shared by two or more components, and δ is the conditional probability that the cause of component failure which is shared by two or more components will be shared by all components. It should be noted that the definition of β used here differs from that in the conventional beta factor method.

Humphreys and Johnston (1987 SRD R418) state that typical values of β, γ and δ are 0.1, 0.76 and 0.82, and that system failure probabilities computed by this method typically differ from those obtained by the conventional beta factor method by a factor of about 1.5.

9.8.12 Binomial failure rate method

In the binomial failure rate model developed by Vesely (1977a) and modified by Attwood (1991), multiple dependent failures in a redundant subsystem are treated as being either probabilistic (binomial) or deterministic (global) in nature. In this model the external event is referred to as a 'shock'. Shocks are divided into non-lethal and lethal, the former being those which have a certain probability of causing subsystem failure and the latter being those which are certain to do so:

$$\lambda = \lambda_{si} + p\mu + \omega \qquad [9.8.10]$$

where p is the conditional probability that the external event will cause subsystem failure, λ is the overall failure rate of the subsystem, λ_{si} is the overall failure rate of the subsystem for independent failures, μ is the occurrence rate of non-lethal shocks, and ω is the occurrence rate of lethal shocks.

9.8.13 Revealed and unrevealed failure models

In the account given so far the emphasis has been on dependent failures in protective systems where failures are unrevealed when they occur and are detected only by proof testing. The probability of failure is a function of the failure rate and of the proof test interval.

In most plant systems, however, where an equipment is required to perform some active function on a continuous basis, failures are detected by the fact that this function is no longer being performed. In this case the failures are revealed. The probability of being in a failure state is a function of the failure rate and of the repair time, taking this as the total time for detection and repair. An example of a system in which failures are revealed is two pumps each of 50% capacity operating in parallel.

The models for the probability of failure in systems with revealed failures and in systems with unrevealed failures are different. Treatments are given in Chapters 7 and 13, respectively. Stand-by systems are an intermediate case. In the stand-by mode the failures are unrevealed, whilst in the operating mode they are revealed.

9.8.14 Diversity and its modelling

Diversity is recognized as one of the principal defences against dependent failures. A system with diversity is much less vulnerable to many of the events which can cause dependent failures in redundant systems. The modelling of diversity is, however, much less well developed than that of redundancy, and it is therefore less easy to obtain quantitative results for subsystems embodying diversity.

9.8.15 Fault tree analysis

One of the principal tools for the analysis of high reliability systems is fault tree analysis. The type of system to which fault tree analysis is typically applied is one in which undesirable events are protected against by the use of control loops and trip systems. In systems where very high reliability is required, use is generally made of redundancy in some of the trip systems. The benefit of redundancy may be negated if there is dependency between failures.

In fault tree analysis dependent failures may be handled in two different ways. One is to construct the tree as normal, without explicit regard to dependency, but to flag each group of base events where a dependency exists and to deal with this dependency at the cut set stage. The alternative approach is to represent the dependent failure part of the subsystem failure as an explicit failure in its own right.

The two approaches are illustrated in Figure 9.17, which shows a fault tree for a system in which a measurement is made using a half-redundant system. The top event X is the failure of the measurement function. In Figure 9.17(a) the events A and B are defined as failure of measurement channel 1 and failure of channel 2, respectively. These failures include both independent and dependent failure contributions. If the dependency of A and B is to be dealt with at the cut set stage, the events would be flagged in some way.

In Figure 9.17(b) the failure which causes failure of the two channels, i.e. the dependent failures, is isolated as event C. The events A and B are now redefined as, respectively, the failures of channels 1 and 2, excluding the dependent failure contribution. In this tree the event C is shown accompanying both event A and event B. In the alternative, but equivalent, configuration shown in Figure 9.17(c), event C is separated out and shown only once.

Computer codes for the analysis of dependent failures have been described by Humphreys and Johnston (1987 SRD R418) and by B.D. Johnston (1987a,b). An account of these codes is given in Chapter 29.

9.8.16 Effect on reliability

The effect of dependent failures is seen most markedly in protective systems with a high degree of redundancy, where it can dominate. This may be illustrated by considering the half-redundant protective system

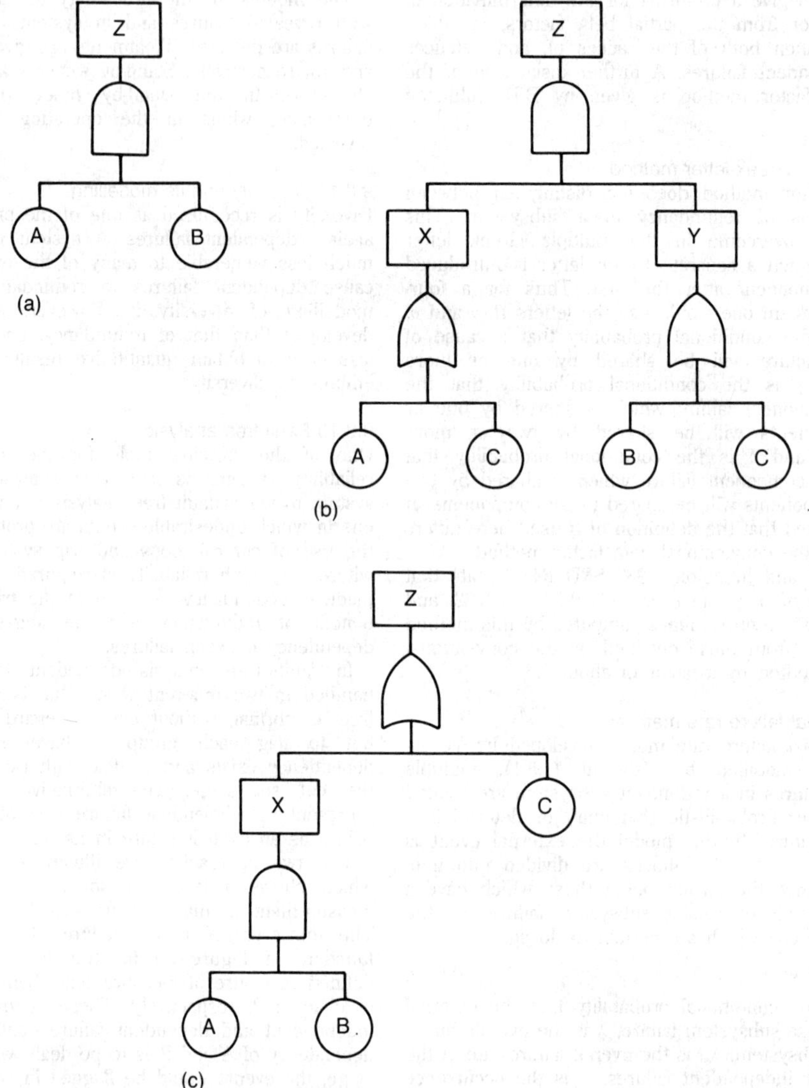

Figure 9.17 *Treatment of dependent failures in fault tree analysis: (a) tree with events A and B treated as independent events; (b) tree with events A and B treated as dependent on event C; (c) alternative form of tree in (b)*

described earlier. The expression for the fractional dead time ϕ of such a system is Equation 9.8.8. Typical values of the parameters are given by Humphreys and Johnston (1987 SRD R418) as follows: $\lambda = 10^{-6}$ failures/h; $\tau_p = 10^3$ h; and $\beta = 0.1$. The fractional dead time is then:

$$\phi = 0.27 \times 10^{-6} + 0.5 \times 10^{-4}$$

where the first term is the contribution of the system for independent failures and the second that for dependent failures. The dominant contribution is that of the dependent failures.

The impact of dependent failures can be even greater. In the example just given the redundancy was minimal. Systems with higher degrees of redundancy are relatively even more vulnerable to dependent failures. One early study, by Epler (1969), stated that dependent failures may be dominant by a factor as high as 10^5.

9.8.17 Benchmark study

A benchmark study of common cause failure is described in *Common-cause Failure Reliability Benchmark Exercise* (Johnston and Crackett, 1985 SRD R383). The study is one of a series of benchmark exercises conducted by the Commission of the European Communities (CEC) and organized by JRC-Ispra.

A summary of this study is given by Humphreys and Johnston (1987 SRD R418). The situation investigated was dependent failures on the feedwater system of a pressurized water reactor (PWR) under emergency conditions. Of the ten teams whose studies are listed, five used the multiple Greek letter method and four the Marshall–Olkin method. The system is characterized by redundancy rather than diversity, and for this situation agreement was reasonably good. The three most structured presentations estimated the probability of failure as lying in the range 10^{-4} to 5×10^{-5}. The treatment of the rather limited amount of diversity in the system exhibited greater discrepancies.

9.8.18 Cascade failures

Cascade failures occur when a fault propagates through the plant and induces further failures. The conditions created by the fault may include abnormal process conditions, equipment overload, loss of potential correction on control loops, activation of trip systems and actions by the operator in response. The cascade failure problem appears to have received relatively little attention.

9.8.19 Defences

The defences available against dependent failures are described by Edwards and Watson (1981 SRD R146) and are further developed by Bourne *et al.* (1981 SRD R196) and Humphreys and Johnston (1987 SRD R418). The headings of the list given by Edwards and Watson are:

1.0 Design
- 1.1 Administration
- 1.2 Functional diversity
- 1.3 Equipment diversity
- 1.4 Fail-safe design
- 1.5 Redundancy and voting logic
- 1.6 Protection and segregation
- 1.7 Proven design and standardization
- 1.8 Equipment derating and simplicity
- 1.9 Operational interfaces
- 1.10 Quality control
- 1.11 Design review

2.0 Operation
- 2.1 Administration
- 2.2 Maintenance procedures
- 2.3 Proof testing
- 2.4 Operating procedures
- 2.5 Reliability performance monitoring

3.0 Reliability assessment

Detailed accounts of the elements of each of these defences are given by the three sets of authors quoted.

Bourne *et al.* (1981 SRD R196) describe an overall defence strategy in terms of (1) general strategy, (2) management aspects and (3) technical aspects. The general strategy should be based on good administrative controls of both design and operations. There should be a specification for the reliability requirements of the system with particular reference to dependent failures (DFs). The design should be subject to reviews of DFs at appropriate stages. There should be an independent assessment of reliability in which DFs are considered. During the operational life of the plant, the continued effectiveness of the defences against DFs should be periodically reviewed.

Management should ensure that there is awareness of the problem of DFs among all personnel who have a contribution to make in minimizing it, and should give a lead. It should make defence against DFs an explicit objective in both design and operations stages. It should make it clear where the need for such defence imposes requirements additional to those of the regular standards and codes. It should assure the necessary reviews and independent assessment.

Technical aspects cover a wide range of measures. Probably the most effective is functional diversity, which is generally used only at system level. This defence may be enhanced by equipment diversity. Within subsystems fail-safe design may be exploited. Other measures are: redundancy and voting logic; protection and segregation; proven design and standardization; and equipment derating and simplicity. These design measures need to be complemented by adequate systems and procedures for defence during operations.

Bourne *et al.* suggest the following priorities. As the most significant contributors to DFs are errors in design and in maintenance and testing, the first priority should be the management of the design process and of design and operational tasks in relation to DFs. Of the technical measures, fail-safe characteristics offer the greatest potential for reliability enhancement. Next come measures to minimize human error at the interfaces of tasks such as maintenance and testing.

The defences most commonly applied are protection and segregation. Protection guards against influences from the external environment and segregation can be effective against most causes of DF. Both have a contribution to make, but it is not a predominant one. The authors state that other design defences considered (redundancy and voting, proven design and standardization, derating and simplicity) are of less significance, but have a role to play.

9.9 Expert Judgement

It is sometimes difficult to obtain data on event frequency or probability. In such cases a possible solution is the use of expert judgement to obtain the required estimates. This approach has been used particularly to obtain estimates for human error rates, where it is especially difficult to obtain field data, and for equipment failure rates, but it also has other applications.

An account of expert judgement is given in *Eliciting and Analysing Expert Judgement: A Practical Guide* (M.A. Meyer and Booker, 1990). A review of five areas of application of expert judgement methods is given by Mosleh, Bier and Apostolakis (1987). Further guidance on knowledge representation and elicitation is available in the literature on expert systems, though there the emphasis is generally more on expert rules rather than on quantification. Expert systems are discussed in Chapter 30.

9.9.1 Scope

Some purposes for which expert judgement has been used are given by M.A. Meyer and Booker (1990) as:

(1) to provide estimates on new, rare, complex or otherwise poorly understood phenomena;
(2) to forecast future events;
(3) to integrate or interpret existing data;
(4) to learn an expert's problem-solving processes or a group's decision-making processes;
(5) to determine what is currently known, what is not known, and what is worth learning in a field of knowledge.

A study in which expert judgement has been used for all these purposes in the context of the NUREG-1150 project is *Analysis of Core Damage Frequency from Internal Events: Expert Judgement Elicitation* by Wheeler *et al.* (1989).

9.9.2 Expert problem-solving

In solving a problem the tasks which an expert has to perform are essentially to (1) understand the question, (2) retrieve the relevant information, (3) make a judgement and (4) formulate and report an answer.

It is to be expected that the results of an expert judgement exercise will be affected by bias. One form of bias is that arising from the elicitation process and from the expert's involvement with the project. Another form is a discrepancy between the expert's knowledge and some sort of norm. These two forms of bias are sometimes termed 'motivational bias' and 'cognitive bias', respectively.

Some forms of motivational bias are social pressure, misinterpretation or misrepresentation by the interviewer, and overoptimism by the expert. Forms of cognitive bias include inconsistency, anchoring, availability and underestimation of uncertainty. Anchoring is failure to make sufficient adjustment to an initial opinion in the light of further evidence. Availability is the difference in ease of retrieval of different types of information, infrequent catastrophic events being more readily retrieved than more frequent mundane ones.

The handling of bias is not well developed, but there are certain measures which can be taken. The elicitation process can be planned so as to minimize it. The expert can be made aware of the potential for bias. The elication process can be monitored for bias and, if necessary, adjustments made to counter it. The results can be analysed for the effect of bias.

9.9.3 Formulation of questions

The elicitation process involves asking the expert a number of questions. The quality of the results obtained depends in large part on the formulation of these questions. This involves defining the objectives of the project, selecting the general question areas and formulating the specific questions. Closely related to this is the selection of the experts who are to try to answer the questions. These may be in-house and/or external experts. Different options exist on the extent to which the experts are involved in the selection of the question areas and the formulation of the questions.

The questions are then refined by structuring them. Some aims of the refinement process are: to focus attention on what is required and to minimize misinterpretation; to present the question in a form that is assimilable by the expert; and to make the question acceptable to him. Some ways in which questions may be refined include decomposition, phrasing, and use of representational aids such as diagrams.

There are certain circumstances in which the involvement of external experts in question refinement is critical. These are given by M.A. Meyer and Booker (1990) as those where the purpose is to capture the experts' problem-solving process, where there is evidence that the experts may not accept the questions, and where outside reviewers are likely to be concerned about bias in the questions.

9.9.4 Selection of experts

A basic distinction in the selection of the experts is between applications where the output is to be the experts' answers and those where the output is to be insight into the experts' problem-solving process. In most cases it is the experts' answers which are of interest. In order to select an expert it is necessary to have some idea of what is to be regarded as constituting expertise. Two types of expertise are distinguished. One is the expert's experience in the domain, or substantive expertise. The other, or normatise, expertise is his knowledge of the response mode, or form in which he is to be asked to give his answers, e.g. as probabilities, odds or rankings. Both types of expertise are needed for satisfactory results. For certain projects the credibility of the expert in the world at large may be a factor.

It is normally advisable to use multiple and diverse experts. Diversity of experts guards against the excessive influence of a single individual, and is particularly effective in face-to-face meetings. The number of experts is typically five to nine, the lower figure being about the minimum needed to obtain diversity and the upper one about the maximum which can be readily handled.

The diversity of expertise achieved depends on the selection of experts from the expert population; this is essentially a sampling problem. It is necessary to guard against over-representation of any particular tendency.

The success of the exercise depends on the motivation as well as the selection of the experts, and measures should be taken to motivate those participating.

9.9.5 Elicitation methods and planning

In designing the elicitation it is necessary to consider (1) the elicitation method, (2) the response mode and dispersion measures, (3) the aggregation of response, (4) the problem-solving process and (5) the documentation.

There are three principal methods of elicitation: the individual interview, the interactive group, and the response of experts operating in isolation (the Delphi method). Each method has its advantages and disadvantages. The individual interview is the main method for obtaining detailed responses and insight into the expert's problem-solving process, but does not give interaction between experts and is time-consuming. The interactive group is claimed to give more accurate responses and to generate more ideas, but it can suffer from group bias. The Delphi method is freer from group bias. It can be framed so as to offer some insight into problem-solving or interaction.

Some response modes which may be used are: single probability values; odds ratios; sets of probability values, effectively a probability distribution; location on a continuous scale; location on a discrete scale, with ranking or rating; paired comparisons; and revisions of estimates. The expert may provide an indication of the uncertainty in the response in the form of a dispersion measures such as a range, or standard deviation.

Methods available for eliciting the expert's problem-solving process include verbal protocol, verbal probe and the ethnographic method. The first requires the expert to think aloud, the second is the questioning of the expert about his problem-solving process immediately after he has given his answer and the third involves transposing the expert's answers into questions, which draw from the expert elaboration of his answers. The use of verbal protocol avoids interviewer bias but is liable to disturb the expert's skilled problem-solving behaviour.

The answers given by a group of experts are likely to differ and need to be aggregated in order to obtain an overall response. This may be done by the experts themselves or by a mathematical technique.

The documentation required from the exercise should be specified. It may record the answers only, the answers and the problem-solving process or something intermediate such as the answers with a short explanation of the underlying reasoning.

The planning of the elicitation includes the logistics of involving the experts, the structuring of the elicitation process, the precautions against bias and the specification of the documentation to guide and record the elicitation. There will normally need to be some training and practice, probably with pilot testing, prior to the main elicitation exercise.

9.9.6 Analysis of responses

The data obtained from the elicitation normally consist of (1) answers to questions and (2) ancillary information. The latter is essentially information about (1) the experts and (2) the expertise.

The data obtained should be examined for two features in particular. One is the degree of detail, or granularity. Variations in the level of granularity can adversely affect the results. It is desirable in principle to operate in the analysis with a single level of granularity. There is, however, a loss of information in passing from the particular to the general and this may act as a constraint.

The other feature is conditionality. Answers are conditioned by the path by which they have been reached and it is desirable to make this explicit. Meyer and Booker comment that: 'Two experts could arrive at exactly the same final answer but for very different reasons, or two experts could arrive at different answers for exactly the same reasons.'

The analysis of the data is facilitated if they can be converted to quantitative form. There are a number of methods of doing this. One approach is to convert qualitative descriptors into values on some kind of scale. Another is to merge separate categories into a small number of broader ones. Another is the combination of quantitative data using weighting.

The data may be analysed to obtain relationships between different answers, between different ancillary variables and/or between answers and ancillary variables. One major aspect of such analysis is that concerned with lack of independence between experts' answers, or correlation, and with bias in these answers. There are various factors which experts may have in common, such as shared training, similar work experience and exposure to the same data sources. The biases which may be present include motivational and cognitive bias, as described earlier.

Another aspect of data analysis is the modelling of the experts' answers in terms of the ancillary data on the expert and his expertise.

A third aspect of the analysis is the application of aggregation methods to combine the answers of a number of experts into a single value or distribution of values. One particular application of aggregation occurs where there is a decision-maker and a single expert. The elements in this situation are the decision-maker's prior estimate, his choice of the expert, the aggregation of the prior estimate with the expert's information, and the inference. There are several quite different potential outcomes. Thus the decision-maker may consider both items of information as valid and combine them, he may prefer one or the other or he may decide to adopt a range of values. Another situation occurs where there is a decision-maker and *n* experts. The potential outcomes are similar, but it is more likely that the decision-maker will not simply prefer his own prior.

A fourth aspect of the analysis is the handling of uncertainties in the data. Some sources of uncertainty are definitions, errors associated with small samples, non-sampling errors such as missing data, and deficiencies in scientific and modelling techniques. Methods are available which assist in the measurement, modelling and control of such uncertainties.

The purpose of the analysis is to reach conclusions by making inferences from the responses and analysis of these responses. What an expert judgement exercise does is to sample the world of experts and their expert knowledge. It does not sample the real world and the conclusions drawn may or may not represent the true state of nature. The validity of these conclusions depends very much on the quality of expert knowledge about the domain in question.

Critiques of expert judgement refer to the subjective nature of such judgement. The validity of subjective judgement is considered in more detail in Chapter 7.

9.9.7 Interviews

One of the main techniques of elicitation used in expert judgement work is the individual interview. There is a large literature on interview and questionnaire techniques. Guidance on interviewing is given by Kahn (1957), by Oppenheim (1966) and by Gorden (1987) and also in texts on expert systems.

9.9.8 Delphi method

Another of the principal techniques of expert judgement is the Delphi method. As already described, this method involves the use of multiple experts operating in isolation. This method is described in *The Delphi Method* (B. Brown, 1964), and by Linstone and Turoff (1975) and Dalkey (1969). The Delphi method was used for the estimation of failure rates in IEEE Standard 500: 1984, which gives an appendix on the procedure used.

9.9.9 Ranking and scaling

Work on ranking and scaling of expert judgement has been described by Thurstone (1927, 1931) and by Kendall (1948, 1955, 1970). An account of scaling methods has been given in *Theory and Method of Scaling* (Torgerson, 1967).

In general, the overall approach in expert judgement work is first to rank the items. From this ranking the items are then located on a numerical scale. If absolute values are known for some of the items on the scale, these may be used to calibrate the scale and thus obtain absolute values for the other items.

A distinction is made between psychological and probability scales. The responses from the judges lie on a psychological scale and have to be converted to a probability scale. The theory underlying the conversion of rankings to locations on the psychological scale is complex, but essentially it is based on the concept that the responses for a particular item exhibit a Gaussian distribution about some point on the scale. Then if two items are located close together, there will be an overlap between the two distributions and hence a higher frequency of disagreement between judges, while if they are located far apart there will be little or no overlap between distributions and little or no disagreement between judges. The key result obtained by Thurstone is that

$$\mu_{k-j} \approx x_{kj} \qquad [9.9.1]$$

where μ_{k-j} is the mean separation on the scale between item j and item k, and x_{kj} is the unit normal deviation corresponding to the probability that k will be ranked above j. This is the basis for the conversion of the rankings of the bench of judges into locations on the psychological scale.

9.9.10 Method of paired comparisons

A major difficulty in asking an expert to rank a number of items is that his preferences may not lie on a linear scale. If the interviewer insists on a simple ranking, the responses may be forced into too rigid a framework. This problem can be mitigated by asking the expert to express preferences between pairs of items only. This is the basis of the method of paired comparisons.

The method of paired comparisons was proposed by Thurstone (1927) and developed by Kendall (1948, 1955, 1970). Further accounts of the method are given by Guilford (1954), Keats (1971) and Hunns (1980, 1982).

In the method of paired comparisons a bench of m judges is asked to rank a set of n items. The response of a single judge is considered first. For each item pair, say A and B, the judge is asked to express a preference. A preference for A over B is denoted by A → B or B ← A. A $n \times n$ matrix of responses is produced, as shown in the following illustrative example given by Kendall (1970):

	A	*B*	*C*	*D*	*E*	*F*
A	–	1	1	0	1	1
B	0	–	0	1	1	0
C	0	1	–	1	1	1
D	1	0	0	–	0	0
E	0	0	0	1	–	1
F	0	1	0	1	0	–

The diagonals are blocked out. An entry 1 in column Y, row X signifies X → Y and is therefore accompanied by the entry 0 in column X, row Y. In this example, A → B, A → C and A ← D. This $n \times n$ matrix provides the basic data from a single judge.

The responses of a judge often exhibit some degree of inconsistency. This is seen in its simplest form where a set of three pairs, or triad, is inconsistent, or circular. For example, the responses F → G → H → F are inconsistent.

In making paired comparisons on n items a judge takes $n(n-1)/2$ decisions. The maximum number of circular triads is $(n^3 - n)/24$ if n is odd and $(n^3 - 4n)/24$ if n is even. The minimum number is zero. A coefficient of consistency ζ for a single judge may then be defined as

$$\zeta = 1 - 24d/(n^3 - n) \qquad n \text{ odd} \qquad [9.9.2a]$$

$$= 1 - 24d/(n^3 - 4n) \qquad n \text{ even} \qquad [9.9.2b]$$

where d is the observed number of circular triads. This may be obtained from the relation

$$d = \frac{n}{12}(n-1)2n - 1 - \frac{\Sigma a_i^2}{2} \qquad [9.9.3]$$

where a_i is the sum of entries in row i.

The number of triads may also be analysed, as described by Kendall, using the chi-square distribution to determine the probability that the responses are random and could have arisen by chance.

The measure of agreement between the bench of judges is the coefficient of agreement u:

$$u = \frac{8\Sigma}{m(m-1)n(n-1)} - 1 \qquad [9.9.4]$$

where Σ is the sum of agreements between pairs of judges. The maximum possible value of u is 1. The minimum possible value is not -1 but $-1/m$ for m odd and $-1/(m-1)$ for m even.

9.9.11 Saaty's method

One method for paired comparisons is that of Saaty (1977, 1980). His method is part of a methodology described in his book *The Analytic Hierarchy Process* (AHP) (Saaty, 1980). The problem is cast in hierarchical form and the paired comparison technique is then used at the successive levels of the hierarchy.

The following scheme is used in making the paired comparisons:

Number	*Description*
1	The two items are of equal importance or equally likely
3	A slight favouring of the first item over the second
5	A strong favouring of the first item over the second
7	A demonstrated dominance of the first over the second
9	An absolute affirmation of the first over the second

Values 2, 4, 6, 8 are used where the comparison is assessed as a case intermediate between the above descriptions. The comparisons are expressed as the number 1, for no preference, or the fractions $\frac{1}{3}$, $\frac{1}{5}$, $\frac{1}{7}$ and $\frac{1}{9}$ for positive preferences. Thus, for example, an entry $\frac{1}{3}$ means that there is a slight preference for the second item over the first.

The expert proceeds by making all the possible comparisons and recording them in a $n \times n$ matrix with 1s down the diagonal and the comparisons in the upper triangular portion of the matrix. The reciprocals of the entries in the upper triangular portion are entered in the lower triangular portion. The relative weights of the n items are then obtained. These weights are the normalized eigenvectors of the principal eigenvalue of the matrix.

An index of consistency is obtained based on the deviation of the principal eigenvalue from the theoretical eigenvalue of a perfectly consistent matrix. If the consistency ratio is greater than 0.10, inconsistency is indicated.

As an illustration of Saaty's method, consider the example given by M.A. Meyer and Booker (1990). It is required to rank the following meteorological conditions in respect of the likelihood that they will cause a loss of off-site power to a plant:

(1) flash flooding at plant site with 0.5 to 2 in. of water;
(2) flash flooding with 2 to 4 in.;
(3) flash flooding with more than 4 in.;
(4) direct hit of lightning on power lines;
(5) direct hit by tornado;
(6) winds between 20 and 40 mph;
(7) winds higher than 40 mph.

The expert makes the following judgements:

1 vs 2 . . . . 1/3
1 vs 3 . . . . 1/4
1 vs 4 . . . . 1/5
1 vs 5 . . . . 1/5
1 vs 6 . . . . 1/3
1 vs 7 . . . . 1/4

2 vs 3 . . . . 1/2
2 vs 4 . . . . 1/3
2 vs 5 . . . . 1/3
2 vs 6 . . . . 1/2
2 vs 7 . . . . 1/3

3 vs 4 . . . . 1/2
3 vs 5 . . . . 1/2
3 vs 6 1
3 vs 7 . . . . 1/2

4 vs 5 . . . . 1/3
4 vs 6 . . . . 1/5
4 vs 7 . . . . 1/4

5 vs 6 5
5 vs 7 4

6 vs 7 . . . . 1/2

and produces the following matrix:

	1	*2*	*3*	*4*	*5*	*6*	*7*
1	1	1/3	1/4	1/5	1/5	1/3	1/4
2	3	1	1/2	1/3	1/3	1/2	1/3
3	4	2	1	1/2	1/2	1	1/2
4	5	3	2	1	1/3	1/5	1/4
5	5	3	2	3	1	5	4
6	3	2	1	5	1/5	1	1/2
7	4	3	2	4	1/4	2	1

Eigenvalue analysis of this matrix gives the principal eigenvalue as 8.001. The weights for the seven items are obtained by normalizing the seven terms in the corresponding eigenvector. For this example the respective weights of the seven items are: 0.03, 0.06, 0.11, 0.11, 0.35, 0.15, and 0.19.

However, the consistency check gives a value of 0.13, indicating a degree of inconsistency. Examination of the initial judgements shows that the inconsistencies include the following: 1 vs 4 is the same as 1 vs 5; 2 vs 4 same as 2 vs 5; 3 vs 4 same as 3 vs 5; 6 vs 4 same as 6 vs 5; and 7 vs 4 same as 7 vs 5. These judgements imply that 4 and 5 are the same, but 4 vs 5 is given as 1/3. Furthermore they imply that 6 < 7, but the judgements 4 vs 6 and 4 vs 7 imply that 6 > 7. There may be other inconsistencies, but these three are major. The following corrections are made:

Comparison	*Correction*	*Objective*
4 vs 5	1	To make 4 and 5 the same
4 vs 6	5	To match 5 vs 6
4 vs 7	4	To match 5 vs 7

This now gives a consistency ratio of 0.06, which is much more acceptable, and yields the following revised weights: 0.03, 0.07, 0.11, 0.28, 0.28, 0.08, and 0.15.

These weights are to be interpreted solely as indicating the relative ranking. Thus the items judged most likely to cause loss of power are 4 and 5, whilst that judged least likely to do so is item 1. The weights must not be used to draw quantitative conclusions. It is not correct to infer that item 7 (weight 0.15) is five times as likely as item 1 (0.03).

Saaty's method has the advantages that it is easily used by an expert and that it monitors consistency. It is used primarily where the overall problem can be formulated to have a hierarchical structure. It has been widely used in decision analysis.

Many problems, however, are not readily cast in hierarchical form. Moreover, the forcing of a problem into such form is contrary to a philosophy of letting the data suggest the methods of analysis and the models to be used.

9.9.12 Hunns' method

A method for paired comparisons which allows items to be not only ranked but also scaled has been given by Hunns (1980, 1982). In order to calibrate the scale in terms of probability it is necessary to know the relationship between the scale units and the relative probability. Hunns suggests that there is evidence that this is a logarithmic one. Some support for the existence of a logarithmic scale has been given by Pontecorvo (1965) in a study involving comparison of expert judgements on repair times with field data.

The relation may be expressed as:

$$\log_{10} P_x = k_1 + k_2 S_x \quad [9.9.5]$$

where P is the relative probability of an item, S the scale value of the item, k_1 and k_2 are constants and the subscript x refers to item x.

Taking one scale unit to correspond to a ratio r of probabilities the following relations then apply:

$$\frac{P_u}{P_l} = r^{(S_u - S_l)} \quad [9.9.6]$$

$$\frac{P_x}{P_l} = r^{(S_x - S_l)} \quad [9.9.7]$$

$$= \left[\frac{P_u}{P_l}\right]^{(S_x - S_l)/(S_u - S_l)} \quad [9.9.8]$$

Hence

$$r = \left[\frac{P_u}{P_l}\right]^{1/(S_u - S_l)} \quad [9.9.9]$$

and

$$k_1 = \log_{10}(P_l r^{-S_l}) \quad [9.9.10]$$

$$k_2 = \log_{10} r \quad [9.9.11]$$

where subscripts l and u are to lower and upper.

If absolute probability values are known for two items on the scale, Equations 9.9.6–9.9.11 may be used to obtain absolute values for the others.

The application of Hunns' method is now described using the illustrative example given by him. A bench of 20 judges is asked to rank four items A, B, C and D. The results obtained are recorded in the raw frequency matrix **F**, which for this example is as follows:

	A	B	C	D
A	–	6	7	11
B	14	–	12	20
C	13	8	–	14
D	9	0	6	–

The entries give the number of judges out of 20 expressing the preference in question. An entry n in column Y, row X signifies that n judges prefer X to Y. Thus in this example all 20 judges prefer B to D. The entries on the diagonal are left blank. For an entry n in the upper triangular portion of the matrix there is an entry $20 - n$ in the corresponding location in the lower triangular portion.

Next the probability matrix **P** is obtained by dividing the entries in the **F** matrix by 20 and thus normalizing them:

	A	B	C	D
A	–	0.30	0.35	0.55
B	0.70	–	0.60	1.0
C	0.65	0.40	–	0.70
D	0.45	0	0.30	–
Σ	1.80	0.70	1.25	2.25

The entries in each column are then summed as shown in the bottom additional row.

The transformation matrix **X** is then obtained as follows. The order of the columns is rearranged in the numerical order of the values of Σ in the **P** matrix, starting with the highest value. The entries in the **X** matrix are obtained by replacing the probability values from the **P** matrix by the values of the unit normal deviates. Thus for each entry value P the probability P' is obtained as follows:

$$P' = P \qquad P < 0.5$$

$$= P - 0.5 \qquad P > 0.5$$

From the value of P' the unit normal deviate is then determined. For this use is made of a table of the cumulative normal distribution function (e.g. A.E. Green and Bourne, 1972, p. 554). Thus for the entry in column A, row B in the **P** matrix, which has the value 0.70, the value of P' is 0.20 (= 0.70 – 0.50) and that of the unit normal deviate is 0.53. The **X** matrix is:

	D	A	C	B
A	0.13	–	–0.38	–0.53
B	?	0.53	0.25	–
C	0.53	0.38	–	–0.25
D	–	–0.13	–0.53	?

The query signs occur where the judges were unanimous, namely in preferring B to D.

The column difference matrix Z is then obtained by taking the differences between entries in the adjacent columns of the X matrix:

	D-A	A-C	C-B
A	0.13	0.38	0.15
B	?	0.28	0.25
C	0.15	0.38	0.25
D	0.13	0.40	?
Σ/n	0.14	0.36	0.22

The entries in each column are then averaged as shown in the bottom additional row. These latter entries are the scale values which are used to locate the items on the ranking scale.

The range of the scale in this example, as given by the bottom additional row of the **Z** matrix, is 0.72 (= 0.14 + 0.36 + 0.22). Item D has the lowest preference rating and is at the bottom of the scale and item B is at the top of the scale. The intervals between D and A, A and C, and C and B are, respectively, 0.14, 0.36 and 0.22.

This psychological scale may then be calibrated as a relative probability scale, as described above, and if the necessary absolute probability values are available for some of the items, absolute probabilities for the others may be calculated as described above. In performing this conversion, it is necessary to ensure that the upper limit of the scale is unity and to set some lower limit. For the latter Hunns uses a value of 10^{-6}. Hunns describes an application of his method but does not claim it to be well tested. It is, however, one of the methods referred to in the literature.

9.9.13 Applications

The use of expert judgement is now a well established approach. A critical review of some principal applications has been given by Mosleh, Bier and Apostolakis (1987). Five types of application are reviewed.

The first is the assessment of component failure rates and other reliability parameters. Examples of the estimation of failure rates considered are those in the *Rasmussen Report* (AEC 1975), in IEEE Standard 500: 1984, and in the Seabrook Station PRA (Pickard, Lowe and Garrick, 1983).

The second type is the assessment of seismic hazard rates as exemplified in the studies by the Electrical

Power Research Institute (EPRI, 1986) and Veneziano, Cornell and O'Hara (1984).

The third type is the assessment of nuclear reactor containment phenomenology as instanced by the work of the Steam Explosion Review Group of the NRC (1985 NUREG/CR 1116) and the Surry Station PRA by Benjamin *et al.* (1986).

The fourth type is modelling for the assessment of human error rates (HERs). Examples considered here are: the approach described in the *Handbook of Human Reliability Analysis* (Swain and Guttman, 1983); the human cognitive reliability (HCR) model of Hannaman et al. (1986); the SLIM-MAUD method of estimating HERs of Embrey (1984); and, again, the Seabrook Station PRA.

The fifth type is studies of precursor events, which have been carried out by Minarick and co-workers (Minarick and Kukielka, 1982; Cottrell, Minarick *et al.*, 1984; Minarick *et al.*, 1985).

These applications covered cases with a wide range of characteristics, including the following:

(1) cases involving single, as well as multiple, experts;
(2) cases involving structured group processes as well as unstructured ones;
(3) cases with a wide range of substantive problem areas;
(4) cases with and without availability of relevant empirical data;
(5) cases of varying degrees of complexity;
(6) cases involving varying degrees of mathematical sophistication in the aggregation of responses;

Some of the conclusions of the study are as follows. (1) An improvement in the quality of the results may often be obtained by decomposition of the problem into a set of more elementary problems. (2) Where experts are used to make estimates without the aid of formal methods, they tend to produce under- or over-estimates and to underestimate the degree of uncertainty in these estimates. (3) Two effective techniques for reducing overconfidence are calibration training of the experts and exercises in which the experts are required to identify contrary evidence. (4) A structured group meeting tends to give better performance than an unstructured one. (5) Aggregation of multiple opinions tends to yield a more accurate result than the opinion of a single expert. (6) For aggregation of the opinions of multiple experts, mathematical methods of aggregation are to be preferred to behavioural methods of reaching a consensus.

Decomposition is identified as an effective strategy. The approach suggested by the authors is to enlist the experts in the decomposition process.

The use of multiple experts as opposed to a single expert is well established. But the methods of combining the responses of multiple experts need more development. Combination by an unstructured group does not give accurate results. On the other hand, the highly structured group used in the Delphi method seems cumbersome. The authors suggest as promising the multiple team approach used by EPRI in the seismic hazard work.

9.9.14 Failure rates

As just described, one of the principal applications of expert judgement has been the estimation of failure and event rates. Experience in applications of this sort has been discussed by Apostolakis (1986). He distinguishes three general situations in which judgement is very prominent. These are where: (1) the probability distribution is developed solely from non-statistical knowledge, e.g. human error distributions; (2) the advice of experts is incorporated, e.g. generic failure rates in the *Rasmussen Report* and fragility curves in seismic work; and (3) the statistical evidence if available but subject to different interpretations. The author discusses some of the detailed features of the generic failure rate estimates in the *Rasmussen Report* and IEEE Standard 500: 1984.

Expert judgement may also be used in the utilization of such generic data. A common situation is where plant data are available to supplement the generic data and it is then necessary to aggregate them. This may be done using Bayes' theorem. The generic data constitute the prior information which is adjusted using the plant data to obtain the posterior distribution.

Apostolakis draws attention to the fact that in some cases of failure rate estimation when this Bayesian approach is used, it is found that the posterior distribution lies in the tail region of the prior distribution on the high side, indicating bias in the prior distribution. He suggests that this points to the need to broaden the prior distribution, but acknowledges that this is a matter of debate (e.g. Apostolakis, 1982, 1985a; Martz, 1984).

9.10 Rare Events and External Threats

One of the most difficult problems in event frequency estimation is that of rare events. Such events come under the following headings:

(1) equipment failure;
(2) external events
  (a) natural,
  (b) man-made.

Usually it is necessary to take into account rare events only where the potential consequences may be very serious. The rare event problem is therefore essentially associated with major hazards.

9.10.1 Equipment failure

The failure rate of equipment is usually estimated from historical data. The difficulty in using this approach is that for rare events such data are sparse. This may render it necessary to resort to making an alternative estimate based on engineering principles.

It is not uncommon that the historical record appears to show there to be some cases of the event of interest, but that it is judged that many, or even all, of these do not apply to the particular situation of interest. Discarding of failures which are considered inapplicable reduces the data set further.

In the process, and nuclear, industries the rare event which has received most attention is almost certainly pressure vessel failure. The historical record shows only a handful of failures and some of these have been judged inapplicable for vessels designed to and operated at high standards.

Statistics exist which allow an estimate to be made of the failure rate of a device even if the data are limited to just a few, or even one, failure. The confidence limits, however, are a function of the number of failures, and if there are only one or two of these the bounds are wide. It is also possible to make an estimate of the failure rate, or rather its upper bound, even if there have been no failures, based on the number of failure-free years of the equipment. For a rare but high hazard event, however, it is generally necessary to accumulate a rather large number of years without failure before the failure rate estimate begins to approach an acceptable value, and often the number of equipment-years recorded is insufficient. Similarly, statistics exist which allow estimates to be made of failure probability on demand for a device, both with and without recorded failures. The appropriate statistics are given in Chapter 7.

If a statistical approach is considered inappropriate, it is necessary to resort to an attempt to make an estimate based on engineering principles. For example, for a pressure vessel use is made of estimates based on fracture mechanics and inspection considerations.

A prime example of equipment failure as a rare event is the failure of the nuclear reactor pressure vessel considered at the Sizewell inquiry, which is described in Appendix 26.

9.10.2 Natural hazards

There are various natural hazards which pose a threat to plant. Some of these are listed in Table 9.9. The relative importance of a particular natural hazard varies between different countries and between different locations within a country.

For some natural hazards, particularly those due to weather, there are historical records which for certain sites at least may allow a frequency estimate to be made. Others such as those involving surface instability tend to be one-off events which are not amenable to prediction. A third class are those involving subterranean stress where the probability of the event may increase over time until the stress is relieved and where the prediction techniques are still developing.

Information on recurring natural hazards is generally of two kinds. The hazard may be a discrete event which either occurs or does not occur and in this case the data required are the frequency of occurrence. For such events the usual default assumption is that occurrence is random and that the data fit a Poisson distribution.

The other situation is where the hazard is an event that has associated with it a variable which has a range of values. In this case the data required are the maximum values of the variable. For example, for wave hazard the required data are maximum wave heights. Data of this latter kind may relate either to maxima of events occurring at random or, where a number of such events always occur within a year, to annual maxima.

The statistical treatment of natural hazards is given in *Statistics of Extremes* (Gumbel, 1958). Descriptions are available of the application of such statistics to various natural hazards such as floods, hurricanes, waves and earthquakes. Further treatments are given by Perry (1981) and Schueller (1982).

In considering such natural phenomena the point of interest is to know the probability that a given value x of the variable, say wind speed or wave height, will be equalled or exceeded, – this is the 'exceedance probability'. If the event occurs at random, then for an event rate λ and time period t the number of trials v in which excedance may occur is

Table 9.9 *Some natural hazards which pose a threat to plant*

Subterranean stress:
Earthquakes
Volcanoes
Tsunamis
Surface instability:
Landslides and avalanches
Ground surface collapse
Weather:
Wind, storm
Tornadoes
Hurricanes
Floods
Fires

$$v = \lambda t \qquad [9.10.1]$$

If the event is the annual event for which the variable is the annual maximum, the number of trials v equals the number of years in the time period t.

Considering first a general treatment, define $P(x)$ as the distribution of the probability that excedance does not occur and p as the probability that in a single trial it does occur. Then,

$$p = 1 - P(x) \qquad [9.10.2]$$

Furthermore,

$$q = 1 - p \qquad [9.10.3]$$

$$= P(x) \qquad [9.10.4]$$

where q is the complement of p.

The probability $W(v)$ of exceedance in v trials is

$$W(v) = 1 - q^v \qquad [9.10.5]$$

The mean number of trials $\bar{v}$ at which exceedance occurs is

$$\bar{v} = 1/p \qquad [9.10.6]$$

From Equations 9.10.3, 9.10.5 and 9.10.6:

$$W(v) = 1 - (1 - 1/\bar{v})^v \qquad [9.10.7]$$

$$= 1 - \exp(-v/\bar{v}) \qquad 1/\bar{v} = p << 1 \qquad [9.10.8]$$

A return period $T(x)$ may be defined as the mean time between exceedances, so that

$$T = \bar{v} \qquad [9.10.9]$$

Then, from Equations [9.10.8] and [9.10.9]

$$W(v) = 1 - \exp(-v/T) \qquad 1/T << 1 \qquad [9.10.10]$$

$$\approx v/T \qquad v/T << 1 \qquad [9.10.11]$$

Equation 9.10.10 is independent of the form of the distribution $P(x)$.

Turning now to the assignment of a specific distribution to the event of interest, Gumbel gives a number of distributions, but that most commonly referred to as the 'Gumbel distribution' is as follows. For $P(x)$ he gives

$$P(x) = \exp\{-\exp[-\alpha(x - u)]\} \quad [9.10.12]$$

where u is a location parameter and α is a scale parameter. This is the distribution of the probability that exceedance does not occur. The distribution of the probability that exceedance does occur is

$$F(x) = 1 - \exp\{-\exp[-\alpha(x - u)]\} \quad [9.10.13]$$

It should be noted that Gumbel uses $F(x)$ for the distribution which is here denoted by $P(x)$. Since exceedance is akin to failure, $F(x)$ has been reserved here for the complement of $P(x)$ so that it is akin to the failure distribution functions given in Chapter 7.

Equation 9.10.13 is a form of the extreme value distribution which is described in Chapter 7. This distribution is the appropriate one to use for extreme values of a variable which follows the exponential distribution.

Observed values of the variable of interest may be plotted on special extreme value distribution graph paper, or 'Gumbel paper'. The vertical axis is a linear scale for the values of the variable. The horizontal axis is the probability scale. The observed values are first ordered in ascending order of magnitude. The ranks are then determined from

$$r = m/(n + 1) \quad [9.10.14]$$

where m is the order number, n the number of readings, and r the rank. The rank r is then equivalent to $P(x)$.

As an illustrative example, consider the determination of the distribution parameters for the wind gust data given in Table 9.10, where the data are already ordered and the rank calculated. The data are shown plotted in Figure 9.18. From this graph, at the 0.10 and 0.99 probability points (equivalent to $P(x)$) the values x of the wind speeds are 51 and 95 knots, respectively. Then the parameters u and α are 57.9 and 0.124, respectively.

The return period T of a particular wind speed is obtained as the reciprocal of $F(x)$ $(= 1 - P(x))$. Consider the return period of a wind speed of 95 knots. Since for this wind speed $P(x) = 0.99$, the probability of excedance $F(x)$ is 0.01 and the wind speed has a return period of 100 years. This value is extrapolated from 23 data points in which the maximum observed wind speed is 81 knots and is about the limit of extrapolation.

If Gumbel paper is not available, the horizontal scale may be constructed using the reduced variate $y = -\ln(-\ln P(x))$, as shown in Figure 9.18.

9.10.3 Man-made hazards

Man-made hazards also pose a threat to plant. Some of these are listed in Table 9.11. The estimation of the frequency of events of this type is usually based on the use of historical data together with detailed consideration of the application of the data to the plant in question.

For example, an estimate of the frequency of aircraft crash on a plant would be made by using general data on aircraft crash frequency in combination with information for the particular location about such factors as aircraft flight paths at any nearby airport, deviations from such paths, etc.

Table 9.10 *Annual maximum wind gust velocities at Cardington 1932–54 (after Perry, 1981) (Courtesy of George Allen & Unwin)*

Order No. (m)	*Highest gust* (knots)	*Year*	*Rank* (r)
1	48	1953	0.042
2	51	1950	0.083
3	52	1941	0.125
4	53	1951	0.167
5	54	1952	0.208
6	55	1937	0.250
7	55	1939	0.292
8	56	1942	0.333
9	57	1933	0.375
10	58	1949	0.417
11	59	1948	0.458
12	60	1945	0.500
13	62	1940	0.542
14	63	1934	0.583
15	63	1944	0.625
16	66	1954	0.667
17	68	1943	0.708
18	68	1946	0.750
19	71	1932	0.792
20	72	1936	0.833
21	75	1938	0.875
22	77	1935	0.917
23	81	1947	0.958

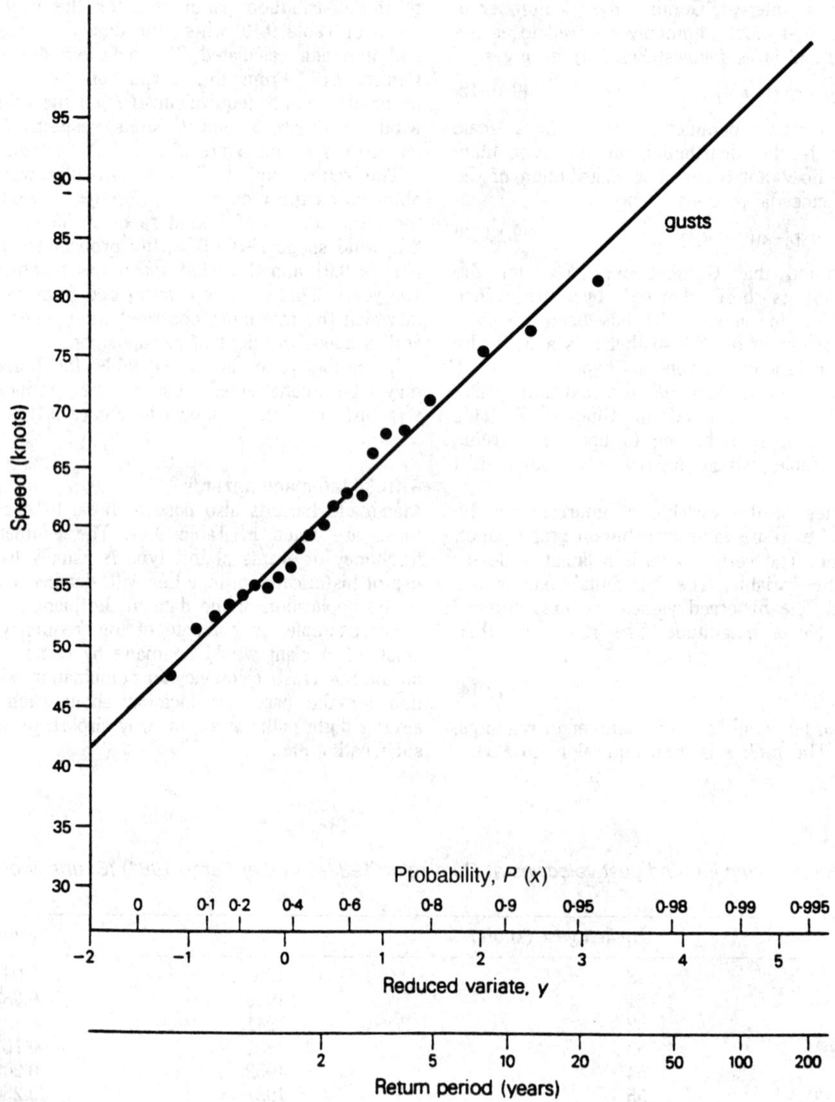

Figure 9.18 *Annual maximum gust velocities given in Table 9.10 plotted on extreme probability (Gumbel) paper (after Perry, 1981)*

Table 9.11 *Some man-made hazards which pose a threat to plant*

Dam bursts
Vehicle crashes
Aircraft crashes
Fires
Explosions

9.11 Human Factors and Human Error

Human error and, more generally, human factors is a wide and complex topic and the treatment here is limited to the bare essentials necessary for a balanced treatment of hazard assessment. A more detailed account is given in Chapter 14. The account given there includes details of available techniques. The discussion in this section tries to set these techniques in the context of hazard assessment.

9.11.1 Human error in hazard assessment

Analyses of failures in technological systems generally conclude that in the vast majority of cases human error has played a predominant role. Insofar as any failure can be attributed to some form of human failing, human error in its broadest sense becomes, by definition, the basic cause of failures. An exception to this can occur in the case where a risk has been assessed and, after evaluation, accepted, but where the hazard nevertheless materializes, as statistically it must in a proportion of instances.

It is well appreciated by hazard analysts that it is necessary to take human error into account in hazard assessment work. The problem has been how to achieve this. As with other aspects of hazard assessment, there are two main motivations for investigating human error. One is to reduce it by identifying defects in the human factors environment and effecting improvements. The other is to obtain the data required to be able to quantify the human aspects of the assessment.

The implications of these two motivations for the hazard assessment as a whole are different. The purpose of hazard assessment is to assist decision-making. There is a danger that the second approach will furnish data for quantification, but that it may be relatively ineffective in indicating improvements in the human factors area.

It is fair to say that interest in human error in the process industries has been driven by the perceived need to address human error in hazard assessment, and in particular to obtain the human error data required for quantification.

9.11.2 Human error in operation

Historically, in hazard assessment most attention has been focused on human error in process operation. In the operation of a plant human error may contribute to an incident either as an initiating or enabling cause or as a failure to achieve some form of prevention or mitigation.

Much of the early work on human error was concerned with estimating error rates in relatively well-defined activities such as detecting a signal, pressing a button or opening a valve. Work then moved on to human error in executing a whole task using a relatively well-defined plan. Both activities and plans may be relevant to either causation or prevention of an incident.

One method was to break a task down into its constituent elements, to collect data on the probability of error in the performance of these elements and to store these data in a human error data bank. This approach, however, tends to do less than justice to the operator's information processing and decision-making which are generally critical to the performance of the task. An alternative method, therefore, is to treat the task as a whole and to collect data on the probability of error in the performance of the task.

Both approaches make allowance for the influence of the work situation in which the task is performed by identifying performance shaping factors. The application of such factors requires the development for each factor of a measure of the strength of the factor and a relationship between the factor and the probability of error.

The most widely quoted method is the THERP technique developed by Swain and Guttman (1983). This technique makes widespread use of estimates of the probability of error in individual actions. A method which gives error estimates for whole tasks is the SLIM technique described by Embrey (1983a,b). These and other methods are described Chapter 14.

Human error often occurs in the development of a branch of a fault tree. The typical application of the methods just described is to obtain an estimate of the probability of failure, or error, for use at this point.

Another application of these methods is in the development of the branches of an event tree. Again the methods described have been used to estimate the probability of error at the branch points of an event tree. In particular, interest has centred on the prevention by the operator of escalation of the incident. The operator action event tree (OAET) is a particular form of event tree in which the branch points are defined in terms of failure of operator action.

As so far described, the assessment of operator error occurs within essentially predefined structures such as fault trees and event trees. In some cases the situation is more complex. One such case is where the operator makes some fundamental error in assessing the situation and performs a series of actions, which may be quite complex, and which cause an incident. This is probably the most difficult case to handle and only limited progress has been made. One area where there has been some progress is in respect of communication errors, as described below.

Another case is where the operator is, in principle, able to prevent an incident but only by making a correct assessment of the situation and performing a series of actions, which may be quite complex and may involve initiative. This situation is sometimes dealt with by making a global estimate of the probability of no effective action. The probability is usually taken as a function which decreases quite strongly with time.

9.11.3 Human error in maintenance

Human error in maintenance has received rather less attention, although there is growing interest in it. Some maintenance errors result in an incident during plant operation. Insofar as the operations function has overall responsibility for safe operation, these may generally be regarded as operational failures. Other maintenance errors lead to an incident while the plant is shut down. Maintenance errors are also an influencing factor on the failure rate of plant equipment.

9.11.4 Human error in communication

Analysis of incidents shows that an important contribution comes from errors of communication. It is fair to say that so far this work has contributed more to defining good practice in communications that to quantifying this aspect of human error.

9.12 Management Aspects

The prime determinant of the control over a hazard is the quality of management. The influence of management is all-pervasive. In other words it is a powerful cause of failure and has many of the features of a common cause. It is clearly desirable, therefore, to take the quality of management into account in any hazard assessment.

The first point to be made is that if the quality of management falls below a certain level, then there is generally little point in doing a hazard assessment at all. Hazard assessment should be able to start from the assumption that management meets at least some minimum standard of quality, particularly for major hazard installations. If this is not so, the effort would be better employed in remedying the defects than in estimating the probable results of those defects.

A second important point is that the use of management quality as an explicit input in a hazard assessment is a sensitive issue. This is so whether the assessment is being conducted by analysts within the organization concerned or by an outside body.

In general, the realization of a hazard involves the occurrence of an initial event followed by escalation of that event. Management influences both the frequency of the initial event and the effectiveness of measures taken to prevent its escalation.

9.12.1 Quantification of management influence

The attempt to make allowance for the effect of management quality involves both devising a measure of quality and a model for the impact of that quality on event frequency and escalation. Discussions of allowance for management quality in hazard assessment has been given by R.A. Cox and Comer (1982), Pitblado, Williams and Slater (1990) and J.C. Williams and Hurst (1992).

Some limited progress has been made in this area. Aspects of the quality of management are considered in Chapter 6. Indices of quality may be devised for this specific purpose or, alternatively, use may be made of other indices such as the International Safety Rating System.

The most straightforward application of such indices is to failure rates. Generic equipment failure rates may be regarded as being derived from organizations with 'average' quality management. These base failure rates may be adjusted according to the assessment made of management quality.

9.12.2 MANAGER model

The MANAGER model developed by DnV Technica and described by Pitblado, Williams and Slater (1990) exemplifies this approach.

A review was conducted of systems which might be adapted to provide modifiers for generic failure data to take into account management factors. Systems reviewed included the International Safety Rating System (ISRS) (Bird and Germain, 1985), the Management Oversight and Risk Tree (MORT) (W.G. Johnson, 1980) and the Technique for Human Error Rate Prediction (THERP) (Swain and Gutmann, 1983). The first two are described in Chapter 28 and the last in Chapter 14. It was concluded that ISRS would be relatively slow and costly to use and that it would not fit well with QRA, and that the strength of MORT is in accident investigation and that of THERP is in assessing the human error contribution in fault tree analysis. A requirement was defined for a modification of risk (MOR) method which would be simpler to use than human reliability assessment.

The properties which such a method should have were identified as the following. It should: (1) be based on a review of the role of safety management in actual accident causation; (2) address all areas shown to be important in accident causation, including human factors; (3) confirm that widely accepted management principles are suitably embedded in all key elements of the safety management system; and (4) provide both a qualitative overview of safety management and an indication of quantitative modification to generic failure rates.

Table 9.12 Twelve topic areas of MANAGER questionnaire (after Pitblado, Williams and Slater, 1990)

1. Written procedures
2. Incident and accident reporting
3. Safety policy
4. Formal safety studies
5. Organization factors
6. Maintenance
7. Emergency resources and procedures
8. Training
9. Management of change
10. Control room instrumentation and alarms
11. Other human factors influences
12. Fire protection systems

MANAGER is based on a questionnaire which utilizes the concept of performance shaping factors, which are described in Chapter 14. The four most significant for this purpose were identified as (1) resources (manning, instrumentation), (2) system norms (incident reporting, safety policy, training), (3) communications (information flow, written documentation) and (4) pressures (stress, boredom). The questions are structured to conform with loss prevention principles as well as with these human factors influences. The questionnaire is structured into 12 broad topics, as shown in Table 9.12.

The responses to the questionnaire are scored as the proportion P which are rated as industry average (P_A), better than average (P_G) or worse than average (P_B). Use is made of a triangular diagram with these three proportions P_A, P_G and P_B at its corners. Values of a management factor (MF) are located on the triangular diagram. A completely average plant ($P_A = 1$) has a value of MF = 1. This same value of the MF is also obtained by a plant with only good and bad responses ($P_A = 0$) at $P_G = 0.74$ and $P_B = 0.26$. A completely good plant has MF = 0.1 and for a completely bad plant MF = 100. In some 30 plant audits values of MF ranged from 0.5 to 8.0.

The MF may then be used in QRA as a modifier for generic failure rates:

$$F_{est} = \text{MF} \times F_{gen} \qquad [9.12.1]$$

where F_{est} is the modified failure frequency and F_{gen} is the generic failure frequency.

A case study of the application of MANAGER has been described by J.C. Williams and Hurst (1992). A comparative investigation was made of two technically similar major hazard sites. The comparison was between the number of incidents reported under the RIDDOR regulations and the computed MF. The numbers of RIDDOR incidents per 1000 employees over the periods April 1986–March 1987, April 1987–March 1988 and January–March 1991 were 8.7, 7.3 and 5.9 for company

A and 12.5, 10.3 and 8.0 for company B. The computed MFs were 0.9 for company A and 1.7 for company B. Thus at site B the occupational incident rate was almost twice that at site A, and likewise the risk at site B as given by the MF was nearly twice that at site A. QRA showed that an individual risk of 10^{-5}/year occurred at site B at a distance of 400 m but at site A at 290 m. Further studies would be required to confirm the generality of the correlation.

As the study just described indicates, MANAGER may also be used for audit purposes.

9.12.3 STATAS model

Another model which is relevant here is STATAS. This has been developed for the purposes of audit. However, just as MANAGER may also be applied to audit, so STATAS has a potential use in making allowance for management factors in QRA. It is described in Chapter 6.

9.13 Hazard Models

The estimation of the physical effects of a particular release scenario involves the use of a series of hazard, or consequence, models. Some of the models required are shown in Table 9.13. Most of the entries are straightforward, but the references to air entrainment and source term require some further explanation.

The models for emission allow an estimate to be made of the amount of material which is released or the rate at which it is released, but they do not in themselves necessarily give any information on the extent to which the emission entrains and mixes with the air. Dispersion models can be sensitive to the amount of air which mixes with the initial emission. It is necessary, therefore, to make some estimate of this.

The emission constitutes the source term for the dispersion models. A real emission does not usually correspond exactly to idealized source terms such as pure instantaneous or continuous sources, and a source term has to be selected which represents the situation as closely as possible but is suitable as input to the dispersion model.

Consequence models are described mainly in Chapters 15–17.

9.14 Domino Effects

Another aspect which needs to be taken into account in hazard assessment is the possibility of knock-on or 'domino' effects, leading to escalation. An event at one unit may be the cause of a further event at another unit, and so on.

The possibility of domino effects was one of the principal concerns which led to the hazard assessment of the complex of installations at Canvey, which was the subject of the two *Canvey Reports* described in Appendix 7.

The incident at Feyzin in 1968 (Case History A38) in which a fire on one storage sphere led to a series of fires and explosions involving other spheres illustrates the danger. Other subsequent incidents in which a domino effect has played an important role include Mexico City in 1985, Pasadena in 1988 and Piper Alpha in 1988. These incidents are described in Appendices 4, 6 and 19, respectively.

Table 9.13 *Some principal hazard models*

Emission
Holes:
  Gas flow
  Liquid flow
Two-phase flow
Vessel rupture
Pipeline rupture
Vents
Relief valves

Vaporization
Flashing liquids
On land:
  Spreading liquid
  Volatile liquid
  Cryogenic liquid
On sea:
  Spreading of immiscible liquid
  Mixing of miscible liquid
  Volatile liquid
  Cryogenic liquid

Air entrainment
(for each emission situation)

Source term

Gas cloud dispersion
Neutral density clouds
Heavy gas clouds
Buoyant gas clouds

Plumes
Neutral plumes
Heavy gas plumes
Buoyant plumes

Jets
Momentum jets

Fires
Pool fires
Fireballs
Flash fires
Jet flames
Engulfing fires

Explosions
Physical explosions of plant
Combustion explosions inside plant
Vapour cloud explosions
Explosions in buildings
Boiling liquid expanding vapour explosions

A method of handling potential domino effects in hazard assessment has been described by Bagster and Pitblado (1991). Table 9.14 shows a matrix of interactions between the process areas of two different operators obtained from the *Canvey Report* and quoted by these authors. A domino event may be treated either as an

Table 9.14 *Matrix of potential interactions of process area explosions for two installations at Canvey: frequency of massive release of LPG per 10^6 years (Bagster and Pitblado, 1991) (Courtesy of the Institution of Chemical Engineers)*

	Fire spreading to LPG storage at	
Explosion at	*Occidental*	*URL*
Occidental process area	0.12	0.075
URL process area	0.09	0.065

LPG, liquefied petroleum gas; URL, United Refineries Ltd

increase in the consequences of the event at the initiating unit, or as an increase in the frequency of events at the victim unit. The method described in this work adopts the latter approach.

Five events are included in the treatment: pool fire, explosion, boiling liquid expanding vapour explosion (BLEVE) giving rise to a fragment, jet fire and delayed explosion of a vapour cloud. The escalation is described in terms of a loss of containment at the victim unit. The following relation is used to describe loss of containment at unit j due to an event at the initiating unit i:

$$p_{\text{loc};j;i} = 1 - \left[1 - \frac{r}{r_{\text{lim};i}}\right]^2 \qquad [9.14.1]$$

where $p_{\text{loc};j;i}$ is the probability of loss of containment at unit j due to an event at unit i, r is the distance of unit j from unit i, and $r_{\text{lim};i}$ is the maximum distance at which the event at unit i can cause damage.

The basic relation for the frequency of loss of containment at the unit j due to an event at unit i is

$$F_{\text{loc};j;i} = \text{PI}_i\left(F_{\text{cat};i}\sum_{m=1}^{3} p_{\text{loc};j;i;m}P_{m;i}M_{m;i} + F_{\text{leak};i}\sum_{m=4}^{5} p_{\text{loc};j;i;m}P_{m;i}M_{m;i}\right) \qquad [9.14.2]$$

with

$$M_{1;i} = M_{2;i} = 1 \qquad [9.14.3]$$

where $F_{\text{cat};i}$ is the frequency of a catastrophic failure at unit i, $F_{\text{leak};i}$ is the frequency of a leak failure at unit i, $F_{\text{loc};j;i}$ is the frequency of loss of containment at unit j due to an event at unit i, $M_{m;i}$ is the factor for mitigation of the event at unit i due to its directionality, $p_{\text{loc};j;i;m}$ is the probability of loss of containment at unit j due to an event at unit i of type m, $P_{m;i}$ is the probability that the event at unit i is an event of type m and PI_i is the probability of ignition following failure at unit i. The five events are pool fire (PF; $m = 1$), explosion (EX; $m = 2$), BLEVE fragment (BF; $m = 3$), jet fire (JF; $m = 4$) and delayed explosion (DEX; $m = 5$). The same probability of ignition is used for both types of failure and for all types of event.

In addition to this escalation of the primary event at unit i into a secondary event at unit j, the method also takes into account escalation of the latter into a tertiary event at unit k. In order to do this, it is necessary to make some assumption about the probability distribution of the failures induced at unit j. The assumption made is that the distribution between catastrophic failures and leak failures at unit j is the same as that at unit i. The relation for the probability of loss of containment at unit k due to an event at unit j, itself due to an event at unit i, is then

$$P_{\text{loc};k;j;i} = PI_{j\text{loc};j;i}\left[\phi_j\sum_{m=1}^{3} p_{\text{loc};k;j;m}P_{m;j}M_{m;j} + (1-\phi_j)\sum_{m=4}^{5} p_{\text{loc};k;\,j;m}P_{m;j}M_{m;j}\right] \qquad [9.14.4]$$

with

$$\phi_j = \frac{F_{\text{cat},j}}{F_{\text{cat},j} + F_{\text{leak},j}} \qquad [9.14.5]$$

$$M_{1;j} = M_{2;j} = 1 \qquad [9.14.6]$$

where $M_{m;j}$ is the factor for mitigation of the event at unit j due to its directionality, $P_{\text{loc};k;j;i}$ is the probability of loss of containment at unit k due to an event at unit j, itself due to an event at unit i, $p_{\text{loc};k;j;m}$ is the probability of loss of containment at unit k due to an event at unit j of type m, $P_{m;j}$ is the probability that the event at unit j is of type m, PI_j is the probability of ignition following failure at unit j, and ϕ_j is the probability that the failure at unit j is catastrophic.

It is also necessary to take into account the fact that the damage zone of the event at unit j may be partially or totally within the damage zone of the event at unit i, or vice versa. This is taken into account by utilizing for each type of event a correction factor S_m. This correction factor is applied to the frequency of loss of containment at unit j for an event at unit i of type m. In effect, this is a correction to allow for overlap of the damage zones and thus to avoid double counting of the damage.

The three situations which are possible if overlap occurs are shown in Figure 9.19. An overlap factor q_m is defined which has the values

$q_m = 1$ damage zone of event i lies entirely within that of event j

$q_m = \frac{1}{2}$ damage zones of events i and j partially overlap

$q_m = 0$ damage zone of event j lies entirely within that of event i

The overlap correction factor S_m is then

$$S_m = \frac{\sum_{m=1}^{5} P_m M_m q_m}{\sum_{m=1}^{5} P_m M_m} \qquad [9.14.7]$$

where M_m is the probability of mitigation of an event at unit j of type m, P_m is the probability that the event at unit j is an event of type m, and q_m is the overlap factor for an event at unit j of type m.

The authors give the following illustrative example. They investigate the domino effects for two alternative layouts of a plant consisting of four items – an

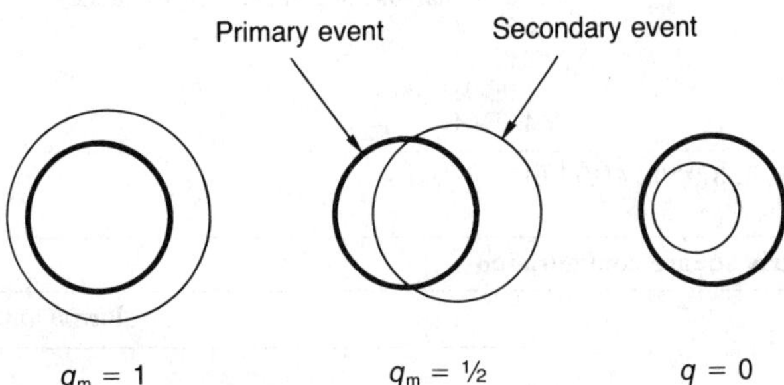

Figure 9.19 *Relationship between primary and secondary events in a domino incident (Bagster and Pitblado, 1991) (Courtesy of the Institution of Chemical Engineers)*

Table 9.15 *Illustrative example of potential interactions for four plant items on a single site (after Bagster and Pitblado, 1991) (Courtesy of the Institution of Chemical Engineers)*

A Data

Plant item number	1	2	3	4
Plant type	Atmospheric storage tank	Sphere	Sphere	Compressor
Toxic?	No	No	No	No
F_{CAT} (per million years)	1.0	1.0	1.0	50
F_{LEAK} (per million years)	5.0	1.0	1.0	100
Probability ignition, *PI*	0.1	0.2	0.2	0.1
Probability of pool fire, P_{PF}	0.3	0.25	0.25	0.1
Probability of explosion, P_{EX}	0.2	0.25	0.25	0.1
Probability of BLEVE, fragments, P_{BF}	0	0.1	0.1	0
Probability of jet fire, P_{JF}	0.05	0.2	0.2	0.1
Probability of delayed explosion, P_{DEX}	0.1	0.1	0.1	0.05
Mitigation factor for BLEVE fragments, M_{MBF}	0	0.005	0.005	0.05
Mitigation factor for jet, M_{JF}	0.01	0.01	0.01	0.05
Mitigation factor for drift of explosion, M_{DEX}	0.005	0.005	0.005	0.05
Pool fire range, r_{pf} (m)	25	20	15	25
Explosion range, r_{EX} (m)	30	40	25	30
Fragment range, r_{BF} (m)	0	100	100	0
Jet fire range, r_{JF} (m)	10	20	25	20
Delayed explosion range, r_{DEX} (m)	50	50	35	50

B Loss of containment frequencies: square configuration

Primary events

Item	Primary loss of containment frequency
1	6
2	2
3	2
4	150

First order knock-ons

Item	Additional loss of containment frequency
1	0.89
2	0.48
3	0.36
4	0.11

Second order knock-ons

Item	Additional loss of containment frequency
1	2.8E-02[a]
2	7.8E-03
3	8.1E-03
4	2.2E-02

Third order knock-ons

Item	Additional loss of containment frequency
1	6.3E-04
2	2.6E-04
3	1.9E-04
4	4.7E-04

(Coordinates (m): 1(0,0), 2(10,0), 3(0,10), 4(10,10))
[a] $E\text{-}n = 10^{-n}$

C First-order domino matrix square configuration

	Victim unit			
Causing unit	*1*	*2*	*3*	*4*
1	6	0.021	0.015	0.026
2	0.061	2	0.019	0.038
3	0.070	0.024	2	0.051
4	0.762	0.432	0.326	150
First order domino frequency (column sum minus major diagonal entry)	0.893[a]	0.477	0.360	0.115

Coordinates of units (m): 1 (0,0), 2 (10,0), 3 (0,10), 4 (10,10)
[a] 0.061 + 0.070 + 0.762.

atmospheric storage tank, two storage spheres and a compressor. In one configuration the items are laid out in a square and in the other they are in a line. Details of the units are given in Table 9.15, Section A. The frequencies of loss of containment for three levels of escalation in the square layout case are shown in Table 9.15, Section B, and the corresponding first order domino matrix in Section C. The results given in the domino matrix show, for example, that for unit 1 the base failure frequency of 6×10^{-6}/year is augmented by 0.893×10^{-6}/year due to domino effects.

9.15 Hazard Model Systems

There are a number of hazard model collections and complete hazard assessment systems. These include:

(1) the vulnerability model;
(2) the 'Yellow Book';
(3) plant layout model;
(4) SAFETI system;
(5) WHAZAN system.

These are now described in turn.

9.15.1 Vulnerability model system

One major hazard assessment system is the vulnerability model (VM), later renamed the population vulnerability model (PVM), developed under contract for the US Coast Guard to investigate the possible effects of hazardous material spills. The basic VM is described by Eisenberg, Lynch and Breeding (1975) in a report entitled *Vulnerability Model: A Simulation System for Assessing Damage Resulting from Marine Spills.* Specific models developed in support of the VM are given by Raj and Kalelkar (1974). The second stage of the development of the VM is described by Rausch, Eisenberg and Lynch (1977) and the third stage by Rausch, Tsao and Rowley (1977). Modifications to the VM are described by Tsao and Perry (1979) and by Perry and Articola (1980) and a users' guide is given by Rowley and Rausch (1977). The VM and its applications have been reviewed by Parnarouskis, Perry and Articola (1980).

The VM is a consequence assessment system. It consists of a set of hazard models and injury/damage relations which are used in conjunction with meteorological and population data for a given site to assess the effects of specified releases of hazardous materials. The overall structure of the VM is shown in Figure 9.20. Within this structure there are a number of decision trees for the selection of the scenarios and the corresponding hazard models. The decision tree for the behaviour of the cargo on release is shown in Figure 9.21. The information required by and obtained from the model is shown in Table 9.16. Figure 9.22 illustrates the simulation of a vapour cloud from a marine spill.

Some of the principal elements in the VM are shown in Table 9.17. As the table indicates, the VM is a prime source both of hazard models and of injury relations for both flammable and toxic releases. Many of these are considered in later chapters.

The injury relations in the VM, given in the form of probit equations, have proved of particular interest, owing to the paucity of information on this aspect. The VM studies did pioneering work in defining the form of the causative factor and in deriving constants for the probit equations. Inevitably some of these require modification in the light of more recent work. Table 9.18 lists some of the original probit equations in the model. In the later work revisions and additions were made to the probits for toxic gases, as described in Chapter 18.

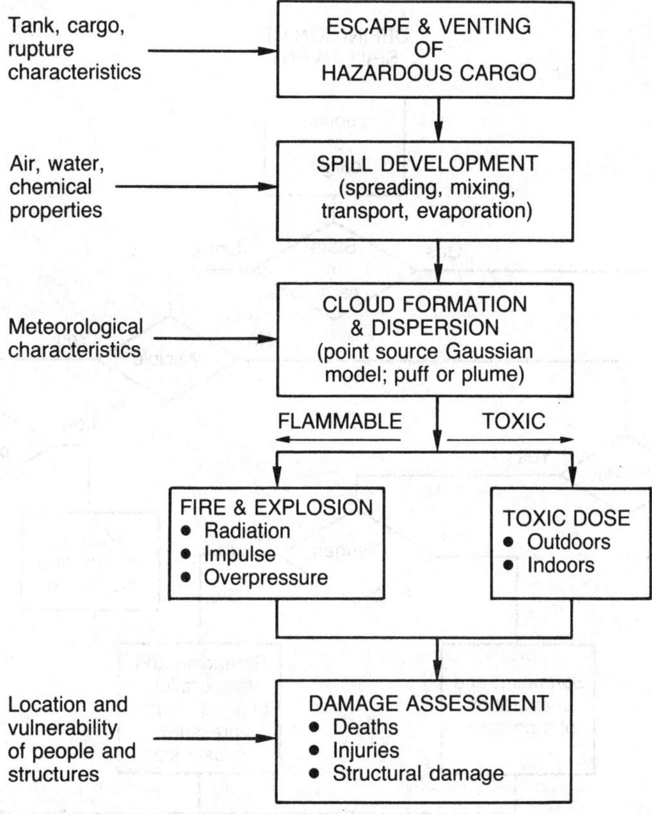

Figure 9.20 *Vulnerability model: structure of the model (Parnarouskis, Perry and Articola, 1980)*

The principal application of the VM is consequence assessment for planning purposes. It is used to estimate for a particular site the effects of hazardous releases, to determine whether particular releases would cause casualties, and to give a hazard ranking. The VM is also used to compare the hazards of specified releases at different sites. For example, comparisons have been made of the relative hazards of liquefied natural gas (LNG) terminals at Point Conception and Oxnard in California.

9.15.2 'Yellow Book' models

A major collection of hazard models is given in *Methods for the Calculation of the Physical Effects Resulting from Release of Hazardous Material* by the Committee for the Prevention of Disasters (CPD, 1992a) in the Netherlands. The report is published by the Dutch Labour Inspectorate and the work was done by Toegepast-Natuurwetenschappelijk Onderzoek (TNO). It is usually referred to as the 'Yellow Book'. The principal contents of the Yellow Book are shown in Table 9.19.

9.15.3 Plant layout models

A set of hazard models for use in plant layout has been given in *Process Plant Layout* (Mecklenburgh, 1985). These models are listed in Table 9.20.

9.15.4 SAFETI computer code

Another major collection of hazard models is that in the SAFETI computer code for risk assessment developed by Technica. The hazard models used, as described by Pitblado and Nalpanis (1989), are as follows. For gas dispersion, four regimes are recognized. These are (1) turbulent jet dispersion, (2) hydrid dispersion with both turbulent jet and dense gas behaviour, (3) dense gas cloud, and (4) passive dispersion. The dense gas dispersion model used is that of R.A. Cox and Carpenter (1980).

The other models cover: (1) vaporization from a pool using the model of O.G. Sutton (1953) or, for liquids with boiling points below ambient, that of Drake and Reid (1975); (2) vapour cloud explosion using the TNO correlation model by Opschoor (1979); (3) fireball, or BLEVE, based on the model of Moorhouse and Pritchard

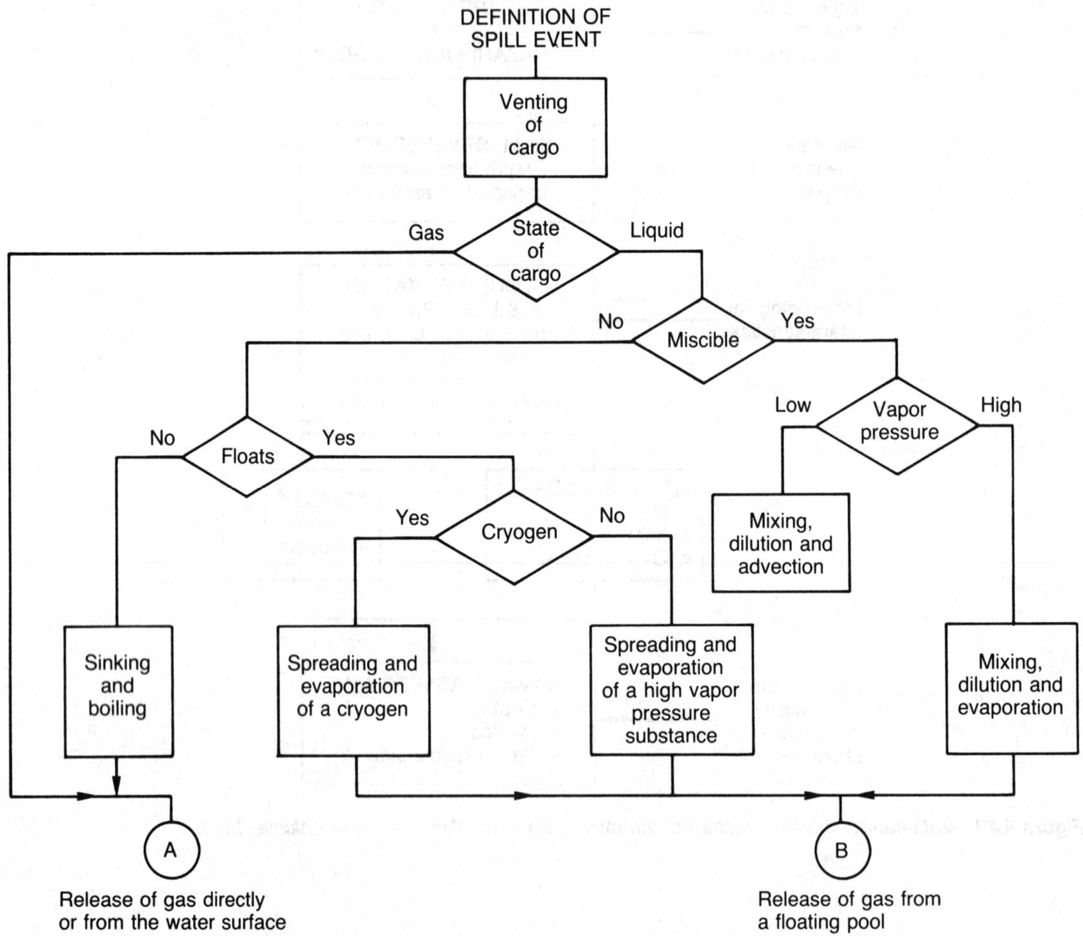

Figure 9.21 *Vulnerability model: decision tree for behaviour of cargo on release (Eisenberg, Lynch and Breeding, 1975)*

(1982); (4) pool fire, using an in-house model; and (5) jet flame, using the TNO free jet model to obtain the jet length combined with the American Petroleum Institute (API) model for thermal radiation. There is no flash fire model as such; it is assumed that the fire is co-terminous with the cloud and in determining injury a small additional increment is added to this contour to allow for the thermal radiation close to the cloud. Probit equations are used for injury and damage effects.

9.15.5 WHAZAN computer code

A set of hazard models is the core of another computer code WHAZAN developed by Technica for the World Bank. The package is described in *Manual of Industrial Hazard Assessment Techniques* (World Bank, 1985) This system allows the user to explore interactively the consequences of a set of release scenarios. The frequency of occurrence of the scenarios is not estimated and the hazard models are relatively simple. The package is not intended for probabilistic risk assessment.

The models used are given in Table 9.21 and a typical event tree which the user can explore using WHAZAN is illustrated in Figure 9.23. These relate to the initial version of the program. WHAZAN is therefore a much simpler package than SAFETI. It is described further in Chapter 29.

9.16 Population Characteristics

Information on the characteristics of the exposed population is frequently required in a hazard assessment. The people exposed consist of two groups, those working on the site and those living and/or working in the area around the site.

Table 9.16 *Vulnerability model: outline of information required by and obtained from the model (after Perry and Articola, 1980)*

A Information required
Characteristics of the cargo (e.g. chemical composition, size of tank, temperature of cargo)
Size and location of the rupture
Characteristics of the spill environment (e.g. marine characteristics and weather conditions)
Geographical location of the spill
Location and characteristics of the vulnerable resource (people and property) in the vicinity of the spill
B Information obtained
Size and characteristics of the spill
Disposition of the hazardous materials (e.g. mixing, sinking, dilution, vaporization, diffusion, dispersion)
Concentrations and hazardous effects of spilled material as a function of position and time (e.g. toxic concentration and dose, thermal intensity and dose, overpressure)
Number of people killed and injured and amount and value of property damage

Table 9.17 *Vulnerability model (VM): some principal elements of the model*

A Reports on VM	
Eisenberg, Lynch and Breeding (1975)	ELB
Raj and Kalelkar (1974)	RK
Rausch, Eisenberg and Lynch (1977)	REL
Rausch, Tsao and Rowley (1977)	RTR
Tsao and Perry (1979)	TP
Perry and Articola (1980)	PA
Rowley and Rausch (1977)	RR

B Elements of VM	*Reference/ page number*
Overview	ELB
Computer programs	ELB; REL, 211; TP, 30
Emission and vaporization	
Overview	ELB, xi, xii
Spreading of liquid on land	
Spreading of liquid on water or other liquids	ELB, 20; RK, 19, 85, 113, 139, 228, 232, 234, 235
Mixing and dilution of soluble liquid in water	ELB, 21; RK, 29, 115, 228, 236
Boiling and sinking of heavy liquid in water	ELB, 22; RK, 171, 237
Gas/vapour dispersion	
Overview	ELB, xi, 23
Instantaneous release (puff)	ELB, 33; RTR, 16
Continuous release (plume)	ELB, 33; RTR, 15
Model selection	ELB, 195
Dispersion coefficients	ELB, 42
High concentration sources	ELB, 185
Area sources	RK, 55
Plume width	RK, 57
Meandering plume	ELB, 42
Fire	
Overview	ELB, xi, xiv, xviii, 92
Ignition, ignition sources	ELB, xiii, 49, 207; REL, 35
Flammable cloud size	ELB, 211
Flash fire	ELB, 59; REL, 31; RTR, 13; TP, 17
Pool fire	ELB, xiv, 66
Flame dimensions	ELB, xiv; RK, 63, 65, 230, 231; RTR, 12
Thermal radiation	ELB, xiv
View factor	ELB, xiv
Burning time	ELB, xiv
Fireball	RTR, 5
Jet flame	RK, 64, 231
Thermal radiation	
View factor	
Structure ignition	REL, 49
Secondary fires	REL, 5
Toxic combustion products	REL, 91, 114
Explosion	
Overview	ELB, xi
Explosion energy	ELB, 221
Explosion	ELB, xiii, xviii, 49, 90, 229; RTR, 18
Toxic release	
Overview	ELB, xi, xii, xviii, 49
Infiltration	REL, 137
Dosage: outdoor dosage	TP, 5
indoor dosage	TP, 6
Injury and damage relations	
Injury outdoors	
Fire	ELB, xi; TP, 12
Explosion	ELB, xi
Toxic gas	TP, 6
Injury indoors	REL, 137
Fire	REL, 202
Explosion	REL, 163
Toxic gas	REL, 137
Damage	ELB, 252
Structure ignition	ELB, 251
Fire	ELB, xi
Explosion	ELB, xi, 247;

Table 9.17 Continued

	Reference/ page number
	RTR, 23
Confined explosion	ELB, 226
Gas toxicity	
Overview	ELB, xvii, 83; REL, 73
Acrolein	REL, 76, 88; PA, C-2, C-3
Acrylonitrile	PA, C-2, C-3
Ammonia	ELB, 257; PA, C-2, C-4
Carbon monoxide	REL, 114
Carbon tetrachloride	REL, 81, 88; PA, C-2, C-5
Chlorine	ELB, xviii, 83, 257; PA, C-2, C-5
Hydrogen chloride	REL, 83, 88; PA, C-2, C-6
Hydrogen cyanide	PA, C-2, C-6
Hydrogen fluoride	RTR, 33; PA, C-2, C-7
Hydrogen sulphide	PA, C-2, C-7
Methane, propane	REL, 78
Methyl bromide	REL, 84, 89; PA, C-2, C-9
Phosgene	REL, 86, 89; PA, C-2, C-9
Propylene oxide	PA, C-2, C-9
Sulphur dioxide	PA, C-2, C-10
Toluene	PA, C-2, C-11
Hazard assessment	
Overview	
Double counting	ELB, 325
Case histories	ELB, 137, 269

It is usually a straightforward matter to obtain the necessary information on the on-site workforce, but this is not so for the off-site population. Here the data required include

(1) population density;
(2) population composition;
(3) population changes by time of day;
(4) vulnerable population;
(5) population outdoors.

For estimation of societal risk information is required on all these items, but even for estimation of individual risk only it is still necessary to have data on the vulnerable population and the population outdoors.

The accuracy of estimation of the density and other characteristics of the population should broadly match that of the other stages of the hazard assessment. A very refined estimate of the population characteristics will often not be justified. On the other hand it should be borne in mind that some of the factors which characterize the population are not independent. Thus the number of people at home during the day is less than at night, but the proportion of vulnerable people is greater.

A method for the determination of the population characteristics for Britain has been given by Petts, Withers and Lees (1987). The method is intended primarily for use in estimating the societal risk of fatalities for sets of scenarios where typically in a large proportion of cases the fatal effects may not extend sufficiently far for more approximate methods of estimating population density to give sufficient accuracy.

9.16.1 Population density

Estimates of off-site population density have been made in the two *Canvey Reports* and in the *Rijnmond Report*.

The *Canvey Reports* are mainly concerned with the off-site population, but the *Rijnmond Report* gives estimates of the on-site population density. For the on-site population, data on the number of workers at each site were obtained and the numbers present during the working day N_d and at other times N_n were determined. Then assuming three-shift, 7-day working, the number of man-shifts worked per week is $5N_d + 16N_n$, and assuming that each employee has n_h weeks off for holidays and sickness, etc., and therefore works $(52 - n_h)$ weeks per year, the average number N_s of employees on site is

$$N_s = ((5N_d + 16N_n) \times 52)/(5 \times (52 - n_h)) \qquad (9.16.1)$$

The average number on site in the daytime for the six sites in this study was found to be 200 persons/km^2. The site area used in this computation appears from the map of the area to be filled with process plant and storage and does not contain much open space.

For the off-site population in the First *Canvey Report* the general approach is to define the built-up areas on the map and to use for these areas a uniform population density of 4000 person/km^2. The population profile was obtained in some cases up to 32 km from the hazard source. In some of the studies in the report, however, there are variations. In Appendix 3 use is made of population densities of 5000 and 100 persons/km^2 for urban and rural areas, respectively. In Appendix 14 use is made of census data on the number of people in 100 m squares. In the Second *Canvey Report* it is this latter method which is used.

The *Rijnmond Report* also makes use of census data on the number of people in grid squares, in this case 500 m squares. This grid covered an area of some 75 km^2. Uniform population densities were assumed outside the grid.

The method for the estimation of the population density for British conditions given by Petts, Withers and Lees is as follows. Use is made of national census population data. In Britain there is a full census every decade, the then most recent being in 1971 and 1981. The 1981 Census (OPCS, 1981a) is available as is the official user guide (OPCS, 1981b) and a further guide (Rhind, 1983). Other useful documents are the 1981 Labour Force Survey (OPCS, 1981c) and the Annual Abstract of Statistics (CSO, 1985). The basic small geographical unit for the census is the enumeration district (ED), which is an area of land defined in terms of the number of households representing a suitable census workload. The average population of an ED in

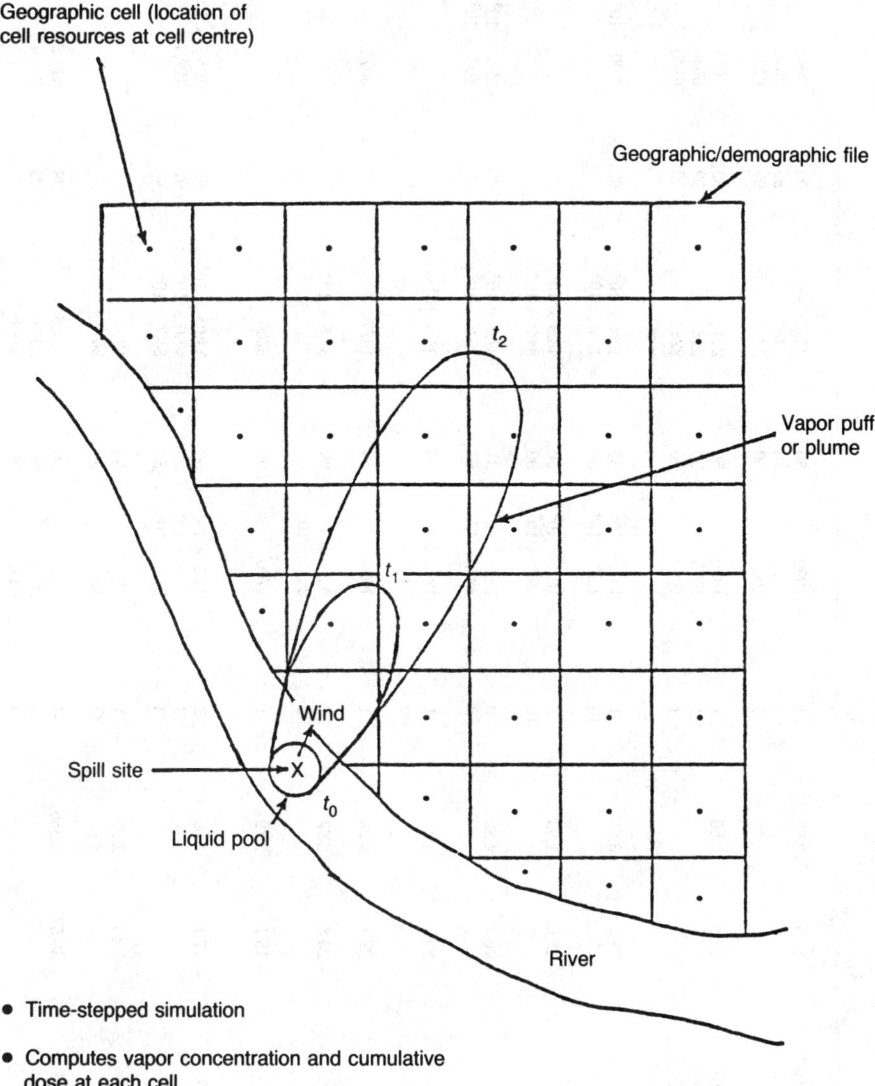
Geographic cell (location of cell resources at cell centre)
Geographic/demographic file
t_2
Vapor puff or plume
t_1
Wind
Spill site
X
t_0
Liquid pool
River
• Time-stepped simulation
• Computes vapor concentration and cumulative dose at each cell
• For flammable vapors, simulation stops when ignition occurs, and fire or explosion damage to each cell is computed at that time
• Ignition occurs if cell contains ignition source and concentration is above lower flammable limit

Figure 9.22 *Vulnerability model; simulation of vapour release from a marine spill (Parnarouskis, Perry and Articola, 1980)*

England and Wales is about 500 in urban areas and 150 in rural areas. The local geography of the census in Scotland is slightly different in that EDs are built up from the Post Office postcode areas. The main small unit statistical output from the census is the small area statistics (SAS), which are supplied by the Census Office on magnetic tape, on microfilm or as paper copy. There are delays of several years in the publication of some census data. Naturally the data become out of date, and this needs to be borne in mind.

Table 9.18 *Vulnerability model: some probit equations in the original model (Eisenberg, Lynch and Breeding, 1975)*

Phenomenon and type of injury or damage	*Causative variable*	*Probit equation parameters*		*Data from which the probit equation was derived*					
		k_1	k_2	*Per cent affected*	*Value of variable*	*Per cent affected*	*Value of variable*	*Per cent affected*	*Value of variable*
Fire:									
Burn deaths from flash fire	$t_e I_e^{4/3}/10^4$	– 14.9	2.56	1	1099	50	2417	99	7008
				1	1073	50	2264	99	6546
				1	1000	50	2210	99	6149
Burn deaths from pool burning	$tI^{4/3}/10^4$	– 14.9	2.56	1	1099	50	2417	99	7008
				1	1073	50	2264	99	6546
				1	1000	50	2210	99	6149
Explosion:									
Deaths from lung haemorrhage	p^o	– 77.1	6.91	1	1.00×10^5	50	1.41×10^5	99	2.00×10^5
				10	1.20×10^5	90	1.76×10^5		
Eardrum ruptures	p^o	– 15.6	1.93	1	16.5×10^3	50	43.5×10^3		
				10	19.3×10^3	90	84.3×10^3		
Deaths from impact	J	– 46.1	4.82	0	18.0×10^3	31	37.3×10^3	96	49.7×10^3
				8	28.6×10^3	63	45.2×10^3	100	60.7×10^3
Injuries from impact	J	– 39.1	4.45	1	13×10^3	90	28×10^3		
				50	20×10^3				
Injuries from flying fragments	J	– 27.1	4.26	1	1024	50	1877	99	3071
Structural damage	p^o	– 23.8	2.92	1	6.2×10^3	99	34.5×10^3		
				50	20.7×10^3				
Glass breakage	p^o	– 18.1	2.79	1	1700	90	6200		
Toxic release:									
Chlorine deaths	$\Sigma C^{2.75}T$	– 17.1	1.69	3	14.1×10^4	50	34.05×10^4	97	105.8×10^4
				3	17.0×10^4	50	47.0×10^4	97	129.4×10^4
				3	21.5×10^4	50	64.7×10^4		
Chlorine injuries	C	– 2.40	2.90	1	6	50	13		
				25	10	90	20		
Ammonia deaths	$\Sigma C^{2.75}T$	– 30.57	1.385	3	37.3	50	74.6	99	411.8
				3	90.9	50	204.6	99	334.4
				3	44.6	50	148.6		

t_e = effective time duration (s)
I_e = effective radiation intensity (W/m^2)
t = time duration of pool burning (s)
I = radiation intensity from pool burning (W/m^2)

p^o = peak overpressure (N/m^2)
J = impulse (N s/m^2)
C = concentration (ppm)
T = time interval (min)

Table 9.19 *CPD models: principal contents of Yellow Book and Green Book (Committee for the Prevention of Disasters, 1992a,b)*

A Physical effects (Yellow Book)
Outflow
Turbulent free jet
Spray release
Evaporation
Heat radiation
Dispersion
Vapour cloud explosion
Consequences of rupture of vessels
B Damage effects (Green Book)
Damage caused by heat radiation
Consequences of explosion effects on structures
Consequences of explosion effects on humans
Survey study of the products which can be released during a fire
Damage caused by acute intoxication
Protection against toxic substances by remaining indoors
Population data

Table 9.20 *Plant layout models: principal models (after Mecklenburgh, 1985)*

1. Introduction
2. Instantaneous release of gas or vapour
   2.1 Size of cloud
   2.2 Dispersion of cloud to lower flammability limit
   2.3 Explosion overpressure
   2.4 Fireball size
   2.5 Dispersion of toxic cloud
3. Steady leak of gas or vapour
   3.1 Leakage rates
   3.2 Dispersion of jet to lower flammability limit
   3.3 Size of jet flame
   3.4 Dispersion of toxic plume
4. Loss of liquid
   4.1 Leakage rate, pool size and evaporation rate
   4.2 Pool and tank fires
5. Fire damage and protection
6. Implications for plant layout
   6.1 Risk criteria
   6.2 On-site and off-site effects
7. Blast effects
   7.1 Human and building tolerances
   7.2 Plant components
   7.3 Control rooms
8. Hazardous area classification
   8.1 Small continuous release of gas or vapour in the open
   8.2 Small release of liquid in the open
   8.3 Small continuous release of gas or vapour in a building
   8.4 Small release of liquid in a building

Table 9.21 *WHAZAN: principal models (Technica, 1985)*

Emission	
Liquid	
Gas	
Two-phase	Fauske–Cude model
Behaviour on release	
Spreading liquid	
Jet	
Flashing liquid	
Dispersion	
Heavy gas	Cox and Carpenter model
Neutral density gas	Pasquill–Gifford model
Buoyant plume	Briggs model
Fire	
Pool fire	
Jet fire	Modified API model
Fireball	API model
Flash fire	Gas dispersion model
Fire damage	
Explosion	
Explosion damage	DSM model
Toxic release	Gas dispersion model + probit equations

The estimation of the population density off site requires the use of Ordnance Survey maps. Four principal map sizes relevant here are the 1:25 000, 1:10 000, 1:2500 and 1:1250 scale series. The 1:10 000 scale is that normally used by local authorities. It is also the size on which the census EDs are recorded and maps showing the boundaries of the EDs are available from the Census Office as paper copies of microfilmed 1:10 000 scale maps. The 1:2500 and 1:1250 scale maps can be useful in locating more accurately the boundaries between EDs; unfortunately the whole country is not covered by these series. These maps give the most detailed record of buildings. In the 1:10 000 and 1:25 000 series there is some loss of detail and accuracy with respect to buildings.

The method which gives the most accurate results for population density is the use of the census data. This should therefore be used where it is practical to do so. There may, however, be reasons for not using census data for the whole area of interest. One is that it can be time-consuming. Another is that the census may have become out of date.

In these circumstances a rapid estimate of population may be made using a map to identify the built-up areas and assuming population densities of 4000 persons/km^2 for large built-up areas, 100 persons/km^2 for other inhabited areas, and zero population for uninhabited areas.

For more accurate work a more detailed approach is necessary. It is recommended by Petts, Withers and Lees that three zones be defined around the hazard source, delimited by distances of 400 and 1000 m. For the outer zone (>1000 m) the rapid estimation method

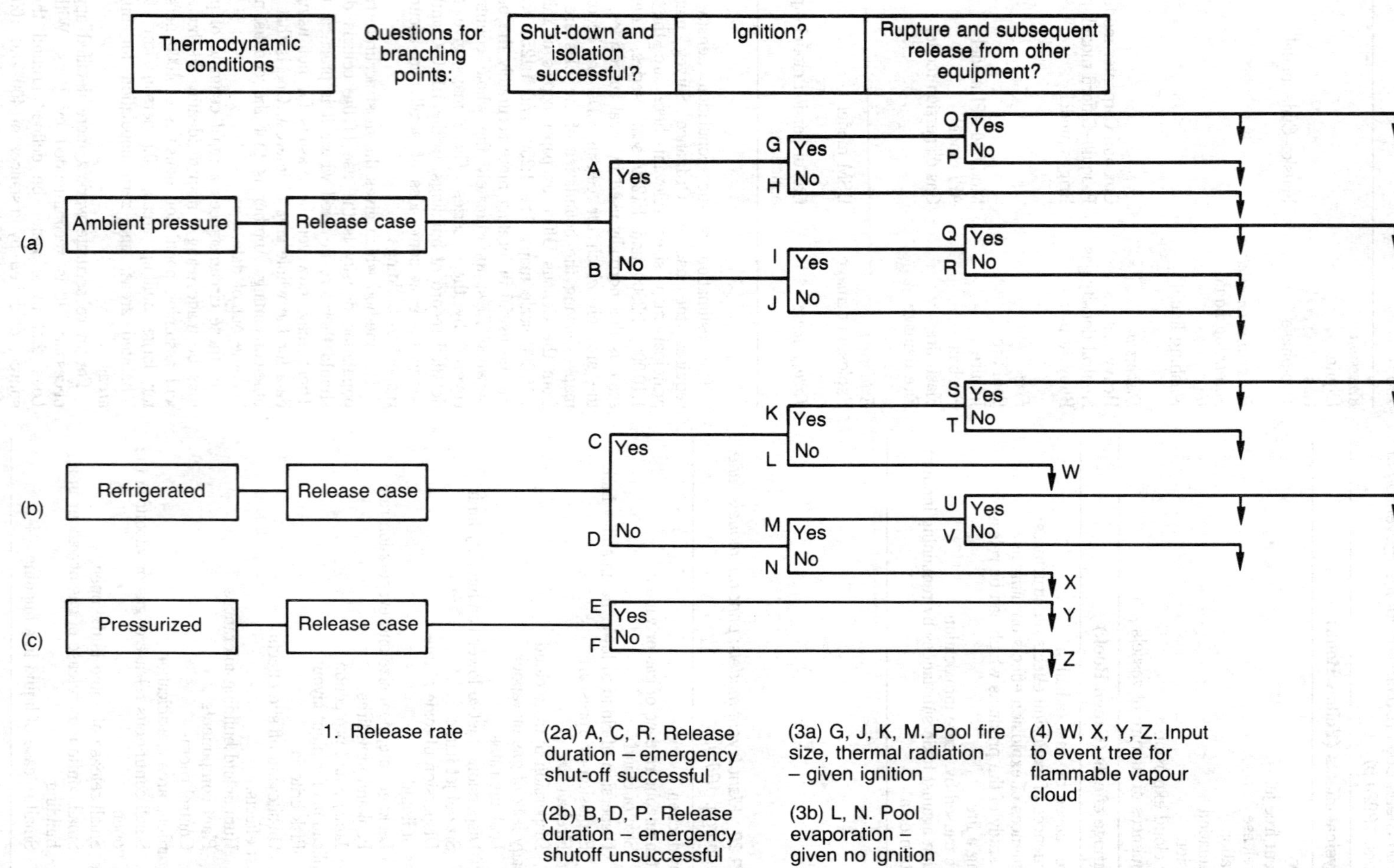

Figure 9.23 *WHAZAN: event trees for the models (after Technica, 1985): (a) tree for flammable gas release; (b) tree for flammable or toxic liquid release; (c) tree for toxic gas release*

Table 9.22 *Methodology for estimation of density of population around a hazard source (after Petts, Lees and Withers, 1987) (Courtesy of Elsevier Science Publishers)*

Distance from hazard source (m)	*Method*	*Population density* (persons/km^2)
<400	Use of Census data	From enumeration district data
400–1000	Use of Ordnance Survey maps	Dense terrace housing: 15 000 Semi-detached housing: 10 000 Sparse detached housing: 1000 Uninhabited areas: 0
>1000	Use of Ordnance Survey maps	Built-up areas: 4000 Other inhabited areas: 100 Uninhabited areas: 0

should be used. For the intermediate zone (400–1000 m) the map should be used to identify three types of residential area and the uninhabited area. These three types of residential development are: dense, usually in-town, terraced housing (and high rise flats); semi-detached housing, usually suburban; and sparse, detached housing. Population densities of 15 000, 10 000 and 1000 persons/km^2, respectively, should be used.

For the inner zone (<400 m) the population should be estimated from census ED data. It is recommended, however, that for this inner zone this estimate should be checked against visual inspection and local inquiry. The radius of the inner zone was originally set at 250 m, but site inspection indicated that this may be too short a distance for acceptable accuracy. The distance of 400 m is therefore preferred as the boundary of the inner zone. This method for the estimation of the density of the population around a hazard source is summarized in Table 9.22.

It is necessary to define the point from which the circles defining these zones are drawn. There appears at present to be some difference of practice. In some cases this point is taken to be the plant itself, in others the site boundary. The former makes more sense for a plant at a specific point, but the latter allows for relocation of the plant within the site boundary.

The accuracy of the method described for the estimation of the population density in the middle zone was checked against the 1981 Census data for three hazardous sites. The average difference in the estimates of the numbers of people was 14%, one estimate being low and two high.

The use of the census data is not entirely straightforward. The approach used by the authors was first to identify the codes of the required EDs from 1:10 000 scale maps provided by the Census Office and then to obtain the ED data by interrogating the Census computer data bank.

There are in existence several computer programs which process population data. These use data on the numbers of people in each 100 m square. This information was provided in the 1971 Census, but is not available in the 1981 Census.

The basic case for the off-site population is the night time population. It is this which is given by the census data and by the method just described. An approximate estimate of the number of people over the whole 24 h is 80–85% of this value. However, the probability of some hazards may be a function of time of day and it then becomes necessary to estimate population by time of day also. The daytime population may be estimated more accurately, as described below.

Attention is drawn by Petts, Lees and Withers to the fact that errors in drawing the boundaries between areas of different population densities can easily lead to errors which are as great or greater than those which are due to errors in estimating those densities.

9.16.2 Population composition

In order to obtain a more detailed picture of the population it is necessary to define the population composition. This is relevant to the probability that an individual is at his home base, that he is a member of a more vulnerable population and/or that he is outdoors.

Information on population categories is available in the 1981 Labour Force Survey (OPCS, 1981c) as shown in Table 9.23. From these data Petts, Withers and Lees have derived the population composition model given in Table 9.24.

In order to use this population composition model it is necessary to define the times of day. The set of times of day used by the authors is shown in Table 9.25, Section A. These are to be used with the population categories at home by time of day given in Table 9.25, Section B.

9.16.3 Population change by time of day

Using the definitions of population composition and of times of day just given, estimates may be made of the population changes by time of day. The estimates based on the data given in Table 9.25, Sections A and B, are given in Section C of the table. It was found that for random sections of the Canvey and Rijnmond populations, the proportions of the night-time population who were at home during the day were 42% and 46%, respectively.

Table 9.23 *Population categories and composition based 1981 Labour Force Survey (after Petts, Lees and Withers, 1987) (Courtesy of Elsevier Science Publishers)*

	Number (thousands)	*No. per household*
Adults in full-time employment	16 595	0.85
Adults in part-time employment	4042	0.21
Self-employed	2164	0.11
Unemployed	2447	0.13
Housewives	7092	0.36
Children: 0–4 years	3222	0.17
Children: school age	8753	0.45
Students	1415	0.07
Retired	6266	0.32
Others (including permanently sick and disabled)	1100	0.06
Total	53 096	2.73
Total No. of households	9442	

Table 9.24 *Population composition model (after Petts, Lees and Withers, 1987) (Courtesy of Elsevier Science Publishers)*

A Unemployment 5%

	No. per household	*Proportion (%)*
1. Adults in full time employment:		
(a) at work (including self-employed)	0.89	32.9
(b) sick, on holiday, working from home	0.10	3.7
2. Adults in part time employment:		
(a) at work	0.21	7.7
(b) sick, on holiday	0.02	0.7
3. Unemployed	0.07	2.6
4. Homekeepers	0.35	12.9
5. Children of school age:		
(a) at school	0.31	11.4
(b) sick, on holiday	0.14	5.2
6. Students:		
(a) at college	0.04	1.5
(b) sick at home, on vacation	0.03	1.1
7. Children under school age	0.17	6.3
8. Retired	0.32	11.8
9. Others (including permanently sick and disabled)	0.06	2.2
Total	2.71	100.0

B Unemployment 10%

	No. per household	*Proportion (%)*
1. Adults in full time employment:		
(a) at work (including self-employed)	0.84	31.0
(b) sick, on holiday, working from home	0.10	3.7
2. Adults in part time employment:		
(a) at work	0.19	7.0
(b) sick, on holiday	0.02	0.7
3. Unemployed	0.14	5.2
4. Homekeepers	0.35	12.9
5. Children of school age:		
(a) at school	0.31	11.4

(b) sick, on holiday	0.14	5.2
6. Students:		
(a) at college	0.04	1.5
(b) sick at home, on vacation	0.03	1.1
7. Children under school age	0.17	6.3
8. Retired	0.32	11.8
9. Others (including permanently sick and disabled)	0.06	2.2
Total	2.71	100.0

Table 9.25 *Population composition at home by time of day (after Petts, Lees and Withers, 1987) (Courtesy of Elsevier Science Publishers)*

A Time-of-day categories

	Time of day	*Duration*	
		(h)	(%)
School day	08.00 – 16.00	8	33
Work day	08.00 – 18.30	10.5	44
Night	18.30 – 08.00	13.5	56

B Population categories at home

	Categories
School day	1(b), 2(b), 3, 4(b), 5(b), 6(b), 7–9
Work day	All except 1(a)
Night	All

C Proportion of population at home

	Proportion at home (%)	
	Unemployment 5%	*Unemployment 10%*
School day	46.5	49.1
Work day	67.2	69.0
Night	100.0	100.0

9.16.4 Vulnerable population

Some members of the population are likely to be more vulnerable to the hazard than others and it may be necessary to take this into account. In general, it is children, old people and infirm people who tend to be most vulnerable, and the proportion of vulnerable people may be estimated as a first approximation by determining the proportion in these categories. However, vulnerability must be a function of the particular hazard. For example, children may actually recover better from some burns than adults, and persons with respiratory disease are likely to be more susceptible to irritant toxic gas, but not necessarily to thermal radiation.

As a first approximation, therefore, the population may be divided into two broad groups: (1) adults of working age and older children and (2) young children and old people. The first group is some 75% and the second some 25% of the population. In general, the latter is the more vulnerable group, although for some hazards it may be necessary to have a more specific definition. In this case the proportion of vulnerable people may be assessed in relation to the particular hazard considered using the population composition model given in Table 9.24.

Mention may be made here of estimates of the proportion of vulnerable people given by other workers. Table 9.26 shows some estimates made by Hewitt (1976) for the proportion of people vulnerable to a toxic irritant gas such as chlorine.

9.16.5 Population outdoors

Very little information was found by Petts, Withers and Lees on the proportion of the population which is outdoors at different times of the day. In the *Rijnmond Report* the proportion of people indoors was taken in the context of toxic gas hazard as 99%, allowing for the fact that some people would seek shelter from this hazard indoors.

The population location model derived by the authors is shown in Table 9.27. It was assumed that the regular and vulnerable groups spend 1 and $\frac{1}{2}$h/day outdoors, respectively, that the proportion of the total population

Table 9.26 Vulnerable members of population (after Hewitt, 1976)

	No. per 1000 people
Children:	
<6 months	8
<12 months	8
12 months – 5 years	75
5 – 9 years	82
Old people (>70 years)	85
People with chronic heart trouble	5
People with respiratory diseases	9
People with restricted mobility	4
Blind people	2
Healthy youngsters and adults	722

outdoors at night (18.30 – 08.00 h) is 1%, and that those outdoors are drawn exclusively from the regular population. Then the time outdoors and the proportions outdoors by day, by night and overall may be calculated for the regular, vulnerable and total populations. A partial cross-check on the model or, more specifically, a cross-check on the lower bound of the proportion of the population outdoors, was obtained from wartime data on V-2 rocket bomb casualties.

9.17 Modification of Exposure

The characteristics of the population at risk may need to be modified to take account of changes in the exposure of the population which occur if there is warning of the hazard. There are a number of ways in which an individual who becomes aware of a hazard may modify his exposure. He may seek shelter indoors or he may try to distance himself from the hazard, either by individual escape or by participating in an organized evacuation.

While these are the principal actions, they are by no means the only ones. There are others which are more specific to the hazard concerned. For example, an individual may seek shelter from the thermal radiation of a fireball behind some structure. Or he may try to reduce his vulnerability to the blast from an explosion by adopting a different body posture.

Some aspects of modification of exposure by escape and evacuation are considered here. Other aspects such as modification by shelter are treated in the chapters on particular hazards. Organized evacuation as part of emergency planning is discussed in Chapter 24.

9.17.1 Reaction

The most basic feature bearing on modification of exposure is the response time of humans.

Estimates of human response time in emergencies have been made by the British Compressed Gases Association (BCGA, 1984) in relation to response to ignition of clothing in an oxygen enhanced atmosphere. Work on human factors (e.g. Denison and Tonkins, 1967) is quoted, to the effect that the time taken by humans to respond to an unexpected situation varies between 5 and 20 s. The BCGA obtain from this the following response model:

(1) 5% of the population (i.e. babies and disabled persons) are incapable of any reaction
(2) 95% of the population are capable of reaction, of whom it is assumed:
80% can react within 15 s,
25% can react within 10 s,
10% can react within 7.5 s.

It is taken that none of the population can react within 5 seconds.

9.17.2 Escape

A systematic treatment of individual escape is difficult. The principal case where this has been attempted is for escape from toxic gas by walking out of the cloud. The treatment of this problem given in the First *Canvey Report* is described in Appendix 7.

9.17.3 Evacuation

Organized evacuation is more readily analysed. A simple model for evacuation is:

$$\phi(t) = 1 \qquad t < t_d \qquad [9.17.1a]$$

$$= \phi(\infty) + [1 - \phi(\infty)]\exp[-\lambda(t - t_d)] \qquad t \geq t_d \qquad [9.17.1b]$$

where t_d is the time lag before any evacuation starts, λ is a constant, $\phi(t)$ is the fraction of population which has not evacuated at time t, and $\phi(\infty)$ is the fraction of population which does not evacuate at all. The constant λ is related to the evacuation half-life $t_{1/2}$ as:

$$\lambda = 0.693/t_{1/2} \qquad [9.17.2]$$

This model has been used by Solomon, Rubin and Okrent (1976) to allow for the effect of evacuation in hazard assessments of large toxic releases. Most other workers appear to have used a similar model.

A study of evacuation effectiveness in actual emergencies has been made by Hans and Sell (1974), who give details of evacuation from transportation, hurricane and flood emergencies. The transportation emergencies

Table 9.27 Population outdoor exposure model (after Petts, Lees and Withers, 1987) (Courtesy of Elsevier Science Publishers)

Population	*Time outdoors* (h/week)			*Proportion outdoors* (%)		
	Day	*Night*	*Overall*	*Day*	*Night*	*Overall*
Regular	0.81	0.19	1.0	7.70	1.33	4.17
Vulnerable	0.5	0	0.5	4.8	0	2.08
Total	0.74	0.14	0.88	7.05	1.0	3.67

Table 9.28 *Evacuation in some major transport emergencies (after Hans and Sell, 1974)*

Date	*Location*	*Incident*	*Number evacuated*	*Area evacuated* ($mile^2$)	*Evacuation distance* (mile)	*Evacuation period* (h)
1965	Baton Rouge, LA	Chlorine barge, no chlorine escape	150 000	8	30	2
1969	Glendora, MI[a]	Vinyl chloride escape from rail tankers	35 000	1200	20	4
1972	Louisville, KY	Chlorine barge, no chlorine escape	4 000	0.35	1	3
1973	Morgan City, LA	Chlorine barge, no chlorine escape	3 000	1.8	2	4

[a] The Glendora incident is described in Case History A43.

include the incidents shown in Table 9.28. The overall conclusion of these investigators is that behaviour in such emergencies is less irrational and more effective than is often supposed.

The data presented in this work were analysed in the *Rasmussen Report* (AEC, 1975). The evacuation speed and time for each type of incident were found to be correlated by the equations:

$$v = kd^n \qquad [9.17.3]$$

$$t = d/v \qquad [9.17.4]$$

where d is the evacuation distance (mile), t is the evacuation time (h), v is the evacuation speed (mile/h), k is a constant and n is an index.

The type of incident which is most relevant in the present context appears to be the transportation emergency. For such incidents

$$v = 0.30d^{1.02} \qquad [9.17.5]$$

The evacuation speeds were found to fit a log–normal distribution with the density function

$$f = \frac{1}{\sigma v(2\pi)^{1/2}} \exp\left[-\frac{(\ln v - m^*)^2}{2\sigma^2}\right] \qquad [9.17.6]$$

where m^* and σ are parameters in the log–normal distribution.

For transportation incidents $m^* = 0.20$ and $\sigma = 1.64$ and hence the modal and mean values of v are 0.08 and 4.7 mile/h, respectively.

There have been a number of more recent major evacuations. One of the largest was at Mississauga, Toronto, where a leaking chlorine rail tank car led to the evacuation of some 223 000 people (Case History A97). Another large evacuation took place at Bhopal, but in this case there was no chance of averting a major disaster. Bhopal is described in Appendix 5. At Chernobyl, after a delay in deciding to evacuate, some 40 000 people were evacuated in about 2 h. Chernobyl is described in Appendix 22.

The value of evacuation in modifying exposure was investigated in the First *Canvey Report*, as described in Appendix 7. Opportunity to evacuate depends on the hazard and on the way in which the incident develops. Although some hazards, such as a single, unexpected explosion may give little warning, in other cases there may be a substantial degree of warning. One example is an initial explosion in a works followed by a fire which may lead to further explosions and fires. Another is derailment of a rail tank car with the possibility of fireballs and toxic release. In both such cases evacuation may be beneficial.

9.17.4 Incident control

Avoidance of casualties also has another aspect. Besides getting people who are exposed out of the danger zone, it is often necessary also to keep others from entering the danger zone. In particular, an incident may tend to attract spectators who thereby put themselves at risk. In 1947, ammonium nitrate explosions aboard the *Grandcamp* and the *High Flyer* in the harbour in Texas City killed 552 people (Case History A16). The explosion on the *Grandcamp* killed everyone in the dock, many of whom were spectators. Spectators also constituted a large proportion of the death toll of more than 150 at Caracas in 1982, when burning oil from a storage tank on fire at the top of a hill flowed down into a crowd below (Case History A102).

9.18 Injury Relations

The estimation of the injury or damage caused by a physical effect such as heat radiation intensity, overpressure or toxic concentration requires the use of injury or damage relations.

The normal method of formulating such relations is in three stages. The first is the determination of the causative (or injury or damage) factor which best correlates the data. The second is the determination of the probability distribution for this factor. The third is the conversion of this distribution into the more convenient form of a probit equation.

These steps are now described. In the account given reference is primarily to injury rather than damage relations, but the principles are the same for the latter.

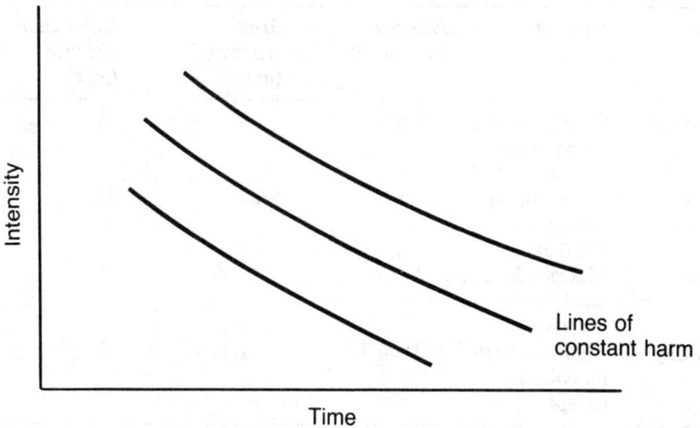

Figure 9.24 *Determination of the form of the injury factor*

Injury and damage relations may be expressed in forms other than probit equations and these also are briefly mentioned.

9.18.1 Injury factors

The starting point for the derivation of an injury relation is a set of data giving the probability of a specified degree of injury as a function of some physical effect. Some typical physical effects are:

Fire	Thermal radiation intensity	I
Explosion	Overpressure	p^o
	Impulse	J
Toxic gas	Concentration	C
	Dosage	Ct

where t is time.

The probability of injury may or may not correlate directly with the physical effect. In some cases the correlation is with some power or time function of the effect. For example, probability of eardrum rupture does correlate with explosion overpressure p^o, but probability of burn death correlates with the radiation intensity time function $I^{\frac{4}{3}}t$ The first step, therefore, is to determine the appropriate injury factor. It may be helpful to plot a graph of the form shown in Figure 9.24.

9.18.2 Injury distributions

Next it is necessary to determine the probability distribution of the injury factor. The distribution usually considered first is the log–normal distribution. The fit of the data to this distribution may be tested by plotting on log–probability paper. A plot of this type is shown in Figure 9.25.

The reason that data on probability of injury tend to give a log–normal distribution is that this distribution fits the case where the population contains a proportion of individuals who are unusually resistant. For example, the distribution is widely used to correlate the toxicity of an insecticide, where the dosage required to kill off the last few individuals is relatively high.

9.18.3 Probit equations

The injury distribution just described gives the information required on the probability of injury as a function of the injury factor. It is possible, however, to carry out a transformation which casts it in the more convenient

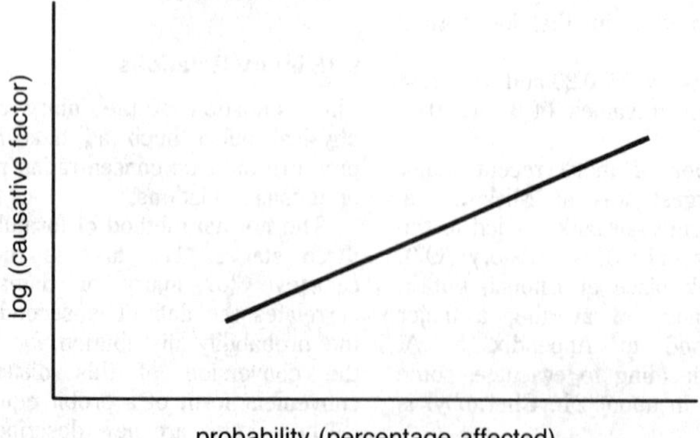

Figure 9.25 *Determination of the distribution of the injury factor*

Table 9.29 Transformation of percentages to probits (Finney, 1971) (Courtesy of Cambridge University Press)

%	0	1	2	3	4	5	6	7	8	9
0	–	2.67	2.95	3.12	3.25	3.36	3.45	3.52	3.59	3.66
10	3.72	3.77	3.82	3.87	3.92	3.96	4.01	4.05	4.08	4.12
20	4.16	4.19	4.23	4.26	4.29	4.33	4.36	4.39	4.42	4.45
30	4.48	4.50	4.53	4.56	4.59	4.61	4.64	4.67	4.69	4.72
40	4.75	4.77	4.80	4.82	4.85	4.87	4.90	4.92	4.95	4.97
50	5.00	5.03	5.05	5.08	5.10	5.13	5.15	5.18	5.20	5.23
60	5.25	5.28	5.31	5.33	5.36	5.39	5.41	5.44	5.47	5.50
70	5.52	5.55	5.58	5.61	5.64	5.67	5.71	5.74	5.77	5.81
80	5.84	5.88	5.92	5.95	5.99	6.04	6.08	6.13	6.18	6.23
90	6.28	6.34	6.41	6.48	6.55	6.64	6.75	6.88	7.05	7.33
–	0.0	0.1	0.2	0.3	0.4	0.5	0.6	0.7	0.8	0.9
99	7.33	7.37	7.41	7.46	7.51	7.58	7.65	7.75	7.88	8.09

form of a probit equation. An account of the probit method is given in *Probit Analysis* (Finney, 1971). One of the first major applications of the method to hazard assessment in the process industries was in the vulnerability model of Eisenberg, Lynch and Breeding (1975).

The probit Y is an alternative way of expressing the probability P of injury. It is defined by the relation:

$$P = \frac{1}{(2\pi)^{1/2}} \int_{-\infty}^{Y-5} \exp(-u^2/2)\mathrm{d}u \qquad [9.18.1]$$

The probit is a random variable with a mean 5 and variance 1. The probability range (0–1) is generally replaced in probit work by a percentage (range 0–100). The relationship between percentages and probits is shown in Table 9.29 and in Figure 9.26.

It is shown below that for an injury factor x which fits the log-normal distribution the probit equation has the form:

$$Y = k_1 + k_2 \ln x \qquad [9.18.2]$$

A typical probit transformation is illustrated in Figure 9.27. The sigmoidal curve of the log-normal distribution is transformed into the straight line of the probit. This makes it easier both to correlate the data in the first place and then to use the correlation.

The derivation of an equation of the general type given in Equation 9.18.2 is conveniently shown by considering first a normal distribution. The density function f is

$$f = \frac{1}{(2\pi)^{1/2}\sigma} \exp\left[-\frac{(x-m)^2}{2\sigma^2}\right] \qquad [9.18.3]$$

The distribution function F is

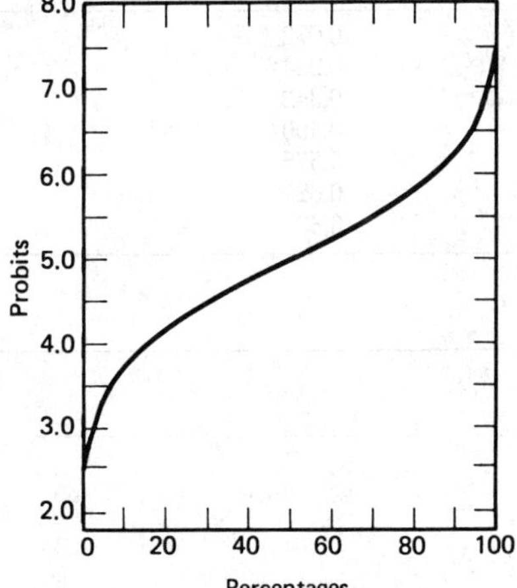

Figure 9.26 *Relationship between percentages and probits (Finney, 1971) (Courtesy of Cambridge University Press)*

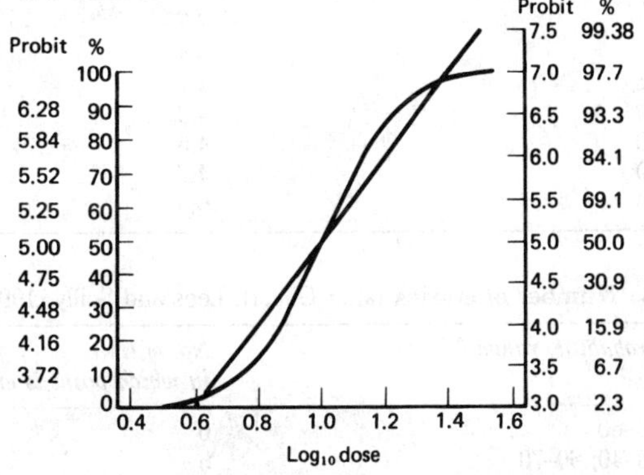

Figure 9.27 *Effect of the probit transformation (Finney, 1971). The figure illustrates a typical experiment on the toxicity of an insecticide. The logarithm of the lethal dose fits a normal distribution. The figure shows the distribution function and the corresponding probit function (Courtesy of Cambridge University Press)*

$$F = \int_{-\infty}^{x} f(x)\mathrm{d}x \qquad [9.18.4]$$

But the distribution function F is the same as the probability P of injury, so that entering Equation 9.18.3 in Equation 9.18.4 gives:

$$P = \frac{1}{(2\pi)^{1/2}\sigma}\int_{-\infty}^{x} \exp\left[-\frac{(x-m)^2}{2\sigma^2}\right]\mathrm{d}x \qquad [9.18.5]$$

Then equating Equation 9.18.5 with Equation 9.18.1 gives

$$u = (x - m)/\sigma \qquad [9.18.6]$$

and

$$Y - 5 = u \qquad [9.18.7]$$

Hence

$$Y = (5 - m/\sigma) + (1/\sigma)x \qquad [9.18.8]$$

$$= k_1' + k_2'x \qquad [9.18.9]$$

Similarly, for the log-normal distribution:

$$Y = k_1 + k_2 \ln x \qquad [9.18.10]$$

with

$$k_1 = 5 - m^*/\sigma \qquad [9.18.11a]$$

$$k_2 = 1/\sigma \qquad [9.18.11b]$$

Equation 9.18.10 is the usual form of the probit equation.

The derivation of a probit equation from a set of data giving the probability of a specified injury as a function of the injury factor may be illustrated using the following data for eardrum rupture due to overpressure given by Eisenberg, Lynch and Breeding:

Percentage affected	*Probit*	*Peak overpressure* (N/m2)
1	2.67	16.5×10^3
10	3.72	19.3×10^3
50	5.00	43.5×10^3
90	6.28	84.3×10^3

The corresponding probit equation given by these authors is

$$Y = -15.6 + 1.93 \ln p^{\circ} \qquad [9.18.12]$$

Probit equations for some major hazards given by Eisenberg, Lynch and Breeding are shown in Table 9.13. The derivation of these equations is discussed in more detail in Chapters 16–18.

It is emphasized that the probit equations given in Table 9.13 were early examples and were avowedly tentative and approximate, as will be clear from their derivations. They were developed essentially for use in the vulnerability model. Some have been superseded by improved equations. They are given here primarily as an illustration of the general form of injury factors and probit equations.

Probit equations were used in the *Canvey Reports* and the *Rijnmond Report* and have been widely used in many hazard assessments since then.

With respect to probit equations, there are several points which need to be borne in mind. First, any injury or damage correlation is only as good as the original data. Second, a check should be made that the data do

Table 9.30 *Weighting factors for construction of probit equations*

A Weighting coefficient (after R.A. Fisher and Yates, 1957)

Probability (%)[a]	*Probit*	*Weighting coefficient*
1	2.7	0.072
5	3.4	0.224
10	3.7	0.342
20	4.2	0.490
30	4.5	0.575
40	4.7	0.621
50	5.0	0.637

B Number of entries (after Gilbert, Lees and Scilly, 1994a)

Probability range (%)	*No. of times in which point is entered*
40–60	6
30–40; 60–70	5
20–30; 70–80	4
10–20; 80–90	3
5–10; 90–95	2
<5; >95	0

[a] The weighting coefficent for a percentage x greater than 50% may be obtained from that for $100 - x$.

actually fit the distribution implied, generally the log-normal distribution. Third, it is desirable to allow for the fact that data for probabilities in the middle of the probability range are likely to be more accurate than those for probabilities at the extremes of the range.

9.18.4 Probit equations: data weighting

Taking up this latter point, it is the case that given data for probabilities P near the centre of the range (say $P = 0.5$) and at an extreme of the range (say $P = 0.01$ or 0.99) obtained from samples of the same size, the level of confidence will be higher in the former. Or, to put the matter another way, for the same level of confidence it is necessary to have a larger sample size at an extreme of the range than near the centre.

This problem arises particularly in laboratory work on the toxicity of, say, pesticides. It has been addressed by Fisher and Yates (1957), who give a set of weighting coefficients for use in probit analysis. Table 9.30, Section A, gives a set of weighting coefficients interpolated from their values.

A technique described by Gilbert, Lees and Scilly (1994) which is convenient in dealing with a set of data points and which gives an essentially similar weighting is to multiply each point by the weighting factor shown in Section B of Table 9.30. The point is discussed further in Chapter 18.

9.18.5 Other relations

Not all correlations of injury and damage are in terms of probability distributions and probit equations. In particular, certain effects of explosions are often correlated in other ways. Among these effects are damage to plant and buildings, injury due to building collapse and damage and injury due to missiles. The treatment of explosion effects is described in Chapter 17.

9.18.6 Multiple injury and double counting

It is appropriate at this point to draw attention to some of the problems associated with multiple injury and double counting of injuries.

A treatment of double counting is given in the *Green Book* CPD (1992b), which identifies three basic types of double counting. In double counting of the first type, a single injury mechanism due to a single event creates victims with several, successively more severe, degrees of injury. A problem arises where the number of victims with a lower degree of injury is also included in the number with the higher degree(s) of injury. The correct procedure is to subtract the number of victims with the higher degree(s) of injury from the number with the lower degree of injury.

In double counting of the second type, several injury mechanisms due to a single event create victims with the same degree of injury. The problem here is that a proportion of victims are injured by more than one of these mechanisms and that, mathematically, it is incorrect to obtain the probabilities of injury by simply summing the probabilities for each mechanism. The correct procedure mathematically is to utilize the expressions for the probability of a joint event, as given by Equations 7.5.5–7.5.8.

In double counting of the third type, several events, occurring in succession, each theoretically injure to a given degree the same person, whereas in reality the first event produces such injury that any subsequent event has no additional injurious effect. In this case, the correct procedure mathematically is to subtract the number of persons suffering this degree of injury after the first event from the population at risk for subsequent events.

This approach to the avoidance of double counting is essentially a mathematical one. It is valid for double counting of the first type and also for double counting of the second and third types where the degree of injury concerned is fatal injury. However, for degrees of injury which are less than fatal it fails for the second and third types to make sufficient allowance for physiological as opposed to mathematical factors.

As far as physiological factors are concerned, it is difficult to generalize. Each case needs to be considered on its merits. It might be thought that from a physiological rather than a mathematical viewpoint the overall probability of a given degree of injury from more than one event capable of causing that degree of injury might be greater, rather than less, than the sum of the individual probabilities, and also that there might be escalation to a more severe degree of injury, but this may not always be so.

Thus on the one hand there is some evidence from battle casualties that the physiological effect of fragment wounds is not necessarily proportional to the number of fragments involved. On the other hand, the severity of burn wounds increases with the proportion of the body surface which is affected. A further discussion of combined injuries is given by R.L. Jones (1971).

9.18.7 Overview

It is peculiarly difficult to obtain injury relations for humans, since the normal method of establishing a correlation, i.e. direct experimentation, is ruled out. The approaches taken have often been opportunistic, seeking data wherever it may be found. For example, the correlation for burn injury which has long held the field is based on data on injuries caused by the atomic bombs at Hiroshima and Nagasaki. Not surprisingly, the derivation and use of injury relations has been subject to some criticism. Critiques include those of V.C. Marshall (1989b), Turner and Fairhurst (1989 HSE SIR 21) and R.F. Griffiths (1991b).

A general criticism is the relatively high uncertainty in injury relations. This is inherent in the problem and there is not very much that can be done about it. The degree of uncertainty may be quantified if it is possible to determine confidence bounds, but often it is not obvious how this is to be done. It may well be possible to give bounds for, say, the concentrations of a toxic gas which are lethal to a particular animal species, but these data have then to be extrapolated to humans, an operation which involves a good deal of judgement.

Another criticism concerns extrapolation from the region of 50% probability to those of 5% and or even 1%. Here there are two distinct problems. One is the confidence which can be placed in the slope of the probit relation on which such extrapolation is based. The other is the possibility that at low values of the physical effect the injury mechanisms which come into play may be qualitatively different. Since the level of physical effect which applies to the 5% or 1% probabilities is relatively low and thus applies to the far field, the area affected is large and differences in the value of the effect for these probability levels can have a significant impact on the estimate of the number injured.

Although in principle a similar problem of extrapolation exists for the region of 95% or 99% probability, in practice this is of less importance, because these probability values apply in the near field where the area affected is relatively small and because in consequence the difference in numbers of injured between estimated values of, say, 90% and 95% is not great.

9.19 Presentation of Results

There are a number of different outputs which may be obtained from a hazard assessment. These outputs may be, and often are, in the form of risk estimates, but other outputs may also be used. The results are sometimes presented in terms of the distance at which the hazard will cause a defined effect. Such a hazard range may be based on a worst-case scenario or on some less serious scenario. In the latter case there is an implied cut-off for the frequency of the event.

Alternatively, the results are presented in terms of risk. One form of risk is the frequency of a defined effect. In this case no explicit computation is made of harm to people. If the risk to people is determined, the exercise becomes a full risk assessment. The risk may be expressed in terms of fatalities only, or of fatalities and injuries.

The risk of damage to property may also be determined, although such assessment appears to be less common as an end in itself than as a means of estimating the risk to people.

9.19.1 Types of risk to people

As just described, the results presented from a risk assessment may be confined to risk of death or may embrace risk of injury. Relying on the concept of a more or less fixed ratio of injuries to fatalities, the engineer may well consider that a risk profile of the plant formulated in terms of fatalities is usually sufficient as an aid to decision-making. As described below, however, risk of injury is also taken into account. In particular, the HSE are interested in the risk of receiving a 'dangerous dose'. This dose translates into different risks of death for vulnerable and non-vulnerable people.

9.19.2 Forms of presentation

There are a number of ways in which the results of a risk assessment may be presented. Some of the forms of presentation are shown in Table 9.31. For a fixed installation the risk will relate to some event which might occur on the site in question. The risk is generally measured from a particular point within the site. In some cases the risk may be presented simply as the frequency of a defined event, usually a major accident, at the installation.

For individual risk to the workforce the usual form is either an annual risk or the fatal accident rate (FAR). There is a straight conversion between the two. It is not usual to derive for the workforce any form of collective risk equivalent to the societal risk for the public described below.

The forms of presentation for risk to the public are more varied. For land use planning, contours and

***Table 9.31** Forms of presentation of results of a risk assessment*

A Risk to workforce
Annual risk
Fatal accident rate (FAR)

B Risk to public
Physical effects:
  Contours on site map
  Transects on site map
Individual risk:
  Contours on site map
  Transects on site map
  Annual risk at fixed location
Societal risk
  FN table
  FN curve
Equivalent annual fatalities

Note: Risk may be risk of fatality or injury, or other risk such as that of HSE dangerous dose.

transects (plan and elevation, respectively) may be given. These contours and transects may represent the frequency of a given intensity of a physical effect, or alternatively the frequency of individual fatality, or individual risk. Such contours and transects may relate to a single hazard or to the combined results of several hazards.

Individual risk to members of the public may be presented as a function of distance from the site, but is more usually given as the average risk to groups at different locations around the site.

The main form in which societal risk is presented is as a relation between incidents which cause some number N or more fatalities and the frequency F of such incidents. These data may be given in tabular form, but are also often plotted as a frequency–number (FN) curve. The basic FN data may be integrated to obtain a value for the equivalent annual fatalities. One use of this statistic is to compare the risks of different activities, e.g. coal mining vs nuclear energy. It is generally regarded as less appropriate for a hazard on a particular site, since it lumps together small incidents with relatively high frequency and large incidents with relatively low frequency. However, for such a hazard it is of some value in assessing the relative contribution of high frequency and low frequency events to the overall risk.

9.19.3 Risk contours and transects

The use of risk contours and transects is illustrated in Figures 9.28 and 9.29. Figure 9.28 gives overpressure risk contours for a natural gas liquids (NGL) plant and refrigeration facility. Figure 9.29 gives a risk transect for interaction between a vapour cloud from a pipeline and the public.

9.19.4 FN tables and curves

Table 9.32 shows some results of the risk assessment given in the first *Canvey Report*. Figure 9.30 shows the corresponding FN curve. This was one of the first FN curves published for the process industries. Also shown

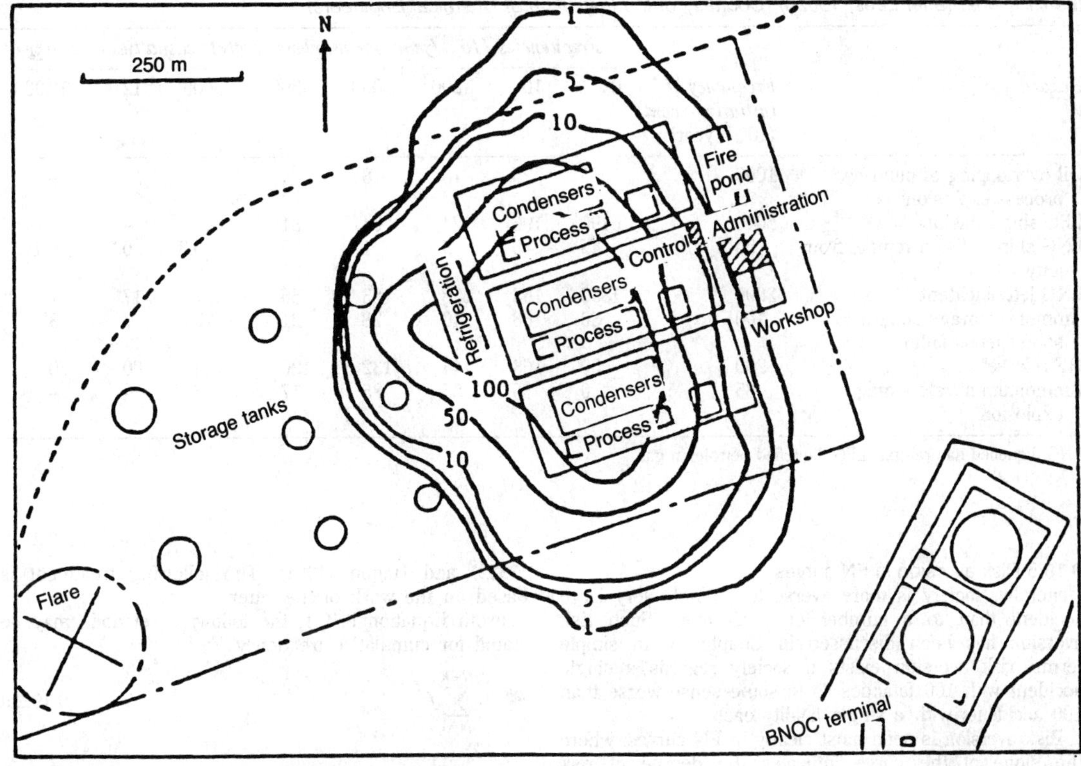

Figure 9.28 *Risk contours for a NGL plant and refrigeration facility (Ramsay, Sylvester-Evans and English, 1982) (Courtesy of the Institution of Chemical Engineers)*

in Figure 9.30 are other curves given in the two *Canvey Reports*.

Figure 9.31 shows a set of FN curves given by the HSE for accidents, in the chemical and petrochemical industries, in the UK and world-wide.

If the FN relation is a straight line, it may be written as:

$$fN^{\alpha} = r = \text{Constant} \tag{9.19.1}$$

where f is the frequency, N is the number of fatalities, r is a constant and α is an index.

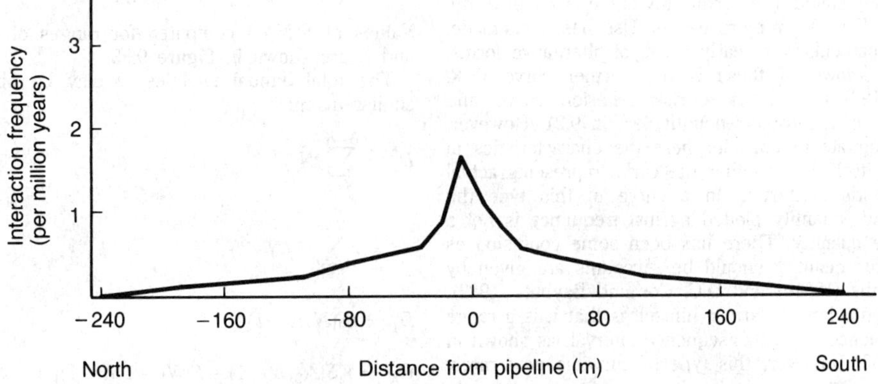

Figure 9.29 *Risk transect for interaction between a vapour cloud from a pipeline and the public (Ramsay, Sylvester-Evans and English, 1982) (Courtesy of the Institution of Chemical Engineers)*

Table 9.32 Assessed risks for some principal hazards given in the First Canvey Report (before proposed modifications) (after Lees, 1982b) (Courtesy of the Institution of Chemical Engineers)

Hazard	Frequency of initiating event (10^{-6}/year)	Frequencies (10^{-6}/year) for numbers of offsite casualties exceeding							
		0	10	1500	3000	4500	6000	12 000	18 000
Oil overtopping of bund by process explosion[a]	1000	975	25	18	8	4	–	–	–
LPG ship collision	6640	6490	196	124	64	31	–	–	–
LNG ship collision remote from jetty	50	45	5	3	1	0.5	0.3	0.2	0.1
LNG jetty incident	2000	1830	168	118	83	56	37	17	7
Ammonia storage sphere spontaneous failure	100	30	68	40	28	21	15	7	3
HF release[a]	200	30	168	144	132	120	114	80	70
Ammonium nitrate storage explosion	85	0	85	85	85	17	17	–	–

LNG, liquefied natural gas; LPG, liquefied petroleum gas.
[a] Occidental.

9.19.5 Risk aversion in FN curves

Generally, society is more averse to a single very large accident than to a number of small ones. Such risk aversion has been discussed in Chapter 4. In simple terms, risk aversion exists if society regards a single accident with 100 fatalities as in some sense worse than 100 accidents with a single fatality each.

Risk aversion is seen most clearly in FN curves, where the slope of the curve indicates the degree of risk aversion. If the FN relation is in fact a straight line, its slope is directly related to the parameter α, which is therefore known as the 'risk aversion index'.

9.19.6 Characteristics of FN curves

In order to make full use of the information given in FN curves and also to avoid certain misunderstandings which have arisen in their use, it is necessary to consider some of the characteristics of such curves. Accounts of FN curves include those given by R.F. Griffiths (1981e), the Royal Society Study Group (1983) and Hagon (1984).

As already stated, the normal FN curve is a plot of the frequency F of N or more events. Use has been made, however, particularly in early work, of alternative forms. The best known of these is the Farmer curve (F.R. Farmer, 1967b). This is a risk criterion curve and description of it is deferred until Section 9.21. However, it is appropriate to consider here the characteristics of the curve itself, i.e. whether the curve represents actual risk or a risk criterion. In a curve of this type the consequence quantity plotted against frequency is not a cumulative quantity. There has been some confusion as to what this quantity should be. Accounts are given by R.F. Griffiths (1981e) and D.C. Cox and Baybutt (1982). The interpretation of Farmer himself is that it is a range of consequences, or a consequence interval, as shown in Figure 9.32. However, this type of curve is now rarely used.

Accounts of the relations underlying FN curves include those by R.F. Griffiths (1981e), D.C. Cox and Baybutt (1982) and Hagon (1984). The following treatment is based on the work of the latter.

From Equation 9.19.1, the following relations may be stated for cumulative frequency F:

$$F_1^K = \sum_{N=1}^{N=K} f \qquad [9.19.2a]$$

$$= r\sum_{N=1}^{N=K} (1/N^{\alpha}) \qquad [9.19.2b]$$

Defining

$$S(K,\alpha) = \sum_{N=1}^{N=K} (1/N^{\alpha}) \qquad [9.19.3]$$

gives

$$F_1^N = rS(N,\alpha) \qquad [9.19.4]$$

Also

$$F_{N_1}^{N_2} = r[S(N_2,\alpha) - S(N_1 - 1,\alpha)] \qquad [9.19.5]$$

Values of $S(N,\alpha)$ computed for ranges of values of N and α are shown in Figure 9.33.

The total annual fatalities D may be obtained in a similar manner:

$$D_1^K = \sum_{N=1}^{N=K} fN \qquad [9.19.6a]$$

$$= r\sum_{N=1}^{N=K} (1/N^{\alpha-1}) \qquad [9.19.6b]$$

$$D_1^N = rS(N,\alpha - 1) \qquad [9.19.7]$$

$$D_{N_1}^{N_2} = r[S(N_2,\alpha - 1) - S(N_1 - 1,\alpha - 1)] \qquad [9.19.8]$$

If f is treated as a continuous function of N, the following relations apply:

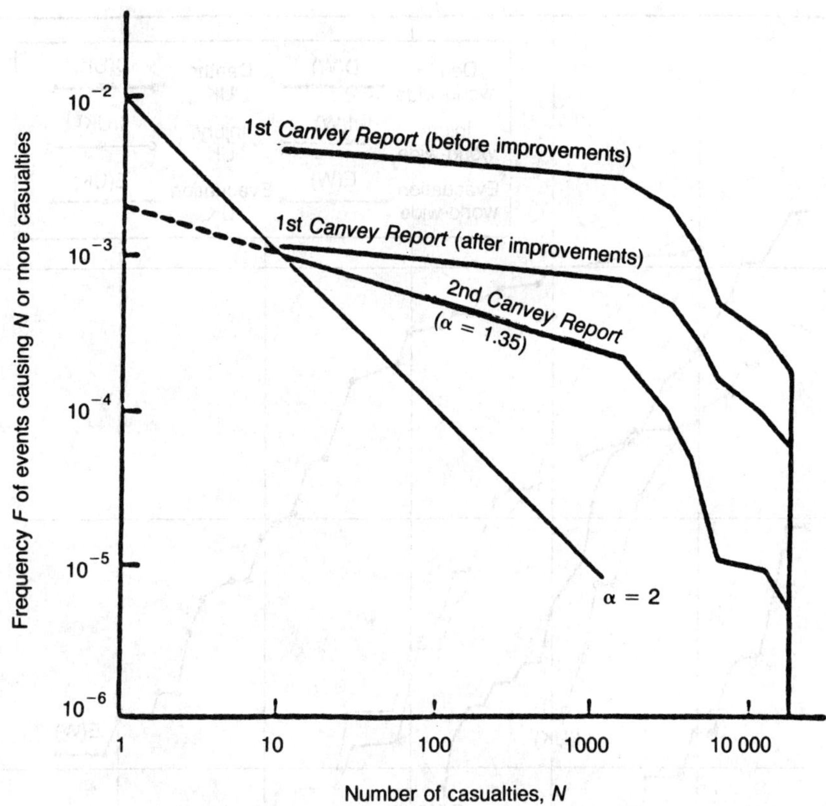

Figure 9.30 *FN curve for some principal hazards given in the first Canvey Report, (Hagon, 1984) (Reproduced by permission of the Institution of Chemical Engineers)*

$$F_1^K = \int_1^K f\,dN \qquad [9.19.9]$$

$$D_1^K = \int_1^K fN\,dN \qquad [9.19.10]$$

The relations for these integrations are shown in Table 9.33. The relations change at $\alpha = 2$. The values based on the assumption that N is continuous are not accurate for N less than approximately 30.

The application of these equations to a set of FN curves given in an actual risk assessment has also been described by Hagon. Figure 9.34 shows the FN curves concerned which are for the ammonia storage sphere in the *Rijnmond Report*.

Also shown in Figure 9.34 is a dotted line which corresponds to

$\alpha = 2$

$F_1^N = 10^{-3}$

Then for r taking some large value of N, say 5000,

$r = F_1^N / S(N, \alpha)$

From Figure 9.30, for $N = 5000$

$S(5000,\ 2) \approx 1.65$

Hence

$r = 10^{-3}/1.65 = 0.61 \times 10^{-3}$

Then the cumulative frequency of accidents and expected annual fatalities may be estimated. Values of these quantities derived from (a) the exact relations and (b) the approximate relations given in Table 9.33 are shown in Table 9.34. These equations are considered further in Section 9.21 in relation to risk criteria.

9.20 Confidence in Results

The results of a hazard assessment are subject to uncertainty and it is necessary therefore to be able to estimate the extent of this uncertainty. Discussions of uncertainty in hazard assessment include those by Nussey (1983), Goosens, Cooke and van Steen (1989) and Chhibber, Apostolakis and Okrent (1991)

9.20.1 Sources of uncertainty

Sources of uncertainty in a hazard assessment may be classed under the following broad headings:

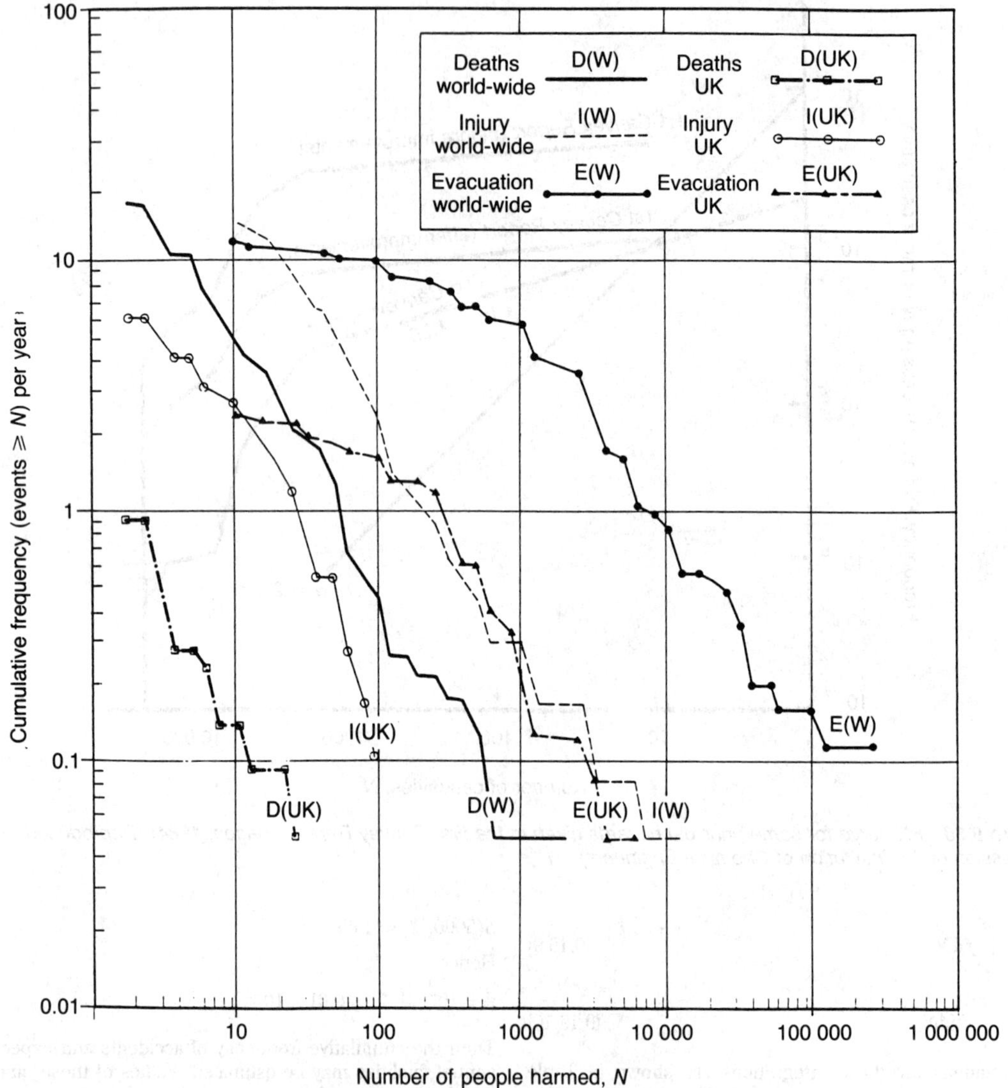

Figure 9.31 *Some FN curves for accidents in the chemical and petrochemical industries, in the UK and world-wide (HSE, 1989e) (Courtesy of the HM Stationery Office)*

(1) scenarios for initial event;
(2) scenarios for escalation;
(3) event frequency, probability;
(4) models for physical events;
(5) models for injury;
(6) models for mitigation.

These sources of uncertainty are amplified in Table 9.35. An account of sources of uncertainty with special reference to toxic release has been given by Nussey (1983).

9.20.2 Characterization of uncertainty

Uncertainty may attach to a single event or to a chain of events. The approach taken to the characterization of uncertainty depends on the complexity of the problem. The conventional analytical approach to uncertainty in the frequency, or probability, of a single event is to utilize not a point value but a distribution of values. A distribution which is widely utilized is the log–normal distribution, described below.

Continuing with an analytical approach, the propagation of uncertainty is handled by utilizing relationships

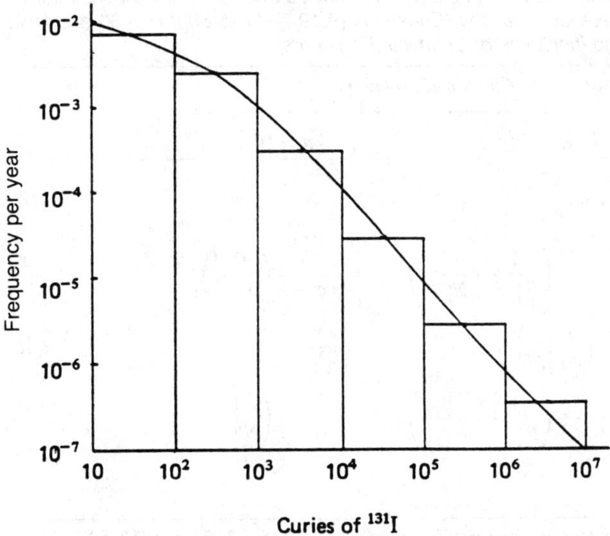

Figure 9.32 *Interpretation of the Farmer curve (R.F. Griffiths, 1981e) (Courtesy of Manchester University Press)*

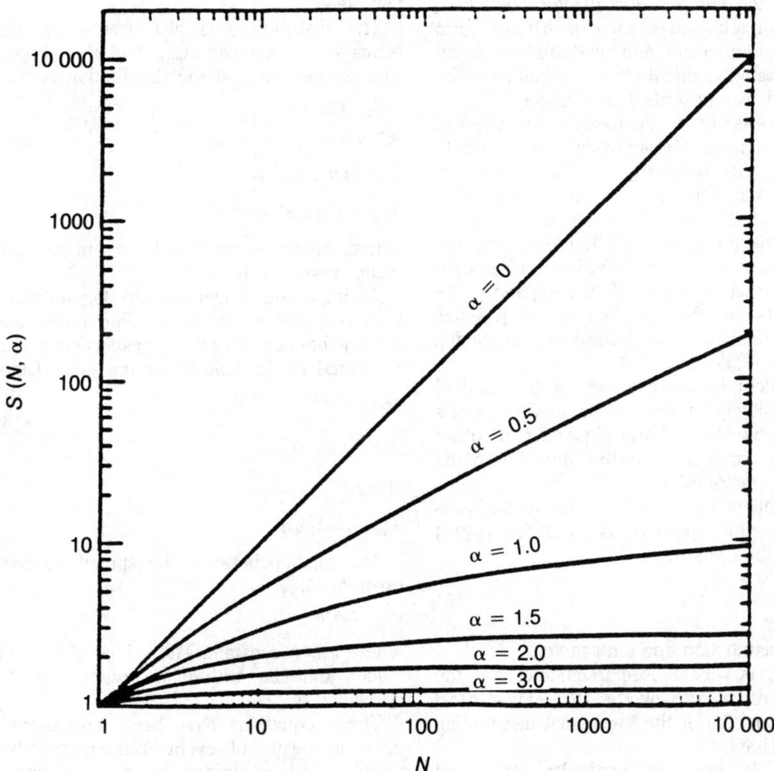

Figure 9.33 *Values of parameter S(N,α) (Hagon, 1984) (Courtesy of the Institution of Chemical Engineers)*

Table 9.33 *Approximate relations for cumulative frequencies and expected annual fatalities (Hagon, 1984). These relations approximate to the exact relations (Equations (9.19.3)–(9.19.6)) if N > 30 and limits are taken as $N_2 + 0.5$, N – 1 – 0.5 (Courtesy of the Institution of Chemical Engineers)*

Aversion index	*Cumulative frequency*		*Total deaths/year*
	$(F)_{N_1}^{N_2}$	$F_N = (F)_N^{\alpha}$	$(D)_{N_1}^{N_2}$
$\alpha = 1$	$r \ln \frac{N_2}{N_1}$	∞	$r(N_2 - N_1)$
$1 < \alpha < 2$	$\frac{r}{\alpha - 1}\left(\frac{1}{N_1^{\alpha-1}} - \frac{1}{N_2^{\alpha-1}}\right)$	$\frac{r}{\alpha - 1}\left(\frac{1}{N^{\alpha-1}}\right)$	$\frac{r}{2-\alpha}(N_2^{2-\alpha} - N_1^{2-d})$
$\alpha = 2$	$r\left(\frac{1}{N_1} - \frac{1}{N_2}\right)$	$\frac{r}{N}$	$r \ln \frac{N_2}{N_1}$
$\alpha > 2$	$\frac{r}{\alpha - 1}\left(\frac{1}{N_1^{\alpha-1}} - \frac{1}{N_2^{\alpha-1}}\right)$	$\frac{r}{\alpha - 1}\left(\frac{1}{N^{\alpha-1}}\right)$	$\frac{r}{\alpha-2}\left(\frac{1}{N_1^{\alpha-2}} - \frac{1}{N_2^{\alpha-2}}\right)$ $= \frac{r}{\alpha - 2}$ for $N_2 = \infty$, $N_2 = 1$

Note: The above relations approximate to the exact Equations 9.20.4, 9.20.5, 9.20.6 and 9.20.7 if $N > 30$ and limits are taken to be $N_2 + 0.5$, $N_1 - 0.5$.

for the combination of distributions, principally relations for the sum and product of distributions. The result obtained is a distribution of the frequency, or probability, of the event of interest.

If the problem is too complex for analytical solution, use may be made of techniques such as Monte Carlo simulation to obtain frequencies and probabilities. Again a frequency, or probability, distribution is obtained. The Monte Carlo method was described in Chapter 7.

Moving on to uncertainty in the models, for physical effects, injury and mitigation, the analytical approach is to derive relations for the sensitivity of the results to changes in the models. One method is to determine the first derivative of the output of the model to its inputs, in other words the gains. Another is to express the parameters of the model as distributions rather than point values. A formal method of investigating the sensitivity of results to the parameters of physical models and injury relations is the hazard impact model, described in Section 9.23.

If again the problem is too complex to be handled analytically, the sensitivity of the results to the models may be investigated by Monte Carlo simulation or other methods. Again the parameters in the model may be given in the form of distributions.

The characterization of uncertainty in the scenarios is less well developed. One approach is to utilize expert judgement, as described below.

9.20.3 Log–normal distribution and error factor

A distribution which is widely used to characterize the spread of values of a variable is the log–normal distribution. The properties of the lognormal distribution were described in Chapter 7.

This distribution is used in particular for event frequency λ and probability p. For a failure rate λ, the log–normal distribution may be written as

$$f(\lambda) = \frac{1}{(2\pi)^{\frac{1}{2}}\sigma\lambda} \exp\left[-\frac{(\ln\lambda - m^*)^2}{2\sigma^2}\right] \qquad [9.20.1]$$

where m^* is the location parameter and σ the shape parameter.

The distribution is also frequently characterized in terms of the median λ_{med} and the shape parameter σ. The median λ_{med} of the distribution is

$$\lambda_{med} = \exp m^* \qquad [9.20.2a]$$

$$m^* = \ln\lambda_{med} \qquad [9.20.2b]$$

The mean λ_{mn} is

$$\lambda_{mn} = \exp(m^* + \sigma^2/2) \qquad [9.20.3]$$

where subscripts med and mn indicate the median and mean, respectively.

An important property of the log–normal distribution is its associated error factor. For lower and upper confidence bounds λ_l and λ_u, respectively, the median λ_{med} is related to the bounds by the error factor ϕ:

$$\lambda_{med} = \phi\lambda_l \qquad [9.20.4a]$$

$$= \lambda_u/\phi \qquad [9.20.4b]$$

Hence

$$\lambda_{med} = (\lambda_l\lambda_u)^{\frac{1}{2}} \qquad [9.20.5]$$

The relation between the spread parameter σ and the error factor ϕ is

$$\sigma = \ln\phi/k \qquad [9.20.6]$$

where k is a constant. The values of k for 90% and 95% of values that lie within the range are 1.64 and 1.96, respectively.

These equations have been written for further reference in terms of event frequency λ but they could equally well be written in terms of any variable x. In particular, they are also applicable to event probability p, and are so utilized below.

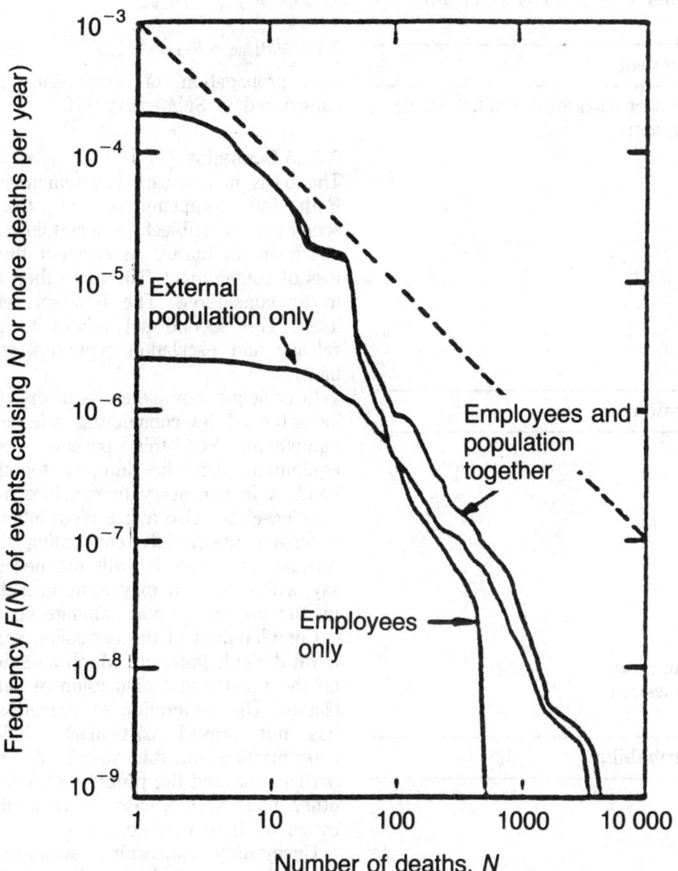

Figure 9.34 *FN curves for an ammonia storage sphere given in the* Rijnmond Report *(Hagon, 1984) (Courtesy of the Institution of Chemical Engineers)*

Table 9.34 *Cumulative frequencies and expected annual fatalities for an ammonia storage sphere (after Hagon, 1984) (Courtesy of the Institution of Chemical Engineers)*

	Approximate relation (per year)	*Exact relation* (per year)
F_3^{30}	1.8×10^{-4}	2.2×10^{-3}
F_{30}^{300}	1.8×10^{-5}	2.2×10^{-5}
D_1^{30}	3.5×10^{-3}	2.4×10^{-3}
D_1^{3000}	4.8×10^{-3}	5.2×10^{-3}

9.20.4 Propagation of uncertainty

A broad outline of approaches to the propagation of uncertainty was given above. An account is given here of the propagation of uncertainty in event rates.

Since a combination of events is considered, it is more appropriate to work in terms of probability p rather than frequency λ. For a variable p_3 which is the product of two independent variables p_1 and p_2 so that

$$p_3 = p_1 p_2 \qquad [9.20.7]$$

it can be shown that if variables p_1 and p_2 are lognormally distributed, the distribution of p_3 is also log–normal and the relations between the location parameters and the variances are

$$m_3^* = m_1^* + m_2^* \qquad [9.20.8]$$

$$\sigma_3^2 = \sigma_1^2 + \sigma_2^2 \qquad [9.20.9]$$

where subscripts 1–3 indicated events 1–3, respectively.

Then it follows from Equations 9.20.2, 9.20.3, 9.20.8 and 9.20.9 that the median and mean of p_3 are given by:

Table 9.35 Some sources of uncertainty in hazard assessment

A Scenarios for initial event

Some uncertainties in the scenarios for the initial event, and thus the source term, are:

Release sources

and, for each source:

Size
Orientation
Flash, rainout
Initial air entrainment
Duration

B Scenarios for escalation

Pool vaporization
Gas dispersion
Ignition
Jet flames
Pool fires
Fireballs
Flowing, burning liquid
BLEVEs
Missiles
Failure of emergency shut-down
Failure of fire protection system
Structural collapse

C Event frequency, probability

Failure frequency
Leak frequency
Ignition probability
Explosion probability

D Models for physical events

Flash fires
Pool fires
Fireballs
Jet flames
BLEVEs
Missiles

E Models for injury

Heat radiation
Explosion effects:
  Direct blast effects
  Housing collapse
  Missiles
Toxic effects

F Models for mitigation

Human behaviour
Escape
Shelter

$$p_{3,\mathrm{med}} = p_{1,\mathrm{med}} \times p_{2,\mathrm{med}} \quad [9.20.10]$$

$$p_{3,\mathrm{mn}} = p_{1,\mathrm{mn}} \times p_{2,\mathrm{mn}} \quad [9.20.11]$$

The propagation of uncertainty in a fault tree is considered in Subsection 9.20.7.

9.20.5 Scenarios

The basis of a hazard assessment is a set of scenarios. Both the completeness and the realism of these scenarios are subject to uncertainty.

Usually in hazard assessment the scenarios relate to loss of containment. There are then two sets of scenarios to be considered. The first set relates to the release itself. The second set relates to the escalation. Both release and escalation scenarios are subject to uncertainty.

In principle, completeness of the release scenarios may be achieved by considering release from each item of equipment. For this purpose certain categories of equipment may be lumped together. On the other hand, it is necessary to consider not just a single size of release, but also a size distribution. There is, however, a further uncertainty concerning the geometry of the release. The material will not necessarily come out as, say, a free jet, but may issue at a downwards angle and hit the ground or may impinge on other equipment.

Completeness of the escalation scenarios is probably a more difficult problem. Much attention been has focused on the release and dispersion of flammable or toxic gas clouds. The generation of scenarios for such a hazard has not proved particularly difficult. However, the concentration on this rather stereotyped situation has perhaps masked the problem of defining the escalation in other, more varied, cases. Such a situation is exemplified by an offshore module.

Uncertainty concerning scenarios shades over into uncertainty concerning models, as described below.

9.20.6 Event rate

For an event rate, a frequency or probability, uncertainty may be taken into account by treating the event rate as a random variable with a distribution of values. This approach was used in the *Rasmussen Report* and since then has been widely adopted. In this study data for both failure frequencies and probabilities were correlated using the log–normal distribution, as described in Section 9.20.3.

The procedure adopted was to consider for each equipment the available data on failure frequency or probability and to determine by expert judgement the lower and upper bounds for the range of values covering 90% of cases. The tables of failure data quote the lower and upper bounds, the median and the error factor.

9.20.7 Fault trees

It is also necessary to be able to obtain the probability distribution of the frequency of an event which has been synthesized using a fault tree. Where a measure of the uncertainty in the frequency estimate is made in a risk assessment, this is generally done numerically using Monte Carlo simulation. The frequencies of the base events in the tree are expressed in terms of probability distributions of frequency, such as the log–normal distribution as just described. Each simulation trial generates a set of frequencies for the base events. These are combined according to the tree structure to obtain a frequency of the top event. The results of a series of such trials is the probability density function for the top event. The principle of the method is illustrated in Figure 9.5.

Analytical treatments of the propagation of uncertainty through the logic gates of a fault tree are also available for certain defined cases. This approach is illustrated here by the work of Keey and Smith (1984). The problem considered is the combination at OR and AND gates of inputs which are normally distributed. Taking first an OR gate, consider two inputs whose frequencies take random values λ_1 and λ_2 with corresponding mean values $\bar{\lambda}_1$ and $\bar{\lambda}_2$ and standard deviations σ_1 and σ_2. The frequency λ of the output from the OR gate is

$$\lambda = \lambda_1 + \lambda_2 \quad [9.20.12]$$

and the corresponding mean frequency $\bar{\lambda}$ is

$$\bar{\lambda} = \bar{\lambda}_1 + \bar{\lambda}_2 \quad [9.20.13]$$

But mean values can be expressed as expected values. Then Equation 9.20.13 can be rewritten as

$$E[\lambda] = E[\lambda_1] + E[\lambda_2] \quad [9.20.14]$$

The variance of the output becomes

$$\sigma^2 = E[\lambda^2] - (E[\lambda])^2 \quad [9.20.15]$$

Then, from the properties of expected values,

$$E[\lambda^2] = E[\lambda_1^2 + 2\lambda_1\lambda_2 + \lambda_2^2] \quad [9.20.16]$$

$$= E[\lambda_1^2] + E[2\lambda_1\lambda_2] + E[\lambda_2^2] \quad [9.20.17]$$

Also

$$(E[\lambda])^2 = \bar{\lambda}^2 = \bar{\lambda}_1^2 + 2\bar{\lambda}_1\bar{\lambda}_2 + \bar{\lambda}_2^2 \quad [9.20.18]$$

From Equations 9.20.15 and 9.20.17

$$E[\lambda^2] = E[\bar{\lambda}_1^2 + \sigma_1^2] + E[\bar{\lambda}_2^2 + \sigma_2^2] + 2\bar{\lambda}_1\bar{\lambda}_2 \quad [9.20.19]$$

Substitution of Equations 9.20.18 and 9.20.19 into Equation 9.20.15 gives

$$\sigma^2 = (\bar{\lambda}_1 + \sigma_1^2) + (\bar{\lambda}_2 + \sigma_2^2) - (\bar{\lambda}_1 + \bar{\lambda}_2) \quad [9.20.20]$$

$$= (\sigma_1^2 + \sigma_2^2)^{\frac{1}{2}} \quad [9.20.21]$$

A measure of spread of the frequency is the coefficient of variation C:

$$C_i = \sigma_i/\lambda_i \quad [9.20.22]$$

The coefficient of variation of the output is

$$C = \sigma/\bar{\lambda} = (\sigma_1^2 + \sigma_2^2)^{\frac{1}{2}}/(\bar{\lambda}_1 + \bar{\lambda}_2) \quad [9.20.23]$$

One limiting case is where the two inputs have the same mean value ($\lambda_0 = \lambda_1 = \lambda_2$) and same standard deviation ($\sigma_0 = \sigma_1 = \sigma_2$). Then

$$C = (2\sigma_0^2)^{\frac{1}{2}}/2\bar{\lambda}_0 = 2^{\frac{1}{2}}\sigma_0/2\bar{\lambda}_0 \quad [9.20.24]$$

or

$$C = C_0/2^{\frac{1}{2}} \quad [9.20.25]$$

Another limiting case is where one input becomes single valued ($\sigma_1 = 0$) and very small ($\lambda_1 \ll \lambda_2$):

$$C = (\sigma_2^2)^{\frac{1}{2}}/(\bar{\lambda}_1 + \bar{\lambda}_2) = \sigma_2/\lambda^2 \quad [9.20.26]$$

or

$$C = C_2 \quad [9.20.27]$$

Hence the output coefficient of variation has the range:

$$(1/2^{\frac{1}{2}}) \min(C_1, C_2) < C < \max(C_1, C_2) \quad [9.20.28]$$

Equation 9.20.18 shows that for an OR gate the coefficient of variation of the output is less than the larger of the coefficients of variation of the inputs, and that in this sense this gate reduces the uncertainty in the input data.

For an AND gate with an input of frequency λ_1 and another input of probability p_2, with corresponding standard deviations σ_1 and s_2, respectively, the frequency λ of the output is

$$\lambda = \lambda_1 p_2 \quad [9.20.29]$$

The corresponding mean frequency $\bar{\lambda}$ is related to the mean frequency $\bar{\lambda}_1$ and probability $\bar{p}_2$ of the inputs:

$$\bar{\lambda} = \bar{\lambda}_1\bar{p}_2 \quad [9.20.30]$$

Then

$$E[\lambda^2] = E[\lambda_1^2 p_2^2] \quad [9.20.31a]$$

$$= E[\lambda_1^2] \cdot E[p_2^2] \quad [9.20.31b]$$

$$= (\bar{\lambda}_1^2 + \sigma_1^2)(\bar{p}_2^2 + s_2^2) \quad [9.20.31c]$$

Also

$$(E[\lambda])^2 = \bar{\lambda}^2 = \bar{\lambda}_1^2\bar{p}_2^2 \quad [9.20.32]$$

Then, from Equation 9.20.15, the standard deviation of the output σ is

$$\sigma^2 = (\bar{\lambda}_1^2 + \sigma_1^2)(\bar{p}_2^2 + s_2^2) - \bar{\lambda}_1^2\bar{p}_2^2 \quad [9.20.33]$$

Hence

$$\sigma = (\sigma_1^2 s_2^2 + \bar{\lambda}_1^2 s_2^2 + \sigma_1^2\bar{p}_2^2)^{\frac{1}{2}} \quad [9.20.34]$$

The output coefficient of variation is then

$$C = \sigma/\bar{\lambda} = (\sigma_1^2 s_2^2 + \bar{\lambda}_1^2 s_2^2 + \sigma_1^2\bar{p}_2^2)^{\frac{1}{2}}\bar{\lambda}_1^2\bar{p}_2 \quad [9.20.35]$$

Noting that

$$C_2 = s_2/\bar{p}_2 \quad [9.20.36]$$

Equation 9.20.35 can be recast in terms of the input coefficients of variation

$$C = (C_1^2 + C_2^2 + C_1^2 C_2^2)^{\frac{1}{2}} \quad [9.20.37]$$

One limiting case is where both input coefficients have the same value ($C_0 = C_1 = C_2$)

$$C = C_0(2 + C_0^2)^{\frac{1}{2}} \quad [9.20.38]$$

Another limiting case is where one input coefficient becomes very small. Hence the output coefficient of variation has the range

$$C_m < C < C_0(2 + C_0^2)^{\frac{1}{2}} \quad [9.20.39]$$

with

$$C_m = \max(C_1, C_2) \quad [9.20.40]$$

Relation 9.20.39 shows that for an AND gate the coefficient of variation of the output is more than the larger of the coefficients of variation of the inputs, and that in this sense this gate increases the uncertainty in the input data.

These results show that strictly in terms of error propagation for a sizeable fault tree a large number of initiating events may produce an outcome for which the uncertainty is relatively low, while a large number of mitigating or protective features may produce one for which the uncertainty is relatively high.

9.20.8 Event trees

An event tree consists of a series of binary splits. Each branch of the tree constitutes a separate escalation scenario. Uncertainty attaches both to the set of escalation scenarios and then, assuming that the set is correct, to the probability of each branch. The question of uncertainty in the escalation scenarios was discussed in Section 9.20.5. The probability distribution for the occurrence of a given branch may be obtained from probability distributions for each of the branch points.

9.20.9 Scenarios, models and parameters

The uncertainty in models has been discussed by Chhibber, Apostolakis and Okrent (1991), in the context of modelling of the fate of chemicals in the environment, particularly in geohydrochemical systems. They consider especially the following areas of uncertainty: (1) completeness of scenarios, (2) completeness of chains of consequences, (3) validity of individual models and (4) parameters in these models, with particular reference to expert judgement.

The formulation of scenarios by experts has been the subject of study. *The Reactor Risk Reference Document* of the Nuclear Regulatory Commission (NRC) (1989 NUREG-1150) gives an account of the expert treatment of scenarios in terms of 'issues' and 'levels' for a nuclear incident. The issues concern the scenarios which are credible. The levels represent different sets of assumptions about these scenarios.

Likewise, the models, whether for physical phenomena or for injury, may be based on a wide variety of assumptions and may thus be very disparate. Up to a certain point, variation may be handled by assigning different values to the parameters in the model, but beyond this point the model can be stretched no more, and a different model has to be used. There can be a thin dividing line between models and parameters.

Similarly, the value assigned to a parameter may depend on the model used. A well-known case is particle diameter, which may be characterized in different ways, depending on the situation being modelled. The relationships between models are a problem area. The models are not necessarily exclusive, exhaustive or independent.

9.20.10 Physical models

In general, the question of uncertainty in the models of physical phenomena is a relatively neglected one. Whereas treatments are available for uncertainty in event data and fault trees, there is no real equivalent for models. The problem is considered here first in terms of certain general principles. The role of expert judgement in selecting models is then described.

It is possible to identify certain features of models which bear on the degree of uncertainty. Models may be

(1) theoretical;
(2) empirical;
(3) semi-empirical.

They may also be

(1) deterministic;
(2) probabilistic.

The characteristics of these different types of model are well known. A theoretical model may have the weakness that it fails to take into account some relevant phenomenon. It may contain parameters to which it is difficult to assign accurate values or which it is impractical to quantify at all. The weakness of an empirical model may be that it has a limited, and poorly defined, range of applicability. A semi-empirical model, incorporating some combination of theory and experiment, may go some way to overcoming these weaknesses. Alternatively, a straight theoretical model may be validated by experiments.

A comprehensive model incorporates knowledge of the relevant regimes, the ranges of these regimes, and, in each regime, the mechanisms and the values of the parameters. Some factors which bear on the confidence which can be placed in a model include:

(1) theoretical basis;
(2) experimental validation;
(3) range of applicability;
(4) variety of tests;
(5) origin of model;
(6) life of, and interest in, model;
(7) number and convergence of competing models.

Confidence in a model must be greater if it has a theoretical basis. Equally, experimental validation increases confidence. It is, however, the quality of these two factors which matters. It is not uncommon that a theoretical model with some experimental support later proves inadequate.

The range of applicability is therefore important. Greater confidence can be placed in a model which covers a wide range and has been tested over this range. Testing using a variety of methods also increases confidence.

In assessing quality the origin of the model is a crucial factor. Notwithstanding the challenge to conventional thinking which sometimes occurs from a previously unknown author, confidence is clearly greater if the workers(s) concerned are well established.

Where there is a single model, confidence should in principle increase with time. This will be true, however, only if there is some interest in the model so that the passage of time does actually result in activities which provide some degree of support for it.

A more encouraging situation is where there is a number of competing models that over a period of time yield results which converge. Whilst this by no means gives a guarantee of correctness, it does nevertheless build confidence.

These points are illustrated by the history of the development of models for heavy gas dispersion. In the early days the models used for gas dispersion were neutral density models. These had both a strong theoretical basis and extensive experimental support. They originated largely from the Meteorological Office and had a good scientific pedigree. It was shown by experiment, however, that the behaviour of a heavy gas cloud is quite different from that of a neutral density gas. It has some of the characteristics of the flow of a liquid, but differs in that air is entrained at the top and the edge of the cloud. This led to the development of a number of models for heavy gas dispersion.

Estimates of the travel distance of a flammable cloud were given in an early study, *Transportation of Liquefied Natural Gas* by the Office of Technology Assessment (OTA) of the US Congress (1977). The following estimates were quoted for the maximum downwind distance for a cloud from a 25 000 m^3 instantaneous spill of LNG onto water in a 5 mile/h wind for dilution to 5% concentration:

Travel distance (miles)	*Source*
0.75	Federal Power Commission
1.2[a]	Science Applications Inc.
5.2	American Petroleum Institute
16.3	US Coast Guard
25.2–50.3	US Bureau of Mines

[a]37 500 m^3 release in a 6.7 mile/h wind.

The figures illustrate the wide spread given by the different early models. As the field has become more mature, this wide spread has disappeared. There are now available heavy gas dispersion models which have emerged from a large field of competitive models and for which there is experimental underpinning from both large-scale field trials and wind tunnel tests.

9.20.11 Injury relations

Many of the same general considerations that apply to physical models apply also to injury relations, but in addition the latter have some characteristics of their own. In general, an injury relation is obtained either from laboratory work on animals or from incident data affecting humans. In the former case confidence limits are usually given for the results. Uncertainty arises, however, concerning the applicability of the results to man. If the data relate to an incident, there may be uncertainty concerning both the physical phenomenon and the exposure of humans to it. It is difficult to provide confidence limits. The uncertainty tends to be greatest at the two extremes of very low and very high probability of the injury effect and least at the 50% probability level. Often the injury relations are expressed as probit equations. The remarks just made apply equally to these.

9.20.12 Expert judgement

The choice of models to be used is normally made by a single expert. One of the main methods available to assess the uncertainty in the models is that of expert judgement, using a panel of experts.

Several studies of expert judgement in assessment of models have been made. One is the NRC study already mentioned (NUREG-1150).

Another is the work of Goossens, Cooke and van Steen (1989). This addresses uncertainty in the relative likelihood of different phenomena and in the values of parameters. The areas of application include failure rates, physical models and injury effects.

Models are developed by a domain expert. The uncertainty in the models may be assessed by a normative expert, whose expertise resides in such assessment.

One way of quantifying the uncertainty in a model is the error factor method. The error factor was described in Section 9.20.3. Consider a deterministic reference model (DRM) which yields some variable T. Then using an error factor E,T may be written as

$$T = ET_{\mathrm{DRM}} \quad [9.20.41]$$

where T_{DRM} is the value of given by the DRM. The error factor is external to the model. It may be updated whilst the model remains unchanged. This error factor approach has been applied by Siu and Apostolakis (1985) in the context of fire growth in a building. It is described further by Chhibber, Apostolakis and Okrent (1991).

It can occur that experimental data become available which might be used to modify a model. In principle, such updating may be effected using Bayesian methods, but this may not be straightforward. In order to apply the Bayesian approach it is necessary to formulate a likelihood function and for a complex model it may not be easy to do this. In this case one approach may be to use expert judgement methods to formulate a suitable likelihood function.

9.20.13 Results

The results of a hazard assessment should in principle include not only the best estimate of the risk but also a measure of the uncertainty which attaches to this estimate. There are, however, problems associated with providing such measures of uncertainty. These include: the difficulties of principle in assigning uncertainty, particularly to models; the effort required to compute the measures, which may be comparable with that of making the basic estimate; and an inability to make effective use of this additional information in decision-making.

9.21 Risk Criteria

In order to evaluate a risk assessment it is necessary to have appropriate risk criteria. Some of the more philosophical aspects of risk are considered in Chapter 4. The treatment here is confined to a consideration of some of the principles which should underlie the choice of risk criteria and of some of the types of criterion, systems and numerical criterion values used or proposed.

Accounts of risk criteria include those given in *Dealing with Risk* (R.F. Griffiths, 1981c), *Risk Assessment. A Report of a Royal Society Study Group* (Royal Society Study Group, 1983) (the RSSG Report), the *Third Report of the Advisory Committee on Major Hazards* (Harvey, 1984), *The Tolerability of Risk from Nuclear Power Stations* (HSE, 1988c) and *Risk Criteria for Land-use Planning in the Vicinity of Major Industrial Hazards* (HSE, 1989c). Selected references on risk criteria are given in Table 9.36.

Table 9.36 *Selected references on toleration of risk and risk criteria*

Kletz (1971, 1976d, 1977c, 1980a,b, 1980–, 1981d,f,k,m, 1982e, 1983d,g); Grosser *et al.* (1964); Sowby (1964, 1965); Crowe (1969); Starr (1969, 1970, 1972, 1985); R.M. Stewart (1971); NAE (1972); Robens (1972); Sinclair (1972); Sinclair, Marstrand and Newick (1972a,b); Tudor-Hart (1972); Weinberg (1972a); Siccama (1973); Baldewicz (1974); Brook (1974); Hirschleifer, Bergstrom and Rappaport (1974); Koshland (1974); Lawless (1974); Sather (1974, 1976); AEC (1975); Chicken (1975); Linnerooth (1975); Otway (1975); R. Wilson (1975); Ash, Baverstock and Vennart (1976); J.H. Bowen (1976); Gibson (1976a,c,d,f, 1977a,b); C.H. Green and Brown (1976/77, 1978, 1980a,b); Hammond and Adelman (1976); Harvey (1976, 1979b, 1984); Kastenburg, McKone and Okrent (1976); Lowrance (1976); Slovic, Fischoff and Lichtenstein (1976, 1981, 1982); Solomon, Rubin and Okrent (1976); Starr, Rudman and Whipple (1976); Council for Science and Society (1977); F.R. Farmer (1977b); Glaser (1977); Mcginty and Atherley (1977); McLean (1977a, 1982); Mark and Stewart (1977); Rowe (1977); Adcock (1978); Boe (1978); Critchley (1978); HSC (1978c); HSE (1978b, 1981a); Rothschild (1978); Throdahl (1978); Dunster and Vinck (1979); C.H. Green (1979); Nelkin and Pollak (1979); Atallah (1980, 1981b–e); Brett-Crowther (1980a); Conrad (1980); Dierkes, Edwards and Coppock (1980); Dowie and Lefrere (1980); Giarini (1980); Goodin (1980); R.F. Griffiths (1980); Houston (1980); Risk Research Committee (1980); Schwing and Albers (1980); Starr and Ritterbush (1980); F. Warner (1980, 1981a,b); Weaver (1980); Zuckerman (1980); Berg and Maillie (1981); A.V. Cohen (1981); Cotgrove (1981); Fischhoff *et al.* (1981); Harvey (1981); Hildyard (1981); T.R. Lee (1981); McLoughlin (1981); Okrent (1981); Payne (1981); F. Warner and Slater (1981); von Winterfeldt, John and Borcherding (1981); Wu-Chien and Apostolakis (1981); Hohenemser and Kaspersen (1982); Lagadec (1982); Lind (1982); Kinchin (1982); Macgill (1982); W.W. May (1982); O'Riordan (1982); Otway and Thomas (1982); Covello *et al.* (1983); Giannini and Galluzzo (1983); The Royal Society (1983, 1992); Covello (1984); Jouhar (1984); Pierson (1984); Sass (1986); BMA (1987); Ashmore and Shama (1988); Corbett (1988 LPB 82); Cumo and Naviglio (1989); H.J.S. Petersen (1989); Smithson (1989); V.C. Marshall (1991a); Vlek and Keren (1991); Alder and Ashurst (1992); Engineering Council (1992); Philley (1992a); Council (1993); Irish (1993); Melchers (1993b); Melchers and Steward (1993); Song, Black and Dunne (1993)

Risk perception
Douglas (1966, 1982, 1985, 1987, 1990, 1992); Ashby (1977); ICRP (1977a,b); Fischhoff *et al.* (1978); C.H. Green and Brown (1978); Lichtenstein *et al.* (1978); Ravetz (1979); Slovic, Fischhoff and Lichtenstein (1980); M. Thompson (1980); Wynne (1980, 1982, 1989, 1992); Fischhoff *et al.* (1981); Watson (1981); Cotgrove (1982); Douglas and Wildavsky (1982a,b); Otway and Winterfeldt (1982); Fischhoff and Macgregor (1983); E.J. Johnson and Teversky (1984); Slovic, Lichtenstein and Fischhoff (1984); J.G.U. Adams (1985); Pidgeon, Blockley and Turner (1986); Slovic (1986); B.B. Johnson and Covello (1987); R. Wilson and Crouch (1987); Rip (1988); Wildavsky (1988); J. Brown (1989); Sharlin (1989); M. Thompson, Ellis and Wildavsky (1990); Beder (1991); C.H. Green, Tunstall and Fordham (1991); Helms (1981); Slovic, Flynn and Layman (1991); Krimsky and Goldring (1992)

Risk communication
MCA (n.d./17); Wynne (1982, 1992); Dewhurst (1986); Plough and Krimsky (1987); Corbett (1988); Covello and Allen (1988); Covello, Sandman and Slovic (1988); Kasperson *et al.* (1988); Hadden (1989a,b); F.R. Johnson and Fisher (1989); NRC (1989); Otway and Wynne (1989); Siegel (1989); Bord and O'Connor (1990); Handmer and Penning-Rowsell (1990); Covello (1991); A. Fisher (1991); Kasperson and Stallen (1991); B.B. Johnson (1992); AIHA (1993/28)

Natural and man-made hazards
Macdonald (1972); Blume (1978); Burton, Kates and White (1978); M.G. Cooper (1985); C.H. Green, Tunstall and Fordham (1991)
Aircraft: North (1949); M. Hill (1971); Warren (1977)
Buildings: Flint (1981); Holdgate (1981)
Chemicals: McLean (1979, 1981)
Chronic hazards: Travis *et al.* (1987); J.B. Cox (1989)
Flooding (Dutch dikes): van Dantzig (1960); Turkenburg (1974); Harvey (1976)
Foodstuffs: McLean (1977b)
Leisure: J. Wilkinson (1981)
Medical: Roach (1970); Leach (1972); H. Miller (1973); Pochin (1975, 1981a,b)
Nuclear radiation: Burhop (1977); Pochin (1983)
Roads: Reynolds (1956); Dawson (1967); Hayzelden (1968); A.D. Little (1968); Jones-Lee (1969); Thedie and Abraham (1969)

Major hazards
Harvey (1976, 1979b, 1984); S.B. Gibson (1978); Lowe (1980); Helsby and White (1985)

Nuclear industry
F.R. Farmer (1967b, 1971); AEC (1975); Weinberg (1976); Orr (1977); Pochin (1983); HSC (1978c); J. Hill (1981); W.L. Wilkinson (1981); Okrent and Baldewicz (1982); T.R. Lee, Brown and Henderson (1984); Franklin (1985); C. Tayler (1985f); Gunning (1987); HSE (1988c)

Conventional energy sources
Orr (1977); HSC (1978c); HSE (1978/4, 1980 RP 11); A.V. Cohen and Pritchard (1980); S. Russell and Ferguson (1980); Ferguson (1981, 1982); Inhaber (1981a,b); K. Thomas (1981); Dunster (1982); IAEA (1984); Fremlin (1985)

Fire, explosion
Melinek (1972 FRS Fire Research Note 950, 1973 FRS Fire Research Note 978, 1974 BRE CP 88/74); North (1973 FRS Fire Research Note 981); Coward (FRS 1973 Fire Research Note 982)

Chemical industry
F.R. Farmer (1971); Kletz (1971, 1972a, 1977a,f, 1980a,b, 1981a); Bulloch (1974); S.B. Gibson (1976a,c,d,f, 1977a,b); Harvey (1976); HSE (1978b, 1981a); de Heer,

Kortlandt and Hansen (1980); Livingstone (1979); N.C. Harris (1982, 1986)

Gas terminals
Keeney (1980); Kunreuther, Linnerooth and Starnes (1981)

Offshore (*see* Table A18.1)

Environment
Sewell (1971); Programmes Analysis Unit (1972); Ashby (1976); Barker (1977); Fischoff *et al.* (1978); Kletz (1981b)

Fatality rates
Pochin (1973); Fryer and Griffiths (1978 SRD R110, 1979 SRD R149); Grist (1978 SRD R125); Anon. (1980n); C. Wright (1986)

Risk criteria
NRC (Appendix 28 *Risk Criteria, Safety Goals*); Kletz (n.d.b, 1971, 1977a,b, 1980a,b); Provinciale Waterstaat Groningen (1979); S.B. Gibson (1981); Lees (1980a); Lowe (1980); Rasbash (1980b, 1985); R.F. Griffiths (1981b–d); Petkar (1981); D.C. Cox and Baybutt (1982); Matthews (1982); Okrent and Baldewicz (1982); Hagon (1984); Holden (1984); Helsby and White (1985); Ministry of Housing (1985); R. Wilson and Crouch (1987); Directorate General for Environ. Prot. (1988–89); NSW Govt (1990); ACDS (1991); Cameron and Corran (1993)
Fatal accident rate (FAR): Kletz (1971, 1978a); B.J. Wilson and Myers (1979); A.F.C. Wallace (1980); Harrod (1981); Lees (1981b)
Value of a life: Hayzelden (1968); Schelling (1968); FRS (1972 Fire Research Note 950, 1973 Fire Research Note 978, 982); Melinek (1974); Linnerooth (1975); Jones-Lee (1976, 1982, 1989); Mooney (1977); BRE (1978 CP52/78); Marin and Psacharoploulos (1982); Slovic *et al.* (1984); Jones-Lee, Hammerton and Philips (1985); Kletz (1985c); Moore and Viscusi (1988); A. Fisher, Chestnut and Violette (1989); Nawar and Salter (1993)
HSE criteria: D.A. Jones (1989)
Delayed fatalities: R.F. Griffiths (1994a)
Ring of igniters: Fallows (1982)

Cost–benefit analysis
Jones-Lee (1969, 1976, 1982, 1989); Marin and Psacharopoulos (1982); May (1982); Jones-Lee, Hammerton and Philips (1985); Marin (1992)

9.21.1 Royal Society Study Group

The *RSSG Report* sets out some possible quantitative guidelines for risk. It prefaces these by a number of qualifications. The risks considered are primarily risks of death, but in addition some suggestions are also made for criteria for non-fatal risks.

The RSSG define an upper bound of risk above which the risk is 'unacceptable in essentially all circumstances', except perhaps for activities entered into voluntarily or in wartime. It also defines a lower bound below which risks may 'legitimately be regarded as trivial by the decision-maker'. Between these two bounds there lies a region in which the risk needs to be assessed and all reasonably practicable steps taken to reduce it. This is referred to here as the 'three-region approach', as opposed to the point value, approach.

The RSSG suggest that most people regard as insignificant a risk of death below about 10^{-6}/year. This therefore constitutes a lower bound. For the upper bound the RSSG state that few would dissent from the proposition that imposing a continuing risk of 10^{-2}/year is unacceptable. The position is less clear with regard to a risk of 10^{-3}/year. For males only, those between 1 and 20 years of age have an annual risk of death markedly less than this. Therefore pre-existing levels of risk are more likely to cause death than a new imposed risk. The RSSG argue that an imposed risk of 10^{-3}/year can therefore hardly be called totally unacceptable provided that the individual is aware of the situation. The RSSG finally take this value of 10^{-3}/year as the upper bound. The central region contained within these bounds is therefore that where the risk lies between 10^{-6}/year and 10^{-3}/year. It is in this region that the concept of reducing risk 'as far as reasonably practicable' applies. As indicated, the upper bound of 10^{-3}/year is evidently associated with situations in which there is a degree of voluntary acceptance.

The risk criteria described are those which may be used by the decision-maker, particularly the central decision-maker in government. The categorization of a risk as insignificant does not necessarily imply that it will be perceived as such by those affected, but relates rather to the action which the decision-maker should require.

9.21.2 Advisory Committee on Major Hazards

In its First *Report* (Harvey, 1976), the ACHM made the statement

> If, for instance, such tentative conclusions indicated with reasonable confidence that in a particular plant a serious accident was unlikely to occur more often than once in 10,000 years (or – to put it another way – a 1 in 10,000 chance in one year), this might perhaps be regarded as just on the borderline of acceptability. . . .

Insofar as the committee was set up following the Flixborough disaster, in which 28 people were killed, it may be assumed to have had in mind a similar accident, say one involving 30 fatalities. The above statement has been taken by a number of parties as a starting point for the development of criteria for societal risk. Otherwise, however, the ACHM proposed little by way of quantitative risk criteria. It did, however, in its Third *Report* (Harvey, 1984) state a number of principles, including the following:

(1) The risk from a major hazard to an individual employee or member of the public should not be significant when compared with other risks to which a person is exposed in everyday life.
(2) The risk from any major hazard should, whenever reasonably practicable, be reduced.
(3) Where there is a risk from a major hazard, additional hazardous development should not add significantly to the existing risk.

(4) If the possible harm from an incident is high, the risk that the incident might actually happen should be made very low indeed.

9.21.3 HSE nuclear installations risk criteria

The HSE have published two documents outlining their approach to the development of risk criteria. The first of these is *The Tolerability of Risk from Nuclear Power Stations* (HSE, 1988c). This document gives the following treatment of risk of death to workers and to members of the public. A distinction is made between risk from normal operation, due essentially to radiation dose, and risk from a nuclear accident.

At nuclear installations the actual risk to workers from the average levels of radiation dose lies between about 10^{-4}/year and 2.5×10^{-4}/year. A very small number of workers are exposed to a risk some ten times greater. The figure of 2.5×10^{-4}/year happens to correspond to the risk to workers in heavy manufacturing and mineral extraction and that of 10^{-4}/year to the average for manufacturing industry. The HSE conclude that, broadly, the limit of tolerable risk to a worker is 10^{-3}/year.

On the basis that the risk to a member of the public should be at least an order of magnitude lower than that to a worker, the limit of tolerable risk to a member of the public is taken as 10^{-4}/year. The risk to a member of the public which might be regarded as acceptable, as opposed to tolerable, is then taken 10^{-6}/year. This appears to be a level at which in general a risk causes little concern. It happens to be approximately the risk of being electrocuted at home and one hundredth of the risk of dying in a traffic accident.

The actual risk to a member of the public from normal operation of a nuclear installation is well below 10^{-5}/year and that from a nuclear accident at an installation designed to Nuclear Installation Inspectorate (NII) principles is about 10^{-6}/year. The risk to a member of the public from both causes is assessed as 10^{-6}/year for most people in the vicinity of the installation, but for a small number it is about 10^{-5}/year and for a very few it may exceed this.

The document does not given a criterion for societal risk in terms of FN curves or the equivalent. Instead it takes as a measure of societal risk three scenarios: (1) a limiting design basis accident, (2) an uncontrolled release, and (3) an uncontrolled release large enough to produce doses of 100 mSv within 3 km. It suggests that the limit of tolerable risk of a considerable uncontrolled release anywhere in the UK might be about 10^{-4}/year. The assessed risk quoted for such a release is in fact 10^{-6}/year.

9.21.4 The ALARP principle

The HSE nuclear risk document also deals with the question of reducing the risk to a level that is as low as reasonably practicable (ALARP). The reduction of risk in accordance with the ALARP principle is a basic requirement of the Health and Safety at Work etc. Act 1974.

The ALARP principle is illustrated in Figure 9.35. There is some level above which the risk is intolerable. Above this level the risk cannot be justified on any

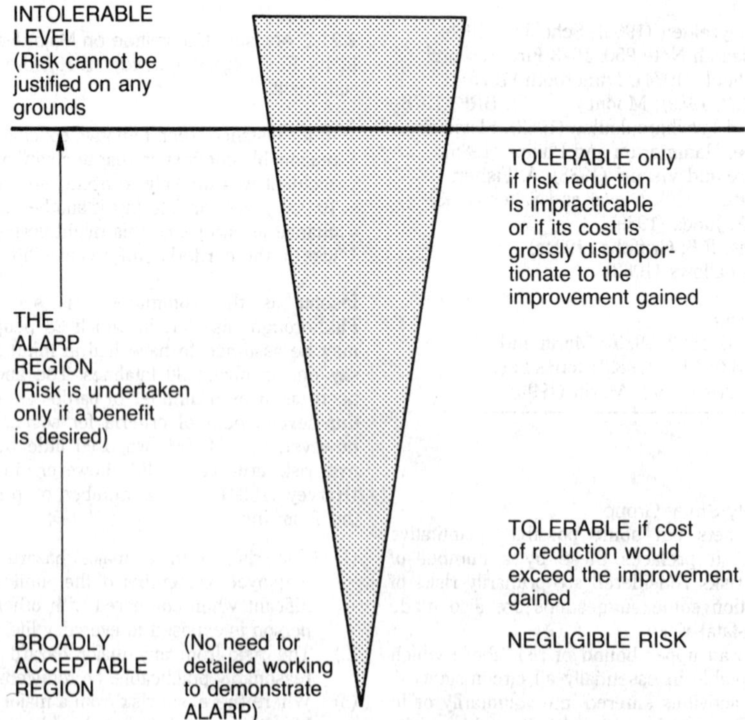
INTOLERABLE
LEVEL
(Risk cannot be
justified on any
grounds
TOLERABLE only
if risk reduction
is impracticable
or if its cost is
grossly dispropor-
tionate to the
improvement gained
THE
ALARP
REGION
(Risk is undertaken
only if a benefit
is desired)
TOLERABLE if cost
of reduction would
exceed the improvement
gained
BROADLY
ACCEPTABLE
REGION
(No need for
detailed working
to demonstrate
ALARP)
NEGLIGIBLE RISK

Figure 9.35 *The ALARP principle (HSE, 1988c) (Courtesy of HM Stationery Office. Copyright All rights reserved)*

grounds and must be reduced. There is another level below which the risk is negligible and no action is required, not even demonstration that the ALARP principle has been applied. Between these two levels lies the region in which there is a requirement to apply the ALARP principle.

9.21.5 HSE land use planning risk criteria

The other HSE publication on risk criteria is *Risk Criteria for Land-use Planning in the Vicinity of Major Industrial Hazards* (HSE, 1989c). The risk criteria described in this document are strictly for land use planning of new developments around major hazard installations. They are not intended to be used for (1) siting of a new major hazard installation, (2) new activities on an existing site, or (3) an existing major hazard installation and existing development.

As described in Chapter 4, the basic approach to land use planning adopted by the HSE is to divide the area around a hazardous installation into three zones. In the top zone risk is a major factor, in the bottom zone it is treated as insignificant, and in the middle zone it is a consideration and further assessment is indicated.

The risk criteria given are largely determined by the use to which they are put in land use planning and differ from the conventional criteria, which tend to be for risk of death. Instead the HSE have developed the concept of a 'dangerous dose'. This dangerous dose is effectively an injurious load imposed by some physical phenomenon such as thermal radiation, explosion overpressure or toxic concentration. The dangerous dose is that which will result in the death of a small proportion of the exposed population. It is assumed that the risk to a vulnerable person is an order of magnitude greater than that to the average person.

For the exposed population as a whole, the HSE take as the upper bound for the risk of a dangerous dose or worse 10^{-5}/year and as the lower bound 10^{-6}/year. This implies that these figures are also approximately the risk of death for vulnerable people. For the case where the exposed population contains a high proportion of vulnerable people the lower bound is taken instead as 0.33×10^{-6}/year. The upper bound risk is about one-tenth of the risk of death in a road traffic accident. The lower bound risk is about ten times the risk of death from lightning.

With regard to societal risk, the HSE rehearse the problems of risk aversion and of FN curves. Instead they elect to use a 'judgmental approach', in which qualitative factors are more prominent. The starting point is the individual risk. The element of societal risk is taken into account by applying a harsher judgement to larger developments in the middle zone. These risk criteria are intended to be used in conjunction with a risk assessment based on the 'cautious best estimate' approach.

The application of these criteria to various types of development, which is given in the document, is described in Chapter 4.

9.21.6 Factors restricting choice of individual risk criteria

A difficulty in the selection of risk criteria is that the choice of criteria is liable to appear arbitrary. The description just given of some systems of risk criteria hints at the fact that when account is taken of the various relevant factors, the scope for arbitrary choice is quite severely restricted. This aspect is now explored in more detail.

Considering risk of death, the average risk to a worker in manufacturing industry is about 3×10^{-5}/year. On the assumption that a new plant should be three times safer than the current average, the risk criterion for a worker on a new plant might be taken as 10^{-5}/year.

It is reasonable to assume that the risk to a worker on a potentially affected plant other than the plant under consideration should be less than that to a worker on the latter, and an order of magnitude reduction is often used. On this assumption the risk criterion for a worker on another plant would be 10^{-6}/year.

Adopting the usual assumption that the risk to a member of the public should be less than that to a worker, the figure of 3×10^{-5}/year might also be taken as an upper bound for a member of the public. The risk of being killed by lightning, which is 10^{-7}/year, is often taken as one which causes little concern. On this basis, the lower bound for risk to a member of the public might be taken as 10^{-6}/year.

A further factor which needs to be taken into account is the risk from a complex of plants. It is assumed that where there is such a complex the risk is increased by a factor of 3.

Then, taking into account (1) new vs existing plants and (2) single plants vs a complex of plants, the risks obtained for a member of the public are as follows:

Single plant		*Complex*	
New plant	*Existing plant*	*New plants*	*Existing plant*
10^{-6}/year	3×10^{-6}/year	10^{-5}/year	3×10^{-5}/year

The highest and lowest risks here correspond exactly to the upper and lower bounds given above.

The foregoing account is given not primarily to suggest a further set of risk criteria, although the values given do not differ greatly from criteria often proposed, but rather to illustrate the fact that by the time account is taken of the relevant factors, the scope for arbitrary choice narrows appreciably.

9.21.7 FN curve criteria of societal risk

Given that a formal risk criterion is used at all for societal risk, the criterion most commonly used is the FN curve. The basic characteristics of FN curves were described in Section 9.19. The account given here is concerned with the parameters to be used in a risk criterion curve.

Like other forms of risk criterion, the FN curve may be cast in the form of a single criterion curve or of two criteria curves dividing the space into three regions – where risk is unacceptable, where it is negligible and where it requires further assessment. The latter approach corresponds to application to societal risk of the ALARP principle and is much to be preferred.

The characteristics of a FN relation are most easily appreciated by considering an FN line. The two defining parameters of such a line are the frequency of a single fatality accident, or intercept on the F axis, and the slope of the line. A characterization of FN curves, whether for actual risks or as risk criteria, has been given by Hagon (1984).

9.21.8 Farmer curve

Before considering the more usual type of FN curve, it is appropriate to refer to the risk criterion curve given by F.R. Farmer (1967b), which pioneered this form of criterion. Figure 9.36 shows the Farmer curve.

As explained in Section 9.19, the quantity plotted on the ordinate of this curve is accident intervals rather than accidents equal to or greater than the defined size.

The argument supporting this criterion (as developed in 1967) is roughly on the following lines. A release of a few thousand curies of ^{131}I constitutes an accident which could cause fatalities and would probably provoke public outcry. Programmes of reactor installation in several countries will accumulate about 100 reactor-years operation before the turn of the century. The risk of one such event during this period is probably the limit of acceptability. This therefore gives one point on the graph, at the frequency of 10^{-3} accidents per reactor year. The slope of the main part of the graph, for large accidents, is -1.5 (on the log–log plot). This slope implies that if the severity of an event increases by two orders of magnitude, its frequency decreases by three orders of magnitude. This represents a weighting to reduce the probability of large-scale accidents which may be expected to give rise to severe public reaction. The curve on the graph in the smaller accident region represents a reduction in the frequency of small incidents, since numerous small incidents are also likely to cause public concern.

9.21.9 Criteria expressing risk aversion

Reference has already been made to risk aversion in respect of societal risk. The Farmer curve just described is an early illustration of such risk aversion. As mentioned above, a measure of risk aversion is the parameter α defined in Equation 9.19.1. This quantity is known as the 'risk aversion index'. Considering a FN line on a log–log plot, if the slope of the line is $-x$ the risk aversion index α is $1+x$. A number of authors have discussed the appropriate value of this index, including

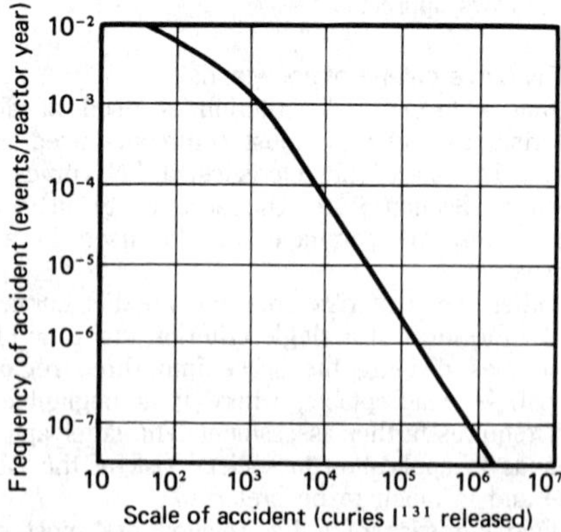

***Figure 9.36** Accident scale–frequency criterion for nuclear power reactors (the Farmer curve) (after F.R. Farmer, 1967b) (Courtesy of the British Nuclear Energy Society)*

Okrent (1981), D.C. Cox and Baybutt (1982) and Hagon (1984).

It is argued by Okrent that it is unlikely that society demands a value of $\alpha \geq 1.5$, that one of $\alpha = 2$ is certainly excessive and that even one of $\alpha = 1.2$ may not be of general applicability. On the other hand, as described below, Hagon has shown that at values of α of less than 2 the contribution of large accidents to the average risk tends to become relatively large, or even predominant.

9.21.10 Groningen FN curve risk criterion

An early societal risk criterion utilizing FN curves was that promulgated by the Provinciale Waterstaat Groningen (PWG, 1979), the 'Groningen criterion'. The Groningen criterion is illustrated in Figure 9.37. There are three zones and two bounding curves. In one zone the risk is unacceptable, in one it is acceptable and in one further assessment is required. For accidents with, on average, more than one fatality ($N > 1$) the slope of the two bounding curves corresponds to a degree of risk aversion in which the consequence is raised to the power 2 (N^2).

Some further characteristics of the Groningen criterion have been given by Hagon (1984), using his relations given in Section 9.19. For the upper curve over the range $N = 1$ to $N = 1000$

$$F_1 = 10^{-2}$$

$$F_{1000} = 10^{-8}$$

The risk aversion is therefore given by

$$FN^2 = \text{Constant}$$

$$\alpha = 3$$

Beyond $N = 1000$ the curve is vertical so that $\alpha = \infty$. For the lower curve over the range $N = 1$ to $N = 10$

$$F_1 = 10^{-6}$$

$$F_{10} = 10^{-8}$$

with the same degree of risk aversion. Beyond $N = 10$, the curve is again vertical.

9.21.11 Hagon FN curve risk criteria

Hagon has used his FN curve relations to describe the characteristics of several well known FN curves. His treatment of the Groningen criterion has just been described. For those given in the Second *Canvey Report*, shown in Figure 9.30 which gives a curve approximating to a straight line, for events up to $N > 1000$ he obtains:

$$F_{10} = 10^{-3}/\text{year}$$

$$F_{1000} = 2 \times 10^{-4}/\text{year}$$

$$\alpha = 1.35$$

This is a low degree of risk aversion.

For the ammonia storage sphere in the *Rijnmond Report*, the FN curves for which are shown in Figure 9.34, Hagon sets a bounding line characterized by $F_1 = 10^{-3}$/year and $\alpha = 2$. He then obtains:

$$F_3^{30} = 2.2 \times 10^{-3}$$

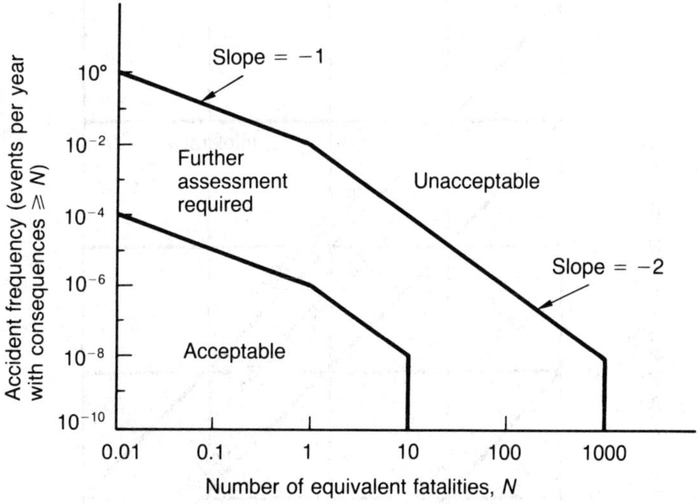

Figure 9.37 *Groningen FN curve risk criterion (Hagon, 1984; after PWG, 1979) (Courtesy of the Institution of Chemical Engineers)*

$F_{30}^{300} = 2.2 \times 10^{-5}$

and

$D_1^{30} = 2.4 \times 10^{-3}$

$D_1^{3000} = 5.2 \times 10^{-3}$

This indicates that large accidents are a significant contributor to the average risk of death.

Hagon takes as a prime determinant of the risk aversion index α the parameter F_N^M, which is the frequency of all accidents between one with N fatalities and one with the maximum number M of fatalities. Figure 9.23 shows that F_N^M increases only slowly with N, provided that $\alpha > 1.5$.

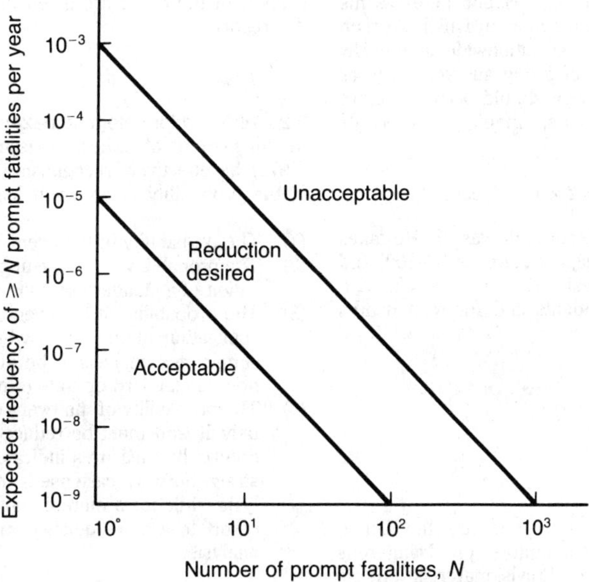

Figure 9.38 *Some risk criteria for the Netherlands (after Ale, 1991; Pasman, Duxbury and Bjordahl, 1992) (Courtesy of Elsevier Science Publishers)*

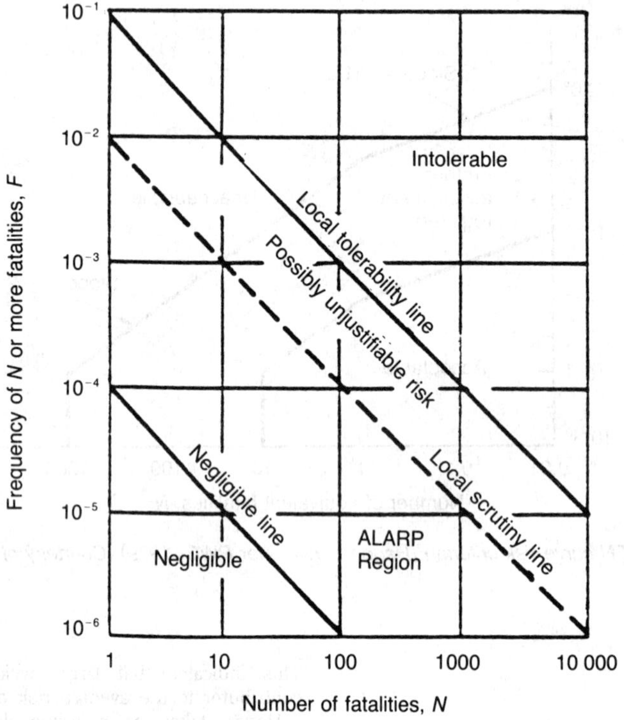

***Figure 9.39** ACDS FN curve risk criterion (after Advisory Committee on Transport of Dangerous Substances, 1991)*

Hagon takes as another determinant of α the parameter D. From Figure 9.33 the increase of D with N is slow, provided that $\alpha > 2$.

With regard to the choice of F_N^M, Hagon takes as his starting point the Flixborough disaster, in which 28 men died. If there are assumed to be nationwide in the UK some 500 installations capable of giving an event such as Flixborough, and if such an event should occur no more often than once in 100 years, then for a single installation:

$$F_N^M = \frac{1}{500 \times 100} = 2 \times 10^{-5}/\text{year}$$

The number of fatalities at Flixborough was 28. He takes the range of fatalities for a major hazard as 30–300, and thus sets $F_{30}^{300} = 2 \times 10^{-5}$/year. He sets $\alpha = 2$, and obtains for single fatality accidents and for total deaths due to the major hazard risk:

$$F_1 = 6.5 \times 10^{-4}/\text{year}$$

and

$$D_1^{300} = 4.1 \times 10^{-3}/\text{year}$$

9.21.12 Dutch risk criteria

Two more recent sets of risk criteria are the Dutch criteria and the Advisory Committee on Dangerous Substances (ACDS) FN curve. Provisional risk criteria according to Dutch law are quoted by Ale (1991). They are shown in Figure 9.38.

9.21.13 ACDS FN curve

The ACDS *Transport Hazards Report* (1991) also gives a FN curve, shown in Figure 9.39. Arguments in support of the curve are given in detail in Appendix 6 of the report.

9.21.14 Need for criteria in hazard assessment

In the method of hazard assessment described by Kletz (1971) he effectively recognizes four situations in respect of the probability of an event leading to a hazard:

(1) The probability of the event is low.
(2) The probability of the event can be made low by application of a standard or code of practice.
(3) The probability of the event can be made low by the application of measures which can be shown by a simple risk analysis to be of equivalent safety to the normal standard or code of practice.
(4) The probability of the event must be assessed quantitatively and must be reduced to conform to the risk criteria by measures indicated by the analysis. For a straightforward case use is make of a simple risk analysis while for a more complex one it is necessary to resort to a more detailed method such as fault tree analysis.

Risk criteria are needed mainly for the fourth case. The second and third cases are considered first.

9.21.15 Standards and codes

If the probability of the event which may lead to realization of the hazard is not intrinsically low, nevertheless it may often be reduced to a low level by the use of an appropriate standard or code of practice.

An example given by Kletz is that there is some probability of overpressure in a pressure vessel, but the use of pressure relief valves as specified in standards and codes for pressure vessels reduces this probability to a low level. Another example is that electrical equipment has some probability of acting as a source of ignition, but the application of codes for hazardous area classification and equipment safeguarding results in a low value of this probability.

The main *caveat* to be made here is that where major hazards are concerned it should not be assumed that the use of the standards and codes is sufficient.

9.21.16 Equivalent risk

It is sometimes not appropriate, or even possible, to use a standard or code. In this case it may be possible to devise an alternative method which can be shown to have a degree of safety at least equivalent to the standard or code. A simple risk analysis is carried out to demonstrate this. The analysis by Kletz (1974a) of the use of trips instead of pressure relief valves described in Chapter 13 is a case in point.

9.21.17 Individual and average risk

Before considering the evaluation of the more complex analyses mentioned using suitable risk criteria, there are some further preliminary points to be made.

In considering risk, it is necessary to distinguish between the average risk run by a group of people, whether employees or public, and the higher risk which may be run by some individuals within the group. Thus some employees may be exposed to a higher risk because the plant on which they work is more hazardous and some members of the public may be exposed to a higher risk because they live nearer the works. There is fairly general agreement that there should be a limit to the risk run by any individual. Thus it is not sufficient to achieve a low average risk; it is necessary also to reduce to a minimal level the risk to the most exposed individual. In other words the risk to that person should not be 'lost' by combining it with the risk to a larger group most of whom are much less exposed.

9.21.18 Death and injury

Most treatments of risk deal primarily with the risk of death. This may appear somewhat arbitrary, but there is justification for this approach. Data on fatalities are recorded and are relatively straightforward. For other levels of injury, however, there can be problems. Often the data are not available, but even if they are, there tend to be difficulties in interpreting them. Moreover, for any given activity in industry there tends to be a relationship between fatalities and other injuries. The work of Heinrich (1959) on this ratio has been described in Chapter 1.

Measures which reduce deaths from a particular hazard tend to reduce injuries in proportion and it is this reduction which is important. The use of death rate as the main practical criterion does not therefore imply any disregard of the personal tragedy arising from serious injury.

Several indices of industrial injury risk are used in the literature. The lost time accident rate is expressed as the number of accidents per 10^5 exposed hours; this period corresponds approximately to a working lifetime. Fatal accidents are much less frequent and in order to avoid inconveniently small numbers the fatal accident rate (FAR) is defined as the number of deaths per 10^8 exposed hours; this roughly corresponds to the number of deaths over a working lifetime of 1000 persons. Another index which is useful in relation to general fatality risks is the death rate per annum.

It should be recognized, however, that the use of fatality risk is not universally accepted as the most appropriate approach, and there is considerable literature on the subject. Some hazards are particularly likely to give rise to delayed, as opposed to prompt, deaths or to serious injury. A release of radioactivity from a nuclear reactor may have the potential to cause many deaths which are predominantly delayed.

9.21.19 Engineering feasibility

There is obviously a limit to the degree of plant reliability which can be achieved even by the best engineering practice and, equally important, to the degree of confidence which can be placed in estimates of that reliability.

A figure frequently quoted for the reliability to which plants can be engineered is a hazard rate for a major accident of 10^{-5} events/year. This is the figure given by J.H. Bowen (1976) from a nuclear industry perspective. The reason that it is difficult to achieve lower hazard rates is that at this level the risk begins to be affected by rather improbable failures and by dependent failures.

In this respect, a distinction may be made between a risk which is assessed as 10^{-5}/year as a single figure and a one which is also assessed as 10^{-5}/year but as the product of two separate risks of 10^{-2}/year and 10^{-3}/year; in so far as the two latter values represent figures which are known to be achievable in engineering terms, the second of the two risks of 10^{-5}/year may be considered to be a more robust estimate.

9.21.20 Value of a life

One criterion for evaluating measures available to reduce risk is the value of a life. A general discussion of this concept, and its difficulties, has been given in Chapter 4. The account here is confined to an overview of the principal approaches and of the values which emerge from them.

The cost of accidents and the cost effectiveness of safety measures were major concerns of the Robens Committee (Robens, 1972). The ***Robens Report*** itself attempted to assess the cost of industrial accidents to the nation. The committee also supported work by Sinclair, Marstrand and Newick (1972a, b) on the cost effectiveness of approaches to industrial safety. The problem is an important one and there is a large literature on the subject (e.g.; National Academy of Engineering, 1972).

Some of the methods which can be used to estimate the value of a life have been reviewed by Melinek (1974 BRE CP88/74) and Kletz (1976d). The following account is based on their work and the figures quoted from it are taken unchanged and reflect the value of money at that time. The methods which they describe may be summarized as (1) future production, (2) administrative decisions, (3) consumer preference, (4) court awards and (5) life assurance.

Thus one method of assessing the value of a life is to calculate the future production or services which an individual may be expected to give to the community. A variant of this is to estimate future earnings, although these are not synonymous with the value of the work done. Most estimates of the value of future production lie in the range £10 000–£100 000. The Road Research Laboratory has given an estimate of £15 000 (R.F.F. Dawson, 1967).

There are many administrative decisions made by government which effectively set a particular value on life. Mellinek quotes:

Worker safety	£15 000–£20 m.
Consumer safety	£0–£20 m
Medical care	up to £20 000

Similarly Klutz quotes:

Agriculture	£2000
Steel handling	£200 000
Pharmaceuticals	£5 m.

The value set in medical work seems to be particularly low. Leach (1972) has shown that lives can be saved for very small expenditures:

Lung X-ray machines for older smokers	£400
Cervical cancer screening	£1400
Artificial kidney	£9500

Another approach is based on consideration of consumer preference. The following illustration is given by Melinek for the implicit valuation put on his own life by a pedestrian who crosses the road on the surface instead of using the safer subway:

Probability of being killed crossing a road	$\approx 10^{-8}$
Extra time taken by using subway	$\approx 15\ \text{s} = 0.004\ \text{h}$
Value which people put on own time	$\approx$ £0.25/h
Value of a life	$\approx \dfrac{0.25 \times 0.004}{10^{-8}} = £100\,000$

The value which people put on their own time is taken from transport studies. The value of a life as obtained from more extensive studies by the Fire Research Station (FRS) is, as described by Melinek, about £50 000.

The damages which courts award for death or injury often include a sum for the reduction of life expectancy and are thus another means of estimating the value of a life. It has also been suggested that the value of a life may be determined from the sums covered by life assurance.

There are quite serious objections to some of these approaches. In the case of the valuation of life in terms of the value of the person's future production, the *reductio ad absurdam* is that this criterion puts a zero value on the life of a retired person. Society, however, is prepared to pay to preserve that life. Consumer preference seeks an objective basis in what people actually do, but hardly appears to reflect the way they actually think. Court awards set a notoriously low value on life expectancy and are thus a poor guide. Life assurance is taken out to protect dependants or as an investment and is not intended to compensate for death. The values implied in administrative decisions show a very wide spread and again are far from satisfactory as a guide.

A rather different, and more relevant, approach is described by Kletz (1976d). This is to consider the cost of achieving the risk target which he uses for process plants. He estimates this cost as equivalent to about £1 m. per life and suggests that in general this is a reasonable figure to take as the value of a life. A valuation less than £100 000 would hardly be acceptable, while one more than £10 m. would seem extravagant.

The value of a life criterion is applied as follows:

$$v = \frac{c}{h} \quad [9.21.1]$$

where c is the annual cost of eliminating the hazard (£/year), h is the hazard rate (deaths/year) and v is the value of a life (£/death avoided). Thus, for example, if in a given case a particular hazard is assessed as presenting a hazard rate of 10^{-4} deaths/year, if the value of a life is taken as £1 m. and if £1 m. capital expenditure is equivalent to £200 000/year recurrent expenditure, then the application of the criterion indicates that it is appropriate to spend up to £20/year to eliminate the hazard.

9.21.21 Comparisons of process and other risks

Whereas the risk to which a member of the public is exposed from a hazardous installation is an involuntary one, it is generally considered that to some degree at least, an employee accepts voluntarily the risk associated with work on the plant.

Writers on hazard analysis such as Kletz (1971) and S.B. Gibson (1976a) have compared the risks to employees on process plant with other risks accepted voluntarily or borne involuntarily. Following Kletz (1976d) some data on voluntary and involuntary risk are

	Fatality rate (deaths/person-year)
Voluntary risk	
Taking a contraceptive pill[a]	2×10^{-5}
Playing football[b]	4×10^{-5}
Rock climbing[b]	4×10^{-5}
Car driving[c]	17×10^{-5}
Smoking (20 cigarettes/day)[b]	500×10^{-5}
Involuntary risk	
Meteorite[d]	6×10^{-11}
Transport of petrol and chemicals (UK)	0.2×10^{-7}
Aircraft crash (UK)[a]	0.2×10^{-7}
Explosion of pressure vessel (USA)[d]	0.5×10^{-7}
Lightning (UK)[e]	1×10^{-7}

	Fatality rate (deaths/person-year)
Flooding of dikes (The Netherlands)[f]	1×10^{-7}
Release from nuclear power station (at 1 km) (UK)	1×10^{-7}
Fire[g]	150×10^{-7}
Run over by vehicle (UK)	600×10^{-7}
Leukaemia[a]	800×10^{-7}

[a] S.B. Gibson (1976c); [b] Pochin (1975); [c] Roach (1970); [d] Wall (1976); [e] Bulloch (1974); [f] Turkenburg (1974); [g] Melinek (1974 BRE CP 88/74)

The validity of such comparisons has been discussed in Chapter 4.

9.21.22 Computation of risk

The risk of an individual may be formulated in simple terms as:

$$r = \frac{1}{N}\sum_{i=1}^{n} x_i f_i \quad [9.21.2]$$

where f_i is the frequency of accident type i, r is the individual risk of death, x_i is the number of deaths for accident type i, n is the number of types of accident and N the total number of persons at risk.

The calculation is concerned with the risk to the individual. All accidents which might have a significant effect on this individual risk should be taken into account. The estimate should not be distorted by inclusion of numbers of people who have a significantly lower exposure.

Often the relationship for individual risk is more complex than Equation 9.21.2. The ACDS *Transport Hazards Report* (1991) contains a number of examples of the more complex formulations necessary in certain cases.

9.21.23 Application of risk criteria

The foregoing treatment has outlined some approaches to the setting of individual and societal risk criteria. An account is now given of some of the practicalities of engineering design utilizing these and other criteria.

The treatment for risk to employees essentially follows the systems developed by Kletz (1971) and S.B. Gibson (1976a). Their approach is based on the use of hazard analysis and of the fatal accident rate (FAR) criterion, as described in Section 9.2. For, say, an overall FAR of 3.5 the target value for a given plant hazard is 0.35, or one tenth of the overall value. If some particular jobs on a process are more subject to this hazard than others, then the FAR is applied to the more hazardous ones; it is not averaged over all the jobs. Thus no one is exposed to a risk higher than the target value. This *primary criterion* of maximum risk must normally be met and any expenditure required to meet it must be incurred.

It is not necessarily enough, however, to meet the FAR criterion. It still remains to determine whether all that is 'reasonably practicable' has been done to ensure safety. For this, use is made as a *secondary criterion* of the concept of the value of a life. The value of a life criterion is applied to evaluate additional expenditure aimed at eliminating hazards and reducing risk.

Various figures have been given for the value of a life for use in this context. As described in Section 9.21.20, the value suggested for the chemical industry by Kletz when he wrote in the mid-1970s was £1 million. Publications in the recent past, say 1990, have quoted for a wider range of activities a value of about £3 million.

This approach represents the practical application of the APARP principle.

9.21.24 Safety improvement

It is often suggested that it would be unfortunate if the adoption of particular risk criteria was to lead to the situation where there ceased to be any improvement in safety standards. In fact the system described, based on the FAR, does tend towards a continuous improvement in safety standards provided the system is universally applied, that factors other than the plant design, particularly the standard of management, do not deteriorate and that the target FAR used is the moving average for the industry. For while it is then not accepted that any plant have an FAR worse than the current average, some will have one which is better, so that the average will gradually improve. Moreover, the use of the value of a life as a secondary criterion injects a further degree of improvement. There is therefore a ratchet effect which tends to raise standards.

9.21.25 Allocation of resources

The problem of expenditure on safety measures is one of allocation of resources and of cost effectiveness. As already indicated, the administrative decisions made by government and industry imply a wide range of valuations of human life.

Some of the examples quoted do at least prompt the question whether the chemical industry perhaps spends too much on safety. This may be so, but the hazards of the chemical industry tend to provoke a strong reaction and it is doubtful if much relaxation would be tolerated. This is doubly so following disasters such as that at Bhopal.

Moreover, it is probably of value to the community as a whole to have an industry which despite the high intrinsic hazards of its materials and processes is able to pioneer methods of improving control of hazards and to achieve high levels of safety.

Nevertheless, the allocation of resources to safety is a legitimate subject for public debate. The chemical industry can only try to respond responsibly to this.

9.22 Guide Assessments

The arrangements for the control of major hazards in the UK were described in Chapter 4. It was mentioned there that in order to give advice to local planning authorities the HSE has developed assessment methods for LPG and for chlorine installations.

An outline approach to assessment for emergency planning for a chlorine release has been given by the UK Chlorine Producers under the auspices of the CIA. The CIA has also issued a Control of Industrial Major Accident Hazards (CIMAH) guide for ammonia. Other CIMAH guides include those on LPG by the Liquefied Petroleum Gas Industry Technical Association (LPGITA),

Table 9.37 HSE guide assessment on LPG installations: hazardous events (after Crossthwaite, 1984)

Event	*Consequences*	*Likelihood*
Small flange leak	Local effects only. No off-site hazard	Considerable number each year
Severe pipe leak (guillotine rupture)	If ignition occurs on site, then off-site consequences unlikely to be serious. If ignition occurs off site, a flash fire would injure persons in and near the cloud	Average of about 1/year in UK
BLEVE	High levels of thermal radiation at substantial distances from point of release	Two incidents known to have occurred in UK
Vessel rupture	Flash fire: persons within the (quasi-instantaneous) cloud likely to be killed. Explosion: this will have serious effects at substantial distances from release	No history in UK

on liquid oxygen by the British Compressed Gases Association (BCGA) and on ammonium nitrate by the Fertilizer Manufacturers Association (FMA). These guide assessments are now described in turn. Some of the features mentioned draw on material described in Chapters 15–18.

9.22.1 HSE LPG methodology

The HSE methodology for the hazard assessment of LPG has been described by Crossthwaite (1984, 1986) and its further development has been described by Clay *et al.* (1988). It is embodied in the program RISKAT. The assessment has been carried out as part of the work done by HSE on their advice to local planning authorities on consultation distances. The overall approach taken is that described in the *Third Report* of the ACMH which states:

> It seems reasonable to aim for a separation which gives almost complete protection for lesser but more probable accidents and worthwhile protection for major but less probable accidents.

In applying this principle the HSE has judged that releases due to pipework failure fall in the first category and fireballs and vapour cloud explosions fall in the second.

The events considered by Crossthwaite (1984) are shown in Table 9.37. For thermal radiation from a fireball the effects of different levels of received radiation are taken to be:

	kJ/m^2
50% fatalities	700
Blistering of exposed skin	200
Blistering of skin (threshold)	100

and of explosion overpressure:

	psi
5% fatalities	5
Injury due to flying glass	1
Injury but very unlikely to be serious	0.7

The distances at which and the areas over which these levels of effect are estimated to occur for the more serious accidents such as a release of 100 te of LPG are shown in Table 9.38.

The guidelines on consultation distances issued by the HSE on the basis of such assessments are described in Chapter 4.

9.22.2 HSE chlorine methodology

The HSE methodology for the hazard assessment of chlorine has been described by Pape and Nussey (1985). It is embodied in the program Risk Assessment Tool (RAT), now RISKAT. The type of installation considered is shown in Figure 9.40. The listing of hazard scenarios given is shown in Table 9.39. These releases were developed as shown in Figure 9.41 and the consequences analysed as shown in Figure 9.42. Some of the principal assumptions used in the analysis are given in Table 9.40.

The hazard scenarios to be considered were simplified by reducing the pipe failures to two cases: guillotine

Table 9.38 HSE guide assessment on LPG installations: physical effects from a 100 te release of propane (after Crossthwaite, 1984)

A Fireball from a BLEVE

Thermal radiation (kJ/m^2)	*Distance from installation* (m)	*Area of land* (hectares)
(Fireball radius)	107	3.6
700	200	12.6
200	380	45
100	530	88

B Explosion of a vapour cloud

Overpressure (psi)	*Distance from installation* (m)	*Area of land* (hectares)
5	176	9.7
1	513	83
0.7	675	143

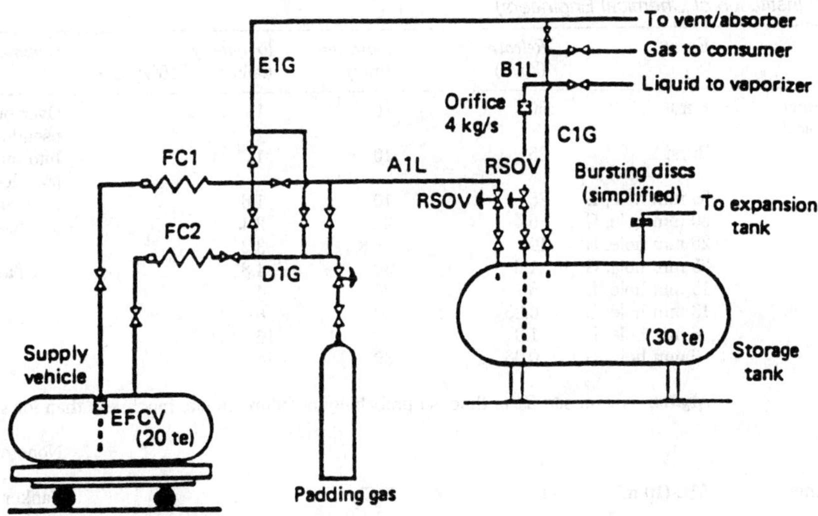

Figure 9.40 HSE guide assessment on chlorine installations: typical installation (after Pape and Nussey, 1985) (Courtesy of the Institution of Chemical Engineers)

fracture with both ends open, and a split equivalent to a hole of half the pipe size. Each release was expressed as a vapour flow rate. It was assumed that small releases from pipes would vaporize completely. The justification for this was that for unbunded releases the SPILL computer code for vaporization predicts that the vaporization rate quickly reaches the release rate.

For convenience, large instantaneous releases were treated as pseudo-continuous. A rule of thumb was used that a release over 10 te is equivalent to a continuous release of 1.5 times the actual release with a duration of 10 min, and that smaller releases have an effective duration of 5 min. The heavy gas dispersion was modelled using the CRUNCH computer code for dispersion of a continuous release of heavy gas.

The results for the base case studied are shown in Table 9.41. They show that very close to the plant the principal risks are from gasket failures, pipe splits,

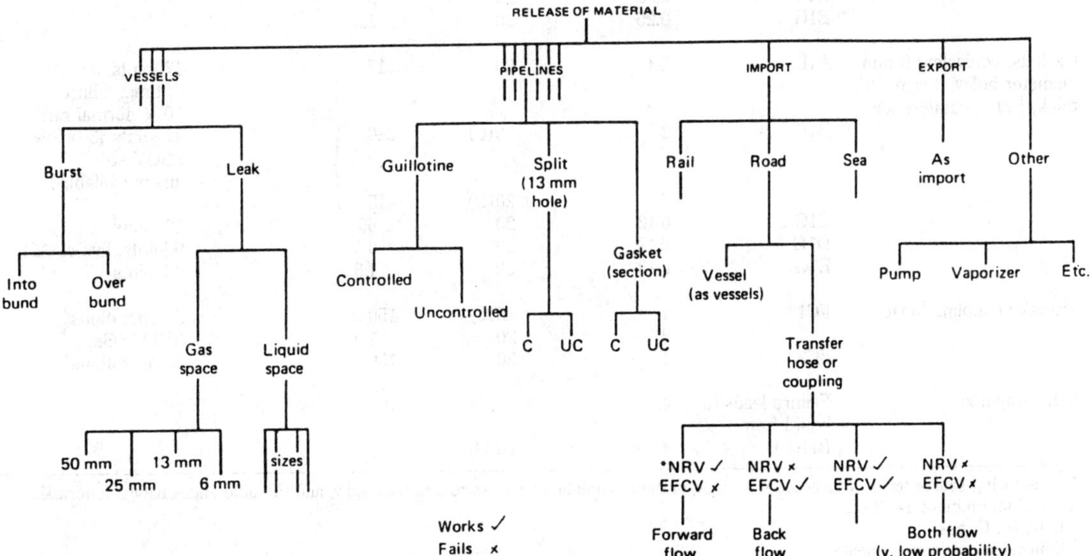

Figure 9.41 HSE guide assessment on chlorine installations: releases (after Pape and Nussey, 1985) (Courtesy of the Institution of Chemical Engineers)

Table 9.39 HSE guide assessment on chlorine installations: hazard scenarios[a] (after Pape and Nussey, 1985) (Courtesy of the Institution of Chemical Engineers)

Item	*Event*[b]	*Release* (kg/s)	*Duration*[d] (min)	*Frequency* (failures × 10^6/year)	*Comments*
Storage vessels; only 1 live at once; typical stock 20te	Burst	50[c]	10	1	Over bund, pseudo-plume
	Burst	25[c]	10	1	Into bund, pseudo-plume
	50 mm hole, L	25[c]	10	1.6	Pseudo-plume
	50 mm hole, G	6.4	20	2.4	2 × flash
	25 mm hole, L	19	8.8	3.2	
	25 mm hole, G	1.6	30	4.8	2 × flash
	13 mm hole, L	5	30	4	
	13 mm hole, G	0.25	30	6	
	6 mm hole, L	1.3	30	16	
	6 mm hole, G	0.06	30	24	
Tanker vessels	*Neglect*: only on site 2% of time, so probability of failure on site much less than for static tanks				
Other vessels					None on site
Pipelines, guillotine fractures	A1L (10 m)	1	5(C)	0.6	Tanker EFVC works, live 2% of time
		9	20	0.006	Tanker EFVC fails, failure 10 × B1L/m[e]
	B1L (40 m)	4	5(C)	12 }	Normally live, limited to 4 kg/s by orifice plate
		4	20	0.12 }	
	C1G (20 m)	1	20	6	Normally live
	D1G (20 m)	1.25	20	0.3	Live as A1L[f]
	E1G	1.25	20	0.15	Live 0.1%; 10 × failure rate
Pipe splits	A1L	5	20	6	EFVC on tanker not actuated
	B1L	4	5(C)	120	
	C1G	0.25	20	60	
	D1G	0.25	20	3	
	E1G	0.25	20	1.5	
Gaskets, equivalent 9 mm diameter holes, 3 mm thick, $\frac{1}{4}$ of circumference	A1L	2.4	20	17	17 joints, live 2% of time, failure 10 × normal rate
	B1L	2	5(C)	220 }	47 joints (3 below RSOV so 'uncontrollable')
		2	20(U)	15 }	
	C1G	0.13	20	60	12 joints
	D1G	0.13	20	9	9 joints, live as A1L
	E1G	0.13	20	1.3	26 joints
Transfer coupling/hose	FC1	1	5(C)	150	50 operations[g]
		9	20	1.5	EFVC fails
	FC2	1.25	20	150	50 operations[g]
Other vaporizer	Failure leads to liquid from	4	5(C)	100	
	B1L	4	20(U)	1	

[a] In deducing source terms, due account is taken of the possibilities for forward and back flow, and the differences between normally live and intermittent use items.
[b] L, liquid; G, gas.
[c] Equivalent continuous release.
[d] C, controlled by remotely operated shut-off valve (RSOV); U, RSOV fails or absent.
[e] Pipe used intermittently. Failure rate assumed to be 10× that of normal pipeline, then multiplied by fractional use.
[f] Pipe live when A1L is live.
[g] 50 operations per annum.
[h] EFVC, excess flow valve cut off.

Table 9.40 *HSE guide assessment on chlorine installations: key assumptions (after Pape and Nussey, 1985) (Courtesy of the Institution of Chemical Engineers)*

A Failure frequency[a]

Vessels:[b]
Frequency of near instantaneous release of whole contents = 2×10^{-6}/year

Frequency of lesser releases = 6×10^{-5}/year partitioned as follows:

Equivalent hole diameter (mm)	*Frequency*
50	4×10^{-6}/year
25	8
13	10
6	40

and

Gas space 60%, liquid space 40%

Pipework:
Frequency of guillotine fracture on 25 mm pipework = 0.3×10^{-6}/m years
Frequency of lesser releases (equivalent to 13 mm hole) = 0.3×10^{-5}/m years

Gaskets:[c]
Frequency of failure for 0.6 mm thick gaskets = 3×10^{-6}/year
Frequency of failure for 3 mm thick gaskets = 5×10^{-6}/year

Tankers:
Probability of failure of coupling/hose = 3×10^{-6}/operation
Probability of failure of excess flow valve = 0.01/demand
Frequency of failure of tanker vessels same as vessels given above, but with allowance made for time on site

B Release duration

Vessels:
For lesser releases duration is 30 min or time taken to release all the contents, whichever is less[d]

Pipework:
For automatic shut-off duration[e] = 1 min
For remote manual shut-off duration[e] = 5 min
For local manual shut-off duration = 20 min

C Release rate

Vessels:
For vessel bursts over bund, 100% of release vaporizes
For vessel bursts directed into bund, 50% vaporizes

Pipeline:
For pipeline release, 100% vaporizes
For two-phase flashing flow from pipeline guillotine fracture in a 25 mm internal diameter pipe with flow driven by chlorine vapour pressure, flow rate = 4 kg/s
For single phase liquid flow from tanker coupling failure in 25 mm pipe driven by padding pressure, flow rate = 9 kg/s
For single phase liquid flow from pipework in 13 mm hole in pipe, flow rate = 4 kg/s

D Release dilution

For releases from pressurized containment initial release is diluted by a factor of 10

E Weather conditions

Pasquill category/ wind speed (m/s)	*Probability*
D/2.4	0.30
D/4.3	0.24
D/6.7	0.29
F/2.4	0.17

F Gas dispersion

Gas dispersion calculated using CRUNCH heavy gas dispersion model. Concentrations within the plume predicted by this model are assumed to be Gaussian, concentrations outside the plume are not considered. During the time of passage of the plume, the concentration at a particular location is assumed to be uniform for the duration of the release

G Modification of exposure

Probability of being initially outdoors = 0.1 Pasquill D
= 0.01 F

For person initially outdoors, probability of escape indoors after receiving significant dose:

Concentration outdoors (ppm)	*Probability of escape indoors*
>1000	0
570–1000	0.2
140–570	0.8
<140	1

Evacuation occurs 30 min after arrival of cloud, or later if cloud persists more than 30 min

H Effect of shelter

Concentrations indoors calculated using a single exponential stage model with following values for the product of the ventilation rate constant λ and the mixing efficiency factor k[f]:

Pasquill category/ wind speed (m/s)	$k\lambda$ (h^{-1})
D/2.4	0.7
D/4.3	1.0
D/6.7	1.5
F/2.4	0.5

I Gas toxicity

Significant exposure is $C^{1.67}t > 20\,000$, equivalent to the Dicken 'fatal' dose

Threshold of significant dose is 140 ppm for 5 min

[a] Failure rates are based on aggregated data from various sources modified by judgement. Rates used are for sudden failures, i.e. leaks which can develop into major failures before preventive action can be taken.
[b] Vessel failure includes events up to and including the first flange on any nozzle or penetration.
[c] Gasket failure means loss of one section between two adjacent bolts. It should be checked whether the gasket internal diameter equals that of the pipeline. Actual frequency may depend on inspection and replacement practices.
[d] For leaks in gas space, the available contents for 50 and 25 mm holes are 2 × the flash fraction and, for smaller holes, the flash fraction.
[e] Fractional dead time of automatic or remote shut-off is 0.01. Such a failure places a demand on the manual shut-off leading to a release duration of 20 min.
[f] These values are based on data modified by judgement and make some allowance for the possibility of a few windows being open.

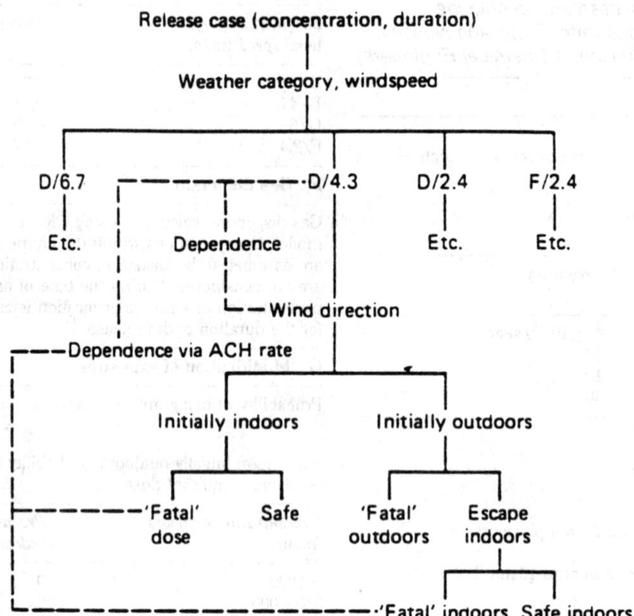

Figure 9.42 *HSE guide assessment on chlorine installations: consequences of release (after Pape and Nussey, 1985) (Courtesy of the Institution of Chemical Engineers)*

coupling/hose failures and releases from the vaporizer unit. At 200 m it is the last three which are the main risks, while at 300 m and beyond the risk of major vessel failures becomes dominant, with a contribution from uncontrolled gasket failures and pipe splits. A large proportion of the risk beyond 200 m occurs with Pasquill category F weather.

The assessment includes a sensitivity analysis on the following features:

(1) plant size;
(2) vessel failure rate;
(3) gasket size;
(4) proportion of time outdoors;
(5) ventilation rate;
(6) evacuation time;
(7) gas toxicity.

A plant larger than the base case was studied. The increase in risk was roughly proportional to the change in the numbers of components. A 10-fold increase in the vessel failure rate had a strong effect on the risks in the far field, doubling the range of the 10^{-7}/year risk. The effect of reducing gasket thickness was to cut release rates by a factor of 4 and to reduce risks at short ranges. The proportion of time spent outdoors was varied from zero to 100%. Zero time outdoors gave results little different from the base case, while 100% time outdoors increased the risk by a factor of 2 at intermediate distances. Ventilation rate increases by factors of 4 in weather category D and 2 in category *F* resulted in risks at intermediate distances which were 2 or 3 times the base case, while decreases of the rate to 0.5 in D/2.4 weather and a halving of the rate in *F* weather gave reductions in the risk of up to 4 at intermediate distances. The effect of increasing to 60 min the time taken to evacuate the building after passage of the cloud was to increase the risk at intermediate distances by a factor of about 2.5. For gas toxicity several relations were investigated. The use of a toxicity estimate by ten Berge and van Heemst instead of the Dicken value had a dramatic effect on risks, reducing the range of the 10^{-7}/year risk from 750 to 300 m. In addition to the results given in Table 9.41 the results were also presented in the form of risk contours on a typical site plan and of a FN curve.

9.22.3 UKCP chlorine emergency planning guidelines

Guidelines for chlorine emergency planning are given in *General Guidance on Emergency Planning with the CIMAH Regulations for Chlorine Installations* by the UK Chlorine Producers (UKCP) (CIA, 1986) (the UKCP *Chlorine Guide*); an extract (CIA, 1989) is given in Lees and Ang (1989b). Although the guidelines are intended primarily for emergency planning, it is explicitly stated that the methods given may also be helpful in producing a safety case.

The type of installation considered is shown in Figure 9.43. Typical equipment and inventories at the installation and the process activities are shown in Table 9.42, Sections A and B, respectively. The steps in the hazard assessment are given in Section C of the table. A list of typical incidents to assist in hazard identification is given in Section D. The summary of possible incidents given in the guide and the interpretation of the terminology are given in Table 9.43, Sections A and B, respectively.

Table 9.41 *HSE guide assessment on chlorine installations: risks (after Pape and Nussey, 1985) (Courtesy of the Institution of Chemical Engineers)*

			Risks (× 10^6/year)			*Distance* (m)				
Rate (kg/s)	*Duration* (min)	*Frequency* (× 10^6/year)	*50*	*100*	*200*	*300*	*500*	*750*	*1000*	*1500*
50	10	1	0.29 (23)	0.23 (24)	0.17 (26)	0.13 (29)	0.08 (39)	0.04 (57)	0.01 (94)	0
22	10	5	1.29 (24)	0.99 (25)	0.68 (30)	0.46 (38)	0.18 (62)	0.06 (95)	0.02 (92)	0
6.4	20	2	0.5 (26)	0.35 (30)	0.18 (42)	0.08 (59)	0.02 (94)	0	0	0
5	30	4	0.82 (27)	0.56 (31)	0.28 (44)	0.11 (63)	0.03 (93)	0	0	0
1.25	20	156	23.8 (34)	9 (52)	1.46 (93)	0.04 (8)	0	0	0	0
4	5	232	44.6 (29)	22.4 (42)	4.33 (89)	0.13 (22)	0	0	0	0
5	20	7	1.44 (27)	0.98 (31)	0.43 (48)	0.17 (69)	0.04 (94)	0 0	0 0	0 0
0.2	20	141	3.04 (94)	0.04 (8)	0	0	0	0	0	0
2.4	20	17	3.09 (30)	1.73 (40)	0.42 (72)	0.14 (93)	0	0	0	0
2	5	220	33.05 (36)	7.36 (86)	0.15 (19)	0.04 (23)	0	0	0	0
1	5	150	13.64 (49)	1.36 (82)	0.03 (20)	0	0	0	0	0
9	20	1	0.33 (26)	0.24 (29)	0.14 (36)	0.07 (48)	0.01 (94)	0	0	0
1.4	30	20	3.38 (33)	1.55 (46)	0.26 (94)	0.08 (92)	0	0	0	0
2	20	15	2.63 (31)	1.37 (43)	0.23 (94)	0.09 (93)	0	0	0	0
Total			131.5 (35)	48.1 (51)	8.8 (76)	1.6 (53)	0.4 (66)	0.1 (80)	0	0

Note: The values in parentheses show the percentage contributions to risk levels due to accidents in stable weather conditions.

The guide gives, in addition, methods of estimating the emission rate and gas dispersion, using for the latter overlays which can be placed on a map of the site, and data on gas toxicity.

9.22.4 CIA ammonia CIMAH guidelines

The CIMAH Safety Case: Ammonia by the CIA (1988 PA9) (the CIA *Ammonia Guide*) gives guidance for that substance. The main body of the CIA *Ammonia Guide* is largely concerned with the safety report itself. It says relatively little about the frequency of events, but it gives in Chapter 5 a set of release scenarios for different storage conditions and some guide to toxic concentration estimates.

The guide contains a detailed table (15 pages) of causes of release. It gives five main release scenarios as follows: Scenario 1, release of liquid from a hole in pressurized storage; Scenario 2, release of liquid from a hole in refrigerated storage; Scenario 3, release of vapour from a relief valve; Scenario 4, failure of a refrigerated bunded storage tank; and Scenario 5, a spill of refrigerated liquid ammonia onto water. For the first two continuous releases, two flows (10 and 40 kg/s) are considered, and for the relief valve a flow of 6 kg/s is considered. In the fourth scenario, two subscenarios are considered: (a) a failure near the top of the tank and (b) a failure near the bottom; in the latter case a flash evaporation of 5 te is estimated. In the fifth scenario the evaporation rate is taken as 50 kg/s during the 2 min of

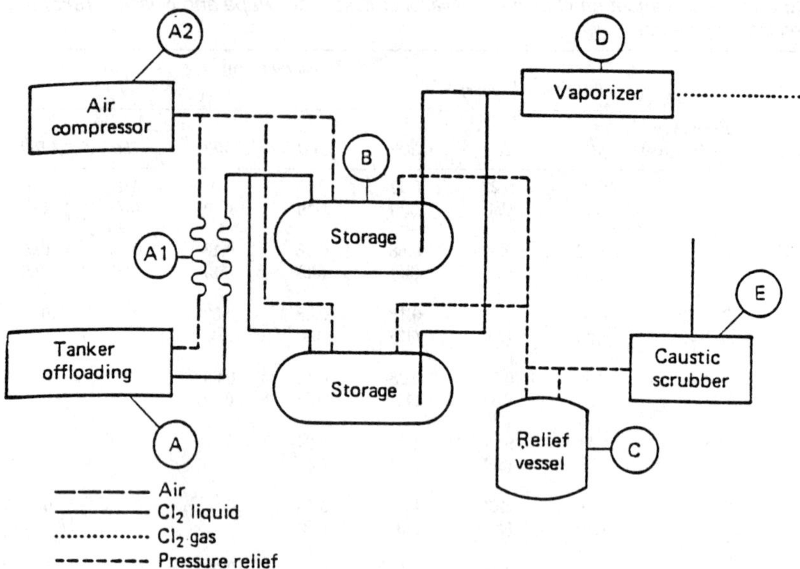

Figure 9.43 *UKCP guide on emergency planning for chlorine installations: typical installation (after Chemical Industries Association, Chlorine Sector Group, 1989)*

the spill. Results are given without and with allowance for plume rise. The guide gives estimates of the ground level concentration of ammonia for the five scenarios, including the results given here in Table 9.44.

9.22.5 LPGITA LPG CIMAH guidelines

Guidance for LPG is given in *A guide to the writing of LPG safety reports* by the LPGITA (1988 GN1) (the LPGITA *LPG Guide*).

The LPGITA LPG *Guide* deals in the main body largely with the safety report. It has relatively little on the frequency of events, but it gives in Appendix 2 a set of graphs based on simplified hazard models for propane, butane and LPG.

These models appear broadly similar to those given in the Second *Canvey Report* and by Considine and Grint (1985) and the graphs are similar in form to those given by Grint (1989). The graphical correlations cover gas flow and two-phase flow from holes, distance to the LFL, mass of gas in cloud formed, distance to given levels of thermal radiation from a fireball, length of a jet flame and distance to given levels of thermal radiation from such a flame and distance to given overpressures from a vapour cloud explosion. There are also tables listing the injury effects at these levels of thermal radiation and overpressure.

9.22.6 BCGA liquid oxygen CIMAH guidelines

A method for estimating the Off-site Risks from Bulk Storage of Liquid Oxygen (LOX), by the British Compressed Gases Association (BCGA, 1984) (the BCGA *Guide*), gives guidance for that substance. The hazard of liquid oxygen is quite different from that for conventional flammables or toxics. Essentially it is the hazard due to enhancement of flammability in an oxygen enriched atmosphere. This effect is discussed in Chapter 16.

The BGCA *Guide* deals with (1) the consequences of oxygen enrichment, (2) the chance of being injured, (3) the dispersion of oxygen vapour, and (4) the probability of the failure event, and gives (5) examples of release and their range of hazard. It contains appendices on (1) the effect of oxygen enrichment on the burning characteristics of cloth materials, (2) the potential for low temperature and wind chill to cause injury, (3) the factors affecting the number of casualties, (4) the principles for validation of storage tank design and (5) release rate calculations and dispersion estimates.

The *Guide* treats three representative scenarios: (1) release of liquid from a hole in a 2 in. pipe, (2) release of liquid from a hole in a 6 in. pipe into a bund and (3) failure of a 135 te storage vessel. It gives for different air entrainment factors estimates of the distances to specified oxygen concentrations.

In determining the effect of enhanced oxygen concentration, the BCGA *Guide* assumes that clothing fires are the only type of fire which would increase casualties. It concentrates largely on the potential for persons who are smoking or using matches to ignite their clothing, the source of fuel closest to them.

9.22.7 FMA ammonium nitrate CIMAH guidelines

Guidance for ammonium nitrate is given in *Safety Case for Ammonium Nitrate Required by Regulation 7 of CIMAH* by the Fertilizer Manufacturers Association (the FMA *Ammonium Nitrate Guide*) (1989); an extract (FMA, 1989) is given in Lees and Ang (1989b). The FMA *Ammonium Nitrate Guide* is mainly concerned with the safety report and, in particular, with the properties of ammonium nitrate.

The *Guide* discusses the hazard of explosion of ammonium nitrate and considers the consequences of deflagration of a stack of 300 te of ammonium nitrate. It

Table 9.42 *UKCP guidance on emergency planning for chlorine installations: typical installation (after Chemical Industries Association, Chlorine Sector Group, 1989)*

A Typical equipment and inventories

Equipment	*Pressure,* (bar g)		*Size*	*Quantity*
	Normal	*Design*		
Stock tanks	7	13	20–200 te	2 × 35 te, typically
Expanse tanks	0	13	10% of stock tank	1 × 5 te, typically
Vaporizers	3	14	0.2–2 te/h	1, typically
Scrubbers	Pressurized			1
Air compressors	9	11	–	1
Piping:				
gas	2–9			<20 m
liquid	9 ASA 150 1 in. diam.			<20 m
vent	13			<20 m
Valves	Primary – globe		1 in. Table D	<20
	Isolation – ball		1 in. ASA 150	<20
'Flexibles' (liquid/vent)	2–9	25	1 in. ASA 150	<10 m

B Process activities

Air compression
Pressurization of tanker for offloading
Pressurization of storage tanks for chlorine transfer
Vaporization of chlorine
Controlled venting of tanker and storage tanks after transfer or for maintenance
Pressure relief
Possible interaction between the chlorine and its point of application

C Steps in hazard assessment

Examination of piping and instrument diagram
Systematic identification and listing of hazards
Assessment of credible size of leak
Definition of operating pressure
Assessment of extent of vaporization
Estimation of extent of gas concentrations at different locations
Estimation of potential duration of leak

D List of typical incidents[a]

1. Loss of containment of chlorine from pipework
   (a) Corrosion due to freeze/thaw conditions (particularly under lagging)
   (b) Pipe flange leaks (e.g. expansion of trapped liquid)
   (c) Flange gasket failure (due to incorrect gasket material)
   (d) Leak from pipework (e.g. spool fabricated/fitted to wrong specification, section of pipework not subject to routine inspection)

2. Loss of containment from the offloading installation
   (a) Leaks from flexibles (e.g. pinhole on weld)
   (b) Damage to filling/discharge connection due to transport container movement (e.g. collision)
   (c) Liquid chlorine ingress into vent/compressed air systems/absorption systems overloaded

3. Incidents involving bulk storage and relief systems
   (a) Contamination of stock tank contents (e.g. water ingress/incomplete drying out procedure)
   (b) Leaks from valves (incorrect valve specification)
   (c) Leaks from relief system/expanse vessel (valve maloperation, water ingress, etc.)
   (d) Total loss of containment from bulk storage needs to be considered in each case to estimate the likelihood of this event. In particular, the systems and procedures used in the operation and maintenance of the plant must be carefully reviewed

4. Loss of containment from the system supplying chlorine to the point of application
   (a) Internal chlorine iron fires (e.g. vaporizers)
   (b) Nitrogen trichloride explosion (e.g. drains from vaporizer)
   (c) Interaction in the chlorine delivery pipework, backflow of water/reagents

[a] These examples are of typical incidents but do not cover all possibilities, particularly those associated with site specific features.

***Table 9.43** UCKP guidance on emergency planning for chlorine installations: hazard scenarios[a] (after Chemical Industries Association, Chlorine Sector Group, 1989)*

						Consequence *Downwind distance affected* (m)			
			Hole size (diameter)	*Leak rate* (kg/s)		*Neutral D* 5 m/s		*Inversion F* 2 m/s	
Event	*Event likelihood*	*Pressure* (bar g)	(mm)	*Liquid*	*Gas*	140 ppm	15 ppm	140 ppm	15 ppm
Valves									
Leak from a packed gland				0.1 NF[b]		–	200	130	400
Leak on a bolted tongue and groove flange	Quite likely	7	<2						
Porosity in the valve body or bonnet									
Leak from flange (hole in 1.6 mm gasket)					0.01	–	–	–	100
Piping									
Small leak from liquid flange (1.6 mm gasket)					0.8 NF	200	600	400	1100
Small hole in liquid pipework wall (corrosion/erosion/defect)	Likely	7	5						
Small hole in gas pipework					0.05	–	150	100	300
Hole in pipework due to corrosion/erosion/defect. A full segment of 1.6 mm gasket expelled between adjacent bolts	Unlikely	7 sustained	12	3.5 NF	–	400	1300	800	>1500
Expansion of trapped liquid between closed valves (full segment of 1.6 mm gasket expelled between adjacent bolts)	Unlikely	7 reducing	12	1.2	–	200	700	500	1400
Hole in gas pipework (e.g. flange leak)	Likely	7 sustained	12	–	0.3	100	400	200	700
Storage									
T&G flange leak on liquid branch	Likely	7 reducing slowly	<2	0.1 NF		–	200	130	400
Flange leak on gas branch					0.01	–	–	–	100
Small leak from liquid branch	Unlikely	7 reducing slowly	5	0.8 NF		200	600	400	1100
Small leak from gas branch					0.05	–	150	100	300
Transfer equipment									
Flexible hose	Very likely		Pinhole	0.01	–	–	–	–	100
	Likely		2	0.1	–	–	200	130	400

B Interpretation of the qualitative frequency terms

Extremely unlikely	$<10^{-6}$/year
Very unlikely	10^{-6}–10^{-5}/year
Unlikely	10^{-5}–10^{-4}/year
Quite unlikely	10^{-4}–10^{-3}/year
Somewhat unlikely	10^{-3}–10^{-2}/year
Fairly probable	10^{-2}–10^{-1}/year
Probable	$>10^{-1}$/year

[a] This table is Appendix IV of the original document. The table is by way of example only.
[b] NF, non-flashing flow

***Table 9.44** Ground level concentrations (ppm) for scenarios 1–5 in the CIA Ammonia Guide (CIA, 1988 PA49)*

	D5			*F2*		
Scenario	500 m	1000 m	2000 m	500 m	1000 m	2000 m
1: 10 kg/s	500–1000		50–100	1000–5000		200–600
40 kg/s	600–4000		150–300	3000–10^4		1000–3000
2	200–500		20–50	500–1000		100–200
3	150–250	80–150	30–70	50–100	200–300	250–350 (no plume rise)
	40–80	40–80	20–40	0	0	10–20 (plume rise)
4(b)	300–500	200–400	100–200	200–300	400–600	600–800 (no plume rise)
	50–100	60–120	30–60	0	0	10–20 (plume rise)
	Same order of magnitude as Scenario 1					

obtains for such an explosion a trinitrotoluene (TNT) equivalent of 41 te, based on an ammonium nitrate TNT equivalent of 55% and a maximum explosion efficiency of 25%, and hence at 600 m a scaled distance of 175 m. It reproduces the ACMH correlation for the overpressure from a TNT explosion (Harvey, 1979b) and utilizes this correlation to determine that for this explosion an overpressure of 1 psi occurs at 600 m.

9.23 Hazard Impact Model

The impact of a hazard on the surrounding area may be modelled analytically by making certain simplifying assumptions. The hazard impact model so derived may be used for various purposes, including estimating the number of injured and assessing the error in the estimates.

9.23.1 Approximate model

An approximate model which may be used to estimate the number of people who suffer injury is

$$N_i = \pi r_{50}^2 d_p \qquad [9.23.1]$$

where d_p is the population density (persons/m^2), N_i is the total number of injured and r_{50} is the radius (metres) at which the probability of injury is 50%.

In this model the assumption is made that the number of people inside the circle of radius r_{50} who escape injury is equal to the number outside the circle who suffer injury. It is also assumed that the population density is uniform. The use of the radius r_{50} has some justification apart from computational convenience in that, although injury relations are usually subject to much uncertainty, the uncertainty tends to be least at the 50% injury level.

It can readily be appreciated that the degree of error in this approximate model depends on the relation between the distance and the probability of injury. Three possible cases are illustrated in Figure 9.44. In case (a), the transition from a probability of injury of unity to one of zero is very gradual, in case (b) more sharp, and in case (c) immediate. The error in the estimate reduces to zero for this latter case.

If the population density is not uniform, the approximate model given by Equation 9.23.1 cannot itself be used, but use may still be made of the radius r_{50} in conjunction with the site population density to obtain an approximate estimate of the number of injured, although in this case the number of injured inside the circle who escape injury and the numbers outside who suffer injury may not balance. As before the error in the estimate reduces to zero for case (c).

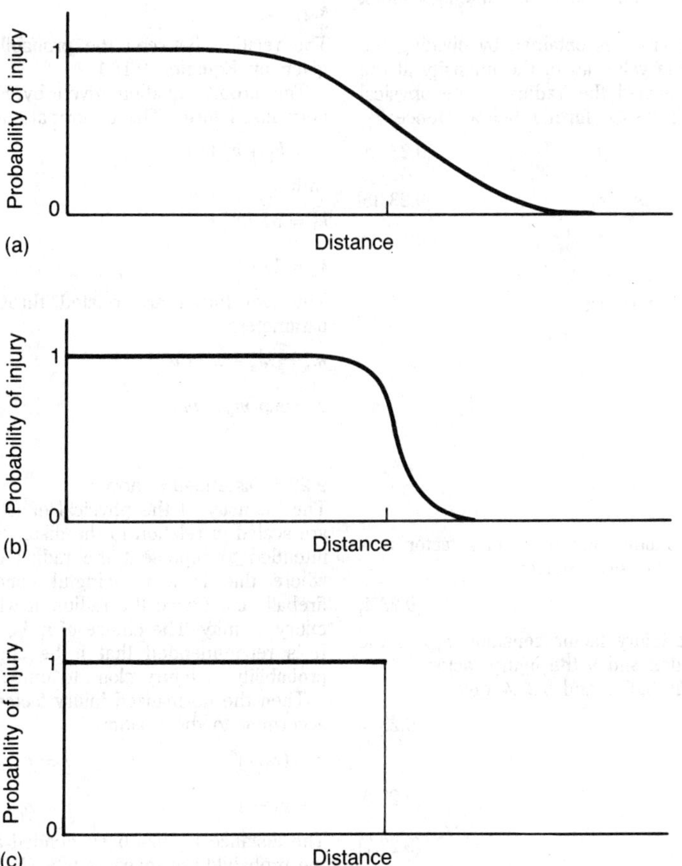

Figure 9.44 *Some hypothetical relations between distance and probability of injury*

It has been shown, however, by Poblete, Lees and Simpson (1984) and Lees, Poblete and Simpson (1986) that it is possible to derive an alternative, more accurate model.

9.23.2 Intensity of physical effect

Some principal physical effects are:

Fire	
Thermal radiation	I
Thermal radiation dose	It
Thermal radiation-time function	$f(I,t)$
Explosion	
Overpressure	p^0
Impulse	I_p
Toxic release	
Concentration	C
Dose	Ct
Concentration-time function	$f(C,t)$

It is assumed in the model that the intensity of the physical effect decays with distance according to an inverse power law. Hence:

$$w = k_w/r^{n_w} \qquad [9.23.2]$$

where k_w is the intensity constant, n_w the intensity index and w the intensity.

The normalized intensity i is obtained by dividing the actual intensity w by the value w_0 of the intensity at the distance r_0, which is termed the 'radius of the physical effect', and which is discussed further below. Hence:

$$i = w/w_0 \qquad r = r \qquad [9.23.3a]$$

$$i = 1 \qquad r = r_0 \qquad [9.23.3b]$$

9.23.3 Injury factor

Some principal injury factors are

Fire	$I^{\frac{4}{3}}t$
Explosion	p^o
	I_p
Toxic release	C
	Ct
	C^2t

It is assumed in the model that the injury factor is a power function of the intensity. Hence:

$$v = k_{vw}w^{n_{vw}} \qquad [9.23.4]$$

where k_{vw} is the first injury factor constant, n_{vw} is the injury factor power index and v the injury factor.

Combining Equations 9.23.2 and 9.23.4 gives

$$v = k_v/r^n \qquad [9.23.5]$$

with

$$k_v = k_{vw}k_w^{n_{vw}} \qquad [9.23.6]$$

$$n = n_w n_{vw} \qquad [9.23.7]$$

where k_v is the second injury factor constant and n is the injury factor decay index.

The normalized injury factor x is obtained by dividing the actual injury factor v by the value v_0 of the injury factor at the distance r_0:

$$x = v/v_0 \qquad r = r \qquad [9.23.8a]$$

$$x = 1 \qquad r = r_0 \qquad [9.23.8b]$$

9.23.4 Probability of injury

The relationship between the injury factor and the probability of injury is assumed to be the log-normal distribution:

$$P = \frac{1}{(2\pi)^{\frac{1}{2}}\sigma}\int_0^x \frac{1}{x}\exp\left[-\frac{\ln(x - m_n^*)^2}{2\sigma^2}\right]dx \qquad [9.23.9]$$

where P is the probability of injury, m_n^* is the normalized location parameter of the distribution, and σ is the spread parameter of the distribution.

An alternative way of expressing this relation is in the form of a probit equation:

$$Y = k_{n1} + k_{n2}\ln x \qquad [9.23.10]$$

with

$$k_{n1} = 5 - m_n^*/\sigma \qquad [9.23.11a]$$

$$k_{n2} = 1/\sigma \qquad [9.23.11b]$$

The relation between the probability and the probit is given by Equation 9.18.1.

The probit equation given by Equation 9.23.10 is in normalized form. The unnormalized equation is

$$Y = k_1 + k_2 \ln v \qquad [9.23.12]$$

with

$$k_1 = 5 - m_u^*\sigma \qquad [9.23.13a]$$

$$k_2 = 1/\sigma \qquad [9.23.13b]$$

The two forms are related through the two location parameters:

$$m_n^* = m_u^* - n \ln v \qquad [9.23.14a]$$

$$v = \exp(m_u^* - m_n^*) \qquad [9.23.14b]$$

9.23.5 Distances r_0 and r_{50}

The intensity of the physical effect and the injury factor are scaled in relation to the distance r_0. In principle, r_0 is intended to represent the radius of the physical effect, where this is a meaningful concept as with, say, a fireball, and hence the radius at which the probability of injury is unity. The choice of r_0 is, however, unrestricted. It is recommended that it be chosen so as to give a probability of injury close to unity.

Then the normalized injury factor decays with distance according to the relation

$$x = (r_0/r)^n \qquad r = r \qquad [9.23.15a]$$

$$x = x_0 = 1 \qquad r = r_0 \qquad [9.23.15b]$$

The distance r_{50} has been defined as the radius at which the probability of injury is 50% ($P = 0.5$). The normalized injury factor x_{50} at this distance is then:

$$x_{50} = (r_0/r_{50})^n \qquad [9.23.16]$$

Also, putting $x = x_{50}$ and $Y = 5$ $(P = 0.5)$ in Equation 9.23.10, gives

$$x_{50} = \exp(m_n^*) \quad [9.23.17]$$

Hence from Equations 9.23.16 and 9.23.17

$$r_{50} = r_0/x_{50}{}^{\frac{1}{2}} \quad [9.23.18a]$$

$$= r_0/\exp(m_n^*/n) \quad [9.23.18b]$$

9.23.6 Number of injured

It is assumed that the density d_p of the population around the hazard source is uniform. This assumption can be relaxed in certain specific ways as described below.

The number N_i of people injured is:

$$N_i = \int_0^{\infty} 2\pi d_p P(r) r \mathrm{d}r \quad [9.23.19]$$

It can be shown that the solution of Equation 9.23.19 is

$$N_i = \pi r_{50}^2 d_p \phi \quad [9.23.20]$$

with

$$\phi = \exp(2\sigma^2/n^2) \quad [9.23.21]$$

where ϕ is the correction factor for variance and decay.

Equation 9.23.20 together with Equation 9.23.21 constitutes the hazard impact model derived by Lees, Poblete and Simpson.

9.23.7 Practical decay relations

An important feature of the model is the assumption that the intensity of the physical effect decays according to an inverse power law. It has been shown by Poblete, Lees and Simpson (1984) and Lees, Poblete and Simpson (1986) that a number of the simpler hazard models give such a decay relation. There are others, however, particularly some gas cloud models, which do not.

The question of the decay of the physical effect is considered in more detail in Section 9.25.

9.23.8 Sensitivity estimates

It has been shown by Lees (1987) that there can be derived from Equation 9.23.20 a set of sensitivity coefficients which can be used to determine the effect of errors in the various physical models and injury relations utilized on the estimate of the number of injured given by that model. The parameters of interest are: d_p, k_w, n_w, k_{vw}, n_{vw}, m_u^* and σ. The normalized partial derivatives, or sensitivity coefficients, of N_i with respect to these parameters are given in Table 9.45.

Table 9.45 *Hazard impact model sensitivity coefficients (after Lees, 1987) (Courtesy of Elsevier Science Publishers)*

$(\partial N_i/N_i)/(\partial d_p/d_p) = 1$
$(\partial N_i/N_i)/(\partial k_w/k_w) = 2/n_w$
$(\partial N_i/N_i)/(\partial n_w/n_w) = -(2/n)\ln k_v + (2/n)m_u^* - 4\sigma^2/n^2$
$(\partial N_i/N_i)/(\partial k_{vw}/k_{vw}) = 2/n$
$(\partial N_i/N_i)/(\partial n_{vw}/n_{vw}) = -(2/n)\ln k_{vw} + (2/n)m_u^* - 4\sigma^2/n^2$
$(\partial N_i/N_i)/(\partial m_u^*/m_u^*) = -2m_u^*/n$
$(\partial N_i/N_i)/(\partial \sigma/\sigma) = 4\sigma^2/n^2$

9.23.9 Applications of model

The hazard impact model presented in Equation 9.23.20 is more accurate than that given in Equation 9.23.1 and, if the assumptions hold exactly, the results given by the model are exact, but in most practical cases there is likely to be a degree of approximation in the assumptions and hence in the results.

One application of the model is to make a rapid estimate of the number of injured. The model may be used to make a quick, ranging estimate, usually prior to a more detailed assessment. The use of the model is straightforward, but there are two common situations where some modification is necessary. These are for population density and directional effects.

In many cases the population density is not uniform. Two specific deviations can readily be taken into account. One is the case where there is a different population density, perhaps zero, up to some distance r_w from the hazard source. If this is such that the probability of injury up to this point is virtually unity, it is straightforward to calculate the number of injured inside the circle radius r_w and to adjust the value of N_i accordingly. The other case is where there are different population densities in different sectors around the hazard source. Again it is straightforward to adjust the value of N_i. Combination of these two cases can also be handled and, taken together, these adjustments extend appreciably the scope of the model.

Some physical phenomena are directional and affect not a circular area around the hazard source but an area which is a sector with its apex at the source. The most important case here is a gas cloud.

Directional effects can be handled by the model by considering only the sector affected and, where necessary, dividing it into a sufficient number of subsectors such that the conditions along any arc in a subsector are essentially uniform.

Another, and perhaps more significant, application of the model is to determine the effect of errors in the physical models and injury relations on the estimate of the number of injured. This may be of value not only in the conduct of hazard assessments but also in the identification for research purposes of areas where improved models and correlations would, or would not, give significant improvements in the accuracy of the assessment.

9.24 Simplified assessment methods

A full hazard assessment is a major exercise. It is useful, therefore, to have short-cut methods. The short-cut methods described may be used to obtain an approximate estimate with economy of effort. The use of such a method may improve understanding of the hazards and may highlight unexpected features. But short-cut methods have their limitations. By definition they tend to be based on generalized relations and thus are much less effective in revealing critical features and assumptions in a specific design.

A study of simplified methods of hazard assessment has been described by R.A. Cox and Comer (1982). The study was commissioned following the *Rijnmond Report*. The aim of the work was to devise a method of hazard assessment which would yield risk contours, FN curves, and individual and group risks. It was accepted that such a method would not be able to give individual risk for employees, since the degree of detail required to do this was not consistent with the simplicity sought. The method was required to yield intermediate as well as final results to permit checking and to improve confidence in the latter.

Two methods were devised and compared:

(1) Simplified classical method
(2) Parametric correlation method.

These are now described in turn.

9.24.1 Short-cut classical method

The short-cut classical method (SCM) is a simplified form of the full classical method in which individual failure scenarios are defined and estimates are made of their frequency and consequences.

The SCM approach involves three stages: failure case selection and frequency estimation; application of consequence submodels; and summarization of risks.

Crucial to the method is the selection of the failure cases. The original intent was to pick a set of representative cases. But it was concluded that it is necessary to start from the full set of failure cases. These cases are then reduced to a more limited set of equivalent discrete failures (EDFs). One method is to cluster together 'similar' failures, another is to compare each case with a standard EDF (SEDF). It was the latter approach which was adopted.

A standard EDF is defined by standard physical consequences such as the distance travelled by a vapour cloud in standard weather conditions. An SEDF list was created, giving for each EDF the distances for defined intensities of the relevant physical effects such as toxic concentration. For each EDF the consequences are determined by a once-and-for-all assessment. Then, in applying the method to a particular plant the actual EDF for a particular failure case is equated to one of the standard EDFs using consequence submodels. This is the only use of these submodels in the method.

Some of the principal elements in the method are shown in Table 9.46. Section A of the table lists the data schedules required to define the failure cases. They are schedules of the plant units, of the items on the plant and of the connections between items. Section B shows the six main categories of release considered, and section C gives the consequence submodels used. Figure 9.45 is a flow diagram of the algorithm for pipe failure and Figure 9.46 one for application of the method to an actual plant.

Up to this point the analysis is independent of the environment. Factors defining this environment, such as the weather, ignition sources, etc., are now brought in. The effect of introducing these factors is to alter not consequences, but the frequencies for each standard.

From this information on the frequency and consequences of each standard EDF at the particular plant it is possible to obtain estimates of risk. The frequencies of the various physical outcomes, the areas affected and the number of casualties are determined, and risk estimates made in the usual way.

Table 9.46 *Some elements of the short-cut classical method (SCM) of hazard assessment (after R.A. Cox and Comer, 1982) (Courtesy of the Institution of Chemical Engineers)*

A Schedule of item interconnections
Each item to item connection defined by:
Identification of items of either end of pipe Pipe diameter Location of connection at either end of pipe – in vapour space or below liquid level Presence of particular types of valving along pipe (excess flow, non-return and shut-off valves)
B Schedule of items
For each plant item:
Item identification Material and inventory contained Temperature and pressure
The item location is specified by the item identification which identifies the plant unit on which the item is located
C Schedule of plant units
For each plant unit:
Set of parameters to define failure frequencies Location coordinates

9.24.2 Parametric correlation method

The alternative method considered was the parametric correlation method (PCM). With this method there are two principal features which must be specified: the form of the correlation function and the method of determining its parameters.

One approach to selecting the form of the correlation function is to run a large number of full hazard assessments and derive a correlation function from these. Another is to use expert judgement to select suitable functional forms. The correlation function should be carefully chosen to give an appropriate number of parameters. There is a potential problem of escalation of the number of parameters required.

A possible form of correlation function might be:

$$R(d) = \frac{R(0)}{2}[1 + \cos(\pi d/d_{max})] \quad [9.24.1]$$

where d is the distance from the unit, R is the risk and the subscript max indicates the maximum at which any risk exists.

The correlation function given in Equation 9.24.1 has two parameters, R(0) and d_{max}. Its characteristics are shown in Figure 9.47. Since $R(0)$ is the risk at the unit itself, it is in effect a measure of frequency, while d_{max} is a measure of consequences. Thus the two parameters are independent. This correlation function therefore has the attraction of simplicity.

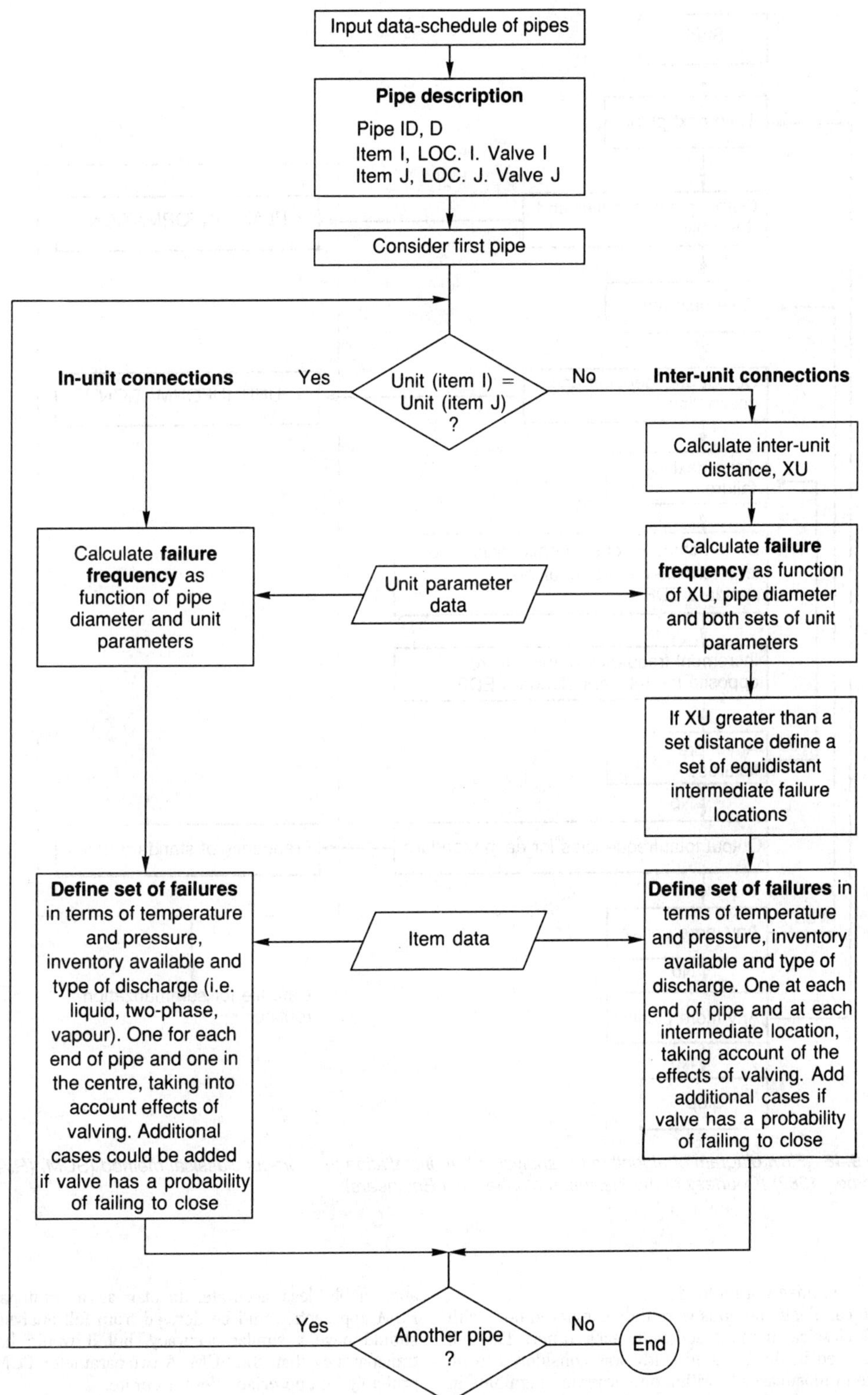

Figure 9.45 *Flow diagram of algorithm for pipe failure in short-cut classical method (SCM) (R.A. Cox and Comer, 1982) (Courtesy of the Institution of Chemical Engineers)*

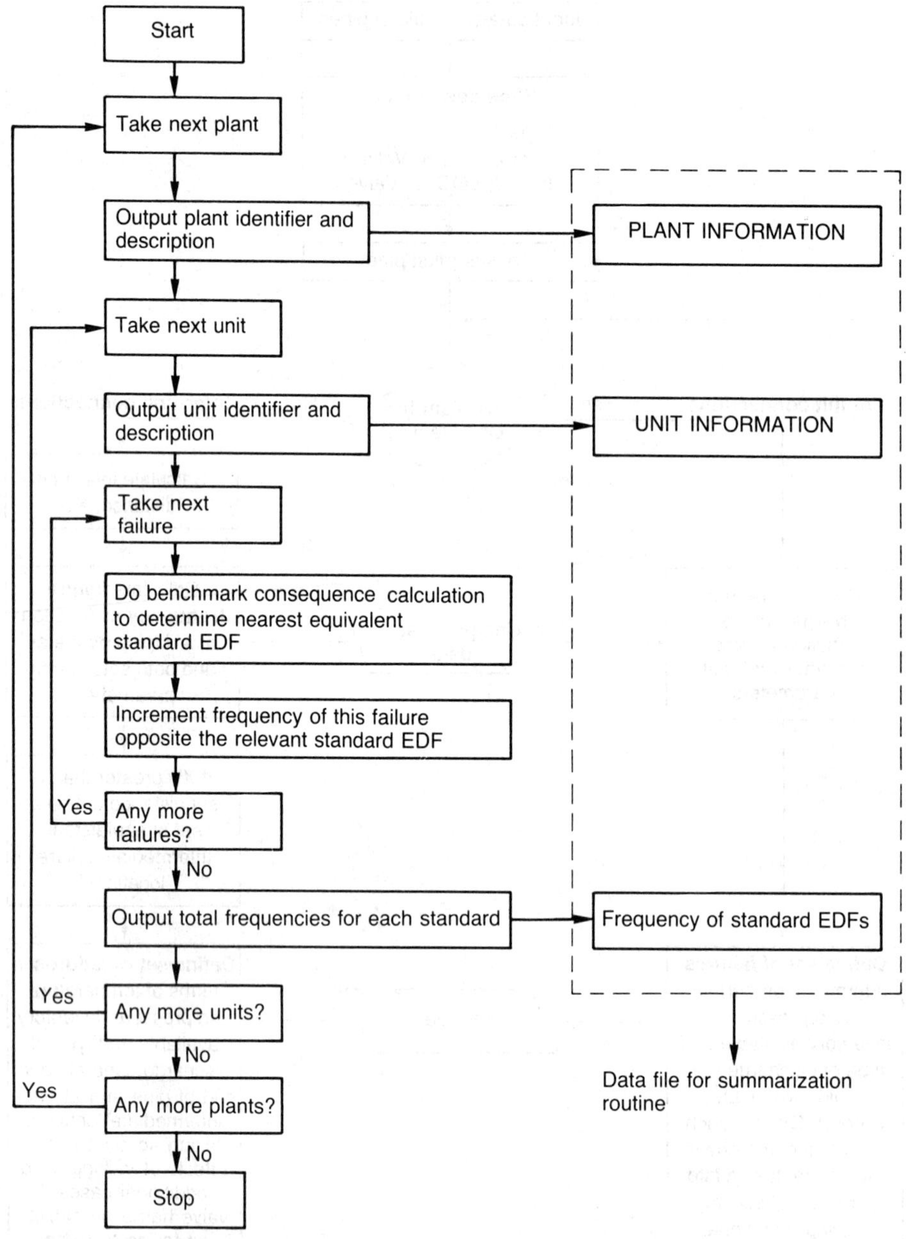

Figure 9.46 *Flow diagram of algorithm for analysis of an installation in short-cut classical method (SCM) (R.A. Cox and Comer, 1982) (Courtesy of the Institution of Chemical Engineers)*

9.24.3 Comparison of methods

The two candidate methods were then compared both with the full classical method and with each other. The full method used in the Rijmond study was considered to be accurate in most cases to within one order of magnitude in each direction, but in some cases only to within two orders of magnitude. The SCM approach was not expected to be appreciably less accurate. Insofar as a multi-parameter PCM approach would be derived from full analyses it too should have a similar accuracy, but it would have less transparency than the SCM. A two-parameter PCM would probably be appreciably less accurate.

With regard to cost, the capital cost of the SCM and two-parameter PCM were assessed as comparable and

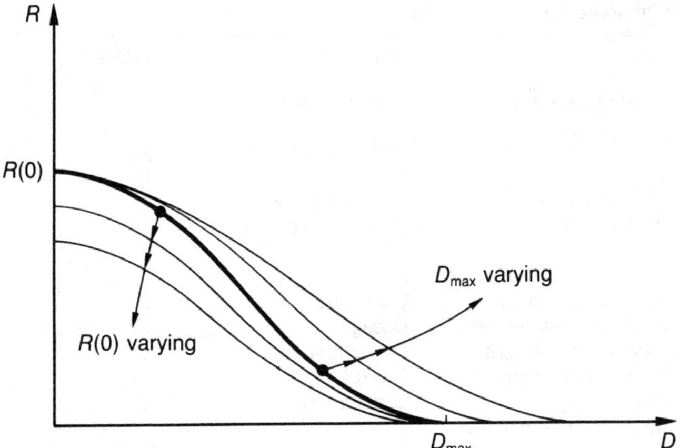

Figure 9.47 *Characteristics of the decay function R(d) used in the parametric correlation method (R.A. Cox and Comer, 1982) (Courtesy of the Institution of Chemical Engineers)*

that of a multi-parameter PCM as about double. On the other hand, the operating costs of the two-parameter PCM were estimated as half those of the SCM, with the multi-parameter PCM costs lying in between.

9.24.4 Hazard impact model

The hazard impact model may be used as a short-cut method for hazard assessment. The method depends on the availability of decay and injury relations and on certain simplifying assumptions. The use of the model for this application is described in Section 9.23.

9.25 Decay Relations

Several of the methods described make use of decay relations. These may be generic relations intended to be applicable to any hazard with suitable choice of parameters or they may be specific to a particular hazard.

For the simpler hazard models decay may often be derived analytically, while for the more complex models embodied in computer codes it is necessary to obtain and then correlate numerical results. Similarly, correlations may be derived from experimental data.

9.25.1 Generic decay relations

Several decay relations have been given in the literature. They include: Equation 9.24.1, used by R.A. Cox and Comer (1982) for the parametric correlation model; Equation 9.23.20 used by Lees, Poblete and Simpson (1986) for the hazard impact model; and Equation 9.14.1, used by Bagster and Pitblado (1991) for the domino effect model.

9.25.2 Decay relations for specific hazards

Decay relations can be obtained from the simpler hazard models for fire, explosion and toxic release, as described by Poblete, Lees and Simpson (1984) and Lees, Poblete and Simpson (1986).

The intensity of heat radiation from a fireball is:

$$F = E/4\pi r^2 \quad [9.25.1]$$

where E is the heat radiated (kW), F is the heat radiated on the target (kW/m^2), and r is the distance (m) from the centre of the fireball to the target. Similar equations apply to other types of fire, such as pool fires and flares.

The peak overpressure and impulse from the explosion of a high explosive are functions of scaled distance:

$$p^o = f(z) \quad [9.25.2]$$

$$I_p = f(z) \quad [9.25.3]$$

with

$$z = r/W^{\frac{1}{3}} \quad [9.25.4]$$

where I_p is the impulse (N s/m^2), p^o is the peak incident overpressure (N/m^2), W is the mass of explosive (kg) and z is the scaled distance (m/kg$^{1/3}$). The function in Equation 9.25.2 is usually provided in graphical form, such as the curves given by W.E. Baker *et al.* (1983). However, over a limited overpressure range it can be represented by the relation

$$p^o = 1/r^{n_1} \quad [9.25.5]$$

where n_1 is an index. The Baker curves correspond over the overpressure range 1.0–0.1 bar to decay indices of 1.7 and 0.9 for peak overpressure and impulse, respectively.

For vapour cloud explosions (VCEs) there is no established model. Some theoretical models give a decay index of approximately unity. On the other hand, the decay curve established for the Flixborough explosion gives over the overpressure range 1.0–0.1 a decay index of 1.7.

Decay indices for dispersion of neutral density gas may be derived from the Sutton equations. For an

***Table 9.47** Decay indices for some principal hazards*

Hazard	*Model*	*Physical effect*	*Intensity*	*Decay index*
Fire	Fireball, pool fire	Thermal radiation	I	2
Explosion	TNT	Peak overpressure	p^o	1.7[a]
		Impulse	I_p	0.9[a]
	VCE (Flixborough)	Peak overpressure	p^o	1.7[b]
	VCE (models)	Peak overpressure	p^o	1.0[c]
				1.1[d]
Toxic release	Neutral density gas: instantaneous release	Concentration	C	2.6
		Dosage	Ct	1.75
	Neutral density gas: continuous release	Concentration	C	1.75
		Dosage	Ct	1.75
	Heavy gas: instantaneous release	Concentration	C	1.4[e]
	Heavy gas: continuous release	Concentration	C	1.7[f]

[a] Over the range 1.0–0.1 bar (W.E. Baker *et al.*, 1983).
[b] Over the range 1.0–0.1 bar (Sadee, Samuels and O'Brien, 1976–77).
[c] Model of Wiekema (1980).
[d] Model of Ebert and Becker (1982).
[e] Specific example of 200 te ammonia release using DENZ computer code (Fryer and Kaiser, 1979 SRD R152).
[f] Specific example of 23.9 kg/s chlorine release using CRUNCH computer code (Jagger, 1983 SRD R229).

instantaneous release the concentration at ground level on the centre line of the cloud and at cloud centre is

$$\chi = \frac{2Q^*}{\pi^{\frac{3}{2}}C^3(ut)^{\frac{3}{2}(2-n)}} \qquad [9.25.6a]$$

where C is the diffusion parameter $(m^{n/2})$, n is the diffusion index, Q^* is the mass released (kg), t is the time (s), u is the wind velocity (m/s) and χ is the concentration (kg/m^3). For neutral conditions, $n = 0.25$:

$$\chi \propto 1/x^{2.6} \qquad [9.25.6b]$$

The total integrated dosage D_{tid} ((kg/m^3) s) is

$$D_{tid} = \frac{2Q^*}{\pi C^2 u(ut)^{2-n}} \qquad [9.25.7a]$$

Setting $n = 0.25$:

$$D_{tid} \propto 1/x^{1.75} \qquad [9.25.7b]$$

For a continuous release the concentration at ground level on the centre line of the cloud is

$$\chi = \frac{2Q}{\pi C^2 u x^{2-n}} \qquad [9.25.8a]$$

where Q is the mass rate of release (kg/s) setting $n = 0.25$:

$$\chi \propto 1/x^{1.75} \qquad [9.25.8b]$$

The total integrated dosage is obtained by assuming that the continuous release lasts for some finite time. Then the total integrated dosage is:

$$D_{tid} \propto 1/x^{1.75} \qquad [9.25.9]$$

Models of heavy gas dispersion tend to be too complex to permit the analytical derivation of general decay indices, although an index may sometimes be derived for particular cases. Some decay indices for some principal hazards are given in Table 9.47.

The decay index for the physical effect can be combined with the power index for the injury factor to give an overall decay index for injury, as described the discussion of the hazard impact model in Section 9.23.

Another decay relation used is:

$$r = kM^{n_2} \qquad [9.25.10]$$

where M is mass released (kg), n_2 is a decay index, r is the distance for some physical effect (m) and k is a constant.

Equation 9.25.10 has been used to correlate results from numerical computations or empirical data. In particular, it has been used to give a correlation for the decay of the concentration of a flammable gas cloud to its lower flammability limit (LFL) or some fraction of this. For example, it was used in the Second *Canvey Report* to correlate distance to 0.5 LFL.

9.26 Hazard Warning

The use of hazard assessment of the kind so far described to provide assurance that a hazard is under control has certain weaknesses. One is the problem of credibility. The most difficult situation arises where there is a major hazard. Industry may attempt to argue qualitatively and then to demonstrate quantitatively that although in the most unfavourable circumstances the hazard has the potential to kill a large number of people, the probability that the hazard will be realized during the lifetime of the plant is very low. Frequently such arguments do not satisfy the objectors, who tend to concentrate exclusively on the hazard potential and to emphasize that the major accident could occur out of the blue at any time.

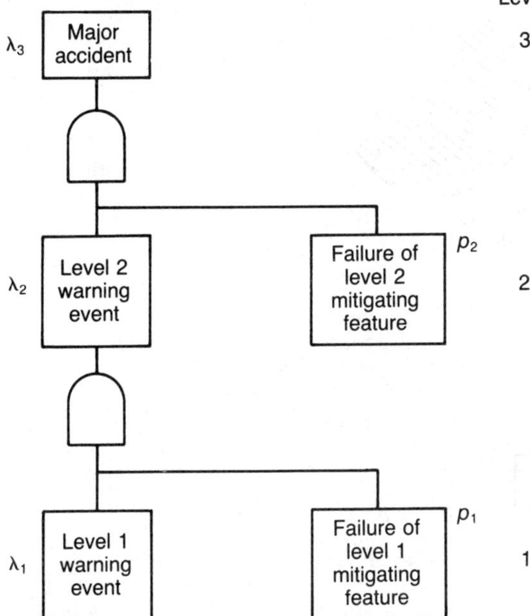

Figure 9.48 *A hazard warning tree (Lees, 1983b) (Courtesy of the Institution of Chemical Engineers)*

Another weakness is that when the plant comes to be operated, usually relatively little use is made of the information generated during the design, particularly the fault trees, for the control of hazards and the analysis of accidents and failures. These problems may be tackled using the concept of hazard warning structure described by Lees (1982b, 1983b, 1985). This is now outlined.

9.26.1 Hazard warning structure

If the structure of a major accident is analysed by constructing a fault tree, it is generally found that this event occurs as a result of a minor accident and of failure of a mitigating feature which is normally effective in preventing such escalation. In turn the minor accident may occur only if there is a lesser accident and failure of a second mitigating feature. Typically, there is a series of events which result in a major accident only if all the mitigating features fail. Generally, therefore, there are many more lesser accidents than major accidents. This high ratio of lesser to major accidents is the concept underlying the accident pyramid discussed in Chapter 1 and shown in Figure 1.4.

The major accident is likely therefore to be preceded by a number of lesser events, some of which may be regarded as 'near misses'. Thus the time order of the events is also important. The exploitation of the time order of major and lesser accidents underlies the concept of learning from near misses.

These ideas are well understood. What does not appear to have been sufficiently appreciated and exploited is the potential for using these concepts in conjunction with a formal analysis of accident structure such as the fault tree. A hazard possesses a characteristic warning structure. The hazard warning structure may be laid bare by casting the fault tree for the hazard in a particular way.

9.26.2 Hazard warning tree

A typical hazard warning tree is shown in Figure 9.48. An event at level 1, which has a frequency λ_1, escalates into a level 2 event only if there is failure of a level 1 mitigating feature, which has a probability p_1. Similarly, the level 2 event with frequency λ_2 escalates into a level 3 event with frequency λ_3 only if there is failure of the level 2 mitigating feature with probability p_2. The probabilities p_1 and p_2 are attentuation fractions associated with the mitigating features.

A simple illustration is a release from a large refrigerated atmospheric storage tank holding toxic liquefied gas towards a housing estate as shown in Figure 9.49. In this case the first level event might be defined as a release of unspecified size which might have a number of possible causes. The first level feature might then be the size of release, since only a proportion of such releases would be large ones. Further mitigating features at higher levels in the tree would be the wind direction, the stability condition, the population exposure and the injury relations.

9.26.3 Events before and after release

For a major hazard the critical event is usually loss of containment, and consideration here is directed mainly to this.

Generally in hazard assessment the frequency of the critical event is estimated either directly from failure data or else the structure of the events leading up to the release is analysed and the frequency obtained indirectly using a fault tree, while the structure of the events following the release and the frequency of particular outcomes is determined using an event tree. For a particular outcome the event tree can then be recast in the form of a fault tree. In some cases, therefore, there are available, in principle, both pre-release and post-release fault trees, whilst in other cases only the latter are available.

Even where it is not normally considered necessary to construct a pre-release fault tree in order to estimate release frequency, it may nevertheless by worth doing so for the purpose of analysing hazard warning structure. If both pre- and post-release fault trees are available, the two trees may be consolidated to form a single fault tree.

The concept of hazard warning was originally described primarily in terms of the critical event and the post-release tree, using the illustrative example of the release of refrigerated toxic liquefied gas as just described. The reason for this was simple. After the release the event/mitigating feature pairs are readily defined. They are features such as leak size, wind direction, stability condition, and so on. The hazard warning analysis is therefore relatively straightforward.

However, the application of hazard warning analysis to the situation prior to release is equally valid and may actually be of more value, since it deals with the features over which management has some control. Management can do little about the weather, but it can do something about plant design and operation. Pre-release analysis may therefore be more valuable, but it is usually more difficult.

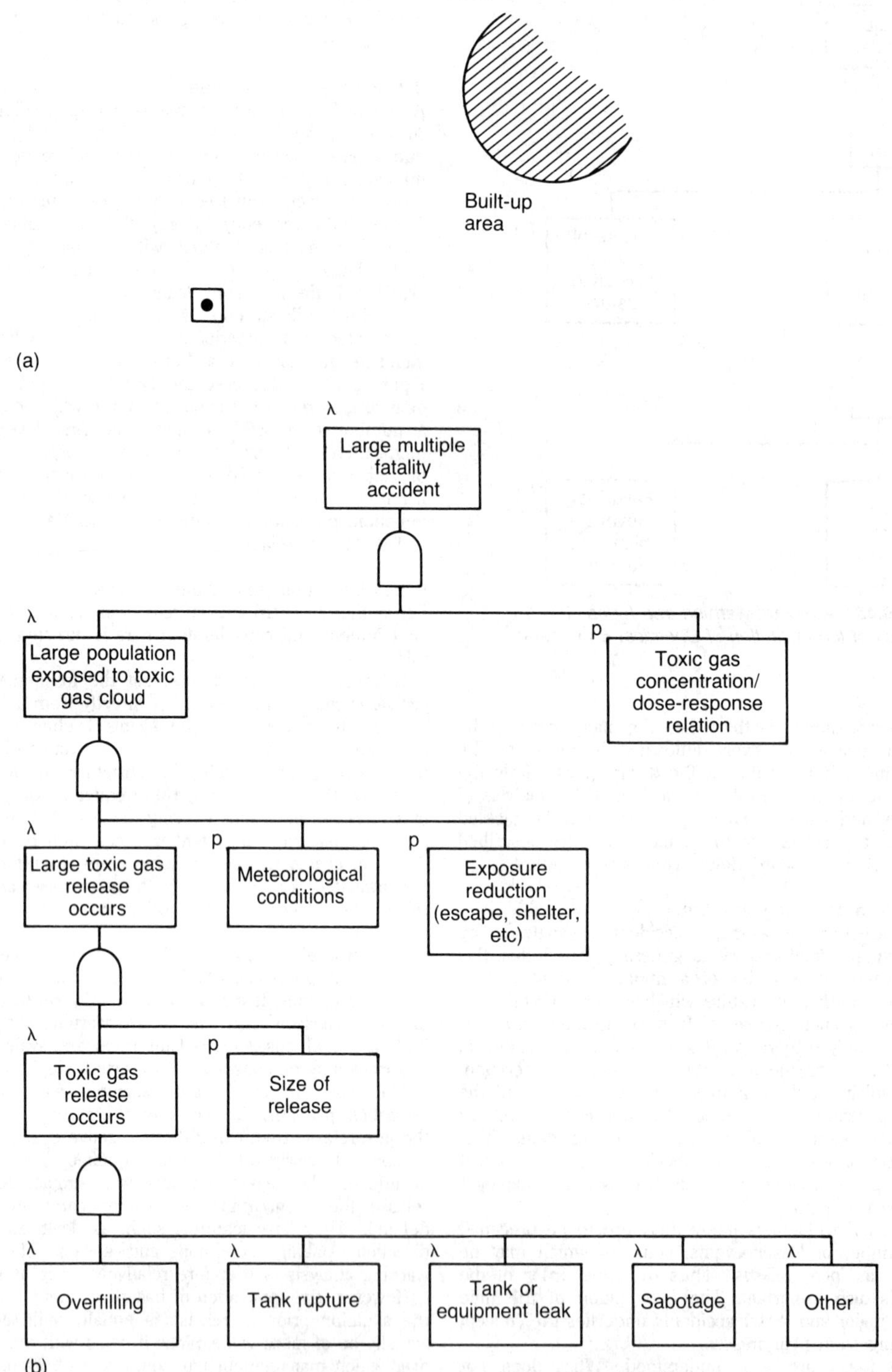

***Figure 9.49** Hazard warning tree for large multiple fatality accident due to large release of toxic liquefied gas: (a) storage tank with built-up area distant and in one sector only; (b) hazard warning tree (Lees, 1982b) (Courtesy of the Institution of Chemical Engineers)*

9.26.4 Events, mitigating features and attenuation factors

The essential feature of a hazard warning analysis is that an event above the lowest level is defined as the outcome of a lower level event and of failure of a mitigating feature. The quality of the analysis depends largely on this first stage in which events, or in effect event/mitigating feature pairs, are defined.

In defining an event the object should be to give a definition which corresponds to an observable, and recordable, event. This is an essential condition both for the use of historical event data in conventional hazard assessment and for the use of event data from the operating plant. The definition of an event implies that of the associated mitigating feature. The term 'mitigating feature' is interpreted very broadly. For example, hole size for a noxious release is a mitigating feature on the basis that only a proportion of holes are large ones.

As in constructing a fault tree, with a hazard warning tree it is important that events be defined in such a way as to make the tree comprehensive and to guard against omissions. Such robustness of event definition may be achieved by working, at least in the first instance and at higher levels of the tree, in terms of generic failure modes rather than specific causes. For example, blockage may represent a generic mode of which solids deposition, freezing, crystallization and polymerization are specific causes. This approach is a safeguard against omissions insofar as it is much less probable that a generic mode will be overlooked than a specific, and perhaps low probability, cause.

Some events and mitigating features are under the influence of management, while others are not. Broadly speaking, management has relatively good control up to the point where loss of containment occurs, but much less thereafter. Management decides the hardware and software protection which is to be provided. The former includes both passive protection such as plant layout and active protection such as trip systems, while the latter covers a range of measures such as formal systems and procedures and training. Management is also able to influence the occurrence of initiating events. Once emission has occurred, the features which come into play are those such as weather conditions and exposure of people, factors over which management has little or no control.

Another important distinction is between predictable and unpredictable event rates and attenuation factors. It is much more difficult to make quantitative estimates for some event/mitigating feature pairs than for others.

Since there is a potential ambiguity it is necessary to define the sense in which 'attenuation' and associated terms are used here. The frequency of a lower level event is divided by the attenuation, or attenuation factor, associated with the mitigating feature in order to obtain the frequency of the higher level event. The attenuation is therefore greater than unity. The attenuation fraction, which is the probability of failure of the mitigating feature, is the reciprocal of the attenuation factor and is less than unity.

A hazard has a high or low warning structure depending on the total attenuation between the major accident and the lowest level of event which can serve as a warning. For a hazard to be classed as 'high warning', which is the desirable case, the total attenuation should ideally be several orders of magnitude. Generally, this means that the attenuation attributable to a single feature should be an order or magnitude or more. Attentuations of 2 or 3 are of limited value, although not completely worthless.

Fortunately, for many hazards there are mitigating features which do tend to be associated with quite large attenuations. For example, the proportion of large leaks to total leaks, the probability of explosion rather than fire, the probability of particular meteorological conditions may each be quite low.

If a mitigating feature represents a significant barrier to escalation and if it is one over which management has some control, it is important that it be identified and that measures be taken to maintain or even improve its effectiveness.

Hazard warning is best viewed not so much as a property inherent in an installation but rather as one conferred on it by the observer. The property of hazard warning is conferred partly by recognition of features capable of giving warning and partly by the creation of such features. The principal means of creating a warning

Table 9.48 *Some mitigating features for post-release hazard warning*

Size of leak

Employee exposure: workforce location
  Distance to population
  Direction of population
  Density of population

Population exposure: siting
  Distance to population
  Direction of population
  Density of population

Modification of exposure (employees and public):
  Time of day
  Escape, evacuation
  Shelter
  Protective clothing

Meteorology:
  Wind direction
  Wind speed
  Stability category

Nature and intensity of failure

Nature and size of equipment failure:
  Size of hole
  Size of inventory

Flammable cloud behaviour

Ignition:
  Probability
  Time delay

Explosion:
  Probability
  Directionality

feature are, in the pre-release phase, the use of monitoring and inspection in some form and, in the post-release phase, the use of separation distances.

9.26.5 Post-release analysis

The starting point for the post-release analysis is the critical event, which is here taken as a loss of containment. In constructing the post-release tree it is generally convenient to include as a mitigating feature the size of release, even though this could be regarded strictly as part of the pre-release tree, provided that double counting is avoided in the final assessment. The justification for this practice is that leak size is nearly always an important mitigating feature with a high attenuation factor, and that inclusion in the post-release tree ensures that it is taken into account even if no pre-release tree is developed.

Some principal mitigating features in a post-release tree are given in Table 9.48. The list is illustrative rather than exhaustive, and is self-explanatory. A post-release hazard warning analysis of the St Fergus to Moss Morran pipeline has been given by Lees (1985). A hazard assessment for this pipeline has been made by the HSE, as described in Chapter 23.

The hazard considered in the HSE report is a release of NGL from the pipeline, resulting in interactions with housing and hospitals. For this hazard there is a single base event which has 11 potential bad outcomes. The base event is a release of NGL from the pipeline. The pipeline is 220 km long. The frequency of the base event is:

Frequency of release of NGL = 2.32×10^{-4}/ km-year

$= 220 \times 2.32 \times 10^{-4} = 5.1 \times 10^{-2}$/years

The bad outcomes are interactions with housing at eight locations and with hospitals at three locations. The frequencies of these bad outcomes are:

Location	*Frequency* (10^6 interactions/ year)	*Location*	*Frequency* (10^6 interactions/ year)
1	2.5	7	1.0
2	2.0	8	1.2
3	3.3	9	1.0
4	4.0	10	1.3
5	1.5	11	1.4
6	2.8		
Total			22.0

The frequency of all bad outcomes is 22×10^{-6}/year. The overall attenuation factor for all bad outcomes is:

Overall attenuation factor = $(5.1 \times 10^{-2})/(22 \times 10^{-6}) = 2300$

The individual attenuation factors which make up this overall attenuation factor may be derived from the report as follows. For each location there is an estimated length of pipeline which is capable of giving rise to an interaction. The total length for all locations is 32.5 km. Hence:

Attenuation factor for location = 220/32.5 = 6.8

There is no interaction for ruptures with hole sizes up to 20 mm, but there can be interactions for ruptures with hole sizes in the range 20–80 mm. The estimated frequencies for these two cases are:

Frequency of rupture with hole size up to 20 mm $= 2.0 \times 10^{-4}$/km-year

Frequency of rupture with hole size 20-80 mm $= 2.2 \times 10^{-5}$/km-year

and hence

Attenuation factor for leak size =

$$(2.0 \times 10^{-4} + 2.2 \times 10^{-5})/(2.2 \times 10^{-5}) = 10.1$$

Interaction can occur only if the wind direction allows. The probability of such wind direction, weighted in accordance with the interaction frequencies, is 0.240. Hence:

Attenuation factor for wind direction = 1/0.240 = 4.2

Also, interaction can occur only if the stability condition allows. Interactions are possible at all locations for Pasquill stability category F and at eight locations for category E, giving for the latter a weighting factor of 0.72 based on the length of line at each location over which interaction can occur. The probabilities of categories E and F are 0.06 and 0.08, respectively. Hence:

Attenuation factor for stability conditions = $1/[(0.06 \times 0.72) + 0.08] = 8.1$

The overall attenuation factor from these four separate attenuation factors is 2300, which agrees with the figure derived earlier. In this case, therefore the overall attenuation factor is a very large one.

9.26.6 Pre-release analysis

It is generally much more difficult to carry out a pre-release analysis and the methods for doing this are not well developed. As before, the approach involves the definition of mitigating features. Some principal mitigating features are shown in Table 9.49. These features are rather less straightforward and require some explanation. The pre-release tree may be envisaged as consisting of the main potential hazard tree which has as its top event the hazard, namely a leak, which will be realized unless protection acts to prevent it, and as having grafted onto it two branches, one of hardware protection and one of software protection. These two latter modes of protection are effected by protective devices such as pressure relief valves, trips and interlocks and by the process operator, respectively. Within the potential hazard tree there are two main mitigating features. These are inspection and procedures. Inspection is the means of detecting and correcting mechanical failures. Procedures are the means of preventing human error. Both types of failure can constitute initiating events. The construction of a pre-release tree requires a degree of inventiveness. There is wide scope for creative definition of mitigating features.

A pre-release hazard warning analysis of a hydrocarbon sweetening plant has been described by Lees (1985). The analysis is based on an account given by Kletz (1972a) of the use in the design of the plant of several trip systems to protect against the hazard of explosion.

Table 9.49 Some release modes and mitigating features for pre-release hazard warning

A Release modes
Pressure envelope failure:
Equipment defect
External threat
Maintenance/modification activity
Process parameter deviation
B Mitigating features
Leak size
Protective devices
Operator action
Inspection procedures

The plant consists of a reactor which is fed with both hydrocarbon and air. The air is supplied from an air receiver, which is fed by an air compressor. There is a pressure relief valve on the air receiver and a non-return valve between the air receiver and the reactor. An explosion is assumed to occur if an explosive mixture develops, since an ignition source is assumed always to be present. An explosive mixture can occur if an air pocket develops and the feed is above its flashpoint or if backflow occurs from the reactor into the air receiver, which is at a temperature above the flashpoint. The combination of an air pocket and of a feed above its flashpoint can result from a high feed temperature, which is a common cause for both these conditions, or from other causes. The fault tree for the system is given by Kletz in his paper.

The hazard warning tree derived by Lees from this fault tree is shown in Figure 9.50. The release mode in this case is a process parameter deviation. The initiating and enabling events are:

(1) relief valves fails open;
(2) compressor fault;
(3) mixer fault;
(4) supply fault;
(5) instrument fault;
(6) operator error.

The protective device failures are:

(1) non-return valve fails;
(2) high temperature trip fails;
(3) air pocket trip fails.

The procedure failures are

(1) purging procedure fails;
(2) flash point control procedure fails.

The mitigating features of protective devices and of procedures are underlined in Figure 9.50. Events which might be reduced in frequency by mitigating features such as inspection (equipment failures) and procedures (operator errors) are shown with broken underlining. This indicates points at which additional mitigating features might in some cases be created to give reduced frequency of occurrence and increased warning if these are judged desirable.

This hazard warning tree shows that there is one mitigating feature, the non-return valve and the high temperature trip, on the first and second branches, respectively, but two mitigating features, the air pocket trip and the flashpoint control procedure, on the third branch. In the event, it was decided to accept low flashpoint feeds as well, so that with this policy there was then one mitigating feature on the third branch also.

In general, the process operator appears both as a source of initiating events and as a form of protection. In the tree considered the protection afforded by the operator is limited to formal checking procedures. In principle, the operator might also intervene to effect protection at other points. For example, he may act as a further layer of protection against high feed temperature. However, for the operator to be able to give protection against a condition he must be able to observe or at least infer it. Observability is one of the principal aspects of operability, which in turn is one of the two main concerns of hazard and operability (hazop) studies.

Another illustrative example of pre-release analysis is shown in Figure 9.51 for the hazard of rupture of a batch reactor due to reaction runaway. Figure 9.51(a) gives the master tree and Figures 9.51(b) and 9.51(c) show the subsidiary trees. The reactor is one for which only a proportion of batches are prone to reaction runaway, which gives an additional mitigating feature.

Of particular interest in this tree is the fact that two of the initiating events may also cause failure of a mitigating feature. Thus loss of agitation is not only an initiating event but also prevents the use of shortstop, since without agitation distribution of the shortstop is poor. Similarly, if there is a single temperature measuring instrument both for the control loop and for display to the operator, failure of this instrument is not only an initiating event but may also cause a failure of operator intervention.

In the first example of pre-release analysis given above the problem was of the type to which fault tree analysis is typically applied. There is a basic hazard and trip, and other protective systems are used to guard against it. The problem in the second example possesses a somewhat similar structure. It is also possible to apply hazard warning analysis to less explicitly structured problems, but the investigator must then impart much of the structure himself.

Of particular importance is loss of containment from the plant. Figure 9.52 gives an outline hazard warning tree for a major release of flammables from a plant, referred to as Plant EB. Figure 9.52(a) gives the master tree and Figures 9.52(b)–9.52(g) show the subsidiary trees. For example, for vessel corrosion (Figure 9.52(b)), the initiating event is vessel deterioration. Only if the mitigating feature of inspection fails does this escalate into a vessel leak. In turn, only in a proportion of cases will the leak be a major one, so that hole size is another mitigating feature.

For pipework rupture by maintenance or modification (Figure 9.52(c)), the initiating event is non-adherence to

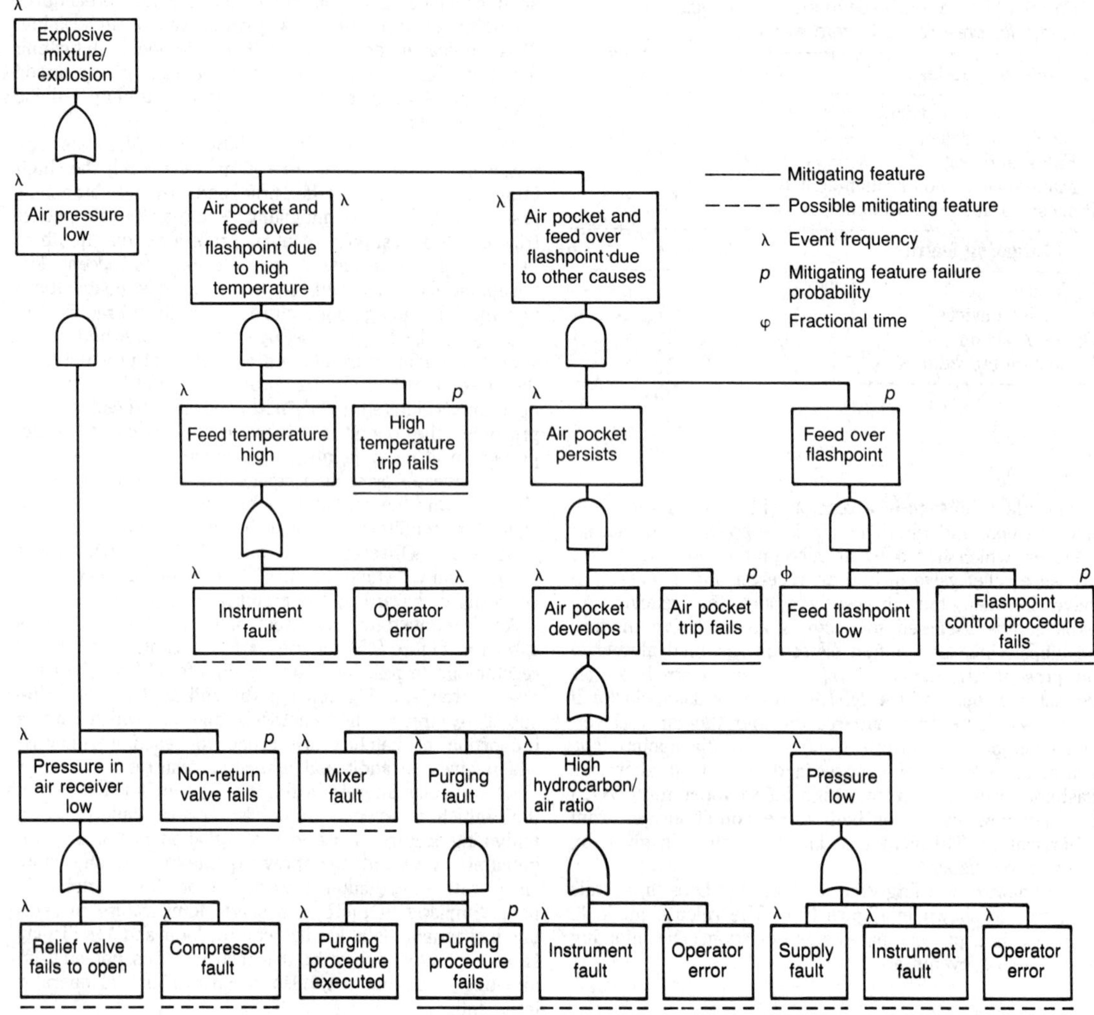

***Figure 9.50** Hazard warning tree for explosion in hydrocarbon sweetening plant (Lees, 1985) (Courtesy of the Institution of Chemical Engineers)*

procedures anywhere in the works. Only in a proportion of cases will this lead to a leak, which gives a mitigating feature. Furthermore, only in a proportion of cases will such a leak occur on Plant X, as opposed to elsewhere in the works, which gives a further mitigating feature. Finally, only in a proportion of cases will the leak be a major one, so that again hole size is another mitigating feature.

The other subsidiary trees for vehicle impact, missile impact and fire on the same plant and on another plant shown (Figures 9.52(d)–9.52(g)) follow the same principles. In some of the trees not all the branches are fully developed.

In this example, therefore, the lowest level warnings even include events which occur elsewhere on the works. It is quite legitimate to include these as warnings if they are indicators, for example, of non-adherence to procedures. The example illustrates the fact that it is up to the analyst to impose the hazard warning structure on the problem.

Another illustrative example of a similar type is that shown in Figure 9.53 for the hazard of failure of a materials testing system. Such failure may lead to a major leak from a pipework component. The initiating event is receipt of an unsuitable component. The materials testing system allows through only a proportion of defective components and acts as a mitigating feature. Only in a proportion of cases will the leak be a major one, which is a further mitigating feature. The failure of the mitigating feature of detection can occur as a result of failure of the detection procedure, but less obviously it can also be a result of failure to select the component for inspection in the first place.

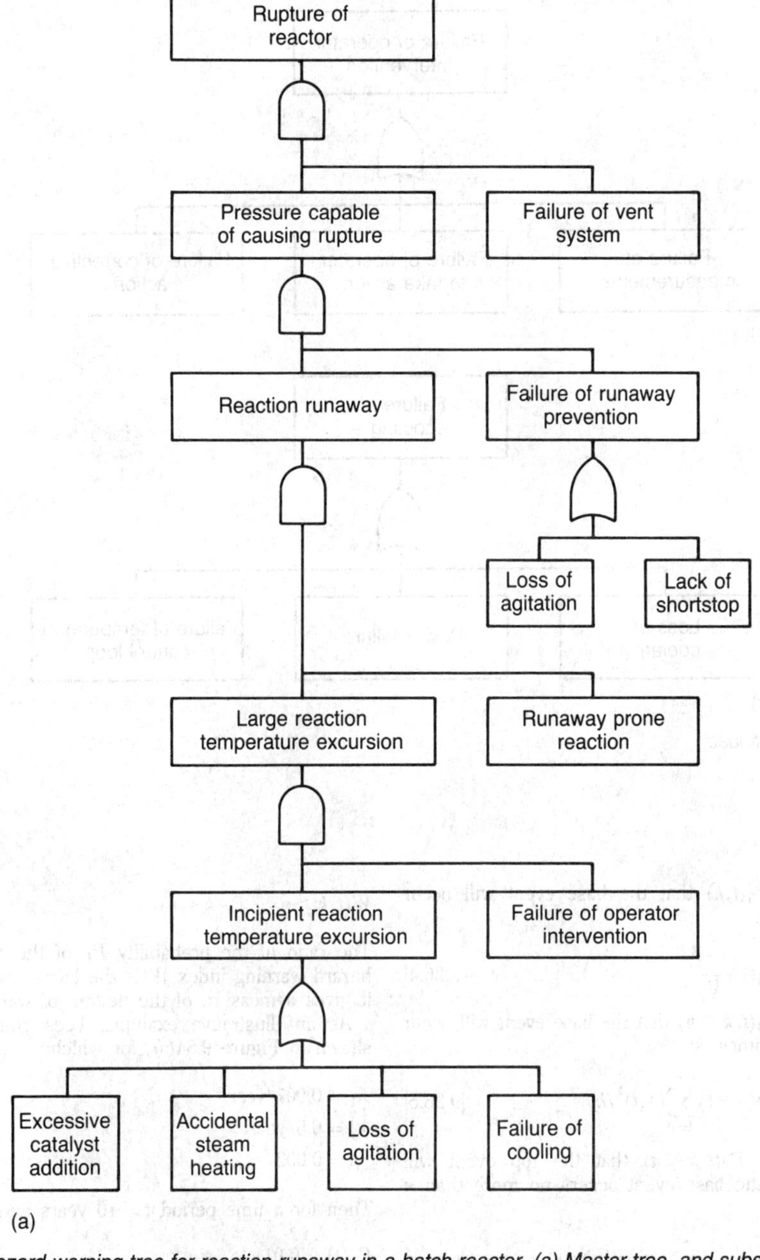

Figure 9.51 *Hazard warning tree for reaction runaway in a batch reactor. (a) Master tree, and subsidiary trees for (b) failure of cooling; (c) failure of operator intervention*

9.26.7 Failure of warning

It is of interest to be able to estimate the probability that there may be a failure of hazard warning.

For a top event T of frequency λ_1 which is caused on a proportion p of occasions by a base event B of frequency λ_2:

$$\lambda_1 = p_1\lambda_2 \quad [9.26.1]$$

Over a given time interval t the probability $P_T(t)$ that the top event will occur is:

$$P_T(t) = 1 - \exp(-\lambda_1 t) \quad [9.26.2]$$

The probability $P_{\bar{T}}(t)$ that the top event will not occur is:

$$P_{\bar{T}}(t) = \exp(-\lambda_1 t) \quad [9.26.3]$$

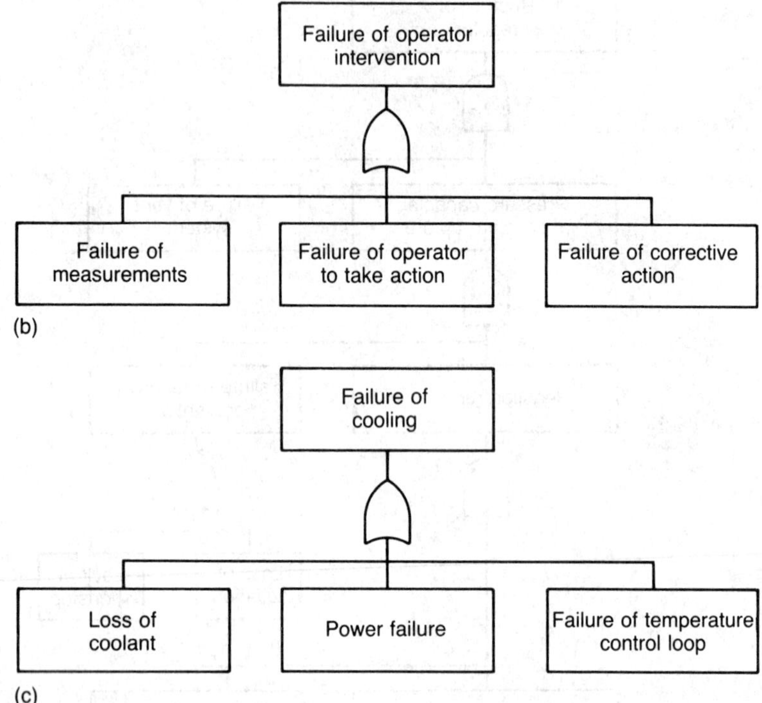

***Figure 9.51** continued*

The probability $P_B(t,k)$ that the base event will occur exactly k times is:

$$P_B(t,k) = \exp(-\lambda_2 t)\frac{(\lambda_2 t)^k}{k!} \qquad [9.26.4]$$

The probability $P_B(t, k \leq n)$ that the base event will occur no more than n times is:

$$P_B(t,\ k \leq n) = \exp(-\lambda_2 t)\sum_{k=0}^{n}(\lambda_2 t)^k/k! \qquad [9.26.5]$$

The probability $P_T(t, k \leq n)$ that the top event will occur given that the base event occurs no more than n times is:

$$P_T(t, k \leq n) = \exp(-\lambda_2 t)\sum_{k=1}^{n}p(k)(\lambda_2 t)^k/k! \qquad [9.26.6]$$

where

$$p(k) = \sum_{j=1}^{k}(-1)^{j-1}\binom{k}{j}p^j \qquad [9.26.7]$$

Then the probability $W(t,\ k \geq n)$ that there will be a failure of hazard warning, in other words that the top event will occur and that the number of base events will be no more than n when the top event occurs so that there will not be n clear warnings, is:

$$W(t, k \geq n) = P_T(k \leq n) \qquad [9.26.8]$$

The ratio of the probability P_T of the top event to the hazard warning index W is the hazard warning ratio and it gives a measure of the degree of warning available.

As an illustrative example, Lees considers the case shown in Figure 9.54(a), for which:

$$\lambda_1 = 0.001/\text{year}$$
$$\lambda_2 = 0.5/\text{year}$$
$$p = 0.002$$

Then for a time period $t = 10$ years and $n = 2$

$$P_T(t) = 0.01$$

$$W(t, k \geq n) = 0.0004$$

Pitblado and Lake (1987) have derived an expression equivalent to Equation 9.26.8 and have explored the numerical implications using Lees' example. Their results are plotted in Figure 9.54(b). They show that the value of the hazard warning ratio is lower in the early years and that it is reduced if the number of warnings specified is excessive, but that otherwise a worthwhile degree of warning is obtained.

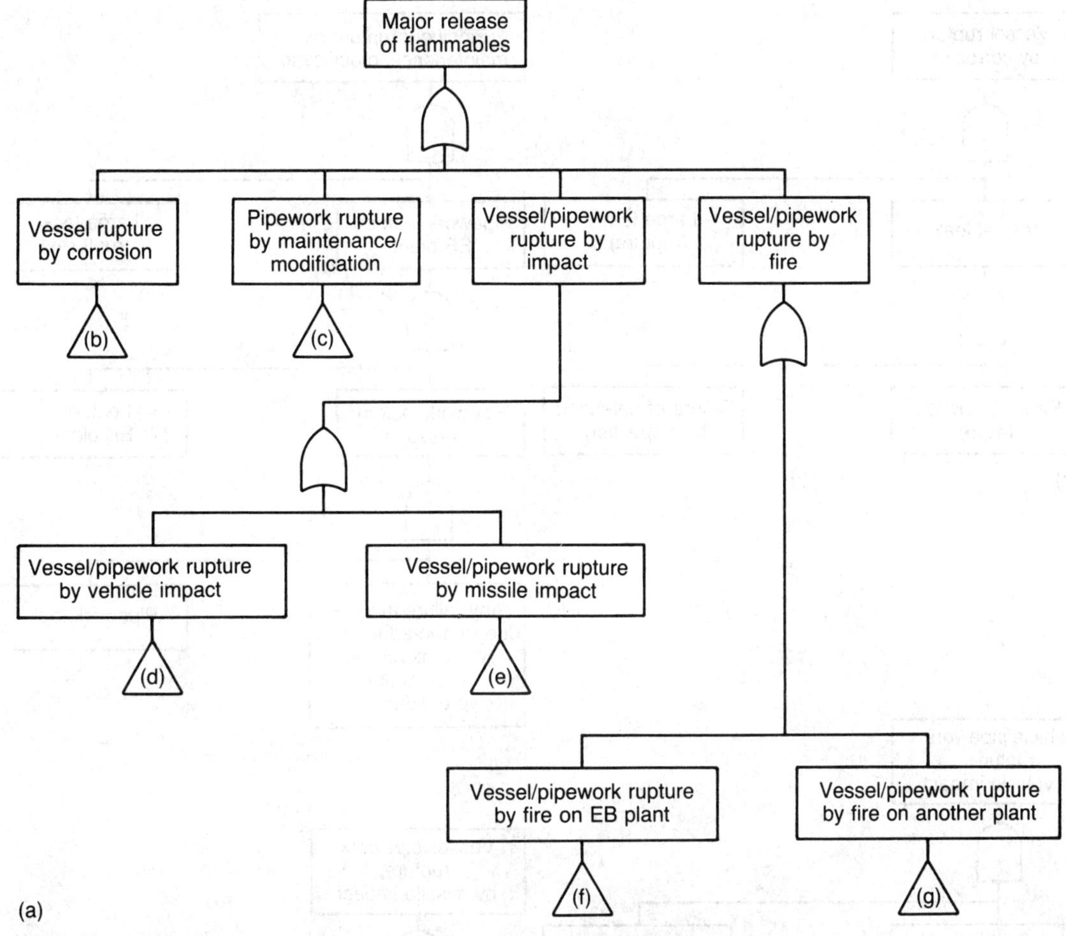

Figure 9.52 *Hazard warning tree for major release of flammables from a pressure system. (a) Master tree, and subsidiary trees for (b) vessel rupture by corrosion, (c) pipework rupture by maintenance/modification, (d) vessel/pipework rupture by vehicle impact, (e) vessel/pipework rupture by missile impact, (f) vessel/pipework rupture by a fire on same plant; and (g) vessel/pipework rupture by fire on another plant. EB, ethyl benzene*

9.26.8 High and low warning hazards

For many hazards the expected number of warnings is large, but for some it is not. The intention behind hazard warning analysis is to exploit the warnings, but analysis is not wasted if it reveals a low warning hazard. It is important to be aware of such a hazard, because it is unforgiving and needs special care.

The degree of warning may be defined as follows (Lees, 1983b):

Degree of warning	*Attenuation factor*
Zero	1
Low	10
Medium	100
High	1000

An example of a low warning hazard was given in the First *Canvey Report*, where an ammonium nitrate explosion was identified as a hazard which, if it occurred at all, would cause casualties.

9.27 Computer Aids

Hazard assessment is an active field for the development of computer aids. Computer aiding has been applied to the following:

(1) fault tree analysis;
(2) fault tree synthesis;
(3) hazard models;
(4) hazard model systems;
(5) risk assessment.

These developments are described in Chapter 29.

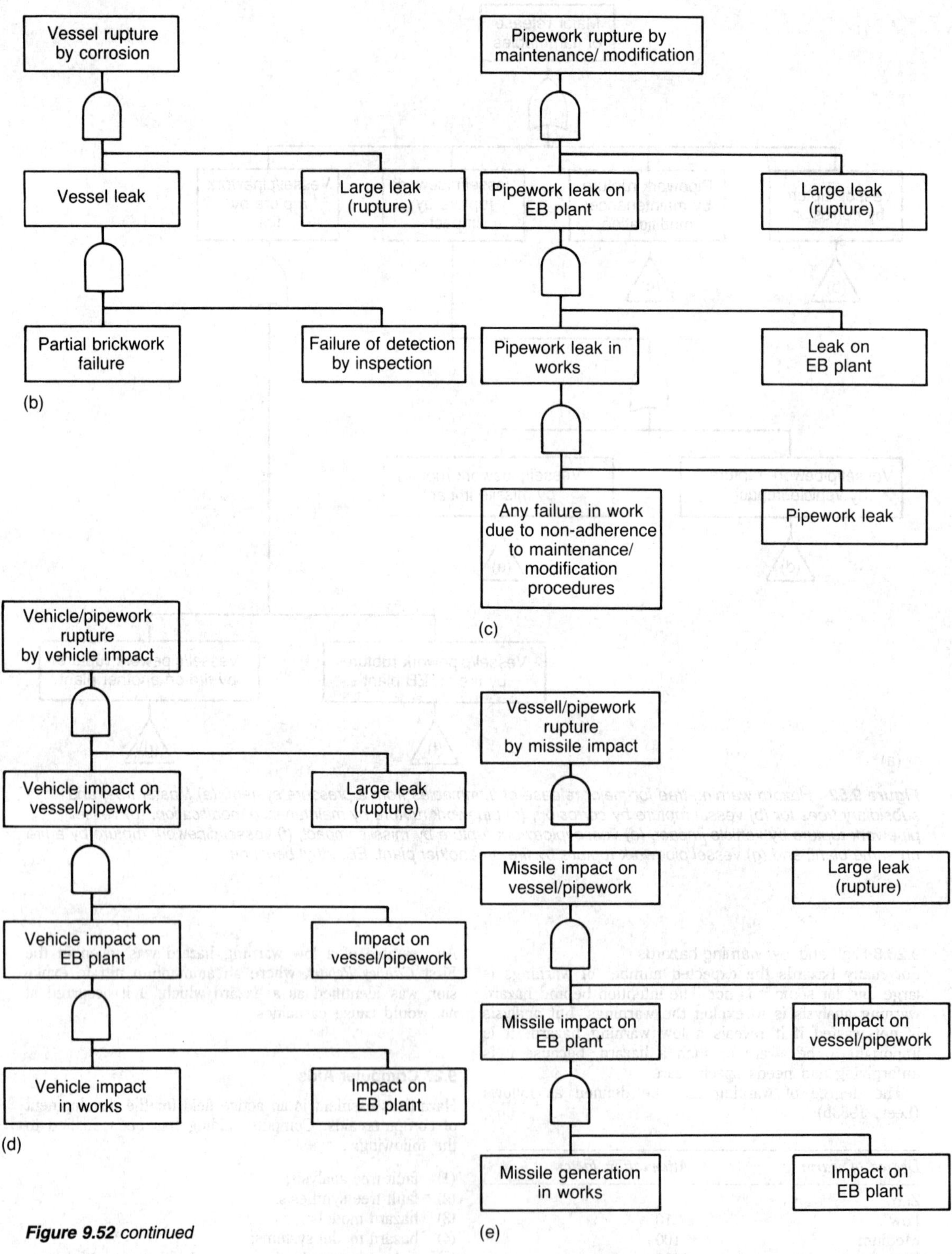

Figure 9.52 *continued*

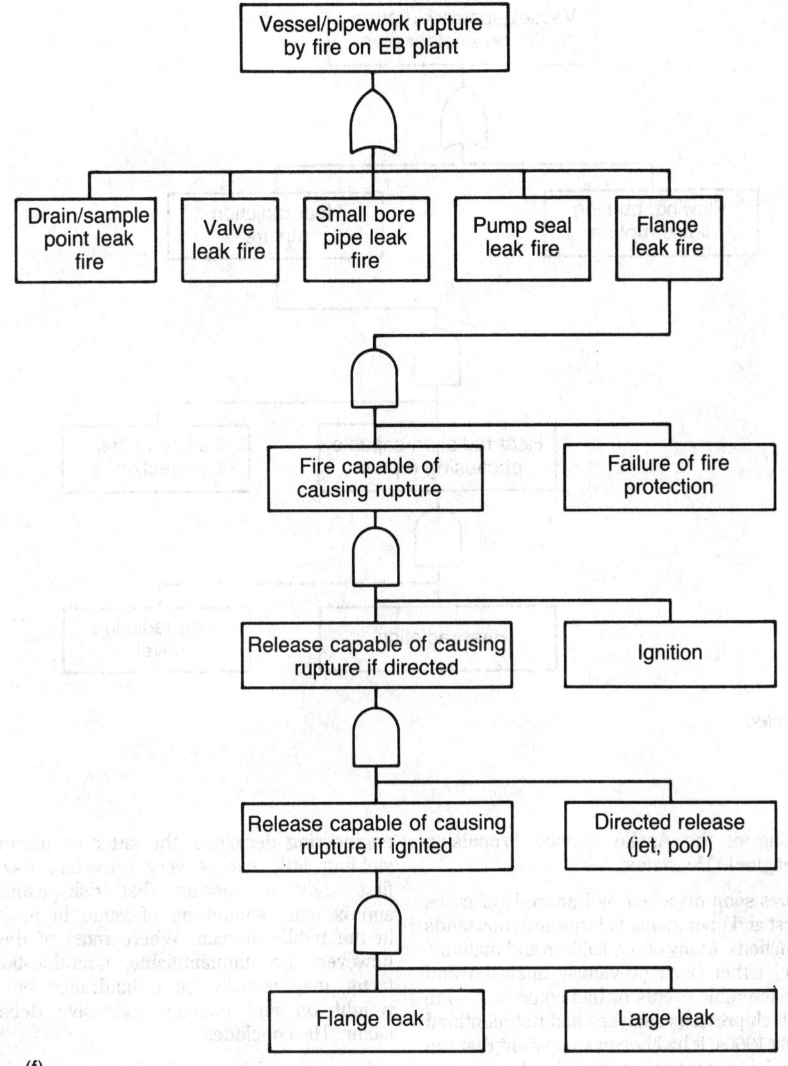

(f)

Figure 9.52 continued

Also important in hazard assessment are computer data banks, which are described in Appendix 14.

9.28 Risk Assessment Debate

Almost from their inception, both the philosophy and methodologies of risk assessment have attracted criticism. The whole status of risk assessment as a scientific activity has been questioned by Weinberg (1972a,b). He argues that the assessed risks for very rare events are likely always to be derived by a process involving much subjective judgement and intractable for peer review and, therefore, to be subject to much uncertainty. In effect, it is possible to pose the problem in scientific terms, but the answer which can be given is not based on science but 'trans-science'. He quotes a major nuclear reactor accident as an example of this type of problem. He argues further that, insofar as in this area truth may be unattainable, the search should be rather for wisdom and that laymen have a correspondingly larger part to play in making the decision. Other critiques of risk assessment as applied to nuclear hazards include those by Hanauer and Morris (1971) and Critchley (1976).

Criticisms of specific techniques were also not slow to appear. Much of the initial work on the use of fault trees was done in the aerospace industry. The effectiveness of fault tree estimates in this work has been criticized by Bryan (1976), who was in charge of reliability assess-

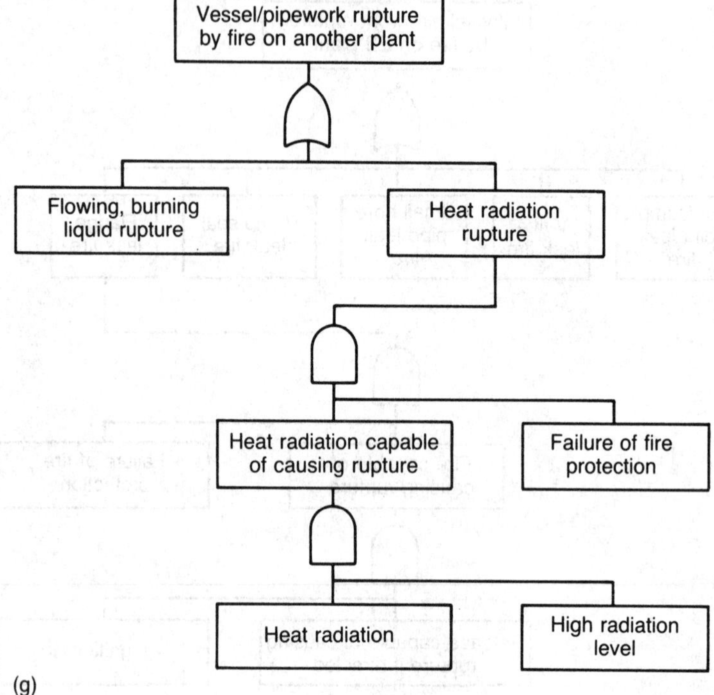

Figure 9.52 *continued*

ment during testing of the Apollo Service Propulsion System (or SPS engine). He states:

> This optimism was soon dispelled by hundreds of cases of unexpected test and operational failures and thousands of system malfunctions. Many of the failures and malfunctions modes had either been previously analysed and seemed to be noncredible events or had come as a complete surprise which previous analyses had not identified at all. By the early 1960s, it had become apparent that the traditional method of identifying potential failure events and assigning historical probabilities of occurrence to these events . . . had consistently led to overly optimistic conclusions. Consequently, the failure rates were consistently underestimated.

Further discussion of experience in reliability assessment of the SPS engine is given by the Union of Concerned Scientists (1977).

The debate on the *Rasmussen Report,* described in Appendix 23, was in large part concerned with the validity of risk assessment. The use of risk assessment in the process industries has also had its critics, including Bjordal (1980), Pilz (1980a,b), Joschek (1983) and Lowe (1984). This criticism comes from experienced practitioners and is directed not at the use of quantitative methods but rather at what is considered to be their misuse.

It is argued by Lowe that hazard analysis and risk analysis are quite different activities. Whereas the former is action oriented, being undertaken to assist in making engineering decisions, the latter is information oriented, yielding little except very uncertain risk estimates. At first sight it appears that risk estimates, even if approximate, should be of value in assisting decisions in the public domain. Where most of the other factors, however, are unquantifiable, quantification of a single factor may actually be a hindrance by putting undue weight on and creating excessive debate about that factor. He concludes:

> In my view risk analysis of the type considered here is to safety what the merry-go-round is to transport. We can spend a lot of time and money on it, only to go round in circles without really getting anywhere.

Bjordal's criticism is directed particularly to risk assessments which concentrate exclusively on fatalities. He draws attention to the wide variety of other factors which may concern the public, ranging from pollution to anxiety. He emphasizes the importance of defining beforehand the use to which the results are to be put, and cites one case where the point which the public were most anxious to have answered was not the risks as usually assessed, but the question: 'Is it more dangerous to live along the fjord than before?'

Pilz describes the many difficulties in the way of obtaining a meaningful risk assessment for a process plant. The value of the assessment is crucially dependent on the insight the analyst has regarding the system. Failure rates are usually assumed to be constant over time, but may well not be so. Failure data are often

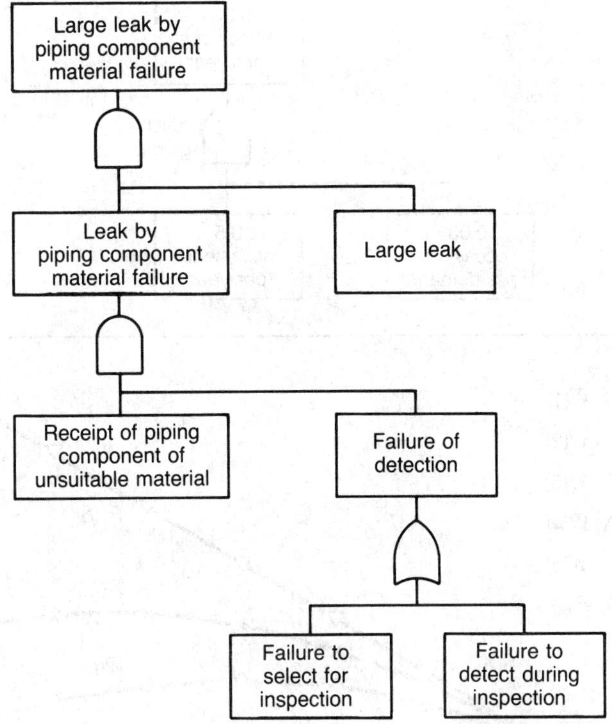

Figure 9.53 *Hazard warning tree for major leak by piping component due to failure of materials inspection*

lacking and are strongly affected by the differences in aggressiveness of the chemicals handled and in operating and maintenance practices. Fault tree methods are based on classifying states which are essentially a continuum as discrete states of failure or success. Human error plays a critical role. It is necessary to consider not simply the steady-state operating condition but also other conditions such as start-up, shut-down, etc. He concludes that it is a virtually hopeless task to obtain a risk assessment which is meaningful in terms of the uses for which it is usually advocated and quotes the wide confidence bounds given by the authors of the *Nuclear Reactor Risk Study* in Germany.

In Pilz's view, fault tree analysis can nevertheless play a useful role. It is a valid technique for comparing different protective systems. It should be used to check different safety concepts and to discover weak points, but it should not be made to bear the burden of risk assessment. He points out that in Germany a risk assessment is not done for every nuclear reactor, and that indeed the *Reactor Risk Study* dismisses this as impossible. What is done is to study a number of typical plants and to determine the average risk from these.

The debate on risk assessment is reflected in the two guides *Methodologies for Hazard Analysis and Risk Assessment in the Petroleum Refining and Storage Industry* (CONCAWE, 1982 10/82) and *Risk Analysis in the Process Industries* by the International Study Group on Risk Assessment (ISGRA, 1985; Pitblado, 1994b). Both publications give guidance on the methodologies of hazard analysis and risk assessment. Neither recommend the universal use of risk assessment. The CONCAWE guide emphasizes the uncertainties inherent in risk assessment. The ISGRA guide explicitly states: 'Whilst a large number of companies have found benefit from the use of quantified risk analysis, it must be recognised that others in the process industries have not found it necessary.'

One aspect of quantification which has attracted less criticism is the modelling of consequences. Certainly there are many comments on the wide disparities in the results from different models of particular physical phenomena, but there is also perhaps a feeling that there is good prospect of resolving these differences in due course and that the models are already becoming sufficiently accurate to give useful results.

9.29 Overview

There are three principal reasons why hazard assessment may be undertaken:

(1) engineering decisions;
(2) public acceptance;
(3) regulatory requirements.

There are important differences between these three situations. The first two are voluntary, while the third is

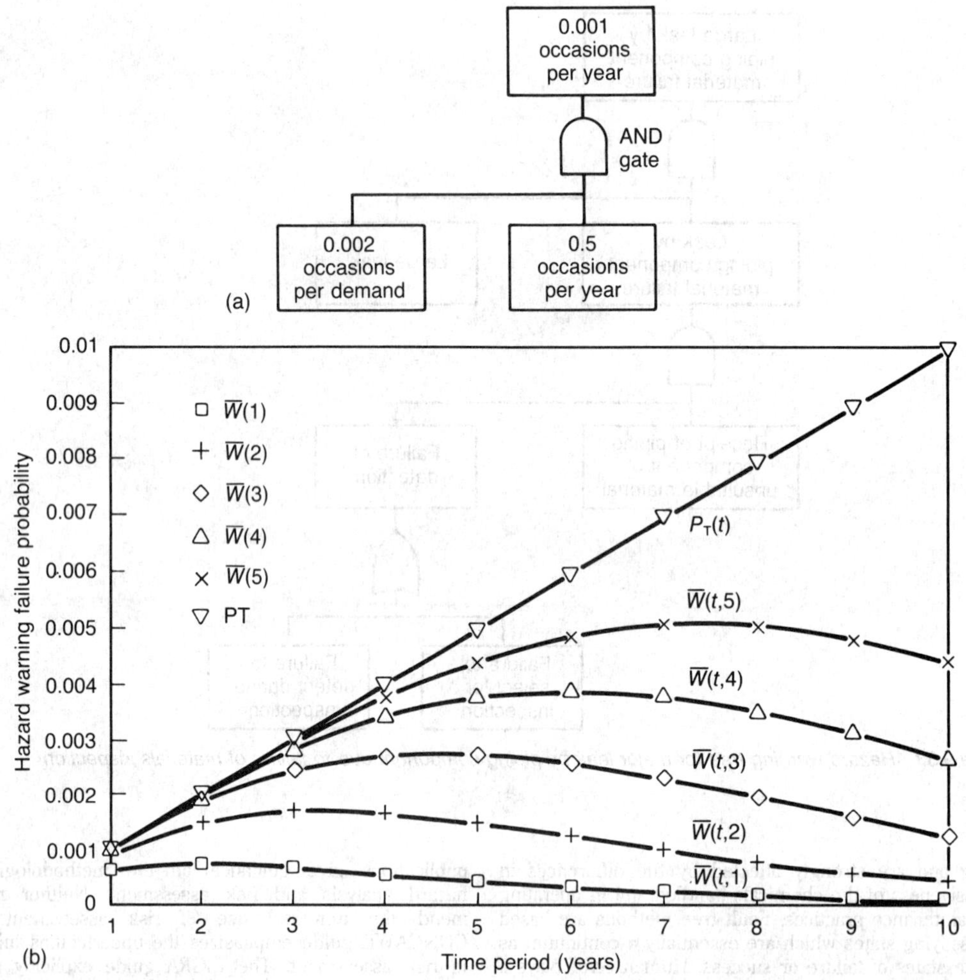

Figure 9.54 *Effectiveness of hazard warning (after Pitblado and Lake, 1987): (a) simple hazard warning tree; (b) probability of failure of warning (Courtesy of the Institution of Chemical Engineers). Note: In these authors' notation P_T is the probability of the top event, or hazard realization; $\bar{W}$ is the probability of failure of hazard warning and is thus equivalent to Lees' W*

imposed. The first is in-company, while the last two are in the public domain.

Hazard assessment is undertaken by industry as an aid to decision-making. It assists in identifying the significant hazard, in choosing cost-effective countermeasures, and in bringing out the underlying assumptions and the conditions which must be met if the hazard is to be controlled. There is little dispute over its value for these purposes. Applied in this way hazard assessment has the aspect of hazard analysis (HAZAN).

Debate centres on the value of hazard assessment in the form of probabilistic risk assessment. Here the common features of risk assessments in respect of consequence modelling appear to have obscured the fact that there are major differences between assessments in the treatment of frequency estimation. Assessments tend to vary in nature according to the system studied. As already pointed out, the *Canvey Reports* contain little in the way of fault trees, whereas the *Rasmussen Report* contains large numbers.

For a process plant it is possible to conduct a risk assessment which is based on a release source review and uses generic failure data. This overcomes many of the problems associated with completeness of hazard identification and with hazard assessment using fault tree methods. On the other hand, the information obtained from such a study is of limited use. Essentially what it reveals is the risks to be expected if the plant is designed and operated to some average standard. A study of this kind appears relevant primarily to decisions about the siting of and development around a major hazard.

A risk assessment for a process plant which starts from a release source review but also includes extensive hazard identification and assessment studies on the particular plant using methods such as hazop and fault trees and utilizing, as far as possible, plant specific data is another matter. Such a study involves much more work, but it also yields much more information and is of much greater assistance in engineering decision-making. It is convenient to refer to these two types of risk assessment as 'generic' and 'specific' assessments, respectively.

The use of risk assessments in the public domain arises from the demand that the operator of a major hazard plant should demonstrate that the hazards on its plant are under control. In making this demand it seems probable that the public is seeking assurance in respect both of the magnitude of the hazard and of the risk of its realization. What the public gains from a generic assessment is a general picture of the hazard and the associated risks, the latter based on the assumption that the plant standards are average.

The value to the public of a specific risk assessment is rather different. In this case what the public gains is the assurance that the company has made a detailed study of the hazards of its plant and that, as a result, the standards should not be below average. The difference between generic and specific risk assessments, therefore, lies less in the assessment of the consequences than in that of the frequency.

A full risk assessment of a major hazard plant is a substantial undertaking. Much of the effort, however, lies in the assembly of data and models and is in some degree transferable to other assessments. Moreover, computer aids are now available which besides carrying out the actual calculations, also provide a framework and set of default data and models.

Some aspects of risk assessment, therefore, may become virtually routine, but care is needed in identifying those features for which in a given case a relatively routine treatment may be appropriate and those which require detailed study. The aim should be to free the engineer to concentrate on the latter. It is such detailed studies that are most likely to yield useful results.

A risk assessment should be realistic rather than conservative. In engineering design it is not uncommon for a number of safety factors to be introduced at the various stages. This approach is not suitable in risk assessment. The use of excessively conservative assumptions leads to overly pessimistic and unrealistic risk estimates which are of little value for decision-making. The assessed risks should be 'best estimates'. The decision-makers can then treat these results with whatever degree of conservatism they think appropriate.

Risk assessment has undergone substantial development in recent years and the pace of change is still rapid. The methodology is still only barely adequate for the tasks often required of it. This should be borne in mind when evaluating the assessed risks.

The numerical risk results are only one of the outputs from a risk assessment. Equally important, and perhaps more so, are the other findings such as the identification and ranking of hazards, the identification and assessment of engineering measures to avoid, protect against and mitigate hazards, and an appreciation of the conditions necessary to keep hazards well controlled.

10 Plant Siting and Layout

Contents

10.1 Plant Siting

Safety is a prime consideration in plant siting. Other important factors include: access to raw materials and to markets; availability of land, labour and cooling water; means of effluent disposal; interlinking with other plants; and government policies, including planning permission and investment incentives. It is only safety aspects which are considered here.

As far as safety of the public is concerned, the most important feature of siting is the distance between the site and built-up areas. Sites range from rural to urban, with population densities varying from virtually zero to high. Separation between a hazard and the public is beneficial in mitigating the effects of a major accident. An area of low population density around the site will help to reduce casualties. In the ideal case the works is surrounded by fields or waste land forming a complete *cordon sanitaire*. In many situations, however, it is unattractive to 'sterilize' a large amount of land in this way, particularly in an urban area, where land is generally at a premium.

The physical effects of a major accident tend to decay quite rapidly with distance. Models for fire give an inverse square law decay, as do many of the simpler models for explosion and toxic release, though other explosion and toxic release models give different decay relations, some with less rapid decay. Decay laws were discussed in Chapter 9 and further treatments are given in Chapters 15–17.

Information on the potential effects of a major accident on the surrounding area is one of the main results obtained from a hazard assessment and such an assessment is of assistance in making decisions on plant siting.

Siting is not a substitute for high standards of design and operation of the plant. It should never be forgotten that the people most at risk are the people on site, and standards should be such as to safeguard this workforce. It is sometimes argued in fact that standards should be sufficiently high that separation between site and public is not necessary. Such standards, however, are essentially a form of active protection, which depends crucially on the quality of management. In most countries, including the UK, the view is taken that it is prudent nevertheless to have a degree of separation. The provision of a separation distance is a form of passive protection which provides a further mitigating factor and which is relatively robust in the event of deterioration in the plant management.

In terms of hazard warning, separation tends to create a hazard, which will give more warnings and which is therefore less unforgiving.

Topography is another relevant feature. It is desirable to avoid terrain where hazardous fluids, whether liquids or dense gases, can flow down into populated areas. Another consideration to be taken into account is contamination of water courses by liquid spills.

In selecting a site, allowance should be made for site emergencies. One factor is the availability of emergency utilities such as electrical power and water. Another is the availability and experience of outside emergency services, particularly the fire service. A third is access for these services.

Table 10.1 *Selected references on plant siting*

NRC (Appendix 28: Siting); Cremer (1945); Mohlman (1950); Bierwert and Krone (1955); Greenhut (1956); von Allmen (1960); J.A. Gray (1960); Anon. (1964b); Risinger (1964i); Liston (1965); Fryer (1966); Farmer (1967a,b, 1969a,b); R. Reed (1967); Fowler and Spiegelman (1968); Kaltenecker (1968); G.D. Bell (1970); Otway and Erdmann (1970); Speir (1970); Tucker and Cline (1970); Yocom, Collins and Bowne (1971); Gronow and Gausden (1973); Balemans *et al.* (1974); Cross and Simons (1975); Roskill (1976); Weismantel (1977); Cremer and Warner (1978); Slater (1979); Dalal (1980); Kletz (1980h); Granger (1981); Considine, Grint and Holden (1982); Lovett, Swiggett and Cobb (1982); Ramsay, Sylvester-Evans and English (1982); Landphair and Motloch (1985)

A discussion of siting for high toxic hazard materials (HTHMs) is given in the CCPS HTHM *Storage Guidelines* (1988/2). Siting policy for major hazard plants in the UK was discussed in Chapter 4. Selected references on plant siting are given in Table 10.1.

10.2 Plant Layout

Plant layout is a crucial factor in the economics and safety of process plant. Some of the ways in which plant layout contributes to safety and loss prevention are:

(1) segregation of different risks;
(2) minimization of vulnerable pipework;
(3) containment of accidents;
(4) limitation of exposure;
(5) efficient and safe construction;
(6) efficient and safe operation
(7) efficient and safe maintenance;
(8) safe control room design;
(9) emergency control facilities;
(10) fire fighting facilities;
(11) access for emergency services;
(12) security.

Plant layout can have a large impact on plant economics. Additional space tends to increase safety, but is expensive in terms of land and also in additional pipework and operating costs. Space needs to be provided where it is necessary for safety, but not wasted.

The topics considered under the heading of 'plant layout' are traditionally rather wide ranging. Many of these subjects are treated here in separate chapters and only a brief treatment is given in this one. This applies in particular to such topics as hazard assessment, emission and dispersion, fire and fire protection, explosion and explosion protection, storage, and emergency planning.

A general guide to the subject is given in *Process Plant Layout* (Mecklenburgh, 1985). This is based on the work of an Institution of Chemical Engineers (IChemE) working party and expands an earlier guide *Plant Layout* (Mecklenburgh, 1973). The treatment of hazard assessment in particular is much expanded in this later volume. The loss prevention aspects of plant layout have also been considered specifically by Mecklenburgh (1976).

Table 10.2 *Selected references on plant layout*

Cremer (1945); Mallick and Gaudreau (1951); Shubin and Madeheim (1951); Muther (1955, 1961, 1973); McGarry (1958); Armistead (1959); R. Reed (1961, 1967); EEUA (1962 Document 12, 1973 Hndbook 7); J.M. Moore (1962); ABCM (1964/3); Dow Chemical Co. (1964, 1966a,b, 1976, 1980, 1987, 1994); Duggan (1964a); Jenett (1964c); Landy (1964a–c); Risinger (1964i); R. Wilson (1964b); Liston (1965, 1982); IP (1980 Eur. MCSP Pt 2, 1981 MCSP Pt 3, 1987 MCSP Pt 9, 1990 MCSP Pt 15); R. Kern (1966, 1977 series, 1978b): M.W. Kellogg Co. (1967); BCISC (1968/7); Fowler and Spiegelman (1968); Kaltenecker (1968); House (1969); Proctor (1969); British Cryogenics Council (1970); J.R. Hughes (1970); ICI/RoSPA (1970 IS/74); Kaess (1970); Sachs (1970); Tucker and Cline (1970); Bush and Wells (1971, 1972); Simpson (1971); Guill (1973); Mecklenburgh (1973, 1976, 1982, 1985); Pemberton (1974); R.B. Robertson (1974a,b, 1976a,b); Unwin, Robins and Page (1974); Falconer and Drury (1975); Beddows (1976); Harvey (1976, 1979b); Spitzgo (1976); Rigby (1977); Kaura (1980b); Kletz (1980h, 1987c); F.V. Anderson (1982); O'Shea (1982); Goodfellow and Berry (1986); Brandt *et al.* (1992); Meissner and Shelton (1992); Bausbacher and Hunt (1993); Madden (1993); Briggs (1994) ANSI A, A10, A37 and D series, BS 5930: 1981

Layout techniques
Mecklenburgh (1973, 1976, 1985); Sproesser (1981); Nolan and Bradley (1987); Madden, Pulford and Shadbolt (1990); Madden (1993).
Virtual reality: IEE (1992 College Digest 92/93)

Civil engineering, including foundations
ASCE (Appendix 27, 28); Urquhart (1959); Biggs (1964); ASTM (1967); MacNeish (1968); Benjamin and Cornell (1970); Tomlinson (1980); Carmichael (1982); M. Schwartz (1982a–c, 1983a–e, 1984); Pathak and Rattan (1985); Blenkinsop (1992); BS (Appendix 27 *Civil Engineering, Construction*), BS 6031: 1981, BS 8004: 1986, BS COP 2010: 1970–, BS COP 2012: 1974–.
Equipment weights: El-Rifai (1979)

Hazardous area classification (*see* Table 16.2)

Materials handling
Woodley (1964); Smego (1966); R. Reed (1969); Department of Employment and Productivity (1970); Brook (1971); DTI (1974); Pemberton (1974); Sussams (1977); *Chemical Engineering* (1978b)

In-works transport, roads
HSE (1973 TDN 44); Mecklenburgh (1973, 1985); HSE (1985 IND(G) 22(L); 1992 GS 9)

Separation distances
C.W.J. Bradley (n.d., 1985); Armistead (1959); Dow Chemical Co. (1964, 1966a, 1976, 1980, 1987, 1994); Scharle (1965); Home Office (1968/1, 1971/2, 1973/4); Masso and Rudd (1968); Goller (1970); J.R. Hughes (1970); ICI/RoSPA (1970 IS/74); Laska (1970); Simpson (1971); OIA (1972 Publication 631); HSE (1973 HSW Booklet 30); Mecklenburgh (1973, 1976, 1985); Unwin, Robins and Page (1974); Butragueno and Costello (1978); IP (1980 Eur. MCSP Pt 2, 1981 MCSP Pt 3, 1987 MCSP Pt 9); API (1981 Refinery Inspection Guide Chapter 13, 1990 Std 620, 1993 Std 650); Nolan and Bradley (1987); D.J. Lewis (1989b); Martinsen, Johnsen and Millsap (1989); NFPA (1989 NFPA 50A, 50B, 1992 NFPA 58, 59); IRI (1991, 1992); LPGITA (1991– LPG Code)

Pipework
R. Kern (1966); Mecklenburgh (1973, 1976, 1985); Clarke (1966 BRE/1)

Corrosion
Mears (1960); ABCM (1964/3)

Buildings
BRE (Appendix 28, 1983 CP2/83, IP8/83); Beigler (1983); Crossthwaite and Crowther (1992)
BS (Appendix 27 *Buildings, CoP Buildings*), BS Handbook 20:1985

Structures and access
EEUA (1962 Document 12, 1973 Handbook 7); Mecklenburgh (1973, 1976, 1985)

Floors, walkways
ABCM (1964/3); Steinberg (1964); Pierce (1968); Friedrich (1974); ASTM (1978 649); EEMUA (1983 Publication 105)

Escape and rescue
FPA (CFSD FPDG 4); HSE (HSW Booklet 40); EEUA (1962 Document 12); Webber and Hallman (1988)

Lighting
Illuminating Engineering Society (n.d.); Electricity Council and British Lighting Council (1967); Mixon (1967); Rowe (1973); HSE (1987 HS(G) 38); BS 5266: 1981–; UL 844–1990, UL 924–1990, UL 781–1992
Emergency lighting: UL (1990 924)

Ventilation (*see* Table 25.1)

Wind resistance
BRE (1972 BR 9, 1975 CP 16/75, SO 8, 1978 CP 25/78, 1986 EP1); Simiu and Scanlan (1978); ASCE (1980/10, 1986/12, 1987/34, 35)

Resistance to flood, hurricane
Fulton (1960); Labine (1961); Neill and Bethel (1962); Weismantel (1969a); Marlar (1971)

Blast resistance (*see* Table 17.38)

Earthquake resistance (*see* Table A15.1)

Compressor houses
D.H.A. Morris (1974); Prentice, Smith and Virtue (1974)

Control rooms
Bradford and Culbertson (1967); Burns (1967); Prescott (1967); Schmidt (1971); E. Edwards and Lees (1973); Mecklenburgh (1973, 1976, 1985); V.C. Marshall (1974, 1976a,c,d); Kletz (1975e); Anon. (1976 LPB 11, p. 16);

Gugan (1976); Harvey (1976, 1979b); Langeveld (1976); Anon. (1977 LPB 16, p. 24); Balemans and van de Putte (1977); CIA (1979); Cannalire *et al.* (1993)

Emergency shelters
Johnston (1968); Lynskey (1985)

Indoor plants
R. Kern (1978a); Munson (1980)

Storage
FPA (1964/1); IP (1980 Eur. MCSP Pt 2, 1981 MCSP Pt 3, 1987 MCSP Pt 9); Home Office (1968/1, 1971/2, 1973/4); J.R. Hughes (1970); ICI/RoSPA (1970 IS/74); HSE (1973 HSW Booklet 30); Wirth (1975); Hrycek (1978); D.W. Johnson and Welker (1978); Aarts and Morrison (1981); NFPA (1986 NFPA 43C, 1989 NFPA 50A, 50B, 1990 NFPA 43A, 50, 59A, 1992 NFPA 58, 59, 1993 NFPA 43B); LPGITA (1991 LPG Code 1 Pt 1)

Fire prevention and protection
FPA (CFSD FPDG 2); IRI (1964/5); BCISC (1968/7); IP (1980 Eur. MCSP Pt 2, 1981 MCSP Pt 3, 1987 MCSP Pt 9, 1993 MCSP Pt 19); Home Office (1974– Manual of Firemanship); J.R. Hughes (1970); ICI/RoSPA (1970 IS/74); Simpson (1971); Mecklenburgh (1973, 1976, 1985); R.B. Robertson (1974a,b, 1976a,b); Klootwijk (1976); Kaura (1980a)

Chimneys
BS 4076: 1989

Drains
J.D. Brown and Shannon (1963a,b); Seppa (1964); ICE (1969); Mecklenburgh (1973, 1976, 1985); Klootwijk (1976); Anon. (1978 LPB 19, p. 10); Elton (1980); Gallagher (1980); D. Stephenson (1981b); Easterbrook and Gagliardi (1984); Mason and Arnold (1984); Chieu and Foster (1993); Crawley (1993 LPB 111); BS 8005: 1987–

Earthing, grounding
IEEE (1982 IEEE 142); UL (1984 UL 467); BS 7430: 1991

Winterisation
J.C. Davis (1979); Facer and Rich (1984); Fisch (1984)

Modular plants
Armstrong (1972); Glaser, Kramer and Causey (1979); IMechE (1980); Saltz (1980); Bolt and Arzymanow (1982); H.R. James (1982); Marcin and Schulte (1982); Parkinson, Short and Ushio (1982); Zambon and Hull (1982); Glaser and Kramer (1983); Hulme and La Trobe Bateman (1983); Kliewer (1983); Tan, Kumar and Kuilanoff (1984); Whitaker (1984); Tarakad, Durr and Hunt (1987); Clement (1989); Hesler (1990); Shelley (1990); Duty, Fisher and Lewis (1993)

Barge mounted and ocean-borne plants
Birkeland *et al.* (1979); Charpentier (1979); Glaser, Kramer and Causey (1979); J.L. Howard and Andersen (1979); Jackson (1979); Jansson *et al.* (1979); Shimpo (1979); Ricci (1981); H.R. James (1982); de Vilder (1982)

Plant identification
NFPA (1990 NFPA 901); API (1993 RP 1109); BS (Appendix 27 *Identification of Equipment*), BS 1710: 1984, BS 5378: 1980–

Hazard assessment
Mecklenburgh (1982, 1985)

Other work on plant layout, and in particular safety and loss prevention (SLP), includes that of: Armistead (1959), R. Kern (1977a–f; 1978a–f), on general aspects and spacing recommendations; Simpson (1971) and R.B. Robertson (1974a, 1976b), on fire protection; Fowler and Spiegelman (1968), the Manufacturing Chemists Association (MCA, 1970/18), Balemans *et al.* (1974) and Drewitt (1975), on checklists; and Madden (1993), on synthesis techniques.

Plant layout is one of the principal aspects treated in various versions of the Dow *Guide* by the Dow Chemical Company (1994b). It is also dealt with in the *Engineering Design Guidelines* of the Center for Chemical Process Safety (CCPS, 1993/13). There are also a large number of codes relevant to plant layout, and particularly separation distances and area classification. These are described below.

The treatment given here for the most part follows that of Mecklenburgh, except where otherwise indicated. It is appropriate to repeat here his caution that the practice described should be regarded only as typical and that it may need to be modified in the light of local conditions, legislation and established safe practices. In particular, the account given generally assumes a 'green-field' site, and some compromise is normally necessary for an existing site. Selected references on plant layout are given in Table 10.2.

10.3 Layout Generation

10.3.1 Factory layout
For factories generally there are a number of different principles on which plant layout may be based (Muther, 1961). Thus in light engineering use is made of layouts in which the material fabricated remains in a fixed position and others in which a particular process or function is performed at a fixed point.

10.3.2 Flow principle
For process plants, however, the most appropriate method is generally to lay the plant out so that the material flow follows the process flow diagram. This is the process flow principle. This arrangement minimizes the transfer of materials, which is desirable both for economics and safety. It is difficult to overemphasize the importance of efficient materials handling. It has been estimated by the Department of Trade and Industry (DTI, 1974) that about a quarter of the production costs of manufacturing industry generally are for materials handling, an activity which in itself is totally unproductive.

Likewise, long runs of pipework with vulnerable features are an undesirable addition to the hazards of the plant. There are features which can lead to the layout sequence diverging from the process sequence. They relate particularly to: requirements for gravity flow; equipment needing specially strong foundations; access for construction, commissioning, operation and maintenance; future extension; operator protection; escape and fire fighting; containment of accidents; and environmental impact.

10.3.3 Correlation and compatibility

There are certain other layout approaches which are used for factory layouts generally and which merit mention. Correlation and compatibility techniques are used for the elimination of layout arrangements which are incompatible or impossible, and also for the preliminary formulation of compatible arrangements.

In the correlation chart method, for example, the procedure is as follows. The constraints and objectives are listed. The floor space is subdivided into a grid and for each item the grid divisions which violate the constraints are deleted. The permissible layouts are then determined. There is a corresponding algebraic method.

Proximity and sequencing techniques are available for the determination of the costs of material transfer with different layouts.

These general factory layout techniques are described in more detail by Mecklenburgh (1973), but he states that they appear to have found little application in process plant layout.

10.3.4 Process plant layout

As with design generally, the design of a process plant layout involves first synthesis and then analysis. Despite its importance, there is relatively little written on the generation of the layout. An indication of some of the principles which guide the designer has been given by Madden (1993). He describes a structured approach to the generation of the layout which has four stages: (1) three-dimensional model, (2) flow, (3) relationships and (4) groups.

10.3.5 Three-dimensional model

The first step is to produce a three dimensional (3D) model of the space occupied by each item of equipment. This 3D envelope should include space for (1) operations access, (2) maintenance access, and (3) piping connections. The effect of allowing for these aspects is generally to increase several-fold the volume of the envelope.

10.3.6 Flow

The concept of 'flow' as used by Madden has two meanings: (1) progression of materials towards a higher degree of completion, and (2) mass flows of process or utility materials. Often the two coincide, but where there is a feature such as a recycle the relationship is less straightforward.

10.3.7 Relationships

A relationship exists between two items when they share some common factor. Relationships may be identified by considering the plant from the viewpoint of each discipline in turn. Broad classes of relationship are (1) process, (2) operations, (3) mechanical, (4) electrical, (5) structural and (6) safety.

Process relationships are exemplified by: direct flow diagram connectivity between items; gravity flow; hydraulics and net positive suction head (NPSH) requirements; and heat interchange and conservation. Operations relationships include multiple items with similar features, e.g. batch reactors and centrifuges. Examples of mechanical relationships are the space needed between items for piping and transmission or isolation of vibrations. Electrical relationships may be associated with electrical area classification and with high voltage or power features. Structural features include the grouping together of heavy items and the location of heavy items on good ground. Some safety features are separation between potential leak sources and ignition sources and the provision of a sterile area such as that around a flare.

10.3.8 Groups

From the relationships identified it is then necessary to select those which are to be given priority. It is then possible to arrange the items into groups. It is found by experience that a group size of about seven items is the largest which a layout designer can readily handle; above this number the arrangement of items within the group becomes excessively complex. A typical group is a distillation column group consisting of the column itself and its associated heat exchangers, etc.

10.3.9 Segregation

A relationship of particular importance in plant layout is that between a hazard and a potential target of that hazard. The minimization of the risk to the target is effected by segregating the hazard from the target. The requirement for segregation therefore places constraints on the layout.

10.4 Layout Techniques and Aids

There are a number of methods available for layout design. These are generally more applicable to the analysis rather than the synthesis of layouts, but some have elements of both. They include:

(1) classification, rating and ranking;
(2) critical examination;
(3) hazard assessment;
(4) economic optimization.

There are also various aids, including:

(5) visualization aids;
(6) computer aids.

10.4.1 Classification, rating and ranking

There are several methods of classification, rating and ranking which are used in layout design. The main techniques are those used for the classification of (1) hazardous areas, (2) storage, (3) fire fighting facilities and (4) access zones, together with methods based on hazard indices.

Hazardous area classification is aimed at the exclusion of ignition sources from the vicinity of potential leak sources and involves the definition of zones in which control of ignition sources is exercised to differing degrees. It is described in Section 10.12 and Chapter 16.

Storage classification is based on the classification of the liquids stored. Accounts are given in Section 10.10 and Chapter 22.

Closely related is classification based on fire fighting requirements, since this is applicable particularly to storage.

Restriction of access may be required near major hazard plants or commercially sensitive processes. Areas are therefore classified by the need to control access.

Ranking methods such as those of the Dow Index and Mond Index may also be used as a means of grouping similar hazards together.

10.4.2 Critical examination

Critical examination, which is part of the technique of method study (Currie, 1960), may be applied to plant layout. This application has been described by Elliott and Owen (1968). In critical examination of plant layout, typical questions asked are: Where is the plant equipment placed? Why is it placed there? Where else could it go?

The technique therefore starts with and involves analysis of a proposed layout, but insofar as other possible solutions are suggested it may be regarded also as a method for the generation of alternatives which can then be evaluated.

As already mentioned in Chapter 8, the working document in an early hazard study is a plant layout diagram, and to this extent such a hazard study may be regarded as a form of critical examination of layout.

10.4.3 Hazard assessment

Hazard assessment of plant layout is practised both in respect of major hazards which affect the whole site, and of lesser hazards, notably leaks, and their escalation. The traditional method of dealing with the latter has been the use of minimum safe separation distances, but there has been an increasing trend to supplement the latter with hazard assessment. An account of hazard assessment is given in Chapter 9 and its role in plant layout is discussed in Sections 10.5 and 10.13.

10.4.4 Economic optimization

The process of layout development generates alternative candidate layouts and economic optimization is a principal method of selection from among these. The points at which such economic optimization is performed are described in Section 10.5. Some factors which are of importance for the cost of a plant layout include foundations, structures, piping and pipetracks, and pumps and power consumption.

10.4.5 Visualization aids

There are various methods of representing the plant to assist in layout design. These include drawings, cutouts, block models and piping models. Cutouts are a two-dimensional (2D) layout aid consisting of sheets of paper, cardboard or plastic which represent items in plan, whether whole plots or items of equipment, and are overlaid on the site or plot plan, as the case may be. The other main physical aids are 3D. Block models are very simple models made from wood blocks or the like which show the main items of equipment and are used to develop plot and floor plans and elevations. Piping models include the pipework, are more elaborate, and can constitute up to 0.5% of the total installed plant cost. They are useful as an aid to: doing layout drawings; determining piping layout and avoiding pipe fouls; positioning valves, instruments, etc.; checking access for operation and maintenance; planning construction and executing it; and operator training.

10.4.6 Computer aids

Plant layout is one of the areas in which computer aided design (CAD) methods are now widely used. One type of code gives visualization of the layout. This may take a number of forms. One is a 2D layout visualization equivalent to cutouts. Another is a 3D visualization equivalent to either a block model or a piping model, but much more powerful. The visualization packages available have become very sophisticated and it is possible in effect for the user to sit at the display and take a 'walk' through the plant. A recent development is enhancement by the use of the techniques of virtual reality. Typically such CAD packages not only give 3D display but hold a large amount of information about the plant such as the co-ordinates of the main items and branches, the piping routes, the materials list, etc.

A particular application of 3D visualization codes is as input to other computer programs such as computational fluid dynamics codes for explosion simulation. The 3D layout required for the latter is provided by the 3D visualization code, which then forms the front end of the total package.

Another type of code tackles the synthesis of layouts. The general approach is to define a priority sequence for locating items of equipment inside a block and then for the location of the block. The pipework is then added and costed. Such a method has been described by Shocair (1978).

A third type of code deals with the analysis of layouts to obtain an economic optimum. Typical factors taken into account in such programs include the costs of piping, space and buildings. A program of this type has been described by Gunn (1970).

The extent to which computer aids are used in the design of plant layout is not great, but some visualization packages are very powerful and are likely to find increasing application. Computer techniques for plant layout are described in more detail in Chapter 29.

10.5 Layout Planning and Development

10.5.1 Layout activities and stages

Plant layout is usually divided into the following activities:

(1) site layout;
(2) plot layout;
(3) equipment layout.

The layout developed typically goes through three stages:

(1) Stage One layout;
(2) Stage Two layout;
(3) Final layout.

The sequence of layout development described by Mecklenburgh is:

Table 10.3 Typical stages in the development of a plant layout (after Mecklenburgh, 1985) (Courtesy of the Institution of Chemical Engineers)

A	**Stage One plot layout**
1	Initial plot data
2	First plot layout
3	Elevation
4	Plot plan
5	Plot buildings
6	Second plot plan
7	Hazard assessment of plot layout
8	Layout of piping and other connections
9	Critical examination of plot layout
B	**Stage One site layout**
10	Initial site data
11	First site layout
12	Hazard assessment of site layout
13	Site layout optimization
14	Critical examination of site layout
15	Site selection
C	**Stage Two site layout**
16	Stage Two site data
17	Stage Two site layout
D	**Stage Two plot layout**
18	Stage Two plot layout data
19	Stage Two plot layout

(1) Stage One plot layout;
(2) Stage One site layout;
(3) Stage Two site layout;
(4) Stage Two plot layout.

Typical stages in the development of a plant layout are given in Table 10.3. The Stage Two and Final Stage design network is shown in simplified form in Figure 10.1. The process of layout development makes considerable use of guidelines for separation distances. These are described in Section 10.11.

Stage One is the preliminary layout, also known as the conceptual, definition, proposal or front end layout. In this stage consideration is given to the various factors which are important in the layout, which may threaten the viability of the project if they are not satisfactorily resolved and which are relevant to site selection.

10.5.2 Stage One plot layout

In the Stage One plot layout, the information available should include preliminary flow sheets showing the major items of equipment and major pipework, with an indication of equipment elevations, and process engineering designs for the equipment. The plot layouts are then developed following the process flow principle and using guidance on preliminary separation distances. The plot size generally recommended is 100 m × 200 m with plots separated from each other by roads 15 m wide.

For each plot layout the elevation and plan are further developed. The proposed elevation layouts are subjected to a review such as critical examination which generates alternatives, and these alternatives are costed. Similarly, alternative plan layouts are generated accommodating the main items of equipment, pipework, buildings and cable runs, and are reviewed to ensure that they meet the principal constraints. These include construction, operation, maintenance, safety, environment and effluents. The plan layouts are then costed. The justification for the use of buildings is examined. The civil engineering aspects are then considered, including foundations and support and access structures.

The outcome of this process for each plot is a set of candidate layouts. These are then presented for view as layout models in block model or computer graphics form. The different disciplines can then be invited to comment. These plot layouts are then costed again and a short list is selected, preferably of one.

The plot layouts are then subjected to hazard assessment. This assessment is concerned largely with the smaller, more frequent leaks which may occur and with sources of ignition for such leaks. The process of hazardous area classification is also performed. Hazard assessment and hazardous area classification are described in Sections 10.12 and 10.13.

Studies are carried out to firm up on piping and piping routes and on electrical mains routes. Finally, each plot layout is subjected to a critical examination, typically using a model and following a checklist.

10.5.3 Stage One site layout

The Stage One plot layouts provide the information necessary for the Stage One site layout. These include the size and shape of each plot, the desirable separation distances, the access requirements and traffic characteristics. The flow of materials and utilities on the site are represented in the form of site flowsheets.

The site layout is now developed to accommodate not only the process plots but also storage and terminals, utilities, process and control buildings, non-process buildings and car parks, and the road and rail systems. The flow principle is again followed in laying out the plots, but may need to be modified to meet constraints. Guidance is available on separation distances for this preliminary site layout.

Hazard assessment is then performed on the site layout with particular reference to escalation of incidents and to vulnerable features such as service buildings and buildings just over the site boundary.

If alternative site layouts have been generated, they are then costed and the most economic identified. The site layout is then subjected to a critical examination. If there is a choice of site, the selection is made at this point.

10.5.4 Stage Two site layout

Stage Two layout is the secondary, intermediate or sanction layout. As the latter term implies, it is carried out to provide a layout which is sufficiently detailed for sanction purposes. It starts with the site layout and then proceeds to the plot layout.

At this stage information on the specific characteristics of the site is brought to bear, such as the legal requirements, the soil and drainage, the meteorological conditions, the environs, the environmental aspects and

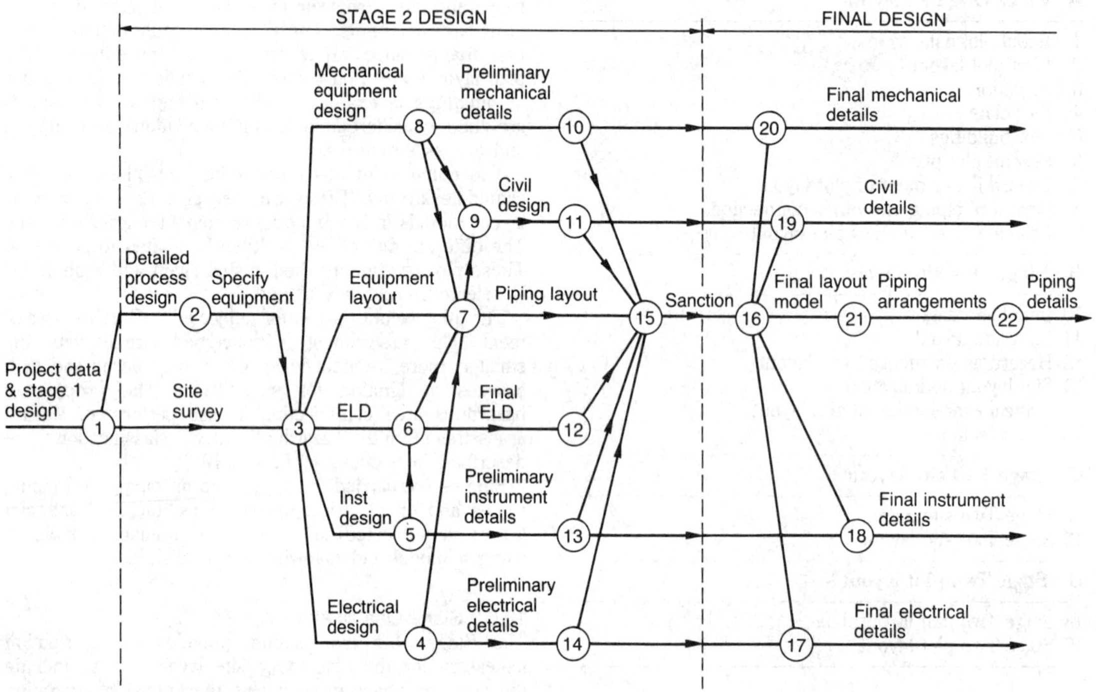

Figure 10.1 Simplified Stage Two and Final Stage design network (Mecklenburgh, 1985). ELD, engineering line diagram (Courtesy of the Institution of Chemical Engineers)

the services. Site standards are set for building lines and finishes, service corridors, pipetracks and roads.

Stage Two layout involves reworking the Stage One site layout in more detail and for the specific site, and repeating the hazard assessment, economic optimization and critical examination.

Features of the specific site which may well influence this stage are: planning matters; environmental aspects; neighbouring plants, which may constitute hazards and/ or targets; other targets such as public buildings; and road, rail and service access points.

At this stage there should be full consultation with the various regulatory authorities, insurers and emergency services, including the police and fire services.

A final site plan is drawn up in the form of drawings and models, both physical and computer-based ones, showing in particular the layout of the plots within the site, the main buildings and roads, railways, service corridors, pipetracks and drainage.

10.5.5 Stage Two plot layout

There then follows the Stage Two plot layout. The information available for this phase includes (1) standards, (2) site data, (3) Stage Two site layout, (4) process engineering design and (5) Stage One plot layout. The standards include international and national standards and codes of practice, company standards and contractor standards. The process engineering design data include the flowsheets, flow diagrams, equipment lists and drawings, process design data sheets and pipework line lists.

The Stage Two plot layout involves reworking in more detail and subject to the site constraints the plot plans and layouts and repeating the hazard assessment, piping layout and critical examination. The reworking of the plot layout, which occurs at node 7 in Figure 10.1, is a critical phase, requiring good co-ordination between the various disciplines.

By the end of Stage 2 an assessment should have been carried out of hazard and environmental problems. This assessment is used to obtain detailed planning permission.

10.6 Site Layout Features

10.6.1 Site constraints and standards

Once a site has been selected the next step is to establish the site constraints and standards. The constraints include:

(1) topography and geology;
(2) weather;
(3) environment;
(4) transport;
(5) services;
(6) legal constraints.

Topographical and geological features are those such as the lie of the land and its load-bearing capabilities. Weather includes temperatures, wind conditions, solar radiation, and thunderstorms. Environment covers people, activities and buildings in the vicinity. Services are power, water and effluents. Legal constraints include planning and building, effluent and pollution, traffic, fire and other safety laws, bylaws and regulations.

Site standards should also be established covering such matters as:

(1) separation distances;
(2) building lines;
(3) building construction, finish;
(4) road dimensions;
(5) service corridors;
(6) pipebridges.

Road dimensions include width, radius and gradient.

10.6.2 Site services

The site central services such as the boiler house, power station, switch station, pumping stations, etc., should be placed in suitable locations. This means that they should not be put out of action by such events as fire or flood and, if possible, not by other accidents such as explosion, and that they should not constitute sources of ignition for flammables.

Electrical substations, pumping stations, etc., should be located in areas where non-flameproof equipment can be used, except where they are an integral part of the plant.

Factors in siting the boiler house are that it should not constitute a source of ignition, that emissions from the stack should not give rise to nuisance and that there should be ready access for fuel supplies.

10.6.3 Use of buildings

Some plant may need to be located inside a building, but the use of a building is always expensive and it can create hazards and needs to be justified. Typically a building is used where the process, the plant, the materials processed and/or the associated activities are sensitive to exposure. Thus the process may need a stable environment not subject to extremes of heat or cold or it may need to be sterile. The plant may contain vulnerable items such as high speed or precision machinery. The process material may need to be protected against contamination or damage, including rain. The activities which the operators have to undertake may be delicate or skilled, or simply very frequent. Thus a building may be used to encourage more frequent inspection of the plant. Similarly, there may be maintenance activities which are delicate or skilled or simply frequent. In some cases where there are high elevations an indoor structure may be, or may feel, safer. The need to satisfy customers of the product and to keep unsightly plant out of view are other reasons. Examples of the use of buildings are the housing of batch reactors, centrifuges and analysis instruments.

Since ventilation in a building is generally less than that outdoors, a leak of flammable or toxic material tends to disperse more slowly and a hazardous concentration is more likely to build up. Moreover, if an explosion of a flammable gas or dust occurs the overpressure generated tends to be much higher. These are major disadvantages of the use of a building.

10.6.4 Location of buildings

Buildings which are the work base for a number of people should be located so as to limit their exposure to hazards.

Analytical laboratories should be in a safe area, but otherwise as close as possible to the plants served. So should workshops and general stores. The latter also require ready access for stores materials.

Administration buildings should be situated in a safe area on the public side of the security point. The main office block should always be near the main entrance and other administration buildings should be near this entrance if possible. Other buildings such as medical centres, canteens, etc., should also be in a safe area and the latter should have ready access for food supplies.

All buildings should be upwind of plants which may give rise to objectionable features.

Water drift from cooling towers can restrict visibility and cause corrosion or ice formation on plants or transport routes, and towers should be sited to minimize this. Another problem is recycling of air from the discharge of one tower to the suction of another, which is countered by placing towers cross-wise to the prevailing wind. The entrainment of effluents from stacks and of corrosive vapours from plants into the cooling towers should be avoided, as should the siting of buildings near the tower intakes. The positioning of natural draught cooling towers should also take into account resonance caused by wind between the towers. The problem of air recirculation should also be borne in mind in siting air-cooled heat exchangers.

10.6.5 Limitation of exposure

An aspect of segregation which is of particular importance is the limitation of exposure of people to the hazards. The measures required to effect such limitation are location of the workbase outside, and control of entry to, the high hazard zone. The contribution of plant layout to limitation of exposure therefore lies largely in workbase location. Limitation of exposure is considered more fully in Chapter 20.

10.6.6 Segregation

Although a layout which is economical in respect of land, piping and transport is in general desirable, in process plants it is usually necessary to provide some additional space and to practise a degree of segregation. The site layout should aim to contain an accident at source, to prevent escalation and avoid hazarding vulnerable targets. A block layout is appropriate with each plot containing similar and compatible types of hazard and with different types segregated in separate plots.

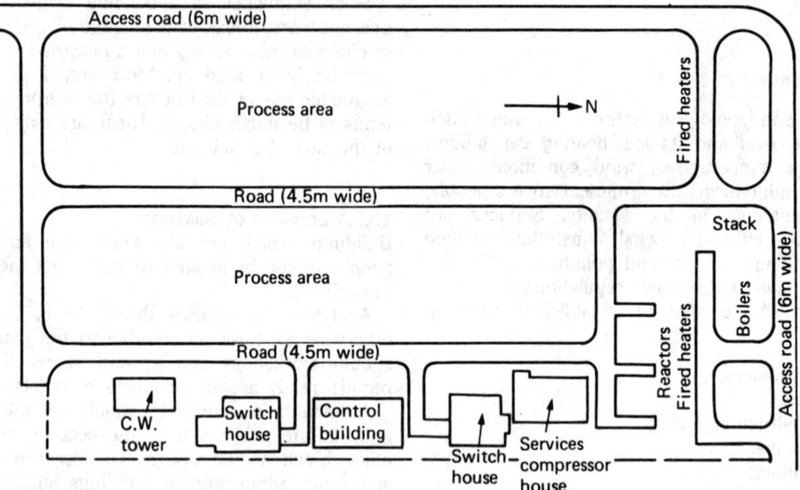

Figure 10.2 Compact block layout system in the process area of a petrochemical plant with 4.5 m roads (Simpson, 1971) (Courtesy of the Institution of Chemical Engineers)

10.6.7 Fire containment

The site layout should contribute to the containment of any fire which may occur and to combating the fire.

Features of the site layout relevant to fire hazard are illustrated in Figures 10.2 and 10.3 (Simpson, 1971). Figure 10.2 shows a compact layout, which minimizes land usage and pipework, for a petrochemical plant consisting of a major process with several stages and a number of subsidiary processes. There are two main process areas and at right angles to these is an area with a row of fired heaters, and associated reactors, steam boilers and a stack.

This layout has several weaknesses. The lack of firebreaks in the main process blocks would allow a

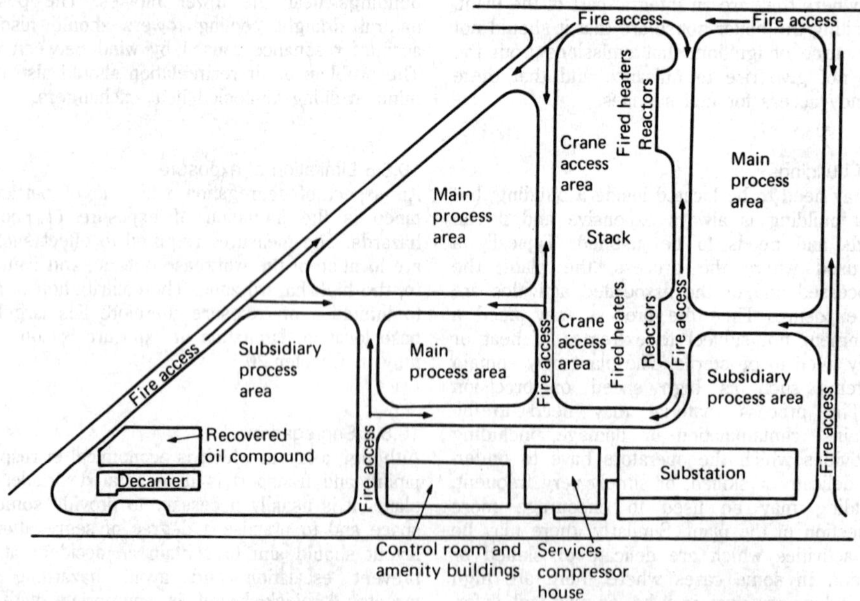

Figure 10.3 Block layout system in the process area of a petrochemical plant with 6 m roads (Simpson, 1971) (Courtesy of the Institution of Chemical Engineers)

fire to propagate right along these, particularly if the wind is blowing along them, which, on the site considered it does for 13% of the year. There is entry to the plant area from the 6m roads from opposite corners, which allows for all wind directions. But the only access for vehicles to the process plots is the 4.5m roads. In the case of a major fire, appliances might well get trapped by an escalation of the fire. The 4.5m roads give a total clearance of about 10m after allowance for equipment being set back from the road, but this is barely adequate as a firebreak. The layout is also likely to cause difficulties in maintenance work.

The alternative layout shown in Figure 10.3 avoids these problems. The process areas are divided by firebreaks. There are more entry points on the site and dead ends are eliminated. The roads are 6m wide with an effective clearance of 15m. The crane access areas provide additional clearances for the fired heaters.

Other aspects of fire protection are described in Section 10.15 and Chapter 16.

10.6.8 Effluents

The site layout must accommodate the systems for handling the effluents – gaseous, liquid and solid – and storm water and fire water. The effluent systems are considered in Section 10.16 and the drain system in Section 10.17.

10.6.9 Transport

It is a prime aim of plant layout to minimize the distances travelled by materials. This is generally achieved by following the flow principle, modified as necessary to minimize hazards.

Access is required to plots for transport of materials and equipment, maintenance operations and emergencies. Works roads should be laid out to provide this to plant plots, ideally on all four sides. Roads should be suitable for the largest vehicles which may have to use them in respect of width, radius, gradient, bridges and pipe-bridges. Recommended dimensions for works roads are given by Mecklenburgh (1973). Road widths of 10m and 7.5m are suggested for works' main and side roads, respectively. Standard road signs should be used. A road width of 7.5m with the addition of free space and/or a pipe trench on the verges may be used to give a separation distance of 15m between units.

There are various types of traffic in a works, including materials, fuel, wastes, stores, food and personnel. These traffic flows should be estimated and their routes planned. Incompatible types of traffic should be segregated as far as possible.

Road and rail traffic should not go through process areas except to its destination and even then should not violate hazardous area classifications. In this connection, it should be borne in mind that some countries still use open firebox engines. Railway lines should not cross the main entrance and should not box plants in. There should be as few railway crossings, crossroads, right angle bends, dead ends, etc., as possible.

There should be adequate road tanker parking and rail tanker sidings at the unloading and loading terminals, so that vehicles can wait their turn at the loading gantry or weighbridge without causing congestion at entrances, or on works or public roads.

Pedestrian pathways should be provided alongside roads where there are many people and much traffic. Bridges may need to be provided at busy intersections. Car and bus parks and access roads to these should be situated in a safe area and outside security points. The park for nightshift workers should be observable by the gatekeeper. There should be gates sited so that the effect of shift change on outside traffic is minimized.

10.6.10 Emergencies

There should be an emergency plan for the site. This is discussed in detail in Chapter 24. Here consideration is limited to aspects of layout relevant to emergencies.

The first step in emergency planning is to study the scenarios of the potential hazards and of their development. Plant layout diagrams are essential for such studies. Emergency arrangements should include an emergency control centre. This should be a specially designated and signed room in a safe area, accessible from the public roads and with space around it for emergency service vehicles.

Assembly points should also be designated and signed in safe areas at least 100m from the plants. In some cases it may be appropriate to build refuge rooms as assembly points. A control room should not be used either as the emergency control centre or as a refuge room.

The maintenance of road access to all points in the site is important in an emergency. The site should have a road round the periphery with access to the public roads at two points at least. The vulnerability of the works road system to blockage should be as low as possible. Data on typical fire services appliance dimensions and weights are given by Mecklenburgh (1985). For several of these the turning circle exceeds 15m.

Arrangements should be made to safeguard supplies of services such as electricity, water and steam to plants in an emergency. Electricity cables are particularly vulnerable to fire and, if possible, important equipment should be provided with alternative supplies run through the plant by separate routes.

10.6.11 Security

The site should be provided with a boundary fence and all entrances should have a gatehouse. The number of entrances should be kept to a minimum. If construction work is going on in part of the works, this building site should have its own boundary fence and a separate entrance and gatehouse. If the works boundary fence is used as part of this enclosure, movement between the building site and the works should be through an entrance with its own gatehouse.

10.7 Plot Layout Considerations

Some considerations which bear upon plot layout are:

(1) process considerations;
(2) economic considerations;
(3) construction;
(4) operations;
(5) maintenance;
(6) hazards;

(7) fire fighting;
(8) escape.

10.7.1 Process considerations

Process considerations include some of the relationships already mentioned, such as gravity flow and availability of head for pump suctions, control valves and reflux returns. Under this heading come also limitations of pressure drop in pipes and heat exchangers and across control valves and of temperature drop in pipes, the provision of straight runs for orifice meters, the length of instrument transmission lines and arrangements for manual operations such as dosing with additives, sampling, etc.

10.7.2 Economic considerations

As already mentioned, some features which have a particularly strong influence on costs are foundations, structures, piping and electrical cabling. This creates the incentive to locate items on the ground, to group items so that they can share a foundation or a structure, and to keep pipe and cable runs to a minimum.

10.7.3 Construction

Additional requirements are imposed by the needs of construction and maintenance. The installation of large and heavy plant items requires space and perhaps access for cranes. Such items tend to have long delivery times and may arrive late; the layout may need to take this into account.

Construction work may require an area in which the construction materials and items can be laid out. On large, single-stream plants major items can often be fabricated only on site. There needs to be access to move large items into place on the plant. If the plot is close to the site boundary, it should be checked that there will be space available for cranes and other lifting gear.

10.7.4 Operation

Access and operability are important to plant operation. Mention has already been made of the development of the 3D envelope of the main items of equipment to allow for operation. Hazop studies, described in Chapter 8, may be used to highlight operating difficulties in the layout.

The routine activities performed by the operator should be studied with a view to providing the shortest and most direct routes from the control room to items requiring most frequent attention. Clear routes should be allowed for the operator, avoiding kerbs and other awkward level changes.

General access ways should be 0.7 m and 1.2 m wide for one and two persons, respectively. Routes should be able to carry the maximum load, which often occurs during maintenance.

Stairways rather than ladders should be provided for main access, the latter being reserved for escape routes on outside structures and access to isolated points which are only visited infrequently. Recommended dimensions for stairways are an angle of 35–40° and overall width of 1 m with railings 0.85 m high and clearances 2.1 m. The height of single flights without a landing should not exceed 4.5 m. No workplace should be more than 45 m from an exit.

Ladders should be positioned so that the person using it faces the structure and does not look into space. A ladder should not be attached to supports for hot pipes, since forces can be transmitted which can distort the ladder. Recommended dimensions for ladders are given by Mecklenburgh (1973).

If plant items require operation or maintenance at elevated levels, platforms should be provided. The levels are defined as 3.5 m above grade for vessels, 2 m above grade for instruments or 2 m above another platform. Platform floors are normally not less than 3 m apart. Headroom under vessels, pipes, cable racks, etc., should be 2.25 m minimum, reducing to 2.1 m vertically over stairways.

Good lighting on the plant is important, particularly on access routes, near hazards and for instrument reading.

Operations involving manipulation of an equipment while observing an indicator should be considered so that the layout permits this. Similarly, it is helpful when operating controls to start or stop equipment to be able to see or hear that the equipment has obeyed the signal.

Hand valves need good access, particularly large valves which may require considerable physical effort to turn. Valves which have to be operated in an emergency should be situated so that access is not prevented by the accident through fire or other occurrences. For emergency isolation, however, it may be preferable to install remotely operated isolation valves, as described in Chapter 12.

Batch equipment such as batch reactors, centrifuges, filters and driers, tends to require more manual operation, so that particular attention should be paid to layout for such items.

Insulation is sometimes required on pipework to protect operating and maintenance personnel rather than for process reasons.

10.7.5 Maintenance

Plant items from which the internals need to be removed for maintenance should have the necessary space and lifting arrangements. Examples are tube bundles from heat exchangers, agitators from stirred vessels and spent catalyst from reactors.

10.7.6 Hazards

The hazards on the plant should be identified and allowed for in the plot layout. This is discussed in other sections, but some general comments may be made at this stage.

Plot layout can make a large contribution to safety. It should be designed and checked with a view to reducing the magnitude and frequency of the hazards and assisting preventive measures. The principle of segregation of hazards applies also to plot layout.

Hazardous areas should be defined. They should not extend beyond the plot boundaries or to railway lines. It is economic to minimize the extent of hazardous areas and to group together in them items which give the same hazard classification.

Plants which may leak flammables should generally be built in the open or, if necessary, in a structure with a roof but no walls. If a closed building cannot be avoided, it should have explosion relief panels in the walls or roof with relief venting to a safe area. Open air construction ventilates plants and disperses flammables but, as already indicated, scenarios of leakage and dispersion should be investigated for the plant concerned.

Fire spread in buildings should be limited by design, as should fire spread on open structures. Sprinklers and other protective systems should be provided as appropriate. A more detailed consideration of fire hazards and precautions is given in Chapter 16.

Plants which may leak toxics should also generally be built in the open air. The hazardous concentrations for toxics are much lower than those for flammables, however, and it cannot be assumed that an open structure is always sufficiently ventilated. A wind of at least 8 km/h is needed to disperse most toxic vapours safely before they reach the next plant. Some toxic plants, however, require a building and in some cases there has to be isolation of the toxic area through the use of connecting rooms in which clothing is changed.

Ventilation is necessary for buildings housing plants processing flammables or toxics. Air inlets should be sited so that they do not draw in contaminated air. The relative position of air inlets and outlets should be such that short circuiting does not occur. Exhaust air may need to be treated before discharge by washing or filtering.

Plants which are liable to leak liquids should stand on impervious ground with suitable slopes to drain spillages away. The equipment should be on raised areas which slope down to valleys and to an appropriate collecting point. Suitable slopes are about 1 in 40 to 1 in 60. Valleys should not coincide with walkways, and kerbs may be needed to keep liquid off these. The collecting points should be away from equipment so that this is less exposed to any fire in the liquid collected. The amount of liquid which may collect should be estimated and the collecting point should be designed to take away this amount. The heat generated if the liquid catches fire should be determined and vulnerable items relocated if necessary.

The use of pervious ground, such as pebbles, to absorb leaks of flammable liquids should be avoided. Such liquid may remain on the water table and may be brought up again by water from fire fighting. Other hazards which are prevalent mainly outside the UK include earthquakes and severe thunderstorms. These require special measures.

Personal safety should not be overlooked in the plot layout. Measures should be taken to minimize injury due to trips and falls, bumping of the head, exposure to drips of noxious substances and contact with very hot or cold surfaces. Where such hazards exist they generally present a threat not just on occasion but for the whole time.

10.7.7 Fire fighting

Access is essential for fire fighting. This is provided by the suggested plot size of 100 m × 200 m with approaches preferably on all four sides and by spacing between plots and buildings of 15 m.

Fire water should be available from hydrants on a main between the road and the plant. Hydrant points should be positioned so that any fire on the plot can be reached by the hoses. Hydrant spacings of 48, 65 and 95 m are suitable for high, medium and low risk plots, respectively. Plants over 18 m high should be provided with dry riser mains and those over 60 m high or of high risk should be equipped with wet riser mains. The inlets on the ground floor to dry riser mains and the outlets on all floors to both types of main should be accessible.

Pipes for fire water supply should be protected against explosion damage. Isolation valves should be provided to prevent loss of fire water from damaged lines and, if these valves are above ground level, they should be protected by concrete blast barriers.

Fire extinguishers of the appropriate type and fire blankets should be placed at strategic points. There should be at least two extinguishers at each point. Some extinguishers should be located on escape or access routes, so that a person who decides to fight the fire using the extinguisher has a route behind him for escape. The location of other fire fighting equipment such as sprinklers and foam sprays is a matter for experts.

Fire equipment should be located so that it is not likely to be disabled by the accident itself. It should be accessible and should be conspicuously marked. The main switchgear and emergency controls should have good access, preferably on an escape route, so that the operator does not have to risk his life to effect shutdown.

There are numerous legal requirements concerning fire, fire construction and fire fighting. There should be full consultation on this at an early stage with the works safety officer and with other parties such as the local authority services, the Factory Inspectorate and the insurers.

10.7.8 Escape

A minimum of two escape routes should be provided for any workspace, except where the fire risk is very small, and the two routes should be genuine alternatives. No workplace should be more than 12–45 m, depending on the degree of risk, from an exit, and a dead end should not exceed 8 m.

Escape routes across open mesh areas should have solid flooring. Escape stairways should be in straight flights. They should preferably be put on the outside of buildings. Fixed ladders may be used for escape from structures if the number of people does not exceed 10. Doors on escape routes should be limited to hinged or sliding types and hinged doors should open in the direction of escape. Handrails should be provided on escape routes across flat roofs. Escape routes should be signposted, if there is any danger of confusion, as in large buildings. They should be at least 0.7 m and preferably 1.2 m wide to allow the passage of 40 persons per minute on the flat and 20 persons per minute down stairways. Good lighting should be provided on escape routes and arrangements made to ensure a power supply in an emergency. The escape times of personnel should be estimated, paying particular attention to people on tall items such as distillation columns or cranes. Bridges between columns may be used.

10.8 Equipment Layout

10.8.1 General considerations

Furnaces and fired heaters are very important. Furnace location is governed by a number of factors, including the location of other furnaces, the use of common facilities such as stacks, the minimization of the length of transfer lines, the disposal of the gaseous and liquid effluents, the potential of the furnace as an ignition source and the fire fighting arrangements. Furnaces should be sited at least 15 m away from plant which could leak flammables.

No trenches or pits which might hold flammables should extend under a furnace, and connections with underground drains should be sealed over an area 12 m from the furnace wall. The working area of the furnace should be provided with ventilation, particularly where high temperatures and high sulphur fuels are involved.

On wall-fired furnaces there should be an escape route at least 1 m wide at each end and on top-fired furnaces there should also be an escape route at each end, one of which should be a stairway. The provision of peepholes and observation doors should be kept to a minimum. Access to these may be by fixed ladder for heights less than 4 m above ground, but platforms should be provided for greater heights.

Incinerators and burning areas for waste disposal should be treated as fired equipment. Waste in burning areas should be lit by remote ignition and, if it is an explosion hazard, blast walls should be provided.

Chemical reactors in which a violent reaction can occur may need to be segregated by firebreaks or even enclosed behind blast walls.

Heat exchangers should have connecting pipework kept to a minimum, consistent with provision of pipe lengths and bends to allow for pipe stresses and with access for maintenance.

Equipments which have to be opened for cleaning, emptying, charging, etc., may need ventilation.

Driers in which volatile materials are driven off solids will generally need ventilation of the drying area and probably of the drier itself. If the materials are noxious, detraying booths may be necessary.

Dust-handling equipment such as driers, cyclones and ducts may constitute an explosion hazard, but tends to be rather weak. It should be separated from other plant by a wall and vented. Vents should be short and should go through the roof. Some equipment such as cyclones is often placed outside the building and this is preferable to ducting a vent to the outside. Vents should pass to a safe area. Mills are relatively strong and are not usually provided with explosion relief. Dust should be transferred through chokes to prevent the transmission of fire or explosion. Surfaces which might collect dust should be kept to a minimum. Dust hazards are considered further in Chapter 17.

Pumps handling liquids which are hot (>60°C) should be separated from those handling liquids which are flammable and volatile (boiling point <40°C) or from compressors handling flammable gases. In the open, separation may be effected by a spacing of at least 7.5 m and in a pump room by a vapour-tight wall.

Hazards associated with particular plant equipment are also considered in Chapter 11.

10.8.2 Corrosive materials

If the process materials are corrosive, this aspect should be taken into account in the plant layout. The layout of plants for corrosive materials is discussed in *Safety and Management* by the Association of British Chemical Manufacturers (ABCM, 1964/3).

Corrosive materials are responsible for an appreciable proportion of accidents on chemical plants and of damage done to the plant. The presence of corrosive materials creates two particular hazards: (1) corrosion of materials of construction, and (2) contact of persons with corrosive materials. On a plant handling corrosive chemicals the materials of construction should be chosen with particular care, should be protected by regular painting and should be checked by regular inspection.

Some features of plant layout which are particularly important in relation to corrosive chemicals are:

(1) foundations;
(2) floors;
(3) walkways;
(4) staging;
(5) stairs;
(6) handrails;
(7) drains;
(8) ventilation.

Foundations of both buildings and machines, especially those constructed in concrete, may be attacked by leakage of corrosive materials, including leakage from drains. If the corrosion is expected to be mild, it may be allowed for by the use of additional thickness of concrete, but if it may be more severe, other measures are necessary. These include the use of corrosion resistant asphalt, bricks and plastics.

Floors should be sloped so that spillages are drained away from vulnerable equipment and from walkways and traffic lanes. The latter should generally be laid across the direction of fall and should as far as possible be at the high points of the slopes.

Severe corrosion of steel stanchions can occur between a concrete subfloor and a brick floor surface, and this possibility should be considered.

Pipe flanges which may drip corrosive substances should not be located over walkways. There should be guardrails around vessels or pits containing corrosive liquids. The floors and walkways should be of the 'non-skid' type.

Staging should not be located over an open vessel which may emit corrosive vapours.

Staircases and handrails should be designed to minimize corrosion. Stairheads should be located at the high points of sloping floors. Handrails tend to corrode internally and may collapse suddenly. They should be made of a suitable material. Aluminium is suitable, if it is not corroded by the atmosphere of the plant. Metal protectors of the vapour type are also available. These are put inside the pipe and the ends sealed. Alternatively, solid rails can be used. The rails should be protected against external corrosion by regular painting or other means. Regular inspection and maintenance is particularly important for staging, staircases and handrails on plants containing corrosive materials.

Drains should be designed to handle the corrosive materials, and mixtures of materials, which may be discharged into them.

Ventilation should be provided and maintained as appropriate. This requires as a minimum the circulation of fresh air. It may also involve local exhaust ventilation.

10.9 Pipework Layout

In general, it is desirable both for economic and safety reasons to keep the pipe runs to a minimum. Additional pipework costs more both in capital and operation, the latter through factors such as heat loss/gain and pumping costs. It is also an extra hazard not only from the pipe itself but more particularly from the joints and fittings.

The application of the flow principle is effective in minimizing pipe runs, but it is also necessary to practise segregation and this will sometimes lead to an unavoidable increase in the length of pipe runs. The design therefore involves a compromise between these two factors.

Piping for fluids servicing a number of points may be in the form of a ring main, which permits supply to most points, even if part of the main is disabled. Ring mains are used for steam, cooling water, process water, fire water, process air, instrument air, nitrogen and even chlorine.

Services such as steam and water mains and electricity and telephone cables should generally be run alongside the road and should not pass through plant or service areas.

Pipes may be buried, run at ground level, run on supports or laid in an open pipe trench. Open pipe trenches may be used where there is no risk of accumulation of flammable vapours, of the material freezing or of flooding.

Water mains should be buried below the frost line or to a minimum depth of 0.75 m to avoid freezing. If they run under roads or concreted areas, they should be laid in ducts or solidly encased in concrete.

Steam mains may be laid on the surface on sleepers. They should be run on the outside edge of the pipeway to allow the expansion loops to have the greatest width and to facilitate nesting of the loops. Steam mains may also be run in open pipe trenches.

Electrical power and telephone cables should be run in sand-filled trenches covered by concrete tiles or a coloured concrete mix. If possible, the cables should be run at the high point of paving leaving room for draw boxes. If use is made of underground piping and cabling, it should be put in position at the same time as the foundation work is being done. Alternatively, cables can be run overhead. Overhead cables are less affected by spillages and are easier to extend, but may require fire protection.

Electrical lines can give rise to fields of sufficient intensity to cause local overheating of adjacent metalwork or to induce static electricity in plant nearby, and this should be taken into account in positioning them. Pipes which are hot or carry solvents should be laid as far as possible from electrical cables.

Piping may be run as a double layer, but triple layers should be avoided. Double layer piping should be run with service lines on the upper and process lines on the lower deck.

Piping may require a continuous slope to permit complete drainage for process, corrosion or safety reasons; other pipes should not be sloped. Sloped lines should be supported on extensions of the steel structure. The slope arrangement should not create a low point from which liquid cannot be drained.

Overhead clearances below the underside of the pipe, flange, lagging or support should have the following minimum values:

Above roads and areas with access for crane	7 m
Plant areas where truck access required	4 m
Plant areas in general	3 m
Above access floors and walkways within buildings	2.25 m
Above railway lines (from top of rail)	4.6 m

Pipe flanges should be positioned so as to minimize the hazards from small leaks and drips. Flanges on pipework crossing roads on pipebridges should be avoided. Pipebridges over roads should be as few as possible. Every precaution should be taken to prevent damage from vehicles, particularly cranes and forklift trucks.

Attention should be paid to the compatibility of adjacent pipework, the cardinal principle being to avoid loss of containment of hazardous materials. Thus it is undesirable, for example, to put a pipe carrying corrosive material above one carrying flammables or toxics at high pressure.

Emergency isolation valves should be used to allow flows of flammable materials to be shut off. Valves may be manually or power operated and controls for the latter may be sited locally or remotely. The use of such valves is described in more detail in Chapter 12.

If a manual valve is used for isolation, it should be mounted in an accessible position. Emergency operation of valves from ladders should be avoided. If the valve is horizontally mounted and its spindle is more than 2.1 m above the operating level, a chain wheel should be provided. A valve should not be mounted in the inverted position, since solids may deposit in the gland and cause seizure.

Discharges from pressure relief valves and bursting discs are normally piped away in a closed system. In particular, a closed system is necessary for hydrocarbon vapours with a molecular weight greater than 60, flammable liquids and toxic vapours and liquids. Pressure relief and flare systems are considered more fully in Chapter 12.

Liquid drains from drainage should also be taken to a safe point. Liquids which are not flammable or toxic may be discharged to grade.

Sample points should be 1 m above the floor and not at eye level.

Flexible piping should be kept to a minimum. Where such piping is used on vehicles, use may be made of devices which shut off flow if the vehicle moves away.

Instruments incorporating glass tubing, such as sight glasses and rotameters, are a source of weakness. In some cases the policy is adopted of avoiding the use of such devices altogether. If this type of instrument is used, however, it should be enclosed in a transparent protective case.

The layout for piping and cabling should allow for future plant expansion. An allowance for 30% additional pipework is typical. Full documentation should be kept on all piping and cabling.

10.10 Storage Layout

Treatments of plant layout frequently cover all aspects of storage, including bunding, venting, etc. In this book storage is dealt with separately in Chapter 22, and only those features which are directly relevant to the layout of the plant as a whole are dealt with at this point.

The principal kinds of storage are bulk storage of fluids, bulk storage of solids and warehouse storage. The storage of main interest in the present context is storage of fluids, particularly flammable fluids. The types of storage include:

(1) liquid at atmospheric pressure and temperature;
(2) liquefied gas under pressure and at atmospheric temperature (pressure storage);
(3) liquefied gas at atmospheric pressure and at low temperature (refrigerated storage);
(4) gas under pressure.

There are also intermediate types such as semi-refrigerated storage.

For liquid storage it is common to segregate the liquids stored according to their class. The current classification, given in the *Refining Safety Code* of the Institute of Petroleum (IP, 1981 MCSP Part 3) and used in BS 5908: 1990, is

Class I	Liquids with flashpoint below 21°C
Class II (1)	Liquids with flashpoint from 21°C up to and including 55°C, handled below flashpoint
Class II (2)	Liquids with flashpoint from 21°C up to and including 55°C, handled at or above flashpoint
Class III (1)	Liquids with flashpoint above 55°C up to and including 100°C, handled below flashpoint
Class III (2)	Liquids with flashpoint above 55°C up to and including 100°C, handled at or above flashpoint

An earlier classification, given in the former BS CP 3013: 1974, was as follows:

Class A	Liquids with flashpoint below 22.8°C (73°F)
Class B	Liquids with flashpoint between 22.8 and 66°C (73 and 150°F)
Class C	Liquid with flashpoint above 66°C (150°F)

The classification given in the National Fire Protection Association's (NFPA 321: 1987) *Basic Classification of Flammable and Combustible Liquids* is:

Class I	Liquids with flashpoint below 37.8°C (100°F)
Class IA	Liquids with flashpoint below 22.8°C (73°F) and boiling point below 37.8°C (100°F)
Class IB	Liquids with flashpoint below 22.8°C (73°F) and boiling point at or above 37.8°C (100°F)
Class IC	Liquids with flashpoint at or above 22.8°C (73°F) and below 37.8°C (100°F)
Class II	Liquids with flashpoint at or above 37.8°C (100°F) and below 60°C (140°F)
Class III	Liquids with flashpoint at or above 60°C (140°F)
Class IIIA	Liquids with flashpoint at or above 60°C (140°F) and below 93.4°C (200°F)
Class IIIB	Liquids with flashpoint at or above 93.4°C (200°F)

NFPA 321 distinguishes between flammable and combustible liquids. It defines a flammable liquid as one having a flashpoint below 37.8°C (100°F) and having a vapour pressure not exceeding 40 psia at 37.8°C (100°F), and a combustible liquid as one having a flashpoint at or above 37.8°C (100°F).

Quantities in storage are almost invariably much greater than those in process. Typical orders of magnitude for a large plant are several hundred tonnes in process and ten thousand tonnes in storage.

Storage is usually built in the open, since this is cheaper and allows dispersion of leaks. The site chosen should have good load-bearing characteristics, since tanks or vessels full of liquid represent a very heavy load. The design of foundations for storage tanks is a specialist matter.

The storage site should be such that the contour of the ground does not allow flammable liquid or heavy vapour to collect in a depression or to flow down to an area where it may find an ignition source. The prevailing wind should be considered in relation to the spread of flammables to ignition sources or of toxics to the site boundary.

Storage should be segregated from process. A fire or explosion in the latter may put at risk the very large inventory in storage. And a small fire in storage which is otherwise easily dealt with may jeopardize the process. The storage area should be placed on one or at most on two sides of the process and well away from it. This gives segregation and allows room for expansion of the process and/or the storage. The separation distance between process and storage has been discussed above. It should not be less than 15 m.

It is also necessary to keep terminals away from the process, since they are sources of accidents. A suitable layout is therefore to interpose the storage between the process and the terminals. The separation distance between storage and terminals should be not less than 15 m.

The storage tanks should be arranged in groups. The grouping should be such as to allow common bunding, if bunds are appropriate, and common fire fighting equipment for each group. There should be access on all four sides of each bund area and roads should be linked to minimize the effect if one road is cut off during a fire.

It is not essential that there be only one storage area, one unloading terminal or one loading terminal. There may well be several, depending on the materials and process, and the principle of segregation. The raw material unloading and the product loading terminals should be separate. Normally both should be at the site boundary near the entrance. If the materials are hazardous or noxious, however, the terminal should not be near the entrance, although it may be near the site boundary, provided it does not affect a neighbour's installation.

10.11 Separation Distances

Plant layout is largely constrained by the need to observe minimum separation distances. For hazards, there are basically three approaches to determining a suitable separation distance. The first and most traditional one is to use standard distances developed by the industry. The second is to apply a ranking method to decide the separation required. The third is to estimate a suitable separation based on an engineering calculation for the particular case. Not all separation distances relate to hazards. Construction, access and maintenance are other relevant factors. The first two methods of determining separation distances are considered in this section, and the third is considered in Section 10.14.

10.11.1 Types of separation

The types of separation which need to be taken into consideration are illustrated by the set of tables of separation distances given by Mecklenburgh (1985) and include:

(1) site areas and sizes;
(2) preliminary spacing for equipment:
  (a) spacing between equipment,
  (b) access requirements at equipment,
  (c) minimum clearances at equipment;
(3) preliminary spacings for storage layout:
  (a) tank farms,
  (b) petroleum products,
  (c) liquefied flammable gas,
  (d) liquid oxygen;
(4) preliminary distances for electrical area classification;
(5) size of storage piles.

Further types of separation used by D.J. Lewis (1980b) are given in Section 10.11.4.

10.11.2 Standard distances

There are a large number of standards, codes of practice and other publications which give minimum safe separation distances. The guidance available relates mainly to separation distances for storage, either of petroleum products, of flammable liquids, of liquefied petroleum gas (LPG) or of liquefied flammable gas (LFG).

Recommendations for separation distances are given in: for petroleum products, *The Storage of Flammable Liquids in Fixed Tanks Exceeding 10 000 m^3 Total Capacity* (HSE, 1991 HS(G) 52), the *Refining Safety Code* (IP, 1981 MCSP Pt 3), the American Petroleum Institute (API) standards API Std 620: 1990 and API Std 650: 1988 and NFPA 30: 1990 *Flammable and Combustible Liquids Code*; for LPG, in *The Storage of LPG at Fixed Installations* by the HSE (1987 HS(G) 34), *Liquefied Petroleum Gas* by the IP (1987 MCSP Pt 9), the *Code of Practice, Part 1, Installation and Maintenance of Fixed Bulk LPG Storage at Consumers' Premises* by the Liquefied Petroleum Gas Industry Technical Association (LPGITA) (1991 LPG Code 1 Pt 1), API Std 2510: 1989 and 2510A: 1989 and NFPA 58: 1989 *Storage and Handling of Liquefied Petroleum Gases*; and for LFG, the ICI *Liquefied Flammable Gases, Storage and Handling Code* (the ICI LFG Code) (ICI/RoSPA 1970 IS/74). Another relevant code is BS 5908: 1990 *Fire precautions in the Chemical and Allied Industries*. Separation distances are also given in many of the NFPA codes.

Further guidance on separation distances and clearances is given by Armistead (1959), House (1969), the Oil Insurance Association (OIA) (1972/6), Backhurst and Harker (1973), Mecklenburgh (1973, 1985), Kaura (1980b), F.V. Anderson (1982) and Industrial Risk Insurers (IRI) (1991, 1992).

Separation distances are specified in the *Fire and Explosion Index. Hazard Classification Guide* (the *Dow Guide*) (Dow Chemical Company, 1976) as a function of the Fire and Explosion Index (F&EI) and the maximum probable property damage (MPPD). These do not appear as such in the current edition of the *Guide* (Dow Chemical Company, 1994b), which is described in Chapter 8. Some tables of separation distances for storage of flammable liquids, for LPG and for LFG are reproduced in Chapter 22.

Separation distances for process units are usually given as the distances between two units or as the distance between a single unit and an ignition source. It is normal to quote distances between the edges of units and not centre to centre. There is generally little explanation given of the basis of the separation distances recommended.

The separation distances for liquids which have a lower vapour pressure, including the bulk of petroleum products and flammable liquids, tend to be less than those for liquids which have a high vapour pressure and so flash off readily, such as LPG and LFG. It is frequently stated that for LPG a smaller separation distance may be allowed if there is provision of adequate radiation walls and/or water drench systems.

There is naturally some tendency for separation distances to be reproduced from one publication to another. In general, however, there are differences between the various codes and guidelines, so that the overall situation is rather confused. This problem has been discussed by Simpson (1971).

Typical separation distances for preliminary site layouts are given by Mecklenburgh (1985). The table of spacings which he gives is shown in Table 10.4. Some interunit and interequipment separation distances given by IRI (1991, 1992) are shown in Table 10.5.

10.11.3 Rating and ranking methods

An alternative to the use of standard separation distances is the utilization of some form of rating or ranking method. The most widely applied method of this kind is that used in hazardous area classification. This method ranks items by their leak potential. An outline of the method is given in Section 10.12. Another such method is the Mond Index, which is now described.

10.11.4 Mond Index

The Mond Index is one of the hazard indices described in Chapter 8. A particular application of this index is the determination of separation distances as described by D.J. Lewis (1980b, 1989b). In the Mond Index method two values are calculated for the overall risk rating (ORR), those before and after allowance is made for offsetting factors. It is the latter rating R_2 which is used in applying the technique to plant layout. The ORR assigns categories ranging from mild to very extreme.

Table 10.4 *Preliminary areas and spacings for site layout (Mecklenburgh, 1985) (Courtesy of the Institution of Chemical Engineers)*

Administration	10 m^2	per administration employee
Workshop	20 m^2	per workshop employee
Laboratory	20 m^2	per laboratory employee
Canteen	1 m^2	per dining space
	3.5 m^2	per place including kitchen and store
Medical centre:	0.1–0.15 m^2	per employee depending on complexity of service
Minimum	10 m^2	
Fire-station (housing 1 fire, 1 crash, 1 foam, 1 generator and 1 security vehicle)	500 m^2	per site
Garage (including maintenance)	100 m^2	per vehicle
Main perimeter roads	10 m	wide
Primary access roads	6 m	wide
Secondary access roads	3.5 m	wide
Pump access roads	3.0 m	wide
Pathways	1.2 m	wide up to 10 people/min
	2.0 m	wide over 10 people/min (e.g. near offices, canteens, bus stops)
Stairways	1.0 m	wide including stringers
Landings (in direction of stairway)	1.0 m	wide including stringers
Platforms	1.0 m	wide including stringers
Road turning circles (90° turn and T-junctions)		radius equal to width of road
Minimum railway curve	56 m	inside curve radius
Cooling towers per tower	0.04 m^2/kW	mechanical draught
	to 0.08 m^2/kW	natural draught
Boiler (excluding house)	0.002 m^3/kW	(Height = 4 × Side)

Property boundary	*Control room (non-pressurized)*	*Control room (pressurized)*	*Administration building*	*Main substation*	*Shippings, buildings, warehouses*	*Loading facilities, road, rail, water*	*Fire pumphouse*	*Cooling towers*	*Process fired heaters*	*Gas compressors*	*Reactors*	*High pressure storage spheres, bullets*	*Atmospheric flammable liquid storage tanks*	*Aircoolers*	*Low pressure storage spheres or tanks <1 bar g*	*Plot limits*	*Process control station*	*Process unit substation*	*Process equipment (low flashpoint)*	*Process equipment (high flashpoint)*	*Cryogenic O_2* plant*
NA																					
30	NA																				
8	NA	NA																			
8	8	8	NA																		
8	30	15	8	NA																	
8	15	NM	30	15	NA																
30	60	30	60	60	60	NA															
8	8	8	8	30	30	45	NA														
30	30	30	30	60	30	45	30	7.5													
30	30	30	75	60	75	60	60	30	7.5												
30	30	30	60	60	60	60	60	30	15	2											
60	30	30	60	60	60	60	60	30	15	10	2										
CP	60	30	75	75	75	CP	75	30	CP	75	60	CP									
CP	60	30	60	60	60	CP	60	30	CP	60	60	CP	CP								
60	30	30	60	60	60	60	60	30	15	7.5	5	60	60	NM							
CP	60	30	75	60	60	CP	60	30	CP	60	60	CP	CP	60	CP						
60	60	60	60	60	60	45	60	30	NA	NA	NA	CP	CP	NA	CP	15					
30	NA	NA	NA	30	NA	60	NA	15	15	15	15	60	60	15	60	NA	NA				
NM	NM	NA	NA	NM	NA	45	NM	15	15	15	15	NM	NM	15	NM	NM	NM	NA			
15	30	NM	60	60	60	60	60	30	30	7.5	5	CP	CP	5	CP	NA	15	15	2		
15	30	NM	60	60	60	60	60	30	15	7.5	5	CP	CP	5	CP	NA	15	15	2	2	
CP	30	NM	30	60	60	CP	45	30	CP	45	60	CP	CP	30	CP	CP	60	50	CP	CP	30

NA, not applicable since no measureable distance can be determined; NM, no minimum spacing established – use engineering judgement; CP, reference must be made to relevant Codes of Practice but see section C.6 of the original reference.

[a] See also section C.6 for minimum clearances.

Notes:

(1) Flare spacing should be based on heat intensity with a minimum space of 60 m from equipment containing hydrocarbons.

(2) The minimum spacings can be down to one-quarter these typical spacings when properly assessed.

Table 10.5 *Some separation distances for oil and chemical plants. The spacings given are applicable for items with potential for fire and vessel explosion. Spacings for items with potential for vapour cloud explosion should be obtained by other means*

A Interunit spacings (Industrial Risk Insurers, 1991)

Service buildings	Motor control centres and electrical substations	Utilities areas	Cooling towers	Control rooms	Compressor buildings	Large pump houses	Process units moderate hazard	Process units intermediate hazard	Process units high hazard	Atmospheric storage tanks	Pressure storage tanks	Refrigerated storage tanks dome roof	Flares	Unloading and loading racks	Fire water pumps	Fire stations
/																
/	/															
50	50	/														
50	50	100	50													
/	/	100	100	/												
100	100	100	100	100	30											
100	100	100	100	100	30	30										
100	100	100	100	100	30	30	50									
200	100	100	100	200	50	50	100	100								
400	200	200	200	300	100	100	200	200	200							
250	250	250	250	250	250	250	250	300	350	*						
350	350	350	350	350	350	350	350	350	350	*	*					
350	350	350	350	350	350	350	350	350	350	*	*	*				
300	300	300	300	300	300	300	300	300	300	300	400	400	/			
200	200	200	200	200	200	200	200	200	300	250	350	350	300	50		
50	50	50	50	50	200	200	200	300	300	350	350	350	300	200	/	
50	50	50	50	50	200	200	200	300	300	350	350	350	300	200	/	/

1 ft = 0.305 m; /, No spacing requirements; * Spacing given in Table 3 of the original reference.

B Interequipment spacings (Industrial Risk Insurers, 1992)

Compressors	Intermediate hazard pumps	High hazard pumps	High hazard reactors	Intermediate hazard reactors	Moderate hazard reactors	Columns, accumulators, drums	Rundown tanks	Fired heaters	Air cooled heat exchanger	Heat exchangers	Pipe racks	Emergency controls	Unit block valves	Analyser rooms
30														
30	5													
50	5	5												
50	10	15	25											
50	10	15	25	15										
50	10	15	25	15	15									
50	10	15	50	25	25	15								
100	100	100	100	100	100	100	100							
50	50	50	50	50	50	50	100	25						
30	15	15	25	15	15	15	100	50	/					
30	10	15	25	15	10	10	100	50	15	5				
30	10	15	25	15	10	10	100	50	/	10	/			
50	50	50	100	50	50	50	100	50	50	50	50	/		
50	50	50	100	50	50	50	100	50	50	50	50	/	/	
50	50	50	50	50	50	50	100	50	50	50	50	/	/	/

1 ft = 0.305 m; /, No spacing requirements.

The objectives of layout are: to minimize risk to personnel; to minimize escalation, both within the plant and to adjacent plants; to ensure adequate access for fire fighting and rescue; and to allow flexibility in combining together units of similar hazard potential.

Lewis enumerates the basic concepts underlying the initial layout. In addition to general layout principles, he includes several applications of the ORR. Control and other occupied buildings should be adjacent to low or medium risk units, the latter being acceptable only if a low risk unit is not available and if the R_2 value is just inside the medium risk band. Units with the highest value of the aerial explosion index A_2 should not be located near to the plant boundary but should be separated by areas occupied by low risk activities and with low population densities (up to 25 persons/acre).

Major pipebridges with medium to high R_2 should be located to reduce their vulnerability to incidents from tall process units and from transport accidents arising from normal vehicle traffic.

Units separately assessed can be combined into a single unit, providing that the hazards are compatible and the risks similar, the potential direct and consequential losses do not become excessive and the reassessed R_2 value is acceptable.

The initial layout is based on a nominal interunit spacing of 10 m. It includes pipebridges and vehicle routes. The nominal interunit distances are then replaced by those established by engineering considerations, including the use of guidance on minimum separation distances and of the ORR.

Lewis states that the minimum separation distances given in the relevant codes are absolute minimum distances and are not necessarily good practice for new installations. Some situations for which separation distances are required are given by Lewis as follows:

(1) distances between a unit of a particular degree of hazard and
 (a) another unit of the same degree,
 (b) another unit of lower or higher degree;
(2) distances between a process unit and
 (a) a storage unit,
 (b) the bund of a storage unit;
(3) distances between adjacent storage units containing materials of different flammability;
(4) distances between a unit and
 (a) occupied buildings,
 (b) potential ignition sources,
 (c) a plant boundary,
 (d) the works boundary.

For units, the relevant distance for the determination of separation is taken as that between the nearest wall, structural frame or free-standing equipment of the two units.

Separation distances for pipebridges receive particular attention. For a pipebridge between two units, the separation distance is between one side of the pipebridge and the adjacent unit. The distance should not include the plan area occupied by the pipebridge itself, but it is not normally necessary to provide two separation distances, one on each side of the pipebridge. A pipebridge which itself has significant potential for a hazardous release should not be located alongside a unit without a separation distance unless assessment shows that the hazard level of the combination of unit and pipebridge is acceptable. If it is not, there should be a separation between the pipebridge and all units, using the pipebridge separation distances given.

The spacings for storage units given in the initial treatment (D.J. Lewis, 1980b) were subsequently revised (D.J. Lewis, 1989b). The principal changes are considerable increases in separation distances for the extreme and very extreme values of R_2, exclusion of units which have potential for 'frothover' or for 'boilover' in a fire, and restriction to units which are on level ground.

The recommendations for separation distances for process units are shown in Table 10.6 and those for storage units in Tables 10.7 and 10.8. D.J. Lewis (1989b) gives further recommendations for dealing with storage

Table 10.6 *Separation distances for process units: spacings between process units and other features obtained using the Mond Index (D.J. Lewis, 1980b) (Courtesy of the American Institute of Chemical Engineers)*

A Minimum spacings between one process unit A and another process unit B (m)

Overall risk rating R_2 of process unit A	*Overall risk rating R_2 of process unit B*						
	Mild	*Low*	*Medium*	*High*	*Very high*	*Extreme*	*Very extreme*
Mild	0	6	9	12	17	20	30
Low	6	8	10	15	20	25	40
Medium	9	10	15	18	25	30	50
High	12	15	18	20	30	40	60
Very high	17	20	25	30	40	50	80
Extreme	20	25	30	40	50	65	100
Very extreme	30	40	50	60	80	100	150

B Minimum spacings between a process unit and another feature (m)

Overall risk rating R_2 of storage unit	*Feature*							
	Works boundary	*Plant boundary works main road, works main railway*	*Control room*	*Offices, amenity, buildings, workshops, laboratories, etc.*	*Electrical switchgear, instrument houses*	*Electrical power lines and transformers*	*Process furnaces and similar ignition sources*	*Forced draught cooling towers*
Mild	20	15	9	12	5	0	7	10
Low	27	20	10	15	10	5	12	17
Medium	35	27	15	20	15	10	17	25
High	50	35	18	27	20	15	25	30
Very high	70	50	25	40	25	20	30	35
Extreme	120	75	30	60	30	25	40	40
Very extreme	200	100	50	75	40	30	60	50

Table 10.7 Separation distances for process units: spacings between storage units and process units or other storage units obtained using the Mond Index (D.J. Lewis, 1989b) (Courtesy of the Norwegian Society of Chartered Engineers)

A Minimum spacings between a storage unit and a process unit: spacing to tank wall (m)

Overall risk rating R_2 of storage unit	*Overall risk rating R_2 of process unit*						
	Mild	*Low*	*Medium*	*High*	*Very high*	*Extreme*	*Very extreme*
Mild	3	7	10	13	18	23	38
Low	6	9	12	17	23	30	50
Moderate	9	12	17	21	31	44	66
High	12	17	21	28	43	56	84
Very high	17	23	31	43	56	72	110
Extreme	23	30	44	56	72	97	145
Very extreme	38	50	66	84	110	145	197

B Minimum spacings between a storage unit and a process unit: spacing to bund wall (m)

Overall risk rating R_2 of storage unit	*Overall risk rating R_2 of process unit*						
	Mild	*Low*	*Medium*	*High*	*Very high*	*Extreme*	*Very extreme*
Mild	2	4	5	7	9	10	15
Low	3	5	6	8	10	13	20
Moderate	4	6	8	9	13	16	26
High	6	8	9	12	16	22	33
Very high	8	10	13	16	22	28	45
Extreme	10	13	16	22	28	36	58
Very extreme	15	20	26	33	45	58	90

C Spacings between two storage units: spacing between one tank wall and the other tank wall

Overall risk rating R_2 of storage unit A	*Overall risk rating R_2 of storage unit B*						
	Mild	*Low*	*Medium*	*High*	*Very high*	*Extreme*	*Very extreme*
Mild	5	7	10	13	18	25	43
Low	7	10	13	19	26	36	55
Moderate	10	13	19	26	36	56	80
High	13	19	26	36	56	72	110
Very high	18	26	36	56	72	97	145
Extreme	25	36	56	72	97	130	185
Very extreme	43	60	80	110	145	185	225

units with frothover or boilover potential and with units located on sloping ground.

10.11.5 Hazard models

Another approach to the determination of separation distances is to use hazard models to determine the separation distance at which the concentration from a vapour escape or the thermal radiation from a fire fall to an acceptable level. This is the other side of the coin to hazard assessment of a proposed layout. Early accounts of the use of hazard models to determine separation distances include those of Hearfield (1970) and Simpson (1971).

Two principal factors considered as determining separation are (1) heat from burning liquid and (2) ignition of a vapour escape.

Permissible heat fluxes are discussed by Simpson, who distinguishes three levels of heat flux: 12.5 kW/m^2 (4000 BHU/ft^2 h), 4.7 kW/m^2 (1500 BTU/ft^2 h) and 1.6 kW/m^2 (500 BTU/ft^2 h). The first value is the limit given in the Building Regulations 1965 and is suggested as a suitable limit for buildings such as control rooms or workshops; the second is the threshold of pain after a short time and is suggested as the limit for workers out on the plant who must continue doing essential tasks; and the third is the level of minor discomfort and is suggested as the limit for people in adjoining areas. A more detailed discussion of thermal radiation criteria is given in Chapter 16.

Simpson also considers separation distances based on the dilution of a vapour leak to a concentration below the lower flammability limit. The estimates are based on calculations of leak emission flows, pool vaporization rate and vapour cloud dispersion, as described in Chapter 15. One problem which he discusses is the separation between storage and an ignition source for petroleum spirit and other flammable liquids of similar volatility. For this case he concludes that in most instances a separation distance of 15 m is adequate.

Another problem is the separation distance between a petrochemical unit and an ignition source. The typical scenarios which he discusses give separation distances as high as 88 m, this being for a necked-off branch on a C_2 fractionator. A further discussion of the basis for separation distances has been given by R.B. Robertson (1976b).

Table 10.8 *Separation distances for process units: spacings between storage units and other features obtained using the Mond Index (D.J. Lewis, 1989b) (Courtesy of the Norwegian Society of Chartered Engineers)*

A Spacings between a storage unit and another feature: spacing to tank wall (m)

Overall risk rating R_2 of storage unit	*Feature*[a]					
	Works boundary	*Plant boundary, works main road, works main railway*	*Control room*	*Offices, amenity, buildings, workshops, laboratories, etc.*	*Process furnaces, other ground level ignition sources, electrical switchgear, instrument houses*	*Flare stacks, of tip height H m above ground*[b]
Mold	20	15	7	12	10	$1.25H+6$
Low	27	20	12	16	15	$1.25H+15$
Moderate	35	25	20	24	22	$1.25H+15$
High	55	41	28	36	33	$1.25H+22$
Very high	81	70	41	58	52	$1.25H+35$
Extreme	125	95	53	72	66	$1.25H+45$
Very extreme	175	130	75	100	90	$1.25H+60$

B Spacings between a storage unit and another feature: spacing to bund wall (m)

Overall risk rating R_2 of storage unit	*Feature*[a]					
	Works boundary	*Plant boundary, works main road, works main railway*	*Control room*	*Offices, amenity, buildings, workshops, laboratories, etc.*	*Process furnaces, other ground level ignition sources, electrical switchgear, instrument houses*	*Flare stacks, of tip height H m above ground*[b]
Mild	15	10	5	8	7	$H+6$
Low	20	12	6	11	10	$H+8$
Moderate	25	15	7	13	12	$H+10$
High	38	22	9	20	18	$H+16$
Very high	46	29	12	25	23	$H+20$
Extreme	54	36	15	30	26	$H+23$
Very extreme	65	45	20	40	32	$H+28$

[a] In the case of a buried tank, the tank wall distance is measured to the position on the plan of the tank wall or other items not more than 10 m below ground level.
[b] In the case of a flare stack, the distances is a function of the flare stack tip height H, as shown in the last column.

The principle of the use of hazard models to set separation distances is now recognized in codes. *Liquefied Petroleum Gas* (IP, 1987 MCSP Part 9) gives separation distances for liquid storage units based on hazard models, as described in Section 10.14.

10.11.6 Liquefied flammable gas

A separation distance of 15 m frequently occurs in codes for the storage of petroleum products, excluding LPG. For LPG and LFG, the separation distances are generally greater. Thus in the ICI *LFG Code* the separation distances recommended between a storage and an ignition source are, for ethylene, 60 m for pressure storage and 90 m for refrigerated storage, and for C_3 compounds, 45 m for both types of storage. The general approach there taken is that there is significant risk of failure for a refrigerated storage but negligible risk for a pressure storage vessel.

Separation distances are also implied in the ICI *Electrical Installations in Flammable Atmospheres Code* (the ICI *Electrical Installations Code*) (ICI/RoSPA 1972 IS/91) in that the code gives guidance on the radius of the electrical area classification zone from potential leak points. For a pump with a mechanical seal containing a flammable liquid with a flashpoint below 32°C the radius of Zone 2 depends on the liquid temperature as follows:

Liquid temperature (°C)	*Radius of Zone 2* (m)
$\leqslant 100$	6
100–200	20
> 200	30

10.12 Hazardous Area Classification

Plant layout has a major role to play in preventing the ignition of any flammable release which may occur. This aspect of layout is known as 'area classification'. One principal type of ignition source is electric motors, and area classification has its origins in the need to specify motors with different degrees of safeguard against ignition. As such, the practice was known as 'electrical area classification' and was usually performed by electrical engineers. The extension of this practice to cover the exclusion of all sources of ignitions is known as 'hazardous area classification' and is generally performed by chemical engineers.

Hazardous area classification is dealt with in BS 5345: 1977 *Code of Practice for the Selection, Installation and Maintenance of Electrical Apparatus for Use in Potentially Explosive Atmospheres (Other than Mining Applications or Explosive Processing and Manufacture)*, and in a number of industry codes, including the *Area Classification Code for Petroleum Installations* (IP, 1990 MCSP Pt 15).

The process of hazardous area classification involves assigning areas of the site to one of four categories. The international definition of these by the International Electrotechnical Commission (IEC), given in BS 5345: 1977, is:

Zone 0	A zone in which a flammable atmosphere is continuously present or present for long periods.
Zone 1	A zone in which a flammable atmosphere is likely to occur for short period in normal operation.
Zone 2	A zone in which a flammable atmosphere is not likely to occur in normal operation and if it occurs only exist for a short time.

A non-hazardous area is an area not classified as Zone 0, 1 or 2. In the UK, this classification system replaces an earlier system based on three divisions: 0, 1 and 2.

In the USA, hazardous area classification is covered in Article 500 of NPFA 70: 1993 *National Electrical Code* and in API RP 500: 1991 *Recommended Practice for Classification of Locations for Electrical Installations at Petroleum Facilities.*

The purpose of hazardous area classification is to minimize the probability of ignition of small leaks. It is not concerned with massive releases, which are very rare. This distinction is a necessary one, but the difference can sometimes be blurred. Mecklenburgh instances a pump seal which, if it leaks, will generally give a rather small release, but which may on occasion give a leak greater than that from the rupture of a small pipe. Because the leaks considered are small and because small flammable releases burn rather than explode, it is fire rather than explosion with which hazardous area classification is concerned.

Since it is difficult to specify leaks fully in terms of size, frequency and duration, the following grades of leak are defined:

(1) *Continuous grade*: release is continuous or nearly so.
(2) *Primary grade*: release is likely to happen regularly or at random times during normal operation.
(3) *Secondary grade*: release is unlikely to happen in normal operation and in any event will be of limited duration.

Broadly speaking, continuous, primary and secondary grade releases equate to Zones 0, 1 and 2, respectively.

Hazardous area classification proceeds by identifying the sources of hazard, or potential leak points, and the sources of ignition. Typical leak points include flanges, seals, sample points and temporary connections; typical ignition sources include electric motors, burners and furnaces, engines and vehicles.

There are three main strategies available for the control of ignition sources: prevention, separation and protection. The approach to hazardous area classification based on these strategies is broadly as follows. First the potential leak sources are identified. The characteristics of the leak are defined, for start-up, shut-down and emergency conditions as well as normal operation, and the grade of leak assigned. For each leak point consideration is given to reducing or eliminating any leak. Guidance on separation distances is then used to determine the area around the leak source from which ignition sources should be excluded. Next the ignition sources near the leak point are identified. For each ignition source in turn, consideration is given to the possibility that it can be eliminated or moved. Where this is not applicable, the zone is specified and appropriate protection of the ignition source is determined. For electrical equipment this means specifying the type of safeguarding appropriate to the zone.

A check may be made on the separation distances used and on the degree of protection required by modelling the dispersion of the leak. Consideration should also be given to the effect of any pool fire arising from flammable liquid released at the leak point. In some instances this may require an increase in the separation distance or the use of protection measures such as insulation or water sprays.

The control of ignition sources reduces the risk of injury to personnel and the risk of property damage. The extent to which the plant design is modified for reasons of hazardous area classification is governed for personal injury by the usual risk criteria and considerations of what is reasonably practicable, and for property damage by economic considerations. In cases where it is property damage which is the issue, it may be preferable to accept a certain risk rather than to undertake unduly expensive countermeasures.

The outcome of this exercise for all the ignition sources identified is the definition of the zones in three dimensions for the whole plant. Drawings are produced showing these zones in plan and elevation, both for individual items of equipment and for the plant as a whole. A typical plan drawing is illustrated in Figure 10.4.

Hazardous area classification provides the basis for the control of ignition sources both in design and in operation. A further discussion of hazardous area classification is given in Chapter 16.

10.13 Hazard Assessment

In the methodology for plant layout described by Mecklenburgh (1985) hazard assessment is used at several points in the development of the layout. In each case the procedure is essentially an iterative one in which hazards are identified and assessed, modifications are made to the design and the hazards are reassessed. The nature of the hazard assessment will vary depending on whether it is done in support of site location, site layout, Stage One plot layout or Stage Two plot layout.

Hazard assessment in support of site location is essentially some form of quantitative risk assessment.

Hazard assessment for site layout concentrates on major events. It provides guidance on the separation distances required to minimize fire, explosion and toxic effects and on the location of features such as utilities and office buildings.

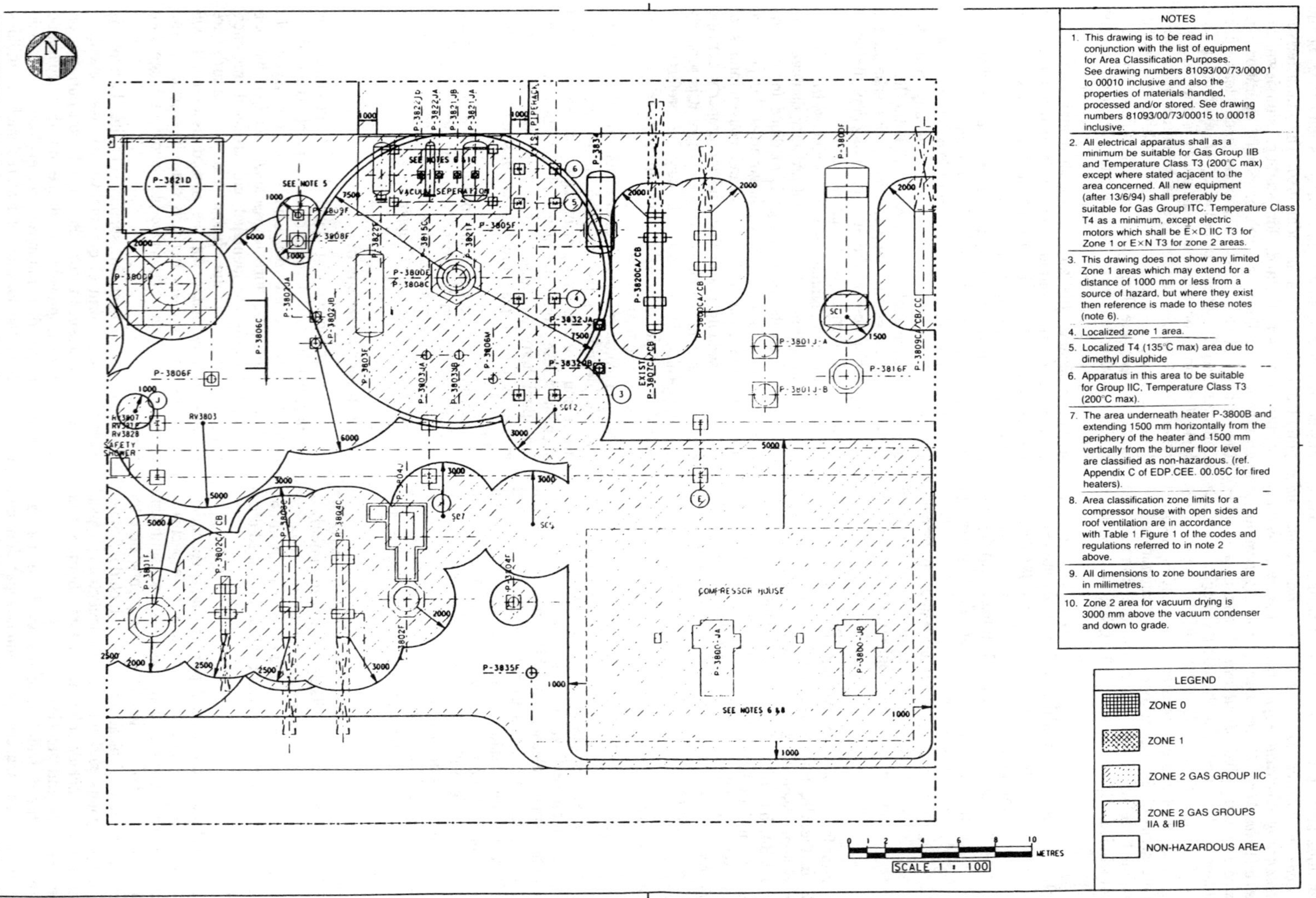

Figure 10.4 *Hazardous area classification drawing (Reproduced with permission from Foster and Wheeler Energy Ltd)*

Hazard assessment for plot layout deals with lesser events and with avoidance of the escalation of such events. It is used as part of the hazardous area classification process and it provides guidance on separation distances to prevent fire spread and for control building location.

At the plot layout level hazard assessment is concerned mainly with flammable releases. It is not usually possible at this level to do much about explosions and toxic releases.

10.14 Hazard Models

10.14.1 Early models

An account has already been given of the early work of Simpson (1971) on the use of models for plant layout purposes. The hazard models described by him include models for two-phase flow and for vapour dispersion and criteria for thermal radiation, as described in Section 10.11.

10.14.2 Mecklenburgh system

A set of hazard models specifically for use in plant layout has been given by Mecklenburgh (1985). A summary of the models in this hazard model system is given in Table 10.9. Some of the individual models are described in Chapters 15, 16 and 17. Although the modelling of some of the phenomena has undergone further development, this hazard model system remains one of the most comprehensive available for its purpose.

10.14.3 IP system

Another, more limited, set of hazard models for plant layout is that given in *Liquefied Petroleum Gas* (IP, 1987 MCSP Pt 9). The models cover:

(1) emission;
(a) pressurized liquid;
(b) refrigerated liquid;
(2) pool fire;
(3) jet flame.

The models include view factors for thermal radiation from cylinders at a range of angles to the vertical and of positions of the target. The requirements for separation distances between storage units are based on thermal radiation flux criteria. These are given in Chapter 22. The code gives worked examples.

10.14.4 Injury and damage criteria

Criteria for injury and damage, principally the latter, are given by Mecklenburgh as part of his hazard model system.

For the heat flux from a flame or fire the tolerable intensities are given as follows:

	Heat flux (kW/m^2)
Drenched storage tanks	38
Special buildings	25
Normal buildings	14
Vegetation	12
Escape routes	6
Personnel in emergencies	3
Plastic cables	2
Stationary personnel	1.5

***Table 10.9** Hazard assessment in support of plant layout: Mecklenburgh hazard model system*

Table No.[a]	
B1	Source term: instantaneous release from storage of flashing liquid (catastrophic failure of vessel) Flash fraction Mass in, and volume of, vapour cloud
B2	Dispersion of flammable vapour from instantaneous release Distance to lower flammability limit (LFL)
B3	Explosion of flammable vapour cloud from instantaneous release Explosion overpressure Damage (as function of overpressure)
B4	Fireball of flammable vapour cloud from instantaneous release Fireball diameter, duration, thermal radiation
B5	Dispersion of toxic vapour from instantaneous release Peak concentration, time of passage Distance to safe concentration, outdoors and indoors
B6	Source term: continuous release of fluid (a) Gas (subsonic) (b) Gas (sonic) (c) Flashing liquid (not choked) (d) Flashing liquid (choked)
B7	Dispersion of flammable vapour jet Jet length, diameter (to LFL)
B8	Jet flame from flammable vapour jet Flame length, diameter Flame temperature, surface heat flux, distance to given heat flux
B9	Dispersion of toxic vapour plume Distance to given concentration Distance to safe concentration, outdoors and indoors
B10	Growth of, and evaporation from, a pool Pool diameter Evaporation rate
B11	Pool or tank fire Flame height Regression rate, surface heat flux View factor
B12	Effect of heat flux on targets Tolerable heat fluxes
B13	Risk criteria Individual risk to employees (as a range) Individual risk to public (as a range) Multiple fatality accident
B14	Explosion overpressure Damage (as function of overpressure) (see also B3)
B15	Dispersion of flammable vapour from small continuous release Jet dispersion: distance to given concentration, to LFL (see B7) Passive dispersion: distance to given concentration (see B9) Jet flame: distance to given heat flux (see B8)
B16	Evaporation and dispersion from small liquid pool Distance to given concentration

B17 Dispersion of flammable vapour from small continuous release in a building
  Jet dispersion: distance to given concentration
  Passive dispersion: distance to given concentration
  Jet flame: distance to given heat flux (see B8)

B18 Evaporation and dispersion from small liquid pool in a building
  Evaporation rate
  Mean space concentration
  Other parameters for (a) horizontal air flow and (b) vertical air flow

[a] In Mecklenburgh (1985).

For a fireball the safe dose is given as $It_B^{2/3} < 47$, where I is the heat flux (kW/m^2) and t_B is the duration of the fireball (s).

For the peak incident overpressure from an explosion the limits which should not be exceeded are given as follows:

	Peak incident overpressure (bar)
Schools	0.02
Housing	0.04
Public roads	0.05
Offices	0.07
Shatter-resistant windows	0.10
Site roads, utilities	0.20
Hazardous plants	0.30–0.40
Protected control room	0.7

10.14.5 Illustrative example

Mecklenburgh illustrates the application of his hazard models by giving for each a scenario and worked example, and for some of the outdoor cases he combines these into an assessment of the effects on site and off site. For this latter assessment he considers a set of scenarios which may be summarized as follows:

(1) Instantaneous release of flashing liquid from storage tank
  (a) Flammable liquid giving rise to unignited vapour cloud, or fireball, or vapour cloud explosion
  (b) Toxic liquid giving rise to toxic gas cloud, in open and around building

(2) Residual liquid in tank
  (a) Flammable liquid giving rise to unignited vapour cloud, pool fire
  (b) Toxic liquid giving rise to toxic gas cloud

(3) Liquid pool from 10 cm leak in tank base following instantaneous release
  (a) Flammable liquid giving rise to unignited vapour cloud, or pool fire
  (b) Toxic liquid giving rise to toxic gas cloud

(4) Continuous release of pressurized liquid from 2.5 cm hole
  (a) Flammable fluid giving rise to passively dispersing unignited vapour cloud, or jet fire
  (b) Toxic fluid giving rise to passively dispersing toxic gas cloud

(5) Continuous release of pressurized liquid from 10 cm hole
  (a) Flammable fluid giving rise to passively dispersing unignited vapour cloud, or jet fire
  (b) Toxic fluid giving rise to passively dispersing toxic gas cloud.

The overall results are summarized in Table 10.10. These results are discussed by Mecklenburgh in relation to both on-site and off-site effects and to the counter-measures which might be taken.

10.15 Fire Protection

Plant layout can make a major contribution to the fire protection of the plant. This has a number of aspects. Plant layout for fire protection is covered in BS 5908: 1990 *Fire Precautions in the Chemical and Allied Industries.* Also relevant are BS 5306:1976 *Fire Extinguishing Installations and Equipment on Premises,* particularly Part 1 on fire hydrants, and BS 5041: 1987 *Fire Hydrant Systems Equipment.* An important earlier code, BS CP 3013: 1974 *Fire Precautions in Chemical Plant,* is now withdrawn. The coverage of BS 5908: 1990 is indicated by the list of contents given in Table 10.11. Accounts of the fire protection aspects of plant layout include those by Simpson (1971), Hearfield (1970) and Kaura (1980a).

Some aspects of plant layout for fire protection may be classed as passive and others as active measures. The former include (1) separation of hazards and targets, (2) measures to prevent fire spread and (3) provision of access for fire fighting; the latter include provision of (4) fire water and (5) fire protection systems. The segregation of hazards and targets and the containment of fire are important aspects of site layout and are considered in Section 10.6. The provision and location of fire water hydrants and fire protection equipment are prominent features of plot layout and are discussed in Section 10.7. This section deals primarily with access, fire water and fire protection equipment. Fire protection is discussed further in Chapter 16.

There are numerous legal requirements concerning fire, fire construction and fire fighting. There should be full consultation at an early stage with the works safety officer and with other parties such as the local authority services, the Health and Safety Executive (HSE) and the insurers.

10.15.1 Fire fighting access

Access is essential for fire fighting. Some basic principles are that it should be possible to get fire fighting equipment sufficiently close to the site of the fire and that there should be access from more than one side. Access should be provided within 18 to 45 m of a hazard and there should be water supplies and hard standing at these access points.

The site should have a peripheral road connected at not less than two points with the public road system. It may be necessary to provide a waiting area for fire fighting vehicles near each main gate. Site roads should be arranged to allow approach to a major fire from two directions. Major process or storage units should be accessible from at least two sides. Access is assisted by a plot size of 100 m × 200 m with approaches preferably

Table 10.10 *Hazard assessment in support of plant layout: illustrative example (after Mecklenburgh, 1985) (Courtesy of the Institution of Chemical Engineers)*

A Off-site effects – summary of distances (m)

	All built-up area	*100 m built-up, then country*	*All country*
1(a) Instantaneous release			
LFL	341	363	377
Fireball, safe dose	463	463	463
Blast, schools	500	500	500
housing	290	290	290
roads	240	240	240
Safe toxic, open	941	1034	1048
Safe toxic, building	35	–	54
1(b) Open tank after instantaneous release			
LFL	At tank	–	At tank
Fire, 1.5 kW/m^2	17	17	17
Safe toxic	Near tank	–	Near tank
1(c) Unconfined pool from 10 cm leak after instantaneous release			
Fire, 1.5 kW/m^2	83	–	83
Pool radius (fire)	9	–	9
LFL	37	–	41
Safe toxic	67	–	96
Pool radius (evap.)	33	–	33 (concrete)
2 2.5 cm steady release under pressure			
LFL (no jet)[a]	16 (28)	–	16 (38)
Fire, 1.5 kW/m^2	43	–	43
Safe toxic (no jet)	60 (72)	–	68 (92)
3 10 cm steady release under pressure			
LFL (no jet)	62 (159)	62 (172)	62 (200)
Fire, 1.5 kW/m^2	172	172	172
Safe toxic in open (no jet)	290 (403)	324 (461)	324 (489)
Safe toxic in building (no jet)	87 (126)	87 (131)	87 (159)

LFL, lower flammability limit.

B On-site effects – summary of distances (m)

1(a) Instantaneous release	
LFL	341
Fireball, safe dose	463
Fireball radius	73
Blast-resistant control rooms	50
Hazardous plants	60–75
Shatter-resistant windows	150
Offices	180
Safe toxic in buildings	35
Safe toxic in open	941

1(b,c) After instantaneous release

	Open tank (m)	*Unconfined pool from 10 cm leak* (m)
LFL	Close	37
Pool radius (fire)	–	9
Fire, drenched tanks	8	30
Special buildings	9	33
Normal buildings	10	39
Vegetation	10	41
Escape routes	12	50
Personnel in emergencies	14	62

Table 10.11 *Continued*

1(b,c) After instantaneous release		
	Open tank (m)	*Unconfined pool from 10 cm leak* (m)
Plastic cables	15	73
Stationary personnel (1.5 kW/m^2)	17	83
Safe toxic limit	3	67
Pool radius (evaporation)	–	33
2 and 3 Steady releases under pressure		
	2.5 cm	*10 cm*
LFL (no jet)	16 (28)	63 (159)
Fire, drenched tanks	29	116
Special buildings	30	120
Normal buildings	32	128
Vegetation	32	128
Escape routes	35	140
Personnel in emergencies	39	156
Plastic cables	41	164
Stationary personnel	43	172
Safe toxic limit in open (no jet)	60 (72)	290 (403)
Safe toxic limit in building (no jet)	60 (72)	87 (126)

LFL, lower flammability limit.
Note Values in brackets are for the case where the release does not take the form of a jet

on all four sides and by spacing between plots and buildings of 15 m.

Access for fire fighting vehicles should be over firm ground, should have sufficient road and gate widths, should give adequate clearance heights and should allow for the necessary turning and manoeuvring. The vehicles requiring access may include heavy bulk foam or carbon dioxide carriers.

10.15.2 Fire water

In a fire, water is required for extinguishing the fire, for cooling tanks and vessels and for foam blanketing systems. The quantities of water required can be large, both in terms of the instantaneous values involved and of the duration for which they may be needed. In some fires water sprinkler systems have been required to operate for several days.

The design of a fire water system requires the determination of the maximum fire water flow which the system should deliver. Some order of magnitude figures are given in the *Refining Safety Code* (IP, 1981 MCSP Pt 3). This states that for a major process fire the fire water flows required might be of the order of 750–1500 m^3/h. The code also quotes for a major fire on a 50 m diameter storage tank a fire water flow of 830 m^3/h for the application of foam to the burning tank and for the cooling of the adjacent tanks. R.B. Robertson (1974a) refers to investigation of the fire water actually used in major process plant fires and quotes water flows in the range of 900–2700 m^3/h.

It is also necessary to specify the length of time for which such fire water flows should be sustained. This specification also may be obtained using the design basis fire approach. Typically this length of time is recommended to be 2–3 hours. Robertson states that study of the time taken to control fires points to a duration of 3 hours.

The fire water requirement may be based on the specification of a design basis fire. Kaura (1980a) suggests that this might be two simultaneous fires, one on a major process unit and the other at the storage tanks. The IP *Refining Safety Code* states, on the other hand, that it is usual to assume that there will be only one major fire at a time. The practical difference between these approaches depends on how generous an allowance is made for the single fire.

The fire water main may be fed from the public water supply, but for large works the public supply may well

Table 10.11 *Principal contents of BS 5908: 1990*

1. General
2. Legal background
3. Principles of initiation, spread and extinction of fire
4. Site selection and layout
5. Buildings and structures
6. Storage and movement of materials
7. Design of process plant
8. Operation of process plant
9. Maintenance of process plant
10. Fire prevention
11. Fire defence
12. Works fire brigades
13. Classification of fires and selection of extinguishing media
14. Fixed fire extinguishing systems
15. Portable and transportable appliances
16. Organization of emergency procedures

not be adequate to provide the quantities of water required. Additional water supplies may be drawn by the fire brigades from rivers, canals, reservoirs or static tanks, but such sources should be near enough to allow suction to be obtained directly, since reliance on relays is likely to involve undue delay. Cooling water should not be used, because loss of cooling on other plants is itself a hazard. Where the public supply is to be used, this should be done in accordance with BS 6700: 1987. The water and fire authorities should be consulted about fire water supplies.

Water supplies for water sprays and sprinklers may be provided in the form of elevated static water tanks. A typical capacity might be such as to supply water for 1 hour. Pumps should be provided to replenish the supplies.

Fire water should be available from hydrants adjacent to the fire hazards on a ring main running alongside the road and located between the road and the plant. The main should preferably be buried under ground. The installation should be generally in accordance with BS 5306: 1976– and fire water hydrants with BS 750: 1984 and BS 5306: Part 1: 1976. The main should take the form of a ring main encircling the plant, with cross-connections and with isolation valves to allow shut-off if a section of the main is damaged.

Hydrant intervals should be 45 m for high risk areas but may be up to 100 m for low risk ones. The distance between the hydrant the plant structure or storage area should be not less than 18 m and may be up to 45 m. The hydrants should be provided with a hard standing and with signs in accordance with BS 5499: Part 1: 1990. The signs should indicate the quantity of water available.

Rising mains should be installed in a building or structure on any floor exceeding 18 m above ground level. Dry rising mains are suitable for heights up to 60 m, but above this height a wet main may be preferable. The inlets on the ground floor to dry riser mains and the outlets on all floors to both types of main should be accessible.

Fire water for sprinkler and water spray systems should be in accordance with BS 5908.

Pipes for fire water supply should be protected against explosion damage. Isolation valves should be provided to prevent loss of fire water from damaged lines and, if these valves are above ground level, they should be protected by concrete blast barriers. The fire water is normally pumped through the main by fixed fire pumps. There should be at least two full capacity pumps with separate power supplies. Cabling for electric pumps should not run through high risk areas or, if this is unavoidable, it should be protected. The location of the pumps is usually determined by that of the source of supply, but they should not be in a high risk area. The fire pumps should be housed to protect them from the weather. In a fire, mobile pumps may sometimes be used to boost the fire main pressure, though their principal use is to supply fire hoses from the main or other sources.

The fire main should be kept pressurized by jockey pumps. A fall in mains pressure should result in automatic start-up of the main fire pump(s) with an indication of this at a manned control point. With regard to the fire water pressure, the IP *Refining Safety Code* states that the system should be able to deliver fire water at the most remote location at a pressure suitable for the fire fighting equipment, which is usually 10 bar. At this pressure the reaction forces on hoses and nozzles are high and make special care necessary.

The quantities of water used in a fire can easily overload the drainage system unless adequate provision is made. This is discussed in Section 10.17.

10.15.3 Fire protection equipment

The other aspects of fire protection of plant, including fire containment by layout, gas, smoke and fire detection, passive fire protection such as fire insulation, and active fire protection such as the use of fixed, mobile and portable fire fighting equipment, are considered in Chapter 16.

It is appropriate to mention here, however, the provision of certain minimal equipment which is generally treated as an aspect of plant layout. Fire extinguishers of the appropriate type and fire blankets should be placed at strategic points on the plant. There should be at least two extinguishers at each point. Some extinguishers should be located on escape or access routes, so that a person who decides to fight the fire using the extinguisher has a route behind him for escape.

Fire equipment should be located so that it is not likely to be disabled by the accident itself, should be accessible and should be conspicuously marked. The main switchgear and emergency controls should have good access, preferably on an escape route, so that the operator does not have to risk his life to effect shut-down.

10.16 Effluents

General arrangements for dealing with effluents are discussed by Mecklenburgh (1973, 1985). Pollution of any kind is a sensitive issue and attracts a growing degree of public control. There should be the fullest consultation with the local and water authorities and the Inspectorate of Pollution in all matters concerned with effluents.

Hazard identification methods should be used to identify situations which may give rise to acute pollution incidents and measures similar to those used to control other hazards should be used to ensure that this type of hazard also is under control.

10.16.1 Liquid effluents

Liquid effluents include soil, domestic and process effluents, and cooling, storm and fire water. Harmless aqueous effluents and clean stormwater may be run away in open sewers, but obnoxious effluents require a closed sewer. One arrangement is to have three separate systems: an open sewer system for clean stormwater and two closed sewer systems, one for domestic sewage and one for aqueous effluent from the plant and for contaminated stormwater.

There are a number of hazards associated with liquid effluent disposal systems such as drains and sewers. One is the generation of a noxious gas by the mixing of incompatible chemicals. Another hazard is that a flammable gas may flow through the drains, become distributed around the plant and then find a source of ignition. This can give rise to a quite violent explosion, or even detonation. A flammable liquid which is immiscible with water flowing through the drains constitutes another hazard. Again it may become distributed around the plant and find an ignition source. If the liquid is already on fire, its entry into the drains may cause the fire to be distributed around the plant.

Other problems with sewers include overloading, blockage and back flow, each of which can be hazardous. Overloading or blockage can result in a liquid fire being floated across to other parts of the site. Some case histories of problems in sewers are given by Anon. (1978 LPB 19, p.10). There are also environmental factors to consider. It is necessary to avoid the discharge of untreated contaminated liquid.

As stated above, process effluents, essentially aqueous, and contaminated stormwater are collected in a common sewer. The liquids discharged to this sewer should be closely controlled. If different effluents are to be mixed together, it should be checked that this can be done safely. Chemical works effluents are quite prone, for example, to generate obnoxious gases.

Water-immiscible flammable liquids should not be allowed to enter the sewers, where they create the hazard of fire or explosion. In particular, open sewers with solvent floating on the water may transmit fires over long distances. There should be arrangements to prevent the entry of such liquids into the sewers. Runoff from the plant area should be routed to interceptors located at the edge of the fire risk area. In order to avoid overloading, use is made of primary interceptors to effect a preliminary separation. Measures may need to be taken to prevent sedimentation in, and freezing of, the interceptors.

It may be necessary to take measures to avoid flooding on process and storage plots. There is need for care to avoid the flooding by effluents of vulnerable points such as pump pits. Flooding of bunds can cause the tanks inside to float. Effluents should not be permitted to run off plant areas onto adjacent sites, or vice versa. If the site slopes or contains a natural water course, additional precautions are needed.

The traditional sewer is the gravity flow type. This should have a gradient and be self-cleaning. Sewer boxes should be used as interconnections with a liquid seal to prevent the transmission of gases and vapours and reduce the hazard of fire/explosion. Where noxious vapours might collect, the sewer box lids should be closed, sealed and vented to a safe place. A suitable point is above grade 3 m, horizontally 4.5 m from platforms and 12 m from furnaces walls. The routing of sewers should be parallel to the road system. They can go under the road, but for preference should be alongside it.

Sewers are considered in more detail in Section 10.17. The sewer system should be settled at an early stage. It is usually not practical to increase the capacity once the plant is built.

10.16.2 Gaseous effluents

Gaseous effluents should be burned or discharged from a tall stack so that the fumes are not obnoxious to the site or the public. The local Industrial Pollution Inspector is able to advise on suitable stack heights and should be consulted. It is also necessary to check whether a high stack constitutes an aerial hazard and needs to be fitted with warning lights.

Flare stacks are a particular problem, because they radiate intense heat and can be very noisy. Quite a large area of ground beneath a flare stack is unusable and is effectively 'sterilized'. A flare stack may have to be relegated to a distant site. A further discussion of flare stacks is given in Chapter 12.

The behaviour of airborne emissions of all types should be carefully considered. Although the prevailing wind is the main factor, other possible troublesome wind conditions should be taken into account. The effect of other weather conditions such as inversions should also be considered.

10.16.3 Solid wastes

Solid waste should preferably be transferred directly from the process to transport. If intermediate storage is unavoidable, care should be taken that it does not constitute a hazard or a nuisance. If combustible solid and solvent wastes are burnt, the incinerator should be convenient to the process.

10.17 Drain Systems

The main plant sewers are of particular importance and merit further description. As already stated, it is common to have an open clean stormwater sewer and a closed contaminated stormwater sewer. These sewers also carry firewater runoff during fire fighting.

Accounts of sewer systems include those by J.D. Brown and Shannon (1963a,b), Seppa (1964), D'Alessandro and Cobb (1976a) and Anon. (1978 LPB 19, p.10). These systems are also considered by Mecklenburgh (1973, 1985). Stormwater systems are discussed by Elton (1980), W.E. Gallagher (1980) and G.S. Mason and Arnold (1984).

10.17.1 Clean stormwater system

Clean stormwater is usually collected in an open sewer. The discharge may be to water courses or the sea, or to a holding pond. On large sites it is generally not practical to discharge it to the public system, due to overload of the latter.

10.17.2 Contaminated stormwater system

The contaminated stormwater system consists of the contaminated stormwater sewers together with an impounding basin to hold the contaminated water prior to treatment and discharge. The design of the impounding basin is discussed by Elton (1980) and W.E. Gallagher (1980) and that of the contaminated stormwater sewers themselves by G.S. Mason and Arnold (1984).

First it is necessary to determine the catchment area, or watershed, from which the stormwater will flow onto the plant site. The next step is to characterize the rainfall. A suitable starting point is a rainfall atlas such as the *Rainfall Frequency Atlas* in the USA. The available data may be used to make an estimate of the maximum 24 hour rainfall. For some locations information is available from which the recurrence interval of particular levels of 24 hour rainfall may be determined. Recurrence intervals for rainfall at Houston have been described by Elton (1980) and W.E. Gallagaher (1980).

In principle only a proportion of this rainfall becomes runoff, this proportion being termed the 'runoff coefficient'. Elton quotes typical values of the runoff coefficient of 1.0 for impervious, 0.7 for semi-pervious and 0.4 for pervious surfaces, respectively. The first group includes process pads, paved areas and impervious clays; the second group includes enclosed, sloping, quickly drained shell and gravel paving; and the third group includes sand or gravel beds, flat open fields.

Correlations may be developed, as described by Elton, for the cumulative volumetric flow per unit area as a function of the recurrence interval and the concentration time. The latter is the time from the start of rainfall until the entire area under consideration is contributing.

Relations are available for the concentration time. Elton quotes:

$$T_c = 5 \times 10^{-5} D \qquad [10.17.1]$$

where D is the distance between the point where the rain falls and the location in question (ft) and T_c the time the water takes to reach the latter (days). He recommends that for steeply sloped areas the constant be reduced by a factor of at least 2.

Contaminated stormwater is usually collected in an impoundment basin and then treated before it is discharged. One of the principal problems in the design of the stormwater system is the sizing of this basin. As already indicated, not all the stormwater will necessarily be contaminated. Investigations may be carried out to determine the degree of contamination of stormwater from various parts of the site. These may show that it is sufficient to collect into the impounding basin only the initial fraction of the runoff. If in a particular area contamination does not fall after the first few inches of runoff, this may be an indication that there is a continuous leak. Mason and Arnold suggest that the amount of rainfall which will typically need treatment is the first 0.5 – 1 inches.

For a given recurrence interval, a curve may be constructed for the cumulative runoff over a period of days. The impoundment basin may then be sized as a function of the capacity of the treatment plant. This exercise may be repeated for other recurrence intervals. This then gives the size of the basin required to prevent discharge of untreated stormwater for different recurrence intervals or, alternatively, the frequency of such discharge for a given basin size. The procedure is described by Elton and by Gallagher.

The design of the contaminated stormwater sewers is described by Mason and Arnold. There are two main systems, the gravity flow and the fully flooded systems. In a gravity flow system there is a network of lines running to a collection sump. The lines are sized to run about three-quarters full at the design flow. The process areas have curbs which direct the water to a catchbasin with a sand trap and liquid seal. Liquid flows from the catchbasins to the collection sump. Sand-trap-type manholes are provided for inspection, cleaning and maintenance.

The water flows under gravity and a minimum slope is required. The authors quote a slope of 0.6–0.8% for a 6 inch line. Given the need for a minimum soil cover, this may involve excavation to some depth, which can reach 2 m at the collection sump. This can cause problems, particularly if the water table is high. In some cases it is necessary to resort to lift pumps at intermediate points.

A gravity flow stormwater system allows nearly immiscible liquids such as chlorinated hydrocarbons to accumulate and to contaminate water passing through until they are gradually dissolved. It may also allow a light, nearly immiscible flammable liquid to float on the top of the water and pass through unless liquid seals are installed to prevent this.

The alternative type of system is the fully flooded system. The system is flooded by a dam at the entrance to the collection sump. As water enters, the sewer becomes fully flooded. The catchbasins and manholes used in this case are of the dry-box type. A fully flooded system prevents the passage of flammable vapours and of burning liquids. There is no accumulation of nearly immiscible liquids and thus no contamination of the stormwater by such liquids.

A flooded stormwater system may not be justified if the liquids handled on the plant are not flammable. Such a system may be impractical in a location sufficiently dusty to cause clogging.

The selection of the materials of construction for a fully flooded system is important and is considered by the authors. These need to withstand both corrosion and thermal shock. They also discuss the conversion of a gravity flow system to fully flooded system and give cost comparisons.

10.17.3 Firewater disposal

There should be arrangements for the disposal of fire water, but it is expensive to provide sewers for the very large quantities of water involved, and different views have been expressed on the necessity for this (e.g. Simpson, 1971; Mecklenburgh, 1976). A practical compromise is to design the sewers to take at any rate the initial 'first aid' fire fighting water (R.B. Robertson, 1974a).

There are also different estimates given of the quantities of fire water likely to be involved. Mecklenburgh (1985) states that the allowance for fire water is about five times the volume allowed for the stormwater. Presumably this refers to UK conditions. A different ratio may well apply in other parts of the world.

Consideration should be given to the fire water flow in all sections of the sewer system. The main trunk sewer usually receives water from a relatively large watershed, but branch sewers may well be prone to overloading from large fire water usage on particular parts of the site.

In view of the large quantities of fire water which can be generated, it may well not be practical to design the sewers for these flows, and other methods of disposal may be needed. These include measures to pump it away or to run it off onto other land.

10.18 Shock-Resistant Structures

It is sometimes necessary in the design of structures such as plant and buildings to allow for the effect of shocks from explosions and/or earthquakes. In both these cases there is a strong probabilistic element in the design in that it is not possible to define the precise load to which the plant structure may be subjected. The starting point is therefore the definition of the design load in terms of the relation between the magnitude of the load and the frequency of occurrence.

The full design of shock-resistant structures is beyond the scope of this book, but some limited comments are made here. Further accounts of explosion-resistant structures are given in Chapter 17 and of earthquake-resistant structures in Appendix 15.

The simpler methods of shock-resistant design are those which assume a static load. This approach builds on the expertise in civil engineering on the design of structures for wind loads. Accounts of such design are given in *Wind Forces in Engineering* (Sachs, 1978) and *Wind Engineering* (Cerkmak, 1980).

10.18.1 Explosion-resistant structures

The methods for the design of a structure to withstand explosion, or blast, shocks start from this point in that one of these methods is to design for the equivalent static pressure exerted by the blast wave. This is a rather simplified approach, however, and the alternative method of dynamic analysis may be preferred.

The type of structure which has received most attention in explosion-resistant design is a rectangular building. Accounts of methods of analysing such a structure have been given in *Explosion Hazards and Evaluation* (W.E. Baker *et al.*, 1983), and by Forbes (1982). Guidance on the design of a explosion-resistant control building has been given by the Chemical Industries Association (CIA, 1979). This is described in Section 10.19.

The analysis of tall structures such as distillation columns has not been as fully treated. Work in support of a method for this type of structure has been described by A.F. Roberts and Pritchard (1982) and D.M. Brown and Nolan (1985).

Accounts of explosion-resistant design frequently assume that the blast profile to be considered is that from an explosion of a condensed phase explosive such as trinitrotoluene (TNT). In many process plant applications the event of interest is a vapour cloud explosion, which has a different blast profile. In the dynamic analysis of a structure the blast profile is in effect the forcing function exciting the dynamic system. The shape of this function, therefore, influences the response of the structure.

An important question is the degree of explosion resistance possessed by plant which is designed to normal codes but which is not designed specifically for blast resistance. Experimentally, such plant has withstood an overpressure of some 0.3 bar (5 psi), except where pipework lacked flexibility. The explosion resistance of plant is considered further in Chapter 17.

10.18.2 Earthquake-resistant structures

As stated in Chapter 9, where the earthquake hazard is briefly treated as one of the natural but rare events which may threaten a plant, this phenomenon is not readily handled either at that point or in this section and is therefore relegated to Appendix 15. The account given here is confined to a limited treatment of earthquake-resistant structures.

Accounts of earthquake-resistant design are given in *Fundamentals of Earthquake Engineering* (Newmark and Rosenblueth, 1971) and *Earthquake Resistant Design* (Dowrick, 1977, 1987) and by Alderson (1982 SRD R246). UK conditions are treated in *Earthquake Engineering in Britain* by the Institution of Civil Engineers (ICE, 1985) and by Alderson.

For regions of high seismicity, such as the USA and Japan, the importance of earthquake-resistant design is clear. The earthquake hazard should not, however, be neglected in other regions. Although earthquakes are often associated with fault lines, they are not confined to such zones. Within a given region of relatively low overall seismicity, there will generally be zones of higher and lower seismicity, but quite severe earthquakes may still occur even in the latter, albeit with lower frequency.

Earthquake-resistant design involves consideration of the whole system of soil and structure, and not simply the latter. Bad ground can reduce markedly the resistance to earthquakes. Some principal problems related to soil are soil–structure interaction, soil amplification of the earthquake and soil liquefaction.

For structures such as buildings there are two principal approaches to earthquake-resistant design. The traditional approach is the use of a suitable building code. Perhaps the best known of these is the *Uniform Building Code* (UBC) of the International Conference of Building Officials (ICBO, 1991) in the USA. This code gives an equation for the total lateral shear at the base of, or base shear on, the structure. The equation contains coefficients for the various influencing factors. It gives the horizontal acceleration of the structure, and hence the force to which it is subjected.

The other, more fundamental, approach is to use some form of dynamic analysis. Design of an earthquake-resistant structure by dynamic analysis starts with the definition of a design basis earthquake. This in turn involves deciding on the severity, and hence recurrence interval, of the earthquake. The ground motion characteristics of the earthquake are then defined, utilizing either profiles from real earthquakes of similar severity or standard reference profiles.

Earthquake-resistant design requirements relevant to plant are most advanced in the nuclear industry. In the USA the Nuclear Regulatory Commission requires earthquake-resistant design. It has issued standard earthquake profiles for seismic design and it has had an extensive programme for the seismic qualification of plant.

Design requirements for the UK nuclear industry have been given by the Nuclear Installations Inspectorate (HSE, 1979d). It is required that there be determined for each site two levels of ground motion, that for the operating basis earthquake (OBE) and that for the safe shut-down earthquake (SSE). The OBE is the most severe earthquake which would be expected to occur at least once in the life of the plant and the SSE the most severe which might be expected to occur based on seismological data. The design is required to ensure that the plant is not impaired by the repeated occurrence of ground motions of the OBE level and that it can shut down safely in the face of those at the SSE level.

The earthquake-resistant design of major hazard plants in the UK has been investigated by Alderson (1982 SRD R 246). Essentially, he proposes that the approach adopted should follow broadly that adopted in the UK nuclear industry. Some US codes for process plant contain seismic design requirements. An example is NFPA 59A: 1990 for liquefied natural gas.

As for explosions, so for earthquakes an important question is the degree of resistance possessed by plant which is designed to normal codes. Evidence from earthquake incidents is that failures which do occur are mainly due to overturning moments, causing yielding of anchor bolts and buckling of storage tank shells and to lack of flexibility in pipework, and that fracture of mains may occur. Failure of storage spheres at the Paloma Cycling Plant in the earthquake at Kern County, California, in 1952 led to a major vapour cloud explosion and fire (Case History A20). Generic studies of the seismic resistance of storage tanks and spheres have shown certain vulnerabilities in larger earthquakes. The earthquake resistance of plant is considered further in Appendix 15.

10.19 Control Buildings

Until the mid-1970s there were few generally accepted principles, and many variations in practice, in the design of control buildings. Frequently the control buildings constructed were rather vulnerable, being in or close to the plant and built of brick with large picture windows.

10.19.1 Flixborough

The Flixborough disaster, in which 18 of the 28 deaths occurred in the control building, caused the Court of Inquiry to call for a fundamental reassessment of practice in this area.

The control building at Flixborough has been described by V.C. Marshall (1976a). It was constructed with a reinforced concrete frame, brick panels and considerable window area. It was $2\frac{1}{2}$ storeys high in its middle section, the $1\frac{1}{2}$ storeys over the control room consisting of a half-storey cable duct and a full-storey electrical switchgear room. The control room was part of a complex of buildings some 160 m long, which also housed managers' offices, a model room, the control laboratory, an amenities building and a production block.

This building complex was 100 m from the assumed epicentre of the explosion and was subjected to an estimated overpressure of 0.7 bar. The complex lay with its long axis at right angles to the direction of the blast. It was completely demolished by the blast and at the control room the roof fell in. The occupants of the control room were presumably killed mainly by the collapse of the roof, but some had been severely injured by window glass or wired glass from the internal doors. It took mine rescue teams 19 days to complete the recovery of the bodies.

The main office block, which was a 3-storey building, again constructed with a reinforced concrete frame, brick panels and windows, was only 40 m from the assumed epicentre and was also totally demolished.

The implications of the Flixborough disaster for control building location and design have been discussed by V.C. Marshall (1974, 1976a,c,d) and by Kletz (1975e).

10.19.2 Building function

The control building should protect its occupants against the hazards of fire, explosion and toxic release. Much the most common hazard is fire, and this should receive particular attention. There are several reasons for seeking to make control buildings safer. One is to reduce to a minimum level the risk to which operators and other personnel are exposed. Another is to allow control to be maintained in the early stages of an incident and so reduce the probability of escalation into a disaster. A third is to protect plant records, including those of the period immediately before an accident.

It is sometimes suggested that another aim should be to equalize the risks to those inside and outside. Since those outside tend to be less at risk in an explosion, this means in effect reducing the risk to those inside. This objective is not self-evident, however. The philosophy of risk described earlier is that no one should be subjected to more than a specified risk, not that the risk should be equal for all. In any event, before designing a control room, it is necessary to be clear as to what the objectives are.

10.19.3 ACMH recommendations

Control building location and design is one of the topics raised by the Court of Inquiry on the Flixborough disaster and considered in the First and Second *Reports* of the Advisory Committee on Major Hazards (ACMH) (Harvey 1976, 1979b). The recommendations of the committee are that control rooms which may be subject to explosion should not be built in brick with large picture windows, but in reinforced concrete with small, protected windows.

10.19.4 Control facilities

There is a tendency for control rooms to become part of a complex of facilities, as the buildings at Flixborough illustrate. As a result, more people are exposed to hazard than is necessary and/or the buildings must be of more elaborate and expensive construction. Some of the additional rooms often associated with the control room include computer room, locker room, mess room, toilets, supervisors' offices, analytical laboratories, test rooms, instrument workshops, electrical relay and switchgear rooms.

The proper policy is to build a secure control room in which the functions performed are limited to those essential for the control of the plant and to remove all other functions to a distance where a less elaborate construction is permissible. The essential functions which are required in the control room are those of process control. There are other types of control which are required for the operation of the plant, such as analytical control and management control, but they need not be exercised from the control room. Thus other facilities such as analytical laboratories, amenities rooms, etc., should be located separately from the control room.

The control room should not be used as a centre to control emergencies. There should be a separate emergency control centre, as described in Chapter 24. The control room should also not be used as an emergency assembly point or refuge room.

10.19.5 Location

The ability of a control building to give protection against a hazard such as an explosion depends not only on its design but also on its location. The siting of a control building can therefore be as important as its construction.

It is good practice to lay plant out in blocks with a standard separation distance. The control building should be situated on the edge of the plant to allow an escape route. Recommended minimum distances between the plant and the control building tend to lie in the range 20–30 m. If hazard studies indicate, however, that the standard separation distance may not be adequate, the distance between the control building and the plant should, of course, be increased.

The control building should not be so near the plant that its occupants are at once put at risk by a serious leak of flammable, toxic or corrosive materials. On the other hand, increasing the distance from the process may make the operators less willing to get out on the plant. Managers are generally opposed to control rooms which are too remote. This is important, because active patrolling by the operator is one of the main safeguards against plant failures.

A control building should not be sited in a hazardous area as defined in BS 5345: 1977–. Further guidance on location of the control building is given below.

10.19.6 Basic principles

Arising from the experience of Flixborough, V.C. Marshall (1976a) has suggested certain principles for control building design which may be summarized as follows:

(1) The control room should contain only the essential process control functions.
(2) There should be only one storey above ground.
(3) There should be only the roof above the operator's head. The roof should not carry machinery or cabling.
(4) The building should have cellars built to withstand earthshock and to exclude process leaks and should have ventilation from an uncontaminated intake.
(5) The building should be oriented to present minimum area to probable centres of explosion.
(6) There should be no structures which can fall on the building.
(7) Windows should be minimal or non-existent and glass in internal doors should be avoided.
(8) Construction should be strong enough to avoid spalling of the concrete, but it is acceptable that, if necessary, the building be written off after a major explosion.

The control building should be constructed in ductile rather than brittle materials. Ductile materials include steel and reinforced concrete. Brick and masonry are brittle materials.

The standards of construction of control rooms subject to the hazard of an explosion, and particularly that of a vapour cloud explosion, have been discussed by a number of workers, including W.J. Bradford and Culbertson (1967), Kletz (1975e, 1980h), Langeveld (1976), Balemans and van de Putte (1977), the CIA (1979), Forbes (1982), Beigler (1983) and Crossthwaite and Crowther (1992). The various approaches proposed are now described.

10.19.7 Bradford and Culbertson method

Bradford and Culbertson (1967), of Esso, in an early paper recommended that the control building be located 100 ft (30 m) from sources of hazard and that it should be in reinforced concrete and should be designed for a 3 psi (0.2 bar) static overpressure. This was based on a 1 te TNT equivalent explosion which would give a peak overpressure of 15 psi (1 bar) at 30 m, combined with an analysis showing that a building designed for 3 psi static pressure would resist a diffraction overpressure of 15 psi and a reflected overpressure of up to 45 psi with only light to moderate structural damage.

10.19.8 Langeveld method

Langeveld (1976) has described the evolution of control building design at Shell, which has been influenced not only by the disaster at Flixborough, but also by the earlier one at Pernis. The explosion at Pernis was estimated to have been equivalent to 20 ton of TNT. He emphasizes that the explosion pressure which the control building must withstand cannot be defined with any precision in the current state of knowledge and that the important thing is to have a good and practical design. In the design described the control building is a reinforced concrete structure capable of withstanding an equivalent static pressure of 10 ton/m^2 (1 bar) on the walls and 2.5–5.0 ton/m^2 (0.25–0.5 bar) on the roof slabs. The purpose of quoting the static pressures is to give the engineer a basis on which to design a building of reasonable dimensions rather than to withstand any particular expected overpressure. The front elevation of a typical Shell control centre is shown in Figure 10.5.

There is perhaps rather less agreement on control room windows. One view is that there should be no windows at all. There were in fact a number of windowless control rooms before Flixborough. But the more common view among plant managers is that it is highly desirable to see the plant from the control room. Langeveld states that this view has been supported by ergonomists.

In the design described by Langeveld, the total window area does not exceed 7% of the front wall and the individual frame sizes are not larger than 0.25 m^2 (e.g. 0.3 m × 0.8 m). A special laminated glass is used consisting of two layers of normal glass, each at least 3 mm thick and a polyvinyl butyral intermediate layer of 1.9 mm. The glass pane is held in a strong flexible way in a window frame with rebates at least 30 mm high, and a catch bar is installed inside the building in the centre of the window behind the glass pane to minimize the effects if the glass is blown inside.

10.19.9 Baalemans and van de Putte method

A guideline for control building design has been described by Balemans and van de Putte (1977) of the Ministry of Social Affairs in the Netherlands. The

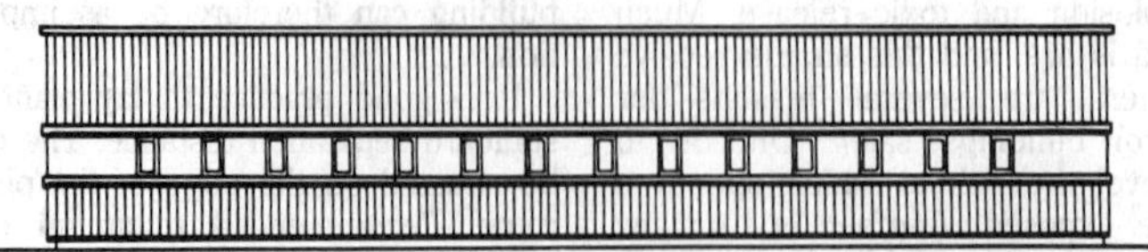

Figure 10.5 *A typical Shell control centre (Langeveld, 1976) (Courtesy of the Institution of Chemical Engineers)*

requirement is that the external walls of the building be capable of withstanding an external static load of 0.3 bar and the roof one of 0.2 bar.

10.19.10 CIA method

An Approach to the Categorisation of Process Plant Hazard and Control Building Design by the CIA (1979) (the CIA *Control Building Guide*) gives a method for the assessment of the explosion hazard from, and for the categorization of, a plant, and guidance on the location of the control building and on the design of the building to resist blast and also to provide protection against toxic release.

Protection of the control building is necessary for the safety of the personnel, the maintenance of control of the plant and the preservation of plant records. The *Guide* starts from the premise that neither the modelling of vapour cloud explosions nor the technique of hazard assessment are mature enough to utilize. It emphasizes that there is no justification for assuming that all plants are subject to a vapour cloud explosion hazard.

The approach suggested is to examine the plant and to identify the points where a major leak may occur. Such a leak is improbable from pressure vessels or large diameter pipes; smaller pipes are more likely sources. The *Guide* enumerates the design techniques which can be used to reduce the size of any leak, such as limitation of inventory, reduction of pressure and temperature, use of high standards of design and construction for flanges, bellows and fittings and of appropriate materials of construction, and devices for leak detection and isolation.

The *Guide* proposes that the duration of the leak be taken as 5 minutes, based on the assumption that this is the time required for detection, diagnosis, decision and action. It recommends that measures be taken to ensure that the leak does not last longer than this, including cessation of heat input, depressurization and isolation. Installation of means for remotely operated isolation allows the release duration to be reduced to 3 minutes. In some cases, exhaustion of the inventory may determine the duration.

The categorization is based on the likely sources of emission, the mass of the release and the probability of a vapour cloud explosion. Three categories of plant are defined. Category I plants are high hazard plants. Category II plants are other plants handling flammables. Most medium-sided, moderate pressure plants containing flammables fall in this category. Category III plants are those handling materials which cannot produce a flammable vapour cloud.

Allocation to category is effected by dividing the plant into sections, making a qualitative assessment of the hazard, estimating the mass of the flammable vapour in the cloud from a potential release and then applying the following categorization:

	Mass of flammable vapour (te)
Category I	>15
Category II	2–15
Category III	<2

These values may be varied for materials of high or low reactivities.

For Category I situations the control building should be located as far as practical, but in any case not less than 30 m, from the nearest source of hazard with a release potential of 15 te, preferably at the edge of the plot and positioned to avoid funnel effects which could give rise to rapid flame acceleration. The number of personnel using it as their workbase should be kept to a minimum, consistent with operational requirements. Heavy equipment should not be located on or over the main roof. The building should be designed to withstand one explosion at or near ground level. This means that the building should be in working condition after the explosion, although the structure may need to be rebuilt. It should have appropriate fire protection.

The *Guide* states that analysis of incidents indicates that the approximate parameters of a typical vapour cloud explosion are a peak overpressure of 0.7 bar and a duration of 20 ms, but that some theoretical studies point to a peak overpressure of 0.2 bar and duration of 100 ms. For partially confined explosions these parameters may be 1.0 bar and 30 ms, but the evidence is conflicting and the mechanisms poorly understood.

The design criteria for a control building in the *Guide* are intended to give a building which will withstand peak overpressure and duration combinations of either 0.7 bar and 20 ms or 0.2 bar and 100 ms. The design for these conditions is conservative and the building should in fact withstand an explosion where the combination is 1.0 bar and 30 ms. This statement is qualified where the material used is other than reinforced concrete, there are long spans or elements with very short natural period.

The outline guidance for the detailed design of the control building is as follows. The building should normally have a single storey. The materials used should be ductile, and brick, masonry or unreinforced concrete should not be used. The building should conform to the normal building codes.

For normal loadings the *Guide* states that the building should conform to BS CP 3 Chapter V: Parts 1 and 2 for dead and imposed loads and wind loads, respectively. It should be noted that, since the *Guide* was written, Part 1 of this standard has been replaced by BS 6399: Part 1: 1984. The loading combinations to be considered are (1) dead + imposed + wind load and (2) dead + imposed + blast load.

The building should be designed for blast loadings 1 (0.7 bar, 20 ms) and 2 (0.2. bar and 100 ms) and checked that it will withstand blast loading 3 (1.0 bar and 30 ms). The walls should be designed with allowance for the reflected pressure and the roof for the incident pressure. Thus for blast load 1 the roof should be designed for 0.7 bar and the walls for 1.75 bar overpressure, both with 20 ms duration, and for blast load 2 the corresponding figures are 0.2 bar, 0.3 bar and 100 ms. The suction phase may be ignored, provided structural rebound is taken into account.

The *Guide* describes a design method based on dynamic analysis. It gives the following relation for the dynamic resistance of a structural element:

$$R = P/\eta \qquad [10.19.1]$$

with

$$\eta = \frac{\tau(2\delta - 1)^{\frac{1}{2}}}{\pi t_0} = \frac{(2\delta - 1)t_0}{2\delta(t_0 + 0.77\tau)} \quad [10.19.2]$$

$$\delta = X_m/X_y \quad [10.19.3]$$

where P is the peak value of the applied blast load, R is the dynamic resistance, t_0 is the duration of the blast load, X_m is the maximum allowable dynamic displacement, X_y is the effective yield displacement, τ is the fundamental period of vibration, and δ is a parameter. X_y is based on the equivalent elastic–plastic load deformation relationship and is the effective displacement at which plastic deformation begins. The *Guide* gives the limits on the ratio δ to be used for steel and reinforced concrete. It also gives guidance on the standards and strengths to be used for steel and reinforced concrete, on foundations, on additional structural requirements and on external doors and openings.

For Category II situations the design philosophy given in the *Guide* is to follow normal building standards but to minimize sensitivity to blast and to arrange structural details so that large plastic deformations occur before collapse. The building should generally be single storey. The materials used should be ductile. Guidance is given on structural, external and internal details.

10.19.11 Kletz method

Kletz (1975e, 1980g), of ICI, has given guidance on control building design. The later guidance (Kletz, 1980g) applies to buildings in general as well as control buildings and is based on the distance between the source of hazard and the building as shown in Figure 10.6. This figure is itself based on a correlation of peak overpressure as a function of distance and mass of hydrocarbon released. The upper boundaries of zones B, C, D and E correspond to peak overpressures of 0.35, 0.2, 0.1 and 0.03 bar, respectively.

Kletz recommends that no building should be nearer the plant than 20 m, but also that a control building be no further than 35 m from the plant. His recommendations for building strength relate to occupied buildings, which he equates to those occupied by at least one person for 20 h/week or more.

Within zone B a building should be designed for a peak incident overpressure of 0.7 bar and a duration of 20 ms. This design allows for the building being within the cloud, since the peak incident overpressure is unlikely to exceed 0.7 bar even in the cloud. No other hazardous plant should be located within this zone and there should be no site roads, though there may be plant roads.

Within zone C a building should be designed for the peak incident overpressure which might occur at the point where it is located. The overpressure range given is between 0.70 and 0.20 bar. There should be no low pressure storage tank in this zone unless it is specially designed or the contents are harmless.

Within zone D a building should be designed for the peak incident overpressure which might occur at the point where it is located, which means for overpressures between 0.20 and 0.10 bar. There should be no public roads in this zone.

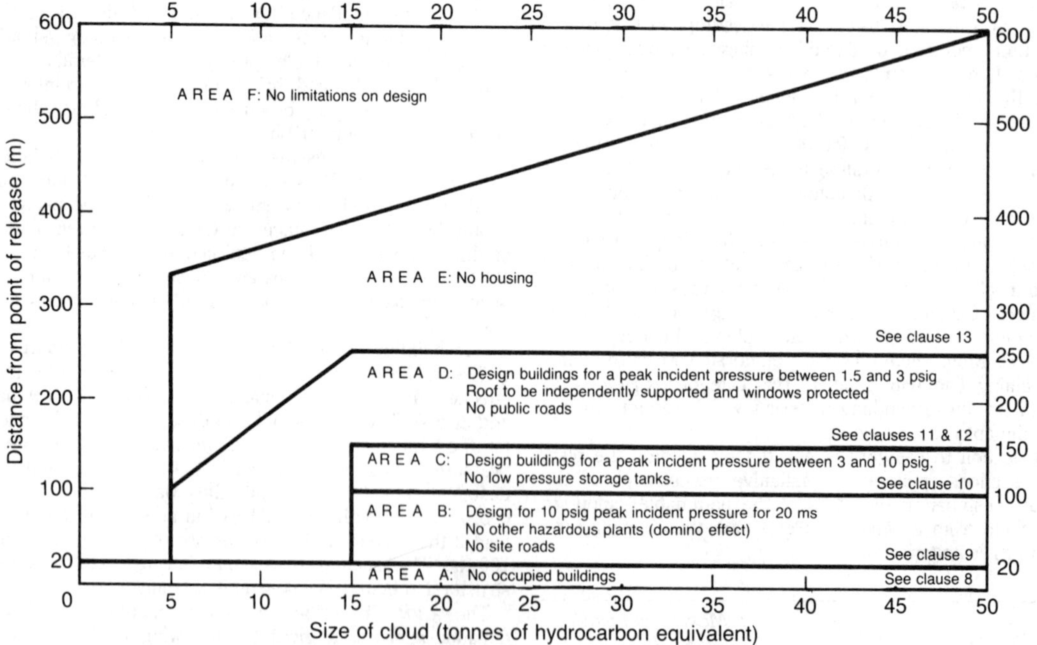

Note: Area E limitations apply in areas D–A, and so on.

Figure 10.6 *Guidelines for location of buildings where a vapour cloud explosion hazard exists (Kletz, 1980h). The clause numbering refers to the appendix of the original paper. (Courtesy of the American Institute of Chemical Engineers)*

Housing should be excluded from zone E.

10.19.12 Forbes method

Forbes (1982) describes an approach to the design of a blast resistant control building based on dynamic analysis. He takes as his design explosion one equivalent to 1 ton of TNT exploding at a distance of 100 ft (30 m) and designs for slight to moderate damage. He gives a table showing the degree of damage to be expected both with the blast-resistant design and with conventional design as a function of mass of TNT and distance from the explosion.

The design peak pressures on the building are then taken as 10 psi (0.7 bar) on the roof and 25 psi (1.7 bar) on the walls, both with a duration of 20 ms. The author also tabulates values for other structural elements. The dynamic design approach described by Forbes is broadly similar to that given in the CIA Guide.

10.19.13 Beigler method

Beigler (1983) has described a method developed in Sweden for the location and construction of buildings based on the energy conversion in the explosion, as shown in Table 10.12. Only essential buildings should be located in Zone B; this is likely to include the control building. In Zone A buildings are to be designed to the normal building code, but with additional static load strength. In Zone Z conventional design applies.

For a control building the requirement is that it be designed to withstand a static load of 0.8 bar or an impulse load of 20 mbar-s, and also that it meet a pressure impulse curve criterion as described by the author.

10.19.14 Crossthwaite and Crowther method

Crossthwaite and Crowther (1992) argue that with the improved understanding of material reactivity and of the effects of confinement of the cloud and with the vapour cloud explosion models now available, an approach to control building design based simply on the mass released has become questionable, and propose instead one based on hazard assessment. They describe an essentially conventional hazard assessment method with identification of release sources and estimation of the frequency and consequences of releases. For the latter, the relevant part of the vapour cloud is taken as that part which is confined within plant structures. The procedure yields a set of site plans with frequency contours for different levels of peak overpressure.

The construction of the control building is governed in a vapour cloud explosion by the impulse of the blast, which the authors obtain by assuming a constant duration of 50–100 ms. Three levels of construction are considered. A conventional brick building is unlikely to withstand a peak overpressure in excess of 0.15 bar. A fully blast resistant building to the CIA Category I standard should withstand 0.7 bar. Between these levels of overpressure they suggest a building strengthened to resist between 0.3 and 0.5 bar, which is somewhat stronger than the CIA Category II building.

The method involves making certain assumptions, which are not described, about the relation between building strength and probability of fatality. The location and construction of the control building are based on criteria for risk of fatality to employees, both individual risk and risk to groups. The risk criteria which the authors suggest are a limit of 10^{-4}/year for individual risk and of 10^{-5}/year for the risk of 10 deaths.

The authors give an illustrative example in which for a given location the individual risks are tabulated for control buildings of different strengths.

10.19.15 Detail design

Several of the methods just described cover also the detailed design of features of the control building such as the foundation, the additional structural requirements, the external doors and openings and the internal parts.

Some other aspects of control room construction have been described by Mecklenburgh (1973, 1976). Windows and doors should be positioned so as to minimize the probability of debris from them striking people. The incidence of direct sunlight on the instrumentation should be avoided if possible, as this can make it difficult to read the instruments; in this connection it is an advantage if the windows face north.

A control room should have forced ventilation with an air intake from a clean area. There should also be emergency air supplies to deal with situations such as entry of foul gases through broken windows.

There are different views on the provision of rooms underground in the control centre. An underground room offers protection against explosion, but has some serious drawbacks. Process leaks of liquid or heavy vapour may get in. It discourages an active patrolling policy and it is less pleasant to work in.

The switch room is often located under the control room to save cabling, but if there is a danger of heavy vapours collecting in it, it may be preferable to put the room at ground level. The integrity of cables is important. These should survive about the same level of accident as the control room itself.

Table 10.12 *Zone distances for building location and construction (after Beigler, 1983)*

Energy conversion in explosion (GJ)	*Distance*		
	Zone B (m)	*Zone C* (m)	*Zone Z* (m)
2	15	125	325
5	20	150	400
10	30	200	500
20	40	250	650
50	60	300	800
100	90	400	1000

10.19.16 Control room layout

The instruments in the control panel are normally laid out in groups by process units and areas. Other grouping criteria include instruments for related variables and instruments used in sequential operations. Control panel displays are discussed in more detail in Chapter 14.

It is recommended (Mecklenburgh, 1973) that the section of the panel carrying recorders and controllers should be no lower than 1.2 m and no higher than 2.1 m with the space up to 2.4 m reserved for indicators, alarms and similar instruments, and that there be a space of 1 m behind the panel and 3 m in front of it.

10.20 Ventilation

Where process plant is located inside a building, ventilation is required to provide a suitable atmosphere for personnel. The plant may generate heat which has to be removed. Any leaks of flammable or toxic materials need to be diluted.

Ventilation is the subject of BS 5925:1990 *Design of Buildings: Ventilation Principles and Designing for Natural Ventilation*. The previous version, BS 5925: 1980, is that referred to in much of the ventilation literature; the differences are not great.

The problem of leaks of gases which are lighter or heavier than air is considered by Leach and Bloomfield (1973) and M.R. Marshall (1983).

10.20.1 Legal requirements

Legal requirements for ventilation include those of the Building Regulations, the Factories Act 1961 and the Offices, Shops and Railway Premises Act 1963.

10.20.2 Ventilation functions

The main function of ventilation is foremost to maintain a suitable atmosphere for personnel. This has a number of aspects, including control of the ambient air in respect of (1) respiration, (2) humidity, (3) thermal comfort and (4) contaminants.

For respiration it is necessary to maintain a minimum oxygen content in the expired air and a maximum carbon dioxide content in the room. The threshold limit value for the latter is 0.5% and this is the governing factor, since the air flows required to maintain this concentration greatly exceed those needed for the oxygen criterion, which is an oxygen concentration of 16.3% in the expired air. The air flows required for an adult male to maintain the carbon dioxide concentration are 0.8 l/s when seated quietly and 2.6–3.9 l/s when performing moderate work; the corresponding air flows to maintain the oxygen concentration are 0.1 l/s and 0.3–0.35 l/s.

Low relative humidity can cause respiratory discomfort, and high relative humidity can cause condensation and mould growth.

There are various types of contaminant, such as odours, which may be present and which need to be removed by ventilation. There may also be smoke resulting from smoking where this is permitted. There may be fugitive emissions from the plant of flammable or toxic chemicals. In accident conditions, there may be a leak of a flammable or toxic material or smoke from a fire. Where there are fuel burning appliances, ventilation is required to supply air for these.

10.20.3 Ventilation systems

Ventilation may be provided either by natural or mechanical means. In deciding between the two means, the main factors to be considered are the quantity and the quality of the air and the control of the air flow.

Natural ventilation can, in theory, supply any required quantity of air, but there are practical limitations. It can supply air of good quality provided it can draw from a source of clean air, but if the air has to be filtered it is necessary to resort to mechanical ventilation. Natural ventilation systems give air flow rates which vary with the weather conditions and can be designed only on a probabilistic basis. Thus where large quantities of air are required, where it is necessary to filter the inlet air and/or where control of the air flow is needed, it is necessary to use mechanical ventilation.

Natural ventilation has the limitations that: neither the air flow for ventilation nor the conditions within the space ventilated can be controlled at all closely; that users may close off air inlets; and that in some cases building layout inhibits good ventilation.

BS 5925: 1990 lists various situations where mechanical ventilation is an absolute necessity and others where it is desirable. Included in the former are factories where it is essential to remove dust, toxic or other noxious contaminants near their source and in the latter factories where it is necessary to remove hot air, moisture and contaminants generally.

The main driving forces for natural ventilation are the pressure differences caused by wind against the side of the building and the temperature difference between the ambient air and the air in the building. Use may also be made of the pressure difference of a column of gas in a chimney.

Locations of the air inlets and outlets are illustrated in BS 5925: 1990 for the two types of ventilation and are as follows. For natural ventilation by wind or temperature difference, the air inlet is set low in a wall on one side of the ventilated space and the outlet high in an opposite wall, but for temperature difference there is also shown a combined inlet/outlet system in one wall with the inlet just below the outlet. For mechanical ventilation the air inlet is low in one wall and the outlet high in the opposite one, but there is also a combined inlet/outlet system shown set in the roof.

10.20.4 Ventilation rates

Ventilation rates may be expressed in several ways. They include volumetric flow per person, volumetric flow per unit floor area and number of air changes per unit time. BS 5925: 1990 gives recommended ventilation rates for various occupancies. For factories the recommended rate is 0.8 l/s per m^2 of floor area. Where the ventilation rate is set by the need to remove a contaminant, the number of air changes per unit time is an appropriate measure. The control of contaminants is considered below

In setting the ventilation rate the effect of air movement on comfort should be considered. Criteria are given in BS 5925: 1990. An air velocity of 0.1 m/s is about the lower limit of perceptibility and one of 0.3 m/s about the upper limit of acceptability, except perhaps in summer. Allowance for air movement may be made be increasing the temperature in the ventilated space using the correlation given in the standard.

10.20.5 Natural ventilation

Natural ventilation is based mainly on wind or temperature difference. The openings through which the air flow takes place are classified in BS 5925: 1990 as (1) cracks or small openings with a typical dimension less than 10 mm and (2) larger openings. For the latter the air flow is given by:

$$Q = C_d A(2\Delta p/\rho)^{\frac{1}{2}} \quad [10.20.1]$$

where A is the area of opening, C_d is the coefficient of discharge, Δp is the pressure difference and ρ is the density of air. The value of C_d conventionally used is 0.61, which is that for a sharp-edged orifice at high Reynolds numbers.

BS 5925: 1990 gives a method for determining the wind pressure on the surface of the building. This pressure is a function of the shape of the building, the wind speed and direction relative to the building and the presence of other structures which affect the flow. The pressure p at a particular point is:

$$p = p_0 + C_p(0.5\rho u_r^2) \quad [10.20.2]$$

and the mean pressure

$$\bar{p} = p_0 + \bar{C}_p(0.5\rho u_r^2) \quad [10.20.3]$$

where C_p is the surface pressure coefficient, p is the pressure at a particular point (Pa), p_0 is the static pressure in the free wind (Pa), u_r is the reference wind speed (m/s), ρ is the density of the air (kg/m^3) and the overbar indicates the mean value. The reference wind speed is conventionally taken as the speed of the undisturbed wind at a height equal to that of the building.

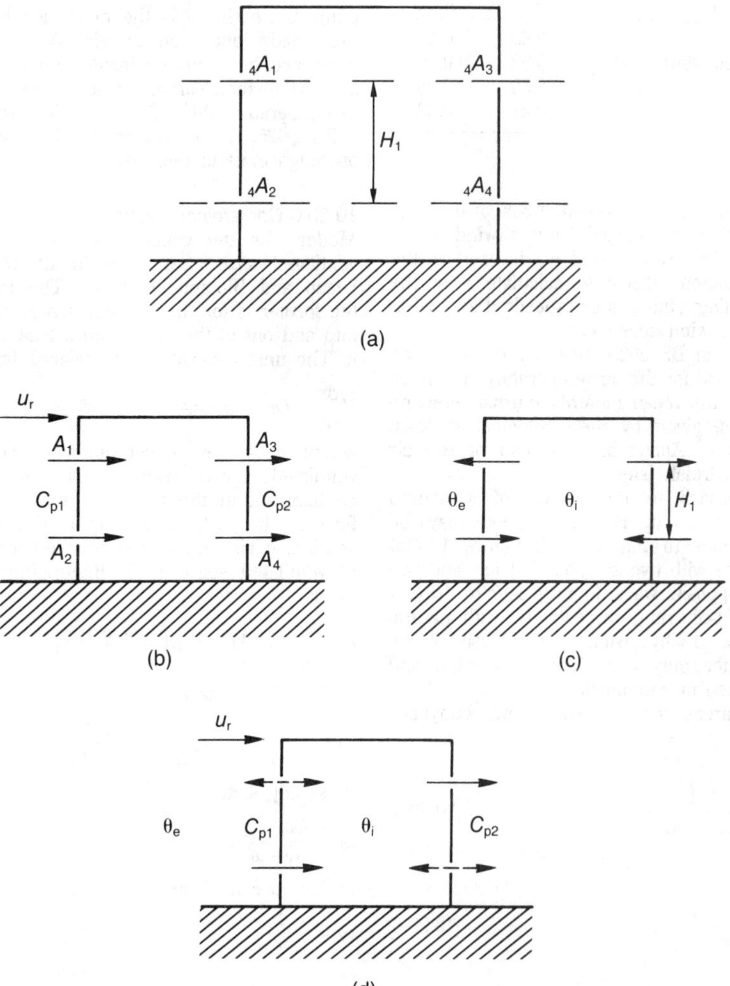

Figure 10.7 Some modes of natural ventilation of a simple building (BS 5925: 1990): (a) ventilated space in a building; (b) wind only; (c) temperature difference only; and (d) wind and temperature difference only (Courtesy of the British Standards Institution)

Values of the surface pressure coefficient C_p are given in BS 5925: 1990, Table 13, based on those in BS CP3 Chapter V, Part 2: 1972. BS 5925: 1990 gives illustrative values of the pressure difference $(p - p_0)$ for values of C_p in the range 0.1–1.0.

For the wind speed at a particular height the relation given is:

$$\frac{u}{u_m} = Kz^a \qquad [10.20.4]$$

where u is the wind speed at height z, u_m is the mean wind speed at 10 m height in open terrain, K is a constant, and a is an index. The values given for the two latter are:

	K	a
Open flat country	0.68	0.17
Country with scattered wind breaks	0.52	0.20
Urban	0.35	0.25
City	0.21	0.33

The wind speed u_{50} which is exceeded 50% of the time at the site in question is obtained from a wind speed map of the country. This wind speed u_{50} is used as the value of u_m in Equation 10.20.4 to calculate u at the heitht of the building, this latter speed then being termed the reference wind speed u_r.

The method given in BS 5925: 1990 for the temperature difference is to use for the air temperature the mean monthly values, and the mean monthly diurnal temperature variation values, given by meteorological stations. The standard gives in Appendix E a map of the air temperature for the British Isles.

The general approach to the design of a natural ventilation system given in BS 5925: 1990 may be illustrated by reference to Figure 10.7. Figure 10.7(a) shows a simple space with two air inlets set low and two outlets set high in opposite walls. Figures 10.7(b)–10.7(d) show the air flows occurring, respectively, with natural ventilation in Case 1 by wind only, Case 2 by temperature difference only and Case 3 by wind and temperature difference in combination.

The equivalent areas for the wind and buoyancy mechanisms are:

$$\frac{1}{A_w^2} = \frac{1}{(A_1 + A_2)^2} + \frac{1}{(A_3 + A_4)^2} \qquad [10.20.5]$$

$$\frac{1}{A_b^2} = \frac{1}{(A_1 + A_3)^2} + \frac{1}{(A_2 + A_4)^2} \qquad [10.20.6]$$

where A_b is the equivalent area for ventilation by temperature difference only (m^2), A_w is the equivalent area for ventilation by wind only (m^2) and areas A_1–A_4 (m^2) are as shown in Figure 10.7.

The relations for ventilation are:

$$Q_w = C_d A_w u_r (\Delta C_p)^{\frac{1}{2}} \qquad \text{wind only} \qquad [10.20.7]$$

$$Q_b = C_d A_b \left(\frac{2\Delta\theta g H_1}{\bar{\theta}}\right)^{\frac{1}{2}} \qquad \text{temperature difference only} \qquad [10.20.8]$$

$$Q = Q_b \qquad \phi < 0.26 \qquad \text{wind and temperature difference combined} \qquad [10.20.9]$$

$$Q = Q_w \qquad \phi > 0.26$$

with

$$\phi = \frac{u_r/(\Delta\theta)^{\frac{1}{2}}}{(A_b/A_w)^{\frac{1}{2}}(H_1/\Delta C_p)^{\frac{1}{2}}} \qquad [10.20.10]$$

where ΔC_p is the differential pressure coefficient, Q is the volumetric air flow (m^3/s), θ is the absolute temperature (K), $\bar{\theta}$ is the mean absolute temperature of the inside and outside air, $\Delta\theta$ is the temperature difference between the inside and outside air (K), ϕ is a discrimination parameter, and subscripts b and w refer to temperature difference and wind, respectively.

BS 5925: 1990 also treats the case where the air openings exist in one wall only.

10.20.6 Contaminant control

Models for the concentration of a contaminant in a ventilated space are given in BS 5925: 1990 and by Leach and Bloomfield (1973). The model described in the former is for the situation where there is an air flow into and out of the space and a leak of contaminant into it. The unsteady-state mass balance is:

$$V\frac{dc}{dt} = Qc_e + q - Qc \qquad [10.20.11]$$

where c is the concentration of contaminant in the ventilated space (v/v), c_e is the concentration of contaminant in the inlet air, (v/v), q is the volumetric flow of the leak (m^3/s), Q is the volumetric flow of ventilation air (m^3/s), t is time (s) and V is the volume of ventilated space (m^3). Integrating Equation 10.20.11 yields:

$$c = \frac{Qc_e + q}{Q + q}\left[1 - \exp\left(-\frac{Q+q}{V}t\right)\right] \qquad [10.20.12]$$

If the inlet air is pure:

$$c = \frac{q}{Q + q}\left[1 - \exp\left(-\frac{Q+q}{V}t\right)\right] \qquad [10.20.13]$$

At steady state

$$c_E = \frac{Qc_e + q}{Q + q} \qquad [10.20.14]$$

or for pure inlet air

$$c_E = \frac{q}{Q + q} \qquad [10.20.15a]$$

$$c_E = \frac{q}{Q} \qquad q << Q \qquad [10.20.15b]$$

where c_E is the steady-state concentration of contaminant in the ventilated space (v/v).

The ventilation air requirement assuming pure air is then from Equation 10.20.15a

$$Q = q\left(\frac{1-c_E}{c_E}\right) \quad [10.20.16]$$

If there is no leak, but an initial concentration of contaminant

$$c = c_0 \exp\left(-\frac{Q}{V}t\right) \quad [10.20.17]$$

where c_0 is the initial concentration of contaminant (v/v).

Leach and Bloomfield treat this case and also two others. One is the case where there is a leak flow into the space and a corresponding flow out of it, but no ventilation air flow. The unsteady-state mass balance is:

$$V\frac{dc}{dt} = q - qc \quad [10.20.18]$$

which on integration gives

$$c = 1 - \exp\left(-\frac{q}{V}t\right) \quad [10.20.19]$$

At steady state $c_E = 1$.

The other case is where there is a leak into a sealed space. The unsteady-state mass balance is:

$$V\frac{dc}{dt} = q \quad [10.20.20]$$

which integrates to give

$$c = \frac{q}{V}t \quad [10.20.21]$$

10.20.7 Buoyant or dense gas

The model just described is based on the assumption of perfect mixing. This assumption is not valid for the case where the leak is that of a gas which is buoyant or dense.

The buoyant gas case has been treated by Leach and Bloomfield (1973). The situation which they consider is a leak of such gas into a room with the leak point in the ceiling and with ventilation air coming in through a low inlet and leaving through a high outlet in the opposite wall. Under these conditions the authors postulate the formation of a stratified layer of buoyant gas between the ceiling and the air outlet.

They argue that mixing between two such layers can be almost totally suppressed, even though there is turbulent mixing in both the gas and air layers. The mixing is governed by the Richardson number which is the ratio of work done against gravity to work done by turbulent stresses. If the Richardson number is large, only a small fraction of the energy is available for turbulent mixing.

The authors give the following model for diffusion at steady state over the cross-sectional area of the room:

$$qc_0 = qc + DA\left(-\frac{dc}{dy}\right) \quad [10.20.22]$$

where A is the cross-sectional area of the room (m^2), c is the concentration of contaminant in the gas layer (v/v), c_0 is the concentration of contaminant in the leak gas (v/v), D is the molecular diffusion coefficient (m^2/s), q is the volumetric flow of leak gas (m^3/s) and y is the vertical distance from the ceiling (m). With the boundary conditions $c = 0$; $y = y_0$, where y_0 is the vertical distance from the ceiling of the interface between the two layers (m), Equation 10.20.22 integrates to give

$$\frac{c}{c_0} = 1 - \exp\left[-\frac{q}{AD}(y_0 - y)\right] \quad [10.20.23]$$

The authors describe experiments which were actually done inverted, using a dense gas, nitrous oxide, introduced through the floor, with the air inlet in the roof and the outlet close to the floor. At low air flows the concentration profile was close to the theoretical one, but at higher flows the concentrations in the 'gas' layer fell and some contaminant appeared in the 'air' layer.

Leach and Bloomfield also present a theoretical investigation of the concentrations associated with a leak of buoyant gas from a source low down in the room. They use the buoyant plume model of B.R. Morton, Taylor and Turner (1956). They point out that the pure plume would exist only for a short time and that the situation soon becomes more complex, with the plume then entraining not pure air, but a mixture of air and gas. This will lead to an increase in the concentration of the buoyant gas in the plume and hence in its concentration below the ceiling and throughout the room. This situation has been studied by Baines and Turner (1969), but the treatment is complex.

Further work on this problem has been described by M.R. Marshall (1983), who performed a series of experiments in a 20 m^3 cubical space and also in an 8 m^3 rectangular space and in buildings, with and without ventilation. In the unventilated situation the dominant factor is the density of the gas released. With the leak source of buoyant gas part way up one wall, a gas-rich mixture is formed in the volume above the source and this volume is well mixed. If the leak source is near the ceiling, a shallow layer of high concentration is formed, whilst if the source is near the floor a deep layer of lower concentration is formed. In both cases the concentration increases with time. This behaviour is shown for natural gas, a buoyant gas, in Figure 10.8. Figures 10.8(a) and (b) show instantaneous concentration profiles with the leak source near the ceiling and near the floor, respectively, and Figure 10.8(c) shows the development of the concentration profile with time for a leak source that is relatively high up. Likewise, with a leak of dense gas, a well mixed gas-rich mixture is formed in the volume below the source.

Where there is ventilation, on the other hand, this may well be the dominant effect. In this case, however, the situation is more complex, because there are various combinations of ventilation pattern and leak source location. In this case the results are presented in terms of the steady-state concentration profiles. Marshall gives results for work with a buoyant gas, natural gas, with upward, downward and cross-flow ventilation patterns. For upward ventilation flow the momentum and buoyancy forces reinforce each other. Figure 10.9 shows the steady-state concentration profiles for upward ventilation flow with a leak source of buoyant gas located at three different heights. The profiles are similar in shape to the transient profiles for the unventilated case. A gas-rich mixture is formed in the volume above the source and this volume is well mixed. At steady state the concentration in this volume is close to that calculated from Equation 10.20.15b.

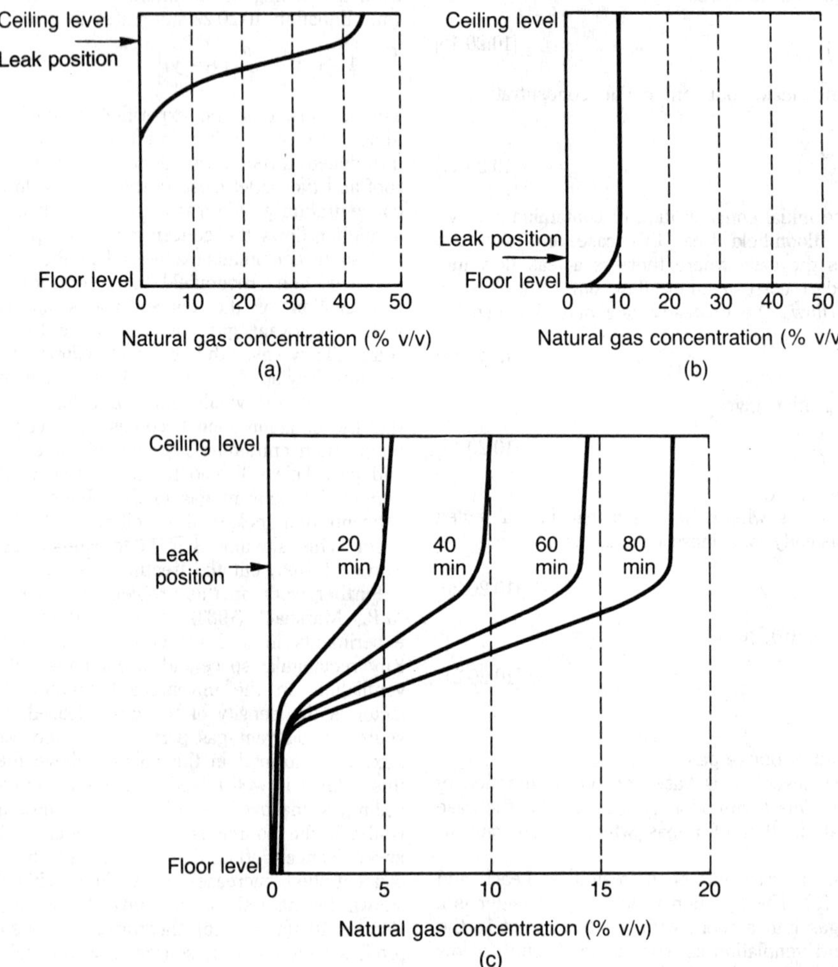

Figure 10.8 *Effect of gas density on mixing in an unventilated space (M.R. Marshall, 1983): (a) instantaneous concentration profile for leak of a buoyant gas located in a side wall near the ceiling; (b) instantaneous concentration profile for leak of a buoyant gas located in a side wall near the floor; (c) development of the concentration profile for leak of a buoyant gas located in the upper part of a side wall (Courtesy of the Institution of Chemical Engineers)*

For a buoyant gas with downward ventilation flow, the momentum and buoyancy forces are opposed. The overall effect is to increase mixing, which results in the formation of a high concentration in the volume above the leak source and a lower concentration in the volume below it. At steady state the concentration in the volume below the source is close to that calculated from Equation 10.20.15b, whilst that in the volume above it is slightly higher. For a buoyant gas with cross-flow ventilation with two pairs of inlets and outlets, one pair at low level and one at high level, the low level pair were found to have little effect, and the concentration profiles were broadly similar to those for the unventilated case. With the appropriate inversion, these results are applicable also to a dense gas.

The practical implication of this work is that for the usual case of upward flow ventilation there is at steady state a volume within which there is a well mixed gas/air mixture, that for a buoyant gas this volume is the volume above the leak source and that for a dense gas it is the whole volume of the ventilated space.

10.20.8 Fire ventilation

Ventilation may also be required to remove smoke generated in fire conditions. This aspect is considered in Chapter 16.

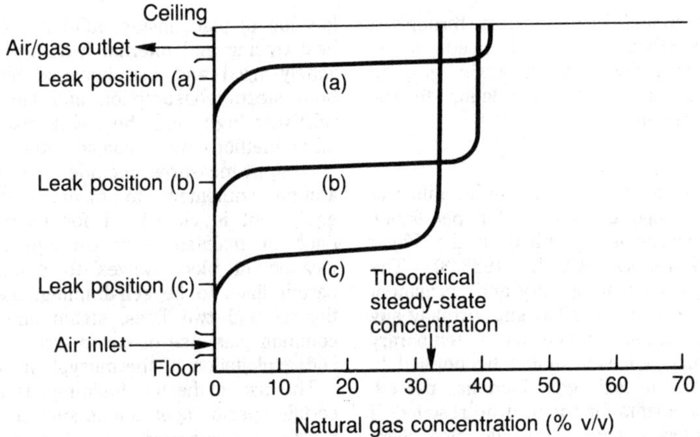

Figure 10.9 *Effect of gas density on mixing in a space with natural ventilation (M.R. Marshall, 1983): steady-state concentration profiles for leaks of a buoyant gas located at different heights in a side wall (Courtesy of the Institution of Chemical Engineers)*

10.21 Toxics Protection

Another hazard against which protection may be required is that posed by a release of toxic gas. In general, ordinary buildings off site and even on site can afford an appreciable degree of protection against a transient toxic gas release, but for certain functions enhanced protection is required. It is also necessary to ensure that the protection potentially available is not defeated. Buildings of particular interest here are (1) the control building, (2) the emergency control centre and (3) any temporary refuges.

10.21.1 Control room

The design of a control room for protection against toxic release is discussed in the *CIA Control Building Guide* (CIA, 1979). The design should start by identifying the release scenarios against which protection is required and by making some quantitative assessment of the dispersion of the gas. If persons outside exposed to the gas release would be incapacitated or unable to escape, protection is needed. The time for which protection is needed should also be defined. This will normally be governed by the time required to shut the plant down or the time needed to control the emergency.

The control building should be located at the edge of the plant, and its siting should take into account both fire/explosion and toxic gas hazards. There should be at least two escape routes, which should be chosen bearing in mind that they may need to be used by partially incapacitated people and by rescue teams wearing heavy breathing apparatus. They should be free from obstructions and well lit.

The construction of the building should be gas-tight. This means, among other things, that windows should be non-opening and that door and window frames should be designed and maintained to minimize entry of gas. There should be no more than two doors, each with an air lock and each gas-tight. The wedging open of these doors should not be tolerated.

The occupants of the control building should be provided with a supply of air sufficient for the time for which protection is required. Generally, it will be necessary to shut off the normal ventilation and the control building will then lose any overpressure and may become contaminated. Self-contained breathing apparatus should be provided for each occupant.

In some cases it may be possible to supply air from a source sufficiently far from the control building that the air from it is clean. The air supply should maintain within the building a positive pressure of 0.5–1.0 in. water gauge, which requires 2 to 15 air changes per hour, depending on building size, construction and gas-tightness.

There are various ways in which toxic gas may enter such a building. One is via service trenches and cellars. These should be avoided, but if used should incorporate sealed barriers and should be subject to a permit-to-work system. Another mode of entry is via instrument air lines and appropriate precautions should be taken. Instrument sample lines should not bring toxic process fluids into the control building.

Where available, gas detectors should be used to give warning of a toxic gas escape by activating an alarm in the control room, where there should be means of activating the toxic gas alarm. There should be in the control room an indication of wind direction.

It may be possible to provide protection to the control building using a water spray system. The gas detector signal may be used to activate such a water spray and to shut off the normal ventilation air.

The control building should have a priority communication link, other than the normal telephone system, to the emergency control centre.

Appropriate breathing apparatus or respirators should be provided in the control building to assist the escape of any persons who have to be evacuated in the emergency.

10.21.2 Emergency control centre

Protection of the emergency control centre from a toxic release will normally be in large part by location. It is necessary for the emergency controllers to gain access to the centre at the start of the emergency and it is undesirable that they should have to pass through a toxic gas cloud. Nevertheless, it may need to be designed to afford protection against toxic gas, in which case the points just made in relation to the control building are pertinent.

10.21.3 Refuges

A temporary refuge, or haven, has the quite different function of providing temporary shelter for personnel. The design of such havens is described in the *Vapor Release Mitigation Guidelines* (CCPS, 1988/3). The *Guidelines* distinguish between temporary and permanent havens, or more effective temporary havens. Virtually any weather-tight building should suffice as a temporary haven. Personnel in such a haven should be notified to leave the building when the toxic gas cloud has passed. There should be arrangements for them to be rescued, if necessary, by well-equipped teams. It would also seem necessary that there be means whereby the emergency controllers know the location of personnel needing rescue.

For permanent havens the CCPS *Guidelines* refer to the arrangements for control buildings as described in the CIA *Control Building Guide*. They also give a method of estimating the capacity of a haven. The conditions in the haven should not exceed the following limits: minimum oxygen concentration 18%; maximum carbon dioxide concentration 3%; maximum temperature 33°C; and maximum 100% relative humidity (RH).

Relations are given for all four of the features and it is shown that humidity is the limiting factor. Thus, starting with an initial temperature of 20°C and 50% RH (8.7 mm Hg) and rising to a temperature of 33°C and 100% RH (37.7 mm Hg), and allowing for a production of water vapour of 2.3 l/min person, the minimum volume of space required per person is calculated as:

$$0.011V + 2.3t = 0.049V \qquad [10.21.1a]$$

or

$$V/t = 60.8 \qquad [10.21.1b]$$

where t is the shelter period (min) and V is the volume of space required per person (l). This factor has the highest value of V/t of the four factors considered, and is therefore the limiting one.

The volume of the human body is 2.65 ft^3 and converting from 60.8 l to 2.1 ft^3 gives, for the required capacity of the haven:

$$V_{tot} = (2.1t + 2.65)N \qquad [10.21.2]$$

where N is the number of people to be sheltered and V_{tot} is the total volume of the haven (ft^3). The corresponding floor area may be obtained assuming a ceiling height of 8 ft.

10.22 Winterization

It is convenient to deal here with the protection of plant against severe winter conditions, or winterization. The winterization of process plants has been described by J.C. Davis (1979) and Fisch (1984), and the shut-down winterization of an ammonia plant has been described by Facer and Rich (1984).

There are five basic techniques for winterization: design and operating methods which avoid freezing, location of plant inside buildings, and use of insulation, heat tracing and internal heating coils. Steam tracing is widely used and has become fairly standardized, but both steam consumption and labour requirements are relatively high, and the rising costs of both mean that other methods merit consideration.

Design measures include: the use of bypass lines around equipment to maintain circulation when the equipment is closed off for maintenance; recirculation lines to maintain flow through non-operating pumps; location of block valves to eliminate dead legs and permit lines to be self-draining; use of common insulation around two lines, steam and condensate being a common pair; use of steam traps to remove condensate; and exploitation of thermosyphon circulation.

The use of heated buildings is generally confined to certain specific applications such as the use of centrifuge rooms and analyser rooms. It is normal to bury fire water lines, but not process lines, because of the hazards of corrosion and of leakage. Where local conditions are suitable, insulation is an economical way of preventing freezing. Tables of times to freezing in a stagnant line have been given by House (1967). For some plant an internal heating coil may be used, but this creates the risk of a leak from the coil which may not be readily detected. Finally, some method of heat tracing may be used to give protection. Methods include steam tracing, circulating medium tracing and electrical tracing.

The winterization protection should be designed as a system in its own right. The weather conditions against which protection is to be provided should be defined and a review conducted of the protection requirements. For each part of the system a method of heat tracing should be selected and the tracing requirements defined, including heat inputs and maximum allowable temperatures. The alternatives to heat tracing should be considered.

The design weather conditions should be combinations of minimum temperature and wind velocity. The consequences of a freeze-up may be sufficiently severe that it is appropriate to design using a formal recurrence interval approach.

There are several factors which may determine the maximum temperature limit for the tracing. One is the maximum tracing temperature which the process fluid can withstand. Another is the maximum process fluid temperature which the tracing can tolerate, allowing for operations such as hot water flushing or steaming out on the process side.

Flow diagrams should be produced for the heat tracing system which show the lines requiring heat make-up in normal operation, those needing freeze protection during shut-down and those normally stagnant. The diagrams should also show where alternatives to heat tracing are to be used, such as self-draining lines, minimum flow bypasses and recirculation arrangements.

The first choice of tracing is usually steam tracing. The minimum practical steam pressure is 25 psig. Steam tracing has been successfully used at temperatures down to −35°C, though in this case the recommended minimum steam pressure is 50 psig. Steam tracing systems are reliable in providing protection, but their installation requires skill and their maintenance requirements are high. Although, in theory, steam tracing can be turned on and off with the weather, this is not always practical and it may be possible to do no more than shut it off between spring and autumn.

Circulating medium tracing systems utilize hot oil or antifreeze and are used mainly where steam tracing is not practical. They are more expensive and are vulnerable to failure of the circulating pump.

Electrical tracing comes in the form of tapes, cable, blankets and custom-built shapes. It can be used over a wide range of temperatures, from below those at which steam is suitable to above those at which hot oil can be utilized. Its main advantage is that it can be thermostatically controlled, either by ambient thermostats or thermostats in contact with pipe surfaces. An electrical tracing system should be provided with alarms to signal failure.

An electrical tracing system may be based on series resistance or parallel resistance cable. The latter is rather more flexible in the range of heat inputs which it can provide and is useful particularly for protecting valves and instruments. Both types are subject to burnout. A third type is temperature-self-limiting tape, which can be provided with a range of cut-off temperatures and is virtually immune to burnout.

Tracing systems can be used in areas subject to hazardous area classification. Electrical tracing systems are available for such areas. But for all types of system consideration should be given to the surface temperature. There have been cases where heat tracing has caused the allowable hot surface temperature to be exceeded. Hazards may also arise when electrical tracing is disconnected for maintenance.

There are also various other devices available. These include preformed and preinsulated accessories such as instrument enclosures and pretraced tubing as well as preapplied heat transfer cement placed over tracer tubing. Further details of winterization system design are given by House and by Fisch.

The winterization measures taken at an Alaskan refinery have been described by J.C. Davis (1979). He highlights in particular the avoidance of water in process streams and the reduction of steam lines to an absolute minimum. The water vapour generated is dispatched quickly to a tall stack. Use is made of air cooling. Heat tracing is by hot oil.

Facer and Rich (1984) describe the shut-down winterization of an ammonia plant. Their account deals with: the aims of the winterization; the measures taken to protect catalyst, furnaces and burners, and rotating equipment such as gas turbines, compressor and pumps; the steam and cooling water systems; and the restart.

10.23 Modular Plants

During the late 1950s and early 1960s there was introduced a type of plant consisting of a number of modules and mounted on skids which could be transported by road from the fabrication to the operating site. The processes were straightforward and the plants were simple and cheap. From these early skid-mounted plants there has developed a whole range of modular and barge-mounted plants, some of which are large and complex.

Accounts of modular plants have been given by Glaser, Kramer and Causey (1979), Zambon and Hull (1982), Glaser and Kramer (1983), Hulme and La Trode-Bateman (1983), Kliewer (1983), Clement (1989), Hesler (1990) and Shelley (1990), and accounts of barge-mounted plants have been given by Birkeland *et al.* (1979), Charpentier (1979), J.L. Howard and Anderson (1979), R.G. Jackson (1979), Jansson *et al.* (1979), Shimpo (1979), Ricci (1981), Bolt and Arzymanow (1982), de Vilder (1982) and Glaser and Kramer (1983). Both types of plant are treated by Mecklenburgh (1985).

10.23.1 Skid-mounted plants

The early skid-mounted plants were typically natural gas processing plants and pipeline compressor stations mounted on skids. The plants had a quite small number of modules of limited dimensions. They were transported by truck from the fabrication works to the operating site. The plants were simple and were equipped to shut down if an operating problem arose. The plant operator typically lived in a house close by. These plants were designed for a relatively short life and had low capital and running costs. A description of such skid-mounted plants is given by Kliewer (1983).

10.23.2 Modular plants

The late 1970s saw a significant extension of the scale and complexity of modular plants. Such plants were seen as offering benefits where site construction was unusually difficult, particularly on remote sites. Factors favouring modular plants include problems associated with (1) access difficulties, (2) severe weather and (3) the labour force.

Advantages of modular construction are those associated with (1) access for equipment suppliers, (2) work in sheltered conditions and (3) availability of a skilled workforce. Arising from these are (4) easier construction and testing, (5) improved quality assurance and (6) shorter project time-scale. Construction of the plant at a dedicated fabrication site minimizes access difficulties for equipment suppliers, and allows the work to be done under cover and by a skilled workforce. Main items of equipment, pipework, supports, instrumentation and cabling can be installed and tested under essentially factory conditions. The project timetable can be shorter, both because work on foundations and on plant construction can proceed in parallel and because construction can be done in more favourable conditions.

Disadvantages of modular construction include those associated with (1) engineering design, (2) modifications, (3) steelwork and (4) transport. Modular construction necessitates high quality and more expensive engineering design. It is relatively unforgiving of modifications, which can therefore be disruptive and expensive. There are additional costs for steelwork but, because steel is relatively cheap, these may be modest. There are additional transport costs which vary depending on the site and the plant, and which can be considerable.

Modular construction requires its own design approach. It is not effective to design a plant by conventional methods and then divide it into modules. It is necessary to design for a modular layout from the start. It is also necessary to accept that the main features affecting layout have to be frozen earlier than is often the case is normal design.

Plants have been constructed with some 200 modules and with modules stacked as high as 50 m. With regard to module dimensions and weight, in one project described by Kliewer (1983) the maximum module dimensions were set at 6.7 m wide × 4.0 m high × 30.5 m long. The weight limit was 125 ton, though most modules did not exceed 50 ton. Shelley (1990) describes rubber-track crawlers and trailers with up to 360 wheels capable of transporting 3000–4000 ton and cranes with lifting capacity of 5000 ton.

Advocates of modular design typically claim savings on project cost and time. Accounts of cost benefits include those of Kliewer (1983) and Shelley (1990). Broadly, capital costs are less, but design, steelwork and transportation cost more. Shelley quotes construction times shortened by up to 50% and capital cost savings of up to 20%.

Progress in modular construction has been reviewed by Shelley (1990). The image of modular plants has tended to be that of skid-mounted plants and plants shipped to remote locations. Modularization has generally been considered only for remote locations where the weather is hostile or skilled labour unavailable. The author discerns a trend towards increasing use of modular construction for regular projects, arising from its advantages of cutting capital costs and shortening construction times. Another stereotype which is somewhat outdated is that modular plants necessarily involve a cramped layout.

An account of five projects involving modular plants has been given by Zambon and Hull (1982). These are a petrochemical complex in the Middle East, a large synthetic crude oil project in Alberta, a plant to convert natural gas to gasoline in New Zealand, and two projects in Alaska, one a gas separation plant and one a seawater treatment facility for oil well water injection. The authors give details of project profiles, listing key factors such as: access, weather and labour availability; schedule and cost data; execution strategies; modules contracts; and module timetables. Labour considerations were important in all five projects and the weather was important in four. Four of the projects were barge mounted.

The modular construction of a large gas processing plant in Wyoming is described by Kliewer (1983). The plant consisted of some 175 modules, some assembled by the vendors and some by a construction company. There were some 390 items of equipment, of which 250 were preassembled in some way, leaving the residue of some 140 units to be site installed.

Glaser and Kramer (1983) describe four modular plant projects: a refinery at Calgary, a crude oil processing facility in Saudi Arabia, a visbreaking unit at Killingholme and a methane recovery unit in New York City. They present a detailed account of the Calgary project showing the items which could and could not be modularized and giving dimensions and weights of modules. Typical reasons for not modularizing are that the item was too tall or was delivered too late.

These authors also described the modular construction of the large crude oil stabilization unit at Sullom Voe in the Shetland Islands, as do Bolt and Arzymanow (1982). The modules include 36 process units, weighing 150–500 ton, 17 compressor units weighing 90 ton each, and 37 pipe rack units weighing 35–350 ton.

10.23.3 Barge-mounted plants

A particular type of modular plant is that mounted on a barge or other vessel. The development of such plants has received impetus from the need for shipyards to diversify. Features of the operating site which favour the use of a barge-mounted plant include: a seaboard inaccessible from the hinterland; a navigable, if shallow, river; or a delta unsuitable for land traffic.

A plant transported by sea may in fact be truly barge mounted or it may be self-floating. In the latter case it is effectively a sea-going object in its own right, must be fully seaworthy and must meet the requirements of the classification societies. The direct costs of transport of a barge-mounted plant may well be modest, but those of providing the stiff framing for, and the measures to counter stresses developed in, a sea voyage can be appreciable. A barge-mounted plant is a sea-going object so that it must be seaworthy and must meet the classification society specifications, which can be expensive.

One solution is to use a vessel designed specifically for the transport of modular plants. The Wijsmuller semi-submersible heavy lift vessel *Super Servant*, described by de Vilder (1982), is of this type. This is a development from the semi-submersible barges which have been in use for some decades, either unpowered or with auxiliary propulsion only.

The options for installation at the operating site have been discussed by Charpentier (1979), who lists four. One is a barge floating at sea or anchored. This means, in effect, a factory ship with its own propulsion and mooring systems. Another is a barge which floats but is moored along a quay, accessible from the sea on one side and from the land on the other. A third is a barge grounded on a dredged bed in a shelter site, possessing connections similar to those in the previous case but not subject to water movement. The fourth option is a barge grounded on a foundation sill and protected by some form of dike or dam.

The use of prestressed concrete hulls for barge-mounted plants has been described by Birkeland *et al.* (1979). They outline three options for installation at the operating site: a self-floating plant may be permanently floating or permanently grounded; a plant delivered by a barge is off-loaded and floated into position and then permanently grounded.

Reviews of projects on barge-mounted plants include those by Charpentier (1979) and Ricci (1981). Birkeland *et al.* (1979) describe several barge-mounted projects. They include a self-floating LPG refrigeration and storage barge, the *Ardjuna Sakti*, sited near Jakarta and permanently floating. Charpentier (1979) describes a number of projects involving barge-mounted plants. They include a refinery, a natural gas liquefaction plant, an ammonia plant and a methanol plant.

The design of a barge-mounted liquefied natural gas (LNG) liquefaction and storage plant, the marine LNG system (MLS), has been described by J.L. Howard and Andersen (1979); the project was intended for the Pars gas field off Kangan, Iran, but was interrupted by political factors. The authors give details of the process flow diagram, the LNG storage spheres and the fire

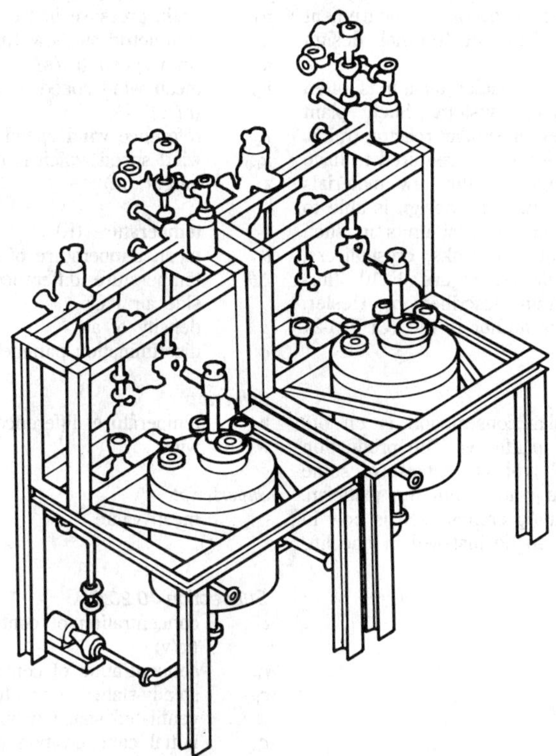

Figure 10.10 *Modular two-unit reactor train (Hesler, 1990) (Courtesy of the American Institute of Chemical Engineers)*

protection and emergency shutdown systems. The design was done according to the requirements of the International Maritime Consultative Organization (IMCO) gas carrier codes. The installation was of the dredged basin type.

In the 1970s the conversion of LNG to methanol prior to transport appeared to be a potentially attractive way of transporting energy on long hauls, and studies of barge-mounted plants for such conversion were carried out (R.G. Jackson, 1979). One application envisaged for such units was the exploitation for smaller, shorter life fields.

A somewhat similar motivation underlies the use of barge-mounted plants to process gas from fields for which a pipeline would be uneconomic and at which the gas would therefore be flared (Jansson *et al.*, 1979). These fields may include subsea completions where there is no production platform. The main design described is for an ammonia plant with the platform a flat, broad barge moored at a single point mooring, but variations include barge-mounted urea, methanol, natural gas liquids (NGL) and LNG plants and beaching of the plant.

Ricci (1981) gives an account of a barge-mounted low density polyethylene (LDPE) plant for Bahia Blanca, Argentina. This plant was transported by the heavy lift semi-submersible described by de Vilder (1982) and referred to above. This author describes in detail the planning of the voyage in respect of the wind and acceleration forces and of the mechanical stresses to which the load would be subjected.

An account of a barge-mounted pulp plant installed in the upper reaches of the Amazon in Brazil using the industrial platform system has been given by Shimpo (1979). The site was one with no roads and accessible only by plane. There were two platforms, one for the pulp plant and one for the power plant. The platforms had to be designed for structural strength whilst being towed and during operation. Platform construction posed various difficulties. It proved impossible to set up a longitudinal bulkhead and there were few straight transverse bulkheads. There were many large irregular openings in the main deck, especially close to the side. At the site the design was for the platform to be set on piles. There were problems arising from unbalanced soil strength and uneven live load on the platform. The project yielded much information on motions and stresses during the voyage and at the site.

10.23.4 Modular design

It is possible to adopt a modular approach to the design of plant, even if modular construction is not intended. An account of such modular design is given by Hesler (1990). A modular approach not only saves on design costs but also allows the design to be optimized and defects eliminated and, by offering equipment suppliers repeat runs, reduces equipment costs and procurement times. For some types of plant the normal design consists of replicated modules.

One type of plant for which modular design is often appropriate is a batch reactor system. Such plant generally consists of a number of similar reactor trains. Furthermore, these trains are frequently required to have the flexibility to permit changes in the raw materials used and products made and a modular design is able to accommodate such modifications. Typical units in such plant are reactors, columns, quench tanks, crystallizers, liquid–solid separators, and driers. Figure 10.10 illustrates the two-unit reactor train described by Hesler. Another example given by this author is the ICI FM-21 membrane chlorine cell.

10.23.5 Offshore modules

Another application of modular construction is on offshore oil and gas production platforms. The production deck of such a platform will typically consist of some four modules which are lifted whole onto the platform. The lifting capacity of the floating cranes used is now in fact such that a whole deck can be installed in one lift.

10.24 Notation

Section 10.14

I heat flux (kW/m^2)
t_B duration of fireball (s)

Section 10.19

P peak value of applied blast load
R dynamic resistance
t_0 duration of blast load
X_m maximum allowable dynamic displacement
X_y effective yield displacement

δ parameter
η variable defined by Equation 10.19.2
τ fundamental period of vibration

Section 10.20

Subsection 10.20.5

a index
A area of opening
A_b equivalent area for ventilation by temperature difference only (m^2)
A_w equivalent area for ventilation by wind only (m^2)
A_{1-4} areas defined by Figure 10.7
C_d coefficient of discharge
C_p surface pressure coefficient
ΔC_p differential pressure coefficient
H_1 vertical distance defined in Figure 10.7
K constant
p pressure at a particular point (Pa)
p_0 static pressure in free wind (Pa)
Q volumetric air flow (m^3/s)
u wind speed (m/s)
u_m mean wind speed at 10 m height in open terrain (m/s)
u_r reference wind speed (m/s)
u_{50} wind speed which is exceeded 50% of time (m/s)
z height (m)

θ temperature (K)
$\bar{\theta}$ mean temperature of inside and outside air (K)
$\Delta\theta$ temperature difference between inside and outside air (°C)
ρ density of air
ϕ discrimination parameter

Subscripts:
b temperature difference
w wind

Superscript:
‾ mean value

Subsection 10.20.6

c concentration of contaminant in ventilated space (v/v)
c_e concentration of contaminant in inlet air (v/v)
c_E steady-state concentration of contaminant in ventilated space (v/v)
c_o initial concentration of contaminant in ventilated space (v/v)
q volumetric flow of leak (m^3/s)
Q volumetric flow of ventilation air (m^3/s)
t time (s)
V volume of ventilated space (m^3)

Subsection 10.20.7

A cross-sectional area of room (m^2)
c concentration of contaminant in gas layer (v/v)
c_0 concentration of contaminant in leak gas (v/v)
D molecular diffusion coefficient (m^2/s)
q volumetric flow of leak gas (m^3/s)
y vertical distance from ceiling (m)
y_0 vertical distance from ceiling of interface between two layers (m)

Section 10.21

N number of people to be sheltered
t shelter period (min)
V volume of space required per person (l)
V_{tot} total volume of haven (ft^3)

11 Process Design

Contents

11.1 The Design Process

The design of a large process plant or plant extension is a quite complex activity which is carried out in stages over a period of time and involves people of many disciplines. The design process normally involves other parties beside the operating company. This aspect is considered in Section 11.2.

The project evolves under the influence of: research and development, which define the technical possibilities and constraints; safety, health and environmental studies, which indicate further constraints; economic studies, which indicate the productions costs and sales returns together with their sensitivity to the various factors; and the financial approvals, which allow the project to proceed to the next stage.

The decisions which are made in the early stages, particularly those concerning the process route, the plant output and the plant location are crucial. Thereafter many options are foreclosed, so that fundamental changes are difficult to make or are simply impractical.

From the safety and loss prevention (SLP) viewpoint it is essential to try to get the process fundamentals right from the start. The aim should be to eliminate hazard rather than to devise measures to control it. This aspect is discussed in Section 11.3. The safety of the plant is determined primarily by the quality of the basic design rather than by the addition of special safety features. It is difficult to overemphasize this point. Nevertheless, it is necessary to build into the design process some quite specific safety checks on safety and to carry out hazard identification and assessment studies.

The design experience and know-how of the organization may be generalized in terms of its knowledge of the characteristics of particular chemicals, unit operations and unit processes, and so on. Full use should be made of this information both for process design and for hazard identification.

Accounts of process design are given in *Applied Process Design for Chemical and Petrochemical Plants* (Ludwig, 1964–), *Equipment Design Handbook for Refineries and Chemical Plants* (Evans, 1971), *Applied Chemical Process Design* (Aerstin and Street, 1978), *Introduction to Chemical Process Technology* (van den Berg and DeJong, 1980), *Chemical Process Synthesis and Engineering Design* (Kumar, 1981), *Process Analysis and Design for Chemical Engineers* (Resnick, 1981), *Process Design for Reliable Operation* (Lieberman, 1983), *Introduction to Material and Energy Balances* (Reklaitis, 1983), *Scaleup of Chemical Processes* (Bisio and Kabel, 1985), *Chemical Process Computations* (Raman, 1985), *The Art of Chemical Process Design* (Wells and Rose, 1986), *Process Modelling* (Dean, 1987), *Conceptual Design of Chemical Processes* (Douglas, 1988), *Foundations of Computer Aided Process Operations* (Reklaitis and Spriggs, 1988), *Chemical Process Equipment* (Walas, 1990) and *Dimensional Analysis and Scaleup in Chemical Engineering* (Zlokarnik, 1991).

Economic aspects of process design are treated in *A Guide to Chemical Engineering Process Design and Economics* (Ullrich, 1984), *Economic Evaluation of Projects* (D.H. Allen, 1988) and *A Guide to Capital Cost Estimating* (IChemE, 1982).

***Table 11.1** Selected references on process design*

Design
Asimow (1962); T.K. Sherwood (1963); W. Hughes and Gaylord (1964); Pye (1964); Dixon (1966); Krick (1969); Middendorf (1969); Leech (1972); Geiger (1975); Simon (1975); AIChE (1982/78); Waldheim, Finneran and Whittington (1983); Landis and Hamilton (1984); Kubic and Stein (1988)

Information sources
Subramanyan (1981); Archbold, Laidlaw and McKechnie (1984); Anthony (1985); Wasserman, Smith and Mottu (1989)

Chemical industry
CIA (n.d./1, 1991 CE5); Kirk and Othmer (1963–, 1978–, 1991–); Beynon (1982); Sharp and West (1982); Harvey-Jones (1983); G. Allen (1984); Heaton (1985, 1993); Chenier (1986); H.L. White (1986); E. Johnson (1989b,c, 1990); Redman *et al.* (1990); Redman and Smith (1991); Redman *et al.* (1991); Goldsmith and John (1992); Osborne, Morgan and Varey (1992)

Biotechnology industry, food industry
Jowitt (1980); Sillett (1988); Collins and Beale (1992); Cottam (1991); Horsley and Parkinson (1990); Cumming and Brown (1991)

Physical, chemical and thermodynamic properties
API (n.d./12, 13, 1983 Publ. 999); NIOSH (Crit. Doc. Series, 1990/18); Washburn (1926–); Perry (1934–); NBS (1947, 1952, 1969); Rossini *et al.* (1947); J.B. Maxwell (1950); Mellan (1950, 1957–, 1977); Timmermans (1950–); MCA (1952– SD series, 1972/21); Rossini (1952); DECHEMA (1953–); Wagman (1953); T.E. Jordan (1954); Dreisbach (1955–); Gambill (1957-); T.K. Ross and Freshwater (1957); Friend and Adler (1958); Reid and Sherwood (1958); Canjar, Manning *et al.* (1962); Marsden (1963); Nesmeyanov (1963); McKelvey (1964); Anon. (1965a); Gallant (1965–); Canjar, Manning *et al.* (1966–); Chase (1966, 1984); Rompp (1966); L.W. Ross (1966); Canjar and Manning (1967); Touloukian, Gerritsen and Moore (1967); Gold (1968–); Home Office (1968); Ullman (1969); Wagman *et al.* (1969); Nabert and Schön (1970); Riddick and Bunger (1970); Staples, Procopio and Su (1970); Starling *et al.* (1971-); Stull and Prophet (1971); NASA (1972–); Perry and Chilton (1973); Wichterle, Linek and Hala (1973–); Bretherick (1974, 1975, 1985); *Chemical Engineering* (1974a); Yaws *et al.* (1974–); Allen (1975); Hirata, Ohe and Nagahama (1975); Horvath (1975, 1982); International Technical Information Institute (Japan) (1975); NFPA (1972/9, 1991 NFPA 49, 321, 325M, 491M, 1991/27, 1992/32); Sax (1957); Anon. (1976 LPB 9, p. 26); Anon. (1976 LPB 12, p. 12); DoE and Chemical Society (1976); Kletz (1976h, 1977i); Barner and Scheuerman (1977); Ganapathy *et al.* (1977); Hawley (1977); Reid, Prausnitz and Sherwood (1977); Verscheuren (1977); Dean (1978); Gmehling and Onken (1978); L.R. Nielsen (1978); Ohe (1978); Wisniak and Tamir (1978, 1980–); Duhne (1979); Koolhaas, Ramdas and Putnam (1979); R. Nelson (1979); R.H. Powell (1979); Sittig (1979, 1981, 1984); Thibodeaux (1979);

Barry (1980); Isman and Carlson (1980); Lin *et al.* (1980); Mansouri and Heywood (1980); J.H. Weber (1980a–d, 1982); Kaye (1981); Chaney *et al.* (1982); Christensen, Hanks and Izatt (1982); C.H. Fisher (1982); AIChE (1983/79, 1984/90, 1985/91, 1986/92); McGarry (1983); Rao (1983); Boublik, Fried and Hala (1984); Buck and Frankl (1984); Edmister and Lee (1984); Perry and Green (1984); Wisniak and Kerskowitz (1984); Adler and Lin (1985); Anthony (1985); Daubert and Danner (1985–); H.H. Fawcett (1985, 1988); Majer and Svoboda (1985); Wagle (1985); Weast (1985); Pedley, Naylor and Kirby (1986); B.D. Smith and Srivastava (1986); Tamas (1986); Wooley (1986); J. Grant and Grant (1987); Reid, Prausnitz and Poling (1987); Ensminger, Lu and Oen (1988); Tayler (1988a); Valenzuela and Myers (1988); Yaws, Ni and Chiang (1988); Anon (1989d); Barin (1989); Möller, Redeker and Schultz (1989); Viswanath and Natarajan (1989); ACGIH (1990/49, 1991/53, 55, 59, 60, 62, 1992/82); Lyman, Reehl and Rosenblatt (1990); Yaws *et al.* (1990a,b); Yaws, Yang and Cawley (1990); Cardozo (1991); Stanley-Wood and Lines (1992); Yaws, Huar-Chung Yang and Xiang Pan (1991); HSE (1992 CRR 39); Sass and Eckermann (1992); Yaws and Xian Pan (1992); Chakravarty (1993); Yaws, Xian Pan and Xiaoyan Lin (1993)
Material safety data sheets: NSC (1952– Safety Data Sheets)
Design Institute for Physical Property Data: AIChE (1985–/157–159, 1987/156); Selover (1990)

Process design, scale-up, unit operations
MCA (SG-14); Bridgman (1931); Perry (1934–); Edmister (1947–); Lyle (1947); Peters (1948); C.S. Robinson and Gilliland (1950); Treybal (1951, 1955); T.K. Sherwood and Pigford (1952); Coulson and Richardson (1955–, 1977–); Cremer (1956); Johnstone and Thring (1957); T.K. Ross and Freshwater (1957); Jeffreys (1961); Norman (1961); L. Clarke and Davidson (1962); Jenson and Jeffreys (1963); Kirk and Othmer (1963–, 1978–, 1991–); Ludwig (1964–); Hudson (1965); Bourton (1967); Landau (1967, 1981); McCabe and Smith (1967); H.S. Robinson (1967b); Henglein (1968); R.G. Hill (1968); D.G. Jordan (1968); Lark, Craven and Bosworth (1968); Rudd and Watson (1968); J.F. Smith (1968); *Chemical Engineering* (1969a,b, 1974b, 1978a,b, 1979b); Grassman (1971); Backhurst and Harker (1973); Perry and Chilton (1973); Rudd, Powers and Siirola (1973); AIChE (1974/75, 1987/86); Considine (1974); R.A. Freeman and Gaddy (1975); Warren (1975b); Zudkevitch (1975); Ramirez (1976); Riggs (1976); Schwartzman and Wiese (1976); Zanker (1976); Mahalec and Motard (1977); Aerstin and Street (1978); Billet (1979); Buck (1978); Clift, Grace and Weber (1978); Economopoulos (1978); Lapedes (1978); Suckling, Suckling and Suckling (1978); D.G. Austin (1979); F.L. Evans (1979); IChemE (1979/118, 1982/124); J.C. Johnson (1979); Malpas (1979, 1983); Reed and Narayan (1979); van den Berg and de Jong (1980); Foust *et al.* (1980); Kauders (1980, 1984); C.J. King (1980); Anon. (1981 LPB 42, p. 21); Kumar (1981); Lamit and Engineering Model Associates (1981); Resnick (1981); Diab and Maddox (1982); Hodgson (1982); IMechE (1983/68); Reklaitis (1983); Waldheim, Finneran and Whittington (1983); Beddow (1984); O'Brien and Porter (1984); Perry and Green (1984); Ruthven (1984); Bisio and Kabel (1985); Raman (1985); Ulrich (1984); Blass (1985); Davidson, Clift and Harrison (1985); Hines and Maddox (1985); McCabe, Smith and Harriott (1985); Seader (1985); McHugh and Krukonis (1986); Stephanopoulos and Townsend (1986); Wells and Rose (1986); Dean (1987); Liu, McGee and Epperley (1987); Rousseau (1987); Sandler and Luckiewicz (1987); Yang (1987); Liang-Shih Fan (1989); Rumpf (1990); Svarovsky (1990); Turney (1990a,b); Walas (1990); Garside, Davey and Jones (1991); Kottowski and Kashiki (1991); Kunii and Levenspiel (1991); Deshotels and Goyal (1992); Kister (1992); Paret (1992); D. Scott and Crawley (1992); Barnes (1993); Biach and Watt (1993); Capps and Thompson (1993); Fabian *et al.* (1993); Kerridge (1993); Koshal (1993); Oliver (1993)
AIChE Design Institutes: Kemp (1983); Prugh (1987b); Selover and Buck (1987); Anon. (1988a); Selover (1990)

Model design projects
Jeffreys (1961); D.G. Austin and Jeffrey (1979); Ray and Johnston (1990); R. Scott and McLeod (1991)

Design error
Kerns (1972); Haastrup (1983); ASCE (1986/27)

Design documentation
Anon. (1991 LPB 98, p. 13); Jowell (1994a)

Thermodynamics
Zemansky (1937); Keenan (1941, 1970); Dodge (1944); Hougen and Watson (1947); Lewitt (1953); Hougen, Watson and Ragatz (1954–); Kiefner, Kinney and Stuart (1954); G.C.F. Rogers and Mayhew (1957); Eashop and McConkey (1963); Redlich (1976); David (1990)

Transfer processes, fluid flow, heat transfer
NRC (Appendix 28 *Fluid Flow, Thermal Hydraulic Analysis*); Lander (1942); L.F. Moody (1944); D.Q. Kern (1950); Streeter (1951); Lewitt (1952); McAdams (1954); Treybal (1955); Crank (1956); G.C.F. Rogers and Mayhew (1957); Emmons (1958); Boucher and Alves (1959–); Landau and Lifshitz (1959); Bird, Stewart and Lightfoot (1960); Duncan, Thomas and Young (1960); C.O. Bennett and Myers (1962); Kutateladze and Borishanskii (1966); Streeter and Wylie (1967, 1975); Massey (1968); Holland *et al.* (1971); D.S. Miller (1971); Collier (1972); D.Q. Kern and Kraus (1972); Siegel and Howell (1972); Slattery (1972); J.A. Adams and Rogers (1973); Churchill (1974); IMechE (1974/9, 1977/38, 1983/70, 1988/101); Clarke and McChesney (1975); T.K. Sherwood, Pigford and Wilke (1975); Hsu and Graham (1976); M.J. Moore and Severding (1976); Welty, Wicks and Wilson (1976); Butterworth (1977); Butterworth and Hewitt (1977); Hahne and Grigull (1977); Karlekar and Desmond (1977); *Chemical Engineering* (1978a); Raju and Rathan (1979); Ricci (1978a); Cavasene and *Chemical Engineering* Staff (1979); Lydersen (1979); Schweitzer (1979); Kollman (1980); L. Thomas (1980); Hausen (1981); IChemE (1981/122); Lienhard (1981); Singh (1981); D. Stephenson (1981a,b, 1984); Vennard and Street (1982); Coulter (1984); Nicholls (1984); Perry and Green (1984); Hines and Maddox (1985); Hobson and Day (1985); Kay and Nedderman (1985); Rohsenow,

Hartnett and Ganic (1985); Cheremisinoff (1986); Crane Co. (1986); Idelchik (1986); Hsi-Jen Chen (1987); Churchill (1989); ASCE (1990/41); Goyal (1990); Coker (1991a,b, 1992); Barnes (1993)

Dimensionless numbers: Boucher and Alves (1959–); Catchpole and Fulford (1966); Fulford and Catchpole (1968); Zlokarnik (1991)

Computational fluid dynamics
Abbott and Cunge (1982); Brebbia and Ferrante (1983); Abbott and Basco (1989); Sharratt (1990); IMeChE (1991/127, 1993/149); Dombrowski, Foumeny and Riza (1993); Foumeny and Benyahia (1993?)

Fuels
Ministry of Power (1944); AGA (1965); D.S. Wilson (1969); Bell (1971a); Lom (1974); Dryden (1975); Goodger (1976); Considine (1977); J.W. Rose and Cooper (1977); Berkowitz (1979); Feeley (1979); Andrew (1981); IMechE (1983/67); Hamilton (1984); Institute of Materials (1985 B367); Melvin (1988); Redman (1990); Speight (1990); Bartok and Sarofim (1991)

Energy
Boyen (1975); Bunton and Buckland (1975); Weston (1975, 1980); Huckins (1978); Linnhoff and Flower (1978); Shaner (1978); Barnwell and Derbyshire (1979); C.D. Grant (1979); Kantyka (1979); Linnhoff, Mason and Wardle (1979); Moon and Tasker (1979); Robnett (1979); Short (1979); Summerfeld and White (1979); Troop (1979); Weston (1979); Linhoff and Turner (1980, 1981); Linhoff, Turner and Boland (1980); Townsend (1980); Yen-Hsiung Kiang (1981); Colbert (1982); Linnhoff *et al.* (1982); Jonas (1982); Townsend and Linhoff (1982); Linnhoff and Hindmarsh (1983); Santoleri (1983); Anon. (1984bb); Boland and Hindmarsh (1984); Chyuan-Chung Chen (1984); Linhoff and Vredeveld (1984); Ganapathy (1985); Kotas (1985, 1986); Skinner (1985); IMechE (1976/28); Linnhoff and Kotjabasakis (1986); O'Reilly (1986); Tjoe and Linnhoff (1986); Anon. (1987a); Kemp and Hart (1987); Linnhoff and Eastwood (1987); G. Smith and Patel (1987); Trivedi, Roach and O'Neill (1987); Linnhoff and Polley (1988); Linnhoff, Polley and Sahder (1988); R. Smith and Linnhoff (1988); Szargut, Morris and Steward (1988); Vogler and Weissman (1988); ASME (1989/202); Kemp and Deakin (1989); Pethe, Singh and Knopf (1989); Linnhoff, Smith and Williams (1990); Tomlinson, Finn and Limb (1990a,b); ASCE (1991/45); Kemp (1991); Polley and Shahi (1991); Polley, Shahi and Nunez (1991); Suaysompol and Wood (1991); S.W. Morgan (1992); Linnhoff (1993); IChemE (1994/15)

Data management
Waligora and Motard (1977); Winter and Newell (1977); Salkovitch (1982); Power (1983)

Data correlation
D.S. Davies (1962); Rowe (1963); J.R. Campbell and Alonso (1978); Heller (1978); Deliquet (1979); Volk (1979a,b); Hughson (1987); Tao and Watson (1987, 1988); Chakravarty (1993)

Trace quantities, impurities
Fair, Crocker and Null (1972a,b); Grollier-Baron (1992a,b); Joshi and Douglas (1992); Tagoe and Ramharry (1993)

Water removal, gas and liquid drying
Gondo and Kusunoki (1969); K.G. Davis and Manchanda (1974); Lees (1974d); Weiner (1974); Sigales (1975); Abernathy (1977); R.A. Johnson (1980); G.S. Mason (1982); Grilc, Golob and Modic (1984); Holmes and Chen (1984); B.W. Bradley (1985); Joshi and Fair (1991)

Safety in process design
MCA (SG-14); Cronan (1960b,c); Jenett *et al.* (1964); Weatherby (1964); G.T. Austin (1965a, 1982a); H.H. Fawcett and Wood (1965, 1982); D.T. Smith (1965a); Leeah (1968–); Burklin (1972); Corley (1972); Kerns (1972); Anon. (1976 LPB 8, p. 1); Coffee (1976); Fitt (1976a); Wells, Seagrave and Whiteway (1976, 1977); H.D. Williams (1976); Kletz (1977i,k, 1979c,e,k, 1981e,h, 1985k, 1990g,h, 1993d); Meeks and Campion (1977); Nakano (1978); Burklin (1979); Lord and Hirst (1979); K.Palmer (1979); Stockburger and Kühner (1979); Goldfarb *et al.* (1981); Collings and Luxon (1982); Husmann and van de Putte (1982); London (1982); Goodier and Cece (1983); Lieberman (1983); Hamm (1984); Burgoyne (1985c); Heard (1986); Bond and Bryans (1987 LPB 75); S.E. Dale (1987); Hendershot (1987); T. Martin (1989c); Trask (1990)

Inherently safer design
N.A.R. Bell (1971); Anon. (1979 LPB 30); Hearfield (1979, 1980a); Kletz (1978c, 1979e,f,n, 1980j,k, 1983 LPB 51, 1984d, 1985h,k, 1988k, 1989c, 1990f–h, 1991f,h,i, 1992c); Middleton and Revill (1983); Anon. (1985t); Anon. (1986q,r); Goodfellow and Berry (1986); Zanetti (1986a); Caputo (1987); S.E. Dale (1987); Anon. (1988f); Barrell (1988a); Gerritsen and van't Land (1988); Lihou (1988); Anon. (1989 LPB 87, p. 21); Hathi, Sengupta and Puranik (1990); Schaller (1990); Englund (1991); Jacob (1991a); R.L. Rogers and Hallam (1991); Butcher (1992c); Frohlich and Rosen (1992a); P. Jones (1992); Villermaux (1993); Mansfield and Cassidy (1994)
Process intensification: N.A.R. Bell (1971); Waldron, Erstfield and Criswell (1979); Anon (1983s); Anon. (1985t); H. Short (1983b); Ramshaw (1983, 1985, 1987); Drinkenburg (1988); Fowler (1989); Kletz (1991h,i); Whiting (1992); Benson and Ponton (1993)
Limitation of inventory: Lofthouse (1969); Tucker and Cline (1970, 1971); Kletz (1975e, 1978c, 1979e, 1985j); R.J. Parker (1975); Harvey (1976, 1979b); Orrell and Cryan (1987); Wade (1987); Malina (1988)
Friendly plants: Kletz (1989c, 1990g,h)

Licensors, vendors, contractors
N.H. Parker (1964); Bresler and Hertz (1965); Spitz (1965); Anon. (1967a, 1970b); R.L. Miller (1970); Gersumsky (1977); Williamson, Hackel and Wright (1978); R.P. Willis (1981); Carmody (1983); MacFie (1983); V. Parker (1991)

Technology transfer
Yawwak (1978)

Project engineering and management
Rase and Barrow (1957); Bergtraun (1978); J.C. Rose, Wells, Yeats (1978); Clough and Sears (1979); Kerridge (1979, 1981); Kimmons (1979); Krishnaswami (1979); Roth (1979); Bush (1980); Kurzawa (1980); Shanmugam (1980); Dinger (1981); Holt and Russell (1981); Cleland and King (1983); Henriksen (1983); Braye (1985); Fawbert (1985); Kharbanda and Stallworthy (1985, 1986a); BS 6046: 1981–
Specification: Elliott (1986); Turney (1994 LPB 118)
Engineering construction contracts, model forms of contract: AIChE (Appendix 28 *Engineering Construction Contracts*); IChemE (1968/46, 1992a–c); IBC (1993/97); D. Wright (1993)
Critical path methods: Glaser and Young (1961); Mauchly (1962); Martino (1963, 1964); Munro (1967); Jenett (1969); Wiest and Levy (1969); Moder and Phillips (1970); Kerridge (1978)

Value engineering
Wolstenholme (1962); Gage (1967); Pegram (1990)

Design documentation
Schwartz and Koslov (1984); Anon. (1991 LPB 98, p. 13)
Symbols: ISA (Appendix 27); IEEE (1991 IEEE 91); BS (Appendix 27 *Symbols*)

Package plants
Kletz (1986h)

Batch plants
E.R. Robinson (1975); Armstrong (1983); Mehta (1983); Parakrama (1985); Love (1987a–c, 1988); A. Wilson (1987)

Multiproduct plants
Flatz (1981); Armstrong (1983); Kirchoff *et al.* (1983); Schramek (1984); Grist (1985)

Pipeless plants
Shimatani and Okuda (1992); Tadao (1993)

Multiple units
Quigley (1966); Kletz (1985h)

Miniplants
Robbins (1979)

Plant construction
H.W.Russell (1989)

Case studies of process design include *Process Design Case Studies* (R. Scott and McLeod, 1991) and the Institution of Chemical Engineers (IChemE) model design projects described by Jeffreys (1961), Austin and Jeffreys (1979) and Ray and Johnston (1990).

Treatment of the SLP aspects of process design is given in *Hazard Survey of the Chemical and Allied Industries* (Fowler and Spiegelman, 1968), *Flowsheeting for Safety* (Wells, Seagrave and Whiteway, 1976), *Industrial Hazard and Safety Handbook* (R. King and Magid, 1979), *Safety in Process Plant Design* (Wells, 1980), *Chemical Process Safety* (Crowl and Louvar, 1990), *Safety in the Process Industries* (R. King (1990), *Safety at Work* (Ridley, 1994) and by Austin (1965a, 1982a), Hudson (1965) and Fitt (1976a).

Selected references on process design are given in Table 11.1; selected references on chemical reactors, unit processes, unit operations and equipments, extreme operating conditions, utilities, particular chemicals, particular processes and plants, ammonia, urea and ammonium nitrate plants, and liquefied petroleum and natural gases (LPG and LNG) are given in further tables in this chapter.

11.1.1 Design responsibility

It is the responsibility of senior management to define policy and to indicate how this may be translated into systems and procedures, to ensure that the policy is carried out and to advise on divergences between policy and legal requirements. Middle management is responsible for preparing detailed systems and procedures, for monitoring their operation and for giving expert advice. The responsibility of engineers who do the detailed work is to follow the systems and procedures and to use standards and codes of practice, but to draw attention to any problems which arise.

Process design is carried out by a team of people of different disciplines. There are several essential conditions which must be observed if this activity is to be carried out properly. The responsibility of individuals should be clearly defined, the nature of their work should not take them outside their sphere of competence and their workload should not be excessive. There should be proper systems and procedures to support them. There should be a suitable proportion of experienced people who can supervise the work of others.

The occurrence of errors in design, as in all other human activities, should be expressly recognized and measures taken to minimize the problem. There should be a system for the cross-checking of designs. Such checks should cover not only the detailed calculations but also the assumptions and information on which the design is based. For critical features the check should be done by personnel who are not intimately involved in the design and are to that extent independent.

An individual involved in the design should advise if he considers a particular aspect is outside his competence. The code of the professional engineer requires him to do this. Its importance is illustrated by the Flixborough disaster. It is also the responsibility of the individual to draw attention to any feature which he thinks is important, but which has apparently been overlooked. He should not assume it is not his job to do so or that someone else has it in hand. It emerged at the Tay Bridge disaster that the man painting the bridge knew about many of the defects, but did not think it was his responsibility to tell anyone.

11.1.2 Design stages

Some important stages of the design of a process plant are

(1) research and development;
(2) process design:
   (a) process flowsheet;
   (b) detailed process design;
(3) engineering design and equipment selection.

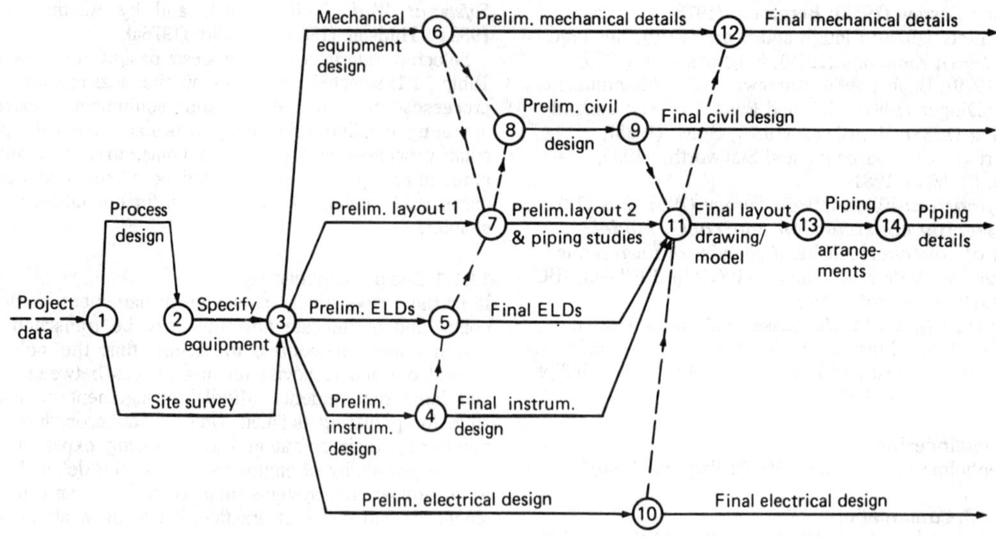

Figure 11.1 Typical plant layout and design network (Mecklenburgh, 1973) (Courtesy of the Institution of Chemical Engineers)

The design of a plant is an iterative process. Modifications are made as more information becomes available, as constraints or opportunities are recognized, and as the situation changes. A typical plant layout and design network is shown in Figure 11.1.

It is necessary to pay close attention to the scheduling and co-ordination of the project, if extensive delays are to be avoided. Much use is made of critical path scheduling (CPS) and of the project evaluation and review technique (PERT).

11.1.3 Design information

Process design can be properly done only if there is adequate and correct design information. This should include as a minimum:

(1) the physical and chemical properties of the chemicals;
(2) the reaction characteristics, including mechanism, kinetics and thermal data for all likely reactions;
(3) fire, explosion and toxic hazards;
(4) the effect of impurities.

These aspects have been considered in more detail in Chapter 8 in relation to hazard identification.

Some of the data particularly relevant to SLP is now given in material safety data sheets (MSDSs). Some principal compilations are those by Keith and Walters (1985–), A. Allen (1988), Walsh (1988), and Kluwer Publishers (1992). The data used in the design should be not only correct but also consistent throughout the design team. It is necessary, therefore, to devote some effort to the provision and documentation of these data.

11.1.4 Design standards and codes

The standards and codes which are to be used in the design should be specified from the start. Standards and codes give design requirements and guidance. They constitute both an aid to the designer and a means of communicating design requirements to other parties, with certain contractual implications. They embody the lessons which industry has learned from past incidents and are therefore a store of experience which the designer should draw on.

Consistency should be maintained both in the initial choice of standards and codes and in their subsequent use. A particular code or set of codes embodies a particular design philosophy. It is not good practice to use an indiscriminate mix of codes or to use a different code for a particular problem simply because its requirements are less onerous. This can cause difficulties not only at the design stage but also later in plant operation and maintenance when confusion may arise as to the design philosophy.

There is often some degree of overlap in the topics covered by different codes. Consideration should be given to areas where codes may conflict or where there may be gaps which are not covered.

An account of standards and codes is given in Appendix 27.

11.1.5 Design experience

It is useful to consider the forms in which process design experience and know-how are available. The information may relate to

(1) chemical reactors;
(2) units processes;
(3) unit operations and equipment;
(4) operating conditions;
(5) utilities;
(6) particular chemicals;
(7) particular processes and plants.

Thus much information is available on: the characteristics of chemical reactors, including reactor protection; unit processes, such as oxidation; unit operations, such as distillation; operating conditions, such as high pressures and low temperature; utilities, such as electricity, steam and nitrogen; on particular chemicals such as ethylene or chlorine; and particular processes and plants, such as air and ammonia plants.

In particular, information is available in the form of national standards and codes of practice, in-house standards, procedures and reports, and checklists. This documentation is the principal means whereby experience won at considerable expense of money, effort and sometimes life is made available for design of new plants. This information should be used in design, and in the reviews of the design.

11.1.6 Design communication and documentation

It is essential in the design process that there be effective communication. Much of this communication is done on an individual basis or in design committees. The most important channel of communication, however, is the documentation. Details of this have been given in Chapter 6 and are not repeated here. Of particular importance are the process flow sheet, the process flow diagram and the engineering line diagram. Examples of the two latter are shown in Figures 11.2 and 11.3.

One item of information which is often required for SLP work and which has frequently not been shown on flow diagrams is the inventory of the principal vessels. It is desirable that flow diagrams show the normal and maximum design inventories.

It is no mean task to ensure that all members of the design team have the up-to-date information which they need. The problem of 'design disclosure' has been the subject of study, notably by the armed forces, which have developed a number of techniques to assist in this.

11.1.7 Design changes

The design process is one of flux in which changes are continuously being made at all levels. It is necessary, therefore, to have a system for the management of change during the design. Elements of such a system include the declaration, checking, authorization and communication of changes. A similar system is required for the control of changes in the construction and commissioning stages. The control of plant modifications is considered further in Chapter 21.

11.1.8 Overdesign

Overdesign in engineering is often equivalent to the incorporation of an extra factor of safety, but this is by no means always so. In some cases such overdesign can reduce safety. There is an inherent tendency to overdesign in a project as the various individuals in the chain introduce factors of safety. In this context overdesign is taken to cover the purchasing as well as the design decisions. What matters is the item which is finally installed.

This is illustrated by two common items of process equipment, pumps and control valves. If a pump is required which has a certain characteristics relating to flow and pressure and if the pump installed is capable of a higher flow or pressure, this may introduce a hazard insofar as the rest of the plant is not designed for these conditions. Similarly, a control valve may be oversized. If reliance has been placed on a valve of a certain size to limit flow to a maximum value and a larger valve is installed, this may create a hazardous situation.

11.1.9 Design error propagation

It is sometimes appropriate to check the error associated with a design calculation, particularly if this has important safety aspects. If a calculated variable z is a function of two other variables x and y

$$z = f(x, y) \qquad [11.1.1]$$

then the calculated value of z will have an error δz which results from the errors δx and δy in x and y, respectively.

There are two errors which are of interest. The maximum error is:

$$\delta z = \left|\frac{\partial f}{\partial x}\right|\delta x + \left|\frac{\partial f}{\partial y}\right|\delta y \qquad [11.1.2]$$

A more realistic estimate is the probable error:

$$\delta z = \left[\left(\frac{\partial f}{\partial x}\right)^2 \delta x^2 + \left(\frac{\partial f}{\partial y}\right)^2 \delta y^2\right]^{\frac{1}{2}} \qquad [11.1.3]$$

Further information is given in standard texts (e.g. Jenson and Jeffreys, 1963). A discussion of error propagation is given by Park and Himmelblau (1980).

11.1.10 Design and terotechnology

The need for a systems approach to the design, operation and maintenance of equipment is the underlying theme of terotechnology, which is defined as: 'A combination of management, financial, engineering and other practices applied to physical assets in pursuit of economic life cycle cost'. And further: 'Its practice is concerned with the specification and design for reliability and maintainability of plant, machinery, equipment, buildings and structures, with their installation, commissioning, maintenance, modification and replacement, and with feedback of information on design, performance and costs.'

The origins of terotechnology are in the problem of maintenance costs and its emphasis is on economics. But the concerns of terotechnology and loss prevention are clearly related. It is a main aim of both to reduce failures by effective feedback of information on the performance of plant equipment. The principles of terotechnology are described in the *Terotechnology Handbook* (Parkes, 1978).

11.1.11 Computer aided design

Modern process design involves considerable use of computer-aided design (CAD) techniques, particularly for flowsheeting, equipment design, plant layout, piping and instrument diagrams. Other applications continue to be developed. In particular, extensions of computer design are developing in the area of information flow in plants, leading to automatic hazard identification, fault tree analysis and reliability assessment.

In the design context, computer systems constitute a powerful tool not only for design but also for information storage and retrieval, and thus for communication. Here their effectiveness depends on the data base. The design

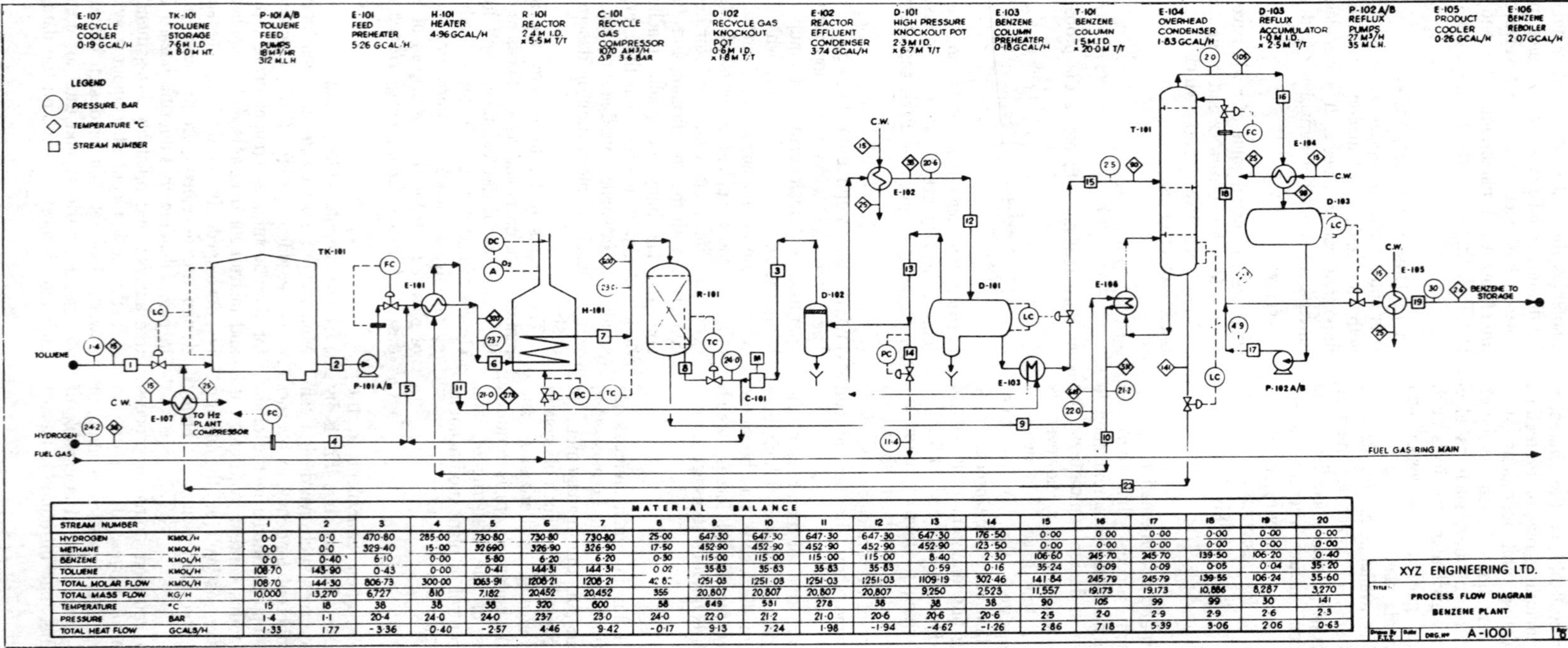

MATERIAL BALANCE

STREAM NUMBER		1	2	3	4	5	6	7	8	9	10
HYDROGEN	KMOL/H	0·0	0·0	470·80	285·00	730·80	730·80	730·80	25·00	647·30	647·30
METHANE	KMOL/H	0·0	0·0	329·40	15·00	326·90	326·90	326·90	17·50	452·90	452·90
BENZENE	KMOL/H	0·0	0·40	6·10	0·00	5·80	6·20	6·20	0·30	115·00	115·00
TOLUENE	KMOL/H	108·70	143·90	0·43	0·00	0·41	144·31	144·31	0·02	35·83	35·83
TOTAL MOLAR FLOW	KMOL/H	108·70	144·30	806·73	300·00	1063·91	1208·21	1208·21	42·8[illegible]	1251·03	1251·03
TOTAL MASS FLOW	KG/H	10,000	13,270	6,727	810	7,182	20,452	20,452	355	20,807	20,807
TEMPERATURE	°C	15	18	38	38	38	320	600	58	649	531
PRESSURE	BAR	1·4	1·1	20·4	24·0	24·0	23·7	23·0	24·0	22·0	21·2
TOTAL HEAT FLOW	GCALS/H	1·33	1·77	−3·36	0·40	−2·57	4·46	9·42	−0·17	9·13	7·24

STREAM NUMBER		11	12	13	14	15	16	17	18	19	20
HYDROGEN	KMOL/H	647·30	647·30	647·30	176·50	0·00	0·00	0·00	0·00	0·00	0·00
METHANE	KMOL/H	452·90	452·90	452·90	123·50	0·00	0·00	0·00	0·00	0·00	0·00
BENZENE	KMOL/H	115·00	115·00	8·40	2·30	106·60	245·70	245·70	139·50	106·20	0·40
TOLUENE	KMOL/H	35·83	35·83	0·59	0·16	35·24	0·09	0·09	0·05	0·04	35·20
TOTAL MOLAR FLOW	KMOL/H	1251·03	1251·03	1109·19	302·46	141·84	245·79	245·79	139·55	106·24	35·60
TOTAL MASS FLOW	KG/H	20,807	20,807	9,250	2,523	11,557	19,173	19,173	10,886	8,287	3,270
TEMPERATURE	°C	278	38	38	38	90	105	99	99	30	141
PRESSURE	BAR	21·0	20·6	20·6	20·6	2·5	2·0	2·9	2·9	2·6	2·3
TOTAL HEAT FLOW	GCALS/H	1·98	−1·94	−4·62	−1·26	2·86	7·18	5·39	3·06	2·06	0·63

Figure 11.2 *Process flow diagram of a benzene plant (Wells, Seagrave and Whiteway, 1976) (Courtesy of the Institution of Chemical Engineers)*

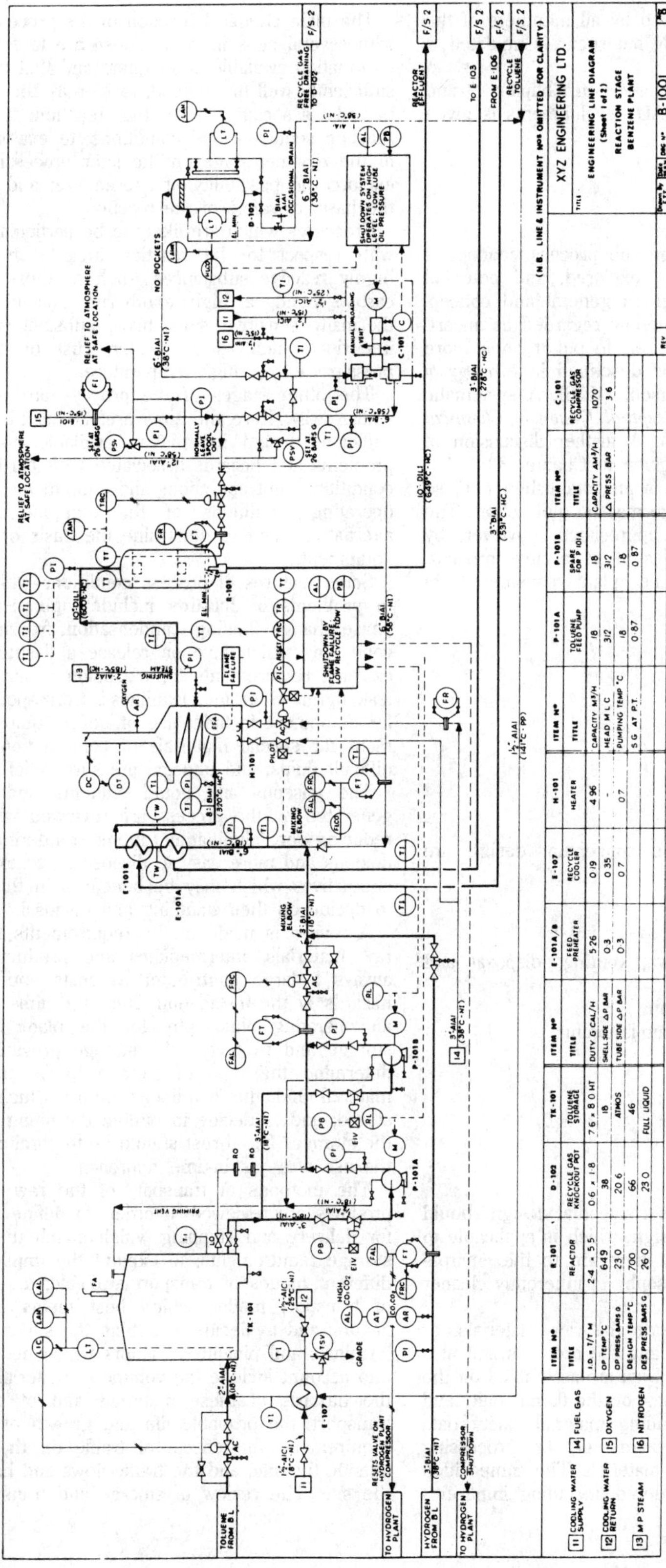

Figure 11.3 *Part of engineering line diagram of a benzene plant (Wells, Seagrave and Whiteway, 1976) (Courtesy of the Institution of Chemical Engineers)*

of data bases which can be used by all members of the team and which can be quickly but securely up-dated is another area of development.

A fuller treatment of CAD is given in Chapter 29 and treatment of more advanced CAD developments is given in Chapter 30.

11.2 Conceptual Design

At the conceptual design stage the process concept is developed, its implications are explored, and potential problems are identified. Design in general and conceptual design in particular is generally regarded as an art. Much work is going on, however, to put it on a more systematic basis. Early work is described in *Strategy of Process Design* (Rudd and Watson, (1968). A systematic approach is stated in *The Conceptual Design of Chemical Processes* (J.M. Douglas, 1988). A further discussion of fundamental developments is given in Chapter 30.

In most cases, the process is an established one, so that the conceptual design stage may be quite short. The nature of this stage is best appreciated, however, by considering the conceptual design of a new process. Elements of the process and plant considered in conceptual design are:

(1) process materials;
(2) chemical reaction;
(3) overall process;
(4) effluents;
(5) storage;
(6) transport;
(7) utilities
(8) siting and layout.

The topics addressed in the conceptual design are principally:

(9) process design;
(10) mechanical design;
(11) pressure relief, blowdown, venting, disposal and drains;
(12) control and instrumentation;
(13) plant construction and commissioning;
(14) plant operation;
(15) plant maintenance;
(16) health and safety;
(17) environment;
(18) costing;
(19) project engineering.

The philosophy underlying the conceptual design should be that of inherently safer design, which is applicable to virtually all aspects. Likewise, in regard to the environment, there should be a philosophy of inherently cleaner design.

The process materials include the raw materials, or feedstocks, the intermediates, and the products and also catalysts, additives and so on. Data are assembled on the physical and chemical properties, on the flammability and the health and toxicity, including material safety data sheets, and any special hazards of, or processing problems associated with, the materials. The composition of the raw materials is considered, including impurities which they may contain.

The main chemical reaction of the process is reviewed with several aims in view. These are to ensure that the information available is adequate and that the reaction is sufficiently well understood; to identify the problems and hazards associated with the reaction; to define the reaction sequence and conditions; to examine the effect of the reaction stage on the later processing stages; to explore the possibility of alternatives; and to determine the basis of safety of the reaction.

Processes which are likely to be particularly hazardous with respect to the reaction include those involving: highly reactive substances, such as those with a triple carbon bond; a high exotherm; unstable substances; thermally sensitive substances; substances sensitive to impurities such as air, water, rust or oil; and high pressure and/or high temperature.

The other stages of the process are reviewed in a similar way. Here the aims are: to ensure that there is sufficient design information available; to identify the problems and hazards associated with particular process conditions, unit operations and equipments; to define the operating conditions of the equipment; to explore alternatives; and to determine the basis of safety of the equipment.

Some features of process conditions which may point to problems or hazards include: operation close to a phase change (boiling, condensation, freezing); operating conditions which give on release a flashing liquid; and extreme requirements for exclusion of air or water, for leak tightness or for cleanliness. Corresponding features for equipment include: use of novel equipment; requirement for special materials of construction; pumping of difficult fluids; and unusual pressure relief requirements.

The gaseous and liquid effluents and solid wastes generated by the process are reviewed with a view to: reducing or eliminating them; rendering them less noxious and more easy to dispose of; to explore process alternatives which may be beneficial in this regard; and to decide on their handling and disposal.

A review is made of the requirements for storage of raw materials, intermediates and products. Storage is always a large contributor to costs and often to the hazards of the installation. Here the aims are: to define an operating philosophy for the plant in respect of storage and to match the storage provided to this; to determine the type of storage to be used for each material and the conditions under which it is to be stored; and to decide in outline the siting and layout of the storage. The thrust should be to eliminate storage by the use of a 'just-in-time' approach.

The methods of transport of the raw materials and products are reviewed in order: to define material flows for delivery and shipping which match the process and storage requirements; to explore the implications of the different modes of transport available; to select the mix of transport modes which best meets the governing factors; and to decide in outline the siting and layout of terminals and pipelines. Factors which need to be taken into account include the volume of materials transported, the hazards of these materials and of the modes of transport, the probable life and growth of the transport requirement, the effects of traffic on the environment outside the site, and the traffic flows and handling within the site. The review of storage and transport generally involves interactions particularly between these two but also with the process.

The utilities for the process are reviewed in order: to define the requirements not simply in terms of the nominal quantities but also of the quality and dependability of the supply; to identify any problems and hazards, including implications for existing plants; to explore alternatives; to determine the method of supply for each utility; and to decide in outline on siting and layout. On the basis of the foregoing information, decisions may then be made concerning the siting and layout. An account has been given in Chapter 10 of the principles applicable at the conceptual design stage.

In the foregoing the conceptual design has been treated in terms of the process and the plant. In the following it is considered essentially in relation to the design disciplines. The first of these is process design. Some account of this has already been given above. In essence, process design at this stage involves consideration of the process and plant from a number of different viewpoints: the process flow sheet and process conditions; the unit processes; and the unit operations and equipments. Two of the principal aims at this stage should be the minimization of hazards by inherently safer design and the minimization of effluents by inherently cleaner design. Regard should also be given to the location where the plant is to be built. Features such as the nature of the available workforce or the access for transport of plant equipment may be relevant. In some cases such factors may point to a more robust design.

The reliability and availability aspects of the plant need to be given some thought at this stage. Different philosophies may be adopted. Options are to accept loss of a function for a period or to minimize its downtime by use of a single high reliability unit or to use several units, with parallel redundancy or stand-by configurations. For some types of plant a single-stream philosophy is the norm. Even with such plants, however, there are certain types of unit, such as furnaces, which are commonly multiple and others for which stand-by configurations are widely used.

The process designer is rarely a completely free agent. In most cases the process in view is an established one, but even where this is not the case there tend to be numerous constraints. There is commonly within the company itself a preference for established technology and an aversion to innovation and the risk perceived to be associated with it. Generally, there are requirements imposed by outside parties. A licensor of a process usually imposes certain requirements which are a condition of the guarantees provided. Likewise, an equipment vendor generally makes his guarantees conditional on requirements such as use of certain ancillary equipment or of alarm and trip instrumentation. An insurer often has preferred arrangements for fire protection. Acceptance of various constraints may be a condition for raising the necessary finance.

Mechanical design at this stage is fairly limited, but consideration is given to materials of construction and to the principal vessels and other items of equipment.

The arrangements for pressure relief, blowdown and venting, for relief and vent disposal, and for drains are also addressed at this stage. These generally have a number of significant implications which it pays to explore at the conceptual stage, because changes later in the design tend to be either disruptive and costly or simply not possible. Thus a decision to contain rather than to relieve a high pressure may entail a stronger vessel, but such vessels have to be ordered early. A decision to dispose of a relief flow by venting or flaring may require a sterile area which needs to be incorporated early in the layout.

Control and instrument system design is concerned at this stage with the philosophy in respect of allocation of function between the human operator and automatic systems and of the means of providing the latter. The process may be reviewed to assess unsteady-state features, including the dynamics of continuous processes, the sequential operation of batch processes, and start-up and shut-down, and to identify any particular control and instrument problems, including difficult measurements.

The conceptual design stage also covers consideration of the construction and commissioning, operation and maintenance of the plant. Construction creates requirements for space on site, for additional traffic and for lifting equipment, and may affect existing plants. Commissioning also may effect existing plants as the new plant is tied in and manned. The operation of the plant too has implications for other linked plants and for storage requirements.

The start-up and shut-down of the plant are considered at the conceptual design stage. Start-up involves taking the plant through a series of discrete stages. In general, some of these stages can be held indefinitely, whilst others are essentially transient, with no choice but to proceed to the next stage or revert to an earlier one. These stages need to be defined and holding states identified. Likewise, normal shut-down may be effected in a series of discrete stages and these need to be defined. It is also prudent to identify holding states, or fall-back positions, short of shut-down, to which the operator can move if abnormal conditions develop, thus taking pressure off him and making the plant more friendly. Definition of emergency shut-down is necessary since it is liable to involve abnormally high flows which have to be catered for.

The plant needs to be reviewed in respect of health, safety and environmental factors. Each of these aspects needs to be assessed using a formal method to identify the problems. Health factors may affect features such as the degree of leak-tightness required, the use of buildings, and the need for particular operating practices.

The factors related to safety are discussed throughout this text and are not rehearsed here, but one particular aspect merits mention at this point. This is the major hazard potential of the plant, which needs to be addressed at the conceptual design stage. This evaluation is necessary because the information gained has a bearing on many fundamental aspects of the design. Methods of hazard identification and assessment appropriate to the conceptual design stage have been described in Chapters 8 and 9. Relevant techniques include the various preliminary hazard study and hazard ranking methods. The formal requirement for the use of such methods as part of the management system has been discussed in Chapter 6.

The approach to environmental aspects is essentially similar, with a formal requirement for the assessment of the major environmental impact potential using specified methods. There are now regulatory requirements for a safety case for major hazard installations and for an environmental impact statement for certain developments, as described in Chapters 3 and 4.

The conceptual design includes, as part of an overall economic assessment, an estimate of the costs of the various elements described.

Finally, there is a project engineering element in the planning carried out at this stage for the various aspects of the project.

11.3 Detailed Design

The detailed design stage involves the detailed process and mechanical design together with detailed design from a large number of supporting disciplines. The treatment given here is confined to a broad outline, with particular reference to features bearing on SLP.

Elements of the detailed design of the process and plant include:

(1) process design;
(2) mechanical design;
(3) storage;
(4) transport;
(5) utilities;
(6) layout;
(7) pressure relief, venting and disposal;
(8) control and instrumentation;
(9) fire protection;
(10) explosion protection;
(11) toxic emission protection;
(12) personnel protection;
(13) plant failures;
(14) plant operation;
(15) plant maintenance;
(16) plant reliability, availability and maintainability;
(17) equipment specification, selection and procurement;
(18) health and safety;
(19) environment.

Central to the process design is the chemical reactor. The design should minimize the probability of a hazardous excursion and provide means for dealing with one should it occur. Reactors and their protection are considered in Sections 11.8–11.13.

The process design should take full account of the experience available, both within the company and elsewhere, in respect of the unit processes, the unit operations and equipments, the operating conditions, the utilities, the particular chemicals and the particular processes and plants, as described in Sections 11.14–11.19, respectively.

The process design sets the operating conditions throughout the process for normal operation and defines the envelope of conditions within which the process can operate safely. This information governs the mechanical design. Consideration should be given in the process design to the various operational deviations which may occur and to the impurities which may be present, as described in Sections 11.20 and 11.21, respectively.

In the mechanical design, the materials of construction need to be compatible with the process materials not only at flow sheet conditions, but within the whole envelope of operating conditions, and compatible also with each other. Attention needs to be paid to impurities which may cause greatly increased corrosion. Consideration has to be given also to erosion.

The design should ensure effective containment of the process fluids within the pressure system of vessels, pipework and rotating machinery. Most losses of containment occur from pipework and fittings. Detailed attention needs to be given to pipework, pipework joints and pipework supports, and to the numerous features which can cause damage, including vibration, thermal expansion and contraction, shut-in fluids, water hammer and external events. Likewise, weak points on rotating machinery such as pump seals require consideration.

In addition to avoidance of loss of containment, it is also necessary to design to minimize much smaller but continuous leaks, known as fugitive emissions. These are considered in Chapter 15.

There often arise situations where there is a trade-off to be made between flexibility and complexity. It may be possible to design manifolds which allow a more flexible use of certain items of equipment but at the price of a considerable increase in complexity of the pipework. Often complexity may be too high a price to pay for the additional flexibility. The mechanical design of the plant is dealt with in Chapter 12.

Features of the design of storage are the choice between pressure and refrigerated storage for liquefied gases, the specification of the design pressure and temperature, the venting and the operating and fire relief arrangements, the fire protection, the bunding or drain-off arrangements, and other aspects of storage layout.

For the loading and unloading terminals for the transport of raw materials and products, features are the fluid transfer arrangements, the fire protection, the emergency isolation, the control of ignition sources, including precautions against static electricity and hazardous area classification, and other aspects of terminal layout.

The design of the utilities should aim to ensure a dependable supply of any critical utility. This includes consideration of protection against events which may threaten the main supply and of the provision of a back-up supply, particularly an emergency power supply. Utilities such as inert gas are often vulnerable to contamination by substances from process or storage and it may be necessary to take measures to prevent this. A common cause is backflow, which may be minimized by the use of utility pressure greater than that of the potential source of contamination. The utilities are considered in Section 11.17.

Plant layout is considered in Chapter 10. Features which merit mention here are: the provision of separation to minimize both ignition of flammable leaks and the effects of fire and explosion, and hence domino effects; the control of ignition sources through hazardous area classification; the provision of a suitable drainage system; and the labelling of equipment to ensure positive identification in plant operation and maintenance; as well as the aspects of the layout of storage and terminals already mentioned.

The arrangements for venting, pressure relief and blowdown are a significant feature and have a pervasive influence on the overall design. It is normally necessary to cater: for venting from the process, either continuously or periodically; for pressure relief for operational conditions; and for pressure relief for fire conditions. Features are: the relief scenarios; the location, setting and capacity of relief devices; the selection of these devices; the relief collection system; the choice between relief to atmosphere or to a disposal system; and the selection of the disposal system. It is also necessary to cater for the relief and blowdown which occurs during emergency shut-down. Venting and pressure relief are treated in Chapters 12 and 15.

Aspects of the control and instrumentation system which are particularly relevant here are the provision of: protective systems in the form of trips and interlocks; a fire and gas detection system; and an emergency shut-down system. Control and instrumentation is considered in Chapter 13.

The fire protection system will generally rely on a combination of passive features such as layout and fire insulation and active ones such as fire and gas detection, emergency blowdown, and fire fighting and equipment cooling systems using water spray, foam and other measures. Fire and fire protection are discussed in Chapter 16; layout aspects are treated in Chapter 10.

Protection may also be required against explosion of flammable vapours or dusts within the plant or inside buildings. Explosions and explosion protection are considered in Chapter 17; layout aspects are treated in Chapter 10.

If the substances handled are particularly toxic, it may well be necessary to design for an enhanced degree of protection against them. The first line of defence is a more leak-tight plant. Dilution and removal of any leaks which occur is assisted if the plant is located in the open, but if it is in a building ventilation may be necessary. Toxic releases and occupational health are dealt with in Chapters 18 and 25; fugitive emissions are considered in Chapter 15.

Personnel protection should not be neglected. Aspects include: access to equipment, particularly hand valves; layout to minimize contact with corrosive substances; guarding of moving machinery; insulation and other measures to prevent contact with hot surfaces; and safeguards to prevent injury due to the operation of fire extinguishing systems. Chapters 10, 16 and 25 in particular deal with such topics.

The design should cater for the requirements of plant operation, discussed in Chapter 20. The requirement to isolate items of equipment or sections of the plant should be studied and features such as high frequency of isolation or need for positive isolation identified and, where appropriate, measures taken or facilities provided. Thus flexibility may be built into the pipework to assist the insertion of slip plates, or isolation facilities such as a spectacle plate or double block and bleed valves, may be provided.

The requirements for plant start-up, normal shut-down and emergency shut-down should be studied and facilities provided as necessary.

Facilities need to be available to dispose of any material which is off specification, contaminated, or otherwise unsatisfactory, which may be produced in the course of plant operation, whether in the plant itself or in storage.

Inspection is practised on a range of plant equipment and using a variety of techniques, and for some of these it may be necessary to provide facilities, even if this amounts to no more than suitable access. Likewise, facilities may be required for various types of test on equipment. Plant inspection and testing is considered in Chapter 19.

For plant maintenance reference has already been made to isolation arrangements. Maintenance often involves emptying the plant of fluids and suitable vents and drains need to be provided for this purpose. It may also be necessary to clean the inside of the plant, and this too may require particular facilities. Plant maintenance is discussed in Chapter 21.

The failures that may occur on the plant should be considered. The items of plant equipment should be identified which are particularly vulnerable to frequent failure and/or failure of which could have serious consequences. A technique particularly suitable for investigation of potential equipment failures is failure modes and effects analysis (FMEA).

Consideration should be given to hazardous passages of fluid through isolation, which include leaks through valves which are not leak-tight and opening of bypasses around equipment.

The effect on the plant of failure of utilities such as electrical power or cooling water should be reviewed. Techniques include FMEA and event trees. The need for emergency back-up supplies should considered, as described earlier.

The effect of abnormal conditions and of start-up, normal shut-down and emergency shut-down on plant under design on other, linked plants should receive attention. Links exist by virtue of the transfer of materials between the plants or common use of a utility. Likewise, the effect on the plant in question of such conditions on other plants should be considered.

For critical functions the reliability and availability required should be specified. Steps can then be taken to ensure that the specification is met by the elimination of causes of failure, the use of single, high reliability items, or the use of suitable configurations of multiple items, such as installed spares. This approach should also be applied to the utilities, including the emergency back-up supplies.

The specification, selection and procurement of equipment should receive its share of attention. The equipment specification should match the design of the plant into which it is to be incorporated. The initial specification may be defective in neglecting some important aspect. Alternatively, it may be unnecessarily restrictive. It may well pay to discuss the specification with other parties to the design and with the prospective manufacturer. In selecting the equipment, the experience of the manufacturer in supplying equipment for similar duties should be taken into account. It should be borne in mind, however, that whilst a manufacturer supplies a range of equipments which are suitable for a generic duty, he cannot be expected to know about special features or requirements which may exist on the plant in question.

The detailed design needs to be subjected to review in respect of health, safety and environment. The design assessments made of these and other aspects are now considered.

11.4 Design Assessments

As the design progresses, it is subject to various assessments, which draw on a number of specific techniques. Those considered here are (1) critical examination, (2) value engineering assessment, (3) energy efficiency assessment, (4) reliability and availability assessment, (5) hazard identification and assessment, (6) occupational health assessment and (7) environmental assessment.

11.4.1 Critical examination

Critical examination, which is a structured method for asking basic questions about the plant, is an effective method for fundamental review, exposing assumptions and generating alternative options. There are in effect a family of techniques which have grown out of critical examination. One is hazop, which is directed to hazard identification, as described in Chapter 8. Another is value engineering, which is now considered.

11.4.2 Value engineering assessment

The review of a plant to ensure that its hazards are identified and are under control has already been described, and in particular the use of hazop studies. Another form of critical examination is that conducted for value engineering. Accounts of value engineering are given in *Value Analysis* (Gage, 1967) and, in its application to process plants, is described by Pegram (1990).

Value engineering has found wide application in manufacturing industry generally, but rather less in the process industries. The central concern of value engineering is with the ratio

$$V = F/C \qquad [11.4.1]$$

where C is cost, F is a function and V is a value. The starting point, therefore, is the definition of the function which is to be performed. Once the function has been established, one option may be to eliminate it. If this is not appropriate, methods are then considered for minimizing the cost of providing it. Value engineering differs from simple cost reduction in the greater attention which it gives to the various options for eliminating or providing the function in question.

Value engineering is a formal technique. The principal stages are planning, function identification, speculation, evaluation, development and implementation. Like hazop, value engineering utilizes a structured team approach. The team is led by a facilitator.

The central activity is the identification and review of the given function. Analysis of the function includes identifying what it is supposed to do, what it actually does, what features are missing and what are undesirable, whether it is necessary at all, and if it is, how else it might be done. Key questions are 'How?' and 'Why?' A significant aspect of the analysis relates to the customer requirements, where this concept is meaningful.

For the costing Pegler gives as an example the function of a distillation column bypass. Costs associated with this would include: those of the valve, piping, piping supports, insulation and foundations; process design, engineering design and project management; valve and piping support purchasing and expediting; construction and construction management; inspection; and documentation.

Value engineering utilizes the techniques of reliability engineering and hazard identification such as Pareto analysis and FMEA. A large value engineering study may take approximately 300–400 hours, a medium one 100–300 hours and a small one 20–100 hours.

11.4.3 Energy efficiency assessment

There is an increasing emphasis in plant design on energy efficiency and on the use of formal methods of analysis to achieve it. This particular analysis needs to be done early, at the stage where it can influence the conceptual design. A fundamental concept here is that of exergy. Accounts of this concept are given in *The Exergy Method of Thermal Plant Analysis* (Kotas, 1985) and by Tomlinson, Finn and Limb (1990a,b) and Linnhoff and co-workers (see below).

According to the Carnot principle the maximum work which can be delivered by a reversible heat engine is:

$$W_{max} = Q(1 - T_c/T_h) \qquad [11.4.2]$$

where Q is the heat flow (W), T_c is the temperature of the cold sink (K), T_h is the temperature of the hot source (K) and W_{max} is the maximum work (W). Similarly, the exergy, or maximum work which can be extracted from a heat flow via a hypothetical reversible process where the heat sink is the environment is

$$E = Q(1 - T_0/T) \qquad [11.4.3]$$

where E is the maximum work, or exergy (W), T is the temperature (K) and T_0 the temperature of the surroundings (K).

Exergy values may be computed for the different streams in the plant. For a process which has inlet and outlet streams and exchanges energy with the environment, the work supplied may be determined using the Guoy–Stodola equation. One of the terms in this equation gives the 'lost work'. A conventional exergy analysis proceeds by determining the exergy of the different streams and the lost work in the units and identifies the potential for thermodynamic improvement. The overall approach and its application to processes for the recovery of natural gas liquids (NGL) is described by Tomlinson, Finn and Limb.

Such an analysis gives essentially an indication of the theoretical thermodynamic potential. The practical exploitation is assisted by the development of further concepts. This is the field which has come to be known as 'process synthesis'. Major developments in this area have followed from the concept of pinch technology, which sets a practical target for energy efficiency. These developments are described in *User Guide to Process Integration for the Efficient Use of Energy* by Linnhoff *et al.* (1982) and by Linnhoff and co-workers (e.g. Linnhoff and Flower, 1978; Linnhoff and Turner, 1980; Linnhoff *et al.*, 1982; Linnhoff and Eastwood, 1987; Linnhoff and Polley, 1988; Linnhoff and Dohle, 1992; Linnhoff, 1993).

The technology of process synthesis is characterized by a more formal analysis of energy efficiency and a more global optimization of this efficiency, not just within a single plant but across a site and by more extensive use of heat exchange, closer temperature approaches and greater interconnection between plants. The increased connectivity and complexity which may result have implications for SLP.

11.4.4 Reliability and availability assessment

The methods of reliability engineering, described in Chapter 7, should be used to assess the reliability and availability features of the plant. Reliability methods may be used to identify features which are critical in this respect and those which are not. This may well involve consideration of the interaction between the plant and its associated storage, since a principal purpose of storage is to provide a buffer on flows entering or leaving the plant to cope with plant downtime.

For critical functions, reliability targets should be formulated. Setting the appropriate targets is as important as devising means of achieving them. The methods of satisfying these objectives may then be considered. Options include the use of high reliability equipment, redundancy and diversity.

The application of reliability engineering does not simply consist of the use of reliability expressions, such as those for simple series, parallel and stand-by systems, but involves consideration of the engineering aspects. In particular, it is necessary to take into account such features as interaction and dependent effects. Thus, for example, the reliability figures calculated for a system of pumps operating in parallel may not be achieved in practice, because the failures of the individual pumps may not be truly independent. If one pump has already failed, there may be a greater probability of failure of the other pumps due to such features as electric motor overload or steam turbine overspeed.

11.4.5 Hazard identification and assessment

It is clearly essential to have a set of hazard identification and assessment techniques which are matched to the stages of the project. The subjects of hazard identification and assessment were discussed in Chapters 8 and 9. For hazard identification the most widely used method is the hazop study, but there is a variety of other techniques. The methods and their application are described in detail in Chapter 8.

The hazard identification stage usually highlights a number of problem areas which require further study and it is here that techniques such as fault tree analysis are used. The application of fault tree analysis is generally relatively restricted, but it becomes more important when the plant has a high degree of instrumented protection.

For some plants it is necessary to carry out a full quantitative risk assessment, as described in Chapter 9. In particular, some form of quantitative assessment is usually necessary for a safety case under the CIMAH Regulations.

It is important, however, that the use of such techniques should not in any way weaken the responsibility of the designer to get the design right first time. If the discovery of fundamental design errors occurs frequently in hazop or other studies, then there is something wrong.

11.4.6 Occupational health assessment

There should be a parallel set of formal methods for the identification and assessment of health hazards. Here a principal problem is the contamination of the working atmosphere. This tends to arise partly from continuous low level, or fugitive, emissions and partly from operations such as opening up of equipment, sampling, etc. Health hazards and their assessment and control are discussed in Chapters 18 and 25.

11.4.7 Environmental assessment

Another area which requires the use of formal methods of identification and assessment is the environment. The principal features are: the gaseous and liquid effluents and solid wastes, together with material relieved or vented under emergency conditions and contaminated material; noise and, sometimes, light; and the traffic entering and leaving the site. Environmental aspects and their assessment and control are considered briefly in Appendix 11 but essentially they are beyond the scope of this text.

11.5 Licensors, Vendors and Contractors

The description of the design process which has just been given is based essentially on that which takes place in an operating company which is designing its own plant. The ultimate responsibility for the safe design and operation of plant lies with the operating company. It should take the appropriate steps to ensure that the processes designed and the equipment supplied by the other parties are safe.

Some design responsibility resides, however, with other parties. These may be (1) licensor, (2) vendor, and (3) contractor. A treatment of licensing is given in *Licensing Technology and Patents* (V. Parker, 1991). Contractual arrangements are the subject of the series *Model Form of Conditions of Contract for Process Plants* in three volumes – (1) subcontracts, (2) lump sum contract and (3) reimbursable contracts – by the IChemE (1992a–c) (the *'Yellow Book'*, *'Red Book'* and *'Green Book'*, respectively) with a guide by D. Wright (1993) (the *'Purple Book'*).

If the process has been bought under licence, then the licensor has some design responsibility. The extent of this depends on how much information the licensor releases as a result of the licence agreement. Normally the licensor is responsible for the basic process flowsheet, but he may also specify some detailed process design and safety features. It is important to ensure that the division of responsibilities is specified in the licence and that adequate information and documentation are made available.

There is a responsibility on the vendor to supply equipment which is safe. The equipment should conform to specified standards and codes and it should have adequate documentation such as fabrication records or operating instructions. The purchaser is responsible for specifying standards and codes to be used, for ensuring that the equipment delivered is that specified and for seeing that it is used as intended. If changes are required to a licensed process or to equipment, consideration should be given as to whether these constitute modifications on which consultation is required.

The responsibilities of the contractor depend on whether he is responsible for the whole process and engineering design on a turnkey basis or whether he is acting as an extension of the client's own organization and is thus undertaking detailed engineering only. In the first case the responsibility of the contractor is obvious, but even in the second he retains a residual responsibility. Again it is important for the contract to state clearly the division of responsibility. If the design responsibility lies primarily with the contractor, he should subject the design to safety checks using methods of hazard identification and assessment, as already described.

It may happen in some cases that the client proposes a feature which the contractor considers unsafe. If agreement cannot be reached, the contractor should not undertake this feature. Another point of difficulty can arise if the client wishes to make a modification. The contractor should make all reasonable checks that the modification does not introduce a hazard.

11.6 Project Management

It is important for safety not only that the plant should be well designed, but that the project as a whole should be well managed. Accounts of project management are given in *A Guide to Project Procedure* (Rose, Wells and Yeats, 1978), *Project Management Handbook* (Cleland and King, 1983), *Effective Project Cost Control* (Kharbanda and Stallworthy, 1985) and *A Guide to Project Implementation* (Kharbanda and Stallworthy, 1985)

The guide by Rose, Wells and Yeats (1978) covers project preliminaries, project planning, capital cost estimation, process design, basic and detailed engineering design, statutory approvals, design and safety reviews, procurement, installation, commissioning, project records, and production, and gives a case study which includes a safety review and sample documentation.

Many of these aspects are treated in other chapters, in particular Chapter 6 on management systems, Chapter 8 on hazard identification, Chapter 12 on pressure systems and Chapter 19 on plant commissioning and inspection.

11.7 Inherently Safer Design

The best way of dealing with a hazard is to remove it completely. The provision of means to control the hazard is very much the second best solution. In other words, the aim should be to design the process and plant so that they are inherently safer. The importance of limitation of the inventory of hazardous materials in the plant was one of the principal lessons drawn from the Flixborough disaster, as described below. In due course this specific aspect was subsumed in the more general principle of inherently safer design.

Inherently safer design is particularly important for major hazard plants and the concept is a recurring theme in the three reports of the Advisory Committee on Major Hazards (ACMH) and receives more detailed treatment in the Third *Report* (Harvey, 1984). While the basic principle of inherently safer design is generally accepted, it is not always easy to put it into practice. An exploration of the concept and numerous practical examples are given in *Cheaper, Safer Plants* (Kletz, 1984d).

'Inherently safer design' is preferred to the alternative term 'intrinsically safer design', since the latter is already associated with electrical safety, and the comparative term 'safer' is preferred to the absolute 'safe', since no plant is totally safe.

11.7.1 Limitation of inventory

One of the principal ways in which a process may be made inherently safer is to limit the inventory of hazardous material. The scale of the Flixborough disaster was due to the fact that the holdup of flammable liquid at high pressure and temperature in the reactors was large. The inventory in the five reactors and one after-reactor operating at the time of the explosions was about 120 te. The *Flixborough Report* recommended that consideration be given to reducing the inventories on process plants.

It is a normal objective in design to minimize the volume of process vessels, as this saves on the cost both of the vessels themselves and of the supporting structures. But the reduction of holdup, though recognized as a generally desirable aim from the safety viewpoint, had not been particularly emphasized as an explicit aim in safe design.

This situation has been changing, largely as a result of Flixborough. The Second *Report* of the ACMH (Harvey, 1979b) states that limitation of inventory should be a specific design objective. The philosophy of limitation of inventory has been memorably captured by the motto of Kletz (1978c): 'What you don't have, can't leak'. It is better to have only a small inventory of hazardous material than a large one which can be rendered relatively safe only by highly engineered safety systems. In other words, it is better to keep a lamb than a caged lion. Some examples of limitation of inventory are given in the following sections.

11.7.2 Some basic principles

Some basic principles of inherently safer design are

(1) intensification;
(2) substitution;
(3) attenuation;
(4) simplicity;
(5) operability;
(6) fail-safe design;
(7) second chance design.

A plant which embodies these principles is described by Kletz (1989c, 1990f) as a 'friendly plant'. Some features of friendly plants are given in Table 11.2.

11.7.3 Intensification

A principal route to limitation of inventory is intensification of the process. This means carrying out the reaction or unit operation in question in a smaller volume. Process intensification is applicable to a wide range of chemical engineering operations, including reactors, mass transfer operations such as distillation and gas absorption, and to heat exchange, as described below.

11.7.4 Substitution

Another principle of wide applicability is that of substitution, in which a hazardous feature is replaced by a less hazardous one. Applications described below include substitution in the main process reaction and in heat transfer media.

Table 11.2 *Some features of friendly plants (Kletz, 1990f) (Courtesy of Elsevier Publishing Company)*

Characteristic		*Friendliness*	*Hostility*
1. Intensification	Distillation	Higee	Conventional
	Heat transfer	Miniaturized	Conventional
	Nitroglycerine manufacture	NAB process	Batch process
	Intermediate storage	Small or nil	Large
	Reaction	Vapour phase	Liquid phase
		Tubular reactor	Pot reactor
2. Substitution	Heat transfer media	Non-flammable	Flammable
	Solvents	Non-flammable	Flammable
	Chlorine manufacture	Membrane cells	Mercury and asbestos cells
	Carbaryl production	Israeli process	Bhopal process
3. Attenuation	Liquefied gases	Refrigerated	Under pressure
	Explosive powders	Slurried	Dry
	Runaway reactants	Diluted	Neat
	Any material	Vapour	Liquid
4. Simpler design with fewer leakage points or opportunities for error		Hazards avoided	Hazards controlled by added equipment
		Single stream	Multi-stream with many cross-overs
		Dedicated plant	Multi-purpose plant
		One big plant	Many small plants
	Spares	Uninstalled	Installed
5. No knock-on effects		Open construction	Closed buildings
		Firebreaks	No firebreaks
	Tank roof	Weak seam	Strong seam
	Horizontal cylinder	Pointing away from other equipment	Pointing at other equipment
6. Incorrect assembly impossible	Compressor valves	Non-interchangeable	Interchangeable
	Device for adding water to oil	Cannot point upstream	Can point upstream
7. Status clear		Rising spindle valve or ball valve with fixed handle	Non-rising spindle valve
		Spectacle plate[a]	Slip plate
8. Tolerant of maloperation or poor maintenance		Continuous plant	Batch plant
		Spiral wound gasket	CAF gasket
		Expansion loop	Bellows
		Fixed pipe	Hose
		Articulated arm	Hose
9. Low leak rate		Spiral wound gasket	CAF gasket
		Tubular reactor	Pot reactor
		Vapour phase reactor	Liquid phase reactor
10. Easier to control	Response to change	Flat	Steep
	Negative temperature coefficient	Processes in which rise in T produces reaction stopper	Most processes
		Most nuclear reactors	Chernobyl reactor
	Slow response	AGR	PWR
	Less dependent on added-on safety systems	AGR, FBR, HTGR PIUS	PWR

Table 11.2 *Continued*

Characteristic		*Friendliness*	*Hostility*
11. Software	Errors easy to detect and correct	Some PES	Some PES
	Training and instructions	Some	Most
	Gaskets, nuts, bolts, etc.	Few types stocked	Many types stocked
12. Other industries	Continuous movement[b]	Rotating engine	Reciprocating engine
	Helicopters with two rotors	Cannot touch	Can touch
	Chloroform dispenser	Reverse connection possible	Reverse connection impossible
13. Analogies		Lamb	Lion
		Bungalow	Staircase
		Tricycle	Bicycle
	Marble on saucer	Concave up	Convex up
	Boiled egg	Pointed end up, hard-boiled, medieval egg-cup	Blunt end up, soft-boiled, standard egg-cup

[a] A spectacle plate is easier to fit (in rigid piping) and easier to find.
[b] In practice reciprocating internal combustion engines are not less friendly than rotating engines, though one might expect that equipment which continually starts and stops would be less reliable.
AGR, advanced gas-cooled reactor; CAF, compressed asbestos fibre; HTGR, high temperature gas reactor; FBR, fast breeder reactor; PES, programmable electronic system; PIUS, process inherent ultimate safety reactor; PWR, pressurised water reactor.

11.7.5 Attenuation

A third principle, that of attenuation, involves the use of less hazardous process conditions. Examples of its application are given below. What constitutes a less hazardous process condition, however, is not always obvious. The problem is a multi-dimensional one. This aspect is described below in the context of the hazards of compromise.

11.7.6 Selection of process

The starting point for inherently safer design is the selection of the process with a view to eliminating particularly hazardous chemicals and/or to operating under less hazardous conditions. Some examples of selection of inherently safer processes are given by Kletz. In dyestuffs production use is no longer made of benzidene and certain other intermediates because they are carcinogenic.

In the production of KA, a mixture of cyclohexanone and cyclohexanol, used in the manufacture of nylon, two routes are available. One is by air oxidation of cyclohexane, the process used at Flixborough. The other involves the hydrogenation of phenol, which occurs in the vapour phase and is less hazardous. That the choice of route cannot be considered in isolation, however, is illustrated by the fact that the usual process for the production of phenol involves oxidation of cumene, a process generally regarded as at least as hazardous as cyclohexane oxidation.

The compounds 4,4-diphenylmethane diisocyanate (MDI) and toluene diisocyanate (TDI) are made using phosgene as an intermediate, which is highly toxic. Efforts have been made to devise an alternative, but these have not been entirely successful. However, manufacturers using the original route have been able to reduce drastically the amount of phosgene stored, or even to eliminate phosgene storage completely by passing the gas direct from the production to the consumer unit.

Generation and immediate consumption, rather than storage, of the intermediate methylisocyanate (MIC) is one of the proposals made for inherently safer design of the process used at Bhopal. Other manufacturers are known to make use of much smaller storages of the intermediate.

11.7.7 Hazards of compromise

It is as well at this point to draw attention to the fact that, as far as operating conditions are concerned, a compromise solution may turn out to be the most hazardous. The hazard may be relatively low for a process with a large inventory but operating at low pressure and temperature such that if an escape occurs only a small amount of material will flash off. At the other extreme a process operating at high temperature and pressure may present only a low hazard, because the use of these operating conditions allows the inventory to be kept low. The compromise solution of moderate inventory at moderate pressure and temperature may actually be the most hazardous, if the conditions are such that on release a large fraction of the material will flash off and the inventory is such that the quantity escaping is likely to be large.

11.7.8 Design of reactors

The reactor is the heart of most processes and it determines the degree of inherent safety, not only because the vessel(s) in which the reaction is carried out usually constitute a hazard but also because the efficiency of the reaction determines the separation processes required downstream. If the conversion is low, large recycles will be necessary and if side reactions produce unwanted by-products, additional separation stages will be required. In the Flixborough

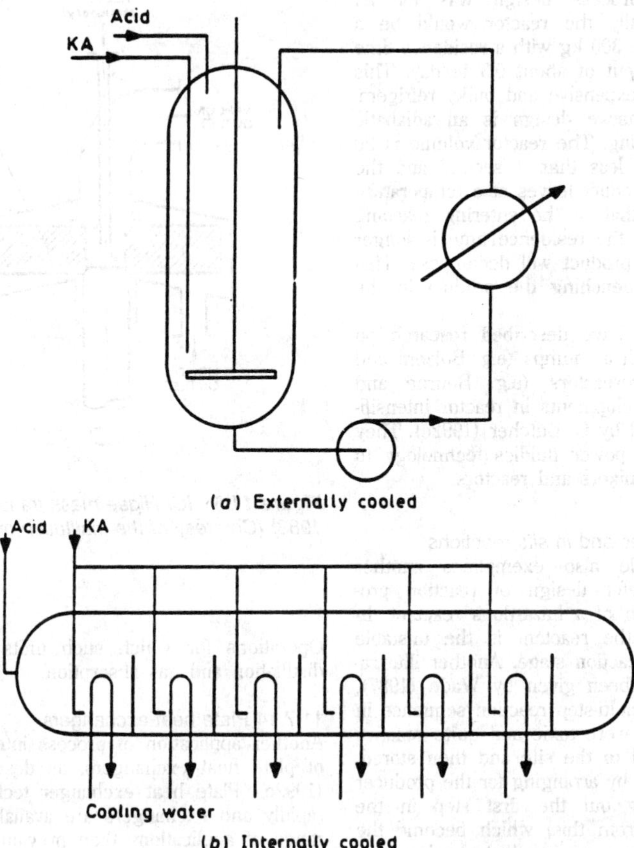

Figure 11.4 Alternative designs of adipic acid reactors (Hearfield, 1980a): (a) original design with external cooling; (b) evolved design with internal cooling (Courtesy of the Institution of Chemical Engineers)

cyclohexane reactors the reaction product still contained approximately 94% of unreacted cyclohexane. The holdup in the reactors and the circulation of unreacted material were both large.

Limitation of inventory in the reactor is illustrated by the evolution of the processes for the manufacture of nitroglycerine, which has been described by N.A.R. Bell (1971). The first stage of development was a batch process with a holdup of about 1000 kg. The second was a continuous process with a 200–300 kg inventory. The third, the NAB process, is again continuous, with reaction taking place in a nozzle and with a holdup of only about 5 kg.

Another example of inherently safer design of a reactor is the development of an alternative design of adipic acid reactor as described by Hearfield (1980a). The original design and the design finally adopted in 1980 are shown in Figure 11.4. In the latter the inventory is reduced and many of the sources of leaks – the agitator, pump, external cooler and connecting pipework – are eliminated.

11.7.9 Semi-batch reactors

In some batch reactors, all the reactants are charged to the reactor at the start and the reaction then proceeds to completion. If a reaction runaway occurs, it is often difficult to control. By contrast, in a semi-batch reactor one of the reactants is fed continuously. In this case, if a runaway reaction starts it is possible to shut off the feed of that reactant. This design is inherently much safer. A more detailed treatment is given below.

11.7.10 High intensity reactors

There is considerable potential for the use of high intensity mixing devices such as pipes, nozzles and pumps as reactors. The NAB process mentioned above is one example. Another example is the process for the production and use of Caro's acid described by Whiting (1992). Caro's acid is an equilibrium mixture of sulphuric acid, water and peroxymonosulphuric acid and is a powerful oxidizing agent. It decomposes exothermically in the presence of transition metals or at elevated temperature. The recommended storage temperature of the concentrated acid is below 0°C. Due to the acid's instability the original process design was for an isothermal reactor. Typically the reactor would be a 30 l vessel containing some 300 kg with a residence time of 30 minutes and an output of about 0.3 te/day. This design involved relatively expensive and bulky refrigeration equipment. An alternative design is an adiabatic reactor with turbulent mixing. The reactor volume is 20 ml, the residence time is less than 1 second and the output is 1 te/day. The product leaves at a temperature of some 50–70°C above that of the entering reactant. Under these conditions, if the residence time is longer than about 1 second, the product will decompose. This problem is handled by quenching the product in the solution to be treated.

Bourne and co-workers have described research on high intensity reactors such as pumps (e.g. Bolzern and Bourne, 1985) and microreactors (e.g. Bourne and Tovstiga, 1988). Other developments in reactor intensification have been described by C. Butcher (1992c). They include the application of power fluidics technology to the design of a range of mixers and reactors.

11.7.11 Reaction sequences and *in situ* reactions

The Caro's acid example also exemplifies another principle in inherently safer design of reaction processes: the ***in situ*** reaction of a hazardous reactant. In the case just described, the reactant is the unstable product from a previous reaction stage. Another illustration of this principle has been given by Wade (1987). The case in point was a multi-step reaction sequence in which the initial reactants were toxic and quite reactive and had to be transported to the site and then stored. This problem was resolved by arranging for the producer of the reactants to carry out the first step in the sequence. The products from this, which become the reactants for the second step, are relatively benign.

11.7.12 Long pipe reactors

Kletz (1991h, 1992c) has drawn attention to the potential of a long, small diameter pipe as a reactor, quoting work by Middleton and Revill (1983). In the event of a rupture the rate of release would be low, and isolation by emergency isolation valves relatively simple.

He points out that, broadly, a flow of 20 000 te/year can pass through a 2 in. (5 cm) pipe at a velocity of 0.5 m/s. He defines the ratio

$$\phi = \frac{800d^2}{W} \qquad [11.7.1]$$

where d is the pipe diameter (cm), W is the throughput (te/year) and ϕ is a ratio. The ideal value of this ratio is unity, but on existing plants it is typically an order of magnitude higher.

11.7.13 Higee process

Returning to applications of intensification other than reactors, an outstanding example in separation operations is the ICI Higee process (Ramshaw, 1983). Mass transfer units based on this principle make use of enhanced acceleration (or g) forces to achieve separations in volumes which are very small compared with conventional equipment. A Higee unit is shown in Figure 11.5.

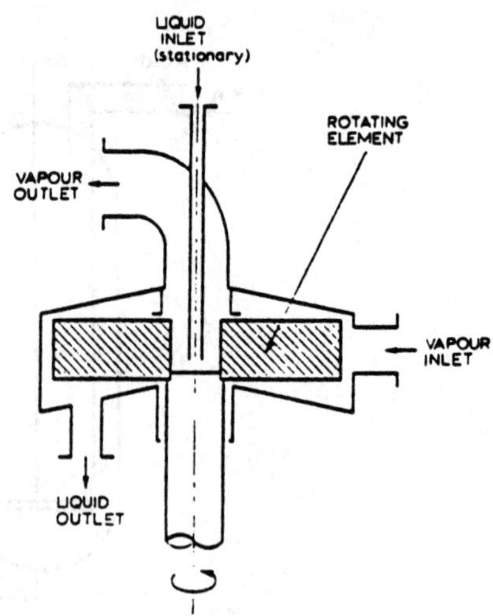

Figure 11.5 *ICI Higee mass transfer unit (Ramshaw, 1983) (Courtesy of the Institution of Chemical Engineers)*

Operations for which such units are suitable include distillation and gas absorption.

11.7.14 Plate heat exchangers

Another application of process intensification is the use of plate heat exchangers, as described by C. Butcher (1992c). Plate heat exchanger technology is developing rapidly and exchangers are available for a much wider range of applications than previously. For example, the pressure range has been extended from about 80 to 1000 bar.

The potential of plate heat exchangers goes beyond conventional heat transfer applications. One concept is the use of plate heat exchangers coated with catalyst as methane reformers. It is estimated that a plate heat exchanger reformer with capacity sufficient for a 1000 te/day methanol plant would be about 4 m^3; this contrasts with conventional reformers which are about 20 m high.

11.7.15 Combination of unit operations

Another form of intensification is the conduct of several unit operations in one equipment. Kletz (1992c) cites the example of a design in which the operations of drying, heat exchange and granulation are combined.

11.7.16 Applications of substitution

Some applications of substitution in the main reaction of the process have been described by S.E. Dale (1987). One is the process for acrylonitrile, which was originally manufactured from acetylene and hydrogen cyanide, both hazardous chemicals, but is now made from propylene, ammonia and air, all less hazardous. Another example is the production of ethylene glycol from ethylene rather than from ethylene oxide, another hazardous material. These are good examples of inherently safer design, even though they have been made for essentially economic reasons.

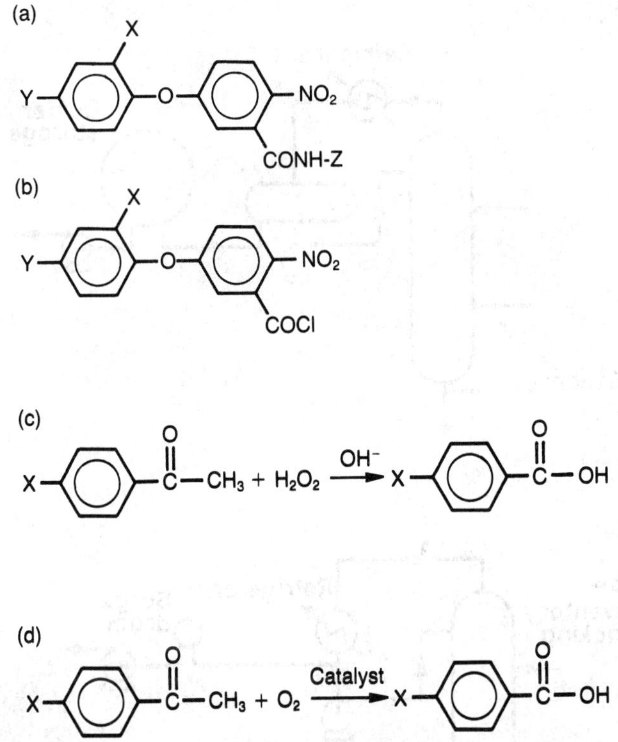

(a)

(b)

(c)

(d)

Figure 11.6 Some reaction routes of varying degree of hazard (R.L. Rogers and Hallam, 1991): (a) substituted nitrocarbamide; (b) nitrobenzoyl chloride intermediate; (c) oxidation of a substituted acetophenone to a carboxylic acid using hydrogen peroxide; and (d) oxidation of the substituted acetophenone to the carboxylic acid using air and catalyst (Courtesy of the Institution of Chemical Engineers)

Further illustrations of the substitution of less hazardous reactions have been given by R.L. Rogers and Hallam (1991). One case quoted is the manufacture of the synthetic nitrocarbamate shown in Figure 11.6(a). One of the routes considered involved the preparation and isolation of the nitrobenzoyl chloride intermediate shown in Figure 11.6(b). Experience indicated that this was likely to be thermally unstable. Further work identified a route in which the nitration is effected at a later stage. Another example involves the oxidation of a substituted acetophenone to a carboxylic acid. The reaction shown in Figure 11.6(c) involves hydrogen peroxide. This was replaced by that shown in Figure 11.6(d) with air and a catalyst.

11.7.17 Applications of attenuation

S.E. Dale (1987) has also described some applications of attenuation. He cites a number of processes which now operate at lower pressures, often due to improvements in the catalysts. Processes which he mentions include methanol synthesis, polyethylene and polypropylene processes, and the Oxo process.

Another process which he quotes is the production of butyl lithium, which is pyrophoric and thus liable to ignite when exposed to air. This hazard is minimized by producing the butyl lithium in dilute solution.

11.7.18 Design of unit operations

Distillation units may contain appreciable inventories of hazardous materials. The quantities held up in a large modern unit are of the same order as those previously held in storage on smaller plants. Substantial reductions in the inventory in distillation units by better design have been demonstrated.

Column inventory may be reduced by the use of low holdup internals. The estimate given by Kletz (1975c) of the holdup per theoretical plate is:

Conventional trays (sieve, valve, etc.)	40–100 mm liquid
Packing	30–70 mm liquid
Film trays	10–20 mm liquid

Holdup in the column base may be reduced by the use of a narrow bottom section. Thermosiphon reboilers have a lower inventory than kettle reboilers. It may be possible to eliminate intermediate storages.

Original and revised designs of a distillation unit for separation of LPG described by Kletz are shown in Figure 11.7. The inventories in the two designs are:

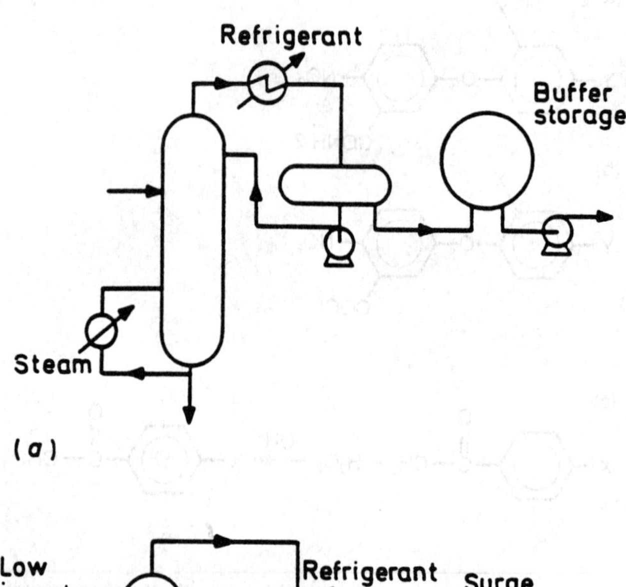

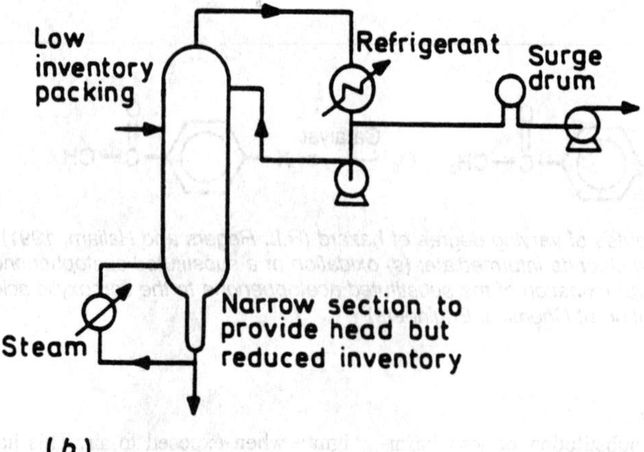

Figure 11.7 *Alternative designs of LPG distillation unit (Kletz, 1984d): (a) original design with high inventory; (b) revised design with low inventory (Courtesy of the Institution of Chemical Engineers)*

	Original inventory (te)		*Revised inventory* (te)	
	Working	*Maximum*	*Working*	*Maximum*
Plant	85	150	50	80
Storage	425	850	nil	nil

In addition to reductions of inventory due to the use of low holdup packing and of a narrow column base, the revised design achieves even larger reductions by the eliminating intermediate storages. The reflux pump suction is taken from the condenser and the reflux drum is omitted. So also are the raw materials and product buffer storage.

Heat exchangers also offer scope for reduction of inventory. In the revised distillation unit design just described the fluids in the condenser were reversed so that the LPG and refrigerant are on the shell and tube sides, respectively. This has the effect of reducing the inventory of coolant and the total flammable inventory. A table of surface compactness of heat exchangers as a guide to inventories is given by Kletz.

11.7.19 Selection of heat transfer media

The refrigeration systems on olefins plants contain large inventories of flammables such as ethylene and propylene. Ammonia is a possible substitute but, in view of the toxic risk seems no safer. A viable alternative refrigerant, however, would make a real contribution to inventory reduction. Some alternative refrigerants have been discussed by Jacob (1991a).

Large inventories of flammables occur in heat transfer systems. For example, in an ethylene oxide plant using kerosene boiling under pressure as the heat transfer medium this may well constitute a greater hazard than the ethylene oxide mixture, which is very closely controlled. Use has been made of steam as an alternative, inherently safer heat transfer fluid.

11.7.20 Design of storage

In general, storage capacity is of great assistance to plant management in the operation of the plant, but this has to be balanced against the cost and the hazard of storage. Application of inherently safer design to storage may involve:

(1) elimination of intermediate storage;
(2) reduction in storage inventory;
(3) storage under less hazardous conditions.

If the plant producing an intermediate can be located near the consumer plant, it may be possible to eliminate intermediate storage altogether. Mention has already been made of the elimination of intermediate storage of phosgene in the manufacture of MDI and TDI.

Even if it is not practical to eliminate storage it may be possible to reduce it. Kletz quotes the case of a company which used to hold several thousand tonnes of chlorine, but found it was in fact able to operate with only a few hundred tonnes.

A third approach is to store the material in a safer form. Examples given by Kletz include the storage and transport of acetylene in acetone, of organic peroxides as solutions and of hydrogen in the form of ammonia which is 'cracked' when the hydrogen is required.

The issue of inherently safer storage conditions arises most often, however, in relation to pressure and refrigerated storage. Considering the storage alone, for a liquefied gas refrigerated storage is safer than pressure storage. Large storages of such gases tend to be refrigerated. On the other hand, if the fluid is required as a gas and it is necessary to provide vaporizers and other equipment, it is the whole system which then needs to be considered and in this case refrigerated storage may be assessed as no safer. Pressure and refrigerated storage are considered further in Chapter 22.

11.7.21 Major hazard storages

An account of the reductions in hazardous storage inventories at one works in the context of the CIMAH Regulations has been given by Orrell and Cryan (1987). The principal hazardous chemicals involved were ethylene oxide, propylene oxide and sulphur trioxide. Disregarding the former, the use of which was discontinued on that site, the reductions achieved between 1980 and 1987 were as follows:

	Inventory (te)				
	1980		*1987*		*1987* CIMAH
	Process	*Storage*	*Process*	*Storage*	*level*
Propylene oxide	51	246	<0.2	<50	50
Sulphur trioxide	<0.5	75	<0.2	<50	75

11.7.22 Toxic storages

Following the Bhopal incident, many companies reviewed their inventories of toxic chemicals and achieved appreciable reductions. An account of one such exercise has been given by Wade (1987), who cites five examples of major reductions of inventory.

In one case, a large acrylonitrile plant, the purification section had a shorter time interval between scheduled shut-downs than the reaction section, and intermediate storage was used to allow the reaction section to continue running whilst the purification section was shut down. Practice was changed so that the reaction section ran at a lower, but safe, throughput whilst the purification section was down and until the latter had run the storage down again, thus effecting a large reduction in the average intermediate storage inventory.

In the second case, by-product hydrogen cyanide from a large acrylonitrile plant was sent to storage before being distributed to satellite plants. The intermediate storage was eliminated, the hydrogen cyanide being passed direct to the user plants. Elimination of the storage created some problems of composition control and of flow smoothing as user plants started up and shut down, but these were met.

A third case involved the elimination of chlorine horizontal storage bullets and their replacement with smaller storage containers. In the fourth case, changes were made to the practice of using as storage rail tank cars hooked up to the plant. One approach was to work more closely with the supplier to ensure that deliveries were made when needed. The other was to work with the railway company to locate areas where such tank cars could be held with minimal public exposure.

The fifth case, described earlier, involved the transfer of the first stage of a multi-step reaction process to the reactant supplier, thus giving more benign reactants for the remainder of the reaction sequence.

11.7.23 On-site production

Another way of eliminating a storage inventory is on-site production of the chemical. There are now available skid-mounted phosgene plants with a capacity of 2–15 te/day which can be installed at the user's work and obviate the need for phosgene storage. On-site production has the further advantage of eliminating the need to transport the chemical produced, although it does of course involve transport of the raw materials.

11.7.24 Design against overpressure

As mentioned earlier, hazards may be introduced by overdesign. Two examples given were installation of an oversize pump or of an oversize valve trim. Either of these may result in overpressure of the downstream plant. Limitation of items such as pump size or valve trim in order to prevent overpressure of the plant is a valid method of achieving inherently safer design, but where it is used it is necessary to ensure that the design intention is not defeated by installation of oversized items either at initial construction or subsequently.

11.7.25 Design for containment

Another aspect of inherently safer design is the prevention and reduction of loss of containment of the more hazardous materials. Three typical leak situations are illustrated in Figure 11.8. Figure 11.8(a) shows a leak of

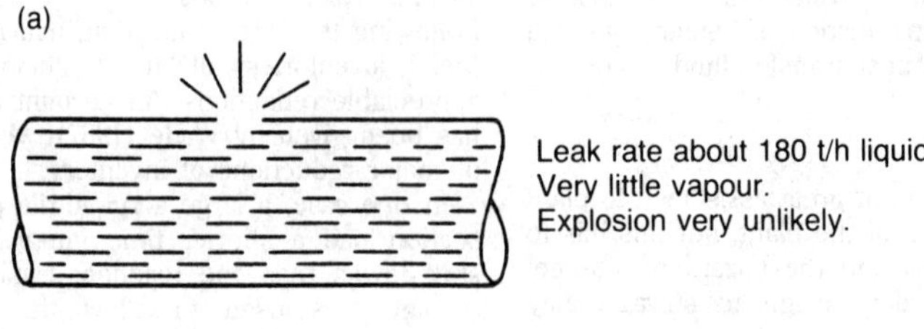

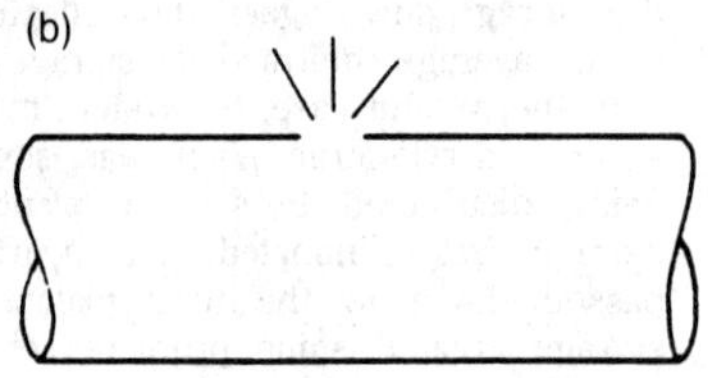

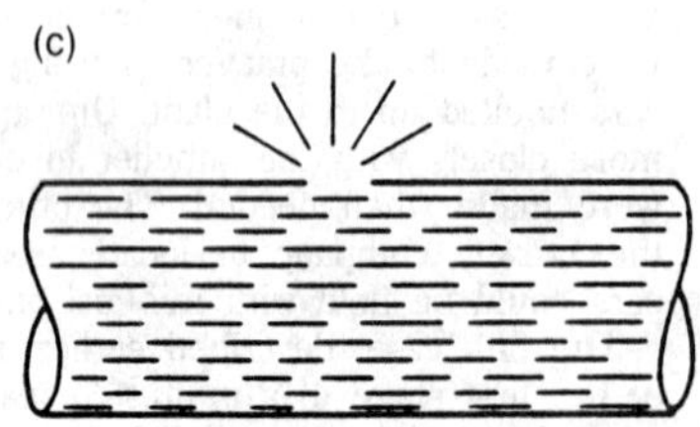

May be dispersed by jet mixing.
Explosion possible but

(c)
Leak rate about 180 t/h liquid
Very little vapour.
Explosion very unlikely.
much of which turns to
vapour and spray.
Explosion much more likely.

Figure 11.8 *Some typical leak situations (Kletz, 1984d): (a) petrol at 7 bar and 10°C; (b) propane gas at 7 bar and 100°C; (c) petrol at 7 bar and 100°C (Courtesy of the Institution of Chemical Engineers)*

liquid petrol at high pressure but atmospheric temperature. For this case the fraction flashing off and the amount of vapour formed are small. Figure 11.8(b) shows a leak of propane gas at high pressure and temperature. The mass flow is much smaller but all the material is in gaseous form. The release may well be rapidly dispersed below its lower flammability limit by the momentum of the escaping jet. Figure 11.8(c) shows a leak of liquid petrol at high pressure and temperature. This is the most hazardous of the three cases. The fluid is a superheated liquid so that the mass flow is high and the flash fraction, and thus the vapour cloud formed, are large. It is therefore this third case which should receive particular attention. One important method of reducing the size of potential leaks is to limit the diameter of nozzle connections on the vessels.

11.7.26 Simplicity in design

A fundamental principle of inherently safer design is simplicity. Many actual plant designs are extremely complex. That shown in Figure 11.9 is used by Kletz (1984d) to illustrate the problem of complexity. Aspects of simpler design instanced by Kletz include:

(1) design for full overpressure;
(2) design modification to avoid instrumentation;
(3) use of resistant materials of construction;
(4) use of simple alternatives to instrumentation.

Where a vessel may be subject to overpressure, the simplest solution may be to design it to withstand this overpressure. This alternative to the provision of a pressure relief system is always worth consideration, though it is generally too expensive.

Sometimes instrumentation is provided to overcome a problem for which the alternative solution is a design change. Kletz gives as an example a liquid phase oxidation plant. The oxidizing air was fed to the plant mixed with the hydrocarbon recycle and instrumentation was provided to keep the mixture outside the flammable range. An alternative design was to add the air directly to the liquid so that a flammable mixture could not form.

Another reason for installing instrumentation may be to prevent attack on materials of construction. Use of more resistant materials may avoid the need for the instruments. Whether this is the best solution depends on the particular case. A contrary example is given in Chapter 9, where hazard assessment showed that use of

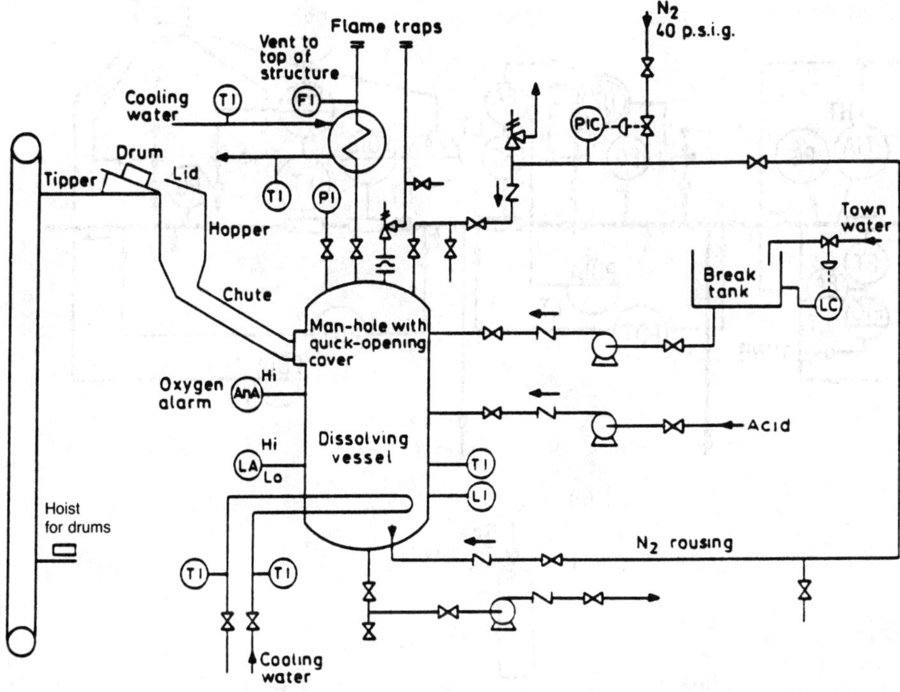

Figure 11.9 *A complex plant design involving instrumentation (Kletz, 1984d) (Courtesy of the Institution of Chemical Engineers)*

a trip was preferable to constructing a pipe in special steel.

Where a control function is to be carried out, simple alternatives to instrumentation should be considered. For example, level control may be effected by a stand-pipe. Some examples of means of avoiding reverse flow are shown in Figure 11.10.

11.7.27 Flexibility of plant

Flexibility is a valuable feature on a plant, but it can be taken too far. Not only can the interconnections needed to give such flexibility become very complex and costly but also they introduce further potential sources of leaks and human errors.

11.7.28 Modification chains

Another source of complexity is the modification chain: an initial modification is made which leads to others. This is a problem not only in design, but perhaps even more on existing plant, where a single *ad hoc* modification can lead to a whole series.

A simple modification chain is shown in Figure 11.11. The manhole cover in Figure 11.11(a) is not leak-tight and traces of flammable vapour escape into the plant. In order to reduce the probability that they will ignite or be inhaled, a vent 4 m high is fitted to the cover, as shown in Figure 11.11(b). In turn, as shown in Figures 11.11(c)–(e), a flame trap is fitted because the gas leaving the vent might be ignited by lightning, an access platform is provided to allow regular cleaning of the flame trap, and handrails and toe boards are fitted because it is realized that the platform is now a 'place of work'.

A more complex example is the system for transfer of slurry under pressure between the two vessels shown in Figure 11.12. Figure 11.12(a) shows the original design with a single transfer line and arrangements to blow the line with steam to clear the line of chokes and to clean it. The plant was the first version to be continuous, earlier plants being batch. In order to avoid down time due to blockages in the transfer line, a second transfer line, Figure 11.12(b), was added. The relief and blowdown review revealed that the two transfer lines could be operated simultaneously and that in this situation the downstream vessels would require twice the relief capacity. In order to avoid this, the valves on the two lines were interlocked, Figure 11.12(c). This created two more dead legs in each line so it was necessary to add four more steam connections to allow steam flushing, Figure 11.12(d). It was then realized that in order to be sure that the stand-by line was ready for use it was necessary to keep steam flowing through it. The transfer line was protected against the pressure of the process but not against that of the steam. Whereas it was considered acceptable not to provide relief for occasional use of steam, it was necessary to do so for steam continuously in the line, and so a pressure relief valve was fitted to the steam line, Figure 11.12(e).

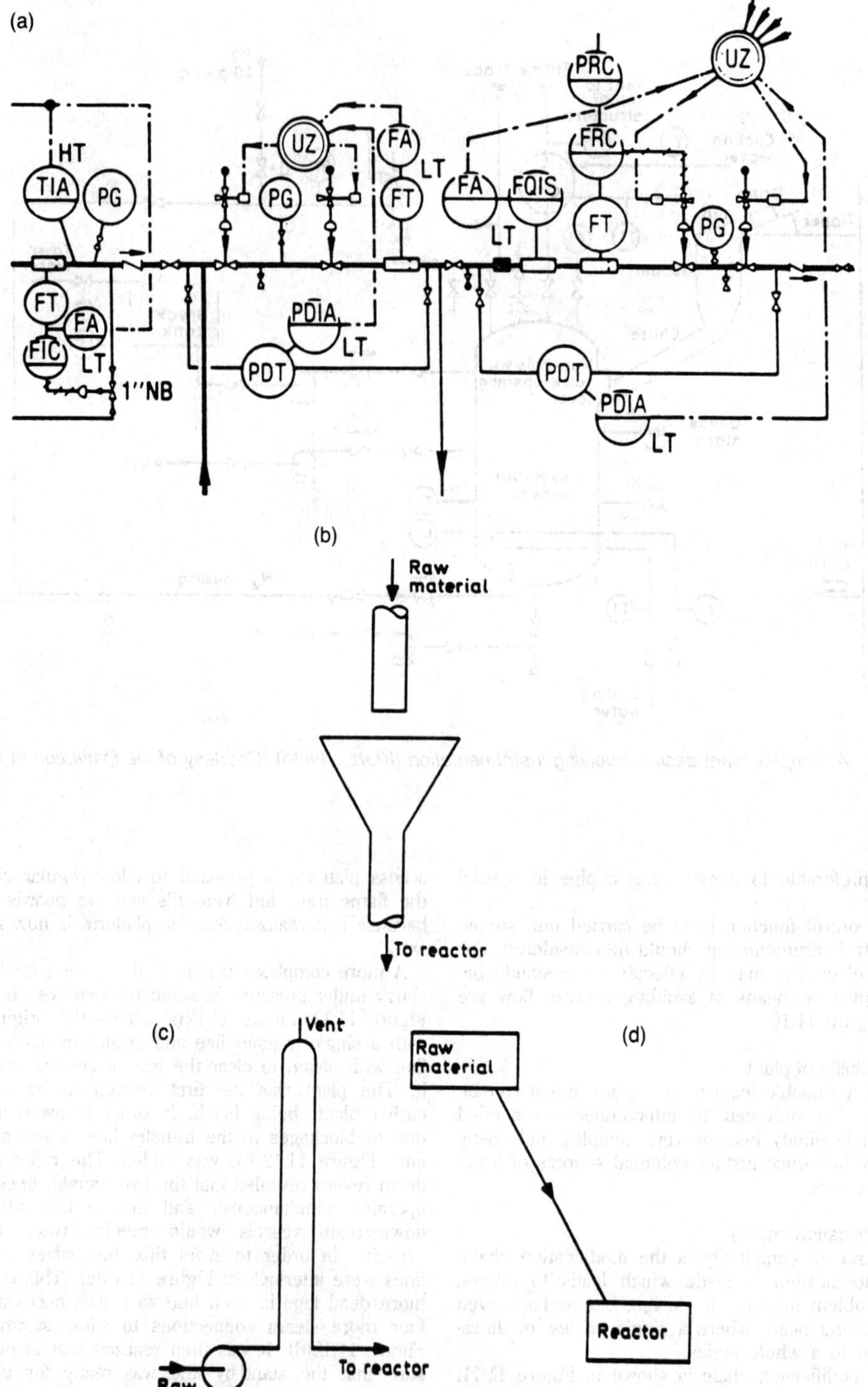

Figure 11.10 *Some simple devices for preventing reverse flow (Kletz, 1984d). (a) Complex instrumented system. FA, flow alarm; FIC, flow indicator controller; FQIS, flow meter (integrating); FRC, flow recorder controller; FT, flow transmitter; HT, high; LT, low; NB, nominal bore; PDIA, pressure difference indicator alarm; PDT, pressure difference transmitter; PG, pressure gauge; PRC, pressure recorder controller; TIA, temperature indicator alarm; UZ, logic system (b) simple tundish system, (c) simple pipe loop system, (d) gravity flow system (Courtesy of the Institution of Chemical Engineers)*

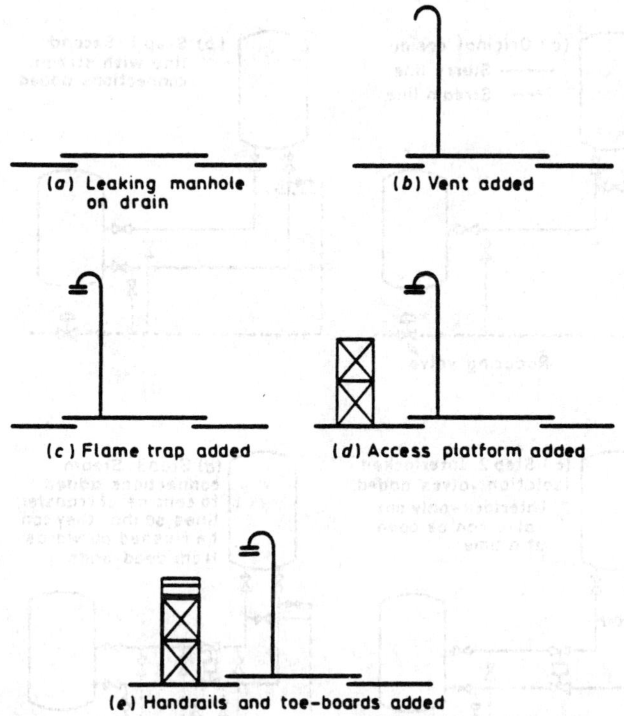

Figure 11.11 *A modification chain for a manhole cover (Kletz, 1984d): (a) manhole cover; (b) addition of vent; (c) addition of flame trap; (d) addition of platform; (e) addition of handrail and toe boards (Courtesy of the Institution of Chemical Engineers)*

11.7.29 Process operability

Some processes are inherently more operable than others. This aspect should be borne in mind in the selection of the process. In particular, a process which has no 'fallback' positions and which in the extreme case presents the operator with a stark choice of continuing to run at a given set of conditions or of shutting down completely is not easy to operate. The operability of process and plant is one of the two main aspects which are examined in hazop studies.

11.7.30 Fail-safe design

The concept of fail-safe design is well established in the process industries. It refers to the design of equipment such as control and solenoid valves so that in the event of failure of a utility such as electricity or instrument air the plant moves to a safe state.

The decision of whether a valve should be open or closed on utility failure is made in the light of the consequences for the process. The fail-safe position, assuming there is one, is then determined. Thus pneumatic control valves, for example, are available as 'air-to-open' or 'air-to-close'; the former close and the latter open on air failure.

The principle of fail-safe design may be extended to include the action of whole control loops. This aspect is considered in Chapter 13.

11.7.31 Second chance design

Another relevant concept is that of 'second chance design'. This means the provision of a second line of defence to guard against an initial hazard or failure. Some features which are prominent in second chance design are:

(1) plant layout;
(2) pressure system design;
(3) materials of construction;
(4) isolation arrangements;
(5) alarms and trips;
(6) operating and maintenance procedures.

Second chance design is illustrated by: many of the normal features of plant layout such as bunding and drainaway arrangements; many normal features of pressure systems such as the installation on pressure vessels of pressure relief and blowdown systems and on pumps of double mechanical seals with monitoring of the pressure sealing system; the use of materials of construction which can withstand deviations from normal operating conditions such as a slug of low temperature liquid; the provision of arrangements for isolation if there is loss of containment; the use of alarms which warn of, and trips which take action against, hazardous conditions; and adherence to operating and maintenance procedures

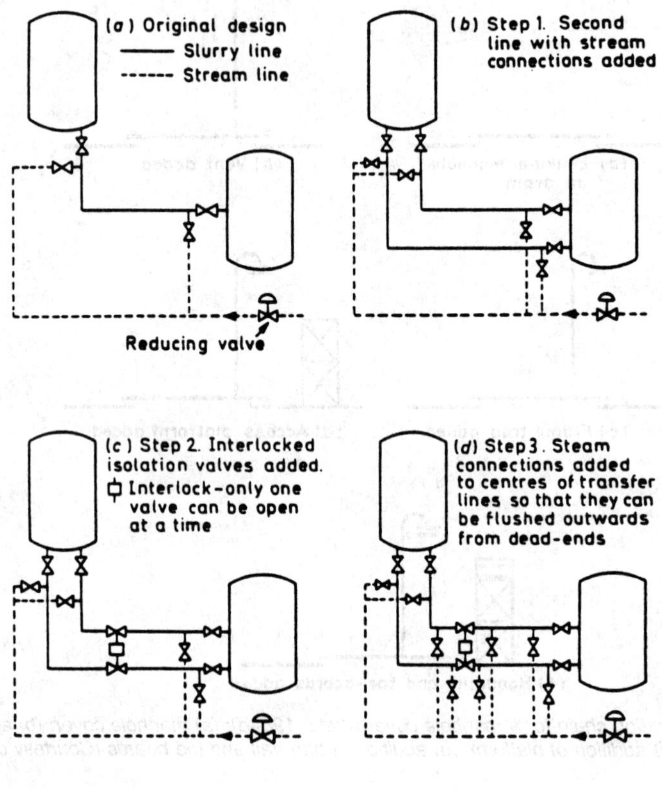

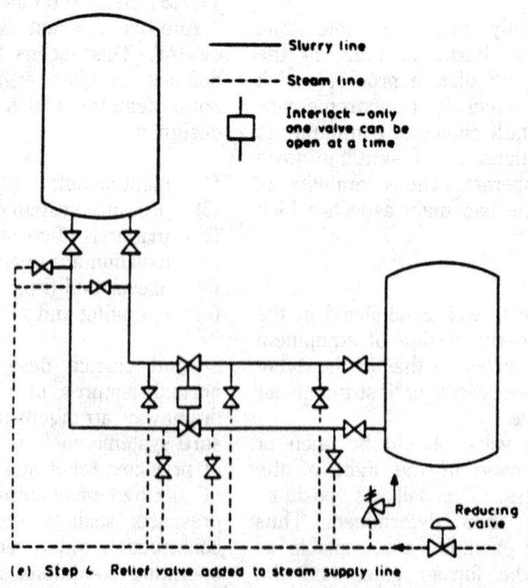

***Figure 11.12** A modification chain for a slurry transfer system (Kletz, 1984d): (a) slurry transfer system; (b) addition of second transfer line; (c) addition of interlock between valves on two transfer lines; (d) addition of extra steam connections; (e) addition of pressure relief valve on steam line (Courtesy of the Institution of Chemical Engineers)*

which reduce the probability of the hazard, such as purging again after a delay in lighting a furnace.

It will be apparent from these examples that second chance design is a rather general concept, embracing both preventive and mitigating features, but is a very valuable one.

11.7.32 Single large vs multiple small systems

A question which occurs in the context of inherently safer design is whether it is preferable from this aspect to have a single large unit, plant or storage, or several smaller ones. Generally it is economically attractive to choose the single large unit. Many consider that this is usually the best choice for safety also. If several small units are built, the number of weak points such as pumps, flanges, etc., increases roughly in proportion to the number of units, and the management effort is more thinly spread. If a single large unit is built, the number of weak points is much reduced and the management can concentrate on this unit. For both these reasons the frequency of incidents with a single large unit should be less than with several smaller ones. The argument for the single large unit is thus, in part, that if one puts all one's eggs in one basket, one can afford a good basket. The drawback is that the scale of any incident which does occur is likely to be larger with the single unit. There is probably no general solution to this problem. Each case should be considered on its merits.

11.7.33 Limitation of exposure

A rather different aspect of inherently safer design is the limitation of exposure of personnel. This is another concept which received impetus from the Flixborough incident, in which 18 of the 28 deaths occurred in the control room. The question was asked whether so many people really needed to be exposed to the hazard in this way. Limitation of exposure of personnel may be achieved by location of the workbase and by control of access to the high hazard zones. These aspects are considered in Chapters 10 and 20.

11.7.34 Inhibiting factors

Progress in the realization of inherently safer designs has not been as rapid as its advocates would wish. Some reasons for this have been examined by Kletz (1991h, 1992c) and Mansfield and Cassidy (1994).

The arguments advanced by Kletz have application particularly where an inherently safer alternative design involves a significant degree of innovation, which affects only a part of the whole plant, and where the scale of production is large. Since, in general, technical and economic risk is increased by novel features, opponents will argue that a large project is being put at risk for relatively small gain.

Essentially, part of the problem is that a single project is being made to bear the 'cost' of an innovation which has the potential eventually to make a significant contribution to the reduction of costs and enhancement of safety over a range of processes operated by the company.

Another feature of the problem is that once plant design is under way, there is often not sufficient time to accommodate such innovation, and there is pressure to follow the previous design. This aspect has been addressed by Malpas (1983), who argues that whilst the design of the next plant is in progress thought should be given to the improvements which have been identified, which it is not practical to incorporate in the design but which, with forethought, could be implemented in 'the plant after next'.

The problem of furthering innovation is exacerbated, moreover, by the move towards business-centred rather than functional organization of companies. However, as Mansfield and Cassidy (1994) point out, inherently safer design is not necessarily radical but may involve incremental change. Among the reasons which they give for lack of progress are lack of awareness, lack of a recognized tool and failure to build the practice into company procedures. Moreover, even where what is in question is a radical change, the proposed process may be such as to constitute a minimal risk. The Caro's acid example described in Section 11.7.10 appears to be such a case.

11.7.35 HSE initiatives

The Health and Safety Executive (HSE) now has an explicit policy of encouraging the application of inherently safer design concepts. Its philosophy and activity in this area have been described by Barrell (1988a) and P. Jones (1992). The work described by Mansfield and Cassidy (1994) is one aspect of this programme.

11.7.36 Inherently safer design methodology

The work of Mansfield and Cassidy is aimed at furthering the practice of inherently safer design by the development of suitable tools. The authors argue that there is need for a technique which will do for inherently safer design what hazop has done for hazard identification. Such a tool needs to address particularly the conceptual design stage. Concept design is generally done quite rapidly. The tendency is that thereafter the basic concept does not undergo any great change.

These authors suggest four broad categories of tool for inherently safer design: (1) brainstorming with a degree of structure; (2) a more highly structured hazop-style examination of flow sheets and process diagrams; (3) examination, using checklists, of the plant layout; and (4) indices of inherent safety.

11.7.37 Inherently safer design index

The development of an index for inherently safer design has been described by D.W. Edwards and Lawrence (1993). This index is intended for use early in the design, at the stage of process selection. The authors list a number of variables relevant to inherent safety, including: properties of the chemicals such as corrosiveness, flammability, explosibility and toxicity; operating conditions such as pressure and temperature; reaction conditions such as phase, rate, heat release, yield and side reactions; and effluents and wastes. At this early stage information on inventory, for example, is lacking and has to be represented by surrogates such as yield.

The method involves assigning scores to the individual parameters. These are then used to compute a chemical score, a process score and an overall inherent safety index. A low value of this index means a high degree of inherent safety. The inherent safety index is obtained from the sum of the scores of the individual process stages. Hence the index penalizes processes with a large number of stages.

Table 11.3 *Some characteristics of miniaturized and of large scale plants (Benson and Ponton, 1993) (Courtesy of the Institution of Chemical Engineers)*

Distributed plants	*Large site plants*
Feedstocks:	
Readily available at the customer's site (e.g. air, methane, water, electricity)	Available from other processes on the site (e.g. chlorine)
Safe to transport (e.g. CO_2, salt, lime, most solids)	Dangerous to transport (e.g. ethylene oxide)
Processes:	
That build new molecules from simple feedstocks (e.g. phosgene)	That involve cracking complex molecules to low molecular weight products (e.g. oil refining, olefins from naphtha)
Catalytic processes (e.g. ammonia from methane)	Low specificity, multi-product processes
Low pressure and temperature (e.g. bioprocesses)	High temperature or pressure (e.g. polyethylene)
Stand-alone processes (e.g. air liquefaction)	Highly integrated processes
Physical change processes (e.g. formulation, shredding)	
Customer products:	
Considered to be dangerous to transport (e.g. phosgene, bromine)	
Production state matches customer requirement, but would require change for transport (e.g. molten polymer, liquefied gases, many powders)	Involving a hazardous processing step (e.g. highly exothermic reaction)
Produced and required in dilute state, but normally concentrated for transport (e.g. aqueous HF)	
Waste products:	
Suitable, after treatment, for returning to the immediate environment (e.g. oxygen, water)	Requiring controlled thermal oxidation (e.g. tars, chlorinated wastes)
May be reacted with other customer waste to produce benign products (e.g. chlorine, caustic)	Treatment requires other plants on the site
Biodegradable waste (e.g. from biological processes)	

Edwards and Lawrence state that one of their aims is to test the hypothesis that inherently safer design means cheaper plants, and give as an illustrative example a comparison of six routes to methylmethacrylate in respect (1) of the estimated costs and (2) of the inherent safety index. Perhaps the most salient point from this single example is the benefit of a smaller number of stages. They also refer to work on comparison of the index with expert rankings.

11.7.38 Miniaturized and distributed plants

In a speculative treatment, Benson and Ponton (1993) suggest that one future trend may well be towards distributed manufacture of chemicals using miniaturized plants at the users' sites. Such an approach chimes with demands for plants which are environmental friendly and deliver their products on a 'just-in-time' basis.

Development on these lines requires that the plants be fully automated and highly reliable. The authors envisage plants that operate at moderate pressure and temperature, are self-cleaning and are sealed for life. There is also an implication that there is no insuperable effluent problem.

Benson and Ponton list a number of types of equipment which have been the subject of much development in recent years but which have not realized their full potential because they come into their own in small scale rather than large scale production. Table 11.3 contrasts some of the characteristics of distributed plants and large scale plants.

11.8 Chemical Reactors: General

The chemical reactor is the heart of the process and is of great importance both in its own right and through its influence on the rest of the process. There are a number of accounts of chemical reactors including *Chemical Engineering Kinetics* (J.M. Smith, 1956), *Chemical Reaction Engineering* (Levenspiel, 1962), *Elements of Chemical Reactor Design and Operation* (Kramers and

Table 11.4 Selected references on chemical reactors and their hazards

Chemical reactivity, chemical reactions
Kirk and Othmer (1963–, 1978–, 1991–); IChemE (1965/41); NFPA (1991 NFPA 491M)

Chemical reactor design, stability, control
E.E. Wilson (1915); Semenov (1935); K.B. Wilson (1946); van Heerden (1953); Bilous and Amundson (1955); J.M. Smith (1956); Aris and Amundson (1957, 1958); Cannon and Denbigh (1957); Boynton, Nichols and Spurlin (1959); Deason, Koerner and Munch (1959); Foss (1959); Walas (1959, 1985); J.E Dawson (1960b); Frost and Pearson (1961); Harriott (1961, 1962, 1964); Levenspiel (1962); Kramers and Westerterp (1963); Berger and Perlmutter (1964); G.T. Austin (1965b, 1982b); Chapman and Holland (1965a,b); Denbigh (1965); Luecke and McGuire (1965); McGuire and Lapidus (1965); Perlmutter (1965); Luyben (1966, 1968); Reilly and Schmitz (1966–); Sabo and Dranoff (1966, 1970); Kneale and Forster (1967); Boudart (1968); Luss and Amundson (1968); Vanderveen, Luss and Amundson (1968); Wang and Perlmutter (1968a–c); Dammers (1969); Gaitonde and Douglas (1969); Ramirez and Turner (1969); Root and Schmitz (1969); Baccaro, Gaitonde and Douglas (1970); Buckley (1970); Ervin and Luss (1970); Satterfield (1970); Vejtasa and Schmitz (1970); Denbigh and Turner (1971, 1984); Luus (1971); Eschenbrenner and Wagner (1972); Barona and Prengle (1973); Cooper and Jeffrey (1973); Guill (1973); Watson (1974); Denn (1975); Carberry (1976); Cresswell (1976); Schafer (1976); C.G. Hill (1977); R. Jackson (1977); P. Johnson (1977); Juvekar and Sharma (1977); Lapidus and Amundson (1977); Rase (1977, 1990); Luyben and Melcic (1978); Rowe (1978); Anon. (1979 LPB 28, p. 91); Corpstein, Dove and Dickey (1979); Froment and Bischoff (1979, 1990); Holcomb and Laschober (1979); Holland and Anthony (1979); Pickett (1979); Y.T. Shah (1978); Trambouze (1979); Butt (1980); J.R. Anderson and Boudart (1981); Espenson (1981); Fogler (1981, 1986); L.M. Rose (1981); Bolliger (1982); Finkelstein (1982); Marzi (1982); Rosenhof (1982a,b); Albright, Crynes and Corcoran (1983); R. Davies (1983); Stiles (1983); Tarhan (1983); Anon. (1984ii); Bartholomew (1984); Berty (1984); Hegedus and McCabe (1984); Hoose (1984); R. Hughes (1984); Morrison (1984); Westerterp, van Swaaij and Beenackers (1984); Augustine (1985); H.H. Lee (1985); Gianetto and Silveston (1986); de Lasa (1986); Bourne *et al.* (1987); Nauman (1987); Rosenhof and Ghosh (1987); Brummel (1989); Trambouze, van Landeghem and Wauquier (1988); AIChE (1989/67); Aris (1990); S.M. Jackson and Dreiblatt (1990); Dickey (1991); HSE (1991a); Nowicki (1991); Bashir *et al.* (1992); Elvers, Hawkins and Schulz (1992); Gordon (1992); Tamir (1994)

Catalysts
Trimm (1980); Stiles (1983); Augustine (1985); Oudar and Wise (1985); Kolb *et al.* (1993)

Low inventory reactors
Marks and Crosby (1975); Middleton and Revill (1983); Bolzern and Bourne (1985); Bourne and Tovstiga (1988); Whiting (1992)

Chemical reactor hazards and safe operation
Anon. (LPB 0 1975, p. 2); Schäfer (1976); Schleich (1976); Devia and Luyben (1978); Knies (1979); Hugo (1980, 1981); Hugo, Konczalla and Mauser (1980); Townsend and Tou (1980); G.T. Austin (1982b); Gilles and Schuler (1982); IBC (1982/25, 1986/66); Welding (1984); Chakrabarti *et al.* (1985); Fauske and Leung (1985); Pilz (1986); Weir, Gravenstone and Hoppe (1986); Eigenberg and Schubler (1987); Grewer and Klais (1987); Steinbach (1987); Fauske (1988a); Verhoeff (1988); Beever and Griffiths (1989); Gusciora and Foss (1989); Gygax (1989, 1990); Marrs and Lees (1989); Marrs *et al.* (1989); Steele *et al.* (1989); Kauffman and Chen (1990); R.L. Rogers and Hallam (1991); J. Singh and Boey (1991); Angelin *et al.* (1992); Spaar and Suter (1992); Barton and Rogers (1993); B.F. Gray, Coppersthwaite and Griffiths (1993); Levy and Larkin (1993); Zaldivar *et al.* (1993); Etchells (1994); Rowe, Starkie and Nolan (1994)
Overall strategies for safe operation: Snyder (1965, 1982); Kunzi (1980); ABPI (1981, 1989); Coates and Riddell (1981a,b); MacNab (1981); Gordon *et al.* (1982); O'Brien *et al.* (1982); Fierz *et al.* (1983, 1984); Schulz, Pilz and Schacke (1983); Regenass (1984); Regenass, Ostwalder and Brogli (1984); Brannegan (1985); Duval (1985); Hoffmann (1985); van Roeckel (1985); N. Gibson (1986b, 1990a); Heemskerk (1986); Pilz (1986); Schofield (1976); Berkey and Workman (1987); Cronin, Nolan and Barton (1987, 1989); N. Gibson, Rogers and Wright (1987); Nolan and Barton (1987); N. Atkinson (1988c); Benuzzi *et al.* (1989); Grewer *et al.* (1989); Lambert and Amery (1989); Pantony, Scilly and Barton (1989); R.L. Rogers (1989b); Gygax (1990); J. Singh (1990a,b, 1993); HSE (1991a); J. Singh and Boey (1991); Raghaven (1992); Barton and Rogers (1993)

Particular reactions
Gupta (1949); Ale (1980); Hearfield (1980a); Heemskerk and Fortuin (1980); Scali, Spinazzola and Zanelli (1980); Storey (1980, 1981); Guenkel, Prime and Rae (1981); Silverstein, Wood and Leshaw (1981); Arabito *et al.* (1983); Roy, Rose and Parvin (1984); Grewer and Hessemer (1987); Tamura *et al.* (1987); Stoessel (1989); Cardillo (1988); Freeder and Snee (1988); Arendt and Marra (1989); Brummel (I989); Cronin, Nolan and Barton (1989); Hoppe and Bruderer (1989); M.C. Jones (1989); Kotoyori (1989b); Levy and Penrod (1989); Ahmed and Lavin (1991); Friedel and Wehmeier (1991); Nicolson (1991); Andreozzi *et al.* (1992); Bergroth (1992); Gustin and Vandermarlieve (1992); Gustin and Vidal (1992); Iizuka and Fujita (1992); Maschio and Zanelli (1992)

Hazard assessment
Arendt and Marra (1989); Marrs and Lees (1989); Marrs et al. (1989)

Westerterp, 1963), *Chemical Reactor Theory* (Denbigh and Turner, 1971, 1984), *Chemical Kinetics and Reactor Design* (A.R. Cooper and Jeffreys, 1973), *Chemical and Catalytic Reaction Engineering* (Carberry, 1976), *An Introduction to*

Chemical Engineering Kinetics and Reactor Design (C.G. Hill, 1977), *Chemical Reactor Design for Process Plants* (Rase, 1977), *Gas-Liquid-Solid Reactor Design* (Shah, 1978), *Fundamentals of Chemical Reaction Engineering* (Holland and Anthony, 1979), *Electrochemical Reactor Design* (Pickett, 1979), *Reaction Kinetics and Reactor Design* (Butt, 1980), *Chemical Reactors* (Fogler, 1981), *Chemical Reactor Design in Practice* (Rose, 1981), *Catalytic Reactor Design* (Tarhan, 1983), *Chemical Reactor Design and Operation* (Westerterp, van Swaaij and Beenackers, 1984), *Elements of Chemical Reaction Engineering* (Fogler, 1986), *Multiphase Chemical Reactors* (Gianetto and Silveston, 1986), *Chemical Reactor Design and Technology* (de Lasa, 1986), *Chemical Reactor Design* (Nauman, 1987), *Chemical Reactors: Design/Engineering/Operation* (Trambouze, van Landeghem and Wauquier, 1988), *Elementary Chemical Reactor Analysis* (Aris, 1990), *Chemical Reactor Analysis and Design* (Froment and Bischoff, 1990) and *Fixed Bed Reactor Design and Diagnostics* (Rase, 1990).

Texts on catalysis and catalysts include *Mass Transfer in Heterogeneous Catalysis* (Satterfield, 1970), *Design of Industrial Catalysts* (Trimm, 1980), *Catalysis* (J.R. Anderson and Boudart, 1981), *Catalyst Poisoning* (Hegedus and McCabe, 1984), *Deactivation of Catalysts* (R. Hughes, 1984) and *Deactivation and Poisoning of Catalysts* (Oudar and Wise, 1985).

Selected references on chemical reactors are given in Table 11.4.

In this section an account is given of reactors in general, with particular reference to continuous stirred tank reactors (CSTRs). In Sections 11.9–11.13 a more detailed treatment is given of batch reactors and their protection.

It is convenient to consider reactors under the following headings:

(1) general considerations;
(2) inventory;
(3) kinetics and heat effects;
(4) stability;
(5) control.

The reactor problem is discussed here primarily in terms of the hazard of strongly exothermic reactions.

11.8.1 General considerations

The most important feature of a reactor is the reaction itself. This may be described in terms of:

(1) phase;
(2) catalysis;
(3) kinetics;
(4) equilibrium;
(5) heat effects.

The reaction may take place in the gas, liquid or even solid phase. It may be catalytic or non-catalytic. It may have simple first or second order kinetics. It may be irreversible or reversible with various degrees of conversion. It may be endothermic or exothermic or have little heat effect.

Important distinctions between reactor types are that between batch and continuous reactors and that between perfectly mixed and plug flow reactors. Some of the main types of reactor are illustrated in Figure 11.13(a–e) which shows, respectively:

(1) batch reactor;
(2) continuous reactor

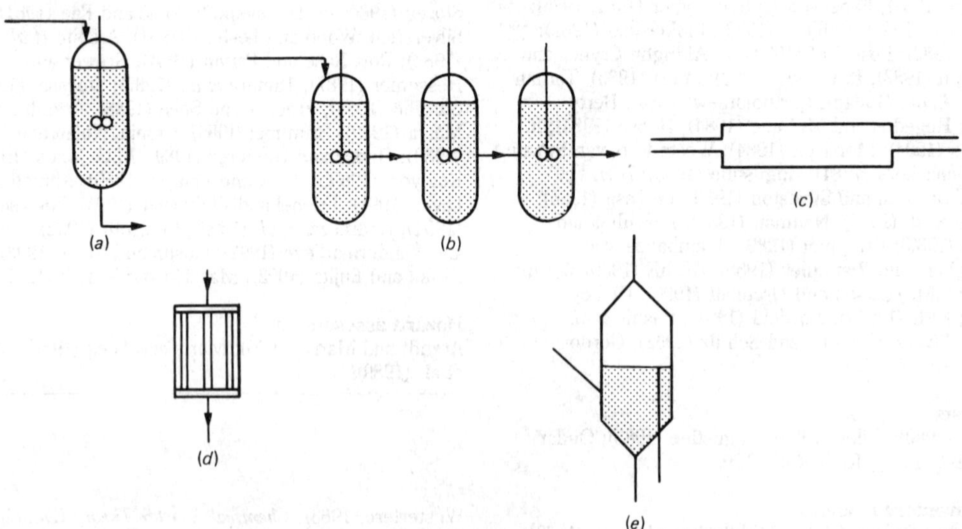

Figure 11.13 Some principal types of chemical reactor: (a) batch reactor; (b) continuous stirred tank reactor (CSTR); (c) tubular reactor (single pipe); (d) tubular reactor (heat exchanger); (e) fluidized bed reactor

(a) CSTR;
(b) tubular reactor (single tube);
(c) tubular reactor (heat exchanger);
(d) fluidized bed reactor.

The tubular reactors may be empty or may contain a packed bed.

The batch and continuous stirred tank reactors approximate to perfect mixing and the tubular reactors to plug flow. The flow through a continuous reactor may be:

(1) single pass;
(2) recycle:
  (a) reactant,
  (b) product,
  (c) inert material.

If conversion per pass is high, a single pass may be sufficient, but if it is low it may be necessary to recycle the reactant. Recycle of product or inert material is often used in order to moderate the reaction. Moderation by product recycle is by dilution or by shifting the equilibrium, that by inert material recycle is by dilution.

There are various ways in which heat may be removed from a reactor. Some heat may be removed in the product stream leaving the reactor, but normally special heat removal arrangements are necessary. Some of these are illustrated in Figures 11.14(a–e), which shows for CSTRs:

(1) recycle cooling;
(2) jacket cooling
  (a) cooling water;
  (b) vaporizing liquid;
(3) tube cooling;
(4) liquid vaporization cooling.

Liquid vaporization cooling involves the vaporization of a liquid in the reactor itself. This liquid may be a reactant, the product or some other material.

The size of the reactor affects the cooling problem. If cooling is applied to the reactor surface, for example, then whereas the reactor volume, and hence the heat generated, increase as the cube of the linear dimension, the heat transfer surface increases only as the square.

A reactor may be operated at low or high conversion. If conversion is low, then there is a potential for further conversion to take place, which in the case of an exothermic reaction may give a greatly increased heat release.

If the reactor is designed to operate with a low difference between the reactor temperature and the coolant temperature, then the amount of heat removed is very sensitive to the temperature difference.

The existence of hot spots may make it difficult to achieve temperature control of the reactor and it is

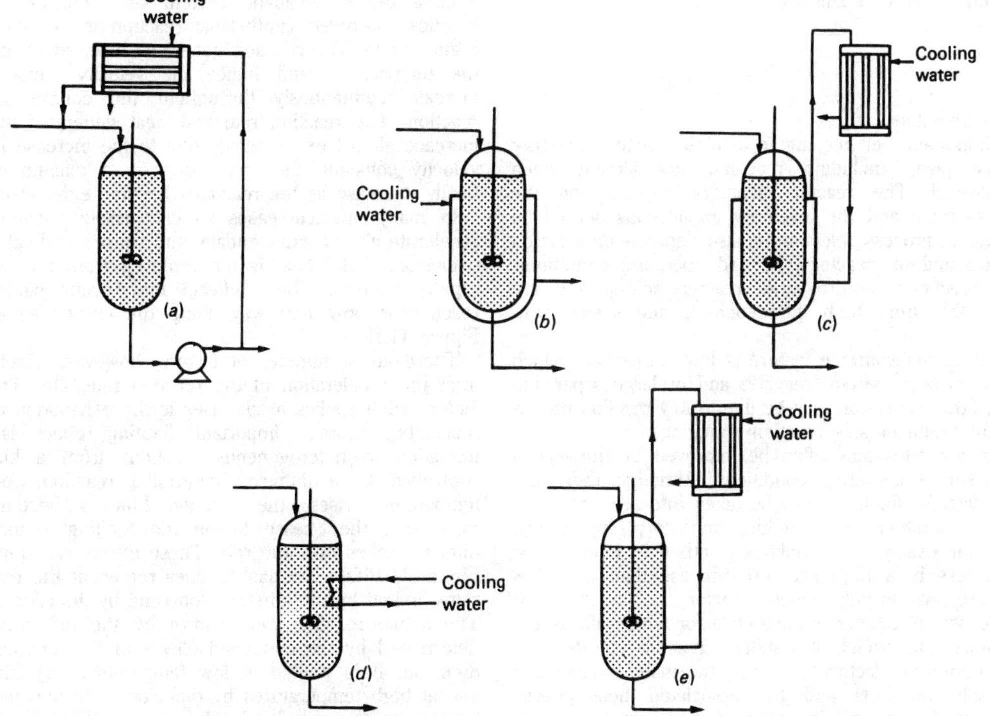

Figure 11.14 *Some methods of removing heat from a continuous stirred tank reactor: (a) recycle cooling; (b) jacket cooling using cooling water; (c) jacket cooling using vaporizing liquid; (d) tube cooling; (e) liquid vaporization cooling*

important to eliminate these. They occur particularly in reactors with solid catalyst. The existence of temperature gradients in a packed catalyst bed, although a normal condition, may also make temperature control difficult if they are too large.

It may be necessary to purge the reactor in order to avoid accumulation of undesirable impurities.

The feasibility of controlling the reactor is highly dependent on the unsteady state characteristics. Usually the reaction temperature is controlled by manipulating a coolant variable. One important characteristic is then the time constant for the response of the temperature to the coolant variable. But other transfer lags and dead time can also greatly affect controllability.

The reactor system should be designed as a pressure system. It should have sufficient combination of mechanical strength and pressure relief to withstand any overpressure to which it may be subjected. Pressure relief on reactors is frequently provided by bursting discs. Design of pressure systems and venting of chemical reactors are discussed in Chapters 12 and 17, respectively.

The abnormal conditions which may occur in the reactor should be reviewed. There may be loss of feed, coolant or utilities. Inability to remove heat or to agitate the reactor may be particularly serious. Alternatively, there may be a temporary interruption with a possible hazard when the facility is restored. Undesired flows may occur from one part of the system to another. In particular, the reactor contents may get into the feed pipes or the coolant may leak into the reactor. Effects may occur due to impurities or catalyst. There may be instrument error or failures.

11.8.2 Inventory

The importance of keeping to a minimum the inventory of the plant, including reactors, has already been emphasized. The reactor inventory depends on the process route and the inventory implications should be a factor in process selection. It also depends on reaction kinetics and on reactor type and operating conditions. Slow reactions require large reactor holdup and are undesirable from both the economic and safety viewpoints.

Another unfavourable feature is low conversion, which results in large reactor recycles and/or large separation units. Low conversion may be due to low reaction rate or to equilibrium or side reaction limitations.

Reaction rates may often be improved by the use of more severe operating conditions. Although there are other factors which have to be taken into account, more severe conditions may reduce inventory appreciably. Thus, for example, the holdup of ethylene tends to be much less in a high pressure process than in a low pressure process polyethylene reactor.

The type of reactor is another factor which affects the inventory. The choice of a suitable reactor type depends on a number of factors, including the main reaction and the side reactions and the associated heat effects. However, in general, inventory is lower in continuous flow than in batch reactors, and lower in vapour phase than in liquid phase reactors.

11.8.3 Kinetics and heat effects

The rate of a chemical reaction is controlled by the reaction kinetics and/or diffusion processes. The reaction considered here is the simple irreversible first-order reaction:

$$\mathrm{A} \rightarrow \mathrm{B} \qquad [11.8.1]$$

with reaction kinetics which may be expressed as

$$r = \rho k x \qquad [11.8.2]$$

where k is the velocity constant at reaction temperature (s^{-1}), r is the reaction rate (kmol/m^3 s), x is the concentration of reactant A (mole fraction), and ρ is the molar density of feed and of reaction mass (kmol/m^3).

The variation of velocity constant with temperature is generally expressed in terms of the Arrhenius equation:

$$k = k_0 \exp(-E/RT) \qquad [11.8.3]$$

where E is the activation energy (kJ/kmol), k_0 is the velocity constant at the reference temperature (s^{-1}), R is the universal gas constant (kJ/kmol K), and T is the absolute temperature (K).

The reaction is accompanied by a heat change so that the heat generated is:

$$Q_s = Vr(-\Delta H) \qquad [11.8.4]$$

where ΔH is the heat of reaction, which is negative for an exothermic reaction (kJ/kmol), Q_s is the heat generated by the reaction (kW), and V is the volume of the reactor (m^3). In considering reactor stability it is exothermic reactions which are of prime importance.

In a closed adiabatic system the behaviour of a kinetics controlled exothermic reaction is as shown in Figure 11.15. The reaction generates heat, which causes the temperature and hence the velocity constant to increase continuously throughout the course of the reaction. The reaction rate and heat generated at first increase almost exponentially due to the increase in the velocity constant, then pass through a maximum and finally decrease as the reactants become exhausted.

In many practical cases an exothermic reaction can accelerate almost exponentially and cause overheating or explosion, if the heat is not removed from the system rapidly enough. These effects may occur while the reaction is only part way along the curves shown in Figure 11.15.

There are a number of factors, however, which can limit the acceleration of the reaction rate. One limiting factor which applies in all cases is the exhaustion of the reactants. Another important limiting effect is the transition in heterogeneous reactions from a kinetics controlled to a diffusion controlled reaction. As the temperature rises, the reaction kinetics become so rapid that they cease to be the limiting factor and diffusion takes over the role. These effects are shown in Figure 11.16(a). The dashed lines represent the reaction rates limited by the kinetics alone and by diffusion alone. The actual reaction rate, shown by the full curve, is determined by the combined effects of the kinetics and diffusion. It is limited at low temperatures by kinetics and at high temperatures by diffusion. Diffusion may be limited by the velocity in the reactor. The change of reaction rate with temperature and velocity for such a case is illustrated in Figure 11.16(b).

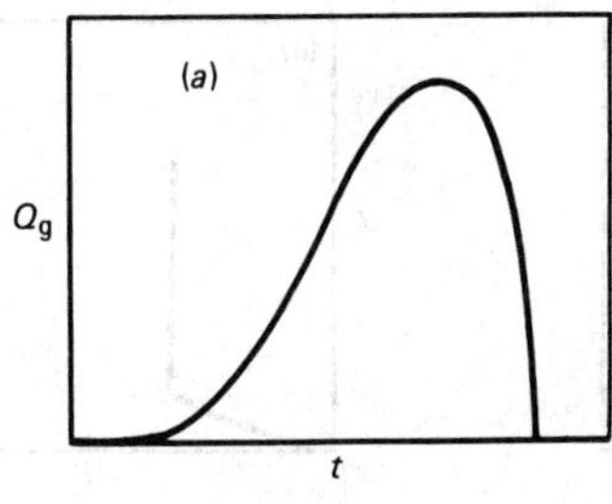

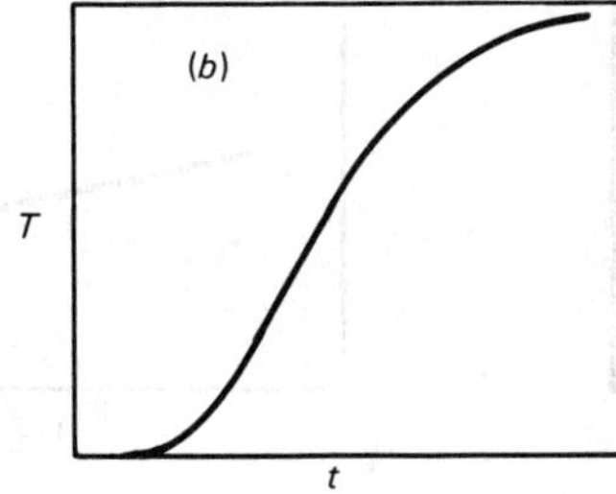

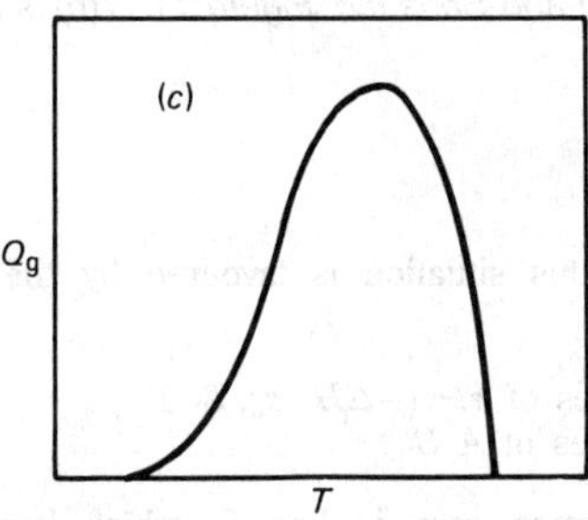

Figure 11.15 *Course of a kinetic controlled exothermic reaction in a closed adiabatic system: (a) heat generated vs time; (b) temperature vs time; (c) heat generated vs temperature*

Another limiting factor is the relation between the reaction rate and temperature. Some of the relations which occur are shown in Figure 11.17. Figure 11.17(a) gives the normal curve of rapid rise in reaction rate with temperature; Figure 11.17(b) shows diffusion control with a slow increase of reaction rate with temperature; Figure 11.17(c) illustrates an explosion with a sudden increase in reaction rate at the ignition point; Figure 11.17(d) is for a catalytic reaction controlled by the rate of adsorption; Figure 11.17(e) corresponds to the occurrence of a side reaction which becomes significant as temperature increases; and Figure 11.17(f) indicates a reaction in which conversion is limited by thermodynamic equilibrium effects.

11.8.4 CSTRs: stability

The factors which determine the stability of a chemical reactor are most easily illustrated by considering a

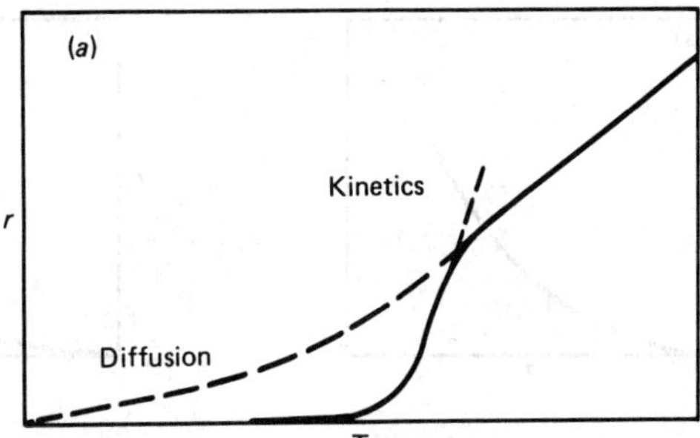

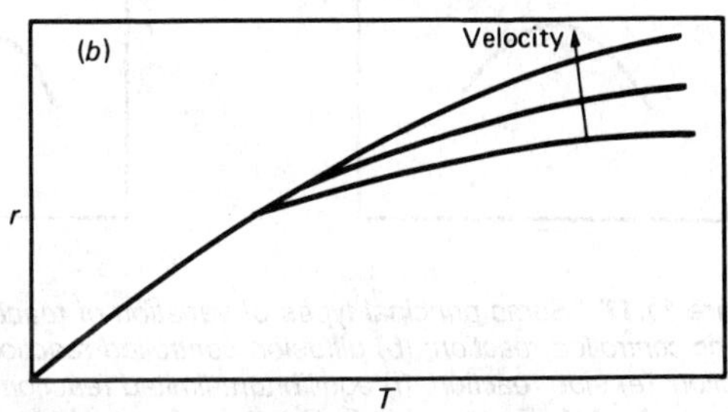

Figure 11.16 *Kinetic and diffusion control of a heterogeneous reaction: (a) transition from kinetic to diffusion control as temperature increases (Denbigh and Turner, 1971) (Courtesy of Cambridge University Press); (b) increase in rate of a diffusion controlled reaction with reactor velocity*

continuous stirred tank reactor (CSTR). For such a reactor with perfect mixing and with the simple irreversible first-order exothermic reaction given by Equation 11.8.1 and by Equations 11.8.2–11.8.4, the following simplified model may be derived.

The steady-state mass balance on the reactor is

$$F\rho x_0 = F\rho x + V\rho kx \quad [11.8.5]$$

$$= F\rho(1 + k\tau)x \quad [11.8.6]$$

with

$$\tau = V/F \quad [11.8.7]$$

where F is the volumetric flow of the feed (m^3/s), x is the mole fraction of reactant in the reactor, x_0 is the mole fraction of reactant in the feed, and τ is the volume-throughput ratio of the reactor.

The heat generated is:

$$Q_g = V\rho kx(-\Delta H) \quad [11.8.8]$$

where Q_g is the heat generated in the reactor (kW). Then, from Equation 11.8.6,

$$x = \frac{x_0}{1 + k\tau} \quad [11.8.9]$$

and from Equations 11.8.3 and 11.8.7–11.8.9

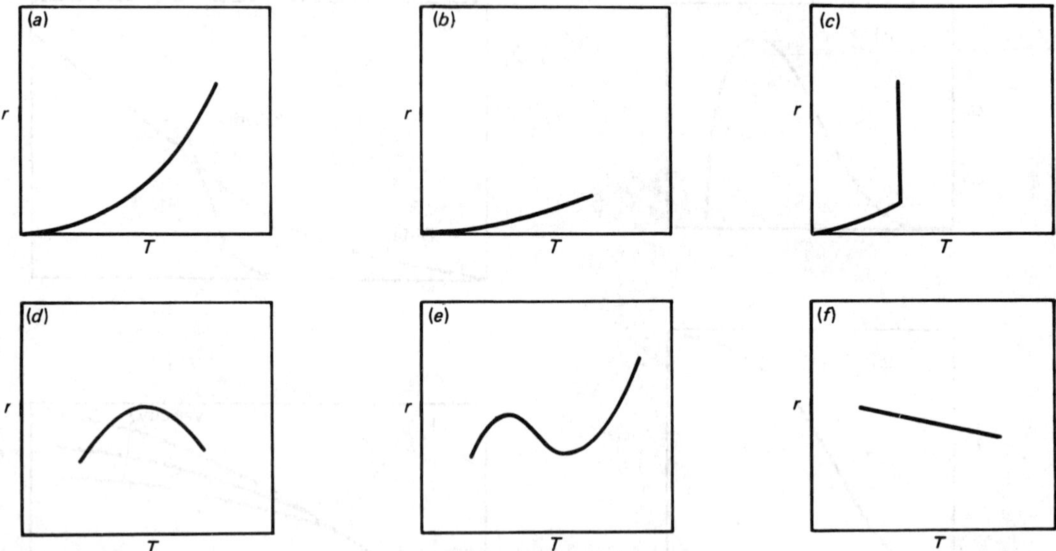

Figure 11.17 Some principal types of variation of reaction rate with temperature (after Frost and Pearson, 1961): (a) kinetic controlled reaction; (b) diffusion controlled reaction; (c) sudden explosive reaction; (d) adsorption controlled reaction; (e) side reaction; (f) equilibrium-limited reaction (Reproduced with permission from Kinetics and Mechanisms, 2nd edn, by A.A. Frost and R.G. Pearson, Copyright ©, 1961, John Wiley and Sons Inc. Figure 11.17(b) is additional to the figures given by these authors)

$$Q_g = \frac{F\rho k\tau x_0(-\Delta H)}{1+k\tau} \qquad [11.8.10]$$

$$= \frac{F\rho k_0\tau x_0(-\Delta H)}{k_0\tau + \exp(-E/RT)} \qquad [11.8.11]$$

The heat removed is:

$$Q_r = F\rho c_p(T - T_0) + UA(T - T_c) \qquad [11.8.12]$$

where A is the heat transfer area (m^2), c_p is the molar specific heat of the reaction mass (kJ/mol °C), Q_r is the heat removed from the reactor (kW), T_c is the absolute temperature of the coolant (K), T_0 is the absolute temperature of the feed (K), and U is the overall heat transfer coefficient (kW/m^2 °C).

The relations between the heats generated and removed and the reaction temperature are shown in Figure 11.18. Equation 11.8.11 for Q_g yields a sigmoidal curve. Equation 11.8.12 for Q_r, which may be rewritten as

$$Q_r = -(F\rho c_p T_0 + UAT_c) + (F\rho c_p + UA)T \qquad [11.8.13]$$

gives a straight line.

The system has stationary states at points where the heat generated Q_g and the heat removed Q_r are equal. These stationary states correspond, therefore, to the points of intersection of the curves for heat generation and heat removal. One extreme case is line A, which has a single stable stationary state at point a at a low reaction rate. Inspection of Equations 11.8.11 and 11.8.13 shows that this situation is favoured by the following conditions:

Small values of $k, \tau, (-\Delta H), x_0, T_0, T$
Large values of A, U

Another extreme case is line C, which has a single stable stationary state at point b, but this time at a high

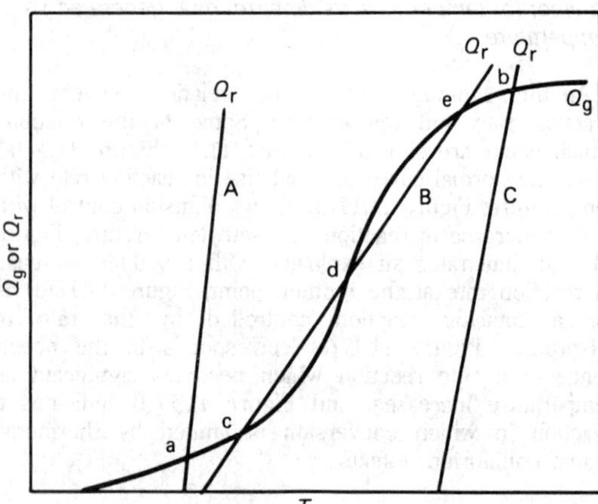

Figure 11.18 Heat generation and heat removal in a continuous stirred tank reactor (after van Heerden, 1953)

reaction rate. This situation is favoured by the converse conditions.

Line B represents a more complex case. Points c and e are stable points similar to points a and b, respectively, and require no further comment except that if it is desired to operate at point e, it is necessary to bypass point c by such means as initial heating up. By contrast, point d is an unstable stationary state. The system tends to move to point c or point e, depending on whether the fluctuation of temperature is downwards or upwards.

The reactor may operate at point d and its stability at this point is now discussed. A full stability analysis should take into account both steady- and unsteady-state conditions.

Consider first a simplified analysis. If the reaction temperature is suddenly increased, there will be a corresponding increase in reaction rate and heat generation. This increase $(\partial Q_g/\partial T)_\phi$ will be greater than the steady-state increase $(\partial Q_g/\partial T)_{ss}$, because the concentration x will not have had time to fall to the new steady-state value. It can be shown that the relation is:

$$\left(\frac{\partial Q_g}{\partial T}\right)_\phi = \left(\frac{1}{1-\phi}\right)\left(\frac{\partial Q_g}{\partial T}\right)_{ss} \qquad [11.8.14]$$

with

$$x = x_0(1-\phi) \qquad [11.8.15]$$

where ϕ is the degree of conversion. But from Equations 11.8.8 and 11.8.15

$$\left(\frac{\partial Q_g}{\partial T}\right)_\phi = V\rho x_0(1-\phi)(-\Delta H)\frac{\partial k}{\partial T} \qquad [11.8.16]$$

and from Equation 11.8.3

$$\left(\frac{\partial Q_g}{\partial T}\right)_\phi = V\rho x_0(1-\phi)(-\Delta H)k_0(E/RT^2)\exp(-E/RT) \qquad [11.8.17]$$

$$= \frac{Q_{gav}}{(RT^2/E)} \qquad [11.8.18]$$

where Q_{gav} is the average heat generated (kW). Also from Equation 11.8.13

$$\frac{\partial Q_r}{\partial T} = UA + F\rho c_p \qquad [11.8.19]$$

The condition for stability is

$$\frac{\partial Q_r}{\partial T} > \left(\frac{\partial Q_g}{\partial T}\right)_\phi \qquad [11.8.20]$$

or

$$UA + F\rho c_p > \frac{Q_{gav}}{RT^2/E} \qquad [11.8.21]$$

For the special, but common, case where almost all the heat is removed by the coolant, Equation 11.8.21 reduces to

$$UA > \frac{Q_{gav}}{RT^2/E} \qquad [11.8.22]$$

$$UA > \frac{UA(T-T_c)}{RT^2/E} \qquad [11.8.23]$$

$$RT^2/E > T - T_c \qquad [11.8.24]$$

A critical temperature difference ΔT_c is defined as:

$$\Delta T_c = RT^2/E \qquad [11.8.25]$$

For stability the difference between the reaction and coolant temperatures should not exceed the critical temperature difference. In other words, the condition for stability is:

$$\Delta T_c > T - T_c \qquad [11.8.26]$$

A fuller steady-state analysis gives the following conditions of stability:

$$UA + F\rho c_p(2 + k\tau) > \left(\frac{\partial Q_g}{\partial T}\right)_\phi = \frac{Q_{gav}}{RT^2/E} \qquad [11.8.27]$$

$$UA + F\rho c_p > \left(\frac{\partial Q_g}{\partial T}\right)_{ss} = \frac{Q_{gav}}{RT^2/E} \qquad [11.8.28]$$

Equation 11.8.28 is simply the steady-state condition that the heat removal must increase with temperature more rapidly than the heat generation. Relation 11.8.27 approximates to relation 11.8.21 if the second term on the left-hand side is relatively insignificant. If condition 11.8.28 is satisfied, but condition 11.8.27 is not, the reactor may exhibit a limit cycle in which the concentration and temperature show constant amplitude but non-sinusoidal fluctuations.

Limit cycles in reactors are discussed by Aris and Amundson (1957, 1958), Harriott (1964) and Coughanowr and Koppel (1965).

11.8.5 CSTRs: control

There are a number of reasons why reactor control is particularly important. The most obvious one is that with strongly exothermic reactions there is the hazard of a runaway reaction. But, in addition, disturbances in the reaction section tend to be propagated throughout the rest of the plant.

The control of reactors is the subject of a considerable literature. Accounts of reactor control are given in *Process Control* (Harriott, 1964), *Process Systems Analysis and Control* (Coughanowr and Koppel, 1965), *Introduction to Chemical Process Control* (Perlmutter, 1965) and *Process Control Systems* (Shinskey, 1967, 1979).

In a CSTR the control variable is normally the temperature of the reaction mass. The most common manipulated variables are probably the temperature or flow of a liquid coolant. Other manipulated variables may be the temperature or flow of the feed or of a recycle, pressure above a vaporizing liquid coolant or pressure above the reactor where this contains a vaporizing liquid. It is less easy, and in some cases theoretically impossible, to control reactor concentration directly.

Control is critically dependent on measurement. This has two principal aspects. One is that it is essential to obtain good measurements of the variables of interest.

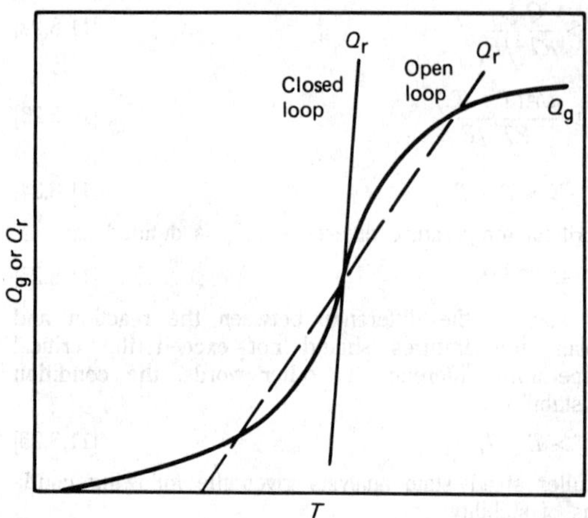

***Figure 11.19** Heat generation and heat removal in a CSTR under closed-loop control*

This may be difficult, for example, with reactor temperature, if there are temperature gradients or hot spots. The dynamic response of the instrument may also be important, particularly where reliance is placed on temperature to detect a runaway reaction. The other aspect is the instrument reliability. Here it may be necessary to resort to a high integrity system.

The conversion in the reactor is often measured, usually as a means of additional monitoring. Sometimes this may be done directly by measurement of reactant and product stream flows and concentrations, and sometimes indirectly from the heat removed by measurement of coolant flows and temperatures.

A full treatment of reactor control is beyond the scope of this book, but it is appropriate to give a brief account of some of the principles involved in a typical reactor control problem. The problem considered is the control of a CSTR at a point where the reactor is unstable under open-loop conditions. The reaction and the reactor system are those described in the previous section. The reaction is a first-order irreversible exothermic reaction. The reactor is a continuous stirred tank reactor with jacket cooling. The reactor temperature is controlled by a controller with proportional action which manipulates the coolant temperature. The graph of heat generation and heat removal under closed-loop conditions is shown in Figure 11.19. A full discussion of the problem has been given by Aris and Amundson (1957, 1958) and it is also described by Coughanowr and Koppel (1965).

The unsteady-state mass and heat balance equations for the open-loop reactor are:

$$V\rho\frac{dx}{dt} = F\rho x_0 - F\rho x - V\rho kx \qquad [11.8.29a]$$

$$\frac{dx}{dt} = \frac{F}{V}(x_0 - x) - kx \qquad [11.8.29b]$$

$$V\rho c_p\frac{dT}{dt} = -F\rho c_p(T - T_0) - UA(T - T_c) + V\rho kx(-\Delta H) \qquad [11.8.30a]$$

$$\frac{dT}{dt} = -\frac{F}{V}(T - T_0) - \frac{UA(T - T_c)}{V\rho c_p} + \frac{Q_g(T)}{V\rho c_p} \qquad [11.8.30b]$$

where t is the time (s). Equations 11.8.29–11.8.30 may be rewritten as:

$$\frac{d\xi}{d\tau*} = 1 - \xi - r(\xi, \eta) \qquad [11.8.31]$$

$$\frac{d\eta}{d\tau*} = -(\eta - \eta_0) + r(\xi, \eta) - q(\eta) \qquad [11.8.32]$$

with

$$\tau* = Ft/V \qquad [11.8.33]$$

$$\xi = x/x_0 \qquad [11.8.34]$$

$$\eta = c_pT/x_0(-\Delta H) \qquad [11.8.35]$$

$$r(\xi, \eta) = \frac{V}{F}k_0\xi \exp[-Ec_p/Rx_0(-\Delta H)\eta] \qquad [11.8.36]$$

$$q(\eta) = \frac{UA(T - T_c)}{F\rho x_0(-\Delta H)} \qquad [11.8.37a]$$

$$= \frac{UA}{F\rho c_p}(\eta - \eta_c) \qquad [11.8.37b]$$

The stability of the reactor at the control point may be determined by linearization. Thus Equations 11.8.29 and 11.8.30 may be linearized with respect to x and T, or Equations 11.8.31 and 11.8.32 with respect to ξ and η, and the limiting controller gain determined.

As an illustration, consider a simplified analysis of the control of a reactor which is open-loop unstable. Equation 11.8.30b may be rewritten in linearized form in terms of a temperature deviation θ from steady state:

$$\frac{d\theta}{dt} = \frac{F}{V}(\theta - \theta_0) - \frac{UA}{V\rho c_p}(\theta - \theta_c) + \frac{1}{V\rho c_p}\left(\frac{\partial Q_g}{\partial T}\right)_\phi \theta \qquad [11.8.38]$$

Then the plant transfer function relating the reaction temperature to the coolant temperature is:

$$\frac{\bar{\theta}}{\bar{\theta}_c} = \frac{K_p}{1 + \tau_p s} \qquad [11.8.39]$$

$$K_p = \frac{UA}{F\rho c_p + UA - (\partial Q_g/\partial\theta)_\phi} \qquad [11.8.40]$$

$$\tau_p = \frac{V\rho c_p}{F\rho c_p + UA - (\partial Q_g/\partial\theta)_\phi} \qquad [11.8.41]$$

where K_p is the plant gain, s is the Laplace operator, τ_p is the plant time constant (s), and the overbar indicates the Laplace transform.

The open-loop transfer function $G(s)$ for a controller gain of K_c is:

$$G(s) = \frac{K_cK_p}{1 + \tau_p s} \qquad [11.8.42]$$

$$= \frac{K}{1 + \tau_p s} \quad [11.8.43]$$

with

$$K = K_c K_p \quad [11.8.44]$$

For the closed loop

$$\frac{\bar{\theta}}{\bar{\theta}_c} = \frac{G}{1 + G} \quad [11.8.45]$$

$$= \frac{K}{1 + K} \frac{1}{1 + \tau_p s/(1 + K)} \quad [11.8.46]$$

The gains K_p and K and the term $\tau_p s$ are negative, since at the control point the system is open-loop unstable with $(F\rho c_p + UA) < (\partial Q_g/\partial\theta)_\phi$, but provided $|K| > 1$ the terms $K/(1+K)$ and $\tau_p/(1+K)$ are positive and the system is stable.

The treatment just given is highly simplified and neglects other lags in the control loop. In fact if other lags, particularly dead time, are present, control may be much more difficult. This point is discussed by Harriott and by Shinskey.

Although linearization gives the conditions for stability, it does not indicate the size of the stable region. This may be determined by phase plane analysis. Eliminating the normalized time $\tau*$ from Equations 11.8.31 and 11.8.32 gives

$$\frac{d\xi}{d\eta} = \frac{1 - \xi - r(\xi, \eta)}{-(\eta - \eta_0) + r(\xi, \eta) - q(\eta)} \quad [11.8.47]$$

The stability of the reactor has been investigated by Aris and Amundson in the phase plane using the following equation for heat removal:

$$q = k_1(\eta - \eta_c)[1 + K_c(\eta - \eta_{ss})] \quad [11.8.48]$$

where k_1 is a constant.

The behaviour of the reactor in Aris and Amundson's work is illustrated in Figure 11.20, which shows a phase plane plot of χ vs η for a stable controller setting. There are both stable and unstable limit cycles. If the reactor

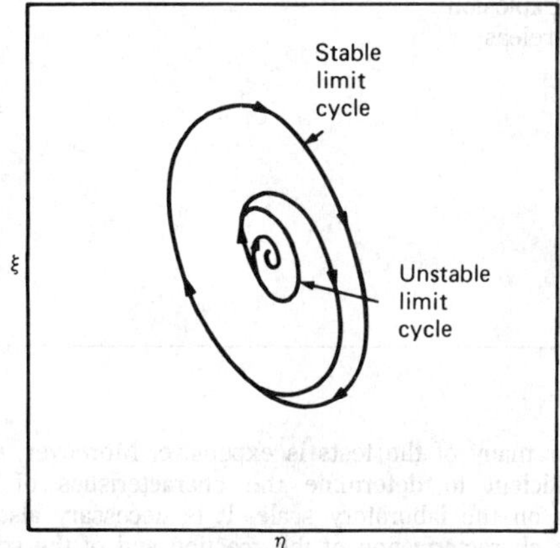

Figure 11.20 *Phase plane plot of limit cycles in a CSTR under closed loop control (after Aris and Amundson, 1958)*

initial conditions are within the envelope of the unstable limit cycle, the system moves towards the stable operating point. If the initial conditions are outside the envelope of the unstable limit cycle, but inside that of the stable limit cycle, the system moves into a stable limit cycle.

11.9 Batch Reactors

The safe operation of a batch chemical reactor requires an overall strategy which covers the whole cycle of development, design and operation. An account of such a strategy is given in *Chemical Reaction Hazards* (Barton and Rogers, 1993) (the IChemE *Reaction Hazards Guide*). This builds on earlier guidance given by the Association of the British Pharmaceutical Industry (ABPI), first in *Guidance Notes on Chemical Reaction Hazard Analysis* (1981) and then *Guidelines for Chemical Reaction Hazard Evaluation* (1989) (the ABPI First and Second *Reaction Hazard Guidelines*). Some accounts of strategies for safe operation of batch reactors given in these guides and by other authors are listed in Table 11.5.

11.9.1 Background

Batch reactors are operated by a wide variety of firms, many of which are quite small. On the other hand, the number of chemicals made by a given firm may be quite

Table 11.5 *Some accounts of strategies for safe reactor operation*

Organization[a]	*Reference*
HSE	Nolan and Barton (1987); Cronin, Nolan and Barton (1987); Pantony, Scilly and Barton (1989)
ABPI	ABPI (1981, 1989)
Bayer	Pilz (1986)
Ciba-Geigy	Fierz *et al.* (1983, 1984); Regenass (1984)
DIERS	Fisher (1991)
Dow	Kohlbrand (1985, 1987a, 1990)
Dupont	Gordon *et al.* (1982); O'Brien *et al.* (1982); Hoffmann (1985)
Hercules	Berkey and Workman (1987)
Hoffman-Laroche	Kunzi (1980)
IChemE	Barton and Rogers (1993)
ICI	McNab (1981); N. Gibson (1984, 1986b)
Lonza	Christen (1980)
Merck	Snyder (1965, 1982)
Pfizer	Brannegan (1985)
Sandoz	Duval (1985)
Sterling Organics	Coates and Riddell (1981a,b); Lambert and Amery (1989)

[a] This is the affiliation of the authors, and is not to be interpreted as meaning that the organization necessarily endorses the scheme which they describe. ABPI, Association of the British Pharmaceutical Industry; DIERS, Design Institute for Emergency Relief Systems; HSE, Health and Safety Executive; IChemE, Institution of Chemical Engineers.

Table 11.6 Taxonomy for reactor overpressure analysis (Marrs et al., 1989) (Courtesy of the Institution of Chemical Engineers)

A System description
Chemical reaction
Reactants, intermediates, products
Heat of reaction
Known exotherms

Reactor pressure
Reactor temperature

Charging
Cooling
Heating
Agitation
Control system
Trip/interlock system
Batch cycle/control
Operating vents
Relief system

B Reaction type
B1 Alcoholysis
B2 Amination
B3 Condensation
B4 Cyclization
B5 Diazotization
B6 Esterification
B7 Halogenation
B8 Hydrogenation
B9 Hydrolysis
B10 Isomerization
B11 Methylation
B12 Nitration
B13 Oxidation
B14 Polymerization
B15 Sulphonation

C Pressurizing fluid
C1 Vaporized liquid
C2 Decomposition gas
C3 Water vapour
C4 Flammable gas
C5 Other gas/vapour

D Pressurizing event
D1 Regular reaction exotherm
D2 Impurity reaction exotherm
D3 Heat of mixing, dilution
D4 Regular reaction decomposition
D5 Impurity decomposition
D6 Water ingress and vaporization
D7 Air ingress and combustion
D8 High pressure gas ingress

E Process deviation
E1 Regular reaction inadequate information
E2 Regular reactant unknown decomposition
E3 Impurity reaction exotherm
E4 Impurity decomposition
E5 Incorrect charging
E6 Inadequate cooling
E7 Excessive heating
E8 Incorrect agitation
E9 Inadequate batch control
E10 Undesired catalysis
E11 High pressure gas connection

F Initiating fault
F1 Inadequate reaction screening
F2 Incorrect design
F3 Mechanical failure
F4 Utilities failure
F5 Control system failure
F6 Operator error

G Overpressure effect
G1 Open vessel
G2 Excursion only
G3 Bursting disc operated
G4 Vessel ruptured
G5 Bursting disc operated but inadequate, vessel ruptured

H Bursting disc failure cause
H1 Undersized for design conditions
H2 Demand greater than design conditions
H3 Installation error
H4 Modification error
H5 Maintenance error
H6 Disc failure
H7 Vent piping failure

I Release effects
I1 Flammable release
1.1 No ignition
1.2 Fire
1.3 Explosion
I2 Toxic release

large. A works may produce over 1000 different chemicals using a range of types of reaction.

The principal hazard in a batch reactor is instability of an individual substance or an uncontrolled exothermic reaction, or reaction runaway. There now exist a bewildering multiplicity of tests for the detection and characterization of substances and reactions. The apparatus for many of the tests is expensive. Moreover, it is not sufficient to determine the characteristics of the reaction on the laboratory scale. It is necessary also to obtain a characterization of the reaction and of the scale-up features which allows a safe design to be developed.

The problem is particularly severe where it is a question of confirming that an existing process is safe,

***Table 11.7** Analysis of reactor overpressure: incident modes (Marrs et al., 1989) (Courtesy of the Institution of Chemical Engineers)*

A Detailed Beakdown

	All reactions	*Nitration*	*Sulphonation*	*Polymerization*
Case 1: C1 Overpressure by vaporized liquid				
D1 Regular reaction exotherm				
E1 Regular reaction inadequate information (7)				
E1.1 Unknown exotherm	4	1		
E1.2 Inadequate definition of operation	3			2
E5 Incorrect charging (35)				
E5.1 Excess of reactant	10			1
E5.2 Deficiency of reactant	9		1	3
E5.3 Too fast addition of reactant (see below)				
E5.4 Too slow addition of reactant				
E5.5 Addition of wrong reactant	1		1	
E5.6 Modification of reactant	4			
E5.7 Incorrect order of reactant addition	2			
E5.8 Too slow reaction of solid (coarse particles)	1			
E5.9 Too fast reaction of solid (fine particles)				
E5.3 Too fast addition of reactant				
E5.3.1 Automatic control failure	1			
E5.3.2 Manual control failure:				
Measurement/alarm				
Operator error	7		2	
E6 Inadequate cooling (26)				
E6.1 Underdesign (esp. scale-up)	5			3
E6.2 Coolant circulation fault (see below)				
E6.3 Inadequate agitation (for heat transfer)	2			
E6.4 Internal fouling				
E6.5 External fouling				
E6.6 Evaporative coolant fault	1			
E6.7 Condenser fault (see below)				
E6.8 Moderating solvent fault	1			
E6.9 Steam jacket cooling inadequate	1			
E6.10 Other causes	2			
E6.2 Coolant circulation fault				
E6.2.1 Coolant source failure				
E6.2.2 Power failure				
E6.2.3 Pump failure	1			
E6.2.4 Coolant turned off	3[a]		1	
E6.2.5 Coolant leak/loss				
E6.2.6 Blockage	1			
E6.2.7 Freezing				
E6.2.8 Automatic control failure	3[b]		1	1
E6.2.9 Manual control failure:				
Measurement/alarm				
Operator error	3			2
E6.7 Condenser fault				
E6.7.1 Condenser vapour inlet blockage	1			1
E6.7.2 Condenser flooding	1			1
E6.7.3 Condenser frozen	1			
E7 Excessive heating (19)				
E7.1 Initial overheating	3		1	2
E7.2 Heating/cooling changeover fault	3		1	1
E7.3 Unintended heating or heating instead of cooling	2[c]			
E7.4 Pump energy	1		1	
E7.5 Agitator energy	2		1	
E7.6 Steam leak	2	1	1	
E7.7 Live steam	1			
E7.8 Automatic control failure	2			1
E7.9 Manual control failure:				
Measurement/alarm	1			
Operator error	1			

Table 11.7 *(Continued)*

A Detailed Beakdown	*All reactions*	*Nitration*	*Sulphonation*	*Polymerization*
E7.10 Overheating in flange joints	1			
E8 Inadequate agitation (for mixing) (20)	20	4	4	2
E9 Incorrect batch control (18)				
E9.1 Initital temperature low	2		1	
E9.2 Initial temperature high				
E9.3 Too fast reactant addition relative to temperature	4		2	
E9.4 Incorrect cycle	3			
E9.5 Inadequate chemical moderation	1			1
E9.6 Stewing	4		1	
E9.7 Other causes	4			
E10 Undesired catalysis (5)				
E10.1 Excess, or too rapid addition, of catalyst	2			1
E10.2 More active catalyst				
E10.3 Catalyst maldistribution				
E10.4 Catalyst impurity	2			
E10.5 Catalyst left over from previous batches	1			
D2 Impurity reaction exotherm (21)				
D2.1 Water	11	3	1	
D2.2 Air				
D2.3 Materials left in reactor	1			
D2.4 Heat transfer fluid	1			
D2.5 Other impurities	8			
D3 Heat of mixing, dilution				
Case 2: C2 Overpressure by decomposition gas				
D4 Regular reactants decomposition (16)				
E2 Regular reactant unknown decomposition	11	2	2	
Ditto (side reaction)	5		1	
D5 Impurity reaction decomposition				
E4 Impurity decomposition				
Case 3: C3 Overpressure by water vapour				
D6 Water ingress and vaporization	1			
Case 4: C4 Overpressure by flammable gas (ignition of explosive mixture)				
D7 Air ingress and combustion (7)	7		1	
Case 5: C5 Overpressure by other gas/vapour				
D8 High pressure gas ingress	1			
Miscellaneous cases				
X1 Overpressure following unknown exotherm, where it is unclear if exotherm is a decomposition	7		3	
X2 Overpressure following operator attempts to recover from fault conditions	4		2	1
Cases where cause is unknown				
Z1 Cause unknown	12			5

B Summary

	Mode (%)			
	All reactions[d]	*Nitration*	*Sulphonation*	*Polymerization*[d]
Regular reaction inadequate information	3.5 (3.7)	9.1	0	7.1 (8.6)
Incorrect charging	17.2 (18.3)	0	13.8	14.3 (17.4)
Inadequate cooling	13.1 (14.0)	0	6.9	28.6 (34.8)
Excessive heating	9.6 (10.2)	9.1	17.2	14.3 (17.4)
Inadequate agitation	10.1 (10.8)	36.4	13.8	7.1 (8.6)
Incorrect batch control	9.1 (9.7)	0	13.8	3.6 (4.4)
Undesired catalysis	2.5 (2.7)	0	0	3.6 (4.4)
Impurity reaction exotherm	10.6 (11.3)	27.3	3.4	0

Table 11.7 (Continued)

B Summary

	Mode (%)			
	All reactions[d]	*Nitration*	*Sulphonation*	*Polymerization*[d]
Regular reactant unknown decomposition	8.1 (8.6)	18.2	10.3	0
Water ingress and vaporization	0.5 (0.5)	0	3.4	0
Air ingress and combustion	3.5 (3.7)	0	3.4	0
High pressure gas ingress	0.5 (0.5)	0	0	0
Exotherm of unknown type	3.5 (3.7)	0	10.3	0
Recovery from fault conditions	2.0 (2.2)	0	6.9	3.6 (4.4)
Cause unknown	6.1	0	0	17.9
Total	100	100	100	100
No. of applicable incidents = 199				

[a] In one case cause is assigned to lack of cooling, although there had also been a change in the quality of the reactants.
[b] In one case cause is assigned to a failure of a temperature controller, although a new more active catalyst was also in use.
[c] In one case cause is assigned to gross overheating by steam, although it was also subsequently found that the reaction mixture exhibited instability.
[d] Values in brackets are based on redistribution of 'Causes unknown' among known causes in proportion to relative frequency of latter.

Table 11.8 Analysis of reactor overpressure: relief arrangements (Marrs et al., 1989) (Courtesy of the Institution of Chemical Engineers)

Relief arrangement[a]	*No.*
BD(s)	11
RV	8
RV + BD	2
RV, vent part closed	1
BD, recommend larger BD	2
RV, recommend BD	1
BD, vessel open	1
No BD	4
Recommend BD	4
No relief	1
Manual vent valve	1
Small vent(s)	1
Vent, recommend BD	1
Small holes	1
Vessel open	15
Vessel open, recommend BD	1
Unknown	13
Total	68

[a] BD, bursting disc; RV, relief valve.

because in this case it is not possible to incorporate the safety margins which might be used in a new process.

11.9.2 Runaway reactions

In a runaway reaction there is a reaction exotherm in the reactor, the heat evolved is not removed sufficiently rapidly, the temperature of the reaction mass rises and the reaction accelerates. The exotherm may be in the main, intended process reaction or in an undesired side reaction. A high pressure may then be generated by the product gases or by the vapour pressure of liquids in the reactor. If this pressure is not relieved, the reactor may suffer overpressure.

11.9.3 Reactor incidents

Analyses of batch reactor incidents in the UK reported to the HSE have been given by Townsend and Pantony (1979), Barton and Nolan (1984), Nolan and Barton (1987) and by Marrs *et al.* (1989).

Marrs *et al.* based their study on two sets of data. The first was a set of UK data from the HSE's own records for the period 1970–81 (the UK national data set) and the second a collection of world-wide incidents made for the HSE (the world data set). The two sets contained 68 and 199 incidents, respectively.

The overall taxonomy used by Marrs *et al.* is given in Table 11.6. The analysis of reactor overpressure incidents by incident mode is shown in Table 11.7, in which the taxonomy is developed in finer detail and which utilizes the world-wide set. The table includes more detailed analyses for (a) sulphonation, (b) nitration and (c) polymerization reactions, for which incidents are relatively common.

Information on reactor relief was obtainable only from the national set. Tables 11.8, 11.9 and 11.10 give information on overpressure relief arrangements, overpressure incident and relief system behaviour.

These workers also carried out an expert judgement field survey the results of which are shown in Table 11.11. The comparison of the ranking of the relative importance, or frequency, of the incident modes (1 = most important, etc.) is shown in Section A. There is reasonable agreement overall, with two exceptions. Thus in both rankings incorrect charging and inadequate cooling are ranked high, undesired catalysis low and excessive heating, incorrect agitation and exotherm from impurity, moderate. The exceptions are unknown exotherm/decomposition, which is ranked much lower by the experts than in the case histories and inadequate batch control. The experts were from companies with expected good practice and the relatively low ranking of

Table 11.9 *Analysis of reactor overpressure: overpressures (Marrs et al., 1989) (Courtesy of the Institution of Chemical Engineers)*

Overpressure Incident	*Proportion* No.	(%)
Vessel open, hazardous release	18[a]	27.3
Glasswork shattered, hazardous release	16[b]	24.2
Vessel ruptured, hazardous release	19	28.8
Vessel ruptured	1	1.5
Explosion	5	7.6
Hazardous release	5	7.6
Catchpot ruptured	1	1.5
Catchpot fire	1	1.5
Total	66	100.0

[a] One case where overpressure was vented through two holes; one where manhole was opened; and one where operator disconnected hose.
[b] One case where reflux divider ruptured – glassware not specifically mentioned.

Table 11.10 *Analysis of reactor overpressure: relief system behaviour (Marrs et al., 1989) (Courtesy of the Institution of Chemical Engineers)*

Relief system behaviour	*Proportion* No.	(%)
Relief operated:		
Glassware ruptured	3	6.3
Vessel ruptured	1	2.1
Explosion	1	2.1
Catchpot ruptured	1	2.1
Catchpot fire	1	2.1
Hazardous release	3	6.3
Relief fitted, but failed:		
Glassware ruptured	3[a,b]	6.3
Vessel ruptured	8	16.7
Explosion	2	4.2
Glassware ruptured	10[c]	20.8
Vessel ruptured	11[d]	22.9
Explosion	2	4.2
Hazardous release	2	4.2
Total	48	100

[a] One case where relief valve (RV) fitted, but vent part closed; and one where bursting disc (BD) fitted, but larger BD recommended.
[b] One case where BD failed to rupture.
[c] Five cases where relief arrangements unknown; three cases where BD recommended; and one where no relief was fitted.
[d] Six cases where relief arrangements unknown; three cases where there was no BD; one case where BD recommended; and one where there was a manual vent valve.

unknown exotherms probably reflects their view that they have taken steps to control this hazard. Sections B–E of Table 11.11 give the corresponding rankings of particular incident submodes.

11.9.4 Overall strategy

A strategy for safe reactor design is given in the IChemE *Guide*. This largely follows that outlined by N. Gibson, Rogers and Wright (1987), who describe the strategy developed at ICI. It involves:

(1) definition of the process conditions and plant design;
(2) characterization of the chemical reaction and its hazards;
(3) selection and specification of safety measures;
(4) implementation and maintenance of safety measures.

The stages in and strategies for the assessment and control of the hazards of chemical reactors are shown in Figures 11.21 and 11.22, respectively.

Table 11.11 *Analysis of reactor overpressure: comparison of rankings from case histories and from expert judgement field study (Marrs et al., 1989) (Courtesy of the Institution of Chemical Engineers)*

	Case histories %	Rank	*Field study* Rank
A Incident modes			
Unknown exotherm/decomposition	15.1	2	6
Incorrect charging	17.2	1	1
Inadequate cooling	13.1	3	3
Excessive heating	9.6	6	4
Incorrect agitation	10.1	5	5
Inadequate batch control	9.1	7	2
Undesired catalyst	2.5	8	8
Exotherm from impurity	10.6	4	7
B Incorrect charging			
Excess of reactant	29.4	1	1
Deficiency of reactant	26.5	2	3
Too fast addition of reactant	23.5	3	2
Modification of reactant	11.8	4	4
Incorrect order of reactant addition	5.9	5	5
C Inadequate cooling			
Coolant source/power failure	0	5	5
Coolant pump set failure	3.8	4	4
Coolant turned off	11.5	= 1	2
Automatic control failure	11.5	= 1	1
Condenser fault	11.5	= 1	3
D Excessive heating			
Initial overheating	15.8	= 1	2
Heating/cooling changeover fault	15.8	= 1	3
Undesired heating	10.5	= 3	5
Automatic control failure	10.5	= 3	4
Manual control failure	10.5	= 3	1
E Incorrect batch control			
Initial temperature too low	11.1	4	3
Initial temperature too high	0	5	5
Too fast addition of reactant relative to temperature	22.2	= 1	1
Incorrect cycle	16.6	3	4
Excessive holding	22.2	= 1	2

(1) Initial chemistry

(a) Characterization of materials/process
(b) Suitability of production

(2) Pilot plant

(a) Chemical reaction hazards
(b) Influence of plant on hazard
(c) Definition of safe procedures

(3) Full-scale production

(a) Re-evaluation of chemical reaction hazards
(b) Effect of expected variations in process conditions
(c) Hazards from plant operations
(d) Definition of safe procedures
(e) Interaction of technical safety with engineering, production, economic and commercial aspects of process

Figure 11.21 *Stages in control of hazards of chemical reactions (N. Gibson, Rogers and Wright, 1987) (Courtesy of the Institution of Chemical Engineers)*

The characterization of the chemical reaction and its hazards involves:

(1) hazard evaluation of the reaction;
(2) characterization of reaction for scale-up;

and safety aspects of the design of the reactor include:

(3) inherently safer design;
(4) control and trip systems;
(5) emergency safety measures;
(6) overpressure relief.

The strategy will be influenced by the types of reaction operated in the company, by evidence from previous incidents and by the options available for safety measures.

11.10 Batch Reactors: Reaction Hazard Evaluation

11.10.1 Principal factors

A review of the factors which should be taken into account in the reaction assessment has been given in the two ABPI *Guidelines*. The components covered should include: the reactants, the reaction intermediates (whether isolated or not) and the reaction products; the by-products; the solvent; the still residues; and the wastes and effluents.

The review should consider the effect of contaminants on the stability of components and on the reaction rate. Contaminants may include: impurities in the reactants; materials of construction; corrosion products; water, steam, refrigerant and other heat transfer fluids; and materials from elsewhere. Other factors mentioned include:

(1) incorrect use of catalysts;
(2) addition of extra quantities of reagents;
(3) sequence of addition or omission of reagents;
(4) addition of wrong reagent;
(5) drift of pH and sensitivity to buffering in normal or abnormal operation;
(6) effect of process errors on reaction conditions.

11.10.2 Overall evaluation strategy

The strategy for reaction hazard evaluation described in the IChemE *Guide* involves the following:

(1) desk screening;
(2) explosibility tests;
(3) preliminary screening tests;
(4) characterization of normal reaction;
(5) characterization of runaway reaction.

An account has been given in Chapter 8 of tests for detecting the instability, reactivity and explosibility of individual substances. Many of these tests are applicable to reaction mixtures also. In particular, the initial screening tests are broadly similar. But reaction hazard evaluation is broader in scope and involves a wider variety of tests.

11.10.3 Desk screening and explosibility tests

Desk screening of the reaction should start with a literature search. The references given in Chapter 8 for the screening of individual components, e.g. *Handbook of Reactive Chemical Hazards* (Bretherick, 1990b) and *Encyclopaedia of Chemical Technology* (Kirk and Othmer, 1963–, 1978–, 1991–), are again a suitable starting point, but for an established commercial reaction it is often helpful to consult the original papers. In certain journals, papers on chemical reactions regularly include information on potential hazards.

It is usual to classify unit processes on the basis of the reaction involved and, as described in Section 11.14, these unit processes have certain generic characteristics. A classification of unit processes in terms of their exothermicity has been given by Hoffmann (1985). He classes halogenation, oxidation, organometallic reactions and some polymerizations as highly exothermic; nitration as exothermic; amination, sulphonation, condensation and some hydrogenations as moderately exothermic; and alkylation, hydrolysis and some hydrogenation as mildly exothermic. Thermodynamic schemes such as CHETAH may be used to evaluate the thermodynamic characteristics of the reaction.

The substances involved in the reaction, whether as reactants, intermediates or products should be screened for explosibility. Basic methods, which have been described in Chapter 8, are examination of reactive chemical groups, computation of the oxygen balance and experimental explosibility testing.

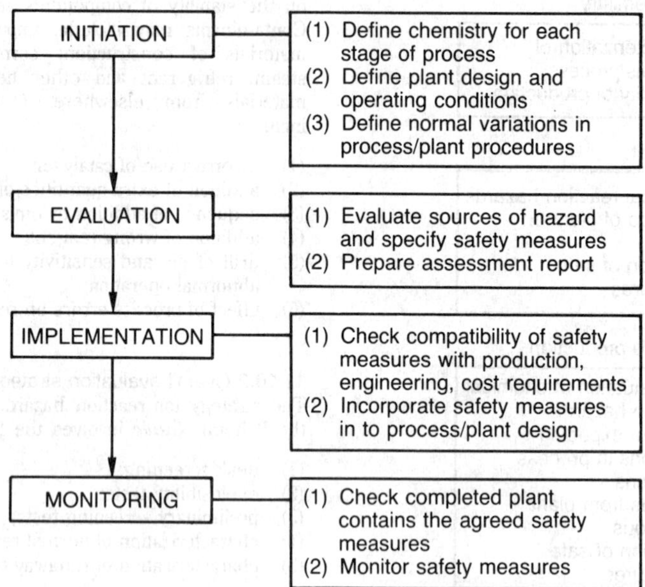
INITIATION
(1) Define chemistry for each stage of process
(2) Define plant design and operating conditions
(3) Define normal variations in process/plant procedures
EVALUATION
(1) Evaluate sources of hazard and specify safety measures
(2) Prepare assessment report
IMPLEMENTATION
(1) Check compatibility of safety measures with production, engineering, cost requirements
(2) Incorporate safety measures in to process/plant design
MONITORING
(1) Check completed plant contains the agreed safety measures
(2) Monitor safety measures

Figure 11.22 *Strategy for control of hazards of chemical reactions (N. Gibson, Rogers and Wright, 1987) (Courtesy of the Institution of Chemical Engineers)*

11.10.4 Overview of tests

The test methods used for reactions are broadly similar to those used for single components, but there are some important differences. In particular, it is necessary for the design of the reactor system to have much more information. There are two main reasons for this. One is that the design involves selecting not only an operating temperature but also a number of other parameters and features such as a feed addition rate and a pressure relief system. The other is that the design involves scale-up to full scale.

The information sought from tests includes data on the presence and quantitative features of exotherms, rate of heat evolution and rate of gas evolution. One group of tests uses a Carius tube or similar equipment. These include the closed tube test and the delayed onset detection test. Another group utilizes a Dewar flask. This includes tests involving: ramped and stepped heating of the reaction sample in a Dewar flask; ramped heating of the sample and a reference material in a Dewar flask (the simple exotherm test, or dynamic heating test); and the hot storage test using a Dewar flask, or heat accumulation test.

In many such tests the sample is held in an apparatus placed in an oven. Tests employ various temperature regimes. The temperature may be held constant, ramped or stepped. The sample may be held under adiabatic conditions by arranging for its temperature to be tracked by that of the oven. Alternatively, the sample conditions may be made isothermal by holding the temperature of the oven constant at a value slightly below that of the sample which is heated just sufficiently to maintain its temperature.

Other tests use instrumentation. In this group are differential thermal analysis (DTA), differential scanning calorimetry (DSC) and accelerating rate calorimetry (ARC) and tests using various types of reaction calorimeter such as the bench scale calorimeter (BSC) and the SEDEX and SIKAREX calorimeters. Use of the SIKAREX apparatus has been described by Hub (1976, 1981).

Numerous other specialized instruments for the investigation of reactions have been described. One such is the TZP described by Kunzi (1980). This allows a component or reaction mixture to be investigated in the same oven under isobaric and isochoric conditions and under identical thermal conditions. The author mentions particularly the investigation of decomposition reactions.

One important test carried out using a reaction calorimeter is the determination of the adiabatic temperature rise. This single test provides a large amount of information on the thermodynamic and kinetic and characteristics of the reaction. The analysis of such as test is considered below.

Heat evolution may be studied by conducting the reaction isothermally in a heat flow calorimeter and measuring the heat evolved. A test of this type is described by Coates and Riddell (1981a).

Gas evolution is another important parameter. One method utilizes a tube and another an autoclave. These tests are also described by Coates and Riddell (1981a).

In addition to the recognized tests, another type of test involves the use of laboratory equipment, particularly the Dewar flask, to simulate and explore the conduct of the reaction in the plant reactor.

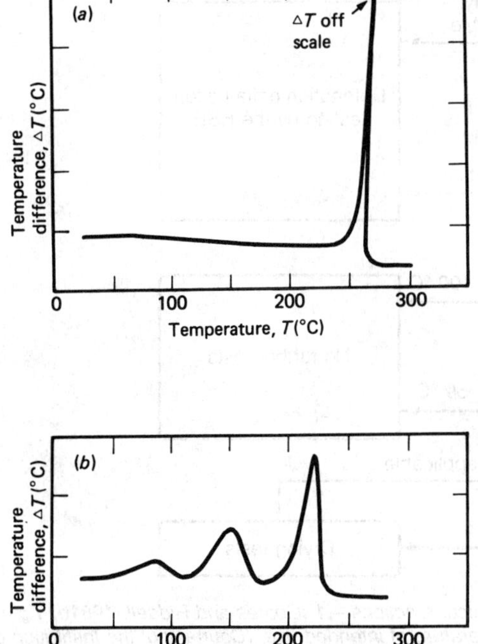

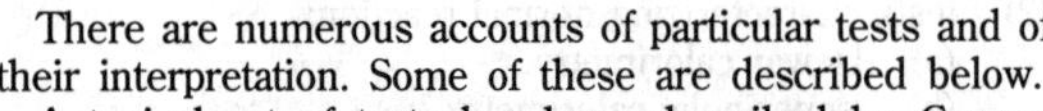

Figure 11.23 *Experimental determination of thermal stability of reaction mixtures – 1 (Grewer, 1974): (a) differential thermal analysis of sulphonation of nitrobenzene; (b) differential thermal analysis of diazotization of 2-cyano-4-nitroaniline in H_2SO_4 (Courtesy of Elsevier Publishing Company)*

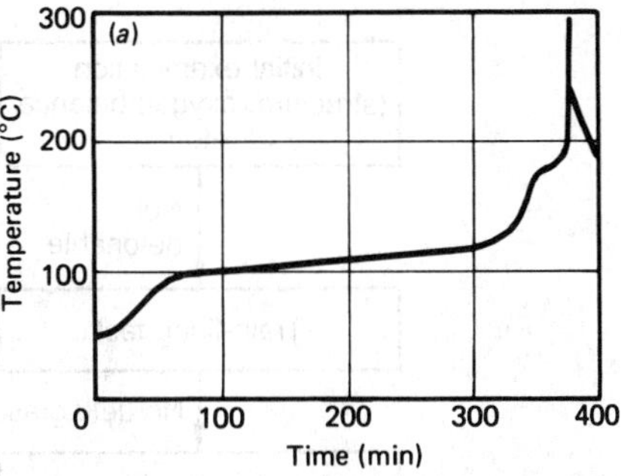

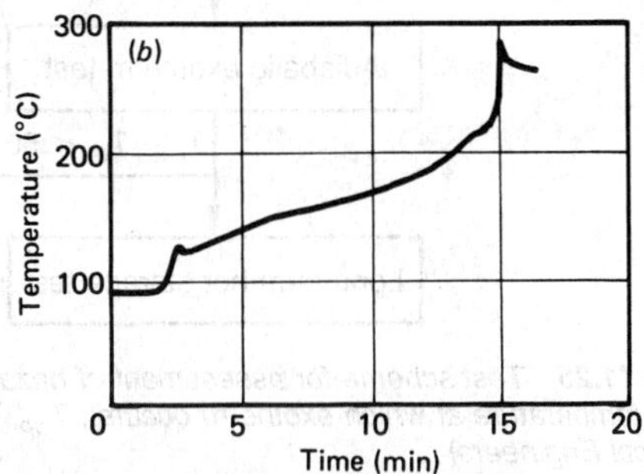

Figure 11.24 *Experimental determination of thermal stability of reaction mixtures – 2 (Grewer, 1974): (a) adiabatic storage test of diazotization solution of 2-cyano-4-nitroaniline in H_2SO_4; (b) adiabatic reaction test of solution of p-nitro-N-methylaniline and carbyl sulphate in nitrobenzene (Courtesy of Elsevier Publishing Company)*

There are numerous accounts of particular tests and of their interpretation. Some of these are described below.

A typical set of tests has been described by Grewer (1974). These tests are illustrated in Figures 11.23 and 11.24. Figure 11.23(a) shows a DTA plot for the sulphonation of nitrobenzene in sulphuric acid in which the product has decomposed and Figure 11.23(b) one for the diazotization of 2-cyano-4-nitroaniline in sulphuric acid in which the three peaks correspond to the reactions associated with the CN, N_2 and NO_2 groups. Figure 11.24(a) gives a plot for an adiabatic storage test of diazotization solution of 2-cyano-4-nitroaniline in which again there are three steps in the curve corresponding to the exotherm peaks in the DTA test. Figure 11.24(b) shows a plot for an adiabatic reaction test, similar to the adiabatic storage test but conducted at the operating temperature, for the reaction of *p*-nitro-*N*-methylaniline with carbyl sulphate in nitrobenzene under pressure.

11.10.5 Test schemes

A systematic approach to testing requires that it be performed within the framework of a scheme for the acquisition of the complete set of information necessary for reactor design. There are a number of test schemes described in the literature.

The scheme described in the ABPI First *Guidelines* comprised the following tests:

(1) preliminary tests –
 (a) safety screen test,
 (b) combustibility test,
 (c) explosion and detonation potential tests;
(2) dynamic heating test;
(3) adiabatic heating test –
 (a) reaction exotherm,
 (b) hot storage;
(4) special tests –
 (a) drying stability tests.

In this scheme the safety screen test is the melting point tube test; the combustibility test is the train firing test; the adiabatic heating test for the main process reaction is similar to the stepped heating test using a Dewar flask.

The test scheme described by Coates and Riddell (1981a,b) is very similar. Figure 11.25 shows the decision tree given by these workers.

The test scheme given by O'Brien *et al.* (1982) is shown in Figure 11.26. Other test schemes include those of Cronin, Nolan and Barton (1987) and N. Gibson, Rogers and Wright (1987), described below.

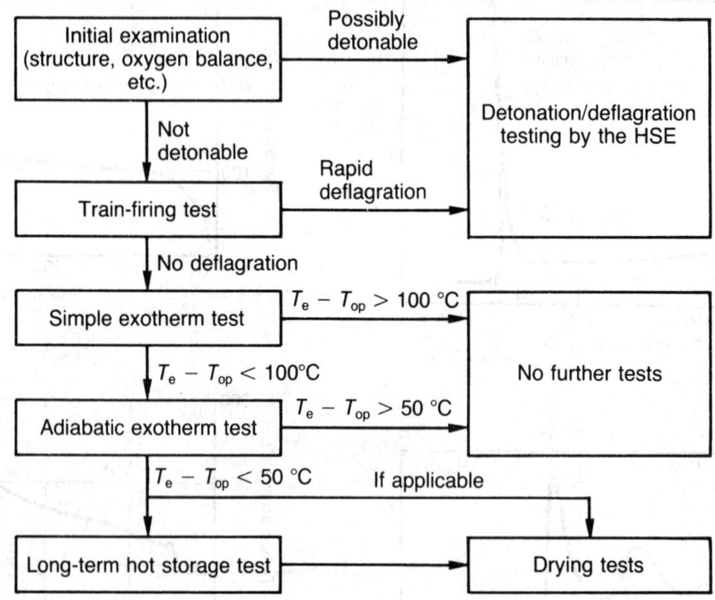

***Figure 11.25** Test scheme for assessment of hazards of chemical reactions – 1 (Coates and Riddell, 1981b) T_e, lowest temperature at which exotherm occurs ; T_{op}, mean temperature of intended use. (Courtesy of the Institution of Chemical Engineers)*

Another test scheme is that described by Kunzi (1980), who states that the problem may come in various forms, and outlines the way in which the scheme is used. It may be required to evaluate a complete process, or one stage of a process such as distillation or storage, or a single component. He gives a decision tree for approaching the general problem. Essentially, steps 1 and 2 are desk screening; step 3 is the standard laboratory tests; step 6 involves tests to characterize the process main reaction; step 8 tests to explore undesired reactions; and step 10 tests to study storage. The actual scheme used is based on the availability in the test centre of a range of instruments including the TZP as described above.

The tests in the scheme given in the ABPI *Second Guidelines* are:

(1) DSC;
(2) combustibility test;
(3) closed tube test;
(4) delayed onset detection test;
(5) adiabatic Dewar test;
(6) ARC;
(7) heat flow calorimeter test;
(8) gas evolution measurement, using automated gas burette;
(9) gas evolution measurement, using thermal mass flowmeter.

A flowchart is given in Figure 8.12.

The ICI test scheme described by N. Gibson, Rogers and Wright is shown in Figure 11.27.

The IChemE *Guide* gives the test scheme shown in Figure 11.27 and lists the following tests:

(1) initial screening tests:
  (a) DSC,
  (b) insulated exotherm test,
  (c) decomposition pressure test,
  (d) 10 g closed tube test,
(2) tests characterizing normal reaction:
  (a) Dewar calorimetry,
  (b) isoperibolic calorimetry,
  (c) power compensation calorimetry,
  (d) heat flow calorimetry,
  (e) gas evolution measurement, using automated gas burette,
  (f) gas evolution measurement, using thermal mass flowmeter,
(3) tests characterizing runaway reaction:
  (a) adiabatic Dewar calorimetry,
  (b) other types of calorimetry,
  (c) Vent Sizing Package (VSP),
  (d) Reactive Systems Screening Tool (RSST),
  (e) Phi-Tec adiabatic calorimeter.

11.10.6 Preliminary or initial screening tests

In the ABPI 1981 scheme preliminary tests the safety screen test is the melting point tube test and the combustibility test is the train firing test, as described in Chapter 8. The explosion and detonation potential tests are of the type described in that chapter.

In the ABPI 1989 scheme, shown in Figure 8.12, the tests are not classified in terms as preliminary or screening tests, but the first six of the tests listed are

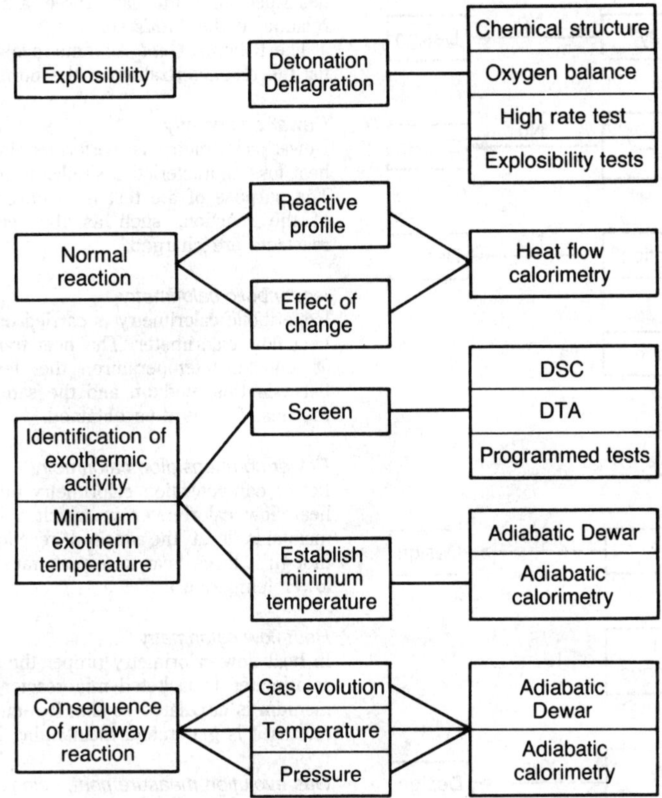

Figure 11.26 *Test scheme for assessment of hazards of chemical reactions – 2 (N. Gibson, Rogers and Wright, 1987) (Courtesy of the Institution of Chemical Engineers)*

grouped under the heading of tests for unexpected decomposition. The combustibility test is again the train firing test. The DSC and closed tube tests are ramped temperature tests for the detection of an exotherm. The delayed onset detection test is an essentially isothermal test for the detection of delayed onset of any exotherm. The adiabatic Dewar test, which simulates the reactor under adiabatic conditions, gives information on the rate of any exothermic reaction. The ARC test also provides this information. Details of the tests not already described are given below.

Closed tube test
The closed tube test uses a closed Carius tube. The oven temperature is ramped at 2°C/min. Measurements are made of the onset temperature of any exotherm and of gas evolution. The sample size is 10 g and the test, developed in ICI, is also known as the 'ICI 10 g sealed tube test'.

Delayed onset detection test
The delayed onset detection test again uses a Carius tube. The oven temperature is held constant. The sample is maintained by a small heater at about 10°C above the oven temperature and thus under essentially isothermal conditions. Any exotherm is detected by a drop in heater power.

Adiabatic Dewar test
The adiabatic Dewar test uses a Dewar flask. The flask simulates the reactor and the process as carried out on the plant. The oven temperature tracks that of the sample and any exotherm is measured.

The IChemE *Guide* scheme involves DSC and the 10 g closed tube test and the following other initial screening tests.

Insulated exotherm test
The insulated exotherm test (IET) is a form of DTA.

Decomposition pressure test
The decomposition pressure test (DPT) is carried out in a pressure vessel. The oven temperature is ramped and the sample temperature and vessel pressure are measured.

11.10.7 Tests characterizing the normal reaction

The ABPI 1981 scheme utilizes as tests characterizing the normal reaction the dynamic heating and the adiabatic heating tests.

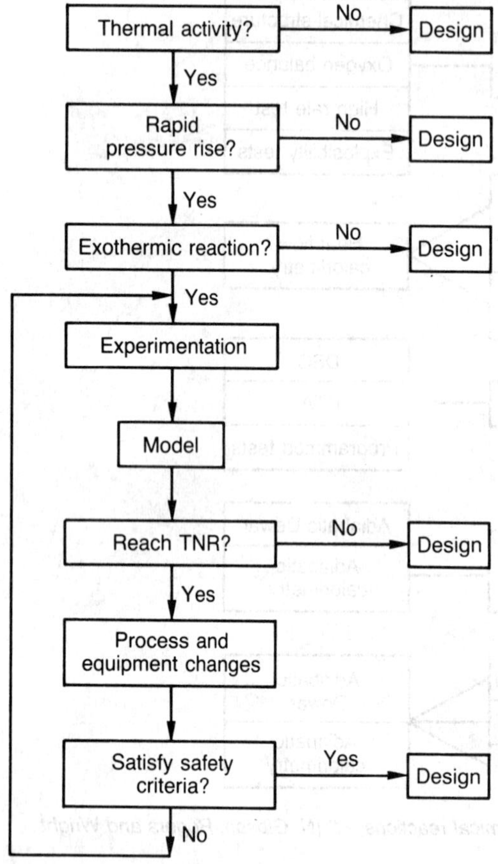

Figure 11.27 Test scheme for assessment of hazards of chemical reactions – 3 (O'Brien et al., 1987). TNR, temperature of no return (Courtesy of the American Institute of Chemical Engineers)

Dynamic heating test
The dynamic heating test has open and closed tube versions. The oven temperature is ramped and the temperature difference between the sample and a reference material is measured to detect any exotherm. The test is also known as the 'simple exotherm test'.

Adiabatic heating test
The adiabatic heating test uses a Dewar flask. The oven temperature is held constant and any exotherm is measured. The test period is at least 8 and preferably 24 hours. This test is also known as the 'heat accumulation test'.

In addition to this storage version of the test, there is a reactor version in which the flask simulates the reactor and the process is carried out as on the plant.

The ABPI 1989 scheme utilizes as tests characterizing the normal reaction the adiabatic Dewar test, ARC, the heat flow calorimeter test and the two gas evolution measurement tests. The first two have already been described and the last three are described below in relation to the *Guide* scheme.

The IChemE *Guide* scheme involves the following tests for the characterization of the normal reaction.

Dewar calorimetry
Dewar calorimetry is performed in a Dewar flask with heat loss characteristics similar to those of the real plant. The purpose of the test is to observe the characteristics of the reaction, such as the temperature rise when reactants are charged.

Isoperibolic calorimetry
Isoperibolic calorimetry is carried out in a simple form of heat flow calorimeter. The heat transfer medium is held at constant temperature, the temperature difference between this medium and the sample is measured, and the heat flow is thus obtained.

Power compensation calorimetry
Power compensation calorimetry utilizes another form of heat flow calorimeter in which a heater in the sample maintains it at the desired reaction temperature, even though it loses heat to a heat transfer medium held at a lower temperature.

Heat flow calorimetry
In heat flow calorimetry proper the sample is held in the calorimeter (a jacketed mini-reactor) and a heat transfer medium is circulated so that it removes heat at the same rate as it is generated, maintaining isothermal conditions.

Gas evolution measurement, using automated gas burette
Gas evolution measurement, using an automated gas burette, is carried out with a burette in which the gas evolved builds up and is then released in a cycle which is then repeated, the total gas evolution being determined from the number of releases made. The releases and the counting are automated.

Gas evolution measurement, using thermal mass flowmeter
Gas evolution measurement, using a thermal mass flowmeter, utilizes this alternative type of flowmeter to measure the flow of gas evolved.

11.10.8 Tests characterizing the runaway reaction

For the design of a reactor vent system it is also necessary to have information characterizing any runway reaction which may occur. A major research programme on reactor venting has been carried out by the Design Institute for Emergency Relief Systems (DIERS). This work is described in Chapter 17. The IChemE *Guide* scheme includes the following tests for the characterization of the runaway reaction.

Adiabatic Dewar calorimetry
Adiabatic Dewar calorimetry is a form of Dewar calorimetry but performed under adiabatic conditions and using a pressure Dewar flask, again with heat loss characteristics similar to those of the real plant. The purpose of the test is to observe the characteristics of the reaction, including any runaway reaction.

Vent Sizing Package and other tests

The DIERS work included the development of a bench scale apparatus for vent sizing, available in commercial form as the Vent Sizing Package (VSP). In addition to the VSP, the Reactive Systems Screening Tool (RSST) and the Phi-Tec adiabatic calorimeter are test methods applicable to the sizing of vents for runaway reactions. Vent sizing and associated tests are discussed in Chapter 17.

11.10.9 Tests characterizing abnormal reactor conditions

A set of tests relating to abnormal reactor conditions, largely but not exclusively concerned with venting, has been described by J. Singh (1988a). They include tests for: liquid and vapour deflagration; pressure attained in an unvented reaction runaway; the vapour–liquid ratio in rapid blowdown; and heat and hence vapour generation for vent sizing.

Certain liquids such as peroxides and hydroperoxides are susceptible to violent decomposition. In contrast to process reactions in which the reaction rate depends on the temperature according to the Arrhenius equation, these decompositions show no simple temperature dependence but occur at about 2000°C and are characterized by Singh as deflagrations. Whether a particular mixture will decompose in this way depends on the concentration, temperature and the presence of an ignition source such as a hot spot, spark or flame. Liquid phase deflagration tests are carried out in a small cylindrical vessel fitted with rupture discs in which the liquid sample is ignited by a heated wire. The maximum pressure resulting from a liquid decomposition is such as to rule out simple containment as a design option.

Deflagration may also occur in the vapour phase. Some compounds such as peroxides and hydroperoxides are sufficiently unstable that they can decompose in the absence of air and give a quite violent deflagration. Vapour phase deflagration tests are carried out in a spherical vessel in which the vapour sample is ignited by a fused wire. In this case containment may be an option, although venting is also widely used.

Singh describes the use of a reaction calorimeter to determine the maximum pressure resulting from runaway of the main process reaction and thus to assess the option of containment instead of venting.

In the blowdown test the reaction mixture is brought to the reaction runaway temperature in a test cell and when the vent pressure is reached a vent valve is opened full. Venting down to atmospheric pressure occurs, typically over a period of some 6 seconds. The residual liquid in the cell is measured. If the cell remains over 60% full of liquid, the vent fluid is taken as all vapour. If the cell contains less than about 5% of liquid, the venting is taken as homogeneous, i.e. the vent fluid is representative of the vapour–liquid mixture in the vessel. For residual fills between these two limits, the venting is taken as two-phase but not homogeneous, i.e. vapour disengagement and preferential venting of vapour have occurred.

The vent sizing test involves bringing the reaction mixture to the reaction runaway temperature and opening the vent valve, but this time the vapour is vented at a slow, controlled rate. The temperature of the reaction mass levels off and the reaction is tempered, the vapour flow being such as to give thermal equilibrium. The vent valve is then closed momentarily, the temperature rises and the valve is opened again. The rate of temperature rise can then be used to size the vent. The author gives an example of the calculation of vent size.

11.10.10 Interpretation of initial screening tests

Initial screening tests can provide information on the occurrence of a thermal excursion, or exotherm, and on the following of its features: (1) onset temperature, (2) heat release and (3) rate of heat release. The data from screening tests need to be interpreted with care. In particular, any onset temperature determined will be a function of the sample size. The IChemE *Guide* gives a method of correcting temperature for sample size.

11.10.11 Interpretation of exotherm onset temperature test

One of the principal parameters of particular interest in characterizing an undesired reaction is the temperature for the onset of that reaction, usually known as the 'onset temperature' or 'exotherm temperature', T_e. Although the onset temperature is a convenient concept, it needs to be borne in mind that even at lower temperatures there is still a finite reaction rate.

A quoted onset temperature is one measured by a particular test and is highly dependent on sample size, degree of adiabaticity, instrument sensitivity, time–temperature history, and reaction kinetic features, notably activation energy.

A traditional approach to reactor safety has been to set the temperature of reactor operation at some fixed temperature interval below the exotherm onset temperature measured in the laboratory test. A widely quoted value of this safety margin is 100°C, the '100 degree rule'.

Various authors discuss this safety margin. Coates and Riddell (1981a) state that any material with an exotherm onset temperature within 100°C of the maximum operating temperature as determined by the simple exotherm test is considered a potential hazard. In such a case the material is subjected to the adiabatic exotherm test. If the exotherm onset temperature in this test is 50°C or more above the operating temperature it is not considered a hazard. If it is within this 50°C safety margin, it is subjected to a long-term hot storage test at 20°C above the operating temperature.

The IChemE *Guide*, which rehearses the factors affecting a measured onset temperature, as just described, states that the 100 degree rule, or a similar one, should not be used as the basis of safety unless previous experience has shown the rule to be valid for the type of reaction in question.

11.10.12 Interpretation of tests characterizing the normal reaction

The interpretation of the tests characterizing the normal reaction is discussed in the 1981 and 1989 ABPI *Guidelines* and the IChemE *Guide*. These give a number of illustrations of test results together with commentaries.

Dewar calorimetry

The IChemE *Guide* describes Dewar calorimetry tests, showing a variety of temperature profiles, including those for adiabatic, isothermal and natural cooling conditions, and for the last two conditions with self-heating. It

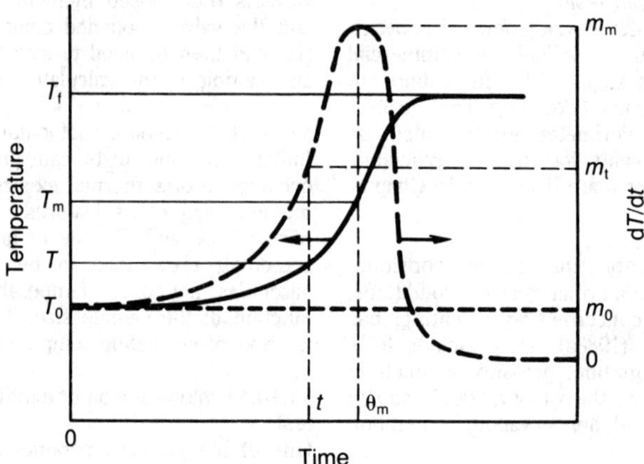

Figure 11.28 *Temperature and self-heat rate of an adiabatic reaction (Townsend and Tou, 1980) (Courtesy of Elsevier Publishing Company)*

describes the use of these tests to determine the adiabatic temperature rise, the heat of reaction and reaction kinetics. It also illustrates the employment of Dewar calorimetry to investigate semi-batch reactions.

Heat flow calorimetry

Another type of test interpreted in the *Guide* is heat flow calorimetry. It shows how data from this test may be used to obtain the heat capacity of the reaction mass, the heat of reaction and the rate of heat production.

11.10.13 Characterization of the normal reaction

The reaction system may be characterized in terms of the following fundamental quantities:

(1) physical and chemical properties of the components;
(2) heat of reaction;
(3) quantity of gas evolved;
(4) kinetics of reaction;
(5) heat transfer parameters.

For design it is necessary to know

(6) rate of heat evolution;
(7) rate of gas evolution;
(8) rate of heat removal.

The rate of heat evolution is a function of the reaction kinetics and the heat of reaction. Likewise, the rate of gas evolution is a function of the kinetics and the quantity of gas evolved. The rate of heat removal is a function of the reaction temperature attained and of the heat transfer characteristics.

The rate of heat evolution depends on the reaction mode, the reaction temperature, the reactant concentration, and any thermal events which may occur. The reaction mode may be batch or semi-batch. The reactant concentration is determined by the reaction mode and the initial temperature. The reaction temperature is a design parameter, but an excursion may occur due to an exotherm. Thermal events are those such as phase changes, gas evolution, decomposition, and so on.

The rate of heat removal is governed by the cooling, arrangements. Assuming jacket cooling it depends primarily on the reaction and coolant temperatures, the effective heat transfer area, and the heat transfer coefficients.

It is necessary to have full information on the gas evolution and decomposition characteristics. If the reaction is subject to gas evolution, data are needed on the rate of gas evolution; if the reaction is subject to decomposition data, information is required on the temperature range in which it occurs and on any tendency to autocatalysis.

11.10.14 Parameters characterizing the normal reaction

There are a number of parameters which are widely used to characterize a reaction. They include:

(1) exotherm onset temperature;
(2) adiabatic temperature rise;
(3) heat of reaction;
(4) rate of temperature rise;
(5) rate of heat generation;
(6) rate of pressure rise;
(7) velocity constant, activation energy and pre-exponential constant;
(8) adiabatic induction time.

The rate of temperature rise is often expressed as the self-heat rate.

11.10.15 Adiabatic temperature rise test

One test which yields a large amount of information on the reaction parameters is an adiabatic test in which the reaction is taken to completion. Such a test gives a sigmoidal temperature profile, with the temperature initially rising rapidly and then levelling off as the reactants are consumed. A simple analysis of such a test is given by the 1981 ABPI *Guidelines*. The analysis yields the adiabatic temperature rise, the self-heat rate, the velocity constant and the activation energy.

A fuller theoretical treatment has been given by Townsend and co-workers (Townsend, 1977, 1981;

Townsend and Tou, 1980). The treatment is for an *n*th-order reaction, but utilizes a pseudo-first-order reaction approach. For an *n*th-order reaction:

$$\frac{dC}{dt} = -kC^n \quad [11.10.1]$$

with

$$k = A\exp(-E/RT) \quad [11.10.2]$$

where A is the pre-exponential factor, C is the concentration of reactant, E is the activation energy, k is the reaction velocity constant, R is the universal gas constant, t is the time, T is the absolute temperature, and n is the order of the reaction.

Consider an adiabatic reaction as shown in Figure 11.28. The adiabatic temperature rise ΔT_{ab} is:

$$\Delta T_{ab} = T_f - T_0 \quad [11.10.3]$$

where T_f is the final absolute temperature and T_0 is the initial absolute temperature. The heat of reaction is:

$$(-\Delta H_r) = Mc_v\Delta T_{ab} \quad [11.10.4]$$

where c_v is the specific heat of the reaction mixture, ΔH_r the heat of reaction, and M is the mass of the reaction mixture.

The appropriate concentration–temperature relationship is approximately

$$\frac{C}{C_0} = \frac{T_f - T}{T_f - T_0} \quad [11.10.5a]$$

and hence

$$C = \frac{(T_f - T)}{\Delta T_{ab}} C_0 \quad [11.10.5b]$$

where C_0 is the initial concentration of reactant.

The self-heat rate m_T is obtained from Equations 11.10.1 and 11.10.5b:

$$\frac{dT}{dt} = k\left(\frac{T_f - T}{\Delta T_{ab}}\right)^n \Delta T_{ab} C_0^{n-1} \quad [11.10.6a]$$

$$= m_T \quad [11.10.6b]$$

A pseudo-reaction velocity constant k^* is defined as:

$$k^* = C_0^{n-1} k \quad [11.10.7a]$$

$$= \frac{m_T}{\left(\frac{T_f - T}{\Delta T_{ab}}\right)^n \Delta T_{ab}} \quad [11.10.7b]$$

From Equations 11.10.2 and 11.10.7a:

$$\ln k^* = \ln(C_0^{n-1}A) - E/RT \quad [11.10.8]$$

Equation 11.10.8 may be used to obtain the pre-exponential factor A and the activation energy E.

The initial self-heat rate m_0 is obtained from Equation 11.10.6b and is

$$m_0 = k_0 \Delta T_{ab} C_0^{n-1} \quad [11.10.9]$$

where k_0 is the reaction velocity constant at the initial temperature. From Equations 11.10.6 and 11.10.9:

$$m_T = m_o \frac{k}{k_0}\left(\frac{T_f - T}{\Delta T_{ab}}\right)^n \quad [11.10.10a]$$

$$= m_o \left(\frac{T_f - T}{\Delta T_{ab}}\right)^n \exp\left[-\frac{E}{R}\left(\frac{1}{T} - \frac{1}{T_0}\right)\right] \quad [11.10.10b]$$

At the maximum self-heat rate:

$$\frac{d^2T}{dt^2} = 0 \quad [11.10.11]$$

and hence it can be shown that

$$nRT_m^2 + ET_m - ET_f = 0 \quad [11.10.12]$$

where T_m is the absolute temperature at the maximum rate. Then from Equation 11.10.11:

$$T_m = \frac{E}{2nR}[(1 + 4nRT_f/E)^{1/2} - 1] \quad [11.10.13]$$

The maximum self-heat rate m_m is obtained by substituting T_m in Equation 11.10.10b.

The time to the maximum rate is obtained as follows. Integrating Equation 11.10.6a

$$\theta_m = t_m - t \quad [11.10.14a]$$

$$= \int_t^{t_m} dt \quad [11.10.14b]$$

$$= \int_T^{T_m} \frac{dT}{k\left(\frac{T_f - T}{\Delta T_{ab}}\right)^n \Delta T_{ab} C_0^{n-1}} \quad [11.10.14c]$$

where t_m is the time for the temperature to reach the maximum rate value T_m and θ_m is the time to the maximum rate from temperature T.

It is possible to obtain an analytical solution of Equation 11.10.14c if it is assumed that

$$A = \alpha T^j \quad [11.10.15]$$

where α is a constant and j is an integer. It can be shown that for a second-order reaction

$$\theta_m = \frac{RT^2}{m_T E} - \frac{RT_m^2}{m_m E} \quad [11.10.16]$$

The same equation applies for other *n*th-order reactions.

If the activation energy E is high, the second term in Equation 11.10.16 is much less than the first, and hence

$$\theta_m = \frac{RT^2}{m_T E} \quad [11.10.17]$$

Then from Equations 11.10.6b and 11.10.17

$$\theta_m = \frac{RT^2}{k\left(\frac{T_f - T}{\Delta T_{ab}}\right)^n \Delta T_{ab} C_0^{n-1} E} \quad [11.10.18a]$$

$$\ln\theta_m = \ln\left[\frac{RT^2}{\left(\frac{T_f - T}{\Delta T_{ab}}\right)^n \Delta T_{ab} C_0^{n-1} E}\right] - \ln A + E/RT \quad [11.10.18b]$$

But, under the same assumption of a high activation energy, the second term in Equation 11.10.18b is much greater than the first, and hence

$$\ln\theta_m = -\ln A + E/RT \quad [11.10.19a]$$

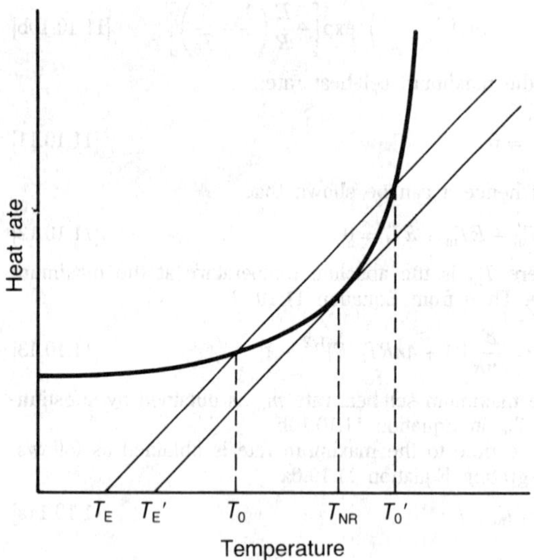

***Figure 11.29** Self-heat rate and heat transfer of a reaction with cooling (Townsend and Tou, 1980) T_E, temperature of heat exchanger; T_{NR}, temperature of no return; T_o, initial temperature; dash ′ indicates alternative set of values (Courtesy of Elsevier Publishing Company)*

or, from Equation 11.10.2,

$$\ln\theta_m = 1/k \qquad [11.10.19b]$$

The time θ_{m0} to maximum rate at temperature T_0, and thus from the start of the reaction, is obtained from Equation 11.10.18a by substituting k_0 and T_0 for k and T:

$$\theta_{m0} = \frac{RT^2}{k_0 \Delta T_{ab} C_0^{n-1} E} \qquad [11.10.20a]$$

Similarly, substituting T_0 for T in Equation 11.10.19a

$$\ln\theta_{m0} = -\ln A - E/RT_0 \qquad [11.10.20b]$$

Now consider a system with external cooling. The heat generation and heat removal curves for such a system are shown in Figure 11.29. As the reaction rate and hence the heat evolution increase, a temperature may be reached at which the rate of heat evolution exceeds the heat removal capacity of the system; this is the temperature of no return T_{nr}. From Equation 11.10.9 the rate of heat evolution Q_e is:

$$Q_e = Mc_v k_0 \Delta T_{ab} C_0^{n-1} \qquad [11.10.21]$$

The rate of heat removal Q_r is:

$$Q_r = UA_c(T_0 - T_c) \qquad [11.10.22]$$

where A_c is the surface area for heat transfer, T_c is the absolute temperature of the coolant, and U is the overall heat transfer coefficient. Equating the heat evolution and removal rates given in Equations 11.10.21 and 11.10.22

$$Mc_v A\exp\left(-\frac{E}{RT_0}\right)\Delta T_{ab} C_0^{n-1} = UA_c(T_0 - T_c) \qquad [11.10.23]$$

It can be seen from Figure 11.29 that Equation 11.10.23 has two solutions. No runaway reaction should occur if the initial temperature of the reaction mass T_0 is below a critical value T_0' and the temperature of the coolant T_c is below a critical value T_c'. But as T_0 increases to the temperature of no return T_{nr}, there is a single solution. At this point the slope of the line of heat evolution equals that of the line of heat removal. Then from Equations 11.10.2, 11.10.21 and 11.10.22:

$$Mc_v k_0 \left[\frac{E}{RT_0^2}\right]\Delta T_{ab} C_0^{n-1} = UA_c \qquad [11.10.24]$$

The time θ_{Tnr} to the maximum rate at the temperature of no return T_{nr} is, from Equations 11.10.20a and 11.10.24:

$$\theta_{Tnr} = \frac{RT_0^2}{k_0 \Delta T_{ab} C_0^{n-1} E} \qquad [11.10.25a]$$

$$= \frac{Mc_v}{UA_c} \qquad [11.10.25b]$$

The group (Mc_v/UA_c) is known as the 'equipment time line'.

The case where the maximum self-heat rate coincides with the initial self-heat rate is of interest. For this:

$$m_m = m_0 \qquad [11.10.26]$$

and

$$T_m = T_0 = T^* \qquad [11.10.27]$$

say, where T^* is the minimum initial absolute temperature above which only deceleration of the reaction occurs.

Then from Equations 11.10.12 and 11.10.27 and noting that

$$T_f = \Delta T_{ab} + T^* \qquad [11.10.28]$$

$$T^* = (E\Delta T_{ab}/nR)^{1/2} \qquad [11.10.29]$$

Thus, theoretically, there exists a temperature above which the effect of concentration depletion on the reaction rate is greater than that of acceleration due to temperature.

The parameters used in the foregoing analysis are the true parameters of the reaction. In practice, they have to be calculated from experimental measurements which are affected by the thermal capacity of the container. The true adiabatic temperature rise ΔT_{ab} is related to the adiabatic temperature rise ΔT_{abs} of the sample:

$$Mc_v \Delta T_{ab} = (Mc_v + M_b c_{vb})\Delta T_{abs} \qquad [11.10.30]$$

or

$$\Delta T_{ab} = \phi \Delta T_{abs} \qquad [11.10.31]$$

with

$$\phi = 1 + \frac{M_b c_{vb}}{Mc_v} \qquad [11.10.32]$$

where c_{vb} is the specific heat of the sample container and M_b is the mass of the container. The quantity ϕ is the thermal inertia and its reciprocal is a measure of the degree of adiabaticity. Similar corrections apply to the other measured quantities:

$$m_0 = \phi m_{os} \qquad [11.10.33]$$

$$\theta_m = \theta_{ms}/\phi \qquad [11.10.34]$$

where m_{os} is the measured self-heat rate at temperature T_0 and θ_{ms} is the measured time to maximum rate at temperature T.

11.10.16 AZT 24

Further methods of characterizing the reaction and the reactor have been described by Grewer *et al.* (1989). It is common for a product to be brought to an elevated temperature for some 24 hours. The adiabatic storage test may be used to determine a suitable temperature limit. The most useful information yielded by the adiabatic storage test is the adiabatic induction time τ_{ad}. Usually this correlates with temperature according to the relation $\log\tau_{ab}$ vs $1/T$. This correlation can be used to determine the temperature at which the adiabatic induction time is 24 hours. This temperature is known as AZT 24.

11.10.17 Mixing and heat transfer

The reaction and heat transfer behaviour of the reactor is strongly affected by the degree of mixing. There may well be departures from near-perfect mixing, and with highly viscous reaction masses these may be considerable.

The IChemE *Guide* gives two models of the profile of the temperature between the centre of the reaction mass and the wall. The first is a uniform temperature, the Semenov model, applicable where the reaction mass is well mixed. The second is a sigmoidal temperature profile, the Frank–Kamenetsky model, applicable where the mass is unmixed. This latter model is fundamental to thermal explosion, or self-heating, theory and is treated in Chapter 16. Reactor situations where the temperature in the reaction mass is liable to rise may be treated using thermal explosion models.

The conventional treatment for the case where the mass is poorly mixed is the Frank–Kamenetsky thermal explosion model, as described in Chapter 16. Specifically, the equations generally quoted are those used in relation to the experimental determination of the critical value of the ignition parameter δ_c, namely Equations 16.6.107–16.6.109.

The *Guide* also quotes the simpler treatment by Leuschke (1981) which yields the relation:

$$\ln\frac{V}{S} \propto \frac{1}{T_d} \quad [11.10.35]$$

where S is the surface area of the mass, T_d is the minimum decomposition temperature, and V is the volume of the mass.

With regard to heat transfer coefficients, correlations are available for the heat transfer coefficients on the inside of a stirred reactor and on the outside of the vessel in the jacket. Accounts of reactor heat transfer coefficients are given by Chapman and Holland (1965a,b). A method widely used for the film heat transfer coefficient on the inside of the vessel is that of E.E. Wilson (1915). The effect of agitator geometry on this coefficient may be taken into account by the use of a geometric factor and a correlation is available for this factor also. Details are given in the IChemE *Guide*, which also discusses cases which are non-ideal due to non-Newtonian behaviour, and so on. The *Guide* also gives a treatment of reactors operated under reflux conditions.

11.10.18 Reaction simulation testing

The testing described thus far is fairly formal, employing specific laboratory tests within the framework of a test scheme. One of the recognized tests, however, is the Dewar flask test in which the conduct of the reaction is explored.

There are a number of accounts of various laboratory tests conducted to explore the conduct of the reaction. Coates and Riddell (1981a,b) have described a 'worst case' analysis of a nitration reaction in a semi-batch reactor. The heat of reaction and the rate of heat evolution were determined in a calorimeter and used to select the time over which the fed reactant should be added, or the 'addition time'. Non-normal conditions were investigated such as accumulation of unreacted nitric acid in the reactor and the effect of this on the temperature of the reaction mass when reaction did occur. These authors also describe two other reaction studies.

The IChemE *Guide* describes the use of Dewar calorimetry to mimic the plant reactor and explore reactor configurations, agitation, cooling and heating systems, addition of gases, liquids and solids, and so on.

11.11 Batch Reactors: Basic Design

11.11.1 Inherently safer design

The application of the principle of inherently safer design to the choice of reaction route was described in Section 11.7. An account is given here of its application to batch reactors, which has been advocated particularly by Regenass (1984). The approach described by this author depends essentially on the avoidance of high concentrations of the reactants. One method of limiting reactant concentration is the use of a semi-batch, or 'fed-batch', reactor in preference to an 'all-in' batch reactor. In the latter all the reactants are added in the initial charge, in the former one of the reactants is added continuously. The hazard in the all-in design is that conditions may occur in which there is a sudden massive reaction of the unreacted reactants. This is liable to occur, for example, on resumption of agitation after an agitator failure. In the semi-batch reactor the continuous feed arrangement, combined with a suitable reactor temperature, keeps the concentration of one of the reactants relatively low and it is possible to effect prompt shut-off of the feed if a potentially hazardous operating deviation occurs.

The full benefit of the semi-batch reaction mode does depend, as just indicated, on the adoption of a suitable reaction temperature. If this temperature, and therefore the reaction rate, is too low, accumulation of the fed reactant can occur. The reaction temperature should be high enough to ensure that the reaction proceeds sufficiently rapidly that accumulation is avoided.

An example of the application of this principle has been given by Fierz *et al.* (1983, 1984), who describe the design of a semi-batch reactor for the further nitration of a substituted nitrobenzene. Two operating temperatures were investigated, 80°C and 100°C, using a bench scale calorimeter (BSC). At 80°C shut-off of the feed at the worst instant, when the reactants are present in equimolar proportion, resulted in heat release before and after shut-off of 90 and 180 kJ/kg of final reaction mass, respectively, whilst at 100°C the corresponding figures were 190 and 80 kJ/kg. At the lower operating

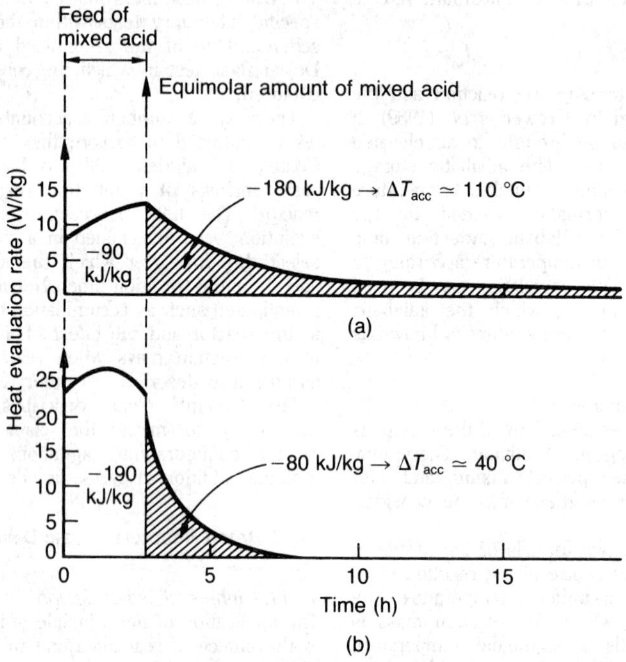

Figure 11.30 Heat effects in a semi-batch reactor at reaction temperatures of (a) 80°C and (b) 100°C (Fierz et al., 1984) (Courtesy of the Institution of Chemical Engineers)

temperature the heat release was sufficient under adiabatic conditions to give a temperature rise of 110°C, taking the temperature to 190°C, at which temperature an undesired exotherm would occur, whilst at the higher operating temperature the adiabatic temperature rise was only 40°C, taking the temperature only to 140°C. Figure 11.30 shows results obtained from the BSC work.

11.11.2 Safe operation criteria

Methods are available for the prediction of safer operating regimes for both batch and semi-batch reactor modes. One approach is based on the application of the Semenov thermal explosion model to the case of a well mixed reactor. This model applies to a liquid mass with agitation and convection. An account is given by Grewer *et al.* (1989).

A limiting condition for the reactor with cooling but without reaction can be derived from the adiabatic induction time. This is:

$$\frac{d\theta}{dt} = -\kappa\theta \qquad [11.11.1]$$

with

$$\kappa = \frac{UA_c}{Vc_p\rho} \qquad [11.11.2]$$

where A_c is the surface area for heat transfer, c_p is the specific heat of the reaction mixture, t is the time, U is the overall heat transfer coefficient, V is the volume of the reaction mixture, κ is the cooling parameter, θ is a dimensionless temperature difference, and ρ is the density of the reaction mixture. The critical condition for thermal explosion is:

$$\tau_{ad}\kappa = e \qquad [11.11.3]$$

where e is the base of the natural logarithm.

The authors define the degree of conversion X in terms of the fraction of heat released. They state that experience shows that for a batch reaction the following equation may be used:

$$r = r_0(1 - X)^n \qquad [11.11.4]$$

with

$$r_0 = kc_{A0}c_{B0} \qquad [11.11.5]$$

where c is the concentration, k is the reaction velocity constant, r is the rate of reaction, n is the effective order of the reaction and subscripts A, B and 0 indicate reactant A, reactant B and initial conditions, respectively. For a semi-batch reaction a relation which fits most cases is:

$$r = r_0\frac{(1 - X)(t/\tau_D - X)}{1 + V_Dt/V_0\tau_D} \qquad [11.11.6]$$

with

$$r_0 = kc_{A0}c_{BD} \qquad [11.11.7]$$

where V_D is the volume of addition, V_0 is the initial volume, τ_D is the time of addition and subscript D indicates addition.

The authors define a thermal reaction parameter B:

$$B = \frac{E\Delta T_{ad}}{RT^2} \quad [11.11.8]$$

where E is the activation energy, R is the universal gas constant, T is the absolute temperature, and ΔT_{ad} is the adiabatic temperature rise. They state that in relative terms a reaction can be regarded as non-critical provided that for batch reactions $B < 5$ and for semi-batch reactions $B < 10$.

Another important parameter is the Damköhler number Da, which is a measure of the rate of reaction:

$$\mathrm{Da} = \frac{(-\nu_u)r_0 t}{c_0} \quad [11.11.9]$$

where c_0 is the initial concentration of the limiting minor component, and ν_u is the stoichiometric coefficient of that component.

For a batch reaction the aim is to have a slow or only moderately fast reaction so as to avoid overloading the cooling capacity. For this Da < 1 is a suitable condition. By contrast, for a semi-batch reaction the aim is to maintain a reaction rate sufficiently fast to prevent the accumulation of reactants. This is assured provided that Da >100.

A measure of the cooling capacity is given by the Stanton number St:

$$\mathrm{St} = \kappa t \quad [11.11.10]$$

where for a batch reactor t is the reaction time and for a semi-batch reactor it is the time of addition τ_D.

For reactors with jacket cooling only a further criterion is the ratio of the Damköhler and Stanton numbers Da/St, known as the 'stability'. For a batch reactor the ratio Da/St should be $\ll 1$ and for a semi-batch reactor it should be >1. For a multi-phase reaction an effective Damköhler number is used which takes into account additional mass transfer phenomena.

The foregoing applies to the case of a well mixed reactor at more or less constant temperature. Other conditions, particularly stoppage and restart, may be allowed for by using in the Damköhler number a critical temperature. For a batch reactor this critical temperature is the highest temperature which can occur in the reactor, and for a semi-batch reactor it is the lowest which can occur. This latter is the reference temperature T_{ref} defined as:

$$T_{ref} = \frac{T_f + \mathrm{St}T_c}{1 + \mathrm{St}} \quad [11.11.11]$$

where subscripts c, f and ref indicate coolant, feed and reference, respectively. The reference temperature is the temperature produced in the reactor by feed addition and by cooling, but without reaction, and is the lowest which can occur, because the feed temperature is usually less than the jacket temperature. At this temperature the rate of reaction is lowest and the danger of accumulation of reactants highest.

11.11.3 Reactor modelling

The most fundamental understanding of the reaction system is obtained by modelling it. An account of such modelling has been given by Gordon *et al.* (1982). These authors describe the modelling of the reaction of the amination of *o*-nitrochlorobenzene (ONCB) with aqueous ammonia under pressure to produce *o*-nitroaniline (ONAN). The study included the determination of the heat of reaction and the reaction kinetics for the amination and the heat of reaction and reaction kinetics for the decomposition of ONAN. The model was validated by comparing model predictions with ARC results and with temperature and pressure profiles of the plant reactors in normal operation and from a reactor incident.

11.11.4 Difficult reactions

One type of reaction which may be difficult to handle is one with a low activation energy, since this means that the reaction rate increases very rapidly with temperature. Another difficult type of reaction is one in which the vapour pressure of the reaction mixture is low. In this case high temperatures can be reached before the pressure relief system responds. This problem is discussed by Regenass (1984).

11.11.5 Secondary reactions

In addition to the primary reaction, there may be one or more secondary reactions. If such a secondary, or side, reaction is exothermic, it may constitute a hazard. This problem has been discussed by Grewer *et al.* (1989).

Whether a secondary reaction occurs may be established using DTA. If it does, the temperature T_s at which the secondary reaction takes place is estimated by subtracting 100 K from the start of the reaction as given by the DTA. This estimate of T_s is compared with the reaction temperature T_r or the maximum possible reactor temperature T_{max}:

$$T_{max} = T_r + \Delta T_{ad} \quad [11.11.12]$$

where ΔT_{ad} is the adiabatic temperature rise (K), T_{max} is the maximum absolute temperature (K), T_r is the reactor temperature (K), and T_s is the absolute temperature of the secondary reaction. If the estimate of T_s is greater than T_{max} there is no hazard and further investigation is not required.

Use may also be made of the reaction energy of the secondary reaction. This may be obtained from the DTA by integration. A small reaction energy of <110–220 kJ/kg or an adiabatic temperature rise of <50–100 K indicates there is no hazard.

If the DTA does not yield clear-cut conclusions, it is necessary to obtain T_s more precisely. The AZT 24 may be determined. If this temperature is greater than T_{max}, again there is no hazard.

If T_s lies between T_r and T_{max} there may be an appreciable hazard, but it may still be possible to carry out the reaction provided it can be ensured that the reactor temperature does not reach T_s during the primary reaction. Alternatively, if this cannot be guaranteed, use may be made of appropriate protective measures. Operation in this region is a specialist matter.

If T_s for a highly exothermic reaction is equal to or less than T_r, it is rarely possible to carry out the primary reaction safely. These guidelines are summarized in Figure 11.31.

11.11.6 Decompositions

Grewer *et al.* (1989) give the following guidance on decomposition reactions. For single phase decomposition of organic substances, if the decomposition energy based

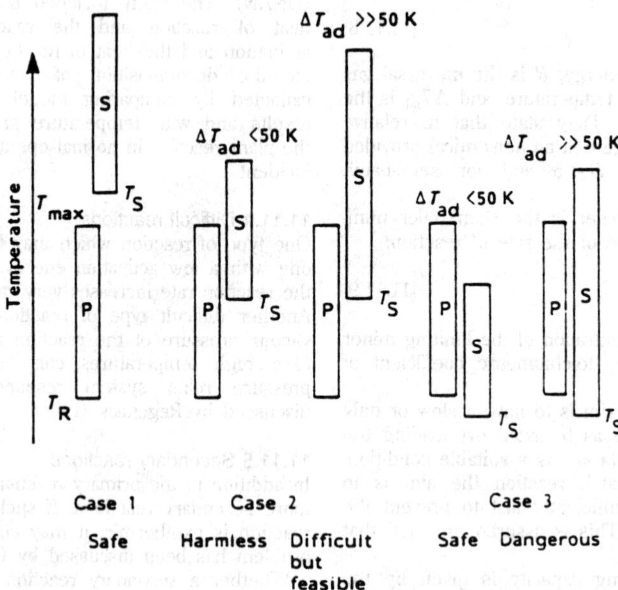

Figure 11.31 *Hazards of side reactions (Grewer et al., 1989) P, primary reaction; S, secondary reaction; T_{max}, maximum reactor temperature; T_R, temperature of reactor; T_S, temperature at which secondary reaction occurs; ΔT_{ad}, adiabatic temperature rise*

on total reaction mass is <100–200 kJ/kg, which corresponds roughly to $\Delta t_{ad} \approx 50$–100 K, the effect of the decomposition is generally not critical. Even for a decomposition energy above the range just quoted, the effect may not be critical if there are other compensating effects such as endothermic reactions or high heat removal capacity. For inorganic substances the criterion decomposition energy should be lower, due to the lower specific heat, whilst for substances containing water it should be higher. The authors state that the decisive criteria are the adiabatic temperature rise ΔT_{ad} or the thermal reaction parameter B, which should not exceed about 8.

11.11.7 Spontaneous ignition

Another hazard of batch reactors is spontaneous ignition of vapour–air mixtures in the vapour space of the reactor. Work on this has been described by Snee and Griffiths (1989). The problem arises particularly in resin manufacture. Experiments were performed using cyclohexane as a substrate related to the components of resins, employing vapour-rich mixtures with molar ratios of cyclohexane/air of 1:2, in closed vessels in the size range 0.2–20 dm^3.

The work showed that fuel-rich mixtures undergo autocatalytic isothermal oxidation, that in this situation the temperature may be increased by self-heating and that as a result spontaneous ignition can occur. Self-heating was found to set in at temperatures just below the autoignition temperature. The minimum temperature for spontaneous ignition decreased as vessel size increased. Using the Semenov thermal explosion model, the authors were able to make comparative predictions of the trend of the minimum temperature with vessel size, but state that absolute predictions would require extremely accurate kinetic data.

11.11.8 Heat loss from reactors

Information on the rate of heat loss from a batch reactor has been given by T.K. Wright and Rogers (1986). They give data on cooling, mainly over the temperature range 96–80°C, for a range of reactors, of different sizes, stirred and unstirred, with different thicknesses of insulation and with none, and with jackets full and empty. The data are for reactors in the full condition.

They correlate the data by plotting the rate of cooling against the cube root of the volume, on the assumption that the rate is proportional to the surface/volume ratio of the reactor, which for a sphere is proportional to the reciprocal of the radius or of the cube root of the volume. Their results are shown in Figure 11.32. Also shown in the figure are the rates of heat loss from 250 and 500 ml Dewar flasks. These entries show that these flasks have heat losses comparable, respectively, to 0.5 and 2.5 m^3 reactors and will not give a satisfactory simulation above these reactor sizes.

11.11.9 Instrumentation and control

Safe operation of a batch reactor requires close monitoring and control of reactant and additive flows to the reactor, of the operating temperature and of the agitation.

For the operating temperature, the temperature at which an uncontrolled exotherm will occur under plant conditions is first defined. A safety margin is set between the operating temperature and this plant exotherm temperature. The operating temperature is displayed and provided with an alarm. The temperature is controlled by a coolant.

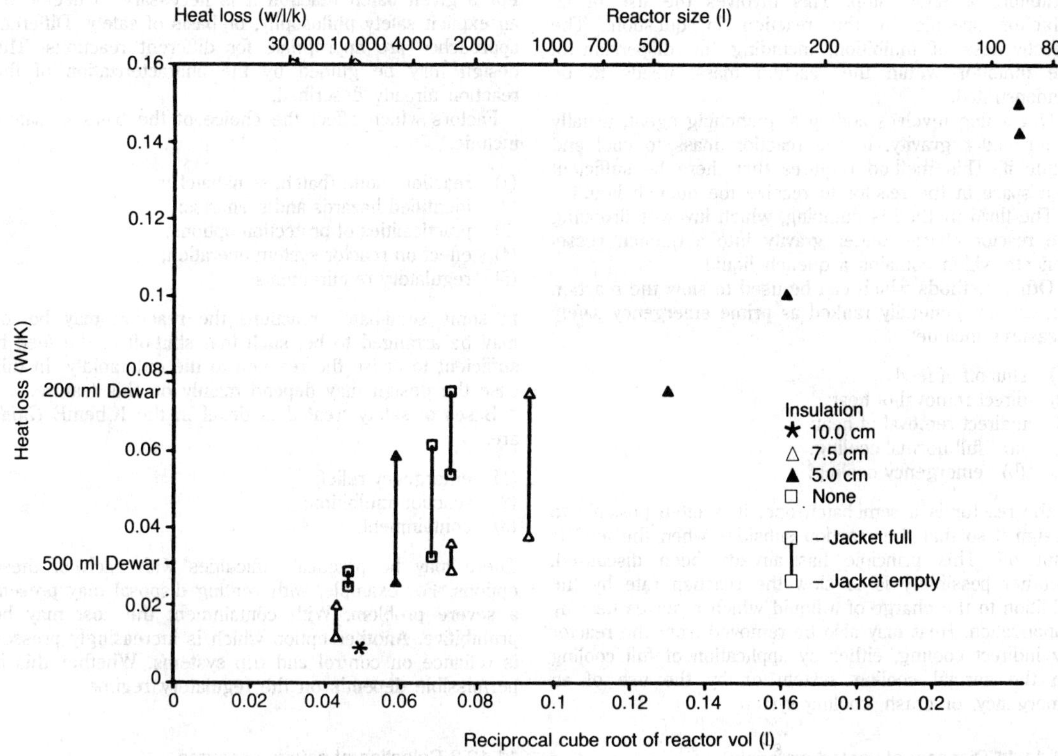

***Figure 11.32** Natural heat loss from reactors (T.K. Wright and Rogers, 1986) (Courtesy of the Institution of Chemical Engineers)*

The measurement of the temperature of the reaction mass should receive careful attention. The instrument should not allow a hot spot to develop undetected and it should not have a large time lag or a high failure rate. It is good practice to have separate measurements for the temperature control loop and the temperature alarm.

Events which may cause the temperature to go out of control, such as agitator or coolant failure, are identified. Actions which can be taken to counter an increase in operating temperature such as shut-off of feed or use of full cooling are identified and appropriate actions are selected to be activated either by operator intervention or by trip.

Too low an operating temperature may also pose a hazard in that it allows reactants to accumulate. A lower temperature limit should be specified and measures taken to maintain the temperature above this limit.

Loss of agitation has two effects, poor mixing and poor heat transfer. Poor mixing tends to lead to the accumulation of reactants. These reactants are then liable to react at some later time, particularly if the agitation is restored. Poor heat transfer may result in inadequate cooling. Either effect may lead to a runaway reaction. Provision should be made to detect loss of agitation, which may occur due to agitator stoppage or due to loss of the agitator paddle. The first condition may be detected by measuring the rotational speed, the latter by measuring power consumption. The action to be taken on loss of agitation depends on the reaction. In semi-batch reactors it is common to provide a trip to shut off the feed.

The importance of matching the controls to the reaction and the danger of overenthusiastic use of generic controls has been highlighted by Brannegan (1985). More sophisticated measurement may be provided in the form of an on-line reaction calorimeter. Such a calorimeter has been described by Hub (1977c).

Increasingly reactor control is based on computers or programmable electronic systems (PESs). The HSE guidelines on PESs give as an illustrative example the control of a nitration reactor. The example includes a full fault tree and failure data for the tree. An account is given in Chapter 13.

11.11.10 Emergency safety measures

There are a number of emergency measures which can be taken if a process deviation occurs which threatens to lead to a reaction runaway. The prime measures are:

(1) inhibition of reaction;
(2) quenching of reaction;
(3) dumping.

The reaction may be stopped by the addition of an inhibitor, or short stop. This involves the use of an inhibitor specific to the reaction in question. The effectiveness of inhibition, including the dispersion of the inhibitor within the reaction mass, needs to be demonstrated.

Quenching involves adding a quenching agent, usually water under gravity, to the reaction mass to cool and dilute it. This method requires that there be sufficient free space in the reactor to receive the quench liquid.

The third method is dumping, which involves dropping the reactor charge under gravity into a quench vessel beneath which contains a quench liquid.

Other methods which can be used to slow the reaction but are not generally ranked as prime emergency safety measures include:

(1) shut-off of feed;
(2) direct removal of heat;
(3) indirect removal of heat:
  (a) full normal cooling,
  (b) emergency cooling.

If the reactor is a semi-batch one, it is often possible to design it so that the reaction subsides when the feed is shut off. This principle has already been discussed. Another possibility is to slow the reaction rate by the addition to the charge of a liquid which removes heat by vaporization. Heat may also be removed from the reactor by indirect cooling, either by application of full cooling on the normal cooling system or by the use of an emergency, or crash, cooling system.

11.11.11 Disposal of vented materials

The material vented from a reactor has to be safely disposed of. This aspect is intimately related to the venting of reactors and is therefore dealt with in Chapter 17.

11.12 Batch Reactors: Reactor Safety

11.12.1 Hazard identification

The hazard identification methods described in Chapter 8 may be applied to batch chemical reactors. This has been described by Pilz (1986), who refers to the use of a variety of techniques including a cause–consequence matrix, a hazop study, a FMEA, a fault tree, an event tree and an incident sequence diagram (similar to a cause–consequence diagram).

A number of generic fault trees for reactors have been published. Figure 11.33 shows a fault tree given by the British Plastics Federation (BPF, 1979) and Figure 11.34 one given by Marrs *et al.* (1989).

Although for batch reactor plants the hazard of reaction runaway receives most attention, there are various other potential hazards which should not be neglected. Some of these have been discussed by Brannegan (1985).

The reaction stage of a process tends to involve various operations additional to the actual reaction, such as quenching, scrubbing and disposal. These operations also may involve scale-up problems and should receive careful attention.

11.12.2 Basis of safety

For a given batch reaction it is necessary to decide on an explicit safety philosophy, or basis of safety. Different approaches are appropriate for different reactions. The design may be guided by the characterization of the reaction already described.

Factors which affect the choice of the basis of safety include:

(1) reaction mode (batch, semi-batch);
(2) identified hazards and scenarios;
(3) practicalities of protection options;
(4) effect on reactor system operation;
(5) regulatory requirements.

In some semi-batch reactions the reaction may be, or may be arranged to be, such that shut-off of the feed is sufficient to cause the reaction to die out rapidly. In this case the design may depend mainly on this feature.

Bases of safety treated in detail in the IChemE *Guide* are:

(1) emergency relief;
(2) reaction inhibition;
(3) containment.

There may be practical difficulties with each of these options. For example, with venting disposal may present a severe problem. With containment the cost may be prohibitive. Another option which is increasingly pressed is reliance on control and trip systems. Whether this is permissible depends on the regulatory regime.

11.12.3 Selection of safety measures

Accounts of the measures for the safe design of a batch reactor include those by Christen (1980), Brannegan (1985), Berkey and Workman (1987) and N. Gibson, Rogers and Wright (1987), and the IChemE *Guide*. The application of the principle of inherently safer design has already been described with special reference to the use of a semi-batch reactor and to reaction temperature.

The approach described by N. Gibson, Rogers and Wright (1987) is based on formal process definition. Four levels of the process are defined:

(1) Level 1: fixed operating parameters;
(2) Level 2: normal variations in operating parameters;
(3) Level 3: generic fault conditions;
(4) Level 4: abnormal situations.

In Level 2 consideration is given to normal variation in the operating conditions. Some of the deviations are well recognized, others less so. Thus there may be an expectation that the temperature of a batch reaction could vary by, say, ±10°C. It may be less well appreciated that the holding time at a particular step may extend from a normal one hour to, say, 12 hours over a weekend.

The effects of certain failures which are generic in that they apply to many reactions, such as loss of coolant, rupture of an internal coil, or agitator stoppage, is assessed in Level 3. The authors regard Level 3 as the minimum standard which leads to an acceptable level of safety in the majority of processes.

Level 4 is concerned with the large number of abnormal conditions which could conceivably lead to

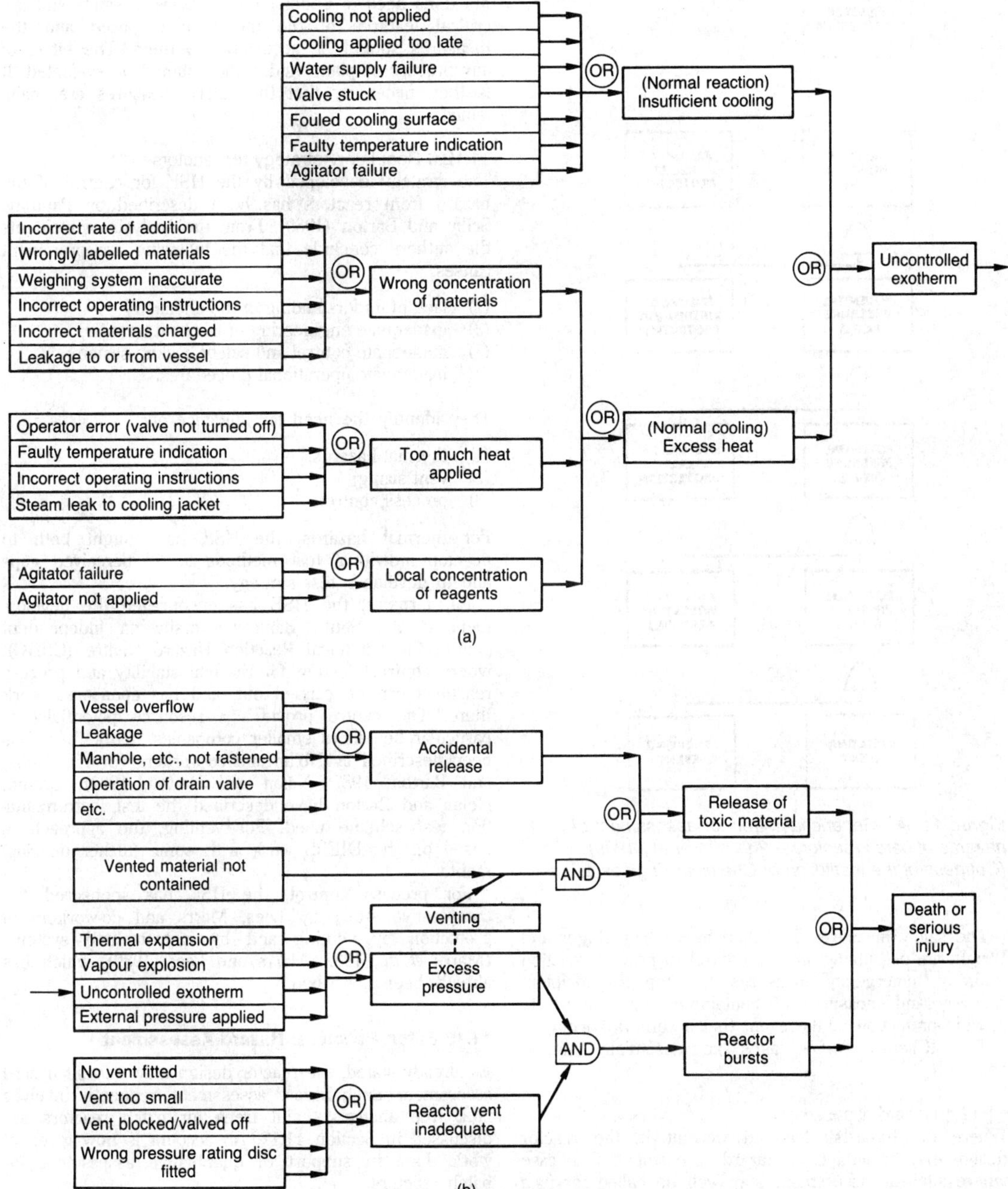

***Figure 11.33** Generic fault trees for assessment of hazards of batch reactors – 1 (BPF, 1979) (a) fault tree for top event 'uncontrolled exotherm'; (b) fault tree for top event 'Death or serious injury'*

reaction runaway. These may be explored using a method such as hazop.

The process definition should cover the operating parameters and the expected variations of these parameters (e.g. concentrations, temperature, hold times) and the details of the operations (e.g. pumping, agitation, cooling) which do not have assured protection by trips.

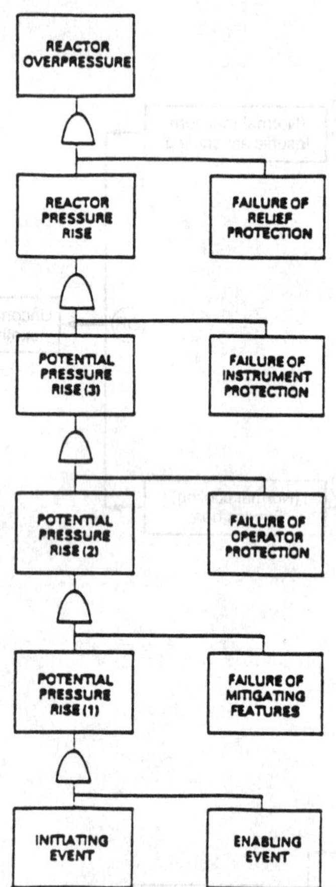

Figure 11.34 *Generic fault tree for assessment of hazards of batch reactors – 2 (Marrs et al., 1989) (Courtesy of the Institution of Chemical Engineers)*

The safety measures described by Gibson, Roger and Wright are essentially process control to prevent reaction runaway, emergency measures to stop an incipient runaway and measures of containment or venting to provide protection in the event that a runaway occurs.

The IChemE *Guide* takes an essentially similar approach.

11.12.4 Hazard assessment

Where the hazards identified warrant it, the reactor design may be subject to hazard assessment. One case where such an assessment may well be called for is a proposal to provide reactor protection by means of trips rather than venting. Hazard assessment of batch reactors in considered in Section 11.13. An illustrative example of such an assessment is given in Appendix 1 of the IChemE *Guide*.

11.12.5 Implementation of safety measures

The implementation of the safety measures selected is discussed by N. Gibson (1986b). His account is in effect an illustration of good practice in this aspect of the development of a process. The design and operating functions need to understand the basis of safety and the critical features of the measures proposed and the measures should be acceptable to them. The effect of any process or plant modification should be evaluated. It is then necessary that the safety measures are maintained.

11.12.6 Regulatory strategy for reactors

The strategy developed by the HSE for control of the hazard from reactors has been described by Pantony, Scilly and Barton (1989). From the analysis of incidents the authors conclude that the following are prominent causes:

(1) lack of understanding of the chemistry;
(2) inadequate engineering of the heat transfer;
(3) inadequate control and safety backup system;
(4) inadequate operational procedures.

They identify the need for guidance on:

(1) thermal hazards;
(2) vent sizing;
(3) process control.

For thermal hazards, the HSE has sought both to develop individual test methods to fill perceived gaps and to develop a test strategy.

Furthermore, the HSE has encouraged the development at the South Bank University an independent centre, the Chemical Reaction Hazard Centre (CRHC), where contract testing for thermal stability and process reactions can be carried out, and has sponsored work there. The centre provides a resource potentially of particular benefit to smaller companies. Work there has been described by Nolan and co-workers (Cronin, Nolan and Barton, 1987; Nolan and Barton, 1987). Cronin, Nolan and Barton have described the test instrumentation test scheme used. For venting, the approach is based on the DIERS work with some further development.

For process control, the HSE has sponsored the comparative study by Lees, Marrs and co-workers of protection by venting and by instrumented systems (Marrs *et al.*, 1989; Marrs and Lees, 1989), which has already been described.

11.13 Batch Reactors: Hazard Assessment

As already stated, the reactor design may warrant hazard assessment. Such hazard assessment frequently involves fault tree analysis. Fault trees for batch reactors are discussed in Section 11.12. An account is now given of work done in support of quantitative assessment for batch reactors.

11.13.1 National reactor inventory

This work also included estimation of the frequency of batch reactor overpressure. In order to make such an estimate a study was made by Marrs and Lees (1989) of the national reactor inventory in the UK. The estimate was based on two independent methods, the first being the number of reactors insured and the second being the number sold by the manufacturers. Table 11.12 gives the

Table 11.12 *UK national reactor inventory (Marrs and Lees, 1989) (Courtesy of the Norwegian Society of Chartered Engineers)*

A Reactors classified by age

Year	*Proportion* (%)
1914–50	2.5
1951–60	12.2
1961–70	25.7
1971–80	45.1
1981–90	14.5
Total	100.0

B Reactors classified by type

Type	*Proportion* (%)
Autoclave	2.7
Reaction vessel	11.7
Jacketed reactor	18.0
Jacketed pan	42.8
Jacketed vessel	11.3
Limpet coil reactor	6.3
Limpet coil vessel	1.2
Mixing vessel	0.4
Mixing reactor	0.2
Jacketed mixer	5.3
Limpet coil blending vessel	0.1
Total	100.0

C Reactors classified by material of construction

Material of construction	*Proportion* (%)
Mild steel	25.6
Mild steel, glass lined	44.9
Mild steel, enamel lined	0.1
Mild steel, stainless steel lined	4.3
Cast iron	0.1
Stainless steel	23.0
Stainless steel, glass lined	1.9
Aluminium	0.1
Total	100.0

D Reactors classified by working pressure

Reactor vessel pressure (psig)	*Proportion* (%)	*Jacket pressure*	*Proportion* (%)
−15–0	4.5	0–15	4.1
0–15	14.0	16–30	7.7
16–30	8.5	31–45	7.5
31–45	44.6	46–60	13.5
46–60	6.7	61–75	40.0
61–75	1.9	76–90	10.0
76–90	2.0	91–105	8.3
91–105	9.1	106–120	0.3
106–120	0.3	121–150	4.0
121–150	3.9	151–180	0.2
151–180	0.2	181–210	0.2
181–210	0.1	211–300	1.7
211–300	1.0	301–355	2.0
500–751	3.2	356–550	0.5
Total	100.0		100.0

inventory by age, by type, by material of construction and by working pressure.

11.13.2 Frequency of reactor excursions and incidents

An attempt was made by Marrs *et al.* to estimate from the national incident record the frequency of reactor overpressure excursions and overpressure incidents. The incidents are the small proportion of the excursions which resulted in a reportable accident. The basis data for this were as follows:

Period of study = 12 years
Estimated inventory of reactors = 2100
No. of applicable incidents = 66
Estimated no. of cycles = 250/year
Estimated batch time = 16 h

Hence

$$\text{Frequency of incidents} = 66/(12 \times 2100) = 2.6 \times 10^{-3}/\text{reactor-year}$$

From information obtained the authors also give the following estimate for the escalation of an excursion into an incident:

Proportion of excursions which become incidents = 0.05

This is consistent with the statement by Prugh (1981) that the probability of success of a pressure relief valve in venting a reaction runaway is 95%. Then,

$$\text{Frequency of excursions} = 2.6 \times 10^{-3}/0.05 = 5.2 \times 10^{-2}/\text{reactor-year}$$

Marrs *et al.* also obtained data from which they were able to make estimates of the failure rates and fractional dead times of bursting disc assemblies and of vent pipework. These are given in Appendix 14.

11.13.3 Frequency of reactor excursion modes

Marrs and Lees have attempted to predict the frequency of reactor excursions due to one of the modes, namely inadequate cooling. Table 11.13 shows their estimates for several of the submodes of inadequate cooling. The observed values are obtained from the overall frequency of excursions just given and from the contribution of the submodes as given in Table 11.7. There is reasonable agreement for some submodes but not for others. The numbers of the observed submodes are, of course, small.

11.13.4 Risk from reactor incidents

Marrs *et al.* have also assessed the risk to reactor operators in the UK. In the period 1967–81, which is slightly different from that used in the rest of the study, there were in the UK two fatalities. The national reactor inventory for this period was estimated as 1987. Assuming an average of 10 reactors and 2 operators per plant operating round the clock for the whole year, the risks obtained were as follows. In terms of incidents:

$$\text{Probability of fatality given an incident} = 2/69 = 0.03 \text{ deaths/incident}$$

It may be noted that this is a much lower probability of fatality than that which is obtained from the study of world-wide incidents, carried out by Nolan and quoted by the authors, in which there were 199 applicable incidents

Table 11.13 Analysis of reactor overpressure: frequency of inadequate cooling (Marrs and Lees, 1989) (Courtesy of the Norwegian Society of Chartered Engineers)

A Estimates used

Coolant source/power failure:
Frequency of coolant source/power failure = 0.1 failure/year
Probability that failure is sufficiently serious to give total loss of power = 0.1
Probability that reactor is in critical condition = 0.2
Frequency of excursion due to this cause
$= 0.1 \times 0.1 \times 0.2 = 2 \times 10^{-3}$/year

Coolant pump set failure:
Frequency of pump failure (complete failure to pump) = 0.1 failure/year
Assume one pump operating and one on standby
Length of batch cycle = 16 h = 0.00183 year
Probability of successful pump changeover = 0.95
Probability of failure during one batch
$= 1 - \exp(-0.1 \times 0.00183)(1 + 0.95 \times 0.1 \times 0.00183) = 9 \times 10^{-6}$
Probability that reactor is in critical condition = 0.2
No. of cycles = 250/year
Frequency of excursion due to this cause
$= 250 \times 0.2 \times 9 \times 10^{-6} = 4.5 \times 10^{-4}$/year

Coolant turned off:
Frequency of manual isolation valve wrongly directed closed = 0.05/year
Probability that operator fails to detect lack of cooling = 0.01
Probability that reactor is in critical condition = 0.5
Frequency of excursion due to this cause
$= 0.05 \times 0.01 \times 0.5 = 2.5 \times 10^{-4}$/year

Automatic control failure:
Frequency of failure of control loop in fail-to-danger mode
= 0.25 failure/year
Probability that operator fails to detect loss of control = 0.01
Probability that reactor is in critical condition = 0.2
Frequency of excursion due to this cause
$= 0.25 \times 0.01 \times 0.2 = 5 \times 10^{-4}$/year

Inadequate agitation:
Frequency of agitator failure = 0.5 failure/year
Frequency of operator failure to start agitator = 0.5 failure/year
Probability that agitator failure is critical = 0.01
Frequency of excursion due to this cause = $(0.5 + 0.5) \times 0.01 = 10^{-2}$/year

B Observed and predicted values

	Observed failures		*Predicted failure*
	No.	*Frequency* (failures/year)	*Frequency* (failures/year)
Coolant source/power failure	0	0	20×10^{-4}
Coolant pump set failure	1	2.8×10^{-4}	4.5×10^{-4}
Coolant turned off	3	8.4×10^{-4}	2.5×10^{-4}
Automatic control failure	3	8.4×10^{-4}	5×10^{-4}
Inadequate agitation (for cooling)	2	5.7×10^{-4}	100×10^{-4}

and 45 fatalities, giving 0.23 deaths/incident. This higher figure reflects the fact that these incidents are biased towards those involving fatalities.

Marrs and Lees express the risk in terms of reactor operating time:

$$\text{Frequency of fatality} = 6.7 \times 10^{-5} \text{ deaths/reactor-year}$$

The fatal accident rate (FAR) from this cause alone was estimated as 3.8.

11.14 Unit Processes

11.14.1 Chemical industry

The chemical and petrochemical industries are described in *The Petroleum Chemicals Industry* (Goldstein and Waddams, 1967), *The Petrochemical Industry* (Hahn and Williams, 1970), *Structure of the Chemical Processing Industries* (Wei, Fraser and Swartzlander, 1978), *The*

Chemical Industry (Sharp and West, 1982), *Shreve's Chemical Process Industries* (G.T. Austin, 1984), *The Chemical Industry* (Heaton (1985–), *Handbook of Chemicals Production Processes* (Meyers, 1986a) and *Petrochemicals* (Spitz, 1988). Further references are given in Section 11.18.

Accounts of the European chemical industry are given in the following: France, E. Johnson (1989c) and Redman *et al.* (1991); Germany, E. Johnson (1990); Italy, Redman *et al.* (1990); and the Netherlands, Redman and Smith (1991).

11.14.2 Principal processes

Descriptions of unit processes and of their thermodynamics and kinetics are given in *Unit Processes in Organic Synthesis* (Groggins, 1952), *Industrial Chemicals* (Faith, Keyes and Clark, 1965), *Chemical Process Industries* (Shreve, 1967), *Organic Chemicals Manufacturing Hazards* (Goldfarb *et al.*, 1981), *Shreve's Chemical Process Industries* (G.T. Austin, 1984), *Survey of Industrial Chemistry* (Chenie, 1986) and *Introduction to Industrial Chemistry* (H.L. White, 1986). Further information is given in other standard texts, including encyclopaedias such as those by Kirk and Othmer (1963–, 1978–, 1991–) and Ullman (1969–, 1985–). Selected references on unit processes and their hazards are given in Table 11.14.

Unit processes are listed by Fowler and Spiegelman (1968) as follows: acylation, alkaline fusion, alkylation, amination, aromatization, calcination, carboxylation, causticization, combustion, condensation, coupling, cracking, diazotization, double decomposition, electrolysis, esterification, fermentation, halogenation, hydration, hydroforming, hydrogenation, hydrolysis, ion exchange, isomerization, neutralization, nitration, oxidation/reduction, polymerization, pyrolysis, and sulphonation. The unit processes which are generally more hazardous are given by Fowler and Spiegelman as alkylation, amination, aromatization, combustion, condensation, diazotization, halogenation, hydrogenation, nitration, oxidation, and polymerization. The unit processes considered here by way of illustration are (1) oxidation, (2) hydrogenation, (3) chlorination and (4) nitration.

A particular unit process tends to have certain characteristic features. These often relate to:

(1) reactant and product;
(2) reactor phase;
(3) main reaction:
  (a) thermodynamic equilibrium,
  (b) heat of reaction,
  (c) velocity constant,
  (d) activation energy;
(4) side reactions;
(5) materials of construction.

11.14.3 Oxidation

The oxidation reactions considered here are those of an organic compound with molecular oxygen.

Some principal industrial oxidation reactions include:

Vapour phase
Ethylene → ethylene oxide
Propylene → acrylic acid
Methanol → formaldehyde

***Table 11.14** Selected references on unit processes and their hazards*

Groggins (1952); Kirk and Othmer (1963–, 1978–, 1991–); Faith, Keyes and Clark (1965); Goldstein and Waddams (1967); Shreve (1967); Considine (1974); NFPA (1991 NFPA 491M)

Biotechnology, food processing
Tong (1978); Frommer and Krämer (1989); Palazzi *et al.* (1989); Jowitt (1980)

Alkylation
B. Scott (1992)

Carbonylation
Goldfarb *et al.* (1981)

Chlorine production
Eichelberger, Smura and Bergenn (1961); Pennell (1963); Chlorine Institute (1981 CPP, Pmphlt 67, 1986 Pmphlt 1); Sommers (1965); Schwab and Doyle (1967); Bunge and Honigh (1969); Elliott (1969); J.L. Wood (1969); Kuhn (1971); Stephens and Livingston (1973); C. Jackson and Kelham (1984); Means and Beck (1984); Coulter (1986); C. Jackson (1986); Wall (1986); I.F. White *et al.* (1986)

Chlorination
Groggins (1952); Daniel (1973); Goldfarb *et al.* (1981)

Electrochemical
Dotson (1978); AIChE (1981/76, 1983/80); Fleischmann and Overstall (1983); Pletcher (1984)

Hydrogenation
Groggins (1952); Augustine (1965); Gant (1978); Rylander (1985); Klais (1987)

Nitration
Groggins (1952); Albright (1966b,e,g); Dubar and Calzia (1968); Albright and Hanson (1969, 1976); Fritz (1969); Ventrone (1969); N.A.R. Bell (1971); Biasutti and Camera (1974); Biasutti (1976, 1979); F.W. Evans, Meyer and Oppliger (1977)

Oxidation
Burgoyne and Kapur (1952); Groggins (1952); Burgoyne and Cox (1953); Sittig (1961-,1962); Salooja (1964a); Shtern (1964); Emanuel (1965); Papenfuss (1965); Zeelenberg and de Bruijn (1965); Berezin, Denisov and Emmanuel (1966); Albright (1967b,e); Farkas (1970); Prengle and Barona (1970); Häberle (1971); Pickles (1971); Talmage (1971); Alexander (1974, 1990a,b); Dumas and Bulani (1974); Gugan (1974b); Hucknall (1974); Dragoset (1976); Nemes *et al.* (1976); Saunby and Kiff (1976); R. Davies (1979); Gans (1979); Weismantel and Ricci (1979); Hobbs (1980); Lyons (1980); Goldfarb *et al.* (1981); Crescitelli *et al.* (1982); Krzyrztoforski *et al.* (1986); Snee and Griffiths (1989); Franz and Sheldon (1991); Memedlyaev, Glikin and Tyulpinov (1992)

Oxychlorination
Quant *et al.* (1963); Illidge and Wolstenholme (1978); Goldfarb *et al.* (1981)

Oxyhydrochlorination
Illidge and Wolstenholme (1978); McNaughton (1983)

Oxo synthesis
Weber and Falbe (1970)

Plasma processing
Szekely and Apelian (1984)

Polymerization
Goldfarb *et al.* (1981)

Naphthalene → phthalic anhydride
Benzene, butenes → maleic anhydride

Liquid phase
Acetaldehyde → acetic acid
Butane → acetic acid and related products
i-Butane → t-butyl hydroperoxide (and hence propylene oxide)
Cumene → cumene hydroperoxide (and hence phenol, acetone)
Cyclohexane → cyclohexanone and cyclohexanol (and hence adipic acid and caprolactam)
Ethylbenzene → ethylbenzene hydroperoxide (and hence propylene oxide)
Naphtha → acetic acid
Toluene → benzoic acid (and hence phenol)
p-Xylene → terephthalic acid
p-Xylene → monomethylterephthalate
Higher hydrocarbons → primary hydroperoxides (and hence primary alcohols, aldehydes and acids)
Higher hydrocarbons → secondary hydroperoxides (and hence secondary alcohols and ketones)

The reactions are generally exothermic and sometimes highly so. There is frequently, therefore, a problem of heat removal and of temperature control.

Oxidation reactions tend not to be limited by thermodynamic equilibrium. There can be problems, however, of the reaction of complete combustion and of side reactions to unwanted by-products. Frequently it is necessary to employ catalysts and to use only moderate temperatures in order to avoid these undesired reactions.

The heat generated in such oxidation processes may considerably exceed the theoretical heat of reaction. Thus the heat of oxidation of napthalene to phthalic anhydride is about 5460 BTU/lb of naphthalene, but for some reactor conditions it is reckoned that the occurrence of the complete combustion reaction can raise the effective heat of reaction to 10 000 BTU/lb of naphthalene fed.

Vapour phase oxidations are carried out in various types of reactor, such as a reactor vessel with catalyst on a tray, a tubular packed bed reactor or a fluidized bed reactor. Tubular and fluidized bed reactors are suitable if the heat release is large. Thus the oxidation of napthalene to phthalic anhydride is highly exothermic. Initially this reaction was carried out in a tubular reactor, but is now usually done in a fluidized bed.

In vapour phase oxidation the reactor feed is generally kept outside the flammable range. This requires the use of well designed and highly reliable trip systems. The use of a high integrity protective system to maintain the ethylene and oxygen feeds in the ethylene oxide process on the ethylene-rich side has been described by R.M. Stewart (1971) and is considered in more detail in Chapter 13.

Another approach to the control of the feed composition has been outlined by Pickles (1971), who describes a method used in the production of formaldehyde by oxidation of methanol. The methanol-air feed is obtained by bubbling air through the methanol vaporizer. The feed is maintained on the methanol-rich side, mainly by ensuring that the temperature of the methanol in the vaporizer is sufficiently high.

The hazard of a flammable mixture is much less if the burning velocity is low, which it is near the flammability limits. In the methanol-air system the burning velocity is low not only at the upper flammability limit but also some way inside it. In the system described by Pickles this fact is exploited.

If the risk of ignition is judged sufficiently low, the reaction may be carried out within the flammable limits. R. Davies (1979) cites benzene and propylene oxidation as two processes which have been operated in this way.

Liquid phase oxidations are typically carried out in tank reactors at a pressure in the range 10–50 bar and a temperature in the range 100–200°C. Air is bubbled into the reactor by a sparge pipe.

One of the main liquid phase processes is the oxidation of cyclohexane to cyclohexanone and cylcohexanol. This was the process at Flixborough. It has been described in *The Oxidation of Cyclohexane* (Berezin, Denison and Emmanuel, 1966) and by Haberle (1971). In this process, which is carried out industrially at pressures in the range 10–25 bar and temperatures in the range 140–170°C, the conversion per pass has to be kept to a few per cent in order to minimize undesirable side reactions. Therefore heat removal is fairly simple. The principal hazard is the large amount of cyclohexane liquid which is held in and circulated around the reactors.

The air inlet line can be a hazard in a liquid phase oxidation process. It is known that in the manufacture of KA, a mixture of cyclohexanone and cyclohexanol, by air oxidation of cyclohexane a blackened product is occasionally made. The conditions under which this 'black KA' is produced are always those of start-up. The cause is that liquid from the reactor flows back up the line, stays there until the plant is started up and then catches fire when the air flow resumes. The temperatures reached are sufficient to melt through the air pipe. There is a hazard, therefore, that the reactor contents may escape through this line. The phenomenon has been discussed by Alexander (1974).

The hazards of oxidation processes have been reviewed by R. Davies (1979). He describes the transition in liquid phase oxidations from batch reactors to large-scale continuous reactors. This growth was accompanied by an increasing number of incidents.

Davies identifies three principal hazards of oxidation processes, in the liquid or vapour phase: flammable release, internal ignition and unstable or reactive substances. The majority of liquid phase oxidation processes require a large inventory held at a temperature above the boiling point. This creates the potential for a large release of flashing flammable liquid. Moreover, the operating conditions are quite difficult, which increases the risk that such a release will occur.

A second hazard of liquid phase oxidations is internal ignition. This can occur if for any reason the concentration of oxygen builds up in the vapour space. Such build-up may happen if the dispersion of the air in the liquid is poor or if the oxygen is not immediately absorbed and reacted. In some processes a flammable mixture can occur where recycle offgas is mixed with the air prior to injection into the liquid.

Vapour phase oxidation processes have a low inventory of process fluids, but a flammable release hazard may still exist where there is a large inventory of flammable heat transfer fluid. Single-pass, atmospheric processes may be protected against internal ignition by explosion venting, but this is more difficult to do where the process is at higher pressure or involves recycle.

The products of oxidation processes tend to contain unstable substances, which in some cases can constitute a severe hazard. These include cumene hydroperoxide and ethylene oxide. Another is acetaldehyde, though this is more readily controlled. Davies gives a detailed account of the characteristics of, and precautions against, these substances.

Davies identifies four oxidation processes which have been responsible for a relatively large number of incidents. They are the liquid phase oxidations of cyclohexane and of cumene, the vapour phase oxidations of ethylene and of naphthalene or *o*-xylene. He gives a list of incidents and identifies several factors which have contributed to such incidents. One is the use of a series of liquid phase reactors in cascade with insufficient attention paid to the interconnecting pipework. Another is modifications which are intended to increase yield or to improve thermal economy but which also increase inventory and complexity. A third is deficiencies of various kinds in the trip systems.

An account of the continuing problem of liquid phase oxidation processes has been given by Alexander (1990b,c). The loss potential in an oxidation process can be large and the margin between a minor and a major incident small. He deals in particular with internal ignition. Such incidents are relatively frequent, but tend not to be reported.

The author suggests that most internal ignitions are due to a common cause and puts forward the following hypothesis. Oxidizable material is exposed to air at a temperature high enough to initiate liquid phase oxidation. Poor dispersion of the air allows a high concentration of oxygen to build up close to the liquid–air interface and a flammable mixture forms. Free radicals from the oxidation reaction in the liquid phase evaporate into the flammable mixture near the interface and ignite it. He cites the work of Sokolik (1960) in support of the ability of free radicals to ignite a flammable mixture well below the autoignition temperature. Alexander argues that this proposed mechanism is consistent with the wide variety of conditions in which ignitions occur. He refers also to other proposed mechanisms of ignition, including deposits of peroxides, of pyrophoric materials, and of catalyst, and friction between metal surfaces.

Alexander describes some of the situations in which internal ignitions occur, and gives explanations based on his hypothesis. These situations involve the formation of a flammable mixture, followed by ignition by free radicals from the liquid, present as bulk liquid or as liquid droplets. One is backflow of liquid into the air feed line. This can occur if the air flow is shut off but the isolation is not good enough to prevent the pressure of the air in the pipe in the reactor falling below that of the liquid. It can occur if the air flow is interrupted momentarily. If the air temperature is above the critical value, burning will occur. This may result only in a very hot pipe, but can cause pipe failure followed by release of the reactor contents. If under very unfavourable circumstances a flammable mixture occurs along a sufficient length of the air pipe, there is potential for detonation.

Another situation with potential for internal ignition occurs where there is a recycle of fuel-rich offgas back into the air pipe, designed so that under normal operation the gas mixture downstream of the mixing point is also fuel-rich. If there is an upset, the mixture may enter the flammable range. Another such situation is where a temporary liquid diversion is made to flush out blockages in the air line.

Alexander also describes precautions which can be taken. One is the design of the air line, which should be such that liquid is not likely to be held up if there is an interruption of the air flow. The line should not rise up through the bottom of the vessel. Nor should it enter through the top, since a leak from the pipe in the vapour space could give a flammable mixture there. The preferred configuration is for the air line to come down from a height, slope gently down to the vessel and then enter its side. Another precaution is positive isolation of the air line by a double block and bleed valve arrangement. Another is the use of cool air. Other precautions relate to recycle gas, liquid diversion, offgas analysis and explosion containment.

11.14.4 Hydrogenation

The hydrogenation reactions considered here are those of an organic compound with molecular hydrogen in the presence of a catalyst. Some principal hydrogenation reactions include:

Vapour phase

Carbon monoxide → methanol
Olefins → paraffins

Liquid phase

Coal → oil

The thermodynamic equilibrium in hydrogenation reactions is almost invariably such that the yield falls as temperature rises and therefore the reaction temperature is a compromise between high reaction yield and high reaction rate.

Hydrogenation reactions are generally exothermic. The reaction usually occurs only on the catalyst, so that the local surface temperature can be very high. This may result in undesirable side reactions and catalyst deterioration, and may have implications for reactor materials of construction and for control.

The activation energy of hydrogenation reactions is normally low. For most hydrogenation reactions an increase in temperature of 50°C or more is required to double the reaction rate. There are few for which an increase of 10°C is sufficient to do this.

Vapour phase hydrogenations are carried out typically in the pressure range from atmospheric to several hundred bar and in the temperature range 100–400°C. Most catalysts used in hydrogenations are also capable of promoting oxidation. Also catalysts can be rendered more pyrophoric by the adsorption of hydrogen. Hydrogenation involves the hazards of the use of hydrogen under pressure. These are considered elsewhere, but particularly in Section 11.16.

11.14.5 Chlorination

The chlorination reactions considered here are those of an organic compound with molecular chlorine. Some principal industrial chlorination reactions include:

Vapour phase
Methanol → chloromethanes

Liquid phase
Ethylene → ethylene dichloride

Chlorination reactions have many similarities with oxidation reactions. They tend not to be limited by thermodynamic equilibrium, but there can be problems of the reaction of complete chlorination and of side reactions to unwanted by-products.

Again the reactions are generally exothermic and often highly so, with the result that there tends to be a problem of heat removal and temperature control. For the substitution reaction

$$RH + Cl_2 \rightarrow RCl + HCl$$

the heat of reaction is about 24 000 kcal/kmol of chlorine.

The vapour phase reaction of chlorine with paraffinic hydrocarbons is apt to become too violent and it may be necessary to take measures to prevent this. These include limiting the proportion of chlorine or admitting it in stages and using diluents such as nitrogen or recycle gases.

Vapour phase reactors are typically empty vessels or packed beds.

Chlorine, like oxygen, forms flammable mixtures with organic compounds. Thus, for example, the following flammability limits have been quoted for methane:

	Oxygen	*Chlorine*
Lower flammability limit (%)	5.2	5.6
Upper flammability limit (%)	60.5	70.0

Further details are given by Daniel (1973).

The reactor feed in vapour phase chlorinations is kept outside the flammable range. Liquid phase reactions are frequently carried out by blowing the chlorine and organic reactants into liquid product in a tank reactor.

11.14.6 Nitration

The nitration reactions considered here are those of an organic compound with a nitrating agent such as nitric acid or mixed acid. Mixed acid is nitric acid admixed with a dehydrating acid such as sulphuric acid. Nitration accidents are among not only the most frequent but also the most destructive in the chemical industry. The processes for the manufacture of nitroglycerine (NG) and trinitrotoluene (TNT) are particularly sensitive.

The typical nitration is a batch process carried out in the liquid phase and it is this which is discussed first. Nitric acid is not only a nitrating agent but also an oxidizing agent. The nitration and the oxidation reactions both constitute a hazard. Nitric acid is itself a hazardous chemical, since it is not only a corrosive substance but also a strong oxidant. As indicated, it may react explosively with some organic liquids.

Both the nitration reaction itself and the oxidation reaction are strongly exothermic. Since nitration mixtures are often very sensitive and liable to explode, unless conditions are closely controlled, this strong heat generation is a problem. The heat evolved in a nitration reaction includes not only the heat of nitration but also the heat of dilution. Thus the heat of nitration of benzene is 761 BTU/lb, but under certain conditions the heat released by reaction and dilution is 895 BTU/lb.

Nitration reactions may constitute a hazard due to a reaction exotherm or a sensitive product. In some incidents the phenomena appear not to have been recorded previously, but in others the hazard was known in general terms, but its full implications have not been appreciated. There have been a number of explosions in nitration reactors which have subsequently been shown to have been due to the fact that the temperature of the reaction was too close to that at which a runaway reaction could start. The temperature differences between the main reaction and the runaway reactions have been typically of the order of 10–50°C. Similarly, explosions have occurred due to the sensitivity of the product of the nitration. Some nitro compounds have a quite low decomposition temperature, in the range 100–150°C, and are thus relatively hazardous. Information is available on this aspect (e.g. T.L. Davis, 1943; Kirk and Othmer, 1963-, 1987-, 1991-).

Rapid autocatalytic decompositions or 'fume-offs' of nitration products sometimes occur. Often they do not involve an explosion, but are nevertheless quite violent. Factors which have contributed to explosions due to runaway reactions or sensitive materials have included lack of knowledge and lack of appreciation of the need for particular precautions and for a wide margin of safety.

Nitrations are frequently carried out in stirred batch reactors. This type of reactor has certain characteristic hazards. These include the charging of reactants in incorrect quantities or sequence and the accumulation of unmixed reactants which react violently when the agitator is switched on.

Another hazard in nitration reactions is the ingress of water. The addition of water to the nitration mixture may result in a large release of heat which causes an explosion either directly or by initiating some other effect.

Whereas the other unit processes considered above are carried out in plants dedicated to that single process, nitrations are frequently performed in plants which make a number of products on a campaign basis. This method of manufacture introduces some additional opportunities for error.

Since nitration often presents problems both in respect of runaway exotherms and heat sensitive products, it is especially important to have full information about the reaction and the products, particularly for a new reaction. Many of the problems of nitration are alleviated by the use of a continuous nitration process. The nitration is carried out in several vessels, each of which has a smaller holdup and time-invariant operating conditions. Many stages such as charging and discharging, precooling and preheating are eliminated, as are some items of equipment such as pumps. The process is more readily adaptable to automatic control. There is less human intervention and thus both less human error and less exposure of personnel. The evolution of nitroglycerine manufacture from a process using a batch reactor to a continuous process carried out in a nozzle has already been described.

Nitration is also carried out in the vapour phase. A further discussion of safety in nitration processes is given by Biasutti and Camera (1974). An accident in a nitration process is described in Case History B22.

11.15 Unit Operations and Equipments

Unit operations include: mixing, dispersion; distillation, gas absorption, liquid–liquid extraction; leaching, ion exchange; precipitation, crystallization; centrifugation, filtration, sedimentation; classification, screening, sieving; crushing, grinding; compacting, granulation, pelletizing; gas cleaning; heat transfer; and drying and dehydration. A further listing is given by Fowler and Spiegelman (1968).

Characteristic types of equipment are associated with these different operations. Accounts of unit operations are given in *Chemical Engineering* (Coulson and Richardson, 1955–, 1977–), *Unit Operations of Chemical Engineering* (McCabe and Smith, 1967), *Handbook of Separation Techniques for Chemical Engineers* (Schweitzer, 1979), *Principles of Unit Operations* (Foust *et al.*, 1980), *Chemical Engineers Handbook* (Perry and Green, 1984) and *Handbook of Separation Processes* (Rousseau, 1987) and by McCabe, Smith and Harriott (1985).

The unit operations and equipments considered here by way of illustration are (1) mixers, (2) centrifuges, (3) driers, (4) distillation columns and (5) activated carbon adsorbers. All five feature in the CCPS *Engineering Design Guidelines* (1993/13) and the first three are the subject of IChemE guides.

There is a large group of unit operations in which dusts are handled with the associated hazard of dust explosion. This includes driers, as described below. Further treatment of this and other operations involving dusts is given in Chapter 17.

Other equipments such as fired heaters, heat exchangers, pumps and compressors are treated in Chapter 12 as parts of the pressure system. Selected references on unit operations and equipments and their hazards are given in Table 11.15.

11.15.1 Mixers

Mixers present a variety of hazards. An account of hazards of, and precautions for, mixers is given in the *Guide to Safety in Mixing* (Schofield, 1982) (the IChemE *Mixer Guide*). The body of the *Mixer Guide* is in the form of a checklist of hazards and precautions under the

Table 11.15 *Selected references on unit operations and equipments and their hazards*

API (1979/1); McCabe, Smith and Harriott (1985); Elvers *et al.* (1988)

Adsorbers
Chapman and Field (1979); Anon. (1986 LPB 69, p. 257); CCPS (1993/13)

Agitation, mixing
Garrison (1981); Schofield (1982, 1983); Harnby, Edwards and Nienow (1985); Kneale (1985 LPB 62); Sandler and Luckiewicz (1987); Ulbrecht and Patterson (1985)

Autoclaves
HSE (1990 PM 73)

Centrifuges
Thrush and Honeychurch (1969); E.B. Price (1970); P. Peterson and Cutler (1973); Butterwick (1976); Funke (1976); Simon (1976); O'Shea (1983); Lindley (1985, 1987); IChemE (1987); Bange and Osman (1993); BS 767: 1983

Conveyors
Anon. (1977i); Ewart Chainbelt Co. Ltd (1977); Conveyor Equipment Manufacturers Association (1979); BS 490: 1975–, BS 2890: 1989

Crushing and grinding equipment
Anon. (1962b); Stern (1962); K.N. Palmer (1973a); Anon. (1979 LPB 26, p. 48); Moir (1984)

Distillation
Guerreri (1969); McLaren and Upchurch (1970); Frank (1977); R.G. Hill (1977); Shah (1978); G.F.P. Harris, Harrison and Macdermott (1981); Nisenfeld and Seeman (1981); Anon. (1983s, 1987g); L.M. Rose (1985); M.E. Harrison and France (1989); Ströfer and Nickel (1989); Hower and Kister (1991); Love (1992)

Driers
Sloan (1967); Keey (1973, 1978); K.N. Palmer (1973a); Glatt (1977); Reay (1977); Grafton (1983); van't Land (1984); Walsh (1984); N. Gibson, Harper and Rogers (1985); Schacke and Falcke (1986); IChemE (1989/134); Abbott (1990); Papagiannes (1992)

Dust-handling plant
Merrill and Velentine (1942); Sargent (1969); K.N. Palmer (1973a); Schafer (1973b); Batel (1976); Horzella (1978); IChemE (1978/116, 1992/32); M.N. Kraus (1979a,b); Anon. (1981b); M.A. Maxwell (1981); P. Swift (1984); D.M. Muir (1985); Mody and Jackhete (1988); Eggerstadt, Zievers and Zievers (1993); Opila (1993); VDI 2263: 1990–

Electrostatic coating
Luderer (1987)

Electrostatic precipitators
H.J. White (1963); H.E. Rose and Wood (1966); Hanson and Wilke (1969); K.N. Palmer (1973a); Schneider *et al.*

(1975a,b); Bump (1977); Frenkel (1978); Lewandowski (1978); Oglesby and Nicholls (1978); Böhm (1982); Coleman (1984); Jaasund (1987)

Evaporating ovens
HSE (1974 HSW Bklt 46, 1981 HS(G) 16)

Gas absorption
Astarita, Savage and Bisio (1983)

Heat treatment baths
HSE (1971 HSW Bklt 27, 1975 SHW 849); FPA (1974 S8)

Oil-water separators
Chambers (1978)

Polymer processing
Sullivan (1989)

Solids handling
Mechanical Handling Engineers Association (1962); Weisselberg (1967); Notman, Gerrard and de la Mare (1981); IChemE (1985/126); Grossel (1988)

Pneumatic conveying
EEUA (1963 Hndbk 15); M.N. Kraus (1965, 1966a,b, 1986); Gluck (1968); D. Smith (1970); Stoess (1970); K.N. Palmer (1973a); Weston (1974a); Gerchow (1975); Caldwell (1976); Barker (1981); O.A. Williams (1983); Maunder (1985); D. Mills (1990); Dahn (1993)

Spray driers
Belcher, Smith and Cook (1963); Sloan (1967); Masters (1972, 1979); N. Gibson and Schofield (1977); Christiansen (1978); Long (1978); Bartknecht (1981a)

Vaporizers
G.T. Wright (1961); W.H. Doyle (1972a); E.R. Peterson and Maddox (1986); R.A. Smith (1987)

Vapour-liquid separators
Gerunda (1981); Purarelli (1982); Wu (1984); Tsai (1985)

Water treatment
Gilwood (1963); Boby and Solt (1965); Arden (1968); Seels (1973); Webb (1974); Betz Laboratories Inc. (1976); D.R. Smith (1976); Rue (1977)

Ancillary equipment
Anon. (1978 LPB 24, p. 167); API (1980/2); Klein (1993)

headings (1) equipment, (2) operations, (3) substances and (4) plant.

The equipment should be suitable for its purpose. Dead spaces where reactions may occur or material may solidify should be avoided. The mixer body may be subjected to pressure or vacuum and to fatigue due to any vibration. It may experience out-of-balance loads such as the dropping in or the shifting of the charge. It should be capable of withstanding these different effects.

The rotating parts may experience inertia loads on start-up and out-of-balance loads during operation. Loads may be imposed due to high liquid viscosity or high powder shear resistance. Other hazards include entry of foreign material such as tramp metal and contact between the impeller and the body of the vessel due to bending of the shaft. There is a critical speed for an impeller. It should normally not be run between 70 and 100% of this critical speed.

There are various features that may constitute weak points on a mixer. They include closures, sight glasses, instruments, sample points on the body, and bearings and seals on the agitator. The moving parts and the charge may present a hazard to personnel. The moving parts may include the mixer body or other features such as counterbalanced closures. Other hazards include unguarded impellers, splashing or even ejected material, or ejected objects.

Failure of utilities can cause problems for mixers. Some failures include loss of electrical power to the agitator, loss of cooling, loss of inerting, and loss of exhaust ventilation. If there is loss of agitation, reactions may occur or material may solidify. Another hazard of failure of electrical power is unexpected restart of the mixer. It should be arranged that the mixer does not restart automatically following restitution of power after a power failure. Measures should also be taken to prevent other inadvertent start-up of the mixer.

Mixers sometimes handle flammable materials. Sources of ignition characteristic of mixers are hot surfaces, friction and impact, and static electricity. Hot surfaces include electric motor casings, drive components, bearings and lamps. Friction and impact can occur due to breakage of rotating parts, entry of foreign material or contact between the impeller and the vessel. Friction may also occur due to bearings or seals running dry. Static electricity may be generated during the loading or during mixing of liquids or powders. A principal precaution against ignition is inerting.

The plant layout should provide good access to the mixer for operation and maintenance and should cater for foreseeable leaks and spillages. Operational hazards include: overfilling, leading to overpressure or spillage; underfilling, leading to splashing or, if the mixer is nearly empty, mechanical damage; and overmixing, leading to thickening, overheating or overreaction of the mixture.

The *Mixer Guide* describes some of the principal types of mixer such as the rotating shape (double cone), orbiting screw, 'U' trough, high speed rotor, Muller, top entry agitator (stirred tank), agitated bead mill (attrition), change can, planetary, heavy paste (Z-blade), internal (Banbury), roll mills and mixer/extruder types, and gives a matrix of mixer types vs relevant hazards.

Two precautions which find widespread use on mixers are fixed guards and interlocks. Fixed guards include annular covers around agitators and bars to prevent entry of a limb into a running nip. These are complemented by non-manual methods of loading and discharging such as chutes and feed belts.

Interlocks are used to prevent a mixer from being started up or approached unless it is in a safe condition. One application is to ensure that the cover is closed, another to ensure that the agitator is properly centred in the body and another to prevent access either to the agitator shaft or to the mixer body while it is moving. The *Mixer Guide* gives a number of examples.

A further treatment of mixers is given in the CCPS *Engineering Design Guidelines.*

11.15.2 Centrifuges

Centrifuges likewise present a variety of hazards. The hazards of, and the precautions for, centrifuges are considered in the *User Guide for the Safe Operation of Centrifuges with Particular Reference to Hazardous Atmospheres* (Butterwick, 1976) and the later publication *User Guide for the Safe Operation of Centrifuges* (J. Lindley, 1987) (the IChemE *Centrifuge Guide*).

Centrifuges often handle volatile flammable liquids and, unless special precautions are taken, are almost certain to contain a flammable mixture at some stage in the operational cycle. The probability of ignition of the flammable mixture is quite high. Main sources of ignition are mechanical friction, hot surfaces and static electricity. A centrifuge rotates at high speed and a mechanical fault, leading to a spark, or a hot surface, can cause ignition.

The movement of the slurry in an operating centrifuge favours the generation of static electricity, particularly if the liquid has a high resistivity. Alternatively, if the centrifuge is stopped and open, a static electricity hazard may occur from an operator who has too high an insulation path to earth.

There is also a mechanical hazard from the high kinetic energy of the rotating centrifuge bowl. The main causes of mechanical failure are basket imbalance, incorrect assembly, corrosion and bearing failure.

Since centrifuge operations involve frequent opening of the centrifuge, there is with some substances a toxic hazard. This may be an immediate, acute hazard or a long-term cumulative one.

If there is an explosion hazard in a centrifuge, it should be blanketed with inert gas, usually nitrogen. The maintenance of an inert atmosphere should be monitored. The three monitoring systems commonly used are based on measurement of (1) flow of inert gas into centrifuge, (2) pressure of gas in centrifuge and (3) oxygen concentration of gas in centrifuge. The most suitable system depends on the degree of hazard. But the system based on the measurement of oxygen concentration, which is the variable of direct interest, is the most positive and should be used for high risk situations.

If there is a toxic hazard, the centrifuge system should be enclosed as far as practicable. Local exhaust ventilation should be provided. Alternatively, it is possible to extract from the casing and operate under a negative pressure, provided that the liquid is not flammable. Other measures which may be required include provision of a forced air supply or of breathing equipment for the operator.

Centrifuges are treated further in the CCPS *Engineering Design Guidelines.*

11.15.3 Driers

The principal hazards presented by driers are those of fire and explosion. These hazards and the corresponding precautions are discussed in *Prevention of Fires and Explosions in Dryers* (Abbott, 1990) (the IChemE *Drier Guide*). This is the second edition of, and thus supersedes, the *User Guide to Fire and Explosion Hazards in the Drying of Particulate Materials* (Reay, 1977).

The *Drier Guide* states that each year in the UK driers are responsible for 30 fires and one explosion serious enough to be attended by the local fire brigade. A detailed account of the self-heating of combustible materials is given in Chapter 16 and of combustibility and explosibility of dusts and of dust fires and explosions and dust explosion venting in Chapter 17, and the treatment given here is correspondingly limited.

Differences between the first and second editions of the *Drier Guide* are in large part accounted for by the up-dating of the methods for explosibility testing and for explosion suppression and venting. The identification of hazards in powders which are to be dried is described in Chapter 8. The *Drier Guide* assumes that the materials processed are essentially materials which may be combustible but are not either explosive or liable to decomposition. The following discussion is concerned with such combustible materials.

With regard to the oxygen concentration to prevent ignition, the *Drier Guide* states that if the published values are unacceptably low for efficient plant operation, the minimum oxygen concentration to support combustion should be established by tests performed using the modified vertical tube apparatus described by Field (1982).

There are three forms in which a combustible material being processed in a drier may be ignited: (1) dust cloud, (2) dust layer and (3) bulk dust.

A dust cloud of explosible concentration will normally occur only in driers where the material is dispersed such as in a fluidized bed drier or a spray drier. In other types of drier such as tray and band driers there should be no dispersed dust, provided the air velocity is kept sufficiently low. If there is a dust cloud, it will almost certainly be within the explosible range. The dust cloud may be ignited by an ignition source. Alternatively, ignition may occur if the cloud is raised above its minimum ignition temperature.

Dust may also form in layers inside the drier. A dust layer may be ignited by an ignition source. Alternatively ignition by self-heating may occur if the dust layer is exposed to an ambient temperature above its minimum ambient temperature for self-ignition.

Ignition may also occur in the bulk dust, for which there is a separate minimum ignition temperature. Since the heat loss from the bulk material is less than from a layer, this minimum ignition temperature is lower than that for a dust layer.

A dust explosion may be prevented by the use of inerting. The concentration of inert gas should be such as to reduce the oxygen concentration below the minimum oxygen concentration required to support combustion.

Dust properties relevant to drier design therefore include in particular the minimum ignition temperature of the dust cloud, the minimum ambient temperature of ignition of a dust layer, the minimum ignition temperature for the bulk dust, and the minimum oxygen concentration to support combustion. The minimum ignition energy of the dust cloud is also relevant. The design of dust explosion venting systems utilizes information on the dust K_{st} class. Particle size is a strong determinant of these properties. Account should also be taken of the fact that if the dust contains any significant amount of flammable liquid, there is a hybrid mixture, with enhanced explosibility.

Many of the regular ignition sources apply to driers. These include hot work, hot surfaces, electrical equipment, and friction and impact. A direct heating system may give rise to hot particles which can act as an ignition source and precautions are taken against this. Air for combustion should be drawn from a source free of dust. Avoidance of dust in this air should be a factor taken into account when considering recirculation of air from the drier. Burners should be cleaned regularly and operated so as to give complete combustion. Erratic operation and flame blow-off should be remedied without delay. Before entering the drier, the hot gases should be passed through a 3 mm mesh to screen out hot particles. The mesh should be able to withstand the high temperatures and should be cleaned regularly. A direct heating system should not be used to evaporate flammable vapours.

Precautions against explosion may be based on prevention or protection. The main preventive measure given in the *Drier Guide* is inerting. The guide gives three main protective measures: venting, suppression and containment.

On the basis of the hazard evaluation a basis of safety should be chosen. The *Drier Guide* states that one measure is sufficient. If inerting is used, there should be continuous on-line monitoring of the oxygen concentration. With some driers there is a facility to recirculate part of the exhaust air to improve thermal efficiency. The gas entering the drier will then have a higher concentration of water vapour and, if it is direct fired, a higher concentration of carbon dioxide also. This provides a degree of inerting and such driers are described as 'self-inertizing'.

The *Drier Guide* gives a detailed discussion of the application of explosion suppression and explosion venting methods to driers. It also discusses the applicability of explosion containment, including pressure shock containment, to which it adopts a cautious approach. It points out, however, that containment may well be a practical option for a vacuum drier, where the initial pressure is much lower.

Measures which protect against a dust explosion, whether explosion suppression, venting or containment, do not necessarily offer protection against a sudden decomposition of the material, with a massive evolution of gas. Measures should also be taken to prevent the spread of dust fire or explosion to other parts of the plant. The process should be specified so as to minimize the hazard. The inlet air temperature is a particularly important variable and the *Drier Guide* discusses in detail the relevant factors and safety margins for different types of drier.

There are various items of equipment which are common to drying plants. They include heating systems, feed systems, dust conveying and recovery systems, and product storage vessels. Dust recovery systems include cyclones, filters, electrostatic precipitators, and scrubbers and washers.

The heating system should be designed in accordance with the codes for gas or oil fired heaters, as the case may be. It should have the appropriate instrumentation, including alarms, trips and interlocks.

Protection should be provided against explosion of a fuel–air mixture in the firing space. Suitable relief should be provided against such an explosion, unless the drier is designed for an explosible dust and is thus already provided with a dust explosion relief which can serve both purposes; where such a dust explosion relief exists, it will generally be adequate for the flammable gas explosion.

Self-heating of material passing hot from a drier to product storage is a hazard. The maximum safe temperature for discharge to storage may be determined for many situations from thermal ignition theory in combination with empirical tests. Measures to reduce the risk of self-heating include limitation of the size of storage vessel and monitoring of the temperature of the material in the vessel.

The operation and maintenance of driers also have certain common features. The *Drier Guide* discusses the operational aspects under the headings of pre-start-up checks, start-up, normal operation, normal shut-down and emergency shut-down. The guide gives what amount to checklists for pre-start-up checks and maintenance.

The *Drier Guide* describes some of the principal types of drier such as spray driers, pneumatic conveying driers, fluidized bed driers, rotary driers, band driers, batch atmospheric tray driers, batch vacuum driers and trough driers. For each type of drier it discusses the principle of operation, hazards, ignition sources, preventive and protective measures, process specification, equipment specification, and operation and maintenance. The CCPS *Engineering Design Guidelines* provide a further treatment of driers.

11.15.4 Distillation columns

Distillation columns present a hazard in that they contain large inventories of flammable boiling liquid, usually under pressure. There are a number of situations which may lead to loss of containment of this liquid.

The conditions of operation of the equipment associated with the distillation column, particularly the reboiler and bottoms pump, are severe, so that failure is more probable.

The reduction of hazard in distillation columns by the limitation of inventory has been discussed above. A distillation column has a large input of heat at the reboiler and a large output at the condenser. If cooling at the condenser is lost, the column may suffer overpressure. It is necessary, therefore, to protect against this by pressure relief devices. On the other hand, loss of steam at the reboiler can cause underpressure in the column. On columns operating at or near atmospheric pressure, inert gas injection is needed to provide protection against this.

Another hazard is overpressure due to heat radiation from fire. Again pressure relief devices are required to provide protection.

The protection of distillation columns is one of the topics treated in detail in codes for pressure relief such as API RP 521. Likewise it is one of the principal applications of trip systems.

Another quite different hazard in a distillation column is the ingress of water. The rapid expansion of the water as it flashes to steam can create very damaging overpressures.

11.15.5 Activated carbon adsorbers

Experience with activated carbon adsorbers indicates that they are another potentially troublesome operation. An account of this type of adsorber is given in the CCPS *Engineering Design Guidelines.*

Activated carbon adsorbers are prone to fires in the adsorbent bed. This is so particularly where the substance being adsorbed is an oxidizable organic such as an aldehyde, ketone or organic acid. Fires are especially likely to occur during a shut-down.

Countermeasures evolved to minimize the risk of fire include: paying attention to adsorbent purity; maintaining the moisture content of the adsorbent; operating at less than 25% of the lower flammability limit; and avoiding dead spots. The adsorbers should be located outdoors and provided with pressure relief and fire protection.

Among the measures used to prevent fire during a shut-down are maintenance of a gas flow 75% of normal, use of water or steam spray to keep the bed moist, inerting of the bed and removal of the bed.

The *Guidelines* give a checklist of precautions for activated carbon adsorbers.

11.16 Operating Conditions

Operation at extremes of pressure and temperature has its own characteristic problems and hazards. Extreme conditions may occur at (1) high pressure, (2) low pressure, (3) high temperature and (4) low temperature.

Some general characteristics of these operating conditions are described below, together with some of the associated literature. In relation to extreme operating conditions in general, mention may be made of *Chemical Engineering under Extreme Conditions* (IChemE, 1965/40) and of the series *Safety in Air and Ammonia Plants* (1960–69/17–26), *Operating Practices in Air and Ammonia Plants* (1961–62/27–28) and *Ammonia Plant Safety and Related Facilities* (1970–94/31–53) by the American Institute of Chemical Engineers (AIChE).

A newer development in extreme operating conditions is the use of plasma processing. Selected references on extreme operating conditions and their hazards are given in Table 11.16.

11.16.1 High pressure

The range of pressures handled in high pressure technology is wide. Three broad ranges of pressure may be distinguished: (1) pressures up to about 250 bar, (2) pressures of several thousand bar and (3) pressures greater than 8000 bar. Most process plant operates at pressures below 250 bar, but certain processes such as high pressure polyethylene plants operate at pressures up to about 3000 bar. Ultra-high pressures begin at a pressure of about 8000 bar, which is the probable limit of the simple elastic cylinder and at which it becomes necessary to use other types of construction such as wire-wound vessels.

Normal high pressure plant is dealt with in numerous standard texts. Mention may be made here of *Theory and Design of Modern Pressure Vessels* (Harvey, 1974) and *Pressure Component Construction* (Harvey, 1980).

Table 11.16 *Selected references on extreme operating conditions and their hazards*

IChemE (1965/40); Zabetakis (1965); Pilborough (1971, 1989)

High pressure

ASME (Appendix 28 *Pressure Vessels and Piping*, PVP 192, 1982/154, 1986 PVP 110, 1987 PVP 125, 1988 PVP 148, 1989 PVP 165); ASTM (STP 374, 1982 STP 755); Tongue (1934); Newitt (1940); Comings (1956); Stevenston (1959); AIChE (1960–69/3–12, 1961–62/13,14, 1967/71, 1970–94/17–38, 1973/62, 1974/63, 1978/64); SCI (1962); F.W. Wilson (1962); E.L. Clark (1963); Bett and Burns (1967); Buchter (1967); Weale (1967); Albright, van Munster and Forman (1968); Manning and Labrow (1971); Munday (1971a); Karl (1973); Witschakowski (1974); Sykes and Brown (1975); Gilbert and Eagle (1977); IMechE (1977/39); Spain and Paauwe (1977); Vodar and Marteau (1980); Homan, MacCrone and Whalley (1983); Boyer (1984); Livingstone (1984); Sherman and Stadtmuller (1987); Prugh, Howard and Windsor (1993)

High temperature

ASME (Appendix 28 *Materials, Pressure Vessels and Piping*, 1975/71, 1979/137, 1982/156, 1983/164, 1984/172, 1989 PVP 163); I.E. Campbell (1956); Goldberger (1966); Kamptner, Krause and Schilken (1966b); I.E. Campbell and Sherwood (1967); Clauss (1969); Albrecht and Seifert (1970); ASME (1971/65); IChemE (1971/50, 1975/62); Zeis and Lancaster (1974); AIChE (1986/84)

Low pressure, vacuum

Avery (1961); Davy (1951); EEUA (1961 Hndbk 11); Pirani and Yarwood (1961); Steinherz (1962); Spinks (1963); van Atta (1967); Roth (1967, 1990); Dennis and Heppel (1968); Carpenter (1970); Mangnall (1971); AIChE (1972/74); O'Hanlon (1980); Ryans and Croll (1981); Patton (1983); R.V. Parker (1984); Ryans (1984); Ryans and Roper (1986); EEMUA (1987 Publ. 152); IChemE (1987/130); Sandler and Luckiewicz (1987); N.S. Harris (1989); Wutz, Adam and Walcher (1989); Eckles and Benz (1992); R.E. Sanders (1993b)

Low temperature

ASME (Appendix 28 *Materials, Pressure Vessels and Piping*, 1980/49; 1984/195); British Cryogenics Council (n.d., 1970, 1975); M. Davies (1949); Ruhemann (1949); Anon. (1960e); J.D. Jackson (1961); Burgoyne (1965b); Krolikowski (1965); Barron (1966–); AIChE (1968/72, 1972/74, 1982/77, 1986/83); Codlin (1968); Harton (1968); Ligi (1969); IChemE (1970/49, 1972/53, 1975/59); V.C. Williams (1970); Haselden (1971); R.W. Miller and Caserta (1971); Edeskuty and Williamson (1983); Springmann (1985); CGA (1987 P-12)

Supercritical conditions

Körner (1985); Penninger *et al.* (1985); Randhava and Calderone (1985); McHugh and Krukonis (1986)

Plant for higher pressure is a more specialized technology. It is treated in *The Design and Construction of High Pressure Chemical Plant* (Tongue, 1934), *The Design of High Pressure Plant and Properties of Fluids at High Pressure* (Newitt, 1940), *High Pressure Technology* (Comings, 1956), *Apparate und Armaturen der Chemischen Hochdrucktechnik* (Buchter, 1967), *High Pressure Engineering* (Manning and Labrow, 1971), *High Pressure Technology* (Spain and Paauwe, 1977) *High Pressure in Science and Technology* (Vodar and Marteau, 1980) and *Experimental Techniques in High Pressure Research* (Sherman and Stadtmuller, 1987), and by Homan, Maccrone and Whalley (1983).

Further accounts of high pressure technology and its hazards are given by Stevenston (1959), Munday (1971a), Sykes and Brown (1975), B.G. Cox and Saville (1975), Boyer (1984), Livingstone (1984) and Prugh, Howard and Windsor (1993).

The relevant standard for high pressure plant is BS 5500: 1976 *Unfired Fusion Welded Pressure Vessels*. There is no UK standard for very high pressure plant.

High pressures are usually associated with high and/or low temperatures.

The use of high pressure greatly increases the amount of energy available in the plant. Whereas in an atmospheric plant stored energy is mainly chemical, in a high pressure plant there is in addition the energy of compressed permanent gases and of fluids kept in the liquid state only by the pressure.

Since the economies of scale apply very strongly to compression, the number of compressors used is generally small. Often there is a single compressor for a particular duty. The machines themselves are large and their technology is complex.

Although high pressures in themselves do not pose serious problems in materials of construction, the use of high temperatures, low temperatures or aggressive materials does. Thus the problem is to obtain the material strength required by high pressure operation despite these factors.

With high pressure operation the problem of leaks becomes much more serious. The amount of fluid which can leak out through a given hole is greater on account of the pressure difference. Moreover, the fluid may be a liquid which flashes off as the pressure is reduced.

11.16.2 Low pressure

The degrees of vacuum handled in low pressure or vacuum technology span many orders of magnitude. In the present context it is relatively low vacuum from 760 torr (atmospheric pressure) to 1 torr and medium vacuum from $1\text{–}10^{-3}$ torr which are the ranges of prime interest.

Normal vacuum technology is treated in *Industrial High Vacuum* (Davy, 1951), *Principles of Vacuum Technology* (Pirani and Yarwood, 1961), *Vacuum Technology* (Spinks, 1963), *Vacuum Science and Engineering* (van Atta, 1967), *Vacuum Sealing Techniques* (Roth, 1967), *Vacuum System Design* (Dennis and Heppel, 1968), *A User's Guide to Vacuum Technology* (O'Hanlon, 1980), *Process Vacuum System Design and Operation* (Ryans and Roper, 1986), *Modern Vacuum Practice* (N.S. Harris, 1989), *Theory and Practice of Vacuum Technology* (Wutz, Adam and Walcher, 1989) and *Vacuum Technology* (Roth, 1990).

Low pressures are not in general as hazardous as the other extreme operating conditions. But a hazard which does exist in low pressure plant handling flammables is the ingress of air, with consequent formation of a flammable mixture.

A major part of vacuum technology is concerned with detecting and preventing leaks. The techniques developed in this field may be relevant, therefore, to plants under positive pressure, particularly those handling very toxic materials.

11.16.3 High temperature

The literature on high temperature technology is relatively diffuse and deals mainly with particular types of plant, e.g. ammonia plants, or particular types of equipment, e.g. fired heaters. Mention may be made, however, of *High Temperature Materials and Technology* (I.E. Campbell and Sherwood, 1967).

As already mentioned, the use of high temperatures in combination with high pressures greatly increases the amount of energy stored in the plant. The heat required to obtain a high temperature is often provided by fired heaters. These have a number of hazards, including explosions in the firing space and rupture of the tubes carrying the process fluid.

There are severe problems with materials of construction in high temperature plants. The main problem is that of creep at high temperatures. But there are other problems also, such as hydrogen embrittlement. The use of high temperatures implies that the plant is put under thermal stresses, particularly during start-up and shutdown. It is essential to take these stresses fully into account in the design.

It is equally necessary to operate the plant with regard to the temperature effects. Its life can be much shortened by excursions inside the high creep range or by transients which cause high thermal stresses.

11.16.4 Low temperature

The temperatures handled in low temperature technology are those below 0°C. There is a somewhat distinct technology dealing with ultra-low temperatures below 20 K, but this is not of prime interest in the present context.

Normal low temperature technology is dealt with in *The Physical Principles of Gas Liquefaction and Low Temperature Rectification* (M. Davies, 1949), *The Separation of Gases* (Ruhemann, 1949), *Cryogenic Systems* (Barron, 1966–), *Cryogenic Fundamentals* (Haselden, 1971) and *Liquid Cryogens* (Edeskuty and Williamson, 1983). Mention should also be made of the *Cryogenic Safety Manual* by the British Cryogenic Council (BCC, 1970).

Low temperature plants contain large amounts of fluids kept in the liquid state only by pressure and temperature. If for any reason it is not possible to keep the plant cold, then the liquids will begin to vaporize.

Impurities in the fluids in low temperature plant are liable to come out of solution as solids. This is particularly likely to happen if parts of the plant are allowed to boil dry. Deposited solids may be the cause not only of blockage but also, in some cases, of explosion. It is necessary, therefore, to ensure that the fluids entering a low temperature plant are thoroughly purified.

A severe materials of construction problem in low temperature plants is low temperature embrittlement. The materials requirements, however, are well understood. The main difficulties arise from aspects such as installation of incorrect materials or flow of low temperature fluids into sections of the plant constructed in mild steel.

In low temperature as in high temperature operations the plant is subject to thermal stresses, especially during start-up and shut-down. These stresses need to be allowed for and, as far as possible, avoided.

The insulation on low temperature plant frequently takes the form of a cold box. It is not easy to gain access to equipment inside the cold box. It is important, therefore, for the equipment to be as reliable and the fluids processed as pure as possible.

In low temperature plant the temperature approaches used in the heat exchangers are very close, say 3°C. This is done to achieve maximum thermodynamic efficiency, since any inefficiency is paid for directly in the power required for compression. The use of such close temperature approaches is another reason for avoidance of fouling by keeping the fluids free of impurities.

The scale of operation in low temperature technology can be very large. In particular, the largest storages of hazardous liquids, both flammables and toxics, tend to be refrigerated storages.

11.17 Utilities

The process plant is dependent on its utilities. These include in particular:

(1) electricity:
   (a) general,
   (b) uninterrupted power supplies,
   (c) electrical heating;
(2) fuels;
(3) steam;
(4) compressed air:
   (a) plant air,
   (b) instrument air,
   (c) process air,
   (d) breathing air;
(5) inert gas;
(6) water
   (a) cooling water:
   (b) process water,
   (c) hot water,
   (d) fire water;
(7) heat transfer media:
   (a) hot fluids,
   (b) refrigerants.

There may also be other special services on particular plants. Selected references on utilities are given in Table 11.17.

Important features of a utility are (1) its security, (2) its quality, (3) its economy, (4) its safety and (5) its environmental effects.

One of the main causes of failure of supply of a utility is loss of part of the generating capacity followed by rapid overload of the rest, leading to total loss. In order to avoid this, measures may be taken to effect load shedding at the less essential points of consumption.

***Table 11.17** Selected references on utilities and their hazards*

NRC (Appendix 28 *Utilities*); D.T. Smith (1965c, 1982); Butikofer (1974); Monroe (1970); W.B. Thomas (1981); API (1981/4); Kletz (1988m); Broughton (1993)

Electricity
American Oil Co. (n.d./5); BG (Appendix 27 *Electrical*); EEUA (Doc. 37D, 1964 Doc. 17); HSE (HSW Bklt 31, 1970 HSW Bklt 24, 1985 OP 10, 1991 GS 47); ICI (n.d.b); IEEE (Appendix 27); IEE (1966 Conf. Publ. 16, 1971 Conf. Publ. 74, 93, 1974 Conf. Publ. 108, 110, 1975 Conf. Publ. 134, 1977 Conf. Publ. 148, 1982 Conf. Publ. 281, 1988 Conf. Publ. 296, 1992 Conf. Publ. 361); MCA (SG-8); NRC (Appendix 28 *Power Supply*); RoSPA (IS/73); Swann (1959); Abbey (1964); Jenett (1964a); Silverman (1964a–d); Tenneco Oil Co. (1964a); R.Y. Levine (1965a); D.T. Smith (1965c, 1982); Edison Electric Institute (1966); Chopey (1967a); Kerkofs (1967); FPA (1968/6, 1974/25); House (1968a); Kullerd (1968); Haigh (1969); Hoorman (1969); R.H. Lee (1969); J.C. Moore (1969); Vaccaro (1969); Nailen (1970a, 1973a–c, 1974, 1975); Tucker and Cline (1970, 1971); Yuen (1970); Yurkanin and Claussen (1970); A.H. Moore and Elonka (1971); API (1982 Refinery Inspection Guide Ch.14, 1991 RP 540, 2003); Consumers Association (1972); Jolls and Reidihger (1972); Stover (1972); Anon. (1973i); Butler (1973); Huey (1973); Imhof (1973); IEEE (1973, 1975a,b); Mattson (1974); Needle (1974); Iammartino (1975); Crom (1977); Fink and Beaty (1978); Margolis (1978); Morrison (1978); Roe (1978); Bos and Williams (1979); T. Brown and Cadick (1979–); Mueller (1980); Macpherson (1981); Wildi (1981); Cohn (1983); A. Jackson (1984); Nicholls (1984); Peate (1984); Goodchild (1985); Hobson and Day (1985); Fordham Cooper (1986); Sandler and Luckiewicz (1987); EEMUA (1988 Publ. 133); Kletz (1988m); UL (1988 UL 1012); Bro and Levy (1990); Anon. (1991 LPB 102, p. 21; 1991 LPB 102, p. 23; 1992 LPB 108, p. 28; 1994 LPB 118, p. 15) Kowalczyk (1992); J.A. McLean (1992); Broughton (1993); McLean (1993 LPB 110); NFPA (1993 NFPA 70, 496) ANSI (Appendix 27), ANSI C series, ANSI C2-1993 BS (Appendix 27 *Electrical*), BS CP 1003: 1964–, BS CP 1013: 1965, BS PD 2379: 1982

Emergency generators
NRC (Appendix 28 *Diesel Generators*); HSE (1985 PM 53); Kletz (1988m); NFPA (1989 NFPA 110A, 1993 NFPA 110, 111)

Uninterruptible power supply
Cullen (1990); UL (1991 UL 1778)

Electrical pipe tracing
Butz (1966); Ando and Othmer (1970); Ando and Kawahara (1976)

Electrical area classification
BASEEFA (n.d./1, SFA 3004, 3006, 3009, 3012); MCA (SG-19); IEE (1962 Conf. Pub. 3, 1971 Conf. Pub. 74, 1975 Conf. Pub. 134, 1982 Conf. Publ 218, 1988 Conf. Publ. 296, 1992 Conf. Publ. 361); API (1973 RP 500B,

1982 RP 500A, 1984 RP 500 C, 1991 RP 500); IP (1990 MCSP Pt 15, 1991 MCSP Pt 1); J.R. Hughes (1970); ICI/RoSPA (1970 IS/74, 1972 IS/91); EEUA (1973 Doc. 47D); LPGITA (1991 LPG Code 1 Pt 1); NFPA (1993 NFPA 70, 496); ANSI (Appendix 27), BS (Appendix 27), BS 4683: 1871–, BS 5345: 1977–, BS 5501: 1977–

Gas (*see* Table 11.20)

Steam, condensate, feedwater
American Oil Co. (n.d./6); Lyle (1947); NRC (Appendix 28 *Steam Supply*); EEUA (1962 Hndbk 8); Monroe (1970); Buffington (1973); IMechE (1974/12, 13, 1981/58, 1984/76, 1987/96, 98, 90/124); Anon. (1976 LPB 11, p. 5); Danekind (1976, 1979); Pitts and Gowan (1977); Reid and Renshaw (1977); Gambhir, Heil and Schuelke (1978); Bos and Williams (1979); J.K. Clark and Helmick (1980); J.J. Jackson (1980); Ward, Labine and Redfield (1980); Pybus (1981); Blackwell (1982b); Andrade, Gates and McCarthy (1983); ASME (1983/193); Freedman (1983); Monroe (1983a); J.F. Peterson and Mann (1985); T.J. Kelly (1986); Plummer (1986); A. Atkinson (1988); Heaton and Handley (1988); Gunn and Horton (1989); National Fuel Efficiency Service (1989); Istre (1992); Babcock and Wilcox (1993); Broughton (1993); C. Butcher (1993b); Ganapathy (1993); Spirax-Sarco (1993); Hahn (1994) BS 759: 1984–, BS 1113: 1992

Cooling water
American Oil Co. (n.d./1); Partridge and Paulson (1963); M. Brooke (1970); Gazzi and Pasero (1970); Troscinski and Watson (1970); Silverstern and Curtis (1971); Klen and Grier (1978); Krisher (1978); Ward, Lee and Freymark (1978); Conger (1979); Veazey (1979); J.W. Lee (1980a,b); Burger (1982, 1983); Holiday (1982); W.J. Scott (1982); D.T. Smith (1982b); G.B. Hill (1983); Willa and Campbell (1983); IChemE (1987/131); Reidenbach (1988); Ellis (1990); G. Parkinson and Basta (1991); NFPA (1992 NFPA 214); Broughton (1993)

Process water
Holiday (1982)

Drinking water
BS 6700: 1987

Compressed air
American Oil Co. (n.d./2); Weiner (1966); La Cerda (1968); A.G. Paterson (1969); McAllister (1970, 1973); ISA (1975 S7.3, 1984 RP 7.7); Allan (1986); Johnston (1990); Broughton (1993); O'Dell (1993)

Instrument air
Nielsen, Platz and Kongso (1977); Broughton (1993)

Inert gas
Hotchkiss and Weber (1953); P.A.F. White and Smith (1962); Funk (1963); Husa (1964); Sittig (1966); Penland (1967); Rosenberg (1968); H.A. Price and McAllister (1970); Loeb (1974); Anon. (1976 LPB 7, p. 11); Simon (1976); Kletz (1980f); Mehra (1982); Metzger *et al.* (1985); Hardenburger (1992); Broughton (1993)

Refrigeration
ASHRAE (Appendix 27, 28, n.d./2, 1972, 1992 ASHRAE 15, 34); Tanzer (1963); Ballou, Lyons and Tacquard (1967); Spencer (1967); Zafft (1967); Kaiser, Salhi and Pocini (1978); Mehra (1978, 1979a–c, 1982); Marsh and Olivo (1979); D.K. Miller (1979); Baggio and Saintherant (1980); Sibley (1983); Jacob (1991a); R.W. James (1994)

Heat transfer media
Geiringer (1962); Seifert, Jackson and Sech (1972); Frikken, Rosenberg and Steinmeyer (1975); Anon. (1978i); Sheehan (1986); Dotiwala (1991)

Water treatment
Bellew (1978); Day (1978); Gasper (1978); Setaro (1980); W.B.Thomas (1980); Saffell (1992)

Specific application of this principle is discussed below for electricity and steam.

It is often attractive to use a ring main layout to distribute a service. This allows the supply to a given point to be provided from either side and thus increases reliability. It is necessary, however, to give careful consideration to the flows which can occur in ring mains, particularly those handling steam.

Since service pipes and trunking pass through the plant they are vulnerable to damage from accidents. The routing of services should be planned so as to minimize such damage.

In particular, cabling for control instruments and other key services should be routed as safely as possible. For critical control functions requiring replication the cables should go by different routes. A severe accident may well cause damage to certain key services. There may be serious consequences if there is loss of electrical power to machines or control instrumentation, of cooling water or of fire water.

11.17.1 IChemE *Utilities Guide*

A treatment of the principal utilities is given in *Process Utility Systems* (Broughton, 1993) (the IChemE *Utilities Guide*). The *Utilities Guide* deals with the following topics: (1) efficient use of utilities, (2) fuel, (3) compressed air, (4) inert gases, (5) thermal fluid systems, (6) water preparation, (7) the boiler house, (8) steam distribution, (9) electricity use and distribution, (10) air and water cooling, (11) refrigeration, (12) fire protection system, (13) building services, and (14) pipework and safety.

The *Guide* starts by outlining five broad principles of general applicability to process utilities. One is the need for critical examination of all aspects, both the requirements and the means of meeting them, with full exploration of alternatives. Another is integration of utilities not just with individual plants but across the site. The third is the appropriate sizing and location of the utilities units. The fourth is the use of modern, efficient systems. And the fifth the need for performance targets and monitoring.

11.17.2 Electricity

Electricity is used for a number of different purposes on process plants. These include (1) machinery, (2) heating and (3) instrumentation. The use of electricity in factories is governed by the Electricity at Work Regulations 1989, which supersede the former Electricity (Factories Act) Special Regulations 1980 and 1944.

UK standards on electrical equipment include: BS 2771: 1986– *Electrical Equipment for Industrial Machines*, replacing BS CP 1015: 1967; BS 4683: 1971– *Specification of Electrical Apparatus for Use in Explosive Atmospheres*; BS 5345: 1977– *Code of Practice for the Selection, Installation and Maintenance of Electrical Apparatus for Use in Potentially Explosive Atmospheres (Other than Mining Applications or Explosive Processing or Manufacture)*; BS 5501: 1977– *Electrical Apparatus for Potentially Explosive Atmospheres*, replacing BS CP 1003; and BS 7430: 1991 *Code of Practice for Earthing*, replacing Bs CP 1013: 1965. Relevant UK codes of practice include the IP *Electrical Safety Code* (1991 MCSP Pt 1) and the IP *Area Classification Code for Petroleum Installations* (1990 MCSP Pt 15). The ICI *Electrical Installations in Flammable Atmospheres Code* (ICI/RoSPA 1972 IS/91) has also been widely quoted in the safety and loss prevention literature.

Relevant US codes are: NFFA 70: 1993 *National Electrical Code*; API RP 540: 1991 *Recommended Practice for Electrical Installations in Petroleum Processing Plants*; and API RP 500: 1991 *Recommended Practice for Classification of Locations for Electrical Installations at Petroleum Facilities*, replacing API RP 500A, 500B and 500C.

Safety aspects of electricity and electrical equipment are dealt with in *Electrical Safety* (Swann, 1959), *Electrical Safety Engineering* (Fordham Cooper, 1986), and *Hazard of Electricity* by the American Oil Company (Amoco) (Amoco/5).

Electrical power in a chemical works may be provided from the National Grid or from a power station within the works. Since steam is also required, a large works usually has its own power station. There are frequently mutual back-up arrangements between the Grid and the works station.

On some processes loss of electrical power for machinery can be serious. This applies particularly where materials may solidify in the plant, liquefied gases may warm up and vaporize, and refractories may cool and collapse. It may be necessary in such situations to provide a stand-by power supply. This may be done by setting up mutual back-up arrangements, by providing stand-by generators, or by installing engine-driven equipment and/or batteries.

Fluctuations of power supply in the form of voltage dips and outages can also occur and cause motors to cut out, which may result in process upsets. Time delay relays are available which can be used to overcome brief fluctuations.

It is particularly important to ensure the removal of heat from the process, as heat exchange depends primarily on the continued operation of pumps, both those for the process fluid and those for the cooling water. The fans of air-cooled heat exchangers are also dependent on electrical power. The fans on fired heaters are important. Loss of power on these fans can lead to the formation of a flammable atmosphere.

In order to maintain power supply to essential equipment when partial loss of generating capacity occurs, it may be necessary to practice load shedding on other equipment. Load may be shed automatically, using load shedding relays which detect deterioration of the supply frequency and operate to shed the load according to a prearranged sequence.

Where electric motors are used to drive pumps, it is not uncommon for sudden overloads to occur on a pump and for the motor overload protection to cut the motor out. This situation tends to occur particularly during a transient condition such as the loss of one pump in a set. In consequence, the reliability of the redundant pumps is not always as good as theoretical reliability predictions suggest. It is necessary, therefore, to give careful consideration to possible causes of pump overload.

Electrical apparatus is a potential source of ignition. It is essential, therefore, to have a system of electrical area classification and to use appropriate equipment in each zone. This is a most important aspect and it is considered in Chapters 10, 16 and 22 in relation to plant layout, fire and storage, respectively.

Stray electrical currents may also be a problem. They can cause corrosion and can act as sources of ignition. The power supplies to instruments as well as to machinery should be such as not to constitute a source of ignition.

11.17.3 Electricity: uninterrupted power supplies

Instrumentation requires a power supply of high quality. Voltage fluctuations, for example, can cause upsets to process computers and programmable electronic systems and to some instrumentation. It is often necessary to provide for such equipment an uninterrupted power supply (UPS), also frequently called an 'uninterruptible power supply'. These terms are used to describe a supply which possesses both high reliability and high quality in the sense that it does not have interruptions even for a few milliseconds. Designs of UPS system are described in the IChemE *Guide*.

The UPS may utilize rotating or static equipment. One method of providing a UPS is a motor–alternator set with diesel engine back-up, other components of the system being an electromagnetic clutch and a flywheel. In normal operation the alternator is driven by the motor. On mains failure the diesel engine starts and the clutch engages, connecting the engine to the alternator, additional energy being supplied by the flywheel until the engine is up to full power.

An alternative method of providing a UPS is a static system using batteries. In one arrangement the AC supply passes to the load via a rectifier and an inverter, with the batteries connected between these two. This design is intended to ensure that in normal operation the batteries are kept charged and that on mains failure the batteries maintain the supply. A further discussion of UPSs is given in the CCPS *Safe Automation Guidelines* (1993/14).

11.17.4 Electricity: electrical heating

Electricity may also be used for electrical heating, or electroheat. Although the efficiency of electrical heating is limited by thermodynamics, this factor may be more than offset by the ability of electroheat systems to deliver heat with greater flexibility and precision. Types of electrical heating include immersion heating, resistance heating of pipes, induction heating of vessel walls, and the various forms of surface heating using jackets, mantles, panels, tapes, etc. An account is given in the IChemE *Guide*. Electrical heating is also used instead of a fired heater in smaller heat transfer fluid systems.

11.17.5 Fuels

Process units use both gaseous fuels such as LPG and liquid fuels in the form of fuel oil. Fuel gas is often available as part of an integrated process, as in a refinery, but it is proper to treat it as a utility. A typical fuel gas supply system is illustrated in Figure 11.35, which shows supplies taken from a primary process source, a secondary source and an LPG vaporizer.

Fuel oil is used in burners on the process, in boilers and in fired heaters. The heavier fuel oils are cheaper but are viscous and require heated storage and preheaters so that for smaller systems it is often more economic to use light fuel oils.

11.17.6 Steam

Steam is used for (1) machinery, (2) heating, (3) process, (4) purging and inerting and (5) snuffing. Steam may provide motive power in steam turbines and also in ejectors. It is extensively used for heating. It may be injected directly into the process as in operations such as steam stripping. It is a principal fluid used for purging and cleaning operations. It is used as a means of snuffing out the flame in fired heaters.

The main source of steam is usually the works power station, but there may be process sources such as waste heat boilers or chemical reactors. On smaller works steam may be supplied by boilers.

Steam systems are described in *The Efficient Use of Steam* (Lyle, 1947), *Industrial Boilers* (Gunn and Horton, 1989) and *Steam, Its Generation and Use* (Babcock and Wilcox, 1993). Hazards are reviewed in *Hazards of Steam* by the American Oil Company (Amoco/6).

An overview of the boiler house on a process plant site is given in the IChemE *Guide*, which describes the common types of boiler and burner and the ancillary plant required, and gives a checklist for boiler selection.

Boilers are either shell boilers or water tube boilers. In firetube boilers the water is in the shell, the hot gases are burned in a combustion chamber inside the shell and then pass through the fire tubes, giving up more of their heat. In one arrangement the boiler is set in brickwork, the Lancashire and Cornish boilers being early workhorses of this type with single and double internal flues, respectively. The economic boiler, by contrast, is self-contained. Package firetube boilers are widely used. Electrical boilers are of two types. One is heated by an immersion element akin to that used in domestic systems. The electrode boiler is heated by passage of current directly through the water between immersed electrodes. Both types come as package units. Finally, water tube boilers are available in a variety of designs and in sizes from package units to the large field assembled units used in power stations.

All boilers, including waste heat boilers, require feed water treated to prevent scaling. The feed water and boiler water qualities required for different steam pressures are listed in the IChemE *Guide*. Boiler feed water treatment is an important aspect of boiler operation. The reliability of the feedwater treatment plant, including the feedwater pumps, is therefore an issue.

The quantities of steam needed for purging can be large. It is necessary, therefore, to provide for this requirement, otherwise there may be a tendency to use insufficient steam in purging operations.

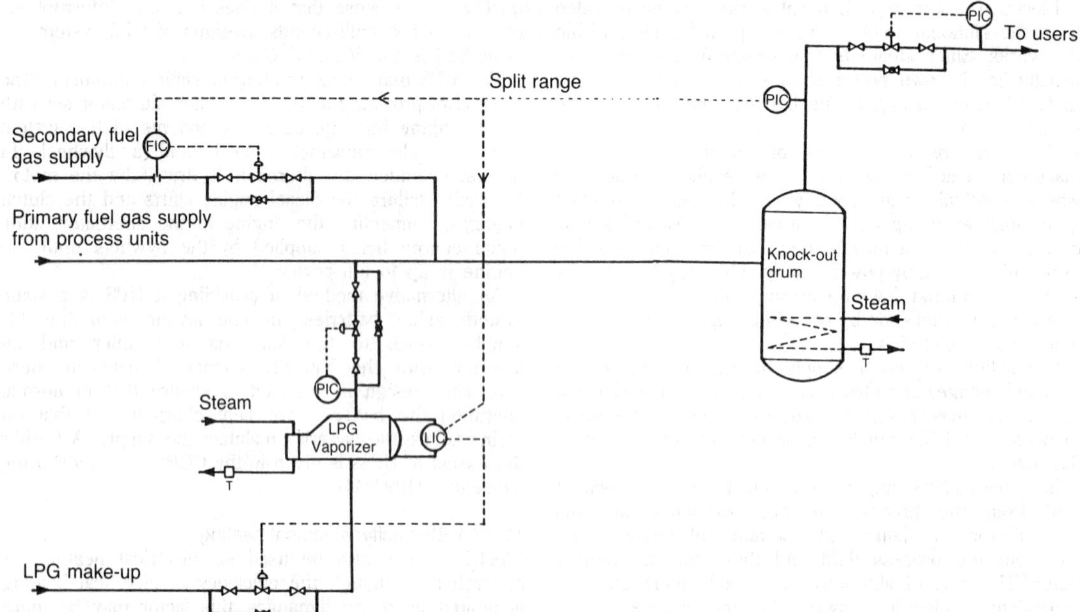

Figure 11.35 *Schematic diagram of a composite gaseous fuel supply system (Broughton, 1993) FIC, flow indicator controller; LIC, level indicator controller; PIC, pressure indicator controller (Courtesy of the Institution of Chemical Engineers)*

As with electrical power, it may be necessary to practice load shedding to protect the supply of steam to essential equipment on loss of part of the generating capacity.

Where there is a set of boilers and some are lost, the resultant demand on the remaining boilers is liable to cause fan overload and loss of these boilers also. A limit control system may be used to prevent such overload.

Loss of steam can normally be tolerated and it is not usual to have a stand-by steam supply. The steam system has some capacity, however, and can usually supply steam for a period. For more critical units the period can be prolonged if steam to other units is cut off.

It is appropriate, however, to consider the effect on a unit if the heating steam is lost. The resultant cooling may lead to condensation of vapours and to underpressure of the unit, and it may be necessary to provide nitrogen injection to counteract this.

Steam from the main boilers, and any waste heat boilers, is distributed to user plants, by a distribution system which may provide steam at low, intermediate and high pressure, pressure being let down at let-down stations.

A serious hazard in steam systems is water hammer. A slug of condensate which has collected in the main may be flung by the steam against the pipework with a very destructive effect.

Steam distribution systems should be designed to effect efficient drainage of condensate. This means that: the lines should be sloped; drainage points should placed at suitable locations and provided with drain pockets so that the condensate collects and does not skim over the top of the drain point; and efficient steam traps should be installed for take-off of the condensate.

Guidance on steam traps is given in *Steam and steam trapping* by Spirax-Sarco Ltd (1993). Further treatments of the design and operation of steam distribution systems, with particular reference to the avoidance of condensate hammer, are given in Chapters 12 and 20, respectively.

Erosion is a common feature in steam systems. Steam can be a surprisingly erosive fluid, especially if it is wet.

Leaking steam can cause a build-up of static electricity and can thus constitute a source of ignition. This possibility should be considered where leaks of flammable vapours may also be present.

Air may be present in a steam system on start-up and manual vents located at high points should be provided to allow it to be removed. The IChemE *Guide* suggests this should be the limit of the problem.

In some cases, as discussed by Plummer (1986), the presence of air and other gases in steam does cause operating problems and needs to be addressed. There are various sources of air or other gas in plant. Air dissolved in the boiler feedwater may not be completely removed by deaeration. Chemical reactions in the boiler, such as the release of carbon dioxide from bicarbonates, can generate gas. Air may be drawn into spaces where steam is condensing at a pressure below atmospheric.

These permanent gases can then create problems. Temperature control is often based on pressure control and if, due to the presence of gas, the relation between pressure and temperature departs from the value corresponding to the steam tables, control is upset. Heat transfer across heat exchange surfaces may deteriorate due to blanketing of these surfaces by gas. Corrosion may be promoted by the presence of oxygen or carbon dioxide.

Unwanted gas may be removed by suitable venting. Vent valves are available which operate on temperature, opening when the temperature is below saturation and seating when saturation temperature is reached. Points where such valves may be located are at the boiler, at remote ends of lines, upstream of isolation valves and on heat exchangers.

11.17.7 Compressed air: plant air

Compressed air is required on process plants as plant air for general applications, instrument air, breathing air and process air. General applications include purging and driving pneumatic tools.

Typically plant air is filtered, compressed to 6–7 barg and then cooled in a heat exchanger with condensation of water, but is not dried. The air enters an air receiver which supplies the service stations on the compressed air main. Compressed air, therefore, should be free of solid matter but contains some oil and is saturated with water.

The demand for plant air is subject to large fluctuations and the supply should be able to cope with such peaks, but it is not generally necessary to install stand-by compressors.

11.17.8 Compressed air: instrument air

The requirements for instrument air are somewhat different. Instrument air needs to be dry and its supply reliable. It is important that the instrument air be of good quality and, in particular, that it be free from solids, oil and water. Dirty, oily or wet air is liable to cause instrument error and failure. It follows that it is not good practice to use ordinary plant air for instrumentation.

The IChemE *Guide* states that instrument air should be free of solids, oil, corrosive substances and noxious gases. The solids content should be less than 0.1 $g/N\,m^3$ with particle sizes less than 3μm. With regard to dryness, the *Guide* quotes the requirement for a dewpoint of -40°C frequently suggested for UK sites, but adds that this can reduce the flexibility of vendor packages and proposes that for temperate climates such as those of central and southern Europe a dewpoint 10°C below the minimum ambient temperature suffices. The compressors used for instrument air should be of the oil-free type.

Instrument air is typically supplied at about 7 barg from an air receiver. The *Guide* states that a pressure of 4 barg is the minimum for satisfactory operation of the instrumentation and that it should not be allowed to fall below this. The supply needs to be adequate to cope with fluctuations. The *Guide* gives typical air consumption figures for instrumentation.

In most applications, the instrument air supply should have high reliability, which will normally mean the provision of stand-by equipment, separate power sources, and so on.

Avoidance of contamination, discussed in Section 11.17.18, is particularly important for instrument air. Connection of another system to the instrument air system creates the possibility of such contamination. It was contamination of the instrument air which initiated the train of events at Three Mile Island, as described in Appendix 21.

11.17.9 Compressed air: breathing air

If the extent of use of air-supplied breathing apparatus justifies it, a dedicated supply of air may be provided for breathing. This is considered in Chapter 25.

11.17.10 Compressed air: process air

Many processes use air as a raw material. Such air is usually obtained by compression of atmospheric air. Generally loss of process air is not critical and special measures are usually not required to ensure security of supply.

Air drawn by the process air compressors necessarily comes from some point in the works. There may be a degree of contamination due to leaks from plants, and in some cases the impurities may have an adverse effect on the process.

Hazards can arise from the build-up of oil from the compressors in the process air lines. Hazards are reviewed in *Hazards of Air* by the American Oil Company (Amoco/2).

11.17.11 Inert gas

The inert gases used in the process industries are mainly nitrogen and carbon dioxide. Alternatively, flue gas from an inert gas generator may be used. The inert gas principally considered here is nitrogen. This is described in *Nitrogen for Industry* (Sittig, 1966) and in the IChemE *Guide.*

Nitrogen is used for a large number of purposes. These include (1) inerting, (2) process, (3) purging and (4) pressurizing. One main use of nitrogen is inerting. Inert atmospheres are often required in processes such as polymerization, catalyst preparation and drier regeneration, and in plant such as centrifuges and storage tanks. Storage tank atmospheres are frequently inerted with nitrogen.

Nitrogen has a number of process uses. In particular, it may be used to dilute the concentration of a reactant such as oxygen in an oxidation process. In high pressure processes the use of nitrogen permits the use of such pressures while limiting the partial pressure of oxygen.

Another use of nitrogen is as a purge gas. Purging of equipment is discussed in Chapters 16, 20 and 21 in relation to fire protection, plant operation and maintenance, respectively. Where purging is carried out to prevent entry of flammable gases into equipment which may spark, nitrogen is not always the most suitable gas to use. Often air is both cheaper and safer. The choice of purge gas for such applications is considered further in Chapter 16.

There are certain situations which can lead to the plant being underpressured so that it may collapse inwards. This may be countered by nitrogen injection. Another application of pressurizing is the use of nitrogen pressure to transfer liquids in order to avoid having to pump them.

The quality of the nitrogen required depends on the application. The IChemE *Guide* distinguishes between general purging (GP) nitrogen and high purity inert medium (HPIM) nitrogen. Supplies of nitrogen may be obtained from gas cylinders or ammonia dissociation; from an inert gas generator in which nitrogen is obtained from combustion of air and removal of other products of combustion; from air separation by pressure swing adsorption (PSA); from air separation by permeation through a membrane; from bulk gas or liquid storage or gas pipeline; or from a low temperature air separation plant on site. Cylinders are economic only for very small quantities and a low temperature air separation plant only for very large ones.

The choice of nitrogen supply method is discussed in the *Guide*. It is essentially a function of the purity and scale. A breakpoint occurs at a demand of about 300 $\mathrm{N\,m^3/h}$. Broadly, the *Guide* indicates that the demands below this can be met using a liquid nitrogen storage/vaporizer system and those above using a low temperature separation plant, with PSA and membrane methods an option for smaller demands of GP nitrogen. Thus in contrast to the other services nitrogen is not always generated on site. Instead it may be brought in and held in storage. In this case it is essential to have a sufficient supply.

The Flixborough Inquiry revealed that the nitrogen required for inerting on the plant was sometimes in short supply and that as a result routine procedures were interrupted. The *Flixborough Report* drew attention to the importance of ensuring an adequate supply of nitrogen when plant safety is dependent on it.

11.17.12 Water: cooling water

Cooling water in the process industries is provided using one of three basic systems. One is the 'once-through' system, which is self-explanatory. The other two involve recirculation, either in an open system in which the water is cooled in cooling towers, or in a closed system in which it is cooled in heat exchangers using another fluid. The use of cooling towers is the normal method except where the site is such as to allow the once-through use of river or seawater.

Other cooling facilities may be provided by air-cooled heat exchangers, which depend on electric fans for air circulation.

Since it is essential to assure removal of heat from the process, the security of the cooling water supply is important. This may be achieved by making sure that the power supply is secure, by providing stand-by pumping arrangements and/or by the use of head tanks.

It is not good practice to use cooling water as fire water, as this may result in the loss of ability to ensure essential heat removal from process units. The cooling water system should be operated so that corrosion of process equipment by cooling water is kept to a minimum.

11.17.13 Water: process water

Water is an important raw material for many processes. It is usually taken from the public water supply or from company sources such as reservoirs and springs. Generally, loss of process water is not serious and a stand-by supply is usually not necessary. High security may be needed, however, for the boiler feedwater supply.

The water purity required depends upon the process. In some cases special purification plant may be needed. Such purification is necessary in particular for boiler feedwater. It is not good practice to draw the process water from the cooling water system.

Hazards are reviewed in *Hazard of Water* by the American Oil Company (Amoco/1).

11.17.14 Water: hot water

A supply of hot water may be required for washing down plants. If this is generated by direct mixing of water and steam, a specially designed mixing nozzle system should be used which shuts off on loss of cold water. The water temperature should be controlled at a temperature which will not injure personnel using it. A water temperature exceeding 57°C can give first degree burns.

11.17.15 Water: fire water

The provision of fire water is treated in Chapters 10 and 16 in relation to plant layout and fire protection, respectively.

11.17.16 Heat transfer media: hot fluids

In addition to steam and cooling water, use may be made of other heat transfer fluids (HTFs) for heating or cooling. HTF, or thermal fluid, systems are used to supply heat to process fluids at temperatures up to about 300°C. Such HTF systems exhibit a number of problems which need to be addressed if they are to operate satisfactorily. Accounts are given in the IChemE *Guide* and the CCPS *Engineering Design Guidelines*; both devote a complete chapter to these systems.

The HTFs used are mainly mineral oils, synthetic aromatics, diphenyl–diphenyl oxide mixtures and silicones. Examples of mineral oils are BP's Transcal 65; of synthetic aromatics, Dow's Dowtherm Q and BP's Transcal SA; of diphenyl–diphenyl oxide mixtures, Dow's Dowtherm A and ICI's Thermex; and of silicone, Dow's Syltherm 800.

An HTF generally has a boiling point in the range 260–340°C, a maximum fluid temperature of 350°C and a vapour pressure at the operating temperature of 1–2 barg. Many are subject to autoignition on escape to atmosphere.

Since the fluid circulates continuously, it is subject to degradation, minimization of which is an important feature of the design and operation of HTF systems.

A typical HTF system consists of a heater, heat exchangers to transfer heat to the process fluid(s), a storage reservoir and an expanse tank. The storage vessel may double as a discharge vessel or there may be a separate discharge vessel. There may be an emergency cooler. Flow diagrams of HTF systems are given in the IChemE *Guide* and the CCPS *Guidelines.*

The heat input is provided by a heater, except for smaller units where electrical heating is used. It follows that an HTF system is subject to the various problems of fired heaters and that the precautions applicable to fired heaters in general, described in Chapter 12, are required for those in HTF systems. They include on the heater tubes trips on high temperature and low flow.

In contrast to most fired heaters, the fluid in an HTF system is recirculated. The conditions created by this combination of fired heating and recirculation are severe and fluid degradation is characteristic of HTF systems. One consequence can be coking and plugging of the heater tubes.

Heat transfer fluids tend to be 'searching' and HTF plants have a history of leaks, particularly at flanges. Ignition occurs where the leaking fluid is above its autognition temperature. Another problem due to leaks is saturation of the lagging, resulting in lagging fires.

Pressure relief valves on HTF systems have not always been well located, so that fluid has been discharged and has then found a source of ignition. Experience also shows that both out-leak of the HFT and in-leak of another fluid may occur. Precautions include the installation of a low level trip and a high level alarm on the expanse tank.

The need for security of supply of the hot fluid varies. A reliable system may be necessary, for example, in an application where heat is required to prevent a process fluid going solid in the plant.

11.17.17 Heat transfer media: refrigerants

Refrigerants are used to cool process fluids to temperatures lower than those achievable using cooling water. Accounts of refrigeration are given in the *ASHRAE Refrigeration Handbook* by the American Society of Heating, Refrigerating and Air-Conditioning Engineers (ASHRAE, 1990/3). Other relevant ASHRAE publications are the *ASHRAE Thermodynamic Properties of Refrigerants* (ASHRAE/2) and the *ASHRAE Handbook of Fundamentals* (ASHRAE, 1993/7).

Basic methods of obtaining refrigeration are Joule–Thomson expansion, work expansion, evaporation refrigeration and absorption refrigeration. Joule–Thompson and work expansion may be used within the process to effect direct cooling of process fluids. Evaporation and absorption refrigeration sets are used to provide indirect cooling.

A variety of refrigerants are available, including: ammonia (R717); carbon dioxide; ethylene (R170), ethane (R1150), propylene (R1270) and propane (R290); and chlorofluorohydrocarbons (CFCs) such as dichlorodifluoromethane (R12) and chlordifluoromethane (R22). In selecting a refrigerant it is necessary to consider the general properties, the safety features and the environmental features.

Relevant general properties are the normal boiling point and refrigerant service temperature, the latent heat and the coefficient of performance (COP). Safety features include flammability and toxicity, in conjunction with the inventory required. The main environmental feature is the ozone depletion potential.

The replacement of materials with noxious properties such as ammonia (toxic) and ethylene (flammable) with non-toxic, non-flammable CFCs has been arrested by the environmental harm which the latter cause. The production and use of CFCs is restricted by the Montreal Protocols. New refrigerants to replace CFCs comprise an active area of development.

In some applications it may be necessary to consider other properties of the refrigerant, such as its interaction with the process fluid in the event of a pinhole leak.

A summary of general and environmental properties of refrigerants is given in the IChemE *Guide* and more detailed information in the ASHRAE handbooks. These publications also deal with refrigeration cycles and with package refrigeration sets.

Use is sometimes made in the process industries of systems in which a refrigerated heat transfer fluid is circulated between a refrigeration set and the process. A typical refrigerated fluid used in such applications is ethylene glycol.

11.17.18 Contamination hazards

Many of the services described are fluids. Hazards can arise from contamination of process materials by service fluids or contamination of service fluids by process materials. The contamination may occur through pipes and valves or by leakage in equipment.

Accidental entry of water, including steam, into the plant can have serious consequences. The water may vaporize, giving a large increase in pressure, or it may cause an oil–water 'slopover'. It may react violently with the process materials, it may ruin catalysts or adsorbents, it may give rise to very rapid corrosion and it may freeze and cause blockages. With accidental air ingress the main hazard is the formation of a flammable mixture.

In order to prevent unwanted entry of water and air into the plant through the pipework, it is necessary to provide positive isolation. This may require the use of double block and bleed valves rather than single valves. There need to be equally effective measures to prevent operational errors.

Contamination can also occur due to leakage of equipment. In particular, steam vaporizers and reboilers often develop pinhole leaks. If the consequences of such a leak might be serious, consideration should be given to the use of an alternative means of heating such as a heat transfer fluid.

The entry of process materials into lines carrying service fluids can have equally serious results. Toxic materials may enter the water supply, flammable vapours may pass into the air lines, or air may get into the nitrogen supply. In all these cases the hazard is obvious. It is essential, therefore, to check whether flow reversals into the service fluids could occur which might create a hazardous situation and to take any appropriate measures.

There is a particular obligation to ensure that the public water supply is not contaminated. If water is drawn from this for process use, positive means such as a break tank should be provided to prevent backflow and possible contamination.

11.18 Particular Chemicals

Accounts of the properties, manufacture and use of chemicals are given in the *Kirk–Othmer Encyclopaedia of Chemical Technology* (Kirk and Othmer, 1963–, 1978–, 1991–) and in *Ullmann's Encyclopaedia of Chemical Technology* (Ullmann, 1969–, 1985–). Selected references on particular chemicals and their hazards are given in Table 11.18.

The number of chemicals used in the process industries is very large. The few principal chemicals considered here are selected because they are important in their own right and because they illustrate the type of factors which have to be taken into account. The chemicals are:

(1) hydrocarbons;
(2) acetylene;
(3) acrylonitrile;
(4) ammonia;
(5) chlorine;
(6) ethylene dichloride;
(7) ethylene oxide;
(8) hydrogen;

Table 11.18 *Selected references on particular chemicals and their hazards*

Notes:
(a) The former MCA SD series, given in the first edition of this book, is not included.
(b) The FPA *Compendium of Fire Safety Data*, vol.4, H series is denoted by the H number.
(c) Company materials safety data sheets are now a much improved source of detailed information.

General
Cloyd and Murphy (1965); DOT, CG (1974a,b); Goldfarb *et al.* (1981); Chissick *et al.* (1984); Keith and Walters (1985–); H.H. Fawcett (1988); Spitz (1988); Agam (1994)

Air sensitive compounds
Shriver (1969)

Acetaldehyde
Gemmill (1961a); Jira, Blau and Grimm (1976); Goldfarb *et al.* (1981); FPA (1986 H29)

Acetic acid
Claydon (1967); Ellwood (1969a); Schwerdtel (1970); Lowry and Aguilo (1974); Goldfarb *et al.* (1981); Aquilo *et al.* (1983); FPA (1988 H120)

Acetylene
Reppe (1952); Penny (1956); H. Watts (1956); Sargent (1957); *Chemical Engineering* Staff (1960); S.A. Miller and Penny (1960); S.A. Miller (1964); Hardie (1965); Mayes and Yallop (1965); Kamptner, Krause and Schilken (1966a); Stobaugh (1966a); Tedeschi *et al.* (1967); Zieger (1969); CGA (1970 G-1.3, 1990 G-1.1, SB-4); Schmidt (1971); Carver, Smith and Webster (1972); Sutherland and Wegert (1973); Holman, Rokstad and Solbakken (1976–); NIOSH (1976 Crit Doc. 76-195); Stork, Hanisian and Bac (1976); Anon. (1978a); Manyik (1978); Tedeschi (1982); P.J.T. Morris (1983); Pässler *et al.* (1985); FPA (1987 H6); Ashmore (1988 LPB 88); Hort and Taylor (1991); Conrad, Dietlen and Schendler (1992)

Acetylenic alcohol
Lorentz (1967, 1973)

Acids and alkalis
Anon. (1977 LPB 16, p. 3)

Acrylic acid
FPA (1986 H95); Levy (1987)

Acrylic monomers
Kirch *et al.* (1988)

Acrylonitrile
Guccione (1965a); S.G.M. Clark and Camirand (1971); Dalin, Kolchin and Serebryakov (1971); Caporali (1972); Pujado, Vora and Krueding (1977); CIA (1978 PA11); CIHSC (1978/4); NIOSH (1978 Crit. Doc. 78-116); Langveldt (1985); FPA (1986 H51); Brazdil (1991)

Air (*see* Table 11.20)

Alcohols
Courty *et al.* (1984); FPA (1986 H2)

Aluminium alkyls
Heck and Johnson (1962); Governale, Ruhlin and Silvus (1965); Albright (1967f); Schmit *et al.* (1978)

Ammonia (*see also* Tables 11.21, 18.1, 18.2, 22.1 and 23.1)
American Oil Co. (n.d./9); Slack and James (1973); NIOSH (1974 Crit Doc. 74-136); CIA (1975/8, 11, PA1, 1988 PA9, 1990 RC3); Blanken (1978); HSE (1978b, 1981a); Strelzoff (1981); CGA (1984 G-2, 1989 G-2.1, TB-2); Bakemeier *et al.* (1985); FPA (1987 H13); Kletz (1988a)

Ammonium nitrate (*see also* Table 11.21)
American Oil Co. (n.d./9); Burns *et al.* (1953 BM RI 5476); Sykes, Johnson and Hainer (1963); van Dolah *et al.* (1966 BM RI 6773); Watchorn (1966); HSE (1978b, 1981a); Keleti (1985); Zapp (1985); FPA (1987 H33); FMA (1989); NFPA (1990 NFC 490); ACDS (1991); Miyake and Ogawa (1992); Weston (1992)

Ammonium perchlorate
Bond (1990 LPB 93)

Aniline
HSE (TDN 10, 1979 EH 4); Gans (1976b); FPA (1986 H80)

Arsine
HSE (1975 TDN 6, 1990 EH11)

Asbestos (*see* Tables 18.1 and 25.1)

Benzene
Stobaugh (1965a); Remirez (1968b); Ockerbloom (1972); Hancock (1975); NIOSH (1976 Crit. Doc. 137); Kohn (1978c); Cheremisinoff and Morresi (1979); CIA (1980 PA3); Malow (1980); HSE (1982 TR4); FPA (1986 H18)

Bischloromethyl ether
Ress (1982)

Bitumen
Swindells, Nolan and Pratt (1986); IP (1990 MCSP Pt 11)

Bromine
Jolles (1966); FPA (1986 H61); CIA (1989 RC4)

Butadiene
D.A. Scott (1940); Stobaugh (1967a); FPA (1987 H63); Glass (1987)

Calcium hypochlorite
V.J.Clancey (1975b,c, 1975/76, 1987); Uehara, Uematsu and Saito (1978); FPA (1987 H46)

Caprolactam
H.F. Steward (1974)

Carbon dioxide
CGA (1984 G-6, 1985 G-6.2)

Carbon disulphide
Thacker (1970); NIOSH (1977 Crit. Doc. 77-152); HSE (1981 SHW 932, TR3); FPA (1986 H43)

Carbon monoxide
HSE (HSW Bklt 29); NIOSH (1972 Crit. Doc. 73-11000); Anon. (1975a); Ribovitch, Murphy and Watson (1977); FPA (1987 H57); CGA (1989 P-13)

Chlorine (*see also* Tables 18.1, 18.2, 22.1 and 23.1)
Chlorine Institute (Appendix 27, 28, 1971 BIBLL, 1986 Pamphlet 1); DoEm (Det. Bklt 10); HSE (HSW Bklt 37, 1978b); Brian, Vivian and Habib (1962); Johnsen and Yahnke (1962, 1973); Payne (1964); Brian, Vivian and Piazza (1966); de Nora and Gallone (1968); W.W. Lawrence and Cook (1970); Sconce (1972); Statesir (1973); BCISC (1975/1); NIOSH (1976 Crit. Doc. 76-110); H.E. Schwarz (1976); CIA (1980/10, 1986 PA 48); Meinhardt (1981); HSE (1986 HS(G) 28); Schmittinger (1986); FPA (1987 H39); J.L. Woodward and Silvestro (1988); CIA, Chlorine Sector Group (1989); Gustin (1989a); Klug (1989); Somerville (1990); Anon. (1991 LPB 98, p. 25); Curlin, Bommaraja and Hansson (1991)
Nitrogen trichloride: Chlorine Institute (1975 MIR-2)

Chlorine dioxide
Cowley (1993 LPB 113)

Chloromethanes
Akiyama, Hisamoto and Mochizuki (1981); FPA (1986 H68)

Cryogenic gases
Blakey (1981); Williamson and Edeskuty (1984)

Cyanuric chloride
Anon. (1979 LPB 25, p. 19)

Cyclohexane
Haines (1962); Dufau *et al.* (1964); Stobaugh (1965b); Berezin, Denisov and Emmanuel (1966); Alagy *et al.* (1968); Ishimoto, Sasano and Kawamura (1968); Craig (1970); Taverna and Chiti (1970); Alagy, Trambouze and van Landeghem (1974); Dragoset (1976); McCorkle (1980); Ciborowski and Krysztoforski (1981); FPA (1986 H25); Krysztoforski *et al.* (1986, 1992)

Diethyl ether
Redeker and Schebsdat (1977)

Dimethylsulphate
Anon (1979 LPB 25, p. 24)

Dimethylsulphoxide (DMSO)
Gostelow (1983); Hall (1993 LPB 114)

Dimethylterephthalate (DMT)
Ueda (1980)

Ethyl acetate
Bond (1994 LPB 119)

Ethylene
Stobaugh (1966c); W.W. Lawrence and Cook (1967); S.A. Miller (1969); Strelzoff (1970); Fiumara and Cardillo

(1976); MITI (1976); Ribovitch, Murphy and Watson (1977); Kniel, Winter and Chung-Hu Tsai (1980); Kniel, Winter and Stork (1980); Pike (1981); Tayler (1984a); C. Britton, Taylor and Wobser (1986); FPA (1987 H17); Glass (1987); Granton and Roger (1987)

Ethylene cyanide
Berthold and Löffler (1983)

Ethylene dichloride
CIA (1975 PA13); CISHC (1975/1); V.L. Stevens (1979); McNaughton (1983); FPA (1986 H77); Rossberg *et al.* (1986)

Ethylene oxide
Burgoyne and Burden (1948, 1949); Burden and Burgoyne (1949); Burgoyne and Cox (1953); Burgoyne, Bett and Muir (1960); Burgoyne, Bett and Lee (1967); Troyan and Levine (1968); Ray, Spinek and Stobaugh (1970); CISHC (1975/2); de Maglie (1976); Gans and Ozero (1976); S.C. Johnson (1976); Kiguchi, Kumazawa and Nakai (1976); NIOSH (1978 Crit. Doc. 77-200); CIA (1979 RC 14); Kuhn (1979, 1980b); Cause *et al.* (1980); Pesetsky and Best (1980); Pesetsky, Cawse and Vyn (1980); Chen and Faeth (1981b); Ozero and Procelli (1984); FPA (1987 H79); Rebsdat and Mayer (1987); Kletz (1988m); de Groot and Heemskerk (1989); Siwek and Rosenberg (1989); Britton (1990b); Brockwell (1990); Grumbles (1990); June and Dye (1990); Conrad, Diellen and Schendler (1992); Ondrey (1992)

Fluorine
McGuffy, Paluzelle and Muldrew (1962)

Formaldehyde
NIOSH (1976 Crit. Doc. 77-126); HSE (1981 TR2); CIA (1983 PA21); Clary, Gibson and Waritz (1983); Dunn *et al.* (1983); FPA (1986 H54)

Formic acid
Aguido and Horlenko (1980); Czaikowski and Bayne (1980); Anon. (1984 LPB 56, p. 24); FPA (1986 H86)

Grignard reagents
Rakita, Aultman and Stapleton (1990)

Halogenated hydrocarbons
Santon and Wrightson (1989)

Hydrocarbons
WHO (EHC20); Wade (1963); Stobaugh (1966b); A. Brown (1964); Burgoyne (1965b); ICI/RoSPA (1970 IS/74); Binns (1978); Boesinger, Nielsen and Albright (1980); Maisel (1980); IP (1991 PUB 60, 61); API (1983/6)

Hydrogen
American Oil Co. (n.d./9); Cronan (1960a); G.R. James (1960); Labine (1960); G.J. Lewis (1961); Vander Arend (1961); Zabetakis and Burgess (BM 1961 RI 5707); Zabetakis, Furno and Perlee (BM 1963 RI 6309); R.B. Scott, Denton and Nicholls (1964); Reiff (1965); Scharle (1965); Stoll (1965); Voogd and Tielrooy (1967); Chopey (1972); CGA (1974 G-5, 1985 P-6, 1990 G-5.3); Anon. (1977 LPB 15, p. 2; 1978 LPB 21, p. 85); K.E.Cox and Williamson (1977); Bassett and Natarajan (1980); W.N. Smith and Santangelo (1980); Donakowski (1981); Mandelik and Newsome (1981); Angus (1984); Mahmood *et al.* (1984); J.D. Martin (1984); Shields, Udengaard and Berzins (1984); FPA (1986 H20); Haussinger, Lohmüller and Watson (1989); NFPA (1989 NFPA 50A, 50B); Johansen, Raghuraman and Hackett (1992); Ondrey, Hoffmann and Moore (1992)

Hydrogen chloride, hydrochloric acid
BCISC (1975/2); CIA (1975 PA12); HSE (1981 BPM 5); FPA (1987 H41)

Hydrogen cyanide
Koberstein (1973); NIOSH (1976 Crit. Doc. 77-108); HSE (1983 SHW 385); FPA (1986 H94)

Hydrogen fluoride
Muehlberger (1928); Simons (1931); K.M. Hill and Knott (1960); NIOSH (1976 Crit. Doc. 76-143); HSE (1978b, 1980 BPM 4); Gall (1980); CIA (1987 PA14); Aigueperse *et al.* (1988); G. Parkinson (1990); van Zele and Diener (1990); Diener (1991)

Hydrogen peroxide
HSE (SIR 19); Schumb, Satterfield and Wentworth (1955); Shell Chemical Corp. (1959); Rawsthorne and Williams (1961); Monger *et al.* (1964); G.A. Campbell and Rutledge (1972); Berthold and Löffler (1980); FPA (1987 H3); Mackenzie (1990, 1991)

Hydrogen sulphide
NIOSH (1977 Crit. Doc. 77-158); API (1981 RP 55); CGA (1981 G-12); Maclachlan (1985); FPA (1986 H78)

Insecticides, pesticides
Presidents Advisory Committee (1963); Brooks (1974)

Iron sulphide
Anon. (1976 LPB 12, p. 1)

Isocyanates
OSHA (OSHA 2248); Corbett (1963); HSE (1975 TDN 41, 1983 MS 8, 1984 EH 16); British Rubber Manufacturers Association (1977); NIOSH (1978 Crit. Doc. 78-215); Chironna and Voelpel (1983); FPA (1986 H5); CIA (1989 RC29)

Isopropyl nitrate
Beeley, Griffiths and Gray (1980)

Lead (*see also* Table 18.1)
HSE (TDN 16); DoE (1974 Poll. Paper 2); NIOSH (1978 Crit. Doc. 78-158); IP (1985)

Lead additives
Associated Octel Co. (Appendix 28, n.d./1); HSE (1978b)

LNG (*see* Table 11.23)

LPG (*see* Table 11.22)

Maleic anhydride
Trivedi and Culbertson (1982)

Mercury
HSE (1975 TDN 21, 1977 ED 17); DoE (1976 Poll. Paper 10); McAuliffe (1977); D. Taylor (1978); Okouchi and Sasaki (1984)

Methanol
Ferris (1974); Kugler and Steffgen (1980); Sherwin (1981); D.L. King, Ushiba and Whyte (1982); Chang (1983); FPA (1986 H42); Fielder *et al.* (1990)

Methyl methacrylate
FPA (1986 H67); Porcelli and Juran (1986)

Molten salts
HSE (1971 HSW Bklt 27); C.B. Allen and Janz (1980)

Monomethylamine nitrate
Miron (1980)

Naphthalene
Stobaugh (1966d); FPA (1987 H69)

NGL
Collins, Chen and Elliott (1985)

Nitric acid
Bingham (1966); van Dolah (1969a); Mandelik and Turner (1977); Bolme and Horton (1980); Calmon (1980); Harvin, Leray and Roudier (1980); Ohrui, Ohkubo and Imai (1980); Keleti (1985); FPA (1987 H23)

Nitrogen
CGA (1980 P-9, 1985 G-10.1); Anon. (1984 LPB 59, p. 30); Bond (1985 LPB 63); Hempseed (1991 LPB 97)

Nitrogen oxides
NIOSH (1976 Crit. Doc. 76-149); Ribovitch, Murphy and Watson (1977); Bretherick (1989 LPB 86); Currie (1989 LPB 88)

Nitrotoluenes
G.F.P. Harris, Harrison and Macdermott (1981)

Organic nitrates
Rebsoch (1976)

Organic peroxides
ASTM (STP 394); D.A. Scott (1940); National Board of Fire Underwriters (1956); Armitage and Strauss (1964); Castrantas, Banerjee and Noller (1965); Fine and Gray (1967); Bowes (1968); Hupkens van der Elst (1969); Swern (1970–); A.G. Davies (1972); Donaldson (1973); Home Office (1974/5); Interox Chemicals Ltd (1975); Wagle *et al.* (1978); de Groot, Groothuizen and Verhoeff (1980); de Groot and Hupkens van der Elst (1981); McCloskey (1989); Britton (1990b); T.A. Roberts, Merrifield and Tharmalingham (1990); Tognotti and Petarca (1992)

Organic phosphorus compounds
MacDonald (1960)

Oxidising substances
Uehara and Nakajima (1985)

Oxygen (*see* also Table 11.20)
CGA (1980 G-4.4, 1983 P-14, SB-2, 1985 G-4.1, 1987 G-4, 1988 G-4.3); NASA (1972–); HSE (1977a); Newton (1979); A.H. Taylor (1981); D.A. Jones (1983); FPA (1987 H12); Kirschner (1991)

Perchlorates
Schumacher (1960)

Perchloric acid
Graf (1967); FPA (1987 H53)

Peroxidizable compounds
H.L. Jackson *et al.* (1970)

Pharmaceuticals
Dickson and Teather (1982); Handley (1985)

Phenol
Richmann (1964); Stobaugh (1966e); Fleming, Lambrix and Nixon (1976); Gelbein and Nislick (1978); HSE (1984 SHW 29)

Phosgene
CISHC (1975/3); NIOSH (1976 Crit. Doc. 76-137); Hardy (1982); Alspach and Bianchi (1984); Anon. (1986q); FPA (1988 H105); Somerville (1990); Schneider and Diller (1991)

Phosphorus
Lemay and Metcalfe (1964); FPA (1987 H19)

Phosphoric acid
Blumrich, Koening and Schwehr (1978)

Phosphorous compounds, chlorides
Anon. (1976 LPB 12, p. 11; 1977 LPB 13, p. 21); HSE (1994 TR30)

Phthalic anhydride
B. Shaw (1961); Ellwood (1969b); J.J. Graham (1970); Schwab and Doyle (1970); G.F.P. Harris and Macdermott (1980); FPA (1988 H130)

Polychlorinated biphenyls (PCBs)
Hutzinger (1974); Higuchi (1976); George *et al.* (1988); Derks (1991); Sundin (1991)

Polymers, including inhibitors
Mark *et al.* (1989); Redman (1991); Levy (1993); Vasile and Seymour (1993)
PVC: BCISC (1974/14); Nass (1976); Terwiesch (1976); Sittig (1978); HSE (1979 BPM 3); Nass and Heiberger (1986)

Propargyl bromide
Coffee and Wheeler (1967)

Propylene
Haines (1963); Stobaugh (1967b); Strelzoff (1970); Hancock (1973); FPA (1986 H75); Glass (1987); C. Butcher (1989b)

Propylene oxide
Jefferson Chemical Co. (1963); Stobaugh *et al.* (1973); Simmrock (1978); Kuhn (1979, 1980b); FPA (1986 H88)

Silane, silane chlorides
Britton (1990a)

Sodium
Bulmer (1972); Zinsstag (1973); Parida, Rao and Mitragotri (1985); FPA (1987 H4)

Sodium borohydride
Duggan, Johnson and Rogers (1994)

Sodium chlorate
V.J. Clancey (1975b,c); Anon. (1980r); HSE (1985 CS 3); FPA (1987 H7)

Solvents
Collings and Luxon (1982); de Renzo (1986)

Styrene
Stobaugh (1965c); HSE (1981 TR1); Short and Bolton (1985); FPA (1986 H44); Anon. (1987 LPB 78, p. 23); Ayers (1988 LPB 84)

Sulphur
J.R. Donovan (1962); Palm (1972); Chao (1980); Parnell (1981); FPA (1986 H14)

Sulphur dioxide, sulphur trioxide
B.O. Davies and Royce (1961); NIOSH (1977 Crit. Doc. 74-11); CGA (1988 G-3)

Sulphuric acid
Biarnes (1982)

Terephthalic acid (TPA)
Derbyshire (1960); FPA (1991 H165)

t-Butylhydroperoxide (TBHP)
Verhoeff (1981)

Toluene
Stobaugh (1966f); Hancock (1982); FPA (1986 H11); HSE (1989 TR20)

Trichloroethylene
Institut National de Sécurité (1967); Tsuda (1970); HSE (1973 TDN 17, 1985 EH 5); FPA (1990 H157)

Vinyl acetate
Reis (1966); Remirez (1968a); Stobaugh, Allen and van Sternbergh (1972); R.D. Brown and Bennett (1978); NIOSH (1978 Crit. Doc. 78-205); Goldfarb *et al.* (1981); Douglas, von Bramer and Jenkins (1982); Ehrler and Juran (1982); FPA (1986 H36)

Vinyl chloride (*see* also Table 18.1)
Gomi (1964); Buckley (1966); Albright (1967d,g); Arne (1967); Keane, Stobaugh and Townsend (1973); NIOSH (1974 Crit. Doc. 78-205); Z.G. Bell *et al.* (1975); CIA (1975/8, 1978 PA15, 1978/9); Reich (1976); Vervalin (1976a); Wimer (1976); Mukerji (1977); Sittig (1978); Mcpherson, Starks and Fryar (1979); Goldfarb *et al.* (1981); Cowfer and Magistro (1983); FPA (1990 H20)

Xylene
Stobaugh (1966g); Atkins (1970); Hancock (1982); FPA (1986 H76); HSE (1992 TR26)

Other chemicals
Bond (1987); L.E. Brown, Johnson and Martinsen (1987); Fitzer *et al.* (1988); Midgley (1989)

Deterioration of chemicals
Crowl and Louvar (1990); Kletz (1992a)

(9) hydrogen fluoride;
(10) oxygen;
(11) phosgene
(12) vinyl chloride.

The emission and dispersion characteristics of materials are treated in Chapter 15 and flammability, explosibility and toxicity in Chapters 16–18, whilst storage, transport and emergency planning are treated in Chapters 22–24, respectively.

The accounts of particular chemicals given below are based mainly on the codes of practice and other guidance documents of the Chemical Industries Association (CIA) and the Health and Safety Executive (HSE). An overview of some principal topics covered in these publications is given in Table 11.19. Storage is dealt with in all the publications to some degree and in several it is the main content. It is convenient, if somewhat arbitrary, to cover it separately in Chapter 22. The treatment of the individual chemicals given here is intended to highlight their characteristic features; it is not a detailed summary of the codes and guidance. For LPG, there are codes issued by the Liquefied Petroleum Gas Association (LPGA) in the UK and the National Fire Protection Association (NFPA) in the USA.

Toxic limits are given in *Occupational exposure limits* (HSE, 1994 EH 40/94). The normal limit is the long-term time weighted average (TWA), but the limit given is in some cases the maximum exposure limit or the short-term exposure limit (STEL). Alternatively, the substance may be classed as an asphyxiant. The HSE has also given guidance on toxicity in relation to major hazards for several of the materials, namely acrylonitrile, ammonia, chlorine and hydrogen fluoride, as described in Chapter 18.

One topic treated in most of the publications mentioned is pressure relief and relief of thermal expansion of liquids. Overall, the treatments are broadly similar, but contain differences of detail, the reasons for which are not always clear. In the treatments given below some mention is made of such reliefs, but only to highlight certain points particular to the chemical in question. An account of pressure relief and thermal relief for chlorine systems, which contains many typical features, is given in Chapter 12.

Guidance on safety cases under the Control of Industrial Major Accident Hazards Regulations 1984 (CIMAH) and on quantitative risk assessment is avail-

Table 11.19 *Some topics covered in certain Codes of Practice, and guidance for particular chemicals*[a]

	Acrylonitrile	*Ammonia*	*Chlorine*	*Ethylene dichloride*	*Ethylene oxide*	*Hydrogen fluoride*	*Phosgene*	*Vinyl chloride*
Physical properties	2, 19	5	34	14, 15	7, 28	15, 16	5	App. I, II
Flammability	2	5			7		6	2
Toxicity, health effects	2	5, 6	36		8, 11	2	8	2
Special properties[b]		17			7, 8, 10		6, 18	7
Phase change: boiling, condensation				7		9		11
Plant design:	3				12	4		2
Siting and layout	4	6	1	3	12	4	3, 10	5
Ignition sources	4, 7	13		8	9			5, 13
Ventilation		13		8				3, 13, 14
Materials of construction	5	7, 12	35	3	13	3	12, 13	3
Material transfer, pumping	5	11	20, 22	8	14	10	9, 18	12
Pipework, pipelines	5	11	5	5	13	6	9, 13, 14	8
Valves	6		8	6		7	14	9
Heat exchange					15		13	
Instrumentation	8							
Backflow prevention					14			
Pressure relief and vents	6		7, 28	6	13	7	9, 12, 19	8, 10, 13
Explosion relief				6				11, 14
Disposal of vented material, absorbers and scrubbers			27	8		10	9, 20	
Disposal of contaminated liquids, drains	6, 15				25		11, 21	15
Vaporizers			23		17			
Plant operation:	10	15, 17	30	11	21	11	21	16
Protective clothing	3, 9, 11				22			
Isolation	12			7		9		11
Maintenance	11		30		23		22	
Emergency planning and procedures:	16	21	31	13	24	13	4, 23	20
Protective equipment	9		29	13		13	23	20
Emergency equipment	17		29	13		13		20
Leaks and spillages	15		29	12	25	12	27	18
Fire	16			12	24			19
First aid	13, 22		33		11	14, 19	28	

[a] The publications referred to are as follows: acrylonitrile (CIA, 1978 PA11); ammonia (HSE, 1986 HS(G) 30); chlorine (HSE, 1986 HS(G) 28); ethylene dichloride (CIA, 1975 PA13); ethylene oxide (CIA, 1992 RC14); hydrogen fluoride (CIA, 1978 PA14); phosgene (CIA, 1975 CPP5); vinyl chloride (CIA, 1978 PA15). The numbers in the body of the table refer to the page numbers of these publications. Storage is considered separately in Chapter 22.
[b] Special properties: ammonia (oxygen); chlorine (nitrogen trichloride); ethylene oxide (decomposition, polymerization); phosgene (decomposition in water, reaction with lubricating oils); vinyl chloride (polyperoxides).

able, from different sources, for several of the materials, as described in Chapter 9. These include LPG, ammonia, chlorine and oxygen.

11.18.1 Hydrocarbons, LPG and LNG

Some of the principal hydrocarbons processed in the petrochemical industry are as follows:

	Normal boiling point (°C)
Methane	−161.4
Ethane	−88.6
Propane	−42.2
n-Butane	−0.6
Ethylene	−103.9
Propylene	−47.7

Methane is the main component of natural gas, which in liquefied form is known as liquefied natural gas (LNG). Propane and butane are referred to as liquefied petroleum gas (LPG). LNG and LPG are treated more fully in Section 11.19.

The hydrocarbons listed above are all gases at normal temperature and pressure, but can all be liquefied with a greater or lesser degree of difficulty. All are handled industrially in the liquid phase by the use of low temperatures, high pressures or both.

In many cases these materials are held not only under pressure but also at the high temperatures necessary for processing. The handling of hydrocarbons in this way means that if containment is lost, dispersion may be quite rapid. In particular, a material held under pressure and above its normal boiling point will flash off when let down to atmospheric pressure.

The use of extremes of temperature, both high and low, creates difficulties in materials of construction. Some of these are discussed in Chapter 12. In particular, there are problems of brittle fracture at low temperature and of creep at high temperature. The use of low temperatures also means that there is continual heat in-leak into the plant, which can become a problem if there is an interruption to processing or if insulation is lost. The hydrocarbons are flammable and can therefore give rise to the hazards both of fire and of explosion.

In addition to their use in the process industries, several of these hydrocarbons are widely used as fuels. In particular, propane and butane enjoy widespread use as fuels in the form of LPG supplied from self-contained storage installations. Natural gas, which is mainly methane, is transmitted by high pressure pipeline to the public distribution system for industrial and domestic users. It is also transported and stored as LNG. Natural gas liquids (NGL), which are mainly ethane, propane and butane, are also transported by pipeline. Again there are hazards of fire and explosion associated with these materials. Fire and explosion hazards of hydrocarbons are considered in detail in Chapters 16 and 17, respectively.

11.18.2 Acetylene

Accounts of acetylene are given in *Acetylene – Its Properties, Manufacture and Use* (S.A. Miller, 1964), *Acetylene Manufacture and Uses* (Hardie, 1965), *Acetylene* by the Compressed Gas Association (CGA, 1990 G-1) and *Acetylene Transmission for Chemical Synthesis* (CGA, 1970 G-1.3) and by Sargent (1957), S.A. Miller and Penny (1960), Schmidt (1971), Manyik (1978), P.J.T. Morris (1983), Passler *et al.* (1985) and Hort and Taylor (1991).

Codes for the handling of acetylene in the UK are issued by the British Compressed Gases Association (BCGA, 1986 CP 5, 1986 CP 6), the successor to the British Acetylene Association.

Acetylene is produced from calcium carbide or by the cracking or oxidation of hydrocarbons. Acetylene is highly reactive due to its triple carbon–carbon bond and was for many years the basis of a whole branch of the chemical industry, including in particular products derived from the intermediate tetrachloroethane, and of vinyl chloride. Acetylene chemistry has always been particularly strong in Germany. Thus in 1958 the production of calcium carbide in West Germany, East Germany, the USA and the UK was 979 000, 815 000, 805 000 and 145 000 ton, respectively. Since the 1960s acetylene has been largely replaced by ethylene as the raw material for these products. Production peaked in the USA in 1960 at 480 000 ton/year and in Germany in the 1970s at 350 000 ton/year. New processes have been developed for the manufacture of acetylene, but these have not arrested its decline. However, there is some evidence that this decline has slowed, partly because acetylene has always been the main route for some chemicals such as 1,4-butanediol and special vinyl esters, and partly because in Europe the relative cost of oil to natural gas feedstock rose.

In the UK acetylene was brought under the provisions of the Explosives Act 1875 by an Order in Council in 1897. For pressures exceeding 24 psig the installation came under the control of the Explosives Inspectorate. The legal requirements associated with this regime are described in *The Law Relating to Petroleum Mixtures, Acetylene, Calcium Carbide, etc.* (H. Watts, 1956). Legislation on acetylene is discussed in Chapter 3.

Acetylene is a gas. It is a flammable and explosively unstable material. For occupational exposure it is classed as an asphyxiant. Acetylene undergoes explosive decomposition at any pressure and temperature and even without the presence of oxygen. Acetylene decomposition may be initiated by shock, temperature or reactive substances. The decomposition may range from a harmless puff of flame to a violent explosion.

In particular, acetylene reacts violently with substances such as chlorine. Thus in the production of vinyl chloride by the reaction of acetylene with hydrogen chloride, where the latter is produced by burning hydrogen and chlorine, it is essential to ensure that there is no free chlorine in the hydrogen chloride. On the other hand, acetylene and chlorine are bubbled into liquid tetrachloroethane in the process for the manufacture of the latter product.

The explosion hazard with pure acetylene is most severe in pipelines. Here a deflagration may travel down a pipe until it becomes a detonation, which is much more destructive. This is a well-known effect in pipeline explosions, but acetylene presents a particularly severe problem in this regard. Further information on acetylene decomposition in equipment and pipelines is given by Schmidt (1971) and by Carver, Smith and Webster (1972). Despite this hazard, long-distance acetylene pipelines have been developed in Germany. Precautions against explosions include the earthing of pipes against static electricity and the use of flame arresters. There are no long-distance acetylene pipelines in the UK.

The usual material of construction for plant handling acetylene is mild steel. It is necessary in such plant to eliminate copper and brass, which can give rise to the formation of the highly explosive compound copper acetylide.

In some processes traces of acetylene may be present in the effluent gases. These gases are sometimes cooled to condense out other components. In such cases consideration should be given to the possibility of forming solid acetylene. Although the literature suggests that pure solid acetylene is not hazardous, this may not be the case if the material is impure.

Bulk acetylene is stored in gasholders. Holders with a capacity of 1000 m^3 have been used in the UK. Acetylene is stored and distributed in cylinders containing acetone and is widely used in this form, particularly for oxyacetylene welding.

11.18.3 Acrylonitrile

A relevant code for acrylonitrile is the *Codes of Practice for Chemicals with Major Hazards: Acrylonitrile* (the *Acrylotnitrile Code*) by the CIA (1978). Accounts are also given in *Acrylonitrile* (Dalin, Kolchin and Serebryakov, 1971) and by Langveldt (1985) and Brazdil (1991).

Acrylonitrile is a liquid with a normal boiling point of 77°C. It has moderate solubility in water. It is flammable, the flammable range being 3–17% and the flashpoint −1°C. It is toxic and has a long-term maximum exposure limit of 2 ppm.

Acrylonitrile is liable to polymerize and thus in storage needs to be stabilized. Stabilizers used are methyl ether of hydroquinone (MEHQ), ammonia and water.

Materials of construction for acrylonitrile plants are carbon steel, and for polymerization plants stainless steel. Copper and its alloys should not be used, since they cause contamination which inhibits subsequent polymerization.

Pumps used for transfer may be of the centrifugal or positive displacement types. For centrifugal pumps located outdoors a single mechanical seal is usually adequate, but indoors consideration should be given to double mechanical seals. At normal pumping rates of about 3 m/s the generation of static electricity by acrylonitrile should be minimal, but plant handling this chemical should be earthed.

In relief systems for acrylonitrile, consideration should be given to possible condensation in the relief line.

11.18.4 Ammonia

There is no single code of practice applicable to ammonia, but there are several codes covering particular aspects. These include *Storage of Anhydrous Ammonia under Pressure in the United Kingdom* (HSE, 1986 HS(G) 30), *Anhydrous Ammonia – Guidance for the Large Scale Storage, Full Refrigerated, in the United Kingdom* (CIA, 1993 RC40) and *Code of Practice for the Safe Handling and Transport of Anhydrous Ammonia by Rail* (CIA, 1975 PA1). Accounts are also given in *Ammonia* (Slack and James, 1973) and by Bakemeier *et al.* (1985) and Czuppon, Knez and Rounes (1992).

Ammonia is produced by the reaction of hydrogen and nitrogen in converters at high pressure, as described in Section 11.19. Much the largest use is for fertilizers, in the form of ammonium sulphate and ammonium nitrate and in anhydrous liquid form. It is used to make nitric acid and, in part via this, a range of plastics and fibres such as nylon, urea–formaldehyde resins, urethane, acrylonitrile and melamine. It is also used in the manufacture of explosives, hydrazine, amines, amides, nitriles and dyestuff intermediates and of urea, sodium cyanide and sodium carbonate.

Ammonia has a normal boiling point of −33.4°C and is therefore a gas. The vapour pressure at 20°C is 8.6 bar. It is frequently handled as a liquefied gas, either under pressure or refrigerated. Pressurized liquefied ammonia gives on release a flashing liquid.

Ammonia is flammable, but not readily so. The flammability range is 15–28%. The minimum ignition energy, however, is 100 mJ, which is high. The flammability and explosibility of ammonia is discussed by Kletz (1988a). The incident at Jonova (Case History A124) illustrates the flammability of ammonia.

Ammonia is a toxic irritant. Its long-term occupational exposure standard is 25 ppm. A concentration of about 1700 ppm can be fatal for a half-hour exposure. Its odour is detectable by most people at about 5 ppm.

After a large release of chlorine, a release of ammonia is potentially one of the worst toxic hazards from the common bulk chemicals in the chemical industry.

Ammonia is fully miscible with water. This creates the hazard that if there is water in a vessel containing ammonia vapour, the latter may dissolve, so that a vacuum is formed and the vessel collapses inwards.

Materials of construction for ammonia depend on the operating temperature. While mild steel may be used at ambient temperatures, special steels are necessary at low temperatures to avoid embrittlement. Materials for ammonia processes are discussed in Section 11.19.

Ammonia is corrosive even in trace amounts to copper, zinc, silver and many of their alloys. It is necessary, therefore, in handling ammonia to avoid the use of valves and other fittings which contain these metals. Under certain conditions ammonia can react with mercury to form explosive compounds (Comings, 1956):

> Under pressure, mercury forms a compound with ammonia consisting of several molecules of ammonia per atom of mercury. This compound is apparently not explosive. However, as the pressure is lowered, the ratio of ammonia to mercury decreases and a compound similar to a fulminate is formed. This has been known to detonate and is a serious hazard. The hazard is greatest when a system containing mercury and ammonia is being depressured.

Impurities in liquid anhydrous ammonia, such as air or carbon dioxide, can cause stress corrosion cracking of mild steel. This is largely inhibited, however, if the ammonia contains 0.2% water and this water content is therefore specified for some applications. This aspect is considered further in Chapters 22 and 23.

The production, storage and transport of ammonia are considered in Section 11.19, and Chapters 22 and 23, respectively, the toxic release hazard is discussed in Chapter 18 and emergency planning in Chapter 24.

11.18.5 Chlorine

Chlorine is treated in *Chlorine* (Sconce, 1972) and it is the subject of *Safety Advice for Bulk Chlorine Installations* (HSE, 1986 HS(G) 28). This code is based on an earlier CIA code and is the code now listed by the CIA; the earlier code *Codes of Practice for Chemicals with Major Hazards: Chlorine* (the *Chlorine Code*) by its predecessor, the British Chemical Industry Safety Council, (BCISC, 1975/1) is no longer listed. Further information is given in the *Chlorine Manual* (Chlorine Institute, 1986 Pamphlet 1). Accounts are also given by Schmittinger (1986) and Curlin, Bommaraju and Hansson (1991).

Chlorine is produced by the electrolysis of brine in mercury or diaphragm cells, with the co-products sodium hydroxide and hydrogen. It is used mainly for the manufacture of a wide range of organic and inorganic chlorine compounds and for the chlorination of water and for bleaching. Some principal organic compounds are chlorinated paraffins such as methyl chloride, methylene chloride, chloroform and carbon tetrachloride from methane and hexachlorobutadiene and hexachloropentadiene from butane and pentane, respectively. The main single organic chloride product is ethylene dichloride which is obtained by chlorination of ethylene and is used mainly to make vinyl chloride. Chlorination of ethylene also gives trichlorethylene and perchlorethylene. Products obtained via chlorination of propylene include propylene oxide, propylene glycol, carbon tetrachloride, glycerine and epoxy resins and synthetic rubber. Aromatic products of chlorination include monochlorbenzene, dichlorbenzene and hexachlorobenzene.

Chlorine has a normal boiling point of −34°C and is therefore a gas. The vapour pressure at 20°C is 6.7 bar. It has a very low solubility in water. Chlorine is frequently handled as a liquefied gas under pressure. In this condition liquefied chlorine gives on release a flashing liquid. Chlorine is not flammable in air, but is itself an oxidant. Chlorine is a toxic irritant. Its long-term occupational exposure standard is 0.5 ppm.

A large chlorine release is potentially one of the most severe hazards from a common bulk chemical presented by the chemical industry. As an oxidant chlorine is in many ways comparable with oxygen. Organic compounds may have flammability limits in chlorine rather similar to those which they have in oxygen. Reactions between organic compounds and chlorine are generally highly exothermic and tend to go to complete chlorination, often with some violence.

Chlorine also forms flammable mixtures with hydrogen. This is important in chlorine production, where both gases are evolved from the electrolysis cells. Hydrogen and chlorine may also be reacted together to produce hydrogen chloride.

Usually, liquid chlorine contains the impurity nitrogen trichloride, which decomposes explosively. Therefore it is necessary when handling chlorine to prevent the accumulation of this substance. For example, a chlorine vaporizer should not be allowed to boil dry so that the nitrogen trichloride is concentrated and may explode.

Mild steel is a suitable material of construction for use with chlorine unless the water content or temperature renders it unsuitable. Wet chlorine corrodes mild steel rapidly. Mild steel should only be used if the chlorine is dry. For wet chlorine gas, ebonite-lined mild steel is commonly used.

There are both low and high temperature limitations on the use of mild steel with chlorine. Special steels are required at low temperatures to avoid embrittlement. At high temperatures mild steel burns in chlorine and again special materials have to be used. Chlorine attacks mild steel significantly above 200°C and rapidly above 230°C. In order to prevent reaction, it is normal to limit the operating temperatures with mild steel. The *Chlorine Code* quotes a limit of 120°C.

The design of pressure systems for handling chlorine incorporates a number of special features. Many of these are generally applicable to the handling of a very toxic substance. They are discussed in Chapter 12. In view of the toxicity of chlorine, it is not appropriate to vent large quantities to atmosphere. Plants producing chlorine therefore require emergency gas absorption facilities. Chlorine from pressure relief devices should go to an expanse tank or to gas absorption plant.

Storage and transport of chlorine are treated in Chapters 22 and 23, respectively, the toxic release hazard in Chapter 18, and emergency planning in Chapter 24.

11.18.6 Ethylene dichloride

A relevant code for ethylene dichloride, or 1,2-dichloroethane, is the *Code of Practice for Chemicals with Major Hazards: Ethylene Dichloride* (CIA, 1975 PA13). Accounts are also given by V.L. Stevens (1979) and Rossberg *et al.* (1986).

Ethylene dichloride is made by the liquid phase chlorination of ethylene. It is used principally for the manufacture of vinyl chloride by cracking the ethylene dichloride to vinyl chloride and hydrogen chloride. Usually, the latter is then reacted in an oxychlorination process to produce more vinyl chloride.

Ethylene dichloride is a liquid with a normal boiling point of 84.4°C. It has a low solubility in water. Ethylene dichloride is flammable, the flammable range being 6–16%. The flash point is 13°C (closed cup) or 18°C (open cup).

Ethylene dichloride is toxic. It has a long-term maximum exposure limit of 5 ppm and carries the 'Skin' notation. Concentrations of about 3% can produce nausea, drowsiness and stupor. It is detectable by odour at a concentration of 50–100 ppm. The decomposition products of ethylene dichloride in a fire include hydrogen chloride, which is toxic.

Plant handling ethylene dichloride is usually constructed in mild steel. Mild steel is suitable for dry ethylene dichloride and also for wet saturated ethylene dichloride below 50°C, provided the aqueous phase is alkaline. But at temperatures above 80°C wet ethylene dichloride undergoes hydrolysis, forms acid and rapidly corrodes mild steel. Transfer of ethylene dichloride may be effected by pumping or inert gas padding.

11.18.7 Ethylene oxide

Ethylene oxide is the subject of the *Guidelines for Bulk Handling of Ethylene Oxide* (CIA, 1992 RC14) (the CIA *Ethylene Oxide Guidelines*), the earlier *Code of Practice for Chemicals with Major Hazards: Ethylene Oxide* by the Chemical Industry Safety and Health Council (CIHSC, 1975/1) (the *Ethylene Oxide Code*) is no longer listed. Accounts are also given by Cause *et al.* (1980), Ozero and Procelli (1984), Rebsdat and Mayer (1987), Ondrey (1992) and Siwek and Rosenberg (1993).

Ethylene oxide is produced by the vapour phase oxidation of ethylene. There are various processes; some use air and some use oxygen, but otherwise the processes are broadly similar. It is used primarily for the production of ethylene glycol, which in turn is used in roughly equal proportions as antifreeze and for production of polyesters. Other products obtained via ethylene oxide are ethanolamines, glycol ethers and surfactants.

Ethylene oxide is a toxic, flammable and explosively unstable material. Therefore it exhibits a rather unusual combination of hazards. The properties of ethylene oxide have been the subject of a series of investigations by Burgoyne and co-workers (e.g. Burgoyne and Burden 1948, 1949; Burden and Burgoyne, 1949; Burgoyne, Bett and Muir, 1960).

Ethylene oxide has a normal boiling point of 10.5°C. At atmospheric pressure it may therefore be a gas or a liquid, depending on the temperature, and in handling it may be treated as a liquefied gas. At higher ambient temperatures a liquid leak undergoes a degree of flashing to vapour. This feature of a normal boiling point close to ambient temperature tends to increase the hazard.

Ethylene oxide is fully miscible with water. In contrast to other flammables such as hydrocarbons, therefore, liquid spills can readily be diluted with water to reduce the hazard of fire.

The lower flammability limit of ethylene oxide is 3%. There is no upper flammability limit as normally conceived, since at high concentrations up to pure ethylene oxide combustion is replaced by explosive decomposition. The flashpoint (open cup) is −17.8°C. The autoignition temperature in air is 429°C.

Ethylene oxide is toxic and has a long-term maximum exposure limit of 5 ppm. The odour of pure ethylene oxide is not detectable by many people below about 700 ppm, and whilst impurities present in the industrial material reduce the odour threshold considerably, odour is not a reliable guide to its presence, and artificial means are necessary. Ethylene oxide liquid or aqueous solution affect the skin.

Ethylene oxide vapour decomposes explosively even in the absence of air, if ignited or heated above about 560°C. The liquid is stable to decomposition.

Liquid ethylene oxide is very susceptible to polymerization. This may be initiated at ambient temperatures by acids, bases and anhydrous chlorides of iron and of some other metals. Also, iron rust is a moderate initiator, so that equipment should be substantially rust free. At around 100°C purely thermal initiation starts, and once this happens iron becomes a promoter. The polymerization is highly exothermic and can result in explosive decomposition. Slow polymerization can also occur, producing a solid polymer which does not decompose explosively.

Mild steel and stainless steel are used as materials of construction for plant handling ethylene oxide, but mild steel must be rust free. The internal surfaces of plant equipment for handling ethylene oxide should be carefully prepared before use and free from foreign matter which could cause slow polymerization such as welding slag, rust and debris. Cleaning may be by shot blasting or by chemical means. If shot blasting is used, it must be possible to remove the dust and debris created; otherwise chemical methods must be used. The stages of acid cleaning are: (1) degreasing; (2) acid pickling, using hydrochloric acid for mild steel and citric acid for stainless steel; and (3) for mild steel, atmospheric passivation.

Flanged joints should have stainless steel, spiral wound, polytetrafluoroethene (PTFE) filled joint rings or trapped Fluon rings. Natural rubber joint rings should not be used. Compressed asbestos fibre (CAF) is suitable for use only below 25°C and is therefore not generally recommended. Joints should be kept to a minimum consistent with enabling lines to be opened for polymer clearance.

In order to avoid polymerization, pipework for handling liquid ethylene oxide should be laid out so that it can be completely drained during a shut-down. There should be facilities for blowing through with nitrogen. Lines of less than 25 mm (1 in) bore should not be used. Pipework should be designed to minimize the trapping of ethylene oxide between closed valves or in stagnant pockets. Relief is not normally required on in-plant pipework, but each case should be considered individually.

Ethylene oxide liquid has a high electrical conductivity, of the order of 10^6 pS/m ($1S = 1\Omega^{-1}$). Since experience shows that buildup of electrostatic charge does not occur in a liquid with a conductivity greater than 300 pS/m, it is not necessary to take the usual precautions against the static electricity hazard, such as limitation of pipeline velocities and avoidance of splash filling. However, in order to maintain consistency with practice for flammable liquids generally, transport containers should be earthed during loading and discharge.

Transfer of liquid ethylene oxide may be by gravity, inert gas padding or pumping. For small quantities it may be preferable to use the first two methods. If a pump is used, it is essential for there to be no conditions which can cause an abnormal temperature rise. This requirement can be met by the use of a recycle system from the pump delivery to the suction when the pump is operating, with cooling of the recycle. The recycle system should be sized to handle all the heat which can be generated by the pump when operating against a closed delivery. If the recycle is to the suction tank, the latter acts as a thermal reservoir. If dead-heading can occur, the pump should be fitted with a high temperature trip, which should shut the pump down if the temperature reaches 10°C above normal.

An in-line centrifugal pump with mechanical seals is the preferred type. It should be fitted with a single seal with a restrictor brush or a double seal flushed with water, which should be analysed regularly to detect leaks. A canned pump should not be used. If a reciprocating pump has to be used, it should be fitted with a double diaphragm seal with suitable sealant and rupture indication.

Refrigerant fluids used for direct heat exchange with ethylene oxide should not react with it and should not contain polymerization initiators.

Vaporizers for liquid ethylene oxide should be designed as once-through systems so as to prevent the accumulation of hazardous residues. The heating medium should be such as to keep the temperature well below that of decomposition. Direct flame or electrical heating should not be used.

Measures are required to prevent contamination of ethylene oxide in storage by backflow from consumer plants, by contamination of inert gas supplies or by flow from the loading terminal. Protection against backflow from the plant should be effected by means of two separate trip systems, the first based on the pressure difference between the ethylene oxide storage outlet and the supply line to the consumer plant and the second on that between the supply line and the inlet of the plant.

Prevention of contamination of the supply of inert gas, typically nitrogen, is ideally effected by the use of separate supplies for storage and process. The measures necessary where the supply is common are detailed in the *Ethylene Oxide Guidelines*.

Ethylene oxide should be stored separately from other flammable liquids. Terminals for loading and unloading ethylene oxide should also be segregated. Storage and transport of ethylene oxide are considered in Chapters 22 and 23, respectively, the toxic release hazard in Chapter 18 and emergency planning in Chapter 24.

11.18.8 Hydrogen

Hydrogen is the subject of the codes NPFA 50A: 1989 *Gaseous Hydrogen Systems at Consumer Sites* and NFPA 50B: 1989 *Liquefied Hydrogen Systems at Consumer Sites* and of *Hydrogen* by the CGA (1974 G-3). Accounts are also given in *Hydrogen* (K.E. Cox and Williamson, 1977–) and *Hydrogen* (W.N. Smith and Santangelo, 1980) and by Donakowski (1981), Mandelik and Newsome (1981), J.D. Martin (1984), Haussinger, Lohmüller and Watson (1989) and Johansen, Raghuraman and Hackett (1992).

Hydrogen is produced by steam reforming or partial oxidation of hydrocarbons or from a hydrogen-rich stream from a low temperature section of a petroleum plant or a coke oven plant or from a chlorine cellroom. It is used in many large-scale processes, notably in ammonia synthesis, refinery hydrogenation processes, and methanol synthesis, and in hydrogenation of coal and hydrogasification of coal to synthetic natural gas (SNG). It is also used in: hydroformylation of olefins, the Oxo synthesis; organic hydrogenations; and hydrogenation of fats and oils.

Hydrogen has a normal boiling point of −252.5°C and is therefore a gas. Hydrogen has a wide flammable range of 4 – 75%, a low minimum ignition energy of about 0.019 mJ and a high burning velocity. It is therefore easily ignited and burns rapidly. For occupational exposure hydrogen is classed as an asphyxiant.

In consequence of its flammability, hydrogen presents an explosion hazard and it is necessary to take suitable precautions. As far as the hazard of an open air explosion is concerned, hydrogen gas has a low density and tends to rise and dissipate rapidly unless it is very cold. Nevertheless, vapour cloud explosions of hydrogen have occurred. They are discussed in more detail in Chapter 17.

Hydrogen gas may burn in air, but hydrogen flames have low heat radiation, about one-tenth that of propane, and tend, therefore, to be less hazardous. The gap through which a hydrogen flame can travel is much smaller than with most other gases. It is therefore more difficult to make electric motors sufficiently flameproof for operation in atmospheres which may contain a flammable hydrogen–air mixture.

Hydrogen at elevated pressures and temperatures attacks mild steel severely, causing hydrogen decarburization and embrittlement, and it is necessary, therefore, to use special alloy steels in hydrogen service.

Leakage of hydrogen constitutes a problem. The pressures in plant handling hydrogen are commonly high and the gas diffuses readily through small holes. Hydrogen exhibits a reverse Joule–Thomson effect, so that leaking gas heats up and may ignite. Since the flame is non-luminous, it may not be visible. An operator may walk unawares into a hydrogen leak flame.

Hydrogen can also diffuse through solid metal. It can pass, for example, into a thermocouple pocket and thence through the connecting leads into the control room. Where this risk exists, therefore, it is good practice to arrange the connections so that there is no flow path for the hydrogen.

Liquid hydrogen is used in certain special applications, notably aerospace work. This is effectively a separate technology. It is described in *Handbook for Liquid Hydrogen Handling Equipment* (A.D. Little Inc., 1960), and by Cassut, Madocks and Sawyer (1964), Stoll (1965) and Scharle (1965).

11.18.9 Hydrogen fluoride

A relevant code for hydrogen fluoride is *Guide to Safe Practice in the Use and Handling of Hydrogen Fluoride* (CIA, 1978 PA14) (the CIA *Hydrogen Fluoride Guide*). Accounts are also given by Gall (1980) and Aigueperse *et al.* (1988).

Hydrogen fluoride is used industrially both as anhydrous hydrogen fluoride and as solution of hydrogen fluoride in water; the *Guide* recommends that the term 'hydrofluoric acid' be reserved for the latter. It is the former, anhydrous hydrogen fluoride, which is of interest here.

Hydrogen fluoride is produced by the reaction of sulphuric acid on fluorspar in a rotary furnace. There are a number of processes, broadly similar but with variations, involving mainly pre-heaters, pre-reactors and recycle. The main uses are for the manufacture of fluorocarbons and of aluminium fluoride and synthetic cryolite used in the aluminium industry. It is also used in refinery processes.

Anhydrous hydrogen fluoride has a normal boiling point of 19.5°C. At atmospheric pressure it may therefore be a gas or a liquid, depending on the temperature and in handling it is conventionally treated as a liquefied gas. At higher ambient temperatures a liquid leak undergoes a degree of flashing to vapour.

Hydrogen fluoride is fully miscible with water. This creates the hazard that if there is water in a vessel containing hydrogen fluoride vapour, the latter may dissolve, so that a vacuum is formed and the vessel collapses inwards; the like hazard with ammonia has already been mentioned.

Hydrogen fluoride is toxic. It has a short-term occupational exposure standard of 3 ppm (as fluorine). It is detectable by odour at about 2–3 ppm, but reliance should not be placed on this. Aqueous solutions of hydrogen fluoride, anhydrous hydrogen fluoride liquid or, to a lesser extent vapour, affect the skin and eyes.

Hydrogen fluoride is highly corrosive, but its corrosivity depends on the water content. Mild steel is resistant to corrosion down to concentrations in water of about 70% hydrogen fluoride. Plant for the manufacture of hydrogen fluoride is largely made of mild steel, but is designed to allow for the foreseeable presence of water. Mild steel is also the material used for the storage and transport of anhydrous hydrogen fluoride. Mild steel becomes decreasingly suitable as the temperature increases. The *Guide* states that the upper limit is usually given as 50–65°C and recommends that this be observed. Silica and silica-containing materials, including metal slags, are attacked by hydrogen fluoride and welds should be such as to avoid slag inclusions.

Special process conditions, such as occur in the manufacture of hydrogen fluoride, may require the use of other materials of construction. The *Guide* gives a list of other materials which may be used for hydrogen fluoride. Pipework for hydrogen fluoride should have flanged or butt-welded joints. The range of materials which can be used for joint rings is narrow and measures are needed to ensure that only suitable materials are used.

Transfer of liquid hydrogen fluoride may be by gas padding or pumping. If pumping is used, measures are needed to prevent cavitation and leaks. A pump with double diaphragm or a canned pump may be used. Padding gas is usually dry air or nitrogen. Unless it is from a source which is inherently dry, the gas should be monitored for water content. The gas supply should have its own pressure control and pressure relief systems, and should not be subject to contamination by hydrogen fluoride.

A vent system is required to handle both normal process and emergency vent flows. The vent gases should pass to a gas absorption system. Water is generally used for absorption, with neutralization by lime slurry.

Methods of relief disposal mentioned in the *Guide* include: returning the relief flow back to a suitable point in the process, which in hydrogen fluoride production is often possible; the use of an expanse tank; and the use of a gas absorption system.

Since with hydrogen fluoride a pressure relief valve is liable to become corroded, the primary relief device used is a platinum bursting disc. A further relief device may be located after the disc to control the relief flow. The design of the relief system ducting should allow for the erosive effect of high velocity hydrogen fluoride gas on mild steel.

Storage and transport of hydrogen fluoride are considered in Chapters 22 and 23, respectively, the toxic release hazard in Chapter 18 and emergency planning in Chapter 24.

11.18.10 Oxygen

Relevant codes for oxygen are *Bulk Liquid Oxygen at Production Plants* (BCGA, 1990 Code 20) and *Bulk Liquid Oxygen at Users' Premises* (BCGA, 1992 Code 19). Oxygen is the subject of NFPA 50: 1990 *Bulk Oxygen Systems at Consumer Sites.* Accounts are also given by A.H. Taylor (1981) and Kirschner (1991).

Oxygen is produced by the liquefaction and separation of air. It is used in bulk quantities in iron and steel making and in chemical processes. It is also used in oxyacetylene welding.

Oxygen is the constituent in the air which supports combustion of fuels and other materials, and in the pure form it supports combustion even more effectively. Combustible materials of all kinds, ranging from hydrocarbons in chemical reactors to workmen's clothing, undergo combustion much more readily in oxygen-enriched air or in pure oxygen. The enhancement of flammability in oxygen-enriched atmospheres is described in Chapter 16.

Oxygen has a normal boiling point of −183°C (90.2 K) and is therefore a gas. It is frequently handled as a liquefied gas under pressure. In this condition liquefied oxygen gives on release a flashing liquid.

Oxygen is odourless and therefore gives no indication of its presence. Short exposure to air with some degree of oxygen enrichment does not appear to be harmful. Oxygen is used in breathing apparatus. The main hazard of an oxygen-enriched atmosphere is the enhancement of flammability.

The effect of a large release of oxygen would be to increase the combustibility of materials in the oxygen cloud. Such materials might include not only people's clothing but even steel equipment on the plant, since steel will burn in oxygen. A possible source of ignition in such circumstances might be a lagging fire. There appears, however, to be no recorded case of any such incident.

Air separation plants are discussed in Section 11.19.

11.18.11 Phosgene

A relevant code for phosgene is *Code of Practice for Chemicals with Major Hazards: Phosgene* (CIHSC, 1975/3) (the CIA *Phosgene Code*), but this code is no longer listed by the CIA. Accounts are also given by Hardy (1982) and Schneider and Diller (1991).

Phosgene is manufactured by passing carbon monoxide and chlorine over activated carbon. It is used mainly to make isocyanates, particularly toluene diisocyanate. It is also used for chloroformic esters, urea azo dyes, carbonate esters and Friedel Crafts acylations. There are available small tonnage phosgene plants which may be installed at a user's site and which avoid the need to transport and store large quantities of the material.

Phosgene has a normal boiling point of 8.2°C. It is not flammable. It is highly toxic and has a long-term occupational exposure standard of 0.02 ppm, which is a five-fold downwards revision from the previous value of 1 ppm and is very low indeed.

In contact with water phosgene undergoes slow decomposition to hydrochloric acid and carbon dioxide. It is essential, therefore, to keep plant handling phosgene dry in order to prevent such decomposition. Phosgene also reacts with most lubricating oils to form a black sludge.

For dry phosgene, carbon steel has proved a satisfactory material of construction, for both vessels and pipework. Where other substances are also present it may be necessary to use other materials. The *Code* cautions against the use of non-metallic materials for pipework.

The *Code* gives detailed requirements for the design of storage vessels for liquid phosgene. Process vessels handling more than one tonne of liquid phosgene in normal operation should be designed to a standard similar to that for storage vessels.

For the pipework on a phosgene plant seamless tubes are preferred to seam welded ones. The pipework should take into account both the maximum and minimum temperatures which may occur. Stress relieving heat treatment may be required. The pipework arrangement should have a good degree of natural flexibility, but bellows should not be used. Joints should preferably be welded, but otherwise flanged. Butt-welded joints should be radiographed. Compressed asbestos fibre gaskets have been used satisfactorily.

Liquid phosgene may be transferred by pumping or inert padding gas. For pumping, glandless pumps should be used. These are suitable for phosgene concentrations down to 5% and for lower concentrations also. Particular attention should be paid to avoidance of cavitation and provision of sufficient net positive suction head (NPSH).

Due to its high solubility in liquid phosgene, nitrogen may be unsuitable as a padding gas. Any gas used should be dry, with a dewpoint less than −30°C. Measures should be taken to prevent contamination of the padding gas supply, preferably by the use of a supply exclusive to the phosgene plant.

Reciprocating compressors are used to compress phosgene vapour, but it is necessary to take special measures to prevent contact between phosgene and lubricating oil. Use is also made in phosgene systems of fans and vacuum pumps.

Relief disposal should be to an expanse tank or to a gas absorption system, but not to atmosphere.

A phosgene plant should normally have two separate gas absorption systems, one to handle process vents and one emergency reliefs. The former should be designed for flows with high concentrations of inerts but low concentrations of phosgene and the latter for flows with higher concentrations of phosgene. The use of a dedicated absorption system for relief flows guards against corrosion and blockage of the relief lines by materials vented from the process. The phosgene content of gas leaving the gas absorption system should be monitored.

With regard to relief devices, pressure relief valves and carbon bursting discs both have some disadvantages for phosgene duties, but either is suitable with good design. A bursting disc alone risks imposing a sudden high load on the gas absorption system. An acceptable arrangement is a bursting disc followed by a pressure relief valve with an alarm on the intermediate space to detect disc failure.

For pipelines an alternative relief device is an expansion bottle. These bottles are fitted to the top of the pipeline at regular intervals and each is isolated from it by a bursting disc. There should be an alarm on the bottle space to detect disc failure.

The use of water or steam in direct heat exchange with phosgene should be avoided due to the possibility that water will leak into the phosgene stream.

Liquid phosgene may be stored under pressure or refrigerated. The *Code* states that plant for handling and storage of phosgene should be kept simple, even if this means storing liquid phosgene under pressure, that storage vessels for liquid phosgene should be pressure vessels and that the contents of a pressure storage vessel should be kept below 8°C in order to reduce the risk of overloading the gas absorption system due to flash-off if pressure has to be reduced.

11.18.12 Vinyl chloride

Vinyl chloride (VC), or vinyl chloride monomer (VCM), is the subject of *Precautions against Fire and Explosion: Vinyl Chloride* (CIA, 1978 PA15) (the CIA *Vinyl Chloride Guide*). The toxicity of vinyl chloride is treated in *Vinyl Chloride*: Toxic Hazards and Precautions (HSE, 1992 EH 63). Accounts are also given in *Vinyl Chloride and PVC Manufacture* (Sittig, 1978) and by Cowfer and Magistro (1983).

Vinyl chloride was originally produced by the gas phase reaction of acetylene and hydrogen chloride. It is now made almost entirely by the cracking of ethylene dichloride and oxychlorination of ethylene. These two processes are generally operated in combination, with the hydrogen chloride from the cracking of ethylene dichloride being reacted with air or oxygen in the oxychlorination process to produce more vinyl chloride. Vinyl chloride is polymerized to polyvinyl chloride (PVC) and also used to make copolymers.

Vinyl chloride has a normal boiling point of −13.9°C. It is frequently handled as a liquefied gas under pressure. In this condition liquefied vinyl chloride gives on release a flashing liquid. Vinyl chloride is flammable, the flammable range being 3.6–33%.

Vinyl chloride is toxic and has a long-term maximum exposure limit of 7 ppm. The toxicity of vinyl chloride was not fully appreciated until the mid 1970s. The discovery of its toxicity, and in particular its carcinogenic properties, presented the industry with a problem which, although quite different in nature, was of the same order of magnitude as that caused by the vapour cloud explosion at Flixborough. The decomposition products of vinyl chloride in a fire include hydrogen chloride and phosgene, which are both toxic, particularly the latter. As its full title indicates, the *Vinyl Chloride Guide* deals with the fire and explosion but not the toxic hazards of vinyl chloride.

In the handling of vinyl chloride it is necessary to ensure that air is excluded. Air can cause the formation of vinyl chloride polyperoxides, which appear as a yellow oil or rubbery solid and which detonate on mild impact. Oxygen also catalyses the polymerization of vinyl chloride. Vinyl chloride can be stored unstabilized if air is excluded, but addition of inhibitor should be considered if this cannot be assured. On the other hand in the absence of other substances, water in vinyl chloride does not constitute any special hazard, although it may cause contamination of the product with rust.

Change of phase, either condensation or boiling, of vinyl chloride can be hazardous. One hazard is a sudden pressure rise, another concentration of impurities and a third polymerization. The *Code* details measures for safeguarding against these hazards.

Materials of construction for vinyl chloride plants are mild steel and for polymerization plants stainless steel. Other materials which may be used where other substances are present in addition are described in the *Code*. Joints on pipework should be welded or flanged.

Liquid vinyl chloride may be transferred by pumping or inert padding gas. Pumps with fine clearances should be avoided as these can give trouble due to polymerization. Particular attention should be paid to avoidance of cavitation and provision of sufficient NPSH and to the pump seals. For padding, the gas normally used is nitrogen.

Compressors used for the compression of vinyl chloride vapour should be such as to ensure exclusion of air.

Relief flows may be sent to a gas absorption system, a flare or an atmospheric vent, if permitted. Both combustion at a flare and combustion following accidental ignition of a vent generate hydrogen chloride, which is toxic.

A suitable relief device for vinyl chloride is a bursting disc followed by a pressure relief valve, or possibly a second bursting disc. The design should allow for the possibility of polymerization in the pipework leading to the relief device and at the device itself.

Plant handling vinyl chloride should preferably be located outdoors, but where it has to be indoors with the attendant hazard of an indoor release of flammable material, consideration should be given to requirements for ventilation and explosion relief of the building.

Storage of vinyl chloride may be in pressure or refrigerated storage.

11.18.13 Some other chemicals

Some other leading inorganic chemicals are covered in *Bromine and Its Compounds* (Jolles, 1966) and *Nitric Acid and Fertiliser Nitrates* (Keleti, 1985). Accounts of some principal organic chemicals include *Propylene Oxide* (Jefferson Chemical Company, 1963), *Propylene and its Industrial Derivatives* (Hancock, 1973), *Benzene and its Industrial Derivatives* (Hancock, 1975), *Benzene* (Cheremisinoff and Moresi, 1979), *Toluene, the Xylenes and their Industrial Derivatives* (Hancock, 1982), *Maleic Anhydride* (Trivedi and Culbertson, 1982), and *Formaldehyde* (Clary, Gibson and Waritz, 1983).

Polymers are treated in *Encyclopaedia of PVC* (Nass, 1976–), *Vinyl Chloride and PVC Manufacture* (Sittig, 1978), *Encyclopaedia of PVC* (Nass and Heiberger, 1986), *Encyclopaedia of Polymer Science and Engineering* (Mark, *et al.*, 1989) and *Handbook of Polyolefins* (Vasile and Seymour, 1993).

11.19 Particular Processes and Plants

The chemical and petrochemical industries operate a great variety of processes and plants, many of them highly interconnected. There are certain processes and plant which have particular hazards, which feature prominently in the safety and loss prevention literature and which it is appropriate to consider here. These are:

(1) air and oxygen plants;
(2) ammonia plants;
(3) ammonium nitrate plants;
(4) methanol plants;
(5) olefins plants;
(6) LPG and LNG installations;
(7) petroleum refineries and gas processing plants.

Descriptions of the first three types of plant and the associated hazards and precautionary measures are given in *Operating Practices in Air and Ammonia Plants* (1961–62/27–28), in the series *Safety in Air and Ammonia Plants* (1960–69/17–26) and *Ammonia Plant Safety and Related Facilities* (1970–1994/31–52), in *Survey of Operating Practices in Thirty-One Ammonium Nitrate Plants* (AIChE, (n.d./15), and in *Safe Operation of Ammonia and Ammonium Nitrate Plants* by the American Oil Company (Amoco/9).

Selected references on particular processes and plants and their hazards are given in Table 11.20 and selected references on ammonia, urea and ammonium nitrate plants and their hazards are given in Table 11.21.

11.19.1 Air separation plants

In addition to the references just given, there are accounts of air separation plants by Newton (1979), A.H. Taylor (1981) and Kirschner (1991).

Air separation plants can produce both oxygen and nitrogen. Each of these products may be in either gaseous or liquid form. Large amounts of oxygen are used in the chemical industry in oxidation processes and

***Table 11.20** Selected references on particular processes and plants and their hazards*

Chemical processes
Groggins (1952); Santini (1961); Kirk and Othmer (1963–, 1978–, 1991–); Faith, Keyes and Clark (1965); Gait (1967); Goldstein and Waddams (1967); Long (1967); Shreve (1967); Hahn and Williams (1970); Kuhn (1971); Considine (1974); Hobson and Pohl (1975); Lowenheim and Moran (1975); British Petroleum Co. (1977); *Financial Times* (1979); Wei, Fraser and Swartzlander (1978); E.R. Kane (1979); Benn Publications Ltd (1980); G.T. Austin (1984); Meyers (1986a)

Air and oxygen plants
American Oil Co. (n.d./9); Anon. (1960d); AIChE (1960–69/3–12, 1961–62/13, 14); C.P. Anderson (1960); Rotzler *et al.* (1960); Gardner (1961); Matthews (1961); G.T. Wright (1961); Brink (1962); Lang (1962, 1965); T.R. McMurray (1962); Ball (1963, 1964, 1965, 1966a,b, 1968a); Guccione (1963); Matthews and Owen (1963); Rendos (1963, 1964, 1967); Calvert (1967); Eschenbrenner, Thielsch and Clark (1968); van der Ende (1969); Tanne (1970); NASA (1972–); Booth (1973); Göller (1974); L'Air Liquide (1976); Honti (1976); Springman (1977); Newton (1979); Wolff, Eyre and Grenier (1979); C. Butcher (1989c); CGA (1989 P-8); NFPA (1990 NFPA 50); Dunrobbin, Werley and Hansel (1991); Kirschner (1991); Shelley (1991)
Impurities: Bollen (1960); Coleman (1961, 1962); Karwat (1961a,b, 1963); Kerry and Hugill (1961); Ball (1962, 1965, 1966b); M.H. Jones and Sefton (1962); Schilly (1962); Hofmaier (1963); Parks and Hinkle (1963); McDonnell, Glass and Daues (1967)

Petrochemical plants, olefins plants
OIA (Loss Inf. Bull. 1, Publ. 301, 1972 Loss Inf. Bull. 2, 1974 Publ. 101); *Hydrocarbon Processing* (1963–); Prescott (1966); Stobaugh (1966c, 1967a,b, 1988); R.F. Phillips (1967); Feldman and Grossel (1968); Lofthouse (1969); J.F. Tucker (1969); Loftus (1970); Sugai (1970); de Blieck, Cijfer and Jungerhans (1971); Silsby and Ockerbloom (1971); Gambro, Muenz and Abrahams (1972); Wilkson and Dengler (1972); Dewitt *et al.* (1974); Stork, Abrahams and Rhoe (1974); Zdonik, Bassler and Hallee (1974); Barlow (1975); Fuge and Sohns (1976); Barker, Kletz and Knight (1977); *Hydrocarbon Processing* (1977, 1981, 1985, 1987b, 1991b, 1993c); Barnwell (1979); Maples and Adler (1978); Boyett (1979); Geihsler (1979); Picciotti (1978, 1980); Wilkinson (1979); Burchell (1980); Bockmann, Ingebrigtsen and Hakstad (1981); Ahearne (1982); Albright, Crynes and Corcoran (1983); Ng, Eng and Zack (1983); Walter (1983); Zack and Skamser (1983); Zdonik and Meilun (1983); Eng and Barnes (1984); Gillett (1984); Kister and Townsend (1984); Maddock (1984); R.V. Parker (1984); C. Tayler (1984a); API (1985 Std 2508); van Camp *et al.* (1985); Wett (1985); Anon. (1986o); List (1986); Redman (1986b); Short (1986); Kister and Hower (1987); Plehiers, Reyniers and Froment (1990); Wiedeman (1992); Olivo (1994a,b); Shelley (1994)

Gas, gas plants, including industrial utilization
AGA (Appendix 28); BG (Appendix 27, 28, 1978 Comm. 1110, 1989 Comm. 1404); IGasE (Appendix 28, n.d./6,

1978 IGE/TM/2, 1985 IGE/TM/2A); Gas Council (1960); Kintz and Hill (1960 BM Bull. 588); Tiratsoo (1967); D.S. Wilson (1969); Anon. (1972i); Bresler and Ireland (1972); Crossland (1972); Hart, Baker and Williams (1972); Thornton, Ward and Erickson (1972); BP Ltd (1973); J.A. Gray (1973); Jockel and Triebskorn (1973); Wall (1973); Walters (1973); Medici (1974); API (1975 RP 50); *Hydrocarbon Processing* (1975, 1984a, 1986b, 1988b, 1990b, 1992b); Franzen and Goeke (1976); IChemE (1976/65); Sharpe (1976); Detman (1977); Caldwell, Eifers and Fankhanel (1978); Pritchard, Guy and Connor (1978); UN (1978); J.A. Gray (1980); Meyers (1984); Røren (1987); Melvin (1988); NFPA (1992 NFPA 54, 1992/32)

Coal-based plants

Hoffman (1978); O'Hara *et al.* (1978); J.P. Leonard and Frank (1979); Pitt and Millward (1979); C.L. Reed and Kuhre (1979); IBC (1981/12); J.M. Evans and Verden (1982); Merrick (1983); Norton (1984)

Methanol plants

Royal and Nimmo (1969); AIChE (1970/73); Hedley, Powers and Stobaugh (1970); Hiller and Marschner (1970); Mehta and Ross (1970); Prescott (1971); Pettman and Humphreys (1975); Supp (1981); Wade *et al.* (1981); Meinhold and Laading (1983); Kobayashi, Kobayashi and Coombs (1986)

Oil, oil refineries

API (Appendix 27, 28, 1983 Publ. 999); ASME (Appendix 28 *Hydrocarbon Processing*); HSE (Appendix 28 *Oil Industry*); IP (Appendix 27, 28, Oil Data Shts, TP series, 1968–/1, 1980 Eur. MCSP, 1981 MCSP Pt 3, 1984/2); OIA (see Appendix 28); Shell International Petroleum Co. (1933–); W.L. Nelson (1958); *Hydrocarbon Processing* (1962–, 1984b, 1986c, 1988c, 1990c, 1992); J.R. Hughes (1967, 1970); Adelman (1972); Anon. (1975m); J.A. Price (1962); Vervalin (1964a, 1973a); Crowe and Marysiuk (1965); Albright (1966a,c); Forry and Schrage (1966); Bland and Davidson (1967); Good (1967); Dosher (1970); Hobson and Pohl (1975); Verde, Moreno and Riccardi (1977); Cantrell (1982); M. Sherwood (1982); Shell (1983); Gary and Handwerk (1984); IBC (1985/62, 1993/105); Loubet (1986); Meyers (1986b)

Polyethylene plants

Albright (1966d,f, 1967a,c); Forsman (1972); AIChE (1973/62, 1974/63, 1978/64); Fitzpatrick (1973, 1974); Guill (1973); Hatfield (1973); Marzais (1973); O'Hara and Poole (1973); O'Neal, Elwonger and Hughes (1973); Royalty and Woosley (1973); Ziefle (1973); Afzal and Livingstone (1974); Mauck and Tomita (1974); R.F. Murphy (1974); Prentice, Smith and Virtue (1974); Traversari and Beni (1974); Watson (1974); Beret, Muhle and Villamil (1978); Olivier and Scheuber (1978); H.S. Robinson (1978); Sandner (1978); Siegel (1978); Yasuhara, Kita and Hiki (1978); Chriswell (1983); Redman (1991)

in the steel industry in both the blast furnace and steelmaking processes. Nitrogen is used in large quantities as a synthesis gas and for gas washing, inerting and purging.

An air separation plant is an integral part of an ammonia plant where the hydrogen for the synthesis gas is obtained from a hydrogen-rich stream or by partial oxidation of hydrocarbons. Hydrogen from partial oxidation or from processes such as coke ovens, catalytic reformers or electrolysis plants is purified by washing in liquid nitrogen, which removes hydrocarbons and carbon monoxide. Nitrogen is then added to the hydrogen to form synthesis gas. The oxygen and nitrogen for these processes are obtained from an air separation plant.

Air separation is carried out by compression, liquefaction and distillation of air. If the products are gases near ambient temperature, the refrigeration required is small and is principally to cover heat inefficiencies and losses, but if they are liquids, there is a much larger refrigeration requirement. This refrigeration is produced by compressing the air and then expanding it through a valve, using the Joule–Thomson effect, or in an expansion engine. The plant is contained in an insulated cold box.

Low pressure plants operate in the approximate pressure range 5–10 bar and intermediate pressure range 10–45 bar. The temperatures are low with nitrogen being separated at about −190°C.

There are two main hazards in an air separation plant. These are that (1) liquid oxygen reacts with any combustible material with explosive violence and (2) hydrocarbons may accumulate in the liquid oxygen and give an explosive reaction. Combustible materials with which liquid oxygen may react violently include wood, lubricating oils, greases and graphite.

Impurities which may enter the plant and accumulate in the liquid oxygen in the distillation column reboiler include methane, ethane, ethylene and acetylene. These hydrocarbons have a very low solubility in oxygen and therefore tend to precipitate out as solids. The solubility of acetylene in liquid oxygen, for example, is only 7 ppm. The solid hydrocarbons are then liable to react explosively with the oxygen. If nitrogen oxides, especially nitric oxides, are present, they catalyse this reaction.

Precautions which are necessary in liquid air plants, therefore, include the elimination of combustibles, both in the construction and operation of the plant, use of clean air and purification of this air, avoidance of build-up of contaminants in the liquid oxygen and safe disposal of the liquid oxygen.

In order to eliminate the combustible materials, equipment is degreased before installation. Grease and graphite materials are not used for pipe joints and gaskets containing graphite are not used for pipe flanges. Every effort is made to ensure that items such as wood, clothing or rags are not left inside the plant. Measures are taken to reduce the risk of hydrocarbon lubricating oils being carried into the low temperature section from the main compressors. Non-flammable lubricants are used for applications such as liquid oxygen pump seals.

Air is taken from an area which is not contaminated by hydrocarbons or nitrogen oxides. Methods of purifying the air in the air intake include catalytic combustion of

Table 11.21 *Selected references on ammonia, urea and ammonium nitrate plants and their hazards*

Ammonia plants
American Oil Co. (n.d./9); Anon. (1959b), 1960d); Harding (1959); AIChE (1960–69/3–12, 1961–62, 13, 14, 1964/15, 1970–77/17–38); Bresler and James (1965); Guccione (1965b); Wrotnowski (1965); B. Powell (1967); Badger (1968); Eschenbrenner, Thielsch and Clark (1968); Finneran, Sweeney and Hutchinson (1968); Quartulli, Fleming and Finneran (1968); Vancini (1971); Finneran, Buividas and Walen (1972); Sawyer, Williams and Clegg (1972); K. Wright (1973); J.F. Anderson (1974); Sawyer and Williams (1974); Webb (1974); Atwood (1976); Butzert (1976); Hess (1976); Ostroot (1976a); Partridge (1976); Wheeler (1976); K.A. Carter *et al.* (1977); Gilbert and Eagle (1977); R.W. James (1977); Livingstone (1977, 1989); Akitsune, Takahashi and Jojima (1978); Attwood, Lombard and Merriam (1978); Banks (1978); Gadsby and Livingstone (1978); Hager, Long and Hempenstall (1978); Kusha (1978); Nakano (1978); Wakabayashi (1978); G.P. Williams (1978); Bishop and Mudahar (1979); Czuppon and Buividas (1979); Farinola and Langana (1979); Ricci (1979b); Archambault, Baldini and Train (1980); Butwell, Kubek and Sigmund (1980); Colby, White and Notwick (1980); Handley (1980); Kokemor (1980); Setoyama, Wadar and Funakoshi (1980); Swanson (1980); Buividas (1981); Moon, Mundy and Rich (1981); Saviano, Lagana and Bisi (1981); C.P.P. Singh and Saraf (1981); Strelzoff (1981); Hodgson (1982); A. Nielsen *et al.* (1982); Reddy and Husan (1982); Anon. (1983q); Livingstone and Pinto (1983); G.P. Williams and Hoehing (1983); Atwood (1984); Facer and Rich (1984); Kolff and Mertens (1984); Prijatel (1984); Rohlfing (1984); Grotz, Gosnell and Grisolia (1985); Ruziska *et al.* (1985); Erskine (1986); Song (1986); Josefson (1987); Madhavan and Sathe (1987); W.K. Taylor and Pinto (1987); G.B. Williams, Hoehing and Byington (1987); Epps (1988); Madhavan and Kirsten (1988, 1989); Madhavan and Landry (1988); Mall (1988); Kershaw and Cullen (1989); Short (1989); Armitage *et al.* (1992); Czuppon, Knez and Rounes (1992)

Converters
W.D. Clark and Mantle (1967); Eschenbrenner and Wagner (1972); Appl, Feind and Liebe (1976); Casey (1976); Kusha and Lloyd (1976); W.D. Clark (1977); Dye (1977); J.A. Lawrence (1977); Patterson (1977); R.L. Thompson and Brooks (1977); Wahl and Neeb (1977); B.R. Phillips (1979); Rao, Wiltzen and Jacobs (1986); Karkhanis and van Moorsel (1987); Mack and Shultz (1987); Shimagaki *et al.* (1987); Veazey and Winget (1989)

Catalysts
A. Nielsen (1964); Comley (1968); D.W. Allen (1969); Cromeans and Knight (1969); J.W. Marshall (1969); Delong (1971); Fleming and Cromeans (1971); Scharle, Salot and Hardy (1971); Cromeans and Fleming (1972, 1976); K. Wright and Haney (1972); Salot (1973); R.L. Thompson (1973); P.W. Young and Clark (1973); Bridger (1976); Collard (1976); Lundberg (1979); Dybkjaer *et al.* (1980); Prince and Odinga (1986); Mukherjee *et al.* (1988)

Reformers
Francis (1964); Adams (1967); Pennington (1968); Kobrin (1969, 1978); Bongiorno, Connor and Walton (1970); Ballantyne (1972); Salot (1972, 1975); Lombard and Culberson (1973); Fuchs (1974); Demarest (1975); Fuchs and Rubinstein (1975); E.R. Johnson (1975, 1977); Leyel (1975); Sterling and Moon (1975); MacMillan (1976); Tendolkar, Sitaraman and Ponnuswamy (1976); Blackburn (1977); Hundtofte (1978); Thuillier and Pons (1978); Hasaballa (1981); Connaughton and Clark (1984); Ennis and Le Blanc (1984); G.M. Lawrence (1984); Roney and Persson (1984); Schlichtharle (1984); Scherf and Novacek (1985); Schuchart and Schlichthärle (1985); Gupton, Stal and Stockwell (1986); McCoy, Dillenback and Traux (1986); Vick (1987); Orbons (1988); Cromarty (1992)

Materials of construction
Speed (1966); W.D. Clark and Mantle (1967); Dial (1968); G.R. James (1968); A. Nielsen (1971); Hutchings *et al.* (1972); van der Horst (1972); Atkins, Fyfe and Rankin (1974); Zeis (1975); Appl and Feind (1976); Tendolkar, Sitaraman and Ponnuswamy (1976); W.D. Clark and Cracknell (1977); Jordan and Rohlfing (1979); Moniz (1985); Barkley (1986); Krishr (1986); Shibasaki *et al.* (1987); van Grieken (1989); Vilkus and Severin (1989)

Pipework
Coats (1967); Chaffee (1970); Appl (1975); Luddeke (1975); Mitcalf (1975); Wicher (1975); Janssen (1976); Osman and Ruziska (1976); Hakansson (1977); Isbell (1977); Pebworth (1977); Batterham (1985); El Ganainy (1985); G.M. Lawrence (1986); Legendre and Solomon (1986); Prescott, Blommaeth and Grisolia (1986); Sötebier and Rall (1986); Schwarz and Spähn (1988); Veazey and Winget (1989); Nightingale (1990)

Compressors
Deminski and Hunter (1962); Penrod (1962); Deminski (1964a,b); Morain (1964); Ball (1965); Wrotnowski (1965); Hile (1967); Stafford (1971); Kusha (1977); Tipler (1979); Verduijn (1979); A.G. Smith *et al.* (1985); Fromm and Rall (1987a); G.R. Thomas (1987); Whiteside (1987); Kumar and Grewal (1988); J.B. Smith and Paulson (1989); Bergenthal, Fromm and Liebe (1990); Iliadis (1990); Leingang and Vick (1992)

Other equipment
Linton and Brink (1967); Roney and Acree (1973); W.D. Clark (1976a); Edmondson (1977); Rall and Fromm (1986); Karkhanis and van Moorsel (1987); Cagnolatti (1988); Mukherjee, Ghosh and Chatterjee (1992)

Control and instrumentation
Daigre and Nieman (1974); van Eijk (1975); Pebworth and Wickes (1975); Prijatel (1984)

Spillages
Kalelkar and Cece (1974); Feind (1975); O'Driscoll (1975a); Raj *et al.* (1975); W.D. Clark (1976b)

Urea plants
Reynolds and Trimarke (1962); Croysdale, Samuels and Wagner (1965); Walton (1965, 1966, 1967); J.F. Anderson

(1967); Mavrovic (1974); Dooyeweerd (1975); W.D. Clark and Dunmore (1976); Otsuka, Inoue and Jojima (1976); Bress and Packbier (1977); Cabrini and Cusmai (1977); Jojima (1980); Khadra (1980); P.C. Campbell (1981); Luetzow (1989)

Ammonium nitrate plants
American Oil Co. (n.d./9); Commentz *et al.* (1921); Cronan (1960c); AIChE (1966/16); van Dolah *et al.* (1966 BM RI 6773); Huijgen and Perbal (1969); Hansen and Berthold (1972); Perbal (1983); FMA (1989); NFPA (1993 NFPA 490)

---

hydrocarbons and adsorption of hydrocarbons and nitrogen oxides.

Build-up of contaminants in the liquid oxygen is minimized by the avoidance of pockets where liquid oxygen accumulates, by running oxygen reboiler tube plates always flooded and by continuous blowdown of liquid oxygen into an auxiliary vaporizer.

If the air supply becomes contaminated, it may be necessary to stop air supply to the plant. Under these conditions heat leaks into the plant and causes liquid oxygen to evaporate and contaminants to concentrate. There is therefore a certain time period after which hazardous concentrations of contaminants may accumulate. Thus there need to be arrangements for the discharge of the plant liquids.

Contaminants which accumulate in the plant may be removed by regular cleaning. This involves heating up the plant with warm air, which vaporizes the contaminants. The accumulation of contaminants is reduced if the insulation is efficient. The cold box is kept at a positive pressure by air, so that water vapour does not enter and freeze in the insulation, thus making it less effective. The air used has to be free of hydrocarbons.

The small bore instrument and sampling pipework inside the cold box is vulnerable to fracture when insulation is being put in. A suitable precaution is to do this work with the lines under a low nitrogen pressure so that a fracture can be detected by a fall in pressure. A further hazard is the twisting of the tubing when work is being done on tubing which extends through the cold box wall. The likelihood of this may be reduced by the use of back-up wrenches.

Materials of construction used in air separation plants include aluminium, copper and stainless steel. Mild steel is not used in applications where it may undergo low temperature embrittlement or may oxidize as a result of a liquid oxygen leak. This applies to its use not only in vessels and pipework but in supports also.

Tarmacadam road surfaces are avoided in areas where liquid oxygen may spill. Oily or greasy clothing is also avoided.

11.19.2 Ammonia plants

Treatments of ammonia and ammonia plants have been referenced in Section 11.18. Additional accounts of ammonia manufacture are given in *Technology and Manufacture of Ammonia* (Strelzoff, 1981) and by Bakemeier *et al.* (1985) and Czuppon, Knez and Rounes (1992).

Large tonnages of ammonia are used directly in liquid form as a fertilizer and as a chemical raw material for the production of other fertilizers such as ammonium sulphate, ammonium nitrate and ammonium phosphate, of explosives such as ammonium nitrate, of inorganic chemicals such as nitric acid and hydrogen cyanide and of organic chemicals such as acrylonitrile. Other uses are described in Section 11.18.

The basic ammonia synthesis reaction is:

$$N_2 + 3H_2 \rightarrow 2NH_3$$

There are a number of processes for ammonia synthesis. The differences between these are partly in the source of hydrogen and partly in the operating conditions and synthesis converter design.

As already described, hydrogen is generally produced by steam reforming of natural gas or naphtha or partial oxidation of fuel oil or from a hydrogen-rich stream from the low temperature separation section of a petroleum or coke oven plant, or from a chlorine cell-room. Nitrogen is normally obtained either from an air separation plant or from air which has been burned to remove the oxygen. Some 75–80% of ammonia production world-wide uses steam reforming and of this some 65–70% is based on a natural gas feedstock.

In the steam reforming of natural gas the principal process steps are feedstock purification, primary and secondary reforming, shift conversion, carbon dioxide removal, synthesis gas purification, synthesis gas compression, ammonia synthesis and recovery.

Following purification the natural gas is reacted with steam in a primary reformer, which consists of a tubular reactor placed vertically in the furnace and packed with nickel catalyst. The reaction, which takes place in the temperature range approximately 700–800°C, is strongly endothermic, requires a substantial heat input and produces a mixture of hydrogen, carbon dioxide and carbon monoxide:

$$2CH_4 + 3H_2O \rightarrow 7H_2 + CO_2 + CO$$

Air is injected into this gas as it enters the secondary reformer, which is an internally insulated vessel filled with nickel catalyst operating at a temperature above 1100°C. Combustion of the air provides the heat for completion of the reaction and nitrogen to form the synthesis mixture. The air flow is adjusted to give this correct synthesis gas composition. The gas is then passed through a shift converter to assist the reaction of steam and carbon monoxide to hydrogen and carbon dioxide:

$$H_2O + CO \rightarrow H_2 + CO_2$$

Carbon dioxide and carbon monoxide are removed from the gas by various purification processes. The synthesis gas then passes to the compression section.

In the ICI LCA process a variation is introduced in which the primary reformer can be greatly reduced in size or even eliminated by waste heat reforming, in which part of the reforming is carried out in a reforming heat exchanger utilizing waste heat from the secondary reformer.

In partial oxidation a similar hydrocarbon feed and oxygen from an air separation plant are preheated separately and reacted in non-catalytic gas generators at a temperature above 1100°C. The reaction is:

$$2CH_4 + O_2 \rightarrow 4H_2 + 2CO$$

The gas is cooled, followed by the shift reaction and purified of carbon dioxide and carbon monoxide by processes similar to those used in steam reforming. It is then cooled and washed with liquid nitrogen to remove traces of impurities. The gas is then adjusted in composition by addition of further nitrogen and passes to the compressors.

If the hydrogen is the off-gas from a refinery reformer, it is purified, dried and then washed with liquid nitrogen. After adjustment of the gas composition by further nitrogen addition, this synthesis gas enters the compressors.

The synthesis gas is then compressed. Whereas reciprocating compressors were a feature of earlier ammonia plants, on modern plants centrifugal compressors are used for compression of the synthesis gas. They are usually driven by steam turbines.

Following compression, the synthesis gas enters the ammonia converters, which on large plants are multiple-bed. There are a variety of designs which differ in respect of the flow pattern, the method of temperature control and the reaction heat recovery. The reaction takes place on an iron catalyst, is exothermic and gives a conversion of about 25%. The ammonia formed is condensed out by cooling the exit gases, which are recycled to the converter.

The ammonia synthesis is one of the classic high pressure processes, but the pressure at which ammonia converters operate has tended to fall. In plants designed before the mid-1960s synthesis converters operated at pressures up to about 340 bar (5000 psi), but the operating pressure in large plants is now of the order of 150 bar (2200 psi).

Up to about 1960 ammonia plants were of relatively low capacity operating at high pressures obtained using reciprocating machines. The early 1960s saw the introduction of the Kellogg single-stream ammonia plant using centrifugal compression and with an output of 544 t/day. This involved a step change in the scale and technology of ammonia plants and was one of the first examples of the large, single-stream plant. By the 1970s single-stream ammonia plants with a capacity of 1500 t/day were becoming common.

The hazards which are encountered in an ammonia plant depend somewhat on the process. But they include those of the chemicals (ammonia and hydrogen), of the operating conditions (high pressure, high temperature and low temperature), and of the pressure system components (furnaces and compressors).

The hazards of and the precautions necessary with ammonia and hydrogen were considered in Section 11.18. Those associated with extreme operating conditions were discussed in Section 11.16 and those associated with pressure systems are dealt with in Chapter 12.

11.19.3 Ammonium nitrate plants

Ammonium nitrate is the subject of NFPA 490: 1986 *Storage of Ammonium Nitrate*. Accounts of ammonium nitrate plants are given by Zapp (1985) and Weston (1992).

Ammonium nitrate is used in large quantities as an agricultural fertilizer. It is also used as an explosive. Some of the worst disasters in the chemical industry have been ammonium nitrate explosions. These include the explosion at Oppau in 1921, which killed 561 people, and that in Texas City in 1947, in which 522 died. Further details are given in Case Histories A5 and A16, respectively.

Despite these occurrences, ammonium nitrate is widely used. It is manufactured by neutralizing nitric acid with ammonia. The nitric acid is made by oxidation of anhydrous ammonia by air on a platinum–rhodium gauze catalyst. The reaction gases are cooled and absorbed in water to give about 56–60% nitric acid. The temperature of the ammonia entering this converter needs to be kept above 60°C in order to prevent liquid ammonia reaching the gauze and generating high temperatures, which may cause an explosion. The chloride content of the water entering the absorber has to be kept low to prevent formation of hydrochloric acid, and thus aqua regia, which is highly corrosive to stainless steel, and to avoid the catalysis by chloride of thermal decomposition of ammonium nitrate. Nitric acid is extremely corrosive and may react explosively with carbonaceous materials, so leaks require to be repaired promptly.

The nitric acid is neutralized with ammonia vapour. The reaction is exothermic and the heat liberated vaporizes water to produce 83–86% ammonium nitrate solution. Solutions of higher concentrations may be produced by further vaporization of water in evaporators. Concentrated ammonium nitrate solutions are blended with ammonia, additional water or urea as required to give ammonium nitrate–ammonia solutions. Alternatively, solid ammonium nitrate is produced by spray drying in a prilling tower or by concentrating to dryness and spraying the molten nitrate.

It is necessary to control closely the temperatures in ammonium nitrate neutralizers and evaporators. Although pure ammonium nitrate does not decompose rapidly below 200°C, impurities can cause a significant reduction in the decomposition temperature. The temperature in the neutralizers and evaporators is therefore generally kept below 135°C. In operating this plant it is important not to produce dry ammonium nitrate, which can become hazardous if then heated further.

Limitation of impurities is equally essential, since they increase the tendency of the ammonium nitrate to decompose. Impurities which sensitize decomposition include copper and zinc as well as nitric acid and chlorides. Solid ammonium nitrate is very hygroscopic and tends to cake into a large mass which is difficult to handle. This is overcome by the use of a surface coating such as diatomaceous earth.

Addition of carbonaceous materials to ammonium nitrate can turn it into a blasting agent. It is necessary, therefore, to keep materials such as hydrocarbons away from ammonium nitrate. The sensitizing effect of hydrocarbon surface coatings used to prevent caking is partly responsible for some of the worst explosions which have occurred in the past. Other serious accidents have occurred from the former use of explosives to loosen caked ammonium nitrate, as at Oppau. This is no longer done.

The extent to which ammonium nitrate should be regarded as a high explosive was for some time a matter of debate. The matter was brought to a head when NFPA 490 was first adopted in 1963. A distinction was drawn between fertilizer grade and explosive grade ammonium nitrate, the latter containing some 1% of organic material. In effect the former was treated in the code as not being a high explosive (W.H. Doyle, 1980). Furthermore, under certain circumstances ammonium nitrate exposed to fire can explode. For some time the conditions necessary for an explosion were not well defined. Accounts of work on both these problems have been given by van Dolah (1966 BM RI 6903) and van Dolah *et al.* (1966 BM RI 6743, 1966 BM RI 6773).

If ammonium nitrate is heated to about 200°C, it begins to decompose and give off heat. This temperature may be lower if there are impurities present. Up to about 260°C the reaction is well behaved, but about this temperature decomposition becomes much more rapid. In a fire ammonium nitrate burns freely and does not explode. Thus over one 3-year period records showed that 13 freight cars containing ammonium nitrate had burnt without explosion. Similarly, there have been several fires which have destroyed without explosion some 5 million pounds of ammonium nitrate stored in warehouses. But an explosion may occur if there is substantial confinement, so that high temperature and pressure can develop. The risk may be minimized by providing adequate spacing and ventilation in storages.

There are certain precautions which should be taken in the storage, transport and handling of ammonium nitrate. For aqueous solutions the hazard is mainly that of evaporation of water to give a more concentrated solution. Precautions include limits on the strength of solutions stored and facilities for water addition. For bagged solids the precautions required include spacing and ventilation, elimination of explosives and oxidizing materials, and avoidance of heat and ignition. In both cases there should be prompt clearing up of spills.

Ammonium nitrate fires should be fought by the application of large volumes of water. Steam should not be used on an ammonium nitrate fire. The nitrate itself can supply enough oxygen, so that steam does not smother the fire, but only increases the temperature and the degree of hazard. Water sprinklers, fog nozzles and fire extinguishants such as foam or dry powder tend to be ineffective. In contrast to the situation with other fires in confined spaces, it may be advisable with ammonium nitrate fires to provide ventilation, so as the reduce the chance of pressure and temperature build-up.

11.19.4 Methanol plants

Accounts of methanol production processes are given by Wade *et al.* (1981) and Fielder *et al.* (1990). Methanol is used in the production of a large range of organic chemicals, notably formaldehyde, methyl-t-butyl ether (MTBE), acetic acid, methyl methacrylate (MM) and dimethyl terephthalate. It is also finding use as a fuel, for which it has considerable potential.

The basic methanol synthesis reactions are:

$$CO + 2H_2 \rightarrow CH_3OH$$

$$CO_2 + 3H_2 \rightarrow CH_3OH + H_2O$$

There are a number of processes for methanol synthesis. All are based on the conversion of hydrogen, carbon monoxide and carbon dioxide to methanol in the presence of a catalyst. The main differences lie in the source of hydrogen and in the catalyst and operating pressure of the synthesis stage.

Sources of hydrogen were discussed above in the context of ammonia synthesis. For methanol, hydrogen is obtained mainly from natural gas, petroleum residues and naphtha. In 1980 some 70% of methanol production world-wide was based on steam reforming of natural gas.

The principal reactions occurring in steam reforming are for methane:

$$CH_4 + H_2O \rightarrow CO + 3H_2$$

and for heavy hydrocarbons

$$C_nH_{(2n+2)} + (n - 1)H_2 \rightarrow nCH_4$$

and the water gas shift reaction

$$CO + H_2O \rightarrow CO_2 + H_2$$

The reaction is strongly endothermic, requires a substantial heat input and produces a mixture of hydrogen, carbon monoxide and carbon dioxide.

Steam reforming gives a hydrogen content higher than that required for the methanol reaction and processes differ in the way this feature is handled. In some the hydrogen is burned whilst in others an addition of carbon dioxide is made. The synthesis gas is then compressed by centrifugal compressors.

At the synthesis stage the pressure required depends on the activity of the catalyst used and these are the main sources of difference between processes. Three pressure ranges are distinguished: low pressure (50–100 bar), medium pressure (100–250 bar) and high pressure (250–350 bar).

Up to the mid-1960s, methanol processes utilized relatively high pressures of the order of 350 bar obtained using reciprocating compressors. In 1966, ICI began operation of a low pressure methanol plant operating at 50 bar with an output of 400 t/day using all-rotating compression equipment. This plant set the pattern for subsequent developments. By 1980, a typical methanol plant was a single-stream plant, operating in the low–medium pressure range, using all-rotating compression equipment and with an output of 1000–2000 t/day.

11.19.5 Olefins plants

An account of olefins production processes is given in *Ethylene* (Kniel, Winter and Stock, 1980) and by Kniel, Winter and Chung-Hu Tsai (1980), Pike (1981) and Granton and Roger (1987). Olefins plants produce not only ethylene but other major co-products, including propylene and butadiene, and are the key units of a petrochemical works.

The main hazard on an olefins plant is that of the flammable hydrocarbons. This hazard is simply stated, but has many aspects. The hydrocarbons are processed under pressure at both low and high temperatures, which increase the risk both of loss of containment and of formation of a vapour cloud which may cause a fire or explosion.

Many aspects of olefins plants are discussed in Chapters 10, 12, 20 and 22, which deal with plant layout, pressure system design, plant operation and storage, respectively. A review of the evolution of safety aspects of olefins plants has been given by Barker, Kletz and Knight (1977), who consider in particular the following aspects: (1) plant layout, (2) plant equipment, (3) pressure relief, and (4) leak control. The authors discuss these features in relation to differences of practice on ICI's successive olefins plants, particularly the then most recent plants, Nos 4 and 5. In the account given below, 'policy' refers to thinking described as 'current' in the paper by Barker, Kletz and Knight.

On plant layout the policy is to space units so that an expected leak does not find a source of ignition (rather than to use rules-of-thumb), to increase the number of fire breaks and to put more equipment at grade level, where it is more accessible to inspection and to fire fighting.

Electrical and instrument cables are vulnerable to flash fires and damage to cabling can cause extensive downtime. Whereas on No. 4 plant no special attention was paid to this aspect, on No. 5 plant in high risk areas cables are run in asbestos/galvanized steel troughs on the cable racks. Policy is to develop preformed reinforced concrete troughs.

Defects in the drainage systems of early olefins plants were that insufficient care was taken in siting low points and gullies and that drains did not have sufficient capacity to carry away fire water. These deficiencies were rectified on the No. 5 plant, in which the high and low points and the drains are located so as to avoid pools of liquid under equipment and the drains are sized to take the quantities of water used to cool major vessels in and near a fire area. The plant, however, has conventional luted drains and some drains from roads and pebbled areas do not pass through the oil–water separator. The policy is to use fully flooded drains with no vapour space in which explosion can occur and with no lutes which can choke, to make extensive use of paving, and to take all drainage from roads and pebbled areas to the oil–water separator.

With regard to process equipment, the early olefins plants had multiple reciprocating or spared centrifugal compressors. Current plants have single unspared machines. This requires the highest standard of design and operation. But it is very effective in reducing leaks.

Another area where leaks can occur is on cold pump glands. The use of O-rings and the elimination of routine pump changeovers have greatly reduced such leaks. Leakage has been further reduced by the use of brazed aluminium plate fin heat exchangers instead of tubular heat exchangers on cold duties. It is necessary, however, to provide these finned tube exchangers with special protection against fire damage.

Developments in pressure relief include the reduction of the total amount of material to be released, reduction of the probability that hydrocarbons will be discharged to the atmosphere unburnt and improvements in the safety of unburnt discharges.

Measures which can be taken to reduce discharge to the flare system include the more extensive use of trip systems and of higher design pressures for vessels such as low pressure catchpots.

The No. 5 plant has two separate flare systems, and this practice would be repeated. One system is for wet streams and the other for cold, dry streams. Process gas can be dumped to either flare or via a process gas transfer system to the fuel gas system or to other flare systems.

Various measures have been developed to reduce and control leaks. Both plants No. 4 and 5 have a leak detection system using Sieger detectors. There is a total of 140 detectors on the two units. Both plants are also provided with a steam curtain to dilute any vapour cloud which may escape from the plant. Extensive use is made of emergency isolation valves. The valves on No. 5 plant have been individually listed and justified by Kletz (1975b).

The possibility of hydrocarbons entering steam or condensate systems is investigated and potential routes are classed as high or low risk. On high risk routes the risk is either eliminated completely or at least is reduced. Thus a tubular heat exchanger is high risk if it may be subjected to conditions such as low temperature, high pressure, vibration or intermittent use. If, for example, a steam-heated heat exchanger might reach subzero temperatures during plant upsets, the practice would be to fit the steam supply with a check valve and the condensate system with a trip valve closed as appropriate by low level and/or low temperature.

The authors list the following safety reviews which are made on olefins plants: (1) piping and instrument diagram, (2) electrical distribution, (3) relief and blowdown, (4) layout, (5) paving and drainage, (6) cable routing and protection, (7) means of escape from structures and buildings, (8) fire fighting, and (9) fire protection.

The foregoing is a necessarily brief summary of the hazards of olefins plants and of some of the measures developed to deal with them. Many of the aspects mentioned are considered in more detail in other chapters.

A further account of loss prevention in an ethylene plant has been given by Olivo (1994a)

11.19.6 LPG and LNG installations

Propane and butane are stored in liquid form as liquefied petroleum gas (LPG). They play an important part in the oil and chemical industries, which have large LPG storages.

LPG is treated in *An Introduction to Liquefied Petroleum Gas* by the LPG Industry Technical Association (LPGITA, n.d.) and are the subject of a set of codes of practice, notably *Installation and Maintenance of Fixed Bulk LPG Storage at Consumer Premises* (LPGITA, 1991 LPG Code pt1); the LPGITA is now renamed the Liquefied Petroleum Gas Association (LPGA). LPG is also covered in *Storage of LPG at Fixed Installations* (HSE, 1987 HS(G) 34) and in the *Liquefied Petroleum Gas Safety Code* (IP, 1987 MCSP Pt 9). It is the subject of NFPA 58: 1989 *Storage and Handling of Liquefied Petroleum Gases* (also published as the *Liquefied Petroleum Gases Handbook*) and NFPA 59: 1989 *Storage and Handling of Liquefied Petroleum Gases at Utility Plants*. Further accounts of LPG are given by Selim (1981) and S.M. Thompson (1990).

LPG is widely used as a fuel in general manufacturing industry, which therefore has a large number of LPG storage installations of various sizes. In general, these industries are not as well versed in the hazards of flammable materials as the chemical industry.

Natural gas, which is mainly methane, has replaced coal gas as the gas supplied to industrial and domestic users by British Gas. Most of the gas is obtained from the North Sea, but some is imported as liquefied natural gas (LNG). In contrast to LPG, the number of LNG

***Table 11.22** Selected references on LPG and its hazards*

IGasE (n.d./5); LPGITA (Appendix 27; n.d., 1974/1, 1991 Code Pt 1); Bray (1964); FPA (1964/1); van Fossan (1965); Eckhart (1969); Andrews (1970); Home Office (1971/2, 1973/4); T.H. Taylor (1971); HSE (1973 HSW Bklt 30); A.F. Williams and Lom (1974); Heng-Joo Ng and Robinson (1976); Skillern (1976); Wesson (1976); Jensen (1978); Rasbash (1978/79); Steinkirchner (1978); Anon. (1979c); Craig and White (1980); Hogan, Ermak and Koopman (1981); C. Jones and Sands (1981); Luks and Kohn (1981a); Selim (1981); BG (1983 BGC/PS/DAT24); Desteese and Rhoads (1983); Chowdhury (1984); Moodie, Billinge and Cutler (1985); Rammah (1985); Jenkins and Martin (1983); UL (1985 UL 144, 1986 UL 21); Avgerinos (1987); IP (1987 MCSP Pt 9); API (1989 Std 2510, Publ. 2510); Dunne and Higgins (1989); Pantony and Fullam (1989); B.M. Lee (1989); Vermeiren (1989); S.M. Thompson (1990); NFPA (1992 NFPA 58, 59)

Storage
ICI/RoSPA (1970 IS/74); Lafave and Wilson (1983); Fuvel and Claude (1986); Lagron, Boulanger and Luyten (1986); Blomquist (1988); Droste and Mallon (1989); ILO (1989)

Transport
Sea: Gosden, Smith and Elkington (1982); Lyon, Pyman and Slater (1982); Lakey and Thomas (1983); Buret, Hervo and Tessier (1985); Chauvin and Bonjour (1985); Pakleppa (1991)

Projects, installations, operations
Borseth, Huse and Olsen (1980); Shtayieh *et al.* (1982); Bonnafous and Divine (1986); Branchereau and Bonjour (1986); Bendani (1989); Pangestu (1991)

Fire protection (*see also* Tables 16.2 and 22.1)
J.M. Wright and Fryer (1982); Fullam (1987); S. Stephenson and Coward (1987); API (1989 Publ. 2510A); B.M. Lee (1989); NFPA (1992/31)

Hazard assessment and control
SAI (1974); van der Schaaf and Opschoor (1980); Romano, Dosi and Bello (1983); Wicks (1983); Crossthwaite (1986); Crocker and Napier (1988a); Droste and Mallon (1989)

***Table 11.23** Selected references on LNG and its hazards*

Abadie (1966); Gram (1968); IGU (1968–); P.C. Johnson (1968); Stark and Washie (1968); Bloebaum and Lewis (1969); Barron (1970); Cribb and Hildrew (1970); Haselden (1970, 1971); Potter and Bouch (1970); R. Shepherd (1970); IGasE (1971 IGE/SR/11); AGA (1973); Thorogood, Davey and Hendry (1973); Wall (1973, 1975); AGA (1974/18, 1979/5, 1981/6, 1984/10, 1986/11); Battison (1974); A.D. Little (1974c); Lom (1974); Sarkes and Mann (1974); ASCE (1976/6); Peebles (1977); Proes (1977); US Congress, OTA (1977); Whitlock (1977); P.J. Anderson and Daniels (1978); Ediger (1978); HSE (1978b); US Senate (1978); Arnoni (1979); Yamanouchi and Nagasawa (1979); Ait Laoussine (1980); Atallah (1982, 1983); Faridany (1982); Geist (1985); Gilbert *et al.* (1985); Rideout *et al.* (1985); Leray, Petit and Paradowski (1986a,b); McGuire and White (1986); Anon (1987r); Melvin (1988); Hammer *et al.* (1991)

Physical and chemical properties
Gonzalez *et al.* (1968); Klosek and McKinley (1968); Huebler, Eakin and Lee (1970); Nakanishi and Reid (1971); Luks and Kohn (1981b); FPA (1987 CFSD H30)

Materials of construction
Burkinshaw (1968); Duffy and Dainora (1968); Lake, DeMorey and Eiber (1968); Cordea, Frisby and Kampschaefer (1972); ASTM (1975 STP 579); Khenat and Hasni (1977); NBS, Cryogenics Div. (1978); Leeper (1980); Charleux and Huther (1983); Duffaut (1983); Krause (1986); Minoda *et al.* (1986); Ohsaki *et al.* (1986); Vercamer, Sauve and Lootvoet (1987)

Instrumentation
Blanchard (1975); Blanchard and Sherburne (1985); Broomhead (1986); Flesch and Dourche (1986); Saiga *et al.* (1991)

Heat exchangers
Merte and Clark (1964); Gaumer *et al.* (1972); O'Neill and Terbot (1972); Tarakad, Durr and Hunt (1987)

Process machinery
Bourguet (1968); Linhart (1970); Schlatter and Noel (1972); L.R. Smith (1972); Guguen and Cherifi (1974); Kato (1977)

Storage
IGasE (n.d./2–4); Tutton (1965); E.L. Smith (1966); Closner (1968, 1970); Ward and Egan (1969); Berge (1970); G.H. Gibson and Walters (1970); Hashemi and Wesson (1971); Geist and Chatterjee (1972); Gondouin and Murat (1972); Kümper (1972); Lusk and Dorney (1972); Mansillon (1972); US Congress (1973); Schuller, Murphy and Glasser (1974); K.A. Smith and Germeles (1974); Stone, Hill and Needels (1974); Walters, Dean and Carne (1974); Ferguson (1975); Bellus *et al.* (1977); Fujita and Raj (1977); Haddenhorst and Lorenzen (1977); Ishimasa and Umemura (1977); US Congress, OTA (1977); Dinapoli (1978); HSE (1978b); Bakke and Andersen (1982); Beevers (1982); Brumshagen (1982); van Hoof and Ofrenchk (1982); Yamakawa (1982); Zick and LaFave (1982); Boulanger and Luyten (1983a,b); Cheyrezy (1983); Closner and Wesson (1983); Collins *et al.* (1983); Lafave and Wilson (1983); de la Reguera *et al.* (1983); Steel, Faridany and Ffooks (1983); Steimer (1983); Capdevielle and Goy (1985); Flüggen, Nüssbaumer and Reuter (1985); Huther, Zehri and Anslot (1985); Marchaj (1985); Speidel (1985); Vater (1985); Crawford, Durr and Handman (1986); Lagron, Boulanger and Luyten (1986); Leray, Petit and Paradowski (1986a,b); Acketts (1987); Bomhard (1987); Morrison (1987); Beese, Trollux and Jean (1989); Itoyama *et al.* (1989); A.L. Marshall (1989); NFPA (1990 NFPA 59A); Meratla (1990); Carre and Bre (1991);

Neville and White (1991); Chen-Hwa Chiu and Murray (1992)
Rollover: J.S. Turner (1965); Hashemi and Wesson (1971); Chatterjee and Geist (1972); Maher and van Gelder (1972a,b); Sarsten (1972); Drake, Geist and Smith (1973); Germeles (1975a,b); K.A. Smith *et al.* (1975); Drake (1976); Takao and Narusawa (1980); Nakano *et al.* (1983); Takao and Suzuki (1983); Lechat and Caudron (1987); Benazzouz and Lasnami (1989); Marcel (1991)

Transport

Dyer (1969); A.E. Gibson and Pitkin (1972); Gondouin and Murat (1972); US Congress, OTA (1977); HSE (1978b)
Road: Eifel (1968); Stahl (1968); Montet, Przydrozny and Inquimbert (1977)
Rail: Eifel (1968); Backhaus and Jannsen (1974)
Waterway: Kober and Martin (1972); Backhaus and Jannsen (1974); Backhaus (1982)
Pipelines: C.J. Gibson (1968); Gineste and Lecomte (1970); Hoover (1970); Ivantzov, Livshits and Rozhdestvensky (1970); Katz and Hashemi (1971); Dimentberg (1972); Ivantzov (1972); Walker, Coulter and Norrie (1972); Dumay (1982); Backhaus (1983); van Tuyen and Regnaud (1983); Kostering and Becker (1987)
Sea: Tutton (1965); Pilloy and Richard (1968); Ward and Hildrew (1968a); F.S. Atkinson (1970); Filstead and Rook (1970); Guilhem and Richard (1970); Rook and Filstead (1970); Gilles (1972); Kober and Martin (1972); Booz-Allen Applied Research Inc. (1973); H.D. Williams (1973); Soesan and Ffooks (1973); Hansen and Vedeler (1974); R.C. Hill (1974); Kniel (1974); AGA (1975/24); D.S. Allan, Brown and Athens (1975); Corkhill (1975); Eisenberg, Lynch and Breeding (1975); J.P. Johnson and Jamison (1975); Authen, Skramstad and Nylund (1976); Prew (1976); Department of Commerce, Maritime Administration (1977); DOT, CG (1977a); Eke and Gibson (1977); Ffooks (1977); Findlater and Prew (1977); Mathiesen *et al.* (1977); Schwendtner (1977); US Congress, OTA (1977); Vrancken and McHugh (1977); HSE (1978b, 1981a); Glasfeld (1980); F.S. Harris (1980); Böckenhauer (1980, 1981, 1985, 1987); Shumaker (1980); Bourguet (1981); Angas (1982, 1985); Armand (1982); Beevers (1982); Benoit (1982); Böjrkman (1982); Brumshagen (1982); Edinberg *et al.* (1982); Hillberg (1982); Holdsworth (1982, 1983, 1985); Lyon, Pyman and Slater (1982); Mabileau (1982); Mankabady (1982); Masaitis and Tornay (1982); van Mater *et al.* (1982); P.R. Mitchell (1982); Nagamoto *et al.* (1982); Nassopoules (1982); Peck and Jean (1982); Riou and Zermati (1982); Veliotis (1982); Aprea (1983); Berger (1983); Fujitani *et al.* (1983); Huther and Benoit (1983); Jean and Lootvoet (1983); Lakey and Thomas (1983); Murata *et al.* (1983); Nagamoto *et al.* (1983); van Tassel (1983); Vogth-Eriksen (1983); B. White and Cooke (1983); Aldwinckle and McLean (1985); Fujitani *et al.* (1985); Jean and Bourgeois (1985); Jenkin, Singleton and Woodward (1985); Latreille (1985); Tanaka and Umekawa (1985); Bakke (1986a); Betille and Lebreton (1986); J. Bradley (1986); J.A. Carter (1986); Itoyama *et al.* (1986); Lootvoet (1986); McLean and Cripps (1986); Ogawa *et al.* (1986); Ogiwara *et al.* (1986); Ölschlager (1986); Rowek and Cook (1986); Schrader and Mowinckel (1986); Ackerman and Hutmacher (1987); Flesch and Lootvoet (1987); Huther, Anslot and Zehri (1987); Fujitani *et al.* (1987); Ferguson, McLean and Sakai (1989); Fujitani, Okumura and Ando (1989, 1991); Jean, Lootvoet and Bennett (1989); Meratla (1990); Claude and Etienne (1991); Itoyama *et al.* (1991); Mathiesen *et al.* (1991); Ogiwara *et al.* (1991); Pakleppa (1991); Tornay, Gilmore and Feskos (1991); Wayne, Böckenhauer and Gray (1991)
Shipping statistics: J.R. Evans (1987); Kolb and Boltzer (1987)
Ships engines: Engesser *et al.* (1987); Grøne and Pedersen (1987); Terashima *et al.* (1987)
Ship–shore transfer: Whitmore and Gray (1987)

Training

Blogg (1987); Karim *et al.* (1987)

Projects, installations, operations

Guccione (1964); Filstead (1965); IGasE (1965 Comm. 696); Culbertson and Horn (1968); Engler (1968); P.C. Johnson (1968); R.T. Miller (1968); Naeve (1968); Pierot (1968); Rerolle (1968); Stanfill (1968); Ward and Hildrew (1968b); Bourguet, Garnaud and Grenier (1970); Gineste and Lecomte (1970); Laur (1970); Asselineau *et al.* (1972); Berge and Poll (1972); Bourguet (1972); C. Gibson (1973); Eke, Graham and Malyn (1974); Horn *et al.* (1974); Jenkins, Frieseman and Prew (1974); Purvin, Withington and Smith (1974); Anon. (1976c,e); Dolle and Gilbourne (1976); Daniels and Anderson (1977); Khenat and Hasni (1977); Ploum (1977); Seurath, Hostache and Gros (1978); Anspach, Baseler and Glasfeld (1979); Kime, Boylston and van Dyke (1980); Rust and Gratton (1980); K.W. Edwards *et al.* (1982); Zermati (1982); Chauvin and Bonjour (1985); McKinney and Oerlemans (1985); Sellers, Luck and Pantony (1985); de Sola (1985); Branchereau and Bonjour (1986); Colonna, Lecomte and Caudron (1986); Craker, Scott and Dutton (1986); Leray, Petit and Paradowski (1986); White-Stevens and Elliott (1986); Benazzouz and Albou (1987); Tarakad, Durr and Hunt (1987); Vik and Kjersem (1987); Bendani (1989); Dassonville and Lechat (1989)

Safety and environmental aspects

Copp (1971); Kober and Martin (1972); Mansillon (1972); Napier (1972); J. Davis (1973); P.J. Anderson and Bodle (1974); FPC (1974); Stone, Hills and Needels (1974); Walters, Dean and Carne (1974); F.H. Warren *et al.* (1974); California State Legislature (1976, 1977); von Ludwig (1975); F.W. Murray, Jaquette and King (1976); FERC (1977); Comptroller General (1978); General Accounting Office (1978); US Senate (1978); L.N. Davis (1979); Ahern (1980); R.A. Cox *et al.* (1980); McGuire (1980); Rust and Gratton (1980); Atallah (1982, 1983); Kotcharian and Simon (1982); Navaz (1987)

Leaks and spillages

R.O. Parker and Spata (1968); Humbert-Basset and Montet (1972); AGA (1974); Battelle Columbus Laboratories (1974); Fay and Lewis (1975); H.H. West, Brown and Welker (1975); Neff, Meroney and Cermak (1976); R.A.Cox and Roe (1977); HSE (1978b, 1981a)
Marine spillages: Burgess, Biordi and Murphy (1970 BM S4105); Burgess, Murphy and Zabetakis (1970 BM RI 7448); Enger and Hartman (1972a,b); Kneebone and Prew (1974); Raj and Kalelkar (1974); Eisenberg, Lynch and Breeding (1975); Germeles and Drake (1975);

Havens (1977, 1978); US Congress, OTA (1977); HSE (1978b, 1981a)

Fires (*see* Tables 16.1 and 16.2)

Fire protection (*see also* Tables 16.2 and 22.1)
van Dyke and Kawaller (1980); Rudnicki and vander Wall (1980); H.H. West, Pfenning and Brown (1980); Sonley (1982); S. Stephenson and Coward (1987); B.M. Lee (1989); NFPA (1990 NFPA 59A).
Inerting: Johannessen (1987); Oellrich (1987); Tepper (1987)

Hazard assessment and control
F.H. Atkinson (1970); Nassikas (1970); Brubaker, Koerner and Mathura (1972); Burgess, Biordi and Murphy (1972); Crouch and Hillyer (1972); Mansillon (1972); Fay (1973); McKinley (1973); R.C. Hill (1974); Horner (1974); Horner and Ecosystems Inc. (1974); Raj and Kalelkar (1974); SAI (1974); Allan, Brown and Athens (1975); Eisenberg, Lynch and Breeding. (1975); Fedor, Parsons and de Coutinho (1975); Gratt and McGrath (1975a,b, 1976); DOT, CG (1976, 1977b); van Horn and Wilson (1976); California Assembly (1977); California Energy Resources Conservation and Development Communication (1977); R.A.Cox and Roe (1977); DOT, OPSO (1977a); Fairley (1977); FPC, Bureau of Natural Gas (1977); Kopecek (1977); Mathiesen *et al.* (1977); John J. McMullen Associates Inc. (1977); Resource Planning Associates (1977); Snellink (1977); Socio-Economic Systems (1977); US Congress, OTA (1977); Wesson and Associates Inc. (1977); BGC (1978); California Control Communication (1978); FERA (1978); HSE (1978b, 1981a); Nikodem (1978, 1980); Philipson (1978a,b, 1980); Rigard and Vadot (1979); R.A. Cox *et al.* (1980); Feely *et al.* (1980); Lautkaski and Fieandt (1980); Summer *et al.* (1980); Kunreuther and Lathrop (1981); Lyon, Pyman and Slater (1982); Roopchand (1983); Solberg and Skramstad (1982); Valckenauers (1983); Wicks (1983); Dale and Croce (1985); D.A. Jones (1985); Sellers, Luck and Pantony (1985); Valk and Sylvester-Evans (1985); Sellers and Luck (1986); Schrader and Mowinckel (1986); Zehri *et al.* (1986); Navaz (1987)

installations is very small, but the quantities stored are large. There is sophisticated management of the hazard.

LNG is the subject of NFPA 59A: 1990 *Production, Storage and Handling of Liquefied Natural Gas and of LNG Materials and Fluids* by the National Bureau of Standards (NBS, 1978), *Liquefied Energy Gases Safety* by the General Accounting Office (GAO,(1978), *Liquefied Natural Gas: Safety, Siting and Policy Concerns* (US Senate, 1978) and *Liquefied Gas Handling Principles on Ships and in Terminals* (McGuire and White, 1986). Further accounts of LNG are given by Boesinger, Nielsen and Albright (1980) and Hammer *et al.* (1991). Selected references on LPG and on LNG and their hazards are given in Tables 11.22 and 11.23, respectively.

LPG and LNG installations present the hazards of fire and of explosion. In view of the scale involved, the siting of a large LNG installation is a major planning matter, which is likely to involve a public inquiry.

LNG is shipped to the storage terminals by specially designed vessels. The shipping of LNG involves the hazard that LNG may be spilled on the sea, may evaporate rapidly and may give a vapour cloud fire or explosion. This hazard has been a matter of much concern and has been the subject of considerable investigation. It is considered further in Chapters 16 and 17.

Storages of LPG and LNG are among the installations which have been of particular concern in relation to interactions between hazardous plants. Thus, for example, LPG storages, if poorly sited, are a possible threat to nuclear reactors and LNG storages may be hazarded by other plants. The British Gas LNG terminal at Canvey was one of the installations investigated in the *Canvey Report*, which is described in Appendix 7.

Closely related are synthetic fuels, or synfuels, including synthetic natural gas (SNG). An account is given in *Handbook of Synfuels Technology* (Meyers, 1984).

11.19.7 Petroleum refineries and gas processing plants

Petroleum refining and gas processing involve the handling of hydrocarbons in very large quantities. This industry has faced many of the classic problems in SLP and has contributed much to their solutions.

Accounts of petroleum refining include *The Petroleum Handbook* (Shell International Petroleum Company, 1933–), *Modern Petroleum Technology* (Hobson and Pohl, 1975), *Our Industry Petroleum* (British Petroleum Company, 1977), *Petroleum Refining Technology and Economics* (Gary and Handwerk, 1984) and *Handbook of Petroleum Refining Processes* (Meyers, 1986b).

Gas processing is treated in *Gas Encyclopaedia* (L'Air Liquide, 1976), *Handbook of Industrial Gas Utilization* (Pritchard, Guy and Connor, 1978) and *Natural Gas* (Melvin, 1988).

Reference to petroleum refining and gas processing is made throughout this book

11.20 Operational Deviations

An important aspect of process design is consideration of possible deviations of operating parameters from their design values. Some of these deviations are listed in Table 11.24. A further list of deviations is given in the checklist for hazard and operability (hazop) studies shown in Figure 8.24.

The causes of deviation of a process variable are mostly specific to that variable, but there are some general causes of deviation. One is the deterioration or failure of equipment. Another is the maloperation of the control system, including in this the process operator.

The effects of deviations and the measures which may be taken to prevent them are also specific. But there are certain devices which are generally provided to give protection against extreme deviations, particularly of pressure and temperature. These are considered in the discussion of pressure systems and of control systems in Chapters 12 and 13, respectively. A further discussion of operating deviations is given by Wells, Seagrave and Whiteway (1976).

When a plant has been designed, it should be checked by some method such as a hazop study, as described in

***Table 11.24** Some deviations of operating parameters from design conditions (after Wells, Seagrave and Whiteway, 1976) (Courtesy of the Institution of Chemical Engineers)*

Process variables	Pressure, temperature, flow, level, concentration
Pressure system	Mechanical stress, loading, expansion, contraction, cycling effects, vibration, cavitation, resonance, hammer; corrosion, erosion, fouling
Chemical reactions	Reactions in reactors: nature and rate of main reactions and side reactions Catalyst behaviour: reaction, regeneration, poisoning, fouling, disintegration Unintended reactions elsewhere: explosion, heating, polymerization, corrosion
Material characteristics	Vapour density; liquid density, viscosity; melting point, boiling point; latent heat; phase change; critical point effects; solids physical state, particle size, water content
Impurities	Contaminants; corrosion products; air; water
Localized effects	Mixing effects, maldistribution; adhesion, separation, vapour lock, surging, siphoning, vortex generation, sedimentation, fouling, blockage, hot spots
Time aspects	Contact time, control lags, sequential order
Process disturbances	Operating point changes, changes in linked plants, start-up, shut-down, utilities failure, equipment failure, control disturbance, operator disturbance, blockage, leakage, climatic effect, fire
Constructional defects	Plant not complete, not aligned, not level, not supported, not clean, not leak-tight; materials of construction incorrect or defective
Loss of containment	Leakage, spillage

Chapter 8. Again this technique is based on consideration of operational deviations.

11.20.1 Pressure deviations

Pressure deviations can occur as a result of: changes in action of pumping equipment, e.g. failure of a pump or compressor; changes in flow, e.g. closure of a valve or pump outlet; changes of heat input, e.g. heat from sun or fire; changes in heat output, e.g. loss of condenser cooling; contacting of materials, e.g. hot oil and water; thermal expansion and contraction, e.g. liquid density changes in a pipeline; and chemical reaction and explosion, e.g. runaway reaction.

Effects of pressure deviations include overpressure and underpressure of equipment and changes of temperature, flow and level. Measures should be taken as appropriate to reduce the pressure deviations. In addition, it is essential to provide protective devices which prevent overpressure and underpressure.

11.20.2 Temperature deviations

Some causes of temperature deviations are: changes in heat input, e.g. loss of fuel; changes in heat output, e.g. loss of cooling; changes in heat transfer, e.g. fouling of heat exchangers; generation of heat, e.g. runaway reaction; changes of flow, e.g. flow to a reactor; changes of pressure, e.g. pressure reduction causing liquid flashing; thermal lags, e.g. lags in heat exchangers; hot spots, e.g. machinery in distress.

Temperature deviations in reactors can occur as a result of errors in the charging sequence, delays in initiating agitation, feed flow or temperature variations, maldistribution of reactants, development of catalyst hot spots, loss of cooling and fouling of heat transfer surfaces.

Effects of temperature deviations include overtemperature and undertemperature of equipment, changes of pressure and flow, and runaway reactions. The appropriate measures should be taken to reduce the deviations. In particular, attention should be paid to the heat input and output from the plant and to the control of reactors.

11.20.3 Flow deviations

Flow deviations may be too high, too low, zero, reverse or fluctuating. Changes in flow are generally the result of pressure changes and should be considered in conjunction with these.

Some causes of no flow are lack of feed material, lack of pressure difference, vapour effects and equipment failures. Vapour effects which can cause loss of flow include vapour locks and gassing up of pumps. Equipment failures include the stoppage of pumping equipment, the fracture of equipment resulting in leakage, and the blockage of equipment.

Blockage may occur in a number of ways. Items may be left in the plant after construction and maintenance work. Deposits may build up from liquid impurities, from solid particles, from corrosion or erosion products. The fluid may polymerize or solidify. Cessation of flow for other reasons can cause solidification or polymerization in a heat exchanger. Low ambient temperatures increase the risk of freezing, particularly of liquids which are solids under ambient conditions.

Equipment which is prone to blockage includes pipes, heat exchangers, packed beds and filters. Control valves may block or jam shut.

Reverse flow occurs if there is a reversal of the pressure differential. It should therefore be considered in conjunction with pressure changes. Failure of pumping equipment is a main cause of reverse flow.

Flow deviation can cause deviations of pressure, temperature and level. It can result in materials entering the wrong parts of the plant, can cause erosion, cavitation and hammer blow, and can result in pump overload and trip.

Complete stoppage of flow may cause disturbances to other parts of the plant. Restart after temporary interruption of flow can be hazardous also, as in the case of the extinction of a flame followed by a resumption of flow of unignited fuel.

The manipulation of fluid flows is the principal means by which control is exercised. There are many hazards which can arise if there is a loss of control. A particularly serious one occurs if it is no longer possible to effect necessary heat removal. Low flow may be as hazardous as zero flow, if the deviation is not sufficient to activate the protective devices.

Leakage flow may cause contamination between process streams or emission to the atmosphere. Contamination may result in blockages, corrosion, phase changes, chemical reaction and explosion. Water is the most common substance used in process plants and can give rise to all of these effects. It may enter other streams through pinholes or other tube failures, or through leaking shut-off valves.

The measures required to reduce flow deviations depend on the specific cause, but include attention to the reliability of pumping equipment, the provision of adequate storage and the elimination of blockages and of phenomena which have a similar effect.

Arrangements should be made for the easy removal of blockages at points where these are reasonably foreseeable. Provision should be made for cleaning pipes and heat exchangers by such methods as rodding out, chemical cleaning and burning of carbon deposits. Surfaces of packed beds may be protected from blinding by putting a filter layer such as ceramic packing on top. It may be necessary to install bypasses on equipments which are liable to blockage, but these may introduce their own hazards, which should be considered. Where fluids may freeze, it may be necessary to provide trace steam or electrical heating to prevent this.

11.20.4 Level deviations

Levels may be too high, too low, zero or fluctuating. Changes in level result from flow changes and should be considered in conjunction with these. Level deviations often occur as a result of changes in the flow in or out of a vessel. These may be controlled changes which result in overfilling or emptying of the vessel, or they may be caused by failure of equipment, such as the off-take pump, by vapour effects such as gassing up of that pump, or by blockages.

Changes in the pressure above the liquid, such as loss of vacuum, also affect the level. Some of the other causes of level deviations are surging, foaming and thermal expansion, siphoning and blown lutes. Excessive agitation or vibration of the vessel can also cause level deviations.

The measurement of liquid level is often difficult if there are solids, two liquid phases or foam, or if inerts, water or sediment accumulate. If the level is too high, liquid may overflow, the vessel may be overpressured or the weight of the system may become excessive. Where there is a gas flow through the vessel, a high level may cause liquid entrainment. If the level is zero, the outlet liquid flow is lost. Loss of liquid level can also result in breakthrough of high pressure gas to a part of the plant which is not designed for it. Measures to reduce level deviations depend on the specific case. In particular, attention should be given to the problems of level measurement.

It is not always the aim to eliminate level deviations completely. For some vessels, such as a chemical reactor, the level should normally be held constant, but other vessels are specifically used as surge vessels to smooth fluctuations of flow between units. Such vessels can fulfil their function only if the level in them does vary. They are usually provided with a low gain control loop which permits this.

In the case of level, control of the upper level limit may be achieved without resort to automatic controls by the simple and reliable device of an overflow pipe, and this should always be considered.

In addition to the levels in the main vessels, those in other equipments should be considered. These include levels in heat exchangers and on distillation trays.

11.20.5 Inhomogeneities and accumulations

There are many ways in which inhomogeneities of conditions and accumulations of material can arise. Poor mixing in reactors or other equipment can result in: side reactions and reaction runaway; hot spots, overheating and thermal degradation; and fouling. The accumulation of substances which are relatively innocuous at high dilution but which when concentrated in some way, such as by vaporization with an insufficient bleed-off or by recycling of material, may present a hazard. Material which is held up in the system in a dead leg or elsewhere, for an abnormally long time may undergo degradation and become hazardous.

11.20.6 Control disturbances

The control system, including the process operator, may be a cause of operating deviations. One cause of this is errors, faults or lags in measurement. A level measurement based on liquid head may be in error due to frothing. The reading of a flow measuring instrument may be plausible, but the instrument may nevertheless have failed. There may be appreciable lags in the measurement of temperature.

Deviations may also be caused by the action of the automatic control system. The unsteady-state characteristics of the process such as the nature of the disturbances and of the time lags, the presence of dead time and loop interactions, and a requirement to operate at different throughputs, may make control inherently difficult. As a result, controller settings are sometimes inadequate. Other deviations may be introduced by the actions of the process operator, either through the control system or in other ways.

It is important for the process design to provide sufficient potential correction for control. If a variable is to be controlled, then there must be other variables which can be manipulated in order to achieve that control. This may seem obvious, but it is often not appreciated in practice. If, for example, a cement kiln is operated at maximum air flow, it is not possible to have close control of the outlet gas oxygen content by means of the air flow, since this requires that it be possible both to increase and decrease that flow.

11.21 Impurities

In the foregoing account frequent reference has been made to impurities. These merit some attention in their own right. The problem has been considered by Grollier-Baron (1992a,b), who gives a number of examples of incidents due to impurities, involving mainly explosion or corrosion. The use of nitrogen from ammonia cracking to inert a storage tank containing ethylene oxide led to decomposition of the latter due to traces of ammonia in the nitrogen. The presence of traces of acetylene in air passing over bronze fittings resulted in the formation of cuprous acetylide which then exploded. In a storage facility, replacement of demineralized water used in a heat exchanger for cooling acrolein with water from another source combined with a small leak in the exchanger allowed mineral ions to enter the acrolein, which polymerized exothermically so that the tank exploded. In catalytic crackers and methane reformers unsaturated hydrocarbons and NO_x are produced and when cooled to low temperature can form compounds, some of which are explosive. Mercury present in some natural gas condenses in the aluminium heat exchangers of the liquefaction units and forms an amalgam which corrodes the exchangers. Explosions can occur due to the accumulation of nitrogen trichloride in chlorine vaporizers.

Grollier-Baron outlines a formal approach to dealing with the problem of impurities based essentially on the following measures: (1) identification of the species present in trace quantities, (2) review of the fate of these species, (3) review of the hazards posed by the species, (4) formulation of corrective measures, and (5) review to ensure that these countermeasures do not themselves create new hazards.

Identification involves consideration of the trace species in the raw materials, those created in the process and those which could be introduced accidentally. It also requires a review of the sensitivity of the raw materials, intermediates and products to these species. It cannot be assumed that the composition of the raw materials will necessarily remain constant with respect to trace species. There are many factors which may give rise to a change in composition of a raw material, ranging from minor changes in natural or quarried products or in inputs or process at the supplier plant to a complete change of supplier or of process.

There are a number of ways in which impurities may be formed in the process, including by side reactions and by corrosion. Impurities may also be introduced accidentally. One significant source is leaks. Another is deposits left behind in equipment or following maintenance.

It is desirable to have a material balance on, and to know the fate of, any impurities which may be significant. The hazard caused by impurities is increased by accumulation. This accumulation occurs because the impurity has some property different from that of the fluid in which it is present. Hydrogen from corrosion processes is light and may collect at the top of a tank or in a vent system. Nitrogen trichloride is less volatile than chlorine and can build up in a chlorine vaporizer. A difference in relative volatility can cause a compound to build up to quite a high concentration on the tray of a distillation column. Accumulation may also occur due to special circumstances. Cold weather can cause trace quantities to freeze out and cause blockages or worse.

Further treatment of contaminants is given by Tagoe and Ramharry (1993).

11.22 CCPS *Engineering Design Guidelines*

11.22.1 Guidelines for Engineering Design for Process Safety

Process design is one of the principal topics covered in the *Guidelines for Engineering Design for Process Safety* (CCPS, 1993/13) (the CCPS *Engineering Design Guidelines*). The scope of the *Engineering Design Guidelines* is, however, much broader. Table 11.25 gives the principal topics covered together with the location in this book where they are treated.

As far as process design is concerned, the *Guidelines* deal in particular with (1) inherently safer design, (2) process equipment (3) utilities, (4) heat transfer fluids, (5) effluent disposal and (6) documentation. The treatment of inherently safer design is organized around the themes of (1) intensification, (2) substitution, (3) attenuation, (4) limitation of effects and (5) simplification and error tolerance, and includes an inherent safety

Table 11.25 *Some principal topics treated in the CCPS* Engineering Design Guidelines *and their treatment in this text*

	Chapter[a]
1. Overview	–
2. Inherently safer plant	11
3. Plant design	10, 11
4. Equipment design	11, 12, 22
5. Material selection	12
6. Piping systems	12
7. Heat transfer fluid systems	11
8. Thermal insulation	12
9. Process monitoring and control	13, 14
10. Documentation	6, 11
11. Sources of ignition	10, 16
12. Electrical system hazards	10, 11, 16, 25
13. Deflagration and detonation flame arresters	17
14. Pressure relief systems	12, 15, 17
15. Effluent disposal systems	10, 11, 17
16. Fire protection	10, 16
17. Explosion protection	17

[a] Chapter in this book in which this topic is principally addressed.

checklist. The general approach is broadly similar to that taken in Section 11.7.

The items of process equipment treated in the *Guidelines* are (1) chemical reactors, (2) columns, (3) heat exchangers, (4) furnaces and boilers, (5) filters, (6) centrifuges, (7) process vessels, (8) gas/liquid separators, (9) driers, (10) solids handling equipment, (11) pumps and compressors, (12) vacuum equipment and (13) activated carbon adsorbers.

The *Guidelines* give a table showing the common causes of loss of containment for different types of process equipment. They also include checklists for safe operation of fired equipment and activated carbon adsorbers. The treatment covers all the main utilities, with particular attention to reliability of electrical power supplies, and includes a checklist of possible utility failures and equipment liable to be affected.

The account of heat transfer fluid systems describes the fluids available and their applications, discusses the relative merits of steam, liquid and vapour–liquid systems, outlines the system design considerations and describes the system components, and considers the safety issues.

With regard to effluent disposal, the systems considered are (1) flares, (2) blowdown systems, (3) incineration systems and (4) vapour control systems. The treatment does not extend to disposal of materials discharged in reactor venting.

The section on documentation covers (1) design documentation, (2) operations documentation, (3) maintenance documentation and (4) record keeping.

11.23 Notation

Section 11.4

E	maximum work, or exergy (W)
Q_T	heat flow (W)
T	temperature (K)
T_c	temperature of cold sink (K)
T_h	temperature of heat source (K)
T_0	temperature of surroundings (K)
W_{max}	maximum work (W)

Section 11.7

d	pipe diameter (cm)
W	throughput (te/year)
ϕ	ratio defined by Equation 11.7.1

Section 11.8

A	heat transfer area (m^2)
c_p	molar specific heat of feed and of reaction mass (kJ/kmol °C)
E	activation energy (kJ/kmol)
F	volumetric flow of feed (m^3/s)
$G(s)$	open-loop transfer function
ΔH	heat of reaction (negative for exothermic reaction) (kJ/kmol)
k	velocity constant at reaction temperature (s^{-1})
k_0	velocity constant at reference temperature (s^{-1})
k_1	constant
K	open-loop gain
K_c	controller gain
K_p	plant gain
q	variable defined by Equation 11.8.37
Q_g	heat generated in reactor (kW)
Q_r	heat removed from reactor (kW)
Q_s	heat generated by reaction (kW)
r	reaction rate (kmol/m^3 s)
R	universal gas constant (kJ/kmol K)
s	Laplace operator
t	time (s)
T	absolute temperature (K)
T_c	absolute temperature of coolant (K)
ΔT_c	critical temperature difference (°C)
T_0	absolute temperature of feed (K)
U	overall heat transfer coefficient (kW/m^2 °C)
V	volume of reactor (m^3)
x	mole fraction of reactant in reactor
x_0	mole fraction of reactant in feed
η	normalized temperature ($= c_p T/x_0(-\Delta H)$)
θ	temperature deviation (°C)
θ_c	temperature deviation of coolant (°C)
ξ	normalized concentration ($= x/x_0$)
ρ	molar density of feed and of reaction mass (kmol/m^3)
τ	volume-throughput ratio of reactor
τ^*	normalized time ($= Ft/v$)
τ_p	plant time constant (s)
ϕ	degree of conversion

Subscripts:

av	average
c	coolant
o	feed
ss	steady state
ϕ	at conversion ϕ

Superscipt:

–	Laplace transform

Section 11.10

A_c	surface area for heat transfer
E	activation energy
k	reaction velocity constant
R	universal gas constant
t	time
T	absolute temperature
U	overall heat transfer coefficient

Subsection 11.10.15

A	pre-exponential factor
c_v	specific heat of reaction mixture
c_{vb}	specific heat of sample container
C	concentration of reactant
C_0	initial concentration of reactant
ΔH_r	heat of reaction
j	integer
k^*	pseudo-reaction velocity constant
k_0	reaction velocity constant at initial temperature T_0
m	self-heat rate
m_m	self-heat rate at temperature T_m
m_0	self-heat rate at initial temperature T_0
m_T	self-heat rate at temperature T
M	mass of reaction mixture
M_b	mass of sample container

n order of reaction
Q_e rate of heat evolution
Q_r rate of heat removal
t time
t_m time for temperature to reach maximum rate value T_m
T^* minimum initial absolute temperature above which only deceleration of reaction occurs
ΔT_{ab} adiabatic temperature rise
T_c absolute temperature of coolant
T_f final absolute temperature
T_m absolute temperature at maximum rate
T_{nr} absolute temperature of no return
T_0 initial absolute temperature
V volume of reaction mass

α constant
θ_m time to maximum rate at temperature T
θ_{m0} time to maximum rate at temperature T_0
θ_{Tnr} time to maximum rate at temperature T_{nr}
ϕ thermal inertia

Subscript:
s value measured for sample

Superscript
$'$ critical value

Subsections 11.10.16-11.10.17
S surface area of reaction mass
T_d minimum absolute decomposition temperature
V volume of reaction mass
τ_{ad} adiabatic induction time

Section 11.11
B thermal reaction parameter

c concentration
c_0 initial concentration of limiting minor component
c_p specific heat of reaction mixture
e base of natural logarithms
n effective order of reaction
ΔT_{ad} adiabatic temperature rise
T_c absolute temperature of coolant
T_f absolute temperature of feed
r rate of reaction
V volume of reaction mixture
V_D volume of addition
X degree of conversion
θ dimensionless temperature difference
κ cooling parameter
ν_u stoichiometric coefficient of limiting minor component
ρ density of reaction mixture
τ_{ad} adiabatic induction time
τ_D time of addition

Subscripts:
A, B reactant A, B
c coolant
D addition
f feed
0 initial value
ref reference

Subsection 11.11.5
B reaction parameter
T_{max} maximum absolute temperature (K)
T_r absolute temperature of reactor (K)
T_s absolute temperature at which secondary reaction occurs (K)
ΔT_{ad} adiabatic temperature rise °C

12 Pressure System Design

Process materials are normally contained within a pressure system. The main problem in loss prevention is the avoidance of loss of containment from this system. Thus the *First Report* of the ACMH (Harvey, 1976) states:

> Containment is the very essence of the problem of control of dangerous materials, and therefore we regard the integrity of pressure systems as of the highest importance. (paragraph 64)

The report continues:

> Scrutiny of incidents suggests that the outright failure of properly designed, constructed, operated and maintained pressure vessels is rare, perhaps because of the lessons learned over many years from steam boiler practice. It is pipework, valves, pumps, etc., which are vulnerable and much more prone to failure. (paragraph 65)

Some of the principal features of pressure systems are briefly discussed in this chapter. It cannot be too strongly emphasized, however, that the discussion is limited to the background information necessary for the appreciation of the problems which occur in loss prevention and that many of the topics touched on, such as materials of construction, pressure vessels, piping, process machinery and overpressure protection, are complex matters which require specialist knowledge.

The operation of pressure systems is at least as important as their design. This is considered in Chapter 6 on management systems and in Chapter 20 on plant operation. Selected references on pressure systems and components are given in Table 12.1 and on materials of construction for and corrosion in pressure systems in Table 12.2.

***Table 12.1** Selected references on pressure systems and components*

ASME (Appendix 27, *General and Safety Standards*, 1987– *Code for Pressure Piping B31*, 1992 *Boiler and Pressure Vessel Code*; Appendix 28 *Pressure Vessels and Piping*); Associated Offices Technical Committee (n.d.); British Gas (Appendix 27 *Pressure Vessels*); IMechE (Appendix 28, 1970/3, 1975/ 16, 18, 1976/27, 1977/33, 1979/50, 1981/57, 1991/130, 1993/148); Lloyds Register of Shipping (n.d.); NRC (Appendix 28 *Containment, Equipment Qualification, Pressure Systems, Pressure Vessels*); NSC (Safe Practice Pmphlt 68); Welding Inst. (Appendix 28, 1972/20); den Hartog (1952, 1956); Chuse (1954–); D.A.R. Clark (1956); Timoshenko and Young (1956); Manning (1957, 1960, 1978); Arnold (1959); Brownell and Young (1959); AIChE (1960–69/3–12, 1967/71, 1970–94/17–38, 1973/62, 1974/13, 1978/14); MacCary (1960a,b); Hughson (1961d, 1969); Begg (1963–64); Maynard (1963); Faupel (1964); Jenett (1964b); Voelker (1964); Blick (1965); IP (1980 Eur. MCSP Pt 2, 1981 MCSP Pt 3, 1987 MCSP Pt 7, 1993 MCSP Pt 13); Johns (1965); MacDermod (1965, 1982); Norden (1965); Thielsch (1965); Kemper *et al.* (1966–67); API (1967– Refinery Inspection Guide, 1992 RP 510, Publ. 910); Bickell and Ruiz (1967); Canavan (1967); Canham (1967); Spence and Carlson (1967–68); Titze (1967); Eschenbrenner, Honigsberger and Impagliazzo (1968); Strelzoff and Pan (1968); Strelzoff, Pan and Miller (1968); T.E. Taylor (1968–69); Witkin (1968); Anon. (1969f); Fowler (1969); McCabe and Hickey (1969); McLeod (1969); Nichols (1969, 1976b, 1979a,b, 1980, 1983, 1987); Warwick (1969); J.S. Clarke (1970); Le Coff (1970); Franzel (1970); Gill (1970); Hearfield (1970); Kemp (1970); Losasso (1970); MacFarland (1970, 1974); Pugh (1970); Dall'Ora (1971); F.L. Evans (1971); Markovitz (1971, 1977); J.R. Palmer (1971); Pilborough (1971, 1989); Toogood (1972); Whenray (1972); Ford *et al.* (1973); Guill (1973); Jaeger (1973, 1975); Karl (1973); Timoshenko and Gere (1973); Dimoplon (1974); L. Evans (1974); Fitzpatrick (1974); Harvey (1974, 1980); M.R. Johnson *et al.* (1974); Stokes, Holly and Mayer (1974); Anon. (1975c,g,l); Berglund (1975, 1978); B.G.Cox and Saville (1975); ICI/RoSPA (1975 IS/107); Koike (1975a–c, 1978a–c); Mazzoncini (1975, 1978); Roark and Young (1975); van Rossen (1975); Steffen (1975, 1978); Anon. (1976j); Bhattacharyya (1976); Dickenson (1976); Logan (1976); Warburton (1976); Bacon and Stephen (1977); Gilbert and Eagle (1977); Heyman (1977); HSC (1977/1); Mahajan (1977b); Pludek (1977); Shigley (1977); Anon. (1978c); Baumeister (1978); Chambard (1978); Megyesy (1978, 1983); Strawson (1978); Witkin and Mraz (1978); Blevins (1979); Heinze (1979); Koenig (1979); Mraz and Nisbett (1979); Puzak and Loss (1979); Unrug (1979); Vreedenburgh (1979); Widera and Logan (1979); D.J.D. White and Wells (1979); Buhrow (1980); Facer (1980); Gerlach (1980); HSE (1980 EM 7); Kletz (1980i, 1984i,k); Wells and White (1980); Zeis and Eschenbrenner (1980); Azbel and Chereminisoff (1981); Burr (1981); R. Cook and Guha (1981); Moy (1981); Collier, Davies and Garne (1982); W. Marshall *et al.* (1982); Polak (1982); Smolen and Mase (1982); IAEA (1983); Rosaler and Rice (1983); Wicks (1983 LPB 53); Boyer (1984); Chuse and Eber (1984); Hurst (1984); Jawad and Farr (1984); Murray (1984); Arnold, Mueller and Ross (1986); Bednar (1986); Escoe (1986, 1992); J.P. Gupta (1986); Kutz (1986); Rogerson (1986); Yokell (1986); S.J. Brown (1987); Grosshandler (1987a,b); Kirkpatrick (1987); Kohan (1987); D.R. Moss (1987); Prugh (1987b); Sandler and Luckiewicz (1987); Coleman (1989 LPB 85); Trbojevic and Gjerstad (1989); K.P. Singh (1990); Factory Mutual Int. (1991b); LPGITA (1991 LPG Code 1 Pt 1); Snow (1991); EEMUA (1992 Publ. 162); HSE (1992 GS 4); Kassatly (1992); Batra *et al.* (1993); Chuse and Carson (1993); E.H. Smith (1993); Norton, Pilkington and Carr (1994); Pilkington, Platt and Norton (1994); Spence and Tooth (1994)

ANSI B series, BS (Appendix 27 *Pressure and Other Vessels*, PD series), BS 5500: 1991, VDI 2224: 1988

Prestressed concrete pressure vessels: IOCI (1968)

Composites, FRP

ASME (PVP 115, 1977/131, 1987 PVP 121, 1990 PVP 196, 1992 Boiler and Pressure Vessel Code Pt X, 1992 AMD 150); ASTM (1975 580, 1976 503, 1986 D4012, 1988 D4097, 1991 D3517); Puckett (1976); IMechE (1977/35, 1984/77, 1990/116, 118); I.R. Miller (1979); Fasano and Eberhart (1980); IChemE (1980/119); Weatherhead (1980); Anon. (1986 LPB 70, p. 7); BG (1986 BGC/PS/ PL2); Institute of Materials (1986 B366); Maddison (1987 LPB 76); HSE (1991 PM 75); Britt (1993)

Foundations, structures
Brownell (1963); Deghetto and Long (1966); Pridgen and Garcia (1967); Tang (1968); A.A. Brown (1969, 1971, 1973, 1974); Czerniak (1969); G.B. Moody (1969, 1972); Youness (1970); Eichmann (1971); Molnar (1971); Arya, Drewer and Pincus (1975, 1977); Mahajan (1975, 1977a); Raynesford (1975); K.P. Singh (1976); Simiu and Scanlon (1978); Faber and Alsop (1979); Faber and Johnson (1979); Moy (1981); Cowan (1982); Mosley and Bungey (1982); M. Schwartz (1982a–c, 1983a–e, 1984); Kong *et al.* (1983); HSE (1991 GS 49)
BS (Appendix 27 *Civil Engineering*), BS CP 2012: 1974–, BS 8004: 1986

Boilers
American Oil Co. (n.d./10); Institute of Fuel (1963); Csathy (1967); FMEC (1967); van Loosen (1968); Attebery (1970); Impagliazzo and Murphy (1970); OIA (1971 Publ. 502); Horsler and Lucas (1972); API (1974 Refinery Inspection Guide Ch. 8); Vodsedalek and Bielak (1977); BRE (1978 CP59/78); Wilcox (1978); Needle (1979); A. Gibson (1980); W.S. Robertson (1981); Andrade, Gates and McCarthy (1983); Wilcox and Baker (1986); Harmsworth (1987); HSE (1987 PM 60, 1989 PM 5); Kauffman (1987); Hewett (1988); Kingshott (1988); Prescott, Podhorsky and Blommaert (1988); Gunn and Horton (1989); NFPA (1989 NFPA 85H); Kakac (1991); Anon (1992a); ASME (1984/174, 1988/201, 1992 Boiler and Pressure Vessel Inspection Code); Colannino (1993); Ganapathy (1994)
BS (Appendix 27 *Boilers*), BS 759: 1984–, BS 1113: 1992

Waste heat boilers, process heaters
Csathy (1967, 1981); Streich and Feeley (1972); Gupton and Krisher (1973); Din (1975); T.B. Gibson (1975); Hinchley (1975, 1977, 1979a,b); W.P. Knight (1978); O'Sullivan, McChesney and Pollock (1978); Ozmore (1978); Salot (1978, 1982); Subrahmanyam, Pandian and Ganapathy (1979); Fuchs and Blanken (1986); Pariag, Welch and Kerns (1986); Sitaraman, Santoso and Sathe (1988)

Heat exchangers
ASME (Appendix 28 *Heat Transfer*, 1990 PVP 194); NRC (Appendix 28 *Steam Generators*); D.Q. Kern (1950, 1966); Drake and Carp (1960); Rubin (1960, 1961, 1980); Bohlken (1961); D.S. Morton (1962); E.M. Cook (1964); Fraas and Ozosik (1965); Gilmour (1965, 1967); P.R. Owen (1965); API (1967 Refinery Inspection Guide Ch. 7, 1982 Std 660, 1992 Std 661); R.B. Moore (1967); Y.N. Chen (1968); Gainsboro (1968); Hargis (1968); Small (1968); A.A. Thompson (1969); Lord, Minton and Slusser (1970a,b); Thongren (1970); EEMUA (1971 Publ. 135); Stuhlbarg (1971); D.Q. Kern and Kraus (1972); Knulle (1972); Simpson (1972); Taborek *et al.* (1972a,b); Barrington (1973, 1978); Doyle and Benkly (1973); Eilers and Small (1973); J.A. Moore (1973); Song and Unruh (1974); P.M.M. Brown and France (1975); Char (1975b); Grossman (1975); Osman (1975); Piehl (1975); Anon. (1976 LPB 10, p. 1); Fanaritis and Bevevino (1976); G.W. Schwarz (1976); Blevins (1977); Cabrini and Cusmai (1977); R. Brown (1978); Ganapathy (1978, 1992, 1994); Glass (1978); Sueyama and Takami (1978); Rubin and Gainsboro (1979); Standiford (1979); Subrahmanyam *et al.* (1979); Needle (1979); Triggs (1979); W.J. Baker (1980); Butterworth (1980, 1992); Devore, Vago and Picozzi (1980); Greene (1980); Gutterman (1980); Landrum and Watson (1980); Lufti, El-Migharbil and Hasaballah (1980); Malone (1980); Saunders (1980); Scaccia and Theoclitus (1980); Vukadinovic (1980); Kakac, Bergler and Mayinger (1981); Lopinto (1982); Schlunder (1982); TEMA (1982 B78.1); Wassom (1982); K. Bell (1983); Cizmar (1983); Crane and Gregg (1983); Mehra (1983); Shipes (1983); Yokell (1983); Knudsen (1984); Anon. (1985m); Rall and Spaehn (1985); J.P. Gupta (1986); Nieh and Zengyan (1986); R.G. Thomas (1986); Crittenden, Kolaczkowski and Hout (1987); Sandler and Luckiewicz (1987); Tammani (1987); Weaver (1987); Zakauskas, Ulinskas and Katinas (1988); Fraas (1989); Bott (1990); Mir and Siddiqui (1990); Crisi (1992); Love (1992); Lian, Kawaji and Chan (1993)
BS (Appendix 27 *Heat Exchangers*), BS 3274: 1960

Pipework
ASME (Appendix 27, 1987– *Code for Pressure Piping B31*, 1992 *Boiler and Pressure Vessel Code*; Appendix 28 *Pressure Vessels and Piping*); British Cryogenics Council (n.d.); IGasE (Appendix 28, n.d./7, 1967/8, 1976/10); NRC (Appendix 28 *Pipework, Pipe Whip*); M.W. Kellogg Co. (1956); G.N. Smith (1959); Bagnard (1960); R. Kern (1960, 1966, 1969, 1971, 1972a,b, 1974–, 1975b,c); Hilker (1962); Littleton (1962); Stubenrauch (1962); Whalen (1962); EEUA (1963 Hndbk 18, 1965 Doc. 23., 1968 Hndbk 23); Rase (1963); Francois (1964); Ingels and Powers (1964); Jenett (1964c); Masek (1964b, 1968); Surdi and Romaine (1964); Chapman and Holland (1965–); W.H. Doyle (1965); Thielsch (1965); Canham (1966, 1981); Judson (1966); Kiven (1966); Lancaster and Hoyt (1966); Koch (1966); Mallinson (1966a,b); W.C. Turner (1966); Cherrington and Ciuffreda (1967); R.C. King and Cracker (1967); Avery and Valentine (1968); Prescott (1968); Simpson (1968, 1969); Ward (1968); C.E. Wright (1968); L. Wright (1968); Yoder (1968); Anon. (1969c); Meador and Shah (1969); Aslam (1970); G.B. Moody (1970); Phelps (1970); Wills (1970); F.L. Evans (1971); Simmon (1972); Benson (1973); FPA (1973/21); Guill (1973); O'Neal, Elwonger and Hughes (1973); Royalty and Woosley (1973); Holmes (1973); Styer and Weir (1973); Weaver (1973); API (1974 Refinery Inspection Guide Ch. 11); Mauck and Tomita (1974); W.W. Russell (1974); Ruziska and Worley (1974); Char (1975a, 1979); Lancaster (1976); Pothanikat (1976); Wachel and Bates (1976); D.W. Moody (1977); Surtees and Rooney (1977); Getz (1978); G.R. Kent (1978a); R. Kern (1978e); Constance (1979); J.D. Dawson (1979); Dilworth (1979); Lazzeri (1979); Marks (1979); Peng (1979); Stevens and Littlewood (1979); API (1980 Std 605, 1994 Bull. 6F2); Bedson (1980); Facer (1980); G. Montgomery (1980); Anon. (1981v); R.J. Cook (1981); D. Stephenson (1981a); Chlorine Institute (1982 Publ. 60); Hooper (1982); Kannappan (1982); Kentish (1982a,b); Rao (1982); Gardner (1983); Hills (1983); Anon. (1984n); Mikasinovic and Marcucci (1984); de Nevers (1984a); Broyles (1985); Hodge (1985); IGasE (1985 IGE/TD/12); IMechE (1985/86, 89, 1989/112, 1993/158, 159); Hansen (1986); Hanson (1986); Helguero (1986); Prescott, Blommaert and Grisalia (1986); Sötebier and Rall (1986); G.Parkinson (1987); Sandler and Luckiewicz (1987); P.R. Smith and van Laan (1987); Towndrow (1987 LPB 73); Anon. (1988p); J.K. Rogers (1988); Droste and Mallon

(1989); Geyer *et al.* (1990); Hancock (1990a,b); Coker (1991a,b); Couch (1991); Hurst *et al.* (1991); Hartman (1993); Sixsmith (1993), Sadler and Matusz (1994); ANSI B18 series; BS (Appendix 27 *Pipework*)
Reaction forces: F.J. Moody (1969, 1973); Stikvoort (1986); Hansen (1991)
Coaxial pipes, jacketed pipes: Anon. (1977 LPB 13, p. 17; 1977 LPB 18, p. 2); Stubblefield (1993)
Plastic lined pipes: J.R. Ward (1968); Chasis (1976); Spencer (1978); Anon. (1981s); Castro (1982); Carroll (1985); M.E. Jones (1990); Jeglic and Lindley (1992, 1994)
Glass-lined pipes: Cowley, Dent and Morris (1978)
Plastic pipes: British Gas (1985 Comm. 1277); AGA (1989/12)
Gaskets: Rothman (1973); Stevens-Guille and Crago (1975); ASTM (1977 620); G.R. Kent (1978b); Payne (1980); Granek and Heckenkampf (1981); Payne and Bazergui (1981); API (1982 Std 601); G. Parkinson (1986); B. Singh (1991); ASME (1992 B16.2); Childs (1992); Crowley (1993)
Bellows, expansion joints: Anon. (1975 LPB 5, p. 19); Kobatake *et al.* (1975); British Gas (1977 TIN1); Engineering Appliances Ltd (1977); ASME (1981/147, 1984/175, 1989 PVP 168); Jetta, Brown and Pamidi (1981); McCulloch (1981); P.E. Smith (1983); C. Taylor (1985); Smith (1988 LPB 83); D.J. Peterson (1991), BS 6129: 1981
Hoses: C.W. Evans (1974)
Filters: Uberoi (1992); Artus (1976); Anon. (1977 LPB 13, p. 5; 1977 LPB 15, p. 35); Carriker (1979)
Sightglasses, rotameters: Anon. (1988 LPB 80, p. 19)
Thermowells : Masek (1964a, 1972, 1978)

Pipeline transients, water hammer
NRC (Appendix 28 *Water Hammer*); ASME (1933/21, 1961/22, 1984 ASME A112.26M, 1984/183); Thomson (1951); Lewitt (1952); Lupton (1953); Parmakian (1955); Marchal and Duc (1959); Worster (1959); Hayashi and Ransford (1960); Karplus (1961); Harding (1964, 1965–66); Streeter (1964); Pearsall (1965–66); Fabic (1967); Streeter and Wylie (1967); Pickford (1969); Casto (1973); Ludwig and Ruijterman (1974); Anon. (1975 LPB 0, p. 6; 1975 LPB 1, p. 7); J.A. Fox (1977); Watters (1979); Mukaddam (1982); Collier (1983); R. King (1983); Kremers (1983); K. Austin (1984); Crawford and Santos (1986); D. Clarke (1988a,b); Swierzawski and Griffith (1990); Kletz (1994 LPB 115)

Pipe tracing, steam tracing, steam traps
Bower and Petersen (1963); Eland (1966); House (1968b); Northcroft and Barber (1968); Bertram, Desai and Interess (1972); Mathur (1973); Anon. (1974a); Monroe (1975, 1976a–c, 1985); Cronenwett (1976); Mikasinovic and Dautovich (1977); Beatty and Kruger (1978); McWhorter (1978); Kohli (1979); Anon. (1981v); Blackwell (1982a,b); Lonsdale and Mundy (1982); Russo, Haydel and Epton (1984); HSE (1988 SIR 5); Haas (1990); David (1991); Kenny (1991, 1992); Lam and Sandberg (1992); Mackay (1992); O'Dell (1992); Radle (1992); C. Butcher (1993b); Morran (1993)

Insulation
British Gas (GBE/DAT30); Hughson (1961b); EEUA (1963 Handbk 12); Lawson (1964); Mathay and MacKnight (1966); W.C. Turner (1966, 1974); Hoffman (1967); House (1968b); Isaacs (1968); Ellis (1969); Malloy (1969); Marks and Holton (1974); ASTM (1975 581, 1980 718, 1983 789, 1984 826, 1985 880); Paros (1976); M.R. Harrison and Pelanne (1977); M.R. Harrison (1979); Webber (1979); Lang, Moorhouse and Paul (1980); McChesney and McChesney (1981, 1982); W.C. Turner and Malloy (1981); Nagl (1982); Schroder (1982); Kletz (1984m); Laxton (1985); Sandler and Luckiewicz (1987); Irwin (1991a,b); McMarlin and Gerrish (1992); Reddi (1992); Sloane (1992); Britton and Clem (1991); Gamboa (1993)
BS (Appendix 27 *Insulation*)

Valves (*see also* Table 13.2)
British Gas (Appendix 27 *Valves*); NRC (Appendix 28 *Valves*); Holmberg (1960); F.C. Price (1961); Antrim (1963); Calef (1964); Liptak (1964); Burger and Hoogendam (1965); Ciancia and Steymann (1965); Holmes and Ramaswami (1966); API (1974 Refinery Inspection Guide Ch. 11, 1984 Std 526, 1988 Std 599, 1989 Spec. 6A, Std 608, 1990 Std 598, 1991 Spec. 6D, Std 600); Boger (1969); Canon (1969); Driskell (1969, 1983); Glickman and Hehn (1969); Simon and Whelan (1970); Brodgesell (1971); *Chemical Engineering* (1971); Lawson and Denkowski (1971); Lovett (1971); Templeton (1971); British Valve Manufacturers Association (1972); G.W. Brown (1974); F.L. Evans (1974b); R. Kern (1975a); Wicher (1975); Wier (1975); Anon. (1976 LPB 10, p. 3); Bertrem (1976); Hays and Berggren (1976); Pikulik (1976); Babbidge, Partridge and d'Angelo (1977); Karcher and Ball (1977); Clayton and Johnson (1978); Constance (1979); Farley (1979); IMechE (1979/51, 1980/54, 1989/114, 1993/161, 1994/168); Anon. (1980 LPB 31, p. 17); Kaplan (1981a); Whitaker (1981); D.T. Cook (1982); Anon. (1984 LPB 59, p. 28); Ball and Howarth (1984); Stacey (1984); C. Tayler (1984c); Wallbridge and Gates (1984); Chowdhury (1985a); Greene *et al.* (1985b); Merrifield (1985); Pittman (1985); ASME (1986 PVP 109, 1989 PVP 180); Bond (1986 LPB 69); Merrick (1986); Anon. (1987n); Latty (1987); Morley and Heasman (1987); Warring (1987); Anon. (1988p); Hunter (1988); T.M. Rogers (1988); Royce (1988b); Pinnington (1989); McGuinness (1990); Whitehouse (1990); Newby and Forth (1991); Hotchkiss (1991); Miles (1991); Ridey (1991); Zappe (1991); Anon. (1992 LPB 103, p. 25); Beasely (1992); Grumstrup (1992); Dana (1993); Fruci (1993); Graczyk and Hannon (1993); Kroupa (1993); T. Robinson (1993); J.B. Wright (1993); Hingoraney (1994); Peters (1994); BS (Appendix 27 *Valves*)
Pressure regulators and reducers, restriction orifices: Fadel (1987); Liptak (1987); Baumann (1992); Khandelwal (1994)
Non-return valves and devices, blackflow protection: Malleck (1969); Fitt (1974); Nicholson (1974); Kletz (1976b); Cherry (1980); Emery (1983); K. Austin (1984); Thorley (1984); Fluid Controls Institute (1985); Tomfohrde (1985); Anon. (1986 LPB 87, p. 1); Ellis and Mualla (1986); Anon. (1987b; 1988 LPB 83, p.9); Zappe (1991); Englund, Mallory and Grinwis (1992); Anon. (1993 LPB 111, p. 25)
Emergency isolation valves: Roney and Acree (1973); Kletz (1975b); Tomfohrde (1985); C. Butcher (1991b)

Excess flow valves: Anon. (1977 LPB 14, p. 15; 1977 LPB 18, p. 22); R.A. Freeman and Shaw (1988)
Fire safe valves: J.B. Wright (1981); Cory and Riccioli (1985); API (1993 Std 589, 607, 1994 Spec. 6FA)
Fluidic valves: J. Grant and Marshall (1976, 1977)

Lutes, seals
Anon. (1989 LPB 87, p. 1; 1992 LPB 104, p. 27)

Joining
Manning (1960); Hamm (1964); Matley (1965); Graves (1966); Canham and Hagerman (1970); Isaacs and Setterlund (1971); Vossbrinck (1973); Briscoe (1976); Cloudt (1978); Girin (1978); Pengelly (1978); Strawson (1978); ASME (1979– Code for Pressure Piping, 1988 B16.6, 1989 PVP 158); Stippick (1979); M.G. Murray (1980b); Anon. (1981 LPB 38, p. 23); Institute of Materials (1981 B349); Zelnick (1981); G.Thompson (1994)
ANSI B18 series, BS (Appendix 27 *Pipework*)

Welding (*see also* Table 25.1)
AGA (Appendix 28); AWS (Appendix 27, 28); ASTM (STP 11, 494); British Gas (Appendix 27); HSE (HSW Bklt 38); IMetall (Item/1); Welding Institute (Appendix 28); Voelcker (1964, 1965, 1973); ASME (1968/64, 1984/93, 1989 PVP 173, 1992 Boiler and Pressure Vessel Code Pt IX); Maukonen and Vest (1973); Assini (1974); Houldcroft (1975); Schofield (1975); API (1978 Refinery Inspection Guide App., 1982 RP 942, 1988 Std 1104); Cary (1979); M.M. Schwartz (1979); A.C. Davies (1984); Faltus (1985); J.E. Jones and Olsen (1986); D. Smith (1986); Hicks (1987); NFPA (1989 51B); W. Lucas (1991); Stippick (1992)
BS (Appendix 27 *Welding*)

Equipment identification
D.F. Allen (1976); API (1993 RP 1109)
ANSI A13.1–1981, BS (Appendix 27 *Identification of Equipment*), BS 1710: 1984

Mechanical failure (*see also* Table 7.1)
SMRE (*Engineering Metallurgy* 4, 6); Thielsch (1965, 1968); Stokoe, Potts and Marron (1969); Weeks and Hodges (1969–70); Anon. (1970a); IMechE (1970/3, 1984/79, 1988/105, 1994/172); Loescher (1971); Pilborough (1971, 1989); Streich and Feeley (1972); Collins and Monak (1973); Colangelo and Heiser (1974); R.W. Wilson (1974); Lancaster (1975); Melville and Forster (1975); Hutchings (1976); Kletz (1980i, 1984k); Mischiatti and Ripamonti (1985); S.J. Brown (1987); Nishida (1991); Chapman and Lloyd (1992)

Pressure vessel failure, failure rates
Kellerman (1966); Kellerman and Seipel (1967); Phillips and Warwick (1968 UKAEA AHSB(S) R162); Slopianka and Mieze (1968); Butler (1974); Engel (1974); T.A. Smith and Warwick (1974, 1978, 1981 SRD R203); Boesebeck (1975); Bush (1975); Solomon, Okrent and Kastenberg (1975a,b); W. Marshall *et al.* (1976); Boesebeck and Homke (1977); Cottrell (1977); H.M. Thomas (1977); Arulanantham and Lees (1981); Harrop (1982 SRD R217); W. Marshall (1982); W. Marshall *et al.* (1982); Smith (1986 SRD R314, 1987 SRD R353); Kavianian, Rao and Brown (1990); Davenport (1991); Hurst (1991); Medhekar, Bley and Gekler (1993); Ardillon and Bouchacourt (1994); Crombie and Green (1994); Hurst, Davies *et al.* (1994)

Pipework failure
Janzen (n.d.); Gibbons and Hackney (1964); British Gas (1978 Comm. 1103); AGA (1981/34, 35, 1983/38); K.E. Petersen (1983); Cannon and Lewis (1987 NCSR/GR/71); Bellamy, Geyer and Astley (1989); Hancock (1990a,b); Geyer *et al.* (1990); Geyer and Bellamy (1991); Hurst *et al.* (1991); Medhekar, Bley and Gekler (1993); Hurst, Davies *et al.* (1994): Strutt, Allsop and Ouchet (1994); G. Thompson (1994)

Table 12.2 *Selected references on materials of construction for and corrosion in pressure systems*

Materials science
ASTM (Appendix 27, STP 289, 325, 1970 STP 466, 1981 STP 736, 1983 STP 806, 814); Institute of Materials (Appendix 28, B345); Godfrey (1959); Guy (1959, 1976); Jastrzebski (1959); Dieter (1961); W. Johnson and Meller (1962); Grossman and Bain (1964); Tweeddale (1964); Hull (1966); Prince (1966); Cottrell (1967); Honeycombe (1967); Spencer (1968); Gillam (1969); Gregory (1970); Timoshenko and Goodier (1970); van Vlack (1970); IMechE (1971/4); Chadwick (1972); Gabe (1972); Benham and Warnock (1973); J.D. Campbell (1973); C.T. Lynch (1974); Popov (1976); Summitt and Sliker (1980); Crane and Charles (1984); Wulpi (1985)
ANSI Z178 series

Process plant materials
ASME (Appendix 27, 1992 Boiler and Pressure Vessel Code Pt II; Appendix 28 *Materials, Pressure Vessel and Piping*, 1986 PVP 111); ASTM (Appendix 28, STP 15C); EEMUA (Appendix 28); NRC (Appendix 28 *Materials of Construction*); DECHEMA (1953–); Greathouse and Wessel (1954); Aldrich (1960c); McConnell and Brady (1960); Norden (1960a,c); Rumford (1960); Chelius (1962); *Chemical Engineering* (1962–, 1970–, 1980, 1984); Heckler *et al.* (1962); Wyma (1962); Brauweiler (1963); Gleekman (1963, 1970); Kane and Horst (1963); J.F. Mason (1963); Samans (1963, 1966); E.M. Sherwood (1963); Tracy (1963); Bulow (1964); Fenner (1964, 1967, 1968, 1970); Hughson and Labine (1964); Jaffee (1964); Juniere and Sigwalt (1964); Klouman (1964); Kuli (1964); Lancaster (1964, 1969, 1970, 1971); Renshaw (1964); Funk (1965); G.A.Nelson (1965); Anon. (1966a); Carmichael *et al.* (1966); Halbig (1966); IChemE (1966/42, 1978/71); C.M. Parker (1966); Petsinger and Marsh (1966); Speed (1966); Duhl (1967); Mara (1967); Sheets, O'Hara and Snyder (1967); Dukes and Schwarting (1968); R. Miller (1968); Leonard (1969); Schweitzer (1969); Skaudahl and Zebroski (1969); Anon. (1970c); Benzer (1970); Briton, Declerck and Vorhis (1970); Desenby (1970); Fontana (1970); Gregory (1970); Gulya and Marshall (1970); C.A. Robertson (1970, 1972); Skabo (1970); Tarlas (1970); Tator (1970); Tucker and Cline (1970, 1971); Zolin (1970); Anon. (1971b); IMechE (1971/4, 1982/63, 1989/107, 1992/139); F.E. Lawrence (1973); Smithells (1973); Tesmen (1973); L. Evans

(1974); W. Lee (1974); McDowell (1974); Hughson (1976); Kerr (1976); McCandless and Ingram (1976); Mack (1976); Menzies (1976); Wyatt (1976); Bonner (1977); Grafen, Gerischer and Gramberg (1977); Anon (1978 LPB 23, p. 136); Brady and Clauser (1978); Cangi (1978); Chambard (1978); Hornbostel (1978); R.E. Moore (1979); Schillmoller (1979); Weiner and Rogers (1979); Ashby and Jones (1980); Bro and Pillsbury (1980); Gallagher (1980); Gramberg, Gunther and Grafon (1980); Greene (1980); Hagel and Miska (1980); Nichols (1980); Tvrdy *et al.* (1980); Beer and Johnstone (1981); W.W. Marshall (1981); Pollock (1981–); Weismantel (1981); Schiefer and Pape (1982); Crook and Asphahani (1983); Gackenbach (1983); Martino (1983); Murali (1983); Shuker (1983); Baker-Counsell (1984a); Fensom and Clark (1984); Horner (1984); Bayer and Khandros (1985); Kirby (1985b); Asphahani (1986); Declerck and Patarcity (1986); R. King (1986); Cornish (1987); Sandler and Luckiewicz (1987); M. Turner (1987); Dean (1989); Setterlund (1991); Kane (1992); Pollock (1992); Puyear (1992)
ANSI H series: BS (Appendix 27 *Concrete, Steels and Tubes, Test Methods*), BS EN series
Stainless steel: ASTM (STP 369, 1952 DS 5, 1965 DS 5–S1, 1969 DS 5–S2, 1982 STP 756); Hughson (1961b); Luce and Peacock (1962, 1963); Bates (1963); Edstrom and Ljungberg (1964, 1965); Scharfstein (1964); Krisher (1965); McDowell (1965); Merrick and Mantell (1965); Keating (1968); Long (1968); Husen and Samans (1969); Kies, Franson and Coad (1970); Knoth, Lasko and Matejka (1970); Vandelinder (1970); Gaugh (1972, 1976); Pitcher *et al.* (1976); British Gas (1977 TIN2); Peckner and Bernstein (1977); Fowle (1978); Anon. (1979 LPB 28, p. 118); Davison and Miska (1979); Sedricks (1979); Truman and Haigh (1980); R. Brown (1981); R. Cook and Guha (1981); J.D. Redmond and Miska (1982, 1983); J.B. Wright (1982); Martenson and Supko (1983); J.R. Fletcher (1984); P.Marshall (1984); Schillmoller and Althoff (1984); J.D. Redmond (1986); Smith (1987 SRD R353); Institute of Materials (1988 B426); Henderson, King and Stone (1990); Avery (1991); Debold (1991); Warde (1991); Underwood (1992); Whitcraft (1992)
Titanium: R.S. Sheppard and Gegner (1965); Cotton (1970); Feige and Kane (1970); C.P. Williams (1970); W.D. Clark and Dunmore (1976); Covington, Shutz and Franson (1978); Seagle and Bannon (1982)
Clad materials: Beckwith (1987); Lednicky and Lindley (1991); Lerman and Carrabotta (1991); Henthorn and Lednicky (1992); B. Singh (1993)
Plastics including fibre reinforced plastics: Norden (1960c); P. Morgan (1961); Costello, Rhodes and Yovino (1965); Parkyn (1970); Parratt (1972); Catherall (1977); Cheremisinoff and Cheremisinoff (1978); Fasano and Eberhart (1980); N.J. Kraus (1980); McBride (1980); Piggott (1980); Puyear and Conlisk (1980); Rolston (1980); Margus (1982); Baines (1984); Dibbo (1984); Watt and Perov (1985); Rubin (1990); M. Schwartz (1991); M.A. Clark (1992); Goldsmith (1992b); Currieo (1993); W.A. Miller (1993)
Timber: Desch (1981); T. Smith (1987)
Glass: Lofberg (1965); Rawson (1980); Anon. (1981 LPB 41, p. 1; 1981 LPB 42, p. 11; 1982 LPB 44, p. 35); Bucsko (1983); Hoult (1983); Lerman and Carrabotta (1991)
Paints: Charlton (1963); British Gas (1983 BGC/PS/DAT12); Baker-Counsell (1985a); Foscante (1990)

High temperature materials
ASME (Appendix 28 *Materials, Pressure Vessels and Piping*, 1975/71, 1979/137, 1982/156, 1983/164, 1984/172, 1989 PVP 163); ASTM (Appendix 28, DS series); IMechE (Appendix 28, EGF and ESIS series); Institute of Materials (Appendix 28); Schley (1960); E.W. Ross and McHenry (1963); R. Miller (1968); Gaugh (1976); Mack (1976); Hasselman and Heller (1980); IMechE (1983/71, 1987/95, 1988/104); Gooch *et al.* (1983); Skelton (1983, 1987); Marriott *et al.* (1988)

Low temperature materials
ASTM (STP 47, 63, 78, 158, DS 22); E.W. Johnson (1960); R.J. Johnson (1960); Rote and Proctor (1960); Vanderbeck (1960); Zenner (1960); Hwoschinsky (1962); Hurlich (1963); C.M. Parker and Sullivan (1963); R.W. Campbell (1967); Wigley (1979); IMechE (1982/60); R.P. Reed and Clark (1983); British Gas (1990 TIN26)

Materials for particular applications
Acids: Falcke and Lorentz (1985); Ireland (1985)
Ammonia: W.L. Ball (1968b); Inkofer (1969); Rohleder (1969); Kobrin and Kopecki (1978); van Grieken (1979); Ishimaru and Takegawa (1980); Prescott and Badger (1980); Prescott (1982); El Ganainy (1985)
Chlorine: Dukes and Schwarting (1968); BCISC (1975/1); Horowitz (1981); Chlorine Institute (1982 Pmphlt 60, 1985 Pmphlt 6, 1986 Pmphlt 1); Hamminck and Westen (1986); Royce (1988b)
Hydrogen: C.M.Cooper (1965, 1972a); API (1967 Publ. 940, 1971 Publ. 942, 1975 Publ. 945, 1978 Publ. 956, 1990 Publ. 941); G.A.Nelson (1965, 1966); Molstad and Gunther (1967); Gorman (1962); Cracknell (1976); Bonner (1977); C.C. Clark (1978); Treseder (1981); ASME (1982/88); Webb and Gupta (1984)
Hydrogen sulphide: Sivalls (1985); Schwinn and Streisselberger (1993)
Oxygen: Lowrie (1987)

Materials control, identification
W.D. Clark and Sutton (1974); Duff (1976); Ostrofsky (1980–); Baker-Counsell (1985d)

Corrosion
ASTM (STP 179, 567, 1972 STP 516, 1974 STP 534, 558); CONCAWE (23/70); Institute of Materials (Appendix 28); NRC (Appendix 28 *Corrosion*); Uhlig (1951); Greathouse and Wessel (1954); Chlorine Institute (1956 Publ. 18); Fontana (1957, 1986); Jelinek (1958–); J.E. Dawson (1960a); U.R. Evans (1960); Gladis (1960); Norden (1960b); Wachter (1960); Hughson (1963); J.D. Jackson (1961); Brooke (1962); Draley (1962); Swandby (1962); Tracy (1962a,b); Bergstrom and Ladd (1963); Charlton (1963); Fochtmann, Langion and Howard (1963); Hinst (1963); Krebs (1963); Wilder (1963); Wilten (1963, 1965); Ashbaugh (1965a); Menzies *et al.* (1965); Sorell (1965, 1968, 1970); E.H. Anderson (1966); Bates (1966); A.S. Cooper and McConomy (1966); T.A. Lees (1966); Sculley (1966); Canavan (1967); Dingman (1967); Falck-Muus (1967); Iverson (1968); Rabald (1968); D. Stewart and Kulloch (1968); Landrum (1969, 1970); Maylor (1969); Schweitzer (1969, 1976, 1989); J.M. Brooke (1970); Dunlop (1970); Fenner (1970); Husen and Samans

(1970); Thornton (1970); Diamaui (1971); Henthorne (1971–); Hoar (1971); Lochmann (1971); Anon. (1972h); Bonar (1972); C.M. Cooper (1972b); P.D. Thomas (1972); Cantwell and Bryant (1973); IChemE (1973/54); Lux (1973); McDowell (1973); J.A. Richardson and Templeton (1973); F.L. Evans (1974a); G.A. Nelson (1974); Rodgers (1974); DoI (1975); Appl and Feind (1976); Dragoset (1976); Lancaster (1976); Shreir (1976); Berger (1977, 1982); Brautigam (1977); Chadwick and Jamie (1977); Hawk (1977); Pludek (1977); Cangi (1978); Edeleanu (1978); Fontana and Greene (1978); Fryer (1978); Layton (1978); Leach (1978); Ueda *et al.* (1978); Chase (1979); Schumacher (1979); Sedriks (1979); Braunton (1980); J.M. West (1980); Danilov (1981); Rozenfeld (1981); A.E. Wallace and Webb (1981); Ailor (1982); McIntyre (1982); Mallinson (1982); Parkins (1982); Bernie (1983); Stafford and Whittle (1983); Cihal (1984); Elliot (1984); NACE (1984); Baker-Counsell (1985b); Holmes (1985); Kyte (1985); Pelosi and Cappabianca (1985); R. Stevens (1985); Twigg (1985); Anon. (1986f); Dillon (1986); AGA (1987/48, 49, 1988/51); Anon. (1987z); DECHEMA (1987–); R.W. Green (1987); Kofstad (1988); Munn (1988); Royce (1988a); Sathe and O'Connor (1988); M. Turner (1988, 1989a,b, 1990a–c); Avery (1991a); R.D. Kane (1991); Wilhelm (1991); Priest (1992); R.D. Kane *et al.* (1993)
Erosion: Thiruvengadam (1966); ASTM (1979 STP 664); Craig (1985)
External corrosion: Hughson (1961c); Figg (1979); SCI (1979); Anon. (1984 LPB 56, p. 1); Dorsey (1984); Batterham (1985); Liss (1987); Pollack and Steeley (1990)
Corrosion of steel in concrete: ASTM (1977 629, 1980 713, 1984 818)
Concrete: Figg (1979, 1983); Sheppard (1984); Closner (1987a)

Materials and corrosion problems
Field (1963); D.T. Williams (1963); Gleekman (1964); C.F. Lewis (1964); Heckler (1969); Capel and van der Horst (1970); Krystow (1971); Schwab (1971); J.L. Cook (1972); R. Lee (1972); Rollins (1972); Spangler (1972); Butwell, Hawkes and Mago (1973); W.D. Clark and George (1973); Kussmaul and Kregeloh (1973); Lancaster (1973); Anon. (1975 LPB 6, p. 1); R.J. Parker (1975); Anon. (1976 LPB 9, p. 26); R.P. Lee (1976a,b, 1977a,b); Livsey and Junejo (1976); Anon. (1977 LPB 15, p. 23); B. Turner (1977); Anon. (1978 LPB 24, p. 172); Bognar, Peters and Schatzmayr (1978); Schmeal, MacNab and Rhodes (1978); K. Brown (1982); Hare (1982); Quraidis (1982); R.W. Clarke and Connaughton (1984); Sheilan and Smith (1984); W.L. Sheppard (1984); R. Stevens (1985); Anon. (1986f); Kolff (1986); A. Atkinson (1988); Schofield (1988); Schofield and King (1988); Cantwell (1989 LPB 89); Nightingale (1989a,b, 1990); Anon. (1990 LPB 92, p. 11); Linstroth (1991)

Fracture, fracture mechanics
ASME (Appendix 28 *Applied Mechanics, Materials, Pressure Vessels and Piping*, 1975/72); ASTM (Appendix 27); IMechE (Appendix 28 EGF, ESIS series); Institute of Materials (Appendix 28); NRC (Appendix 28 *Crack Growth, Fatigue, Fracture Mechanics*); A.A. Griffith (1920–21); Neuber (1937); Westergaard (1939); Sneddon (1946, 1973); Irwin (1948, 1957, 1958, 1962, 1968); Dryden, Rhode and Kuhn (1952); E.R. Parker (1957); Dugdale (1960, 1968); A.A. Wells (1961, 1969); Bilby, Cottrell and Swinden (1963); Paris and Erdogan (1963); Hahn and Rosenfield (1965, 1967, 1968); R.E. Johnson (1965); Paris and Sih (1965); Burdekin and Stone (1966); Pratt (1966); Sih (1966, 1973a,b); Irwin *et al.* (1967); Tetelman and McEvily (1967); Liebowitz (1968); Rice (1968a,b, 1976); Sneddon and Lowengrub (1969); Pellini (1971); ASTM (1972 STP 513, 514, 1974 STP 556, 559, 560); Battelle Columbus Labs (1972); Begley and Landes (1972); Bravenec (1972); Bucci *et al.* (1972); Eftis and Liebowitz (1972); Heald, Spink and Worthington (1972); Formby, Kirby and Ratcliffe (1973); Irvine (1973 SRD R21, 1974 SRD R26, 1977 SRD R48); Kiefner *et al.* (1973); Kihara and Ikeda (1973); Knott (1973); Nichols (1973); Tada, Paris and Irwin (1973); Broek (1974); Fearnehough (1974); Harvey (1974); Hood (1974); J.N. Robinson and Tetelman (1974); Shannon (1974a,b); Tanaka (1974); Cartwright and Rooke (1975); Crossley and Ripling (1975); Dumm and Fortmann (1975); Gibbons, Andrews and Clarke (1975); MacCary (1975); Pook (1975a,b, 1977, 1979); Underwood (1975); Dragoset (1976); Erdogan (1976); Rooke and Cartwright (1976); Eftis, Subramanian and Liebowitz (1977); Engineering Sciences Data Unit (1977); T.G.F. Gray (1977); Knowles, Tweedle and van der Post (1977); Rolfe and Barsom (1977, 1987); Sorenson and Besuner (1977); Stanley (1977); H.M. Thomas (1977, 1981); AGA (1978/28, 1988/50, 1991/72, 76, 1992/79, 84, 85); Chell (1978); Hudson and Seward (1978); Paris *et al.* (1978); Vosikovsky and Cooke (1978); Cesari and Hellen (1979); R.P. Harrison, Loosemore and Milne (1979); Nichols (1979a); A.P. Parker (1979, 1981); Suzuki, Takahashi and Saito (1979); CEC (1980 EUR 6371 EN); HSE (1980 EM 4, EM 5); G.O. Johnson (1980); O'Neil (1980); Ponton (1980); Wiberg (1980); R.P. Harrison and Milne (1981); Kastner *et al.* (1981); Iorio and Crespi (1981); Shih, German and Kumar (1981); Ziebs *et al.* (1981); Johnston (1982); Ainsworth and Goodall (1983); Temple (1983, 1985); Haines (1983); Tomkins (1983); Arimochi *et al.* (1984); Ewalds and Gielisse (1984); Ichikawa (1984c, 1985, 1987); Imai *et al.* (1984); Latzko *et al.* (1984); Lidiard (1984); Saldanha Peres and Rogerson (1984); Funderburg (1985); Milne *et al.* (1986); IMechE (1987/95); Jutla (1987); Thomson (1987); Milne *et al.* (1988a,b); J.K.W. Davies (1989b); HSE (1989 Nuclear Installations 12); Tomkins, Lidbury and Harrop (1989); Schulz and Braun (1992); Medhekar, Bley and Gekler (1993)

Fatigue
ASME (Appendix 28 *Materials*, 1981/11, 84, 1984/92, 94); ASTM (STP 91, 237, 1972 513, 514, 1973 STP 520, 1974 STP 556, 559, 560); IMechE (Appendix 28 EGF, ESIS series, 1975/15, 1977/40); NRC (Appendix 28 *Crack Growth, Fatigue, Fracture Mechanics*); Dryden, Rhode and Kuhn (1952); Kooistra and Lemcoe (1962); Yao and Munse (1962); Manson (1966); Avery (1972b); Tomkins (1973); Harvey (1974); Frost, Marsh and Pook (1975); Pook (1975a,b, 1977); W.J. Harris (1976); Duggan and Byrne (1977); Linhart and Jelinek (1977); Bongers, Diols and Linssen (1978); Cowley and Wylde (1978); Michelini (1978); Klesnil and Lukas (1980); Skelton (1983, 1987); Larsson (1984); Younas and Sheikh (1987); Fromm, Liebe and Siegel (1988); J.K.W. Davies (1989)

Ageing
NRC (Appendix 28 *Ageing*)

Embrittlement, brittle fracture
E.R. Parker (1957); Biggs (1960); Tipper (1962); Ladd (1966); W.D. Clark and Mantle (1967); W.J. Hall *et al.* (1967); Sorell and Zeis (1967); Harnby (1968); Madayag (1969); Boyd (1970); Karinen (1971); Moisio (1972); Anon. (1974i); ASTM (1974 STP 543, 1984 STP 844); Afzal and Livingston (1974); Harvey (1974); Lamberton and Vaughan (1974); Lonsdale (1975); Watanabe and Murakami (1981); Murza, Gentner and McMahon (1981); B.J. Shaw (1981); AGA (1983/38); Merrick and Ciuffreda (1983); Maxey *et al.* (1985); Wilkie (1985a); Snyder (1988); API (1990 Publ. 920); Burke and Moore (1990)

Stress corrosion cracking
ASTM (STP 64, 264, 397, 1967 STP 425, 1972 STP 518, 1976 STP 610, 1979 STP 665); Copson and Cheng (1957); Hughson (1961a); Loginow and Phelps (1962); Wilten (1962); Ashbaugh (1965b, 1970); Truman and Kirkby (1965); Logan (1966); Anon. (1972j); van der Horst (1972); Hutchings *et al.* (1972); Phelps (1972, 1974); Collins and Monack (1973); Creamer (1974); Harvey (1974); Zeis (1975); Zeis and Paul (1975); van Grieken (1976, 1979); Harvey (1976); HSE (1976 TDN 53/2); Arup (1977); W.D. Clark and Cracknell (1977); Anon. (1979 LPB 28, p. 104); Karpenko and Vasilenko (1979); Cracknell (1980); Ishemaru and Takegawa (1980); Macintyre (1980); Takemura, Shibasaki and Kawai (1981); Gossett (1982); Anon. (1983r); AGA (1984/40, 1988/56, 1990/68, 70, 1991/71, 1992/87, 88); Blanken (1984); L. Lunde (1984); Orbons and Huurdeman (1985); Lemoine *et al.* (1986); Loginow (1986); Schillmoller (1986); L. Lunde and Nyborg (1987, 1989, 1990); Stephens and Vidalin (1988); Anon. (1989 LPB 89, p. 27); Böckenhauer (1989); Byrne, Moir and Williams (1989); Parkins (1989); Appl *et al.* (1990); Selva and Heuser (1990); Conley, Angelsen and Williams (1991); Crawley (1992 LPB 104)

Creep
ASME (Appendix 28 *Materials*, 1981/84, 1984/92, 94, 1988 PVP 35, 151);ASTM (STP 26, 37, 107, 325, 391); Institute of Materials (Appendix 28); Finnie and Heller (1959); Anon. (1969e); Lochmann (1972); Harvey (1974); IMechE (1974/7, 1975/15, 1977/36, 1978/47, 1994/162); HSE (1977 TDN 53/3); Imoto, Terada and Maki (1982); Konoki, Shionohara and Shibata (1982); Ashby and Brown (1983); Tomkins (1983); H.E. Evans (1984); Kawai *et al.* (1984); Larsson (1984); Anon. (1986e)

Hydrogen corrosion
Anon. (1962d); Cooper (1965, 1972a); McDowell and Milligan (1965); McDowell (1967); Ciuffreda and Hopkinson (1968); Ciuffreda and Greene (1972); AGA (1974/19); Harvey (1974); K.L. Moore and Bird (1965); Bonner (1977); A.W. Thompson (1978); Truax (1978); Tvrdy *et al.* (1981); Timmins (1983, 1984); Webb and Gupta (1984); Genet and Perdrix (1987)

Liquid metal attack
British Gas (1982 TIN16)

Zinc embrittlement
Ball (1975a,b, 1976); Cottrell and Swann (1975a,b, 1976); Harvey (1976); HSE (1976 TDN 53/1, 1977 PM 13); Anon. (1979 LPB 29, p. 149); Anon. (1979 LPB 30, p. 175); British Gas (1983 BGC/PS/DAT11); SETE Consultants and Services Ltd (1984 LPB 55)

Tribology, wear
ASTM (STP 30, 567, 1969 STP 446); Burwell (1957–58); Norden (1960a); MacGregor (1964); Furey (1969); Summers-Smith (1969); M.C. Shaw (1971); Avery (1972a); M.J. Neale (1973); Halling (1975); IMechE (1975/17, 22, 1977/45, 1980/53, 1994/167); Engel (1976); P.R. Williams (1976); ASME (1977/74, 1979/80, 1980/10, 1981/87, 1983/91, 1985/96); Fenton (1977) Schumacher (1977); Sorenson and Besuner (1977); Sayles and Macpherson (1980); Suh and Saka (1980); Szeri (1980); Buckley (1981); Briscoe (1982)

12.1 Pressure Systems

Any system in which the pressure departs at all significantly from atmospheric, and which is therefore of rigid construction, needs to be considered as a pressure system. In the UK, pressure systems are covered by the Pressure Systems and Transportable Gas Containers Regulations 1989 (the Pressure Systems Regulations). For fixed plant the associated code is COP 37 *Safety of Pressure Systems* (HSE, 1990).

The definition of a 'pressure system' under the Pressure Systems Regulations is given in Regulation 2 and is: '(a) a system comprising one or more pressure vessels of rigid construction, any associated pipework and protective devices; (b) the pipework with protective devices to which a transportable gas container is, or is intended to be, connected; or (c) a pipeline and its protective devices; which contains or is liable to contain a relevant fluid, but does not include a transportable gas container.'

Regulation 2 defines 'relevant fluid' as: '(a) steam; (b) any fluid or mixture of fluids which is at a greater pressure than 0.5 bar above atmospheric pressure, and which fluid or mixture of fluids is – (i) a gas, or (ii) a liquid which would have a vapour pressure greater than 0.5 bar above atmospheric pressure when in equilibrium with its vapour at either the actual temperature of the liquid or 17.5 degrees Celsius; or (c) a gas dissolved under pressure in a solvent contained in a porous substance at ambient temperature and which could be released from the solvent without application of heat.'

Regulation 4 deals with the design, construction, repair and modification of pressure systems. Regulations 5–13 cover essentially operation, inspection and maintenance. Regulation 7 requires there to be safe operating limits. Regulation 5 requires marking of equipment, Regulation 8 a written scheme of examination, Regulation 9 examination in accordance with this scheme and Regulation 13 the keeping of records. Regulation 11 covers operation and Regulation 12 maintenance, whilst Regulation 10 deals with action in the case of imminent danger.

Requirements for the provision of protective devices are contained in Regulation 4 and also in Regulations 14 and 15. This aspect is discussed in more detail in Section 12.12.

12.2 Pressure System Components

The main components of pressure systems have been described by Dickenson (1976). They are:

(1) pressure vessels (reactors, distillation columns, storage drums and vessels);
(2) piping system components (pipes, bends, tees, reducers, flanges, valves, nozzles, nipples);
(3) means of adding, controlling or removing heat (fired heaters, reboilers, vaporizers, condensers, coolers, heat exchangers generally);
(4) means of increasing, controlling or reducing pressure (pumps, compressors, fans, letdown turbines, control valves);
(5) means of adding or removing fluids or solids to or from the process system (pumps, compressors, dump valves);
(6) measurement and control devices and systems (instrumentation);
(7) utilities and services (electricity, steam, water, air).

Item (7) was considered in Chapter 11 and item (6) is described in Chapter 13. The other items are dealt with below.

12.3 Steels and Their Properties

The main material of construction used in pressure systems is steel and it is necessary to consider some of the properties of steel which are particularly important in relation to pressure systems and which are the basis of pressure systems standards and codes of practice.

Steels and other materials are discussed in *Selecting Materials for Chemical and Process Plant* (L. Evans, 1974), *Handbook of Materials Science* (C.T. Lynch, 1974–), *Materials Handbook* (Brady and Clauser, 1978), *Construction Materials* (Hornbostel, 1978), *Materials for Low Temperature Use* (Wigley, 1979), *Mechanics of Materials* (Beer and Johnstone, 1981), *Physical Properties of Materials for Engineers* (Pollock, 1981–) and *Selection and Use of Engineering Materials* (Crane and Charles, 1984).

Accounts of the general properties of steels in relation to their use in pressure vessels are given in *Theory and Design of Modern Pressure Vessels* (Harvey, 1974), *Pressure Component Construction* (Harvey, 1980) and *Pressure Vessel Systems* (Kohan, 1987). The main steels used in process plant and their relation to the standards and codes are described by L. Evans (1974).

12.3.1 Stress, strain and elasticity

If a specimen test bar of a material is subjected to a tensile load in a tensile testing machine, stress is induced in the material and strain occurs.

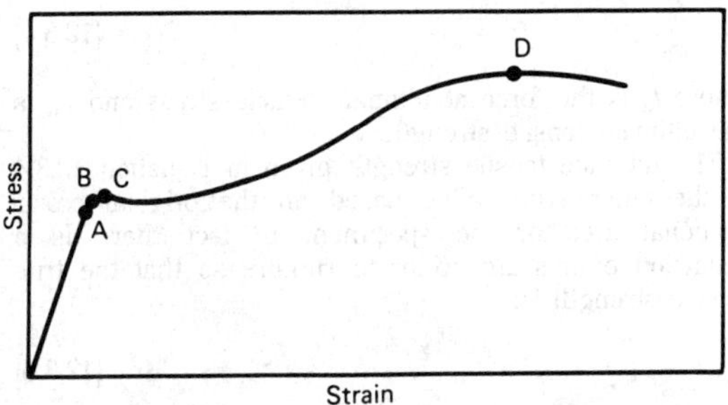

***Figure 12.1** Typical load–extension diagram for mild steel*

The stress is:

$$\sigma = \frac{f}{a_0} \qquad [12.3.1]$$

where a_0 is the original cross-sectional area, f is the force, and σ is the stress.

The strain, or fractional elongation, is:

$$e = \frac{\delta}{l_0} \qquad [12.3.2]$$

where e is the strain, l_0 is the original length, and δ is the elongation.

The typical load–extension, or stress–strain, diagram obtained for many metals, including mild steel, is shown in Figure 12.1. At low loads the material is elastic, returning to its original dimensions when the stress is removed, and the strain is proportional to the stress:

$$\frac{e}{\sigma} = \frac{1}{E} \qquad [12.3.3]$$

where E is the elastic modulus. Equation 12.3.3 is Hooke's law and the constant E is Young's modulus.

As the load increases, there is a change in material behaviour indicated by points A and B in Figure 12.1. The proportionality limit at A is the stress beyond which the strain is no longer proportional to the stress. The elastic limit at B is the stress beyond which the material no longer returns to its original dimensions, but undergoes some permanent deformation. For mild steel and many other materials the proportionality and elastic limits coincide, but for some, such as rolled aluminium, they are separate.

12.3.2 Yield and tensile strengths

The yield point C, which is close to points A and B, is the stress at which there occurs a marked increase in strain without an increase in stress. For many engineering applications it is not acceptable that the material should yield. The yield point represents the limit of strength in such applications.

As the load increases further, there is continued extension until point D is reached at which fracture occurs. The stress at fracture is the ultimate tensile stress, or strength:

$$\tau_u = \frac{f_u}{a_0} \quad [12.3.4]$$

where f_u is the force at ultimate tensile stress and τ_u is the ultimate tensile strength.

The ultimate tensile strength given in Equation 12.3.4 is the engineering value based on the original cross-sectional area of the specimen. In fact there is a reduction of this area prior to rupture so that the true tensile strength is:

$$\tau_{u,t} = \frac{f}{a} \quad [12.3.5]$$

where a is the actual cross-sectional area and $\tau_{u,t}$ is the true ultimate tensile strength. The ultimate tensile stress represents the limit of strength in applications where yielding is acceptable, but rupture is not.

12.3.3 Ductility

Another important property of a material is its ductility, or its ability to undergo deformation. Ductility may be expressed as fractional elongation:

$$d_e = \frac{l_r - l_0}{l_0} \quad [12.3.6]$$

or fractional reduction in area

$$d_a = \frac{a_0 - a_r}{a_0} \quad [12.3.7]$$

where a_r is the cross-sectional area at rupture, d_a is the fractional reduction in area at rupture, d_e is the fractional elongation and l_r is the length at rupture.

Ductility is a very desirable property in a steel and is important for design and for fabrication. In pressure vessels, features such as joints, openings and nozzles create additional load stresses which may not be fully allowed for in the design. Ductility permits local yielding so that the stress is adjusted. Similarly, ductility allows material to undergo fabrication operations, such as rolling, forging, drawing and extruding, without fracturing.

12.3.4 Toughness and impact strength

A further property which is important in a material is its toughness, or ability to resist impact. Toughness is measured by the energy absorbed in stressing to fracture, which is the area under the stress–strain curve, and thus the modulus of toughness is given by the area:

$$T = \int_0^{e_r} \sigma \, de \quad [12.3.8]$$

where e_r is the strain at rupture and T is the toughness.

A material which has both high tensile strength and high ductility has high toughness. Toughness is another property which is very desirable in a steel, and loss of toughness, or embrittlement, is a serious defect.

12.3.5 Low temperature strength

At low temperatures mild and other steels tend to suffer a loss of toughness as shown in Figure 12.2. For mild steel the ductile/brittle transition occurs at about 0°C; the precise transition temperature depends on the type and quality of the steel. An account of brittle fracture is given in *The Brittle Fracture of Steel* (Biggs, 1960). Since much process plant operates at low temperatures, such embrittlement is important. Low temperature embrittlement has been the cause of many failures, and is discussed further in Section 12.27.

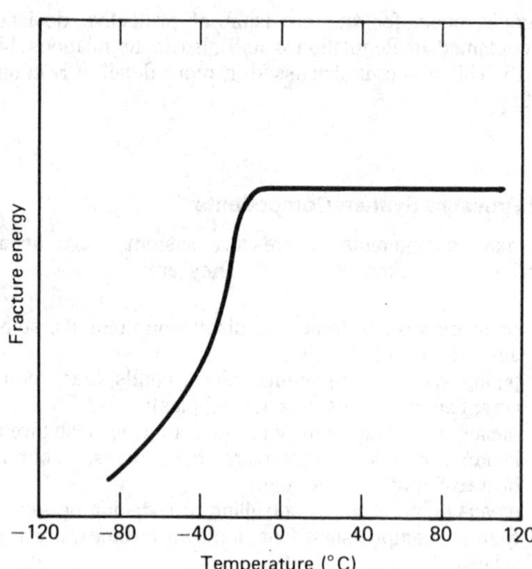

Figure 12.2 *Typical impact strength diagram for mild steel*

12.3.6 High temperature strength

At high temperatures the strength of many materials, including mild steel, tends to decrease, as illustrated in Figure 12.3. The yield point falls and becomes less pronounced, and the ultimate tensile strength also falls. Moreover, a further factor begins to limit strength at high temperature. This is creep.

12.3.7 Creep strength

Creep is an extension which occurs in metals, including mild steel, at high temperatures under constant load over

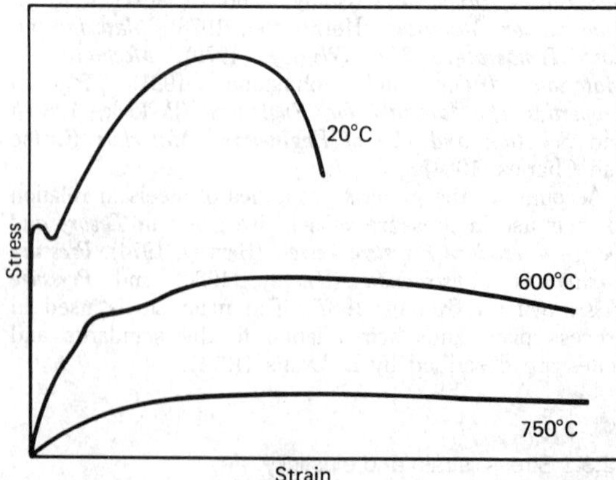

Figure 12.3 *Typical load–Nextension diagram at high temperatures for mild steel*

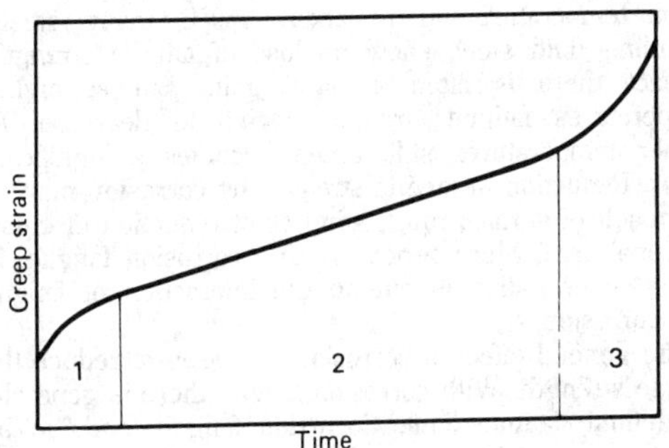

Figure 12.4 *Typical creep test at high temperatures for mild steel*

a period of time. Accounts of creep are given in *Perspectives in Creep Fracture* (Ashby and Brown, 1983) and *Mechanisms of Creep Fracture* (H.E. Evans, 1984).

The typical results of a creep test on mild steel carried out in a tensile testing machine at constant load and constant temperature are illustrated in Figure 12.4. There is a short primary phase (1) when the creep rate is fairly rapid, a long secondary phase (2) when it is approximately constant, and a short tertiary phase (3) when it accelerates to rupture.

The typical variation of creep rate with loading and temperature is illustrated by Figure 12.5. The creep strength of a material at a particular temperature is usually expressed either as the stress for 1% extension in 100 000 h or as the stress for rupture in 100 000 h. The former is normally taken as not less than two-thirds of the latter.

Frequently, a steel is subjected to a number of different temperatures and/or stresses. For such cases it is common practice to calculate the fraction of creep life used up as follows. For time at different stresses and temperatures, the condition for exhaustion of the creep life is given by the 'life fraction' rule:

$$\sum_{i=1}^{n} \frac{t_i}{t_{ri}} = 1 \qquad [12.3.9]$$

where t_i is the time at stress and temperature conditions i, and t_{ri} is the creep life at stress and temperature conditions i. Similarly, for strain at different stresses and temperatures the condition is given by the 'strain fraction' rule:

$$\sum_{i=1}^{n} \frac{e_i}{e_{ri}} = 1 \qquad [12.3.10]$$

where e_i is the strain at stress and temperature conditions i, and e_{ri} is the strain for rupture at stress and temperature conditions i.

Use is also made of the following equation which utilizes the geometric mean of these two approaches:

$$\sum_{i=1}^{n} \left(\frac{t_i}{t_{ri}} \frac{e_i}{e_{ri}} \right)^{\frac{1}{2}} = 1 \qquad [12.3.11]$$

It is emphasized that Equations 12.3.9–12.3.11 are approximate. The applicability of these equations is discussed in more detail by Harvey.

Since high temperatures are common in process plant, creep is very important. It is considered again in Section 12.27.

12.3.8 Fatigue strength

Many ductile metals, including mild steel, can under certain conditions suffer fatigue failure at a stress well below the ultimate tensile strength if subjected to repeated cycles of stress. Accounts of fatigue are given in *Fatigue and Fracture of Metals* (Dryden, Rhode and Kuhn, 1952), *Fatigue of Metallic Materials* (Klesnil and Lukas, 1980), *The Mechanics of Fatigue and Fracture* (A.P. Parker, 1981), *Fatigue at High Temperatures* (Skelton, 1983) and *High Temperature Fatigue* (Skelton, 1987).

Whether fatigue failure occurs depends on the magnitude of the stress range which is applied and on the number of repetitions of the stress. The relationship between these two factors is shown in Figure 12.6, which illustrates the results of a typical fatigue test on mild

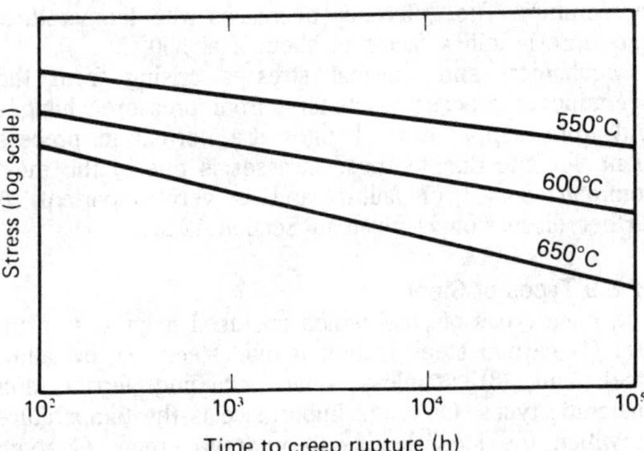

Figure 12.5 *Typical variation of creep rate with loading and temperature for mild steel*

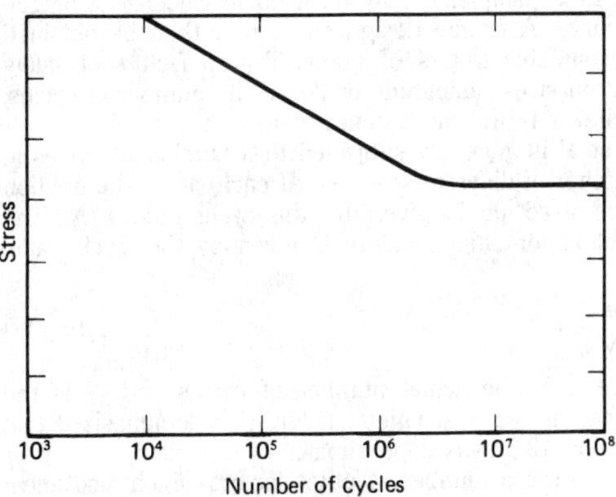

Figure 12.6 *Typical fatigue test for mild steel*

steel. Fatigue testing is usually done using the reversed bending test, in which a load is applied at the end of a cantilever test specimen which is then rotated at constant speed, so that the full reverse bending stresses are created each revolution. The data from such tests may be plotted as maximum stress or as stress range. The stress range is the fundamental parameter, but the maximum stress is often plotted.

The curve given in Figure 12.6 has two distinct regimes. In the first region to the left of the break point, failure occurs at decreasing stresses but increasing number of cycles, while in the second region to its right there is a stress level below which failure does not occur irrespective of the number of cycles. This stress level is the fatigue limit, or endurance limit, and is an important property. For most steels at room temperature a good estimate of the fatigue limit may be obtained from the ultimate tensile stress using the relation:

$$\sigma_E = d_a \sigma_u \qquad [12.3.12]$$

where σ_{E} is the fatigue limit and σ_{u} is the ultimate tensile stress. The fatigue limit for steels other than high strength steels is generally 40–55% of the ultimate tensile strength.

The number of cycles at which the fatigue limit occurs at room temperatures is of the order of 10^6–10^7. It is often convenient, therefore, to take the fracture stress at 10^8 cycles as the practical limit.

Although it is high cycle fatigue which has been most studied, low cycle fatigue is also important. This occurs at higher stresses and smaller number of cycles and corresponds to the first region in Figure 12.6. In this region, if the data are expressed as stress there is usually considerable scatter, but the scatter is reduced if the data are expressed as strain. A strain is often multiplied by one-half of the elastic modulus to give a pseudo-stress amplitude. Low cycle fatigue is generally considered to cover the region below 10^5 cycles.

It can be shown for many metals, including mild steel at room temperature, that the relation between the stress amplitude and the number of cycles N at fatigue failure can be described approximately by the equation:

$$\sigma = \frac{e}{4N^{\frac{1}{2}}} \ln\left(\frac{1}{1-d_a}\right) + \sigma_E \qquad [12.3.13]$$

This equation covers the range of both low cycle and high cycle fatigue. It may be used to calculate a fatigue life curve. A fatigue design curve may then be obtained using suitable factors of safety. Typical factors of safety are 2 on stress amplitude or 20 on the number of cycles, whichever is the more conservative.

A steel is generally subjected to a number of cycles at a number of different stresses. At each stress the fraction of life used up is given by the cycle ratio n/N. The condition for fatigue failure is given by the 'cycle ratio' rule:

$$\sum \frac{n}{N} = 1 \qquad [12.3.14]$$

where n is the actual number of cycles and N is the number of cycles at fatigue failure. It is emphasized that Equation 12.3.14 is approximate.

There are a number of other factors which also affect fatigue strength. They include (1) temperature, (2) corrosion and (3) surface condition. In the range from room temperature up to about 350°C many steels, including mild steel, show no loss of fatigue strength. In fact there is often a slight gain. But at higher temperatures fatigue strength begins to decrease. At higher temperatures still, creep becomes a significant factor. Reduction of fatigue strength by corrosion may be the result of surface roughening or of reduction of cross-sectional area. More serious is true corrosion fatigue, in which deterioration is due to the interaction of fatigue and corrosion.

The general effect of corrosion fatigue is to reduce the fatigue strength. With corrosion fatigue there is generally no definite fatigue limit. Corrosion fatigue tends to be worse if oxygen is present. Oxygen normally reduces the low cycle fatigue strength. Whether the fatigue limit is reduced depends on the surface condition. A reduction in fatigue strength due to oxygen tends to occur if very sharp cracks are present.

Fatigue strength is in general greatly reduced if there are cracks or similar surface defects. A high stress concentration occurs at the extremities of the crack and the crack grows a finite amount each cycle. The growth of fatigue cracks is the subject of fracture mechanics, which is considered in Section 12.28.

Another important factor in fatigue is the temperature changes which occur and which cause thermal stress fatigue. This is an effect quite separate from that of temperature on fatigue strength. Thermal changes are a source of cycling and tend to be difficult to predict, but are often appreciable. If the thermal stress is large, it may cause the metal to deform. This deformation results in a more favourable stress distribution, and to this extent thermal stress is self-correcting, but there may also be some cracking. Cracks arising during thermal stress may grow as a result of mechanical stress.

An index of susceptibility to thermal stress is given by the thermal shock parameter

$$\phi = \frac{E\alpha}{k} \qquad [12.3.15]$$

where k is the thermal conductivity, α is the coefficient of thermal expansion, and ϕ is the thermal shock parameter.

The higher the thermal shock parameter the more susceptible the metal is to thermal stress. The thermal shock parameter of mild steel is less than that of austenitic stainless steel by a factor of about 3 at room temperature. The difference decreases with temperature, but there is still a factor of about 2 at 350°C.

Mechanical and thermal stresses arising from the operation of process machinery, from pressure changes and from temperature changes are normal in process plant. Fatigue due to these stresses is one of the most common causes of failure and is very important. A further discussion is given in Section 12.27.

12.3.9 Types of Steel

The main types of steel which are used in process plant are: (1) carbon steel, including mild steel; (2) low allow steel; and (3) stainless steel, including ferritic and austenitic types. Of prime importance is the temperature at which the steel begins to undergo creep. At room temperature the strength of the steel is limited by the yield and/or tensile strengths. As the operating temperature increases, however, the creep strength becomes the

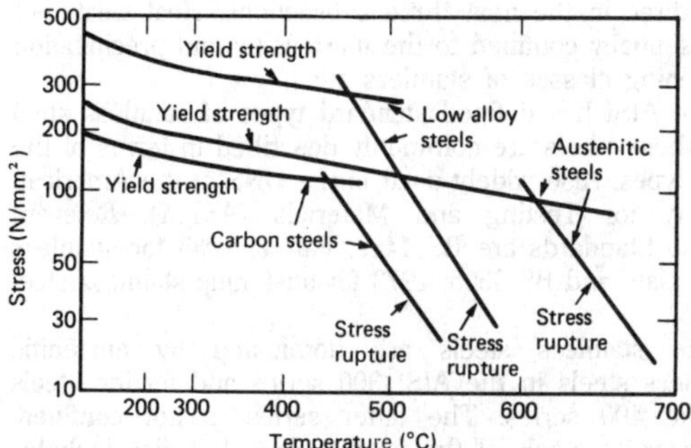

***Figure 12.7** Typical yield stress and creep rupture stress curves for mild, low alloy and austenitic steels. (L. Evans, 1974) (Courtesy of Business Books)*

limiting factor. Thus the operating temperatures may be divided into (1) temperatures below the creep range and (2) temperatures above the creep range. The basis of design is different for these two cases.

The approximate threshold temperatures at which onset of creep occurs are:

Mild steel	400°C
Low allow steel	500°C
Austenitic stainless steel	600°C

The change with temperature in the limiting strengths of these three types of steel is shown in Figure 12.7. The shape of the curve is broadly similar for all three types.

12.3.10 Carbon steel

Carbon steels are generally defined as those containing iron, carbon and manganese. Most steels have a carbon content of 0.1–0.9% carbon and 0.5–2.0% manganese.

The most common type of carbon steel is low carbon, or mild, steel with a carbon content up to 0.25%. Mild steel is described in the following subsection. High carbon steels are those with a carbon content above this figure, up to about 0.9%.

Steels containing more than 0.3% carbon are difficult to weld. High carbon steel is produced mainly as bar and forgings for uses such as shafts, bolts, etc. These steels require heat treatment by way of quenching and tempering to give their optimum properties.

12.3.11 Mild steel

Mild steel is low carbon steel with, as just stated, a carbon content less than 0.25%. It is the cheapest and commonest form of steel. The properties of mild steel may be illustrated by the room temperature properties given in BS 4360: 1990 for plate to grade 43EE:

Tensile strength	430/580 N/mm^2
Yield strength	265 N/mm^2
Elongation	20%

where the yield strength applies to thicknesses in the range 16–40 mm and the elongation applies over 200 mm.

Mild steel is ductile and weldable, but it has certain limitations which are related to:

(1) corrosion resistance;
(2) strength at
 (a) room temperature;
 (b) high temperature;
 (c) low temperature.

There are many chemicals which corrode mild steel. Examples are dilute acids, such as sulphuric, nitric and hydrochloric acids, and chemicals containing water, such as wet chlorine.

For pressure vessels it is often attractive to use a material which has a higher strength than mild steel and so to have a thinner walled vessel. Low alloy steel is generally used instead of mild steel in such applications.

At high temperatures, above about 400°C, mild steel suffers quite severe creep and it becomes necessary to use other steels. At low temperatures, below about 10°C, the ductility of mild steel falls off sharply, and brittle fracture can occur, as described above. Again, other steels have to be used.

Mild steel can be modified by the incorporation of small amounts of alloying elements. Addition of up to 0.1% niobium can increase yield strength and low temperature ductility. Addition of up to 1.5% manganese can give substantial increases in yield strength such as from 230 N/mm^2 to 400 N/mm^2.

12.3.12 Low alloy steel

The low alloy steels which are important in process plant are mainly those which have a carbon content less than 0.2% and contain a total $< 12\%$ alloying elements (Ni, Cr, Mo, V, B, W or Cu).

Many low alloy steels are given a heat treatment of normalizing and tempering by the manufacturer, but there is an increasing tendency to a quenching and tempering treatment. Low alloy steels are all weldable, but for some steels pre- or post-weld heat treatment is necessary in order to avoid weld zone cracking.

Some principal low alloy steels are:

0.5 Mo	12 CrMoVW
1.25 CrMo	0.25 Cr 0.25 Mo 1.5 Ni
2.25 CrMo	

Some significant advantages of low alloy steels over mild steel are:

(1) yield strength;
(2) high temperature properties:
 (a) creep strength;
 (b) oxidation resistance;
 (c) hydrogen resistance;
(3) low temperature ductility.

The improvement in yield strength and in creep strength was illustrated in Figure 12.7. Yield strength is the main design parameter used in advanced codes for pressure vessels and the gain in yield strength is valuable. Low alloy steels make it possible to have thinner walled pressure vessels. Low alloy steels such as 0.5 Mo, 1.25 CrMo, 2.25 CrMo and 12 CrMoVW are used for their creep properties in applications such as steam boilers and refinery crackers and reformers. The upper temperature limit for low alloy steels is about 600°C.

Oxidation resistance is another important property of low alloy steels. The principal alloying element which imparts oxidation resistance is chromium. Low alloy steels also offer greater resistance to hydrogen attack in such operations as cracking and reforming. Mild steel ceases to be suitable above 250°C, 0.5 Mo steel can used up to about 350°C, while 2.25 CrMo steel can go up to about 550–600°C. Apart from resistance to oxidation and hydrogen attack, low alloy steels do not offer much greater resistance to corrosion than does mild steel.

An increase in low temperature ductility down to − 50°C can be obtained by addition of 1.5 Ni to 0.25 Cr 0.25 Mo steel. This is a most valuable modification, since low temperature embrittlement is a serious problem. Nickel is the main alloying element used to give low temperature ductility.

12.3.13 Stainless steel

The main alloying elements in high alloy steels are chromium and nickel. A stainless steel is one which contains 11% or more of chromium. The chromium forms an oxide film giving a passive surface, rendering it generally more resistant to corrosion than that on the lower alloy or carbon steels.

Accounts of stainless steels are given in *Handbook of Stainless Steels* (Peckner and Bernstein, 1977), *Corrosion of Stainless Steels* (Sedriks, 1979), *Austenitic Stainless Steels* (P. Marshall, 1984) and *High Temperature Alloys* (Marriott *et al*., 1988) and by Redmond and Miska (1982), J.R. Fletcher (1984), Henderson, King and Stone (1990), and Warde (1991).

The review by Redmond and Miska (1982) covers: the role of the alloying elements; the classes of stainless steel; their corrosion resistance; their physical and mechanical properties; and their fabrication. It also gives numerous tables listing the American Iron and Steel Institute (AISI) types and Unified Numbering System (UNS) numbers for the different classes of stainless steel.

The principal alloying elements in stainless steel are chromium, nickel and molybdenum. Each has a number of effects, and mention is made here of only a few. As already stated, chromium gives a passive oxide layer which increases the corrosion resistance. It also promotes the formation of the ferrite phase. Nickel, by contrast, promotes the formation of the austenite phase and enhances toughness, ductility and weldability. It also increases resistance to strong acids. A principal function of molybdenum is to enhance the resistance of the stainless steel to pitting and crevice corrosion.

Broadly, stainless steels have four main advantages: (1) higher corrosion resistance, (2) higher resistance to heat, (3) higher creep and stress rupture resistance and (4) higher strength at ambient and intermediate temperatures.

For a long time the development of stainless steels was hampered by the presence of carbon and nitrogen, which can have deleterious effects, but the use of techniques such as argon–oxygen decarburization (AOD), vacuum–oxygen decarburization (VOD) and vacuum induction melting (VIM), which allow the production of steels with very low carbon and nitrogen contents, has largely obviated these difficulties.

Redmond and Miska distinguish five classes of stainless steel: (1) austenitic, (2) ferritic, (3) duplex, (4) martensitic and (5) precipitation hardening. The last two classes are considered later in this subsection and the first three in the next three subsections. Heat treatment is essentially confined to the martensitic and precipitation hardening classes of stainless steel.

The AISI has defined standard types of stainless steel and these steels are commonly described in terms of the AISI types. Also widely used is the UNS of the American Society for Testing and Materials (ASTM). Relevant British Standards are BS 1449: Part 2: 1983 for stainless steel plate and BS 3605: 1973 for austenitic stainless steel pipe.

The stainless steels are dominated by austenitic stainless steels in the AISI 300 series and ferritic steels in the 400 series. The latter series is not confined, however, to steels of the ferritic class, but also includes some in other classes such as martensitic and superduplexes.

Martensitic stainless steels generally contain 11–18% chromium but little or no nickel, and have a structure which can be transformed by suitable cooling from austenite at high temperature to martensite at room temperature. They include a low carbon subclass and a high carbon one, the division occurring at a carbon content of about 0.15%. Martensitic steels, given suitable heat treatment, can have very high strength and hardness. They have a ductile–brittle transition temperature which is usually close to room temperature. A general purpose low carbon martensitic steel is type 410. Martensitic steels are used for such items as bolts, shafts and bearings.

Precipitation-hardening (PH) stainless steels contain both chromium and nickel, and other alloying elements such as copper or aluminium, which impart the precipitation hardening property, allowing the steel to be hardened to high strength for use in such applications as gears.

12.3.14 Austenitic stainless steel

The basic austenitic stainless steel may be regarded as containing 18% chromium and 8% nickel, but there is a whole family of such steels with variations about these values. These steels are described in *Chromium Nickel Austenitic Steels* by Keating (1968) and Redmond and Miska (1982).

The workhorse austenitic steels include AISI types 304, 304L, 316, 316L, 321 and 347. The suffix L denotes low carbon. As already mentioned, austenitic steels are covered in BS 1449: Part 2: 1983 and BS 3605: 1973.

Type 304 is a 19/10 (19 Cr 10 Ni) steel. It contains the minimum amount of alloying elements to give a stable austenitic structure under all fabrication conditions. Type 304L has a lower carbon content than type 304 and is used for applications involving the welding of plate thicker than about 6.5 mm in order to avoid intergranular corrosion, or weld decay.

Type 316 is a 17/12 steel with 2.5 Mo addition to improve resistance to reducing conditions such as brine and dilute sulphuric acid. Type 316L has a lower carbon content than type 316 and is used in applications where the heat input during fabrication will exceed the incubation period of the 316 grade, such as the welding of plate thicker than about 10 mm.

Type 321 is an 18/10 steel with titanium addition to prevent intergranular corrosion. Type 347 is an 18/11 steel with niobium addition to improve welding properties.

Standard austenitic stainless steels are subject to chloride-induced stress corrosion cracking, although certain types exist such as 904L and, particularly 825, which are resistant.

More highly alloyed austenitic stainless steels include the HyResist 317LM and 94L steels, with enhanced corrosion resistance, and the HyProof 3126L and 317L steels with higher proof strength.

The absence of ferrite means that austenitic stainless steels do not have a ductile-brittle transition temperature.

12.3.15 Ferritic stainless steel

Ferritic stainless steels contain chromium but little or no nickel. The class may be subdivided into two main subclasses: those with 11% Cr and those with 17% Cr. Ferritic stainless steels can be rendered brittle by even quite low contents of carbon and nitrogen (say 0.1% C + N), and for a long time this inhibited their use, but in manufacture this problem has been largely obviated by the techniques of AOD, VOD and VIM mentioned above.

The workhorse ferritic stainless steels include AISI types 409, 430, 434 and 444. Type 409 has 11% Cr and type 430 17% Cr.

The welding of ferritic steels poses a problem in that carbon or nitrogen pick-up can have serious effects. In the 11% Cr steels an improvement in weldability, and also an extension of the size of section which can be produced, has been achieved with the use of the 3CR 12 type, which is so balanced that some austenite forms above about 800°C, which is beneficial both in welding and rolling.

Ferritic, and also martensitic, stainless steels are immune to chloride-induced stress corrosion cracking but are susceptible to sulphide-induced stress corrosion cracking. Ferritic stainless steels have a ductile–brittle transition temperature. There are also superferritic stainless steels with a high alloy content, which imparts enhanced resistance to pitting corrosion and crevice corrosion.

12.3.16 Duplex stainless steel

Duplex stainless steels contain both ferrite and austenite, in approximately equal amounts. Their properties reflect the fact that they contain both phases. The presence of ferrite means that duplex stainless steels have a ductile–brittle transition temperature.

The duplexes differ from standard austenitic stainless steels in that they have higher resistance to stress corrosion cracking and higher yield strength. They differ from the ferritic stainless steels in having higher low temperature toughness and in being more weldable. Some types of duplex stainless steels are 'lean' duplex, 22/5 and '25 Cr'

For some time the use of duplex stainless steels was inhibited by fabrication difficulties. The addition of nitrogen substantially eased these constraints and reduced the need for post-weld heat treatment to restore corrosion resistance.

Recent years have seen the introduction of a new breed of duplex, the superduplex stainless steels. Standard duplexes have limited resistance to pitting and crevice corrosion. The superduplexes include Ferralium 255, Zeron 100 and Uranus 52N+. They are not yet widely used but are starting to find application offshore. They are described by Warde (1991).

12.3.17 Low temperature steel

For low temperature applications use is made of alloy steels or aluminium alloys. Two principal applications are in storage tanks and in liquefied gas carriers. Some of the materials used are illustrated by the following list taken from the IMO *International Gas Carrier* Code (1983 IMO 782):

Minimum design temperature	*Chemical composition and heat treatment* (°C)	*Impact test temperature* (°C)
−60	1.5% nickel steel, normalized	−65
−65	2.25% nickel steel, normalized or normalized and tempered	−70
−90	3.5% nickel steel, normalized or normalized and tempered	−95
−105	5% nickel steel, normalized or normalized and tempered	−110
−165	9% nickel steel, double normalized and tempered or quenched and tempered	−196
−165	Austenitic steels such as types 304, 304L, 316, 316L, 321 and 347 solution treated	−196
−165	Aluminium alloys such as type 5083 annealed	Not required
−165	Austenitic Fe–Ni alloys (36% nickel), heat treatment as agreed	Not required

The qualifications on the use of these materials are given in the code.

12.3.18 Heat treatment of steel

There are a number of processes to which steel may be subjected in order to obtain a better microstructure and enhanced mechanical and corrosion properties. An account oriented to pressure vessels is given by Kohan (1987).

Critical Temperature and Transformation Range

A basic concept in heat treatment of steel is that of the critical temperature at which a transformation occurs in the steel. Actually changes occur over a range of temperature (the transformation range). There are separate transformation ranges for heating and for cooling.

Quenching

Quenching is rapid cooling of the steel by immersion in a medium such as oil or water.

Hardening

Hardening is the heating and quenching of the steel to produce an increase in hardness.

Normalizing

Normalizing involves heating the steel to about 100°F (38°C) above its critical temperature and cooling in still air to room temperature.

Annealing
Annealing involves heating and controlled cooling of the steel in order to refine its structure, modify its properties and/or remove stresses. There is a variety of annealing processes. Full annealing involves heating above the critical temperature and then slow cooling.

Tempering
Tempering is an operation performed after hardening of the steel, and involves heating to a temperature below the critical value and then cooling.

Stress Relief Heat Treatment
Stress relief heat treatment involves uniform heating of all or part of a component to a temperature sufficient to relieve the major part of the residual stresses and then uniform cooling.

Post-weld Heat Treatment
Post-weld heat treatment (PWHT) refers to any heat treatment carried out following welding.

12.3.19 Steel coding systems

There are a number of systems for the coding of types of steel. These include British, European and American systems.

BS 5500: 1991 for pressure vessels refers to BS 1501–1504 and BS 3059 and BS 3601–3605. These standards designate steels by type and grade, e.g. 224 grade 490. Another relevant standard is BS 4360: 1990 *Weldable Structural Steels*. This refers to grades, e.g. 50EE. The 1990 edition gives cross-referencing to nomenclature used in earlier editions of this standard. The standards also use the following classification of steels:

Grade	
M0	Carbon steels
M1	Carbon and carbon manganese steels
M2–M10	Low alloy steels
–	High alloy steels

Details are given in Appendix H of BS 5500.

Currently, in the UK transition is taking place to the European system. This is described in BS EN 10027 *Designation Systems for Steel*; this and BS 10028 *Flat Products Made of Steels for Pressure Purposes* give cross-referencing.

For the USA the ASTM has established a unified numbering system (UNS) and issues standards for steel, e.g. ASTM A-515/A515M-90. ASME utilizes these standards adding the prefix S, e.g. SA-515. Principal steels for process plant are given in the ASME *Boiler and Pressure Vessel Code*, Section II: 1990 *Materials Specifications*. Reference has already been made to the AISI system, particularly in relation to stainless steels. A further discussion of steel coding is given by Pilborough (1989).

12.4 Pressure Vessel Design

Pressure vessels are subject to a variety of loads and other conditions which stress them and, in certain cases, may cause serious failure.

Accounts of pressure vessel design are given in *Pressure Vessels* (Chuse, 1954–), *Pressure Vessel Design and Analysis* (Bickell and Ruiz, 1967), *The Stress Analysis of Pressure Vessels and Pressure Vessel Components* (Gill, 1970), *Theory and Design of Modern Pressure Vessels* (Harvey, 1974), *Pressure Component Construction* (Harvey, 1980), *Chemical and Process Equipment Design: Vessel Design and Selection* (Azbel and Cheremisinoff, 1981), *Pressure Vessels* (Chuse and Eber, 1984), *Pressure Vessel Design Handbook* (Bednar, 1986), *Mechanical Design of Process Systems* (Escoe, 1986), *Fundamentals of Pressure Vessel and Heat Exchanger Technology* (J.P. Gupta, 1986), *Pressure Vessel Systems* (Kohan, 1987), *Pressure Vessel Design Manual* (D.R. Moss, 1987), *Pressure Vessels* (Chuse and Carson, 1993) and *Pressure Vessel Design* (Spence and Tooth, 1994).

In general, structures are subject to two types of loading: (1) static loading, and (2) dynamic loading. For pressure vessels this loading is normally due to pressure. The load caused by the pressure creates stresses in the vessel. There may also be other stresses, which include (1) residual stress, (2) local stress, and (3) thermal stress.

12.4.1 Design basis

The basis of the design of pressure vessels is the use of appropriate formulae for vessel dimensions in conjunction with suitable values of design strength. In determining the strength of materials a basic distinction is drawn between (1) temperatures below the creep range, and (2) temperatures inside the creep range.

For temperatures below the creep range two important properties are (1) tensile strength, R; and (2) yield strength, E. Different design methods may use these properties as measured at room temperature (generally 20°C) or at the operating temperature (t°C). Use has been made, therefore, of all the four properties:

(1) tensile strength at room temperature, R_{20};
(2) tensile strength at operating temperature, R_t;
(3) yield strength at room temperature, E_{20};
(4) yield strength at operating temperature, E_t.

For temperatures inside the creep range two further important properties are (1) stress for 1% extension in 100 000 h at the operating temperature, S_c and (2) stress for rupture in 100 000 h at the operating temperature, S_r.

These material strengths are divided by a factor of safety to obtain the design strengths for use in the design.

12.4.2 Static pressure loading

Pressure vessels may often be treated as thin-walled cylinders or spheres. For a thin-walled cylinder, a simple force balance gives the longitudinal stress:

$$p\pi r^2 = \sigma_1 t 2\pi r \tag{12.4.1}$$

$$\sigma_1 = \frac{pr}{2t} \tag{12.4.2}$$

where p is the pressure, r is the radius of the vessel, t is the wall thickness, and σ_1 is the longitudinal stress.

The circumferential, or hoop, stress is:

$$\sigma_2 = \frac{pr}{t} \tag{12.4.3}$$

where σ_2 is the circumferential stress.

Similarly, for a sphere the longitudinal and circumferential stresses are:

$$\sigma_1 = \sigma_2 = \frac{pr}{2t} \qquad [12.4.4]$$

Other methods are required for thick-walled vessels.

12.4.3 Dilation

If the pressure vessel is subject to stress, it exhibits dilation, or radial growth. For an element which is subject to tensile stress in two perpendicular directions, the strain in one direction depends not only on the stress in this direction but also on that in the perpendicular direction and is a function of Poisson's ratio μ:

$$e_1 = \frac{\sigma_1}{E} - \frac{\mu\sigma_2}{E} \qquad [12.4.5]$$

$$e_2 = \frac{\sigma_2}{E} - \frac{\mu\sigma_1}{E} \qquad [12.4.6]$$

where e_1 is the longitudinal strain, e_2 the circumferential strain and E is the elastic modulus.

The dilation, or radial growth, of a cylindrical pressure vessel may be determined by integrating the hoop strain in the wall from an axis through the centre of rotation and parallel to a radius. If the co-ordinates of a point in the wall are r,ϕ, the dilation δ is:

$$\delta = \int_0^{\pi/2} e_2 r \cos\phi \, d\phi \qquad [12.4.7a]$$

$$\delta = e_2 r \qquad [12.4.7b]$$

Then from Equation 12.4.6

$$\delta = r\left(\frac{\sigma_2}{E} - \frac{\mu\sigma_1}{E}\right) \qquad [12.4.8]$$

and from Equations 12.4.2 and 12.4.3

$$\delta = \frac{pr^2}{2tE}(2-\mu) \qquad [12.4.9]$$

Similarly, for a spherical vessel the dilation is

$$\delta = \frac{pr^2}{2tE}(1-\mu) \qquad [12.4.10]$$

12.4.4 Dynamic pressure loading

The behaviour of a material such as mild steel under dynamic loading differs considerably from its behaviour under static loading. The shape of the stress–strain diagram alters with the rate of loading. At very high strain rates (say $de/dt = 300\ s^{-1}$) the yield point more than doubles. There are also large increases in the ultimate tensile strength and elongation. In consequence of this characteristic, the loads which can be applied to the material without permanent deformation or rupture are considerably greater with dynamic loading than with static loading.

12.4.5 Residual stress

The basic design equations for pressure vessels assume that the stresses are caused only by external loads. They do not take into account residual stresses resulting from fabrication and construction processes such as heat treatment, weld shrinkage or casting cooling. These stresses are of great importance in brittle materials. They are usually rather less significant in ductile materials, which tend to yield and relieve the stress, but they can contribute to fatigue failure.

Moreover, in thick sections residual stresses can combine with load stresses to create a three-dimensional stress pattern which restricts the redistribution of the localized stresses by yielding. This is the reason why stress relieving is particularly important for thick vessels.

12.4.6 Local stress

Another assumption of the basic design equations is that there is continuity of stress. The localized stresses in a region where there is a discontinuity are therefore greater than those predicted. These localized stresses are more important in brittle than in ductile materials, but again they are significant for fatigue failure.

12.4.7 Thermal stress

Most materials undergo expansion if their temperature is raised and contraction if it is lowered. If an unrestrained body is subjected to non-uniform heating, the different parts experience different degrees of strain and these differential strains give rise to corresponding stresses. Such stresses do not develop if the body is heated up uniformly. If the body is restrained, however, stress is developed even though the heating is uniform. There are, therefore two basic causes of thermal stress, non-uniform temperature and restraint.

With a ductile material thermal stresses do not on first application cause failure by rupture, because the material tends to yield and relieve the stress, but their repeated application can cause fatigue failure. In addition, the deflection or distortion due to normal stresses can render equipment unserviceable.

For a body which is restrained in one dimension only, the thermal stress σ is

$$\sigma = -E\alpha\Delta T \qquad [12.4.11]$$

where ΔT is the temperature difference and α is the coefficient of thermal expansion. The product $E\alpha$ is the thermal stress modulus. The minus sign indicates that the body is in compression. For a body restrained in two dimensions

$$\sigma = -\frac{E\alpha\Delta T}{(1-\mu)} \qquad [12.4.12]$$

and for one restrained in three dimensions

$$\sigma = -\frac{E\alpha\Delta T}{(1-2\mu)} \qquad [12.4.13]$$

Pressure vessel conditions mainly correspond to two-dimensional constraint.

Non-uniform temperature distribution may be either steady or unsteady state. The thermal stresses caused by thermal gradients in a long thin-walled cylinder or pipe may be shown to be as follows. For a steady-state, linear temperature gradient through the pipe wall the stress on the inside of the pipe is:

$$\sigma = -\frac{E\alpha\Delta T}{2(1-\mu)} \qquad [12.4.14]$$

For a transient in which a hot fluid is suddenly contacted with the pipe wall, which previously was at a uniform temperature, the initial stress is:

$$\sigma = -\frac{E\alpha\Delta T}{(1-\mu)} \qquad [12.4.15]$$

12.4.8 Design construction features

There are numerous design features which tend to create local stresses. These include:

(1) discontinuities:
  (a) vessel ends,
  (b) changes of cross-section,
  (c) changes of thickness;
(2) joints
  (a) bolted joints,
  (b) welded joints;
(3) bimetallic joints;
(4) holes and openings;
(5) flanges;
(6) nozzles and connections;
(7) bolt seating and tightening;
(8) supports and lugs.

The minimization of the local stresses caused by these stress raisers is a major aspect of pressure vessel design. Some of the principal methods of minimizing local stresses are as follows. Discontinuities are made as gradual as possible and sudden discontinuities are avoided. Weld defects such as overfill are kept to a minimum. Holes are spaced relative to each other so as not to cause serious weakening. The same applies to the spacing of nozzles and other connections. Flanges and nozzles are reinforced so as to minimize stress. The reinforcement relies on good design rather than on massive addition of metal. Self-adjusting nut seats are provided for bolts which have to be tightened on sloping surfaces. Thermal stresses at supports are reduced by measures to allow heat to flow between the vessel and the support, and so avoid large temperature gradients.

12.4.9 Vessels in composite materials

Pressure vessels may also be constructed in composite materials, notably glass reinforced plastic (GRP), or fibre reinforced plastic (FRP). The main use of GRP in the process industries is for atmospheric, or near-atmospheric, storage tanks, but vessels can be built to withstand higher pressures.

Accounts of the use of composite materials are given in *Glass Reinforced Plastic* (Parkyn, 1970), *Fibre Reinforced Materials Technology* (Parratt, 1972), *Fibre Reinforcement* (Catherall, 1977), *FRP Technology* (Weatherhead, 1980), *Materials and the Designer* (Cornish, 1987), *Composite Materials Handbook* (M. Schwartz, 1991) and by Puckett (1976), I.R. Miller (1979) and Britt (1993).

The use of GRP for storage tanks is discussed in Chapter 22.

12.5 Joining, Fastening and Welding

There are two main methods by which materials may be joined: (1) mechanical fastening, and (2) physical bonding. For pressure systems the main fastening methods are riveting and bolts and nuts, while the methods of bonding are soldering and brazing and fusion welding. Although riveting was used as a method of joining in older pressure vessels, some of which are still in use, the modern method of joining is by fusion welding. Fastening by bolts and nuts, however, is a widely used method of joining, particularly on flanged joints. Flanged joints are used mainly in pipework and are considered in Section 12.7.

Soldering and brazing are methods of bonding in which the joint is made by introducing molten metal between the two parts to be joined without deliberate fusion of the parent metal. With fusion welding, by contrast, the parts of the parent metal adjacent to the joint are brought to molten temperature and caused to fuse together, generally with the addition of molten filler material. Fusion welding is the principal method of making permanent joints in pressure vessels and pipework.

An account of methods of joining, fastening and welding is given by Pilborough (1971, 1989).

12.5.1 Fusion welding

In fusion welding the parts to be welded are heated so that they are molten along the line of the joint, and then generally a filler rod, also raised to molten temperature, is brought over the joint. This forms a pool of molten metal which is made to run along the joint and fuse the two parts together so that when they cool and solidify a homogeneous joint is made.

Some of the main methods of fusion welding described are:

(1) oxyacetylene welding;
(2) metallic arc welding;
(3) inert gas shielded welding:
  (a) argon arc welding,
  (b) CO_2 gas shielded welding,
  (c) pulsed arc welding;
(4) electrical resistance welding:
  (a) spot welding,
  (b) seam welding,
  (c) projection welding,
  (d) butt welding,
  (e) flash welding.

Oxyacetylene welding is cheap and is widely used. It is suitable where high quality welds are not required and where corrosion is not a serious problem. In oxyacetylene welding heat is generated by combustion of acetylene in oxygen using a blowtorch and the weld is made using a filler rod. The flame should be kept as nearly neutral as possible. Excess acetylene tends to cause carburization of the weld, leading to lower corrosion resistance and possibly embrittlement, while excess oxygen generally gives an oxidized and porous weld.

Metallic arc welding is perhaps the most commonly used method. It is suitable for welding a wide range of plate thicknesses, from thin plate up to plate many centimetres thick, and a wide range of materials including stainless steel. In metallic arc welding an electrode of the same metal as the work is used and an arc is struck between the two. The heat of the arc melts both the tip of the electrode and the parent metal beneath it, and metal drops across the arc from this electrode to the parent metal. Both a.c. and d.c. are used, the former being cheaper, but the latter being more suitable for some high quality work and for stainless steel.

Inert gas shielded welding is widely used to obtain a high quality weld. In the UK the inert gas used is generally argon and the method is also known as 'argon arc welding'. Helium is normally used in the USA. The two main methods used are tungsten inert gas (TIG) and metal inert gas (MIG) welding. In the TIG technique an arc is struck between the tungsten electrode and the work in an inert atmosphere. Filler is added as necessary in the form of a bare wire. In the MIG process the arc is struck and a bare wire filler and inert gas are fed in through a feeder gun or head.

An alternative gas used in inert gas shielded welding is CO_2, which is much cheaper than argon. CO_2 gas shielded welding tends to be used for high quality work on ferrous metals and argon shielded welding for high quality work on stainless steel and on other materials.

The quality of the weld in MIG welding is affected by the mode of transfer of the molten metal to the work. The most desirable mode is spray transfer, in which the metal passes from the wire to the weld pool as small drops, but there occur also dip transfer, in which metal is deposited when the wire is dipped in the pool and which causes irregularity, and globular transfer, in which large drops are released. Spray transfer is favoured by high current, but the latter tends to burn through on thin material and a compromise is necessary.

Pulsed arc welding overcomes this problem and gives spray transfer without burn-through by using a high pulsed current to effect transfer, while maintaining a relatively low mean welding current.

There are a number of electrical resistance welding processes in which an electric current is passed through the work to be joined; the material is heated and rendered molten by the electrical resistance at the joint and is thus welded together.

In spot welding the points to be joined are clamped together between two shaped electrodes. A heavy surge of current is passed between these electrodes and welds the work together by a small spot.

In seam welding a series of spot welds is produced by a pulsed current from a rotating disc electrode. This method does not give a completely continuous weld. There is some tendency, therefore, for seam welds to leak and to entrap materials and corrode.

In projection welding projections are raised on the parts to be joined, the parts are pressed together between the electrodes and the projections are fused by the weld current.

In butt welding special projections are not used. The parts are pressed together between the electrodes and the adjoining surfaces are fused by the welding current. The adjoining faces in butt welding should be flat and parallel. In flash, or flash-butt, welding the parts are again pressed together between electrodes, a current is passed which causes the slight roughnesses of finish to melt and flash-off, the parts are pressed closer and more flashing occurs and, at the right moment, further pressure is applied, the parts fuse together and the current is switched off. Butt welding is used to join smaller components and flash welding large ones.

The terminology used in arc welding in the USA is described in the account by J.E. Jones and Olson (1986). They classify arc welding processes by type of shielding and by whether or not the electrode is consumed; they distinguish the following principal types:

Method	*Shielding*	*Electrode type*	*Usual thickness* (in.)
1. Shielded-metal arc welding (SMAW)	Flux plus coating; some generated gas	Solid wire, coated with flux	0.25–4
2. Submerged arc welding (SAW)	Granulated flux	Solid wire, may be copper flashed	0.1–2
3. Gas-metal arc welding (GMAW)	Inert gas or gas mixture	Solid wire, may be copper flashed	0.05–2
4. Fluxed-core arc welding (FCAW)	Flux and/or gas; may be self-shielding	Hollow electrode with core of metal powder	0.05–1.5
5. Gas-tungsten arc welding (GTAW)	Inert gas or gas mixture	Solid tungsten wire, water or air cooled	0.05–0.4
6. Plasma arc welding (PAW)	Inert gas or gas mixture	Solid tungsten wire, water cooled	0.04–0.24

In the first four of these methods the electrode is consumed, whilst in the last two it is not. The authors give details of the materials for which the different methods are suitable and of typical applications.

The American Welding Society (AWS) issues standards for most of these methods, with individual standards covering the arc welding of different materials such as carbon steel, stainless steel, etc. Thus, for example, AWS B2.1.001–90, B2.1.002–90 and B2.1.008–90, and B2.1004–90 cover welding of carbon steel by the SMAW, GTAW and GMAW methods, whilst B2.1.0013–91 and B2.1.005–90 cover welding of austenitic stainless steel by the SMAW and GMAW methods, respectively.

The qualifications of welders, the inspection of welds and defects in welds are considered in Chapter 19.

12.6 Pressure Vessel Standards and Codes

The design, fabrication and construction of pressure vessels is a subject which is particularly well covered by standards and codes of practice. Pressure vessel codes fall into two main types. The first type is the conventional codes, the second the so-called 'advanced codes'. The differences between the two types of code are considered below.

In the UK the main pressure vessel codes until 1976 were BS 1500: 1958 *Fusion Welded Pressure Vessels for General Purposes* and BS 1515: 1965 *Fusion Welded Pressure Vessels for Use in the Chemical, Petroleum and Allied Industries*. BS 1500 was a conventional code and BS 1515 was an advanced code. In 1976 both these codes were superseded by a new pressure vessel code BS 5500. The current version of this code is BS 5500: 1991 *Specification for Unfired Fusion Welded Pressure Vessels*.

The corresponding US code is the ASME *Boiler and Pressure Vessel Code*: 1992; this is described in Subsection 12.6.5.

Another code is ISO R831:1968 *Recommendations for Stationary Boilers*, which is applicable to pressure vessels also.

There is an EC Directive 77/767/EEC on pressure vessels.

12.6.1 Conventional and advanced codes

The maximum design stresses allowed in a code either are tabulated in the code or are calculated from materials properties using factors of safety given in the code. The difference between the conventional and the advanced codes lies mainly in the design stress considered. These differences may be illustrated by considering on the one hand BS 1500: 1958 and ASME Section VIII, Division 1, and on the other BS 1515: 1965 and ASME Division 2.

At temperatures below the creep range the conventional codes BS 1500: 1958 and ASME Division 1 use as the design stress the tensile strength divided by a factor of safety, which is 4 in both codes, whilst the advanced codes BS 1515: 1965 and ASME Division 2 use the yield stress at design temperature divided by a factor of safety, which in BS 1515: 1965 is 1.5. This different design criterion favours the use of steel with high yield strength such as low alloy steels. The use of low alloy steels makes possible the design of thinner walled pressure vessels.

At temperatures within the creep range the design stress used is the creep strength, again in conjunction with a suitable factor of safety. The difference between the two types of code may be seen in the comparisons of design strengths given in the earlier BS codes by L. Evans (1974) and in the ASME code by Pilborough (1989). For a particular material at a particular temperature the maximum permissible design strength is the lowest strength obtained by dividing the specified properties by the specified factor of safety. For BS 1500 and BS 1515 Evans gives the design strengths in terms of the material tensile strengths R_{20} and R_t, the yield strengths E_{20} and E_t and the creep strengths S_r and S_c (as defined in Section 12.4), and an appropriate factor of safety.

BS 1500: 1958	$\frac{R_{20}}{4}$	$\frac{R_t}{4}$				$\frac{S_c}{1}$
BS 1515: 1965	$\frac{R_{20}}{2.35}$			$\frac{E_t}{1.5}$	$\frac{S_r}{1.5}$	$\frac{S_c}{1}$

The comparison given by Pilborough (1989) for the two divisions of the ASME code and also for ISO R831 may be summarized as follows. For each code the design strength is the lowest of:

ASME Section VIII Division 1	$\frac{R_{20}}{4}$	$\frac{R_t}{4}$	$\frac{E_{20}}{1.6}$	$\frac{E_t}{1.6}$	$\frac{S_r}{1.25}$	$\frac{S_c}{1}$
ASME Section VIII Division 2		$\frac{R_t}{3}$		$\frac{E_t}{1.5}$		$\frac{S_c}{1.5}$
ISO R831		$\frac{R_t}{2.7}$		$\frac{E_t}{1.5}$	$\frac{S_r}{1.5}$	$\frac{S_c}{1}$

The design strengths laid down in BS 5500: 1991 and the ASME code are given below.

12.6.2 Pressure vessel classification

BS 5500: 1991 gives a classification of pressure vessels. An earlier classification was given in BS 1500: 1958. The latter classified pressure vessels as Class I, II or III. For a Class I vessel the requirements include full radiography of the main seam welds, while Classes II and III have less stringent requirements. A Class I vessel also requires PWHT for stress relief.

BS 5500: 1991 places pressure vessels in three construction categories. For a Category 1 vessel full non-destructive testing of the main seam welds is required, while for Categories 2 and 3 the requirements are limited to spot non-destructive testing and visual examination, respectively. The PWHT requirements are related to steel type and thickness rather than to vessel category.

12.6.3 BS 5500

The coverage of BS 5500: 1991 *Specification for Unfired Fusion Welded Pressure Vessels* is indicated by the list of contents given in Table 12.3. The standard applies to the design, construction, inspection, testing and certification of pressure vessels. It does not cover certain defined types of vessel, which may be summarized as atmospheric storage tanks, vessels for very high pressure (e.g. strip wound compound vessels), vessels in which the stresses are less than 10% of the permitted design stress, vessels for specific applications covered by other British Standards and pressure vessels used in transport.

Specific recognition is given in the standard to the Inspecting Authority, which acts on behalf of the purchaser, and the Regulating Authority, which enforces the laws and regulations in the country concerned.

With respect to materials of construction the standard allows the use either of materials specified in British Standards or of other materials, provided these meet specifications agreed between the manufacturer and the user for certain specified properties as tested by specified methods listed in the standard. The latter should cover as a minimum the manufacturing process, compositional limits for all constituents, deoxidation practice, heat treatment and appropriate mechanical properties.

BS 5500: 1991 lists in its Table 2.3 and Appendix K the other standards which it uses for the design strength of steels. These include BS 1501–1504, BS 3059 and BS 3601–3605. It gives in Tables 2.3(a)–(j) extensive tabulations of design strengths from these standards.

The basis for the determination of design strengths in the standards just listed is described in Appendix K of BS 5500. A distinction is made between (1) the time-independent design strength and (2) the time-dependent design strength, in the creep region. For the former, distinctions are made between materials with and without specified elevated temperature properties and between carbon, carbon manganese and low alloy steels and austenitic stainless steels. Thus, for the time-independent strength, relations are given for the following four cases:

(1) carbon, carbon manganese and low alloy steels:
   (a) materials with specified elevated temperature properties,
   (b) materials without specified elevated temperature properties;

(2) austenitic stainless steels:

Table 12.3 Principal contents of BS 5500: 1991

1. General
   - 1.1 Scope
   - 1.2 Interpretation
   - 1.3 Definitions
   - 1.4 Responsibilities
   - 1.5 Information and requirements to be agreed and documented
2. Materials
   - 2.1 Selection of materials
   - 2.2 Materials for low temperature applications
   - 2.3 Carbon, carbon manganese and alloy steels
3. Design
   - 3.1 General
   - 3.2 Application
   - 3.3 Corrosion, erosion and protection
   - 3.4 Construction categories and design stresses
   - 3.5 Vessels under internal pressure
   - 3.6 Vessels under external pressure
   - 3.7 Supports, attachment and internal structures
   - 3.8 Bolted flanged connections
   - 3.9 Flat heat exchanger tubesheets
   - 3.10 Design of welds
   - 3.11 Jacket construction
   - 3.12 Manholes and inspection openings
   - 3.13 Protective devices for excessive pressure or vacuum
4. Manufacture and workmanship
   - 4.1 General aspects of construction
   - 4.2 Cutting, forming and tolerances
   - 4.3 Welded joints
   - 4.4 Heat treatment
   - 4.5 Surface finish
5. Inspection and testing
   - 5.1 General
   - 5.2 Approval testing of fusion welding procedures
   - 5.3 Welder and operator approval
   - 5.4 Production control test plates
   - 5.5 Destructive testing
   - 5.6 Non-destructive testing
   - 5.7 Acceptance criteria for weld defects revealed by visual examination and non-destructive testing
   - 5.8 Pressure tests

Also various appendices including:

- B Recommendations for cylindrical, spherical and conical shells under combined loadings, including wind and earthquakes
- C Recommendations for the assessment of vessels subject to fatigue
- D Requirements for ferritic steels in bands M0 to M4 inclusive for vessels required to operate below 0°C
- E Recommendations for welded connections of pressure vessels
- G Recommendations for methods of calculation of stresses from local lads, thermal gradients, etc.
- H Recommendations for post-weld heat treatment of dissimilar ferritic steel joints
- J Recommendations for pressure relief protective devices
- K Requirements for the derivation of material nominal design strengths for construction Category 1 and 2 vessels
- U Guidance on use of fracture mechanics analysis

(a) materials with specified elevated temperature properties,
(b) materials without specified elevated temperature properties.

The nominal design strength f is taken as the lesser of the nominal design strength f_e corresponding to the short-time tensile strength characteristics and the nominal design strength f_F corresponding to the creep characteristics. The material strengths used are as follows. R_m is the minimum tensile strength at room temperature. R_e is the minimum yield strength at room temperature. Where a standard specifies minimum strength values of R_{eL} or $R_{p,\ 0.2}$, these values are taken as corresponding to R_e. $R_{e(T)}$ corresponds to R_{eL} or $R_{p,\ 0.2}$ at temperature T. S_{Rt} is the mean stress required to produce rupture at time t at temperature T. The relations given in Appendix K of the standard for the determination of the nominal design strength f are shown in Table 12.4. The design strength for a typical steel, BS 1501 26B, in plate form up to a thickness of 32 mm, which has a minimum tensile strength of 402 N/mm^2, is given in Table 12.5.

The equations given in the standard for minimum shell thickness for pressure loading only are as follows:

Cylinder:

$$e = \frac{pD_i}{2f - p} \qquad [12.6.1a]$$

or

$$e = \frac{pD_o}{2f + p} \qquad [12.6.1b]$$

Sphere:

$$e = \frac{pD_i}{4f - 1.2p} \qquad [12.6.2]$$

or

$$e = \frac{pD_o}{4f + 0.8p} \qquad [12.6.3]$$

where D_i is the internal diameter of the shell, D_o is the external diameter of the shell, e is the minimum thickness of the shell, f is the nominal design stress, and p is the design pressure. The dimensions do not include corrosion allowances.

The standard gives guidance on the design pressures and temperatures to be used. For the maximum design pressure it states that the design pressure, or the pressure to be used in the equations for the purposes of calculation, should not be less than:

(a) the pressure which will exist in the vessel when the pressure relieving device starts to relieve, or the set pressure of the pressure relieving device, whichever is the higher (see Appendix J of standard);
(b) the maximum pressure which can be attained in service where this pressure is not limited by a relieving device.

The static head should be taken into account.

For the minimum design pressure, the standard recommends that vessels subject to vacuum should be designed to a full negative pressure of 1 bar.

Table 12.4 *Design strengths given in BS 5500: 1991*

Material	*Regime*	*Temperature range (°C)*	*Design strength lowest of*
Carbon, carbon manganese and low alloy steels	Time-independent[a]	≤50	$f_E = \frac{R_e}{1.5}$ or $\frac{R_m}{2.35}$
		>150[b]	$f_E = \frac{R_{e(T)}}{1.5}^c$ or $\frac{R_m}{2.35}$
Austenitic stainless steel	Time-independent	≤50	$f_E = \frac{R_e}{1.5}$ or $\frac{R_m}{2.5}$
		>150[b]	$f_E = \frac{R_{e(T)}}{1.35}^d$ or $\frac{R_m}{2.5}$
Both types of steel	Time-dependent		$f_F = \frac{S_{Rt}}{1.3}$

[a] Values given for time-independent nominal design strengths are for material with specified elevated temperature values.
[b] For temperatures between 50 and 150°C values are interpolated.
[c] For materials without specified elevated temperature values, value is $R_e(T)/1.6$.
[d] For materials without specified elevated temperature values, values is $R_e(T)/1.45$.
[e] See detailed notes in Appendix K of the standard.

Table 12.5 *Design strengths (f, N/mm²) for steel 224, 490 M1 (A and B) for plate up to 40 mm thick, as given in BS 5500: 1991 (Courtesy of the British Standards Institution)*

R_m (N/mm²)	R_e (N/mm²)	*Thickness* (mm)	*Design temperatures (°C) not exceeding* 50	100	150	200	...	480	*Design lifetime* (h)
490	325	3 to 16	208	192	177	160	...	42	100 000
	315	40	208	190	173	158	...	38	150 000
	305	63	204	187	171	156	...	34	200 000
	281	100	187	176	164	153	...	34	250 000
	250	150	167	162	157	150			

The maximum design temperature should not be less the actual metal temperature expected in service, allowing adequate margin for uncertainties in predicting this temperature. An appropriate design lifetime should be agreed between purchaser and manufacturer. No vessel designed on this basis should remain in service beyond the agreed lifetime without a review as specified in the standard.

The minimum design temperature, which is used to assess the suitability of the material to resist brittle fracture, should be the lowest metal temperature expected in service. In the case of components with external thermal insulation, this temperature is that of the contents of the vessel under the appropriate loading conditions, whilst in the uninsulated case the minimum temperature is a matter for agreement.

Some of the topics on which the standard gives guidance are (1) local stresses, (2) low temperature properties, (3) corrosion and (4) fatigue. The calculation of local stresses is covered rather comprehensively in the standard, which gives a large number of charts for particular cases.

The standard deals in Appendix D with the suitability of steels in bands M0 to M4 for low temperature service, below 0°C. A design reference temperature θ_R is defined such that

$$\theta_R \leq \theta_D + \theta_S + \theta_C + \theta_H \quad [12.6.4]$$

where θ_D is the minimum design temperature (°C), and θ_R is the design reference temperature (°C), and θ_C, θ_H and θ_S are temperature adjustments. θ_C is an adjustment which depends on the construction category and has the values 0, – 10 and – 20°C for Category 1, 2 and 3 vessels, respectively. θ_H is an adjustment in applications where all plates incorporating subassemblies undergo PWHT before they are butt welded together, but the main seams do not subsequently undergo PWHT. In these applications θ_H is +15°C. θ_S is an adjustment depending on the calculated membrane stress which has the value 0°C when this stress is equal to or exceeds 2f/3, +10°C when the stress is equal to or exceeds 50 N/mm² but does not exceed 2f/3; and +50°C when the stress does not exceed 50N/mm². In the latter case the membrane stress should take into account internal and external pressure, static head and self-weight.

Where the calculated membrane stress can vary with the minimum design temperature, for example in autorefrigeration during depressurization, the standard requires that the coincident values of θ_D and θ_S be evaluated, where appropriate, allowing for the possibility of repressurization whilst still cold, and gives detailed guidance.

The standard gives graphs showing the permissible design reference temperature/reference thickness/material impact test temperature relationships for (1) as-welded and (2) post-weld heat treated components.

Guidance on corrosion given in the standard is fairly general. It is recommended that the possible forms of corrosion such as chemical attack, rusting, erosion and high temperature oxidation be reviewed, that particular attention be paid to impurities and to fluid velocities, and that where doubt exists corrosion tests be done. If corrosion is expected to be negligible, no corrosion allowance is required, but where this is not the case, the recommended minimum allowance is 1 mm.

The standard deals in Appendix C with fatigue, for which it lists the following causes: (1) periodic temperature transients; (2) restrictions of expansion or contraction during normal temperature variations; (3) applications or fluctuations of pressure; (4) forced vibrations; and (5) variations in external loads. Attention is drawn to the influence on fatigue of corrosion and of creep.

The standard requires that the design stress be kept reasonably below the fatigue limit if the expected number of cycles during the service life of the vessel might otherwise lead to fatigue failure. A detailed fatigue analysis is not required, however, where the design is based on previous and satisfactory experience of strictly comparable service or where certain alternative conditions are satisfied. A generalized chart for a design fatigue curve is given in the standard. It also gives a recommended method of constructing a design fatigue curve from test data.

The standard gives requirements for protective devices and for inspection. These are described here in Section 12.12 and Chapter 19, respectively.

12.6.4 Standards committees

Support for BS 5500: 1991 is provided by the Pressure Vessels Technical Committee (PVTC) (PVE/1) of the British Standards Institute (BSI). There is also a Pressure Vessel Quality Assurance Board (PVQAB) set up under the sponsorship of the Institution of Mechanical Engineers (IMechE).

12.6.5 ASME *Boiler and Pressure Vessel Code*

Another principal code is the American Society of Mechanical Engineers (ASME) *Boiler and Pressure Vessel Code*: 1992. The 11 sections of the code are shown in Table 12.6. Of particular relevance here are Sections II and VIII which deal, respectively, with materials specifications and with pressure vessels themselves. Section VIII has two parts: *Pressure Vessels, Division 1* and *Pressure Vessels, Division 2—Alternative Rules*. Division 1 is a conventional code and Division 2 an advanced code.

Table 12.6 *Principal contents of ASME* Boiler and Pressure Vessel Code

Section No.	
I	Power boilers
II	Material specification
III	Nuclear power plant components
IV	Heating boilers
V	Non-destructive examination
VI	Recommended rules for care and operation of boilers
VII	Recommended rules for care of power boilers
VIII	Pressure vessels: Division 1 Division 2 – Alternative rules
IX	Welding and brazing qualifications
X	Fiber-glass reinforced plastic pressure vessels
XI	Rules for in-service inspection of nuclear power plant components

12.6.6 Fibre reinforced plastic vessels

There are also a number of standards and codes for the design of vessels and tanks in fibre reinforced plastic. Relevant standards and codes are BS 4994: 1987 *Specification for Design and Construction of Vessels and Tanks in Reinforced Plastic*; the ASME *Boiler and Pressure Vessel Code*, Part X, *Fiber-Glass Reinforced Plastic Pressure Vessels* (1992); ASTM D 4021–86 *Standard Specification for Contact Moulded Glass-fiber-reinforced Thermosetting Resin Underground Petroleum Storage Tanks* and ASTM D 4097–88 *Standard Specification for Contact Moulded Glass-fiber-reinforced Thermosetting Resin Chemical Resistant Tanks*. HSE guidance is given in PM 75 *Glass Reinforced Plastic Vessels and Tanks: Advice to Users* (1991).

12.6.7 Limitations of standards

It is important to appreciate that standards and codes of practice have certain inevitable limitations. They are by their nature generalized and cannot readily cover all situations that may arise. They are prepared by committees and may represent the minimum standard on which agreement can be obtained. Their revision is a time-consuming task and they can become out of date. A fuller discussion of standards and codes is given in Appendix 27. These points have particular relevance to plants which have major hazards and/or novel technology. In these cases it should be appreciated that the standards represent a minimum requirement. Nevertheless, standards and codes of practice have an essential role to play in the design and operation of process plant.

12.7 Pipework and Valves

Loss of containment from a pressure system generally occurs not from pressure vessels but from pipework and associated fittings. At least as much attention should be paid to the pipework and fittings as to the vessels.

The cause of the Flixborough disaster was a modification to a 28 in. pipe connection between two reactors. The modification involved the installation of a temporary 20 in. pipe with bellows at each end. The design of the pipe system was defective in that it did not take into account the bending moments on the pipe due to the pressure in it. The bellows were not installed in accordance with the manufacturer's instructions. The pipework assembly was not adequately supported. The relevant British Standards, notably BS 3351 and 3974, were not followed. Further details are given in Appendix 2.

12.7.1 Pipework

The plant pipework and fittings include the piping itself, flanges and joints, and fittings, such as the many types of valves, bellows, etc., together with the pipe supports.

Accounts of pipework on process plants include those given in *Handbook of Industrial Pipework Engineering* (Holmes, 1973), *Process Piping Design* (Weaver, 1973), *Handbook of Pipeline Engineering Computations* (Marks, 1979), *Piping Stress Handbook* (Helguero, 1986), *Piping and Pipe Support Systems* (P.R. Smith and van Laan, 1987) and the *Engineering Design Guidelines* of the Center for Chemical Process Safety (CCPS, 1993/13).

Pipework in process plant in the UK has been covered by BS 3351: 1971 *Piping Systems for Petroleum Refineries and Petrochemical Plants* and by BS 3974: 1966 *Pipe Supports,* together with numerous other British Standards for pipework and fittings, some of which are listed in Appendix 27.

BS 3351: 1971 is now withdrawn. The code now commonly used is the ASME B31 *Code for Pressure Piping.* Also relevant until its withdrawal was the American Petroleum Institute *Guide for Inspection of Refinery Equipment,* Chapter 11, *Pipes, Valves and Fittings* (API, 1974). ASME B31.3 is described below, but first there are some preliminary points to be made.

A large proportion of failures of containment in process plants occur on the pipework and fittings. Some suggestions for reducing pipework failures have been given by Kletz (1984k) as part of a survey of such failures, which is described in Section 12.30. The design of pipework should be done by a fully integrated design organization working in a structured manner. There should be a relatively small number of designers of high quality making full use of computer aids. Similar principles should apply to the fabrication and construction stages.

Kletz recommends detailed design of even small bore pipework, though he recognizes that some organizations consider this impractical. He states that efforts should be made to reduce the number of grades of steel required so as to reduce the chance of installation of the incorrect grade, and instances restriction of steam temperatures so as to avoid the need for creep resistant steel. His survey highlights the high proportion of failures which are attributable to the construction phase and he makes suggestions for improved inspection during and after construction. Kletz states that the incidents which he lists suggest that of all the measures proposed this would be the most effective.

The pipework should be designed for ease of maintenance. If a joint may have to be broken, there should be adequate access and sufficient 'spring' in the pipework. If the insertion of a slip plate into a joint is likely to be a frequent operation, consideration should be given to installing a slip ring or spectacle plate.

Work on safe piping systems has been the subject of a study by the Institution of Chemical Engineers (IChemE), as described by Hancock (1990a). The principal features considered were (1) layout, (2) quality control, (3) construction, (4) pipe supports, (5) dead ends and (6) vibration and the principal causes of failure (1) vibration, (2) external corrosion, (3) temporary supports, (4) blocked in liquids, (5) water hammer, (6) steam hammer, (7) cavitation and (8) pressure surge.

12.7.2 ASME B31.3

The ASME B31 *Code for Pressure Piping* includes, in particular, Sections B31.1: 1992 *Power Piping,* B31.3: 1990 *Chemical Plant and Petroleum Refinery Piping,* B31.4 *Liquid Transportation Systems for Hydrocarbons, Liquid Petroleum Gas, Anhydrous Ammonia, and Alcohol,* B31.5: 1987 *Refrigeration Piping,* B31.8 *Gas Transmission and Distribution Systems,* and B31.9: 1988 *Building Services Piping.* In this chapter it is B31.3, the ASME *Chemical Plant and Refinery Piping Code,* which is the most relevant. The principal contents of ASME B31.3 are shown in Table 12.7.

Table 12.7 *Principal contents of ASME B31.3*

Chapter		
I	Scope and definition	
II	Design	
	Part 1	Conditions and criteria
	Part 2	Pressure design of piping components
	Part 3	Fluid service requirements for piping components
	Part 4	Fluid service requirements for piping joints
	Part 5	Flexibility and support
	Part 6	Systems
III	Materials	
IV	Standards for piping components	
V	Fabrication, assembly and erection	
VI	Inspection, examination and testing	
VII	Non-metallic piping and piping lined with non-metals[a]	
VIII	Piping for Category M fluid service[a]	
IX	High pressure piping[a]	

Also various appendices, including:

A	Allowable stresses and quality factors for metallic piping and bolting materials
B	Stress tables and allowable pressure tables for non-metals
C	Physical properties of piping materials
E	Reference standards
F	Precautionary considerations
G	Safeguarding
K	Allowable stresses for high pressure piping
M	Owner's guide to classifying fluid services
X	Metallic bellows expansion joints

[a] Chapters VII–IX have a structure broadly similar to, but in some cases more detailed than, that of Chapter II.

B31.3 defines the following categories of fluid service: (1) Category D, fluid service; (2) Category M, fluid service; (3) high pressure fluid service; and (4) normal fluid service. Category D applies where: (1) the fluid handled is non-flammable, non-toxic and not damaging to human tissues; (2) the design pressure does not exceed 150 psig (10.3 barg); and (3) the design temperature is in the range − 20 to 366°F (− 29 to 186°C). The term 'damaging to human tissues' describes a fluid such that 'exposure to the fluid, caused by leakage under expected operating conditions, can harm skin, eyes, or exposed mucous membranes so that irreversible damage may result unless prompt restorative measures are taken.'

Category M, which is particularly relevant here, is 'a fluid service in which the potential for personnel exposure is judged to be significant and in which a single exposure to a very small quantity of a toxic fluid, caused by leakage, can produce irreversible harm to persons on breathing or bodily contact, even when prompt restorative measures are taken.'

High pressure fluid service is one for which the owner specifies the use of Chapter IX of the code. Normal fluid service is fluid service not covered by the other three categories and not subject to severe cyclic conditions, and is the most commonly occurring service.

Appendix M of the code gives a guide to classification of fluid services (Figure M-1).

The code utilizes a system of listed materials and components, which are those conforming to a specification given in Appendices A, B or K or to certain listed standards (Tables 326.1, A326.1 and K326.1).

It also utilizes certain prefixes. These include A for entries in Chapter VII on non-metallic piping, M for those in Chapter VIII on Category M fluid service piping, and K for those in Chapter IX on high pressure piping.

Dealing first with the provisions for metallic pipework, the code gives guidance in Chapter II on design pressures and temperatures. Essentially, the design pressure of a component should not be less than the pressure at the most severe condition to which the equipment may be subject, the most severe condition being that combination of pressure and temperature which requires the greatest component thickness and highest component rating.

For a straight pipe under internal pressure the code gives for the pipe thickness:

$$t_m = t + c \qquad [12.7.1]$$

where c is the sum of mechanical, corrosion and erosion allowances, t is the pressure design thickness and t_m is the minimum required thickness. The pressure design thickness t is given by:

$$t = \frac{PD}{2(SE + PY)} \qquad [12.7.2]$$

It also allows three alternative equations, one of which is

$$t = \frac{PD}{2SE} \qquad [12.7.3]$$

where D is the outside diameter of the pipe, E is the quality factor, P is the internal design gauge pressure, S is the stress value for the material, and Y is a coefficient. Values for the quality factor E are given in Appendix A of the code and those for the coefficient Y are given in a table. These equations apply for $t < D/6$. For $t \geq D/6$ or $P/SE > 0.385$ the design is more complex and requires account to be taken of factors such as thermal stress and fatigue, and application of failure theory.

The design temperature of a component should be that at which, in combination with the relevant pressure, the greatest component thickness and highest component rating are required. In determining the design temperatures, factors to be considered include fluid temperatures, ambient temperatures, solar radiation, and heating and cooling medium temperatures.

The code gives guidance on the component temperatures to be assumed for uninsulated, externally insulated and internally insulated components. For the former, for fluid temperatures below 150°F (65°C) the component temperature is taken as that of the fluid, unless there are factors such as solar radiation which indicate that a higher value should be taken. For fluid temperatures above 150°F the component temperature is taken as a given fraction of the fluid temperature, as follows: (1) 95% of the fluid temperature for pipes, valves, lapped ends, welding fittings, and other components of wall thickness comparable to that of the pipe; (2) 90% for flanges, except lap joint flanges; (3) 85% for lap joint flanges; and (4) 80% for bolting. For externally insulated pipework the component temperature should be taken as the fluid temperature unless some other temperature can be justified by calculation, test or experience.

The code deal with materials in Chapter III, in Appendices A–C and also in other parts. For severe cyclic conditions only certain types of pipe may be used, as listed in Chapter II (paragraph 305.2.3). The code gives guidance on materials suitable for use at high temperatures and those suitable for low temperatures. It contains a table of requirements for low temperature toughness tests for metals (Table 323.2.2). This table gives design minimum temperatures.

Flanges, valves and other fittings are dealt with in Chapter II of the code and also in Chapter IV and Appendix F. Chapter IV gives for flanges, valves and other fittings, listed components, each with its appropriate standard (Table 326.1), together with guidance on the use of unlisted components.

Appendix F of the code on precautionary considerations provides guidance on certain potential hazards and weak points in pipework. It highlights certain hazards such as expansion of trapped liquids and fluid transients including geysering and bowing during cooldown, but is mainly concerned with: (1) valves; (2) flanged joints, including flanges, gaskets and bolting; and (3) materials.

The integrity of a flanged joint depends on the flanges, the gasket and the bolting. For the flanges themselves, the code gives guidance on flange rating, type, facing and facing finish. It also gives guidance on selection of gaskets.

Bolting covers bolts, bolt studs, studs, cap screws, nuts and washers. The codes gives listed bolting. It also gives guidance for conditions subject to high pressure, to high and low temperatures and temperature cycling and to vibration. For high, low and cycling temperatures and for vibration conditions consideration should be given to the use of controlled bolting procedures to reduce (1) joint leakage due to differential thermal expansion and (2) stress relaxation and loss of bolt tension.

With respect to valves, the use of extended bonnet valves is recommended where appropriate to establish a differential between the temperature of the valve stem packing and that of the fluid. This can be an effective way of avoiding problems of packing shrinkage, leakage and icing. Attention is also drawn to the potential for entrapment of liquids in valves such as double-seated valves, with its attendant liquid expansion problem.

Metallic bellows expansion joints are treated in the code in Appendix X, which delineates the relative responsibilities of the pipework designer and the expansion joint manufacturer, and gives guidance on: expansion joint design conditions; joint design, including fatigue analysis; joint support and restraint; and fabrication, examination and pressure and leak testing. Reference is made to the standards of the Expansion Joint Manufacturers Association (EJMA). The use of expanded joints is not allowed in Category M (fluid service) or high pressure piping.

Appendix G of the code on safeguarding deals with measures to protect pipework against accidental damage and to minimize the consequences of pipework failure. Normal safeguards listed include: (1) plant layout; (2) equipment, such as instrumentation, ventilation and fire protection; (3) systems of access and of work; (4) means for fluid containment, recovery or disposal; and (5) operating procedures. Whilst in most cases additional engineered safeguards are not necessary, in some they may be. Engineered safeguards listed include: (1) temperature protection by instruments, thermal insulation or thermal shields; (2) mechanical protection by shields, barricades, etc.; (3) vibration protection; and (4) measures to mitigate pipework failure.

The code addresses the question of the use of isolation valves, or stop valves, in pressure relief piping. Stop valves between the piping being protected and the protective device(s) and between the latter and the point of discharge are not allowed, except subject to conditions. Essentially these are: (1) the stop valves are so constructed and subject to such positive control that any closure made does not reduce the pressure relief capacity; (2) the stop valves are either full-area valves or are such that they will not affect either the operation or the capacity of the pressure relief device(s); and (3) the stop valves are so arranged that they can be locked or sealed in both the open and closed positions. The procedures to be followed if stop valves installed according to these principles are to be closed whilst the plant is operating are given in Appendix F of the code. This requires the presence of an authorized person who has the means to observe the pressure in the equipment and to relieve it in the event of overpressure. Before leaving the scene, this person should lock or seal open the stop valves.

The requirements for examination of the pipework are given in Chapter VI of the code. For piping for normal fluid service the minimum requirements are certain specified visual examinations, covering materials, fabrication, longitudinal welds, assemblies and erection and examination by radiographic or ultrasonic methods of a random sample of 5% of circumferential butt welds and mitre groove welds.

Chapter VI also contains the provisions for testing. For normal fluid service piping the essential requirement is for a hydrostatic test. The test is performed using water and the test pressure is 1.5 times the design pressure, but with an adjustment for the difference between the test and design temperatures and with a limitation to ensure that the yield strength is not exceeded. There are various alternative provisions such as those for the use of a pneumatic test and of hydraulic test fluids other than water.

High pressure pipework is dealt with in Chapter IX of the code. The basic approach is similar, but the requirements are more stringent in respect of a number of features. There is a different formula for the thickness of straight pipe under internal pressure.

Piping for Category M fluid service is treated in Chapter VIII. Again, the requirements for certain features are more stringent.

12.7.3 Pipe supports

Pipe supports are dealt with in BS 3974: 1974. This standard lists the following loads for which supports are required: (1) mass of the pipe, including operating or testing medium, insulation and associated equipment; (2) expansion and contraction; (3) reaction due to pumping effects and discharge to atmosphere; and (4) wind, snow or ice loads.

Equations are given for the calculation of maximum bending stresses and deflections based on a single span simply supported or a continuous beam. Pipes are usually assumed to be simply supported or unrestrained rather than supported as beams with fixed ends. Concentrated loads, direction changes and pipe joints require special attention. Increase in pipe diameter is sometimes used to obtain a greater distance between supports.

Some devices used to support horizontal pipes include pipe hangers, slider supports and roller supports. Spring and turnbuckle hangers are used for suspending hot insulated pipes. For vertical pipes methods of support include trunnions supported on guide lugs, and sling rods supported by pipe hangers.

Pipe clips or hangers should be in direct contact with the pipe or insulation, but should not be too tight around them. Distance pieces are generally used to prevent this. The insulating material should be capable of bearing the compression load imposed. Pipe guides are used to restrain sideways movement of pipes. Anchors are used where it is necessary to provide fixed points for pipe bends and loops.

Pipework supports are also dealt with in ASME B31.3. This code gives separate treatments for pipe supports for piping for normal fluid service, Category M fluid service and high pressure service, and for non-metallic piping.

12.7.4 Gaskets

Flanged joints in process plant are sealed mainly by gaskets or ring seals. Gaskets are considered in this subsection and ring seals in the next.

Two commonly used types of gasket are compressed asbestos fibre (CAF) gaskets and metallic spiral wound gaskets. These are covered in BS 1832: 1972 *Specification for Oil Resistant Compressed Asbestos Fibre Jointing* and BS 3381: 1989 *Specification for Spiral Wound Gaskets for Steel Flanges to BS 1560*. There is a trend towards the use of metallic spiral wound gaskets instead of CAF gaskets for many of the more severe and hazardous duties. A metallic spiral wound gasket takes the form of a ring made out of metal windings, with a V-shaped cross-section, wrapped in a continuous spiral with soft filler between the metal plies, all encased in an outer wrap.

The features of metallic spiral wound gaskets are discussed by Granek and Heckenkamp (1981). Such spiral wound gaskets have many merits. They provide a relatively simple and low cost joint. They offer the elasticity and material flow required for good sealing. They are well suited to high pressures. They are resistant to high temperature and corrosive conditions. They are resilient to thermal shock and vibration. The method of construction and the permutations of metal and filler support a family of gaskets with shapes compliant to different flange configurations and with properties such as the stiffness matched to the application. On the other hand they have certain disadvantages. They are not readily tightened up to stop leaks and they are not resusable. They are also relatively sensitive to the quality control on the gasket itself, the application in which it is used and the manner in which it is bolted up. These authors also discuss the extent of confinement of the gasket, favouring the practice of confining it so that it has no freedom to move and immobilizing it to prevent leak paths when pressurized. Metallic spiral wound gaskets are resistant to blowout, and this is often an important factor in their selection.

12.7.5 Ring seals

The other main kind of seal for flanged joints is the ring seal, also known as a lens or ring type joint (RTJ). This type of seal is used particularly for high pressure plant. A ring seal is a ring with an approximately ovoidal cross-section which fits into two grooves, one on each of the two flanges. The seal is made at the edges rather than at the bottom of the groove. The ring is capable of a degree of deformation to make a good seal. It is good practice to use a new ring seal when remaking a joint. Ring seals are another form of blowout-resistant sealing.

12.7.6 Bellows

Pipework needs to incorporate means of accommodating expansion and contraction. One method of doing this is by utilizing bellows. A bellows assembly may be used, for example, between two pipe sections, between two vessels or between a pipe and a vessel.

After the Flixborough disaster in 1974, referred to above, the use of bellows came under a cloud. It is now subject to much stricter control. Typically, proposals to use bellows in pipework with potential for hazardous release are reviewed individually during design, and in some cases there may be restrictions on their use. An account of one company's policy on bellows is given by Boyett (1979).

Following Flixborough, a British Standard on bellows was issued. This is BS 6129: 1981 *Code of Practice for the Selection and Application of Bellows Expansion Joints for Use in Pressure Systems*, which has a single part, Part 1: 1981 *Metallic Bellows Expansion Joints*. In the USA, standards for bellows are issued by the EJMA.

An account of the design and application of bellows as expansion joints is given by C. Taylor (1985). The bellows itself consists of a convolution. The convolution may be 'thick-walled' or 'thin-walled', the latter term covering multi-wall construction which may be thicker than a 'thick-walled' unit. Its stiffness is expressed as a spring rate.

The design of a bellows assembly has to take into account: the forces on, and stresses induced in, the bellows, and its movement; the forces and moments on equipment connected to it; the external forces such as wind loads on it; its stability; and its fatigue performance. The design requires also to cover the anchors and guides.

The EJMA method gives relations for determining the spring rate of the convolution. It treats the bellows as an elastic shell with stresses induced by pressure and deflection. It deals with the stability of the bellows and gives a fatigue curve.

Taylor describes three types of system used to compensate for expansion: (1) compression systems, (2) tension systems and (3) pressure balanced systems. The compression system is the simplest: the bellows is placed between the two pipe sections and, when expansion occurs, the bellows is compressed. The fixed pipe section is held by an anchor which resists the compressive forces and the moving section may be provided with a guide.

In a tension system the pressure end load is taken up not by anchors but in some other way. The arrangement exploits angulation, or slight off-axis movement of the bellows. Three tensile systems are hinged assemblies, gimbal assemblies and tied double assemblies. Hinged units give movements in one plane by angulating the bellows; the pressure end load is contained by the hinge parts. Gimbal assemblies allow angular movements. Tied double assemblies permit large movements in a plane and, using two tie bars, angulation.

In a pressure balanced system, the force induced by the pressure in the bellows is balanced so that only the spring rate needs to be accommodated by external means. Two commonly used systems are the in-line assembly and the elbow-type assembly. Systems of this kind tend to be relatively bulky and expensive, but find application where movements are small and space is restricted, and also on high pipework where anchors are impractical. They are widely used on compressors and turbines.

Hydrostatic testing requirements can present a difficulty in bellows design. Whereas with most pipework fittings there is no fundamental problem in increasing the strength, a bellows is unique in that there is a trade-off to be made between strength and deflection. An increase in thickness of the bellows gives an increase in strength, but a decrease in allowable deflection, the limiting factor being squirm. The point is discussed by D.J. Peterson (1991), who instances design of bellows for piping on a fluid catalytic cracker unit with gas at 45 psig and 1300°F. He argues that in practice the metal temperature of such bellows would be 1000°F. Design for a test pressure of 76 psig and the actual operating temperature of 1000°F yields a bellows thickness of 0.048 in. However, if the test pressure is taken as that for the nominal operating temperature of 1300°F and, furthermore, as that applicable to the material of construction of the pipework rather that of the bellows, this results in a test pressure of 343 psig and a corresponding bellows thickness of 0.150 in. The spring rate is increased by a factor of 28 and the cycle life is reduced by a factor of 317.

12.7.7 High integrity pipework

Where the consequences of pipework failure could be especially serious, the pipework should be of high integrity. Measures to improve pipework integrity are discussed by numerous authors. A convenient summary of many of these is given for highly toxic materials in the CCPS *HTHM Guidelines* (1988/3).

Measures should be taken to reduce the stresses to which the pipework is liable to be subjected and to give it robustness in the face of such stresses. Sources of vibration and of severe cycling should be minimized and, if necessary, countermeasures taken. A flexibility analysis of the pipework should be carried out.

Joints on pipework should be welded or flanged, and threaded joints should be avoided. Features which may prove to be weak points should be minimized; these include expansion bellows, hoses, and sight glasses.

Gaskets should be suitable for the fluid handled and the temperatures, and should be resistant to blow-out. Blow-out resistant gaskets include metal reinforced, spiral wound and ring joint types.

Fittings, particularly valves, are potential weak points and should receive attention both in respect of fugitive emissions and outright failure. Leakage can be reduced by the use of double packing boxes or bellows seals.

The design should seek to limit the size of any release which may occur. One measure is the avoidance on vessels of nozzles larger than is strictly necessary. Another is the use of flow restrictor orifices in pipes. The approach taken here is to make the pipe of larger diameter, and thicker walled, with a lower normal flow velocity.

The pipework should be protected against overtemperature, particularly from fire. Methods of protection include fire insulation and water sprays used as alternatives or in combination. The fire insulation should not only possess a suitable nominal fire resistance but also should remain in place under fire conditions, which include the use of water jets for fire fighting. With respect to fire insulation there is a balance to be struck between fire protection which it affords and the threat of external corrosion beneath it.

Where the pipework may be subjected to low temperature, whether this be the normal operating temperature or one resulting from a deviation such as sudden depressurization, the constructional materials should have sufficient impact resistance to prevent low temperature brittle fracture.

If the material in the pipe may freeze at the lowest temperatures expected, countermeasures such as trace heating are necessary. Freezing poses two distinct threats. One is expansion on freezing. The other is uneven thawing and expansion of the unfrozen liquid.

The pipework should be protected against overpressure, using a suitable combination of protective measures. These may include trips, expansion chambers and pressure relief devices.

With respect to physical abuse the CCPS *Guidelines* state that a properly supported NPS $1\frac{1}{2}$in. steel pipe is generally accepted as being resistant to loads such as being stepped on or struck by a heavy pipe wrench. It is desirable that piping of other materials be similarly resistant, whether by materials selection, pipe thickness or protection.

Impurities may constitute a threat to pipework integrity, particularly where the materials handled are highly reactive, which includes those which are strong oxidizers. The system may therefore need to be cleaned after construction. This can be assisted by a design which avoids dead spaces and provides adequate vents and drains.

In supporting pipework carrying a corrosive fluid, the hangers which support the pipe from above by rods are liable to suffer corrosion of these rods, and the arrangement is generally to be avoided. An arrangement which supports the pipe from below is less liable to fail. In more critical duties, it may be necessary to consider protection of the pipe support themselves against very low or high temperatures, including those occurring in fire conditions.

The routing of pipes should take account of the threat from tanks and vessels, pumps and pipework holding flammables and, generally, of the threat from fire. Routes should pass through protected areas and away from roads. It may be appropriate to furnish protection against vehicle impact and against physical abuse. In some cases piping is armoured and extra protection is provided for nozzles and small bore pipework.

The pipework should be designed with a view to maintenance of the valves, important aspects being access and ease of slip plating. Attention should also be paid to the avoidance of unbolting errors in the maintenance of valves, so the bolts necessary to containment of the pressure are not removed in the mistaken belief that they are associated only with, say, the actuator or with a spool piece.

Tanks and vessels should be provided where necessary with remotely operated emergency isolation valves, as described below. Open-ended external valves should be fitted with a blind flange or plug.

Another potential weak point is the pumps. These should be designed to withstand the highest pressures and most extreme temperatures to which they may be subject.

Construction of the pipework should also be to high standards. This aspect is dealt with in Chapter 19. Aspects mentioned in the CCPS *Guidelines* include: colour coding for identification; cleaning after construction, and design to facilitate this; procedures for controlled bolting; and access for inspection.

12.7.8 Fibre reinforced plastic pipework

There are also a number of standards and codes for the design of pipework in fibre reinforced plastic. Relevant standards and codes are BS 6464: 1984 *Specification for Reinforced Plastics Pipe, Fittings and Joints for Process Plants*, BS 7159: 1989 *Code of Practice for Design and Construction of Glass-reinforced Plastic (GRP) Piping Systems for Individual Plants or Sites*, and ASTM D 3517–91 *Standard Specification for 'Fiberglass' (Glass-fiber-reinforced Thermosetting Resin) Plastic Pipe*. ASME B31.3, the *Chemical Plant and Refinery Piping Code*, deals with non-metallic piping.

12.7.9 Valves

Valves constitute another important class of component in pipework systems. There is a wide variety of valves in use, and only a few types, i.e. those with an essentially protective function, are considered here. They are (1) non-return valves, (2) excess flow valves and (3) emergency isolation valves. Regular control valves are dealt with in Chapter 13. Another aspect of protection is the interlocking of valves to prevent their being operated in an incorrect sequence. This is discussed in Subsection 12.7.13.

12.7.10 Non-return valves

Non-return valves (NRVs), or check valves, are used to prevent undesired reverse flows. Reverse flow can have serious consequences and some form of protection is often necessary. Some situations in which reverse flow occurs have been described by Kletz (1976b). They include: (1) flow into plant from storage vessels or blowdown lines, (2) flow from plant into service lines, (3) reverse flow through a pump and (4) reverse flow from reactors. A non-return valve is a prime means of preventing reverse flow in many of these situations, although it is not always an adequate, or even the most appropriate, method in all cases.

Process materials can flow back into service lines with disastrous results if a pressure reversal occurs. A non-return valve should be installed on service lines, but other measures are also necessary. If the service line is in continuous use, it may be necessary to have low service pressure and high process pressure alarms/trips. If the service line is used intermittently, it should be isolated when not in use by positive means such as hose disconnection or double block and bleed valves.

Non-return valves are widely used on the discharge of pumps to prevent reverse flow through the pump. If reverse flow occurs, it can disintegrate the impeller and damage the motor. Consideration should also be given to reverse rotation locks on pumps.

Reverse flow of one reactant from a reactor into the feed pipe of another reactant is likely in many cases to result in an explosion. Non-return valves may be appropriate on the feed pipes. Another precaution is the use of a small inventory feed tank. This limits the scale of any possible explosion, although it also lessens the degree of dilution of a contaminant and may thus increase the probability of a reaction.

Non-return valves are not a fully reliable means of eliminating reverse flow. Numerous instances have occurred in which materials have travelled back not just through one but several non-return valves in series. According to Ellis and Mualla (1986), a survey of stoppages in Central Electricity Generating Board (CEGB) power stations showed that some 16% were connected with non-return valves.

Non-return valves on critical duties are important pressure system components and should be included in the pressure system register and regularly tested and maintained.

Leakage from storage vessels or blowdown lines into the plant tends to occur when the latter is shut down, and can have serious consequences. It is usually due to leaks through closed but defective valves. The remedy is more positive isolation by means such as blanking off.

Increasingly, environmental considerations lead to requirements for vents from storage tanks and other equipment to be routed to a disposal system. It is necessary to ensure that reverse flow does not occur in such a system. However, reverse flow protection for storage tank vents is not readily effected by the use of

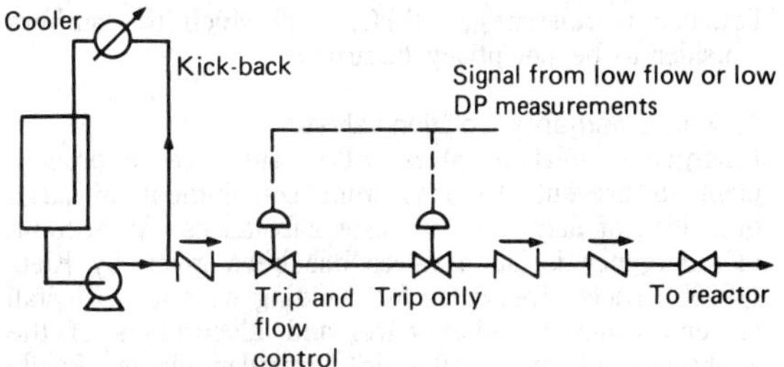

Figure 12.8 *Ethylene oxide feed system to a reactor (Kletz, 1976b) (Courtesy of Hydrocarbon Processing)*

non-return valves, because the storage tank pressures are low and the flows are often very low and intermittent. The problem is discussed by Englund, Mallory and Grinwis (1992), who describe instrument systems for this duty. These authors also describe reverse flow protection arrangements, including non-return valves, for a number of other situations, such as pump and reactor systems.

In some cases it is necessary to take quite elaborate precautions to prevent reverse flow. The importance of avoiding reverse flow in ethylene oxide plants was emphasized in Chapter 11. Figure 12.8 shows a reverse flow protection system for the ethylene oxide feed to a reactor. There is a kick-back line around the pump to prevent overheating and a non-return valve on the pump delivery. There are two trip initiators and two shut-down valves with both initiators and valves arranged as 1-out-of-2 systems, as described in Chapter 13. One of the valves is also used as a control valve. There are two further non-return valves. Diversity of the types of non-return valve is used to reduce the probability of common cause failure.

Reverse flow of ammonia into an ethylene oxide feed tank on a reactor was responsible for the disaster in 1962 at Doe Run, Kentucky (Case History A31). Appendix 1 also describes various other incidents involving reverse flow and non-return valves. Further case histories are given by Kletz (1976b).

12.7.11 Excess flow valves

Another type of valve with a protective function is the excess flow valve (EFV). Such a valve is used to shut off flow in a situation where the flow has suddenly risen far in excess of its normal value. The rise in flow which triggers the operation of an excess flow valve is generally 50% or more above the normal value.

Excess flow valves are used at loading facilities to give flow shut-off in the event of rupture of a filling line hose. The excess flow valve on a filling line may be installed internally in the storage vessel.

Where the fluid is held as a superheated liquid under pressure and will flash when released to atmosphere, the flow will become two-phase. The sizing of the excess flow valve thus involves a two-phase flow calculation. A calculation method for two-phase flow in this application has been described by R.A. Freeman and Shaw (1988). The method is presented as an alternative to testing of excess flow valves, which is a code recommendation for liquefied petroleum gas (LPG), but which the authors consider to be potentially hazardous.

12.7.12 Emergency isolation valves

Emergency isolation valves (EIVs) are used in process plant to prevent the loss from containment of large quantities of flammable or toxic substances. An account of emergency isolation valves has been given by Kletz (1975b). Kletz gives rules for deciding whether to install an emergency isolation valve, and illustrations of the application of these rules in particular plants. Points where large escapes of material are liable to occur include (1) pumps, (2) drain points and (3) hose connections.

Large leakages can occur from pump glands and seals. Some cases of pump leakage are quoted by Kletz. In one instance there was a 3 ton escape of ethylene over a period of 20 min.

Drain points are also liable to large escapes. Again a number of examples are given by Kletz. The drainage of water from hydrocarbons can take some minutes, so that an operator tends to go and do something else. Then, if there is less water than he expected, or if he forgets to return, there may be a leak of hydrocarbons.

Another problem is the blockage of drain lines and drain valves by ice or hydrate formation. In some cases a drain valve freezes open and the operator is unable to close it. In others a blockage occurs and the operator clears it, but is then unable to approach to close the valve on account of the material escaping. The Feyzin disaster in 1966 (Case History A38) began with a blockage in the drain line from a propane storage sphere, due almost certainly to ice or hydrate.

Emergency isolation valves may be installed between the inventory and the point of expected leakage, but it is unnecessary to install an emergency isolation valve between every inventory and every leakage point. In fact it is undesirable to do so, since every such device itself introduces further chances of leakage. The decision as to whether an emergency isolation valve is required in a given case depends, therefore, on the expected scale and effects of a leakage and on its probability. Inventory is an important factor, but it is not the only one.

In many cases a rough estimate of the probability of a leak may be sufficient to show that an emergency isolation valve is or is not needed. In cases where the decision is not clear-cut it may be helpful to do a hazard assessment as described in Chapter 9. Relevant to such assessment is the probability of pump fires. Some estimates given by Kletz (1971) are quoted in Table 16.55.

The condition of the material should also be taken into account in determining the need for an emergency isolation valve. A cold liquid which is well below its atmospheric boiling point is less hazardous than one which is superheated and will flash off when let down to atmospheric pressure. Other factors which may be taken into account include the possibility of isolating the inventory by other means, such as on the feed to the unit, or of pumping the inventory away.

Kletz (1975b) gives some examples of the need for emergency isolation valves, which illustrate the fact that inventory is not the only criterion. An emergency isolation valve is recommended for a 10 ton inventory on an ethylene sidestream pump which in the past has leaked and ignited, whereas one is not recommended for a 70 ton inventory on a naphtha feed pump which contains cold material and has not leaked.

An emergency isolation valve may be installed as a new valve or by motorization of an existing valve. Figure 12.9 illustrates the two methods for two pumps piped up in parallel. In this case the use of existing valves requires the motorization of two valves. But there are some advantages in this method in that the amount which can leak out after the valves are closed is much reduced and that the valves are more easily tested.

The valve may be arranged so that air, hydraulic or electrical power is required to keep it open or, alternatively, to close it. The normal arrangement is closure on power failure, unless this is undesirable for some reason. Valves which are very large and existing valves which have to be motorized, however, require power to move them either to open or close.

If the emergency isolation valve closes on loss of power, it is not generally necessary to provide fire protection on the air or electrical power lines to it. But if power is needed to close the valve, the power lines should be provided with 15-minute protection and a latching device installed to keep the valve closed if the power lines are destroyed. The 15-minute protection allows the operator time to decide whether to close the valve.

There appears to be little evidence to suggest that either pneumatic or electrical operation is more reliable, but it is desirable to ensure that the valve actuator is

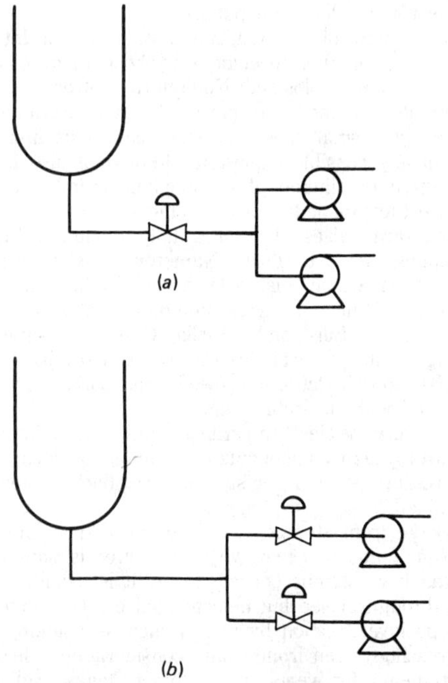

Figure 12.9 *Emergency isolation valves between a vessel and pumps: (a) new isolation valve; (b) existing isolation valves motorized. (after Kletz, 1975b) (Courtesy of the American Institute of Chemical Engineers)*

sufficiently powerful, particularly on dirty fluids, slurries, and so on. An oversized actuator may well be worthwhile.

Operation of the emergency isolation valve should be sufficiently remote that the operator can close it readily in an emergency. He should not have to approach a gas cloud or a fire or to go up on higher level platforms or to stand on a ladder. The most suitable arrangement is often remote operation from the control room. Emergency isolation valves which are controlled remotely are also known as remotely operated valves (ROVs) and remotely operated block valves (RBVs).

For emergency isolation valves it is recommended by Kletz that two switches should be provided at a safe distance from potential leak points and separated by at least 30 ft. Where there is more than one device to be operated remotely, such as an emergency isolation valve and a pump shut-off, it will generally be desirable to group the switches for the different devices together.

In some older plants the handles of emergency isolation valves are taken through a wall, which thus provides protection for the operator. This method is not recommended, however, since walls interfere with ventilation and hinder the dispersal of leaks. Remote operation is better.

Where an emergency isolation valve is fitted on the suction of a pump, operation of the valve should cause the pump motor to trip so that the pump does not overheat and ignite the leak.

The closure of an emergency isolation valve is not instantaneous. An electrically operated valve typically takes about 1 minute to close, a pneumatically operated ball valve somewhat less and a pneumatically operated gate valve somewhat more.

In some cases an automatic control valve is used for emergency isolation. Such a valve is not an ideal isolation valve, because a characterized valve often does not give a tight shut-off, especially after a period of operation. But where a control valve is installed, it may be difficult to justify a separate emergency isolation valve. If a control valve is used for this purpose, there should be a separate manual control for isolation; the manual mode of the controller should not be used.

Emergency isolation valves have not been widely used on compressors, but their installation on a new compressor plant has been described by Kletz. He makes the point that, since the valves are very large and are motorized, they are useful for plant operation as well as for emergencies, and this makes their installation easier to justify.

The Flixborough disaster involved the escape of large quantities of superheated flammable liquid from a train of high inventory reactors in series. It has been suggested that it would have been desirable to have an emergency isolation valve between each reactor. Conversely, it has also been argued that the probability of an escape from a pipe between two reactors in the original, as opposed to the modified, plant was remote and that the introduction of emergency isolation valves might have increased rather than decreased the hazard; the latter is probably the more widely held view.

Emergency isolation valves may be used to prevent low temperature fluids getting into mild steel plant and causing low temperature embrittlement, but in general where this hazard exists it is better to use materials which can withstand the lower temperatures.

The most positive method of ensuring isolation of the bottom offtake from a storage vessel is to have a single offtake pipe with an emergency isolation valve on it and with all other devices such as sample points or drain points downstream of this.

Self-closing valves using the dead-man's handle principle may be used for drains, but they are liable to be wired open by operators. Drain points which are required only for maintenance should be blanked off.

With regard to the type of valve used as an emergency isolation valve, relevant features are the torque required to operate it, the propensity to trap liquid and the leak tightness. Valves should be provided with power operated actuators, but should be capable of being operated manually. Valves which do not trap liquid include globe, ball and high pressure butterfly types. Leak-tightness cannot be assumed in an emergency isolation valve and the pipework arrangements should be such that positive isolation can be effected by methods such as double block and bleed systems or slip plates.

Emergency isolation valves with potential exposure to flammables, and hence fire, should be fire resistant or should have fire protection in the form of fire shields and insulation or water deluge.

Emergency isolation valves are also installed on pipelines as described in Chapter 23.

Emergency isolation valves are important components of the pressure system. They should be included in the pressure system register and should be regularly tested and maintained.

12.7.13 Valve interlocks

In some cases it is critical that operations in a pressure system take place in a specified sequence. This applies particularly to the opening and closing of valves. The control of such operations may be effected using interlocks. This topic is treated in Chapter 13, but it is convenient here to consider the more specialized matter of mechanical interlocks, particularly for valves.

Some principal forms of mechanical interlock are those used for (1) supply, (2) access and (3) exchange. In supply there is a choice between two mutually exclusive conditions. An example might be where two different fluids may be admitted to a system, but where only one is to be admitted at a time. One interlock system that can ensure this is to fit each of the two valves with a single lock and to provide a single key. The two states of the lock are (1) switch on with the key turned and trapped in the lock and (2) switch off with the key removable. When the key is turned and trapped the valve is closed and when it is removed the valve is open. Another example might be two power supplies that are configured in parallel but only one of which should be connected at a given time.

With regard to access, there is to be no access unless there is a safe condition. The safe condition generally depends on the isolation of a power supply or other source of energy or pressure. One interlock system that can ensure this is to fit both the access point and the power supply point each with a lock and to provide a single key. The basic principle is similar to that in the case of the two-valve system just described.

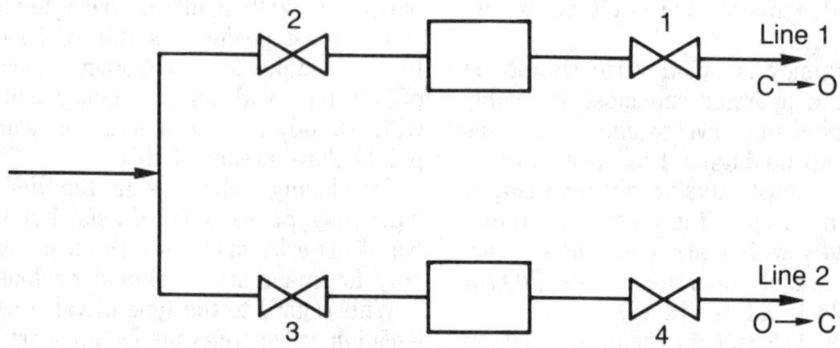

Figure 12.10 *Valve system controlled by key interlocks. C, closed; O, open*

An alternative method, which is applicable to the systems just described but comes into its own particularly in more complex systems, is exchange. Some interlock systems based on key exchange may be illustrated by reference to the system shown in Figure 12.10 in which there are two lines, only one of which is operational at a given time.

The particular interlock system to be used for such a system will depend on whether it is critical that one line always be open or that one line always be closed. It is the first case which is considered here for illustrative purposes. One interlock system is: to fit each of the four valves with a single lock, valves 1 and 2 being fitted by key type A, and valves 3 and 4 by key type B such that when a key is turned and trapped the valve is locked and when the key is removed the valve is open; to provide two keys of each type, A and B; and to provide also a key exchange box with four locks which contains at any one time the pair of keys of one type such that the pair of keys in residence (say type B) are removable only when the pair of keys of the other type (A) are inserted, turned and trapped. For the case where the initial state is line 1 closed and line 2 open and the system is to be moved under interlock to the final state of line 1 open and line 2 closed, the procedure is to open Valves 1 and 2, remove the two A keys, insert, turn and trap them in their locks in the key exchange, thus allowing the B keys to be removed from the exchange and inserted, turned and trapped in valves 3 and 4, thus closing it. The sequence is then as follows:

	Initial state		*Final state*	
	Key	*Valve*	*Key*	*Valve*
Key exchange	B, B	–	A, A	
Valve 1	A	Closed	–	Open
Valve 2	A	Closed	–	Open
Valve 3	–	Open	B	Closed
Valve 4	–	Open	B	Closed

Another interlock system applicable to the same valve system is as follows. Each of the four valves is fitted with a double lock with two different keys such that when the first key is turned and trapped the valve is open and the second key is removable, and when the second key is turned and trapped the valve is closed and the first key is removable. Thus for valve 1 the first key might be A and the second B. Five keys are used for the four valves. The keys overlap around the system so that for valve 2 the two keys are B and C, for valve 3 keys C and D, and for valve 4 keys D and E; key A would then be held in the key box. For the case where the initial state is line 1 closed and line 2 open with keys B and C turned and trapped to hold closed valves 1 and 2, respectively, with keys D and E turned and trapped to hold open valves 3 and 4, respectively, and with key A in the key box, and the system is to be moved under interlock to the final state of line 1 open and line 2 closed, the procedure is to insert, turn and trap key A in valve 1, thus opening it; this allows key B to be removed and inserted, turned and trapped in valve 2, thus opening it; this allows key C to be removed and inserted, turned and trapped in valve 3, thus closing it; this allows key D to be removed and inserted, turned and trapped in valve 4, thus closing it; and this allows key E to be removed and returned to the key box. The sequence is then as follows:

	Initial state		*Final state*	
	Key	*Valve*	*Key*	*Valve*
Key exchange	A	–	E	–
Valve 1	B	Closed	A	Open
Valve 2	C	Closed	B	Open
Valve 3	D	Open	C	Closed
Valve 4	E	Open	D	Closed

12.7.14 Fluid transients

There are a number of flow phenomena which can constitute a threat to the piping system. They include various forms of hammer, two-phase flow, and geysering. Some of these fluid transients are referred to in codes such as ASME B31.3: 1990 and API RP 521: 1990.

The best known of the hammer phenomena is water hammer, which occurs when the closure of a valve at the end of a pipe is too rapid. Steam hammer is another rather similar effect in high pressure steam mains. Steam flow may also accelerate condensate causing another hammer phenomenon which is referred to here as 'condensate hammer'.

The terminology used to describe these effects reflects the fact that, in a many cases, the fluids concerned are water and/or steam, but similar effects can occur with other substances. Thus, for example, water hammer may more generally be termed 'hammerblow'. Another form of hammer which occurs when a non-return valve is slow to close, so that a return velocity builds up and the fluid movement is then arrested. It is referred to here as 'reverse flow hammer'.

12.7.15 Water hammer

If a valve on a liquid pipeline is closed quickly, a large change in momentum of the liquid column occurs in a short period of time and a large force is exerted on the valve. This is known as 'water hammer' and it can have a very destructive effect, resulting in the shattering of the valve and/or the line.

Accounts of water hammer are given in *Hydraulics* (Lewitt, 1952), *Hydraulic Transients* (Streeter and Wylie, 1967), *Analysis of Surge* (Pickford, 1969), *Hydraulic Analysis of Unsteady State Flow in Pipe Networks* (J.A. Fox, 1977) and by Hayashi and Ransford (1960), Streeter (1964), Pearsall (1965–66), Ludwig and Ruiterman (1974), Ardron, Baum and Lee (1977), Kremers (1983) and D. Clarke (1988a,b).

Following Lewitt (1952), a simple treatment of hammer blow in a pipelines is as follows. Consider the pipeline system shown in Figure 12.11. If the valve at A is closed suddenly, there will be a rapid rise in pressure at this point due to the change in momentum of the column of liquid in the pipe. This sudden pressure rise at A causes a pressure wave to pass back down the pipe to B at the velocity of sound in the liquid. When this pressure wave reaches B, the liquid in the pipe surges back in the direction of B and causes the pressure at A to fall. The time for this to happen is

$$t = \frac{l}{u_s} \quad [12.7.4]$$

where l is the length of column of liquid, t is the time for the pressure wave to travel the length of the column of liquid, and u_s is the velocity of sound in the liquid. For any closure time less than this the pressure at A has its maximum value. Thereafter the pressure at A oscillates as the pressure waves pass back and forth between A and B, the oscillation being damped by friction in the pipe.

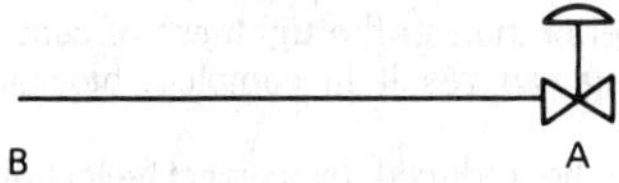

Figure 12.11 *Simple pipeline system*

For an inelastic pipe the pressure on the valve may be obtained by equating the loss of kinetic energy of the liquid to the gain in the strain energy of the liquid. The latter is

$$W = \frac{1}{2}PV \quad [12.7.5]$$

where P is the pressure of the liquid, V is the volume of the liquid, and W is the strain energy of the liquid.

The bulk modulus K of the liquid is defined as:

$$K = \frac{P}{(dV/V)} \quad [12.7.6]$$

Hence:

$$W = \frac{1}{2}\frac{P^2 V}{K} \quad [12.7.7]$$

Then, equating the kinetic energy loss to the strain energy gain for a differential element:

$$\frac{\rho a dl}{2}u^2 = \frac{1}{2}\frac{P^2}{K}a dl \quad [12.7.8]$$

$$P = u(K\rho)^{\frac{1}{2}} \quad [12.7.9]$$

where a is the cross-sectional area of the pipe, l is the length of the column of liquid, u is the velocity of the liquid, and ρ is the density of the liquid.

For an elastic pipe it is necessary to equate the loss of kinetic energy of the liquid to the gain in the strain energy not only of the liquid but also of the pipe:

$$\frac{\rho a dl}{2}u^2 = \frac{1}{2}\frac{P^2}{K}a dl + \left(\frac{1}{2}\frac{\sigma_1^2}{E} + \frac{1}{2}\frac{\sigma_2^2}{E} - \frac{\mu\sigma_1\sigma_2}{E}\right)a dl \quad [12.7.10]$$

with

$$\sigma_1 = \frac{Pr}{2h} \quad [12.7.11]$$

$$\sigma_2 = \frac{Pr}{h} \quad [12.7.12]$$

where E is the elastic modulus of the pipe, h is the thickness of the pipe wall, r is the radius of the pipe, μ is the Poisson's ratio of the pipe, σ_1 is the longitudinal stress in the pipe, and σ_2 is the circumferential stress in the pipe.

Hence, assuming $\mu = 1/4$:

$$\frac{\rho u^2}{2} = \frac{1}{2}\frac{P^2}{K} + \frac{P^2 r}{Eh} \quad [12.7.13]$$

$$P = u\left[\frac{\rho}{(1/K + 2r/Eh)}\right]^{\frac{1}{2}} \quad [12.7.14]$$

Equation 12.7.14 reduces to Equation 12.7.9 as the elastic modulus E goes to infinity. A treatment of the case where the fluid flow is two-phase bubbly flow has been given by Karplus (1961).

A principal means of avoiding water hammer is the limitation of the rate of closure of the valve in question. There are available computer codes for the analysis of liquid flows in networks which can be used to analyse the system and determine the safe rate of closure of the valve.

12.7.16 Steam hammer

A treatment of transients in lines containing a steam–water mixture is given by Karplus (1961). An account of the occurrence in high pressure steam lines of a steam hammer effect on rapid closure of the stop valves to a steam turbine has been given by Crawford and Santos (1986). The closure time of the valves was very short, some 50–100 ms.

12.7.17 Condensate hammer

Another hammer effect can occur when steam is admitted, typically after a shut-down, into a line which contains condensate. The condensate can be propelled along the line with tremendous and destructive force.

This effect is treated in *Avoiding Water Hammer in Steam Systems* by Mortimer and Edwards (1988 HSE SIR 5). They distinguish three cases: (1) condensate driven by steam, (2) condensate drawn by vacuum and (3) flash steam effects. In the first case, condensate collects in a pipe or at fittings. When steam is admitted and flows past, at velocities which can be up to 40 m/s, waves are set up on the condensate surface and, in due course, slugs of condensate are formed. When a liquid slug encounters an obstruction, such as a sharp bend or valve, it strikes it with a 'hammer blow'. The second case is where the steam is admitted to a space where it is cooled in some way and condenses, drawing condensate at high speed towards the space. The third case occurs in a pipe containing superheated water which is let down in pressure so that some vaporizes as flash steam and drives the water away. The flash steam then condenses, creating a vacuum with the effects described in the second case. Frequently the damage done is slight, but in some instances it is severe. It does not necessarily manifest itself at the point of hammer, but sometimes a metre or more away.

The principal measure taken to avoid condensate hammer is design to prevent the build-up of condensate. Provision of adequate fall in the pipe is one aspect, avoidance of fittings which collect condensate another and provision of means to remove condensate a third. Fittings such as globe valves and strainers are liable to collect condensate. With a globe valve, for example, the depth of water collecting behind the weir even when the valve is open can exceed half the pipe diameter. In this particular case the problem can be avoided by mounting the valve with its spindle horizontal. Manual drain points may be appropriate at some locations, but automatic steam traps are the principal means of draining condensate.

There are also operational measures which can be taken. One is to drain condensate before starting up. Another is to exercise care in the admission of steam. It should be appreciated that condensate may form from leaks through closed steam valves, even several closed valves in series, while the pipe is shut down.

12.7.18 Reverse flow hammer

Another form of hammer is that which occurs when a non-return valve does not close rapidly enough. This is referred to here as 'reverse flow hammer'. The effect is described by Ellis and Mualla (1986). If the closure of a non-return valve is not sufficiently rapid there is an appreciable 'reverse velocity', the arrest of which, when the valve does close, gives rise to a high pressure transient on one side of the valve and a low pressure one on the other. Incorrect selection of non-return valves has for a long time been a source of troublesome pressure transients. In order to make the correct choice, it may be necessary to conduct tests and to model valve behaviour. The trend is to present the valve characteristics as a plot of reverse velocity on closure vs the rate of velocity reversal. The lack of a standard definition of the latter has caused some variability in treatments.

12.7.19 Two-phase flow effects

Two-phase flow can often give rise to pressure oscillla-tions and surges. Another phenomenon associated with such flow is 'bowing', which occurs when a refrigerated liquid is introduced at such a rate that two-phase stratified flow occurs, with attendant large temperature differences.

12.7.20 Geysering

Geysering is a phenomenon which occurs when a liquid near its boiling point is handled in a vertical, or sometimes in an inclined, pipe. It involves a rapid evolution of vapour, expelling liquid from the pipe and causing a destructive pressure surge.

12.8 Heat Exchangers

Heat exchangers are components of importance in pressure systems. They include (1) vaporizers, (2) reboilers, (3) condensers, and waste heat boilers, as well as heat exchangers on general heat interchange duties. The design of heat exchangers is dealt with in *Process Heat Transfer* (D.Q. Kern, 1950) and in *Process Heat Transfer* (Hewitt *et al.*, 1993).

Relevant standards are BS 3274: 1960 *Tubular Heat Exchangers for General Purposes*, API Std 660: 1973 *Heat Exchangers for General Refinery Services* and the Tubular Exchangers Manufacturers Association (TEMA) *Standards of the Tubular Heat Exchangers Manufacturers Association* (n.d.). Until its withdrawal, heat exchangers were dealt with in the API *Guide for Inspection of Refinery equipment*, Chapter 7, Heat Exchangers, Condensers and Cooler Boxes.

Some problems that occur in heat exchangers and which may affect safe operation are (1) fouling, (2) polymerization and solidification, (3) leakage, (4) tube vibration and (5) tube rupture.

12.8.1 Fouling

If fouling occurs in a heat exchanger, it may affect (1) heat transfer and (2) pressure drop. Reduction of heat transfer capacity may be particularly serious if the heat exchanger is a cooler on a critical duty such as removal of heat from a reactor in which an exothermic reaction is carried out. Effects of increasing pressure drop which may be serious are reduction in the maximum flows which can be passed through the exchanger and reduction in the pressure drop across the control valve and hence deterioration in the tightness of control. In the extreme, fouling can result in complete blockage of the exchanger.

Fouling may be reduced by correct selection of tube and shell side fluids and by the use of appropriate fluid velocities. Thus, for example, D.Q. Kern (1950) gives separate tabulations of fouling factors for water velocities below and above 3 ft/s. An allowance may be made for fouling by the use of a fouling factor which is equivalent to a heat transfer coefficient. This may be a general value taken from the literature or a specific value determined on the plant.

Generally, the so-called 'fouling factor' measures all the deficiencies of the heat exchanger, not just fouling. Often these deficiencies are much more significant than fouling itself. Thus, for example, a symmetrical positioning of the inlet and outlet on a two-pass heat exchanger can cause bypassing of the tubes below and above the level of these connections and a consequent reduction in heat transfer. Or again, a vertical condenser tends to have an inferior performance if it is updraft rather than downdraft.

It is claimed, however, that it is possible by good design virtually to eliminate the features which tend to cause serious reductions in heat transfer. Detailed design aspects have been described by Gilmour (1965). He states that where heat transfer deficiencies do still occur, they are usually on the shell side.

The use of a large fouling factor is not necessarily conservative. It may give a degree of overdesign which results in higher than expected temperatures and greater corrosion or coke deposition in the equipment. An example is discussed by Small (1968).

The use of very compact equipment can give rise to fouling due to lack of turbulence. Gilmour quotes a case in which reboiler tubes were very tightly packed and tended to foul by polymerization with bridging of polymer between tubes. The original tubes were replaced by half the number of finned tubes so that turbulence was increased and satisfactory heat transfer restored. This point is clearly relevant to design for limitation of inventory.

Fouling can also occur due to operational factors such as stoppage of a pump, omission of a filter, transfer of debris from other equipment, and so on. If this can have a serious effect, it may be necessary to take appropriate precautions, but the incorporation of a fouling factor in the design is of little use in this situation.

12.8.2 Tube vibration

The use of high capacity, compact heat exchangers with long tubes and high fluid velocities has intensified the problem of tube vibration. Tube vibration can cause damage to the tubes themselves and possibly to the exchanger shell and its supports or to the pipework connected to it, can transmit destructive pressure fluctuations from the exchanger and can generate considerable noise. Vibration can be sufficiently severe to lead to shut-down with multiple tube failure within 1 or 2 days. Two such cases are quoted by Gainsboro (1968) and by Eilers and Small (1973).

A heat exchanger tube may be regarded as a beam clamped at both ends and with distributed load. For such a system the natural frequency f_n is:

$$f_n = k\frac{\pi}{l^2}\left(\frac{EI}{w}\right)^{\frac{1}{2}} \qquad [12.8.1]$$

with k = 1.136 and where E is the elastic modulus, I is the moment of inertia, l is the length of the beam, and w is the mass per unit length. In a more detailed treatment A.A. Thompson (1969) gives a value of the constant k for heat exchanger tubes in the range 0.781–1.125.

The tube may be excited to resonance by vortex shedding. The vortex shedding frequency f_v has been studied by Y.N. Chen (1968), who obtained for gases a correlation in terms of the Strouhal number S:

$$f_v = \frac{Su}{d} \qquad [12.8.2]$$

where d is the diameter of the tube and u is the fluid velocity (based on the minimum flow area between tubes of a row).

The Strouhal number was correlated by a relation of the following form:

$$S = f(x_l, x_t) \qquad [12.8.3]$$

where x_l is the longitudinal spacing ratio, and x_t is the transverse spacing ratio.

The problem of tube vibration in heat exchangers has been investigated by P.R. Owen (1965), Y.N. Chen (1968), Gainsboro (1968), A.A. Thompson (1969) and Eilers and Small (1973).

Also relevant is the acoustic frequency. The acoustic resonant frequency f_a in a cavity is

$$f_a = \frac{nc}{2z} \qquad [12.8.4]$$

where c is the velocity of sound, n is the wave mode (integer), and z is the characteristic dimension.

If the vortex shedding frequency approaches the acoustic resonant frequency, loud noise may be generated. Acoustic resonance in heat exchangers has been described by Barrington (1973). Resonance may be avoided by alteration of the parameters in Equations 12.8.1–12.8.4.

In one case quoted by Barrington a heat exchanger with an operating weight of 125 000 lb suffered displacements of 1 mil (0.001 in.) at the anchor bolts with accelerations up to 175 ft/s. A force of 340 ton was applied to the foundations at a frequency of about 350 Hz. He found that few foundations can withstand such forces without damage. The monitoring of heat exchanger vibration is discussed in Chapter 19.

12.8.3 Tube rupture

In a heat exchanger with high pressure gas or vapour in the tubes and low pressure liquid in the shell, tube rupture can lead to overpressure of the shell. The need for protection of the heat exchanger shell is recognized in pressure vessel codes such as BS 5500 and the ASME *Boiler and Pressure Vessel Code*. The latter states: 'Heat exchangers and similar vessels shall be protected with a relieving device of sufficient capacity to avoid overpressure in the case of an internal failure.'

More specific guidelines are given in API RP 520: 1990. This recommends that it be assumed that the area available for flow of gas or vapour from the high pressure tube into the shell be taken as twice the cross-sectional area of the tube, which is the worst case for failure of a single tube.

The pressure rise in the shell caused by a tube rupture is not a simple matter, however. The problem has been investigated by Simpson (1972), who distinguished two effects: (1) local pressure rise at point of rupture, and (2) overall peak pressure rise in shell. The time-scale of these two phenomena differs by an order of magnitude; the first occurs in about 1 ms, and the second in about 10 ms. But these times are very dependent on the precise conditions in the shell, particularly the amount of gas or vapour present initially. As little as 1% of inert gas in the shell can increase by a factor of 3 the time required to reach peak shell pressure.

Simpson presents a calculation of the bubble pressure created by the rupture in 1 ms of a tube at 10 000 psia. The peak bubble surface pressures reached at 1 tube diameter and 5 tube diameters from the rupture were about 900 and 550 psia, respectively. In other words, the peak bubble pressure was less than 10% of the tube pressure. The pressure experienced by the shell wall is the reflected pressure P_r:

$$P_r \approx 2P_i - P_0 \qquad [12.8.5]$$

where P_i is the incident pressure and P_0 is the initial liquid pressure. Here the effective incident pressure is that of the bubble surface. If the rupture time was halved, the peak bubble pressure was increased by about 50%.

Although the localized overpressure is less than the overall peak pressure rise in the shell which occurs later, the latter is normally catered for by a pressure relief device, while the former is not. The effect of local overpressure on the shell and its implications for the positioning of tubes relative to the shell should be considered as a separate matter. The pressure relief of heat exchangers is discussed in Section 12.12.

12.8.4 Waste heat boilers

Process heat recovery boilers, or waste heat boilers, are widely used, particularly on ethylene and ammonia plants and on sulphuric acid and nitric acid plants. This type of boiler has developed as a cross between a conventional shell-and-tube heat exchanger and a firetube boiler. Its original function was to cool high temperature process gas and, as a by-product, to generate low pressure steam. The steam generation aspect, however, has grown in importance and the operating conditions have become increasingly severe. Moreover, as with heat recovery schemes in general, the use of waste heat boilers increases the degree of interlinking of plant and its vulnerability to failure.

Waste heat boilers and their problems have been discussed by Streich and Feeley (1972). A comparison given by these authors of the typical process waste heat boiler with the traditional Scotch marine boiler is instructive:

Typical conditions	*Process waste heat boiler*	*Scotch boiler*
Heat load (BTU/ft^2 h)	110 000	15 000
Steam pressure (psig)	$\leq$2300	200–225

Waste heat boilers do not have a raw flame licking at the tube seats, but otherwise experience conditions at least as severe.

Firetube boilers tend to leak at the tube inlets and this problem occurs in waste heat boilers also. Tube size is a compromise between the conflicting requirements for low velocity at the high temperature inlet and high velocity at the low temperature outlet.

12.9 Fired Heaters and Furnaces

Fired heaters are another important component in pressure systems. They include pipe stills on crude oil units, furnaces on olefins plants, and heaters for reactors and for heat transfer media.

General treatments of fired heaters are given in *Safe Furnace Firing* by the American Oil Company (Amoco/3), *Handbook of Industrial Loss Prevention* by the Factory Mutual Engineering Corporation (FMEC, 1967), *Heaters for Chemical Reactors* by Lihou (IChemE 1975 /61) and the CCPS *Engineering Design Guidelines* (1993/13).

Relevant codes include BS: 799: 1981– *Oil Burning Equipment* and 5410: 1976– *Code of Practice for Oil Firing*, and the British Gas (BG) *Code of Practice for Large Gas and Dual Fuel Burners* (the BG *Burner Code*) (1976/1). National Fire Protection Association (NFPA) codes include the NFPA 31, 85 and 86 series.

Selected references on furnaces, fired heaters and flare systems are given in Table 12.8.

Fired heaters are a prime source of hazards. These hazards are principally (1) explosion in firing space and (2) rupture of tubes. Explosion in the firing space occurs mainly either during lighting up or as the result of flame failure. Rupture of tubes is usually caused by loss of feed or by overheating. In addition, heaters are a source of ignition for escapes of flammable materials from other parts of the plant.

***Table 12.8** Selected references on furnaces, fired heaters and flare systems*

American Oil Co. (n.d./3, 9); ASME (Appendix 28 *Heat Transfer*, 1984/55, 1985/56, 57, 1992 CSD-1); Griswold (1946); Thring (1952); Krebs (1962); Finnie (1963); Wimpress (1963); Lichtenstein (1964); Charlton (1965); Runes and Kaminsky (1965, 1973); F.A. Williams (1965); Backensto, Prior and Porter (1966); Ellwood and Danatos (1966); FMEC (1967); Loftus, Schutt and Sarofim (1967); Maddock (1967); Lenoir (1969); C. Davies (1970); von Wiesenthal and Cooper (1970); Demarest (1971); Mack (1971, 1976); OIA (1971 Publ. 501); Siegel and Howell (1972); J. Chen and Maddock (1973); Fitzsimmons and Hancock (1973); R.D. Reed (1973); Ulrich (1973); Chambers and Potter (1974); Mol and Westenbrink (1974); Taube (1974); Fertilio and Princip (1975); IChemE (1975/61); Mitcalf (1975); Ashbaugh (1976); Coulter and Tuttle (1976); Krikke, Hoving and Smit (1976); Anon. (1977l); Hoffman (1977); Lihou (1977); Tuttle and Coulter (1977); Berman (1978); Hougland (1978); Jensen (1978); R. Kern (1978f); Ministry of Social Affairs (1979b); Goyal (1980); API (1981 Refinery Inspection Guide Ch. 9, 1986 Std 560, 1988 RP 530); Lovejoy and Clark (1983); Schillmoller and van den Bruck (1984); Gomes (1985); Institute of Materials (1985 B367); Kauffman (1987); T.O. Gibson (1988); NFPA (1985 NFPA 85E, 1987 NFPA 85A, 85G, 1988 85F, 1989 NFPA 85B, 85D, 1990 NFPA 86, 1991 NFPA 86C, 1992 NFPA 31, 54); Ghosh (1992); L.Thomas (1992)

Reformers

Francis (1964); Axelrod and Finneran (1966); Avery and Valentine (1968); Kratsios and Long (1968); Pennington (1968); Jacobowitz and Zeis (1969); Kobrin (1969); Bongiorno, Connor and Walton (1970); F.W.S. Jones (1970); Zeis and Heinz (1970); Holloway (1971); Nisbet (1971); Attebery and Thompson (1972); Salot (1972,

1975); Strashok and Unruh (1972); A.J.P. Tucker (1972); Fuchs (1974); Ruziska and Bagnoli (1974); Demarest (1975); Fuchs and Rubinstein (1975); van der Horst and Sloan (1975); E.R. Johnson (1975); Leyel (1975); Sterling and Moon (1975); Sparrow (1986)

Burners
British Gas (Appendix 27, 28); Anon. (1957–58); P.G. Atkinson, Grimsey and Hancock (1967); FMEC (1967); Gas Council (1969 GC166); Brook (1978); Elias (1978); J.A.Wagner (1979); Murphy (1988); NFPA (1992 NFPA 31, 52)
ANSI Z83; BS (Appendix 27 *Burners*)

Refractories
Chesters (1961, Institute of Materials 1974 B135); Burst and Spieckerman (1967); Crowley (1968); McGreavy and Newmann (1974)

Flare systems
Zink and Reed (n.d.); P.D. Miller, Hibshman and Connell (1958); Hajek and Ludwig (1960); Bluhm (1961, 1964b); Husa (1964); G.R. Kent (1964, 1968, 1972); Kevil (1967); P. Peterson (1967); Tan (1967a,b); Kilby (1968); R.D. Reed (1968, 1972); Grumer *et al.* (1970 BM RI 7457); F.R. Steward (1970); Escudier (1972); Lauderback (1972); Seebold (1972a,b, 1984); Swithenbank (1972); Brzustowski (1973, 1976, 1977); Brzustowski and Sommer (1973); Peters (1973); H.M. Chief Alkali Inspector (1974 annual report); Kletz (1974b); Vanderlinde (1974); Brzustowski *et al.* (1975); Klooster *et al.* (1975); Ito and Sawada (1976); Bonham (1977); Jenkins, Kelly and Cobb (1977); Schmidt (1977a,b); R. Schwartz and Keller (1977); Straitz (1977); Straitz and O'Leary (1977); Straitz *et al.* (1977); Agar (1978); Bonilla (1978); McGill and McGill (1978); Anon. (1979 LPB 28, p. 97); Boeye (1979); Paruit and Kimmel (1979); Anon. (1980 LPB 31, p. 27); Oenbring and Sifferman (1980a,b); Kandell (1981); E.B. Harrison (1982); McMurray (1982); P.J. Turner and Chesters (1982); Anon. (1983n); G.D. Allen, Wey and Chan (1983); Beardall (1983); Blanken and Groefsma (1983); Chung-You Wu (1983); B.C. Davis (1983, 1985); Fumarola *et al.* (1983); Lützow and Hemmer (1983); Max and Jones (1983); Romano (1983); Alcazar and Amillo (1984); Coulthard (1984); Vaughan (1984); Banerjee, Cheremisinoff and Cheremisinoff (1985); Boix (1985); Chunghu Tsai (1985); Cindric (1985a); Corbett (1985); de Faveri *et al.* (1985); Herbert and Rawlings (1985); Narasimhan (1986); Pohl *et al.* (1986); Swander and Potts (1986, 1989); British Gas (1987 Comm. 1349); D.K. Cook, Fairweather, Hammonds and Hughes (1987); D.K. Cook, Fairweather, Hankinson and O'Brien (1987); Straitz (1987); Leite (1988, 1990); de Silva (1988); Tite, Greening and Sutton (1989); API (1990 RP 521); Crawley (1993 LPB 111); Niemeyer and Livingston (1993); Bryce and Fryer-Taylor (1994)

Explosions (see Table 17.1)

The elimination of hazards in fired heaters is a good illustration of the need for measures both in design and operation. The design aspects are considered here and the operating aspects in Chapter 20.

Many explosions in the firing space take place during start-up when an attempt is made to light a burner. The usual situation is that fuel leaks and forms a flammable atmosphere in the firing space while the plant is shut down and an explosion occurs when an attempt is made to light a burner. There are a number of measures which should be taken to prevent this hazard. The most important is to eliminate the admission of unburnt fuel into the firing space. This is principally a matter of achieving a positive isolation on the main fuel feed pipe to the burner. The importance of this is strongly emphasized.

For isolation it is not sufficient to rely on a single valve. Use should be made either of double block and bleed valves or of slip plates. There are different views on which of these methods is preferable and in any case the choice is affected by the fuel. For gas, double block and bleed valves should be used and for oil either double block and bleed valves or slip plates.

Where oil is used, the configuration of the pipe between the shut-off valve and the burner should be such that oil does not flow from this section of pipe into the firing space after the valve has been shut. In addition, the entry of fuel from the main feed pipe during the few seconds while the burner is shutting down after flame failure or from the subsidiary feed pipe to the pilot burner should be avoided.

Before a burner is started up, the firing space should be purged with air. The atmosphere in the space should then be sampled to confirm that it is not flammable. The purging of the firing space is essential. It is emphasized, however, that it is in no way a substitute for elimination of leakage of fuel into this space.

Means for lighting a burner include a poker, a spark igniter, a permanent pilot burner and an interrupted pilot burner. An interrupted pilot burner is extinguished a few seconds after main flame ignition. The use of a poker is not a good method of ignition. An interrupted pilot burner may be more satisfactory than a permanent pilot burner, since the latter tends to go out.

The ignition period is a critical one. The approach adopted is illustrated by the procedures given in the BG *Burner Code*. The burner is started up on a start-gas flow which is less than the normal flow. The establishment of the start flame consists of two periods: an initial period of not more than 5 seconds in which ignition is checked and a second period of at least 5 seconds during which flame stability is checked.

It is essential that if the flame fails, the burner should be shut down immediately. A flame failure detector is used to monitor the flame and to initiate the trip. Loss of combustion air also necessitates immediate shut-down. Methods of detecting this condition are measurement of static pressure or of air flow.

The sequence of operations for the start-up of a burner should be carefully specified. In particular, the following operations should be carried out in sequence and the start-up should not proceed unless this has been done: establishment of air flow; purging of firing space; testing of atmosphere in space; and establishment of flame. Some detailed start-up sequences are given in the BG *Burner Code* and by the FMEC (1967). It is necessary to ensure that any start-up sequence specified is rigidly adhered to. Generally, the reliability of execution of these sequences can be improved by the use of automatic operation or of interlocks.

Many cases of tube rupture in fired heaters occur because the flow of process fluid through the tubes either falls too low or ceases altogether. If this happens, it is essential for the flame to be extinguished immediately. It should be appreciated that in a furnace where at normal throughput the ratio of the heat required to vaporize liquid to that required to superheat vapour is high, a given reduction in feed flow can result in a disproportionate increase in outlet tube temperature. Overheating of the tubes is another cause of rupture. It is necessary, therefore, to have measurements of tube temperature.

Instrumentation should be provided to measure and control the main operating parameters and, if necessary, to trip the furnace. Measurements should normally include: fuel pressure and flow; combustion air flow and fuel/air ratio; combustion gas oxygen content; process fluid pressure, flow and outlet temperature; flame state; and tube temperatures. There should be appropriate alarms based on these measurements.

The fail-safe principle should be applied in the design of the control system. Thus there is frequently a control loop which manipulates the feed flow in order to control the temperature of the process fluid leaving the furnace. If the instrument measuring this outlet temperature fails, the control loop may increase the fuel flow, which is a hazardous condition. It is essential, therefore, that the measurement on which any alarm or trip is based be separate from that used for control.

There should be trip systems which shut off the fuel flow to the furnace and inject snuffing steam (1) if the electrical power fails, (2) if the fuel pressure and/or flow is low, (3) if the flame goes out, (4) if the combustion air pressure and/or flow is low, or (5) if the process fluid pressure and/or flow is low or its outlet temperature high.

The provision of automatic sequence controls or interlocks for burner start-up has already been mentioned. The safe operation of burners depends critically on the reliability of the instrumentation and it is essential for this to be high.

Difficulties are sometimes experienced with the flame failure device. One problem is instrument failure. Flame failure devices are available with some degree of self-checking. Another is sighting of a pilot burner flame instead of the main flame. If this situation might arise, the flame failure detector should be fitted in such a way that it cannot 'see' the pilot flame.

The foregoing indicates some of the general safety aspects of the design of fired heater systems, but it should be emphasized that this is a specialist matter.

12.10 Process Machinery

Process machines such as compressors and pumps are particularly important items in pressure systems. Not only are they themselves potential sources of loss of containment, but also they affect the rest of the plant by imposing pressure and/or flow fluctuations and by causing vibrations.

The size, complexity and severe operating conditions of many process machines create numerous problems. Much attention is devoted to critical machinery, however,

Table 12.9 *Selected references on process machinery*

ASME (Appendix 28 *Applied Mechanics, Design Engineering, Fluids Engineering*, 1981/85); ASTM (STP 231); IMechE (Appendix 28, 1987); Routh (1882, 1905); Riegel (1953); EEUA (1960 Doc. 15); Meyer (1961); Everett (1964); Axelrod, Daze and Wickham (1968); Sternlicht, Lewis and Rieger (1968); Blubaugh and Watts (1969); Scheel (1970); Tucker and Cline (1970, 1971); Carrier (1971); Perkins and Stuhlbarg (1971); Bultzo (1972); Weaver (1972); Abraham (1973); Dziewulski (1973); J.E. Ross (1973); Barnes (1974); Ryder (1975); API (1976 Refinery Inspection Code Ch. 10); HSE (1978b); Sohre (1977, 1979, 1981); C. Jackson (1976a,b, 1981); Bloch and Geitner (1983, 1990); Bloch (1988, 1989); IEE (1991/5); Oberg *et al.* (1992); Rutan (1993)
BS (Appendix 27 *Machinery*), BS 5304: 1988; VDI 2224: 1988

Turbines, turbomachinery
Keith (1965); Shield (1967, 1973); Naughton (1968); Purcell (1968); Swearingen (1970, 1972); Farrow (1971); ASME (1972/28, 1977/189, 1982/43); H. Cohen, Rogers and Saravanamuttoo (1972); Wachel (1973); B. Turner (1974); Bloch (1976); IMechE (1976/29); Leonard (1976); Millar (1978); Simmons (1978); Athearn (1979); Bergmann and Mafi (1979); C. Jackson and Leader (1979); Molich (1980); Neerken (1980b); Fielding and Mondy (1981); Harman (1981); Nippes (1981); Sohre (1981); Nicholas (1983); L.L. Fisher and Feeney (1984); Turton (1984); Campagne (1985); Rall and Fromm (1986); API (1987 Std 612, 1988 Std 611, 1992 Std 616); Lake (1988); AGA (1989/13)
BS 132: 1983, BS 3863: 1992

Compressors
American Oil Co. (n.d./9); Horlock (1958); Coopey (1961); Troyan (1961c); Younger and Ruiter (1961); Esplund and Schildwachter (1962); Borgmann (1963); Anon. (1964a); Chlumsy (1966); Bultzo (1968); Telesmanic (1968); Kauffmann (1969); Bresler (1970); R.N .Brown (1972, 1974); Hallock, Farber and Davis (1972); Mehta (1972); Dwyer (1973); D.F Neale (1973, 1976); H.M. Davis (1974a,b); P. Lewis (1974); Bauermeister (1975); C. Jackson (1975, 1978); Neerken (1975); Sayyed (1976, 1978); Bryson and Dickert (1977); Cordes (1977); IMechE (1977/34, 1989/108); Kusha (1977); W.E.Nelson (1977); Dimoplan (1978); Winters (1978); Matley and *Chemical Engineering* Staff (1979–); van Ormer (1979); Stokes (1979); Broekmate (1980); Haselden (1980); Burke (1982); AGA (1983/39, 1988/52, 58, 1989/12); G. Bowen (1984); Greene *et al.* (1985a); Zafar (1986); Ablitt (1987): EEMUA (1987 Publ. 152); Sandler and Luckiewicz (1987); ASME (1990 19.1, 1991 19.3); Nissler (1991); B.C. Price (1991); Bloch and Noack (1992); Hallam (1992); Livingstone (1993); O'Neill (1993)
Centrifugal compressors: Schildwachter (1961); Church (1962); Sedille (1965–66); Hile (1967); Badger (1968); Chodnowsky (1968); Morrow (1968); Moschini and Schroeder (1968); Scheel (1968, 1971); Zech (1968); Sohre (1970, 1977); ASME (1971/25, 1976/36, 1984/46, 1990 19.1, 1991 B19.3); Burns (1971); H.M. Davis (1971); Dwyer (1971a,b, 1974); C.Jackson (1971b, 1975); D.L.E. Jacobs (1971); D.F. Neale (1971); Rassman (1971); Schirm (1971); Stafford (1971); Wachel (1973); Cameron

and Danowski (1974); B. Turner (1974); Lapina (1975, 1982); Strub and Matile (1975); Leonard (1976); Sayyed (1976, 1985); Boyce (1978); Gupta and Jeffrey (1979); Rehrig (1981); H. Davis (1983); API (1988 Std 617)

Reciprocating compressors: Deminski and Hunter (1962); Deminski (1964a,b); Morain (1964); Chlumsy (1966); Scheel (1967a,b); Gallier (1968); F.G. Jones (1971); Bultzo (1972); Dziewulski (1973); Prentice, Smith and Virtue (1974); Traversari and Beni (1974); ASME (1975/33); Whittaker (1975); Barnes (1976); Messer (1979); Schiffhauer (1984); API (1986 Std 618); Dube, Eckhardt and Smalley (1991); Woollatt (1993)
BS 7322: 1990

Rotary compressors: Scheel (1969b); van Ormer (1980); API (1990 Std 619); IMechE (1994/163)

Compressor control: Claude (1959); Hagler (1960); R.N. Brown (1964); Daze (1965); Marton (1965); Hatton (1967); Magliozzi (1967); Gallier (1968); Labrow (1968); Spence (1972); Staroselsky and Carter (1990)

Fans

NRC (Appendix 28 *Fans and Blowers*); Woods Ltd (1960); Scheel (1969a); Pollak (1973); Eck (1974); IMechE (1974/8, 1975/24, 1977/34, 1984/78, 1990/119, 1993/151); Martz and Pfahler (1975); J.P. Lee and Chockshi (1978); Matley and *Chemical Engineering* Staff (1979); Summerell (1981); J.E. Thompson and Trickler (1983); Frings and Kasthuri (1986); ASHRAE (1992 ASHRAE 87.1)

Pumps

ASME (Appendix 28 *Fluids Engineering*, 1991 B73.1, B73.2M); ASTM (STP 307, 408); IMechE (Appendix 28, 1974 Item 11, 1975 Items 14, 24, 1976 Item 25, 1977 Item 34); NRC (Appendix 28 *Pumps*); Church (1944); Hicks (1957, 1958); Stepanoff (1957); Minami *et al.* (1960); Doolin (1961, 1963, 1978, 1984, 1990); Korzuch (1961); Younger and Ruiter (1961); Troyan (1961c); Anon. (1962g); Niemkiewicz (1962); T.E. Johnson (1963); Pollak (1963); C. Jackson (1965, 1972a, 1973); J.K. Jacobs (1965); Thurlow (1965, 1971); Addison (1966); Chapman and Holland (1966); Holland and Chapman (1966a,b); Hummer (1966); R. Montgomery (1967); I. Taylor (1967); EEUA (1968 Hndbk 26); Hernandez (1968); Rost and Visich (1969); Hattiangadi (1970b); *Chemical Engineering* (1971); Glikman (1971); R. Kern (1971); Knoll and Tinney (1971); J.A. Reynolds (1971); Stindt (1971); Anon. (1972h); H.E. Doyle (1972); D'Innocenzio (1972); Karassik (1972, 1977, 1982, 1993); Platt (1972); Ramsden (1972); C.A. Robertson (1972); M.G.Murray (1973); Simo (1973); Black *et al.* (1974); van Blarcom (1974, 1980); British Pump Manufacturers Association (1974); Hancock (1974); Neerken (1974, 1980a, 1987); Waring (1974); Yedidiah (1974); Buse (1975, 1985, 1992); Karassik and Krotsch (1975); Anon. (1976d); R. James (1976); Makay (1976); de Santis (1976); Bloch (1977a, 1978, 1980, 1982, 1983a, 1989); R.J. Meyer (1977); Rinard and Stone (1977); Sparks and Wachel (1977); W.H. James (1978); McLean (1982); Morlock (1978); Panesar (1978); Penney (1978); Tinney (1978); Tsai (1982); Anon. (1979 LPB 29, p. 139); Grohmann (1979); Henshaw (1979, 1981); Lightle and Hohman (1979); Panesar (1979a); Poynton (1979); A.P.Smith (1979a,b); Webster (1979); Mattley and *Chemical Engineering* Staff (1979–); API (1980 Std 674, 676, 1989 Std 610); IChemE (1980/121, 1985/127); Krienberg (1980); Vetter and Hering (1980); Anon. (1981f); Bristol (1981); Ekstrum (1981); Fraser (1981); J.D. Johnson (1981); Lapp (1981); Mikasinovic and Tung (1981); W.V. Adams (1982); Hallam (1983); F.J. Hill (1983); R. King (1983); Krutzsch (1983); R.R. Ross (1983); Talwar (1983); Baker-Counsell (1984b); Dobrowolski (1984); Hornsby (1984); Vlaming (1984); Bloch and Johnson (1985); Cody, Vandell and Spratt (1985); Etheridge (1985); S. Hughes (1985, 1987); Lobanoff and Ross (1985); Nevill (1985); Rattan and Pathak (1985); Ionel (1986); Karassik *et al.* (1986); Nasr (1986, 1992); R. Edwards (1987); EEMUA (1987 Publ. 157); Reeves (1987); Sandler and Luckiewicz (1987); I. Taylor (1987); T. Martin (1988a,b, 1989a); Rendell (1988b); API (1990 Std 610); Dufour (1989); Garbers and Wasfi (1990); Gravenstine (1989); Gülich and Rösch (1989); Lahr (1989); Luetzow (1989); McCaul (1989); G. Parkinson and Johnson (1989); J.A. Reynolds (1989); H. Davis (1990); Doolin and Teasdale (1990); Flyght Pumps Ltd (1990); Mabe and Mulholland (1990); Anon. (1991i); C. Butcher (1991d); Doolin *et al.* (1991); Jaskiewicz (1991); Margus (1991); Newby and Forth (1991); Wild (1991); Hawks (1992); G. Parkinson and Ondrey (1992); Schiavello (1992); UL (1992 UL 51); Bitterman (1993); Blair (1993); Buck (1993); Chyuan-Cheng Chen (1993); Cleary (1993); J.R. Peterson and Davidse (1993); Vandell and Foerg (1993); Vetter *et al.* (1993)
BS 4082: 1969–, BS 5257: 1975

Agitators

N.H. Parker (1960); Troyan (1961c); EEUA (1962 Hndbk 9); Penney (1970); Gates, Hicks and Dickey (1976); Hicks and Dickey (1976); R.S. Hill and Kime (1976); Ramsey and Zoller (1976); Ketron (1980)

Centrifuges

IChemE (1976/67, 1987/81); Crosby (1979)
BS 767: 1983

Conveyors

EEUA (1965 Doc. 30); ASME (1990 B20.1–1990)
BS 490: 1975–, BS 2890: 1989

Drives

NRC (Appendix 28 *Electric Motors*); Gillett (1960); H.H. Meyer (1961); Cunningham (1962); Lane and Holzbock (1962); Olson (1963, 1979); ASME (1966/122); Caplow (1967); Olson and McKelvy (1967); EEUA (1968 Hndbk 29); Roe (1969); Nailen (1970a,b, 1973a–c, 1974, 1975, 1978); E.F. Cooke (1971, 1973); P. Bell (1971b); Cates (1972); Hallock, Faber and Davis (1972); IEE (1972 Conf. Publ. 93); Platt (1972); Pritchett (1972); J.C. Moore (1975); Plappert (1975); Bloch (1977a); M.G. Murray (1977); Constance (1978); Albright (1979); Deutsch (1979); Fishel and Howe (1979); Kohn (1979b); Panesar (1979b); Basta (1980); Finn (1980); M.N. Kraus (1980); Pollard (1980); Anon. (1981e); IMechE (1981/55, 1983/69, 1985/85); Antony and Gajjar (1982); Doll (1982); Hartmann (1984); Long (1984); Beevers (1985b); API (1987 Std 541); Feldman (1987); Sandler and Luckiewicz (1987); EEMUA (1988 Publ. 132); I. Evans (1988); Ranade, Robert and Zapate-Suarez (1988); UL (1988 UL 1004, 1991 UL 1247); Anon. (1989a); Burton (1989); Mabe and Mulholland (1990); Basso (1992); Patzler (1992); Thibault (1993)

BS (Appendix 27 *Electrical*)

Power transmission
North and Parr (1968); Anon. (1973h); Beard (1973); E.F. Cooke (1973); Dubner (1975); Finney (1973); Shipley (1973); Thoma (1973); Bloch (1974); Kraemer (1974); Arndt and Kiddoo (1975); Calistrat (1975, 1978–); Roney (1975); Wattner (1976); Vanlaningham (1977); Coupland (1980); Renold Ltd (1980); J.D. Smith (1983); ASME (1985 B106.1M); C.M. Johnson (1985); Polk (1987); API (1988 Std 613, 1989 Std 677); Mancuso (1988); Timmerman (1989); Horrell, Neal and Needham (1991); IMechE (1993/153); Thibault (1993)
ANSI B93 series, BS (Appendix 27 *Machinery*)

Bearings, seals, lubrication
ASME (Appendix 28 *Applied Mechanics, Fluids Engineering*); ASTM (STP 77, 84, 88, 437); IMechE (Appendix 28); Whalen (1963); Anon (1964d); Coopey (1965, 1967, 1969); EEUA (1965 Hndbk 10, 1971 Hndbk 27); L.H. Price (1965); Koch (1966); Stock (1966); Lindsey (1967); C. Jackson (1968, 1970, 1972a, 1975); N.H. Miller (1968); Samans (1968); Samoiloff (1968); Furey (1969); Yaki and Carpenter (1970); Battilana (1971, 1989); C.P. Shaw (1971, 1977); E. Meyer (1972); Ruckstuhl (1972); EEMUA (1973 Publ. 115); Bushar (1973); Kellum (1973); M.J. Neale (1973); Rothman (1973); J.O.S. MacDonald (1973); Tinney, Knoll and Diehl (1973); B. Turner (1973); R.M. Austin and Nau (1974); Miannay (1974); Nisbet (1974); Anon. (1975k); Pattinson (1975); J. Wright (1975); Anon. (1976 LPB 9, p. 12); Cameron (1976); Fern and Nau (1976); Houghton (1976); M.G. Murray (1976); Ramsey and Zoller (1976); Summers-Smith and Livingstone (1976); Czichos (1978); Hills and Neely (1978); Hoyle (1978); Ramsden (1978); Barwell (1979); McNally (1979); J. Phillips (1979); Tipler (1979); Ummarino (1979); Hawk (1980); P.T. Jones (1980); Mendenhall (1980); Panesar (1980); P. Rogers (1980); Sangerhausen (1981); Warring (1981); Kerklo (1982); Lansdown (1982); Summers-Smith (1982, 1988); Anon. (1983g); W.V. Adams (1983, 1987); Bloch (1983b, 1989); Booser (1983); Cameron and McEttles (1983); Cleaver (1983); Higham (1983); M.H. Jones and Scott (1983); BOHS (1984 TG3); Fuller (1984); W.S. Robertson (1983); Dunhill (1984b); Wallis (1984); Wong and Ansley-Watson (1985); Baker-Counsell (1985h); Buse (1985); Wallace and David (1985); Flitney (1986, 1987); Trade and Technical Press (1986); Anon. (1987e); Chynoweth (1987); Martel, Botte and Regazzacci (1987); Abrams and Olson (1988); Ferland (1988); Fort and Jehl (1988); Newby (1988); API (1992 Std 614); Gregory (1993); Wells (1993)
BS 1399: 1970

Vibration (*see also* Table 19.9)
ASME (Appendix 28 *Applied Mechanics, Design Engineering, Dynamic Systems and Control, Flow Induced Vibration and Noise, Fluids Engineering, Heat Transfer, Pressure Vessels and Piping*, 1988/197); Church (1963); Summers-Smith (1969); Ker Wilson (1970–71); R.H. Wallace (1970); Anon. (1973h); IMechE (1975/24, 1977/43, 1979/49, 1984/84, 1988/100, 106, 1991/131, 1992/145); J.M. Baker (1976); R.J. Meyer (1977); Sallenbach (1980); M.H. Jones and Scott (1983); J.D. Smith (1983)

Alignment and balancing
Blake (1967); C. Jackson (1971a,b, 1976a,b); Essinger (1973, 1974); M.G. Murray (1974, 1979); Ryman and Steenbergen (1976); Kirlan (1977); AGA (1985/44); Horrell (1991); IMechE (1991/128)

as much for economic as for safety reasons. In consequence, the level of reliability attained is generally rather high. Nevertheless, a failure of a process machine can be serious.

An account of some of the problems associated with process machinery is given in *Safe Operation of Air, Ammonia and Ammonium Nitrate Plants* by the American Oil Company (Amoco/9). Selected references on process machinery are given in Table 12.9.

12.10.1 Process compressors

The process industries use both positive displacement and centrifugal compressors, some of which are very large machines with high throughputs and energy. An account of compressors is given in *Industrial Compressors* (O'Neill, 1993).

The three principal types of compressor are centrifugal compressors, reciprocating compressors and screw compressors. Both the last two are positive displacement machines, but with reciprocating and rotary motions, respectively.

Reciprocating compressors have long been used for high pressure duties. There has been a trend, however, towards centrifugal compressors and, although these machines were originally used mainly for relatively low compressions, the pressures for which they are used have tended to increase. Some centrifugal compressors are now very big. A large ethylene plant may depend on single large compressors not only for main gas compression but also for refrigeration compression. Another trend is the increase in the application of screw compressors. The range of throughput and compression of these machines has extended so that they now constitute a significant third option.

Compressors are complex machines and their reliability is crucial. It is essential, therefore, to put much effort into the specification of the machine and to liaise with the manufacturer. It is equally necessary to maintain high standards in the operation of compressors. Some general features of compressors which are important are (1) lubrication, (2) protection, (3) isolation, (4) purging, (5) liquid slugs, (6) housekeeping and (7) observation.

The lubrication systems are an integral part of the compressor. It is essential for them to be well engineered and reliable if they are not to be a weak link.

Compressors are provided with protective devices such as pressure relief valves and pressure, temperature and vibration measuring instruments which activate alarms and trips. These are particularly important instruments and they should be reliable.

There need to be arrangements for the isolation of the compressor. Isolation may be by a single block valve and blind or by double block and bleed valves. It may also be appropriate to have an emergency isolation valve on the compressor suction.

Facilities should be provided to allow the compressor to be purged. A permanent purge connection is often considered undesirable because of the hazard of reverse flow from the compressor into the purge gas main. If a permanent connection is installed, it should be shut off when not in use by a blind.

A slug of liquid entering a compressor can be as destructive as a lump of metal. It is important, therefore, for the system to be designed so that this does not happen. It is also necessary before start-up to check that there is no liquid in the inlet or interstage piping, intercoolers, knockout drums or pulsation bottles.

A high standard of housekeeping around compressors is very desirable. Leaks of oil or water should be eliminated and cleaned up promptly.

The process operators should, and usually do, learn to detect changes in external appearance, noise and vibration as well as in the pressure and temperature measurements. This monitoring by the operators often gives early detection of malfunction and allows corrective action to be taken in time. For large compressors, however, it is now usual to monitor condition and performance using instrumentation.

It is essential to have the highest standard of maintenance on compressors. The maintenance methods need to be thought out thoroughly and the maintenance personnel well trained in them. The conditions under which the maintenance is done should be suitable both with respect to weather and to cleanliness.

Prime movers for compressors include electric motors, turbines and gas engines.

12.10.2 Reciprocating compressors

Reciprocating compressors are utilized for high compressions. They have long been the machines used for compression to the high pressures required in ammonia plants and on offshore platforms. Accounts of reciprocating compressors are given by Whittaker (1975), Barnes (1976), Messer (1979) and Woollat (1993). Relevant standards are BS 7322: 1990 *Specification for the Design and Construction of Reciprocating Type Compressors for the Process Industry* and API Std 618: 1986 *Reciprocating Compressors for General Refinery Services.*

Reciprocating compressors can be provided with capacity control, or turndown, by the use of volume pockets and by drive speed control. Some principal malfunctions on reciprocating compressors are (1) valve leakage, (2) cylinder/piston scoring, (3) piston ring leakage, (4) gasket failure, (5) tail rod failure and (6) vibration, as well as the general compressor failures such as those caused by liquid slugs or loss of lubricating oil or cooling water.

Leakage of the suction or discharge valves is one of the commonest failures. There are a number of symptoms of valve malfunction. The valve may become unusually hot, the cylinder capacity may fall, the discharge temperature may rise and the interstage pressures may be abnormal. The suction pressure may rise and the discharge pressure fall unless automatically controlled.

On some machines there is nothing to prevent a suction valve being fitted on the discharge, or vice versa, by mistake. If this happens, it is possible to create very high pressures in the cylinder, particularly the high pressure cylinder, and so cause failure of the cylinder, the piston or the drive system. Preferably this feature should be designed out. Where this is not the case, there should be procedures to minimize the probability of error.

The piston rod in high pressure cylinders is sometimes balanced by a tail rod on the other side of the piston. There have been some serious accidents in which the tail rod has broken off and flown out like a projectile. There should be a 'catcher' of sufficient strength to prevent escape of the tail rod if it does break. Tail rods are frequently surface hardened and it is important for them to be free of surface cracks. They should be regularly inspected. A tail rod failure allows the escape of high pressure gas.

The reciprocating movement of the piston inevitably causes some degree of vibration. This vibration may be transmitted to, and cause failures in, the process pipework. Small auxiliary piping on the machine tends to be particularly vulnerable to fracture from this vibration. It should be anchored to reduce vibration and inspected regularly.

Changes in the discharge temperature are often a sign of malfunction on a reciprocating compressor. High discharge temperature may be associated with valve failure, piston ring leakage, increased compression ratio, gas composition change or loss of cooling water.

On air compressors lubricated with oil a high discharge temperature can result in an explosion. Such air compressor explosions, and the discharge temperature limits necessary to avoid them, are discussed in Chapter 17. Some compressors are required to produce compressed air which is free of oil. It is very desirable, for example, that instrument air be oil-free. Carbon ring compressors are often used for this purpose.

The reciprocating compressors were one of the potential sources of the gas leak investigated in the Piper Alpha Inquiry (Cullen, 1990). The questions of the tolerance of such machines to ingestion of liquid slugs and of the bolt tightening practices used were considered in evidence by Bett (1989).

12.10.3 Centrifugal compressors

Centrifugal compressors are the main workhorse machines in the process industries. They can be built for very high throughputs. Although the compression obtained has been lower than that given by reciprocating machines, the range of pressures attainable has gradually been extended. As mentioned, centrifugal compressors are used for the main duties on ethylene plants, both for process gas and for refrigeration.

Accounts of centrifugal compressors are given by Rehrig (1981), H. Davis (1983) and Sayyed (1985). A relevant standard is API Std 617: 1988 *Centrifugal Compressors for General Refinery Services.*

Centrifugal compressors have relatively limited turndown. On centrifugal compressors some of the main malfunctions are (1) rotor or shaft failure, (2) bearing failure, (3) vibration and (4) surge, as well as the general compressor failures mentioned earlier.

In some duties the rotor suffers relatively little attack, but rotors sometimes have flaws or suffer embrittlement or debris deposition. Rotors can become unbalanced or displaced axially. Shafts may contain flaws.

The various kinds of bearing failure are one of the commonest faults in centrifugal compressors. Shaft or casing misalignment and rotor imbalance are frequent causes of bearing failure.

Vibration can cause damage to parts of the compressor itself, such as the bearings. It may also be transmitted to and induce failure in the pipework.

Surging can cause serious damage on a centrifugal compressor. Surge is a condition which occurs when the machine is operating on too low a load, generally in the range 50–85% of normal capacity, which gives rise to violent rapid flow changes and vibration. It is usually prevented by automatic controls which detect the near-surge condition and bypass gas from the discharge to the suction of the compressor. A further discussion of surge control is given in Chapter 13.

12.10.4 Screw compressors

Screw compressors are positive displacement machines with rotary motion, and are also known as 'helical screw' or 'spiral lobe' compressors. They are relatively simple and low in capital cost. They have gradually extended their range of application to emerge as a third force in process industry compression. They are now in wide-spread use for lower compression and refrigeration duties.

Accounts of screw compressors are given by B.C. Price (1991) and Bloch and Noack (1992). A relevant standard is API Std 619: 1985 *Rotary Type Positive Displacement Compressors for General Refinery Services.*

The original screw compressors were oil-free, or dry, machines. They ran at high speed to minimize internal gas bypassing and were very noisy. In the 1950s, oil-injected machines came in and found application in natural gas compression and refrigeration.

Capacity control on screw compressors is effected by the use of a slide valve which moves axially along the housing. They have good turndown, being able to operate at loads as low as 10% of the normal throughput. Another feature of screw compressors is that they can tolerate relatively large changes in suction pressure, a characteristic useful in refrigeration duties.

Screw compressors have the advantage over reciprocating compressors that they require no suction or discharge valves. By comparison with centrifugal compressors, the spares problem with screw compressors tends to be less severe. Since screw compressors are designed for oil-injection, they have some tolerance to a liquid slug, although continuous liquid slugging can cause problems.

A large proportion of the heat of compression in a screw compressor is absorbed by the oil. This characteristic may be advantageous if it is desirable to avoid high gas compression temperatures, as in cases where the gas may polymerize or explode if overheated.

12.10.5 Gas engines

Gas engines are used to drive compressors. Some principal failures which occur are explosions of (1) starting air line, (2) fuel line and (3) crankcase. A number of explosions have occurred in the starting line to gas engines. Generally, these have been the result of leakage from the power cylinder through a defective non-return valve. Preventive measures include regular inspection and maintenance of the non-return valve on the starting line, elimination of oil from this line and venting of the header when the engine is in normal operation.

When the engine is shut down, it is desirable that the fuel be shut off also. If this is not done, fuel may collect in the engine and the exhaust system, and may explode when the engine is restarted. There have been a number of explosions of combustible mixtures of oil or gas and air in crankcases during operation. Crankcase explosions may be minimized by ventilating the crankcase or purging it with inert gas and fitting an explosion relief device. A further discussion of crankcase explosions is given in Chapter 17.

12.10.6 Process pumps

Most process pumps are centrifugal machines, although reciprocating machines are used in some cases. Where the application is severe and/or critical, a degree of redundancy may be provided, using one or more stand-by pumps or using several pumps operating at less than full capacity. Thus, for example, the equipment to perform the duty of a single pump may be one pump operating at 100% throughput and a similar pump on stand-by or two pumps each rated for 100% capacity, but both operating at 50% throughput. Diversity of the power supply may also be provided using as drives steam turbines as well as electric motors, thus reducing dependence on electrical power.

If a stand-by pump is provided, it is necessary to ensure that it can be started up rapidly and dependably. The reliability of the pump set can be seriously degraded by pump changeover failure. The arrangements for pump shut-down also need to be reliable. This is particularly the case where there is a potential weak point in the discharge piping such as a flexible hose at a loading facility.

As described earlier, it will often be appropriate to install an emergency isolation valve between a pump and its feed vessel. It is good practice to group together controls for both items.

12.10.7 Centrifugal pumps

The principal malfunctions on centrifugal process pumps are:

(1) bearing failure;
(2) gland/seal failure;
(3) maloperation damage:
  (a) cavitation,
  (b) deadheading,
  (c) dry running.

The most common pump faults are failures of bearings or of glands or seals. A common cause of failure is shaft misalignment. Also, a bearing failure can induce a gland or seal failure. If seal failure is particularly undesirable, a type of pump may be used which has a more reliable sealing arrangement. Such pumps include:

(1) pumps with mechanical seals:
  (a) single mechanical seals,
  (b) double mechanical seals;
(2) seal-less pumps.

The pump sealing arrangements should both minimize fugitive emissions and have a low probability of catastrophic failure. Gland seals and single mechanical seals leak to some degree. Leaks may be minimized by the use of double mechanical seals or of seal-less pumps. As a precaution against catastrophic failure use may be made of a double mechanical seal with monitoring of the atmosphere between the two seals. Another arrangement with the same purpose is a single mechanical seal with a packing gland backup.

Seal-less pumps are also referred to as 'glandless pumps' or 'canned pumps'. In a canned pump the impeller of the pump and the rotor of the motor are mounted on an integral shaft, the rotor and stator being encased, or canned, so that the process fluid can circulate in the space which is normally the air gap of the motor. The bearings are lubricated by the process fluid. Pumps with mechanical seals and seal-less pumps are less liable to leakage, but are also more expensive.

Pumps are particularly liable to be damaged by maloperation. Cavitation, deadheading and dry running can have destructive effects and failure can be catastrophic. Not infrequently pumps are allowed to run whilst cavitating. Cavitation occurs if the liquid is close to its boiling point and involves incipient vaporization in the pump, the bubbles formed collapsing on the pump impeller due to the pressure induced condensation. This causes pitting and, eventually, more serious damage to the impeller, and can be very destructive. It is avoided by provision of adequate net positive suction head. In order to ensure this it is helpful to keep the suction lines to the pump short. Another measure which may be taken is the provision of a bypass.

Deadheading involves pumping against a closed outlet, or possibly a closed outlet with a closed inlet also. The pump may then generate enough heat and/or pressure to rupture. Cases have occurred of a runaway reaction in the pump or a reaction between the pump material and the fluid. Although avoidance of deadheading is principally an operations matter, there are countermeasures which can be taken in the design. One is a trip system, based on rise in temperature or pressure in the pump. Another is a bypass line free of valves from the pump delivery to the feed tank, or 'kickback' line.

The provision of bypasses on centrifugal pumps is discussed by I. Taylor (1987). Apart from boiler feedwater pumps, centrifugal pumps have traditionally not been equipped with a bypass. Over the last 20 years, however, this has changed. It remains true that most centrifugal pumps do not need a bypass. In order to decide whether one is needed, it is necessary to establish the expected frequency and extent of low flow operation, including shut-off of a downstream control valve, and the extent of any hazard if failure occurs. Centrifugal pumps can generally tolerate '1 minute' periods of low flow immediately after start-up. The cases where bypasses are required tend to be those involving pumps which are of large size, high horsepower and high head, and are in continuous low flow operation. The author gives some ten reasons for installing a bypass, the two principal ones being to prevent excessive temperature rise and to avoid unstable flow conditions. For pump thermal control a bypass flow of 10% is generally sufficient.

Taylor discusses the net positive suction head (NPSH) required to prevent cavitation. In the 1960s and early 1970s a spate of problems occurred on high head and high horsepower pumps, involving excessive vibration and pulsation at low flows. Work by Minami *et al.* (1960) and others showed that, whereas traditionally a 3% reduction of first stage head had been taken as an indication of impairment of pump performance by cavitation due to lack of NPSH, at low flows incipient cavitation can occur even at a much higher suction head. In fact the NSPH required to suppress 'low flow' cavitation in such cases could be 2–5 times as much as the conventional figure. Subsequently, Fraser (1981) presented a method of estimating the minimum continuous flow based on the suction specific speed index number N_{ss}.

Incorrect design of a bypass, however, can itself be a cause of failure at pumps. Taylor recommends that, with certain exceptions, an automatic bypass should branch off on each pump upstream of the first valve, whether this be an isolation valve, non-return valve or control valve. Whilst it is true that in many cases involving two motor driven pumps, a common bypass has often been satisfactory, there are two dangers. One is that if one pump stops or trips out and the other, on stand-by, cannot be started up, the bypass opens and connects the high pressure of the delivery side to the suction side. The other danger is that if both pumps are running but with one at a higher speed, perhaps because one is driven by a motor and the other by a steam turbine, the difference in speed may be sufficient to hold closed the non-return valve of the lower speed pump, so that it overheats.

Pumps are sometimes allowed to run completely dry, often due to loss of head in a feed tank. Under these conditions the pump can be wrecked very quickly. Pumps which are not operating can suffer damage due to vibration of the plant, which causes 'brinelling' at one particular point. Such pumps may be turned over periodically to prevent this.

12.10.8 Positive displacement pumps

Positive displacement pumps are less common in process plant than are centrifugal pumps, and, apart from metering applications, are used mainly for high pressure work. In particular, reciprocating pumps are used to obtain high pressures. The discharge side of a positive displacement pump requires protection against overpressure. The usual arrangement is the provision of a pressure relief valve with its discharge returned to the suction side.

12.11 Insulation

Another element of the pressure system on process plants is the insulation. Insulation is employed to control heat transfer both in normal operation and in fire conditions. It has more implications for safety than might at first appear. Accounts of insulation are given in *Thermal Insulation* (Malloy, 1969), in the CPPS *Engineering Design Guidelines* (1993/13) and by Britton (1991), Britton and Clem (1991), Irwin (1991a,b), McMarlin and Gerrish (1992), Reddi (1992) and Gamboa (1993).

There are two basic types of insulation: thermal insulation and fire insulation. There is potential for confusion in that many types of thermal insulation have fire-resistant properties. Thermal insulation is used to (1) reduce heat loss from plant operating at temperatures above ambient, (2) reduce heat gain from plant at temperatures below ambient, (3) protect personnel from hot or cold surfaces on the plant and (4) attenuate sound from the plant. Two particular reasons for reducing heat loss are to prevent (1) freezing of liquid and (2) condensation of vapour.

12.11.1 Safety aspects of insulation

Some of the principal safety aspects of insulation are (1) corrosion beneath insulation, (2) self-heating in insulation, (3) insulation against fire and (4) effect on the process of inadequacies, defects or failures of insulation. Corrosion beneath insulation is considered in Section 21.11.4, self-heating in insulation in Section 12.11.6, fire insulation in Section 12.11.7 and fire properties of thermal insulation in Subsection 12.11.8. The effects on process or storage of defects in insulation depend on the particular case, but may include hazards from freezing, condensation, rollover, etc.

12.11.2 Materials for thermal insulation

The thermal conductivity is by no means the sole criterion in selecting an insulation. Materials with similar thermal conductivities may differ widely in their other relevant properties. Categories of non-combustible insulation material are (1) calcium silicate, (2) expanded perlite, (3) expanded vermiculite, (4) mineral fibre and (5) cellular glass.

Some properties of thermal insulation which are important in the present context, in addition to its thermal conductivity, are its characteristics in respect of fire resistance, liquid absorption, fabrication, durability and damage resistance. The liquid absorptivity and fire resistance of insulation are considered in Sections 12.11.3 and 12.11.8, respectively.

The insulation should provide complete cover of the areas of the equipment which it is intended to insulate. Properties which bear on this are dimensional stability and shrinkage. Furthermore, it should be sufficiently easy to fabricate that gaps do not occur as a result of fabrication difficulties. Insulation is often subject to conditions which lead to its being crushed or torn, damage from feet being quite common. Vibration may lead to degradation of short-fibre insulation.

12.11.3 Absorption of liquids in insulation

An important property of an insulation is the extent to which it retains liquid that it already contains or that leaks into it. This property is relevant to the insulation in respect of its (1) thermal performance, (2) mechanical performance and (3) weight, and to (4) corrosion beneath the insulation and (5) self-heating in the insulation.

Water absorption is greatest for calcium silicate and least for cellular glass, whilst for the other types it is variable, depending in part on any added water repellent. Water may enter the insulation as rain or washing water or in other ways such as a steam leak from a trace heating system. Water present in insulation can cause a marked increase in the effective thermal conductivity of the latter, as spaces which should be filled with air become filled with water, with a thermal conductivity more than 20 times that of air. Water also has the potential to cause disruption of the insulation, a typical situation being that which occurs when water freezes to ice and then melts. Since some insulation materials can absorb twice their mass in liquid, water absorption can cause a significant increase in the weight of the insulated system.

Another effect of water is to cause external corrosion of the equipment beneath the insulation, as discussed in Subsection 12.11.4.

With some insulations the moisture can be driven out and thermal performance restored, but with others this is not so. Calcium silicate, which has a high propensity to absorb water, falls in this latter group.

Process liquids may enter from pipework as may heat transfer fluids. Possible sources of in-leak are flanges and drain and sample points. Some liquids react with the binder in the insulation and promote its disintegration. Flammable liquid entering insulation can give rise to self-heating, as described in Section 12.11.6.

12.11.4 External corrosion beneath insulation

Factors which determine the extent of external corrosion beneath insulation include (1) the insulation material, (2) the equipment material, (3) the equipment configuration, (4) the equipment coating, (5) the equipment stress, (6) the service temperature, (7) any temperature transients, and (8) the climate and location.

One important factor governing external corrosion is the insulation material. An insulation which absorbs a large amount of water is likely to cause increased corrosion. One which contains chloride is liable to cause corrosion of stainless steel. The extent to which water is excluded by efficient cladding is another significant factor in external corrosion.

The material of construction of the equipment insulated is another factor. In particular, stainless steel is prone to corrosion by chloride in the insulation. A further factor is the equipment configuration. There are a number of equipment features where corrosion is especially likely to occur. On a vessel these might include nozzles, manholes, supports, brackets, lifting lugs and small bore and instrumentation piping. There are also features such as flanges and drain and sample points which may allow entry of liquids into the insulation. The extent of external corrosion under insulation also depends on the degree of protection afforded by any coating such as paint. External corrosion may be increased by residual stress in the equipment. This is a factor particularly in the stress corrosion cracking of stainless steel.

External corrosion under insulation occurs mainly on equipment operating in the range −5 to 105°C, and especially in the range 60 to 80°C. At lower temperatures the reaction rate is slower, whilst at higher temperatures water tends to be driven off. Much equipment is subject to temperature transients, which can increase external corrosion. These may be major changes occurring during the start-up and shut-down of high or low temperature plants, or they may take the form of temperature cycling. High plant temperatures can cause concentration of salts which then cause severe corrosion when rewetted.

Corrosion occurs on low temperature plant at thawing zones which tend to remain wet and corrosive.

External corrosion can be aggravated by the presence of salt, as at a coastal location, especially in combination with water precipitating, say, from cooling towers, or condensing, say, from low temperature plant. Pollack and Steely (1990) have described a case in which a large steel column had to be replaced due to corrosion underneath the insulation.

One of the proposals put to the Piper Alpha Inquiry (Cullen, 1990) was that gas risers on production platforms should be provided with passive fire protection in the form of lagging. However, the Inquiry heard evidence that there were also potential problems with this approach – a riser could be put at risk by corrosion beneath the lagging. It made no recommendation on the fire insulation of risers.

It is also convenient to mention at this point the tendency of some insulation to cause abrasive wear of equipment.

12.11.5 External corrosion beneath insulation: protective measures

Measures to prevent external corrosion under insulation relate broadly to (1) the insulation material, (2) the cladding, (3) the equipment coating and (4) the equipment design features. The insulation material should match the equipment to be insulated. For example, an insulation for stainless steel should contain minimal amounts of chloride. The insulation should be one with low water content and absorption capacity.

The insulation should be designed to exclude water. It should be provided with weatherproof cladding. This should permit water entering to be drained and water in the insulation to vaporize and exit when the insulation is heated. The equipment should be protected by coating with paint or other suitable covering.

The design should assist inspection for external corrosion. Inspection ports are available which are specifically designed for this purpose. With regard to abrasive wear of the equipment, there are available anti-abrasive coatings which can be used to reduce this.

12.11.6 Self-heating in insulation

Another insulation issue related to fire is self-heating in insulation soaked with a flammable liquid. Above a critical temperature, which depends on the liquid–substrate system and on the geometry, self-heating initiates and smouldering occurs, limited both by heat transfer from the combustion zone and mass transfer of oxygen into the zone. The explosive destruction of an ethylene oxide column at Antwerp in 1989 was found to be due to self-heating of a leak of the fluid into the insulation (Case History A122).

Accounts of self-heating in insulation are given by Gugan (1974a), Britton (1991) and Britton and Clem (1991). The latter authors discuss the spontaneous ignition temperature (SIT) of liquid in insulation. Values of the SIT range from above 180°C down to near ambient temperature.

Precautions against self-heating include (1) use of a non-porous insulation, (2) exclusion of flammable liquids and (3) use of a geometry less favourable to its occurrence. One non-porous insulation is cellular glass. However, its fire resistance properties leave something to be desired. Measures for the exclusion of flammable liquids have been discussed in Section 12.11.3. Both self-heating in general and lagging fires in particular are discussed in more detail in Chapter 16.

12.11.7 Fire insulation

Fire insulation proper is provided on equipment which requires fire protection but not thermal insulation. For fire insulation use is made of material which is generally cementitious, such as vermiculite cement. A further discussion of fire insulation, including criteria and tests, is given in Chapter 16.

12.11.8 Fire performance of thermal insulation

On equipment which does require thermal insulation, fire protection may be provided by a fire resistant thermal insulation system. This phrasing is used advisedly, since the effectiveness of the protection is a function of the whole system rather than of the insulation alone.

From the fire protection viewpoint, the ideal thermal insulation is one which does not burn or melt and is non-absorbent. To the extent that it does undergo combustion, it should not give off toxic fumes. Some insulating materials such as polyisocyanurate foam are not suitable. The main categories of non-combustible insulation are listed in Subsection 12.11.2.

Standard tests of fire properties such as those for flame spread and smoke development are applicable to the insulation materials. There are also standard fire tests such as ASTM E-119 and other tests which seek to simulate more closely hydrocarbon fires. Tests for the insulation material itself and for the insulation system are discussed in the CCPS *Guidelines* and by Britton and Clem (1991). A review of insulation materials for fire protection has been given by Wright and Fryer (1982).

The insulation needs to be protected against the weather. Methods of weatherproofing include the use of caulking, of mastic or of metal jacketing. Metal jacketing is to be preferred. Stainless steel provides a quality jacket, with galvanized steel being a less expensive alternative. There is some experimental evidence of good fire performance of galvanized steel jackets, but it is uncertain how far this can be generalized. Aluminium jackets melt on exposure to fire, in some cases within minutes.

The jacket should have points through which any water entering can drain away and should allow for the egress of any water vaporized when the insulation is heated. In a fire the insulation system may be subject to quite severe conditions, including impingement by flame and/or water jets. These need to be borne in mind in the design.

Britton and Clem (1991) describe the measures used in Union Carbide for certain situations where fire insulation is regarded as particularly critical. One such case is fire insulation of equipment handling reactive chemicals. An example of such a chemical is acetylene. For such reactive chemicals a double layer of insulation is used with staggered joints. Stainless steel is specified for the metal jacketing.

12.11.9 Personnel protection by insulation

One function of thermal insulation is to prevent injury to personnel. Hazard exists not only on high temperature plants but also on those operating at low temperatures which can cause 'cold burns'. The usual method of protection is to make the insulation sufficiently thick to prevent injury. For high temperatures, a criterion commonly used is that the external surface of the insulation should not exceed 60°C. Britton and Clem (1991) discuss the effect on the outer surface temperature of the insulation of its emissivity and absorptivity. An alternative means of protecting personnel from hot or cold equipment is the use of guards to prevent access to the hazardous surface.

12.11.10 Design of insulation systems

As already emphasized, design must address the total insulation system rather than just the insulation itself. It needs to have regard to the several goals of thermal insulation, fire protection and personnel protection and to the avoidance of corrosion beneath the insulation, self-heating and hazardous effects on the process. It is likely to involve a number of trade-offs. Attention should be paid to any special situations. These include, as discussed above, the insulation of stainless steel equipment and of equipment handling reactive chemicals.

Some of the errors commonly made in insulation, not all in design, are rehearsed by Irwin (1991a,b). They include (1) using the wrong insulation material, (2) installing the wrong thickness, (3) struggling with difficult calculations, (4) specifying the wrong thickness for jacketing, (5) laying insulation over jacketing, (6) poor workmanship, (7) hiding damage and (8) no insulation at all.

12.12 Overpressure Protection

Pressure systems need to be provided with protection against failure, particularly from overpressure. Standards and codes which deal with overpressure include: BS 5500: 1991; the ASME *Boiler and Pressure Vessel Code* 1992; the ASME B31 *Code for Pressure Piping*, particularly B31.3: 1990 *Chemical Plant and Petroleum Refinery Piping*; API RP 520: 1990 *Recommended Practice for the Design and Installation of Pressure Relieving Systems in Refineries*; API RP 521: 1990 *Guide for Pressure Relief and Depressuring Systems*; and API Std 2000: 1992 *Venting Atmospheric and Low Pressure Storage Tanks*

Accounts of pressure relief include those given in *Relief Systems Handbook* (Parry, 1992) (the IChemE *Relief Systems Guide*) and by Isaacs (1971), Fitt (1974), Duxbury (1976), Crawley and Scott (1984), A. Moore (1984), Cunningham (1985), Tomfohrde (1985), Crooks (1989), Duckworth and McGregor (1989), Parry (1989) and Crowl and Louvar (1990).

The overpressure protection considered here applies to situations in which the pressure rise is relatively gradual. Protection against explosion overpressure is treated in Chapter 17. A number of equations given in the standards and codes are quoted in the following sections. Particular symbols may have different meanings in different equations, but in all cases the symbols are fully defined with the equation and in the notation at the end of the chapter. Selected references on overpressure protection, pressure relief and blowdown are given in Table 12.10.

12.12.1 Requirements for protection

The statutory requirements for protection of pressure vessels and the legislative background to these were outlined in Chapter 3. As described in Section 12.1, the

***Table 12.10** Selected references on overpressure protection, pressure relief and blowdown*

Overpressure protection
Sylvander and Katz (1948); C.G. Weber (1955); Conison (1960, 1963); Block (1962); Jenett (1963a–c); Loudon (1963); Driskell (1964, 1976); Ruleo (1964); Rearick (1969); Warwick (1969); Hattiangadi (1970a); Heitner (1970); ICI/RoSPA (1970 IS/74); Wittig (1970); Isaacs (1971); Pilborough (1971, 1989); Fitt (1974, 1976b, 1983); Kletz (1974a,d,e, 1989); BCISC (1975/1); Whelan and Thomson (1975); Anon. (1976m); Duxbury (1976); Henderson and Kletz (1976); Huff (1977a, 1988); Jenkins, Kelly and Cobb (1977); Frankland (1978); Wia-Bu Cheng and Mah (1976); Richter (1978a,b); Sengupta and Staats (1978a,b); IMechE (1979, 1984); J.S. Parkinson (1979); Sallet (1979a); Anon. (1980 LPB 36, p. 1); Chambard (1980); IP (1980 Eur. MCSP Pt 2, 1981 MCSP Pt 3, 1987 MCSP Pt 9); D. Scott (1980, 1980 LPB 34, 36); Gerardu (1981); P.L. Jones (1981); Kauders (1980–, 1984); Chen-Hwa Chiu (1982); Doelp and Brian (1982); McKinley (1982); Uchiyama (1982); Badami (1983); IMechE (1984/81); Knox (1984); Middleton and Lloyd (1984); A. Moore (1984); Cunningham (1985); Francis and Shackleton (1985); Poole (1985); Tomfohrde (1985); Valdes and Svoboda (1985); Politz (1988); Crooks (1989); Duckworth and McGregor (1989); IChemE (1989/136); Parry (1989, 1992); Crowl and Louvar (1990); LPGITA (1991 LPG Code Pt 1); Ridey (1991); Aarebrot and Svenes (1992); Prugh (1992b); Cassata, Dasgupta and Gandhi (1993); Nichols (1994); Schiappa and Winegardner (1994); Tanner (1994)
BS 5500: 1991

Fire relief
Fitt (1974); A.F. Roberts, Cutler and Billinge (1983); Grolmes and Epstein (1985); Moodie *et al.* (1988); Morris (1988b); Wilday (1988); Epstein, Fauske and Hauser (1989): Venart (1990b)

Containment
Bartknecht (1981a); Wilday (1991)

Particular equipment
Gas filled vessels: Heitner, Trautmanis and Morrissey (1983a,b)
Heat exchangers: Case (1970); P.M.M. Brown and France (1975); Sumaria *et al.* (1976); W.Y. Wong (1992a, b)
Storage vessels: Kutateladze (1972)
Offshore: Crawley and Scott (1984)

Pressure relief valves, liquid relief valves, relief pipework
NRC (Appendix 28 *Safety Relief Valves*); Burgoyne and Wilson (1957); Conison (1960); Missen (1962); Porter (1962); Boyle (1967); Rearick (1969); Hattiangadi (1970a); Wissmiller (1970); Klaassen (1971); K. Wood (1971); Anon. (1972k); G.F. Bright (1972); Beck and Raidl (1973); Bodurtha, Palmer and Walsh (1973); API (1974 Refinery Inspection Guide Ch. 16, 1984 Std 526, 1988– RP 520, 1990 RP 521, 1991 Std 527); Forrester (1974); Kletz (1974a,d,e, 1985l, 1986a,g, 1987j, 1989a); Pasman, Groothuizen and de Gooijer (1974); Anon. (1975 LPB 6, p. 12); ASME (1975/130, 1983/168); Chlorine

Institute (1975 Pmphlt 41, 1982 Pmphlt 39); Schampel and Steen (1975, 1976); Anon. (1976i); Anon. (1976 LPB 10, p. 15); F.E. Anderson (1976); Anon. (1977 LPB 17, p. 5); R. Kern (1977f); D. Martin (1977); Welsh (1977); Frankland (1978); Sallet (1978, 1979a,b, 1990a–c); Sengupta and Staats (1978a,b); Bourdelon and Lai (1979); Coffman and Bernstein (1979); Ezekoye (1979); Haupt and Meyer (1979); Moody, Wheeler and Ward (1979); Sallet, Weske and Gühler (1979); Semprucci and Holbrook (1979); Strong and Baschiere (1979); Wheeler and Moody (1979); CGA (1980 S-1.1, S-1.3, 1989 S-1.2); Copigneux (1980, 1982); Crozier (1980, 1985); Dockendorff (1980); Mukerji (1980); Weighell (1980); O.J. Cox and Weirick (1981); Scully (1981); Aird (1982, 1983); van Boskirk (1982); AGA (1983/39, 1988/58); Heller (1983); Huff (1983); Papa (1983a,b, 1991); Bradford and Durrett (1984); Brahmbhatt (1984); J.A. Fuller (1984); S.F. Harrison (1984); Chester and Phillips (1985); Emerson (1985, 1988); I.D. Pearce (1985); Sallett and Somers (1985); Anon. (1986t); Bayliss (1987); DIERS (1987); DnV (1987 RP C202); Maher *et al.* (1988); Chambers and Fisher (1989); Desaedeleer *et al.* (1989); Morley (1989a,b); W.W. Wong (1989); Anon. (1990 LPB 92, p. 1); Kast (1990 LPB 95); Gavrila and Sethi (1991); Glinos and Miers (1991); Ridey (1991); Zappe (1991); Coker (1992); Lai (1992); Leung (1992a); W.Y. Wong (1992a–c); Bravo and Beatty (1993); W.E. Nelson (1993); British Gas (1994 BGES/DAT44); Crombie and Green (1994); Hanks (1994); D.W. Thompson (1994); BS 1123: 1987–, BS 5500: 1991.
Relief pipework: Duxbury (1976); Friedel and Schmidt (1993); S.M.Hall (1993); Perbal (1993)

Bursting discs
CGA (S-3); Fire Metal Products Corp. (n.d.); Bonyun (1935, 1945); Creech (1941); Murphy (1944a,b); Bigham (1958); J.F.W. Brown (1958); Lowenstein (1958); Luker and Leibson (1959); Diss, Karam and Jones (1961); Franks (1961); Block (1962); Solter, Fike and Hansen (1963a,b); Ruleo (1964); Liptak (1965); Myers and Wood (1965); Sestak (1965); L.E. Wood (1965); Boyle (1967); Alba (1970); C.R.N. Clark (1970); Harmon and Martin (1970); Kayser (1972); Brodie (1973); Huff (1973, 1977a); Anon. (1975 LPB 6, p. 12); Anon. (1976i); Zook (1976); Ganapathy (1976); Cockram (1977); Anon. (1978 LPB 23, p. 125); Fitzsimmons and Cockram (1979); Beese, Organ and Wade (1980); Hoffman, Hansen and Doelling (1980); L. Wood (1981); Anon. (1982 LPB 45, p. 1, 7); Falconer *et al.* (1982 LPB 45); L.R. Harris (1983a,b); Mathews (1983); Phadke (1983); British Gas (1984 TIN20); Prickett (1984); Watton and Brodie (1984); Zanetti (1984b); Beveridge (1985); Walker (1985); Anon. (1989 LPB 90, p. 30); Zappe (1991); Brazier (1993); Anon. (1994a)
BS 2915: 1990

Fusible plugs
Warwick (1969)

Venting and blowdown
Rudinger (1959); API (1988– RP 520, 1990 RP 521); Grote (1967); Klaassen (1971); Craven (1972); Simon and Thomson (1972); Fitt (1974, 1976b); BCISC (1975/1); Wia-Bu Cheng and Mah (1976); Pilz (1977); von Boskirk (1987); Haque *et al.* (1990); Haque, Richardson and Saville (1992); Haque *et al.* (1992); Perbal (1992); Mahgerefteh, Giri and Wong (1993); N.E. Stewart and McVey (1994)
Vents: Cindric (1985a,b); P. Watts (1985); Burgoyne (1986a); P.F. Thorne (1986)
Depressuring: Chiu (1982); Sonti (1984)

Relief disposal
Kneale (1984, 1989, 1984 LPB 59); Grossel (1986); DnV (1987 RP C202)

Trip systems
Kletz (1974a,d,e); Lawley and Kletz (1975)

principal legislation in the UK covering pressure systems is the Pressure Systems and Transportable Gas Containers Regulations 1989 (the Pressure Systems Regulations), with, for fixed plant, the associated code COP 37 *Safety of Pressure Systems* (HSE, 1990).

The requirement for protection is given in Regulation 4(5) which states 'The pressure system and transportable gas container shall be provided with such protective devices as may be necessary for preventing danger; and any such device designed to release contents shall do so safely, so far as is practicable.' It is also required by Regulations 14 and 15 that a permanent outlet to the atmosphere, or to a space at atmospheric pressure, be kept open and free of obstruction at all times.

It has been a requirement of the Factories Act 1961 in Sections 32, 35 and 36 that steam boilers and receivers and air receivers have a safety valve to prevent overpressure. These sections of the act were revoked by the Pressure Systems Regulations 1989.

Relevant to disposal is the legislation dealing with pollution. The Control of Pollution Act 1974 requires the use of best practicable means for preventing emission of noxious or offensive substances and for rendering harmless and inoffensive any which may be emitted There is also the general duty to provide safe equipment, which is reinforced by the Health and Safety at Work, etc. Act 1974.

The statutory controls are not the only external influence to which industry is subject. Provision of protection similar to the statutory requirements is normally a condition of obtaining insurance for other vessels.

It is necessary to define both the conditions against which protection is required and the nature of the system to be protected. BS 5500: 1991 requires that every pressure vessel be protected from excessive pressure or vacuum. If the source of pressure or temperature, however, is external to the vessel and is controlled so that overpressure of the vessel cannot occur, then the standard does not require the use of a protective device. Thus an overpressure protection is not required if the source of pressure is a pump which has a maximum delivery pressure less than the design pressure of the vessel.

For a vessel which is subdivided, each compartment should be treated as a separate vessel and suitably connected to a protective device. On the other hand, BS 5500 states: 'Vessels connected together in a system by piping of adequate capacity, free from potential blockages and which does not contain any valve that can isolate any vessel may be considered as a system of vessels for the application of pressure relief.' The protective device should have adequate capacity to prevent overpressure in the face of failure of any heating coil or other similar element in the vessel.

The protective devices envisaged in the standard are primarily safety valves and bursting discs complying with BS 6759: 1984 and BS 2915: 1990, respectively, but it also allows other devices provided they are suitable and reliable. It requires the maintenance of a register of all protective devices fitted to the vessel. The standard states 'Where the total capacity of the devices necessary to protect an installation from overpressure requires appropriate account to be taken of operating and fault conditions, the register should also include a record of the relevant calculations.'

12.12.2 API RP 521

Guidance on protection of plant against overpressure is given in API RP 521. The contents are (1) general, (2) causes of overpressure, (3) determination of individual relieving rates, (4) selection of disposal systems, (5) disposal systems and (6) bibliography. The appendices are: A, determination of fire relief requirements; B, principal causes of overpressure; C, sample calculations for sizing a flare stack; D, typical details and sketches; E, design of relief manifolds; and F, special system design considerations. This is supplemented by the information on the relief devices themselves given in API RP 520.

12.12.3 IChemE *Relief Systems Guide*

Further guidance is available in the *Relief Systems Handbook* by Parry (1992) (IChemE *Relief Systems Guide*). The contents are (1) general, (2) relief devices, (3) determination of set pressure and bursting pressure, (4) total relief systems, (5) vacuum relief, (6) thermal relief, (7) fire relief, (8) reliability of relief systems, (9) performance of relief systems, (10) installation of relief devices, (11) operation and maintenance and (12) current and future developments. The *Guide* contains some 19 appendices covering the legislation and (1) cause of relief situations, (2) relief rates, (3) minimum discharge velocity for dispersion, (4) Mach number, (5) atmospheric dispersion from a relief, (6) heat radiation at ground level from an elevated flare, (7) steam supply for smokeless flare operation, (8) flare purge gas velocity, (9) knockout drum droplet drag coefficient, (10) ground flares, (11) elevated flares, (12) thermal relief rate, (13) classification of liquids for thermal relief, (14) reliability data and (15) discharge reaction force. The *Guide* gives a collection of pressure relief case histories.

12.12.4 Sources of overpressure

API RP 521 gives detailed guidance on causes of overpressure. The principal events may be summarized as follows: (1) connection to a high pressure source, (2) disconnection from a low pressure sink, (3) increased heat input, (4) decreased heat output, (5) vapour evolution, (6) absorbent failure, (7) heat exchanger tube failure, (8) expansion of blocked-in liquid, (9) reverse flow, (10) fluid transients and (11) plant fire.

Connection to a high pressure source can occur if a valve is opened in error. Likewise, loss of connection to a low pressure sink can occur if a valve is erroneously closed.

Increased heat input can occur due to malfunction of heating equipment or to chemical reaction. A heat exchanger such as a reboiler when just cleaned may temporarily have a high heat transfer capacity.

Decreased heat output, or loss of cooling, can occur in numerous ways. Malfunction of cooling equipment such as a heat exchanger is an obvious case. Distillation columns particularly can lose cooling in a number of ways. In a single column loss of cooling can occur not only due to malfunction of the condenser but also due to loss of reflux or of subcooled feed. Where there is a set of columns in series, loss of heat input to one column can cause overpressure of the next column, due to carry forward of light ends in the bottoms from the first column which then overload the second column.

Overpressure can be caused by admission of water or light hydrocarbons to hot oil resulting in rapid evolution of vapour. Failure to remove sufficient gas due to loss of flow of absorbent can also cause overpressure. Failure of a tube in a heat exchanger so that the low pressure shell is exposed to the high pressure tube fluid can cause overpressure. This has been discussed in Section 12.8. Overpressure can be caused by expansion of liquid in a section of line between two closed block valves.

A particular case of increased heat input is a fire on the plant. In this case the potential heat input is so great that it is accorded special treatment, as described in Section 12.14. Failure of a pressure raiser and reverse flow from the high pressure discharge to the low pressure suction can cause overpressure of the latter. Fluid transients, such as those described in Section 12.7, can be another source of overpressure.

Some of these events may themselves be caused by failure of a utility such as electrical power to equipment or cooling water, or other coolant, or by failure of air or electrical power to instruments. The actions of control loops and those of the process operator are further causes. API RP 521 gives a table (Table 1) of possible utility failures and the equipment affected.

12.12.5 Identification of relief requirements

An account of the identification of the nature and magnitude of the sources of overpressure has been given by Fitt (1974), with emphasis on practical interpretation. The first step in dealing with hazards is to ensure that they have been fully identified. Hazard identification, which has been described in Chapter 8, is therefore an essential part of the design of protection for pressure systems. In particular, the technique of hazard and operability (hazop) studies is effective in defining protection requirements.

Once the hazards have been identified, it is appropriate to consider whether measures other than overpressure protection are more suitable. Some alternatives are to make the plant inherently safer by means such as increasing vessel strength or limiting the delivery pressure of pumps, or to install trip systems.

Full consideration should be given to the action of the process operator. He is an integral part of the hazard situation. It is usually reasonable to place some reliance on operator action in averting a hazard, but the extent of such reliance depends on such factors as ease of recognition of the existence of the hazard, instructions and training for dealing with it, and difficulty of and time available for this task. The possibility that the operator would do something positively harmful has to be allowed for, although this is less probable that that he will fail to take some beneficial action. Problems associated with action by the process operator are discussed in Chapter 14.

Some hazards may have such a low probability that they can be disregarded. The cut-off level is determined by the consequences of a failure, which in turn depend on factors such as the number of people exposed to the hazard. Overpressure protection should be considered at an early stage in the design so that unnecessarily severe demands can be identified and averted, but the main work has to be done at a relatively late stage.

Pressure relief becomes necessary when plant conditions are abnormal. Thus flow sheet quantities may be an inadequate guide to the relief capacities required. Some cases discussed by Fitt to illustrate the practical application of pressure relief are (1) shut-down pumps and compressors, (2) pressure let-down, (3) continuous distillation columns, (4) isolable equipment, (5) low pressure storage tanks, (6) control loop failures, (7) power turbines and let-down engines, (8) heat exchanger tube bursts and (9) batch operations.

For a pump discharging into a receiver the pressure relief valve should pass the volume of fluid which the pump could deliver if the feed valve is open and the receiver discharge is blocked. Except in certain special cases, flow sheet suction conditions should be assumed.

Failure of pressure let-down devices can cause a large and sudden rise in downstream pressure, particularly where a liquid line is blown down by gas. Therefore, consideration should always be given to making the downstream section capable of withstanding the upstream pressure. An alternative policy is to size the pressure relief valve to handle the combination of let-down valve and bypass valve both fully open. Neither of these valves should be oversized and bypasses should be avoided wherever practicable.

On a continuous distillation column overpressure may be created by an increase in heat input. Some causes of increased heat input include increase in temperature difference in the reboiler, loss of cooling, loss of reflux and loss of subcooled feed. Loss of cooling is often, but not always, the worst case. In some columns a subcooled feed provides as much cooling as the condenser.

The case of a column with a subcooled feed also provides an illustration of the need to consider the action of the process operator. If there is a loss of reflux on the column, he may well shut off the feed until he has re-established conditions. This then represents a worse case for pressure relief than loss of either reflux or feed alone.

Pressure relief should be provided between isolations if the equipment is subject to pressure from a source of high pressure or of process heat and, generally, if the equipment can be isolated when the plant is operating and it is in a fire zone. Relief need not be provided, however, between isolations if there is no such high pressure or process heat source and if either there is no significant fire risk or the equipment cannot be isolated while the plant is on line and there is adequate provision of vents and drains between the isolations. This philosophy applies to all fluids, including steam and water. It is particularly applicable to valve-spared heat exchanger sets.

Low pressure storage tanks are capable of withstanding only a very low pressure. It is important, therefore, to avoid exposing them to high pressure. The problem is discussed in Chapter 20 on plant operation and in Chapter 22 on storage.

Failure in a control loop may occur in the measuring element, in the controller, in the control valve or its actuator, in the transmission lines or by operator action. The possible failures and combinations of failure should be reviewed and the worst case considered. It should be assumed that both the control valve and the bypass may be fully open, if this constitutes a worst case. It need not be assumed that all control valves on an equipment will fail to danger simultaneously, but neither should it be assumed that all control loop actions will be favourable. In considering control loops, it should be assumed that flow may not be fully controlled by the control valve, either because the bypass is open or because the controller is on manual setting.

The system downstream of a power turbine or let-down engine is liable to be subjected to the upstream pressure. Such machines often have a low resistance to flow when stopped. Flow may also occur through an open bypass. The pressure relief valve on the downstream system should be sized to handle the maximum flow from the combination of let-down engine and open bypass. Alternatively there should be protection by trips. Alarms such as 'sentinel valves' are not adequate.

For the bursting of a high pressure tube in a heat exchanger, the pressure relief valve on the low pressure side should be sized to handle the flow from twice the cross-sectional area of the tube. Bursting of a furnace tube gives rise to a large increase in heat release in the furnace and causes rapid rise in pressure on the process side. One consequence of this may be abnormally high rates of steam generation in process heaters, or waste heat boilers. Adequate pressure relief capacity should be installed to relieve steam from these boilers, and arrangements should be made to provide boiler feed-water at the abnormally high rates necessary.

Batch operations require special consideration, because the conditions change throughout the cycle.

12.13 Overpressure Protection: Pressure Relief Devices

12.13.1 Pressure relief valves

Valves for the relief of pressure are referred to by a number of names and there is no universally agreed terminology. Two names commonly used as generic terms are 'pressure relief valve' and 'safety valve'.

A set of definitions is given in the successive editions of API 520. API RP 520: 1990 makes the following distinctions. A relief valve is a spring-loaded pressure relief valve actuated by the static pressure upstream of the valve. Normally it opens in proportion to the increase in pressure over the opening pressure. It is used mainly for incompressible fluids. A safety valve is a spring-loaded pressure relief valve actuated by the static pressure upstream of the valve and characterized by rapid opening or pop action. Normally it is used for compressible fluids. A safety relief valve is a spring-

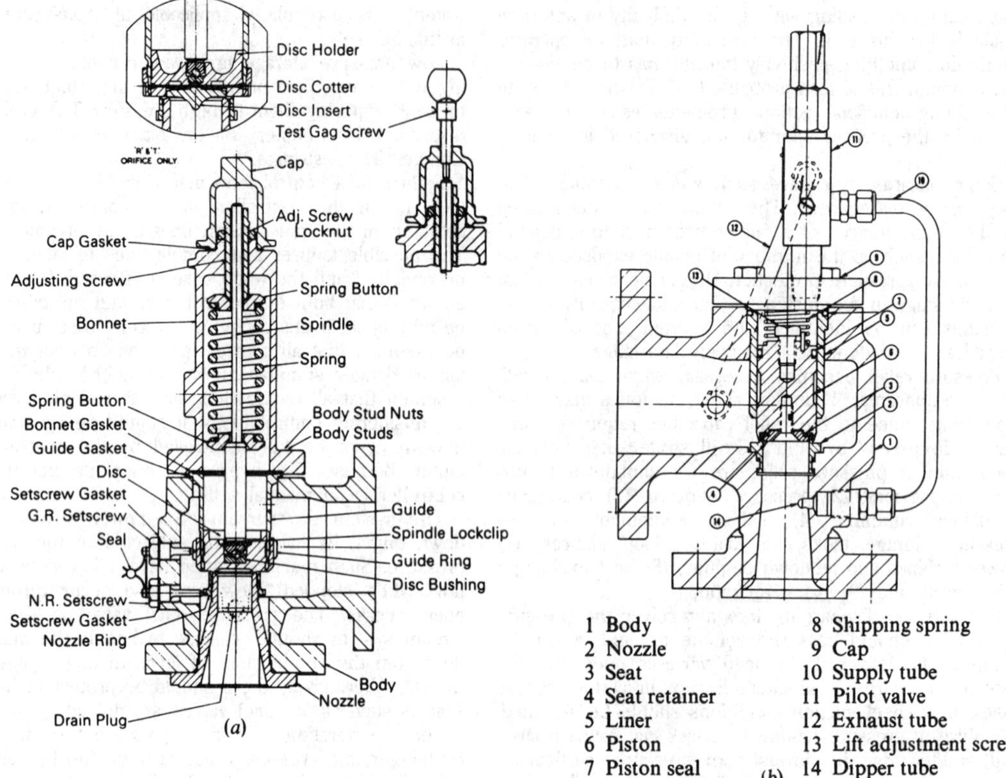

Figure 12.12 *Some typical pressure relief valves: (a) conventional pressure relief valve (Crosby Valve and Engineering Company Ltd); (b) pilot-operated relief valve (Anderson Greenwood Company)*

loaded pressure relief valve which can be used as either a safety or a relief valve, depending on the application.

According to the IChemE *Relief Systems Guide*, the BSI has decided that the use of these definitions is impossible without ambiguity. The relevant standard, BS 6759: 1984, describes a valve for the relief of pressure as a safety valve. Similarly, BS 5500: 1990 refers to safety valve. This term is also that adopted in the *Guide*. The term used here is 'pressure relief valve'.

BS 5500 requires that safety valves comply with BS 6759 *Safety Valves*. BS 6759 has three parts, the most relevant here being BS 6759: Part 3: 1984 *Specification for Safety Valves for Process Fluids*. Part 1: 1984 deals with steam and hot water and Part 2: 1984 with compressed air or inert gases. The installation is required to comply where appropriate with these standards and with BS 1123: 1961 *Safety Valves, Gauges and Other Safety Fittings for Air Receivers, and Compressed Air Installations*, which has one part, Part 1: 1987 *Code of Practice for Installation*.

There are a number of different types of pressure relief valve. Three broad categories of valve are (1) the conventional, direct-loaded valve, (2) the balanced valve and (3) the pilot-operated, indirect-loaded valve.

A conventional pressure relief valve, the simplest type, is a spring-loaded valve. The pressure at which the valve relieves is affected by the back pressure.

The direct-loaded pressure relief valve has a number of variants. One is the assisted-opening valve, which has power-assisted opening. One use of this type of valve is to give depressurization down to a predetermined level. In a supplementary-loaded pressure relief valve an external power source is used to impose an additional sealing force, which is released automatically when the set pressure is reached. This arrangement gives an improved degree of leak-tightness.

A third variant is the pilot-assisted valve, which is a different type from the pilot-operated valve. A pilot-assisted valve is a direct-loaded valve in which some three-quarters of the load is due to the spring and the rest to the pressure of the fluid from the pilot valve. When the set pressure is reached, the pilot opens and that part of the load contributed by it is removed.

A balanced pressure relief valve is one which incorporates means of minimizing the effect of back pressure on performance. With this type of valve the back pressure has little effect on the pressure at which the valve relieves. A balanced relief valve is used to accommodate the back pressure in a relief header.

The principal type of balanced pressure relief valve is the balanced bellows valve. The valve incorporates a bellows with an effective area equal to that of the valve seat so that it counteracts the effect of back pressure on

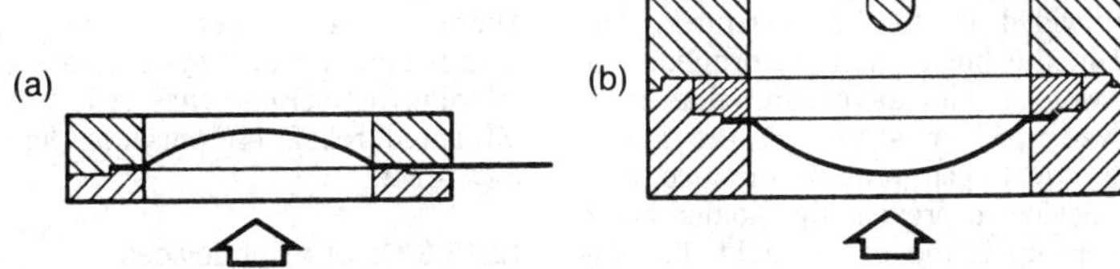

Figure 12.13 *Some typical bursting discs: (a) conventional bursting disc; and (b) reverse buckling bursting disc*

the set pressure. The balanced pressure relief valve also comes in a number of variants. They include the balanced piston valve, in which the counteracting balance is effected by a piston, and the balanced bellows valve with auxiliary piston.

A pilot-operated, indirect-loaded pressure relief valve is one in which the main valve is combined with and controlled by an auxiliary direct-loaded pressure relief valve. The whole load on the main valve is provided by the fluid pressure from the pilot valve. When the set pressure is reached, the pilot valve opens and releases the loading pressure.

A pilot-operated pressure relief valve may be used where the margin between the operating and set pressures is narrow or where the set pressure is low. It is also used when discharging direct to atmosphere to ensure high discharge velocity and hence good jet mixing. The principal drawback of a pilot-operated pressure relief valve is that the small passages of the pilot valve are liable to become blocked if the process fluid is dirty, so that is suitable only for relatively clean duties.

Some pressure relief valves are illustrated in Figure 12.12. Figure 12.12(a) shows a conventional pressure relief valve and Figure 12.12(b) a pilot-operated pressure relief valve. Further accounts of pressure relief valves are given in API RP 520 and the IChemE *Guide*. The latter gives diagrams of the various types.

12.13.2 Bursting discs

The conventional bursting disc is the simple disc illustrated in Figure 12.13(a). The disc has its dome with the direction of the bursting pressure. Conventional bursting discs tend to be very thin. Many are less than 0.05 mm in thickness.

A conventional bursting disc should be carefully installed. Manufacturers provide disc mountings which have a number of foolproofing features of which the following are typical. The ring holding the disc on the vent side is made thicker than the dome in order to protect the latter. The disc has an identification tag with full details stamped on. The tag neck serves to centre the disc in place and is notched on one side so that, if correctly installed, the disc does not seat properly and, if not readjusted, will vent at a low pressure. Discs with different pressures have pegs located at different points so that a disc for one pressure will not fit into a holder for a different pressure. Conventional bursting discs are normally made of metal, but graphite discs are also used.

Some other types of disc which are used in general applications include:

(1) composite slotted disc;
(2) reverse buckling disc.

A composite slotted disc is a variation on the conventional disc. It consists of the main disc, which is slotted to burst at the rated pressure, and a protective membrane, in plastic or metal. The use of slotting allows the disc to be made of thicker material so that it is less liable to fatigue, while the use of the membrane gives protection against corrosion by the process fluid. There are also grooved discs which are less deeply scored.

A reverse buckling disc has its dome against the direction of the bursting pressure, as shown in Figure 12.13(b). The principle of operation is that at the burst pressure the dome of the disc reverses and is cut by a knife on the downstream side. The use of reverse buckling allows the disc to be made thicker, generally some 3–5 times as thick as a conventional disc, and thus less liable to fatigue, and also eliminates fragmentation. An account of reverse buckling discs is given by Watton and Brodie (1984), who list the following types:

(1) disc which, on reversal, bursts by being sheared radially by a knife or circumferentially by a serrated cutters mounted on the vent side;
(2) disc with preformed diametral grooves which, on reversal, opens at the grooves;
(3) disc which, on reversal, slips out of the holder;
(4) disc which, on reversal, shears around the dome.

The latter type is available in metal or graphite.

A reverse buckling disc is thicker than a conventional disc for the same duty. Watton and Brodie quote for 50 mm diameter discs rated for a burst pressure of 20 bar at 20°C thicknesses for stainless steel of 0.04 and 0.3 mm for conventional and reverse buckling discs, respectively. The latter is thus 7.5 times thicker than the former. The greater thickness of a reversed buckling disc makes it less liable to failure by fatigue, creep or corrosion.

The reversal time of a reverse buckling disc bursting as designed is very short. For 450 mm diameter discs Watton and Brodie quote reversal times of 40 ms at 0.035 bar and 5 ms at 10 bar. In order for the disc to be cut by the knife there needs to be sufficient energy. The disc may fail to burst properly if the pressure rise is too slow. If the disc is damaged, even by a quite small dent, in the dome, it may 'roll through' and lie on the knife without being cut. The disc may then fail to burst at the design burst pressure. If the disc slips sideways, the dome shape and burst pressure may be altered.

A reverse buckling disc should be carefully installed and securely mounted. The following foolproofing features are typically provided by the manufacturer. The disc is preassembled in the holder and pretorqued. The holder is provided with a pin to ensure that it is mounted the right way up. The screws in the holder are of a special type which can require the use of a special tool. Another device to prevent the holder being installed the wrong way up is the use of a J-bolt. Only when the bolt is in can the holder be centred and the flange bolts tightened. This arrangement is particularly useful where frequent cleaning of discs to remove blockages is necessary.

In addition to its use as the primary relief device a bursting disc may also be used mounted in series below a pressure relief valve to protect the latter from corrosion. This arrangement is effective only as long as there are no pinhole leaks in the disc. A pressure gauge is usually mounted on the space between the valve and the disc to detect any rise of pressure due to leakage. This arrangement is suitable only if the bursting of the disc will not cause obstruction of the relief valve.

There are also available disc assemblies based on the reverse buckling principle which give protection against both overpressure and underpressure, or vacuum. An arrangement described by Watton and Brodie consists of a reverse buckling metal disc perforated with a number of holes and sealed with a graphite membrane. If overpressure occurs, the metal disc and graphite membrane burst in the normal reverse buckling mode; in effect the system has the characteristics of the metal disc. If underpressure occurs, the graphite membrane bursts, acting like a simple domed disc.

Bursting discs can be made in quite large diameters. Manufacturers are able to supply discs up to some 1.2 m. There are a number of other applications for which special bursting disc systems are available. They include (1) high temperature gas relief, (2) general liquid relief, (3) liquid relief, (4) pulsating liquid relief and (5) two-way relief.

12.13.3 Other relief devices

In addition to pressure relief valves and bursting discs, there are a number of other devices which may be used to relieve pressure. The simplest of these devices is the atmospheric vent, typically a short pipe. An atmospheric vent is used on atmospheric storage tanks where there is minimal hazard from the ingress of air or egress of vapour.

On an atmospheric storage tank where the liquid held is more volatile, use is generally made of pressure-vacuum valves which allow the tank to 'breathe'. Another device which can be used to provide pressure relief on low pressure tanks is the liquid seal. Fire relief for an atmospheric storage tank may be provided in the form of a weak roof-to-shell seam, or rupture seam. For a pressure vessel a device which provides fire relief is the fusible plug.

12.14 Overpressure Protection: Relief System Design

12.14.1 Location of relief devices

In some cases the provision of a pressure relief device on a particular kind of equipment is a legal or code

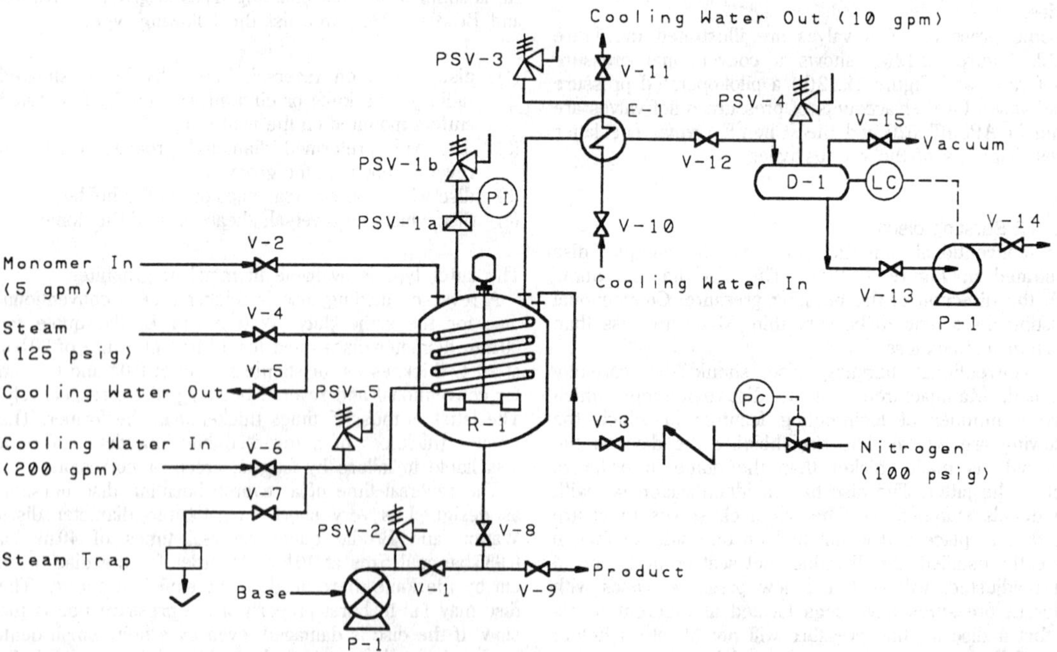

Figure 12.14 Polymerization reactor with pressure safety valves (Daniel A. Crowl/Joseph F. Louvar, Chemical Process Safety Fundamentals with Applications, © 1990, pp. 246–247) D–1, 100 gallon drum; R–1, 1000 gallon reactor; P–1, gear pump; P–2, centrifugal pump. (Reprinted by permission of Prentice-Hall, Englewood Cliffs, New Jersey)

requirement. An example is the requirement in the Factories Act 1961 for a pressure relief device on a steam boiler, steam receiver or air receiver. Generally, a process vessel is provided with a pressure relief device, but, as described below, there are exceptions.

Guidance on the location of pressure relief devices has been given by Crowl and Louvar (1990) drawing on the work of Isaacs (1971). They give guidelines for specifying relief locations which may be summarized as follows: (1) all vessels, including reactors, storage tanks, towers and drums; (2) storage vessels; (3) positive displacement pumps discharge; (4) blocked-in sections of cool liquid-filled lines which are exposed to heat (like the sun) or refrigeration; and (5) vessel steam jackets.

Storage vessels also require vacuum protection. Some of the conditions which create the need for both overpressure and vacuum relief are described in Chapter 22. Vessel steam jackets are often designed for a low pressure, which explains the need for relief in this case. A comment on this guidance is that an alternative policy for blocked-in liquid sections is to assess the particular case, and to provide relief only where the assessment shows this to be necessary.

An illustrative example of the location of pressure relief devices is given by Crowl and Louvar. The plant considered is shown in Figure 12.14. The design involves the provision of pressure safety valves (PSVs) at five locations, at one of which there are two such valves. The justification for these reliefs is given by the authors as follows. PSV-1 on reactor R-1 and PSV-4 on drum D-1 are in accordance with the practice of installing a relief on each process vessel. PSV-2 provides relief on a positive displacement pump discharge. PSV-3 on heat exchanger E-1 provides protection against tube rupture from overpressure if the water is blocked in (valves V-10 and V-11 closed) and the exchanger is heated, for example by steam. PSV-5 on the coil of reactor R-1 provides protection against tube rupture from overpressure if water is blocked in (valves V-4, V-5, V-6 and V-7 closed) and the coil is heated. There is actually a more positive reason for installing PSV-1 on reactor R-1. This is that the reactor is designed for 50 psig but is connected to a nitrogen line at 100 psig.

The selection of pressure relief devices given by the authors for this example is as follows. PSV-1a is a safety relief valve sized for two-phase flow, liquid and vapour, and protected by a bursting disc PSV-1b. PSV-4 is again a safety relief valve, also sized for liquid and vapour flow. The other PSVs are relief valves sized for liquid flows.

Conditions for the protection of several components in a pressure system by a single pressure relief device are given in API RP 521: 1990 in Appendix F. Four criteria are given. The first is that there should not exist any means for blocking in any of the components so protected. The other three criteria, given in detail in the code, have to do with the set pressure, the accumulation pressure and the operating pressure. The code gives an example of a hydrotreater–reactor–recycle gas loop system which has a compressor and six process vessels/units but is protected by a single pressure relief device on the last vessel in the train, and lists the conditions for the acceptability of such protection in this case.

In many ways the location of pressure relief devices is best approached by starting with the identification of sources of overpressure. Full identification of such sources is necessary in any case in order to size the reliefs, and it is logical to start with this. The pressure reliefs should be installed not simply to protect the particular points at which they are located, but to provide protection for all vulnerable points.

12.14.2 Selection of relief devices and systems

Guidance on the selection of the pressure relief devices to be used for a particular duty is given in the IChemE *Relief Systems Guide*. The first decision is whether to make use of a bursting disc, either alone or in combination with a pressure relief valve. It advises the use of a bursting disc if there is a completely free choice and if this is the more economic option or if either (1) the pressure rise is too rapid for a pressure relief valve, or (2) the process fluid properties make a pressure relief valve unsuitable (toxicity, corrosiveness, blockage-forming components, aggressiveness).

If a system based on pressure relief valves alone is indicated, this may be a single valve or a set of multiple valves in parallel, depending on whether a single valve can provide the necessary capacity. The principal types of pressure relief valve are (1) conventional valves, (2) balanced valves, (3) pilot-operated valves, (4) supplementary-loaded valves and (5) assisted-opening valves. The characteristics and applications of these different valves are described in Section 12.13.

One fundamental factor which affects the choice is the sink to which the valve will discharge. If this is a relief header and therefore exerts a back pressure, a balanced valve may be indicated. On the other hand, if discharge is to atmosphere, a pilot-operated valve may be selected for its high discharge velocity. Other factors which affect the choice are (1) the margin between operating and the set pressure, (2) the required speed of opening and (3) the required valve tightness. The first case favours a pilot-operated valve.

If the discharge is into a header where the back pressure is variable or exceeds 10% of the absolute set pressure, a balanced pressure relief valve is needed, the workhorse valve being of the balanced bellows type. In this case consideration should be given to the effects of bellows failure in altering the set pressure or leaking process fluid into the valve bonnet.

If the system is to incorporate a bursting disc, it is necessary to consider whether, following operation of the relief, (1) the loss of the plant contents is acceptable and (2) the plant can be shut down to replace the disc. If neither is acceptable, a system incorporating a bursting disc in series with a pressure relief valve is indicated. If only the first is acceptable, use may be made of two bursting disc systems in parallel. If both are acceptable, a bursting disc system alone suffices. For a bursting disc system, use may be made of a single bursting disc or, if the process conditions are aggressive, two bursting discs in series.

For a bursting disc/pressure relief valve system, the valve may have a bursting disc (1) upstream, (2) downstream or (3) on both sides, depending on whether it is the process-side fluid, the discharge-side fluid or both fluids which are aggressive.

The principal types of bursting disc are (1) conventional domed disc, (2) reverse domed disc and (3) composite slotted disc. Use of the conventional type is

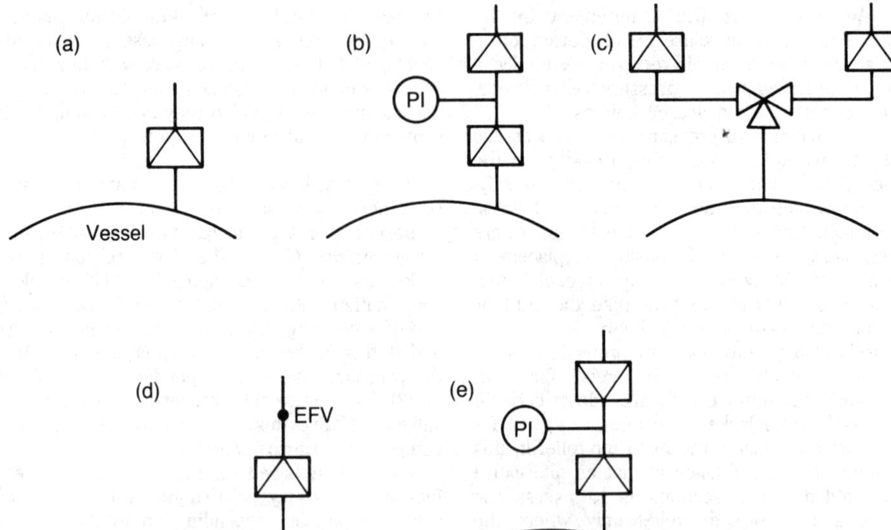

Figure 12.15 Some typical bursting disc arrangements: (a) single bursting disc; (b) two bursting discs in series with pressure indicator (PI) on intermediate space; (c) two bursting discs in parallel with three-way valve to allow replacement without plant shut-down; (d) bursting disc in series with excess flow valve (EFV, not shown) to prevent back pressure; and (e) opposed bursting discs for vacuum service

suitable provided that there is a wide margin, say 30%, between the operating and design pressures and that the pressure does not pulsate. If these conditions are not met, if a long life is required or if the disc is to be used in series with a safety valve, consideration should be given to the other two types.

12.14.3 Relief device configurations

Figure 12.15 illustrates some relief configurations based on bursting discs only. Figure 12.15(a) shows a single bursting disc. Figure 12.15(b) shows two bursting discs in series with a pressure indicator on the space between to detect pinholes leaks. Figure 12.15(c) shows two bursting disc in parallel with a three-way valve. Figure 12.15(d) shows a bursting disc in series with an excess flow valve. Figure 12.15(e) shows a pressure and a vacuum bursting disc in series.

Figure 12.16 illustrates some pressure relief valve + bursting disc configurations. Figure 12.16(a) shows a pressure relief valve and bursting disc in parallel and Figure 12.16(b) a pressure relief valve with an upstream bursting disc to protect against aggressive process conditions. Other arrangements (not shown) are a pressure relief valve with a downstream bursting disc to protect against aggressive discharge conditions, and one with both upstream and downstream bursting discs to protect against aggressive conditions on both sides.

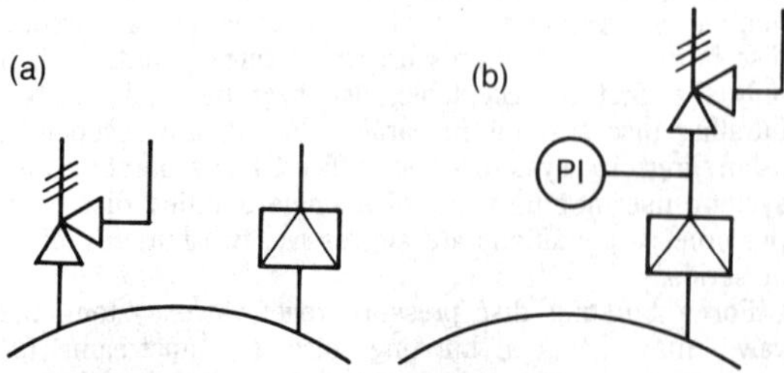

Figure 12.16 Some typical pressure relief valve + bursting disc arrangements: (a) pressure relief valve and bursting disc in parallel; and (b) pressure relief valve in series with bursting disc upstream and with pressure indicator on intermediate space

12.14.4 Setting of relief devices

Once the location of the relief system and the type of system to be used have been decided, it is possible to determine the setting and capacity of the device. Guidance on the set pressure for pressure relief valves and for bursting discs is given in the codes. Discussion of this aspect is deferred to Sections 12.19 and 12.20, respectively.

12.14.5 Sizing of relief devices

Guidance on the sizing of individual pressure relief valves in given in API RP 521, which considers in detail the assumptions to be made in relation to each of the sources of overpressure listed in Section 12.12. The situations treated are similar to those discussed in Section 12.12. The approach taken is to describe the scenario and to define a basis for the determination of the relief capacity, indicating whether credits can be taken for particular for mitigating factors and discussing features to be taken into account. The account given in Section 12.15 of the treatment of external fire illustrates the general approach. An essentially similar philosophy is applicable to the sizing of bursting discs.

12.15 Overpressure Protection: Fire Relief

The pressure relief requirements just discussed are those for relief of conditions which may arise during operation. In addition to such operational relief, there is a requirement for fire relief. Except in storage areas, operational upsets are more likely to give rise to a requirement for pressure relief than are fires. But fires tend to require a larger pressure relief capacity. In cases where fire can affect a number of vessels, the requirement for fire relief is very much greater than that for operational relief.

Equipment for which fire relief may be required includes atmospheric storage tanks, pressure storage vessels and process systems. For pressure vessels there are two distinct cases: vessels containing gas only and vessels containing liquids. Guidance on fire relief for pressure storage vessels has been given in the successive editions of API RP 520 and RP 521. There are also a number of other codes which give recommendations on fire relief. A critique of these codes and summary of the current situation is given in the IChemE *Relief Systems Guide*, as described below.

The account of fire relief given here is supplemented by the further discussion in Chapter 22 of fire relief for storage.

12.15.1 Scenarios for fire relief

The first step in design for fire relief is to define the scenarios for which such relief is to be provided. A discussion of this aspect is given by Fitt (1974). The amount of fire relief required depends on the surfaces which may be exposed to fire. The requirement is greatest for groups of process vessels and of storage vessels. If the scenario indicates that the fire would develop slowly and would cause overpressure only after a long period, say several hours, it may normally be assumed that the fire would be brought under control by fire fighting before overpressure occurs.

12.15.2 Fire protection measures

Some methods of protection against fire in addition to pressure relief are (1) inventory reduction, (2) fire insulation and (3) water sprays. Inventory reduction is discussed in the following section. Fire insulation and water sprays are discussed in Chapter 16 and fire insulation and water sprays for storage vessels in Chapter 22.

12.15.3 Inventory reduction in fire

Consideration also needs to be given to measures for limitation of developed pressure and for reduction of inventory in fire conditions. One method of limiting the overpressure which can develop is vapour depressurization, or blowdown. Vapour is blown down to a suitable disposal system through a depressurization valve which is separate from any pressure relief device and which is normally operated before the pressure reaches the set pressure of that device. The depressurisation valve needs to be capable of remote operation.

Provision may also be made for the removal of the liquid inventory. One reason for doing this is to limit the amount of vapour which has to be handled. Another is to prevent release of material which could then feed the fire. The preferred means of removal is the normal liquid withdrawal system, but a separate liquid pulldown system may be used. However, removal of the liquid inventory is not necessarily the best policy. Liquid remaining in the vessel provides effective cooling for the portion of the walls which it wets. Each case needs to be considered on its merits.

12.15.4 Fire relief of pressure vessels containing liquid: API RP 520

A method for determining of the requirement for fire relief of a pressure vessel is given in Appendix D of API RP 520. A text apparently identical, except for a less extensive treatment of heat absorption rates, is given in Appendix A of API RP 521.

API RP 520 gives a treatment for a vessel containing liquid and one for a vessel containing gas when engulfed in fire. For the former, it is the wetted surface which is effective in generating vapour. The code states that the portion of the vessel to be considered is that which is wetted by the liquid and which is at a height equal to or less than 25 ft (7.6 m) above the source of the flame, usually, but not necessarily, the ground surface. For the heat absorbed, two cases are considered. For the case where there is drainage of flammable liquid away from the vessel and fire fighting is prompt, the relation given is:

$$Q = 21\,000\,FA^{0.82} \qquad [12.15.1]$$

where A is the total wetted surface (ft^2), F is the environment factor and Q is the total heat absorption to the wetted surface (BTU/h). For the case where these conditions are not met:

$$Q = 34\,500\,FA^{0.82} \qquad [12.15.2]$$

The heat absorbed per unit area q is therefore proportional to $A^{-0.18}$, which is termed the 'area exposure factor'; it takes account of the fact that a large vessel is less likely than a small one to be completely exposed.

From Equation 12.15.1 or 12.15.2 the flow of vapour generated may be obtained and hence the capacity of the pressure relief valve determined. The two important parameters of this model, therefore, are the wetted area A and the environment factor F.

For the wetted area the standard gives guidance (Table D-2) on the degree of fill to be assumed for various vessels, including storage vessels, process vessels, distillation columns, etc.

The environment factor F is, in principle, a function of a number of variables related to fire protection: (1) insulation, (2) water application and (3) depressurizing and emptying facilities. The factor F has a value of unity for bare metal and 0.3, 0.15, 0.075 and 0.03 for insulations with conductances of 4, 2, 1 and 0.4 BTU/ft^{2}h °F, respectively. A full set of values is given in the code (Table D-3). No credit is given by way of reduction of this factor, so that F is unity, for depressurizing and emptying facilities or for water sprays on a bare vessel.

As described below, differences between API RP 520 and other codes in respect of fire relief are in large part associated with differences in the credit given for fire protection by way of the environment factor F.

12.15.5 Fire relief of pressure vessels containing gas: API RP 520

For a vessel containing gas engulfed in fire API RP 520 gives the following simplified method, which in this case gives directly the required discharge area of the pressure relief valve:

$$A = F'A'/P_1^{\frac{1}{2}} \qquad [12.15.3]$$

with

$$F' = \frac{0.1406}{CK_d}\left[\frac{(T_w - T)_1^{1.25}}{T_1^{0.6506}}\right] \qquad [12.15.4]$$

$$T_1 = T_\eta(P_1/P_\eta) \qquad [12.15.5]$$

where A is the effective discharge area of the valve (in.^{2}), A' is the exposed surface area of the vessel (ft^2), C is a coefficient, F' is a parameter, K_d is the coefficient of discharge, P_1 is the upstream relieving pressure (psia), P_η is the normal operating pressure (psia), T_1 is the upstream relieving temperature (°R), T_w is the vessel wall temperature (°R) and T_η is the normal operating temperature (°R). The code states that the recommended minimum value of F' is 0.01 and when the minimum value is unknown a value of 0.045 should be used. Then, from Equations 12.15.3 and 12.15.4 together with Equation 12.19.15 given below the relief load is:

$$W = 0.1406(MP_1)^{\frac{1}{2}}\left[A'\frac{(T_w - T_1)^{1.25}}{T_1^{1.1506}}\right] \qquad [12.15.6]$$

where M is the molecular weight of the gas and W is the required mass flow (lb/h).

This simple method is based on the assumption that the gas behaves as an ideal gas, that the vessel is uninsulated and has negligible mass, that the gas temperature does not change and that the vessel wall temperature does not reach its rupture value. The code states that these assumptions should be reviewed and, if necessary, use made of a more rigorous method, on which it gives some further guidance.

12.15.6 Fire relief of pressure vessels: other codes

There are several other codes which give guidance on fire relief. A critical review is given in the IChemE *Guide*. The other major code discussed in the *Guide* is NFPA 30: 1981, together with the Occupational Safety and Health Administration code (OSHA) 1910.106: 1981, which is essentially similar. It states that for pressure vessels with wetted areas below 2000 ft^2, NFPA 30 appears to give heat fluxes twice those of API RP 520, but that this difference gradually reduces for wetted areas between 2000 and 2800 ft^2 so that for wetted areas above the latter value the two coincide.

It is also necessary, however, to consider the environmental factor. For good drainage NFPA 30 gives credit in the form of a value of 0.5 for this factor. As stated above, API RP 520 gives no explicit credit for good drainage, the existence of good drainage being a condition of the use of Equation 12.15.1. Thus for wetted areas of less than 2000 ft^2 the heat fluxes given by the method of API RP 520 and that of NFPA 30 are the same. The *Guide* concludes that the method of API RP 520 is to be preferred A further discussion of heat fluxes to pressure vessels is given in Chapter 16.

The *Guide* also considers the Compressed Gas Association code CGA S-1.3: 1980 for gas cylinders, together with the Chlorine Institute code CI 5.3.3: 1977, which is essentially similar. CGS S-1.3 utilizes for gas cylinders the more conservative Equation 12.15.2. The *Guide* argues that cylinders may well be exposed to higher heat fluxes so that a case can be made for this approach.

12.15.7 Fire relief of atmospheric storage tanks: API Std 2000

Guidance on fire relief for atmospheric and low pressure storage tanks is given in API Std 2000: 1992. This standard distinguishes between tanks with a weak roof-to-shell attachment, or rupture seam, and those without. For tanks with a rupture seam, the standard does not require any further emergency venting for fire relief. For tanks without a rupture seam, the standard provides a table of relieving requirements given as a function of the wetted area of the tank.

For a vertical tank the wetted area is taken as the total cylindrical surface area of the shell up to height of 30 ft above grade. For a horizontal tank it is the greater of 75% of the total surface area or the surface area to a height of 30 ft (9.14 m) above grade. For a sphere or spheriod it is the greater of 55% of the total surface area or the surface area to a height of 30 ft above grade.

The value of the wetted area used in API Std 2000 reflects the fact that such an atmospheric storage tank normally stands in a bund in which there could be a liquid fire. This contrasts with the values of the wetted area in API RP 520, which reflect the fact that the ground under a pressure storage vessel is normally sloped to drain any liquid away. Fire relief for atmospheric storage is discussed further in Chapter 22.

12.15.8 Effect of fire on relief devices

It is also necessary to consider the effect of fire on pressure relief devices. On a pressure relief valve there are two main effects of fire. One is to cause thermal expansion of the spring, which reduces the force on the valve and is a deviation in the safe direction. The other is to cause thermal expansion of the valve spindle which can lead to jamming. However, if the valve is immersed in the fire so that the temperatures in it are equalized, this effect may be minimal. On a bursting disc the effect of fire is to reduce the bursting pressure or to cause rupture, both of which are effects in the safe direction.

12.15.9 Heating of unwetted surfaces

Provision of fire relief does not give full protection against fire. Unwetted surfaces which are exposed to fire may experience overtemperature and may rupture, even though overpressure of the vessel has not occurred.

API RP 520 gives a graph for the time taken by steel plate of different thicknesses to reach particular temperatures when exposed to an open gasoline fire on one side, and another graph for the time to rupture at different rupture stresses and temperatures. These two graphs are shown in Figures 12.17(a) and (b), respectively. The code quotes the example of an unwetted steel vessel of plate thickness 1 in. and with a stress of 15 000 psi in an open fire. The time for the vessel wall to heat up to 1300°F is 17 minutes and the time for rupture to occur at that temperature is 2.5 minutes.

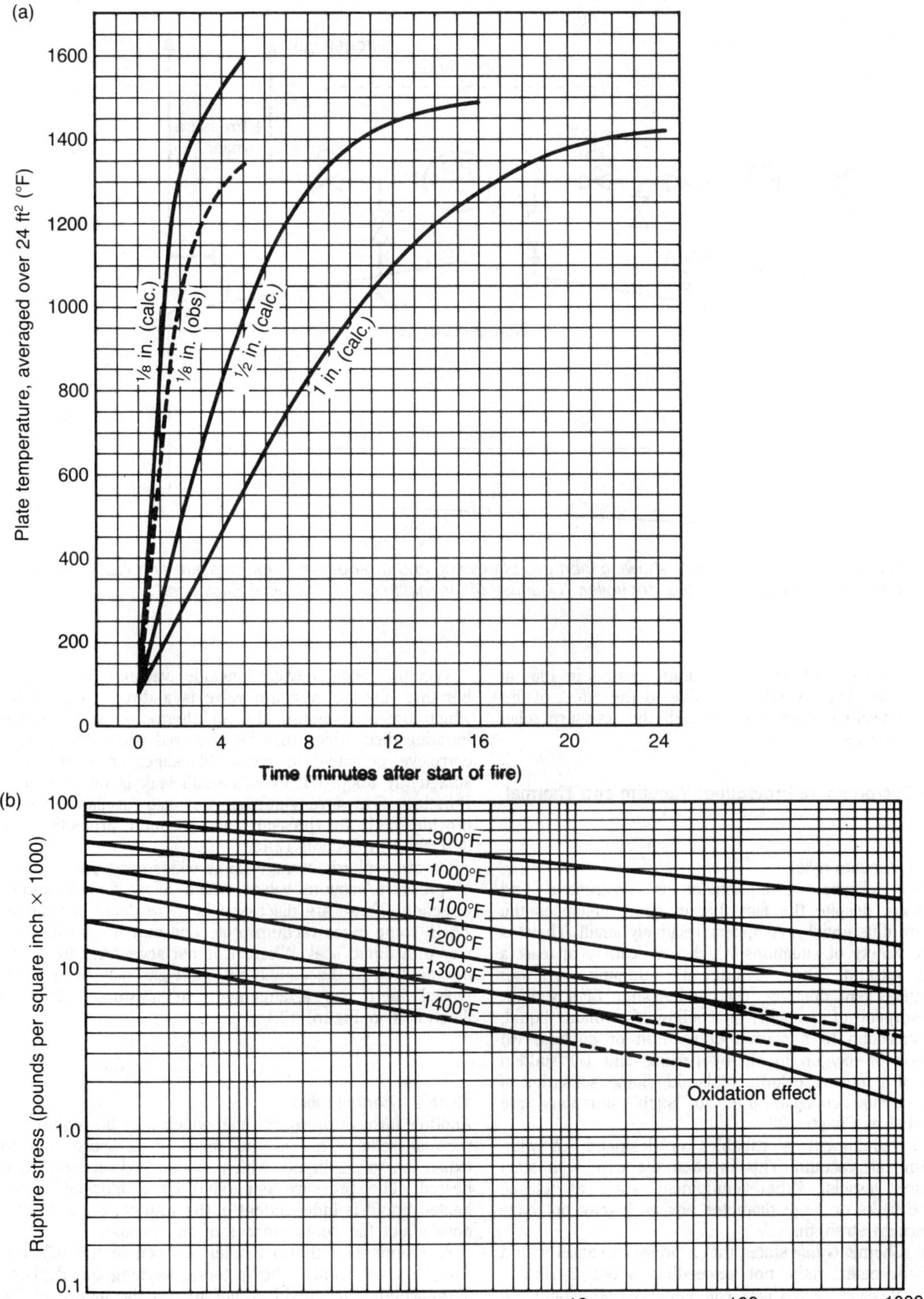

Figure 12.17 *Effect of fire engulfment on steel plate (American Petroleum Institute, 1990 API RP 520): (a) average rate of heating of steel plates of different thickness exposed to an open gasoline fire on one side; and (b) time to rupture of steel plate (ASTM A515 grade) as a function of temperature and stress (Courtesy of the American Petroleum Institute)*

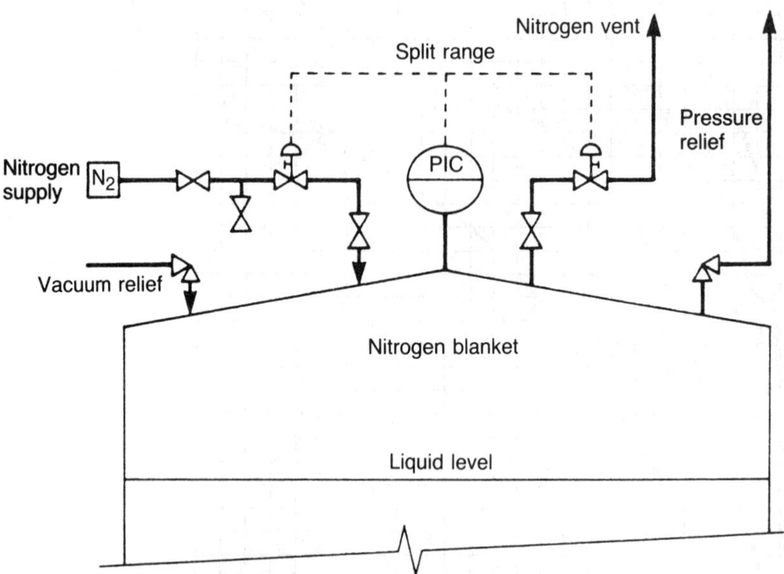

Figure 12.18 *Storage tank system showing inert gas blanketing and arrangements for pressure and vacuum relief (Parry, 1992). PIC, pressure indicator controller. (Courtesy of the Institution of Chemical Engineers)*

The failure of LPG pressure storage vessels in 1966 at Feyzin (Case History A38) was due to the effect of fire on the unwetted walls, even though the pressure relief valves operated.

12.16 Overpressure Protection: Vacuum and Thermal Relief

12.16.1 Vacuum relief

Vacuum collapse of equipment is destructive and hazardous, despite the fact that in some instances the pressure differential may appear relatively small. There is a wide variety of situations which can lead to at least a partial vacuum. They include (1) pumping out with inadequate vent opening, (2) condensation of a vapour, (3) absorption of a vapour, (4) cooling of a volatile liquid, (5) connection to a source of vacuum or suction, (6) depletion of oxygen in air by rusting and (7) sudden arrest of a moving column of liquid. Some scenarios of storage tank collapse due to such situations are illustrated in Chapter 22.

Some equipment is capable, as designed, of withstanding full vacuum. This is often the case with steel pressure vessels. Other equipment such as plastic storage tanks or large diameter pipework may not have full vacuum strength.

The IChemE *Guide* states that a pressure vessel with a length/diameter ratio not exceeding about 3 and a design pressure of 3.5 barg can generally withstand full vacuum, but that it is prudent to make the check. An increase in the vacuum strength of a vessel can often be obtained by a quite small increment in wall thickness or addition of a stiffening ring.

Vacuum relief devices include vacuum valves and bursting discs. A vacuum valve is a direct-loaded valve which opens to admit air. An alternative is a vacuum bursting disc, which may be preferred where the fluid is corrosive or liable to create blockage, or where it is sufficiently toxic that even a small leak is unacceptable. Bursting disc arrangements are also available which provide both overpressure and vacuum protection, as described in Section 12.13.

On atmospheric storage tanks widespread use is made of pressure/vacuum valves which allow flow in either direction. These are discussed in more detail in Chapter 22. In some cases vacuum protection may be provided by way of a liquid seal. Where it is not acceptable for air to be admitted to the equipment, inert gas may be used. A typical inert gas arrangement on a storage tank is illustrated in Figure 12.18.

12.16.2 Thermal relief

Another form of pressure relief which may be required is thermal relief. This is the relief of pressure due to expansion of a liquid which is blocked in and then heated. The pressure generated by a blocked-in and heated liquid is independent of the volume, but the latter does affect the consequences of any escape.

A treatment of thermal relief is given in the IChemE *Relief Systems Guide*. The factors governing the decision as to whether to provide relief are (1) the nature of the liquid, (2) the heat source, (3) the volume of the plant section or system, (4) the tightness of the closure, (5) the probability of blocking in, (6) the location of the system and (7) the availability of other means of relief.

The *Guide* divides liquids into the following categories: (1) cryogenic, (2) low boiling point, (3) volatile, (4) flammable, (5) toxic, (6) other liquids and (7) water.

Sources of heat are classified as (1) process heat, (2) heat tracing, (3) solar radiation and (4) ambient temperature. Subject to the volume exemption described below, the first of these cases should be provided with relief, as should the second, unless the controls are such as to make it unnecessary, whilst the third and fourth cases require the exercise of judgement.

For liquids in categories 1–5 relief should generally be provided except for insignificant volumes, i.e. volumes $< 0.1\,m^3$. The location of the system may strengthen the case for relief, which is indicated if the system is located in an unmanned offsite area or on public property.

Turning to category 6 liquids, these require judgement. Account may be taken of the frequency with which the system may be blocked in and the degree of control. If the frequency is judged sufficiently low, it may be possible to dispense with relief.

The *Guide* advises that in borderline cases account may be taken of the volume of the system. The volume which may be regarded as insignificant is related to the tightness of closure. The following tightness categories are given:

Construction	*Valves*		
	Bubble tight	*Soft seated*	*Metal seated*
All welded	A	A	B
Class 300 flanges or higher	A	B	C
Class 150 flanges	B	C	D

It also gives the following relationships between tightness category and limiting volume:

Tightness category	*Limiting volume* (m^3)
A	0.1
B	1.0
C	5.0
D	10.0

Separate thermal relief may not be needed if other means of relief exist. These may include a small hole in the valve or a small bore bypass around it, bellows, or relief provided for other purposes.

The following equation is given in the IChemE *Relief Systems Guide* for the thermal relief requirement:

$$W = Q\beta/S \qquad [12.16.1]$$

where Q is the heat input (W), S is the specific heat of the fluid (J/kg °C), W is the required relief rate (kg/s) and β is the coefficient of volumetric expansion (°C^{-1}).

12.17 Overpressure Protection: Special Situations

There are a number of special situations in pressure relief which require separate consideration. They include (1) extreme operating conditions, (2) aggressive fluids and (3) multiphase fluids.

12.17.1 Extreme operating conditions

Extreme operating conditions which can affect pressure relief include (1) high pressure, (2) high temperature and (3) low temperature.

With high pressure relief valves factors to be considered are valve tightness and precision of lift. Increased tightness may be obtained by the use of a supplementary-loaded valve. In high pressure processes it is usual to operate close to the design pressure and precision of lift becomes important. A pilot-operated valve may meet the case for clean fluids. Advice on selection of valves for dirty fluids is given in the IChemE *Guide*. There are also available bursting discs for high pressure duties.

A feature of relief flows at high pressure is that the gas behaviour may not be ideal. For a high temperature pressure relief valve relevant factors are materials of construction, spring relaxation, the hot set pressure and the difference between the operating temperature and the relieving temperature.

A valve spring subject to high temperature may suffer relaxation. The problem, and work on it, have been described by Aird (1982, 1983). A method of minimizing the effect by the hot set procedure has been described by Weighell (1980). The hot set procedure is specified by the valve manufacturer and should be followed.

If the process fluid has to travel any significant distance before reaching the valve, there may be a difference between the operating temperature and the relieving temperature. It is the latter which is to be used in performing the sizing calculations.

For a pressure relief valve on low temperature duty, factors to consider are material of construction, the low temperature set pressure, the difference between the operating temperature and the relieving temperature and icing. The low temperature set pressure may be obtained by using a correction on the ambient set pressure but it is also practice to adjust a valve in place at the operating temperature. Icing is countered in various ways. Use may be made of insulation or trace heating. If the plant is in a cold box, the valve may be mounted outside it so that it is at a temperature close to ambient. For a bursting disc on high or low temperature duty the important feature is the material of construction.

12.17.2 Aggressive fluids

Where a fluid is corrosive, it may be possible to use a pressure relief valve on its own if suitable materials of construction are available. Otherwise, options are the use of a bursting disc upstream of the valve, a bursting disc on its own, or, if the fluid is especially corrosive, two bursting discs in series.

Where the fluid is liable to cause blockage, a pressure relief valve alone may not be suitable. One option is to install a bursting disc upstream of the valve. Another is to use a bursting disc on its own.

If the fouling could be severe, further steps may need to be taken to keep the disc clear. Methods include the use of a coating or anti-fouling compound on the disc, or even water sprays to wash the disc clean.

12.17.3 Multiphase fluids

In some applications the fluid flowing through the pressure relief valve will be a two-phase vapour–liquid mixture, either because the fluid is two-phase at the valve inlet or because it becomes so in passing through the valve. Two-phase flow through a pressure relief valve is a specialist topic. It is discussed further in Chapter 15.

Another type of two-phase system is a liquid containing solid particles. There are two potential effects of such solid matter. One is blockage, which has just been discussed. The other is any effect on the flow properties of the fluid.

12.18 Overpressure Protection: Disposal

12.18.1 Disposal and disposal systems

Pressure relief necessarily involves the provision of arrangements for the disposal of the material vented. The selection of the method of relief disposal and the design of any relief disposal system are integral to the overall design.

Using the scenarios of relief demands, the nature, temperature, phase and flow of the each fluid stream is defined. Fluids may be flammable, toxic and/or corrosive. Some may be at high or low temperature. Some flows may be two-phase or liable to condense on release.

For each relief location there will be a flow profile with time. There may also be contributions to the total flow from other devices such as depressuring valves. If the overall profile shows too high a peak, measures may be taken to shave the peaks by means such as the staggering of pressure relief valve settings.

Discharge of gas or vapour to atmosphere is relatively simple and dependable, but its use may be constrained by regulatory and other requirements. API RP 521 recognizes this method for use where it is acceptable. The code also deals with some of the problems which can arise with atmospheric discharge, including formation of flammable plumes, ignition of such plumes, toxic or corrosive plumes, noise and pollution. Its guidance on avoidance of flammable plumes is discussed in the next subsection.

Another option for gas or vapour which is applicable in some cases is disposal to a lower pressure system. Minimum requirements in this case are that the unit have the capacity to receive the relief safely and that it be available to do so, bearing in mind that this may be affected by the same conditions that created the need for relief in the first place.

If these methods of disposal are not appropriate, other methods available include: scrubbing, followed by discharge of the scrubbed gas; discharge from a vent stack; or burning from a flare stack. In these cases, the necessary pipework has to be provided in the form of a relief header, or headers. Here a basic distinction is between single-valve and multiple-valve disposal systems. In the design of the latter there are a number of features which need to be taken into account.

It is common practice to use a common header both for normal vents and for emergency relief. In this case the normal vent flows should be added to the emergency relief flows in designing the header. Alternatively, separate headers may be provided for normal vents, for gas/vapour relief and for liquid blowdown. Separate headers for gas/vapour and for liquid permit separate disposal arrangements such as a flare stack and blowdown drums, respectively. It may be necessary to consider segregation of streams which are in some way incompatible. Such incompatibility may have to do with fluids which react together, with a fluid at low temperature which requires a more expensive material of construction, or with a fluid which is susceptible to freezing by another, colder fluid.

As far as liquids are concerned, some may be suitable for disposal to the drains. Others may need to be routed to a liquid blowdown receiver or disposed of in an earth burning pit.

12.18.2 Load on disposal system

If the relief flows are discharged for closed system disposal rather than to atmosphere, it becomes necessary to define the load which the system has to handle. The determination of the flow for which the relief header should be designed is not a simple matter. Its capacity must lie somewhere between the largest single relief flow and the sum of all relief flows. Except for very small plants the latter solution is too conservative and uneconomic.

Guidance on the load on such a disposal system is given in API RP 521. The essential principle is to design for the set of individual contingencies, rather than for coincident occurrence of multiple independent contingencies. However, each contingency should be carefully examined, because the consequences of certain events can be very widespread.

The code gives guidance particularly on loss of utilities. For failure of electrical power to equipment the design is commonly based on failure of one bus, or possibly of the distribution centre or the incoming line. Failure of instrument air or of electrical power to instruments is typically assumed to be plant-wide.

Definition of the load resulting from fire requires a specific study taking into account the plant layout, including location of potential sources of release, natural barriers and drainage, etc. The fire on a plant handling only gas can generally be assumed to be more localized than on one handling liquids. The code states that in the absence of special factors the ground area of fire is often taken as being limited to 2500–5000 ft^2 (230–460 m^2).

Fitt (1974) describes an approach to this problem based on certain 'standard hazards'. These are (1) fire, (2) electrical power failure, (3) instrument air failure and (4) cooling water failure.

Fire may affect a group of vessels. It should be assumed, therefore, that there would be simultaneous relief on these vessels. It is usually not necessary to assume a process upset at the same time.

Failures of electrical power, instrument air or cooling water are all obvious causes of simultaneous relief. Total power failure can cause failures on a large number of equipments and all electronic instrumentation. Partial power failure can be even more hazardous. Thus, for example, reboiler fuel pumps might continue to operate after reflux pumps have stopped.

Total instrument air failure disables all pneumatic instrumentation. It is normal practice, however, to design instrument systems so that the control valves 'fail safe' on air failure. Thus again, it may be more hazardous if the air failure is partial. In this case the movement of valves may be in the directions opposite to those on total air failure. In addition, a control valve with an open bypass will not shut off flow completely even if it does close.

Failure of cooling water is usually a consequence of power failure and should be assumed unless the supply can be maintained by pumps with prime movers which have an independent power supply. Then for a given standard hazard the flows occurring over the three time periods (1) less than 10 min, (2) 10–30 min, and (3) up to 1 h (4 h for fire), are determined.

12.18.3 Disposal to atmosphere

Disposal of gas or vapour by direct discharge to atmosphere is relatively simple and dependable, and may be used where it is acceptable. A method for the design of atmospheric discharge arrangements is given in API RP 521. The case considered is that of a safety relief valve discharging a flammable vapour upwards through a vertical tail pipe. The vapour is discharged at a velocity sufficiently high that it is rapidly diluted below the lower flammability limit (LFL).

The criterion for the vapour to be diluted below the LFL whilst the mixing is still dominated by the momentum of the jet is that the Reynolds' number Re exceed following value:

$$\mathrm{Re} > 1.54 \times 10^4 \rho_j/\rho_\infty \qquad [12.18.1]$$

where ρ_j is the density of the vapour at the vent outlet and ρ_∞ is the density of the air.

The code quotes the work of J.F. Taylor, Grimmett and Comings (1951) and gives for the mass flow from the valve the relation:

$$\frac{W}{W_0} = 0.264\frac{y}{D} \qquad [12.18.2]$$

where D is the diameter of the nozzle (ft), W is the mass flow of vapour/air mixture at distance y from the tail pipe (lb/h), W_0 is the mass flow of vapour from the safety relief valve (lb/h), and y is the distance from the end of the tail pipe (ft).

On this basis it is calculated that, provided that the velocity of the discharged material leaving the tail pipe is more than 500 ft/s (150 m/s) so that sufficient energy is supplied for mixing to occur, a hydrocarbon discharging from a safety relief valve into the atmosphere will entrain enough air to become diluted below its LFL at a distance of approximately 120 diameters from the end of the tail pipe.

The code recognizes that, whilst this is fine as long as the valve discharges at this high rate, it cannot be assumed that the flow through a safety relief valve is the full capacity flow. Once a spring-loaded valve has opened, the force exerted on it by the flowing fluid is sufficient to prevent reclosure until the flow has fallen to about 25% of capacity. However, the code quotes a study by Hoehne, Luce and Miga (1970) which has shown that even at flows of 25% of capacity the vapour is safely dispersed.

API RP 521 also refers to studies on venting of oil tankers (ICS, 1978) which indicate that dilution is effective at release velocities exceeding 100 ft/s (30 m/s). It gives a set of graphs for the estimation, for petroleum gases, of the maximum vertical and downwind horizontal distances from the jet exit to the LFL and of the axial distances to the upper flammability limit (UFL) and LFL. For the vertical distance the correlation is:

$$\frac{y}{d_j(\rho_j/\rho_\infty)^{\frac{1}{2}}} \text{ vs } U_\infty/U_j \qquad [12.18.3]$$

where d_j is the inside diameter of the tip, or jet exit diameter, U_j is the jet exit velocity, U_∞ is the wind speed, and y is the vertical distance to the LFL. The correlations for horizontal distance x and axial distance S are of similar form, with y replaced by x and S, respectively.

The code comments that the studies demonstrate the adequacy of the industry practice of locating the discharge of a safety relief valve at a horizontal distance of at least 50 ft (15 m) from any equipment or structure above the discharge level.

API RP 521 also gives consideration to the case where the vapour discharged condenses to form a mist. In principle, condensation can occur if the ambient temperature is below the dewpoint of the vapour. However, there are two factors which may mitigate this situation. One is that condensation does not necessarily occur and the other that, even if it does, the droplets may coalesce and fall out.

Vapour let down in pressure through a safety relief valve becomes superheated and this may prevent condensation. Furthermore, the vapour may be diluted so rapidly that condensation is avoided. A method of determining whether condensation occurs has been given by Loudon (1963). It shows that condensation does not occur with most hydrocarbons, though it may with some of high molecular weight, a finding which is in accord with industrial experience.

The code states that, whilst conclusive data are lacking, condensation at a hydrocarbon partial pressure of 5 psi (0.34 bar) or less should be assumed to result in a fine mist, the dispersion and flammability of which are similar to those of a vapour. A further discussion of discharge to atmosphere and from vents is given in Chapter 15.

12.18.4 Design of a relief header

The use of a closed system involves the design of the relief discharge pipework. In certain cases use may be made of individual discharge lines, but commonly there is a relief header taking discharges from a number of points.

A discussion of relief headers is given by Fitt (1974) in the study mentioned. As stated above, he considers for a given standard hazard the flows occurring over the three time periods (1) less than 10 min, (2) 10–30 min, and (3) up to 1 h (4 h for fire). The relief header is designed to handle the sum of the flows of the largest of these groups. The header may be required to handle both gas/vapour and liquid flows.

If conventional relief valves are used, relief headers should be so sized that the back pressure developed is limited to 10% of the set pressure. This requirement can be relaxed, however, if balanced relief valves are used. The latter thus permit a more economical design of the relief header.

The pressure drop in the header is sometimes high and the changes in gas density and velocity are correspondingly large. Not infrequently, sonic flow conditions occur. The design of relief headers is therefore relatively complex.

The gas flow in the header may be treated either as isothermal or as adiabatic. For an ideal gas the flow is greater under adiabatic than under isothermal conditions, but for a ratio of specific heats $\gamma < 1.8$ this difference does not exceed 23%. Frequently calculations are done on the basis of isothermal flow.

API RP 521 gives in Appendix E guidance on the design basis for a relief header. Essentially, the flows are determined for each contingency and the design is based on the most severe flow condition. This is the condition which gives the highest pressure drop, which is not necessarily that with the highest mass flow; the volumetric flow is relevant, and hence so are molecular weight and temperature.

Once the design loading is determined, then for each contingency the maximum allowable back pressure on each pressure relief valve should be calculated. The code gives detailed guidance on these back pressures.

In addition to the flows from the pressure relief valves there may be flows from depressuring valves. Normally, a depressuring valve is treated as fully open or fully closed and the maximum flow from it taken as the flow capacity of the valve at the maximum pressure of the equipment that it protects, which may be the maximum accumulated pressure. However, the code advises that where the same equipment is protected by both a pressure relief valve and a depressuring valve, only the larger of the two flows is to be considered.

API RP 521 gives a method for the design of the relief header, which is based on the method of Lapple (1943) for isothermal flow of gas through pipes at high pressure drop. The design of relief headers is a specialist matter. It is also an economically important one, since relief headers can be expensive items, particularly if it is necessary to use special materials of construction. Some of the specialist topics in the sizing of relief headers, including non-ideal gas flow and pipe section changes, are discussed by Duxbury (1976).

12.18.5 Disposal to a flare stack

Guidance on the design of a flare for the disposal of gas or vapour is given in API RP 521. The code gives methods for the determination of the diameter and height of the flare stack, the fuel gas supply to the flare, and the steam requirement for a smokeless flare.

The selection of flare stack diameter and height is treated in Appendix C of the code. As described there, the diameter is chosen to give a particular Mach number at the flare tip and the height is chosen to limit thermal radiation at ground level. Appendix D of the code gives typical sketches and details of a complete flare system, and of seal drums and knockout drums.

The code also deals with thermal radiation and pollution from the flare. Appendix C gives two models for the flame on the flare, a simple model and the model of Brzustowski and Sommer. These API flare models are described in Chapter 16.

API RP 521 gives information on the exposure times necessary to reach the threshold of pain (Table 2) and recommended flare radiation levels (Table 3). These and other thermal radiation limits are discussed in Chapter 16.

With regard to the estimation of ground level concentrations of combustion products from the flare, the code refers to the method of Gifford (1960a). Methods for dispersion from elevated stacks are considered in Chapter 15.

12.18.6 Disposal to a vent stack

API RP 521 also contains guidance on the design of a vent stack for the disposal of gas or vapour. It gives methods for the determination of the diameter and height of the stack.

The diameter is chosen to give a high exit velocity. The code states that excellent dispersion is obtained with a velocity of 500 ft/s (150 m/s). However, sonic velocity should be avoided, or else allowance made for the effect on the pressure drop of the pressure discontinuity at the discharge. The choice of the height of the stack is governed by the need to ensure acceptable concentrations at the points of interest. The code gives guidance on the level of noise from such a stack.

An equation for the maximum ground concentration of a gas downwind of a vent stack is:

$$C_{max} = \frac{W}{VH^2M} \qquad [12.18.4]$$

where C_{max} is the maximum ground concentration (% v/v), H is the height of the stack (ft), M is the molecular weight, V is the wind velocity (mile/h), and W is the mass flow of gas (lb/h).

There are a number of restrictions on the use of Equation 12.18.4. It assumes that the air velocity increases with height above the ground and that the gas–air mixture is not much heavier than air itself. It is not applicable if the wind velocity is zero or very low, or if the discharge velocity is much below 500 ft/s.

12.18.7 Selection of disposal system

Guidance on the selection of a disposal system is given in the IChemE *Relief Systems Guide*. The disposals considered in the guide are to (1) atmosphere, (2) quench vessel, (3) process, (4) storage, (5) dump tank, (6) sewer, (7) elevated flare, (8) ground flare, (9) incinerator or (10) scrubber.

The selection scheme passes through four broad stages. If the fluid to be disposed of is mainly vapour and if it is lighter than air or can be discharged at high velocity, disposal to atmosphere is a technical option. If the fluid contains both vapour and liquid and needs to be cooled, disposal to a quench vessel is indicated. Otherwise, if a sink for the fluid is available in process or in storage, it may be disposed of to that sink.

If the fluid is mainly liquid and it is suitable for discharge to a plant sewer, it may be so disposed of. Otherwise it should be discharged to a dump tank.

If the fluid is both vapour and liquid and cannot be disposed of as described above, other options must be considered. Flammable vapours with some liquid may be sent to a flare. If the fluid contains toxic and/or particulate matter which can be scrubbed out, treatment in a scrubber is suitable. Otherwise the fluid should be sent to an incinerator.

12.19 Overpressure Protection: Pressure Relief Valves

The basics of pressure relief values have been outlined in Section 12.13. In this section consideration is given to valve set pressure, sizing and failure.

12.19.1 Valve set pressure

A pressure vessel has a design pressure. The definition of design pressure given in BS 5500: 1991 was quoted in Section 12.6. The set pressure of a safety valve is defined in codes by reference to a specific terminology. The pressure relationships used in BS 5500: 1991 for a safety valve on a vessel are indicated in Figure 12.19(a), which shows the case of a safety valve on gas or vapour service with a 10% overpressure. The pressures on the left-hand side refer to the vessel, whilst those on the right-hand side refer to the safety valve. For the vessel, the normal operating pressure lies between 90 and 95% of its design pressure. The maximum permitted regulated pressure is 110% of the design pressure, the difference between the two being the accumulation. For the safety valve, the set pressure is equal to the design pressure. The relieving pressure is 110% of the set pressure, the difference between the two being the overpressure. Where the set pressure of the valve is the design pressure of the vessel, the overpressure equals the accumulation.

BS 5500: 1991 states that a safety valve should normally be set to operate at a nominal pressure not exceeding the design pressure of the vessel at the operating temperature. However, if the capacity is provided by more than one safety valve, it is permissible for only one of the valves to be set to operate in this way and for the additional valve(s) to be set to operate at a pressure not more than 5% high. The pressure relationships given in BS 5500: 1991 for a safety valve for liquid service with a 25% overpressure are shown in Figure 12.19(b).

Figure 12.19(c) gives the pressure relations for pressure relief valves on a vessel used in API RP 520: 1990. For the vessel, the maximum expected operating pressure is 90% of the maximum allowable working pressure (MAWP), or design pressure. The figure shows several values of the maximum allowable accumulation, for the following three cases: (1) non-fire exposure with a single valve; (2) fire exposure with multiple valves; and (3) fire exposure. The accumulation values are 110%, 116% and 121%, respectively. For the pressure relief valve, a similar set of distinctions apply. The set pressure and maximum relieving pressure are equal, respectively, to 100% and 110% of the design pressure for case 1; 105% and 116% for case 2; and 110% and 121% for case 3. Where the set pressure of the valve is the design pressure of the vessel, the overpressure equals the accumulation. The code states that Figure 12.19(c) conforms to the ASME *Boiler and Pressure Vessel Code*, Section VIII.

For further details of, and qualifications and variations on, the above pressure relief requirements, such as the use of multiple pressure relief valves, reference should be made to the appropriate codes, including the codes just mentioned, and other codes such as API RP 521 and Std 2000.

12.19.2 Valve sizing: BS 5500

Equations for pressure relief valve sizing were given in BS 5500: 1976. BS 5500: 1991 simply refers to the standards for pressure relief valves BS 6759: 1984. The equations given in BS 5500: 1976 are widely quoted and are given here for reference.

For a gas or vapour under critical flow conditions:

$$W = \frac{KAp}{C}\left(\frac{M}{T}\right)^{\frac{1}{2}} \qquad [12.19.1]$$

where A is the actual discharge area (m^2), K is the coefficient of discharge for gas flow, M is the molecular weight, p is the absolute accumulation pressure (N/m^2), T is the inlet temperature (K), W is the rated capacity (kg/s) and C is a constant.

For saturated steam:

$$W_s = \frac{KAp}{686} \qquad [12.19.2]$$

and for air at 20°C

$$W_a = \frac{KAp}{423} \qquad [12.19.3]$$

where W_a is the rated capacity for air at 20°C (kg/s) and W_s is the rated capacity for saturated steam (kg/s).

The coefficient of discharge K may vary. The standard quotes a value of 0.25 for a parallel inlet guided wing type of high lift valve and 0.97 for a nozzle inlet type of flat disc valve. The value of the coefficient should be provided by the manufacturer. The value of C is 133 for a ratio of specific heats of 1.4.

For liquid flow:

$$W = KA(2\Delta pG)^{\frac{1}{2}} \qquad [12.19.4]$$

where G is the relative density at the inlet temperature (kg/m^3), K is the coefficient of discharge for liquid flow and Δp is the pressure drop (N/m^2).

12.19.3 Valve sizing: BS 6759: 1984

The relations given for pressure relief valve sizing in BS 6759: Part 3: 1984 are as follows. For air or any gas at critical flow conditions:

$$q_{mg} = pC\left(\frac{M}{ZT}\right)^{\frac{1}{2}} \qquad [12.19.5]$$

$$= 0.2883\, C\left(\frac{p}{v}\right)^{\frac{1}{2}} \qquad [12.19.6]$$

with

$$C = 3.984\left[k\left(\frac{2}{k+1}\right)^{(k+1)/(k-1)}\right]^{\frac{1}{2}} \qquad [12.19.7]$$

where M is the molecular mass of gas (kg/kmol), p is the actual relieving pressure (bara), q_{mg} is the theoretical flowing capacity (kg/h mm^2 of flow area), T is the actual relieving temperature (K), v is the specific volume at the actual relieving pressure and temperature (m^3/kg), Z is the compressibility factor, k is the isentropic coefficient at the relieving inlet conditions and C is a parameter which is a function of that coefficient.

For air or any gas at subcritical conditions:

$$q_{mg} = pCK_b\left(\frac{M}{ZT}\right)^{\frac{1}{2}} \qquad [12.19.8]$$

$$= 0.2883\, CK_b\left(\frac{p}{v}\right)^{\frac{1}{2}} \qquad [12.19.9]$$

with

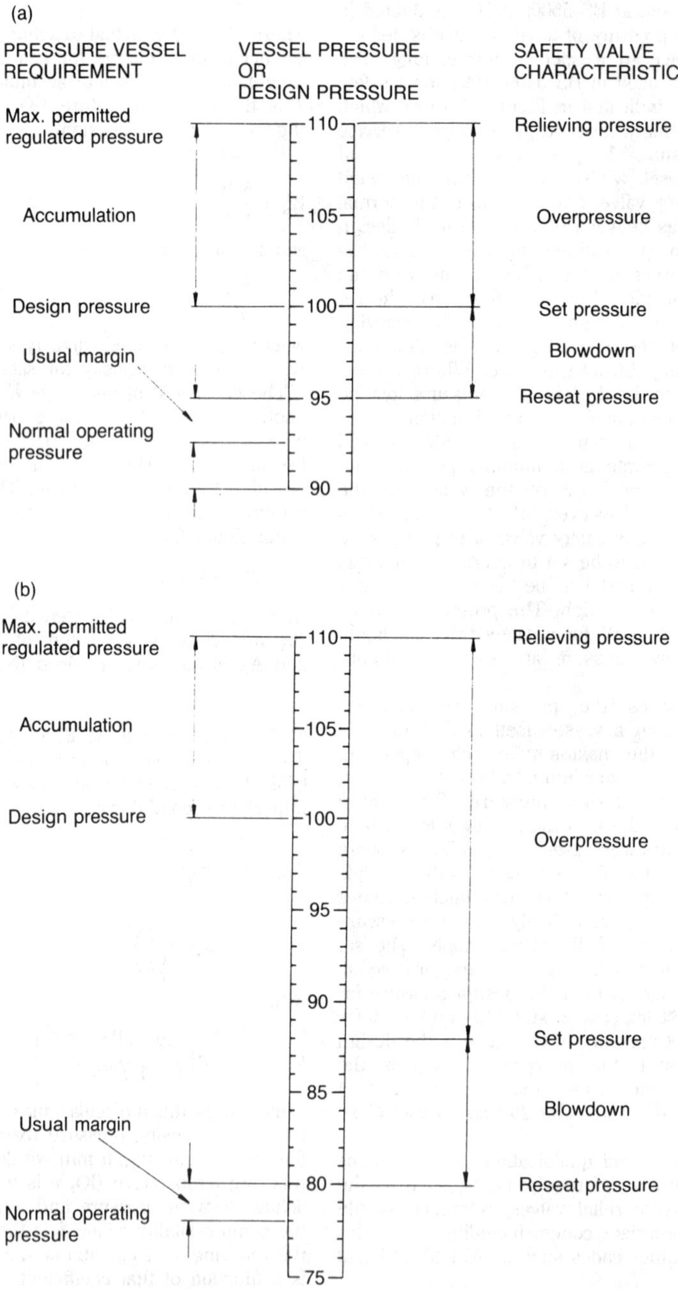

Figure 12.19 *Settings for pressure relief valves: (a) safety valve with 10% overpressure for gas or vapour service (BS 5500: 1991) (Courtesy of the British Standards Institution); (b) safety valve with chosen 25% overpressure for liquid service (BS 5500: 1991) (Courtesy of the British Standards Institution); and (c) safety relief valve for vapour service (API RP 520: 1990) (Courtesy of the American Petroleum Institute)*

PRESSURE VESSEL REQUIREMENT | VESSEL PRESSURE | TYPICAL CHARACTERISTICS OF SAFETY RELIEF VALVES

Maximum allowable accumulated pressure (fire exposure only) — 121 — Maximum relieving pressure for fire sizing

120

Maximum allowable accumulated pressure for multiple-valve installation (other than fire exposure) — 116 — Maximum relieving pressure for process sizing

115 — Margin of safety due to orifice selection (varies) — Multiple valves — Single valve

Maximum allowable accumulated pressure for single-valve installation (other than fire exposure) — 110 — Maximum allowable set pressure for supplemental valves (fire exposure)

Overpressure (maximum)

105 — Maximum allowable set pressure for supplemental valves (process)

Overpressure (typical)

Maximum allowable working pressure or design pressure (hydrotest at 150) — 100 — Maximum allowable set pressure for single valve (average)

Simmer (typical)

Start to open

Blowdown (typical)

95 — Seat clamping force

Reseat pressure (typical) for single valve

Usual maximum normal operating pressure — 90 — Standard leak test pressure

Tolerances — Setting: ±3%; Blowdown, simmer: Not specified by ASME code, Section VIIII; Tightness: API Std 527

85

Percent of maximum allowable working pressure (gauge)

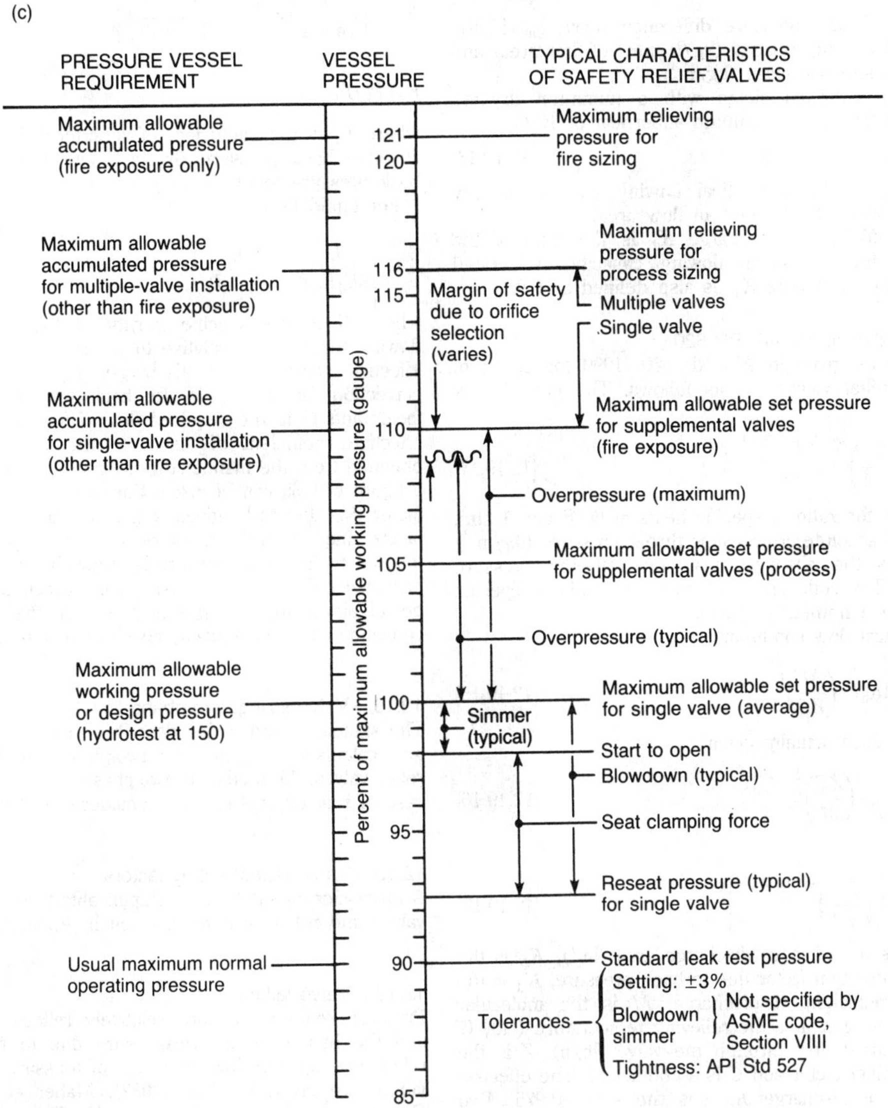

Notes:
1. The operating pressure may be any lower pressure required.
2. The set pressure and all other values related to it may be moved downward if the operating pressure permits.
3. This figure conforms with the requirements of Section VIII, Division 1, of the ASME Code.
4. The pressure conditions shown are for safety relief valves installed on a pressure vessel (vapor phase).

Figure 12.19 continued

$$K_b = \left\{ \frac{\dfrac{2k}{k-1}\left[\left(\dfrac{p_b}{p}\right)^{2/k} - \left(\dfrac{p_b}{p}\right)^{(k+1)/k}\right]}{k\left(\dfrac{2}{k+1}\right)^{(k+1)/(k-1)}} \right\}^{\frac{1}{2}} \quad [12.19.10]$$

where K_b is the capacity correction factor for back pressure, and p_b is the back pressure (bara).

For liquids:

$$q_{ml} = 1.61(\rho \Delta p)^{\frac{1}{2}} \quad [12.19.11]$$

with

$$\Delta p = p - p_b \qquad [12.19.12]$$

where Δp is the pressure difference (bar), q_{ml} is the theoretical flowing capacity (kg/h mm^2 of flow area) and ρ is the volumetric mass (kg/m^3).

For dry saturated steam with a minimum dryness fraction of 0.98 or a maximum superheat of 10°C:

$$q_{ms} = 0.525p \qquad p \leq 110 \qquad [12.19.13]$$

where q_{ms} is the theoretical flowing capacity of dry saturated steam (kg/h mm^2 of flow area).

The coefficient of discharge K_d is the ratio of the actual to the theoretical flowing capacity. A derated coefficient of discharge K_{dr} is also defined as $0.9K_d$

12.19.4 Valve sizing: API RP 520

The relations given in API RP 520: 1990 for sizing of pressure relief valves are as follows. The critical flow criterion is:

$$\frac{P_{CF}}{P_1} = \left(\frac{2}{k+1}\right)^{k/(k-1)} \qquad [12.19.14]$$

where k is the ratio of specific heats at 60°F and 1 atm, P_{CF} is the absolute critical flow throat pressure (lb$_f$/in.^{2}) and P_1 is the absolute upstream relieving pressure (lb$_f$/in.^{2}). The code gives values of the ratio of specific heats k for a number of gases.

For critical flow conditions:

$$W = CK_dAK_bP_1\left(\frac{M}{ZT}\right)^{\frac{1}{2}} \qquad [12.19.15]$$

or, in the form actually given

$$A = \frac{W}{CK_dP_1K_b}\left(\frac{ZT}{M}\right)^{\frac{1}{2}} \qquad [12.19.16]$$

with

$$C = 520\left[k\left(\frac{2}{k+1}\right)^{(k+1)/(k-1)}\right]^{\frac{1}{2}} \qquad [12.19.17]$$

where A is the effective discharge area (in.^{2}), K_b is the capacity correction factor due to back pressure, K_d is the effective coefficient of discharge, M is the molecular weight of the gas, T is the relieving temperature (°R), W is the required flow through the valve (lb/h), Z is the compressibility factor and C is a coefficient. The effective coefficient of discharge K_d has the value 0.975. Two further equations are also given for volumetric flows, one for air and one for gas.

The capacity correction factor due to back pressure is obtained from manufacturer's literature or from a graph given in the code; the graph is for balanced bellows pressure relief valves. The correction factor is not a means of extending Equation 12.19.15 to subcritical flow, for which separate equations are given.

For subcritical flow:

$$W = 735F_2K_dA\left[\frac{MP_1(P_1 - P_2)}{ZT}\right]^{\frac{1}{2}} \qquad [12.19.18]$$

or, in the form given

$$A = \frac{W}{735F_2K_d}\left[\frac{ZT}{MP_1(P_1 - P_2)}\right]^{\frac{1}{2}} \qquad [12.19.19]$$

with

$$F_2 = \left[\left(\frac{k}{k-1}\right)r^{2/k}\left(\frac{1 - r^{(k-1)/k}}{1-r}\right)\right]^{\frac{1}{2}} \qquad [12.19.20]$$

$$r = P_2/P_1 \qquad [12.19.21]$$

where F_2 is the coefficient of subcritical flow, P_2 is the absolute back pressure (lb$_f$/in.^{2}) and r is the ratio of back pressure to upstream pressure.

For liquid flow:

$$A = \frac{Q}{38K_dK_wK_v}\left(\frac{G}{P_1 - P_2}\right)^{\frac{1}{2}} \qquad [12.19.22]$$

where G is the specific gravity of the liquid at the flowing temperature relative to water at 70°F, K_d is the effective coefficient of discharge, K_v is the viscosity correction factor, K_w is the back pressure correction factor and Q is the required flow (USgal/min). For the effective coefficient of discharge the value should be obtained from the manufacturer; for preliminary estimates a figure of 0.65 can be used. For the viscosity correction factor K_v the code gives a graph. For the correction factor due to back pressure K_w the approach is as follows. If the back pressure is atmospheric $K_w = 1$. If a balanced bellows valve is used, the correction factor may be obtained from a graph given in the code. If a conventional valve is used, no correction is necessary.

12.19.5 Valve sizing: two-phase flow

The standards and codes described refer to, but do not give calculation methods for two-phase flow in pressure relief valves. Methods for two-phase flow, including the case of reactor venting, are considered in Chapter 15.

12.19.6 Valve sizing: safety factors

A discussion of safety factors applicable to pressure relief valves and relief systems is given in Appendix 13.

12.19.7 Valve failure

Pressure relief valves are relatively reliable equipment and the incidence of overpressure due to failure of a relief valve appears low. Accounts of pressure relief valve failure are given by Aird (1982), Maher *et al.* (1988), Crombie and Green (1994) and D.W. Thompson (1994).

Modes of failure of pressure relief valves include:

(1) failure to open on demand:
   (a) total failure to open,
   (b) opening at a pressure higher than set pressure;
(2) opening in the absence of demand:
   (a) opening at a pressure lower than set pressure (lifting light),
   (b) chattering,
   (c) failure to reseat after a correct opening on demand,
   (d) leakage;
(3) disabling of valve inlet or outlet.

Failure to open is a functional failure and opening in the absence of demand an operational, or spurious, failure.

Information on the failure rates of pressure relief valves is given in Appendix 14.

12.20 Overpressure Protection: Bursting Discs

An alternative pressure relief device is a bursting disc. The circumstances under which a bursting disc may be applicable are given in BS 5500: 1991 as those where the pressure rise may be so rapid that the inertia of a relief valve may be a disadvantage, where even minute leakage cannot be tolerated or where blockage may render a valve inoperative.

The relevant standard is BS 2915: 1990 *Specification for Bursting Discs and Bursting Disc Devices*. The basics of bursting discs have been outlined in Section 12.13. In this section consideration is given to disc set pressure, sizing and failure.

12.20.1 Bursting disc sizing: BS 2915

For a bursting disc it is usual to specify a range of pressures within which the disc will burst. Therefore, a disc has minimum, mean and maximum bursting pressures.

Equations for the sizing of bursting discs are given in BS 2915: 1990. Earlier editions of this standard contained the following treatment, given here for reference. For gas or vapour:

$$W = CKAP\left(\frac{M}{T}\right)^{\frac{1}{2}} \quad [12.20.1]$$

where A is the actual orifice area (mm^2), K is the coefficient of discharge, M is the molecular weight of the gas, P is the absolute vessel pressure (bara), T is the inlet temperature (K), W is the rated capacity (kg/h) and C is a constant.

For saturated steam

$$W_s = 0.535KAP \quad [12.20.2]$$

and for air at 20°C

$$W_a = 0.85KAP \quad [12.20.3]$$

where W_a is the rated capacity for air at 20°C (kg/h) and W_s is the rated capacity for saturated steam (kg/h). The coefficient of discharge is usually taken as 0.6 and the value of C is 2.7 for a ratio of specific heats of 1.4.

These equations have similarities with Equations 12.19.1–12.19.3.

BS 2915: 1990 gives the following relations for bursting disc sizing. For compressible fluids:

$$Q_{mg} = A_0 pF\alpha(M/ZT)^{\frac{1}{2}} \quad [12.20.4]$$

where A_0 is the required minimum cross-sectional area (mm^2), M is the molecular mass (kg/kmol), p is the relieving pressure (bara), Q_{mg} is the required flow of gas (kg/h), T is the relieving temperature (K), Z is the compressibility factor, α is the coefficient of discharge of the branch/nozzle type and bursting disc combined, and F is a correction factor (described below).

For steam

$$Q_{mw} = 0.2883A_0(F/x)\alpha(p/v)^{\frac{1}{2}} \qquad 0.9 \le x < 1.0 \quad [12.20.5]$$

and for saturated or superheated steam ($x = 1$)

$$Q_{ms} = 0.2883A_0F\alpha(p/v)^{\frac{1}{2}} \quad [12.20.6]$$

where Q_{mw} is the required flow of wet steam (kg/h), Q_{ms} is the required flow of saturated or superheated steam (kg/h), x is the dryness fraction and v is the specific volume at the relieving pressure and temperature (m^3/kg).

The standard gives a graph for the coefficient of discharge α. The correction factor F is:

$$F = 3.984\left\{\frac{2k}{k-1}\left[(p_b/p)^{2/k} - (p_b/p)^{(k+1)/k}\right]\right\}^{\frac{1}{2}} \quad [12.20.7]$$

where k is the isentropic exponent and p_b is the back pressure (bara). For steam the standard gives a graph for the isentropic coefficient k.

For incompressible fluids:

$$Q_{ml} = 1.610A_0f_\mu\alpha(\Delta p \cdot \rho)^{\frac{1}{2}} \quad [12.20.8]$$

where f_μ is the correction factor for liquid viscosity, Q_{ml} is the required flow of liquid (kg/h), Δp is the pressure difference (bar) and ρ is the volumetric mass (kg/m^3). The pressure difference term should take account of any static head. The standard gives a graph for the correction factor for liquid viscosity f_μ. The coefficient of discharge has the value of 0.5 or 0.62, depending on the type of branch/nozzle; details are given in the standard.

12.20.2 Bursting disc sizing: API RP 520

For the sizing of a bursting disc used independently API RP 520: 1990 states that Equations 12.19.16, 12.19.19 and 12.19.22 apply but with a different value of the coefficient of discharge K_d. The code gives for a bursting disc the value $K_d = 0.62$.

12.20.3 Bursting disc sizing: two-phase flow

The standards and codes described refer to, but do not give calculation methods for, two-phase flow in bursting discs. Methods for two-phase flow, including the case of reactor venting, are considered in Chapter 15.

12.20.4 Bursting disc sizing: safety factors

A discussion of safety factors applicable to bursting discs and relief systems is given in Appendix 13.

12.20.5 Bursting disc failure

Like a relief valve, a bursting disc may suffer functional or operational failures. In other words, it may fail to burst at the set burst pressure and thus fail to danger, or it may burst below that pressure and thus fail prematurely but, generally, be safe.

Some of the principal modes of failure of a conventional bursting disc are fatigue, creep and corrosion. Pulsation of the fluid pressure induces fatigue, while high temperature causes creep and contributes to fatigue.

Another type of failure is caused by blockage. Blockage may occur on the process side and may be caused by corrosion, crystallization or polymerization. It can also occur on the vent side.

Malinstallation constitutes a third type of failure. The disc may be installed upside down. Another example of malinstallation is putting more than one disc in the holder. Discs are sometimes supplied in stacks and in this case duplication may be a simple error. In other cases the use of more than one disc is done deliberately to avoid frequent bursting, particularly during commissioning.

Reverse buckling discs are thicker and less prone to fatigue, creep and corrosion, but they have their own characteristic failures. One of these is 'roll through'. The dome of the disc becomes dented and it rolls through onto the knife but with insufficient energy to cause bursting.

The condition of the knife in a reverse buckling disc is critical. It may suffer corrosion, cracking or blunting so that it is no longer capable of cutting. Failure of a reverse buckling disc may also occur if there is insufficient energy to cause cutting of the disc. This is particularly liable to occur on liquid systems.

Malinstallation is another type of failure with a reverse buckling disc. Again the disc may be put in upside down. This may occur where the fitter is accustomed to installing conventional discs and is insufficiently alert to the difference between the two.

Information on the failure rate bursting discs is given in Appendix 14.

12.21 Overpressure Protection: Installation of Relief Devices

In addition to the factors already considered, it is necessary in the design of a relief system to pay close attention to the installation of the relief devices. The installation of relief devices is treated in API RP 520 Pt II, the CCPS *Engineering Design Guidelines* and the IChemE *Relief Systems Guide.*

12.21.1 API RP 520: Part II

A prime source of guidance on the installation of relief devices is API RP 520 Pt II: 1988 *Sizing, Selection and Installation of Pressure-relieving Devices,* Part II, *Installation.* This deals with the following topics: (1) general, (2) inlet piping, (3) discharge piping, (4) bonnet or pilot vent piping, (5) drain piping, (6) valve location and position, (7) bolting and gasketing, (8) multiple pressure-relieving devices with staggered settings and (9) pre-installation handling and inspection.

12.21.2 Inlet piping to pressure relief valves

A pressure relief valve should be mounted directly on the vessel which it protects at a dedicated nozzle unless there is good reason to do otherwise. Similar considerations apply to bursting discs.

Cases where a pressure relief valve may be positioned at some short separation from the vessel include those where this improves access, allows the valve to be put outside the insulation, reduces the valve temperature or minimizes the vibration to which it is subject or permits improved drainage of the discharge pipework.

Where a pressure relief valve is connected to the vessel by piping, the desirable features of this piping are that it should be short, full bore, vertical and self-draining with low pressure drop and exerting minimum stress on the equipment.

12.21.3 Reaction force from pressure relief valve discharge

The reaction forces due to discharge from a pressure relief valve can be appreciable and need to be allowed for in the design. Accounts of reaction forces are given in API RP 521 Pt II: 1988 and the IChemE *Relief Systems Guide* and by Moody (1969, 1973), Stikvoort (1986) and Hansen (1991).

API RP 521 Pt II gives a relationship for the reaction force for the case where the pressure relief valve is mounted vertically with its discharge entering a pipe elbow which runs a short distance horizontally, then round an elbow, and vertically upwards. The equation is:

$$F = \frac{W}{366}\left(\frac{kT}{(k+1)M}\right)^{\frac{1}{2}} + A_0P_2 \qquad [12.21.1]$$

where A_0 is the area of the outlet at the point of discharge (in.^{2}), F is the reaction force (lb$_f$), k is the ratio of specific heats of the gas, M is the molecular weight of the gas, P_2 is the static pressure at the point of discharge (lb$_f$/in.^{2}), T is the absolute temperature (°R) and W is the mass flow of gas (lb/h). The reaction force is exerted downwards on the horizontal section. The code shows a support between this section and the vessel.

12.21.4 Isolation of relief devices

There are certain circumstances where it would be convenient to be able to isolate a relief device on an operating vessel. This is particularly the case where the device needs to be removed for maintenance. Any such isolation is restricted by regulatory requirements and by good practice.

In British legislation, as described in Section 12.12, steam boilers and air receivers have long been subject to specific requirements. It has not been permitted even to install an isolation valve between a steam boiler and its safety valve. For air receivers, BS 1123: Part 1: 1987 permits the installation of an isolation valve between the vessel and its safety valve, provided (1) that the source of supply is also isolated by the same valve and (2) the receiver is fitted with a fusible plug. For process vessels the installation of a isolation valve between the vessel and its relief valve without further precaution is regarded as bad practice.

BS 5500: 1991 does, however, provide in Appendix J for installation of isolation valves subject to the following conditions: 'Intervening stop valves or cocks may be installed provided that they are so constructed and controlled by mechanical interlocks that a limited number only can be closed at any one time and that those stop valves or cocks which remain open are adequately sized to permit the unaffected pressure relieving devices to discharge at the required capacity for the vessel.' Thus there must be at least duplicate pressure relief valve systems and there must be a mechanical interlock between the two systems.

The fitting of an isolation valve on a single relief device is permitted in some companies, provided that there are strict administrative controls. A case in point might be the fitting of an isolation between a liquid relief device and the space protected. It may be a condition of such exemption that it be shown by hazard assessment that the probability of a demand during the period of valve isolation is acceptably low.

The requirements of the Pressure Systems Regulations 1989 given in Regulation 4(5) have been quoted in Section 12.12. Guidance is given in COP 37: 1990. This states: 'Every plant item in which the pressure can exceed the safe operating limit for pressure should be protected, whenever operational, by at least one pressure relieving or pressure limiting device.'

12.22 Flare and Vent Systems

A principal method of disposal of gases and vapours discharged from the process is flaring. The function of a flare system is generally to handle materials vented during (1) normal operations, particularly start-up, and (2) emergency conditions.

Guidance on flare system design is given in API RP 521: 1990, on which much of the other literature draws heavily. Accounts of flare systems are given in *Flare Gas Systems Pocket Handbook* (Bannerjee, Cheremisinoff and Cheremisinoff, 1985), the CCPS *Engineering Design Guidelines* and the IChemE *Relief Systems Handbook*, and by Bluhm (1961, 1964b), Husa (1964), G.R. Kent (1964, 1968, 1972), P. Peterson (1967), Tan (1967a,b), Kilby (1968), Seebold (1984), Straitz (1987), Swander and Potts (1989) and Tite, Greening and Sutton (1989).

12.22.1 Types of flare system

There are three main types of flare system: (1) elevated flare, (2) ground flare and (3) low pressure flare. The last two are described first.

12.22.2 Ground flares

A ground flare generally consists of a battery of tubular burners contained in a short refractory-lined stack, which serves as combustion chamber and windshield. The bottom of the stack is above grade so that the combustion air can enter. Typically, this air inlet is provided with a louvred shield to maintain flame stability and masks the glare from the flame; the stack is circular or square in cross-section and it is set inside an octagonal windshield. The elements of a ground flare system are knockout drum, seal drum, burner system, ignition system, stack and windshield.

In one design of ground flare systems the height H of the stack is about $0.3D$, where D is the diameter of the combustion chamber, and the bottom of the stack wall is also about $0.3D$ from grade. Other designs of ground flare are described by Swander and Potts (1989), who give illustrations of designs ranging from a completely open flare to one with a quite high stack. There are also variations in the number of burners, some low volume ground flares having just a single burner. Details of ground flare systems are given in the IChemE *Guide*.

A ground flare may be favoured where there is a limitation on the height of the flare which can be installed. Insofar as the products of combustion are discharged relatively close to ground level, a ground flare is not suitable if these gases are noxious or pollutant. Flare glow and noise may also be limiting factors.

12.22.3 Low pressure flares

There are in process plants various low pressure units which may produce an off-gas stream. These include storage tanks, and oil–water separator and wastewater treatment units. These off-gases may be routed to a low pressure flare. The tightening of controls on the emission of volatile organic compounds increases the number of situations where a low pressure flare may be needed. There are available a number of designs of low pressure flare such as small 'stick' flares.

The principal problems and hazards of a low pressure flare arise from the fact that the pressure in the discharging unit may be barely sufficient to maintain the flow. The pressure drop in the low pressure header should be calculated with care and the header adequately sized. Not all computer programs used for header design are suitable for this case, and it may be necessary to resort to hand calculation.

Where the discharge source pressure is insufficient, use may be made of a blower installed in the off-gas line. The use of an off-gas blower, however, is not an ideal arrangement in that it introduces rotating equipment which may fail and which may have the potential to act as a source of ignition. Designs based on off-gas blowers are discussed in the CCPS *Guidelines*.

12.22.4 Elevated flares

Elevated flares are used to handle both normal and emergency loads and are generally the system of choice except where environmental considerations prevent their use. An elevated flare is a normal feature of a refinery or a petrochemical plant.

A flare system consists of a flare stack and of gathering pipes which collect the gases to be vented. Other features include the flare tip, which typically has steam nozzles to assist entrainment of air into the flare, seals installed on the stack to prevent flashback of the flame and a knockout drum at the base of the stack to remove the liquid from the gases passing to the flare.

A typical elevated flare system is illustrated in the widely quoted diagram from API RP 521 shown in Figure 12.20. The elements of this system are gas collection pipework, knockout drum, seal drum, flare stack, flare tip, molecular seal, ignition system, steam injection system and purge system.

12.22.5 Flaring policy

The type of flare system required and the associated hazards depend very much on the venting and flaring policies adopted. Thus, for example, the decision to vent certain cold columns of an ethylene plant direct to the atmosphere can greatly reduce the size of the system and avoid the need to use the special steels required to handle cold gases.

The emergency load on the flare is highly dependent on factors such as the extent to which trip systems are used to reduce the frequency of discharges from pressure relief valves and on the staggering of such discharges.

Descriptions of practice in flaring include those given in the CCPS *Guidelines*, and the IChemE *Guide*, and by Barker, Kletz and Knight (1977), Bonham (1977), R. Schwartz and Keller (1977) and Straitz *et al.* (1977).

12.22.6 Flare system hazards

There are a number of hazards associated with a flare system. These include (1) failure of the collection system pipework due to low temperature embrittlement or corrosion, (2) obstruction in the flare system, (3) explosion in the flare system, (4) heat radiation from the flare, (5) liquid carryover from the flare and (6) emission of toxic materials from the flare.

Even if these features are not such as to present a hazard, the latter three may be environmentally objectionable. Other features which cause environmental problems are (1) smoke in the flare, (2) glare from the flare and (3) noise from the flare.

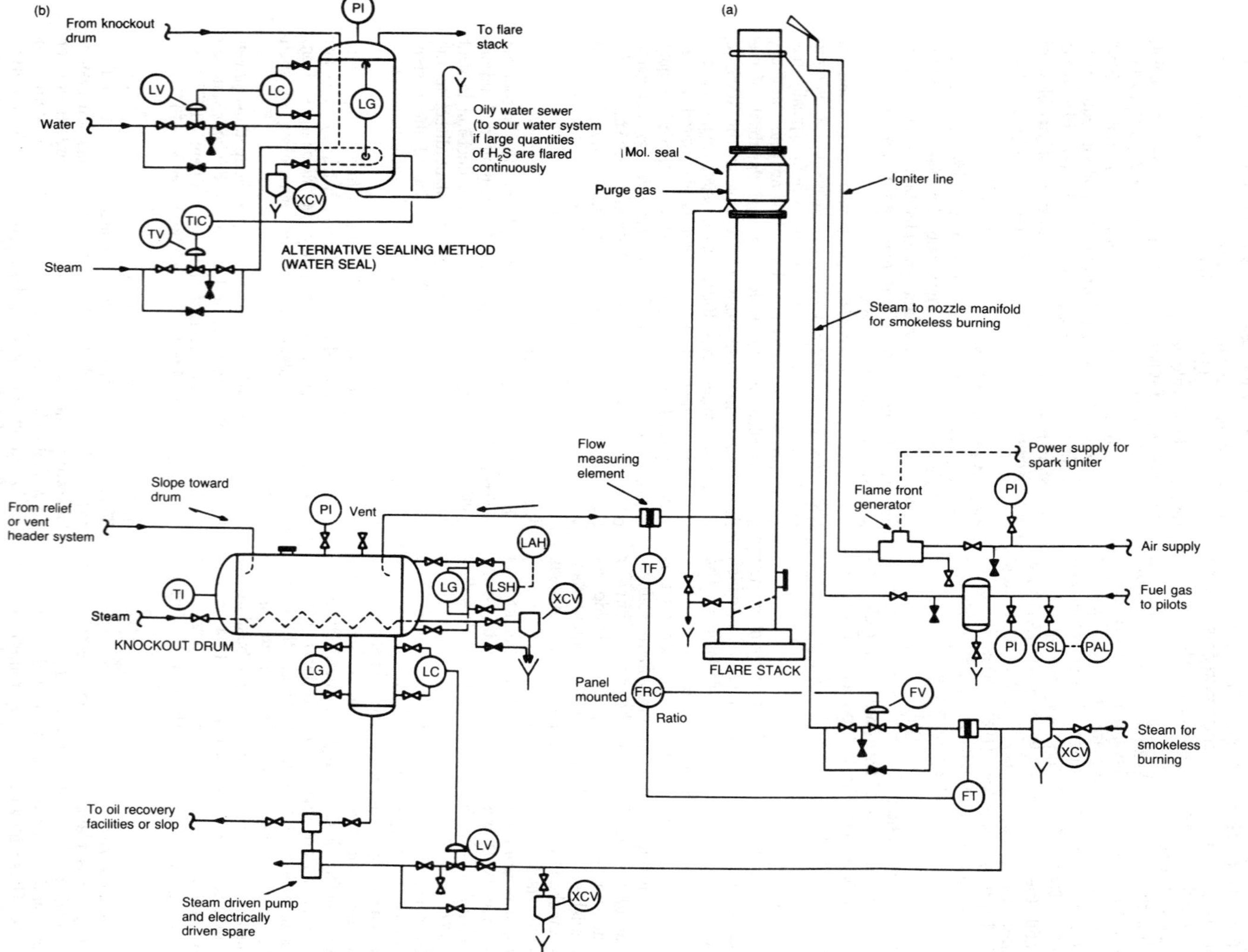

Figure 12.20 *Typical flare system (American Petroleum Institute, 1990 API RP 521): (a) flare system; and (b) alternative sealing system. Note: this figure represents an operable system arrangement and its components. The arrangement of the system will vary with the performance required. Correspondingly, the selection of types and quantities of components, as well as their applications, should match the needs of the particular plant and its specifications. FRC, flow recorder controller; FT, flow transmitter; FV, flow valve; LAH, level alarm (high); LC, level controller; LG, level gauge; LSH, level switch (high); LV, level valve; PAL, pressure alarm (low); PI, pressure indicator; PSL, pressure switch (low); TIC, temperature indicator controller; TV, temperature valve; XCV, steam trap*

12.22.7 Flare collection system

The collection system which gathers the gas and vapour to be flared may comprise a single system or a set of systems in which different types of stream are segregated. Brittle fracture of the pipework in the gathering system is a hazard if the temperature of the steel is taken below its transition value. It may be necessary to use stainless steel in parts of the pipework where this condition may occur. Stainless steel may also be required for highly corrosive gases. The use of such steel is expensive, however, and the amount used is kept to a minimum. The extent to which stainless steel is required depends on the discharge and segregation policies adopted.

The CCPS *Guidelines* distinguish the following cases: (1) cold gas, (2) intermediate gas, (3) hot gas, (4) sour gas. Typical cold gas is ethane and lighter hydrocarbons flashing at or below −45°C. The material of construction used for the header is commonly austenitic stainless steel. Most gaseous effluents are hot wet gases above 0°C which are generally collected in a header made of carbon steel. Liquid droplets are separated in the knockout drum. A third class of gaseous effluent is cold dirty gases at a temperature intermediate between 0 and −45°C. For these use is frequently made of a header constructed in killed carbon steel.

Certain processes give rise to sour gas streams containing appreciable concentrations of hydrogen sulphide which are toxic and corrosive. A sour gas stream is often provided with its own header, made of stainless steel. This segregation avoids the use of an expensive steel for the whole gas collection system. Frequently a separate flare is also used, which gives better control of the flaring.

Separate collection systems may also be provided for low and high pressure gas streams. The size of the system is generally governed by the low pressure streams, but if the system is to take all the discharges, it must be designed for the highest pressure which it could experience. It can be more economic to use a separate high pressure collection system with smaller piping.

Elements of the collection system are the various relief headers, the piping from the individual valves discharging into these headers and the flare header connecting these relief headers to the flare itself.

12.22.8 Flare system

The flare system consists of the collection system, the liquid knockout and seal drum system, the flare assembly and the ancillary equipment. The collection system is described in the next section. The flare assembly comprises the flare stack itself and the associated components, notably (1) the flare tip, (2) the pilot burners, (3) the pilot igniters and (4) the steam injection system. Between the collection system and the flare stack comes the knockout and seal drum system. This typically includes a knockout drum and a seal drum, and may also have a quench drum.

12.22.9 Flare stack

The flare stack itself is tyically self-supporting or supported by guy wires. Associated components are (1) the flare tip, (2) the pilot burners, (3) the pilot igniters, (4) the steam injection system and (5) the flame monitoring instrumentation.

The process design of the flare stack is essentially the determination of the stack height and diameter. API RP 521 gives one design method. The stack height is basically governed by the heat radiation from the flare, which is described below.

The stack diameter is set by the gas velocity. API RP 521 suggests that the design aim for a flare tip Mach number of 0.2 at normal gas flows and 0.5 for short, infrequent peak flows. It also states that a smokeless flare should be sized for the conditions under which it will operate smokelessly.

Flare stack conditions are discussed further in Chapter 16. Discussion of the other elements in the flare system is deferred until the flare stack obstruction and explosion hazards have been considered.

12.22.10 Flare system obstruction

Obstruction of the flare system is a hazard which can occur in a number of ways. One is the blockage of devices such as flame arresters and molecular seals. Another is the freezing of water seals in cold weather.

In some flare systems steam is injected at the base of the stack. The combination of contact with a cold gas such as ethylene and of cold weather may cause the water injected to freeze. There may also be other fluids in the flare system which are liable to freeze in cold weather. Freezing points of two common hydrocarbons are: benzene 5°C and cyclohexane 6.5°C.

The solidification of heavy oils can create blockage. Blockage can also be caused by refractory debris at the base of the flare stack, unless this is catered for in the design.

12.22.11 Flare system explosion

In a flare system there is a hazard that air will enter and form a flammable mixture. The hazard is particularly serious, because a source of ignition is always present in the form of the flame on the flare tip. Air may enter the flare system due to factors such as open valves or corrosion, or by diffusion down from the flare tip when the flame is not operating.

There are three specific conditions which are conducive to entry of air into the flare stack. One is when flaring has just ceased and cooling and shrinkage of the gas column in the stack draws air in. Another is when the gas being flared is light, especially if it is hydrogen, so that the pressure at the bottom of the stack is lower than atmospheric and air may enter through any apertures. The third is when the flare stack creates a natural draught and air leaks in at the bottom.

The precautions taken against the hazard of an explosion in a flare system divide into those aimed at avoiding a flammable mixture, such as (1) use of purge gas, (2) elimination of leaks, (3) monitoring the oxygen concentration, and those directed to mitigating the results of any ignition which may occur, such as (4) the use of flame arresters. Associated measures include the use of (5) molecular seals and (6) seal drums.

The discharge of waste gas to the flare system is inevitably somewhat erratic. If the gas flow falls too low and if there are air leaks into the system, a flammable mixture may form. Also if the gas flow is too low, the flame may go out, flash back down the stack or start pulsating. It is common practice, therefore, to maintain the gas flow in the flare system. This purge is usually a fuel gas. A method of calculating the purge gas rate is given by Husa (1964).

The best approach to the prevention of explosions in flare systems, however, is to prevent air from getting into the system in the first place and to monitor the oxygen content to check that this has been achieved. The elimination of air inleak into the pipework should be a specific objective. The prevention of air inleak through open valves is particularly important.

Air may also enter the system by diffusion down the stack when the flare is not operating. A molecular seal is often installed to prevent this. The use of molecular seals (described below) effects a large reduction in the purge gas flow required to prevent air diffusion down the stack.

The oxygen concentration in the flare system should be monitored. This is necessary in order to ensure that a large air inleak does not develop undetected.

The elimination of air inleak by the measures just described is preferable to the use of large purge gas flows to dilute such inleak. Devices which are used to guard against flashback of the flame down the stack are flame arresters and water seals.

12.22.12 Seal drums

A device widely used to prevent ingress of air from the flare stack into the gas collection system is a seal drum in which the gas passes through a water seal. API RP 521 gives one design for a seal drum together with the method of sizing. The IChemE *Guide* gives further design details.

A seal drum is generally designed for an internal pressure of at least 50 psig in order to withstand an internal explosion. Two problems with water seals are the creation of an uninterrupted gas passage through the water at high gas flows, which can render the seal ineffective, and the tendency to surge, which can affect operation of the flare. Proprietary water seals are available which are designed to overcome both these difficulties.

Another difficulty with water seals is the loss of the water from the seals. A further disadvantage of water seals is that they tend to freeze in cold weather. This problem is sufficiently serious that some operators prefer not to use them at all.

12.22.13 Molecular seals

A molecular seal is a device designed to limit the rate at which air can flow back into the flare stack. The principle of operation is to make the air flow through a set of concentric tubes in which it suffers reversals of direction so that its flow is reduced. A molecular seal is best located close to the top of the stack.

The benefit of a molecular seal is economy of purge gas. Molecular seals have the disadvantage that they tend to block up. One cause of blockage is corrosion products from the stack. Another is blockage by water or ice. In cold climates prevention of ice formation requires the use of heating.

12.22.14 Flashback arresters

One common device for preventing passage of a flame along a pipe or stack is the flame, or flashback, arrester. A disadvantage of flame arresters is the tendency for the small passages to become obstructed. This problem is sufficiently serious that some companies do not use them.

12.22.15 Flare combustion and smoke control

Brzustowski and co-workers (Brzustowski, 1973, 1977; Brzustowski and Sommer, 1973; Brzustowski *et al.*, 1975) have studied the fundamentals of combustion in flare systems. Efficient combustion in the flame depends on achieving good mixing between the fuel gas and the air, and avoidance of a pure diffusion flame. Failure to achieve efficient combustion results in a smoky flame.

There are various methods which are used to promote efficient combustion. One of the principal methods is steam injection. The main function of the steam is to increase the entrainment and mixing of the air. The steam also takes part in the water gas shift reaction

$$C + H_2O = CO + H_2$$

and the steam reforming reaction

$$C_xH_y + H_2O = xCO + zH_2$$

Alternatively, low pressure air may be used to assist the entrainment and mixing of the atmospheric air.

There are available a number of proprietary flare tips for steam injection. One which is widely used is the Zink S type, which has a ring of small jets around the tip of the flare.

The equation given in API RP 521 for the amount of steam injection required to give a smokeless flare is that of Tan (1967):

$$W_{st} = W_{HC}\left(0.68 - \frac{10.8}{M}\right) \qquad M > 16 \qquad [12.22.1]$$

where M is the molecular weight of the gas, W_{HC} is the mass flow of hydrocarbons (kg/s) and W_{st} is the mass flow of steam (kg/s). This is based on maintaining a steam/CO_2 mass ratio of about 0.7.

The control of the state of the flame by the manipulation of the steam supply has traditionally been done by the process operator relying on visual observation. Automatic control is difficult, because it is not easy to obtain a good measure of the state of the flame. One automatic system now in use is based on the measurement of the heat radiation at the root of the flame, where the difference in the heat radiated by a smoky and by a non-smoky flame is particularly pronounced.

There are a number of problems involved in achieving positive ignition and in retaining a flame on the flare tip as well as in preventing flashback. These problems are particularly difficult where the turndown ratio is high, which is often the case.

12.22.16 Flare heat radiation

The flare stack radiates intense heat which constitutes a potential hazard. The flame on a flare stack is often several hundred feet long and has a heat release of the order of 10^7 BTU/h. There is intense heat radiation from a flare. It is generally necessary, therefore, to have an area around the flare in which people do not normally work. A large flare can thus sterilize a sizeable area of land. It may be acceptable in some cases, however, to locate certain types of equipment within the flare compound, provided that the time which personnel have to spend in the compound is strictly limited.

Methods of caculating the heat radiated from a flare are given in API RP 521: 1990. This code gives the following equation based on the work of Hajek and Ludwig (1960):

$$D = \left(\frac{\tau FQ}{4\pi K}\right)^{\frac{1}{2}} \qquad [12.22.2]$$

where D is the distance from the centre of the flame, F is the fraction of heat released which is radiated, K is the allowable intensity of heat radiation, Q is the net heat release rate, and τ is the atmospheric transmissivity.

Combustion in the flare is a complex process. In Equation 12.22.2 this complexity is subsumed in the factor F. This factor has traditionally been taken as a property of the fuel only. Some typical values of F are:

Gas	F	*Reference*
Methane	0.16	Brzustowski and Sommer (1973)
Propane	0.33	G.R. Kent (1964)
Butane	0.30	Brzustowski and Sommer (1973)
Ethylene	0.38	Brzustowski and Sommer (1973)

It has been shown, however, that the F factor depends also on other features of the flame such as the Reynolds number. The F factor is discussed by Brzustowski (1977) and Straitz *et al.* (1977).

Models of the flame on a flare are considered in Chapter 16.

Guidance on the permissible level of heat radiation from a flare in different situations is given in API RP 521 and BS 5908: 1990. Further discussions of acceptable levels of heat radiation are given by G.R. Kent (1964), R. Schwartz and Keller (1977) and Straitz *et al.* (1977). This aspect is considered in Chapter 16.

12.22.17 Liquid carryover in flares

Liquid carryover from the flare stack may result in a more smoky flame, in dispersion of burning drops of flammable material or dispersion of drops of toxic material. Liquid drops as small as 15 μm can negate the devices used for smokeless operation and give a smoky flame. The principal means used to prevent liquid drops reaching the flame is the use of a knockout drum at the base of the stack. It is sometimes difficult, however, to eliminate completely spray and condensation.

12.22.18 Toxics in flares

In some flares gases containing materials such as chlorine and sulphur are burned, giving compounds such as HCl and SO_2. In such cases the flare stack should be sufficiently high to prevent unacceptable ground level concentrations of these toxic gases.

12.22.19 Flare environmental problems

Smoke, glare and noise from a flare are environmentally objectionable. The elimination of smoke has already been discussed. There is relatively little which can be done to eliminate light from the flame. If this is a serious problem, it may be necessary to use a low level enclosed burning system instead of a flare. Flare noise is discussed in Appendix 12.

12.22.20 Combined flare systems

In some installations the flaring arrangements consist of an elevated flare combined with a ground flare. Normal operating and start-up loads are handled by the ground flare, while both flares are used to handle the relatively infrequent high volume emergency loads. In this way the environmental impact of the flaring is kept to a minimum.

12.23 Blowdown and Depressuring Systems

The response to a plant emergency may require the disposal of hazardous inventories. This is effected by the emergency blowdown and depressuring systems.

12.23.1 Blowdown systems

Disposal of liquid streams in an emergency is through the blowdown system. An emergency blowdown system has two basic features: treatment and disposal. An account of blowdown systems is given in the CCPS *Engineering Design Guidelines.*

Fluids which may require disposal are, in general, streams containing volatile organics or mixtures of volatile organics and water. More specific cases are reaction fluids vented from a reactor and cooling water contaminated with organics following a tube burst in a heat exchanger where the process fluid is at the higher pressure. The streams to be disposed of may contain flammable, toxic, corrosive or pollutant substances.

Blowdown streams may be unsuitable for immediate disposal, and may therefore require treatment. Principal types of treatment are disengagement of gases and vapours from liquid, and quenching to cool and partially condense hot vapours.

Disengagement is effected in a disengagement drum. The CCPS *Guidelines* refer to the API RP 521 knockout drum design. In a typical disengagement operation, involving a mixture of organics and water, the vapour is sent to a flare and the organic and water phases are separated by gravity in the drum.

Quenching is carried out in a quench nozzle, a quench drum or a blowdown drum. In a quench nozzle water is injected to effect at least partial condensation of the blowdown fluid. In a quench drum condensation of the blowdown fluid is effected by spraying water or other liquid onto it. The CCPS *Guidelines* refer to the API RP 521 quench drum design (Figure D-2).

A blowdown tank is charged with an inventory of solvent through which the blowdown fluid is sparged, effecting condensation. The design of blowdown tanks for high rate releases of short duration has been discussed by Fauske and Grolmes (1992).

The sinks to which the liquid effluent part of a blowdown stream may be sent include a clean water sewer, an oily water sewer, a closed drain header and a tank for further treatment. The CCPS *Guidelines* give a detailed discussion.

12.23.2 Depressuring systems

Disposal of vapours from equipment at pressures below those of the pressure relief devices is by means of the emergency depressuring system. Depresuring is used to reduce hazard by removing vapour or gas from an equipment. This may be necessary where the equipment may fail at a pressure below the set pressure of the pressure relief valve, as in fire engulfment of a vessel which is filled with gas or one which contains some liquid but in which part of the walks is unwetted. Another situation where depressuring may be appropriate is where there may be benefit in anticipating the operation of the relief, as in the depressuring of a reaction runaway. A third case is where the equipment pressure is very high (say, 1000 psig) where depressuring reduces both the inventory and the stress in the equipment.

12.24 Pressure Containment

An alternative to pressure relief is pressure containment. This may be effected by ensuring that the system is sufficiently strong that any pressure generated is contained. This approach is currently applied only to a relatively limited extent. A discussion of the potential for wider application is given by Wilday (1991).

12.24.1 Problems of pressure relief

Protection against overpressure by the provision of pressure relief has a number of disadvantages. In general, pressure relief involves the use of an active system which is subject to failure and has to be inspected and maintained, and which may operate spuriously.

Pressure relief creates the problem of relief disposal. Environmental pressures make this an increasingly onerous matter. The basis for the design of the disposal train is often not well defined. The demands on the disposal train will be rare and its reliability in responding to them uncertain.

In some cases pressure relief is not a practical option. This is the case, for example, where the vent area on a reactor is simply too large to fit on the vessel.

12.24.2 Inherently safer design

The containment approach to overpressure protection requires that all the potential sources of overpressure be identified. Generic sources include (1) flow in, (2) momentum effects, (3) heat input, and (4) chemical reaction.

Likewise, in principle it is necessary to identify all the potential sources of underpressure. Generic causes of underpressure include (1) flow out, (2) heat removal, and (3) chemical reaction. However, vacuum-causing events, such as vapour condensation, are often difficult to quantify. Thus in the case of underpressure the practical approach is often to design for full underpressure. The pressure deviation scenarios, including combinations, should then be considered and a credible worst case determined as the basis of design.

Design for full overpressure is assisted if sources of overpressure can be designed out. Wilday (1991) gives as examples (1) heat exchanger tube failure, (2) process heating and (3) runaway reactions. A heat exchanger with high pressure gas in the tubes and liquid in the shell can present a difficult case for pressure relief. If the design is to be based on the common assumption of full bore rupture of a single tube, a large two-phase flow may be generated and the pressure relief requirement is severe. One solution may be to eliminate the pressure differential by raising the liquid pressure.

Other aspects of heat exchanger tube failure considered by Wilday include the use of a non-reactive cooling medium to eliminate reaction between the material cooled and cooling water and the merits of using air-cooled heat exchangers.

In the heating of a liquid the total pressure due to the vapour pressure of the liquid and the partial pressure of non-condensable gases can be high. Methods of dealing with this problem include the use of a lower vapour pressure liquid, the provision of adequate gas space, the limitation of the heat input, the prevention of isolation of the equipment, and the use of a sufficiently high design pressure.

The measures suggested for avoidance of high pressure in reactors include laboratory screening to identify reaction exotherms and decompositions, use of a low vapour pressure solvent, avoidance of accumulation of reactants, use of a semi-batch configuration with one component fed continuously, and provision of a trip to cut off the feed.

12.24.3 Instrumented protective systems

One approach to the elimination of sources of overpressure is the use of instrumented protective systems. The use of instrumented protective systems as an alternative to pressure relief is relatively limited and generally subject to strict conditions.

The use of an instrumented protective system requires that all the sources of high pressure be identified and protected against. Another problem is that such a system is an active one, with the disadvantages already mentioned. Nevertheless, where the pressure relief option also poses severe difficulties, as in some reactor relief situations, there has been considerable interest in the use of instrumented protective systems. Work on the options for overpressure protection of reactors is described in Chapter 11.

Examples of the use of instrumented protection given by Wilday are (1) the protection of plant downstream of a pressure letdown valve by a trip which shuts a trip valve and (2) the protection of a reactor by a trip which shuts off the feed.

12.24.2 Problems of, and scope for, containment

Containment also has its problems. One general problem is protection against external fire. If a vessel is engulfed in a fire it is liable to fail due to overtemperature of the metal, even though the normal failure pressure has not been attained. There may also be additional hazards such as a liability of the material held to decompose.

There are certain measures which may be taken in some cases to limit the degree of potential fire exposure such as elimination or reduction of flammable materials near the vessel, provision of fire insulation and, possibly, location of the vessel above the level at which flame impingement could occur, but the overall conclusion is that it is very difficult to protect against external fire by containment.

There are certain types of equipment, such as atmospheric storage tanks, to which overpressure protection by containment is not applicable. The same applies to certain reactors. Wilday suggests that a plant protected solely by containment would be likely to be designed for a relatively high pressure and for full vacuum and would require to be engineered so as effectively to eliminate the external fire hazard or the need for fire relief. However, even if a policy of full containment is not adopted, the principles described may be used to limit the extent of pressure relief needed.

12.25 Containment of Toxic Materials

Some plants handle substances which rank as high toxic hazard materials (HTHMs) and the pressure system must ensure their containment. Guidance is given in the CCPS *Guidelines for Safe Storage and Handling of*

High Toxic Hazard Materials (1988/3) (the CCPS *HTHM Guidelines*), which covers a wide range of topics. These guidelines are considered in Chapter 22. Many aspects are dealt with in other parts of the present text. Table 22.10 gives an indication of the chapter or section where relevant material is to be found.

Containments for HTHMs should be designed and constructed to standards which are higher than average. This applies to, among other things, the vessel wall thickness and connections, the materials of construction, the welds and the quality control.

12.25.1 Storage tanks and vessels

US standards and codes applicable to the storage of HTHMs include API Std 620: 1990, API Std 650: 1988 and the ASME *Boiler and Pressure Vessel Code* 1992. Also relevant is the ASME B31.3: 1990 *Chemical Plant and Petroleum Refinery Piping* code.

API Std 650 is the standard for large oil storage tanks and has limited relevance to HTHMs. The safety factors and welding standards are less stringent and the tank relief is to atmosphere. The more relevant standard is API Std 620. For storage of larger quantities of HTHMs the preferred method will often be the use of refrigerated storage. This is dealt with in the standard in Appendices R and Q which treat, respectively, tanks for refrigerated liquids stored in the temperature range 40 to −60°F and those stored at temperatures in the range down to −270°F.

Tank designs for refrigerated storage given in API Std 620 include both single and double wall metal tanks, but in the latter the outer wall is not intended to withstand major failure of the inner tank, either in respect of the hydrostatic head or of the temperature of the liquid stored. A more suitable design is one incorporating a high bund close to the metal tank. The CCPS *Guidelines* refer to that given the CIA *Refrigerated Ammonia Storage Code,* which is considered in Chapter 22. Materials of construction for refrigerated storage of HTHMs need to possess both the necessary impact toughness and corrosion resistance.

In design of storage tanks for HTHMs particular attention should be given to features such as the foundations, the weld radiography and the pressure testing. API Std 620 contains stress relief requirements which are given in Appendices H and I.

With a storage tank designed to API Std 620 the pressure relief is generally designed for normal 'breathing;' and for operational variations. These arrangements are less suitable for HTHMs, for which there are more severe restrictions on release to atmosphere. It may be necessary to use a closed relief and vent system. If direct relief to atmosphere is permissible, a rupture disc should not be used alone, as it does not reclose after opening.

API Std 620 contains provisions by which fire relief may be provided by a rupture seam. The CCPS *Guidelines* state that a rupture seam is not normally acceptable for HTHMs. The implication of this is that there need to be specific alternative arrangements for fire relief. Whatever arrangements are adopted, it is essential to ensure that the tank does not rupture at the wall–floor seam, which would allow discharge of its entire contents.

For storage vessels, the ASME *Boiler and Pressure Vessel Code* recognizes 'lethal service'. It states that 'by lethal substances are meant poisonous gases or liquids of such a nature that a very small amount of gas or of the vapour of the liquid, mixed or unmixed with air, is dangerous to life when inhaled... this class includes substances of this nature which are stored under pressure or which may generate a pressure if stored in a closed vessel.'

The minimum wall thicknesses given in pressure vessel codes may yield for small vessels at low pressure relatively thin walls, and it may be prudent for certain HTHMs to use a thicker walled container. A suggestion made in the CCPS *Guidelines* is the use of compatible DOT cylinders.

The nozzles are frequently the weak points on a vessel and for containment of HTHMs they need to be of high integrity. Measures to ensure this include high nozzle specifications, use of sufficient thickness and of suitable welds and avoidance of small nozzles and of excessive projection. The CCPS *Guidelines* recommend that: nozzles should preferably be to Class 300 flanged welding necks; all nozzle-to-shell welds be full penetration welds; nozzles of less than 1–1½ in. nominal bore (NB) be avoided, or at least protected; and nozzles less than 2 in. NB do not project more than 6 in., or be provided with additional strength or protection.

Since for HTHMs avoidance of catastrophic failure is of particular importance, use should be made of fracture mechanics so that, where practical, the vessel operates in the lower risk 'leak-before-break' regime. This is discussed in Section 12.28.

12.25.2 Pipework

As for tanks and vessels, so for piping – design and construction for HTHM service should be to high standards. Some features of high integrity pipework have been discussed in Section 12.7. Only those aspects are considered here where the toxicity of the material is of particular relevance.

ASME B31.3 recognizes a Category M Fluid Service in which the potential for personnel exposure is judged to be significiant, and in which a single exposure to a very small quantity of a toxic fluid, caused by leakage, can produce serious irreversible harm to persons upon breathing or bodily contact, even when prompt restorative measures are taken. Chapter VIII of the code deals with Category M Fluid Service.

It is preferable that filling and discharge lines enter a storage tank through the top, thus avoiding piping penetration below the liquid level. However, for discharge this requires the use of submerged pumps, which may constitute a maintenance problem.

The pipework should have an adequate degree of fire protection. Whereas in general the emphasis in fire protection is avoidance of further releases which could feed the fire, in the case of HTHMs it shifts rather to the avoidance of serious toxic release. It follows that for HTHMs it may be necessary to provide pipework with protection which would otherwise be considered unnecessary. The CCPS *Guidelines* recommend a fire resistance of at least 30 minutes at 1100°F.

12.25.3 Limitation of release

There are a number of devices which may be used to shut off or reduce releases of HTHMs. These include emergency isolation valves, non-return valves, excess flow valves, restrictor orifices and pump trips.

Storage tanks and vessels for HTHMs should be provided with emergency isolation valves or control valves with separate shut-off facilities. Other situations which may merit an emergency isolation valve are a large vapour line on a pressurized storage vessel and a long pipeline.

Non-return valves, or check valves, can be used with the purpose of preventing back flow out of a tank in the event of rupture of a liquid line on it, but are notoriously unreliable and are not suitable as the sole device in a critical duty. The flow from a line may be reduced by the use of an excess flow valve or a restrictor orifice. Another measure is remote shut-off of pumps providing the pressure behind the release.

12.25.4 Secondary containment

There are a number of arrangements which can be used to contain a release and prevent its escape to atmosphere. Such secondary containments include double-walled equipment, enclosures and bunds. Pressure relief and vent systems are also sometimes so classified. Common to all these arrangements is the need to provide in addition some form of disposal system.

Double-walled equipment is used particularly for storage tanks. In particular, the principle is exploited in the design of large refrigerated ammonia storage tanks. This is described in Chapter 22.

The double-wall principle is also used to a limited extent for pipework. One system involves the enclosure of the pipe carrying the HTHM in a second, concentric pipe. The annular space between the two pipes is filled with inert gas either at a pressure higher than that in the inner pipe or at a lower pressure with monitoring to detect leakage of the toxic fluid. An alternative system involves the use of double seals on flanges and of bellows seals on valves, both purged with monitored inert gas.

The use of an enclosure building provides another form of secondary containment but allows accumulation of small leaks. It therefore requires the provision of toxic detectors and alarms and of ventilation and disposal systems. Where toxic liquids are handled, it is also necessary to provide for their retention, collection and disposal. The sump should be of small cross-sectional area but deep, in order to minimize vaporization. If there is a possibility of an internal overpressure sufficient to cause structural damage, consideration should be given to this.

Material from the pressure relief and depressurizing, or blowdown, devices should be collected in a closed relief or vent header system. This is usually a common header, though in some instances this aproach may need to be modified in order to avoid a problem arising from the interconnection of systems. The header typically passes to a knockout drum or catchpot and then to the disposal system, usually a flare or scrubber. Disposal systems are considered in Chapters 11 and 22.

12.26 Pressure Systems for Chlorine

As just described, the design of pressure systems for hazardous materials requires special consideration. This is illustrated by the requirements for chlorine, which is highly toxic and, if wet, very corrosive, and which is produced in large tonnages.

The handling of chlorine is treated in the publications of the Chlorine Institute, including the *Chlorine Manual* (1986 Pmphlt 1). The generally used code has been the Chemical Industries Association (CIA) *Code of Practice for Chemicals with Major Hazards: Chlorine* (BCISC, 1975/1) (the CIA *Chlorine Code*). Another CIA publication has been *Guidelines for Bulk Handling of Chlorine at Customer Installations* (1980/9). These are both now superseded by HS(G) 28 *Safety Advice for Bulk Chlorine Installations* published by the Health and Safety Executive (HSE, 1986) in co-operation with the CIA; it is this document which now appears in the CIA publication list.

An account of chlorine as a chemical is given in Chapter 11. The storage of chlorine is considered in Chapter 22 and its transport, including pipeline transport, is discussed in Chapter 23. The principles of chlorine handling are, in large part, applicable to the handling of many other chemicals which are toxic, irritant and non-flammable, and are made on a large scale.

12.26.1 Chlorine and chlorine plants

Typically, chlorine is produced in integrated systems in which part of the production is liquefied and transported away and part is consumed as gas in user plants within the works. Such a system is illustrated in Figure 18.13.

The quantity of chlorine handled in a producer plant, or cell room, is generally much greater than in a user plant. In a producer plant the important thing is to be able in an emergency to stop the production of the gas quickly and so avoid a large release to atmosphere. This requires that means be provided to detect this need, to communicate it and to shut down rapidly and safely. There should be facilities for emergency venting of the chlorine to gas absorption plant.

In a user plant, the size of potential chlorine release is usually less, but the operating conditions tend to be more arduous. In such a works chlorine is handled both as gas and as liquid (see Chapter 11). Particularly important in chlorine handling are the facts that wet chlorine attacks mild steel and that chlorine forms the unstable explosive compound nitrogen trichloride.

12.26.2 Materials of construction

The usual material of construction for dry chlorine is mild steel. If the chlorine is wet, however, it attacks mild steel severely. It is essential, therefore, for water to be excluded if mild steel is to be used for handling chlorine. Corrosion is particularly severe with liquid chlorine if there is a separate water phase. The corrosiveness of wet chlorine also depends on the temperature.

The *Chlorine Code* states that the normal dissolved water content of chlorine is 20–60 ppm and that this is satisfactory for the common materials of construction. In conformity with BS 3947, commercial chlorine should have a maximum water content of 100 ppm.

Mild steel equipment for use with dry chlorine itself needs to be dried out before commissioning. HS(G) 28 recommends purging with air or inert gas down to a dewpoint of −40°C.

Corrosion of mild steel by chlorine is also affected by temperature. The *Chlorine Code* states that attack is significant at about 200°C and rapid at about 230°C and that it is normal practice to impose a limit of 120°C. This is the normal limit advised by HS(G) 28, for example, for steel tubes in chlorine vaporizers.

Materials widely used for handling wet or dry chlorine gas include ebonite-lined mild steel or reinforced polyester resin. Glass and stoneware may also be used in appropriate applications. With certain exceptions, plastic materials are unsatisfactory for liquid chlorine. Another material which can be used for chlorine gas or liquid up to about 100°C is titanium, but only provided that the fluid is wet; titanium is attacked by dry chlorine. If this material is used for wet chlorine, therefore, there must be assurance that it will not be contacted with dry chlorine, even under fault conditions, otherwise some alternative material should be selected. There are available carbon steels which are suitable for the low temperatures found in the normal handling of chlorine.

12.26.3 Vessels, pipework and equipment

Pressure systems for handling chlorine should be of a particularly high standard, with appropriate attention paid to control of impurities, isolation arrangements, protective devices, pressure relief arrangements and protection against external damage.

Chlorine vessels are designed to normal pressure vessel codes but with the additional requirements stated in HS(G) 28. Storage tanks and vessels for chlorine are discussed in Chapter 22. For chlorine process vessels particular consideration should be given to possible impurities in the chlorine, which are a more severe problem in process than in storage.

Chlorine pipework may carry chlorine gas in which the presence of liquid is excluded, chlorine gas which may contain some liquid, or liquid chlorine. Pipelines are designed for dry gas, wet gas and dry liquid. HS(G) 28 gives guidance on pipework for liquid chlorine. The pipework should be designed, fabricated, inspected and tested to a recognized code. The design pressure should be in accordance with the code, but in any case not less than 12 barg which corresponds to a design temperature of 45°C.

Seamless steel tubing is preferred, but resistance seam welded tubing that has been welded and stress relieved automatically during manufacture is acceptable. The number of flanged joints should be kept to a minimum. All butt welds should be fully radiographed or ultrasonically examined.

Gaskets used for liquid chlorine are of compressed asbestos fibre and should be to BS 2815 Grade A. The use of an incorrect gasket for chlorine can be hazardous and gaskets should be tabbed if there is any possibility of misidentification.

Pipelines should be sited so as to be as safe as possible from impact, fire or other threats. They should be protected from external corrosion, but accessible for inspection and maintenance.

HS(G) 28 gives less guidance on lines for chlorine gas. The following guidance is given in the *Chlorine Code*, but it should be borne in mind that it is somewhat dated. Gas pipelines in which it is known with confidence that there will be no liquid chlorine are usually designed so that they are sufficiently strong to withstand the highest pressure which can be applied or so that relief is inherent in the design, and do not normally require relief devices. Pipelines for gas which may contain liquid are operated at rather higher pressures than the previous ones, but are not necessarily designed for the full pressure which could arise from the vaporization of trapped liquid chlorine. Usually some relief capacity is necessary to cater for any liquid present and this should be considered in relation to the gas absorption facilities. Often such pipelines contain vaporized chlorine. It may be appropriate to provide trace heating. This is necessary if the lowest winter temperature to which the pipeline is exposed could cause formation of the liquid. The gaskets used for dry gas are compressed asbestos fibre, whilst those for wet gas are a special rubber. Although the latter is suitable for low pressure dry gas, it is quite unsuitable for liquid or high pressure gas. It is essential, therefore, that means such as tabbing be provided to ensure that the correct gasket is used.

Liquid pipelines may be subject to overpressure caused by hammer blow resulting from rapid closure of valves and by thermal expansion of the liquid. Slow-acting valves and surge vessels may be used to protect against hammer blow from liquid chlorine. Good communication between operators is also helpful in reducing this problem.

Liquid chlorine has a high coefficient of thermal expansion and the possible effects of liquid thermal expansion should always be considered in liquid chlorine pipework systems. The trapping of liquid between two valves should be minimized by a combination of design and operations measures. Liquid trapping is particularly likely to occur if there is a proliferation of valves which can close automatically. This may be avoided by the judicious use of a proportion of valves which are manually operated.

Traditionally, for small lines the industry has placed some reliance on liquid thermal relief by the springing of flanged joints. Thus the *Chlorine Code* states that on smaller systems the release may be relatively small and harmless, and indicates that in order to provide such relief flanged joints should not be eliminated completely.

HS(G) 28 states that if the capacity of the system is such that a release could have serious consequences, automatic means of relief should be provided and that reliance on the springing of flanged joints for thermal relief is not acceptable. It refers to two means of relief: (1) a bursting disc discharging to a suitable disposal system and (2) an expansion tank designed to accommodate the gas phase.

The *Chlorine Code* states that an expansion bottle or other relief device is normally used if:

(1) the system is fully welded and includes no flanged joints;
(2) there is a substantial length of pipework involved such that the volume of chlorine released in easing a potential pressure would be significant;
(3) the two valves, the closure of which traps the liquid chlorine, are under the control of different operators and there is a reasonable chance of this occurring without either being aware of it;
(4) overpressure due to external causes can be foreseen.

However, it also states that an expansion bottle is effective only provided it contains non-condensable inert gas and does not fill up with liquid. HS(G) 28 states that, in view of the difficulty in determining whether the gas space contains chlorine or inert gas, the use of such a device is not recommended for new installations.

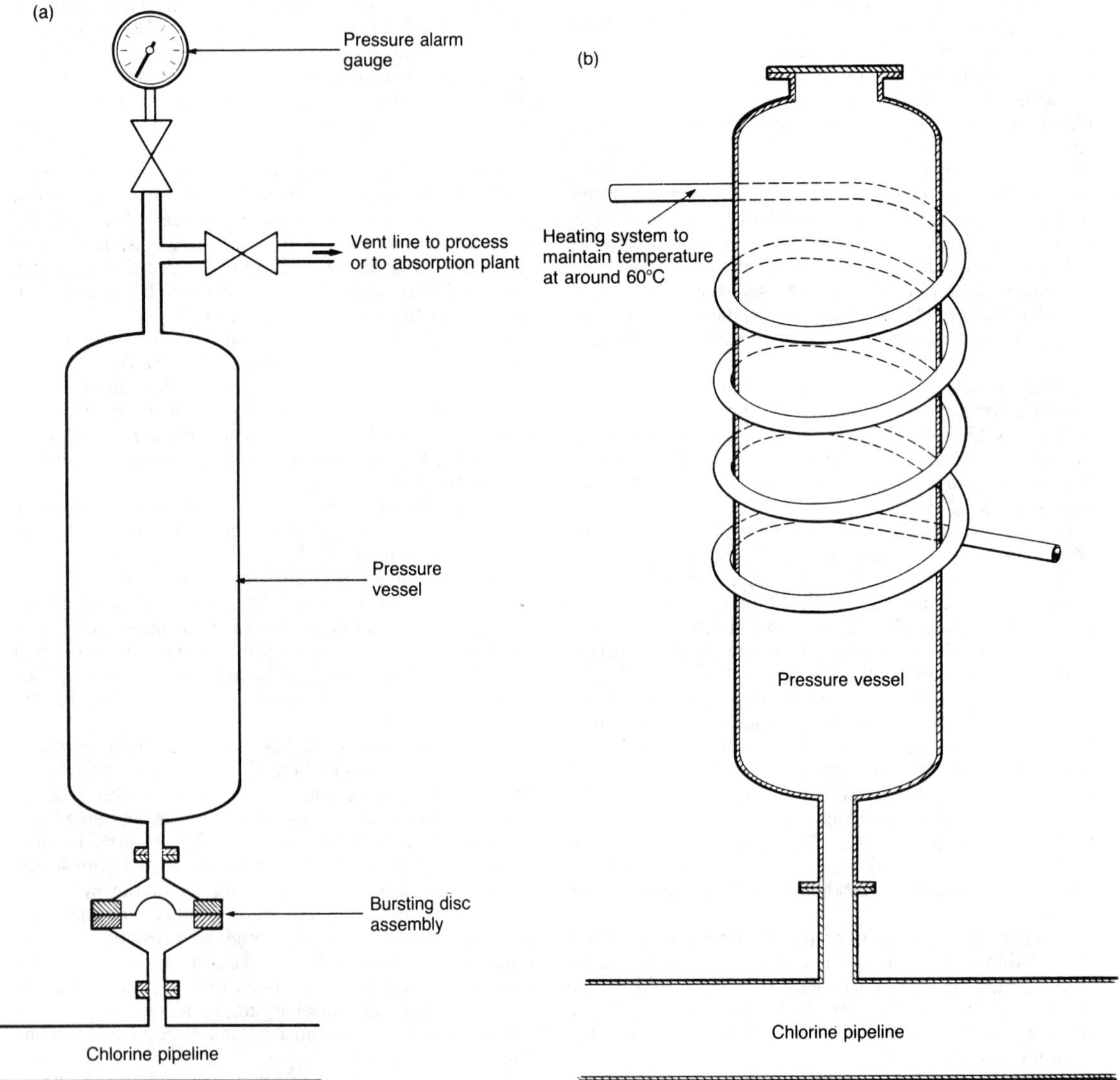

Figure 12.21 *Overpressure protection arrangements for chlorine pipelines (Health and Safety Executive, 1986 HS(G) 28): (a) bursting disc system; and (b) expansion vessel system (Courtesy of HM Stationery Office. Copyright. All rights reserved)*

Arrangements for pressure relief of chlorine pipelines using these two systems are given in HS(G) 28. In both cases the relief volume needs to be at least 20% of the line volume. Figure 12.21 shows the two systems. Figure 12.21(a) gives the bursting disc system. The disc relieves into a pressure vessel fitted with a pressure alarm and vents from there to the disposal system. Figure 12.21(b) gives the expansion vessel system. There is a heating system to maintain the temperature at around 60°C. There are a number of further design requirements, which are given in the guide. In each case the vessel used should be a pressure vessel and entered as such in the pressure vessel register.

The possibility of low temperature embrittlement should be considered carefully for pipelines carrying liquid chlorine. The line may be subjected to temperatures down to the normal boiling point of chlorine (−34°C) or even lower. Pipelines carrying liquid chlorine are liable to ice up on the outside and to corrode under the ice, and require to be protected by suitable wrapping or painting.

Valves for chlorine pipelines should be carefully chosen, particularly in respect of corrosion resistance, strength against possible overpressures, leak-tightness of the gland and ease of removal and maintenance. Some requirements for valves in liquid chlorine systems are given in HS(G) 28. Such valves should preferably be in forged steel, and in any event not in cast iron. If the valve is of a type in which liquid chlorine may become trapped, measures need to be taken to prevent the development of excessive pressure due to temperature rise. The types of valve referred to in HS(G) 28 as suitable for liquid chlorine or dry chlorine gas are (1) vertical globe valves, (2) conical plug valves and (3) ball valves. Detailed guidance is given on each type.

Isolation arrangements are particularly important in chlorine systems. A single valve should not be relied on to give complete isolation. More positive isolation is required using means such as the provision of spool pieces which can be removed and replaced by blanks and the use of slip plates. Where the latter are to be utilized, the layout and piping should provide for this. Large low pressure chlorine gas mains are particularly difficult to isolate effectively due to the size of the line and the dirtiness of the gas. Ease of removal of valves and of insertion of slip plates is especially important in such systems.

Every item of equipment which contains, and could release, a significant quantity of chlorine should have a means of isolation and a means of venting to a gas absorption unit. Stop valves at inlet and outlet are the normal means provided for effecting immediate isolation, but the arrangements should facilitate the insertion of slip plates.

Chlorine liquefiers are subject to the general requirements for chlorine vessels, but involve some special considerations. The operating temperatures are low and require suitable materials of construction. The chlorine gas, even after drying, is not completely clean and tends to cause blockage and corrosion, so that ease of cleaning is important. The refrigerant used should not react with the chlorine if there is a leak.

Likewise, there are some particular considerations in chlorine vaporizers. There is a potential for creating overpressure. The higher temperatures are conducive to increased corrosion. The impurities, particularly nitrogen trichloride, may concentrate and explode. There is a possibility of failure of the heating surfaces. Materials of construction for chlorine vaporizers should withstand both the temperatures and impurities. There should be effective control on pressure, temperature and level. Pressure relief should be provided and should take account of the possibility of heating surface failure. There should be specific arrangements to prevent concentration of nitrogen trichloride. The effects of heating surface failure should be thoroughly reviewed and means provided for its prompt detection. Chlorine vaporizers are utilized by users of chlorine which are not necessarily large chemical works. A robust design is therefore important. A further discussion of chlorine vaporizers is given in Chapter 22.

12.26.4 Compressors, pumps and padding

Compressors are used to re-compress chlorine vapour and need to be suitable for intermittent operation. Dry carbon ring and diaphragm types are used, the latter having double stainless steel diaphragms with inert fluid in between. The shaft of a carbon ring compressor needs to be sealed so as to prevent escape of chlorine and ingress of moist air. The shaft glands should be pressurized during operation with dry inert gas. After use the compressor should be purged with dry air to prevent leakage of residual chlorine.

The temperature rise in the compressor may need to be controlled. The cooling arrangements should take account of the possibility of water leaks. Direct water cooling should be avoided. It is preferable to use air cooling. The compressor should have a bypass allowing the chlorine to be recycled so that its temperature is high enough to avoid liquefaction in the delivery lines. There should, however, be an alarm which will activate if the temperature exceeds 90°C.

In general, transfer of liquid chlorine may be effected by the use of (1) chlorine vapour at the vapour pressure of the liquid, (2) dry compressed chlorine vapour, (3) dry compressed padding gas and (4) pumping. The choice of method for different applications such as transfer from storage to consuming units and unloading from tankers to storage is discussed in HS(G) 28. For the former, all four methods are listed, but with the first two preferred, whilst for the latter the third method is the preferred one, with the second as a possible alternative.

In many cases transfer of liquid chlorine from storage may be effected under the vapour pressure of the liquid. If this method is used factors to be taken into account are the effect on the vapour pressure of low external temperatures and the need to maintain in the vessel a positive pressure, which should be specified. Alternatively, liquid chlorine may be transferred by pressure of gas, such as dry air or nitrogen, or of chlorine vapour from a vaporizer or compressor. HS(G) 28 gives the following guidance.

Gas used for transfer of liquid should be a dedicated supply. Nitrogen at pressure may be obtained from a liquid nitrogen vaporizer. Compressed air can be provided by an air compressor. It is highly desirable that this be an oil-free machine; if oil lubrication is used, the machine should have an oil filter which is regularly maintained. Measures should be taken to ensure that the air is dried to a dewpoint of −40°C. The padding gas, nitrogen or air, should have its own pressure control and pressure relief arrangements. It should be held in a gas receiver with a relief valve set to operate at the safe working pressure of the chlorine plant or 150 psig, whichever is the lesser. Further detailed requirements are given in the guide.

Chlorine vapour for padding may be obtained by recompression of vapour from a storage vessel or by vaporization of liquid chlorine. The storage vessel from which transfer is made should have a design pressure and pressure relief arrangements which protect against pressures which the padding gas or vapour can develop. It should also have a relief to the gas absorption facilities.

It may be necessary to pump liquid chlorine if it has to be delivered at a higher pressure (say, 100 psig) or if the use of dry padding gas is for some reason inappropriate. Some of the problems in pumping chlorine are discussed in the *Chlorine Code*. Pump cavitation can be a problem in pumping chlorine, which tends to vaporize. It is essential to provide sufficient liquid head on the suction side to avoid liquid flashing. Severe cavitation can damage the pump or even cause the casing to fail. Another problem in pumping chlorine is seal leakage. This virtually rules out the use of rotating shaft seals. Liquid chlorine pumps are generally canned, diaphragm or submersible types. The code recommends that centrifugal pumps should be provided with a bypass to a pumping tank to prevent liquid boiling and that a positive displacement pump should have a pressure relief valve on the outlet bypassing to the suction.

HS(G) 28 describes the use of an arrangement in which liquid chlorine is transferred from a storage vessel using a canned pump. There should be sufficient NPSH and a remotely operated emergency isolation valve installed either inside the vessel or between it and the pump. The guide also discusses the use of submersible pumps.

12.26.5 Pressure relief and gas absorption

The philosophy of pressure relief for chlorine has traditionally been somewhat different from that for most other materials. It is stated in the *Chlorine Code* thus:

(1) Any relief system shall be regarded as a safeguard of last resort; its existence shall not be regarded as an alternative to providing protective systems and operating instructions such that process conditions cannot, in the absence of human or equipment failure, lead to the relief conditions arising.
(2) Relief to atmosphere shall be sanctioned only as a last resort when the consequences of not doing so would be worse.
(3) A relief device shall not be isolated unless approved alternative arrangements to protect the vessel have been made.

Likewise, the philosophy in HS(G) 28 for chlorine systems lays emphasis on prevention of overpressure with pressure relief as a last resort. These measures include (1) prevention of overfilling of storage tanks by use of weighing systems or of ullage pipes giving warning of entry of liquid into vent lines, combined in both cases with alarms and (2) prevention of overpressure by pressure control and relief on the padding gas system and storage vessel high pressure alarms.

As a last line of defence, a chlorine storage vessel should be provided with pressure relief. This pressure relief should be provided by a bursting disc rather than a relief valve, because chlorine tends to corrode the latter. The relief should normally pass to an expansion vessel. For a simple system consisting of a single storage tank and expansion tank, a single bursting disc system may suffice, installed directly on the storage vessel without isolation valves. A system commonly found on much existing plant is two bursting discs in series with a pressure alarm on the intervening space and with isolation valves upstream and downstream of this assembly, with both valves locked open. The preferred arrangement is two such assemblies of bursting discs and isolation valves in parallel. In this case one arrangement is that the valves are interlocked and another that the upstream valves and at least one downstream valve are locked open.

Where there are several storage vessels, they may each be provided with an expansion vessel or there may be a common expansion vessel. An expansion vessel should have a capacity at least 10% of the largest storage vessel. It should be capable of being vented manually to an absorption system. It should be fitted with a high pressure alarm to indicate build-up of chlorine. Any necessary measures should be taken to ensure that the expansion vessel is not itself overpressurized.

Arrangements should be made to ensure that facilities exist for absorption of chlorine gas vented by the pressure relief system. In some installations there are chlorine consuming plants to which the relief gas can be routed, and a separate gas absorption facility may be deemed unnecessary. In this case it is essential to ensure that the absorption capability is available whenever it may be required. Usually it is necessary to provide a separate gas absorption facility. This should be a high reliability plant with adequate redundancy and comprehensive instrumentation. HS(G) 28 gives detailed recommendations.

The pressure relief arrangements should cater for fire relief as well as operational relief. In some cases it may be necessary to provide protection against the risk of explosion of chlorine with other gases, particularly hydrogen. The design of such explosion relief is a specialist matter. The *Chlorine Code* refers to the possible need for explosion relief to atmosphere in unavoidable cases, and states that this a matter for expert advice. HS(G) 28 does not consider this eventuality.

12.27 Failure in Pressure Systems

Maintenance of the integrity of the pressure system and avoidance of loss of containment is the essence of the loss prevention problem. It is necessary, therefore, to consider the failures which occur in pressure systems. Of particular importance are catastrophic failures in service.

Catastrophic failure of a properly designed, constructed and operated pressure vessel is comparatively rare. The most common cause of such failure is inadequate operational procedures. Most failures in pressure systems occur, however, not in pressure vessels but in the rest of the system, which includes pipework, valves and fittings and equipment such as heat exchangers and pumps.

The problem of failure in pressure systems is treated here by first describing some principal causes of failure, particularly sudden failures, and then giving some historical data on failure. In the present context it is service failures which are of prime importance. Service failures of pressure systems are generally caused by exposure to operating conditions more severe than those for which the system was designed.

It is usual to classify service failures as (1) mechanical failure, through stress and fatigue and (2) corrosion failure, although many failures have an element of both of these. Accounts of causes of failure in pressure systems are given in *Defects and Failures in Pressure Vessels and Piping* (Thielsch, 1965 and *Inspection of Chemical Plant* (Pilborough, 1971, 1989).

12.27.1 Materials identification errors

Mistakes in identifying materials of construction which result in the construction of plant using the wrong materials are a significant problem. Such errors are particularly likely to occur where the materials have a somewhat similar appearance, e.g. low allow steel and mild steel, or stainless steel and aluminium-painted mild steel. It is necessary, therefore, to exercise careful control of materials. Methods of reducing errors involve marking, segregation and instrument spot checks. For critical applications is may be necessary to carry out a full 100% *in situ* check on materials using an instrument such as a Metascope.

Misidentification of materials and some accidents caused by it have been described by W.D. Clark and Sutton (1974). In 1984 a leak due to erosion failure on a hydrocarbon liquid line at Fort McMurray, Alberta, resulted in a major fire (Case History A109). The investigation found that inadvertently an 18 in. long section of carbon steel had been inserted into an alloy steel line.

12.27.2 Mechanical failure

Some common causes of mechanical failure in process plant are:

(1) excessive stress;
(2) external loading;
(3) overpressure;
(4) overheating;
(5) mechanical fatigue and shock;
(6) thermal fatigue and shock;
(7) brittle fracture;
(8) creep;
(9) hydrogen attack.

Excessive stress in equipment may be caused by factors such as stress raisers or by malpractices such as uneven tightening of flanges.

Equipment may be externally loaded by loads carried on additions such as lugs, supports, brackets and hangers. Common sources of external loads are platforms, stairs and ladders. Movement of foundations may cause external loads, as may restraint on thermal expansion of pipework passing through concrete walls. Wind and occasionally ice may contribute a significant external load.

Overpressure may occur due to failure of a pressure relief valve to fulfil its function. Such failures may be due to poor design or incorrect operation, or to relief valve failure. Freezing of liquid, particularly in pipework, is another cause of overpressure.

Gross overheating may cause rapid failure due to operation well outside the temperature range of the material. In particular, direct flame impingement can give rise to sudden failure. Less severe overheating can cause creep as discussed below.

The other causes of mechanical failure are now considered.

12.27.3 Mechanical fatigue and shock

Conditions which give rise to mechanical fatigue include (1) pressure variations, (2) flow variations, (3) expansion effects and (4) imposed vibrations. Stress cycling can be caused by normal changes of pressure in the process. Normal flow fluctuations or effects such as hammer blow or cavitation also give rise to stress cycles. So do differential expansions and contractions of the process plant. Other stress cycling is imposed by vibrations from such equipment as compressors, pumps or valves.

Although there is often some degree of vibration in pressure vessels and pipework, it is usually at a sufficiently low level as not to cause fatigue failure. In some cases, however, fatigue can progress very rapidly. Thielsch quotes the case of the fatigue failure of a carbon monoxide generator which was subjected in a period of only 17 days to several hundred million stress cycles.

Mechanical shock differs from mechanical fatigue in that the load is greater and causes failure within only one or a few cycles. Shock is often caused by hammer blow or liquid slugs. In 1975 at Antwerp, Belgium, a leak of ethylene at high pressure due to fatigue failure of a vent connection on the suction of a compressor led to an explosion and fire at a low density polyethylene plant (Case History A74). Six persons were killed and extensive damage was done.

12.27.4 Thermal fatigue and shock

Similarly, the distinction between thermal fatigue and thermal shock is that in the latter the applied temperature difference and the rate of change of temperature are greater and cause failure in one, or a few, cycles. A given temperature cycle may give rise to thermal fatigue or shock, depending on the material. It may constitute thermal fatigue for a ductile material, but shock for a brittle one. Thermal cycling is caused by (1) intermittent operation and (2) particular equipment conditions.

There are many types of intermittent operation ranging from batch processes to throughput changes and including forced shutdowns. These all create thermal stress cycles. This source of thermal fatigue is much reduced if the plant is operated continuously, even though the operating temperatures may be relatively high.

Thermal fatigue is also caused by the conditions in particular types of equipment. Steam desuperheaters using direct water spray cooling may suffer thermal fatigue failures. Thermal fatigue occurs in reducing valves due to repeated flow changes. Slugs of condensate can cause thermal fatigue in steam lines.

12.27.5 Brittle fracture

The materials used in plant handling low temperature fluids should have a ductile–brittle transition temperature below not only the normal operating temperature but also the minimum temperature which may be expected to occur under abnormal conditions. If this requirement is not observed and cold fluid contacts metal below the transition temperature, brittle fracture may occur. Brittle fracture is catastrophic, since the fracture can propagate at a velocity close to that of sound.

In 1973 at Potchefstroom, South Africa, an ammonia tank suffered brittle fracture, which resulted in the release of some 30 ton of ammonia (Case History A65). The fracture occurred in a dished end which was fabricated in carbon steel and which had not been stress relieved after manufacture. The minimum transition temperatures obtained by subsequent testing were 20°C for the fragment and 115°C for the remaining part of the dished end. Thus the metal was below its transition temperature under normal operating conditions.

The possibility of accidental entry of cold fluids into parts of the plant where they are not normally present and the resultant hazard of brittle fracture present a difficult problem. In general, the best practice is to use suitable materials in any part of the plant where it is realistic to expect that low temperature fluids could enter under abnormal conditions. But alternative means of protection such as trip systems may be used in some cases. An example of the application of hazard analysis to this problem was discussed in Chapter 9.

Where a liquefied gas is held under pressure, sudden depressurization to atmospheric pressure will cause the temperature of the gas to fall to its normal boiling point. This could be below the transition temperature of the metal unless this has been taken into account in the design.

It is not only the vessels and pipework which may suffer low temperature embrittlement, but also features such as vessel supports and flange bolts. It is essential, therefore, for these latter also to be made of the appropriate material.

12.27.6 Creep

The problem of creep at high temperatures has already been discussed in Section 12.3. Since design for high temperatures is based on creep strength, failure due to creep is usually unlikely under normal operating conditions and is generally caused by abnormal conditions such as maloperation or fire. It is appropriate to emphasize here the very rapid increase in creep rate with temperature. For mild steel in the creep range a temperature increase of about 10–15°C is sufficient to double the creep rate. A stress which gives an 11-year life at 500°C can result in rupture in 1 hour at 700°C.

Creep involves plastic deformation and, eventually, rupture. However, some metals have a very low ductility in the normal creep range, although they are more ductile at low and at very high temperatures. Low ductility fractures with an elongation of less than 1% are frequently associated with the formation of cavities at the grain boundaries. The development of cavities can occur at low stress over long time periods. The application of higher temperatures and higher strain rates, as in a fire, can accelerate cavitation and cause low ductility failure.

Creep cavitation of stainless steel occurred on the plant at Flixborough, as described in Appendix 2. The 8 in. pipe, which was made of 316L stainless steel, had a 50 in. rupture which was initiated by creep cavitation. It was estimated that for failure in 20 minutes a temperature of 900–950°C would have been needed.

If there has been overheating, there may be creep effects such as deformation and distortion which can be detected by visual examination and by measurement. Often there are also oxidation and scaling. But, since creep can occur with very little deformation, it may not be readily detectable. Expert advice should be sought if this is a possibility.

Fire on a process plant may result in local overheating and hence creep. Equipment should be protected against overheating by fixed protection such as fire insulation and by fire fighting. If equipment has been exposed to fire so that creep may have occurred, expert advice should be sought before it is put back into service.

The process should be operated at the temperatures specified. If an alteration is proposed, its effect on creep should be checked. This includes such alterations as change in lagging thickness, since this affects equipment temperature.

Further information on creep is given in TDN 53/3 *Creep of Metals at Elevated Temperatures* (HSE, 1977).

12.27.7 Hydrogen attack

Mild steel is subject to the following types of hydrogen attack: (1) hydrogen blistering and (2) hydrogen embrittlement and damage.

Hydrogen blistering can occur when atomic hydrogen diffuses into steel. Normally the atomic hydrogen diffuses right through, but if it encounters voids it forms molecular hydrogen, which can generate pressures of several hundred bar. Internal splitting of the steel occurs and gives surface blisters. Steels which are dirty and contain numerous voids, laminations and inclusions are especially prone to hydrogen blistering. Hydrogen blistering has been a quite common problem in catalytic reformers and has occurred at temperatures as low as low as 315°C. It has also been reported in butane storage tanks with butane run down from a gas separation treatment plant at 21°.C. The butane was sometimes contaminated with small amounts of water with traces of hydrogen sulphide. When corrosion occurs, the latter favours the evolution of atomic rather than molecular hydrogen.

At high temperatures and high hydrogen partial pressures steels can suffer hydrogen embrittlement and damage. The steel suffers decarburization by the hydrogen, which causes a loss of ductility, and development of microfissures at the grain boundaries, which leads to loss of strength. The decarburization reaction is reversible, but the formation of cracks is not. These two effects are sometimes distinguished as hydrogen embrittlement and hydrogen damage, but often one of these terms is used to describe both phenomena. Both effects are referred to here as 'hydrogen attack'.

Hydrogen attack has been a relatively common problem in equipment handling hydrogen at high temperatures such as refinery boilers and catalytic reformers. Failures usually occur in the temperature range 300–550°C. The resistance of steel to hydrogen attack can be greatly increased by the addition of suitable alloying elements, and CrMo steels are widely used for hydrogen service. Operating limits for steels in hydrogen service have been given by a number of workers, notably G.A. Nelson (1966). The Nelson curves are published with periodic amendments in *Steels for Hydrogen Service at Elevated Temperatures and Pressure in Petroleum Refineries and Petrochemical Plants* (API, 1991 Publ. 941). Typical Nelson limit curves are illustrated in Figure 12.22. These are used to select a suitable material. The basis of design for equipment in which hydrogen attack may occur is therefore selection of a suitable material and use of a design stress less than the lower critical stress.

In 1984 at Romeoville, Illinois, an absorption column came apart at the weld and rocketed up, initiating a vapour cloud explosion and a series of fires and boiling liquid expanding vapour explosions (BLEVEs), in which 17 people were killed (Case History A111). The vessel had a history of hydrogen attack problems.

12.27.8 Corrosion failure

A large proportion of failures in process plant are due to corrosion Accounts of corrosion include *Deterioration of Materials* (Greathouse and Wellel, 1954), *Seawater Corrosion Handbook* (Schumacher, 1979), *Basic Corrosion and Oxidation* (J.M. West, 1980), *Corrosion Inhibitors* (Rozenfeld, 1981), *Corrosion Processes* (R.N. Parkins, 1982, *Intergranular Corrosion of Steels and Alloys* (Cihal, 1984), *Corrosion Control in the Chemical Process Industries* (Dillon, 1986), *Corrosion Engineering* (Fontana, 1986), *DECHEMA Corrosion Handbook*

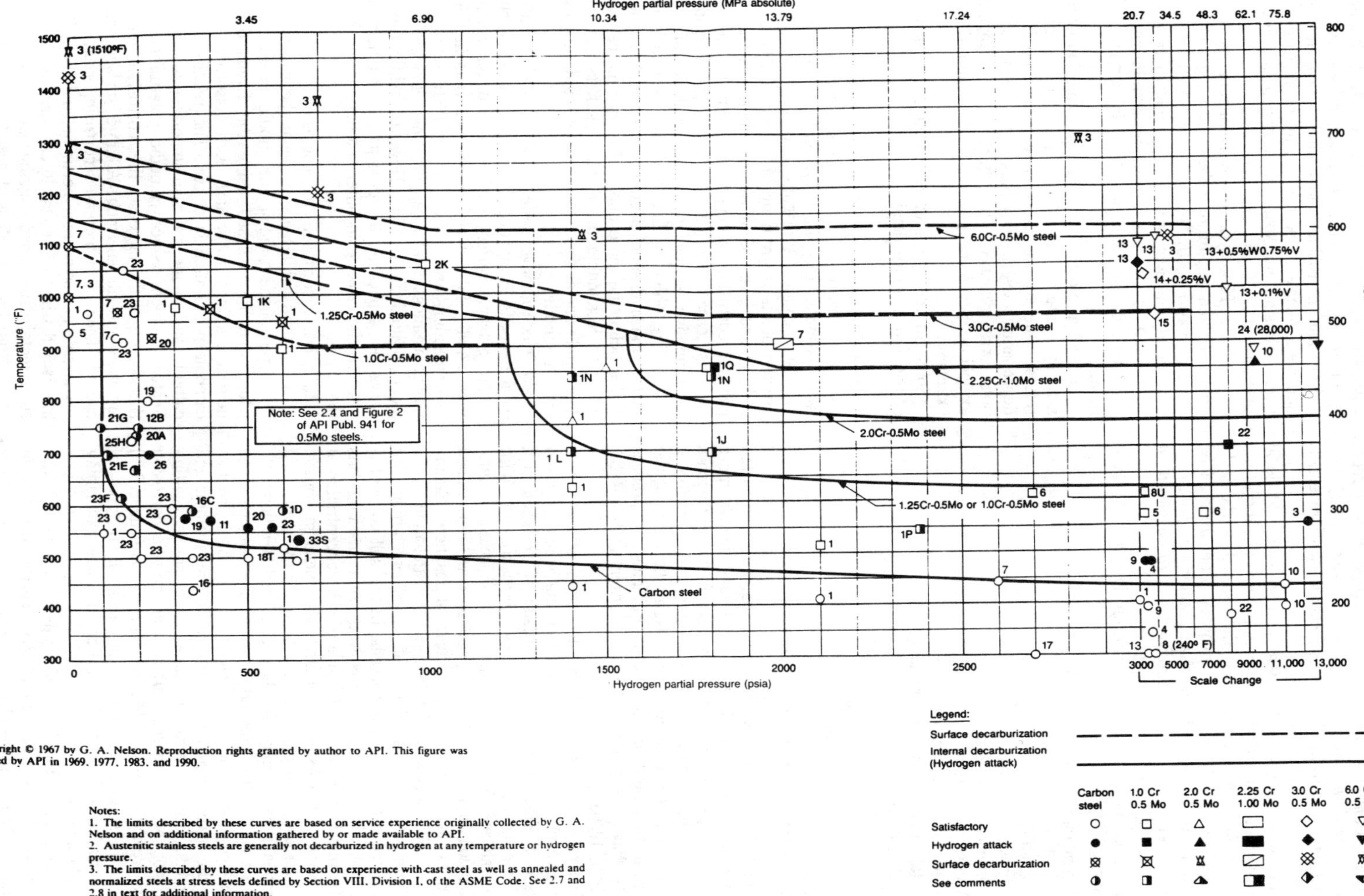

Figure 12.22 *Recommended operating limits of steel for hydrogen service: the Nelson curves (American Petroleum Institute, 1991 Publ. 941) (Courtesy of the American Petroleum Institute)*

(DECHEMA, 1987–), the *Chemical Engineering Guide to Corrosion Control in the Process Industries* (R.W. Green, 1987) *High Temperature Corrosion* (Kofstad, 1988) and *Corrosion and Corrosion Protection Handbook* (Schweitzer, 1989).

Corrosion occurs as (1) general corrosion, (2) local corrosion and (3) erosion. In general corrosion there is a fairly uniform deterioration of the overall surface. In contrast, localized corrosion involves little generalized corrosion but severe local attack, often at points of surface defects or stress. Erosion is also localized at points of high velocity or impact.

Some common types of corrosion in process plant, following Thielsch (1965), are

(1) general corrosion;
(2) scaling;
(3) exfoliation;
(4) galvanic corrosion;
(5) crevice corrosion;
(6) corrosion pitting;
(7) stress-related corrosion:
  (a) stress corrosion cracking,
  (b) corrosion fatigue,
  (c) stress-enhanced corrosion;
(8) intergranular corrosion;
(9) knife-line corrosion;
(10) erosion;
(11) external corrosion.

It should be emphasized, however, that many corrosion failures are not easily classified and may involve more than one type of corrosion. Corrosion frequently occurs at welds rather than in the parent metal and corrosion at welds ranks as a topic in its own right. Accounts of the practical treatment of corrosion problems in process plants have been given by M. Turner (1987, 1988, 1989a,b, 1990a,b).

General corrosion often results from attack by a corrosive chemical or impurity over the whole exposed surface. Scaling corrosion is the result of oxidation in a gaseous atmosphere caused by exposure to high temperatures. It occurs particularly in steam boilers. Exfoliation is another type of scaling corrosion caused again by oxidation, but in a steam atmosphere. Steam-heated boiler feedwater heaters exhibit exfoliation.

There are several kinds of corrosion which are associated with concentration cells. In the salt concentration cell, which involves a dissolved salt containing ions of the metal, there is a difference of salt concentration at two parts of the exposed surface. The metal in contact with the lower salt concentration becomes the anode and corrodes preferentially. In the differential aeration cell, two parts of the exposed surface are subject to a difference of oxygen concentration. The metal in contact with the lower oxygen concentration becomes anodic and corrodes. Such corrosion takes two forms: pitting corrosion of an exposed surface, and crevice corrosion in a crevice. The detailed mechanisms are described further in relation to crevice corrosion. Both types of concentration cell corrosion occur in water containing chloride ions.

Galvanic corrosion is considered in Section 12.27.9; crevice corrosion in Section 12.27.10; corrosion pitting in Section 12.27.11; stress-related corrosion, including stress corrosion cracking, corrosion fatigue and stress-enhanced corrosion, in Section 12.27.12; particular forms of stress corrosion in Sections 12.27.13 and 12.27.14; and intergranular corrosion, or weld decay, in Section 12.27.18.

Knife-line corrosion takes place between the parent and weld metals. It occurs mainly in austenitic stainless steels. Erosion and external corrosion are discussed in Sections 12.27.16 and 12.27.17.

In 1970 in Brooklyn, New York, a road tanker carrying liquid oxygen ruptured violently causing a number of fires, with two deaths (Case History A48). The investigation found, amongst other things, that there had been a appreciable loss of metal from the aluminium tank shell, even though the tank had been in service for only a month.

In 1988 at Norco, Louisiana, an explosion occurred on a catalytic cracker (Case History A121). The cause was a release due to internal corrosion of an elbow in the depropanizer overhead piping. Some 9 te of hydrocarbons was released and there was a vapour cloud explosion. Seven people were killed and damage was extensive.

Some of the types of corrosion mentioned are now considered in more detail.

12.27.9 Galvanic corrosion

Galvanic corrosion is caused by current flowing between two dissimilar metals which form a galvanic cell, or bimetallic couple. A discussion of galvanic corrosion in process plants is given by M. Turner (1990c).

A typical pair of dissimilar metals which exhibits galvanic corrosion is copper and iron, typically in steel. In this case it is the steel which corrodes. The electromotive series gives the ranking of metals in descending order of their electrode potential relative to that of the hydrogen electrode, the metals higher in the series being the more noble. If a galvanic cell is set up, it is the less noble metal which suffers the corrosion. The noble metal forms the cathode and the less noble one the anode.

The electromotive series is an imperfect indicator of what happens in practice. For this a more relevant guide is the galvanic series which ranks metals according to their nobility in seawater. Some materials in this series, in descending order of nobility, are 316 stainless steel > ferritic stainless steel > copper > low allow steel > carbon steel > aluminium > zinc. However, while the galvanic series is applicable to an aqueous medium of wide interest, the ranking in other solutions is not necessarily the same.

In galvanic corrosion, the corrosion rate is a function of the potential difference between the two surfaces and the slope of the cathodic and anodic polarization curves, which may be analysed using the Evans diagram. In neutral solutions, particularly with dissolved oxygen, the cathodic polarization is crucial. The relative sizes of the cathode and anode are also important in this situation. Features which reduce cathodic polarizations, and can therefore greatly increase the corrosion rate, include agitation and air sparging. A large surface area of the cathode favours an increased corrosion rate.

Galvanic corrosion may occur in equipment incorporating two dissimilar metals. It can also occur when two such metals are joined together at a weld. It may also result from the utilization of a second metal as a plated surface. But such corrosion can also occur with minor dissimilarities in composition such as different metallurgical phases.

Measures which can be taken against galvanic corrosion include the selection of compatible materials and the insertion in the circuit of an ohmic resistance, either in the metal path or in the ionic path. Examples of the insertion of a resistance for these two paths are the use of an insulated gasket in pipework and the coating of one of the metal surfaces, respectively.

The purpose of such coating is to protect the less noble metal, but it does not follow that it is this metal which should be coated. The coating is liable to contain holes, or 'holidays'. A couple of coats of paint may still leave gaps of perhaps 1% of the area. If the coating is put on the less noble metal and gaps exist amounting to 1% of the area, the corrosion at these gaps will simply be enhanced by a factor of about 100.

12.27.10 Crevice corrosion

Crevice corrosion of steel occurs in crevices sufficiently narrow for an oxygen deficiency to occur which leads in due course to a build-up of acidity. Its occurrence in process plant is described by M. Turner (1989a).

It is exemplified by the corrosion of carbon steel in salt water. In general corrosion of an exposed surface, part of the surface acts an anode and part as a cathode. The corrosion can be represented as:

$Fe \rightarrow Fe^{2+} + 2e^-$	Anodic reaction
$O_2 + 2H_2O + 4e^- \rightarrow 4OH^-$	Cathodic reaction

The location of the anodes and cathodes shifts with time.

Where the metal surface is within a crevice the situation is quite different. The cathodic reaction consumes oxygen and the concentration falls, since the rate of diffusion into the crevice is slow. A differential aeration cell is set up. The metal surface in the crevice becomes anodic relative to the exposed surface. The anodic reaction creates ferrous ions, which combine with chloride ions to give ferrous chloride; this oxidizes to ferric chloride, which combines with hydroxyl ions from the cathodic reaction to form ferric hydroxide. This insoluble hydroxide in turn forms ferric oxide, or rust, and hydrochloric acid. The chloride ions are attracted to the anode. The water in the crevice can become so acidic that corrosion then occurs simply by acid attack.

From this point on the rate of corrosion is determined by the acidity, which depends on the particular solution. In the example just given the solution contained chloride. There is a rank order of aggressiveness, headed by chlorides, then sulphates, phosphates, etc. With crevice corrosion, the corrosion rate can be orders of magnitude greater than with general corrosion.

Crevice corrosion can occur where the two surfaces forming the crevice are of the same metal, where they are of different metals or where one is non-metallic. In the second case there may be bimetallic corrosion as well.

Many of the situations where crevice corrosion occurs involve single-sided fillet welds. It occurs, for example, on lap joints with such a weld. On process plants use is often made of a support ring welded around the inside of a vessel. The use of a full penetration weld instead of a single-sided fillet weld can minimize such corrosion. Another support detail in plants is a pad seal welded to the inside of a thin walled vessel to support a heavy internal member. In this case crevice corrosion may be reduced by making the seal weld a continuous one. In some cases a 'sentinel hole' is drilled from the outside through the vessel wall to leak test the weld; it also serves to provide warning of any leak during service. Crevice corrosion occurs at the tube–tube plate joints of heat exchangers. It is fairly common in threaded joints.

12.27.11 Corrosion pitting

Corrosion pitting also involves concentration cell corrosion by a differential aeration cell, but in this case without crevice effects. The anode and cathode move over the surface of the exposed metal, with the oxygen-lean area acting as the anode and suffering preferential corrosion. Pitting is favoured by no-flow conditions and by surface defects. It occurs in boiler feedwater heaters during shut-down if the oxygen content of the feedwater is high. This type of corrosion may be prevented by chemical treatment.

The resistance of steels to pitting and crevice corrosion can be characterized by the pitting index (Henderson, King and Stone, 1990)

$$PRE_N = \%Cr + 3.3(\%Mo) + 16(\%N) \quad [12.27.1]$$

Broadly, alloy steels with an index value below 32 are susceptible to pitting and crevice corrosion in salt water, whilst those with a value of 36 are resistant. A minimum value of 40 is quoted for resistance to hot seawater.

12.27.12 Stress-related corrosion

Often failure results from a combination of corrosion and stress. Stress corrosion cracking (SCC) is non-ductile failure caused by the combination of corrosion and static tensile stresses. Corrosion fatigue is caused by corrosion and stress cycling. An account of stress corrosion is given in *The Stress Corrosion of Metals* (Logan, 1966).

Many types of steel are liable to SCC, including mild steel and austenitic stainless steel. The stresses necessary to cause SCC may be internal or due to externally applied loads. Generally, normal operating pressures are not alone sufficient to cause such cracking and appreciable internal stress must be present. SCC can occur in the parent metal of vessels and pipes, in welds and on nozzle connections and support attachments. In welds the cracks may be transverse or longitudinal.

Chlorides are a common cause of SCC. They are a very widespread impurity and are rather difficult to eliminate. SCC due to chlorides is discussed in Section 12.27.13. There is also a form of SCC due to nitrates, which is considered in Section 12.27.14.

SCC caused by an alkaline solution is known as 'caustic embrittlement', which has been a frequent cause of failure in boilers. It is normal to treat boiler feedwater to reduce the risk of caustic embrittlement. SCC, particularly by chlorides, can occur externally as well as internally. A common cause is the leaching of chlorides from lagging. Some failures caused by SCC have been described by Ashbaugh (1970) and M. Turner (1989b).

Corrosion fatigue involves a mutual interaction between corrosion and fatigue. Corrosion occurs in the fatigue cracks. Alternation of stress prevents the build-up of films which protect against corrosion. Thus corrosion reduces the fatigue limit and fatigue speeds up the corrosion. In consequence, corrosion tends to occur with a wider range of chemicals and concentrations than is the case in the absence of fatigue.

Stress-enhanced corrosion describes corrosion at features where there are high residual stresses, such as welds, nozzles and attachments. Measures to counter stress-related corrosion depend on the particular case, but include selection of suitable materials, operation in a suitable temperature range, elimination of corrodants, reduction of residual stresses, reduction of vibrations, and close control of operations.

12.27.13 Chloride stress corrosion cracking

Chloride SCC occurs in austenitic stainless steels. A discussion of this form of SCC is given by M. Turner (1989b). When a dislocation propagates and a slip plane forms, there is a rupture of the protective oxide layer. Although the passive film usually reforms on the surface thus exposed, in an aqueous solution of chloride ions the repair is imperfect and the film contains chloride ions, so that the renewed surface has a different composition. The repaired surface constitutes an anode and, since it is small relative to the rest of the surface which becomes the cathode, rapid corrosion occurs. This in turn creates a crack, which propagates along the slip plane. The corrosion is further promoted by the build-up of acidity at the anodic crack and now becomes well established.

There are four main factors which influence SCC: (1) the temperature, (2) the chloride ion concentration, (3) the stress and (4) the grade of steel. SCC rarely occurs at temperatures below 60°C, but above 200°C it can progress rapidly, the relevant temperature being that of the metal rather than of the liquid. The rate at which SCC occurs is a function of the chloride ion concentration, the relevant concentration being that in the liquid film at the metal surface. The stress causing SCC may sometimes be the tensile stress in the metal surface at the operating pressure, but a more common cause is residual stresses. Stainless steels such as the commonly used 304, 321 and 347 types are susceptible, and the 316 and 317 types are only marginally less so.

Where SCC is detected, specialist advice is required. If the cracks are sufficiently deep, immediate shut-down is necessary. If they are more shallow, the advice may be that it is possible to continue in operation with appropriate monitoring and perhaps that the cracks may be dealt with in due course by grinding out.

Prevention of SCC by elimination of stress is generally difficult. Stress relief may be practised, although the temperatures required are high and can cause distortions. On the other hand, operation with a metal temperature below 60°C eliminates the problem. Control of chloride ion concentration is more problematic, particularly if chlorides can concentrate by evaporation. Selection of an appropriate stainless steel is an effective measure. The steels which show resistance to SCC are those with higher nickel contents: the 904L grade is resistant and the 825 grade highly so.

Other measures of protection include exclusion of oxygen, use of inhibitors and cathodic protection. A critique is given by M. Turner (1989b), who rehearses some drawbacks of these methods. The quantity of oxygen required to initiate SCC is very small. Inhibitors have not proved universally effective. Cathodic protection requires very precise control. Failures due to SCC are mostly not catastrophic, though they can be. They tend to exhibit leak-before-break behaviour.

12.27.14 Nitrate stress corrosion cracking

A particular type of SCC is the nitrate stress corrosion cracking of mild steel. It was nitrate stress corrosion which caused the crack in No. 5 Reactor at Flixborough and led to the removal of the reactor and the installation of the 20 in pipe. The cracking occurred because cooling water treated with nitrite had been played on the reactor to dilute small leakages of cyclohexane.

The conditions for nitrate cracking of mild steel to occur are a high concentration of nitrate, a low pH, a temperature above 80°C and high stress. This combination of features is not often present and nitrate cracking is not common outside plants handling nitric acid and nitrates. Most water contains some nitrates. In particular, water treatment of industrial cooling water results in an appreciable nitrate concentration. But such cooling water does not normally cause nitrate cracking. It is concentration of nitrates which presents the main hazard of nitrate cracking. Concentration may occur on a nitric acid plant or a plant producing nitrates, inside a crack which has been created by some other mechanism such as fatigue, or due to evaporation.

Contamination with nitrates should be avoided as far as practicable. In particular, it is not good practice to let water penetrate thermal insulation. This creates a risk of various kinds of corrosion apart from nitrate cracking. If heavy contamination may occur, equipment should be stress relieved. This is not a full answer for all fittings, since some items such as bolts have a high working stress. These need to be checked by regular inspection.

Where it is not possible to rely on either avoidance of contamination or stress relief, some protection may be obtained by the use of certain paint coatings. On plants handling nitric acid and nitrates, nitrate cracking is a well known problem and special procedures have been evolved.

Further information on nitrate cracking is given in TDN 53/2 *Nitrate Stress Corrosion of Mild Steel* (HSE, 1976).

12.27.15 Zinc embrittlement

At high stresses and temperatures, traces of other metals such as copper or zinc can cause rapid and severe embrittlement of some types of steel. The effects of zinc embrittlement of austenitic stainless steel were illustrated at Flixborough. Many of the stainless pipes found on the site had suffered zinc embrittlement.

For zinc embrittlement of stainless steel to occur it is necessary for the material to be under high stress and at a high temperature, but given these conditions the quantity of molten zinc required to cause embrittlement is very small and failure can occur in seconds. The features which determine zinc embrittlement of stainless steel are (1) temperature, (2) applied stress, (3) wetting, (4) type of steel, and (5) time.

Austenitic stainless steel is not used in process plant above 750° except in certain specialized applications. Zinc embrittlement is not likely to occur below this temperature. Zinc has a melting point of 419°C and molten zinc penetrates stainless steel above 450°C, but embrittlement is improbable below 750°C. Investigations by Cottrell and Swann (1976), after Flixborough, showed that the most favourable metal temperature for rapid attack is 800–900°C. Thus zinc embrittlement is unlikely to occur under normal operating conditions, but requires temperatures such as usually occur only in a fire.

High stress is also a condition for zinc embrittlement to occur. In Cottrell and Swann's work zinc embrittlement did not occur at low stress, even though the temperature was 1050°C and the specimen was coated on all sides by a pool of zinc.

Wetting of the steel by the molten zinc is the condition most favourable to zinc embrittlement. Surface layers such as metal oxides can prevent penetration, but if the layer is broken in some way, such as by abrasion or a reducing atmosphere, wetting can occur. There is some evidence that embrittlement can be caused by contamination from zinc vapour, but this is unlikely to occur, unless the molten zinc is at a distance of no more than an inch or so.

The type of stainless steel affects its susceptibility to zinc embrittlement. The latter is believed to be the result of interaction between zinc and nickel in the steel. It is the austenitic chromium nickel steels which are particularly affected.

Given conditions favourable to zinc embrittlement failure can be rapid. Cottrell and Swann obtained failures in a matter of seconds.

There are a number of possible sources of zinc on process plants. Zinc is used in galvanized, sprayed and painted coatings. Typical zinc-coated items are galvanized fittings, walkways, wire and finned tubes. Paints which contain zinc compounds but not metallic zinc are not a serious risk in this respect.

The principal hazard posed by zinc embrittlement is rapid and catastrophic failure as a result of a fierce local fire on the plant. Zinc embrittlement also creates the problem that, after a general plant fire, stainless steel equipment which has a nearby source of zinc is suspect. The problem of zinc embrittlement, however, should be kept in perspective. It is significant mainly as a secondary effect which can increase the severity of a local fire.

On stainless steel plant where zinc embrittlement could have serious consequences in the event of fire, zinc should be eliminated as far as possible. In particular, zinc-coated items should not be placed in direct contact with stainless steel or in positions where they could drip molten zinc onto it. Thus, for example, galvanized wire netting used in insulation should not be in direct contact with stainless steel pipe. Similar considerations apply to stainless steel plant which has a normal operating temperature above 400°C. Care should also be taken to prevent zinc contamination of stainless steel in welding and other fabrication or maintenance activities.

If plant made of stainless steel is subjected to fire, it should be examined by experts to check that it has not suffered zinc embrittlement. Deterioration is difficult to determine by normal inspection methods and requires metallurgical examination. Further information on zinc embrittlement is given in TDN 53/1 *Zinc Embrittlement of Austenitic Stainless Steel* (HSE, 1976).

12.27.16 Erosion

Erosion is a common form of corrosion and takes many forms. It occurs particularly at sites where there is a flow restriction or change of direction. These included nozzles and valves, elbows, tees and baffles, and points opposite to inlet nozzles. It is enhanced by the presence of solid particles in gas or liquid, by drops in vapours or by bubbles in liquids and by two-phase flow. Conditions which can cause severe erosion include pneumatic conveying, wet steam flow, flashing flow and pump cavitation.

12.27.17 External corrosion

External corrosion can be caused by components in insulation and in fireproofing. The leaching of chloride salts from insulation by dripping water corrodes pipework. An account of corrosion beneath lagging has been given in Section 12.11. External corrosion is also considered in relation to maintenance in Chapter 21.

Underground pipework can be corroded by the soil. This form of corrosion is often electrochemical, and cathodic protection is used to combat it.

12.27.18 Corrosion at welds

In much process equipment the welds constitute both a potential fault line and a line of common cause failure, in that if there is corrosion at one point in the weld it is quite likely to be occurring along much of it. It is this combination, with its potential for catastrophic failure, which makes corrosion at welds so serious. An account of corrosion at welds is given by M. Turner (1990a), who distinguishes three basic causes of such corrosion: (1) hydrodynamics, (2) differences in composition and (3) differences in metallurgical structure.

Corrosion due to hydrodynamic effects is seen in butt welds on pipes. One cause is lack of root fusion, such that there is a small hollow at the root of the weld on the inner surface of the pipe, which then becomes the site of turbulence and impingement corrosion. Another cause is protrusion of a weld bead, which gives rise to impingement pitting in the pipe surface beyond the protrusion. In both cases attack can be rapid and can accelerate. If the defect extends sufficiently far round the pipe, it may result in pipe rupture. In addition to control of welding to eliminate such defects, countermeasures may include the use of a backing ring; for a small bore connection, the use of an undersized connection which is then drilled out to size; and the limitation of fluid velocity.

A weld has the structure of cast metal and, if made of the same material as the parent metal, will generally be weaker. In order to compensate for this, use is commonly made of a weld metal of different composition. However, this difference in composition can be another cause of weld attack, by galvanic cell, or bimetallic couple, corrosion. The corrosion effects are governed by the ionic conductivity of the liquid which determines the extent of the pipe which participates as an electrode. If the weld metal is more noble, it is cathodic towards the parent metal, and it is principally the latter which will corrode, but even if the corrosion zone is relatively narrow, the rate of corrosion will tend to be limited by the small area of the cathode. If the weld metal is less noble and is therefore anodic, corrosion will occur mainly at the weld, with the rate of corrosion being dependent on the area of the pipe acting as the cathode. If the ratio of this area to that of the weld is high, corrosion can be rapid.

During welding a zone of the parent metal is affected by the heat. This heat affected zone (HAZ) can be some 20 mm wide and usually it is more susceptible to corrosion. Early stainless steels suffered serious problems of grain boundary corrosion, or weld decay, which led to brittle failure. The cause was depletion of chromium at the grain boundaries, due to the fact that under these conditions the carbon in solid solution migrates to the boundaries and forms chromium carbide, which precipitates, whilst chromium migrates more slowly and thus the chromium deficiency is not made good. The solution was the production of stainless steels with low carbon contents.

In steels generally, the high temperatures involved in welding cause a number of metallurgical changes in the HAZ, many of which make it more susceptible to corrosion. One mechanism of corrosion is a bimetallic couple. Control of the welding, including the heat input and interpass temperatures, can minimize, but generally not totally eliminate, the corrosion susceptibility. This susceptibility is, however, little affected by post-weld heat treatment, which is a form of tempering aimed at partial stress relief of the weld to reduce its brittleness, and is conducted at temperatures below those required to reverse changes in the metallurgical structure of the heat affected zone.

In 1984, an oil line fracture at Las Piedras, Venezuela, sprayed hot oil across a roadway onto hydrogen units, causing a major fire (Case History A110). The pipe failure was a circumferential fracture in the parent metal in the heat affected zone about 1.5 in. from the weld.

12.27.19 Corrosion testing

Corrosion is generally a function of a number of variables and can be very sensitive to the particular process conditions. It is not always easy to predict, and in some cases corrosion testing is necessary. As described by M. Turner (1988), there are essentially two distinctly different types of corrosion test. The first type is standard tests, which are characterized by the fact that they apply essentially to a set of relatively straightforward conditions and are precise, reproducible, rapid and cheap, and require neither skilled supervision and interpretation nor input from the client. Plant simulation tests, on the other hand, which simulate actual plant conditions, are not very precise or reproducible, can be prolonged and expensive, and require close supervision, skilled interpretation and client input.

Since in a candidate design corrosion is likely to proceed relatively slowly, testing can take a long time. It is therefore not uncommon to use some form of accelerated test. As with any form of accelerated testing, this can be misleading. An alternative is the use of corrosion rate monitoring for which instrumentation now exists. This allows information to be obtained in tests which simulate the plant conditions. However, since the corrosion rate can vary with time, if this method is used it is necessary to ensure that the steady-state rate has been reached. It is desirable that the design of corrosion tests be a joint exercise between the chemical engineer and the corrosion specialist.

12.27.20 Corrosion policy

In establishing a policy on corrosion, there is a need to strike a balance between neglect and excessive caution. Process plants live with corrosion. The important thing is to identify, and act on, those cases where corrosion can be especially hazardous or costly. Some of the factors to be considered are discussed, with illustrations, by M. Turner (1987, 1990b). Most experienced corrosion engineers will themselves admit to cases where they have been surprised by the rapid corrosion which has occurred and others where they anticipated corrosion which never materialized.

The *Report of the Committee on Corrosion and Protection* (Hoar, 1971) (the *Hoar Report*) reviewed national policy on corrosion. It made the point that there were substantial gains to be made simply by applying existing knowledge; in other words, there is a need for 'corrosion awareness' on the part of engineers. However, there has to be an economic balance. Corrosion technologists for their part need a keen appreciation of the practicalities of the economics of process plants. For example, in many applications the choice of carbon steel will be the right one, even though it will corrode faster than an alternative, more expensive material.

As an instance of living with corrosion, M. Turner (1987) cites the case of the steam reforming process for hydrogen. Theoretically this should have been severely inhibited by the problem of creep rupture of the catalyst tubes, which, since creep is a probabilistic process, was liable to cause disruption due to premature failures. The solution was the use of 'pigtails' which could be nipped closed, allowing the furnace to continue in operation.

Corrosion is another area in which it is important that engineers know enough to recognize those situations in which specialist advice is necessary.

12.28 Fracture Mechanics

A very powerful tool for the avoidance of catastrophic failure in pressure equipment is fracture mechanics. In this section an account is given of deterministic fracture mechanics, which deals essentially with the propagation of cracks, whilst the following section treats probabilistic fracture mechanics, which deals with the probabilistic aspects of equipment failure due to cracks such as the probabilities of prior existence of defects, of failure to detect them, and so on.

Accounts of fracture mechanics are given in *Fracture of Structural Materials* (Tetelman and McElivy, 1967), *Elements of Elasticity* (Dugdale, 1968), *Fracture – An Advanced Treatise* (Liebowitz, 1968), *Fundamentals of Fracture Mechanics* (Knott, 1973), *Elementary Engineering Fracture Mechanics* (Broek, 1974), *Fracture and Fatigue Control in Structures* (Rolfe and Barsom, 1977, 1987, *The Mechanics of Fracture and Fatigue* (A.P.Parker, 1981) and *Engineering Safety Assessment* (Thomson, 1987). Background treatments of elasticity and plasticity include those by Godfrey (1959), W. Johnson and Meller (1962), Spencer (1968), Timoshenko and Goodier (1970) and Benham and Warnock (1973),

Fracture mechanics deals with the initiation and growth of cracks to a critical size and with failure due to cracks. It identifies different regimes of crack growth, and the effect of material properties, plate thickness, and so on. It provides the basic for strategies for the avoidance, control and mitigation of crack propagation.

As such it provides a tool not only for the avoidance of catastrophic failure, but also for discrimination between those defects which require prompt action and those which do not. The failure mechanism mainly treated in fracture mechanics is fatigue, but it is applicable to other mechanisms also.

12.28.1 Loading regimes

The account of fracture mechanics given here is concerned essentially with steel plate such as is used in pressure vessels. An applied load gives rise in a body to stress σ and strain ϵ. The general problem of determining the stresses and strains within a loaded body is a complex one. A considerable simplification can be obtained if for a body of co-ordinates x, y, z it can be assumed that there is no change in the distribution of stress in the z direction. Then for deformation in the x, y plane two cases may be distinguished: plane stress and plane strain.

Plane stress is applicable to the case of a plate sufficiently thin that it can be assumed incapable of supporting stress σ_z through the thickness (the z direction) so that $\sigma_z = 0$. Plane strain applies in the case of a plate sufficiently thick that it prevents strain ϵ_z through the thickness so that $\epsilon_z = 0$.

In most engineering applications the combination of the dimensions of the structure and the steels used is such that at normal service temperatures the thickness of the steel is not sufficient to ensure plane strain under slow loading conditions. The material exhibits a behaviour which is not purely elastic but is to some degree elastic–plastic. Another distinction is in the mode of crack surface displacement. Three such modes are commonly treated. The modes involve opening (mode I), shearing (Mode II) and tearing (Mode III). It is the first mode which is of interest here.

12.28.2 Modelling of crack behaviour

In describing the modelling of the stresses and strains caused by an applied load, and of the effect on cracks, it is usual to start by considering purely elastic behaviour. Models of loading are based on equations of equilibrium and compatibility. There are a number of stress functions which can be shown to satisfy the equilibrium conditions. One of these is the Airy stress finction F. There are also a number of complex stress functions which satisfy both the equilibrium and compatibility conditions. It can be shown that both sets of conditions are satisfied by the solution

$$F + iB = \bar{z}\phi(z) + \int \psi(z)dz \qquad [12.28.1]$$

where F is the Airy function, B is some other function, ϕ and ψ are analytic functions, and $z = x + iy$ and $\bar{z} = x - iy$.

12.28.3 Stored energy

When a body is loaded the movement caused by the applied load does work which is stored as strain energy. For uniaxial tension this work is:

$$U = \sigma^2/2E \qquad [12.28.2]$$

where E is the modulus of elasticity, or Young's modulus, and U is the energy stored per unit volume of material.

12.28.4 Griffith crack model

Early work on fracture faced the apparent paradox that a sharp notch is capable of producing very high stress concentration, yet an infinitely sharp one does not necessarily lead to failure. This problem was addressed by Griffith (1920–21), who for the case of a single crack of length $2a$ in a brittle material considered the energy changes associated with an incremental extension of that crack and showed that

$$U = \frac{\sigma^2}{2E}\pi a^2 \qquad \text{plane stress} \qquad [12.28.3]$$

$$U = \frac{\sigma^2}{2E}\pi a^2(1 - v^2) \qquad \text{plane strain} \qquad [12.28.4]$$

since for plain strain

$$\epsilon = (1 - v^2)\sigma/E \qquad [12.28.5]$$

where a is the crack half-length and v is Poisson's ratio.

The energy release rate for crack extension G is defined as

$$G = \partial U/\partial a \qquad [12.28.6]$$

Hence from Equations 12.28.3, 12.28.4 and 12.28.6

$$G = \frac{\pi\sigma^2 a}{E} \qquad \text{plane stress} \qquad [12.28.7]$$

$$G = \frac{\pi\sigma^2 a}{E}(1 - v^2) \qquad \text{plane strain} \qquad [12.28.8]$$

The energy absorption rate for crack extension R is

$$R = \partial W/\partial a \qquad [12.28.9]$$

where W is the energy absorbed in producing crack extension.

At the threshold condition for unstable crack growth $\partial U/\partial a = \partial W/\partial a$ and hence $G = R$. But since R is a constant, there is a critical value G_{cr} at which such growth will occur and this value is a constant. Then from Equations 12.28.7 and 12.28.8

$$G_{cr} = G_c = \frac{\pi{\sigma_c}^2 a}{E} \qquad \text{plane stress} \qquad [12.28.10]$$

$$G_{cr} = G_{IC} = \frac{\pi{\sigma_c}^2 a}{E}(1 - v^2) \qquad \text{plane stress} \qquad [12.28.11]$$

where G_c and G_{IC} are the critical values of G for plane stress and plane strain, respectively, and σ_c is the critical value of the stress.

Equations 12.28.10 and 12.28.11 give

$${\sigma_c}^2 = \frac{EG_c}{\pi a} \qquad \text{plane stress} \qquad [12.28.12]$$

and

$${\sigma_c}^2 = \frac{EG_{IC}}{\pi a(1 - v^2)} \qquad \text{plane stress} \qquad [12.28.13]$$

Equation 12.28.12 is sometimes written as

$$\sigma_c = \left(\frac{2\gamma E}{\pi a}\right)^{\frac{1}{2}} \qquad [12.28.14]$$

with

$$G_c = 2\gamma \qquad [12.28.15]$$

where γ is the surface energy per unit area.

12.28.5 Stress intensity factor

For a body containing a crack the magnitude of the stress field at the crack tip may be characterized by the stress intensity factor K. For the first Mode of crack surface displacement, Mode I, which is that of interest here, the stress intensity factor is K_I.

This stress intensity factor K_I has a critical value K_c at which unstable crack propagation occurs. As the plate thickness is increased, the value of K_c decreases asymptotically to a minimum value K_{IC}. In other words, under plain strain conditions K_c has a minimum value K_{IC}, the plane strain fracture toughness or, simply the fracture toughness.

For biaxial loading an expression for the stress intensity factor is

$$K_I = \sigma(\pi a)^{\frac{1}{2}} \qquad [12.28.16]$$

Then from Equations 12.28.7, 12.28.8 and 12.28.16:

$$K_I = (GE)^{\frac{1}{2}} \quad \text{plane stress} \qquad [12.28.17]$$

$$K_I = \left(\frac{GE}{1-\upsilon^2}\right)^{\frac{1}{2}} \quad \text{plane strain} \qquad [12.28.18]$$

Compilations of stress intensity factors are given by Sih (1973a), Rooke and Cartwright (1976) and Hudson and Seward (1978).

The fracture toughness is a property of the material, and plays a crucial role in fracture mechanics. It is discussed further below.

12.28.6 Westergaard stress function

Mention has already been made of the Airy function. Another stress function, or rather family of functions, which has proved fruitful is the Westergaard stress function Z_I, where the subscript I refers to the x axis of symmetry. The stress intensity factor K_I may be expressed in terms of a Westergaard function:

$$K_I = (2\pi)^{\frac{1}{2}} \lim_{z\to a} (z-a)^{\frac{1}{2}} Z_I(z) \qquad [12.28.19]$$

It can be shown that using the appropriate Westergaard function the stress intensity factor for a biaxially loaded crack in a plate is given by

$$K_I = \sigma(\pi a)^{\frac{1}{2}} \qquad [12.28.20]$$

as already given.

Another situation of interest is that of an array of cracks aligned in the same direction. Then, using the Westergaard function appropriate to this case, it can be shown that

$$K_I = \sigma(\pi a)^{\frac{1}{2}} \left[\frac{2b}{\pi a}\tan\left(\frac{\pi a}{2b}\right)\right]^{\frac{1}{2}} \qquad [12.28.21]$$

where the distance between the centres of the cracks is $2b$. Equation 12.28.21 reduces to Equation 12.28.20 as $b/a \to \infty$.

12.28.7 Geometric factor

Equation 12.28.20 may be generalized to cover other cases such as that given in Equation 12.28.21 by the introduction of a configuration correction factor, or geometric factor Q, so that

$$K_I = Q\sigma(\pi a)^{\frac{1}{2}} \qquad [12.28.22]$$

Then, for example, for the case of an array of cracks given by Equation 12.28.21

$$Q = \left[\frac{2b}{\pi a}\tan\left(\frac{\pi a}{2b}\right)\right]^{\frac{1}{2}} \qquad [12.28.23]$$

The geometric factor Q has the value unity for a central through crack in a thick plate. Values for other configurations are given by Harvey (1974).

12.28.8 Fracture toughness

The fracture toughness K_{IC} is a property of the material and there are standard tests for it. They include BS 5447: 1977 and ASTM Test Method E-399: 1990. There are also relations between fracture toughness and other parameters such as the Charpy V-notch (CVN) toughness and the crack opening displacement (COD). Both tests and relationships are described by Rolfe and Barsom (1987).

As described earlier, below the transition temperature there is a marked fall in the toughness of a ferritic steel. Typically the fracture toughness is reduced by a factor of about four. Hydrogen embrittlement and impurities can also cause a pronounced reduction in fracture toughness.

12.28.9 Modelling of crack behaviour: plastic deformation

In the account given so far the assumption has been made that the behaviour of the material is elastic. In practice, most materials exhibit at some critical combination of stresses a degree of plastic deformation, or yielding.

There are two principal criteria for the onset of yielding. These are the Tresca criterion

$$|\sigma_1 - \sigma_3| = \sigma_{ys} \qquad [12.28.24]$$

and the von Mises criterion

$$(\sigma_1-\sigma_2)^2 + (\sigma_2-\sigma_3)^2 + (\sigma_1-\sigma_1)^2 = 2\sigma_{ys}{}^2 \qquad [12.28.25]$$

where σ_{ys} is the yield stress and $\sigma_1 > \sigma_2 > \sigma_3$.

Some principal models of crack tip plastic behaviour are those given by Irwin (1958) and Dugdale (1960).

12.28.10 Irwin model

A model of crack tip plastic behaviour has been given by Irwin (1958). If the plastic zone is treated as a circle of radius λ, the model gives

$$\lambda = \frac{1}{2\pi}\left(\frac{K_I}{\sigma_{ys}}\right)^2 \qquad [12.28.26]$$

The crack may be envisaged as having a notional length $(a+\lambda)$. Then, following Equation 12.28.22, an alternative stress intensity factor K^* may be defined as

$$K^* = Q\sigma[\pi(\alpha+\lambda)]^{\frac{1}{2}} \qquad [12.28.27]$$

12.28.11 Dugdale model

Another model for crack tip plasticity is that of Dugdale (1960). In this model the notional length of the crack is taken as $(a+\rho)$, where ρ is the length of the plastic zone. The model yields the relation:

$$\rho = \frac{\pi}{8}\left(\frac{K_I}{\sigma_{ys}}\right)^2 \qquad [12.28.28]$$

which may be compared with Equation 12.28.26 in the Irwin model.

12.28.12 Modelling of critical crack length

There are a number of models for the critical length of crack for fracture. They include the following: the linear elastic fracture mechanics (LEFM) model, the stress concentration theory (SCT) model, the Bilby, Cottrell and Swinden (BCS) model, and the J integral model.

An early model for failure at the crack tip was that of Neuber (1937). In this model failure is considered to occur when, over some characteristic distance b from the crack tip, the average stress, or stress integrated over the distance from a to $(a+b)$, reaches a critical value.

The characteristic distance b has a critical value b_{cr}, such that at shorter distances the stress concentration at the crack becomes critical. Following the treatment of Thomson (1987), itself based on that of Irvine (1977 SRD R48), when the average stress over the characteristic distance reaches a critical value

$$\frac{\sigma_u}{\sigma_c} = \left(\frac{2a}{b_{cr}}+1\right)^{\frac{1}{2}} \qquad [12.28.29]$$

where b is a 'spreading length', b_{cr} its critical value and σ_u is the ultimate tensile stress.

Use is also made of another parameter S, effectively an alternative critical spreading length, where

$$b_{cr} = \frac{2}{3}S \qquad [12.28.30]$$

An account is now given of the first three models mentioned earlier. The fourth is the J integral model of Rice (1968a,b).

12.28.13 LEFM model

The linear elastic fracture mechanics (LEFM) model follows directly from Equation 12.28.29. If the spreading length is much less than the crack size ($b_{cr} \ll a$), then from that equation:

$$\frac{\sigma_u}{\sigma_c} = \left(\frac{2a}{b_{cr}}\right)^{\frac{1}{2}} \qquad [12.28.31]$$

and from Equations 12.28.30 and 12.28.31

$$\frac{a}{S} = \frac{1}{3}\left(\frac{\sigma_u}{\sigma}\right)^2 \qquad [12.28.32]$$

12.28.14 SCT model

The stress concentration theory (SCT) model is a modification of the LEFM model. It may be expressed in the form of Equation 12.28.29 or, using equation 12.28.30, as:

$$\frac{a}{S} = \frac{1}{3}\left[\left(\frac{\sigma_u}{\sigma}\right)^2 - 1\right] \qquad [12.28.33]$$

12.28.15 BCS model

The third model, that of Bilby, Cottrell and Swinden (1963), or the BCS model, may be stated as:

$$\frac{a}{S} = \frac{\pi}{8}\left[\ln \sec\left(\frac{\pi}{2}\frac{\sigma}{\sigma_u}\right)\right] - 1 \qquad [12.28.34]$$

12.28.16 Crack opening displacement

In the extension of the modelling of fracture into the region of elesto-plastic behaviour, an important concept is the crack opening displacement (COD), or crack tip opening displacement (CTOD), proposed by Wells (1961). In this region, if a crack is envisaged as a triangular wedge extending some distance λ into a circular plastic zone, such that the effective half-length is $(a+\lambda)$, then at the distance a from its centre, where it enters the plastic zone, it must have a certain width. This width is the crack opening displacement.

The COD is a property of the material, is the subject of standard tests and may be related to other properties such as the fracture toughness. Standard tests include BS 5762: 1979 and ASTM Test Method E1290: 1990 and relations exist between the COD and other parameters; the tests and relationships are discussed by Rolfe and Barsom (1987).

Expressions for the crack opening displacement δ may be obtained from the models of crack tip plastic behaviour. For the model of Dugdale (1960):

$$\delta = \frac{8}{\pi}\frac{\sigma_{ys}}{E} a \ln\left(\frac{a+\rho}{a}\right) \qquad [12.28.35]$$

and also

$$\delta = \frac{8}{\pi}\frac{\sigma_{ys}}{E} a \ln\left[\sec\left(\frac{\pi a}{2\sigma_{ys}}\right)\right] \qquad [12.28.36]$$

This last equation may be approximated by

$$\delta \approx \frac{\pi\sigma^2 a}{E\sigma_{ys}} \qquad \sigma < 0.7\sigma_{ys} \qquad [12.28.37]$$

Then from Equations 12.28.16 and 12.28.37

$$\delta = K_I^2/E\sigma_{ys} \qquad [12.28.38]$$

The crack opening displacement δ has a critical value δ_c corresponding to the critical stress σ_c.

In line with the foregoing, a general relation for the critical crack opening displacement has been given by J.N. Robinson and Tetelman (1974)

$$\delta_c = \frac{K_{IC}{}^2}{\lambda' E\sigma_{ys}} \qquad \text{plane stress} \qquad [12.28.39]$$

$$\delta_c = \frac{K_{IC}{}^2(1-\upsilon^2)}{\lambda' E\sigma_{ys}} \qquad \text{plane strain} \qquad [12.28.40]$$

where λ' is a constraint factor.

12.28.17 Fatigue crack growth

A principal mechanism of crack propagation is fatigue. If a crack is subjected to a cyclic stress of amplitude $\Delta\sigma$

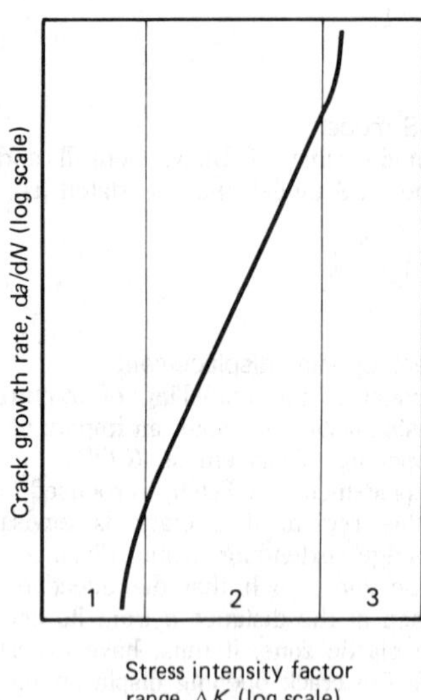

Figure 12.23 Typical crack growth rate for mild steel

this induces a corresponding range of the stress intensity factor ΔK. Then from Equation 12.28.22

$$\Delta K = Q\Delta\sigma(\pi a)^{\frac{1}{2}} \quad [12.28.41]$$

If the rate of increase da/dN of the crack length is plotted against the stress intensity factor range ΔK on a log–log plot, then as shown in Figure 12.23, the curve has three regions. In the first region (1) there is rapid crack growth; in the second region (2) the crack growth is slower and more predictable; and in the third region (3) the crack growth accelerates again to failure. Over the second region the curve is given by the Paris equation:

$$\frac{da}{dN} = C(\Delta K)^m \quad [12.28.42]$$

where N is the number of loading cycles, C is a constant, and m is an index. For most ferrous and non-ferrous metals $m \approx 4$.

Equation 12.28.42 may be integrated as follows:

$$N = \int_{a_i}^{a_f} \frac{1}{C(\Delta K)^m} da \quad [12.28.43]$$

where subscripts f and i indicate final and initial, respectively. Fatigue is not, however, the only mechanism of crack propagation. At high temperatures crack growth can occur by creep. In this case the rate of growth is expressed not in terms of the number N of cycles but of time t, hence as da/dt. Treatments have been given by Tomkins (1983) and by Ainsworth and Goodall (1983).

12.28.18 Failure assessment diagram

A method for the assessment of the regime in which a particular equipment is operating has been developed at the CEGB, and is described by R.P. Harrison and Milne (1981).

Linear elastic failure and fully plastic failure represent two limiting cases. Two parameters are defined as follows. For linear elastic failure

$$K_r = K_I(a)/K_{IC} \quad [12.28.44]$$

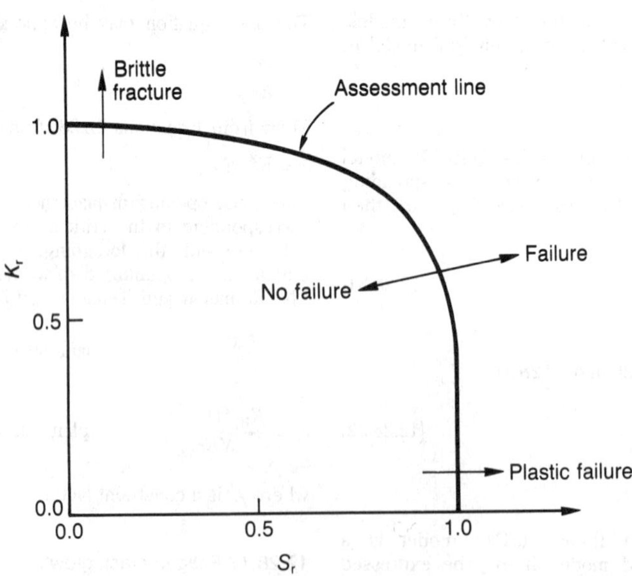

Figure 12.24 Failure assessment diagram (after R.P. Harrison and Milne, 1981) (Courtesy of The Royal Society)

and for fully plastic failure

$$S_r = \sigma/\sigma_1(a) \quad [12.28.45]$$

where K_r is a measure of the proximity to linear elastic failure, S_r is a measure of the proximity to fully plastic failure and $\sigma_1(a)$ is the plastic collapse stress. As indicated, both K_I and σ_1 are functions of the crack half-length.

Figure 12.24 illustrates the failure assessment diagram (FAD) given by the authors and based on these two parameters. The failure assessment line in the diagram is constructed using an elastic–plastic model from the work of Dugdale (1960), Bilby, Cottrell and Swinden (1963) and Heald, Spink and Worthington (1972) to interpolate between the two limiting behaviours. It has been shown by Chell (1978) that this approach is a reasonable approximation to the failure curve for any structural geometry. It is supported by tests on laboratory specimens and on structures as described by R.P. Harrison, Loosemore and Milne (1979). Failure is conceded if the assessment point lies outside (to the right) or on the curve. If lower bound data are used, the curve may be regarded as a failure avoidance line.

The method has been developed to support a comprehensive system of fracture mechanics used in the CEGB and elsewhere. This system is referred to as the 'R6 Procedure'. Its third revision is given in a CEGB report by Milne *et al.* (1986) and in the open literature by Milne *et al.* (1988b).

12.28.19 Fracture analysis diagram

Another widely used diagram is the fracture analysis diagram (also FAD). This was developed by Pellini (1971) as a means of providing guidance on steels which exhibit a transition temperature. The diagram is a plot of stress vs temperature and has four reference points: the null ductility transition (NDT), fracture transition elastic (FTE) and fracture transition plastic (FTP) points and midrange point between NDT and FTE.

As described by the author, restricting the service temperature to just above the NDT provides fracture initiation protection for most common failures; restricting it to the midrange between NDT and FTE provides fracture arrest protection for $\sigma \leq 0.5\sigma_{ys}$; restricting it to above the FTE provides fracture arrest protection for $\sigma \leq \sigma_{ys}$; and restricting it above the FTP ensures that the fracture is fully ductile.

12.28.20 Through-thickness yielding

It is of interest to be able to establish in a given case whether through-thickness yielding will occur before fracture. Such yielding is a function of plate thickness. As already described, the two limiting conditions are those of plane stress in thin plates and plane strain in thick plates. Practical situations to lie between these two extremes.

Through-thickness yielding is generally treated in terms of the dimensionless parameter β

$$\beta = \frac{1}{B}\left(\frac{K_{IC}}{\sigma_{ys}}\right)^2 \quad [12.28.46]$$

where B is the plate thickness

A discussion of the values given by various workers for the parameter β and their interpretation is given by Rolfe and Barsom (1987). They quote ASTM Committee 24 on Fracture Testing as advising that to ensure plain strain behaviour it is necessary to have $\beta \geq 0.4$. Hahn and Rosenfield (1968) found a significant increase in the rate at which through-thickness deformation occurs for $\beta \geq 1$. Irwin (1962) used a value of $\beta = 1.4$ in his leak-before-break model, described below.

12.28.21 Leak-before-break behaviour

If a crack grows to the point where it will cause failure, it is obviously desirable that it give warning of impending failure by leaking. It is possible using fracture mechanics to assess whether or not in a given situation the crack will exhibit such leak-before-break (LBB) behaviour. This behaviour is also sometimes called leak-before-fracture (LBF) and its obverse as fracture-before-leak (FBL).

The LBB criterion was proposed by Irwin *et al.* (1967) with the aim of predicting the toughness required in a pressure vessel so that before fracture occurs a surface crack will grow through the wall and give a leak. Another way of putting this is that the critical crack size is greater than the wall thickness.

An account of the LBB criterion is given by Rolfe and Barsom (1977). It is assumed that a surface crack might grow through a wall of thickness B into a through-thickness crack of length $2B$. Hence the LBB criterion is $2a \approx 2B$. The authors use for the stress intensity factor K_I the relation

$$K_I^2 = \frac{\sigma^2 \pi a}{1 - 0.5(\sigma/\sigma_{ys})^2} \quad [12.28.47]$$

or, for low stress

$$K_I = \sigma(\pi a)^{\frac{1}{2}} \quad [12.28.48]$$

as given earlier. Then for the critical stress $K_I = K_c$. The critical stress intensity factor K_c is obtained using the following relation:

$$K_c{}^2 = K_{IC}{}^2(1 + 1.4\beta_{IC}{}^2) \quad [12.28.49]$$

with

$$\beta_{IC} = \frac{1}{B}\left(\frac{K_{IC}}{\sigma_{ys}}\right)^2 \quad [12.28.50]$$

where B is the wall thickness and β_{IC} a dimensionless parameter.

Another treatment of LBB is given by Harvey (1974), who deals with practical aspects such as allowance for residual stress.

One use of the LBB method is in the design of a vessel to ensure that it operates in the LBB region. Another is to calculate the allowable stress in a vessel where a crack has been observed. Another is to check that the size of crack which requires a reduction in operating stress is detectable by the inspection methods used.

12.28.22 Some design features

The fracture characteristics just described have a number of implications for the design of pressure systems. They include: the effects of plate thickness and through-thickness yielding; the effect of the use of advanced design codes; and the LBB principle.

If a plate is sufficiently thin, fracture is ductile; if it is sufficiently thick, fracture is brittle. Much equipment in pressure systems will lie somewhere in between. The relations for through-thickness yielding provide a means of quantifying this behaviour.

Advanced design codes such as BS 5500 and ASME *Boiler and Pressure Vessel Code*, Section VIII, Division 2, permit the use of a stress which constitutes a higher proportion of the ultimate tensile stress; in other words the ratio σ_u/σ is lower, so that instead of being 4 it lies in the range 2–3. Whilst this gives economies in equipment design, it means that the length of the critical crack is appreciably reduced, thus making it more difficult to ensure detection of such a crack.

For critical equipment, such as a vessel containing a large inventory of a hazardous material, it is possible to identify the region in which a fracture will exhibit LBB behaviour, with a view to designing to ensure that the equipment operates always in that region.

There are also implications for inspection. Fracture mechanics can indicate the size of crack which needs to be detected by the inspection procedures and the inspection interval necessary to ensure that no crack grows to its critical value before the next inspection.

12.29 Probabilistic Fracture Mechanics

A comprehensive approach to equipment failure from crack propagation requires that the deterministic treatment just described be complemented by probabilistic methods. Probabilistic fracture mechanics addresses such questions as the probability that a crack will be present, that it is not detected on a given inspection and that it grows to a critical size before the next inspection, causing failure. Accounts of probabilistic fracture mechanics include those by O'Neil (1980), Temple (1983, 1985), Haines (1983), Lidiard (1984) and Thomson (1987).

The probability that a pressure vessel, or similar equipment, will fail due to crack propagation may be expressed as follows:

$$P_f = P_{x0}(x)P_{nf}P_{gc} \qquad [12.29.1]$$

where P_f is the probability that the vessel fails, P_{gc} is the probability that the defect grows to critical size before the next inspection, P_{nf} is the probability that the defect is not detected by non-destructive examination (NDE) at a given inspection, $P_{x0}(x)$ is the probability that the vessel has an initial defect of depth $> x$, and x is the crack depth (mm).

Work in this area has focused particularly on pressure vessels for nuclear pressurized water reactors (PWRs). In a study of this problem, Lidiard (1984) obtained the relation

$$P_{x0}(x) = \frac{A}{\Theta}\exp(-\Theta x) \qquad (12.29.2)$$

where A and Θ are constants (both mm^{-1}). He obtained for such vessels values of A and Θ of 0.59 and 0.16, respectively. These values imply, for example, that the probability of such a vessel having a crack of depth exceeding 25 mm is 0.067 and the probability of one exceeding 50 mm is 1.24×10^{-3}.

The reliability of detection by NDE was the subject of a study by the European Community Plate Inspection Steering Committee (PISC). Information obtained on defect detection probability (DDP) is described by O'Neil (1980) and an analysis is given by Haines (1983).

The data on the probability of detection, or DDP, may be fitted to an appropriate distribution. That used by W. Marshall (1982) is the exponential distribution. Thomson (1987) proposes, for a simplified analysis, that the probability of detection by NDE, P_n, may be fitted to a log–normal distribution, as given by Equation 7.7.20. He gives for the parameters of the distribution m^* and σ the values 3.135 and 0.667 mm, respectively. The probability P_{nf} of failure to detect is the complement of P_n.

The probability that a crack will grow to critical size may be obtained deterministically by the method described in the previous section. Alternatively, it too may be treated probabilistically. This approach has been described by Temple (1985). The probability P_{gc} that the crack grows to critical size given that it has an initial depth x_0 and stress intensity factor range ΔK and is subject to N loading cycles may be written as $P_{gc}(x \mid x_0, \Delta K, N)$. Then, taking for this probability a log–normal distribution, Temple obtains for PWR vessels:

$$m^* = x_0 + 6.56 \times 10^{-5}\Delta K \cdot N \,\text{mm}$$

$$\sigma = 2.93 \times 10^{-5}\Delta K \,\text{mm}$$

where ΔK is the stress intensity factor range ($\text{MPa.m}^{\frac{1}{2}}$).

The application of the relations just described to obtain the probability of vessel failure is discussed, with illustrative examples, by Thomson. He casts Equation 12.29.1 in the form

$$P_f = \int_0^{x_c} P_{x0}(x)P_n(x)p_{gc}(x)\,dx \qquad [12.29.3]$$

where p_{gc} is the probability density function corresponding to the probability distribution P_{gc} and x_c is the critical crack length.

H.M. Thomas (1981) describes an approach to the determination of the probability P_C of catastrophic failure and P_L of leak which starts with generic estimates but involves a learning process using data from the actual plant to obtain more refined estimates.

12.30 Failure of Vessels, Equipment and Machinery

There is a certain amount of statistical information available on the failure of pressure system components.

12.30.1 Failure of pressure vessels

Work on the failure rates and modes of pressure vessels has been driven initially by the needs of the nuclear industry.

Information on pressure vessel failure is given in *A Survey of Defects in Pressure Vessels Built to High Standards of Construction and its Relevance to Nuclear Primary Circuit Envelopes* by Phillips and Warwick (1968 UKAEA AHSB(S) R162). The survey deals with pressure vessels built to Class 1 requirements of BS 1500, BS 1515 and comparable standards. It classifies failures as catastrophic or potentially dangerous. The former are disruptions of the vessel which require major repair or scrapping; the latter are defects which might deteriorate under the working conditions and which require remedial action. The following information is given on failures prior to service and in service:

	Sample size	*Failure rate* (failures/year)	
		Potentially dangerous failures	*Catastrophic failures*
Failure in construction	12 700 vessels	5.5×10^{-4}	2.3×10^{-4}
Failure in service	100 300 vessel-years	1.25×10^{-3}	0.7×10^{-4}

For all the service failures the causes were classified. The results are shown in Table 12.11. The vast majority of failures, some 89.3%, were due to cracks. The causes of cracks were therefore analysed separately as shown in the table. The table also gives the methods by which the failures were detected.

In addition, Phillips and Warwick analysed the effects of other variables such as materials of construction, design conditions, fluid handled, component state and age. The results are given in Table 12.12. As indicated in Table 12.12, there were 7 catastrophic failures. The causes of these failures were: maloperation (4 cases), fatigue (2 cases), and pre-existing from manufacture (1 case). The authors comment: 'Most of these catastrophic failures are due to inadequate operational procedures and highlight the need for more consideration of control techniques.'

The authors then analyse the data to assess the failure frequency of the pressure vessels used in nuclear reactor primary circuit envelopes. They conclude that many of the failures given in the survey would not apply in such vessels, e.g. cracks in boiler furnaces, and reduce the total number of applicable service failures from 132 to 62 and the number of applicable catastrophic failures from 7 to 2. Then the failure frequencies become:

	Failure rate (failures/year)	
	Potentially dangerous failures	*Catastrophic failures*
Failure in service	6×10^{-4}	2×10^{-5}

Table 12.11 *Causes and methods of detection of service failure in pressure vessels (after Phillips and Warwick, 1968 UKAEA AHSB(S) R162) (Courtesy of the UK Atomic Energy Authority)*

	No. of cases	*Percentage of total cases*
Causes of failures:		
Cracks	118	89.3
Maloperation	8	6.1
Pre-existing from manufacture	3	2.3
Corrosion	2	1.5
Creep	1	0.8
	132	100.0
Causes of cracks:		
Fatigue	47	35.6
Corrosion	24	18.2
Pre-existing from manufacture	10	7.6
Miscellaneous	2	1.5
Not ascertained	35	26.5
	118	89.4
Method of detection:		
Visual examination	75	56.9
Leakage	38	28.8
Non-destructive testing	10	7.5
Hydraulic tests	2	1.5
Catastrophic failure	7	5.3
	132	100.0

Table 12.12 *Other variables relevant to service failure in pressure vessels (after Phillips and Warwick, 1968 UKAEA AHSB(S) R162) (Courtesy of the UK Atomic Energy Authority)*

Variables	*Failures*			
	Potentially dangerous failures[a]		*Catastrophic failures*[a]	
	No.	*(%)*	*No.*	*(%)*
Materials of construction:				
Mild steel	105	79.5	5	3.9
Alloy steel	20	15.1	2	1.5
Design pressure:				
<500 lb/in.^{2}	41	31.1	4	3.1
≥ 500 lb/in.^{2}	84	63.5	3	2.3
Design temperature:				
$<600°F$	15	11.4	1	0.7
$\geq 600°F$	69	52.5	1	0.7
Not stated	41	31.1	5	3.9
Age:				
<10 years	59	45.0	6	5.0
≥ 10 years	64	48.5	1	0.7
Not stated	2	1.5		
Component state:				
Fired	29	22.0	4	3.1
Unfired	96	72.6	3	2.3
Fluid handled:				
Gas (including steam)	69	52.2	4	3.1
Liquid	14	10.6		
Mixture	42	31.8	3	2.3

[a] These columns give alternative breakdowns of the 125 potentially dangerous and 7 catastrophic failures.

These results agree reasonably well with the catastrophic failure frequencies estimated for nuclear vessels by Kellerman and Seipel (1967). This study provides a good illustration of the determination of failure data and of the derivation from crude data of failure data applicable to the case which is of interest. The survey just described has been updated by a further survey by Smith and Warwick (1974 SRD R30). Again this survey dealt only with pressure vessels built to Class 1 requirements. The period covered by the survey was from the latter half of 1967 to the end of 1972. The information shown in the table below was obtained on failures prior to service and in service:

	Sample size	*Failure rate* (failures/year)	
		Potentially dangerous failures	*Catastrophic failures*
Failure in construction	8823 vessels	2.1×10^{-3}	–
Failure in service	105 402 vessel-years	1.1×10^{-3}	1.5×10^{-4}

In this latter survey, therefore, there were no catastrophic failures in construction, but the failure rates of potentially dangerous failures in construction and catastrophic failures in service are appreciably greater than in the earlier survey. The failure rate of potentially dangerous failures in service is similar. The survey includes further tables giving the classifications fo service failures and the variables relevant to service failures similar to Table 12.11 and 12.12.

A further analysis of pressure vessel failure statistics made by the Atomic Energy Commission (AEC) Advisory Committee on Reactor Safety (ACRS) has been described by Engel (1974). This work refers to further large collections of pressure vessel failure data by the Edison Electric Institute-Tennessee Valley Authority (EEI-TVA), by the American Boiler Manufacturers Association (ABMA) and by the Institut für Reactorsicherheit der Technischen Überwachungs-Verein (TUV). The EEI-TVA data are effectively a subset of the ABMA data. Studies based on the TUV data have been published by Kellerman (1966), Kellerman and Seipel (1967) and Slopianka and Mieze (1968).

The most severe category of failure used by Engel is a disruptive failure, which is breaching of the vessel accompanied by the release of a large volume of contained fluid. He treats this as a more restrictive category other than of catastrophic failure used by Phillips and Warwick. Engel considers failures of non-nuclear vessels applicable to nuclear reactor vessels. The largest data collections are as tabulated below:

The work by Kellerman and Seipel does not indicate how many of the 49 failures reported might be classified as disruptive and therefore the number of failures and the upper bound on this failure rate are shown in the table as ranges. The upper bound of the failure rate in the table is determined by the statistical methods described in Chapter 7 and for zero disruptive failures is equal at the 99% confidence level to 4.6 (9.2/2) divided by the number of vessel-years.

Engel concludes that the ABMA data are the most useful and significant. These data are for pressure vessels designed to the ASME *Boiler and Pressure Vessel Code*, Section I or Section VIII, version 1956 or later, for pressures and temperatures equal to or exceeding those in light water reactors (LWRs) and with wall thicknesses exceeding 1.5 in. He states that the data support the conclusion that for failure of pressure vessels applicable to nuclear reactors the disruptive failure rate is less than 1×10^{-5} failures/year.

In a further survey, Smith and Warwick (1981 SRD R203) obtained for the period 1962–78 the following failure rates for pressure vessel failures in service. For some 20 000 vessel they obtained the results given at the top of the next page. Of the 229 failures, 206 were cracks.

A survey of pressure vessel failures in process plants has been reported by Arulanantham and Lees (1981). The work covered process pressure vessels, pressure storage vessels, non-pressure storage vessels, heat exchangers and fired heaters. The plants studied were four olefins plants and one plant handling a toxic and corrosive material, in a single works. Two of the olefins plants were commissioned in the 1950s and closed down in the 1960s, one was commissioned in the 1960s and closed in the 1970s and one was commissioned in the 1960s and was still operating. The toxics plant was also commissioned in the 1960s and was still operating. The olefins plants operated at about 3.5 barg and 35 barg at the hot and cold ends, respectively, except that at the hot end there was steam generation at about 85 barg. The temperatures ranged from −100° to 800°C. The toxics plant operated at relatively low pressures and mild temperatures. Excluded from the survey were failures in

Data source	*Sample size* (vessel-years)	*Disruptive failures* (failures/year)	
		No. of failures	*Upper bound of failure rate at 99% confidence*
Phillips and Warwick (1968)	100 300	0	4.6×10^{-5}
ABMA (Marx, 1973)	723 000	0	0.64×10^{-5}
Kellerman and Seipel (1967)	1 700 000	0–49	0.27×10^{-5} to 4.0×10^{-5}

Sample size (vessel-years)	No. of failures: Potentially dangerous failures	No. of failures: Catastrophic failures	Failure rate (failures/year): Potentially dangerous failures	Failure rate (failures/year): Catastrophic failures
310 000	216	13	6.9×10^{-4}	4.2×10^{-5}

steam boilers and steam and air receivers and in pipework and also heat exchanger tube failures. The authors define failure as a condition in which a crack, leak or other defect has developed in the equipment so that repair or replacement is required, a definition which they treat as including some of the potentially dangerous as well as all the catastrophic failures in that of Phillips and Warwick. They report failure rates classified by type of vessel, operating conditions and causes of failure.

For pressure vessels other than fired heaters, i.e. process pressure vessels, pressure storage vessels and heat exchangers they obtain:

Sample size (vessels)	Sample size (vessel-years)	No. of failures	Failure rate (failures/year)
1216	16417	70	4.3×10^{-3}

Other failure data given are as follows. For the olefins plants:

Vessel	Sample size (vessels)	Sample size (vessel-years)	No. of failures	Failure rate (failures/year)
Process pressure vessel	415	5535	15	2.7×10^{-3}
Pressure storage vessel	129	2220	4	1.8×10^{-3}
Heat exchanger	446	5950	10	1.7×10^{-3}
Fired heaters	36	447	181	405×10^{-3}
High temperature vessel, except fired heater	58	809	6	7.4×10^{-3}
Low temperature vessel	147	1941	3	1.5×10^{-3}

Here the pressure storage vessel data are for the whole works and the fired heater data are for the toxics plant also. For the toxics plant:

Vessel	Sample size (vessels)	Sample size (vessel-years)	No. of failures	Failure rate (failures/year)
Process pressure vessel	131	1572	15	26×10^{-3}
High temperature vessel, except fired heater	16	192	7	36×10^{-3}
Vessel in corrosive duty	45	540	21	39×10^{-3}
Vessel subject to stress corrosion	49	588	12	20×10^{-3}

Davenport (1991) has reported a survey involving some 360 000 vessels and 1.8×10^6 vessel-years over the period 1983–88, which is a much larger sample than that obtained by Smith and Warwick. The survey was not confined to Class 1 vessels, but took in a much larger range, and in fact is dominated by relatively thin-walled air receivers, so that the two surveys are not comparable. In the survey by Davenport there were 92 failures, of which 60 were cracks.

The distribution of pressure vessel disruptive failure rates from a number of sources has been analysed by Hurst, Davies *et al.* (1994). In this work the authors give various sets of failure rate data as plots of $\log_{10}\lambda$ vs normal quantiles, where λ is the failure rate of the exponential distribution, in order to check whether the failure rates are log–normally distributed. Their plot for pressure vessels is shown in Figure 12.25(a), the straight line confirming that the distribution of the failure rates is log–normal. The mean pressure vessel disruptive failure rate obtained from this plot is $\lambda \approx 10^{-5}$/year. The failure rates vary, however, by some five orders of magnitude at the 95% confidence level.

With regard to pressure vessel failure rates used in hazard assessments, the First *Canvey Report* gives for spontaneous failure of a pressure vessel a failure rate of some $10^{-5} - 10^{-4}$ failures/year. A failure rate of 10^{-5} failures/year is used in the study, but the sensitivity of the results to the higher failure rate is also given. The treatment of pressure vessel failure rates in the report is described in Appendix 7.

The *Rijnmond Report* gives for catastrophic failure and for serious leakage of a pressure vessel failure rates of 10^{-6} and 10^{-5} failures/year, respectively.

The estimated failure rate of a pressure vessel is given by Batstone and Tomi (1980) as 10^{-6} failures/year.

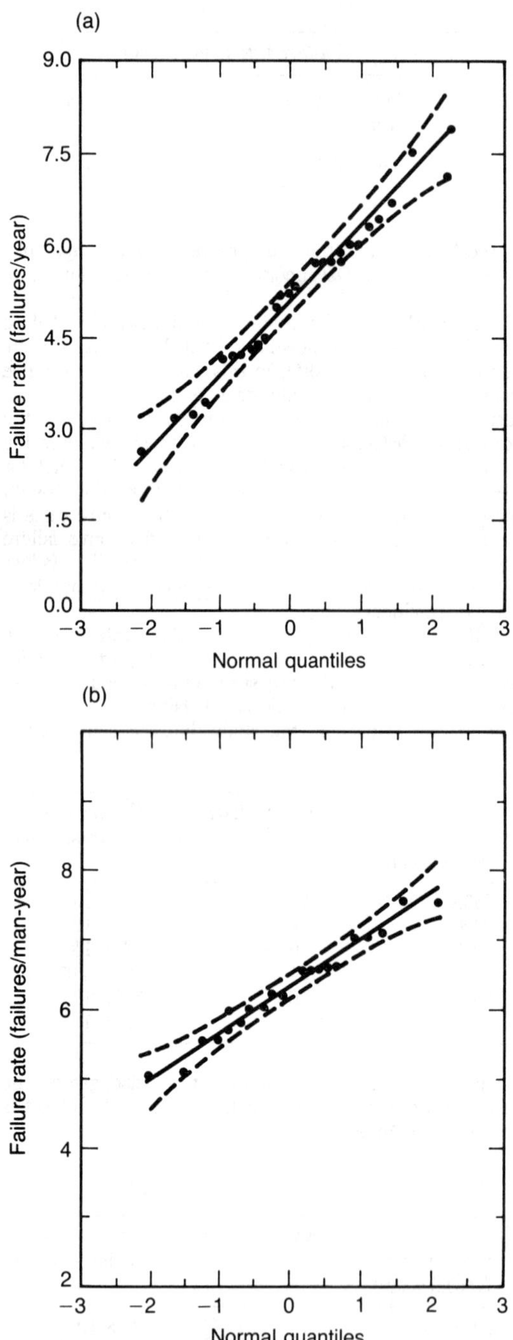

Figure 12.25 Distribution of disruptive failure rates of pressure vessels and pipework (Hurst Davies et al., 1994): (a) pressure vessel disruptive failure rates (failures/year); and (b) pipework guillotine fracture failure rates (failures/m-year). In these graphs the variable of interest is the logarithm of the exponential failure rate, which is shown plotted against the quantiles of the normal distribution. A straight line through the points indicates that this variable is log–normally distributed. The value corresponding to the zero value of the normal quantile is the mean of the failure rate values.

12.30.2 Failure of pipework

There is a considerable amount of data available on pipework failures, but the range of values quoted is wide and tends to be confusing. There are several important distinctions to be made. One concerns the type of failure. Complete pipe breaks, or guillotine fractures, constitute only a small proportion of failures. Another relates to the pipe size. The failure rate tends to be higher for small than for large diameter pipes. A further difference in the figures arises from the basis on which figures are quoted, which may be per unit length of pipe run, per pipework section or per plant.

Even so, the quoted failure rates of pipework show considerable variation. Table 12.13 shows some estimates of the failure rate of pipework on nuclear plants quoted in the *Rasmussen Report*. Similarly, Figure 12.26 has been given by Hawksley (1984) to illustrate the variability of pipework failure rates.

As with pressure vessels, so with piping, early work on failure rates and modes was directed to nuclear industry needs. Data on pipework failure have been given by a number of authors, particularly those concerned with nuclear systems, such as Smith and Warwick (1974 SRD R30) and Bush (1977).

Smith and Warwick (1974 SRD R30), applying definitions of catastrophic and potentially dangerous failures, similar to those which they used for pressure vessels as described in the previous section, obtained for reactor systems the data contained in the table at the top of the next page.

The reactor system included reactors and other equipment such as heat exchangers as well as pipework, but the failures were predominantly in the latter.

The failure rate of pipework in nuclear plants is given by Bush (1977) as 4.3×10^{-5} failures/man-year derived from eight failures, which are individually described and of which four were due to water hammer.

A survey of pipework failure in nuclear plants to provide information more up-to-date than that given in *Rasmussen Report* has been described by R.E. Wright, Steverson and Zuroff (1987). They distinguish between pipework for which a failure would result in a loss-of-coolant accident (LOCA), which they term 'LOCA-sensitive', and other pipework. Their summary of the pipework failure rates used in the *Rasmussen Report* is shown in Table 12.14.

The authors obtained data for US nuclear reactors with a total operating time of nearly 800 reactor-years, covering both PWRs and boiling water reactors (BWRs). They consider three pipe ranges: $\frac{1}{2}$–2 in., > 2–6 in. and > 6 in. For LOCA-sensitive systems they defined a failure as a leak flow $\geq$ 50 USgal/min for PWRs and $\geq$ 500 US gal/min for BWRs. For non-LOCA-sensitive systems all pipe break failures were collected which had a leak flow $\geq$ 1 US gal/min for pipes $\geq$ 2 in. diameter or which had a leak flow of $\geq$50 US gal/min for all pipe sizes.

Fir PWRs with a total period of operation of 485 years and BWRs with a period 313 years there were no failures of LOCA-sensitive pipework. There were, therefore, 798 years of operation of the two types of (LWR) free of failure. The point estimates of the failure rate based on this information were 0.0005, 0.0007 and 0.0003 failures/reactor–year for PWRs, BWRs and LWRs, respectively. The value for LWRs is evidently not the average of the

Sample size (reactor system-years	*No. of failures*		*Failure rate* (failures/year)	
	Potentially dangerous failures	*Catastrophic failures*	*Potentially dangerous failures*	*Catastrophic failures*
2397	35	6	1.5×10^{-2}	2.5×10^{-3}

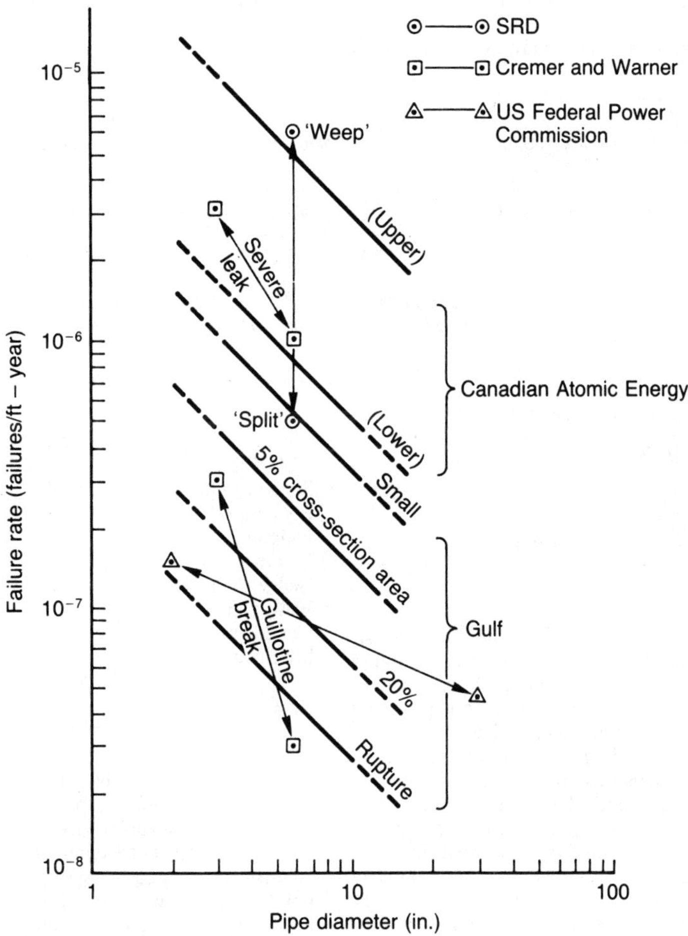

Figure 12.26 *Some data on pipework failure rates (Hawksley, 1984)*

values for the two types of LWR, but a value determined in its own right and based on the total of 798 years accumulated by both reactor types.

For the non-LOCA-sensitive pipework there were 9 failures in PWRs and 10 in BWRs, making a total of 19 failures for LWRs. Pipe length and weld population data were available for 18 plants. The authors used these data to give the pipe length and weld populations of a nuclear plant. The mean pipe lengths and weld numbers obtained are given in Table 12.15, Sections A and B. Breakdowns of the failures by plant type and pipe size, by leak flow and by operational mode are given in Sections C, D and E of Table 12.15, respectively. The causes of failure were mainly vibration (10 cases) and water hammer (3 cases). The other cases were attributed one each to pump seizure, corrosion, fatigue, impact, operator error and unknown cause.

In the survey of failure rates of pressure equipment in process plants described in the previous section, Arulanantham and Lees (1981) also report data on major failures of pipework in one of the olefins plants. Over a 9.5-year period there were 7 failures, giving a

Table 12.13 *Estimates of failure rates of pipework in nuclear plants given in the* Rasmussen Report *(after Atomic Energy Commission, 1975)*

	Failure rate
'Probability of large scale rupture of primary coolant system' (A.E. Green and Bourne) (1968)	$3 \times 10^{-6} - 2 \times 10^{-3}$/plant-year
'Catastrophic rupture of primary system pipes' (Salvatory) (1970)	1×10^{-4}/plant-year
'Pipe rupture' (Erdmann) (1973):	1.5×10^{-6}/section-year
Which corresponds roughly to	$1 \times 10^{-4} - 1 \times 10^{-2}$/plant-year
'Pessimistic probability for catastrophic failure of primary system of PWR' (Otway)	1.7×10^{-7}/plant-year
'Total probability of severence anywhere in primary system piping' (General Electric Report) (1970):	
Without ultrasonic testing	1×10^{-3}/plant-year
With ultrasonic testing	1×10^{-4}/plant-year
'Failure rate for rupture of primary coolant system piping' (Wells–Knecht) (1965)	1×10^{-7}/plant-year

Note: The references given in this table refer to those in the *Rasmussen Report*, not this book.

Table 12.14 *Failure rates for LOCA-initiating ruptures used in the* Rasmussen Report *(after R.E. Wright, Steverson and Zuroff, 1987)*

Pipe rupture size (in.)	*LOCA-initiating rupture rates* (failures/plant-year)	
	Median	*Range*
$\frac{1}{2}$–2	1×10^{-3}	$1 \times 10^{-4} - 1 \times 10^{-2}$
2–6	3×10^{-4}	$3 \times 10–5 - 3 \times 10^{-3}$
>6	1×10^{-4}	$1 \times 10^{-5} - 1 \times 10^{-3}$

failure rate of 0.74 failures/plant. The failures are described individually: two were due to water hammer, two to vibration-induced fatigue, one to brittle fracture resulting from admission of freezing hydrocarbon and one to overpressure of a pump suction line.

A survey of pipework failure in plants in the nuclear, chemical and other industries had been described by Blything and Parry (1988 SRD R411). The data sources were as follows. For chemical plants use was made of data from a medium-sized plant and of incident data from four separate sources; for refineries data were obtained from the plants of an oil company and from incidents; for nuclear plants data sources were a study by Riso National Laboratory and a presentation by the Nuclear Installations Inspectorate (NII); for steam plants the data were taken from T.A. Smith and Warwick (1974) and Gibbons and Hackney (1964).

The data were analysed by 'failure cause' and 'root cause'. Essentially, failure causes are the mechanical causes, such as corrosion, fatigue and water hammer, and root causes are activities such as error in design, operation and maintenance. Their results are summarized in Table 12.16, which gives the failure causes vs root causes for chemical plants and refineries (Section A) and for nuclear plants and steam plants (Section B).

The severity of failure expressed as leak flow is shown in Table 12.17 for the medium-sized chemical plant in Section A and for the GEC steam plant survey in Section B.

The mechanical failures for the five chemical plant sources were as follows:

	No.	*Proportion* (%)
Weld failure	45	33.1
Stress rupture	30	22.1
Bending stress	2	1.5
Fatigue	3	2.2
External load	4	2.9
Bellows failure	3	2.2
Valve failure	27	19.9
Seal failure	15	11.0
Miscellaneous	7	5.1
Total	136	100.0

Kletz (1984k) has given information on some 50 major pipe failures in process plants as shown in Table 12.18. Section A of the table gives general failures, Section B gives failures in dead ends, and Section C gives bellows failures. The suggestions which Kletz makes for prevention of each failure are given in Table 12.19. He also makes proposals for reduction of pipework failures by improved design and describes some points to look for in inspection of pipework. These aspects are described in Section 12.7 and Chapter 19, respectively.

A study of pipework failures in process plants has been described by Bellamy and co-workers (Bellamy, Geyer and Astley 1989; Geyer *et al.*, 1990; Geyer and Bellamy, 1991; Hurst *et al.*, 1991). This work was concerned particularly with human factors as a cause of failure. This study reviewed 921 incidents from incident data bases such as the HSE MARCODE, the Safety and Reliability Directorate (SRD) MHIDAS and (the Toegepast-Natuurwetenschappelijk Onderzoek (TNO) FACTS data bases. Some analysis was undertaken on all 921 incidents, and some 500 of these incidents were selected as suitable for further analysis. Pipelines and flexible hoses were excluded.

Table 12.15 *Failure rates for rupture in non-LOCA-sensitive pipework in US nuclear plants (after R.E. Wright, Steverson and Zuroff, 1987)*

A Pipe length populations (ft)[a]

	Pipe size (in.)				
	2	*>2–6*	*>6*	*No. of plants*	*Total length (in.)*
PWRs:					
LOCA-sensitive	212	338	209	6	1834
Non-LOCA-sensitive[b]	6312	11 883	15 021	2–8 (av. 4.7)	33 239
BWRs:					
LOCA-sensitive	2603	3024	4427	1–2 (av. 1.7)	5803
Non-LOCA-sensitive[b]	–	5906	6577	0–2 (av. 1.0)	3045

B Pipe weld populations (no of welds)[a]

	Pipe size (in.)				
	2	*>2–6*	*>6*	*No. of plants*	*Total*
PWRs:					
LOCA-sensitive	123	161	100	8	579
Non-LOCA-sensitive[b]	2914	3862	3800	3–10 (av. 5.2)	12 289
BWRs:					
LOCA-sensitive	870	448	622	2	1078
Non-LOCA-sensitive[b]	–	1915	2114	1–2 (av. 1.7)	861

C Failures by plant type and pipe size

Pipe size (in.)	*No. of failures*	*No. of reactor years*	*Failure rate* (failures/reactor-year)
PWRs:			
$\frac{1}{2}$–2	2	484	0.0041
>2–6	4	484	0.0083
>6	3	484	0.0062
Total			0.0186
BRWs:			
$\frac{1}{2}$–2	3	313	0.0096
>2–6	2	313	0.0064
>6	5	313	0.0160
Total			0.0320
LWRs:			
Total	19	798	0.0238

D Failures by leak flow

Leak flow (US gal/min)	*No. of failures*	*No. of reactor-years*	*Leak frequency* (leaks/reactor-year)
PWRs:			
≥1–<15	5	484	0.0103
≥15	4	484	0.0083
BWRs:			
≥1–<15	1	313	0.0032
≥15	9	313	0.0287

E Failures by operational mode

Plant type	*No. of failures*			
	Starting up	*Normal operation*	*Whilst shut down*	*Total*
PWR	2	5	2	9
BWR	0	8	2	10

[a] The figures in Sections A and B of the table are for background information only. They give an indication of the population of pipe lengths and welds on which data were available. They were not used as such in determining failure rates, which are given as failures/ reactor-year. The total in the final columns is not necessarily the sum of those in the earlier columns. A full explanation of these data is given by the authors.

[b] For PWRs this is primary circuit pipework, for BWRs it is recirculation, main steam and main feed (interpreted as feedwater) pipework

Table 12.16 *Failure of pipework in chemical, refinery, nuclear and steam plants: failure cause vs root cause (after Blything and Parry, 1986 SRD R441) (Courtesy of the UKAEA Safety and Reliability Directorate)*

A Failures in chemical plants and refineries – 'failure cause' vs 'root cause'

	Design	*Installation*	*Design/ installation*	*Operation*	*Maintenance*	*Manufacture*	*Unknown*	*Unspecified*	*Total*
Corrosion:									
External	18	8	–	2	4	–	–	1	33
Internal	56	1	2	1	1	1	–	3	65
Stress	15	–	1	–	–	–	–	–	16
Erosion	2	1	–	–	1	–	–	–	4
Restraint	1	2	4	–	–	–	–	–	7
Vibration	9	1	3	1	–	–	–	1	15
Mechanical	28	10	5	11	12	18	2	21	107
Material	5	7	10	–	4	2	–	21	49
Freezing	13	1	–	2	–	–	–	1	17
Thermal fatigue	2	1	–	2	–	1	–	1	7
Water hammer	2	1	1	4	–	–	–	–	8
Work systems	6	4	36	47	49	–	–	2	144
Unknown	–	–	–	–	–	–	29	1	30
Unspecified	1	1	13	3	3	–	–	33	54
Total	158	38	75	73	74	22	31	85	556

B Failures in steam plants -'failure cause' vs 'root cause'

Corrosion:									
External	–	–	–	–	–	–	–	–	–
Internal	16	–	–	14	–	1	–	–	31
Stress	5	–	–	–	–	1	–	–	6
Erosion	13	–	–	53	–	1	–	–	67
Restraint	2	–	2	–	–	–	–	–	4
Vibration	7	1	–	–	–	–	–	–	8
Mechanical	11	6	–	4	1	22	1	–	45
Material	3	–	–	–	–	14	–	–	17
Freezing	1	–	–	–	–	–	–	–	1
Thermal fatigue	7	–	–	–	–	3	–	–	10
Water hammer	–	–	–	2	–	–	–	–	2
Work systems	–	2	1	–	–	–	–	–	3
Unknown	–	–	–	–	–	–	4	–	4
Unspecified	–	–	–	–	–	4	–	–	4
Total	65	9	3	73	1	46	5	–	202

Table 12.17 *Failure of pipework in chemical, refinery, nuclear and steam plants: leak flows (after Blything and Parry, 1986 SRD 441) (Courtesy of the UKAEA Safety and Reliability Directorate)*

A Failures in a chemical plant

Pipe size (in.)	*No. of leaks*	*No. of ruptures*
$d \leq 6$	29	2
$10 < d < 15$	12	–
Total	41	2

B Failures of steam plant (GEC study)

Pipe size (in.)	*No. of leaks*	*No. of ruptures*
$d \leq 6$	115	9
$10 < d \leq 15$	26	7
$d > 15$	6	1
Unspecified	18	2
Total	165	19

The data were biased towards failures in larger pipes, as the following breakdown of pipework failures shows:

Pipe size (mm)	*Frequency of failure*
>150	70
51–149.9	31
26–50.9	19
13.1–25.9	22
≤13	3
Total	145

This reflects the fact that the study was based on incidents held in data bases.

Incidents were classified under the three headings: (1) direct cause, (2) origin of failure or underlying cause, and (3) recovery from failure or preventive mechanism. Detailed accounts of these definitions are given by Bellamy and Geyer (1989) and by Hurst *et al.* (1991).

Table 12.20 gives the direct causes of failure and Table 12.21 the underlying causes vs the recovery failure. Table 12.22 gives a breakdown of the failures of pipes, valves and other equipment, for those incidents where the information was available. Table 12.23 shows the state of the equipment, again for those cases where there was information. The survey also yielded information on various aspects of the releases, some of which is described in other chapters.

In the work already described, Hurst Davies *et al.* (1994) also plotted the distribution of pipework guillotine failure rates from a number of sources, in order to check whether the failure rates are log–normally distributed. Their plot for pipework guillotine failure rates is shown in Figure 12.25(b), the straight line confirming that the distribution of the failure rates is log–normal. The mean pipework guillotine failure rate obtained from this plot is $\lambda \approx 4.6 \times 10^{-7}$/m-year. The failure rates vary, however,

Table 12.18 *Some pipework failures in process plants: individual cases (after Kletz, 1984k) (Courtesy of the American Institute of Chemical Engineers)*

A General failures

1. A pipe was secured too rigidly by welding to supports; vibration caused a section to be torn out
2. Thermal expansion of a pipe caused a $\frac{3}{4}$-in. branch to press against a girder on which the pipe rested; the branch was torn off
3. A batch of pipe-hangers were too hard and many cracked
4. A crane was used to move a live line slightly so that a joint could be remade
5. An old pipe was reused after being on corrosive/erosive service; it failed
6. A pipe was laid on the ground and corroded badly. Construction team may have used a pipe which was already corroded
7. Water injection caused excessive corrosion/erosion because the mixing arrangements were poor
8. Water injection caused excessive corrosion/erosion because the mixing arrangements were poor
9. A temporary support was left in position
10. The exit pipe from a converter was made of carbon steel instead of $\frac{1}{2}$% Mo; the pipe fragmented and the reaction force caused the converter to fall over
11. A crane hit an overhead pipeline
12. Vibration caused fatigue failure of a 2-in. long pipe
13. The wrong valve was opened and liquid nitrogen entered a mild-steel line causing it to disintegrate
14. Vibration caused fatigue failure of a 1-in. long pipe
15. An underground pipeline corroded and leaked
16. Vibration caused a fatigue failure at a badly designed joint
17. A construction worker cut a hole in a pipeline in the wrong place and, discovering his error, patched the pipe. The repair was substandard, corroded, and leaked badly
18. A heat transfer oil line failed by fatigue as the result of repeated expansion and contraction. There should have been more expansion bends in the line
19. Water froze in an LPG drain line. A screwed joint fractured. Screwed joints should not be used for LPG
20. A section of steam tracing was isolated, causing a blockage. Expansion of the liquid in the rest of the pipe caused it to burst
21. A level controller fractured at a weld, the result of poor workmanship. According to the report, the failure 'emphasizes the need for clear instructions on all drawings and adequate inspection during manufacture'
22. A 1-in. screwed nipple blew out of a hot oil line. It was installed 20 years before, during construction, for pressure testing and was not shown on any drawing
23. Water injection caused corrosion/erosion because a properly designed mix nozzle was installed pointing in the wrong direction (see (7))
24. A Be/Cu circlip was used in an articulated arm carrying ammonia instead of a stainless steel one. The joint blew wide open
25. Decomposition of the contents caused a pipeline to fail
26. An LPG line corroded because it passed through a pit full of water contaminated with acid

27. The tail pipe from a relief valve came down to the ground and dipped into a pool of water which froze
28. Underground propane and oxygen lines leaked causing an underground explosion. the report states, 'During execution of the pipework, doubts were expressed by the works management as to the quality of the workmanship and the qualifications of those employed'
29. A little-used line was left full of water. Frost split it
30. A portable hand-held compressed air grinder being used on a new pipeline was left resting between two live lines. When the air compressor was started the grinder, which had been left switched on, started to turn and ground away part of a live line
31. A new line had the wrong slope so the contractors cut and welded some hangers. They failed. Other hangers failed due to incorrect assembly and absence of lubrication
32. The space between a reinforcement pad and the pipe was not vented. Water in the space vaporized causing collapse of a pipe
33. An old pipe was reused after use on duty which used up most of its creep life (see (5))
34. Two pipe-ends which were to be welded together did not fit exactly and were welded with a step between them
35. An ice/hydrate plug blocked a blowdown line. It was cleared by external steaming. When the choke cleared the pressure above it caused it to move with such force that the line fractured at a T
36. A carbon steel line was installed instead of $1\frac{1}{4}$ Cr $\frac{1}{2}$ Mo and ruptured after 16 years by H_2 attack (see (10))

B Failures in dead ends

1. Water collected in a dead end branch 10 ft long and caused corrosion. Five men were killed when the branch failed and the escaping gas ignited
2. Water collected in a dead end branch and froze, breaking the branch
3. Water collected in the dead end branch leading to a flowmeter which had been removed. It froze and damaged a valve
4. Corrosion products collected in the branch leading to a spare pump which was never used; the branch failed
5. A stainless steel line operating at 360°C was fitted with a branch leading to a relief valve. The branch was made of stainless steel for 1 m and then mild steel. The temperature of the mild steel exceeded the 100°C estimate and the line failed by hydrogen attack
6. Corrosive by-products collected in a blanked branch; the branch failed
7. Water collected in a branch on the feed line to a furnace; the branch was permanently connected to a steam supply. The water froze and fractured the line
8. Water collected in an open-ended branch which was welded onto a pipeline as an instrument support. After 4 years the process line had corroded right through

C Bellows failures

1. A bellows was damaged before delivery
2. A Fluon bellows failed because the pipe was free to move sideways
3. Two pins in a hinged bellows failed by fatigue and the bellows became distorted
4. A bellows was designed for normal operating conditions but distorted under abnormal, but foreseeable, conditions
5. A bellows blew apart a few hours after installation. The split rings around the convolutions which support them and equalize expansion were slack
6. Flixborough – two 28-in. bellows failed completely

Table 12.19 *Some pipework failures in process plants: preventive measures (after Kletz, 1984k) (Courtesy of the American Institute of Chemical Engineers)*

A Points to look for during construction

1. Equipment is made of the grade of steel specified and has received the right heat treatment
2. Old pipe is not being reused without checking that it is suitable
3. Pipes are not laid underground
4. Workmanship is of the quality specified and tests are carried out as specified
5. Purchased equipment is undamaged

B Points to look for after construction

1. Pipes are not secured too rigidly (by welding or clamping) so that they are not free to expand
2. Pipes will not foul supports or other fixtures when they expand
3. Pipes are not in contact with the ground
4. Temporary supports have been removed
5. Pipes are free to expand
6. Screwed joints have not been used
7. Steam tracing cannot be isolated on different sections of the same process line
8. Temporary branches, nipples, and plugs have been removed and replaced by properly designed welded plugs
9. Equipment has not been assembled wrongly (First identify equipment which can be assembled wrongly)
10. Pipes do not pass through pits or depressions which will fill with water
11. Relief-valve tail pipes are not so close to the ground that they may be blocked by ice or dirt
12. Lines which may contain water can be drained
13. The slope of lines is correct (for example blowdown line should slope towards the blowdown drum)
14. There is no 'bodging'
15. Reinforcement pads are vented
16. There are no dead ends in which water or corrosive materials can collect (Note: dead ends include little used branches as well as blanked lines)
17. There are no water traps formed by brackets, etc., fixed to equipment
18. Bellows are not bent because the two pipe ends are not in line
19. The support rings on bellows are not loose

C Points to look for after start-up

1. Pipes are not vibrating

by some 2.5 orders of magnitude at the 95% confidence level.

Further values for pipework failure are quoted in work on hazard assessment. Estimated failure rates of pipework quoted by Batstone and Tomi (1980) are given in Table 12.24.

Pape and Nussey (1985) in a hazard assessment of a chlorine installation have used for 25 mm diameter pipe the following values:

Frequency of guillotine fracture
$= 3 \times 10^{-7}$ failures/m-year
Frequency of lesser failure
$= 3 \times 10^{-6}$ failures/m-year
Frequency of gasket failure
Gaskets 0.6 mm thick $= 3 \times 10^{-6}$ failures/year
Gaskets 3 mm thick $= 5 \times 10^{-6}$ failures/year

The authors state that these data include valve leaks.

The failure rates given by Hawksley (1984) in Figure 12.26 yield the values given in Table 12.25. Comparison of the frequency of a guillotine rupture for pipe diameters $\geq$ 3 in. quoted by some of these sources gives:

	Failure frequency (failures/year)
Gulf	3×10^{-7}
Pape and Nussey	3×10^{-7}
Cremer and Warner	$\approx 1 \times 10^{-6}$
Batstone and Tomi	3×10^{-6}

In converting the data of Batstone and Tomi, it has been assumed that a connection is a pipe section 10 m long. From these data it might be estimated that for regular pipework the frequency of guillotine rupture for a pipe $\geq$ 3 in. diameter is 1×10^{-6} failures/m-year and that for more critical pipework it is some 3 times less.

Table 12.20 *Failures of pipework in incidents: direct causes (after Bellamy, Geyer and Astley, 1989) (Courtesy of the Health and Safety Executive)*

	No. of incidents	*Contribution*	
		Normalized No.[a]	*Proportion* (%)
Corrosion	92	85.5	9.3
Erosion	11	7.3	0.8
External load	35	27.5	3.0
Impact	49	43.8	4.8
Overpressure	129	111.8	12.1
Vibration	16	14.0	1.5
Temperature (high or low)	44	34.8	3.8
Wrong or incorrectly located in-line equipment	44	36.8	4.0
Operator error	190	167.8	18.2
Defective pipe or equipment	303	293.5	31.9
Other	17	14.0	1.5
Unknown	84	84.0	9.1
Total	1014	921.0	100.0

[a] Some incidents were assigned to more than one direct cause so that the total number of entries in the first column is 1014. The second column gives the number of incidents normalized to a total of 921.

12.30.3 Failure of process equipment

A survey of the causes of service failure in process equipment generally has been given by Collins and Monack and this has been further analysed by Lancaster (1975). Some 685 failures were recorded. The results of the survey are shown in Table 12.26. Causes of general service failure which are emphasized by this work are general corrosion, stress corrosion cracking and mechanical failure. The number of failures caused by brittle fracture is small.

12.30.4 Failure of process machinery

The results of a survey of causes of service failure in process machinery are shown in Table 12.27. The importance of misalignment as a cause of failure is brought out strongly.

Table 12.21 *Failures of pipework in incidents: underlying cause vs recovery failure (after Geyer et al., 1990)*

Underlying cause	*Recovery failure (%)*						
	Not recoverable	*Hazard study*	*Human factors review*	*Task checking*	*Routine checking*	*Unknown recovery*	*Total*
Natural causes	1.8	—	—	0.2	—	—	2.0
Design	—	24.5	2.0	—	0.2	—	26.7
Manufacture	—	—	—	2.4	—	—	2.4
Construction	0.1	0.2	1.9	7.5	0.2	0.4	10.3
Operation	—	0.1	11.0	1.6	0.2	0.8	13.7
Maintenance	—	0.4	14.5	12.7	10.3	0.8	38.7
Sabotage	1.2	—	—	—	—	—	1.2
Domino	4.5	0.2	—	—	0.3	—	5.0
Total	7.6	25.4	29.5	24.4	11.1	2.0	100.0

Table 12.22 *Failures of pipework in incidents: failures of pipes, valves and other equipment (after Bellamy, Geyer and Astley, 1989) (Courtesy of the Health and Safety Executive)*

A Pipe failures

	Frequency
Full bore release:	
Spontaneous rupture	165
Release during line opening or other human activity	25
Leak:	
Spontaneous leak	85
Leak during line opening or other human activity	5
Failure	99
Total	379

B Valve failures

	Frequency
Valve operation mode:	
Manual	7
Automatic	4
Remotely operated	2
Valve function:	
Flow control	20
Isolation	7
Drain	5
Emergency shut-down	2
Safety/relief	6
Other	3
Valve position:	
Valve failed open	2
Valve failed closed	25

C Failure of other equipment[a]

	Frequency
Bellows	5
Coupling	6
Drum	3
Elbow	2
Gasket	10
Joints	20
Compressor	2
Filter	4
Flange	48
Fitting	2
Gasket	10
Packing	14
Gland (gland packing)	2
Pump	10
Rupture/bursting disc	5
Seal	9
Shell	5
Sightglass	2
T pieces	3
Weld	17
Total	179

[a] Only equipment with more than one reported failure is listed

Table 12.23 *Failures of pipework in incidents: equipment in incorrect status (after Bellamy, Geyer and Astley, 1989) (Courtesy of the Health and Safety Executive)*

A Type of equipment

	Frequency
Pipe	76
Valve	53
Flange	5
Pump	4
Other	9
Total	147

B Nature of incorrect status

	Frequency
Disconnected/connected	4
Removed	10
Not bled/drained/cleaned	35
Not effectively isolated	25
Open/on	50
Closed/off	8
Other/unknown	8
Total	140

Table 12.24 *Some failure rates of pipework used in hazard assessment (after Batstone and Tomi, 1980)*

Pipe diameter (mm)	*Failure rate per connection* (failures/10^6 years)
≤25	30
40	10
50	7.5
80	5
100	4
≥150	3

Table 12.25 *Some failure rates (failures/ft-year) of pipework given in the literature (after Hawksley, 1984)*

Source	*Type of failure*	*Pipe diameter (in.)* 3	6	10
Canadian Atomic Energy	Upper line	10^{-5}		3×10^{-6}
	Lower line	1.7×10^{-6}		4×10^{-7}
Gulf	Small	10^{-6}		3×10^{-7}
	5%	5×10^{-7}		1.3×10^{-7}
	20%	2×10^{-7}		6×10^{-8}
	Rupture	9×10^{-8}		2.5×10^{-8}
Cremer and Warner	Severe leak	3×10^{-6}	10^{-6}	
	Guillotine break	3×10^{-7}		3×10^{-8}
SRD	Weap		6×10^{-6}	
	Split			5×10^{-7}
FPC	Undefined	1.3×10^{-7}		10^{-7}

FPC, Federal Power Commission; SRD, Safety and Reliability Directorate.

Table 12.26 *Causes of service failure in metal equipment and piping in chemical plants (Collins and Monack, 1973) (Courtesy of Materials Protection and Performance)*

Corrosion	(%)
Cavitation	0.3
Cold wall	0.4
Cracking, corrosion fatigue	1.5
Cracking, stress corrosion	13.1
Crevice	0.9
Demetallification	0.6
End grain	0.4
Erosion–corrosion	3.8
Fretting	0.3
Galvanic	0.4
General	15.2
Graphitization	0.1
High temperature	1.3
Hot wall	0.1
Hydrogen blistering	0.1
Hydrogen embrittlement	0.4
Hydrogen grooving	0.3
Intergranular	5.6
Pitting	7.9
Weld corrosion	2.5
Subtotal	55.2
Mechanical failure	
Abrasion, erosion or wear	5.4
Blisters, plating	0.1
Brinelling	0.1
Brittle fracture	1.2
Cracking, heat treatment	1.9
Cracking, liquid metal pen	0.1
Cracking, plating	0.6
Cracking, thermal	3.1
Cracking, weld	0.6
Creep or stress rupture	1.9
Defective material	1.6
Embrittlement, sigma	0.3
Embrittlement, strain age	0.4
Fatigue	14.8
Galling	0.1
Impact	0.1
Leaking through defects	0.4
Overheating	1.9
Overload	5.4
Poor welds	4.4
Warpage	0.4
Subtotal	44.8

No. of failures: 685
Period: 1968–71

Table 12.27 *Causes of service failure of rotating machinery in the process industries (after Anon., 1970a)*

Cause	*(%)*
Misalignment between machines	>50
Ingestion of solid materials	10
Portions of rotating element thrown	10
Vibration causing seal loss, causing thrust bearing failure	10
Various other sources: lubrication failures, internal misalignment, piping vibration, overspeed, slugging with liquid, chemical attack, surge	10
Design and manufacturing errors	<10

13 Control System Design

The operation of the plant according to specified conditions is an important aspect of loss prevention. This is very largely a matter of keeping the system under control and preventing deviations. The control system, which includes both the process instrumentation and the process operator, therefore has a crucial role to play. Selected references on process control are given in Table 13.1.

Traditionally, control systems have tended to grow by a process of accretion as further functions are added. One of the thrusts of current work is to move towards a more systematic design approach in which there is a more formal statement of the control objectives, hierarchy, systems and subsystems.

Once the objectives have been defined, the functions of the systems and subsystems can be specified. Typical subsystems are those concerned with measurement, alarm detection, loop control, trip action, etc. The next step is the allocation of function between man and machine – in this case the instrumentation and the operator. This allocation of function and the human factors aspects of process control are discussed in Chapter 14.

It is convenient to distinguish several broad categories of function that the control system has to perform: these are (1) information collection, (2) normal control and (3) fault administration.

A control system is usually also an information collection system. In addition to that required for immediate control of the process, other information is collected and transmitted. Much of this is used in the longer term control of the process. Another category which is somewhat distinct from normal control is the administration of fault conditions which represent disturbances more severe than the control loops can handle.

Table 13.1 *Selected references on process control*

NRC (Appendix 28 *Control Systems*); A.J. Young (1955); Ceaglske (1956); D.F. Campbell (1958); Grabbe, Ramo and Wooldrige (1958); Macmillan (1962); Buckley (1964); R.J. Carter (1964, 1982); Harriott (1964); Hengstenberg, Sturm and Winkler (1964); Coughanowr and Koppel (1965); Perlmutter (1965); Franks (1967); IChemE (1967/45); C.D. Johnson (1967); E.F. Johnson (1977); H.S. Robinson (1967b); Shinskey (1967, 1977, 1978, 1983); Himmelblau and Bischoff (1968); *Chemical Engineering* (1969c); Gould (1969); McCoy (1969); Soule (1969–); Himmelblau (1970); Considine (1971); Hartmann (1971); Pollard (1971); Luyben (1973); C.A.J. Young (1973); C.L. Smith and Brodman (1976); Lees (1977a); R.E. Young (1977, 1982); C.L Smith (1979); Dorf (1980); L.A. Kane (1980); Basta (1981d); Frankland (1981); Auffret, Boulvert and Thibault (1983); Stephanopoulos (1984); *Hydrocarbon Processing* (1986a–); Tsai, Lane and Lin (1986); Benson (1987); Prett and Morari (1987); W.R. Fisher, Doherty and Douglas (1988); Prett and Garcia (1988); Asbjornsen (1989); T. Martin (1989b); K. Pritchard (1989); R. Hill (1991); Ayral and Melville (1992); Y.Z. Friedman (1992); T. Palmer (1992); C. Butcher (1993c); Holden and Hodgson (1993); Ponton and Laing (1993); Roberson, O'Hearne and Harkins (1993)

Sequence control, batch control, including computer control
Kochhar (1979); Thome, Cline and Grillo (1979); Ghosh (1980); Rosenof (1982b); Armstrong and Coe (1983); Severns and Hedrick (1983); Anon. (1984ii); M. Henry, Bailey and Abou-Loukh (1984); Bristol (1985); Cherry, Preston and Frank (1985); E.M. Cohen and Fehervari (1985); Krigman (1985); Namur Committee (1985); Preston and Frank (1985); Egli and Rippin (1986); Love (1987a,b, 1988); Rosenof and Ghosh (1987); ISA (1988); Kondili, Pantiledes and Sargent (1988); Cott and Macchietto (1989); IChemE (1989/135); T.G. Fisher (1990); Crooks, Kuriyna and Macchietto (1992); Wilkins (1992); Sawyer (1992a,b, 1993a,b); Hedrik (1993)

Reactor control (*see also* Table 11.4)
Aris and Amundson (1957, 1958); Harriott (1961, 1964); Levenspiel (1962); Dassau and Wolfgang (1964); Coughanowr and Koppel (1965); Denbigh (1965); Perlmutter (1965); Shinskey (1967); Buckley (1970); Schöttle and Hader (1977); Rosenhof (1982a,b); R. King and Gilles (1986); Rosenof and Ghosh (1987); Craig (1989)

Compressor control, turbine control
Claude (1959); Hagler (1960); Tezekjian (1963); R.N. Brown (1964); Daze (1965); Marton (1965); Hatton (1967); Magliozzi (1967); Hougen (1968); Labrow (1968); M.H. White (1972); Nisenfeld *et al.* (1975); Sweet (1976); IEE (1977 Coll. Dig. 77/38); Nisenfeld and Cho (1978); Staroselsky and Ladin (1979); D.F. Baker (1982); Bass (1982); Gaston (1982); Maceyka (1983); B. Fisher (1984); Rana (1985); AGA (1988/52)

Process instrument and control systems
Isaac (1960); Anon. (1962a); Fusco and Sharshon (1962); Richmond (1965); Fowler and Spiegelman (1968); Byrne (1969); Frey and Finneran (1969); Klaassen (1971); Hix (1972); Nisenfeld (1972); Jervis (1973); K. Wright (1973); Calabrese and Krejci (1974); Wilmot and Leong (1976); Gremillion (1979); Mosig (1977); Redding (1977); Shinskey (1978); Kumamoto and Hensley (1979); Rinard (1982); Cocheo (1983); Rindfleisch and Schecker (1983); Swanson (1983); Galuzzo and Andow (1984); Love (1984); E.M. Cohen (1985); B. Davis (1985); S.J. Brown (1987); Cluley (1993); I.H.A. Johnston (1993)

13.1 Process Characteristics

The control system required depends very much on the process characteristics (E. Edwards and Lees, 1973). Important characteristics include those relating to the disturbances and the feedback and sequential features. A review of the process characteristics under these headings assists in understanding the nature of the control problem on a particular process and of the control system required to handle it.

Processes are subject to disturbances due to unavoidable fluctuations and to management decisions. The disturbances include:

(1) raw materials quality and availability;
(2) services quality and availability;
(3) product quality and throughput;
(4) plant equipment availability;
(5) environmental conditions;

and due to

(6) links with other plants;
(7) drifting and decaying factors;
(8) process materials behaviour;
(9) plant equipment malfunction;
(10) control system malfunction.

Quality may relate to any relevant parameter such as the composition or particle size of the material, the voltage level of a power supply or the specification of a product. Plant equipment may be taken off or brought back into service. Links with other plant may require changes in the operation of the process. Typical drifting and decaying factors are fouling of a heat exchanger and decay of catalyst. Process materials introduce disturbances through such behaviour as the clogging of solids on weighbelts or the blocking of pipes. Plant equipment failures constitute disturbances, as do those of the control system such as instrument faults, measurement noise, control loop instability or operator error.

Certain trends in modern plants tend to intensify the process disturbances. They include use of continuous, high throughput processes, existence of recycles, elimination of storage and interlinking of plants.

Some process characteristics which tend to make feedback control more difficult include:

(1) measurement problems;
(2) dead time;
(3) very short time constants;
(4) very long time constants;
(5) recycle;
(6) non-linearity;
(7) inherent instability;
(8) limit cycles;
(9) strong interactions;
(10) high sensitivity;
(11) high penalties;
(12) parameter changes;
(13) constraint changes.

Measurement has always been one of the principal problems in process control. A measurement may be difficult to make; it may be inaccurate, noisy, or unreliable; or it may be available in sampled form only. Even if the measurement itself is satisfactory, it may not be the quantity of prime interest. An 'indirect' or 'inferred' measurement may have to be computed or otherwise obtained from the actual plant measurement(s). Feedback control is totally dependent on measurement.

Dead time or time delay arises in various ways in processes. It may be introduced by the distance–velocity lag in pipework, the nature of distributed parameter systems or the time to obtain a sample or laboratory analysis. Dead time makes feedback control more difficult, owing to the delay before any error is measured and corrective action is initiated.

Processes with very short time constants are obviously difficult to control, because the speed of response required for control decisions and actions is rapid. But so also are processes with very long time constants, where the problems have to do with the increased chance of disturbances and other control interactions upsetting the control action taken and with the difficulty of remembering all the relevant factors.

Recycle takes a number of forms, including recycle of a process stream to an earlier point in the process and internal recycle within a vessel.

If a process is very non-linear, its behaviour tends to vary with throughput, its responses to disturbances and corrective actions differ, and it becomes difficult to find satisfactory controller settings.

Some processes, notably certain chemical reactors, are inherently unstable over a certain range of operation. If the process enters the unstable region, variables such as temperature and pressure may increase exponentially, leading to an explosion. In other cases the process enters a limit cycle and oscillates between definite limits.

The relationships between the input and output variables of a process are often complex and there may be strong interactions. One input may change several outputs and one output may be changed by several inputs. Where the output variables are controlled by single loops, severe interactions may occur between these loops.

Some processes are very sensitive and this clearly intensifies the difficulty of control. So also does the existence of very high penalties for excursions outside the control limits.

Process parameter changes tend to reduce the effectiveness of controller settings and may make the process inherently more difficult to control. Constraint changes alter the envelope within which the process is to be controlled.

The sequential control characteristics of a process include:

(1) plant start-up;
(2) plant shut-down;
(2) batch operation;
(4) equipment changeover;
(5) product quality changes;
(6) product throughput changes;
(7) equipment availability changes;
(8) mechanical handling operations.

The sequential element in the start-up and shut-down of continuous processes and in batch processes is obvious, but there are other operations with sequential features. Continuous processes often contain semi-continuous equipment, particularly where regeneration is necessary. Deliberate changes in product quality or throughput or in equipment status involves sequential operation. In general, a sequence consists of a series of stages, of which some are initiated by events occurring in the process and others are initiated after the lapse of a specified time.

Some other process characteristics which may be significant include requirements for:

(1) monitoring;
(2) feedforward control;
(3) optimization;

(4) scheduling;
(5) process investigation;
(6) plant commissioning.

Monitoring is usually a very important function in the control system. The monitoring requirements posed by a process vary, but in cases such as multiple identical units or batch operations they can be very large.

Feedforward control may be appropriate if there are difficulties in feedback control due to measurement problems or process lags. It is applicable where the disturbances can be measured but not eliminated, and where a model exists which makes possible the prediction of the effect on the controlled variable of both the disturbing and correcting variables.

If the plant has a time-varying operating point, continuous optimization may be appropriate. Although optimization is carried out normally for economic reasons, it is characterized by adherence to a set of constraints. Operation within the envelope of constraints contributes to process safety.

Some processes pose a scheduling requirement, particularly where batch operations are concerned. There is normally some element of novelty, in the process or plant equipment and this may give rise to a requirement for process investigation and collection of information which is not otherwise needed for control.

The investigative element is particularly important during plant commissioning. So also is the need for facilities which assist in bypassing problems on plant equipment or control instrumentation, while solutions are sought or equipment ordered.

13.2 Control System Characteristics

The characteristics of process control systems have passed through three broad phases: (1) manual control, (2) analogue control and (3) computer control (covering all forms of programmable electronic system). However, such a classification can be misleading, because it does not bring out the importance of measuring instrumentation and displays, because neither analogue nor computer control is a homogeneous stage and because it says very little about the quality of control engineering and reliability engineering and the human factors involved.

The sophistication of the measuring instrumentation greatly affects the nature of the control system even at the manual control stage. This covers instruments for measuring the whole range of chemical and physical properties. The displays provided can also vary widely. These are discussed in more detail in the next chapter.

The stage of analogue control implies the use of simple analogue controllers, but may also involve the use of other special purpose equipment. Most of this equipment serves to facilitate one of the following functions: (1) measurement, (2) information reduction and (3) sequential control.

The first two functions, therefore, improve the information available to the operator and assist him to digest it, but leave the control to him. The equipment typically includes data loggers and alarm scanners. The third function does relieve the operator of a control function. Batch sequential controllers exemplify this sort of equipment.

Another crucial distinction is in the provision of protective or trip systems. In some cases the safety shut-down function is assigned primarily to automatic systems; in others it is left to the operator. Similarly, computer control is not a homogeneous stage of development. In some early systems the function of the computer was limited to the execution of direct digital control (DDC). The real control of the plant was then carried out by the operator with the computer as a rather powerful tool at his disposal. In other systems the computer had a complex supervisory program which took most of the control decisions and altered the control loop set points, leaving the operator a largely monitoring function. The two types of system are very different.

The quality of the theoretical control engineering is another factor which distinguishes a system and largely determines its effectiveness in coping with problems such as throughput changes, dead time and loop interactions.

Equally important is the reliability engineering. Unless good reliability is achieved nominally automated functions will be degraded so that they have to be done manually or not at all. Control loops on manual setting are the typical result.

The extent to which human factors has been applied is another distinguishing feature. This aspect is considered further in Chapter 14. The general trend in control systems is an increase in the degree of automation and a change in the operator's role from control to monitoring.

Computer control itself has progressed from control by a single computer, or possibly several such computers, to distributed control by programmable electronic systems (PESs). These are described further in Section 13.4.

13.3 Instrument System Design

The design of process instrument systems, like most kinds of design, is largely based on previous practice. The control panel instrumentation and the control systems on particular operations tend to become fairly standardized. Selected references on process instrumentation are given in Table 13.2.

13.3.1 Some design principles

There are some basic principles which are important for control and instrument systems on hazardous processes. The following account has been given by Lees (1976b):

(1) There should be a clear design philosophy and proper performance and reliability specifications for the control and instrumentation. The design philosophy should deal among other things with the characteristics of the process and of the disturbances to which it is subject, the constraints within which the plant must operate, the definition of the functions which the control system has to perform, the allocation of the function of these between the automatic equipment and the process operator, the requirements of the operator and the administration of fault conditions. The philosophy and specification should cover: measurements, displays, alarms and control loops; protective systems; interlocks; special valves (e.g. pressure relief, non-return, emergency isolation); the special purpose equipment; and the process computer(s).

Table 13.2 Selected references on process instrumentation

British Gas (Appendix 27 *Instrumentation*); IEEE (Appendix 27); ISA (Appendix 27); Gillings (1958); Howe, Drinker and Green (1961); Jenett (1964a); J.T. Miller (1964); O.J. Palmer (1965); Richmond (1965, 1982); Holstein (1966); Liptak (1967, 1970, 1993); Regenczuk (1967); Considine (1968, 1971, 1985); Fowler and Spiegelman (1968); EEUA (1969 Doc. 32, 1970 Doc. 37D, 1973 Hndbk 34); HSE (1970 HSW Bklt 24); Tully (1972); Whitaker (1972); Zientara (1972); EEMUA (1973 Publ. 120); Perry and Chilton (1973); Weston (1974a,b); Anon. (1975i); Andrew (1975); Doebelin (1975); Anon. (1976 LPB 7, p. 1); J. Knight (1976); Benedict (1977); Hayward (1977, 1979); C.D. Johnson (1977); Yothers (1977); Anon. (1978 LPB 21, p. 68); Cavaseno (1978b); C. Tayler (1987a); Verstoep and Schlunk (1978); Cheremisinoff (1979, 1981); B.E. Cook (1979); Hayward (1979); Hougen (1979); Marcovitch (1979); Ottmers *et al.* (1979); Andrew and Williams (1980); *Chemical Engineering* Staff (1980); Coppack (1980); IChemE (1980/73); Medlock (1980); Messniaeff (1980); Hewson (1981); Cramp (1982); Liptak and Venczel (1982); R.J. Smith (1982); R.H. Kennedy (1983); Anon. (1984gg); IBC (1984/51); Klaassen (1984); Perry and Green (1984); Atkinson (1985); Borer (1985); Cahners Exhibitions Ltd (1985); Demorest (1985); M.J. Hauser, McKeever and Stull (1985); Higham (1985a,b); Langdon (1983); Leigh (1985); Challoner (1986); A. Moore (1986); A. Morris (1986); Tily (1986); Leigh (1987); Sinnott (1988); C. Butcher (1990b, 1991c); Bosworth (1991); Burchart (1991); Bond (1992 LPB 106); Krohn (1992); Nimmo (1992); K. Petersen (1992); API (1993 RP 551); Goodner (1993); Chilton Book Co. (1994); McClure (1994)
BS (Appendix 27 *Instrumentation*), VDI (see Appendix 27)

Symbols
ISA (1976, 1982)

Measurement
Flow: IBC (1982/26, 1984/54); IMechE (1989/100)
Level: IBC (1982/28)
Pressure, vacuum: Waters (1978); Pressure Gauge Manufacturers Association (1980); Masek (1981, 1982, 1983); Demorest (1985); Liptak (1987); Roper and Ryans (1989)
Temperature: ASTM (1974 STP 470A)
Process analysers: Huyten (1979); Verdin (1973, 1980); Huskins (1977); Carr-Brion (1986); Clevett (1986); EEMUA (1988 Publ. 138); Dailey (1993)

Non-invasive instruments
Asher (1982)

Intelligent and self-checking instruments
Hasler and Martin (1971, 1973, 1974); J.O. Green (1978); R.E. Martin (1979, 1980); Barney (1985); Dent (1988); Anon. (1994b)

Control valves
ISA (Appendix 27); Charlton (1960); Liptak (1964, 1983); EEUA (1969 Hndbk 32); Driskell (1969, 1983, 1987); Baumann (1971, 1981); Baumann and Villier (1974); Hays and Berggren (1976); Hutchison (1976); Forman (1978); R.T. Wilson (1978); Kawamura (1980); Perry (1980); Royle and Boucher (1980); Whitaker (1981); Langford (1983); M. Adams (1984); Kerry (1985); Kohan (1985); Vivian (1988); Barnes and Doak (1990); Bhasin (1990); Fitzgerald (1990); Luyben (1990); B.A. White (1993); Anon. (1994b)
BS 5793: 1979–

Fluidics
J. Grant and Marshall (1976, 1977); Grant and Rimmer (1980); Anon. (1981 LPB 40, p. 7)

Sampling
Cornish, Jepson and Smurthwaite (1981); Strauss (1985)

Signal transmission, cabling
Berry (1978); Garrett (1979); Kaufman and Perz (1978); Boxhorn (1979); Anon. (1984cc); K. Hale (1985); Higham (1985a,b); Mann (1985); C. Tayler (1986e); P. Reeves (1987); Fuller (1989)

Sneak circuits
McAlister (1984); Rankin (1984)

Intrinsic safety (*see* Table 16.2)

Fail-safe philosophy
Fusco and Sharshon (1962); Axelrod and Finneran (1965); Hix (1972); Nisenfeld (1972); Bryant (1976); Ida (1983)

Instrument commissioning
Gans and Benge (1974); Spearing (1974); Shanmugam (1981); Meier (1982)

Instrument maintenance
Upfold (1971); Skala (1974); Denoux (1975); van Eijk (1975); R. Kern (1978d)

Instrument failure (see also Appendix 14)
SIRA (1970); Anyakora, Engel and Lees (1971); A.E. Green and Bourne (1972); Lees (1976b); Cornish (1978a–c); English and Bosworth (1978); H.S. Wilson (1978); Mahood and Martin (1979); Kletz (1981i); Perkins (1980); Weir (1980); R.I. Wright (1980); Vannah and Calder (1981); Rooney (1983); Prijatel (1984); May (1985)

Logic systems
Hodge and Mantey (1967); F.J. Hill and Peterson (1968); Maley (1970); Steve (1971); D. King (1973); E.P. Lynch (1973, 1974, 1980); Zissos (1976); Kampel (1986); S.B. Friedman (1990)

Protective systems, trip systems
Bowen and Masters (1959); Obermesser (1960); Eames (1965 UKAEA AHSB(S) R99, 1966 UKAEA AHSB(S) R119, 1967 UKAEA AHSB(S) R122, R131); A.E. Green and Bourne (1965 UKAEA AHSB(S) R91, 1966 UKAEA AHSB(S) R117, 1972); L.A.J. Lawrence (1965–66); Bourne (1966 UKAEA AHSB(S) R110, 1967); A.E. Green (1966 UKAEA AHSB(S) R113, 1968, 1969 UKAEA AHSB(S) R172, 1970); Hensley (1967 UKAEA AHSB(S) R136, 1968, 1971); Hettig (1967); Vaccaro (1969); Schillings (1970); M.R. Gibson and Knowles (1971, 1982

LPB 44); Kletz (1971, 1972a, 1985n, 1987j, 1991n); R.M. Stewart (1971, 1974a,b); Stewart and Hensley (1971); Tucker and Cline (1971); Wood (1971); Bennet (1972); R.L. Browning (1972); Herrmann (1972); Nisenfeld (1972); Ruziska (1972); J.T. Fisher (1973); de Heer (1973, 1974, 1975); J.R. Taylor (1973, 1976c); AEC (1975); van Eijk (1975); Lawley and Kletz (1975); E.J. Rasmussen (1975); Hullah (1976); B.R.W. Wilson (1976); Giugioiu (1977); Quenne and Signoret (1977); B.W. Robinson (1977); Süss (1977); Troxler (1977); M.R. Gibson (1978); Kumamoto and Henley (1978); Verde and Levy (1979); Chamany, Murty and Ray (1981); Wheatley and Hunns (1981); Aitken (1982); Lees (1982a); Rhodes (1982); Ciambarino, Merla and Messina (1983); Jonstad (1983); Yip, Weller and Allan (1984); Enzina (1985); Lihou and Kabir (1985); Hill and Kohan (1986); Onderdank (1986); C. Tayler (1986c); Zohrul Abir (1987); Barclay (1988); R. Hill (1988, 1991); Kumar, Chidambaram and Gopalan (1989); Oser (1990); Papazoglu and Koopman (1990); Rushton (1991a,b, 1992); Argent, Cook and Goldstone (1992); Beckman (1992a,b, 1993); Englund and Grinwis (1992); S.B. Gibson (1992); Gruhn (1992a,b); Kobyakov (1993); R.A. Freeman (1994); VDI 2180 (1967)

Interlocks
D. Hughes (n.d.); Richmond (1965, 1982); E.G. Williams (1965); Platt (1966); Holmes (1971); Rivas and Rudd (1974); Rivas, Rudd and Kelly (1974); Becker (1979); Becker and Hill (1979); E.P. Lynch (1980); Kohan (1984); Rhoads (1985)

Control system classification
W.S. Black (1989); EEMUA (1989 Publ. 160)

Emergency shut-down systems
DoEn (1984); AGA (1988/52); Cullen (1990); HSE (1990b); J. Pearson (1992)

Leak detection
ISA (1982 S67.03)

Gas, smoke and fire detectors (*see* Table 16.2)

Toxics detectors (*see* Table 18.1)

Reaction runaway detectors
Hub (1977c); Wu (1985)

Fracture detectors
Ponton (1980); Wilkie (1985a)

Instrument air (*see* Table 11.17)

---

(2) The process should be subjected to a critical examination such as a hazop study to discover potential hazards and operating difficulties.
(3) If a process contains serious hazards and requires an elaborate instrument system, it should be re-examined to determine whether the hazards can be reduced at source.
(4) If the process continues to contain serious hazards, these should be assessed and protective systems provided as appropriate. If necessary, these should be high integrity protective systems.
(5) For pressure systems it is necessary to provide protection not just against overpressure, but also against other conditions such as underpressure, overtemperature, undertemperature, overfilling, etc.
(6) The measurements should be as far as possible on the variable of direct interest. If this variable has to be inferred from some other measurement, this fact should be made clear. It is also important that the measurement should be at the right location.
(7) If the variable is critical for process safety, the same measurement should not be used for control and for an alarm or trip.
(8) If the variable is critical for operator comprehension, it may be desirable to provide additional integrity.
(9) The alarm system should have a properly thought out philosophy, which relates the variables alarmed, the number, types and degrees of alarm, and the alarm displays and priorities to factors such as instrument failure and operator confidence, the information load on the operator, the distinction between alarms and statuses, and the action which the operator has to take.
(10) The control loops should have fail-safe action as far as possible, particularly on loss of instrument air or electrical power to the control valves. The action for other equipment should also be fail-safe where applicable.
(11) Those control loops which can add material or energy to the process are particularly critical and it may be desirable to provide additional integrity.
(12) The control system as a whole and the individual instruments should have the 'rangeability' necessary to maintain good measurement and control at low throughputs.
(13) The control system should be designed for off-normal as well as normal conditions, e.g. start-up and shut-down.
(14) Restart situations, such as restarting after a trip or restarting an agitator, tend to be particularly hazardous.
(15) Manual stations should be provided which allow the operator to manipulate control valves in situations such as the failure of the automatic controls.
(16) The fact of instrument failure should be fully taken into account. The reliability of critical instrumentation should be assessed quantitatively where possible.
(17) The ways in which dependent failures can occur and the ways in which the instrument designer's intentions may be frustrated should be carefully considered.
(18) Instrumentation which is intended to deal with a fault should not be disabled by the fault itself. And if the process operator has to manipulate the instrumentation during the fault, he should not be prevented from doing so by the condition arising from the fault.
(19) The services (instrument air, electrical power, inert gas) on which instruments depend should have an appropriate degree of integrity.
(20) The instrument system should be checked regularly and faults repaired promptly. It should not be allowed to deteriorate, even though the process operator compensates for this. The process operator should be

trained not to accept instrumentation unrepaired over long periods.

(21) Ease of detection of instrument faults should be an objective in the design of the instrument system. The process operator should be trained to regard detection of malfunction in instruments as an integral part of his job.

(22) Instruments which are required to operate only under fault conditions, and which may therefore have an unrevealed fault, require special consideration.

(23) Important instruments should be checked regularly. The proof test interval should, where possible, be determined from a reliability assessment. The checks should not be limited to protective systems and pressure relief valves, but should include non-return valves, emergency isolation valves, etc., and often also measurements, alarms, control loops, etc.

(24) Tests should correspond as nearly as possible to the expected plant conditions. It should be borne in mind that an instrument may pass a workshop test, but still not perform satisfactorily on the plant.

(25) Valves, whether control or isolation valves, are liable to pass fluid even when closed. Characterized control valves in particular tend not to give a tight shut-off. More positive isolation may require measures such as the use of double block and bleed valves or of slip plates.

(26) Valves, particularly control valves, also tend to stick. This can give rise to conditions which do not always emerge from a simple application of fail-safe philosophy. Jamming in the open position is often particularly dangerous.

(27) Practices which process operators tend to develop in their use of the instrumentation should be borne in mind, so that these practices do not invalidate the assumptions made in the reliability assessments.

(28) The fact of human error should be fully taken into account. To the extent that is practical, human factors principles should be applied to reduce human error, and the reliability of the process operator should be assessed quantitatively.

It is also necessary to pay careful attention to the details of the individual instruments used. Some features which are important are as follows:

(1) Instruments are a potential source of failure, either through a functional fault on the instrument or through loss of containment at the instrument.

(2) Use of inappropriate materials of construction can lead to both kinds of failure. Materials should be checked carefully in relation to the application, bearing in mind the possible impurities as well as the bulk chemicals. It should be remembered that the instrument supplier usually has only a very general idea of the application.

(3) Instruments containing glass, such as sight glasses or rotameters, can break and give rise to serious leaks and should be avoided if such leaks could be hazardous.

(4) Instruments may need protection against the process fluid due to its corrosiveness. Examples of protection are the use of inert liquids in the impulse lines on pressure transmitters or of chemical diaphragm seals on pressure gauges.

(5) Sampling and impulse lines should be given careful attention. Purge systems are often used to overcome blockages in impulse lines. Freezing is another common problem, which can be overcome by the use of steam or electrical trace heating.

(6) Temperature measuring elements should not normally be installed bare, but should be protected by a thermowell. A thermowell is frequently exposed to quite severe conditions such as erosion/corrosion or vibration and should be carefully designed.

(7) Pulsating flow is a problem in flowmeters such as orifice plate devices and can give rise to serious inaccuracies. This is a good example of a situation where replication of identical instruments is no help.

(8) Pressure transmitters and regulators are easily damaged by overpressure and this needs to be borne in mind.

(9) Complex instruments such as analysers, speed controllers, vibration monitors and solids weighers are generally less reliable than other instruments. This requires not only that such instruments should receive special attention but that the consequences of failure should be analysed with particular care.

(10) Different types of pressure regulator are often confused, with perhaps a pressure reducing valve being used instead of a non-return valve, or vice versa. It is specially necessary with these devices to check that the right one has been used. Also, bypasses should not be installed across pressure regulators.

(11) Selection of control valves is very important. A control valve should have not only the right nominal capacity but also appropriate rangeability and control characteristics. It should have any fail-safe features required, which may include not only action on loss of power but also a suitable limit to flow when fully open. It should have any necessary temperature protection, e.g. cooling fins. Bellows seals may need to be provided to prevent leaks. The valve should have a proper mechanical balance for the application, so that it is capable of shutting off against the process pressure. It should be borne in mind that any valve, but particularly a characterized valve, may not give completely tight shut-off, and also that a badly adjusted valve positioner can prevent shut-off.

(12) Instruments should not be potential sources of ignition and should conform with the hazardous area classification requirements.

Further discussions of the SLP aspects of instrument systems are given by Hix (1972) and the Center for Chemical Process Safety (CCPS, 1993/14).

13.3.2 Instrument distribution

A feel for the distribution of types of instrument on a process plant may be obtained from the following figures given by Tayler (1987a):

	Overall (%)	*Monitoring* (%)	*Control* (%)
Pressure	40	26	21
Temperature	32	56	15
Flow	20	8	47
Level	8	4	8
Analysis		3	4
Miscellaneous		3	5

The first column evidently refers only to the four main types. It can be seen that, whereas temperature is dominant for monitoring, it is flow which predominates in control.

13.3.3 Instrument accuracy

Most process plant instrumentation is quite accurate provided it is working properly. Information on the expected error limits of commercially available instrumentation has been given by Andrew and Williams (1980), who list limits for over 100 generic types of instrument. Some ranges of total error quoted by these authors are:

Pressure:
  Bellows transmitter ±0.5%
Temperature:
  Thermocouple ±0.25–5%
  Resistance thermometer ±0.2–0.5
Flow:
  Orifice meter ±0.5–1%
Level:
  Differential pressure ±0.5–2%
Analysis:
  Gas chromatograph ±0.5–1%

13.3.4 Instrument signal transmission

Pneumatic instrument signals are transmitted by tubing, but several means are available for the transmission of electrical signals: wire, fibre optics and radio waves. The signals from measuring instruments can become corrupted in transmission. Pneumatic signals may be affected by poor quality instrument air, while electrical signals are liable to be subject to electromagnetic interference.

Both pneumatic and electrical instrument signals utilize live zero, standard ranges being 3–15 psig for pneumatic instruments and 4–20 mA for electronic ones. This avoids the situation where a zero signal is ambiguous, meaning either that the measured variable actually has a zero value or that the instrument signal has simply gone dead.

13.3.5 Instrument utilities

Instrument systems require high quality and high reliability utilities. A general account of instrument utilities has been given in Chapter 11. As far as quality is concerned, pneumatic systems require instrument air which is free of dirt and oil. Many electronic instrument systems can operate from an electrical feed which does not constitute an uninterruptible power supply (UPS). But computers and PESs are intolerant of even millisecond interruptions, unless they have their own in-built means of eliminating them. A further treatment of instrument utilities is given by the CCPS (1993/14).

13.3.6 Valve leak-tightness

In many situations on process plants the leak-tightness of a valve is of some importance. The leak-tightness of valves is discussed by Hutchison (1976) in the ISA *Handbook of Control Valves.*

Terms used to describe leak-tightness of a valve trim are (1) drop tight, (2) bubble tight or (3) zero leakage. Drop tightness should be specified in terms of the maximum number of drops of liquid of defined size per unit time and bubble tightness in terms of the maximum number of bubbles of gas of defined size per minute.

Zero leakage is defined as a helium leak rate not exceeding about 0.3 cm^3/year. A specification of zero leakage is confined to special applications. It is practical only for smaller sizes of valves and may last for only a few cycles of opening and closing. Liquid leak-tightness is strongly affected by surface tension.

Specifications for leak tightness of a stop, or isolation, valve are given in SP-61 by the US Valve Manufacturers Standardization Society, and are quoted in the *ISA Handbook*. In respect of control valves, the *Handbook* states:

> Properly designed control valves can achieve stop valve tightness and maintain it throughout a long service life before trim replacement; particularly with cage guided, balanced trim having elastomer plug-to-cage seals. The control valve, however, is expected to throttle and often shuts off much more frequently than stop valves. For example, some dump valves may have from 4000 to 7000 opening and closing cycles per day, handling high pressure and erosive fluids at 1000 to 4000 psi pressure drop. Few stop valves could match this performance and remain tight.

It is normal to assume a slight degree of leakage for control valves. It is possible to specify a tight shut-off control valve, but this tends to be an expensive option.

A specification for leak-tightness should cover the test fluid, temperature, pressure, pressure drop, seating force and test duration. For a single-seated globe valve with extra tight shut-off the *Handbook* states that the maximum leakage rate may be specified as 0.0005 cm^3 of water per minute per inch of valve seat orifice diameter (not the pipe size of the valve end) per pound per square inch pressure drop. Thus a valve with a 4 in. seat orifice tested at 2000 psi differential pressure would have a maximum water leakage rate of 4 cm^3/min.

13.3.7 Hazardous area compatibility

The instrument system, including the links to the control computers, should be compatible with the hazardous area classification. Hazardous area classification involves first zoning the plant and then installing in each zone instrumentation with a degree of safeguarding appropriate to that zone. Since much instrumentation is of low power, an approach based on inherent safety is often practical. These various aspects of hazardous area classification are dealt with in Chapter 16.

13.3.8 Multi-functional vs dedicated systems

An aspect of basic design philosophy which occurs repeatedly in different guises is the choice which has to be made between a multi-functional and a dedicated system. Some basic functions which are typically required are (1) monitoring, (2) control, (3) trips and interlocks, (4) fire and gas detection, (5) emergency shut-down (ESD) and (6) communication. The trip system may well be separate from the monitoring and control system and the ESD system trips separate from the other trips.

The situation which develops is illustrated in Figure 13.1(a) which shows a traditional design for an offshore production platform system (A. Morris, 1986). The

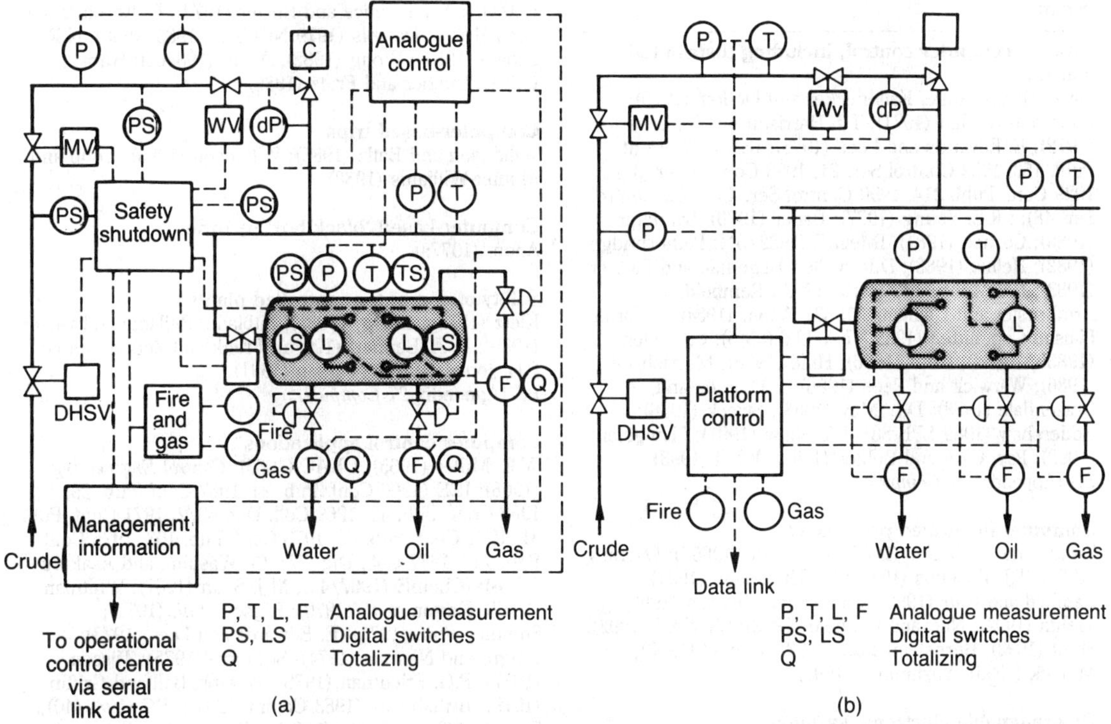

Figure 13.1 *Instrumentation for a system on an offshore production platform (A. Morris, 1986): (a) conventional system; and (b) alternative system (Courtesy of Process Engineering)*

alternative design which he proposes for consideration is shown in Figure 13.1(b). To the objection that this latter design puts all its eggs in one basket, the author puts two arguments. First, the overall reliability has been improved to such an extent that the frequency of a complete system failure will be very low. Second, in the majority of cases the process should be able to survive such failure because it can be brought to a safe state by simple measures, notably by shutting off the heat input and depressurizing.

A particular but common example of the multi-functional vs dedicated system problem is the choice between a computer-based and a hardwired trip system. This aspect is discussed further in Sections 13.9, 13.12 and 13.15.

13.4 Process Computer Control

The use of computers in control systems began in the late 1950s and is now a mature technology. Process control computer systems and applications are described in *Computer Control of Industrial Processes* (Savas, 1965), *Computer Control of Industrial Processes* (Lowe and Hidden, 1971), *Handbook of Industrial Control Computers* (Harrison, 1972), *Understanding Distributed Process Control* (Moore and Herb, 1983), *Computer Systems for Process Control* (Güth, 1986) and *Industrial Digital Control Systems* (Warwick and Rees, 1986), while a description of computer control and its relation to operator control has been given in *Man and Computer in Process Control* (E. Edwards and Lees, 1973). Selected references on process computer control are given in Table 13.3.

The inclusion of a process control computer greatly extends the capabilities, but also affects the reliability, of the control system. These two aspects are now considered.

13.4.1 Computer configurations and reliability

There are several ways in which a computer may be incorporated in a process control system. The approaches originally used are illustrated in Figure 13.2. If there is no computer, then the loops are controlled by analogue controllers as shown in Figure 13.2(a).

The configuration given in Figure 13.2(b) is set-point control. The computer takes in signals from measuring instruments and sends signals to the set points of analogue controllers. If there is a computer failure, control is still maintained by the analogue controllers. Figure 13.2(c) shows direct digital control (DDC). The computer again takes in signals from measuring instruments, but now sends signals direct to the control valves;

Table 13.3 *Selected references on process computer control*

Process computer control, including distributed control
Savas (1965); Anke, Kaltenecker and Oetker (1970); Lowe and Hidden (1971); T.J. Harrison (1972); Lees (1972); E. Edwards and Lees (1973); IEE (1977 Conf. Publ. 153, 1982 Control Ser. 21, 1988 Control Ser. 37, 1989 Conf. Publ. 314, 1990 Control Ser. 44, 1993 Control Ser. 48); ; R.E. Young (1977); Bader (1979); Sandefur (1980); Cocheo (1981); IMechE (1982/61); Petherbridge (1982); Helms (1983); D.R. Miller, Begeman and Lintner (1983); J.A. Moore and Herb (1983); Rembold, Armbruster and Ülzmann (1983); Anon. (1984rr); Nordic Liaison Committee (1985 NKA/LIT (85)5); C. Tayler (1985b, 1986d); Güth (1986); Hide (1986); Morrish (1986); Warwick and Rees (1986); J. Pearson and Brazendale (1988); D.L. May (1988); Strock (1988); Eddershaw (1989 LPB 88); J.A. Shaw (1991); Livingston (1992); Ray, Cary and Belger (1992); Wadi (1993)
BS (Appendix 27 *Computers*)

Computer integrated processing
Zwaga and Veldkamp (1984); C. Tayler (1985d); O'Grady (1986); T.J. Williams (1989); W. Thompson (1991); Canfield and Nair (1992); Conley and Clerrico (1992); Mehta (1992); Nair and Canfield (1992); Sheffield (1992); Stout (1992); Bernstein *et al.* (1993); Koppel (1993); Mullick (1993); Yoshimura (1993)

Programmable electronic systems
Zielinski (1978); Bristol (1980); Sargent (1980); EEMUA (1981 Publ. 123); HSE (1981 OP 2, 1987/21, 22); Dartt (1982); IBC (1982/39); Devries (1983); Martinovic (1983); Martel (1984); Lihou (1985b, 1987); Skinner (1985 LPB 62); Weiner (1985); Wilkinson and Balls (1985); R. Bell (1986); Daniels (1986); Fulton and Barrett (1986); Holsche and Rader (1986); Margetts (1986a,b, 1987); Pinkney (1986); Wilkinson (1986); Anon. (1987u); Pinkney and Hignett (1987); Wilby (1987); Bellamy and Geyer (1988); Clatworthy (1988); D.K. Wilson (1988); Deja (1989); IGasE (1989 IGE/SR/15); Max-Lino (1989); Oser (1990); British Gas (1991 Comm. 1456); Borer (1991); J. Pearson (1991); Sawyer (1991a); Gruhn (1992b); Prugh (1992d)

Control rooms, computer displays
Bernard and Wujkowski (1965); Wolff (1970); IEE (1971 Conf. Pub. 80, 1977 Conf. Pub. 150); Dallimonti (1972, 1973); E. Edwards and Lees (1973); Strader (1973); Lees (1976d); Bonney and Williams (1977); Jervis and Pope (1977); Hammett (1980); Burton (1981); Lieber (1982); C.M. Mitchell and Miller (1983); Banks and Cerven (1984); Jansen (1984); Mecklenburgh (1985); C. Tayler (1986a); Gilmore, Gertman and Blackman (1989)

Computer system reliability, including safety critical systems, fault tolerant systems, computer system security (*see also* Table 7.1)
Hendrie and Sonnenfeldt (1963); R.J. Carter (1964); Sonnenfeldt (1964); Burkitt (1965); A. Thompson (1965); Lombardo (1967); Regenczuk (1967); Amrehn (1969); Stott (1969); Anon. (1970d); Barton *et al.* (1970); Hubbe (1970); Luke and Golz (1970); H.F. Moore and Ballinger (1970); Parsons, Oglesby and Smith (1970); J. Grant (1971); J.A. Lawrence and Buster (1972); E. Edwards and Lees (1973); Daniels (1979 NCRS 17, 1983, 1986); N.R. Brown (1981); Wong (1982); Anon. (1984cc); Hura (1984); Bucher and Fretz (1986)

Computer-based trips
Wilkinson and Balls (1985); Wilkinson (1986); Cobb and Monier-Williams (1988)

Computer-based 'black box' recorder
Anon. (1977a)

Safety of computer controlled plants
Kletz (1982g, 1991g, 1993a); Pitblado, Bellamy and Geyer (1989); P. A. Bennett (1991a); Frank and Zodeh (1991); P.G. Jones (1991); Pearson (1991)
BS (Appendix 27 *Computers*)

Computer control applications
W.E. Miller (1965); UKAC (1965); *Control Engineering* (1966); IEE (1966 Conf. Pub. 24, 1967 Conf. Pub. 29, 1968 Conf. Pub. 43, 1969 Coll. Dig. 69/2, 1971 Conf. Pub. 81, 1972 Conf. Pub. 83, 1973 Conf. Pub. 103, 1975 Conf. Pub. 127, 1977 Coll. Dig. 77/30); Washimi and Asakura (1966); IChemE (1967/45); M.J. Shah (1967); Whitman (1967); Barton *et al.* (1970); Higson *et al.* (1971); Sommer *et al.* (1971); E. Edwards and Lees (1973); Daigre and Nieman (1974); St Pierre (1975); Tijssen (1977); P.G. Friedman (1978); Weems, Ball and Griffin (1979); British Gas (1983 Comm. 1224); IBC (1983/40); Seitz (1983); C. Tayler (1984b); Tatham, Jennings and Klahn (1986)

there are no analogue controllers. If there is a computer failure, control is lost on all loops, unless stand-by arrangements have been made. Although set-point control developed first, it was followed quickly by DDC, and both methods came into use.

The first large DDC installation on a chemical plant was on the ammonia soda plant of Imperial Chemical Industries (ICI) at Fleetwood (Burkitt, 1965; A. Thompson, 1965). The computer carried out DDC on 98 loops and achieved an availability of about 99.8%. Further accounts of DDC systems have been given by Barton *et al.* (1970) and by Higson *et al.* (1971).

Although the initial intention was for DDC to save the cost of analogue controllers, it soon became apparent that many other factors were involved in the choice between set-point control and DDC. Since, with DDC, computer failure leads to loss of control, it may be necessary to achieve a much higher reliability than with set-point control. The effort required to implement a DDC installation tends, therefore, to be much greater. It is necessary to pay very careful attention to details of the computer, the power supply and the environment, the input–output equipment and the programming. Usually DDC does not reduce the cost of adding computer control to the control system much below that for set-point control. Savings in costs per loop tend to be slight,

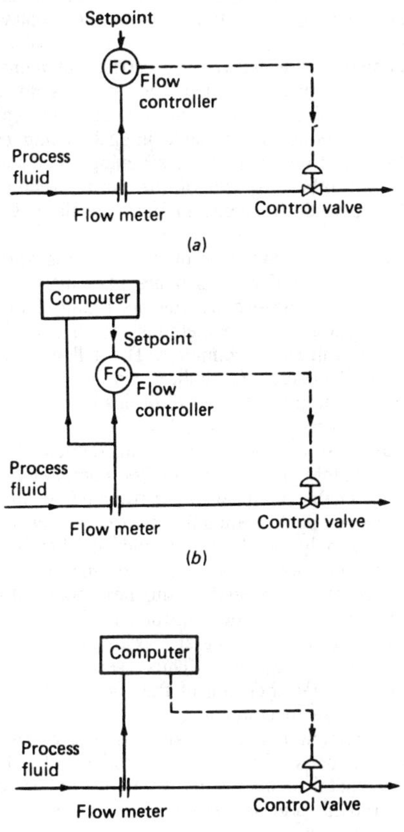

Figure 13.2 *Process computer control systems: set-point and direct digital control: (a) analogue control; (b) set-point control by computer; (c) direct digital control by computer*

because the equipment needed to get measurements into the computer and to position the control valves from it is quite expensive. It is necessary to provide stand-by analogue controllers for critical control loops and change-over equipment to transfer between computer and analogue control. The extra general effort required to assure integrity in DDC is also significant.

On the other hand, DDC does offer some advantages, not only over conventional control but also over set-point control. The advantages derive from the fact that the computer takes in signals from the measuring instruments and can process them in all sorts of ways before sending out the results as signals to the control valves. It makes it possible to: carry out operations on the measurements, such as calculation of indirect measurements and filtering of measurement signals; ensure that the control algorithm is truly proportional, integral and derivative without the inaccuracies and interactions which tend to occur in analogue controllers; use different control algorithms such as non-linear or asymmetrical algorithms or algorithms with some logic in them; eliminate features such as integral saturation and derivative kick; position valves more accurately; alter the control configuration; and so on.

There are several ways in which the reliability of DDC systems can be improved. One of these, as mentioned earlier, is the use of stand-by controllers on critical loops. But this is by no means a complete answer to the problem. The system may still be upset by intermittent faults, there may be difficulties in keeping the stand-by instrumentation maintained and avoiding degradation, and the operator is faced with a different interface to use on loss of computer control. Another approach is the use of duplication. In this case it is necessary not only to use dual computers, but also to duplicate other parts of the system such as power supplies and input–output equipment. Various configurations are possible and in normal operation the work may be divided either on a parallel or a hierarchical basis, but in all cases the essential principle is that the surviving computer takes over the critical control functions. The reliability of dual computer systems is undoubtedly higher, but it can still be affected by factors such as intermittent failures, data link troubles, hardware faults in common, such as earthing, and software faults in common, such as programming errors. With regard to reliability, for the type of system just described, the most reliable systems achieved a mean time between failures (MTBF) and an availability of not less than 2000 hours and 99.9%, respectively.

Advances in process control systems, and particularly the trend towards distributed PESs, have largely resolved the dilemmas described and have gone far towards solving the reliability problems. Figure 13.3 shows schematically a system configuration typical of these developments. The backbone of the system is a data highway to which various devices are connected. The individual PES controllers are capable of operating as DDC controllers in the stand-alone or set-point control

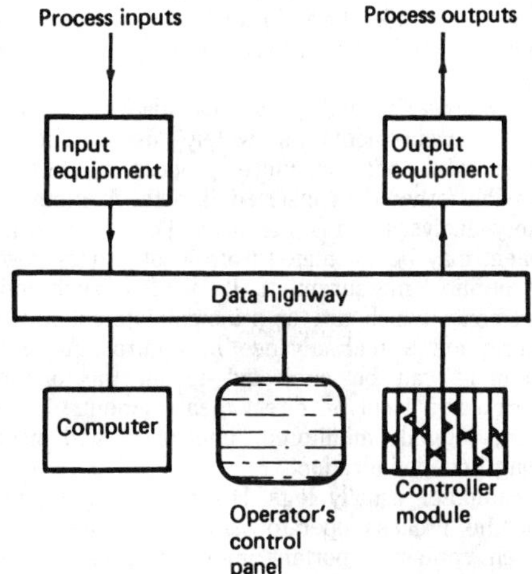

Figure 13.3 *Process computer control systems: distributed control system*

modes. The VDU display can also operate independently of the computer. Thus the system allows the full facilities of DDC if the computer is working, but on computer failure the controllers maintain control and the VDU display continues to provide the operator with the usual interface.

Various configurations may be used to obtain back-up control of critical loops. Where a loop is backed up it is desirable to ensure 'bumpless' transfer when the stand-by equipment assumes control. This involves a process of initialization before control is transferred.

Accounts of computer-based and PES-based process control systems based on these principles include those by E. Johnson (1983), Tatham, Jennings and Klahn (1986), Cobb and Monier-Williams (1988) and the CCPS (1993/14). Programmable electronic systems for process control are considered further in Section 13.12. Data on the reliability of computer systems are given in Appendix 14.

13.4.2 Computer functions

If the computer carries out DDC, then this is its most important function. The facilities and flexibility which DDC offers have already been described. However, as just described, modern process control systems are generally based on distributed PESs.

The other main functions which a process control computer or PES performs are:

(1) measurement;
(2) data processing and handling;
(3) monitoring;
(4) other control;
(5) sequential and logical control;
(6) optimization;
(7) scheduling;
(8) communication.

Several of these functions are important in relation to safety and loss prevention (SLP).

The measurements on which control depends are critical. The computer is often used to carry out certain checks on the measurements as described in Chapter 30. It can also upgrade them in various ways such as by extraction of non-linearities, zero or range correction, or filtering.

The computer's ability to calculate 'indirect' or 'inferred' measurements is widely used. These are calculated from one or more process measurements and possibly other data inserted into the computer, e.g. laboratory analyses. Thus the mass flow of a particular component may be calculated from a total mass flow and a concentration measurement. It is often such indirect measurements which are of principal interest and their use represents a real advance in control. An indirect measurement can be subjected to all the operations which are carried out on direct measurements: it can be displayed, logged, monitored, controlled and used in modelling and optimization.

The computer usually logs data and provides summaries for the process operator and management. These logs often contain important information on equipment faults, operator interventions, etc. Arrangements are also sometimes made for a *post-mortem* log in the event of a serious incident on the process. This usually involves holding a continuously up-dated set of data on process instrument readings so that it can be replayed if necessary.

The computer almost invariably carries out monitoring of the process measurements and statuses to detect abnormal conditions. This constant scanning of the operating conditions is invaluable in maintaining control of the process. Computer alarm scanning is considered, together with other aspects of the alarm system with which the operator interacts, in more detail in Chapter 14.

Frequently, there are one or two process variables, equipments or operations which are particularly difficult to control and for these more advanced control methods may be appropriate. These methods are usually difficult to implement without a computer. The following appear to be especially useful: (1) indirect variable control, (2) automatic loop tuning, (3) control of dead time processes and (4) non-interacting control.

The execution by the computer of sequential operations in a reliable manner is another common function which is invaluable in maintaining trouble-free operation of the process. Such sequential control involves much more than simply sending out control signals. It is essential for checks to be made to ensure that the process is ready to proceed to the next stage, that the equipment has obeyed the control signals, and so on. There is therefore a liberal sprinkling of checks throughout the sequence. Thus sequential control involves continuous checking of the state of the process and the operation of equipment.

Using a computer it is possible to carry out more complex sequences with greater reproducibility. This is particularly useful in operations where it is necessary to follow a rather precise schedule in order to avoid damage to the equipment.

On some processes where there is a time-varying optimum, the computer carries out continuous optimization. Optimization is usually performed with a set of constraints. Computer optimization therefore provides as by-product a more formal definition of, and adherence to, process constraints.

There are several other computer functions which are particularly relevant to SLP. These include computer alarm analysis, valve sequencing and malfunction detection. These are dealt with in Chapter 30.

For many years there was very little use of computers to carry out the protective function of tripping plant when a hazardous condition occurs. The protective system has almost invariably been a system separate from the control system, whether or not the latter contains a computer, and engineered for a greater degree of integrity. There is now movement towards the use of PESs for the trip functions also, but only where it can be demonstrated that the system has a reliability at least equal to that of a conventional hardwired system.

13.4.3 Computer displays and alarms

Process computers, as just indicated, are powerful tools for the support of information display and alarm systems. The design of such systems is intimately bound up with the needs of the process operator, and discussion is therefore deferred to Chapter 14.

13.4.4 Fault-tolerant computer systems

To the extent practical, process computer systems should be fault tolerant. A fault-tolerant system is one which continues to perform its function in the face of one or more faults. Accounts of fault-tolerant design of computer systems, including process computer systems, are given by Shrivastava (1991), the CCPS (1993/14) and Johnston (1993).

The creation of a fault-tolerant system involves a combination of approaches. A necessary preliminary is effort to obtain high reliability and thus to eliminate faults. The methods of reliability engineering may be used to model the system and to identify weak points. The use of redundancy and diversity is a common strategy. Dependent failures and methods of combating them should receive particular attention.

Prompt detection and repair of faults is an important part of a strategy for a fault-tolerant system. A fault-tolerant system should degrade gracefully, and safely. One important aspect is the fail-safe action of the system.

13.4.5 Computer power supplies

Process computers and PESs require a high reliability and high quality power supply. A general account of power supplies is given in Chapter 11. The operation of such equipments can be upset by millisecond interruptions, unless they have in-built means of dealing with them. They therefore generally require an uninterruptible power supply (UPS). Devices used to provide a UPS include motor generators, DC/AC inverters and batteries.

The power supply also needs to be uninterruptible in the sense that it has high reliability. One option is the use of batteries, another is some form of redundancy or diversity of supply.

A treatment of power supplies for PESs is given by the CCPS (1993/14). A relevant code for UPSs is IEEE 446.

13.4.6 Computer system protection

Process computers and PESs require suitable protection against fire and other hazards.

For fire protection relevant codes are BS 6266: 1992 *Code of Practice for Fire Protection of Electronic Data Processing Installations,* NFPA 75: 1992 *Protection of Electronic Computer/Data Processing Equipment* and NFPA 232: 1991 *Protection of Records.*

Lightning protection is covered in NFPA 78: 1989 *Lightning Protection Code.*

Codes for earthing are BS 1013: 1965 *Earthing* and IEEE 142: 1982 *Grounding of Industrial and Commercial Power Systems* (the IEEE *Green Book*).

These hazards and protection against them are treated by the CCPS (1993/14).

13.5 Control of Batch Processes

The control of batch processes involves a considerable technology over and above that required for the control of continuous processes. Accounts of batch process control are given in *Batch Process Automation* (Rosenof and Ghosh, 1987), *Batch Control Systems* (T.G. Fisher, 1990) and *Computer-Controlled Batch Processing* (Sawyer, 1993a) and by Love (1987a–c, 1988).

Batch processes constitute a large proportion of those in the process industries. Sawyer (1993a) gives the following figures:

Industry sector	*Mode of operation*	
	Batch	*Continuous*
Chemical	45%	55%
Pharmaceutical	80%	20%

Many batch plants are multi-purpose and can make multiple products. Their outstanding characteristic is their flexibility. They differ from continuous plants in that: the operations are sequential rather than continuous; the environment in which they operate is often subject to major variability; and the intervention of the operator is to a much greater extent part of their normal operation rather than a response to abnormal conditions. A typical batch plant is shown in Figure 13.4.

13.5.1 Models of batch processing

There are a number of models which have been developed to represent batch processing. Three described by T.G. Fisher (1990) are (1) the recipe model, (2) the procedure model and (3) the unit model.

The recipe model centres on the recipe required to make a particular product. Its elements are the procedures, the formula, the equipment requirements and the 'header'. The procedure is the generic method of processing required to make a class of product. The formula is the raw materials and operating conditions for the particular product. The equipment requirements cover the equipment required to execute the formula, including materials of construction. The header is the identification of the batch in terms of product, version, recipe and so on.

The procedure model has the form:

Procedure → Operation → Phase → Control step

The overall procedure consists of a number of operations, akin to the unit operations of continuous processes, except that they may be carried out by the same equipment. The phase is a grouping of actions within an operation. The control step is the lowest level of action, typically involving the movement of a small number of final control elements.

The concept of phase is a crucial one in batch processing. A phase is a set of actions which it is logical to group together and which ends at a point where it is logical and safe for further intervention to take place. It is closely connected, therefore, with the concept of 'hold' states at which it is safe for the process to be held. The possibility that other facilities on which the progress of the batch depends may not be immediately available makes such hold states essential.

The unit model is equipment-oriented and has the form

Unit → Equipment module → Device/loop → Element

The unit is broken down into functional equipment modules such as vessels and columns. These in turn are decomposed into devices and loops which are groupings of elements such as sensors and control valves.

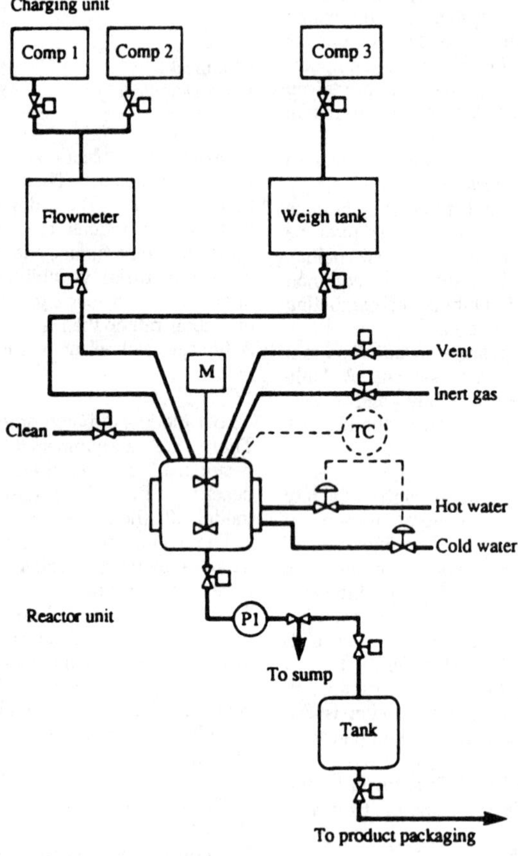

Figure 13.4 *A typical batch plant (Sawyer, 1993) (Courtesy of the Institution of Chemical Engineers)*

13.5.2 Representation of sequential operations

The control of a batch process is a form of sequential control. A typical sequential control procedure, expressed in terms of the procedure model, is shown in Table 13.4. Various methods are available for the specification of sequences. They include (1) flowcharts, (2) sequential function charts and (3) structured plain language.

The flowchart is a common method of representing sequences, but its successful use requires that: a consistent style be adopted; that the method cater for the procedure hierarchy by the use of a hierarchy of charts for operations, phases and control steps; it also allows for parallel activities and for actions prompted by alarms and failures; and is supplemented by information on recipes, units, etc., and by other representations such as structured language. Computer-based drafting aids are invaluable in creating flowcharts.

The sequential flowchart has been developed expressly to describe sequential control and has three basic features: (1) steps, (2) transitions, and (3) directed links. A step is an action and ends with a conditional transition. If the condition is satisfied, control passes to the next step. This latter step then becomes active and the previous step inactive. A directed link creates a sequence from steps and transitions. Figure 13.5 shows a sequential flowchart together with the standard symbols used in the creation of such charts.

With regard to the use of structured language, Rosenof and Ghosh advise that: (1) simple statements should be used; (2) the required function should be clearly defined in a statement; (3) the plant hardware addressed should, where possible, be identified; (4) text should be indented where necessary; (5) negative logic should be avoided; and (6) excessive nested logic should be avoided.

13.5.3 Structure of batch processing

The overall structure of batch processing is commonly represented as a hierarchy. The following structure and terminology by Rosenof and Ghosh (1987) is widely used:

Production planning → Production scheduling → Recipe management → Batch management → Sequential control → Discrete/regulatory control → Process interlocks → Safety interlocks

Table 13.4 *Typical sequential control procedure (Sawyer, 1993a) (Courtesy of the Institution of Chemical Engineers)*

Operations	*Phases*	*Control steps*
Initialize	Initialize	Start jacket circulation pump. Put reactor temperature controller in SECONDARY AUTO mode with set point of 120°C
Weigh tank	Weigh Component 3	Initialize (tare-off weigh tank). Open outlet valve from head tank. When weight of Component 3 equals preset, close outlet valve from head tank
Charge	Add Component 3	Open outlet valve from weigh tank. When enough of Component 3 has been added, start the agitator. When weigh tank is empty, close outlet valve
	Add Component 1	Initialize (reset flow totalizer to zero). Open outlet valves from head tank to flowmeter and from flowmeter to reactor. When volume of Component 1 charged equals preset, close outlet valves
React	Heat	Initialize (put reactor temperature controller in CASCADE mode with set point of 120°C)
	Hold	Initialize (reset timer). Start timer
	Sample	
Discharge	Cool	Initialize (set reactor temperature set point to 35°C)
	Transfer	Initialize (set reactor outlet valves to correct destination, ie storage tank). Start discharge pump. Set reactor temperature controller to MANUAL mode with output at zero (full cooling). Before agitator blades are uncovered, stop agitator. When reactor is empty, close reactor outlet valves, stop discharge pump, stop jacket circulation pump

A treatment of batch processing as a form of computer integrated manufacturing is given in Section 13.7.

13.5.4 Batch control systems

Batch processing may be controlled by the process operator, by a system of single controllers or by a programmable logic control (PLC) system, a distributed control logic system (DCL) or a centralized control system (CCS). The selection of the system architecture and hardware is discussed by Sawyer (1993).

Recommendations for batch control have been made in Europe by the NAMUR committee (1985), which addresses particularly the need for standard terminology and for a hierarchical structure of the control system which reflects that of batch processing itself.

In the USA, guidance is available in the form of ISA SP88: 1988 *Batch Control Systems.*

13.6 Control of Particular Units

The safe operation of process units is critically dependent on their control systems. Two particularly important features of control in process plant are (1) compressor control and (2) chemical reactor control. These are now considered in turn.

13.6.1 Compressor control

Centrifugal and axial compressors are subject to the phenomenon of surging. Surging occurs when flow through the compressor falls to a critical value so that a momentary reversal of flow occurs. This reversal of flow tends to lower the discharge pressure and normal flow resumes. The surge cycle is then repeated. Severe surging causes violent mechanical shock and noise, and can result in complete destruction of parts of the compressor such as the rotor blades.

A typical centrifugal compressor characteristic showing the surge limit is illustrated in Figure 13.6(a). A centrifugal compressor is usually fitted with anti-surge controls which detect any approach to the surge conditions and open the bypass from the delivery to the suction of the machine, thus increasing the flow through the machine and moving it away from the surge conditions.

The compressor delivery and suction pressures P_d and P_s are related to the gas flow Q as follows:

$$P_d - P_s \propto Q^2 \qquad [13.6.1]$$

The shape of the surge curve is therefore parabolic as shown in Figure 13.6(a). An expression of this form is generally inconvenient in instrumentation, for which linear relations are preferred. A linear relation can be

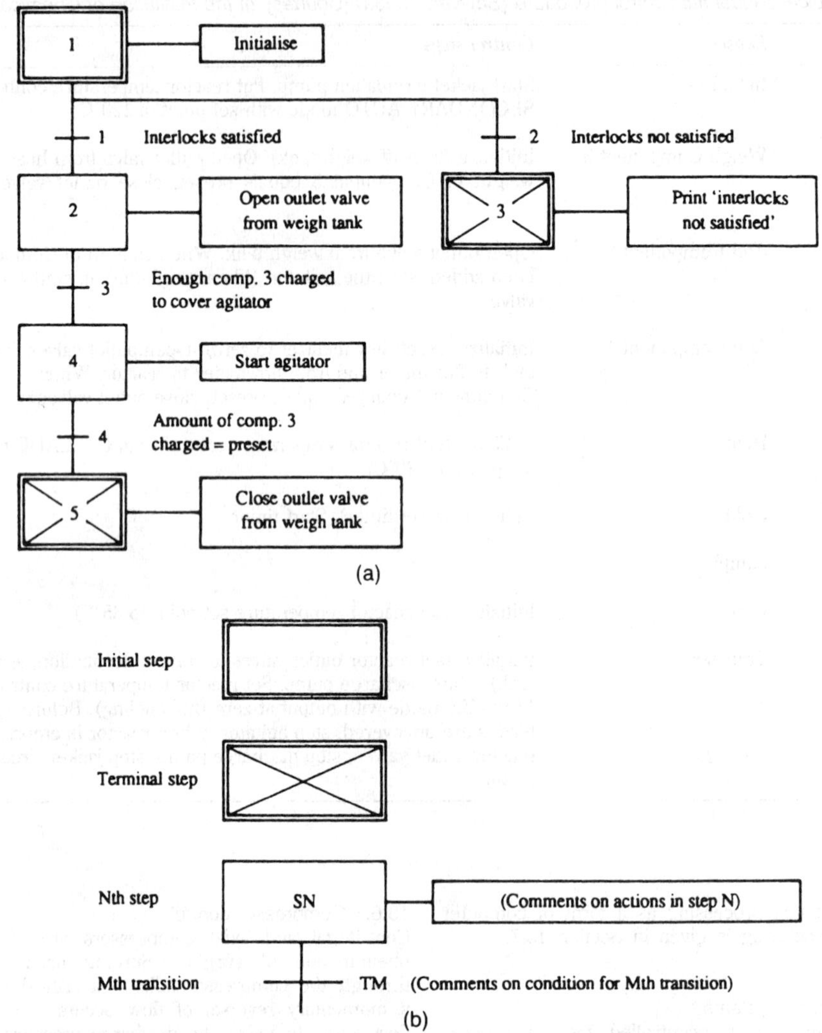

Figure 13.5 Sequential function chart (Sawyer, 1993): (a) chart for control steps ADD COMPONENT; and (b) basic symbols (Courtesy of the Institution of Chemical Engineers)

obtained by making use of the relation for pressure drop ΔP across the orifice flowmeter on the compressor suction:

$$Q^2 \propto \Delta P \qquad [13.6.2]$$

Hence from relations 13.6.1 and 13.6.2

$$P_d - P_s \propto \Delta P \qquad [13.6.3]$$

Figure 13.6(b) shows the compressor characteristics redrawn in terms of this pressure drop. The surge condition is now given by a straight line. The antisurge control system is set to operate on a line somewhat in advance of the surge limit, as shown in Figure 13.6.(b). The anti-surge controller is usually a $P+I$ controller and, since it operates only intermittently, it needs to have arrangements to counteract integral saturation.

Accounts of centrifugal and axial compressor control are given by Claude (1959), R.N. Brown (1964), Daze (1965), Hatton (1967) and Magliozzi (1967), and accounts of reciprocating compressor control are given by Hagler (1960) and Labrow (1968). Multi-stage compressor control is discussed by D.F. Baker (1982), Maceyka (1983) and Rana (1985), and control of compressors in parallel by Nisenfeld and Cho (1978) and B. Fisher (1984).

13.6.2 Chemical reactor control

The basic characteristics of chemical reactors have already been described in Chapter 11, in which, in particular, an account was given of the stability and control of a continuous stirred tank reactor. It is

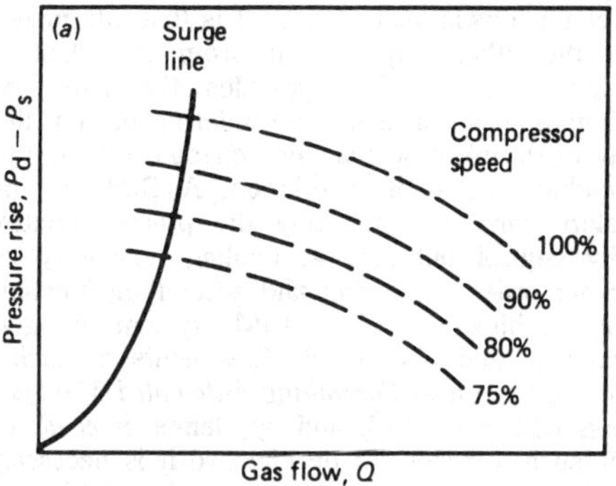

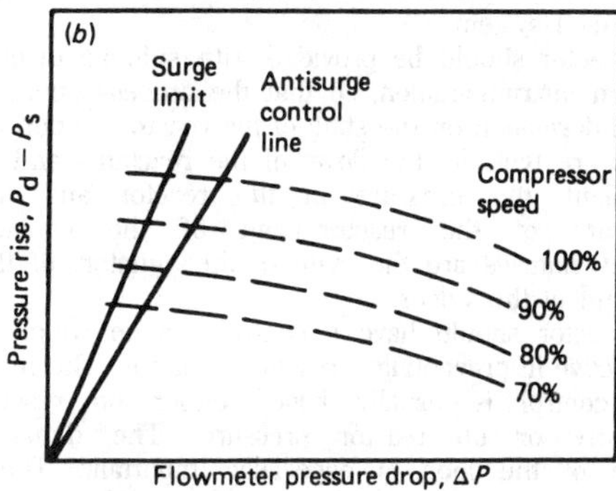

Figure 13.6 *Centrifugal compressor characteristics illustrating surge and antisurge control: (a) conventional characteristic; (b) characteristic for antisurge control*

appropriate here to consider some additional aspects of reactor control.

A continuous stirred tank reactor is generally stable under open-loop conditions, but in some cases a reactor may be unstable under open-loop but stable under closed-loop conditions. Some polymerization reactors and some fluidized bed reactors may be open-loop unstable under certain conditions.

The reactor should be designed so that it is open-loop stable unless there is good reason to the contrary. One method of achieving this is to use jacket cooling with a large heat transfer area. Another is to cool by vaporization of the liquid in the reactor. This latter method gives a virtually isothermal reactor.

If the reactor is or may be open-loop unstable, the control system should be very carefully designed. The responses of the controls should be fast. One method of achieving this is the use of cascade control for the reactor temperature to the coolant temperature. The dead time should be minimized. A high coolant flow assists in reducing dead time.

Continuous stirred tank reactors and batch reactors have their own characteristic control problems. Some of the control problems of continuous stirred tank reactors are as follows. A reaction in a continuous reactor is often carried out in a single phase in one pass. This requires accurate control of the feed flows to the reactor. Failure to achieve such control may have effects such as unconverted reactant leaving the reactor, undesirable side reactions or rapid corrosion.

It is often possible for impurities to build up in a continuous reactor. Where this is the case, arrangements should be made to purge the impurities. If the reactants to a continuous reactor need to be preheated, this should be done before they are mixed, unless the reaction requires a catalyst. A continuous reactor is sometimes provided with regenerative preheating. It should be borne in mind that such preheating constitutes a form of positive feedback.

As described in Chapter 11, batch reactors are of two broad types. In the first, the 'all-up' batch reactor, the main reactants are all charged at the start. In the semi-batch reactor one reactant is not charged initially but is fed continuously.

The reaction mass in a batch reactor cannot necessarily be assumed to be completely mixed. It is not uncommon for there to be inhomogeneities, hot spots and so on. This has obvious implications for reactor control.

Some of the control problems of batch reactors are as follows. In a typical all-up batch cycle the reactants and catalyst are charged, the charge is heated to reaction temperature, and the reaction mass is then cooled and discharged. In some cases the reaction stage is followed by a curing stage which may be at a temperature below or above the reaction temperature.

In the initial heating up period the temperature of the charge should be brought up to the operating point rapidly, but it should not overshoot. If the reactor temperature is controlled by an ordinary three-term controller, integral saturation in the controller will cause overshoot. It is necessary, therefore, to employ a controller which is modified to avoid this. Alternatively, the heating up may be controlled in some other way which avoids overshoot. Once the reaction is under way in a batch reactor the initial heat release is large. The cooling system should be adequate for this peak heat release.

Semi-batch reactors have different problems. The addition of the continuously fed reactant before the batch is up to temperature should be avoided, otherwise it is liable to accumulate and then to react rapidly when the operating temperature is reached.

If agitation is interrupted and then resumed, there may be a sudden and violent reaction of reactants which have accumulated. There should be suitable alarms, trips and interlocks to signal loss of agitation, to cut off feed of reactant, and to ensure an appropriate restart sequence.

In both types of reactor there should be arrangements to prevent material from the reactor passing back into reactant storage tanks where this could constitute a hazard. The control of flows in the reactant feed pipes is important. It is necessary to ensure tight shut-off of the reactants and to prevent flow from the reactor into the reactant feed system.

The reactor should be provided with suitable display and alarm instrumentation, so that the process operator has full information on the state of the reactor. Important variables are typically the flows of the reactants and of the coolant, the pressure in the reactor and the temperature of the reactor and of the coolant. Important statuses are the state of the agitator, of the pumps and of the valves.

The reactor should have a control system which is fully effective in preventing a reaction runaway. The main reactor control is usually based either on reactor temperature or on reactor pressure. The dynamic response of the loop is especially important. There should be adequate potential correction on the control loops. In other words, the steady-state gain between the manipulated variable and the controlled variable should be high enough to ensure that control of the latter is physically possible.

The instrumentation should possess both capability and reliability for the duty. Important aspects of capability are accuracy and dynamic response. The effects of instrument failure should be fully considered. In particular, failure in the measurement and control of the main variable, which is usually temperature, should be assessed. The ease of detection of instrument malfunction by the process operator should be considered. Factors which assist in malfunction detection include the use of measuring instruments with a continuous range rather than a binary output and the provision of recorders and of indications of valve position.

Trip systems should be provided to deal with potentially hazardous conditions. These typically include loss of feed, loss of coolant, loss of agitation and rise in reactor temperature. Emergency shut-down arrangements for reactors are discussed in Chapter 11.

Use should be made of interlocks to ensure that critical sequences which have to be carried out on the reactor are executed safely and to prevent actions which are not permissible. Many of these control functions are facilitated by the use of a process control computer. A fuller discussion of instrumentation is given in Section 13.8.

13.7 Computer Integrated Manufacturing

There is now a strong trend in the process industries to integrate the business and plant control functions in a total system of computer integrated manufacturing (CIM). Accounts of CIM are given by T.J. Williams (1989), Canfield and Nair (1992), Conley and Clerico (1992), Mehta (1992), Nair and Canfield (1992), Bernstein *et al.* (1993) and Koppel (1993).

The aim of CIM is essentially to obtain a flexible and optimal response to changes in market demand, on the one hand, and to plant capabilities on the other. It has been common practice for many years for production plans to be formulated and production schedules to be produced by computer and for these schedules to be passed down to the plant. In refineries, use of large scheduling programs is widespread. In addition to flexibility, other benefits claimed are improved product quality, higher throughputs, lower costs and greater safety.

A characteristic feature of CIM is that information also flows the other way, i.e. up from the plant to the planning function. This provides the latter with a continuous flow of up-to-date information on the capability of the plant so that the schedule can be modified to produce the optimal solution. A CIM system may therefore carry out not only the process control and quality control but also scheduling, inventory control, customer order processing and accounting functions.

The architecture of a CIM system is generally hierarchical and distributed. Treatments of such architecture are given in *Controlling Automated Manufacturing Systems* (O'Grady, 1986) and by Dempster *et al.* (1981).

For such a system to be effective it is necessary that the data passing up from the plant be of high quality. The system needs to have a full model of the plant, including the mass and energy balances and the states and capabilities of the equipment. This involves various forms of model-based control, which is of such prominence in CIM that the two are sometimes treated as if they are equivalent.

Plant data are corrupted by noise and errors of various kinds, and in order to obtain a consistent data set it is necessary to perform data reconciliation. Methods based on estimation theory and other techniques are used to achieve this. Complete and rigorous model-based reconciliation (CRMR) is therefore a feature of CIM. Data reconciliation is discussed further in Chapter 30. One implication of CIM is the plant is run under much tighter control, which should be beneficial to safety.

13.7.1 Batch plants

Batch processing involves not only sequential operations but also a high degree of variability of equipment states and is particularly suited to CIM. Accounts of integrated batch processing include those by Rosenof (1982b), Armstrong and Coe (1983), Rippin (1983), Severns and Hedrick (1983), Bristol (1985), Krigman (1985), Egli and Rippin (1986), Kondili, Pantelides and Sargent (1988), Cott and Macchietto (1989) and Crooks, Kuriyna and Macchietto (1992).

In the system described by Cott and Macchietto (1989), use is made of three levels of control, which are, in descending order: plant level control, batch level control and resource level control, operating respectively on typical time-scales of days, minutes and seconds. A comprehensive approach to batch processing requires the integration of tools for plant design, automation and operating procedures.

13.8 Instrument Failure

Process plants are dependent on complex control systems and instrument failures may have serious effects. It is helpful to consider first the ways in which instruments are used. These may be summarized as follows:

Instrument	*System application*
Measuring instrument	Input to: Display system – measurement/status/alarm Control loop Trip system Computer model
Control element	Output from: Control loop Trip system

Measuring instruments are taken to include digital as well as analogue outputs. Control elements are normally control valves, but can include power cylinders, motors, etc.

The important point is that some of these applications constitute a more severe test of the instrumentation than others. The accuracy of a flowmeter may be sufficient for flow control, but it may not be good enough for an input to a mass balance model in a computer. The dynamic response of a thermocouple may be adequate for a panel display, but it may be quite unacceptable in a trip system.

This leads directly, of course, to the question of the definition of failure. In the following sections various kinds of failure are considered. It is sufficient here to emphasize that the reliability of an instrument depends on the definition of failure, and may vary depending on the application.

13.8.1 Overall failure rates

There are more data on the failure rates of instrumentation than on most other types of plant equipment. It is now usually possible to obtain sufficient data for assessment purposes, though there are inevitably some gaps. There are two types of failure data on instruments. The first relates to performance in standard instrument tests and the second to performance on process plant. It is the latter which is of primary interest here.

Many of the data on the failure rates of instruments on process plants derive from the work of the UK Atomic Energy Authority (UKAEA). Table 13.5 gives data quoted in early investigations by UKAEA workers. Table 13.6 shows data in another early survey by Anyakora, Engel and Lees (1971) in three works in the chemical industry. The first (works A) was a large works producing a wide range of heavy organic chemicals. The second (works B) made heavy inorganic chemicals. The third (works C) was two sites in a glass works. The failures were defined as and derived from job requests from the process operators. The failure rates were calculated on the assumption of a constant failure rate. The environment factor quoted in the table is explained below.

The failure rates given in Table 13.6 are in broad agreement with other work published about the same time, such as that of Skala (1974). It should be

Table 13.5 *Some data on instrument failure rates published by the UKAEA*

Instrument	*Failure rate* (faults/year)		*Reference*[a]
	Observed	*Assumed/predicted*	
Control valve[b]	0.25		1–4
		0.26	5, 6
Solenoid valve		0.26	5, 6
Pressure relief valve		0.022	5, 6
Hand valve		0.13	5, 6
Differential pressure transmitter[b]	0.76		1–4, 7, 8
Variable area flowmeter transmitter[b]	0.68		1–4, 8
Thermocouple		0.088	5, 6
Temperature trip amplifier:			
type A	2.6		7
type B	1.7		7
Pressure switch	0.14		1–4, 8
Pressure gauge		0.088	5, 6
O_2 analyser	2.5		1, 2, 4, 8
Controller[b]	0.38		7
Indicator (moving coil meter)		0.026	5, 6, 8
Recorder (strip chart)		0.22	5, 6, 8
Lamp (indicator)		0.044	5, 6
Photoelectric cell		0.13	5, 6
Tachometer		0.044	5, 6
Stepper motor		0.044	5, 6
Relay[b]	0.17		7
Relay (Post Office)		0.018	5, 6

[a] (1) Hensley (1967 UKAEA AHSB(S) R136); (2) Hensley (1968); (3) Hensley (1969 UKAEA AHSB(S) R178); (4) Hensley (1973 SRS/GR/1); (5) Green and Bourne (1966 UKAEA AHSB(S) R117); (6) A.E. Green and Bourne (1972); (7) Eames (1966); (8) Green (1966 UKAEA AHSB(S) R113)

[b] Pneumatic.

Table 13.6 *Some instrument failure rate data from three chemical works (Anyakora, Engel and Lees, 1971) (Courtesy of the Institution of Chemical Engineers)*

Instrument	*No. at risk*	*Instrument years*	*Environment factor*	*No. of faults*	*Failure* (faults/year)
Control valve	1531	747	2	447	0.60
Power cylinder	98	39.9	2	31	0.78
Valve positioner	334	158	1	69	0.44
Solenoid valve	252	113	1	48	0.42
Current/pressure transducer	200	87.3	1	43	0.49
Pressure measurement	233	87.9	3	124	1.41
Flow measurement (fluids):	1942	943	3	1069	1.14
Differential pressure transducer	636	324	3	559	1.73
Transmitting variable area flowmeter	100	47.7	3	48	1.01
Indicating variable area flowmeter	857	409	3	137	0.34
Magnetic flowmeter	15	5.98	4	13	2.18
Flow measurement (solids):					
Load cell	45	17.9	–	67	3.75
Belt speed measurement and control	19	7.58	–	116	15.3
Level measurement (liquids):	421	193	4	327	1.70
Differential pressure transducer	130	62	4	106	1.71
Float-type level transducer	158	75.3	4	124	1.64
Capacitance-type level transducer	28	13.4	4	3	0.22
Electrical conductivity probes	100	39.8	4	94	2.36
Level measurement (solids)	11	4.38	–	30	6.86
Temperature measurement (excluding pyrometers):	2579	1225	3	425	0.35
Thermocouple	772	369	3	191	0.52
Resistance thermometer	479	227	3	92	0.41
Mercury-in-steel thermometer	1001	477	2	13	0.027
Vapour pressure bulb	27	10.7	4	4	0.37
Temperature transducer	300	142	3	124	0.88
Radiation pyrometer	43	30.9	4	67	2.17
Optical pyrometer	4	3.4	4	33	9.70
Controller	1192	575	1	164	0.29
Pressure switch	549	259	2	87	0.34
Flow switch	9	3.59	–	4	1.12
Speed switch	6	2.39	–	0	–
Monitor switch	16	6.38	–	0	–
Flame failure detector	45	21.3	3	36	1.69
Millivolt-current transducer	12	4.78	–	8	1.67
Analyser:	86	39.0	–	331	8.49
pH meter	34	15.8	–	93	5.88
Gas–liquid chromatograph	8	3.43	–	105	30.6
O_2 analyser	12	5.67	–	32	5.65
CO_2 analyser	4	1.90	–	20	10.5
H_2 analyser	11	5.04	–	5	0.99
H_2O analyser (in gases)	3	1.38	–	11	8.00
Infrared liquid analyser	3	1.43	–	2	1.40
Electrical conductivity meter (for liquids)	5	1.99	–	33	16.70
Electrical conductivity meter (for water in solids)	3	1.20	–	17	14.2
Water hardness meter	3	1.20	–	13	10.9
Impulse lines	1099	539	3	416	0.77
Controller settings	1231	609	–	84	0.14

emphasized that of the failures given in the tables only a very small proportion resulted in a serious plant condition. In most cases the failures were detected by the process operator, who then called in the instrument maintenance personnel.

The failure rates quoted are those for normal commercial instruments in the process industries. In certain other applications where higher instrument costs are acceptable, the failure rates are lower. Thus instruments used in some defence applications are an order of

Table 13.7 Effect of environment on instrument reliability: instruments in contact with and not in contact with process fluids (Anyakora, Engel and Lees, 1971) (Courtesy of the Institution of Chemical Engineers)

Instrument	*No. at risk*	*No. of faults*	*Failure rate* (faults/year)
Instruments in contact with process fluids:	2285	1252	1.15
Pressure measurement	193	89	0.97
Level measurement	316	233	1.55
Flow measurement	1733	902	1.09
Flame failure device	43	28	1.37
Instruments not in contact with process fluids:	2179	317	0.31
Valve positioner	320	62	0.41
Solenoid valve	168	24	0.30
Current-pressure transducer	89	23	0.54
Controller	1083	133	0.26
Pressure switch	519	75	0.30
Control valve	1330	359	0.57
Temperature measurement	2391	326	0.29

Table 13.8 Effect of environment on instrument reliability: instruments in contact with clean and dirty fluids (Anyakora, Engel and Lees, 1971) (Courtesy of the Institution of Chemical Engineers)

Instrument	*No. at risk*	*No. of faults*	*Failure rate* (faults/year)
Control valve:			
Clean fluids	214	17	0.17
Dirty fluids	167	71	0.89
Differential pressure transmitter:			
Clean fluids	27	5	0.39
Dirty fluids	90	82	1.91

magnitude more expensive, but have a much higher reliability. Further data on instrument failure rates are given in Appendix 14.

13.8.2 Factors affecting failure

Some of the factors which affect instrument failure are listed below.

(1) System context:
 (a) application (display, control, etc.);
 (b) specification (accuracy, response, etc.);
 (c) definition of failure.
(2) Installation practices.
(3) Environmental factors – process materials:
 (a) degree of contact (control room, plant);
 (b) material phase (gas, liquid, solid);
 (c) cleanliness;
 (d) temperature;
 (e) pressure;
 (f) corrosion;
 (g) erosion.
(4) Environmental factors – ambient and plant conditions:
 (a) temperature;
 (b) humidity;
 (c) dust;
 (d) frost exposure;
 (e) vibration;
 (f) impact exposure.
(5) Operating factors:
 (a) movement, cycling.
(6) Maintenance practices.

There is little information available on which to assess the effect of these factors. In the survey by Anyakora, Engel and Lees an attempt was made to assess the effect of environment, defining this rather loosely in terms of both ambient conditions and process materials. Two approaches were tried. One was to compare the effect of being or not being in contact with process fluids. Table 13.7 shows this effect for two groups of instruments, one consisting of those which are in contact and one consisting of those which are not. The instruments which are not in contact with process fluids show a much lower failure rate, although control valves and temperature measurements are exceptions.

The other approach, which was applied to instruments which are in contact with process fluids, was to distinguish between 'clean' and 'dirty' fluids. A fluid was regarded as dirty if it contained 'gunk', polymerized, corroded, etc. Table 13.8 gives data for instruments in these two cases.

From this work it was concluded, as a first approximation and for the instruments considered, that the severity of the environment of an instrument depends on the aggressiveness of any process materials with which it is in contact and that other factors are generally of secondary importance.

If the failure rate is taken to be a product of a base failure rate and of an environment factor, then Tables 13.7 and 13.8 suggest that a maximum value of about 4 is appropriate. Environment factors are given in Table 13.6; the failure rates given in the table are the original data and should be divided by the environment factor to give the base failure rate. It should also be noted that the sampling/impulse line failure rates given in Table 13.6 should be added to the failure rates of the instruments themselves to obtain the failure rates of installations.

13.8.3 Failure modes

The overall failure rate of an instrument gives only limited information. It is often necessary to know its failure modes. Failure modes can be classified in several ways. Some important categories are (1) condition, (2) performance, (3) safety, and (4) detection.

In a failure classification based on conditions, a failure mode is exemplified by a faulty bellows on a flowmeter or a broken diaphragm in a control valve. In a classification by performance illustrations of failure are

Table 13.9 *Failure modes of some instruments (Lees, 1973b) (Courtesy of the Institution of Chemical Engineers)*

Instrument failure mode	*No. of faults*
Control valve:	
Leakage	54
Failure to move freely:	
sticking (but moving)	28
seized up	7
not opening	5
not seating	3
Blockage	27
Failure to shut off flow	14
Glands repacked/tightened	12
Diaphragm fault	6
Valve greased	5
General faults	27
Thermocouple:	
Thermocouple element faults	24
Pocket faults	11
General faults	20

Table 13.10 *Failure modes of some instruments defined by performance*

		Reference
Level measurement and alarm:	*Failure rate* (faults/year)	Lawley (1974b)
Level indicator fails to danger	2	
High level alarm fails	0.2	
	Probability	
Operator fails to observe level indicator or take action	0.04	
Operator fails to observe level alarm or take action	0.03	
Flow measurement and control:	*Failure rate*	S.B. Gibson (1977b)
For an FRC where high flow is undesirable:	(faults/year)	
Flow element fails giving low reading	0.1	
Flow transmitter fails giving low reading	0.5	
Flow recorder controller fails calling for more flow	0.4	
Flow control valve fails towards open position	0.1	
	Fractional dead time	
High flow trip fails to operate	0.01	
For an FRC where low flow is undesirable:	*Failure rate* (faults/year)	
Flow element fails giving high reading	0.2	
Flow transmitter fails giving high reading	0.4	
Flow recorder controller fails calling for less flow	0.4	
Flow control valve fails towards closed position	0.2	
	Fractional dead time	
Low flow trip fails to operate	0.01	
Low flow trip left aborted after start-up	0.01	
Manual and control valves:	*Failure rate* (faults/year)	Lawley (1974b)
Manual isolation valve wrongly closed	0.05 and 0.1	
Control valve fails open or misdirected open	0.5	
Control valve fails shut or misdirected shut	0.5	

FRC, flow recorder controller

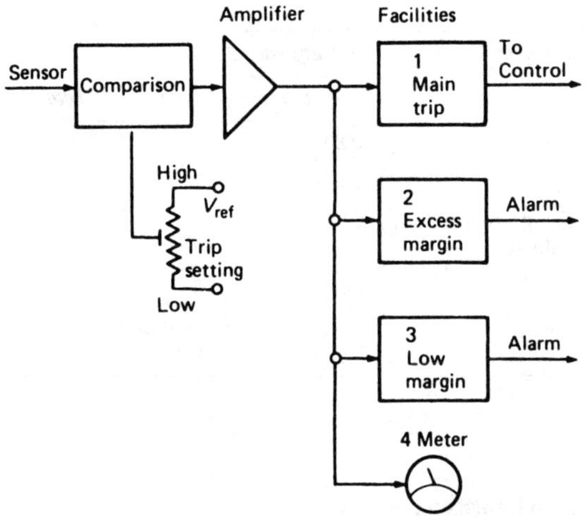

Figure 13.7 *Temperature trip amplifier (Eames, 1973 UKAEA SRS/GR/12) (Courtesy of the UK Atomic Energy Authority, Systems Reliability Directorate)*

a zero error in a flowmeter or the passing of fluid when shut by a control valve. A performance classification emphasizes effects and a condition classification emphasizes causes, but the distinction is not rigid: a blockage in a control valve could reasonably be classed either way. The safety classification divides faults into fail-safe and fail-dangerous. The detection classification distinguishes between revealed and unrevealed faults: a revealed fault signals its presence and is at once detectable, an unrevealed fault is not immediately detectable, but is usually detected by a proof check.

Condition and performance may be regarded as the primary types of failure. Safety and detection modes may be obtained from these and from the system context of the instrument.

Table 13.9 shows some data obtained by Lees (1973b) in the survey already described on failure modes in thermocouples and control valves. These failure modes are essentially classified by condition, although the condition is often revealed by a performance failure. Similar data for other instruments are given in the original paper. Some data used in reliability studies described in the literature on failure modes of instruments and control loops are shown in Table 13.10.

If information is available on overall but not on mode failure rates, it is sometimes assumed that about one-third of the faults are in the fail-dangerous mode. The safety and detection failure modes of the temperature trip amplifier shown in Figure 13.7 have been analysed by Eames (UKAEA 1973 SRS/GR/12) as shown in Table 13.11. The fault is described by a four-letter code. The first indicates that it is fail-safe (S), fail-dangerous (D) or a calibration shift in the dangerous direction (C). The second is the number of the equipment adversely affected by the fault. The number 1–5 refers to, respectively, the main trip, the excess margin alarm, the low margin alarm, and the indicating meter shown in the figure and the indicating lamp (not shown). The third letter indicates that the fault is revealed (r) or unrevealed (u). The fourth is the number of the equipment which reveals the fault; the numbering code is as before. The various failure rates of the equipment are as follows:

Faults	*Failure rate*	
	(faults/10^6 h)	(faults/year)
Fail-dangerous (D1)	9.85	0.086
Fail-dangerous, unrevealed (D1u)	4.6	0.040
Total	145.5	1.27

Thus the total fail-dangerous and fail-dangerous unrevealed faults are, respectively, 6.8% and 3.2% of the total faults, which is one measure of the success of the fail-safe design of the equipment.

13.8.4 Prediction of failure rates

It is sometimes necessary to know the failure rate of an instrument for which field data are not available. To meet this situation methods have been developed for estimating the failure rate of an instrument from those of its constituent parts. Table 13.12 shows part of a prediction by Hensley (UKAEA 1969 AHSB(s) R178) of the failure rate of a pressure switch.

A comparison of some observed and predicted failure rates is given in Table 13.13. It can be seen that the agreement is quite good. A more quantitative measure of the effectiveness of the technique is Figure 13.8, which

Table 13.11 *Failure modes and rates of a temperature trip amplifier (Eames, 1973 UKAEA SRS/GR/12) (Courtesy of the UK Atomic Energy Authority, Systems Reliability Directorate)*

				Failure rate (faults/10^6 h)			
Sr1	53.89	D1r2	4.85	D2r3	9.95	C1u	3.15
Sr2	2.1	D1r3	0.4	D2r4	0.15	C2u	6.79
Sr3	7.4	D1u	4.6	D2u	6.0	C3u	5.77
Sr5	5.0			D3r2	0.35		
Su	15.84			D3u	3.3		
				D4r4	5.8		
				D5u	10.2		
Total S	84.23	Total D1	9.85	Total other D	35.75	Total C	15.71

Table 13.12 Predicted failure rates of a pressure switch (Hensley, 1969 UKAEA AHSB(S) R178) (Courtesy of the UK Atomic Energy Authority, Systems Reliability Directorate)

Component	*Fault*	*Category*	*Failure rate* (faults/10^6 h)		
			Dangerous	*Safe*	*Total*
Spring	Fracture	Dangerous	0.2		
Bellows	Rupture	Safe		5.0	
Screws – pivot (2 items)	Loosen	Dangerous	1.0		
Microswitch	Random	Dangerous 25%	0.5		
		Safe 75%		1.5	
Total of above 5	–	–	1.7	6.5	
Total for 30 components in instrument			2.9	11.7	14.6
				(faults/year)	
Total for 30 components in instrument			0.025	0.10	0.13

is given by A.E. Green and Bourne (1972) and shows the ratio of observed and predicted failure rates for a number of equipments. The median value of this ratio is 0.76 and the probabilities of the ratio being within factors of 2 and 4 of this value are 70% and 96%, respectively.

13.8.5 Loop failure rates

Data on failure rates of complete control loops have been given by Skala (1974) and are shown in Table 13.14. Loop failure rates can be calculated from the failure rates of the constituent instruments. The failure rates of a loop with a pneumatic flow indicator controller, as calculated from the data in Table 13.5 (UKAEA), as calculated from the data in Table 13.6 (Anyakora, Engel and Lees), and as given by Skala, are shown in Table 13.15.

Again it should be emphasized that of the failure rates for loops given in these tables only a very small proportion results in a serious plant upset or trip. In one study of the control loop failures on a large chemical plant quoted by M.R. Gibson (1978), it was found that there had been three control loop failures which resulted in plant trips and that the frequency of such failures was one failure every 20 years per loop.

13.8.6 Detection of failure

If instrument failure occurs, it is important for it to be detected. The ease of detection of an instrument failure depends very much on whether the fault is revealed or unrevealed. Unrevealed faults are generally detectable only by proof testing.

An instrument fault which is revealed is usually detected by the process operator either from the behaviour of the instrument itself or from the effect of the failure on the control system. There are, however, developments in the use of the process computer to detect instrument faults. Fault detection by the operator and by the computer is discussed in Chapters 14 and 30, respectively.

The detection of failure in instruments which have a binary output such as pressure or level switches is particularly difficult, because the fault is generally unrevealed, but is particularly important, because such instruments are frequently part of an alarm or trip system. One approach to the problem is to use an instrument with a continuous range output rather than a binary output. Thus a level measuring instrument may be used instead of a level switch. In this way may of the faults on the instrument which would otherwise be unrevealed become revealed.

Table 13.13 Observed and predicted instrument failure rates

Instrument	*Failure rate* (faults/year)		*Reference*[a]
	Observed	*Predicted*	
Control valve[b]	0.25	0.19	1–3
Differential pressure transmitter[b]	0.76	0.45	1–4
Variable area flowmeter transmitter[b]	0.68	0.7	1–3
Temperature trip amplifier:			
type A	2.6	2.8	1, 2, 4
type B	1.7	2.1	4
Controller	0.38	0.87	4
Pressure switch	0.14	0.13	1–3
Gas analyser	2.5	3.3	1, 2
Relay[b]	0.17	0.35	4

[a] (1) Hensley (1967 UKAEA AHSB(S) R136); (2) Hensley (1968); (3) Hensley (1969 UKAEA AHSB(S) R178); (4) Eames (1966).
[b] Pneumatic.

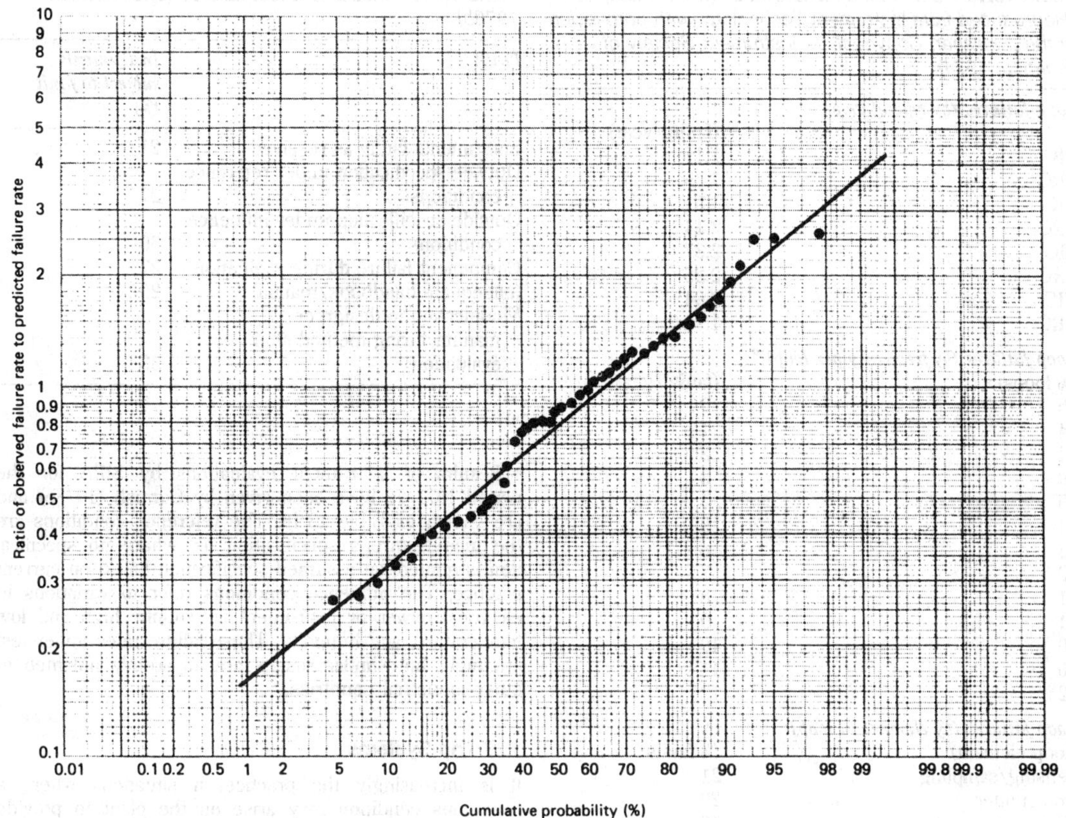

Figure 13.8 Ratio of observed to predicted equipment failure rates (A.E. Green and Bourne, 1972) (Reproduced with permission from Reliability Technology *by A.E. Green and J.R. Bourne, Copyright ©, 1972, John Wiley and Sons Inc.)*

13.8.7 Self-checking instruments

Developments are also occurring in instruments which have a self-checking capability. Principles on which such instruments are designed include (1) multiple binary outputs and (2) electrical sensor check.

A self-checking level measuring instrument which uses multiple binary outputs has been described by Hasler and Martin (1971). The instrument has a series of binary output points, which measure the liquid level at different heights. These points provide a mutual check. Thus, for example, if there are 10 points and the liquid level is up to point 5, so that this point gives a positive output, the absence of a positive output from point 4 indicates a failure on that point.

A self-checking level switch in which electronic signals are used to check the state of the sensor has been described by J.O. Green (1978).

Increasingly, instruments are also being provided with the enhanced capabilities available from the incorporation of microprocessors. Self-checking is one such capability.

A general account of such instruments is given in *Intelligent Instrumentation* (Barney, 1985). A further discussion of intelligent, or smart, instruments is given by the CCPS (1993/14).

13.8.8 Fault-tolerant instrumentation

Instrument systems should have a degree of fault tolerance. The need for fault-tolerant systems has already been mentioned in relation to computer systems, where certain basic principles were outlined. These principles are equally applicable to the design of fault-tolerant instrument systems. Two main features of such instrumentation are redundancy and/or diversity and fail-safe operation. Fault-tolerant design of instrument systems is discussed by Bryant (1976), Ida (1983), Frederickson and Beckman (1990) and the CCPS (1993/14).

13.8.9 Instrument testing

Information is also available on the performance of instruments when subjected to a battery of standard tests. The evaluation of instruments is carried out both by special testing organizations and by major users. In the UK the main organization concerned with instrument evaluation is the Scientific Instrument Research Association (SIRA). Some of the tests carried out by SIRA have been described by Cornish (1978a). The instruments tested are normal production models.

Table 13.14 *Control loop failure rates (after Skala, 1974) (Reproduced from* Instrument Technology *with permission of the publisher, Copyright ©, Instrument Society of America, 1974)*

Loop failures (by type of loop):	
	(faults/year)
PIC	1.15
PRC	1.29
FIC	1.51
FRC	2.14
LIC	2.37
LRC	2.25
TIC	0.94
TRC	1.99
Loop failures (by frequency per loop):[a]	
(% loops)	(faults/year)
25	0
34	1
14	2
9	3
5	4
4	5
3	6
2	7
1	8
1	9
0	10
0	11
2	12
Loop failures (by element in loop):	
(loop element)	(% faults)
Sensing/sampling	21
Transmitter	20
Transmission	10
Receiver[b]	18
Controller	7
Control valve	7
Other	17

[a] These data have been read from Figure 2b of the original paper.
[b] Presumably indicators, recorders.

Table 13.15 *Failure rates for a pneumatic flow indicator control loop*

UKAEA data:	(faults/year)
Differential pressure transmitter	0.76
Controller	0.38
Control valve	0.25
	1.39
Anyakora, Engel and Lees' data:	
Impulse lines	0.26
Differential pressure transmitter[a]	0.58
Controller[a]	0.29
Control valve[a]	0.30
Valve positioner[a]	0.09 (0.2×0.44)[b]
	1.52
Skala's data:	
FIC loop[c]	1.51

[a] Pneumatic
[b] It is assumed that 20% of the valves have positioners.
[c] FIC, Flow indicator controller.

Table 13.16 *Instrument test failures (after Cornish, 1978b)*

Fault	*Instruments subject to fault* (%)
Instruments faulty as received	21
Outside specification under reference conditions	27
Outside specification under influence conditions	30
Component failure during evaluation	27
Inadequate handbook/manual	26
Modification to design or manufacturing method after evaluation	33

Results of instrument evaluations by SIRA for the period 1971–76 have been given by Cornish (1978b) and are shown in Table 13.16. The reference conditions are the manufacturer's specification or, where no specification is quoted, an assumed specification based on current practice. The influence conditions refer to variations in electrical power or instrument air supply, high and low temperature, and humidity. These failure rates under test are high, but similar results are apparently obtained in other industrial countries.

13.9 Trip Systems

It is increasingly the practice in situations where a hazardous condition may arise on the plant to provide some form of automatic protective system. One of the principal types of protective system is the trip system, which shuts down the plant, or part of it, if a hazardous condition is detected. Another important type of protective system is the interlock system, which prevents the operator or the automatic control system from following a hazardous sequence of control actions. Interlock systems are discussed in Section 13.10.

Accounts of trip systems are given in *Reliability Technology* (A.E. Green and Bourne, 1972) and by Hensley (1968), R.M. Stewart (1971), Kletz (1972a), de Heer (1974), Lawley and Kletz (1975), Wells (1980), Barclay (1988), Rushton (1991a,b) and Englund and Grinwis (1992).

The existence of a hazard which may require a protective system is usually revealed either during the design process, which includes, as routine, consideration of protective features, or by hazard identification techniques such as hazop studies.

The decision as to whether a trip system is necessary in a given case depends on the design philosophy. There are quite wide variations in practice on the use of trip systems. There is no doubt, however, about the general trend, which is towards the provision of a more comprehensive coverage by trip systems. The problems are considered further in Chapter 14. The decision as to whether to install a trip system can be put on a less subjective basis by making a quantitative assessment of the hazard and of the reliability of the operator in preventing it.

13.9.1 Single-channel trip system

A typical, single-channel trip system is shown in Figure 13.9. It consists of a sensor, a trip switch and a trip valve. The configuration of a trip loop is therefore not dissimilar to that of a control loop. The difference is that, whereas the action of a control loop is continuous, that of a trip loop is discrete.

The trip switch may be of a general type, being capable of taking an electronic or pneumatic signal from any type of sensor. Thus in a pneumatic system a pressure switch would serve as the trip switch. Alternatively, the trip switch and the sensor may be combined to give a switch dedicated to a particular variable. Thus common types of trip switch include flow, pressure, temperature, level and limit switches.

13.9.2 Dependability of trip systems

Since a trip system is used to protect against a hazardous condition, it is essential for the system itself to be dependable. The dependability of a trip system depends on (1) capability and (2) reliability. Thus it is necessary both for the system to have the capability of carrying out its function in terms of features such as accuracy, dynamic response, etc., and for it to be reliable in doing so.

The reliability of the trip system may be improved by the use of (1) redundancy and (2) diversity. Thus one approach is to use multiple redundant instruments, which generally give a reliability greater than that of a single instrument. But redundancy is not always the full answer, because there are some dependent failures which may disable the whole set of redundant instruments. This difficulty can be overcome by the use of diversity, which is exemplified by the use of different measurements to detect the same hazard and by the use of different instruments to measure the same variable.

Most trip systems consist of a single channel comprising a sensor, a switch and a shut-off valve, but where the integrity required is higher than that which can be obtained from a single channel, redundancy is generally used.

A trip system should be reliable against functional failure, i.e. failure which prevents the system shutting the plant down when a hazardous condition occurs. Such a condition is not normally present and its rate of occurrence, which is the demand rate on the trip system, is usually very low. Thus functional failures of the system are generally unrevealed failures. The trip system should also be reliable against operational failure, i.e. failure which causes the system to shut the plant down when no hazardous condition exists. Thus operational failures of the system are always revealed failures.

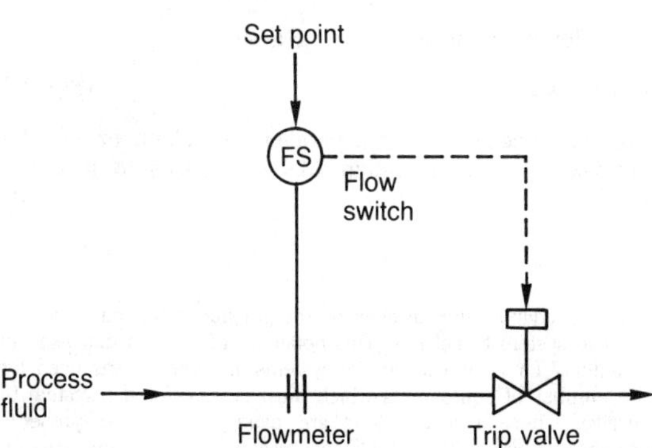

Figure 13.9 *A trip system*

It is the object of trip system design and operation to avoid both loss of protection against the hazardous condition due to functional failure and plant shut-down due to operational failure, or spurious trip. Since functional failure of the system is generally unrevealed, it is necessary to carry out periodic proof testing to detect such failure.

The simpler theoretical treatments of trip systems usually assume that the functional failures are unrevealed and the operational failures revealed and that the failure rates are constant; this approach is followed here. The treatment draws particularly on the work of A.E. Green and Bourne (1966 UKAEA AHSB(S) R117), some of which was later published by the same authors in *Reliability Technology* (1972).

13.9.3 Fractional dead time

The fractional dead time (FDT) of an equipment or system gives the probability that it is in a failed state. If the failure of an equipment is revealed with a revealed failure rate λ, the FDT ϕ depends on the failure rate and the repair time τ_r:

$$\phi = \lambda\tau_r \qquad \tau_r \ll 1 \qquad [13.9.1]$$

For a series system with revealed failure, the FDT ϕ of the system is related to the FDT ϕ_i of the constituent equipments as follows:

$$\phi = \sum_{i=1}^{n} \phi_i \qquad \lambda_i\tau_{ri} \ll 1 \qquad [13.9.2]$$

For a parallel system with revealed failure, the FDT ϕ of the system is related to the FDT ϕ_i of the equipments as follows:

$$\phi = \prod_{i=1}^{n} \phi_i \qquad \lambda_i\tau_{ri} \ll 1 \qquad [13.9.3]$$

For a parallel redundant, or $1/n$ (1-out-of-n), system with revealed failure, the FDT $\phi_{1/n}$ of the system is related to the FDT $\phi_{1/1}$ of a single equipment as follows:

$$\phi_{1/n} = \phi_{1/1}^{n} \qquad [13.9.4]$$

If, however, the failure of an equipment is unrevealed with an unrevealed failure rate λ, the FDT depends on this failure rate and on the proof test interval τ_p. The probability q of failure within time period t is:

$$q = 1 - \exp(-\lambda t) \qquad [13.9.5]$$

or, for small values of λt

$$q = \lambda t \qquad \lambda t \ll 1 \qquad [13.9.6]$$

Then the FDT is:

$$\phi = \frac{1}{\tau_p}\int_0^{\tau_p} q\mathrm{d}t \qquad [13.9.7]$$

For a $1/n$ system with unrevealed failure the FDT $\phi_{1/n}$ of the system is obtained from the probability $q_{1/n}$ of failure of the system within the time period t:

$$\phi_{1/n} = \frac{1}{\tau_p}\int_0^{\tau_p} q_{1/n}\mathrm{d}t \qquad [13.9.8]$$

A detailed account of fractional dead times is given by A.E. Green and Bourne (UKAEA 1966 AHSB(S) R117).

13.9.4 Functional reliability of trip systems

Functional failure of a trip system is here assumed to be unrevealed. The failure rate λ used in the equations in this section is that applicable to these unrevealed fail-dangerous faults.

For a simple trip system consisting of a single channel 1/1 (1-out-of-one) system with a failure rate λ the probability q of failure within proof test interval τ_p is:

$$q = 1 - \exp(-\lambda\tau_p) \qquad [13.9.9]$$

For small values of $\lambda\tau_p$:

$$q = \lambda\tau_p \qquad \lambda\tau_p \ll 1 \qquad [13.9.10]$$

The trip system is required to operate only if a hazardous plant condition occurs. The probability p_δ that such a plant demand, which has a demand rate δ, will occur during the dead time τ_0 after the failure is:

$$p_\delta = 1 - \exp(-\delta\tau_0) \qquad [13.9.11]$$

But, on average, the dead time τ_0 is half the proof test interval τ_p:

$$\tau_0 = \frac{\tau_p}{2} \qquad [13.9.12]$$

Hence:

$$p_\delta = 1 - \exp(-\delta\tau_p/2) \qquad [13.9.13]$$

For small values of $\delta\tau_p$:

$$p_\delta = \frac{\delta\tau_p}{2} \qquad \delta\tau_p \ll 1 \qquad [13.9.14]$$

The probability p_η that a plant hazard will be realized can be written in terms of the plant hazard rate η:

$$p_\eta = 1 - \exp(-\eta\tau_p) \qquad [13.9.15]$$

For small values of $\eta\tau_p$:

$$p_\eta = \eta\tau_p \qquad \eta\tau_p \ll 1 \qquad [13.9.16]$$

Frequently, some or all of the approximations of Equations 13.9.10, 13.9.14 and 13.9.16 apply. If all do, then taking

$$p_\eta = qp_\delta \qquad [13.9.17]$$

gives

$$\eta = \frac{\delta\lambda\tau_p}{2} \qquad \lambda\tau_p \ll 1; \delta\tau_p \ll 1; \eta\tau_p \ll 1 \qquad [13.9.18]$$

If the assumptions underlying Equation 13.9.18 are not valid, an expression which has been commonly used is:

$$\eta = \lambda p_\delta \qquad [13.9.19]$$

Alternatively, the plant hazard rate can be expressed in terms of the FDT ϕ of the system:

$$\eta = \delta\phi \qquad [13.9.20]$$

As given earlier, the probability q of failure of the simple trip system within the time period t is:

$$q = 1 - \exp(-\lambda t) \qquad [13.9.21]$$

For small values of λt:

$$q = \lambda t \qquad \lambda t \ll 1 \qquad [13.9.22]$$

Then the FDT ϕ of the simple trip system is:

$$\phi = \frac{1}{\tau_p}\int_0^{\tau_p} q\mathrm{d}t \qquad [13.9.23]$$

Hence:

$$\phi = \frac{\lambda\tau_p}{2} \qquad [13.9.24]$$

and

$$\eta = \frac{\delta\lambda\tau_p}{2} \qquad [13.9.25]$$

as before.

For a parallel redundant, or $1/n$ (1-out-of-n), system, or for an m/n (m-out-of-n) system, which may be a majority voting system, the following treatment applies. If there are n equipments of which m must survive for the system to survive and r must fail for the system to fail*:

$$r = n - m + 1 \qquad [13.9.26]$$

The probability $q_{m/n}$ of failure of this system within the proof test interval is:

$$q_{m/n} = \sum_{k=r}^{n}\binom{n}{k}q^k(1-q)^{n-k} \qquad [13.9.27]$$

For small values of q:

$$q_{m/n} = \binom{n}{r}q^r \qquad q \ll 1 \qquad [13.9.28]$$

The FDT $\phi_{m/n}$ of the system is:

$$\phi_{m/n} = \frac{1}{\tau_p}\int_0^{\tau_p} q_{m/n}\mathrm{d}t \qquad [13.9.29]$$

Then from Equations 13.9.22, 13.9.28 and 13.9.29 the FDT $\phi_{m/n}$ of the system is:

$$\phi_{m/n} = \binom{n}{r}\frac{(\lambda\tau_p)^r}{r+1} \qquad [13.9.30]$$

Thus for a 2/3 majority voting system:

$$\phi_{2/3} = (\lambda\tau_p)^2 \qquad [13.9.31]$$

For the special case of a parallel redundant, or $1/n$ (1-out-of-n), system, Equation 13.9.30 reduces to give the FDT $\phi_{1/n}$:

* In this chapter the number of equipments which must fail for the trip system to fail is r. This notation differs from that used in Chapter 7 for r-out-of-n parallel systems, in which r was used for the number of equipments which must survive for the system to survive. These two notations are used in order to preserve correspondence with established usage in texts on general reliability (e.g. Shooman, 1968) and on trip systems (e.g. A.E. Green and Bourne, 1972).

***Table 13.17** Fractional dead times for trip systems with simultaneous proof testing*

System	n	m	r	ϕ	
1/1	1	1	1	$\frac{\lambda\tau_p}{2}$	$\phi_{1/1}$
1/2	2	1	2	$\frac{(\lambda\tau_p)^2}{3}$	$\frac{4}{3}\phi^2_{1/1}$
1/3	3	1	3	$\frac{(\lambda\tau_p)^3}{4}$	$2\phi^3_{1/1}$
2/2	2	2	1	$\lambda\tau_p$	$2\phi_{1/1}$
2/3	3	2	2	$(\lambda\tau_p)^2$	$4\phi^2_{1/1}$

***Table 13.18** Fractional dead times for trip systems with staggered proof testing (A.E. Green and Bourne, 1972) (Reproduced with permission from Reliability Technology by A.E. Green and J.R. Bourne, Copyright ©, 1972, John Wiley and Sons Inc.)*

System	ϕ^*	$\frac{\phi}{\phi^*}$
1/1	$\frac{\lambda\tau_p}{2}$	1
1/2	$\frac{5}{24}(\lambda\tau_p)^2$	1.6
1/3	$\frac{1}{12}(\lambda\tau_p)^3$	3.0
2/2	$\lambda\tau_p$	1
2/3	$\frac{2}{3}(\lambda\tau_p)^2$	1.5

$$\phi_{1/n} = \frac{(\lambda\tau_p)^n}{n+1} \qquad [13.9.32]$$

Thus for a 1/2 parallel system

$$\phi_{1/2} = \frac{(\lambda\tau_p)^2}{3} \qquad [13.9.33]$$

and for a 1/3 system

$$\phi_{1/3} = \frac{(\lambda\tau_p)^3}{4} \qquad [13.9.34]$$

Fractional dead times calculated from Equation 13.9.30 are shown in Table 13.17, both as functions of $\lambda\tau_p$ and of the FDT $\phi_{1/1}$ for a single channel. Table 13.17 assumes that the instruments are tested simultaneously at the end of the proof test interval. Some improvement can be obtained by staggered testing, as shown by the data in Table 13.18, which are taken from A.E. Green and Bourne (1972).

13.9.5 Operational reliability of trip systems

It is also necessary to consider operational failure of trip systems. Operational failure is here assumed to be revealed. The failure rate λ used in the equations in this section is that applicable to these revealed fail-safe, or fail-spurious, faults.

For a simple trip system consisting of a single channel 1/1 system with a failure rate λ the operational failure, or spurious trip, rate γ is:

$$\gamma = \lambda \qquad [13.9.35]$$

For a parallel redundant, or $1/n$, system the operational failure rate $\gamma_{1/n}$ is:

$$\gamma_{1/n} = n\lambda \qquad [13.9.36]$$

For an m/n system, which may be a majority voting system, the following treatment applies. The rate at which the first operational failure of a single channel occurs is $n\lambda$. This first failure only results in a system trip if further operational failure of single channels sufficient to trip the system occurs within the repair time τ_r. The probability q_γ that this will occur is:

$$q_\gamma = \binom{n-1}{m-1}(\lambda\tau_r)^{m-1} \qquad \lambda\tau_r \ll 1 \qquad [13.9.37]$$

Then the operational failure rate $\gamma_{m/n}$ of the system is:

$$\gamma_{m/n} = n\lambda q_\gamma \qquad [13.9.38]$$

$$\gamma_{m/n} = n\lambda\binom{n-1}{m-1}(\lambda\tau_r)^{m-1} \qquad [13.9.39]$$

Thus for a 2/3 majority voting system:

$$\gamma_{2/3} = (3\lambda)\cdot(2\lambda\tau_r) \qquad [13.9.40]$$

$$\gamma_{2/3} = 6\lambda^2\tau_r \qquad [13.9.41]$$

Operational failure, or spurious trip, rates calculated from Equations 13.9.36 and 13.9.39 are shown in Table 13.19.

It is emphasized again that the foregoing treatment is a simplified one. The expressions derived here appear, however, to be those in general use (e.g. Hensley, 1968; Kletz, 1972a; de Heer, 1974; Lawley and Kletz, 1975). Full theoretical treatments of trip systems are been given by A.E. Green and Bourne (1966 UKAEA AHSB(S) R117) and Wheatley and Hunns (1981). The latter give expressions for a wide variety of trip systems.

13.9.6 Proof testing of trip systems

The treatment of the functional reliability of trip systems which has just been given demonstrates clearly the importance of the proof test interval. The expressions derived show that the condition for high functional reliability is

***Table 13.19** Spurious trip rates for trip systems*

System	γ
1/1	λ
1/2	2λ
1/3	3λ
2/2	$2^2\tau_r$
2/3	$6\lambda^2\tau_r$

$$\lambda\tau_p \ll 1 \qquad [13.9.42]$$

As an illustration of the effect of the proof test interval, consider a simple trip system which has a failure rate of 0.67 faults/year on a duty where the demand rate is 1 demand/year:

$\delta = 1$ demand/year
$\lambda = 0.67$ faults/year

If the proof test interval is 1 week:

$\tau_p = 0.0192$ year

Then the fractional dead time is:

$$\phi = \frac{\lambda\tau_p}{2}$$

$$= 0.0064$$

and the plant hazard rate is

$$\eta = \delta\phi$$

$$= 0.0064 \text{ hazards/year}$$

The plant hazard rate for a range of proof test intervals is:

τ_p	η (hazards/year)
1 week	0.0064
1 month	0.027
1 year	0.26

For the longer proof test intervals the approximate Equation 13.9.18 is not valid and Equation 13.9.19 was used.

Some additional factors which affect the choice of proof test interval are the facts that, while it is being tested, a trip is disarmed and that each test is an opportunity for an error which disables the trip, such as leaving it isolated after testing.

Thus for a simple trip system with a trip disarmed period τ_d and an isolation dead time ϕ_{is} the FDT becomes:

$$\phi = \frac{\lambda\tau_p}{2} + \frac{\tau_d}{\tau_p} + \phi_{is} \qquad [13.9.43]$$

This expression has a minimum at:

$$(\tau_p)_{min} = \left(\frac{2\tau_d}{\lambda}\right)^{1/2} \qquad [13.9.44]$$

The effect of these factors can be illustrated by considering the trip system described in the previous example. If the trip disarmed period is 1 h and the isolation dead time is 0.001:

$$\tau_d = 1.14 \times 10^{-4} \text{ year}$$

$$\phi_{is} = 0.001$$

$$(\tau_p)_{min} = 0.0184 \text{ year} = 0.96 \text{ weeks}$$

Assume a proof test interval of 1 week is chosen:

$\tau_p = 0.0192$ year

Then the fractional dead time is

$\phi = 0.013$

and the plant hazard rate is

$\eta = 0.013$ hazards/year

In some instances it is not possible to test all parts of the trip system every time a proof test is carried out. For example, it is often not permissible to close the shut-off valve completely. In such cases a partial test is done, checking out the system to demonstrate valve movement but not valve shut-off.

If for a simple trip system the functional failure rate and proof test interval of the first part of the system are λ_A and τ_{pA}, respectively, and those of the second part are λ_B and τ_{pB}, respectively, then:

$$q_A = 1 - \exp(-\lambda_A t) \qquad [13.9.45]$$

$$q_A = \lambda_A t \qquad \lambda_A t \ll 1 \qquad [13.9.46]$$

Similarly

$$q_B = \lambda_B t \qquad \lambda_B t \ll 1 \qquad [13.9.47]$$

Then the FDT is:

$$\phi = \frac{1}{\tau_{pA}}\int_0^{\tau_{pA}} q_A \mathrm{d}t + \frac{1}{\tau_{pB}}\int_0^{\tau_{pB}} q_B \mathrm{d}t \qquad [13.9.48]$$

$$\phi = \frac{1}{2}(\lambda_A\tau_{pA} + \lambda_B\tau_{pB}) \qquad [13.9.49]$$

There appears to be some variability in industrial practice with respect to the proof test interval. Generally a particular firm tends to have one longer interval, which is the standard one, and a shorter interval which is used for more critical cases. One such pair of intervals is 3 months and 1 month. Another is 1 month and 1 week. Thus 1 month and 1 week proof test intervals are mentioned in many of the trip system applications described by Lawley and Kletz. In some cases the policy is adopted that if analysis shows that the proof test interval for a single trip is short, a redundant trip system is used.

13.9.7 Some other trip system characteristics

There are certain general characteristics which are desirable in any instrument system, but which are particularly important in a trip system. A trip system should possess not only reliability but also capability. In other words, when functional, it should be capable of carrying out its function. If it is not, then no amount of redundancy will help. The measuring instrument of the trip system should be accurate. This is particularly important where the safety margin is relatively fine.

The trip system should have a good dynamic response. What matters here is the ability of the trip to give rapid detection of the sensed variable and to effect rapid correction of that variable or rapid plant shut-down. This response therefore depends on the trip system itself, but also on the dynamic response of the plant. This aspect is considered further in Section 13.9.16.

The trip system should have sufficient rangeability to maintain accuracy at different plant throughputs. Another important property of a trip system is the ease and completeness with which it can be checked. It is obviously desirable to be able to check all the elements in a trip system, but this is not always easy to arrange.

13.9.8 Trip system applications

Some illustrations of the specification and design of protective systems have been given by Kletz (1972a, 1974a) and by Lawley and Kletz (1975). As already mentioned in Chapter 12, the use of trip systems instead of pressure relief valves is sometimes an attractive proposition, particularly where the relief valve solution involves large flare or toxic scrubbing systems. This is considered by Kletz (1974a) and by Lawley and Kletz (1975), who suggest that if a trip system is used instead of a relief valve, it should be designed for a reliability 10 times that of the latter. The reason for this is the uncertainty in the figures and the difference in the modes of failure; a relief valve which fails to operate at the set pressure may nevertheless operate at a higher pressure, whereas a trip is more likely to fail completely.

The fail-dangerous failure rate of a pressure relief valve and of a simple 1/1 trip system are quoted by these workers as 0.01 and 0.67 faults/year, respectively (Kletz, 1974a). Then, using these failure rates and assuming a demand rate of 1 demand/year, the plant hazard rates shown in Table 13.20 are obtained. For the longer proof test interval $\delta\tau_p \ll 1$ and $\lambda\tau_p \ll 1$ do not apply and for this case the data in the table are obtained not from the approximate equation 13.9.18, but from Equation 13.9.19. Thus to meet the design criterion suggested for functional reliability a 1/2 trip system with weekly testing is required. The spurious trip rate for this system, however, might be unacceptable, leading to a requirement for a 2/3 system.

Another example of the use of trip systems is the hydrocarbon sweetening plant shown in Figure 13.10 (Kletz, 1972a). The hydrocarbon is sweetened with small quantities of air which normally remain completely dissolved, but conditions can arise in which an explosive mixture may be formed. Initially it was assumed that the principal problem lay in a change in the air/hydrocarbon ratio but a hazop study revealed that a hazard could arise in a number of ways. One is for an air pocket to be formed, which can occur as follows:

(1) the temperature can be so high that the amount of air normally used will not dissolve;
(2) an air pocket can be left behind when a filter is recommissioned;
(3) the pressure can fall, allowing air to come out of solution;
(4) a fault in the mixer can prevent the air being mixed with hydrocarbon, so that the pockets of air can be carried forward;
(5) a fault in the air/hydrocarbon ratio controller can result in the admission of excess air.

In this case it is also necessary for the feed to be above its flashpoint, which can occur in a number of ways: (1) the temperature can be above the normal flashpoint and (2) the feed can contain low flashpoint material. Alternatively, if there is both a loss of pressure in the receiver and a failure of the non-return valve, hydrocarbon may find its way into the air receiver.

The fault tree for the hazard is shown in Figure 13.11. The probabilities of the various fault paths were evaluated from this and the trip requirements were identified. The trip initiators considered, shown by the circles in Figures 13.10 and 13.11, were

(1) a device for detecting a pocket of air in the reactor;
(2) a pressure switch for detecting a low pressure in the receiver;
(3) a temperature measurement device for detecting a high temperature in the feed;
(4) laboratory analysis for detecting a low flashpoint feed;
(5) a device for detecting a high air/hydrocarbon ratio.

As the latter condition is detected by the trip initiator 1 anyway, trip initiator 5 was not used. The shut-down arrangement was that any of the trip initiators 1–3 shuts a valve in the air line and shuts down the compressor. The use of the laboratory analysis of feed flashpoint was restricted to ensuring that low flashpoint feeds were only present a sufficiently small fraction of the time to meet the system specification.

A third example is the distillation column heating system shown in Figure 13.12 (Lawley and Kletz, 1975). Heat was supplied to the distillation column from an existing steam-heated reboiler and a hot-water-heated feed vaporizer. The plant was to be uprated by the addition of another reboiler and vaporizer.

The existing pressure relief valve was a 2/2 system and was adequate to handle overpressure from the

Table 13.20 *Plant hazard rates for pressure relief valves and for trip systems (after Kletz, 1974a) (Courtesy of Chemical Processing)*

System	***Plant hazard rate,*** η (hazards/year)	(years/hazard)
No relief valve or trip	1	say, 1
Relief valve:		
annual testing	0.004	250
Single 1/1 trip:		
annual testing	0.264	4
monthly testing	0.027	36
weekly testing	0.0064	180
Duplicate 1/2 trip:		
weekly testing	5.5×10^{-5}	18000

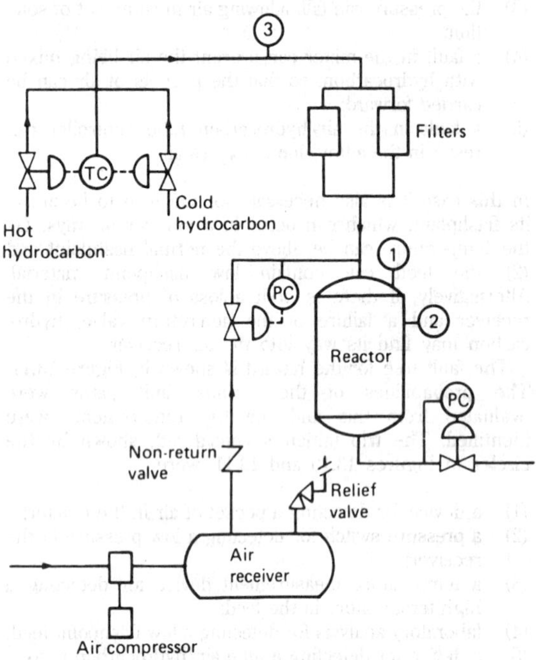

Figure 13.10 Hydrocarbon sweetening plant (Kletz, 1972a) (Courtesy of the American Institute of Chemical Engineers)

existing reboiler and vaporizer. The problem was to cope with the overpressure to which the new reboiler and vaporizer might give rise. Both re-sizing of the existing relief valves and the addition of a third were unattractive in the particular situation. There was therefore a requirement for a trip system which would shut down both the steam to the new reboiler and the hot water pump on the new vaporizer.

Both 1/1 and 1/2 trip systems were considered. The 1/1 system consisted of a pressure switch for detecting a high pressure on the overhead vapour line, a relay and a contact on the power supply to the hot water pumps and a relay and contact on that to the solenoid-operated shut-off valve on the reboiler steam supply. The 1/2 system consisted of a duplication of the 1/1 system. The summary of the failure rates of a simple trip system for this case is shown in Table 13.21.

The functional reliability of the existing 2/2 relief valve system was calculated as follows:

$$\lambda = 0.005 \text{ faults/year}$$

$$\tau_p = 2 \text{ years}$$

$$\phi = \lambda\tau_p = 0.01$$

The target FDT for the new trip system was taken as a factor of 10 less than this, namely 0.001, for the reasons already explained. The functional reliability of a 1/1 trip system was calculated as follows:

$$\lambda = 0.42 \text{ faults/year (from Table 13.21)}$$

$$\phi = \frac{\lambda\tau_p}{2}$$

$$= 0.21\tau_p$$

Then, taking into account also the disarmed time and the isolation dead time:

$$\tau_d = 1 \text{ h/test}$$

$$\approx 0.0001 \text{ years/test}$$

$$\phi_{is} = 0.001$$

$$\phi = 0.21\tau_p + \frac{0.0001}{\tau_p} + 0.001$$

The minimum FDT is at a proof test interval of

$$(\tau_p)_{min} = (2\tau_p/\lambda)^{1/2}$$

$$= 0.022 \text{ years}$$

$$= 8 \text{ days}$$

Assuming a proof test interval of 1 week, the FDT is

$$\phi = 0.01$$

This does not meet the target.

The authors therefore consider a 1/2 trip system. The analysis for this system is more complex and takes into account the fact that the complete shut-off of the valves can be checked only on a proportion of the tests. It is concluded that a 1/2 system does just meet the target set.

13.9.9 High integrity trip systems

The trip system applications described so far have been 1/1 or 1/2 systems. The more complex system with 2/3 majority voting is now considered.

A major system of this kind, which appears to be the most sophisticated on a chemical plant and which has been influential in the general development of protective systems in the industry, is that on the ethylene oxide process of ICI described by Stewart and co-workers (R.M. Stewart, 1971, 1974a; R.M. Stewart and Hensley, 1971). The ethylene oxide process is potentially extraordinarily hazardous: it operates with a reaction mixture very close to the explosive limit, there is a fire/explosion hazard and a toxic release hazard.

The design of the protective system followed the methods already outlined. The risk criterion was set at a probability of one fatality of 3×10^{-5} per year. The hazards were assessed by means of a fault tree, part of which is shown in Figure 13.13.

The system devised is a high integrity protective system (HIPS), consisting of the high integrity trip initiators (HITIs), the high integrity voting equipment (HIVE) and the high integrity shut-down system (HISS). A schematic system diagram, which omits replicated signal connections, is shown in Figure 13.14.

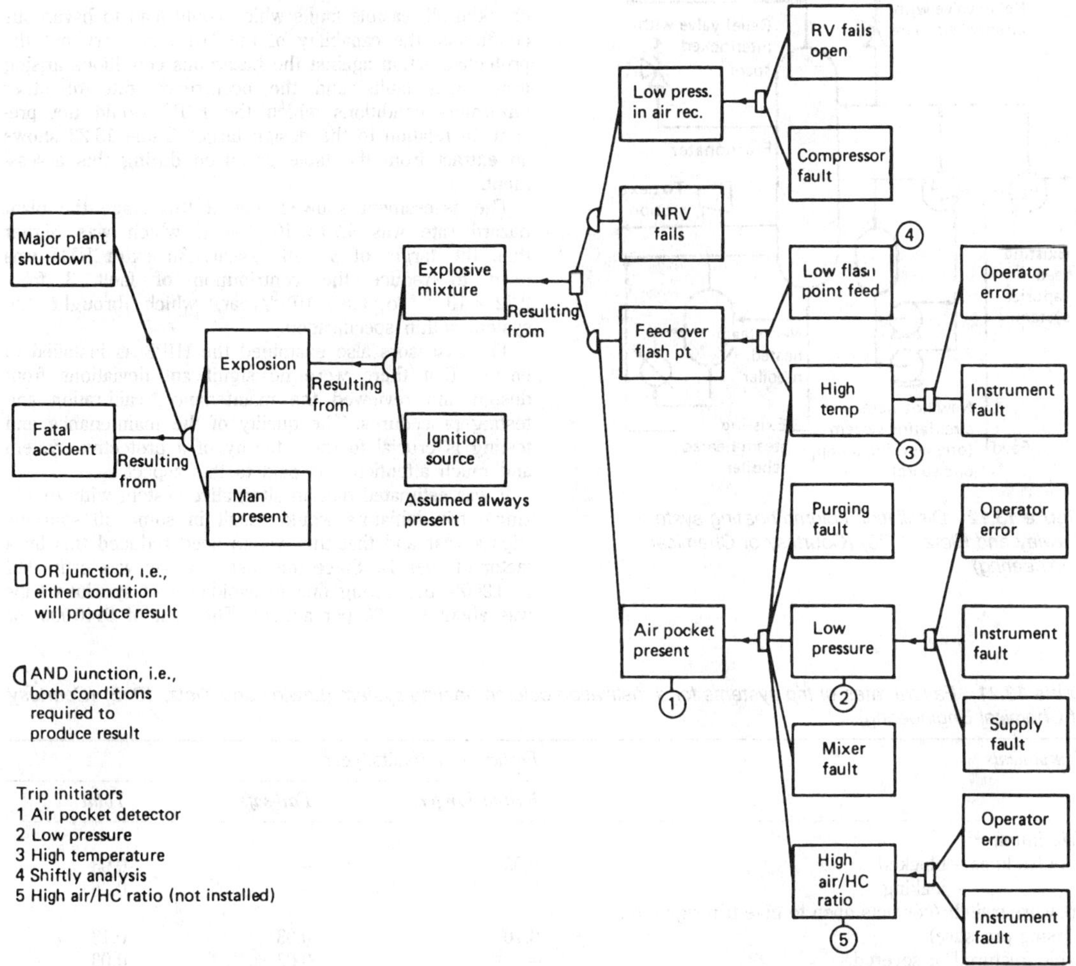

Figure 13.11 *Fault tree for explosion on a hydrocarbon sweetening plant (after Kletz, 1972a) (Courtesy of the American Institute of Chemical Engineers)*

Redundancy is fully exploited throughout the system. Against each logic path to fire/explosion in the fault tree at least one parameter was selected initially to be a trip initiator. The integrity specified in fact required the use of at least two parameters. The choice of the trip initiating parameters is important but difficult. Some are obvious such as high oxygen concentration, high reactor temperature, low recycle gas flow. Others are far less obvious, but are needed to guard against combinations of faults or to substitute for other parameters. This latter occurs, for example, where the measurement response is too slow, e.g. oxygen concentration, or where the trip could result in a hazardous condition, e.g. recycle compressor trip.

The measuring instruments used are carefully selected and, if necessary, modified to ensure high reliability. Each parameter is measured by triplicate instruments. The cables from each trip initiator on a parameter go by different routes so that there is less chance of all three being disabled by an incident such as a flash fire; the equipments have separate power supplies; and so on. The arrangement of the shut-down valves in the oxygen line illustrates further the use of redundancy. There are two lines each with three valves. A single line represents a 1/3 shut-down system. Duplication is provided to permit complete testing without disarming.

The advantages of the system are that the failure of one trip initiator in the fail-safe mode does not cause the plant to be tripped spuriously, the failure of one trip initiator in the fail-dangerous mode does not prevent the plant from being tripped, and the proof testing can be done without disarming the system.

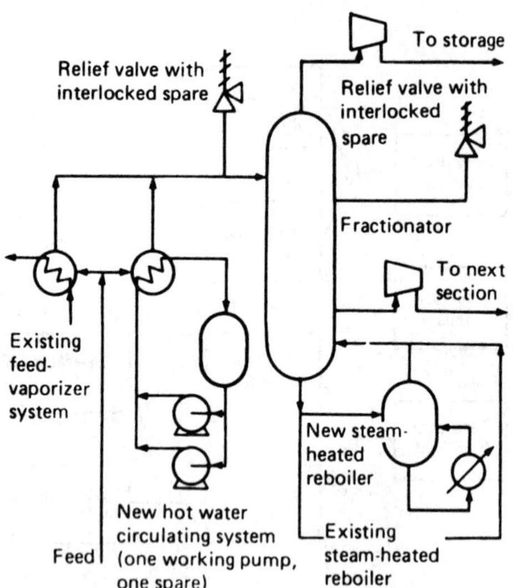

Figure 13.12 *Distillation column heating system (Lawley and Kletz, 1975) (Courtesy of Chemical Engineering)*

The design of the system was subjected to an independent assessment by assessors within the company who were advised by the UKAEA. The assessors checked all feasible faults which could lead to hazardous conditions, the capability of the HIPS to carry out the protective action against the hazardous conditions arising from such faults and the occurrence rate of other hazardous conditions which the HIPS would not prevent, in relation to the design target. Table 13.22 shows an extract from the table produced during this assessment.

The assessment showed that at this stage the plant hazard rate was 4.79×10^{-5}/year, which was higher than the target of 3×10^{-5}/year. An extra HITI was used to reduce the contribution of fault 3 from 2.72×10^{-5} to 0.8×10^{-5}/year, which brought the system within specification.

The assessors also examined the HIPS as installed to ensure that there were no significant deviations from design and reviewed the maintenance, calibration and testing procedures. The quality of the maintenance and testing is crucial to the integrity of a protective system and much attention was paid to this aspect.

It was estimated that an alternative system with 70 1/1 single trip initiators would result in some 30 spurious trips a year and that the system used reduced this by a factor of over 12. Since the cost of a trip was estimated as £2000, the saving due to avoidance of spurious trips was about £55 000 per annum. The cost of £140 000 for

Table 13.21 *Failure rates of trip systems for a distillation column heating system (Lawley and Kletz, 1975) (Courtesy of Chemical Engineering)*

Components	*Failure rate* (faults/year)		
	Fail-to-danger	*Fail-safe*	*Total*
Trip initiator:			
Impulse lines – blocked	0.03	–	0.03
– leaking	0.06	–	0.06
Pressure switch (contacts open to give trip signal on rising pressure)	0.10	0.03	0.13
Cable fractured or severed	–	0.03	0.03
Loss of electrical supply	–	0.05	0.05
Total	0.19	0.11	0.30
Steam shut-off system:			
Relay coil (de-energize to trip)	–	0.05	0.05
Relay contact	0.01	0.01	0.02
Relay terminals and wire	–	0.01	0.01
Solenoid valve (de-energize to trip)	0.10	0.20	0.30
Loss of electrical supply (to solenoid valve)	–	0.05	0.05
Trip valve (closes on air failure)	0.10	0.15	0.25
Air supply line – blocked or crushed	0.01	–	0.01
– fractured or holed	–	0.01	0.01
Loss of air supply	–	0.05	0.05
Total	0.22	0.53	0.75
Pump shut-off system:			
Relay coil, contact, terminals and wire (as above)	0.01	0.07	0.08
	0.01	0.07	0.08

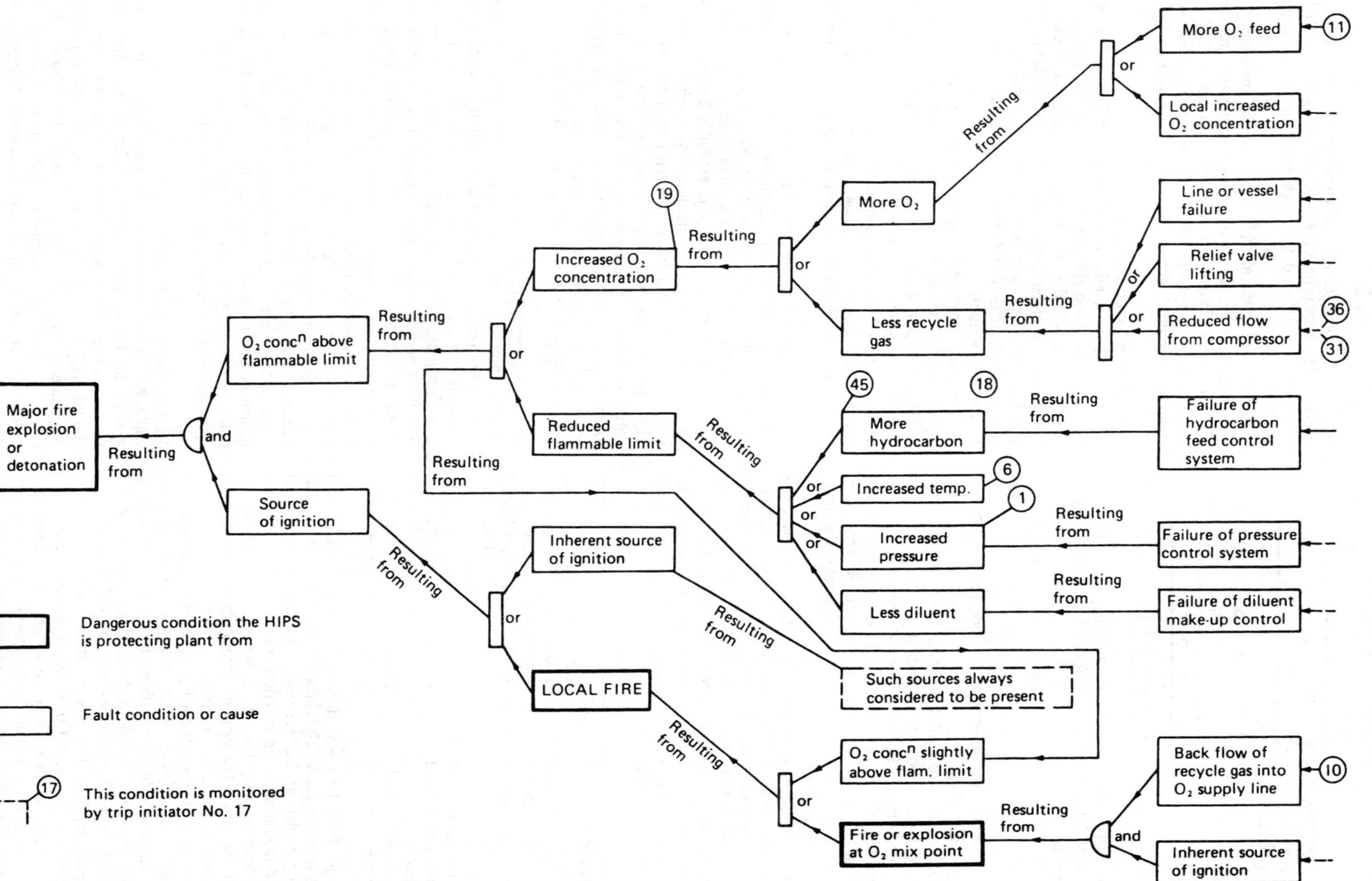

Figure 13.13 Fault tree for fire/explosion on an ethylene oxide plant (after R.M. Stewart, 1971) (Courtesy of the Institution of Chemical Engineers)

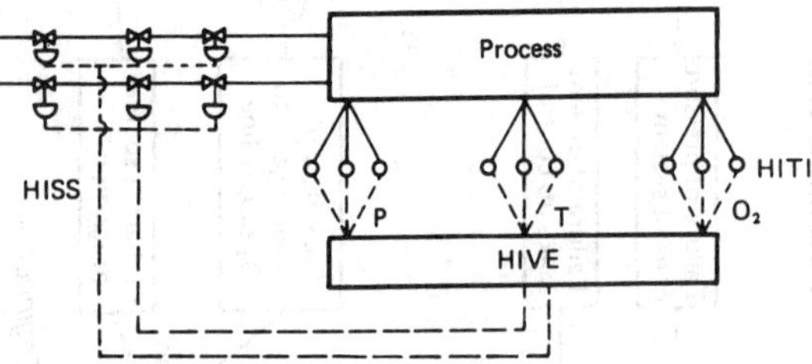

Figure 13.14 *High integrity protective system (after Stewart, 1971) (Courtesy of the Institution of Chemical Engineers)*

the installation was therefore considered justified on these grounds alone.

A coda to this account has been given by A. Taylor (1981), who describes the operation of the trip system over the period 1971–80. The information on the performance of the trip system is of two kinds: operations of and tests on the system. Events were classified as spurious, genuine and deliberate, the latter being initiations by the operators.

Analysis of these events revealed that in a few cases the demand frequency was greater than that originally estimated by orders of magnitude. The author gives a table listing seven fault conditions, with the event numbering rising to 53, which exemplify the worst discrepancies. The two fault conditions which show the greatest discrepancies are the opening of a certain relief valve and loss of reaction:

Fault condition	*Fault condition*		
	Predicted frequency (events/year)	*Actual frequency* (events/year)	*Ratio of actual/ predicted frequency*
15 A certain relief valve opens	0.001	1.68	1680
48b Loss of reaction	0.01	1.16	116

The relief valve fault was due to 'feathering', which had not been anticipated. The loss of reaction fault is not explicitly explained, but references by the author to the effect of modifications in reaction conditions may bear on this. It is noteworthy that of the seven fault conditions it is those for which the original frequency estimates were lowest which are most in error.

With regard to instrument failure rates, in the case of magnetic float switches three different failure mechanisms, and three different failure rates, were observed. Switches operating submerged in clean lubricating oil recorded no failures; those operating in recycled gas with occasional slugs of dirty water choked up; and a new type of switch was found to suffer from corrosion.

The author quotes three examples of dependent failure. One of these relates to the choked level switches just mentioned, which were all on one vessel. On four occasions, testing of the switches revealed that all were choked. The test procedure was altered to require that if one switch was found to be choked, the others should be tested.

There were also mistakes made in the installation of the instruments. In one case pneumatic pressure switches, of a flameproof type which is not waterproof, were located downwind of a low pressure steam vent pipe, and suffered water ingress and corrosion.

13.9.10 PES-based trip systems

A trip system needs to be highly reliable. For this reason, it has been the practice to design trip systems as separate, hardwired systems. The acceptability of using a programmable electronic system (PES) to implement the trip functions has long been a matter of debate. There has been a marked reluctance to abandon dedicated, hardwired systems.

The most constructive approach to the problem is to try to define the conditions which must be met by a PES-based trip system. As described in Section 13.12, the Health and Safety Executive (HSE, 1987b) has issued guidance based on this approach. Further guidance is given in the CCPS *Safe Automation Guidelines* described in Section 13.15, which are largely concerned with this topic.

An account of a computer-based trip system on an ammonia plant has been given by Cobb and Monier-Williams (1988). The reason given for moving to such a system is the avoidance of spurious trips. Design options were considered based on programmable logic controller and computer systems. The latter was selected largely because it offered a better interface with the operator. The system uses two computers operating in parallel. Some features of the system are: the ability to use inferred measurements; improved reliability of the trips; decreased defeating, or disarming, of the trips; and better control of any disarming which does occur.

13.9.11 Disarming of trip systems

It may sometimes be necessary to disarm a trip. This need arises particularly where a transition is being made between one state and another such as during start-up. The disarming of a trip should be assessed to ensure that it does not negate the design intent, whether this check is made at the time of the original design or subsequently. Such disarming should be the subject of a formal authorization procedure. This may be supplemented by hardware measures such as a key interlock.

If a trip proves troublesome, it is liable to be disarmed without such authorization. This is particularly likely to occur if there are frequent spurious trips, due to sensor failure or other causes. In order to disarm a trip it may not be necessary to interfere with the hardware. It is often sufficient simply to alter to set-point.

13.9.12 Restart after a trip

Once a trip has operated, it is necessary to reset the system so that a safer restart can be made. Therefore, the trip action which has driven the plant to a safe state should not simply be cancelled, but instead a planned sequence of actions taken to effect the restart. One situation which has frequently led to incidents is the restart of agitation in a batch reactor following an interruption of agitation.

Table 13.22 *Assessment of reliability of a high integrity trip system (after R.M. Stewart, 1971) (Courtesy of the Institution of Chemical Engineers)*

Description	Fault condition					Fractional dead time				Hazard
	Occasions per year	*Probability that it leads to rupture*	*Probability that operator's intervention fails*	*Demand rate (demands/year)*	*Relevant trip initiator No.*	*HITI*	*HIVE*	*HISS*	*Overall*	*rate* ($\times 10^5$) (hazards/ year)
	a	b	c	$(d = a \times b \times c)$		e	f	g	$(h = e + f + g)$	$(i = d \times h)$
1 Feed filters blocked	0.001	0.2	0.1	0.000 02	10 & 12	10^{-4}	10^{-5}	10^{-5}	1.2×10^{-4}	0.000 24
2 Oxygen supply failure	2.0	0.2	0.1	0.04	10 & 38	10^{-4}	10^{-5}	10^{-5}	1.2×10^{-4}	0.48
3 PCV fails open	0.25	0.1	1.0	0.025	11	10^{-3}	8.3×10^{-5}	10^{-5}	1.09×10^{-3}	2.72
4 Compressor antisurge bypass fails open	0.2	1.0	0.1	0.02	18 & 36	10^{-4}	10^{-5}	10^{-5}	1.2×10^{-4}	0.24
5 Gross carryover from absorber	0.1	1.0	0.1	0.01	18, 24	10^{-4}	10^{-5}	10^{-5}	1.2×10^{-4}	0.12
6 etc.										etc.

PCV, pressure control valve.

A case history caused by restart after a trip has been described by Kletz (1979a). A cumene oxidation reactor was fitted with a high temperature trip for which the trip action was to shut off the air and dump the contents of the reactor into a water tank. A spurious trip occurred, the air valve closed and the dump valve opened. The trip condition cleared itself, the dump valve remained open, but the air valve reopened. Air passed into the reactor, creating a flammable mixture.

13.9.13 Restart after a depressurization

A particular case of restart after a trip is the repressurization of a vessel following emergency depressurization. The effect of rapid reduction of pressure in a vessel containing a material such as liquefied gas may be to chill the vessel below the transition temperature, thus creating the hazard of brittle fracture. Too prompt a repressurization, before the vessel has warmed up sufficiently, can result in realization of this hazard.

Cases where this has occurred are mentioned by Valk and Sylvester-Evans (1985). Treatment of the problem using a model of blowdown has been described by S.M. Richardson and Saville (1992).

13.9.14 Hazard rate of a single-channel trip system

In the relations for the functional reliability of a single-channel trip given in Section 13.9.3:

$$\eta = \frac{\delta\lambda\tau_p}{2} \qquad [13.9.50]$$

The assumptions made are that $\lambda\tau_p \ll 1; \delta\tau_p \ll 1; \eta\tau_p \ll 1$, as stated.

If Equation 13.9.50 is used outside its range of validity, the results obtained can be not only incorrect but nonsensical. Consider the case where the failure rate is λ=0.01 failures/year, the demand rate is $\delta = 3$ demands/year and the proof test interval is $\tau_p = 1$ year. Then Equation 13.9.50 gives for the hazard rate η a value of 0.015 hazards/year, which is actually greater than the failure rate λ.

A treatment is now given for the more general case, based on the work of Lees (1982a) as extended by de Oliveira and Do Amaral Netto (1987). For a single-channel trip one formulation of the possible states is: (1) trip operational; (2) trip failed but failure undetected; and (3) trip failed, failure detected and trip under repair. The corresponding Markov model is:

$$\dot{P}_1(t) = -\lambda P_1(t) + \mu P_3(t) \qquad [13.9.51a]$$

$$\dot{P}_2(t) = \lambda P_1(t) - \delta P_2(t) \qquad [13.9.51b]$$

$$\dot{P}_3(t) = \delta P_2(t) - \mu P_3(t) \qquad [13.9.51c]$$

where P_n is the probability that the trip is in state n.

With the initial condition that the trip is operational, the solution of Equation 13.9.51 is as follows:

$$P_1(t) = \frac{\mu\delta}{r_1 r_2} + \frac{(r_1+\mu+\delta)+\mu\delta}{r_1(r_1-r_2)}\exp(r_1 t) - \frac{r_2(r_2+\mu+\delta)+\mu\delta}{r_2(r_1-r_2)}\exp(r_2 t) \qquad [13.9.52a]$$

$$P_2(t) = \frac{\lambda\mu}{r_1 r_2} + \frac{\lambda(r_1+\mu)}{r_1(r_1-r_2)}\exp(r_1 t) - \frac{\lambda(r_2+\mu)}{r_2(r_1-r_2)}\exp(r_2 t) \qquad [13.9.52b]$$

$$P_3(t) = \frac{\lambda\delta}{r_1 r_2} + \frac{\lambda\delta}{r_1(r_1-r_2)}\exp(r_1 t) - \frac{\lambda\delta}{r_2(r_1-r_2})\exp(r_2 t) \qquad [13.9.52c]$$

with

$$r_1 = \frac{-(\mu+\lambda+\delta) - [(\lambda+\delta-\mu)^2 - 4\lambda\delta]^{1/2}}{2} \qquad [13.9.53a]$$

$$r_2 = \frac{-(\mu+\lambda+\delta) + [(\lambda+\delta-\mu)^2 - 4\lambda\delta]^{1/2}}{2} \qquad [13.9.53b]$$

Then the fractional dead time and hazard rate obtained from Equations 13.9.52a and 13.9.52b are instantaneous values, and are:

$$\phi(t) = P_2(t) + P_3(t) \qquad [13.9.54]$$

$$\eta(t) = \delta[P_2(t) + P_3(t)] \qquad [13.9.55]$$

The fractional dead time and the hazard rate given in Equations 13.9.54 and 13.9.55 are functions of time. The average value of the hazard rate over the proof test interval is:

$$\eta = \frac{1}{\tau_p}\int_0^{\tau_p} \eta(t)\mathrm{d}t \qquad [13.9.56]$$

Then, substituting Equation 13.9.55 in Equation 13.9.56 and integrating gives for the average hazard rate:

$$\eta = \frac{\lambda\delta(\mu+\delta)}{r_1 r_2} + \frac{\lambda\delta(r_1+\mu+\delta)}{r_1^2\tau_p(r_1-r_2)}[\exp(r_1\tau_p) - 1] - \frac{\lambda\delta(r_2+\mu+\delta)}{r_2^2\tau_p(r_1-r_2)}[\exp(r_2\tau_p) - 1] \qquad [13.9.57]$$

The foregoing treatment is based on the assumptions that the trip is always operational after a proof test is performed and that the test duration is negligible compared with the proof test interval.

Although this model has a high degree of generality, it is based on the assumption that, following detection of a trip failure, the plant continues to operate while the trip is repaired. If in fact the policy is that the plant operation does not continue while the trip is being repaired, different expressions apply. If the state $P_3(t)$ is dropped from the instantaneous hazard rate $\eta(t)$ in Equation 13.9.55 and the repair rate μ=0, Equation 13.9.57 for the average hazard rate then becomes:

$$\eta = \frac{1}{\tau_p}\left\{1 - \frac{1}{\lambda-\delta}[\lambda\exp(-\delta\tau_p) - \delta\exp(-\lambda\tau_p)]\right\} \quad \lambda \neq \delta \qquad [13.9.58a]$$

$$\eta = \frac{1}{\tau_p}[1 - (1+\lambda\tau_p)\exp(-\lambda\tau_p)] \quad \lambda = \delta \qquad [13.9.58b]$$

This case is essentially that considered by Lees (1982a), who used the joint density function method. Two of the relations which he gives, for the failure density function f_η and the probability p_η of realization of the hazard, are also of interest and are

$$f_\eta = \frac{\lambda\delta}{\lambda-\delta}[\exp(-\delta\tau_p) - \exp(-\lambda\tau_p)] \qquad \lambda \neq \delta \qquad [13.9.59a]$$

$$f_\eta = \lambda^2\tau_p\exp(-\lambda\tau_p) \qquad \lambda = \delta \qquad [13.9.59b]$$

and

$$p_\eta = 1 - \frac{1}{\lambda-\delta}[\lambda\exp(-\delta\tau_p) - \delta\exp(-\lambda\tau_p)] \qquad \lambda \neq \delta \qquad [13.9.60a]$$

$$p_\eta = 1 - (1+\lambda\tau_p)\exp(-\lambda\tau_p) \qquad \lambda = \delta \qquad [13.9.60b]$$

A number of other relations have been given in the literature for situations where Equation 13.9.50 is not valid. An expression given by Kletz and by Lawley (Kletz, 1972a; Lawley and Kletz, 1975; Lawley, 1976) is

$$\eta = \lambda[1 - \exp(-\delta\tau_p/2)] \qquad [13.9.61]$$

This is in effect Equation 13.9.19. It may be derived from Equation 13.9.17 together with Equations 13.9.10, 13.9.13 and 13.9.16. It is applicable for small $\lambda\tau_p$ and $\eta\tau_p$, but higher $\delta\tau_p$.

Lawley (1981) has subsequently given the more accurate Equation 13.9.58. The assumptions underlying this equation have just been described.

Wells (1980) has given an expression

$$\eta = \frac{\delta\lambda}{\delta+\lambda} \qquad [13.9.62]$$

as an upper bound on the hazard rate for higher values of $\delta\tau_p(>2)$. This expression is equivalent to taking the fractional dead time as $\phi = \lambda/(\lambda+\delta)$.

De Oliveira and Do Amaral Netto give the relation:

$$\eta = \delta\left\{1 - \frac{1}{\lambda\tau_p}[1 - \exp(-\lambda\tau_p)]\right\} \qquad [13.9.63]$$

for low values of δ but higher values of $\lambda\tau_p$.

Numerical results for some of these expressions have been given by Lees and by de Oliveira and Do Amaral Netto. Table 13.23 shows some comparative results obtained, mainly by the latter workers.

13.9.15 Frequency of events in a trip system

A method of determining for a trip system the frequency of the events of principal interest has been described by Kumamoto, Inoue and Henley (1981). These events are the demand, the functional failure and the operational failure of the trip. The method is implemented in the program PROTECT.

The procedure is to designate each of these events in turn as the top event of a fault tree, to create the fault tree and to determine its cut sets. These cut sets together with the proof test interval for the trip are the inputs used by the model to provide estimates of the frequency of the events mentioned.

The application of this program to determine the expected frequency of these events for an ammonia–air mixing plant as a function of the proof test interval has been described by Kumar, Chidambaram and Gopalan (1989).

***Table 13.23** Some numerical values given by expressions for the average hazard rate of a single-channel trip system (after de Oliveira and Do Amaral Netto, 1987) (Courtesy of Elsevier Science Publishers)*

Parameter			*Equation*				
τ_p (year)	λ (year^{-1})	δ (year^{-1})	*13.9.57*	*13.9.58*	*13.9.61*	*13.9.63*	*13.9.18*
0.0192	0.1	0.1	0.958×10^{-4}	0.958×10^{-4}	0.958×10^{-4}	0.958×10^{-4}	0.959×10^{-4}
		1.0	0.954×10^{-3}	0.952×10^{-3}	0.954×10^{-3}	0.958×10^{-3}	0.959×10^{-3}
		10	0.919×10^{-2}	0.900×10^{-2}	0.914×10^{-4}	0.958×10^{-2}	0.959×10^{-2}
	1.0	0.1	0.952×10^{-3}	0.952×10^{-3}	0.958×10^{-3}	0.953×10^{-3}	0.959×10^{-3}
		1.0	0.949×10^{-2}	0.947×10^{-2}	0.954×10^{-2}	0.953×10^{-2}	0.959×10^{-2}
		10	0.913×10^{-1}	0.895×10^{-1}	0.914×10^{-1}	0.953×10^{-1}	0.959×10^{-1}
	10	0.1	0.900×10^{-2}	0.900×10^{-2}	0.958×10^{-2}	0.900×10^{-2}	0.959×10^{-2}
		1.0	0.897×10^{-1}	0.895×10^{-1}	0.954×10^{-1}	0.900×10^{-1}	0.959×10^{-1}
		10	0.864	0.845	0.914	0.900	0.959
1.0	0.1	0.1	0.468×10^{-2}	0.468×10^{-2}	0.488×10^{-2}	0.484×10^{-2}	0.5×10^{-2}
		1.0	0.359×10^{-1}	0.355×10^{-1}	0.393×10^{-1}	0.484×10^{-1}	0.5×10^{-1}
		10	0.916×10^{-1}	0.860×10^{-1}	0.993×10^{-1}	0.484	0.5
	1.0	0.1	0.358×10^{-1}	0.355×10^{-1}	0.488×10^{-1}	0.368×10^{-1}	0.5×10^{-1}
		1.0	0.284	0.264	0.393	0.368	0.50
		10	0.847	0.591	0.993	3.68	5.0
	10	0.1	0.892×10^{-1}	0.860×10^{-1}	0.488	0.900×10^{-1}	0.5
		1.0	0.827	0.591	3.93	0.900	5.0
		10	4.82	1.00	9.93	9.00	50

[a] In equations where μ is used, $\mu = 365$/year.

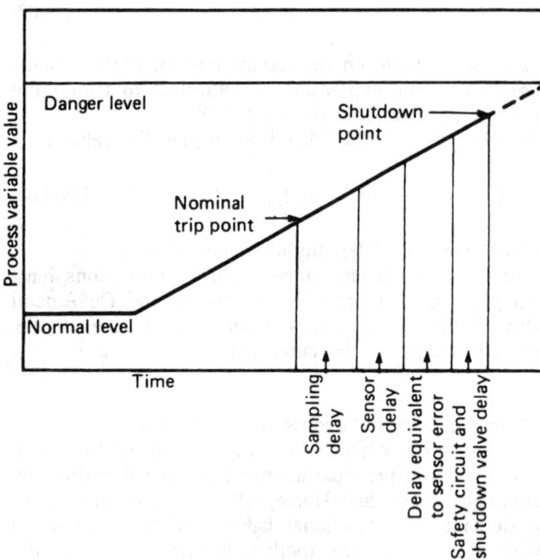

Figure 13.15 Effect of instrument error and dynamic response on the safety margin in a trip system (after Hensley, 1968) (Courtesy of the Institute of Measurement and Control)

13.9.16 Time response of a trip system

The point has already been made that the dependability of a trip system is a function not only of its reliability but also of its capability. An important aspect of capability is the dynamic response. The effect of the dynamic response of the instrument is illustrated in Figure 13.15. It is assumed in the figure that, when a fault occurs, the variable increases linearly from its normal level to the danger level. The nominal trip point is set part way up the ramp, but the trip will not usually occur at the point in time corresponding to this level of the variable. There will normally be delays due to sampling and the dynamic response of the measuring instrument and there may be an instrument error. After the measuring instrument has responded, there will be delays in the safety circuitry and the shut-down valve. There will be a further delay in the process itself, before the effect of the shut-off is felt on the variable measured. All these factors, delays and errors, erode the nominal safety margin and should be considered carefully. The original assumptions concerning the maximum rate of rise of the variable are clearly critical also.

Further reduction of the nominal trip point may be appropriate, but the setting should not be put so low that noise on the variable at its normal level activates the trip. A spurious trip can arise from too low a level of the trip setting as well as from instrument unreliability.

The dynamic response of the complete situation against which the trip system is designed to protect may be modelled using standard methods. An account of unsteady-state modelling of plant is given in *Mathematical Modeling in Chemical Engineering* (Franks, 1967) and the modelling of instrumentation is treated in texts on process control such as those by Harriott (1964) and Coughanowr and Koppel (1965).

The following treatment is confined to the dynamic response of the measuring instrument, or sensor. The inputs to a sensor are generally characterized by a set of idealized forcing functions, of which the main types relevant here are (1) the step function, (2) the ramp function and (3) the impulse function. The unit step function changes suddenly at time zero from a value of zero to one of unity. The unit ramp function increases linearly with time and has a slope t. The unit impulse is a function which is infinitely large at time zero and zero elsewhere, but which also has an area which is unity. These three forcing functions are shown in Figure 13.16(a)–(c).

The instrument itself is typically modelled as either a first- or second-order system. Thus a temperature sensor might be modelled as a first-order system:

$$Mc_p \frac{dT}{dt} = UA(T_i - T) \quad [13.9.64]$$

where A is the area for heat transfer to the sensor, c_p is the specific heat of the sensor, M is the mass of the sensor, t is time, T is the temperature of the sensor, U is the overall heat transfer coefficient, and the subscript i indicates input, or forcing. Thus T_i is the temperature of the surrounding fluid.

Equation 13.9.64 may be written in the more general form for a first-order system as:

$$\tau \frac{dT}{dt} = T_i - T \quad [13.9.65]$$

where

$$\tau = Mc_p/UA \quad [13.9.66]$$

and τ is a time constant. A lag of the form given by Equation 13.9.65 is known as a 'transfer lag'.

In this case the model obtained is a linear one. If the model obtained is non-linear, it needs first to be linearized. A non-linear model is obtained, for example, if the mode of heat transfer to the sensor is radiation rather than conduction.

The normal approach is then to express each term in the linear model as the sum of the steady-state value and of a transient component. Equation 13.9.65 then becomes:

$$\tau \frac{d(T_{ss} + \theta)}{dt} = (T_{i,ss} + \theta_i) + (T_{ss} + \theta) \quad [13.9.67]$$

The corresponding steady-state equation is:

$$0 = T_{i,ss} - T_{ss} \quad [13.9.68]$$

Subtracting Equation 13.9.68 from Equation 13.9.67 gives:

$$\tau \frac{d\theta}{dt} = \theta_i - \theta \quad [13.9.69]$$

where θ is the transient component of temperature and the subscript ss indicates steady state.

Equation 13.9.69 is then transformed into the Laplace, or s, domain by taking the Laplace transform:

$$s\bar{\theta} - \theta(0) = \bar{\theta}_i - \bar{\theta} \quad [13.9.70]$$

Taking the initial condition as the steady state with zero deviation gives $\theta(0) = 0$ and hence the ratio of the output to the input, or the transfer function, is:

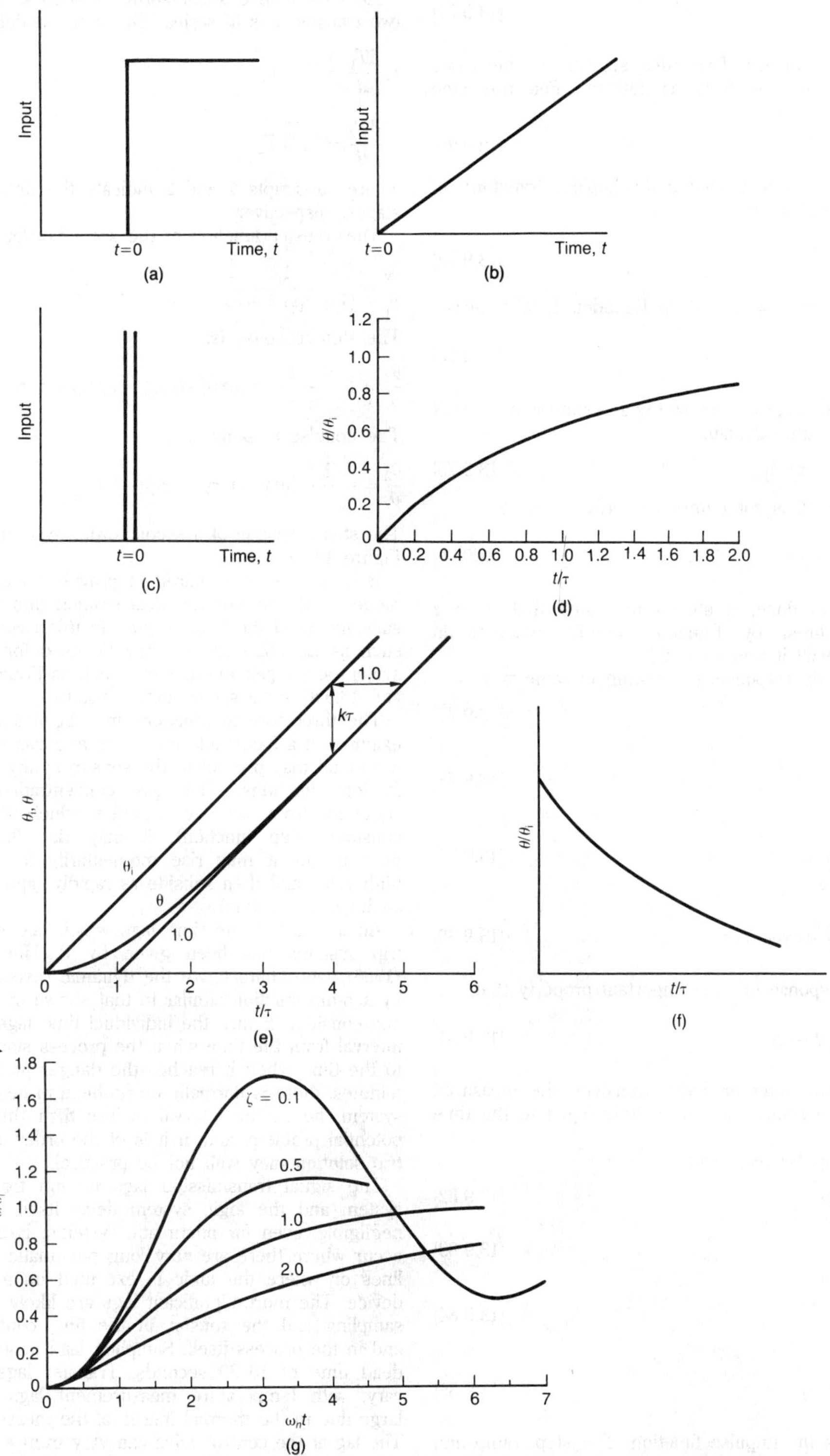

***Figure 13.16** Dynamic response of sensor systems: (a) step forcing function; (b) ramp forcing function; (c) impulse forcing function; (d) response of first-order system to step forcing function; (e) response to first-order system to ramp forcing function; (f) response to first-order system to impulse forcing function; (g) response of second-order system to step forcing function*

$$\frac{\bar{\theta}}{\bar{\theta}_i} = \frac{1}{1 + \tau s} \qquad [13.9.71]$$

The response of the first-order system to the three forcing functions is then as follows. For the step response:

$$\theta_i = k \qquad [13.9.72]$$

where k is a constant. Taking the Laplace transform of Equation 13.9.72 gives

$$\bar{\theta}_i = \frac{k}{s} \qquad [13.9.73]$$

Substituting Equation 13.9.73 in Equation 13.9.71 gives:

$$\bar{\theta} = \frac{k}{(1 + \tau s)s} \qquad [13.9.74]$$

Inverting the Laplace transformed expression 13.9.74 back into the time domain:

$$\theta = k[1 - \exp(-t/\tau)] \qquad [13.9.75]$$

Equation 13.9.75 is sometimes written as:

$$\frac{\theta}{\theta_i} = 1 - \exp(-t/\tau) \qquad [13.9.76]$$

Where this is done, it should be noted that θ_i is a constant, defined by Equation 13.9.72, whereas in Equation 13.9.69 it was a variable.

For the ramp response, proceeding in same way:

$$\theta_i = kt, \text{ say} \qquad [13.9.77]$$

$$\bar{\theta}_i = \frac{k}{s^2} \qquad [13.9.78]$$

$$\bar{\theta} = \frac{k}{(1 + \tau s)s^2} \qquad [13.9.79]$$

$$\frac{\theta}{\theta_i} = \tau\left\{\frac{t}{\tau} - [1 - \exp(-t/\tau)]\right\} \qquad [13.9.80]$$

The ramp response has the important property that

$$\frac{\theta}{\theta_i} = t - \tau \qquad t \to \infty \qquad [13.9.81]$$

In other words, after an initial transient, the measured value lags the actual value by a time equal to the time constant τ.

For the impulse response:

$$\theta_i = k\delta(t), \text{ say} \qquad [13.9.82]$$

$$\bar{\theta}_i = k \qquad [13.9.83]$$

$$\bar{\theta} = \frac{k}{1 + \tau s} \qquad [13.9.84]$$

$$\frac{\theta}{\theta_i} = \frac{1}{\tau}\exp(-t/\tau) \qquad [13.9.85]$$

where $\delta(t)$ is the impulse function. The step, ramp and impulse responses of a first-order system are shown in Figure 13.16(d)–(f).

An overdamped second-order system is equivalent to two transfer lags in series. The basic model is therefore

$$\tau_1 \frac{dT_1}{dt} = T_i - T_1 \qquad [13.9.86]$$

$$\tau_2 \frac{dT_2}{dt} = T_1 - T_2 \qquad [13.9.87]$$

where subscripts 1 and 2 indicate the first and second stages, respectively.

The transfer function of the second-order system is:

$$\frac{\bar{\theta}_2}{\bar{\theta}_i} = \frac{1}{(1 + \tau_1 s)(1 +_2 s)} \qquad [13.9.88]$$

The step response is:

$$\frac{\theta_2}{\theta_i} = 1 - \frac{1}{\tau_1 - \tau_2}[\tau_1 \exp(-t/\tau_1) - \tau_2 \exp(-t/\tau_2)] \qquad [13.9.89]$$

The impulse response is:

$$\frac{\theta_2}{\theta_i} = \frac{1}{\tau_1 - \tau_2}[\exp(-t/\tau_1) - exp(-t/\tau_2)] \qquad [13.9.90]$$

The step response of a second-order system is shown in Figure 13.16(g).

It is sometimes required to provide an unsteady-state model of the sensor for incorporation into an unsteady-state model of the total system. In this case an equation such as Equation 13.9.65 may be used for a first-order system and a pair of equations such as Equations 13.9.86 and 13.9.87 for a second-order system.

The three forcing functions may be illustrated by the example of a flammable gas cloud at a gas detector. The gas cloud may present to the sensor as any one of these forcing functions. The gas concentration may rise suddenly from zero to a value which then remains constant (step function), it may rise linearly (ramp function) or it may rise momentarily from zero to a high value and then subside as rapidly (approximated by an impulse function).

An account of the time lags which occur in practical trip systems has been given by R. Hill and Kohan (1986), who characterize the dynamic response of a trip by a ramp function similar to that shown in Figure 13.15 and consider in turn the individual time lags. If the total interval from the time when the process starts to deviate to the time when it reaches the danger point exceeds 2 minutes, there is normally no problem in designing a trip system, but if the interval is less than this there is a potential problem, and, if it is of the order of seconds, a trip solution may well not be practical.

The signal transmission lags to and from the logic system and the logic system delay itself are normally negligible, even for pneumatic systems. Exceptions may occur where there are very long pneumatic transmission lines or where the logic is executed on a time-shared device. The more significant lags are likely to be in the sampling and the sensor, in the final control element, and in the process itself. Sampling lags may amount to a dead time of 10–30 seconds. Transfer lags in sensors vary, with temperature measurement lags often being large due to the thermal inertia of the measuring pocket. The lag at the control valve can vary from a fraction of a second up to several minutes, depending on the valve size. The lag in the process is also highly variable.

***Table 13.24** Ranking of trip system configurations with respect to functional and operational failure (Rushton, 1991b) (Courtesy of the Institution of Chemical Engineers)*

Functional ranking		*Operational ranking*	
$\phi_{m/n}$	*System*	$\gamma_{m/n}$	*System*
Low	1/3	Low	3/3
	1/2		2/2
	2/3		2/3
↓	1/1	↓	1/1
	2/2		1/2
High	3/3	High	1/3

13.9.17 Configuration of trip systems

Inspection of Tables 13.17 and 13.19 indicates that the use of parallel redundant, or $1/n$, systems gives an increase in functional reliability, but a decrease in operational reliability compared with a 1/1 system. Better overall reliability characteristics can be obtained by the use of a majority voting system, of which the 2/3 system is the simplest. A comparison of a 2/3 system with a 1/1 system shows that the 2/3 system has a high functional and operational reliability, while a comparison with a 1/2 system shows that the 2/3 system has a slightly lower functional reliability, but a much higher operational one. The 2/2 system is little used for trip systems but has some interesting characteristics. It is effectively a series system which has a rather lower functional reliability than a 1/1 system, but its operational reliability exceeds not only that of the 1/1 but also that of the 2/3 system.

The configuration of trip systems has been discussed by Rushton (1991a,b), who describes a formal approach. According to Rushton, for typical systems the ranking of trip systems with respect to their functional and operational reliability is invariant and is as shown in Table 13.24. The trip systems most commonly used are the 1/1, 1/2 and 2/3 systems. The requirement for functional reliability is rarely such as to justify a 1/3 system and that for operational reliability rarely such as to justify a 2/2 or 3/3 system.

The criterion given by Rushton for selection of the trip system configuration is an economic one and is:

$$V = nC + \delta\phi_s H + \gamma_s S + \delta(1 - \phi_s)G \quad [13.9.91]$$

where C is the annualized cost of a single channel trip, G is the cost of a genuine trip, H is the cost of realization of a hazard, S is the cost of a spurious trip, V is the overall annual cost, and the subscript s indicates the trip system (as opposed to a single channel). The most economic solution is that which minimizes V.

For a genuine trip there is an element of loss related to the process failure which causes the demand. If, for purposes of comparison, this element (which will occur in all cases) is neglected, the cost of a genuine trip is approximately the same as that of a spurious one ($G = S$), so that Equation 13.9.91 becomes:

$$V = nC + \delta\phi_s H + [\gamma_s + \delta(1 - \phi_s)]S \quad [13.9.92]$$

If the basic parameters of a particular application are known, namely the demand rate δ, the fail-to-danger and spurious failure rates λ_s and γ_s, the proof test interval τ_p, the repair time τ_r and the costs of hazard realization H and spurious trip S, then a plot of H/C vs S/C gives a map showing the regions where a particular configuration is optimal. The boundaries of the regions are curves of constant V/C.

As an illustration, consider the case given by Rushton where the application is characterized by $\delta = 0.01$ demands/year, $\lambda = 0.2$ failures/year, $\gamma = 0.5$ failures/year, $\tau_p = 1/12$ and $\tau_r = 1/52$. The map giving the optimal configurations for this case is shown in Figure 13.17(a). If δ is increased to 1.0 demands/year, the map becomes that shown in Figure 13.17(b).

Rushton also treats the case where there is an element of common cause failure (CCF) and uses for this the beta method described in Chapter 9. He considers the simplest trip configuration to which such failure applies, the 1/2 system. For such a system:

$$\phi_{1/2} = \frac{[\lambda(1 - \beta_1)\tau_p]^2}{3} + \frac{\beta_1 \lambda \tau_p}{2} \quad [13.9.93]$$

$$\gamma_{1/2} = 2 \times (1 - \beta_2)\gamma + \beta_2 \gamma \quad [13.9.94a]$$

$$= (2 - \beta_2)\gamma \quad [13.9.94b]$$

where β_1 is the fraction of the functional failure rate which is common cause, or the beta value for that failure rate, and β_2 is the operational beta value.

The effect of CCF may be illustrated by considering the extension given by Rushton of his example to the case of a 1/2 trip system where $\delta = 1$ demand/year, $\lambda = 0.2$ failures/year, $\gamma = 0.5$ failures/year, $\tau_p = 1/12$, $\tau_r = 1/52$ and where $\beta_1 = \beta_2 = \beta$. Maps of the configuration space for this case are shown in Figure 13.17(c) and (d) for different values of β. Table 13.25 gives expressions for the fractional dead time and spurious trip rate for different trip configurations.

This cost-based approach allows the different trip system configurations to be put on a common basis for purposes of comparison. Where the hazard includes one to human life, there will be a certain level of functional reliability which must be achieved and this should be a factor in the choice of configuration. The approach described may still be applicable with adaptation in judging which configurations are reasonably practicable.

13.9.18 Integration of trip systems

As already described, a trip system is normally dormant and comes to life only when a demand occurs. An element of the trip system such as a sensor or a valve may experience failure and such a failure will lie unrevealed unless detected by proof testing or some other means. By contrast, equivalent elements in a control system are exercised continuously, and failure in such an element is liable to cause an operational excursion of some kind. The failure in this case is a revealed one. Yet the actual physical fault in the two cases may well be identical. A sensor may fail giving a low/zero or high reading, or a valve may jam open or shut. The concept of trip integration, which has been described by Rushton (1992), is based on this contrast

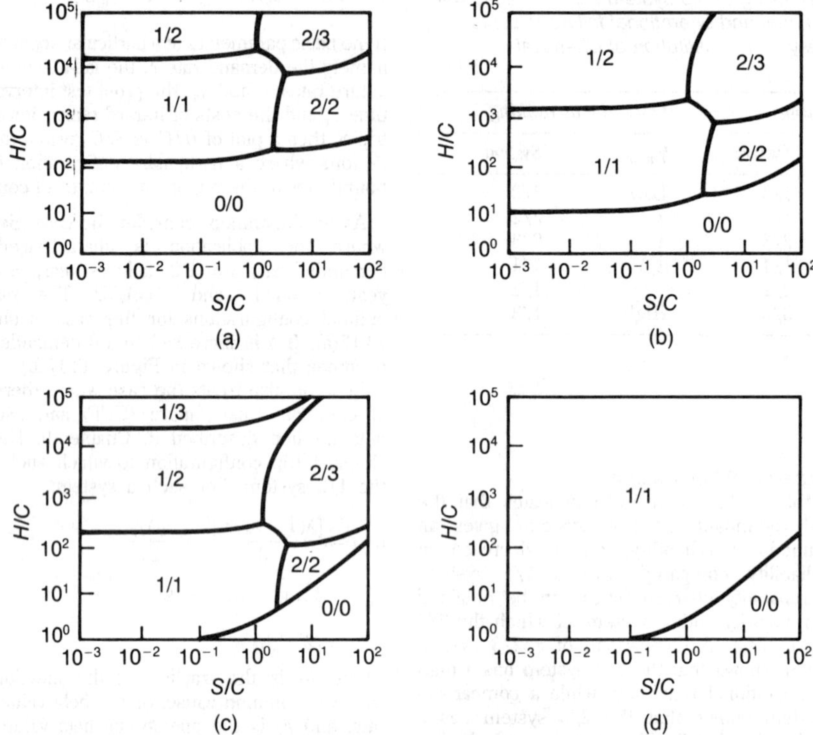

***Figure 13.17** Configuration selection map for trip systems: illustrative examples (Rushton, 1991b): (a) case with $\delta = 0.01$, $\beta = 0$; (b) case with $\delta = 0.1$, $\beta = 0$; (c) case with $\delta = 1$, $\beta = 0.3$; and (d) case with $\delta = 1$, $\beta = 1$. See text for further details (Courtesy of the Institution of Chemical Engineers)*

between a fault which lies unrevealed in a trip system but is revealed in a measurement and control system. The principle applies to any system which has a protective function. The system is regarded as integrated provided it is regularly exercised, which generally means that it is in use during the normal operation of the plant.

As an illustration, Rushton describes a refrigerated storage tank for a toxic liquid, equipped with a cooling system and a pressure relief valve. Both are protective systems, but the cooling system is in more or less continuous use and is thus integrated, whereas the relief valve is not.

In this case the integration is benign, but it can also be malign. As an example of the latter, the author cites the case of a sensor which is common to both a level control loop and a high level alarm. Failure of the sensor results in failure not only of the control loop but also of the alarm.

It should be an aim of trip system design to convert unrevealed failures into revealed failures, and hence to enhance reliability, by the judicious exploitation of benign integration.

13.9.19 Maintenance of trip systems

The foregoing account of trip systems has brought out the importance of proof testing. This testing and, more generally, the maintenance of trips needs to be of a high standard if the design reliabilities are to be achieved. Accounts of the testing and maintenance of trip systems have been given by R.M. Stewart (1971), A. Taylor (1981) and Barclay (1988).

The system described by Barclay is broadly as follows. The trips on a plant are covered by a testing schedule which specifies a test interval for each trip system. A common test interval is 12 months, but the interval is established for each trip individually. A change to the test interval, or complete removal of the trip, are governed by formal procedures which involve consultation with the interested parties.

There is a written procedure for the test which details the actions to be taken. This is necessary because the procedure can be quite complex, because the individual performing the test may not be familiar with the particular trip and because in many cases the test is one of the last tasks done prior to a start-up when there may be considerable pressure. This procedure can be changed only after formal consultations.

The test should cover the whole trip from initial to final element. From the point of view of testing, the preferred method is an actual test in which the procedure is to take the process to the trip point and verify the trip action. The alternative is a simulated test which is performed by simulating process conditions using test equipment.

Table 13.25 Fractional dead times and spurious trip rates for trip systems with simultaneous proof testing and common cause failures accounted for by the beta method (Rushton, 1991b) (Courtesy of the Institution of Chemical Engineers)

System	$\phi_{m/n}$	$\gamma_{m/n}$
1/1	$\frac{\lambda\tau_p}{2}$	γ
1/2	$\frac{[\lambda\tau_p(1-\beta_1)]^2}{3}+\frac{\beta_1\lambda\tau_p}{2}$	$(2-\beta_2)\gamma$
2/2	$\lambda\tau_p(1-\beta_1/2)$	$2\gamma^2(1-\beta_2)^2\tau_r+\beta_2\gamma$
1/3	$\frac{(\lambda\tau_p)^3}{4}(1-\beta_1)^3+\frac{\beta_1\lambda\tau_p}{2}$	$(3-2\beta_2)\gamma$
2/3	$(\lambda\tau_p)^2(1-\beta_1)^2+\frac{\beta_1\lambda\tau_p}{2}$	$6\gamma^2(1-\beta_2)^2\tau_r+\beta_2\gamma$
3/3	$\frac{\lambda\tau_p}{2}(3-2\beta_1)$	$3\gamma^3(1-\beta_2)^3\tau_r^2+\beta_2\gamma$
m/n	$\binom{n}{r}\frac{[\lambda\tau_p(1-\beta_1)]^r}{r+1}+\frac{\beta_1\lambda\tau_p}{2}$	$n\gamma(1-\beta_2)\binom{n-1}{m-1}[\gamma(1-\beta_2)\tau_r]^{m-1}+\beta_2\gamma$

In many cases it is impractical to carry out an actual test. In the case of a hazardous process, the reasons are obvious. But even for a less hazardous process the number of trips may be such that repeated shut-down and start-up is not practical. On a plant with 30 or 40 trips the equipment may be worn out just by testing.

It can be misleading to rely on a single-point trip as a sufficient test. And particularly where there is complex logic it is necessary to exercise all the steps in the chain; omission of intermediate steps can be misleading.

Instruments which are part of a trip system are provided with identification, both on circuit diagrams and in the field by a tag. This helps avoid shut-downs caused by work on such instruments.

The trip system is maintained in good condition by preventive maintenance. Equipment is inspected for deterioration. Critical equipment is classified as such and subject to periodic overhaul. It is required that following maintenance work a function check be carried out on the equipment. A trip which is out of service or fails to operate is not tolerated. It is classed as a hazard and action is taken.

At the site described by Barclay, there are some 35 000 to 40 000 instruments with more than 5000 trips and interlocks. Trip maintenance is handled by a computerized system. The responsibility for testing in this works lies with the operations rather than the mechanical function. Essentially similar considerations apply to the maintenance of interlocks.

13.10 Interlock Systems

Interlocks are another important type of protective device. They are used to control operations which must take place in a specified sequence and equipments which must have specified relations between their states. This definition of an interlock differs from that often used in the American literature, where the term 'interlock' tends to be applied to both trip and interlock systems (as defined here).

Accounts of interlock systems are given in *Applied Symbolic Logic* (E.P. Lynch, 1980) and *Logical Design of Automation Systems* (V.B. Friedman, 1990) and by D. Richmond (1965), E.G. Williams (1965), Becker (1979), Becker and Hill (1979), Kohan (1984) and the CCPS (1993/14).

There are various kinds of interlock. The original type is a mechanical device such as a padlock and chain on a hand valve. Another common type is the key interlock. Increasing use is made of software interlocks based on process computers.

Some typical applications of interlocks are in such areas as:

(1) electrical switchgear;
(2) test cubicles;
(3) machinery guards;
(4) vehicle loading;
(5) conveyor systems;
(6) machine start-up and shut-down;
(7) valve systems;
(8) instrument systems;
(9) fire protection systems;
(10) plant maintenance.

An interlock is often used to prevent access as long as an equipment is operating. Thus electrical switchgear may be installed in a room where an interlock prevents the door opening until there is electrical isolation. Similarly, an interlock prevents access to a test cubicle for operations involving high pressure or explosive materials until safe conditions pertain. An interlock may be used to stop access to a machine or entry into a vessel unless the associated machinery cannot move. In vehicle loading, interlocks are used to prevent a tanker moving away while it is still connected to the discharge point.

Where synchronized operation of equipment is necessary, as in a conveyor system, interlocks are used to ensure this. Interlocks are used for the start-up of machinery to ensure that all the prestart conditions are met, that the correct sequence is followed and that conditions for transition from stage to stage are met. For large rotating machinery key factors are process conditions and oil pressures.

Pressure relief valves have interlocks to prevent all the valves being shut off simultaneously. There may be interlocks on other critical valve systems. Interlocks are also a part of instrument systems. An interlock may be used to prevent the disarming of a trip system unless certain conditions are met. Fire protection systems are provided with interlocks as a safeguard against leaving the system disabled, particularly after testing or maintenance. Plant maintenance operations make much use of interlocks to prevent valves being opened or machinery started up while work is in progress.

Some features of a good hardware interlock are that it (1) controls operations positively, (2) is incapable of defeat, (3) is simple, robust and inexpensive, (4) is readily and securely attachable to engineering devices and (5) is regularly tested and maintained.

Some interlocks are quite simple, but some interlock systems are quite complex. Such systems are often not confined to interlocks, but incorporate other logic functions. Interlock systems therefore shade over into general logic control systems. In particular, there are some very large interlock systems on boilers and gas turbines.

An especially important type of logic control is the control of sequential operations. Sequential control systems usually have numerous checks which must be satisfied before the next stage is initiated and checks that equipment has obeyed the control signals. These checks constitute a form of interlock.

Since an interlock can bring the process to a halt, it is important to provide adequate status and alarm signals to indicate which feature is responsible for the stoppage. It will be apparent that some interlocks are effectively trips. The distinction between the two is often blurred.

The interlocks described so far are simple rather than high integrity systems, but the latter can, of course, be used, if the situation warrants it. The general approach is similar to that described for trip systems.

13.10.1 Interlock diagrams

As with protective systems so with interlock systems the design may involve a number of parties and a common language is needed. Unfortunately, this is an area of some difficulty, for three reasons. The description of interlock systems involves the use of several different types of diagram; there appears to be considerable variability in the types of diagram employed and in the nomenclature used to describe them. The symbols for use in these diagrams are given in standards; however, not only are these standards subject to continuous revision, but also the symbols given are often not those in common use. Interlock systems are not well served with textbooks. In particular, there is in electrical engineering a large literature on switching systems, but very little of this addresses process interlock systems as such.

Three types of diagram commonly used in the design of interlock systems in the process industries are (1) the process flowchart, (2) the logic diagram and (3) the ladder diagram. The last two are sometimes referred to as the 'attached logic diagram' and the 'detached logic diagram', respectively.

The starting point for design of an interlock system is a description of a sequence of operations. A diagram showing this is a process chart. Process chart symbols have been given in *Work Study* (Currie, 1960) and are shown in Table 13.26, Section A.

The logic required to implement this sequence may be shown in a logic diagram. This utilizes standard symbols for functions such as OR, AND and NOT, similar to those used in fault tree work, as described in Chapter 9. Standard symbols for fault trees are given in BS 5760 *Reliability of Systems, Equipment and Components*, Part 7: 1991 *Guide to Fault Tree Analysis*. For some functions, two sets of symbols are given, the preferred and the alternative. It is the latter which are commonly used in the process industries and which are used here. The logic symbols used here are the alternative symbols given in BS 5760 and are shown in Table 13.26, Section B.

The logic diagram may then be converted to a ladder diagram. Standard symbols for protective logic systems are given in BS 3939: 1985 *Graphical Symbols for Electrical Power, Telecommunications and Electronic Diagrams*. The relevant IEC standard is IEC 617 *Graphical Symbols for Diagrams*. BS 3939: Part 7: 1985 *Switchgear, Controlgear and Protective Devices*, which is identical to IEC 617-7, gives relevant symbols. Other sets of symbols include those given by E.G. Williams (1965) and those of E.P. Lynch (1980). An account of the evolution of logic symbols is given in *An Introduction to the New Logic Symbols* (Kampel, 1986). Table 13.26, Section C, shows a selection of symbols, including those used here, from those given by Lynch.

13.10.2 Some basic systems

Some of the basic building blocks of interlock systems are illustrated in Figure 13.18. Figure 13.18(a) shows a simple starting circuit. Activation of the circuit occurs if there is a signal due to depression of the start pushbutton AND a signal due to non-depression of the stop pushbutton. Since the signal from the start pushbutton will disappear when it is no longer being depressed by the operator, it is necessary to provide the feedback signal shown, which ensures that there continues to be an output signal. If the stop pushbutton is depressed, the output signal is extinguished.

Figure 13.18(b) shows a time delayed holding circuit. If following activation by the start pushbutton, the signal X does not appear within the time interval specified, the output signal disappears. A typical application of this circuit is start-up of a motor-driven pump which is supplied with lubricating oil by a lube oil pump driven from the same motor. If after the time interval specified the lubricating oil pressure signal is still absent, the pump is shut down.

Figure 13.18(c) shows a self-extinguishing circuit. Activation of the pushbutton gives an output signal which continues until the time interval specified has

Table 13.26 *Interlock logic symbols*

A Work study symbols[a]

Symbol	*Activity*	*Predominant result*
	Operation	Produces, accomplishes changes further the process
	Inspection	Verifies quantity or quality
	Transport	Moves or carries
	Delay	Interferes or delays
	Storage	Holds, retains or stores

B Logic symbols[b]

Symbol	
	AND
	OR
	NOT
	Delay

C Ladder diagram symbols[c]

Symbol	
PB start	Pushbutton start
PB stop	Pushbutton stop
	Position, or limit, switch
	Relay or solenoid contacts, normally open, closed when relay or solenoid is energized
	Relay or solenoid contacts, normally closed, opened when relay or solenoid is energized

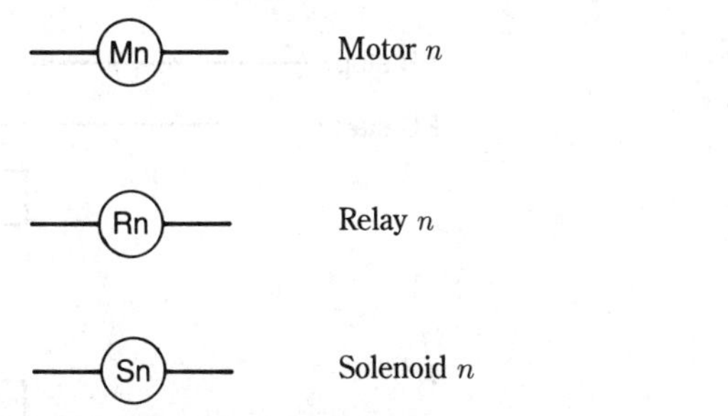

[a] These symbols are given by Currie (1960), who attributes them without reference to the American Society of Mechanical Engineers.
[b] These symbols are given in BS 5670: Part 7: 1991. The alternative symbol for NOT is a common alternative and is that used by E.P. Lynch (1980).
[c] These symbols are those used by E.P. Lynch (1980).

elapsed, when the output signal is extinguished. This circuit might typically be used to have a motor-driven equipment run for a period and then shut down.

13.10.3 Illustrative example: conveyor system

As an illustration of an interlock system, consider the conveyor system described by Lynch. A screw conveyor A feeds material from a car vibrator to an elevator which discharges to screw conveyor B above two storage bins A and B. There is a slide gate on the pipe between conveyor B and each bin, with a limit switch on each gate. Material is fed from a bin by a star feeder into screw conveyor C. The loading equipment can fill the bins at several times the rate at which it can be withdrawn.

Figure 13.19(a) shows a logic diagram for the interlocks for manual operation of this system. Conveyor B can be started only if either A or B slide gate is open. The elevator can be started only if conveyor B is running. Conveyor A can be started only if the elevator is running. The diagram also shows the simple non-interlocked starting circuit for the car vibrator.

The corresponding ladder diagram is shown in Figure 13.19(b). The diagram shows six circuits A–F. Certain relays occur in more than one circuit, e.g. relay R1 in circuits A and D, and it is this which imparts the interlocking feature. Circuit A is the starting circuit for conveyor B. This circuit can be activated only if either relay R2 or R3, the relays for the slide gates limit switches (LS), is closed. If this condition is met, depression of the start pushbutton (PB) energizes relays R1 and M1 and causes R1 to close and M1 to operate a relay in the power circuit. When the stop pushbutton is pressed, the circuit is de-energized and R1 opens.

In circuit B closure of the slide valve limit switch LS1 energizes relay R2 and causes it to close, and opening of the switch causes R2 to open. Circuit C implements a similar relationship between limit switch LS2 and relay R3. Circuit D is the starting circuit for the elevator. The circuit can be activated only if relay R1 is closed. If this condition is met, depression of the start button energizes relays R4 and M2 and causes R4 to close and M2 to

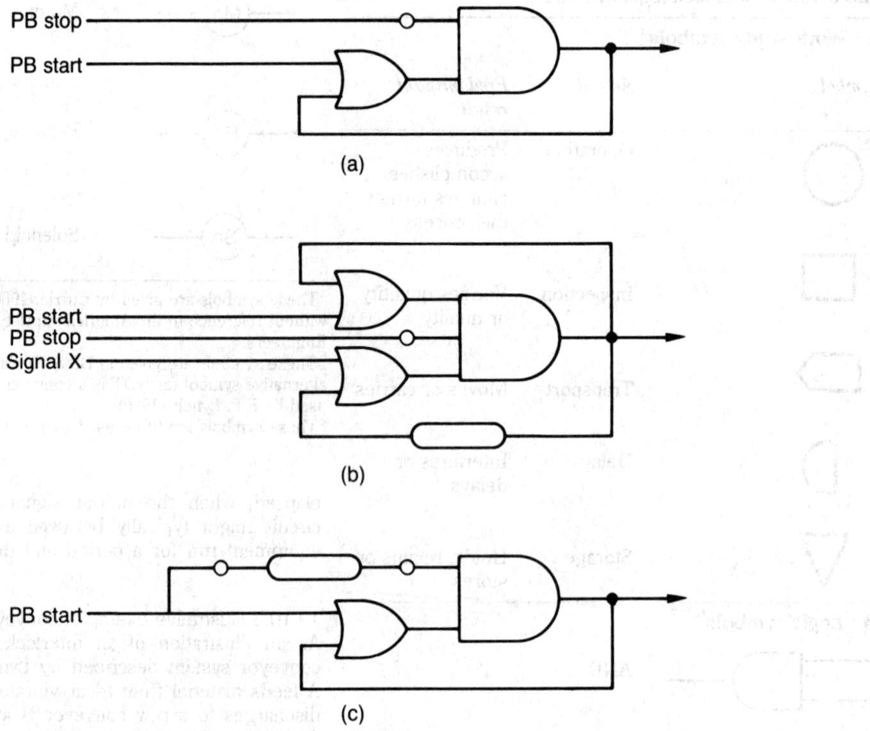

***Figure 13.18** Some basic interlock system logic diagrams: (a) simple starting circuit; (b) time delayed holding circuit; and (c) self-extinguishing circuit. PB, pushbutton*

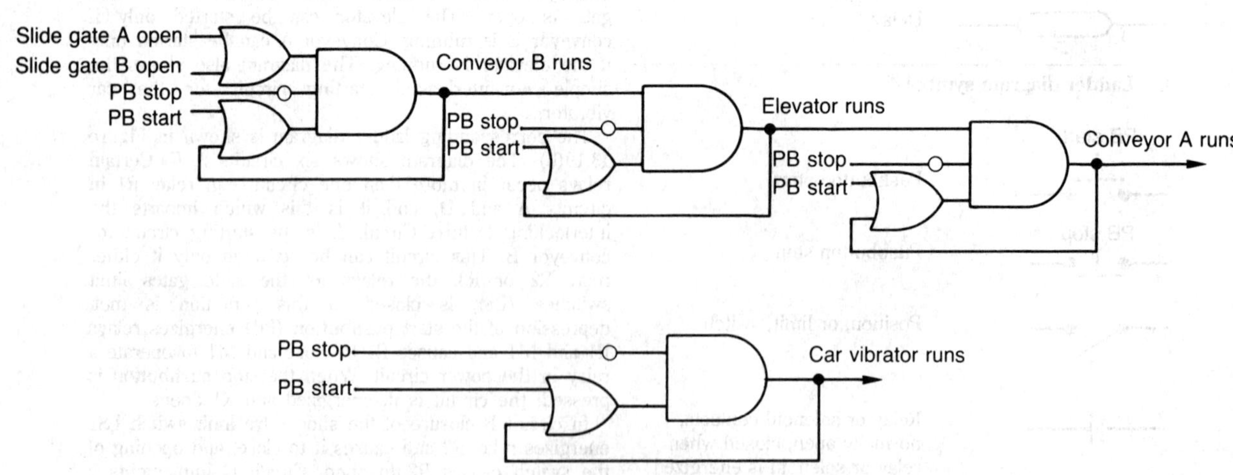

***Figure 13.19** Conveyor interlock system diagrams: (a) logic diagram; and (b) ladder diagram (E.P. Lynch, 1980) (Reproduced with permission from Applied Symbolic Logic by E.P. Lynch, Copyright ©, 1980, John Wiley and Sons Inc.)*

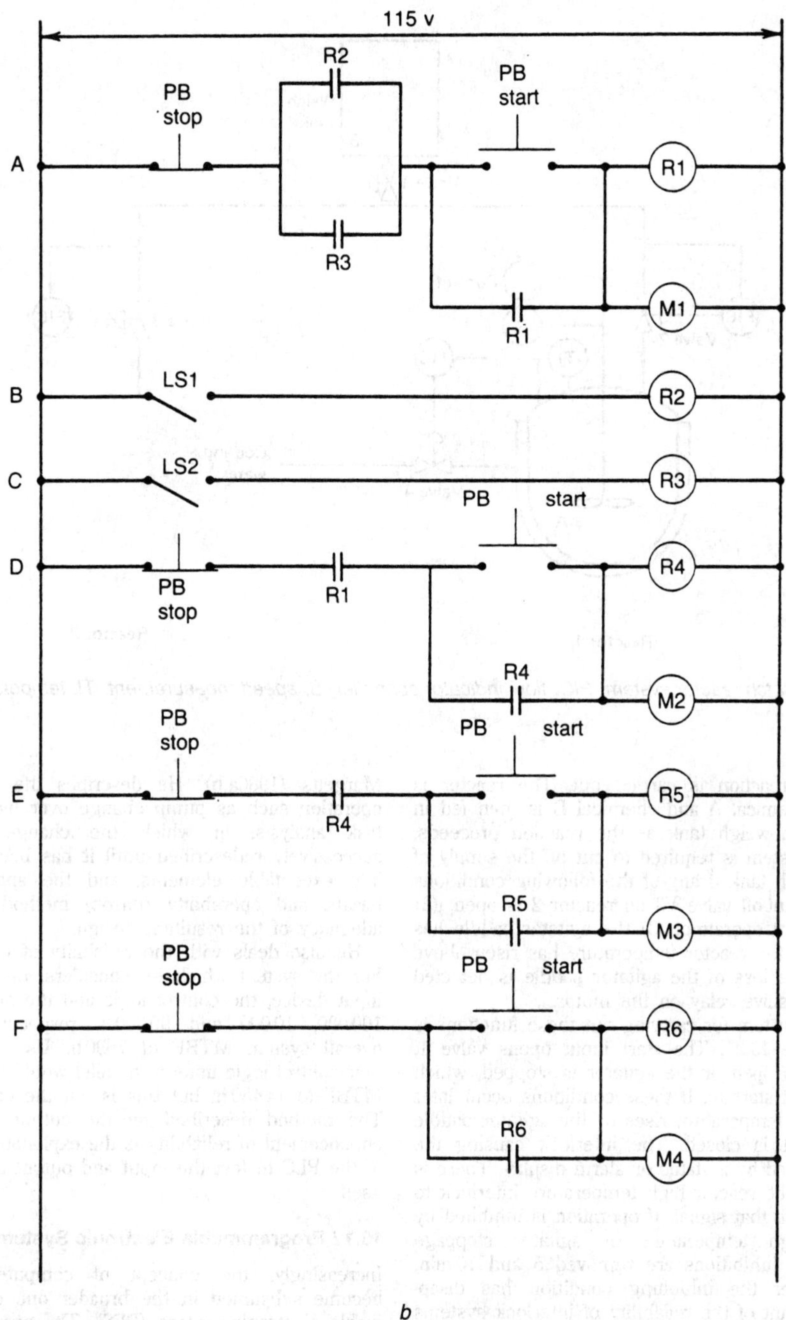

Figure 13.19 continued

operate. Circuit E is the starting circuit for conveyor A, and is similar to circuit D. The circuit can be activated only if relay R4 is closed. Circuit F is a simple starting circuit and is not interlocked.

13.10.4 Illustrative example: reactor system

Another example of a simple interlock system is illustrated in Figures 13.20 and 13.21. Figure 13.20 shows a plant consisting of a water-cooled reactor in

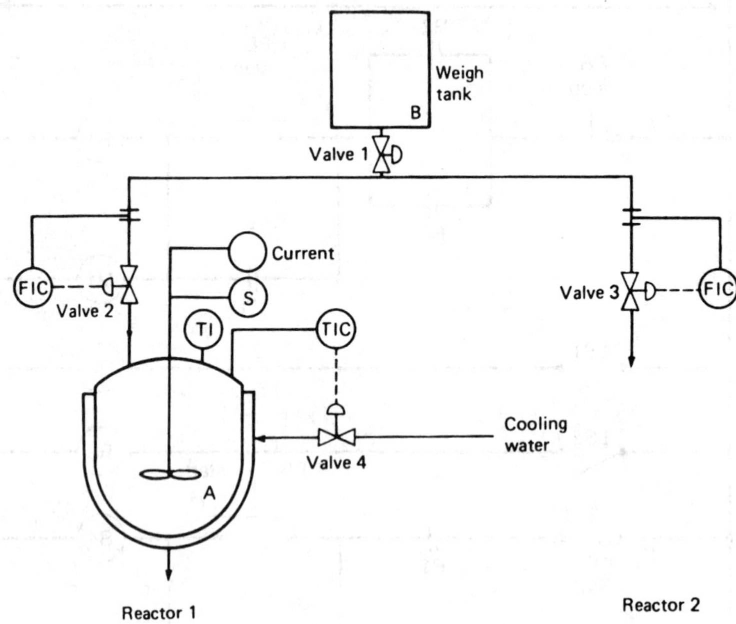

Figure 13.20 *Batch reactor system. FIC, flow indicator controller; S, speed measurement; TI, temperature indicator controller*

which a batch reaction is carried out. The reactor is charged with chemical A and chemical B is then fed in gradually from a weigh tank as the reaction proceeds. The interlock system is required to cut off the supply of B from the weigh tank if any of the following conditions apply: (1) the shut-off valve V3 on reactor 2 is open; (2) the agitator is not operating; (3) the agitator paddle has fallen off; or (4) the reactor temperature has risen above a fixed limit. The loss of the agitator paddle is detected by a current-sensitive relay on the motor.

An interlock system for carrying out these functions is shown in Figure 13.21. The start input opens valve 1, unless valve 3 is open or the agitator is stopped, which conditions inhibit start-up. If these conditions occur later or if the reactor temperature rises or the agitator paddle falls off, valve 1 is closed. The interlock causing the closure is signalled by a status or alarm display. There is a 10 s delay on the reactor high temperature interlock to allow for noise on that signal. If operation is inhibited by the reactor high temperature or agitator stoppage interlocks, these inhibitions are removed 5 and 10 min, respectively, after the inhibiting condition has disappeared. An account of the reliability of interlock systems is given by R.A. Freeman (1994).

13.11 Programmable Logic Systems

As already indicated, increasing use is made in process control systems of programmable logic controllers (PLCs). An account of the application of PLCs to functions such as pump change over, fire and gas detection and emergency shut-down has been given by Margetts (1986a,b). He describes the planning of an operation such as pump change over using hierarchical task analysis, in which the change over task is successively redescribed until it has been broken down into executable elements, and the application of the hazard and operability (hazop) method to assess the adequacy of the resultant design.

He also deals with the reliability of the PLC system. For the system which he considers, the MTBFs of the input device, the control logic and the output device are 100 000, 10 000 and 50 000 h, respectively, giving an overall system MTBF of 7690 h. Use of as many as four control logic units in parallel would raise the system MTBF to 14 480 h, but this is not the complete answer. The method described by the author for the further enhancement of reliability is the exploitation of the ability of the PLC to test the input and output devices and also itself.

13.12 Programmable Electronic Systems

Increasingly, the concept of computer control has become subsumed in the broader one of the programmable electronic system (PES). The account given here is confined to the safety aspects of PESs and is based on the HSE *PES Guide*. The treatment in the CCPS *Safe Automation Guidelines* is discussed in Section 13.15.

13.12.1 HSE *PES Guide*

An account of programmable electronic systems and their safety implications is given in ***Programmable Electronic Systems in Safety Related Applications*** (HSE, 1987b) (the HSE *PES Guide*), of which Part 1 is an

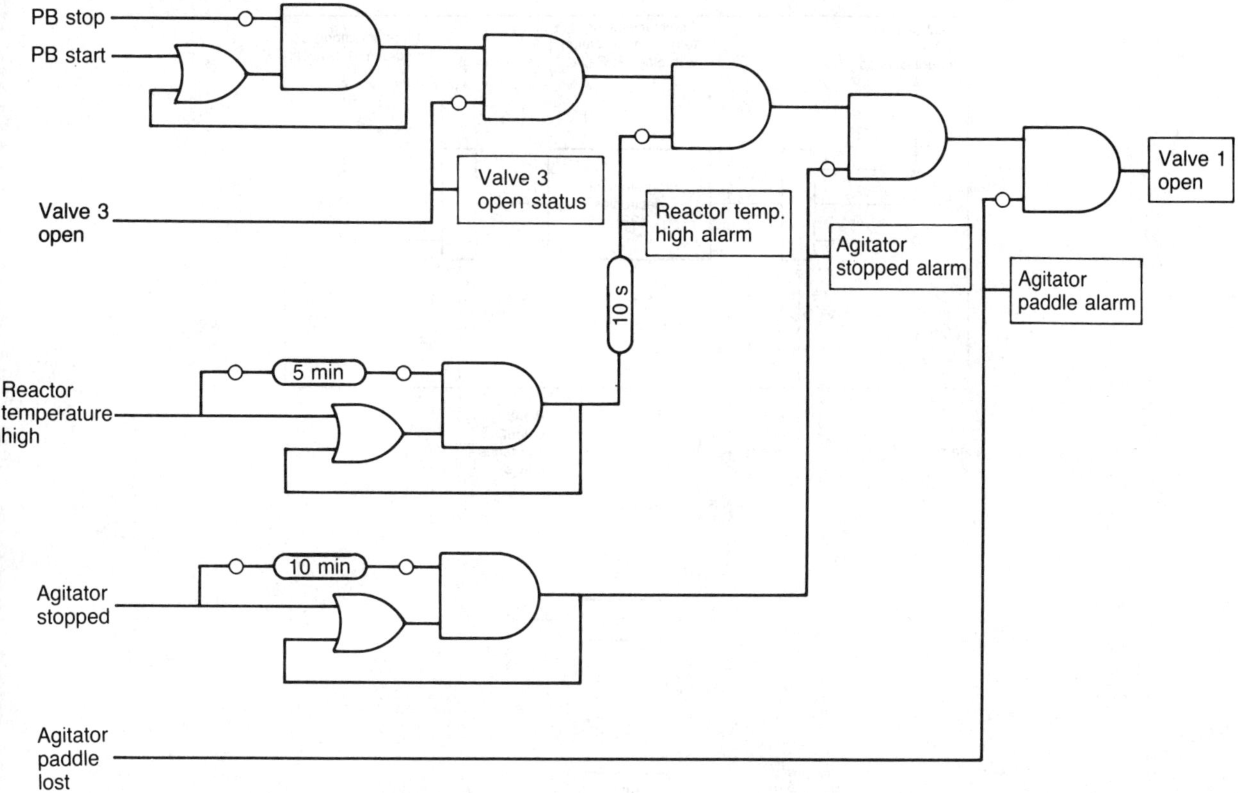

Figure 13.21 *Batch reactor interlock system logic diagram*

Introductory Guide (*PES 1*) and Part 2 the *General Technical Guidelines* (*PES 2*). The general configuration of a PES is shown in Figure 13.22.

Whereas in a safety-related system the use of conventional hardwired equipment is routine, the use of a PES in such an application has been relatively unknown territory. The approach taken, therefore, has been to assess the level of integrity required in the PES by reference to that obtained with a conventional system based on good practice. This level of integrity is referred to as 'conventional safety integrity'.

PES 2 gives three system elements which should be taken into account in the design and analysis of safety-related systems:

(1) configuration;
(2) reliability;
(3) overall quality.

Safety integrity criteria for the system should be specified which cover all three of these system elements.

13.12.2 Configuration

The configuration of the system should be such as to protect against failures, both random and systematic. The former are associated particularly with hardware and the latter with software. *PES 2* lays down three principles which should govern the configuration:

(1) the combined number of PES and non-PES safety related systems which are capable, independently, of maintaining the plant in a safe condition, or bringing it to a safe state, should not be less than the number of conventional systems which have provided conventional safety integrity;
(2) no failure of a single channel of programmable electronic (PE) hardware should cause a dangerous mode of failure of the total configuration of safety-related systems;
(3) faults within the software associated with a single channel of PE should not cause a dangerous mode of failure of the total configuration of safety-related systems.

Observance of the second principle may require that, in addition to the single channel of PE hardware, there should be at least one additional means of achieving the required level of safety integrity. Three such means might be:

(1) additional non-programmable hardware;
(2) additional programmable hardware of diverse design;
(3) additional PE hardware of same design.

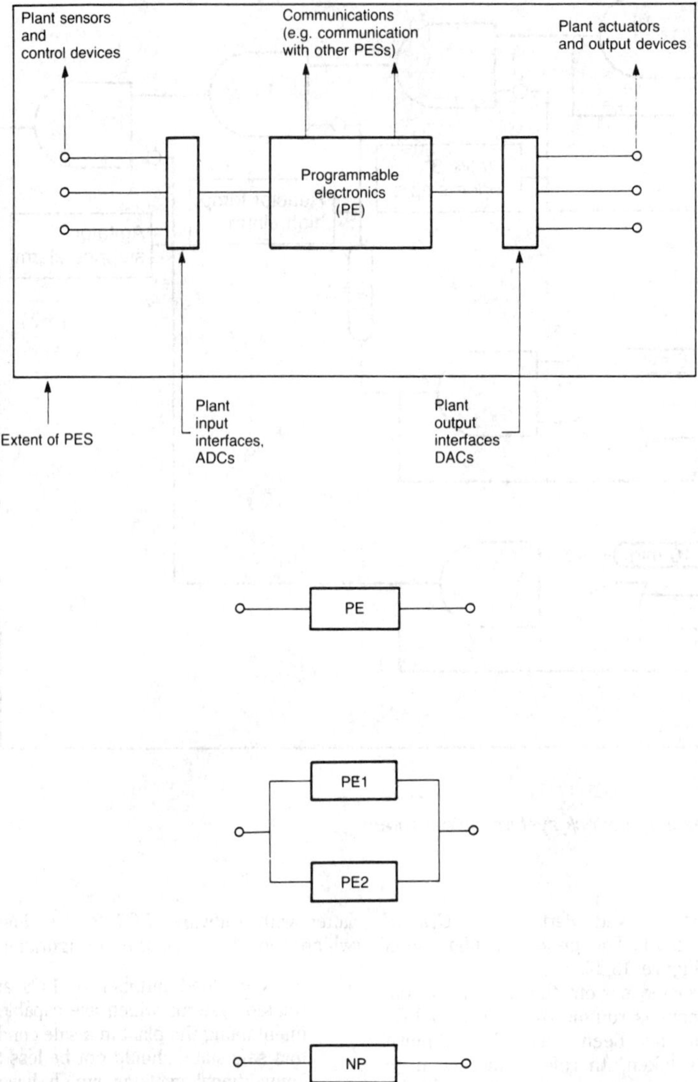

Figure 13.22 *A programmable electronic system (HSE, 1987b). ADC, analogue-to-digital converter; DAC, digital-to-analogue converter; NP, non-programmable hardware; PE, programmable electronics (Courtesy of HM Stationery Office)*

The latter is applicable only where the design is well established and there is a record of reliable operation in an environment similar to that under consideration.

Observance of the third principle may require that where a single design of software is used, there should be an additional means of achieving safety integrity. Such means may be:

(1) additional software of diverse design;
(2) additional non-programmable hardware.

Diversity of software is required only where:

(1) PES safety-related systems are the sole means of achieving the required level of safety integrity;
(2) faults in the software of a single channel of PE might cause a dangerous mode of failure of the total configuration of safety-related systems.

This strategy is intended to protect against systematic failures and, in particular: (a) software errors in the embedded or applications software; (b) differences in the detailed operation of microprocessor and other large-scale integrated circuits from that specified; (c) incompatibility between original and replacement hardware modules; and (d) incompatibility of updated or replacement embedded software with original software or hardware.

The extent to which it is necessary to have diversity of software depends on the application. As a minimum, the safety-related function should use diverse applications software. For higher reliability it may be necessary to consider also diverse embedded software. The safety requirements specification is necessarily a common feature of the diverse software implementations. It is therefore important that it be correct.

PES 2 recognizes that there may exist other ways of providing against failure. In some applications it may be possible to achieve the required level of safety integrity by adopting a formal approach to the software design and testing. Furthermore, for situations where a relatively low level of reliability is acceptable, the use of a single PE channel may be acceptable provided there is extensive self-monitoring of the hardware and automatic safety action on detection of failure.

13.12.3 Reliability

The governing principle for reliability of the hardware is that the overall failure rate in a dangerous mode of failure, or, for a protection system, the probability of failure to operate on demand, should meet the standard of conventional safety integrity. *PES 2* specifies three means of meeting this criterion:

(1) a qualitative appraisal of the safety-related systems, using engineering judgement;
(2) a quantified assessment of the safety-related systems;
(3) a quantified assessment of the safety of the plant.

Essentially, the level of reliability should be governed by the conventional safety integrity principle. Where the acceptable level of reliability is relatively low, the first method may suffice, but where a higher reliability is required the second and third methods will be appropriate.

13.12.4 Overall quality

Whilst the foregoing measures concerning configuration and reliability are a necessary framework, they are not in themselves sufficient. In particular, systematic errors creep in due to deficiencies in features such as the safety requirements specification and software faults. They need, therefore, to be supplemented by the third system element, overall quality. Overall quality is concerned essentially with high quality procedures and engineering. These should cover the quality of the specification, design, construction, testing, commissioning, operation, maintenance and modification of the hardware and software.

In determining the level of overall quality to be aimed for, regard should be paid to the level which would be appropriate for conventional safety integrity and to the level determined for the system elements of configuration and reliability. As a minimum, attention should be paid to (a) the quality of manufacture and (b) the quality of implementation. For overall quality to match a higher level of reliability (c) each procedural and engineering aspect should be reviewed.

Qualitative assessment checklists in support of such a review are included in *PES 2* in Appendix 7. Three sets of checklists are given for (1) a control computer, (2) programmable logic control (PLC) and (3) common cause failure (CCF). The headings of these checklists are (1) safety requirements specification, (2) hardware specification, (3) hardware design, (4) hardware manufacture, (5) hardware test, (6) installation, (7) system test, (8) operations, (9) hardware maintenance and modification, (10) software specification, (11) software design, (12) software coding, (13) software test, (14) embedded software, (15) application programming and (16) software maintenance and modification. The applicable headings are: for a control computer, all except (14)–(15); for a PLC, all except (10)–(13); and for CCF, all.

13.12.5 Design considerations

PES 2 describes a number of design considerations which are particularly relevant to the safety integrity of PESs. The replacement of a control chain in which the sensor sends a signal directly to the actuator by one which involves analogue-to-digital (A/D) converters and PEs may reduce the safety integrity. Unless there is a positive contribution to safety by doing otherwise, the direct route between sensor and actuator should normally be retained. An additional signal may be taken from the sensor with suitable isolation to the PEs.

For the execution of safety functions, it may be necessary to have a shorter sampling interval than is required for normal measurement and control functions.

There should be hardware or software to ensure that on switch-on or on restart after a power failure the resetting of the system is complete and the point in the program at which entry occurs is a safe one. Interruptions of the power supply should be catered for and should not lead to unidentified or unsafe conditions.

As far as practicable, safety critical functions should be automatically monitored or should be self-checking. The emergency shut-down systems should be proof checked at appropriate intervals to discover unrevealed failures.

PES 2 also gives detailed guidance on the environmental aspects of PESs, particularly in respect of electrical interference and of electrostatic sensitive devices.

13.12.6 Software considerations

The software for use in safety-related applications needs to be of high quality and *PES 2* gives an account of some of the measures which may be taken to achieve this. These include:

(1) safety requirement specification;
(2) software specification;
(3) software design, coding and test;
(4) system test.

They also include:

(5) software modification procedures.

For all these aspects there should be

(6) formal documentation.

PES 2 puts considerable emphasis on the safety requirements specification, as already described. It also devotes a good deal of space to the control of software changes.

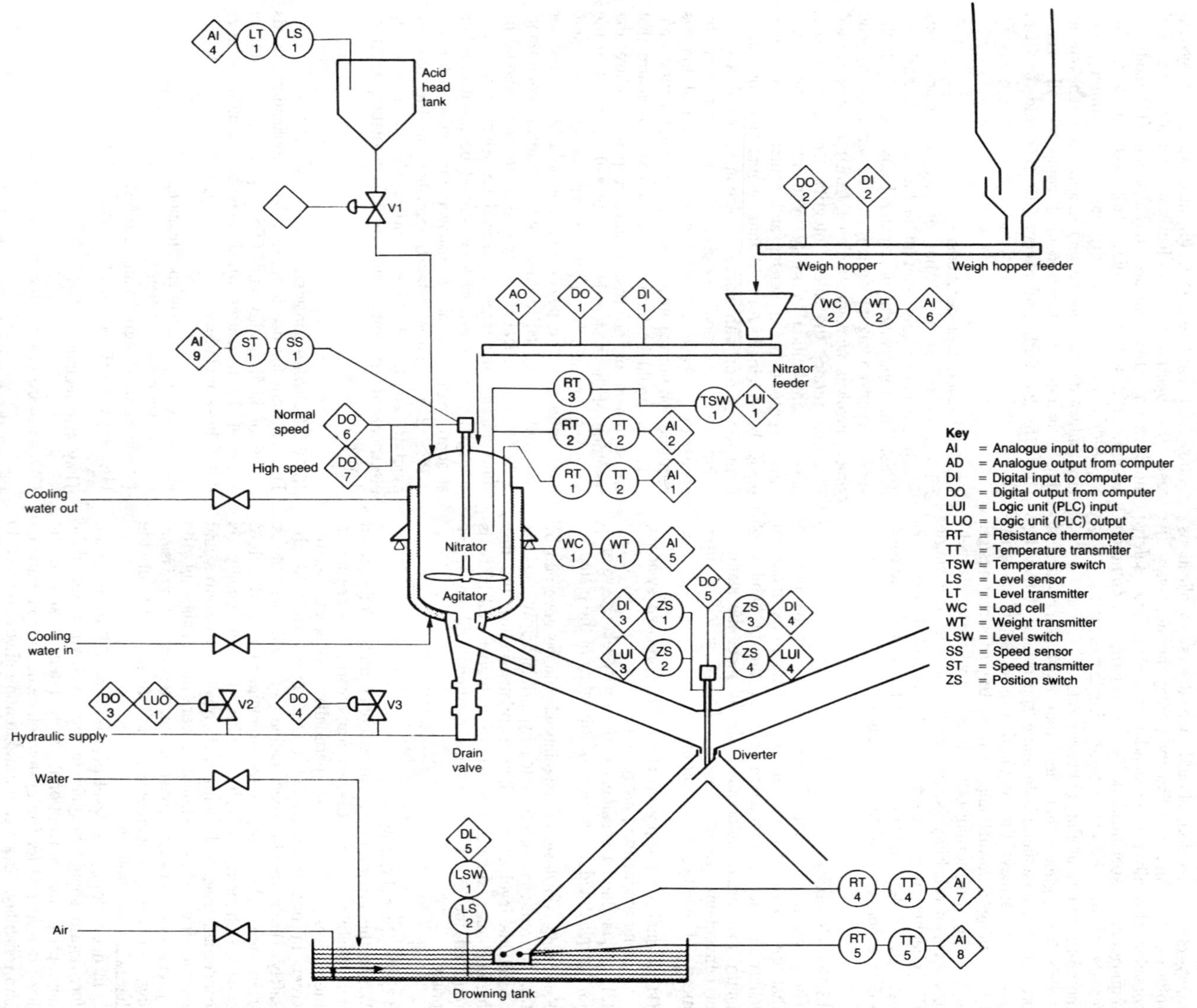

Figure 13.23 *A nitrator unit under control of a PES (HSE, 1987b) (Courtesy of HM Stationery Office)*

A further account of software reliability is given in Section 13.13.

13.12.7 Illustrative example

PES 2 gives as an illustrative example part of the safety integrity assessment of a plant for the manufacture of the explosive pentaerythritol tetranitrate (PETN). Figure 13.23 shows a schematic diagram of one of the nitrators.

A particularly critical parameter is high temperature in the nitrator, the limit being 35°C. For this, protection is provided in the form of a dump valve, which opens to dump the reactor contents to a drowning tank. The conventional control and protection system for such a plant is a control system incorporating some protection features and a single dedicated protection system.

In the design considered, the plant is controlled by a control computer which performs the basic control of the operating sequence. At each stage of the sequence the computer performs checks to ensure that the previous stage is complete and that the plant is in the correct state and ready to proceed to the next stage.

On each of the critical parameters there are duplicate sensors. The signal from one sensor goes to the computer and that from the other to a PLC. The computer and the PLC operate their own relays in the appropriate interlock system. An attempt by the computer to take an action is inhibited if: (1) the PLC relay contact is not closed in agreement with the computer; or (2) the combination of permissives being sent to the computer is correct but indicates that the action is to be inhibited; or (3) the combination of permissives being sent to the computer is incompatible with the inputs to the PLC. The latter occurrence is assessed by the computer. Thus the control of critical parameters and sequences is by the computer monitored by the PLC, which is in turn monitored by the computer.

For the critical parameter of high temperature in the nitrator, the computer and PLC act in effect as a 1/2 protective system. If either detects a high temperature, it acts to open the dump valve on the reactor.

A hazard analysis of the nitrator system was performed and a fault tree developed for the top event 'Decomposition', as shown in Figure 13.24(a). This event occurs if there is a demand in the form of high temperature and a failure of the top level of protection ('protection fails'). The guide includes the further fault trees for the events C1–C4 and B4–B8.

Figure 13.24(b) shows one of these constituent fault trees, subtree B5 for the event 'no dump signal'. This has several interesting features. The top event in the

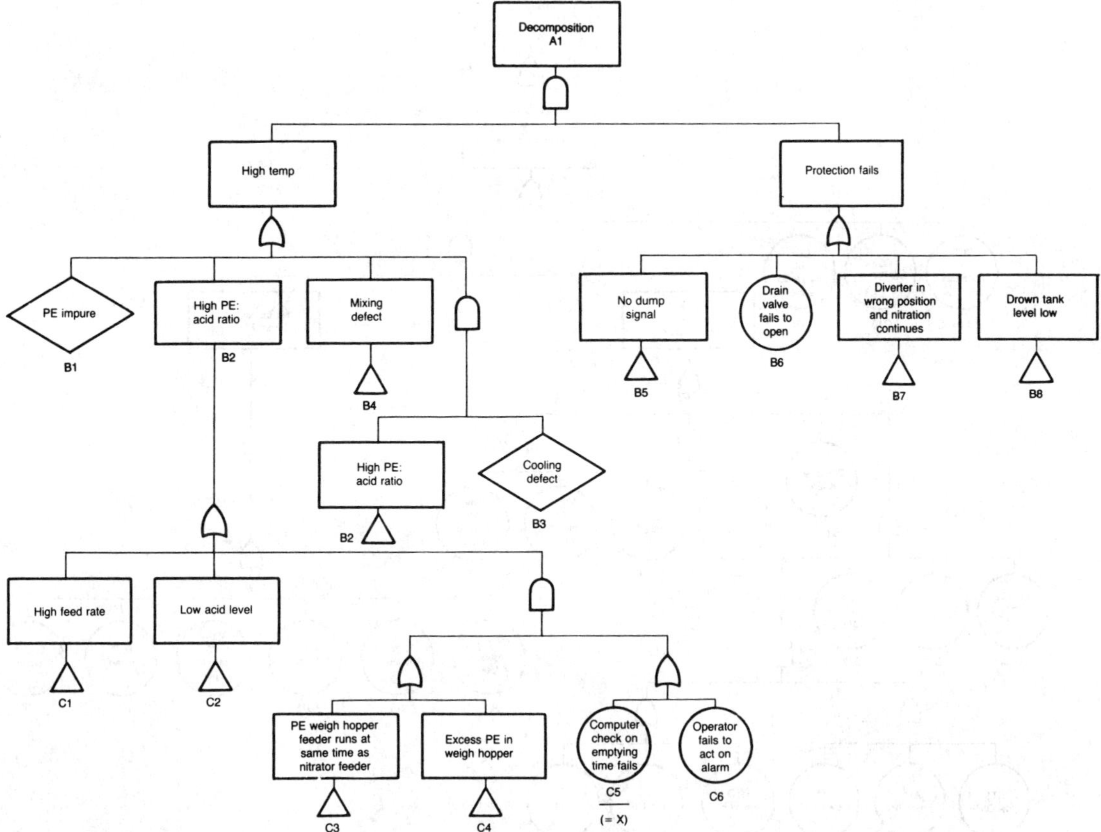

Figure 13.24 *Sections of fault tree for the top event 'Decomposition' for the nitrator unit shown in Figure 13.23 (HSE, 1987b): (a) top section of fault tree; and (b) subtree for event B5 (Courtesy of HM Stationery Office)*

subtree will occur if both the computer and the PLC fail to send a trip signal. Among the causes for the computer failing to send this trip signal are various combinations of instrument failure, including common cause failure of all the instruments in one group, e.g. resistance thermometers RT1 and RT2. In the tree the failures of RT1 and RT2 are regular random failures, whilst the failure 'CCF of RTs' is the CCF for this group. The separation of the CCF in this way both highlights it as a specific event and assists in assigning to it a numerical value.

Other CCFs occur higher up the subtree just beneath the top event. These are the CCFs of: (a) all three resistance thermometers, RT1 and RT2 on the computer and RT3 on the PLC; (b) the hydraulic valves; and (c) both the computer and the PLC.

A safety integrity analysis of the system is given in which each of the three system elements (configuration, reliability and overall quality) are examined. For configuration, a check is made against each of the three principles given in Section 13.12.2. In respect of criterion 1, the combined number of PES and non-PES systems is two, as in the conventional system, so this criterion is met. For criterion 2, failure of no single channel, computer or PLC, will cause loss of protection, so criterion 2 is met. For criterion 3, failure of no single set of software, on the computer or on the PLC, will cause loss of protection, so criterion 3 is met.

For reliability, the fault tree is analysed to produce the cutsets and it is shown that the Boolean relation for the top event A1 is:

$$A1 = D10 + Z + X(B6 + J13 + J14 + J15 + J16 + J18 + J20 + L1 + L2 + Y) + Y(D11 + D12) + (B1 + D1 + D2 + D11 + D12 + H1 + H2)(J20 + B6) \quad [13.12.1]$$

The frequency of the top event A1 was then estimated by applying data on the frequency of failures and, for protective features, by utilizing Equation 13.9.18 with data on proof test intervals. Table 13.27 shows the numerical values obtained for the events given in Equation 13.12.1. The frequency of the top event was found to be 8.8×10^{-4}/year. This was based on a number of pessimistic assumptions and on this basis was deemed an acceptable frequency.

For overall quality, it was considered necessary to examine not only the quality of manufacture and implementation, but also the procedures and engineering. For this the checklists given in Appendix 7 of *PES 2* were used and the results obtained for each item in this example are shown in that appendix. This check led to consideration of the following modifications: (a) addition of software limits on programmable alarm and trip levels; and (b) provision of test signal injection and monitoring

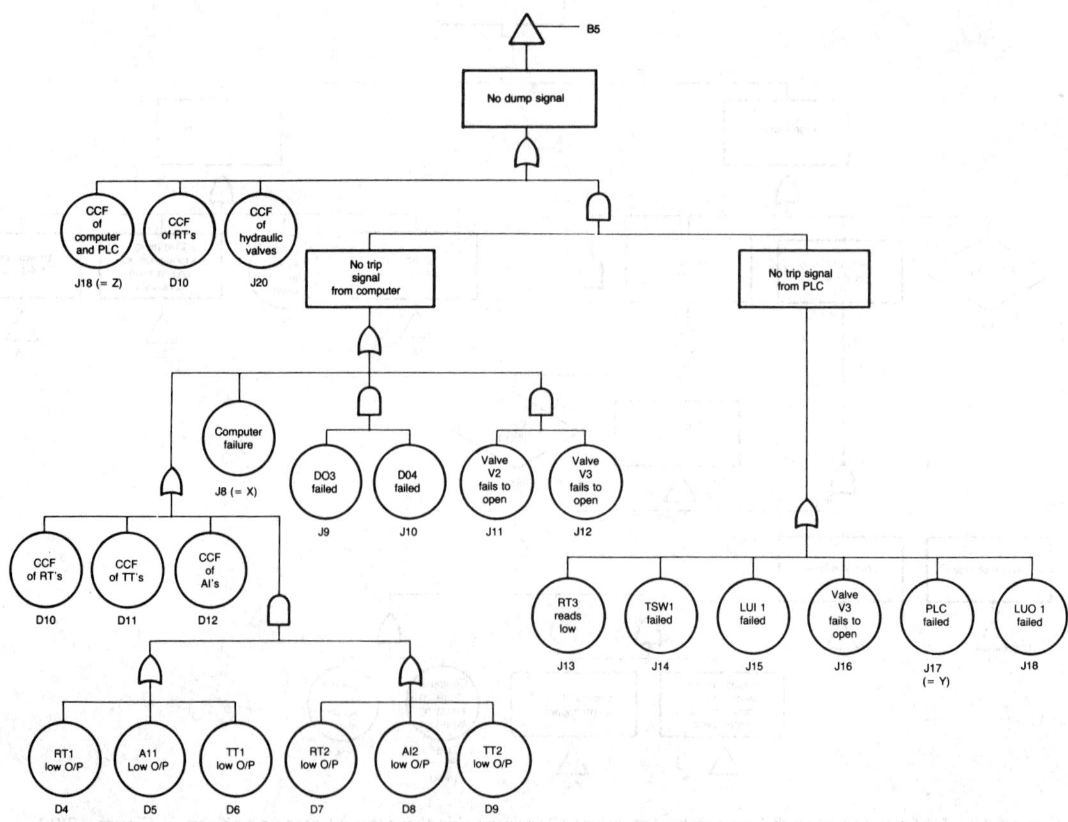

Figure 13.24 continued

Table 13.27 *Some results obtained in the estimation of the hazard rate of the nitrator shown in Figure 13.23 (HSE, 1987b) (Courtesy of HM Stationery Office)*

Event reference	*Event description*	*Failure rate* (failures/10^6 h)	*Test interval*	*Probability of failure on demand*
D10	Common cause failure of resistance thermometers RT1, RT2 and RT3 giving low output	0.06		
Z	Common cause failure of computer and PLC	Not quantified		
X	Failure in a dangerous mode of control computer	7.5		
B6	Drain valve fails to open			3.8×10^{-4}
J13	Failure of resistance thermometer RT3 giving low output	2.0	1 month	7.3×10^{-4}
J14	Temperature switch TSW1 fails to operate on high temperature	3.4	1 month	1.2×10^{-3}
J15	Logic unit (PLC) input LUI 1 failed; does not respond to TSW1	1.0	1 month	3.7×10^{-4}
J16	Valve V3 fails to open			7.5×10^{-4}
J18	Logic unit (PLC) output LUO 1 failed; does not de-energize	1.0	1 month	3.7×10^{-4}
L1	Drowning tank leaking			Negligible
L2	Drowning tank drain valve opened			1×10^{-3}
Y	PLC failed in dangerous mode	2.5	1 month	9.1×10^{-4}
$\overline{\Sigma_1}$				5.7×10^{-3}
X. Σ_1		0.042		
Y	PLC failed in dangerous mode	2.5	1 month	9.1×10^{-4}
D11	Common cause failure of temperature transmitters TT1, TT2 giving low output	0.06		
D12	Common cause failure of control computer analogue inputs A11 and A12 giving low reading	0.03		
Σ_2		0.09		
Y. Σ_2		8.2×10^{-5}		
B1	PE impure	Not quantified		
D1	Feeder fails at high speed	2.6		
D2	Control computer analogue output A01 fails to high O/P	1.0		
D11	Common cause failure of temperature transmitters TT1, TT2	0.06		
D12	Common cause failure of control computer analogue inputs A11, A12	0.03		
H1	Hydraulic failure of agitator causing high speed	Negligible		
H2	Stirrer breaks from shaft	Negligible		
Σ_3		3.7		
J20	Common cause failure of hydraulic valves V2 and V3	2.3×10^{-6}		
B6	Drain valve fails to open	3.8×10^{-4}		
Σ_4		3.8×10^{-4}		
$\Sigma_3 \cdot \Sigma_4$		1.4×10^{-3}		
Total	$(D10 + Z + X\Sigma_1 + Y\Sigma_2 + \Sigma_3\Sigma_4)$	0.1		

points, particularly on the resistance thermometers measuring high temperature in the nitrator.

13.13 Software Engineering

The use of various types of computer aid in process plant design and operation is now routine. The dependability of these aids is determined by the quality of the computer programs. The dependability of this software is therefore important and may be critical. This is especially the case in real-time, on-line computer-based systems.

The dependability of software, particularly in safety critical systems, is a major topic in software engineering and is beyond the scope of this book. However, it cannot be neglected, and therefore a brief description is given of some of the principal issues of which engineers in the process industries should be aware.

Accounts of software engineering and software reliability are given in *Software Engineering* (Bauer, 1975a), *Software Reliability, Principles and Practice* (Meyers, 1976), *Quality Assurance for Computer Software* (Dunn and Ullman, 1982), *Program Verification Using ADA* (McGettrick, 1982), *Software Engineering* (Shooman, 1983), *Software Defect Removal* (Dunn, 1984), *Program Construction and Verification* (Backhouse, 1986), *Systematic Software Development Using VDM* (C.B. Jones, 1986), *The Spine of Software: Designing Probably Correct Software – Theory and Practice* (Baber, 1987), *Achieving Safety and Reliability with Computer Systems* (Daniels, 1987), *The Craft of Software Engineering* (Macro and Buxton, 1987), *Software Reliability* (Littlewood, 1987b), *Software Reliability* (Musa *et al.*, 1987), *Handbook of Software Quality Assurance* (Schulmeyer and MacManus, 1987), *Software Diversity in Computerised Control Systems* (Voges, 1987), *Managing the Software Process* (Humphrey, 1989), *High Integrity Software* (Sennett, 1989), *Software Engineering* (Somerville, 1989), *Case Studies in Systematic Software Development* (C.B. Jones and Shaw, 1990), *Deriving Programs from Specifications* (C. Morgan, 1990), *Software Quality and Reliability* (Ince, 1991a), *Software Engineers Reference Book* (McDermid, 1991), *Developing Safety Systems: A Guide Using ADA* (Pyle, 1991), *Reliability in Instrumentation and Control* (Cluley, 1993) and *Safety Aspects of Computer Control* (P. Bennett, 1993).

13.13.1 Software dependability

The software provided should be dependable in serving the purposes of the system. The dependability of the software has two aspects: (1) specification and (2) reliability. The requirements of the system need to be defined and then converted into a specification. Both the formulation of the requirements and the conversion into a specification are critical features. It is then necessary to ensure that the software conforms with the specification to a high degree of reliability. One of the recurring themes in discussions of software dependability is that reliability alone is not enough. If the specification is defective, the software will be so too, however high its reliability.

13.13.2 Some software problems

There are some persistent problems associated with software. A review of these problems by Bauer (1975b) cites the following tendencies: (1) software is produced in a relatively amateurish and undisciplined way, (2) it is developed in the research environment by tinkering or in industry by a human wave approach, (3) it is unreliable and needs permanent maintenance, (4) it is messy, opaque and difficult to modify or extend and (5) it arrives later, costs more and performs less well than expected.

13.13.3 Software error rates

There are a number of rules-of-thumb used in the software industry for the error rates which occur in programming. An account is given by Cluley (1993). For software, an important distinction is that made between a fault and failure. A fault is an error in the program. A failure occurs when the program is run and produces an incorrect result for software reasons. It is a common occurrence that a program which contains a fault may be run many times before a failure occurs.

A rule-of-thumb widely used in the industry is that a program typically contains 1 fault per 1000 instructions. This is supported by data given by Musa *et al.* (1987) to the effect shown that for programs of some 100 000 lines of source code when first operational the incidence of faults varies between 1.4 and 3.9 faults per 1000 lines. For programs when first written the number of faults is much higher.

Faults may be corrected, but correction is not always straightforward and the potential exists to introduce other faults. The data of Musa *et al.* indicate that between 1 and 6 new faults are introduced for every 100 faults corrected.

Musa *et al.* also quote data for the number of failures per fault for a single run of a program. The average value of this ratio is 4.2×10^{-7} failures/fault. In other words, this implies that in order to detect a fault by triggering a failure, it is necessary on average to run a program 2.4×10^6 times.

In real-time applications of safety critical systems another metric of concern is the failure intensity, or frequency of failure per unit time, or per mission. For passenger aircraft an error rate used has been 10^{-9} per mission, where the mission is a flight of 1 to 10 hours.

The progress of a debugging task may be monitored by 'seeding' the program with deliberate errors which are not known to the team engaged in the work. Thus if 35 faults have been introduced deliberately and 25 genuine and 7 deliberate faults are found, the estimated number of original faults is 125 ($=25 \times 35/7$).

There exist reliability growth models for software which may be used by management to estimate the time necessary to debug a program. One such model is described by Cluley.

13.13.4 Software management

Management commitment is crucial in achieving dependability in software, as in other fields. Management needs to create a culture which gives priority to, and so ensures, dependability of the software. The management of a software project includes the following aspects:

(1) project management;

(2) software quality assurance;
(3) software standards;
(4) system requirements and software specifications;
(5) software development;
(6) software documentation;
(7) software verification;
(8) software modification control;
(9) software validation and testing;
(10) software maintenance.

Accounts of project management are given by Tsichritzis (1975a) and P.A.V. Hall (1991). The other aspects are considered below.

13.13.5 Software quality assurance

There should by a system of quality assurance (QA) for the software. The extent of this system will depend on the scale of the operation, and in some cases will be governed by standards and/or user requirements, but as a minimum there should be a formal system and an independent QA function. Some of the methods of assuring quality are described below.

13.13.6 Software standards

Use has long been made in software development of the traditional quality standards such as BS 5750 and ISO 9000, but there are an increasing number of standards specific to software. Accounts of developments in these standards are given by P. Bennett (1991a), the CCPS (1993/14) and Rata (1993). In the UK, standards and guidance include: BS 5887: 1980 *Code of Practice for Testing of Computer-Based Systems*, BS 6238: 1982 *Code of Practice for Performance Monitoring of Computer-based Systems*, BS 5515: 1984 *Code of Practice for Documentation of Computer-based Systems* and BS 6719: 1986 *Guide to Specifying using Requirements for a Computer-based Systems; Programmable Electronic Systems in Safety Related Applications* (HSE, 1987b) (the HSE *PES Guide*); the Ministry of Defence (MoD) Interim Defence Standards 0055: 1989 *Requirements for the Procurement of Safety Critical Software in Defence Equipment* (MoD, 1989c) and 0056: 1989 *Requirements for the Analysis of Safety Critical Hazards* (MoD, 1989b).

Relevant US standards are IEEE 1058-1987 *Software Project Management Plans*, IEEE 1012-1987 *Software Verification and Validation Plans*, IEEE 1028-1988 *Software Reviews and Audits*, IEEE 730-1989 *Software Quality Assurance Plans* and IEEE 1063-1989 *Software User Documentation*, as well as the guides IEEE 830-1984 *Guide to Software Requirement Specifications* and IEEE 1042-1987 *Guide to Software Configuration Management*. An international standard is IEC SC65A WG9: 1991 *Software for Computers in the Application of Industrial Safety-related Systems.*

The *PES Guide* has been described in Section 13.12.

MOD 0055 is in three main sections. The first deals with the project management, the parties involved and the documentation; the second with the software engineering; and the third relates the requirements of these two sections to the life cycle of the project.

MOD 0056 gives requirements for the hazard analysis of safety critical systems.

There are also two IEC working groups, WG9 and WG10, which deal with software for safety-related applications and with generic safety aspects, respectively. WG9 is responsible for IEC SG65A.

13.13.7 Software development

The process of software development is generally described broadly in the following terms:

(1) requirements specification;
(2) system specification;
(3) program specification;
(4) program design;
(5) program production;
(6) program verification;
(7) program validation and testing;
(8) system integration and testing.

In software development two terms widely used are 'verification' and 'validation' (V&V). Verification is the process of determining whether the product of a given phase of development meets the requirements established in the previous phase. Validation is the process of evaluating software at the end of the software development process to ensure compliance with software requirements.

It is good practice to verify the software produced in each phase of the project before proceeding to the next phase. Another aspect of good practice is the production of good documentation.

One method of software development which is found useful in many cases is prototyping. There is more than one kind of prototype. An account of prototyping is given by Ince (1991b).

13.13.8 Software specification

The conversion of the user's requirements into an unambiguous specification for the system and then for the software is one of the most important, but difficult, tasks in software development. There is a high degree of formality in the approach taken to the specification of the software and a number of formal methods have been developed. An account is given by Webb (1991). Use is made of mathematically based languages such as VDM, Z and OBT and of mathematically based methodologies such as JSD, EPOS and Yourdan Structural Development. For many safety critical systems such formal methods are a requirement.

13.13.9 Software design, production and verification

There are a number of basic principles governing software design. They include (1) modularity and (2) hierarchy.

The computer program required for even a moderately sized project may be large. A large program needs to be subdivided into manageable parts, or modules. Whilst subdivision into modules is necessary, problems arise if the interfaces between modules are poorly defined. The specification of the interfaces between modules requires careful attention. In some applications, it may be possible to exploit the use of verified modules and of a module library.

The program will generally have a hierarchical structure, with the higher level modules controlling the lower level ones. Structured programming involves the use of a hierarchy of conceptual layers and provides a formal approach to the creation of hierarchical software.

As already mentioned, verification of the programs produced at each phase should not be relegated until the end but should be performed before proceeding to the next phase.

13.13.10 Software modification control

A major software project will generally be subject to modifications. Demands to make modifications may occur at any level, starting with the system requirements. There should be a system for the control of such modifications. The ease with which such a system can be created and operated depends very much on the quality of the software design and documentation.

13.13.11 Software reliability

The point has already been made that 'software reliability' is not the same as 'software dependability'. It is nevertheless an essential feature. Accounts of software reliability are given in the texts quoted and by Tsichritzis (1975b). Some aspects of software reliability are:

(1) programming language;
(2) programming practice;
(3) software design;
(4) measurement of reliability;
(5) assessment of reliability.

The programming language used can influence the reliability of the software produced. A number of examples of differences between languages in this respect are given by Tsichritzis.

Likewise, the programming style can affect reliability. One aspect is the naming of items. It is usually recommended that semantic naming be practised in which the name is a meaningful one. Another aspect is the length of sections of the program. Here the recommendation is to keep the verification length short. A practice which tends to increase the verification length is the use of GO TO statements.

One aspect of good design practice which contributes to software reliability is a strong structure. Another is transparency of the programs. A third is well-defined interfaces between modules.

In principle, improvement of reliability depends on the ability to measure it. Traditionally, metrics have been concerned primarily with aspects of performance such as execution time rather than with reliability. Measures for reliability are discussed by Tsichritzis. Software protection contributes to reliability by providing barriers to the transmission of errors between different features of the system.

13.13.12 Software testing and debugging

The traditional way of dealing with errors in a program is testing and debugging. An account of this aspect is given by Poole (1975). Debugging and testing are greatly facilitated if they are planned for in the design phase. Another feature which can make a major contribution is documentation written with this requirement in mind.

Debugging tends to be a difficult task and various aids are available. One is the system dump, activated by a call in, or by catastrophic failure of, the program. Another is the snapshot, similar to a dump, but occurring during execution. The trace mode of program execution causes an output to be made for each statement in the section traced. The traceback facility shows how control reached the point in the program where the error has occurred.

It is helpful to debugging if key quantities in the program are made parameters which the user can alter. This permits a fuller exploration of the program characteristics. Debugging is also assisted by the incorporation of debugging code in the program. The use of such code is discussed by Poole.

Testing is assisted by subdivision of the program into modules. It is not, however, a straightforward matter to devise test beds and test strategies for modules.

13.13.13 Software protection

Software protection may be regarded as an aspect of software reliability. The aim of software protection is to guard against error and malice. There are a variety of items, such as files and programs, which need to be protected and a corresponding variety of means of achieving protection.

Protection establishes barriers to the transmission of an error between one part of the system and another. It therefore contributes to reliability by limiting the effect of an error. Protection contributes to reliability in another way. The occurrence of an error usually results in an attempt to violate a protection barrier. This can be used as a means of error detection.

Closely related to software protection is software security. The aim of software security is to guard against unauthorized use.

13.13.14 Software assessment

There are a number of methods available for the assessment of the software reliability. Accounts of these techniques are given in the texts mentioned and by Tsichritzis (1975b), Fergus *et al.* (1991), Webb (1991) and M.R. Woodward (1991). Three main approaches are:

(1) auditing;
(2) static analysis;
(3) dynamic analysis.

Auditing particularly addresses aspects such as the quality assurance and standards, the comprehensibility and readability of the program, and the documentation.

Static analysis involves analysing the program without running it. Some methods which may be used include:

(1) semantic checking;
(2) control flow analysis;
(3) data use analysis;
(4) information flow analysis;
(5) semantic analysis;
(6) compliance analysis.

The program compiler is generally utilized to perform checks on statements in the program, or semantic checks. The power of this facility depends on the programming language used.

The control flow of the program may be analysed to reveal its structure and to detect undesirable features such as multiple starts, multiple ends, unreachable code, etc. One method of doing this is to represent the program as a graph of nodes joined by arcs, where initially each node represents a statement. A process of reduction is then applied whereby nodes are successively eliminated to reveal the underlying structure.

The data use of the program may be analysed to identify incorrect uses of data such as attempts to read data which are not available or failure to utilize data which have been generated.

The information flow in the program may be analysed to identify the dependence of output variables on input variables.

Semantic analysis determines the mathematical relationship between the input and output variables for each semantically feasible path. It can be used to determine the outputs for the whole input space, including unexpected inputs.

Compliance analysis compares the program with the specification and reveals discrepancies. The specification is expressed as a statement in the predicate calculus of the pre-conditions and post-conditions to be satisfied by the program. For a complex program assertions may be provided about the functionality of the program for intermediate stages.

Use may also be made of diagrams showing the logic of the program, such as fault trees, event trees, Petri nets, and state transition diagrams, as described by P. Bennett (1991a). The application of fault tree analysis to programs has been developed by Leveson and co-workers (Leveson and Harvey, 1983; Leveson and Stolzy, 1983). Figure 13.25 shows the analysis of an IF...THEN...ELSE statement by fault tree, Petri net and event tree methods.

Fault trees and event trees are described in Chapter 9, but the Petri net representation requires brief explanation. A Petri net consists of the quintuple C:

$$C = (P, T, I, O, u) \qquad [13.13.1]$$

where P is a place, T a transition, I an input, O an output and u an initial condition. Initialization of a Petri net is called 'marking' it. A transition are said to 'fire'. Assigning a value is referred to as 'passing' a 'token' to a place.

Software tools have been developed to assist in the static analysis of software. Some of the tools available are described by Fergus *et al.* (1991). They include MALPAS, SPADE and the LDRA testbed. An account of MALPAS, which includes control flow, data use, information flow, semantic and compliance analysers, is given by Webb (1991).

Dynamic analysis, or testing, involves running the program and analysing the results. The basic technique is to force situations where errors are revealed. There are two main approaches. One is 'black box' testing and the other 'white box' testing. The distinction is that the latter relies on knowledge of the structure of the program, whilst the former has no such knowledge but tests the performance against the requirements and relies essentially on knowledge of the application domain. Dynamic testing may be control flow or data flow driven. A discussion is given by M.R. Woodward (1991).

One aim of testing is to remove whole classes of error. A technique for doing this is mutation testing. An account is given by Woodward. The basic concept is to make a small change to the program in the expectation that this will make an observable difference in its performance.

It should be appreciated that good results from a validation test do not necessarily indicate high reliability. This is so only if the exercise of the control path in the validation test corresponds to that which will occur in practice.

The level of assessment should be matched to the application. This aspect is discussed by P. Bennett (1991a). He lists five classes of assessment:

0 System overview.
1 System structure analysis.
2 System hazard analysis.
3 Rigorous analysis.
4 Formal mathematical methods.

13.13.15 Software correctness

The use of formal methods to prove the correctness of the program has already been mentioned. This is a major area of research. There are differing views as to the feasibility of such proof.

The methods used to prove correctness may be informal or formal. The informal method derives from work of Naur (1966), Floyd (1967) and London (1968), following von Neumann. Points are selected on all the control paths at which assertions can be made about the variables. Then if A is an assertion at one point in a control path and B an assertion at the following point, the approach taken is to prove that the code is such that if A is true, B is true. If this verification is performed for all adjacent pairs of assertions and for all control paths, the partial correctness of the program is proved. Proof of complete correctness requires a separate proof of halting. It tends to be a substantial task, however, to develop the assertions and to perform the proofs.

The formal method of proving correctness is based on the demonstration by Floyd (1967) that proof of partial correctness is equivalent to proving corresponding theorems in the first order-predicate calculus. Manna and Pnueli (1969) extended such proof to include halting. The approach taken is to formulate the problem so that it is possible to apply automatic theorem-proving techniques.

13.13.16 Software maintenance

Software generally requires a good deal of maintenance. This is particularly true of safety related software, especially real-time software. The project management should make suitable provisions for software maintenance. The quality of the software, and the associated documentation, largely determines the ease of maintenance.

13.13.17 Software for real-time systems

Real-time, on-line systems controlling process plants place even more stringent demands on software. An account of the software aspects is given by Fergus *et al.* (1991).

The characteristics of real-time systems have been described by Quirk (1985). In such systems the demands are driven in timing and sequencing by the real world, they may occur in parallel and they may be unexpected and even conflicting. The software must satisfy time constraints and it must continue to operate. Moreover, the software is part of a total system and is difficult to validate in isolation.

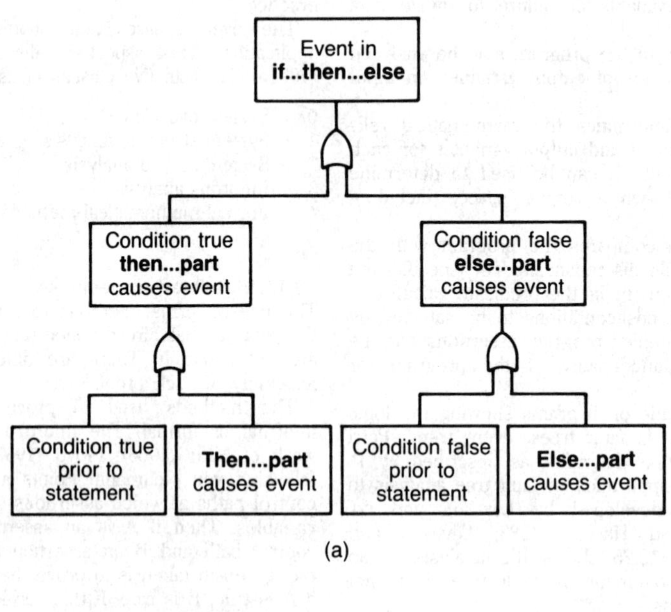

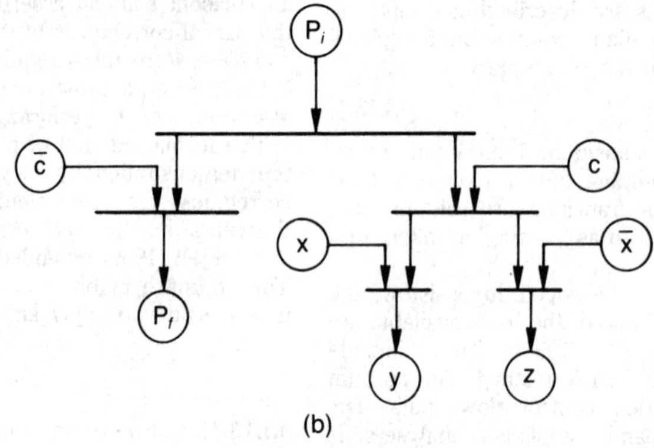

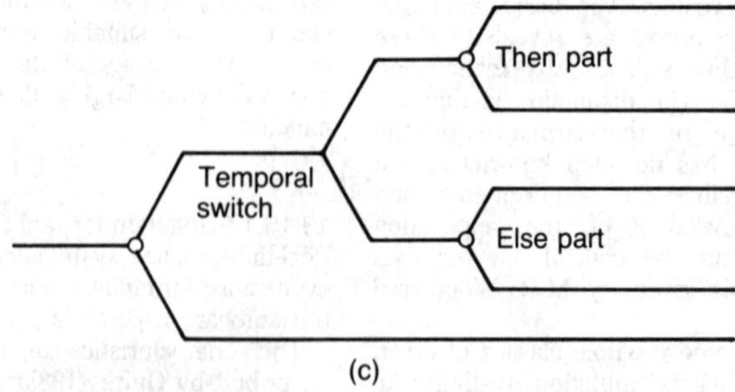

***Figure 13.25** Some representations used in error analysis of an IF… THEN… ELSE statement in a computer program (after P. Bennett, 1991a): (a) fault tree; (b) Petri net; and (c) event tree (Courtesy of Butterworth-Heinemann)*

Work on the methods of describing the behaviour of real-time systems typically deals with issues such as concurrency and synchronization, resource scheduling, and liveness and deadlock.

The dynamic testing of a real-time program may be carried out using an off-line host machine. Such testing is described by Fergus *et al.*

13.13.18 Safety critical systems

If the consequences of failure of a real-time computer system are sufficiently serious, the system is a safety critical system (SCS). SCSs are of particular concern in the military, aerospace and nuclear fields, but are of growing importance in the process industries.

SCSs are treated in *Safety Aspects of Computer Control* (P. Bennett, 1993). Other accounts are given in the texts cited at the start of this section and by P. Bennett (1991a,b), Bologna (1993), Ehrenberger (1993), Malcolm (1993), McDermid (1993) and Pyle (1993). Standards are particularly important for SCSs. Some relevant standards and guidance are detailed in Section 13.13.6. Practical guidance is available in the HSE *PES Guidelines* and the CCPS *Safe Automation Guidelines*, described in Sections 13.12 and 13.15, respectively.

There are a number of real-time languages and environments which have special safety related features. One such is ADA. Accounts are given in *Program Verification Using ADA* (McGettrick, 1982), *ADA for Specification and Design* (Goldsack, 1985), *ADA in Industry* (Heilbrunner, 1988) and *Developing Safety Systems: A Guide Using ADA* (Pyle, 1991). Pyle (1993) discusses the guidance given in the HSE *PSE Guide* in the context of ADA.

Where the process is dependent on a computer or PES, methods are required to identify the associated hazards. The application of hazop to process computers (chazop) is described in Chapter 8

13.14 Safety-related Instrument Systems

It will be apparent from the foregoing that it is necessary to adopt a systematic approach to the whole system of instrumentation, control and protection.

13.14.1 EEMUA *Safety-related Instrument System Guide*

A scheme for this is described in *Safety-related Instrument Systems for the Process Industries* by the Engineering Equipment and Materials Users Association (EEMUA) (1989 Publ. 160) (the EEMUA *Safety-related Instrument System Guide*). This document complements the HSE *PES Guide* by proving additional guidance specific to the process industries. The background to, and an account of, the scheme is given by W.S. Black (1989).

The starting point is the practice in conventional systems of separating the protective functions from the control functions. Whereas in such systems control functions may be performed by a PES, it has been almost universal practice to use hardwired systems for protective functions.

13.14.2 Categories of system

Four categories of system are defined:

0 Self-acting devices.
1 Non-self-acting devices.
2 System which protects against damage to environment.

Table 13.28 *Categories of control and protective system given in the EEMUA Guide (after EEMUA, 1989) (Courtesy of the Engineering Equipment Manufacturers and Users Association)*

Category	*Type of system*	*Purpose*	*Consequence of failure*	*Requirements*
0[a]	Self-acting device such as PRV, BD, or containment	Safety	Hazard to persons	Relevant BSs
1[b]	Instrumented safety system	Safety	Hazard to persons	*PES Guide;* EEMUA *Guide*
2[c]	Protective system	Economic or environmental	Loss of production or harm to environment	Reliability comparable to conventional analogue systems so that demands on protective devices are limited
3[c]	Control system	Operational	Loss of production and possible demand on Category 0, 1 or 2 system	

[a] Where Category 0 devices are installed and their capability and integrity alone are adequate to ensure safety, Category 1 systems will be unnecessary.
[b] Where mechanical devices cannot be used or are not adequate alone to ensure safety, Category 1 systems will be necessary.
[c] If programmable systems are used for Category 2 or 3 systems, a full assessment of the system according to the *PES Guide* or the EEMUA *Guide* will be unnecessary.

Table 13.29 Selection scheme for control and protective systems given in EEMUA Guide (EEMUA, 1989) (Courtesy of the Engineering Equipment Manufacturers and Users Association)

System	*Self-acting*	*Non-programmable*	*Fixed program*	*Limited variability*	*Full variability*
Ultimate safety, Category 0	Preferred	–	–	–	–
Ultimate safety, Category 1	–	Preferred	Acceptable	Acceptable	Avoid
Protection, Category 2	–	Preferred	Preferred	Acceptable	Avoid
Regulatory, Category 3	–	Acceptable	Acceptable	Preferred	Avoid
Supervisory control	–	–	Avoid	Preferred	Acceptable
Information	–	–	–	Acceptable	Preferred

3 System which ensures reliable production and keeps plant operation with operational limits.

These categories are amplified in Table 13.28.

13.14.3 Categorization process

Assignment of systems to these categories should be made on the basis of a review, involving consideration of the plant line by line. A schedule should be prepared of all the failures which result in excursions outside normal process operating limits. The process conditions after failure should be determined. Cases where the process conditions are unacceptable with respect to safety should be identified. The option of making a modification to eliminate the unacceptable condition should be considered. If it is decided to rely on the instrument system to prevent the unacceptable condition, the system should be listed together with the potential hazard.

The review may be part of a hazop study or it may be separate. A separate review has the advantage that any rethinking can be done outside the hazop. If a separate review is undertaken, the results should be considered in the hazop.

13.14.4 Selection of systems

In selecting a system for a Category 1, 2 or 3 duty, consideration should be given as to whether the system should be programmable or non-programmable. The Institution of Electrical Engineers (IEE) classification of programmable systems recognizes three types:

(1) fixed program system;
(2) limited variability system;
(3) full variability system.

Examples of these three types are a three-term controller which emulates its analogue equivalent, a programmable logic controller (PLC) and a minicomputer. Table 13.29 gives the selection scheme presented in the EEMUA *Guide*.

13.14.5 Review of systems

Once the systems have been selected, the arrangements should be subjected to a review by a team including process engineers, control engineers and operations managers. It should be established that the requirements given in the *PES Guide* for configuration, reliability and quality are met. The EEMUA *Guide* refers to the checklists in the *PES Guide* and gives its own checklists.

13.14.6 Implementation of systems

A Category 0 or 1 system should have the capability and reliability to deal with the foreseeable failure modes and failure frequency of the plant itself and of the Category 2 and 3 systems.

Where a Category 1 system is used, the system should be engineered in accordance with the *PES Guide* and the EEMUA *Guide*. The requirements in these documents relating to hardware, quality and reliability are applicable both to programmable and non-programmable systems. For the latter, however, the requirements relating to software are not applicable.

Failure of a Category 2 or 3 system may put a demand on a Category 0 or 1 system. Where a Category 0 or 1 system is used which is based on a PES, the failure rate should not exceed that of the equivalent conventional system.

13.14.7 Failures in systems

The EEMUA *Guide* gives an account of, and guidance on, the failures which occur in conventional and programmable systems. In conventional systems a single output failure is usually to the zero or low states. This is the mode of failure on loss of air or power. Systems are designed so that the plant goes to a safe state on this failure mode. In such systems the usual assumption is that multiple failures are in the zero or low mode. This is commonly the basis on which relief capacity is sized.

In programmable systems a single output failure will be due to failure in an input or output channel. The failure rate to the high state is unlikely to exceed that in a conventional system. There is potential, however, in a programmable system for multiple failures to the high state due to random hardware failure or systematic software failure. An assessment should be made of the system to ensure that the probability of multiple failure to the high state due to random hardware failure is low. Failures of software may be failures of system software or of applications software. It is rare for system software to be fault free, although a mature system can be expected to contain fewer faults than a new one. The system software for any control system to be used in a safety related application should be evaluated. The alternative means given are formal evaluation and user experience.

Failure in the applications software should be minimized by good software engineering. The EEMUA *Guide* gives guidance on software development and testing.

13.14.8 Loop allocation strategies

There are two basic strategies for loop allocation: (1) outputs distributed and (2) outputs grouped. The principles are that, in the first case, the outputs from a single PES unit are distributed around a number of process units, whereas in the second they are concentrated at a single process unit or at least at a minimum number of units.

If the outputs are distributed, loops which may fail simultaneously are not concentrated on the same process unit. The resultant problem at any given unit will therefore be less severe. In particular, this policy allows the pressure relief system to be designed for a single failure on each unit. On the other hand there may then be a quite large number of process units with some degree of problem. Alternatively, the outputs may be grouped. The problem of multiple failures of loops is then concentrated on one process unit.

The choice between these strategies depends on the characteristics of the process and the probability of multiple failure. A distributed strategy may be suitable for a simple, slow-responding process, but a grouped strategy for a fast-acting process. With a well-designed instrument system, particularly with redundancy, the probability of multiple failure may be small compared with failures of the process unit for other causes.

13.15 CCPS *Safe Automation Guidelines*

13.15.1 *Guidelines for Safe Automation of Chemical Processes*

The safety aspects of process control systems are the subject of *Guidelines for Safe Automation of Chemical Processes* (CCPS, 1993/14) (the CCPS *Safe Automation Guidelines*). The *Safe Automation Guidelines* cover the safety aspects of the whole process control system, including the basic process control system (BPCS), the safety interlock system (SIS) and the human operator. Two types of interlock are distinguished: (1) failure interlocks and (2) permissive interlocks. The distinction corresponds to that used here between trips and interlocks proper.

The headings of the *Guidelines* are: (1) overview; (2) the place of automation in chemical plant safety – a design philosophy; (3) techniques for evaluating integrity of process control systems; (4) safety considerations in the selection and design of BPCSs; (5) safety considerations in the selection and design of SISs; (6) administrative controls to ensure control system integrity; (7) an example involving a batch polymerization reactor; and (8) the path forward. Appendices deal with SIS technologies, separation of the BPCS and SIS, watchdog timer circuits, communications, sensor fail-safe considerations, SIS equipment selection, PES failure modes and factory acceptance test guidelines.

The *Guidelines* are concerned particularly with PES-based SISs. As described earlier, at least until recently the normal approach has been to use for the safety interlock a hardwired system separate from the rest of the control system, whether or not this be computer based. The *Guidelines* describe a design philosophy in which the system of choice for an SIS is a PES-based system. In large part the guidance is concerned with ensuring that a PES-based system has the availability and reliability required for this duty.

This section gives an outline of the *Guidelines*. The latter contain a wealth of practical guidance on the various topics which are touched on here.

13.15.2 Basic design method

The design requirements for the SIS arise out of the process hazard analysis (PHA). The *Guidelines* require that the SIS should be designed by a formal method, but are flexible with respect to the method used. They give a basic design method, which includes what they term a 'qualitative approach to specification' of the safety interlocks required but which allows for the use of alternative quantitative approaches.

The basic design method given in the *Guidelines* is based on the following features:

(1) independent protection layers;
(2) process risk ranking;
(3) safety interlock integrity level specification;
(4) safety interlock integrity level implementation.

This philosophy is outlined in Table 13.30. Section A of the table lists the features which are treated as layers of protection and Section B gives the criteria for a layer or combination of layers to constitute an independent layer of protection (IPL). An IPL protects against a particular type of hazardous event. The event severity and event likelihood are obtained from the process hazard analysis as shown in Section C. The scheme given in Section D indicates the integrity level (IL) required for any safety interlock (SI). There are three integrity levels: Levels 1, 2 and 3 (IL1, IL2 and IL3). As stated in the footnotes, the number of IPLs to be used in the table is the total number of IPLs, including the safety interlock being classified. The implementation of a safety interlock system of specified integrity level is indicated in Section E and an example of the determination of the integrity level of a safety interlock is given in Section F.

13.15.3 Evaluation of control system integrity

The *Guidelines* review the various safety and integrity evaluation techniques applicable to process control systems, including under the qualitative techniques operating experience, standards and codes, design guidelines, checklists, What-If analysis, failure modes and effects analysis (FMEA) and hazop and under the quantitative techniques trip capability analysis, fault tree analysis, event tree analysis, reliability block diagrams, Markov models, Monte Carlo simulation, non-destructive fault insertion testing and QRA.

The BPCS and SIS should both be certified, either by self-certification or third party certification. For PES devices three maturity levels are recognized: user-approved (for BPCS), user-approved safety (UAS) (for SIS) and user-obsolete. The *Guidelines* give criteria for user approvals.

Table 13.30 Basic design philosophy of the safety interlock system in the CCPS Safe Automation Guidelines (CCPS, 1993/14) (Courtesy of the American Institute of Chemical Engineers)

A Layers of protection

1. Process design
2. Basic controls, process alarms, operator supervision
3. Critical alarms, operator supervision and manual intervention
4. Automatic SIS
5. Physical protection (relief devices)
6. Physical protection (containment dikes)
7. Plant emergency response
8. Community emergency response

B Criteria for independent layers of protection

The criteria for a protection layer or a combination of protection layers to qualify as a independent protection layer (IPL) are:

1. The protection provided reduces the identified risk by a large amount, that is, at least by a 100-fold reduction
2. The protective function is provided with a high degree of availability – 0.99 or greater
3. The protection has the following characteristics:
   – Specificity. An IPL is designed solely to prevent or to mitigate the consequences of one potentially hazardous event (e.g. a runaway reaction, release of toxic material, a loss of containment, or a fire). Multiple causes may lead to the same hazardous event and therefore multiple event scenarios may initiate action of one IPL
   – Independence. An IPL is independent of the other protection layers associated with the identified danger
   – Dependability. It can be counted on to do what it was designed to do. Both random and systematic failure modes are addressed in the design
   – Auditability. It is designed to facilitate regular validation of the protective functions. Functional testing and maintenance of the safety system is necessary

C Process risk ranking

An event is assigned a severity and a likelihood:

Event severity

Minor incident	Impact initially limited to local area of the event with potential for broader consequences if corrective action is not taken
Serious incident	One that could cause: – Any serious injury or fatality on site or off site – Property damage of \$1 million offsite or \$5 million onsite
Extensive incident	One that is five more times worse than a serious incident

Event likelihood

Low	A failure or series of failures with a very low probability of occurrence within the expected lifetime of the plant ($<10^{-4}$ failures/year). Examples: (1) three or more simultaneous instrument, valve or human failures; (2) spontaneous failure of single tanks or process vessels
Moderate	A failure or series of failures with a very low probability of occurrence within the expected lifetime of the plant (10^{-4} to 10^{-2} failures/year). Examples: (1) dual instrument failures; (2) combination of instrument failures and operator errors; (3) single failures of small process lines or fittings
High	A failure can reasonably be expected to occur within the expected lifetime of the plant ($>10^{-2}$ failures/year). Examples: (1) process leaks; (2) single instrument or valve failures; (3) human errors that could result in material releases

The event is designated low risk for any of the following combinations: (1) severity low, likelihood low; (2) severity serious, likelihood low; (3) severity low, likelihood moderate. It is designated high risk for any of the following combinations: (1) severity extensive, likelihood high; (2) severity serious, likelihood high; (3) severity extensive, likelihood moderate. It is designated moderate risk for the other three combinations

D Safety interlock integrity level specification[a]

	Event severity								
	Minor			*Serious*			*Extensive*		
Event likelihood	Low	Moderate	High	Low	Moderate	High	Low	Moderate	High
No. of IPLs[b] 3	(5)	(5)	(5)	(5)	(5)	(5)	(5)	1	1
2	(5)	(5)	1	(5)	1	2	1	2	3 (2)
1	1	1	3	1	2	3 (2)	3 (2)	3 (2)	3 (1)

[a] The values in the table without brackets refer to the integrity level (IL) required; the values in brackets refer to the number of the note given below.
[b] Total number of IPLs, including the safety interlock being classified.

Notes:
1. One Level 3 safety interlock does not provide sufficient risk reduction at this risk level. Additional PHA modifications are required.
2. One Level 3 safety interlock may not provide sufficient risk reduction at this risk level. Additional PHA review is required.
3. Event likelihood – likelihood that the hazardous event occurs without any of the IPLs in service (i.e. the frequency of demand).
4. Event likelihood and total number of IPLs are defined as part of the PHA team work.
5. SIS IPL is probably not needed.

Integrity level availability

	Availability (%)
Level 1	about 99
Level 2	99–99.9
Level 3	up to 99.9–99.99

E Safety interlock integrity level implementation

Integrity level (IL)	Minimum interlock design structure
1	Non-redundant: best single path design
2	Partially redundant: redundant independent paths for elements with lower availability
3	Totally redundant: redundant, independent paths for total interlock system. Diversity should be considered and used where appropriate. A single fault of an SIS component is highly unlikely to result in a loss of process protection

F Illustrative example

Event severity	Extensive
Event likelihood without benefit of either IPL	Moderate
Total number of IPLs (non-SIS IPL + SIS interlock)	2
Required SIS interlock integrity level	2

13.15.4 Basic process control system

The basic process control system is not usually an IPL. It is, nevertheless, the next line of defence after the process design and has an important part to play. The *Guidelines* therefore deal with the safety considerations in the selection and design of the BPCS. The account given covers (1) the technology selection, (2) the signals, (3) the field measurements, (4) the final control elements, (5) the process controllers, (6) the operator/control interfaces, (7) communication considerations, (8) electrical power distribution systems, (9) control system grounding, (10) batch control, (11) software design and data structures and (12) advanced computer control strategies, and contain much practical material on these features.

The *Guidelines* advise that use of a supervisory computer should be subject to a discipline which restricts it to manipulation of loop set points. It should not normally be able to change the operational mode of the loops except for transfer to the back-up mode on computer failure or to computer mode on initialization. It should not compromise the integrity of the back-up controls.

The design philosophy of the *Guidelines* requires that the BPCS and the SIS should be separate systems. The BPCS should not be relied on to protect against unsafe process conditions. The integrity of the SIS should not be compromised by the BPCS. Appendix B of the *Guidelines* gives detailed guidance on separation.

13.15.5 Safety interlock system

As far as concerns safety considerations in the selection and design of the SIS, the *Guidelines* cover (1) the design issues, (2) the requirements analysis, (3) the technology selection, (4) the architecture selection, (5) the equipment selection and (6) the system design. The necessary preliminaries are to determine the need for safety interlocks and to establish their integrity levels.

Design issues

Design issues are of two main kinds: function and integrity. Issues concerning function include the parameters to be monitored, the trip actions to be taken, the testing facilities and policy. Among those bearing on integrity are the number of integrity levels required, which affects the choice of technology.

Some specific design issues addressed in the *Guidelines* are (1) the fail-safe characteristics, (2) logic structures, (3) fault prevention and mitigation, (4) separation of the BPCS and SIS, (5) diversity, (6) software considerations, (7) diagnostics, (8) the human/machine interface and (9) communications.

The fail-safe issue involves the choices de-energize-to-trip vs energize-to-trip. There is also the question of the failures modes of PES-based devices. Even at the chip level probable states are equally likely to be on or off. The problem is even more severe at the level of a PES-based device. Effectively the *Guidelines* suggest alternative approaches based on use of equipment of proven reliability, capable of self-diagnosis and of proof-testing, with judicious use of redundancy.

There should be a separation between the SIS and BPCS such as to ensure the integrity of the former. Conventional SISs have long utilized separate sensors and power supplies. The *Guidelines* bring into consideration also the input/output system, the software and the human/machine interface.

Diagnostics may be used to detect fail-to-danger failures in the safety interlock equipment including the sensor, the logic solvers, the final control elements and the energy sources. The *Guidelines* distinguish between passive and active diagnostics. In passive diagnostics the failure is revealed only when a demand is imposed, either by the system or by a user test. In active diagnostics the device is subjected continuously to testing by input of out-of-range conditions and its response monitored, but over a time interval short enough not to upset the safety interlock loop. The example quoted is the perturbation of a solenoid valve on a control valve with sufficient rapidity that the control valve is not affected.

Requirements analysis

The requirements analysis determines the targets for availability and reliability (or functional and operational reliability).

Technology selection

The SIS technologies given in the *Guidelines* include: (1) fluid logic (pneumatic, hydraulic); (2) electrical logic, including direct-wired systems, electromechanical devices (relays, timers), solid state relays, solid state logic and motor-drive timers; (3) PES technology, involving programmable logic controllers (PLCs) and distributed control systems (DCSs); and (4) hybrid systems. The technologies are detailed in Appendix A of the *Guidelines*.

The hardware of a typical SIS as envisaged in the *Guidelines* might consist of a logic solver with input modules receiving sensor signals, output modules sending out signals to final control elements, a BPCS interface, a human/machine interface and an engineer's interface.

Architecture selection

Under architecture selection the *Guidelines* discuss the various ways of achieving an integrity appropriate for the integrity level determined. Thus for IL1 redundancy is usually not necessary, though it may be appropriate for a lower reliability element. For IL3, on the other hand, there should be full redundancy. Other features mentioned for IL3 are use of analogue sensors so that active diagnostics can be practised, monitoring of the logic solver outputs by the BPCS and consideration of the use of diversity in the sensors. For both high availability and high reliability use may be made of a triple modular redundant (TMR) system, or 2/3 voting system.

Equipment selection

The equipment selected for a PES-based SIS should be of user-approved safety.

System design

The basic design method for the SIS has already been described, but the design involves more than this. The design should allow for the special features of PES-based systems. One of these, the difficulty of determining fail-safe states, has already been mentioned. Another feature is false 'turn-ons' of inputs or outputs.

Another problem in PES-based systems is that the life of a given version of the software is relatively short so that the version initially used is liable to become out of date and, after a time, no longer supported by the vendor. The problem then arises that insertion of an updated version constitutes a software modification, with all that that entails. There are various approaches to the problem, none entirely satisfactory.

The design should take into account the potential impacts of the SIS on the other components of the process control system, including the alarm system, the communications system and the human/machine interfaces.

Most process control systems involve some sequential control even if it is largely limited to start-up and shut-down. The sequential logic should operate in such a way as not to cause any safety problems. Its operation should be tested against the safety interlock logic to ensure that normal operation of the sequential control does not trigger interlock action.

The documentation for the SIS specified in the *Guidelines* includes (1) the operational description, (2) the schematic diagrams, (3) the binary logic diagrams and (4) the single line diagrams. Examples are given of these different types of diagram.

13.15.6 Administrative actions

In order to ensure the control system integrity the design process just described needs to be supported by administrative actions The *Guidelines* outline minimum procedural requirements, the scope of which includes (1) operating procedures, (2) maintenance facilities, (3) testing of the BPCS, (4) testing of the SIS and alarms, (5) test frequency requirements, (6) testing facilities, (7) operations training, (8) documentation, and (9) auditing of maintenance and documentation.

The test frequency indicated in the *Guidelines* for SIS functional testing is, for minimal risk systems, testing once every 2 years or at major turnarounds, whichever is more frequent, and for high risk systems, testing at least once a year or on major maintenance, whichever is more frequent.

13.16 Emergency Shut-down Systems

In a quite large proportion of cases, the plant is provided not just with individual trips but with a complete automatic emergency shut-down (ESD) system. There is relatively little written about ESD systems. One of the principal accounts is that given in *Offshore Installations: Guidance on Design and Construction, Guidance Notes to the Offshore Installations (Construction and Use) Regulations 1974* issued by the Department of Energy (1984) followed now by *Offshore Installations: Guidance on Design, Construction and Certification* (HSE, 1990b) (the HSE *Design, Construction and Certification Guidance Notes*).

13.16.1 Conceptual design of ESD

The function of an ESD system is to detect a condition or event sufficiently hazardous or undesirable as to require shut-down and then to effect transition to a safe state. The potential hazards are determined by a method of hazard identification such as hazop. Estimates are then made of the frequency and consequences of these hazards. The hazards against which the ESD system is to protect are then defined.

This protection is effected by identifying the operating parameters which must be kept within limits if realization of the hazards is to be avoided and selecting shut-down actions which will achieve this. A shut-down sequence is determined and the shut-down logic formulated. It is not always necessary to shut down the whole plant and there are different levels of ESD which fall short of this, such as shut-down of an individual unit or of a section of plant.

13.16.2 Initiation of ESD

The arrangements for initiation of the ESD are critical. If these are defective, so that the system is not activated when it should be, all the rest of the design goes for nothing. There is a balance to be struck between the functional and operational reliability of the ESD system. It should act when a hazard arises, but should not cause unnecessary shut-downs or other hazards.

One factor which affects this balance is the fact that usually the plant is safest in the normal operating mode and that transitions such as shut-down and start-up tend to be rather more prone to hazards and are to be avoided unless really necessary. Another, related factor is that shut-down of one plant may impose shut-down on other, linked plants.

Initiation may be manual, automatic or, more usually, both. The usual arrangement is a manual initiation point, or shut-down button, in the control centre, other manual initiation points located strategically throughout the plant and initiation by instrumentation. Such automatic initiation may be effected by the fire and gas system and/or by process instruments. Measures should be taken to avoid inadvertent activation, including activation during maintenance and testing.

13.16.3 Action on ESD

There are a variety of actions which an ESD system may take. Three principal types are:

(1) flow shut-off;
(2) energy reduction;
(3) material transfer.

Flow shut-off includes shut-off of feed and other flows. It often involves shut-down of machinery and may include isolation of units. Energy reduction covers shut-off of heat input and initiation of additional cooling. Material transfer refers to pressure reduction, venting and blow-down.

A fundamental principle in ESD is failure to a safe state. The overall aim is failure to a safe state for the system as a whole. This is normally effected by applying the principle to individual units, but there may be exceptions, and cases should be considered individually. Each required action of the ESD system should be effected by positive means. Reliance should not be placed on the cascading effect of other trip actions.

13.16.4 Detail design of ESD system

It is a fundamental principle that protective systems be independent of the rest of the instrument and control system, and this applies equally to an ESD system. The design of the ESD system should follow the principles which apply to trip systems generally, as described in Section 13.9. There should be a balance between functional and operational reliability. Dependent failures should be considered. The reliability may be assessed using fault tree and other methods. The techniques of diversity and redundancy should be used as appropriate. Use may be made of majority voting systems.

The emergency shutdown valves (ESVs) should have a high degree of integrity. Such valves are frequently provided with pneumatic or hydraulic power supplies in addition to electrical power supply. An ESV should be located so as that it is unlikely to be disabled by the type of incident against which it is intended to protect.

The ESD system should be provided with power supplies which have a high degree of integrity. The normal approach is to provide an uninterruptible power supply. This supply should be designed and located so that it is unlikely to be disabled by the incident itself. The cables from the power supply to the final shut-down elements should be routed and protected to avoid damage by the incident.

13.16.5 Operation of ESD system

The status of the ESD system should be clear at all times. There should be a separate display showing this status in the control centre. This display should give the status of any part of the EDS system which is under test or maintenance and of any part which is disarmed. Initiation of ESD should activate audible and visual

Table 13.31 *Elements of three candidate protective systems for a large steam boiler plant (after Hunns, 1981) (Courtesy of Elsevier Science Publishers)*

Control and protective features	*Protective system design*[a]		
	Manual	*Medium automated*	*Highly automated*
Trip parameters[b]	A	A	A
Boiler purge sequence	M	A	A
Burner flame failure detection	M	A	A
Gas valves leak test	M	M	A
Ignition burner control	M	M	A
Burner fuel valves operation	M	A	A

[a] A, automated; M, manual.
[b] Low boiler drum level; low combustion air flow; low instrument air pressure; low fuel oil pressure; low atomizing steam pressure; low fuel gas pressure; high fuel gas pressure; high knockout drum level; and loss of 110 V DC supplies.

alarms in the control centre. There should be an indication of the source of the initiation, whether manual or instrument. ESD should also be signalled by an alarm which is part of the general alarm system.

It may be necessary in certain situations such as start-up, changeover or maintenance to disarm at least part of the ESD system, but such disarming should be governed by formal arrangements. The principles are essentially similar to those which apply to trip systems generally, as described in Section 13.9.

13.16.6 Testing and maintenance of ESD system

The ESD system should be subject to periodic proof testing and such testing should be governed by a formal system. The principles of proof testing were discussed in Section 13.9. As far as is practical, the test should cover the complete system from initiation to shut-down condition.

The need for proof testing and, more generally, for the detection of unrevealed failure should be taken into account in the design. The equipment should be designed for ease of testing. It should be segregated and clearly identified. Techniques for detection of instrument malfunction should be exploited. In voting systems, the failure of a single channel should be signalled.

13.16.7 Documentation of an ESD system

The ESD system should be fully documented. The HSE *Design, Construction and Certification Guidance Notes* give details of recommended documentation.

13.16.8 ESD of a gas terminal

The design of systems for ESD and emergency depressurization (EDP) of a gas terminal has been described by Valk and Sylvester-Evans (1985). The design philosophy described is that the ESD system should operate only in an extreme emergency, that the ESD and EDP systems are separate from the control, trip and relief systems, and that the systems should be simple and reliable.

The preliminary design of the ESD and EDP systems was reviewed by means of a hazop study. Potential operational failures were studied using general reliability engineering methods and functional failures were studied using, in particular, fault tree analysis. A further hazop was conducted on the final design for the ESD and EDP system.

Design studies showed that a totally fail-safe concept would result in a relief and flare system of exceptional size. Alternatives considered were to allow an increase in the depressurization time for certain critical equipment beyond that recommended in the codes and to control the peak depressurization flow in the relief and flare system. In the design adopted, the plant was divided into sections such that the depressurization of each section could be done independently and the operation of the sections was interlocked. The depressurization time of certain items was extended to 30 minutes as opposed to the 15 minutes recommended in API 521, but the design compensated for this extension by provision of additional fireproofing and water cooling arrangements.

The authors highlight the differences of philosophy between companies on whether the ESD and EDP systems should be used for normal shut-down and depressurization or reserved as systems dedicated for emergency use. This project reaffirmed the need to consider the ESD and EDP systems at an early stage and to avoid treating them as an 'add-on' feature to be dealt with late in the design.

13.16.9 ESD on Piper Alpha

The ESD system on Piper Alpha illustrates a number of the points just made. Overall, the system was largely effective in achieving shut-down and venting and blow-down, but there were a number of features which are of interest.

The main button for the initiation of the ESD caused closure of the ESV on the main oil pipeline but not on the three gas pipelines. One reason for this was that closure of these latter ESVs would impose a forced shut-down on the linked platforms. There were three separate shut-down buttons, one for each of these valves, and shut-down depended on manual action by the control room operator – he was thrown across the control room by the explosion.

The ESVs on the risers of the gas pipelines were so located that they were vulnerable to the fires that developed. This defect was widespread throughout the North Sea and regulations were introduced without delay to require such valves to be relocated. There was also evidence that some of the ESVs did not achieve tight

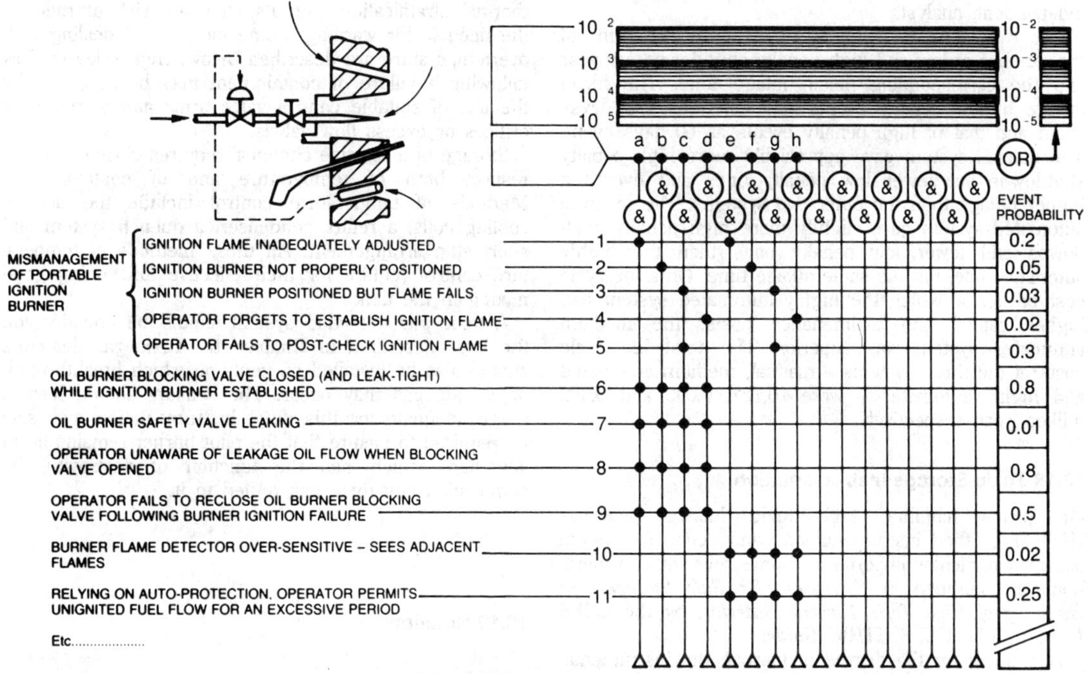

Figure 13.26 *Matrix logic diagram for a steam boiler protective system: event 'excessive unignited fuel release' (Hunns, 1981) (Courtesy of Elsevier Science Publishers)*

shut-off. The explosion damaged power supplies, and in some cases closure of ESVs occurred, not due to survival of the intended power supply, but fortuitously. Further details are given in Appendix 19.

13.17 Level of Automation

The allocation of function between man and machine is a principal theme in human factors and is discussed in the next chapter. Of particular interest here is the allocation of control and protective functions to the process operator or the instrument system. This was touched on above and is now considered in more detail by means of a industrial example.

13.17.1 Illustrative example: steam boiler protective system

A case study of the optimum level of automation has been described by Hunns (1981). The system investigated was the protective system of a large steam plant. The plant consisted of a 100 MW boiler operating at 1500–2000 psi and producing 500 ton/h of steam, the boiler being dual fired with oil and gas. The principal relevant features of the three candidate control and protective systems are shown in Table 13.31.

Each system was assessed for its reliability in the start-up and operational phases of the plant. A variety of start-up sequences were considered, each related to the event which had caused the previous shut-down. Some 200 logic diagrams were produced.

The criterion used to determine the optimum system was a function of the expected shut-downs. These were classified as low penalty and high penalty. Low penalty shut-downs were unwanted shut-downs due to spurious trips and correct shut-downs in response to a demand, whilst high penalty, or catastrophic, shut-downs were those caused by a demand to which the protective system did not respond.

One such case is an excessive release of unignited fuel into the combustion chamber. Figure 13.26 shows a section of the logic, which the authors term 'matrix logic'. Events envisaged by the analyst are shown in the left-hand column. A particular event sequence is shown by a vertical column of the matrix containing one or more dots. The circle enclosing the & symbol at the head of the column indicates that the events are ANDed together. The set of event sequences is collected under the OR symbol, which indicates that these event sequences are related to the top event, the unignited release, by OR logic.

The equipment failure data were taken from the Safety and Reliability Directorate SYREL data bank, whilst the human error estimates were obtained by expert judgement. There were some 180 elements in the latter list. Estimates of the values in the list were made by two experienced analysts. Use was made of performance shaping factors such as 'time to react', 'prior expectancy', 'conspicuity of task' and 'perception of consequence'. Good agreement was obtained between the two. Those estimates which were particularly critical or where a divergence had emerged were mediated by a third, independent, analyst.

The results of the study were expressed in terms of the number of low and high penalty shut-downs per year and the corresponding mean outage time. The mean outage of low penalty events was taken as 1/3 days/event and that of high penalty events as 60 days/event. The manual system gave appreciably more high penalty shutdowns but fewer low penalty ones and overall a higher outage than the other two systems. The medium automated system gave slightly more high penalty shutdowns and fewer low penalty ones than the highly automated one, but the same outage time. On a life cycle cost basis, for which the highly automated system had higher capital and maintenance costs, the medium automated system was superior. The total life cycle costs of the three systems – manual, medium automated and highly automated – were £0.122, 0.095 and 0.099 million/year, respectively.

13.18 Toxic Storage Instrumentation

On plants handling high toxic hazard materials (HTHMs), the instrumentation and control system assumes particular importance. Some relevant considerations are outlined in *Guidelines for Safe Storage and Handling of High Toxic Hazard Materials* by the CCPS (1988/2) (the CCPS HTHM *Guide*).

Depending on the degree of hazard, the instrumentation and control system should be a high integrity one. This requires adherence to the various principles already described for high integrity design, including the application of principles such as fail-safe and second chance design and the use, as appropriate, of high reliability instrumentation, instrument diversity and redundancy and high quality maintenance. It also involves the application of the techniques of hazard identification and assessment to the design.

In respect of measurement, principal considerations are that the potential for release from the instrument, or its fittings, should be minimized, that the instrument be reliable and that the measurement be accurate. For flow measurement this favours the use of non-invasive sensors such as magnetic flowmeters and avoidance of glass in instruments such as rotameters. Orifice flowmeters also have the disadvantage of an extra flange and associated piping. For pressure measurement, diaphragm pressure sensors are preferred to direct-connected gauges of the Bourdon tube type. Precautions to be taken where the latter have to be used include protection by inert liquid filling in corrosive service and installation of shut-off valves and, possibly, flow limiters in the form of restriction orifices. For level measurement, weighing methods have advantages, but use of sightglasses should be avoided. For temperature measurement, particular care should be taken in the design of the thermowell, which can be a weak point.

The arrangements for control and protection should address the hazards of particular importance for the storage of toxics. These include (1) overpressure, (2) overfilling, (3) overtemperature and (4) high release flow. For overpressure, the main requirements are the provision of overpressure protection and of means of disposal for the relief flows. For overfilling, a significant role is likely to be played by trip systems. For temperature deviations, which may indicate reaction runaway or thermal stratification with its attendant risk of rollover, the need is for warning. Some methods of dealing with overtemperature are described below. High releases flow following a failure of containment may be mitigated by the use of suitable control valve trims and of restrictor orifices or excess flow valves.

Storage of a reactive chemical requires close control in respect both of temperature and of contamination. Methods of temperature control include the use of cooling coils, a reflux condenser, a quench system and short stop arrangement. All these methods of temperature control require for their effective functioning good mixing in the tank.

A toxic gas detection system should be provided, on the lines described in Chapter 18. Toxic gas detectors should also be installed on vents on which breakthrough of a toxic gas may occur. The sensors should have a range adequate for this duty. Instrumentation may also be required to ensure that the pilot burner remains lit on any flare which has the function of destroying by combustion any toxic gas routed to it.

13.19 Notation

Section 13.6

P_d	delivery pressure
P_s	suction pressure
ΔP	pressure drop
Q	gas flow

Section 13.9

f_η	density function for a plant hazard occurring
m	number of equipments which must survive for trip system to survive
n	number of identical equipments
p_η	probability of a plant hazard occurring
r	number of equipments which must fail for trip system to fail
t	time
γ	spurious trip rate
δ	plant demand rate
η	plant hazard rate
λ	equipment failure rate
τ_p	proof test interval
τ_r	repair time
ϕ	fractional dead time (with simultaneous testing)

Subsections 13.9.1–13.9.13

p_δ	probability of a plant demand occurring
q	probability of a single channel failing
q_γ	probability defined by Equation 13.9.37
τ_d	disarmed time
τ_0	dead time
ϕ^*	fractional dead time (with staggered testing)
ϕ_{is}	isolation dead time

Subscripts:

m/n	for an *m/n* (*m*-out-of-*n*) system
min	minimum

Subsection 13.9.14

P_n	probability that system is in state n
r_1, r_2	terms defined by Equation 13.9.53

Subsection 13.9.16

A	heat transfer area of sensor
c_p	specific heat of sensor
k	constant
M	mass of sensor
s	Laplace operator
t	time
T	temperature
T_i	input temperature
U	overall heat transfer coefficient of sensor
$\delta(t)$	delta function
ζ	damping factor
θ	temperature deviation
θ_i	input temperature deviation
τ	time constant of sensor
ω_n	natural frequency

Subscripts:

i	input
ss	steady-state
1,2	first,second stage

Superscript:

–	Laplace transform

Subsection 13.9.17

C	annualized cost of a single channel trip
G	cost of genuine trip
H	cost of realization of hazard
S	cost of spurious trip
V	overall annual cost
β_1	functional beta value
β_2	operational beta value

Subscripts:

s	trip system
1/2	1/2 system

14 Human Factors and Human Error

It is appropriate at this point to deal with the topic of human factors, and one important aspect of that, human error. Human factors considerations are relevant to all aspects of the design and operation of process plants. The topic is, however, a vast one and the account given here is necessarily limited. The approach taken is to consider in particular the process operator and then to touch on certain other aspects such as communications, maintenance and construction.

Overviews of human factors are given in *Biotechnology* (Fogel, 1963), *Human Factors Engineering* (McCormick, 1957b), *Human Engineering Guide to Equipment Design* (C.T. Morgan *et al.*, 1963), *Human Engineering Guide for Equipment Designers* (Woodson and Conover, 1964), *Human Factors Evaluation in System Development* (Meister and Rabideau, 1965), *Ergonomics* (Murrell, 1965a), *Human Performance in Industry* (Murrell, 1965b), *Human Factors* (Meister, 1971), *Handbook of Human Factors* (Salvendy, 1987), *Applied Ergonomics Handbook* (Burke, 1992 ACGIH/76) and *Human Factors in Design and Engineering* (Sanders, 1993). Accounts of human factors with specific reference to safety are given in *Ergonomics Guides* (AIHA, 1970–/1), *Human Aspects of Safety* (Singleton, 1976b) and *Human Factors in Process Operations* (Mill, 1992). HSE guidance is given in HS(G) 48 *Human Factors in Industrial Safety* (HSE, 1989).

Selected references on human factors are given in Table 14.1 and references on the process operator are given in Table 14.2.

Table 14.1 *Selected references on human factors*

IOHSI (Inf. Sht 15); NRC (Appendix 28 *Human Factors*); Chapanis, Garner and Morgan (1949); Crossman (1956); McCormick (1957, 1976); Institute of Personnel Management (1961); Bennett, Degan and Spiegel (1963); Fogel (1963); C.T. Morgan *et al.* (1963); E. Edwards (1964, 1973); Woodson and Conover (1964); Chapanis (1965); Meister and Rabideau (1965); Murrell (1965a,b); Edholm (1967); Sackman (1967, 1970); Singleton, Easterby and Whitfield (1967); Kelley (1968); Hands (1969); Hurst (1969); Ragsdale (1969); Siegel and Wolf (1969); Bongard (1970); De Greene (1970, 1974); Meister (1971, 1987); Poulton (1971); Sayers (1971 SRS/GR/9); Singleton, Fox and Whitfield (1971); E. Edwards (1973); E. Edwards and Lees (1973, 1974); R.G. Mills and Hatfield (1974); Christensen (1976); Singleton (1976a,b); Towill (1976); Welford (1976); S. Brown and Martin (1977); Jennings and Chiles (1977); Rouse and Gopher (1977); Craig (1978); Jagacinski and Miller (1978); Stammers (1978); V.R. Hunt (1979); Kiguchi and Sheridan (1979); Anon. (1980i); Shackel (1980); McCormick and Sanders (1982); M.S. Sanders (1982–, 1993); J.C. Williams (1982); S. Cox and Cox (1984); Helmreich (1984); Nordic Liaison Committee (1986 NKA/LIT(85)1); L.E. Davis and Wacker (1987); Goldstein (1987); Lehner and Zirk (1987); Parasuraman (1987); Salvendy (1987); Sanders and McCormick (1987); Swain (1987c); de Vries-Griever and Meijman (1987); Wiener (1987); Woodson (1987); Holloway (1988); Grollier-Barron (1989); HSE (1989b); Bond (1990 LPB 92); Broadbent, Reason and Baddeley (1990); J.R. Wilson and Corlett (1983); A.F. Sanders (1991); Mill (1992); Moraal (1992); Needham (1992)

Allocation of function
Fitts (1962); Ephrath and Young (1981); Bainbridge (1983); H.E. Price (1985); Kantowitz and Sorkin (1987); Kletz (1987b); Swain (1987b); J. Lee and Moray (1992)

Human information processing, including decision-making, control tasks
Wald (1947); Heider (1958); Newell, Shaw and Simon (1960); Sinaiko (1961); Senders (1964); W. Edwards (1965); Zadeh (1965); Dreyfus (1972); Newell and Simon (1972); Kelley (1973); Ince and Williges (1974); Levine and Samet (1974); Pew (1974); Sheridan and Ferrell (1974); Tversky and Kahneman (1974); McLeod and McCallum (1975); J. Anderson (1976); W. Edwards and Tversky (1976); Gaines (1976); Keeney and Raiffa (1976); Senders and Posner (1976); Gaines and Kohout (1977); Janis and Mann (1977); Rouse (1977); Schank and Abelson (1977); Jagacinski and Miller (1978); Kochhard and Ali (1979); Feigenbaum (1979); Ringle (1979); Hammond, McClelland and Mumpower (1980); Cuny and Boy (1981); Barnett (1982); Baron *et al.* (1982); Rasmussen and Lind (1982); Rasmussen (1983); Eberts (1985); Friedman, Howell and Jensen (1985); Sayers (1988); Moray *et al.* (1991); Thimbleby (1991); Bainbridge (1993b); Bainbridge *et al.* (1993); van der Schaaf (1993); Sünderstrom (1993)

Vigilance
Levine and Samet (1974); Wiener (1974, 1987); Kvalseth (1979); Mackie (1977); Loeb (1978); Craig (1979, 1980, 1981, 1987); Curry (1981); Fisk and Schneider (1981); Kessel and Wickens (1982); Anon. (1984s); Wiener (1987); Wogalter *et al.* (1987)

Inspection tasks
Mackenzie (1958); Drury and Addison (1973); Drury and Fox (1975); Yao-Chung Tsao, Drury and Morawski (1979); Gallwey (1982); Geyer and Perry (1982); Drury and Sinclair (1983)

Displays, controls, display–control relations
Sleight (1948); Murrell (1952a,b, 1965a); Sinaiko (1961); Loveless (1962); Fogel (1963); Ziebolz (1964); Carbonell (1966); H.H. Bowen (1967); Meister (1967); Kelley (1968); Luxenberg and Kuehn (1968); Singleton (1969); Bainbridge (1971); Bernotat and Gartner (1972); Ince and Williges (1974); Jacob, Egeth and Bevan (1976); Seeberger and Wierwille (1976); Sheridan and Johanssen (1976); Curry, Kleinman and Hoofman (1977); Potash (1977); Demaio, Parkinson and Crosby (1978); B. Gibson and Laios (1978); Cakir, Hart and Stewart (1980); Penniall (1980); Petropoulos and Brehner (1981); Goodstein (1984); Kautto *et al.* (1984); Bullinger, Kern and Muntzinger (1987); Downing and Sanders (1987); Helander (1987); Triggs (1988); Whalley (1989); Buttigieg and Sanderson (1991); Dillon (1992)

Diagnostic tasks
Dale (1958, 1964, 1968); Shriver, Fink and Trexler (1964); Tilley (1967); Rasmussen and Jensen (1973, 1974); Gai and Curry (1976); Brooke and Duncan (1980a,b, 1981, 1983a,b); Bond (1981); Brehmer (1981); Brooke (1981); Freedy and Lucaccini (1981); Gaddes and Brady (1981); R.M. Hunt and Rouse (1981); Leplat (1981); Moray (1981); Patrick and Stammers (1981);

Sheridan (1981); Syrbe (1981); W.B. Johnson and Rouse (1982); Rouse and Hunt (1982); Brooke, Cook and Duncan (1983); N.M. Morris and Rouse (1985); Patrick *et al.* (1986); Toms and Patrick (1987, 1989); Carlson, Sullivan and Schneider (1989); Brinkman (1993); Patrick (1993); Reinartz (1993); Schaafstal (1993)

Emergency actions
Haas and Bott (1982); Lester and Bombaci (1984)

Shiftwork
HSE (1978 Research Review 1, 1992 CRR 31); Folkard and Monk (1979); Akerstedt and Torsvall (1981); Tilley *et al.* (1982); Folkard and Condon (1987); Costa *et al.* (1989); Folkard (1992); Wedderburn (1992); Wilkinson (1992a,b); Rosa and Bonnet (1993)

Workload, mental load, stress, fatigue, boredom
Bartlett (1943); Berkun *et al.* (1962); Berkun (1964); Bainbridge (1978); Goldstein and Dorfman (1978); Leplat (1978); Welford (1978); Moray (1979); R.P. Smith (1981); Sharit and Salvendy (1982); Anon. (1984ss); Braby, Harris and Muir (1993); Dörner and Pfeifer (1993); Gaillard (1993); HSE (1993 CRR 61)

Drink, drugs
NTSB (annual reports); HSE (1981 OP 1)

Task sharing
Goldstein and Dorfman (1978); Cellier and Eyrolle (1992)

Teamwork
D.P. Baker and Salas (1992); Driskell and Sals (1992); Reinartz (1993); Rogalski and Samurcay (1993)

Personnel selection
Mallamad, Levine and Fleishman (1980); Gallwey (1982); Osburn (1987); HSE (1993 CRR 58)

Training
Holding (1965, 1987); Boydell (1970, 1976); Patrick (1975, 1992); API (1977 Publ. 756, 1979 Publ. 757); S.L. Johnson (1981); Nawrocki (1981); Wickens and Kessel (1981); Svanes and Delaney (1981); Towne (1981); W.B. Johnson and Rouse (1982); Kessel and Wickens (1982); Buch and Diehl (1984); Gagné (1985); Cannon-Bowers *et al.* (1991); Bainbridge (1993a); Kozak *et al.* (1993)

Task analysis
Annett *et al.* (1971); Leplat (1981); Piso (1981); Drury (1983); Kirwan and Ainsworth (1992)

Operating instructions
S. Jones (1968); Booher (1975); Kammann (1975); Anon. (1976 LPB 8, p. 19); Duffy, Curran and Sass (1983); Krohn (1983); Berkovitch (1985); Hartley (1985); DTI (1989); R.C. Parker (1988); Chung-Chiang Peng and Sheue-Ling Hwang (1994)

Computer aids
Mitter (1991); Marmaras, Lioukas and Laios (1992); Adelman *et al.* (1993); Kirlik (1993)

Air traffic control systems
Whitfield, Ball and Ord (1980); Wiener and Curry (1980)

Formal safety assessment (FSA)
Bellamy, Kirwan and Cox (1986); Bellamy and Geyer (1991)

Evacuation
Bellamy and Harrison (1988)

Offshore (see *Table A18.1*)

Table 14.2 *Selected references on the process operator, including fault administration and computer aids (see also Table A14.1)*

NRC (Appendix 28 *Operating Personnel, Simulators, Training*); Riso National Laboratory (Appendix 28); Crossman (1960); W.R. King (1965); Rasmussen (1968a–c, 1969, 1971, 1973, 1974, 1976a,b, 1977, 1981a,b, 1983, 1986, 1988); Beishon (1969); Lees (1970, 1974c); E. Edwards and Lees (1971a,b, 1973, 1974); E. Edwards (1973); IEE (1975 Coll. Dig. 75/12); Purdue Europe (1977); A. Shepherd (1979, 1993); SN (1979 123); Lenior, Rijnsdorp and Verhagen (1980); Bainbridge (1981); Ergonomics Society (1983); Boel and Daniellou (1984); Brouwers (1984); Daniellou and Boel (1984); EPRI (1984); IChemE (1984/77); Kragt and Daniels (1984); S.B. Gibson (1987); Hoyos (1987); Bellamy and Geyer (1988); Sanderson, Verhage and Fuld (1989); J.R. Wilson and Rutherford (1989); ACSNI (1990, 1991, 1993); Mill (1992); Hoc (1993)

Manual vs automatic systems
Hunns (1981); Bessant and Dickson (1982); Bainbridge (1983); Rijnsdorp (1986); Visick (1986); Bodsberg and Ingstad (1989)

Operator activities
Troyan (1963); Lees (1970); E. Edwards and Lees (1973, 1974); Skans (1980)

Control tasks
E. Edwards and Lees (1973, 1974); Patternotte (1978); Umbers (1979, 1981)

Information display
Rasmussen (1968a–c, 1969, 1971, 1973, 1976b); Rasmussen and Goodstein (1972); E. Edwards and Lees (1973, 1974); Pedersen (1974); A.Shepherd (1979); Pew, Miller and Feehler (1981); Buttigieg and Sanderson (1991)

Fault administration
Annett *et al.* (1971); CAPITB (1971 Inf. Pap. 8); E. Edwards and Lees (1973, 1974); Rasmussen and Jensen (1973, 1974); Duncan (1974); Goodstein *et al.* (1974); Lees (1974c); Duncan and Gray (1975a,b); Duncan and Shepherd (1975a,b); A. Shepherd (1976); Rouse (1977, 1979a,b, 1981); A. Shepherd *et al.* (1977); Himmelblau (1978); Landeweerd (1979); Lihou (1979, 1981); A. Shepherd and Duncan (1980); Christeansen and Howard (1981); Duncan (1981); Johannsen (1981); E.C. Marshall, Duncan and Baker (1981); E.C. Marshall and Shepherd (1981); E.C. Marshall *et al.* (1981); Patrick and Stammers

(1981); Process Control Training (1981); Rasmussen and Rouse (1981); D.A. Thompson (1981); Rouse and Hunt (1982); Embrey (1986); P.J. Smith *et al.* (1986); Toms and Patrick (1987, 1989); Vermeulen (1987); Malaterre *et al.* (1988); Morrison and Duncan (1988); Patrick and Haines (1988); Moray and Rotenberg (1989); Patrick *et al.* (1989); Yukimachi, Nagasaka and Sasou (1992); Decortis (1993); Hukki and Norros (1993); Patrick (1993); Chung-Chiang Peng and Sheue-Ling Hwang (1994)

Alarm systems
Andow and Lees (1974); Hanes (1978); Kortlandt and Kragt (1980a,b); E. Edwards (1979); Hanes (1980); British Gas (1986 Comm 1296); Benel *et al.* (1981); Dellner (1981); Andow (1982, 1985a,b); Kragt (1982, 1983, 1984a,b); Kragt and Bonten (1983); Renton (1984); Schellekens (1984); J.A. Shaw (1985); Arnold and Darius (1988); Sorkin, Kantowitz and Kantowitz (1988); Fort (1989); Edworthy, Loxley and Dennis (1991); Fordestrommen and Haugset (1991)

Batch processes
Lihou and Jackson (1985); Rayment (1986)

Operator-computer interaction
Rasmussen (1968b, 1981a); Rasmussen and Goodstein (1972); Goodstein (1981, 1982, 1984); Kletz (1982g, 1991g); Rasmussen and Lind (1982); Goodstein *et al.* (1983); Hollnagel (1984); J.C. Shaw (1985); Rayment (1986); Gilmore, Gertman and Blackman (1989); Hockey *et al.* (1989)

Control room design
NRC (Appendix 28 *Control Rooms*); Ergonomics Society (1983); Stumpe (1984); Nordic Liaison Committee (1985 NKA/LIT(85)4); Singleton (1986); Wanner (1986); Pikaar *et al.* (1990); HSE (1993 CRR 60); Ainsworth and Pendlebury (1995); Umbers and Rierson (1995); Whitfield (1995)

Operator training, including simulators
MCA (SG-15); NRC (Appendix 28 Operating Personnel, Simulators, Training); Pontius, van Tassel and Field (1959); Crossman (1960); S.D.M. King (1960, 1964); Crossman and Cooke (1962); Stapleton (1962); Crossman, Cooke and Beishon (1964); Weltge and Clement (1964); BCISC (1965/5); Whitesell and Bowles (1965); Carmody and Staffin (1970); Annett *et al.* (1971); Atherton (1971); CAPITB (1971 Inf. Paps 6–10, 1972 Inf. Pap. 13); PITB (1971/1, 2, 1972/3, 1975/6); Biggers and Smith (1972); City and Guilds Institute (1972); E. Edwards and Lees (1973, 1974); Duncan (1974, 1981); Goodstein *et al.* (1974); Lees (1974c); Duncan and Gray (1975a, b); Duncan and Shepherd (1975a,b); Patrick (1975, 1992); Barber and Tibbets (1976); A. Shepherd (1976, 1982b, 1986, 1992); J. Davies (1977); Doig (1977); A. Shepherd *et al.* (1977); Shindo and Umeda (1977); Stephens (1977); Crawford and Crawford (1978); Goldstein (1978, 1987); Duncan, Gruneberg and Willis (1980); Landeweerd, Seegers and Praagma (1981); Process Control Training (1981); Vervalin (1981c, 1984); Sorotzkin and Lock (1983); Demena *et al.* (1984); Madhavan (1984); Clymer (1985); Pathe (1985); Tomlinson (1985); Embrey (1986); Marcille (1986); Nordic Liaison Cttee (1986 NKA/LIT(85)6); Wetherill and Wallsgrove (1986); Drury *et al.* (1987); Elshout and Wetherill (1987); Flexman and Stark (1987); Avisse (1989); IAEA (1989); Patrick *et al.* (1989); ACSNI (1990); Mani, Shoor and Petersen (1990); Ferney (1991); Grossman and Dejaeger (1992); IEE (1992 Coll. Dig. 92/123); Sanquist (1992)

Operating procedures, instructions
Oriolo (1958); Troyan (1961a); Minich (1979); Lopinto (1983); S.T. Wood (1984); Kujawski (1985); Bardsley and Jenkins (1991 SRDA R1); Connelly (1992); McIntyre (1992); I.S. Sutton (1992); Swander and Vail (1992)

Three Mile Island (see also *Table A21.1*)
Livingston (1980); Malone *et al.* (1980); Kletz (1982l)

14.1 Human Factors in Process Control

In the previous chapter the importance of plant operation was emphasized and the automatic control system was described. Consideration is now given to the other element in the overall control system – the process operator.

Although modern control systems achieve a high degree of automation, the process operator still has the overall immediate responsibility for safe and economic operation of the process. There are different philosophies on the extent to which the function of safety shut-down should be removed from the operator and assigned to automatic trip system. In general, the greater the hazards, the stronger the argument for protective instrumentation. This question is considered in more detail later. But, whatever approach is adopted, the operator still has the vital function of running the plant so that shut-down conditions are avoided.

The job of the process operator is therefore a crucial one, but it is also rather elusive in that it presents the engineer with a type of problem with which he is not normally required to deal. The study of industrial jobs and work situations is the province of ergonomics or, its American near-equivalent, human factors. It is appropriate, therefore, to consider the contribution which this discipline can make to the problems associated with the work of the process operator.

It should be said, however, that the chemical industry in general appears to make little use of human factors in this area. There is considerable willingness to do so among engineers, but perhaps also some lack of appreciation of the scope of human factors as a discipline.

14.2 Human Factors in System Design

The pace of technological change and the scale of systems have now become so great that it is often not possible to rely on evolutionary trial and error to achieve the proper adaptation of human tasks. Instead it is

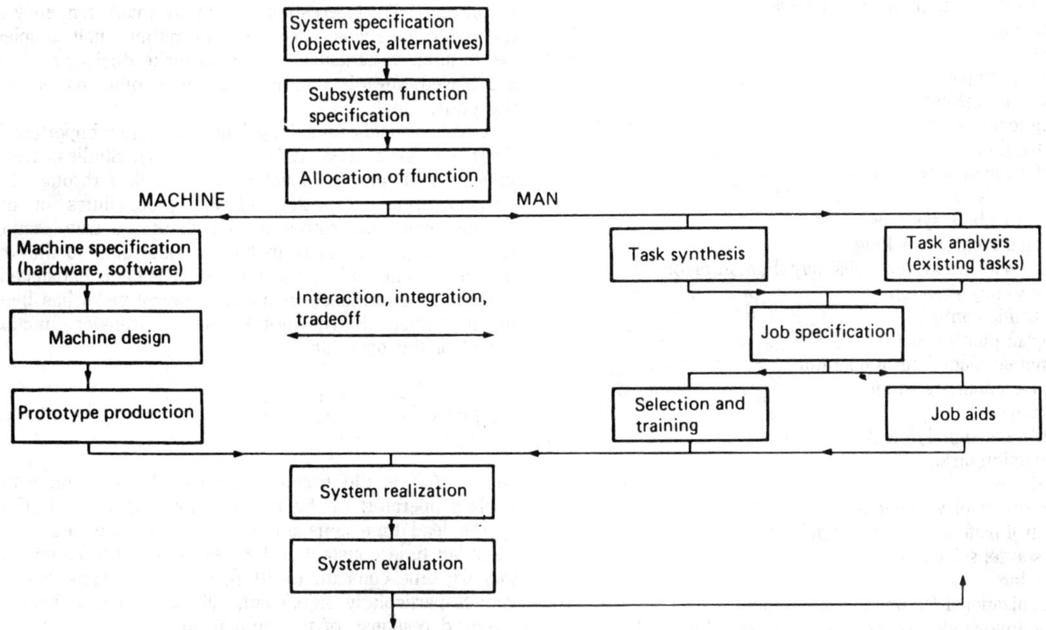

Figure 14.1 *Human factors activities in system design (Lees, 1974c) (Courtesy of Taylor & Francis Ltd)*

necessary to try to foresee the problems and to design to overcome them. The discipline which is concerned with this on the human side of systems is human factors.

The development of human factors has been strongly influenced by the problems of large, complex man–machine systems which occur in the fields of defence, aerospace and computers. Much of the fundamental research and design experience are in these areas.

An important area of study in the early work on human factors was the compatibility of man and machine, with its emphasis on 'knobs and dials'. More recent work has laid a greater stress on system design. In consequence the ergonomist has concerned himself increasingly with all stages of the design process, particularly the early stages where the crucial decisions are made.

Human factors is now established, therefore, as an aspect of systems engineering. An outline of the human factors activities in system design is shown in Figure 14.1. Two main points may be noted: (1) human factors play a role at all stages of the design; and (2) the decisions taken early on, such as those on allocation of function, are especially important and the design process is a highly interactive and iterative one. Only one iteration loop is shown, but in fact iteration occurs at all stages of the design process. Important emphases in human factors, therefore, are the system criteria and the system design process.

The engineer unfortunately frequently misunderstands human factors. The view of the subject as being concerned with knobs and dials is entrenched. Human factors is too often abused by being called in late in the day to fulfil a rescue or cosmetic function. Its greatest contribution should in fact come earlier, particularly at the allocation of function stage.

Like the engineer, the ergonomist is concerned with solving problems, particularly those arising in design. He draws on the work of psychologists and others, much as the engineer draws on that of physicists.

14.3 Themes in Human Factors

It is appropriate at this point to give a very brief account of some of the themes in human factors work which have obvious relevance to process control. Some of these are shown in Table 14.3. This is an abstract from a fuller table, which gives detailed references, in an account of the development of human factors in this context by Lees (1974c). Much of the early work on human factors was concerned with physical tasks, but in more recent years the emphasis has been increasingly on mental tasks. This is certainly more relevant as far as the process operator is concerned, since his job is essentially decision-making.

The question of the sampling and processing of information by the human operator is therefore of great importance. Work in this area largely evolved from study of the skills involved in physical tasks to investigation of those required for perceptual and decision-making tasks (Crossman, 1956). The model of man as an information processor has proved fruitful, although the application of information theory to the problem has not been entirely successful. Work in the area has emphasised: the ability

***Table 14.3** Selected topics in human factors*

Information sampling and processing
Learning
Skill
Stress, fatigue
Decision-making
Diagnostic tasks
Motivation
Performance assessment
Task analysis
Man–machine systems
Manual control, tracking
Man–machine system reliability, human error
Emergency situations
Air traffic control
Aircraft pilot's task
Dynamic modelling of operator
Man–computer systems
Displays
Vigilance, signal detection
Inspection tasks
Controls
Control–display relations
Control panels, computer consoles
Personnel selection
Training
Organizational factors, job enrichment
Repetitive work, boredom, rest pauses, shift work

of man to accept information coming through many sensory channels and coded in many different ways, and his ability to compensate for errors in the information; the differences in the amounts of information which can be handled by the various channels; the sampling of information and the updating of his mental model of the environment; the effect of information overload, resulting in selective omission of parts of the task; and the characteristics of memory, particularly short-term memory such as is exercised in remembering a telephone number to make a phone call.

The ergonomist's approach to a particular job tends to be to enquire into the skill involved, including the nature of the skill, its acquisition through the learning process and its disintegration under stress. Skills differ greatly in their amenability to study; some skills, such as that of the process operator, are particularly inaccessible. Nevertheless, skilled performance does exhibit certain common characteristics. Skill seems to lie largely in the timing and co-ordination of activities to give a smooth and effortless performance. It is highly learned and barely accessible to consciousness, as indicated by the fact that the attempt to describe it, as in instructing a novice, often leads to actual degradation in performance.

The effects on skill of various forms of stress such as fatigue, work load and anxiety, have been investigated both on account of the importance of these effects in themselves and of the light which they throw on the nature of skill as it degrades under stress. An important finding is that skilled performance tends to improve with moderate stress, but that beyond a certain threshold, which varies greatly with the individual, it deteriorates rapidly.

The characteristics of human as opposed to mathematically optimal decision-making have been studied (W. Edwards, 1965). Interesting results are man's tendency to make decisions which are based on rather small samples, i.e. to jump to conclusions, and to make decisions which are biased towards optimism, i.e. to gamble on beating the odds.

A type of decision-making which is rather important is diagnosis (Dale 1958, 1964; Tilley, 1967). Studies of this indicate that man does not so much follow through the decision tree irrespective of the probabilities of the various paths, but rather moves about the tree, testing first the high probability paths, and only goes to the low probability ones when the former have been exhausted.

Human performance in manual control tasks has been much studied. Early work evolved a transfer function model of the operator:

$$H(s) = K\frac{\exp(-\tau_d s)(1+\tau_L s)}{(1+\tau_N s)(1+\tau_I s)} \qquad [14.3.1]$$

where $H(s)$ is the transfer function; K the gain; s the Laplace operator; τ_d the reaction time (usually 0.1–03 s); τ_L the lead time constant (0.2–2.5 s); τ_N the neuromuscular lag time constant (0.1–0.2 s); and τ_I the compensatory lag time constant (5–20 s). Of these parameters the gain is particularly significant, alteration of gain being a favoured response of the human operator to a changing situation. This approach has been developed by later workers to take account of non-linearities, sampled data features, etc.

Work in this area shows clearly the increased difficulty which the human operator has in controlling processes with an increasing number of transfer lags or integrations or with dead time. Certain systems are virtually uncontrollable by the operator, unless he is provided with specially processed information. In particular, systems with more than three integrations in series tend to be beyond the limits of manual control. For systems such as submarines, which have this feature, the technique of quickening has been developed, in which the signal displayed to the operator is a weighted sum of signals from various points in the series of integrations. Another feature of man as a controller is his ability to carry out predictive and feedforward control functions. This is a rather characteristic feature of operator control.

Much work has been done on displays, in terms both of detailed design of dials, etc., and of the display layout, and it is perhaps this aspect which the engineer most readily identifies as human factors. Particularly relevant here is the classification of the uses of displays (Murrell, 1965a). These are

(1) Indicating, i.e. the operator perceives one of two binary states.
(2) Quantitative reading, i.e. the operator requires a precise numerical value.
(3) Check reading, i.e. the operator requires confirmation that value lies within an acceptable range.
(4) Setting, i.e. the operator manipulates his machine controls in order to achieve a predetermined display state.
(5) Tracking, i.e. the operator carries out an on-going control task in order to achieve certain display conditions which may vary as a function of time.

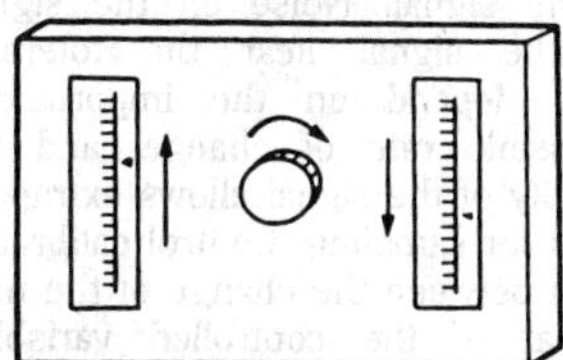

***Figure 14.2** Expected relationships between control and display movements (E. Edwards and Lees, 1973) (Courtesy of the Institution of Chemical Engineers)*

These are very different uses and a display which is optimal for one is not necessarily so for another. There is some work which suggests that some 75% of industrial applications is accounted for either by check reading alone or by check reading and setting combined (Murrell, 1952b).

The acquisition of information from large display layouts is another important problem. Work in this area tends to emphasize the value to the operator of being familiar with the position of dials which give particular readings. Such spatial coding is lost, for example, if the same instrument is used to display different variables at different times. The operator then has to devote more effort to finding out which variable is on a particular display and he is less able to recognize patterns.

Control–display relations can be important. In a given culture there tend to be expectations of particular relations between control movements and display readings. A typical stereotype of a control–display relation is shown in Figure 14.2. Although an operator can be trained to use equipment which embodies faulty control–display relations in its design, he may tend under stress to revert to the expected relation. Violation of the stereotype can result in severe penalties.

Monitoring, signal detection and vigilance is another related and important area on which a large amount of work has been done. Some of this has been concerned with the fall-off in attention over the watch-keeping period, i.e. the vigilance effect. The application of this to process control is doubtful, but perhaps more relevant is the well-established relation between the frequency of a signal and the probability of its detection. The probability of detection of a rare signal is rather low. This has been studied both in dial monitoring and in inspection tasks.

The problems of man–computer systems have been studied quite extensively. These include allocation of function between man and machine, man–computer interaction and man–computer problem-solving. For many functions the emphasis has moved away from early attempts at complete automation towards computer-assisted operator decision-making.

This rather brief survey is intended to show that the problems with which human factors deals are very relevant to process control by the human operator. These and other topics such as learning and training, organizational and social factors, and human error are now considered more specifically in relation to process control.

14.3.1 Human factors in process control

The task of process control is of interest to workers in human factors as an example of a task involving cognitive rather than manipulative skills and was the subject of a series of early classic studies by Crossman. Since then there have been numerous investigations of the process control task. This work is described below.

A review of the control of processes by operators and computers is given in *Man and Computer in Process Control* and some of the classic papers on the process operator are collected in *The Human Operator in Process Control*, both by E. Edwards and Lees (1973, 1974).

14.4 Process Operator Functions

The job of the process operator has developed over the years from one based largely on manual work to one consisting primarily of decision-making. The physical work content is now frequently vestigial.

The process operator is part of the control system. The primary functions of the control system, and therefore of the operator, depend on the nature of the process. In a chlorine cellroom it may be the monitoring of alarm conditions, and in a batch reactor plant the conduct of sequential operations.

The nature of the control system provided also influences strongly the operator's functions. The stages of development of control systems have already been described in Chapter 13. The job of the operator is generally different in systems based on analogue and on computer control. The job of the process operator, at least in the control room, is essentially decision-making in a rather artificial situation, involving the manipulation of symbolic displays.

If the process control task as a whole is considered, a number of distinct operator functions may be identified (E. Edwards and Lees, 1973; Lees, 1974c):

(1) goal formulation;
(2) measurement;
(3) data processing and handling;
(4) monitoring;
(5) single variable control;
(6) sequential control;
(7) other control;
(8) optimization;
(9) communication;
(10) scheduling;
(11) manual operations.

These functions have been discussed in detail by E. Edwards and Lees (1973). Although most process control tasks have elements of all these functions, the relative importance varies widely.

The state of the process also affects the operator's task. If process conditions are abnormal, he has the crucial function of fault administration. In simple terms this may be regarded as having three stages: (1) fault detection, (2) fault diagnosis and (3) fault correction. But this can be an oversimplification, in that following fault detection the priority is often to move to a safe condition rather than to diagnose the fault. As control systems achieve increasing automation, the function of fault administration tends to grow in importance. It is a crucial one in relation to loss prevention.

14.5 Process Operator Studies

In a review in 1974 of studies of the process operator, Lees (1974c) listed some 140 items. This list was limited to work directly concerned with the process operator and did not include work on more general but related problems such as vigilance or manual control. There exists, therefore, a quite substantial body of work on the process operator and his problems, although it is not well known among engineers. Selected studies of the process operator are given in Table 14.4. Some of the salient points from these studies are now considered.

It may be mentioned at this point that the compilation of this list appears to have coincided with the end of that phase in the development of human factors in which studies were conducted of the process operator *per se*. Subsequent research has been directed rather to specific tasks, particularly fault diagnosis, to methodologies such as task analysis and to training.

Much of the pioneering work on the process operator was initiated by Crossman and his co-workers. His interest in process control was that it is a good example of an industrial skill which is predominantly cognitive rather than physical. Crossman (1960) did an investigation of a number of process control tasks and concluded that it is particularly difficult to control the following processes:

(1) where several display and control variables depend on one another;
(2) where the process has a long time constant;
(3) where important variables have to be estimated by the operator rather than measured by an instrument;
(4) where the readings of instruments at widely separated points have to be collated, and the operator has to remember one while going to another ('short-term memory');
(5) where the operator gets imperfect knowledge of the results of his performance, or where the knowledge arrives late;
(6) where the basic process is either difficult to visualize, for example chemical reactions, or contradicts 'common-sense' assumptions, or is too complicated to be held in mind at one time.

One aspect of process control skill which was investigated by Crossman and his colleagues is information sampling (Crossman, Cooke and Beishon, 1964). The information sampling behaviour of the operator was studied in the laboratory task of controlling the temperature of a water bath and the industrial one of controlling the basis weight on a paper machine. It was found that the Shannon–Wiener sampling theorem did give a basic minimum sampling rate, provided a system bandwidth modified for error tolerance was used. But many other factors were identified which tend to increase the sampling rate. Sampling behaviour depends on the operator's uncertainty and its growth over time and on the cost of sampling, and it cannot be separated from the control problem, which raises questions of the control accuracy required, the penalty for error, the operator's understanding of the system, the nature and predictability of the disturbances and the lags in the process. Five factors were identified as governing the sampling rate: (1) bandwidth, (2) noise, (3) tolerance, (4) predictability and (5) control calibration.

Bandwidth is a function of the maximum possible rate of change of the signal. Noise on the signal causes excursions of the signal near the tolerance limit. Tolerance limits depend on the importance of the variable, its possible rate of change, and the signal noise. Predictability of the signal allows extrapolation and reduces the need for sampling. Control calibration, which gives the relation between the change of the manipulated variable and that of the controlled variable, assists predictability.

In the light of this work the authors criticized the use of displays which show only a deviation between the set point and a measured value. Such displays greatly reduce the operator's ability to learn the signal characteristics such as noise, predictability and control calibration. This problem has also been considered by de Jong and Köster (1971) in terms of the sampling of information from chart recorders.

Crossman and his colleagues also studied manual control of the laboratory water bath and the paper machine (Crossman and Cooke, 1962; Beishon, 1967). The control action required in both cases was to bring the system to a new operating point. The work showed that a good subject often begins with a closed-loop strategy which tends to produce the oscillatory response shown in Figure 14.3(a), but soon learns to use an open-loop approach, limiting the use of feedback to fine tuning as in Figure 14.3(b).

Similar work has been done by Attwood (1970), also on a laboratory water bath and a paper machine, by Kragt and Landeweerd (1974) on a laboratory heater and by B. West and Clark (1974) on a distillation column in a computer-controlled pilot plant. This work further confirms the importance of the open-loop strategy.

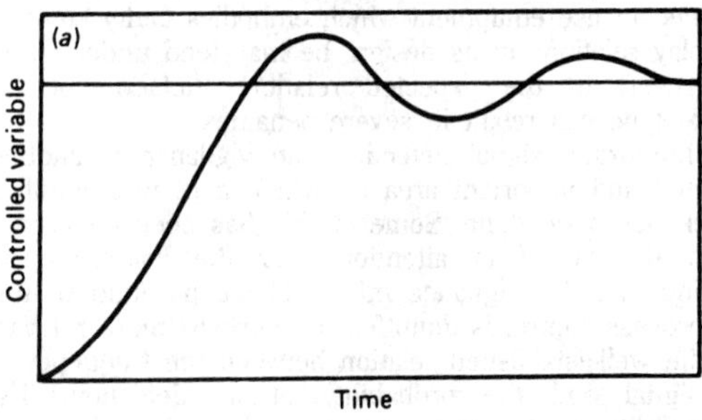

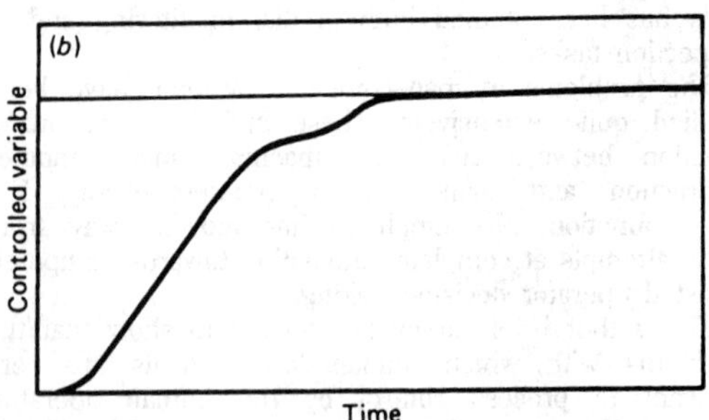

Figure 14.3 Change of operating point in a manually controlled system: (a) closed-loop behaviour; (b) open-loop behaviour

Table 14.4 *Selected studies of the process operator*

Author	*Subject of study*	*Type of study*
Hiscock (1938)	Selection tests for process operators	Industrial study
Crossman (1960)	Process control task and skill in plants of different types	Industrial study
Kitchen and Graham (1961)	Decision-making and mental load of process operators in 50 chemical plants	Industrial study
Crossman and Cooke (1962)	Manual control by process operator	Experimental laboratory study
Sell, Crossman and Box (1962)	Evaluation of human factors aspects of control system of hot strip mill	Industrial study
Spencer (1962)	Manual control by process operator on liquid washing plant	Experimental in-plant study
Crossman, Cooke and Beishon (1964)	Information sampling by process operator	Experimental laboratory and in-plant studies
Vander Schraaf and Strauss (1964)	Acceptability of process computer to process operator	Industry study
Bernard and Wujkowski (1965)	Acceptability of process computer to process operator	Industry study
CEGB (1966)	Evaluation of human factors aspects of control room of nuclear reactors	Industry study
Sinclair *et al.* (1966)	Evaluation of human factors aspects of control room of Linz-Donan (LD) converter waste heat boilers	Industrial study
Beishon (1967)	Manual control by process operator on paper machine	Experimental in-plant study
Davies (1967)	Selection tests for process operators	Industrial study
Bainbridge *et al.* (1968); Bainbridge (1971, 1972, 1974)	Decision-making by process operator in scheduling electric arc furnaces	Experimental simulation study
Crawley (1968)	Allocation of function between process operator and process computer and computer graphic displays for process operator on LD converter	Industrial study
Munro, Martin and Roberts (1968)	Workload of process operator	Experimental simulation study
Rasmussen (1968a–c, 1971, 1973, 1974, 1976a,b, 1977, 1978); Rasmussen and Goodstein (1972); Rasmussen and Jensen (1973); Goodstein *et al.* (1974); Rasmussen and Taylor (1976)	Control and surveillance by process operator, displays for process operator, fault diagnosis by process operator, reliability of process operator	Experimental, industrial and general studies
Beishon (1969)	Manual control and decision-making by process operator on baking ovens	Experimental in-plant studies
Whitfield (1969)	Evaluation of human factors aspects of control room of nuclear reactor	Industrial study
Attwood (1970)	Manual control by process operator on paper machine	Experimental laboratory and in-plant studies
Engelstad (1970)	Social and organizational features in process control in paper mill	Experimental in-plant study
Ketteringham, O'Brien and Cole (1970); Ketteringham and O'Brien (1974)	Computer-aided scheduling by process operator in steel mill soaking pits	Experimental simulation study
Annett *et al.* (1970); Duncan (1974); Duncan and Gray (1975); Duncan and Shepherd (1975a,b)	Task analysis and training for process control tasks	Experimental in-plant studies
Daniel, Puffler and Strizenec (1971)	Mental load of process operator in four chemical plants	Experimental in-plant studies
de Jong and Köster (1971)	Process operator in computer-controlled refinery	Review
Anyakora and Lees (1972a)	Detection of instrument malfunction by process operator in chemical plants	Experimental in-plant study
Dallimonti (1972, 1973)	Design of man–computer interface and acceptability of process computer to process operator	Industrial study
E. Edwards and Lees (1973, 1974)	The human operator in process control, including process computer systems	Review

Lees (1973a, 1976a)	Design for human reliability and human reliability assessment in process control	Review
Duncan (1974); Duncan and Gray (1975); Duncan and Shepherd (1975a,b); Shepherd (1976); Shepherd *et al.* (1977)	Training of process operator in fault diagnosis	Experimental in-plant studies
Kragt and Landeweerd (1974)	Manual control and surveillance by process operator	Experimental laboratory and in-plant studies
West and Clark (1974)	Information display for and manual control and fault administration by process operator in computer-controlled pilot plant	Experimental in-plant study
Brigham and Laios (1975)	Manual control by process operator	Experimental laboratory study
Lees and Sayers (1976)	Emergency behaviour of process operator	Experimental simulation study

Crossman also raised the question of the operator's mental model of the process and investigated the difference in performance between subjects who were given an account of the physics of the plant and those who were simply told to control calibration settings. The performance of the former was much less effective. He concluded that the results cast some doubt on the practice of instructing operators in the physics and chemistry of the process. Similar results were obtained by Attwood and by Kragt and Landeweerd.

Brigham and Laois (1975) have studied manual control of level in a laboratory rig consisting of three tanks in series, with interaction between the levels. The tanks were glass, so that the operator also had feedback of information from intermediate points in the process. Cross-correlation of his control manipulations with level error showed little correlation between the two, indicating that the control was based on predictive rather than feedback strategy.

The effect of the number and types of lag in the process has also been studied. Crossman and Cooke added further lags to their laboratory apparatus by putting additional sheaths round the thermometer and found that control became more difficult and response was more oscillatory. Attwood introduced dead time into his water bath and found a similar effect. Control behaviour on paper machines observed by Crossman, Cooke and Beishon and by Attwood shows similar oscillatory responses.

Manual control of a liquid washing plant has been studied by Spencer (1962). This work highlighted the difficulties of control in the absence of feedback of information on the results of control actions and the wide differences between operators in terms of the control gain which they employ.

The mental load on the operator has been studied by Kitchin and Graham (1961) in an investigation covering some 50 plants. The decision-taking load was analysed under the following five headings: (1) number of factors in the situation; (2) complexity of comprehension of each factor; (3) memory; (4) interdependence of factors; and (5) delay characteristics of situation. This is somewhat similar to Crossman's description of the factors which make process control difficult, as described earlier.

Another study which involved sampling of the operator's activities and assessment of the mental workload imposed by the work is that done by Daniel, Puffler and Strizenec (1971) on four chemical plants. The Monte Carlo simulation of the workload on the operator has been described by Munro, Martin and Roberts (1968).

There have been a number of studies by ergonomists of industrial control rooms. Examples are those of a hot strip mill by Sell, Crossman and Box (1962), of an Linz-Donan (LD) converter waste heat boiler control room by Sinclair *et al.* (1966), and of nuclear power station control rooms by the Central Electricity Generating Board (1966) and by Whitfield (1969).

Decision-making by the operator has been investigated by Bainbridge *et al.* (1968) in a simulation study of electric arc furnace scheduling. A running commentary, or verbal protocol, was given by the subjects as they performed the task and has since been analysed in detail by Bainbridge (1971, 1972, 1974). This analysis identifies subroutines which the operator uses and attempts to describe the executive program which organizes these.

An important aspect of this work is the light which it throws on the way in which the operator keeps track of the state of the process and so updates his mental model of it. In general, he tends to predict the future of the process state and then subsequently samples only enough readings to confirm that his prediction was correct. This is in line with other work such as that of Crossman, Cooke and Beishon (1964) referred to earlier. This work has significant implications for displays. Good display facilities allow the operator to survey the state of the process as a whole with minimum effort.

The advent of the process computer has brought profound changes in the work of the operator. A number of investigations have been done (e.g. Vander Schraaf and Strauss, 1964; Bernard and Wujkowski, 1965) on the reaction of process operators to computers. Another such study is that by Dallimonti (1972, 1973), which was particularly concerned with the design of computer facilities.

The problem of system objectives and allocation of function in computer control of a basic oxygen furnace has been discussed by Crawley (1968). He gives as an example of a function which is more appropriately allocated to man the interpretation of the noise signal given by a bomb thermocouple. The point is also made, however, that the optimal allocation of function changes continuously as technology progresses.

The difficulty of automating certain functions is further illustrated by the work of Ketteringham, O'Brien and Cole (Ketteringham, O'Brien and Cole, 1970; Ketteringham and O'Brien, 1974) on the scheduling of soaking pits using a man–computer interactive system. Problems of providing all the necessary information to the computer and of complexity in the decision-making make it difficult to automate this function, but it is possible to provide computer facilities which, by storage of large amounts of information, the use of predictive models and the provision of appropriate displays, greatly assist the operator to make decisions. The work involved simulation using a realistic interface and actual steel works schedulers. The problem is very similar to that of air traffic control, which has been extensively investigated in human factors.

The importance of organizational and social factors is shown by the work of Engelstad (1970), who investigated these in a paper mill. The work revealed that individuals tended to operate too much in isolation and to have goals which were not necessarily optimal as far as the system was concerned. It sought to encourage communication by treating the control room as an information and control centre, which all concerned should use, and by designing jobs to give greater variety, responsibility, learning opportunity and wholeness. Other workers too have commented on the losses which can occur in continuous flow processes, if there are poor relationships between the men controlling the process at different points.

Other studies have been concerned with task analysis, fault administration, displays, selection and training, and human error. These topics are considered below.

14.6 Allocation of Function

As already emphasized, human factors has at least as important a role to play in matters of system design such as allocation of function as in those of detailed design. The classic approach to allocation of function is to list the functions which machines perform well and those which men perform well and to use this as a guide. One of the original lists was compiled by Fitts (1962) and such a list is often referred to as a 'Fitts' list'. However, this approach needs some qualification. As Fitts pointed out later, the functions which should be allocated to man are not so much those which he is good at as those which it is best from the system point of view that he perform, which is slightly different. The question of motivation is also important. De Jong and Köster (1971) have given a similar type of list showing the functions which man is motivated to perform. Further accounts of allocation of function is given by H.E. Price (1985) and Kantowitz and Sorkin (1987). Particularly important in relation to loss prevention are functions concerned with fault administration, fault diagnosis, plant shut-down and malfunction detection.

14.7 Information Display

Once the task has been defined, it is possible to consider the design of displays. Information display is an important problem, which is intensified by the increasing density of information in modern control rooms. The traditional display is the conventional control panel. Computer graphics now present the engineer with a more versatile display facility, offering scope for all kinds of display for the operator, but it is probably fair to say that he is somewhat uncertain what to do with it.

The first thing which should be emphasized is that a display is only a means to an end, the end being improved performance by the operator in executing some control function. The proper design of this function in its human factors aspects is more important than the details of the display itself. Some types of display which may be provided are listed in Table 14.5.

Table 14.5 *Some displays for the process operator*

Displays of flow and mimic diagrams
Displays of current measurements, other variables, statuses (other variables include indirect measurements, valve positions, etc.)
Displays of trends of measurements, other variables, statuses
Displays of control loop parameters
Displays of alarms
Displays of reduced data (e.g. histograms, quality control charts, statistical parameters)
Displays of system state (e.g. mimic diagrams, 'status array', 'surface' and polar plots)
Displays for manual control (e.g. predictive displays)
Displays for alarm analysis
Displays for sequential control
Displays for scheduling and game-playing
Displays for valve sequencing
Displays for protective system checking
Displays of maloperation
Displays for malfunction detection
Command displays

The provision of displays which the operator deliberately samples with a specific object in view is only part of the problem. It is important also for the display system to cater both for his characteristic of acquiring information 'at a glance' and for his requirement for information redundancy. There is a need, therefore, for the development of displays which allow the operator to make a quick and effortless survey of the state of the system. As already described, the operator updates his knowledge of system state by predicting forward, using a mental model of the process and sampling key readings to check that he is on the right lines. He needs a survey display to enable him to do this.

This need still exists even where other facilities are provided. Facilities such as alarm systems are based on a 'management by exception' approach, which is essential if the large amounts of information are to be handled. But when the exceptional condition has been detected, the operator must deal with it and for this he needs knowledge of the state of the process, which a survey display provides. A display of system state also allows the operator to use his ability to recognize patterns. This aspect is considered below.

Figure 14.4 Control panel in a chemical plant (courtesy of Kent Instruments Ltd)

14.7.1 Regular instrumentation

It will be apparent from the foregoing that the conventional control panel has certain virtues. The panel shown in Figure 14.4 is typical of a modern control panel. The conventional panel does constitute a survey display, in which the instruments have spatial coding, from which the operator can obtain information at a glance and on which he can recognize patterns. These are solid advantages not to be discarded lightly. This is only true, however, if the density of information in the panel is not allowed to become too great. The advantages are very largely lost if it becomes necessary to use dense blocks of instruments which are difficult to distinguish individually.

An important individual display is the chart recorder. A trend record has many advantages over an instantaneous display. As the work of Crossman, Cooke and Beishon (1964) shows, it assists the operator to learn the signal characteristics and facilitates his information sampling. Both Attwood (1970) and B. West and Clark (1974) found that recorders are useful to the operator in making coarse adjustments of operating point, while the latter authors also noted the operator's use of recorders in handling fault conditions. Anyakora and Lees (1972a) have pointed out the value of recorders in enabling the operator to learn the signal characteristics and so recognize instrument malfunctions.

14.7.2 Computer consoles

The computer console presents a marked contrast to conventional instrumentation. This is illustrated with particular starkness in Figure 14.5(a), which shows the console of the original ICI direct digital control (DDC) computer. Some of the features of ergonomic importance in this panel are: (1) specific action is required to obtain a display; (2) there is no spatial coding and the coding of the information required has to be remembered or looked up; (3) only one variable is displayed at a time; (4) only the instantaneous value of the variable is displayed; and (5) the presentation is digital rather than analogue.

This is a revolutionary change in the operator's interface. While there is, among engineers, a general awareness that the change is significant, its detailed human factors implications are not so well appreciated. The subsequent refinement of process computer consoles and the introduction of computer graphics have somewhat mitigated these features. A more modern computer console is shown in Figure 14.5(b). Moreover, the facilities which the computer offers in functions such as

Figure 14.5(a) *Control panel of the direct digital control (DDC) computer on ICI's ammonia soda plant at Fleetwood, 1961 (courtesy of Ferranti Ltd and Imperial Chemical Industries Ltd)*

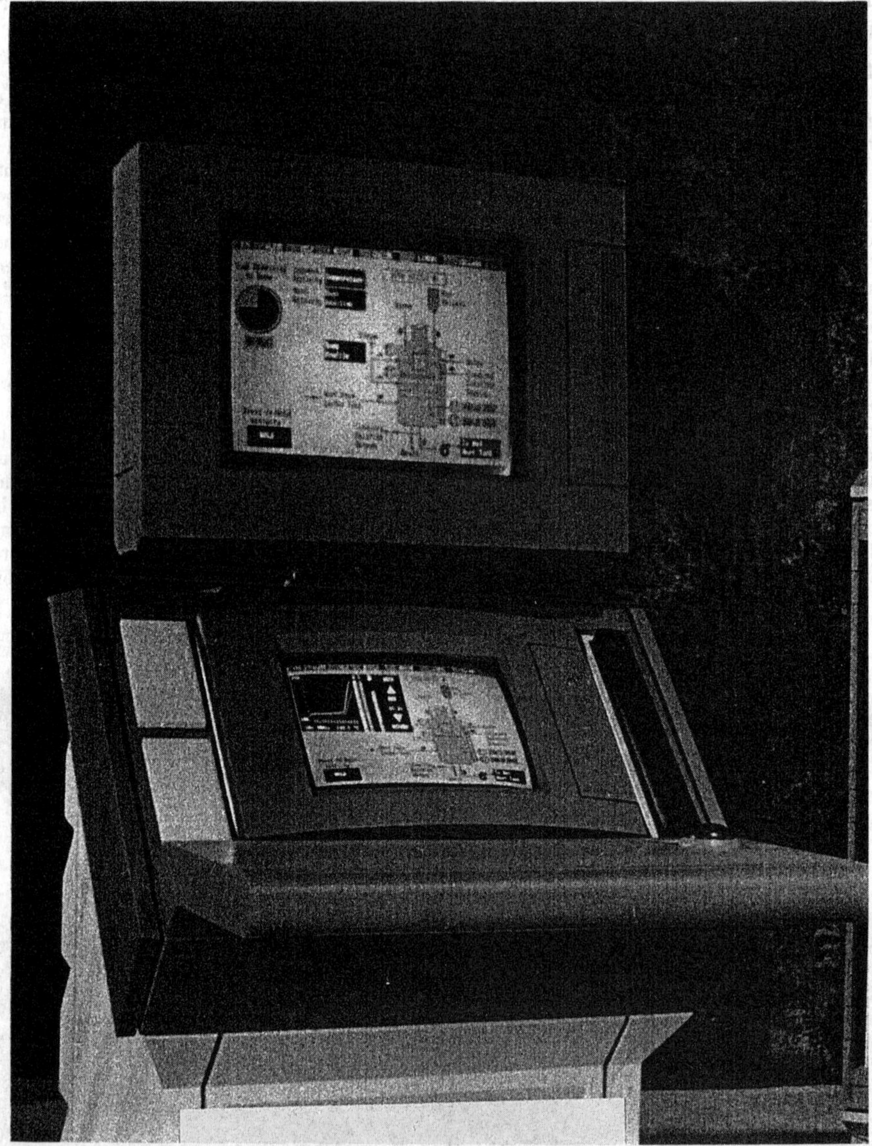

Figure 14.5(b) Control panel and display of Foxboro Intelligent Automation computer system (Foxboro Company Ltd)

alarm monitoring or sequential control are powerful new aids to the operator.

It remains true, however, that the transition from the conventional panel to the computer console involves some serious human factors losses, particularly in information display. This does not necessarily mean that the change should not be made. Conventional panels are expensive and lose much of their advantage if the information density becomes too high. Computers offer some very worthwhile additional facilities. But the change should at least be made in the full awareness of its implications and with every effort to restore in the new system the characteristic advantages of the old.

14.8 Alarm Systems

As stated in the previous chapter, the progress made in the automation of processes under normal conditions has focused attention on the monitoring and handling of abnormal or fault conditions. It is the function of the control system to prevent, if possible, the development of conditions which will lead to plant shut-down, but to carry out the shut-down if necessary. The responsibility for averting shut-down conditions falls largely to the operator. The principal automatic aid provided to assist him is the alarm system. Alarm systems are an extremely important but curiously neglected and often unsatisfactory aspect of process control.

An alarm system is a normal feature of conventional control systems. If a process variable exceeds specified limits or if an equipment is not in a specified state, an alarm is signalled. Both audible and visual signals are used. Accounts of alarm systems include those by E. Edwards and Lees (1973), Andow and Lees (1974), Hanes (1978), Swain and Guttman (1983), Schellekens (1984), Shaw (1985) and the Center for Chemical Process Safety (CCPS) 1993/14). There have also been investigations of the operation of alarm systems, as described below.

14.8.1 Basic alarm systems

The traditional equipment used for process alarms is a lightbox annunciator. This consists of a fascia, or array, of separate small rectangular coloured glass panels, behind each of which there is a lamp which lights up if the alarm is active. Each panel is inscribed with an alarm message. The panels are colour coded, usually with red being assigned to the highest priority. This visual display is complemented by a hooter, When a new alarm occurs, the hooter sounds and the fascia light flashes until the operator acknowledges receipt by pressing a button. The panel then remains lit until the alarm condition is cleared by operator action or otherwise, as described below.

There are a number of variations on this basic system. An account of some of these is given by Kortland and Kragt (1980a). One is a two-level hierarchical arrangement in which there is a central fascia and a number of local fascias, the light on the former indicating the number of the local fascia on which the alarm has occurred. These authors also describe hierarchical systems with additional levels. Another arrangement is a mimic panel with alarms located at the relevant points on the flow diagram.

In computer-controlled systems alarms are generally printed out on hard copy in the time sequence in which they occur and are also displayed on the VDU. On the latter there are numerous display options, some of which are detailed by Kortlandt and Kragt. One is the use of a dedicated VDU in which the alarms come up in time sequence, as on the printer. Another is a group display in which all the alarms on an item of equipment are shown with the active alarms(s) highlighted. Another is a mimic display akin to the hardwired mimic panel.

The occurrence of an alarm is normally accompanied by an audible warning in the form of the sounding of the hooter and a visual warning in the form of a flashing light. On a computer system the sound of the printer serves as another form of audible signal. The operator 'accepts' the alarm by depressing the appropriate pushbutton.

Annunciator sequences are given in ISA Std S18.1: 1975 *Annunciator Sequences and Specifications,* which recognizes three types: (1) automatic reset (A), (2) manual reset (M) and (3) ringback (R). Automatic reset returns to the normal state automatically once the alarm has been acknowledged and the process variable has returned to its normal state. Manual reset is similar except that return to the normal state requires operation of the manual reset pushbutton. Ringback gives a warning, audible and/or visual, that the process condition has returned to normal.

There are alarms on the trips in the Safety Interlock System (SIS). Operation of a trip is signalled by an alarm. The SIS alarm system, like the rest of the SIS, should be separate from the basic process control system (BPCS).

In a conventional instrument system the hardware used to generate an alarm consists of a sensor, a logic module and the visual display. The sensor and the logic elements may be separate or may be combined in an alarm switch. Such combined switches are referred to as flow switches, level switches, and so on.

In a computer-based system there are two approaches which may be used. In one the computer receives from the sensor an analogue signal to which the program applies logic to generate the alarm. In the other the signal enters in digital form. The latter can provide a cheaper system, but reduces the scope for detection of instrument malfunction.

Some basic requirements for an alarm are that it attracts the attention of the operator, that it be readily identifiable and that it indicate unambiguously the variable which has gone out of limit.

14.8.2 Alarm system features

As just described, an alarm system is a normal feature of process computer control. The scanning of large numbers of process variables for alarm conditions is a function very suitable for a computer. Usually the alarm system represents a fairly straightforward translation of a conventional system on to the computer. Specified limits of process variables and states of equipment are scanned and resulting alarms are displayed on a typewriter, necessarily in time order, and also on a VDU, where time order is one of a number of alarm display options.

The process computer has enormous potential for the development of improved alarm systems, but also brings with it the danger of excess. There is first the choice of variables which are to be monitored. It is no longer necessary for these to be confined to the process variables measured by the plant sensors. In addition, 'indirect' or 'inferred' measurements calculated from one or more process measurements may be utilized. This considerably increases the power of the alarm system.

Then there are a number of different types of alarm which may be used. These include absolute alarms, relative, deviation or set-point alarms, instrument alarms and rate-of-change alarms, in which the alarm limits are, respectively, absolute values of the process variable, absolute or proportional deviations of the process variable from the loop set-point, zero or full-scale readings of the instrument and rate of change of the process variable.

The level at which the alarm limits are set is another important feature. Several sets of alarm limits may be put on a single variable to give different degrees of alarms such as early warning, action or danger alarms. The alarms so generated may be ordered and displayed in various ways, particularly in respect of the importance of the variable and the degree of alarm.

The conventional alarm system, therefore, is severely limited by hardware considerations and is relatively inflexible. The type of alarm is usually restricted to an absolute alarm. The computer-based alarm system is potentially much more versatile.

The alarm system, however, is frequently one of the least satisfactory features of the control system. The most common defect is that there are too many alarms and that they stay active for too long. As a result, the system tends to become discredited with the operator, who comes to disregard many of the alarm signals and may even disable the devices which signal the alarms.

Computer-based alarm systems also have some faults peculiarly their own. It is fatally easy with a computer to have a proliferation of types and degrees of alarm. Moreover, the most easily implemented displays, such as time-ordered alarms on a typewriter or a VDU, are inferior to conventional fascias in respect of aspects such as pattern recognition.

The main problem in alarm systems is the lack of a clear design philosophy. Ideally, the alarm system should be designed on the basis of the information flow in the plant and the alarm instrumentation selected and located to maximize the information available for control, bearing in mind instrumentation reliability considerations. In fact an alarm system is often a collection of subsystems specified by designers of particular equipment with the addition of some further alarms. An alarm system is an aid for the operator. An important but often neglected question is therefore what action he is required to take when an alarm is signalled.

There are also specific problems which cause alarms to be numerous and persistent. One is the confusion of alarms and statuses. A status merely indicates that an equipment is in a particular state, e.g. that an agitator is not running. An alarm, by contrast, indicates that an equipment is in a particular state and should be in a different one, e.g. that an agitator is not running but should be. On most plants there are a number of statuses which need to be displayed, but there is frequently no separate status display and so the alarm display has to be used. As a result, if a whole section of plant is not in use, a complete block of alarms may be permanently up on a display, even though these are not strictly alarm conditions. The problem can be overcome by the use of separate types of display, e.g. yellow for alarms, white for statuses, in conventional systems and by similar separate displays in computer systems, but often this is not done.

A somewhat similar problem is the relation of the alarms to the state of the process. A process often has a number of different states and a signal which is an alarm in one state, e.g. normal operation, is not a genuine alarm in another, as with, say, start-up or maintenance. It may be desirable to suppress certain alarms during particular states. This can be done relatively easily with a computer but not with a conventional system. It should be added, however, that suppression of alarms needs to be done with care, each case being considered on its merits.

On most plants there is an element of sequential control, e.g. start-up. As long as no fault occurs, control of the sequence is usually straightforward, but the need to allow for faults at each stage of the sequence can make sequential control quite complex. With sequential operations, therefore, the sequential control and alarm systems are scarcely separable. A computer is particularly suitable for performing sequential control.

The state of the art in alarm system design is not satisfactory, therefore. In conventional systems this may be ascribed largely to the inflexibility of the hardware, but the continuance of the problem in computer systems suggests that there are also deficiencies in design philosophy.

The process computer provides the basis for better alarm systems. It makes it possible to monitor indirect measurements and to generate different types and degrees of alarm, to distinguish between alarms and statuses, and to adapt the alarms to the process state. It is also possible to provide more sophisticated facilities such as analysis of alarms, as described below. However, there is scope for great improvement in alarm systems even without the use of such advanced facilities.

14.8.3 Alarm management

It will be apparent from the foregoing that in many systems some form of alarm management is desirable. Alarm management is discussed by the CCPS (1993/14). Approaches to the problem include (1) alarm prioritization and segregation, (2) alarm suppression and (3) alarm handling in sequential operations.

Alarms may be ranked in priority. The CCPS suggests a four level system of prioritization in which critical alarms are assigned to Level 1 and important but non-critical alarms to Level 2, and so on. The alarms are then segregated by level.

With regard to alarm suppression, the CCPS describes two methods. One is conditional suppression, which may be used where an alarm does not indicate a dangerous situation and where it is a symptom which can readily be deduced from the active alarms. The other is flash suppression, which involves omitting the first stage in the alarm annunciation sequence, namely the sounding of the hooter and flashing of the fascia lamp. Instead, the alarm is shown illuminated, as if already acknowledged.

As already discussed, sequential operations such as plant start-up tend to activate some alarms. With a computer-based system arrangements may be made for appropriate suppression of alarms during such sequences.

Alarm management techniques need to be approached with caution, taking account of the overall information needs of the operator, of the findings of research on the operation of alarm systems and of other factors such as the fact that an alarm which is less important in one situation may be more so in another.

14.8.4 Alarm system operation

Studies of the operation of process alarm systems have been conducted by Seminara, Gonzalez and Parsons (1976) and Kragt and co-workers (Kortlandt and Kragt, 1980a,b; Kragt, 1983, 1984a,b; Kragt and Bonten, 1983; Kragt and Daniels, 1984).

The work of Seminara, Gonzalez and Parsons (1976) was a wide-ranging study of various aspects of control room design. On alarms, two features found by these authors are of particular interest. One is that in some cases the number of alarms was some 50–100 per shift, and in one case 100 in an hour. The other is the proportion of false alarms, for which operator estimates ranged from 'occasional', through 15% and 30%, up to 50%.

Kortland and Kragt (1980a,b) studied five different control room situations, by methods including questionnaire and observation, two of the principal investigations being in the control rooms of a fertilizer plant and a high pressure polyethylene plant. On both plants the authors identified two confusing features of the alarms. One was the occurrence of oscillations in which the measured values moved back and forth across the alarm limit. The other was the occurrence of clusters of alarms. The number of alarms registered on the two plants was as follows:

	Fertilizer plant	*Polyethylene plant*
Observation period (h)	63	70
Single alarms, not occurring during clusters or oscillations	816	1714
Single alarms during clusters or oscillations	280[a]	325[a]
Signals during oscillations	410[a]	NA
Signals from clusters	1288[a]	1300[a]
Total	2794	3339

[a] Estimated from analysis of the data.

The authors suggest that oscillations can be overcome by building in hysterisis, which is indeed the normal approach, and that clusters may be treated by grouping and suppression of alarms. The intervals between successive signals, disregarding oscillations and clusters, were analysed and found to fit a log–normal distribution. The response of the operators to the alarms was as follows:

	Fertilizer plant	*Polyethylene plant*
Signal followed by action (%)	47	43
Action followed by signal (%)	46	50
No action (%)	7	7

Thus about half the alarms were actually feedback cues on the effects of action taken by the operators, who in many cases would have been disturbed not to receive such a signal. Further evidence for this is given by the fact that on the fertilizer plant 55% of the alarm signals were anticipated. On this plant, the operators judged the importance of the signals as follows: important 13%, less important 36%, not important 43%, and unknown 8%.

Kragt (1984a) has described an investigation of the operator's use of alarms in a computer-controlled plant. The main finding was that sequential information presentation is markedly inferior to simultaneous presentation.

14.9 Fault Administration

As already stated, it is the function of the control system to avert the development of conditions which may lead to shut-down, but if necessary to execute shut-down. Generally, there are automatic trip systems to shut the plant down, but the responsibility of avoiding this situation if at all possible falls to the operator.

Fault administration can be divided into three stages: fault detection, fault diagnosis and fault correction or shut-down. For the first of these functions the operator has a job aid in the form of the alarm system, while fault correction in the form of shut-down is also frequently automatic, but for the other two functions, fault diagnosis and fault correction less drastic than shut-down, he is largely on his own.

14.9.1 Fault detection

The alarm system represents a partial automation of fault detection. The operator still has much to do, however, in detecting faults. This is partly a matter of the additional sensory inputs such as vision, sound and vibration which the operator possesses. But it is also partly due to his ability to interpret information, to recognize patterns and to detect instrument errors.

14.9.2 Fault diagnosis

Once the existence of some kind of fault has been detected, the action taken depends on the state of the plant. If it is in a safe condition, the next step is diagnosis of the cause. This is usually left to the operator. The extent of the diagnosis problem may vary considerably with the type of unit. It has been suggested, for example, that whereas on a crude distillation unit the problem is quite complex, on a hydrotreater it is relatively simple (Duncan and Gray, 1975a).

There are various ways in which the operator may approach fault diagnosis. Several workers have observed that an operator frequently seems to respond only to the first alarm which comes up. He associates this with a particular fault and responds using a rule-of-thumb. This is an incomplete strategy, although it may be successful in quite a high proportion of cases, especially where a particular fault occurs repeatedly.

An alternative approach is pattern recognition from the displays on the control panel. The pattern may be static or dynamic. The static pattern is obtained by instantaneous observation of the displays, like a still photograph. The operator then tries to match this pattern with model patterns or templates for different faults. Duncan and Shepherd (1975b) have developed a technique for training in fault diagnosis in which some operators use this method. The alternative, and more complex, dynamic pattern recognition involves matching the development of the fault over a period of time.

Another approach is the use of some kind of mental decision tree in which the operator works down the paths of the tree, taking particular branches, depending on the instrument readings. Duncan and Gray (1975a) have used this as the basis of an alternative training technique. Yet another method is the active manipulation of the controls and observation of the displays to determine the reaction of the plant to certain signals. Closely related to this is the situation where no fault has been detected, but the operator is already controlling the process when he observes some unusual feature and continues his manipulation to explore this condition.

Whatever approach is adopted by the operator, fault diagnosis in the control room is very dependent on the instrument readings. It is therefore necessary for the operator to check whether the instruments are correct. The problem of checking to detect malfunctions is considered later, but attention is drawn here to its importance in relation to fault diagnosis.

These different methods of fault diagnosis have important implications for aspects such as displays and training. The conventional panel assists the recognition of static patterns, whereas computer consoles generally do not. Chart recorders aid the recognition of dynamic patterns and instrument faults. The question of training for fault diagnosis is considered later.

Fault diagnosis is not an easy task for the operator. There is scope, therefore, for computer aids, if these can be devised. Some developments on these lines are described below.

14.9.3 Fault correction and shutdown

When, or possibly before, a fault has been diagnosed, it is usually possible to take some corrective action which does not involve shutting the plant down. In some cases the fault correction is trivial, but in others, such as operating a complex sequence of valves, it is not. Operating instructions are written for many of these activities, but otherwise this is a relatively unexplored area. Fault correction is one of the activities for which interlocks may be provided. Conventional interlocks were described in the previous chapter and developments in computer software interlocks are outlined below.

Some fault conditions, however, require plant shut-down. Although fault administration has been described in terms of successive stages of detection, diagnosis and correction, in emergency shut-down usually little diagnosis is involved. The shut-down action is triggered directly when it is detected that a critical process limit has been passed.

There are differing philosophies on the problem of allocation of responsibility for shutting down the plant under fault conditions. In some plants the operator deals with fault conditions with few automatic aids and is thus required to assure both safety and economic operation. In others, automatic protective systems are provided to shut the plant down if it is moving close to an unsafe condition and the operator thus has the economic role of preventing the development of conditions which will cause shut-down.

In a plant without protective systems the operator is effectively given the duty of keeping the plant running if he can, but shutting it down if he must. This tends to create in his mind a conflict of priorities. Usually he will try to keep the plant running if he possibly can and, if shut-down becomes necessary, he may tend to take action too late. Mention has already been made of the human tendency to gamble on beating the odds. There are numerous case histories which show the dangers inherent in this situation (Lees, 1976b).

The alternative approach is the use of automatic protective systems to guard against serious hazards in the plant. The choice is made on the basis of quantitative assessment of the hazards. This philosophy assigns to the operator the essentially economic role of keeping the plant running.

Although the use of protective systems is rapidly increasing, the process operator usually retains some responsibility for safe plant shut-down. There are a number of reasons for this. In the first place, although high integrity protective systems with 2/3 voting are used on particularly hazardous processes (R.M. Stewart, 1971), the majority of trip systems do not have this degree of integrity. The failure rate of a this simple 1/1 trip system has been quoted as 0.67 faults/year (Kletz, 1974d).

Protective systems have other limitations which apply even to high integrity systems. One is that it is very difficult to foresee and design for all possible faults, particularly those arising from combinations of events. It is true, of course, that even if a process condition arises from an unexpected source a protective system will usually handle it safely. But there remains a residual of events, usually of low probability, against which there is no protection, either because they were unforeseen or because their probability was estimated as below the designer's cut-off level.

Another problem is that a protective system is only partially effective against certain types of fault, particularly failure of containment. In such an event the instrumentation can initiate blowdown, shut-off and shut-down sequences, but while this may reduce the hazardous escape of materials, it does not eliminate it.

Yet another difficulty is that many hazards occur not during steady running but during normal start-up and shut-down or during the period after a trip and start-up from that condition. A well designed protective system caters, of course, for these transitional regimes as well as continuous operations. Nevertheless, this remains something of a problem area. Even with automatic protective systems, therefore, the process operator tends to retain a residual safety function. His effectiveness in performing this is discussed later.

14.10 Malfunction Detection

Another aspect of the administration of fault conditions by the control system is the detection of malfunctions, particularly incipient malfunctions in plant equipment and instruments. These malfunctions are distinguished from alarms in that although, they constitute a fault condition of some kind, they have not as yet given rise to a formal alarm.

Malfunction detection activities are not confined to the control system, of course. Monitoring of plant equipment by engineers, as described in Chapter 19, is a major area of work, usually independent of the control system. Insofar as the control system does monitor malfunctions, however, this function is primarily performed by the operator. The contribution of the computer to malfunction detection is considered later.

Detection of instrument malfunction by the operator has been investigated by Anyakora and Lees (1972a). In general, malfunction may be detected either from the condition of an instrument or its performance. Detection from condition is illustrated by observation on the plant of a leak on the impulse line to a differential pressure transducer or of stickiness on a control valve. Detection from performance is exemplified by observation in the

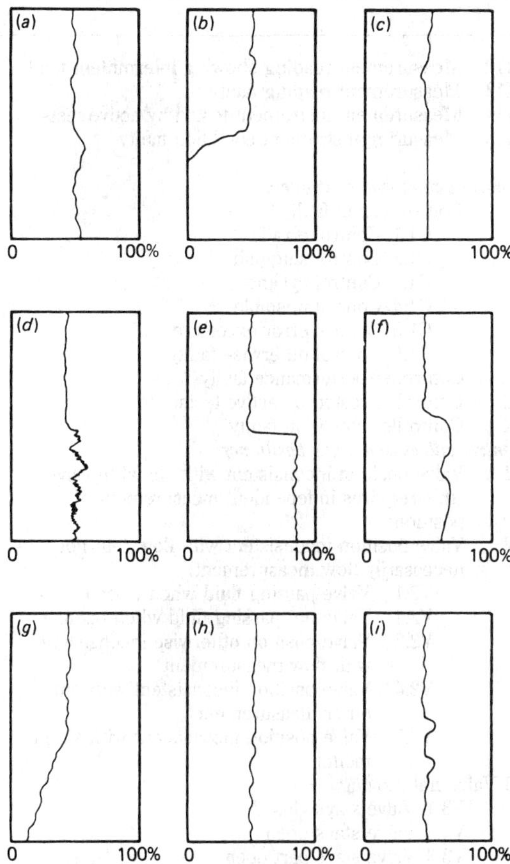

***Figure 14.6** Typical chart recorder displays of measurement signals (Anyakora and Lees, 1972a): (a) normal reading; (b) reading zero; (c) reading constant; (d) reading erratic; (e) reading suddenly displaced; (f) reading limited below full scale; (g) reading drifting; (h) reading cycling; (i) reading with unusual features (Courtesy of the Institution of Chemical Engineers)*

control room of an excessively noise-free signal from the transducer or of an inconsistency between the position of a valve stem and the measured flow for the valve. Most checks on instrument condition require the operator to visit the equipment and use one of his senses to detect the fault. Most performance checks can be made from the control room by using instrument displays and are based on information redundancy.

Some of the ways in which the operator detects malfunction in instruments are illustrated by considering the way in which he uses for this purpose one of his principal detection aids – the chart recorder. Some typical chart records are shown in Figure 14.6. The operator detects error in such signals by utilizing some form of redundant information and making a comparison. Some types of redundant information are (1) *a priori* expectations, (2) past signals of instrument, (3) duplicate instruments, (4) other instruments, and (5) control valve position.

Thus it may be expected *a priori* that an instrument reading will not go 'hardover' to zero or full scale, that it will give a 'live' rather than a 'dead' zero, that it will exhibit a certain noise level, that its rate of change will not exceed a certain value and that it is free to move within the full scale of the instrument. On the basis of such expectations, the operator might diagnose malfunction in the signals shown in Figures 14.6(b–f). However, the firmness of such expectations may vary with the plant operating conditions. For example, during start-up, zero readings on some instruments may be correct.

It may not be possible to decide *a priori* what constitutes a reasonable expectation. The level of noise, for example, tends to vary with the individual measurement. In this case, the operator must use his knowledge of the range of variation of the noise on a particular instrument in the past. Thus Figures 14.6(c) and 14.6(d) might or might not indicate malfunction.

If there is a duplicate instrument, then detection of the fact that one of the instruments is wrong is straightforward, although it may not be possible to say which. However, duplication is not usual in the normal instrument systems which are of primary concern here. On the other hand, near duplication is quite common. For example, the flow of a reactant leaving a vaporizer and entering two parallel reactors may be measured at the exit of both the vaporizers and the reactors, and the flow measurement systems provide a check on each other.

What constitutes a reasonable signal may depend on the signals given by other instruments. Thus, although a signal which exhibits drift, such as that in Figure 14.6(g), may appear incorrect, a check on other instruments may show that it is not.

Some types of variation in a signal, such as a change in the noise level, appear easy to detect automatically. Others, such as that shown in Figure 14.6(i), are probably more difficult, especially if their form is not known in advance. Here the human operator with his well developed ability to recognize visual patterns has the advantage.

There are a number of ways in which the readings of other instruments can serve as a check. Some of these are (1) near-duplication, (2) mass and heat balances, (3) flow–pressure drop relations, and (4) consistent states. This last check is based on the fact that certain variables are related to each other and at a given state of operation must lie within certain ranges of values. The position of control valves also provides a means of checking measurements. This is most obvious for flow measurement but it is by no means limited to this.

These remarks apply essentially to measuring instruments, but checks can also be developed for controllers and control valves. A general classification of instrument malfunction diagnoses by the operator is shown in Table 14.6.

Many of the checks described do not show unambiguously that a particular instrument is not working properly; often they indicate merely that there is an inconsistency which needs to be explored further. However, this information is very important. Another kind of information which the operator also uses in checking instruments is his knowledge of the probabil-

***Table 14.6** General classification of methods of instrument malfunction detection (Anyakora and Lees, 1972a) (Courtesy of the Institution of Chemical Engineers)*

Measuring instruments
M1 Measurement reading zero or full scale
- M1.1 Reading zero
- M1.2 Reading full scale

M2 Measurement reading noise or dynamic response faulty
- M2.1 Reading constant
- M2.2 Reading erratic
- M2.3 Reading sluggish

M3 Measurements reading displaced suddenly
- M3.1 Reading fell suddenly
- M3.2 Reading rose suddenly

M4 Measurement reading limited within full scale
- M4.1 Reading limited above zero
- M4.2 Reading limited below full scale

M5 Measurement reading drifting
- M5.1 Reading falling
- M5.2 Reading rising

M6 Measurement reading inconsistent with duplicate measurement
M7 Measurement reading inconsistent with one other measurement
- M7.1 Reading inconsistent with near-duplicate measurement
- M7.2 Reading inconsistent with level-flow integration
- M7.3 Reading otherwise inconsistent

M8 Measurement reading inconsistent with simple model
- M8.1 Reading inconsistent with mass balance
- M8.2 Reading inconsistent with heat balance
- M8.3 Reading inconsistent with flow–pressure drop relations
- M8.4 Reading otherwise inconsistent

M9 Measurement reading inconsistent with plant operating state
- M9.1 Reading zero but variable not zero
- M9.2 Reading not zero but variable zero
- M9.3 Reading low
- M9.4 Reading high
- M9.5 Reading otherwise inconsistent

M10 Measurement reading inconsistent with control valve position
- M10.1 Reading (flow) zero with valve open
- M10.2 Reading (flow) not zero with valve closed
- M10.3 Reading otherwise inconsistent

M11 Measurement reading periodic or cycling
M12 Measurement reading showing intermittent fault
M13 Measurement reading faulty
M14 Measurement instrument tested by active tests
M15 Measuring instrument condition faulty

Control action and controllers
C1 Control action faulty
- C1.1 Control erratic
- C1.2 Control sluggish
- C1.3 Control cycling
- C1.4 Control unstable
- C1.5 Control error excessive
- C1.6 Control otherwise faulty

C2 Controller performance faulty
C3 Controller tested by active tests
C4 Controller condition faulty

Control valves and valve positioners
V1 Valve position inconsistent with signal to valve (this requires independent measurement of position)
V2 Valve position inconsistent with flow (but not necessarily flow measurement)
- V2.1 Valve passing fluid when closed
- V2.2 Valve not passing fluid when open
- V2.3 Valve position otherwise inconsistent with flow measurement
- V2.4 Valve position inconsistent with one other measurement
- V2.5 Valve position inconsistent with simple model

V3 Valve not moving
- V3.1 Valve stays closed
- V3.2 Valve stays open
- V3.3 Valve stays part open

V4 Valve movement less than full travel
- V4.1 Valve not closing fully
- V4.2 Valve not opening fully
- V4.3 Valve travel otherwise limited

V5 Valve movement faulty
- V5.1 Valve movement erratic
- V5.2 Valve movement sluggish
- V5.3 Valve movement otherwise faulty

V6 Valve movement cycling
V7 Valve or positioner performance faulty
V8 Valve or positioner tested by active tests
V9 Valve or positioner condition faulty

ities of failure of different instruments. He usually knows which have proved troublesome in the past.

The detection of instrument malfunction by the process operator is important for a number of reasons. Instrument malfunctions tend to degrade the alarm system and introduce difficulties into loop control and fault diagnosis. Their detection is usually left to the operator and it is essential for him to have the facilities to do this. This includes appropriate displays and may extend to computer aids.

14.11 Computer-based Aids

There are some functions which the computer can perform automatically, whereas there are others which at present are performed by the operator, but for which computer aids have been, or may be, developed to assist him. Some computer-based aids to assist the operator which have been described include:

(1) system state display;

(2) alarm diagnosis;
(3) valve sequencing;
(4) malfunction detection.

The provision of such an aid is not solely a matter of engineering. There is an essential human factors angle also. A facility which is intended to assist the operator to perform a particular function, as opposed to replacing him by executing that function automatically, must conform to the requirements of the operator.

The main account of computer-based aids is given in Chapter 30. The treatment at this point is confined to consideration of one particular aid, computer-based alarm diagnosis. This, is in its own right, one of the most significant aids, but in addition it also illustrates the human factors problems which can arise.

14.11.1 Alarm analysis

In some process systems the number of alarms which can be generated is large and it seems desirable to help the operator to assimilate these by an analysis of the alarms using a process computer. The problem is most acute in nuclear power stations and alarm analysis has been pioneered by the nuclear industry. Computer-based alarm systems, which include alarm analysis, were installed on the nuclear reactors at Oldbury (Kay, 1966; Kay and Heywood, 1966; Patterson, 1968) and at Wylfa (Jervis and Maddock, 1965; Welbourne, 1965, 1968).

The number of alarms in the systems described is very large. At Wylfa, for example, there are two reactors with some 6000 fuel channels, 2700 mixed analogue inputs and 1900 contacts on each reactor. These systems, therefore, represent an extreme form of the problem of potential information overload which is always encountered when a wide span of control is concentrated in one centralized control station. Some kind of information reduction is required if overload is to be avoided. Thus alarm analysis was undertaken not as a desirable optional facility but rather as a matter of necessity. There are two possible objectives of alarm analysis which are related but distinct. The weaker one is to interpret fresh alarms as they appear in the real-time situation, whilst the stronger one is to identify the original, usually mechanical, cause.

On the Oldbury system, as described by Patterson (1968), the method of analysis is 'alarm tree analysis' in which the propagation of a fault is followed up through successively higher levels of the tree. A prime cause alarm is an active alarm at the lowest level. All important alarms are displayed together, with the cause alarm at the head of the group and the effect alarms beneath it. Other associated effect alarms which are not considered important for the operator's understanding of the fault are not displayed but are 'suppressed' or 'inhibited' by 'darkening'.

Initially, the alarms displayed were the prime cause alarms and the uninhibited alarms; the highest level alarm reached was not shown, unless it was an uninhibited alarm. It was found, however, that this system led to an excessive demand for more uninhibited alarms. It was therefore modified so that the highest alarm reached is always displayed. The operator thus knows how far the fault has propagated.

The analysis method at Wylfa, as described by Welbourne (1965, 1968), is rather similar. When a fresh alarm occurs, it is classed as a prime cause alarm, unless it could have been caused by an existing prime cause alarm, in which case it is classed merely as a new alarm. There are separate VDU displays for prime cause and new alarms. There is, however, no suppression of intermediate alarms.

At Oldbury, in addition to real process alarms based on plant sensors, messages may be generated based on such alarms. If the message relates to a fault, it is in effect a 'deduced alarm'. Similarly, at Wylfa there are 'synthetic' alarms. The number of alarms to be displayed at Oldbury proved to be rather larger than expected. It was necessary to increase the number of 32-line pages of alarms from 3 to 9. The number of alarms is also large at Wylfa, where there is no suppression of alarms.

It is apparent that there are a number of problems in the provision and use of an alarm analysis facility. They fall into two broad categories: those which are concerned with the engineering work involved in conducting a comprehensive analysis, and those which concern the use of the analysis by the operator.

The method of conducting the analysis in the systems described is that a team of experienced engineers studies systematically and in detail the various situations which can occur on the plant and the alarms to which these give rise. The engineering effort in the system described by Patterson was a team of five for 2 years.

As far as concerns process plants, an early exploration was undertaken by Barth and Marleveld (1967), but was not apparently followed through.

One of the main obstacles to the use of an aid such as alarm analysis is the large effort required to create the alarm data structure. It is clear that progress in its application to process plants would be greatly assisted by the development of a systematic and economical method of creating the alarm data structure. By the mid-1970s there were indications of work in this area (Andow, 1973; Andow and Lees, 1975; Powers and Tompkins, 1974b).

Since that time, as described in Chapter 30, the development of real-time aids to assist the operator in handling fault conditions, including expert system aids, has been an active area of investigation, but progress is perhaps best described as sporadic.

Turning to the human factors aspects, alarm analysis is a facility to assist the operator. It is essential, therefore, to give the fullest consideration to the human factors problems which arise. Otherwise, the engineering effort is largely wasted.

The objective of alarm analysis is principally to assist the operator in handling rarely occurring but seriously hazardous situations. These occur typically as a result not of a single fault but of a combination of faults. If an analysis is to be useful, it must be sufficiently comprehensive to include these low probability but hazardous conditions, and it must do this in such a way as to convince the operator, who may be disinclined initially to believe in the existence of a low probability event or combination of events until he has exhausted the more familiar high probability causes. The point is well put by Rasmussen (1968b):

> It may be extremely difficult, if not impossible, for the designer of a large plant to carry through an analysis that takes into account not only all failures in the plant

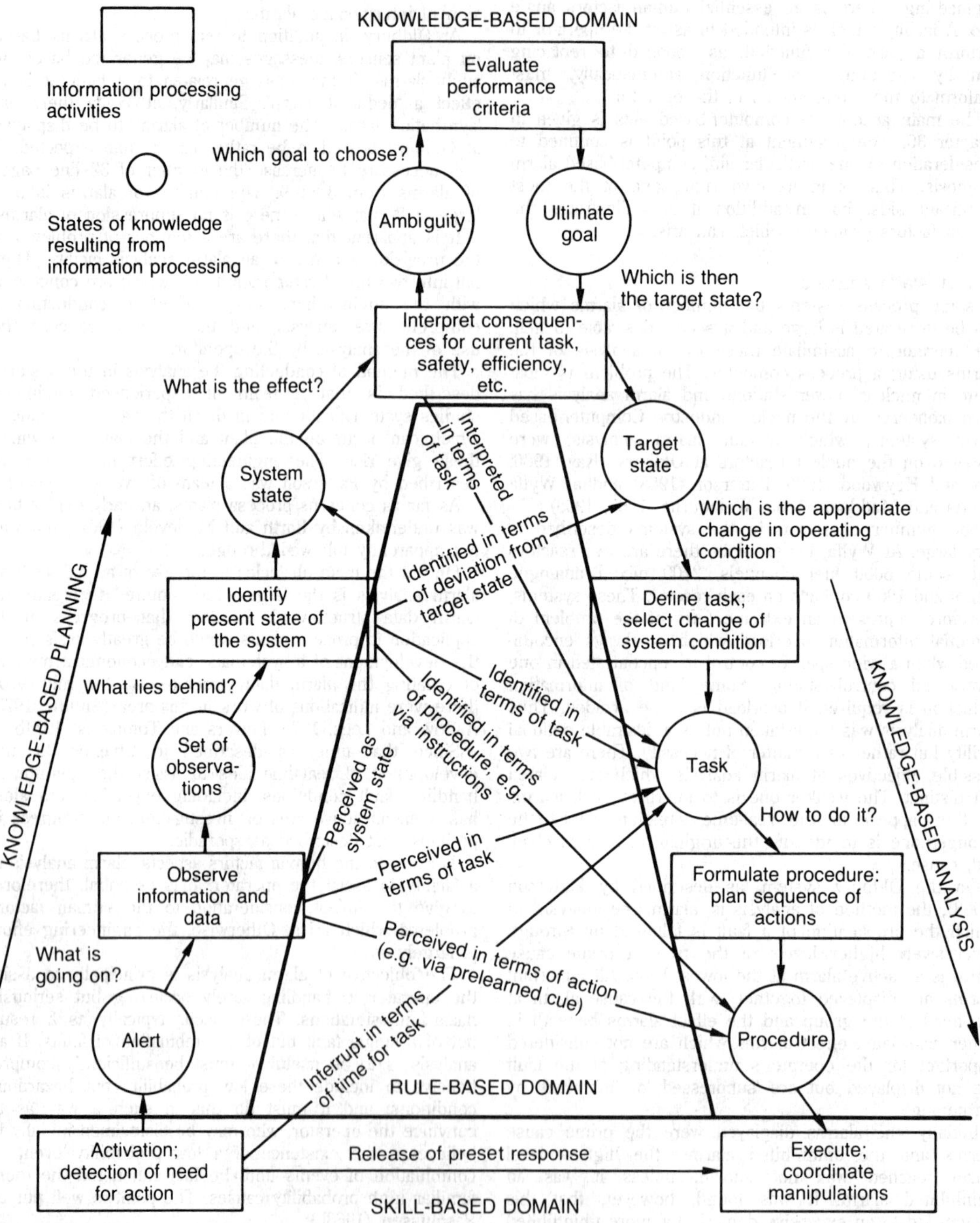

Figure 14.7 *The information processing activities of the operator in making a control decision represented as a ladder diagram (Rasmussen, 1986). The diagram also shows (a) bypassing of certain stages and (b) the domains of skill-based, rule-based and knowledge-based behaviour (Reproduced with permission of John Wiley & Sons Ltd from* Information Processing and Human–Machine Interaction *by J. Rasmussen, 1986, Copyright ©)*

itself and the instrumentation, but also the combination of failures...

One has to realize that direct automatic identification of the primary fault based on a not completely comprehensive analysis, which assumes that the operator critically evaluates the results of the analysis, involves a great risk of further decreasing that probability that the operator takes into consideration very improbable but hazardous failures not dealt with in the simplified analysis of the designer and thus of the instrumentation.

If one utilizes a simplified analysis of failure conditions in this way, one may therefore be in the paradoxical situation that it is as risky for the operator to trust the analysis too much when it indicates a probable cause of failure as to incline towards distrusting it when it indicates an *a priori* improbable cause.

In addition to this problem of the confidence which the operator can place in the analysis, there is the question of the way in which the results are to be displayed. Moreover, the display of the analysis itself is not the whole of the display problem. A knowledge of the current state of the process is also necessary to the operator when a fault is detected. Both his own diagnosis and his corrective action depend very much on this background information. It is not satisfactory that he should have to begin his evaluation of the process state *ab initio* on detection of a fault before he can take action.

An additional aspect is the phenomenon of operator indecision. It has been found by several workers that an operator may have difficulty in bringing himself to actually take a control action. Thus Bainbridge (1974) describes this situation, which in her work occurred even in a simulation study:

> Another type of poor performance, independent of ability to think about the task, is shown by the subjects who can choose the action to take, considering more and more refined dimensions of this choice, but have difficulty in committing themselves to making it.

It seems reasonable to assume that this inability to come to the point of decision is largely a matter of the confidence which the operator has in his assessment of the situation and in his intended remedial action. This in turn depends to a considerable extent on his knowledge of the state of the process, on information redundancy in the displays, and so on.

14.12 Human Information Processing

At this point it is necessary to devote some consideration to the topic of human information processing, some appreciation of which is essential to the understanding of modern developments in areas such as training and human error. The task of the process operator is largely one of processing information. An understanding of the characteristics of human information processing in such a task is therefore crucial to the design of his work situation. In particular, it can contribute to the design of the interface and to assessment of human error. The process operator as an information processor is described in *Information Processing and Human-Machine Interaction* (Rasmussen, 1986). The following is a simplified account of this work.

14.12.1 Cognitive task analysis

An account has already been given of the use of task analysis as an aid to understanding the task of the process operator and its role in design of the system and provision of assistance to the operator. Figure 14.7 shows a generalized framework for task analysis in process operation in the form of information processing activities required to pass from one state of knowledge to the next. The left-hand, upward ladder gives the activities leading to evaluation of some plant state which requires action. The activities are (1) detection, (2) observation, (3) identification, (4) interpretation and (5) evaluation. The right-hand, downward ladder gives the activities leading to action in response to that state. They are (1) evaluation, (2) interpretation, (3) definition of task, (4) formulation of procedure and (5) execution of procedure. Also shown in the diagram are some of the links which allow bypassing of intermediate stages.

Several different types of information processing are involved. There is analysis of the current state of the system, evaluation both of the current state and of the ultimate goal, and planning of the task to be undertaken in response to that state.

14.12.2 Level of abstraction

The operator may think about the system at several different levels of abstraction: (1) physical form, (2) physical function, (3) generic function, (4) abstract function and (5) functional purpose. Consider the example of a pump. The level of physical form corresponds to the appearance of the pump. A typical representation at this level might be the symbol for a pump on the flow diagram. The level of physical function corresponds to the pumping function of the pump. The level of generic function corresponds to the maintenance of flow and that of abstract function to the maintenance of the mass and/or heat balance. These functions in turn contribute to the ultimate functional purpose of the plant. The levels of abstraction constitute a means–ends hierarchy, with the functional end at the top and the physical means of fulfilling this at the bottom.

In making decisions about the system the operator may need to move between the different levels of abstraction. In an emergency situation, for example, he may have to step back and consider the ultimate functional purpose. If he decides that this is best served by maintaining the heat balance, he may consider whether to reconfigure the system. This is a decision at a lower level of abstraction. He may then decide to effect this by altering a flow and may alter this flow by making a physical change to a pump system. This takes the decision-making to still lower levels.

14.12.3 Human behaviour in diagnostic tasks

One of the principal functions of the process operator is diagnosis. On a process plant, however, diagnosis is not entirely straightforward. If an equipment is known to have a fault, the response required from the maintenance technician is relatively simple. It is to identify the fault and repair it. With a fault on a process plant fault matters are not quite so simple. The identification of the fault is typically not the first priority. The priority is generally to evaluate the situation, decide whether the ultimate goal is to be modified, decide the new target state of the plant and define the tasks, and formulate and execute the procedures required to bring the plant to that state. Only if diagnosis of the fault is necessary to this sequence need it be undertaken at this stage.

A study of trouble-shooting in electronic equipment was undertaken by Rasmussen and Jensen (1973, 1974) as an example of human behaviour in a diagnostic task. They identified two broad search strategies:

(1) functional search;
(2) topographic search.

Functional search involves detailed observation of the specific characteristics of the fault, followed by interpretation in the light of a model of the system. It requires relatively few items of data but a possibly quite complex model. It is characteristic of the expert designer.

Topographic search is quite different. It involves a search through the system making a rapid sequence of good/bad checks until the fault is located. It requires a large number of items of data, on average, but only simple good/bad decisions. It is characteristic of the expert maintenance technician.

The maintenance technician is able to operate successfully a topographic search strategy using a rather general search procedure which is not dependent on the specific equipment or fault. The task is viewed not as one of solving a problem but of locating a faulty item. Aspects of the search strategy include search along the main flow path and hierarchical ordering of the search. The technician is also able to make do with a very generalized model of the system. In electronic trouble-shooting the circuit diagram is used for simple purposes such as identification of the main signal paths rather than to obtain a full functional understanding.

In performing topographic search, the maintenance technician follows a strategy which he finds to be broadly optimal in terms of time and trouble. It may not appear optimal in terms of the conventional measures. He will tend to follow the line of least resistance and to take apparently impulsive decisions.

Behaviour in topographic search may sometimes appear to display a fixation. The search may come to an impasse. Even in these circumstances there is relatively little tendency to resort to the use of functional reasoning. The technician is likely to take a break and then to resume using a topographic search strategy again.

Functional search is often used to identify the subsystem which is faulty, prior to the use of topographic search on the subsystem. A common example is trouble-shooting on a TV set where the behaviour of the 'picture' may provide an indication of the type of fault and the associated faulty subsystem.

This work provides background to the diagnostic tasks performed by the process operator. The operator may utilize one of two broad strategies:

(1) topographic search;
(2) symptomatic search –
  (a) pattern recognition,
  (b) sequential decision-making.

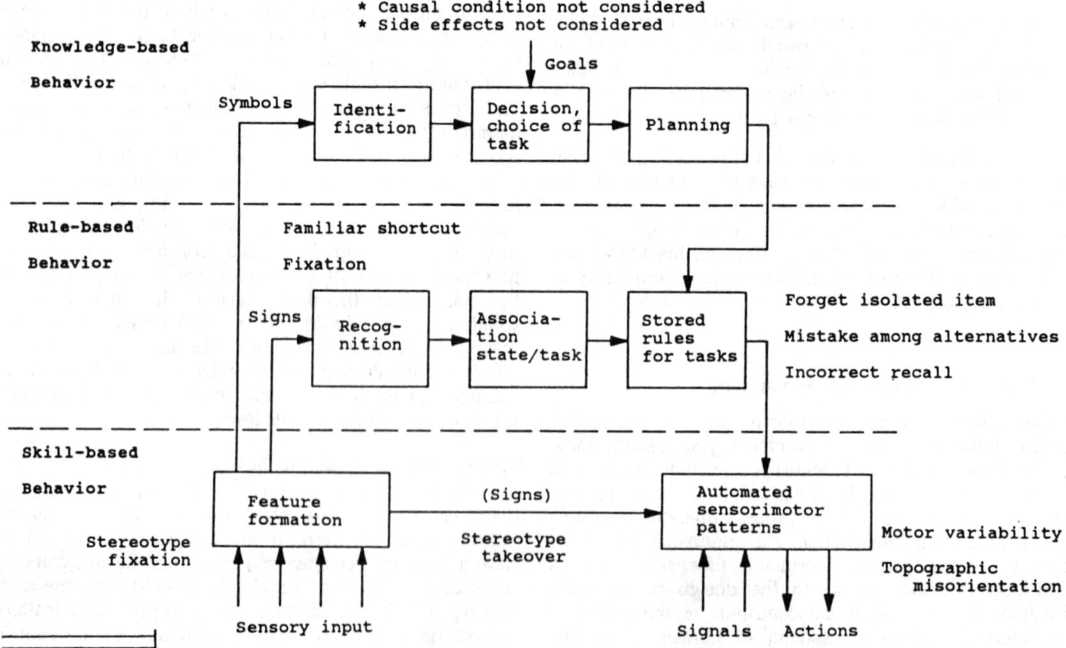

Figure 14.8 The three levels of control of human actions: skill-based, rule-based and knowledge-based behaviour (Rasmussen, 1986) (Reproduced with permission of John Wiley & Sons Ltd from Information Processing and Human–Machine Interaction *by J. Rasmussen, 1986, Copyright ©)*

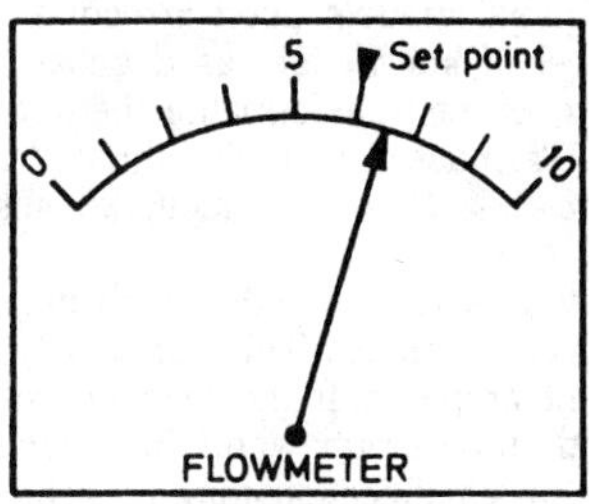

Signal

- Keep at set point
- Use deviation as error signal
- Track continuously

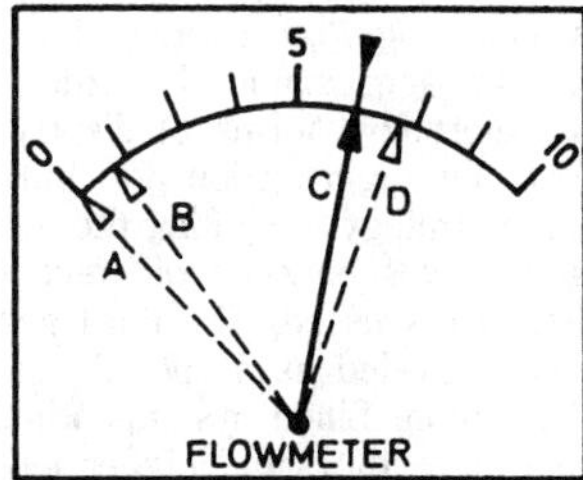

Sign

Stereotype acts:

If valve open, and
  if indication C: ok
  if indication D: adjust flow

If valve closed, and
  if indication A: ok
  if indication B: recalibrate meter

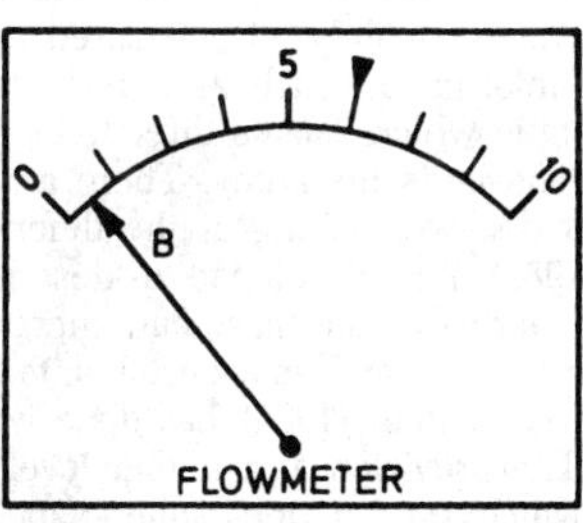

Symbol

If, after calibration, indication is still B, read flow and think functionally (could be a leak)

Figure 14.9 *Three perceptions of the same indication (Rasmussen, 1983): as a signal; as a sign; and as a symbol (Courtesy of the Institute of Electrical and Electronic Engineers)*

The topographic search is based on a map of the system which shows the location of items the state of which can be observed and subjected to a good/bad check.

The symptomatic search may take two forms. It may involve assessment of the state of the system by some form of pattern recognition. Or it may involve a sequential process of decision-making equivalent to logical forms such as truth tables, fault trees and event trees. An alternative sequential approach is search by hypothesis and test.

These considerations have implications for various aspects affecting the operator such as information display, computer aiding and training.

14.12.4 Design of overall control system

Design of the overall control system for a process plant is seen by Rasmussen as involving three principal stages:

(1) definition of control requirements;
(2) analysis of the decision tasks;
(3) cognitive task analysis and design.

The designer defines the control requirements by moving down through the levels of abstraction, defining at each level the context of the control task, including the task specification from the level above and the resources available at the level below. The decision tasks implied in this scheme of control are then analysed. The decision ladder diagram given in Figure 14.7 provides a framework for such analysis.

In the cognitive task analysis the designer formulates the possible information-processing strategies which the operator may use. System design does not require knowledge of the detailed mental models used by the operator. Rather it can be based instead on knowledge of the higher level, more generic, models which operators can and do use. The fact that different operators may use different detailed models also argues for a higher level approach.

Rasmussen gives several ladder diagrams illustrating the decision task analysis for different tasks and in different systems. These all have the same broad structure but differ in detail. For example, different diagrams apply for fully and partially automated systems.

14.12.5 Human behaviour in cognitive tasks

Rasmussen distinguishes three types, or levels, of human behaviour in tasks such as process control. Such behaviour may be:

(1) skill-based;
(2) rule-based;
(3) knowledge-based.

The relation between the three types is shown in Figure 14.8.

This description of operator behaviour by Rasmussen is frequently referred to as the skill–rule–knowledge, or SRK model. Skill-based, or skilled, behaviour occurs without conscious attention and is data-driven. Rule-based behaviour is consciously controlled and goal-oriented. Knowledge-based behaviour is also conscious and involves reasoning. These three types of behaviour apply to both parts of the decision ladder diagram, as shown in Figure 14.7.

Another, closely related distinction is that between (1) a signal, (2) a sign and (3) a symbol, illustrated in Figure 14.9. The distinction lies in the way in which the input is perceived. A signal is an input perceived simply as a continuous quantitative indicator of system state. A sign is perceived as an indication of some discrete state and often of the need for action. A symbol relates to some functional property of the system. A sign cannot be used for functional reasoning about the system, which is the province of the symbol. Signs belong essentially to the domain of rule-based behaviour, and symbols to that of knowledge-based behaviour, as shown in Figure 14.8.

14.12.6 Human as control system component

Much work has been done on human behaviour in control tasks generally and in process control tasks in particular, and various characterizations have been developed of man as a component in a control system.

Rasmussen discusses the behaviour of the human as a component in a control loop, and his needs for aiding, in relation to the successively more difficult tasks of (1) direct manipulation, (2) indirect manipulation, (3) remote manipulation and, finally, (4) remote process control. In the first three cases the operator receives information in analogue form and exercises a basic sensorimotor skill in a relatively simple space–time loop. This may not be so in the fourth case, where the signals on both the input, or sensory, channel and the output, or manipulative, channel, are likely to be symbolic and therefore to require translation. Such translation is a relatively high-level activity and therefore breaks the simple space–time loop.

In this light, the aim of the designer should be to restore to the operator the ability to apply to the task the basic sensorimotor skill. This requires that the designer do at least two things. On the sensory channel the information presented should be based on symbols directly related to the function to be controlled. On the manipulative channel the information presented should provide symbols and structure which allow the symbolic representation of the space–time feature to be directly manipulated.

14.12.7 Mental models used in cognitive tasks

At the level of knowledge-based behaviour, the operator requires to effect transformation between different types of model. Rasmussen identifies three strategies which he uses to do this: (1) aggregation, (2) abstraction and (3) analogy.

If a system is viewed with a high degree of resolution, it may appear complex. This complexity, however, is not an inherent property. If the elements of the system are aggregated into a larger whole by reducing the degree of resolution, a simplification may be achieved.

The apparent complexity of a typical control system is largely due to the one-sensor–one indicator technology. Given this technology, the operator has to transform the information by aggregation in order to reduce the complexity. Alternative forms of display can be devised which effect this aggregation for him. Another basic transformation strategy is shifting the level of abstraction. The operator moves between different levels of abstraction, as already discussed. The third strategy is analogy. This may be regarded as a special case of shifting the level of abstraction. Since systems which are physically different may have the same higher level model, a shift to this higher level may provide a useful analogy.

These considerations have implications for the decision support systems provided for the operator. Traditionally, displays consist mainly of measured values and some structural information such as mimic diagrams. There is relatively little which relates directly to the various levels of the means–ends hierarchy. There is need for displays to support decision-making at the different levels.

Lind (1981) has developed a description of process systems in terms of the mass and energy balances which has several levels and maps well onto the means–ends hierarchy. Goodstein (1984) has done work on multilevel displays. The provision at a higher level of abstraction of displays which give a one-to-one mapping between the appearance of the display and the properties of the process to be controlled allows rule- and skill-based behaviour to be extended to this higher level.

A functional representation can serve as a set of prescriptive signs. Cuny and Boy (1981) have shown that an electrical circuit diagram can be analysed as a set of signs controlling activities in design, installation and repair. A closely associated topic is co-operative decision-making where the computer supports human decision-making by undertaking parts of the information processing.

With regard to the form of mental models, the model used by man tends to be based on common-sense reasoning, or causal reasoning. As a model at the purely physical level this does not have the precision of a formal mathematical model such as a set of differential equations, but it is serviceable and can be extended to higher levels of reasoning.

14.12.8 Human error in cognitive tasks

The causes of unsatisfactory system performance are technical faults and human error. The tracing of the causes of poor performance by an investigator is arbitrary to a degree. Rasmussen points out that there is a strong tendency for the trace to terminate if it reaches a human; it does not often pass through.

He discusses human error mainly in terms of human variability and of human–machine mismatch. This variability is a desirable characteristic. It is an important ingredient in human adaptability. In particular, it plays an important part in the learning processes. Small excursions outside limits are needed in order to learn where limits lie. Similar experimentation underlies the development of rules-of-thumb.

If the environment is unforgiving, however, so that if such excursions prove to have bad effects it is not possible to recover, error becomes a problem. Overall, it is generally more fruitful both for error prevention and error prediction to concentrate on recovery from error than on the initial error.

At the level of knowledge-based behaviour error may occur due to selection of an inappropriate goal or selection of an appropriate goal followed by incorrect implementation.

At the level of rule-based behaviour a prime cause of error is a change in the environment. Practice of a task leads to a tendency for knowledge-based behaviour to be replaced by rule-based behaviour and use of signs. If the environment changes so that the signs do not alter but the rules are no longer appropriate, error is liable to occur. This may be compounded by the tendency to utilize convenient, informal signs instead of formal signs and to modify rules to give more convenient sequences.

Human error may occur in a normal situation in the execution of familiar tasks. Some of the mechanisms include (1) motor variability, (2) topographic misorientation, (3) stereotype take-over, (4) forgetting an isolated item and (5) selecting an incorrect alternative.

Motor variability is exemplified by an operator applying different degrees of force to a set of valves so that one of them is not leak-tight. Selecting the wrong pump in a set of pumps is an example of topographic misorientation. Stereotype take-over is illustrated by switching during an emergency shut-down sequence to the regular shut-down sequence.

Forgetting an isolated item can take many forms. A typical form is omission of an isolated act. An analysis of nuclear plant test and calibration reports by Rasmussen (1980b) found that this category accounted for 50% of errors. Precisely because the act is an isolated one, the probability of initial error tends to be high and that of recovery low. Another form of forgetting an isolated item is incorrect recall of a number. Insertion by an operator of a wrong set-point is an example. An example of incorrect selection between alternatives is adding to a figure a correction factor which should in fact be subtracted.

Other types of human error occur in off-normal situations. Skill-based behaviour is appropriate to the normal situation, but if the situation changes it may no longer be appropriate. Generally, however, adaptation tends not to take place until a mismatch has occurred.

Rule-based behaviour is liable to two types of mismatch: (1) stereotype fixation and (2) stereotype take-over. An example of stereotype fixation is execution of a sequence of operations appropriate to normal dusts when in fact the dust being handled is radioactive and calls for additional precautions. Stereotype take-over, by contrast, is exemplified by an operator initially conscious of the need to vary a sequence of operations to suit the particular circumstances, such as the dust-handling task just described, but then relapsing into the normal sequence. Such interference is more likely to occur in situations where the mind is occupied by forward planning of other activities before the action concerned has been executed.

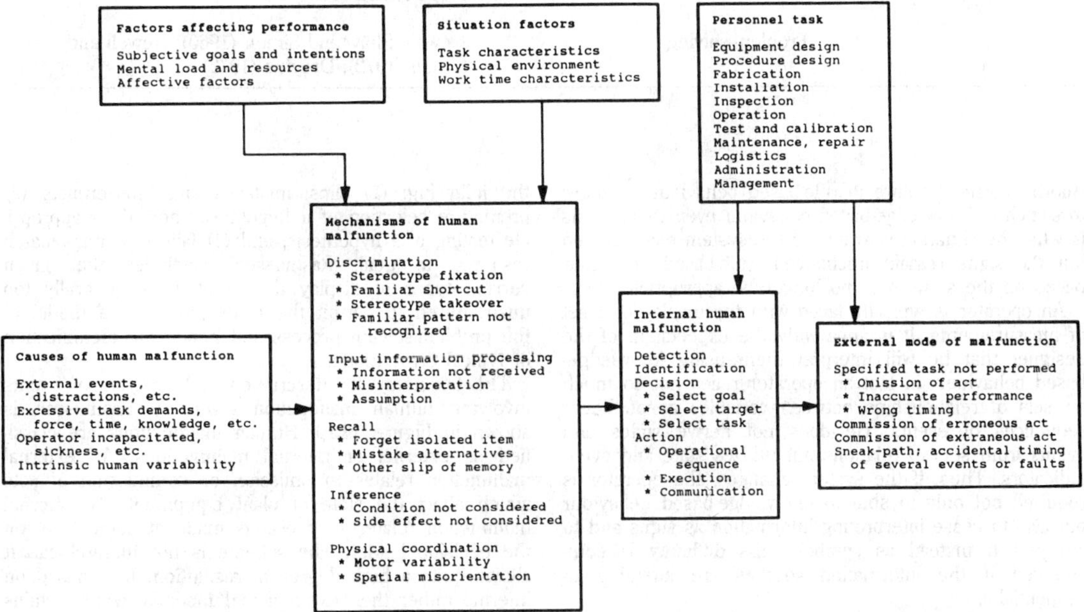

Figure 14.10 *A taxonomy for description and analysis of events involving human malfunction (Rasmussen, 1982b) (Reproduced with permission of John Wiley & Sons Ltd from* High Risk Safety Technology *by A.E. Green, 1982, Copyright ©)*

Table 14.7 *Some models of human information processing.*

Model	*Further description*	*References*
Attention allocation	Queueing theory	Carbonell (1966); Senders and Posner (1976); Rouse (1977)
	Sampling theory	Senders (1964)
Detection	Signal detection Estimation theory	Wald (1947); Sheridan and Ferrell (1974) Gai and Curry (1976)
Manual control		Crossman and Cooke (1962); Sheridan and Ferrell (1974); Pew (1974); Baron *et al.* (1982)
Human judgement		Hammond, McClelland and Mumpower (1980)
Decision theory		Keeney and Raiffa (1976)
Behavioural decision theory	Subjective probability and its revision by Bayesian approach	W. Edwards and Tversky (1976)
Psychological decision theory	Subjective probability	Tversky and Kahneman (1974)
Social judgement	Expert judgement	Hammond, McClelland and Mumpower (1980); Brehmer (1981)
Information integration theory	Cognitive algebra of stimulus–response	N. Anderson (1974)
Attribution theory		Heider (1958); Kelley (1973)
Fuzzy set theory		Zadeh (1965); Gaines (1976); Gaines and Kohout (1977)
Artificial intelligence models	Expert systems	Ringle (1979); Schank and Abelson (1977); Feigenbaum (1979)
	Problem-solving	Newell, Shaw and Simon (1960); Newell and Simon (1972); Dreyfus (1972)

Another form of failure in rule-based behaviour is failure to switch to knowledge-based behaviour even though this is what the situation warrants. If the system has changed but the signs remain unchanged, rule-based behaviour based on the signs may no longer be appropriate.

An operator is typically faced with the display of a set of measurements. It is apparently the expectation of the designer that he will interpret them using knowledge-based behaviour. In fact an operator relies not so much on sets of relationships between variables as on linear sequences of events. He does not derive states and events from sets of relations but utilizes state and event indicators. Thus, if the system changes, the operator is required not only to shift to knowledge-based behaviour but also to cease interpreting information as signs and to interpret it instead as symbols. His difficulty is compounded if the information sources are attended to sequentially.

Once the domain of knowledge-based behaviour is entered, it becomes much more difficult to characterize the mechanisms leading to mismatch. Some types of human error which may occur in this situation include the following: (1) slips, mistakes and interference; (2) premature selection of a hypothesis; and (3) inappropriate testing of a hypothesis, and (4) failure to map/match resource to goal. Rasmussen concludes that given current types of display the situation is generally too unstructured to permit the development of a model of the problem-solving process and hence the identification of typical error modes.

The taxonomy for description and analysis of events involving human malfunction given by Rasmussen is shown in Figure 14.10. Human malfunctions are classified as external or internal malfunctions. An external malfunction relates to omission or commission of acts which affect the state of plant equipment. An internal malfunction relates to decisions made at some level on the decision ladder. The scheme is not hierarchical. It allows quite a high degree of resolution. It is based on internal rather then external malfunctions, which retains the internal structure of the error, allows a reasonably economical description of it and avoids the combinatorial explosion which is liable to occur in a scheme based on external malfunctions.

One of the factors which affects the probability of error is stress. However, although much work has been done on the effects of stress, relatively little is applicable to cognitive tasks. An abnormal situation tends to result in a modification of behaviour even if there is little stress. But frequently in an abnormal situation there is an increase in stress. Apart from the effect of anxiety, there are some specific effects related to the cognitive task itself. The need to apply functional reasoning to a disturbed function increases the workload and reduces the time available for general monitoring of the state of the system. At the same time the existence of abnormal conditions means that familiar indicators may be less reliable as guides to the state of the system. There is a tendency for the operator to focus on the disturbed function and thus to exhibit the cognitive tunnel effect. The need for shifts of strategy is another factor increasing stress.

Work on pilot performance by Bartlett (1943) indicates that under stress the tendency is for skilled subroutines to be retained but for the higher level co-ordination of these routines to deteriorate.

14.12.9 Models of human information processing

There exists a variety of models of human information processing activities which may have a contribution to make in the present context. These are reviewed by Rasmussen in respect of the characteristics of each model and of its relationships with the other models. The models are summarized in Table 14.7.

14.13 Task Analysis

The analysis of the task to be done logically precedes other stages of the design process, such as interface or training design, as shown in Figure 14.1. Insofar as any analysis of the operator's task is made in the process industries, this has traditionally tended to be done implicitly in the course of writing the plant operating instructions. Increasingly, however, use is being made of more fundamental approaches, mainly based on the various forms of task analysis.

14.13.1 Hierarchical task analysis

Early work was given in *Task Analysis* (Annett *et al.*, 1971). The method described by these authors is hierarchical task analysis (HTA), in which the task is broken down into a hierarchy of task elements. The hierarchical task analysis method of Annett *et al.* has been developed by Duncan and co-workers (Duncan, 1974; Duncan and Gray, 1975a,b; Duncan and Shepherd, 1975a,b).

The method adopted is to break the task down into a hierarchy. The elements of the hierarchy are a goal, plans and operations. The task involves a goal and this is then redescribed in terms of the plans and operations necessary to achieve it. Operations are units of behaviour, typically with an action–information feedback structure. The main operations in the task of controlling an acid purification plant are shown in Figure 14.11, the full hierarchy of operations in Figure 14.12 and representative subordinate operations in a start-up procedure in Figure 14.13.

One of the difficulties in any task analysis is to know when to stop. Unless there is a suitable stopping rule, the redescription gets out of hand. The rule used is that redescription stops when the product of the probability p and the cost c of failure is acceptably low. The application of the rule is shown by the underlining in Figure 14.11. Double underlining indicates that this

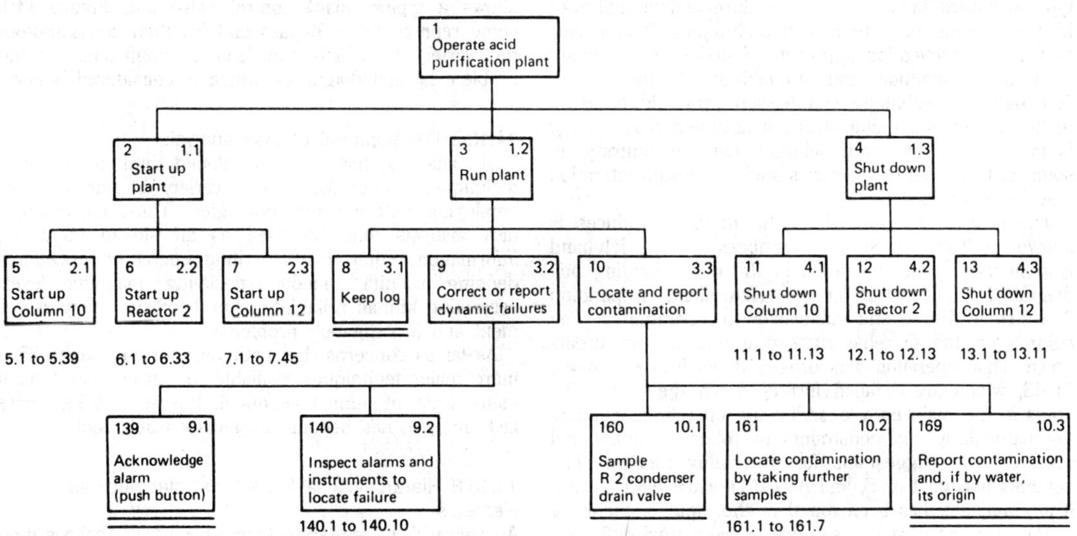

Figure 14.11 *Main operations in an acid purification task (Duncan, 1974) (Courtesy of Taylor & Francis Ltd)*

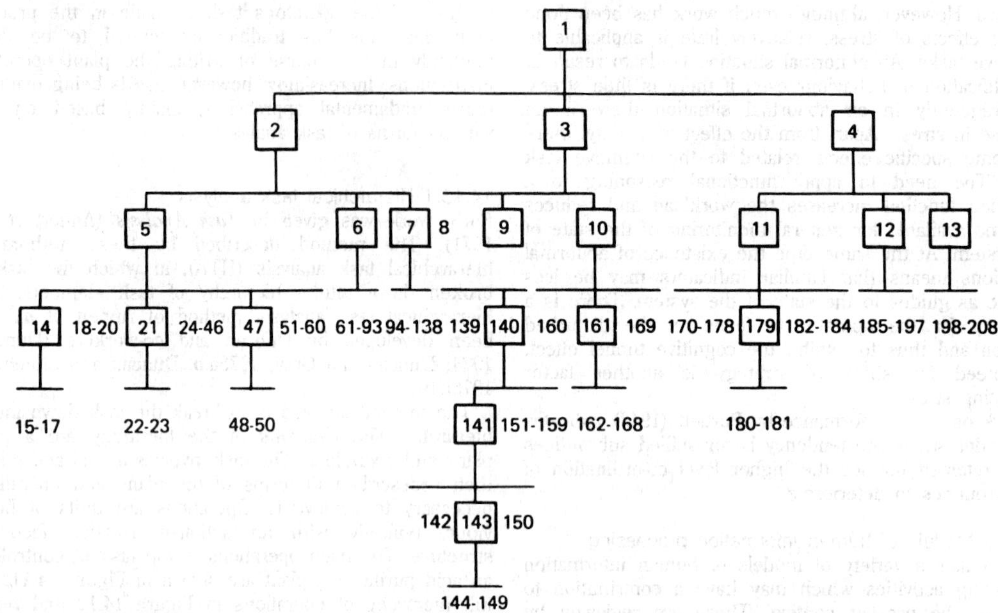

Figure 14.12 *Hierarchy of operations in an acid purification task (Duncan, 1974) (Courtesy of Taylor & Francis Ltd)*

product has an acceptably low value. This is usually because the action required is trivial and does not call for training. Single underlining denotes a possible difficulty and perhaps need for training.

Thus, as mentioned in Chapter 8, this method is a type of hazard identification procedure and should take its place along with other such techniques. It conforms to the loss prevention approach of taking into account both the magnitude and probability of the hazard. Although the technique is intended primarily to assist in the design of training, it is not assumed that training is in all cases the right solution. On the contrary, in some instances other measures such as design alteration may be more appropriate.

The type of material which the method produces is shown in Table 14.8. The numbers in the left-hand column identify the operation in the corresponding box diagram of Figure 14.12. Thus $\frac{3}{1.2}$ (Run plant) is operation 3, which is the second subdivision of operation 1. The column on the extreme right refers to further breakdown. Thus operation 4 is broken down into operations 11–13, which are detailed further down the table. The notes in the main body describe the operation, mentioning particularly any constraints or tolerance limits and also tentative suggestions for a training method. The columns headed I or F, and A, are checked with a cross if any difficulty has been found in the input or feedback to the operator (that is a sensory or perceptual difficulty) or if an action difficulty (inability to perform the motor act) has been found. The letter R indicates that the operation is redescribed elsewhere in the table.

The task analysis breaks the task down into operations which are carried out according to a plan. The simplest plan is a fixed sequence, but variable sequences may be handled, and training may be particularly important for these. One problem that task analysis studies reveal is the identification of equipment, such as bypass and isolation valves around control valves. Figure 14.14 shows a typical plant control valve and Figure 14.15 some representative bypass and isolation configurations, the variety of which can lead to confusion. Another problem is fault diagnosis, which is considered below.

14.13.2 Development of task analysis

Task analysis has now developed into a family of techniques, undertaken for a variety of purposes and employing different methodologies. Thus, for example, task analysis may be used as an aid to identifying information requirements, writing operating procedures, defining training needs, specifying manning levels, estimating human reliability in probabilistic safety assessment and investigating problems.

As far as concerns the methods, there are some 25 or more major techniques available. In effect, therefore, in many areas of human factors in the process industries, task analysis has become an indispensable tool.

14.13.3 Hierarchical task analysis: standard task elements

An approach to rendering hierarchical task analysis more systematic has been described by A. Shepherd (1993), with special reference to the information requirements of the process operator, as described below. In HTA the task is decomposed into a set of subgoals for each of which there is a set of information requirements, which

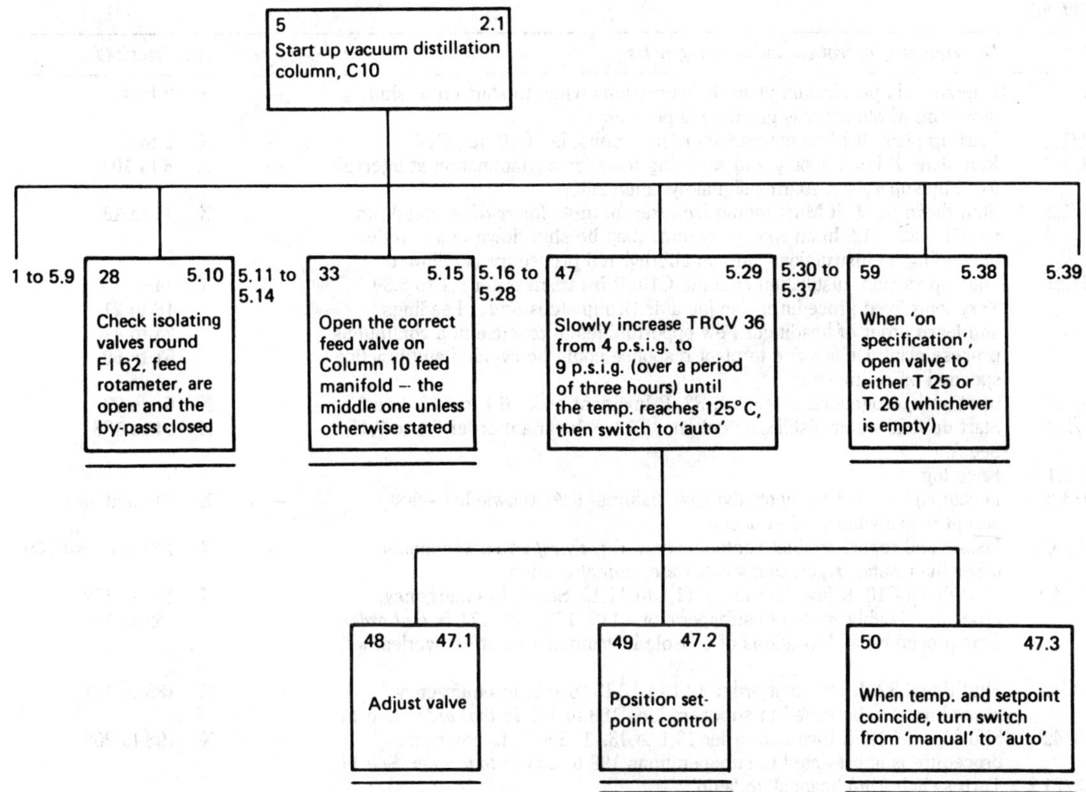

Figure 14.13 *Representative subordinate operations in a start-up procedure (Duncan, 1974) (Courtesy of Taylor & Francis Ltd)*

the author calls 'subgoal templates' (SGTs). He identifies five broad classes of task element:

(1) sequence elements (S);
(2) action elements (A);
(3) communications elements (C);
(4) monitoring elements (M);
(5) decision-making elements (D).

He further divides each class of element into subclasses to give the scheme shown in Table 14.9.

Shepherd illustrates the translation of a conventional HTA into his standard form by reference to the simple example shown in Figure 14.16. The original plan is:

> *Upon instructions from supervisor, do 1, then do 2, then when temperature = 100 degrees, do 3. Then if pH is within specification, do 4, otherwise do 5.*

This plan is translated into standard task elements as follows, where the left-hand side shows the plan and the right-hand side the standard elements:

Upon instruction from supervisor	– receive instruction (C4)
do	– then do (S1)
1 (Establish feeds)	– redescribed
then do	– then do (S1)
2 (Open steam valve to 7 psi)	– redescribed
then	– then do (S1)
when temperature = 100 degrees	– monitor to anticipate change (M2)
do	– then do (S1)
3 (Switch steam to 'automatic')	– activate (A1)
then	– then do (S1)
If pH is within specification	– monitor to anticipate change (M2)
do	– then do ... either (S2)
4 (Switch feeds to 'automatic')	– activate (A1)
otherwise do	– then do ... or (S2)
5 (Adjust feeds)	– redescribed

In this method sequence elements need to be introduced between each of the other elements. This serves to clarify for each subgoal in the task the information required and the time when it is needed. Another feature is that monitoring elements and decision-making elements need to be introduced within sequences to handle contingencies that may arise. Frequently in task analyses these elements are not made explicit.

Table 14.8 *Extracts from the analysis of an acid purification task (Annett et al., 1971) (Courtesy of HM Stationery Office)*

No.	*Description of operation and training notes*	*I or F*	*A*	*redescribed*
1	Operate acid purification plant. R Instructions when to start up or shut down the whole process given by supervisor	—	X	2 to 4
2/1,1	Start up plant. R Must memorize order of units, i.e. C10, R2, C12	—	X	5 to 7
3/1,2	Run plant. R Log keeping and sampling tests for contamination at intervals fixed by supervisor. Alarms signal dynamic failure	—	X	8 to 10
4/1,3	Shut down plant. R Must memorize order of units for ***routine*** shut-down, i.e. C10, R2, C12. In an ***emergency*** units may be shut down in any order depending on instructions, and an abbreviated procedure is followed	—	X	11 to 13
5/2,1	Start up vacuum distillation column, C10. R Invariant order 5,1 to 5,39. Very long fixed procedure. Use job aid? Group steps under headings and learn order of headings? Few opportunities to practise in a continuous process plant. Delays due to plant response could be avoided and practice speeded on a simulator	—	X	14 18 to 21 24 to 47 51 to 60
6/2,2	Start up hydrogenerator reactor, R2. R Invariant order 6,1 to 6,35. See 5	—	X	61 to 93
7/2,3	Start up final water distillation column, C12. R Invariant order 7,1 to 7,45. See 5	—	X	94 to 138
8/3,1	Keep log			—
9/3,2	Locate and correct or report dynamic failures. R Acknowledge—first accept responsibility—then locate	—	X	139 and 140
10/3,3	Locate and report product contamination. R ***Only report*** contaminants other than water; ***report and locate*** water contamination	—	X	160, 161 and 169
11/4,1	Shut down C10. R Invariant order 11,1 to 11,13. See 5. In emergency, procedure is abbreviated to suboperations 172, 170, 171, 174 ***in that order.*** This procedure and locations of console instruments must be overlearned	—	X	170 to 179 182 to 184
12/4,2	Shut down R2. R Invariant order 12,1 to 12,13. See 5. In emergency, procedure is abbreviated to suboperations 188 to 193 ***in that order.*** See 11	—	X	185 to 197
13/4,3	Shut down C12. R Invariant order 13,1 to 13,11. See 5. In emergency, procedure is abbreviated to suboperations 198 to 203 ***in that order.*** See 11	—	X	198 to 208
23/21,2	Turn switch from 'manual' to 'auto'			
25/5,7	Open inlet valve to reflux pump in use and check outlet, drain and valves about stand-by pump are closed (P66/67). I Must find way through plant to pump plinth 66/67—'landmarks' for learning plant geography? Pumps are labelled but inlet valve is one of six unlabelled hand valves on pump lines. Rote learn which is which, OR learn generalization and discrimination of classes of valves, i.e. ***inlet, outlet*** or ***discharge,*** and ***drain?*** Sketch to indicate relevant valves and lines against cluttered plant background?	X	—	—
48/47,1	Adjust valve. A Compensatory tracking of pen movements to produce an acceptable slope on recorder. Plant control law might be simulated, but expense justifiable? Practice on the job under supervision feasible, but would be infrequent. F Examples of 'acceptable' and 'unacceptable' slopes could easily be provided	X	X	—
78/6,18	Set up a pressure of 1 psig on the gauge after the block valve. Then check that there is a purge through the system by putting a hand over the outlet vent. F Must find way through plant to outlet valve. Once pointed out, vent is easily identified	X	—	—
161/10,2	Locate origin of water contamination by taking further samples from other points on the plant. R The number and order of samples taken, i.e. suboperations, varies. Given (a) a relative cost index for each of 7 sampling points and (b) for each of 8 plant components, estimated probability of failure (i.e. contaminating product), the operator must apply a search strategy which minimizes sampling costs. Need flow diagram showing only relevant information, i.e. (a) and (b). Simulation inevitable—in this case instructor might provide 'lab reports' for a series of 'faults' (occurring with a frequency proportional to their probabilities). Should trainee (1), be told efficient strategy, e.g. decision tree-prescribing branching sequence of samples? or (2), attempt unaided to minimize his sampling costs?	—	X	162 to 168

***Figure 14.14** A slave valve and sensing device with bypass and isolating configurations (Duncan, 1974) (Courtesy of Taylor & Francis Ltd)*

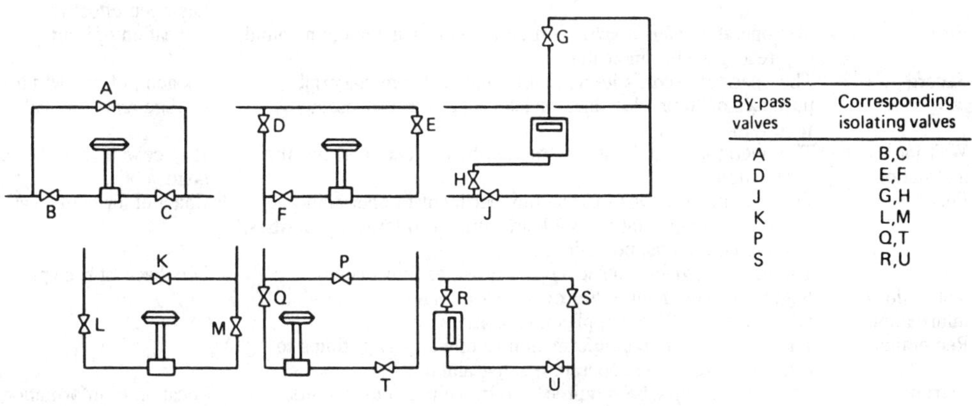

By-pass valves	Corresponding isolating valves
A	B,C
D	E,F
J	G,H
K	L,M
P	Q,T
S	R,U

***Figure 14.15** Some bypass and isolating configurations (Duncan, 1974) (Courtesy of Taylor & Francis Ltd)*

14.13.4 Applications of task analysis: operator information requirements

A methodology for the definition of the information requirements of the process operator where he is part of a control loop has been described by A. Shepherd (1993) in the work just referred to. He describes a number of operator information acquisition problems found in industrial studies. One is 'breaking the loop'. In one case the display of information on the effect of control actions was distant from the controls to be manipulated, being two floors up from the controls. In another the problem was that the control actions on a rather sensitive distillation process on a batch plant had to be performed by interrupting work on preparation of batches and going to another part of the plant, with the result that the operators tended to adopt a rather conservative and uneconomic control strategy. In a third example the target information and the controls required to effect changes were shown on different VDUs because the VDUs were arranged to display information by geographical area on the plant and the equipments were in different areas, with the result that in order to make a change two operators, each viewing one screen, had to shout across the control room to each other.

Another problem identified was 'information fragmentation'. This can arise where the operator has to examine several items of information before making a decision. In

Table 14.9 *Standard task elements for hierarchical task analysis (A. Shepherd, 1993) (Courtesy of Taylor & Francis Ltd)*

Code	*Label*	*Description*	*Information suggestion*
S1	then do	Fixed sequence—where the first operation of REMAINDER is carried out upon completion of the previous operation	
S2	then do either ... or ...	Choice—where the first operation of REMAINDER varies in accordance with a specified condition detected in the first operation	
S3	then do together ... and	Time sharing—where the first operation of REMAINDER is carried out at the same time as the previous operation	
S4	do in any order ...	Sometimes it does not matter which order suboperations are carried out in. This is the case in setting-up operations, e.g. prior to plant start-up	
A1	Prepare equipment	The operator sets items of equipment, e.g. valves, in readiness for subsequent operations	Indication of alternative operating states. Feedback that equipment is set to required state
A2	Activate	The operator carries out an act locally on plant or on a central or local control panel to bring an item of equipment on-line or take it off-line, e.g. open valve, switch on centrifuge	Auditory or mechanical feedback that the action has been effective
A3	Adjust	The operator carries out an act locally or on a central or local control panel to adjust an item of equipment to modify operating rates, e.g. increase amps, adjust valve controller, decrease feed-pump rate	Feedback confirming controller position
A4	De-activate	The converse of A1	Auditory or mechanical feedback that the action has been effective
C1	Read	The operator reads a value of a parameter or listens to a sound, e.g. read pressure indicator	Indication of item
C2	Record	The operator records items of information for managerial purposes or to provide data for later stages in processing (e.g. remember, record)	Location of record for storage and retrieval
C3	Wait for instruction	The operator waits for an instruction before commencing the next action	Project wait time. Contact point
C4	Receive	The operator receives instructions pertinent to operation, e.g. to proceed or not proceed, adjustment to targets, priorities, running conditions, potential hazards	Channel for confirmation
C5	Give instruction/ information	The operator gives instructions or provides information (e.g. to management, other operators in plant, operators on other plants outside, suppliers, customers, services)	Feedback of receipt
C6	Remember	The operator commits information to memory (e.g. times to act, operating values, operating constraints)	
C7	Retrieve	The operator may be required to retrieve previously stored information to provide current operating information	Location of information for retrieval
M1	Monitor to detect deviance	Target values of particular parameters are to be maintained through specified phases of processing, usually as a precautionary action. This form of monitoring is usually intermittent, with low expectancy of deviation, unless certain preconditions have occurred. Often this form of monitoring is dealt with by attention grabbing alarms	Key parameters to monitor
M2	Monitor to anticipate change	The operator anticipates change to detect a cue for a subsequent action. In contrast to M1, the operator expects change, usually after the passing of some time or the occurrence of an intermediate event	Key parameters
M3	Monitor rate of change	Where the current plant status is changing over time (e.g. during a plant start-up or other change of operating conditions), the operator may have to ensure that rates of change are within tolerance	Key parameters to monitor Change rates
M4	Inspect plant and equipment	The satisfactory state of equipment may only be determined by human inspection via the human senses	Access to symptoms
D1	Diagnose	Operators must diagnose situations judged out of tolerance	Information requirements

	process problems	to determine a course of compensatory or corrective action	must be identified in conjunction with training strategies
D2	Plan adjustment	The operator must observe conditions and work out a sequence of steps to move to a desired (safe) state in an acceptable fashion. Planning must anticipate the consequence of choices of intermediate actions and select only those which lead to acceptable conditions	Planning information deriving from analysis of typical cases
D3	Locate contaminant	In plants where there may be problems associated with contaminants entering pipework, through impure feeds or leaks in heat exchangers, there may be a need for examination of samples at various points in order to detect the site of the problem	Sample points in the system enabling the problem, eventually to be bracketed between a clean input and a contaminated output
D4	Judge adjustment	Often operators are required to make an adjustment to an item of equipment to affect processing conditions. The degree of adjustment must be judged prior to the action	The target indicator Adjustment values

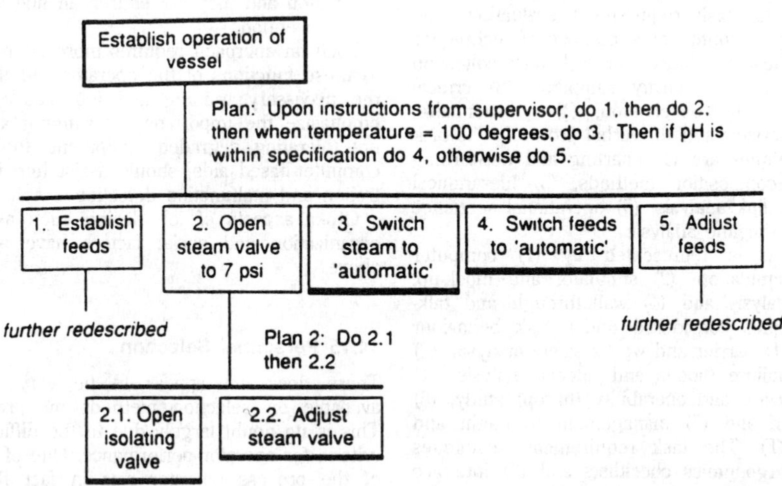

Figure 14.16 *Hierarchical task analysis of part of a process control operation (A. Shepherd, 1993) (Courtesy of Taylor & Francis Ltd)*

one case the operator was required to deal with disturbances in a particular part of the plant but in order to do so required information from points throughout the plant. This was not too difficult to do using the original conventional control panel but with a new VDU system it became very difficult, because the menu hierarchy was organized geographically and considerable search activity was involved in order to locate the necessary data.

14.13.5 Applications of task analysis: emergency stand-by system

Umbers and Reiersen (1994) have described the application of task analysis to assessment of an emergency stand-by system. This arose as part of the long-term safety review (LTSR) of the UK Magnox nuclear reactors, which includes the installation of a new, tertiary feed water system as stand-by for the two existing systems. Several task analysis techniques were used, including hierarchical task analysis and timeline analysis.

Information for the work was drawn from a wide variety of sources. The documentation consulted included the outline design specification, summary of safety arguments, fault study results, station operating procedures, site drawings, and plant system drawings. Discussions were held with operations staff, the feed system designers and fire service teams. A timed walk-through was done.

Also of interest are the topics on which assumptions had to be made, including: plant status following a fault; system design aspects; maintenance and testing aspects; and the number, training and responsibilities of personnel.

The study had a number of outcomes. It gave estimates of the time to commission the tertiary feed water system in an emergency. It highlighted the importance of avoiding delay in establishing loss of the primary and secondary feeds and identified the indications of such loss, some of which varied as between one plant and another. It showed up potential errors and simple design enhancements to avoid them.

The work included a comparison of the commissioning times predicted by the task analysis with those found in demonstration exercises. The differences were attributable mainly to pessimistic assumptions made by the analysis.

14.13.6 *Task Analysis Guide*

A full treatment of the topic is given in *A Guide to Task Analysis* (Kirwan and Ainsworth, 1992) (the *Task Analysis Guide*). Other accounts include those by Drury *et al.* (1987) and A. Shepherd (1992).

The *Task Analysis Guide* divides task analysis methods into five broad categories: (1) task data collection, (2) task description, (3) task simulation, (4) task behaviour assessment and (5) task requirement evaluation. The *Guide* gives short accounts of a number of techniques under each of these headings. For task data collection the techniques are (1) activity sampling, (2) critical incident technique, (3) observation, (4) questionnaire, (5) structured interview and (6) verbal protocol. The task description techniques are (1) charting and networking methods, (2) decomposition methods, (3) hierarchical task analysis, (4) link analysis, (5) operational sequence diagram and (6) timeline analysis.

Task simulation is represented by (1) computer modelling and simulation, (2) simulator and mock-up, (3) table top analysis and (4) walk-through and talk-through. Techniques considered under task behaviour assessment are (1) barrier and works safety analysis, (2) event tree, (3) failure modes and effects analysis, (4) fault tree, (5) hazard and operability (hazop) study, (6) influence diagram and (7) management oversight and risk tree (MORT). The task requirement techniques treated are (1) ergonomics checklists and (2) interface surveys.

The *Guide* also gives some 10 case histories of the application of task analysis to (1) allocation of function between humans and automatic systems, (2) preliminary communications system assessment, (3) plant local panel review, (4) staffing assessment for a local control room, (5) operator workload assessment in a command system, (6) analysis of operator safety actions, (7) maintenance training, (8) quantification of effectiveness of ultrasonic inspection, (9) operations safety review and (10) a task analysis programme for a large plant.

14.13.7 CCPS *Guidelines*

Other task analysis methods are described in the *Human Error Prevention Guidelines* (CCPS, 1994/17), described in Section 14.38. Under the heading of 'action oriented techniques' they give an account of hierarchical task analysis, operator action event trees, decision/action flow diagrams, operational sequence diagrams and signal flow graph analysis. The two cognitive task analysis techniques given are the critical action and decision evaluation technique (CADET) and the influence modelling and assessment system (IMAS).

14.14 Job Design

Job design involves the arrangement of the individual tasks which the man has to do into a job which he is capable of doing and from which he obtains satisfaction. An account of job design is given by L.E. Davis and Wacker (1987).

As the control system becomes more automatic, the active control work of the operator is reduced and his function is increasingly one of monitoring. Such passive monitoring, however, is not a function to which man is well suited.

There is, therefore, a potential problem of job design. An entry to this problem is to consider why man is part of the control system at all. To a large extent he is there as the component used by the designer to give the system a self-repairing capability. An approach to job design which starts from this fact is to develop explicitly those functions of the operator which are concerned with handling faults and keeping the plant running, notably fault administration and malfunction detection. These are tasks for which, in general, man has the ability and the motivation and they are entirely in line with the aims of loss prevention.

Such an approach requires more attention to be given to these functions of the operator and that he be given the necessary training and job aids. Training should emphasize the importance of running a tight ship and not tolerating degraded equipment. Job aids, including computer-based aids, should assist him in fault administration and malfunction detection.

Other aspects of job design such as workload and organization and social factors have been considered above.

14.15 Personnel Selection

There does not appear to be very much guidance available on selection methods for process operators. This is no doubt largely due to the difficulty of defining criteria for operator performance. One of the first studies of the process operator was in fact that by Hiscock (1938), who described the development of a set of selection tests, but subsequent work by D.G. Davies (1967) showed little correlation between performance assessed by a selection test battery and by the judgement of supervisors.

It is probably true that enough is now known about process control skill to offer a better prospect of success in developing selection tests. The abilities which an operator needs are better understood. They include signal detection, signal filtering, probability estimation, system state evaluation, manual control and fault diagnosis. Tests may be devised to measure these abilities, using perhaps a computer-based simulator.

It is also important, however, to take into account in selection personal qualities. Crossman (1960) suggested that it is desirable that a process operator should be:

(1) Responsible – able to make satisfactory judgements on matters of discretion, so that his work does not need frequent checking by superiors.
(2) Conscientious – ready to take extra trouble and care, without direct instructions, when the situation demands it.

(3) Reliable – never making mistakes, forgetting instructions, overlooking important indications, etc., or otherwise failing in his prescribed duties.
(4) Trustworthy – honest and truthful in reporting to superiors; not concealing the facts when his own actions may have had adverse effects.

Other important factors are temperament, motivation and social skills. Temperament includes response to monotony and to stress. Motivation is partly an individual matter, but is also influenced by job design. Social skills cover the wide range of communication activities that the work entails.

The operator's job requires a reasonably high level of intelligence, particularly on large plants, but it is not necessary to look for a high intelligence score or university degree. The graduate sometimes makes a rather poor operator. The question of education is considered below together with that of training. A further account of personnel selection is given by Osburn (1987).

Selection for process control tasks is touched on in a study by the Advisory Committee on the Safety of Nuclear Installations (ACSNI, 1990) in relation to testing for trainability. The authors emphasize the need for the formulation of criteria for operator performance applicable to both training and selection, and thus imply that full development of selection tests must wait on the availability of such criteria. They do, however, consider it practical to select out individuals who have chronic stress problems or are unable to cope with high levels of stress. This aspect is also addressed by Weisaeth (1992), who describes the variability of individual response to the stress of emergency conditions.

The elimination of completely unsuitable individuals is probably the most useful feature of selection tests in other fields, such as those for aircraft pilots, which might appear to be a suitable model for some process situations.

A review of the state of knowledge and current practice on operation selection has been published by the HSE (1993 CRR 58).

14.16 Training

The training of process operators is an area in which more can be achieved and, in general, industry devotes considerable effort to this. An overview of training applicable to the process industries is given in *Training* (Patrick, 1992).

14.16.1 Training and education

There is a distinction to be made between training and education: the former is specific to a particular task or job, the latter is more general. The difference may be seen clearly in the evolution of training for electronic maintenance technicians in the US armed services (Shriver, Fink and Trexler, 1964). Initially this involved a general education in electronics, but this did not prove very effective in training people to repair equipment. Latterly the emphasis has been on specific training for the diagnostic tasks involved.

The question of the operator's mental model of the process and of his need for a scientific understanding of it was raised in the work of Crossman and Cooke (1962) on manual control and has been extensively discussed by subsequent workers (Attwood, 1970; Kragt and Landeweerd, 1974). Investigations in fairly simple manual control tasks do tend to suggest that it is better to train the operator in control strategies, controller settings, etc., but it is not clear how far this can be generalized to other tasks such as fault diagnosis.

In the UK, the City and Guilds Institute runs a Chemical Technicians Certificate (1972) which constitutes a substantial education in chemical processes. Such an educational background probably is beneficial, but it is not a substitute for specific training in process tasks.

General guidance on the training of process operators is given in the publications of the former Industry Training Boards (ITBs), in particular the Chemical and Allied Products ITB (CAPITB) and the Petroleum ITB (PITB). For example, CAPITB *Training Recommendation 12* (1971/9) contains *Information Papers* which represent outline approaches to job analysis, training needs, training programmes and fault analysis which the user should apply to his own plant, while other *Information Papers* such as that on distillation (1972/6) represent outline subject syllabuses. Another *Information Paper* (1971/5) deals with safety training. The work of Duncan and his colleagues has been supported by the CAPITB and has been published partly as reports of that body (Duncan and Shepherd, 1975a).

14.16.2 Training principles

> The job of the trainer is to observer and analyse, and to arrange to supply the right amount of the right kind of information to the learner at the right time. He must know a great deal about the task being trained, but the kind of knowledge needed is what comes from careful, objective analysis of the job and of the necessary skill rather than from the experience of becoming personally proficient. His task is to find out what factors affect the learning of the skill with which he is concerned, to watch the effects of varying them and to try to arrange the best combination.

This account by Holding (1965) highlights some of the important features in training. It brings out clearly the importance of the prior task analysis to determine where the operator may have difficulty and where training may be necessary. The content of training should be appropriate, which in effect means it should be related to these difficulties. The training should be at a suitable pace. It should provide feedback of results, since this is essential for learning.

Motivation of the trainee is another important factor and this can be strengthened by recognizing successful performance and treating failures objectively by explaining the cause in a non-condemnatory manner.

There are a number of classic problems in training. The training period may be spaced out or massed together. The task may be learned whole or in parts. The task may have to be done under some form of stress and this may need to be taken into account.

A particularly important question is the transfer of skill from one task to another. It is possible, for example, to achieve good performance in a particular task using methods such as decision trees which leave relatively little to the operator, but the penalty tends to be that he cannot transfer the skill to other tasks.

Many of these problems are discussed in the specific context of process control in the work of Duncan and his colleagues (Duncan, 1974; Duncan and Shepherd, 1975a,b). A further account of concepts of training is given by Holding (1987). The relationship between training goals and training systems is described by Goldstein (1987) and the use of training simulators by Flexman and Stark (1987).

14.16.3 Process operator training

The process operator needs to have the safety training received by other employees also, but in addition has certain specific needs of his own. In this section consideration is given to the content of training for process operators. A further account of operator training is given in Section 14.17.5. Safety training is considered in Chapter 28. Some topics for training are listed in Table 14.10.

In general, it is usually a mistake to provide numerous lectures on areas such as details of the process design or of the process computer program. It is better to stick fairly closely to the operator's own problems. Nevertheless, he does need to understand the process flow diagram, the unit operations which compose it, and the control system. In addition, a basic understanding of the goals and constraints with which the plant manager operates and the possible changes of priorities which may occur is necessary.

The operator also requires some knowledge of the plant equipment and the instrumentation. In particular, he needs to be able to identify items and to carry out the manipulations for which he is responsible.

Table 14.10 *Some aspects of the training of process operators*

Process goals, economics, constraints and priorities
Process flow diagram
Unit operations
Process reactions, thermal effects
Control systems
Process materials quality, yields
Process effluents and wastes
Plant equipment
Instrumentation
Equipment identification
Equipment manipulation
Operating procedures
Equipment maintenance and cleaning
Use of tools
Permit systems
Equipment failure, services failure
Fault administration
Alarm monitoring
Fault diagnosis
Malfunction detection
Emergency procedures
Fire fighting
Malpractices
Communications, record-keeping, reporting

See also Table 28.5

There are numerous operating procedures with which he has to become familiar. These include start-up, shut-down, batch operation and all other sequential routines.

The operator needs to learn to administer faults and, in particular, to interpret the alarm system, to diagnose faults and to detect incipient malfunction. He must also be thoroughly conversant with the emergency procedures.

The system of permits-to-work on a plant is extremely important and the operator requires to have a full grasp of the system used.

Training in fire fighting is a particularly important aspect of emergency work. The best training is provided by realistic fire fighting exercises in a training area dedicated to this purpose.

Operational malpractices develop on most plants. A typical example is the operation of furnaces at temperatures which drastically reduce creep life. If this can be foreseen, training may be given to counter this.

The operator has quite an important role to play in communicating information about plant operation to other people and it may be appropriate to include this in training.

14.16.4 Fault diagnosis

Fault diagnosis is a particularly important activity of the process operator. It has been studied in detail by Duncan and his co-workers (Duncan, 1974; Duncan and Shepherd, 1975a,b; Duncan and Gray, 1975a,b) and is considered here as an example of training in a specific task.

Fault diagnosis is usually carried out by the operator at the control panel. Frequently, therefore, use is made of a panel simulator for training. The simulator used at the beginning of Duncan's work is shown in Figure 14.17. It represents a typical panel and the trainer sets up manually combinations of instrument readings corresponding to particular process conditions. The training is interrupted, however, by the rather slow process of resetting the simulator to represent the next set of conditions. Duncan and Shepherd (1975b) have overcome this problem by using a mock-up of the panel, shown in Figure 14.18, and back-projecting this on to a full-size screen in front of which the trainee operator can stand. Different sets of instrument readings can be brought up by projection without delay.

Using this device, Duncan and Shepherd have studied the training of operators to recognize different fault patterns on the panel. A cumulative part-training method was used in which the number of sets of faults was progressively increased. The subjects achieved encouragingly high rates of success in diagnosing faults. Debriefing of the subjects suggested that different diagnosis strategies were used. One was a pattern recognizer, looking first at the alarms and then considering possible failures consistent with these. Another examined first the instruments and then used hueristics based on plant functions.

An alternative approach is the use of decision trees as described by Duncan and Gray (1975a). Figures 14.19 and 14.20 show for a crude distillation column the decision tree and fault–symptom matrix, respectively. The operators were taught to use the decision tree to diagnose faults. Again a high success rate was achieved.

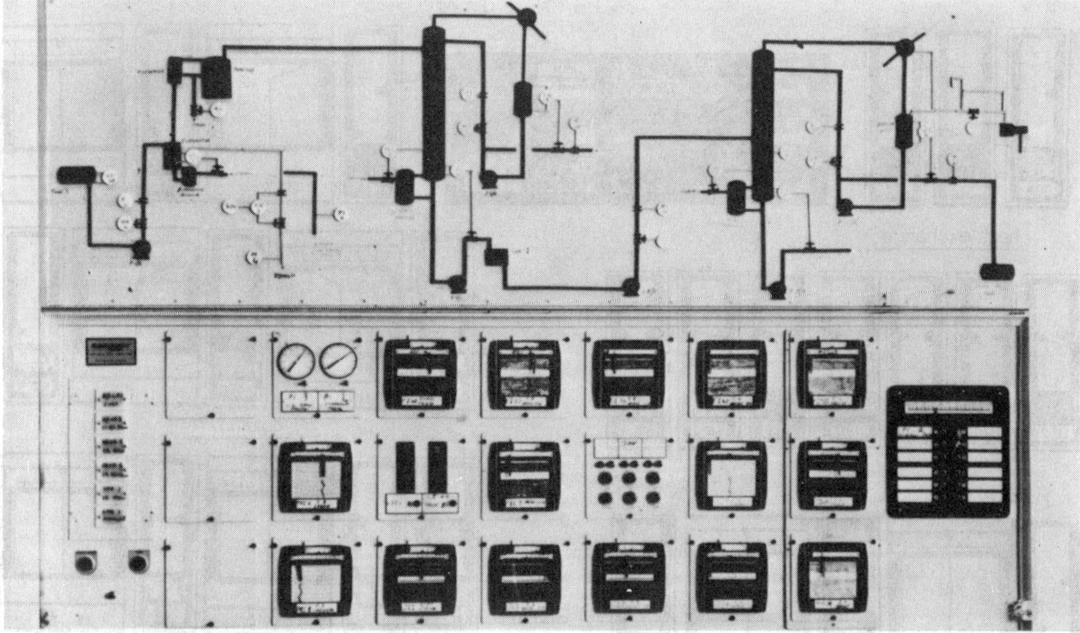

Figure 14.17 The Carmody 'Universal Process Trainer' (Courtesy of Taylor & Francis Ltd; photograph courtesy of BP Chemicals (International) Ltd))

There are various ways in which the information in the decision tree in Figure 14.19 may be presented and learned. The tree as shown gives the faults in the 'natural' grouping. An alternative tree can be constructed in which the maximum number of decisions is reduced from 8 to 3, but the rationale of such a tree is no longer apparent, and the tree would be more difficult to remember. The method actually adopted was to retain the tree in its natural form and to use the 'linked lists' shown in Figure 14.21 to assist in learning it.

Some of the decisions in the tree depend on the instrument readings, which may be in error. Training in the checking of instruments was therefore developed also. The emphasis here was on determining the true value of a parameter rather than on deciding the instrument failure. The following general principles were evolved for checking:

(1) Direct observation – represents truth.
(2) An outside report, such as a sight glass reading, is a direct observation only if the correct procedure for reading is followed.
(3) Two control instruments which agree represent truth.
(4) The costs (time, effort, danger) must be taken into account, and the cheapest route to truth taken.
(5) Only independent indicators of the same parameter can be used for checking.
(6) Where only two indicators of a parameter are available, the more reliable of the two is to be believed.
(7) An individual instrument may show itself to be faulty by deviant behaviour; for example, a straight line on a pen record.

This work brings out the importance of verifying the instrument readings before embarking on the main diagnosis. The need for this in the automated equivalent of computer alarm analysis was mentioned earlier.

This work also included the development of programmed learning texts to assist the operator to learn both the linked lists of the main decision tree and the instrument verification procedures. These developments in training do not of course solve the problem of diagnosing very rare but very hazardous fault conditions. It is doubtful if training is the proper approach to this problem. Such faults are better dealt with by the protective system; since they are very rare, the shutdowns involved are probably acceptable.

14.17 Training: ACSNI Study Group Report

A review of training is given in the *First Report: Training and Related Matters* of the Advisory Committee on the Safety of Nuclear Installations (ACSNI) (1990). Although this is a report to the nuclear industry, it is applicable in large part to the process industries also. The report starts from the point that human error is the major source of accidents and refers to the cases of Three Mile Island and Chernobyl.

The authors treat the topic under the following headings: (1) definition of training, (2) the safety

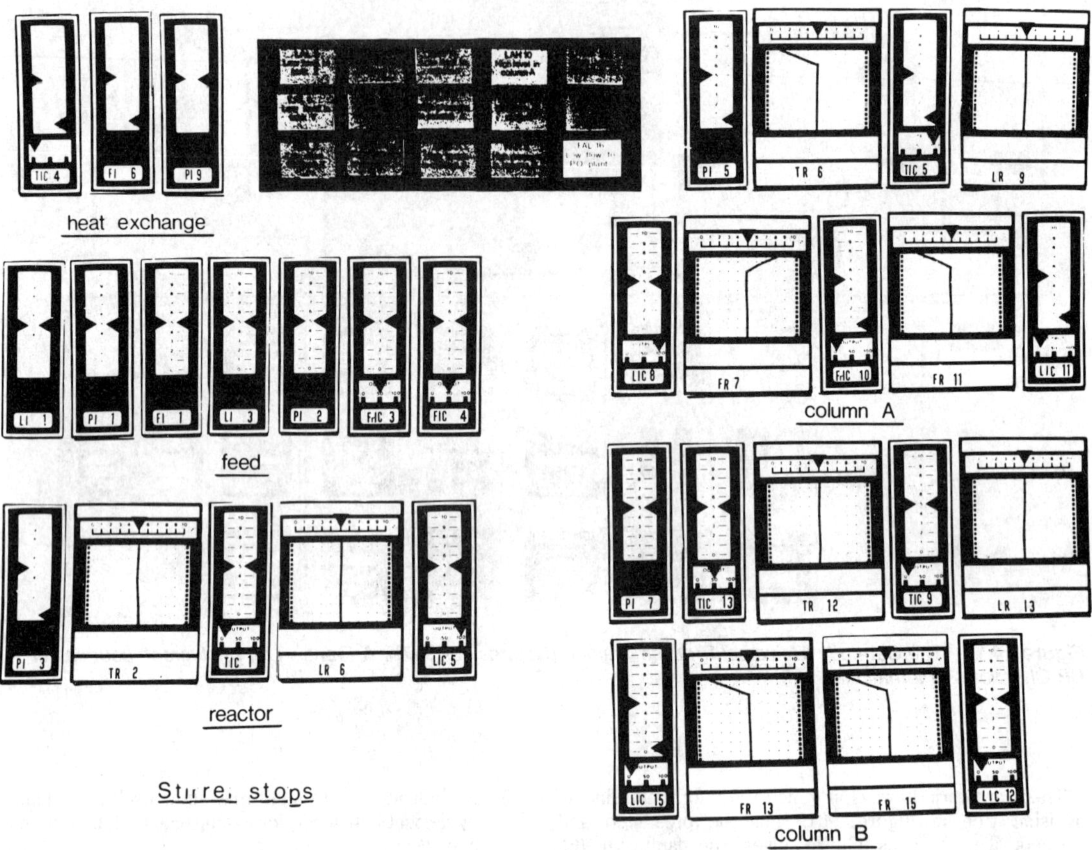

Figure 14.18 *Control panel display used for operator training (Duncan and Shepherd, 1975b) (Courtesy of Taylor & Francis Ltd)*

culture, (3) initiation of training, (4) internal monitoring, (5) training needs analysis, (6) criteria for operator performance, (7) standards for training, (8) methods of training, (9) central vs site-based simulators, (10) individual vs team training, (11) training for stress, (12) training of management, instructors, etc., (13) certification and (14) privatization. They state:

> It is misleading to think of training solely in terms of the transfer of items of verbal knowledge or technical skill from instructor to pupil. Although both knowledge and skill are necessary, they are not on their own sufficient to assure safety. They must be augmented by different qualities: habits of forethought and precaution that place minimization of risk first, and other goals such as short term performance or convenience, second.

The main purpose of training is to create a safety culture. The report rehearses some of the elements of this, which have already been described in Chapter 6. It is necessary that senior management give a lead and take certain specific measures. Line management should have operational responsibility *inter alia* for safety and training. There should be monitoring of the safety culture using objective measures and independent assessment and a policy of continuous improvement with specific targets and with feedback to the workforce. Training should aim to rehearse the individual's experience deliberately so as to reinforce compliance and self-monitoring and awareness and reporting of hazards. Senior management should ensure not only that the organization has a suitable formal training system, but that it is operating effectively in practice.

Training should be monitored. One method of monitoring is the use of objective measures such as the frequency of accidents, incidents, operational deviations, trips, etc. Another method is to sample regularly what people say about the safety culture and about their own attitudes. The monitoring should be done by an assessor who is independent both of the local line management and of the training specialists.

Task analysis should be used to establish training needs. The analysis should specify in concrete terms the

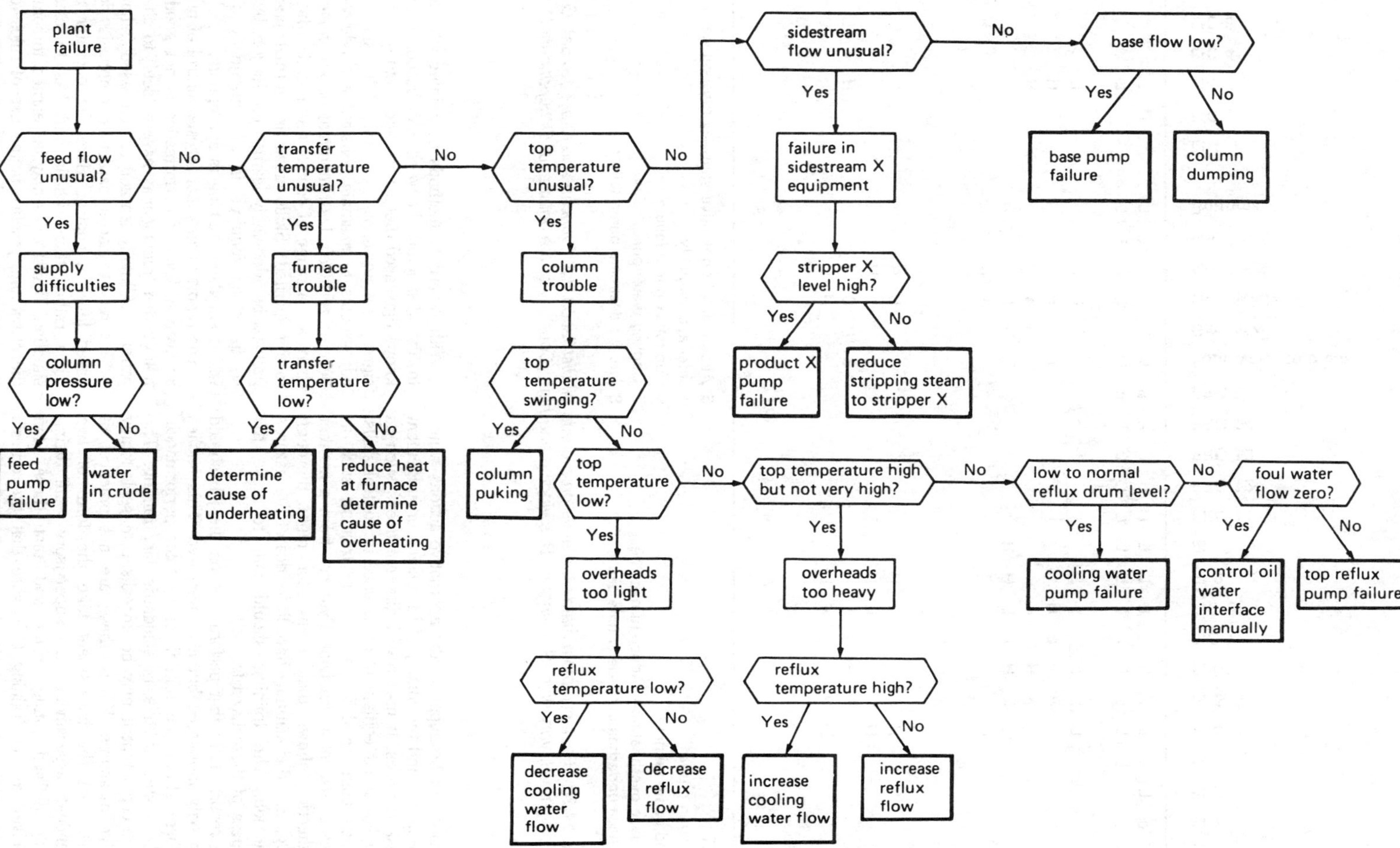

***Figure 14.19** Decision tree for fault diagnosis on a crude oil distillation unit (Duncan and Gray, 1975a) (Courtesy of the Journal of Occupational Psychology)*

FAULT	FI 1	FI 2	FI 3	FI 4	FI 5	FI 6	FIC 7	FIC 8	FI 9	FI 10	FRC 11	FRC 12	FRC 13	FRC 14	FR 15	Foul water to drain	Gas flow	Overheads	No. 1 S.S.	No. 2 S.S.	Bottoms	LI 1	LIC 2	LIC 3	LIC 4	LIC 5	Feed tank sight glass	Column base sight glass	R/X drum sight glass
1	ø	L	L	L	L	L	L	L	L	L	L	L	L	L	ø	L	L	L	L	L	L		H	L				H	L
2	ø	L	L	L	L	L	L	L	L	L	L	L	L	L	ø	L	L	L	L	L	L			L					L
3																													
4				L	L	L		L	L								L	L	L					L					L
5					H	H	ø	H								ø	H	H						H					H
6					ø	H		ø									H	ø						H					H
7					L	H	L	L	H	H						L	H	L											
8				ø															ø							H			
9			ø																	ø					H				
10						H	L	L								L	H												
11								L																					
12				H	L			L	H									L	H					L					L
13				H	L			H	H									L	H										
14					S	H	S	S								S	H	S						H	L	L			H
15				L	H			H	L									H	L							L			
16			L							L										L					L	H			
17		ø																			ø		H					H	
18		S																			S		B					B	

FAULT:
1 Water in crude
2 Feed pump failure
3 Furnace overheating—reduce furnace heat
4 Furnace underheating—determine cause
5 Accumulator boot valve shut—control oil water interface manually
6 Top reflux pump failure
7 Cooling water pump failure
8 No. 1 sidestream pump failure

Figure 14.20 *Fault-symptom matrix for faults on a crude oil distillation unit (Duncan and Gray, 1975a): Ø zero; L, low; H, high; vH, very high; S, swinging; B, swinging and high (Courtesy of the Journal of Occupational Psychology)*

information to be supplied, the alternative actions which can be taken, and the quality of calculation or judgement the task requires. It may reveal different needs such as knowledge about equipment or practice in control skills, and thus point to different types of training such as lectures or use of a simulator. Evidence from a number of industries shows that, even after quite thorough training, an individual may have little idea of the hazards of the job. The training should aim to raise the awareness of these hazards.

The report states that performance on the job should be not only monitored but also measured. This implies that there should be criteria of operator performance. Closely related to this is standards for training. The report accepts that it may be unrealistic to seek totally objective measures. Where these are not practical, it suggests that the assessment take the form of two independent assessments by supervisory staff using carefully defined rating scales and that these staff themselves receive training to ensure that the criteria used are consistent.

With regard to methods of training, the report refers to the distinction between skill-based, rule-based and knowledge-based behaviour and between responses to signal and to signs. Control skills such as making adjustments to maintain constant a single variable have become relatively less important on automated plant, but there are a range of control room skills which still need to be learned. Skills used in normal operation may perhaps be adequately learned on the job, but this is not so for skills required in an emergency situation, for which simulator training is appropriate.

The report argues that for the control of nuclear plants an average level of performance is not good enough. It follows that training may have to seek to change certain patterns of highly learned skilled-based behaviour.

The report places particular emphasis on the rule-based level of control. Here the operator responds to signs rather than signals and has to categorize the situation. Whilst rules may be learned to some degree in the classroom, practice in the use of rules to categorize the situation is most effectively given on a simulator.

No. 1 S.S.S. sight glass	No. 2 S.S.S. sight glass	PI 1	PI 2	PI 3	PI 4	PI 5	PI 6	PI 7	PI 8	PI 9	PRC 10	TI 1	TI 2	TI 3	TI 4	TI 5	TI 6	TI 7	TRC 8	TI 9	TI 10	TI 11	TI 12	TI 13	TI 14	TI 15	TI 16	TI 17	TI 18
					H	H	H	H	H	H	H	L	L	L	H	H	H	H	L	L	L	L	L	L	L	L	L	L	L
		L	L	L	L	L	L	L	L	L	L	L	L	L	H	H	H	H	L	L	L	L	L	L	L	L	L	L	L
															H	H	H	H	H	H	H	H	H	H	H	H	H	H	H
												L	L		L	L	L	L	L	L	L				L	L	L	L	L
									H	H	H	H	H	H										H	H	vH	vH	L	L
									L	L	L	L	L	L										H	H	vH	vH	L	L
									H	H	H	L	L	L										H	H	vH	vH	L	L
H													L	L															
	H													L															
									H	H	H	L	L	L											H	H	H	H	H
										L	L														H	H	H		
										L	L	L													L	L	L	L	L
										H	H	L													L	L	L		
L	L							S	S	S	S													S	S	S	S	S	S
L												H																	
H	L												H																
														L															
								S	S	S	S			L						L	L	L							

9 No. 2 sidestream pump failure
10 Too little cooling water—increase flow
11 Too little reflux—increase flow
12 Too much cooling water—decrease flow
13 Too much reflux—decrease flow
14 Column puking
15 Too much stripping steam (S.S.S.1)—reduce
16 Too much stripping steam (S.S.S.2)—reduce
17 Base pump failure
18 Column dumping

Figure 14.20 continued

The report gives high priority to the provision of simulators for each type of control room, since there are aspects of task which cannot readily be taught without them. It discusses the merits of local vs off-site simulators, but regards this question as less important that clear specification of the skills to be learned and the level of competence to be achieved.

With regard to individual vs team training, the authors suggest that the initial emphasis should be on individual training, with team training introduced later. Simulator training is treated as a form of team training.

Operator performance is affected by the level of stress experienced. It is not desirable that any operator of a nuclear plant should have to carry out the job under an abnormal degree of stress. This has implications for selection and for monitoring of individuals. The individual should be free from chronic stress symptoms and able to tolerate acute stress. Given this, the training should give practice in performing the task in realistic conditions and with a raised level of stress so that the trainee learns to handle it. Factors which affect stress include the level of demand and the degree of control over the task which the person feels he has. The individual's confidence that he is in control may be built up by experience of successful handling of similar problems. This experience can be given using a simulator. The level of demands should be built up until it reaches that experienced in real operation.

Another aspect of stress is the danger inherent in a major incident. The reports acknowledges that this aspect is difficult to simulate in an exercise, but refers to the expectation of possible criticism. In a healthy safety culture criticism of the performance of an individual or team can be made and accepted constructively. The report urges that management at all levels should be periodically assessed for training need and that no-one should be 'above it'.

The report devotes considerable attention to the question of certification, which is supported by some and opposed by others. In order to disentangle the argument, it is necessary to assign a clear meaning to the word 'certification'. The report gives some eight

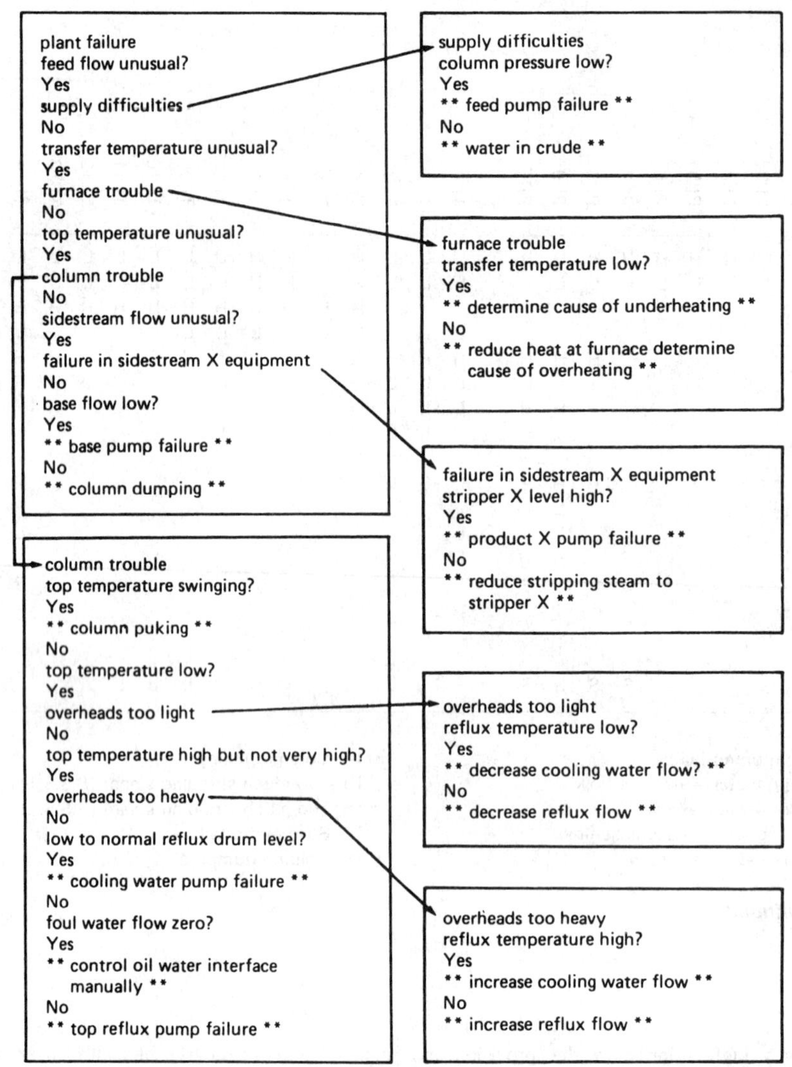

Figure 14.21 *Linked lists for learning fault diagnosis on a crude oil distillation unit (Duncan and Gray, 1975a) (Courtesy of the Journal of Occupational Psychology)*

possible definitions, together with comments. These may be summarized as follows: (1) specification of the content of training and clear responsibility for authorization of an individual to undertake certain actions; (2) separation between individuals who undertake (1) and line management; (3) a requirement that line management accept without further question separate assessment of the individual competence; (4) a requirement that line management authorize only persons from amongst those who have satisfied a separate assessment; (5) a rule that (3) or (4) apply only at the stage common to all reactors with different procedures thereafter; (6) a requirement that separate assessment should not mean simply testing by the training function but by another part of the organization; (7) a requirement that separate assessment should not be carried out by the licensee's organization at all but by a separate body; and (8) a requirement that assessment should include observation of performance on the job and not merely verbal knowledge.

The report suggests that the first two interpretations would attract widespread consent. Opponents of certification may understand it in the third sense, but the authors doubt whether this view has proponents. A more serious argument turns on the fourth definition. Here opponents of certification argue that the trend is towards more involvement by line management in any matter which affects production and that there are dangers if it is not involved in training, whilst proponents emphasize that line management is liable to develop blind spots. The authors suggest that this points to the need for a 'two key' system, in which an operator needs to satisfy both an independent assessment and local line management. They argue that given the wide variety of reactor types in Britain the fifth definition would effectively rule out independent assessment. With regard to the sixth and seventh definitions the report supports an assessment of the trainee performed by assessors who are independent both of the local line management and of the training function, but is unconvinced that the independent assessors need to come from outside the licensee. The authors support the type of assessment given in the eighth definition. With regard to privatization, the report makes the point that, in order to support the type of training envisaged, a certain minimum size of training establishment is required.

The framework for training developed in the report is therefore broadly on the following lines. The starting point is task analysis. On the basis of this a specification for training is formulated by line management and the training is carried out by the training function. It is monitored by assessors independent of both local line management and the training establishment. Assessment of the trainee contains a component of actual or simulated performance. The trainee first has to satisfy the independent assessors and is then authorized by line management. Assessment is a continuing process.

The report deals particularly with the training of process operators. However, it emphasizes that training should not be confined to process operators but should also cover instrument artificers, maintenance personnel, etc.

Under the guise of trainability testing, the report touches on the question of personnel selection. It states:

> The Study Group believes it is important to find a valid way of deciding the relative success of different individuals, and that thereby it will become possible to devise satisfactory selection procedures. In other occupations it has frequently been found that systematic assessment of individuals can significantly improve trainability and subsequent performance. The potential of such methods should be explored further.

14.18 Human Factors and Industrial Safety

An account of the role of human factors in safety is given in *Human Factors in Industrial Safety* (HSE, 1989 HS(G) 48). This study sets out the factors which emerge most frequently from accident investigations and gives numerous case histories to illustrate the points made. Some of these case histories are given in Appendix 1.

It refers to the disaster at Chernobyl as an illustration. The background to the incident was serious defects in the management culture and in the regulatory system. The personnel conducting the test on the reactor were under pressure to complete the test in a short time. They failed to distinguish between small and large risks. They removed layer upon layer of protection and violated operational rules. The incident has many lessons, but the authors warn against treating it as a universal model.

The study starts from the viewpoint that the human operator is basically a positive rather than a negative factor in process plants. The operator gives the system a much enhanced ability to deal with abnormal situations. But it is necessary to design systems which protect against, and are tolerant of, human error and to train operators to improve their decision-making in these situations. Above all, the operator needs the support of a good safety culture.

The authors describe case histories which illustrate three aspects of human factors in particular: inadequate information, inadequate design and minimization of the consequences of human error. Human error is classified, with examples, under the headings of slips and lapses, mistakes, misperceptions, mistaken priorities and violations.

Measures for the prevention of human error, and an overall strategy for this, are considered under the headings of the organization, the job and personal factors. In this strategy the organization is considered in terms of the safety climate, standard setting, monitoring, supervision and incident reports; the job is considered in terms of task analysis, decision-making, man–machine interfaces, procedures and operating instructions, the working environment, tools and equipment, work patterns and communication; and personal factors are considered in terms of personnel selection, training, health assessment and monitoring. Checklists are given under each of these headings.

Many of the examples given in this study are drawn from the process industries. The application of human factors to the process industries specifically is treated in *Human Factors in Process Operations* (Mill, 1992). This is a report of the Human Factors Study Group of the Loss Prevention Panel of the European Federation of Chemical Engineering (CEFIC).

The study covers a number of topics dealt with elsewhere in the present text, notably: management; accident models; control room design; operator control, mental models and workload; operating and emergency procedures; and human error.

A five-step strategy is described for modification of human behaviour in relation to accidents and hazards. These steps are: (1) identification and analysis of accidents and hazards; (2) revision of safety rules concerning working behaviour; (3) development of a plan of measures; (4) implementation of the plan; and (5) follow-up – assessing the effectiveness of the measures taken.

14.19 Human Error

The topic of human error as such is a vast one and beyond the scope of the present work. The treatment here is limited to consideration of the effect of human error on the performance of process systems and on methods of dealing with it.

Accounts of human error include *Handbook of Human Reliability Analysis with Emphasis on Nuclear Power Plant Applications* (Swain and Guttmann, 1983), *Information Processing and Human-Machine Interaction* (Rasmussen, 1986), *Human Reliability* (Park, 1987), *Human Reliability Analysis* (Dougherty and Fragola, 1988), *Human Error* (Reason, 1990) and *An Engineers View of Human Error* (Kletz, 1991e). Selected references on human error and operator error are given in Table 14.11.

***Table 14.11** Selected references on human error, operator error*

NRC (Appendix 28 *Human Error*); Riso National Laboratory (Appendix 28); Flanagan (1954); Lincoln (1960); Rook (1962, 1964); Irwin, Levitz and Ford (1964); Irwin, Levitz and Freed (1964); Leuba (1964); Swain (1964a,b, 1968, 1969, 1972, 1973a,b, 1982); Chase (1965); Pontecorvo (1965); Blanchard *et al.* (1966); Rigby (1967, 1971a,b); Rasmussen (1968a–c, 1969, 1976b, 1978, 1980a,b, 1982a,b, 1985, 1987, 1990); Rigby and Edelman (1968a); Askren and Regulinski (1969); Favergé (1970); Sayers (1971 UKAEA SRS/GR/9); E. Edwards and Lees (1973); Kletz and Whitaker (1973); Lees (1973a, 1976b, 1983a); Regulinski (1973); Finley, Webster and Swain (1974); AEC (1975); Anon. (1976 LPB 7, p.15); Apostolakis and Bansal (1976); R.L. Browning (1976); Embrey (1976a,b, NCSR R10, 1979a,b, 1981, 1983a,b, 1984, 1985, 1992a–c); Rasmussen and Taylor (1976); Hopkin (1977); Reason (1977, 1979, 1986, 1987a,b, 1990); Skans (1978); Bello and Colombari (1980); J. Bowen (1980); Griffon (1980); R.A. Howard and Matheson (1980); W.G. Johnson (1980); Kletz (1980c,e, 1982g, 1985d,f, 1987f,h, 1989d, 1990c, 1991e,g, 1993a); Senders (1980); Swain and Guttman (1980, 1983); B.J. Bell and Swain (1981, 1983); J.A. Adams (1982); Carnino and Griffon (1982); Rasmussen and Pedersen (1982, 1983); Singleton (1982, 1984); Beare *et al.* (1983); Brune, Weinstein and Fitzwater (1983); Embrey and Kirwan (1983); HSE (1983a); Mancini and Amendola (1983); Nieuwhof (1983a); Seaver and Sitwell (1983); J.C. Williams (1983, 1985a,b, 1986, 1988a,b, 1992); Embrey *et al.* (1984); Siegel *et al.* (1984); Willey (1984); Ball (1985 LPB 62); Dhillon and Misra (1985); Leplat (1985); Carnino (1986); Dhillon (1986); Dhillon and Rayapati (1986); Hannaman *et al.* (1986); P. Miller and Swain (1987); Rasmussen, Duncan and Leplat (1987); Whalley (1987); D.D. Woods, O'Brien and Hanes (1987); Bercani, Devooght and Smidts (1988); Bersini, Devooght and Smidth (1988); Dougherty and Fragola (1988); Drager, Soma and Falmyr (1988); Humphreys (1988a,b); Kirwan (1988); D. Lucas and Embrey (1988); Purdy (1988); Whittingham (1988); Worledge *et al.* (1988); Whalley and Kirwan (1989); Park (1987); J.B. Smith (1987); ACSNI (1990, 1991, 1993); CMA (1990); Gall (1990); Kirwan *et al.* (1990); Lorenzo (1990); Ball (1991 SRDA R3); Bellamy (1991); Masson (1991); Hollnagel, Cacciabue and Rouhet (1992); Ishack (1992); Paradies, Unger and Ramey-Smith (1992); Samdal, Kortner and Grammeltvedt (1992); Sten and Ulleberg (1992); Vestrucci (1992); Welch (1992); Yukimachi, Nagasaka and Sasou (1992); Nawar and Samsudin (1993); Wreathall (1993); Wreathall and Reason (1993); CCPS (1994/17)

Organizational factors as cause of human error
Bellamy (1983, 1984, 1986)

Effect of human error on system performance
J. Cooper (1961); Cornell (1968); Ovenu (1969); R.L. Scott (1971); S. Brown and Martin (1977); Burkardt (1986); Gerbert and Kemmler (1986); Hancock (1986); B.J. Bell (1987); Latino (1987); Whitworth (1987); Holloway (1988); Rasmussen (1990); Suokas (1989); Vervalin (1990b); Pyy (1992); Rothweiler (1994 LPB 118)

Data banks
Payne and Altman (1962); Altman (1964); Meister (1964); Rigby (1967); Swain (1970); Topmiller, Eckel and Kozinsky (1982)

Vigilance tasks
Ablitt (1969 UKAEA AHSB(S) R160); A.E. Green (1969 UKAEA AHSB(S) R172, 1970); Kantowitz and Hanson (1981)

Inspection tasks
Mackenzie (1958); McCornack (1961); Rigby and Edelman (1968a); Drury and Addison (1973); Rigby and Swain (1975); Drury and Fox (1978); Pedersen (1984); Murgatroyd *et al.* (1986)

Emergency situations
Fitts and Jones (1947); Ronan (1953); Rigby and Edelman (1968b); Lees (1973a); AEC (1975); Lees and Sayers (1976); Danaher (1980); Apostolakis and Chu (1984); Giffin and Rockwell (1984); Woods (1984); Briggs (1988); Waters (1988b)

Deliberate violations
W.B. Howard (1983, 1984); Zeitlin (1994)

Human reliability assessessment
NRC (Appendix 28 *Human Reliability Assessment*); R.A. Howard and Matheson (1980); Swain and Guttman (1980, 1983); R.G. Brown, von Herrman and Quilliam (1982); R.E. Hall, Fragola and Wreathall (1982); Rasmussen and Pedersen (1982); Hannaman *et al.* (1983); L.D. Phillips *et al.* (1983); Weston (1983); Anon. (1984aa); Watson (1984 LPB 58); White (1984 SRD R254); Hannaman *et al.* (1985); Hayashi (1985); Heslinger (1985); Pedersen (1985); Soon Heung Chang, Myung Ki Kim and Joo Young Park (1985); I.A. Watson (1985); R.F. White (1986); Hannaman and Worledge (1987); Murgatroyd and Tate (1987); Dougherty and Fragola (1988); Humphreys (1988a); Purdy (1988); Kirwan (1990); Oliver and Smith (1990); L.D. Phillips *et al.* (1990); Delboy, Dubnansky and Lapp (1991); Paradies, Unger and Ramey-Smith (1991); Banks and Wells (1992); Bridges, Kirkman and Lorenzo (1992); Zimolong (1992); J.C. Williams (1993); CCPS (1994/17)

Benchmark exercise
Waters (1988a, 1989); Poucet (1989)

Pipework failures
Bellamy, Geyer and Astley (1991); Geyer and Bellamy (1991); Hurst *et al.* (1991)

Computer control (see also Table 8.1)
Bellamy and Geyer (1988); Kletz (1993a)

14.19.1 Engineering interest in human error

Engineering interest in human error derives from two principal sources. One is accident investigation, where the apparent salience of human error gives rise to concern. The other is hazard assessment, where the requirement to quantify the effect of human error on system performance creates a demand for a methodology capable of doing this.

In certain areas, such as an aircraft flightdeck, the prime emphasis has been on the application of human factors in system design to improve performance and reduce human error. In the nuclear field, however, there has been a much greater emphasis on the development of methodology for the assessment of the effect of human error on system performance. This has been the case in the process industries also.

There are therefore two distinct sources of interest in human error in engineering. Design seeks to improve the work situation and should not too readily accept the inevitability of human error. Assessment must to a considerable extent accept the situation as it is and seek to evaluate human error.

Work both on features which cause human error and on the methods of assessing it is a well established aspect of human factors.

14.19.2 Human error as a cause of accidents

Human error often figures as a major factor in analyses of incidents. Such an attribution appears in large part to derive from the nature of the investigative process. Reference has already been made to the comment by Rasmussen that in the investigation of accidents the tracing of the causes is generally terminated when it reaches a human, as if this was a stopping rule of the tracing process.

When a failure occurs, there is frequently an administrative requirement to determine the cause, and allocation to human error is notorious. In fact, the incident has occurred in a specific set of circumstances involving men, machines, systems and procedures, physical and social factors, and their interactions. Often the error is more truly that of another party who is responsible for some aspect of the work situation. This point is important because, whereas assignment to human error suggests little can be done, recognition of the effect of the work situation does tend to indicate the possibilities for improvement.

It is relevant to remember that experimental psychologists frequently conduct experiments in which various aspects of the work situation are altered and the subject's error rate is evaluated. This demonstrates clearly the importance of these factors in determining human error.

The reporting of errors in man–machine systems is often deficient, because reporting systems are frequently designed essentially to give information on equipment failure. The reporting of malfunctions in missile systems was studied by J. Cooper (1961), who found that, whilst the reporting system did reveal failures involving equipment only, it was deficient in discovering those involving man as well. He found that man appeared to be involved in 20–53% of all malfunctions for the systems studied, and this result is probably typical.

Thus other investigations of the contribution of man to system failure have been made—in aerospace by Cornell (1968), on nuclear reactors by R.L. Scott (1971), and on industrial boilers by Ovenu (1969). These studies too attribute a large proportion of failures in the system to human error.

The contribution of human error to major accidents appears to be greater than it is to less serious failures. It has been estimated by Rasmussen (1978) that whereas human error probably contributes about 10% to general failures, it contributes about 50–80% to major accidents.

14.19.3 Approaches to human error

In recent years, the way in which human error is regarded, in the process industries as elsewhere, has undergone a profound change. The traditional approach has been in terms of human behaviour, and its modification by means such as exhortation or discipline. This approach is now being superseded by one based on the concept of the work situation. This work situation contains error-likely situations. The probability of an error occurring is a function of various kinds of influencing factors, or performance shaping factors.

The work situation is under the control of management. It is therefore more constructive to address the features of the work situation which may be causing poor performance. The attitude that an incident is due to 'human error', and that therefore nothing can be done about it, is an indicator of deficient management. It has been characterized by Kletz (1990c) as the 'phlogiston theory of human error'.

There exist situations in which human error is particularly likely to occur. It is a function of management to try to identify such error-likely situations and to rectify them. Human performance is affected by a number of performance shaping factors. Many of these have been identified and studied so that there is available to management some knowledge of the general direction and strength of their effects.

The approach to the work situation has itself undergone development. Three phases may be distinguished. In the first phase the concern was with error-likely situations and performance shaping factors in general and on the application of ergonomic and human factors principles. The second phase saw greater emphasis on cognitive and decision-making aspects of the task. The third phase seeks the root causes in the organizational and, more generally, socio-technical background.

The approaches taken to human error may therefore be summarized as:

(1) behavioural approach;
(2) work situation approach –
  (a) general work situation;
  (b) cognitive features;
  (c) organizational features.

Any approach which takes as its starting point the work situation, but especially that which emphasizes organizational factors, necessarily treats management as part of the problem as well as of the solution. Kipling's words are apt: 'On your own heads, in your own hands, the sin and the saving lies!'

14.20 Models and Classifications of Human Error

A systematic approach to human error must involve the classification of errors and must therefore be based on appropriate models, either explicitly or implicitly. The classification is not necessarily along a single dimension.

Table 14.12 *Classification of operator error in 200 licensee event reports (J.R. Taylor, 1979; after Rasmussen)*

	No.
Task condition	
Routine task on schedule	89
Routine task on demand	11
Special task on schedule	51
Ad hoc, improvization	21
Various, not mentioned	27
Task control	
Paced by system dynamics	9
Paced by program, orders	4
Self-paced	166
Various, not mentioned	21
Error situation	
Spontaneous error in undisturbed task	93
Change in condition of familiar task	27
Operator distracted in task, preoccupied	10
Unfamiliar task	22
Various, not mentioned	48
Task	
Monitoring and inspection	3
Supervisory control	13
Manual operation, manual control	17
Inventory control	30
Test and calibration	47
Repair and modification	60
Administrative, recording	4
Management, staff planning	13
Various, not mentioned	13
Effect from	
Specified act not performed	103
Positive effect of wrong act	65
Extraneous effect	15
Sneak path	12
Various, not mentioned	6
Potential for recovery	
Effect not immediately reversible	29
Effect not immediately observable	137
Various, not mentioned	34
Error categories	
Absent-mindedness	3
Familiar association	6
Capability exceeded	1
Alertness low	10
Manual variability, lack of precision	10
Topographic, spatial orientation inadequate	10
Familiar routine interference	0
Omission of functionally isolated act	56
Omission of administrate act	12
Omission, other	9
Mistake, interchange among alternative possibilities	11
Expect, assume rather than observe	10
System knowledge, insufficient	2
Side-effects of process not adequately considered	15
Latent causal condition or relations not adequately considered	20
Reference data recalled wrongly	1
Sabotage	1
Various, not mentioned	17

Most workers in the field have found it necessary to classify in terms of at least two dimensions: (1) human behaviour and (2) task characteristics.

An early detailed classification of operator error, which was developed for use in analysing licensee event reports (LERs) for nuclear plants by Rasmussen, quoted by J.R. Taylor (1979), is shown in Table 14.12. Also shown in the table is an analysis of 200 such reports. A more comprehensive version of this analysis is given by Rasmussen (1980). This classification includes categories both of behaviour and of task.

14.20.1 Task analysis framework and model

The general framework in which most models and classifications of human error are applied is that of task analysis. That is to say, the task is decomposed into elements such as plans and actions and the errors associated with these are modelled and classified. The extent of the decomposition varies, some approaches being highly decompositional and others more holistic, as discussed below.

In addition to being a general framework within which particular models are applied, task analysis, particularly hierarchical task analysis, may be regarded as a model in its own right. One common classification of human error is in terms of actions. This type of classification refers to acts of omission, acts of commission, and also delays in taking action, and so on. It does not attempt to explain them in terms of any other model, such as one involving skills or absentmindedness, although it make take into account various types of influencing factor. This approach has much in common with hazop in the way in which action errors are classified. The similarity between the task analysis scheme and the human error classification shown in Tables 14.9 and 14.18 respectively, makes the point.

14.20.2 Work situation framework and models

Another general framework in the modelling of human error is the work situation, which covers the task itself and the various influencing factors including the ergonomic, the cognitive and the organizational. Where an approach is adopted to human error which accords a central role to the work situation, or some aspect of it such as communication, the work situation virtually becomes the model. In such a model the emphasis tends to be on identifying, classifying and quantifying the strength of the factors influencing human error rather than on seeking a more fundamental explanation.

14.20.3 Demand-capacity mismatch model
Turning now to other models of human error, or human reliability, one early model views it as a mismatch between the demands of the task and the capacity of the human to perform the task. The mismatch may arise in various ways. It may be due, for example, to physical incapacity. Or it may arise from lack of training.

A particular case, which is properly regarded as involving human reliability rather than human error, is where sudden death or incapacity, such as a heart attack, occurs. This may need to be taken into account in relation to crucial functions.

14.20.4 Tolerance-variability model
Another early model starts from the viewpoint that fundamental to human error is variability, both of people and of tasks. Variability in performance is associated with the desirable human feature of adaptability. Variability is also present in the task. Even a standardized assembly operation involves some variability. More complex industrial tasks are much more variable.

Some variations in the performance of the task can be tolerated. It is only when some limit is overstepped that error is said to occur. The definition of error is therefore critical. The limits may be defined in various ways such as physical barriers, warning signs, operating procedures, production tolerances and standards, process conditions, and social and legal codes.

Swain (1972) distinguishes between random, systematic or sporadic errors. Random and systematic errors can usually be corrected by reducing, respectively, the variance and the bias in the performance of the task, but it may be more difficult to understand and correct sporadic errors, which typically involve sudden and often large excursions out of limits.

14.20.5 Time availability model
Another type of model is based on the premise that human reliability in certain tasks, particularly emergency response, increases with the time available to perform the task. This model might be regarded as a particular case of the demand–capacity mismatch model.

14.20.6 Skills–rules–knowledge model
The skills–rules–knowledge (SRK) model, developed by Rasmussen (1986), treats the human as operating on the three levels of skills, rules and knowledge. Each level has its characteristic types of error. An account of this model has been given in Section 14.12.

14.20.7 Absentmindedness model
The absentmindedness model, developed by Reason (1990), distinguishes between slips, lapses and mistakes. A slip is either (a) an error in implementing a plan, decision or intention, where the plan is correct but the execution is not, or (b) an unintended action. A lapse is an error in which the intended action is not executed due to a failure of memory. A mistake is an error in establishing a course of action such as an error in diagnosis, decision-making or planning. This model is described more fully in Chapter 26.

14.20.8 Organizational model
The organizational, or socio-technical system, model stresses the contribution of organizational and wider socio-technical factors to human error. It may be regarded as a version of the work situation model which puts particular emphasis on these factors.

14.20.9 Violations
It is recognized that some actions are not strictly human errors but outright violations. There appears to be no developed model for violations. Whilst it is now accepted that human error is something which is under the influence of management, this is less so for violations. In this respect the treatment of violations seems to be at the point where that of human error was some two or three decades ago. There is a need for a better understanding of why violations occur and how they may be designed out.

14.20.10 Decompositional and holistic approaches
Along a different dimension, approaches to human error may be characterized either as decompositional or as holistic. In the decompositional approach, also referred to as 'reductionist', 'mechanistic' or 'atomistic', the task is decomposed into its constituent elements, and for each element the probability of failure is estimated. The probability of failure for the task overall is taken as the product of the individual probabilities. Allowance may be made for certain features such as dependent failure, but for the most part other aspects of task failure such as errors in decision-making are not well handled. The holistic approach by contrast takes high level elements of the work and thus covers features such as decision-making.

14.20.11 Classifications of human error
Classifications of human error flow from the above described models. A common distinction, based essentially on the task analysis model, is between errors of omission, transformation or commission. A human may fail to execute a required action, or he may execute it but incorrectly or out of sequence or too slowly, or he may execute an unwanted action. Errors of omission may involve failures of attention or memory, and errors of commission failures in identification, interpretation or operation. Error classifications along these lines have been given by Lincoln (1960) and Swain (1970).

Classifications related to the work situation model are typically in terms of the influencing factors. The SRK model of Rasmussen gives rise to a classification based on distinction between errors arising at the levels of skill, rules and knowledge. The classification deriving from Reasons's absentmindedness model is into slips, lapses and mistakes.

Another classification of human error is based on the relationship in time of the error to the initiating event in an incident sequence. If the error precedes the initiating event, it is a latent, or enabling, error. If it is the direct cause of the event, it is an initiating error. And if it follows the event, it is a response or recovery error.

14.21 Human Error in Process Plants

14.21.1 Human error in general and in plant operation

An overview of human error in process plants is given in *An Engineer's View of Human Error* (Kletz, 1991e). This work provides a useful account of the practicalities of human error in the various activities occurring on such plants. Kletz describes the following types of human error:

(1) simple slips;
(2) errors due to poor training or instructions;
(3) errors due to lack of physical or mental ability;
(4) wrong decisions;
(5) management errors.

He treats human error in:

(6) design;
(7) construction;
(8) operation;
(9) maintenance.

Klete also gives numerous examples.

The basic approach which he adopts is that already described. The engineer should accept people as they are and should seek to counter human error by changing the work situation. In his words: 'To say that accidents are due to human failing is not so much untrue as unhelpful. It does not lead to any constructive action.'

In designing the work situation the aim should be to prevent the occurrence of error, to provide opportunities to observe and recover from error, and to reduce the consequences of error.

Some human errors are simple slips. Kletz makes the point that slips tend to occur not due to lack of skill but rather because of it. Skilled performance of a task may not involve much conscious activity. Slips are one form of human error to which even, or perhaps especially, the well trained and skilled operator is prone. Generally, therefore, additional training is not an appropriate response. The measures which can be taken against slips are to (1) prevent the slip, (2) enhance its observability and (3) mitigate its consequences.

As an illustration of a slip, Kletz quotes a incident where an operator opened a filter before depressurizing it. He was crushed by the door and killed instantly. Measures proposed after the accident included: (1) moving the pressure gauge and vent valve, which were located on the floor above, down to the filter itself; (2) providing an interlock to prevent opening until the pressure had been relieved; (3) instituting a two-stage opening procedure in which the door would be 'cracked open' so that any pressure in the filter would be observed; and (4) modifying the door handle so that it could be opened without the operator's having to stand in front of it. These proposals are a good illustration of the principles for dealing with such errors. The first two are measures to prevent opening while the filter is under pressure; the third ensures that the danger is observable; and the fourth mitigates the effect.

One area of process operation where slips are liable to occur is in the emptying and filling of equipment. It is often necessary to drain a vessel or other equipment such as a pump. Not infrequently, fluid is admitted to the equipment while the drain valve is still open. In some applications an interlock may be the appropriate solution, as described in Chapter 13.

Misidentification of the equipment to be worked on is another fertile source of error. Kletz gives as an example a plant which had an alarm and trip on the same process variable and where the technician who intended to work on the alarm in fact disabled the trip. He also gives a number of examples of the operation of the wrong press button or the wrong valve caused by inconsistent labelling. The identification of equipment for maintenance is considered in more detail in Chapter 21.

Another type of slip is failure to notice. This is particularly liable to occur if some item is similar but not identical. A case in point is where an operator draws from a drum which contains the wrong chemical but where the drums containing the right and the wrong chemicals have similar markings.

The need to calculate some process quantity introduces a further opportunity for error. In some cases the result may be sufficiently unusual as to highlight the error, but in others it may not.

Many human errors in process plants are due to poor training and instructions. In terms of the categories of skill-, rule- and knowledge-based behaviour, instructions provide the basis of the second, whilst training is an aid to the first and the third, and should also provide a motivation for the second. Instructions should be written to assist the user rather than to hold the writer blameless. They should be easy to read and follow, they should be explained to those who have to use them, and they should be kept up to date.

Problems arise if the instructions are contradictory or hard to implement. A case in point is that of a chemical reactor where the instructions were to add a reactant over a period of 60–90 minutes, and to heat it to 45°C as it was added. The operators believed this could not be done as the heater was not powerful enough and took to adding the reactant at a lower temperature. One day there was a runaway reaction. Kletz comments that if operators think they cannot follow instructions, they may well not raise the matter but take what they believe is the nearest equivalent action. In this case their variation was not picked up as it should have been by any management check. If it is necessary in certain circumstances to relax a safety-related feature, this should be explicitly stated in the instructions and the governing procedure spelled out.

Certain features of the operation and maintenance of process plants should be covered in elementary training. Incidents occur because operators do not fully appreciate the basic properties and hazards of flammable and toxic chemicals and the precautions necessary in handling them, the operation and hazards of pressure systems, the reasons for the principal procedures, and the importance of adherence to all procedures. This elementary training is discussed in more detail in Chapter 28.

There are a number of hazards which recur constantly and which should be covered in the training. Examples are the hazard of restarting the agitator of a reactor and that of clearing a choked line with air pressure.

Training should instil some awareness of what the trainee does not know. The modification of pipework which led to the Flixborough disaster is often quoted as an example of failure to recognize that the task exceeded the competence of those undertaking it.

Kletz illustrates the problem of training by reference to the Three Mile Island incident. The reactor operators had a poor understanding of the system, did not recognize the signs of a small loss of water, and they were unable to diagnose the pressure relief valve as the cause of the leak.

Installation errors by contractors are a significant contributor to failure of pipework. Details are given in Chapter 12. Kletz argues that the effect of improved training of contractors' personnel should at least be more seriously tried, even though such a solution attracts some scepticism.

Human error may occur due to physical or mental inability to perform the task. Even if there is not total inability, the task may be very difficult and therefore liable to error. Physical difficulties can arise in the manipulation of controls and valves. The task may require the operator to observe a display whilst manipulating a control, but the display may be too far from the control to do this. Manual valves on the plant may be inaccessible or stiff to turn. The need to use protective equipment such as breathing apparatus may introduce physical difficulties.

There are a number of situations which may create mental difficulties. Mental overload is one such situation which may come in various forms. The operator may suffer overload due to an excessive number of alarms. A supervisor may be overloaded by the number of permits-to-work. Another type of mental difficulty occurs where the operator is required to undertake tasks which are not his *forte*. One case is detection of rare events. Another is taking over in an emergency from a totally automated system.

Another category of human error is the deliberate decision to do something contrary to good practice. Usually it involves failure to follow procedures or taking some other form of short-cut. Kletz terms this a 'wrong decision'. W.B. Howard (1983, 1984) has argued that such decisions are a major contributor to incidents, arguing that often an incident occurs not because the right course of action is not known but because it is not followed: 'We ain't farmin' as good as we know how'. He gives a number of examples of such wrong decisions by management.

Other wrong decisions are taken by operators or maintenance personnel. The use of procedures such as the permit-to-work system or the wearing of protective clothing are typical areas where adherence is liable to seem tedious and where short-cuts may be taken.

A powerful cause of wrong decisions is alienation.

Wrong decisions of the sort described by operating and maintenance personnel may be minimized by making sure that rules and instructions are practical and easy to use, convincing personnel to adhere to them and auditing to check that they are doing so.

Responsibility for creating a culture which minimizes and mitigates human error lies squarely with management. The most serious management failing is lack of commitment. To be effective, however, this management commitment must be demonstrated and made to inform the whole culture of the organization.

There are some particular aspects of management behaviour which can encourage human error. One is insularity, which may apply in relation to other works within the same company, to other companies within the same industry or to other industries and activities. Another failing to which management may succumb is amateurism. People who are experts in one field may be drawn into activities in another, related field in which they have little expertise.

Kletz refers in this context to the management failings revealed in the inquiries into the Kings Cross, *Herald of Free Enterprise* and Clapham Junction disasters. Senior management appeared unaware of the nature of the safety culture required, despite the fact that this exists in other industries.

14.21.2 Human error and automation

The automation of functions previously performed by the human operator can give rise to certain characteristic problems. These ironies of automation have been discussed by Bainbridge (1983), who identifies four areas of concern. One is the potential for deterioration of the skills of the process operator. Another is the need for the operator to monitor the functioning of the automatic equipment, a role to which he is not particularly well suited. The third is the difficulty for the operator of maintaining an up-to-date mental model of the status of the process in case he has to intervene. The fourth is the tendency of automated systems to reduce small variations at the price of occasionally introducing very large ones.

These characteristics of automated systems were discussed in general terms in the early part of this chapter. Here attention is drawn to their potential to induce human error.

14.21.3 Human error in plant maintenance

Much of the literature on human error in process plants is devoted to the process operators. However, a large proportion of serious incidents is attributable to errors in maintenance work. Maintenance errors may endanger those doing the work, the plant, or both.

Several treatments of human error in maintenance work have been published by the Health and Safety Executive (HSE). Accounts illustrated with case histories are given in *Deadly Maintenance* (HSE, 1985b) and *Dangerous Maintenance* (HSE, 1987a). *Human Factors in Industrial Safety* (HSE, 1989 HS(G) 48) gives general guidance on human factors relevant *inter alia* to maintenance work.

Here again it is appropriate to start with the designer. If it is practical to design out the need for maintenance, the problem disappears. One cause of error in maintenance is lack of understanding either of the process and chemicals or of the equipment. This is exemplified by a number of incidents where the wrong bolts have been loosened on a valve, resulting in a release.

Maintenance is an area where poor practices can easily creep in. An example is breaking a flange the wrong way. Maintenance personnel on process plants spend a large proportion of their time dealing with heavy, dirty equipment. It is not easy to make the transition required when they have to work on relatively delicate items. A case in point is leaving flameproof equipment with loose screws and an excessive gap so that it is no longer flameproof.

In some cases defects of workmanship in maintenance are gross. Kletz cites a case where a fitter had to remake a flanged joint using a new spiral wound gasket. Since it

Table 14.13 *Human error as a cause of pipework failure (after Bellamy, Geyer and Astley, 1989) (Courtesy of the Health and Safety Executive)*

A Direct causes

Cause	*Contribution of human error (%)*
Operator error	30.9
Wrong or incorrectly located in-line equipment	4.5
Human-initiated impact	5.6
Total	41.0

B Underlying cause

Cause	*Distribution of human error between underlying causes (%)*
Design	8
Manufacture	2
Construction	8
Operation	22
Maintenance	59
Sabotage	1
Total	100

C Recovery failures

Failure	*Distribution of human error between recovery failures (%)*
Not recoverable	1
Hazard study	6
Human factors review	60
Task checking	31
Routine checking	2
Total	100

was too large, he ground depressions in the outer metal ring of the gasket so that it would fit between the bolts. He made matters worse by making only three depressions so that the gasket did not fit centrally between the bolts.

Maintenance is an activity which can often be frustrating and short-cuts may be taken. Use of protective clothing, isolation procedures and the permit-to-work system are areas where such short-cuts tend to be taken.

Reference has already been made to errors in identification of equipment. Measures need to be taken to ensure effective identification. An overview of the management of maintenance to ensure safe working is given in Chapter 21.

14.21.4 Human error in pipework failure

In construction, errors in the installation of pipework loom large. The problem was highlighted in a survey by Kletz (1984k), who has given a checklist aimed at eliminating such defects. Two of the measures proposed by Kletz to reduce pipework construction errors are improved training of the construction workforce, as already described, and the use of the barrier principle. A more detailed treatment of Kletz' work on pipework errors is given in Chapter 12.

This theme has been taken up again in the study of pipework failure, and human error as cause of such failure, described by Bellamy and co-workers (Bellamy, Geyer and Astley, 1989; Geyer *et al.*, 1990; Geyer and Bellamy, 1991; Hurst *et al.*, 1991). This study has already been referred to in Chapter 12 in dealing with failure of pipework, where a number of tables are given based on this study.

A total of 921 incidents from incident data bases were reviewed, some analysis was done on all 921 incidents and further analysis on some 502 incidents. Incidents were classified in three ways: (1) 'direct cause', (2) 'origin of failure' or underlying cause, and (3) 'recovery failure' or preventive mechanism.

The crude data showed that operator error contributed 18.2% to the direct causes of pipework failure, whilst defective pipe or equipment contributed 31.9% and unknown causes 9.1%. If the last two categories are removed from consideration, operator error then contributes 30.9%. Table 14.13, Section A, shows the contribution to the direct causes of operator error and of two other human errors. Sections B and C of the table give the distribution of human error among the underlying causes and the recovery failures, respectively. For the former the predominant errors are in maintenance and for the latter in human factors review and task checking.

14.21.5 Human error and plant design

Turning to the design of the plant, design offers wide scope for reduction both of the incidence and consequences of human error. It goes without saying that the plant should be designed in accordance with good process and mechanical engineering practice. In addition, however, the designer should seek to envisage errors which may occur and to guard against them.

The designer will do this more effectively if he is aware from the study of past incidents of the sort of things which can go wrong. He is then in a better position to understand, interpret and apply the standards and codes, which are one of the main means of ensuring that new designs take into account, and prevent the repetition of, such incidents.

A significant contribution can be made by the designer to the elimination of errors leading to pipework failures. One is to remove situations vulnerable to an operator error such as opening the wrong valve so that the plant suffers overpressure or undertemperature. Another is to counter errors in construction and maintenance by applying the barrier principle so that it is literally impossible to assemble an item in a manner other than the correct one.

Another area where the designer has a large contribution to make is in facilitating maintenance activities on the plant. This includes: provision of access to equipment, pipework and valves; pipework arrangements which assist isolation of sections of plant; and measures to prevent of misidentification of equipment.

An aspect of operation which provides a new field for human error is computer control of process plants. This is considered in Chapter 13.

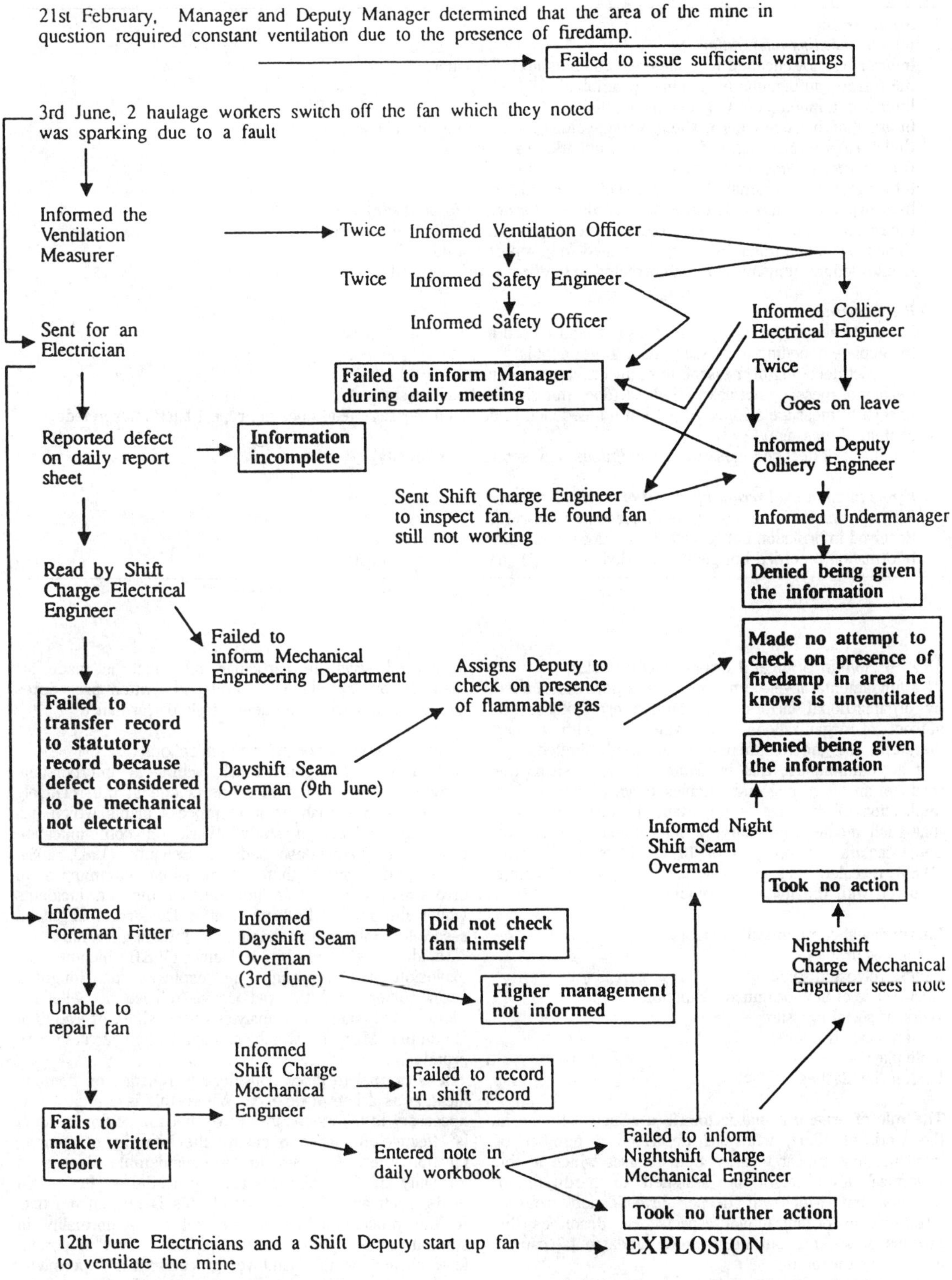

Figure 14.22 Chain of events in the Houghton Main colliery accident illustrating communication errors (Bellamy, 1983. Reproduced by permission.)

Table 14.14 *Some human errors in inter-personal communication (Bellamy, 1983. Reproduced with permission.)*

1. *Sender errors*
   (a) *in encoding information*
   Information not encoded (e.g. message contains no information)
   Information ambiguous (e.g. semantic ambiguity)
   Information incomplete (e.g. details omitted)
   Information or code incorrect (e.g. wrong values given, wrong terms used)
   Code inappropriate (e.g. code foreign to intended receiver)
   (b) *In transmitting information*
   Information not transmitted (e.g. not sent, not recorded)
   Information not transmitted in time (e.g. too late for action to be taken)
   Information transmitted via incorrect channel (e.g. standard channel not used)
   Unacknowledged transmission not repeated (e.g. repeat not possible)
   Acknowledged transmission not corrected (e.g. error in the acknowledgement not noticed)

2. *Receiver errors*
   Failure to acquire message (e.g. does not read record book, ignores message)
   Incomplete decoding of message (e.g. forget details)
   Incorrect decoding of message (e.g. misinterprets meaning)
   Receipt of message not acknowledged (does not give feedback to sender)
   Receipt of message acknowledged but no feedback of decoded message (e.g. does not repeat back interpreted content of message)
   Feedback of decoded message is ambiguous (e.g. semantic ambiguity)

3. *Errors in additional recording of sent or received messages*
   Sent information not recorded (e.g. not written down)
   Received information not recorded (e.g. not written down)
   Information recorded but poorly encoded (see 1(a)) (e.g. in written report)

14.21.6 Human error and organizational factors

At a fundamental level human error is largely determined by organizational factors. Like human error itself, the subject of organizations is a wide one with a vast literature, and the treatment here is strictly limited.

It is commonplace that incidents tend to arise as the result of an often long and complex chain of events. The implication of this fact is important. It means in effect that such incidents are largely determined by organizational factors. An analysis of 10 incidents by Bellamy (1985) revealed that in these incidents certain factors occurred with the following frequency:

Interpersonal communication errors	9
Resources problems	8
Excessively rigid thinking	8
Occurrence of new or unusual situation	7
Work or social pressure	7
Hierarchical structures	7
'Role playing'	6
Personality clashes	4

The role of organizational factors is a principal theme in the work of Kletz, who has described a number of incidents that appear to arise from failures which at the superficial level might be assigned to hardware or humans, but which on deeper investigation can be attributed to organizational aspects. He describes the process of seeking out these more fundamental causes as 'peeling the onion'.

Organizational factors are to a degree taken into account in current methods of human reliability analysis in several ways. In particular, allowance is made for these factors in relation to oral and written procedures and to factors which influence task performance.

14.21.7 Human error in communications

One aspect of organizational factors is interpersonal communication. Communications is one of the relatively few topics in human error in process plants which has been the subject of study. Work on communication errors has been described by Bellamy (1983, 1984, 1985). She found that interpersonal communication errors played a part in nine out of the ten incidents which she studied Bellamy (1985). The errors might be associated with either enabling or initiating events.

Incidents studied by Bellamy (1983) include the explosion at Flixborough, the explosion at Houghton Main colliery and the radioactive release at Sellafield. Figure 14.22 shows her analysis of the chain of events at Houghton Main in which communication errors were prominent.

It is normal to place considerable reliance on communication as a line of defence. Where this is the case it is necessary to take steps to ensure that the communication is effective. In order to ensure the integrity of communication there is need for a considerable degree of formality in the communication procedures. In certain fields, such as air traffic control, this is normal practice. In the process industries the degree of formality in communication tends to be variable. Two important features include the handover procedures and permit-to-work system. Other aspects of communication may be less formal.

Table 14.14 is a list given by Bellamy (1983) of some of the errors which may occur in interpersonal communications. One feature prominent in the incidents which she describes is ambiguous or incomplete messages occurring within a formal system. The existence of a formal system such as handover with associated logbooks may not, therefore, be sufficient if in practice the quality of communication within it is poor.

Two central features of the system of communication on a plant are the permit-to-work system and the handover system. It is not surprising, therefore, that many communications errors are associated in some way with one or other of these systems. Kletz (1991e) gives as an illustration the instance of the failure of a permit-to-work system, described in Chapter 20.

14.22 Prevention and Mitigation of Human Error

There exist a number of strategies for prevention and mitigation of human error. Essentially these aim to:

(1) reduce frequency;
(2) improve observability;
(3) improve recoverability;
(4) reduce impact.

Some of the means used to achieve these ends include:

(1) design-out;
(2) barriers;
(3) hazard studies;
(4) human factors review;
(5) instructions;
(6) training;
(7) formal systems of work;
(8) formal systems of communication;
(9) checking of work;
(10) auditing of systems.

The *Guide to Reducing Human Error in Process Operation* compiled by the Human Factors in Reliability Group (HFRG), (1985 SRD R347) gives guidance in checklist form. The checklists cover (1) operator–process interface, (2) procedures, (3) workplace and working environment, (4) training, and (5) task design and job organization.

14.22.1 Role of engineering measures

There is some danger that emphasis on the apparently dominant role of human error in major incidents may induce in the engineer a feeling of helplessness. This is misplaced. As this chapter indicates, there is a variety of methods, from the managerial to the technical, which can be used to reduce the frequency with which human error gives rise to an initiating event. But it may well be in his contribution to the mitigation of the consequences that the engineer comes into his own.

14.23 Assessment of Human Error

Turning now to the assessment of human error, the growing use of quantitative risk assessment has created a demand for the development of techniques for the assessment of human error to complement those available for the assessment of hardware failures.

In the following sections an account is given of developments in the assessment of human error. First the task analytic approach is described in which a task is decomposed into its elements and error rates are assigned to each of these elements (Section 14.24). Next an overview is given of early attempts to decompose the process control task into elements for which there might be available information on human error probabilities (Section 14.25). The account then reverts to qualitative methods developed to ensure that the analyst has a sound understanding of the problem before embarking on quantification (Section 14.26). There follow descriptions of some of the principal methods of estimating human reliability (Sections 14.27–14.32), of performance shaping factors (Section 14.33) and of human error data (Sections 14.34). Treatments are then given of certain guides and overviews, including a benchmark exercise, and of applications at Sizewell B (Sections 14.35–14.39). The topic of human reliability assessment is replete with acronyms and the account here necessarily reflects this.

14.24 Assessment of Human Error: Task Analytic Approach

Turning then to the methods available for making a quantitative assessment of human error in process plant operation, the starting point is analysis of the task.

14.24.1 Task analysis

Early work on discrete tasks generally took as its starting point some form of task analysis in which the task was broken down into its component parts. The probabilities of error in the performance of these component subtasks were then estimated and the probability of error in the complete task then assessed.

A methodology based on the task analytical approach, the technique for human error rate prediction (THERP), developed by Swain and co-workers is described below. A overview of this type of approach is given here based on an early account by Swain (1972).

Estimates of the probability of success in a planned discrete task may be made at different levels of sophistication. At one extreme an average task error rate of, say, 0.01 is sometimes quoted. This is apparently based on the assumption that the average element error rate is 0.001 and that there are, on average, 10 elements per task. At the other extreme is the collection and application of experimental data on error rates as a function of the performance shaping factors.

An important aspect of task reliability is the possibility of recovery by detection and correction of the error. If an error is retrievable in this way, the error rate for the task may be reduced by orders of magnitude. The probability of recovery depends very much on the cues available concerning the error, whether from displays and controls or from the system generally.

The effect of an error is also relevant, since only a proportion of errors has significant effects. Swain suggests that if the probability that an error will occur is between 10^{-4} and 10^{-3}, that it will be corrected is between 0.02 and 0.2 and that it will have a significant effect is between 0.2 and 0.3, then the probability that an error will occur, will remain undetected and will have a significant effect is of the order 0.4×10^{-6} to 60×10^{-6}.

The application of task reliability calculation to process control is exemplified by the work of Ablitt (1969 UKAEA AHSB(S) R160) on the estimation of the reliability of execution of a schedule of trip and warning tests involving some 55 items.

14.25 Assessment of Human Error: Process Operation

It was recognized early on that the above approach, based as it is on routine tasks such as equipment assembly, was of dubious applicability to the overall task of process control, and attempts were made to classify tasks in process control so as to make the links between these tasks and relevant areas of research in human factors.

One such classification is that given by Lees (1973a, 1980b), who breaks the overall process control tasks down into three categories on which information is available – (1) simple tasks, (2) vigilance tasks and (3) emergency behaviour – and two further more difficult categories – (4) complex tasks and (5) control tasks. In this classification simple tasks are essentially the planned, usually routine, discrete tasks of the type just described, whereas complex tasks are those which are not routine but require some element of decision-making. The control task is essentially the residual of the operator's overall task in supervising the process, after exclusion of the specific simple and complex tasks, the vigilance, or monitoring, task and emergency response. The classification was an early attempt to decompose the process control task into categories which are recognized in work on human factors and on which there is therefore some prospect of obtaining data, or at least guidance.

A more recent but somewhat similar classification is that given by Dougherty and Fragola (1988). They give the following analogy between hardware failures and human error:

Hardware failure	*Human activity*
Demand failure	Plan-directed behaviour
Stand-by failure	Vigilance behaviour
Running failure	Event-driven behaviour

To the above list might be added, for hardware, emergency response failure and, for human activity, emergency response behaviour.

Other techniques which take a more holistic approach to complex tasks are the goal-directed activity (GDA) technique and success likelihood index method (SLIM) of Embrey and the human error assessment and reduction technique (HEART) of Williams, described below.

14.25.1 Routine tasks

As described above, the usual approach to the estimation of human error rates in simple, or routine, tasks is to break the task down into its constituent elements and to derive an estimate of the reliability of the task from that of its constituent elements. The deficiencies of this rather mechanistic approach have long been recognized for any task which involves decision-making, but it may serve to the degree that a routine task is defined as one which does not require decision-making so that errors are confined to slips rather than mistakes. Given this restriction the methods developed in THERP may be used. Alternatively, use may be made of the methods developed for complex, or non-routine tasks, as described below.

14.25.2 Vigilance tasks

A vigilance task involves the detection of signals. There is a large body of work on the performance of vigilance tasks. Some of the factors which affect performance in such tasks are:

(1) sensory modality;
(2) nature of signal;
(3) strength of signal;
(4) frequency of signal;
(5) expectedness of signal;
(6) length of watch;
(7) motivation;
(8) action required.

Sight and hearing are the main channels for receiving the signal. The signal itself may be a simple GO–NO GO one or a complex pattern. In this latter case the out-of-limit condition may not be well defined in advance, but may have to be judged by the subject at the time. The signal may be strong and noise-free or weak and noisy, it may be frequent or infrequent and it may be an expected or unexpected type. The performance of the subject is affected by such factors as length of watch, degree of motivation and the action required on receipt of the signal.

Some general conclusions may be drawn from the work on vigilance. In general, the auditory channel is superior to the visual for simple GO–NO GO signals such as alarms. The probability of detection of a signal varies greatly with signal frequency; detection is much more probable for a frequent than for an infrequent signals. A vigilance effect exists so that performance tends to fall off with time.

There is an appreciable amount of data on vigilance tasks such as scanning displays, responding to alarms, and so on. A typical piece of work on vigilance in process control is that described by A.E. Green (1969 UKAEA AHSB(S) R172, 1970). The equipment used was the Human Response Analyser and Timer for Infrequent Occurrences (HORATIO), which automatically produces signals that are reasonably representative in form and time distribution of certain types of fault indication, and measures the time interval between the onset of the signal and the operator's response to it. Experiments were conducted in nuclear reactor control rooms in which operators were asked to respond to a simultaneous visual indication and an audible alarm signal by pressing a button. The signal rates were between 0.35 and 1.5 events/h. The probability of failure to respond to signals within the allowed response time was about 10^{-3}.

Another related type of task is inspection, which has also been much studied. In general, there is a greater probability of error in an inspection task where it is rather passive, the discrimination required is simple, the defect rate is low and the inspection is continuous. If there are more active features, such as taking measurements to detect the defect, the defect rate is high and inspection is alternated with another task, the error rate is reduced. A survey by McCornack (1961) suggests that with a defect probability of 0.01 the average inspection error is about 0.15, but inspection error probabilities vary widely and Swain (1972) quotes the range 0.01–0.6.

14.25.3 Emergency behaviour

Another definable task is response to an emergency. This is treated here as a situation which requires decision-making under stress. Here a certain amount of information is available from military and aerospace sources. One source of information is the critical incident technique which involves debriefing personnel who have been involved in critical incidents.

Two studies in particular on behaviour in military emergencies have been widely quoted. One is an investigation described by Ronan (1953) in which critical incidents were obtained from US Strategic Air Command aircrews after they had survived emergencies, e.g. loss of engine on take-off, cabin fire or tyre blowout on landing. The probability of a response which either made the situation no better or made it worse was found to be, on average, 0.16.

The other study, described by Berkun (1964), was on army recruits who were subjected to emergencies, which were simulated but which they believed to be real, such as increasing proximity of mortar shells falling near their command posts. As many as one-third of the recruits fled rather than perform the assigned task, which would have resulted in a cessation of the mortar attack.

Information is also available from work on simulators. In process control a typical study is the bounding study on operator response to emergencies done with reactor operators on a process simulator in the UKAEA and described by Lees and Sayers (1976). The simulation involved typical faults, e.g. control rod runout, blower failure and gas temperature rise. The fault rate was high at 10 faults/h. The response was to push a single button. These conditions were favourable and the result may therefore be regarded as an upper limit to the operator reliability which can be expected. For response times in the range 0–30 s the error probability was 0.24, and for those in the range 60–90 s it was zero.

14.25.4 Non-routine tasks

For complex, or non-routine, tasks the mechanistic approach of subdividing the task into its constituent elements is inappropriate. The appropriate approach is to treat the task in a more holistic way. Using this approach it is necessary to obtain estimates for the reliability of the whole task.

One method of doing this is to obtain estimates of task reliability from the judgement of experts. Early work on this was done by Rook (1964), who developed a method in which tasks were ranked by experts in order of their error-likeliness, and ranking techniques were used to obtain error rates.

Embrey (1979b) has developed this line of approach in his work on GDA. Whole tasks which are undertaken in process control are identified and estimates made of task reliability by a combination of data collection and expert judgement.

The other main method of determining the reliability of the whole task is the use of simulation. The information from simulation is generally a correlation between the probability of failure and the time available to perform the task.

14.25.5 Overall control task

There remains an overall control task which is essentially the residual remaining after the above activities have been treated. The effect of progress in the treatment of human error in process control has been to reduce this undefined residue.

14.25.6 Isolated acts of commission

There is a finite possibility that in performing his overall control task the operator will intervene unnecessarily, with unfortunate results. Such an event is not brought out by analyses that consider only activities which the operator is required to perform.

There are also other possibilities, including deliberate damaging action by the operator. One such action is suicide by destruction of the plant which is discussed by Ablitt (1969 UKAEA AHSB(S) R190):

> The probability per annum that a responsible officer will deliberately attempt to drop a fuel element into the reactor is taken as 10^{-3} since in about 1000 reactor operator years there have been two known cases of suicide by reactor operators and at least one case in which suicide by reactor explosion was a suspected possibility. The typical suicide rate for the public in general is about 10^{-4} per year although it does vary somewhat between countries.

The report on the underground tube train crash at Moorgate Station in London (DoE 1976/7) drew attention to the possibility of suicide by the driver. Sabotage is another type of action by the operator which is not unknown in process plants.

14.25.7 Time–reliability correlations

Error in executing a task is, in general, a function of the time available to perform it. There has been a growing tendency to use time as a principal correlator of operator error.

Thus error in vigilance situations depends very much on the response time which is allowed. Ablitt (1969 UKAEA AHSB(S) R190) tentatively proposed the following estimates of the error probability q for operator action in response to an alarm signal as a function of response time t:

t	1 s	10 s	60 s	5 min	10 min	> 10 min
q	1	10^{-1}	10^{-2}	10^{-3}	10^{-4}	10^{-5}–10^{-6}

A similar type of correlation underlies the estimates of the general probability of ineffective behaviour in emergencies used in studies such as the *Rasmussen Report*. In that study the error probability q used for operator action in response to an emergency as a function of response time t is:

t	60 s	5 min	30 min	Several hours
q	1	0.9	10^{-1}	10^{-2}

Table 14.15 General estimates of error probability used in the Rasmussen Report *(Atomic Energy Commission, 1975)*

Estimated error probability	*Activity*
10^{-4}	Selection of a key-operated switch rather than a non-key switch (this value does not include the error of decision where the operator misinterprets situation and believes key switch is correct choice)
10^{-3}	Selection of a switch (or pair of switches) dissimilar in shape or location to the desired switch (or pair of switches), assuming no decision error. For example, operator actuates large-handled switch rather than small switch
3×10^{-3}	General human error of commission, e.g. misreading label and therefore selecting wrong switch
10^{-2}	General human error of omission where there is no display in the control room of the status of the item omitted, e.g. failure to return manually operated test valve to proper configuration after maintenance
3×10^{-3}	Errors of omission, where the items being omitted are embedded in a procedure rather than at the end as above
3×10^{-2}	Simple arithmetic errors with self-checking but without repeating the calculation by re-doing it on another piece of paper
$1/x$	Given that an operator is reaching for an incorrect switch (or pair of switches), he selects a particular similar appearing switch (or pair of switches), where x = the number of incorrect switches (or pairs of switches) adjacent to the desired switch (or pair of switches). The $1/x$ applies up to 5 or 6 items. After that point the error rate would be lower because the operator would take more time to search. With up to 5 or 6 items he does not expect to be wrong and therefore is more likely to do less deliberate searching
10^{-1}	Given that an operator is reaching for a wrong motor operated valve (MOV) switch (or pair of switches), he fails to note from the indicator lamps that the MOV(s) is (are) already in the desired state and merely changes the status of the MOV(s) without recognizing he had selected the wrong switch(es)
~1.0	Same as above, except that the state(s) of the incorrect switch(es) is (are) *not* the desired state
~1.0	If an operator fails to operate correctly one of two closely coupled valves or switches in a procedural step, he also fails to correctly operate the other valve
10^{-1}	Monitor or inspector fails to recognize initial error by operator. *Note*: With continuing feedback of the error on the annunciator panel, this high error rate would not apply
10^{-1}	Personnel on different work shift fail to check condition of hardware unless required by checklist or written directive
5×10^{-1}	Monitor fails to detect undesired position of valves, etc., during general walk-around inspections, assuming no checklist is used
0.2–0.3	General error rate given very high stress levels where dangerous activities are occurring rapidly
$2^{(n-1)}x$	Given severe time stress, as in trying to compensate for an error made in an emergency situation, the initial error rate, x, for an activity doubles for each attempt, n, after a previous incorrect attempt, until the limiting condition of an error rate of 1.0 is reached or until time runs out. This limiting condition corresponds to an individual's becoming completely disorganized or ineffective
~1.0	Operator fails to act correctly in first 60 seconds after the onset of an extremely high stress condition, e.g. a large LOCA
9×10^{-1}	Operator fails to act correctly after the first 5 minutes after the onset of an extremely high stress condition
10^{-1}	Operator fails to act correctly after the first 30 minutes in an extreme stress condition
10^{-2}	Operator fails to act correctly after the first several hours in a high stress condition
x	After 7 days after a large LOCA, there is a complete recovery to the normal error rate, x, for any task

Notes:
(1) Modifications of these underlying (basic) probabilities were made on the basis of individual factors pertaining to the tasks evaluated.
(2) Unless otherwise indicated estimates or error rates assume no undue time pressures or stresses related to accidents.

14.25.8 Human error probabilities: *Rasmussen Report*

One of the earliest sets of estimates for the probability of human error was that given in the *Rasmussen Report* as shown in Table 14.15. These estimates are mostly of the type used in THERP, but they include the time reliability correlation just referred to.

14.25.9 Human error probabilities: hazard analysis

Estimates of human error probability are given in a number of fault trees. Operator error occurs in the tree as errors which either initiate or enable to the fault sequence and as errors which constitute failures of protection. Table 14.16 is a summary by Lees (1983a)

***Table 14.16** Some estimates[a] of operator error used in fault tree analysis (Lees, 1983a; after Lawley, 1974, 1980) (Courtesy of the Institution of Chemical Engineers)*

Crystallizer plant		
		Probability
Operator fails to observe level indicator or take action		0.04
Operator fails to observe level alarm or take action		0.03
		Frequency (events/year)
Manual isolation valve wrongly closed (p)		0.05 and 0.1
Control valve fails to open or misdirected open		0.5
Control valve fails shut or misdirected shut (l)		0.5
Propane pipeline		
Operator fails to take action:	*Time available*	*Probability*
To isolate pipeline at planned shut-down		0.001
To isolate pipeline at emergency shut-down		0.005
Opposite spurious tank blowdown given alarms and flare header signals	30 min	0.002
Opposite tank low level alarm		0.01
Opposite tank level high given alarm with	5–10 min	
(a) controller misdirected or bypassed when on manual		0.025
(b) level measurement failure		0.05
(c) level controller failure		0.05
(d) control valve or valve positioner		0.1
Opposite slowly developing blockage on heat exchanger revealed as heat transfer limitation		0.04
Opposite pipeline fluid low temperature given alarm	5 min	0.05
Opposite level loss in tank supplying heat transfer medium pump given no measurement (p)	5 min	0.2
Opposite tank blowdown without prior pipeline isolation given alarms which operator would not regard as significant and pipework icing	30 min	
(a) emergency blowdown		0.2
(b) planned blowdown		0.6
Opposite pipeline fluid temperature low given alarm		0.4
Opposite pipeline fluid temperature low given alarm	Limited	0.8
Opposite backflow in pipeline given alarm	Extremely short	0.8
Opposite temperature low at outlet of heat exchanger given failure of measuring instrument common to control loop and alarm		1
Misvalving in changeover of two-pump set (stand-by pump left valved open, working pump left valved in)		0.0025/changeover
Pump in single or double operation stopped manually without isolating pipeline		0.01/shut-down
Low pressure steam supply failure by fracture, blockage or isolation error (p)		0.1/year
Misdirection of controller when on manual (assumed small proportion of time)		1/year

Notes:
[a] l, Literature value; p, plant value; other values are assumptions.

of the estimates used in two fault trees published by Lawley (1974b, 1980). Errors which result in failure of protection (expressed as probabilities) predominate over errors which initiate or enable the fault sequence (expressed as frequencies). Initiating and enabling errors tend to be associated with an item of equipment, and protection errors with a process variable. Such a variable may have (1) no measurement, (2) measurement only, or (3) measurement and alarm, and this is an important feature influencing the error rate.

The great majority of figures given are assumed values, with a few values being obtained from the literature and a few from the works. Engineering judgement is used in arriving at these values and the values selected take into account relevant influencing factors such as variable measurement and alarm and time available for action. The way in which allowance is made for these influencing factors is illustrated by the following extracts from the supporting notes:

> There is therefore a very high probability that the operator would be made aware of a spurious blowdown condition by the alarms and this would be augmented by observation of excessive flaring and header noise which would highlight the cause of the problem. Because alarms will be set quite close to normal operating pressure and level, there would be almost 30 min available for action before the pipeline is chilled to −15°C.

and

> The probabilities quoted are based on experience assuming that 5 min would be available for action, and including allowance for failure of the alarm. They take into account factors such as whether or not the operator would be in close attendance at the time of the fault, ease of diagnosis of the problem, whether or not the fault could be corrected from the control room or only by outside action, reluctance to shut down the export pumps until correction of the fault has been attempted because back-up trip protection is provided, etc.

14.26 Assessment of Human Error: Qualitative Methods

The foregoing account has described some early approaches to human reliability assessment in support of probabilistic risk assessment. It is now necessary to backtrack a little and revert to a consideration of qualitative methods for analysis of human error.

Methods are available which may be utilized to identify and so reduce error-likely situations and also to support incident investigation. The elements of such methods are the underlying model of human reliability, the taxonomy of human error and the analysis technique.

These methods may be used in their own right to reduce error or as a stage in a human reliability assessment method. Their significance in the latter application is that they provide a structured approach to gaining understanding of the problem. In the absence of a high quality technique for this essential preliminary stage, quantification is premature.

Task analysis may be regarded as the prime technique, or rather family of techniques, but there are also a number of others. Early work in this area was that of J.R. Taylor (1979), who described a variety of approaches.

A method based on hierarchical task analysis is predictive human error analysis, described below. Another method is the work analysis technique described by Pedersen (1985).

A fuller discussion of human error analysis methods is given by Kirwan (1990) and the CCPS (1994/17).

14.26.1 Some error analysis strategies

Against the background of a long-term programme of work on human error J.R. Taylor (1979) has developed a set of four error analysis strategies:

(1) action error method;
(2) pattern search method;
(3) THERP;
(4) sneak path analysis.

The third of these has already been outlined and is considered further below. The others are now described.

14.26.2 Action error method

The action error method is applicable to a sequence of operator actions which constitute intervention on the plant. The structure of the sequence takes the form: action/effect on plant/action/effect on plant. ... An outline of the structure of the procedure in the form of a cause–consequence diagram is shown in Figure 14.23. The range of errors handled is shown in Table 14.17.

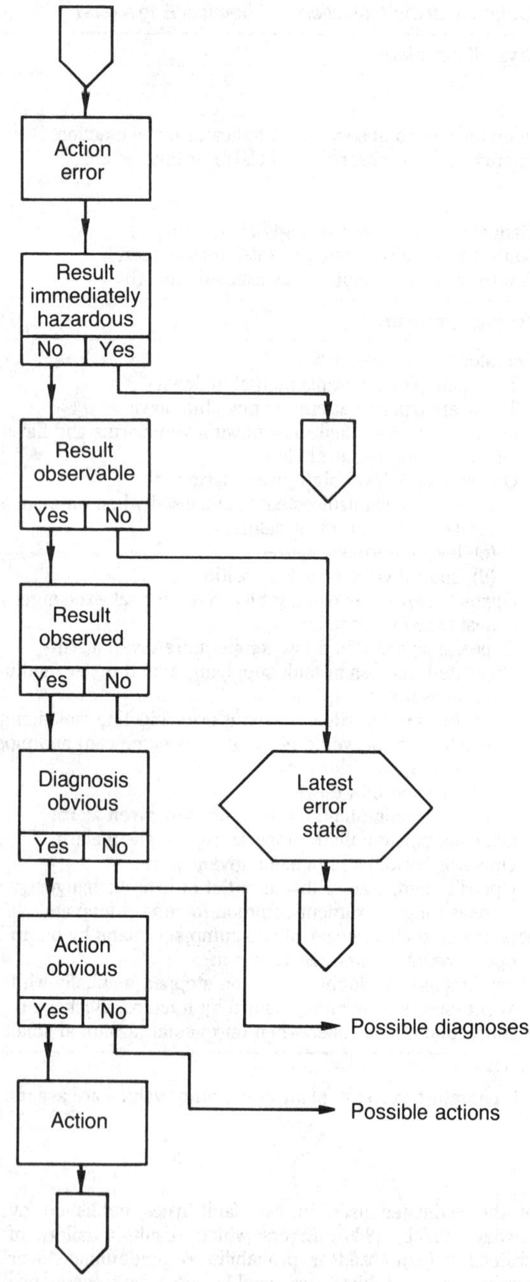

Figure 14.23 Outline structure of action error method (J.R. Taylor, 1979. Reproduced by permission.)

Usually it is found that for any reasonably large operating procedure it is practical to take into account

Table 14.17 Operator errors addressed in action error method (J.R. Taylor, 1979. Reproduced by permission.)

Cessation of a procedure
Excessive delay in carrying out an action or omission of an action
Premature execution of an action—too early
Premature execution of an action—preconditions not fulfilled
Execution on wrong object of action
Single extraneous action
In making a decision explicitly included in a procedure, taking the wrong alternative
In making an adjustment or an instrument reading, an error outside tolerance limits

only single initial errors, although in a few cases it may be possible to use heuristic rules to identify double errors which it is worthwhile to explore. For example, one error may result in material being left in a vessel, while a second error may result in an accident arising from this.

Taylor states that the method is not suitable for quantitative assessment, because the spread of error rates on the individual elements is considered to be too wide. Factors mentioned as influencing these error rates are cues, feedback and type of procedure (freely planned, trained).

14.26.3 Pattern search method

The pattern search method is addressed to the problem that an accident is typically the result of a combination of operator errors. For such cases detailed analysis at the task element level is impractical for two reasons. One is the combinatorial explosion of the number of sequences. The other is that the error rates, and above all the error rate dependencies, are not determinable.

An important feature of such accidents is that they may have a relatively long sequence of errors, say 3 to 5, which have a common cause, such as error in decision-making, in work procedure or in plant state assessment. Often the sequence is associated with an unrevealed plant failure.

The pattern search method is based on identifying a common cause error, developing its consequence, perhaps using an event tree, and using the results to 'steer' the construction of the fault tree.

14.26.4 Sneak path analysis

Sneak path analysis is concerned with the identification of potential accident situations. It is so called by analogy with sneak circuits. It seeks to identify sources of hazard such as energy or toxins and targets such as people, critical equipment or reactive substances. The standpoint of the analysis is similar to the accident process model of Houston (1971) described in Chapter 2.

For an accident to occur it is necessary for there to be some operator action, operator error, equipment failure or technical sequence. A search is made to determine whether any of these necessary events can occur. In examining operator error attention is directed particularly to actions which are 'near' to the necessary error. Nearness may be temporal, spatial or psychological. Often such an action is very near to the normal operator action.

Table 14.18 Error classification for predictive human error analysis (Center for Chemical Process Safety, 1994/X) (Courtesy of the American Institute of Chemical Engineers)

Action	
A1	Action too long/short
A2	Action mistimed
A3	Action in wrong direction
A4	Action too little/too much
A5	Misalign
A6	Right action on wrong object
A7	Wrong action on right object
A8	Action omitted
A9	Action incomplete
A10	Wrong action on wrong object
Checking	
C1	Checking omitted
C2	Check incomplete
C3	Right check on wrong object
C4	Wrong check on right object
C5	Check mistimed
C6	Wrong check on wrong object
Retrieval	
R1	Information not obtained
R2	Wrong information obtained
R3	Information retrieval incomplete
Transmission	
T1	Information not transmitted
T2	Wrong information transmitted
T3	Information transmission incomplete
Selection	
S1	Selection omitted
S2	Wrong selection made
Plan	
P1	Plan preconditions ignored
P2	Incorrect plan executed
P3	Correct but inappropriate plan executed
P4	Correct plan executed too soon/too late
P5	Correct plan executed in wrong order

14.26.5 Predictive human error analysis (PHEA)

Predictive human error analysis (PHEA) is described by Embrey and co-workers (e.g. Murgatroyd and Tait, 1987; Embrey, 1990) and by the CCPS (1994/17). PHEA uses hierarchical task analysis to discover the plan involved in the task, combined with the error classification shown in Table 14.18. The task is then analysed step by step in terms of the task type, error type, task description, consequences, recovery and error reduction strategy.

In a validation study of PHEA, Murgatroyd and Tait (1987) found that the proportion of errors with potentially

Table 14.19 HRA Handbook: *contents (Swain and Guttmann, 1983)*

1. Introduction
2. Explanation of Some Basic Terms
3. Some Performance Shaping Factors Affecting Human Reliability
4. Man–Machine Systems Analysis
5. A Technique for Human Reliability Analysis
6. Sources of Human Performance Estimates
7. Distribution of Human Performance and Uncertainty Bounds
8. Use of Expert Opinion in Probabilistic Risk Assessment
9. Unavailability
10. Dependence
11. Displays
12. Diagnosis of Abnormal Events
13. Manual Controls
14. Locally Operated Valves
15. Oral Instructions and Written Procedures
16. Management and Administrative Control
17. Stress
18. Staffing and Experience Levels
19. Recovery Factors
20. Tables of Estimated Human Error Probabilities
21. Examples and Case Studies
22. Concluding Comments

Appendices

A Methods for Propagating Uncertainty Bounds in a Human Reliability Analysis and for Determining Uncertainty Bounds for Dependent Human Activities
B An Alternative Method for Estimating the Effects of Dependence
C Calculations of Mean and Median Trials to Detection
D Calculations of Basic Walk-around Inspections as a Function of Period between Successive Walk-arounds
E Reviews of the Draft Handbook
F A Comparison of the October 1980 and Present Versions of the Handbook

significant consequences which actually occurred in a equipment calibration task over a 5 year period was 98%.

14.26.6 System for predictive error analysis and reduction (SPEAR)

The system for predictive error analysis and reduction (SPEAR) is a set of qualitative techniques, of which PHEA is one. It is described by the CCPS (1994/17). SPEAR comprises the following techniques: (1) task analysis, (2) performance influencing factor (PIF) analysis, (3) PHEA, (4) consequence analysis and (5) error reduction analysis.

Consequence analysis involves consideration not just of the consequences of failure to perform the task but also of the consequences of any side-effects which may occur whether or not the task is executed. Error reduction analysis is concerned with measures to reduce those errors which do not have a high probability of recovery.

Task analysis and PHEA have already be described and PIF analysis is treated in Section 14.33.

14.27 Assessment of Human Error: *Human Reliability Analysis Handbook*

The first systematic approach to the treatment of human error within a probabilistic risk assessment (PRA) was the Technique for Human Error Rate Prediction (THERP). An early account of THERP was given by Swain (1972). Its origins were work done at Sandia Laboratories on the assessment of human error in assembly tasks. This work was then extended to human error in process control tasks, with particular reference to nuclear reactors. This extension to nuclear reactor control was used in the *Rasmussen Report*, or WASH-1400 (AEC, 1975), which contains generic PRAs for US commercial nuclear reactors.

The accident at Three Mile Island gave impetus to work in this area and led to the publication of the *Handbook of Human Reliability Analysis with Emphasis on Nuclear Power Plant Applications. Final Report (*the *HRA Handbook)* by Swain and Guttmann (1983). This report was widely circulated in draft form in 1980 and many literature references are to the draft. Further work is described in *Accident Sequence Evaluation Program. Human Reliability Analysis Procedure* (Swain, 1987a).

The *HRA Handbook* gives a complete methodology for addressing the human error aspects of a PRA, but THERP is central to the approach, and the methodology as a whole is generally referred to by that acronym. However, the *Handbook* represents a considerable extension of the original THERP methodology, particularly in respect of its adoption of the time reliability correlation method.

A further review of human reliability analysis (HRA) and THERP is given by P. Miller and Swain (1987). Table 14.19 gives the contents of the *HRA Handbook* and Figure 14.24 shows the structure of the principal data tables. An account is now given of THERP. The *HRA Handbook* should be consulted for a more detailed treatment.

14.27.1 Overall approach

The overall approach used in the *Handbook* is shown in Figure 14.25. The tasks to be performed are identified as part of the main PRA. The HRA involves the assessment of the reliability of performance of these tasks.

14.27.2 Technique for human error rate prediction (THERP)

The starting point is a task analysis for each of the tasks to be performed. The method is based on the original THERP technique and uses a task analytic approach in which the task is broken down into its constituent elements along the general lines described above. The basic assumption is that the task being performed is a planned one.

The task is described in terms of an event tree as shown in Figure 14.26. This figure gives the event tree for two tasks A and B which are performed sequentially and which constitute elements of a larger overall task. In reliability terms the relationship between the two constituent tasks to the overall task may be a series or a parallel one.

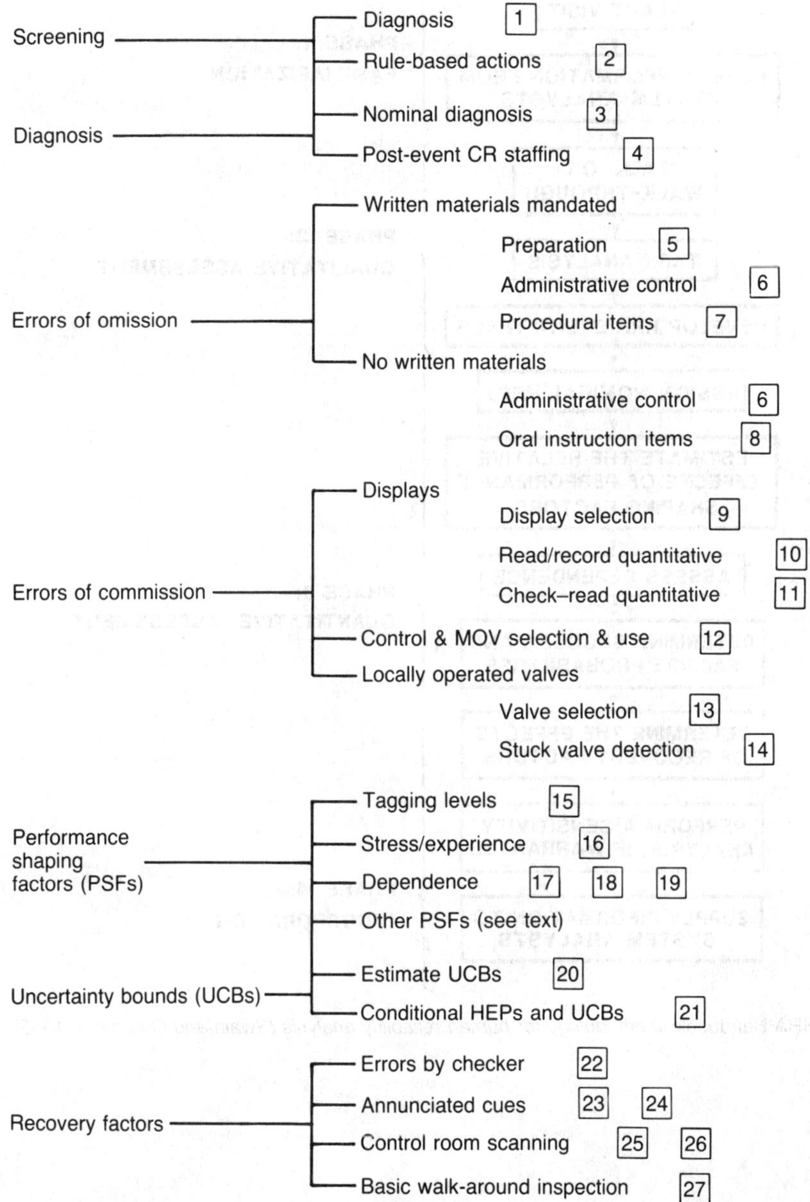

Figure 14.24 *HRA Handbook: structure of principal data tables (Swain and Guttmann, 1983). CR, control room; HEP, human error probability; MOV, motor operated value*

The probability that task A will be performed successfully is a and the probability that it will not be performed successfully is A. Since task A is the first in the sequence it is assumed to be independent of any other task and the probabilities a and A are therefore unconditional probabilities. The probability that task B will be performed successfully is b and the probability that it will not be performed successfully is B. Since task B is performed after task A it is assumed to be dependent in some degree on task A and the probabilities b and B are therefore conditional probabilities. Thus for b it is necessary to distinguish between $b|a$ and $b|A$ and for B between $B|a$ and $B|A$. The probabilities to be used are therefore as shown in Figure 14.26.

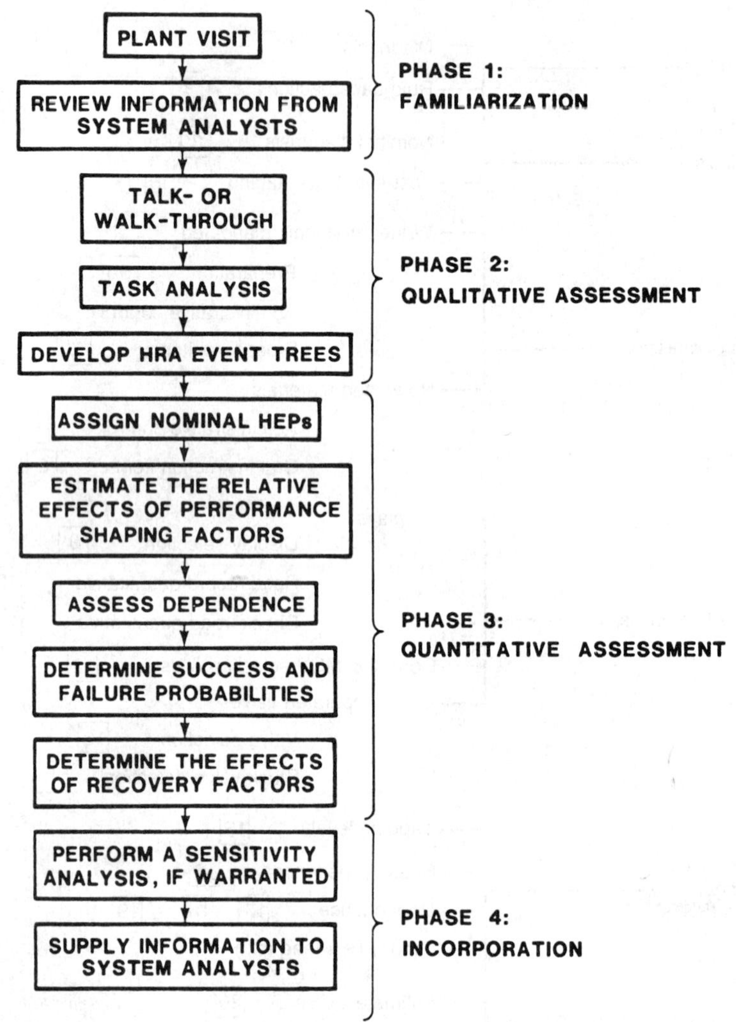

Figure 14.25 HRA Handbook: *methodology for human reliability analysis (Swain and Guttmann, 1983)*

The structure of the tree is the same whether the configuration is a series or parallel one but the status of the outcomes is different. As shown in Figure 14.26 for a series system in which it is necessary for both tasks to be successful, there is only one which rates as a success, whilst for a parallel system in which it is sufficient for only one of the tasks to be successful, there are three outcomes which rate as success.

A distinction is made between step-by-step tasks and dynamic tasks. The latter involve a higher degree of decision-making. The approach just described is most readily justified for step-by-step tasks.

14.27.3 Human error probability

The event tree so produced, and the corresponding equations, are then used to determine the probability of failure for the overall task. For this estimates are required of the human error probability (HEP) for each of the constituent tasks.

Several different human error probabilities are distinguished: nominal, basic, conditional and joint. A nominal HEP is a generic value before application of any performance shaping factors. A basic HEP (BHEP) is the basic unconditional HEP after application of performance shaping factors. A conditional HEP (CHEP) is a

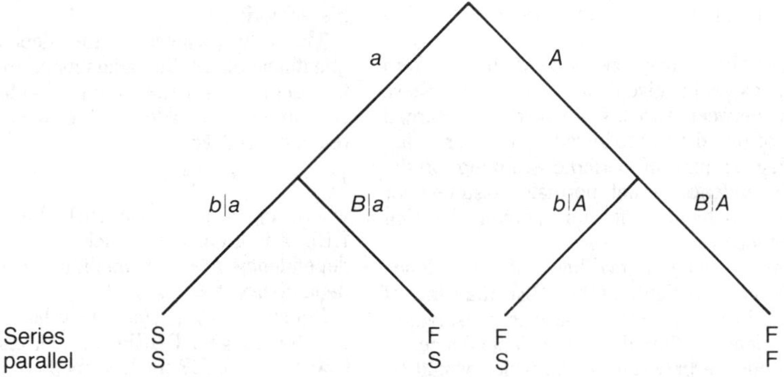

Task A = The first task

Task B = The second task

a = Probability of successful performance of task A

A = Probability of unsuccessful performance of task A

$b|a$ = Probability of successful performance of task B given a

$B|a$ = Probability of unsuccessful performance of task B given a

$b|A$ = Probability of successful performance of task B given A

$B|A$ = Probability of unsuccessful performance of task B given A

FOR THE SERIES SYSTEM:

$\Pr[S] = a(b|a)$

$\Pr[F] = 1 - a(b|a) = a(B|a) + A(b|A) + A(B|A)$

FOR THE PARALLEL SYSTEM:

$\Pr[S] = 1 - A(B|A) = a(b|a) + a(B|a) + A(b|A)$

$\Pr[F] = A(B|A)$

Figure 14.26 HRA Handbook: *event tree for two tasks in sequence illustrating conditional probabilities (Swain and Guttmann, 1983); F, failure; S, success*

basic HEP adjusted to take account of dependency. A joint HEP (JHEP) is the HEP for the overall task.

The simple application of HEP values which make no allowance for dependence tends to give very low probabilities of failure which do not accord with experience and carry little conviction. Allowance for dependence is therefore important. There is in any event some value of the HEP below which the estimate is no longer credible. A cut-off value is therefore applied. Reference is made to a cut-off of about 5×10^{-5}.

Many of the HEPs in the *Handbook* are expressed as log–normal distributions, quoted in terms of the two parameters median and error factor.

14.27.4 Dependence model

The dependence model is an important feature of the methodology. A significant proportion of the *Handbook* is concerned with dependency. There are two basic forms of dependence: dependence between tasks and dependence between people.

Where two tasks are performed in sequence the second task may be influenced by the first. Dependence is likely, for example, where an operator has to change two valves one after the other.

The other situation is where two people are involved in the same task. The form of the involvement may vary. The two persons may be involved in a joint task such as calibrating an instrument. They may perform separate but closely linked functions such as those of two operators sharing a control room but controlling different sections of the plant. The work of one person may be subject to the general supervision of another person. Or it may be formally checked by another.

There are various approaches which may be used to quantify dependence, including data and expert judgement. There are few relevant data. Expert judgement can be used to a limited extent. An example is given of the use of expert estimates of dependence in a calibration task.

The approach used in the *Handbook*, therefore, is the development of a generalized dependence model. The degrees of dependence used are zero, low, medium, high and complete. The determination of the appropriate degree of dependence depends on the situation under consideration. The *Handbook* gives quite extensive guidance. Here it is possible only to give a few examples.

In some cases there may be judged to be zero dependence. An example given of a situation where zero dependence between two tasks would be assumed is check-reading of one display followed by check-reading of another display as part of periodic scanning of the displays. Zero dependence is not normally assumed for persons working as a team or for one person checking another's performance.

The assumption of even a low level of dependence tends to result in an appreciably higher HEP than that of zero dependence. If there is any doubt, it is conservative to use low dependence rather than zero dependence.

A level of dependence between people which would be assessed as low is illustrated by the checking of the work of a new operator by a shift supervisor. In this situation the shift supervisor has an expectation that the new operator may make errors and is more than usually alert.

A moderate level of dependence is usually assessed between the shift supervisor and the operators for tasks where the supervisor is expected to interact with them.

A high level of dependence, or even complete dependence, would be assigned for the case where the shift supervisor takes over a task from a less experienced operator, since the latter may well defer to the supervisor both because of his greater experience and his seniority.

The other aspect of the dependence model is the quantification of the adjustment to be made given that the degree of dependency has been determined. The adjustment is made to the basic HEP (BHEP). The relation used is:

$$P_c = (1 - kP_b)/(k + 1) \qquad [14.27.1]$$

where P_b is the basic HEP and P_c is the conditional HEP. k is a constant which has the following values: low dependency $k = 19$; medium dependency $k = 6$; high dependency $k = 1$.

Equation 14.27.1 and the values of the constant k are selected to give CHEPs of approximately 0.05, 0.15 and 0.50 of the BHEP for low, medium and high dependency, respectively, where BHEP ≤ 0.01. Where BHEP > 0.01 the effective multiplying factor is slightly different. Thus, for example, for a BHEP of 0.1 the values of the CHEP are 0.15, 0.23 and 0.55, respectively, for these three levels of dependency.

14.27.5 Displays, annunciators and controls

Common tasks in process control are obtaining information from displays, responding to annunciators and manipulating controls. This is very much the home ground of human factors and there is a good deal of

Table 14.20 HRA Handbook: *human error probability (HEP) estimates and error factors (EFs) for oral instructions*[a] *(after Swain and Guttmann, 1983)*

Item[b]	Number of oral instruction items or perceptual units	(a) Pr(F) to recall item 'N', order of recall not important		(b) Pr(F) to recall all items, order of recall not important		(c) Pr(F) to recall all items, order of recall is important	
		HEP	EF	HEP	EF	HEP	EF
Oral instructions are detailed:							
(1)	1[c]	0.001	3	0.001	3	0.001	3
(2)	2	0.003	3	0.004	3	0.006	3
(3)	3	0.01	3	0.02	5	0.03	5
(4)	4	0.03	5	0.04	5	0.1	5
(5)	5	0.1	5	0.2	5	0.4	5
Oral instructions are general:							
(6)	1[c].	0.001	3	0.001	3	0.001	3
(7)	2	0.006	3	0.007	3	0.01	3
(8)	3	0.02	5	0.03	5	0.06	5
(9)	4	0.06	5	0.09	5	0.2	5
(10)	5	0.2	5	0.3	5	0.7	5

[a] It is assumed that if more than five oral instruction items or perceptual units are to be remembered, the recipient will write them down. If oral instructions are written down, use Table 20–5 in the *Handbook* for errors in preparation of written procedures and Table 20–7 for errors in their use.
The first column of HEPs (a) is for individual oral instruction items, e.g. the second entry, 0.003 (item 2a), is the Pr(F) to recall the second of two items, given that one item was recalled, and order is not important. The HEPs in the other columns for two or more oral instruction items are joint HEPs, e.g. the second 0.004 in the second column of HEPs is the Pr(F) to recall both of two items to be remembered, when order is not important. The 0.006 in the third column of HEPs is the Pr(F) to recall both of two items to be remembered in the order of performance specified. For all columns, the EFs are taken from Table 20–20.

[b]. The term 'item' for this column is the usual designator for tabled entries and does *not* refer to an oral instruction item.

[c] The Pr(F) values in rows 1 and 6 are the same as the Pr(F) to initiate the task.

Table 14.21 HRA Handbook: *human error probability (HEP) estimates and error factors (EFs) for written procedures (after Swain and Guttmann, 1983)*

A Preparation of written procedures[a]

Item	*Potential errors*	*HEP*	*EF*
(1)	Omitting a step or important instruction from a formal or *ad hoc* procedure[b] or a tag from a set of tags	0.003	5
(2)	Omitting a step or important instruction from written notes taken in response to oral instructions [c]	Negligible	
(3)	Writing an item incorrectly in a formal or *ad hoc* procedure or a tag	0.003	5
(4)	Writing an item incorrectly in written notes made in response to oral instructions [c]	Negligible	

[a] Except for simple reading and writing errors, errors of providing incomplete or misleading technical information are not addressed in the *Handbook*. The estimates are exclusive of recovery factors, which may greatly reduce the nominal HEPs.
[b] Formal written procedures are those intended for long-time use; *ad hoc* written procedures are one-of-a-kind, informally prepared procedures for some special purpose.
[c]. A maximum of five items is assumed. If more than five items are to be written, use 0.001 (EF = 5) for each item in the list.

B Neglect of written procedures

Item	*Task*	*HEP*	*EF*
(1)	Carry out a plant policy or scheduled tasks such as periodic tests or maintenance performed weekly, monthly or at longer intervals	0.01	5
(2)	Initiate a scheduled shiftly checking or inspection function[a]	0.001	3
	Use written operations procedures under:		
(3)	Normal operating conditions	0.01	3
(4)	Abnormal operating conditions	0.005	10
(5)	Use a valve change or restoration list	0.01	3
(6)	Use written test or calibration procedures	0.05	5
(7)	Use written maintenance procedures	0.3	5
(8)	Use a checklist properly[b]	0.5	5

[a] Assumptions for the periodicity and type of control room scans are discussed in Chapter 11 of the *Handbook* in the section, 'A General Display Scanning Model'. Assumptions for the periodicity of the basic walk-around inspection are discussed in Chapter 19 of the *Handbook* in the section, 'Basic Walk-around Inspection'.
[b] Read a single item, perform the task, check off the item on the list. For any item in which a display reading or other entry must be written, assume correct use of the checklist for that item.

C Use of written procedures[a]

Item[b]	*Omission of item*	*HEP*	*EF*
When procedures with check-off provisions are correctly used:[c]			
(1)	Short list, ≤ 10 items	0.001	3
(2)	Long list, > 10 items	0.003	3
When procedures without check-off provisions are used, or when check-off provisions are incorrectly used:[d]			
(3)	Short list ≤ 10 items	0.003	3
(4)	Long list, > 10 items	0.01	3
(5)	When written procedures are available and should be used but are not used[d]	0.05[e]	5

[a] The estimates for each item (or perceptual unit) presume zero dependence among the items (or units) and must be modified by using the dependence model when a nonzero level of dependence is assumed.
[b] The term 'item' for this column is the usual designator for tabled entries and does *not* refer to an item of instruction in a procedure.
[c] Correct use of check-off provisions is assumed for items in which written entries such as numerical values are required for the user.
[d] Table 20–6 in the *Handbook* lists the estimated probabilities of incorrect use of check-off provisions and of non-use of available written procedures.
[e] If the task is judged to be 'second nature', use the lower uncertainty bound for 0.05, i.e. use 0.01 (EF = 5).

Table 14.22 HRA Handbook: *human error probability (HEP) estimates and error factors (EFs) for manipulation and checking of locally operated valves (after Swain and Guttmann, 1983)*

A Selection of valve

Item	*Potential errors*	*HEP*	*EF*
	Making an error of selection in changing or restoring a locally operated valve when the valve to be manipulated is:		
(1)	Clearly and unambiguously labelled, set apart from valves that are similar in *all* of the following: size and shape, state, and presence of tags[a]	0.001	3
(2)	Clearly and unambiguously labelled, part of a group of two or more valves that are similar in *one* of the following: size and shape, state, or presence of tags[a]	0.003	3
(3)	Unclearly or ambiguously labelled, set apart from valves that are similar in *all* of the following: size and shape, state, and presence of tags[a]	0.005	3
(4)	Unclearly or ambiguously labelled, part of a group of two or more valves that are similar in *one* of the following: size and shape, state, or presence of tags[a]	0.008	3
(5)	Unclearly or ambiguously labelled, part of a group of two or more valves that are similar in *all* of the following: size and shape, state, and presence of tags[a]	0.001	3

[a] Unless otherwise specified, level 2 tagging is presumed. If other levels of tagging are assessed, adjust the tabled HEPs according to Table 20–15 in the *Handbook*.

B Detection of stuck valves

Item	*Potential errors*	*HEP*	*EF*
	Given that a locally operated valve sticks as it is being changed or restored,[a] the operator fails to notice the sticking valve, when it has:		
(1)	A position indicator[b] only	0.001	3
(2)	A position indicator[b] and a rising stem	0.002	3
(3)	A rising stem but no position indicator[b]	0.005	3
(4)	Neither rising stem nor position indicator[b]	0.01	3

[a] Equipment reliability specialists have estimated that the probability of a valve's sticking in this manner is approximately 0.001 per manipulation, with an error factor of 10.
[b] A position indicator incorporates a scale that indicates the position of the valve relative to a fully opened or fully closed position. A rising stem qualifies as a position indicator if there is a scale associated with it.

C Checking, including valves[a]

Item	*Potential errors*	*HEP*	*EF*
(1)	Checking routine tasks, checker using written materials (includes over-the-shoulder inspections, verifying position of locally operated valves, switches, circuit breakers, connectors, etc., and checking written lists, tags, or procedures for accuracy)	0.1	5
(2)	Same as above, but without written materials	0.2	5
(3)	Special short-term, one-of-a-kind checking with alerting factors	0.05	5
(4)	Checking that involves active participation, such as special measurements	0.01	5
	Given that the position of a locally operated valve is checked (item (1) above), noticing that it is not completely opened or closed:	0.5	5
(5)	Position indicator[b] only	0.1	5
(6)	Position indicator[b] and a rising stem	0.5	5
(7)	Neither a position indicator[b] nor a rising stem	0.9	5
(8)	Checking by reader/checker of the task performer in a two-man team, *or* checking by a *second* checker, routine task (no credit for more than 2 checkers)	0.5	5
(9)	Checking the status of equipment if that status affects one's safety when performing his tasks	0.001	5
(10)	An operator checks change or restoration tasks performed by a maintainer	Above HEPs ÷2	

[a] This table applies to cases during normal operating conditions in which a person is directed to check the work performed by others either as the work is being performed or after its completion.
[b] A position indicator incorporates a scale that indicates the position of the valve relative to a fully opened or fully closed position. A rising stem qualifies as a position indicator if there is a scale associated with it.

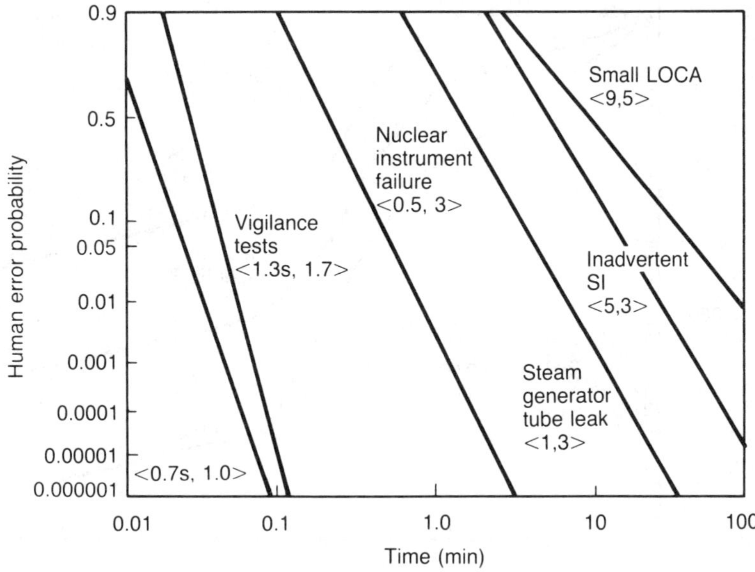

Figure 14.27 *Some time–reliability correlations relevant to nuclear power plants (R.E. Hall, Fragola and Wreathall, 1982). LOCA, loss of coolant accident; SI, safety interlock. Notation <x, y> denotes median m, error factor f of log-normal distribution*

information available on response times and probabilities of error in such tasks. The *Handbook* gives a number of tables for HEPs for these tasks.

For unannunciated displays, the *Handbook* gives the following HEP values for selection of a display (Table 20–9). The HEP depends on the existence of similar adjacent displays. It is assumed to be negligible if the display is dissimilar to adjacent displays and the operator knows the characteristics of the display he requires. The HEP is taken as 0.005 (error factor (EF) 10) if it is from a group of similar displays on a panel with clearly drawn mimic lines which include the displays; as 0.001 (EF 3) if it is from a group of similar displays which are part of a well-delineated functional group on the panel; and as 0.003 (EF 3) if it is from an array of similar displays identified by label only. These HEP values do not include recovery from any error. The probability that this will occur is high if the reading obtained is grossly different from that expected.

For check reading from displays the HEPs given are as follows (Table 20–11). The HEP is taken as 0.001 for digital displays and for analogue meters with easily seen limit marks; as 0.002 for analogue meters with limits marks which are difficult to see and for analogue chart recorders with limits marks; as 0.003 for analogue meters without limit marks; and as 0.006 for analogue chart recorders without limit marks. In all cases the EF is taken as 3. The HEP for confirming a status change on a status lamp and that for misinterpreting the indications an indicator lamp are assumed to be negligible. These HEPs apply to the individual checking of a display for some specific purpose.

14.27.6 Oral instructions and written procedures

Communication in process control includes both oral instructions and written procedures. For oral instructions a distinction is made between general and detailed instructions. Table 14.20 gives some HEP estimates for these two cases.

For written instructions the types of HEP treated include error in the preparation of the instructions, failure to refer to them and error in their use. Table 14.21 gives some HEP estimates for these three cases, respectively.

14.27.7 Locally operated valves

Another common task in process control is the manipulation of locally operated valves (LOVs). This task therefore receives special treatment in the *Handbook*. The valves concerned are manually operated. They include valves with or without a rising stem and with or without position indicators.

Three principal errors are distinguished. One is selection of the wrong valve. Here the base case is a single isolated valve. Where there are other valves present, the possibility exists that the wrong valve will be selected. Another type of error is reversal, or moving the valve in the wrong direction. The operator opens it instead of closing it, or vice versa. One form of this error is to reverse the state of a valve which is in fact already in the desired state. The third type of error is failure to detect that the valve is stuck. A common form of this error is to fail to effect complete closure of a valve. Table 14.22 gives some HEP estimates for selection of a valve, for detection of a stuck valve and for checking a valve.

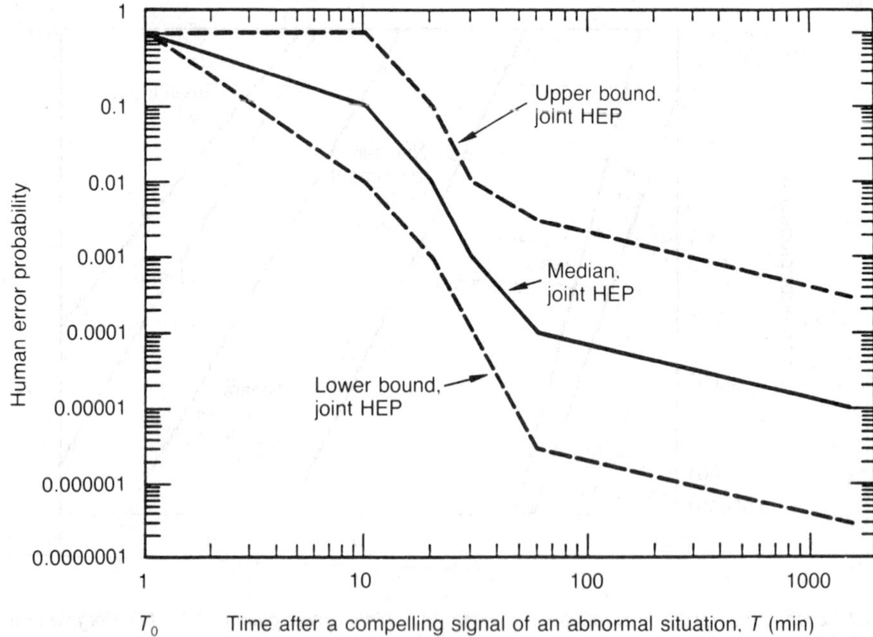

Figure 14.28 HRA Handbook: *nominal diagnosis model (Swain and Guttmann, 1983): human error probability (HEP) of diagnosis of one abnormal event by control room personnel*

The manipulation of valves is a particular case where there may exist a strong dependence, or coupling, between two tasks. This case is one which was taken into account in the *Rasmussen Report*. An operator may be required to cut off flow by closing two valves, closure of either of which is sufficient to stop the flow. If the probability of error in closing one valve is 10^{-2} and there is zero dependence, the probability of error in the overall task is 10^{-4} ($10^{-2} \times 10^{-2}$). On the other hand, if there is complete coupling, the probability of error is 10^{-2} ($10^{-2} \times 1$). These two cases represent the two extremes and constitute lower and upper bounds on the probability of failure. For the more realistic case of loose coupling, the approach used in WASH-1400 was to take for the probability of error the log–normal median or square root of the product of the two bounds: $(10^{-2} \times 10^{-4})^{\frac{1}{2}} = 10^{-3}$. The dependence model given in the *Handbook* allows the use of more levels of dependence.

14.27.8 Time–reliability correlation

As already mentioned, there has been an increasing tendency, associated mainly with experimental work with operators on simulators, to correlate the probability of operator failure with time. In particular, use is made of the time–reliability correlation (TRC) to obtain human error probabilities for complex, or non-routine, tasks, including handling an emergency. The assumption underlying such a TRC is that, although there are in principle other factors which affect operator performance in such tasks, time is the dominant one.

An aspect of operator performance of particular concern to the nuclear industry is behaviour following the initial event. One of the main methods used to study such behaviour is the use of simulators. Work on simulators has shown that the probability of success in post-event behaviour correlates strongly with time. Early work on this was done by R.E. Hall, Fragola and Wreathall (1982). A number of TRC models have since been produced based on simulator results.

Figure 14.27 shows some TRCs given by Hall, Fragola and Wreathall for operator vigilance and for particular events to which a nuclear reactor operator may have to respond.

14.27.9 Nominal diagnosis model

The task analysis approach on which THERP is based is not well adapted to handling the behaviour of the operators in an abnormal situation. For this use is made of the TRC approach. Several TRCs are given in the *Handbook*. Two are used for screening: one for diagnosis and one for post-diagnosis performance. The main TRC is the nominal diagnosis model which is applicable to diagnosis only and not to post-diagnosis performance. This TRC is shown in Figure 14.28. In contrast to the other HEP relations, which refer to individuals, the TRCs refer to the behaviour of a team. The nominal diagnosis model therefore implies a particular manning model. This is described in the next section.

Figure 14.28 includes curves for the upper and lower bounds. The *Handbook* gives guidance on the choice of

Table 14.23 HRA Handbook: *manning model for nominal diagnosis model—illustrative example (after Dougherty and Fragola, 1988, from Swain and Guttmann, 1983)*

Time (min)	*Conditional probability of error*		*Joint probability of error*	*TRC value*
10	Operator 1	0.1 (basic probability)		
	Operator 2	1.0 (complete dependence)		
	Shift supervisor	0.55 (high dependency)		
	Shift technical adviser	1.0 (no credit)	0.055	0.1
20	Operator 1	0.1 (basic probability)		
	Operator 2	0.55 (high dependency)		
	Shift supervisor	0.23 (moderate dependency)		
	Shift technical adviser	1.0 (high dependency)	0.007	0.01
30	Operator 1	0.1 (basic probability)		
	Operator 2	0.55 (high dependency)		
	Shift supervisor	0.15 (low dependency)		
	Shift technical adviser	0.15 (low dependency)	0.0012	0.001

curve. Essentially the lower bound is applicable if the event is a well-recognized one and the operators have practised on a simulator and demonstrate by interview that they know how to handle it. The upper bound is applicable if the event is not covered by training or is covered only in initial training or if the operators demonstrate by interview that they do not know how to handle it. The main, or nominal, curve is applicable if the operators have practised the event only on simulator requalification exercises or if none of the rules for the lower or upper bound apply.

The nominal diagnosis TRC does not itself fit a log–normal distribution, but it may be approximated by such a distribution. The parameters of the log–normal distribution approximating to this TRC have been estimated by Dougherty and Fragola (1988), who obtain values for the median and the error factor of m = 4 and f = 3.2, respectively. The relation between the nominal diagnosis TRC and the approximating log–normal distribution is:

Time (min)	*Human error probability*	
	Nominal diagnosis TRC	*Approximating log–normal distribution*
5	0.9	0.4
10	0.1	0.1
20	0.01	0.01
30	0.001	0.002
60	0.0001	0.00006

14.27.10 Manning model

The nominal diagnosis TRC just described applies to a whole team and is in fact postulated on a particular team composition. In other words there is an implied manning model.

The manning model used is shown in Table 14.23. The team consists of operators 1 and 2, the shift supervisor and the shift technical advisor. At 10 minutes into the incident for operator 1 the BHEP is 0.1. Operator 2 has complete dependency. The shift supervisor has high dependency. At this stage no credit is taken for the shift technical advisor. For operator 1 the BHEP remains constant at 0.1. At 20 minutes for operator 2 the dependency reduces to the high level and for the shift supervisor it reduces to medium level. The shift technical advisor is now taken into account with a high dependency. At 30 minutes for operator 2 the dependency remains at the high level but for the shift supervisor it reduces again to the low level. For the shift technical advisor the dependency reduces to the low level. The CHEPs shown are those given by Equation 14.27.1. The JHEPs shown are the products of the BHEP for operator 1 and the CHEPs for the other members of the team. These JHEPs are then rounded to give the actual values used in the nominal diagnosis TRC.

14.27.11 Recovery model

Some errors are not recoverable, but many are and the recovery model is therefore another important feature of the methodology. The probability of recovery depends on the opportunities for detection, the use made of these opportunities and the effectiveness of the recovery action.

Recovery mechanisms include:

(1) human actions – checking;
(2) plant states – panel indications;
(3) equipment states – inspections.

Recovery is treated under the headings:

(1) human redundancy;
(2) annunciated indications;
(3) active inspections;
(4) passive inspections.

Human redundancy is essentially the checking of one person's work by another person. For checking the

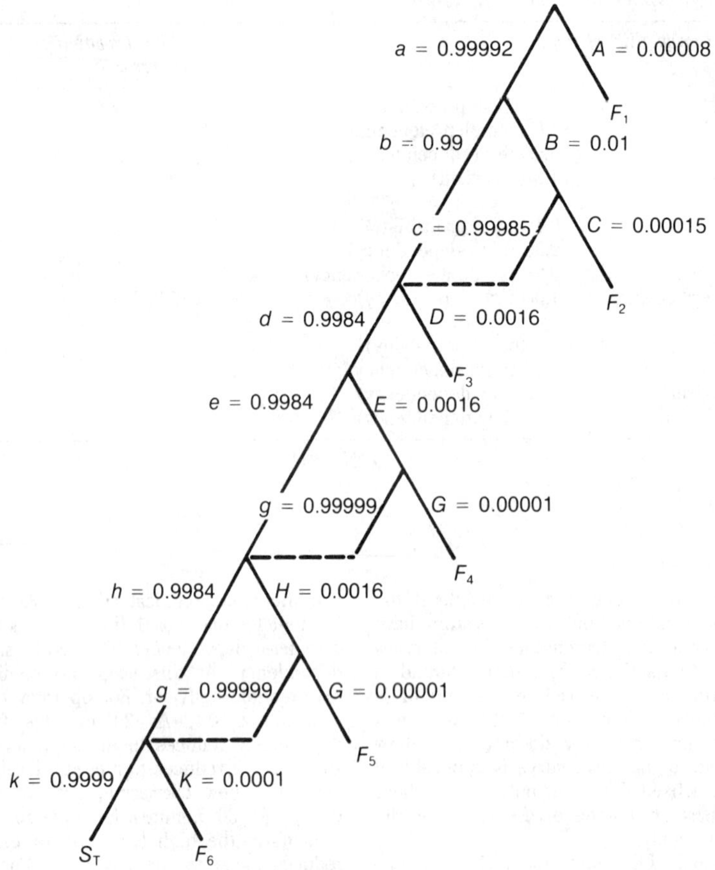

$a = 0.99992$ $A = 0.00008$
F_1
$b = 0.99$ $B = 0.01$
$c = 0.99985$ $C = 0.00015$
F_2
$d = 0.9984$ $D = 0.0016$
F_3
$e = 0.9984$ $E = 0.0016$
$g = 0.99999$ $G = 0.00001$
F_4
$h = 0.9984$ $H = 0.0016$
$g = 0.99999$ $G = 0.00001$
F_5
$k = 0.9999$ $K = 0.0001$
S_T F_6

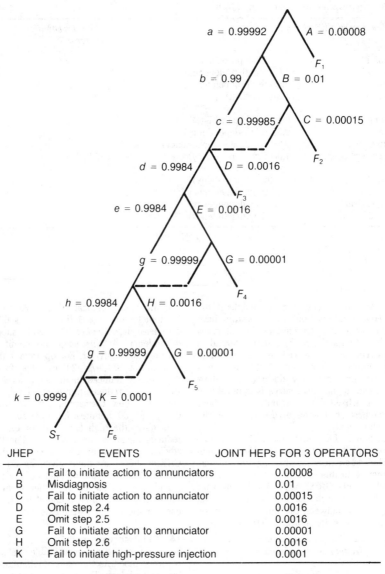

JHEP	EVENTS	JOINT HEPs FOR 3 OPERATORS
A	Fail to initiate action to annunciators	0.00008
B	Misdiagnosis	0.01
C	Fail to initiate action to annunciator	0.00015
D	Omit step 2.4	0.0016
E	Omit step 2.5	0.0016
G	Fail to initiate action to annunciator	0.00001
H	Omit step 2.6	0.0016
K	Fail to initiate high-pressure injection	0.0001

Figure 14.29 HRA Handbook: *event tree for task of handling loss of steam generator feed in a nuclear power plant, illustrating recovery from error (Swain and Guttmann, 1983)*

Handbook gives the following HEP values (Table 20–22). The HEP for checking is taken as being determined by two distinct errors, failure to execute the check at all and error in performing it. The HEP is taken as 0.1 if a written procedure is used; as 0.2 if a written procedure is not used; as 0.05 if the check is a one-off with alerting factors; and as 0.01 if the check involves active participation. The HEP is taken as 0.5 for checking by a second member of a two-man team or by a second checker. It is taken as 0.001 for checking of equipment which affects the safety of the checker. In all cases the EF is taken as 5. These HEPs apply where a person is directed to check the work of others, either as the work is being performed or after its completion. Credit for checking is limited to the use of two checkers. The HEP of the second checker in a routine task is taken as 0.5.

Recognition is given to a number of problems associated with checking. Checking is particularly affected by psychological considerations. There is an expectation that an experienced person will not make errors. Conversely, there is an expectation that an inexperienced person may well do so.

Table 14.24 HRA Handbook: *some performance shaping factors (PSFs)*[a] *(after Swain and Guttmann, 1983)*

External PSFs		*Stressor PSFs*	*Internal PSFs*
Situational characteristics: those PSFs general to one or more jobs in a work situation	*Task and equipment characteristics: those PSFs specific to tasks in a job*	*Psychological stressors: PSFs which directly affect mental stress*	*Organismic factors: characteristics of people resulting from internal and external influences*
Architectural features Quality of environment Temperature, humidity, air quality, and radiation Lighting Noise and vibration Degree of general cleanliness Work hours/work breaks Shift rotation Availability/adequacy of special equipment, tools and supplies Manning parameters Organizational structure (e.g. authority, responsibility, communication channels) Actions by supervisor, co-workers, union representatives, and regulatory personnel Rewards, recognition, benefits *Job and task instructions: single most important tool for most tasks* Procedures required (written or not written) Written or oral communications Cautions and warnings Work methods Plant policies (shop practices)	Perceptual requirements Motor requirements (speed, strength precision Control–display relationships Anticipatory requirements Interpretation Decision-making Complexity (information load) Narrowness of task Frequency and repetitiveness Task criticality Long- and short-term memory Calculational requirements Feedback (knowledge or results) Dynamic vs step-by-step activities Team structure and communication Man-made interface factors: design of prime equipment, test equipment, manufacturing equipment, job aids, tools, fixtures	Suddenness of onset Duration of stress Task speed Task load High jeopardy risk Threats (of failure, loss of job) Monotonous, degrading, or meaningless work Long, uneventful vigilance periods Conflicts of motives about job performance Reinforcement absent or negative Sensory deprivation Distractions (noise, glare, movement, flicker, colour) Inconsistent cueing *Physiological stressors: PSFs which directly affect physical stress* Duration of stress Fatigue Pain or discomfort Hunger or thirst Temperature extremes Radiation G-force extremes Atmospheric pressure extremes Oxygen insufficiency Vibration Movement constriction Lack of physical exercise Disruption of circadian rhythm	Previous training/experience State of current practice or skill Personality and intelligence variables Motivation and attitudes Emotional state Stress (mental or bodily tension) Knowledge of required performance standards Sex differences Physical condition Attitudes based on influence of family and other outside persons or agencies Group identifications

[a] Some of the tabled PSFs are not encountered in present-day nuclear power plants (e.g. *g*-force extremes), but are listed for application to other man–machine systems.

It is often suggested that if a person knows that his work is to be checked he may perform it with less care and that the end result may be a lower task reliability than if checking were not employed. This view is rejected. It is argued that for any credible values of the basic HEP and the conditional HEP for checking the joint HEP will be lower with checking.

The possibility exists that on a particular plant checking may have fallen into disuse. This is one feature in particular which it is prudent for the analyst to observe and check. There is a tendency in some situations for the task and its checking to become elided and for the whole to become a joint operation. Where this occurs, there is no longer an independent check.

Annunciated indicators, or alarms, are treated at two levels. The HEP for taking the prescribed corrective action in response to a single alarm is 0.0001, but this may be drastically modified for other alarm situations.

An annunciator response model is used which applies to multiple alarms and is expressed by two equations. The probability P_i of failure to initiate action in response to the ith alarm in a group of n alarms is:

$$P_i = 10^{-4} \qquad i = 1 \qquad [14.27.2a]$$

$$P_i = 2^{i-2} \times 10^{-3} \qquad 1 < i \leq 10 \qquad [14.27.2b]$$

$$P_i = 0.25 \qquad i > 10 \qquad [14.27.2c]$$

The probability P_r of failure to initiate action in response to a randomly selected alarm in a group of alarms and is:

$$P_r = \sum_{i=1}^{n} P_i/n \qquad [14.27.3]$$

Active inspection is defined as inspection for a specific purpose. The main forms are prescribed periodic logging of readings and prescribed audit of indications with written instructions, both in the control room, and a walk-around inspection on the plant with instructions. The inspection may be based on oral instructions or written instructions. The HEP for an active inspection is that applicable to oral instructions and written procedures, already described.

Passive inspection is defined as a casual type of inspection. There are no written instructions and no instructions to look for any particular feature. The main forms are scanning of the control room displays and a walk-around on the plant.

As described above, HEPs are given for detection of deviant unannunciated displays in a periodic scan.

For passive inspection by walk-around the *Handbook* gives the following HEP values (Table 20–27). The event concerned is failure to detect a particular deviant state within 30 days. It is assumed that there is one inspection per shift. The HEPs are taken as 0.52, 0.25, 0.05, 0.003, 0.002, 0.001 and 0.001 for periods between walk-around of 1, 2, 3, 4, 5, 6 and 7 days, respectively.

For a planned task recovery may be introduced into the task event tree. This is illustrated in the event tree shown in Figure 14.29, which shows recoveries from error represented by tasks C and G. C is a recovery from failure at task B. G is a recovery from failure at task E and also at task H.

14.27.12 Performance shaping factors

The performance shaping factors (PSFs) are divided into the following classes:

(1) external factors –
  (a) situational characteristics;
  (b) task and equipment characteristics;
  (c) job and task instructions;
(2) internal factors;
(3) stressors.

Stressors are treated separately and are considered in the next section.

The performance shaping factors used listed in the *Handbook* are shown in Table 14.24. Each PSF is discussed in some detail. The *Handbook* does not, however, appear to give any simple method of adjusting the nominal HEPs by way of a multiplying factor or otherwise. It is evidently up to the analyst to judge the quality of a particular PSF for the situation concerned and to make a suitable adjustment.

However, some estimates, described as speculative and conservative, are given of the potential benefit of the adoption of good ergonomic practices. The authors state that in nuclear power plants violations of conventional human factors practices are the rule rather than the exception. The *Handbook* indicates that a reduction in HEP by a factor in the range 2–10 might be attained by adoption of good human factors practices in the design or displays and controls, and a similar reduction can be achieved by the use of checklists and well-written procedures instead of narrative procedures (Table 3–8). The error factors given alongside the nominal HEPs also provide further guidance for the analyst.

14.27.13 Stress

As already described, stress is an important determinant of performance and must be taken into account. It is one of the performance shaping factors, but is accorded special treatment. Stress may be caused by both external and internal factors. Some of these stressors are listed in Table 14.24 for stress which could potentially be rather complex.

The approach adopted in the *Handbook* is to simplify and to treat stress as a function of workload. The assumption underlying this approach is that, although there are in principle other factors which affect stress, workload is the dominant one.

At a very low workload performance is less than optimal. There is some higher workload at which it is optimal. At a higher workload yet, performance again deteriorates. Finally, the situation may induce threat stress, which is qualitatively different and is accorded separate treatment.

The four levels of stress are therefore defined as:

(1) very low task load;
(2) optimum task load;
(3) heavy task load;
(4) threat stress.

A heavy task load is one approaching the limit of human capacity.

For the first three levels of stress the *Handbook* gives multiplying factors which are applied to the nominal HEP (Table 20–16). The multiplying factors for an experienced operator are: 2 for a very low task load and for a heavy task load of a step-by-step task; and 5 for a heavy task load of a dynamic task and for a threat stress condition of a step-by-step task. The multiplier for optimum workload is unity. Different factors are given for an inexperienced operator.

A situation which can arise is where an error is made and recognized and an attempt is then made to perform the task correctly. Under conditions of heavy task load the probability of failure tends to rise with each attempt as confidence deteriorates. For this situation the doubling rule is applied. The HEP is doubled for the second attempt and doubled again for each attempt thereafter, until a value of unity is reached. There is some support for this in the work of Siegel and Wolf (1969) described above.

For a dynamic task or for diagnosis under threat stress the approach is different. A multiplier is not used, but instead a HEP value is given. The HEP for these cases is taken as 0.25 for an experienced operator and as 0.5 for an inexperienced one (both EF 5). The *Handbook* gives guidance on the assignment of the levels of workload, and hence stress.

The basis of the HEP value of 0.25 is the work on behaviour in emergencies by Ronan and Berkun already described. The probability of ineffective behaviour from the work on in-flight emergencies is about 0.15. The training of a pilot is particularly intensive. Operators are not expected to perform as well. Hence the HEP value of 0.25 is used for operators.

This HEP for threat stress conditions applies to dynamic tasks and to diagnosis, not to step-by-step tasks, for which, as already stated, a multiplier of 5 is used. A different treatment again is applied to a loss-of-coolant accident (LOCA).

14.27.14 Sources of human performance estimates

Ideally, the data on human performance used in this study would have been obtained from the nuclear industry alone. In fact the data from this source were very limited and a much wider range of sources was used, as follows: nuclear power plants (NPPs); NPP simulators; process plants; other industrial and military sources; experiments and field studies on real tasks; and experiments on artificial tasks. This list is in order of decreasing relevance but, unfortunately, of increasing data availability. Some 29 'experts' on human error were approached for assistance in providing HEP estimates, but virtually none were forthcoming.

For the nuclear industry the main potential source of human error data is licensee events reports (LERs). These contain an entry 'Personnel error'. HEPs were determined from these LERs for tasks involving operation, maintenance and testing of manual isolation valves, motor operated valves (MOVs) and pumps. The HEP values obtained were low. Work by Speaker, Thompson and Luckas (1982) on valves has shown, however, that for every LER classification of 'Personnel error' there were some 5 to 7 additional reportable events which in their judgement involved human error. Multiplication of the original HEP estimates by a correction factor of 6 brought them much closer to those from other sources.

Most studies on NPP and other simulators have not yielded usable human error data. The first systematic study found which does was that of Beare *et al.* (1983). This work came too late for incorporation in the *Handbook* but was used as a cross-check. Extensive use was made of HEP data from the process industries given by Kletz and Whitaker (1973), E. Edwards and Lees (1974) and the *ICI Safety Newsletter* (1979). Other industrial HEP data mentioned are those of Rook (1962), Rigby and Edelman (1968a,b) and Rigby and Swain (1968, 1975) on the production and testing of military systems.

A number of field studies and experiments in industrial settings were conducted and yielded usable data. These were subject, however, to the usual caution that the very fact that an experiment is being conducted tends to distort the results.

There is a large amount of experimental data on artificial tasks such as those conducted in laboratories. This work suffers from the fact that not only is it artificial, but also the limits of acceptable performance are often very narrow. It tends to be a poor indicator of absolute performance in real situations, but is a much more reliable guide to comparative performance. The correction which needs to be applied to such data to allow for the broader tolerances in industrial situations was the subject of a study by Payne and Altman (1962), who obtained an average correction factor of 0.008. The *Handbook* states that using this factor the HEP values obtained are similar to those found in field operations.

Expert judgement was utilized extensively to obtain HEP estimates where hard data were not available. Use was made of scaling techniques to calibrate HEPs for tasks estimated by the experts against known task HEPs. For the HEP in an emergency or highly stressed situation use was made of the work of Ronan (1953) and Berkun and co-workers (Berkun *et al.*, 1962; Berkun, 1964).

The *Handbook* discusses the estimation of HEPs where these are a function of time. For such tasks the three relevant features are: the time to begin the task, essentially the response time, and the time required for diagnosis, if any; the time required to do the task correctly; and the time available to do the task correctly. Data on the time to perform the task were obtained from operating records and from experts. The time available was often determined by the characteristics of the plant.

The *Handbook* also gives an account of the determination of HEPs for: displays and controls; locally operated valves; oral instructions and written procedures; administrative controls; and abnormal events.

14.27.15 Human performance estimates from expert judgement

The estimates of human performance given in the *Handbook*, whether for human error probabilities or performance shaping factors, are based on expert judgement. A discussion of expert judgement techniques applicable to human factors work is given in the *Handbook* by Weston (1983). He discusses the following methods: (1) paired comparisons, (2) ranking and rating procedures, (3) direct numerical estimation and (4) indirect numerical estimation. In dealing with estimates of human error it is particularly important to give a full definition of the task for which the estimate is to be made. If the definition is poor, the estimates obtained are liable to exhibit wide differences.

A study is quoted by Seaver and Stillwell (1983) in which these methods were compared in respect of six criteria. For the three criteria selected by Weston (quality of judgements, difficulty of data collection and empirical support) the rankings obtained by these workers were as follows: (1, best; 4, worst):

Criterion	*Type of procedure*			
	Paired comparisons estimation	*Ranking/ rating estimation*	*Direct numerical*	*Indirect numerical*
Quality of judgement	1	2	4	3
Difficulty of data collection	4	1	2	3
Empirical support	3	4	1	1

14.27.16 Uncertainty bounds and sensitivity analysis

It is normal to include in a hazard assessment a sensitivity analysis and this creates a requirement to express an estimate of human error probability not just as a point value but as a distribution. The distribution which is generally used is the log–normal distribution. As described in Chapter 7, the log–normal distribution is characterized by the two parameters m^* and σ. Alternatively, it may be defined instead in terms of the log–normal median m and the error factor f. Often only a point value is available, and generalized values of σ or f are used to give the spread.

The log–normal distribution is that used in the *Handbook*. It is admitted that the basis for preferring this distribution is not strong. The experimental support that exists relates to distributions of response times. On the other hand, it is argued that the choice of distribution does not appear critical. It is also the case that the log–normal distribution is a convenient one to use. Uncertainty arises from (1) lack of data, (2) deficiencies in the models used, (3) the effects of the PSFs and (4) the variable quality of analysts.

Using the log–normal distribution, characterized by the log–normal median m and the error factor f, the uncertainty bounds (UCBs) are expressed in terms of the error factor. As an illustration, consider the following case:

Nominal HEP $= 0.01$
Lower UCB $= 0.003$
Upper UCB $= 0.03$

Error factor $f = (0.03/0.003)^{\frac{1}{2}} \approx 3$

For the most part symmetrical UCBs are used, but there are exceptions. If the median value is low, the use of symmetrical UCBs may give a lower bound which is below the HEP cut-off, whilst at the other extreme for an HEP ≥ 0.25 it may give an upper bound which exceeds unity. In these cases asymmetrical bounds are used.

The general guidelines given in the *Handbook* for estimating the error factor (Table 20–20) are given in Table 14.25.

14.27.17 Validation

A methodology of the type just described is clearly difficult to validate. There is a good deal of information provided in the *Handbook* in support of individual HEP estimates and some of this is described above. An account is also given of validation exercises carried out in support of the original THERP methodology, but these relate to tasks such as calibration and testing rather than process control.

14.28 Assessment of Human Error: Success Likelihood Index Method (SLIM)

A method of obtaining HEP estimates based on PSFs is the success likelihood index (SLI) which is incorporated in the SLI method (SLIM). Accounts are given in *SLIM–MAUD: An Approach to Assessing Human Error Probabilities using Structured Judgement* by Embrey *et al.* (1984 NUREG/CR-3518) and by Embrey (1983a,b) and Kirwan (1990).

SLIM treats not only the quality of the individual PSFs but also the weighting of these in the task. It is thus a complete method for assessing of human error, and not merely a technique for determining values of the PSFs.

The basic premise of SLIM is that the HEP depends on the combined effects of the PSFs. A systematic approach is used to obtain the quality weightings and relevancy factors for the PSFs, utilizing structured expert judgement. From these PSFs the SLI for the task is obtained.

As defined by Embrey (1983a) the SLI for n PSFs is:

$$\mathrm{SLI} = \sum_{i=1}^{n} r_i w_i \qquad [14.28.1]$$

where r is a relevancy factor and w a quality weighting. Thus the SLI approach makes explicit the distinction between quality and relevance.

The quality weighting is obtained from the judgement of a panel of experts and is assigned a value on the scale 1–9. The relevancy factor, which again is obtained from the judgement of the expert panel, is a measure of the contribution of that PSF, the sum of the relevancy factors being unity. The SLI so obtained is a relative value.

In order to convert an SLI into an HEP it is necessary to calibrate it against tasks for which the HEP is known. The relation used is:

$$\log_{10}(\mathrm{HEP}) = a\,\mathrm{SLI} + b \qquad [14.28.2]$$

where a and b are constants. These constants are obtained from two tasks of known HEP. Due to the logarithmic relationship, SLI values which do not differ greatly (e.g. 5.5 and 5.75) may correspond to very different HEPs.

The SLI methodology has two modules. The first is the multi-attribute utility decomposition (MAUD), usually referred to as SLIM-MAUD. The second is the systematic approach to the reliability assessment of humans (SARAH). The former is used to obtain the SLI and the latter to perform the calibrations. Both are embodied in computer programs. Accounts of work done to validate this approach have been given by Embrey (1983a) and (Embrey and Kirwan, 1983). Illustrative examples of SLIM are given by Dougherty and Fragola (1988), as described below, and Kirwan (1990).

14.29 Assessment of Human Error: Human Error Assessment and Reduction Technique (HEART)

In the human error assessment and reduction technique (HEART), described by J.C. Williams (1986, 1988a,b, 1992), the human error probability of the task is treated as a function of the type of task and of associated error producing conditions (EPCs), effectively PSFs. The method is based on a classification of tasks into the generic types shown in Section A of Table 14.26, which also gives the proposed nominal human unreliabilities for execution of the tasks. There is an associated set of EPCs, for each of which is given an estimate of the maximum predicted normal amount by which the unreliability might change going from 'good' to 'bad'. Section B of the table shows the first 17 EPCs listed, those with the strongest influence; entries 18–38 list further EPCs with a weaker influence.

In applying an EPC use is made of a weighting for the proportion of the EPC which is effective. Thus for a task of type D with a nominal unreliability of 0.09, a single EPC of 4 and a weighting of 0.5, the resultant unreliability is:

$$0.09[(4-1) \times 0.5 + 1] = 0.23$$

Table 14.25 HRA Handbook: *general guidelines on estimation of the error factor (EF)*[a] (after Swain and Guttmann, 1983)

Item	*Task and HEP guidelines*[b]	*EF*[c]
	Task consists of performance of step-by-step procedure[d] conducted under routine circumstances (e.g. a test, maintenance, or calibration task); stress level is optimal:	
(1)	Estimated HEP < 0.001	10
(2)	Estimated HEP 0.001–0.01	3
(3)	Estimated HEP > 0.01	5
	Task consists of performance of step-by-step procedure[d] but carried out in non-routine circumstances such as those involving a potential turbine/reactor trip; stress level is moderately high:	
(4)	Estimated HEP < 0.001	10
(5)	Estimated HEP ≥ 0.001	5
	Task consists of relatively dynamic[d] interplay between operator and system indications, under routine conditions, e.g. increasing or reducing power; stress level is optimal:	
(6)	Estimated HEP < 0.001	10
(7)	Estimated HEP ≥ 0.001	5
(8)	Task consists of relatively dynamic[d] interplay between operator and system indications but carried out in non-routine circumstances; stress level is moderately high	10
(9)	Any task performed under extremely high stress conditions, e.g. large LOCA, conditions in which the status of ESFs is not perfectly clear, or conditions in which the initial operator responses have proved to be inadequate and now severe time pressure is felt (see text of Handbook for rationale for EF = 5)	5

[a] The estimates in this table apply to experienced personnel. The performance of novices is discussed in Chapter 18 of the *Handbook*.
[b] For UCBs for HEPs based on the dependence model, see Table 7–3 of the *Handbook*.
[c] The highest upper UCB is 1.0.
See Appendix A to calculate the UCBs for $Pr(F_T)$, the total-failure term of an HRA event tree.
[d] See Table 18–1 of the *Handbook* for definitions of step-by-step and dynamic procedures.

Table 14.26 *Classification of generic tasks and associated unreliability estimates (J.C. Williams, 1986)*

A Generic classifications

	Generic task	*Proposed nominal human unreliability (5th–95th percentile boundaries)*
A	Totally unfamiliar, performed at speed with no real idea of likely consequences	0.55 (0.35–0.97)
B	Shift or restore system to a new or original state on a single attempt without supervision or procedures	0.26(0.14–0.42)
C	Complex task requiring high level of comprehension and skill	0.16 (0.12–0.28)
D	Fairly simple task performed rapidly or given scant attention	0.09 (0.06–0.13)
E	Routine, highly practised, rapid task involving relatively low level of skill	0.02 (0.007–0.045)
F	Restore or shift a system to original or new state following procedures, with some checking	0.003 (0.0008–0.007)
G	Completely familiar, well-designed, highly practised, routine task occurring several times per hour, performed to highest possible standards by highly motivated, highly trained and experienced person, totally aware of implications of failure, with time to correct potential error, but without the benefit of significant job aids	0.0004 (0.00008–0.009)
H	Respond correctly to system command even when there is an augmented or automated supervisory system providing accurate interpretation of system stage	0.00002 (0.000006–0.00009)

Table 14.26 *Continued*

M	Miscellaneous task for which no description can be found. (Nominal 5th to 95th percentile data spreads were chosen on the basis of experience suggesting log-normality)	0.03 (0.008–0.11)

B Error-producing conditions[b]

	Error-producing condition	*Maximum predicted nominal amount by which unreliability might change going from 'good' conditions to 'bad'*
1.	Unfamiliarity with a situation which is potentially important but which only occurs infrequently or which is novel	× 17
2.	A shortage of time available for error detection and correction	× 11
3.	A low signal-to-noise ratio	× 10
4.	A means of suppressing or overriding information or features which is too easily accessible	× 9
5.	No means of conveying spatial and functional information to operators in a form which they can readily assimilate	× 8
6.	A mismatch between an operator's model of the world and that imagined by the designer	× 8
7.	No obvious means of reversing an unintended action	× 8
8.	A channel capacity overload, particularly one caused by simultaneous presentation of non-redundant information	× 6
9.	A need to unlearn a technique and apply one which requires the application of an opposing philosophy	× 6
10.	The need to transfer specific knowledge from task to task without loss	× 5.5
11.	Ambiguity in the required performance standards	× 5
12.	A mismatch between perceived and real risk	× 4
13.	Poor, ambiguous or ill-matched system feedback	× 4
14.	No clear direct and timely confirmation of an intended action from the portion of the system over which control is to be exerted	× 3
15.	Operator inexperienced (e.g. a newly qualified tradesman, but not an 'expert')	× 3
16.	An impoverished quality of information conveyed by procedures and person–person interaction	× 3
17.	Little or no independent checking or testing of output	× 3
18.	A conflict between immediate and long-term objectives.	× 2.5
19.	No diversity of information input for veracity checks	× 2.5
20.	A mismatch between the educational achievement level of an individual and the requirements of the task	× 2
21.	An incentive to use other more dangerous procedures	× 2
22.	Little opportunity to exercise mind and body outside the immediate confines of the job	× 1.8
23.	Unreliable instrumentation (enough that it is noticed)	× 1.6
24.	A need for absolute judgements which are beyond the capabilities or experience of an operator	× 1.6
25.	Unclear allocation of function and responsibility	× 1.6
26.	No obvious way to keep track of progress during an activity	× 1.4
27.	A danger that finite physical capabilities will be exceeded	× 1.4
28.	Little or no intrinsic meaning in a task	× 1.4
29.	High-level emotional stress	× 1.3
30.	Evidence of ill-health amongst operatives, especially fever	× 1.2
31.	Low workforce morale	× 1.2
32.	Inconsistency of meaning of displays and procedures	× 1.2
33.	A poor or hostile environment (below 75% of health or life-threatening severity)	× 1.15
34.	Prolonged inactivity or highly repetitious cycling of low mental workload tasks	× 1.1 for first half-hour × 1.05 for each hour thereafter
35.	Disruption of normal work-sleep cycles	× 1.1
36.	Task pacing caused by the intervention of others	× 1.06
37.	Additional team members over and above those necessary to perform task normally and satisfactorily	× 1.03 per additional man
38.	Age of personnel performing perceptual tasks	× 1.02

[a] If none of the task descriptions A–H fits the task under consideration, the values given under M may be taken as reference points.
[b] Conditions 18–38 are presented simply because they are frequently mentioned in the human factors literature as being of some importance in human reliability assessment. To a human factors engineer, who is sometimes concerned about performance differences of as little as 3%, all these factors are important, but to engineers who are usually concerned with differences of more than 300%, they are not very significant. The factors are identified so that engineers can decide whether or not to take account of them after the initial screening.

The method also includes a set of associated remedial measures to be applied to improve the reliability.

HEART has been designed as a practical method, and is easy to understand and use. It was on of the principal techniques used in the quantitative risk assessment for Sizewell B, as described in Section 14.39.

14.30 Assessment of Human Error: Method of Dougherty and Fragola

The deficiencies of THERP in respect of non-routine behaviour have led to the development of alternative methods. One of these methods is that described in *Human Reliability Analysis* by Dougherty and Fragola (1988). Like THERP, this method has been developed essentially as an adjunct to fault tree analysis, but the approach taken is rather different.

The earlier approach associated human errors essentially with equipment or procedure failures. One effect was the generation of a large number of human events linked to the equipment failures. Another was the neglect of more significant but complex human errors. The approach taken in the human reliability analysis (HRA) of Dougherty and Fragola is to concentrate on a smaller number of more significant failures and human errors. There is also strong emphasis on integrating the HRA into the probabilistic risk assessment (PRA).

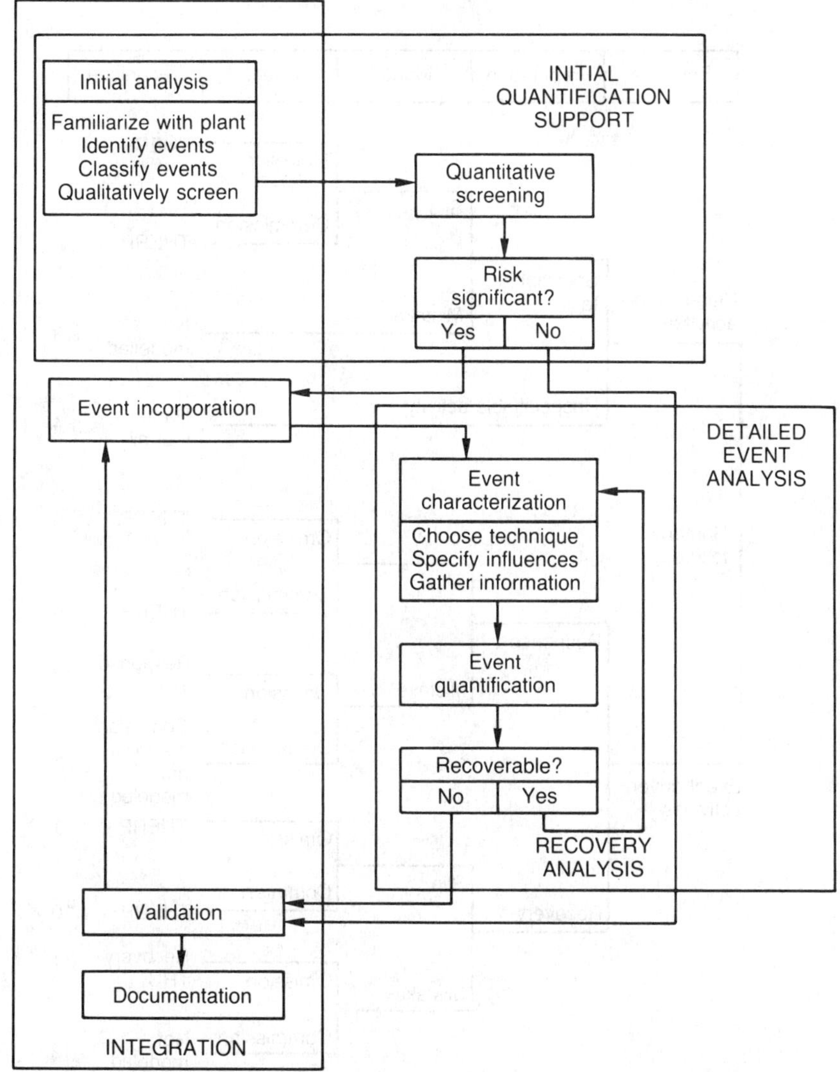

Figure 14.30 Method of Dougherty and Fragola: methodology for human reliability analysis (Dougherty and Fragola, 1988) (Reproduced with permission of John Wiley & Sons Ltd from Human Reliability Analysis *by E.M. Dougherty and J.R. Fragola, 1988, Copyright ©)*

14.30.1 Overall approach

The HRA process is shown in Figure 14.30. The generation of the human events to be considered is part of the PRA. As with physical hazards, identification has a good claim to be the most difficult stage of the process. Error classification schemes have been developed as an aid, but give no guarantee of completeness.

The human events are identified, characterized and quantified according to the scheme shown in Figure 14.31.

Certain principles have been identified to guide the development of the PRA so that there is compatibility with the HRA. In broad outline these guidelines include the following:

(1) The description of the human event should refer to failure of a function rather than to some lower level of abstraction.
(2) The human events should be confined to the three categories of pre-initiator (or latent) events, human-induced initiator events and post-initiator events.
(3) Human events in the latent category should be incorporated in the fault tree at the highest appropriate level.
(4) A human-induced initiator event should be subsumed in the initiator type which includes the human event. The data required should be expressed as a probability and not as a frequency.

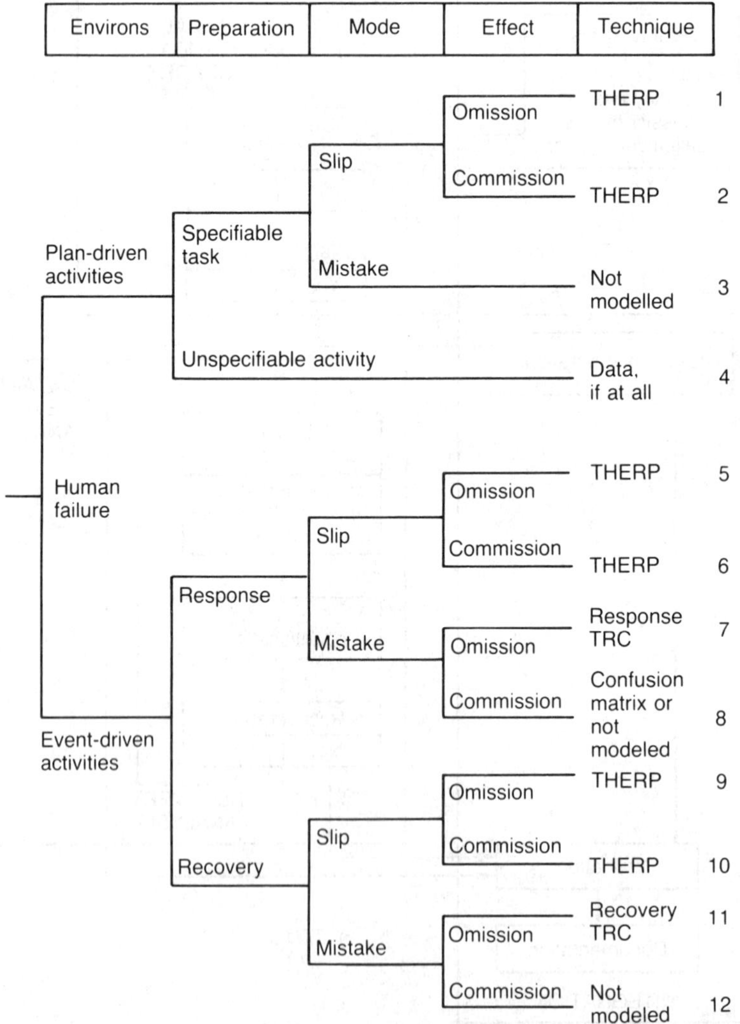

Figure 14.31 *Method of Dougherty and Fragola: classification of human events (Dougherty and Fragola, 1988) (Reproduced with permission of John Wiley & Sons Ltd from* Human Reliability Analysis *by E.M. Dougherty and J.R. Fragola, 1988, Copyright ©)*

(5) A human event in the post-initiator category which relates to failure of a system should be modelled as a single event under the gate below the top event of the fault tree for that system, the gate being (a) an OR gate if the system is manually activated or (b) an AND gate if the system is automatically activated.

14.30.2 Task analysis

The starting point is again a task analysis. A typical task analysis is outlined in Table 14.27.

14.30.3 Response and recovery events

Human events which occur in the post-initiator period are treated as either response or recovery events. The nuclear industry has gone to some pains to provide a rather comprehensive set of emergency response guidelines. Following such guidelines, a planned activity is classed as a response, whilst an unplanned activity is classed as a recovery. Recovery activity is applicable only to those events from which recovery is possible; there are some events for which there is no recovery.

The incorporation of recovery is not undertaken during the development of the main fault tree but is deferred until the tree, and its cut sets, are available. There are two reasons for this. One is that introduction of recovery during the main synthesis tends to result in an undue increase in the size of the tree. The other is that recovery analysis is an iteration through much of the PRA/HRA process, embraces system and human aspects, and tends to be highly specific to the set of events from which the recovery is to be made. It is therefore better to consider the set of events and the associated recovery as a separate exercise at the end.

Table 14.27 *Task analysis in support of a human reliability analysis for a task in nuclear power plant operation (Dougherty and Fragola, 1988) (Reproduced with permission of John Wiley & Sons Ltd from* Human Reliability Analysis *by E.M. Dougherty and J.R. Fragola, 1988, Copyright ©)*

Goal
Obtain reactor core system water make-up and cooling following a small loss of coolant accident (LOCA)

Step	*Means*
Diagnose event	
Detect plant upset condition	Several alarms
Observe RCS level indicator	Pressurizer or reactor level
Observe decreasing RCS pressure	Pressure indicator
Observe sump level increasing	Level indicator
Observe containment pressure increasing	Pressure indicators
Observe no secondary side radiation	Radiation monitors
Isolate the LOCA	
Close PORV block valve 1	One valve control
Close PORV block valve 2	Another valve control
Close letdown line	One valve control
Close RCP seal isolation valve 1	One valve control
Close RCP seal isolation valve 2	Another valve control
Verify safety system actuation	
Observe HPI pump meters	Two flow indicators
Start HPI pumps	Two pump start controls
Observe AFW pump meters	Two flow indicators
Start AFW pumps	Two pump start controls
Obtain long-term cooling	
Await low level tank alarm	One level indicator
Open sump valve 1	One valve control
Open sump valve 2	One valve control
Open tank valve 1	One valve control
Open tank valve 2	One valve control

AFW, auxiliary feed water; HPI, high pressure injection; PORV, power operated relief valve; RCP, reactor coolant pump; RCS, reactor cooling system.

14.30.4 Operator action event trees

Some of the activity of the operator consists of planned tasks, but he may also have to respond to an abnormal condition on the plant. Thus, although some events can be incorporated into the main fault tree of the PRA, post-initiator events involving action by the operator require the introduction of a specific event tree – the operator action event tree (OAET).

The response of the operator to an abnormal condition may be described in terms of an event tree. The use of event trees for this purpose has been formalized as the OAET or, simply, the operator action tree (OAT) technique.

The OAET method has been described by R.E. Hall, Fragola and Wreathall (1982). Three basic features are recognized: (1) perception, (2) diagnosis and (3) response. Figure 14.32 shows a typical OAET for a nuclear reactor coolant pump seal LOCA. The OAET tends to be used in conjunction with a time–reliability correlation.

14.30.5 Modelling of human events

The classification of human events and the associated event models used in the HRA are shown in Figure 14.31. The human activity is divided into three classes which are specifiable and one which is not. The three definable activities are planned tasks, response and recovery. Each of these definable activities is further divided into slips and mistakes, and these in turn are divided into errors of omission or commission.

In each case a slip is modelled using a modified version of THERP. Mistakes of response and recovery are modelled using TRCs. Other mistakes are not modelled, except that response mistakes of commission may be modelled. Recovery mistakes of commission may be estimated as described below.

14.30.6 Filtering of human events

The human events identified are subjected to both a qualitative and a quantitative filter. This process is assisted if the analysis of the hardware and human aspects proceeds simultaneously. This is the most effective way to apply a qualitative filter to the large proportion of the human events identified.

A quantitative filter is then applied. This means that an approximate estimate is made of the probability of each

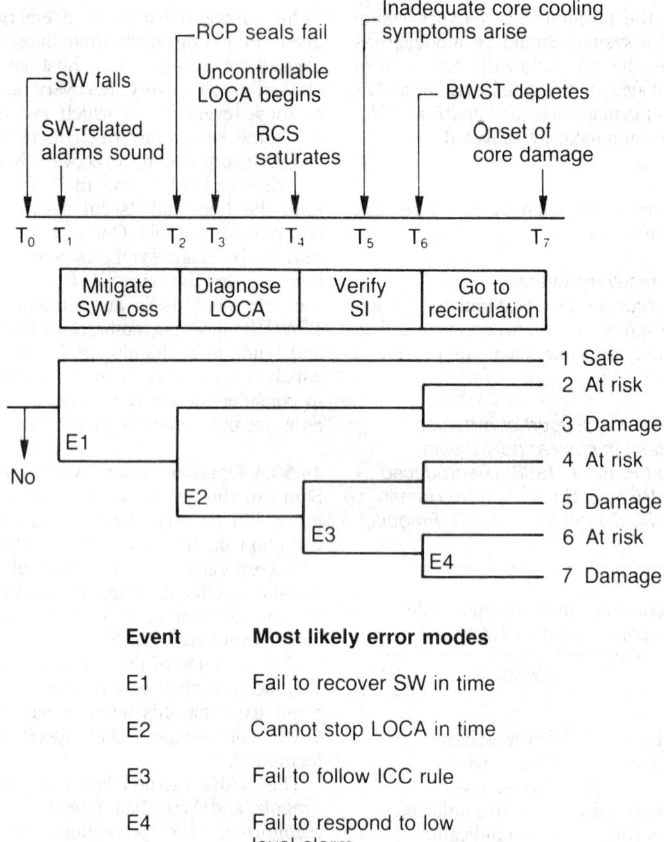

***Figure 14.32** Operator action event tree for a nuclear reactor pump seal loss-of-coolant accident (LOCA) (Dougherty and Fragola, 1988). BWST , borated water storage tank; RCP, reactor coolant pump; RCS, reactor cooling system; SI, safety interlock; SW, seal water (Reproduced with permission of John Wiley & Sons Ltd from* Human Reliability Analysis *by E.M. Dougherty and J.R. Fragola, 1988, Copyright ©*

human event and, if at this level of probability the event has negligible effect on the fault tree, it is not pursued.

The approach used draws on THERP. A probability of 0.001 is used as a basic screening value for the probability of a human event for latent or human-induced initiator failures. If there is redundancy so that a second error must occur for the failure to occur, a conditional probability of value of 0.1 is used, corresponding to moderate dependency. For post-initiator events use is made of the appropriate TRC, taking the probability at 5 minutes as the screening value.

14.30.7 Modelling of slips

Slips are errors in an activity which has to some degree been planned. This applies even to recovery, where the operator first formulates and then executes a plan. The method used to model slips is a modified version of THERP. One modification is to consider only one slip per task. The probability of the slip is taken as a basic value of 0.001 or, if there is redundancy, 0.0001. These values are the same as those used for screening. They are then adjusted using appropriate performance shaping factors. The authors also describe alternative, more complex methods of estimating the probability of slips.

14.30.8 Modelling of mistakes

Response and recovery mistakes are modelled using time–reliability correlations. The underlying assumption in this approach is that, although the probability of success depends, in principle, on many factors, the time available is the dominant one.

Four separate TRCs are used. These are for the cases

(1) response without hesitancy;
(2) response with hesitancy;
(3) recovery without hesitancy;
(4) recovery with hesitancy.

Hesitancy is associated with the burden of the task which in turn depends on a number of factors as described below. Account is also taken of the effect of performance shaping factors. The incorporation of these two features into the TRCs is described below.

14.30.9 Time–reliability correlations

It is found empirically that response times tend to fit a log–normal distribution. The TRC models used are based on the assumption that the log–normal distribution is applicable. They are characterized by a log–normal median m and an error factor f and give a straight line when plotted on log–probability paper.

The basic response time τ is given by:

$$\tau_r = k_1 k_2 \tau \qquad [14.30.1]$$

where τ is the median response time, τ_r is the adjusted median response time and k_1 and k_2 are adjustment factors. The value of m used is the adjusted median response time.

The first factor k_1 takes account of the availability or otherwise of a rule, in other words of the difference between response and recovery. It has the values:

$$k_1 = 1 \quad \text{rule available} \qquad [14.30.2a]$$

$$k_1 = 0 \quad \text{no rule available} \qquad [14.30.2b]$$

The second factor k_2 takes account of the performance shaping factor as measured by the success likelihood index (SLI). It is assumed that at best (SLI = 1) the median response time is halved and at worst (SLI = 0) it is doubled. The factor is defined as:

$$k_2 = 2^{(1-2x)} \qquad [14.30.3]$$

where x is the SLI.

The base TRC is the nominal diagnosis curve given in THERP. This curve has a median m of 4 and an error factor f of 3.2. It therefore has a relatively high median and low error factor. The high median corresponds to a recovery rather than a response and the low error factor to absence of hesitancy. On this basis this curve is taken as that for recovery without hesitancy.

The other TRC curves are then obtained from this basic curve. The curve for response without hesitancy has the parameters m = 2 and f = 3.2; that for response with hesitance m = 2 and f = 6.4; and that for recovery with hesitance m = 4 and f = 6.4.

The TRC curves so derived are shown in Figure 14.33. These curves already incorporate the factor k_1 but not the factor k_2. The response time obtained from the curves should be multiplied by k_2.

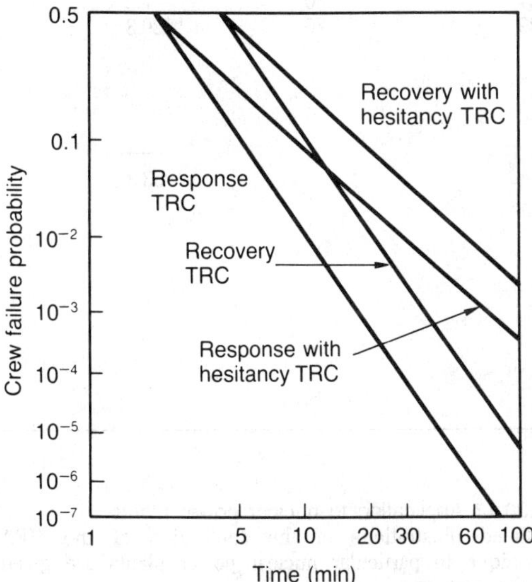

***Figure 14.33** Method of Dougherty and Fragola: time–reliability correlations (Dougherty and Fragola, 1988) (Reproduced with permission of John Wiley & Sons Ltd from* Human Reliability Analysis *by E.M. Dougherty and J.R. Fragola, 1988, Copyright ©)*

14.30.10 Performance shaping factors

Performance shaping factors are treated using SLIM. In the format given by the authors the SLI is defined as:

$$\text{SLI} = \sum_{i=1}^{n} r_{ni} q_i \qquad [14.30.4]$$

with

$$r_{ni} = r_i / \sum_{i=1}^{n} r_i \qquad [14.30.5]$$

where q is the quality, r the rank and r the normalized rank.

In general, q has a value in the range 0–1. This is so if the PSF can have either a bad or a good influence. If it can have only a bad influence, the range of q is restricted to 0–0.5 and if it can have only a good influence the range is restricted to 0.5–1. The determination of the SLI is illustrated in Table 14.28, which gives the SLI for a human error in a recovery mistake. The SLI is can be used to adjust probabilities of slips or mistakes. In the latter case it is applied to the median response time of the TRC, as described below.

14.30.11 Recovery mistakes of commission

In an incident situation the possibility exists that the operator will take some action which actually makes the situation worse. As so far described, the methodology does not include a model for such an action which is called a 'recovery mistake of commission'.

A method is given, however, for estimating the probability that such an error will occur. This probability is the product of three probabilities. The first is the probability of a significant and extended commission error, the second the probability that the emergency response guidelines (ERGs) do not cover the resulting situation, and the third the probability that the senior reactor operator (SRO) or other personnel fail to recover the situation.

The values of these three probabilities are estimated as follows. At the time of writing there had been 10 000 reactor scrams. Two involved misdiagnoses which led to core melt, including Chernobyl. The probability of a significant and extended commission error is thus estimated as 0.0002 (2/10 000). It is assumed that ERGs would effect a reduction in probability of between one and three orders of magnitude and a reduction of two orders of magnitude is selected so that the second probability is estimated as 0.01. The action of the SRO, whose function is to stand back and monitor plant status, is assumed to have only low dependency with that of the crew and the probability that he will fail to effect recovery is estimated as 0.05. From these figures the

Table 14.28 *Success likelihood index calculation for a recirculation event in nuclear power plant operation (Dougherty and Fragola, 1988) (Reproduced with permission of John Wiley & Sons Ltd from* Human Reliability *Analysis by E.M. Dougherty and J.R. Fragola, 1988, Copyright ©)*

Influence	*Type*	*Rank*	*Relative rank*	*Quality*	*Product*	%
Competing resources	bad	10	0.05	0.4	0.02	3
Tank level indication	good	50	0.23	0.9	0.21	31
Size of LOCA	both	10	0.05	0.2	0.01	1
Expectation of failure	both	50	0.23	0.3	0.07	10
Training on contingencies	both	100	0.45	0.8	0.36	54
		220		SLI = 0.67		

Table 14.29 *Proforma for a mistake in a recovery event in nuclear power plant operation (Dougherty and Fragola, 1988) (Reproduced with permission of John Wiley & Sons Ltd from Human Reliability Analysis by E.M. Dougherty and J.R. Fragola, 1988, Copyright ©)*

Event design actors	NDXOVERH	**Event type:**	Recovery	
Event description				
The crew fails to realign equipment following recirculation hardware failures				
Option information:		**Screening value:**	4E-1	
Rule based?	No			
Hesitancy?	No			
SLI calculated?	Yes			
Standard TRC?	Yes			
Influences				
	Rank	*Normed-rank*	*Quality*	*Product*
1. Display adequacy	10	0.06	70	4.2
2. Procedure adequacy	40	0.24	30	7.2
3. Team effectiveness	20	0.12	80	9.6
4. Communication effectiveness	10	0.06	80	4.8
5. Workload	40	0.24	30	7.2
6. Training adequacy	50	0.29	70	20.3
7.				
8.				
9.				
10.				
				53.3
SLI	50.3			
Available time (min)	20			
Mean probability and statistics	1E-2			
Lower bound	4E-4			
Upper bound	4E-2			
Median time (min)	4.0			
Error factor	3.2			

$E - n = 10^{-n}$

authors derive a figure for the frequency of unrecovered mistakes of commission.

14.30.12 Computer aids

The documentation required of the HRA by the PRA is quite extensive and a computer program ORCA (Operator Reliability and Assessment) has been developed to assist in producing it. Table 14.29 shows a typical document for a recovery mistake.

14.30.13 Application to nuclear power plants

Detailed illustrations of the application of the HRA technique to particular nuclear power plants are given by the authors.

14.30.14 Validation

The HRA method just outlined is described by its authors as speculative. In other words, it lacks validation. In this it is on a par with most of the techniques used for quantifying human error in nuclear and process plants. Nevertheless, this general type of approach represents the most systematic method currently available for the treatment of the human error aspects of a PRA.

14.31 Assessment of Human Error: CCPS Method

Another methodology for human reliability assessment is that described in the CCPS *Human Error Prevention Guidelines* (1994/17). The structure of the method is shown in Figure 14.34. The core is the four stages of (1) critical human interaction identification and screening, (2) qualitative prediction of human error, (3) representation of event development and (4) quantification of significant human errors.

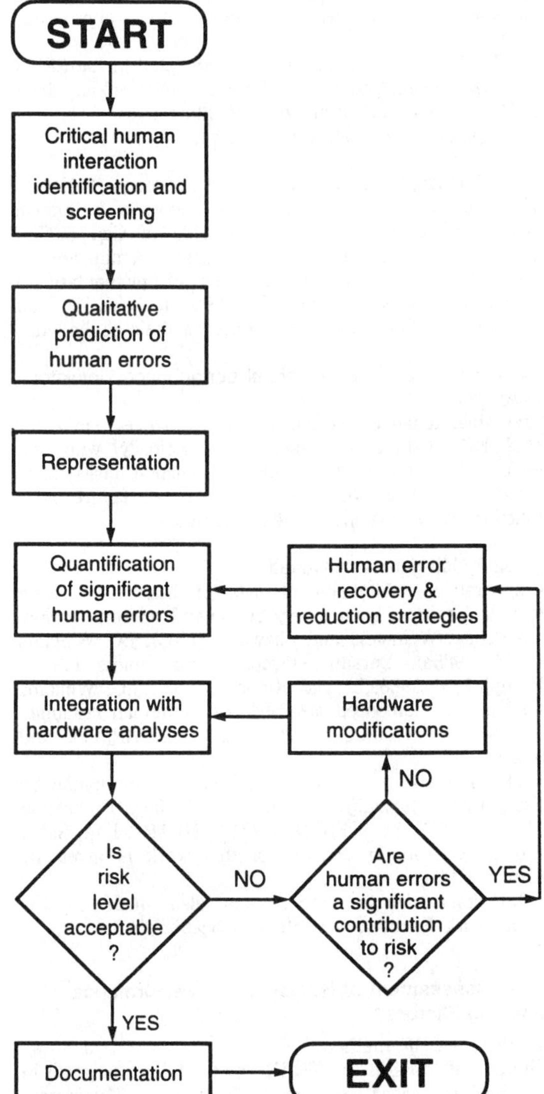

Figure 14.34 A methodology for human reliability assessment (CCPS, 1994/17) (Courtesy of the American Institute of Chemical Engineers)

The qualitative error prediction stage utilizes the SPEAR method involving task analysis, PIF analysis, PHEA, consequence analysis and error reduction analysis. The representation of event development typically takes the form of fault trees and event trees. The techniques described for the quantitative error prediction stage are THERP, SLIM and influence diagram analysis (IDA). The latter is described in the next section.

The CCPS HRA methodology has the important characteristic that it requires the analyst to start by acquiring a thorough understanding of the system first by critical human interaction identification and then by detailed qualitative analysis. Only then does the work progress to the use of the quantitative methods.

14.32 Assessment of Human Error: Other Methods

The following methods also merit mention. A brief account of each is given in the *Second Report* of the Study Group on Human Factors of the ACSNI (1991). Accounts are also given by J.C. Williams (1985), Waters (1989) and Brazendale (1990 SRD R510).

14.32.1 TESEO method

A simple model for the estimation of the probability of operator error is that used in TESEO (Tecnica Empirica Stima Errori Operatori) developed by Bello and Colombari (1980). The probability, q, of error is assumed to be the product of five parameters K_1–K_5 as follows:

$$q = K_1 K_2 K_3 K_4 K_5 \qquad [14.32.1]$$

Definitions and values of the parameters are given in Table 14.30.

An illustration of the use of TESEO has been given by Kletz (1991e). He considers a daily task of filling a tank by watching the level and closing a valve when the tank is full. He suggests that a reasonable estimate of failure is 1 in 1000, or once in 3 years. In practice, men operate such systems without incident for periods of 5 years. Using the TESEO approach he sets K_1 as 0.001, K_2 as 0.5 and the other parameters as unity, obtaining a probability of failure of 1 in 2000, or once every 6 years.

14.32.2 Absolute probability judgement (APJ) method

In the absolute probability judgement (APJ) method, described by Seaver and Sitwell (1983), experts are asked to make direct estimates of human error probabilities for the task.

14.32.3 Method of paired comparisons (PCs)

The use of the method of paired comparisons (PCs) in expert judgement was outlined in Chapter 9. It may be applied to the estimation of human error probabilities, as described by Blanchard *et al.* (1966) and Hunns (1980, 1982).

14.32.4 Influence diagram approach (IDA)

Influence diagram analysis (IDA) is a tool developed in the context of decision analysis (R.A. Howard and Matheson, 1980). It has been adapted for work on human factors by L.D. Phillips, Humphreys and Embrey (1983). Essentially it is a form of logic tree

Table 14.30 Operator error probability parameters used in TESEO (Bello and Colombari, 1980) (Courtesy of Elsevier Applied Science Publishers Ltd)

Type of activity	
	K_1
Simple, routine	0.001
Requiring attention, routine	0.01
Not routine	0.1
Temporary stress factor for routine activities	
Time available (s)	K_2
2	10
10	1
20	0.5
Temporary stress factor for non-routine activities	
Time available (s)	K_2
3	10
30	1
45	0.3
60	0.1
Operator qualities	
	K_3
Carefully selected, expert, well trained	0.5
Average knowledge and training	1
Little knowledge, poorly trained	3
Activity anxiety factor	
	K_4
Situation of grave emergency	3
Situation of potential emergency	2
Normal situation	1
Activity ergonomic factor	
	K_5
Excellent microclimate, excellent interface with plant	0.7
Good microclimate, good interface with plant	1
Discrete microclimate, discrete interface with plant	3
Discrete microclimate, poor interface with plant	7
Worst microclimate, poor interface with plant	10

showing the relations between particular performance shaping factors. As such it may be used to obtain quantitative estimates based on expert judgement of the effects of these factors.

14.32.5 Human Cognitive Reliability (HCR) correlation

The human cognitive reliability (HCR) correlation of Hannaman and co-workers (Hannaman *et al.*, 1985; Hannaman and Worledge, 1987) is a method in which the actions of the operating crew are represented in the form of an extended operator action tree and the probability of failure to respond is assessed using a set of three time–reliability correlations.

14.32.6 Systematic human error reduction and prediction approach (SHERPA)

The systematic human error reduction and prediction approach (SHERPA), and its application to human performance in ultrasonic inspection, has been described by Murgatroyd *et al.* (1986). SHERPA involves the three stages of identification of the set of tasks, hierarchical task analysis and human error analysis. Hierarchical task analysis (HTA) is used to identify the two main types of task element handled: those involving manual skills, and those involving the application of condition–action rules. The human error analysis utilizes a flowchart technique to identify the external error modes of the actions comprising the task and the psychological error mechanisms which give rise to these modes. Other types of task element are dealt with by some other appropriate method.

SHERPA is one of the two methods cited in the SRDA *Operating Procedures Guide* (Bardsley and Jenkins, 1991 SRDA-R1) for the treatment of human error in the development of operating procedures.

14.32.7 Critical action and decision approach (CADA)

The other method quoted by these authors is the critical action and decision approach (CADA) of Gall (1990). This is a technique for systematic examination of decision-making tasks and is thus complementary to SHERPA. CADA utilizes checklists to classify and examine decision errors and to assess their likelihood.

14.32.8 Maintenance personnel performance simulator (MAPPS)

The maintenance personnel performance simulator (MAPPS) method is concerned principally with the effect of manning levels on maintenance tasks. The basic premise is that the probability of failure is a function of the loading on the personnel.

14.32.9 Comparative evaluations

Comparative evaluations of sets of human reliability techniques have been given by several authors, including Brune, Weinstein and Fitzwater (1983), J.C. Williams (1983, 1985a), Bersini, Devooght and Smidts (1988), Humphreys (1988a,b) and Kirwan (1988). J.C. Williams (1985b) has compared six methods of human reliability assessment, including the AIR data bank, APJ, PC and SLIM.

The methods treated in the comparative evaluation by Humphreys (1988a,b), described more fully in Section 14.35, are APJ, PC, TESEO, THERP, HEART, IDA, SLIM and HCR. Another account of this work is given by Kirwan (1988).

A comparative evaluation was also made in the Benchmark Exercise described in Section 14.36.

14.33 Assessment of Human Error: Performance Shaping Factors

14.33.1 THERP method

The use of PSFs in the THERP methodology as given in the *HRA Handbook* has been outlined above. The factors are listed and described but a formal quantitative method of determining for each PSF an adjustment factor to be applied to the human error probability does not appear to be used.

14.33.2 SLIM

PSFs are the basis of SLIM, as described above. In this method the value, or quality, of a PSF is determined by structured expert judgement. The influence diagram technique is used to show the structure of the relationship between PSFs.

14.33.3 HEART

HEART is another method in which PSFs play a central role. The classification of the generic tasks is itself based on PSF-like distinctions and the error producing conditions (EPCs) which are applied to the basic generic task reliability estimates are in effect PSFs.

14.33.4 White method

R.F. White (1984 SRD R254) has described a form of PSF, the observable operational attribute (OOA) in which, as the name implies, the emphasis is on the observability of the attribute. He provides a checklist of attributes, broken down into (1) plant attributes and (2) maintenance attributes. Examples of attributes listed as observable are:

(1) What is the time-scale of a filling operation of one tank (or two at a time)? (e.g. Within one shift? Longer than one shift?)
(2) Is formal training given at instrument fitter level?

14.33.5 Whalley model

A methodology for identifying error causes and making the link between them and the performance shaping factors has been developed by Whalley (1987). The classification structure of the PSFs used is shown in Figure 14.35. A total of 146 PSFs is defined. The PSFs influence the error causes. Each error cause may be affected by several PSFs and each PSF may affect several error causes.

The method utilizes classifications of (1) task types, (2) response types, (3) error types, (4) error mechanisms and (5) error causes. There are seven task types (TTs), which are shown in Table 14.31, Section A. There are also seven response types (RTs) as shown in Section B of the table. The ten error types (ETs) are shown in Section C. The error mechanisms and error causes are shown in Figure 14.36. There are 10 error mechanisms (EMs) and 37 error causes (ECs) (including causes 23a and 30a).

The task types may be summarized as follows. TT 1 is response to a familiar input and requires essentially no decision-making. TT 2 is response to several familiar inputs, matches the mental model and requires no decision-making. TT 3 is interpretation of, and response to, a developing situation using the mental model. TT 4 is a pre-determined response to a recognized situation. TT 5 is a self-determined activity which may involve planning. TT 6 is selection of one of several alternative plans. TT 7 is correction of error, and thus differs from the other types. For the response types the primary

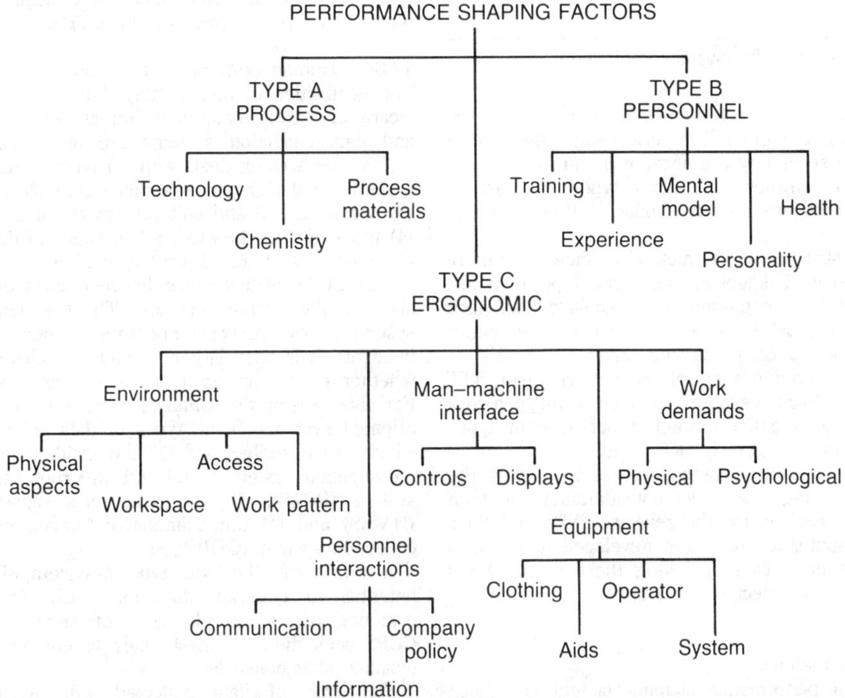

Figure 14.35 Classification structure of performance shaping factors (Whalley, 1987. Reproduced by permission.)

Table 14.31 Some classifications of task, response and error for the human operator (Whalley, 1987. Reproduced with permission.)

A Task types

1. Stimulus
2. Integration
3. Interpretation
4. Requirement
5. Self-generation
6. Choice
7. Correction required

B Response types[a]

	Discrete	*Sequence*
Action	Y	Y
No action	Y	N
Give information	Y	Y
Get information	Y	Y

C Error types

1. Not done
2. Less than
3. More than
4. As well as
5. Other than
6. Repeated

Timing errors

7. Sooner than
8. Later than
9. Misordered
10. Part of

[a] Y, valid combination; N, invalid combination

classification is into discrete or sequence responses, the former involving a single unit of performance, the latter a sequence. The secondary classification is into action and communication activities. The error types are cast as guidewords which are broadly similar to those used in hazop studies.

The overall structure of the method is shown in Figure 14.37. The primary linkages are from task type, response type and error type to possible error causes. Each task type can be mapped to a set of error causes, either through an information processing chain (IPC) (TT → IPC → EC) or through a set of error mechanisms (TT → EM → EC). Each response type can be mapped to a set of error causes either through a set of error types (RT → ET→ EC) or directly (RT → EC).

The analyst uses these various routes to identify the error causes for the task under consideration and then determines for each cause the relevant PSFs and their impact. A computer aid has been developed to assist in identifying the error causes, linking these to the PSFs and quantifying the effect of the latter.

14.33.6 CCPS method

Another set of performance shaping factors are those given in the *Human Error Prevention Guidelines* by the CCPS (1994/17), which refers to them as performance influencing factors (PIFs). The PIF classification structure is shown in Table 14.32. The *Guidelines* give a detailed commentary on each of these factors.

14.34 Assessment of Human Error: Human Error Data

The various methods described for quantitative assessment of human reliability give rise to a demand for data on human error. Some aspects of this are now considered. Human error data and its acquisition has been discussed by a number of workers. The account given in *Human Error Prevention Guidelines* by the CCPS (1994/17) deals with the essentials.

14.34.1 Human error data

Data on human error may be acquired by a number of methods. One approach involves study of the task from documentation, which may be aided by task analysis. Another group of methods are those based on some form of direct observation. This may be informal or may utilize formal techniques such as activity sampling, verbal protocol, or withholding of information.

A third group of methods are based on debriefing. One technique here is the critical incident technique, originally used by Flanagan (1954), in which a person who has experienced a near miss is debriefed. Debriefing may also be used to gain information about more normal tasks and situations. A fourth approach is the elicitation of information from experts.

The situation studied may be that on real plant or on a simulator. Simulation is widely used to present situations representative of real life in a compressed time-scale. A review of these and other methods of acquisition of data on human error is given by the CCPS.

14.34.2 Human error data collection

The acquisition of high quality data on human error is clearly of central importance. Human error data collection and data collection systems are also treated by the CCPS. The account deals with (1) types of data collection system, (2) design principles for data collection systems, (3) organizational and cultural aspects of data collection, (4) types of data collected, (5) methods of data collection and storage and (6) data interpretation.

The CCPS distinguishes between consequence-driven and causally oriented systems. The traditional reporting system in the process industries is one triggered by incidents. It is often largely a matter of chance, however, whether or not an error has significant consequences. For this reason the *Guidelines* concentrate on causally oriented systems. Some types of data collection system which are described are (1) the incident reporting and investigation system (IRIS), (2) the root cause analysis system (RCAS), (3) the near miss reporting system (NMRS) and (4) the quantitative human reliability data collection system (QHRDCS).

It is primarily the latter type of system which has the potential to generate data for HRA. There is little evidence of the development of such systems. The CCPS sees them as most likely to emerge in the first instance at in-house level.

The type of data collected will depend on the perspective dominant in the organization, on the human error model implied and on the associated error

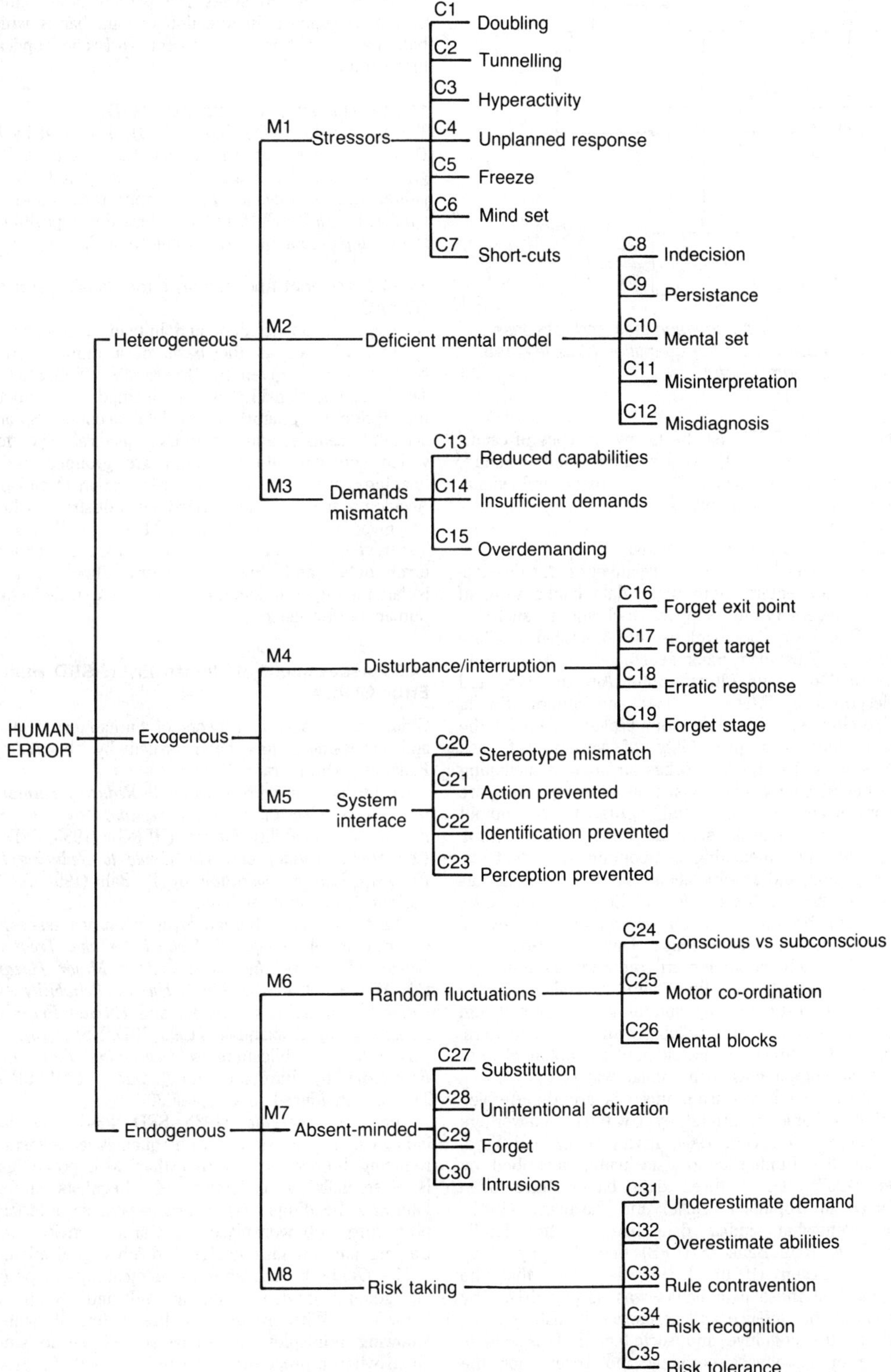

Figure 14.36 *Structure of error mechanisms and error causes (Whalley, 1987. Reproduced by permission.)*

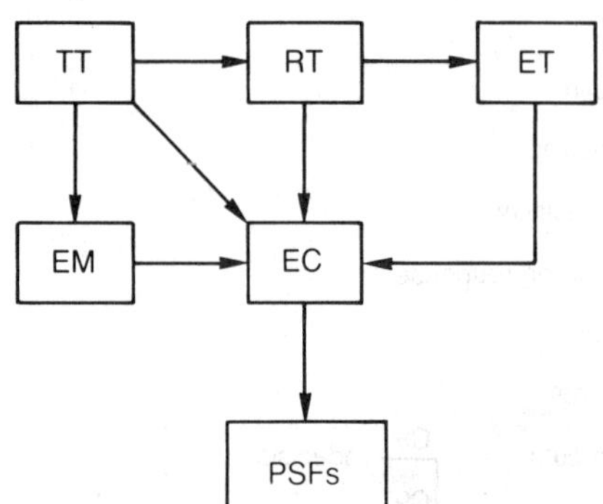

Figure 14.37 *Relations between task and response types, error causes and other quantities (Whalley, 1987. Reproduced by permission.)*

classification. The CCPS list the following types of causal data. Data may refer to (1) event sequence and structure, (2) human error tendencies, (3) performance-influencing factors and (4) organisational issues.

14.34.3 Human reliability data banks

Human error data banks were a relatively early development. The data entered into these data banks were of the type necessary to support techniques such as THERP. One such data bank is the American Institute for Research (AIR) data bank developed by Altman and co-workers (Payne and Altman, 1962; Altman, 1964) and also described by Meister (1964) and others (Swain, 1968; De Greene, 1970). The store provides data for the execution time and probability of success of task 'elements' or 'steps'. Each step has an input, a mediating and an output component. Typical input components are indicators such as scales and lights, typical output components are controls such as knobs and pushbuttons, and the two mediating components are identification/recognition and manipulation. Each component has several parameters. Those for a light include size, number, and brightness. Each parameter has several discrete levels or 'dimensions'. Data on time and reliability for each parameter are recorded as functions of its dimensions. The time and reliability of the component are estimated by summing and multiplying, respectively, the time and reliability of the parameters. The time and reliability of the element are obtained from those of the components in a similar way.

Two other data banks from this era are the Aerojet-General data bank, described by Irwin and co-workers (Irwin, Levitz and Freed, 1964; Irwin, Levitz and Ford, 1964), and the Bunker-Ramo data bank, described by Meister (1967). These three data banks have been reproduced in Topmiller, Eckel and Kozinsky (1982). Another somewhat similar data bank is the Sandia Human Error Rate Bank (SHERB) described by Rigby (1967) and Swain (1970). It will be clear that the development of these data banks was largely driven by the needs of the defence and aerospace industries.

Work on the cognitive and socio-technical aspects of human error is almost certainly too recent for the emergence of data banks utilizing classifications based on these approaches. Nor is it clear whether the needs of the process industries are perceived as sufficiently urgent to support the creation of data banks which are both based on this more recent work and applicable to those industries.

14.34.4 Human error database (HED)

The Human Error Database (HED), described by Kirwan (1988), is based on the human error probability data given in the *Rasmussen Report*, tempered by expert judgement. In that it derives from that report, it has similarities to THERP, but it is less decompositional and is not dependent on any specific model.

14.34.5 Accident and human error classification system (TAXAC)

A design for an accident and human error classification system, TAXAC, as the basis of a human error data bank, has been given by Brazendale (1990 SRD R510). The accident classification is obtained as a function of the accident signature (a skeletal account) S, and the accident causes and conducive factors. The features which contribute to the latter are grouped under the headings: activity A, man M, organization O and plant P. Accidents are also categorized by industry I. Checklists are given for features I, S, A, M, O and P. The author also reviews the requirements and prospects for a human error data bank. Such a venture should not be an isolated entity but should be part of a wide activity on human performance.

14.35 Assessment of Human Error: SRD Human Error Guides

Guidance on several aspects of human error in design and assessment has been issued by the Safety and Reliability Directorate (SRD).

Design is dealt with in *Guide to Reducing Human Error in Process Operation (short version)* by the Human Factors in Reliability Group (HFRG) (1985 SRD R347) (the *HFRG Guide*) and *The Guide to Reducing Human Error in Process Operation* by P. Ball (1991 SRDA-R3), the short and full versions.

The treatment of human error in hazard assessment is covered in *A Suggested Method for the Treatment of Human Error in the Assessment of Major Hazards* by R.F. White (1984 SRD R254), *Human Reliability Assessors Guide* by Humphreys (1988a) and *Human Error in Risk Assessment* by Brazendale (1990 SRD/HSE/510).

A related publication is *Developing Best Operating Procedures* by Bardsley and Jenkins (1991 SRDA-R1). This is considered in Chapter 20.

The *HFRG Guide* (1985 SRD R347), as its title indicates, is concerned with qualitative measures for reducing human error rather than with predicting it. It is essentially a collection of checklists under the following headings: (1) operator–process interface, (2) procedures, (3) workplace and working environment, (4) training and (5) task design and job organization.

The *Guide* distinguishes three meanings of procedure: (1) general guidance, (2) an aid and (3) prescribed behaviour. With regard to the latter, it states the following principles: (1) there should be no ambiguity about when a procedure is to be used, (2) if a procedure is mandatory there should be no incentive to use another

Table 14.32 *Classification structure of performance influencing factors (CCPS, 1994/17) (Courtesy of the American Institute of Chemical Engineers)*

Operating environment	*Task characteristics*
Chemical process environment: Frequency of personnel involvement Complexity of process events Perceived danger Time dependency Suddenness of onset of events Physical work environment: Noise Lighting Thermal conditions Atmospheric conditions Work pattern: Work hours/rest hours Shift rotation	Equipment design: Location/access Labelling Personal protective equipment Control panel design: Content and relevance of information Identification of displays and controls Compatibility with user expectations Grouping of information Overview of critical information and alarms Job aids and procedures: Clarity of instructions Level of description Specification of entry/exit conditions Quality of checks and warnings Degree of fault diagnostic support Compatibility with operational experience Frequency of updating Training: Clarity of safety and production requirements Training in using new equipment Practice with unfamiliar situations Training in using emergency procedures Training in using automatic systems
Operator characteristics	*Organizational and social factors*
Experience: Degree of skill Experience with stressful process events Personality: Motivation Risk taking Risk homeostasis Locus of control Emotional control Type A versus type B Physical condition and age	Teamwork and communications: Distribution of workload Clarity of responsibilities Communications Authority and leadership Group planning and orientation Management policies: General safety policy Systems of work Learning from operational experience Policies for procedures and training Design policies

method, (3) where possible a procedure should support the operator's skill and discretion rather than replace them and (4) a procedure should be easy to understand and to follow.

R.F. White (1984 SRD R254) describes the use of event tree and fault tree methods to identify the points at which a human action occurs which has an effect on the outcomes. As described earlier, he uses a form of PSF, the observable operational attribute (OOA), which is distinguished by the fact that it is observable and for which he provides a checklist. He gives as an illustrative example the analysis of the filling of an liquefied natural gas tank.

The *Human Reliability Assessors Guide* (Humphreys, 1988a), given in overview by Humphrey (1988b), is in two parts, the first of which gives a summary of eight techniques for human reliability assessment and the second evaluation criteria for selection and a comparative evaluation based on these criteria. The eight techniques are APJ, PC, TESEO, THERP, HEART, IDA, SLIM and HCRM. For each technique the *Guide* gives a description and a statement of advantages and disadvantages. The evaluation criteria used relate to (1) accuracy, (2) validity, (3) usefulness, (4) effective use of resources, (5) acceptability and (6) maturity. Accuracy has to do with correspondence with reality and with consistency;

validity with incorporation of human factors knowledge and of the effect of appropriate PSFs.

The report by Brazendale (1990 SRD/HSE/510) presents a taxonomy for an accident classification scheme, TAXAC, intended for use in connection with a human error data bank. The classification is preceded by a discussion of models of human error. An account of this work is given in Section 14.34.

14.36 Assessment of Human Error: Benchmark Exercise

A benchmark exercise of human reliability is described in *HF-RBE: Human Factors Reliability Benchmark Exercise: Summary Contributions of Participants* (Poucet, 1989). The study is one of a series of benchmark exercises conducted by the Commission of the European Communities (CEC) and organized by the Joint Research Centre (JRC) at Ispra, Italy. The exercise was carried out over 1986–88 with 13 participants. The system studied was the emergency feedwater system (EFS) on a nuclear reactor at the Kraftwerk Union (KWU) site at Grohnde. Two studies were performed: a routine test and an operational transient.

The study carried out by SRD is described in Poucet's volume by Waters. The work showed the importance of the qualitative modelling stage prior to any quantification and the value of event trees as a tool for such modelling. Methods investigated included APJ, TESEO, THERP, HEART and SLIM. No single method appeared superior in all applications.

14.37 Assessment of Human Error: ACSNI Study Group Report

The accounts given of attempts to assess human error illustrate the difficulties and raise the question of whether such assessment is even feasible. The use of quantitative risk assessment has created a demand for the development of techniques for the assessment of human error and this demand has been satisfied by the development of methods, which have just been described, but the validation of these methods leaves much to be desired.

This problem is addressed in *Second Report* of the ACSNI (1991) entitled *Human Reliability Assessment - A Critical Review*. The study deals with the control of the process, but emphasizes that the question is wider than this, embracing also maintenance and other activities. Although it is a report to the nuclear industry, its findings are in large part applicable to the process industries also.

As systems become more automated and reliable, but still vulnerable to human error, the relative importance of human error increases.

The report quotes the view expressed to the Sizewell B Inquiry by the Nuclear Installations Inspectorate (HSE, 1983e) that it was 'considered that comprehensive quantification of the reliability of human actions was not, with current knowledge, meaningful or required', and considers how far the situation has changed since that time.

The report describes the basic procedure of human reliability assessment (HRA). This is the breakdown of the task into a series of events, each of which has two alternative outcomes and the assignment to these events of basis error probabilities (HEPs). These human error probabilities are then modified in respect of dependencies and of performance shaping factors. The HEP for the task is then determined. The robustness of the result is assessed by means of a sensitivity analysis.

The variations on this basic procedure lie in the areas of the data sources, the rules for combination and the handling of time. The different techniques vary in the use which they make of field data, data banks, and expert judgement. They also differ in the way in which they combine the influencing factors. In one method some factors may be taken into account by making distinctions in the basic tasks, whilst in another they may be handled as PSFs. The effect of time may be taken into account by treating it simply as another PSF, or, alternatively, it may be accorded a special status. In this latter case the favoured method is the time response correlation, in which time is the principal independent variable determining the HEP, the HEP decreasing as the time available increases.

The report examines some of the areas of HRA in which problems arise or over which care must be taken. The include (1) classification of errors, (2) modelling and auditing, (3) operator error probabilities, (4) maintenance error probabilities, (5) interaction between PSFs, (6) sensitivity analysis and (7) changes in management and organization.

It adopts the distinction between skill-based, rule-based and knowledge-based behaviour and that between slips and mistakes. Essentially it argues that slips, associated with skilled behaviour, are amenable to prediction, but mistakes, associated with rule-based behaviour, are harder to predict, whilst for errors in knowledge-based behaviour there is currently no method available. It also draws attention to wilful actions, or violations, which again are not well covered in current methods.

The first step in an assessment of HEPs is the modelling of the task. There is evidence that for skill-based errors at least, the variability between techniques is less than that between analysts. This points to the need for training of analysts and for an independent check. For auditing it is necessary that the analyst record sufficient detail on his procedures and reasoning.

The report cites estimates of HEPs given in the literature such as the *Rasmussen Report,* described below. It draws attention to the importance of dependency between human actions. It is characteristic of human error that the probability of a further error following an initial one is often not independent but conditional on, and increased by, the occurrence of the first error. On the other hand, humans have the capability to recover from error. It is also necessary to consider the HEP for tasks where more than one operator is involved.

With regard to maintenance error, the report makes the point that hardware failure rate data already include the effects of maintenance errors. Nevertheless, it is prudent to make some assessment of the maintenance error. In particular, there is the possibility that such errors may be a source of dependent failures, as described in Chapter 9.

Most methods make the assumption that the PSFs are independent of each other. Yet the effects of lack of supervision and of training, for example, are likely to be greater when they occur in combination. An exception is SLIM, which does allow for this effect. The HEP estimates should be explored using sensitivity analysis.

The HRA is likely to more affected than the PRA as a whole by changes in the management and organization of the company, but the assessment of such changes on HEPs is not straightforward. It seems likely that certain types of error will be more affected than others and, therefore, that their relative importance will change. Skill-level errors may be less influenced by management changes than by knowledge-level ones. Violations might well be sensitive to such changes.

The report reviews the strengths and limitations of HRA. It is accepted that PRA has an essential role to play in improving reliability; HRA is a logical and vital extension. Its use has been given impetus by the retrospective assessments conducted after the Three Mile Island incident. An HRA contributes to system design in three ways: it provides a benchmark for designs and safety cases; it gives a quantitative assessment of alternative design or organizational solutions; and it provides a means and a justification for searching out the weak points in a system.

The limitations of HRA principally considered are those of HEP data and of validation. There is a wealth of data from various kinds of human factors experiments, but the situations in which they are obtained are usually to a degree artificial and their applicability is questionable. They provide guidance on the relative importance of different PSFs but generally need to be supplemented by information from other sources. The quantity and quality of field data are relatively low. Those data which do exist are predominantly for slips rather than mistakes or violations. Again they are a guide to the relative importance of different PSFs, but need to be supplemented. A third source of data is expert judgement. The methods for maximizing the quality of assessments and avoiding pitfalls are discussed. The importance is emphasized of the experts having full information on the operational context, such as normal operation or abnormal conditions.

The report addresses the question of the lack of validation of HRA methods. It quotes the following critique of J.C. Williams (1985b): 'It must seem quite extraordinary to most scientists engaged in research in other areas of the physical and technological world that there has been little or no attempt made by human reliability experts to validate the human reliability assessment techniques which they so freely propagate, modify and disseminate.'

Four kinds of validity are considered by the authors: (1) predictive validity, (2) convergent validity, (3) content validity and (4) construct validity. Essentially, for the prediction of a given analyst, predictive validity is concerned with agreement with the real situation, convergent validity is concerned with agreement with the predictions of other analysts, content validity is concerned with agreement between the elements of the model and the features which are critical in real life, and construct validity is concerned with agreement between the structure of the model and that of the real-life situation.

A study of predictive validity has been made by Kirwan (1988). He compared six HRA methods in respect of their ability to predict accident data from experimental, simulator and plant sources, and found that on some criteria some methods seemed reasonably successful. The work was limited, however, to the assessment of the HEP for specified errors and did not deal with modelling of tasks. Two studies of convergent validity, by Brune, Weinstein and Fitzwater (1983) and Bersini, Devooght and Smidts (1988), have found that different assessors give widely differing estimates, even when using the same method. Content validity is probably best assessed by peer review. A study by Humphreys (1988a) based on a comparison of eight methods against a checklist concluded that the relative performance of the methods depends on the problem and that few are completely comprehensive. The report states that construct validity does not appear to have been applied in HRA and that it may not be applicable at this stage of development.

Considering the areas where work is required, the report suggests that one important topic is the development of an improved model of operator behaviour. Here it is the high level decision-making processes which are of prime importance, since it is failures at this level which have the most serious consequences. There is a hierarchy of patterns of learned behaviour, highly learned at the lower levels and but less so towards the upper ones. It is the conservatism of these processes which is responsible for behaviour such as 'mind set' and switch to 'automatic pilot' and for slips when the mind 'jumps the points' and then continues with a whole series of inappropriate actions. Another area where work is needed is the creation of a data set which may be used for validation studies.

The report supports the development and use of methods of HRA. It is an essential element of PRA and is beneficial in its own right. But the techniques available for predicting slips are better than those for mistakes, and particular caution should be exercised with the latter. It should be recognized that HRA is still in its infancy.

It rejects the view that HRA in the nuclear industry should be optional and supports a requirement for its use. HRA provides a structured and systematic consideration of human error. It is already a valuable tool and has the potential to become an invaluable one. A requirement for its use is necessary to provide the necessary impetus for this. The report emphasizes, however, that effort should not be concentrated exclusively on assessment; there is equal need for a systematic approach to the reduction of human error.

The report contains appendices (Appendices 1–4) on HRA methods and their evaluation, dependability of human error data, attempts to establish validity of HRA methods, and approaches to reducing human error. Some of the procedures reviewed in Appendix 1 are APJ, IDA, TESEO, THERP, HEART, SLIM, TRCs, MAPPS and HED.

Appendix 4 of the report gives guidance on approaches to the reduction of human error. It discusses accident chains and latent and active (enabling and initiating) failures and gives the accident model shown in Figure 2.5 which shows the 'shells of influence' for the PSFs. The approach is based on the identification by the HRA, including the sensitivity analysis, of the features where human error is critical, the application of human factors methods and the implementation of the improvements indicated by these methods. The role of human factors both in system design and in the design of the man-machine interface and the workplace is described.

14.38 CCPS Human Error Prevention Guidelines

14.38.1 CCPS *Guidelines for Preventing Human Error in Process Safety*

The prevention of human error on process plants is addressed in the *Guidelines for Preventing Human Error in Process Safety* edited by Embrey for the CCPS (1994/X) (the CCPS *Human Error Prevention Guidelines*).

The *Human Error Prevention Guidelines* are arranged under the following headings: (1) the role of human error in chemical process safety, (2) understanding human performance and error, (3) factors affecting human performance in the chemical industry, (4) analytical methods for predicting and reducing human error, (5) qualitative and quantitative prediction of human error in risk assessment, (6) data collection and incident analysis methods, (7) case studies and (8) setting up an error reduction program in the plant.

14.38.2 Approaches to human error

The *Guidelines* distinguish four basic perspectives on human error, which they term: (1) the traditional safety engineering approach, which treats the problem as one of human behaviour and seeks improvement by attempting to modify that behaviour; (2) the human factors engineering and ergonomics (HF/E) approach, which regards human error as arising from the work situation which it therefore seeks to improve; (3) the cognitive engineering approach, which again accords primacy to the work situation, but places its main emphasis on the cognitive aspects; and (4) the socio-technical systems approach, which treats human error as conditioned by social and management factors. Until quite recently it has been the HF/E approach which has made the running. In the last decade, however, the last two approaches, cognitive engineering and socio-technical systems, have emerged strongly.

The stance of the *Guidelines* is one of system-induced error. This is akin to the work situation approach but enhanced to encompass the cognitive and socio-technical approaches.

14.38.3 Performance influencing factors

The factors shaping human performance are referred to in the *Guidelines* as performance influencing factors (PIFs). They give a PIF classification structure and a detailed commentary on each factor.

14.38.4 Methods of predicting and reducing human error

The *Guidelines* deal with methods for predicting and reducing human error in terms of (1) data acquisition techniques, (2) task analysis, (3) human error analysis and (4) ergonomics checklists.

They describe a number of methods of task analysis, grouped into action-oriented techniques and cognitive techniques. The former include hierarchical task analysis (HTA) and operation action event trees (OAETs). The two representatives of the latter are the critical action and decision evaluation technique (CADET) and the influence modelling and assessment system (IMAS). A critical review is given of each method.

Human error analysis is represented by predictive human error analysis (PHEA) and work analysis. The PHEA method is that utilized in the *Guidelines'* methodology for HRA.

14.38.5 Methods of predicting human error probability for risk assessment

The *Guidelines* describe a methodology for HRA, as part of quantitative risk assessment, utilizing both qualitative and quantitative methods for predicting human error. They begin with an illustrative example of fault tree analysis in which the prime contributors are human error events.

The HRA methodology presented in the *Guidelines* is that already described in Section 14.31 and outlined in Figure 14.34. A detailed commentary is given on each of the stages involved. The core of the method is the four stages of (1) critical human interaction identification and screening, (2) qualitative prediction of human error, (3) representation of event development and (4) quantification of significant human errors.

It is a feature of this HRA methodology that it directs the analyst to acquire understanding of the system and its problems by critical human interaction identification and to undertake a detailed qualitative analysis to further enhance this understanding before embarking on the use of the quantitative methods. It is this rather than the introduction of any new quantitative technique which is its most distinctive characteristic of the method.

14.38.6 Methods for data collection and incident analysis

The *Guidelines* review the collection of data on human error. The treatment given is outlined in Section 14.34.

The *Guidelines* give a number of methods of incident analysis. They are (1) the causal tree/variation diagram, (2) the management oversight and risk tree (MORT), (3) the sequentially timed events plotting procedure (STEP), (4) root cause coding, (5) the human performance investigation process (HPIP) and (6) change analysis. They also refer to the CCPS *Incident Investigation Guidelines*, where most of these techniques are described. A relatively detailed account is given of HPIP by Paradies, Unger and Ramey-Smith (1991); HPIP is a hybrid technique combining several of those just mentioned. The CCPS treatment of incident investigation both in these *Guidelines* and in the *Incident Investigation Guidelines* is described Chapter 27.

14.38.7 Case studies

A feature of the *Human Error Prevention Guidelines* is the number of case histories given. The section on case studies gives five such studies: (1) incident analysis of a hydrocarbon leak from a pipe (Piper Alpha); (2) incident investigation of mischarging of solvent in a batch plant; (3) design of standard operating procedures for the task in Case Study 2; (4) design of VDUs for a computer controlled plant; and (5) audit of offshore emergency blowdown operations.

Other case studies occur throughout the text. One illustrates system induced error. Other scene-setting case studies cover (1) errors occurring during plant changes and stressful situations, (2) inadequate human–machine interface design, (3) failures due to false assumptions, (4) poor operating procedures, (4) routine violations, (5) ineffective organization of work, (6) failure explicitly to allocate responsibility and (7) organizational failures. There are case studies illustrating the application of models of human error such as the step ladder and sequential models.

The importance of human error in QRA is illustrated by the case study, already mentioned, dealing with the prevalence of human error in fault trees. Other case studies illustrate HTA, SPEAR, THERP and SLIM.

14.38.8 Error reduction programmes

The *Guidelines* provide guidance on the implementation of an error reduction programme in a process plant. A necessary precondition for such a programme is a management culture which provides the background and support for such initiatives. The general approach is essentially that given in the CCPS *Process Safety Management Guidelines* described in Chapter 6.

Since both safety-related and quality-related errors tend to have the same cause, an error reduction programme may well run in parallel with the quality programme. An error reduction programme should address both existing systems and system design. The tools for such a programme given in the *Human Error Prevention Guidelines* include (1) critical task identification, (2) task analysis, (3) PIF analysis and (4) error analysis, as described in Sections 14.38.3 and 14.38.4. System design should also address allocation of function. Error reduction strategies for these two cases are presented in the *Guidelines.*

14.39 Human Factors at Sizewell B

14.39.1 Sizewell B Inquiry

The potential contribution of human factors to the design and operation of nuclear power stations was urged in evidence to the Sizewell B Inquiry by the Ergonomics Society, as described by Whitfield (1994).

The Society saw this contribution as being in the areas of: (1) setting operational goals; (2) allocation of function between humans, hardware and software; (3) definition of operator tasks; (4) job design; (5) overall performance assessment and monitoring of operational experience; (6) operator–plant interface and workplace conditions; (7) operator support documentation; (8) selection and training of operating staff; (9) human reliability assessment; and (10) construction and quality assurance. The Nuclear Installations Inspectorate also laid emphasis on human factors aspects (HSE, 1983e).

The report of the Inquiry Inspector (Layfield, 1987) states: 'I regard human factors as of outstanding significance in assessing the safety of Sizewell B since they impinge on all stages from design to manufacture, construction, operation and maintenance.' (paragraph 25.90). It recommended the involvement of the discipline.

The plan put forward by the Central Electricity Generating Board in support of its licence application included commitments to an extensive schedule of training and to the use of probabilistic safety assessment, including quantification of human error.

14.39.2 Human factors studies

A review of the applicability of ergonomics at Sizewell B was undertaken by Singleton (1986), who identified applications in control room design, use of VDU displays, documentation, fault diagnosis, maintenance and task analysis. He stated: 'We know that human operators can achieve superb performance if they are given the right conditions. Appropriate conditions in this context can be listed within the four categories: the information presentations, the training, the support systems and the working conditions and environment.' An overview of the application of human factors at Sizewell B has been given by Whitfield (1994).

Human factors specialists are involved in extensive work on training, supported by task analysis, as described below. In the main control room one feature is the use of a plant overview panel, separate from the other displays, for the monitoring of safety critical parameters. Another area of involvement is in operating instructions. Use is made of 'event-based' instructions to diagnose a fault and to initiate recovery. But in addition, for safety critical functions, there are 'function-based' procedures, which assist in restoring the plant to a safe condition if for some reason the event-based instructions are inappropriate. Further details are given by McIntyre (1992).

Human reliability analysis within the PSA utilizes OAETs and the HEART method, with some use of THERP. An account is given by Whitworth (1987). The application of task analysis at Sizewell B is described by Ainsworth (1994). The programme for this comprised five stages:

(1) preliminary task analysis of critical tasks;
(2) task analysis of selected safety critical tasks;
(3) preliminary talk-through/walk-through evaluations of procedures in a control room mock-up;
(4) validation of procedures on a control room simulator;
(5) task analysis of tasks outside the main control room.

One crucial task on which task analysis was performed was the cooling down and depressurization of the reactor. This was a major study involving some 60 task elements and taking some 44 person-weeks. Time line analysis was used to address issues such as manning.

A review of procedures revealed a number of defects. Besides obvious typographical errors, they included: (1) incorrect instrument numbering (in procedures); (2) incorrect instrument labelling (on panels and in procedures); (3) omission of important clarifiers such as 'all', 'or', 'either' and 'if available'; (4) omission of important cautions and warnings; (5) requirements for additional information; and (6) lack of consistency between procedures, panels and VDU displays. Overall, the task analyses identified a number of mismatches between task requirements and man–machine interfaces.

Ainsworth makes the point that the ergonomists were often better at identifying problems than in devising solutions, those which they proposed often being impractical, but that it is possible to achieve a mode of working in which these problems are communicated to the designers who take them on board and come up with effective solutions.

14.40 Notation

Section 14.3

$H(s)$ operator transfer function
K gain

τ_d reaction time (s)
τ_I compensatory lag time constant (s)
τ_L lead time constant (s)
τ_N neuromuscular lag time constant (s)

Section 14.27

Subsection 14.27.4

k constant
P_b basic human error probability
P_c conditional human error probability

Subsection 14.27.11

n number of alarms in group
P_i probability of failure to initiate action in response to ith alarm
P_r probability of failure to initiate action in response to a randomly selected alarm

Section 14.28

a,b constants
HEP human error probability
n number of performance shaping factors
r relevancy factor
SLI Success Likelihood Index
w quality weighting

Section 14.30

f error factor
k_1, k_2 constants
m median
q quality
r rank
r_n normalized rank
x Success Likelihood Index

τ mean response time
τ_r adjusted median response time

Section 14.32

$K_1 - K_5$ parameters
q probability of error

15 Emission and Dispersion

Contents

The three major hazards – fire, explosion and toxic release – usually involve the emission of material from containment followed by vaporization and dispersion of the material. The treatment given here is relevant to all three hazards; in particular, the development of the emission and dispersion phases is relevant to such situations as:

(1) Escape of flammable material, mixing of the material with air, formation of a flammable cloud, drifting of the cloud and finding of a source of ignition, leading to (a) a fire and/or (b) a vapour cloud explosion, affecting the site and possibly populated areas.
(2) Escape of toxic material, formation of a toxic gas cloud and drifting of the cloud, affecting the site and possibly populated areas.

15.1 Emission

There is a wide range of circumstances which can give rise to emission. The situation which appears to be most frequently discussed is a failure of plant integrity, but it is important to consider other occurrences, including escape from valves which have been deliberately opened and forced venting in emergencies. Thus on a low pressure refrigerated liquefied petroleum gas (LPG) installation, for example, a large quantity of flammable vapour may be released by failure of a pressure vessel or of pipework. But loss of refrigeration with resultant forced venting could also give a large release of vapour.

In view of the many different situations which can give rise to emission, it is peculiarly difficult to obtain meaningful estimates of the quantity and duration of emissions. The emission stage is therefore subject to great uncertainty. It is nevertheless very important, because the way in which emission occurs can greatly influence the nature and effect of the release, and particularly of any vapour cloud which is formed.

Emission flows may be determined by the basic relations of fluid mechanics. Fundamentals of fluid flow are treated in such texts as *Hydraulics* (Lewitt, 1952), *Thermodynamics Applied to Heat Engines* (Lewitt, 1953), *The Dynamics and Thermodynamics of Compressible Flow* (Shapiro, 1953–), *Chemical Engineering* (Coulson and Richardson, 1955–, 1977–), *Mechanics of Fluids* (Massey, 1968–) and *Chemical Engineers Handbook* (Perry and Green, 1984). Selected references on emission are given in Table 15.1.

***Table 15.1** Selected references on emission*

Cremer and Warner (1978); HSE (1978b, 1981a); Rosak and Skarka (1980); McQuaid and Roberts (1982); O'Shea (1982); Pikaar (1985); Ramskill (1986 SRD R352)

Liquid, gas and vapour flow
Stanton and Pannell (1914); Lacey (1922/23); Lea (1930); Lapple (1943); Streeter (1951–); Lewitt (1952, 1953); Shapiro (1953–); Coulson and Richardson (1955–,1977–); Bulkley (1967); Grote (1967); Massey (1968–); Tate (1970); Simpson (1971); V.J. Clancey (1974a); Loeb (1975); Levenspiel (1977); HSE (1978a, 1981a); Pham (1979); Nowak and Joye (1981); Gyori (1985); Lipowicz (1985); Belore and Buist (1986); Pirumov and Roslyakov (1986); Leung and Epstein (1988); J.L. Woodward and Mudan (1991)

Coefficient of discharge
Perry (1949); Arnberg (1962); S.D. Morris (1990a)

Friction factor
Colebrook and White (1937); L.F. Moody (1944); Colebrook (1939); Churchill (1973, 1977, 1980); Jain (1976); N.H. Chen (1979–); Serghides (1984)

Slow leaks
D.H. Jackson (1948); Mencher (1967); Pilborough (1971, 1989)

Vessel drainage times
Loiacono (1978); T.C. Foster (1981); Schwartzhoff and Sommerfeld (1988); Sommmerfeld (1990); J.L. Woodward and Mudan (1991); Papas and Sommerfeld (1991); Crowl (1992b); Hart and Sommerfeld (1993); Sommerfeld and Stallybrass (1993); K.S. Lee and Sommerfeld (1994)

Bund overflow
Michels, Richardson and Sharifi (1988); Phelps and Jureidni (1992)

Pipelines
Cronje, Bishnoi and Svrcek (1980); T.B. Morrow (1982a,c); Fayed and Olten (1983); Oranje (1983); Lagiere, Miniscloux and Roux (1984); Battarra *et al.* (1985); Cawkwell and Charles (1985); Maddox and Safti (1985); Schweikert (1986); Picard and Bishnoi (1988, 1989); Clerehugh (1991); Olorunmaiye and Imide (1993)

Two-phase flow (see Table 15.2)

Vessel venting, blowdown
Zuber and Findlay (1965); DnV (1983 83–1317); Fauske, Grolmes and Henry (1983); Swift (1984); Grolmes and Epstein (1985); Fauske (1988a); Evanger *et al.* (1989); Haque *et al.* (1990); Haque, Richardson and Saville (1992); Haque *et al.* (1992); S.M. Richardson and Saville (1992)

Pressure relief valves
F.J. Moody (1966); Richter (1978b); Sallett (1978, 1979a,b, 1990a–c); Lyons (1979); Sale (1979a,b); Sallet, Weske and Gühler (1979); L. Thompson and Buxton (1979a,b); Z. Chen, Govind and Weisman (1983); Heller (1983); Rommel and Traiforos (1983); Cloutier (1985); Forrest (1985); Sallett and Somers (1985); Moodie and Jagger (1987); Wilday (1987); Morley (1989a,b)

Bursting discs
Huff and Shaw (1992)

Vessel rupture
Artingstall (1972); Hardee and Lee (1974, 1975); Hess, Hoffman and Stoeckel (1974); J.D. Reed (1974); Lonsdale (1975); Maurer *et al.* (1977); HSE (1978b); Bongers, ten Brink and Rulkens (1980); R.A. Cox and Comer (1980); Leiber (1980); A.F. Roberts (1982); Drivas, Sabnis and Teuscher (1983); Friedel (1986, 1987a); CEC (1990 EUR 12602 EN); Nolan *et al.* (1990); Schmidli, Bannerjee and

Yadigaroglu (1990); J.L. Woodward (1990); Nolan, Hardy and Pettitt (1991); Bettis and Jagger (1992); Pettit *et al.* (1992)

Pipeline rupture
Inkofer (1969); Westbrook (1974); R.P. Bell (1978); HSE (1978d); D.J. Wilson (1979b, 1981b); Morrow (1982a,c); Morrow, Bass and Lock (1982); Grolmes, Leung and Fauske (1983); Tam (1989); Tam and Higgins (1990); J.R. Chen, Richardson and Saville (1992)

Fugitive emissions (see Table 15.70)

15.1.1 Emission situations

Emission situations may be classified as follows:

(1) Fluid –
  (a) gas/vapour,
  (b) liquid,
  (c) vapour–liquid mixture;
(2) Plant –
  (a) vessel,
  (b) other equipment,
  (c) pipework;
(3) Aperture –
  (a) complete rupture,
  (b) limited aperture;
(4) Enclosure
  (a) in building,
  (b) in open air;
(5) Height –
  (a) below ground level,
  (b) at ground level,
  (c) above ground level;
(6) Fluid momentum –
  (a) low momentum,
  (b) high momentum.

Some typical emission situations are shown in Figure 15.1.

The fluid released may be gas, vapour, liquid or a two-phase vapour–liquid mixture, as shown in Figure 15.1(a). If the escape is from a container holding liquid under pressure, it will normally be liquid if the aperture is below the liquid level and vapour or vapour–liquid mixture if it is above the liquid level. For a given pressure difference the mass of release is usually much greater for a liquid or vapour–liquid mixture than for a gas or vapour.

The plant on which the release occurs may be a vessel, other equipment such as a heat exchanger or a pump, or pipework, as shown in Figure 15.1(b). The maximum amount which can escape depends on the inventory of the container and on the isolation arrangements.

The aperture through which the release occurs may range from a large fraction of the envelope of the container in the case of complete rupture of a vessel to a limited aperture such as a hole, as shown in Figure 15.1(c). An aperture on a vessel may be (1) a sharp-edged orifice, (2) a conventional pipe branch, (3) a rounded nozzle branch or (4) a crack. The flow through a rounded nozzle is greater than through a conventional pipe branch, but it is the latter which is generally used. Other apertures on vessels, equipment and pipework include: (5) drain and sample points; (6) pressure relief devices—(a) pressure relief valves, (b) bursting discs, and (c) liquid relief valves; (7) seals; (8) flanges; and (9) pipe ends.

The relief may take place from plant in a building or in the open air, as shown in Figure 15.1(d). This greatly affects the dispersion of the material. A large proportion of escapes to the open air are dispersed without incident.

The height at which the release occurs also has a strong influence on the dispersion, as shown in Figure 15.1(e). A liquid escape from plant below ground level may be completely contained. An escape of gas or vapour from above ground level may be dispersed over a considerable distance.

Dispersion is further affected by the momentum of the escaping fluid, as shown in Figure 15.1(f). Low and high momentum releases of gas or vapour form a plume and a turbulent momentum jet, respectively. Low and high momentum releases of liquid form a liquid stream and a high 'throw' liquid jet, respectively, both of which then form a liquid pool.

A generalized chart for the calculation of the discharge of liquids and gases under pressure has been given by Pilz and van Herck (1976) and is reproduced in Figure 15.2.

15.1.2 Elementary relations

For flow of a single fluid

$$W = Q\rho \qquad [15.1.1]$$

$$Q = uA \qquad [15.1.2]$$

$$G = W/A \qquad [15.1.3]$$

$$G = u\rho \qquad [15.1.4a]$$

$$G = u/v \qquad [15.1.4b]$$

where A is the cross-sectional area of the pipe, G is the mass flow per unit area, Q is the total volumetric flow, u is the velocity, v is the specific volume of fluid, W is the total mass flow, and ρ is the density of the fluid.

The general differential form of the energy balance for the flow of unit mass of fluid is

$$-\mathrm{d}q + \mathrm{d}W_s + \mathrm{d}H + \mathrm{d}\left(\frac{u^2}{2}\right) + g\mathrm{d}z = 0 \qquad [15.1.5]$$

where g is the acceleration due to gravity, H is the enthalpy, q is the heat absorbed from the surroundings, W_s is the work done on the surroundings (shaft work), and z is the height.

It is often convenient to eliminate H from Equation 15.1.5 using the relation:

$$\mathrm{d}H = \mathrm{d}q + \mathrm{d}F + v\mathrm{d}P \qquad [15.1.6]$$

where F is the mechanical energy irreversibly converted to heat (frictional loss) and P is the absolute pressure.

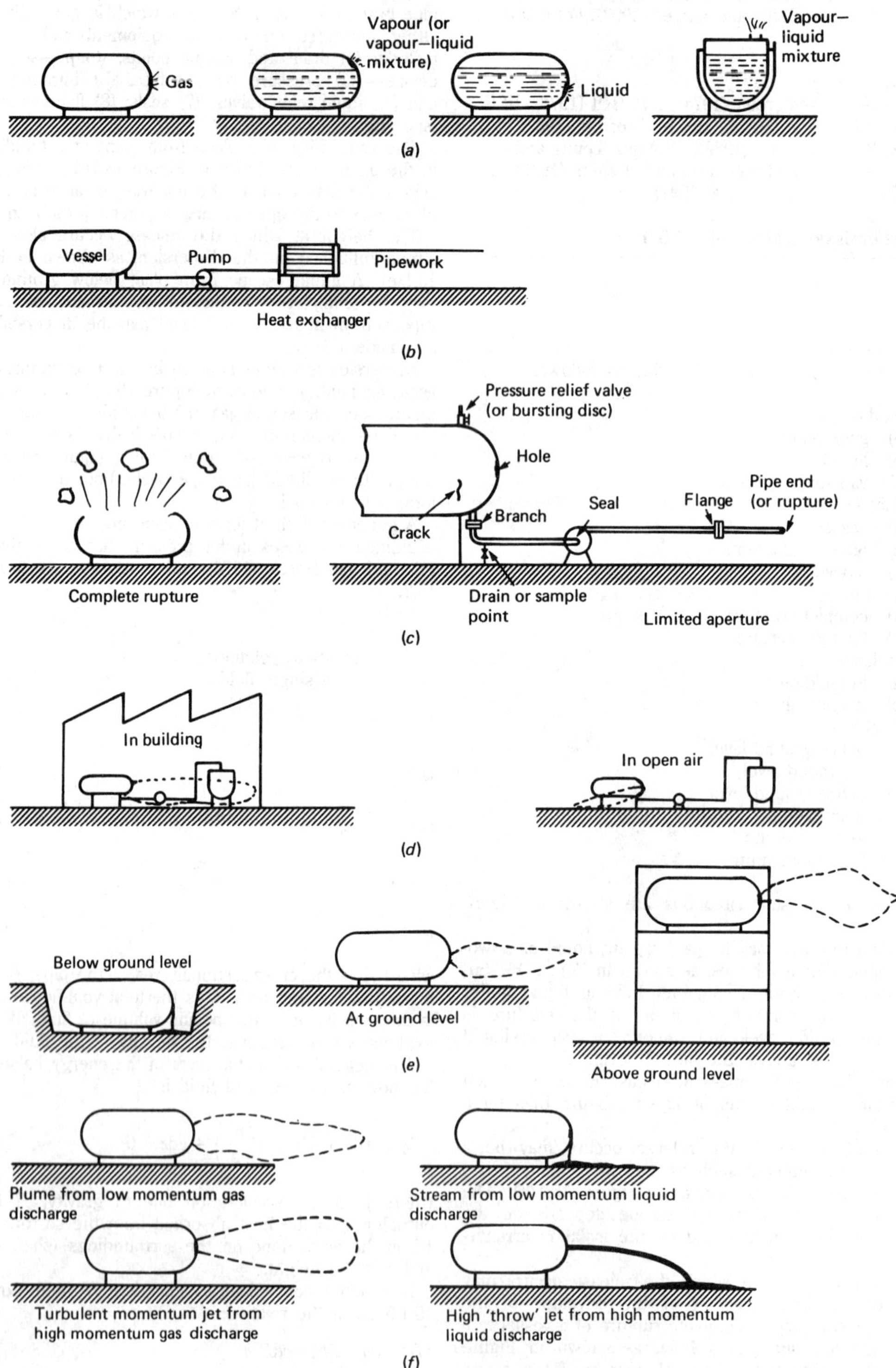

Figure 15.1 *Some emission situations*

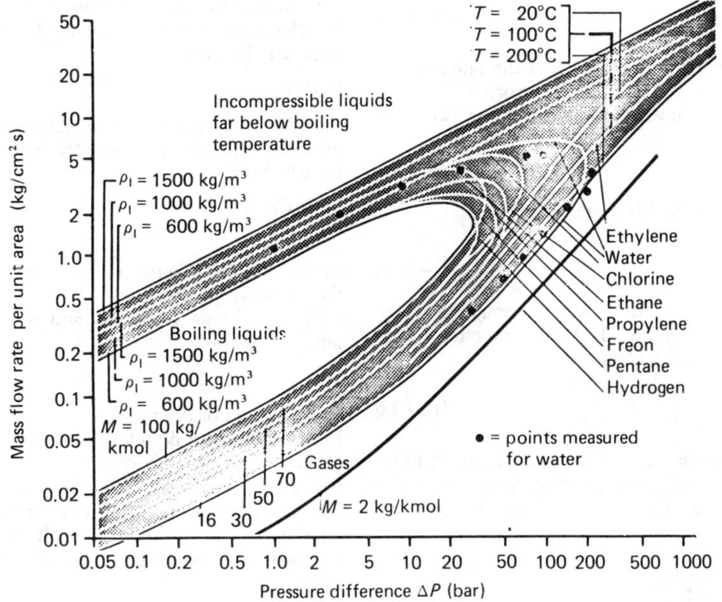

Figure 15.2 *Mass flow vs pressure difference for flow of gases, vapours and liquids through an orifice (Pilz and van Herck, 1976; reproduced by permission)*

Then, combining Equations [15.1.5] and [15.1.6] gives

$$dW_s + v dP + dF + d\left(\frac{u^2}{2}\right) + g dz = 0 \quad [15.1.7]$$

For the discharge processes considered here, the work done on the surroundings W_s is zero. Then Equation 15.1.7 becomes:

$$v dP + dF + d\left(\frac{u^2}{2}\right) + g dz = 0 \quad [15.1.8a]$$

or, alternatively,

$$v dP + dF + u du + g dz = 0 \quad [15.1.8b]$$

In some applications the friction loss and potential energy terms may be neglected, in which case Equation 15.1.8b reduces to:

$$v dP + u du = 0 \quad [15.1.9]$$

Equation 15.1.8b may also be expressed in the form of a pressure drop relation:

$$\frac{dP}{dl} + \frac{1 dF}{v dl} + G\frac{du}{dl} + \frac{g dz}{v dl} = 0 \quad [15.1.10a]$$

or

$$\frac{dP}{dl} + \rho\frac{dF}{dl} + G\frac{du}{dl} + \rho g\frac{dz}{dl} = 0 \quad [15.1.10b]$$

or

$$\frac{dP}{dl} + \frac{1 dF}{v dl} + \frac{u du}{v dl} + \frac{g dz}{v dl} = 0 \quad [15.1.10c]$$

where l is the distance along the pipe.

A relation for pressure drop may also be obtained from the momentum balance:

$$\frac{dP}{dl} + \frac{S}{A}R_0 + G\frac{du}{dl} + \rho g\frac{dz}{dl} = 0 \quad [15.1.11]$$

where R_o is the shear stress at the pipe wall and S is the perimeter of pipe. The term dz/dl in Equations 15.1.10 and 15.1.11 is often replaced by the term $\sin\theta$, where θ is the angle between pipe and horizontal.

Comparing Equations 15.1.10b and 15.1.11, which each give the pressure drop, it follows that:

$$\frac{S}{A}R_0 = \rho\frac{dF}{dl} \quad [15.1.12]$$

The pressure drop dP/dl may also be expressed in terms of its constituent elements:

$$-\frac{dP}{dl} = -\frac{dP_f}{dl} - \frac{dP_a}{dl} - \frac{dP_g}{dl} \quad [15.1.13]$$

where dP_f/dl is the frictional pressure change, dP_a/dl is the accelerational pressure change and dP_g/dl is the gravitational pressure change.

For flow in pipes the friction term dF is:

$$dF = 4\phi\frac{dl}{d}u^2 \quad [15.1.14a]$$

$$= 8\phi\frac{dl}{d}\frac{u^2}{2} \quad [15.1.14b]$$

with

$$\phi = \frac{R}{\rho u^2} \qquad [15.1.15]$$

where d is the diameter of the pipe, R is the shear stress at the pipe wall and ϕ is the friction factor.

The most commonly used friction factor f is twice the friction factor ϕ:

$$f = 2\phi \qquad [15.1.16]$$

Hence from Equations 15.1.14b and 15.1.16

$$\mathrm{d}F = 4f\frac{\mathrm{d}l}{d}\frac{u^2}{2} \qquad [15.1.17]$$

Another form of the momentum balance equation which is derived from Equation 15.1.10b together with Equations 15.1.14 and 15.1.17 is:

$$\frac{\mathrm{d}P}{\mathrm{d}l} + G^2\left(\frac{\mathrm{d}v}{\mathrm{d}l} + \frac{4fv}{2d}\right) + \frac{g}{v}\frac{\mathrm{d}z}{\mathrm{d}l} = 0 \qquad [15.1.18]$$

The equations just given are in differential form. In order to obtain relations useful in engineering, they need to be integrated. Integrating the energy balance Equation (15.1.8a)

$$\int_1^2 v\mathrm{d}P + F + g\Delta z + \Delta\left(\frac{u^2}{2}\right) = 0 \qquad [15.1.19]$$

where limit *1* is the initial state and limit 2 is the final state.

Integration of the first term in Equation 15.1.19 requires a relation between pressure P and specific volume v. Three principal relations are those for incompressible, isothermal and isentropic flow:

v = Constant Incompressible flow [15.1.20a]

Pv = Constant Isothermal flow [15.1.20b]

Pv^γ = Constant Isentropic flow [15.1.20c]

where γ is the ratio of the gas specific heats. Also, for an adiabatic expansion:

Pv^k = Constant Adiabatic flow [15.1.20d]

where k is the expansion index.

Then it can be shown that for these flow regimes:

$$\int_1^2 v\mathrm{d}P = v(P_2 - P_1) \quad \text{Incompressible flow} \qquad [15.1.21a]$$

$$= P_1 v_1 \ln\left(\frac{P_2}{P_1}\right) \quad \text{Isothermal flow} \qquad [15.1.21b]$$

$$= \frac{\gamma}{\gamma - 1}P_1 v_1\left[\left(\frac{P_2}{P_1}\right)^{(\gamma-1)/\gamma} - 1\right]$$

Isentropic flow [15.1.21c]

$$= \frac{k}{k - 1}P_1 v_1\left[\left(\frac{P_2}{P_1}\right)^{(k-1)/k} - 1\right]$$

Adiabatic flow [15.1.21d]

It is frequently convenient to work in terms of the head h rather than the pressure P of the fluid. The relation between the two is:

$$P = h\rho g \qquad [15.1.22]$$

It is generally necessary to define the flow regime. The principal criterion of similarity is the Reynolds number Re which is

$$\mathrm{Re} = \frac{Gd}{\mu} \qquad [15.1.23]$$

where d is the diameter of the orifice or pipe, and μ is the viscosity of the fluid.

The discharge of liquids and of gases and vapours is now considered.

15.1.3 Liquid discharge

For the flow of a liquid, or incompressible fluid, the term $\int_1^2 v\mathrm{d}P$ is given by Equation 15.1.21a. Then, assuming negligible friction loss and flow on a horizontal axis, it can be shown from Equations 15.1.19 and 15.1.21a that for the velocity of liquid through an orifice with a cross-sectional area A which is small relative to that of the vessel

$$u = C_\mathrm{D}[2v(P_1 - P_2)]^{\frac{1}{2}} \qquad [15.1.24a]$$

$$= C_\mathrm{D}\left[\frac{2}{\rho}(P_1 - P_2)\right]^{\frac{1}{2}} \qquad [15.1.24b]$$

$$= C_\mathrm{D}(2gh)^{\frac{1}{2}} \qquad [15.1.24c]$$

or, in terms of the mass velocity G

$$G = \frac{C_\mathrm{D}}{v}[2v(P_1 - P_2)]^{\frac{1}{2}} \qquad [15.1.25a]$$

$$= C_\mathrm{D}[2\rho(P_1 - P_2)]^{\frac{1}{2}} \qquad [15.1.25b]$$

$$= C_\mathrm{D}\rho(2gh)^{\frac{1}{2}} \qquad [15.1.25c]$$

where C_D is the coefficient of discharge and h is the head of liquid.

For the flow of liquid through a pipe the frictional loss F is given by

$$F = F_\mathrm{f} + F_\mathrm{ft} + F_\mathrm{c} \qquad [15.1.26]$$

where F is the total frictional loss, F_c is the frictional loss due to sudden contraction, F_f is the frictional loss due to the pipe, and F_ft is the frictional loss due to fittings.

The frictional loss due to the pipe is:

$$F_\mathrm{f} = \frac{4\phi l u^2}{d} \qquad [15.1.27]$$

The friction loss from pipe fittings is usually expressed as a number of velocity heads:

$$h_\mathrm{ft} = k_\mathrm{h}\frac{u^2}{2g} \qquad [15.1.28]$$

$$F_\mathrm{ft} = g h_\mathrm{ft} \qquad [15.1.29]$$

where h_ft is the head loss due to pipe fittings and k_h is the number of velocity heads.

The friction loss from contraction is:

$$F_c = \frac{u_2^2}{2}\left(\frac{1}{C_c} - 1\right)^2 \qquad [15.1.30a]$$

where C_c is the coefficient of contraction and u_2 is the velocity of the fluid after contraction. The coefficient of contraction C_c increases from about 0.6 to 1.0 as the ratio of pipe diameters increases from 0 to 1. At the normal value of $\frac{2}{3}$:

$$F_c = \frac{u_2^2}{8} \qquad [15.1.30b]$$

The frictional loss from sudden contraction cannot exceed the kinetic energy of the fluid.

The flow through a pipe system can be calculated from Equation 15.1.19 using Equations 15.1.26–15.1.30. For the case of flow through a horizontal pipe where the pressure drop is due predominantly to the friction loss in the pipe itself, Equations 15.1.19, 15.1.26 and 15.1.27 yield

$$G = \left[\frac{\rho d(P_1 - P_2)}{4\phi l}\right]^{\frac{1}{2}} \qquad [15.1.31]$$

or, in terms of the alternative friction factor f

$$G = \left[\frac{\rho d(P_1 - P_2)}{2fl}\right]^{\frac{1}{2}} \qquad [15.1.32]$$

15.1.4 Coefficient of discharge

The coefficient of discharge C_D was defined in Equation 15.1.24. It is normally applied to the discharge from orifices and nozzles. The coefficient of discharge is the ratio of the actual to the theoretical discharge. This actual discharge is the product of the actual cross-sectional area of the jet and of the actual velocity of the jet. Then defining a coefficient of contraction C_c as the ratio of the actual to the theoretical area, and a coefficient of velocity C_v as the ratio of the actual to the theoretical velocity gives

$$C_D = C_c C_v \qquad [15.1.33]$$

The coefficient of contraction C_c may be defined as

$$C_c = \frac{a_c}{a_0} \qquad [15.1.34]$$

where a_0 is the cross-sectional area of the orifice and a_c is the cross-sectional area of the vena contractor. The coefficient of velocity C_v may be defined as

$$C_v = \frac{u}{(2gh)^{\frac{1}{2}}} \qquad [15.1.35]$$

For a sharp-edged orifice in the side of a vessel there are available theoretical expressions for the coefficient of contraction. Thus for a two-dimensional sharp-edged orifice for liquid

$$C_c = \frac{\pi}{\pi + 2} = 0.61 \qquad [15.1.36]$$

and for gas

$$C_c = \frac{\pi}{\pi + 2(\rho_2/\rho_1)} \qquad [15.1.37]$$

where the subscripts 1 and 2 refer to stagnation and outlet conditions, respectively.

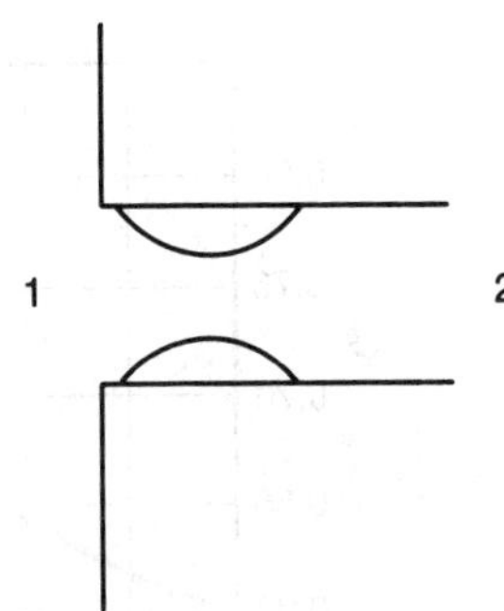

Figure 15.3 *External mouthpiece*

An average experimental value of C_c for sharp-edged orifices for liquids is 0.64. The difference between the actual and theoretical velocities for a sharp-edged orifice for liquid is very small. An average experimental value of C_v for sharp-edged orifices for liquids is 0.97. Thus for a sharp-edged orifice for liquids an average experimental value of the coefficient of discharge C_D is 0.62 (= 0.97 × 0.64).

Another common discharge situation is that of an external mouthpiece on the side of a vessel, such as a short nozzle or pipe stub, as shown in Figure 15.3. For an external mouthpiece the velocity u_c at the vena contractor is

$$u_c = \frac{ua}{a_c} \qquad [15.1.38]$$

where a is the cross-sectional area of the pipe and a_c is the cross-sectional area of the vena contracta. But the coefficient of contraction C_{cv} at the vena contractor is

$$C_{cv} = \frac{a_c}{a} \qquad [15.1.39]$$

Hence from Equations 15.1.38 and 15.1.39

$$u_c = \frac{u}{C_c} \qquad [15.1.40]$$

The head loss h_c at the contraction is

$$h_c = \frac{(u_c - u)^2}{2g} \qquad [15.1.41]$$

From Equations 15.1.40 and 15.1.41

$$h_c = \left(\frac{1}{C_{cv}} - 1\right)^2 \frac{u^2}{2g} \qquad [15.1.42]$$

Allowing for the velocity head lost at the pipe outlet the total head loss h is

$$h = \left[1 + \left(\frac{1}{C_{cv}} - 1\right)^2\right]\frac{u^2}{2g} \qquad [15.1.43]$$

Then from Equations 15.1.35 and 15.1.43

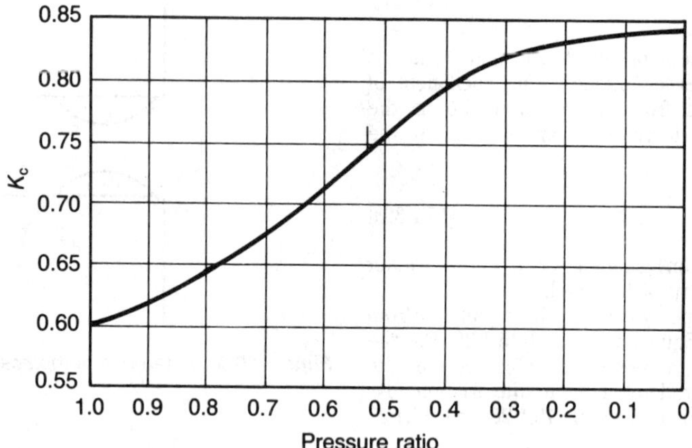

Figure 15.4 *Coefficient of discharge for gas flow through an orifice (after J.A. Perry, 1949) (Courtesy of the American Society of Mechanical Engineers)*

$$C_v = \frac{1}{\left[1 + \left(\frac{1}{C_{cv} - 1}\right)^2\right]^{\frac{1}{2}}} \qquad [15.1.44]$$

Since for an external mouthpiece the coefficient of contraction C_c (as opposed to C_{cv}) is unity, the coefficient of discharge is equal to the coefficient of velocity.

An average experimental value of the coefficient of contraction C_{cv} at the vena contractor of an external mouthpiece for liquids is 0.62. Then the value of the coefficient of velocity C_v obtained from Equation 15.1.44 is 0.85. This is also the value of the coefficient of discharge C_D. An average experimental value of the coefficient of discharge is 0.81. The difference between the theoretical and experimental values is due to friction. The experimental value applies to mouthpieces for which the length/diameter ratio of the pipe is not less than 3.

The concept of the coefficient of discharge may also be applied to discharge from longer sections of pipe where there is appreciable head loss due to friction and fittings. For a pipe the total number K of velocity heads, excluding any exit, is

$$K = K_e + K_f + K_{ft} \qquad [15.1.45]$$

where K_e is the number of velocity heads lost at the entrance, K_f is the number of velocity heads lost due to friction and K_{ft} is the number of velocity heads lost due to fittings.

Thus for an external mouthpiece, from Equation 15.1.24c

$$h = \frac{1}{C_D^2}\frac{u^2}{2g} \qquad [15.1.46]$$

But the head losses in the mouthpiece are the entrance and exit losses, so that

$$h = (1 + K_e)\frac{u^2}{2g} \qquad [15.1.47]$$

Then from Equations 15.1.46 and 15.1.47

$$C_D = \frac{1}{(1 + K_e)^{\frac{1}{2}}} \qquad [15.1.48]$$

Similarly, for a longer pipe an effective coefficient of discharge C can be defined as

$$C = \frac{1}{(1 + K)^{\frac{1}{2}}} \qquad [15.1.49]$$

Studies on the coefficient of discharge from sharp-edged orifices for water have been described by Lea (1930). The coefficient of discharge for sharp-edged orifices for gases has been studied by Perry (1949). He found that the coefficient of discharge is a function of the ratio of the outlet to the stagnation pressure, or pressure ratio, as shown in Figure 15.4. The limiting values of the coefficient are 0.6 and 0.84 at high and low pressure ratios, respectively. Similar results have been obtained for short pipe outlets by Arnberg (1962). Values of the coefficient of discharge used in hazard assessment work range between 0.6 and 1.0. A value of 0.6 is frequently used for orifices and of 0.8 for short nozzles and pipe stubs. Coefficients of discharge are treated in more detail in standard texts (e.g. Lewitt, 1952; Coulson and Richardson, 1977–).

15.1.5 Friction, friction factor and fittings loss

The friction loss in a pipe is given by Equation 15.1.31 or 15.1.32, where the friction factors ϕ or f are obtained from Equations 15.1.15 and 15.1.16, respectively. The relationship between the friction factors ϕ and f is given by Equation 15.1.16.

Relationships between the principal friction factors used are given by Churchill (1977). The friction factor f as defined above is identical to the Fanning friction factor f_F so that

$$f_F = 2\phi \qquad [15.1.50]$$

Another friction factor is the Darcy friction factor f_D. This is related to the other friction factors as follows:

$$f_D = 8\phi \quad [15.1.51a]$$

$$= 4f \quad [15.1.51b]$$

$$= 4f_F \quad [15.1.51c]$$

Graphs of the friction factor are available based primarily on the work of Stanton and Pannell (1914) and of L.F. Moody (1944) and are given in standard texts (e.g. Coulson and Richardson, 1977–, Perry and Green, 1984).

It is also desirable to have equations for the friction factor, and a number of these are available. Some of the equations are implicit, others explicit. Some are more convenient for use if the flow is known, others if the pressure drop is known.

The friction factor depends on the flow regime, the three regimes being those of laminar, transitional and turbulent flow. In the laminar regime the friction factor may be obtained from the Poiseuille equation and may be written as:

$$\phi = \frac{8}{\text{Re}} \quad [15.1.52]$$

In the other regimes the friction factor is, in general, a function of the Reynolds number Re and of the roughness ratio e/d, where e is the pipe roughness.

A correlation of the friction factor for these regimes has been given by Colebrook (1939) as follows:

$$\frac{1}{f_D^{\frac{1}{2}}} = -2\log_{10}\left(\frac{e}{3.7d} + \frac{2.51}{\text{Re}\, f_D^{\frac{1}{2}}}\right) \quad [15.1.53]$$

He states that Equation 15.1.53 reduces for smooth pipes to

$$\frac{1}{f_D^{\frac{1}{2}}} = 2\log_{10}\left(\frac{\text{Re}\, f_D^{\frac{1}{2}}}{2.51}\right) \quad [15.1.54]$$

and for rough pipes to

$$\frac{1}{f_D^{\frac{1}{2}}} = 2\log_{10}\left(\frac{3.7d}{e}\right) \quad [15.1.55]$$

Colebrook also gives as an alternative to Equation 15.1.54 which avoids the implicit formulation by means of the following approximate equation:

$$\frac{1}{f_D^{\frac{1}{2}}} = 1.8\log_{10}(\text{Re}/7) \quad [15.1.56]$$

Coulson and Richardson (1977–) state that the transition regime occurs at 2000 < Re < 3000 and give the following equations for the turbulent regime.

Smooth pipes:

$$\phi = 0.0396\,\text{Re}^{-0.25} \qquad 2.5\times 10^3 < \text{Re} < 10^5 \quad [15.1.57]$$

$$\phi^{-0.5} = 2.5\ln(\text{Re}\,\phi^{0.5}) + 0.3 \qquad 10^5 < \text{Re} < 10^7 \quad (15.1.58)$$

Rough pipes:

$$\phi^{-0.5} = -2.5\ln(0.27e/d + 0.885\text{Re}^{-1}\phi^{-0.5}) \quad [15.1.59]$$

$$\phi^{-0.5} = 3.2 - 2.5\ln(e/d) \qquad (e/d)\,\text{Re}\,\phi^{0.5} \gg 3.3 \quad [15.1.60]$$

Equation 15.1.57 is the Blasius equation.

Churchill (1977) has given an explicit equation for the friction factor ϕ for all three regimes as follows:

$$\phi = \left[\left(\frac{8}{\text{Re}}\right)^{12} + \frac{1}{(\text{A}+\text{B})^{\frac{3}{2}}}\right]^{\frac{1}{12}} \quad [15.1.61]$$

with

$$A = \left[2.457\ln\left(\frac{1}{(7/\text{Re})^{0.9} + 0.27e/d}\right)\right]^{16} \quad [15.1.62]$$

$$B = (37530/\text{Re})^{16} \quad [15.1.63]$$

where A and B are constants. Elsewhere, Churchill (1980) states that the transition regime occurs at 1800 < Re < 4000.

Another explicit equation for the friction factor for all three regimes, expressed this time in terms of the Darcy friction factor, is that of N.H. Chen (1979):

$$\frac{1}{f_D^{\frac{1}{2}}} = -2.0\log\left\{\frac{e}{3.7065d} - \frac{5.0452}{\text{Re}} \times \log\left[\frac{1}{2.8257}\left(\frac{e}{d}\right)^{1.1098} + \frac{5.8506}{\text{Re}^{0.8991}}\right]\right\} \quad [15.1.64]$$

The relative merits of the equations for the friction factor are considered in the discussion on the work of N.H. Chen (1980).

Tabulations are available for the pressure drop across the various types of pipe fitting such as elbows, tees, valves, etc. A prime source is *Flow of Fluids through Valves, Fittings and Pipe* (Crane Company, 1986), which gives the loss in terms of the number of velocity heads. Tables are given in standard texts (e.g. Perry and Green, 1984) and appear mostly to be based on the data given in the various editions of the Crane Company publication.

An alternative formulation of fittings loss is in terms of number of pipe diameters. A table giving for selected fittings both the number of velocity heads and the number of pipe diameters is given by Coulson and Richardson (1977–). For many of the entries in this table the number of pipe diameters equivalent to one velocity head is about 50.

15.1.6 Critical flow

For compressible fluids, including gases and two-phase mixtures, it is necessary to consider critical flow. If, for such a fluid, the upstream pressure is high relative to the downstream pressure, the flow is choked. The flow is then no longer a function of the pressure difference but is a function of the upstream pressure alone and has a maximum or critical value. This is the condition of critical or choked flow. For a gas this flow is then sonic and the condition is also referred to as sonic flow.

The measure of approach to sonic conditions is the Mach number Ma which is

$$\text{Ma} = \frac{u}{c} \quad [15.1.65]$$

where c is the velocity of sound. For sonic flow the Mach number is unity and the velocity u of the fluid equals the velocity of sound c. But the velocity of sound in a gas of absolute pressure P and specific volume v is given by the relation

$$c = (kPv)^{\frac{1}{2}} \quad [15.1.66]$$

where k is the expansion index. Hence under critical conditions

$$u = (kP_2v_2)^{\frac{1}{2}} \quad [15.1.67]$$

where subscript 2 indicates downstream conditions.

A pressure ratio η may be defined

$$\eta = \frac{P_2}{P_1} \quad [15.1.68]$$

Then it can be shown that the critical pressure ratio η_c is

$$\eta_c = \left(\frac{2}{k+1}\right)^{k/(k-1)} \quad [15.1.69]$$

Equation 15.1.69 is often used to determine the downstream, or choke, pressure under critical conditions.

For a given upstream pressure the flow is zero both when the $\eta = 0$ and when $\eta = 1$. At some intermediate value of η, and P, the flow passes through a maximum, which occurs when

$$\frac{dG}{dP} = 0 \quad [15.1.70a]$$

or

$$\frac{dG}{d\eta} = 0 \quad [15.1.70b]$$

Differentiating Equation 15.1.4b with respect to P and applying Equation 15.1.70a gives for the critical flow G_c

$$\frac{du}{dP} = G_c \frac{dv}{dP} \quad [15.1.71]$$

Then from Equations 15.1.71 and 15.1.9

$$G_c{}^2 = -\frac{dP}{dv} \quad [15.1.72]$$

Critical flow is discussed further in the following sections.

15.1.7 Gas and vapour discharge

For the flow of a gas, or compressible fluid, through an orifice it is necessary to define the type of fluid and the type of expansion.

For an ideal gas

$$Pv = \frac{RT}{M} \quad [15.1.73]$$

and for a non-ideal gas

$$Pv = \frac{ZRT}{M} \quad [15.1.74]$$

where M is the molecular weight, R is the universal gas constant, T is the absolute temperature, and Z is the compressibility factor.

The flow of an ideal gas may be calculated using Equation 15.1.21 with a suitable expression for $\int_1^2 v\mathrm{d}P$. The flow of a non-ideal gas may be obtained from Equation 15.1.5 using enthalpy values from thermodynamic tables.

The expansions which take place in discharge situations are normally throttling expansion or free expansion. Throttling expansion occurs when gas issues through a very narrow aperture or crack. There is high frictional resistance and low kinetic energy. When applied to vapour, the process is sometimes called 'wire-drawing'.

The frictional resistance of a fluid passing through a pipe varies inversely with the fifth power of the pipe diameter. Similarly, as an aperture becomes larger a situation is quickly reached where there is free expansion. The frictional resistance is low and the kinetic energy is high. In the present context it is free expansion which is of prime interest and which is considered in the treatment given below.

The free expansion of an ideal gas is usually treated as an approximately adiabatic expansion with some degree of irreversibility. The entropy change dS for the expansion is

$$\mathrm{d}S = \frac{\mathrm{d}q}{T} + \frac{\mathrm{d}F}{T} \quad [15.1.75]$$

If the expansion is adiabatic $\mathrm{d}q = 0$, and if it is reversible $\mathrm{d}F = 0$. If the expansion is reversible adiabatic $\mathrm{d}S = 0$. The free expansion of an ideal gas is usually treated by the use of the Equation 15.1.21d.

If the expansion process is reversible and is adiabatic or intermediate between adiabatic and isothermal, Equation 15.1.21d is applicable. If the expansion is not reversible, it may not be possible to give a continuous function for the relation between pressure and volume, but Equation 15.1.21d is often applied over a limited range of conditions. If the expansion is isentropic, then the index k is equal to the ratio of specific heats γ. If the expansion is not isentropic, then k is less than γ.

If the absolute upstream pressure P_1 is only slightly greater than the absolute downstream pressure P_2, in other words for low values of the pressure ratio P_1/P_2, it may be adequate to assume an isothermal expansion according to Equation 15.1.20b. For isothermal flow of an ideal gas ($Pv = P_1v_1$)

$$\int_1^2 v\mathrm{d}P = P_1v_1 \int_1^2 \frac{1}{P}\mathrm{d}P \quad [15.1.76a]$$

$$= P_1v_1 \ln\frac{P_2}{P_1} \quad [15.1.76b]$$

Then, assuming flow on a horizontal axis ($\Delta z = 0$) and using a coefficient of discharge C_D to take account of the friction term F in Equation 15.1.19, it can be shown from Equations 15.1.4a, 15.1.19 and 15.1.76b that for the flow of an ideal gas through an orifice with a cross-sectional area A which is small relative to that of the vessel

$$u = C_D\left[2P_1v_1 \ln\left(\frac{P_1}{P_2}\right)\right]^{\frac{1}{2}} \quad [15.1.77]$$

$$G = C_D\left[\frac{2P_2^2}{P_1v_1} \ln\left(\frac{P_1}{P_2}\right)\right]^{\frac{1}{2}} \quad [15.1.78]$$

For adiabatic flow of an ideal gas ($Pv^k = P_1v_1{}^k$)

$$\int_1^2 v\mathrm{d}P = \int_1^2 \left(\frac{P_1 v_1^k}{P}\right)^{1/k} \mathrm{d}P \qquad [15.1.79a]$$

$$= \frac{k}{k-1} P_1 v_1 \left[\left(\frac{P_2}{P_1}\right)^{(k-1)/k} - 1\right] \qquad [15.1.79b]$$

$$= \frac{k}{k-1}(P_2 v_2 - P_1 v_1) \qquad [15.1.79c]$$

Following the same approach, it can be shown from Equations 15.1.4a, 15.1.19 and 15.1.79 that for the adiabatic flow case

$$G = \frac{C_d}{v_2}\left\{\frac{2P_1}{v_1} \cdot \frac{k}{k-1}\left[1 - \left(\frac{P_2}{P_1}\right)^{(k-1)/k}\right]\right\}^{\frac{1}{2}} \qquad [15.1.80]$$

Introducing the pressure ratio η from Equation 15.1.68 in Equation 15.1.80 gives

$$G = C_d\left\{\frac{2P_1}{v_1} \cdot \frac{k}{k-1}[\eta^{2/k} - \eta^{(k+1)/k}]\right\}^{\frac{1}{2}} \qquad [15.1.81]$$

If the ratio of the upstream pressure P_1 to that downstream P_2 is sufficiently high, the flow is choked, or sonic. Under these conditions Equation 15.1.69 applies. Then, taking friction into account the maximum discharge G_c under these conditions is

$$G_c = C_d\left[\frac{P_1}{v_1} k\left(\frac{2}{k+1}\right)^{(k+1)/(k-1)}\right]^{\frac{1}{2}} \qquad [15.1.82]$$

Equation 15.1.82 shows that, under sonic conditions, the flow depends on the upstream pressure P_1, but is independent of the downstream pressure P_2.

Equations 15.1.80 and 15.1.82, for subsonic and sonic flow of gas, respectively, are those normally required here for the calculation of high pressure gas discharges both from vents and pressure relief valves and from ruptures.

Equation 15.1.20d and the expressions based on it should be used only within the range of their validity. The equation is based on a modification of the equations for isentropic expansion of an ideal gas, in which the index k is equal to the ratio of specific heats γ. The index k is used instead of γ to take account of a degree of non-ideal behaviour. The value of k is obtained from thermodynamic charts by fitting Equation 15.1.20d to the appropriate constant entropy line. In addition, the index k is also used to take into account a degree of friction energy loss in expansion. In this case the value of k is determined empirically for the system concerned. The use of Equation 15.1.20d is not suitable for gases at very high pressures. These require separate treatment, as discussed in standard texts (e.g. Shapiro, 1953–; Massey, 1968–).

The use of Equation 15.1.20d is less appropriate for vapour which becomes wet when expanded. This case may be treated by the direct use of Equation 15.1.5 assuming an isentropic expansion ($\mathrm{d}S = 0$). Then, neglecting potential energy changes ($\mathrm{d}z = 0$), for finite changes Equation 15.1.5 becomes

$$\Delta H = \Delta\left(\frac{u^2}{2}\right) = 0 \qquad [15.1.83]$$

Hence, neglecting friction, for the flow of a vapour

$$G = \rho_2[2(H_1 - H_2)]^{\frac{1}{2}} \qquad [15.1.84]$$

In order to solve Equation 15.1.84 it is necessary to determine the mass fraction constituted by the vapour phase, or quality, x. This is obtained from the assumption of isentropic expansion which from the entropy balance gives

$$S_1 = S_2 \qquad [15.1.85a]$$

$$S_2 = S_{v2}x + S_{l2}(1 - x) \qquad [15.1.85b]$$

where subscripts l and v refer to the liquid and the vapour, respectively.

The enthalpy after expansion H_2 is obtained from the enthalpy balance:

$$H_1 = H_2 \qquad [15.1.86a]$$

$$H_2 = H_{v2}x + H_{l2}(1 - x) \qquad [15.1.86b]$$

The effect of friction may be taken into account by applying to the enthalpy change (H_1 – H_2) an efficiency factor. Efficiencies of orifices and nozzles are discussed in standard texts (e.g. Lewitt, 1953).

If the vapour flow is sonic, as determined by Equations 15.1.68 and 15.1.69, then, as for a gas, the flow is independent of the downstream pressure. In this case the pressure P_2 for the calculation of the terminal conditions of the fluid is that obtained from Equation 15.1.68 with $\eta = \eta_c$. An illustrative example is given by Lewitt (1953, p. 275).

A vapour undergoing sudden expansion tends, however, to be supersaturated with less than the equilibrium quantity of condensed liquid droplets. It is frequently assumed that such a vapour does in fact follow a path given by Equation 15.1.20d.

A calculation which is sometimes required is the flow of vapour from a liquid vaporizing adiabatically in a vessel. The pressure in the vessel decreases as the temperature of the liquid falls due to the removal of the latent heat of vaporization.

Further discussions of methods of calculating the expansion of gases and vapours are given in the *High Pressure Safety Code* (B.G. Cox and Saville, 1975) and by Lapple (1943) and Duxbury (1976).

15.1.8 Temperature effects

For an adiabatic expansion the absolute temperature after expansion T_2 is

$$\frac{T_2}{T_1} = \left(\frac{P_2}{P_1}\right)^{(k-1)/k} \qquad [15.1.87]$$

For sonic flow the pressure ratio P_2/P_1 is given by Equation 15.1.69. Hence for this case

$$\frac{T_2}{T_1} = \frac{2}{k+1} \qquad [15.1.88]$$

15.1.9 Vessel discharge: gas flow through pipe

It is sometimes necessary to calculate the flow of high pressure gas from a vessel. The pressure in the vessel

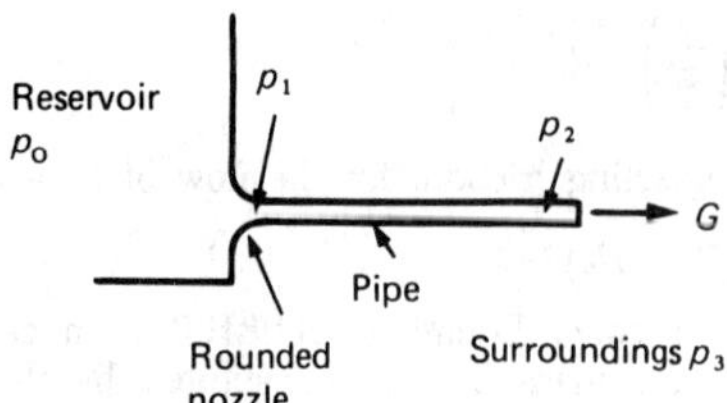

Figure 15.5 Vessel with connected discharge pipe (Levenspiel, 1977) (Courtesy of the American Institute of Chemical Engineers)

falls as the mass of gas decreases. Equation 15.1.82, for sonic flow, can be used until the upstream pressure falls to the value at which flow becomes subsonic, at which point Equation 15.1.80 is used. An illustrative example is given by Coulson and Richardson (1977–, Vol. 1, p. 110).

Another problem which may occur is the flow of gas from a vessel through a pipe, as shown in Figure 15.5. In this case the calculation involves simultaneous solution of the equations for flow through the aperture at the entrance to the pipe and through the pipe itself.

Charts to facilitate the calculation have been constructed by Lapple (1943) and are widely used. It has been pointed out by Levenspiel (1977), however, that this work is in error due to the assumption that sonic flow can be isothermal. Corrected charts given by Levenspiel are reproduced in Figure 15.6. The parameters used in the charts are the pressures shown in Figure 15.5 and

d = diameter of pipe (m)
f_F = Fanning friction factor
G = mass velocity of gas (kg/m^2 s)
G^* = critical mass velocity of gas through an adiabatic nozzle (kg/m^2 s)
k = expansion index (= ratio of gas specific heats)
L = length of pipe (m)

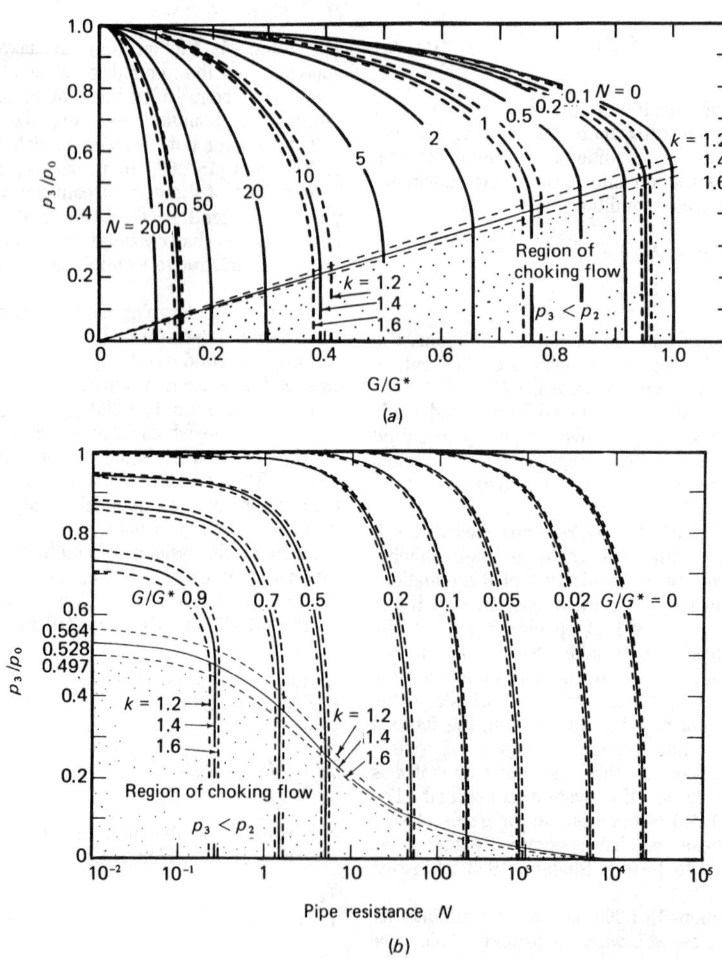

Figure 15.6 Charts for calculation of adiabatic flow of gas from a vessel to atmosphere (Levenspiel, 1977): (a) chart useful for finding the allowable length of pipe for a given flow rate; (b) chart useful for finding the flow rate for a given length of pipe (Courtesy of the American Institute of Chemical Engineers)

N = pipe resistance factor ($= (4f_F L)/d$)
p = absolute pressure (N/m^2)

Figure 15.6(a) is useful for finding the length of pipe for a given flow rate and Figure 15.6(b) for finding the flow rate for a given length of pipe.

Expressions exist for the calculation of the fraction of fluid left in a vessel after a given discharge time. Such relations are given in the *Rijnmond Report* and by Mecklenburgh (1985).

15.1.10 Vessel discharge: liquid discharge time

It is convenient to have expressions for the time taken for a vessel containing liquid to discharge its contents. Treatments of liquid discharge times for vessels have been given in the *Rijnmond Report* and by Lewitt (1952), T.C. Foster (1981), Mecklenburgh (1985), Loiacono (1987), Sommerfeld and co-workers (Schwartzhoff and Sommerfeld, 1988; Sommerfeld, 1990; Papas and Sommerfeld, 1991; Hart and Sommerfeld, 1993; Sommerfeld and Stallybrass, 1993; K.S. Lee and Sommerfeld, 1994), Crowl and Louvar (1990), J.L. Woodward and Mudan (1991) and Crowl (1992b).

In the most elementary case the vessel is of simple geometry and the outflow is through a hole at the bottom and is due solely to the head of liquid. Complexity is introduced by the following features: (1) pressure above the liquid in the vessel, (2) hole at an arbitrary height in the liquid space, and (3) outflow through a pipe rather than an orifice.

The basic relations for liquid outflow, taking account of the pressure on the liquid surface and for a hole of arbitrary height, are

$$-A\frac{dh}{dt} = Q \qquad [15.1.89]$$

$$Q = C_D A_0\{2[g(h-h_0) + \Delta P/\rho]\}^{\frac{1}{2}} \qquad [15.1.90]$$

where A is the cross-sectional area of the vessel at height h (m^2), A_o is the cross-sectional area of the orifice (m^2), C_D is the coefficient of discharge, g is the acceleration due to gravity (m/s^2), h is the height of the liquid above the bottom of the vessel (m), h_o is the height of the orifice above the bottom of the vessel (m), Q is the outflow through the orifice (m^3/s), ΔP is the pressure difference between the liquid surface and atmosphere, or gauge pressure (Pa), t is the time (s) and ρ is the density of the liquid (kg/m^3).

Then, rearranging and integrating gives the general relation for the time for discharge, or efflux time:

$$t = \frac{1}{C_D A_0 (2)^{\frac{1}{2}}} \int_{h_0}^{h} \frac{A(h)}{[g(h-h_0) + \Delta P/\rho]^{\frac{1}{2}}} dh \qquad [15.1.91]$$

The solution of this equation requires the specification of the terms $A(h)$ and $\Delta P/\rho$. For the first of these, the function $A = f(h)$ depends on the geometry of the vessel. In the simple case of a vertical cylindrical vessel, or right cylinder, $A = [(\pi/4)D^2]$, where D is the diameter of the cylinder (m). Other more complex cases are given below.

As for the term $\Delta P/\rho$, this is zero for the case where there is no imposed pressure on the liquid surface. The other simple case is where the pressure is maintained constant, as with an inert gas padding system, so that the term may be expressed as $\Delta P/\rho = h_p$, where h_p is an equivalent liquid head (m).

Some expressions for the vessel cross-sectional area $A(h)$ are:

$$A = \frac{\pi}{4}D^2 \quad \text{vertical cylindrical vessel} \qquad [15.1.92]$$

$$A = 2L(hD - h^2)^{\frac{1}{2}} \quad \text{horizontal cylindrical vessel (flat ends)} \qquad [15.1.93]$$

$$A = \frac{\pi}{4}C^2 \quad \text{spherical vessel} \qquad [15.1.94]$$

with

$$C = 2(hD - h^2)^{\frac{1}{2}} \qquad [15.1.95]$$

$$A = \pi h^2 \tan^2\theta \quad \text{vertical conical vessel, tip down} \qquad [15.1.96]$$

where L is the length of the vessel (m) and θ is the half-angle of the vessel tip (°).

It may sometimes be helpful to derive the cross-sectional area from the volume using the relation

$$A = \frac{dV}{dh} \qquad [15.1.97]$$

where V is the volume of the vessel (m^3).

Vertical cylindrical vessel, finite imposed pressure

A number of specific cases are now considered. Crowl and Louvar (1990) have treated the case of a vertical cylindrical vessel with a finite imposed pressure on the liquid surface. For this case, Equation 15.1.91 yields

$$t = \frac{2A}{C_D A_0 2^{\frac{1}{2}}}[g(h-h_0) + \Delta P/\rho]^{\frac{1}{2}} \qquad [15.1.98]$$

where A is given by Equation 15.1.92. For the case where there is no imposed pressure $\Delta P = 0$:

$$t = \frac{2A}{C_D A_0 2 g^{\frac{1}{2}}}[g(h-h_0)]^{\frac{1}{2}} \qquad [15.1.99]$$

And for the case where the hole is at the bottom of the vessel ($h_o = 0$)

$$t = \frac{2A}{C_D A_0 2^{\frac{1}{2}}}(gh)^{\frac{1}{2}} \qquad [15.1.100]$$

Horizontal cylindrical vessel, no imposed pressure

The alternative case of a horizontal cylindrical vessel has been treated by Sommerfeld and Stallybrass (1993), who consider the case of a vessel with flat ends and no imposed pressure. Then from Equations 15.1.91 and 15.1.93

$$t = \frac{2L}{C_D A_0 (2g)^{\frac{1}{2}}} \int_{h_0}^{h} \frac{[(D-h)h]^{\frac{1}{2}}}{(h-h_0)^{\frac{1}{2}}} dh \qquad [15.1.101]$$

The authors obtain an analytical solution, albeit a somewhat complex one.

Spherical vessel, no imposed pressure

The case of a spherical vessel with no imposed pressure has been described by Hart and Sommerfeld (1993). For this case, from Equations 15.1.91 and 15.1.94

$$t = \frac{2\pi}{15C_DA_0(2g)^{\frac{1}{2}}}(5Dh + 10Dh_0 - 3h^3 - 4hh_0 - 8h_0^2)(h - h_0)^{\frac{1}{2}} \quad [15.1.102]$$

For the case where the hole is at the bottom of the vessel ($h_o = 0$):

$$t = \frac{2\pi}{3C_DA_0(2g)^{\frac{1}{2}}}[(D - 3h/5)h^{\frac{3}{2}}] \quad [15.1.103]$$

Vessels with other geometries

K. S. Lee and Sommerfeld (1993) have obtained analytical solutions for vessels with a number of other geometries with no imposed pressure. These are conical, paraboloid and ellipsoid vessels, both with the tapering end at the bottom and with it at the top.

Vessel drainage times and flows

Expressions have just been given for vessel drainage time. The average drainage flow is then obtained as the initial volume of liquid in the vessel divided by the drainage time. Expression of the drainage relations in dimensionless form allows some more general conclusions to be drawn. K. S. Lee and Sommerfeld (1993) define a dimensionless drainage time τ as the ratio of the time to drain the vessel with the liquid at the actual height h and the hole at the actual height h_o to the time to drain a full vessel with the hole at the bottom (h_o = 0). They also define a dimensionless hole height x as the ratio of the hole height h_o to the vessel diameter D or height H.

They show that for certain vessel geometries the variation of the dimensionless drainage time τ with the dimensionless hole height x is such that it passes through a maximum. The geometries where this occurs are those where the cross-sectional area of the vessel $A(h)$ decreases as the height h_o decreases. They give a tabulation of values of the maximum dimensionless drainage time τ_{max}, which includes the following:

Vessel geometry	*Dimensionless drainage time maximum,* τ_{max}	*Corresponding dimensionless height of hole,* x
Vertical cylindrical	None	–
Horizontal cylindrical	1.161	0.17
Spherical	1.299	0.25
Cone, tip down	1.776	0.683
Paraboloid, tip down	1.414	0.50
Vertical ellipsoid	1.299	0.25
Horizontal ellipsoid	1.299	0.25

The sphere may be considered a special case of the ellipsoid.

The authors state that in no case does the maximum dimensionless drainage time attain a value of 2.

Vessels with pipe attached

Treatments for the case where the outflow is through a pipe attached to the vessel are given by Loiacono (1987), Sommerfeld (1990), Papas and Sommerfeld (1991) and Sommerfeld and Stallybrass (1992).

15.1.11 Model systems

There are available a number of model systems for determining the emission of fluid from vessels. These include those of the two *Canvey Reports* and the *Rijnmond Report* and of the Committee for the Prevention of Disasters (CPD), (1992a), Solberg and Skramstad (1982), Considine and Grint (1985), Mecklenburgh (1985) and Napier and Roopchand (1986). There are also the model systems associated with computer codes for hazard assessment such as WHAZAN and SAFETI.

15.2 Two-phase Flow

In some cases the fluid is neither a pure gas or vapour nor a pure liquid, but is a vapour–liquid mixture, and it is then necessary to use a correlation for two-phase flow.

The behaviour of fluids in two-phase flow is complex and by no means understood. Of the large literature on the topic mention may be made of the accounts in *One Dimensional Two-Phase Flow* (Wallis, 1969), *Two-Phase Flow and Heat Transfer* (Butterworth and Hewitt, 1977), in *Two-Phase Flow in Pipelines and Heat Exchangers* (Chisholm, 1983) and *Emergency Relief Systems for Runaway Reactions and Storage Vessels: A Summary of Multiphase Flow Methods* (AIChE, 1992/149) and of the work of Benjaminsen and Miller (1941), Burnell (1947), Lockhart and Martinelli (1949), Pasqua (1953), Schweppe and Foust (1953), O. Baker (1954, 1958), Isbin, Moy and da Cruz (1957), Zaloudek (1961), Fauske (1962, 1963, 1964), Isbin *et al.* (1962), Fauske and Min (1963), Dukler, Wicks and Cleveland (1964), Levy (1965), F.J. Moody (1965), Baroczy (1966), Chisholm and Watson (1966), Chisholm and Sutherland (1969–70), R. E. Henry and Fauske (1971), M.R.O, Jones and Underwood (1983, 1984), van den Akker, Snoey and Spoelstra (1983), Nyren and Winter (1983), B. Fletcher (1984a,b), B. Fletcher and Johnson (1984), Leung (1986a, 1990a,b, 1992a) and S.D. Morris (1988a,b, 1990a,b).

Work on two-phase flow, particularly in pipelines, has been carried out by the American Institute of Chemical Engineers (AIChE) Design Institute for Multiphase Processing (DIMP). An important distinction in two-phase flow is that between one-component systems, such as steam and water, and two-component systems, such as air and water. It is the former which is of prime concern here and, unless otherwise stated, it is to these

***Table 15.2** Selected references on two-phase flow*

ASME (Appendix 28 *Applied Mechanics*, 1984/53, 183); NRC (Appendix 28 *Two-Phase Flow*); Schiller (1933); Benjaminsen and Miller (1941); Burnell (1947); Lockhart and Martinelli (1949); W.F. Allen (1952–); Pasqua (1953); Schweppe and Foust (1953); O. Baker (1954, 1958); Lottes and Flynn (1956); Brigham, Holstein and Huntington (1957); Isbin, Moy and da Cruz (1957); Chisholm and Laird (1958); Hesson and Peck (1958); Fauske (1961, 1962, 1963, 1964, 1983, 1985a,b, 1986a–c, 1987a,b, 1988a); Zaloudek (1961); R. Brown and York (1962); Friedrich and Vetter (1962); Isbin *et al.* (1962); Fauske and Min (1963); D.S. Scott (1963); Dukler, Wicks and Cleveland (1964); Starkman *et al.* (1964); R.J. Anderson and Russell (1965–); Lacey (1965); Levy

(1965); F.J. Moody (1965); Baroczy (1966); Chisholm and Watson (1966); Gouse (1966); Min, Fauske and Petrick (1966); Romig, Rothfus and Kermode (1966); Uchida and Nariai (1966); Cruver and Moulton (1967); Paige (1967); Collier and Wallis (1968); R.E. Henry (1968, 1970); Hubbard and Dukler (1968); Ogasawara (1969); Simpson (1968, 1991); Chisholm and Sutherland (1969–70); Schicht (1969); Wallis (1969, 1980); Degance and Atherton (1970); R.E. Henry, Fauske and McComas (1970); Flinta, Gernborg and Adesson (1971); Greskovich and Shrier (1971); R.E. Henry and Fauske (1971); Klingebiel and Moulton (1971); Collier (1972); Beggs and Brill (1973); Choe and Weisman (1974); T.W. Russell *et al.* (1974); Cude (1975); Dukler and Hubbard (1975); Ishii (1975); Sozzi and Sutherland (1975); Isii, Chawla and Zuber (1976); Kopalinsky and Bryant (1976); Taitel and Dukler (1976); Ardron, Baum and Lee (1977); Butterworth and Hewitt (1977); Kevorkov, Lutovinov and Tikhonenko (1977); Ardron (1978); Wallis and Richter (1978); Fauske, Grolmes *et al.* (1980); Rassokhin, Kuzevanov and Tsiklauri (1980); Azbel (1981); Bergles, Collier and Delhaye (1981); Bergles and Ishigai (1981); Blackwell (1981); Chisholm (1981, 1983); B. Fletcher (1982, 1984a,b); Hewitt (1982); IBC (1982/31, 1984/55); Landis (1982); S. Levy Inc. (1982); Sabnis, Simmons and Teuscher (1982); Fernandes, Semiat and Dukler (1983); Azzopardi and Gibbons (1983); van den Akker, Snoey and Spoelstra (1983); El-Emam and Mansour (1983); Friedel and Purps (1983, 1984a,b); Grolmes and Leung (1983, 1984); Hutcherson, Henry and Wollersheim (1983); M.R.O. Jones and Underwood (1983, 1984); R. King (1983); Nyren and Winter (1983, 1987); Veziroglu (1983); AGA (1984/43, 1989/66); van den Akker (1984); Botterill, Williams and Woodhead (1984); Duiser (1984); Dukler and Taitel (1984); B. Fletcher and Johnson (1984); S.D. Morris and White (1984); Nicholson, Aziz and Gregory (1978); Soliman (1984); Berryman and Daniels (1985); Grolmes and Epstein (1985); Huff (1985, 1990); D.A. Lewis and Davidson (1985); Olujic (1985); D.A. Carter (1986 LPB 70, 1988); Friedel (1986, 1987a, 1988); S.C. Lee and Bankoff (1986); Leung (1986a, 1990a,b, 1992a); Orell and Rembrand (1986); Yamashiro, Espiell and Farina (1986); Bettis, Nolan and Moodie (1987); Cindric, Gandhi and Williams (1987); Leung and Grolmes (1987–, 1988); Ewan, Moodie and Harper (1988); Hardekopf and Mewes (1988); S.D. Morris (1988a,b, 1990a,b); First and Huff (1989); Leung and Fisher (1989); Nyren (1989); Eggers and Green (1990); Hague and Pepe (1990); Haque *et al.* (1990, 1992); Leung and Epstein (1990a,b, 1991); D.S. Nielsen (1991); Nolan, Pettit and Hardy (1991); Sumipathala, Venart and Steward (1990a,b); J.L. Woodward (1990, 1993); Haque, Richardson and Saville (1992); Lantzy (1992); Giot (1994); Khajehnajafi and Schinde (1994); Seynhaefve *et al.* (1994); Holland and Bragg (1995)
Control valves: Ziegler (1957); Hanssen (1961); Sheldon and Schuder (1965); Romig, Rothfus and Kermode (1966)

Pressure relief valves
Tangren, Dodge and Seifert (1949); Richter (1978); Sallet (1978, 1979a,b, 1984, 1990a–c); Sallet, Weske and Gühler (1979); Simpson, Rooney and Grattan (1979); Fauske *et al.* (1980); O.J. Cox and Wierick (1981); Zahorsky (1983); J.W. Campbell and Medes (1985); Fauske (1985b, 1990); Forrest (1985); Sallet and Somers (1985); DnV (1982 RP C202); Friedel and Kissner (1987, 1988); Banerjee (1988); Friedel (1988); Morris (1988a,b, 1990a,b); M. Epstein, Fauske and Hauser (1989); Friedel and Molter (1989); Morley (1989a,b); Alimonti, Fritte and Giot (1990); API (1990 API RP 520); Bilicki and Kestin (1990); Curtelin (1991); M.R. Davis (1991); Lemonnier *et al.* (1991); Selmer-Olsen (1991, 1992); Simpson (1991); Leung (1992a); Wehmeier, Westphal and Friedel (1994)
ISO 4126: 1991
Computer codes: Nylund (1983 DnV Rep 83–1317, 1984); Middleton and Lloyd (1984); Bayliss (1987); Klein (1987); Evanger *et al.* (1990); Haque, Richardson and Saville (1992); Haque *et al.* (1992)
Benchmark study: Skouloudis (1992)

Reactor venting (see also Table 17.13)
DIERS: Klein (1987); Wilday (1987); Fauske (1988a, 1989a); Leung and Fisher (1989)
Liquid swell: Grolmes (1983); Grolmes and Fauske (1983); Friedel and Purps (1984a,b)

Storage vessel venting
Fauske, Epstein *et al.* (1986)

Leaks through cracks
Pana (1976); Amos *et al.* (1983); Abdollahian *et al.* (1984); Collier (1984); John *et al.* (1986); Kefer *et al.* (1986); Friedel (1987b); Friedel and Westphal (1987, 1988, 1989)

---

systems that the account given here refers. Two situations which are of particular interest in the present context are the escape of a superheated, flashing liquid and the venting of a chemical reactor. Selected references on two-phase flow are given in Table 15.2.

15.2.1 Experimental studies

Much work on two-phase flow has been done on two-component systems, e.g. air and water, but for the two cases mentioned the more relevant work is that on one-component systems, e.g. steam and water. The bulk of the experimental work has been done on small pipes, often 1 in. diameter or less, but more recently work has been done on pipes of larger diameter. Details of some of the experimental work are given in Table 15.3.

15.2.2 Two-phase flow phenomena

Some characteristic features of, and problems in, two-phase flow are:

(1) flow pattern;
(2) phase equilibrium;
(3) phase velocity;
(4) critical flow;
(5) friction factor.

There is a variety of flow patterns in two-phase flow, the pattern depending on the mass flow and on the mass fraction of vapour, or quality, and also on whether the pipe is horizontal or vertical. The flow patterns are considered in the next section.

For very short flow outlets there may be insufficient time for relaxation to occur and equilibrium between the

Table 15.3 *Some experimental studies on two-phase flow*

Investigator(s)	*System fluid*[a]	*Geometry*
W. Schiller (1933)		
Benjaminsen and Miller (1941)	Single component	Orifices
Burnell (1947)	Single component	Orifices, nozzles, pipes
Lockhart and Martinelli (1949)	Two components	Horizontal pipes
Pasqua (1953)	Single component	
Schweppe and Foust (1953)	Single component	Vertical tubes
Brigham, Holstein and Huntington (1957)	Two components	Inclined pipes
Isbin, Moy and da Cruz (1957); Isbin *et al.* (1962)	Single component	
Chisholm and Laird (1958)	Two components	Horizontal pipes
Zaloudek (1961)	Single component	Pipes
Fauske (1962, 1963, 1964)	Single component	Horizontal pipes
Friedrich and Vetter (1962)	Single component	Nozzles, short tubes
Baroczy (1966)	Single and two component systems	
Chisholm and Watson (1966)	Single component	Orifices
Uchida and Nariai (1966)	Single component	Orifices, nozzles, pipes
Flinta, Gernborg and Adesson (1971)	Single component	Horizontal pipes
Beggs and Brill (1973)	Two components	Inclined pipes
Sozzi and Sutherland (1975)	Single component	Horizontal pipes
Kevorkov, Lutovinov and Tikhonenko (1977)	Single component	Horizontal pipes
Sallet (1979b); Sallet, Weske and Gühler (1979); Sallet and Somers (1985)	Single component	Safety valve
van den Akker, Snoey and Spoelstra (1983)	Single component	Vertical then horizontal section
Fernandes, Semiat and Dukler (1983)	Two components	Vertical tubes
Hutcherson, Henry and Wollersheim (1983)	Single component	Horizontal pipes
Nyren and Winter (1983)	Single component	Horizontal pipes
B. Fletcher (1984a,b); B. Fletcher and Johnson (1984)	Single component	Horizontal pipes
M.R.O. Jones and Underwood (1984)	Single component	Nozzles, short tubes
D.A. Lewis and Davidson (1985)	Single component	Orifices, nozzles
Friedel (1988)	Two components	Bursting disc, safety valve

[a] Single component systems are those with a single saturated or superheated liquid, typically water. Two component systems are those with a gas and a liquid, typically air and water.
[b] See also work of Fauske and coworkers (Section 15.3) and work on relaxation length (Table 15.4), on pressure relief valves (Section 15.6), on vessel blowdown (Section 15.7) and in the DIERS project (Chapter 17).

liquid and vapour phase to be established. In this case it is necessary to make some assumption about the degree of equilibrium attained. One limiting assumption is that of frozen flow in which there is no interchange between the phases. The alternative limiting assumption is that equilibrium is attained.

Another principal feature which needs to be defined is the relative velocity between the two phases. One limiting assumption is that of zero slip in which the relative velocity of the two phases is the same.

Critical flow occurs in two-phase flow but it is a much more complex phenomenon than in gas flow and it tends not to be well defined. The essential feature, however, is that, as in gas flow, the flow is independent of the downstream pressure.

For flow in long pipes it becomes necessary to take friction into account. Pressure drop due to friction is characterized in two-phase as in single-phase flow by a friction factor, but again this is somewhat more complex than in single-phase flow.

The types of two-phase flow calculation which are of interest here are primarily releases from plant and venting of reactors and vessels. An assumption which is conservative for one of these cases is not necessarily conservative for the other.

15.2.3 Two-phase flow patterns

There are a number of different flow patterns in two-phase flow. Among the numerous descriptions are those by O. Baker (1954, 1958), D.S. Scott (1963), Schicht (1969), Choe and Weisman (1974), Butterworth and Hewitt (1977), Chisholm (1983), and Cindric, Gandhi and Williams (1987).

The flow patterns for horizontal two-phase flow in pipelines were identified and correlated by O. Baker (1954, 1958). Figure 15.7, after Coulson and Richardson (1977–), shows a set of flow patterns denoted in accordance with Baker's terminology.

Figure 15.8 shows the correlation of flow patterns given by Baker in terms of the gas and liquid mass velocities G and L and the parameters λ and ψ, where the latter are defined as

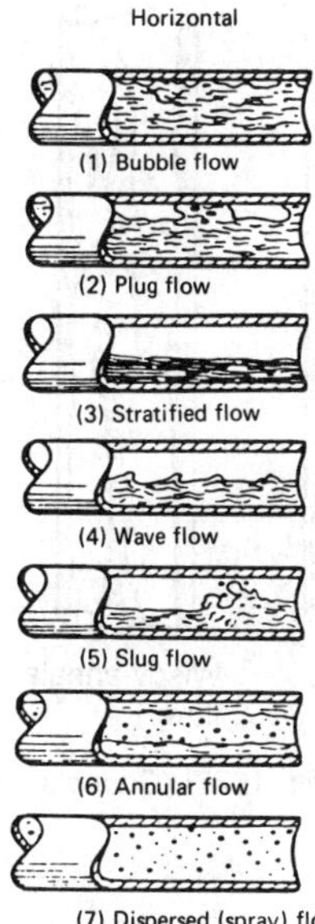

Figure 15.7 *Flow patterns in two-phase flow: horizontal flow (Coulson and Richardson, 1977–) (Courtesy of Pergamon Press)*

$$\lambda = \left[\frac{\rho_g \rho_l}{\rho_a \rho_w}\right]^{0.5} \qquad [15.2.1]$$

$$\psi = \frac{\sigma_w}{\sigma_l}\left[\frac{\mu_l}{\mu_w}\left[\frac{\rho_w}{\rho_l}\right]^2\right]^{0.33}$$

where μ is the viscosity, ρ is the density, σ is the surface tension, and the subscripts a, g, l and w denote air, gas, liquid, and water respectively.

The Baker curves have been widely used to correlate the flow patterns in two-phase flow. A set of equations for the Baker curves has been given by Yamashiro, Espiell and Farina (1986).

The flow patterns for horizontal two-phase flow given by Butterworth and Hewitt are similar except that they omit dispersed flow. Figure 15.9 shows the flow patterns for vertical two-phase flow and Figure 15.10 a flow pattern map given by Butterworth and Hewitt.

There is a considerable variation in the terminology used for the flow patterns in two-phase flow. A summary of the terminology and a classification has been given by Chisholm (1983).

A simplified classification scheme for horizontal two-phase flow has been proposed by Choe and Weisman (1974), following earlier work by Hubbard and Dukler (1968). They identify four flow patterns, which correlate as follows with mass velocity G and with mass fraction of vapour, or quality, x:

Flow pattern	*Mass velocity,* G (kg/m^2 s)	*Quality,* x
Homogeneous	$G > 2700$	0–1.0
Intermittent	$94 < G < 2700$	0–0.8
Annular	$94 < G < 2700$	>0.8 (± 0.1)
Separated	$G < 94$	0–1.0

In terms of the relations given by Chisholm between the usual terms and those of Choe and Weisman, homogeneous flow corresponds to bubble flow, intermittent flow to plug and slug flow and separated flow to stratified flow; annular flow is the same in both cases.

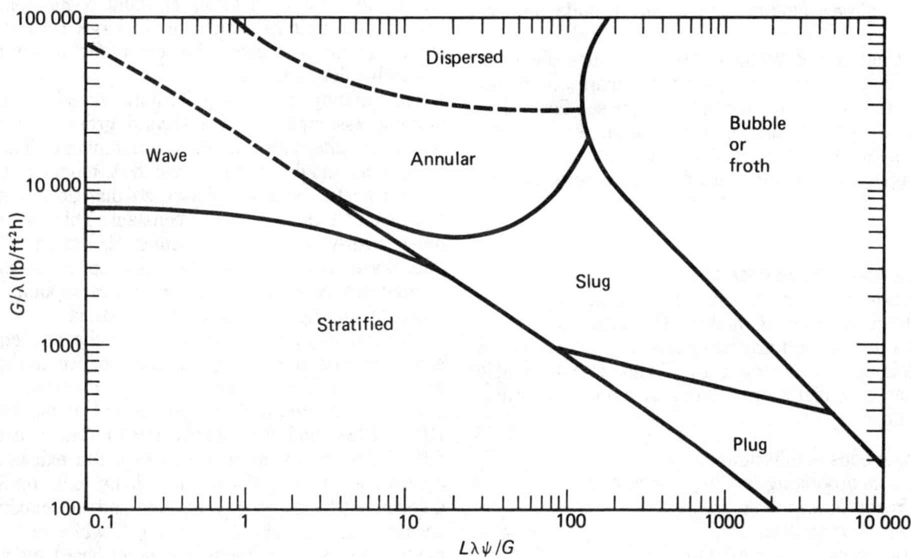

Figure 15.8 *Regimes in two-phase flow: the Baker diagram (O. Baker, 1954) (Courtesy of the Oil and Gas Journal)*

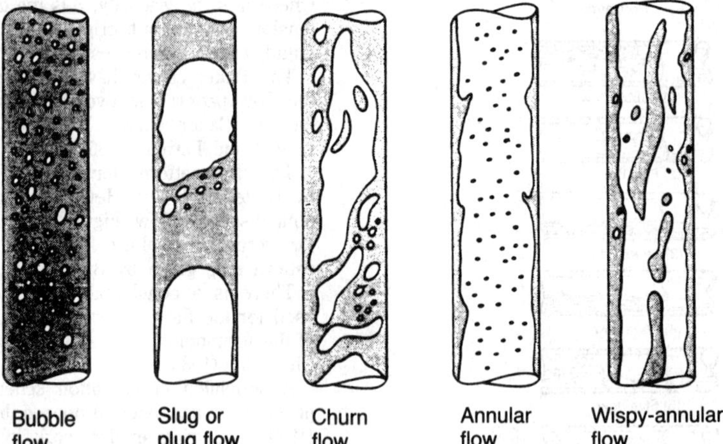

Figure 15.9 *Flow patterns in two-phase flow: vertical flow (Butterworth and Hewitt, 1977) (Courtesy of Oxford University Press)*

Olujic (1985) has given a classification of flow patterns in horizontal two-phase flow based on a distinction between high and low quality. He defines parameters α and β:

$$\alpha = \left(1 + \beta \frac{u_g}{u_l}\right)^{-1} \qquad [15.2.3]$$

$$\beta = \frac{1 - x}{x} \frac{\rho_g}{\rho_l} \qquad [15.2.4]$$

where u is the velocity. In the β regime the velocity of the two phases is approximately equal, while in the α regime that of the gas is much higher. The β regime corresponds to bubble and slug flows, and the α regime to wavy, slug and annular flows. The author gives a map of the two regimes, plotting $1/\beta$ vs the Froude number Fr.

Cindric, Gandhi and Williams (1987) discuss the work of DIMP and present flow pattern maps, one for horizontal and one for vertical two-phase flow. The former is a plot of certain flow parameters vs the Martinelli parameter (described below), and the latter a plot of superficial liquid velocity vs superficial gas velocity.

15.2.4 Modelling of two-phase flow

Two-phase flow is a complex phenomenon and has given rise to a large number of models. Of particular interest here are models for critical two-phase flow. A review of such models has been given by Wallis (1980). Wallis distinguishes essentially three types of model for critical two-phase flow:

(1) homogeneous equilibrium model;
(2) models incorporating limiting assumptions –
  (a) frozen flow models,
  (b) slip flow models,
  (c) isentropic stream tube models;
(3) non-equilibrium models –
  (a) empirical models,
  (b) physically based models for thermal equilibrium,
  (c) two fluid models.

The homogeneous equilibrium model (HEM) is based on the two limiting assumptions: that the flow is homogeneous, i.e. there is no slip between the two phases; and that the flow is in equilibrium, i.e. there is effectively perfect transfer between the two phases.

This is a well-established and widely used model; it is used in a number of computer codes. Wallis states that the model gives reasonably good predictions for the mass velocity under conditions where the pipe is long enough for equilibrium to be established and the flow pattern is such as to suppress relative motion. For short pipes where there is insufficient relaxation time the flow can be in error by a factor of about 5. For longer pipes with a flow pattern which allows large relative velocity, such as annular flow, the error factor tends to be somewhat less than 2.

The homogeneous equilibrium model is based on limiting assumptions. The second group of models are based on alternative limiting assumptions. The limiting assumption made in the frozen flow model is that there is no transfer between phases so that flow is effectively frozen and quality remains constant. This assumption is most nearly met if the outlet is short. While no interphase transfer is the key assumption in this model it may be combined with other assumptions such as those of no slip or isentropic expansion.

The limiting assumption made in the slip equilibrium model is that there is equilibrium between the phases, which is the direct opposite of the previous case. Two principal slip equilibrium models are those by Fauske (1962, 1963) and F.J. Moody (1965). These two models differ in terms of the way in which the exit condition is determined. In the Moody model the exit conditions are determined by an energy balance and the maximum flow occurs at a slip ratio $k = (\rho_l/\rho_g)^{\frac{1}{3}}$, while in the Fauske model the exit conditions are determined by a momentum balance and the maximum flow occurs at $k = (\rho_l/\rho_g)^{\frac{1}{2}}$.

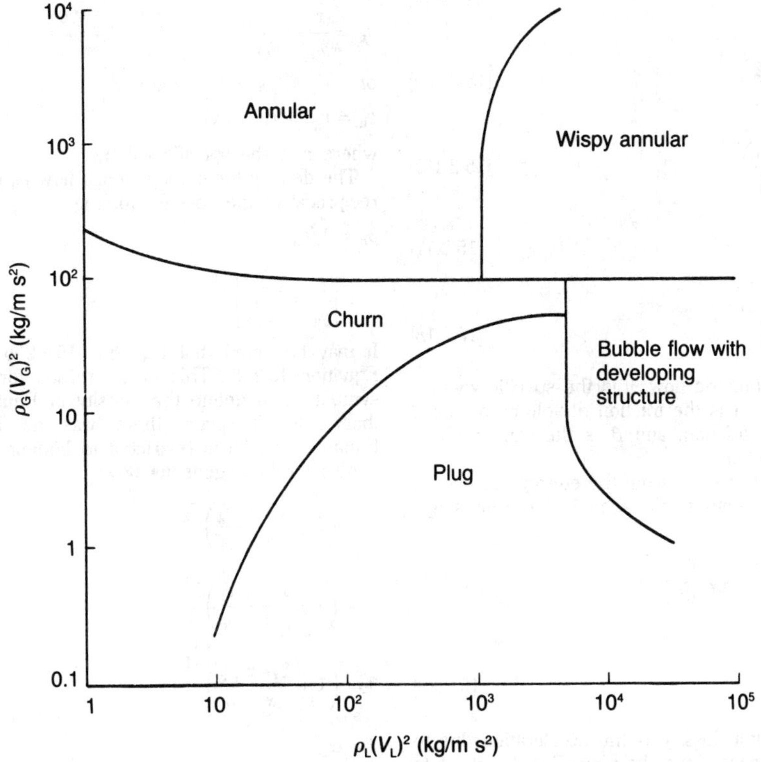

Figure 15.10 *Regimes in two-phase flow: flow patterns map for vertical flow (Butterworth and Hewitt, 1977) V_G, velocity of gas (m/s); V_L, velocity of liquid (m/s); ρ_G, density of gas (kg/m^3); ρ_L, density of liquid (kg/m^3). (Courtesy of Oxford University Press)*

The slip equilibrium models just described treat the velocity ratio k as a parameter to be adjusted to obtain the maximum flow, without indicating how this condition is actually achieved. Wallis and Richter (1978) have described an isentropic stream tube model, based on isentropic expansion of individual stream tubes originating from the vapour–liquid interface, which attempts to address this point.

The third group of models incorporate non-equilibrium effects. One approach used to handle such effects is the use of empirical correction factors. This is the basis of the model by R.E. Henry and Fauske (1971).

Other non-equilibrium models are based on physical descriptions of the processes of bubble nucleation and vapour generation. Wallis concludes that the development of such models is not such as to offer much improvement on the purely empirical models.

Another type of non-equilibrium model is the two fluid, or separated flow, model, in which effects such as interphase drag are taken into account, but again Wallis concludes that the stage of development is not such as to make these models more useful than the empirical ones.

15.2.5 Elementary relations

For a two-phase system the following equations apply for continuity and other relations:

$$W_g = Wx \qquad [15.2.5]$$

$$W_l = W(1 - x) \qquad [15.2.6]$$

$$G_g = u_g \rho_g \qquad [15.2.7]$$

$$G_g = \frac{Gx}{\alpha} \qquad [15.2.8]$$

$$G_l = \frac{G(1 - x)}{1 - \alpha} \qquad [15.2.9]$$

$$Q_g = \frac{Wx}{\rho_g} \qquad [15.2.10]$$

$$Q_l = \frac{W(1 - x)}{\rho_l} \qquad [15.2.11]$$

$$u_g = \frac{Gx}{\alpha \rho_g} \qquad [15.2.12]$$

$$u_l = \frac{G(1 - x)}{(1 - \alpha)\rho_l} \qquad [15.2.13]$$

$$k = \frac{u_g}{u_l} \qquad [15.2.14]$$

$$k = \frac{x}{1-x}\,\frac{1-\alpha}{\alpha}\,\frac{\rho_l}{\rho_g} \qquad [15.2.15]$$

$$\alpha = \left(1 + \frac{1-x}{x}\,k\,\frac{\rho_g}{\rho_l}\right)^{-1} \qquad [15.2.16a]$$

$$\alpha = \left(1 + \frac{1-x}{x}\,k\,\frac{v_l}{v_g}\right)^{-1} \qquad [15.2.16b]$$

$$\beta = \frac{Q_g}{Q_l + Q_g} \qquad [15.2.17]$$

$$\beta = \left(1 + \frac{1-x}{x}\,\frac{\rho_g}{\rho_l}\right)^{-1} \qquad [15.2.18]$$

where Q is the volumetric flow, v is the specific volume, W is the mass flow, α is the fraction of volume occupied by vapour, or void fraction, and β is the vapour phase volumetric flow fraction.

The pressure drop derived from the energy balance for two-phase flow, analogous to Equation 15.1.10b for single phase flow, is

$$\frac{dP}{dl} + \rho_h\,\frac{dF}{dl} + G\,\frac{du}{dl} + \rho_h g\,\frac{dz}{dl} = 0 \qquad [15.2.19]$$

with

$$\rho_h = \left[\frac{x}{\rho_g} + \frac{1-x}{\rho_l}\right]^{-1} \qquad [15.2.20]$$

where F is the friction loss, g is the acceleration due to gravity, l is the distance along the pipe, P is the absolute pressure, z is the vertical distance, and subscript h denotes homogeneous flow.

The pressure drop derived from the momentum balance for two-phase flow, analogous to Equation 15.1.11 for single phase flow, may be written as

$$\frac{dP}{dl} + \frac{S}{A}R_0 + \frac{d[G_g\alpha u_g + G_l(1-\alpha)u_l]}{dl} + g[\alpha\rho_g + (1-\alpha)\rho_l]\frac{dz}{dl} = 0 \qquad [15.2.21]$$

where A is the cross-sectional area of the pipe, R_0 is the shear stress at the pipe wall and S is the perimeter of the pipe.

Comparing Equations 15.2.19 and 15.2.21, which each give the pressure drop, it follows that, in a manner analogous to the derivation of Equation 15.1.12,

$$\frac{S}{A}R_0 = \rho_h\,\frac{dF}{dl} \qquad [15.2.22]$$

As with Equation 15.1.19 for single-phase flow, the pressure drop given in Equation 15.2.19 may also be expressed in terms of its constituent elements, as given in Equation 15.1.13.

15.2.6 Homogeneous flow

If the gas and liquid velocities are equal and the slip ratio k therefore unity, the flow is said to be homogeneous. The specific volume for homogeneous flow v_h is defined as

$$v_h = \frac{Q_g + Q_l}{W} \qquad [15.2.23]$$

and hence

$$v_h = \frac{x}{\rho_g} + \frac{1-x}{\rho_l} \qquad [15.2.24]$$

or

$$v_h = v_g x + v_l(1-x) \qquad [15.2.25]$$

where v is the specific volume.

The density for homogeneous flow ρ_h is defined as the reciprocal of the specific volume

$$\rho_h = 1/v_h \qquad [15.2.26a]$$

$$= \left[\frac{x}{\rho_g} + \frac{1-x}{\rho_l}\right]^{-1} \qquad [15.2.26b]$$

It may be noted that Equation 15.2.26b is the same as Equation 15.2.20. This is the reason for the use of the symbol ρ_h to define the density in Equation 15.2.20. In that case, however, there was no implication that Equation 15.2.20 is restricted to homogeneous flow.

Also for homogeneous flow:

$$\alpha = \left(1 + \frac{1-x}{x}\,\frac{\rho_g}{\rho_l}\right)^{-1} \qquad [15.2.27a]$$

$$\alpha = \left(1 + \frac{1-x}{x}\,\frac{v_l}{v_g}\right)^{-1} \qquad [15.2.27b]$$

$$\beta = \left(1 + \frac{1-x}{x}\,\frac{\rho_g}{\rho_l}\right)^{-1} \qquad [15.2.28]$$

$$\beta = \alpha \qquad [15.2.29]$$

For homogeneous flow, since $\rho_h = 1/v_h$, Equation 15.2.19 may be written as

$$\frac{dP}{dl} + \frac{1}{v_h}\,\frac{dF}{dl} + G\,\frac{du}{dl} + \frac{g}{v_h}\,\frac{dz}{dl} = 0 \qquad [15.2.30]$$

15.2.7 Non-homogeneous flow

If the gas and liquid velocities are different, the flow is non-homogeneous. For non-homogeneous flow the specific volume of the two-phase mixture v_m is not equal to the reciprocal of the density of the two-phase mixture ρ_m:

$$v_m \neq 1/\rho_m \qquad [15.2.31]$$

The specific volume for non-homogeneous flow is often defined as

$$v_m = [\rho_g\alpha + \rho_l(1-\alpha)]^{-1} \qquad [15.2.32]$$

The use of this relationship for specific volume has been criticized by Fauske (1962). His alternative treatment is described in the following section.

15.2.8 Critical flow

As an introduction to critical two-phase flow, the condition is derived for critical flow in homogeneous flow. The acceleration pressure drop term in Equation 15.1.13 may be written as

$$-\frac{dP_a}{dl} = G^2\,\frac{dv_h}{dl} \qquad [15.2.33]$$

But

$$v_h = f(P, x) \qquad [15.2.34]$$

Hence

$$\frac{dv_h}{dl}=\frac{dv_h}{dP}\frac{dP}{dl}+\frac{dv_h}{dx}\frac{dx}{dl} \qquad [15.2.35]$$

But, from Equation (15.2.24)

$$\frac{dv_h}{dx}=v_g-v_l \qquad [15.2.36]$$

Then, from Equations 15.2.33, 15.2.33 and 15.2.36

$$-\frac{dP}{dl}=-\frac{dP_f}{dl}+G^2\left[\frac{dv_h}{dP}\frac{dP}{dl}+(v_g-v_l)\frac{dx}{dl}\right]+\frac{g}{v_h}\frac{dz}{dl} \qquad [15.2.37a]$$

$$=\frac{-\dfrac{dP_f}{dl}+G^2(v_g-v_l)\dfrac{dx}{dl}+\dfrac{g}{v_h}\dfrac{dz}{dl}}{1+G^2\dfrac{dv_h}{dP}} \qquad [15.2.37b]$$

where dP_f/dl is the pressure drop due to friction.

The critical condition occurs where the pressure gradient becomes infinite

$$1+G^2\frac{dv_h}{dP}=0 \qquad [15.2.38a]$$

$$-\frac{dv_h}{dP}=\frac{1}{G^2} \qquad [15.2.38b]$$

15.2.9 Empirical friction correlations

The solution of the equations for pressure drop in two-phase flow requires a relation for the friction loss term. Much effort has been devoted to obtaining a correlation for this. Typically, the two-phase pressure drop has been correlated in terms of some single phase pressure drop. One pair of relations used for correlation is

$$\frac{dP_f}{dl}=\phi_g^2\left(\frac{dP_f}{dl}\right)_g \qquad [15.2.39a]$$

$$\frac{dP_f}{dl}=\phi_l^2\left(\frac{dP_f}{dl}\right)_l \qquad [15.2.39b]$$

where dP_f/dl is the actual two-phase pressure drop, $(dP_f/dl)_g$ is the pressure drop which would occur if the gas phase were flowing alone in the pipe, $(dP_f/dl)_l$ is the pressure drop which would occur if the liquid phase were flowing alone and $\phi_g{}^2$ and $\phi_l{}^2$ are friction parameters.

Another pair of relations are

$$\frac{dP_f}{dl}=\phi_{go}^2\left(\frac{dP_f}{dl}\right)_{go} \qquad [15.2.40a]$$

$$\frac{dP_f}{dl}=\phi_{lo}^2\left(\frac{dP_f}{dl}\right)_{lo} \qquad [15.2.40b]$$

where $(dP_f/dl)_{go}$ is the pressure drop which would occur if the whole fluid were flowing as gas in the pipe, $(dP_f/dl)_{lo}$ is the pressure drop which would occur if the whole fluid were flowing as liquid and $\phi_{go}{}^2$ and $\phi_{lo}{}^2$ are further friction parameters.

Lockhart and Martinelli (1949) introduced the parameter

$$X^2=\frac{(dP_f/dl)_l}{(dP_f/dl)_g} \qquad [15.2.41]$$

Chisholm and Sutherland (1969–70) utilized the parameter

$$\Gamma^2=\frac{(dP_f/dl)_{go}}{(dP_f/dl)_{lo}} \qquad [15.2.42]$$

This parameter was also used by Baroczy (1966) in the form

$$\frac{1}{\Gamma^2}=\frac{\rho_g}{\rho_l}\left(\frac{\mu_l}{\mu_g}\right)^{0.2} \qquad [15.2.43]$$

The friction factor may be expressed in a generalized form of the Blasius equation

$$f \propto \mathrm{Re}^{-n} \qquad [15.2.44]$$

where n is an index. Then the parameters X and Γ are related to the quality x as follows:

$$X^2=\frac{1}{\Gamma^2}\left[\frac{1-x}{x}\right]^{2-n} \qquad [15.2.45]$$

15.2.10 Lockhart–Martinelli model

A method for determining the pressure drop in two-phase, two-component flow has been given by Lockhart and Martinelli (1949), who correlated $\phi_g{}^2$ and $\phi_l{}^2$ with X^2. The Equations of the model are Equations 15.2.39 and 15.2.41. The relations given by Lockhart and Martinelli between the parameters ϕ_g, ϕ_l and X are shown in Figure 15.11.

The method is to calculate the single-phase gas and liquid Reynolds numbers Re_g and Re_l and pressure drops $(dP_f/dl)_g$ and $(dP_f/dl)_l$, to calculate X from Equation 15.2.41, to obtain from Figure 15.11 either ϕ_g or ϕ_l, depending on the value of X, and then to calculate (dP_f/dl) from Equation 15.2.39a or 15.2.39b. The curves given in Figure 15.11 for ϕ_g and ϕ_l depend, as shown, on whether the gas and liquid flow regimes are viscous or turbulent.

Equations for the Lockhart-Martinelli parameters, which may be used instead of Figure 15.11, have been given by Degance and Atherton (1970). Although it is one of the earlier two-phase flow correlations, the method of Lockhart and Martinelli remains one of the most widely used.

15.2.11 Other friction correlation methods

Baroczy (1966) obtained a correlation in terms of ϕ_{lo}, x and the group which is defined as $1/\Gamma^2$ in Equation 15.2.43. A further more economical correlation was derived by Chisholm and Sutherland (1969–70). The correlation depends on the value of n in relation 15.2.44. For the case of $n = 0$ their correlation is

$$\phi_g^2=1+CX+X^2 \qquad [15.2.46a]$$

$$\phi_l^2=1+\frac{C}{X}+\frac{1}{X^2} \qquad [15.2.46b]$$

with

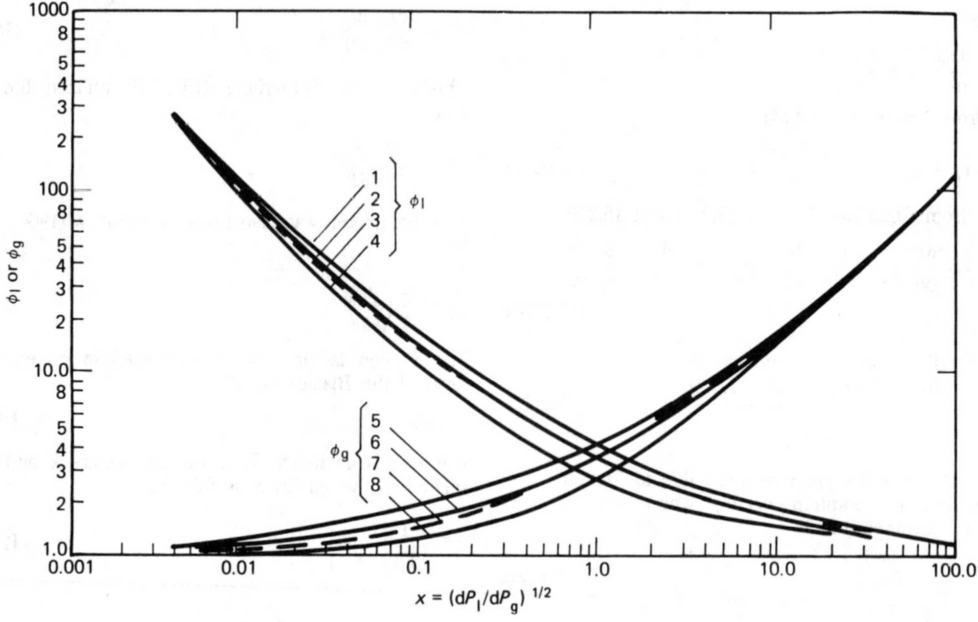

Reynolds no.	*Regime*			
	Turbulent–turbulent	Viscous–turbulent	Turbulent–viscous	Viscous viscous
Re_l	>2000	<1000	>2000	<1000
Re_g	>2000	>2000	<1000	<1000
Curve for ϕ_l	1	2	3	4
Curve for ϕ_g	5	6	7	8

If $X \geqslant 1.0$ use ϕ_l
If $X < 1.0$ use ϕ_g

Figure 15.11 *Pressure gradient in two-phase flow: the Lockhart–Martinelli correlation (after Lockhart and Martinelli, 1949) (Courtesy of the American Institute of Chemical Engineers)*

$$C = \frac{1}{k}\left(\frac{\rho_l}{\rho_g}\right)^{\frac{1}{2}} + k\left(\frac{\rho_g}{\rho_l}\right)^{\frac{1}{2}} \qquad [15.2.47]$$

The parameter C is a function of the mass velocity G and the parameter Γ.

15.2.12 Friction factor

The friction factor to be used in two-phase flow equations depends on the particular correlation. Use is often made of a single-phase flow friction factor, obtained from one of the relations given in Section 15.1. In some cases the approach taken is even simpler. The friction factor recommended by Perry and Green (1984) for use with Equation 15.2.48 below is the single value of 0.003.

15.2.13 DIMP project

A study of non-flashing two-phase flow has been conducted by the Design Institute for Multi-phase Processing (DIMP) of the AIChE. An account of the DIMP work has been given by Cindric, Gandhi and Williams (1987).

15.2.14 Homogeneous flow models

Much of the recent development in two-phase flow which is of interest here relates to homogeneous equilibrium flow models.

15.2.15 Homogeneous equilibrium flow model

The homogeneous equilibrium model has been described by a number of workers. The minimal assumptions in this model are that the phase velocities are equal and that thermal equilibrium exists.

From Equation 15.1.18 but substituting v_h for v it is possible to derive for a horizontal pipe

$$G = \left[\frac{-1}{\dfrac{dv_h}{dP} + \dfrac{4fv_h}{2d(dP/dl)}}\right]^{\frac{1}{2}} \qquad [15.2.48]$$

This appears to be the form most commonly quoted (e.g. Perry and Green, 1984).

For critical flow the following relation is given by Fauske (1962). From Equation 15.2.38b utilizing the definition of v_h in Equation 15.2.25 and assuming isentropic expansion, the critical mass velocity G_c is

$$G_c = \left[\frac{-1}{x\left(\frac{dv_g}{dP}\right)_s + (v_g - v_l)\left(\frac{dx}{dP}\right)_s + (1-x)\left(\frac{dv_l}{dP}\right)_s}\right]^{\frac{1}{2}} \quad [15.2.49]$$

where subscript s indicates constant entropy.

A third form of the model is that given by Starkman *et al.* (1964), as quoted by R.E. Henry and Fauske (1971). Again the additional assumption of isentropic expansion is made:

$$G_c = \frac{\{2[h_0 - (1 - x_E)h_{lE} - x_E h_{gE}]\}^{\frac{1}{2}}}{(1 - x_E)v_{lE} + x_E v_{gE}} \quad [15.2.50]$$

where h is the specific enthalpy and subscripts c, E and o indicate critical, equilibrium and stagnation conditions, respectively.

15.2.16 Homogeneous frozen flow model

The homogeneous frozen flow model has also been described by a number of workers. In the form given by R.E. Henry and Fauske (1971), the model is based on the two necessary assumptions for this type of model, that the phase velocities are equal and that there is no transfer between phases so that thermal equilibrium does not exist, but also on the assumption that the vapour expansion is isentropic and that the critical flow is determined by gas dynamic principles. The expression for the critical mass velocity is

$$G_c = \frac{1}{v}\left[2x_0 v_{g0} P_0 \frac{\gamma}{\gamma - 1}(1 - \eta_c{}^{(\gamma-1)/\gamma})\right]^{\frac{1}{2}} \quad [15.2.51]$$

where γ is the ratio of gas specific heats, η_c is the critical pressure ratio and subscript o indicates the stagnation condition. The critical pressure ratio is the ratio of the throat pressure to the downstream pressure, and the expression given for it is relatively complex.

15.2.17 Two-phase flashing flow

The two-phase flow which is of particular interest here is single-component, two-phase flashing flow. This is the type of flow which occurs when liquefied gas is released to the atmosphere, whether in a controlled manner through a relief device or due to an accident. A number of models have been developed for two-phase flashing flow, particularly through the work of Fauske and co-workers in the DIERS project, as described below.

The type of model characteristic of this work is a homogeneous equilibrium model for a very short pipe. The presence of the pipe makes it possible to assume equilibrium conditions, but the pipe is sufficiently short that frictional effects can still be neglected. Allowance is then made for frictional pressure drop by the use of empirical correlations.

The assumption of negligible pipe length leads to much simpler expressions for the critical flow. Initially, the length of pipe at which equilibrium is established was assumed to be defined by a length/diameter, or l/d, ratio. As described below, it has been shown that equilibrium is reached as a fixed length of some 0.1 m.

Two models, or rather families of models, of the type described are those of Fauske and of Leung. These are described in Sections 15.3 and 15.4, respectively.

15.2.18 Maximum critical flow

The maximum possible flow of a two-phase fluid in a pipe is obtained if it is assumed that friction can be neglected and that the energy released by the drop in pressure is converted to kinetic energy. At the critical flow conditions $dG/dP = 0$ or, alternatively, $dG/d\eta = 0$.

For a horizontal pipe Equation 15.1.77 yields $v dP = u du$ ($dW_s = dF = dz = 0$). Noting that the velocity $u = Gv$, it can readily be shown that at the critical flow conditions ($dG/dP = 0$) the maximum, or critical, mass velocity G_c is given by the relation $G_c^2 = -(dP/dv)$, where the pressure P and the specific volume v refer to the outlet conditions.

Critical flow conditions can occur at quite a high value of the ratio of the downstream pressure to the upstream pressure, or critical pressure ratio. A value as high as 0.7 in some cases is quoted by Coulson and Richardson (1977–, p. 95). The Leung model, described below, includes correlations for the critical pressure ratio.

15.2.19 Relaxation length

As already mentioned, it was initially believed that in two-phase flow the length of pipe over which equilibrium becomes established, or relaxation length, is a function of the length/diameter ratio. However, it has been shown in a number of studies that it is the absolute length which matters.

Work designed specifically to elucidate this point has been described by B. Fletcher (1984a,b) and B. Fletcher and Johnson (1984). Experiments were carried out on discharges of Refrigerant 11 through sharp edged orifices and tubes in the diameter range 3.2–10.8 mm and length/diameter ratio (l/d) range 0.88–200. In this work it was found that the onset of flashing was favoured by higher excess pressure, and hence superheat, and by increased tube length. It was also found that as the tube length was increased the flow at first decreased rapidly but then much more gradually. The transition point marking the end of the rapid decrease in flow with length occurred at a length of about 75 mm. Figure 15.12 shows the jet in some of Fletcher and Johnson's experiments.

At the time when this work was done one of the most widely used models for two-phase flow was the empirical model of Fauske (1964)

$$G = C_d[2\rho_l(P_0 - P_c)]^{\frac{1}{2}} \qquad l/d < 3 \quad [15.2.52]$$

where C_d is the coefficient of discharge, P_c is the absolute critical outlet pressure and P_o is the absolute upstream pressure.

As described below, Fletcher and Johnson are somewhat critical of Equation 15.2.52. The correlation preferred is based on the assumption that for a given length

$$G \propto C_d P_0^{\frac{1}{2}} \quad [15.2.53]$$

Figure 15.13 shows the results of discharge experiments correlated in this way. Figure 15.13(a) shows Fletcher's own results for Refrigerant 11 and

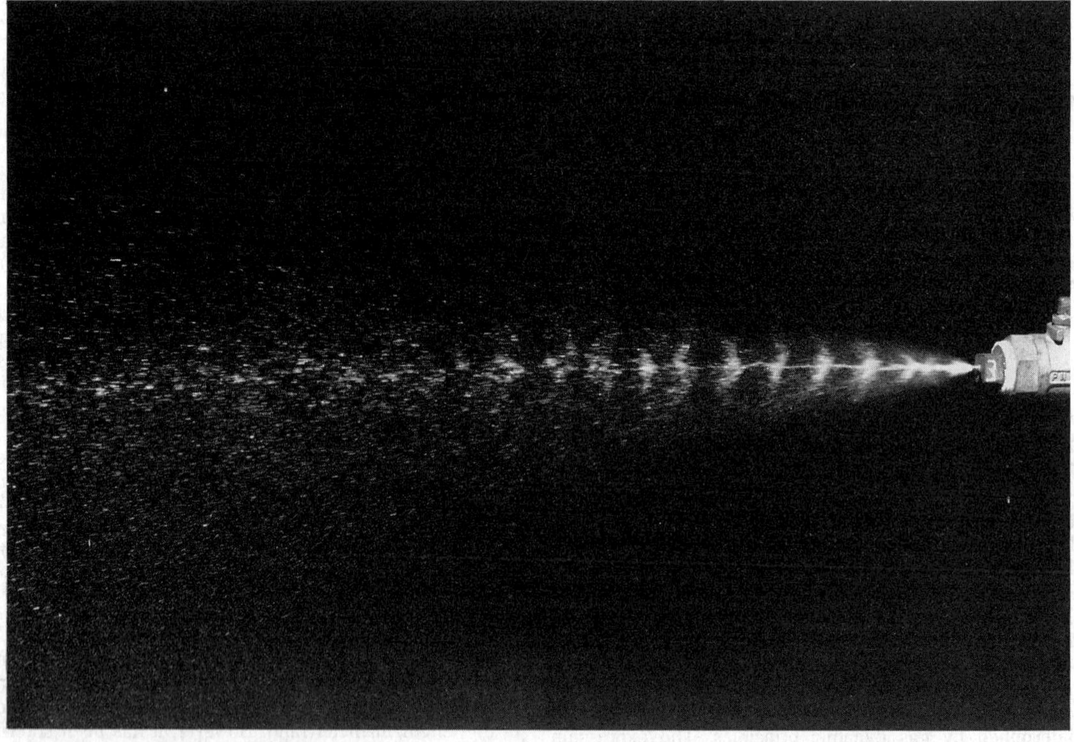

(a)

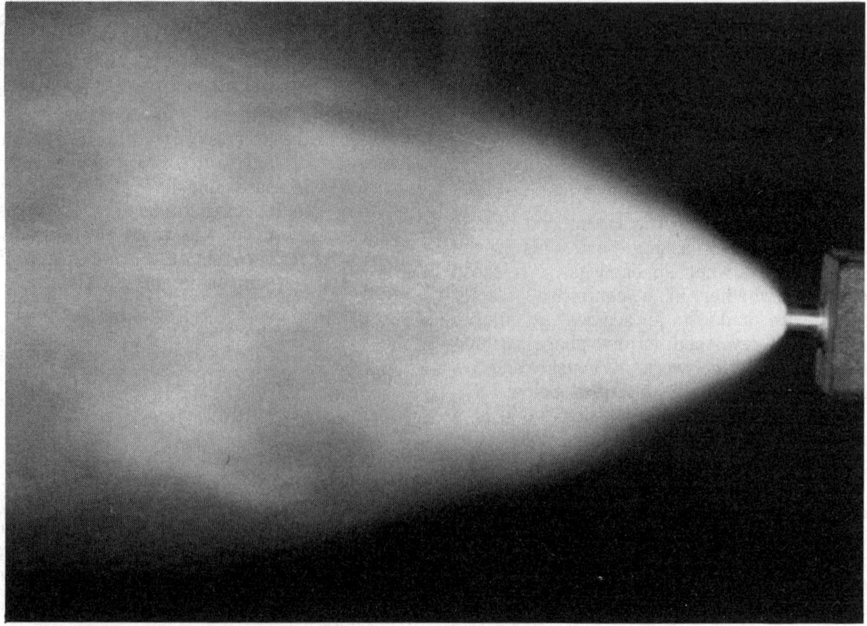

(b)

Figure 15.12 Discharge of Refrigerant 11 through 3.2 mm pipe (B. Fletcher and Johnson, 1984): (a) excess pressure of 0.7 bar; (b) excess pressure of 2.5 bar (Courtesy of HM Stationery Office)

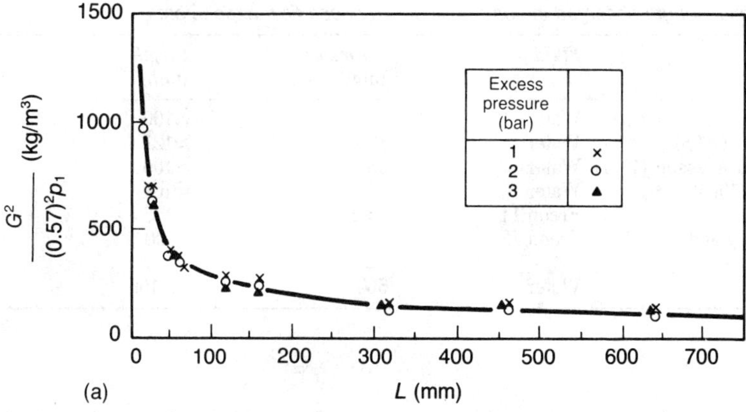

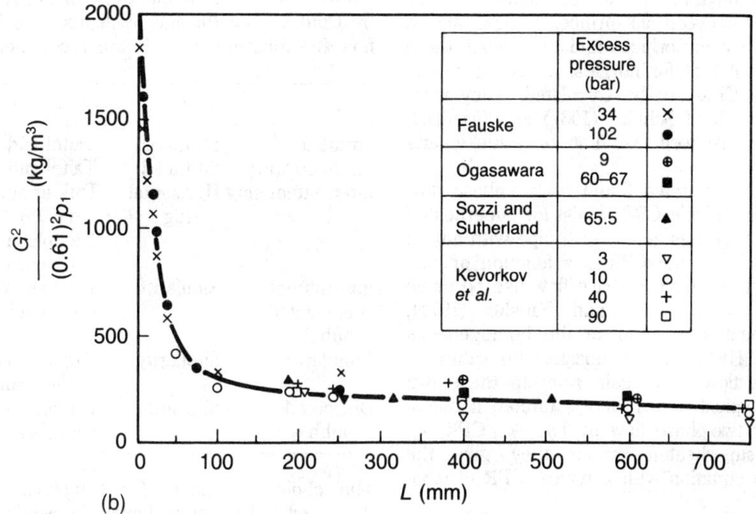

***Figure 15.13** Correlations for discharge of Refrigerant 11 and of saturated water (B. Fletcher and Johnson, 1984): (a) Refrigerant 11; (b) saturated water G, mass velocity (kg/m^{2}s); p_1 upstream pressure (N/m^2) (Courtesy of HM Stationery Office)*

Figure 15.13(b) shows the results of other workers. Thus in this work relation 15.2.53 is used for excess pressures in the range 1–102 bar.

Fletcher points out that the transition distance of 75 mm found in his work corresponds in Fauske's work to a length/diameter ratio of 12, a ratio which Fauske treats as a transition point.

Fletcher discusses the point at which all liquid flow ceases and flashing begins. He refers to the assumption in the *Rijnmond Report* that all liquid flow can be assumed to exist for $l/d \leq 2$ and states that he has observed flashing at $l/d < 1$ if the excess pressure is sufficiently high.

In presenting their own correlation, Fletcher and Johnson state certain criticisms of Equation 15.2.52. Two of these have already been discussed: the use of l/d instead of l as a correlator and the assumption of all liquid flow for $l/d \leq 2$. Another is the assumption that $P_c = 0.55P_o$ for all fluids. They refer to the use in the Second *Canvey Report* of Equation 15.2.52, but point out that this applies strictly only to the all gas, or vapour, condition.

Confirmation that the relaxation length is a function of the absolute length of pipe for large pipe diameters is provided by the work at Marviken by S. Levy Inc. (1982). The available data on relaxation length has been reviewed by Fauske (1984c), who quotes the data shown in Table 15.4. The later models of Fauske incorporate the absolute length method of correlation. This method is now generally adopted. The value of the relaxation length is taken as 0.1 m.

15.2.20 Critical pressure ratio

As the foregoing account indicates, a crucial parameter in two-phase flow is the critical pressure ratio (CPR), defined as the ratio of the critical pressure to the

***Table 15.4** Relaxation length observed in some critical two-phase flow experiments*

Source	*Fluid*	*Diameter* (mm)	*Length* (mm)	*Length/ diameter*
Fauske (1964)	Water	6.35	≈100	≈16
Sozzi and Sutherland (1975)	Water	12.7	≈127	≈10
Flinta, Gernborg and Adesson (1971)	Water	35	≈100	≈3
Uchida and Nariari (1966)	Water	4	≈100	≈25
B. Fletcher (1984)	Freon 11	3.2	≈105	≈33
van den Akker, Snoey and Spoelstra (1983)	Freon 12	4	90	≈22
S. Levy Inc. (1982)[a]	Water	500	< 166	< 0.33

[a] Marviken data.

upstream, or stagnation pressure. A review of the CPR is given by Hardekopf and Mewes (1988).

For the fluid they distinguish between a saturated liquid, a two-phase mixture and a superheated vapour and for the outlet between an orifice, a pipe and a nozzle. They state that for orifices and long pipes there is no single CPR, but that for nozzles, including outlets with short pipes, a CPR can be calculated. They refer particularly to the work of Schiller (1937) and Friedrich and Vetter (1962), as well as that of Fauske and coworkers.

For an orifice with saturated liquid or two-phase flow there is no readily calculable CRP, whilst for superheated vapour they give the CPR as zero. For a pipe with any of the three fluid conditions the CPR is a function of pipe length; they suggest that for two-phase flow use be made of the model by R.E. Henry and Fauske (1971), described in the next section, or of the homogeneous equilibrium model (HEM). For a nozzle with saturated liquid or two-phase flow they again refer to these two models and quote typical values for a saturated liquid of CPR ≈0.85 and for two-phase flow of 0.85 < CPR < 0.55, whilst for superheated vapour they give the adiabatic expansion equation with a typical CPR of 0.55.

15.2.21 Two-phase flow transients

The two-phase flow models described apply to steady-state conditions. The steady state is preceded and followed by transients. A treatment of such transients in pipelines is given by Seynhaefve *et al.* (1994).

15.2.22 DIERS project

Two-phase flow was a major aspect of the project by the Design Institute for Emergency Relief Systems (DIERS) of the AIChE. Although it was not intended that the project should produce new relations for two-phase flow, much work was done to investigate the suitability of the various correlations, and new methods have resulted. This work is described in the DIERS *Technical Summary* and in a series of papers by Fauske, Epstein, Grolmes, Leung and co-workers.

15.2.23 Choice of model

With regard to choice of model, it is relevant to distinguish between one-component, two-phase flashing flow, mainly in controlled or accidental releases from plant, and two-component, two-phase non-flashing flow, mainly of fluids in pipelines. It is the former which is of prime interest here.

Guidance on the choice of model for non-flashing two-phase flow based on the work of DIMP has been given by Cindric, Gandhi and Williams (1987). For horizontal flow the following models are recommended:

Flow	*Model*	
Stratified	Stratified	Taitel and Dukler (1976)
Stratified wavy	Stratified	Taitel and Dukler (1976)
Intermittent slug	Horizontal slug	Dukler and Hubbard (1975) Nicholson, Aziz and Gregory (1978)
Intermittent elongated bubble	Similarity	Duckler, Wicks and Cleveland (1964)
Annular	Similarity	Dukler, Wicks and Cleveland (1964)
Dispersed bubble	Similarity	Dukler, Wicks and Cleveland (1964)

The choice of model for two-phase flashing flow is discussed below, in particular in Sections 15.3 and 15.4 which deal with families of models of Fauke and Leung, Section 15.6 which treats the model given by Leung for pressure relief valves and Chapter 17 which covers venting of reactors.

For reactor venting Fauske (1985b) suggests: for orifices, the non-equilibrium model; for short outlets, the equilibrium rate model (ERM); and for longer outlets, where friction is significant, the homogeneous equilibrium model (HEM).

15.2.24 Design conservatism

In considering what constitutes a conservative assumption, it is necessary to make a number of distinctions. In determining the flow through a relief such as a pressure relief valve or a reactor vent, for the purpose of sizing the relief, it is conservative to use a method which tends to underestimate the flow. However, this approach is not conservative with respect to aspects such as the reaction forces on the vessel, the pressure drop in the relief header or the flow through the relief disposal train, for all of which it is conservative to use a method which tends to overestimate the flow.

For an accidental release to atmosphere it is generally conservative to use a method which tends to overestimate the flow. However, this may not always be true. For example, if the release gives a jet flame, a smaller flame of longer duration could be a worse case.

15.2.25 Two-phase flow in valves

Two-phase flashing flow occurs in valves, both control valves and pressure relief valves. Accounts of methods used to calculate such flow through control valves have been given by W.F. Allen (1952–), Ziegler (1957), Hanssen (1961), Sheldon and Schuder (1965) and Romig, Rothfus and Kermode (1966). Flow through pressure relief valves is considered in Section 15.6.

15.2.26 Two-phase flow through cracks

It is convenient to deal at this point with flow, particularly two-phase flow of a flashing liquid, through cracks in pressure vessels such as may be relevant to the leak-before-break condition. A number of workers have studied the leak rate through such cracks and have modelled it using various two-phase flow models (e.g. Amos *et al.*, 1983; Abdollahian *et al.*, 1984; R. P. Collier, 1984; John *et al.*, 1986; Kefer *et al.*, 1986). The two-phase flow models used by the workers cited are principally those of R.E. Henry (1970), Pana (1976) and Levy (1982).

The work has been reviewed by Friedel (1987b) and Friedel and Westphal (1987, 1988, 1989). They observe that there are a number of features, such as viscosity changes, surface tension and wetting effects, and non-equilibrium conditions, which current models tend not to account for and that no model can be regarded as universally valid, even for water.

15.3 Two-phase Flow: Fauske Models

15.3.1 Fauske empirical model

An early empirical model for two-phase flow was that described by Fauske (1964). The model is relatively simple and readily adaptable, and has been widely used. The general form of the model is

$$G = C_d[2\rho_l(P_0 - P_*)]^{\frac{1}{2}} \qquad [15.3.1]$$

where C_d is the coefficient of discharge, G is the mass velocity, P_o is the absolute stagnation pressure, P_* is the effective downstream pressure and ρ_l is the density of the liquid.

A correlation for the transition from single-phase to two-phase flow in fluid flowing from a vessel through an aperture or short pipe to atmosphere has been given by Min, Fauske and Petrick (1966). The transition is correlated in terms of a modified cavitation number C_a':

$$C_a' = \frac{2\Delta P}{\rho_l U^2}\frac{l}{d} \qquad [15.3.2]$$

where d is the diameter of the pipe, l is the length of the pipe, ΔP is the pressure difference, U is the average velocity of fluid and ρ_l is the density of the saturated or subcooled fluid in the vessel. Flow is single-phase for values of C_a' less than 9 and two-phase for values greater than 15. At values between 9 and 15, unstable transitional flow occurs.

The empirical model given by Fauske (1964) is based on experimental work on the flow of a saturated liquid from a vessel through an aperture or an aperture connected to a short pipe to atmosphere. For a system in which the aperture was a sharp edged orifice, the following equations were obtained:

$$G = C_d[2\rho_l(P_0 - P_b)]^{\frac{1}{2}} \qquad [15.3.3a]$$

$$G = C_d[2\rho_l(P_0 - P_c)]^{\frac{1}{2}} \qquad l/d < 3 \qquad [15.3.3b]$$

where subscripts b and c denote the downstream and critical values, respectively. The value of the coefficient of discharge was 0.61.

Equation 15.3.3a applies to an orifice without a pipe and Equation 15.3.3b to an orifice with a pipe attached. For this latter case the critical exit pressure P_c was found to be a function of the length/diameter ratio l/d of the pipe. Fauske considered separately the regions $3 < l/d < 12$ and $12 < l/d < 40$. He found that in the latter region there was very little decrease in flow with l/d, indicating a relatively small friction effect. His data over the two regions may be represented by the approximate equation

$$\frac{P_c}{P_0} = 0.55[1 - \exp(-l/3d)] \qquad [15.3.4]$$

Equation 15.3.23 was derived from experiments on saturated water flowing through a sharp-edged orifice of 0.25 in. diameter with an upstream pressure of up to 2000 psi. This model has been used extensively in industry to determine the two-phase flow from pipe ruptures. An account of such application has been given by Simpson (1971).

Equation 15.3.3 is that used in the Second *Canvey Report* for estimation of two-phase flow from pipe ruptures. The report uses for the critical pressure ratio equation 15.1.69.

15.3.2 Relaxation length

As described earlier, Fauske (1965) correlated his results in terms of the length/diameter ratio. Subsequent work indicates, however, that it is the absolute length l rather than the length/diameter ratio l/d which is the appropriate correlator and this position has been adopted by Fauske.

15.3.3 Jones and Underwood model

The Fauske empirical model has been widely used and has been subject to various adaptations. M.R.O. Jones and Underwood (1983) refer to Fauske's empirical model and point out that the use of an average density ρ_m implies the assumption of homogeneous equilibrium flow. They present as an alternative relations derived from the theory for critical annular flow given by Fauske (1961):

$$G = C_{DM}\left[\frac{\rho_l \rho_n}{\rho_l - \rho_n}(P_o - P_c)\right]^{\frac{1}{2}} \qquad [15.3.5]$$

with

$$\rho_n = \left[\frac{x_c^2}{\rho_g R_g} + \frac{(1 - x_c)^2}{\rho_l(1 - R_g)}\right]^{-1} \qquad [15.3.6]$$

$$R_g = \left[\frac{(1-x_c)}{x_c}\left(\frac{\rho_g}{\rho_l}\right)^{\frac{1}{2}} + 1\right]^{-1} \quad [15.3.7]$$

where C_{DM} is the modified coefficient of discharge, R_g is the fraction of the cross-sectional area occupied by vapour, x is the mass fraction of the vapour, and subscripts c, g, l, and n denote the pipe exit, vapour, liquid and two-phase annular value, respectively.

The authors describe experimental work on Freons 12, 22 and 114 discharging through pipes of diameter 0.5 and 4 mm and length/diameter ratios between 10 and 617. The authors found that the coefficient of discharge C_{DM} in Equation 15.3.5 was a function of the length/diameter ratio and obtained the relation

$$C_{DM} = 5.46\left(\frac{l}{d}\right)^{-0.77} + 0.29 \quad [15.3.8]$$

Equation 15.3.8 actually gives a coefficient of discharge greater than unity for $l/d \leq 15$ and the coefficient so obtained should be treated as a purely empirical one.

In a further study, M.R.O. Jones and Underwood (1984) investigated releases from apertures rather than full bore ruptures. Experiments were carried out using Freons and propane with pipes 4.3–4.5 mm in diameter and ending in a ball valve followed by one of the following apertures:

(1) circular aperture, d = 0.5–4.4 mm, $l/d \leq 3$;
(2) circular aperture, d = 0.5 mm, l/d = 10;
(3) rectangular aperture, height/width ratio 5.5, 10.2 and 25.6.

It might be expected that for pipes of cross-sectional area A_p and apertures of area A_a the flow would be proportional to the area ratio A_a/A_p, but the correlation obtained was in fact in terms of $(A_a/A_p)^{\frac{1}{2}}$. Three regions were observed:

$0 \leq (A_a/A_p)^{\frac{1}{2}} < 0.25$	Flow effectively metastable.
$0.25 \leq (A_a/A_p)^{\frac{1}{2}} < 0.9$	Ratio of actual flow to maximum flow proportional to $(A_a/A_p)^{\frac{1}{2}}$. Constant of proportionality 1.15.
$0.9 \leq (A_a/A_p)^{\frac{1}{2}} \leq 1.0$	Flow equal, or close, to maximum flow.

The authors warn, however, that their experiments were small scale and that caution should be exercised in extrapolating the results to the large scale.

15.3.4 Henry-Fauske non-equilibrium model

A non-equilibrium model for outlets with short pipes has been given by R.E. Henry and Fauske (1971). The model receives frequent mention. The assumption is made that flow is homogeneous but not that it is in equilibrium. The expansion of the vapour is assumed to be polytropic. The model gives the flow in terms of the stagnation conditions.

For critical two-phase flow the authors derive the relation

$$G_c^2 = \left(-\frac{d}{dP}\left\{\frac{[xk+(1-x)]}{k}[(1-x)kv_l + xv_g]\right\}\right)_t^{-1} \quad [15.3.9]$$

where k is the slip ratio, v is the specific volume and subscript t indicates the throat.

The equations of the model are:

$$G_c^2 = \frac{(1-x_o)v_{lo}(P_o - P_t) + [x_o\gamma/(\gamma-1)](P_o v_{go} - P_t v_{gt})}{[(1-x_o)v_{lo} + x_o v_{gt}]^2/2} \quad [15.3.10]$$

$$\eta = \left[\frac{\frac{(1-\alpha_o)}{\alpha_o}(1-\eta) + \frac{\gamma}{\gamma-1}}{\frac{1}{2\beta\alpha_t^2} + \frac{\gamma}{\gamma-1}}\right]^{\gamma/(\gamma-1)} \quad [15.3.11]$$

$$\beta = \left\{\frac{1}{n} + \left(1 - \frac{v_{lo}}{v_{gt}}\right)\left[\frac{(1-x_o)NP_t}{x_o(s_{gE} - s_{lE})_t}\frac{ds_{lE}}{dP}\right]_t - c_{pg}\frac{1/n - 1/\gamma}{s_{go} - s_{lo}}\right\} \quad [15.3.12]$$

$$\alpha_0 = \frac{x_o v_{go}}{(1-x_o)v_{lo} + x_o v_{go}} \quad [15.3.13]$$

$$\alpha_t = \frac{x_o v_{gt}}{(1-x_o)v_{lo} + x_o v_{gt}} \quad [15.3.14]$$

$$v_{gt} = v_{go}\eta^{-1/\gamma} \quad [15.3.15]$$

$$\eta = \frac{(1-x)c_l/c_{pg} + 1}{(1-x)c_l/c_{pg} + 1/\gamma} \quad [15.3.16]$$

$$N = \frac{x_{Et}}{0.14} \qquad x_{Et} \leq 0.14 \quad [15.3.17a]$$

$$N = 1 \qquad x_{Et} > 0.14 \quad [15.3.17b]$$

$$x_E = \frac{s_o - s_{lE}}{s_{gE} - s_{lE}} \quad [15.3.18]$$

where c is the specific heat, n is the polytropic index, N is a non-equilibrium parameter, s is the specific entropy, α is the void fraction, γ is the isentropic index and the subscript E denotes equilibrium (corresponding to local static pressure).

For N = 1 the model reduces approximately to the homogeneous equilibrium model, while for N = 0 it reduces approximately to the homogeneous frozen model. For inlet quality greater than x_o = 0.1, corresponding to throat quality x_{Et} in the range 0.125–0.155, with an assumed average of 0.14, the homogeneous equilibrium model is assumed, so that N = 1, as given in Equation 15.3.17b. For inlet qualities less than x_o = 0.1, Equation 15.3.17a is used.

15.3.5 Fauske slip equilibrium model

In the models described so far the usual assumptions are thermal equilibrium and homogeneous flow. Fauske (1962, 1963) has developed a slip equilibrium model in which the latter assumption is relaxed. The model is applicable essentially to annular flow, in which the vapour and liquid flows are largely separate so that the vapour velocity is likely to exceed that of the liquid. The starting point of the model is the momentum balance:

$$\mathrm{d}P + \mathrm{d}F' + \mathrm{d}(G_g u_g + G_l u_l) = 0 \quad [15.3.19]$$

where $\mathrm{d}F'$ is the friction term. Differentiating Equation 15.3.19 with respect to the quality x gives

$$\frac{\mathrm{d}P}{\mathrm{d}x} + \frac{\mathrm{d}F'}{\mathrm{d}x} + \frac{\mathrm{d}(G_g u_g + G_l u_l)}{\mathrm{d}x} = 0 \quad [15.3.20]$$

It is assumed that the friction term can be expressed as

$$\mathrm{d}F' = \frac{f^* G^2 v \, \mathrm{d}l}{2d} \quad [15.3.21]$$

where f^* is a fraction factor for two-phase flow.

Substituting for u_g, u_l and $\mathrm{d}F'$ in Equation 15.2.20 from Equations 15.2.12, 15.2.13 and 15.2.21 gives, eventually,

$$\frac{\mathrm{d}P}{\mathrm{d}l} + G^2 \left\{ \frac{\mathrm{d}}{\mathrm{d}l} \left[\frac{x^2 v_g}{\alpha_g} + \frac{(1-x)^2}{1-\alpha_g} \frac{v_l}{\rho_l} \right] + \frac{fv}{2d} \right\} = 0 \quad [15.3.22]$$

But the corresponding equation for single-phase flow is

$$\frac{\mathrm{d}P}{\mathrm{d}l} + G^2 \left[\frac{\mathrm{d}v}{\mathrm{d}l} + \frac{fv}{2d} \right] = 0 \quad [15.3.23]$$

where f is the friction factor.

Equations 15.3.22 and 15.3.23 become identical because the specific volume v is defined as

$$v = \frac{v_g x^2}{\alpha_g} + \frac{v_l (1-x)^2}{1-\alpha_g} \quad [15.3.24]$$

This definition of the specific volume is a crucial feature of the model.

In the model the slip ratio k is treated as a variable to be adjusted to obtain the maximum flow. The specific volume v can be expressed as a function of k by combining Equations 15.2.16b and 15.3.24 to give

$$v = \frac{[(1-x)v_l k + x v_g][1 + x(k-1)]}{k} \quad [15.3.25]$$

Differentiating Equation 15.3.22 with respect to k gives

$$G^2 \left[\frac{\delta}{\delta l} \left(\frac{\delta v}{\delta k} \right) + \frac{f}{2d} \frac{\delta v}{\delta k} + \frac{v}{2d} \frac{\delta f}{\delta k} \right] = 0 \quad [15.3.26]$$

In order to obtain from Equation 15.3.26 an expression for the maximum flow it is necessary to make some limiting assumptions. One is that of isentropic flow ($f = 0$, $\delta v/\delta k = 0$), but this is rejected because of the irreversibilities inherent in different phase velocities. Instead the assumption made is isenthalpic flow ($\delta f/\delta k = 0$, $\delta v/\delta k = 0$).

Differentiating Equation 15.3.25 with respect to k gives

$$\frac{\delta v}{\delta k} = (x - x^2) \left(v_l - \frac{v_g}{k^2} \right) \quad [15.3.27]$$

Then the values of k for which Equation (15.3.27) is equal to zero are

$$k = 1 \qquad x = 0, 1 \quad [15.3.28a]$$

$$k = \left(\frac{v_g}{v_l} \right)^{\frac{1}{2}} \qquad 0 < x < 1 \quad [15.3.28b]$$

Integrating Equation 15.3.22 over the length l of the pipe gives

$$\int_{P_o}^{P} \frac{\mathrm{d}P}{v} + G^2 \left(\ln \frac{v}{v_o} + \frac{f_m l}{2d} \right) = 0 \quad [15.3.29]$$

with

$$f_m = \int_{P_o}^{P} f \frac{\mathrm{d}}{\mathrm{d}P} (1/l) \, \mathrm{d}P \quad [15.3.30]$$

where f_m is the average friction factor.

Differentiating Equation 15.3.30 with respect to P and setting $\mathrm{d}G/\mathrm{d}P = 0$ and then combining the result with Equation 15.3.29 gives

$$G_c = \left[\frac{-1}{v[\mathrm{d}\ln v/\mathrm{d}P + (1/2d) \, \mathrm{d}f_m/\mathrm{d}P]} \right]^{\frac{1}{2}} \quad [15.3.31]$$

But

$$\left(\frac{\delta f_m}{\delta P} \right)_c = \frac{\delta f_m}{\delta k} \frac{\delta k}{\delta P} \quad [15.3.32]$$

and, since $\delta f_m/\delta k = 0$, it follows that $\delta f_m/\delta P = 0$.

Then from Equations 15.3.31 and 15.3.25

$$G_c = \left(\frac{-k}{\frac{\mathrm{d}}{\mathrm{d}P} \{[(1-x)v_l k + x v_g][1 + x(k-1)]\}} \right)^{\frac{1}{2}} \quad [15.3.33]$$

Completing the differentiation of the denominator of Equation (15.3.33) with respect to P gives

$$G_c = (-k/\zeta)^{\frac{1}{2}} \quad [15.3.34]$$

with

$$\zeta = [(1 - x + kx)x] \frac{\mathrm{d}v_g}{\mathrm{d}P} + [v_g(1 + 2kx - 2x) + v_l(2xk - 2k - 2xk^2 + k^2)] \frac{\mathrm{d}x}{\mathrm{d}P} + [k(1 + x(k-2) - x^2(k-1)] \frac{\mathrm{d}v_l}{\mathrm{d}P} \quad [15.3.35]$$

Equation 15.3.34 is the required relation for the critical two-phase flow.

The derivative terms in Equation 15.3.35 are calculated under isenthalpic conditions as follows:

$$\frac{\mathrm{d}v_g}{\mathrm{d}P} \approx \left(\frac{\Delta v_g}{\Delta P} \right)_h \quad [15.3.36]$$

$$\frac{\mathrm{d}v_l}{\mathrm{d}P} \approx \left(\frac{\Delta v_l}{\Delta P} \right)_h \quad [15.3.37]$$

$$\frac{\mathrm{d}x}{\mathrm{d}P} \approx -\frac{1}{h_{fg}} \left(\frac{\mathrm{d}h_f}{\mathrm{d}P} + x \frac{\mathrm{d}h_{fg}}{\mathrm{d}P} \right) \quad [15.3.38a]$$

$$\frac{\mathrm{d}x}{\mathrm{d}P} \approx -\frac{1}{h_{fg}} \left(\frac{\Delta h_f}{\Delta P} + x \frac{\Delta h_{fg}}{\Delta P} \right) \quad [15.3.38b]$$

where the subscripts f, fg and h denote the liquid, the liquid–vapour transition and constant enthalpy, respectively.

The method of calculation is as follows. For a given pressure P and given value of the quality x, the quantities v_g, v_l, h_{fg}, $\Delta v_g/\Delta P$, $\Delta v_l/\Delta P$, $\Delta h_f/\Delta P$ and $\Delta h_{fg}/\Delta P$ are obtained from thermodynamic tables. As an illustration, consider the calculation (in British units) given by Fauske for the maximum mass velocity of a steam–water mixture at a pressure of 600 psi. From steam tables

$$v_g = 0.7698$$

$$v_l = 0.0201$$

$$h_{fg} = 731.6$$

and for $\Delta P = 40$

$$\Delta v_g = 0.7440 - 0.7973$$

$$\Delta v_l = 0.0202 - 0.0201$$

$$\Delta h_f = 727.2 - 736.1$$

$$\Delta h_{fg} = 475.2 - 467.4$$

Hence

$$\frac{dx}{dP} = \frac{10^{-6}}{4.214}(-8.3 + 8.9x)$$

$$\frac{dv_g}{dP} = -9.2535 \times 10^{-6}$$

$$k = 6.19$$

Then the maximum mass velocity G_c for different values of quality x is

x	G_c
0.01	8960
0.05	7605
0.10	6510
0.20	5100
0.40	3570
0.60	2740
0.80	2235

The term (dv_l/dP) can often be neglected. In the calculation given for a value of the pressure P of 600 psi, the error in neglecting the term is 3%. At 400 psi the error reduces to 1%.

A further illustrative calculation has been given by Ramskill (1986 SRD R352) for an ammonia flow problem.

In the Fauske slip-equilibrium model the friction factor defined is relatively complex and is unique to that model, although this does not present a problem since it is eliminated in the manipulations and does not appear in the final equation for the critical mass velocity.

15.3.6 Fauske non-equilibrium model

A model for critical two-phase flow in short outlets which has both non-equilibrium and equilibrium versions has been given by Fauske (1985b). The model is suitable for venting of vessels such as reactors and storage vessels where the outlet is sufficiently short for friction to be neglected.

The general, or non-equilibrium, form of the model is:

$$G \approx \frac{h_{fg}}{v_{fg}}\left(\frac{1}{NTC}\right)^{\frac{1}{2}} \qquad [15.3.39]$$

with

$$N \approx \frac{h_{fg}^2}{2\Delta P \rho_l K^2 v_{fg}^2 TC} + l/l_e \qquad 0 < l/l_e < 1 \qquad [15.3.40]$$

where C is the specific heat of the liquid, K is the coefficient of discharge, l_e is the equilibrium or relaxation length, N is a non-equilibrium parameter and ΔP is the total available pressure drop. The discharge coefficient K is the usual coefficient of discharge; it is given by Fauske variously as 0.6 and 0.61. The relaxation length l_e is 0.1 m.

Equation 15.3.39 is applicable over the range $0 \leq l/l_e \leq 1$. For $l/l_e = 0$ it reduces to

$$G = K(2\Delta P \rho_l)^{\frac{1}{2}} \qquad [15.3.41]$$

with

$$\Delta P = P_o - P_a \qquad [15.3.42]$$

where P_a is the absolute atmospheric pressure.

15.3.7 Fauske equilibrium rate model

If, on the other hand, the pipe length is sufficient for equilibrium to be reached, which occurs at $l/l_e \geq 1$, then in equation [15.3.39] $N = 1$, and hence

$$G \approx \frac{h_{fg}}{v_{fg}}\left(\frac{1}{TC}\right)^{\frac{1}{2}} \qquad [15.3.43]$$

Fauske (1985b) terms Equation 15.3.43 the equilibrium rate model (ERM).

An alternative form of Equation 15.3.43 which is often more convenient is

$$G \approx \frac{dP}{dT}\left(\frac{T}{C}\right)^{\frac{1}{2}} \qquad [15.3.44]$$

Equation 15.3.44 gives a good estimate of two-phase flashing flow, provided the mass fraction x of vapour does not exceed a limiting value. The condition for its validity is given by Fauske and Epstein (1989) as

$$x < \frac{P v_{fg}}{h_{fg}^2} TC \qquad [15.3.45]$$

This condition implies a corresponding value of the void fraction α, since

$$\alpha = \frac{x v_g}{(1-x)v_f + x v_{fg}} \qquad [15.3.46]$$

15.3.8 Flow through a pipe

The equilibrium rate model applies essentially to flow through a nozzle just long enough for equilibrium conditions to be established but sufficiently short that frictional effects can be neglected. The flow is only weakly dependent on the length of the outlet, but if it is necessary to take friction into account, Fauske suggests two approaches. One is the use of the homogeneous equilibrium model.

The other approach given by Fauske (1989a) is the use of a flow correction factor F so that Equation 15.3.44 becomes

$$G = F\frac{dP}{dT}\left(\frac{T}{C}\right)^{\frac{1}{2}} \qquad [15.3.47]$$

with

Table 15.5 *Flow correction factor for Fauske equilibrium rate model (after Fauske, 1989a)*

l/d	F
0	1.0
50	0.85
100	0.75
200	0.65
400	0.55

$$F = f(l/d) \qquad [15.3.48]$$

In the context of venting of reactors with vapour pressure, or tempered systems, Fauske (1989a) has reported the values of F given here in Table 15.5.

15.3.9 Subcooled liquids

For flow of a liquefied gas which is initially subcooled Fauske (1985b) gives the following relations. For a liquid with a large degree of subcooling:

$$G = K[2(P_o - P_a)\rho_l]^{\frac{1}{2}} \qquad [15.3.49]$$

For a liquid with a small degree of subcooling, and thus in the transition zone between a saturated and strongly subcooled liquid:

$$G = \{2[P_o - P_v(T_o)]\rho_l + G_{ERM}^2\}^{\frac{1}{2}} \qquad l/l_e \geq 1 \qquad [15.3.50]$$

with

$$G_{ERM} = \frac{P_v}{P}\frac{dP}{dT}\left(\frac{T}{C}\right)^{\frac{1}{2}} \qquad [15.3.51]$$

where P_v is the absolute vapour pressure.

Equation 15.3.50 reduces to Equation 15.3.44 for zero subcooling and to Equation 15.3.49 for strong subcooling.

15.3.10 Source term models

From this family of models, Fauske (1985b) and Fauske and Epstein (1988) have summarized those which may be used for the source term of an accidental release. They are:

15.4 Two-phase Flow: Leung Models

Another family of models for two-phase flow is that given by Leung, another contributor to the DIERS project. The model of particular interest here is the homogeneous equilibrium model (HEM) for one-component two-phase flashing flow (Leung, 1986a). In addition to the basic model, the author has treated two-phase flashing flow in a horizontal duct (Leung, 1990a; Leung and Grolmes, 1987–), in an inclined duct (Leung, 1990a; Leung and Epstein, 1990a), with a subcooled liquid (Leung and Grolmes, 1988) and with non-condensable gas (Leung and Epstein, 1991), has presented a unified approach for nozzles and pipes (Leung, 1990b) and an overall comparison of flashing flow methods (Leung and Nazario, 1990). He has also given a model for two-phase non-flashing flow (Leung and Epstein, 1990b) and has shown the similarity between the flashing and non-flashing flow models (Leung, 1990a).

Applications of the model given by Leung include its use for: two-phase flow in storage vessels (Leung, 1986b); pressure relief valves (Leung, 1992a); and venting of reactors (Leung and Fauske, 1987; Leung and Fisher, 1989). These applications are described in Sections 15.5, 15.6 and Chapter 17, respectively.

15.4.1 Flow through a nozzle

The homogeneous equilibrium flow model of Leung (1986a) follows on from an earlier model by Grolmes and Leung (1984) which was restricted to an inlet condition of all liquid flow. This has been generalized by Leung to allow the handling of two-phase flow at the inlet.

The basic momentum and energy balance relations are:

$$v\,dP + G^2\left(v dv + 4fv^2\frac{dL}{2D}\right) = 0 \qquad [15.4.1]$$

$$h_o = h + \frac{G^2v^2}{2} \qquad [15.4.2]$$

where D is the diameter of the pipe, f is the friction factor, h is the specific enthalpy, L the length of the pipe and subscript o denotes stagnation. The situation considered is frictionless flow through a nozzle. This is treated as an isentropic process.

Liquid condition		*Equation*
Saturated stagnation	Orifice (non-equilibrium, frictionless, $l/l_e = 0$)	15.3.41
	Nozzle (non-equilibrium, frictionless, $0 < l/l_e < 1$)	15.3.39
	Nozzle/short pipe (equilibrium, frictionless, $l/l_e > 1$)	15.3.44
	Longer pipe (equilibrium, frictional effects, $l/l_e > 1$)	15.3.47
Weak subcooling	Nozzle (non-equilibrium, $l/l_e < 1$)	15.3.50 (with G_{ERM} replaced by G from 15.3.39)
	Longer pipe (equilibrium, $l/l_e > 1$)	15.3.50 (with G_{ERM} from 15.3.51)
Strong subcooling	Nozzle/longer pipe	15.3.49

For a nozzle ($L = 0$) Equation 15.4.1 becomes

$$G = (-\mathrm{d}v/\mathrm{d}P)^{-\frac{1}{2}} \quad [15.4.3]$$

The relations for specific enthalpy h, specific entropy s and specific volume v are:

$$h = h_f + xh_{fg} \quad [15.4.4]$$

$$s = s_f + xs_{fg} \quad [15.4.5]$$

$$v = v_f + xv_{fg} \quad [15.4.6]$$

where subscripts f and fg indicate the liquid and the liquid–vapour transition, respectively.

Equation 15.4.1 for the energy balance can be written as

$$G = \frac{[2(h_o - h)]^{\frac{1}{2}}}{v} \quad [15.4.7]$$

But

$$h_o - h = -\int_{P_o}^{P} v\mathrm{d}P \quad [15.4.8]$$

Then from Equations 15.4.7 and 15.4.8

$$G = \frac{\left(-2\int_{P_o}^{P} v\mathrm{d}P\right)^{\frac{1}{2}}}{v} \quad [15.4.9]$$

In order to solve Equation 15.4.9, use is made of an equation of state proposed by M. Epstein *et al.* (1983) for isentropic expansion from an all liquid stagnation condition:

$$\frac{v}{v_{fo}} = \omega\left(\frac{P_o}{P} - 1\right) + 1 \quad [15.4.10]$$

where ω is a parameter. Grolmes and Leung (1984) utilized Equation 15.4.10 to obtain the flow through a nozzle for an all liquid inlet condition.

For an isentropic expansion the correlating parameter ω in Equation 15.4.10 is

$$\omega = \frac{C_{fo}T_oP_o}{v_{fo}}\left(\frac{v_{fgo}}{h_{fgo}}\right)^2 \quad [15.4.11]$$

where C_f is the specific heat of the liquid and T is the absolute temperature.

The critical mass velocity G_c is

$$G_c = \eta_c\left(\frac{P_o}{\omega v_{fo}}\right)^{\frac{1}{2}} \quad [15.4.12]$$

where η_c is the critical pressure ratio.

The pressure ratio η is defined as

$$\eta = \frac{P}{P_o} \quad [15.4.13]$$

The critical pressure ratio η_c was found to be the value of η satisfying the relation

$$\eta^2 + (\omega^2 - 2\omega)(1-\eta)^2 + 2\omega^2 \ln \eta + 2\omega^2(1-\eta) = 0 \quad [15.4.14]$$

Further, for the all liquid inlet condition

$$\omega = \frac{P_o}{v_{fo}G_{lm}^2} \quad [15.4.15]$$

and hence from Equations 15.4.11 and 15.4.15

$$G_{lm} = \left(\frac{1}{C_{fo}T_o}\right)^{\frac{1}{2}}\frac{h_{fgo}}{v_{fgo}} \quad [15.4.16]$$

where G_{lm} is the mass velocity at this limiting condition.

Starting from these relations, Leung extends the treatment to the case of a two-phase inlet condition. Equation 15.4.10 becomes

$$\frac{v}{v_o} = \omega\left(\frac{P_o}{P} - 1\right) + 1 \quad [15.4.17]$$

which is also given as

$$\omega = \frac{(v/v_o - 1)}{(P_o/P - 1)} \quad [15.4.18]$$

and Equation 15.4.12 becomes

$$G_c = \eta_c\left(\frac{P_o}{\omega v_o}\right)^{\frac{1}{2}} \quad [15.4.19]$$

Introducing a dimensionless mass velocity G^*

$$G^* = \frac{G}{(P_o/v_o)^{\frac{1}{2}}} \quad [15.4.20]$$

the critical value G_c^* becomes

$$G_c^* = \frac{G_c}{(P_o/v_o)^{\frac{1}{2}}} \quad [15.4.21]$$

At this point the assumption is made that the process is isenthalpic. This approximation allows ω to be expressed more readily in terms of the stagnation properties. The relations for the specific enthalpy, specific entropy and specific volumes are:

$$h_o = h_{fo} + x_oh_{fgo} \quad [15.4.22]$$

$$s_o = s_{fo} + x_os_{fgo} \quad [15.4.23]$$

$$v_o = v_{fo} + x_ov_{fgo} \quad [15.4.24]$$

where x is the mass fraction of the vapour.

It is shown that with the isenthalpic assumption and certain other approximations:

$$\omega = \frac{x_ov_{fgo}}{v_o} + \frac{C_{fo}T_oP_o}{v_o}\left(\frac{v_{fgo}}{h_{fgo}}\right)^2 \quad [15.4.25]$$

Starting from Equation 15.4.19 and utilizing data for two-phase flow of water and nine other fluids, Leung obtains the following empirical correlation:

$$\frac{G_c}{(P_o/v_o)^{\frac{1}{2}}} = \frac{\eta_c}{\omega^{0.5}} \qquad \omega \geq 4.0 \quad [15.4.26a]$$

$$= \frac{0.66}{\omega^{0.39}} \qquad \omega < 4.0 \quad [15.4.26b]$$

The critical pressure ratio η_c was found to be the value of η satisfying the following relations. For a stagnation pressure of $P_o = 5$ bara

$$\eta = 0.6055 + 0.1356 \ln \omega - 0.131(\ln \omega)^2 \quad [15.4.27]$$

and for $P_o = 15$ bara

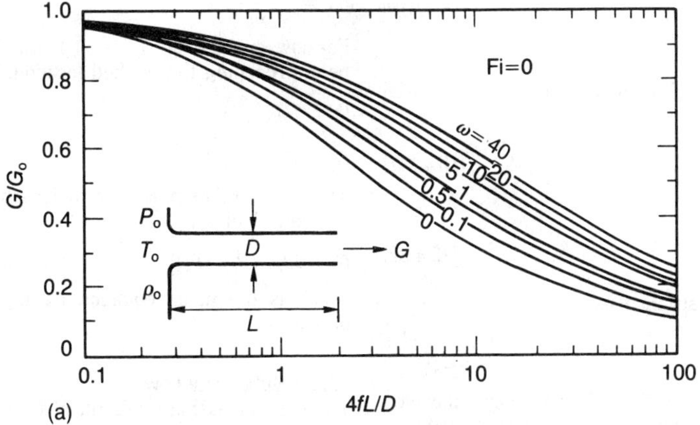

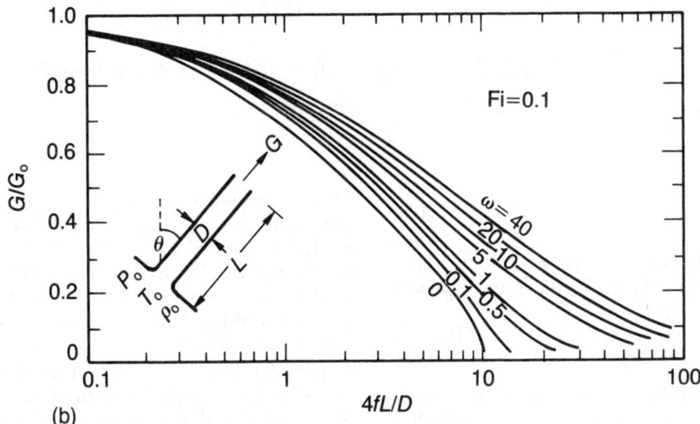

Figure 15.14 *Leung homogeneous equilibrium flow model: critical flow for discharge through pipe (Leung, 1990a): (a) horizontal pipe; and (b) inclined pipe Fi = 0.1 (Courtesy of the American Institute of Chemical Engineers)*

$$\eta = 0.55 + 0.217 \ln \omega - 0.046(\ln \omega)^2 + 0.004(\ln \omega)^3 \qquad [15.4.28]$$

This correlation was found to apply up to a reduced temperature $T_r = 0.9$ and reduced pressure $P_r = 0.5$.

The most general formulation of the parameter ω is:

$$\omega = \alpha_o + (1 - \alpha_o)\rho_{fo} C_{fo} T_o P_o \left(\frac{v_{fgo}}{h_{fgo}}\right)^2 \qquad [15.4.29]$$

where α is the void fraction. For the all liquid condition $\alpha_o = 0$, and for the all vapour condition $\alpha_o = 1$.

The parameter ω takes the following values:

$x_o = 1$	$\alpha_o = 1$ all vapour	$\omega = 1$
$0 < x_o < 1$	$\alpha_o < 1$ but flashing	$\omega > 1$
$0 < x_o < 1$	$\alpha_o < 1$ but non-flashing	$\omega = \alpha_o < 1$

The larger the value of ω, the greater the increase in specific volume upon depressurization and the more readily the flow attains the choking condition.

15.4.2 Flow through a pipe

The case of flow through a horizontal pipe as opposed to a nozzle has been treated by Leung and Grolmes (1987–), whilst Leung and Epstein (1990a) have considered flow through a pipe inclined upwards. The results of this work have been summarized by Leung (1990a).

For the inclined pipe use is made of the flow inclination number Fi defined as:

$$\text{Fi} = \frac{gD \cos \theta}{4fP_o v_o} \qquad [15.4.30]$$

where g is the acceleration due to gravity and θ is the angle between the pipe axis and the vertical.

Figure 15.14 shows the ratio G/G_o of the mass velocity though a pipe G to that through a nozzle G_o, or flow reduction factor. Figure 15.14(a) is for a horizontal pipe (Fi = 0) and Figure 15.14(b) for an inclined pipe with Fi = 0.1.

For the case of all liquid inlet flow ($\omega = 0$):

$$\frac{G}{G_o} = \left(\frac{1 - NFi}{1 + N}\right)^{\frac{1}{2}} \quad [15.4.31a]$$

$$= \frac{1}{(1+N)^{\frac{1}{2}}} \qquad Fi = 0 \text{ (horizontal pipe)} \quad [15.4.31b]$$

$$\rightarrow 0 \qquad NFi = 1 \quad [15.4.31c]$$

with

$$N = 4f\,\frac{L}{D} \quad [15.4.32]$$

where N is the pipe resistance factor.

15.4.3 Subcooled liquids

The case of flashing critical flow of an initially subcooled liquid is considered by Leung and Grolmes (1988). A more general relation which applies to such flow as well as flow of an initially two-phase mixture is

$$G^* = \frac{G}{(P_o/v_o)^{\frac{1}{2}}} = \frac{\{2(1-\eta_s) + 2[\omega\eta_s \ln(\eta_s/\eta) - (\omega - 1)(\eta_s - \eta)]\}^{\frac{1}{2}}}{\omega\left(\frac{\eta_s}{\eta} - 1\right) + 1} \quad [15.4.33]$$

with

$$\eta_s = \frac{P_s}{P_o} \quad [15.4.34]$$

where the subscript s denotes saturation.

Setting $dG^*/d\eta = 0$ gives for the critical mass velocity

$$G_c^* = \frac{\eta_c}{(\omega\eta_s)^{\frac{1}{2}}} \quad [15.4.35]$$

The flow behaviour depends on the degree of subcooling. The critical value η_{sc} separating the two regimes is

$$\eta_{sc} = \frac{2\omega - 1}{2\omega} \quad [15.4.36]$$

For low subcooling ($\eta_s \geq \eta_{sc}$), the fluid attains flashing before reaching the choked location. For high subcooling

$$\eta_c = \eta_s \quad [15.4.37]$$

$$G_c^* = [2(1 - \eta_s)]^{\frac{1}{2}} \quad [15.4.38]$$

or, from a relation similar to Equation 15.4.21 but with v_{fo} instead of v_o,

$$G_c = [2\rho_{fo}(P_o - P_s)]^{\frac{1}{2}} \quad [15.4.39]$$

which is the usual equation for liquid flow through an orifice.

15.4.4 Subcritical flow

The case of flashing subcritical flow is treated by Leung (1992a). A more general relation which applies to such flow as well as critical flow is

$$G^* = \frac{G}{(P_o/v_o)^{\frac{1}{2}}} = \frac{\{-2[\omega \ln \eta + (\omega - 1)(1 - \eta)]\}^{\frac{1}{2}}}{\omega\left(\frac{1}{\eta} - 1\right) + 1} \quad [15.4.40]$$

Setting $dG^*/d\eta = 0$ gives for the critical mass velocity

$$G_c^* = \frac{\eta_c}{\omega^{\frac{1}{2}}} \quad [15.4.41]$$

15.4.5 Non-flashing flow

The case of non-flashing critical flow has been dealt with by Leung and Epstein (1990b). Non-flashing two-phase flow is exemplified by two-phase flow of a two-component mixture such as air and water. A comparison of flashing and non-flashing flows is given by Leung (1990a).

Figure 15.15 shows the correlation given by Leung (1990a) for the dimensionless mass velocity G_c* and the critical pressure ratio η_c in two-phase critical flow in both the flashing and non-flashing regimes.

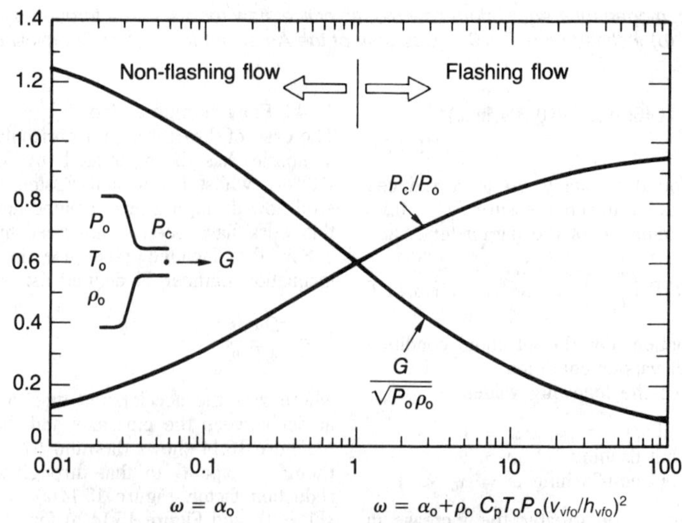

Figure 15.15 *Leung homogeneous equilibrium flow model: critical flow for discharge through a frictionless nozzle (Leung, 1990a) (Courtesy of the American Institute of Chemical Engineers)*

15.5 Vessel Depressurization

Some of the most complex fluid flow problems arise in the depressurization or venting of vessels, particularly reactors and storage vessels. In order to specify the venting arrangements for a vessel, it is necessary to determine the flow, the phase condition of fluid and the vent area required for this flow. The venting of vessels is complicated by the fact that in many situations the flow is likely to be two-phase. It is necessary, therefore, to be able both to estimate the vapour mass fraction, or quality, of the fluid entering the vent and the flow through the vent.

The need to determine the quality of the fluid vented has been highlighted during the work of the DIERS project. The quality of the fluid entering the vent is determined by liquid swell and vapour disengagement. There are two principal regimes which are recognized as occurring when a vessel is depressurized. If the liquid is non-foaming the regime tends to be churn turbulent, whereas if it is foaming the regime is bubbly.

For a non-foaming liquid, methods have been developed to predict the onset of two-phase flow. In the depressurization of a vessel containing such a liquid there will, in general, be a region in which the flow is all vapour and another region in which it is two-phase. A large initial vapour space, or freeboard, will favour all vapour flow. For a foaming liquid it is necessary to assume two-phase flow. Broadly, a pure liquid held in a storage vessel may well be non-foaming. It is to this situation that the methods developed principally apply.

An otherwise non-foaming liquid may be rendered foaming by the presence of impurities. Quite small quantities of a surface active agent may suffice to effect this. Since impurities are likely to be present in the liquid in a reactor, it is usual to treat a reaction mass as foaming.

In this section an account is given of models of liquid swell and vapour disengagement. Models have been given in the DIERS *Technical Summary* and by Fauske, Grolmes and Henry (1983), Swift (1984) and Grolmes and Epstein (1985). The application of these models to venting, particularly of storage vessels, is described in Chapter 17.

15.5.1 Liquid swell

Depressurization of a vessel containing superheated liquid or addition of heat to liquid in a vessel gives rise to the formation of vapour bubbles, which results in liquid swell. If this is severe, the level of the swollen liquid may reach the vent. But even if this is not so, there may be carryover of droplets into the vent. Liquid swell and vapour disengagement therefore affect the vapour mass fraction entering the vent. Some of the flow regimes which can occur on operation of the relief are shown in Figure 15.16.

For a liquid subject to swell, the average void fraction is

$$\bar{\alpha} = 1 - \frac{H_o}{H} \quad [15.5.1]$$

with

$$H_o = \frac{V_l}{A_x} \quad [15.5.2]$$

where A_x is the cross-sectional area of the vessel (m^2), H is the height of the liquid (m), H_o is the height of the liquid without swell (m), V_l is the volume of the liquid (m^3) and $\bar{\alpha}$ the average void fraction. A distinction is made between the average void fraction $\bar{\alpha}$ and the local void fraction α.

The initial void fraction in the vessel, or freeboard, α_o is

$$\alpha_o = 1 - \frac{H_o}{H_x} \quad [15.5.3]$$

where H_x is the height of the vessel (m).

15.5.2 Gas or vapour generation

Gas or vapour may be present in the liquid because gas is injected into the liquid or because gas or vapour is

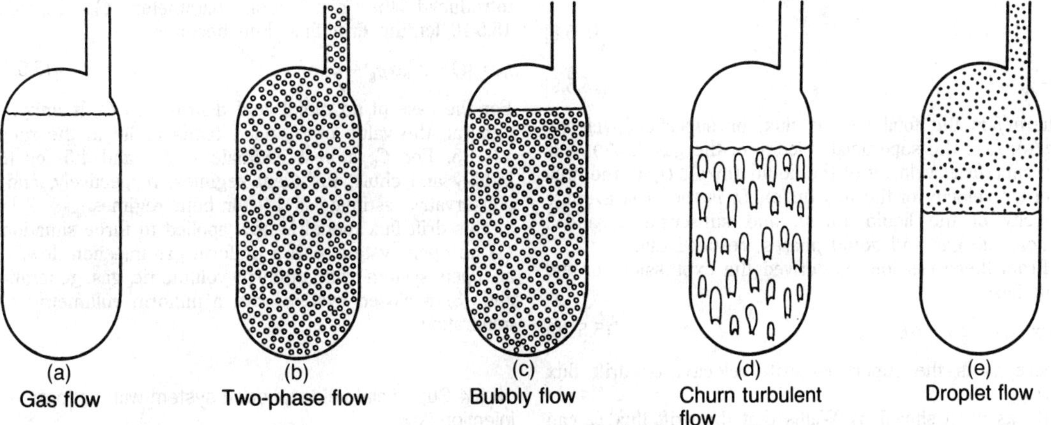

Figure 15.16 *Some flow regimes in venting of vessels (Swift, 1984) (Courtesy of the Institution of Chemical Engineers)*

generated in the liquid, due either to exothermic reaction or to an external heat source. The first case is that of uniform flux (UF) and the second that of a uniform volumetric source (VS). It is the latter which is of principal interest here.

The vapour superficial velocity above the height of the swelled liquid is

$$j_{g\infty} = \frac{S}{A_x} \qquad [15.5.4]$$

where $j_{g\infty}$ is the vapour superficial velocity (m/s) and S is the volumetric source strength (m^3/s).

A dimensionless source strength ψ is defined as

$$\psi = \frac{j_{g\infty}}{U_\infty} \qquad [15.5.5]$$

where U_∞ is the bubble rise velocity (m/s). The parameter ψ is also referred to as the 'dimensionless velocity'.

The volumetric source strength is a function of the heat input. For an external heat source

$$q = \frac{FQ_{ex}}{\rho_l D} \qquad [15.5.6]$$

where D is the vessel diameter (m), q is the heat input per unit mass (kW/kg), Q_{ex} is the external heat input flux (kW/m^2), ρ_l is the density of the liquid (kg/m^3) and F is a geometric factor. For the geometric factor F the *Technical Summary* gives for a sphere $F = 6$, for a vertical cylinder $F = 4$ and for a horizontal cylinder with hemispherical ends $F = (L/D + 1)/(3L/D + 2)$, where L/D is the length/diameter ratio.

15.5.3 Drift flux

A drift flux model for the behaviour of the liquid in a vessel subject to vapour generation has been described in the *Technical Summary*. The concept of drift flux has been described in work by Zuber and Findlay (1965) and Wallis (1969).

The local gas and liquid superficial velocities are:

$$j_g = \alpha U_g \qquad [15.5.7]$$

$$j_l = (1-\alpha)U_l \qquad [15.5.8]$$

$$j = j_g + j_l \qquad [15.5.9]$$

where j is the total volume flux, or superficial velocity (m/s), j_g is the superficial velocity of the gas (m/s), j_l is the superficial velocity of the liquid (m/s), U_g is the area average velocity of the gas (m/s), U_l is the area average velocity of the liquid (m/s), and subscripts g and l denote the gas and liquid phases, respectively.

From these relations is derived the expression for the drift flux:

$$j_{gl} = j_g(1-\alpha) - \alpha j_l \qquad [15.5.10]$$

where j_{gl} is the superficial drift velocity, or drift flux (m/s).

It has been shown by Wallis that the drift flux j_{gl} can be expressed as a function of the bubble rise velocity U_∞. The particular relation adopted in the *Technical Summary* differs slightly from that of Wallis and is

$$j_{gl} = \frac{\alpha(1-\alpha)^n U_\infty}{1-\alpha^m} \qquad [15.5.11]$$

where U_∞ is the bubble velocity (m/s) and m and n are indices. The velocity U_∞ is also variously termed the 'particle velocity' and the 'characteristic velocity'.

The bubble velocity is given by

$$U_\infty = \frac{k[\sigma g(\rho_l - \rho_g)]^{\frac{1}{4}}}{\rho_c^{\frac{1}{2}}} \qquad [15.5.12]$$

where ρ is the density (kg/m^3), σ is the surface tension of the liquid (N/m), k is a constant, and subscript c denotes the continuous phase.

The values of the indices m and n and of the constant k depend on the regime and are as follows:

	n	m	k
Bubbly regime	2	3	1.18
Churn turbulent regime	0	$\to \infty$	1.53

Substituting these values in Equation 15.5.11 gives

$$j_{gl} = \frac{\alpha(1-\alpha)^2}{1-\alpha^3} U_\infty \qquad \text{bubbly regime} \qquad [15.5.13]$$

$$j_{gl} = \alpha U_\infty \qquad \text{churn turbulent regime} \qquad [15.5.14]$$

A further drift flux is defined j_{lg}, complementary to j_{gl}, such that

$$j_{gl} + j_{lg} = 0 \qquad [15.5.15]$$

The two drift fluxes are given by

$$j_{gl} = \alpha(1-\alpha)(U_g - U_l) \qquad [15.5.16a]$$

$$j_{lg} = -\alpha(1-\alpha)(U_g - U_l) \qquad [15.5.16b]$$

where j_{lg} is the superficial drift velocity of the liquid phase relative to the gas phase (m/s).

The foregoing treatment does not take account of the radial distribution of the void fraction. In order to obtain better correlations Zuber and Findlay (1965) have introduced the distribution parameter C_o. Equation 15.5.10 for the drift flux then becomes

$$j_{gl} = (1 - C_o\alpha)j_g - C_o\alpha j_l \qquad [15.5.17]$$

For the case of no radial void distribution C_o is unity. In general, the value of C_o used tends to lie in the range 1.0–1.5. For C_o a best estimate is 1.2 and 1.5 for the bubbly and churn turbulent regimes, respectively, and a conservative estimate is 1.0 for both regimes.

This drift flux model is now applied to three situations: (1) an open system with a uniform gas injection flow, (2) an open system with uniform volumetric gas generation and (3) a closed system with a uniform volumetric gas generation.

15.5.4 Superficial velocity: open system with uniform injection flow

For an open system with uniform gas injection at the bottom of the vessel, the steady state condition is $j_l = 0$ and Equation 15.5.17 with Equation 15.5.11 then gives

$$j_g = \frac{\alpha(1-\alpha)^n U_\infty}{(1-\alpha^m)(1-C_o\alpha)} \quad [15.5.18]$$

But, since the gas flow is constant throughout the height, $j_g = j_{g\infty}$, and hence from Equation 15.5.5:

$$\psi = \frac{\alpha(1-\alpha)^n}{(1-\alpha^m)(1-C_o\alpha)} \quad [15.5.19]$$

Then for the two regimes

$$\psi = \frac{\alpha(1-\alpha)^2}{(1-\alpha^3)(1-C_o\alpha)} \quad \text{bubbly regime} \quad [15.5.20]$$

$$\psi = \frac{\alpha}{1-C_o\alpha} \quad \text{churn turbulent regime} \quad [15.5.21a]$$

$$\psi \to \alpha \quad \text{churn turbulent regime; } \alpha \ll 1 \quad [15.5.21b]$$

15.5.5 Superficial velocity: open system with uniform volume source

For an open system with uniform volumetric gas generation throughout the liquid, the vapour flux varies with height. The energy balance is:

$$\rho_g h_{fg}\, dj_g = q\rho_l(1-\alpha)\, dH \quad [15.5.22]$$

where h_{fg} is the latent heat of vaporization (kJ/kg). Limiting the treatment to the churn turbulent regime, differentiation of Equation 15.5.18 with respect to α and combination with Equation 15.5.22 yields

$$\psi \frac{dH^*}{d\alpha} = (1-\alpha)^{-3} \quad [15.5.23]$$

with

$$H^* = H/H_o \quad [15.5.24]$$

$$\psi = \frac{q(\rho_l/\rho_g - 1)H_o}{h_{fg}U_\infty} \quad [15.5.25]$$

where H^* is the dimensionless liquid height.

Equation 15.5.23 may then be integrated:

$$\psi H^* = \int_0^\alpha (1-\alpha)^{-3}\, d\alpha \quad [15.5.26]$$

Two successive integrations are then performed. The first gives α as a function of H^*:

$$\alpha = 1 - (1 + 2\psi H^*)^{-\frac{1}{2}} \quad [15.5.27]$$

The second gives $\bar{\alpha}$ as a function of H^*, subject to the condition

$$H^*_{max} = 1/(1-\bar{\alpha}) \quad [15.5.28]$$

This second integration yields

$$\bar{\alpha} = \frac{\psi}{2+\psi} \quad [15.5.29a]$$

or

$$\psi = \frac{2\bar{\alpha}}{1-\bar{\alpha}} \quad [15.5.29b]$$

Introducing the distribution parameter C_o, Equation 15.5.29 becomes

$$\bar{\alpha} = \frac{\psi}{2+C_o\psi} \quad [15.5.30a]$$

$$\psi = \frac{2\bar{\alpha}}{1-C_o\bar{\alpha}} \quad [15.5.30b]$$

The corresponding equation for bubbly flow is

$$\psi = \frac{\bar{\alpha}(1-\bar{\alpha})^2}{(1-\bar{\alpha}^3)(1-C_o\bar{\alpha})} \quad [15.5.31]$$

15.5.6. Superficial velocity: closed system with uniform volume source

The case of a closed system is also relevant, since it represents the condition just before the relief device operates. It is used for the analysis of transient conditions. For this case j_l is not equal to zero.

For a closed system with uniform volumetric gas generation throughout the liquid, the treatment is as follows. Continuity of flow gives:

$$j_l = aj_g \quad [15.5.32]$$

with

$$a = \frac{1-x_e}{(\rho_l/\rho_g)x_e} \quad [15.5.33]$$

where a is a parameter and x_e is the mass fraction of vapour at the vent. Then from Equations 15.5.11, 15.5.17 and 15.5.32:

$$j_g = \frac{\epsilon\lambda U_\infty}{1-C_o\lambda_a} \quad [15.5.34]$$

with

$$\epsilon = \frac{(1-\alpha)^n}{1-\alpha^m} \quad [15.5.35]$$

$$\lambda = \frac{\alpha}{1-C_o\alpha} \quad [15.5.36]$$

where ϵ and λ are parameters.

For the two regimes the values of the parameters α, ϵ and λ are as follows. For the bubbly regime:

$$\alpha = \bar{\alpha} \quad [15.5.37]$$

$$\epsilon = \frac{(1-\alpha)^2}{1-\alpha^3} \quad [15.5.38]$$

$$\lambda = \frac{\bar{\alpha}}{1-C_o\bar{\alpha}} \quad [15.5.39]$$

For the churn turbulent regime:

$$\alpha = \frac{2\bar{\alpha}}{1+C_o\bar{\alpha}} \quad [15.5.40]$$

$$\epsilon = 1 \quad [15.5.41]$$

$$\lambda = \frac{2\bar{\alpha}}{1-C_o\bar{\alpha}} \quad [15.5.42]$$

A mass balance on the vapour at the vent gives:

$$j_g = \frac{Gx_e\Omega}{\rho_g} \quad [15.5.43]$$

with

$$\Omega = A/A_x \quad [15.5.44]$$

where A is the area of the vent (m^2) and Ω is the ratio of the vessel area to the vent area. Then from Equations 15.5.34–15.5.43:

$$\frac{Gx_e\Omega}{U_\infty\rho_g} = \frac{\epsilon\lambda}{1 - C_o\lambda a} \qquad [15.5.45]$$

Equation 15.5.45 provides an expression coupling between the vessel and the vent line. As described by Grolmes and Leung (1985), it is used for this purpose in numerical models of vessel venting.

The value of ψ at the all vapour flow condition with $x_e = 1$ may be obtained as follows. Noting that Equation 15.5.43 is for the vent point and hence $j_g = j_{g\infty}$, then Equations 15.5.43 and 15.5.5 give

$$\psi = \frac{G\Omega}{U_\infty\rho_g} \qquad [15.5.46]$$

15.5.7 Vapour generation by external heat

From Equations 15.5.2, 15.5.6 and 15.5.25 it may be shown that for an externally heated vessel

$$\psi = \frac{FQ_{ex}(1-\alpha_o)H_x}{U_\infty h_{fg}\rho_g D} \qquad [15.5.47]$$

15.5.8 Criterion for two-phase flow

These relations may be used to indicate the boundary between all vapour venting and two-phase venting. The criterion for two-phase flow into the vent is that the average void fraction is greater than the vessel freeboard

$$\bar{\alpha} > \alpha_o \qquad [15.5.48]$$

The value of the average void fraction $\bar{\alpha}$ for use in relation 15.5.48 is obtained as follows. The actual value of the dimensionless velocity ψ is determined from the appropriate equation. The corresponding value of $\bar{\alpha}$ is then determined from the appropriate relation between ψ and $\bar{\alpha}$. For example, for an externally heated vessel with a churn turbulent regime, Equation 15.5.47 gives ψ and Equation 15.5.30a the corresponding value of $\bar{\alpha}$.

Alternatively, the criterion for two-phase flow to occur may be formulated in terms of the dimensionless velocity ψ. The criterion is then

$$\psi(\bar{\alpha}) > \psi(\alpha_o) \qquad [15.5.49]$$

For example, for the externally heated vessel with churn turbulent regime $\psi(\bar{\alpha})$ is obtained as before from Equation 15.5.47 and $\psi(\alpha_o)$ from Equation 15.5.30b with α_o instead of $\bar{\alpha}$.

15.5.9 Criterion for two-phase flow: illustrative example

An illustrative example of the determination of the quality of the vapour entering the vent is given in Table 15.6. The case considered is a vertical cylindrical vessel with external heating. The calculation shows that the flow entering the vent is two-phase.

15.5.10 Vapour carryunder

A model based on vapour carryunder has been developed and applied to an externally heated vessel by Grolmes and Epstein (1985). A dimensionless heat flux J_o is defined as

$$J_o = \frac{Q_{ex}}{U_\infty h_{fg}\rho_g} \qquad [15.5.50]$$

The external heating creates a boiling two-phase boundary layer at the vessel walls so that the vapour bubbles there rise. This rise is balanced by liquid downflow in the centre of the vessel, which creates the potential for vapour carryunder. Vapour carryunder is assumed to occur if the downward recirculating velocity U_c exceeds the bubble rise velocity U_∞. The conditions which can occur are shown in Figure 15.17, where Figure 15.17(a) shows no vapour carryunder ($U_c < U_\infty$ throughout) and Figure 15.17(b) shows vapour carryunder to a level H_{BL} ($U_c = U_\infty$).

***Table 15.6** Illustrative calculation for occurrence of two-phase flow of fluid entering the vent of an externally heated storage vessel*

A Scenario

Liquid density, $\rho_l = 750$ kg/m^3
Vapour density, $\rho_g = 4.5$ kg/m^3
Latent heat of vaporization, $h_{fg} = 900$ kJ/kg
Surface tension, $\sigma = 0.02$ N/m
Vessel type: vertical cylindrical
Vessel diameter, $D = 1.0$ m
Vessel height, $H_x = 2.0$ m
Vessel freeboard, $\alpha_o = 0.16$
Vessel geometric factor, $F = 4$
External heat input, $Q_{ex} = 60$ kJ W/m^2
Vessel regime: churn turbulent
$k = 1.53$
$C_o = 1$

B Fluid phase

From Equation [15.5.12]:
$U_\infty = 1.53[0.02 \times 9.81(750 - 4.5)]^{\frac{1}{4}}/750^{\frac{1}{2}} = 0.19$ m/s
From Equation [15.5.47]

$$\psi = \frac{4 \times 60(1 - 0.16) \times 2.0}{0.19 \times 900 \times 4.5 \times 1.0} = 0.51$$

From Equation [15.5.5]

$j_{g\infty} = 0.51 \times 0.19 = 0.097$ m/s

From Equation [15.5.29a]

$$\bar{\alpha} = \frac{0.51}{2 + 0.51} = 0.20$$

But

$\alpha_o = 0.16$

Hence $\bar{\alpha} > \alpha_o$, indicating that flow entering vent is two phase.

Alternatively, from Equation 15.5.29b with α_o instead of $\bar{\alpha}$:

$$\psi(\alpha_0) = \frac{2 \times 0.16}{1 - 0.16} = 0.38$$

Hence $\psi(\bar{\alpha}) > \psi(\alpha_0)$, again indicating that flow entering vent is two phase.

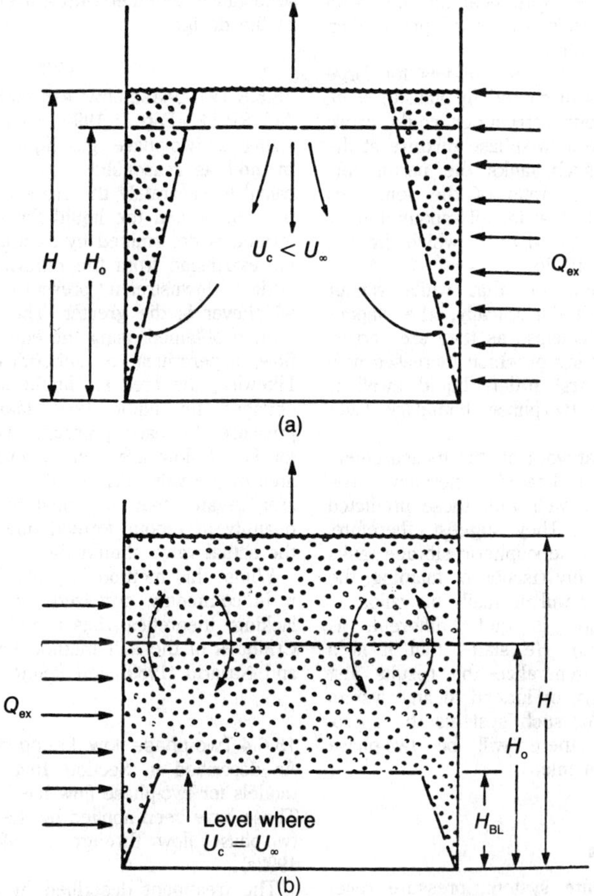

Figure 15.17 *Vapour carryunder in venting of a vessel (Grolmes and Epstein, 1985): (a) case where the recirculating liquid velocity U_c is less than the bubble rise velocity U_∞ throughout the vessel; and (b) case where the recirculating liquid velocity U_c is not less than the bubble rise velocity U_∞ throughout the vessel (Courtesy of the American Institute of Chemical Engineers)*

The downward liquid recirculating velocity U_c is related to the upward liquid superficial velocity U_s, evidently the same as the bubble rise velocity U_∞, by a geometric factor ϵ:

$$U_c = \epsilon U_s \qquad [15.5.51]$$

The geometric factor ϵ is the ratio of the area for downward recirculation flow to the boundary layer area for upward flow. It is a function of the boundary layer thickness $\delta(H)$ measured at the height H of the liquid, which itself is a function of the heat flux:

$$\epsilon = f[\delta(H)] \qquad [15.5.52]$$

$$\delta(H) = f(J_o) \qquad [15.5.53]$$

The interface height H_{BL} below which there is no vapour carryunder is determined as the height at which $U_c = U_s$ or $\epsilon = 1$.

The authors give a graphical relation of the form

$$U_s = f(J_o, z) \qquad [15.5.54]$$

where z is the height (m). Using this relation, the height H_{BL} can be determined.

Using this model the authors derive the following results for the fraction H_{BL}/H of vessel height which there is no vapour carryunder:

Heat flux, J_o	*Vessel diameter* (m)			
	0.61	1.52	3.05	6.1
0.310	0.41	0.32	0.26	0.22
0.062	0.62	0.48	0.39	0.33
0.006	1	0.86	0.69	0.58

The results show that the proportion of the liquid subject to vapour carryunder increases with heat flux and with vessel size. Thus large vessels are more prone than small ones to vapour carryunder.

The potential significance of these models for large atmospheric storage tanks is discussed by Fauske *et al.* (1986). Assuming that vapour carryunder could cause sufficient liquid swell to give a two-phase mixture at the vent inlet, then since in such tanks the permissible overpressure is very low, the ratio of the vent area required for two-phase flow to that for all vapour flow is, to first order, proportional to $(\rho_g/\rho_l)^{\frac{1}{2}}$, which implies augmentation by a factor of 10–30.

The authors suggest, however, that while vapour carryunder is possible hydrodynamically, the vapour bubbles will in fact tend to collapse as they are carried down into liquid, the subcooling of which increases with depth, and state that, to first order, liquid swell is determined by the boiling two-phase boundary layer without vapour carryunder.

They describe experimental work on the measurement of liquid swell in vessels heated externally. The measured liquid swells agree well with those predicted neglecting vapour carryunder. They suggest, therefore, that in the sizing of vents for atmospheric storage tanks for liquids which are not highly viscous or foaming, the assumption of all vapour flow will normally be valid.

For reactors the situation is quite different, as discussed by Fauske (1985a). He states that it is a hopeless task to seek to generalize the results just described to reacting systems, or indeed to any system with traces of impurities. For such systems the appropriate assumption is that there will be two-phase homogenous flow at the vent inlet.

15.6 Pressure Relief Valves

As components of a pressure system, pressure relief valves have been treated in Chapter 12. In this section some further consideration is given in respect of the flow through such devices. Dispersion of the discharge is considered in Section 15.20 and, for dense gases in Section 15.43.

15.6.1 Single-phase flow

Relations for single-phase flow of gases and liquids have been presented in Section 15.1. Formulae derived from these are given in the codes for pressure relief. The API and BS formulae are stated in Chapter 12.

15.6.2 Two-phase flow

Two-phase vapour-liquid flow has been discussed in Sections 15.2–15.4, from which it will be apparent that this is a much more complex topic. The various models for two-phase flow do not lend themselves to formulae for calculating the area of the pressure relief valve orifice that are comparable in simplicity with those for single-phase flow. Moreover, with two-phase flow the determination of the flow through the valve orifice by no means exhausts the problem. The flows in the inlet and outlet lines are also complex.

The treatment here deals primarily with flow through the valve orifice and is confined to an account of the widely used American Petroleum Institute (API) method and of a method given by Leung which has gained some acceptance, together with a overview of the wider aspects of the design.

15.6.3 Two-phase flow: API method

API RP 520 Part 1: 1990 gives the following method for sizing a two-phase gas–liquid relief. It describes the method as reasonably conservative. The method is to calculate separately the cross-sectional areas required for the vapour and the liquid flows. The quantity of vapour formed is determined by assuming an adiabatic, isenthalpic expansion from the relieving condition down to the critical downstream pressure or the back pressure, whichever is the greater. The area for the vapour flow is then obtained using the equations given for all vapour flow, supercritical or subcritical, as the case may be. Likewise, the area for liquid flow is obtained from the equation for liquid flow, taking as the downstream pressure the back pressure. The areas for vapour and for liquid flow are then summed to give the required area of the valve orifice. The valve selected will have an area greater than or equal to this required area. The quantity of vapour formed downstream of the orifice of the actual valve should then be rechecked.

Whilst this method is probably the most widely used, it is commonly criticized by workers in the field as lacking theoretical basis and experimental validation. Critiques of the API method have been given by Leung and Nazario (1989) and Leung (1992a).

15.6.4 Two-phase flow: Leung method

As described in Section 15.4, a comprehensive set of models for two-phase flow has been developed by Leung. These have been applied by the author to the problem of two-phase flow through a safety relief valve (Leung, 1992a).

The treatment described by Leung is essentially the application of his homogenous equilibrium model (HEM) to the safety relief valve (SRV) design requirements of API RP 520: 1990 and RP 521: 1990. The equations used are a subset of those given in Section 15.4, based mainly on Leung (1986a). For clarity, this subset is given again here together with an explanation of their use in SRV system design and with the illustrative examples given by Leung.

In his method Leung distinguishes four initial states of the liquid in the vessel:

(1) saturated liquid;
(2) two-phase gas–liquid mixture;
(3) subcooled liquid –
 (a) low subcooling,
 (b) high subcooling.

Each of these cases is treated differently.

The model utilizes the parameter ω which determines the equation of state

$$\frac{v}{v_o} = \omega\left(\frac{P_o}{P} - 1\right) + 1 \qquad [15.6.1]$$

and is given by

$$\omega = \frac{x_o v_{go}}{v_o} + \frac{C_{fo} T_o P_o}{v_o}\left(\frac{v_{fgo}}{h_{fgo}}\right)^2 \qquad [15.6.2]$$

where C is the specific heat (J/kg K), h is the specific enthalpy (J/kg), P is the absolute pressure (Pa), T is the absolute temperature (K), v is the specific volume (m^3/kg), x is the mass fraction of vapour, and subscripts f, fg, g and o indicate the liquid, the liquid–vapour transition, the vapour and stagnation, respectively. The inlet void fraction α_o is

$$\alpha_o = \frac{x_o v_{go}}{v_o} \qquad [15.6.3]$$

The specific volume of the fluid is

$$v_o = x_o v_{go} + (1 - x_o) v_{fo} \qquad [15.6.4]$$

The following pressures are defined: P the pressure, P_b the back pressure at the orifice, P_c the critical pressure, P_o the stagnation pressure in the vessel, P_s the saturation pressure of the liquid (all pressures absolute in Pa). Subscripts 1 and 2 denote upstream and downstream ends of the outlet line, respectively. The corresponding pressure ratios are:

$$\eta = \frac{P}{P_o} \qquad [15.6.5]$$

$$\eta_c = \frac{P_c}{P_o} \qquad [15.6.6]$$

$$\eta_s = \frac{P_s}{P_o} \qquad [15.6.7]$$

where the subscripts c and s denote critical and saturation, respectively.

For the outlet line the following expression is used for the critical pressure at the downstream end:

$$P_{2c} = \frac{W_{act}}{A_2}\left(\frac{P_o \omega}{\rho_o}\right)^{\frac{1}{2}} \qquad [15.6.8]$$

where A_2 is the cross-sectional area of the outlet line (m^2), W is the mass flow (kg/s), ρ is the density (kg/m^3) and the subscript act indicates actual.

Saturated liquid

For the case where the inlet fluid condition is a saturated liquid, the vapour fraction $x_o = 0$.

A value of the parameter ω is calculated from Equation 15.6.2. This value is then used in the following equation to determine the critical pressure ratio η_c:

$$\eta_c^2 + (\omega^2 - 2\omega)(1 - \eta_c)^2 + 2\omega^2 \ln \eta_c + 2\omega^2(1 - \eta_c) = 0 \qquad [15.6.9]$$

From the value of η_c so obtained the critical pressure P_c is calculated using Equation 15.6.6.

If $P_c < P_b$ the flow is not choked and P is set equal to P_b. The mass flux is then calculated from the following equation:

$$\frac{G}{(P_o/v_o)^{\frac{1}{2}}} = \frac{\left(-2\left\{\omega \ln\left(\frac{P}{P_o}\right) + (\omega - 1)\left[1 - \left(\frac{P}{P_o}\right)\right]\right\}\right)^{\frac{1}{2}}}{\omega\left(\frac{P_o}{P} - 1\right) + 1} \qquad [15.6.10]$$

where G is the mass velocity (kg/m^2 s). If $P_c > P_b$ the flow is choked and the mass flux may be obtained either from Equation 15.6.10 or from

$$\frac{G_c}{(P_o/v_o)^{\frac{1}{2}}} = \frac{\eta_c}{\omega^{\frac{1}{2}}} \qquad [15.6.11]$$

Two-phase mixture

For the case where the inlet fluid condition is a two-phase gas–liquid mixture, the vapour fraction x_o has a finite value. Otherwise the calculation procedure is essentially the same as for the case of a saturated liquid.

Subcooled liquid

For the case where the inlet fluid condition is a subcooled liquid, a distinction is made between a low degree of subcooling and high subcooling.

A form of the parameter ω appropriate to the saturated condition is defined as

$$\omega_s = \frac{C_{fo} T_o P_s}{v_{fo}}\left(\frac{v_{fgo}}{h_{fgo}}\right)^2 \qquad [15.6.12]$$

The condition for low subcooling is

$$\eta_s \geq \frac{2\omega_s}{1 + 2\omega_s} \qquad [15.6.13]$$

If the case is one of low subcooling, the fluid flashes in the throat.

The value of the parameter ω_s is then used to determine the critical pressure ratio η_c:

$$\frac{\omega_s + (1/\omega_s) - 2}{2\eta_s}\eta_c^2 - 2(\omega_s - 1)\eta_c + \omega_s \eta_s \ln\left(\frac{\eta_c}{\eta_s}\right) + \frac{3}{2}\omega_s \eta_s - 1 = 0 \qquad [15.6.14]$$

From the value of η_c so obtained the critical pressure P_c is calculated using Equation 15.6.6.

The mass flux is calculated from the following equation:

$$\frac{G}{(P_o/v_{fo})^{\frac{1}{2}}} = \frac{\left\{2(1 - \eta_s) + 2\left[\omega_s \eta_s \ln\left(\frac{\eta_s}{\eta}\right) - (\omega_s - 1)(\eta_s - \eta)\right]\right\}^{\frac{1}{2}}}{\omega_s\left(\frac{\eta_s}{\eta} - 1\right) + 1} \qquad [15.6.15]$$

As the condition of a saturated liquid is approached $P_o \to P_s$ and $\eta_s \to 1$, so that Equations 15.6.14 and 15.6.15 reduce to Equations 15.6.9 and 15.6.10.

If the case is one of high subcooling, no flashing occurs at the throat. The critical pressure ratio is

$$\eta_c = \frac{P_s}{P_o} \qquad [15.6.16]$$

and the critical pressure is the saturation pressure P_s.

The mass flux is given by Equation 15.6.15, which for this case reduces to

$$\frac{G_c}{(P_o/v_{fo})^{\frac{1}{2}}} = [2(1 - \eta_s)]^{\frac{1}{2}} \qquad [15.6.17]$$

or

$$G_c = [2\rho_{fo}(P_o - P_s)]^{\frac{1}{2}} \qquad [15.6.18]$$

This completes the set of expressions for the mass flow through the valve orifice. It remains to consider the valve sizing and selection and the effect of the inlet and outlet lines.

Valve sizing
The pressure relief requirement is specified as a required mass flow. The actual mass flux is obtained by the methods just described. The required area of the valve orifice is then

$$A_{req} = \frac{W_{req}}{KG_{act}} \quad [15.6.19]$$

where A is the cross-sectional area of the valve (m^2), K is the valve discharge coefficient and the subscript req denotes required.

A valve size is then selected. Since, in general, this will have an orifice cross-sectional area slightly larger than that required, the actual mass flow will also be larger:

$$W_{act} = KA_{act}G_{act} \quad [15.6.20]$$

It is this actual mass flow which is used in the piping calculations which follow.

Inlet line
The treatment of the inlet and outlet piping conforms with the API requirements. For the inlet piping these are that the piping be as short as possible with the recommendation that the line losses be limited to 3% of the gauge set pressure.

The pressure drop in the inlet line is determined from the equation

$$\Delta P_{in} = \frac{1}{2}\bar{v}_{in}\left(\frac{W_{act}}{A_{in}}\right)^2\left[4f\left(\frac{L_{in}}{D_{in}}\right)\right] + \frac{gH}{\bar{v}_{in}} \quad [15.6.21]$$

where A_{in} is the cross-sectional area of the inlet pipe (m^2), D_{in} is the diameter of the inlet pipe (m), f is the Fanning friction factor, g is the acceleration due to gravity (m/s^2), H is the change in elevation (m), L_{in} is the length of the inlet pipe (m), ΔP_{in} is the total pressure difference across the inlet pipe (Pa) and $\bar{v}_{in}$ is the average specific volume in the inlet pipe (m^3/kg). The value of K is taken as 0.95.

The average specific volume in the inlet is evaluated at 0.98 of the stagnation pressure, so that

$$\bar{v}_{in} = v_{fo}\left[\omega\left(\frac{1}{0.98} - 1\right) + 1\right] \quad [15.6.22]$$

However, if the liquid is subcooled and $P_s/P_o < 0.98$, then $\bar{v}_{in} = v_{fo}$.

Outlet piping
For the outlet piping a calculation is made to determine the maximum allowable length of the pipe. This is governed by the maximum allowable back pressure on the valve. This back pressure is specified by the SRV manufacturer and is typically 10% of the differential set pressure for unbalanced valves and up to 50% for balanced valves.

The maximum allowable length of the outlet line is determined from the equation

$$4f\left(\frac{L_2}{D_2}\right) = 2\frac{P_o\rho_o}{(W_{act}/A_2)^2}\left\{\frac{\eta_1 - \eta_2}{1-\omega} - \frac{\omega}{(1-\omega)^2}\right.$$
$$\left.\ln\left[\frac{(1-\omega)\eta_1 + \omega}{(1-\omega)\eta_2 + \omega}\right]\right\} + 2\ln\left[\frac{(1-\omega) - \eta_1 + \omega}{(1-\omega)\eta_2 + \omega}\left(\frac{\eta_2}{\eta_1}\right)\right] \quad [15.6.23]$$

with

$$\eta_1 = \frac{P_1}{P_o} \quad [15.6.24]$$

$$\eta_2 = \frac{P_2}{P_o} \quad [15.6.25]$$

where D_2 is the diameter of the outlet pipe (m) and L_2 is the length of the outlet pipe (m).

Illustrative example
As an illustration of this method, consider the example given by Leung, which is the design of an SRV for the venting of liquefied ammonia. The example is given in two parts, Part A being the determination of the mass flux for the specified liquid condition and Part B the sizing of the valve itself and of the inlet and outlet lines.

Part A: Mass flux
The initial, or stagnation, state of the liquid is

$P_o = 1 \times 10^6$ Pa

$T_o = 298$ K

$v_{fo} = 0.001\,658\ m^3/kg$

$v_{go} = 0.1285\ m^3/kg$

$v_{fgo} = 0.1268\ m^3/kg$

$C_{fo} = 4806$ J/kg K

$h_{fgo} = 1\,165\,430$ J/kg

The mass flux is calculated for two liquid conditions: (1) saturated liquid, $x_o = 0$; and (2) two-phase gas-liquid mixture, $x_o = 0.1$. The latter is included to illustrate how this liquid condition is tackled, but is not strictly relevant to the main problem.

Case 1: Saturated liquid

$x_o = 0$

$v_o = 0.001\,658\ m^3/kg$

$\rho_o = 603\ kg/m^3$

$\alpha_o = 0$

$\omega = 10.2$ (from Equation 15.6.2)

$\eta_c = 0.85$ (from Equation 15.6.9)

$G_c = 6510\ kg/m^2\,s$ (from Equation 15.6.11)

Case 2: Two-phase mixture

$x_o = 0.1$

$v_o = 0.01434\,m^3/kg$ (from Equation 15.6.4)

$\rho_o = 69.7\,kg/m^3$

$\alpha_o = 0.9$ (from Equation 15.6.3)

$\omega = 2.1$ (from Equation 15.6.2)

$\eta_c = 0.7$ (from Equation 15.6.9)

$G_c = 6010\,kg/m^2 s$ (from Equation 15.6.11)

Part B: Sizing of valve and piping
The pressures around the SRV are:

Atmospheric pressure = 0.101 MPa (a)

Back pressure = 0.0101 MPa (a)

Set pressure = 0.915 MPa (a) = 1.016 (g)

Differential set pressure = 0.915 − 0.101 = 0.814 MPa

Relieving pressure = 1.0 MPa (a)

The gauge relieving pressure is 110% of the differential set pressure (1.1 × 0.814 = 0.90) and thus the absolute relieving pressure is 1.0 MPa.
The details of the SRV piping are

Elevation of valve, $H = 1\,m$

Length of inlet pipe, $L_{in} = 1\,m$

Equivalent length of outlet pipe, $L_2 = 20\,m$

Outlet pipe Fanning friction factor, $f = 0.005$

The SRV is sized for the saturated liquid condition. The valve is required to vent a mass flow W_{req} of 1.0 kg/s, the actual mass flux G_{act} is 6510 kg/s and the value of the valve coefficient K is 0.95. Then from Equation 15.6.19 the required area A_{req} of the valve orifice is $1.62 \times 10^{-4}\,m^2$. Consulting the table of standard valve orifice sizes, the next size up is an 'F' size valve with an actual area A_{act} of $1.98 \times 10^{-4}\,m^2$. This is the valve size selected.

Then from Equation 15.6.20 the actual mass flow is $W_{act} = 1.22\,kg/s$.

For the size-F valve the diameter D_{in} and cross-sectional area A_{in} of the inlet are 0.0409 m and $1.313 \times 10^{-3}\,m^2$, respectively. An inlet line of the same diameter is used and a check is made that the inlet line losses do not exceed 3% of the gauge set pressure. Then Equation [15.6.22] with $\omega = 10.2$ and $v_o = 0.001658\,m^3/kg$ yields $\bar{v}_{in} = 0.002\,m^3/kg$.

The pressure drop ΔP_{in} in the inlet line is calculated from Equation 15.6.21. For the vertical line the length L_{in} and the elevation H are identical at 1 m. The value obtained for ΔP_{in} is 5328 Pa. The inlet line losses are thus much less than 3% of the gauge set pressure of 1.016×10^6 Pa.

For the outlet line the calculation required is to establish that the equivalent length of 20 m is within the maximum allowable length of the line. The first step is to determine the values to be used for the pressures P_1 and P_2 at the inlet and outlet of the line, respectively. For P_1 the value is 10% of the gauge set pressure and is thus 0.183 MPa. For P_2, from Equation [15.6.8] $P_{2c} = 5.64 \times 10^4$ Pa. The back pressure P_b is 0.101 MPa, or 10.1×10^4 Pa. Hence $P_{2c} < P_b$ and thus P_2 is set equal to P_b. Then from Equations 15.6.24 and 15.6.25 $\eta_1 = 0.183$ and $\eta_2 = 0.101$.

The line diameter D_2 is taken as that of the valve outlet, which is 0.0629 m, giving a corresponding cross-sectional area A_2 of $3.089 \times 10^{-3}\,m^2$. Then with a value of the Fanning friction factor f of 0.005, Equation [15.6.23] yields a maximum allowable line length of 28 m. The actual equivalent line length is 20 m, which is within the allowable range.

15.6.5 Two-phase flow: overall system

As indicated earlier, there is more to the design of a pressure relief for a two-phase vapour–liquid mixture than the sizing of the valve orifice. Accounts of overall design for pressure relief include those by Crawley and Scott (1984) and S.D. Morris (1988b). An overview, with particular reference to offshore systems, has been given by Selmer-Olsen (1992).

Selmer-Olsen identifies a number of difficulties: (1) there is no accepted method of deciding when two-phase flow needs to be considered; (2) there is no accepted design method for two-phase flow; (3) the widely used API method may be inadequate; (4) the methods available are generally not sufficiently validated by experimental work, particularly on the large scale; and (5) there is a strong coupling between the thermohydraulic behaviour of the fluid system and the flow through the pressure relief device for which the commonly used modular approach may be inappropriate.

The occurrence of two-phase flow can alter completely the characteristics of the relief system, in particular the behaviour of the fluid in the pipework downstream of the pressure relief valve. This can result in increased back pressure or in choking in the relief header, possibly with oscillating location of the choking throat between the valve and the pipe – the multiple choke effect. Potential consequences of such flow effects are prolonged depressurization times, intermittent flows, high header pressures and blowback to lower pressure sources, and high thermal and mechanical loads.

The author identifies the following work relevant to the design of a pressure relief system for two-phase flow. Work on the sizing of the valve itself includes that of Simpson, Rooney and Grattan (1979), Sallet and co-workers (Sallet, 1984, 1990b,c; Sallet and Somers, 1985), Campbell and Medes (1985), Friedel and Kissner (1985, 1987, 1988), S.D. Morris (1988b, 1990a,b), Morley (1989a,b), Alimonti, Fritte and Giot (1990), Curtelin (1991), M.R. Davis (1991), Simpson (1991) and Leung (1992a).

Treatments of the inlet line to the valve are given by O.J. Cox and Weirick (1980), Zahorsky (1983), S.D. Morris (1988b) and Leung (1992a) and treatments of the outline line are given by Richter (1978b), Friedel and Löhr (1982), S.D. Morris (1988b, 1990b) and Leung (1992a).

The author states that the method of Leung (1992a), based on the homogeneous equilibrium model (HEM) of two-phase flow appears to be the increasingly preferred choice for flow through the valve, for which it is said to be reasonably conservative. He comments, however, that it may well not be conservative for the associated inlet and outlet lines and downstream tanks.

The effects of non-equilibrium assumptions are discussed by Bilicki and Kestin (1990) and Lemonnier *et al.* (1991).

Experimental work on the flows through a configuration relevant to pressure relief has been done by Selmer-Olsen (1991, 1992), using a converging–diverging nozzle with inlet and outlet lines. For short nozzles and low pressures the flows exceeded those predicted by the HEM by a factor of up to 2. As the throat length and pressure were increased the predictions of the HEM were approached, but never reached. The concomitant of this is that the flows in the inlet and outlet lines may be higher than predicted by the HEM.

15.6.6 Two-phase flow: computer codes

A number of computer codes have been developed for pressure relief flows. An account is given by Selmer-Olsen (1992).

One of these codes is BLOW-DOWN by Nylund (1983 DnV Rep. 83–1317, 1984). Another is the BLOWDOWN code of Haque, Richardson and Saville (1992), described in the next section. Other codes include those by Middleton and Lloyd (1984), Bayliss (1987), Klein (1987), Evanger *et al.* (1990) and the NEL code PIPE3.

A benchmark exercise on vessel depressurization methods is described by Skouloudis (1990 CEC EUR 12602 EN).

15.7 Vessel Blowdown

Emergency shut-down of a hydrocarbon processing plant may involve the depressurization of major process vessels by blowdown to the flare. On offshore production platforms blowdown is a normal part of emergency shut-down.

In many cases, blowdown will be two-phase. One aspect of the treatment of blowdown is therefore the estimation of the flow. The account of flow in relief systems given in the previous section is applicable to blowdown. The other aspect is the hazards associated with blowdown, particularly the fall in the temperature of the gas.

15.7.1 BLOWDOWN

A treatment which covers both aspects has been given by Richardson, Saville and co-workers (S.M. Richardson, 1989; Haque *et al.*, 1990; Haque, Richardson and Saville, 1992; S.M. Richardson and Saville, 1992).

The hazards arise essentially due to the large temperature drop associated with the sudden depressurization of a vessel containing gas or vapour. Thus, for example, if nitrogen at 150 bar and atmospheric temperature is expanded adiabatically down to atmospheric pressure the isentropic expansion gives a final gas temperature of 78 K.

The type of vessel of interest is one which contains, under pressure, a vapour space, a layer of liquid hydrocarbon and beneath that a layer of water. The potential effects of the temperature drop associated with sudden depressurization are chilling of the vessel walls, condensation of vapour to form droplets with carryover of these droplets into the flare header and formation of hydrates in the hydrocarbon liquid. It is therefore of practical interest to be able to predict the temperature profiles of the fluid phases and of the vessel walls.

An account of a model for vessel blowdown has been given by Haque, Richardson and Saville (1992). The model is embodied in the code BLOWDOWN. For a non-condensable gas, an acceptable estimate of the time profiles of pressure in and of flow from the vessel can generally be obtained assuming choked adiabatic flow. For this case the authors give:

$$P = P_o\left[1+\frac{\gamma-1}{2}\frac{A}{V}t\left(\frac{\gamma RT_o}{M}\right)^{\frac{1}{2}}\left(\frac{2}{\gamma+1}\right)^{(\gamma+1)/(2\gamma-2)}\right]^{-2\gamma/(\gamma-1)} \quad [15.7.1]$$

$$W = A\rho_o\left(\frac{\gamma RT_o}{M}\right)^{\frac{1}{2}}\left(\frac{2}{\gamma+1}\right)^{(\gamma+1)/(2\gamma-2)}\left(\frac{P}{P_o}\right)^{(\gamma+1)/2\gamma} \quad [15.7.2]$$

where A is the cross-sectional area of the choke (m^2), M is the molecular weight of the gas, P is the absolute pressure in the vessel (Pa), R is the universal gas constant (J/kg mole K), T is the absolute temperature of the gas (K), V is the volume of the vessel (m^3), W is the mass flow from the vessel (kg/s), γ is the ratio of the gas specific heats, ρ is the density of the gas (kg/m^3), and the subscript o indicates initial conditions in the vessel.

Since in this model the conditions are adiabatic, it excludes heat transfer between the gas and the wall of the vessel. By definition, therefore, it cannot be used to determine the temperature of the vessel wall. On these grounds alone a fuller model is needed. Moreover, it turns out that this model is also inadequate for the representation of the more complex vessel blowdown scenarios of the kind described earlier. The BLOWDOWN model has been developed to meet this requirement. It consists of a thermophysical model in combination with relationships for heat, mass and momentum and for flow through the choke. The thermophysical properties are calculated using the computer package PREPROP.

The vessel is modelled as a set of three fluid zones (gas, liquid hydrocarbon and water), three corresponding vessel wall zones, and the outflow system consisting of the pipework between the vessel and the choke, the choke itself and the pipework downstream of the choke.

The outflow from the vessel is modelled as follows. It is assumed that the pressure drop at the orifice is such that the flow is critical. For a gas the flow is choked and is determined by its velocity through the choke which is equal to the velocity of sound in the gas:

$$a = \left[\left(\frac{\partial P}{\partial \rho}\right)_s\right]^{\frac{1}{2}} \quad [15.7.3]$$

where a is the velocity of sound in the gas at the choke (m/s), P is the absolute pressure at the choke (Pa), ρ is the density of the gas at the choke (kg/m^3) and subscript s denotes isentropic conditions. The expansion is isentropic so that

$$S_i = S_c \qquad [15.7.4]$$

where S is the entropy of the gas (kJ/kmol K) and the subscripts c and i denote choke and vessel, respectively. The energy balance is

$$H_i = H_c + a^2/2 \qquad [15.7.5]$$

where H is the enthalpy of the gas (J/kg). These three equations are solved simultaneously using the thermophysical package. The mass flow is then obtained as

$$W = C_D A a \rho_c \qquad [15.7.6]$$

where C_D is the coefficient of discharge. The value of C_D at the choke is generally taken as 0.8.

For a gas–liquid, thus two-phase, fluid the flow is again assumed to be critical, but in this case the velocity does not equal the speed of sound. Consider changes in the overall pressure difference caused by, say, decreasing the downstream pressure. At small overall pressure differences the liquid is in the metastable condition and does not flash as it passes through the orifice. As the overall pressure difference increases, the limit of metastability is reached. There is then no further change either in the flow through the orifice or the pressure at the orifice. The fluid downstream of the orifice undergoes flashing. The behaviour of the fluid is therefore analogous to that of a gas in that a critical condition is reached, but differs from it in that the speed of sound plays no role.

For the pressure at which metastability can no longer be supported, the authors give the following relation, generalized from the work of Abuaf, Jones and Wu (1983):

$$P_s - P_{ms} = 1.1 \times 10^{11} \frac{\sigma^{1.5}(T/T_c)^8(1 + 2.2 \times 10^{-8}\Pi^{0.8})^{0.5}}{T_c^{0.5}[1 - (\rho_g/\rho_l)]}$$

$$T < 0.85T_c \qquad [15.7.7]$$

where P_{ms} is the absolute pressure at which metastability is no longer supported (Pa), P_s is the absolute saturated vapour pressure of the liquid at temperature T (Pa), T is the absolute temperature of the liquid upstream of the orifice (K), T_c is the absolute critical temperature of the liquid (K), Π is the depressurization rate (N/m^{2}s), ρ_g is the density of the gas (kg/m^3), ρ_l is the density of the liquid (kg/m^3), σ is the surface tension of the liquid (N/m), and subscripts g and l denote the gas and liquid, respectively. The depressurization rate Π is approximated as the pressure difference across the orifice divided by the time required to traverse twice the diameter of the orifice.

The flow conditions are then as follows. For the range $P_u > P_s$, $P_d < P_s$:

$P_d > P_{ms}$ flow unchoked

$P_d \leq P_{ms}$ flow choked

where subscripts d and u denote downstream and upstream, respectively. The mass flow through the orifice is given by

$$W = C_D A[2\rho_l(P_u - P_x)]^{\frac{1}{2}} \qquad [15.7.8]$$

where C_D is the coefficient of discharge and P_x is an appropriate absolute orifice pressure (Pa). The pressure P_x is the greater of P_{ms} and P_d. The coefficient of discharge is taken as approximately 0.65 for an orifice and unity for a properly formed nozzle.

The situations treated in this work are complex and it is not practical to construct a model which represents in equal detail all the relationships involved. Rather, the authors proceeded by seeking to identify the more significant phenomena and to model these accurately.

Experiments done to validate the model are described by Haque *et al.* (1990, 1992) and S.M. Richardson and Saville (1992). Two vessels were utilized, one 1.52 m long and the other 3.24 m long. In some tests the fluid used was nitrogen, in others it was a mixture of light hydrocarbons. The papers cited present for selected experiments a number of time profiles of the pressure in the vessel and of the temperatures of the gas and the vessel walls.

Some of these time profiles for tests of blowdown through a top outlet are shown in Figure 15.18. Figures 15.18(a) and 15.18(b) are for a vessel containing nitrogen; the first shows the profiles of the vessel pressure, the second those of the temperatures of the gas and of the vessel wall. Figures 15.18(c) and 15.18(d) are for a vessel containing hydrocarbons with the composition C_1 66.5%, C_2 3.5% and C_3 30.0%; the first shows the profiles of the gas and the liquid and the second those of the vessel wall in contact with these two phases.

Case studies of the application of the model are also described. One is for the blowdown of the suction scrubber of a gas compressor and another for the blowdown of a gas–condensate separator. The model is also applicable to the prediction of an accidental release from a vessel.

15.8 Vessel Rupture

In certain circumstances a vessel may rupture completely. If vessel rupture occurs, a large vapour cloud can be formed very rapidly. Accounts of vessel rupture include those by Hardee and Lee (1974, 1975), Hess, Hoffmann and Stoeckel (1974), J.D. Reed (1974), Maurer *et al.* (1977), A.F. Roberts (1981/82) and B. Fletcher (1982).

15.8.1 Vaporization

If a vessel containing a superheated liquid under pressure ruptures, a proportion of the liquid vaporizes. This initial flash fraction is determined by the heat balance, the latent heat of vaporization being supplied by the fall in the sensible heat of the liquid. The rapid formation of vapour bubbles also generates a spray of liquid drops so that typically most or all of the remaining liquid becomes airborne, leaving little or no residue in the vessel.

This effect has been demonstrated by J.D. Reed (1974), who carried out experiments on sudden vessel depressurization. In one series of experiments 3.5 kg of liquid ammonia contained at an absolute pressure of 3 bar and a temperature of −9°C in a vessel 15 cm diameter and 45 cm high was released using a quick release lid. One of the experiments is shown in Figure 15.19, in which the time interval between the first and

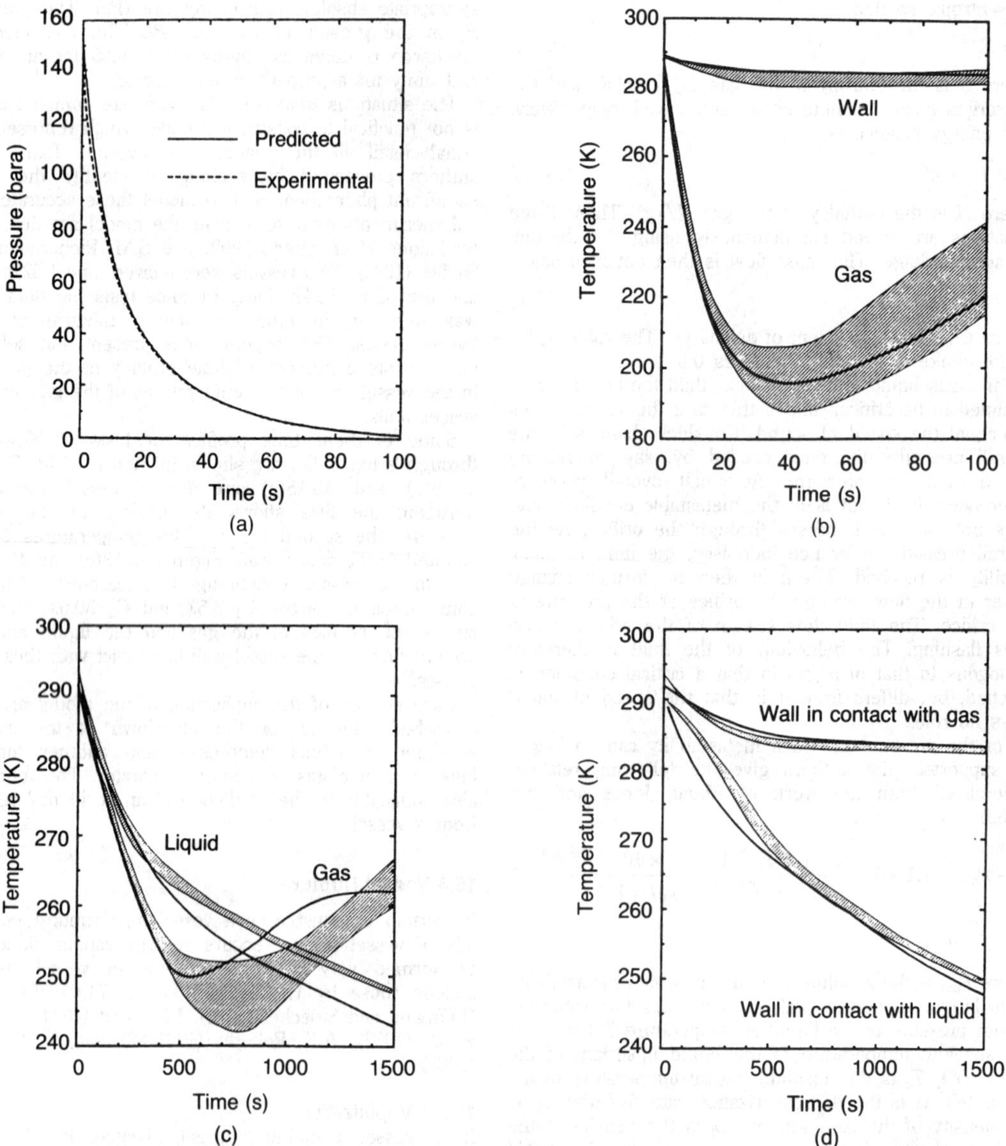

Figure 15.18 *Profiles of pressure and temperature in blowdown of a vessel (Haque et al., 1992): (a) pressure in vessel for nitrogen; (b) temperatures of bulk gas and of vessel wall for nitrogen; (c) temperatures of bulk gas and of bulk liquid for hydrocarbon mixture (C_1, 66.6%; C_2, 3.5%; and C_3, 30.0%); (d) temperature of sections of vessel inside wall in contact with (1) gas and (2) liquid for hydrocarbon mixture (C_1, 66.6%; C_2, 3.5%; and C_3, 30.0%). Vertical vessels, top outlets. Hatched regions span experimental measurements; solid lines are predictions (Courtesy of the Institution of Chemical Engineers)*

last frame is one-sixth of a second. In all the experiments at least 90% of the liquid ammonia was vaporized.

Similarly, Maurer *et al.* (1977) have carried out experiments (described below) on the rapid release of propylene held at pressures of 22–39 bar and temperatures of 50–80°C. The flash fraction of vapour was 50–65% and the remaining liquid formed spray.

An investigation of the extent of vapour and spray formation and of retained liquid has been made by B. Fletcher (1982), who carried out experiments, principally in a vertical vessel of 127 × 47 mm cross-section, in which a charge of superheated Refrigerant 11 was depressurized and the liquid residue was determined. The results are correlated in terms of the ratio (h_b/h_o) of

Figure 15.19 *Sudden depressurization of a vessel containing liquid ammonia under pressure (J.D. Reed, 1974) (Courtesy of Elsevier Publishing Company)*

the height reached by the liquid on depressurization h_b to the initial height of the liquid h_o, as shown in Figure 15.20. The parameter in the figure is the ratio K of the vent area to the vessel cross-sectional area.

The mass of retained liquid is given by the relation

$$\frac{m_r}{m_o} = \frac{1}{K}\left[\frac{m_v \rho_l}{m_o \rho_v} + \left(1 - \frac{m_v}{m_o}\right)\right]^{-1} \qquad [15.8.1]$$

where m is the mass of fluid, ρ is the density, and subscripts l, o, r and v denote the liquid, initial, retained and vapour, respectively. For a full bore release the value of the constant K is unity.

Further correlation of the results is given in terms of the superheat ΔT_{onb} required for vapour nucleation, and hence nucleate boiling. The expression for superheat is

$$\Delta T \approx \frac{4RT_{sat}^2 \sigma}{\Delta H_v M \delta P_l} \qquad \rho_l \gg \rho_v;\ 4\sigma/P_l\delta \ll 1 \qquad [15.8.2]$$

where ΔH_v is the latent heat of vaporization, M is the molecular weight, P is the absolute pressure, R is the universal gas constant, T is the absolute temperature, ΔT is the superheat, δ is the diameter of the vapour nucleus, σ is the surface tension and subscripts onb and sat denote onset of nucleate boiling and saturation, respectively.

The proportion of retained liquid was expressed in terms of the ratio (h/h_v) of the height h of the liquid residue to the height h_v of the vessel. The limiting value of the superheat, or value at the onset of nucleate boiling (ΔT_{onb}), corresponds to the situation where all the liquid is retained and thus $h/h_v = 1$.

For Refrigerant 11 the value of ΔT_{onb} is about 1.9 K. Using this value in Equation 15.8.2 with the appropriate physical value for Refrigerant 11 a value of δ can be obtained. Then, utilizing this value, the superheat ΔT_{onb} for other substances can be obtained.

The correlation given by Fletcher for the superheat effect is shown in Figure 15.21. The proportion of retained liquid h/h_v falls off rapidly with increase in the ratio $\Delta T/\Delta T_{onb}$ and for values in excess of about 12 falls to less than 5%.

A comparison of the proportion of liquid retained in the tank cars in several transport incidents with that estimated from Equation [15.8.1] is given as:

Incident	*Liquid*	*Mass fraction retained* (%)	
		Reported	*Estimated*
Pensacola	Ammonia	50	40
Mississauga	Chlorine	10	13
Youngstown	Chlorine	44	37

Further work on vaporization following vessel rupture has been described by Schmidli, Bannerjee and Yadigaroglu (1990).

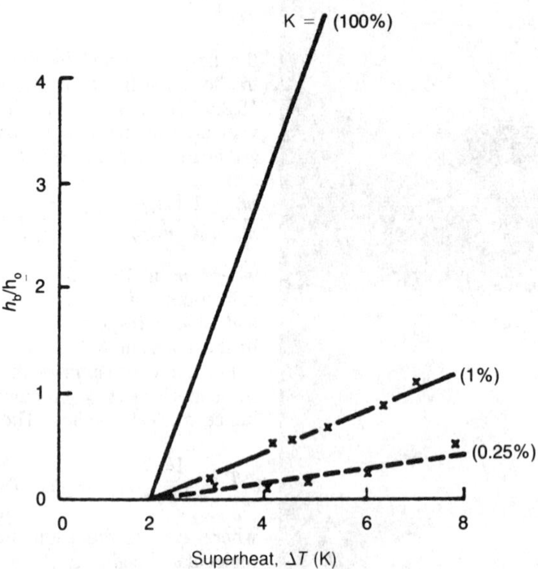

Figure 15.20 Sudden depressurization of a vessel containing superheated liquid: effect of a restriction on the height to which the liquid rises (B. Fletcher, 1982) (Courtesy of the Institution of Chemical Engineers)

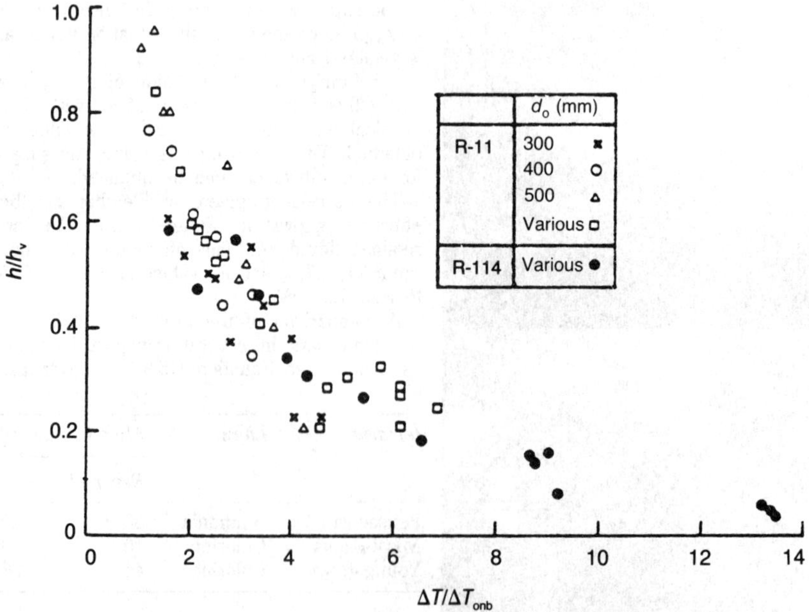

Figure 15.21 Sudden depressurization of a vessel containing superheated liquid: depth of liquid remaining (B. Fletcher, 1982) (Courtesy of the Institution of Chemical Engineers)

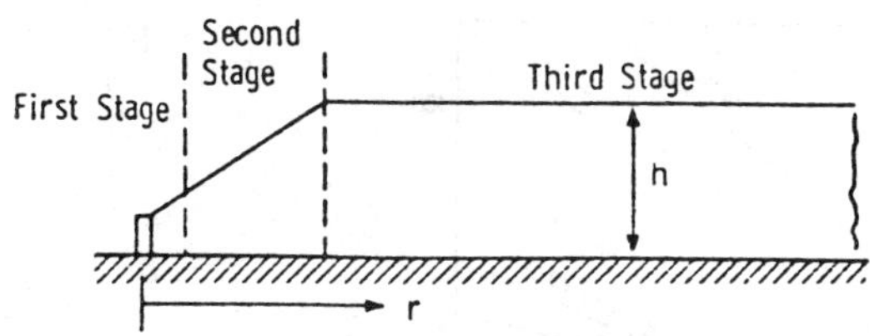

Figure 15.22 Rupture of a vessel containing superheated liquid: cloud expansion process (Hardee and Lee, 1975) (Courtesy of Pergamon Press)

15.8.2 Hardee and Lee model

An investigation of the rapid depressurization of a vessel containing a superheated liquid has been described by Hardee and Lee (1975). The growth of the cloud is considered to occur in three stages, as shown in Figure 15.22. The first stage is the expansion of the fluid from the original vessel pressure to atmospheric pressure, the second the entrainment of air, and the third the dispersion of the cloud.

For the first stage the expansion is assumed to be isentropic and the flash fraction at the end of the expansion is given by the entropy balance

$$x_1 s_{v1} + (1 - x_1)s_{f1} = x_2 s_{v2} + (1 - x_2)s_{f2} \qquad [15.8.3]$$

where s is the specific entropy, x is the mass fraction of vapour and subscripts f and v denote liquid and vapour and 1 and 2 initial and final states, respectively.

The velocity of the expanding fluid may then be obtained from the energy balance. Taking the initial velocity u_1 of the liquid as zero,

$$x_1 h_{v1} + (1 - x_1)h_{f1} = x_2 h_{v2} + (1 - x_2)h_{f2} + u_2^2/2 \qquad [15.8.4]$$

where h is the specific enthalpy and u is the velocity.

The first stage is evidently to be regarded as virtually instantaneous. At the end of this stage the fluid consists of vapour and spray. In the second stage air is entrained, the spray is vaporized and the cloud grows. At the end of this stage the cloud has attained a height which thereafter does not increase. In the third stage the cloud grows, but only in the radial direction.

In the second stage the momentum of the cloud increases linearly with time until the end of the stage is reached at the depressurization, or dump, time t_d. In the third stage the momentum remains constant. For the second stage the volume of the expanding fluid is

$$V_2 = u_2 A_2 t \qquad [15.8.5]$$

where A_2 is the area of cloud perpendicular to direction of expansion at end of the second stage, t is the time and V is the volume of fluid; subscript 2 denotes the value at the end of expansion down to atmospheric pressure. At the end of the second stage, and hence at time t_d,

$$V_2 = u_2 A_2 t_d \qquad [15.8.6]$$

The initial mass of fluid W may be expressed in terms of the volume given in Equation 15.8.6 as follows:

$$W = \rho_2 V_2 \qquad [15.8.7a]$$

$$= \rho_2 u_2 A_2 t_d \qquad [15.8.7b]$$

where ρ is the density.

Then, during the second stage, the momentum ψ is

$$\psi = \rho_2 u_2^2 A_2 t \qquad t < t_d \qquad [15.8.8a]$$

$$= \rho_2 u_2^2 A_2 t_d \qquad t > t_d \qquad [15.8.8b]$$

$$= W u_2 \qquad [15.8.8c]$$

The momentum may be obtained via the momentum balance from the initial pressure P_1 and final pressure P_2 as follows:

$$(P_1 - P_2)A_1 = \rho_2 u_2^2 A_2 \qquad [15.8.9]$$

where A_1 is the area of the vessel aperture.

Also, from Equations 15.8.8 and 15.8.9

$$\frac{A_1 t_d}{W} = \frac{u_2}{P_1 - P_2} \qquad [15.8.10]$$

For the momentum of a general moving vortex

$$\psi = c\rho u V \qquad [15.8.11]$$

where c is a constant. The value of the constant c is shown by the authors to be approximately 3/2. Then, from Equation 15.8.11:

$$\frac{dr}{dt} = u \qquad [15.8.12]$$

$$= \frac{2\psi}{3\rho V} \qquad [15.8.13]$$

where r is the radius of the cloud.

For a hemispherical cloud

$$V = \frac{2}{3}\pi r^3 \qquad [15.8.14]$$

Hence substituting from Equations 15.8.8a and 15.8.14 in Equation 15.8.13 and integrating

$$r = \left(\frac{2\rho_2 u_2^2 A_2}{\rho\pi}\right)^{\frac{1}{4}} t^{\frac{1}{2}} \qquad [15.8.15]$$

or, from Equation 15.8.8b

$$r = \left(\frac{2\psi}{\rho\pi t_d}\right)^{\frac{1}{4}} t^{\frac{1}{2}} \qquad [15.8.16]$$

At the end of this stage

$$r = \left(\frac{2\psi}{\rho\pi t_d}\right)^{\frac{1}{4}} t_d^{\frac{1}{2}} \qquad [15.8.17]$$

For the third stage the height h of the cloud is constant at the value of the radius r reached at the end of the second stage. Then for a cylindrical cloud

$$V = \pi r^2 h \qquad [15.8.18]$$

Since the cloud is now relatively dilute it is possible to set $\rho \approx \rho_a$, where ρ_a is the density of air. Then from Equations 15.8.13 and 15.8.18:

$$\frac{dr}{dt} = \frac{2\psi}{3\rho_a \pi r^2 h} \qquad [15.8.19]$$

and integrating

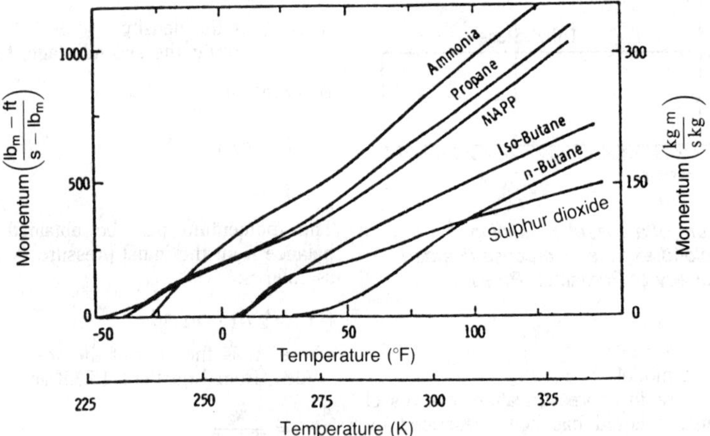

Figure 15.23 *Rupture of a vessel containing superheated liquid: momentum release per pound of fuel (Hardee and Lee, 1975) (Courtesy of Pergamon Press)*

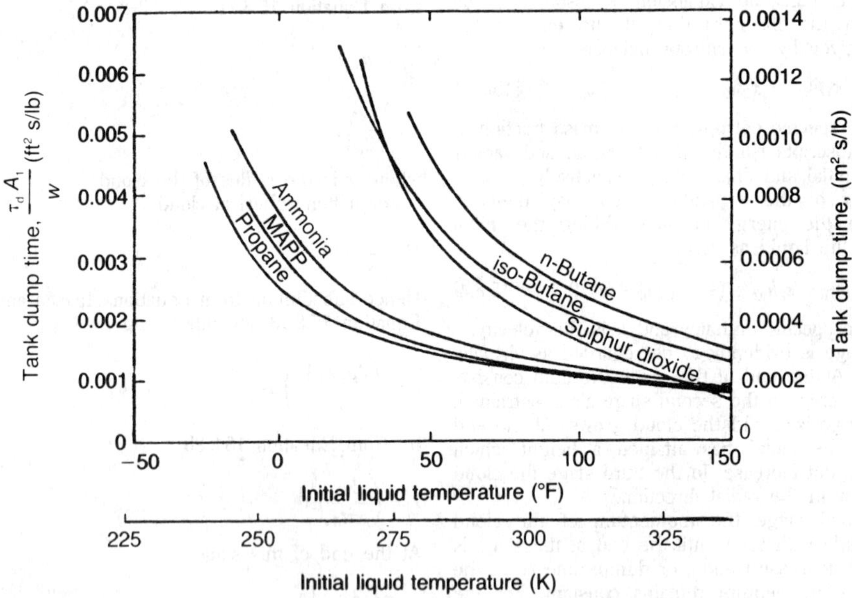

Figure 15.24 *Rupture of a vessel containing superheated liquid: vessel dump time (Hardee and Lee, 1975) (Courtesy of Pergamon Press)*

$$r = \left(\frac{2\psi}{\rho_a \pi h}\right)^{\frac{1}{3}} t^{\frac{1}{3}} \qquad [15.8.20]$$

The use of the model is assisted by the graphs given in Figures 15.23 and 15.24, which give the momentum and the dump time, respectively.

For a hazardous release there will be some concentration, corresponding to a cloud volume V, below which the cloud no longer presents a hazard. The model may be used to determine the range r and time t at which the concentration falls below this critical value.

The authors carried out experiments in which vessels containing up to 422 kg of propane and 436 kg of a methyl acetylene-propylene-propadiene (MAPP) mixture were suddenly depressurized and obtained good agreement for the increase of the cloud radius with time and for the cloud height.

Thus for a release of 29 kg of MAPP at a temperature of 284 K through an aperture of 0.728 m^2 the relation for the growth of the cloud radius, derived from Equation (15.8.20), is

$$r = 8.84t^{\frac{1}{3}} \qquad [15.8.21]$$

where r is the radius (m) and t the time (s). The cloud height is 2.74 m.

Hardee and Lee also describe the use of the model to determine cloud growth for the incidents of propane release at Lynchburg, Virginia, and of ammonia at Crete, Nebraska.

15.8.3 Model of Hess, Hoffman and Stoeckel

Another model for vessel rupture has been given by Hess, Hoffmann and Stoeckel (1974). The situation modelled is again the depressurization of a superheated liquid.

After the vessel burst the growth of the cloud is considered to occur in three stages. The first stage is expansion down to atmospheric pressure with flash-off of vapour and formation and partial evaporation of spray and some admixture of air. The second stage is the evaporation of the remaining spray and entrainment of air into the cloud. The third stage is the further entrainment of air and, if the fluid is flammable, the formation of a flammable mixture.

The model deals primarily with this third stage. For this stage, using spherical symmetry, the basic equation describing the variation of concentration c with radial distance r and time t is:

$$\frac{dc}{dt} = \epsilon\left[\frac{d^2c}{dr^2} + \frac{2}{r}\frac{dc}{dr}\right] \qquad [15.8.22]$$

where ϵ is eddy diffusion coefficient. Time zero is taken as the start of the third stage. The boundary conditions used are as follows: $0 < r < a$, $t = 0$, $c = c_o$; $r > a$, $t = 0$, $c = 0$. Then from a standard solution given by Carslaw and Jaeger (1959, p. 257), Equation 15.8.22 can be integrated to give

$$\frac{c}{c_o} = \frac{a^3}{6(\epsilon t)^{\frac{3}{2}}\pi^{\frac{1}{2}}}\exp\left(\frac{-r^2}{4\epsilon t}\right) \qquad [15.8.23]$$

where a is the radius of the initial cloud formed on completion of the bursting process and c_o is the concentration of vapour in the initial cloud.

The authors carried out experiments in which 40 mm and 60 mm diameter cylindrical vessels with a length/diameter ratio of 3.5:1, filled with liquid propane, were heated until they burst. The bursting of the vessels was recorded by high speed photography. They describe in detail the bursting of a 40 mm cylinder. The time for the vessel itself to burst was less than 1 ms. Following bursting the vapour cloud formed in two phases. In the first phase the propane expanded to ambient pressure and about half the liquid flashed off as vapour with the temperature dropping to about 230 K. It was estimated that without admixture of air the vapour would form a hemisphere with a radius of about 20 cm, but the actual radius was somewhat larger, implying that some air had already been mixed in. This phase was complete within about 3 ms. In the second phase the remaining liquid propane vaporized on mixing with air, from which it obtained the necessary heat of vaporization. It was estimated that from the energy balance the vapour–air mixture would form a hemisphere with a radius of about 46 cm. The average concentration of propane was then 20%. This phase was complete in about 8 ms.

The concentrations of vapour obtained following this bursting process were compared with those predicted by using Equation 15.8.23. The radius of and concentration in the initial cloud were obtained from the experiment. The eddy diffusion coefficient ϵ was obtained from the Prandtl mixing length formula:

$$\epsilon = lu' \qquad [15.8.24]$$

where l is the mixing length and u' the fluctuating velocity. The mixing length l was taken as the average eddy diameter and the fluctuating velocity u' was taken as the speed at which the eddies moved, both quantities being obtained from photographs. The parameters required for solution of Equation 15.8.23 were then

$$c_o = 20\%$$

$$a = 46\,\text{cm}$$

$$\epsilon = 0.52\,\text{m}^2/\text{s}$$

Using these values in the model, good agreement was obtained with the results of this experiment.

15.8.4 Model of Maurer *et al.*

A development of the model of Hess, Hoffman and Stoeckel has been described by Maurer *et al.* (1977). In the second model it is again assumed that there is a central highly mixed core at uniform concentration and that the concentration outside this core decays in a Gaussian manner, but the second model differs in that it allows for the variability of the eddy diffusion coefficient, which in the first model is assumed to be constant.

For the air entrainment stage, using spherical symmetry, the basic equations given are:

$$\frac{c}{c_c} = \exp\left\{X_c^2\left[1 - \left(\frac{r}{r_g}\right)^2\right]\right\} \qquad r > r_g \qquad [15.8.25]$$

$$c_c = f(X_c)\frac{V_g}{(4\epsilon t)^{1.5}} \qquad [15.8.26]$$

$$r_g = X_c(4\epsilon t)^{\frac{1}{2}} \qquad [15.8.27]$$

where V_g is the effective volume of vapour, X_c is a parameter characterizing the mass fraction of gas in the core, and subscripts c and g denote the core concentration and the core radius, respectively. The volume V_g is taken as twice the mass divided by the density at standard pressure and temperature; this allows for the use of spherical symmetry to model a hemispherical cloud.

The parameter $f(X_c)$ is obtained experimentally. The value used is 1.36, which corresponds to a fraction of vapour in the core of 50%.

In addition, as observed by A.F. Roberts (1981/82), the following additional relation for the eddy diffusion coefficient may be obtained from a graph of experimental results given by the authors:

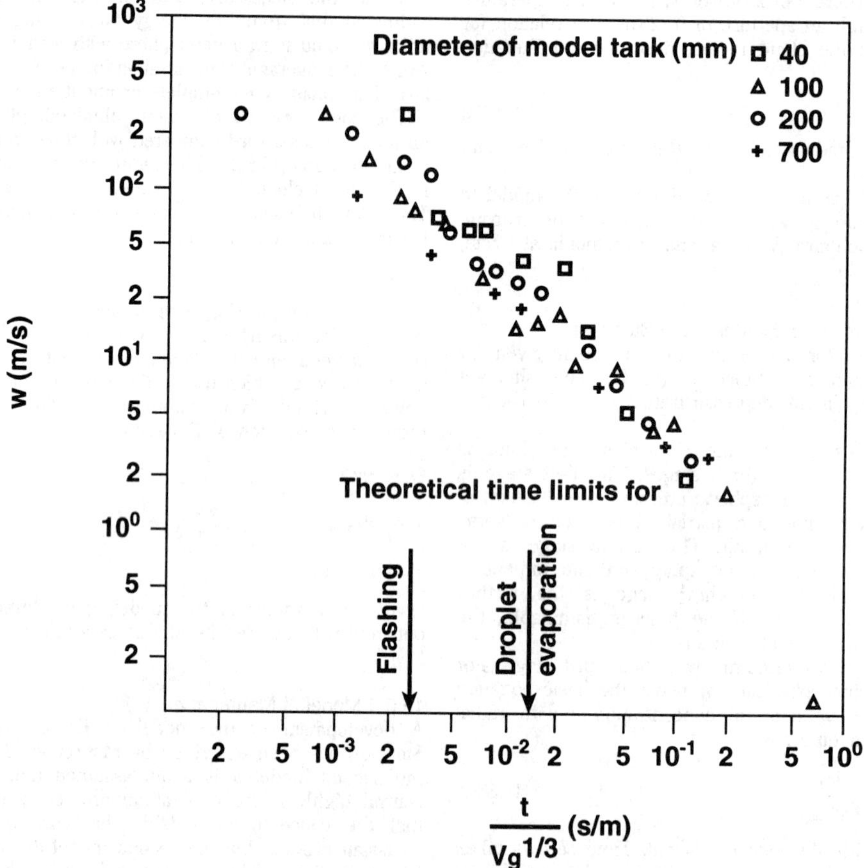

Figure 15.25 Rupture of a vessel containing superheated liquid (Maurer et al., 1977): propagation velocity of vapour cloud boundary due to condensation from humid air (Courtesy of DECHEMA)

$$\epsilon = 0.75 V_g^{\frac{1}{3}} \left(\frac{t}{V_g^{\frac{1}{3}}} \right)^{-\frac{1}{4}} \quad [15.8.28]$$

Experiments were carried out in which cylindrical vessels containing propylene were heated up and then made to burst by mechanical means or by a small explosive charge. The quantities of propylene used ranged from 0.124 to 452 kg, the vessel in this latter case being 0.7 m diameter × 2.8 m long. The propylene was heated to temperatures of 50–80°C corresponding to pressures of 22–39 bar. After rupture and flash evaporation the vapour was ignited and overpressure generated was recorded. Some 50–65% of the liquid vaporized and the rest formed spray. The propagation velocity of the cloud boundary was followed by the condensation contours, which correspond to vapour concentrations of 1–3%.

The expansion velocities obtained, and the turbulence generated, were high. Only a very small proportion of the mechanical energy available was required to generate this turbulence.

The authors correlate their results in terms of a reduced time $t/V_g^{\frac{1}{3}}$, a dimensionless time $\tau = \epsilon t/V_g^{\frac{2}{3}}$ and a scaled distance $r/V_g^{\frac{1}{3}}$. Some of these results are shown in Figures 15.25 and 15.26. Figure 15.25 shows the propagation velocity w of the cloud boundary. The plateau at short times corresponds to the expansion velocities predicted thermodynamically, which ranged from 240 to 370 m/s for vessel preheats of 35 and 80°C, respectively. Figure 15.26 gives the overpressures measured due to flash evaporation.

Figure 15.27 shows the concentration profiles and contours predicted by the model. Figure 15.28 shows the ratio $V_{g,ign}/V_g$ of the flammable volume $V_{g,ign}$ to the total volume V_g as a function of t. The maximum flammable fraction is 70% and the period during which a flammable mixture exists at all is $0.15 < t < 0.4$ s.

The flammable mixture showed considerable unmixedness, as evidenced by the afterburning of the cloud subsequent to the pressure-generating combustion. Investigation revealed that only some 30–50% of the macromixed gas participated in the pressure-generating combustion. Then taking the maximum flammable fraction as 70% and the fraction of the latter participating in pressure-generating combustion as 40%, the maximum fraction of the original contents participating in pressure-

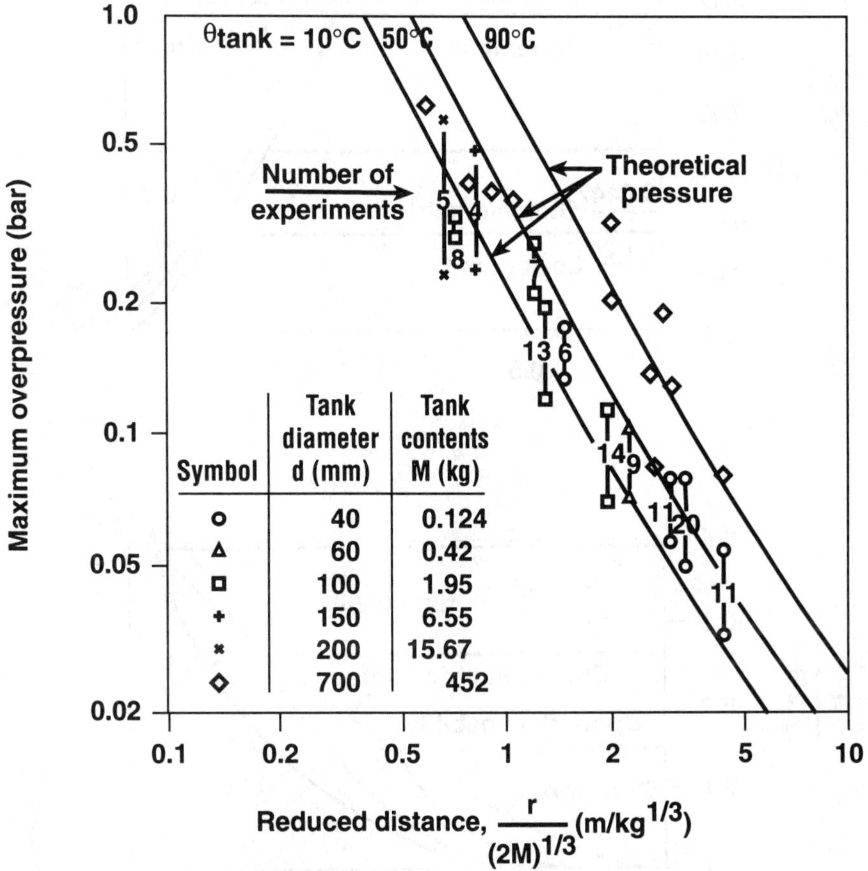

Figure 15.26 Rupture of a vessel containing superheated liquid: blast wave overpressure due to flash expansion (Maurer et al., 1977) (Courtesy of DECHEMA)

generating combustion is 0.28 (= 0.4 × 0.7). The authors therefore quote a round figure for this fraction of 30%.

15.8.5 Roberts model

A discussion of vessel rupture in the context of the fireballs has been given by A.F. Roberts (1981/82). Roberts compares the models of Hardee and Lee and of Maurer *et al.* in respect of the predictions of the transition from control by diffusion to control by gravity slumping, using the transition criterion – proposed by Jagger and Kaiser (1981):

$$N = \frac{gr\Delta}{(\mathrm{d}r/\mathrm{d}t)^2} \qquad [15.8.29]$$

with

$$\Delta = \frac{\rho_\mathrm{v} - \rho_\mathrm{a}}{\rho_\mathrm{a}} \qquad [15.8.30]$$

where g is the acceleration due to gravity, r is the radial distance, Δ is a relative density, ρ is the density, and subscripts a and v denote air and vapour, respectively. The value of N at the transition point is taken to be unity.

For the Hardee and Lee model Roberts obtains

$$N = \frac{6g\Delta \cdot t}{\alpha} \qquad [15.8.31]$$

and for the model of Maurer *et al.*

$$N = 0.6\frac{\rho_\mathrm{v}^{\frac{5}{6}}}{\rho_\mathrm{a}}\left(1 - \frac{\rho_\mathrm{a}}{\rho_\mathrm{v}}\right)M^{\frac{1}{6}}t^{\frac{1}{2}} \qquad [15.8.32]$$

where M is the initial mass of liquid and α is the momentum per unit mass.

Then, for a transition with $N = 1$, Roberts obtains for propane for the time t_g to transition for the Hardee and Lee model

$$t_\mathrm{g} = 0.05\alpha \qquad [15.8.33]$$

and for that of Maurer *et al.*

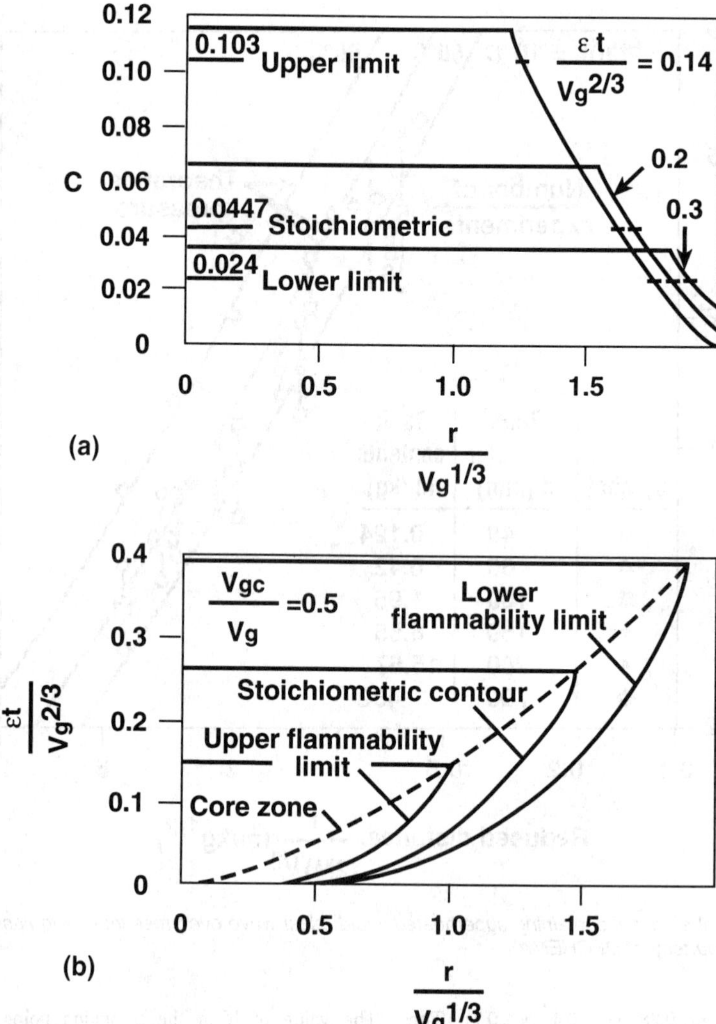

Figure 15.27 Rupture of a vessel containing superheated liquid – model predictions (Maurer et al., 1977): (a) concentration profiles; (b) concentration contours (Courtesy of DECHEMA)

$$t_g = 10M^{-\frac{1}{3}} \qquad [15.8.34]$$

Also in the Hardee and Lee model the concentration c at transition is

$$c = 22M^{\frac{1}{4}}\alpha^{-\frac{3}{2}} \qquad [15.8.35]$$

Thus in both models gravity slumping becomes more important as the release size increases. With increase in release size the concentration at transition to gravity slumping also increases.

Roberts also gives the following expression for the radial velocity w in the Hardee and Lee model:

$$w = \left(\frac{\alpha\rho_v}{128\pi\rho_a}\right)^{\frac{1}{4}}\left(\frac{t}{V_g^{\frac{1}{3}}}\right)^{-\frac{3}{4}} \qquad [15.8.36]$$

with

$$V_g = 2M/\rho_v \qquad [15.8.37]$$

where V_g is the volume of vapour released and the factor of 2 is included to allow for the hemispherical symmetry. Substituting values for propane of $\rho_v/\rho_a = 2$ and $\alpha = 220$ m/s gives

$$w = 1.0\left(\frac{t}{V_g}\right)^{-\frac{3}{4}} \qquad [15.8.38]$$

He comments that this also agrees well with the data of Maurer *et al.*

15.8.6 Other treatments

A further discussion of vessel rupture has been given by Appleton (1984 SRD R303).

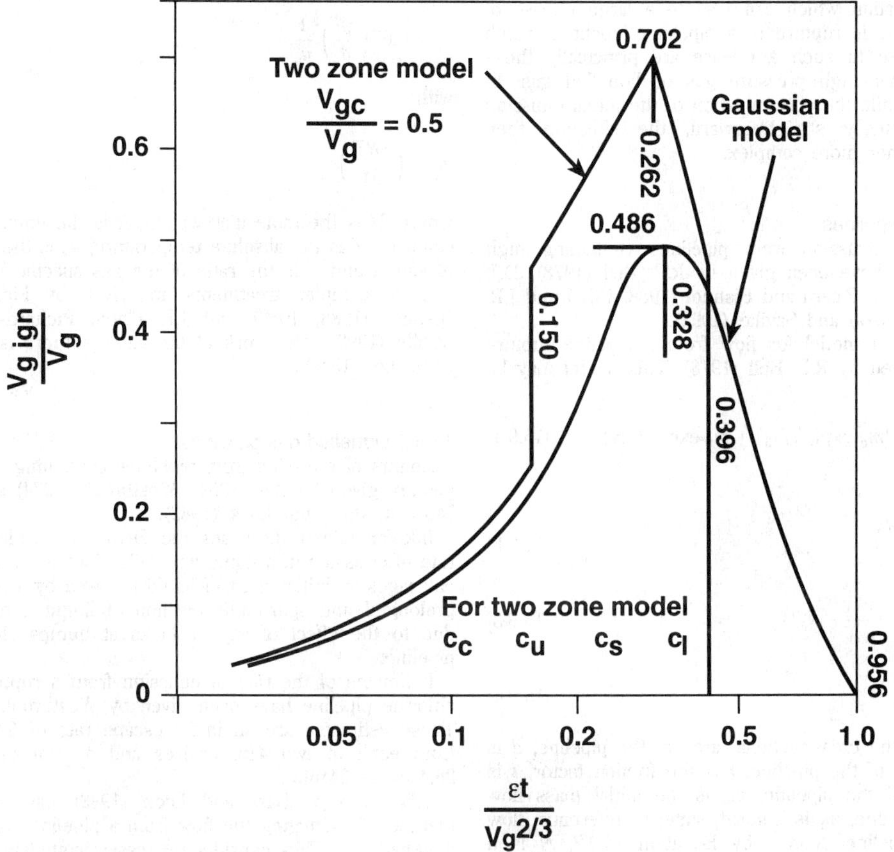

Figure 15.28 Rupture of a vessel containing superheated liquid – model predictions (Maurer et al., 1977): fractions of flammable and reacting gas (Courtesy of DECHEMA)

15.8.7 Entrainment of air

The cloud formed from a vessel rupture entrains air. The determination of the amount of air entrained is important, because the cloud constitutes the source term for dispersion models and these models are rather sensitive to the degree of air entrainment. The air entrained may be estimated using the models just described. Alternatively, use may be made of empirical relations.

Air entrainment has been discussed by Griffiths and Kaiser (1979 SRD R154) in relation to rupture of vessels containing liquefied ammonia. They propose for ammonia the rule-of-thumb that in such a release the ammonia mixes with ten times as much air by mass. They also state: 'to the accuracy that is possible here, this is also equivalent to mixing in about ten times as much air by volume'.

This estimate is based partly on the incident at Potchefstroom in which a horizontal pressure vessel burst, releasing 38 te of ammonia through a hole some 7 m^2. From eyewitness accounts the immediate resulting gas cloud was 'about 150 m in diameter and about 20 m in depth'. They also refer to the work of van Ulden (1974) in which 1 te of Freon 12 was poured onto water, giving immediate vigorous boiling and rapid formation of a gas cloud in which the volume of air was about ten times that of the Freon.

The rule-of-thumb that for a vessel rupture the volume of air entrained is some ten times that of the gas released has been widely used in the modelling of heavy gas dispersion.

Griffiths and Kaiser give an example of a release of 20 te of liquefied ammonia, giving 4 te of vapour and 16 te of spray. If sufficient dry air is entrained at 20°C just to evaporate the drops, this requires 400 te of air, or a factor of 20 on the mass of ammonia released. This gives a cloud at a temperature of − 33°C and a relative density of 1.18. The density of the cloud is greatly in excess of that of air and the cloud will exhibit heavy gas behaviour.

15.8.8 Storage tank rupture

A quite different problem arises when rupture of an atmospheric storage tank occurs. If the rupture is sufficiently sudden and complete, a wave of liquid surges outwards and may overflow the bund. This hazard is considered further in Chapter 22.

15.9 Pipeline Rupture

Another situation which can lead to a large release of gas or vapour is rupture of a pipeline. Pipelines which may give rise to such a release are principally those carrying either high pressure gas or liquefied gas. In each case, while the determination of the initial emission rate is relatively straightforward, the situation then becomes rather more complex.

15.9.1 Gas pipelines

Accounts of emission from pipelines containing high pressure gas have been given by R.P. Bell (1978), D.J. Wilson (1979b), Picard and Bishnoi (1988, 1989) and J.R. Chen, Richardson and Saville (1992).

An empirical model for flow from a pipeline rupture has been given by R.P. Bell (1978). This model may be written as

$$m = \frac{m_o}{m_o + m_r}[m_o \exp(-t/\tau_2) + m_r \exp(-t/\tau_1)] \quad [15.9.1]$$

with

$$m_r = A\left(\frac{2P\rho d}{flN}\right)^{\frac{1}{2}} \quad [15.9.2]$$

$$\tau_1 = \frac{W_o}{m_r} \quad [15.9.3]$$

$$\tau_2 = \frac{W_o m_r}{m_o^2} \quad [15.9.4]$$

where A is the cross-sectional area of the pipeline, d is the diameter of the pipeline, f is the friction factor, l is the length of the pipeline, m_o is the initial mass flow from the pipeline, m_r is a steady-state, or reference, flow from the pipeline defined by Equation 15.9.2, N is a correction factor, P is the absolute pressure, W_o is the initial mass holdup in the pipeline, ρ is the density and τ_1 and τ_2 are time constants. The friction factor f is evidently the Darcy friction factor f_D (= 8ϕ).

For the correction factor N, Bell gives the empirical formula

$$N = 8[1 - \exp(-26400d/l)] \quad [15.9.5]$$

He also states that he used a value of 0.02 for the friction factor f.

Bell also discusses the dispersion from the pipeline.

D.J. Wilson (1979b) has derived a rather more complex model, but also states that Bell's model compares quite well.

A formulation of the Bell model has been given by the CCPS (1987/2) and this may be put in the following form:

$$m = \frac{m_o}{1+\psi}[\exp(-t/\tau_2) + \psi \exp(-t/\tau_1)] \quad [15.9.6]$$

with

$$\psi = m_r/m_o \quad [15.9.7a]$$

$$\psi = W_o/m_o\tau_1 \quad [15.9.7b]$$

$$\tau_1 = W_o/m_r \quad [15.9.8]$$

$$\tau_2 = \psi^2\tau_1 \quad [15.9.9]$$

where ψ is a parameter. Furthermore,

$$\tau_1 = 0.67\left(\frac{\gamma fl}{d}\right)^{\frac{1}{2}}\frac{1}{u_s} \quad [15.9.10]$$

with

$$u_s = \left(\frac{\gamma RT}{M}\right)^{\frac{1}{2}} \quad [15.9.11]$$

where M is the molecular weight, R is the universal gas constant, T is the absolute temperature, u_s is the velocity of sound and γ is the ratio of the gas specific heats.

More complex treatments are given by Picard and Bishnoi (1988, 1989) and J.R. Chen, Richardson and Saville (1992). The work of the latter authors is treated in Section 15.9.3.

15.9.2 Liquefied gas pipelines

Accounts of emission from pipelines containing liquefied gas are given Inkofer (1969), Westbrook (1974) and T.B. Morrow, Bass and Lock (1982).

Inkofer (1969) discusses the factors determining the rate of emission in a liquid ammonia pipeline rupture. He envisages an initial spurt of liquid followed by a period of prolonged and spasmodic ejection of liquid and vapour due to the effect of vapour locks at humps along the pipeline.

Estimates of the rate of emission from a rupture in a chlorine pipeline have been given by Westbrook (1974). These estimates are an initial escape rate of 60.3 ton/h from each of two 4 in. orifices and a total escape of 28.4 ton in 24 min.

T.B. Morrow, Bass and Lock (1982) have given a method of estimating the flow from a pipeline containing liquefied gas. They consider two cases: complete rupture and partial rupture. For the two-phase critical flow at the rupture point the method utilizes Fauske's slip equilibrium model. Upstream of the rupture point it is assumed that there is a transition, or interface, point at which the flow changes from liquid flow to two-phase flow and that this point is that at which the pressure corresponds to the bubble point of the liquid. The basic equation for the two-phase pressure drop is

$$\frac{dP}{dz} = \frac{\phi_g^2 4fu_{fs}}{2dv_f} \quad [15.9.12]$$

where f is the friction factor, u is the velocity, v is the specific volume, z is the distance along the pipe from the rupture point, ϕ_g is a parameter, and subscripts f and fs denote liquid and superficial value for liquid, respectively. The friction factor f is the Fanning friction factor (= 2ϕ).

The actual liquid velocity u_f is

$$u_f = \frac{u_{fs}}{1-Y} \quad [15.9.13]$$

where Y is the void fraction. The parameter ϕ_g is

$$\phi_g^2 = \frac{1}{(1-Y)^2} \quad [15.9.14]$$

Then from Equations 15.9.12–15.9.14

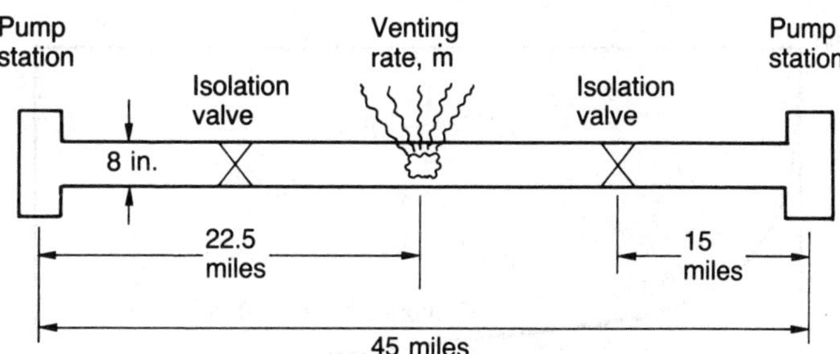

- Pipeline flow rate — 40 000 bbl/day
- Isolation valve spacing — 45, 15 and 5 miles
- Pump stations — 45 miles apart
- Pump shut-off time — 5 min and 20 min after rupture
- Valve closure time — 5 min after pump shut-off time
- Break sizes — Complete break (venting from both ends), and partial break (initial venting rate 10 000 bbl/day)

Figure 15.29 *Emission from an LPG pipeline: configuration of pipeline studied (T.B. Morrow, Bass and Lock, 1982) (Courtesy of the American Society of Mechanical Engineers)*

$$\frac{dP}{dz} = \frac{2fu_f^2}{dv_f} \qquad [15.9.15]$$

Using the same assumption for slip as made by Fauske in the derivation of the slip equilibrium model, namely

$$\frac{u_g}{u_f} = \left(\frac{v_g}{v_f}\right)^{\frac{1}{2}} \qquad [15.9.16]$$

where subscript g denotes vapour. Expressing the quality in terms of the fluid enthalpies, and hence of pressure, the authors obtain

$$\left(\frac{mv_f}{Au_f}\right)^2 = f(P) \qquad [15.9.17]$$

Hence

$$\frac{dP}{dz} = \frac{2fm^2 v_f}{A^2 d f(P)} \qquad [15.9.18]$$

In order to integrate Equation 15.9.18 the mass flow m is expressed as a function of the distance z from the rupture point:

$$m = m_e\left(\frac{m_e - m_i}{z_i}\right)z \qquad [15.9.19]$$

where subscript e denotes the rupture point, or exit, and i denotes the transition, or interface, point. Expressions are also given for the volume of the vapour space, and hence of the liquid removed, between the transition and rupture points. The model allows the mass flow from the rupture point and the position of the transition point to be determined as a function of time.

The authors have used the model to study ruptures in the propane pipeline system shown in Figure 15.29. They investigated different pumping station distances, valve spacings and shut-down times of the upstream pump, with valves shut down 5 min after the pump. Typical results for the flow profiles given by ruptures are shown in Figure 15.30. Curve A shows the flow for a complete rupture. This flow is for some time unaffected by pump shut-down and valve closure, since the two-phase interface moves relatively slowly. For example, for a spacing between pumps of 15 miles the interface is estimated to reach the isolation valves at $7\frac{1}{2}$ miles only after some 25 min. Curve B in Figure 15.30(b) shows the flow for a partial rupture. In this case the pressure is maintained at the value before rupture until the pump shuts off, then falls, and finally reaches a new steady value corresponding to the bubble point. The fall may approximate to a ramp or may change to a steeper slope as the isolation valve shuts.

A description of the actual pipeline break at Port Hudson has been given by the NTSB (1972 PAR-72–01) and Burgess and Zabetakis (1973 BM RI 7752). The event is described in Case History A52.

15.9.3 BLOWDOWN

An account has been given in Section 15.8 of the model BLOWDOWN developed for the prediction of conditions during blowdown of, or release from, a vessel. This model has also been adapted for use in the determination of conditions arising during outflow of fluid from a pipeline, whether as intentional blowdown or accidental release.

One of the topics in which the Piper Alpha Inquiry was interested was the explanation of the change in pressure which occurred during the accident in the gas pipelines connected to the platform. Evidence on this topic, based on the use of the BLOWDOWN code, was presented to the inquiry by Richardson (1989b).

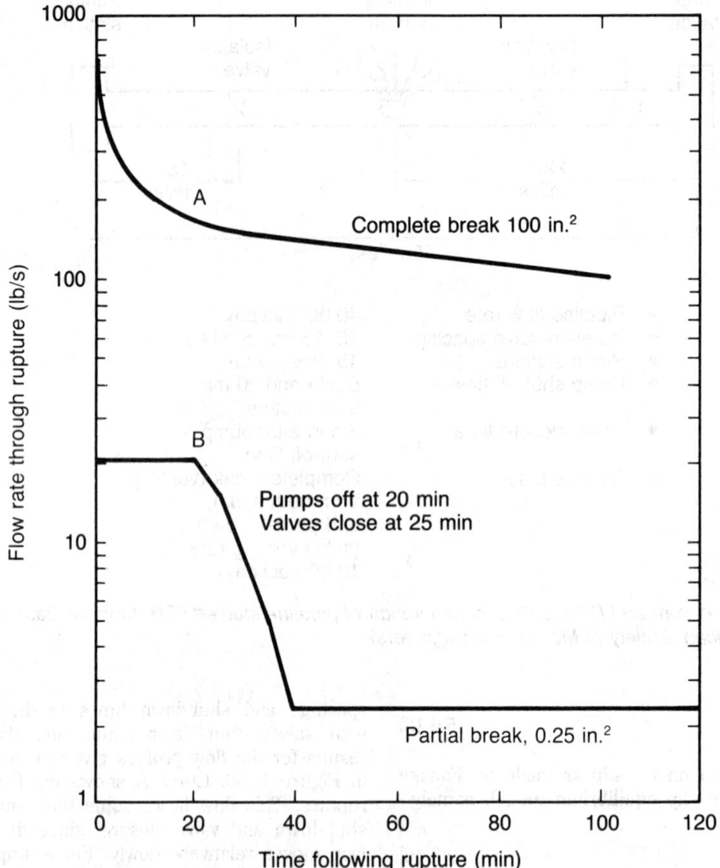

Figure 15.30 *Emission from an LPG pipeline: flow of propane for partial and complete ruptures. Isolation valve spacing = 45 miles (T.B. Morrow, Bass and Lock, 1982) (Courtesy of the American Society of Mechanical Engineers)*

An account of the extensions made to the BLOWDOWN model to allow it to be applied to a gas pipeline has been given by J.R. Chen, Richardson and Saville (1992). The basic model comprises the unsteady one-dimensional Euler equations supplemented by friction and heat transfer relations. Chen *et al.* describe the investigation of a number of methods for the solution of these equations, including finite difference methods (FDMs), the method of characteristics (MOC), a hybrid method and a wave tracing method. The technique found to be most efficient and accurate was the multiple wave tracing method.

The investigation made use of the test problem of Picard and Bishnoi (1989) and of data on the depressurization of the pipeline between Piper Alpha and MCP-01 on the night of the accident. Results are presented which include values of the intact end pressure, the open end pressure and the release rate.

15.9.4 Pipeline isolation

The quantity of material released from a pipeline rupture depends on whether the section of line is isolated or not. The arrangements for line break detection and isolation are therefore of the greatest importance in minimizing escapes.

15.10 Vaporization

If the fluid which escapes from containment is a liquid, then vaporization must occur before a vapour cloud is formed. The process of vaporization determines the rate at which material enters the cloud. It also determines the amount of air entrained into the cloud. Both aspects are important for the subsequent dispersion. Selected references on vaporization are given in Table 15.7.

15.10.1 Vaporization situations

In considering the generation of a vapour cloud from the liquid spillage, the following situations can be distinguished:

(1) A volatile liquid at atmospheric temperature and pressure, e.g. acetone.
(2) A superheated liquid –
  (a) at ambient temperature and under pressure, e.g. butane;

Table 15.7 Selected references on vaporization

Liquid spreading
Stoker (1957); Abbott (1961); Fay (1969); Hoult (1969, 1972b); Fannelop and Waldman (1972); Webber and Brighton (1986 SRD R317)

Spillage and evaporation
Hinchley and Himus (1924); O.G. Sutton (1934); R.W. Powell and Griffiths (1935); Lurie and Michailoff (1936); R.W. Powell (1940); T.K. Sherwood (1940); Wade (1942); Pasquill (1943); Linton and Sherwood (1950); Langhaar (1953); Burgoyne (1965b); Humbert-Basset and Montet (1972); Pancharatnam (1972a,b); Mackay and Matsugu (1973); AGA (1974); V.J. Clancey (1974a, 1977c); Kalelkar and Cece (1974); Drake and Reid (1975); Feind (1975); Opschoor (1975b, 1978); Japan Gas Association (1976); Meadows (1976); Bellus, Vincent *et al.* (1977); Deacon (1977); HSE (1978b); Reid and Wang (1978); Shaw and Briscoe (1978 SRD R100); Flothmann, Heudorfer and Langbein (1980); N.C. Harris (1980, 1982); Reijnhart *et al.* (1980); Reijnhart and Rose (1980); Raj (1981, 1991); Brutsaert (1982); Drivas (1982); Hilder (1982); O'Shea (1982); Jensen (1983); Kunkel (1983); Webber and Brighton (1986, 1987 SRD R390); Anon. (1985i); Brighton (1985b, 1986 SRD R371, 1987 SRD R375, 1990); Pikaar (1985); Prince (1985 SRD R324); Lebuser and Schecker (1986, 1987); Moorhouse and Carpenter (1986); Kawamura and Mackay (1987); Webber (1987 SRD R421, 1988 SRD R404, 1989); Webber and Jones (1987); Studer, Cooper and Doelp (1988); Deaves (1989); Lantzy *et al.* (1990); Raj and Morris (1990); J. Singh and McBride (1990); J.L. Woodward (1990); Alp and Matthias (1991); Mikesell *et al.* (1991); Angle, Brennan and Sandhu (1992); Frie *et al.* (1992); Stramigioli and Spadoni (1992); J. Cook and Woodward (1993a,b); Leonelli, Stramigioli and Spadoni (1994); Takeno *et al.* (1994)

Solar radiation, solar constant
Fritz (1954); C.O. Bennett and Myers (1962); N. Robinson (1966); Coulson (1975); L.B. Nielsen *et al.* (1981); Multer (1982); Lide (1994)

Burning spills
Opschoor (1975a,b, 1980)

Spillage and vaporization on water
Blokker (1964); Burgess, Murphy and Zabetakis (1970 RI 7448); Enger and Hartman (1972a,b); Hoult (1972a); Kneebone and Boyle (1973); Mackay and Matsugu (1973); Raj and Kalelkar (1973, 1974); Raj, Hagopian and Kalelkar (1974); Drake, Jeje and Reid (1975); Eisenberg, Lynch and Breeding (1975); Opschoor (1975a, 1977, 1980); Raj *et al.* (1975); Dincer, Drake and Reid (1977); Griffiths (1977 SRD R67); Havens (1977, 1980); HSE (1978b); Shaw and Briscoe (1978 SRD R100); Raj and Reid (1978a,b); Reid and Smith (1978); Raj (1979, 1981, 1991); Shell Research Ltd (1980); Anon. (1982f); Chang and Reid (1982); Dodge *et al.* (1983); Hirst and Eyre (1983); Waite *et al.* (1983); Prince (1985 SRD R324); Zumsteg and Fannelop (1991); J. Cook and Woodward (1993a)

Flash-off, flash fraction, aerosol
NRC (Appendix 28 *Aerosols*); Pinto and Davis (1971); A.R. Edwards (1978); Mesler (1985); API (1986 Publ. 4456); Ramsdale (1986 SRD R382, 1987 SRD R401); Emerson (1987b); Bache, Lawson and Uk (1988); Britter and McQuaid (1988); I. Cook and Unwin (1989); Lantzy *et al.* (1990); Schmidli, Bannerjee and Yadigaroglu (1990); D.W. Johnson (1991); D.W. Johnson and Diener (1991); Raj (1991); Kukkonen and Vesala (1991, 1992); J.L. Woodward and Papadourakis (1991)
Vaporization in jets (see *Table 15.40*)

Spill control
BDH (1970); Anon. (1972b); May, McQueen and Whipp (1973); AIChE (1974/99, 1988/100, 1989/70); May and Perumal (1974); University Engineers Inc. (1974); Weismantel (1974); Welker, Wesson and Brown (1974); Lindsey (1975); Otterman (1975); Wirth (1975); L.E. Brown *et al.* (1976); W.D. Clark (1976b); Martinsen and Muhlenkamp (1977); D.P. Brown (1978); Harsh (1978a,b); Norman and Dowell (1978, 1980); Whiting and Shaffer (1978); Kletz (1982i); Jeulink (1983)

Vaporization suppression
R.H. Hiltz (1982, 1987); Dilwali and Mudan (1986, 1987); Norman (1987); ASTM (1988 F1129); Dimaio and Norman (1988, 1990); Norman and Dimaio (1989); Leone (1990); Norman and Swihart (1990); Martinsen (1992); Howell (1993); Scheffler, Greene and Frurip (1993)

(b) at high temperature and under pressure, e.g. hot cyclohexane.

(3) A refrigerated liquefied gas at low temperature but at atmospheric pressure, e.g. cold methane.

The vaporization of the liquid is different for these three cases. In the first case the liquid after spillage is approximately at equilibrium and evaporates relatively slowly. In the second case the liquid is superheated and flashes off when spilt, and then undergoes slower evaporation. The first category of a superheated liquid, at ambient temperature but under pressure, is that of a liquefied gas, while the second, at high temperature and under pressure, is that of a liquid heated above its normal boiling point. The third case is that of a refrigerated liquefied gas which on spillage evaporates rapidly at first and then more slowly.

15.10.2 Vaporization of a superheated liquid

If the liquid released from containment is superheated, a proportion flashes off as vapour. The remaining liquid is cooled by the removal of the latent heat of vaporization and falls to its atmospheric boiling point.

The theoretical adiabatic flash fraction (TAFF) of vapour so formed is usually determined by the simple heat balance

$$\phi = \frac{c_{pl}}{\Delta H_v}(T_i - T_b) \qquad T_i > T_b \qquad [15.10.1]$$

where c_p is the specific heat, T is the absolute temperature, ΔH_v is the latent heat of vaporization, ϕ is the fraction of liquid vaporized and subscripts b, i and l denote boiling point, initial and liquid, respectively.

An alternative expression which takes account of the differential nature of the vaporization is

Table 15.8 Flow characteristics of discharge of liquid ammonia (after Wheatley, 1986) (Courtesy of the Institution of Chemical Engineers)

Storage temperature (°C)	*Flow type*	*Axial velocity* (m/s)	*Flow diameter* (cm)	*Flash fraction* (%)	*Spray characteristics* *Drop size* (μm)	*Settling velocity* (m/s)	*Inclination* (°)
20	Liquid	154	13	17.6	25	0.013	0.0048
20	Two-phase	37.7	61	18.7	420	1.4	2.1
-34	Liquid				1900	5.2	16

$$(1-\phi)c_{\mathrm{pl}}(-\mathrm{d}T) = \Delta H_{\mathrm{v}}\mathrm{d}\phi \quad [15.10.2]$$

and hence

$$\phi = 1 - \exp\left[-\frac{c_{\mathrm{pl}}}{\Delta H_{\mathrm{v}}}(T_{\mathrm{i}} - T_{\mathrm{b}})\right] \qquad T_{\mathrm{i}} > T_{\mathrm{b}} \quad [15.10.3]$$

Equations 15.10.1 and 15.10.3 give the theoretical fraction of vapour formed. The sudden growth and release of vapour bubbles also results in the formation of liquid droplets, or spray. The mass of liquid in spray form is generally of the same order as that in the initial vapour flash and may exceed it.

This spray then either vaporizes, increasing the vapour cloud, or rains out as liquid, forming a pool on the ground. The total amount of vapour formed, both that from the initial flash and that from the evaporation of spray, constitutes the ultimate flash.

For the fraction of liquid which forms spray a rule-of-thumb frequently used is that it is equal to the initial vapour flash (Kletz, 1977j). Another rule-of-thumb is that if the fraction flashing off is small it may be appropriate to assume that the spray fraction is two or three times the initial vapour flash-off (W.G. High, 1976).

A discussion of spray and rainout following the discharge of a flashing liquid, with particular reference to liquid ammonia, has been given by Wheatley (1986). Two discharge situations may be distinguished: metastable flow of a superheated liquid and choked two-phase flow. These two cases have been modelled by Wheatley. As shown in Table 15.8, the flow characteristics obtained are significantly different.

In the non-choked liquid flow case there is a large increase in diameter of the flow at the outlet. Such large increases have been observed in the Desert Tortoise tests (Koopman *et al.*, 1986). In the choked two-phase flow case the flow velocity is much higher. The flash fraction is similar in the two cases.

The drops formed are subject to shear stress and there is a maximum size of drop given by the drop Weber number:

$$\mathrm{We}_{\mathrm{d}} = \frac{u_{\mathrm{d}}^2 d\rho_{\mathrm{d}}}{\sigma_{\mathrm{d}}} \quad [15.10.4]$$

where d is the diameter of the drop, u is the velocity, ρ is the density, σ is the surface tension and subscript d denotes drop. Table 15.8 shows the drop sizes obtained for the two cases.

The extent to which rainout of the drops occurs depends on a balance between the inertial and gravity forces. The drops will have higher inertia than the vapour but will be more affected by gravity. Wheatley treats rainout in terms of the inclination of the bounding drop trajectory, which is obtained from the ratio of the settling velocity to the horizontal velocity of the drops. The settling velocity and inclination of the drops for the cases considered are shown in Table 15.8. Both for the superheated liquid flow and the choked two-phase flow the inclinations are too small to give significant rainout, but for the refrigerated liquid flow the inclination is much larger and rainout could be significant.

Following flash-off the residual liquid is at its normal boiling point. Vaporization then continues as a rate-limited process. This secondary stage of rate-limited vaporization is usually regarded as relatively less important compared with the initial flash-off, particularly with respect to the formation of flammable gas clouds.

The spray fraction from a ruptured vessel was considered in Section 15.7. Frequently, very little liquid is left in the vessel. The spray fraction in fireballs is discussed in Chapter 16.

15.10.3 Mass and heat transfer for a pool

In general, vaporization from a pool is a mass and heat transfer limited process, but in a specific case there may be one dominant transfer mode. Thus for vaporization of a cryogenic liquid it is the heat transfer from the ground to the pool which governs the rate of vaporization.

For a pool there is only one mode of mass transfer, that between the liquid surface and the atmosphere. By contrast, for heat transfer there are several modes, namely heat transfer by convection between the liquid surface and the atmosphere, by conduction between the liquid and the ground, and by radiation, both solar and between the liquid surface and the atmosphere.

The heat balance on the pool is thus

$$wc_{\mathrm{pl}}\frac{\mathrm{d}T}{\mathrm{d}t} = -A_{\mathrm{p}}(q_{\mathrm{v}} - q_{\mathrm{cn}} - q_{\mathrm{cd}} - q_{\mathrm{r}}) \quad [15.10.5]$$

with

$$q_{\mathrm{r}} = q_{\mathrm{rs}} + q_{\mathrm{ra}} - q_{\mathrm{rl}} \quad [15.10.6]$$

$$q_{\mathrm{v}} = m_{\mathrm{v}}\Delta H_{\mathrm{v}} \quad [15.10.7]$$

where A_{p} is the area of the pool, c_{pl} is the specific heat of the liquid, m_{v} is the mass vaporization rate per unit area, q_{cd} is the heat flow per unit area by conduction from the ground to the pool, q_{cn} is the heat flow per unit area by convection from the atmosphere to the pool, q_{r} is the net heat flow per unit area by radiation to the pool, q_{ra} is the heat flow per unit area by radiation from the atmosphere to the pool, q_{rl} is the heat flow per unit area by radiation from the pool to the atmosphere, q_{rs} is the heat flow per unit area by solar radiation to the pool, q_{v}

is the heat required per unit area for vaporization, t is the time and w is the mass of liquid.

The mass balance on the pool is

$$-\frac{dw}{dt} = A_p m_v \qquad [15.10.8]$$

with

$$m_v = \frac{k_m M(p^o - p_\infty)}{RT} \qquad [15.10.9]$$

where k_m is the mass transfer coefficient, p^o is the vapour pressure of the liquid and p_∞ is the partial pressure of the liquid outside the influence of the pool. The partial pressure p_∞ can usually be set equal to zero.

For a multi-component liquid Drivas (1982) has modified Equation 15.10.9 as follows:

$$m_v = \frac{k_m M_T^o}{RTn_T} \frac{\Sigma x_i^o M_i p_{is}}{\Sigma x_i^o M_i} \exp(-kp_{is}t) \qquad [15.10.10]$$

with

$$k = \frac{k_m A_p}{RTn_T} \qquad [15.10.11]$$

where M_i is the molecular weight of component i, M_T^o is the initial mass of the liquid, n_T is the number of moles of the liquid, p_{is} the vapour pressure of component i, x_i^o is the initial mole fraction of component i in the liquid and k is a constant.

In general, heat transfer may be correlated by the factor j_h of Chilton and Colburn (1934) and mass transfer by the corresponding factor j_m. For a geometry where the characteristic dimension is the diameter d these j factors are defined as

$$j_h = \mathrm{St}\,\mathrm{Pr}^{0.67} \qquad [15.10.12]$$

$$j_m = \frac{k_m}{u}\frac{C_{Bm}}{C_T}\mathrm{Sc}^{0.67} \qquad [15.10.13]$$

where C_{Bm} is the log mean concentration difference, C_T is the total concentration and k_m is the mass transfer coefficient. The term C_{Bm}/C_T is the drift factor.

The definitions of, and relations between, the Reynolds number Re, the Prandtl number Pr, the Nusselt number Nu, the Schmidt number Sc and the Stanton number St are

$$\mathrm{Re} = \frac{ud\rho}{\mu} \qquad [15.10.14]$$

$$\mathrm{Pr} = \frac{c_p\mu}{k} \qquad [15.10.15]$$

$$\mathrm{Nu} = \frac{hd}{k} \qquad [15.10.16]$$

$$\mathrm{Sc} = \frac{\mu}{\rho D} \qquad [15.10.17]$$

$$\mathrm{St} = \frac{h}{c_p\rho} \qquad [15.10.18]$$

$$\mathrm{St} = \frac{\mathrm{Nu}}{\mathrm{Re}\,\mathrm{Pr}} \qquad [15.10.19]$$

where d is the pool diameter, D is the diffusion coefficient, k is the thermal conductivity and μ is the viscosity.

Given a correlation for j_h, the heat transfer coefficient h may be obtained from the definition of j_h in Equation 15.10.12. Then, using the approximate equality,

$$j_m \approx j_h \qquad [15.10.20]$$

and using the definitions of j_h and j_m in Equations [15.10.12] and [15.10.13]

$$k_m = \frac{hC_T}{c_p\rho C_{Bm}}\left(\frac{\mathrm{Pr}}{\mathrm{Sc}}\right)^{0.67} \qquad [15.10.21]$$

Alternatively, if the correlation available is for j_m, this may be used to obtain j_h. For other geometries the diameter d is replaced by the characteristic length.

A number of workers have measured mass transfer from a plane liquid surface to a gas stream above, particularly in the evaporation of water. An account is given by Coulson and Richardson (1977–). Their results have been correlated by T.K. Sherwood (1940) in terms of a point Reynolds number Re_x, where the characteristic dimension is the distance across the liquid surface to the point considered. His results may be expressed as

$$j_m = 0.0415\mathrm{Re}_x^{-0.21} \qquad [15.10.22]$$

over the approximate range $10^4 < \mathrm{Re}_x < 10^5$. Equation [15.10.22] may be used, in conjunction with Equation [15.10.20], to obtain both the mass and heat transfer coefficients.

The mass transfer coefficient may also be correlated directly in terms of the Sherwood number Sh:

$$\mathrm{Sh} = \frac{k_m d}{D} \qquad [15.10.23]$$

Fleischer (1980) has given the following correlation for the Sherwood number over a plane liquid surface:

$$\mathrm{Sh} = 0.037\mathrm{Sc}^{\frac{1}{3}}(\mathrm{Re}^{0.8} - 15200) \qquad [15.10.24]$$

The transition from the laminar to the turbulent regime is taken here to occur at Re = 320 000.

In the case of the plane liquid surface there is also available for the mass transfer a specific correlation for the evaporation rate derived by Sutton and Pasquill. This is described in Section 15.10.4.

Heat transfer from the ground is complex, because it involves an unsteady-state process. An account of unsteady-state heat transfer, including many standard cases, is given by Carslaw and Jaeger (1959). The basic equation is

$$\frac{d^2\theta}{dz^2} = \frac{1}{\alpha_s}\frac{d\theta}{dt} \qquad [15.10.25]$$

with

$$\theta = T - T_s \qquad [15.10.26]$$

$$\alpha_s = \frac{k_s}{\rho_s c_{ps}} \qquad [15.10.27]$$

where α is the thermal diffusivity, θ the temperature above the datum, or soil, temperature and subscript s denotes soil.

For the case of constant liquid temperature, Equation 15.10.25 may be solved with the boundary conditions $z = 0, \theta = \theta_l$; $z = \infty, \theta = \theta_s$. For temperature this gives

$$\frac{\theta}{\theta_1} = \text{erfc}\left[\frac{z}{2(\alpha_s t)^{\frac{1}{2}}}\right] \qquad [15.10.28]$$

where z is the vertical distance down into the ground and subscript l denotes liquid.

For the heat flow per unit area q

$$q = -k\frac{\mathrm{d}\theta}{\mathrm{d}z} \qquad [15.10.29]$$

Then from Equations [15.10.28] and [15.10.29]

$$q = \left(\frac{k_s \rho_s c_{ps}}{\pi}\right)^{\frac{1}{2}} \frac{\theta_1}{t^{\frac{1}{2}}} \qquad [15.10.30]$$

For the total heat transferred per unit area Q

$$Q = \int_0^t q\mathrm{d}t \qquad [15.10.31]$$

and from Equations [15.10.30] and [15.10.31]

$$Q = 2\left(\frac{k_s \rho_s c_{ps}}{\pi}\right)\theta_1 t^{\frac{1}{2}} \qquad [15.10.32]$$

For the more complex case of varying liquid temperature, Equation 15.10.25 may be solved with the boundary conditions $t = 0, \theta = 0$; $z = 0, \theta = \theta_1(t)$; $z = \infty, \theta = 0$. For temperature this gives

$$\theta = \frac{z}{2(\pi\alpha_s)^{\frac{1}{2}}}\int_0^t \frac{\theta_1}{(t-\tau)^{\frac{3}{2}}}\exp\left[-\frac{z^2}{4\alpha_s(t-\tau)}\right]\mathrm{d}\tau \qquad [15.10.33]$$

The decrease in temperature of a vaporizing pool will be approximately exponential. An analytical solution of Equation 15.10.33 for the case $\theta_l = \exp(-\lambda t)$, where λ is a constant, is given by Carslaw and Jaeger.

For heat transfer by radiation the net heat radiated to the pool is given by Equation [15.10.6]. The individual terms in the equation are given by

$$q_{rs} = \psi(1-a) \qquad [15.10.34]$$

$$q_{ra} = \epsilon_a \sigma T_a^4 \qquad [15.10.35]$$

$$q_{rl} = \epsilon_l \sigma T_l^4 \qquad [15.10.36]$$

where a is the albedo of the liquid surface, ϵ is the emissivity, σ is the Stefan–Boltzmann constant, ψ is the solar constant and subscript a denotes atmosphere. The value of the solar constant is 1373 W/m^2 (Lide, 1994).

The albedo and emissivities have been discussed by Mackay and Matsugu (1973), who propose for the albedo a value of 0.14 and for the emissivities of the atmosphere and of the liquid values of 0.75 and 0.95, respectively.

The albedo of a plane water surface has also been discussed by Shaw and Briscoe (1978 SRD R100). The albedo is actually a function of the sun's elevation. They give the following values:

Elevation (°)	90	50	40	20	10	0
Albedo	0.02	0.025	0.034	0.134	0.348	1.0

Data on the albedo for different surfaces are given by Oke (1978). A further discussion of albedo is given by Byrne *et al.* (1992 SRD R553).

15.10.4 Sutton–Pasquill model

A model for mass transfer in evaporation from a liquid surface at constant temperature has been derived by O.G. Sutton (1934). Sutton's model was subsequently modified by Pasquill (1943) and the modification was accepted by O.G. Sutton (1953). In the form given by Pasquill, the relations for evaporation from a rectangular and a circular pool are:

$$E = Ku_1^{(2-n)/(2+n)} x_o^{2/(2+n)} y_o \quad \text{rectangular pool} \qquad [15.10.37]$$

$$E = K'u_1^{(2-n)/(2+n)} r^{(4+n)/(2+n)} \quad \text{circular pool} \qquad [15.10.38]$$

with

$$K = \chi_o\left(\frac{2+n}{2-n}\right)^{(2-n)/(2+n)}\left(\frac{2+n}{2\pi}\right)\sin\left(\frac{2\pi}{2+n}\right) \Gamma\left(\frac{2}{2+n}\right) \times a^{2/(2+n)} z_1^{-n^2/(4-n^2)} \qquad [15.10.39]$$

$$K' = \frac{2^{2+n}\pi^{\frac{1}{2}}\Gamma\left(\frac{3+n}{2+n}\right)K}{\Gamma\left(\frac{8+3n}{2(2+n)}\right)} \qquad [15.10.40]$$

$$a = \left|\frac{[(\pi/2)]^{1-n}(2-n)^{1-n}n^{1-n}}{(1-n)(2n-2)^{2(1-n)}}\right| k^{2(1-n)}\lambda^n z_1^{(n^2-n)/(2-n)} \qquad [15.10.41]$$

$$\chi_o = \frac{Mp^o}{RT} \qquad [15.10.42]$$

where E is the evaporation rate (g/s), k is the von Karman constant, M is the molecular weight, n is the diffusion index, p^o is the vapour pressure of the liquid (dyn/cm^2), r is the radius of the pool (cm), R is the universal gas constant (erg/g mol K), T is the absolute temperature (K), u_1 is the wind speed at height z_1 (cm/s), x_o is the downwind length of the pool (cm), y_o is the crosswind width of the pool (cm), z_1 is the height at which the wind speed u_1 is measured (cm), λ is a parameter (cm^2/s), χ_o is the concentration (g/cm^3) and a, K and K' are constants.

The constant K may be written as:

$$K = \chi_o f_1(n) a^{2/(2+n)} z_1^{-n^2/(4-n^2)} \qquad [15.10.43]$$

$$= \chi_o f_1(n)^{2/(2+n)} f_2(n)^{2/(2+n)} k^{4(1-n)/(2+n)} \chi\lambda^{2n/(2+n)} z_1^{-n/(2+n)} \qquad [15.10.44]$$

with

$$f_1(n) = \left[\frac{2+n}{2-n}\right]^{(2-n)/(2+n)} \left[\frac{2+n}{2\pi}\right] \sin\left[\frac{2\pi}{2+n}\right] \Gamma\left[\frac{2}{2+n}\right] \qquad [15.10.45]$$

$$f_2(n) = \frac{(\pi/2)^{1-n}(2-n)^{1-n}n^{1-n}}{(1-n)(2-2n)^{2-2n}} \qquad [15.10.46]$$

Sutton's treatment is based on the wind velocity profile:

$$\frac{u}{u_1} = \left[\frac{z}{z_1}\right]^{n/(2-n)} \qquad [15.10.47]$$

where u is the wind speed (cm/s) and z is the height (cm).

Sutton identified the parameter λ with the kinematic viscosity of the air ν. Pasquill gives a table of values for the constants K and K' for the following case:

$$k = 0.4$$

$$\lambda = \nu = 0.147 \text{ cm}^2/\text{s}$$

$$\chi_o = 1 \text{ g/cm}^3$$

$$z_1 = 1 \text{ cm}$$

$$u_1 = 500 \text{ cm/s}$$

Hence

n	K	K'
0.20	0.0094	0.0283
0.25	0.0180	0.0537
0.30	0.0313	0.0926

He also gives the results of an experiment for which

$$M = 157$$

$$T = 290 \text{ K}$$

$$x_o = 10 \text{ cm}$$

$$y_o = 20 \text{ cm}$$

$$n = 0.219$$

Hence

$$K = 0.0122\chi_o$$

and

$$\frac{E}{p^o} = 1.89 \times 10^{-6} (\text{g/s})/(\text{dyn/cm}^2)$$

However, Pasquill argued that the parameter λ should properly be the diffusion coefficient D of the vapour in air and this argument was accepted by O.G. Sutton (1953). Hence in Equations 15.10.39–15.10.44 λ should be set equal to D.

Equations for the Sutton–Pasquill model have been given by V.J. Clancey (1974a) in the following form. For a rectangular pool set square to the wind direction:

$$E = 1.2 \times 10^{-10} \frac{Mp^o}{T} u^{0.78} x_o^{0.89} y_o \quad \text{rectangular pool} \qquad [15.10.48]$$

$$E = 3.6 \times 10^{-10} \frac{Mp^o}{T} u^{0.78} r^{1.89} \quad \text{circular pool} \qquad [15.10.49]$$

Clancey states that in deriving Equations 15.10.48 and 15.10.49 he has used values $\lambda = 0.147 \text{ cm}^2/\text{s}$ and $D \approx 0.075 \text{ cm}^2/\text{s}$, the latter being typical for hydrocarbons. In determining K from Equation 15.10.39 he appears to have included both the terms $\lambda^{2n/(2+n)}$ and $D^{2n/(2+n)}$, which for $n = 0.25$ have the values 0.653 and 0.56, respectively. Equations 15.10.48 and 15.10.49 may be corrected for this by dividing by 0.653.

Relations for the Sutton–Pasquill model in SI units are given in the *Rijnmond Report*. The report gives Equations [15.10.37] and [15.10.38] together with the equations

$$K = a\chi_o D^{2n/(2+n)} z_1^{-n/(2+n)} \qquad [15.10.50]$$

$$K' = a'\chi_o D^{2n/(2+n)} z_1^{-n/(2+n)} \qquad [15.10.51]$$

where D is the diffusion coefficient of vapour in air (m²/s), E is the evaporation rate (kg/s), r is the radius of the pool (m), x_o is the downwind length of the pool (m), y_o is the crosswind width of the pool (m), χ_o is the concentration of the vapour in the air (kg/m³), u is the wind speed at height z_1 (m/s) z_1 is the height at which the wind speed is measured (m) and a and a' are constants. The constant a is different from that given in Equation 15.10.41. The following values are given for the groups K/χ_o and K'/χ_o:

n	K/χ_o	K'/χ_o
0.20	1.278×10^{-3}	3.846×10^{-3}
0.25	1.579×10^{-3}	4.685×10^{-3}
0.30	1.786×10^{-3}	5.285×10^{-3}

15.10.5 Vaporization of a volatile liquid

Vaporization of a volatile liquid is governed by the mass and heat transfer rate processes described in Section 15.10.3. If the vaporization rate is low so that the heat transfer to the pool is sufficient to prevent a fall in the temperature of the liquid, the vaporization is a mass transfer limited process which depends on the vapour pressure of the liquid and the wind flow across the pool. At higher vaporization rates, where the heat transfer to the pool is insufficient to prevent chilling, the liquid temperature will fall, approaching asymptotically to a steady state value. The approach will generally be approximately exponential.

It has been shown theoretically by Flothmann, Heudorfer and Langbein (1980) that for pools of liquids such as hydrogen cyanide and acrolein, neglect of heat transfer from the ground can lead to appreciable error in the estimation of the vaporization rate, particularly under conditions of low solar radiation and/or low wind speeds. Vaporization from a pool of a volatile liquid constitutes a steady, continuous source of vapour.

Higher vaporization rates still occur where the liquid is at or close to its boiling point. There is relatively little information available on vaporization under these conditions. This situation has been discussed by Shaw and

Briscoe (1978 SRD R100), who consider the vaporization of a liquid such as butane (b.p. −0.5°C) on a cold surface at 0°C. The liquid boils and the vaporization is limited by heat transfer, the heat transfer being by convection from the air and by radiation, that from the ground being minimal. They propose the use of the relations of Mackay and Matsugu (1973), given in Section 15.10.6, both for convective heat transfer from the air to the pool and for the net radiative heat transfer. For the former the heat transfer coefficient may be obtained from the expression for the mass transfer coefficient. They also discuss the question of the concentration and temperature at which the necessary physical properties should be determined.

A correlation for the vaporization rate for a liquid at a temperature above ambient has been given by the Centre for Chemical Process Safety (CCPS, 1987/2):

$$m_v = C_D u(\rho_{gs} - \rho_g) \qquad [15.10.52]$$

where C_D is the drag coefficient, u is the wind speed, ρ_g is the density of gas 10 m above the pool and ρ_{gs} is the saturation density of the gas at ambient conditions. The value of C_D is about 10^{-3} if u is measured 10 m above the pool.

15.10.6 Mackay and Matsugu model

A model for the vaporization of a volatile liquid from a circular pool has been given by Mackay and Matsugu (1973). The authors carried out experiments on the evaporation of water, gasoline and cumene from 4 × 4 ft and 4 × 8 ft pans.

In their model the heat balance on the pool is given by Equation 15.10.5, except that the term for heat conduction from the ground is omitted. For mass and heat transfer between the pool and the atmosphere:

$$N = \frac{k_m(p - p_\infty)}{RT_l} \qquad [15.10.53]$$

$$q_{cn} = h(T_a - T_l) \qquad [15.10.54]$$

with

$$k_m = Cu^{0.78}x^{-0.11} \qquad [15.10.55]$$

$$h = k_m \rho_v c_v (\mathrm{Sc}/\mathrm{Pr})^{0.67} \qquad [15.10.56]$$

$$C = k\mathrm{Sc}^{-0.67} \qquad [15.10.57]$$

where c is the specific heat, h is the heat transfer coefficient, k_m is the mass transfer coefficient, N is the mass transferred per unit area, p is the partial pressure of the vapour, q is the heat flow per unit area by convection from the atmosphere to the pool, u is the wind speed, x is the diameter of the pool, C and k are constants and subscripts a, l, v and ∞ are air, liquid, vapour and atmosphere beyond influence of pool, respectively. In SI units the value of k is 0.00482 $m^{0.33}/s^{0.22}$.

Equation 15.10.55 is based on the Sutton–Pasquill model with the diffusion index n = 0.25. The authors state that in terms of the factor j_m their correlation for the mass transfer coefficient is equivalent to

$$j_m = 0.0565\mathrm{Re}^{-0.22} \qquad [15.10.58]$$

where Re is in the range 7×10^4 to 4.6×10^5.

For the net radiation the authors use Equations 15.10.34–15.10.36 with the values of the parameters quoted in Section 15.10.3.

As stated earlier, the model given by these authors does not include a term for heat transfer from the surface beneath the pool. This term was presumably not significant in their particular experiments, which involved evaporation pans.

15.10.7 Spreading of a liquid

Accounts of the spreading of liquids are given in *Water Waves* (Stoker 1957) and *Oil on Sea* (Hoult, 1969) and by Abbott (1961) and Webber and Brighton (1986). Much of the work is concerned with the spreading of oil slicks, which is treated by Fay (1969, 1973) and by Hoult (1969, 1972b).

If the liquid is not confined within a hollow or a bund, but is free to spread, it is usually necessary to take account of such spreading and to use a treatment based on simultaneous spreading and vaporization. Spreading of a liquid on land and on sea differ somewhat and may require separate treatment.

The volume V of the liquid pool is a function of its radius r and height h

$$V = \pi r^2 h \qquad [15.10.59]$$

The growth of the pool may be expressed in terms either of the velocity u of the pool edge

$$\frac{dr}{dt} = u \qquad [15.10.60]$$

or of its acceleration.

As the pool spreads, it passes through different spreading regimes. Initially the frictional resistance may be negligible so that the gravity force is balanced by the inertia force. At a later stage the inertia may be negligible so that the gravity force is balanced by the resistance force.

The pool growth has most commonly been modelled by the energy balance relation

$$\frac{dr}{dt} = (2g'h)^{\frac{1}{2}} \qquad [15.10.61a]$$

or, more generally,

$$\frac{dr}{dt} = c(g'h)^{\frac{1}{2}} \qquad [15.10.61b]$$

with

$$g' = g \quad \text{on land} \qquad [15.10.62a]$$

$$g' = g\Delta \quad \text{on water} \qquad [15.10.62b]$$

$$\Delta = \frac{\rho_w - \rho_l}{\rho_w} \qquad [15.10.63]$$

where g is the acceleration due to gravity, g' is the reduced gravity, Δ is the reduced density, c is a constant and subscripts l and w denote liquid and water, respectively.

The value of the constant c in Equation 15.10.61b has been extensively treated in the literature. A value of c of 1.4 is often used. This value is close to the value of $2^{\frac{1}{2}}$ implied in Equation 15.10.61a.

It may be noted that Equations 15.10.59 and 15.10.61 give

$$\frac{dr}{dt} \propto r^{-1} \qquad [15.10.64]$$

A treatment which takes account of both the gravity–inertia and gravity–viscous regimes has been given by Raj and Kalelkar (1974). In the former the gravity force F_g is opposed by the inertial force F_i:

$$F_g = \pi\rho g' h^2 r \qquad [15.10.65]$$

$$F_i = -C\pi\rho h r^2 \frac{d^2 r}{dt^2} \qquad [15.10.66]$$

where C is a constant. The constant C represents the ratio of the inertia of the liquid system to the inertia which would exist if all the liquid were moving at the acceleration of the spill edge. The value given for C is 0.754.

Then, equating the two forces F_g and F_i gives

$$r\frac{d^2 r}{dt^2} = -\frac{g'h}{C} \qquad [15.10.67]$$

The authors also give a treatment of the gravity–viscous regime.

A relation which takes account of the frictional resistance has been given by Webber and Brighton (1986). They suggest that the natural model to use for the spreading of a thin layer of liquid is the shallow water, or shallow layer, equations (Stoker, 1957). From these equations they derive the relation

$$\frac{d^2 r}{dt^2} = \frac{4g'h(1-s)}{r} - F \qquad [15.10.68]$$

where F is a friction term and s is a shape factor. The friction factor is

$$F = C_L \frac{u^2}{h} + \left(\frac{3}{2}\right)^5 \frac{\nu u}{h^2} \quad \text{on land} \qquad [15.10.69a]$$

$$F = C_W \frac{u^2}{h} + 1.877 \frac{\nu u}{h^2}(1-f) \quad \text{on water} \qquad [15.10.69b]$$

where ν is the kinematic viscosity and C_L and C_W are constants for land and sea, respectively. The factor f allows for the effect on the shear stress on the pool of the radial motion in the water beneath it. In Equation 15.10.69 the right-hand side contains both a turbulent and a laminar term.

In a further treatment, Webber and Brighton (1986 SRD R371) have expressed their results in the form

$$\frac{d^2 r}{dt^2} = -\frac{gV}{\pi}\frac{\beta}{r^3} \qquad [15.10.70]$$

where β is a shape factor. The shape factor β determines the edge height of the pool and its value depends on the pool boundary conditions. Relevant factors include the surface tension of the liquid and the resistance of the ambient medium. For a convex pool $\beta < 0$, for a concave pool $\beta > 0$, for a cylindrical pool $\beta = 0$. A value of $\beta = -4$ corresponds to a pool with zero edge height (zero surface tension) and one of $\beta = 4$ corresponds to a pool with its centre tending to zero thickness. Hence from Equation 15.10.70

$$\frac{d^2 r}{dt^2} \propto r^{-3} \qquad [15.10.71]$$

15.10.8 Vaporization of a cryogenic liquid

In the general case of vaporization of a cryogenic liquid, or refrigerated liquefied gas, it is necessary to consider simultaneous spreading and vaporization. Considering initially just the vaporization, this is governed by heat transfer from the ground to the liquid. There is a short period of very rapid vaporization followed by a relatively steady lower rate of vaporization.

The rate of vaporization is obtained from the heat flux as given by Equation 15.10.30:

$$q = \left(\frac{k_s \rho_s c_{ps}}{\pi}\right)^{\frac{1}{2}} \frac{\theta_1}{t^{\frac{1}{2}}} \qquad [15.10.72]$$

Equation [15.10.72] may be written as

$$q = A_v t^{-\frac{1}{2}} \qquad [15.10.73]$$

with

$$A_v = \left(\frac{k_s \rho_s c_{ps}}{\pi}\right)^{\frac{1}{2}} \theta_1 \qquad [15.10.74]$$

where A_v is the vaporization parameter. Equation [15.10.73] has been confirmed experimentally.

It should be noted that for zero time Equation 15.10.72 gives an infinite heat transfer rate and is not applicable, but the period over which the heat transfer rate is overestimated is very short.

Most studies of the vaporization of cryogenic liquids have been concerned with liquefied natural gas (LNG) or liquefied ammonia. These are considered in the following sections.

Large-scale tests on the vaporization of refrigerated liquid ethylene have been conducted in Japan by a working group headed by Professor Hikita Tsuyoshi under the auspices of the Ministry of International Trade and Industry (MITI) (1976) and other bodies. This work was part of a programme of tests on various aspects of hazards of liquid ethylene.

In one test 392 kg of liquid ethylene at a temperature of −104°C was poured into a 2.5 m diameter bund with a pebble floor. There was violent boiling of the liquid which caused an estimated 66% of it to vaporize or form spray within the first minute. The spillage gave rise to a cloud of ethylene with regions of mist and vapour.

The contours of the mist and of the vapour concentrations corresponding to lower explosive limit were similar, but not identical. The maximum distances reached by vapour concentrations of 2.7% (the lower explosive limit), 2.0% and 1.0% were 60, 80 and 86 m, respectively, all attained after 70 s. The mean wind speed was approximately 2.5 m/s.

The initial vaporization rate of the liquid ethylene over the first minute was fitted by the equation

$$v = 0.625 \exp(-0.0507t) \qquad [15.10.75]$$

where t is the time (s) and v is the regression rate (cm/s). After the first minute the rate of vaporization was much reduced.

A collection of experimental data for the validation of models of spills of cryogenic liquids on land and water has been made by Prince (1985 SRD R324).

Table 15.9 *Vaporization parameters for LNG on different substrates (after Drake and Reid, 1975) (Courtesy of the American Institute of Chemical Engineers)*

	ρ_s (g/cm^3)	c_{ps} (cal/g °C)	k_s (cal/s cm °C)	A_v (cal/s$^{1/2}$ cm^2)
Carslaw and Jaeger (1959, p. 497)				
Soil (average)	2.5	0.2	0.0023	3.41
Soil (sandy, dry)	1.65	0.19	0.00063	1.40
Soil (sandy, 8% moist)	1.75	0.24	0.0014	2.45
Concrete (1:2:4)	2.3	0.23	0.0022	3.39
AGA tests				
Soil (compacted)				8.5
Gaz de France				
Soil (dry, 15°C)				18.2
Soil (wet, 50°C)				12.7
Soil (wet, 15°C)				5.68

Vaporization from a pool of a cryogenic liquid constitutes initially a near-instantaneous source of vapour followed by steady, continuous source.

15.10.9 Spreading and vaporization of LNG

Accounts of the spreading and vaporization of LNG have been given by a number of workers including Burgess, Murphy and Zabetakis (1970 BM RI 7448), Humbert-Basset and Montet (1972), Kneebone and Boyle (1973), Feldbauer *et al.* (1972), Raj and Kalelkar (1973), the American Gas Association (AGA) (1974), Drake and Reid (1975), Opschoor (1977, 1980), Shaw and Briscoe (1978 SRD R100) and Raj (1979, 1981).

A general account of the spreading of a liquid such as LNG has been given in Section 15.10.7. In this section a description is given of work on the vaporization rate of a pool of LNG of fixed area on land or on water and of some simple treatments of simultaneous spreading and vaporization of LNG. Some principal models for the latter are given in the following sections.

Evaporation of liquid methane on common surfaces such as soil has been investigated by Burgess and Zabetakis (1962 BM RI 6099). They found that initially the vaporization rate v was limited by the rate of heat transfer from the substrate and was given by the relation

$$v = \frac{k_s}{\rho_l \Delta H_v} \frac{T_s - T_l}{(\pi \alpha_s t)^{\frac{1}{2}}} \qquad [15.10.76]$$

Equation 15.10.76 applied only to the initial rapid vaporization period, which did not exceed 1 min.

Values of the thermal conductivity and thermal diffusivity for different substrates are given by Carslaw and Jaeger (1959), as described below. There is relatively little difference between the values for average soil and normal concrete.

During the initial period the vaporization rate for liquid methane is high. Thereafter it decays and becomes dependent on the wind speed and is thus convection controlled. It approaches a constant value of approximately 0.02 in./min at low wind speeds after periods of 0.25–2 h. A log–log plot of vaporization rate vs time gives a straight line.

The evaporation of LNG on various surfaces has been investigated by Drake and Reid (1975). These authors describe experimental work at MIT and refer also to other experimental work by the American Gas Association (AGA) (1974) and work done by Gaz de France (Humbert-Basset and Montet, 1972). They found that the initial transfer and vaporization rates are in accordance with Equation 15.10.73, but that they vary considerably, depending on the nature of the substrate.

Drake and Reid quote both calculated and experimental values of the vaporization parameter A_v in Equation [15.10.73]. Values of the parameters ρ_s, c_{ps} and k_s for various substrates are given by Carslaw and Jaeger (1959). From these parameters the corresponding values of A may be calculated. It is assumed that the substrate is at 16°C (60°F) and the LNG is at −162°C (−260°F) so that $(T_s - T_l)$ is 178°C. These calculated values of A_v are compared by Drake and Reed with experimental values for LNG spills obtained in the AGA and Gaz de France tests, as shown in Table 15.9. By contrast the value of A given for insulating concrete is 0.61.

Drake and Reid state that the rate of vaporization of LNG is greater than that of pure methane. It is greater on surfaces such as crushed rock or pebbles than on compacted soil. On soils the vaporization rate is enhanced by percolation if the soil is dry. The extent of percolation in most soils is limited by the formation of a frozen barrier. Foaming is another factor which may increase the vaporization rate of LNG. A marked reduction in the vaporization rate of LNG may be obtained by the use of insulating concrete.

V.J. Clancey (1974a) has developed equations for the vaporization of a cryogenic liquid such as LNG. For the initial rapid vaporization

$$E_i = \frac{k_1 (T_s - T_l)^2}{\Delta H_v} \qquad [15.10.77]$$

and for the steady continuous vaporization

$$E_s = \frac{k_2(T_s - T_l)}{\Delta H_v} \quad [15.10.78]$$

where E_i is the mass vaporized per unit area within 1 min (g/cm^2), E_s is the steady-state mass vaporization rate per unit area (g/cm^2 min), ΔH_v is the latent heat of vaporization (cal/g) and k_1 and k_2 are constants. The values of the constants k_1 and k_2 are as follows:

Substrate	k_1	k_2
Average soil	7.1×10^{-4}	1.5×10^{-2}
Concrete	7.5×10^{-4}	1.5×10^{-2}
Sandstone	1.3×10^{-3}	2.6×10^{-2}

Equation 15.10.77 is applicable where the temperature difference $(T_s - T_l)$ is large, as for a spillage of liquid methane. It is assumed in Equation [15.10.77] that the depth of ground which gives up heat to the pool is proportional to $(T_s - T_l)$; this accounts for the occurrence in Equation 15.10.77 of the term $(T_s - T_l)^2$.

For a large spillage the time which elapses before the vaporization rate reaches a quasi-steady state may be an hour or more. Clancey suggests that the values of the vaporization rate in this intermediate period may be obtained by logarithmic interpolation between the initial and steady vaporization rates.

Vaporization rates of LNG on water are relatively high. Work has been carried out by a number of investigators, including Burgess, Murphy and Zabetakis (1970 BM RI 7448), Kneebone and Boyle (1973), Feldbauer *et al.* (1972), Germeles and Drake (1975) and Opschoor (1977, 1981).

Burgess, Murphy and Zabetakis obtained a mass vaporization rate of 0.18 kg/m^2 s and Feldbauer *et al.* (1972) one of 0.20 kg/m^2 s. The First *Canvey Report* states that the vaporization rate obtained in such work is of the order of 0.19 kg/m^2 s.

For the vaporization rate, Opschoor (1980) quotes the following relation of J.S. Turner (1965):

$$q_w = 0.085 k_w \left(\frac{g \beta \Delta T^4}{\alpha_w \nu_w} \right)^{\frac{1}{3}} \quad [15.10.79]$$

where q_w is the heat flow per unit area by conduction from the water to the spill, ΔT is the temperature difference between the water and the liquid, β is the coefficient of cubical expansion and ν is the kinematic viscosity. For LNG, taking ΔT = 47.4 K, Equation [15.10.79] gives

$$q_w = 2.3 \times 10^4 \ W/m^2$$

which corresponds to a mass vaporization rate of

$$m_v = 0.045 \ \text{kg/m}^2\text{s}$$

This value agrees well with the experimental results of Kneebone and Boyle (1973). Opschoor gives a preferred value of 0.05 kg/m^2 s.

Shaw and Briscoe (1978 SRD R100) use a regression rate based on the Esso work of 4.7×10^{-4} m/s. The regression rate used by Raj (1979) is 6.5×10^{-4} m/s.

Opschoor also gives relations for the estimation of the mass vaporization rate with ice formation beneath the LNG. It is found that an ice layer forms in about 20 s. Before the ice forms, film boiling pertains; once the film has formed heat transfer is by conduction through the ice. He gives for these two regimes the following mass vaporization rates:

$$m_v = 0.008t \qquad 0 \le t \le 25 \quad [15.10.80a]$$

$$m_v = \frac{0.517}{(t-20)^{\frac{1}{2}}} \qquad t > 25 \quad [15.10.80b]$$

where m_v is the mass vaporization rate per unit area (kg/m^2 s) and t is the time (s).

For simultaneous spreading and vaporization of LNG, Burgess, Murphy and Zabetakis (1970) have given the relations

$$r = k_3 t \quad [15.10.81]$$

$$\frac{dm}{dt} = \pi \rho v r^2 \quad [15.10.82]$$

$$m = \pi \rho v \int_0^t r^2(t) \, dt \quad [15.10.83]$$

$$m = \frac{\pi}{3} \rho v k_3^2 t^3 \quad [15.10.84]$$

where m is the total mass vaporized, r is the radial distance and k_3 is a constant. The value of the constant k_3 is 0.381. Then, taking the density of LNG as 416 kg/m^3, Equation 15.10.84 becomes

$$m = 63.2 v t^3 \quad [15.10.85]$$

Lind (1974) has given a similar treatment but uses a value of k_3 of 0.635 based on the Esso experiments described by Feldbauer *et al.* (1972). From Equation 15.10.84 this gives

$$m = 175.7 v t^3 \quad [15.10.86]$$

Fay (1973) has used Equations 15.10.59 and 15.10.61a, integrating the latter with the boundary conditions t = 0 and r = 0 to obtain

$$r = \left(\frac{8g'V}{\pi} \right)^{\frac{1}{4}} t^{\frac{1}{2}} \quad [15.10.87]$$

Equation 15.10.87 may be used with Equation 15.10.83 to determine the mass vaporized. Fay assumes in his treatment that ice forms beneath the spill, and that this determines the vaporization rate.

Raj and Kalelkar (1973) obtained the relation

$$m = 1.712 \pi \rho v A^{\frac{1}{2}} t^2 \quad [15.10.88]$$

where A is the volume of liquid for an instantaneous spill.

These and other models have been reviewed by Shaw and Briscoe (1978 SDR R100).

15.10.10 Shaw and Briscoe model

A series of models for simultaneous spreading and vaporization of continuous and instantaneous spills of LNG on land and on water has been given by Shaw and Briscoe (1978 SRD R100).

For the vaporization of LNG on land in a fixed area such as a bund they give

$$\frac{dm}{dt} = \frac{A_p X k_s (T_s - T_b)}{\Delta H_v (\pi \alpha_s)^{\frac{1}{2}} t^{\frac{1}{2}}} \quad [15.10.89]$$

where A_p is the area of the pool, ΔH_v is the latent heat of vaporization, k is the thermal conductivity, m is the total mass vaporized, α is the thermal diffusivity, and subscripts b and s denote the boiling point and soil, respectively. The factor X takes account of surface roughness. In the tests by the Japan Gas Association, after 1 s the regression rate was some 3 times the theoretical value ($X = 1$), and on this basis the authors suggest a value for X of 3.

In general, for an LNG spill the volume V of the pool is

$$V = A + Bt - m/\rho \quad [15.10.90]$$

where A is the volume of liquid for instantaneous spill and B is the volumetric flow for a continuous spill.

From Equations 15.10.59 and 15.10.61

$$\frac{dr}{dt} = \left(\frac{2g'V}{\pi r^2}\right)^{\frac{1}{2}} \quad [15.10.91]$$

where r is the radius of the pool.

Analytical solutions of Equation 15.10.91 are obtained by neglecting the term m/ρ in Equation 15.10.90.

For an LNG spill on water the heat transfer rate, and hence the vaporization rate, is taken as constant and is characterized by the regression rate v. The authors use Equations 15.10.90 and 15.10.91 together with the relation

$$\frac{dm}{dt} = \pi r^2 \rho v \quad [15.10.92]$$

For an instantaneous spill on water, Equation 15.10.91) is solved with the condition $B = 0$ (and hence $V = A$), $t = 0$ and $r = r_o$, where r_o is the initial radius of the pool. Then, integrating Equation 15.10.91

$$r = \left[\left(\frac{8g'A}{\pi}\right)^{\frac{1}{2}} t + r_o^2\right]^{\frac{1}{2}} \quad [15.10.93]$$

and using Equations 15.10.92 and 15.10.93 and integrating the former

$$m = \pi \rho v \left[\left(\frac{2g'A}{\pi}\right)^{\frac{1}{2}} t^2 + r_o^2 t\right] \quad [15.10.94]$$

For a continuous spill on water, Equation [15.10.91] is solved with the condition $A = 0$ (hence $V = Bt$), $t = 0$ and $r = 0$. Then, as before, integrating Equation 15.10.91 followed by Equation 15.10.92 gives

$$r = \left(\frac{2}{3}\right)^{\frac{1}{2}} \left(\frac{8g'B}{\pi}\right)^{\frac{1}{4}} t^{\frac{3}{4}} \quad [15.10.95]$$

$$m = \frac{4}{15} \pi \rho v \left(\frac{8g'B}{\pi}\right)^{\frac{1}{2}} t^{\frac{5}{2}} \quad [15.10.96]$$

For an LNG spill on land the vaporization rate is determined by the heat transfer, and hence the vaporization rate decreases with time. The relation used for the vaporization rate is

$$\frac{dm}{dt} = 2\pi\theta \int_0^{r_2} \frac{r_1}{(t_2 - t_1)^{\frac{1}{2}}} dr_1 \quad [15.10.97]$$

with

$$\theta = \frac{X k_s (T_s - T_b)}{\Delta H_v (\pi \alpha_s)^{\frac{1}{2}}} \quad [15.10.98]$$

where r_1 is the radius at time t_1, r_2 is the radius at time t_2 ($t_2 > t_1$) and θ is a parameter. Equations 15.10.90 and 15.10.91 are again used.

For an instantaneous spill on land, Equation 15.10.91 is solved with the conditions $B = 0$ (hence $V = A$), $t = 0$ and $r = r_o$. Also on land, $g = g'$. Then, integrating Equation 15.10.91 gives

$$r = \left[\left(\frac{8gA}{\pi}\right)^{\frac{1}{2}} t + r_o^2\right]^{\frac{1}{2}} \quad [15.10.99]$$

For small r_o, Equation [15.10.99] becomes

$$r^2 \approx \left(\frac{8gA}{\pi}\right)^{\frac{1}{2}} t \quad [15.10.100]$$

Hence

$$t = \beta_1 r^2 \quad [15.10.101]$$

with

$$\beta_1 = \left(\frac{\pi}{8gA}\right)^{\frac{1}{2}} \quad [15.10.102]$$

where β_1 is a constant.

Utilizing Equation 15.10.101 and integrating Equation 15.10.97 gives

$$m = \frac{8}{3}\theta (2\pi g A)^{\frac{1}{4}} t^{\frac{3}{4}} \quad [15.10.103]$$

For a continuous spill on land, Equation 15.10.91 is solved with the conditions $A = 0$ (hence $V = Bt$), $t = 0$ and $r = 0$. Then, as before, integrating Equation 15.10.91 gives

$$r = \frac{2}{3}\left(\frac{8gB}{\pi}\right)^{\frac{1}{4}} t^{\frac{3}{4}} \quad [15.10.104]$$

Hence

$$t = \beta^2 r^{\frac{4}{3}} \quad [15.10.105]$$

with

$$\beta^2 = \left(\frac{9}{32}\frac{\pi}{gB}\right)^{\frac{1}{3}} \quad [15.10.106]$$

where β_2 is a constant.

Utilizing Equation 15.10.105 and integrating Equation 15.10.97 gives

$$m = \pi^{\frac{3}{2}} \theta \left(\frac{gB}{2}\right) t^2 \quad [15.10.107]$$

As stated above, the authors use for the regression rate for LNG on water the value from the Esso work of 4.7×10^{-4} m/s.

The authors also performed numerical integration of the full equations and obtained for the radius r_e and time t_e at the end of evaporation:

$$\ln r_e = 1.5 \ln t_e - 2.45 \qquad [15.10.108]$$

$$\ln t_e = 0.25(\ln M_s + 6) \qquad [15.10.109]$$

where M_s is the mass of the spill.

They also confirmed that it is justifiable to neglect the term m/ρ.

15.10.11 Raj model

Another series of models for simultaneous spreading and vaporization of continuous and instantaneous spills of LNG on land and on water has been given by Raj (1981), developing the earlier work of Raj and Kalelkar (1974).

For a continuous spill on land the basic equation is

$$\frac{dq}{dt} = \frac{k_s \Delta T}{(\pi \alpha_s)^{\frac{1}{2}}(t - t_1)^{\frac{1}{2}}} 2\pi r_1 \, d_r \qquad [15.10.110]$$

with

$$\Delta T = T_s - T_l \qquad [15.10.111]$$

where k is the thermal conductivity, q is the heat flux per unit area, r is the radial distance, ΔT is the temperature difference between the liquid and the soil at distance r_1, and subscript 1 denotes at distance r_1.

Then, equating the steady-state vaporization rate per unit area and the volumetric flow of liquid into the spill V_c:

$$V_c = \frac{2\pi k_s \Delta T}{\Delta H_v \rho_l (\pi \alpha_s t)^{\frac{1}{2}}} \int_0^{r(t)} \frac{r_1}{(1 - t_1/t)^{\frac{1}{2}}} dr_1 \qquad [15.10.112]$$

where subscript l denotes the liquid.

Integration of Equation 15.10.112 gives

$$r(t) = \left(\frac{2\Delta H_v \rho_l V_c}{\pi^2 S \Delta T}\right)^{\frac{1}{2}} t^{\frac{1}{4}} \qquad [15.10.113]$$

with

$$S = \left(\frac{k_s \rho_s c_s}{\pi}\right)^{\frac{1}{2}} \qquad [15.10.114]$$

where c is the specific heat and ρ is the density.

Raj also shows that

$$V_c = \frac{\pi r^2 S \Delta T}{(2/\pi)\Delta H_v \rho_l} t^{\frac{1}{2}} \qquad [15.10.115]$$

The regression rate v is

$$v = \frac{V_c}{\pi r^2} \qquad [15.10.116]$$

For a continuous spill of finite size on land, using the result just given in Equation 15.10.115

$$V(t) + \frac{\pi^2 S \Delta T}{2\Delta H_v \rho_l} \int_0^t \frac{r^2}{t^{\frac{1}{2}}} dt = V_c t \qquad t \le t_{sp} \qquad [15.10.117a]$$

$$V(t) = V_c t_{sp} \qquad t > t_{sp} \qquad [15.10.117b]$$

where V is the volume of the pool and the subscript sp denotes the spill duration. The first term on the left-hand side of Equation 15.10.117 is the volume of liquid remaining and the second the volume vaporized.

Raj utilizes the liquid spreading relations 15.10.59 and 15.10.61b to obtain a numerical solution of Equation 15.10.117. For an instantaneous spill of LNG on land the basic equation is

$$V(t) + \int_0^t v(t) \pi r^2(t) \, dt = V_i \qquad [15.10.118]$$

with

$$v(t) = \frac{S \Delta T}{(2/\pi)\Delta H_v \rho_l t^{\frac{1}{2}}} \qquad [15.10.119]$$

where V_i is the volume of liquid in the instantaneous spill.

The relations utilized for the liquid spreading are Equations 15.10.59 and 15.10.67, together with Equation 15.10.118. From these relations are obtained for the final evaporation radius r_e and time t_e:

$$r_e = 1.248\left(\frac{\Delta H_v \rho_l}{S \Delta T}\right)^{\frac{1}{3}} V_i^{\frac{4}{9}} \qquad [15.10.120]$$

$$t_e = 0.639\left(\frac{\Delta H_v \rho_l}{S \Delta T}\right)^{\frac{2}{3}} \left(\frac{1}{g}\right)^{\frac{1}{2}} V_i^{\frac{7}{18}} \qquad [15.10.121]$$

For a continuous spill on water the basic equation is

$$\dot{V}(t) + \pi r^2 v = V_c \qquad [15.10.122]$$

The relations utilized for liquid spreading are Equations 15.10.59 and 15.10.61. The following relation is obtained for the final evaporation radius r_e:

$$r_e = \left(\frac{V_c}{\pi v}\right)^{\frac{1}{2}} \qquad [15.10.123]$$

For an instantaneous spill on water the basic equation is

$$V(t) + \pi \int_0^t v(t) r^2(t) \, dt = V_i \qquad [15.10.124]$$

The relation utilized for spreading is Equation [15.10.67]. The following relations are obtained for the final evaporation radius and time:

$$r_e = \left(\frac{V_i^3 g'}{v^2}\right)^{\frac{1}{8}} \qquad [15.10.125]$$

$$t_e = 0.6743\left(\frac{V_i}{g' v^2}\right)^{\frac{1}{4}} \qquad [15.10.126]$$

Raj (1979) has also given a criterion for classifying a spill of LNG on water as instantaneous or continuous. He defines the following dimensionless quantities:

$$L = V_i^{\frac{1}{3}} \qquad [15.10.127]$$

$$t_{ch} = L/v \qquad [15.10.128]$$

$$\xi = r/L \qquad [15.10.129]$$

$$\tau = t/t_{ch} \qquad [15.10.130]$$

where L is a characteristic length, t_{ch} is a characteristic evaporation time, ξ is a dimensionless maximum spread and τ is a dimensionless time.

Utilizing Equations 15.10.123 and 15.10.125 for the final spill radii for the two types of spill together with the fact that $V_c = V_i/t$, he obtains

$$\xi = \left(\frac{Lg'}{v^2}\right)^{\frac{1}{8}} \quad \text{instantaneous spill} \qquad [15.10.131]$$

$$\xi = \frac{1}{(\pi\tau)^{\frac{1}{2}}} \quad \text{continuous spill} \qquad [15.10.132]$$

Then from Equations 15.10.131 and 15.10.132 a value of τ may be obtained for the point at which the value of ξ is the same for the two equations.

Using a typical value of the regression rate v of LNG on water of 6.5×10^{-4} m/s gives a value of τ of 2×10^{-3} s. Then for the crossover time t_{cr} Raj gives the following values:

Volume spilled (m^3)	1000	10 000	25 000	50 000
Crossover time (s)	31	68	92	116

15.10.12 Opschoor model

A model for the simultaneous spreading and vaporization of an instantaneous spill of LNG on water has been given by Opschoor (1977, 1980).

For an instantaneous spill of LNG on water

$$V(t)\rho_l + \pi \int_0^t m_v r^2(t)\, dt = V_i \rho_l \qquad [15.10.133]$$

where m_v is the mass rate of vaporization per unit area.

Integration of Equation 15.10.133 gives

$$r = \left[0.44 \frac{m_v g'}{\rho_l} t^3 + 1.3(g' V_i)^{\frac{1}{2}} t\right]^{\frac{1}{2}} \qquad [15.10.134]$$

For the final evaporation radius and time

$$r_e = 1.02\left(\frac{g' \rho_l^2 V_i^3}{m_v^2}\right)^{\frac{1}{8}} \qquad [15.10.135]$$

$$t_e = 0.67\left(\frac{\rho_l^2 V_i}{g' m_v^2}\right)^{\frac{1}{4}} \qquad [15.10.136]$$

Opschoor also considers an LNG spill on a water surface of limited area A_o. Using Equation 15.10.134, but neglecting the first term on the right-hand side, the following relation is obtained for the time t_o for a water surface of area A_o to be fully covered:

$$t_o = 0.24\left(\frac{A_o^2}{g' V_i}\right)^{\frac{1}{2}} \qquad [15.10.137]$$

The work of Opschoor on the mass vaporization rate of LNG on water has already been described in Section 15.10.9.

15.10.13 Spreading and vaporization of ammonia

Work on the spreading and vaporization of liquefied ammonia (LNH_3) on water has been described by Raj, Hagopian and Kalelkar (1974). Experiments were carried out in which quantities of ammonia up to 50 USgal were spilled on water.

When ammonia is spilled on water, part of it vaporizes and part goes into solution in the water. Defining a partition coefficient p as the proportion which dissolves, the authors obtained a value of $p \approx 0.6 \pm 0.1$. The remainder $(1 - p)$ vaporizes.

The authors modelled the release assuming that the ammonia was at a temperature of −33°C, there was no initial entrainment of air and no spray formation. Under these conditions the cloud is buoyant. They treated the dispersion in terms of a buoyant plume model.

For the final evaporation radius and time they give:

$$r_e = 2.5V^{0.375} \qquad [15.10.138]$$

and

$$t_e = 0.674\left(\frac{V}{g' v^2}\right)^{\frac{1}{4}} \qquad [15.10.139]$$

where r_e is the final evaporation radius (ft), t_e is the final evaporation time (s), v is the regression rate (ft/s), and V is the volume spilled (USgal). The authors obtained a regression rate of 2.8 in./min.

Using a thermodynamic analysis, Raj and Reid (1978a) have shown that it is possible to determine the point at which further addition of water simply dilutes the solution without causing any further evolution of ammonia and to obtain the value of the partition coefficient. For the illustrative example considered the experimental partition coefficient was 0.73 and the theoretical value 0.715.

Griffiths (1977 SRD R67) has criticized the interpretation given by Raj, Hagopian and Kalelkar and gives an alternative interpretation of the results assuming that spray formation occurs and that the plume is not buoyant.

15.10.14 Computer codes

A number of computer codes have been written for vaporization. A review of these has been given by the CCPS (1987/2). The Safety and Reliability Directorate (SRD) code SPILL is based on the model of Shaw and Briscoe (1978 SRD R100). It has been described by Prince (1981 SRD R210). Another program is the Shell SPILLS code which is described by Fleischer (1980).

15.11 Dispersion

Emission and vaporization are followed by dispersion of the vapour to form a vapour cloud.

Accounts of gas dispersion of particular relevance here include those given in *Micrometeorology* (O.G. Sutton, 1953), *Atmospheric Diffusion* (Pasquill, 1962a, 1974) *Meteorology and Atomic Energy* (Slade, 1968), *An Evaluation of Dispersion Formulas* (Anderson, Hippler and Robinson, 1969), *Recommended Guide for the Prediction of the Dispersion of Airborne Effluents* (ASME, 1969/1, 1973/2, 1979/4), *Workbook of Atmospheric Dispersion Estimates* (D.B. Turner, 1970), *Turbulent Diffusion in the Environment* (Csanady, 1973) and *Handbook on Atmospheric Diffusion* (Hanna, Briggs and Hosker, 1982), and those given by Pasquill and Smith (1983).

Dispersion models used in the nuclear industry are described in the Nuclear Regulatory Commission (NRC) *Regulatory Guide 1.111 Methods for Estimating Atmospheric Transport and Dispersion of Gaseous Effluents in Routine Releases from Light-water-cooled Reactors* (1974) and the NRC *Regulatory Guide 1.145 Atmospheric Dispersion Models for Potential Accident Consequence Assessments at Nuclear Power Plants* (1979a). The NRC has also published an account of the basis for *Regulatory Guide 1.145* (1981 NUREG/CR-2260). Another nuclear industry dispersion model is *A Model for*

Short and Medium Range Dispersion of Radionuclides into the Atmosphere by Clarke (1979 NRPB R91).

Work on dispersion is primarily concerned with the dispersion of pollutants from industrial chimney stacks. Most of the fundamental work on dispersion relates to this problem. There is, however, an increasing amount of work on dispersion of hazardous releases from process plant.

Selected references on dispersion are given in Table 15.10.

15.11.1 Dispersion situations

Dispersion situations may be classified as follows.

The fluid and the source may be classified as:

(1) Fluid buoyancy –
 (a) neutral buoyancy,
 (b) positive buoyancy,
 (c) negative buoyancy.
(2) Momentum –
 (a) low momentum,
 (b) high momentum.
(3) Source geometry –
 (a) point source,
 (b) line source,
 (c) area source.
(4) Source duration –
 (a) instantaneous,
 (b) continuous,
 (c) intermediate.
(5) Source elevation –
 (a) ground level source,
 (b) elevated source.

The dispersion takes place under particular meteorological and topographical conditions. Some principal features of these are as follows:

(1) Meteorology –
 (a) wind,
 (b) stability.
(2) Topography –
 (a) surface roughness,
 (b) near buildings and obstructions,
 (c) over urban areas,
 (d) over coastal zones and sea,
 (e) over complex terrain.

These aspects of the dispersion situation are now considered.

15.11.2 Buoyancy effects

The fluid may have neutral, positive or negative buoyancy. Neutral density is generally the default assumption and applies where the density of the gas–air mixture is close to that of air. This is the case where the density of the gas released is close to that of air or where the concentration of the gas is low. In determining the density of the gas it is necessary to consider not only molecular weight but also the temperature and liquid droplets. Gases with positive buoyancy include those with low molecular weight and hot gases. Many hazardous materials, however, form negatively buoyant gases, or heavy gases.

Much of the fundamental work on dispersion, and the models derived from this work, relate to the dispersion

Table 15.10 *Selected references on dispersion*

Meteorology
NRC (Appendix 28 *Meteorology*, 1972); Lamb (1932); Nikuradse (1933); Prandtl (1933); Schlichtling (1936, 1960); Brunt (1934); Milne-Thomson (1938); O.G. Sutton (1949, 1953, 1962); Geiger (1950); Hewson (1951); Malone (1951); Batchelor (1953, 1956, 1964, 1967); Brooks and Carruthers (1953); Singer and Smith (1953, 1966); Clauser (1954, 1956); G.F. Taylor (1954); AEC (1955); Townsend (1956); Haltiner and Martin (1957); Scorer (1958, 1978); Byers (1959); Priestley (1959); Chandrasekhar (1961); Monin (1962); Ogura and Phillips (1962); Pasquill (1962, 1974); Lumley and Panofsky (1964); Roll (1965); Munn (1966); Wu (1965); Bowne, Ball and Anderson (1968); Meteorological Office (1968, 1972, 1975, 1991, 1994); Slade (1968); Swinbank (1968); Trewartha (1968); ASME (1969/1, 1975/2, 1979/3, 4); McIntosh and Thom (1969); Monin (1970); D.B. Turner (1970); R.J. Taylor, Warner and Bacon (1970); Bradshaw (1971, 1978); Businger and Yaglom (1971); B.J. Mason (1971); Monin and Yaglom (1971); Tsang (1971); Launder and Spalding (1972); McIntosh (1972); Tennekes and Lumley (1972); Zilitinkevich (1972a); Busch (1973); Blackadar (1976); Oke (1978); Caughey, Wyngaard and Kaimal (1979); Sethuraman and Raynor (1979); Cermak (1980); Kreith (1980); Venkatram (1980b, 1981a); A.E. Mitchell (1982); Plate (1982); Tennekes (1982); Volland (1982); Pasquill and Smith (1983); Panofsky and Dutton (1984); Houghton (1985); J.C.R. Hunt (1985); Page and Lebens (1986); Anfossi (1989); Bartzis (1989); A.D. Young (1989); Lindzen (1990); Swaid (1991)
Lagrangian length scale: J.S. Hay and Pasquill (1959); J.D. Reid (1979); Hanna (1979a, 1981b); Hanna, Briggs and Hosker (1982)
Atmospheric boundary layer: Panofsky (1968, 1973, 1974, 1978); Deardorff (1970, 1973, 1974); Pasquill (1972); Zilitinkevich (1972b); Businger (1973, 1982); Csanady (1973); Tennekes (1973a); Wyngaard (1973, 1975, 1982); Wyngaard, Arya and Coté (1974); Wyngaard and Cote (1974); Wyngaard, Coté and Rao (1974); Counihan (1975); Kaimal *et al.* (1976); H.N. Lee (1979); Nichols and Readings (1979); Caughey (1982); Nieuwstadt (1984a,b); Holtslag and Nieuwstadt (1986)
Wind characteristics: Irwin (1979b); Wieringa (1980); Oehlert (1983); Nitz, Endlich and Ludwig (1986); R. Weber (1992)
Roughness length, friction velocity: Paeschke (1937); O.G. Sutton (1953); Plate (1971); Mulhearn (1977); Oke (1978); Wieringa (1981); Brutsaert (1982); Venkatram and Paine (1985); San Jose *et al.* (1986)
Wind velocity profile: Frost (1947); Deacon (1949, 1957); Lettau (1959); Panofsky, Blackadar and McVehil (1960); McVehil (1964); Swinbank (1964); Bowne and Ball (1970); Dyer and Hicks (1970); Paulson (1970); Webb (1970); Businger *et al.* (1971); P.M. Jones, Larrinaga and Wilson (1971); Golder (1972); Tennekes (1973a); Dyer (1974); Sethuraman and Brown (1976); Touma (1977); Yaglom (1977); Carson and Richards (1978); H.N. Lee (1979); Sethuraman and Raynor (1979); Skibin and Businger (1985); Hanafusa, Lee and Lo (1986); Leahey (1987)

Mixed layer height: Holzworth (1967, 1972); Zilitinkevich (1972a,b); Businger and Arya (1974); Tennekes and van Ulden (1974); Gryning *et al.* (1987)
Mixed layer scaling, convective velocity scale: Wyngaard, Coté and Rao (1974); Deardorff and Willis (1975); Wyngaard (1975); Willis and Deardorff (1976, 1978); Panofsky *et al.* (1977); Venkatram (1978); Nieuwstadt (1980b)
Exchange coefficients: O.G. Sutton (1953); Pasquill (1974); Mizuno and Yokoyama (1986)
Stability criteria: L.F. Richardson (1920, 1925, 1926); Monin and Obukhov (1954); Kazanski and Monin (1960); Monin (1970); Obukhov (1971); Willis and Deardorff (1976, 1978)
Stability classification: O.G. Sutton (1953); M.E. Smith (1956); Singer and Smith (1953, 1966); Cramer (1957, 1959a,b, 1976); Pasquill (1961); D.B. Turner (1961, 1964); Klug (1969); Carpenter *et al.* (1971); Golder (1972); Luna and Church (1972); F.B. Smith (1973, 1979); Liu *et al.* (1976); Hanna *et al.* (1977); AMS (1978); Clarke (1979 NRPB R91); Sedefian and Bennett (1980); Schacher, Fairall and Zannetti (1982); Taglizucca and Nanni (1983); Kretschmar and Mertens (1984); Skupniewicz and Schacher (1984, 1986); Hasse and Weber (1985); Larsen and Gryning (1986); Ning and Yap (1986); Draxler (1987)
Vertical heat flux: Caughey and Kaimal (1977); Clarke (1979 NRPB R91); Maul (1980); Holtslag and van Ulden (1983)
Atmospheric visibility: Horvath (1981)

Dispersion
NRC (Appendix 28 *Gas Dispersion*, 1974, 1979a); G.I. Taylor (1915, 1921, 1927); O.F.T. Roberts (1923); L.F. Richardson and Proctor (1925); Schmidt (1925); L.F. Richardson (1926); O.G. Sutton (1932, 1947, 1953); Dryden (1939); Bakhmeteff (1941); Katan (1951); Batchelor (1952, 1953, 1956, 1964); Frenkiel (1952); Chamberlain (1953); AEC (1955); Bodurtha (1955); Gosline, Falk and Helmers (1955); Crank (1956); Cramer (1957, 1959a,b); Gifford (1957, 1960a, 1961, 1962a,b, 1968, 1976a,b, 1977, 1986, 1987); H.L. Green and Lane (1957–); Cramer, Record and Vaughan (1958); Hilst and Simpson (1958); Hinze (1959); Ellison and Turner (1960); Bowne (1961); Friedlander and Topper (1961); Pasquill (1961, 1962, 1965, 1975, 1976a,b); P.H. Roberts (1961); Monin (1962); Beattie (1963 UKAEA AHSB(S) R64); Bryant (1964 UKAEA AHBS(RP) R42); Cramer *et al.* (1964); Panofsky and Prasad (1965); Yih (1965); Kreyzig (1967); Hansen and Shreve (1968); D.O. Martin and Tikvart (1968); R.O. Parker and Spata (1968); Ross (1968); Slade (1968); Anderson, Hippler and Robinson (1969); API (1969 Publ. 4030, 1982 Publ. 4360, 1987 Publ. 4457); ASME (1969/1, 1975/2, 1979/3, 4); Briggs (1969); Csanady (1969, 1973); Fay and Hoult (1969); Larsen (1969); Burgess, Murphy and Zabetakis (1970); A.J. Clarke, Lucas and Ross (1970); Hearfield (1970); MCA (1970/16); D.B. Turner (1970); Pasquill and Smith (1971); Ramsdell and Hinds (1971); Simpson (1971); Beryland (1972, 1975); J.T. Davies (1972); Monji and Businger (1972); Tennekes and Lumley (1972); Yang and Meroney (1972); Calder (1973); Haugen (1973, 1976); Leslie (1973); Mathis and Grose (1973); Ragland (1973); F.B. Smith (1973); J.S. Turner (1973); Winter (1973); V.J. Clancey (1974a, 1976a, 1977a,c); Deardorff and Willis (1974); J.D. Reed (1974); Eisenberg, Lynch and Breeding (1975); R.W. McMullen (1975); Runca and Sardei (1975); Drysdale (1976a); Nappo (1976, 1981); Willis and Deardorff (1976, 1978); Bass, Hoffnagle and Egan (1977); Comer (1977); S.K. Friedlander (1977); Griffiths (1977 SRD R85); Kletz (1977–78); US Congress, OTA (1977); AMS (1978); R.P. Bell (1978); Carson and Richards (1978); HSE (1978b); Nieuwstadt and van Ulden (1978); Slater (1978a); Venkatram (1978, 1980a); Beychok (1979); Bowne and Yocom (1979); Dobbins (1979); S.R. Hanna (1979b); C.J. Harris (1979); Harvey (1979b); Nieuwstadt and van Duuren (1979); Raj (1979); TNO (1979); T.B. Morrow *et al.* (1980); Nieuwstadt (1980a,b); Reijnhart, Piepers and Toneman (1980); Reijnhart and Rose (1980); Rodi (1980); Spencer and Farmer (1980); Arya and Shipman (1981); Arya, Shipman and Courtney (1981); Bower and Sullivan (1981); Dunker (1981); Holtslag, de Bruin and van Ulden (1981); Nappo (1981); Steenkist and Nieuwstadt (1981); Venkatram (1981b, 1988a,b); D.J. Wilson (1981a); de Wispelaere (1981, 1983, 1984, 1985); S.R. Hanna (1982); S.R. Hanna, Briggs and Hosker (1982); N.C. Harris (1982); J.C.R. Hunt (1982); Lamb (1982); Mecklenburgh (1982, 1985); Nieuwstadt and van Dop (1982); O'Shea (1982); Comer *et al.* (1983); C.D. Jones (1983); Ludwig, Liston and Salas (1983); Ludwig and Livingston (1983); Diedronks and Tennekes (1984); Panofsky and Dutton (1984); Pendergrass and Arya (1984); Zeman (1984); Cogan (1985); Hukkoo, Bapat and Shirvaikar (1985); Lupini and Tirabassi (1985); K.R. Peterson (1985); Carson (1986); Davidson (1986); Enger (1986); Hamza and Golay (1986); Jakeman, Taylor and Simpson (1986); Mizuno and Yokoyama (1986); de Wispelaere, Schiermeier and Gillani (1986); Zanetti (1986, 1990); Andren (1987); Irwin *et al.* (1987); Li Zong-Kai and Briggs (1988); Lawson, Snyder and Thompson (1989); Nema and Tare (1989); Underwood (1989 SRD R483); Finch and Serth (1990); S.R. Hanna, Chang and Strimatis (1990); Linden and Simpson (1990); Guinnup (1992); Runca (1992); van Ulden (1992); Verver and de Leeuw (1992); Eckman (1994); Straja (1994)
Lagrangian models: van Dop (1992); Runca (1992)
Large eddy simulation: Henn and Sykes (1992); Nieuwstadt (1992); Sykes and Henn (1992)

Models, codes
Kaiser (1976 SRD R63); NRC (1972); Church (1976); Tennekes (1976); Benson (1979); D.B. Turner and Novak (1978); NRPB (1979 R91); AGA (1980/31, 32); Pierce and Turner (1980); Schulman and Scire (1980); Zanetti (1981, 1990); Hanna *et al.* (1984); Doron and Asculai (1983); Mikkelsen, Larsen and Thykierg-Nielsen (1984); Ames *et al.* (1985); Irwin, Chico and Catalano (1985); Berkowitz, Olesen and Torp (1986); EPA (1986); Layland, McNaughton and Bodner (1986); H.N. Lee (1986); W.B. Petersen (1986); Benarie (1987); API (1988 Publ. 4461, 1989 Publ. 4487); Doury (1988); Freiman and Hill (1992); Seigneur (1992); Bianconi and Tamponi (1993)

Experimental trials
M.E. Smith (1951); Pasquill (1956, 1962); Lettau and Davidson (1957); Barad (1958); Haugen (1959); Islitzer (1961, 1965); MacCready, Smith and Wolf (1961); F.B. Smith and Hays (1961); Barad and Fuquay (1962); Haugen and Fuquay (1963); Haugen and Taylor (1963); Islitzer and Dumbauld (1963); Cramer *et al.* (1964);

Fuquay, Simpson and Hinds (1964); Gartrell *et al.* (1964); Islitzer and Markee (1964); Islitzer and Slade (1964, 1968); T.B. Smith *et al.* (1964); J.H. Taylor (1965); Lapin and Foster (1967); McElroy and Pooler (1968); Seargeant and Robinett (1968); Carpenter *et al.* (1971); Halitsky and Woodward (1974); van der Hoven (1976); Nickola (1977); Nieuwstadt and van Duuren (1979); AGA (1980/31); Defense Technical Information Center (1980); Doury (1981); Dabberdt *et al.* (1982); Dabberdt *et al.* (1983); Yersel, Goble and Merrill (1983); Vanderborght and Kretschmar (1984); Ramsdell, Glantz and Kerns (1985); Briggs *et al.* (1986); Schiermeier, Lavery and Dicristofaro (1986)

Meandering plumes
Falk *et al.* (1953); M.E. Smith (1956); Gifford (1959, 1960b); Csanady (1973); Vogt, Straka and Geiss (1979); Zanetti (1981); S.R. Hanna (1984, 1986); Ride (1988)

Inversion conditions, non-Gaussian models
Gifford (1960b); Sagendorf (1975); van der Hoven (1976); Tennekes (1976); Robson (1983); Hukkoo, Bapat and Shirvaikar (1985); Venkatram and Paine (1985)

Calm conditions
Sagendorf (1975); Gifford (1976a,b); van der Hoven (1976)

Time-varying conditions
Ludwig (1981, 1984, 1986); Skibin (1983)

Puff vs plume criterion
Eisenberg, Lynch and Breeding (1975); Hesse (1991b)

Similarity models
Townsend (1956); Panofsky and Prasad (1965); Gifford (1968); Csanady (1973); Pasquill (1974, 1976b); Scorer (1978); Horst (1979); S.R. Hanna, Briggs and Hosker (1982)

Gradient transfer models, *K* models
O.F.T. Roberts (1923); Nieuwstadt and van Ulden (1978); Dvore and Vaglio-Laurin (1982); S.R. Hanna, Briggs and Hosker (1982); Runca (1982); Gryning, van Ulden and Larsen (1983); Gryning and Larsen (1984); Larsen and Gryning (1986); Gryning *et al.* (1987)

Urban areas
D.B. Turner (1964); Pasquill (1970); S.R. Hanna (1971, 1976); W.B. Johnson *et al.* (1971); Gifford (1972, 1976a,b); Ragland (1973); Bowne (1974); Chang and Weinstock (1974); Dabberdt and Davis (1974); D.S. Johnson and Bornstein (1974); Yersel, Goble and Morrill (1983); Depaul and Sheih (1985, 1986); Oerlemans (1986); Jakeman, Jun and Taylor (1988); Surridge and Goldreich (1988); Bachlin, Plate and Theurer (1989); Gatz (1991); Grimmond, Cleugh and Oke (1991)

Coastal areas, sea
Roll (1965); Sethuraman, Brown and Tichler (1974); Kondo (1975); Gifford (1976a,b); Lyons (1976); Sethuraman, Meyers and Brown (1976); Nichols and Readings (1979); Cooper and Nixon (1984 SRD R307); S.R. Hanna, Paine and Schulman (1984); Manins (1984); Hasse and Weber (1985); Larsen and Gryning (1986); Stunder and Sethuraman (1986); Gryning *et al.* (1987); S.R. Hanna (1987a); Spangler and Johnson (1989); Khoo and Chew (1993)

Model comparisons
Nappo (1974); Gifford (1976a,b); Long and Pepper (1976); Venkatram (1981b); EPA (1984a,b); W.M. Cox and Tikvart (1986); Hayes and Moore (1986); Hinrichsen (1986); Layland, McNaughton and Bodmer (1986); M.E. Smith (1986); Stunder and Sethuraman (1986); Irwin *et al.* (1987); Zannetti (1990); Cirillo and Poli (1992); Dekker and Sliggers (1992); Poli and Cirillo (1993)

Diffusion parameters
O.G. Sutton (1953); Haugen, Barad and Athanaitas (1961); Pasquill (1962, 1974, 1976a); Högstrom (1964); Beals (1971); Eimutis and Konicek (1972); Montgomery *et al.* (1973); Hosker (1974b); Pendergast and Crawford (1974); R.W. McMullen (1975); Draxler (1976, 1979); Gifford (1976a,b, 1980, 1987); A.H. Weber (1976); Doran, Horst and Nickola (1978); Horst, Doran and Nickola (1979); Sedefian and Bennett (1980); D.J. Wilson (1981a); S.R. Hanna (1981a); S.R. Hanna, Briggs and Hosker (1982); Schayes (1982); Comer *et al.* (1983); Pasquill and Smith (1983); Irwin (1984); Venkatram, Strimaitis and Dicristoforo (1984); Atwater and Londergan (1985); Bowling (1985); Henderson-Sellers (1986); San Jose *et al.* (1986); Gryning *et al.* (1987); Wratt (1987); Georgopoulos and Seinfeld (1988); R.F. Griffiths (1994b)
Estimation schemes: AMS (1977); Hanna *et al.* (1977); Irwin (1979a, 1983); Randerson (1979)
Urban areas: D.B. Turner (1964); McElroy (1969); Bowne (1974); Gifford (1976a,b); Santomauro, Maestro and Barberis (1979); Yersel, Goble and Morrill (1983)
Complex terrain: Kamst and Lyons (1982)
Dispersion over water: Hosker (1974a); Kondo (1975); Ching-Ming Sheih (1981); Hanna, Paine and Schulman (1984); Hasse and Weber (1984); Skupniewicz and Schacher (1984); Larsen and Gryning (1986)

Dispersion over short ranges
O.G. Sutton (1950); Long (1963); G.A. Briggs (1974); Hanzevack (1982); Palazzi *et al.* (1982); Yersel, Golle and Morrill (1983); R. Powell (1984); Callander (1986); Kaufman *et al.* (1990)

Dispersion from pipelines
D.J. Wilson (1979b, 1981b)

Concentration fluctuations, peak-mean concentrations
Gosline (1952); Fuquay (1958); Gifford (1960b, 1970); Becker, Hottel and Williams (1965); Hinds (1969); Barry (1971, 1972); Ramsdell and Hinds (1971); Csanady (1973); C.D. Jones (1979, 1983); Fackrell and Robins (1982); D.J. Wilson (1982, 1986, 1991a); D.J. Wilson, Robins and Fackrell (1982, 1985); D.J. Wilson, Fackrell and Robins (1982); Deardorff and Willis (1984); R.F.Griffiths and Megson (1984); S.R. Hanna (1984, 1986); Ride (1984a,b, 1988); R.F. Griffiths and Harper (1985); D.J. Wilson and Sims (1985); Apsimon and Davison (1986); Georgopoulos and Seinfeld (1986); Lewellen and Sykes (1986); S.T. Brown (1987); Derksen and Sullivan (1987); Bara, Wilson and Zelt (1992)

Emergency gas dispersion modelling (see also Table 29.1)
M.E. Smith *et al.* (1983); Mudan (1984b); Lynskey (1985); Nitz, Endlich and Ludwig (1986); McNaughton, Worley and Bodmer (1987); Mulholland and Jury (1987)

Jets and plumes (see Table 15.40)

Obstacles, buildings
Gifford (1960a, 1976a,b); Halitsky (1963, 1968, 1977); Barry (1964); Hinds (1967, 1969); van der Hoven (1968); Slade (1968); Dickson, Start and Markee (1969); J.C.R. Hunt (1971, 1985); Meroney and Yang (1971); AEC (1975); Cagnetti (1975); Huber and Snyder (1976, 1982); D.J. Wilson (1976, 1979a); Vincent (1977, 1978); Brighton (1978, 1986); J.C.R. Hunt, Snyder and Lawson (1978); D.J. Wilson and Netterville (1978); Hosker (1979, 1980); Huber (1979, 1984, 1988, 1989); Ferrara and Cagnetti (1980); J.C.R. Hunt and Snyder (1980); C.D. Barker (1982); Castro and Snyder (1982); Fackrell and Robins (1982); Ogawa and Oikawa (1982); Ogawa *et al.* (1982); D.J. Wilson and Britter (1982); Li and Meroney (1983); Ogawa, Oikawa and Uehara (1983); Fackrell (1984a,b); Snyder and Hunt (1984); C.D. Jones and Griffiths (1984); Ryan, Lamb and Robinson (1984); M.E. Davies and Singh (1985a); Rottman *et al.* (1985); Arya and Gadiyaram (1986); Bachlin and Plate (1986, 1987); Boreham (1986); McQuaid (1986); Maryon, Whitlock and Jenkins (1986); R.S.Thompson and Shipman (1986); M. Epstein (1987); G.A. Briggs *et al.* (1992); R.S. Thompson (1993)

Complex terrain
Egan (1976); Egan, d'Errico and Vaudo (1979); Reible, Shair and Kauper (1981); Lott (1984, 1986); Callander (1986); P.A. Davis *et al.* (1986); Dawson, Lamb and Stock (1986); Dicristofaro and Egan (1986); Horst and Doran (1986); Lavery *et al.* (1986); Massmeyer *et al.* (1986); Rowe and Tas (1986); Schiermeier, Lavery and Dicristofaro (1986); Strimaitis and Snyder (1986); R.S. Thompson and Shipman (1986); Arya, Capuana and Fagen (1987); Snyder and Britter (1987); Castro, Snyder and Lawson (1988); Lawson, Snyder and Thompson (1989); Ohba, Okabayashi and Okamoto (1989); Spangler and Johnson (1989); Ohba *et al.* (1990); Ramsdell (1990); Snyder (1990); Yoshikawa et al (1990)

Physical modelling, wind and water tunnel experiments
Strom and Halitsky (1955); Strom, Hackman and Kaplin (1957); Meroney and Yang (1971); Hoot, Meroney and Peterka (1973); Meroney, Cermak and Neff (1976); Britter (1980); Castro and Snyder (1982); Huber and Snyder (1982); Meroney (1982); Ogawa *et al.* (1982); Bradley and Carpenter (1983); Cheah, Cleaver and Milward (1983a,b); Chea *et al.* (1984); Wighus (1983); Ogawa, Oikawa and Uehara (1983); Meroney and Lohmeyer (1984); T.B. Morrow, Buckingham and Dodge (1984); D.J. Hall and Waters (1985); van Heugten and Duijm (1985); Milhe (1986); Riethmuller (1986); Schatzmann, Lohmeyer and Ortner (1987); Snyder and Britter (1987); Bara, Wilson and Zelt (1992)

Flammable cloud formation
Hess and Stickler (1970); Hess, Leuckel and Stoeckel (1973); Burgess *et al.* (1975); V.J. Clancey (1974a, 1977a,c); Eisenberg, Lynch and Breeding (1975); Sadée, Samuels and O'Brien (1976–77); R.A. Cox (1977); J.G. Marshall (1977, 1980); Harvey (1979b); van Buijtenen (1980); T.B. Morrow (1982b); Hesse (1991a)

Infiltration into buildings
Dick (1949, 1950a,b), Dick and Thomas (1951); BRE (1959); Megaw (1962); Slade (1968); Handley and Barton (1973); AEC (1975); Ministry of Social Affairs (1975); CIBS (1976); Brundrett (1977); HSE (1978b); Guillaume *et al.* (1978); Kronvall (1978); Jackman (1980); Warren and Webb (1980a,b); Terkonda (1983); Haastrup (1984); Pape and Nussey (1985); Sinclair, Psota-Kelty and Weschler (1985); P.C. Davies and Purdy (1986); Pietersen (1986c); British Gas (1987 Comm. 1355); ASHRAE (1988 ASHRAE 119); El-Shobosky and Hussein (1988); Deaves (1989); Jann (1989); van Loo and Opschoor (1989); D.J. Wilson (1990, 1991b); D.J. Wilson and Zelt (1990); Zelt and Wilson (1990); McQuaid (1991); Engelmann (1992); McCaughey and Fletcher (1993); Rosebrook and Worm (1993)
Infiltration into cars: M. Cooke (1988)

Dispersion in buildings
B.R. Morton, Taylor and Turner (1956); Baines and Turner (1969); L.A. Wallace *et al.* (1985); Loughan and Yokomoto (1989); Cleaver, Marshall and Linden (1994)

Safe discharge and dispersion
Bosanquet (1935, 1957); Bosanquet and Pearson (1936); Chesler and Jesser (1952); Bodurtha (1961, 1980, 1988); Long (1963); Loudon (1963); ASME (1969/1); Cairney and Cude (1971); Bodurtha, Palmer and Walsh (1973); Cude (1974a,b); Nonhebel (1975); de Faveri, Hanzevack and Delaney (1982); Hanzevack (1982); Palazzi *et al.* (1982); Jagger and Edmundson (1984); Palazzi *et al.* (1984); Burgoyne (1987); Moodie and Jagger (1987)

Deposition and removal
Gregory (1945); Bosanquet, Carey and Halton (1950); Chamberlain (1953, 1959, 1961, 1966a,b, 1975); Chamberlain and Chadwick (1953, 1966); N.G. Stewart *et al.* (1954); Csanady (1955, 1973); F.G. May (1958); P.A. Sheppard (1958); E.G. Richardson (1960); Leighton (1961); Gifford and Pack (1962); F.B. Smith (1962); Hage (1964); Calvert and Pitts (1966); Engelmann (1968); van der Hoven (1968); Slade (1968); Marble (1970); Beadle and Semonin (1974); Demerjian, Kerr and Clavert (1974); Hosker (1974b); Krey (1974); AEC (1975); McEwen and Phillips (1975); Dana and Hales (1976); Hales (1976); Heicklen (1976); Overcamp (1976); V.J. Clancey (1977b); Horst (1977, 1980); Sehmel and Hodgson (1978, 1980); Slater (1978a); McMahon and Denison (1979); Jensen (1980); NRPB (1980 R101); Sehmel (1980); Corbett (1981); Lodge *et al.* (1981); Meszaros (1981); P.M. Foster (1982); Garland and Cox (1982); S.R. Hanna, Briggs and Hosker (1982); Hosker and Lindberg (1982); Sievering (1982, 1989); R.M. Williams (1982a,b); Bartiz (1983); Ibrahim, Barrie and Fanaki (1983); Murphy and Nelson (1983); Pruppacher, Semonin and Slinn (1983); Cadle, Dasch and Mulawa (1985); Cher (1985); Doran and Horst (1985); El-Shobosky (1985); Schack, Pratsinis and Friedlander (1985); Sinclair, Psota-Kelty and Weschler (1985); Kumar (1986); Bettis, Makhviladze and Nolan (1987); Underwood (1987, 1987 SRD R423, 1988 SRD

442); Bache, Lawson and Uk (1988); M. Bennett (1988); El-Shobosky and Hussain (1988); Joffre (1988); Nicholson (1988, 1993); Venkatram (1988b); Dugstad and Venkatram (1989); Adhikari *et al.* (1990); M.P. Singh, Kumari and Ghosh (1990); A.G.Allen, Harrison and Nicholson (1991); Schorling and Kardel (1993)

Vapour solubility and chemical reactions
Clough and Garland (1970); Murata *et al.* (1974); HSE (1978b); Burton *et al.* (1983); Das Gupta and Shen Dong (1986); Raj (1986); Shi and Seinfeld (1991)

Topics common to passive and dense gas dispersion, including source terms, mitigation systems and hazard assessment (see Table 15.42)

of gas of neutral density, or neutral buoyancy. This work is relevant to dispersion from stacks once buoyancy effects have decayed. There are separate models which treat gases of positive buoyancy, which apply to releases close to stacks, and gases of negative buoyancy. Dispersion of gases which do not exhibit positive or negative buoyancy is generally referred to as 'passive dispersion'.

15.11.3 Momentum effects

A continuous release of material with low kinetic energy forms a plume which tends to billow. If the kinetic energy is high, however, a momentum jet is formed which has a well-defined shape.

The momentum of the release has a marked effect on the extent of air entrainment. If the kinetic energy is high, large quantities of air are entrained. The degree of air entrainment affects the density of the cloud and is important in its further dispersion.

15.11.4 Source terms

The principal types of source used in idealized treatments of dispersion are the point source, the line source and the area source. An escape from a pipe is normally treated as a point source, while vaporization from a pool may be treated as an area source. There may also be some situations which may be modelled as an infinite or semi-infinite line source.

A very short and a prolonged escape may approximate to an instantaneous release and to a continuous release, respectively. An escape of intermediate duration, however, may need to be treated as a quasi-instantaneous or, alternatively, quasi-continuous release.

The most common scenarios considered are an instantaneous release from a point source, or 'puff', and a continuous release from a point source, or 'plume'.

It will be apparent that the source terms described are idealizations of the actual situation.

15.11.5 Source elevation

Another distinction is the elevation of the source. Sources are classed as ground level or elevated. Most hazardous escapes are treated as ground level sources. Stacks are the principal elevated sources.

15.11.6 Meteorology and topography

The two main meteorological conditions which affect the dispersion are the wind direction and speed and the stability conditions.

The stability of the atmosphere determines the degree of mixing. The simplest classification of stability is:

(1) unstable;
(2) neutral;
(3) stable.

Dispersion is greatest in unstable conditions and least in stable conditions.

With respect to topography the default condition may be regarded as dispersion over flat grassland of moderate roughness, but there are many other situations, including dispersion over surfaces with very low or very high roughness, at buildings or other obstacles, over urban areas, over coastal zones and sea, and over other complex terrain.

Meteorology and topography are discussed further in Sections 15.12 and 15.13, respectively.

15.11.7 Dispersion of passive gas plume

As already mentioned, the dispersion situation which has been studied most extensively is the dispersion of a

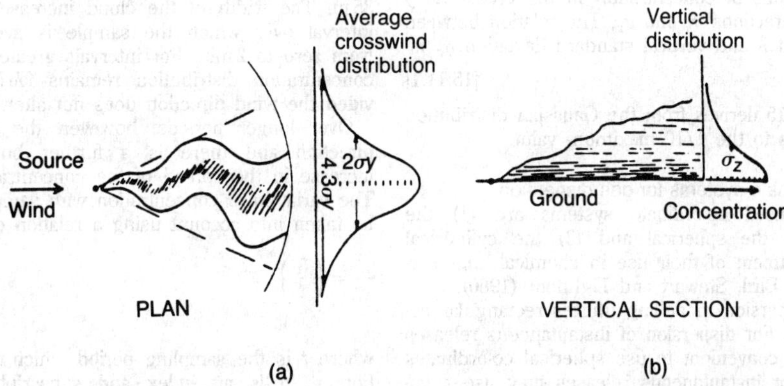

Figure 15.31 Plume from a continuous release: definition of plume (Pasquill and Smith, 1983; reproduced by permission)

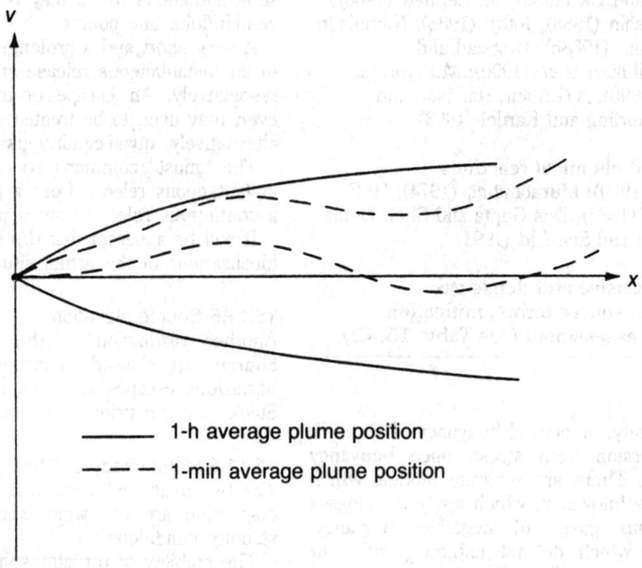

Figure 15.32 Plume from a continuous release: effect of meandering of plume (S.R. Hanna, Briggs and Hosker, 1982)

plume of neutral density gas from a continuous point source. It is convenient at this point to consider some basic features of gas dispersion in relation to this source term.

It has frequently been shown by experiment that for such a plume both the crosswind and vertical concentration distributions are approximately Gaussian, as illustrated in Figure 15.31(a).

By convention, the boundaries of a cloud are defined in dispersion work as the locus of the points at which the concentration has fallen to 1/10 of that at the centre. For a plume at ground level the horizontal and vertical cloud boundaries are given by the envelope at which the concentrations are 1/10 of those down the centre line. This defines the cloud width w, the lateral spread θ and the cloud height h, as shown in Figure 15.31(b).

Since the concentration distribution is Gaussian, the cloud spread may also be defined in terms of the standard deviations of concentration in the crosswind y and vertical z directions σ_y and σ_z. The relation between the cloud height h and vertical standard deviation σ_z is

$$h = 2.15\sigma_z \qquad [15.11.1]$$

The factor of 2.15 derives from the Gaussian distribution and corresponds to the 1/10 maximum value.

15.11.8 Co-ordinate systems for gas dispersion

Three widely used co-ordinate systems are (1) the rectangular, (2) the spherical and (3) the cylindrical systems. A treatment of their use in chemical engineering is given by Bird, Stewart and Lightfoot (1960).

Most gas dispersion modelling utilizes rectangular co-ordinates (x,y,z). For dispersion of instantaneous releases it is sometimes convenient to use spherical co-ordinates (r,θ). Models for instantaneous releases may use a co-ordinate system moving with the cloud. Some use is made of special co-ordinate systems. An example is that used in the DRIFT model described in Section 17.35.

15.11.9 Concentration fluctuations and sampling

It is found experimentally that if the concentration in a gas cloud from a continuous point source under near neutral conditions is measured at points along an arc across wind, the distribution of concentration obtained is Gaussian, but the shape of the distribution depends on the sampling interval. If the sample is virtually instantaneous, the distribution has a relatively narrow spread and a high maximum, while if the sample is averaged over a time period the distribution has a wider spread and a lower maximum. This behaviour is illustrated in Figure 15.32.

Typically, at 100 m the total width of the cloud as measured by instantaneous samples is about 20 m, but the width as measured by time mean samples is about 35 m. The width of the cloud increases rapidly as the interval over which the sample is averaged increases from zero to 2 min. For intervals greater than 3 min the concentration distribution remains fairly constant, provided the wind direction does not alter.

Over longer periods, however, the wind does alter direction and there is a further but more gradual increase in the spread of the concentration distribution. The variation of concentration with sampling period may be taken into account using a relation of the form

$$\frac{\chi}{\chi_r} = \left(\frac{t_r}{t}\right)^p \qquad [15.11.2]$$

where t is the sampling period which yields concentration χ, p is an index and subscript r denotes the reference value.

There can be considerable differences between peak and time-mean concentrations. Values of the peak/mean ratio of 50 or more have been reported in some investigations. The concentrations given by the common dispersion models for passive dispersion gas are time-mean values. In some applications it is necessary also to consider the peak values.

15.11.10 Models for passive gas dispersion

The modelling of dispersion, particularly that of neutral density gas, and passive gas dispersion models are discussed further in Sections 15.15 and 15.16, respectively. Passive gas dispersion over particular surfaces and in particular conditions is considered in Sections 15.17 and 15.18, while dispersion parameters for passive gas dispersion models are described in Section 15.19.

15.11.11 Models for dense gas dispersion

For high concentration releases of many of the hazardous materials of interest in process plants the assumption of neutral density gas behaviour is not valid. In particular, the gas cloud is often heavier than air. In this situation the common neutral density gas models are not applicable. However, the behaviour of dense gases has been the subject of much work in recent years and dense gas dispersion models have been developed. Dispersion of dense gas is discussed further in Section 15.22 and succeeding sections.

15.11.12 Other topics

Other topics in gas dispersion considered below include dispersion of buoyant plumes and momentum jets, concentration fluctuations, dispersion over short distances, transformation and removal mechanisms, flammable and toxic clouds, infiltration into buildings, fugitive emissions, dispersion by fluid curtains and leaks and spillages.

15.12 Meteorology

Gas dispersion depends on meteorology and, in particular, on turbulence. Both the form of the models used to describe dispersion and the values of the parameters in them derive from meteorological considerations.

Accounts of meteorology include those given in *Physical and Dynamical Meteorology* (Brunt, 1934), *The Climate Near the Ground* (Geiger, 1950–), *Micrometeorology* (O.G. Sutton, 1953), *The Challenge of the Atmosphere* (O.G. Sutton, 1962), *Elementary Meteorology* (G.F. Taylor, 1954), *Exploring the Atmosphere's First Mile* (Lettau and Davidson, 1962), *Natural Aerodynamics* (Scorer, 1958), *Environmental Aerodynamics* (Scorer, 1978), *General Meteorology* (Byers, 1959), *Atmospheric Diffusion* (Pasquill, 1962a, 1974; Pasquill and Smith, 1983), *An Introduction to Climate* (Trewartha, 1968), *Essentials of Meteorology* (McIntosh and Thom, 1969), *Evaporation into the Atmosphere* (Brutsaert, 1982), *CRC Handbook of Atmospherics* (Volland, 1982), *Engineering Meteorology* (Plate, 1982) and *Meteorological Glossary* (McIntosh, 1972), *Dynamics in Atmospheric Physics* (Lindzen, 1990), and in the *Meteorological Glossary* (McIntosh, 1972). An account specific to the UK is available in *Climate in the United Kingdom* (Page and Lebens, 1986).

Works on turbulence and related phenomena include *Massenaustausch in freier Luft and verwandte Erscheinungen* (Schmidt, 1925), *Turbulence* (Hinze, 1959), *Turbulent Transfer in the Lower Atmosphere* (Priestley, 1959), *Boundary Layer Theory* (Schlichtling, 1960), *Turbulence: Classic Papers on Statistical Theory* (Friedlander and Topper, 1961), *The Structure of Atmospheric Turbulence* (Lumley and Panofsky, 1964), *Introduction to Turbulence and Its Measurement* (Bradshaw, 1971), *Statistical Fluid Mechanics: Mechanics of Turbulence* (Monin and Yaglom, 1971), *Turbulence Phenomena* (J.T. Davies, 1972), *A First Course in Turbulence* (Tennekes and Lumley, 1972), *Turbulent Diffusion in the Environment* (Csanady, 1973), *Buoyancy Effects in Fluids* (J.S. Turner, 1973), *Mathematical Modelling of Turbulent Diffusion in the Environment* (C.J. Harris, 1979), and *Turbulence and Diffusion in Stable Environments* (J.C.R. Hunt, 1985).

Treatments of gas dispersion, with particular emphasis on air pollution, include those given in *Atmospheric Pollution: Compendium of Meteorology* (Hewson, 1951), *Air Pollution* (Scorer, 1968), *Contemporary Problems of Atmospheric Diffusion and Pollution in the Atmosphere* (Beryland, 1975), *Atmospheric Motion and Air Pollution* (Dobbins, 1979), *Air Pollution Modelling and its Application* (de Wispelaere, 1981, 1983, 1984, 1985; de Wispelaere, Schiermeier and Gillani, 1986), *Atmospheric Diffusion* (Pasquill and Smith, 1983), and *Atmospheric Turbulence and Air Pollution Modelling* (Nieuwstadt and van Dop, 1982).

15.12.1 Atmospheric boundary layer

The meteorological conditions which are of prime relevance here are those within the atmospheric boundary layer (ABL), or planetary boundary layer (PBL). In unstable, or convective, conditions this layer is also known as the convective boundary layer (CBL).

The ABL has an outer and an inner region, as shown in Figure 15.33. The outer region, or defect layer, has a height of some 10^2–10^3 m and the inner region, or surface layer, one of some 10 m. There is within the surface sublayer a dynamic sublayer with a height of 1–10 m. Between the dynamic sublayer and the surface is the interfacial layer, the height of which depends on the surface roughness.

Within the PBL, the surface layer is of particular importance. Within this layer stress is virtually constant and it is also termed the constant stress layer, or surface stress layer. Considering the PBL over land, conditions in the layer depend on the stability conditions. The simplest case is that of neutral conditions where buoyancy forces are negligible. It is to this case that the classical treatments primarily apply.

In the neutral PBL turbulent energy derives from two main sources. One is due to the mechanical drag of the wind over the surface, and the other to the turning of the wind direction with height. In the unstable boundary layer a further source of turbulent energy is that due to buoyancy. Over land, such buoyancy is mainly caused by vertical flux of sensible heat into the air at the surface. Over sea, vertical flux of latent heat associated with water vapour plays a significant role.

Stable conditions occur mainly at night. In the stable boundary layer the surface is cool relative to the air and the vertical heat flux is in the reverse direction so that

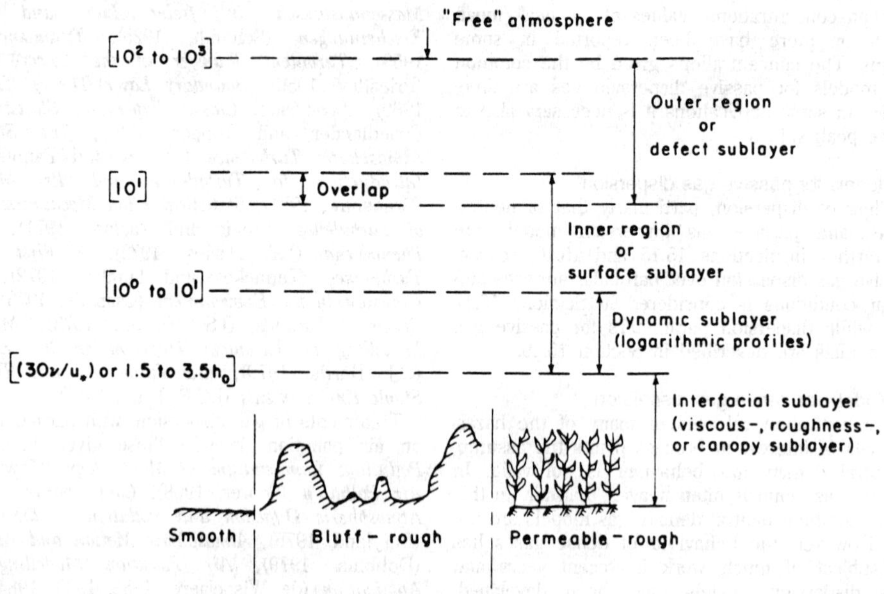

Figure 15.33 *Atmospheric boundary layer: orders of magnitude of the heights of the sublayers (Brutsaert, 1982). The vertical scale (m) is distorted; h_0 is the height of the roughness obstances (Courtesy of Reidel Publishing Company)*

the mechanical and buoyancy forces are opposed. At some height above the surface turbulence becomes very weak and is virtually suppressed.

15.12.2 Wind characteristics

Wind is a main factor in determining dispersion. Some principal wind characteristics are:

(1) direction;
(2) speed –
  (a) at surface,
  (b) above ground;
(3) persistence;
(4) turbulence.

Wind direction is defined as the direction from which the wind is blowing. Information on wind direction and speed at a given location is conveniently summarized in the form of a 'wind rose'. This is a polar diagram in which the length of the sections of the spokes is proportional to the observed frequencies of wind direction and speed. The period for which the wind rose is drawn is typically a month or a year. Some monthly wind roses show a marked degree of seasonal variation, others do not. The degree of symmetry in wind roses also varies considerably. A table of percentage frequency of wind direction and speed and corresponding wind rose for Watnall is shown in Figure 15.34.

The predominant wind direction is usually called the 'prevailing wind'. This wind direction only applies, however, for a relatively limited proportion of the time, and it is usually necessary in dispersion calculations to consider other directions also.

Persistence of the wind direction is important in assessing dispersion. This is often expressed in the form of a persistence table such as that shown for Watnall in Table 15.11, which gives the number of occasions during the given period in which the wind remained within the sectors indicated for the number of hours in sequence indicated.

Alternatively, the persistence of wind direction may be expressed in terms of the constancy, which is based on the ratio of the vector and scalar winds and has a value of 1 for an invariant wind direction and 0 for a completely uniform distribution of wind directions. This measure has the advantage that, unlike persistence, it is relatively little affected by brief excursions outside the 45° sector.

Localized variations of wind direction and speed can occur and may affect dispersion. Features which can give rise to these variations include irregularities in terrain and differences in surface temperature. Where there is a marked slope, a drainage wind can occur. This is a downhill flow of air cooled by radiation at night and it may be in a direction quite different from the gradient wind. In well-defined valleys there are usually complex flow patterns. Typically, at night there is a drainage wind from the sides and down the bottom of the valley and during the day a tendency for wind to flow up the valley bottom. On the coast the land is warmer in daytime and the sea at night. Hence during the day there tends to be a sea breeze blowing onto the land and during the night a land breeze blowing towards the sea.

Turbulence is another feature of wind variation. In the present context turbulence includes wind fluctuations with a frequency of more than 2 cycle/h, the most

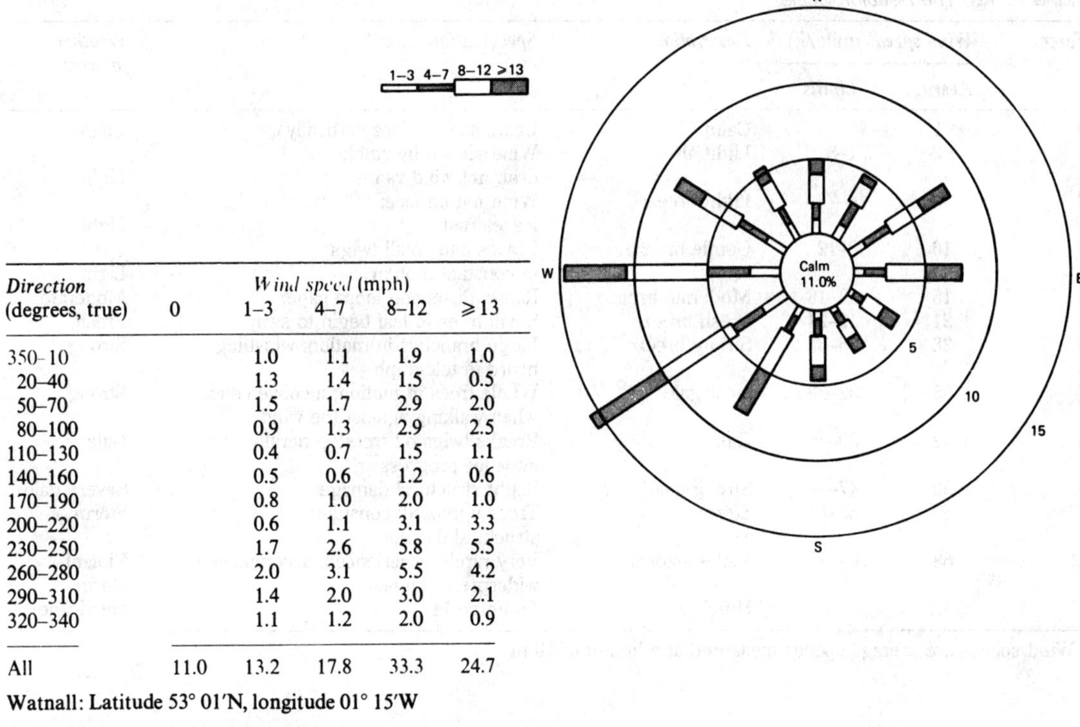

Direction (degrees, true)	0	Wind speed (mph) 1–3	4–7	8–12	≥13
350–10		1.0	1.1	1.9	1.0
20–40		1.3	1.4	1.5	0.5
50–70		1.5	1.7	2.9	2.0
80–100		0.9	1.3	2.9	2.5
110–130		0.4	0.7	1.5	1.1
140–160		0.5	0.6	1.2	0.6
170–190		0.8	1.0	2.0	1.0
200–220		0.6	1.1	3.1	3.3
230–250		1.7	2.6	5.8	5.5
260–280		2.0	3.1	5.5	4.2
290–310		1.4	2.0	3.0	2.1
320–340		1.1	1.2	2.0	0.9
All	11.0	13.2	17.8	33.3	24.7

Watnall: Latitude 53° 01′N, longitude 01° 15′W

Figure 15.34 *Table of percentage frequency of wind direction and speed and corresponding wind rose for Watnall, 1959–68 (Meteorological Office, 1977; reproduced by permission)*

Table 15.11 *Table of wind persistence for Watnall, 1974–76 (Meteorological Office, 1977; reproduced by permission)*

No. of hours in sequence	*Direction* N	*NE*	*E*	*SE*	*S*	*SW*	*W*	*NW*
1	756	565	510	411	587	1043	1087	721
7	91	110	64	29	82	176	202	46
13	27	49	29	4	49	97	72	12
25	14	8	8	2	6	15	9	2
37	2	7	1	0	0	6	5	0
>49	0	5	0	0	0	2	3	0
Total no. of hours	3374	4007	2404	1190	3106	6503	6153	2202

Watnall: latitude 53° 01′ N, longitude 01° 15′ W

important fluctuations lying in the range 0.1–1 cycle/s. The main factors which determine the turbulence are the gradient wind speed and the roughness of the terrain, and the temperature differences between the surface and the air. Turbulence tends to increase as the gradient wind speed increases, or as the temperature of the air close to the surface increases, relative to that of the air aloft. A measure of turbulence is given by the standard deviation, or σ, value of the wind fluctuations over a 1 h period. These wind speed σ values can be related to the dispersion σ values in the models of gas dispersion.

Before leaving wind characteristics, it is convenient to give at this point the method of characterizing high wind speeds. This is the Beaufort scale, developed originally for use at sea. It is shown in Table 15.12. The scale is not used in work on gas dispersion, but is relevant to wind hazard.

15.12.3 Geostrophic wind and Ekman spiral

In the free atmosphere above the PBL it is usual to assume that the wind is horizontal and free from friction. In this situation the wind velocity becomes a function of

Table 15.12 The Beaufort Scale

Force	*Wind speed*[a] (mile/h)		*Description*	*Specification (land)*	*Weather forecast*
	Average	*Limits*			
0	< 1		Calm	Calm; smoke rises vertically	Calm
1	2	1–3	Light air	Wind shown by smoke drift, not wind vanes	Light
2	5	4–7	Light breeze	Wind felt on face; leaves rustle	Light
3	10	8–12	Gentle breeze	Leaves and small twigs in constant motion	Light
4	15	13–18	Moderate breeze	Raises dusts and loose paper	Moderate
5	21	19–24	Fresh breeze	Small trees in leaf begin to sway	Fresh
6	28	25–31	Strong breeze	Large branches in motion; whistling heard in telegraph	Strong
7	35	32–38	Near gale	Whole trees in motion; inconvenience when walking against the wind	Strong
8	42	39–46	Gale	Breaks twigs off trees; generally impedes progress	Gale
9	50	47–54	Strong gale	Slight structural damage	Severe gale
10	59	55–63	Storm	Trees uprooted; considerable structural damage	Storm
11	68	64–72	Violent storm	Very rarely experienced; accompanied widespread damage	Violent storm
12	> 73		Hurricane	As Force 11	Hurricane

[a] Wind speeds are average speeds measured at a height of 10 m

the pressure gradient and of the forces arising from the rotation of the Earth. The wind which satisfies these conditions is known as the gradient level wind.

Where the lines of constant pressure, or isobars, are straight so that any centripetal acceleration is negligible, the gradient level wind is known as the geostrophic wind. The direction of the geostrophic wind is not the same as the surface wind. An idealized representation of the wind velocity vectors at different heights is given by the Ekman spiral, developed originally in relation to ocean currents, as shown in Figure 15.35. The Ekman spiral is the locus of the end points of the wind velocity vectors. These vectors approach the geostrophic wind velocity at the mixed layer height. The angle of a vector is the angle by which it is backed from the direction of the geostrophic wind. The angle of the surface wind vector is backed 45° from this direction. Actual observed angles between the geostrophic and surface winds are typically 5–10°, 15–20° and 30–50° for unstable, neutral and stable conditions, respectively.

Wind direction shear affects dispersion, particularly over large distances, where the top and bottom of a plume can move with a difference of direction of some 40–50°, which gives a much larger plume spread than would be obtained by diffusion alone.

15.12.4 Force balance equation

The balance of forces acting horizontally on the air in free stream conditions is given by the relations

$$\frac{\partial u}{\partial t} + fv = \frac{1}{\rho}\frac{\partial p}{\partial x} + X \qquad [15.12.1a]$$

$$\frac{\partial v}{\partial t} + fu = \frac{1}{\rho}\frac{\partial p}{\partial y} + Y \qquad [15.12.1b]$$

where f is the Coriolis force, p is the pressure, t is the time, u and v are the wind velocity components in the downwind and crosswind directions, respectively, X and Y are accelerations representative of forces not due to pressure and gravity, and ρ is the air density.

The Coriolis force, or parameter, f is

$$f = 2\omega \sin \phi \qquad [15.12.2]$$

where ϕ is the latitude and ω is the angular velocity of the Earth. The Coriolis force has a value of approximately $10^{-4}\,s^{-1}$ at intermediate latitudes and tends to zero at the equator.

15.12.5 Turbulent exchange, momentum flux, eddy viscosity and mixing length

The theory of turbulence derives from the work of Schiller (1932), Nikuradse (1933), Prandtl (1933) and Schlichtling (1936) on the boundary layer for flow in pipes. The extension of this theory for flow over open ground is described by O.G. Sutton (1953) and Pasquill (1962a). The theories of turbulence which describe the mixing that occurs when wind passes over a surface are complex. The treatment given here is limited to the description of a few simple concepts and relations.

The momentum flux is given by the equation

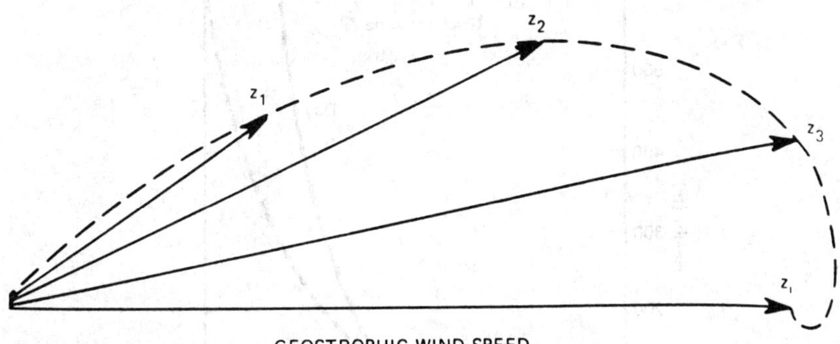

Figure 15.35 *Geostrophic wind and Ekman spiral (S.R. Hanna, Briggs and Hosker, 1982). The wind velocity vectors (z_1, z_2, z_3) approach the geostrophic wind velocity vector at the top of the mixed layer z_i*

$$\frac{\tau}{\rho} = (\nu + K_M)\frac{du}{dz} \qquad [15.12.3a]$$

The eddy viscosity K_M is generally much greater than the kinematic viscosity ν, and hence

$$\frac{\tau}{\rho} = K_M\frac{du}{dz} \qquad \nu \ll K_M \qquad [15.12.3b]$$

where K_M is the eddy viscosity, u is the mean wind speed, z is the height, ν is the kinematic viscosity of air, ρ is the density of air and τ the mean momentum flux per unit area.

The constant K_M is variously described as the coefficient of exchange for momentum, the eddy diffusivity for momentum, or the eddy viscosity. In terms of the mixing length theory of Prandtl,

$$K_M = l^2\frac{du}{dz} \qquad [15.12.4]$$

where l is the mixing length. The momentum flux is then

$$\frac{\tau}{\rho} = l^2\left(\frac{du}{dz}\right)^2 \qquad [15.12.5]$$

On the assumption that the mixing length is proportional to the distance from the surface,

$$l = kz \qquad [15.12.6]$$

where the constant k is von Karman's constant. The value usually quoted for von Karman's constant is 0.4. A value of 0.35 has been given by Businger *et al.* (1971).

From Equation 15.12.3b and the relation $u_*^2 = \tau/\rho$, described below,

$$K_M = \frac{u_*^2}{du/dz} \qquad [15.12.7]$$

Then from Equations 15.12.7 and the 'law of the wall' Equation 15.12.15, given below,

$$K_M = ku_*z \qquad [15.12.8]$$

In terms of mixing length theory, using Equation 15.12.6, Equation 15.12.15 may be rewritten as

$$\frac{l}{u_*}\frac{du}{dz} = \frac{1}{l} \qquad [15.12.9]$$

and Equation 15.12.8 as

$$K_M = u_*l \qquad [15.12.10]$$

15.12.6 Friction velocity

Closely related to the momentum flux is the friction velocity u_* defined by the relations

$$\frac{\tau}{\rho} = u_*^2 \qquad [15.12.11a]$$

or

$$u_* = (\tau/\rho)^{\frac{1}{2}} \qquad [15.12.11b]$$

The friction velocity may be obtained from the geostrophic wind velocity using the relation

$$u_* = c_g u_g \qquad [15.12.12]$$

where c_g is the drag coefficient and u_g is the geostrophic wind speed.

An empirical relation for the drag coefficient c_g in neutral conditions has been given by Lettau (1959) as follows:

$$c_g = \frac{0.16}{\log_{10}\mathrm{Ro} - 1.8} \qquad [15.12.13]$$

with

$$\mathrm{Ro} = \frac{u_g}{fz_o} \qquad [15.12.14]$$

where Ro is the Rossby number and z_o is the roughness length.

For other stability conditions Lettau uses the following approximate ratios of the drag coefficient to that in neutral conditions:

Unstable	1.2
Slightly stable	0.8
Stable	0.6

The value of the friction velocity u_* varies between about 3 and 12% of the mean wind speed, the lower values being associated with smooth surfaces. It is often taken

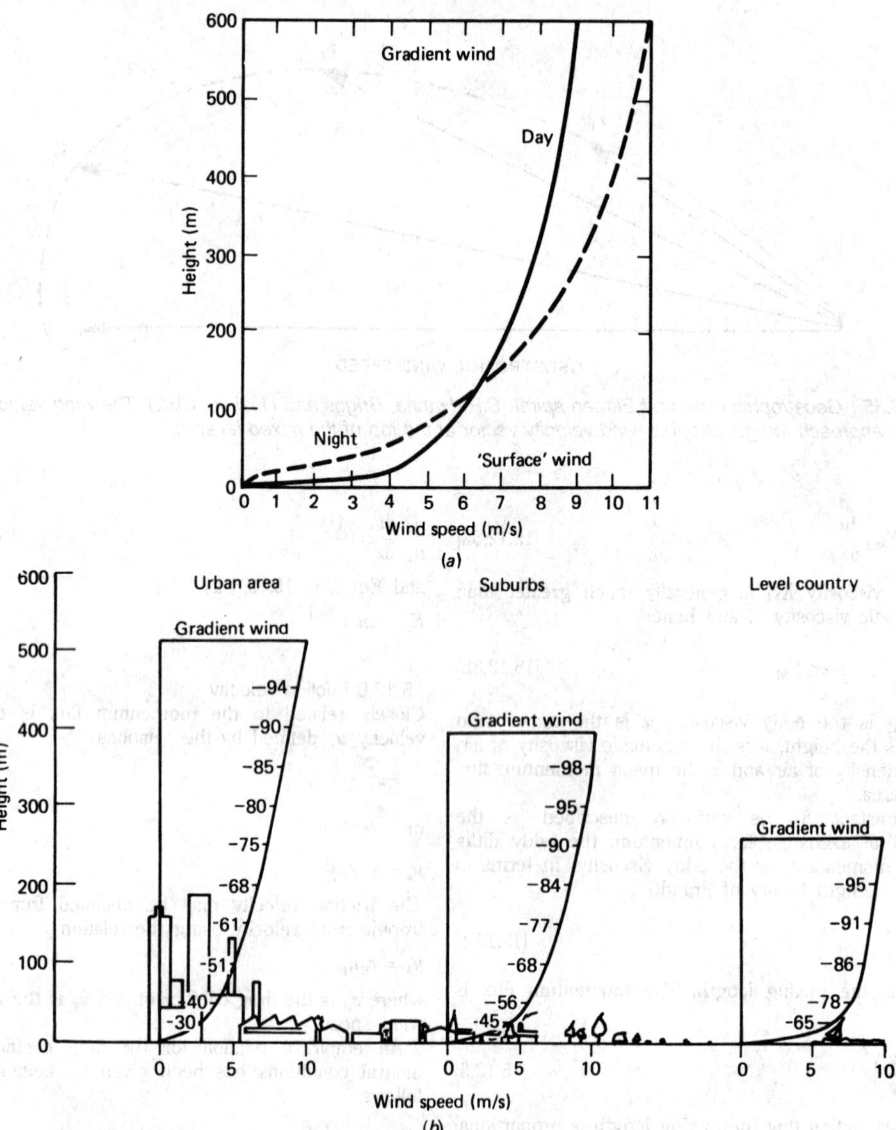

Figure 15.36 Wind speed vs height above ground (American Society of Mechanical Engineers, 1973/2): (a) effect of stability conditions on wind speed; (b) effect of terrain on wind speed

as one-tenth of the wind speed. The friction velocity is generally tabulated together with the surface roughness as a function of the type of surface, as described below.

15.12.7 Law of the wall

On the assumption that the flow at the surface depends only on the shear stress and the distance from the wall

$$\frac{du}{dz} = \frac{u_*}{kz} \qquad [15.12.15]$$

This is the so-called 'law of the wall'. The term $(kzdu/u_*dz)$ is a non-dimensional wind shear. Equation 15.12.15 is applicable to adiabatic, or neutral, conditions.

15.12.8 Empirical vertical wind velocity profile

The gradient wind is, by definition, uninfluenced by friction effects, but at lower heights the wind velocity is reduced by such effects. Figure 15.36 illustrates some vertical profiles of wind speed.

A widely used empirical relation for the vertical wind velocity profile is

$$u = u_r \left(\frac{z}{z_r}\right)^p \quad z < z_r \qquad [15.12.16]$$

where u is the wind speed, u_r is the wind speed at the reference height, z is the height, z_r the reference height and p is an index.

The gradient level reference height z_r lies approximately in the range 300–750 m, being typically 300 m over level ground and 500 m over an urban area.

The value of the index p generally lies in the range $0.11 < p < 0.65$, and invariably in the range $0 < p < 1$. Range values obtained include: $0.145 < p < 0.77$ by Frost (1947); $0.10 < p < 0.63$ by Touma (1977); and $0.11 < p < 0.65$ by Hanafusa, Lee and Lo (1986).

As pointed out by O.G. Sutton (1953), the generally accepted value of the index p for the turbulent boundary layer of a flat plate in a wind tunnel is 1/7 (0.142).

Touma states that in the absence of measured values the wind speed is generally estimated using the 1/7 wind profile and assuming neutral stability conditions. He presents experimental data, however, which indicate that the index p has a value of about 0.10 at Pasquill stability category A, may pass at some sites through a weak minimum and then increases markedly through categories D–F to a value of up to 0.62.

In further experimental work in Japan by Hanafusa, Lee and Lo (1986), the value of the index p varied with the Pasquill stability categories approximately as follows:

Pasquill stability category	*Index, p*
A	0.33
B	0.26
C	0.20
D, E	0.38
F	0.42
G	0.57

The strong increase in the index p with stability categories D–F broadly confirmed that found by Touma, but the values of p for categories A and B were higher than those obtained by the latter. The authors suggest that the difference may be due to the different methods used to define the stability category.

The index p may also be written as

$$p = n/(2 - n) \qquad [15.12.17]$$

where n is an index. This index is one of the parameters in the Sutton model for neutral density gas dispersion, as described below.

15.12.9 Empirical vertical wind velocity profile: Deacon relation

An empirical relation for the vertical wind velocity gradient was formulated by Deacon (1949) as follows:

$$\frac{du}{dz} = az^{-\beta} \qquad [15.12.18]$$

where a is a constant and β an index. He gave the following values of the index β:

$\beta > 1$ superadiabatic conditions
$\beta = 1$ adiabatic conditions, or small temperature gradients
$\beta < 1$ inversion conditions

Then for $\beta = 1$

$$\frac{du}{dz} = \frac{a}{z} \qquad [15.12.19]$$

Equation 15.12.19 is an alternative form of the law of the wall given above as Equation 15.12.15].

15.12.10 Logarithmic vertical wind velocity profile

Another vertical wind velocity profile is the logarithmic profile. This is applicable to adiabatic, or neutral, conditions. The logarithmic vertical wind velocity profile is obtained by integration of Equation 15.12.15 to give

$$u = \frac{u_*}{k} \ln\left(\frac{z}{z_o}\right) \qquad [15.12.20]$$

where z_o is a constant of integration. This constant is termed the 'roughness length'.

15.12.11 Modified logarithmic vertical wind velocity profiles

For non-adiabatic, or diabatic, conditions a different treatment is required. The analysis is complicated by buoyancy effects. For diabatic conditions Equation 15.12.15 may be modified by the inclusion of the similarity factor introduced by Monin and Obukhov (1954). It then becomes

$$\frac{du}{dz} = \frac{u_*}{kz}\phi_M \qquad [15.12.21]$$

where ϕ_M is the similarity factor. These authors take

$$\phi_M = f(z/L) \qquad [15.12.22]$$

They also use the relation

$$\phi_M = 1 + \alpha\frac{z}{L} \qquad [15.12.23]$$

which yields for Equation 15.12.21

$$\frac{du}{dz} = \frac{u_*}{kz}\left(1 + \alpha\frac{z}{L}\right) \qquad [15.12.24]$$

where L is a characteristic length (the Monin–Obukhov length), and α is the Monin–Obukhov coefficient.

Integrating Equation 15.12.24

$$u = \frac{u_*}{k}\left[\ln(z/z_o) + \alpha\frac{z - z_o}{L}\right] \qquad [15.12.25]$$

and taking $z \gg z_o$ yields

$$u = \frac{u_*}{k}\left[\ln(z/z_o) + \alpha\frac{z}{L}\right] \qquad [15.12.26]$$

This is the log–linear vertical wind velocity profile. The log-linear velocity profile is applicable over a limited range of unstable conditions and a wide range of stable conditions.

In a further treatment Ragland (1973) has given

$$\phi_M = 1 \quad \text{neutral conditions} \quad z < (ku_*/f) \qquad [15.12.27a]$$

$$= 1 + \alpha\frac{z}{L} \quad \text{stable conditions} \quad 0 < \frac{z}{L} < 1 \qquad [15.12.27b]$$

Table 15.13 Some values of roughness length

A Values given by O.G. Sutton (1953)

Type of surface	z_o (cm)	u_* (cm/s)
Very smooth (mud flats, ice)	0.001	16
Thin grass up to 10 cm high	0.7	36
Thick grass up to 50 cm high	9	63

B Values given by Pasquill and Smith (1983)

Type of surface	z_o (m)
Grass: closely mown	10^{-3}
short (*c.* 10 cm)	10^{-2}
long	3×10^{-2}
Agricultural–rural complex	0.2
Towns, forests	1

$$= 1 + \alpha \quad \text{neutral conditions} \quad 1 < \frac{z}{L} < 6 \qquad [15.12.27c]$$

$$= \left(1 - 15\frac{z}{L}\right)^{-0.25} \quad \text{stable conditions} \quad -2 < \frac{z}{L} < 0 \qquad [15.12.27d]$$

He quotes for α the value of 5.2 given by Webb (1970).

15.12.12 Surface roughness and roughness length

A surface is aerodynamically rough when the flow is turbulent down to the surface itself. Over such a surface the velocity profile and surface drag are independent of viscosity and depend on the roughness length. In meteorology nearly all surfaces are aerodynamically rough for any significant wind speed. The roughness length depends on the height and spacing of the roughness elements.

From work on pipe roughness, the roughness length z_o and the mean height ϵ of the roughness elements are related approximately as follows:

$$z_o = \epsilon/30 \qquad [15.12.28]$$

Roughness of flow may be defined in terms of the roughness Reynolds number $u_* z/\nu$. For fully rough flow the criterion obtained from Nikuradse's work is

$$\frac{u_*\epsilon}{\nu} > 75 \qquad [15.12.29a]$$

or, using Equation 15.12.28],

$$\frac{u_* z_o}{\nu} > 2.5 \qquad [15.12.29b]$$

Paeschke (1937) correlated roughness length z_o with the mean height ϵ of roughness elements on a larger scale and obtained for various grass and field surfaces

$$z_o = \epsilon/7.35 \qquad [15.12.30]$$

Other workers have obtained similar results, while yet others have found it to be a more complex function. Some typical values of the roughness length given in the literature are shown in Table 15.13.

For very rough surfaces the logarithmic wind profile of Equation 15.12.20 may be extended using the empirical modification

$$u = \frac{u_*}{k} \ln\left(\frac{z - d}{z_o}\right) \qquad z > (d + z_o) \qquad [15.12.31]$$

where d is the zero plane displacement, or displacement height.

The displacement height d is the datum level above which normal turbulent exchange occurs. Using the displacement height concept the base of the roughness elements is at $z = 0$.

15.12.13 Vertical temperature profile

The stability of the atmosphere is essentially the extent to which it allows vertical motion by suppressing or assisting turbulence. One source of turbulence is the mechanical turbulence due to wind movement. Another is the turbulence associated with the vertical temperature gradient. Traditionally stability has been expressed primarily in terms of the latter.

If a small volume of air is taken vertically upward in the atmosphere, it meets lower pressure and therefore expands and cools. The rate of decrease of temperature with height is known as the lapse rate. If the air were dry and the process adiabatic, then the rate of decrease would have a particular value which is known as the dry adiabatic lapse rate of temperature. Although such a process does not occur in the atmosphere, the dry adiabatic lapse rate provides a standard of comparison for real atmospheric conditions.

The rate of change of temperature with height dT/dz under adiabatic, or neutral, conditions is approximately -0.01°C/m. The definition of lapse rate as a rate of decrease of temperature, and thus as a positive quantity under adiabatic conditions, is a potential source of confusion. It is appropriate, therefore, to quote the following statement by O.G. Sutton (1953, p. 9): '$dT/dz \approx -1$°C per 100 m. This particular rate of decrease of temperature with height, known as the dry adiabatic lapse rate and denoted by the symbol Γ (gamma), is one of the fundamental constants of meteorology'.

The potential temperature θ of dry air is the temperature attained if a small volume of the air is taken adiabatically from its existing pressure to a standard pressure, usually that at the surface. Then

$$\theta = T\left(\frac{p_o}{p}\right)^{(\gamma-1)/\gamma} \qquad [15.12.32]$$

where p is the absolute original pressure, p_o is the absolute pressure at the surface, T is the absolute temperature of the volume of air and θ is the potential temperature of this air.

Taking logarithms of Equation 15.12.32 and differentiating with respect to z:

$$\frac{1}{\theta}\frac{d\theta}{dz} = \frac{1}{T}\frac{dT}{dz} - \frac{\gamma - 1}{\gamma}\frac{1}{p}\frac{dp}{dz} \qquad [15.12.33]$$

and noting that

$$dp = g\rho(-dz) \qquad [15.12.34]$$

and

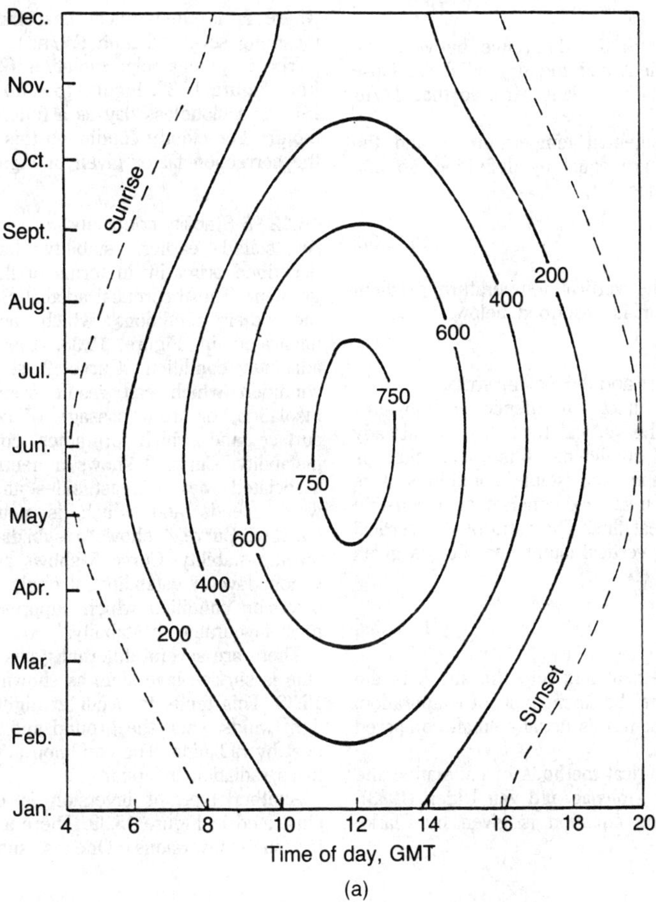

Cloud amount (eighths)	Multiply ISR by
0	1.07
1	0.89
2	0.81
3	0.76
4	0.72
5	0.67
6	0.59
7	0.45
8	0.23

(b)

Figure 15.37 *Incoming solar radiation (ISR) at Cambridge (after Clarke, 1979 NRPB R91; reproduced by permission): (a) ISR on a cloudless day (W/m²); (b) correction factor by which ISR for cloudless day is multiplied to allow for cloudy conditions.*

$$\rho = \frac{p}{RT} \qquad [15.12.35]$$

where g is the acceleration due to gravity, R is the universal gas constant and γ is the ratio of the specific heats of air, the following relation is obtained:

$$\frac{1}{\theta}\frac{\mathrm{d}\theta}{\mathrm{d}z} = \frac{1}{T}\left(\frac{\mathrm{d}T}{\mathrm{d}z} + \Gamma\right) \qquad [15.12.36]$$

with

$$\Gamma = \frac{\gamma - 1}{\gamma} \frac{g}{R} \qquad [15.12.37]$$

The term $(\mathrm{d}T/\mathrm{d}z + \Gamma)$ is the difference between the actual temperature gradient and the dry adiabatic lapse rate. Under adiabatic conditions it is zero, so that $\mathrm{d}T/\mathrm{d}z = -\Gamma$.

At the surface the potential temperature θ and the absolute temperature T are equal by definition, so that Equation 15.12.36 becomes

$$\frac{\mathrm{d}\theta}{\mathrm{d}z} = \frac{\mathrm{d}T}{\mathrm{d}z} + \Gamma \qquad [15.12.38]$$

The relation between the vertical temperature gradient and the stability condition is described below.

15.12.14 Vertical heat flux and the Bowen ratio

Another feature which affects turbulence is buoyancy effects associated with the vertical heat flux. As already mentioned, in neutral conditions this heat flux is negligible, but in unstable and stable conditions it is significant. The vertical heat flux consists of a sensible heat flux and a latent heat flux. The ratio of the vertical sensible heat flux to the vertical latent heat flux is given by the Bowen ratio Bo:

$$\mathrm{Bo} = \frac{H}{L_e E} \qquad [15.12.39]$$

where H is the sensible heat flux into the air, E is the evaporation rate and L_e is the latent heat of evaporation. Over land, the latent heat flux is usually small compared to the sensible heat flux.

There are several empirical methods of estimating the vertical heat flux H (e.g. Holtslag and van Ulden, 1983). For the UK the following equation is given by Clarke (1979 NRPB R91):

$$H = 0.4(S - 100) \qquad [15.12.40]$$

where H is the vertical heat flux (W/m^2) and S the incoming solar radiation (W/m^2).

The incoming solar radiation (ISR) may be estimated from Figure 15.37. Figure 15.37(a) gives the value of the ISR for a cloudless day as a function of time of day and month. For cloudy conditions this value is multiplied by the correction factor given in Figure 15.37(b).

15.12.15 Stability conditions

As stated earlier, stability has traditionally been described primarily in terms of the vertical temperature gradient. The theoretical adiabatic condition and some of the other conditions which occur in practice are illustrated in Figure 15.38. Curve 1 shows the dry adiabatic condition. Curve 2 shows a super-adiabatic condition which can result from strong sunlight, or insolation, or from passage of cold air over a warm surface and which promotes convection and favours instability. Curve 3 shows a neutral condition which is associated with overcast skies and moderate to strong wind speeds and which is neutral with respect to stability. Curve 4 shows a subadiabatic condition which favours stability. Curve 5 shows an isothermal condition which favours stability strongly. Curve 6 shows an inversion condition which suppresses convection and is most favourable to stability.

There are several different types of inversion condition. One is surface inversion, as shown by Curve 6 in Figure 15.38. This tends to occur at night with clear skies and light winds when the ground and the air near to it lose heat by radiation. The condition is therefore also referred to as radiation inversion.

Another type of inversion is elevated inversion, as illustrated in Figure 15.39. There are a number of causes for such inversions. One is subsidence of air from

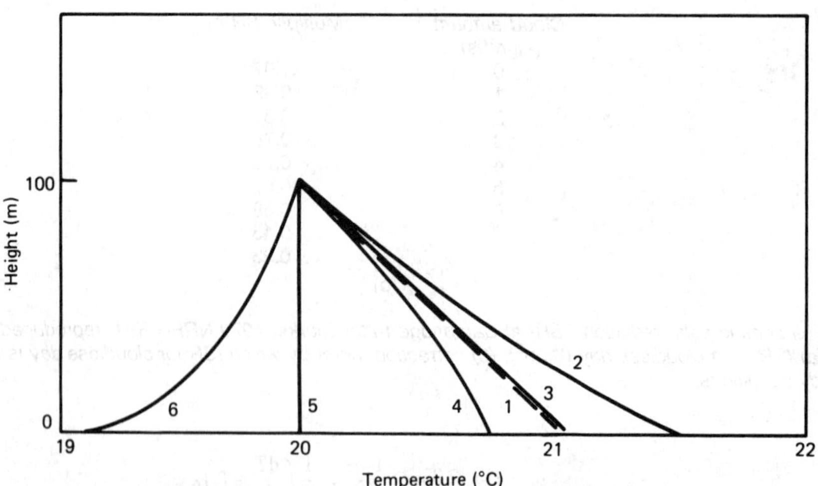

Figure 15.38 *Vertical temperature profiles and lapse rates (American Society of Mechanical Engineers, 1973/2; reproduced by permission): (1) dry adiabatic condition; (2) super-adiabatic condition; (3) neutral condition; (4) subadiabatic condition; (5) isothermal condition; (6) inversion condition*

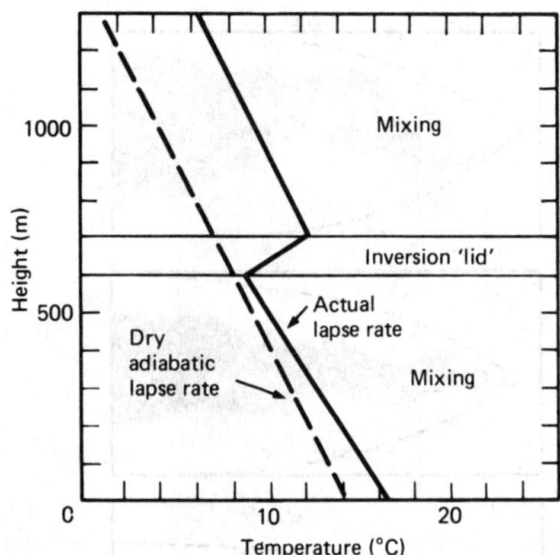

Figure 15.39 Elevated inversion (American Society of Mechanical Engineers, 1973/2; reproduced by permission)

greater heights, which results in compression and hence warming. Another is a sea breeze, which may introduce a layer of cold air beneath a warm air mass. A third is a meteorological front, which also constitutes a boundary between cold air below and warm air above.

An inversion layer inhibits vertical motion. Surface inversion suppresses upward dispersion of a gas release at ground level and also downward dispersion of a release at elevated level. Elevated inversion acts as a 'lid', inhibiting further upwards dispersion. There is a virtually permanent inversion lid on the atmosphere at a height of about 10 000 m.

If in any layer the rate of change of temperature with height is negative, that layer is effectively a 'mixing layer'. There is diurnal variation of stability within the lowest few hundred metres above the ground. The rate of change of temperature with height tends to be negative by day and positive by night giving, respectively, a lapse and an inversion condition. The latter is a surface inversion. Figure 15.40(a) illustrates a typical diurnal variation. An elevated inversion, however, can last for days or even weeks on end.

There are several special features which can affect stability. These include (1) semi-permanent pressure areas, (2) sea–land locations and (3) urban areas. Some areas are subject to a relatively fixed high or low pressure system. The UK, for example, is often under the influence of the Icelandic semi-permanent low pressure area, with cloudy weather and near neutral stability. Stability in coastal locations is strongly influenced by sea–land interactions. One effect may be the complete suppression of surface inversion during winter due to relatively warm sea breezes, as illustrated in Figure 15.40(b). Urban areas affect stability in various ways. One major aspect is the 'heat island' effect, which prevents the development of surface inversion at night.

15.12.16 Stack plume regimes

The effect of stability conditions on dispersion has been studied extensively and is most readily illustrated in relation to the behaviour of the plume from an elevated source such as a factory chimney. Some of the principal types of plume behaviour are shown in Figure 15.41.

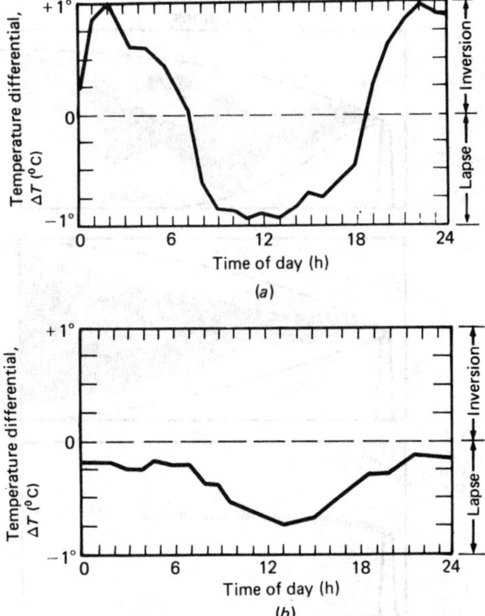

Figure 15.40 Diurnal variation of vertical temperature profile (American Society of Mechanical Engineers, 1973/2; reproduced by permission): (a) open country site all year; (b) coastal site in winter

15.12.17 Stability criteria

A simple stability parameter, the environmental stability s, may be defined:

$$s = \frac{g}{T}\left(\frac{dT}{dz} + \Gamma\right) \qquad [15.12.41a]$$

$$= \frac{g}{\theta}\frac{d\theta}{dz} \qquad [15.12.41b]$$

This parameter is proportional to the rate at which the generation of turbulence is suppressed.

15.12.18 Richardson number

The rate at which mechanical turbulence is generated by wind shear is proportional to $(du/dz)^2$. A criterion for the stability of the atmosphere which is effectively the ratio of the rates of suppression and generation of turbulence, as just described, was proposed by L.F. Richardson (1920, 1925) and is known as the Richardson number.

There are several forms of the Richardson number. The gradient Richardson number Ri is defined as

$$\mathrm{Ri} = \frac{g}{T}\left[\frac{dT/dz + \Gamma}{(du/dz)^2}\right] \qquad [15.12.42a]$$

$$= \frac{g}{\theta}\left[\frac{d\theta/dz}{(du/dz)^2}\right] \qquad [15.12.42b]$$

The bulk Richardson number $\mathrm{Ri_B}$ is defined as

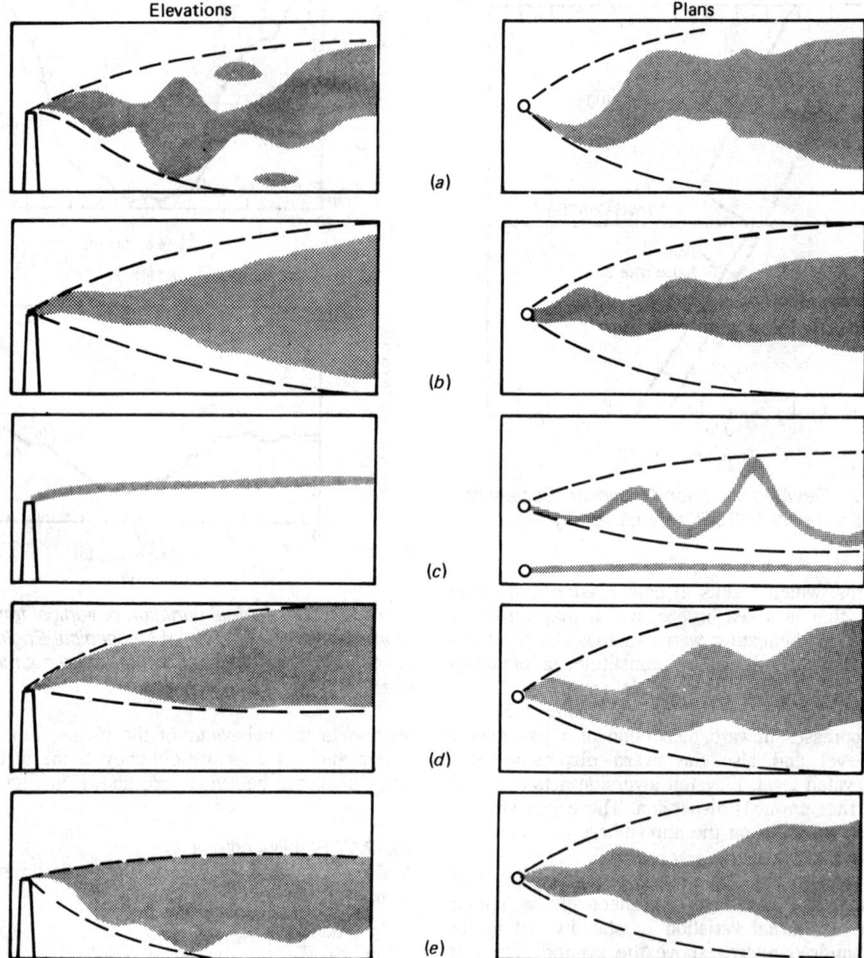

Figure 15.41 *Plume behaviour as a function of atmospheric stability conditions: (a) unstable ('looping'); (b) neutral ('coning'); (c) stable above and below source; surface inversion ('fanning'); (d) stable below source only: surface inversion ('lofting'); (e) stable above source only – elevated inversion ('fumigation')*

$$\mathrm{Ri_B} = \frac{g}{T}\frac{\Delta\theta/\Delta z}{u^2}z^2 \qquad [15.12.43]$$

This form of the Richardson number is convenient for use where the value is determined experimentally from measurements made at two heights. The value of z is usually taken as the geometric mean of these heights and the value of u as that at the upper level.

The flux Richardson number $\mathrm{Ri_F}$ is defined as

$$\mathrm{Ri_F} = -\frac{gH}{Tc_p\tau(\mathrm{d}u/\mathrm{d}z)} \qquad [15.12.44]$$

where c_p is the specific heat of air, H is the vertical heat flux and τ is the Reynolds stress.

A modified Richardson number $\mathrm{Ri_{mod}}$ which is also used is

$$\mathrm{Ri_{mod}} = g\frac{\Delta\rho}{\rho_a}\frac{l}{u^2} \qquad [15.12.45]$$

where l is the mixing length and $\Delta\rho$ is the density difference between the cloud and the air.

The Richardson number Ri is the ratio of the buoyancy to the turbulent stress and is a criterion of similarity for turbulent motion. In his original treatment Richardson suggested that there is a critical value $\mathrm{Ri_{cr}}$ at which motion becomes turbulent and postulated that this value is unity. Then for $\mathrm{Ri} < 1$ motion is laminar and for $\mathrm{Ri} > 1$ it is turbulent.

Subsequent work indicates other values for the critical Richardson number. Obukhov (1971) quotes various values derived from theory including 1/2 (Prandtl), 1/4 (Taylor) and 1/24 (Tollmien) as well as an experimental value of 1/11 (Sverdrup).

Table 15.14 *Pasquill's stability categories (Pasquill, 1961) (Courtesy of HM Stationery Office)*

Surface wind speed at 10 m height (m/s)	*Insolation*			*Night*	
	Strong	*Moderate*	*Slight*	*Thinly overcast or* $\geq$ *4/8 cloud*	$\leq$*3/8 cloud*
<2	A	A–B	B	–	–
2–3	A–B	B	C	E	F
3–5	B	B–C	C	D	E
5–6	C	C–D	D	D	D
>6	C	D	D	D	D

Table 15.15 *Pasquill stability categories for Great Britain (see also Figure 15.42) (Meteorological Office, 1977; reproduced by permission)*

Surface wind speed at 10 m height (mile/h)	*Frequency of stability category (%)*						
	A	*B*	*C*	*D*	*E*	*F*	*G*
16	0	4	11	80	3	2	0
14.5	0	4	13	75	4	3	1
13	0	4	14	70	5	5	2
11.5	1	5	15	65	6	6	2
10	1	6	17	60	7	7	2
9	2	7	19	55	7	7	3
8	2	9	21	50	6	8	4

Since mechanical turbulence decreases quite rapidly with increase in height, the Richardson number is a function of height.

15.12.19 Monin–Obukhov length

The stability criteria just described take no account of two parameters which are now known to be important determinants of similarity in the surface layer. These are the vertical heat flux H and the friction velocity u_*.

These parameters are taken into account in another criterion of stability which is now increasingly used. This is the Monin–Obukhov length scale L (Monin and Obukhov, 1954; Obukhov, 1971), which is defined as

$$L = -\frac{u_*^3 c_p \rho T}{kgH} \quad [15.12.46]$$

This parameter has the dimensions of length. It may be regarded as a measure of the depth of the mechanically mixed layer near the surface.

The Monin–Obukhov length L is negative for unstable and positive for stable conditions and infinite for adiabatic conditions. A dimensionless parameter, the Monin–Obukhov parameter ζ, is obtained by taking the ratio of the height z to the Monin–Obukhov length L:

$$\zeta = z/L \quad [15.12.47]$$

The Monin–Obukhov parameter is thus directly proportional to the height. Thus both the Richardson number and the Monin–Obukhov parameter are functions of the height.

The Richardson numbers and the Monin–Obukhov length and parameter are related, but the relations depend on stability conditions. Golder (1972) gives the following relations. For unstable conditions the Pandolfo–Businger hypothesis gives a good approximation:

$$\frac{z}{L} = \text{Ri} \quad [15.12.48a]$$

For stable conditions use may be made of the empirical relation given by McVehil (1964):

$$\frac{z}{L} = \frac{\text{Ri}}{1 - \beta \text{Ri}} \quad [15.12.48b]$$

Values of the constant β are reviewed by Golder who uses $\beta = 7$.

The relation between the Richardson number and the Monin–Obukhov parameter may also be written in alternative form as given by Tagliazucca and Nanni (1983). For unstable conditions

$$\text{Ri} = \frac{z}{L} \quad [15.12.49a]$$

$$= \frac{z/L}{1 + \gamma z/L} \quad [15.12.49b]$$

where γ is a constant.

The following empirical relation for the Monin–Obukhov length has been obtained by Venkatram (1980b):

$$L = 1100 u_*^2 \quad [15.12.50]$$

15.12.20 Other stability criteria

Another stability parameter is the Kazanski–Monin parameter μ (Kazanski and Monin, 1960). This is defined as

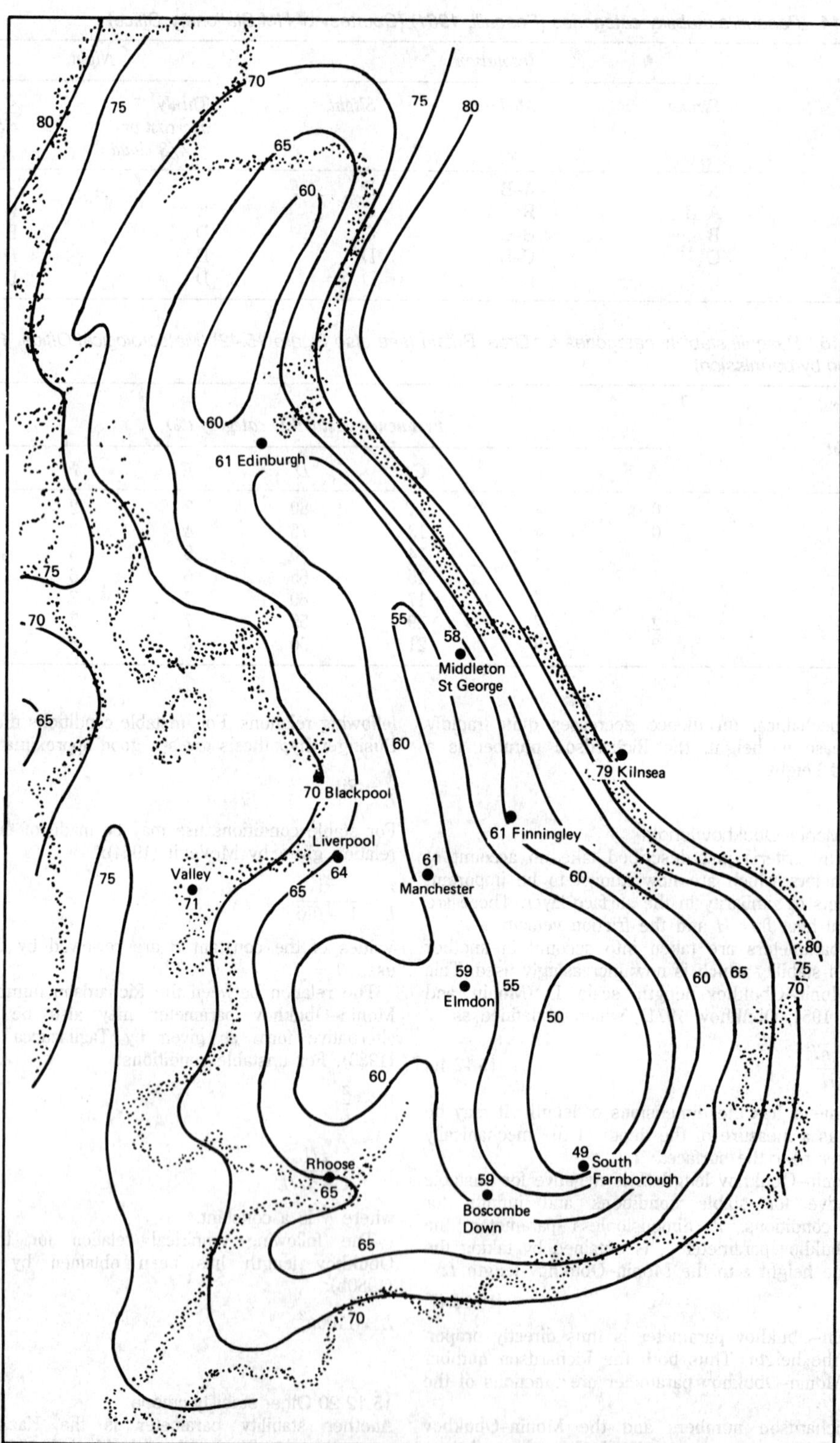

Figure 15.42 *Frequency of occurrence of Pasquill stability categories over Great Britain (Meteorological Office, 1977; reproduced by permission). Contours showing percentage of time during which Pasquill stability category D prevails (see Table 15.15)*

Table 15.16 *Relations between Pasquill stability categories and stability criteria (after Pasquill and Smith, 1971) (Courtesy of Academic Press)*

Pasquill stability category	*Richardson no. Ri*	*Obukhov length L*
A	−1.0 to −0.7	−2 to −3
B	−0.5 to −0.4	−4 to −5
C	−0.17 to −0.13	−12 to −15
D	0	∞
E	0.03 to 0.05	35 to 75
F	0.05 to 0.11	8 to 35

$$\mu = \frac{ku_*}{fL} \quad [15.12.51a]$$

$$= \frac{h'}{L} \quad [15.12.51b]$$

where f is the Coriolis parameter and h' is the scale height of the neutral ABL. This parameter is also sometimes denoted by the symbol S.

A further stability criterion which is sometimes used is the non-dimensional windshear S:

$$S = \frac{kz}{u_*}\frac{du}{dz} \quad [15.12.52]$$

The condition $S = 1$ corresponds to the law of the wall as given in Equation 15.12.15.

15.12.21 Stability classification

The stability conditions have a strong influence on dispersion. The principal dispersion relations contain parameters which are functions of these conditions. Stability is a complex phenomenon, however, and its characterization is a difficult problem. There is a large literature dealing with approaches to turbulence typing or stability classification. Accounts are given by Pasquill (1962a), Gifford (1976b) and Sedefian and Bennett (1980).

A simple set of stability categories is:

(1) unstable conditions – lapse conditions;
(2) neutral conditions;
(3) stable conditions – including inversion conditions.

These categories are used to define the diffusion parameters in the Sutton equations, as described below.

A stability classification based on insolation and wind speed has been given by Pasquill (1961, 1962a) and is shown in Table 15.14. Night refers to the period from one hour before sunset to one hour after sunrise. Strong insolation corresponds to sunny, midday conditions in midsummer England and slight insolation to similar conditions in midwinter. If during day or night, there are overcast conditions then, regardless of wind speed, category D should be assumed.

Stability conditions in terms of the Pasquill stability categories are approximately:

A Unstable conditions
D Neutral conditions
F Stable conditions

Table 15.17 *Some stability classification schemes*

Stability classification scheme	*Categories*	*Reference(s)*
Pasquill	A-F	Pasquill (1961, 1962)
Brookhaven National Laboratory (BNL)	A-D	M.E. Smith (1951), Singer and Smith (1953, 1966)
Turner	1–7	D.B. Turner (1961, 1964)
Klug		Klug (1969)
Cramer	σ_A, σ_E	Cramer (1957, 1959b)
Tennessee Valley Authority (TVA)	A-F	Carpenter *et al.* (1971)

The frequency of the different Pasquill stability categories in Great Britain is shown in Table 15.15. The contours on the map given in Figure 15.42 show the percentage of time during which the neutral category D prevails.

The Pasquill stability categories are used to define the dispersion coefficients in the Pasquill–Gifford equations and the parameters in the Pasquill equations, as described below.

An additional stability category G is used by some workers (e.g. Bryant, 1964 UKAEA AHSB(RP) R42).

Relations between the Pasquill stability categories and the two main stability criteria, the Richardson number Ri and the Obukhov length L have been given by Pasquill and Smith (1971) and are shown in Table 15.16.

There are a number of other stability classifications schemes. A review of these has been given by Gifford (1976b). Some of schemes are shown in Table 15.17.

The scheme of Brookhaven National Laboratory (BNL) (Singer and Smith, 1966) is based on the fluctuations of the horizontal wind direction trace recorded over a 1 hour period. The categories are

A Fluctuations >90° peak to peak
B2 Fluctuations 40–90°
B1 Fluctuations 15–45°
C Fluctuations >15° distinguished by unbroken solid core of trace
D Trace approximates a line, short-term fluctuations <15°

Typical wind direction traces illustrating these wind gustiness categories given by Singer and Smith are shown in Figure 15.43.

The BNL scheme makes it possible to classify turbulence using relatively simple wind direction measurements and to relate the stability categories to experimental dispersion data obtained at a site. Strictly the categories are site specific and those published for BNL apply to the BNL site. The relations between the BNL categories and the wind speed and temperature gradient is shown in Table 15.18.

The scheme proposed by D.B. Turner (1961, 1964) is a version of Pasquill's scheme in which the insolation is

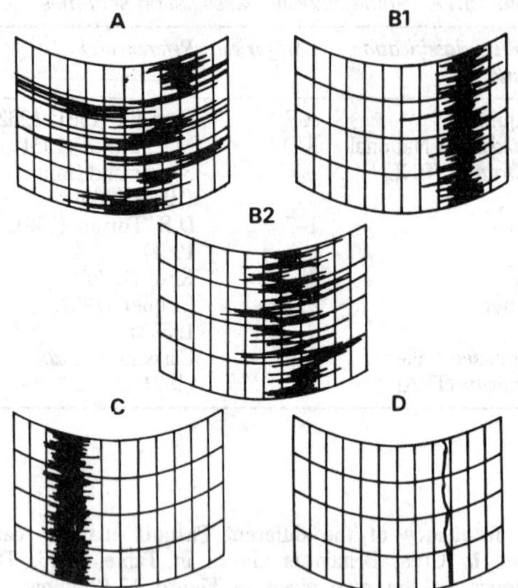

Figure 15.43 *Typical wind direction traces illustrating wind gustiness categories (Singer and Smith, 1966) (Courtesy of Brookhaven National Laboratory)*

classified according to the amount and height of cloud cover and the solar elevation angle.

The scheme developed by Klug (1969) is similar to that of Pasquill, but differs in that the categories are defined using more detailed rules for cloud cover, wind speed, time of day and season.

The scheme of Cramer (1957, 1959b) is based on correlation of stability with the azimuth σ_A and the elevation angle σ_E. Cramer's turbulence categories are shown in Table 15.19.

The Tennessee Valley Authority (TVA) scheme, described by Carpenter *et al.* (1971) is concerned with stability classification for buoyant plumes from tall chimneys. It is based on lapse rate $\Delta\theta/\Delta z$. The TVA stability categories are given in Table 15.20.

Relationships between some of these stability classification, or turbulence typing, schemes have been given by Gifford and are shown in Table 15.21.

The Pasquill scheme has been modified by F.B. Smith (1973) so that the stability is described in terms of a continuous index *P* rather than by discrete categories. The *P* value is a function of the wind speed, the upward heat flux and the incoming solar radiation, and also of the roughness length. *P* values of 0.5 to 6.5 correspond to the Pasquill categories A–G as follows:

	P
A	0.5
B	1.5
C	2.5
D	3.6
E	4.5
F	5.5
G	6.5

Table 15.18 *BNL turbulence categories (after Gifford, 1976b) (Courtesy of Nuclear Safety)*

Turbulence category	*Wind speed* (m/s)		*Temperature gradient* ($\Delta T/\Delta z$)
	9 m	*108 m*	(°C/123 m)
A	–	1.8 ± 1.1[a]	−1.25 ± 7[a]
B2	2.5	3.8 ± 1.8	−1.6 ± 0.5
B1	3.4	7.0 ± 3.1	−1.2 ± 0.65
C	4.7	10.4 ± 3.1	−0.64 ± 0.52
D	1.9	6.4 ± 2.6	2.0 ± 2.6

[a] Standard deviation.

Table 15.19 *Cramer's turbulence categories*

Stability	*Azimuth,* σ_A (°)	*Elevation angle,* σ_E (°)
Extremely unstable	30	10
Near neutral (rough surface; trees, buildings)	15	5
Near neutral (very smooth grass)	6	2
Extremely stable	3	1

Table 15.20 *TVA stability categories (after Carpenter et al., 1971) (Courtesy of the Air Pollution Control Association)*

Stability category	*Lapse rate,* $\Delta\theta/\Delta z$ (K/100 m)
A Neutral	0
B Slightly stable	0.27
C Stable	0.64
D Isothermal	1.00
E Moderate inversion	1.36
F Strong inversion	1.73

Table 15.21 *Relations between turbulence typing schemes (Gifford, 1976b) (Courtesy of Nuclear Safety)*

Stability	*Scheme*			
	Pasquill	*Turner*	*BNL*	σ_A (°)
Very unstable	A	1[a]	B2	25
Moderately unstable	B	2	B1	20
Slightly unstable	C	3	B1	15
Neutral	D	4	C	10
Moderately stable	E	6	–	5
Very stable	F	7	D	2.5

[a] After Golder (1972).

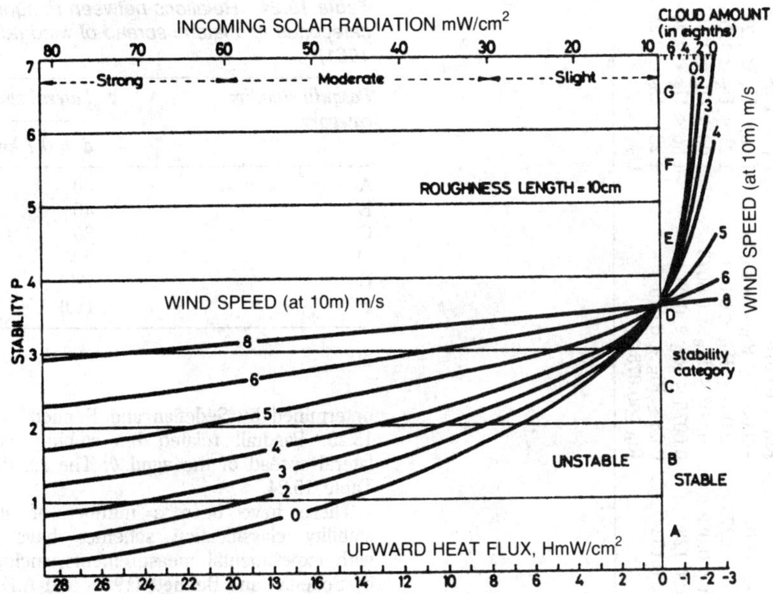

Figure 15.44 Stability parameter, or P value (after F.B. Smith, 1973; Clarke, 1979 NRPB R91; reproduced by permission)

Table 15.22 Some schemes for assigning Pasquill stability categories (after Sedefian and Bennett, 1980)

Scheme	*Comment*
Wind direction standard deviation σ_θ	One of two methods given in NRC *Guide* 1.23
Temperature difference, ΔT	Other method given in NRC *Guide* 1.23
Richardson number, Ri	Authors' own method, utilizing relations given by Golder (1972) between Pasquill categories and Monin–Obukhov length L and roughness length z_0
Bulk Richardson number, Ri_B	Authors' own method
Wind speed ratio U_R	Authors' own method, based on ratio of wind speeds at heights of 10 and 50 m
Stability array (STAR)	Author's own method, based on wind speed at 10 m, solar altitude and cloud cover

The graph given by Smith for the determination of the *P* value is shown in Figure 15.44.

There are a number of other stability classification schemes based on relations between the Pasquill stability categories and particular parameters. A review of several such schemes has been given by Sedefian and Bennett (1980), as shown in Table 15.22. The authors also analysed dispersion data and derived relations for the *P* value in some of the schemes.

In the σ_θ, or sigma theta, method the stability category is taken as a function of the wind direction standard deviation σ_θ. This method is one of two which has been recommended by the Nuclear Regulatory Commission and is widely used.

Sedefian and Bennett give the following relation involving σ_θ:

$$\sigma_\theta(z_2) = \sigma_\theta(z_1)(z_2/z_1)^{P_\theta} \qquad [15.12.53]$$

where P_θ is the *P* value and where the heights z_1 and z_2 are 10 and 50 m, respectively. The values of P_θ corresponding to Pasquill categories A–F are −0.06, −0.15, −0.17, −0.23, −0.38 and −0.53, respectively.

The temperature difference ΔT method is the other method recommended by the NRC.

The wind speed ratio U_R is a parameter devised by Sedefian and Bennett and is the ratio of the wind speeds at 50 and 10 m height. The authors give the following relation involving U_R

$$U_R = (z_2/z_1)^{P_u} \qquad [15.12.54]$$

where P_u is and index and U_R is the wind speed ratio and where the heights z_1 and z_2 are 10 and 50 m, respectively. The values of P_u are related to the Pasquill stability categories.

The stability array (STAR) method is based on cloud cover, solar altitude and wind speed.

The relation between the Pasquill stability categories and the parameters of these various methods as

Table 15.23 *Relations between Pasquill stability categories and parameters of some typing schemes (after Sedefian and Bennett, 1980) (Courtesy of Pergamon Press)*

Pasquill stability	σ_θ (10 and 50 m) (°)	σ_θ (50 m)[a] (°)	$\Delta T/\Delta z$ (°C/100 m)	Ri[b]	Ri_B[b]	U_R[b]
A	$22.5 < \sigma_\theta$	$20.0 < \theta$	$\Delta T/\Delta z < -1.9$	$Ri < 2.51$	$Ri_B < 0.0016$	$U_R < 1.18$
B	$17.5 < \sigma_\theta \leq 22.5$	$13.75 < \sigma_\theta \leq 20.0$	$-1.9 \leq \Delta T/\Delta z < -1.7$	$-2.51 \leq Ri < -1.07$	$-0.0016 \leq Ri_B < -0.009$	$1.18 \leq U_R < 1.21$
C	$12.5 < \sigma_\theta \leq 17.5$	$9.5 < \sigma_\theta \leq 13.75$	$-1.7 \leq \Delta T/\Delta z < -1.5$	$-1.07 \leq Ri < -0.275$	$-0.009 \leq Ri_B < 0.003$	$1.21 \leq U_R < 1.26$
D	$7.5 < \sigma_\theta \leq 12.5$	$5.0 < \sigma_\theta \leq 9.5$	$-1.5 \leq \Delta T/\Delta z < -0.5$	$-0.275 \leq Ri < 0.089$	$-0.003 \leq Ri_B < 0.0075$	$1.26 \leq U_R < 1.56$
E	$3.75 < \sigma_\theta \leq 7.5$	$2.0 < \sigma_\theta \leq 5.0$	$-0.5 \leq \Delta T/\Delta z < 1.5$	$0.089 \leq Ri < 0.128$	$0.0075 \leq Ri_B < 0.05$	$1.56 \leq U_R < 2.28$
F	$2.0 < \sigma_\theta \leq 3.75$	$0.75 < \sigma_\theta \leq 2.0$	$1.5 \leq \Delta T/\Delta z < 4.0$	$0.128 \leq Ri < 0.134$	$0.05 \leq Ri_B < 0.075$	$2.28 \leq U_R < 3.28$
G	$\sigma_\theta \leq 2.0$	$\sigma_\theta \leq 0.75$	$\Delta T/\Delta z \geq 4.0$	$Ri \geq 0.134$	$Ri_B \geq 0.075$	$U_R \geq 3.28$

[a] Corrected for height variation of σ_θ.
[b] Ri, Ri_B and U_R limits based on Businger formulation for 22 m.

Table 15.24 *Relations between Pasquill stability categories and lateral spread of wind (after Pasquill, 1961)*

Pasquill stability category	*Lateral spread* θ (deg)	
	$d = 0.1$ *km*	$d = 100$ *km*
A	60	(20)
B	45	(20)
C	30	10
D	20	10
E	(15)	(5)
F	(10)	(5)

determined by Sedefian and Bennett is shown in Table 15.23. Pasquill related his stability categories to the lateral spread of the wind θ. The relations are given in Table 15.24.

There have been a number of studies in which stability classification schemes have been compared with experimental measurements, including those done by Sedefian and Bennett (1980) and A.E. Mitchell (1982).

An alternative approach to stability is to treat it as primarily a function of the Obukhov length L and the roughness length z_o. Golder (1972) has developed relationships between the Pasquill stability categories and these two parameters. Figure 15.45 shows these relations. Golder also concluded that the most appropriate conversion between the Pasquill and Turner categories is A to 1, B to 2, C to 3, D to 4, E to 6 and F to 7.

Relations between the Turner stability categories and the Monin–Obukhov length, inferred from the work of Golder (1972) have been given by Ning and Yap (1986). These are given in Table 15.25.

Another stability classification scheme is that of Holtslag and van Ulden (1983). This is based on the sensible heat flux.

The stability classifications just described were developed for use over land. They do not necessarily hold over water.

A stability classification scheme for use over water has been developed at the Naval Postgraduate School (NPS), Monterey, by Schacher, Fairall and Zannetti (1982). Following Golder, stability is treated as primarily a function of the Monin–Obukhov length and the roughness length. The Monin–Obukhov length is determined from air–sea temperature difference, wind speed and relative humidity. Figure 15.46, given by Skupniewicz and Schacher (1986), illustrates the classification for 50% relative humidity. A somewhat similar scheme has been given by Hasse and Weber (1985). In that stability is characterized by the Monin–Obukhov length, the derivation and use of this parameter may be regarded as a stability classification 'scheme'.

The selection of a stability classification method scheme for gas dispersion modelling is exemplified by the set of schemes used in the DRIFT dense gas dispersion model described in Section 15.35. The model can accommodate three schemes. One is the use of the Monin–Obukhov length, valid for all locations and all times. The others are the Pasquill scheme, used for day

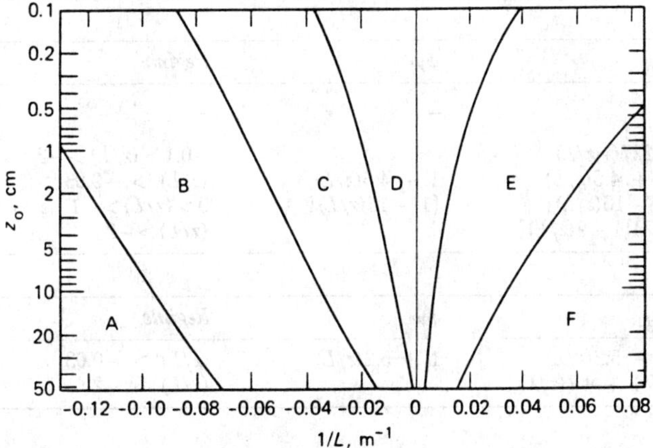

Figure 15.45 *Relations between Pasquill stability categories and Monin–Obukhov length and surface roughness (Golder, 1972) (Courtesy of Kluwer Academic Publishers)*

Table 15.25 *Relations between Turner stability categories and Monin-Obukhov length (after Ning and Yap, 1986) (Courtesy of Pergamon Press)*

Turner stability category[a]	1/L (m^{-1})
A	$1/L \leq -0.087$
B	$-0.087 \leq 1/L < -0.028$
C	$-0.028 \leq 1/L < -0.005$
D	$-0.005 \leq 1/L$

[a] D.B. Turner (1964) denoted his stability categories 1–7. Ning and Yap state that the Turner stability categories A–D are derived from the work of Golder (1972) for a roughness length z_o of 20 cm.

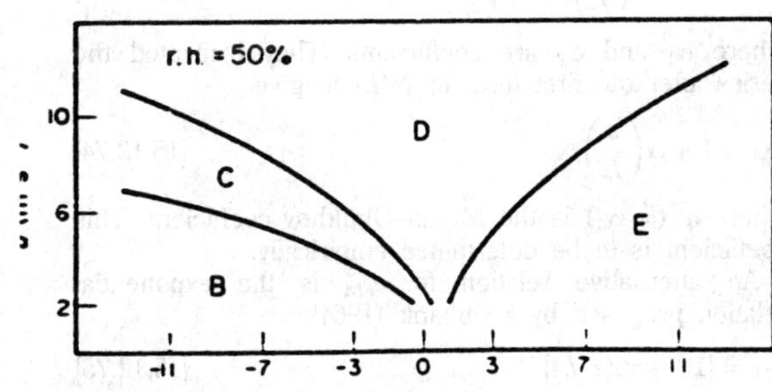

Figure 15.46 *Illustration of the NPS stability classification scheme over water (Skupniewicz and Schacher, 1986) (Courtesy of Pergamon Press)*

on land, and the Holtslag and van Ulden scheme, used for neutral or unstable conditions for day on land.

The stability category is the other main piece of information which is required, in addition to the wind characteristics, for the estimation of dispersion at a given location.

15.12.22 Flux-gradient relations

In the surface layer it is assumed that stress does not vary with height. Similar arguments based on the work of Monin and Obukhov lead to the conclusion that for any transferable property S the relation between vertical flux and vertical gradient is determined by the following parameters: z, ρ, g/T, τ/ρ, and $H/c_p\rho$.

The flux–gradient relations may be expressed in the form:

$$\frac{dS}{dz} = \frac{S_*}{kz}\phi_s\left(\frac{z}{L}\right) \qquad [15.12.55]$$

where S is the mean quantity of the property per unit mass of air, S_* is a scale parameter with the same dimensions as S, ϕ is a similarity factor, and the subscript s denotes the property specified.

The scale parameter S_* is defined by the relation to the vertical flux F_s:

$$S_* = \frac{F_s}{\rho u_*} \qquad [15.12.56]$$

The vertical flux is given by the relation:

$$F_s = -\rho K_s \frac{dS}{dz} \qquad [15.12.57]$$

where K is a turbulent exchange coefficient.

For momentum, heat and water vapour the flux relations are:

Table 15.26 *Some forms of the universal functions (after Dyer, 1974) (Courtesy of Kluwer Academic Publishers)*

A Unstable conditions

ϕ_M	ϕ_H	ϕ_W	*Regime*	*Reference*
$\frac{z}{L}[1-\exp(z/L)^{-1}$	–	–	–	Swinbank (1964)
$0.613(-z/L)^{-0.20}$	$0.227(-z/L)^{-0.44}$	–	$-0.1 > (z/L) > -2$	Swinbank (1968)
$1+4.5(z/L)$	$1+4.5(z/L)$	$1+4.5(z/L)$	$(z/L) > -0.03$	E.K. Webb (1970)
$[1-16(z/L)]^{-\frac{1}{4}}$	$[1-16(z/L)]^{-\frac{1}{2}}$	$[1-16(z/L)]^{-\frac{1}{2}}$	$0 > (z/L) > -1$	Dyer and Hicks (1970)
$[1-15(z/L)]^{-\frac{1}{4}}$	$0.74[1-9(z/L)]^{-\frac{1}{2}}$		$(z/L) > -2$	Businger *et al.* (1971)

B Stable conditions

ϕ_M	ϕ_H	ϕ_W	*Regime*	*Reference*
$1+5.2(z/L)$	$1+5.2(z/L)$	$1+5.2(z/L)$	$(z/L) > -0.03$	E.K. Webb (1970)
$1+4.7(z/L)$	$0.74+4.7(z/L)$		$(z/L) > -2$	Businger *et al.* (1971)

$$\tau = \rho K_M \frac{du}{dz} \qquad [15.12.58]$$

$$H = -\rho c_p K_H \frac{d\theta}{dz} \qquad [15.12.59]$$

$$E = -\rho L_W K_W \frac{dq}{dz} \qquad [15.12.60]$$

where c_p is the specific heat of air, E is the vertical flux of water vapour, H is the vertical heat flux, L_W is the latent heat of water vapour, q is the specific humidity, or concentration of water vapour, u is the wind velocity, θ is the potential temperature, ρ is the density of air, and the subscripts H, M and W denote heat, mass and water vapour, respectively.

The corresponding equations for the turbulent exchange coefficients are then

$$K_M = \frac{\tau}{\rho du/dz} \qquad [15.12.61]$$

$$K_H = -\frac{H}{\rho c_p d\theta/dz} \qquad [15.12.62]$$

$$K_W = -\frac{E}{\rho L_W dq/dz} \qquad [15.12.63]$$

Then, applying Equations 15.12.56 and [15.12.57 to momentum and noting that for this case

$$S = u \qquad [15.12.64]$$

and

$$F_s = -\rho u_*^2 \qquad [15.12.65]$$

gives

$$\frac{du}{dz} = \frac{u_*}{kz}\phi_M \qquad [15.12.66]$$

Similarly, the gradient relations for heat and water vapour are:

$$\frac{d\theta}{dz} = -\frac{H}{\rho c_p k u_* z}\phi_H \qquad [15.12.67]$$

$$\frac{dq}{dz} = -\frac{E}{\rho L_W k u_* z}\phi_W \qquad [15.12.68]$$

The relations for the turbulent exchange coefficients become:

$$K_M = ku_*z/\phi_M \qquad [15.12.69]$$

$$K_H = ku_*z/\phi_H \qquad [15.12.70]$$

$$K_W = ku_*z/\phi_W \qquad [15.12.71]$$

Also from the definitions of the Richardson number Ri and the Monin–Obukhov parameter z/L, as given by Equations 15.12.42b and 15.12.46, and from Equations 15.12.66 and 15.12.67

$$\mathrm{Ri} = \frac{z}{L}\frac{\phi_H}{\phi_M^2} \qquad [15.12.72]$$

15.12.23 Universal functions

Monin and Obukhov expanded the function $\phi_M(z/L)$ as the power series

$$\phi_M = 1 + \alpha_1\left(\frac{z}{L}\right) + \alpha_2\left(\frac{z}{L}\right)^2 \ldots \qquad [15.12.73]$$

where α_1 and α_2 are coefficients. They truncated the series after the first term in (z/L) to give

$$\phi_M = 1 + \alpha\left(\frac{z}{L}\right) \qquad [15.12.74]$$

where α $(= \alpha_1)$ is the Monin–Obukhov coefficient. This coefficient is to be determined empirically.

An alternative relation for ϕ_M is the exponential relation proposed by Swinbank (1964):

$$\phi_M = [1-\exp(z/L)]^{-1} \qquad [15.12.75]$$

Relations for the universal functions have been derived by a number of authors and have been reviewed by Dyer (1974). Some of these relations are given in Table 15.26.

Obukhov (1971) has defined an alternative universal function by the relation

$$K_M = K_{Mo}\phi(\text{Ri}) \quad [15.12.76]$$

where K_{Mo} is the adiabatic value of K_M. He gives the following relations for $\phi(\text{Ri})$:

$$\phi(\text{Ri}) = 1 \qquad \text{Ri} = 0 \quad [15.12.77a]$$

$$= \left(1 - \frac{\text{Ri}}{\text{Ri}_{cr}}\right)^{\frac{1}{2}} \qquad \text{Ri} < \text{Ri}_{cr} \quad [15.12.77b]$$

$$= 0 \qquad \text{Ri} > \text{Ri}_{cr} \quad [15.12.77c]$$

where the subscript cr denotes critical.

Equation 15.12.77a follows from the definition of $\phi(\text{Ri})$ and from the fact that for adiabatic conditions Ri = 0, and Equation 15.12.77c follows from the fact that turbulence is suppressed for $\text{Ri} > \text{Ri}_{cr}$.

As already described, the Monin–Obukhov form of the universal function ϕ_M gives for the wind velocity gradient Equation 15.12.24, which on integration yields Equation 15.12.25 or, for $z \gg z_o$, Equation 15.12.26.

15.12.24 Turbulent exchange coefficients

The turbulent exchange coefficients are:

K_M Turbulent exchange coefficient for momentum; eddy viscosity.
K_H Turbulent exchange coefficient for heat; eddy thermal diffusivity.
K_W Turbulent exchange coefficient for water vapour; eddy diffusivity.

These coefficients are defined by Equations 15.12.58–15.12.60, respectively.

It was assumed in early work by Schmidt (1925) that

$$K_M = K_H = K_W \quad [15.12.78]$$

The identity of the three exchange coefficients is still a common assumption for neutral conditions. According to Dyer (1974), however, there is evidence that for such conditions the ratio K_H/K_M is equal to 1.35.

From Equations 15.12.69–15.12.71 the ratio of the exchange coefficients may be expressed in terms of the ratios of the universal functions ϕ_M, ϕ_H and ϕ_W. For neutral conditions the latter approach unity. For other conditions the ratios K_H/K_M and K_W/K_M will be unity only if the asymptotic approach to unity of the ϕ_H and ϕ_W is identical with that of ϕ_M. For unstable conditions there is evidence that this is not so.

The identity of the exchange coefficients for mass and heat transfer implies, from Equations 15.12.58–15.12.59,

$$\frac{H}{C_p\tau} = -\frac{d\theta/dz}{du/dz} \quad [15.12.79]$$

Equation 15.12.79 is a form of the Reynolds analogy.

15.12.25 Wind direction fluctuations

Wind direction fluctuations may be defined in terms of the standard deviation of the horizontal direction and inclination of the wind:

σ_θ Standard deviation of the horizontal direction of the wind.
σ_ϕ Standard deviation of the inclination to the horizontal of the wind.

Alternative notation for these quantities is for σ_θ the standard deviation of the azimuth σ_A and for σ_ϕ the standard deviation of the elevation σ_E. Distributions of the standard deviations σ_θ and σ_ϕ by Pasquill stability category have been given by Luna and Church (1972) and are shown in Figure 15.47.

As described above, the concentration measured in a plume varies with the sampling time. Similarly, the value measured for σ_θ varies with sampling time. Slade (1968) suggests that a correction for sampling time may be made using the following approximate relation:

$$\frac{\sigma_\theta}{\sigma_{\theta r}} = \left(\frac{t}{t_r}\right)^p \quad [15.12.80]$$

where t is the sampling period which yields σ_θ, p is an index and subscript r denotes the reference value. The value of p is of the order of 0.2.

The standard deviations σ_θ and σ_ϕ are used in the Pasquill model for passive gas dispersion.

15.12.26 Wind velocity fluctuations

The wind velocities u, v and w in the downwind, crosswind and vertical directions may be written as the sum of a mean velocity and a velocity fluctuation:

$$u = \bar{u} + u' \quad [15.12.81a]$$

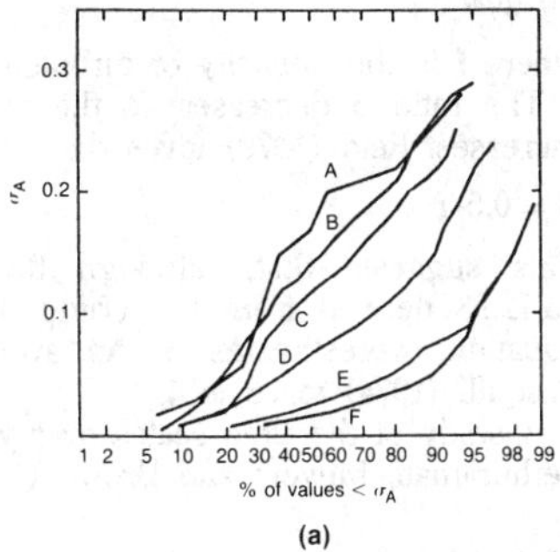

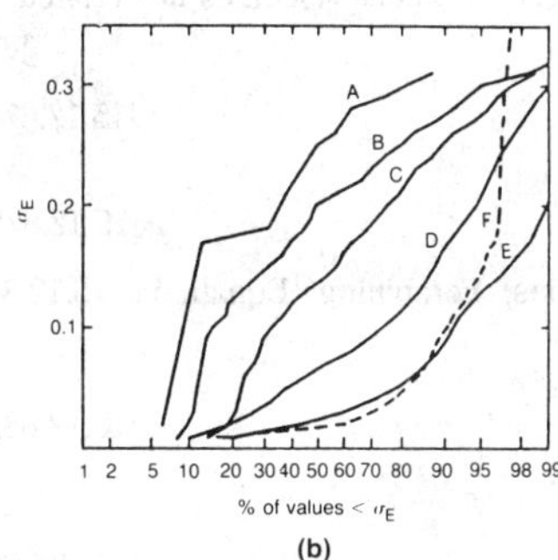

Figure 15.47 *Standard deviations of wind direction fluctuations (Luna and Church, 1972): (a) distribution of σ_A by stability class; (b) distribution of σ_E by stability class. A–F, Pasquill stability categories. Reproduced by permission*

$$v = \bar{v} + v' \quad [15.12.81b]$$

$$w = \bar{w} + w' \quad [15.12.81c]$$

where $\bar{u}$, $\bar{v}$ and $\bar{w}$ are the mean values and u', v' and w' the fluctuations of u, v and w, respectively.

Wind velocity fluctuations may be defined in terms of the standard deviations σ_u, σ_v and σ_w of the velocities as follows:

$$\sigma_u = (\overline{u'^2})^{\frac{1}{2}} \quad [15.12.82a]$$

$$\sigma_v = (\overline{v'^2})^{\frac{1}{2}} \quad [15.12.82b]$$

$$\sigma_w = (\overline{w'^2})^{\frac{1}{2}} \quad [15.12.82c]$$

where $\overline{u'^2}$, $\overline{v'^2}$ and $\overline{w'^2}$ are the variances of u, v and w, respectively.

Relationships for σ_u, σ_v and σ_w derived from the work of Wyngaard, Cote and Rao (1974) and Wyngard (1975) have been given by S.R. Hanna, Briggs and Hosker (1982). For σ_u:

$$\frac{\sigma_u}{u_*} = \left(12 - 0.5\frac{z_i}{L}\right)^{\frac{1}{3}} \quad \text{unstable} \quad [15.12.83]$$

$$= 2.0 \exp\left(-3\frac{fz}{u_*}\right) \quad \text{neutral} \quad [15.12.84]$$

$$= 2.0\left(1 - \frac{z}{z_i}\right) \quad \text{stable} \quad [15.12.85]$$

where f is the Coriolis parameter, L is the Monin–Obukhov length and z_i is the mixed layer height. The authors also give relationships for σ_v and σ_w.

15.12.27 Turbulence intensity

Intensities of turbulence are defined as:

$$i_y = (\overline{v'^2})^{\frac{1}{2}}/\bar{u} \quad [15.12.86a]$$

$$i_z = (\overline{w'^2})^{\frac{1}{2}}/\bar{u} \quad [15.12.86b]$$

where i_y and i_z are the intensities of the lateral and vertical components of turbulence, respectively.

The root-mean-square turbulent velocities are related to the friction velocity:

$$(\overline{v'^2})^{\frac{1}{2}} = 2.2u_* \quad [15.12.87a]$$

$$(\overline{w'^2})^{\frac{1}{2}} = 1.25u_* \quad [15.12.87b]$$

For neutral conditions, combining Equations 15.12.86, 15.12.87 and 15.12.20:

$$i_y = \frac{0.88}{\ln(z/z_o)} \quad [15.12.88a]$$

$$i_z = \frac{0.5}{\ln(z/z_o)} \quad [15.12.88b]$$

Also, combining Equations 15.12.82 and 15.12.86:

$$i_y = \sigma_v/\bar{u} \quad [15.12.89a]$$

$$i_z = \sigma_w/\bar{u} \quad [15.12.89b]$$

15.12.28 Eddy dissipation rate

The eddy dissipation rate ϵ is the rate at which on the small scale turbulence is dissipated into heat.

In the surface layer:

$$\epsilon = \frac{u_*^3}{kz}(\phi_M - z/L) \quad [15.12.90]$$

For neutral conditions:

$$\epsilon = \frac{u_*^3}{kz} \quad [15.12.91]$$

15.12.29 Lagrangian turbulence

Turbulence may be measured at a fixed point or by reference to a particle moving through the turbulent field. The types of turbulence measured in these two different ways are known as Eulerian and Lagrangian turbulence, respectively. The dispersion of a gas in air is a Lagrangian process, but it is generally observed by Eulerian measurements.

The difference between the two types of turbulence may be illustrated by the following simplified argument. Consider a wind speed u carrying a circular eddy radius r and tangential velocity ω. The Eulerian time scale t_E is the time for the eddy to pass a fixed point:

$$t_E = 2r/u \quad [15.12.92]$$

The Lagrangian time scale t_L is the time for a particle on the circumference of the eddy to travel once round the eddy:

$$t_L = 2\pi r/\omega \quad [15.12.93]$$

The ratio β of the Lagrangian to the Eulerian time scale is then

$$\beta = \frac{t_L}{t_E} \quad [15.12.94a]$$

$$= \frac{\pi}{\omega/u} \quad [15.12.94b]$$

$$= \frac{\pi}{i} \quad [15.12.94c]$$

with

$$i = \omega/u \quad [15.12.95]$$

where i is the intensity of turbulence.

The ratio β decreases as the intensity of turbulence increases. Reid (1979) gives the relation

$$\beta = 0.5/i \quad [15.12.96]$$

This suggests that, although the form of Equation 15.12.93 derived from the crude model is correct, the equation overestimates β. An average value given by Pasquill (1974) for β is 4.

A study of the time scales over water is described by Sethuraman, Meyers and Brown (1976).

15.12.30 Autocorrelation function

If in a field of homogenous turbulence measurements are made at two points of the instantaneous components of velocity u'_1 and u'_2 and, if the position of the second point is moved while that of the first is kept fixed, a correlation coefficient $R(x)$ may be defined as

$$R(x) = \overline{u'_1 u'_2} / \overline{u'^2} \qquad [15.12.97]$$

For eddy sizes which are large compared with the distance x this correlation coefficient is high, and vice versa.

The correlation coefficient may also be expressed in terms of time. This time correlation coefficient $R(\tau)$, which is also known as the autocorrelation coefficient, is defined as

$$R(\tau) = \overline{u(t)u(t+\tau)} / \overline{u^2} \qquad [15.12.98]$$

The two correlation coefficients are identical:

$$R(\tau) = R(x) \qquad [15.12.99]$$

with $x = u\tau$.

Properties of the autocorrelation coefficient are:

$$R(\tau) = 1 \qquad \tau = 0 \qquad [15.12.100a]$$

$$= 0 \qquad \tau \rightarrow \infty \qquad [15.12.100b]$$

The form of the autocorrelation coefficient is not fully understood. One model commonly used is the Markov model for which

$$R(\tau) = \exp(-\tau / t_L) \qquad [15.12.101]$$

15.12.31 Taylor's theorem

A relation between the autocorrelation function and the dispersion parameters has been derived by G.I. Taylor (1921). The displacement x of a particle is related to its velocity u as follows:

$$x(t) = \int_0^t u(t')\,dt' \qquad [15.12.102]$$

Then

$$\frac{d[x^2(t)]}{dt} = 2x\frac{dx}{dt} \qquad [15.12.103a]$$

$$= 2\int_0^t u(t)u(t')\,dt' \qquad [15.12.103b]$$

Substituting the autocorrelation coefficient from Equation 15.12.98 in Equation 15.12.103 yields

$$\frac{d\overline{x^2}}{dt} = 2\overline{u^2}\int_0^t R(\tau)\,d\tau \qquad [15.12.104]$$

But it can be shown that

$$\sigma_x^2 = \overline{x^2} \qquad [15.12.105]$$

Hence

$$\frac{d\sigma_x^2}{dt} = 2\overline{u^2}\int_0^t R(\tau)\,d\tau \qquad [15.12.106]$$

and

$$\sigma_x^2 = 2\overline{u^2}\int_0^t\int_0^{t'} R(\tau)\,d\tau\,dt' \qquad [15.12.107]$$

Equation 15.12.107 is Taylor's theorem.

If a form of the autocorrelation coefficient is assumed, such as that given in Equation 15.12.101, it is possible to derive from Equation 15.12.107 an expression for the dispersion coefficient. The application of Taylor's theorem for this purpose is described below.

15.12.32 Mixed layer sealing

Much of the foregoing treatment has been concerned with the properties of, and similarity within, the surface stress layer. In mixed layer scaling similarity is extended to the whole boundary layer.

The initial treatment by Deardorff (1970) was for the convective boundary layer, in other words for the planetary boundary layer in convective, or unstable, conditions, and is also termed convective layer, or convective velocity, scaling.

Scaling has also been extended by other workers such as Caughey, Wyngaard and Kaimal (1979) to stable conditions and for this the term 'mixed layer scaling' is perhaps more appropriate. A parameter widely used in mixed layer scaling is

$$X = \frac{w_* x}{u z_i} \qquad [15.12.108]$$

where w_* is the convective velocity and z_i the mixed layer height.

15.12.33 Mixed layer height

Over land the mixed layer has a height of some 1000–2000 m by day but about an order of magnitude less by night. Over sea the diurnal variation is much less. By day the mixed layer height is usually set by an inversion layer capping the well mixed layer above the ground surface. By night, when there is some degree of inversion at all levels, the height of the mixed height may be taken as that at which surface induced mechanical turbulence dies out. Extensive data on the mixed layer height in the US have been published by Holzworth (1972).

The mixed layer height z_i has been investigated by Zilitinkevich (1972a), who gives the following relations:

$$Z_i = \frac{k u_*}{f} \qquad \text{neutral conditions} \qquad [15.12.109a]$$

$$= c\left(\frac{u_* L}{f}\right)^{\frac{1}{2}} \qquad \text{stable conditions} \qquad [15.12.109b]$$

where c is a constant. Values of the constant c are quoted by Venkatram and Paine (1985) as varying between 0.4 and 0.7.

Another correlation, applicable to stable conditions at night in high wind, is that of Venkatram and Paine (1985):

$$z_i = 2300 u_*^{\frac{3}{2}} \qquad [15.12.110]$$

They state that the constant, given here as 2300, tends to be site specific and quote a value of 1300 at another site.

15.12.34 Convective velocity scale

The convective velocity, or velocity scale, w_* is defined as

$$w_* = \frac{g Q_o z_i}{T_o} \qquad [15.12.111]$$

where Q_o is the surface heat flux and T_o is the absolute average temperature of the mixed layer.

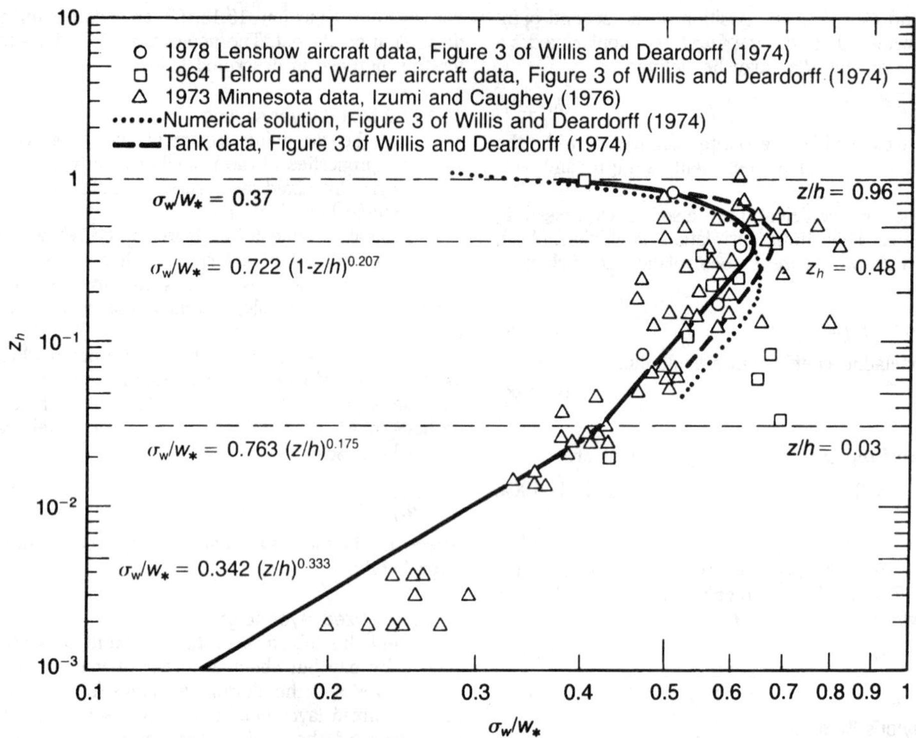

Figure 15.48 Relation between the standard deviations of the vertical velocity fluctuations σ_w and the convective velocity w_* for fully corrective conditions (Irwin, 1979a; reproduced by permission)

The convective velocity w_* may be estimated from correlations for the quantity σ_w/w_* together with correlations for σ_w or, more directly, from correlations for w_* itself. A correlation of the first type has been given by Irwin (1979a) and is shown in Figure 15.48. Correlations for σ_w have been discussed above.

Direct relations for w_* have been given by Venkatram (1978) in terms of the maximum surface heat flux Q_m and the maximum mixed layer height z_{im} as follows:

$$w_* = AQ_m^{\frac{1}{2}} \qquad [15.12.112]$$

with

$$Q_m = Q_o \sin(\pi t/2\tau) \qquad [15.12.113]$$

and

$$w_* = Bz_{im} \qquad [15.12.114]$$

where τ is the time after sunrise when Q_o is a maximum, and A and B are constants. The values of the constants are A = 4.74 and B = 1.12×10^{-3}.

15.12.35 Density effects

In some conditions the density of the air may be assumed to be a function of temperature, and humidity, but not of pressure. This is the Boussinesq approximation. Then the buoyancy of the air, which is a function of the ratio of the density difference to the mean density, may be assumed to be a function of the ratio of the temperature difference to the mean absolute temperature. The Boussinesq approximation is generally valid when the ratio of the density difference to the mean density is 0.1 or less.

15.12.36 Meteorological variables

The meteorological variables of particular interest here in respect of gas dispersion are (1) wind speed and (2) stability category. Other variables which are of interest in hazard assessment generally include: (1) minimum temperature, (2) maximum temperature, (3) maximum wind speed and (4) maximum rainfall. The behaviour of the variable may be expressed in terms of the values averaged over a fixed period such as an hour, day, month or year.

15.12.37 Meteorology of the UK

Information on the meteorological characteristics of the UK is available from the Meteorological Office. A simple introduction is given in *Climate of the British Isles* (Meteorological Office, 1994), which contains a number of maps of wind speed, minimum temperature, maximum temperature, rainfall and snowfall. An account of British meteorology with particular reference to the hazards which it poses is given in ***Environmental Hazards in the British Isles*** (Perry, 1981).

15.13 Topography

In dispersion work the ground surface which may be regarded as the base case is flat grassland such as that on Salisbury Plain, where much of the early work on dispersion was done. Terrain may differ from this base case in a number of ways, particularly in respect of:

(1) surface roughness;
(2) urban areas;
(3) coastal zones and sea;
(4) complex terrain;
(5) buildings and obstructions.

General aspects of such terrain are considered briefly in this section. Models for dispersion over such terrain are treated in Section 15.14 and the corresponding dispersion parameters in Section 15.16.

15.13.1 Surface roughness

There is a wide range of surface roughness which is of interest in dispersion work. At one extreme ice has a very low value of the roughness length, while trees or urban areas have a very high value. There are also considerable differences in the roughness length of fields. For example, that for short grass is much lower than for wheat. The surface roughness of sea is affected by the wave motion and requires special treatment.

15.13.2 Urban areas

Accounts of dispersion over urban areas have been given by Gifford (1976a,b) and by S.R. Hanna (1976). An urban area acts as a heat island. It also constitutes terrain with a high surface roughness. Over an urban area the principal meteorological features such as wind speed, air and surface temperatures, mixing height and stability, are different from those over rural terrain.

Stable air from the surrounding rural area is modified as it approaches an urban area. Stability near a city tends to be nearly neutral, however stable the rural area may be. Hence the more stable conditions associated with Pasquill categories F and G tend not to occur. An urban area also tends to give increased turbulence. Measurements by Bowne, Ball and Anderson (1968) at Fort Wayne showed that the intensity of turbulence was some 40% above that of the surrounding area.

15.13.3 Coastal areas and water

Dispersion over coastal areas and water has been described by Gifford (1976a,b) and Lyons (1976). Water in this context includes both sea and lakes, but unless otherwise stated reference here is to sea. Much of the experimental work described by Lyons, however, relates to the Great Lakes, which constitute a feature well suited to the study of coastal area dispersion.

Conditions at the coastline may exhibit extreme variations of wind pattern, temperature and humidity. In fact radical changes of stability and turbulence are the rule. Dispersion under such conditions is much more complex than over flat rural terrain and Lyons places some emphasis on what is not known as well as what is.

In general, turbulence, and hence dispersion, is often less over sea than over land. In particular, this is so when warm air is advected over colder water. On the other hand, when colder, or drier air, is advected over warmer water, strong turbulence can occur.

Lyons distinguishes three situations:

(1) dispersion over water;
(2) dispersion over the coastline –
    (a) with gradient wind,
    (b) with sea breeze.

The surface of the sea tends to be more uniform than that of land, but sharp differences of water temperature and the sudden discontinuity at the coastline introduce complications not met on a land mass.

For dispersion over water, in unstable and neutral conditions, with a flow of cold air over warmer water, intensity of turbulence is high. Water spouts and steam devils, may occur. In stable conditions, with a flow of warm air over colder water, intense inversion can occur. Conditions in which conduction of heat from the air to the water below is a dominant influence on the over-water air temperature are termed by Lyons 'conduction inversions'.

For dispersion in a gradient onshore flow, without a sea breeze, the low level flow crossing the coastline reacts to the changes in surface roughness, temperature and evaporation. This creates internal boundary layers for momentum, heat and water vapour which do not necessarily coincide.

Two conditions are of particular interest: plume trapping and fumigation. As described above, plume trapping occurs when a plume is trapped beneath an elevated inversion, while fumigation is of several types, that of interest here being coastline fumigation in which cold sea air is heated rapidly by the ground.

Sometimes a sea breeze blows onshore at low level while the gradient wind above blows in the opposite direction. This condition tends to occur in very light gradient winds, strong insolation and daytime air temperatures greater than the sea surface temperature. The condition, known as a 'sea breeze', is therefore something different from a simple onshore wind. The sea breeze inflow layer has a depth of some 100–1000 m, typically 500 m, and the peak wind speed is about 7 m/s.

Dispersion in a sea breeze is often regarded as the least favourable condition for dispersion of pollutants from a source near the coastline, in that the breeze will tend to blow these back on shore. In practice, along shore air movements, wind shear and other effects can cause appreciable dispersion. The effects of sea breeze tend to be varied.

Over the sea, most of the meteorological parameters, and their diurnal variation, tend to differ from those over land. The stability categories required to classify stability are different, as are the roughness length and friction velocity, and the mixed layer height.

15.13.4 Complex terrain

An account of dispersion over complex terrain has been given by Egan (1976) and Gifford (1976a,b). Urban and industrial development often occurs at sites which have some special geographical feature, such as along rivers, in valleys, or at the base of mountains beside a lake or the sea or at the end of a plain. Thus complex terrain is likely to be encountered quite frequently. The type of

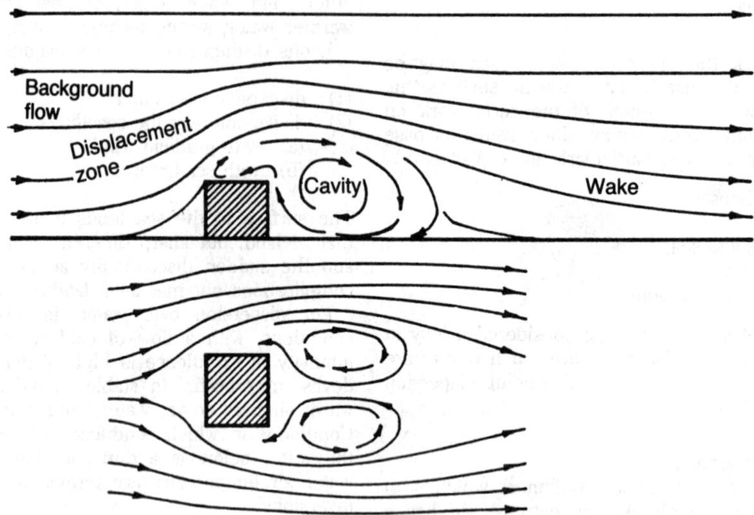

Figure 15.49 Simplified flow pattern around a sharp-edged building (Pasquill and Smith, 1983; after Halitsky, 1968; reproduced by permission)

terrain principally discussed by Egan is mountainous areas.

In complex terrain there are a number of effects which tend to increase dispersion. One is wind shear due to the variation of wind speed and wind direction with height. Another is distortion of the plume.

There are significant distortions of flow in complex terrain. Flow separation, which occurs when the streamlines no longer follow the shape of an obstacle, may take place. Separation tends to be less severe for flow around an isolated three-dimensional object such as a hill than for flow normal to an object which is effectively two dimensional such as a ridge. Separation is more pronounced for unstable or neutral conditions than for stable conditions, which tend to suppress it.

The difference between complex and level terrain in respect of stability is most marked for stable conditions. Stable stratification is liable to produce shearing motions and thus to promote turbulence. The effect of the terrain is less pronounced for unstable and neutral conditions.

Air flow in stable stratified conditions has been the subject of considerable work, since the phenomenon constitutes a hazard to aircraft, but this work tends to concentrate on the two-dimensional aspects, and there is less work available on three-dimensional stratified flow.

In valleys there tend to be local variations of wind direction and velocity. Typically there are upslope winds during the day and downslope winds at night. Two issues in air pollution in complex terrain are the impingement of a plume from an elevated source on the ground and episodes of high pollution in stagnant regions.

15.13.5 Buildings and obstructions

Dispersion at buildings and other obstructions has been described by Halitsky (1968), Gifford (1976a,b) and S.R. Hanna, Briggs and Hosker (1982). The presence of buildings or other obstructions in the path of the gas cloud has a marked influence on its flow and dispersion. Buildings act as obstacles to the flow of gas. They also give rise to flow distortions with local pressure and velocity fluctuations.

In considering the effect of a building in the near field, therefore, there is a distinction to be made between its influence on a release from a source some distance upwind and its influence on one from a source at a point near the building such as on its roof or in its lee. In the far field the influence of a number of buildings may often be treated in terms of their effect on the surface roughness.

Diagrams and photographs showing the flow patterns around buildings have been given by a number of authors, including Halitsky (1968), Hosker (1979) and D.J. Wilson (1979a). The flow pattern for conditions where the wind is perpendicular to the upwind face of a rectangular building are shown in Figure 15.49. This shows several important features. There is a roof cavity at the front edge of the building. Above this cavity, starting at about its maximum height, is a high turbulence layer boundary and a roof wake boundary. In the lee of the building there is a wake cavity.

15.14 Dispersion Modelling

There are a number of different approaches to the modelling of dispersion. These include:

(1) gradient transfer models;
(2) statistical models;
(3) similarity models;
(4) top hat, box and slab models.

These are now described in turn.

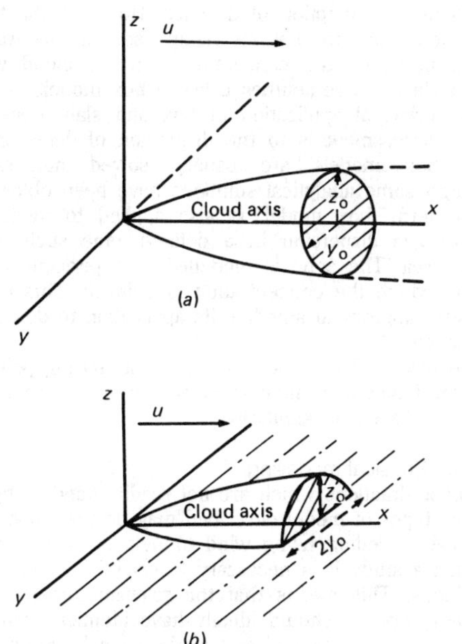

***Figure 15.50** Co-ordinates for dispersion equations: (a) elevated source; (b) ground level source*

15.14.1 Diffusion equation

The fundamental equation for diffusion of a gas, in rectangular co-ordinates, is

$$\frac{d\chi}{dt} + u\frac{d\chi}{dx} + v\frac{d\chi}{dy} + w\frac{d\chi}{dz} = K_x\frac{d^2\chi}{dx^2} + K_y\frac{d^2\chi}{dy^2} + K_z\frac{d^2\chi}{dz^2} \quad [15.14.1]$$

where x, y, z are the rectangular co-ordinates (m), K_x, K_y, K_z are the diffusion coefficients in the x, y, z directions (m^2/s), t is the time (s), u, v, w are the mean wind speeds in the x, y, z directions (m/s) and χ is the concentration (kg/m^3). The co-ordinate system is shown in Figure 15.50(a).

If the wind speed in the y and z directions is zero ($v = w = 0$) and the diffusion coefficients are the same in each direction ($K_x = K_y = K_z = K$), Equation 15.14.1 becomes

$$\frac{d\chi}{dt} + u\frac{d\chi}{dx} = K\left(\frac{d^2\chi}{dx^2} + \frac{d^2\chi}{dy^2} + \frac{d^2\chi}{dz^2}\right) \quad [15.14.2]$$

where x, y, z are the distances in the downwind, crosswind and vertical directions (m) and K is the diffusion coefficient (m^2/s).

If the wind speed in the x direction is also zero ($u = 0$), Equation 15.14.2 reduces to

$$\frac{d\chi}{dt} = K\left(\frac{d^2\chi}{dx^2} + \frac{d^2\chi}{dy^2} + \frac{d^2\chi}{dz^2}\right) \quad [15.14.3]$$

The corresponding equation for a symmetrical spherical system is

$$\frac{d\chi}{dt} = \frac{K}{r^2}\frac{d}{dr}\left(r^2\frac{d\chi}{dr}\right) \quad [15.14.4]$$

with

$$r^2 = x^2 + y^2 + z^2 \quad [15.14.5]$$

where r is the radial co-ordinate (m).

Analytical solutions of the above equations with a constant value of the diffusion coefficient K have been given by O.F.T. Roberts (1923) as described below.

15.14.2 Gradient transfer models

Gradient transfer models, or K models, are solutions of the diffusion equation. Although the assumption of a constant diffusion, or turbulent exchange, coefficient was made in the early work, it is known that this is an oversimplification which yields unsatisfactory results. The approach now adopted is to solve the diffusion equation using relationships for the variation of the individual exchange coefficients K_x, K_y and K_z and for the wind speed u. If the form of these relations is amenable, an analytical solution may be obtained, but usually it is necessary to resort to numerical solution.

15.14.3 Statistical models

Analytical solution of the diffusion equation with a constant exchange coefficient K shows that the concentration profiles obtained for releases such as plumes and puffs are Gaussian in form and may therefore be characterized by standard deviations for dispersion, or dispersion coefficients, σ_x, σ_y and σ_z.

The statistical properties of turbulence may be described in terms of quantities such as the Lagrangian time scale and the autocorrelation function. Moreover, using Taylor's theorem the autocorrelation function may be related to the standard deviations of the wind velocity fluctuations, which in turn may be related to the standard deviations for dispersion.

In the model developed by O.G. Sutton (1953) the diffusion parameter C is related to the dispersion coefficients, while in the Pasquill–Gifford model these dispersion coefficients are used directly. These models therefore possess a statistical basis.

15.14.4 Similarity models

Another approach is the use of dimensional analysis to derive similarity, or self-similar, models. Typically, the basis of a similarity model is an equation, derived from dimensional analysis, for the rate of growth of some characteristic dimension of the cloud, such as the height of a plume. This part of the model, however, gives no information about the concentration distribution within the cloud; for this, additional relations are required, which tend to have an empirical basis.

Similarity models are used particularly for buoyant plumes and momentum jets, but such models also exist for plumes and puffs. The similarity approach is also used to obtain relationships for the dispersion coefficients used in the Gaussian models.

15.14.5 Similarity criteria

There are a number of dimensionless groups which constitute criteria of similarity and which are particularly relevant in dispersion work.

The Reynolds number Re is defined as

$$\mathrm{Re} = \frac{ul\rho}{\mu} \quad [15.14.6a]$$

$$= \frac{ul}{\nu} \quad [15.14.6b]$$

where l is a characteristic dimension, u is the velocity of the fluid, μ its viscosity, ν its kinematic viscosity and ρ its density. The Reynolds number is the ratio of the inertial forces to the viscous forces and is a criterion for similarity of flow regime.

The Peclet number Pe is

$$\mathrm{Pe} = \frac{ul}{D} \quad [15.14.7]$$

where D is the diffusion coefficient, or diffusivity.

The Peclet number characterizes the ratio of the inertial forces to the diffusivity and is a criterion for the relative importance in mass transfer of bulk transport and diffusive transport.

The Schmidt number Sc is

$$\mathrm{Sc} = \frac{\mu/\rho}{D} = \frac{\mu}{\rho D} \quad [15.14.8a]$$

$$= \frac{\nu}{D} \quad [15.14.8b]$$

It is thus the ratio of the Peclet and Reynolds numbers

$$\mathrm{Sc} = \mathrm{Pe}/\mathrm{Re} \quad [15.14.9]$$

The Schmidt number is the ratio of the kinematic viscosity to the diffusivity and is a criterion for similarity of mass transfer.

The Prandtl number Pr is

$$\mathrm{Pr} = \frac{\mu/\rho}{k/c_p\rho} = \frac{c_p\mu}{k} \quad [15.14.10]$$

where c_p is the specific heat of the fluid and k its thermal conductivity. The Prandtl number is the ratio of the kinematic viscosity to the thermal diffusivity and is a criterion of similarity for heat transfer, corresponding to the Schmidt number for mass transfer.

The Froude number Fr is

$$\mathrm{Fr} = \frac{u}{(gl)^{\frac{1}{2}}} \quad [15.14.11]$$

It is the ratio of the inertial force to the gravitational force and is a criterion of similarity where buoyancy is significant.

15.14.6 Top hat, box and slab models

Another family of models is that comprising the models referred to as top hat, box and slab models.

A top hat model has an essentially flat top, although vertical mixing takes place at the top surface. The cloud can therefore be considered to have a defined height.

In a box model the cloud is treated as a vertical cylinder which at a given instant has a uniform concentration throughout. A source cloud is defined, which may or may not contain some initial entrained air, and the subsequent development of the cloud is described in terms of its advection and of the entrainment of air into it. In a slab model the concentration in the cloud is a function of distance. However, the terms 'box' and 'slab' are not invariably used in this way. A model in which the concentration in the cloud varies with distance is sometimes called a box model.

The principal application of box and slab models in hazard assessment is to the dispersion of dense gases. Such box models are usually solved numerically, although some analytical solutions have been obtained.

The term 'box model' is also applied to models of passive gas dispersion in a defined zone such as an urban area. The zone is modelled as a perfectly mixed box in which the concentration is uniform. This use of the term appears to antedate its application to dense gas dispersion.

The box model has also served as a starting point for the development of models in which the concentration profile is based on similarity.

15.14.7 Physical modelling

Complex situations which are not readily handled by the various types of model described may be investigated by physical modelling using wind tunnels or water flumes. In such a study it is necessary to establish similarity of conditions. This means that the relevant dimensionless similarity criteria should ideally have identical values in the scenario investigated and in the tunnel experiments. Since in some cases several such groups may be involved, it may not be possible to achieve complete identity of all the groups.

15.15 Passive Dispersion

The dispersion of gases with neutral buoyancy, or passive dispersion, has been the subject of a very large volume of work. Some of the early work was oriented to gas warfare, but latterly air pollution has been the principal concern.

Neutral buoyancy is commonly due to the low concentration of the contaminant gas released, although it may also occur if the density of the gas is close to that of air. The neutral buoyancy condition may be negated if the gas release causes a large change in the temperature of the resultant cloud.

Hazard assessment utilizes, but has not greatly contributed to, work on neutral density gas dispersion. Most of the work has therefore been concerned with dispersion of continuous releases from an elevated point source, as represented by an industrial chimney stack. The two other main types of release studied which are of industrial relevance are continuous and instantaneous point source releases at ground level. There is some work on continuous line sources at ground level, relevant to gas warfare, but also to industrial area releases.

Early work on the subject includes that of G.I. Taylor (1915) and O.F.T. Roberts (1923). The models which are most widely used, however, are those of O.G. Sutton (1953) and of Pasquill (1961, 1962a) and the Pasquill–Gifford model (Pasquill, 1961; Gifford, 1961).

In this section an account is given of experimental studies and empirical features of passive dispersion. Section 15.16 describes passive dispersion models and Sections 15.17–15.19 describe dispersion over particular surfaces, dispersion in particular conditions and dispersion parameters, respectively.

15.15.1 Experimental studies

Passive gas dispersion has been the subject of a large volume of experimental work. Accounts of this work include those by Islitzer and Slade (1968), Slade (1968), Golder (1972), Gifford (1976a,b) and A.E. Mitchell (1982). Some experimental studies of such dispersion are listed in Table 15.27. More detailed listings are given by Islitzer and Slade (1968).

Work on passive gas dispersion on Salisbury Plain at Porton Down established what may be regarded as a

Table 15.27 *Some experimental studies of passive gas dispersion*

A Continuous elevated sources[a]

Location	*Experiment*	*Reference*
Porton Down	Porton	J.S. Hay and Pasquill (1957)
Idaho Falls, ID	National Reactor Testing Station (NRTS)	Islitzer (1961)
Richland, WA	Hanford	Hilst and Simpson (1958); Nickola (1977)
Harwell	Harwell (BEPO)	Stewart *et al.* (1954)
Long Island, NY	Brookhaven National Laboratory (BNL)	
Tennessee Valley, TN	Tennessee Valley Authority (TVA)	Gartrell *et al.* (1964); Carpenter *et al.* (1971)

B Continuous ground level releases[a]

Location	*Experiment*	*Reference*
Porton Down	Porton	J.S. Hay and Pasquill (1959)
Cardington	Cardington	Pasquill (1962)
O'Neill, NE	Prairie Grass	Cramer (1957); Lettau and Davidson (1957); Barad (1958); Islitzer and Slade (1964)
Round Hill	Round Hill	Islitzer and Slade (1964)
Richland, WA	Green Glow	Barad and Fuquay (1962); Fuquay, Simpson and Hinds (1964)
Cape Kennedy, FL	Ocean Breeze	Haugen and Fuquay (1963); Haugen and Taylor (1963)
Vandenberg AFB, CA	Dry Gulch	Haugen and Fuquay (1963); Haugen and Taylor (1963)
Idaho Falls, ID	NRTS	Islitzer and Dumbault (1963)

C Instantaneous ground level releases[a,b]

Location	*Experiment*	*Reference*
Porton Downs	Porton	F.B. Smith and Hay (1961)
Edwards AFB, CA	Sand Storm	J.H. Taylor (1965)
Agesta and Studsvik, Sweden		Högström (1964)
Dugway Proving Grounds, Utah	Dugway	Cramer *et al.* (1964)
Point Arguello, CA	Naval MissileFacility (NMF)	T.B. Smith *et al.* (1964)
Idaho Falls, ID	NRTS	Islitzer and Markee (1964)
Dallas, TX	Cedar Hill	McCready, Smith and Wolf (1961)

D Releases in urban areas

Location	*Experiment*	*Reference*
Worcester, MA		Yersel, Goble and Morrill (1983)
St Louis, MO		McElroy and Pooler (1968)

E Releases over water

Location	*Experiment*	*Reference*
Cameron, LA		Dabberdt *et al.* (1982)
Pismo Beach, CA		Dabberdt *et al.* (1982)

[a] Many of these experiments are also described by Slade (1968)
[b] These releases are more accurately described as 'quasi-instantaneous'.

base case for terrain. The terrain at Porton is open country with some clumps of trees. The work at Porton provides the experimental background for the work of O.G. Sutton (1953) and Pasquill (1962a).

An important set of experiments carried out on this terrain are those used by Pasquill to derive parameters in his dispersion models. In this work there was a continuous release from a point source at ground level, the distance to which concentration measurements were made was 800 m and the sampling period was 10 min.

15.15.2 Empirical features

Experiments on passive gas dispersion indicate several important empirical features. The fundamental features have been described by O.G. Sutton (1953).

One of the most important features is that, for both continuous and instantaneous releases from a point source at ground level, the concentration profiles are Gaussian. Another basic feature is that for both types of release the spread of the measured concentration increases as the sampling period increases. It is observed that the plume from a continuous point source release tends to meander and that the dispersion due to turbulence is augmented by that due to this meandering.

The concentration downwind of a continuous or instantaneous point source at ground level is found to vary according to the strength of the source, provided that the latter does not itself cause appreciable convection. For a continuous point source the concentration is also inversely proportional to the mean wind speed.

The concentration on the centre line of a continuous point source is

$$\chi \propto x^{-1.76} \quad [15.15.1]$$

and that on the centre plane of a continuous infinite line crosswind source is

$$\chi \propto x^{-1.09} \quad [15.15.2]$$

This information on the variation of concentration with distance has played an important role in guiding the development of dispersion models.

15.16 Passive Dispersion: Models

Some principal models for passive dispersion are:

(1) Roberts model;
(2) Sutton model;
(3) Pasquill model;
(4) Pasquill–Gifford model.

An account is now given of each of these models in turn.

15.16.1 Roberts model

The fundamental diffusion equation has been given in Section 15.14. Solutions of this equation have been given by O.F.T. Roberts (1923), who analysed the behaviour of smoke from various types of release. The following treatment is based on Roberts' work and on modifications of it derived by O.G. Sutton (1953).

For dispersion from an instantaneous point source under windless conditions Equation 15.14.4 is applicable. The relevant boundary conditions are:

$$\chi \to 0 \quad \text{as} \quad t \to 0, \quad r > 0 \quad [15.16.1a]$$

$$\chi \to 0 \quad \text{as} \quad t \to \infty \quad [15.16.1b]$$

The continuity condition is

$$\int_{-\infty}^{\infty}\iint \chi \, dx\, dy\, dz = Q^* \quad [15.16.2]$$

where Q^* is the mass released instantaneously (kg). The solution of Equation 15.14.4 is then

$$\chi(r,t) = \frac{Q^*}{8(\pi K t)^{\frac{3}{2}}} \exp\left(-\frac{r^2}{4Kt}\right) \quad [15.16.3a]$$

$$\chi(x,y,z,t) = \frac{Q^*}{8(\pi K t)^{\frac{3}{2}}} \exp\left[-\frac{(x^2+y^2+z^2)}{4Kt}\right] \quad [15.16.3b]$$

Equation 15.16.3b may also be applied to an instantaneous point source with a wind speed in the x direction by measuring the co-ordinates from an origin moving with the cloud at the mean wind speed.

For dispersion from a continuous point source under windless conditions, Equations 15.16.3, with Equation 15.14.5, may be integrated with respect to time to give

$$\chi(r,t) = \frac{Q}{4\pi K r} \operatorname{erfc}\left[\frac{r}{2(Kt)^{\frac{1}{2}}}\right] \quad [15.16.4a]$$

$$\chi(x,y,z,t) = \frac{Q}{4\pi K(x^2+y^2+z^2)^{\frac{1}{2}}} \operatorname{erfc}\left[\frac{(x^2+y^2+z^2)^{\frac{1}{2}}}{2(Kt)^{\frac{1}{2}}}\right] \quad [15.16.4b]$$

where Q is the continuous mass rate of release (kg/s). At steady state, Equation 15.16.4 becomes

$$\chi(r) = \frac{Q}{4\pi K r} \quad [15.16.5a]$$

$$\chi(x,y,z) = \frac{Q}{4\pi K(x^2+y^2+z^2)^{\frac{1}{2}}} \quad [15.16.5b]$$

Equation 15.16.5b may be applied to a continuous point source with a wind speed u in the x direction by a transformation of co-ordinates. This gives

$$\chi(x,y,z) = \frac{Q}{4\pi K(x^2+y^2+z^2)^{\frac{1}{2}}} \times \exp\left\{-\frac{u}{2K}[(x^2+y^2+z^2)^{\frac{1}{2}} - x]\right\} \quad [15.16.6]$$

If the concentrations considered are those not too far from the x axis so that

$$\frac{y^2+z^2}{x^2} \ll 1 \quad [15.16.7]$$

then for all but the lightest winds Equation 15.16.6 becomes

$$\chi(x,y,z) = \frac{Q}{4\pi K x} \exp\left[-\frac{u}{4Kx}(y^2+z^2)\right] \quad [15.16.8]$$

For dispersion from a continuous infinite line source at right angles to the wind direction, Equation 15.16.6 can be modified and integrated in the y direction to give

$$\chi(x,z) = \frac{Q'}{2\pi K} \exp\left(\frac{ux}{2K}\right) K_o\left[\frac{u(x^2+z^2)^{\frac{1}{2}}}{2K}\right] \qquad [15.16.9]$$

where Q' is the continuous mass rate of release per unit length (kg/m s). K_o is the modified Bessel function of the second kind. Provided the term $u(x^2 + z^2)^{\frac{1}{2}}/2K$ is sufficiently large, Equation 15.16.9 may be approximated by

$$\chi(x,z) = \frac{Q'}{(2\pi Kx)^{\frac{1}{2}}} \exp\left(-\frac{uz^2}{4Kx}\right) \qquad [15.16.10]$$

For the case where dispersion is anisotropic the equations are as follows. For an instantaneous point source:

$$\chi(x,y,z,t) = \frac{Q^*}{8(\pi t)^{\frac{3}{2}}(K_xK_yK_z)^{\frac{1}{2}}} \times \exp\left[-\frac{1}{4t}\left(\frac{x^2}{K_x}+\frac{y^2}{K_y}+\frac{z^2}{K_z}\right)\right] \qquad [15.16.11]$$

For a continuous point source

$$\chi(x,y,z) = \frac{Q}{4\pi x(K_xK_y)^{\frac{1}{2}}} \exp\left[-\frac{u}{4x}\left(\frac{y^2}{K_y}+\frac{z^2}{K_z}\right)\right] \qquad [15.16.12]$$

For a continuous infinite line crosswind source:

$$\chi(x,z) = \frac{Q'}{(2\pi K_zx)^{\frac{1}{2}}} \exp\left(-\frac{uz^2}{4K_zx}\right) \qquad [15.16.13]$$

The equations derived so far apply to an elevated source dispersion which is unaffected by the ground. If the source is on the surface, the ground forms an impervious boundary. The co-ordinate system is shown in Figure 15.50(b). The effect of the ground is to double the concentration. Thus Equations 15.16.11–15.16.13 become: for an instantaneous point source

$$\chi(x,y,z,t) = \frac{Q^*}{4(\pi t)^{\frac{3}{2}}(K_xK_yK_z)^{\frac{1}{2}}} \times \exp\left[-\frac{1}{4t}\left(\frac{x^2}{K_x}+\frac{y^2}{K_y}+\frac{z^2}{K_z}\right)\right] \qquad [15.16.14]$$

for a continuous point source

$$\chi(x,y,z) = \frac{Q}{2\pi x(K_yK_z)^{\frac{1}{2}}} \exp\left[-\frac{u}{4x}\left(\frac{y^2}{K_y}+\frac{z^2}{K_z}\right)\right] \qquad [15.16.15]$$

and for a continuous infinite line crosswind source

$$\chi(x,z) = \frac{2Q'}{(2\pi K_zx)^{\frac{1}{2}}} \exp\left(-\frac{uz^2}{4K_zx}\right) \qquad [15.16.16]$$

An intermediate situation occurs where the ground forms an impervious barrier to the material from an elevated source such as a chimney or high vent. For this case it can be shown that for a continuous elevated point source

$$\chi(x,y,z) = \frac{Q}{4\pi Kx} \exp\left(-\frac{uy^2}{4Kx}\right) \times \left\{\exp\left[-\frac{u}{4Kx}(z-h)^2\right] + \exp\left[-\frac{u}{4Kx}(z+h)^2\right]\right\} \qquad [15.16.17]$$

where h is the height of the source (m). Equation 15.16.17 reduces to Equation 15.16.15 if h is set equal to zero.

The equations for continuous sources are steady-state equations and therefore apply only to fully established plumes. The concentrations given by the equations are applicable only if the duration of the release t is equal to or greater than the ratio of the distance x at the location of interest to the wind speed u ($t \geq x/u$).

Equations 15.16.14–15.16.17 give the concentrations at all distances from the source. For practical purposes it is convenient to define some boundaries for the cloud. The convention adopted is to take the cloud boundaries y_o and z_o as a fixed proportion, usually one-tenth, of the maximum concentration at distance x. If y_o and z_o are the semi-lateral and semi-vertical dimensions of the cloud, as shown in Figure 15.50, then from Equation 15.16.15, for a continuous point source

$$y_o = \left(\frac{4}{u}K_yx \ln 10\right)^{\frac{1}{2}} \qquad [15.16.18]$$

$$z_o = \left(\frac{4}{u}K_zx \ln 10\right)^{\frac{1}{2}} \qquad [15.16.19]$$

Hence

$$y_o \propto z_o \propto x^{\frac{1}{2}} \qquad [15.16.20]$$

The ground level concentration on the centre line ($y = z = 0$) of a continuous point source is given by Equation 15.16.15 as

$$\chi(x) = \frac{Q}{2\pi x(K_yK_y)^{\frac{1}{2}}} \qquad [15.16.21]$$

and that on the centre plane ($z = 0$) of a continuous infinite line crosswind source is given by Equation 15.16.16 as

$$\chi(x) = \frac{2Q'}{(2\pi K_zx)^{\frac{1}{2}}} \qquad [15.16.22]$$

The visible outline of the cloud is not determined as simply, but depends on the theory of opacity.

An important implication of Equations 15.16.15 and 15.16.16 is that for Equation 15.16.15 at a fixed distance x from the source the concentration distribution in both the crosswind and the vertical directions is a Gaussian, or normal, distribution. Similarly, Equation 15.16.16 gives a concentration distribution which is Gaussian in the vertical direction. This feature is important in the further development of dispersion equations, as described below.

It can also be seen from Equations 15.16.15 and 15.16.16 that the ground level concentration downwind along the centre line, or axis, of a continuous point source ($y = z = 0$), is

$$\chi \propto x^{-1} \qquad [15.16.23]$$

and that the ground level concentration downwind over the centre plane of a continuous infinite line crosswind source is

$$\chi \propto x^{-\frac{1}{2}} \qquad [15.16.24]$$

Comparison of relations 15.16.23 and 15.16.24 with the relations 15.15.1 and 15.15.2 obtained by experiment shows that this model is unsatisfactory.

Thus it is not appropriate to model dispersion in the atmosphere using the constant Fickian diffusion coefficient K. Nevertheless, the equations given form the basis for most subsequent developments in work on passive dispersion.

15.16.2 Sutton model

The failure of the simple Fickian diffusion model has prompted the search for more realistic models of dispersion. The first model considered here is that of O.G. Sutton (1953).

The basic equation derived by Sutton for an instantaneous point source at ground level is

$$\chi(x,y,z,t) = \frac{2Q^*}{\pi^{\frac{3}{2}} C_x C_y C_z (ut)^{\frac{3}{2}(2-n)}} \times \exp\left[-(ut)^{n-2}\left(\frac{x^2}{C_x^2}+\frac{y^2}{C_y^2}+\frac{z^2}{C_z^2}\right)\right] \qquad [15.16.25]$$

where C_x, C_y, C_z are diffusion parameters in the downwind, crosswind and vertical (x, y, z) directions ($m^{\frac{1}{2}n}$), n is the diffusion index, and the co-ordinates x, y, z are measured from an origin moving with the cloud at the mean wind speed u.

Equation 15.16.25 is related to Equation 15.16.14 in the Roberts model and becomes identical with it by putting

$$n = 1 \qquad [15.16.26a]$$

$$C^2 = \frac{4K}{u} \qquad [15.16.26b]$$

The equation for a continuous point source at ground level is

$$\chi(x,y,z) = \frac{2Q}{\pi C_y C_z u x^{(2-n)}} \exp\left[-x^{n-2}\left(\frac{y^2}{C_y^2}+\frac{z^2}{C_z^2}\right)\right] \qquad [15.16.27]$$

and that for a continuous infinite line crosswind source at ground level is

$$\chi(x,z) = \frac{2Q'}{\pi^{\frac{1}{2}} C_z u x^{\frac{1}{2}(2-n)}} \times \exp\left[-x^{n-2}\left(\frac{z^2}{C_z^2}\right)\right] \qquad [15.16.28]$$

The equation for a continuous elevated point source is

$$\chi(x,y,z) = \frac{Q}{\pi C_y C_z u x^{2-n}} \exp\left(-x^{n-2}\frac{y^2}{C_y^2}\right) \times \left\{\exp\left[-x^{n-2}\frac{(z-H)^2}{C_z^2}\right] + \exp\left[-x^{n-2}\frac{(z+H)^{\frac{1}{2}}}{C_z^2}\right]\right\} \qquad [15.16.29]$$

where H is the height of the source (m).

As mentioned above, the concentrations given by the equations for continuous sources are applicable only if the duration t of the release is equal to or greater than the ratio of the distance x to the wind speed u ($t \geq x/u$).

Equations 15.16.27–15.16.29 correspond to Equations 15.16.15–15.16.17 in the Roberts model. Equations 15.16.25 and 15.16.27–15.16.29 are frequently written with the assumption of isotropic conditions, for which the diffusion parameters are

$$C_x = C_y = C_z = C \qquad [15.16.30]$$

The index n and the generalized diffusion parameter C are meteorological constants. Values of these constants have been discussed in detail by O.G. Sutton (1947, 1953). The index n is a function of the stability conditions. The limiting values of n are zero and unity under conditions of very high and very low turbulence, respectively. In average conditions n has a value of approximately 1/4. The value of the generalized diffusion parameter C is a function of the height above ground and of the stability conditions. Values of these parameters given by Sutton are shown in Table 15.28. Further values are given in Section 15.17.

15.16.3 Pasquill model

Another general system of dispersion equations has been derived by Pasquill (1961, 1962a, 1965) and is generally referred to as the Pasquill model. The basis of the system is a modification of Equation 15.16.29 for a continuous point source at ground level:

$$C_o = \frac{2.8 \times 10^{-3} Q}{u d h \theta} \qquad [15.16.31]$$

where C_o is the ground level concentration on the axis of the plume (units/m^3), d is the distance in the downwind direction (km), h is the vertical spread (m), Q is the mass rate of release (units/min), u is the mean wind speed (m/s) and θ is the lateral spread (degrees).

Equation 15.16.31 is based on Equation 15.16.29, generalized to allow for relatively slow changes of wind direction. The lateral spread θ and the vertical spread h define an envelope at the edge of which concentrations are one-tenth of the axial or ground level values, respectively, i.e. the envelope of the cloud defined by the usual convention.

Equation 15.16.31 is for a continuous source and hence, as mentioned above, the concentrations given are applicable only if the duration of the release is equal

***Table 15.28** Meteorological parameters for Sutton model[a] (after O.G. Sutton, 1947)*

Source height (m)	n	C_y ($m^{1/8}$)	C_z ($m^{1/8}$)
0	0.25	0.21	0.12
10	0.25	0.21	0.12
25	0.25	0.12	
30	0.25	0.10	
75	0.25	0.09	
100	0.25	0.07	

[a] Values are for small lapse rate and wind speed of 5 m/s.

Table 15.29 *Meteorological parameters for Pasquill model (after Pasquill, 1961)*

A Lateral spread for a short release, θ

Pasquill stability category	*Lateral spread* θ (deg)	
	$d = 0.1$ *km*	$d = 100$ *km*
A	60	(20)
B	45	(20)
C	30	10
D	20	10
E	(15)	(5)
F	(10)	(5)

B Vertical spread, h

Pasquill's values, shown in Figure 15.51, are fitted by the following approximate equations:

A	$\log_{10} h = 2.95 + 2.19 \log_{10} d + 0.723(\log_{10} d)^2$	$0.1 < d < 2$
B	$\log_{10} h = 2.36 + 1.05 \log_{10} d + 0.067(\log_{10} d)^2$	$0.1 < d < 10$
C	$\log_{10} h = 2.14 + 0.919 \log_{10} d - 0.017(\log_{10} d)^2$	$0.1 < d < 30$
D1	$\log_{10} h = 1.85 + 0.835 \log_{10} d - 0.0097(\log_{10} d)^2$	$0.1 < d < 100$
D2	$\log_{10} h = 1.83 + 0.754 \log_{10} d - 0.087(\log_{10} d)^2$	$0.1 < d < 100$
E	$\log_{10} h = 1.66 + 0.670 \log_{10} d - 0.100(\log_{10} d)^2$	$0.1 < d < 100$
F	$\log_{10} h = 1.48 + 0.656 \log_{10} d - 0.122(\log_{10} d)^2$	$0.1 < d < 100$

to or greater than the ratio of the distance to the wind speed u ($t \geq d \times 10^3/u$).

For the lateral spread θ used in Equation 15.16.31, Pasquill recommends that if suitable data are available the following relations should be used:

$$\theta = 4.3\sigma_\theta \qquad [15.16.32]$$

with

$$\sigma_\theta \approx \sigma_y/x \qquad [15.16.33]$$

where x is the downwind distance (m), σ_y is the standard deviation of concentration in the crosswind direction (m) and σ_θ is the standard deviation of wind direction (degrees).

If such data are not available, he suggests that approximate estimates of θ for a long release, lasting 1 h or more, may be made from wind direction traces using the following rules:

$d =$ 0.1 km
$\theta =$ difference between maximum and minimum trace over period of release

$d =$ 100 km
$\theta =$ difference between maximum and minimum '15-min averages' of wind direction

For a short release, lasting a few minutes, Pasquill gives the estimates of θ shown in Table 15.29.

The vertical spread used in Equation 15.16.31 increases with distance d at a rate which depends on the extent of vertical mixing. If, however, vertical convection is suppressed by an isothermal or inversion layer, the height of this layer sets a limit to h. Ultimately, the concentration between the ground and the layer becomes uniform.

For the vertical spread h, Pasquill recommends that if suitable data are available the following relation be used:

$$h = 2150d\sigma_\phi \qquad [15.16.34]$$

where σ_ϕ is the standard deviation of the wind inclination (rad) and ϕ is the wind inclination (rad).

If such data are not available, he suggests that tentative estimates of h be made using the graph shown in Figure 15.51. These estimates are applicable only to open country. The curves given in the graph are fitted by the approximate equations given in Table 15.29. The limits of the equations are as shown in the table.

In the stable conditions of a clear night with very light winds (< 2 m/s), and thus under conditions conducive to sharp ground frost or heavy dew, the vertical spread h may be even less than that given for stability category F.

In unstable conditions the vertical spread h may be estimated by the methods just described only until it reaches a value h' equal to the estimated vertical extent of convection which will occur at some distance d'. Under these conditions at the distance d' there is a concentration profile with a limiting vertical spread h'. For approximate work it may be assumed that over a further interval d', and therefore at a distance $2d'$, a uniform concentration develops. Thus for distances greater than $2d'$ the vertical spread h used in Equation 15.16.31 may be taken as $2h'$.

The wind speed u and the wind direction used in Equation 15.16.31 at distances 0.1 and 1 km are the usual surface values. At distances of 10 and 100 km the wind speed and direction used are the averages of the surface and geostrophic values, but in addition the latter is backed (i.e. moved anticlockwise) by 10°.

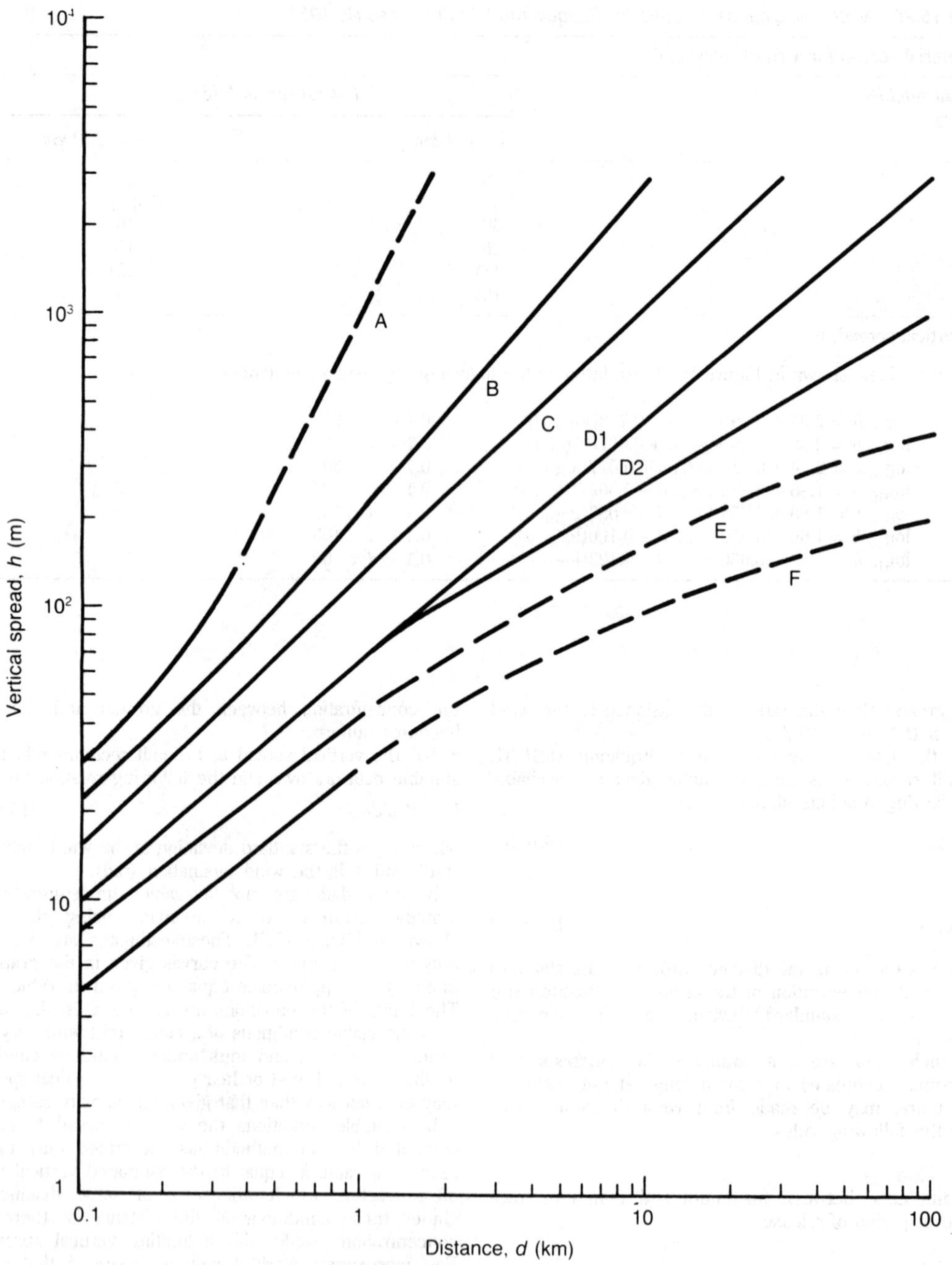

Figure 15.51 *Vertical spread h for Pasquill equation (Pasquill, 1961) (Courtesy of HM Stationery Office)*

For an elevated source, Equation 15.16.31 is modified by the stack correction factor F_1 to give

$$C_c = \frac{2.8 \times 10^{-3} Q F_1}{udh\theta} \qquad [15.16.35]$$

with

$$F_1 = \exp\left[-2.303\left(\frac{H}{h}\right)^2\right] \qquad [15.16.36]$$

where C_c is the ground level concentration on the axis of the plume for the elevated source (units/m^3), and H is the height of the stack (m).

Table 15.30 *Illustrative example of the application of the Pasquill equation for a prolonged release from continuous point sources at and above ground level (after Pasquill, 1961)*

A	***Problem***	
	Effective height of elevated source H	100 m
	Stability category	B–C
	Surface wind speed	4 m/s
	Surface wind direction	275°
	Geostrophic wind speed	8 m/s
	Geostrophic wind direction	325°
	Vertical extent of convection	1000 m
	Source strength	1 unit/min
B	***Solution***	
	Distance at which vertical spread h equals vertical extent of convection h'	5.5 km

1	Ground level source				
	Distance d, km	0.1	1	10	100
	Effective wind speed u, m/s	4	4	6	6
	Effective wind direction, deg	275	275	290	290
	Lateral spread θ,[a] deg	120	93	67	40
	Effective vertical spread h, m	20	170	2000	2000
	Concentration along axis C_0, units/m^3	2.9×10^{-6}	4.4×10^{-8}	3.5×10^{-10}	5.8×10^{-11}

2	Elevated source							
	Arbitrary values of h/H	$\frac{1}{2}$	$\frac{2}{3}$	$\frac{4}{5}$	1	$1\frac{1}{2}$	2	4
	Corresponding values of:							
	Effective vertical spread h, m	50	67	80	100	150	200	400
	Distance d, km	0.28	0.37	0.46	0.59	0.86	1.15	2.20
	Concentration along axis C_0, $\times 10^{-10}$ units/m^3	4600	2700	1800	1200	580	330	82
	Stack correction factor F_1	10^{-4}	5.6×10^{-3}	0.027	0.10	0.36	0.56	0.87
	Corrected concentration C_c, $\times 10^{-10}$ units/m^3	0.46	15	49	120	210	185	71

[a] These values are obtained from wind direction traces.

Equations 15.16.31 and 15.16.35 give the concentration on the axis of the plume. Other concentrations may be calculated by multiplying the values given by these equations by the off-axis correction factor F_2:

$$F_2 = \exp\left[-2.303\left(\frac{2\alpha}{\theta}\right)^2\right] \qquad [15.16.37]$$

where α is the deviation from the axis (degrees). Thus Equation 15.16.31 becomes

$$C = \frac{2.8 \times 10^{-3} QF_1F_2}{udh\theta} \qquad [15.16.38]$$

where C is the ground level concentration (units/m^3).

The parameters θ and h are functions of distance and should be determined for the distance d at which the concentration C_o or C_c is to be calculated. It is convenient to tabulate and plot base values of C_o or C_c at distances d of 0.1, 1, 10 and 100 km and to obtain other values by interpolation. Lines of equal concentration, or isopleths, may be calculated from Equation 15.16.38 by fixing the concentration C and determining the corresponding distance d and deviation α.

As an illustration of the application of the method the example described by Pasquill (1961, 1962a) is given. The problem is the determination of the ground level concentration for a prolonged release from continuous point sources (1) at ground level and (2) above ground level. A statement of the problem is given in Section A of Table 15.30. The results of the calculations for the solution of the problem are shown in Section B of the table and in Figures 15.52 and 15.53.

It may be noted that, although Equation 15.16.31 is for a continuous point source, so that the concentration given is applicable only if the duration t is equal to or greater than the ratio of the distance to the wind speed ($t \geq d \times 10^3/u$), it nevertheless gives the correct value of the total dosage for the total mass released, and may therefore be used to calculate dosage for a nearly instantaneous point source.

Graphical solutions of the Pasquill model have been presented by Bryant (1964 UKAEA AHSB(RP) R42).

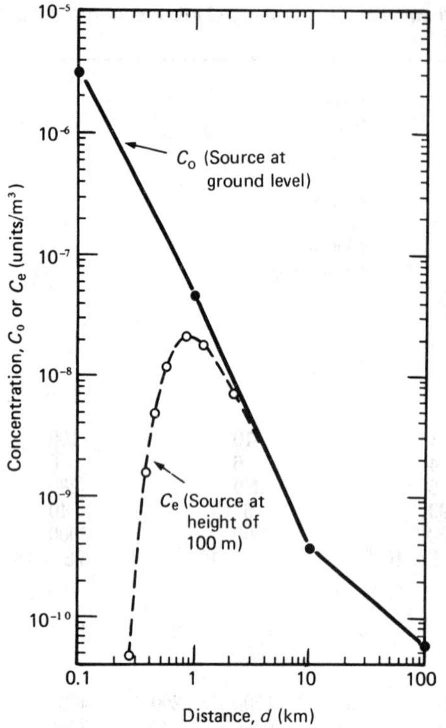

Figure 15.52 *Illustrative example of Pasquill equation: ground level concentrations along axis for a prolonged release from continuous point sources at and above ground level (Pasquill, 1961) (Courtesy of HM Stationery Office)*

15.16.4 Pasquill–Gifford model

An alternative form of the Sutton equations has been presented by Pasquill (1961, 1962a) and values of the dispersion coefficients used in this model have been obtained by Gifford (1961). This alternative form is often referred to as the Pasquill–Gifford model, as described below.

In this formulation use is made of the relations given by O.G. Sutton (1953, p. 286):

$$\sigma_x^2 = \frac{C^2}{2}(ut)^{2-n} \qquad [15.16.39]$$

Similar expressions apply for σ_y and σ_z where σ_x, σ_y and σ_z are the standard deviations, or dispersion coefficients, in the downwind, crosswind and vertical (x, y, z) directions (m).

The equation for an instantaneous point source at ground level then becomes

$$\chi(x,y,z,t) = \frac{2Q^*}{(2\pi)^{\frac{3}{2}}\sigma_x\sigma_y\sigma_z} \times \exp\left[-\frac{1}{2}\left(\frac{x^2}{\sigma_x^2}+\frac{y^2}{\sigma_y^2}+\frac{z^2}{\sigma_z^2}\right)\right] \qquad [15.16.40a]$$

where the co-ordinates x, y, z are measured from an origin moving with the cloud at the mean wind speed u.

Alternatively, if the co-ordinates are measured from the point of release

$$\chi(x,y,z,t) = \frac{2Q^*}{(2\pi)^{\frac{3}{2}}\sigma_x\sigma_y\sigma_z} \times \exp\left\{-\frac{1}{2}\left[\frac{(x-ut)^2}{\sigma_x^2}+\frac{y^2}{\sigma_y^2}+\frac{z^2}{\sigma_z^2}\right]\right\} \qquad [15.16.40b]$$

The equation for a continuous point source at ground level becomes

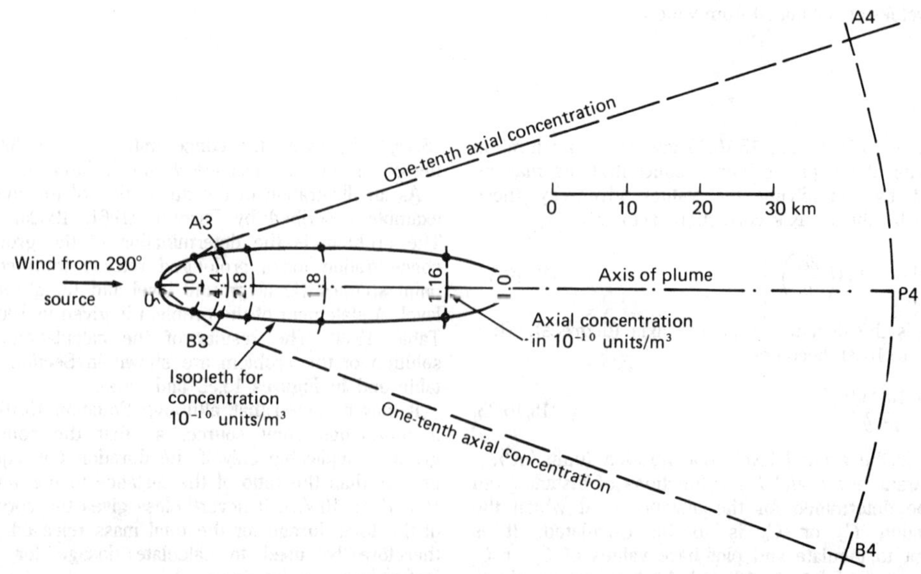

Figure 15.53 *Illustrative example of Pasquill equation: ground level concentration profile for a prolonged release from a continuous point source at ground level (Pasquill, 1961; reproduced by permission of HM Stationery Office)*

Table 15.31 Meteorological parameters for the Pasquill–Gifford model

A Values of σ_y and σ_z for continuous sources given by Turner (1970) are shown in Figure 15.54. Turner's values are fitted by the following approximate equations:

Pasquill stability category	*Dispersion coefficient*		
	σ_y (m)	σ_z (m)	
A	$\sigma_y = 0.493x^{0.88}$	$\sigma_z = 0.087x^{1.10}$	$100 < x < 300$
		$\log_{10}\sigma_z = -1.67 + 0.902\log_{10}x + 0.181(\log_{10}x)^2$	$300 < x < 3000$
B	$\sigma_y = 0.337x^{0.88}$	$\sigma_z = 0.135x^{0.95}$	$100 < x < 500$
		$\log_{10}\sigma_z = -1.25 + 1.09\log_{10}x + 0.0018(\log_{10}x)^2$	$500 < x < 2 \times 10^4$
C	$\sigma_y = 0.195x^{0.90}$	$\sigma_z = 0.112x^{0.91}$	$100 < x < 10^5$
D	$\sigma_y = 0.128x^{0.90}$	$\sigma_z = 0.093x^{0.85}$	$100 < x < 500$
		$\log_{10}\sigma_z = -1.22 + 1.08\log_{10}x - 0.061(\log_{10}x)^2$	$500 < x < 10^5$
E	$\sigma_y = 0.091x^{0.91}$	$\sigma_z = 0.082x^{0.82}$	$100 < x < 500$
		$\log_{10}\sigma_z = -1.19 + 1.04\log_{10}x - 0.070(\log_{10}x)^2$	$500 < x < 10^5$
F	$\sigma_y = 0.067x^{0.90}$	$\sigma_z = 0.057x^{0.80}$	$100 < x < 500$
		$\log_{10}\sigma_z = -1.91 + 1.37\log_{10}x - 0.119(\log_{10}x)^2$	$500 < x < 10^5$

B Values of σ_y and σ_z for instantaneous sources given by Slade (1968)

Stability condition	*x = 100 m*		*x = 4000 m*		*Approximate equation*
	σ_y (m)	σ_z (m)	σ_y (m)	σ_z (m)	
Unstable	10	15	300	220	$\sigma_y = 0.14x^{0.92}$; $\sigma_z = 0.53x^{0.73}$
Neutral	4	3.8	120	50	$\sigma_y = 0.06x^{0.92}$; $\sigma_z = 0.15x^{0.70}$
Very stable	1.3	0.75	35	7	$\sigma_y = 0.024x^{0.89}$; $\sigma_z = 0.05x^{0.61}$

$$\chi(x,y,z) = \frac{Q}{\pi\sigma_y\sigma_z u} \times \exp\left[-\frac{1}{2}\left(\frac{y^2}{\sigma_y^2} + \frac{z^2}{\sigma_z^2}\right)\right] \quad [15.16.41]$$

The equation for a continuous infinite line crosswind source at ground level becomes

$$\chi(x,z) = \frac{2Q'}{(2\pi)^{\frac{1}{2}}\sigma_z u} \times \exp\left(-\frac{1}{2}\frac{z^2}{\sigma_z^2}\right)\Big] \quad [15.16.42]$$

The equation for a continuous elevated point source becomes

$$\chi(x,y,z) = \frac{Q}{2\pi\sigma_y\sigma_z u}\exp\left(-\frac{y^2}{2\sigma_y^2}\right) \times \left\{\exp\left[-\frac{(z-h)^2}{2\sigma_z^2}\right] + \exp\left[-\frac{(z+h)^2}{2\sigma_z^2}\right]\right\} \quad [15.16.43]$$

As mentioned above, the concentrations given by the equations for continuous sources are applicable only if the duration t of the release is equal to or greater than the ratio of the distance x to the wind speed u ($t \geq x/u$).

Equations 15.16.40–15.16.43 correspond to Equations 15.16.25 and 15.16.27–15.16.29, respectively. As stated above, Equations 15.16.40–15.16.43 are often referred to as the Pasquill–Gifford model. Pasquill (1962a, p. 190) himself states that an equation of the particular form of Equation 15.16.40a appears to have been first given by Frenkiel (1952), but that Equations 15.16.41 and 15.16.42 are implicit in the equations given by O.G. Sutton (1947). D.B. Turner (1970, p. 3) states that values of the dispersion coefficient σ have been obtained by Cramer and co-workers (Cramer, Record and Vaughan, 1958; Cramer, 1959a,b) and by Gifford (1961b), the latter's values for continuous releases being converted from Pasquill's values of angular spread and height for similar releases, which were described above. It is the latter which are recommended for use by D.B. Turner (1970, p. 3).

Again the dispersion coefficients σ_x, σ_y and σ_z are meteorological constants. Different values are used for Equations 15.16.40 for an instantaneous point source and for Equations 15.16.41–15.16.43 for a continuous source. The values of the dispersion coefficients σ_y and σ_z for a continuous point source have been discussed in detail by D.B. Turner (1970, p. 7). The values given by the latter, which are based on the work of Gifford (1961b), are shown in Figures 15.54(a) and *15.54*(b). These estimates are applicable only to open country. The curves in the graph are fitted by the approximate equations given in Table 15.31, Section A. The limits of the equations are as shown in the table.

The values given for the dispersion coefficients are best estimates (D.B. Turner, 1970, p. 7). In unstable and in stable conditions several-fold errors may occur in the estimate of σ_z. There are some circumstances, however, when the estimate of σ_z may be expected to be within a factor of 2. These are (1) for all stability conditions up to

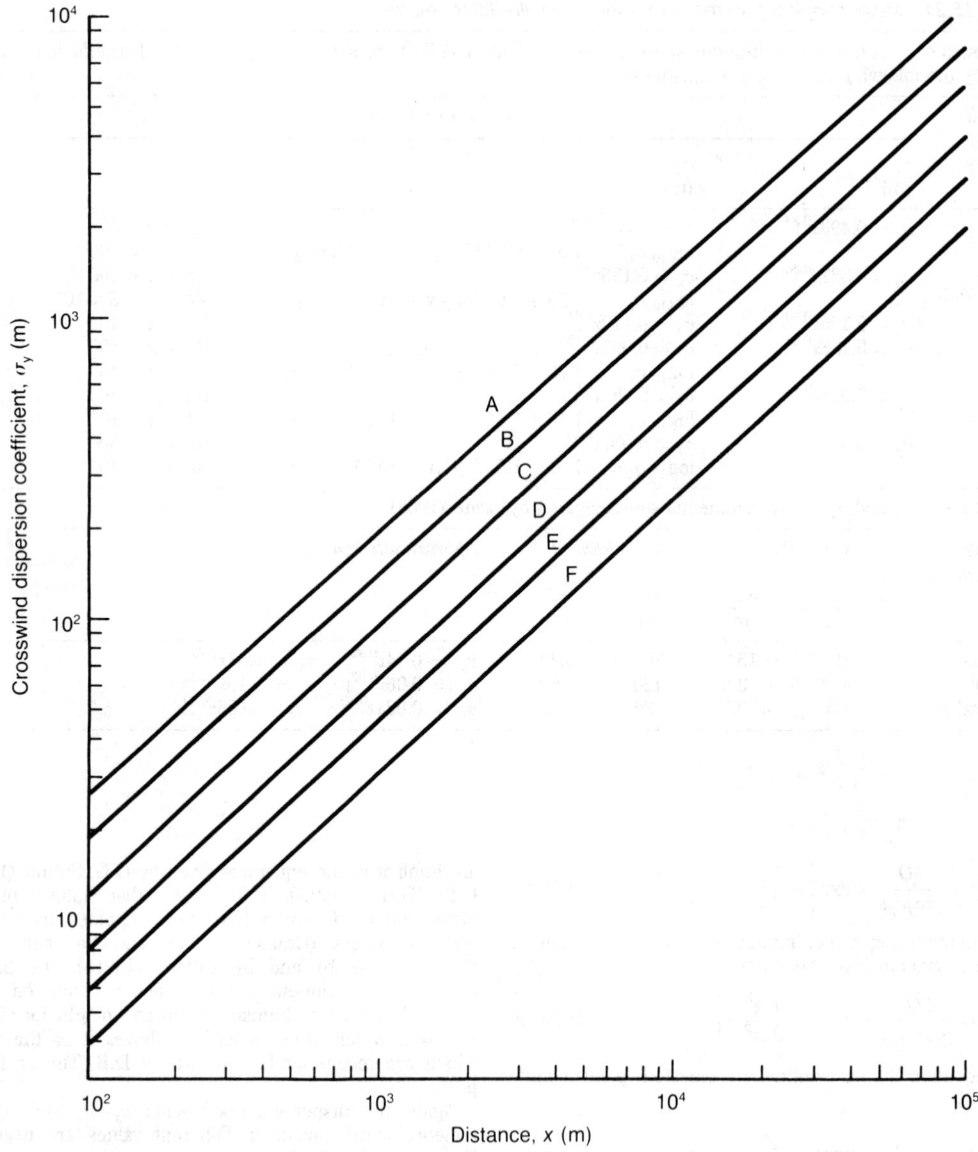

Figure 15.54(a) *Meteorological parameters for Pasquill–Gifford equations for continuous source: crosswind dispersion coefficient σ_y (D.B. Turner, 1970). A–F refer to Pasquill stability categories*

a few hundred metres, (2) for neutral or moderately stable conditions at distances up to a few kilometres and (3) for unstable conditions in the lower 1000 m of the atmosphere with a marked inversion above at distances up to 10 km or more. The uncertainty in the estimation of σ_y is, in general, less than that in the estimation of σ_z.

Equations 15.16.41 and 15.16.43 are applicable only for time periods up to about 10 min. For time periods exceeding 10 min the concentration downwind of the source is somewhat less, due to alteration of the wind direction. This effect can be taken into account for time periods of between 10 min and 2 h using the following approximate relation suggested by D.B. Turner (1970, p. 38):

$$\chi \propto t^{-0.17} \qquad [15.16.44]$$

The concentration from a continuous point source at ground level or at an elevation obtained from Equations

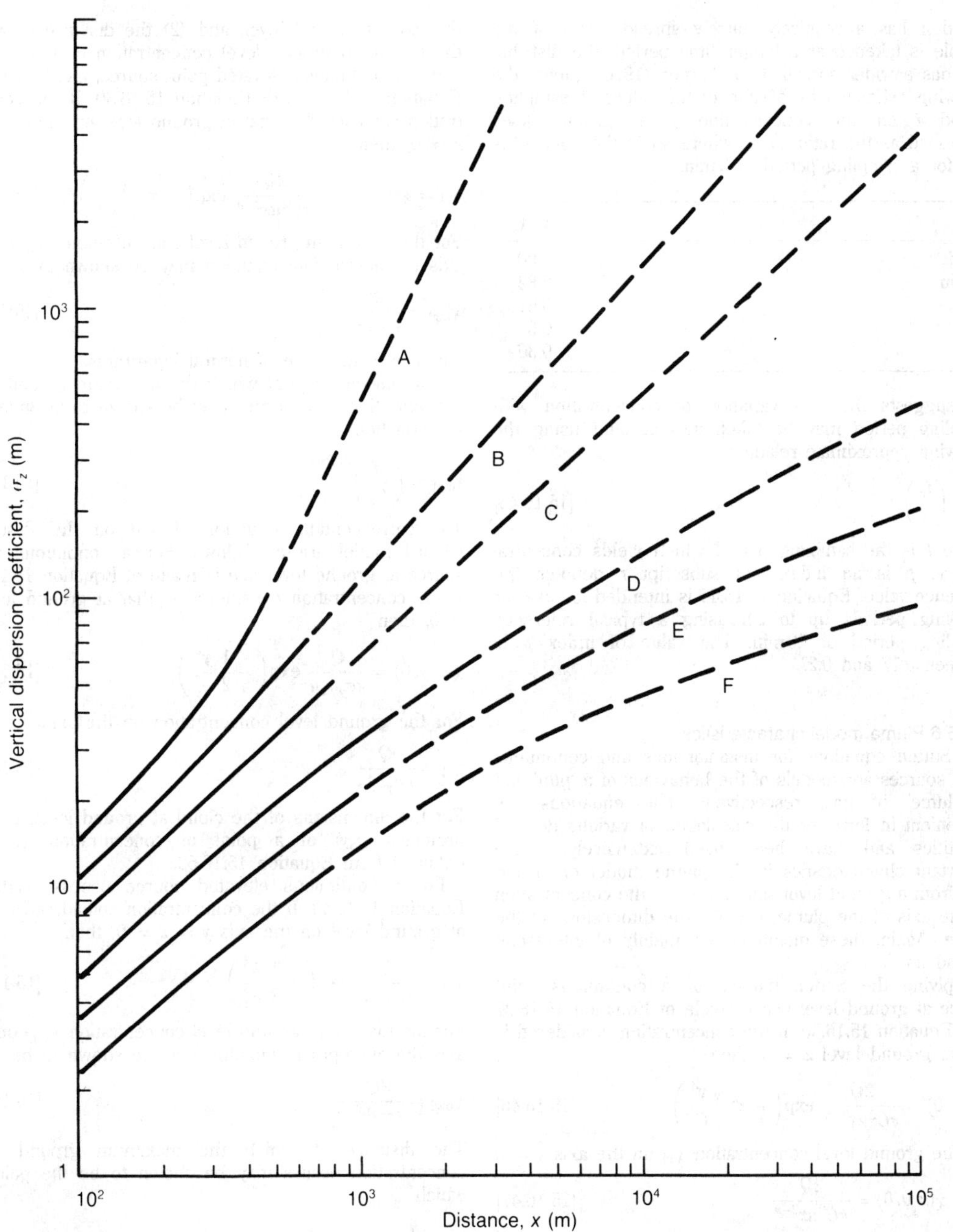

Figure 15.54(b) *Meteorological parameters for Pasquill–Gifford equations for continuous source: vertical dispersion coefficient σ_z (D.B. Turner, 1970). A–F refer to Pasquill stability categories*

15.16.41 and 15.16.43, respectively, is often plotted as $\chi u/Q$ vs distance x.

The values of the dispersion coefficients σ_x, σ_y and σ_z for an instantaneous point source, which differ from those for continuous sources, are discussed by Slade (1968). The values of σ_y and σ_z given by the latter are shown in Table 15.31, Section B. It is normally assumed that σ_x is equal to σ_y.

15.16.5 Effect of sampling period

As described above, if the concentration in a plume is sampled virtually instantaneously, the concentration distribution has a relatively narrow spread, while if the sample is taken over a longer time period, the distribution has a wider spread. D.B. Turner (1970) quotes the following estimates by Meade of the effect of sampling period t on the concentration χ at ground level expressed as the ratio χ/χ_3, where χ_3 is the concentration for a sampling period of 3 min.

t	χ/χ_3
3 min	1.00
15 min	0.82
1 h	0.61
3 h	0.51
24 h	0.36

He suggests that the variation of concentration with sampling period may be taken into account using the following approximate relation:

$$\frac{\chi}{\chi_r} = \left(\frac{t_r}{t}\right)^p \qquad [15.16.45]$$

where t is the sampling period which yields concentration χ, p is an index and subscript r denotes the reference value. Equation 15.16.45 is intended for use for sampling periods up to 2 h using a typical reference sampling period of 10 min. The value of index p is between 0.17 and 0.2.

15.16.6 Plume model characteristics

The Sutton equations for instantaneous and continuous point sources are models of the behaviour of a 'puff' and a 'plume' of gas, respectively. The equations are convenient in form for the calculation of various derived quantities and have been used extensively. Some important characteristics in the plume model of dispersion from a ground level source are (1) the concentration on the axis of the plume, and (2) the dimensions of the plume. Again, these quantities are mainly of interest at ground level.

Applying the Sutton model, for a continuous point source at ground level use is made of Equation 15.16.27 with Equation 15.16.30. If the concentration considered is that at ground level $z = 0$, then

$$\chi(x, y, 0) = \frac{2Q}{\pi C^2 u x^{2-n}} \exp\left(-x^{n-2}\frac{y^2}{C^2}\right) \qquad [15.16.46]$$

For the ground level concentration χ_{cl} on the axis $y = 0$

$$\chi_{cl} = \chi(x, 0, 0) = \frac{2Q}{\pi C^2 u x^{2-n}} \qquad [15.16.47]$$

For $n = 0.25$ and $C = 0.14$

$$\chi_{cl} = \frac{32Q}{u x^{1.75}} \qquad [15.16.48]$$

The cloud boundary concentration χ_{cb} may be defined as either one-tenth of the axial concentration χ_{cl} or as the extinction concentration χ_{ex}. For the dimensions of the cloud at ground level the co-ordinates (x,y) of a point of concentration χ_{cb} are obtained from Equation 15.16.46.

The plume model for dispersion from an elevated source is also relevant. In this case two important characteristics are (1) the maximum concentration on the axis at ground level, and (2) the distance at which the maximum ground level concentration occurs.

For a continuous elevated point source, use is made of Equation 15.16.29 with Equation 15.16.30. If the concentration considered is that at ground level on the axis $y = z = 0$, then

$$\chi_{cl} = \chi(x, 0, 0) = \frac{2Q}{\pi C^2 u x^{2-n}} \exp\left(-x^{n-2}\frac{H^2}{C^2}\right) \qquad [15.16.49]$$

For the maximum ground level concentration χ_{mgl} on the axis the appropriate equation may be shown to be

$$\chi_{mgl} = \frac{2Q}{e\pi u H^2}\frac{C_z}{C_y} \qquad [15.16.50]$$

where e is the base of natural logarithms.

The distance x_{mgl} at which the maximum ground level concentration χ_{mgl} occurs may be shown to be given by the equation

$$x_{mgl} = \left(\frac{H}{C_z}\right)^{2/(2-n)} \qquad [15.16.51]$$

The corresponding equations based on the Pasquill–Gifford model are as follows. For a continuous point source at ground level use is made of Equation 15.16.41. If the concentration considered is that at ground level $z = 0$, then

$$\chi(x, y, 0) = \frac{Q}{\pi\sigma_y\sigma_z u} \exp\left(-\frac{1}{2}\frac{y^2}{\sigma_y^2}\right) \qquad [15.16.52]$$

For the ground level concentration on the axis

$$\chi_{cl} = \frac{Q}{\pi\sigma_y\sigma_z u} \qquad [15.16.53]$$

For the dimensions of the cloud at ground level, the co-ordinates (x,y) of a point of concentration χ_{cb} are obtained from Equation 15.16.52.

For a continuous elevated source use is made of Equation 15.16.43. If the concentration considered is that at ground level on the axis $y = z = 0$, then

$$\chi_{cl} = \frac{Q}{\pi\sigma_y\sigma_z u} \exp\left(-\frac{1}{2}\frac{h^2}{\sigma_z^2}\right) \qquad [15.16.54]$$

For the maximum ground level concentration χ_{mgl} on the axis the appropriate equation may be shown to be

$$\chi_{mgl} = \frac{2Q}{e\pi u h^2}\frac{\sigma_z}{\sigma_y} \qquad [15.16.55]$$

The distance at which the maximum ground level concentration occurs may be shown to be the point at which

$$\sigma_z = \frac{h}{2^{\frac{1}{2}}} \qquad [15.16.56]$$

15.16.7 Puff model characteristics

Some important quantities in the puff model of dispersion from a ground level source are (1) the concentration at the centre of the cloud, (2) the dimensions of the cloud, (3) the maximum distance travelled and (4) the total integrated dosage given by the cloud. These quantities are mainly of interest at ground level.

Applying the Sutton model, for an instantaneous point source at ground level use is made of Equation 15.16.25

with Equation 15.16.30]. If the concentration considered is that at ground level $z = 0$, then

$$\chi(x,y,0,t) = \frac{2Q^*}{\pi^{\frac{3}{2}}C^3(ut)^{\frac{3}{2}(2-n)}} \times \exp\left[-(ut)^{n-2}\left(\frac{x^2}{C^2}+\frac{y^2}{C^2}\right)\right] \qquad [15.16.57]$$

For the ground level concentration χ_{cc} at the centre of the cloud $x = y = 0$

$$\chi_{cc} = \chi(0,0,0,t) = \frac{2Q^*}{\pi^{\frac{3}{2}}C^3(ut)^{\frac{3}{2}(2-n)}} \qquad [15.16.58]$$

For $n = 0.25$ and $C = 0.14$

$$\chi_{cc} = \frac{132Q^*}{(ut)^{2.62}} \qquad [15.16.59]$$

For the other quantities mentioned it is necessary to define the cloud. If the definition used is the convention that the concentration χ_{cb} at the cloud boundary is one-tenth of that at the centre, then

$$\frac{\chi_{cb}}{\chi_{cc}} = 0.1 \qquad [15.16.60]$$

Alternatively, the cloud boundary concentration may be defined as some extinction value χ_{ex} at which the concentration is no longer of interest:

$$\chi_{cb} = \chi_{ex} \qquad [15.16.61]$$

A typical extinction concentration is the lower flammability limit for a flammable gas.

For the dimensions of the cloud at ground level the cloud radius x is determined by calculating χ_{cc} from Equation 15.16.58, χ_{cb} from Equation 15.16.60 or 15.16.61 and x from Equation 15.16.57, setting $\chi = \chi_{cb}$.

For the maximum distance travelled by the cloud the group ut is calculated from Equation 15.16.58 with χ_{cc} set equal to χ_{ex}.

The corresponding equations based on the Pasquill–Gifford model are as follows. For an instantaneous point source at ground level use is made of Equation 15.16.40. If the concentration considered is that at ground level $z = 0$, then

$$\chi(x,y,0,t) = \frac{2Q^*}{(2\pi)^{\frac{3}{2}}\sigma_x\sigma_y\sigma_z} \times \exp\left[-\frac{1}{2}\left(\frac{x^2}{\sigma_x^2}+\frac{y^2}{\sigma_y^2}\right)\right] \qquad [15.16.62]$$

For the ground level concentration χ_{cc} at the centre of the cloud $x = y = 0$

$$\chi_{cc} = \frac{2Q^*}{(2\pi)^{\frac{3}{2}}\sigma_x\sigma_y\sigma_z} \qquad [15.16.63]$$

For the dimensions of the cloud at ground level the cloud radius x is determined by calculating χ_{cc} from Equation 15.16.63, χ_{cb} from Equation 15.16.60 or 15.16.61 and x from Equation 15.16.62, setting $\chi = \chi_{cb}$.

For the maximum distance travelled by the cloud use is made of Equation 15.16.63 with χ_{cc} set equal to χ_{ex}.

15.16.8 Dosage and dosement

The total integrated dosage D_{tid}, or more simply the dosage D, is

$$D = \int_0^\infty \chi \, dt \qquad [15.16.64]$$

In the Sutton model for a puff release the dosage at ground level on the axis may be obtained by integrating Equation 15.16.57 to obtain

$$D(x,0,0) = \frac{2Q^*}{\pi C^2 u(ut)^{2-n}} \qquad [15.16.65]$$

In the Pasquill–Gifford model for a puff release the dosage D at ground level and at ground level on the axis, the appropriate equations may be shown to be, respectively,

$$D(x,y,0) = \frac{Q^*}{\pi\sigma_y\sigma_z u}\exp\left(-\frac{1}{2}\frac{y^2}{\sigma_y^2}\right) \qquad [15.16.66a]$$

and

$$D(x,0,0) = \frac{Q^*}{\pi\sigma_y\sigma_z u} \qquad [15.16.66b]$$

For some materials it is necessary to work in terms not of the dosage D as defined by Equation 15.16.64 but of a load, or dosement, defined as

$$L = \int_0^\infty \chi^n \, dt \qquad [15.16.67]$$

where n is an index.

For the Pasquill–Gifford model it has been shown by Tsao and Perry (1979) that for a puff release the dosement at ground level is

$$L = \left[\frac{2Q^*}{(2\pi)^{\frac{3}{2}}\sigma_x\sigma_y\sigma_z}\frac{\pi^{\frac{1}{2}}}{(2n)^{\frac{1}{2}}}\right]^n \frac{\sigma_x}{u}\left[1+\text{erf}\left(\frac{n^{\frac{1}{2}}x}{2^{\frac{1}{2}}\sigma_x}\right)\right] \times \exp\left(-n\frac{y^2}{2\sigma_y^2}\right) \qquad [15.16.68]$$

For a plume release the load, or dosement, is obtained from the steady-state concentration and the period of exposure to that concentration.

15.16.9 Virtual sources

In the models just described it is assumed that the sources are point sources, instantaneous or continuous. If the source is of finite size, it may be necessary to allow for this using the virtual source method. An account of the method has been given by Mecklenburgh (1985).

The need to allow for finite source size may occur, for example, as the result of the instantaneous flashing off of vapour at the source to form a relatively large cloud. A change of surface roughness, and hence of the dispersion parameter, may also require the use of the virtual source method.

Mecklenburgh treats the determination of the ground level concentration at the centre of the cloud formed from an instantaneous release. For this he utilizes Equation 15.16.63 rewritten in the form

$$\sigma_x\sigma_y\sigma_z = \frac{2V}{(2\pi)^{\frac{3}{2}}C} \qquad [15.16.69]$$

where C is the volumetric concentration (v/v) and V is the volume of gas released (m^3), together with the following relationships for the dispersion parameters:

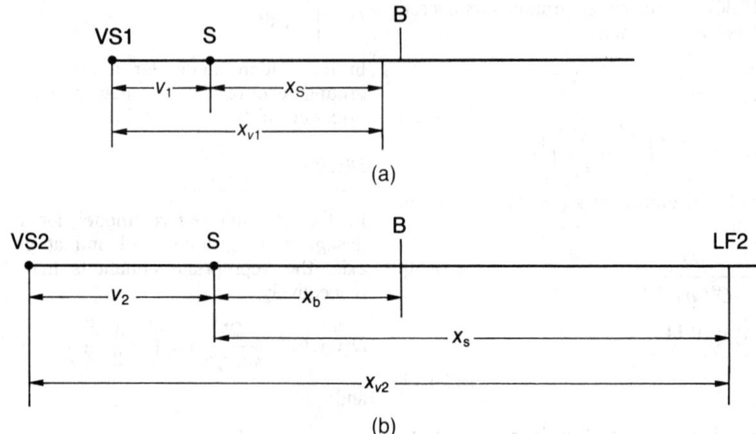

Figure 15.55 Virtual source method: (a) use of virtual source to allow for finite size of initial cloud; and (b) use of virtual source to allow for change in surface roughness

On site:

$$\sigma_x = 0.13x; \quad \sigma_y = 0.064x^{0.905}; \quad \sigma_z = 0.395x^{0.701};$$

$$(\sigma_y \sigma_z)^{\frac{1}{2}} = 0.159x^{0.803}; \quad (\sigma_x \sigma_y \sigma_z)^{\frac{1}{3}} = 0.149x^{0.869}$$

off site:

$$\sigma_x = 0.13x; \quad \sigma_y = 0.064x^{0.905}; \quad \sigma_z = 0.200x^{0.760};$$

$$(\sigma_y \sigma_z)^{\frac{1}{2}} = 0.113x^{0.833}; \quad (\sigma_x \sigma_y \sigma_z)^{\frac{1}{3}} = 0.118x^{0.888}$$

The equations are for Pasquill stability category D and a surface roughness of 0.1 on site and 1.0 (urban) off site. The method involves the repeated use of Equation 15.16.69, sometimes to determine the group $\sigma_x \sigma_y \sigma_z$ from C and sometimes vice versa.

Consider as an illustration the example given by Mecklenburgh, which is the determination of (1) the concentration of gas at the works boundary and (2) the distance to the lower flammability limit for a release of 8240 m^3 of gas from a point where the distance in the downwind direction between that point and the works boundary is 100 m.

The release gives a cloud of finite size. Equation 15.16.69 is strictly applicable to an instantaneous point source. In order to allow for the finite size of the cloud, the release is assumed to originate from a point source VS1 located at a virtual distance v_1 upwind of the actual release point. Then

$$x_{v1} = v_1 + x_s \qquad [15.16.70]$$

where v_1 is the distance between the virtual and actual sources (m), x_s is the distance between the actual source and the point of interest (m) and x_{v1} is the distance between the virtual source and the point of interest (m). These distances are shown in Figure 15.55(a). The concentration of the cloud at the actual source is taken as pure gas so that $C = 1$ v/v.

Then Equation 15.16.70 with $V = 8240$ m^3 and $C = 1$ v/v yields $\sigma_x \sigma_y \sigma_z = 1046$ m^3. The on-site equations for the σ values (with x in these relations equal to v_1) give $v_1 = 129$ m. The virtual source VS1 is thus 129 m upwind of the actual source.

For the concentration at the works boundary, the distance x_{v1} between the virtual source and the boundary is given by Equation 15.16.70 with $x_s = x_b$, so that

$$x_{v1} = v_1 + x_b \qquad [15.16.71]$$

where x_b is the distance between the actual source and the boundary (m). Using the on-site σ equations (with $x = x_{v1}$) yields $\sigma_x \sigma_y \sigma_z = 4696$ m^3. Then from Equation 15.16.69 at the boundary $C = 0.223$ v/v.

Across the boundary the surface roughness changes and it is necessary to utilize a second virtual source VS2, determining its distance upwind from conditions at the works boundary. Utilizing again $C = 0.223$ v/v, and hence $\sigma_x \sigma_y \sigma_z = 4696$ m^3, but this time using the off-site σ equations (with $x = x_{v2}$), gives $x_{v2} = 365$ m. At the boundary

$$x_{v2} = v_2 + x_b \qquad [15.16.72]$$

where v_2 is the distance between the virtual and actual sources (m). Then, from Equation 15.16.72, $v_2 = 385 - 100 = 265$ m.

For the distance to the lower flammability limit, the distances are shown in Figure 15.55(b). Setting $C = 0.0355$ v/v in Equation 15.16.69 yields $\sigma_x \sigma_y \sigma_z = 29\,475$ m^3. Using the off-site σ equations (with $x = x_{v2}$) gives $x_{v2} = 528$ m. But

$$x_{v2} = v_2 + (x_s - x_b) \qquad [15.16.73]$$

Hence $x_s = 528 - (265 - 100) = 363$ m.

15.16.10 K-theory models

An alternative approach to that given in the Sutton, Pasquill and Pasquill–Gifford models is the direct solution of the fundamental diffusion equation in which the dispersion is characterized by the parameter K. As mentioned above, early work on this, in which K was assumed to be a constant, such as that of Roberts, gave unsatisfactory results, but more modern models, which take into account the variation of K, have been more successful and find use alongside the models already described.

Accounts of such gradient transfer, or K-theory, models are given by S.R. Hanna, Briggs and Hosker (1982), by Runca (1982) and Gryning, van Ulden and Larsen (1983). These K-theory models may be solved analytically, if correlations of a suitable form are available for K. Alternatively, they may be solved numerically.

For analytical solutions the profiles of the wind velocity u and of the parameter K may be expressed as power functions of the height z:

$$u = u_r \left(\frac{z}{z_r} \right)^m \quad [15.16.74]$$

$$K = K_r \left(\frac{z}{z_r} \right)^n \quad [15.16.75]$$

where m and n are indices and r denotes the reference value.

A further discussion of values of the parameter K is given in Section 15.16. Various analytical solutions for K-theory models are given by S.R. Hanna, Briggs and Hosker (1982).

Numerical solution of K-theory models tends to involve severe problems both of numerical inaccuracy and numerical instability. The effect of the former is to give a spurious diffusion which arises solely from the numerical method of solution. The effect of the latter is to require very small step lengths and hence very long computation times.

15.16.11 Criterion for type of release

The dispersion situations which occur in practice often correspond not to the puff or the plume model but to the intermediate case. The usual method of dealing with this problem is to calculate dispersion using both models and to take the worst case.

The alternative approach is to select the model on the basis of a suitable criterion. A criterion for deciding whether to use the puff or the plume model, as given by the respective Pasquill–Gifford equations, has been suggested by Eisenberg, Lynch and Breeding (1975). The puff model is preferable if the diffusion in the downwind, or x, direction is large relative to the length of the plume. A measure of diffusion in the downwind direction is the dispersion coefficient σ_x, where this is evaluated using the puff data, and a measure of the length of the plume is ut_e, where u is the wind speed and t_e is the time taken for total discharge or evaporation of the gas. The coefficient σ_x is evaluated using the puff data at $ut_e/2$. The authors suggest the following criteria:

$ut_e < 2\sigma_x$	Use puff model
$ut_e > 5\sigma_x$	Use plume model
$2\sigma_x < ut_e < 5\sigma_x$	Neither model entirely appropriate

In addition, the plume model is undefined for zero wind speeds, is inaccurate for low wind speeds and is therefore not recommended for use at wind speeds of less than 2 m/s. Further criteria for model selection are given in relation to some of the model systems described below.

15.17 Passive Dispersion: Dispersion over Penticular Surfaces

15.17.1 Dispersion over urban areas

Accounts of dispersion over and in urban areas have been given by Gifford (1972, 1976a,b), S.R. Hanna (1976) and S.R. Hanna, Briggs and Hosker (1982). There are two rather different types of model which are used to describe the concentration of pollutant in an urban area. The first type is concerned with the overall concentration of the pollutant in the urban atmosphere as required for pollution studies. The second describes the dispersion and gives concentration profiles.

Models used for dispersion over and within an urban area include the Pasquill–Gifford model and K models. D.B. Turner (1964) carried out a study to validate the use of the Pasquill–Gifford model over and within a city, and various other workers have carried out studies to determine the parameters of this model in the urban situation.

Models used for pollution studies include rollback models, statistical correlations and box models. In a rollback model it is assumed that the concentration is proportional to the strength of the emission source(s). A rollback model therefore focuses attention on the level of emissions. Statistical correlations are used to relate concentrations to meteorological variables such as solar radiation, wind speed and temperature.

In a box model the urban area is treated as a box as shown in Figure 15.56. The basic equation is

$$\Delta x z_i \frac{d\chi}{dt} = \Delta x Q_a + u z_i(\chi_b - \chi) + \Delta x \frac{dz_i}{dt}(\chi_a - \chi) \quad [15.17.1]$$

where Q_a is the mass rate of release per unit area, Δx is an incremental length, z is the mixed layer height, χ is the concentration, χ_a is the concentration above the mixed layer height and χ_b is the concentration upwind of the city, or background concentration.

With zero background concentration and at steady state, Equation 15.17.1 becomes

$$\chi = \frac{\Delta x Q_a}{z_i u} \quad [15.17.2]$$

In some versions of the box model it is assumed that the pollutant is mixed not to the mixed layer height z_i but to some other height. A correlation for this height is given by S.R. Hanna (1976).

Dispersion coefficients for use in the Pasquill–Gifford model in urban areas are discussed in Section 15.19.

15.17.2 Dispersion over coastal areas and water

An account of the meteorology relevant to coast and sea is given in *Physics of the Marine Atmosphere* (Roll, 1965). The modelling of dispersion over coastal areas and water has been described by Gifford (1976a,b) and by Lyons (1976).

Dispersion over coastal areas and water is most often modelled using the Pasquill–Gifford method. It is pointed out by Lyons that this involves use of the model outside the range of conditions for which it was developed, which were flat rural terrain. Moreover, a further weakness is that the assumption of a steady state is less valid in the more variable wind conditions characteristic of coastal areas. Nevertheless, this is the model

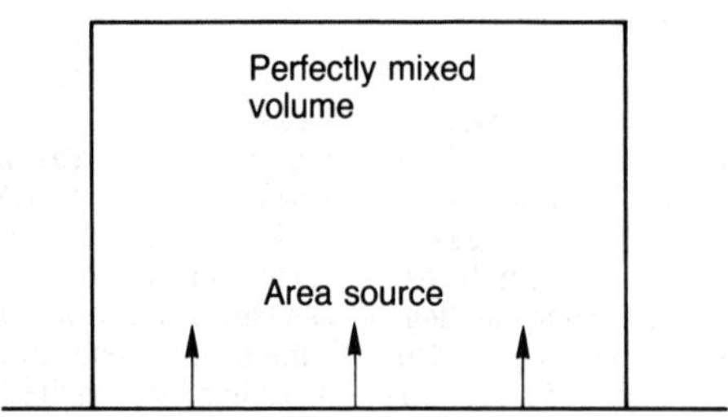

Figure 15.56 *Box model for dispersion over an urban area*

which is most widely used. Dispersion coefficients for this model are described in Section 15.19.

Several other models are described by Lyons. These include a model for conduction inversion and one for shoreline fumigation based on a modification of the nocturnal inversion breakup model of Turner.

15.17.3 Dispersion over complex terrain

A description of the modelling of dispersion over complex terrain has been given by Egan (1976).

A model which has been used for complex terrain is the model of Environmental Research and Technology Inc. (ERT) which is based on the Pasquill–Gifford equation for a release from an elevated continuous point source as given in Equation 15.16.43. The effective stack height h_o is taken as usual as the sum of the actual stack height h and the plume rise Δh. Then an effective plume height, essentially the height of the centre line of the plume, is defined as follows. For a receptor on a hill, the height h_1 of which is lower than the effective stack height h_o, the effective plume height h_p is taken as

$$h_p = h_o - h_1/2 \qquad h_1 < h_o \qquad [15.17.3a]$$

whilst for a receptor on a hill height h_2, which is higher than h_o, the expression used for h_p is

$$h_p = h_o/2 \qquad h_2 > h_o \qquad [15.17.3b]$$

The model has been used for unstable, neutral and stable conditions. For a hill higher than the effective stack height this model is the so-called 'half-height' model.

Flow in complex terrain may be modelled using potential flow theory. Standard cases in potential flow are treated by Lamb (1932) and Milne-Thomson (1938). Two cases relevant to complex terrain are the half circular cylindrical ridge and the hemispherical hill. Potential flow theory may be extended to include diffusion.

Thus the Pasquill–Gifford equation giving the ground level centre line concentration from an elevated point source may be modified to give

$$\chi = \frac{Q}{(\pi\sigma_y\sigma_z z_u)_f(D_yD_z)} \exp\left[-\frac{(\eta h_o)^2}{2\sigma_{zf}^2\zeta^2D_z^2}\right] \qquad [15.17.4]$$

with

$$D_y = \sigma_y/\sigma_{yf} \qquad [15.17.5a]$$

$$D_z = \sigma_z/\sigma_{zf} \qquad [15.17.5b]$$

$$\eta = (h_p - z)/h_o \qquad [15.17.6]$$

$$\zeta = (d\psi/dh_p)_o/(d\psi/dh_p) \qquad [15.17.7]$$

where h_p is the local height of the plume, z is the local height of the terrain, ψ is the stream function and subscripts f and o denote flat terrain and effective source height, respectively. ζ is the ratio of the average vertical gradient of the stream function at the effective source height to the gradient at the plume centre line over the surface.

For flat terrain, D_y, D_z, η and ζ are all unity. For two-dimensional flow, as over a ridge, it can be shown that near the surface η and ζ are approximately equal and, assuming that D_y and D_z are also equal to Equation 15.17.4 reduces to Equation 15.16.43. For three-dimensional flow, as over a hill, the parameters η and ζ have the values

$$\eta = h_o/3a \qquad [15.17.8]$$

$$\zeta = 2h_o/3a \qquad [15.17.9]$$

where a is the radius of the hill. Then, again assuming that D_y and D_z are unity, Equation 15.17.4 becomes

$$\chi = \frac{Q}{(\pi\sigma_y\sigma_z u)_f} \exp\left[-\frac{(h_o/2)^2}{2\sigma_{zf}^2}\right] \qquad [15.17.10]$$

Comparing Equation 15.17.10 with Equation 15.16.43 it can be seen that the effect of the hill is to give an equivalent source height h_e half the effective source height h_o. Thus Equation 15.17.10 is another form of the 'half-height' equation for complex terrain.

Another approach is physical modelling using wind and water tunnels. As described above, Egan discusses the requirements of similarity in such work. He gives several examples of tunnel studies for complex terrain.

15.17.4 Dispersion at buildings and obstacles

Accounts of models available for dispersion at buildings and obstacles are given by Halitsky (1968), Gifford (1976a,b), Vincent (1977), S.R. Hanna, Briggs and Hosker (1982) and Fackrell (1984a). A review of the aerodynamics of flow around buildings has been given by J.C.R. Hunt (1971).

Treatments of the effect of a building on release from an upwind source are mainly concerned with the case of an elevated source. A rule-of-thumb applicable to this situation is that the plume will rise over the building if the distance from the source to the building is greater than 2h, where h is the height of the building, and if the height h_s of the source exceeds two-thirds that of the building ($h_s > 2h/3$).

The overall effect of the building is to enhance the dispersion of the plume. D.J. Wilson and Netterville (1978) found that the main effect is to increase the mixing between the roof and ground levels. Around the building itself aerodynamic effects may give high local concentrations.

For groups of buildings or buildings of less regular shape, the flow patterns tend to be complex. It has been shown by Cagnetti (1975) for a reactor site that

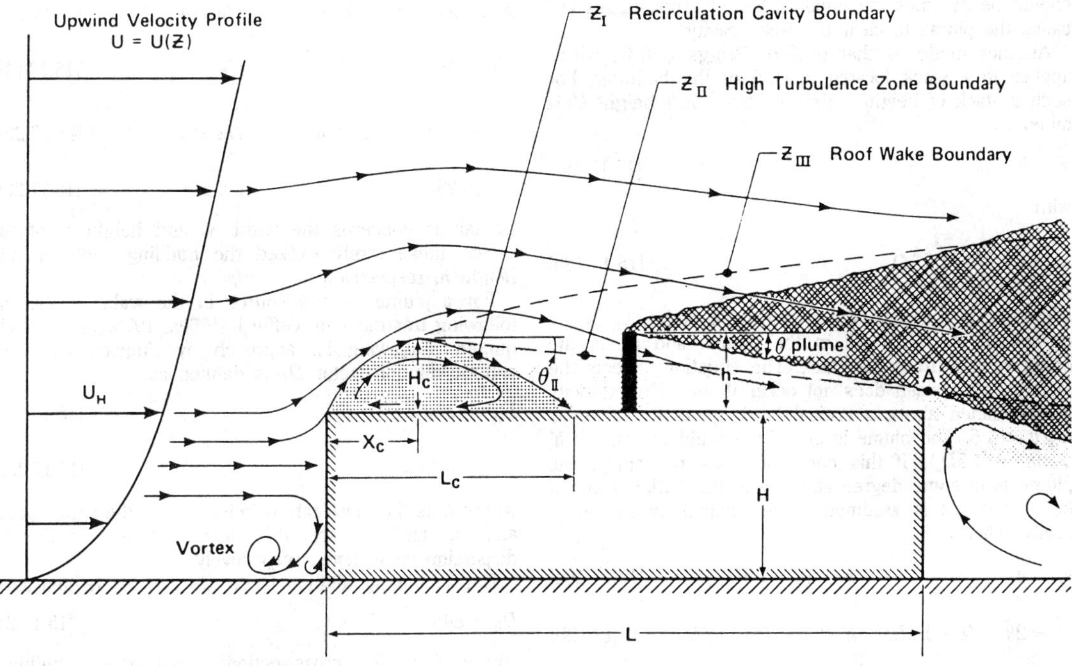

Figure 15.57 Flow pattern and flow zones around the roof of a square edged building for wind perpendicular to the upwind face (D.J. Wilson, 1979a) H, H_C, L, L_C in the figure are denoted in the text by h, h_C, l, l_C. (Courtesy of the American Society of Heating, Refrigerating and Air Conditioning Engineers)

dispersion may vary markedly for relatively small changes in wind direction. Dispersion from sources close to a building has been modelled more extensively, since this is relevant to the siting of vents and stacks.

Rules for the siting of roof stacks for the case of wind impinging perpendicularly on the face of a rectangular building, as shown in Figure 15.57, have been derived by D.J. Wilson (1979a). For a building with height h, crosswind width w and alongwind length l, he defines a scaling factor R such that

$$R = D_s^{0.67} D_l^{0.33} \qquad [15.17.11]$$

where D_s is the smaller dimension of the height h and the width w, and D_l is the larger of these two. He gives the following correlations for the roof cavity length l_c and height h_c:

$$l_c = 0.9R \qquad [15.17.12]$$

$$h_c = 0.22R \qquad [15.17.13]$$

The upper boundary Z_{II} of the high turbulence layer, which starts near the maximum roof cavity height, is given as

$$\frac{Z_{II}}{R} = 0.27 - 0.1\frac{x}{R} \qquad [15.17.14]$$

and that of the roof wake Z_{III} as

$$\frac{Z_{III}}{R} = 0.28\left(\frac{x}{R}\right)^{\frac{1}{3}} \qquad [15.17.15]$$

where x is the downwind distance.

Wilson suggests that for a roof-mounted stack there is little risk of high pollutant concentrations on the roof if plume clears the boundary of the high turbulence zone and virtually no risk if it clears the roof wake boundary.

Concentrations arising from vents on roofs have been studied in wind tunnel investigations as described by Halitsky (1963, 1968) and D.J. Wilson (1976). Results of such studies are plotted in terms of the dimensionless concentration K defined as

$$K = \frac{\chi A u}{Q} \qquad [15.17.16]$$

where A is a characteristic area, Q is the mass rate of release and u is the wind speed. A typical characteristic area is wh.

Wilson gives the following correlation for the maximum value of the concentration $\chi(x)$ from a roof vent:

$$\frac{\chi(x)}{\chi(0)} = 9.1\frac{w_o A_v}{ux^2} \qquad [15.17.17]$$

where A_v is the vent area, u is the wind speed, w_o is the effluent velocity, x is the downwind distance, and $\chi(0)$ is the effluent concentration.

For a roof stack a widely used rule-of-thumb is the '$2\frac{1}{2}$' rule that to avoid downwash problems the stack height should be $2\frac{1}{2}$ times the building height. This height will cause the plume to clear the wake cavity.

Another model is that of G.A. Briggs (1974), which applies to a stack located upwind of the building. For such a stack of height h the effective stack height h' is taken as

$$h' = h - h_d \qquad [15.17.18a]$$

with

$$h_d = 2\left(\frac{w_o}{u} - 1.5\right)D_s \qquad [15.17.18b]$$

where D_s is the diameter of the stack outlet, h_d is the downwash, or initial fall in the plume, and w_o is the efflux velocity from the stack. The equation reflects the fact that downwash does not occur if the efflux velocity is maintained above a critical level, generally taken as $w_o/u = 1.5$. The plume is out of the building wake if $h' > (h + 1.5D_s)$. If this condition does not apply, the plume is to some degree affected by the building and if $h_o < 0.5D_s$, it is assumed to be trapped by the wake cavity. Then

$$h_o = h' \qquad h' > (h + 1.5D_s) \qquad [15.17.19a]$$

$$= 2h' - (h + 1.5D_s) \qquad h' < h' < (h + 1.5D_s) \qquad [15.17.19b]$$

$$= h' - 1.5D_s \qquad h' < h \qquad [15.17.19c]$$

If the plume is trapped it may be treated as a ground level source with initial area D_s^2.

If a plume is entrained into the wake cavity ($h_o < 0.5D_s$), the dimensionless concentration tends to lie in the range 0.2–2. Given that the plume is entrained into the wake cavity, the precise height of the stack makes little difference to the ground concentrations. There is, however, one particular case where stack height can have a marked effect. If the stack height is about the same height as the wake cavity, the effluent may be carried directly to the ground along the boundary of the cavity.

There have been a number of treatments of the wake cavity in the lee of the building. These include studies of the characteristics of the wake cavity itself and of the concentrations resulting from a release within the cavity. The wake cavity approximates a perfectly mixed zone with the concentration being nearly the same throughout.

A review of models for building-influenced dispersion has been given by Fackrell (1984a). These include: the model of Gifford (1960a), described below; the models of D.B. Turner (1970) and Barker (1982), which involve the use of a virtual source; and those of Ferrara and Cagnetti (1980) and Huber and Snyder (1976), which involve modification of the dispersion coefficients. Both the models that use a virtual source and those that use modified dispersion coefficients correlate the parameters in terms of the building width w and height h.

For the wake cavity itself, Hosker (1979, 1980) has developed the following correlations for the length x_r:

$$\frac{x_r}{h} = \frac{A(w/h)}{1 + B(w/h)} \qquad [15.17.20]$$

with

$$A = -2.0 + 3.7(l/h)^{-\frac{1}{3}} \qquad l/h < 1 \qquad [15.17.21a]$$

$$= 1.75 \qquad l/h > 1 \qquad [15.17.21b]$$

$$B = -0.15 + 0.305(l/h)^{-\frac{1}{3}} \qquad l/h < 1 \qquad [15.17.22a]$$

$$= 0.25 \qquad l/h > 1 \qquad [15.17.22b]$$

As far as concerns the width y_r and height z_r of the cavity, these rarely exceed the building width w and height h, respectively.

For a plume from a source in the wake cavity, the following treatment by Gifford (1960a, 1976a,b) is widely quoted. Following an approach by Fuquay, an atmospheric dilution factor D_A is defined as

$$D_A = Q/\chi \qquad [15.17.23a]$$

$$= \pi\sigma_y\sigma_z u \qquad [15.17.23b]$$

where Q is the mass rate of release, u is the wind speed, and σ_y and σ_z are the downwind and crosswind dispersion coefficients, respectively.

But a building dilution factor D_B can be defined as

$$D_B = cAu \qquad [15.17.24]$$

where A is the cross-sectional area of the building normal to the wind (= wh) and c is a constant. The total dilution factor D_{tot} is

$$D_{tot} = D_A + D_B \qquad [15.17.25a]$$

$$= (\pi\sigma_y\sigma_z + cA)u \qquad [15.17.25b]$$

$$= Q/\chi \qquad [15.17.25c]$$

Hence

$$\chi = \frac{Q}{(\pi\sigma_y\sigma_z + cA)u} \qquad [15.17.26]$$

Gifford (1960a) has estimated that the value of c lies in the range 0.5–2. He states (Gifford, 1976b) that this estimate, though widely quoted, was purely intuitive and that there are other studies which support a value of about 0.5.

There is evidence that the dispersion coefficients σ_y and σ_z applicable in Equation 15.17.26 differ somewhat from the regular values. The latter tend to give downwind concentrations which vary with distance to a power lying between −1.3 and −1.6, depending on the stability condition. Laboratory studies by Meroney and Yang (1971) and Huber and Snyder (1976) have found this power to be close to −0.8 to a distance of $50h$, although field studies such as those by Dickson, Start and Markee (1969) have given the power as about −1.3.

Gifford (1976b) draws attention to other conditions in which dispersion deviates from that predicted by Equation 15.17.26.

Barry (1964) has reviewed published expressions for the concentration given by pollutants in the lee of a rectangular block of square cross-section (alongwind length = height = S) and found them to be of the general form

$$\bar{C} = k\left(\frac{Q}{S^2 u}\right) \quad [15.17.27]$$

where C^- is the mean concentration, Q is the source strength, u is the main stream air velocity and k is a constant. The value of the constant k is in the range 0.5–20.

Vincent (1977) has described wind tunnel work on the dispersion of pollutant near buildings, and particularly in building wakes. He uses dimensional analysis to derive the principal parameters governing flow around a rectangular block. These are X/S and

$$H = \frac{ut_d}{S} \quad [15.17.28]$$

$$\Lambda = \frac{l_f k_f^{\frac{1}{2}}}{Su} \quad [15.17.29]$$

where k_f is the characteristic energy of free stream turbulence, l_f is the characteristic length scale of free stream turbulence, t_d is the mean residence time of the scalar entities in the wake bubble, X is the length of the wake 'bubble' and Λ is a free stream turbulence parameter. Vincent gives results of wind tunnel work correlating H and X/S against Λ.

For a pollutant source wholly contained within the bubble he derives the relation

$$\bar{C} \approx \frac{QSH}{Vu} \quad [15.17.30]$$

But

$$V \approx \left(\frac{X}{S}\right)S^3 \quad [15.17.31]$$

where V is the volume of the wake bubble. Then from Equations 15.17.30 and 15.17.31

$$\bar{C} \approx \left(\frac{Q}{S^2 u}\right)\left(\frac{H}{X/S}\right) \quad [15.17.32]$$

Vincent points out that Equation 15.17.32 gives for k in Equation 15.17.27 the explicit expression $H/(X/S)$.

Fackrell (1984b) has derived a model for a building wake and for the concentration in this wake and has used wind tunnel experiments to obtain values of the parameters involved. For flow out of the wake

$$F = \alpha S u_h C_w \quad [15.17.33]$$

where C_w is the average concentration in the near-wake region, F is the flow out of the wake, S is the surface area of the near-wake region, u_h is the wind speed at the building height in undisturbed flow, and α is a flux constant.

If there is a source of strength Q in the wake

$$F = Q \quad [15.17.34]$$

so that from Equations 15.17.33 and 15.17.34

$$Q = \alpha S u_h C_w \quad [15.17.35]$$

The residence time in the wake is

$$t_r = \frac{V}{\alpha S u_h} \quad [15.17.36]$$

where t_r is the residence time in the near-wake region and V is the volume of the near-wake region.

Fackrell defines the following dimensionless quantities for the recirculation, or near-wake, region:

$$\beta = \frac{AL_r}{V} \quad [15.17.37]$$

$$\lambda_r = L_r/h \quad [15.17.38]$$

$$\tau_r = \frac{u_h t_r}{h} \quad [15.17.39]$$

$$\chi_w = \frac{C_w u_h A}{Q} \quad [15.17.40]$$

where A is the frontal area of the building, h is the height of the building, L_r is the length of the recirculation region, β is the wake shape parameter, λ_r is the dimensionless length of the recirculation region, τ_r is the dimensionless residence time in the recirculation region and χ_w is the dimensionless average concentration in the near-wake region.

Then from Equations 15.17.35–15.17.40

$$\chi_w = \frac{A}{\alpha S} \quad [15.17.41]$$

$$= \beta\frac{\tau_r}{\lambda_r} \quad [15.17.42]$$

Fackrell reviews the correlations available for λ_r and τ_r. He modifies Equation 15.17.20 of Hosker to give

$$\lambda_r = \frac{1.8\frac{b}{h}}{\left[\left(\frac{l}{h}\right)^{0.3}\right]\left[1+0.24\frac{b}{h}\right]} \qquad 0.3 \le \frac{l}{h} \le 3 \quad [15.17.43]$$

where b is the crosswind breadth of the building.

He also obtains

$$\tau_r = \frac{11\left[\frac{b}{h}\right]^{1.5}}{\left[1+0.6\left(\frac{b}{h}\right)^{1.5}\right]} \quad [15.17.44]$$

15.18 Passive Dispersion: Dispersion in Particular Conditions

The other type of special situation is determined not by the terrain but by the meteorological conditions. Two such situations are considered here:

(1) dispersion in calm conditions;
(2) dispersion in a uniformly mixed layer.

15.18.1 Dispersion in calm conditions

For a small proportion of the time, calm, near windless, conditions prevail. These occur typically on clear nights with frost or heavy dew.

Conditions which are usually classified as Pasquill category F apply in the UK for some 20% of the time. Included in these are the periods of calm. Beattie (1963 UKAEA AHSB(S) R64) found that the proportion of time for which such conditions apply is about 5–8%. Pasquill actually excluded such conditions from his original categories because dispersion then tends to be irregular. Beattie has classified such conditions as category G, but does not give the associated dispersion coefficients.

It is natural to assume that dispersion is less in G than in F conditions. This assumption is made, for example, in NRC *Regulatory Guide* 1.21. However, there is evidence to show that under G conditions dispersion is ill-defined. Work on such conditions by Sagendorf (1975) has shown that a plume is subject to considerable meander, while work by van der Hoven (1976) has indicated values of dispersion coefficients which correspond to those found across the whole range from A to F conditions.

15.18.2 Dispersion in a uniformly mixed layer

Another meteorological condition which requires special treatment is that of limited mixing fumigation, where there is an elevated inversion which acts as a 'lid'. There is then a well mixed layer of height H in which, beyond a certain distance, the concentration is uniform.

For this condition Gifford (1976a) gives the relation

$$\chi = \frac{Q}{(2\pi)^{\frac{1}{2}}\sigma_y u H} \qquad [15.18.1]$$

where, in this case, H is the height of the 'lid', Q is the mass rate of release, u is the wind speed, σ_y is the crosswind dispersion coefficient and χ is the concentration.

15.19 Passive Dispersion: Dispersion Parameters

The dispersion models described contain parameters which are functions of the meteorological variables and of the distance from the source. The Sutton model contains the diffusion parameter C, the Pasquill model the parameters θ and h, and the Pasquill–Gifford models the dispersion coefficient σ. The values of, and correlations for, these parameters proposed by the authors of the models were given in the overall account of these models in Section 15.16. There is, in addition, a considerable literature on these parameters, especially the Pasquill–Gifford dispersion coefficients, ranging from fundamental relations to empirical correlations, which is now described.

In the estimation of dispersion parameters a crucial aspect is the definition of the stability condition. It is argued by Atwater and Londergan (1985) that the representation of stability is of more importance than the correlation of dispersion coefficients with stability. In some cases there are differences of two or more stability classes between the class based on the meteorology and that inferred from the observed dispersion coefficients and between that inferred from the horizontal dispersion coefficient and that inferred from the vertical dispersion coefficient. The discussion of dispersion parameters is confined initially to those for a plume. Those for a puff are considered in Section 15.19.12.

15.19.1 Fundamental relations

The relations between the turbulent exchange coefficient K, the Sutton diffusion parameter C, the Pasquill parameters θ and h, and the Pasquill–Gifford dispersion coefficient σ may be summarized as follows.

As described above, the condition for the reduction of the Sutton model to the K-model is

$$n = 1 \qquad [15.19.1]$$

$$C_x^2 = \frac{4K_x}{u} \qquad [15.19.2]$$

Similar equations apply relating C_y to K_y and C_z to K_z.

The relation between the Pasquill–Gifford dispersion coefficients and the Sutton diffusion parameters is

$$\sigma_x = \tfrac{1}{2}C_x^2(ut)^{2-n} \qquad [15.19.3]$$

Similar equations apply relating σ_y to C_y and σ_z to C_z.

The relation between the Pasquill–Gifford dispersion coefficients and the turbulent exchange coefficients is

$$\sigma_x^2 = 2K_x t \qquad [15.19.4]$$

Similar equations apply relating σ_y to K_y and σ_z to K_z.

The Pasquill parameters are related to the Pasquill–Gifford dispersion coefficients as follows:

$$\sigma_y = \sigma_\theta x \qquad [15.19.5]$$

with

$$\sigma_\theta = \frac{\theta}{4.3} \qquad [15.19.6]$$

and

$$\sigma_z = \frac{h}{2.15} \qquad [15.19.7]$$

The constants 4.3 and 2.15 in Equations 15.19.6 and 15.19.7 are the values derived from the Gaussian distribution corresponding to the cloud boundary as defined by convention.

The physical meaning of the dispersion coefficient may be interpreted as follows. For diffusion through a vertical plane in the x direction, the dispersion variance σ_x^2 is the second moment of the concentration with respect to distance x, normalized with respect to concentration:

$$\sigma_x^2 = \frac{\int_{-\infty}^{\infty} \chi x^2 \, dx}{\int_{-\infty}^{\infty} \chi \, dx} \qquad [15.19.8]$$

Similar expressions apply to σ_y and σ_z.

Following Csanady (1973), consider the one-dimensional diffusion of a mass Q_{ps}^* of material (kg/m^2) initially concentrated in a thin sheet at $x = 0$, in an instantaneous plane source. Then the diffusion is given by the relationship

$$\frac{\partial \chi}{\partial t} = K_x \frac{\partial^2 \chi}{\partial x^2} \qquad [15.19.9a]$$

The mass released is

$$Q_{ps}^* = \int_{-\infty}^{\infty} \chi \, dx \qquad [15.19.9b]$$

and the solution is

$$\chi = \frac{Q_{ps}^*}{2(\pi K_x t)^{\frac{1}{2}}} \exp\left(\frac{-x^2}{4K_x t}\right) \qquad [15.19.9c]$$

Furthermore, it may be shown that

$$Q_{ps}^* = \frac{1}{2K_x t}\int_{-\infty}^{\infty} \chi x^2 \, dx \qquad [15.19.10]$$

Hence from Equations 15.19.8, 15.19.9b and 15.19.10

$$\sigma_x = (2K_x t)^{\frac{1}{2}} \quad [15.19.11]$$

Moving on, if the exponential form of the autocorrelation coefficient $R(\tau)$ given in Equation 15.12.101 is used in Taylor's theorem as given in Equation 15.12.107, the solution obtained is

$$\sigma_x^2 = 2\sigma_u^2 t_L^2\left[\frac{t}{t_L} - 1 + \exp\left(-\frac{t}{t_L}\right)\right] \quad [15.19.12]$$

where t_L is the Lagrangian time scale.

Then for a short time t ($t \rightarrow 0$)

$$\sigma_x \propto t \quad [15.19.13a]$$

while for a long time ($t \rightarrow \infty$)

$$\sigma_x \propto t^{\frac{1}{2}} \quad [15.19.13b]$$

Since the distance x is proportional to the time t ($x = ut$), at short times

$$\sigma_x \propto x \quad [15.19.14]$$

Similar relations apply for σ_y and σ_z.

The dispersion coefficients may be related to the standard deviations of the wind velocity (σ_v, σw) as follows:

$$\sigma_y = \sigma_v t f_1(t/T_y) \quad [15.19.15a]$$

$$\sigma_z = \sigma_w t f_2(t/T_z) \quad [15.19.15b]$$

where f_1 and f_2 are universal functions and T_y and T_z are the time scales (s) for lateral and vertical dispersion, respectively. But

$$\sigma_v t \approx \sigma_\theta x \quad [15.19.16a]$$

$$\sigma_w t \approx \sigma_\phi x \quad [15.19.16b]$$

Hence from Equations 15.19.15 and 15.19.16

$$\sigma_y = \sigma_\theta x f_1(t/T_y) \quad [15.19.17a]$$

$$\sigma_z = \sigma_\phi x f_2(t/T_z) \quad [15.19.17b]$$

The universal functions f_1 and f_2 may be derived from theoretical considerations or expressed as empirical correlations. Pasquill (1976a) has tabulated values of f_1, which is taken as independent of stability and source height and as a function of distance only. Irwin (1979a) has correlated these values as follows:

$$f_1 = (1 + 0.031x^{0.46})^{-1} \qquad x \leq 10^4 \quad [15.19.18a]$$

$$= 33x^{-\frac{1}{2}} \qquad x > 10^4 \quad [15.19.18b]$$

Correlations for both f_1 and f_2 have been given by Draxler (1976). He obtains for f_1 the relation

$$f_1 = \frac{1}{[1 + 0.9(t/T_i)]^{\frac{1}{2}}} \quad [15.19.19]$$

where T_i is the dispersion time. He states that the function f_2 may be assumed equal to f_1 except for a ground level source in unstable conditions and an elevated source in stable conditions. For the former case it was difficult to obtain a correlation, but for a limited set of data

$$f_2 = \frac{0.3(t/T_i - 0.4)^2}{0.16} + 0.7 \quad [15.19.20a]$$

For the latter case he obtained

$$f_2 = \frac{1}{1 + 0.945(t/T_i)^{0.806}} \quad [15.19.20b]$$

The dispersion time T_i is the time for f_1, or f_2, to become equal to 0.5. Draxler gives the following values for T_i (in seconds)

Source height	*Horizontal dispersion*		*Vertical dispersion*	
	Stable	*Unstable*	*Stable*	*Unstable*
Ground level	300	300	50	100
Elevated	1000	1000	100	500

A further discussion of these universal functions is given by Gryning *et al.* (1987).

15.19.2 Sutton diffusion parameters

Sutton's original values for the diffusion parameters C_y and C_z have been given in Section 15.16. Values have also been given by other workers as shown in Table 15.32.

15.19.3 Pasquill cloud parameters

Pasquill's original values for the cloud parameters θ and h have been given in Section 15.16.

For a long release where wind direction traces are not available, Beattie (1963 UKAEA AHSB(S) R64) has suggested the use of the following values for θ:

$d = 0.1$ km, $\quad \theta = 30°$

$d = 100$ km, $\quad \theta = 15°$

These values of θ are based on neutral weather conditions with Pasquill stability category D and relatively invariant wind direction, but are also applicable to stable weather conditions with Pasquill stability category F if the wind direction is variable. The values quoted are three times greater than those given by Pasquill for a short release in stability category F conditions, as shown in Table 15.29. The difference may be regarded as allowing for the variability of wind direction which tends to apply to this type of weather. If, however, the wind direction is invariant, then the lower values given in Table 15.29 for stability category F should be used.

15.19.4 Pasquill–Gifford dispersion coefficients

The dispersion coefficients most often used with the Pasquill–Gifford model are probably those of D.B. Turner (1970), as given in Section 15.16. There are, however, several other sets.

Correlations of the dispersion coefficients in terms of the Brookhaven National Laboratory (BNL) stability categories have been given by Singer and Smith (1966) as shown in Table 15.33 and in Figure 15.58.

Correlations have also been given by Carpenter *et al.* (1971) of the Tennessee Valley Authority (TVA) in terms of temperature gradient categories as shown in Figure 15.59.

15.19.5 Briggs' dispersion coefficient scheme

It can be seen from comparison of Figure 15.54 and Figures 15.58 and *15.59* that the correlations for the

Table 15.32 *Sutton diffusion parameters*

A Values given by N.G. Stewart, Gale and Crooks (1954)
Source height: 68 m

Downwind distance (m)	*Stability condition*	C_y	C_z
150–1000	Unstable		0.27
	Neutral		0.22
	Stable		0.12
590–620	Unstable, neutral	0.46	
880–1050	Unstable, neutral	0.28	
1200	Unstable, neutral	0.33	
2400–2800	Unstable, neutral	0.26	
6000–9700	Unstable, neutral	0.18	

B Values of Brookhaven National Laboratory (after Slade, 1968)[a]
Source height: Elevated and ground level sources

Wind gustiness category	n	C_y	C_z
B2	0.15	0.48	0.50
B1	0.26	0.44	0.38
C	0.48	0.54	0.32
D	0.57	0.41	< 0.08

C Values of Barad and Fuquay (1962)

Stability condition	*Wind speed* (m/s)	n	C_y	C_z
Source height: elevated				
Unstable	1	0.20	0.30	0.30
	5		0.26	0.26
	10		0.24	0.24
Neutral	1	0.25	0.15	0.15
	5		0.12	0.12
	10		0.11	0.11
Source height: ground level				
Unstable	1	0.20	0.35	0.35
	5		0.30	0.30
	10		0.28	0.28
Neutral	1	0.25	0.21	0.17
	5		0.15	0.14
	10		0.14	0.13

D Values of V.J. Clancey (1974a)

Stability condition	n	C
Lapse	0.17	0.2
Neutral	0.25	0.14
Inversion	0.35	0.09

E Values of Long (1963)

Stability condition	n	C_y ($m^{n/2}$)	C_z ($m^{n/2}$)
Large lapse	1/5	0.37	0.21
Neutral	1/4	0.21	0.12
Moderate inversion	1/3	0.13	0.08
Large inversion	1/2	0.11	0.06

[a] A slightly different set of values are given by Gifford (1976b).

Table 15.33 Pasquill–Gifford dispersion coefficients: BNL formulae (Singer and Smith, 1966) (Courtesy of International Journal of Air and Water Pollution)

BNL gustiness class	Dispersion coefficient (m)	
	σ_y	σ_z
B2	$0.40x^{0.91}$	$0.41x^{0.91}$
B1	$0.36x^{0.86}$	$0.33x^{0.86}$
C	$0.32x^{0.78}$	$0.22x^{0.78}$
D	$0.31x^{0.71}$	$0.06x^{0.71}$

dispersion coefficients differ. An attempt to resolve these differences has been made by G.A. Briggs (1974). His work has been summarized by Gifford (1976b).

The Pasquill–Gifford curves are based on experiments on dispersion of non-buoyant gas over flat ground up to a downwind distance of 800 m. The BNL curves are for dispersion of a non-buoyant plume from an elevated stack (108 m) out to a distance of several kilometres with few measurements taken within 800 m. The TVA curves are for dispersion of a buoyant plume with stack heights of 75–250 m, and with effective stack heights at least twice these values, and to a distance of up to some tens of kilometres. The dispersion of this latter type of plume is a function more of buoyancy and entrainment effects than of atmospheric turbulence.

Briggs therefore proposed a set of interpolation formulae with properties conformable to these three sets of results. They would agree with the Pasquill–Gifford curves for downwind distances over the range 100–10 000 m except that for A and B stability classes and for $\sigma_z > 100$, the σ_z curves would be based on the values given by the American Society of Mechanical Engineers (ASME) (1968/1), which reflect mainly BNL data. The BNL and TVA curves agree reasonably well with each other, except at short distances where the latter are influenced by buoyancy; they also agree reasonably well with the Pasquill–Gifford curves at about 10 km, except for the A and B conditions, as just described.

Briggs' formulae are shown in Table 15.34 and curves given by Hosker (1974b) based on these are shown in Figure 15.60. The formulae are applicable up to 10 km.

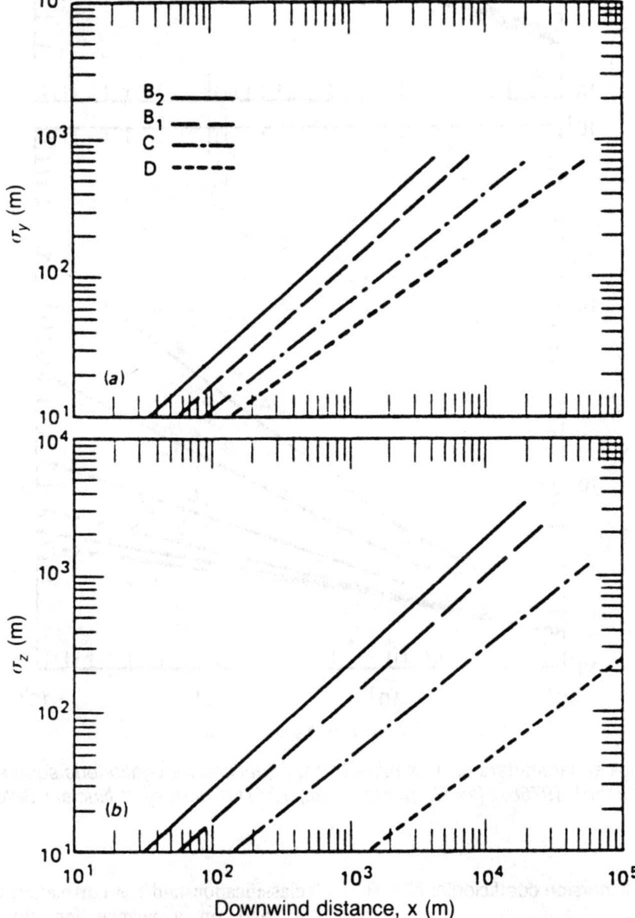

***Figure 15.58** Meteorological parameters for Pasquill–Gifford equations for continuous source: dispersion coefficients based on BNL scheme (Gifford, 1976b; after Singer and Smith, 1966) (Courtesy of Nuclear Safety)*

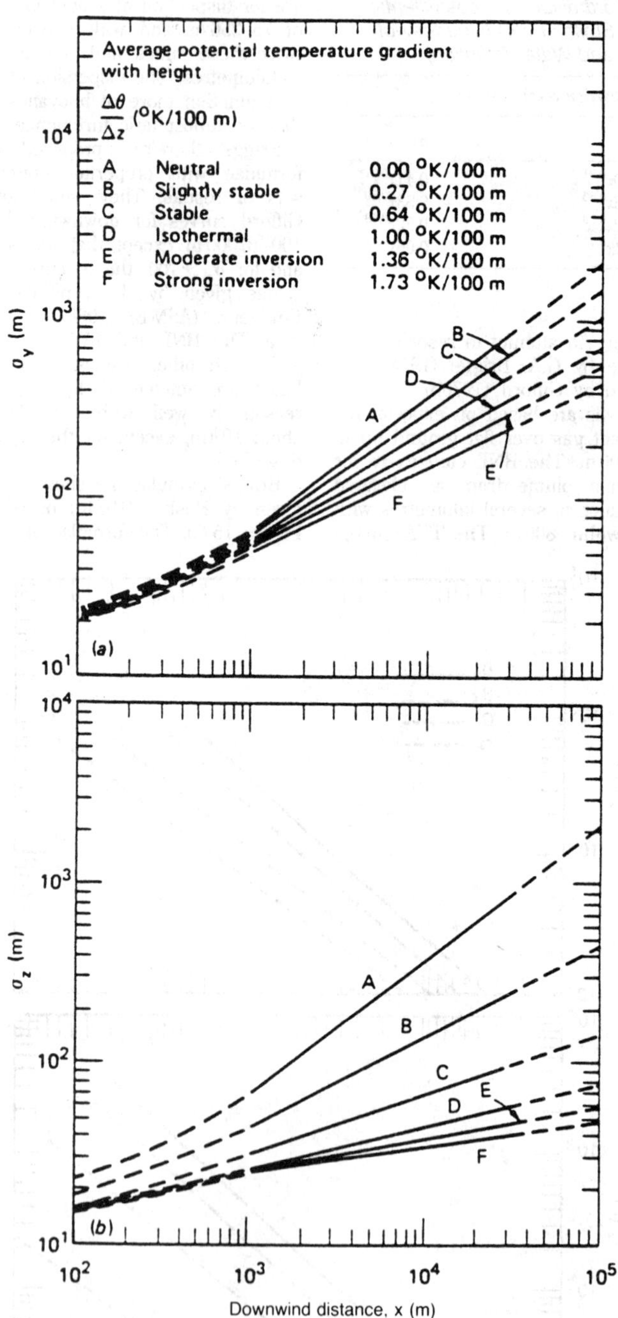

Figure 15.59 *Meteorological parameters for Pasquill–Gifford equations for continuous source: dispersion coefficients based on TVA scheme (Gifford, 1976b; after Carpenter et al., 1971) (Courtesy of Nuclear Safety)*

15.19.6 Pasquill–Gifford dispersion coefficients: NRPB scheme

A scheme for the estimation of the Pasquill–Gifford dispersion coefficients based on F.B. Smith's *P* stability classification and his correlation of the vertical coefficient σ_z and on allowance for plume meandering for the horizontal dispersion σ_y has been developed at the National Radiological Protection Board (NRPB), as

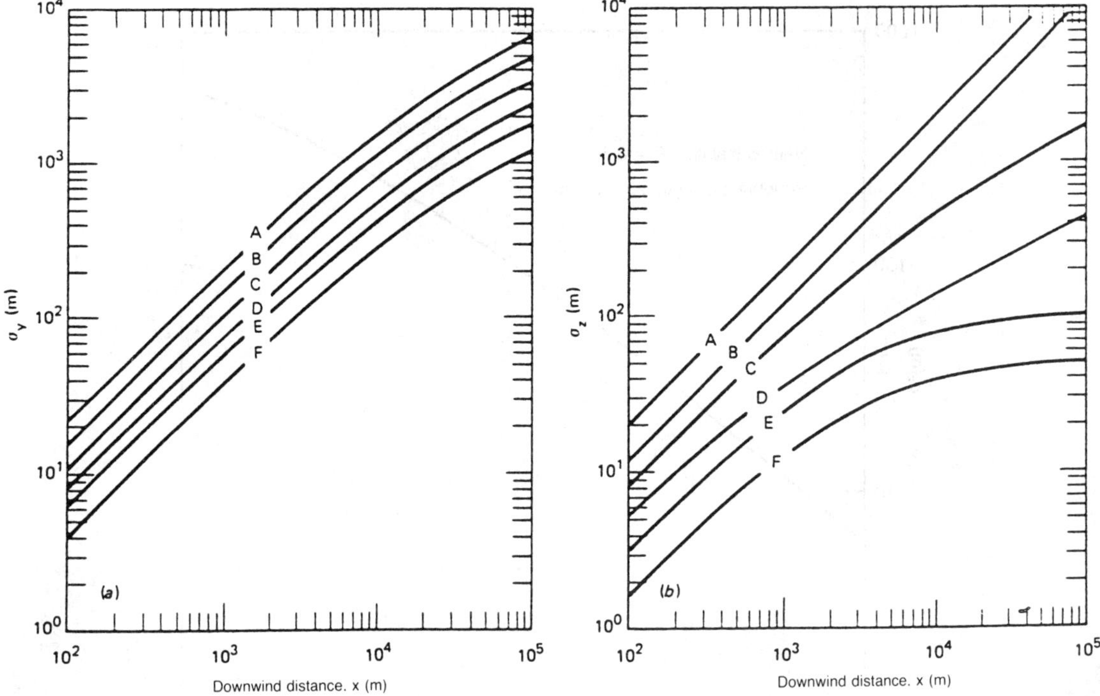

***Figure 15.60** Meteorological parameters for Pasquill–Gifford equations for continuous source: dispersion coefficients based on Briggs' interpolation formulae for open country (Gifford, 1976b; after Hosker, 1974b) (Courtesy of Nuclear Safety)*

***Table 15.34** Pasquill–Gifford dispersion coefficients: Briggs interpolation formulae for open country (G.A. Briggs, 1974)*

Pasquill stability class	*Dispersion coefficient* (m)	
	σ_y	σ_z
A	$0.22x(1 + 0.0001x)^{-1/2}$	$0.20x$
B	$0.16x(1 + 0.0001x)^{-1/2}$	$0.12x$
C	$0.11x(1 + 0.0001x)^{-1/2}$	$0.08x(1 + 0.0002x)^{-1/2}$
D	$0.08x(1 + 0.0001x)^{-1/2}$	$0.06x(1 + 0.0015x)^{-1/2}$
E	$0.06x(1 + 0.0001x)^{-1/2}$	$0.03x(1 + 0.0003x)^{-1}$
F	$0.04x(1 + 0.0001x)^{-1/2}$	$0.016x(1 + 0.0003x)^{-1}$

described by Clarke (1979 NRPB R91).

Following F.B. Smith (1973), the vertical dispersion coefficient σ_z is a function of the distance, the P class and the roughness length z_o. The value of σ_z is obtained from Figures 15.61–15.63. Figure 15.61 gives σ_z as a function of distance for $P = 3.6$ and $z_o = 0.1$ m. Figure 15.62 gives the correction for other P values at the same value of z_o. Figure 15.63 gives the correction for other values of z_o for all P values.

Formulae for σ_z for the discrete Pasquill classes, or P values, and for discrete values of z_o have been given by Hosker (1974b). These are

$$\sigma_z = \frac{ax^b}{1 + cx^d} F(z_o, x) \qquad [15.19.21]$$

with

$$F(z_o, x) = \ln\{fx^g[1 + (hx^j)^{-1}]\} \qquad z_o \geq 0.1 \qquad [15.19.22a]$$

$$= \ln[fx^g(1 + hx^j)^{-1}] \qquad z_o < 0.1 \qquad [15.19.22b]$$

where $F(z_o, x)$ is the roughness length correction factor and a–j are constants or indices. Values of the constants and indices for use in Equations 15.19.21 and 15.19.22 are given in Table 15.35.

Figure 15.64 shows the values of σ_z given by Smith's scheme for roughness length $z_o = 0.1$ m.

For the horizontal dispersion coefficient σ_y the regular Pasquill–Gifford values, which apply for very short releases, typically of 3 min duration, are used for releases of up to 30 min duration. For longer releases, account is taken of meandering using the relation

$$\sigma_y^2 = \sigma_{yt}^2 + \sigma_{yw}^2 \qquad [15.19.23]$$

where σ_{yt} is the turbulent diffusion (or 3 min) term and σ_{yw} is the term due to fluctuations in wind direction.

The values used for σ_{yt} are the regular Pasquill–Gifford values. For σ_{yw} two relations are given. If the standard deviation of the horizontal wind direction σ_θ is known, use is made of the equation

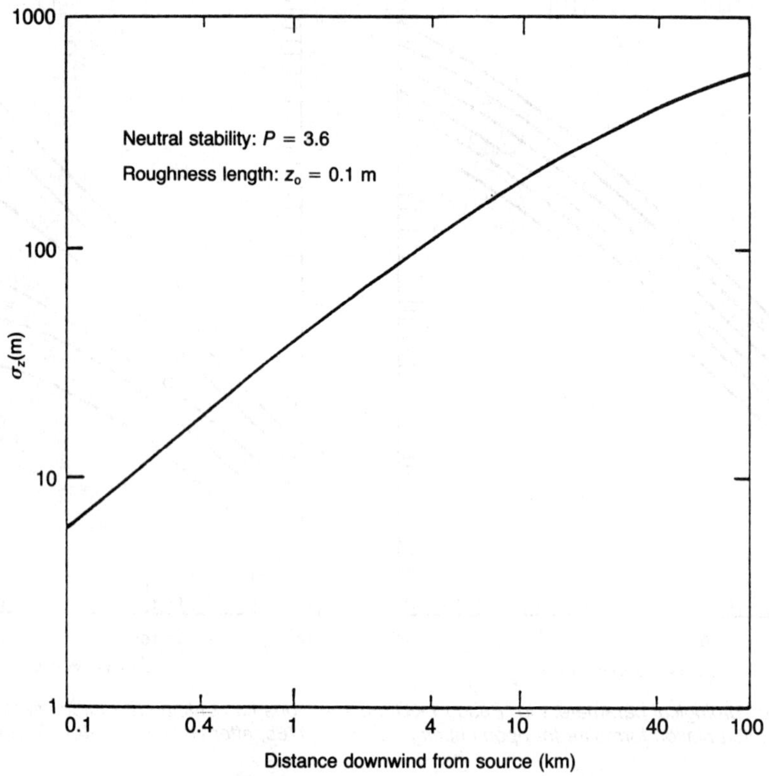

Figure 15.61 *Meteorological parameters for Pasquill–Gifford equations for continuous source: dispersion coefficients based on NRPB scheme (Clarke 1979 NRPB R-91): vertical dispersion coefficient σ_z for stability parameter $P = 3.6$ and surface roughness $z_0 = 0.1$ m*

$$\sigma_{yw} = \sigma_\theta x \qquad [15.19.24a]$$

and if it is not, the equation used is

$$\sigma_{yw} = 0.065\left(\frac{7T}{u_{10}}\right)^{\frac{1}{2}} x \qquad [15.19.24b]$$

where T is the duration of the release (h), x is the downwind distance (m), u_{10} is the wind speed at 10 m height (m/s) and σ_θ is the horizontal standard deviation of the wind direction averaged over 3-min periods and sampled over the duration of the release (rad).

15.19.7 Pasquill–Gifford dispersion coefficients: other correlations

Various authors have given for the Pasquill–Gifford coefficients power law formulae of the form

$$\sigma_y = ax^b \qquad [15.19.25a]$$

$$\sigma_z = cx^d \qquad [15.19.25b]$$

Many of the Pasquill–Gifford curves given in Figures 15.54 and 15.58–15.60 clearly cannot be fitted over their whole range by such formulae, but provided they are used over the range of their validity such equations have their use in making it possible to obtain analytical solutions to dispersion problems.

R.W. McMullen (1975) has given the following formulae for both dispersion coefficients:

$$\sigma = \exp[a + b \ln x + c(\ln x)^2] \qquad [15.19.26]$$

where σ may be either σ_y or σ_z. The values of the constants a–c are given in Table 15.36.

15.19.8 Dispersion coefficients for urban areas

Work on dispersion coefficients for use in the Pasquill–Gifford model over and in urban areas has been described by a number of workers, including D.B. Turner (1964), Bowne, Ball and Anderson (1968), Bowne and Ball (1970) and Bowne (1974), McElroy and Pooler (1968), McElroy (1969) and Yersel, Goble and Morrill (1983). In general, dispersion over or within an urban area is greater than over a flat rural area. Mechanical turbulence is enhanced by the increase surface roughness and thermal turbulence by the heat island effect.

Summaries of experimental work on dispersion over urban areas have been given by Gifford (1972, 1976a,b). The experimental work of McElroy and Pooler (1968)

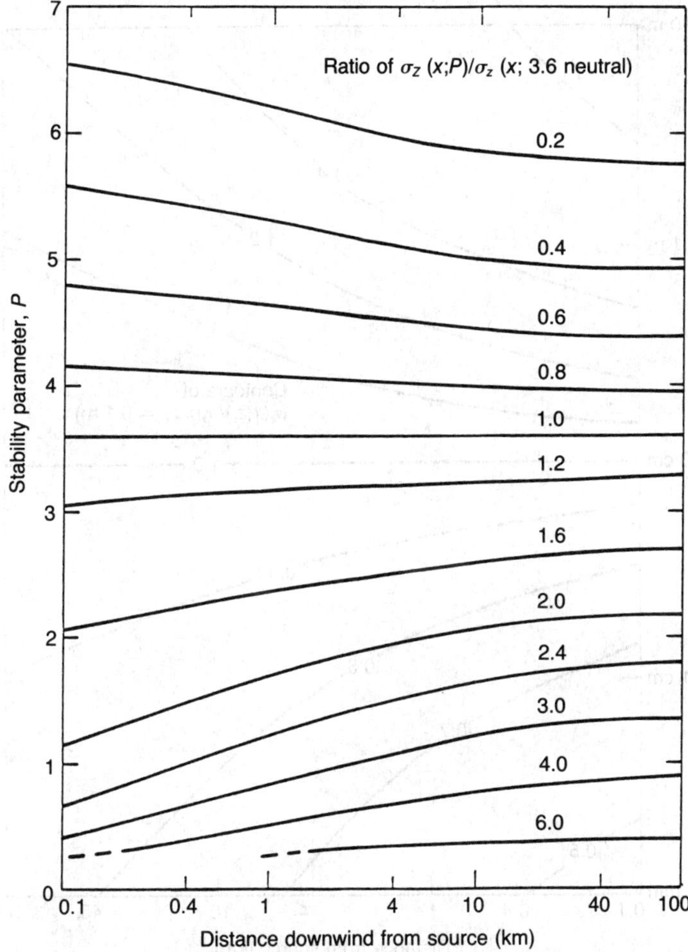

Figure 15.62 *Meteorological parameters for Pasquill-Gifford equations for continuous source: dispersion coefficients based on NRPB scheme (Clarke 1979 NRPB R91): correction factor to vertical dispersion coefficient σ_z for stability parameter P*

over St Louis has provided dispersion data. These data have been used by Pasquill (1970) to compare dispersion over urban and rural terrains. Further data have been analysed by W.B. Johnson *et al.* (1971). On the basis of this work G.A. Briggs (1974) has proposed relations for the dispersion coefficients. Table 15.37 gives information on the vertical dispersion coefficient σ_z and Table 15.38 gives the Briggs' formulae for both the horizontal and the vertical dispersion coefficients σ_y and σ_z. Figure 15.65 shows graphs obtained by Hosker from Briggs' equations.

Experimental work on dispersion over short distances within an urban area at Worcester, Massachusetts, has been conducted by Yersel, Goble and Morrill (1983). They carried out releases in a parking lot, with the nearest building 40 m away, and measured the concentration at distances up to 600 m. Their results for σ_y and σ_z are shown in Figure 15.66. For σ_y the values for stability class B are well described by the regular Pasquill–Gifford curves for class A, a shift of one class, while for classes C and D the values are best described by a shift of two classes. For σ_z, however, the data differ significantly in magnitude and slope and around 200 m are substantially higher than even the class A curve.

15.19.9 Dispersion coefficients for coastal areas and water

A number of authors have described work on dispersion coefficients for the Pasquill–Gifford model for coastal areas and water, including Kondo (1975), Gifford (1976a,b), Lyons (1976), S.R. Hanna, Paine and Schulman (1984), Hasse and Weber (1985), Larsen and Gryning (1986) and Skupniewicz and Schacher (1986). In general, dispersion over water tends to be less than over

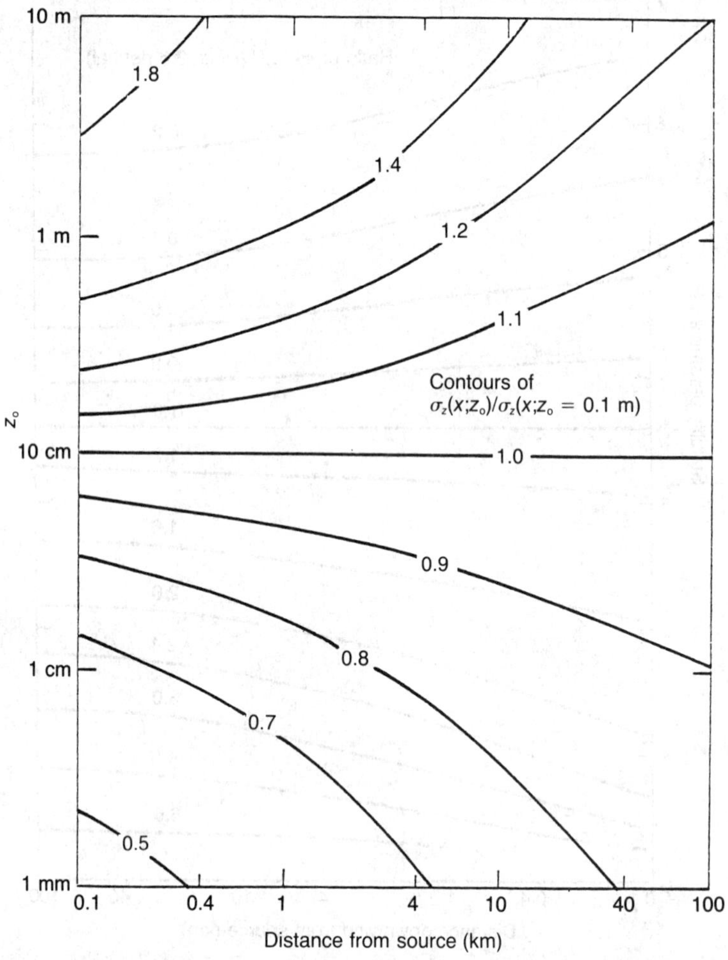

***Figure 15.63** Meteorological parameters for Pasquill–Gifford equations for continuous source: dispersion coefficients based on NRPB scheme (Clarke 1979 NRPB R91): correction factor to vertical dispersion coefficient σ_z for surface roughness z_0*

land but, as described above, this generalization needs heavy qualification.

The surface roughness of the sea may be characterized by a roughness length, but waves do not behave as simple roughness elements. This problem is discussed by Gifford (1976a,b).

In order to use the Pasquill–Gifford model it is necessary to define the stability classes. The Pasquill classes over land are not necessarily applicable without modification over sea. The definition of suitable classes for use over sea was discussed in Section 15.12.

Emperical correlations for the Pasquill–Gifford dispersion coefficients for use over water have been given by Skupniewicz and Schacher (1986) in terms of the NPS stability classification. The relations are

$$\sigma_y = \sigma_y(R)\left(\frac{x}{R}\right)^{\alpha} \qquad [15.19.27a]$$

$$\sigma_z = \sigma_z(R)\left(\frac{x}{R}\right)^{\beta} \qquad [15.19.27b]$$

where σ_y(R) and σ_z(R) are the values of σ_y and σ_z at the reference distance R (= 100 m), and α and β are indices. Values of σ_y and σ_z at 100 m and of the indices α and β are given in Table 15.39.

These equations are empirical correlations. The data for σ_y exhibit considerable scatter. The authors state that stability classification can reasonably be used to characterize σ_z but not σ_y.

Table 15.35 Pasquill–Gifford dispersion coefficients: Hosker's formulae (Hosker, 1974b) (Courtesy of International Atomic Energy Agency)

A Vertical dispersion coefficient

Stability class	*a*	*b*	*c*	*d*
A	0.112	1.06	5.38 x 10^{-4}	0.815
B	0.130	0.950	6.52 x 10^{-4}	0.750
C	0.112	0.920	9.05 x 10^{-4}	0.718
D	0.098	0.889	1.35 x 10^{-3}	0.688
E	0.0609	0.895	1.96 x 10^{-3}	0.684
F	0.0638	0.783	1.36 x 19^{-3}	0.672

B Roughness length correction factor

Roughness length (m)	*f*	*g*	*h*	*j*
0.01	1.56	0.0480	6.25 × 10^{-4}	0.45
0.04	2.02	0.0269	7.76 × 10^{-4}	0.37
0.1	2.72	0	0	0
0.4	5.16	-0.098	18.6	−0.225
1.0	7.37	-0.0957	4.29 × 10^{3}	−0.60
4.0	11.7	-0.128	4.59 × 10^{4}	−0.78

An alternative treatment for the estimation of dispersion coefficients for use over water based on the AMS Workshop scheme has been given by S.R. Hanna, Paine and Schulman (1984).

15.19.10 Dispersion coefficients for complex terrain

Dispersion coefficients for complex terrain are discussed by Egan (1976). In general, dispersion over complex terrain tends to give complex dispersion patterns. The overall effect is generally to increase dispersion, but, as described below, there are two features which may lead to local high concentrations.

There tends to be enhancement of both the dispersion coefficients, but the crosswind dispersion coefficient σ_y is enhanced more than is the vertical dispersion coefficient σ_z. This is often attributed mainly to plume meandering and splitting. For a ground level source the implication is that the ground level centre line concentration will be less than for flat terrain.

For an elevated source the situation is not quite so straightforward. The Pasquill–Gifford Equation 15.16.55 for the maximum ground level concentration for a release from an elevated continuous point source shows that, if there is an increase in the ratio σ_z/σ_y, this maximum concentration increases. Therefore, in principle, if σ_z increases and there is no compensating increase in σ_y, there will be an increase in the maximum ground level concentration and a decrease in the distance at which this occurs. The other feature which can give rise to a high local concentration is a stagnant pocket.

15.19.11 Effect of sampling period

As described above, if the concentration in a plume is sampled virtually instantaneously, the concentration distribution has a relatively narrow spread, while if the sample is taken over a longer time period the distribution has a wider spread. It follows that the dispersion coefficients are a function of the sampling period. The following relation has been proposed by Gifford (1976a) to take account of this variation:

$$\frac{\sigma_y}{\sigma_{yr}} = \left(\frac{t}{t_r}\right)^p \qquad [15.19.28]$$

where t is the sampling period which yields dispersion coefficient σ_y, p is an index, and subscript r denotes the reference value. The value of p is approximately 0.2 for sampling periods between 3 min and 1 h and in the range 0.25–0.3 for periods between 1 and 100 h.

Gifford states that similar considerations apply to σ_z, except that the effect of variations in σ_z should not extend beyond downwind distances of a few kilometres.

The regular Pasquill–Gifford curves correspond to a sampling period of about 10 min.

15.19.12 Puff dispersion coefficients

For a puff, the size of eddy which mainly influences dispersion lies within a limited range, approximately 1/3 to 3 times the size of the puff. A smaller eddy effects little dispersion, while a larger one moves the whole cloud.

Relations for a puff dispersion coefficient σ have been derived by Batchelor (1952) from similarity considerations. It can be shown that

$$\sigma^2 = \sigma_o^2 + c_1(\epsilon\sigma_o)^{\frac{2}{3}}t^2 \qquad \text{short times} \qquad [15.19.29a]$$

$$= c_2\epsilon t^3 \qquad \text{intermediate times} \qquad [15.19.29b]$$

$$= c_3 t \qquad \text{long times} \qquad [15.19.29c]$$

where ϵ is the eddy dissipation rate, σ_o is the initial size of the cloud, and c_1, c_2 and c_3 are constants.

The constant c_2 is generally found to lie in the range 0.5–2 and is often taken as unity.

Then, from Equation 15.19.29,

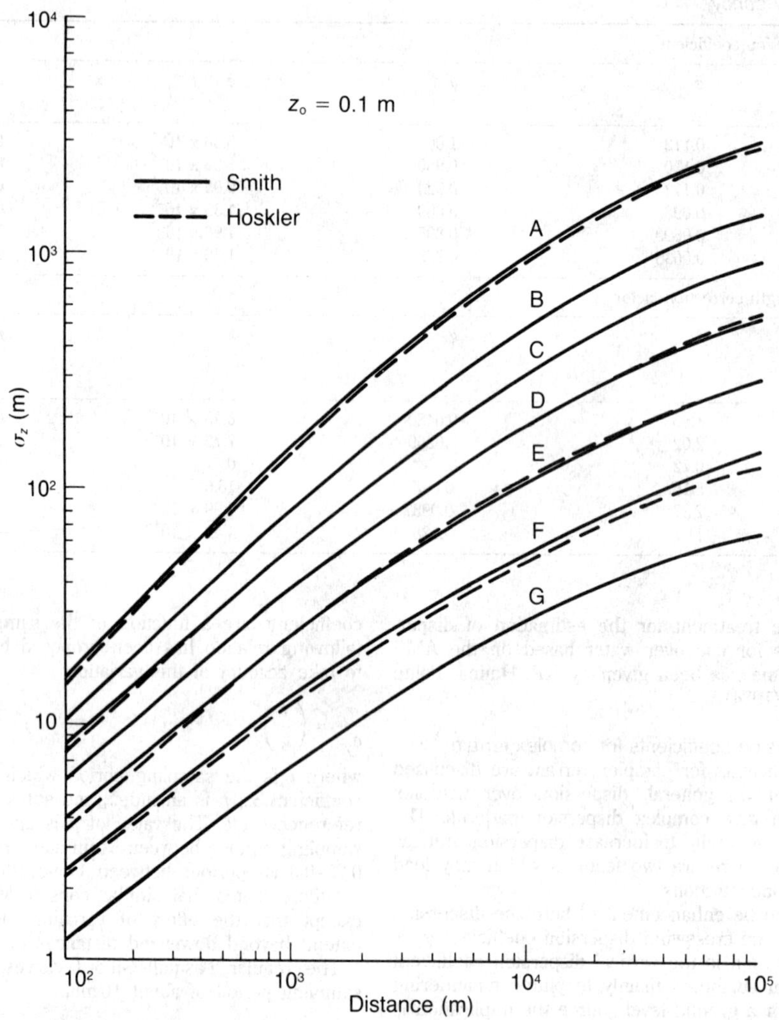

Figure 15.64 *Meteorological parameters for Pasquill–Gifford equations for continuous source: dispersion coefficients based on NRPB scheme (Clarke 1979 NRPB R91): vertical dispersion coefficient σ_z as a function of stability*

Table 15.36 *Pasquill–Gifford dispersion coefficients: McMullen's formulae (R.W. McMullen, 1975) (Courtesy of Air Pollution Control Association)*

Pasquill stability class	*Formulae parameters*					
	σ_y			σ_z		
	a	b	c	a	b	c
A	5.357	0.8828	−0.0076	6.035	2.1097	0.2770
B	5.058	0.9024	−0.0096	4.694	1.0629	0.0136
C	4.651	0.9181	−0.0076	4.110	0.9201	−0.0020
D	4.230	0.9222	−0.0087	3.414	0.737	−0.0316
E	3.922	0.9222	−0.0064	3.057	0.6794	−0.0450
F	3.533	0.9181	−0.0070	2.621	0.6564	−0.0540

Table 15.37 *Pasquill–Gifford dispersion coefficients for urban areas: comparison of vertical dispersion coefficient for rural and urban terrain (after Gifford, 1976b)[a] (Courtesy of Nuclear Safety)*

Downwind distance (km)	Terrain	Ratio of σ_z for following stability classes to value in neutral conditions			
		B	C	D	E–F
1	City[b]	4.5	2.7	1.7	0.7
	City[c]	4.0	2.4	1.5	0.6
	Open country	3.2	1.9	1.0	0.5
10	City[b]	9	3.4	1.0	0.3
	City[c]	11	4.1	1.2	0.4
	Open country	6	2.4	1.0	0.3

[a] Based on work of McElroy and Pooler (1968) at St Louis and analysis by Pasquill (1970).
[b] Based on McElroy and Pooler's Figure 2 using curve for bulk Richardson number $B = \pm 0.01$.
[c] Using data for $B = \pm 0.01$ in evening conditions only.

Table 15.38 *Pasquill–Gifford dispersion coefficients for urban areas: Briggs interpolation formulae (after G.A. Briggs, 1974)*

Pasquill stability class	Dispersion coefficient (m)	
	σ_y	σ_z
A–B	$0.32x(1 + 0.0004x)^{-\frac{1}{2}}$	$0.24x(1 + 0.001x)^{\frac{1}{2}}$
C	$0.22x(1 + 0.0004x)^{-\frac{1}{2}}$	$0.20x$
D	$0.16x(1 + 0.0004x)^{-\frac{1}{2}}$	$0.14x(1 + 0.0003x)^{-\frac{1}{2}}$
E–F	$0.11x(1 + 0.0004x)^{-\frac{1}{2}}$	$0.08(1 + 0.0015x)^{-\frac{1}{2}}$

$$\sigma \propto t \quad \text{short times} \qquad [15.19.30a]$$

$$\propto t^{\frac{3}{2}} \quad \text{intermediate times} \qquad [15.19.30b]$$

$$\propto t^{\frac{1}{2}} \quad \text{long times} \qquad [15.19.30c]$$

Also at long times

$$\sigma^2_{\text{puff}} \rightarrow \sigma^2_{\text{plume}} \qquad t \rightarrow 0 \qquad [15.19.31]$$

Here short times may be interpreted as a few seconds and long times as greater than 10^4 s. Generally it is the intermediate times which are of most interest.

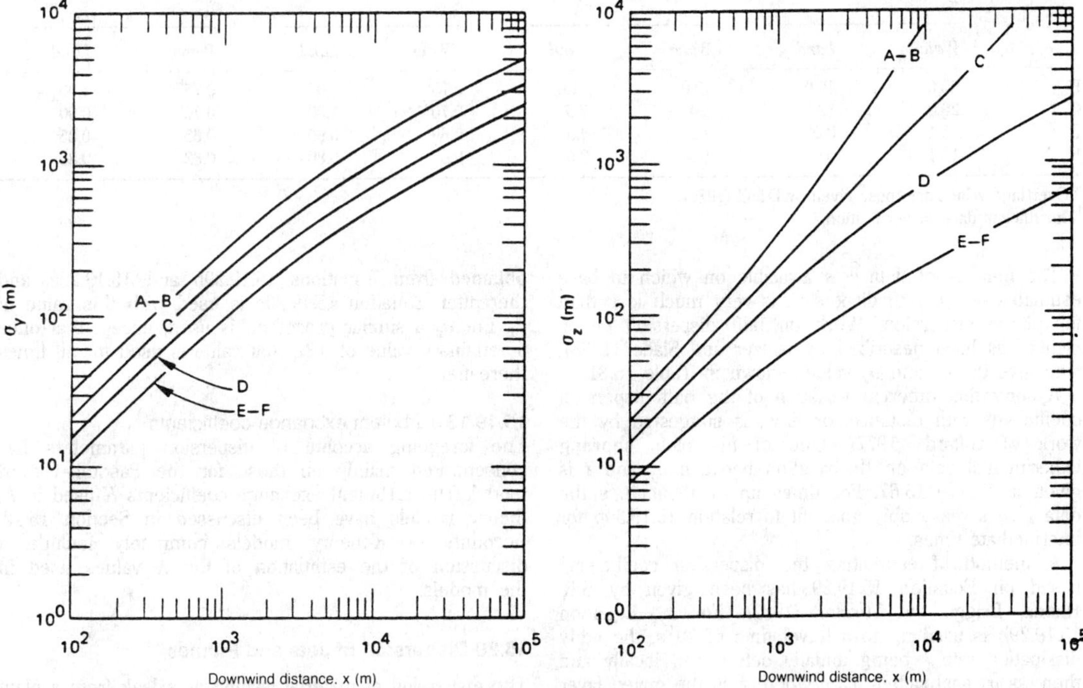

Figure 15.65 *Meteorological parameters for Pasquill–Gifford equations for continuous source: dispersion coefficients based on Briggs' interpolation formulae for urban areas (Gifford, 1976b; after Hosker, 1974b) (Courtesy of Nuclear Safety)*

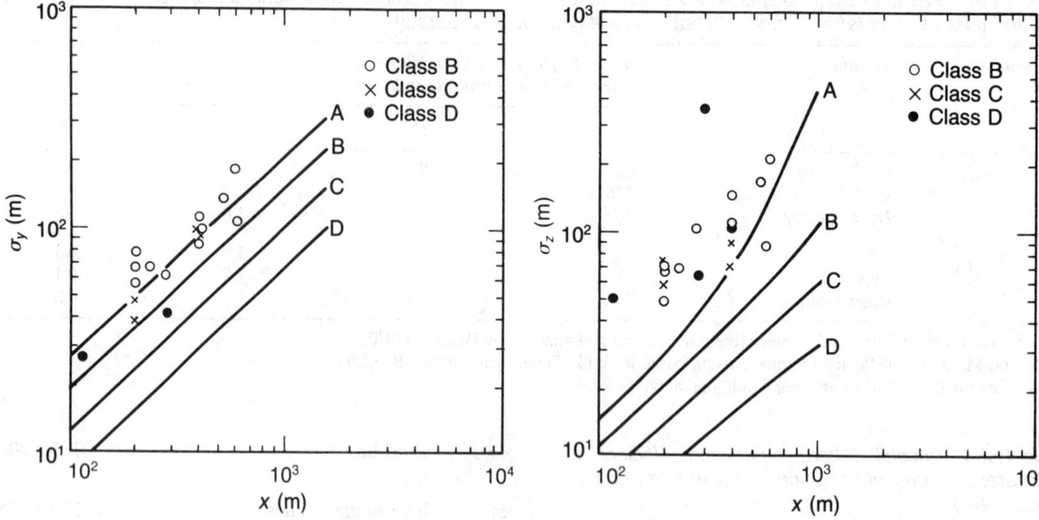

Figure 15.66 Meteorological parameters for Pasquill–Gifford equations for continuous source: dispersion coefficients for urban areas (Yersel, Goble and Morill, 1983) (Courtesy of Pergamon Press)

Table 15.39 Pasquill–Gifford dispersion coefficients over water: empirical formulae (after Skupniewicz and Schacher, 1986) (Courtesy of Pergamon Press)

NPS class	*Dispersion coefficient*[a] (100 m)				*Indices*			
	σ_y		σ_z		α		β	
	Water	*Land*	*Water*	*Land*	*Water*	*Land*	*Water*	*Land*
B	25.0	19.0	10.0	11.0	0.75[b]	1.00	0.75[b]	1.00
C	20.0	12.5	8.0	7.5	0.70[b]	1.00	0.70[b]	0.90
D	15.1	8.0	3.2	4.5	0.69	0.90	0.65	0.85
E	16.1	6.0	1.8	3.5	0.65	0.80	0.62	0.80

[a] Over land values are those given by DTIC (1980).
[b] Insufficient data for verification.

The number of data sets available on which to base estimates of the puff dispersion is very much less than for plume dispersion. Work on puff dispersion coefficients has been described by Islitzer and Slade (1968), who give the σ_y and σ_z values shown in Table 15.31.

A somewhat different variation of the puff dispersion coefficients with distance, or time, is suggested by the work of Gifford (1977). One of his plots showing experimental data on the relation between σ_y and t is given in Figure 15.67. For times up to about 10^4 s the data give a reasonably good fit to relation 15.19.30b for intermediate times.

A method of estimating the dispersion coefficients based on Equation 15.19.29 has been given by S.R. Hanna, Briggs and Hosker (1982). For σ_y, Equation 15.19.29b is used up to a travel time of 10^4 s, the eddy dissipation rate ϵ being initially determined locally and then, as σ_y approaches $0.3z_i$, where z_i is the mixed layer height, at a height midway in the boundary layer. At a travel time equal to 10^4 s, the constant c_3 in Equation 15.19.29c is evaluated by equating the values of σ_y obtained from Equations 15.19.29b and 15.19.29c, and thereafter Equation 15.19.29c is used with this value of c_3. For σ_z a similar procedure is used, except that once σ_z attains a value of $0.3z_i$ that value is used for all times thereafter.

15.19.13 Turbulent exchange coefficients

The foregoing account of dispersion parameters has concentrated mainly on those for the Pasquill–Gifford model. The turbulent exchange coefficients K used in K-theory models have been discussed in Section 15.12. Accounts of K-theory models commonly include a discussion of the estimation of the K values used in the models.

15.20 Dispersion of Jets and Plumes

The dispersion of material issuing as a leak from a plant is determined by its momentum and buoyancy. If momentum forces predominate, the fluid forms a jet, while if buoyancy forces predominate, it forms a plume.

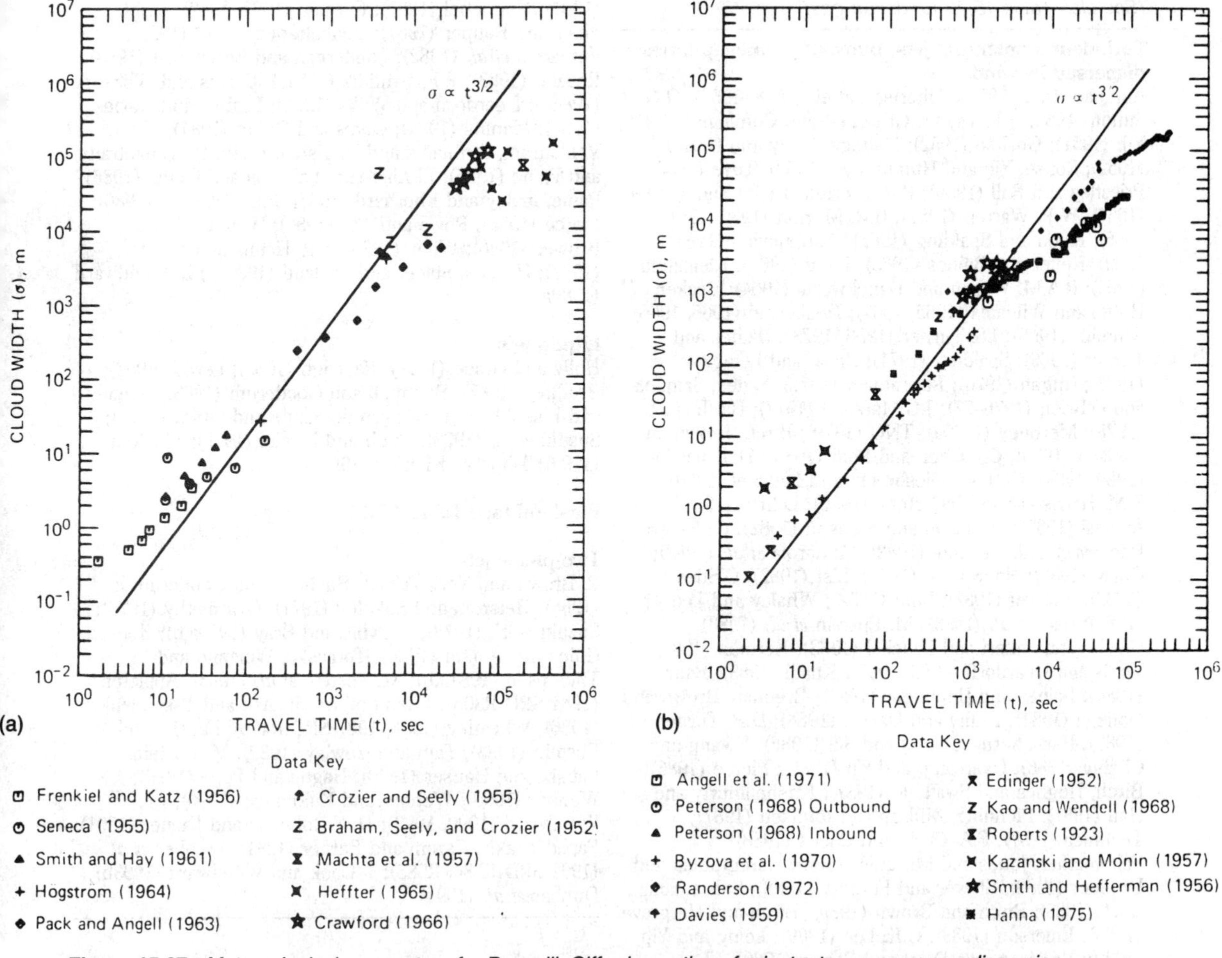

***Figure 15.67** Meteorological parameters for Pasquill–Gifford equations for instantaneous source: dispersion coefficient σ (Gifford, 1977): (a) selection of original data of Crawford (1966); (b) collection of subsequent data by Gifford. The references quoted are those given by Gifford and do not all appear in this text*

Such dispersion contrasts with the dispersion by atmospheric turbulence considered so far. However, once the momentum or buoyancy decay dispersion by atmospheric turbulence becomes the predominant factor for leaks also.

Emission situations were classified in Section 15.1. In respect of jets and plumes the following distinctions may be made:

(1) Fluid –
  (a) gas,
  (b) liquid,
  (c) two-phase vapour–liquid mixture;
(2) Fluid momentum –
  (a) low momentum,
  (b) high momentum;
(3) Fluid buoyancy –
  (a) positive buoyancy,
  (b) neutral buoyancy,
  (c) negative buoyancy;
(4) Atmospheric conditions –
  (a) low turbulence,
  (b) high turbulence.

If the momentum of the material issuing from an orifice on a plant is high, the dispersion in the initial phase at least is due to the momentum, and the emission is described as a momentum jet. If the momentum is low, either because the initial momentum is low or because it has decayed, the dispersion is due to buoyancy and atmospheric turbulence, and, if buoyancy is involved, the emission is described as a buoyant plume: the buoyancy may be positive or negative.

An account is now given of dispersion by momentum jets and buoyant plumes. Both types of emission are sometimes described as plumes, the jet being a forced plume. The forces in a buoyant plume are often of the same order as those in a momentum jet. Selected references on jets and plumes are given in Table 15.40. Some jet and plume dispersion situations are illustrated

Table 15.40 Selected references on jets and plumes (See also Table 15.42 for dense gas)

Turbulent momentum jets, buoyant plumes, plumes dispersed by wind
Abraham (n.d., 1970); Albertson *et al.* (1948, 1950); O.G. Sutton (1950); J.F. Taylor, Grimmett and Comings (1951); Yih (1951); Gosline (1952); Nottage, Slaby and Gojsza (1952); Rouse, Yih and Humphreys (1952); Tuve (1953); Priestley and Ball (1955); B.R. Morton, Taylor and Turner (1956); W.R. Warren (1957); B.R. Morton (1959, 1961, 1965); Ricou and Spalding (1961); Abramovich (1963, 1969); Keffer and Baines (1963); Long (1963); Kleinstein (1964); R.A.M. Wilson and Danckwerts (1964); Becker, Hottel and Williams (1965, 1967); Heskestad (1965, 1966); Schmidt (1965); J.S. Turner (1966, 1973); Baines and Turner (1969); Spalding (1971); Ballal and Lefebvre (1973); Gugan (1976); Rajaratnam (1976); Sadée, Samuels and O'Brien (1976–77); J.G.Marshall (1977); Birch *et al.* (1978); Meroney (1979a); TNO (1979); Birch, Brown and Dodson (1980); C.J. Chen and Rodi (1980); Delichatsios (1980, 1988); Gartner, Giesbrecht and Leuckel (1980); R.M. Harrison and McCartney (1980); Lehrer (1981); Anfossi (1982); Davidson and Slawson (1982); de Faveri, Hanzevack and Delaney (1982); Fumarola *et al.* (1982); Golay (1982); Hanzevack (1982); List (1982); O'Shea (1982); Rittman (1982); Rodi (1982); Whaley and Lee (1982); Barr *et al.* (1983); M. Epstein *et al.* (1983); Giesbrecht, Seifert and Leuckel (1983); Kerman (1983); Willis and Deardorff (1983, 1987); Khalil (1983); Badr (1984); Baines and Hopfinger (1984); Brennan, Brown and Dodson (1984); Leahy and Davies (1984); D.B. Turner (1985); Hsu-Cherng Chiang and Sill (1986); Hwang and Chaing (1986); Overcamp and Ku (1986); Pierce (1986); Birch, Hughes and Swaffield (1987); Krishnamurty and Hall (1987); Pfenning, Millsap and Johnson (1987); Technica (1987); P.A. Clark and Cocks (1988); Fairweather, Jones and Marquis (1988); Haroutunian and Launder (1988); Shaver and Fornery (1988); Subramanian *et al.* (1988); Birch and Brown (1989); Birch and Hargrave (1989); Emerson (1989); G.K. Lee (1989); Loing and Yip (1989); Vergison, van Diest and Basler (1989); J.L. Woodward (1989a); Arya and Lape (1990); Cleaver and Edwards (1990); M. Epstein, Fauske and Hauser (1990); Moodie and Ewan (1990); Netterville (1990); Anfosssi *et al.* (1993); Bagster (1993); Lane-Serff, Linden and Hillel (1993); Webber and Wren (1993 SRD R552); Landis, Linney and Hanley (1994).
Sonic jets: Hess, Leuckel and Stoeckl (1973); Ramskill (1985 SRD R302); Ewan and Moodie (1986)

Elevated plumes, stacks, chimneys
Bosanquet (1935, 1957); Bosanquet and Pearson (1936); Church (1949); Barad (1951); Gosline (1952); N.G. Stewart *et al.* (1954); Strom and Halitsky (1955); F.B. Smith (1956); O.G.Sutton (1956); Best (1957); Hay and Pasquill (1957); Hilst (1957); Strom, Hackman and Kaplin (1957); N.G. Stewart, Gale and Crooks (1958); Strauss and Woodhouse (1958); Nonhebel (1960); Bodurtha (1961, 1988); Bowne (1961); Scorer and Barrett (1962); Briggs (1965, 1968, 1969, 1974, 1976); D.J. Moore (1969, 1975); Slawson and Csanady (1971); Ooms (1972); Cude (1974a,b, 1975); Hoot and Meroney (1974); Ooms, Mahieu and Zelis (1974); Readings, Haugen and Kaimal (1974); Bowne, Cha and Murray (1978); Lamb (1979); Venkatram and Kurtz (1979); Venkatram (1980a,c); P.M. Foster (1981); Nonhebel (1981); Kerman (1981, 1982); Reible, Shair and Kauper (1981); Venkatram and Vet (1981); Fumarola *et al.* (1982); Londergan and Borenstein (1983); Robson (1983); R.F. Griffiths (1984a); Carras and Williams (1984–); Deardorff and Willis (1984); Leahey and Davies (1984); Manins (1984); Ooms and Duijm (1984); Venkatram, Strimaitis and Dicristoforo (1984); Venkatram and Paine (1985); Li Xiao-Yun, Leijdens and Ooms (1986); Paine, Insley and Eberhard (1986); R.L. Petersen (1986); Pierce (1986); Rowe and Tas (1986); Weil, Corio and Brower (1986); Willis (1986); S.R. Hanna and Paine (1987); Havens, Spicer and Layland (1987); J.L. Woodward (1989a)

Liquid jets
Holly and Grace (1921); Rayleigh (1932); Lewitt (1952); Brodkey (1967); Bushnell and Gooderum (1968); Benatt and Eisenklam (1969); van de Sande and Smith (1976); Suzuki *et al.* (1978); Engh and Larsen (1979); Blanken (1990); Tilton and Farley (1990)

Flash-off (see Table 15.7)

Two-phase jets
R. Brown and York (1962); Bushnell and Gooderum (1968); Hetsroni and Sokolov (1971); Gyarmathy (1976); Suzuki *et al.* (1978); Melville and Bray (1979a,b); Koestel, Gido and Lamkin (1980); Tomasko, Weigand and Thompson (1981a,b); M. Epstein *et al.* (1983); Appleton (1984 SRD R303); Kitamura, Morimatsu and Takahashi (1986); Wheatley (1986, 1987 SRD R410); Fauske and Epstein (1988); Tam and Cowley (1989); M. Epstein, Fauske and Hauser (1990); Hague and Pepe (1990); Webber (1990); Webber and Kukkonen (1990); J.L. Woodward (1990, 1993); D.W. Johnson and Diener (1991); Papadourakis, Caram and Barner (1991); Webber *et al.* (1992 SRD R584, R585); J. Cook and Woodward (1993b); Dunbar *et al.* (1994)

in Figure 15.68. The figure shows schematically instances releases issuing as jets and becoming plumes as the influence of buoyancy takes over from that of momentum.

15.20.1 Jets
Accounts of jets are given in *The Theory of Turbulent Jets* (Abramovich, 1963), *Turbulent Jets of Air, Plasma and Real Gas* (Abramovich, 1969), *Turbulent Jets* (Rajaratnam, 1976), *Vertical Turbulent Buoyant Jets* (C.J. Chen and Rodi, 1980) and *Turbulent Buoyant Jets and Plumes* (Rodi, 1982). Work on jets includes that by Albertson *et al.* (1948), O.G. Sutton (1950), J.F. Taylor, Grimmett and Comings (1951), Nottage, Slaby and Gojsza (1952), Tuve (1953), B.R. Morton, Taylor and Turner (1956), B.R. Morton (1959, 1961), Ricou and Spalding (1961), Long (1963), Kleinstein (1964), Heskestad (1965, 1966), J.S. Turner (1966), Spalding (1971), Hess, Leuckel and Stoeckl (1973), Birch *et al.* (1984); Ramskill (1985 SRD R302), Ewan and Moodie (1986), Birch, Hughes and Swaffold (1987), Birch and Brown (1989) and Birch *et al.* (1989).

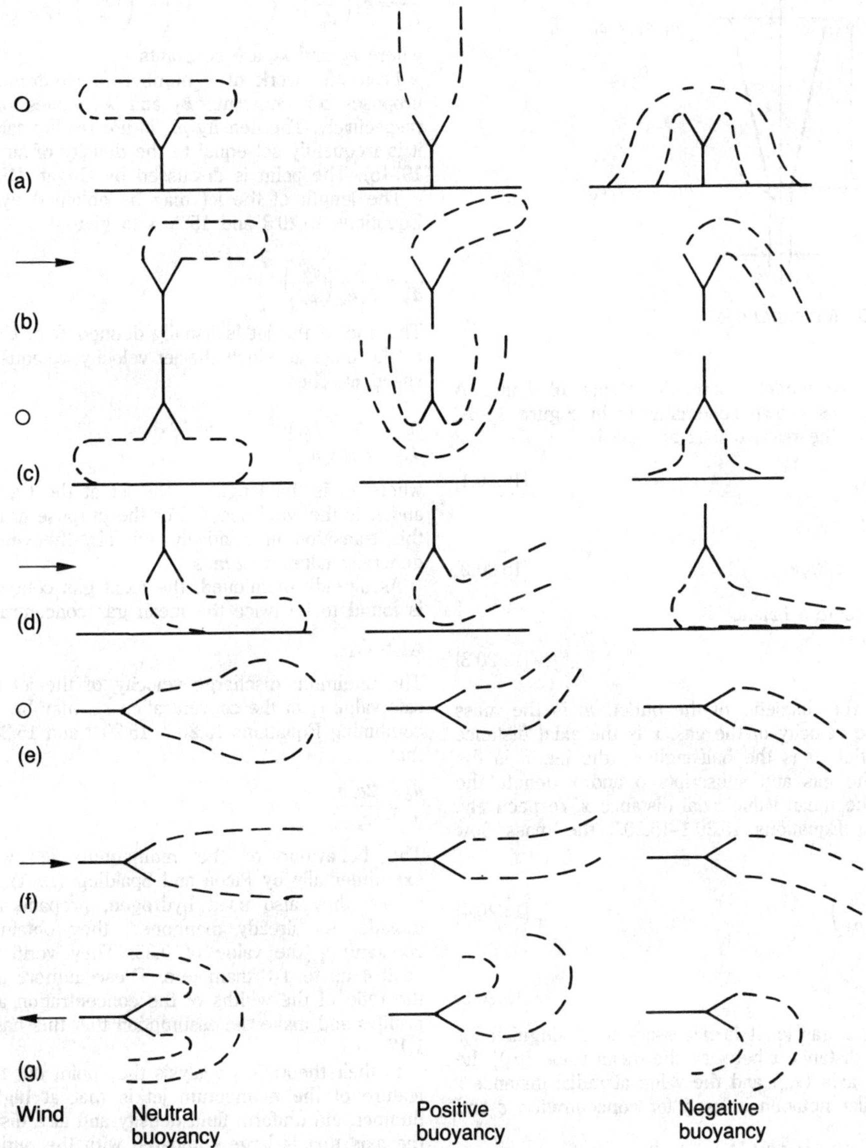

Figure 15.68 *Some jet and plume dispersion situations (after TNO, 1979): (a) upwards vertical release, zero wind speed; (b) upwards vertical release, finite wind speed; (c) downwards vertical release, zero wind speed; (d) upwards vertical release, finite wind speed; (e) horizontal release, zero wind speed; (f) horizontal release, wind speed in direction of release; and (g) horizontal release, wind speed in direction opposed to release*

15.20.2 Gas jets

The behaviour of a circular, or conical, turbulent momentum gas jet has been studied by Ricou and Spalding (1961), Long (1963) and Cude (1974b).

Long summarizes the empirical features of such a jet. The characteristics which he enumerates include the following. The jet is conical and apparently diverges from a virtual point source upstream of the orifice; dilution occurs by turbulent mixing; the time–mean velocity and concentration profiles are similar after 10 diameters and are approximately Gaussian; the concentration profile is wider than the velocity profile; the jet entrains air, but conserves its momentum, so that the momentum flux at any plane normal to the axis is constant.

Models of a momentum jet have been given by Ricou and Spalding, Long and Cude. The following treatment is

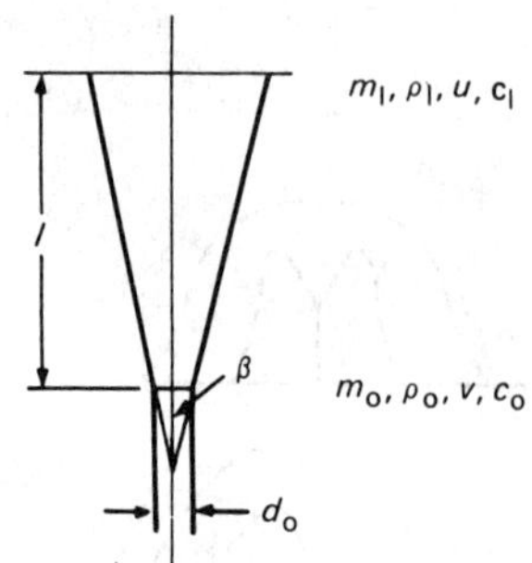

Figure 15.69 Momentum jet

based on these models, particularly that of Long. A momentum jet is shown schematically in Figure 15.69. For such a jet the mass discharge flow is

$$m_o = \frac{\pi}{4} d_o^2 \rho_o u_o \qquad [15.20.1]$$

the mass balance is

$$m_x = \frac{\pi}{4}(2x \tan \beta) 2\rho_x u \qquad [15.20.2]$$

and the momentum balance is

$$\frac{m_x}{m_o} = \frac{u_o}{u} \qquad [15.20.3]$$

where d_o is the diameter of the outlet, m is the mass flow, u is the velocity of the gas, x is the axial distance from the outlet, β is the half-angle of the jet, ρ is the density of the gas and subscripts o and x denote the orifice and the mean value axial distance x, respectively.

Then from Equations 15.20.1–15.20.3 the mass flow ratio is

$$\frac{m_x}{m_o} = k_1 \frac{x}{d_o}\left(\frac{\rho_x}{\rho_o}\right)^{\frac{1}{2}} \qquad [15.20.4]$$

with

$$k_1 = 2 \tan \beta \qquad [15.20.5]$$

In modelling a gas jet it is necessary to distinguish for velocity u at distance x between the mean value (u_x), the value on the axis (u_{xo}) and the value at radial distance r (u_{xr}). A similar notation is used for concentration c and density ρ.

From Equation 15.20.4 the velocity ratio is

$$\frac{u_x}{u_o} = \frac{1}{k_1}\left(\frac{x}{d_o}\right)^{-1}\left(\frac{\rho_x}{\rho_o}\right)^{-\frac{1}{2}} \qquad [15.20.6]$$

The mean volumetric concentration is inversely proportional to the volumetric flow of the gas–air mixture and hence

$$\frac{c_x}{c_o} = \frac{1}{k_1}\left(\frac{x}{d_o}\right)^{-1}\left(\frac{\rho_x}{\rho_o}\right)^{\frac{1}{2}} \qquad [15.20.7]$$

where c is the volumetric concentration. Long proposes for the constant k_1 the value of 0.32 obtained by Ricou and Spalding.

If a Gaussian radial concentration profile is assumed, Equation 15.20.7 may be rewritten as

$$\frac{c_{xr}}{c_o} = k_2\left(\frac{x}{d_o}\right)^{-1}\left(\frac{\rho_{xo}}{\rho_o}\right)^{\frac{1}{2}} \exp\left[-\left(\frac{k_3 r}{x}\right)^2\right] \qquad [15.20.8]$$

where k_2 and k_3 are constants.

From the work of a number of experimenters Long proposes for constants k_2 and k_3 values of 6 and 5, respectively. The density ρ_{xo} is not readily calculated, but it is frequently set equal to the density of air (e.g. Cude, 1974b). The point is discussed by Gugan (1976).

The length of the jet may be obtained by combining Equations 15.20.3 and 15.20.4 to give

$$\frac{x}{d_o} = \frac{u_o}{k_1 u_x}\left(\frac{\rho_x}{\rho_o}\right)^{-\frac{1}{2}} \qquad [15.20.9]$$

The end of the jet is usually defined (e.g. Cude, 1974b) as the point at which the jet velocity u_x equals the wind speed w. Then

$$\frac{x_{tr}}{d_o} = \frac{u_o}{k_1 w}\left(\frac{\rho_x}{\rho_o}\right)^{-\frac{1}{2}} \qquad [15.20.10]$$

where x_{tr} is the length of the jet at the transition point and w is the wind speed. For the purpose of determining this transition in relatively still air, the wind speed is generally taken as 2 m/s.

As already mentioned, the axial gas concentration c_{xo} is found to be twice the mean gas concentration c_x:

$$c_{xo} = 2c_x \qquad [15.20.11]$$

The minimum discharge velocity of the jet to obtain a safe value c_s of the concentration c_{xo} may be obtained by combining Equations 15.20.7, 15.20.9 and 15.20.11 and so that

$$\frac{u_o}{u} = \frac{2c_o \rho_{xo}}{c_{xo}\rho_o} \qquad [15.20.12]$$

The behaviour of the momentum jet was studied experimentally by Ricou and Spalding (1961). In addition to air, they also used hydrogen, propane and carbon dioxide. As already mentioned, they obtained for the constant k_1 the value of 0.32. They verified Equation 15.20.4 up to 418 diameters. These authors also discuss the ratio of the widths of the concentration and velocity profiles and make the assumption that this has a value of 1.17.

In their theoretical analysis they point out that another feature of the momentum jet is that at high Reynolds number and uniform fluid density and at a distance along the axis that is large compared with the outlet diameter the mass flow m of entrained air is proportional to the axial distance x. They give the relations

$$\frac{dm}{dx} = k(M\rho_a)^{\frac{1}{2}} \qquad [15.20.13a]$$

$$\frac{m}{x} = k(M\rho_a)^{\frac{1}{2}} \qquad [15.20.13b]$$

with

$$M = M_o = \frac{\pi}{4} d_o^2 \rho_o u_o^2 \qquad [15.20.14]$$

where M is momentum flux and k is a constant. They give for k a value of 0.282.

For work on jets, Mecklenburgh (1985) introduces the air/fuel mass ratio θ. For this ratio

$$\theta = \left(\frac{1}{c_x} - 1\right)\frac{\rho_a}{\rho_x} \qquad [15.20.15]$$

$$u = \frac{u_o}{1+\theta} \qquad [15.20.16]$$

A set of equations for velocity ratio u_{xo}/u_o and the concentration ratio c_{xo}/c_o for circular and planar jets has been given by C.J. Chen and Rodi (1980). These are, for a circular jet

$$\frac{u_{xo}}{u_o} = 6.2\left(\frac{x}{d_o}\right)^{-1}\left(\frac{\rho_o}{\rho_a}\right)^{\frac{1}{2}} \qquad [15.20.17]$$

$$\frac{c_{xo}}{c_o} = 5\left(\frac{x}{d_o}\right)^{-1}\left(\frac{\rho_o}{\rho_a}\right)^{-\frac{1}{2}} \qquad [15.20.18]$$

and for a planar jet

$$\frac{u_{cl}}{u_o} = 2.4\left(\frac{x}{d}\right)^{-\frac{1}{2}}\left(\frac{\rho_o}{\rho_a}\right)^{\frac{1}{2}} \qquad [15.20.19]$$

$$\frac{c_{cl}}{c_o} = 2\left(\frac{x}{d}\right)^{-\frac{1}{2}}\left(\frac{\rho_o}{\rho_a}\right)^{-\frac{1}{2}} \qquad [15.20.20]$$

where d is the width of the slot and the subscript cl denotes the centre line.

A set of equations for the velocity ratio u_m/u_o only for circular, planar and radial jets for air has been given by Rajaratnam (1976). These are, for a circular jet

$$\frac{u_m}{u_o} = 6.3\left(\frac{x}{d_o}\right)^{-1} \qquad [15.20.21]$$

for a planar jet

$$\frac{u_m}{u_o} = 3.5\left(\frac{x}{b_o}\right)^{-1} \qquad [15.20.22]$$

and for a radial jet

$$\frac{u_m}{u_o} = 3.5\frac{r_p}{b_o}\left(\frac{r}{b_o}\right)^{-1} \qquad [15.20.23]$$

where b_o is the half-width of the slot (for a planar or radial jet), r is the radial distance, r_p is the radius of the pipe or flange in which the outlet exists and the subscript m denotes the centre line, and hence the maximum.

Hess, Leuckel and Stoeckl (1973) have given relations for both subsonic and sonic gas jets. For the subsonic flow the condition is

$$\frac{P_v}{P_a} < \left(\frac{\gamma+1}{2}\right)^{\gamma/(\gamma+1)} \qquad [15.20.24]$$

At the exit plane after isentropic expansion

$$P_E = P_a \qquad [15.20.25]$$

$$T_E = T_v\left(\frac{P_a}{P_v}\right)^{(\gamma-1)/\gamma} \qquad [15.20.26]$$

$$\rho_E = \rho_v\left(\frac{P_a}{P_v}\right)^{1/\gamma} \qquad [15.20.27]$$

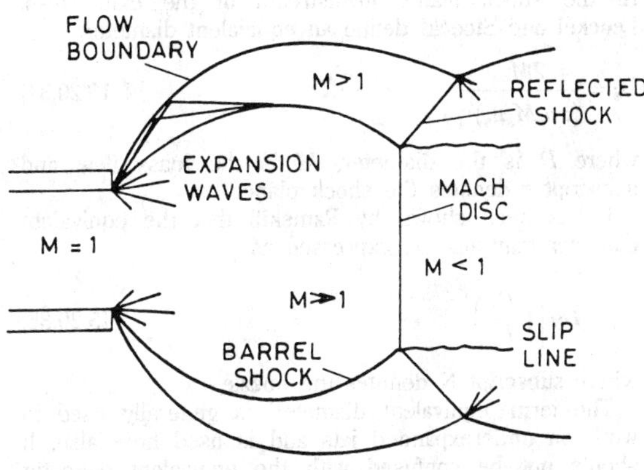

***Figure 15.70** Underexpanded momentum jet (Ewan and Moodie, 1986) M, Mach number (Reproduced by permission of Gordon and Breach Science Publishers, Copyright ©)*

$$u_E = \left\{\frac{2\gamma}{\gamma-1}RT_v\left[1-\left(\frac{P_a}{P_v}\right)^{(\gamma-1)/\gamma}\right]\right\} \qquad [15.20.28]$$

where P is the absolute pressure, R is the universal gas constant, T is the absolute temperature, γ is the ratio of the gas specific heats, and subscripts a, E and v denote atmospheric conditions, exit conditions and vessel conditions, respectively.

Some illustrative calculations on momentum jets have been given by Cude (1974b), Gugan (1976) and Mecklenburgh (1985).

15.20.3 Gas jets: sonic jets

The models just given apply to a gas jet with subsonic flow. A different treatment is necessary for sonic flow conditions. Models for underexpanded sonic jets have been reviewed by Ramskill (1985 SRD R302). The structure of an underexpanded sonic jet is shown in Figure 15.70.

As already mentioned, Hess, Leuckel and Stoeckl (1973) have also given a treatment for circular jets with sonic flow. For sonic flow the condition is

$$\frac{P_v}{P_a} \geq \left(\frac{\gamma+1}{2}\right)^{\gamma/(\gamma-1)} \qquad [15.20.29]$$

At the exit plane, after isentropic expansion $P_E > P_a$, and the conditions are:

$$P_E = P_v\left(\frac{2}{\gamma+1}\right)^{\gamma/(\gamma-1)} \qquad [15.20.30]$$

$$T_E = T_v\left(\frac{2}{\gamma+1}\right) \qquad [15.20.31]$$

$$\rho_E = \rho_v\left(\frac{2}{\gamma+1}\right)^{\gamma/(\gamma-1)} \qquad [15.20.32]$$

$$u_E = \left(\frac{2\gamma}{\gamma+1}RT_v\right)^{\frac{1}{2}} \qquad [15.20.33]$$

At the shock plane downstream of the exit, Hess, Leuckel and Stoeckl define an equivalent diameter:

$$D_s = \frac{2M_s}{(\pi \rho_s M_s u_s)^{\frac{1}{2}}} \qquad [15.20.34]$$

where D is the diameter, M is the mass flow and subscript s denotes the shock plane.

It has been shown by Ramskill that the equivalent diameter may also be expressed as

$$D_s = D_N \left[\left(\frac{P_v}{P_a} \right)^{(\gamma-1)/\gamma} \right]^{\frac{1}{2}} \qquad [15.20.35]$$

where subscript N denotes the nozzle.

The term 'equivalent diameter' is generally used in work on underexpanded jets and is used here also. It should not be confused with the equivalent diameter used to relate non-circular to circular orifices and ducts.

For the velocity at this plane it is assumed that all the excess pressure goes to increase the momentum of the jet:

$$u_s = u_E + \frac{P_E - P_a}{\rho_E u_E} \qquad [15.20.36]$$

Also

$$\rho_s = \frac{\rho_E u_E D_N^2}{u_s D_s^2} \qquad [15.20.37]$$

Another model based on the apparent diameter is that given in the TNO *Yellow Book* (1979). The equations of this model are as follows. In the model use is made of gas densities normalized with respect to the density of air, these densities being denoted by a prime.

The radial velocity and concentration are expressed as

$$\frac{u_{xy}}{u_m} = \exp(-b_1 b_3^2) \qquad [15.20.38]$$

$$\frac{j_{xy}}{j_m} = \exp(-b_2 b_3^2) \qquad [15.20.39]$$

with

$$b_3 = y/x'' \qquad [15.20.40]$$

$$x'' = x + x' \qquad [15.20.41]$$

where j is the volumetric concentration of the gas, u is the velocity of the gas, x is the axial distance from the outlet, x' is the distance between the virtual source and the outlet, y is the radial distance, b_1 and b_2 are constants, b_3 is defined by Equation 15.20.40 and subscript m denotes the centre line, and hence maximum, and subscript xy denotes point (x,y).

The following relations are given for the constants b_1 and b_2:

$$b_1 = 50.5 + 48.2\rho'_{ga} - 9.95(\rho'_{ga})^2 \qquad [15.20.42]$$

$$b_2 = 23 + 41\rho'_{ga} \qquad [15.20.43]$$

where ρ'_{ga} is the normalized density of the gas at atmospheric conditions.

The equivalent diameter d_{eq} is defined as

$$d_{eq} = d_o \left(\frac{\rho'_{go}}{\rho'_{oeq}} \right)^{\frac{1}{2}} \qquad [15.20.44]$$

where ρ'_{go} is the normalized density of the gas at the outlet and ρ'_{oeq} is the normalized density of the gas at the equivalent diameter.

The velocity ratio is

$$\frac{u_m}{u_o} = \frac{b_1}{4} \left(\frac{d_{eq}}{x''} \right)^2 \left[0.32 \frac{x}{d_{eq}} \frac{\rho'_{ga}}{(\rho'_{oeq})^{\frac{1}{2}}} + 1 - \rho'_{ga} \right] \frac{\rho'_{oeq}}{\rho'_{ga}} \qquad [15.20.45]$$

and the concentration ratio is

$$\frac{j_m}{j_o} = \frac{(b_1 + b_2)/b_1}{0.32 \frac{x}{d_{eq}} \frac{\rho'_{ga}}{(\rho'_{oeq})^{\frac{1}{2}}} + 1 - \rho'_{ga}} \qquad [15.20.46]$$

Ewan and Moodie (1986) have given a treatment of underexpanded sonic jets based on the work of Kleinstein (1964). For fully expanded supersonic jets Kleinstein obtained the relation

$$\frac{u_m}{u_n} = 1 - \exp\left[- \frac{1}{k\left(\frac{\rho_a}{\rho_e}\right)^{\frac{1}{2}} \frac{z}{r_n} - x_c} \right] \qquad [15.20.47]$$

where k is a compressible eddy coefficient, r_n is the equivalent radius, u is the velocity of the gas, x_c is a dimensionless core length, z is the distance along the axis from the outlet and subscripts e, m and n denote the exit, the maximum and the nozzle, respectively.

They also quote the following equation given by Warren (1957) and based on a different analysis for the radial variation of velocity in fully developed flow

$$\frac{u}{u_m} = \exp\left\{ -\ln\left[2\left(\frac{r}{r_{0.5}} \right)^2 \right] \right\} \qquad [15.20.48]$$

where $r_{0.5}$ is the radius at which the velocity is half the centre line value.

Ewan and Moodie have adapted the Kleinstein model to an underexpanded jet by introducing the equivalent diameter. For the expansion

$$N = \frac{P_v}{P_a} \qquad [15.20.49]$$

$$P_e = P_v \left(\frac{2}{\gamma + 1} \right)^{\gamma/(\gamma-1)} \qquad [15.20.50]$$

$$\rho_e = \rho_v \left(\frac{2}{\gamma + 1} \right)^{1/(\gamma-1)} \qquad [15.20.51]$$

$$\rho_{eq} = \rho_e \left(\frac{P_a}{P_e} \right) \qquad [15.20.52]$$

where N is the ratio of the vessel pressure to atmospheric pressure and subscripts eq and v denote equivalent and vessel, respectively. Equivalent conditions are based on the sonic jet at ambient pressure, nozzle exit temperature and nozzle mass flow.

They obtain for the velocity ratio

$$\frac{u_m}{u_n} = 1 - \exp\left[-\frac{1}{k\left(\frac{\rho_a}{\rho_{eq}}\right)^{\frac{1}{2}}\frac{z^*}{r_{eq}} - x_c}\right] \quad [15.20.53]$$

with

$$k = 0.08(1 - 0.16\ \mathrm{Ma}) \quad [15.20.54]$$

$$z^* = z - 2z_b \quad [15.20.55]$$

$$z_b = 0.77d_n + 0.068d_n^{1.35}N \quad [15.20.56]$$

where d is the diameter, Ma is the Mach number, z_b is the barrel length and z^* is the modified axial distance. The unit of length in Equations 15.20.55 and 15.20.56 is millimetres.

The equivalent diameter d_{eq} is derived as follows. For continuity

$$\rho_e u_n A_e = \rho_{eq} u_{eq} A_{eq} \quad [15.20.57]$$

where A is the cross-sectional area of flow. But

$$\rho u = \frac{Pu}{RT} \quad [15.20.58a]$$

$$= \frac{Pu}{(\gamma RT)^{\frac{1}{2}}}\left(\frac{\gamma}{RT}\right)^{\frac{1}{2}} \quad [15.20.58b]$$

$$= P\,\mathrm{Ma}\left(\frac{\gamma}{R}\right)^{\frac{1}{2}}\frac{1}{T^{\frac{1}{2}}} \quad [15.20.58c]$$

Then from Equation 15.20.57

$$\frac{\rho_e u_n}{\rho_{eq} u_{eq}} = \frac{A_{eq}}{A_e} = \frac{P_e}{P_a} \quad [15.20.59]$$

Hence

$$d_{eq} = d_n\left(\frac{P_e}{P_a}\right)^{\frac{1}{2}} \quad [15.20.60]$$

In calculating the equivalent diameter the authors apply to the flow a discharge coefficient C_D so that Equation 15.20.60 becomes

$$d_{eq} = d_n\left(\frac{C_D P_e}{P_a}\right)^{\frac{1}{2}} \quad [15.20.61]$$

They quote the value of C_D of 0.85 assumed by Birch *et al.* (1984).

For the concentration ratio they obtain

$$\frac{m_{cl}}{m_e} = 1 - \exp\left[-\frac{1}{k'\left(\frac{\rho_a}{\rho_{eq}}\right)^{\frac{1}{2}}\frac{z^*}{r_{eq}} - x_c}\right] \quad [15.20.62]$$

where k' is another eddy viscosity coefficient and m is the mass fraction. The constant k' has the value 0.104.

15.20.4 Liquid jets

It is also necessary to consider liquid jets. The 'throw' of a liquid jet can be quite large. Such a jet may reach beyond the limit of the area classification zone or may jump over a bund.

The simplest treatment of a liquid jet is based on the assumption that the jet suffers no drag and does not disintegrate. For a jet at ground level, as shown in Figure 15.71(a) the motion in the horizontal direction is given by

$$\frac{dx}{dt} = u \cos \alpha \quad [15.20.63]$$

where t is time, u is the horizontal velocity of the liquid, x is the horizontal distance and α is the angle between the jet and the horizontal. At $t = 0$, $x = 0$ and hence

$$x = ut \cos \alpha \quad [15.20.64]$$

For motion in the vertical direction

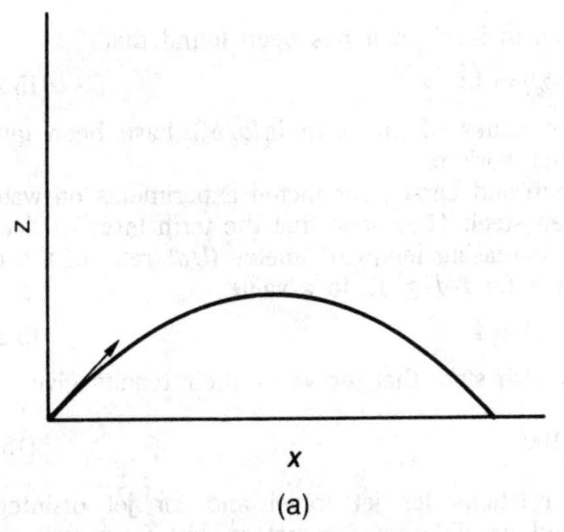

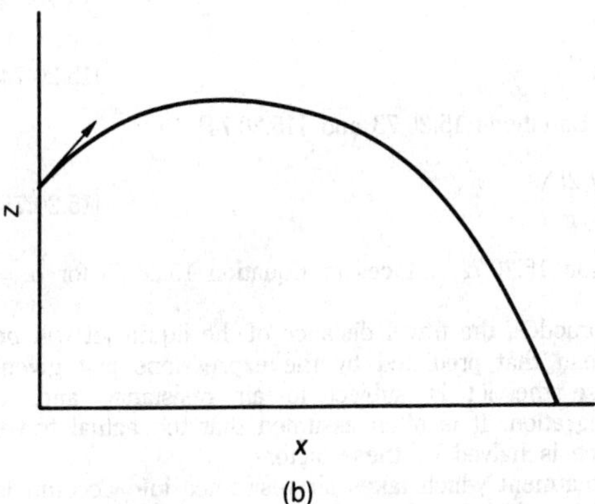

***Figure 15.71** Travel range of a liquid jet, neglecting air resistance and break up: (a) jet from ground level source; (b) jet from elevated source*

$$\frac{d^2z}{dt^2} = -g \quad [15.20.65]$$

Integrating Equation 15.20.65 gives

$$\frac{dz}{dt} = -gt + c \quad [15.20.66]$$

where g is the acceleration due to gravity, z is the vertical distance and c is a constant of integration. At $t = 0$, $dz/dt = u \sin\alpha$ and hence

$$\frac{dz}{dt} = u \sin\alpha - gt \quad [15.20.67]$$

At $t = 0$, $z = 0$ and hence

$$z = ut \sin\alpha - \tfrac{1}{2}gt^2 \quad [15.20.68]$$

The jet returns to the ground at $t = t$, $z = 0$ and hence

$$t = \frac{2u \sin\alpha}{g} \quad [15.20.69]$$

Then from Equations 15.20.64 and 15.20.69 the horizontal distance travelled is

$$x = \frac{u^2 \sin^2\alpha}{g} \quad [15.20.70]$$

The maximum distance of travel x_{max} occurs at $\alpha = 45°$ and is

$$x_{max} = \frac{u^2}{g} \quad [15.20.71]$$

where the subscript max denotes maximum.

For an elevated jet, as shown in Figure 15.71(b), the jet returns to the ground at $t = t$, $z = -l$ and, following the same approach, it may be shown that

$$x = \frac{u^2}{g}\left\{\frac{\sin 2\alpha}{2} + \left[\left(\frac{\sin 2\alpha}{2}\right)^2 + \frac{2lg \cos 2\alpha}{u^2}\right]^{\frac{1}{2}}\right\} \quad [15.20.72]$$

where l is the height above the ground of the outlet and z is the vertical distance above the outlet. Equation 15.20.72 reduces to Equation 15.20.70 for $l = 0$.

For a horizontal elevated jet the following simple treatment is applicable.

$$x = ut \quad [15.20.73]$$

$$l = \tfrac{1}{2}gt^2 \quad [15.20.74]$$

From Equations 15.20.73 and [15.20.74]

$$x = u\left(\frac{2l}{g}\right)^{\frac{1}{2}} \quad [15.20.75]$$

Equation 15.20.72 reduces to Equation 15.20.75 for $\alpha = 0$.

In practice, the travel distance of the liquid jet will be less than that predicted by the expressions just given, because the jet is subject to air resistance and to disintegration. It is often assumed that the actual travel distance is halved by these factors.

A treatment which takes air resistance into account is as follows. It is assumed that the resistance is proportional to the velocity. Then for motion in the horizontal direction

$$\frac{d^2x}{dt^2} = -k\frac{dx}{dt} \quad [15.20.76]$$

Following the same approach as previously, it can be shown

$$x = \frac{u \cos\alpha}{k}[1 - \exp(-kt)] \quad [15.20.77]$$

with

$$t = \frac{(ku \sin\alpha + g)[1 - \exp(-kt)]}{kg} \quad [15.20.78]$$

where k is a constant representing air resistance.

The disintegration of a liquid jet under air resistance has been studied by a number of workers including Rayleigh (1932), van de Sande and Smith (1976) and Engh and Larsen (1979). The following treatment is based on the work of the latter.

The disintegration of a jet is determined by the Weber number We, which gives the ratio of the momentum to the surface tension forces. The Weber number is

$$We = \frac{\rho u^2 d}{2\sigma} \quad [15.20.79]$$

where d is the diameter of the jet, u is the velocity of the jet and σ is the surface tension.

The propagation of a disturbance in a liquid jet leading to its break up was studied by Lord Raleigh. In his model the amplitude δ of a surface disturbance of initial amplitude δ_o increases exponentially with time t:

$$\delta = \delta_o \exp(\alpha t) \quad [15.20.80]$$

where α is a constant. When the amplitude δ of the disturbance is equal to the jet radius a, disintegration into drops begins. Drops are therefore formed at time

$$t = \ln(a/\delta_o)/\alpha \quad [15.20.81]$$

Then the break-up length Z is

$$Z = ut \quad [15.20.82]$$

$$= u \ln(a/\delta_o)/\alpha \quad [15.20.83]$$

From this starting point Engh and Larsen derive the following expression for break-up length:

$$\frac{Z}{d} = 1.03 \ln(a/\delta_o)(2We)^{\frac{1}{2}} \quad [15.20.84]$$

For an inviscid jet it has been found that

$$\ln(a/\delta_o) \approx 12 \quad [15.20.85a]$$

Other values of the term $\ln(a/\delta_o)$ have been given by various workers.

Engh and Larsen conducted experiments on water and molten steel. They state that the term $\ln(a/\delta_o)$ decreases with increasing length/diameter (l/d) ratio of the nozzle, tending for $l/d \geq 13$ to a value

$$\ln(a/\delta_o) \approx 4 \quad [15.20.85b]$$

They also state that for water their results give

$$\frac{Z}{d} \approx 100 \quad [15.20.86]$$

The relations for jet travel and for jet disintegration depend on different parameters, but for many cases of practical interest the jet will start to disintegrate within a short distance of the orifice.

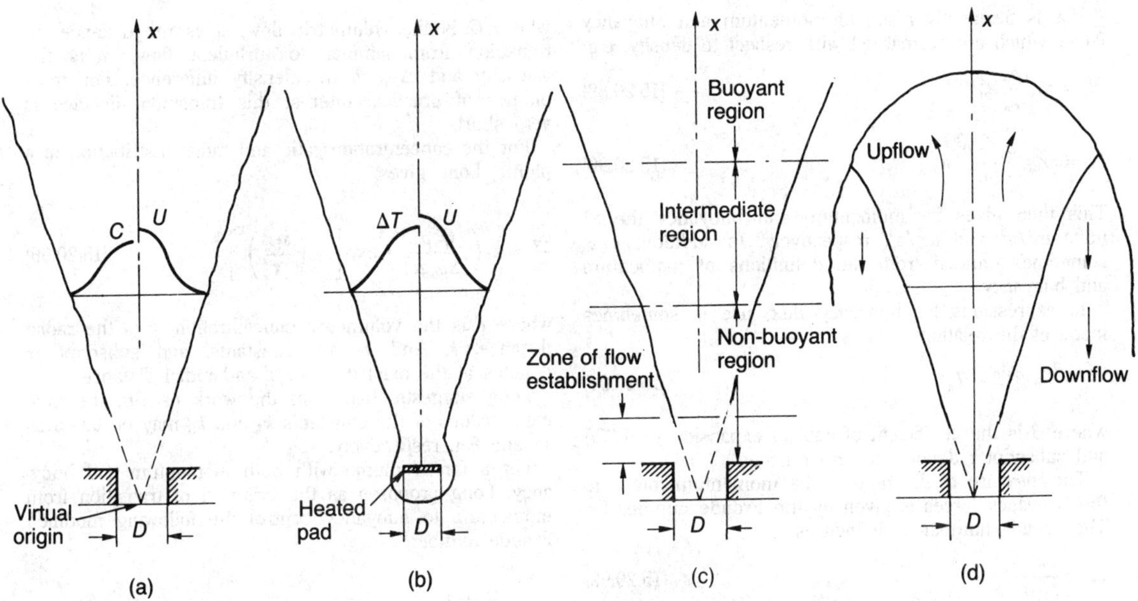

Figure 15.72 *Principal types of jet and plume (after C.J. Chen and Rodi, 1980): (a) pure, non-buoyant jet; (b) pure plume; (c) buoyant jet (forced plume); (d) negatively buoyant jet*

15.20.5 Two-phase jets

Work on two-phase jets includes that by R. Brown and York (1962), Bushnell and Gooderum (1968), Hetsroni and Sokolov (1971), Gyarmathy (1976), Suzuki *et al.* (1978), Melville and Bray (1979a,b), Koestel, Gido and Lamkin (1980), and Tomasko, Weigand and Thompson (1981a,b). A review has been given by Appleton (1984 SRD R303). Two-phase jets may be classified as non-flashing or flashing.

A liquid jet in air tends to disintegrate. Break up may occur due either to air resistance or to vapour bubble formation. The break up of a non-flashing jet due to air resistance was considered in the previous section.

A two-phase non-flashing jet consists of a gas flow with entrained droplets. Characteristics of interest are: for the gas phase, the centre line and the radial velocities; and for the liquid phase, the radial mass flux of the droplets. Relations for these may be found in the work of Hetsroni and Sokolov (1971) and Melville and Bray (1979a,b).

It is two-phase flashing jets, however, which are of prime interest here. An account of such jets is given in Section 15.21.

15.20.6 Plumes

The principal types of jet and plume which occur are illustrated in Figure 15.72. Figure 15.72(a) shows a non-buoyant jet, or pure jet, or simply a jet. Figure 15.72(b) shows a pure plume with no momentum. This figure, therefore, shows as the source not a discharge orifice, but a heated surface. Figure 15.72(c) shows a buoyant plume, with some initial momentum, or forced plume. Figure 15.72(d) shows a negatively buoyant plume.

Accounts of plumes are given in *Plume Rise* (G.A. Briggs, 1969), *Vertical Turbulent Buoyant Jets* (C.J. Chen and Rodi, 1980) and *Handbook on Atmospheric Diffusion* (S.R. Hanna, Briggs and Hosker, 1982). Work on plumes includes that by O.G. Sutton (1950), Yih (1951), Rouse, Yih and Humphreys (1952), B.R. Morton (1959, 1961), Long (1963), Schmidt (1965), J.S. Turner (1966), G.A. Briggs (1965, 1968, 1976), Ooms (1972), Cude (1974a,b), Ooms, Mahieu and Zelis (1974), Davidson and Slawson (1982) and Frick (1984).

Most models of plumes deal with the case of an initial jet in which the momentum decays so that the emission undergoes transition to a buoyant plume. Much of the work on plumes relates to elevated emissions from stacks and is concerned primarily with pollution rather than with hazards. Fundamental quantities in models of dispersion of forced and buoyant plumes are the momentum and buoyancy fluxes.

The momentum flux at the outlet orifice is

$$M_o = \pi r_o^2 \rho_o u_o^2 \quad [15.20.87]$$

where M is the momentum flux, r_o the radius of the outlet, u is the velocity of the gas, ρ is the density and subscript o denotes the outlet.

The buoyancy flux at the outlet is

$$F_o = \pi r_o^2 g(\rho_a - \rho_o)u_o \qquad [15.20.88]$$

where F is the buoyancy flux and subscript a denotes air.

Use is frequently made of momentum and buoyancy fluxes which are normalized with respect to density, e.g.

$$M_o = \pi r_o^2 \frac{\rho_o}{\rho_a} u_o^2 \qquad [15.20.89]$$

$$F_o = \pi r_o^2 g \frac{\rho_a - \rho_o}{\rho_a} u_o \qquad [15.20.90]$$

This then gives for momentum and buoyancy the SI units m^4/s^2 and m^4/s^3, respectively. In addition, π is sometimes omitted from the definitions of momentum and buoyancy.

In expressions for buoyancy flux, use is sometimes made of the relation

$$\frac{\rho_a - \rho}{\rho_r} = \beta(T - T_a) \qquad [15.20.91]$$

where β is the coefficient of cubical expansion ($\approx 1/T_r$) and subscript r denotes the reference.

The measure of the ratio of the momentum forces to the buoyancy forces is given by the Froude number Fr. The Froude number is defined as

$$\mathrm{Fr} = \frac{u}{(gl)^{\frac{1}{2}}} \qquad [15.20.92]$$

where l is a characteristic length.

In the context of plumes the Froude number is more usually defined as

$$\mathrm{Fr} = \frac{u}{\left(gd_o \frac{\rho_a - \rho_o}{\rho_a}\right)^{\frac{1}{2}}} \qquad [15.20.93]$$

where d_o is the diameter of the outlet.

Use is also sometimes made of the following Froude group (e.g. C.J. Chen and Rodi, 1980):

$$\mathrm{Fr} = \frac{u^3}{gd_o\left(\frac{\rho_a - \rho_o}{\rho_a}\right)} \qquad [15.20.94]$$

In some definitions of the Froude number the divisor of the term ($\rho_a - \rho_o$) is ρ_o rather than ρ_a. The Froude number is a criterion for the transition from momentum control to buoyancy control.

Some principal characteristics of a plume have been described by Long (1963). He summarizes the empirical features of a buoyant plume as follows. Under turbulent conditions the plume is conical and apparently diverges from an equivalent point source, the position of the point source depending on the point of transition from laminary to turbulent flow; the time–mean velocity and time–concentration profiles are similar and are approximately Gaussian.

Under adiabatic conditions buoyancy is conserved within the plume and the buoyancy flux remains constant with distance. The momentum flux therefore increases with distance. For the distance within which turbulence sets in for a pure plume Long gives the following relation based on the work of Yih (1951) on plumes from cigarettes:

$$x_{lt} = \left(\frac{\mu^3}{\rho_a^2 Q \Delta\rho_o g}\right)^{\frac{1}{2}} \times 10^5 \qquad [15.20.95]$$

where Q is the volumetric flow, x_{lt} is the distance for transition from laminar to turbulent flow, μ is the viscosity and $\Delta\rho_o$ is the density difference. For most plumes of practical interest this transition distance is very short.

For the concentration ratio and radial distribution of a plume, Long gives:

$$\frac{c_{xr}}{c_o} = k_4\left(\frac{Q^2\rho_a}{\Delta\rho_o g x^5}\right)^{\frac{1}{3}} \exp\left[-\left(\frac{k_5 r}{x}\right)^2\right] \qquad [15.20.96]$$

where c is the volumetric concentration, r is the radial distance, k_4 and k_5 are constants, and subscript xr denotes at the axial distance x and radial distance r.

Long suggests that, from the work of Yih, the time mean values of the constants k_4 and k_5 may be taken as 11 and 8.4, respectively.

For a forced plume with both momentum and buoyancy, Long proposes as the criterion of transition from momentum to buoyancy control the following modified Froude number:

$$\mathrm{Fr} = \frac{\rho_o^{\frac{3}{2}} u^2}{\rho_a^{\frac{1}{2}} \Delta\rho_o g d_o} \qquad [15.20.97]$$

He obtains an approximate value for the transition point by equating Equations 15.20.8 and 15.20.96. Then the transition point occurs at $2.4Fr^{\frac{1}{2}}$ diameters from the source. This criterion is confirmed by the work of Ricou and Spalding (1961) in that the point of equal entrainment in the jet and plume regimes occurs at $2.3Fr^{\frac{1}{2}}$ diameters. The jet persists at least to $0.5Fr^{\frac{1}{2}}$ diameters and the plume is fully developed beyond $3Fr^{\frac{1}{2}}$ diameters.

With the exception of last point, the treatment just given applies to a pure plume in a still atmosphere. In many practical applications the case to be considered is more complex. The emission may have both momentum and buoyancy, starting off as a jet and undergoing transition to a plume. The plume may be affected by the various weather conditions. In addition, the plume may have heavy gas characteristics requiring special treatment.

15.20.7 Plumes in atmospheric turbulence

A different situation occurs where a higher wind speed results in atmospheric turbulence. If conditions are such that the discharge momentum and buoyancy effects are negligible relative to atmospheric turbulent diffusion, the dispersion may be estimated using the methods described in Section 15.10 such as the Sutton and Pasquill–Gifford models.

Using the Sutton model, Long (1963) gives the following equations for an elevated vent, derived from Equation 15.16.29. The full equation is

$$\chi(x,y,z) = \frac{Q}{\pi C_y C_z u x^{2-n}} \exp\left(-\frac{y^2}{C_y^2 x^{2-n}}\right) \times \left[\exp\left(-\frac{z^2}{C_z^2 x^{2-n}}\right) + \exp\left(-\frac{(z+2h)^2}{C_z^2 x^{2-n}}\right)\right] \qquad [15.20.98a]$$

and that for the axial concentration at ground level ($y = z = 0$)

$$\chi(x,0,0) = \frac{Q}{\pi C_y C_z u x^{2-n}} \left[1 + \exp\left(-\frac{4h^2}{C_z^2 x^{2-n}}\right)\right] \qquad [15.20.98b]$$

where x, y and z are the downwind, crosswind and vertical distances, C is the diffusion parameter, h is the height of the outlet, Q is the volumetric flow from the outlet, u is the wind speed, χ is the concentration, n is the diffusion index and subscripts x and y denote downwind and crosswind, respectively. The origin of the co-ordinates in Equation 15.20.98, but in this equation only, is the outlet of the vent.

The maximum ground level concentration χ_{mgl} is

$$\chi_{mgl} = 0.234 \frac{Q}{uh^2} \frac{C_z}{C_y} \qquad [15.20.99]$$

The distance x_{mgl} from the source at which the maximum ground level concentration occurs is

$$x_{mgl} = \left(\frac{h^2}{C_z^2}\right)^{1/(2-n)} \qquad [15.20.100]$$

Values of the index n and of the generalized diffusion parameters C_y and C_z for the first 10 m above ground are given by Long as follows:

Stability condition	n	C_y ($m^{n/2}$)	C_z ($m^{n/2}$)
Large lapse rate	1/5	0.37	0.21
Neutral condition	1/4	0.21	0.12
Moderate inversion	1/3	0.13	0.08
Large inversion	1/2	0.11	0.06

The Sutton equation (15.20.98) was not originally intended for use over the very short distances involved in vent calculations, and its use demands some justification. This point has been discussed by Long, who concludes that the use of the equation is justified. A further discussion of this aspect is given in Section 15.48.

15.20.8 Plumes with negative buoyancy

In many practical situations the emission of interest is that of a dense gas. Plumes with negative buoyancy are therefore of particular interest. A study of a plume with negative buoyancy has been made by J.S. Turner (1966). He carried out experiments in which a dense liquid was injected into a less dense one. The flow pattern obtained is illustrated in Figure 15.72(d).

For this case Turner obtained the following expression for maximum plume rise z_{max}:

$$z_{max} = C\frac{M^{\frac{3}{4}}}{F^{\frac{1}{2}}} \qquad [15.20.101]$$

with

$$F = \pi r_o^2 g\left(\frac{\rho_o - \rho_1}{\rho_1}\right) u_o \qquad [15.20.102]$$

$$M = \pi r_o^2 u_o^2 \qquad [15.20.103]$$

where F is the buoyancy flux, M is the momentum flux, u_o is the outlet velocity of the gas, ρ_o is the density of the outlet gas, ρ_1 is the density of ambient fluid and C is a constant. Turner gives the value of the constant C as 1.85.

15.20.9 Cude model

A model for jets and plumes has been given by Cude (1974b). The Cude model distinguishes between a jet, a buoyant plume, a negatively buoyant plume and a buoyant plume at low wind speed.

Plume in still air

A treatment of the rise or fall of a buoyant plume in low wind speed conditions has been given by Bosanquet (Bosanquet 1935, 1957; Bosanquet and Pearson, 1936). In the simplified form given by Cude, the rise or fall of the plume is

$$i = -wAF_1(x_1) \qquad [15.20.104]$$

with

$$A = 9.42\frac{gN}{w^4}\frac{\rho_g - \rho_a}{\rho_a} \qquad [15.20.105]$$

$$x_1 = t_1/A \qquad [15.20.106]$$

where A is a time parameter, $F_1(x_1)$ is a function which expresses the change in density of the plume as a result of dilution by air, i is the rise or fall of the plume at a distance wt_1 from the source, N is the volumetric flow from the outlet, t_1 is an arbitrarily chosen time interval, w is the wind speed and ρ_g is the density of the gas at the outlet. The time t_1 is taken by Cude as 100 s.

If the gas density is greater than that of air, the plume falls and i has a negative numerical value, while if the gas density is less than that of air the plume rises and i has a positive value.

The function $F_1(x_1)$ is a complicated one. Some values are

x_1	F_1	x_1	F_1
$<10^{-3}$	$1.0543x_1^{0.75}$	10	2.33
10^{-3}	0.0059	10^2	4.50
10^{-2}	0.0323	10^3	6.79
10^{-1}	0.170	10^4	9.09
1	0.767	$>10^4$	$= \ln x_1 - 0.12$

Plume in atmospheric turbulence

For conditions of atmospheric turbulence with higher wind speeds Cude gives the following equations. For the axial concentration p_a from an elevated source

$$p_a = \frac{5.82J}{wx^{1.75}} \qquad [15.20.107]$$

for the maximum ground level concentration p_{mgl}

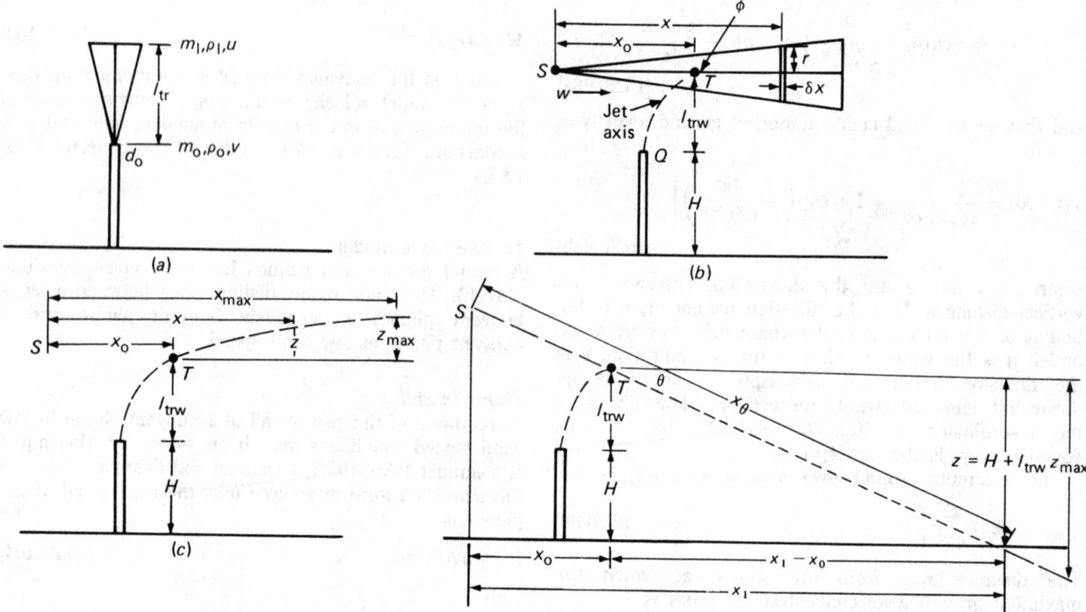

Figure 15.73 *Cude model of dispersion as jet and plumes from a stack (after Cude, 1977): (a) turbulent momentum jet; (b) and (c) buoyant plumes; (d) negatively buoyant plume with steep descent*

$$p_{mgl} = \frac{0.234J}{w(H+l)^2} \quad [15.20.108]$$

and for the distance x_{mgl} at which the maximum ground level concentration occurs

$$x_{mgl} = 5.3(H+l)^{1.14} \quad [15.20.109]$$

where H is the height of the outlet (m), J is the volumetric flow from the outlet referred to the ambient temperature (m^3/s), l is the height of the jet (and of the horizontal axis of the plume above the outlet) (m), p_a is the concentration of gas along the axis at a distance x (mole fraction), p_{mgl} is the maximum ground level concentration of the gas (mole fraction), w is the wind speed (m/s), x is the downwind distance (m) and x_{mgl} is the distance at which the maximum ground level concentration occurs (m).

Momentum jet

For a momentum jet the model given by Cude is that already described in Equations 15.20.1–15.20.5. The height of the jet at which transition occurs is calculated as follows. It is assumed that at the transition point the mass flow of the gas-air mixture m_l can be approximated by the mass flow of air m_a

$$m_l \approx m_a \quad [15.20.110]$$

and that the velocity of the jet u is equal to that of the wind w

$$u \approx w \quad [15.20.111]$$

Then for still air, substituting in Equation 15.20.4 for d_o from Equation 15.20.1 and for m_l from Equation 15.20.110

$$\frac{m_a}{l} = k_1 k_6 (m_o v)^{\frac{1}{2}} \quad [15.20.112a]$$

$$= k_1 k_6 \psi^{\frac{1}{2}} \quad [15.20.112b]$$

with

$$k_6 = \left(\frac{\pi \rho_l}{4}\right)^{\frac{1}{2}} \quad [15.20.113]$$

$$\psi = m_o v \quad [15.20.114]$$

where m_o is the mass flow at the outlet, v is the velocity of the gas at the outlet, ψ is the momentum flux in the still air, ρ_l is the density of the jet at distance l, and k_6 is a constant.

Forced plume in wind

For a momentum jet which undergoes transition to a plume in a wind, Cude gives the following treatment. The situation is illustrated in Figure 15.73. The transition from momentum to buoyancy control is obtained as follows. For windy conditions the momentum flux is a vector quantity. In this treatment it is assumed that

$$\psi_w = [(m_o v)^2 + (m_a w)^2]^{\frac{1}{2}} \qquad [15.20.115]$$

where ψ_w is momentum flux in windy conditions. But from Equations 15.20.3, 15.20.110 and 15.20.111

$$\frac{m_o v}{m_a w} = 1 \qquad [15.20.116]$$

Hence from Equations 15.20.114–15.20.116

$$\psi_w = 2^{\frac{1}{2}}\psi \qquad [15.20.117]$$

Then, since from Equation 15.20.112b the length of the jet is inversely proportional to the root of the momentum flux, the transition heights of the jet in still air conditions l_{tr} and in windy conditions l_{trw} are related as follows:

$$l_{trw} = \frac{l_{tr}}{2^{\frac{1}{4}}} = \frac{l_{tr}}{1.2} \qquad [15.20.118]$$

The following relation is given by Cude for the horizontal displacement of the transition point x_{trw}:

$$x_{trw} \approx 0.5 l_{trw} \qquad [15.20.119]$$

The height reached by the plume from the transition point is calculated as follows. It is assumed that, as shown in Figure 15.73(b), there is a virtual source for the plume at point S which is a distance x_o upwind from the transition point T. This assumption allows for the mixing effect of the jet. The distance x_o is that which would be necessary to achieve the same concentration at the transition point if the mixing were done by the wind. Then, consider the element δx in Figure 15.73(b):

Volume of element $= \pi r^2 \delta x$

Volume of pure gas in element $= Q\delta x/w$

Volume of air in element $= \pi r^2 \delta x - Q\delta x/w$

Mass of pure gas in element $= Q\rho_o \delta x/w$

Mass of air in element $= \pi r^2 \rho_a \delta x - Q\rho_a \delta x/w$

Mass of air displaced by element $= \pi r^2 \rho_a \delta x$

$$\text{Buoyancy force on element} = g[\pi r^2 \rho_a \delta x - (\pi r^2 \rho_a \delta x - Q\rho_a \delta x/w) - Q\rho_o \delta x/w$$

$$= \frac{gQ\delta x}{w}(\rho_a - \rho_o)$$

where Q is the volumetric flow of the gas at the outlet, r is the radius of the plume and x is the downwind distance from the virtual source.

The rising plume experiences a resistance due to the shear stress in the air adjacent to it. Equating the buoyancy force and the shear stress

$$\frac{gQ\delta x}{w}(\rho_a - \rho_o) = 2\pi r \delta x u_b^2 \rho_a D_c \qquad [15.20.120]$$

where D_c is the drag coefficient and u_b is the buoyancy, or upward, velocity of the plume. But

$$r = x \tan \phi \qquad [15.20.121]$$

where ϕ is the angle between the axis and the edge of the plume, or half-angle of plume. Hence Equation 15.20.120 can be rearranged to give

$$u_b = k_7 x^{-\frac{1}{2}} \qquad [15.20.122]$$

with

$$k_7 = \left(\frac{gQ}{2\pi D_c w \tan \phi} \frac{\rho_a - \rho_o}{\rho_a}\right)^{\frac{1}{2}} \qquad [15.20.123]$$

The half-angle of a plume from a stack is generally 5° or 6° so that $\tan \phi \approx 1$. The drag coefficient D_c is estimated by Cude from the work of Bodurtha, Palmer and Walsh (1973) as 0.4.

The treatments given by Cude for rising and for falling plumes are slightly different. The condition for a rising plume is $\rho_o < \rho_a$. The treatment for a rising plume shown in Figure 15.73(c) is as follows. The rise of the plume is given by

$$dz = \frac{u_b dx}{w} \qquad [15.20.124a]$$

$$= \frac{k_7 x^{-1} dx}{w} \qquad [15.20.124b]$$

where z is the vertical distance above the transition point. Integrating Equation 15.20.124 with the boundary conditions

$$z = 0, \qquad x = x_o$$

gives

$$z = \frac{2k_7}{w}(x^{\frac{1}{2}} - x_o^{\frac{1}{2}}) \qquad [15.20.125]$$

The maximum rise z_{max} of a plume occurs at a distance x_{max} where

$$x_{max} = k_8 w + x_o \qquad [15.20.126]$$

where k_8 is a constant. The value of the constant k_8 may be obtained from the observation that a plume generally reaches its maximum rise after 200 s so that $k_8 = 200$. Then from Equation 15.20.125

$$z_{max} = \frac{2k_7}{w}(x_{max}^{\frac{1}{2}} - x_o^{\frac{1}{2}}) \qquad [15.20.127]$$

The distance x_o is obtained from Equation 15.20.107:

$$x_o = \left(\frac{5.82J}{w p_a}\right)^{1/1.75} \qquad [15.20.128]$$

The concentration p_a is, in this case, the concentration on the axis at the transition point and, assuming that the axial concentration is twice the mean concentration in the jet,

$$p_a = \frac{2M_a m_o}{M m_a} \qquad [15.20.129]$$

$$= \frac{2M_a w}{M v} \qquad [15.20.130]$$

where M is the molecular weight of the gas and M_a is the molecular weight of the air.

The total effective height h_{pl} reached by the plume is thus

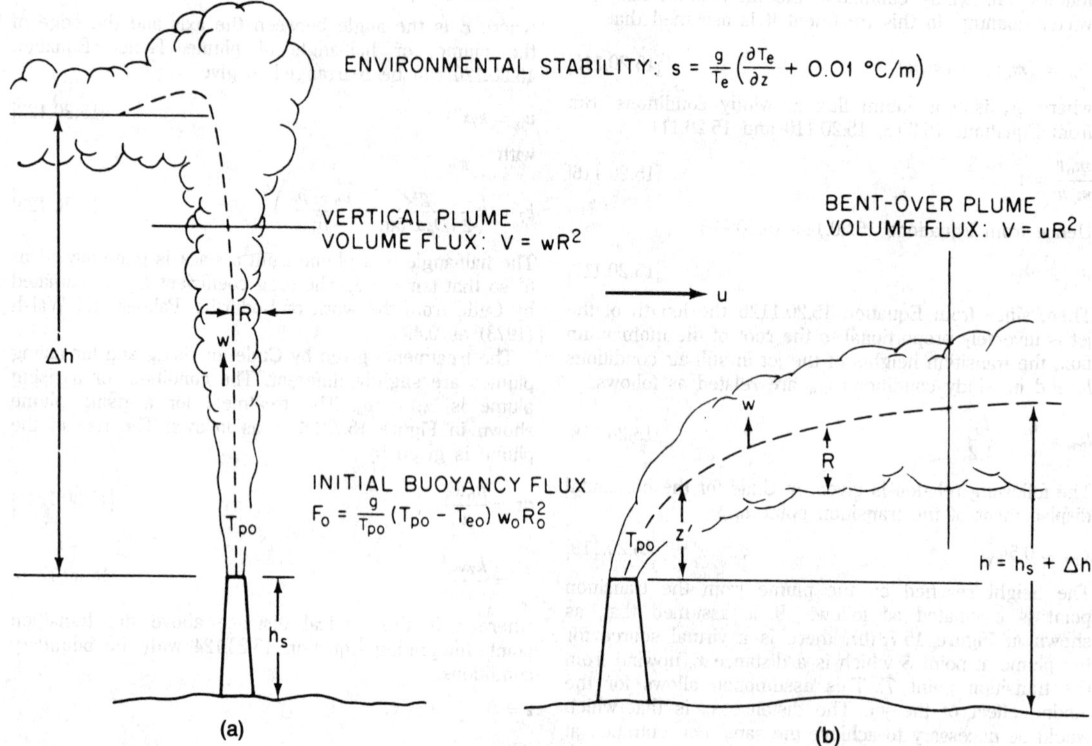

Figure 15.74 Briggs model of dispersion of plumes from a stack (S.R. Hanna, Briggs and Hosker, 1982): (a) vertical plume; and (b) bent-over plume h and R in the figure are denoted in the text by h_e and r respectively

$$h_{pl} = H + l_{trw} + z_{max} \quad [15.20.131]$$

The maximum ground level concentration p_{mgl} is calculated from an equation similar to Equation 15.20.108, using the total effective height of the plume:

$$p_{mgl} = \frac{0.234J}{wh_{pl}^2} \quad [15.20.132]$$

and the distance x_{mgl} at which the maximum ground level concentration occurs is calculated from an equation similar to Equation 15.20.109:

$$x_{mgl} = 5.3h_{pl}^{1.14} \quad [15.20.133]$$

The treatment for a falling plume, which is an approximate one, is as follows. The condition for a falling plume is $\rho_o > \rho_a$. As before, calculations are made of the distances z and z_{max}, but in this case the quantities denote vertical distances below the transition point. In calculating k_7 the term $[(\rho_a - \rho_o)/\rho_a$ is replaced by the term $[(\rho_o - \rho_a)/\rho_a]$. For a falling plume there are two situations which can arise.

If conditions are favourable

$$z_{max} < (H + l_{trw}) \quad [15.20.134]$$

The plume tends to become horizontal as it approaches the ground. In this case the method is to determine the total effective height h_{pl} from Equation 15.20.131 but with the modification that the term z_{max} is subtracted not added. The maximum ground level concentration p_{max} is then calculated as before from Equation 15.20.132.

It should be noted that Equation 15.20.132, which is approximate, gives very large values for p_{mgl} for very small values of h_{pl} and is not valid for such values. A check may be made by substituting x_{max} instead of x_o into Equation 15.20.128 and calculating p_a. If $p_a < p_{mgl}$, the value of p_{mgl} is rejected and the value of p_a is accepted in its place.

If conditions are unfavourable

$$z_{max} > (H + l_{trw}) \quad [15.20.135]$$

The plume approaches the ground at an angle. The maximum ground level concentration is the axial concentration in the plume when it reaches ground level. The situation is shown in Figure 15.73(d). In this case the method is to make the substitution

$$z = H + l_{trw} \quad [15.20.136]$$

The distance x_1 from the virtual source at which the plume axis reaches the ground is then calculated from Equation 15.20.125, using Equation 15.20.136. Then from the geometry of Figure 15.73(d)

$$\tan\theta = \frac{H + l_{trw}}{x_i - x_o} \quad [15.20.137]$$

where θ is angle between axis of plume and horizontal. The horizontal component of the plume velocity is w and the vertical component is $w \tan \theta$. The resultant velocity u_θ is

$$u_\theta = w(1 + \tan^2 \theta)^{\frac{1}{2}} \quad [15.20.138]$$

The distance x_θ from the virtual source S to the point where the plume reaches the ground measured along the axis of the plume is

$$x_\theta = \frac{x_1}{\cos \theta} = x_1(1 + \tan^2 \theta)^{\frac{1}{2}} \quad [15.20.139]$$

The maximum ground level concentration ρ_{mgl} may be estimated by substituting u_θ and x_θ for w and x_o in Equation 15.20.128. The distance x_{mgl} at which the maximum ground level concentration is reached is

$$x_{mgl} = x_1 - x_o \quad [15.20.140]$$

Cude has compared his method with that of Bosanquet and with experimental data on plume paths given by N.G. Stewart *et al.* (1954) and quoted by Pasquill (1962a, p. 221). It predicts a plume rise approximately twice that of Bosanquet's method and apparently more in accordance with the experimental data of Stewart *et al.* It gives a maximum ground level concentration which is approximately half that obtained by Bosanquet's method, although Cude emphasizes that this difference is well within the range of variations which arise as a result of different atmospheric conditions. The method is applicable where the density difference between the plume and the air is zero, a condition where Bosanquet's method is unreliable.

15.20.10 Briggs models

A model for jet and plumes and, in particular, for plume rise has been given by G.A. Briggs (1965, 1968, 1969, 1976), and S.R. Hanna, Briggs and Hosker (1982). The Briggs model distinguishes between a plume dominated by momentum and a plume dominated by buoyancy; between a vertical plume and a bent-over plume; and between a plume near the source and a plume in the three main stability conditions (stable, near neutral and convective, or unstable). The plume scenarios described in Briggs' models are illustrated in Figure 15.74.

In these models the prime concern is with stack gases which have a temperature much greater than that of air but a molecular weight which does not differ greatly from that of air. For such gases Briggs defines the momentum flux as

$$M = wV \quad [15.20.141]$$

and the buoyancy flux as

$$F = g\frac{(T_p - T_{en})}{T_p}V \quad [15.20.142]$$

where F is the buoyancy flux, M is the momentum flux, V is the volumetric flow, w is the velocity in the vertical direction and the subscripts en and p denote environment and plume, respectively.

The volumetric flow V is defined as

$$V = wr^2 \quad \text{Vertical plume} \quad [15.20.143a]$$

$$V = ur^2 \quad \text{Bent-over plume} \quad [15.20.143b]$$

where r is the radius and u is the wind speed.

Thus in definition 15.20.143 Briggs omits the term π and in definitions 15.20.141 and 15.20.142 the term πr^2. He adopts a similar approach to the definition of other quantities.

The initial momentum flux is

$$M_o = w_o V_o \quad [15.20.144]$$

and the initial buoyancy flux is

$$F_o = g\left(\frac{T_{po} - T_{eo}}{T_{po}}\right)V_o \quad [15.20.145a]$$

$$F_o = g\left(1 - \frac{T_{eo}}{T_{po}}\right)V_o \quad [15.20.145b]$$

where subscripts eno, o and po denote the initial environment, the outlet and the initial plume, respectively.

If the molecular weight of the gas emitted differs significantly from that of air, the initial momentum and buoyancy may be modified as follows:

$$M_o = \frac{\rho_o}{\rho_a} w_o V_o \quad [15.20.146]$$

$$F_o = g\left(1 - \frac{T_{eo} m_p}{T_{po} m_o}\right)V_o \quad [15.20.147]$$

where m is the molecular weight.

The basis of the Briggs model is the equations for conservation of momentum and of buoyancy, together with additional relations required for the solution, or closure, of the equations. The closure relation is the Taylor entrainment assumption.

The basic equations are as follows. For conservation of momentum

$$\frac{dM}{dz} = \frac{F}{w} \quad \text{Vertical plume, bent-over plume} \quad [15.20.148]$$

For conservation of buoyancy

$$\frac{dF}{dz} = -sV \quad \text{Vertical plume} \quad [15.20.149a]$$

$$\frac{dF}{dz} = -\frac{sV}{S} \quad \text{Bent-over plume} \quad [15.20.149b]$$

with

$$s = \frac{g}{T_e}\left(\frac{dT_e}{dz} + \Gamma\right) \quad [15.20.150]$$

where s is the environmental stability, S is the ratio of the effective area influenced by the plume momentum to the cross-sectional area of the so-called thermal plume, T_e is the absolute temperature of the environment, z is the vertical distance above the outlet and r is the dry adiabatic lapse rate. The value of S is found to be 2.3. For closure

$$\frac{dV}{dz} = 2r\alpha w \quad \text{Vertical plume} \quad [15.20.151a]$$

$$\frac{dV}{dz} = 2r\beta w \quad \text{Bent-over plume} \quad [15.20.151b]$$

where α and β are constants.

The velocities are

$$w = \frac{dz}{dt} \tag{15.20.152}$$

$$u = \frac{dx}{dt} \tag{15.20.153}$$

An alternative form of the equation for conservation of momentum, utilizing Equations 15.20.141 and 15.20.152, is

$$\frac{d(wV)}{dt} = F \tag{15.20.154}$$

The derivation of the model equations is largely based on solutions of Equation 15.20.154 for the various conditions.

Plume near source: vertical plume

At the source there is a momentum-dominated plume, or jet. Then, taking the buoyancy as zero, or setting $F = 0$, gives

$$wV = \text{Constant} \tag{15.20.155a}$$

Hence

$$(wV)^{\frac{1}{2}} = \text{Constant} \tag{15.20.155b}$$

Then from Equation 15.20.143a]

$$(wV)^{\frac{1}{2}} = wr \tag{15.20.156a}$$

$$= w_o r_o \tag{15.20.156b}$$

But for a jet

$$r = \beta z \tag{15.20.157}$$

The constant β has the value 0.16 for the jet phase. Then

$$wz \propto w_o r_o \tag{15.20.158}$$

From the work of Pai, the centre line velocity w_{cl} of a jet is

$$w_{cl} = 13\frac{w_o r_o}{z} \tag{15.20.159}$$

But the mean velocity is half the centre line velocity and hence

$$w = 6.5\frac{w_o r_o}{z} \tag{15.20.160a}$$

$$= 6.5\frac{M^{\frac{1}{2}}}{z} \tag{15.20.160b}$$

Transition from momentum to buoyancy control occurs at time

$$t = \frac{M}{F_o} \tag{15.20.161}$$

Typically the time to transition is less than 10 s.

After transition there is a buoyancy dominated plume, or buoyant plume. For this

$$w = 2.3\frac{F_o^{\frac{1}{3}}}{z} \tag{15.20.162}$$

The radius r is again given by Equation 15.20.157, but the constant β has a value of 0.15 for the plume phase.

Plume near source: bent-over plume

Near the source, setting F = Constant in Equation 15.20.154 gives on integrating

$$wV = M_o + F_o t \tag{15.20.163}$$

But for a bent-over plume

$$r = \beta z \tag{15.20.164}$$

Then from Equations 15.20.143b, 15.20.152, 15.20.163 and 15.20.164

$$u(\beta z)^2\frac{dz}{dt} = M_o + F_o t \tag{15.20.165}$$

$$\int_0^z z^2 \, dz = \int_0^t \left(\frac{M_o}{\beta^2 u} + \frac{F_o t}{\beta^2 u}\right) dt \tag{15.20.166}$$

$$z = \left(\frac{3M_o t}{\beta^2 u} + \frac{3F_o t^2}{2\beta^2 u}\right)^{\frac{1}{3}} \tag{15.20.167}$$

Alternatively, since

$$u = x/t \tag{15.20.168}$$

Equation 15.20.167 can be written as

$$z = \left(\frac{3M_o x}{\beta^2 u^2} + \frac{3F_o x^2}{2\beta^2 u^3}\right)^{\frac{1}{3}} \tag{15.20.169}$$

For the momentum-dominated plume, setting $F_o = 0$, gives

$$z = \left(\frac{3M_o t}{\beta^2 u}\right)^{\frac{1}{3}} \tag{15.20.170}$$

or from Equation 15.20.168]

$$z = \left(\frac{3M_o x}{\beta^2 u^2}\right)^{\frac{1}{3}} \tag{15.20.171}$$

Equation 15.20.171 is the $\frac{1}{3}$ law for a bent-over momentum plume.

For the buoyancy-dominated plume, setting $M_o = 0$, gives

$$z = \left(\frac{3}{2}\frac{F_o t^2}{\beta^2 u}\right)^{\frac{1}{3}} \tag{15.20.172}$$

and

$$z = \left(\frac{3}{2}\frac{F_o x^2}{\beta^2 u^3}\right)^{\frac{1}{3}} \tag{15.20.173}$$

But for a buoyant plume the constant β has the value 0.6. Substituting in Equation 15.20.173 gives

$$z = C_1\frac{F_o^{\frac{1}{3}}}{u}x^{\frac{2}{3}} \tag{15.20.174}$$

with

$$C_1 = 1.6 \tag{15.20.175}$$

where C_1 is a constant. Equation 15.20.174 is the $\frac{2}{3}$ law for a bent-over buoyant plume. A table of values of the constant C_1 is given by G.A. Briggs (1976). His preferred value is 1.6.

Plume rise in stable conditions
In stable conditions the environmental stability s is constant. Then from Equations 15.20.148, 15.20.149 and 15.20.152

$$\frac{d^2M}{dt^2} = -sM \qquad [15.20.176]$$

This is the equation of a harmonic oscillator.

Briggs introduces a correction to the environmental stability s to allow for the effective vertical momentum flux M_{eff} and defines a modified stability s' such that

$$s' = s \qquad \text{Vertical plume} \qquad [15.20.177a]$$

$$= \frac{M}{M_{eff}} s \qquad \text{Bent-over plume} \qquad [15.20.177b]$$

The ratio M/M_{eff} has the value 2.3. Then from Equations 15.20.141 and 15.20.176, and using s' rather than s,

$$\frac{d^2(wV)}{dt^2} = s'wV \qquad [15.20.178]$$

Integrating Equation 15.20.178 with the initial conditions

$$t = 0; \qquad d(wV)/dt = F_o; \qquad wV = M_o$$

gives

$$wV = M_o \cos(w't) + \frac{F_o}{w'} \sin(w't) \qquad [15.20.179]$$

with

$$w' = s'^{\frac{1}{2}} \qquad [15.20.180]$$

In Briggs' model use is made of several plume heights. One is the maximum rise z_{max}. Another is the 'equilibrium' height z_{eq}, defined as the point at which $F = 0$. A third is the 'falling back' height z_{fb}, defined as the height where the plume at maximum rise would be in equilibrium if no further mixing were to take place. Both z_{eq} and z_{fb} are fractions of z_{max}, as described below.

Plume rise in stable conditions: vertical plume
For the momentum plume, G.A. Briggs (1976) gives

$$z_{max} \approx 4\left(\frac{M_o}{s}\right)^{\frac{1}{4}} \qquad [15.20.181]$$

and also

$$z_{eq} = 0.77z_{max} \qquad [15.20.182a]$$

$$z_{fb} = 0.81z_{max} \qquad [15.20.182b]$$

S.R. Hanna, Briggs and Hosker (1982) also give

$$z_{eq} = 2.44\left(\frac{M_o}{s}\right)^{\frac{1}{4}} \qquad [15.20.183]$$

For the buoyant plume G.A. Briggs (1976) gives

$$z_{max} = 5\frac{F_o^{\frac{1}{4}}}{s^{\frac{3}{8}}} \qquad [15.20.184]$$

S.R. Hanna, Briggs and Hosker (1982) also give

$$z_{eq} = 5.3\frac{M_o^{\frac{1}{4}}}{s^{\frac{3}{8}}} - 6r_o \qquad [15.20.185]$$

The last term in Equation 15.20.185 is a correction to allow for the virtual source.

Plume rise in stable conditions: bent-over plume
For the buoyant plume, Equation 15.20.178 may be reformulated using Equations 15.20.143, 15.20.152 and 15.20.157. Then integrating the equation and taking the maximum value of z gives

$$z_{max} = \left(\frac{3F_o}{\beta^2 us'}\right)^{\frac{1}{3}}\left\{1 + \left[1 + \left(\frac{w'M_o}{F_o}\right)^2\right]^{\frac{1}{2}}\right\}^{\frac{1}{3}} \qquad [15.20.186]$$

But for $F_o \gg (M_o/w')$ Equation 15.20.186 reduces to

$$z_{max} = \left(\frac{6F_o}{\beta^2 us'}\right)^{\frac{1}{3}} \qquad [15.20.187]$$

Taking the constant β as 0.6 in Equation 15.20.187 gives

$$z_{max} = 2.6\left(\frac{F_o}{us'}\right)^{\frac{1}{3}} \qquad [15.20.188]$$

Also

$$z_{eq} = 0.79z_{max} \qquad [15.20.189a]$$

$$z_{fb} = 0.83z_{max} \qquad [15.20.189b]$$

G.A. Briggs (1976) also gives the following equation for plume rise Δh:

$$\Delta h = C_2\left(\frac{F_o}{us}\right)^{\frac{1}{3}} \qquad [15.20.190]$$

where C_2 is a constant. Values of the constant C_2 are tabulated by Briggs (1976). His preferred value is 2.6.

Plume rise limited by atmospheric turbulence
For conditions other than stable conditions, Briggs' treatment utilizes a 'break-up' model in which it is assumed that the plume rise terminates when the internal plume eddy dissipation rate equals the ambient eddy dissipation rate ϵ. Then

$$\frac{\eta w^3}{z} = \epsilon \qquad [15.20.191]$$

where η is a constant. The constant η has the value 1.5. The term on the left-hand side of Equation 15.20.191 is the internal plume eddy dissipation rate.

For a stack, the effective stack height (EHS) is

$$h_e = h_s + \Delta h \qquad [15.20.192]$$

where Δh is the plume rise above the stack outlet, h_e is the effective stack height and h_s is the actual stack height.

The plumes considered are momentum and buoyant plumes, both bent over.

Near neutral conditions
In near neutral conditions the ambient eddy dissipation rate is

$$\epsilon = \frac{u_*^3}{kh_e} \qquad [15.20.193]$$

where u_* is the friction velocity and k is von Karman's constant. The constant k has a value of 0.4.

Near neutral conditions: momentum plume
For a momentum plume for which the first phase of rise is given by Equation 15.20.171

$$\Delta h = \left(\frac{M_o}{\beta^2 u}\right)^{\frac{3}{7}}\left(\frac{\eta}{\epsilon}\right)^{\frac{1}{7}} \qquad [15.20.194]$$

From Equations 15.20.192 and [15.20.193]

$$\Delta h = \left(\frac{M_o}{\beta^2 u u_*}\right)^{\frac{3}{7}}(\eta k)^{\frac{1}{7}}\left(1+\frac{h_s}{\Delta h}\right)^{\frac{1}{7}}\Delta h^{\frac{1}{7}} \qquad [15.20.195]$$

and hence

$$\Delta h = \frac{(\eta k)^{\frac{1}{6}}}{\beta}\left(\frac{u}{u_*}\right)^{\frac{1}{2}}\left(\frac{M_o}{u^2}\right)^{\frac{1}{2}}\left(1+\frac{h_s}{\Delta h}\right)^{\frac{1}{6}} \qquad [15.20.196]$$

Assuming $\Delta h \gg h_s$ so that the last term of Equation 15.20.196 can be neglected and substituting for η and k values of 1.5 and 0.4, respectively, gives

$$\Delta h = \frac{0.9}{\beta}\left(\frac{u}{u^*}\right)^{\frac{1}{2}}\frac{M_o^{\frac{1}{2}}}{u} \qquad [15.20.197]$$

From Equations 15.20.141, 15.20.143a and 15.20.197, taking the constant β as 0.6 and the ratio u/u_* as 15,

$$\Delta h = 2.9\frac{w_o d_o}{u} \qquad [15.20.198]$$

where d_o is the diameter of the source.

G.A. Briggs (1982) also gives

$$\Delta h = 3d_o\left(\frac{w_o}{u} - 1\right) \qquad [15.20.199]$$

The additional term is evidently a virtual source correction.

Near neutral conditions: buoyant plume
For a buoyant plume for which the first phase of rise is given by Equation 15.20.173]

$$\Delta h = \left(\frac{2F_o}{3\beta^2 u}\right)^{\frac{3}{5}}\left(\frac{\eta}{\epsilon}\right)^{\frac{2}{5}} \qquad [15.20.200]$$

Proceeding as before, it can be shown that

$$\Delta h = \frac{2}{3\beta^2}(\eta k)^{\frac{2}{3}}\left(\frac{F_o}{uu_*^2}\right)\left(1+\frac{h_s}{\Delta h}\right)^{\frac{2}{3}} \qquad [15.20.201]$$

Neglecting the last term of Equation 15.20.201 and taking for the constants β, η and k values of 0.6, 1.5 and 0.4, respectively, gives

$$\Delta h = 1.3\left(\frac{F_o}{uu_*^2}\right) \qquad [15.20.202]$$

Briggs also gives as an approximation for Equation 15.20.201 the following equation:

$$\Delta h = 1.54\left(\frac{F_o}{uu^{*2}}\right)^{\frac{2}{3}}h_s^{\frac{1}{3}} \qquad [15.20.203]$$

Convective conditions
For convective, or unstable, conditions Equation 15.20.192 does not apply. Instead, the relation used is

$$\epsilon = 0.25H \qquad [15.20.204]$$

where H is surface buoyancy flux.

The resultant equation for plume rise is

$$\Delta h = 3\left(\frac{F_o}{u}\right)^{\frac{3}{5}}H^{-\frac{2}{5}} \qquad [15.20.205]$$

but this equation is to be regarded as tentative.

Ground level concentration
Once the plume rise Δh and hence the effective stack height h_e have been obtained, the maximum ground level concentration χ_{mgl} may be found in the usual way, using Equation 15.12.57, which may be written in the form

$$\chi_{mgl} = \frac{2Q}{e\pi uh_e^2}\frac{\sigma_z}{\sigma_y} \qquad [15.20.206a]$$

or

$$\chi_{mgl} = 0.234\frac{Q}{uh_e^2}\frac{\sigma_z}{\sigma_y} \qquad [15.20.206b]$$

where e is the base of natural logarithms, Q is the mass rate of release and σ_y and σ_x are the longitudinal and vertical dispersion coefficients, respectively.

The maximum ground level concentration χ_{mgl} is low both at low and at high wind speeds, because in the first case the plume rise is high and in the second dilution is high. It therefore passes through a maximum value χ_{max} at some critical wind speed u_c.

A minimum value of the term uh_e^2 is obtained from Equation 15.20.201 when $\Delta h = h_s/3$. The following relations can then be derived:

$$u_c = 2.15\left(\frac{u}{u_*}\right)^{\frac{2}{3}}\left(\frac{F_o}{h_s}\right)^{\frac{1}{3}} \qquad [15.20.207]$$

$$u_c h_e^2 = 3.8\left(\frac{u}{u_*}\right)^{\frac{2}{3}}F_o^{\frac{1}{3}}h_s^{\frac{5}{3}} \qquad [15.20.208]$$

15.20.11 Jet transients

The jet and plume models described apply to steady-state conditions. The steady state is preceded and followed by transients which are much less well defined. In particular, the decay can result in a plume with a relatively low momentum. A study of gas jet transients has been described by Landis, Linney and Hanley (1994).

15.21 Dispersion of Two-phase Flashing Jets

The type of jet considered here is the jet resulting from a release of superheated liquid at high pressure, with choked liquid or two-phase flow. Such a jet may be divided into a number of zones, as shown in Figure 15.75. One common division is into two zones, a depressurization zone and an entrainment zone. In the

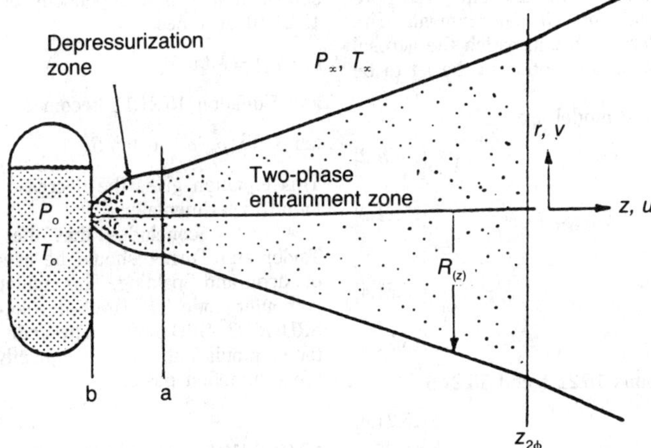

Figure 15.75 Two-phase momentum jet (Fauske and Epstein, 1988). Subscript 2ϕ in the figure is denoted in the text by subscript tp.

depressurization zone, which is some two orifice diameters in length, the pressure falls essentially to atmospheric pressure, but there is little air entrainment. This occurs in the entrainment zone.

A modification of this scheme is to introduce between the depressurization and entrainment zone an intermediate zone in which deposition of liquid droplets, or rainout, is completed but negligible entrainment of air occurs. This is referred to here as the deposition zone.

The jet is commonly considered to end at the point where the jet velocity has fallen to a minimum value or to the wind speed, whichever is higher.

15.21.1 Disintegration of jets

Two-phase jets in general were considered in Section 15.20. The two-phase jets of main interest here are those of superheated liquids, which give flashing jets. A review of such jets has been given by Appleton (1984 SRD R303).

The break-up of a flashing jet will, in general, be due both to air resistance and vapour bubble formation. Appleton states that for most situations of interest in the process industries the latter will usually be the more important effect.

Work by Bushnell and Gooderum (1968) has shown that there is a critical degree of superheat above which the jet will disintegrate due to vapour bubble formation. They define a shattering temperature as follows:

$$\frac{T_{sh} - T_{sat}}{T_{sh}} = \epsilon \qquad [15.21.1a]$$

Hence

$$T_{sh} = \frac{T_{sat}}{1 - \epsilon} \qquad [15.21.1b]$$

where ϵ is a constant and subscripts sat and sh denote saturation and shattering, respectively. They give for the value of the constant ϵ the range 0.07—0.1.

Below the shattering temperature the jet remains well formed, while above it the jet shatters. In their work break-up did no appear to be affected by outlet diameter or liquid velocity. In the work of R. Brown and York (1962), however, for experiments where the Weber number was high, shattering occurred at lower values of the superheat. Mention should also be made of work by Suzuki *et al.* (1978), which has shown that, even when the jet conditions are held constant, large variations can occur in break-up length and pattern.

For a two-phase flashing jet the characteristics of interest are similar, but in this case the mass flow of droplets increases along the axis, so that in addition information is required on the production rate of droplets and on droplet size.

For droplet size relations for minimum and for maximum droplet size have been given by Koestel, Gido and Lamkin (1980) and by Gyarmathy (1976), respectively. Appleton has compared theoretical estimates from the methods of these authors with values obtained by Brown and York as follows:

Brown and York:	
Smooth orifice	34.7 μm
Rough orifice	54.5 μm
Koestel *et al.* (for minimum size)	19 μm
Gyarmathy (for maximum size)	31 μm

Quantitative treatment of the production rate of droplets is sparse. In Brown and York's experiments there was initial flashing and jet expansion some 0.15 m downstream from the outlet and there was then rapid evaporation of the droplets, so that by 1.2–1.5 m downstream practically all the liquid phase had disappeared.

15.21.2 Wheatley model

A model for a two-phase jet of liquefied ammonia has been given by Wheatley (1986, 1987 SRD R410). The model deals with: jet disintegration; gravitational settling of liquid droplets, or rainout; entrainment of air; and thermodynamics of mixing ammonia and moist air. It is not concerned with the effect on the jet of gravity or wind speed.

Wheatley's treatment of jet disintegration and rainout was considered in Section 15.10. The account given here is confined to his model for air entrainment. The situation which he considers is that in which the aerosol particles are sufficiently small for motion in the jet to be taken as homogeneous.

The basic relations of the model are

$$G_d = AuC \quad [15.21.2]$$

$$G_d u_d = Au^2 \rho \quad [15.21.3]$$

and

$$G = G_d + G_a \quad [15.21.4]$$

$$G = Au\rho \quad [15.21.5]$$

Furthermore, from Equations 15.21.3 and 15.21.5

$$G_d u_d = Gu \quad [15.21.6]$$

where A is the cross-sectional area of the jet, C is the concentration (mass per unit volume), G is the mass flow in the jet, u is the velocity of the jet, ρ is the density of the jet and subscripts a and d denote air and the end of deposition zone, respectively.

The relation for the growth of the jet due to air entrainment is

$$\frac{dG_a}{dz} = 2\pi R u \rho_a f(\rho/\rho_a) \quad [15.21.7]$$

where R is the radius of the jet and z is the distance downstream of the source. Here $f(\rho/\rho_a)$ is a function yet to be defined.

Wheatley considers two models of air entrainment. The first is that of Ricou and Spalding (1961). These workers give for an isothermal gas jet over a limited range of initial densities

$$\frac{dG_a}{dz} = \pi^{\frac{1}{2}} \tan \beta_\infty \rho_a^{\frac{1}{2}} u^{\frac{1}{2}} G^{\frac{1}{2}} \quad [15.21.8]$$

where β_∞ is the asymptotic half-angle of the jet.

It is found empirically that $\beta_\infty = 9.1°$ and hence $\tan \beta_\infty = 0.159$. An entrainment coefficient α is sometimes defined as

$$\alpha = \tfrac{1}{2} \tan \beta_\infty \quad [15.21.9]$$

Comparison of Equations 15.21.7 and 15.21.8 implies

$$f(\rho/\rho_a) = \tfrac{1}{2}(\rho/\rho_a)^{\frac{1}{2}} \tan \beta_\infty \quad [15.21.10]$$

This equation indicates that entrainment is affected by the density of the jet.

A density effect is indicated in the following equation obtained for isothermal gas jets:

$$\tan \beta = \frac{1}{2}\left(\frac{\rho_a}{\rho} + 1\right)\left(\frac{\rho}{\rho_a}\right)^{\frac{1}{2}} \tan \beta_\infty \quad [15.21.11]$$

Wheatley quotes experimental work to the effect that, where the density ratio ρ/ρ_a is appreciably greater than unity, β can be smaller than β_∞ by a factor of about 2, which is not consistent with Equation 15.21.11. He notes that in the work of Ricou and Spalding the density ratio ρ/ρ_a was at most about 1.05 and concludes that Equation 15.21.8 is not valid where ρ differs significantly from ρ_a.

Work by B.R. Morton, Taylor and Turner (1956) has shown that f is independent of ρ/ρ_a. Thus Equation 15.21.10 becomes

$$f(\rho/\rho_a) = \tfrac{1}{2} \tan \beta_\infty \quad [15.21.12]$$

and Equation 15.21.11 becomes

$$\tan \beta = \tfrac{1}{2}(\rho_a/\rho + 1) \tan \beta_\infty \quad [15.21.13]$$

This equation gives the decrease of β with increasing ρ found experimentally.

Wheatley concludes that the treatment of Morton, Taylor and Turner should be used in preference to that of Ricou and Spalding. This model forms the basis of the computer code TRAUMA developed by Wheatley (1987 SRD R393, SRD R394). The code was developed initially for ammonia but was subsequently generalized to handle other liquefied gases.

15.21.3 Webber and Kukkonen model

A generalized model of a two-phase jet has been given by Webber and Kukkonen (1990). The model is intended as an exploration of basic principles. Most attempts to model two-phase jets have been partial ones, concentrating on particular aspects and neglecting others. The aim of the work was to determine which are the important phenomena and hence how comprehensive a realistic model needs to be.

In this model the depressurization zone is ignored. For the entrainment zone the basic relations are:

$$\frac{d(CuA)}{dx} = 0 \quad [15.21.14]$$

$$\frac{d(\phi u A)}{dx} = 2\pi R u_E \quad [15.21.15]$$

$$\frac{d(\phi u^2 A)}{dx} = 0 \quad [15.21.16]$$

with

$$A = \pi R^2 \quad [15.21.17]$$

where A is the cross-sectional area of the jet, C is the concentration of the contaminant, R is the radius of the jet, u is the velocity of the jet, u_E is the entrainment velocity, x is the downstream distance and ϕ is the ratio of the densities of the jet and air.

For the density of the jet two cases (A and B) are considered. For Case A the density of the jet is the same as that of the ambient air

$$\phi = 1 \quad [15.21.18]$$

For Case B the jet is an isothermal mixture of contaminant and air:

$$\phi = 1 + C \quad [15.21.19]$$

Two entrainment models are considered. These are the models of B.R. Morton, Taylor and Turner (1956) and Ricou and Spalding (1961), as treated by Wheatley. The authors state these models in the form

$$u_E = \alpha u \quad [15.21.20]$$

and

$$u_E = \alpha \phi^{\frac{1}{2}} u \qquad [15.21.21]$$

where α is an entrainment coefficient. These two entrainment models are identical for Case A, but not for Case B.

For Case A of a jet of ambient density, the solution of Equations 15.21.14–15.21.16 with incorporation of either entrainment model is

$$R = 2\alpha x \qquad [15.21.22]$$

$$u = UL(2\alpha x) \qquad [15.21.23]$$

$$C = L/(2\alpha x) \qquad [15.21.24]$$

where L and U are length and velocity scales determined by conditions at the source, and x is the downstream distance.

The solution for Case B of a dense jet before incorporation of the entrainment model is

$$(\phi - 1)uA = B \qquad [15.21.25]$$

$$\phi u^2 A = f \qquad [15.21.26]$$

where B and f are the buoyancy and momentum flux, respectively. The quantities L and U are defined as

$$L = \left(\frac{B^2}{\pi f}\right)^{\frac{1}{2}} \qquad [15.21.27]$$

$$U = f/B \qquad [15.21.28]$$

With incorporation of the first entrainment model, Equation 15.21.20, Equations 15.21.25 and [15.21.26 yield for Case B

$$R = L(q/p) \qquad [15.21.29]$$

$$1 - \phi = 1/q^2 \qquad [15.21.30]$$

$$\phi = 1/(pq)^2 \qquad [15.21.31]$$

with

$$p^2 = u/U \qquad [15.21.32]$$

$$q^2 = Au/(\pi L^2 U) \qquad [15.21.33]$$

$$p^2 = (1 + q^2)^{-1} \qquad [15.21.34]$$

$$q(1 + q^2)^{\frac{1}{2}} + \ln\,[q + (1 + q^2)^{\frac{1}{2}} = 2\alpha x/L \qquad [15.21.35]$$

where p and q are dimensionless variables.

For the Case B Equations 15.21.25 and 15.21.26 with the second entrainment model, Equation 15.21.21, the result is

$$R = 2\alpha x[1 + L/(2\alpha x)]^{\frac{1}{2}} \qquad [15.21.36]$$

$$u = U[L/(2\alpha x)][1 + L/(2\alpha x)]^{-1} \qquad [15.21.37]$$

$$\phi = 1 + L/(2\alpha x) \qquad [15.21.38]$$

For a two-phase jet which is not dense initially, L is rather smaller than R, and U is larger than u at the end of the depressurization zone. For an initially dense jet, L can be somewhat larger than R, whilst U approximates to u at the end of the zone.

The authors then develop the model to treat the effects of gravity, wind speed and liquid deposition. They give illustrative estimates from this model, utilizing the TRAUMA code to determine the conditions at the end of the deposition zone.

15.21.4 Fauske and Epstein model

Fauske and Epstein (1988) have described a more specific two-phase jet model, shown in Figure 15.75. They point out that, although in laboratory work the shattering temperature exceeds the saturation temperature, in practice due to factors such as vaporization upstream, roughness of the hole and air resistance, the extent of superheat without disintegration is very small. They take it as established that a superheated liquid released to the atmosphere disintegrates into droplets in the size range 10–100 μm, citing the work of Bushnell and Gooderum just described.

The basic relations of the model for the depressurization zone are:

$$u_a = u_b + \frac{P_b - P_a}{u_b \rho_b} \qquad [15.21.39]$$

$$A_a = \frac{A_b u_b \rho_b}{u_a \rho_a} \qquad [15.21.40]$$

$$\rho_a = \frac{1}{x_a/\rho_g + (1 - x_a)/\rho_f} \qquad [15.21.41]$$

$$h_o = h_f + x_a h_{fg} \qquad [15.21.42]$$

where A is the cross-sectional area of the jet, h is the specific enthalpy of the jet, P is the pressure in the jet, u is the velocity of the jet, x is the vapour fraction, or quality, ρ is the density of the jet, subscript b denotes at the orifice, or break, and subscript a denotes the end of the depressurization zone. Subscripts f, fg, g and o denote the liquid, liquid–vapour transition, the vapour and stagnation conditions, respectively. The densities and specific enthalpies at the end of the zone are evaluated at atmospheric pressure.

For the entrainment zone the entrainment relation used is that of Ricou and Spalding (1961):

$$e \propto u\left(\frac{\rho}{\rho_\infty}\right)^{\frac{1}{2}} \qquad [15.21.43a]$$

$$= E_o \rho_\infty u\left(\frac{\rho}{\rho_\infty}\right)^{\frac{1}{2}} \qquad [15.21.43b]$$

where e is the entrainment rate, E_o is the entrainment coefficient and the subscript ∞ denotes atmospheric conditions.

The relations for the conditions at the entrainment zone are:

$$\frac{u}{u_a} = \frac{h - h_\infty}{h_o - h_\infty} \qquad [15.21.44a]$$

$$= \frac{R_a}{R}\frac{\rho_a}{\rho} \qquad [15.21.44b]$$

$$= \left[1 + 2E_o\left(\frac{\rho_\infty}{\rho_a}\right)^{\frac{1}{2}}\frac{z}{R_a}\right]^{-1} \quad [15.21.44c]$$

where R is the radius of the jet, z is the distance along the jet axis and h_∞ is the specific enthalpy of the gaseous phase evaluated at atmospheric temperature T_∞. Equation 15.21.44 is valid at the end of the entrainment zone; it should not be used to determine conditions part way through the zone.

The authors discuss the thermal effects within the entrainment zone, particularly the formation of vapour from evaporating liquid droplets and the condensation of vapour onto droplets, and refer to the work of Weimer, Faeth and Olson (1973) on the latter. They conclude that it is reasonable to proceed on the assumption that the liquid and vapour phases are at the normal boiling point. Then, for a release which is all liquid and is from a vessel at ambient temperature, the specific enthalpies may be taken as

$$h = h_g(T_{bp}) \quad [15.21.45]$$

$$h_o = h_f(T_\infty) \quad [15.21.46]$$

$$h_\infty = h_g(T_\infty) \quad [15.21.47]$$

where the subscript bp denotes boiling point.

Then, rearranging Equation 15.21.44 gives the following equations for the end of the entrainment zone

$$\frac{u_{tp}}{u_a} = \left[1 + 2E_o\left(\frac{\rho_\infty}{\rho_a}\right)^{\frac{1}{2}}\frac{z_{tp}}{R_a}\right]^{-1} \quad [15.21.48]$$

$$\frac{R_{tp}}{R_a} = \left(\frac{\rho_a}{\rho_\infty}\right)^{\frac{1}{2}}\left[1 + 2E_o\left(\frac{\rho_\infty}{\rho_a}\right)^{\frac{1}{2}}\frac{z_{tp}}{R_a}\right] \quad [15.21.49]$$

$$\frac{z_{tp}}{R_a} = \frac{1}{2E_o}\left(\frac{\rho_a}{\rho_\infty}\right)^{\frac{1}{2}}\left[\frac{h_o - h_\infty}{h_g(T_{bp}) - h_\infty} - 1\right] \quad [15.21.50]$$

where the subscript tp denotes two-phase.

Utilizing Equations 15.21.45–15.21.47, the last factor in Equation 15.21.50 is modified to give

$$\frac{z_{tp}}{R_a} = \frac{1}{2E_o}\left(\frac{\rho_a}{\rho_\infty}\right)^{\frac{1}{2}}\frac{(1 - x_a)h_{fg}}{c_{pg}(T_\infty - T_{bp})} \quad [15.21.51]$$

where c_{pg} is the specific heat of the gaseous phase.

As an illustration, consider the example given by the authors of an ammonia jet in one of the Desert Tortoise ammonia release trials, described by Goldwire (1986) and considered in more detail in Section 15.37. This example is summarized in Table 15.41.

***Table 15.41** Illustrative calculation for a two-phase flashing jet of liquefied ammonia from pressure storage (after Fauske and Epstein, 1988) (Courtesy of Butterworth-Heinemann)*

A Scenario

Atmospheric pressure $P_\infty = 1.01 \times 10^5$ Pa
Density of ambient air $\rho_\infty = 1.1$ kg/m^3
Orifice diameter $d_b = 0.0945$ m
Storage temperature $T_o = 297$ K (24°C)
Storage pressure $P_o = 1.4 \times 10^6$ Pa
Discharge velocity $u_f = 22.7$ m/s
Pressure of jet at discharge conditions $P_b = 0.968 \times 10^6$ Pa
Density of jet at discharge conditions $\rho_b = 603$ kg/m^3
Density of gaseous ammonia at discharge conditions (boiling point, atmospheric pressure) $\rho_g = 0.89$ kg/m^3
Specific heat of liquid ammonia $c_{pf} = 4.46 \times 10^3$ J/kg K
Specific heat of gaseous ammonia $c_{pg} = 10^3$ J/kg K
Latent heat of vaporization of ammonia $h_{fg} = 1.37 \times 10^6$ J/kg
Normal boiling point of ammonia $T_{bp} = 240$ K (−33°C)
Entrainment coefficient $E_o = 0.116$

B Jet[a]

Quality at end of depressurization zone:
$x_a = c_{pf}(T_o - T_{bp})/h_{fg} = 4.46 \times 10^6 (297 - 240)/1.37 \times 10^6 = 0.19$
Density of jet at end of depressurization zone:
$\rho_a \approx \rho_g/x_a = 0.89/0.19 = 4.68$ kg/m^3
Velocity of jet at end of depressurization zone:

$$u_a = u_b + (P_b - P_a)/u_b\rho_b = 22.7 + (0.968 \times 10^6 - 0.101 \times 10^6/(22.7 \times 603) = 86 \text{ m/s}$$

Radius of jet at end of depressurization zone:

$$R_a = R_b\left(\frac{u_b\rho_b}{u_a\rho_a}\right)^{\frac{1}{2}} = \frac{0.0945}{2}\left(\frac{22.7 \times 603}{86 \times 4.68}\right)^{\frac{1}{2}} = 0.275 \text{ m}$$

Distance to end of two-phase jet zone:

$$z_{tp} = R_a\frac{1}{2E_o}\left(\frac{\rho_a}{\rho}\right)^{\frac{1}{2}}\frac{h_{fg}(1 - x_a)}{c_{pg}(T_\infty - T_{bp})} = 0.275\frac{1}{2 \times 0.116}\left(\frac{4.68}{1.1}\right)^{\frac{1}{2}}\frac{1.37 \times 10^6(1 - 0.19)}{10^3(306 - 240)} = 39.1 \text{ m}$$

Velocity at end of two-phase jet zone:

$$u_{tp} = u_a\left[1 + 2E_o\left(\frac{\rho_\infty}{\rho_a}\right)^{\frac{1}{2}}\frac{z_{tp}}{R_a}\right]^{-1} = 86\left[1 + 2 \times 0.116\left(\frac{1.1}{4.68}\right)^{\frac{1}{2}}\frac{39.1}{0.275}\right]^{-1} = 5.06 \text{ m/s}$$

Radius at end of two-phase jet zone:

$$R_{tp} = R_a\left(\frac{\rho_a}{\rho_\infty}\right)^{\frac{1}{2}}\left[1 + 2E_o\left(\frac{\rho_\infty}{\rho_a}\right)^{\frac{1}{2}}\frac{z_{tp}}{R_a}\right] = 0.275\left(\frac{4.68}{1.1}\right)^{\frac{1}{2}}\left[1 + 2 \times 0.116\left(\frac{1.1}{4.68}\right)^{\frac{1}{2}}\frac{39.1}{0.275}\right] = 9.6 \text{ m}$$

[a] The numerical values in this example have been revised after consultation with one of the authors (HKF).

15.22 Dense Gas Dispersion

The account given so far of gas dispersion has been confined to the dispersion of gases of neutral buoyancy, or passive dispersion. A large proportion of industrially important gases exhibit negative buoyancy. It is these gases particularly which are prominent in hazard assessment.

An account of the dispersion of dense, or heavy, gases is given in this and following sections. Section 15.23 deals with source terms; Section 15.24 with models and modelling; Section 15.25 with modified conventional models; Sections 15.26–15.28 with some principal box models; Sections 15.29–15.33 with some *K*-theory and other three-dimensional models; Section 15.34 with the *Workbook*; Section 15.35 with the DRIFT three-dimensional model; Section 15.36 with some other models; Sections 15.37 and 15.38 with field trials; Section 15.39 with physical modelling; Section 15.40 with terrain, obstructions and buildings; Section 15.41 with validation and comparison; Section 15.42 with particular gases; and Sections 15.43 and 15.44 with elevated plumes and an elevated plume model. Further aspects of dense gas dispersion are treated in Section 15.45 on concentration and concentration fluctuations; Sections 15.46 and 15.47 on flammable and toxic gas clouds; Section 15.48 on dispersion over short ranges; Section 15.49 on hazard ranges for dispersion; and Sections 15.53 and 15.54 on vapour cloud mitigation. Selected references on dense gas dispersion are given in Table 15.42.

15.22.1 Dense gas

Many of the more hazardous gases such as hydrocarbons, chlorine, ammonia and hydrogen fluoride, and oxygen, are capable of giving a gas cloud which is denser than air. Whether or not a particular gas release forms a gas cloud which is denser than air depends on a number of factors. These are:

(1) the molecular weight of the gas;
(2) the temperature of the gas;
(3) the presence of liquid spray;
(4) the temperature and humidity of the air.

A gas cloud may be dense by virtue of the molecular weight of the gas released. However, even if the molecular weight of the gas is less than that of air, the gas cloud may be rendered dense by other factors. If the boiling point of the gas is low so that the gas cloud is cold, this may be sufficient to make the cloud dense.

The presence of liquid spray is another factor which can render the cloud dense. The effect of vaporization of the droplets of liquid is to remove heat from the gas and so cool it.

The density of the cloud is also affected by the humidity of the ambient air. The effect of the condensation of the droplets of water is to add heat to the gas. In general, the cloud will tend to have a higher density when mixed with dry air than with wet air.

The effect of molecular weight and temperature may be illustrated by considering the densities of gas clouds from releases of chlorine, methane and ammonia into dry air. The molecular weight of chlorine (71) is much greater than that of air (28.9) and thus gives a cloud heavier than air at ambient temperature. The cloud given by chlorine at its boiling point (–34°C) is even heavier.

Table 15.42 *Selected references on dense gas dispersion*

Howerton (n.d., 1969); Abbott (1961); Joyner and Durel (1962); J.S. Turner (1966); Hoult, Fay and Forney (1969); Feldbauer *et al.* (1972); Humbert-Basset and Montet (1972); MacArthur (1972); Yang and Meroney (1972); Bodurtha, Palmer and Walsh (1973); Hoot, Meroney and Peterka (1973); Simmons, Erdmann and Naft (1973, 1974); Battelle Columbus Laboratories (1974); Hoot and Meroney (1974); Kneebone and Prew (1974); Lutzke (1974); Buschmann (1975); Germeles and Drake (1975); Simmons and Erdmann (1975); Anon. (1976 LPB 10, p. 10); McQuaid (1976b, 1979a,b, 1980, 1982a,b, 1984a–c, 1985a); Havens (1977, 1980, 1982b, 1985, 1986, 1987, 1992); Kletz (1977i); US Congress, OTA (1977); HSE (1978b, 1981a); Slater (1978a); Britter (1979, 1980, 1982, 1988a,b); Gutsche, Ludwig and Vahrenholt (1979); Harvey (1979b); R.A. Cox *et al.* (1980); Fiedler and Tangermann-Dlugi (1980); Flothmann and Nikodem (1980); Forster, Kramer and Schon (1980); Hartwig (1980a,b, 1983a,b, 1986a,b); Hartwig and Flothmann (1980); Kinnebrock (1980); Klauser (1980); McNaughton and Berkowitz (1980); Schnatz and Flothmann (1980); Bloom (1980); Fay (1980, 1984); Kaiser (1980, 1982a,b); Shell Research Ltd (1980); Dirkmaat (1981); Fay and Ranck (1981, 1983); Jagger and Kaiser (1981); Jensen (1981, 1984); Snyder (1981); Anon. (1982f); Blackmore, Eyre and Summers (1982); Blackmore, Herman and Woodward (1982); Britter and Griffiths (1982); Farmer (1982 SRD R221); R.F. Griffiths and Kaiser (1982); Gunn (1982, 1984); N.C.Harris (1982, 1983, 1986); Hogan (1982); Jagger (1982, 1983); List (1982); Meroney (1982, 1984a,b, 1985, 1987a,b); Zeman (1982a,b, 1984); Astleford, Morrow and Buckingham (1983); Byggstoyl and Saetran (1983); Emblem and Fannelop (1983); Havens and Spicer (1984); Pikaar (1983, 1985); Schnatz, Kirsch and Heudorfer (1983); Anon. (1984q,y); Emblem, Krogstad and Fannelop (1984); Fanaki (1984); Hartwig, Schatz and Heudorfer (1984); Jacobsen and Fannelop (1984); Jensen and Mikkelsen (1984); A.G. Johnston (1984, 1985); Knox (1984); Webber (1984); Barrell and McQuaid (1985); Ermak and Chan (1985); Fay and Zemba (1985, 1986, 1987); Raj (1985); Spicer (1985); Alp *et al.* (1986); Bachlin and Plate (1986); Brighton (1986, 1988); Duijm *et al.* (1986); Fannelop and Zumsteg (1986); Heinhold *et al.* (1986); Heudorfer (1986); Layland, McNaughton and Bodmer (1986); Riou (1986); Schnatz (1986); M.T.E. Smith *et al.* (1986); Verhagen and Buytenen (1986); Bettis, Makhviladze and Nolan (1987); N.E. Cooke and Khandhadia (1987a,b); Emerson (1987a); S.R. Hanna (1987b); Jacobsen and Magnussen (1987); Raj, Venkataraman and Morris (1987); Redondo (1987); Schreurs and Mewis (1987); Schreurs, Mewis and Havens (1987); Webber and Wheatley (1987 SRD R437); Cave (1988); Doury (1988); C.D.Jones (1988); Mercer (1988); Shaver and Fornery (1988); Kakko (1989); Nielsen and Jensen (1989); Vergison, van Diest and Basler (1989); Askari *et al.* (1990); Chaineaux and Mavrathalassitis (1990); Cleaver and Edwards (1990); Matthias (1990, 1992); P.T. Roberts, Puttock and Blewitt (1990); Schulze (1990); Witlox (1990); J.L. Woodward (1990, 1993); Ayrault, Balint and Morel (1991); Ermak (1991); Huerzeler and Fannelop (1991); Mercer and Porter (1991); J.K.W. Davies (1992); Van Dop (1992);

Anfossi *et al.* (1993); Caulfield *et al.* (1993); Webber and Wren (1993 SRD R552)

Density effects, gravity currents
Penney and Thornhill (1952); Keulegan (1957); NBS (1957 Rep. 5482, 1958 Rep. 5831); Stoker (1957); Ellison and Turner (1959); Lofquist (1960); Middleton (1966); Benjamin (1968); Hoult (1972b); van Ulden (1974); Kantha, Phillips and Azad (1977); Britter and Simpson (1978); J.C.R. Hunt *et al.* (1978); Britter (1979); Fischer *et al.* (1979); Simpson and Britter (1979); Brighton (1980 SRD R444); Britter and Linden (1980); Huppert and Simpson (1980); Yih (1980); Beghin, Hopfinger and Britter (1981); Jagger (1982 SRD R238); Simpson (1982, 1987); Crapper (1984); J.C.R. Hunt, Rottman and Britter (1984); Kranenberg (1984); Rottman and Simpson (1984a,b); Stretch, Britter and Hunt (1984); Zeman (1984); Grundy and Rottman (1985); Stretch (1986); Britter and Snyder (1988); Carruthers and Hunt (1988); Darby and Mobbs (1988); J.C.R. Hunt, Stretch and Britter (1988); J.C. King (1988); Linden and Simpson (1988); McGuirk and Papadimitrou (1988)

Source terms
NRC (Appendix 28 *Source Terms*); HSE (1978b, 1981a); Rijnmond Public Authority (1982); Zeman (1982a,b); Mudan (1984a); Wheatley (1986); Fauske and Epstein (1987, 1988, 1989); C. Harris (1987); Webber and Wheatley (1987); Britter and McQuaid (1988); Puttock (1988a, 1989); Chikhliwala *et al.* (1989); Iannello *et al.* (1989); Kaiser (1989); J.L. Woodward (1989a); Finch and Serth (1990); Matthias (1990, 1992); J. Singh, Cave and McBride (1990); McFarlane (1991); Webber (1991); Lantzy (1992); de Nevers (1992); Spadoni *et al.* (1992).
Release inside buildings: Brighton (1989 SRD R467); Raman (1993)

Heat transfer, thermodynamic effects
Andreiev, Neff and Meroney (1983); Meroney and Neff (1986); Ruff, Zumsteg and Fannelop (1988)

Buildings, obstacles, complex terrain
D.J. Hall, Barrett and Ralph (1976); Brighton (1978, 1986); Kothari and Meroney (1979, 1982); Raupach, Thom and Edwards (1980); Kothari, Meroney and Neff (1981); Meroney (1981, 1992); Britter (1982, 1988a); Deaves (1983a, 1984, 1985, 1987a,b, 1989); Robins and Fackrell (1983); Chan and Ermak (1984); Chan, Rodean and Ermak (1984); Gunn (1984); Jensen (1984); de Nevers (1984b); Guldemond (1986); Krogstad and Pettersen (1986); McQuaid (1986); Riou and Saab (1986); Heinhold, Walker and Paine (1987); Britter and Snyder (1988); Britter and McQuaid (1988); Carissimo *et al.* (1989); Vergison, van Diest and Basler (1989); Briggs, Thompson and Snyder (1990); Deaves and Hall (1990); Snyder (1990); M.Nielsen (1991); Brighton *et al.* (1992 SRD R583); Jones *et al.* (1992 SRD R582); Kukkonen and Nikmo (1992); Nikmo and Kukkonen (1992); Webber, Jones and Martin (1992, 1993); Castro *et al.* (1993); S.J. Jones and Webber (1993); Duijm and Webber (1994); Perdikaris and Mayinger (1994)

Physical modelling, wind tunnels, water flumes
D.J. Hall, Barrett and Ralph (1976); Neff, Meroney and Cermak (1976); Meroney *et al.* (1977); BRE (1978 CP71/78); D.J. Hall (1979a,b); Isyumov and Tanaka (1979); Kothari and Meroney (1979, 1982); Meroney (1979b, 1981, 1982, 1985, 1986, 1987a,b); Meroney and Lohmeyer (1979, 1981, 1982, 1983, 1984); J.C.R. Hunt (1980); Meroney, Lindley and Bowen (1980); Meroney and Neff (1980, 1981, 1984, 1985); T.B. Morrow (1980); Raupach, Thom and Edwards (1980); Hansen (1981); Kothari, Meroney and Neff (1981); Lohmeyer, Meroney and Plate (1981); Neff and Meroney (1981, 1982); D.J. Hall, Hollis and Ishaq (1982, 1984); C.I. Bradley and Carpenter (1983); Cheah, Cleaver and Milward (1983a); Meroney *et al.* (1983); Wighus (1983); Builtjes and Guldemond (1984); Cheah *et al.* (1984); Morrow, Buckingham and Dodge (1984); M.E. Davies (1985); D.J. Hall and Waters (1985, 1989); van Heugten and Duijm (1985); Krogstad and Pettersen (1986); Riethmuller (1986); Schatzmann, Koenig and Lohmeyer (1986); Bachlin and Plate (1987); Knudsen and Krogstad (1987); R.L. Petersen and Ratcliff (1989); Blewitt *et al.* (1991); D.J. Hall, Kukadia *et al.* (1991); D.J. Hall, Waters *et al.* (1991); Moser *et al.* (1991); Seong-Hee Shin, Meroney and Williams (1991); ASME (1992 DSC 36); Heidorn *et al.* (1992); Havens, Walker and Spicer (1994); P.T. Roberts and Hall (1994)

Concentration fluctuations
Chatwin and Sullivan (1980); Chatwin (1982a,b, 1983, 1991); R.F. Griffiths and Megson (1982, 1984); Meroney and Lohmeyer (1983); Carn and Chatwin (1985); Misra (1985); Carn (1987); Mole and Chatwin (1987); Carn, Sherrell and Chatwin (1988)

Models
API (Appendix 28, 1986/16, 1987 Publ. 4459, 1989 Publ. 4491, 4492, DR 229, 1990 Publ. 4522, 4523, 1992 Publ. 4539, 4540, 4545–4547, 1993 Publ. 4559, 4577); Considine (1981); Blackmore, Herman and Woodward (1982); J.L. Woodward *et al.* (1982, 1983); S.R. Hanna and Munger (1983); R.L. Lee *et al.* (1983); Ministry of Environment, Canada (1983, 1986); Scire and Lurmanis (1983); Grint (1984 SRD R315); Kunkel (1985); Whitacre *et al.* (1986); Havens, Spicer and Schreurs (1987a,b); Betts and Haroutunian (1988); Kaiser (1989); Nussey, Mercer and Clay (1990); Gudivaka and Kumar (1990); Zapert, Londergan and Thistle (1990); Goyal and Al-Jurashi (1991); S.R. Hanna, Strimaitis and Chang (1991a); Touma *et al.* (1991); S.R. Hanna *et al.* (1992); S.R. Hanna, Chang and Strimaitis (1993); Kumar, Luo and Bennett (1993); Brighton *et al.* (1994); S.R. Hanna (1994); Kinsman *et al.* (1994).
Eidsvik model: Eidsvik (1978, 1980, 1981a,b); Havens and Spicer (1983)
van Ulden model: van Ulden (1974, 1979, 1984, 1988); van Ulden and de Haan (1983)
BG/C&W model: R.A. Cox and Roe (1977); R.A. Cox and Carpenter (1980); C.I. Bradley *et al.* (1983); Carpenter *et al.* (1987)
TNO models: van den Berg (1978)
VKI models: Foussat (1981); van Dienst *et al.* (1986)
COBRA: Alp (1985); Alp and Mathias (1991)
DENZ, CRUNCH: Fryer and Kaiser (1979 SRD R152); Jagger (1983 SRD R229); Jagger (1985 SRD R277); Wheatley, Brighton and Prince (1986)
DISP2: Fielding, Preston and Sinclair (1986); Preston and Sinclair (1987)

SIGMET: England *et al.* (1978); Havens (1979, 1982a); Su and Patniak (1981); Havens and Spicer (1983); Havens, Spicer and Schreurs (1987a,b)
HEAVYGAS, SLUMP: Deaves (1985, 1987a, 1992)
MARIAH-II: Taft, Rhyne and Weston (1983); Havens, Schreurs and Spicer (1987); Havens, Spicer and Schreurs (1987a,b)
MERCURE-GL: Riou and Saab (1986); Riou (1987, 1988)
SLAB, FEM3: Chan, Gresho and Ermak (1981); Ermak, Chan *et al.* (1982); Zeman (1982a,b); Chan (1983); R.L. Lee *et al.* (1983); D.L. Morgan, Morris and Ermak (1983); Chan and Ermak (1984); Chan, Rodean and Ermak (1984); D.L. Morgan (1984); D.L. Morgan, Kansa and Morris (1984); Rodean *et al.* (1984); Leone, Rodean and Chan (1985); Ermak and Chan (1986, 1988); Rodean and Chan (1986); Chan, Ermak and Morris (1987); Chan, Rodean and Blewitt (1987); Havens, Spicer and Schreurs (1987a,b); Zapert, Londergan and Thistle (1990); S.R. Hanna, Strimatis and Chang (1991b); Touma *et al.* (1991)
HEGADAS: te Riele (1977); Colenbrander (1980); J.L. Woodward *et al.* (1982); Colenbrander and Puttock (1983); Havens and Spicer (1983); Puttock (1987b,c, 1988a, 1989); Zapert, Londergan and Thistle (1990); S.R. Hanna, Strimatis and Chang (1991b); Moser *et al.* (1991)
HEGABOX: Puttock (1987b,c, 1988a); S.R. Hanna, Strimatis and Chang (1991a); Moser *et al.* (1991)
HGSYSTEM: McFarlane *et al.* (1990); Rees (1990); Witlox *et al.* (1990); McFarlane (1991); Puttock *et al.* (1991)
PLUME: McFarlane (1991)
ADREA: Andronopoulos *et al.* (1993); Statharas *et al.* (1993)
DEGADIS: Havens (1985, 1986, 1992); Havens and Spicer (1985); Spicer and Havens (1986, 1987); Spicer *et al.* (1986); Spicer, Havens and Kay (1987); Havens, Spicer and Guinnup (1989); Zapert, Londergan and Thistle (1990); Hanna, Strimatis and Chang (1991a); Havens *et al.* (1991); Moser *et al.* (1991); Touma *et al.* (1991)
CHARM: Fabrick (1982); Radian Corporation (1986); Zapert, Londergan and Thistle (1990); Hanna, Strimatis and Chang (1991a);
HSE Workbook: McQuaid (1982b); Britter and McQuaid (1987, 1988)
DRIFT: Byren *et al.* (1992 SRD R553); Edwards (1992 SRD R563); Webber *et al.* (1992 SRD R586, R587); Webber and Wren (1993 SRD R552)

Experimental trials
AGA (1974/20–23); Buschman (1975); HSE (1978 RP 8); Puttock, Blackmore and Colenbrander (1982); Puttock and Colenbrander (1985a); Lewellen and Sykes (1986); Heinrich, Gerold and Wietfeldt (1988); Havens (1992).
AGA: Feldbaner *et al.* (1972); Drake, Harris and Reid (1973)
China Lake: Ermak *et al.* (1983)
Porton Down: Picknett (1978a–c, 1981); McQuaid (1979a,b); Anon. (1981j); D.J. Hall, Hollis and Ishaq (1982, 1984); Kaiser (1982b); Heudorfer (1986)
Maplin trials: Anon. (1982f); Puttock, Blackmore and Colenbrander (1982); Puttock, Colenbrander and Blackmore (1983, 1984); Colenbrander and Puttock (1984)
Nevada Test Site : Koopman and Thompson (1986); Koopman *et al.* (1986)
LLNL: Hogan (1982); Koopman, Ermak and Chan (1989)

Experimental trials: Thorney Island
McQuaid (1982b, 1984b,c, 1985a–c, 1987); HSE (1984 RP 24); Brighton (1985a, 1985 SRD R319, 1987); Brighton, Prince and Webber (1985); Carn and Chatwin (1985); CEC (1985 EUR 9933 EN, EUR 10029 EN); M.E. Davies and Singh (1985a,b); Deaves (1985); Gotaas (1985); R.F. Griffiths and Harper (1985); D.J. Hall and Waters (1985); N.C. Harris (1985); Hartwig (1985); van Heugten and Duijm (1985); D.R. Johnson (1985); Leck and Lowe (1985); Nussey, Davies and Mercer (1985); Pfenning and Cornwell (1985); Prince, Webber and Brighton (1985 SRD R318); Puttock (1985, 1987a–c); Puttock and Colenbrander (1985b); Riethmuller (1985); Roebuck (1985); Rottman, Hunt and Mercer (1985); Rottman *et al.* (1985); Spicer and Havens (1985); Wheatley, Prince and Brighton (1985, 1985 SRD R355); Wheatley, Brighton and Prince (1986); Brighton and Prince (1987); Cabrol, Roux and Lhomme (1987); Carn (1987); Carpenter *et al.* (1987); Chan, Ermak and Morris (1987); Cornwell and Pfenning (1987); J.K.W. Davies (1987); M.E. Davies and Inman (1987); Havens, Schreurs and Spicer (1987); Knudsen and Krogstad (1987); Mercer and Davies (1987); Mercer and Nussey (1987); Riou (1987, 1988); Sherrell (1987); van Ulden (1987); Wheatley and Prince (1987); Carn, Sherrell and Chatwin (1988); Carissimo *et al.* (1989); Andronopoulos *et al.* (1993)

Particular chemicals
Ammonia: HSE/SRD (HSE/SRD/WP13, 14, 16, 22 and unnumbered); W.L. Ball (1968b); Rohleder (1969); Comeau (1972); MacArthur (1972); J.D. Reed (1974); Lonsdale (1975); Luddeke (1975); HSE (1978b, 1981a); Kaiser and Walker (1978); Slater (1978a); R.F. Griffiths and Kaiser (1979 SRD R154, 1982); Haddock and Williams (1979); Kaiser (1979 SRD R150, 1980, 1981 LPB 38, 1982, 1989); Blanken (1980); Kaiser and Griffiths (1982); R.F. Griffiths and Megson (1984); Goldwire (1985, 1986); Goldwire *et al.* (1985); Guldemond (1986); Koopman *et al.* (1986); Chan, Rodean and Blewitt (1987); Koopman, Ermak and Chan (1989); J. Singh and McBride (1990); S.R. Hanna, Strimaitis and Chang (1991a); Statharas *et al.* (1993)
Antiknock compounds: HSE/SRD (HSE/SRD/WP17)
Chlorine: McClure (1927); Ministry of Social Affairs (1975); R.F. Griffiths and Megson (1984); Riethmuller (1986); Wheatley (1986 SRD R357); Clough and Wheatley (1988 SRD R396); Vergison, van Dienst and Basler (1989); J. Singh and McBride (1990)
Hydrogen fluoride: Beckerdite, Powell and Adams (1968); Bosch and de Kayser (1983); Blewitt *et al.* (1987a,b); Chan, Rodean and Blewitt (1987); Chikhliwala and Hague (1987); Clough, Grist and Wheatley (1987); Schotte (1987, 1988); Koopman, Ermak and Chan (1989); Holve *et al.* (1990); Leone (1990); R.L. Petersen and Diener (1990); Blewitt *et al.* (1991); Fthenakis, Schatz and Zakkay (1991); S.R. Hanna, Strimaitis and Chang (1991a); Meroney (1991); Moser (1991); Puttock, MacFarlane, Prothero and Roberts *et al.* (1991); Antonello and Buzzi (1992); Schatz and Fthenakis (1994); Webber, Mercer and Jones (1994)

LNG : HSE/SRD (HSE/SRD/WP21); Fay (1973); AGA (1974/20–23, 1978/27); Duffy, Gideon and Puttnam (1974); Drake and Puttnam (1975); Fay and Lewis (1975); Meroney, Cermak and Neff (1976); Neff, Meroney and Cermak (1976); Havens (1977, 1978, 1979, 1980, 1982a); Meroney *et al.* (1977); England *et al.* (1978); Koopman, Bowman and Ermak (1980); Meroney and Neff (1980, 1981, 1984, 1985); Hogan, Ermak and Koopman (1981); Kothari, Meroney and Neff (1981); Neff and Meroney (1981, 1982); TNO (1981 Rep. 81–07020); Blackmore, Eyre and Summers (1982); Ermak *et al.* (1982); D.J. Hall, Hollis and Ishaq (1982); Hogan (1982); Koopman *et al.* (1982); Rodean (1982, 1984); Ermak *et al.* (1983); Goldwire *et al.* (1983); Havens and Spicer (1983); Meroney *et al.* (1983); Chan and Ermak (1984); D.L. Morgan *et al.* (1984); Rodean *et al.* (1984); Misra (1985); Havens, Spicer and Schreurs (1987a,b); Leone (1990); Havens *et al.* (1991); Seong-Hee Shin, Meroney and Williams (1991); Chan (1992)
Propane, LPG: Puttock, Colenbrander and Blackmore (1983); Wighus (1983); Meroney and Lohmeyer (1984)
Other gases: McRae *et al.* (1983); McRae (1986); Koopman *et al.* (1986)

Dispersion of aerososls
J.L. Woodward (1989)

Visibility of cloud
CCPS (1988/3); de Nevers (1992)

Mitigation sytems
R.W. Johnson *et al.* (1989); Hague (1992); Prugh (1992b); Buchlin (1994)
Vapour barriers, fences: R.L. Petersen and Diener (1990); Chan (1992)
Gas curtains: Rulkens *et al.* (1983)
Steam curtains: Cairney and Cude (1971); Simpson (1974); Seifert, Maurer and Giesbrecht (1983); Sato (1989); Lopez *et al.* (1990); Barth (1992)
Water curtains: Rasbash and Stark (1962); Buschmann (1975); McQuaid (1975, 1976a, 1977, 1980, 1982b); Ministry of Social Affairs (1975); Eggleston, Herrera and Pish (1976); Heskestad, Kung and Todtenkopf (1976); Vincent *et al.* (1976a,b); J.W. Watts (1976); Benedict (1977); HSE (1978 TP1, 1978b, 1981a); van Doorn and Smith (1980); Beresford (1981); Greenwood (1981); N.C. Harris (1981a); IChemE (1981/123); McQuaid and Fitzpatrick (1981, 1983); Moodie (1981, 1985); P.A.C. Moore and Rees (1981); J.M. Smith and Doorn (1981); Zalosh (1981); McQuaid and Moodie (1982, 1983); Deaves (1983b); de Faveri *et al.* (1984); Kirby and Deroo (1984); Emblem and Madsen (1986); Blewitt *et al.* (1987b); NIOSH (1987/10); Fthenakis and Zakkay (1990); Holve *et al.* (1990); Lopez *et al.* (1990); Schatz and Koopman (1990); Fthenakis, Schatz and Zakkay (1991); Lopez, Lieto and Grollier-Baron (1991); Meroney (1991); Meroney and Seong-Hee Shin (1992); St Georges *et al.* (1992a,b); Fthenakis (1993); Fthenakis and Blewitt (1993); Ratcliff *et al.* (1993); Schatz and Fthenakis (1994)

Jets and plumes
Bodurtha (1961, 1980, 1988); Ooms (1972); J.L. Anderson, Parker and Benedict (1973); Bodurtha, Palmer and Walsh (1973); Hoot, Meroney and Peterka (1973); Ooms, Mahieu and Zelis (1974); Chu (1975); Meroney (1979b); Blooms (1980); Giesbrecht, Seifert and Leuckel (1983); Hirst (1984, 1986); Jagger and Edmundson (1984); Ooms and Duijm (1984); Emerson (1986a,b, 1989); Li Xiao-Yun, Leudens and Ooms (1986); Tam and Cowley (1989); J.L. Woodward (1989a); Epstein, Fauske and Hauser (1990); D.K. Cook (1991b); McFarlane (1991); Schatzmann, Snyder and Lawson (1993)

Hazard assessment
R.A. Cox and Roe (1977); Air Weather Service (1978); HSE (1978b, 1981a); Wu and Schroy (1979); Connell and Church (1980); Doron and Asculai (1983); Barboza, Militana and Haymes (1986); Gebhart and Caldwell (1986); Hart (1986); Kasprak, Vigeant and McBride (1986); Layland, McNaughton and Bodmer (1986); Schewe and Carvitti (1986); Shea and Jelinek (1986); Costanza *et al.* (1987); Fauske and Epstein (1987); Nussey and Pape (1987); Paine, Smith and Egan (1987); Studer, Cooper and Doelp (1988); English and Waite (1989)

Accident simulation
M.P. Singh and Ghosh (1987); Fay (1988); Billeter and Fannelop (1989); M.P. Singh (1990); M.P. Singh, Kumari and Ghosh (1990)

Incident investigation
Vegetation damage: Benedict and Breen (1955); Hindawi (1970); Jacobson and Hill (1970); Booij (1979)

Methane (mol. wt. 16) is much lighter than air at ambient temperature. But the cloud given by methane at its boiling point (−161°C) is heavier than air. Ammonia (mol. wt. 17) is much lighter than air at ambient temperature. The cloud given by ammonia at its boiling point (−33°C) is also lighter than air. But ammonia clouds tend to be denser than air due to presence of liquid spray.

There is now ample evidence, both from experiments and incidents, that releases of liquefied gas such as propane, LNG, ammonia, chlorine and hydrogen fluoride give rise to dense gas clouds.

As the gas cloud is diluted, the concentration, and the relative density, decrease and the mode of dispersion changes from dense gas dispersion to passive dispersion of a neutrally buoyant gas.

15.22.2 van Ulden's experiment

Experimental work on the dispersion of a heavy gas has been described by van Ulden (1974). Some results from this work are shown in Figure 15.76. Figures 15.76(a) and (b) show, respectively, the growth of the height and radius of the dense gas cloud from an instantaneous release at ground level. The first part of the curve in the former figure shows the initial sharp decrease in cloud height due to gravity slumping, followed by a gradual increase in height. The height is much less and the radius much greater than that predicted by the Gaussian model, which overestimates the height and underestimates the radius by factors of 5 and 2.5, respectively.

Van Ulden formulated a model for dense gas dispersion which was influential in the development of the topic. This is described in Section 15.26.

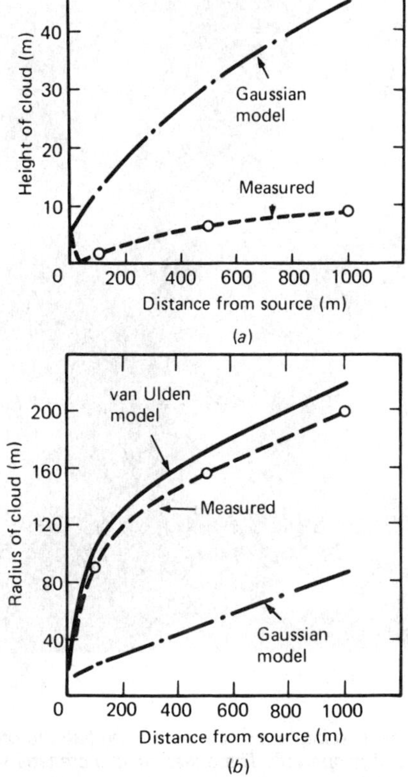

Figure 15.76 *Dutch Freon 12 trial on dense gas dispersion (van Ulden, 1974): (a) growth of height; (b) growth of radius (Courtesy of Elsevier Publishing Company)*

15.22.3 Experimental work

Experimental work bearing on dense gas dispersion includes not only field trials but also physical modelling in wind tunnels and water flumes, the latter being often on the laboratory scale. The work on field trials is described in Sections 15.37 and 15.38 and the physical modelling in Section 15.39. The laboratory experiments are referred to in discussion of the particular aspects to which they relate.

15.22.4 Dense gas behaviour

The behaviour of a dense gas may be described by considering the development of the gas cloud from the bursting of a vessel. The source term for this is usually modelled by assuming that air is entrained in a ratio of some 10–20:1 and that a cloud of unit aspect ratio is formed.

The behaviour of the cloud is influenced by gravity. Three stages may be distinguished:

(1) gravity slumping;
(2) gravity spreading;
(3) passive dispersion.

The flow is gravity driven in both the first two stages. The first stage is normally described as gravity slumping. The second stage is gravity-driven flow with stable stratification. The term 'gravity spreading' is applied sometimes to both the first two stages and sometimes only to the second stage; it is used here in the latter sense. The third stage is one of passive dispersion.

Gravity-driven flow is illustrated in Figure 15.77, which shows shadowgraphs taken in a water tunnel, these illustrate the intrusion of a more dense fluid into a less dense one on surfaces of different slopes. Features are the 'head' at the leading edge of the denser fluid and the mixing at the horizontal interface between the two fluids.

The density difference between the cloud and the atmosphere, or density excess, has four main effects. First, it imparts a strong horizontal velocity to the cloud. Second, it creates velocity shear. Third, it inhibits vertical mixing by atmospheric turbulence through the top of the cloud. Fourth, it affects the inertia of the cloud.

The inhibition of vertical mixing by atmospheric turbulence through the top of the cloud means that the mixing is less sensitive to the stability conditions. Vertical mixing still occurs, however, due to other forms of turbulence in the cloud. Moreover, the inhibition of mixing by atmospheric turbulence is counteracted by the creation of a large top surface area on the cloud due to gravity spreading.

If the ratio of the density difference to the density of the atmosphere, or reduced density, is small, an important approximation may be made. This is to neglect the effect of the density difference on the inertia. This is the Boussinescq approximation and it is widely used in dense gas dispersion modelling.

15.22.5 LNG spill study

Another piece of work which was influential in the development of the topic was the study for the US Coast Guard by Havens (1978) of the hazard range from a spill of 25 000 m^3 of LNG onto water. The work was a comparative study of the predictions of the distance to the lower flammability limit given by seven gas dispersion models. These predictions ranged from 0.75 up to 50 km. Reference to this study was made in Chapter 9 as an illustration of uncertainty in modelling. It is considered again in Section 15.41 in the context of validation of models.

15.23 Dispersion of Dense Gas: Source Terms

The source term can be critical for the modelling of the subsequent dispersion of a dense gas. It is therefore important that the source model should be realistic and complete. If the model for the source term is poor, the results of the whole dense gas dispersion estimate may be seriously in error.

15.23.1 Release scenarios

The modelling of release scenarios is assisted by an appropriate classification. Such classifications have been given by a number of workers, including Fryer and Kaiser (1979 SRD R152), Griffiths and Kaiser (1979 SRD R154) and Kaiser (1989).

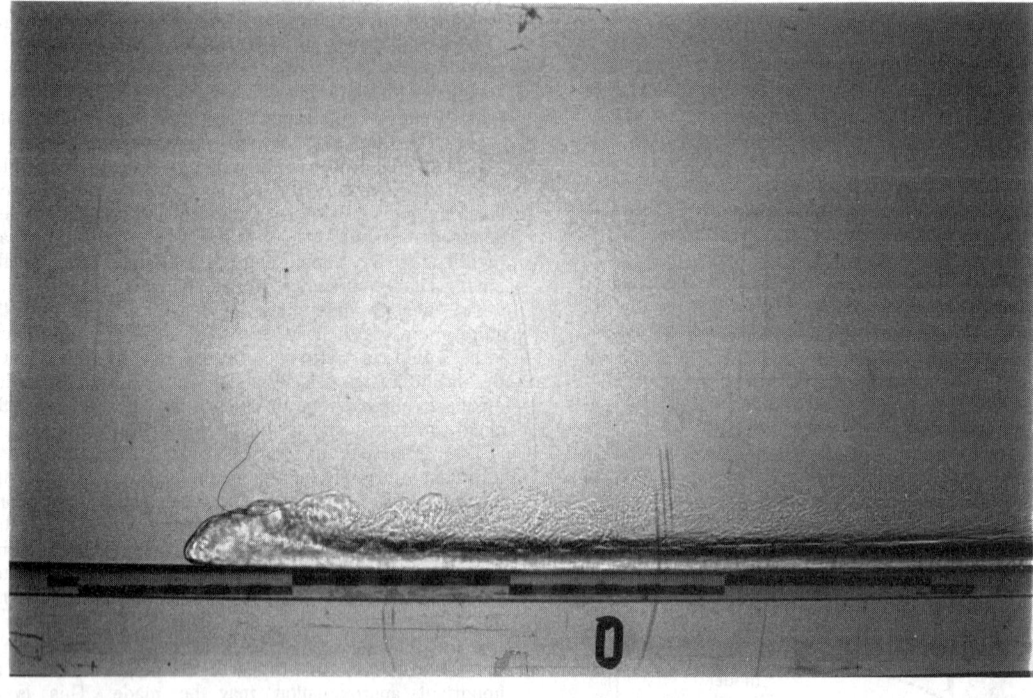

(a)

Figure 15.77 *Water tunnel modelling of the intrusion of a more dense fluid into a less dense one on flat and on sloping surfaces (Britter and Linden, 1980): (a) slope 0°; (b) slope 5°; (c) slope 20°. The entrainment increases with slope both into the head and into the flow behind it (Courtesy of Cambridge University Press)*

The main concern is the modelling of the source term for a release of liquefied gas. The classification given for liquefied gases is

(1) Pressurized release –
  (a) small hole in vapour space,
  (b) large hole in vapour space—catastrophic failure,
  (c) intermediate-sized hole in vapour space,
  (d) hole in liquid space;
(2) Refrigerated release –
  (a) spill onto land,
  (b) spill onto water;
(3) Jet release.

These cases are considered in turn for a dense gas.

15.23.2 Release from pressurized containment

For a pressurized liquefied gas, a release from a small hole in the vapour space (Case 1(a)) gives rise to a vapour jet which loses momentum and turns into a dense gas plume. This is considered further in Section 15.43.

A catastrophic failure of the vessel (Case 1(b)) results in the flash-off of a vapour fraction with the liquid falling to its normal boiling point. The vapour fraction is obtained by heat balance. A proportion of the residual liquid is also ejected as liquid droplets, or spray. The evidence is that in fact substantially all the residual liquid becomes airborne. Such a release generates considerable turbulence and results in the entrainment of large amounts of air. The liquid droplets in the cloud undergo partial evaporation with a further drop in their temperature. Thereafter the heat required for their further evaporation is taken from the gas cloud, which results in cooling of the cloud.

A release from a hole of intermediate size (Case 1(c)) may be presumed to give as spray a liquid fraction intermediate between the above two cases. Work bearing on this is described in Section 15.10. Some simple experiments by A.R. Edwards (1978) suggest that the liquid fraction which becomes airborne is a function of the ratio of the hole size to the vessel area. When this ratio reached about 0.01 the liquid fraction ejected became appreciable, whilst above a ratio of about 0.1 virtually all the liquid become airborne as spray. More detailed work is described in Section 15.10.

A hole in the liquid space (Case 1(d)) gives a two-phase flashing jet, as considered below.

15.23.3 Release from refrigerated containment

For a fully refrigerated liquefied gas, a spill onto land (Case 2(a)) results in a spreading, vaporizing pool. If the spill is into a bund, the size of the pool is constrained by the bund. Otherwise, the pool spreads until the liquid depth is so small that spreading ceases and break-up occurs. If the release is violent, there is a possibility that the wave created will spill over the edge of the bund.

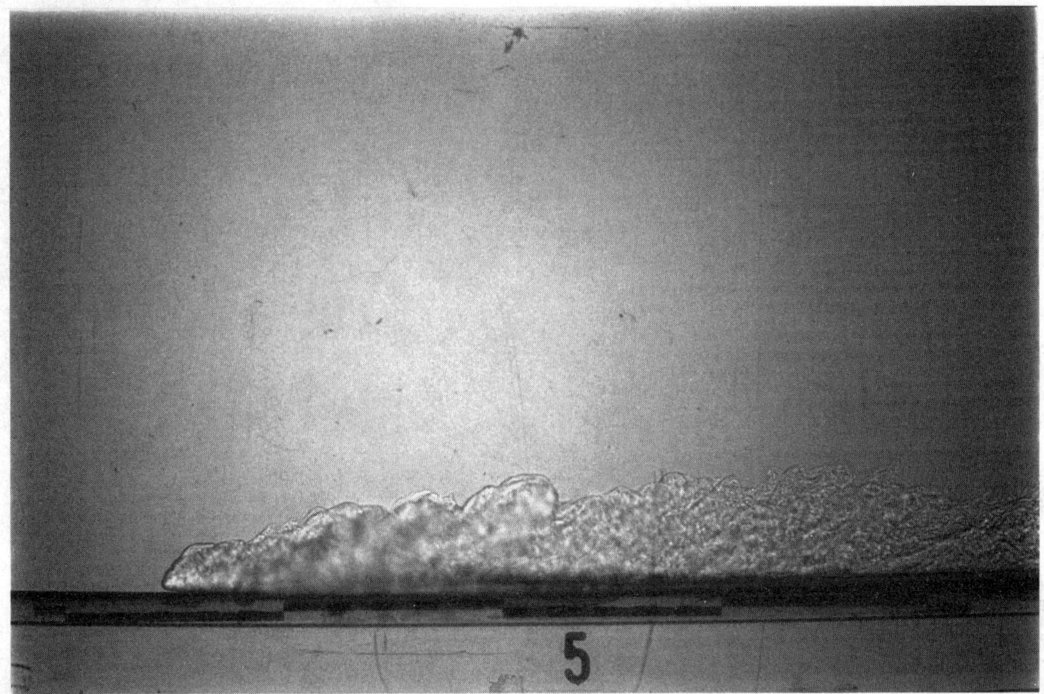

Figure 15.77 continued (b)

The vaporization from the pool is a function of the area of the pool and of the rate of vaporization per unit area. The vaporization process is governed by the heat transfer from the surface to the liquid, which is initially high but rapidly falls off as the surface chills.

For a pool of fixed size filled rapidly, the pattern is one of very high vapour flow for a short initial period, lasting perhaps a minute, followed by rapid and then more gradual decline. For a spreading pool the vapour flow tends to pass through a maximum.

A spill onto water (Case 2(b)) again results in a pool spreading until the liquid depth becomes small and break-up occurs. The vaporization from the pool is a function of the area of pool and of the specific rate of vaporization. In this case, however, renewal of the water surface occurs and the heat transfer rate is constant and high. The vapour flow starts high and increases as the pool spreads, then falls off quite sharply as the pool evaporates.

Another possible scenario is an elevated jet release. In this case the liquid may well vaporize completely before it reaches the surface. Since the heat of vaporization is obtained from the entrained air rather than the surface, the resultant gas cloud is much colder.

15.23.4 Release from semi-refrigerated containment

For a semi-refrigerated liquefied gas, the scenarios are basically similar to those for the fully pressurized case. The liquid is, however, at a lower pressure and a lower temperature. This may be expected to affect the proportion of the residual liquid which becomes airborne and the extent of rainout. The evidence is limited, but suggests that a semi-refrigerated release behaves broadly like a full refrigerated one. The main difference is that some, limited, degree of rainout may occur.

15.23.5 Two-phase jets

For a pressurized liquefied gas, a hole in the liquid space of a vessel or in a pipe gives a two-phase flashing jet. Accounts of such jets include those by Mudan (1984a), Wheatley (1987 SRD R410), Fauske and Epstein (1988), Webber and Kukkonen (1990) and Webber (1991).

The development of such a jet proceeds through three zones. The first is the depressurization, or flashing, zone in which the vapour and liquid droplets form. This extends several hole diameters and increases rapidly in diameter. The second zone is that of the two-phase momentum jet at atmospheric pressure into which air is entrained. The third zone starts where the jet loses momentum and other dispersion mechanisms take over.

The jet disintegrates into a spray of liquid droplets. These tend to remain airborne until they evaporate rather than to rain out.

Two-phase jets are considered in more detail in Section 15.21.

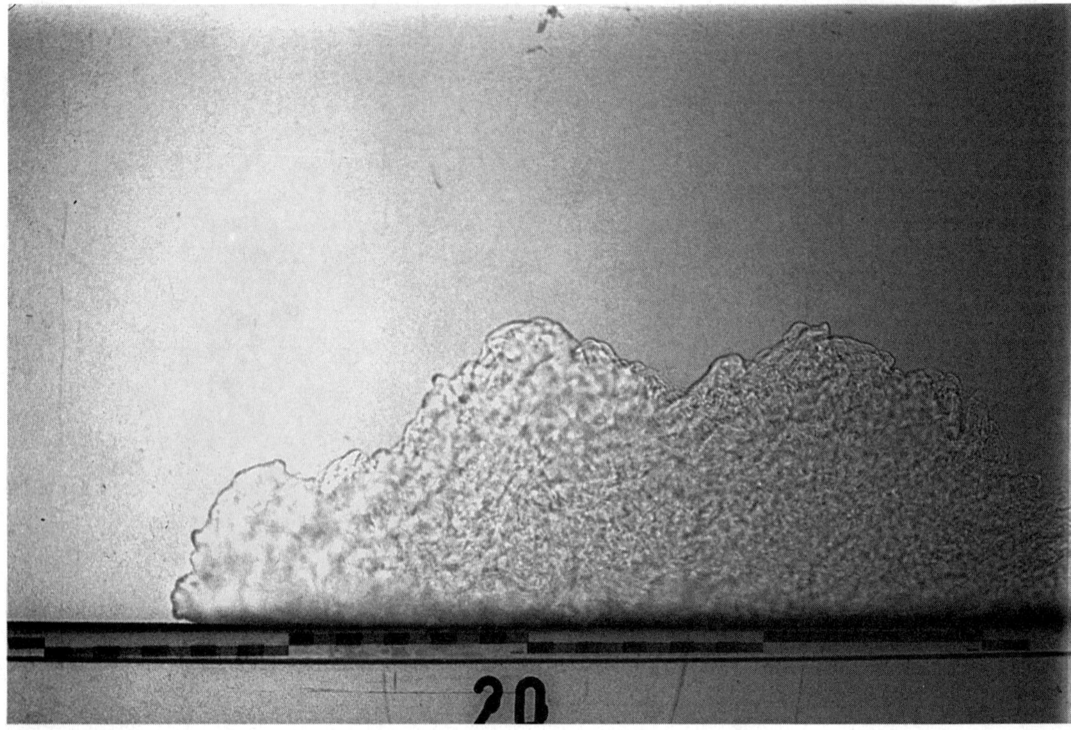

Figure 15.77 continued (c)

15.23.6 Initial density

The density of the gas cloud formed from the release depends on a number of factors. These include the density of the contaminant gas, the normal boiling point, the fraction of liquid present as droplets and the humidity of the ambient air. Typically, the density and temperature of the gas cloud is given as a function of the concentration of contaminant gas with the liquid fraction and humidity as parameters. The factors affecting the density of the gas cloud are discussed in general in Section 15.22 and, for particular gases, in Section 15.42.

15.23.7 Air entrainment

The mass of air entrained into the gas cloud is a crucial feature of the source model. There are available some estimates of the extent of entrainment for some of the scenarios just described.

For vaporization from a pool on land, the vapour is generally taken as pure vapour and the entrainment of air is modelled within the dense gas dispersion model.

For vaporization from a pool on water, the same approach appears often to be used. However, the turbulence generated by the boiling will cause some entrainment of air. Kaiser (1980) has estimated this as an entrainment within 5 s of enough air to dilute the contaminant gas by a factor of 10.

For catastrophic failure of a pressurized containment vessel, estimates have been made by Fryer and Kaiser (1979 SRD R152) of the aspect ratio and of the mass of air entrained for ammonia. The gas cloud is assumed to form a cylinder. From field trials and incidents they suggest that the radius and height of this cloud be taken as equal and the mass ratio of air to contaminant gas as some 20:1. This applies to the failure of a vessel at a quite high pressure, with a correspondingly high degree of induced turbulence; the degree of entrainment may be expected to be rather less if the vessel pressure is lower.

The Second *Canvey Report* takes for the initial cylindrical gas cloud a value of the aspect ratio, or ratio of diameter to height, of unity and this value is adopted in the *Workbook* by Britter and McQuaid (1988) .

Further discussion of such empirical estimates for catastrophic failure is given in Section 15.8, which also gives models of vessel failure from which alternative estimates of the quantity of air entrained may be made.

For two-phase flashing jets and for elevated jets the entraiment of air is part of the model for the jet. An account of these models is given in Section 15.21.

15.23.8 Liquid droplets

Another important aspect of the source term is the initial liquid fraction present in the gas cloud as spray, together with the extent to which this rains out and, if so, evaporates again. The methods of estimating the liquid fraction have been described briefly above and are dealt with in more detail in Section 15.10.

The liquid droplets tend not to rain out but rather to evaporate. The behaviour of liquid droplets in the gas cloud is considered in Section 15.21. If rainout does occur, the liquid in the resulting pool will evaporate and the contribution to the source term of this re-evaporation may be significant.

15.23.9 Matching to dispersion model

For the determination of the dispersion of the material released, the source model has to be matched to the dispersion model. The cases for which such matching is required include:

(1) vaporization from a liquid pool;
(2) catastrophic failure of a vessel;
(3) two-phase flashing jet;
(4) elevated jet.

Vaporization from a liquid pool appears at first sight to constitute a relatively simple source. However, both the area of and the rate of heat transfer to the pool are variable. Therefore either the vapour flow has to be treated as an idealized instantaneous or continuous release or the dense gas dispersion model has to be able to handle a time-varying release. In addition, a gas blanket may form above the pool. A separate submodel of this gas blanket may be included in the dense gas dispersion model.

Catastrophic failure of a vessel is generally modelled by taking as the source term a cylindrical cloud of defined aspect ratio and containing a specified quantity of entrained air. It may then be treated by the dense gas dispersion model as an instantaneous release. A similar approach may be used for a release from an intermediate sized hole.

A two-phase flashing jet model may or may not be readily conformable with the dispersion model. Models of such jets are described in Section 15.21, including some which are designed explicitly to give a smooth transition to a dense gas dispersion model.

For an elevated jet, models have been developed that conform with dense gas dispersion models, but the two-phase aspect of such jets is not well developed. The models for an elevated jet give two main types of output. One type consists simply of the touchdown distance and the concentration at touchdown. The other gives the full profile of the touchdown plume, including dimensions, velocity and concentration. It is this latter type which gives the better match to a dense gas dispersion model. Both types of model for an elevated jet are described in Section 15.43. A jet would generally be expected to constitute a continuous release, unless the time is so short that this is not appropriate.

The matching of emission and vaporization models to dispersion models is treated in the CCPS *Workbook of Test Cases for Vapour Cloud Source Dispersion Models* (1989/8). This is described in Section 15.52.

15.23.10 LNG

Treatments of the source terms for specific gases are also available. Source terms for LNG have been discussed by Kaiser (1980) and Puttock (1987c, 1988a, 1989).

As stated above, for LNG spilled onto water evidence for field trials involving Refrigerant 12 suggests that the violent boiling results within about 5 s in the entrainment of some ten times as much air. It is not known how this effect scales up.

LNG is not normally stored under pressure, but if it were the extent of air entrainment following catastrophic failure would be sufficient to dilute it close to its lower flammability limit.

An elevated jet release of LNG might well disintegrate and evaporate before it reached the surface.

15.23.11 Ammonia

Source terms for ammonia have been discussed by Kaiser and Walker (1978), Fryer and Kaiser (1979 SRD RR152), Griffiths and Kaiser (1979 SRD R154) and Kaiser (1980, 1989). They consider in particular the field trials with Refrigerant 12 and the ammonia release incidents at Potchefstroom in 1973 (Case History A65), Houston in 1976 (Case History A84), and Pensacola in 1977, and deduce for catastrophic failure of a vessel rules of thumb for aspect ratio and air entrainment, as mentioned above and described more fully in Section 15.8. The work of Wheatley (1986) on two-phase flashing jets is concerned specifically with ammonia.

15.24 Dispersion of Dense Gas: Models and Modelling

The mathematical modelling of dense gas dispersion is a complex subject with a large literature.

15.24.1 Types of model

There are three broad classes of model of dense gas dispersion:

(1) modified conventional models;
(2) box and slab models;
(3) K-theory and other three-dimensional models.

15.24.2 Modified conventional models

Once it was realized that some materials exhibit a degree of dense gas behaviour, the initial response was to attempt to modify existing models such as the Sutton and Pasquill–Gifford models, usually by making empirical adjustments to the diffusion parameter C or the dispersion coefficient σ. This approach is described in Section 15.25. It was soon appreciated, however, that the behaviour of a dense gas is radically different from that of a neutrally buoyant gas and that a totally different approach is necessary.

15.24.3 Box models

A new generation of models, usually known as box models, was created. In the simple box model the gas cloud is assumed to be a pancake-shaped cloud with properties uniform in the crosswind and vertical directions. The model contains relations which describe the growth of the radius and height of an instantaneous release, or the crosswind width and height of a continuous release, and the entrainment of air at the top and edges of the cloud. The concentration in the cloud is obtained by mass balance. For an instantaneous release it is uniform in all three directions. For a continuous release it is uniform in the crosswind and vertical directions, but varies in the downwind direction.

The characteristic features of a typical box model are as follows. The cloud shape is determined by gravity-driven flow. Mixing inside the cloud is sufficiently rapid that the cloud has uniform concentration. Air entrainment velocities are specified as a function of cloud advection velocity, density difference and turbulence levels.

Although simple in concept, the box model actually places considerable demands on the modeller to specify correctly the various features just mentioned, such as cloud advection velocity and entrainment rates. In this respect it contrasts with the three-dimensional models described below, which largely take care of these aspects. In this sense, the apparent simplicity of the box model is deceptive. A box model contains a small number of adjustable parameters. These may be determined individually by experiment and confirmed by field trials.

The features of the box model for the investigation of dense gas dispersion have been described by McQuaid (1984a). A box model utilizes a top-hat concentration profile. McQuaid cites a number of other problems in fluid mechanics in which a similar 'black box' approach has been used, including the analysis of the boundary layer by Clauser (1954) and of plumes and of wakes by B.R. Morton, Taylor and Turner (1956) and B.R. Morton (1962).

A box model is based on the principle of similarity. For dense gas dispersion the conditions necessary for similarity are given by McQuaid as that the gas cloud is not so large as to alter the ambient atmosphere and that it is not subject to rapid change of depth.

McQuaid discusses the applicability of box models for the assessment of releases of flammable or toxic materials. He concludes that for toxic releases a box model gives concentrations which are generally sufficiently accurate, given the relatively high uncertainty in relations for toxic injury, but may be insufficiently accurate for some problems which arise in relation to flammability.

Box models are relatively simple, but have a number of drawbacks. One is that they are essentially limited to flat, unobstructed terrain and are not designed to handle terrain with slopes, obstructions, buildings or other special features. Another defect is the assumption of uniform concentration which may be inadequate for assessment of flammability and toxicity. A box model is sometimes termed a bulk property (BP) model.

Box models have been the dominant type of model used for dense gas dispersion and several such models are described below.

15.24.4 Advanced similarity models

Box models are simple similarity models which are based on assumptions about the shape of the gas cloud. A development from these is the advanced similarity models, which incorporate also assumptions about the concentration profile. The HEGADAS model, described below, is an advanced similarity model.

15.24.5 K-theory models

The limitations of box models have led to increasing use of the more fundamental *K*-theory models. A *K*-theory model consists of a set of equations for the conservation of mass, energy, momentum, and so on, together with boundary conditions. The model determines the variation of properties, velocity and concentration along the flow streamlines.

When the basic equations are averaged with respect to time certain second-order terms appear. The number of equations is less than the number of unknowns. It is therefore necessary to specify some of the unknowns in terms of the others. This is known as 'turbulence closure' and much work has been done on methods of effecting it. In a *K*-theory model the closure is a low order one. It is effected by making the gradient transfer assumption.

The model utilizes empirical relationships for the eddy diffusivities and eddy viscosities as a function of space, and sometimes also of stability category or Richardson number. Higher order closures are those in which additional equations are derived for the Reynolds stresses (e.g. $\bar{u}\bar{w}$) and therefore also for the velocity concentration correlations (e.g. $\bar{w}\bar{c}$). The new equations include third-order terms which in turn have to be modelled.

For undisturbed ambient flows *K*-theory models are reasonably satisfactory. But the presence of dense gas or obstructions renders them less so. In such cases the relations used for the eddy diffusivity *K* become inadequate, because they take insufficient account of local variations. In particular, the Reynolds analogy for the equivalence of mass, heat and momentum eddy diffusivities is not valid when the flow has stable density stratification.

The development of *K*-theory models was influenced by the need to obtain analytical solutions. This requirement meant that *K* had to be specified in a way compatible with this aim. The advent of computers has removed this constraint.

In some areas of work on turbulent diffusion the use of eddy diffusivities is regarded as out-dated and use is made of higher order closure models. *K*-theory models were in use before the recognition of the need for models specific to dense gas dispersion and a *K*-theory model, SIGMET, was in fact one of the first models applied to this problem. A *K*-theory model is sometimes termed a turbulence scheme model.

K-theory models are full three-dimensional models, but they may also be used in the two-dimensional mode, and in fact it is often sufficient to use them in this way.

Several of the models described below are *K*-theory models.

15.24.6 κ–ϵ Models

As just mentioned, the presence of a dense gas or of obstructions tends to render inadequate the relations normally used for the eddy diffusivity in a *K*-theory model.

The eddy diffusivity *K* is related to the turbulence energy κ and the energy dissipation rate ϵ as follows:

$$K \propto \frac{\kappa^2}{\epsilon} \qquad [15.24.1]$$

The value of the eddy diffusivity is subject to local variations which are much more significant in the presence of dense gas or obstructions. A region of high density gradient tends to give low values of κ and thus, from Equation [15.24.1], very low values of *K*.

One way of overcoming this difficulty is to determine local values for κ and ϵ and then use them to obtain local values for K. This may be done by formulating the transport equations for κ and ϵ. This creates a two-equation κ–ϵ model. The parameters required for this κ–ϵ model are available and apply over a wide range of flows. This approach retains eddy diffusivity but allows it to vary locally.

κ–ϵ models are used extensively for the study of flows in regions of recirculation. One of the models described below, HEAVYGAS, is a κ–ϵ model.

15.24.7 Algebraic stress models

The κ–ϵ models just described still utilize the gradient transport relationship in regions where the diffusivities are very low and difficult to specify. They still retain the assumption that the eddy diffusivity is isotropic. It is known, however, that stability affects the different components of the Reynolds stresses in different ways.

A development which addresses this problem is the algebraic stress model. This replaces the gradient transport relation with equations which relate the various fluxes and stresses to the velocity and density gradients. This type of model is not, however, developed for dense gas dispersion.

15.24.8 Density difference

The treatment described below makes use of certain common relations and, before describing the principal models, it is convenient to give some of these relations.

For the density intrusion represented by a dense gas spreading in ambient air

$$\Delta\rho = \rho - \rho_a \qquad [15.24.2]$$

$$\Delta = \frac{\Delta\rho}{\rho_a} \qquad [15.24.3]$$

$$g' = g\frac{\rho - \rho_a}{\rho_a} \qquad [15.24.4a]$$

$$= g\Delta \qquad [15.24.4b]$$

where $\Delta\rho$ and Δ are the absolute and reduced density differences between the gas cloud and the ambient air, g and g' are the absolute and reduced accelerations due to gravity, ρ is the density of the cloud and ρ_a is the density of the ambient air. The variable g' is generally referred to simply as the 'reduced gravity'.

An alternative reduced density difference is also used defined as

$$\Delta' = \frac{\Delta\rho}{\rho} \qquad [15.24.5]$$

where Δ' is an alternative reduced density difference.

15.24.9 Wind velocity

The vertical wind velocity profile is frequently obtained from the empirical expression

$$u = u_r\left(\frac{z}{z_r}\right)^p \qquad [15.24.6]$$

where u is the wind velocity, u_r is the wind velocity at the reference height z_r, z is the height, z_o is the roughness length and p is an index.

Use is made of the logarithmic vertical wind velocity profile:

$$u = \frac{u_*}{k}\ln\left(\frac{z}{z_r}\right) \qquad [15.24.7]$$

where u_* is the friction velocity, z_o is the roughness length and k is the von Karman constant.

The logarithmic vertical wind velocity profile is also used in the modified form:

$$u = \frac{u_*}{k}\left[\ln\left(\frac{z+z_o}{z_o}\right) - \psi\left(\frac{z}{L}\right)\right] \qquad [15.24.8]$$

where L is the Monin–Obukhov length and ψ is a profile function.

15.24.10 Heat transfer

Heat transfer from the ground to the cloud will, in general, have contributions from both natural and forced convection. For natural convection use is frequently made of the relation given by McAdams (1954) for heat transfer from an upwards-facing heated plate:

$$\mathrm{Nu} = 0.14(\mathrm{GrPr})^{\frac{1}{3}} \qquad [15.24.9]$$

with

$$\mathrm{Nu} = \frac{hL}{k_f} \qquad [15.24.10]$$

$$\mathrm{Gr} = \frac{L^3\rho_f\beta_f g\Delta T}{\mu_f} \qquad [15.24.11]$$

$$\mathrm{Pr} = \frac{c_{pf}\mu_f}{k_f} \qquad [15.24.12]$$

$$\beta_f = 1/T \qquad [15.24.13]$$

where c_p is the specific heat, Gr is the Grashof number, h is the heat transfer coefficient, k is the thermal conductivity, L is a characteristic linear dimension, Nu is the Nusselt number, Pr is the Prandtl number, T is the absolute temperature, ΔT is the temperature difference for heat transfer, β is the coefficient of thermal expansion, μ is the viscosity, ρ is the density and the subscript f denotes film conditions. Then from Equation 15.24.9

$$h_{gr} = 0.14\left(\frac{\rho_f^2 k_f^3 g}{\mu_f^2}\frac{c_{pf}\mu_f}{k_f}\frac{\Delta T}{T}\right)^{\frac{1}{3}} \qquad [15.24.14]$$

where the subscript gr denotes between the ground and the cloud.

For heat transfer by forced convection use is made of the relation given by Treybal (1955) for heat transfer from an upwards-facing heated plate:

$$j_H = \mathrm{St_H}\mathrm{Pr}^{\frac{2}{3}} = \frac{C_f}{2} \qquad [15.24.15]$$

with

$$\mathrm{St_H} = \frac{h}{c_p G} \qquad [15.24.16]$$

$$C_f = 2\left(\frac{u_*}{u}\right)^2 \qquad [15.24.17]$$

where C_f is a friction factor, G is the mass velocity, j_H is the j factor for heat transfer, $\mathrm{St_H}$ is the Stanton number for heat transfer and u is the velocity of the cloud. Equation 15.24.15 can be rearranged to give, for the case considered,

$$h_{gr} = \left[\left(\frac{c_p \mu}{k}\right)^{-\frac{2}{3}}\left(\frac{u_*}{u}\right)^2\right] u\rho c_p \qquad [15.24.18]$$

15.24.11 Density intrusion models

A dense fluid within a less dense fluid, such as a dense gas cloud in the atmosphere, represents a density intrusion which flows as a gravity current. The modelling of density intrusions and of the associated gravity-driven flow is a distinct area of work with applications in a number of fields. Reviews of gravity currents have been given by Benjamin (1968) and Simpson (1982). Gravity currents occur in many natural situations in the atmosphere and the ocean and they are relevant to industrial accidents such as oil spills and dense gas releases.

Work on gravity-driven flow includes the investigation of: the different phases of the flow (e.g. Huppert and Simpson, 1980); of the flow on a flat surface, and in particular the 'head' which develops at the leading edge (e.g. Benjamin, 1968) and the mixing which occurs (e.g. Britter and Simpson, 1978); of flow on slopes (e.g. Ellison and Turner, 1959; Britter and Linden, 1980; Beghin, Hopfinger and Britter, 1981); and of flow over and around obstacles (e.g. Rottman *et al.*, 1985).

The driving force of the gravity current is the buoyancy force. This is opposed by the inertial force, which is dominant initially, and the viscous force, which becomes dominant later. There is therefore an inertia–buoyancy phase and a viscous–buoyancy phase.

A model of gravity-driven flow which is increasingly utilized is the 'shallow water equations'. These are described in *Water Waves* (Stoker, 1957), *Stratified Flows* (Yih, 1980) and *Introduction to Water Waves* (Crapper, 1984). A derivation of the shallow water equations has been given by Penney and Thornhill (1952).

The non-linear shallow water equations are:

$$\frac{\partial h}{\partial t} + \frac{\partial (uh)}{\partial x} + n\frac{uh}{x} = 0 \qquad [15.24.19]$$

$$\frac{\partial u}{\partial t} + u\frac{\partial u}{\partial x} + g'\frac{\partial h}{\partial x} = 0 \qquad [15.24.20]$$

where g' is the reduced gravity, h is the height of the dense fluid, u is the velocity of the dense fluid, x is the distance and n is a symmetry index. For planar flow n = 0, and for axisymmetric flow n = 1.

From Bernouilli's equation, the velocity of the gravity-driven wave is

$$c = (g'h)^{\frac{1}{2}} \qquad [15.24.21]$$

where c is the velocity of the wave. Then, from Equations [15.24.19]–[15.24.21], for n = 0

$$\frac{\partial c}{\partial t} + \frac{c}{2}\frac{\partial u}{\partial x} + u\frac{\partial c}{\partial x} = 0 \qquad [15.24.22]$$

$$\frac{\partial u}{\partial t} + u\frac{\partial u}{\partial x} + 2c\frac{\partial c}{\partial x} = 0 \qquad [15.24.23]$$

15.24.12 Basic box model relations

The basic equations of the typical box model are:

$$\frac{\mathrm{d}R}{\mathrm{d}t} = c_E (g'H)^{\frac{1}{2}} \qquad [15.24.24]$$

$$\frac{\mathrm{d}V}{\mathrm{d}t} = \pi R^2 u_e + 2\pi R H w_e \qquad [15.24.25]$$

where g' is the reduced gravity, H is the height of the cloud, R is the radius of the cloud, u_e is the top entrainment velocity, V is the volume of the cloud, w_e is the edge entrainment velocity and c_E is the damping coefficient.

Equation 15.24.24 may also be formulated in terms of the negative buoyancy B:

$$B = g'V \qquad [15.24.26a]$$

$$= g'(\pi R^2 H) \qquad [15.24.26b]$$

Thus from Equation [15.24.24] and [15.24.26b]

$$\frac{\mathrm{d}R}{\mathrm{d}t} = c_E\left(\frac{B}{\pi}\right)^{\frac{1}{2}}\frac{1}{R} \qquad [15.24.27]$$

15.24.13 Model development

Since the work on dense gas dispersion in the Netherlands in the early 1970s, notably the work of van Ulden, there has been a high level of interest and work on the subject, including mathematical modelling, physical modelling and field trials. Some early workers, such as the Bureau of Mines and Clancey, attempted to adapt passive dispersion models to dense gas dispersion, as described in Section 15.25. However, in view of the fundamentally different physical phenomena involved, this attempt came to be regarded as inappropriate, although the newer models often incorporate features of, and make transition to, passive dispersion models.

The next generation of models were box models. Of these, the models of van Ulden, of Cox and Roe (developed as the BG/C&W model) and the SRD models (DENZ and CRUNCH) are described in Sections 15.26–15.28, respectively. Even at this stage the box models did not have the field entirely to themselves. An early and influential *K*-theory model was SIGMET, which has already been referred to in connection with LNG spill study and which is described further in Section 15.29.

The more modern systems of models include: SLAB and FEM3; HEGADAS and HGSYSTEM; DEGADIS; and SLUMP and HEAVYGAS. These are described in Sections 15.30–15.33, respectively. Both advanced box models and three-dimensional models are represented here. A different type of modern model is the similarity model given in the *Workbook* and described in Section

15.34. In Section 15.35 an account is given of DRIFT, a model which develops but transcends the box model approach. Some other models are described in Section 15.36.

15.25 Dispersion of Dense Gas: Modified Conventional Models

15.25.1 Bureau of Mines model

Attempts were made in early work to modify the parameters in passive gas dispersion models to take into account heavy gas behaviour. An account is given, although the approach has now been largely superseded by other methods.

Thus, for example, for a neutral gas the vertical dispersion coefficient σ_z in the Pasquill–Gifford equations tends to be about half the value of the horizontal dispersion coefficient, but work at the Bureau of Mines (Burgess and Zabetakis, 1973 BM RI 7752) indicated that for a heavy gas

$$\sigma_z = 0.2\sigma_y \qquad [15.25.1]$$

where σ_y and σ_z are the dispersion coefficients in the crosswind and vertical (y and z) directions, respectively.

Equation [15.25.1] has been used by Burgess and Zabetakis to take into account the density effect in the gas cloud formed in the Port Hudson explosion in 1970 (Case History A52).

15.25.2 Clancey model

V.J. Clancey (1976a) suggested that in Equation [15.16.25] the generalized diffusion parameter in the vertical direction C_z should be taken as

$$C_z = \tfrac{1}{2}C_x = \tfrac{1}{2}C_y \qquad [15.25.2]$$

where C_x, C_y and C_z are the generalized diffusion parameters in the downwind, crosswind and vertical (x, y and z) directions, respectively.

15.25.3 Germeles and Drake model

A modified Gaussian model was presented by Germeles and Drake (1975) to describe the dispersion of LNG vapour clouds. This model has also been described by Ermak, Chan *et al.* (1982)

The Germeles and Drake model for a continuous release contains elements both of a dense gas box model and a passive dispersion model. The source is assumed to be a cylindrical cloud with an initial radius R_i that of the pool and initial height H_i given by

$$H_i = 2R_i W_v/u \qquad [15.25.3]$$

The mass and heat content of the source are

$$m = \pi\rho R^2 H \qquad [15.25.4]$$

$$E = mc_p T \qquad [15.25.5]$$

where c_p is the specific heat of the gas in the source cloud, E is the heat in the source cloud, m is the mass of gas in the source cloud, T is the absolute temperature, u is the wind speed, W_v is the velocity of the vapour source, and ρ is the density. The value of W_v is some 250 times the liquid regression rate.

The development of the source cloud is as follows. For the rate of change of the radius use is made of Equation [15.24.24] and for that of the mass and heat content

$$\frac{dm}{dt} = \pi\rho_a R^2 u_e \qquad [15.25.6]$$

$$\frac{dE}{dt} = \pi\rho_a R^2 c_{pa} T_a u_e + \epsilon_v + \epsilon_w \qquad [15.25.7]$$

where u_e is the top entrainment velocity, ϵ_v is the heat released by the condensation and by freezing of the water vapour, ϵ_w is the surface heat flux, and subscript a denotes air.

It is assumed that the release is from a finite line source of width L located at a virtual source a distance x_v upwind of the true source. Use is then made of a model of the Pasquill–Gifford type to model the dispersion. This gives for the concentration

$$c(x,y,z) = \frac{\dot{V}}{(2\pi)^{\frac{1}{2}} uL\sigma_z} \exp\left(-\frac{z^2}{2\sigma_z^2}\right) \left\{\mathrm{erf}\left[\frac{(L/2)-y}{2^{\frac{1}{2}}\sigma_y}\right] + \mathrm{erf}\left[\frac{(L/2)+y}{2^{\frac{1}{2}}\sigma_y}\right]\right\} \qquad [15.25.8]$$

where c is the volumetric concentration, L is the width of the source, $\dot{V}$ is the volumetric rate of release, and σ_y and σ_z are the dispersion coefficients in the crosswind and vertical directions, respectively. The values used for σ_y and σ_z are the regular Pasquill–Gifford values.

The source model is run until the radial velocity of the cloud is equal to the wind speed. The source length L is then set equal to the diameter $2R$ of the cloud and the virtual source distance x_v is found from the values of $\sigma_y(x_v)$ and $\sigma_z(x_v)$ required to make the concentration at the source centre, as given by Equation [15.25.8], equal to that in the source cloud. Equation [15.25.8] is then used with these parameters to determine the concentrations downwind.

Ermak, Chan *et al.* (1982) have compared the performance of the Germeles and Drake model with that of other models in relation to the Burro trials.

15.26 Dispersion of Dense Gas: Van Ulden Model

15.26.1 Van Ulden model

The experimental work of van Ulden (1974) which demonstrated the marked difference in behaviour between a neutrally buoyant gas and a dense gas has already been described. At the same time van Ulden also presented a theoretical model. The main model was for the dispersion of an instantaneous release of dense gas in still air, but a model for a continuous release with finite wind velocity was also outlined. Van Ulden has subsequently extended his model (van Ulden 1979, 1984, 1987, 1988; van Ulden and de Haan, 1983).

In van Ulden's original work the behaviour of the cloud is modelled as a two-stage process, in which spreading occurs in the first stage by gravity slumping and in the second by Gaussian dispersion. In the first stage the dispersion is treated as a dense fluid flow process but with air entrainment at the edge of the cloud.

The starting point of the model is the following equation for the movement of a front of dense fluid such as that resulting from a dam burst:

$$u_{\mathrm{f}} = c_{\mathrm{E}}(\Delta \cdot gH_{\mathrm{f}})^{\frac{1}{2}} \quad [15.26.1]$$

with

$$\Delta = (\rho_1 - \rho_2)/\rho_1 \quad [15.26.2]$$

where H_{f} is the height of the front, u_{f} is the velocity of the front, Δ is the reduced density difference, ρ_1 is the density of the fluid behind the wavefront, ρ_2 is the density of the fluid ahead of the wavefront, and c_{E} is a slumping constant. Van Ulden states that experiment suggests that the constant c_{E} is approximately unity.

It is assumed that in the first stage the initial gas cloud is a cylinder of radius R, height $H(R)$ and volume $V(R)$ with a front moving at velocity $u_{\mathrm{f}}(R)$ and with changing density $\rho_{\mathrm{c}}(R)$. Then from the model just given

$$\frac{\mathrm{d}R}{\mathrm{d}t} = u_{\mathrm{f}} \quad [15.26.3]$$

$$= c_{\mathrm{E}}(\Delta'_{\mathrm{c}} \cdot gH)^{\frac{1}{2}} \quad [15.26.4]$$

with

$$\Delta'_{\mathrm{c}} = (\rho_{\mathrm{c}} - \rho_{\mathrm{a}})/\rho_{\mathrm{c}} \quad [15.26.5]$$

where t is the time, Δ'_{c} is the reduced density difference of the cloud, ρ_{a} is the density of air and ρ_{c} is the density of the cloud.

The volume V of the cloud is

$$V = \pi R^2 H \quad [15.26.6]$$

Mixing at the front, or edge, of the cloud is assumed to take place at the rate

$$\frac{\mathrm{d}V}{\mathrm{d}t} = \gamma 2\pi RH \frac{\mathrm{d}R}{\mathrm{d}t} \quad [15.26.7]$$

where γ is the edge entrainment coefficient.

Solution of Equations 15.26.3–15.26.7 gives

$$V/V_{\mathrm{o}} = (R/R_{\mathrm{o}})^{2\gamma} \quad [15.26.8]$$

$$H/H_{\mathrm{o}} = (R/R_{\mathrm{o}})^{2\gamma-2} \quad [15.26.9]$$

$$\frac{\mathrm{d}R}{\mathrm{d}t} = \frac{c_{\mathrm{E}}}{R}\left[\frac{gV_{\mathrm{o}}/\pi}{V_{\mathrm{o}}/V + \rho_{\mathrm{a}}/(\rho_{\mathrm{o}} - \rho_{\mathrm{a}})}\right]^{\frac{1}{2}} \quad [15.26.10]$$

$$\Delta = \left[1 + \frac{\rho_{\mathrm{a}} V}{(\rho_{\mathrm{o}} - \rho_{\mathrm{a}})V_{\mathrm{o}}}\right]^{-1} \quad [15.26.11]$$

where the subscript o denotes time zero and hence the source.

Equation [15.26.10] reduces to

$$\frac{\mathrm{d}R}{\mathrm{d}t} = \frac{c_{\mathrm{E}}}{R}\left[\frac{(\rho_{\mathrm{o}} - \rho_{\mathrm{a}})gV_{\mathrm{o}}}{\pi\rho_{\mathrm{o}}}\right] \quad [15.26.12]$$

if the mixing at the front is negligible or if the term $(\rho_{\mathrm{o}} - \rho_{\mathrm{a}})/\rho_{\mathrm{a}}$ is small. Therefore Equation 15.26.12 is an adequate approximation in all circumstances. From Equation 15.26.12 the relation between the cloud radius and the time t is

$$R^2 - R_{\mathrm{o}}^2 = 2c_{\mathrm{E}}\left[\frac{(\rho_{\mathrm{o}} - \rho_{\mathrm{a}})gV_{\mathrm{o}}}{\pi\rho_{\mathrm{o}}}\right]^{\frac{1}{2}} t \quad [15.26.13]$$

The transition from gravity spread to diffusional mixing is taken to occur at the point where the turbulent energy is equal to the average potential energy difference and hence where

$$2\rho_{\mathrm{a}} u_*^2 = \tfrac{1}{2}(\rho_{\mathrm{c}} - \rho_{\mathrm{a}})gH \quad [15.26.14a]$$

$$= \tfrac{1}{2}\rho_{\mathrm{c}} u_{\mathrm{f}}^2 \quad [15.26.14b]$$

The condition for transition with $\rho_{\mathrm{c}} \approx \rho_{\mathrm{a}}$ is then

$$u_{\mathrm{f}} = 2u_* \quad [15.26.15]$$

Van Ulden uses the model just given to calculate the transition radius R_{u} and height H_{u} at which condition 15.26.15 applies and the position of the cloud and then applies an area source Gaussian dispersion model for the further determination of concentrations. The cloud radius and height obtained by van Ulden for the experiment described earlier are shown in Figure 15.76. The fit is much superior to that given by the Gaussian model.

van Ulden also gave a treatment of a continuous release. The mass flux across the cross-section of the plume at right angles to the wind direction is constant. The area of the cross-section is

$$S = 2LH \quad [15.26.16]$$

where L is the half-width of the plume and S is the area of the cross-section of the plume. For the source

$$S_{\mathrm{o}} = 2L_{\mathrm{o}}H_{\mathrm{o}} \quad [15.26.17]$$

and

$$W_{\mathrm{o}} = S_{\mathrm{o}}u \quad [15.26.18]$$

where u is the mean wind speed and W_{o} is the volumetric flow from the source.

Then, by an argument similar to that for the instantaneous release,

$$S/S_{\mathrm{o}} = (L/L_{\mathrm{o}})^{\gamma} \quad [15.26.19]$$

$$H/H_{\mathrm{o}} = (L/L_{\mathrm{o}})^{\gamma-1} \quad [15.26.20]$$

$$u_y = \frac{c_{\mathrm{E}}}{L^{\frac{1}{2}}}\left[\frac{gS_{\mathrm{o}}/2}{S_{\mathrm{o}}/S + \rho_{\mathrm{a}}/(\rho_{\mathrm{o}} - \rho_{\mathrm{a}})}\right]^{\frac{1}{2}} \quad [15.26.21]$$

$$\Delta'_{\mathrm{c}} = \left[1 + \frac{\rho_{\mathrm{a}}}{\rho_{\mathrm{o}} - \rho_{\mathrm{a}}}\left(\frac{L}{L_{\mathrm{a}}}\right)^{\gamma}\right]^{-1} \quad [15.26.22]$$

where u_y is the velocity of the cloud in the crosswind direction.

For the particular case of no mixing ($\gamma = 0$) and hence constant reduced density ($\Delta'_{\mathrm{c}} = \Delta'_{\mathrm{co}}$)

$$u_y = \left(\frac{gS_{\mathrm{o}}\Delta'_{\mathrm{co}}}{2L}\right)^{\frac{1}{2}} \quad [15.26.23]$$

15.26.2 Critique of model

In subsequent work, van Ulden has described certain deficiencies in this model and in models by other workers, and has made modifications to his own model.

Van Ulden and de Haan (1983) analyse the forces which determine the behaviour of an unsteady-state gravity current, notably the buoyant, viscous and inertial forces, and describe the viscous–buoyant and inertia–buoyant phases. They refer to the finding of Huppert and Simpson (1980) in work on flow of a dense fluid under a lighter fluid in a channel that the velocity of the gravity current is not solely a function of $(g'H)^{\frac{1}{2}}$ and that an important parameter is the fractional depth H/D, where D is the total depth of the channel.

Van Ulden (1984) gives a model for the particular case of two-dimensional flow in the inertia–buoyancy regime. In further critiques, van Ulden (1987, 1988) points out that Equation [15.26.4] is deficient in that it takes no account of the horizontal acceleration of the cloud from rest and therefore does not give conservation of momentum.

Various modellers have described air entrainment at the top and at the edge of the cloud by the relation

$$\frac{dV}{dt} = \pi R^2 u_e + 2\pi RHw_e \qquad [15.26.24]$$

where u_e is the top entrainment velocity and w_e is the edge entrainment velocity. However, proportionality between u_e and u_f violates conservation of energy.

15.26.3 van Ulden model 2

Van Ulden (1987, 1988) has presented a second model which corrects some of the weaknesses of the initial model just described. The model is again for an instantaneous release in still air.

The model consists of four simultaneous differential equations for the rate of change with time of the radius R, the volume V, the edge velocity u_f, and the bulk turbulent kinetic energy T_E. It gives a more fundamental treatment in that the relations for u_f and T_E are based on conservation of momentum and of energy, respectively, and the velocity of entrainment u_e through the top surface of the cloud is obtained from the bulk turbulent velocity $\bar{u}_t$ which itself is derived from the turbulent energy T_E.

The equations for the radius R and the volume V are

$$\frac{dR}{dt} = u_f \qquad [15.26.25]$$

$$\frac{dV}{dt} = \pi R^2 u_e \qquad [15.26.26]$$

For the edge velocity u_f, conservation of radial momentum gives

$$\frac{dM_{rd}}{dt} = F_s + F_d + F_a + F_v \qquad [15.26.27]$$

where F_a is the force due to the reaction of the ambient fluid to the outward radial acceleration of the cloud edge, F_d is the drag force on the cloud edge due to the presence of the quiescent ambient fluid, F_s is the static pressure force due to the negative buoyancy of the cloud, F_v is the force due to vertical acceleration in the cloud and the reaction of the ambient fluid to vertical acceleration of the cloud top, and M_{rd} is the radial momentum integral of the cloud. This momentum budget yields a relation of the form

$$\frac{du_f}{dt} = f(g', H, R, u_f, u_e, \rho_c, \rho_a, \bar{\rho}_c, \Delta\bar{\rho}_c) \qquad [15.26.28]$$

where $\bar{\rho}_c$ is the mean density of the cloud and $\Delta\bar{\rho}_c$ is the mean density difference.

The relation for the bulk turbulent kinetic energy T_E is derived from conservation of the total energy in a large control volume containing the cloud and the secondary flows around it. Four types of energy are considered: I_E the internal heat, K_E the kinetic energy of the mean radial and vertical motions, P_E the potential energy and T_E the turbulent kinetic energy. Then

$$\frac{dP_E}{dt} = -G + B \qquad [15.26.29]$$

$$\frac{dK_E}{dt} = G - S \qquad [15.26.30]$$

$$\frac{dT_E}{dt} = S - B - D \qquad [15.26.31]$$

$$\frac{dI_E}{dt} = D \qquad [15.26.32]$$

where B is the rate of buoyant destruction of turbulent energy, which is the conversion of turbulent energy into potential energy by entrainment, D is the rate of dissipation of turbulent energy into internal heat, G is the rate at which gravity transforms potential energy into mean kinetic energy, and S is the rate of shear production of turbulent energy. The relationship between these forms of energy is shown in Figure 15.78(a) and their variation with time is shown in Figure 15.78(b). Conservation of energy gives

$$\frac{d(I_E + K_E + P_E + T_E)}{dt} = 0 \qquad [15.26.33]$$

This energy budget yields a relation of the form

$$\frac{dT_E}{dt} = f(g, H, R, u_f, \bar{u}_t, V, V_o, u_e, \rho_c, \rho_a, \bar{\rho}_c, \Delta\bar{\rho}_o, Ri_t) \qquad [15.26.34]$$

where Ri_t is a bulk turbulent Richardson number and $\Delta\bar{\rho}_o$ is the mean initial density difference for the cloud.

For air entrainment, edge entrainment is neglected and top entrainment is obtained using the entrainment model of Diedronks and Tennekes (1984):

$$u_e = \frac{c_l \bar{u}_t}{c_t + Ri_t} \qquad [15.26.35]$$

with

$$Ri_t = \frac{g\Delta\bar{\rho}_c H}{\bar{\rho}_c \bar{u}_t^2} \qquad [15.26.36]$$

where c_l and c_t are constants. The values of c_l and c_t are 0.2 and 1.5, respectively.

The bulk turbulent velocity $\bar{u}_t$ is a function of the bulk turbulent kinetic energy T_E:

$$\bar{u}_t = (2T_E/\bar{\rho}_c V)^{\frac{1}{2}} \qquad [15.26.37]$$

The average density $\bar{\rho}_c$ and density difference $\Delta\bar{\rho}_c$ of the cloud and average concentration C are

$$\bar{\rho}_c = \rho_a + \Delta\bar{\rho}_c \qquad [15.26.38]$$

$$\Delta\bar{\rho} = \Delta\bar{\rho}_o V_o / V \qquad [15.26.39]$$

$$C = V_o / V \qquad [15.26.40]$$

assuming $C_o = 1$.

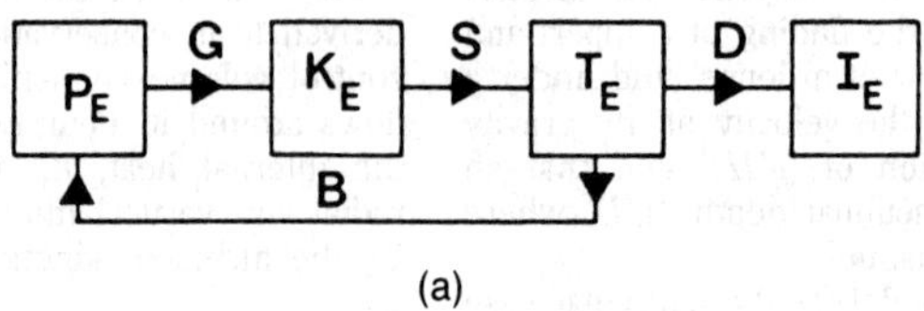

(a)

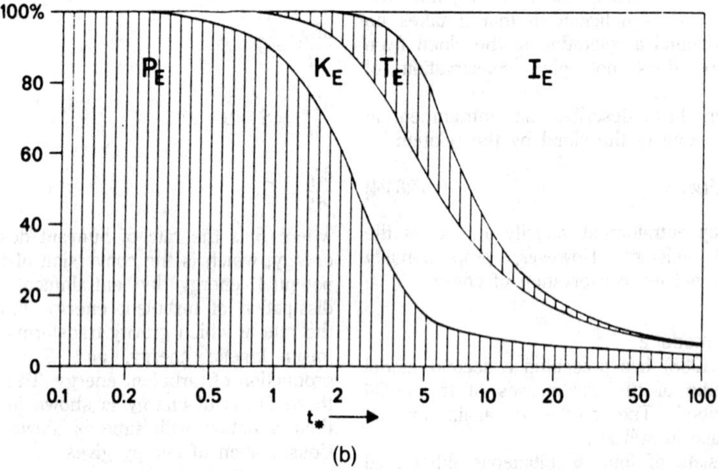

(b)

***Figure 15.78** van Ulden's second model of dense gas dispersion – components of energy balance (van Ulden, 1988): (a) relationship between components; (b) change of components with time, t_* is a scaled time (Courtesy of Clarendon Press)*

The model itself assumes a uniform profile of concentration with height. Van Ulden states that, in general, the concentration may be described by a relation of the form

$$C(z)/\bar{C} = f(z/H) \qquad [15.26.41]$$

where $\bar{C}$ is the mean concentration.

Any such relation should be consistent with conservation of mass:

$$\int_0^\infty f(z/H)\,\mathrm{d}(z/H) = 1 \qquad [15.26.42]$$

and with the definition of H, taken as

$$H = 2\int_0^\infty z(C(z)\,\mathrm{d}z/\int_0^\infty C(z)\,\mathrm{d}z \qquad [15.26.43]$$

Van Ulden quotes as a possible candidate the family of profiles

$$C(z)/\bar{C} = A\exp[-(Bz/H)^s] \qquad [15.26.44]$$

with

$$A = 2s\Gamma(2/s)/[\Gamma(1/s)]^2 \qquad [15.26.45]$$

$$B = 2\Gamma(2/3)/\Gamma(1/s) \qquad [15.26.46]$$

where s is a profile shape factor and A and B are constants. In fitting his model to experimental results he utilizes a shape factor $s = \frac{1}{2}$, giving

$$C(z)/\bar{C} = 6\exp[-(12z/H)^{\frac{1}{2}}] \qquad [15.26.47]$$

15.26.4 Extensions of de Nevers

De Nevers (1984b) has described a model of the type given by van Ulden and has extended it to the case of a continuous release and then to a comparison of instantaneous and continuous releases in windless conditions. The treatment is concerned with the short-term behaviour and ignores entrainment of air.

Following van Ulden, for an instantaneous release

$$\frac{\mathrm{d}R}{\mathrm{d}t} = c_E(g'H)^{\frac{1}{2}} \qquad [15.26.48]$$

$$V_o = \pi R_o^2 H_o \qquad [15.26.49a]$$

$$= \pi R^2 H \qquad [15.26.49b]$$

but with

$$g' = \frac{\rho_g - \rho_a}{\rho_a} \qquad [15.26.50]$$

where g' is the reduced gravity and subscript o denotes time zero. Then, combining Equations 15.26.48 and 15.26.49b

$$\frac{\mathrm{d}R}{\mathrm{d}t} = c_E\left(\frac{g'V_o}{\pi R^2}\right)^{\frac{1}{2}} \qquad [15.26.51]$$

Integrating this equation with the initial condition

$t = t_o; \qquad R = R_o$

and taking the initial aspect ratio as

$$R_o = H_o \qquad [15.26.52]$$

yields

$$\frac{R^2}{2} = c_E\left(\frac{g'V_o}{\pi}\right)^{\frac{1}{2}} t + \frac{1}{2}\left(\frac{V_o}{\pi}\right)^{\frac{2}{3}} \qquad [15.26.53]$$

For a continuous release, Equations 15.26.48 and 15.26.49b are again applicable, but in addition

$$V_o = Qt \qquad [15.26.54a]$$

$$= \pi R^2 H \qquad [15.26.54b]$$

where Q is the volumetric rate of release. Then, combining Equations 15.26.48, 15.26.49b and 15.26.54a

$$\frac{dR}{dt} = c_E\left(\frac{g'Qt}{\pi R^2}\right)^{\frac{1}{2}} \qquad [15.26.55]$$

Integrating this equation with the initial condition

$t = t_o; \qquad R = 0$

yields

$$\frac{R^2}{2} = \frac{2c_E}{3}\left(\frac{g'Q}{\pi}\right)^{\frac{1}{2}} t^{\frac{3}{2}} \qquad [15.26.56]$$

There is a small transient when expansion by bulk flow from the source exceeds that due to gravity-driven flow, but this is so short it can be neglected.

From Equations 15.26.53 and 15.26.56 and setting $V_o = Qt$

$$\frac{R_i}{R_c} = \left[\frac{3}{2} + \frac{3Q^{\frac{1}{6}}}{4c_E(g')^{\frac{1}{2}}\pi^{\frac{1}{6}}t^{\frac{5}{6}}}\right]^{\frac{1}{2}} \qquad [15.26.57a]$$

$$= \left(\frac{3}{2}\right)^{\frac{1}{2}} \qquad t \to \infty \qquad [15.26.57b]$$

where subscripts c and i denote continuous and instantaneous, respectively. The second term on the right-hand side of Equation 15.26.57a falls rapidly to zero, giving Equation 15.26.57b. From Equations 15.26.54b and 15.26.57b

$$\frac{H_i}{H_c} = \frac{2}{3} \qquad [15.26.58]$$

15.26.5 Validation and application

Van Ulden (1974) compared the performance of his first model against the Freon 12 experiment given in his paper, as described in Section 15.22 and shown in Figure 15.76. His work demonstrated that his model was much superior to the Gaussian model in predicting the radius and the height of the cloud, and stimulated work on dense gas dispersion.

15.27 Dispersion of Dense Gas: British Gas/Cremer and Warner Model

The work of van Ulden gave impetus to the development of a number of models of dense gas dispersion. One of the first of these was that by R.A. Cox and Roe (1977) of Cremer and Warner and British Gas, respectively. This model was further developed in treatments by R.A. Cox and Carpenter (1980), C.I. Bradley *et al.* (1983) and Carpenter *et al.* (1987), who refer to it as the British Gas/Cremer and Warner (BG/C&W) model.

15.27.1 Cox and Roe model

The model of R.A. Cox and Roe (1977) is for a continuous release. It follows the same approach as van Ulden in that the cloud is modelled as a sequence of rectangular crosswind slices each of uniform concentration and spreading laterally under gravity. The movement of the cloud edge is

$$\frac{dL}{dt} = (k_E g' H)^{\frac{1}{2}} \qquad [15.27.1]$$

where H is the height of the cloud, L is the half-width of the cloud, g' is the reduced gravity and k_E is a spreading constant. The value of the constant k_E is taken as unity, following van Ulden, although there is theoretical and some experimental basis for a value of 2.

For entrainment of air at the cloud edge, the approach followed is that of van Ulden:

$$Q_e = Hw_e \qquad [15.27.2]$$

with

$$w_e = \gamma \frac{dL}{dt} \qquad [15.27.3]$$

where Q_e is the volumetric flow of air entrained at the edge of the cloud per unit distance crosswind, w_e is the edge entrainment velocity and γ is an edge entrainment coefficient.

In this model account is also taken of entrainment of air at the top of the cloud:

$$Q_t = 2Lu_e \qquad [15.27.4]$$

where Q_t is the volumetric flow of air entrained at the top of the cloud per unit distance downwind and u_e is the top entrainment velocity.

The rate of change of volume of the cloud is then

$$\frac{dV}{dt} = Q_t + Q_e \qquad [15.27.5]$$

where V is the volume of the cloud per unit distance downwind.

For the top entrainment velocity u_e

$$u_e = \frac{\alpha u_1}{\mathrm{Ri}} \qquad [15.27.6]$$

with

$$\mathrm{Ri} = \frac{g'l}{u_1^2} \qquad [15.27.7]$$

where l is a turbulence length scale, Ri is a Richardson number, u_1 is the longitudinal turbulence velocity and α is a top entrainment coefficient. Equation 15.27.6 is valid for moderate values of Ri, but as Ri tends to zero the term (u_e/u_1) tends to a limiting small value.

The longitudinal turbulence velocity u_1 may be expressed in the form

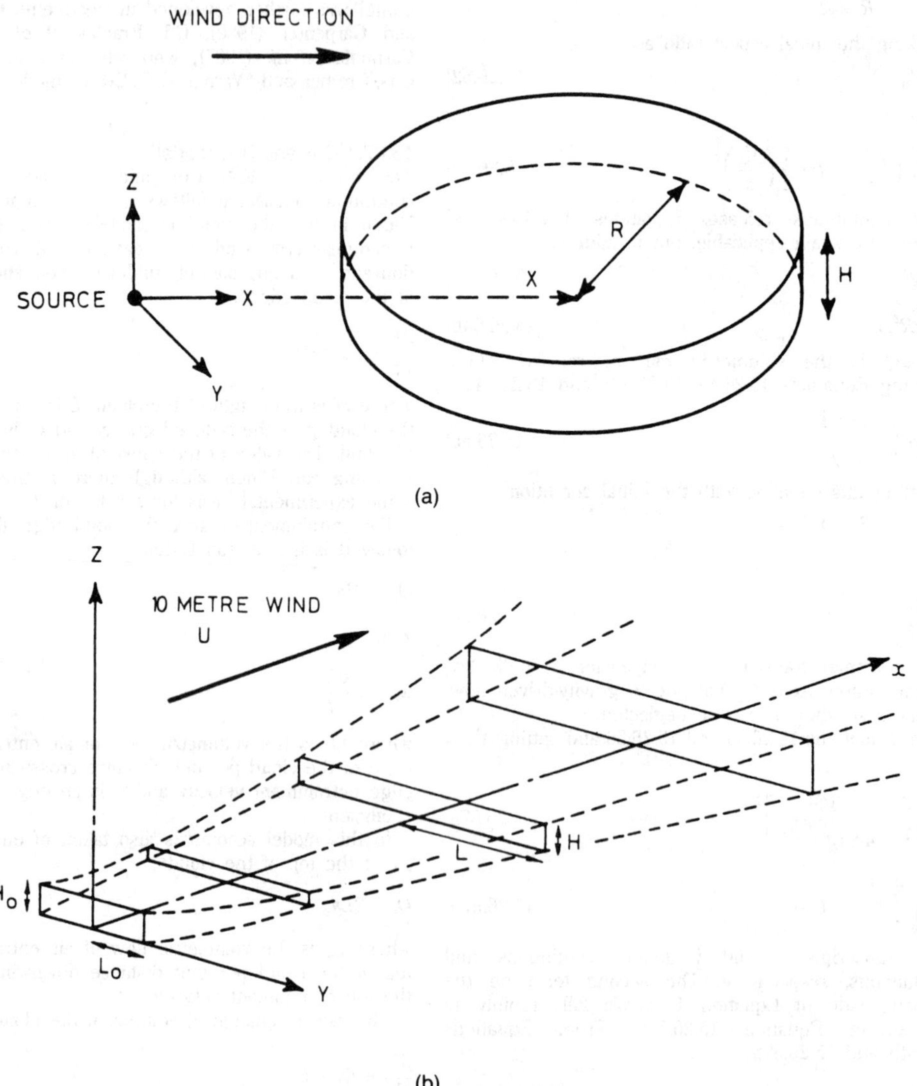

Figure 15.79 *Cox and Carpenter model for dense gas dispersion – idealized cloud shapes (R.A. Cox and Carpenter, 1980): (a) instantaneous release; (b) continuous release (Courtesy of Reidel Publishing Company)*

$$\frac{u_1}{u} = \frac{u_1}{u_*}\frac{u_*}{u} \qquad [15.27.8]$$

where u is the wind speed and u_* is the friction velocity. The ratio (u_1/u_*) has been shown by Monin (1962) to depend primarily on the stability condition. His results indicate the following approximate values:

Stability condition	u_1/u_*
Very stable	3.0
Neutral	2.4
Very unstable	1.6

The ratio (u_*/u) has been shown by O.G. Sutton (1953) to depend primarily on the surface roughness. A typical value for open terrain is 0.1. For the turbulence length scale l, use is made of the results of R.J. Taylor, Warner and Bacon (1970), which show that the parameter varies with the stability condition and the height H of the cloud. For the top entrainment coefficient α work in the literature suggests values ranging from 0.15 to 1.0. A value of 0.5 was obtained by fitting the model to one of the American Gas Association tests for LNG dispersion.

The authors state that in the model the entrained air is assumed to mix adiabatically with the gas and that the effects of water vapour present in the air are taken into account, but give no further details.

15.27.2 Cox and Carpenter model

R.A. Cox and Carpenter (1980) extend the model of Cox and Roe to cover both continuous and instantaneous releases. The situations modelled are shown in Figures 15.79(a) and (b) for the instantaneous and continuous release cases, respectively. The models cover a continuous release in wind and an instantaneous release in still air or in wind, but not a continuous release in still air.

The movement of the cloud edge for the two cases is

$$\frac{dL}{dt} = (k_E g' H)^{\frac{1}{2}} \quad \text{continuous release} \qquad [15.27.9a]$$

$$\frac{dR}{dt} = (k_E g' H)^{\frac{1}{2}} \quad \text{instantaneous release} \qquad [15.27.9b]$$

where R is the radius of the cloud. The value of the constant k_E is unity.

The entrainment of air at the cloud edge is

$$Q_e = Hw_e \quad \text{continuous release} \qquad [15.27.10a]$$

$$Q_e = 2\pi RHw_e \quad \text{instantaneous release} \qquad [15.27.10b]$$

where Q_e is the volumetric flow of air entrained at the edge of the cloud, this being the total value for the instantaneous release case and the value per unit distance crosswind for the continuous release case. The edge entrainment velocity w_e is

$$w_e = \gamma \frac{dL}{dt} \quad \text{continuous release} \qquad [15.27.11a]$$

$$= \gamma \frac{dR}{dt} \quad \text{instantaneous release} \qquad [15.27.11b]$$

The entrainment of air at the cloud top is

$$Q_t = 2Lu_e \quad \text{continuous release} \qquad [15.27.12a]$$

$$Q_t = \pi R^2 u_e \quad \text{instantaneous release} \qquad [15.27.12b]$$

where Q_t is the volumetric flow of air entrained at the top of the cloud, this being the total value for the instantaneous release case and the value per unit distance downwind for the continuous release case.

The treatment of the entrainment velocity u_e at the top of the cloud is essentially that given by Cox and Roe. There is a limiting small value β to which the group (u_e/u_1) tends as Ri tends to zero. The values used for α, β and γ are discussed below.

The rate of change of volume of the cloud is

$$\frac{dV}{dt} = Q_t + Q_e \qquad [15.27.13]$$

where V is the volume of the cloud, this being the total value for the instantaneous release case and the value per unit distance downwind for the continuous release case.

A more detailed description is given of the heat balance used in the model. Vaporization of liquid spray, condensation of water vapour and heat transfer from the ground to the cloud are taken into account. If the source is not one which generates pure vapour but involves a flashing liquid it is assumed that the liquid flash-off is adiabatic, giving vapour and liquid spray. It is also assumed that half the liquid rains out but that this rainout revaporizes instantly. Thereafter the cloud is assumed to be at the saturated condition as long as liquid still remains. The cloud is assumed to be saturated with water.

For heat transfer from the ground by natural convection

$$Q_{gr} = h_{grn}(T_{gr} - T_c)^{\frac{4}{3}} \qquad [15.27.14]$$

and by forced convection

$$Q_{gr} = 0.5 C_f \rho_c c_{pc} u'(T_{gr} - T_c) \qquad [15.27.15]$$

where C_f is a friction factor, c_{pc} is the specific heat of the cloud, h_{grn} is the heat transfer coefficient between the ground and the cloud by natural convection, Q_{gr} is the heat flux from the ground to the cloud, T_c is the temperature of the cloud, T_{gr} is the temperature of the ground, u' is the magnitude of the vector sum of the slumping and wind velocities and ρ_c is the density of the cloud. The friction factor C_f is given by Equation 15.24.17. The value used for the heat flux Q_{gr} is the larger of the values obtained from Equation 15.27.14 for natural convection and Equation 15.27.15 for forced convection.

The heat balance on the cloud is then

$$m_{vo}h_v(T_o) + m_{lo}h_l(T_o) + m_a h_a(T_a) + m_{wv}h_{wv}(T_a) + Q_{gr}$$
$$= m_v h_v(T_c) + m_l h_l(T_c) + m_a h_a(T_c) + m_{wv}h_{wv}(T_c)$$
$$+ m_{wl}h_{wl}(T_c) \qquad [15.27.16]$$

where h is the specific enthalpy of a component, m_i is the mass of component i in the cloud, T_a is the temperature of the air and subscripts a, l, o, v, wl and wv denote air, liquid, source, vapour, liquid water and water vapour, respectively.

The model is extended in the region of passive dispersion. The transition from dense gas to passive dispersion is assumed to occur at the point where the rate of lateral spreading due to turbulence exceeds that due to gravity:

$$\frac{dL}{dt} = \frac{d\sigma_y}{dt} \quad \text{continuous release} \qquad [15.27.17a]$$

$$\frac{dR}{dt} = \frac{d\sigma_y}{dt} \quad \text{instantaneous release} \qquad [15.27.17b]$$

where σ is the dispersion coefficient and subscript y denotes crosswind.

Then from this point a continuous release is modelled using the Gaussian model for a finite line source:

$$\chi(x, y, 0) = \frac{Q}{(2\pi)^{\frac{1}{2}}\sigma_z l_{vs} u}\left[\operatorname{erf}\left(\frac{l_{vs}/2 - y}{2^{\frac{1}{2}}\sigma_y}\right) + \operatorname{erf}\left(\frac{l_{vs}/2 + y}{2^{\frac{1}{2}}\sigma_y}\right)\right] \qquad [15.27.18]$$

where l_{vs} is the width of the virtual source, Q is the continuous mass rate of release and χ is the concentration (mass per unit volume).

The treatment for an instantaneous release is slightly more complex. The virtual source is represented as a set of 253 instantaneous point sources on a rectangular grid:

$$\chi(x, y, 0) = \frac{2Q^* S^*}{253(2\pi)^{\frac{3}{2}}\sigma_y^2 \sigma_z} \qquad [15.27.19]$$

with

$$S^* = \sum_{n=1}^{253}\left\{-\frac{1}{2}\left[\frac{r(n)}{\sigma_y}\right]^2\right\} \quad [15.27.20]$$

where Q^* is the mass released instantaneously and $r(n)$ is the distance downwind from the source n.

The dense phase model gives a uniform cross-sectional concentration profile, whereas the passive model gives a Gaussian concentration profile. The Gaussian model has only two degrees of freedom: the distance of the virtual source upstream of the matching point, and the width of the virtual source. The two cloud parameters selected for matching are the centre line concentration and the cloud width. It is assumed that at the matching point the cloud width is equal to the width at the virtual source plus $2\sigma_y$, where σ_y is the value at the matching point based on its distance from the virtual source.

The authors discuss the calibration of the top entrainment coefficient α, the limiting value β of the group u_e/u_1 and the edge entrainment coefficient γ, and give values of 0.1, 0.15 and 0.6, respectively.

15.27.3 Development of model

Development of the model of Cox and Carpenter has been described by C.I. Bradley *et al.* (1983). A deficiency of that model is the neglect of conservation of momentum and the assumption that the cloud travels at a constant speed equal to that of the wind.

For the cloud from an instantaneous release the rate of change of momentum in the frame of reference of the moving cloud is

$$\frac{d(mu_c)}{dt} = \dot{m}_e u_{av} + \dot{m}_t u_h + D - F \quad [15.27.21]$$

where D is the drag force on the cloud, F is the friction force on the cloud, m is the total mass in the cloud, m_e is the mass of air entrained at the edge of the cloud, m_t is the mass of air entrained at the top of the cloud, u_{av} is the average wind speed over the cloud height, u_c is the velocity of the cloud and u_h is the wind speed at the cloud height H. The drag force D is

$$D = C_d \int_0^H R\rho_a[u(z) - u_c]\,|\,u(z) - u_c\,|\,dz \quad [15.27.22]$$

where C_d is a drag coefficient and $u(z)$ is the wind speed at height z. The value of the drag coefficient C_d is taken as 1.0. The friction force F is obtained by analogy with the shear stress τ in the atmospheric boundary layer

$$F = \tau\pi R^2 \quad [15.27.23]$$

with

$$\tau = \left(\frac{u_*}{u_r}\right)^2 \rho_c u_c^2 \quad [15.27.24]$$

where u_r is the wind speed at the reference height and τ is the shear stress. The reference height is 10 m. Then, from Equation 15.27.21 the acceleration of the cloud relative to a fixed point is

$$\frac{du_c}{dt} = \frac{1}{m}[\dot{m}_e(u_{av} - u_c) + \dot{m}_t(u_h - u_c) + D - F] \quad [15.27.25]$$

For the cloud from a continuous release there is no aerodynamic drag, since the main effect of this force is to reverse the upwind flow and give the plume its initial acceleration. There is, however, an additional term which allows for the interaction of one slice of the plume with another due to the variation of cloud dimensions and density as the plume is carried downwind. The rate of change of momentum is

$$\frac{du_c}{dt} = \frac{1}{\dot{m}}\left\{\frac{d\dot{m}_e}{dt}(u_{av} - u_c) + \frac{d\dot{m}_t}{dt}(u_h - u_c) - 2\tau L u_c - g\frac{d[LH^2(\rho_c - \rho_a)]}{dt}\right\} \quad [15.27.26]$$

A method is also described of handling a transient release which corresponds to neither of the two ideal cases, i.e. to neither a continuous nor an instantaneous release. A case in point is vaporization from a spill of a cryogenic liquid such as LNG. A suitable approach in such a case may be to model the release as an instantaneous release followed by a continuous release. If successively larger masses of vapour are considered, it is found that for a small amount m_1 the time t_1 for vaporization to occur is so short that the trailing edge of the cloud is still upwind of the source. For a large amount m_3 the time t_3 is so long that the trailing edge has moved downwind of the source. There is some intermediate mass m_2 and time t_2 such that the trailing edge is just on the source. This time t_2 is taken as that at which transition occurs. The instantaneous release model is used up to this time and the continuous release model thereafter.

At low wind speeds this method gives a large instantaneous release and the maximum travel distance to a given concentration is determined by this instantaneous part of the release. At high wind speeds and/or for small releases the maximum travel distance is determined by the continuous part of the release.

For very low wind speeds it can occur that the cloud is diluted below its lower flammability limit before it has moved away from the source. This is an indication that the model is inapplicable and it should not be used.

In a further treatment for an instantaneous release, Carpenter *et al.* (1987) express the average wind speed u_{av} over the cloud height as a function of the wind speed u_h at height H:

$$u_{av} = fu_h \quad [15.27.27]$$

where f is the wind speed coefficient.

The authors present a further calibration of the parameters α, β, γ and f based on data from the Thorney Island Phase 1 tests, and give values of 0.08, 0.30, 0.65 and 0.55, respectively.

15.27.4 Validation and application

R.A. Cox and Roe (1977) compared predictions from their model with observations in the Matagordo Bay tests. R.A. Cox and Carpenter (1980) used the SS *Gadila* LNG spill trials to calibrate the parameters in their model. The resultant predicted cloud shape is shown in Figure 15.92(a). Carpenter *et al.* (1987) used the Thorney Island trials to calibrate the parameters in the BG/C&W model.

15.28 Dispersion of Dense Gas: DENZ and CRUNCH

Two models on somewhat similar lines have been developed by the Safety and Reliability Directorate

(SRD). The model for an instantaneous release, DENZ, is described by Fryer and Kaiser (1979 SRD R152) and that for a continuous release, CRUNCH, by Jagger (1983 SRD R229). The two models are incorporated in the computer codes of the same name and have been widely used.

15.28.1 DENZ

Kaiser and Walker (1978) have described a model which was a forerunner of DENZ and applied it to releases of liquefied anhydrous ammonia.

DENZ is a model for an instantaneous release of a dense gas. It is described by Fryer and Kaiser (1979 SRD R152). The model is based on three simultaneous differential equations for the radius R of the cloud, the mass m_a of air in the cloud and the temperature T_c of the cloud.

The cloud is assumed to be advected downwind at an advection velocity which corresponds to the mean wind velocity at the half-height of the cloud. This advection velocity is given by

$$u_c(t) = u_r \frac{\ln(H/2z_o)}{\ln(z_r/z_o)} \quad [15.28.1]$$

where $u_c(t)$ is the mean velocity of the cloud, u_r is the mean wind velocity at the reference height, z_o is the roughness length, z_r is the reference height and H is the height of the cloud. The reference height z_r is 10 m.

The rate of change of the downwind distance $x(t)$ of the centre of the cloud is equal to the advection velocity $u_c(t)$:

$$\frac{dx}{dt} = u_c(t) \quad [15.28.2]$$

For the radius R of the cloud

$$\frac{dR}{dt} = c_E(g'H)^{\frac{1}{2}} \quad [15.28.3]$$

The volume V is

$$V = \pi R^2 H \quad [15.28.4]$$

and c_E is the slumping coefficient.

Equation 15.28.3 is actually cast in the alternative form

$$\frac{dR^2}{dt} = 2c_E\left(\frac{g'V}{\pi}\right)^{\frac{1}{2}} \quad [15.28.5]$$

For the mass m_a of air entrained

$$\frac{dm_a}{dt} = \pi R^2 \rho_a u_e + 2\pi RH\rho_a w_e \quad [15.28.6]$$

The edge entrainment velocity w_e is

$$w_e = \gamma \frac{dR}{dt} \quad [15.28.7]$$

Following van Ulden (1974), the default value of the edge entrainment coefficient γ is taken as zero. Following R.A. Cox and Roe (1977), the top entrainment velocity u_e is obtained from Equation 15.27.6 using the methods given by these authors for relations for the group $(u_1/u_*)$and for the turbulence length scale l.

Heat transfer between the ground and the cloud is assumed to be by natural convection:

$$Q_{gr} = h_{gr}\Delta T_{gr} \quad [15.28.8]$$

with

$$\Delta T_{gr} = T_{gr} - T_c \quad [15.28.9]$$

$$h_{gr} = 0.418\frac{k_f}{L}Z^{\frac{1}{4}}\Delta T_{gr}^{\frac{1}{4}} \quad \text{laminar convection} \quad [15.28.10a]$$

$$= 0.146\frac{k_f}{L}Z^{\frac{1}{3}}\Delta T_{gr}^{\frac{1}{3}} \quad \text{turbulent convection} \quad [15.28.10b]$$

$$Z = \left(\frac{L^3\rho_f^2\beta_f g}{\mu_f^2}\Delta T_{gr}\frac{c_{pf}\mu_f}{k_f}\right)^{\frac{1}{4}} \quad [15.28.11]$$

and with convection turbulent if

$$3 \times 10^{10} > Z\Delta T_{gr} > 2 \times 10^7$$

where c_p is the specific heat, h_{gr} is the heat transfer coefficient between the ground and the cloud, k is the thermal conductivity of the cloud, L is a characteristic dimension of the cloud, Q_{gr} is the heat flux between the ground and the cloud, T_{gr} is the temperature of the ground, ΔT_{gr} is the temperature difference between the ground and the cloud, β is the coefficient of volumetric expansion, μ is the viscosity and subscript f denotes at film temperature (= $(T_c + T_{gr})/2$).

Equation 15.28.8 gives the heat transfer between the ground and the cloud. Then for the temperature T_c of the cloud

$$\frac{dT_c}{dt} = \frac{\dot{m}_a c_{pa}\Delta T_a + h_{gr}(\pi R^2)\Delta T_{gr}}{m_a c_{pa} + m_g c_{pg}} \quad [15.28.12]$$

with

$$\Delta T_a = T_a - T_c \quad [15.28.13]$$

where m_a is the mass of air, m_g is the mass of gas, ΔT_a is the temperature difference between the air and the cloud, and subscripts a, c and g denote the air, cloud and gas, respectively.

Use is also made of the relations

$$V = \frac{m_a + m_g}{\rho_c} \quad [15.28.14]$$

$$\rho_c = \frac{m_a + m_g}{(m_a/\rho_a) + (m_g/\rho_g)}\frac{T_a}{T_c} \quad [15.28.15]$$

For transition from dense gas dispersion to passive dispersion use is made of two alternative criteria. The first criterion is that transition occurs if the following two conditions are satisfied. The first condition is that the rate of increase of the radius due to gravity spreading is less than that due to atmospheric turbulence

$$\frac{dR}{dt} < 2.14\frac{d\sigma_y}{dt} \quad [15.28.16]$$

For the determination of the term $(d\sigma_y/dt)$ use is made of the treatment by Hosker (1974b):

$$\frac{d\sigma_y}{dt} = C^*\frac{dx}{dt} = C^*u_c \quad [15.28.17]$$

where u_c is the velocity of the cloud and C^* is a constant. The constant C^* is a function of the stability category and for categories A, B, C, D, E and F has the values 0.22, 0.16, 0.11, 0.08, 0.06 and 0.04, respectively. The second condition is that

$$u_e > u_1 \qquad [15.28.18]$$

where u_e is determined from Equation 15.27.6 and u_1 is the longitudinal turbulence velocity.

The alternative criterion for transition is based on the density difference $\Delta\rho_c$ for the cloud:

$$\Delta\rho_c < \Delta\rho_r \qquad [15.28.19]$$

where $\Delta\rho_r$ is a reference value. The default value of $\Delta\rho_r$ is 10^{-3} kg/m^3.

Transition occurs at distance x_{tr} with cloud radius R_{tr} and height H_{tr}. Then assuming a Gaussian distribution the dispersion parameters are:

$$\sigma_{ytr} = R_{tr}/2.14 \qquad [15.28.20]$$

$$\sigma_{xtr} = \sigma_{ytr} \qquad [15.28.21]$$

$$\sigma_{ztr} = H_{tr}/2.14 \qquad [15.28.22]$$

Beyond the transition point

$$\sigma_y^2(x) = \sigma_x^2(x) = \sigma_{ytr}^2 + \sigma_{yH}^2(x - x_{tr}) = (R/2.14)^2 \qquad [15.28.23]$$

$$\sigma_z^2(x) = \sigma_{ztr}^2 + \sigma_{zH}^2(x - x_{tr}) = (H/2.14)^2 \qquad [15.28.24]$$

where subscript H denotes the value given by the Hosker scheme.

The concentration distribution in the cloud is assumed to be Gaussian, even though this is not consistent with the gravity spreading treatment. The concentration is

$$\chi(x,y,z,t) = \frac{m_g}{2^{\frac{1}{2}}\pi^{\frac{3}{2}}\sigma_y^2\sigma_z}\exp\left\{-\frac{y^2 + [x - x(t)]^2}{2\sigma_y^2} - \frac{z^2}{2\sigma_z^2}\right\} \qquad [15.28.25]$$

where $x(t)$ is the distance to the centre of the cloud at time t.

A simplified model is also given. The model has four stages: (1) formation of the source cylinder, (2) gravity slumping, (3) 'ground hugging', and (4) passive dispersion. Stage 1 involves the formation of the source cylinder. In Stage 2 this cylinder undergoes gravity slumping with little air entrainment; zero entrainment is assumed. The growth of the cloud radius R is given by Equation 15.28.3. The criterion for the end of this stage is that slumping terminates when the cloud height H is comparable to that of the roughness elements of the surface. In Stage 3, the growth of the cloud radius R is again given by Equation 15.28.3. The growth of the cloud height H is taken for all stability categories as one-third of that which would occur with dispersion by atmospheric turbulence for stability category F. The criterion for the end of this stage is that given by Equation 15.28.19. In Stage 4, passive dispersion applies.

Further developments of DENZ are described by Jagger (1985 SRD R277). It is shown by Jagger in this latter work that, on certain simplifying assumptions, it is possible to obtain a solution for the dense gas dispersion phase. Noting that

$$\Delta \cdot V = \Delta_o \cdot V_o \qquad [15.28.26]$$

Equation 15.28.3 may be integrated to give

$$t^2 = R_o^2 + 2c_E A t \qquad [15.28.27a]$$

with

$$A = \left(\frac{g_o' V_o}{\pi}\right)^{\frac{1}{2}} \qquad [15.28.27b]$$

where Δ is the reduced density difference and the subscript o denotes the initial value. With minimal approximation, Equation 15.28.6 may be integrated and Equation 15.27.6 utilized to give the following solutions:

$$V = \{[B/(10 - C)][R^5 - R_o^5(R/R_o)^{C/2} + V_o^{\frac{1}{2}}(R/R_o)^{C/2}\}^2 \qquad [15.28.28]$$

$$H = \{[B/10 - C][R^4 - R_o^4(R/R_o)^{C/2-1}/\pi^{\frac{1}{2}} + H_o^{\frac{1}{2}}(R/R_o)^{C/2-1}\}^2 \qquad [15.28.29]$$

with

$$B = \alpha u_1^3 \pi^{\frac{1}{2}}/6A^3 c_E \qquad [15.28.30]$$

$$C = 2\gamma \qquad [15.28.31]$$

where B and C are parameters.

In windless conditions there is no top entrainment and $B = 0$, so that Equations 15.28.28 and 15.28.29 reduce to

$$V = V_o(R/R_o)^C \qquad [15.28.32]$$

$$H = H_o(R/R_o)^{C-2} \qquad [15.28.33]$$

15.28.2 CRUNCH

The CRUNCH model for an instantaneous release of a dense gas, described by Jagger (1983 SRD R229), is based on three simultaneous differential equations for the half-width L of the cloud, the mass m_a of air in the cloud and the temperature T_c of the cloud.

The cloud is assumed to be advected downwind at an advection velocity $\bar{u}(t)$ which corresponds to the mean wind velocity at the half-height of the cloud. This advection velocity is given by

$$u_c = u_r \frac{\ln(\dot{V}/4Lu_c z_o)}{\ln(z_r/z_o)} \qquad [15.28.34]$$

where u_c is the mean velocity of the cloud, u_r is the mean wind velocity at the reference height, V is the volume of the cloud and z_r is the reference height. The reference height z_r is 10 m.

The rate of change in the downwind distance x is equal to the advection velocity u_c:

$$\frac{dx}{dt} = u_c \qquad [15.28.35]$$

For the half-width L of the cloud

$$\frac{dL}{dt} = c_E(g'H)^{\frac{1}{2}} \qquad [15.28.36]$$

The rate of change in the volume V is

$$\dot{V} = 2LHu_c \qquad [15.28.37]$$

Equation 15.28.36 is recast, utilizing Equations 15.28.35 and 15.28.37, in terms of downwind distance x:

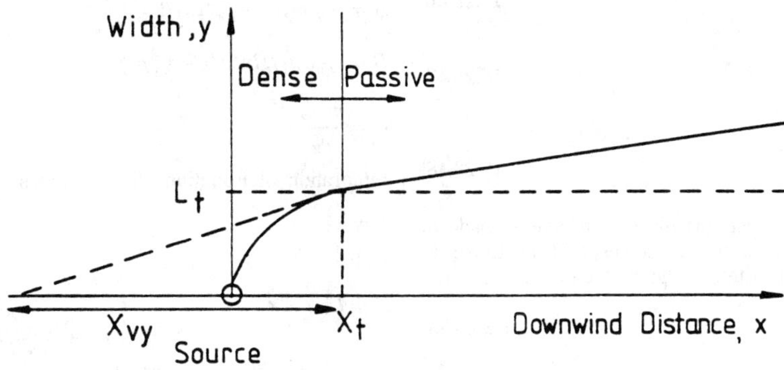

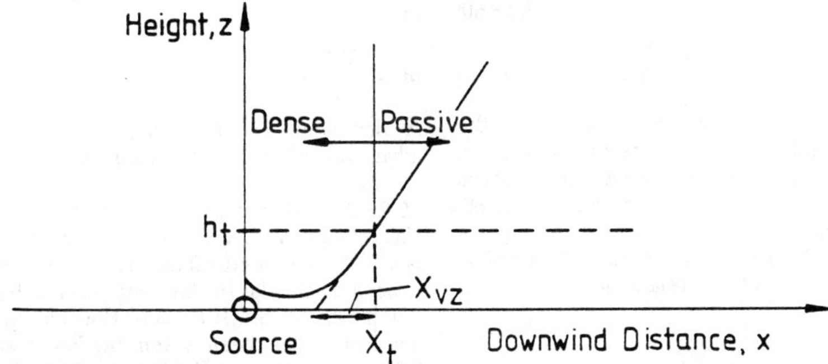

Figure 15.80 *CRUNCH model for dense gas dispersion – transition from dense gas to passive dispersion (Jagger, 1983 SRD R229) h, X and subscript t in the figure are denoted in the text by H, x and subscript tr, respectively (Courtesy of the UKAEA Safety and Reliability Directorate)*

$$\frac{dL}{dx} = c_E\left(\frac{g'V}{2Lu_c^3}\right)^{\frac{1}{2}} \quad [15.28.38]$$

For the mass of air entrained per unit distance downwind

$$\frac{d\dot{m}_a}{dt} = 2L\,dx\rho_a u_e + 2H\,dx\rho_a w_e \quad [15.28.39]$$

where m_a is the mass of air.

Equation 15.28.39 is recast in terms of x

$$\frac{d\dot{m}_a}{dx} = 2L\rho_a u_e + 2H\rho_a w_e \quad [15.28.40]$$

The edge entrainment velocity w_e is

$$w_e = \gamma\frac{dL}{dt} \quad [15.28.41]$$

Following Cox and Roe, the top entrainment velocity u_e is obtained from Equation 15.27.6 using the methods given by these authors for the group (u_1/u_*) and for the turbulence length scale l. For the latter the following relation is quoted:

$$l = 5.88H^{0.48} \quad [15.28.42]$$

Equation 15.28.40 is recast, utilizing Equation 15.28.41, to maintain it as a function of x only:

$$\frac{d\dot{m}_a}{dx} = 2L\rho_a u_e + \gamma\frac{\dot{V}}{L}\rho_a\frac{dL}{dx} \quad [15.28.43]$$

Heat transfer between the ground and the cloud is treated as in DENZ. Then for the temperature T_c of the cloud

$$\frac{dT_c}{dx} = \frac{1}{\dot{m}_a c_{pa} + \dot{m}_g c_{pg}}\left[(T_a - T_c)c_{pa}\frac{d\dot{m}_a}{dx} + 2LQ_{gr}\right] \quad [15.28.44]$$

where m_g is the mass of gas.

Use is also made of the relations

$$\dot{m} = \dot{m}_a + \dot{m}_g \quad [15.28.45]$$

where m is the total mass in the cloud.

$$\dot{m} = 2LH\rho_c u_c \quad [15.28.46]$$

$$\dot{V} = \frac{T_c \dot{m}_a}{\rho_a T_a} + \frac{T_c \dot{m}_g}{\rho_g T_g} \quad [15.28.47]$$

$$\rho_c = \frac{\dot{m}_a + \dot{m}_g}{\dot{V}} \quad [15.28.48]$$

For transition from dense gas dispersion, use is made of two alternative criteria, which correspond to those in DENZ. For the first criterion the first condition is

$$\frac{dL}{dx} < 2.14 \frac{d\sigma_y}{dx} \quad [15.28.49]$$

and the second condition is that given in relation 15.28.18. The alternative criterion is that given in relation 15.28.19.

Transition occurs at downwind distance x_{tr} with cloud half-width L_{tr} and height H_{tr}. Then, assuming a Gaussian profile, the dispersion parameters are

$$\sigma_{ytr} = L_{tr}/2.14 \quad [15.28.50]$$

$$\sigma_{ztr} = H_{tr}/2.14 \quad [15.28.51]$$

For the dispersion beyond the transition point the approach used is to utilize two separate virtual sources at locations x_{vy} and x_{vz} for crosswind and vertical dispersion, respectively, as shown in Figure 15.80, where x_{vy} is upwind of the source, whilst x_{vz} is downwind of it. The determination of these two points is as follows. Use is made of the relations

$$\sigma_{yH}(x_{vy}) = \sigma_{ytr} \quad [15.28.52]$$

$$\sigma_{zH}(x_{vz}) = \sigma_{ztr} \quad [15.28.53]$$

where subscript H is the value given by the Hosker scheme. Since this scheme gives σ values which are a function of x, the solution is a trial-and-error one. Beyond the transition point the dispersion is determined using the virtual source at x_{vy} for the crosswind dispersion and the virtual source at x_{vz} for the vertical dispersion and with σ values from the Hosker scheme.

The concentration distribution in the cloud is assumed to be Gaussian. The concentration χ is

$$\chi(x,y,z) = \frac{\dot{m}_g}{\pi \sigma_y \sigma_z u_c} \exp\left[-\left(\frac{y^2}{2\sigma_y^2} + \frac{z^2}{2\sigma_z^2}\right)\right] \quad [15.28.54]$$

On certain simplifying assumptions, it is possible to obtain a solution for the dense gas dispersion phase. One case considered is that in which it is assumed that the advection velocity of the cloud is constant and that the cloud is at ambient temperature, so that heat effects can be neglected. For this case

$$\frac{dL}{dx} = A(Lu_c^3)^{-\frac{1}{2}} \quad [15.28.55]$$

$$\frac{d\dot{m}_a}{dx} = BL(\dot{m}_a + C) + [\gamma(\dot{m}_a + C)/L\frac{dL}{dx} \quad [15.28.56]$$

with

$$A = c_E[g\dot{m}_g(1 - \rho_a/\rho_g)/2\rho_a]^{\frac{1}{2}} \quad [15.28.57]$$

$$B = 2\rho_a \alpha u_1^3/gl\dot{m}_g(1 - \rho_a/\rho_g) \quad [15.28.58]$$

$$C = \frac{\dot{m}_g \rho_a}{\rho_g} \quad [15.28.59]$$

Integration of Equation 15.28.55 gives

$$L = X^{\frac{2}{3}} \quad [15.28.60]$$

with

$$X = L_o^{\frac{3}{2}} + A'x \quad [15.28.61]$$

$$A' = \frac{3}{2}Au_c^{-\frac{3}{2}} \quad [15.28.62]$$

where L_o is the half-width of the cloud at the source. Substitution of Equation 15.28.60 in Equation 15.28.56 and integration gives

$$\dot{m}_a = C\{X^{\frac{2}{3}\gamma} \exp[3B(X^{\frac{5}{3}} - L_o^{\frac{5}{2}})/5A']/L_o^{\gamma} - 1\} \quad [15.28.63]$$

For short downwind distances Equation 15.28.63 reduces to

$$\dot{m}_a = \frac{CX^{\frac{2}{3}\gamma}}{L_o^{\gamma}} \quad [15.28.64]$$

A derivation is also given for the case where the advection velocity of the cloud is variable.

15.28.3 Validation and application

Kaiser and Walker (1978) have given a rough comparison of the predictions from their model with the observed clouds in the ammonia release incidents at Potchefstroom in 1973 and Houston in 1976; a further discussion has been given by Fryer and Kaiser (1979 SRD R152). Jagger (1985 SRD R277) has discussed the calibration of DENZ against the van Ulden trial and the Porton Down trials. Jagger (1983 SRD R229) has also compared predictions of CRUNCH for the cloud shape in the SS *Gadila* trials as shown in Figure 15.92(b).

DENZ and CRUNCH have been widely used. The Second *Canvey Report* gave hazard ranges computed from DENZ and more detailed relations for hazard ranges based on DENZ and CRUNCH have been given by Considine and Grint (1985), as described in Section 15.49. CRUNCH has been used by McQuaid and Fitzpatrick (1983) to model the dispersion of dense gas by a water spray barrier.

15.29 Dispersion of Dense Gas: SIGMET

Most of the early models were box models. Some of these have just been described. The early models also included, however, several *K*-theory models. Of these, the SIGMET model was particularly influential in the development of the subject.

15.29.1 SIGMET

SIGMET was developed by Science Applications Inc. (SAI) for the US Coast Guard (USCG). It has been described by England *et al.* (1978), and evaluated by Havens, who also gives a description (Havens, 1982a).

The model is intended for investigating the dispersion of a large spill of LNG onto water. The starting point of the model is the set of hydrodynamic equations. In applying these to dense gas dispersion, where the cloud is wide but shallow, a simplification is made which eliminates the equation for the vertical velocity. The governing equations are

$$\frac{\partial(\pi u)}{\partial t}+\frac{\partial(u\pi u)}{\partial x}+\frac{\partial(v\pi u)}{\partial y}+\frac{\partial(\dot{\sigma}\pi u)}{\partial \sigma}+\left(\frac{\partial\phi}{\partial x}+\frac{\sigma}{\rho}\frac{\partial\pi}{\partial x}\right)=0 \quad [15.29.1]$$

$$\frac{\partial(\pi v)}{\partial t}+\frac{\partial(u\pi v)}{\partial x}+\frac{\partial(v\pi v)}{\partial y}+\frac{\partial(\dot{\sigma}\pi v)}{\partial \sigma}+\left(\frac{\partial\phi}{\partial y}+\frac{\sigma}{\rho}\frac{\partial\pi}{\partial y}\right)=0 \quad [15.29.2]$$

$$\frac{\partial(\pi h)}{\partial t}+\frac{\partial(u\pi h)}{\partial x}+\frac{\partial(v\pi h)}{\partial y}+\frac{\partial(\dot{\sigma}\pi h)}{\partial \sigma}+\frac{\pi\omega}{\rho}=0 \quad [15.29.3]$$

$$\frac{\partial(\pi c)}{\partial t}+\frac{\partial(u\pi c)}{\partial x}+\frac{\partial(v\pi c)}{\partial y}+\frac{\partial(\dot{\sigma}\pi c)}{\partial \sigma}=0 \quad [15.29.4]$$

$$\frac{\partial(\pi)}{\partial t}+\frac{\partial(u\pi)}{\partial x}+\frac{\partial(v\pi)}{\partial y}+\frac{\partial(\dot{\sigma}\pi)}{\partial \sigma}=0 \quad [15.29.5]$$

with

$$\sigma=\frac{p-p_t}{\pi} \quad [15.29.6]$$

$$\pi=p_s-p_t \quad [15.29.7]$$

$$\phi=gz \quad [15.29.8]$$

where c is the concentration (mass fraction), h is the specific enthalpy, p is the pressure at a given height, p_s is the pressure at the surface, p_t is the pressure at the top, T is the absolute temperature, ρ is the density, σ is a dimensionless pressure co-ordinate, ϕ is the geopotential height, ω is the substantial pressure derivative, and u, v and w are the components of the velocity in the alongwind, crosswind and vertical directions, respectively. $\dot{\sigma}$ is the substantial derivative in x,y,σ,t co-ordinates. The sigma co-ordinate takes the values $\sigma = 0$ at the top and $\sigma = 1$ at the surface.

The hydrostatic relation is

$$\frac{dz}{d\sigma}=-\frac{\pi}{g\rho} \quad [15.29.9]$$

The following supplementary equations are used:

$$h=[c_{pa}(1-c)+c_{pg}c]T+WL_o(1-c)f(T) \quad [15.29.10]$$

$$\omega=\pi\dot{\sigma}+\sigma\left(\frac{\partial\pi}{\partial t}+u\frac{\partial\pi}{\partial x}+v\frac{\partial\pi}{\partial x}\right) \quad [15.29.11]$$

where L_o is the latent heat of vaporization of water, W is the mass of water per unit mass of air, and the subscripts a and g denote air and gas, respectively.

These equations, which are the set given by England *et al.*, describe the fluid dynamic and thermodynamic processes but do not include the effects of turbulence. Corresponding equations incorporating the turbulence terms, including the eddy diffusivities, are given by Havens. He also gives a diagram showing the boundary conditions for solution of the model equations.

The eddy diffusivities required in the submodel for turbulent mass, momentum and energy transfer, as described by Havens, are obtained as follows. For the horizontal eddy diffusivities the Reynolds analogy is assumed and the eddy diffusivities for turbulent momentum, mass and energy transfer are taken as equal and are denoted by K_h. The vertical diffusivities are likewise assumed equal and are denoted K_v. The relation for the vertical diffusion coefficient is

$$K_v=0.45\sigma_\epsilon ul \qquad u\geq 1\text{ m/s} \quad [15.29.12a]$$

$$=0.45\sigma_\epsilon l \qquad u<1\text{ m/s} \quad [15.29.12b]$$

where l is turbulence length scale and σ_ϵ is the standard deviation of the wind direction. Tabulated values of l and σ_ϵ are given by England *et al.* and a plot for K_h is given by Havens. The diffusion coefficient K_h is obtained from the vertical diffusion coefficient K_v using values of the ratio K_h/K_v. Havens gives for this ratio the values shown in Table 15.43, which evidently supersede the table of values given by England *et al.*

***Table 15.43** Ratio of horizontal to vertical diffusion coefficients (Havens, 1982a) (Courtesy of Elsevier Science Publishers)*

Stability category	K_h/K_v
D	1.0
E	10.0
F	25.0

Other submodels are those for heat transfer and momentum transfer from the surface to the cloud. For heat transfer

$$q=U(T_s-T_c) \quad [15.29.13]$$

where q is the heat flux, T_c is the temperature of cloud, T_s is the temperature of the surface and U is the heat transfer coefficient. The value used for U is 20.4 W/m^{2}K.

For momentum transfer

$$\tau_o=C_d\rho u^2 \quad [15.29.14]$$

where C_d is the drag coefficient, u is the velocity and τ_o is the shear stress at the surface. The velocity u is that at 1 m height. The value of C_d is 0.001.

SIGMET incorporates a source submodel for the spillage of LNG on water. The source is assumed to be a cylindrical pool. The pool radius is determined from the density intrusion model given by Equation 15.24.21. The pool radius is allowed to grow, with a corresponding decrease in height, until a depth is reached at which break-up occurs. After pool break-up, the evaporation rate is determined by the empirical correlation of Feldbauer *et al.* (1972):

$$\dot{m}=\dot{m}_{max}\exp\left[-\frac{0.04}{\rho_l H_{min}}(t-t_{max})\right] \quad [15.29.15]$$

where H_{min} is the height at break up, $\dot{m}$ is the mass rate of evaporation, $\dot{m}_{max}$ is the mass rate of evaporation at break-up, t is time, t_{max} is the time at break up and ρ_l is the density of the liquid.

15.29.2 Validation and application

England *et al.* (1978) give comparisons of the prediction of SIGMET with results of the American Gas Association

LNG dispersion trials. They also use SIGMET to simulate an instantaneous spillage of 30 000 m^3 of LNG onto water.

An assessment of SIGMET has been carried out by Havens (1982a), who studied the somewhat similar case of a 25 000 m^3 LNG spill and performed a sensitivity analysis, investigating the effect of spill size, wind velocity and atmospheric stability, and made comparisons between the distance to the lower flammability limit for this dense gas and for a neutrally buoyant gas.

SIGMET, in the version SIGMET-N, was one of four models in a comparative evaluation of models by Havens and co-workers (Havens, 1986; Havens, Spicer and Schreurs, 1987), the other models being ZEPHYR, MARIAH-II and FEM3. SIGMET-N was discarded because the turbulent mixing model did not scale properly and because of the difficulty of controlling numerical 'diffusion'.

15.30 Dispersion of Dense Gas: SLAB and FEM3

A pair of models have been developed at the Lawrence Livermore National Laboratory (LLNL) by Chan, Ermak and co-workers (Ermak, Chan *et al.*, 1982; S.T. Chan, Rodean and Ermak, 1984; Ermak and Chan, 1986, 1988; S.T. Chan, Ermak and Morris, 1987). SLAB is an advanced similarity model and FEM3 a full three-dimensional model.

15.30.1 SLAB

SLAB was originally formulated by Zeman (1982a,b) and its development has been described by Ermak, Chan *et al.* (1982), D.L. Morgan, Kansa and Morris (1984) and Ermak and Chan (1986, 1988).

The SLAB model is for a continuous release. It is a slab model with properties averaged in the horizontal and vertical directions and is thus one-dimensional. The model is based on a set of six simultaneous differential equations, for the conservation of total mass, conservation of the material released, conservation of momentum and conservation of energy. These equations are given by Ermak, Chan *et al.* (1982) as a set of six partial differential equations, in time t and downwind distance x, but subsequently by Ermak and Chan (1986) as six ordinary differential equations in x. The latter are

$$\frac{d(Bhu_c\rho_c)}{dx} = B_o u_o \rho_o + (Bu_e + hw_e)\rho_a \quad [15.30.1]$$

$$\frac{d(Bhu_c\rho_c\omega)}{dx} = B_o u_o \rho_o \quad [15.30.2]$$

$$\frac{d(Bhu_c\rho_c c_{pc} T_c)}{dx} = B_o u_o \rho_o c_{po} T_o + (Bu_e + hw_e)\rho_a c_{pa} T_a + Q_{gr} \quad [15.30.3]$$

$$\frac{d(Bhu_c^2\rho_c)}{dx} = -\frac{1}{2}[Bh^2(\rho_c - \rho_a)]g + (Bu_e + hw_e)\rho_a u \quad [15.30.4]$$

$$\frac{d(Bhu_c u_y \rho_c)}{dx} = h^2(\rho_c - \rho_a)g \quad [15.30.5]$$

$$\frac{dB}{dx} = \frac{u_y + w_e}{u_c} \quad [15.30.6]$$

where B is a cloud width parameter, c_p is the specific heat, h is a cloud height parameter, Q_{gr} is the heat flux between the ground and the cloud, T is the absolute temperature, u is the wind velocity, u_c is the velocity of the cloud in the downwind direction, u_e is the vertical entrainment velocity, u_o is the velocity of material at the source, u_y is the velocity of the cloud in the crosswind direction, w_e is the horizontal entrainment rate, ρ is the density, ω is the mass fraction of material released, and subscripts a, c and o denote the air, cloud and source, respectively.

The following supplementary equation is used:

$$\rho_c T_c = \frac{\rho_a T_a M_o}{M_o + (M_a - M_o)\omega} \quad [15.30.7]$$

where M is molecular weight.

The vertical entrainment velocity u_e is taken as

$$u_e = \frac{2.7ku^*}{\phi(\mathrm{Ri})} \quad [15.30.8]$$

where k is the von Karman constant and $\phi(\mathrm{Ri})$ is the Monin–Obukhov profile function given by Dyer (1974).

The profile function ϕ is

$$\phi = 1 + 5\mathrm{Ri} \qquad \mathrm{Ri} > 0 \quad [15.30.9a]$$

$$\phi = (1 - 16\mathrm{Ri})^{-p'} \qquad \mathrm{Ri} < 0 \quad [15.30.9b]$$

where Ri is a Richardson number and p' is an index. The index p' is $\frac{1}{4}$ for momentum and $\frac{1}{2}$ for species and energy.

The horizontal entrainment velocity w_e was given in the earlier versions (Ermak, Chan *et al.*, 1982) as

$$w_e = (1.8)^2(h/B)u_e \quad [15.30.10]$$

where the term (h/B) implies that at low values of the height the horizontal entrainment rate is low but that it increases as the height increases. In the later version (Ermak and Chan, 1986, 1988) w_e is described as a function of the ground surface friction coefficient and of the Monin–Obukhov length.

The profile of the volumetric concentration C is represented as

$$C(x,y,z) = C(x)\cdot C_1(y)\cdot C_2(z) \quad [15.30.11]$$

where C is the volumetric concentration and $C(x)$, $C_1(y)$, $C_2(z)$ are concentration functions. These functions are

$$C(x) = \frac{M_a\omega}{M_o + (M_a - M_o)\omega} \quad [15.30.12]$$

$$C_1(y) = \frac{1}{4b}\left[\mathrm{erf}\left(\frac{y+b}{2^{\frac{1}{2}}\beta}\right) - \mathrm{erf}\left(\frac{y-b}{2^{\frac{1}{2}}\beta}\right)\right] \quad [15.30.13]$$

$$C_2(z) = \frac{1}{h}\left(\frac{6}{\pi}\right)^{\frac{1}{2}}\exp\left(-\frac{3z^2}{2h^2}\right) \quad [15.30.14]$$

with

$$B^2 = b^2 + 3\beta^2 \quad [15.30.15]$$

where ω is the mass fraction and b and β are shape parameters. The variation of B with x is given by Equation 15.30.6], and the variation of b with x is given by

$$\frac{db}{dx} = \frac{bu_y}{Bu_c} \quad [15.30.16]$$

The horizontal concentration profile is uniform when $\beta = 0$ and approaches a Gaussian one as $\beta \gg b$. The ratio of β to b remains constant under the influence of the horizontal velocity u_y. It is the horizontal entrainment rate w_e which causes an increase in β and results in a crosswind Gaussian profile.

The cloud shape parameters B, b and h are defined by Equations 15.30.15, 15.30.13 and 15.30.14, and are related to the velocities u_y, w_e and u_e by Equations 15.30.6, 15.30.16 and 15.30.2. The parameters B and h are related to the dispersion parameters as follows:

$$\sigma_y^2 = 3B^2 \quad [15.30.17]$$

$$\sigma_z^2 = 3h^2 \quad [15.30.18]$$

15.30.2 FEM3

Accounts of FEM3 have been given by Ermak, Chan *et al.* (1982), S.T. Chan and Ermak (1984), S.T. Chan, Rodean and Ermak (1984) and Ermak and Chan (1986, 1988).

The FEM3 model is also for a continuous release. It is a full three-dimensional model. The model is based on a set of four simultaneous differential equations, in terms of tensor quantities, for the conservation of total mass, conservation of the material released, conservation of momentum and conservation of energy. The equations are

$$\nabla \cdot (\rho_c \mathbf{u}) = 0 \quad [15.30.19]$$

$$\frac{\partial(\rho_c \mathbf{u})}{\partial t} + \rho_c \mathbf{u} \cdot \nabla \mathbf{u} = -\nabla p + \nabla \cdot (\rho_c \mathbf{K}^m \cdot \nabla \mathbf{u}) + (\rho_c - \rho_h)\mathbf{g} \quad [15.30.20]$$

$$\frac{\partial \theta}{\partial t} + \mathbf{u} \cdot \nabla \theta = \nabla \cdot (\mathbf{K}^\theta \cdot \nabla \theta) + \frac{c_{pg} - c_{pa}}{c_{pc}} (\mathbf{K}^\omega \cdot \nabla \omega) \cdot \nabla \theta + S \quad [15.30.21]$$

$$\frac{\partial \omega}{\partial t} + \mathbf{u} \cdot \nabla \omega = \nabla \cdot (\mathbf{K}^\omega \cdot \nabla \omega) \quad [15.30.22]$$

with

$$\mathbf{u} = (u, v, w) \quad [15.30.23]$$

$$\rho_c = \frac{M_g M_a P}{RT[M_g + (M_a - M_g)\omega]} \quad [15.30.24]$$

$$c_{pc} = c_{pa}(1 - \omega) + c_{pg}\omega \quad [15.30.25]$$

where p is the pressure deviation from an adiabatic atmosphere at rest with corresponding density ρ_h, P is the absolute pressure, S is the temperature source term (e.g. latent heat), θ is the potential temperature deviation from an adiabatic atmosphere, ω is the mass fraction of material released, $\mathbf{g}$ is the acceleration due to gravity, $\mathbf{K}^m$, $\mathbf{K}^\theta$ and $\mathbf{K}^\omega$ are the diagonal eddy diffusion tensors for the momentum, energy and mass fraction, respectively, $\mathbf{u}$ is a velocity tensor, and subscript g denotes material released. The numerical solution of these equations has been described by S.T. Chan, Rodean and Ermak (1984).

FEM3 is based on a generalized anelastic approximation, adapted from Ogura and Phillips (1962), which allows large density changes to be handled whilst precluding sound waves. The approximation is the use of Equation 15.30.19 instead of

$$\nabla \cdot (\rho_c \mathbf{u}) + \frac{\partial \rho_c}{\partial t} = 0 \quad [15.30.26]$$

In FEM3 turbulence is treated using the *K*-theory approach. The eddy diffusion tensors are associated with three diffusion coefficients, two horizontal and one vertical. The latter is particularly important. In the earlier versions (Ermak, Chan *et al.*, 1982) the vertical diffusion coefficient K_v was taken as

$$K_v = K_a(1 - \omega) + K_\rho \omega \quad [15.30.27]$$

where K_a is the ambient vertical diffusion coefficient, K_v is the vertical diffusion coefficient and K_ρ is a dense-layer diffusion coefficient. Two submodels were used for K_ρ, one based on a Richardson number and one on a mixing length.

In the later versions (Ermak and Chan, 1988) three submodels are given for the vertical diffusion coefficient. In the first submodel

$$K_v = \frac{k[(u_{*c}z)^2 + (w_{*c}h)^2]^{\frac{1}{2}}}{\phi(\mathrm{Ri})} \quad [15.30.28]$$

with

$$u_{*c} = u_* \mid u_c/u \mid \quad [15.30.29]$$

where u_* is the friction velocity, u_{*c} is the cloud friction velocity, w_{*c} is a cloud 'convection velocity' and $\phi(\mathrm{Ri})$ is the Monin–Obukhov profile function. The cloud convection velocity is a function of the temperature difference between the ground and the cloud.

The horizontal diffusion coefficient K_h is taken as

$$K_h = \frac{\beta^* k u_{*c} z}{\phi} \quad [15.30.30]$$

where β^* is an empirical coefficient. The value of β^* is taken as 6.5.

15.30.3 Validation and application

Ermak, Chan and co-workers (Ermak, Chan *et al.*, 1982; S.T. Chan, Rodean and Ermak, 1984; Ermak and Chan, 1986; Koopman, Ermak and Chan, 1989) compared predictions from SLAB and FEM3 with observations from the Burro trials. D.L. Morgan, Kansa and Morris (1984) have compared the predictions of SLAB with observations from the Burro and Coyote trials and Blewitt, Yohn and Ermak (1987) have compared its predictions with observations from the Goldfish trials. Touma *et al.* (1991) have compared the performance of this model with that of other models in relation to the Burro, Desert Tortoise and Goldfish trials.

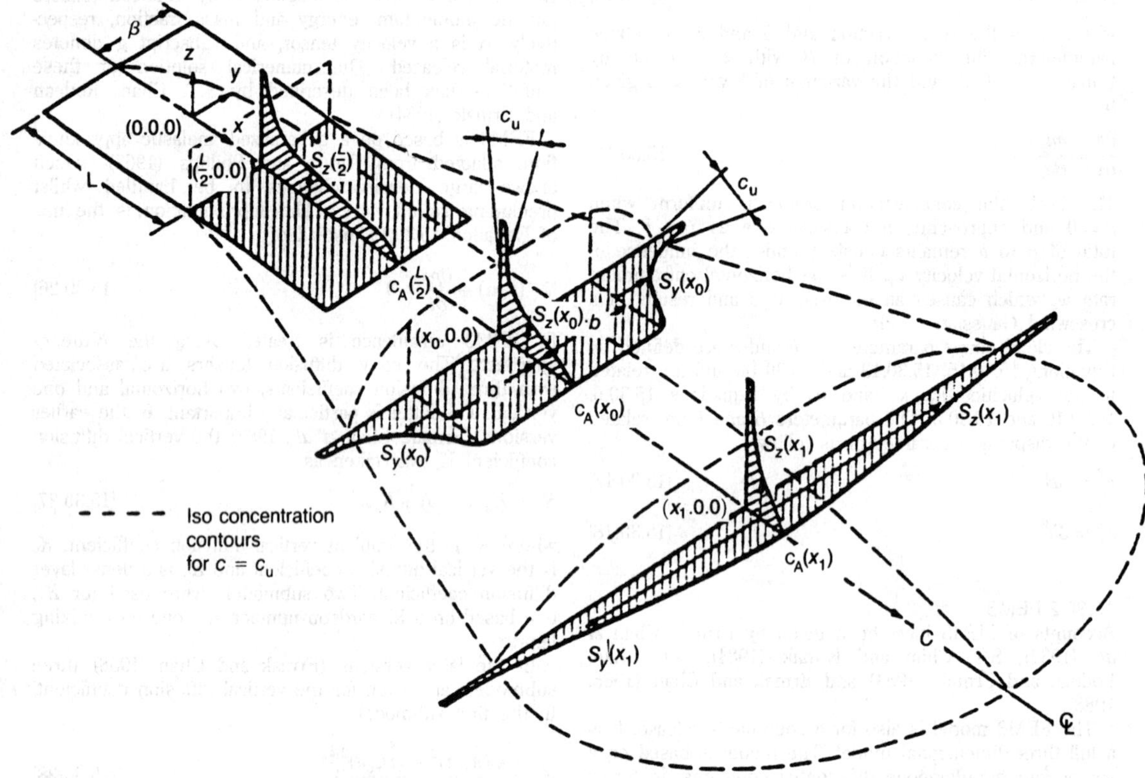

Figure 15.81 HEGADAS model for dense gas dispersion – idealized cloud shapes (Puttock, 1987c) (Courtesy of the American Institute of Chemical Engineers)

15.31 Dispersion of Dense Gas: HEGADAS and Related Models

A series of models have been developed by Shell, leading to the model HEGADAS and then to the model system HGSYSTEM. The first of these models was that of te Riele (1977), which was followed by the model of Colenbrander (1980) and then HEGADAS itself. HEGADAS is complemented by the front-end model HEGABOX.

Another major model, DEGADIS, developed by Havens and co-workers is based on HEGADAS.

15.31.1 te Riele model

The model described by te Riele (1977) is for a continuous release. It is based on two simultaneous partial differential equations for conservation of mass and of momentum. The assumption is made that the take-up of material into the ambient flow field and the dispersion of material within this field are independent. Other assumptions are the Reynolds analogy and the Boussinesq approximation.

The basic equations are:

$$\frac{\partial \rho u}{\partial x} = \frac{\partial}{\partial y}\left(K_y \frac{\partial \rho}{\partial y}\right) + \frac{\partial}{\partial z}\left(K_z \frac{\partial \rho}{\partial z}\right) \qquad [15.31.1]$$

$$\frac{\partial \rho u^2}{\partial x} = \frac{\partial}{\partial y}\left(K_y \frac{\partial \rho u}{\partial y}\right) + \frac{\partial}{\partial z}\left(K_z \frac{\partial \rho u}{\partial z}\right) \qquad [15.31.2]$$

where K is the eddy diffusion coefficient, u is the wind speed, ρ is the density of the cloud and subscripts x and y denote in the downwind and crosswind directions, respectively.

The assumption that the wind speed is independent of the take-up of material, the presence of the dense gas layer and the density gradients in the ambient flow field gives

$$\frac{\partial u}{\partial x} = \frac{\partial u}{\partial y} = 0 \qquad [15.31.3]$$

It can be shown from Equations 15.31.1–15.31.3 that

$$K_z = \frac{\rho_a \tau_o}{\rho^2}\left(\frac{\partial u}{\partial x}\right)^{-1} \qquad [15.31.4]$$

where τ_o is the shear stress acting on the upwind ground surface, and subscript a denotes air. Then from

Equation 15.31.4 it is possible to derive the shear stress acting on the top of the plume, which gives

$$\tau_{gr} = \frac{\rho_a}{\rho_{gr}} \tau_o \qquad [15.31.5]$$

where ρ_{gr} is the density downwind and τ_{gr} is the shear stress acting on the downwind ground surface.

For the relation between density and concentration it is assumed that the ambient air and the dense gas are ideal gases with the same molecular specific heats and that equalization of temperature and density occur only through convection originating in atmospheric turbulence. Then

$$\frac{c}{\rho} = \frac{1}{\rho_a + (1 - \rho_a/\rho_g)} \qquad [15.31.6]$$

where c is the concentration (mass per unit volume) and subscript g denotes material released.

The prime purpose of the model is to determine concentrations downwind of an area source such as that from an evaporating pool of liquefied gas. This source is modelled as a rectangular area source of length L_s and half-width B_s and mass rate of release per unit area Q''. Figure 15.81 shows the general form of the model, in the later version HEGADAS.

For the concentration profile it is assumed that there is a middle part of half-width b in which the dispersion occurs in the vertical direction but not in the horizontal direction. The width of this middle part reduces due to mixing at the crosswind edges of the plume and thus decreases from a value of $2B$ at the source to zero at some point downwind. A concentration profile is assumed which in the vertical direction is Gaussian throughout but which in the crosswind direction has Gaussian features at the crosswind edges, but becomes fully Gaussian only when the middle part disappears:

$$\frac{c}{c_A} = \exp\left[-\left(\frac{z}{\sigma_z}\right)^s\right] \qquad |y| < b \qquad [15.31.7a]$$

$$\frac{c}{c_A} = \exp\left[-\left(\frac{|y| - b}{\sigma_y}\right)^r - \left(\frac{z}{\sigma_z}\right)^s\right] \qquad |y| \geq b \qquad [15.31.7b]$$

where c_A is the ground level concentration on the centre line, and r and s are shape factors for the crosswind and vertical concentration profiles, respectively.

This approach ensures a seamless transfer from the dense gas dispersion regime to the passive dispersion regime and avoids the need to define a transition point and to use separate models for the two regimes.

From Equations 15.31.1–15.31.4 and 15.31.6 the integral form of the mass and momentum balances can be derived:

$$\int_0^{y_{Ls}} \left[\frac{\partial}{\partial x}\left(\int_0^\infty cu\,dz\right)\right] dy = \left(\int_0^\infty K_y \frac{\partial c}{\partial y}\,dz\right)_{y=y_{Ls}} + \int_0^{y_{Ls}} Q''\,dy \qquad [15.31.8]$$

$$\frac{d}{dx}\left(\int_0^\infty \int_0^\infty cu^2\,dy\,dx\right) = \int_0^\infty \frac{c_{gr}}{\rho_{gr}} \tau_o\,dy \qquad [15.31.9]$$

where the subscript gr denotes ground level.

Equation 15.31.8 is used with three different values of the limit y_{Ls}. These are:

Total mass balance: $y_{Ls} \to \infty$

Mass balance on middle part: $y_{Ls} < b$; $b > 0$

Mass balance for determination of σ: $y_{Ls} = b + 0.5(2)^{\frac{1}{2}}\sigma_y$

This therefore gives a set of four equations, the three versions of Equations 15.31.8 and 15.31.9, which are used to determine the four variables c_A, b, σ_y and σ_z.

The parameters necessary for this model include: the wind velocity u; the shape parameters r and s; the vertical eddy diffusion coefficient K_y; the upstream shear stress τ_o; and the mass rate of release per unit area Q''.

Wind velocity

For the wind velocity u use is made of the vertical wind velocity profile given in Equation 15.24.6 in the form

$$u = u_r\left(\frac{z}{z_r}\right)^{\alpha'} \qquad [15.31.10]$$

where u_r is the wind speed at reference height z_r and α' is an index. For the friction velocity use is made of Equation 15.24.7.

Shape parameters and eddy crosswind diffusivity

The parameters r, s and K are obtained by applying Equations 15.31.1, 15.31.4, 15.31.6 and 15.31.10 to the case of a point source with release of a neutrally buoyant gas ($b = 0$, $\rho = \rho_a$). This yields

$$r = 2 \qquad [15.31.11a]$$

$$s = 1 + 2\alpha' \qquad [15.31.11b]$$

$$K_y = K_o^*\left(\frac{W_b}{B_s}\right)^{\gamma'}\left(\frac{z}{z_r}\right)^{\alpha'} \qquad [15.31.12]$$

with

$$K_o^* = \left[\frac{\delta'(\pi)^{\frac{1}{2}}}{2B_s}\right]^{1/\beta'} \frac{2B_s^2 u_r \beta'}{\pi} \qquad [15.31.13]$$

$$W_b = b + \frac{(\pi)^{\frac{1}{2}}}{2}\sigma_y \qquad [15.31.14]$$

$$\gamma' = 2 - 1/\beta' \qquad [15.31.15]$$

where K_o^* is the vertical eddy diffusion coefficient at reference height z_r acting on a concentration profile of width $2B$, W_b is the half-width of the crosswind concentration profile, and β', γ' and δ' are constants. The constants β' and δ' are obtained from the relation for the crosswind dispersion coefficient for a point source:

$$\sigma_y = \delta' x^{\beta'} \qquad [15.31.16]$$

Table 15.44 gives the values of the shape parameter α' for the wind velocity profile given by te Riele. The parameters β' and δ' are obtained from Equation 15.31.16, which corresponds in general form to a number of correlations for the crosswind dispersion coefficient.

Shear stress

For the shear stress τ_o the relation used is

Table 15.44 Wind velocity profile shape parameter (after te Riele, 1977) (Courtesy of DECHEMA)

Stability category	*Shape parameter, α'*	*Stability category*	*Shape parameter, α'*
A	0.02	D	0.14
B	0.05	E	0.20
C	0.09	F	0.28

$$\tau_o = 0.16\frac{\rho_a u_r^2}{\ln^2(z_r/z_o)} \quad [15.31.17]$$

where z_o is the roughness length.

Take-up flux

For the flux of material taken up into the ambient atmosphere, or take-up flux, Q'' there is a certain maximum value Q''_{max}. This maximum value is obtained from Equation 15.31.5 which yields

$$\tau_{gr} = Q''_{max}\frac{\int_0^\infty cudz}{\int_0^\infty cdz} \quad [15.31.18]$$

Then the value of Q'' is given by

$$Q'' = Q''_o \qquad Q''_{max} > Q''_o \quad [15.31.19a]$$

$$= Q''_{max} \qquad Q''_{max} \leq Q''_o \quad [15.31.19b]$$

where Q''_o is the mass flux of material from the source.

The model consists of Equations 15.31.8 and 15.31.9 with the supplementary Equations 15.31.6, 15.31.7 and 15.31.10–15.31.12 together with the three values of the integration limit y_{Ls}. It yields the four parameters c_A, b, σ_y and σ_z.

The dimensionless groups which govern dense gas dispersion are given as:

$$\alpha'; \quad \gamma'; \quad \frac{Q''L_s}{\rho_g u_r z_r}; \quad \frac{\tau_o L_s}{\rho_g u_r^2 z_r}; \quad \frac{K_o^* L_s}{B_s^2 u_r}; \quad \frac{\rho_g}{\rho_a}$$

15.31.2 Colenbrander steady-state model

The further development of the te Riele model has been described by Colenbrander (1980). The principal features which he describes are developments in: the source term; the treatment of the four parameters – centreline concentration c_A, half-width of the middle part b, crosswind dispersion parameter S_y and vertical dispersion parameter S_z; and in the criterion for transition to passive dispersion.

Source term

An evaporating pool may generate above itself a gas blanket. The condition for formation of such a gas blanket is that the rate of vapour evolution per unit area Q''_p exceeds the ambient take-up rate per unit area Q''_{max} of vapour. Where this is the case, the gas blanket grows until the area of the cloud is such that the rate of ambient take-up rate equals the rate of input into the cloud. If the rate of vapour evolution is less than the ambient take-up rate, a gas blanket does not form.

For the case where $Q''_p > Q''_{max}$, the ambient take-up rate Q'' is constrained to be

$$Q'' = Q''_{max} \quad [15.31.20]$$

It is shown by a mass balance that

$$Q''_{max} = \frac{\rho_E u_r S_z^{1+\alpha'}}{(1+\alpha')z_r^{\alpha'} L} \quad [15.31.21]$$

where ρ_E is the density at emission conditions and S_z is evaluated at $x = 0.5L$. The length L and half-width B of the gas blanket are obtained from

$$Q''_p L_b B_p = Q''_{max} LB \quad [15.31.22]$$

where B_p is half-width of the pool and L_p is the length of the pool. The centre line ground level concentration c_A is

$$c_A = \rho_E \quad [15.31.23]$$

For the case where $Q''_p < Q''_{max}$, the ambient take-up rate is

$$Q'' = Q''_p \quad [15.31.24]$$

The length L and half-width B of the gas blanket are

$$L = L_p \quad [15.31.25]$$

$$B = B_p \quad [15.31.26]$$

The centre line ground level concentration c_A is obtained from the relation

$$Q''_p = \frac{c_A u_r S_z^{1+\alpha'}}{(1+\alpha')z_r^{\alpha'} L} \quad [15.31.27]$$

where S_z is evaluated at $x = 0.5L$.

The foregoing applies to a rectangular source. It is convenient in dealing with the effective source constituted by a gas blanket to work in terms of a circular source. Assuming that the rectangle is a square, the radius R of this source is then

$$R = \frac{1}{\pi^{\frac{1}{2}}}L \quad [15.31.28]$$

If there is a gas blanket, the radius $R_b(t)$ of the blanket is taken as having a minimum value $R_1(t)$ and as growing at the rate

$$\frac{dR_b(t)}{dt} = c_E[g' H_b(t)]^{\frac{1}{2}} \quad [15.31.29]$$

where H_b is the height of the 100% gas blanket.

The value of $Q''_{max}(t)$ is obtained from the mass balance

$$\frac{d}{dt}(\pi R_b^2(t) H_b(t)\rho_E) = \pi R_1^2(t) Q''_p - \pi R_b^2(t) Q''_{max}(t) \quad [15.31.30]$$

For a spreading pool of radius $R_1(t)$

$$Q''_p = \frac{Q_p(t)}{\pi R_1^2(t)} \quad [15.31.31]$$

where Q_p is the total mass rate of evaporation.

Vertical dispersion parameter

The vertical concentration distribution given by Equation 15.31.7a satisfies the two-dimensional diffusion equation

$$u\frac{\partial c}{\partial x}=\frac{\partial}{\partial z}\left(K_z\frac{\partial c}{\partial z}\right) \quad [15.31.32]$$

with

$$K_z=\frac{ku_* z}{\phi(\mathrm{Ri}^*)} \quad [15.31.33]$$

$$\mathrm{Ri}_*=\frac{g' H_{\mathrm{eff}}}{u_*^2} \quad [15.31.34]$$

where H_{eff} is the effective height of the cloud, Ri is a Richardson number, K is the von Karman constant and $\phi(\mathrm{Ri}_*)$ is the Monin–Obukhov profile function.

Then from Equations 15.31.7a, 15.31.10, 15.31.32 and 15.31.33

$$\frac{\mathrm{d}}{\mathrm{d}x}\left(\frac{S_z}{Z_r}\right)^{1+\alpha'}=\frac{k}{z_r}\frac{u_*}{u_r}\frac{(1+\alpha')^2}{\phi(\mathrm{Ri}_*)} \quad [15.31.35]$$

The ratio (u_*/u_r) is obtained from the logarithmic wind velocity profile Equation 15.24.7. Equation 15.31.35 describes the vertical growth of a plume for the two-dimensional case of a plume of constant width. It is generalized for a dense gas plume which spreads laterally to yield

$$\frac{\mathrm{d}}{\mathrm{d}x}\left[B_{\mathrm{eff}}\left(\frac{S_z}{z_r}\right)\right]^{1+\alpha'}=\frac{k}{z_r}\frac{u_*}{u_r}\frac{(1+\alpha')^2 B_{\mathrm{eff}}}{\phi(\mathrm{Ri}_*)} \quad [15.31.36]$$

with

$$B_{\mathrm{eff}}=b+\frac{\pi^{\frac{1}{2}}}{2}S_y \quad [15.31.37]$$

where B_{eff} is the effective half-width of the plume.

Crosswind dispersion parameter

The crosswind concentration distribution given by Equation 15.31.7b with $z=0$ satisfies the two-dimensional diffusion equation

$$u\frac{\partial c}{\partial x}=\frac{\partial}{\partial y}\left(K_y\frac{\partial c}{\partial y}\right) \quad [15.31.38]$$

with

$$K_y=K_o u W_b^{\gamma'\prime} \quad [15.31.39]$$

where W_b is the half width of the cloud, K_o is a constant and γ'' is an index. Then, again from Equations 15.31.7b, 15.31.38 and 15.31.39,

$$S_y\frac{\mathrm{d}S_y}{\mathrm{d}x}=\frac{4\beta'}{\pi}W_b^2\left[\frac{\delta'(\pi/2)^{\frac{1}{2}}}{W_b}\right]^{1/\beta'} \quad [15.31.40]$$

For $b=0$, Equation 15.31.40 describes the lateral growth of a plume for the case of a plume where the width of the middle part is zero. In the generalization for a dense gas plume with a finite middle part, Equation 15.31.40 still applies but with finite b in Equation 15.31.37.

Width of middle part

The growth of the effective half-width B_{eff} is

$$\frac{\mathrm{d}B_{\mathrm{eff}}}{\mathrm{d}t}=c_E(g' H_{\mathrm{eff}})^{\frac{1}{2}} \quad [15.31.41]$$

with

$$H_{\mathrm{eff}}=\frac{\Gamma\left(\frac{1}{1+\alpha'}\right)}{1+\alpha'}S_z \quad [15.31.42]$$

$$\frac{\mathrm{d}B_{\mathrm{eff}}}{\mathrm{d}x}=\frac{1}{u_{\mathrm{eff}}}\frac{\mathrm{d}B_{\mathrm{eff}}}{\mathrm{d}t} \quad [15.31.43]$$

with

$$u_{\mathrm{eff}}=\frac{\int_0^\infty cu\,\mathrm{d}z}{\int_0^\infty c\,\mathrm{d}z} \quad [15.31.44]$$

$$=u_r\left(\frac{S_z}{z_r}\right)^{\alpha'}\frac{1}{\Gamma\left(\frac{1}{1+\alpha'}\right)} \quad [15.31.45]$$

where u_{eff} is the effective local cloud velocity.

The half-width b of the middle part is obtained from Equation 15.31.37 with B_{eff} obtained from Equation 15.31.43.

Centre-line concentration

For the centre-line concentration c_A the integral mass balance with substitution of the similarity profiles for c and u yields

$$c_A=\frac{Q(1+\alpha')z_r^{\alpha'}}{u_r S_z^{1+\alpha'}(b+0.5\pi^{\frac{1}{2}}S_y)} \quad [15.31.46]$$

The foregoing treatment does not allow for dispersion in the along-wind direction, which can be significant. In order to allow for this a correction is derived which when applied to the centreline ground level concentration c_A yields a concentration c_A' corrected for this effect.

Transition conditions

Transition to passive dispersion occurs when $b=0$ at downwind distance x_{tr}. A virtual source is assumed to exist at x_v. The dispersion parameter S_y is

$$S_y=2^{\frac{1}{2}}\delta'(x+x_v)^{\beta'} \quad [15.31.47]$$

The downwind distance x_v of the virtual source is obtained from Equation 15.31.47 by setting $x=x_{tr}$.

At transition

$$S_y=2^{\frac{1}{2}}\sigma_y \quad [15.31.48]$$

15.31.3 Colenbrander quasi-steady-state model

Colenbrander has also described a method of accommodating within his model a transient release. The scenario of concern is the evaporation of a liquefied gas from a spreading pool. The author gives a source model which defines as a function of time both the rate of release $Q(t)$ and the radius of the pool $R(t)$.

He then applies to this source the concept of 'observers'. A series of observers is envisaged as passing over the source at fixed time intervals. A given observer i passes over at a velocity $u_i(t)$ which is time dependent. The observer velocity increases with downwind distance and at a given distance x the velocity of all observers is the same $(u_i(x)=u_{i+1}(x))$. It is implied in this that at a given time the velocity of observer i is greater than that of the succeeding observer $(u_i(t)>u_{i+1}(t))$. A detailed derivation of the velocity $u_i(t)$ of passage is given by the author.

There is then derived for each observer i a corresponding set of source parameters. If the times when observer i passes over the upwind and downwind edges of the source are t_{1i} and t_{2i}, respectively, and if $x_i(t)$ is the location of the observer at time t, the local half-width $B'_{si}(t)$ is

$$B'_{si}(t) = [R^2(t) - x_i^2(t)]^{\frac{1}{2}} \quad [15.31.49]$$

Then

$$A_{si} = \int_{t_{1i}}^{t_{2i}} u_i(t) B'_{si}(t)\,dt \quad [15.31.50]$$

$$L_{si} = x_i(t_{2i}) - x_i(t_{1i}) \quad [15.31.51]$$

$$B_{si} = A_{si}/L_{si} \quad [15.31.52]$$

$$Q''_i B_{si} L_{si} = \int_{t_{1i}}^{t_{2i}} Q''_i(t) u_i(t) B'_{si}(t)\,dt \quad [15.31.53]$$

where for observer i, A_i is the half-area of the source, B_{si} is the half-width of the source, L_{si} the length of the source and Q''_i the average take-up flux.

Then for any specified time t_s the concentration distribution in the plume may be determined. At this time for observer i the location $x_i(t_s)$ is calculated and the plume parameters c_A, b, S_y and S_z are obtained as functions of this distance. A steady-state calculation is then performed, utilizing Equation 15.31.7, to obtain the concentration attributable to this observer. The calculation is repeated for all observers.

15.31.4 HEGADAS

As already described, HEGADAS is a development of the models of te Riele (1977) and Colenbrander (1980). Accounts of this development have been given by Colenbrander and Puttock (1983, 1984) and Puttock (1987b,c, 1989). Further enhancements of the model have been described by Witlox (1991).

The basic HEGADAS model is for a continuous release of dense gas. The source is assumed to be an area source. The plume from this source is modelled as consisting of a middle part which has a uniform crosswind concentration and an outer part which has Gaussian features. The middle part initially constitutes the whole plume, but eventually disappears completely so that the plume assumes a fully Gaussian form. The development of the plume is illustrated in Figure 15.81. The model is a prime example of an advanced similarity model.

HEGADAS originally had two versions. The first, HEGADAS-S, is the steady-state model for a continuous release; this is the basic model just described. The other version, HEGADAS-T, is the quasi-steady-state model for a transient release, utilizing the 'observer' concept.

In a subsequent version of the model, described by Colenbrander and Puttock (1983, 1984), enhancements were incorporated to take account of heat transfer between the ground, or water, surface and the cloud. The fifth version, HEGADAS-5, described by Witlox (1991), deals with interfacing with other source models and gives a model for vertical sources and an improved gas blanket model, and contains improved treatments of gravity spreading and of crosswind and along-wind dispersion and of time step selection. It also contains an HF thermodynamics model.

Witlox describes particularly the interface of HEGADAS-5 with the jet/plume model PLUME. Transition is taken as occurring when the jet/plume velocity is close to the wind velocity at centroid height and the air entrainment rate of the plume model is close to that of the dense gas model. There is also an advection criterion which stops the transition if the cloud becomes too buoyant. The PLUME model gives averaged values of the density, concentration, velocity and enthalpy. The matching of these variables between the two models is discussed in detail by the author.

The gas blanket model, as originally formulated, exhibited oscillatory behaviour and has been modified to correct this. A model is also given for a vertical source in terms of a breakpoint. Using this model, breakpoint data are specified for a series of times for the effective half-width and for any two of the following: the effective cloud height, the mass flow of gas and the centre line ground level concentration of gas.

The treatment of crosswind gravity spreading in HEGADAS-5 recognizes that gravity spreading undergoes a collapse beyond which the degree of gravity spreading is much reduced. This effect has been shown in wind tunnel work by R.L. Petersen and Ratcliff (1989) and is dealt with by P.T. Roberts, Puttock and Blewitt (1990). For the crosswind dispersion coefficient S_y use is made of an alternative formula for σ_y due to Briggs.

The HEGADAS-5 model gives the four dependent variables: crosswind dispersion coefficient S_y, vertical dispersion coefficient S_z, effective half-width of cloud B_{eff} and centre line ground level concentration c_A. It contains five ordinary differential equations for the variables S_y, S_z, B_{eff}, H_e and y_{w3}, where H_e is the surface heat transfer and y_{w3} is the mole fraction of surface water vapour. The earlier versions of the program were sensitive to the choice of time interval for solution of the equations. HEGADAS-5 contains automatic time interval selection.

HEGADAS-5 incorporates a model of HF thermodynamics. This is described by Puttock *et al.* (1991a,b).

Vertical dispersion

In this later work the vertical dispersion is expressed in terms of the top entrainment velocity u_e:

$$u_e = \frac{(1+\alpha')ku_*}{\phi(\mathrm{Ri}_*)} \quad [15.31.54]$$

with Ri_* defined by Equation 15.31.34].

Improved relations are also given for the Monin–Obukhov profile function $\phi(\mathrm{Ri}_*)$:

$$\phi(\mathrm{Ri}_*) = 0.74 + 0.25\mathrm{Ri}_*^{0.7} + 1.2 \times 10^{-7}\mathrm{Ri}_*^3 \qquad \mathrm{Ri}_* \geq 0 \quad [15.31.55a]$$

$$= \frac{0.74}{1 + 0.65\,|\,\mathrm{Ri}_*\,|^{0.6}} \qquad \mathrm{Ri}_* < 0 \quad [15.31.55b]$$

Equation 15.31.55a applies to stable conditions and is based on work by McQuaid (1976b) and Kranenburg (1984), and Equation 15.31.55b applies to unstable conditions and is based on fitting data from the Prairie Grass experiments.

Heat transfer

For heat transfer between the ground and the cloud, expressions are given for both the natural and forced convection cases. For natural convection the expression given for this heat flux Q_{gr} is

$$Q_{gr} = 0.14\left[\frac{\alpha_{Tc}^2\beta_c g}{\nu_c}(\rho_c c_{pc})^3\right]^{\frac{1}{3}}(T_{gr} - T_c)^{\frac{4}{3}} \quad [15.31.56]$$

with, for an ideal gas,

$$\beta_c = 1/T_c \quad [15.31.57]$$

where C_f is a friction factor $(= (2u_*/u_c^2))$, u_c is the velocity of the cloud, α_{Tc} is the thermal diffusivity of the cloud, β_c is the coefficient of volumetric expansion and ν_c is the kinematic viscosity of the cloud.

For forced convection the heat flux is given as

$$Q_{gr} = \frac{1}{2}\left(\frac{\alpha_{Tc}}{\nu_c}\right)^{\frac{2}{3}} C_f\rho c_{pc}u_c(T_{gr} - T_c) \quad [15.31.58]$$

15.31.5 HEGABOX

HEGADAS models for a continuous release of dense gas the gravity spreading of the gas in the lateral direction. But for an instantaneous release, or for a non-instantaneous release in low wind, there is also strong gravity spreading along the direction of the wind. This aspect is taken care of by HEGABOX, which is used as a front-end addition to HEGADAS to cater for such cases. HEGABOX has been described by Puttock (1987c, 1988a).

In HEGABOX, the cloud from an instantaneous release is treated as a cylinder of uniform concentration. The gravity spreading and air entrainment are modelled as follows:

$$\frac{dR}{dt} = c_E(g'H)^{\frac{1}{2}} \quad [15.31.59]$$

$$\frac{dV}{dt} = \pi R^2 u_e + 2\pi RHw_e \quad [15.31.60]$$

The value of c_E is taken as 1.15.

For the edge entrainment velocity w_e

$$w_e = \gamma\frac{dr}{dt} \quad [15.31.61]$$

where γ is the edge entrainment coefficient.

For the top entrainment velocity u_e

$$u_e = \frac{ku_{*I}}{\phi(Ri_{*I})} \quad [15.31.62]$$

with

$$u_{*I} = \frac{ku_B}{\ln(H/z_o) - 1} \quad [15.31.63]$$

$$\phi(Ri_{*I}) = (1 + 0.8Ri_{*I})^{\frac{1}{2}} \quad [15.31.64]$$

$$Ri_{*I} = \frac{g'H}{u_{*I}^2} \quad [15.31.65]$$

where u_B is a bulk cloud velocity and u_{*I} is an internal velocity scale. The internal velocity scale u_{*I} is chosen to

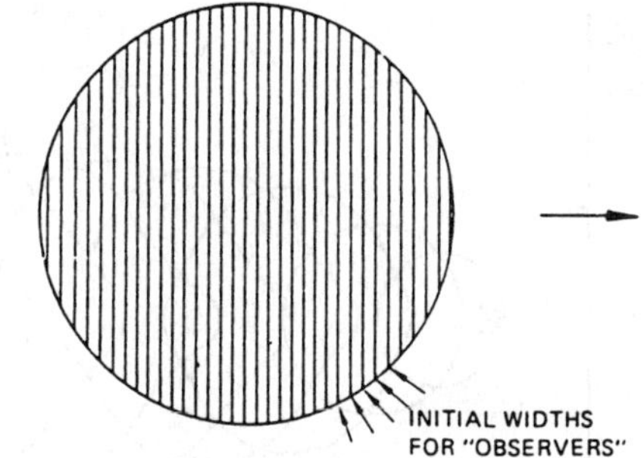

Figure 15.82 *HEGADAS model for dense gas dispersion – initial locations of 'observers' (Puttock, 1987c) (Courtesy of the American Institute of Chemical Engineers)*

be consistent with the uniform density assumed over the height of the cloud.

The bulk cloud velocity u_B is determined as follows. Strictly, this quantity should be obtained from conservation of momentum, but there are a number of aspects such as the effective velocity of the air entrained in the cloud, which are not well understood. The approach taken is therefore a semi-empirical one. The cloud acquires momentum from the entrainment of air passing over it. There is evidence that the effective air velocity u_A 'seen' by the cloud bears a constant ratio to the average air velocity over the height H of the cloud and that this ratio is of the order of 0.7. Furthermore, the cloud must eventually accelerate to the velocity of the ambient air.

These features are taken into account by using the following relations:

$$u_B = f(u_A) \quad [15.31.66]$$

$$u_A = \left(0.7 + \frac{0.3}{1 + Ri_*}\right)\frac{1}{H}\int_0^H u(z)\, dz \quad [15.31.67]$$

with

$$Ri_* = \frac{g'H}{u_*^2} \quad [15.31.68]$$

If HEGABOX is used as a front end for HEGADAS, transition is taken to occur at a transition value of Ri_I^*. This value is taken as 10.

At transition the cylindrical cloud is divided into slices with an observer at each slice, as shown in Figure 15.82. The parameters with which an observer starts are as follows. The half-width b of the middle part is equal to the cloud width at that point. The width parameter S_y is initially zero. The height parameter S_z is obtained from the height H. The centre line ground level concentration c_A is set equal to the uniform concentration c.

The transition from HEGABOX to HEGADAS is illustrated by the example shown in Figure 15.83.

15.31.6 HGSYSTEM

A set of models named HGSYSTEM has been developed, built around HEGADAS. These models were originally

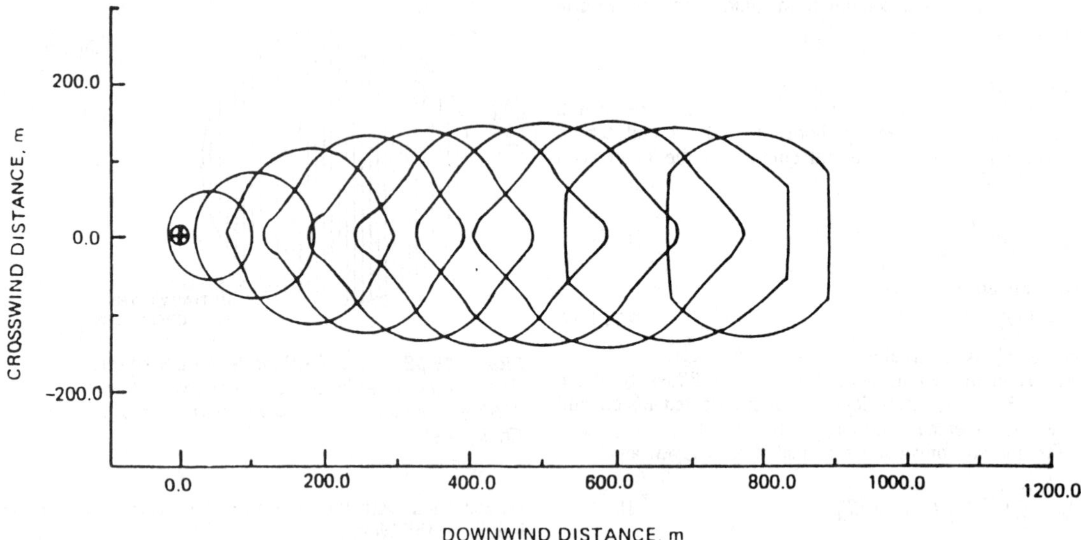

Figure 15.83 HEGABOX/HEGADAS model for dense gas dispersion – prediction of combined model for cloud development for Thorney Island trial 14 (Puttock, 1987c). Contours are for 500 ppm ground level concentration shown at 20 s intervals from time of release. 20 s and 40 s contours are from HEGABOX, remainder from HEGADAS (Courtesy of Reidel Publishing Company)

created specifically for releases of hydrogen fluoride and for this case the model system is called HFSYSTEM and the constituent models mainly bear the prefix HF. The models were subsequently generalized to give a set for an ideal gas. HGSYSTEM has been described in outline by Puttock *et al.* (1991a,b) and in detail by MacFarlane *et al.* (1990), Rees (1990) and Witlox *et al.* (1990). HFSYSTEM contains source models for an unpressurized source and for a pressurized source.

For an unpressurized source the model used is EVAP, which describes evaporation from a pool and which contains two options. The first models evaporation from a pool of fixed size or a spreading pool on land. Evaporation is assumed to be mass transfer limited. The second option models evaporation from a spreading pool on water. Evaporation is assumed to be heat transfer limited.

For a pressurized source at ground level the model used is HFSPILL. This gives the flow of a time-varying release of gas, subcooled liquid or two-phase flashing liquid. The models used are of an accuracy matched to that of the models for the later dispersion phase.

For a pressurized release from an elevated source the model used is HFPLUME. This models the airborne, touchdown and slumping phases of the plume. An account of the generalized model PLUME is given in Section 15.44. Each of these three source models is designed to interface with HEGADAS.

As stated, the models in HFSYSTEM have been generalized to form the set of models for an ideal gas given in HGSYSTEM. This model system also contains an alternative model PGPLUME for passive dispersion. The models in HGSYSTEM are intended to give reasonable accuracy in the near field. One aim is to allow prediction of conditions both at the inlet and outlet of a mitigation system.

15.31.7 Validation and application

Te Riele (1977) compared predictions from his model with observations in the Bureau of Mines, van Ulden, and SS *Gadila* trials. The latter is shown in Figure 15.92(c). Colenbrander (1980) used the Matagordo Bay trials for comparison with predictions from his model.

Validations of the various versions of HEGADAS include comparison of predictions from the model with observations from the Maplin Sands trials by Colenbrander and Puttock (1983) and Puttock (1987c).

Puttock (1987c, 1988a) has compared predictions of HEGABOX/HEGADAS with observations from the Maplin Sands and Thorney Island Phase I trials.

J.L. Woodward *et al.* (1982) have compared the performance of HEGADAS with that of other models in relation to the Matagordo Bay and Porton Down trials, and S.R. Hanna, Strimaitis and Chang (1991b) its performance against other models for nine sets of trials, including the Porton Down, Maplin Sands, Thorney Island Phase I, Burro, Coyote, Desert Tortoise and Goldfish trials.

15.32 Dispersion of Dense Gas: DEGADIS

15.32.1 DEGADIS

As described above, the related model DEGADIS has been developed by Havens and co-workers (Havens, 1985, 1986; Spicer and Havens, 1986, 1987). DEGADIS

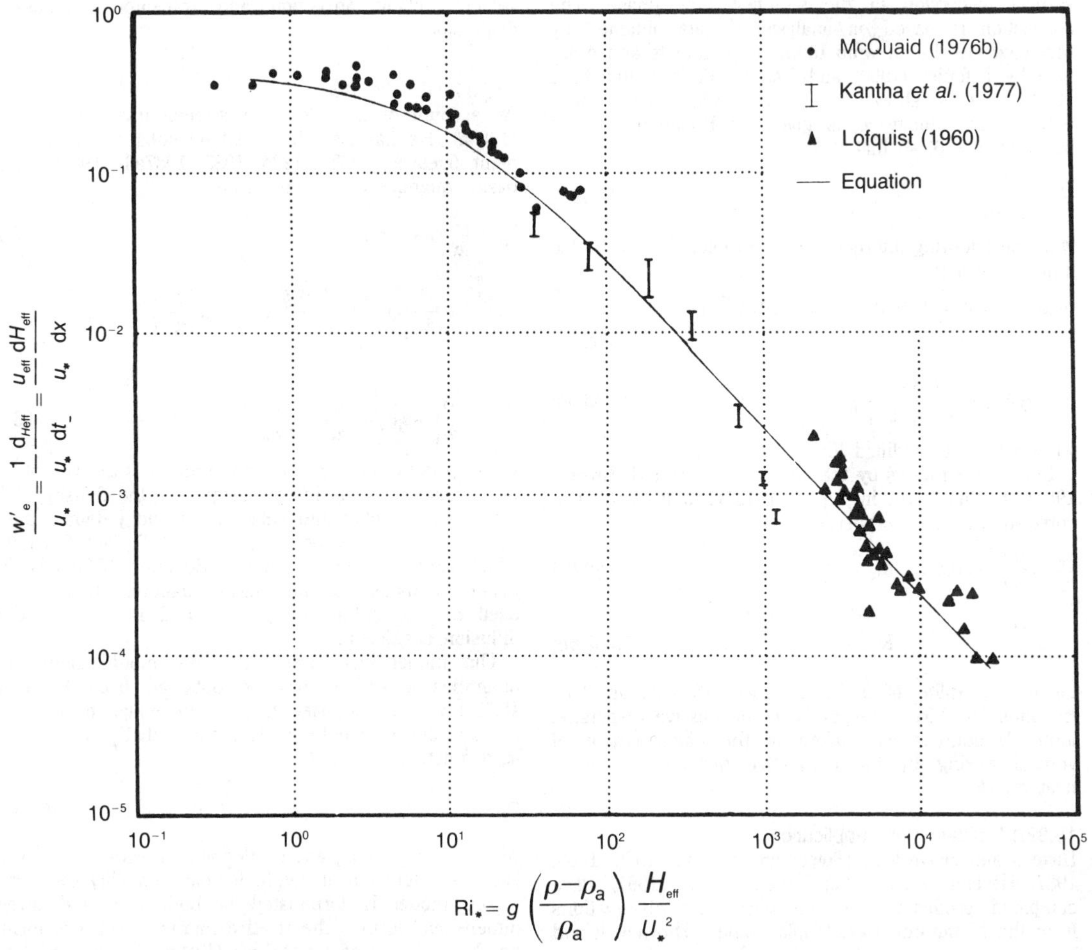

Figure 15.84 *Correlation of vertical entrainment velocity with bulk Richardson number (Spicer and Havens, 1986) (Courtesy of Elsevier Science Publishers) H_{eff}, height at which concentration is one tenth of the maximum value; u_{eff}, effective advection velocity; $w_e' = dH_{eff}/dt$*

is based on HEGADAS, but differs in several respects and, in particular, it incorporates its own source, vertical entrainment and heat transfer models. DEGADIS is for a continuous release but may be adapted to time-varying releases, and an instantaneous release, by treating these as a series of quasi-steady-state releases.

The source model used in DEGADIS has been described by Spicer and Havens (1986). It describes the transition by gravity slumping from a cloud of high aspect ratio to one of low aspect ratio and to gravity spreading.

The source model includes the following treatment of heat transfer from the ground surface to the cloud. Relations are given for both natural convection and forced convection. For natural convection, use is made of Equation 15.24.14. The authors state that for the hydrocarbon gases methane and propane this equation can be written as

$$h_{gr} = 18\left[\left(\frac{\rho}{M}\right)^2 \Delta T\right]^{\frac{1}{3}} \qquad [15.32.1]$$

For heat transfer by forced convection, use is made of Equation 15.24.18. Then utilizing Equation 15.31.10 to give at height H

$$u = u_r\left(\frac{H}{z_r}\right)^{\alpha'} \qquad [15.32.2]$$

and taking Pr = 0.741, Equation 15.32.1 yields

$$h_{gr} = \left[1.22\frac{u_*^2}{u}\left(\frac{z_o}{H}\right)^{\alpha'}\right]\rho c_p \qquad [15.32.3]$$

Another feature of DEGADIS is the correlation used for vertical dispersion in the stratified flow phase. The correlation is based on analysis of data obtained by McQuaid (1976b) in wind tunnel experiments and other data by Lofquist (1960) and Kantha, Phillips and Azad (1977) and is shown in Figure 15.84. The vertical entrainment velocity u_e is given by Equation 15.31.54 but with $\alpha' = 0$, so that

$$\frac{u_e}{u_*} = \frac{k}{\phi(\text{Ri})} \qquad [15.32.4]$$

with the following alternative relation used for the profile function $\phi(\text{Ri}_*)$:

$$\phi(\text{Ri}_*) = 0.88 + 0.099\text{Ri}_*^{1.04} + 1.4 \times 10^{-25}\text{Ri}_*^{5.7} \qquad \text{Ri}^* \geq 0 \qquad [15.32.5a]$$

$$= \frac{0.88}{1 + 0.65 \mid \text{Ri}_* \mid^{0.6}} \qquad \text{Ri}_* < 0 \qquad [15.32.5b]$$

where Ri_* is a modified Richardson number.

In later treatments by the authors (Spicer and Havens, 1987) the last term in Equation 15.32.5a is zero. The correlation has the properties that

$$\frac{u_e}{u_*} \rightarrow \frac{0.35}{0.88} = 0.4 \qquad \text{Ri}_* \rightarrow 0 \qquad [15.32.6a]$$

$$\propto \frac{1}{\text{Ri}_*} \qquad \text{Ri}_* \rightarrow \infty \qquad [15.32.6b]$$

taking the value of k for Equation 15.32.6a as 0.35. Equation 15.32.6a corresponds to the passive dispersion limit. Account is also taken of the enhancement of vertical mixing by the convective turbulence due to heat transfer.

15.32.1 Validation and application

Havens and co-workers (Spicer and Havens, 1985, 1986, 1987; Havens, 1986, 1992; Havens *et al.*, 1991) have compared predictions of DEGADIS with observations from the Matagordo Bay, Maplin Sands, Thorney Island Phase I, Burro, Coyote, Desert Tortoise, Eagle and Goldfish trials, and Blewitt, Yohn and Ermak (1987) have compared its predictions against the Goldfish trials.

Touma *et al.* (1991) have compared the performance of the model with that of other models in relation to the Burro, Desert Tortoise and Goldfish trials, and S.R. Hanna, Strimaitis and Chang (1991b) have compared its performance against other models for nine sets of trials, including the Porton Down, Maplin Sands, Thorney Island Phase I, Burro, Coyote, Desert Tortoise and Goldfish trials.

15.33 Dispersion of Dense Gas: SLUMP and HEAVYGAS

Another pair of models are SLUMP and HEAVYGAS developed at WS Atkins. They have been described by Deaves (1983a,b, 1987a,b). SLUMP is a box model and HEAVYGAS a *K*-theory model with κ–ϵ modification.

15.33.1 SLUMP

SLUMP is described by Deaves (1987a), who refers to the equation set for box models formulated by Wheatley and Webber (1984 CEC EUR EN 9592) and deals particularly with the treatment in his model of slumping velocity, cloud advection and transition to passive dispersion.

15.33.2 HEAVYGAS

An account of HEAVYGAS has been given by Deaves (1983a). He has also described its subsequent development (Deaves, 1983b, 1984, 1985, 1987a,b, 1989b). The basic equations of the model are:

$$\frac{\partial \rho}{\partial t} + \frac{\partial}{\partial x_j}(\rho u_j) = 0 \qquad [15.33.1]$$

$$\frac{\partial(\rho u_i)}{\partial t} + \frac{\partial}{\partial x_j}(\rho u_i u_j) = -\frac{\partial p}{\partial x_i} + \frac{\partial}{\partial x_j}\left(\mu \frac{\partial u_i}{\partial x_j}\right) + \Delta \rho g_i \qquad [15.33.2]$$

$$\frac{\partial \rho \phi}{\partial t} + \frac{\partial}{\partial x_j}(\rho u_j \phi) = \frac{\partial}{\partial x_j}\left(\Gamma_\phi \frac{\partial \phi}{\partial x_j}\right) + S_\phi \qquad [15.33.3]$$

where p is the pressure, t is the time, u is the velocity, x is the distance, μ is the viscosity, ρ is the density, ϕ is any scalar quantity and subscripts i and j denote the i and j directions, respectively. S_ϕ and Γ_ϕ are functions which vary with the quantity ϕ. Equation 15.33.3 is the general equation for any scalar quantity. Thus if it is used for concentration, $S_\phi = 0$ and Γ_ϕ is the eddy diffusion coefficient.

The model incorporates the κ–ϵ modification. The quantities κ and ϵ may be obtained from Equation 15.33.3 with S_ϕ adjusted to give the usual form of the κ–ϵ turbulence model equations and with Γ_ϕ either given as a function of κ and ϵ:

$$\Gamma_\phi = c_\phi \frac{\kappa^2}{\epsilon} \qquad [15.33.4]$$

where c_ϕ is a constant or defined in some other way such as a function of height for each stability category.

The model is formulated in both two- and three-dimensional forms. The relative merits of the two forms has been discussed by Deaves (1987a). In essence, two-dimensional modelling is much simpler and is usually sufficient, but three-dimensional modelling may be required in special cases.

15.33.3 Validation and application

The predictions of HEAVYGAS have been compared with the results of Thorney Island Phase II trials 5 and 21 without and with a solid fence (Deaves, 1987a) and trial 29 with a single building (Deaves, 1984, 1985).

HEAVYGAS has been used to model a number of applications which exploit the ability of this type of model to deal with situations other than flat, unobstructed terrain. It has been used to model the dispersion of chlorine within a building (Deaves, 1983a) and the effect of water spray barriers on the dispersion of a plume of carbon dioxide (Deaves, 1983a,b, 1984). These studies are described in Sections 15.40 and 15.53, respectively

It has also been applied to problems arising in safety cases (Deaves, 1987, 1989a,b). These include: the release of chlorine into and then from a building; the effect of individual buildings on hazard ranges for chlorine; and the effect of a whole plant site on hazard ranges for ammonia. Accounts of this work are given in Section 15.40.

15.34 Dispersion of Dense Gas: *Workbook* Model

A unified methodology for dense gas dispersion is given in the *Workbook on the Dispersion of Dense Gases* (Britter and McQuaid, 1988). The work was supported, though not necessarily endorsed, by the Health and Safety Executive and was closely associated with the Thorney Island trials.

15.34.1 Basic model

The basic model is a similarity relation and consists of a correlation of dimensionless groups. This relation is

$$\frac{C}{C_o} = f\left[\frac{x}{D}, \frac{y}{D}, \frac{z}{D}, \frac{tu_{\mathrm{ref}}}{D}, \frac{q_o(t)}{u_{\mathrm{ref}}D^2} \text{ or } \frac{Q_o^{\frac{1}{3}}}{D}, \frac{u_{\mathrm{ref}}^2}{gD}, \frac{\rho_o}{\rho_a}, \frac{z_o}{D}, \frac{l_i}{D}, G \text{ or } \frac{Q_o^{\frac{1}{3}}}{D}, \text{atmospheric stability}\right] \quad [15.34.1]$$

where C is the volumetric concentration, C_o is the volumetric concentration of the initial release, D is a characteristic dimension of the source, g is the acceleration due to gravity, G is the source geometry, l_i is the length scale of the turbulence in the atmospheric boundary layer, $q_o(t)$ is the volumetric rate of release of material, Q_o is the volume of material released, u_{ref} is a characteristic mean velocity, ρ_a is the density of air, ρ_o is a characteristic density of the material released, and x, y and z are the distances in the downwind, crosswind and vertical directions, respectively.

The influence of some of the items in Equation 15.34.1 is relatively weak and in order to obtain a sufficient data set the following are disregarded: (l_i/D), (z_o/D), G and atmospheric stability. This then yields

$$\frac{C}{C_o} = f\left[\frac{x}{D}, \frac{y}{D}, \frac{z}{D}, \frac{tu_{\mathrm{ref}}}{D}, \frac{q_o(t)}{u_{\mathrm{ref}}D^2} \text{ or } \frac{Q_o^{\frac{1}{3}}}{D}, \frac{u_{\mathrm{ref}}^2}{gD}, \frac{\rho_o}{\rho_a}\right] \quad [15.34.2]$$

15.34.2 Model for continuous release

For a continuous release the concentration of interest is the time mean concentration $\bar{C}$. From Equation 15.34.2]

$$\frac{\bar{C}}{C_o} = f\left[\frac{x}{D}, \frac{y}{D}, \frac{z}{D}, \frac{q_o}{u_{\mathrm{ref}}D^2}, \frac{u_{\mathrm{ref}}^2}{gD}, \frac{\rho_o}{\rho_a}\right] \quad [15.34.3]$$

For the maximum time mean concentration $\bar{C}_m$ on the centre line of the cloud

$$\frac{\bar{C}_m}{C_o} = f\left[\frac{x}{D}, \frac{q_o}{u_{\mathrm{ref}}D^2}, \frac{u_{\mathrm{ref}}^2}{gD}, \frac{\rho_o}{\rho_a}\right] \quad [15.34.4]$$

Since the number of terms in Equation 15.34.4 is still too high for the data set available, two further approximations are called in aid. One is the Boussinesq approximation, in which the density variation is neglected in the inertial terms but is retained in the buoyancy terms. This means in effect that the problem is restricted to the case where formally $[(\rho_o - \rho_a)/\rho_a] \ll 1$ and, in practice, $[(\rho_o - \rho_a)/\rho_a] < 1$; the value of $[(\rho_o - \rho_a)/\rho_a]$ for the initial release tends to approximate to unity or less. The Boussinesq approximation allows the terms (u_{ref}^2/gD) and (ρ_o/ρ_a) to be replaced by $(u_{\mathrm{ref}}^2/g_o')$ with

$$g_o' = \frac{g(\rho_o - \rho_a)}{\rho_a} \quad [15.34.5]$$

where g_o' is the reduced gravity of the initial release.

The other approximation invoked is to neglect the characteristic dimension D of the source. D has progressively less effect as the distance x from the source increases, and becomes insignificant at about $x > 5D$.

With these simplifications, Equation 15.34.4 reduces to

$$\frac{\bar{C}_m}{C_o} = f\left[\frac{x}{(q_o/u_{\mathrm{ref}})^{\frac{1}{2}}}, \frac{g_o' q_o/u_{\mathrm{ref}}^3}{(q_o/u_{\mathrm{ref}})^{\frac{1}{2}}}\right] \quad [15.34.6a]$$

$$= f\left[\frac{x}{(q_o/u_{\mathrm{ref}})^{\frac{1}{2}}}, \phi\right] \quad [15.34.6b]$$

with

$$\phi = \left(\frac{g_o'^2 q_o}{u_{\mathrm{ref}}^5}\right)^{\frac{1}{5}} \quad [15.34.7]$$

where ϕ is a stability parameter. The power 1/5 is introduced solely to give convenient numbers.

Equation 15.34.6b can be recast in the more useful form

$$\frac{x_n}{(q_o/u_{\mathrm{ref}})^{\frac{1}{2}}} = f(\phi) \quad [15.34.8]$$

where x_n is the distance at which the concentration is the fraction n of the initial concentration $\bar{C}_m/C_o$.

Downwind distance to a given concentration

The correlation of the data based on Equation 15.34.8 is shown in Figure 15.85(a). The correlation is intended to apply to the long-term average concentrations. It is based on both laboratory and full-scale trial data. Justifications of the correlation itself and of the limit of effectively passive dispersion are given in Appendices C and E of the report, respectively.

The correlation is valid within the limits

$$0.002 \leq \frac{\bar{C}_m}{C_o} \leq 0.1; \; 0 \leq \phi \leq 4$$

There are two useful relations which can be obtained from Figure 15.85(a). One is

$$\frac{x_n}{(q_o/u_{\mathrm{ref}})^{\frac{1}{2}}} = A\phi^{-\frac{1}{2}} \qquad \phi \geq 1 \quad [15.34.9]$$

or

$$x = Aq_o^{0.4}(g_o')^{-0.2} \quad [15.34.10]$$

where A is given by Figure 15.85(b). This equation shows a quite strong dependence of the distance x to a particular concentration on source strength q_o. It also shows that the distance decreases as the stability parameter ϕ increases. The interpretation is that increased lateral spread more than compensates for the inhibition of vertical mixing.

The other relation is

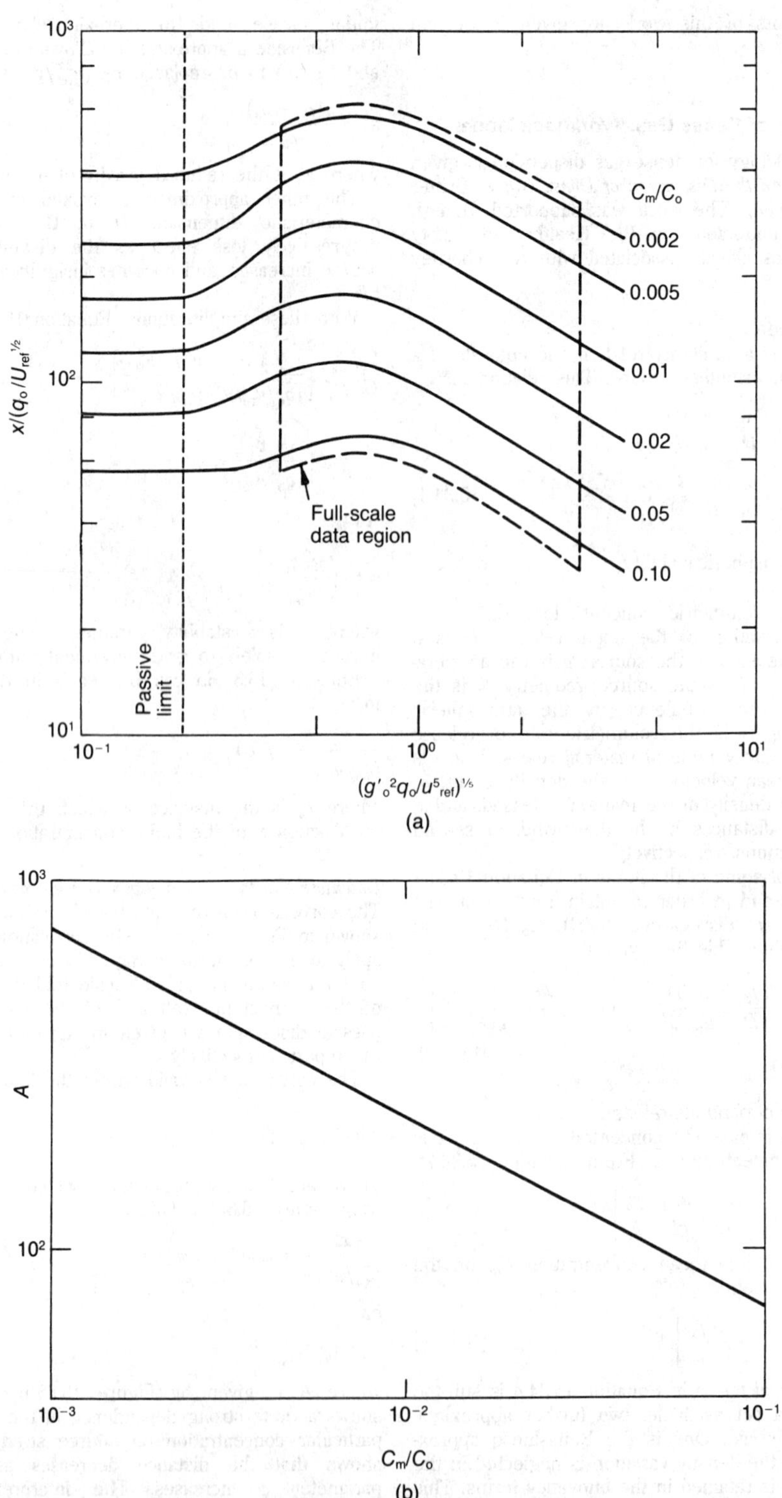

***Figure 15.85** Workbook method for dense gas dispersion – correlation for continuous release (Britter and McQuaid, 1988; reproduced by permission): (a) correlation; (b) parameter A*

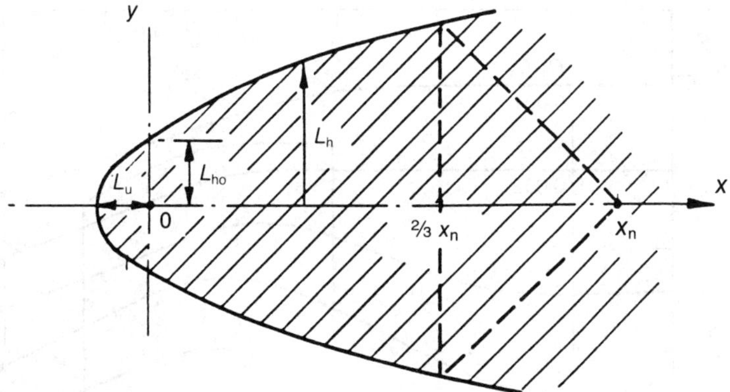

Figure 15.86 *Workbook method for dense gas dispersion – area covered by concentration contour for continuous release (Britter and McQuaid, 1988; reproduced with permission)*

$$x = 17.5\left(\frac{C_m}{C_o}\right)^{-\frac{1}{2}}\left(\frac{q_o}{u_{ref}}\right)^{\frac{1}{2}} \qquad 0 \le \phi \le 3 \qquad [15.34.11]$$

Thus the distance depends only on the length scale $(q_o/u_{ref})^{\frac{1}{2}}$.

For low concentrations outside the range of validity given, the suggestion made is that Equation 15.34.10 may be extrapolated, but still subject to the limits on ϕ.

Area bounded by a given concentration

A method is also given to obtain the area bounded by a given concentration. On the basis of modelling and experiment the dimensions of the cloud are found to be approximately as follows:

$$L_u = D/2 + 2l_b \qquad [15.34.12]$$

$$L_h = L_{ho} + 2.5 l_b^{\frac{1}{3}} x^{\frac{2}{3}} \qquad [15.34.13]$$

with

$$l_b = \frac{q_o g'_o}{u_{ref}^3} \qquad [15.34.14]$$

$$L_{ho} = D + 8l_b \qquad [15.34.15]$$

where l_b is the buoyancy length scale, L_h is the half-width of the plume, L_{ho} is the half-width of the plume at the source and L_u is the upwind extent of the plume.

For the height of the plume, from conservation of mass

$$L_v = \frac{q_o}{u_{ref} L_h} \qquad [15.34.16]$$

where L_v is the height of the plume.

Equation 15.34.16 assumes that the concentration is homogenous throughput the plume. In fact, the concentration varies with height, but the exact vertical profile is uncertain and variable. There is experimental evidence that the bulk of the material is contained within a height of $2(q_o/u_{ref}L_h)$.

Since Equation 15.34.16 is unduly conservative, an alternative approach is also given. This is to locate, for the concentration of interest, the lateral boundaries of the cloud at the distance $\frac{2}{3}x_n$ and to join these boundary points with x_n, as shown in Figure 15.86.

15.34.3 Model for instantaneous release

For an instantaneous release the case considered is a source of unit aspect ratio. This allows the external length scale D to be equated with the internally set length scale $Q_o^{\frac{1}{3}}$.

Then, following an approach similar to that for the continuous case,

$$\frac{C_m}{C_o} = f\left(\frac{x}{Q_o^{\frac{1}{3}}}, \frac{u_{ref}^2/g'_o}{Q_o^{\frac{1}{3}}}\right) \qquad [15.34.17]$$

$$= f\left(\frac{x}{Q_o^{\frac{1}{3}}}, \psi\right)$$

with

$$\psi = \left(\frac{g'_o Q_o^{\frac{1}{3}}}{u_{ref}^2}\right)^{\frac{1}{2}} \qquad [15.34.18]$$

where ψ is a stability parameter. Again the power $\frac{1}{2}$ is used solely to give convenient numbers. Then,

$$\frac{x_n}{Q_o^{\frac{1}{3}}} = f(\psi) \qquad [15.34.19]$$

In this case, since $D = Q_o^{\frac{1}{3}}$, the assumption that the effect of D can be neglected is not necessary.

Downwind distance to a given concentration

The correlation based on Equation 15.34.19 is shown in Figure 15.87(a). The correlation is intended to apply to the ensemble average of the maximum short-time (0.6 s) mean concentrations. It is based on both laboratory and full-scale trial data. Justifications of the correlation itself and of the limit of effectively passive dispersion are given in Appendices D and E of the report, respectively.

The correlation is valid within the limits

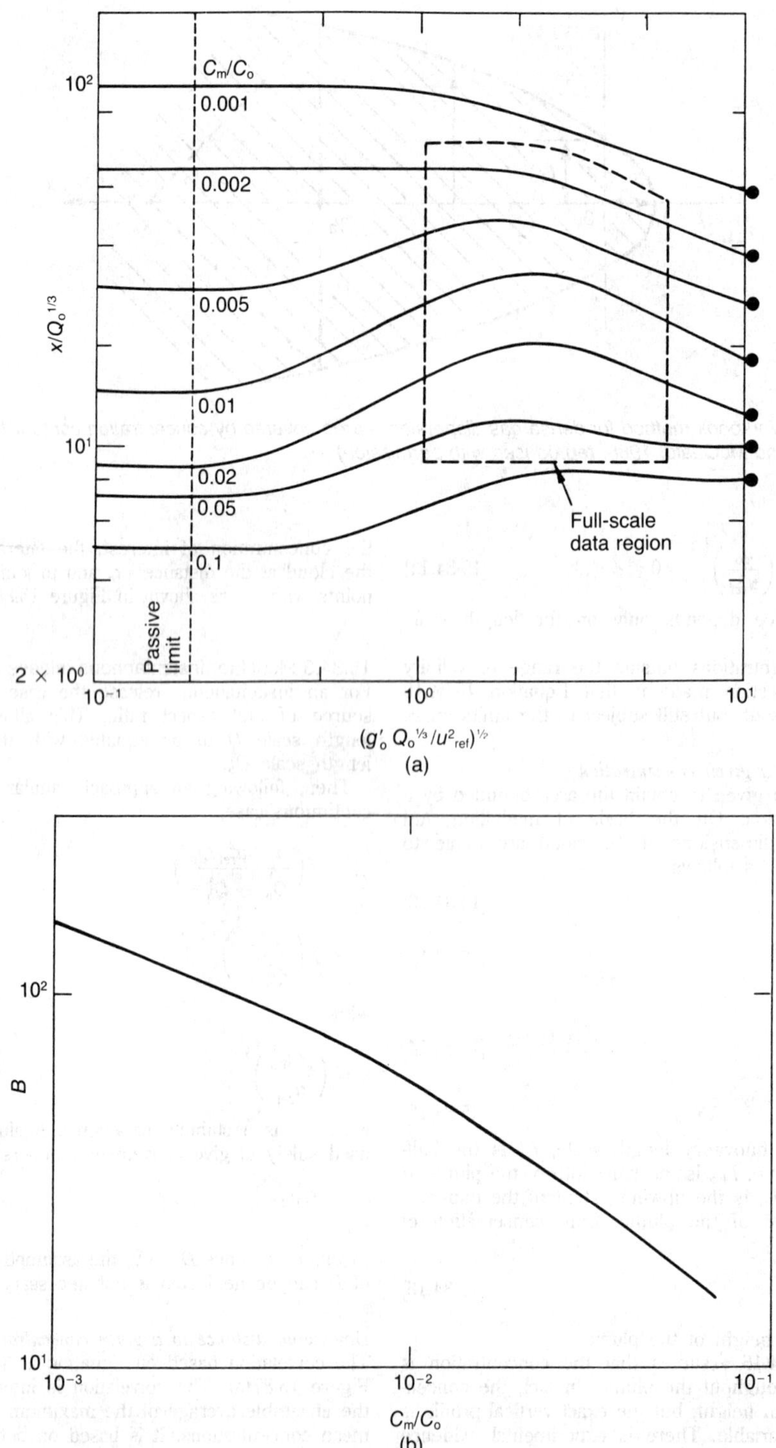

***Figure 15.87** Workbook method for dense gas dispersion – correlation for instantaneous release (Britter and McQuaid, 1988; reproduced with permission): (a) correlation; (b) parameter B; (c) correlation if initial dilution already taken into account*

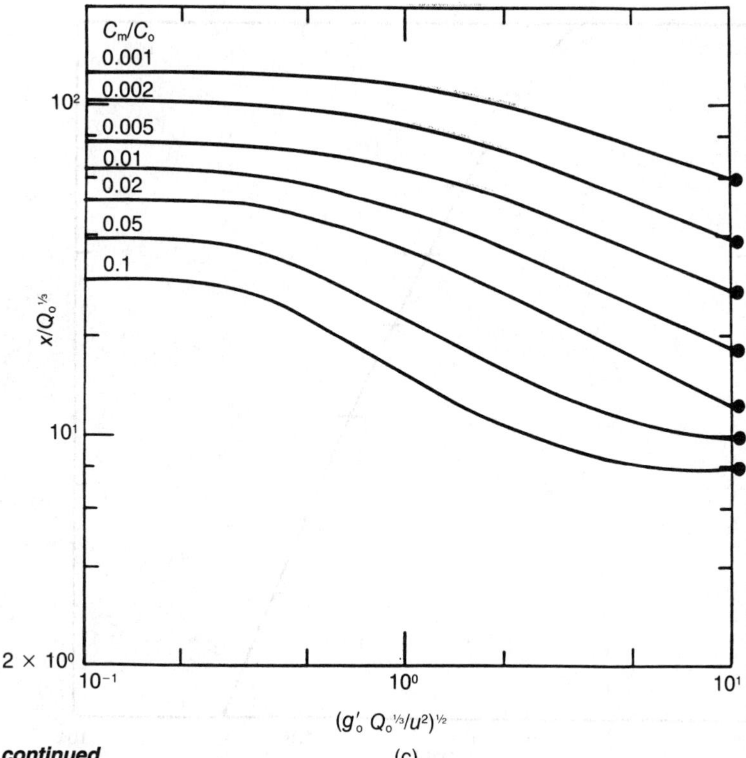

Figure 15.87 continued

$$0.001 \leq \frac{C_m}{C_o} \leq 0.1; \; 0.7 \leq \psi \leq 10$$

There are again two useful relations which can be obtained from Figure 15.87(a). One is

$$\frac{x_n}{Q_o^{\frac{1}{3}}} = B\psi^{-\frac{1}{2}} \qquad 0.001 \leq \frac{C_m}{C_o} \leq 0.02; \; 2 \leq \psi \leq 10 \qquad [15.34.20]$$

or

$$x = BQ_o^{\frac{1}{4}} g_o'^{-\frac{1}{4}} u_{ref}^{\frac{1}{2}} \qquad [15.34.21]$$

where B is given by Figure 15.87(b). This equation shows a rather weaker dependence on source strength of the distance x to a given concentration compared with the continuous case.

The other relation is

$$x = 2.8\left(\frac{C_m}{C_o}\right)^{-\frac{1}{2}} \qquad 1 \leq \psi \leq 5 \qquad [15.34.22a]$$

$$= 1.8\left(\frac{C_m}{C_o}\right)^{-\frac{1}{2}} \qquad \psi \to \infty \qquad [15.34.22b]$$

For low concentrations outside the range of validity given, the suggestion made is that Equation 15.34.21 may be extrapolated, but subject still to the limits on ψ.

For the case where there has already been significant dilution of the cloud and consequent cloud acceleration the *Workbook* proposes the modification shown Figure 15.87(c). The *Workbook* also gives for an instantaneous release the correlation shown in Figure 15.88 obtained from the Thorney Island trials for $1 \leq \psi \leq 5$.

Area swept by a given concentration

For the area swept by a given concentration the method is as follows. A model is used which is based on gravity slumping

$$\frac{dR}{dt} = c_E(g'H)^{\frac{1}{2}} \qquad [15.34.23]$$

with

$$g' = \frac{g(\rho - \rho_a)}{\rho} \qquad [15.34.24]$$

where H is the height of the cloud, R is the radius of the cloud and c_E is the slumping constant. The value of the constant c_E is 1.07. By conservation of buoyancy

$$g'Q = g'_o Q_o \qquad [15.34.25]$$

Also

$$H = Q/\pi R^2 \qquad [15.34.26]$$

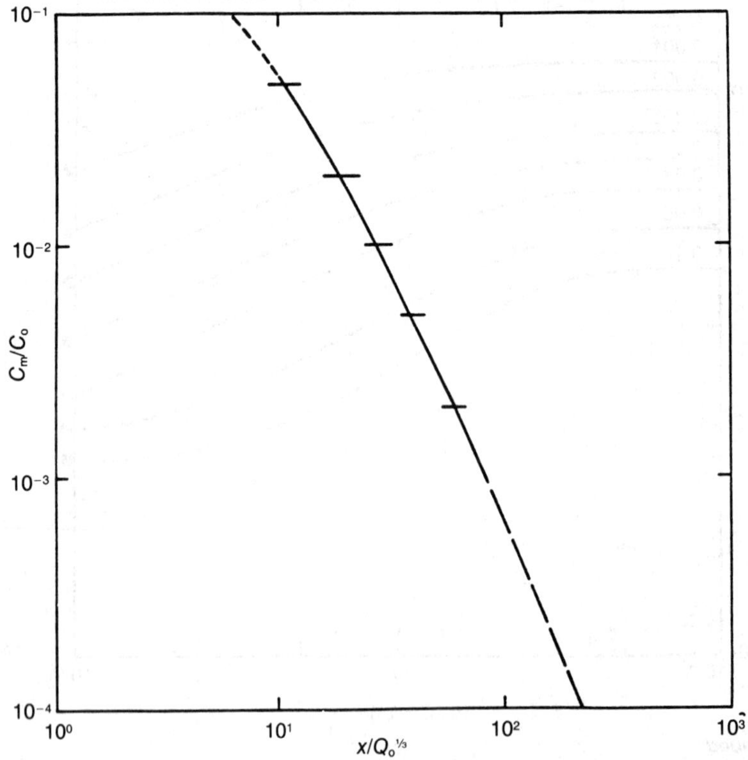

Figure 15.88 Workbook method for dense gas dispersion – variation of maximum concentration with scaled distance. Thorney Island instantaneous release trials (see text) (Britter and McQuaid, 1988; reproduced with permission)

where Q is the volume of the cloud.

Integration of Equation 15.34.23 gives

$$R(t) = [R_o + 1.2(g_o' Q_o)^{\frac{1}{2}} t]^{\frac{1}{2}} \quad [15.34.27]$$

where R_o is the initial radius. The advection velocity of the cloud is taken as $0.4u_{ref}$. At a distance x the arrival time t_a of the leading edge of the cloud is given by

$$x = 0.4u_{ref}t_a + R(t_a) \quad [15.34.28]$$

and the departure time t_d of the trailing edge by

$$x = 0.4u_{ref}t_d - R(t_d) \quad [15.34.29]$$

Now consider any time t_c between the arrival time t_a and the departure time t_d. For this time t_c the distance x_c of the leading edge is

$$x_c = 0.4u_{ref}t_c + R(t_c) \quad [15.34.30]$$

where $x_c = x$, $t_c = t_a$ and $x_c > x$, $t_c > t_a$. Then for a given concentration C of interest, the value of x_c is obtained from the correlation given in Figure 15.87(a), or the equivalent equations, t_c is obtained from Equation 15.34.30 and the cloud half-width y_c is obtained from

$$y_c = R(T_c) \quad [15.34.31]$$

The area required is a parabola enclosing the source and closed off at the downwind end by an arc of radius $R(t_c)$.

By conservation of mass, the cloud height H is

$$H = \frac{C_o Q_o}{\pi R^2 C} \quad [15.34.32]$$

In practice, the bulk of the material is contained within twice this height.

15.34.4 Criterion for type of release

The criterion used to distinguish between a continuous and an instantaneous release is based on the group $(u_{ref}T_o/x)$ or, alternatively, $(u_* T_o/x)$.

An effectively continuous plume may be assumed if

$$\frac{u_{ref}T_o}{x} \geq 2.5 \quad [15.34.33a]$$

or

$$\frac{u_* T_o}{x} \geq 0.15 \quad [15.34.33b]$$

where u_* is the friction velocity. This criterion applies on the centre line of the cloud. Larger values apply off the centre line.

Relation 15.34.33 implies that the distance x at which the release can be regarded as continuous recedes closer to the source as the duration T_o of the release decreases. For distances greater than those given by relation

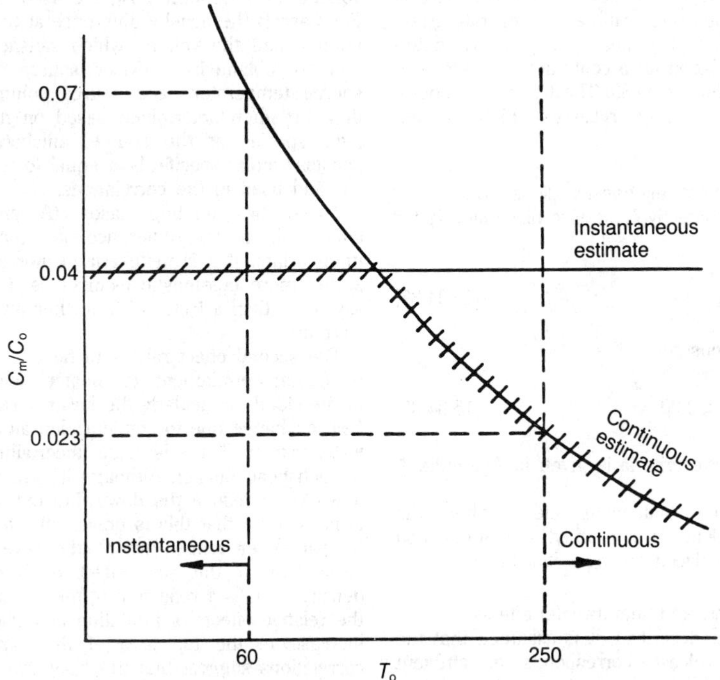

Figure 15.89 Workbook method for dense gas dispersion – treatment for transient release (Britter and McQuaid, 1988; reproduced with permission)

15.34.33, the assumption of a continuous source overestimates the concentration.

An effectively instantaneous release may be assumed if

$$\frac{u_{\mathrm{ref}}T_o}{x} \leq 0.6 \qquad [15.34.34a]$$

or

$$\frac{u_* T_o}{x} \leq 0.04 \qquad [15.34.34b]$$

In the region

$$0.6 \leq \frac{u_{\mathrm{ref}}T_o}{x} \leq 2.5$$

the release is a transient one, being neither effectively continuous nor instantaneous.

15.34.5 Treatment of transient release

A transient release, as just defined, may be treated by considering it first as a quasi-continuous release and then as a quasi-instantaneous one. Taking first the quasi-continuous approach, a release of quantity Q_o over a release duration T_o has an effective release rate q_o. This release rate may be utilized in the continuous release correlation to obtain a value of the concentration. This approach neglects the effect of longitudinal dispersion and will overestimate the concentration. Alternatively, the release may be treated as a quasi-instantaneous one with quantity release Q_o. This approach will again give an overestimate of the concentration. The recommendation is then to take as the upper bound of concentration the lower of the two concentrations obtained by these two methods.

This treatment is based on the assumption that the material is transported downwind by the ambient velocity only. But such transport occurs also due to the buoyancy induced motion. For the ambient velocity the time scale is x/u_{ref} and for buoyancy-driven velocity it is $x^2/(g'_o Q_o)^{\frac{1}{2}}$. Then for an instantaneous release the additional criterion which must be met is

$$\frac{(g'_o Q_o)^{\frac{1}{2}} T_o}{x^2} \leq 0.6 \qquad [15.34.35]$$

By a similar argument for a continuous release the buoyancy-driven time scale is $x^{\frac{4}{3}}/(g'_o q_o)^{\frac{1}{3}}$ and the additional criterion is

$$\frac{(g'_o q_o)^{\frac{1}{3}} T_o}{x^{\frac{4}{3}}} \geq 2.5 \qquad [15.34.36]$$

As an illustration, consider the example given in the *Workbook*. This is the determination of the concentration ratio C_m/C_o at a distance of 200 m for a release of a total of 2000 m^3 of dense gas with initial reduced gravity 10 m/s^2 with a wind speed at reference height of 2 m/s for some release time T_o such that the release is truly neither instantaneous nor continuous. The procedure is as follows. First the release is treated as if it were an instantaneous one. For an instantaneous release $\psi = 5.6$ and $x/Q_o^{\frac{1}{3}} = 15.9$, so that from the correlation for an instantaneous release $C_m/C_o = 0.04$. Then the release is treated as a continuous one, with a release rate q_o of $2000/T_o \, m^3/s$. A series of values of C_m/C_o are determined using the correlation for a continuous release. The results are shown in Figure 15.89. The figure also shows the limits determined from relations 15.34.35 and 15.34.36.

15.34.6 Criterion for effectively passive dispersion

The criterion used for effectively passive dispersion is for a continuous release

$$\left(\frac{g'_o q_o}{u_{ref}^3} \Big/ D\right)^{\frac{1}{3}} \leq 0.15 \qquad [15.34.37]$$

and for an instantaneous release

$$\frac{(g'_o Q_o^{\frac{1}{3}})^{\frac{1}{2}}}{u_{ref}} = \frac{(g'_o Q_o/u_{ref}^2)^{\frac{1}{2}}}{Q_o^{\frac{1}{3}}} \leq 0.20 \qquad [15.34.38]$$

The justification of these criteria is given in Appendix A of the report.

For the case of an instantaneous release where the aspect ratio of the cloud is not unity, the suggestion made is that $Q_o^{\frac{1}{3}}$ be replaced by the initial height.

15.34.7 Thermodynamic and heat transfer effects

In the treatment so far described, it is assumed that the temperature of the release corresponds to ambient temperature, the relevant release temperature being that which pertains after the source effects have subsided and any aerosol has evaporated. If this is not the case, there are two additional effects, one thermodynamic and one of heat transfer, for which allowance needs to be made.

The first effect relates to the non-linear variation of temperature and density with volume fraction (or mole fraction) where the species have different molar specific heats. Considering the initial concentrations, the non-isothermal initial volume is V_o and the isothermal initial volume $V_o T_a/T_o$, where T_a and T_o are the absolute temperature of the air and the initial release, respectively, so that the isothermal concentration C_i and the non-isothermal concentration C_{ni} are

$$C_{ni} = \frac{V_o}{V_o + V_a} \qquad [15.34.39]$$

and

$$C_i = \frac{V_o T_a/T_o}{V_o T_a/T_o + V_a} \qquad [15.34.40]$$

where subscripts a, i and ni denote ambient, isothermal and non-isothermal, respectively. Hence

$$C_i = \frac{C_{ni}}{C_{ni}(1 - C_{ni})(T_a/T_o)} \qquad [15.34.41]$$

For example, for LNG at −162°C (111 K) and with ambient temperature of 17°C (290 K), a non-isothermal concentration of 0.05 in Equation 15.34.41 gives an isothermal concentration of 0.02.

The *Workbook* states that this effect is probably not significant for most hazardous materials, but it is for methane and hydrogen. It suggests that it be handled by determining the following two limiting cases (Cases A and B). For Case A the source density difference and volume, or volumetric flow, are used in the correlations. For Case B the density difference at the concentration of interest and the volume which satisfies mass conservation are obtained. A revised source density difference, source temperature and source volume, or volumetric flow, are then determined based on mass conservation, pure species at the source, adiabatic mixing and a constant molar specific heat equal to that of the ambient air, and used in the correlations.

These two limiting cases (A and B) are then compared. If the difference is small, the effect is unimportant. If it is neither large nor small, say a factor 2, the more pessimistic estimate is used. If it is large, say more than a factor of 2, further investigation may be warranted.

The second effect relates to heat transfer. This occurs by (a) air entrainment, (b) heat transfer at the surfaces of the cloud, particularly the lower surface, and (c) latent heat exchange due to condensation and re-evaporation of water vapour. There is some uncertainty as to the effect of such heat transfer. Although it is usually assumed that it works to reduce the downwind extent of the cloud, it is not certain that this is universally the case; there may be parameter ranges where the reverse could be so, particularly if the sole effect of heat transfer is on density. This is a reflection of the uncertainty concerning the relative effects of inhibition of vertical mixing and of increase in the top area of the cloud. The authors' correlations suggest that this may vary with the stability condition.

This effect is handled in the *Workbook* by determining the following two limiting cases (Cases A and B). For Case A the source density difference and volume, or volumetric flow, are used in the correlations. For Case B the conditions of Case A are taken as the starting point. It is then assumed that heat addition at the source raises the cloud temperature to the ambient temperature. The revised source density difference and volume, or volumetric flow, obtained from this temperature are then used in the correlations. These two limiting cases (A and B) are then compared, the approach being identical to that used for the thermodynamic effect.

15.34.8 Concentration vertical profile

This basic method provides an estimate of the alongwind concentration profile but not of the crosswind or vertical concentration profiles and implies by default that the concentration is uniform in both these directions.

For crosswind concentration, the Thorney Island trials confirm that the assumption of uniform concentration is reasonable near the source where density effects are dominant. However, further from the source the cloud becomes diffuse at the edges. A criterion for the development of these diffuse edges has been given by Britter and Snyder (1988). This is the condition $(g'_m h)^{\frac{1}{2}} \leq u^*$, where g'_m is the reduced gravity corresponding to the maximum concentration and h is a measure of plume height.

For a continuous release, the development of diffuse edges is due partly to true plume dilution and partly to meandering. For an instantaneous release, the equivalent to meandering is the difference between different members of an ensemble of trials at the same nominal conditions. It is necessary to decide whether the frame of reference is to be a fixed point one or a relative one

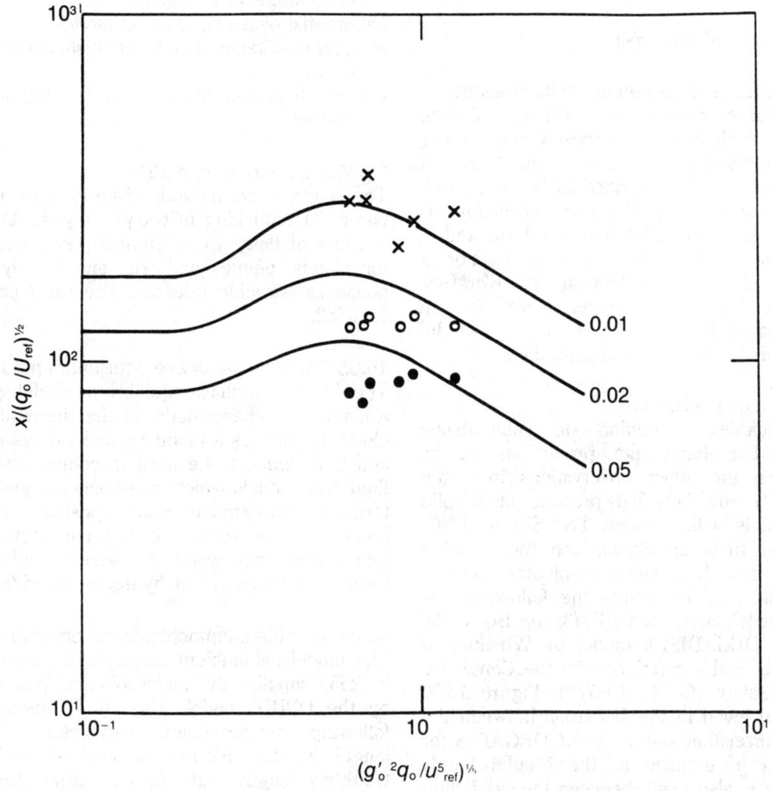

Figure 15.90 *Workbook method for dense gas dispersion – comparison of predictions of* Workbook *correlation with those of DEGADIS for Maplin Sands continuous propane spills (Britter and McQuaid, 1988; reproduced with permission)*

such as that obtained by overlaying the centres of mass of the ensemble members.

For vertical concentration, the trials show clearly that the profile is not uniform. The variation of the vertical concentration can be represented by the relation

$$C \propto \exp(-z^n) \qquad [15.34.42]$$

In the Gaussian models used for passive dispersion the profile is normal with $n = 2$. A value of $n = \infty$ corresponds to a uniform distribution. The *Workbook* suggests that a more appropriate value of n is about 1.5. It also states that where density effects are strong, there is evidence pointing to a still smaller value of n.

In any event, there is in dense gas dispersion a very rapid change in concentration close to the surface. Implications of this are that it is necessary to exercise care in specifying a ground level concentration and that such a concentration may not be that most relevant to hazard assessment.

15.34.9 Dispersion at buildings and obstacles

The treatment of the effect of buildings and obstacles in the *Workbook* is limited to releases upwind and releases in the immediate lee. The *Workbook* treatment of these two cases is described in Section 15.40. Broadly, the effect of a building or obstacle on a release upwind is to reduce the concentration at a given distance. The evidence is that, even close to the face, the upwind concentration is not increased and that downwind it is significantly reduced.

15.34.10 Dispersion in complex terrain

The treatment of complex terrain in the *Workbook* is limited, but consideration is given to two bounding cases. One is where the topographical feature is large compared to the scale of the release and the other where it is small. If the feature is large relative to the scale of the release, the topography will tend to reduce locally at the source to a relatively simple feature, typically a slope.

If the feature is small relative to the scale of the release, the dense gas will tend to flow around it. For the case where the cloud or plume is very wide compared with the feature, the *Workbook* gives the following criterion for the cloud to flow over or around the feature:

Flow over crest of feature:

$$\frac{u^2}{g'h} \gg 2\left(\frac{h_c}{h} - 1\right)$$

Flow around feature:

$$\frac{u^2}{g'h} \ll 2\left(\frac{h_c}{h} - 1\right)$$

where h_c is the height of the crest.

15.34.11 Concentration and concentration fluctuations

The *Workbook* discusses concentration and its definition and use in models, including the *Workbook* correlations, and considers concentration fluctuations. The *Workbook* treatment of concentration is described in Section 15.45. The concentration used in the *Workbook* correlations is defined in Section 15.34.2 for a continuous release and in Section 15.34.3 for an instantaneous release. For a continuous release the concentration in the *Workbook* correlations is the long-term average, whilst for an instantaneous release it is the ensemble average of the maximum short-time (0.6 s) mean concentrations.

15.34.12 Validation and application

The *Workbook* includes a section on comparisons between, on the one hand, the predictions of its correlations and, on the other, observations from the field trials on LNG and liquefied propane at Maplin Sands, the Burro trials at the Nevada Test Site on LNG, the Thorney Island trials on Freon, and the chlorine trials at Lyme Bay, and the predictions of other models for these trials. The models include the following: the Germeles–Drake model; SLAB and FEM3; the BG/C&W model; HEGADAS; DEGADIS; a model by Wheatley *et al.* (1987 SRD R438); and a correlation by the Center for Chemical Process Safety (CCPS, 1987/2). Figure 15.90 shows a comparison given in the *Workbook* between the predictions of its correlation and those of DEGADIS for continuous releases of propane in the Maplin Sands trials. A comparison is also made between the prediction of the correlations and the observed cloud dimensions in the Potchefstroom incident (Lonsdale, 1975).

In estimating the distance to the LFL, the *Workbook* correlations were generally in good agreement. However, the correlations are essentially for gas clouds at ambient temperature and are not intended to handle very cold clouds such as those formed by LNG. The predictions for the Burro trials were conservative, even after utilizing the cold gas correction to the isothermal correlation.

15.35 Dispersion of Dense Gas: DRIFT and Related Models

A family of dense gas dispersion models based on a set of fundamental conservation equations has been developed by the SRD. The work is described in the following reports: *Mathematical Modelling of Two-phase Release Phenomena in Hazard Analysis* (Webber *et al.*, 1992 SRD R584), *Description of Ambient Atmospheric Conditions for the Computer Code DRIFT* (Byrne *et al.*, 1992 SRD R553), *A Differential Phase Equilibrium Model for Clouds, Plumes and Jets* (Webber and Wren, 1993 SRD R552), *A Model of a Dispersing Dense Gas Cloud and the Computer Implementation DRIFT. I, Near-instantaneous Releases* (Webber *et al.*, 1992 SRD R586) and *A Model of a Dispersing Dense Gas Cloud and the Computer Implementation DRIFT. II, Steady Continuous Releases* (Webber *et al.*, 1992 SRD R587)

These reports may be summarized as

Source term	R584
Ambient atmospheric model	R553
Differential phase equilibrium model	R552
Integral dispersion model: near-instantaneous release	R586
Integral dispersion model: steady continuous release	R587

15.35.1 Source term model

The source term model is given in R584, which is in four parts: (1) modelling of two-phase jets; (2) modelling jets in ambient flow; (3) differential phase equilibrium model for clouds, plumes and jets; and (4) dynamics of two-phase catastrophic releases. The third part is the same as R552.

15.35.2 Differential phase equilibrium model

The differential phase equilibrium model given in R552 is a simple two-phase model of the thermodynamics of the cloud. It provides a module which covers this aspect only and is intended to be used in combination with a set of fluid flow models which cover aspects such as the source term, air entrainment and deposition. The model can handle dry or moist air and contaminants which are immiscible with water or water soluble. The report includes a treatment of hydrogen fluoride in moist air.

15.35.3 Ambient atmospheric conditions model

The model for ambient atmospheric conditions described in R553 supplies the meteorological parameters required by the DRIFT model. The DRIFT model requires the following meteorological parameters: the roughness length z_o, the friction velocity u^* and the Monin–Obukhov length scale L_s (or rather the reciprocal $1/L_s$). The two latter quantities are obtained from the ambient atmospheric conditions model.

This model gives three schemes for determining $1/L_s$. The first is the Monin–Obukhov scheme itself and the other two schemes are named after Pasquill and Holtslag. In the Pasquill scheme use is made of the relationship given by Golder (1972) between z_o and $1/L_s$, which has been described in Section 15.12. The graph given in Figure 15.45 by Golder is fitted by the following equation:

$$1/L_s = (1/L_{ref})[\log_{10}(z_o/z_{ref})] \quad [15.35.1]$$

where L_s is the length scale (m), z_o is the roughness length and L_{ref} (m) and z_{ref} (m) are the corresponding reference values. The two latter quantities are functions of the Pasquill stability class:

Pasquill stability category	L_{ref} (m)	z_{ref} (m)
A	33.162	1117
B	32.258	11.46
C	51.787	1.324
D	∞	–
E	−48.330	1.262
F	−31.325	19.36

The Holtslag scheme (Holtslag and van Ulden, 1983) involves the estimation of the sensible heat flux H, which is a function of the solar radiation and the albedo of the surface. The length scale L_s is obtained from Equation 15.12.46.

For all three schemes the friction velocity u^* is obtained from the wind speed at the height of interest using the similarity profile of Businger *et al.* (1971).

15.35.4 DRIFT: near-instantaneous release

The dense gas dispersion model itself is DRIFT (Dense Releases Involving Flammables or Toxics), an integrated model which is described in two reports: the near-instantaneous release version in R586 and the steady continuous release version in R587. The DRIFT models have similarities to the earlier SRD models DENZ and CRUNCH, but represent a significant advance on these. They incorporate both a more comprehensive thermodynamic module and improved models for the dispersion processes.

It is convenient to describe first the near-instantaneous model. The model uses the co-ordinate system (ξ, y, z) of Wheatley (1986 SRD R445), where y is the crosswind co-ordinate, z is the vertical co-ordinate and ξ is a co-ordinate described as the downwind co-ordinate relative to the skewed cloud axis. This allows for the case where the cloud is not a right cylinder but 'leans over'. The variable ξ is defined as

$$\xi = x' - \zeta(z - Z) \qquad [15.35.2]$$

with

$$x' = x - X \qquad [15.35.3]$$

where x' is a relative co-ordinate (m), X is the downwind co-ordinate of the cloud centre (m), Z is the vertical co-ordinate of the cloud centroid (m), ζ is a measure of the skewness of the cloud and ξ is a downwind co-ordinate relative to the skewed cloud axis (m).

A characteristic feature of the model is the way in which the air entrainment is modelled. Use is made of a general profile which comprehends both the tophat and Gaussian approaches. This means that it is possible to dispense with arbitrary definitions of the cloud boundary, such as a concentration which is one-tenth that of the centre line concentration and that there is seamless transfer between the dense gas and passive gas dispersion regimes.

The concentration is expressed as

$$C = C_m F_h(\xi, y) F_v(z) \qquad [15.35.4]$$

with

$$F_h(\xi, y) = \exp\{-[(\xi/a_1)^2 + (y/a_2)^2]^{w/2}\} \qquad [15.35.5]$$

$$F_v(z) = \exp[-(z/a_3)^s] \qquad [15.35.6]$$

where C is the concentration, C_m is the ground level centre concentration and a_1, a_2, a_3, s and w are parameters.

The DRIFT model contains a model of the atmosphere with profiles of the wind speed and the diffusion coefficient and relationships for the moisture content. Essentially, the model may be regarded as a framework which takes care of the basic conservation relations and into which can be slotted models for the cloud dynamics and cloud thermodynamics. The former includes models for lateral spreading, air entrainment and vertical diffusion, downwind travel, and so on.

For the cloud thermodynamics there is a basic homogeneous equilibrium model. Models are also provided covering a number of additional options such as moist air, contaminant immiscible with or miscible in water and for ammonia and hydrogen fluoride. It will be apparent from this description that the model has the ability to accommodate new models both for cloud dynamics and cloud thermodynamics as these may be developed.

15.35.5 DRIFT: steady continuous release

The second DRIFT model, for a steady continuous release, which is described in R587, is essentially similar, but utilizes variables representing mass flux rather than mass, and so on. Apart from this the two versions, and the reports R586 and R587, are very similar.

15.35.6 Validation and application

Accounts of the application and validation of DRIFT do not as yet seem to have appeared.

15.36 Dispersion of Dense Gas: Some Other Models and Reviews

Several other models, or sets of models, and reviews of models merit brief mention. The field trials referred to in this account are described in Sections 15.37 and 15.38.

15.36.1 Eidsvik model

Another early and influential box model was that by Eidsvik (1979, 1980). The model utilizes the basic box model Equations 15.24.24 and 15.24.25. The vertical entrainment velocity is taken to be a function of a characteristic vertical turbulent velocity and the Richardson number.

15.36.2 Fay and Ranck model

Another box model frequently mentioned is that by Fay and Ranck (1983). This again uses the same two basic box model equations. The vertical entrainment velocity is taken as the product of the friction velocity and a function of the Richardson number.

15.36.3 TNO models

Several models for dense gas dispersion have been developed at TNO. One of these is the model described by van Buijtenen (1980). This models the dispersion as an initial phase of slumping without entrainment of air followed by a phase in which gas is 'picked off' into the air from the top of the cloud. Its subsequent dispersion is modelled as passive dispersion from a number of sources on the top of the cloud. A further account of TNO dense gas dispersion models is given by Verhagen and van Buytenen (1986).

15.36.4 TRANSLOC

One of the early *K*-theory models for dense gas dispersion, SIGMET, has already been described. Another early *K*-theory model was TRANSLOC, produced at the Battelle Institute at Frankfurt, and described by Schnatz and Flothman (1980). Further information on TRANSLOC is given in the comparative review by Blackmore, Herman and Woodward (1982).

15.36.5 VKI models

The von Karman Institute (VKI) in Belgium has produced a number of models for dense gas dispersion. Early work was done by Foussat (1981). Van Dienst *et al.* (1986) have described subsequent work. Further development includes the BITC model referred to below.

15.36.6 BITC model

The Bureau International Technique du Chlore (BITC) and the VKI have developed a model for dense gas dispersion, with particular reference to chlorine. An account of the model is given by Vergison, van Dienst and Basler (1989). The BITC model is a three-dimensional one.

15.36.7 ZEPHYR

A further *K*-theory model is ZEPHYR, developed by Energy Resources Co. (ERCO) for Exxon Research and Engineering, as described by McBride (1981). The comparative review by Blackmore, Herman and Woodward (1982) gives further information on ZEPHYR. It is one of the models considered in the comparison of model predictions and results of the field trials at Matagordo Bay and Porton Down described by J.L. Woodward *et al.* (1982).

15.36.8 MARIAH and MARIAH-II

MARIAH is a *K*-theory model developed for Exxon Research and Engineering by Daygon-Ra. Descriptions have been given by Taft (1981) and Taft, Ryne and Weston (1983). MARIAH is one of the models included in the comparative review by Blackmore, Herman and Woodward (1982). The comparison given by J.L. Woodward *et al.* (1982) of model predictions with the results of the field trials at Matagordo Bay and Porton Down includes those of MARIAH.

Havens, Schreuers and Spicer (1987) have described an up-dated version, MARIAH-II, and given a comparison of its predictions with results of the Thorney Island field trials.

15.36.9 DENS20

DENS20 is an advanced slab model which has been described by Meroney (1984a). He gives comparisons of the prediction of the model with results both from wind tunnel experiments and from the Porton Down and Burro trials.

15.36.10 MERCURE-GL

The MERCURE-GL (GL = gaz lourd) model has been developed at Electricité de France (EDF) and described by Riou (1987, 1988) and Riou and Saab (1986). MERCURE-GL, which is a three-dimensional model, is a development of the non-hydrostatic model MERCURE (Caneill *et al.*, 1985). The accounts by Riou (1987, 1988) and Riou and Saab (1986) include comparisons of model predictions with results of the Thorney Island trials.

15.36.11 DISP2

DISP2 is a pair of advanced box models, one for an instantaneous release and one for a continuous release, developed at ICI. Accounts have been given by Fielding, Preston and Sinclair (1986) and Preston and Sinclair (1987). The latter describe the calibration of the model against the results of the Thorney Island trials.

15.36.12 CCPS correlation

The CCPS (1987/2) has obtained expressions for the dilution ratio V/V_o, including the following relation obtained from dimensional analysis:

$$\frac{V}{V_o} = f\left(\frac{x}{V_o^{\frac{1}{3}}}\right) \qquad [15.36.1]$$

where V is the volume of the cloud and subscript o denotes initial. They obtain a fit to the Thorney Island Phase I instantaneous releases with the correlation

$$\frac{V}{V_o} = \left(\frac{x}{V_o^{\frac{1}{3}}}\right)^{1.5} \qquad [15.36.2]$$

or

$$\frac{C}{C_o} = \left(\frac{x}{V_o^{\frac{1}{3}}}\right)^{-1.5} \qquad [15.36.3]$$

where C is the volumetric concentration.

15.36.13 Reviews of model characteristics

A number of reviews have been published in which the characteristics, as opposed to performance, of dense gas dispersion models are compared. These include reviews by Blackmore, Herman and Woodward (1982), Webber (1983 SRD R243), Wheatley and Webber (1984 CEC EUR 9592 EN) and the CCPS (1987/2).

Characteristics typically considered in such reviews include:

(1) Source elevation: ground level, elevated.
(2) Source dimensions: point, line, area, spreading pool.
(3) Source type: jet, plume, evaporating pool.
(4) Type of release: instantaneous, continuous, transient.
(5) Meteorological conditions: wind speed, including calm conditions; stability conditions; atmospheric humidity.
(6) Topography: flat, sloping terrain; open, obstructed terrain.
(7) Type of model: modified Gaussian, box, *K*-theory, other.
(8) Submodels of model: gravity-driven flow; momentum relations; cloud advection; air entrainment; heat transfer.
(9) Model outputs: cloud width, height; concentration profiles vs downwind distance, crosswind distance, time.

Reviews of the performance of dense gas dispersion models are considered in Section 15.41.

15.36.14 Webber generalized box model

Webber (1983 SRD R243) has given a generalized box model in the form of a set of dimensionless equations. The basic box model equations, corresponding to Equations 15.24.24 and 15.24.25, are given as

$$\frac{d\hat{r}}{dt} = \frac{1}{2\hat{r}} \qquad [15.36.4]$$

$$\frac{d\hat{v}}{dt} = \frac{2\hat{v}}{\hat{r}}\left(\frac{u_E t_o}{R_o}\right) + \hat{r}^2\left(\frac{u_T t_o}{H_o}\right) \qquad [15.36.5]$$

with

$$\hat{r} = R/R_o \quad [15.36.6]$$

$$\hat{h} = H/H_o \quad [15.36.7]$$

$$\hat{v} = V/V_o \quad [15.36.8]$$

$$\hat{t} = t/t_o \quad [15.36.9]$$

$$t_o = \frac{R_o}{2u_f(0)} \quad [15.36.10]$$

where H is the height of the cloud, R is the radius of the cloud, t is the time, t_o is an initial time scale, u_f is the cloud front velocity, V is the volume of the cloud, $\hat{h}$, $\hat{r}$, $\hat{t}$ and $\hat{v}$ are normalized values of H, R, t and V, respectively and o denotes initial.

Integration of Equation 15.36.4 gives

$$\hat{r} = (1 + \hat{t})^{\frac{1}{2}} \quad [15.36.11]$$

Dividing Equation 15.36.5 by Equation 15.36.4 yields

$$\frac{\mathrm{d}\hat{v}}{\mathrm{d}\hat{r}} = 2\alpha\hat{v}\hat{u}_E + 2\beta\hat{r}^3\hat{u}_T \quad [15.36.12]$$

with

$$\alpha = \frac{u_E}{u_f(0)} \quad [15.36.13a]$$

$$= 2t_o\frac{u_E(0)}{R_o} \quad [15.36.13b]$$

$$\beta = \frac{1}{2}\frac{u_T(0)}{u_f(0)}\frac{R_o}{H_o} \quad [15.36.14a]$$

$$= t_o\frac{u_T(0)}{H_o} \quad [15.36.14b]$$

$$\hat{u}_E = \frac{u_E}{u_E(0)} \quad [15.36.15]$$

$$\hat{u}_T = \frac{u_T}{u_T(0)} \quad [15.36.16]$$

where u_E is the edge entrainment velocity, u_T is the top entrainment velocity, $\hat{u}_E$ and $\hat{u}_T$ are normalized values of u_E and u_T, respectively, and α and β are parameters. Webber then classifies models according to the dependency of $\hat{u}_E$ and $\hat{u}_T$ on $\hat{v}$.

Table 15.45 *Some field trials on dense gas dispersion*

Location	*Date*	*Gas*	*Code name*	*Organization*[a]	*References*
A Refrigerated liquefied gas					
California	1966–67	Oxygen		Air Products	Lapin and Foster (1967)
	1968	LNG		AGA/TRW	Seargeant and Robinett (1968)
Various	1970–72	LNG		Bureau of Mines	Burgess, Murphy and Zabetakis (1970 BM RI 4105); Burgess, Biordi and Murphy (1972 BM RI 4177)
Matagordo Bay	1971	LNG		Esso/API	Feldbauer *et al.* (1972)
Nantes	1972	LNG		Gaz de France	Humbert-Basset and Montet (1972)
SS *Gadila*	1973	LNG	*Gadila*	Shell	Kneebone and Prew (1974)
Capistrano	1974	LNG		AGA/Battelle	Duffy, Gideon and Putnam (1974)
China Lake	1978	LNG	Avocet	US DOE/LLNL	Koopman, Bowman and Ermak (1980)
China Lake	1980	LNG	Burro	US DOE/LLNL	Koopman *et al.* (1982)
China Lake	1981	LNG	Coyote	US DOE/LLNL	
Maplin Sands	1980	Propane, LNG		Shell/NMI	Puttock, Blackmore and Colenbrander (1982)
NTS[b]	1987	LNG	Falcon	US DOT, GRI/ LLNL	T.C. Brown *et al.* (1989)
B Pressurized liquefied gas					
NTS[c]	1983	Ammonia	Desert Tortoise	USCG, TFI/LLNL	Goldwire (1985, 1986)
NTS[c]	1983	Nitrogen tetroxide	Eagle	USAF/LLNL	McRae *et al.* (1984); McRae (1986)
NTS	1986	Hydrogen fluoride	Goldfish	Amoco/LLNL	Blewitt, Yohn, Koopman and Brown (1987)
NTS	1988	Hydrogen fluoride	Hawk	HMAP	Diener (1991)
C Ambient pressure and temperature gas					
Netherlands	1973	Refrigerant 12		DGL	van Ulden (1974); Buschmann (1975)
Porton Down	1976–78	Refrigerant 12		HSE/CDE	Picknett (1981)
Thorney Island	1980–83	Refrigerant 12		HSE/NMI	McQuaid (1985b, 1987)

[a] AGA, American Gas Association; API, American Petroleum Institute; CDE, Chemical Defence Establishment, Porton Down; DGL, Directorate General of Labour, the Netherlands; GRI, Gas Research Institute; HMAP, Hydrogen Fluoride Mitigation and Assessment Program; HSE, Health and Safety Executive; LLNL, Lawrence Livermore National Laboratory; NMI, National Maritime Institute; TFI, The Fertilizer Institute; TRW, Thomson Ramo Woolridge; USAF, US Air Force; USCG, US Coast Guard; US DOE, US Department of Energy; US DOT, US Department of Transportation.
[b] Nevada Test Site.
[c] Area of NTS known as Frenchman Flat.

Table 15.46 *Experimental conditions in some field trials on dense gas dispersion*

Trial	*Date*	*Gas*	*Surface*	*Topography*	*Release*				
					Type	*No.*	*Volume* (m^3)	*Volumetric flow* (m^3/min)	*Duration* (min)
BM[a]	1970	LNG	Water		I[c]	51	0.04–0.5		
BM	1970	LNG	Water		C	4	0.2–0.3		
Esso/API[a]	1971	LNG	Sea		I	17	0.09–10.2		0.1–0.6
BM	1972	LNG	Water		C	13	0.2–1.3		≤10
Gaz de France	1972	LNG	Soil	Bund	I	40	≤3		
			Soil	Bund	C	1	0.16		4.5
Gadila	1973	LNG	Sea		C	6	27–198	2.7–19.8	10
Avocet	1978	LNG	Water			4	*c.* 4.5	4	
Burro	1980	LNG	Water			8	40	12–18	2.2–3.5
Coyote	1981	LNG	Water			15	3–28	6–19	0.2–2.3
Desert									
Tortoise	1983	Ammonia	Land			4	15–60	7.3–10.3	
Eagle	1983	Nitrogen tetroxide	Land			6	1.3–4.2	0.5–1.8	
Goldfish	1986	Hydrogen fluoride	Land						
Falcon	1987	LNG	Land	Fences			20–66		
Porton Down	See Table 15.47								
Maplin Sands	See Table 15.49								
Thorney Island	See Table 15.52								

Sources: Puttock, Blackmore and Colenbrander (1982); Koopman, Ermak and Chan (1989); Havens (1992); ***inter alia***. See also text.
[a] API, American Petroleum Institute; BM, Bureau of Mines.
[b] See also Table 15.48.
[c] C, continuous; I, instantaneous.

15.37 Dispersion of Dense Gas: Field Trials

As already indicated, there have been a large number of field trials conducted on the dispersion of dense gases. Accounts of such field trials have been given by Puttock, Blackmore and Colenbrander (1982), McQuaid (1984b), Koopman, Ermak and Chan (1989) and Havens (1992), and also by Blackmore, Herman and Woodward (1982) and Koopman *et al.* (1986).

A summary of these trials is given in Table 15.45 and details of some of the principal trials are given in Table 15.46. Details of the Porton Down, Maplin Sands and Thorney Island trials are given separately below.

15.37.1 Early work

A series of experiments on dispersion of chlorine over water were carried out at Lyme Bay in 1927 (McClure, 1927). A summary of these trials has been given by Wheatley *et al.* (1988 SRD R438).

In 1967, some chlorine dispersion experiments were conducted at the Chemical Defence Establishment (CDE), Porton Down, for ICI and BP. Other than this, there appears to have been little work on dense gas dispersion prior to 1970.

In 1966–67, Air Products carried out experiments in which liquid oxygen was spilled into a bund and a steady continuous release maintained using a water spray and the oxygen concentration in the gas plume was measured (Lapin and Foster, 1967).

Some 18 experiments in which about 0.2 m^3 of LNG was spilled into a 2 m^2 bund with three types of surface were carried out in 1968 for the American Gas Association (AGA) by Thomson Ramo Woolridge (TRW) (Seargeant and Robinett, 1968). The vaporization rate decreased over the course of the experiment. Measurements were taken of the methane concentration in the plume. The correlation presented gives the concentration decreasing proportionately with downwind distance.

15.37.2 Bureau of Mines trials

The Bureau of Mines (BM) performed a series of trials in 1970–72 involving the vaporization and dispersion of LNG following spillage on water, as reported by Burgess and co-workers (Burgess, Murphy and Zabetakis, 1970 BM RI 7448, S 4105; Burgess, Biordi and Murphy, 1972 BM PMSRS 4177). In the 1970 work, some 51 trials were conducted in which LNG was spilled instantaneously on water and the pool spread and vaporization were measured; measurements of gas dispersion were not taken. Four trials were also carried out with continuous spillage of LNG and in these the gas dispersion was measured. The Pasquill–Gifford relations were used to model the dispersion, the data being correlated by taking the ratio σ_y/σ_z as 5, indicating a relatively flat cloud. A further 13 continuous spillage experiments were done in 1972, of which six were usable. The ratio σ_y/σ_z was again high.

Figure 15.91 *SS* Gadila *trials on dense gas dispersion at sea – plume from continuous spill of LNG in trial 4 (Kneebone and Prew, 1974) (Courtesy of Gastech)*

15.37.3 Esso/API trials

In 1971, a series of spillages of LNG onto the sea were carried out in Matagordo Bay, Texas, by Esso under the auspices of the AGA (Feldbauer *et al.*, 1972). The concentration of methane in the cloud was measured. There were ten smaller experiments of about 0.9 m^3 and seven larger ones of between 2.5 and 10.2 m^3; the former were virtually instantaneous and the latter nearly so. Again high values of the ratio σ_y/σ_z in the Pasquill–Gifford model were obtained. Downwind concentrations obtained in these tests are shown in Figure A7.3.

15.37.4 Gaz de France trials

A series of 40 experiments involving the instantaneous spillage of LNG into a bund were performed in 1972 at Nantes by Gaz de France to investigate vaporization, dispersion and combustion (Humbert-Basset and Montet, 1972). Most of the tests involved the spillage of 3 m^3 into bunds with areas 9–200 m^3. The vaporization rate decreased with time. There was also a test with a continuous release of 10 m^3/h.

These trials showed several features of interest. Following the spillage the vaporization was initially very rapid but then fell to a lower, plateau value. The initial visible cloud was observed to separate from the steady plume. The flammable region was always contained within the visible cloud. In a number of tests the cloud tended to lift off the ground in its later stages.

15.37.5 SS *Gadila* trials

In 1973, larger scale trials of spillage of LNG onto the sea were performed by Shell from the SS *Gadila*, a LNG carrier (Kneebone and Prew, 1974; te Riele, 1977). The vessel was being commissioned and the purpose of the trials was to confirm recommended procedures for jettisoning cargo in an emergency. The study of gas dispersion was not a main intent, but photographs were taken of the cloud. There were six tests lasting some 10 min and involving releases of up to 200 m^3. The liquid was ejected from a nozzle at the stern 18 m above the water. It is believed that vaporization was complete before the jet reached the water surface. In two of the tests the ship was stationary and in the others it was moving at up to 10.5 knots. The wind speed at 30 m was 1.9–5.1 m/s with a stable atmosphere. The relative humidity was 80–85%.

Two trials, tests 4 and 6, have proved of particular interest. In test 4 the flow was 7.6 m^3/min (53.5 kg/s) of LNG with a wind speed of 5.1 m/s. The visible cloud was 1370 m long, although with considerable inhomogeneity beyond 1000 m. The plume had a maximum width of 300 m at 750 m downwind. Its height was about 8–10 m. This test is shown in Figure 15.91.

In test 6 the flow was 19.3 m^3/h (136.5 kg/s) with a wind speed of 3.9 m/s. The plume was more coherent and had a visible length of 2250 m, which was still increasing at the end of the test. It had a maximum continuous width of 550 m and a height of about 10–12 m. The outline of the plume is shown in Figure 15.92, together with the predictions of the models of Cox and Carpenter, CRUNCH and te Riele.

15.37.6 Dutch Freon 12 trial

The first trial involving the virtually instantaneous release of a dense gas at atmospheric pressure and temperature, as opposed to an evaporating liquefied gas, was conducted in 1973 by the Directorate General of Labour (DGL) of the Dutch Ministry of Social Affairs. This experiment has been described by van Ulden (1974) and Buschmann (1975), and also by te Riele (1977). The release was 1000 kg of Freon 12 which has a relative density of 4.2. The wind speed at 10 m was 3 m/s and the stability was neutral.

On release there was rapid and intensive mixing with air and after some 5 s the estimated volume of the gas–air mixture was 2400 m^3. The shape of the cloud was roughly cylindrical with some elongation along the direction of the wind. Its height fell from an initial value of 5 m to a minimum of 0.2 m before increasing again, reaching 10 m at 1000 m downwind. Its radius increased rapidly with distance at first with the rate of increase then reducing.

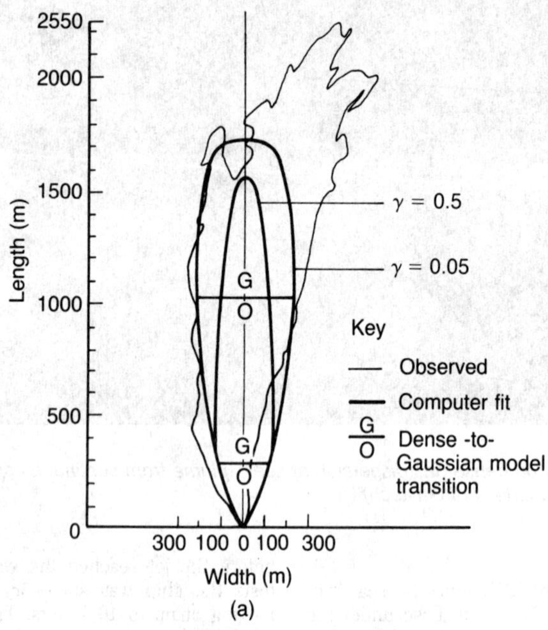

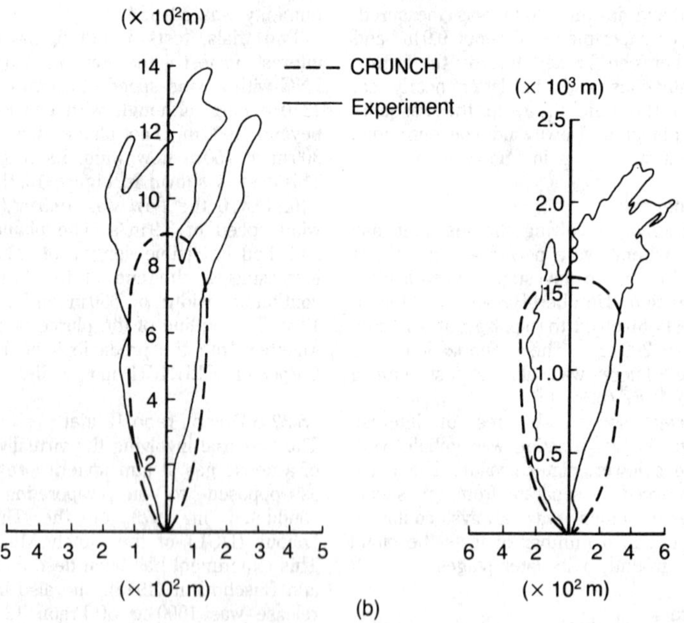

Figure 15.92 *SS* Gadila *trials on dense gas dispersion at sea – plume from continuous spill of LNG and comparison of observed cloud contour with model predictions of contour: (a) prediction of Cox and Carpenter model for trial 6 (R.A. Cox and Carpenter, 1980) Courtesy of Reidel Publishing Company); (b) prediction of CRUNCH for trials 4 and 6 (Jagger 1983 SRD R229) Courtesy of UKAEA Safety and Reliability Directorate); (c) prediction of te Riele model for trials 4 and 6 (te Riele, 1977) (Courtesy of DECHEMA)*

The height and radius of the cloud are shown in Figure 15.76. The radius is much greater and the height much less than predicted for passive dispersion of a cloud of neutral buoyancy.

15.37.7 AGA/Battelle trials

In 1974, further experiments on the spillage of LNG into bunds were conducted near Capistrano, California, by Battelle Columbus Laboratories sponsored by the AGA

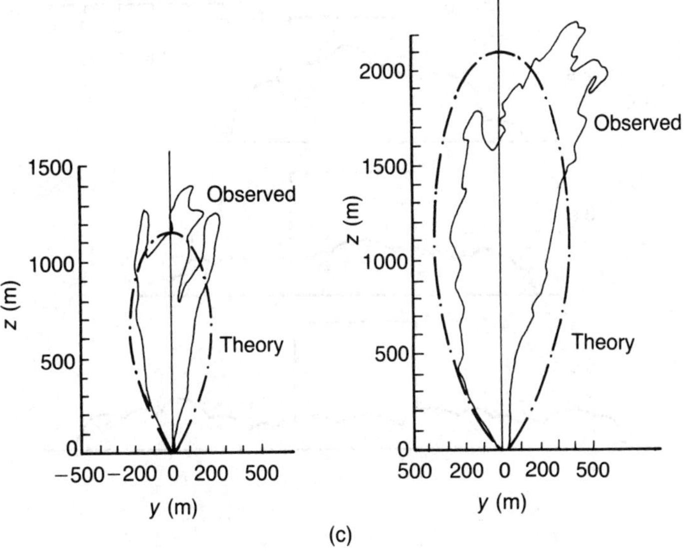

Figure 15.92 continued

***Table 15.47** Experimental conditions in some field trials on dense gas dispersion: Porton Down (after Picknett, 1981)*

Trial No.	*Initial density relative to air*	*Site*	*Roughness length (mm)*	*Wind speed (m/s)*	*Pasquill stability category*	*Comment*
6	1.7	Flat	2	2.8	C–D	
8	2.0	Flat	2	<0.5	F–G	Calm after release. Cloud remained in layer on ground for 20 min
10	2.5	Sloping	2	0.4	F	Calm allowed long period of gravity flow downhill
11	2.4	Sloping	2	2.0	E	
12	2.4	Flat	20	5.7	C	
15	2.0	Sloping	2	6.4	D	Dispersion in uphill wind
24	1.9	Flat	150	2.5	B	Dispersion over very long grass
29	4.2	Flat	2	4.3	C	Initially 100% dense vapour
36	1.5	Flat	10	4.7	C–D	
39	1.03	Flat	10	1.8	E–F	Initial density near neutral

(Duffy, Gideon and Puttnam, 1974). There were 42 trials, of which 14 involved ignition and 28 dispersion. Of the latter, 17 were with a 2 m bund, 9 with a 6 m bund and 2 with a 24 m bund. The test approximated an instantaneous spill, the release being made over 20–30 s. The results were marred by a tendency for the liquid to spill outside the bund. A correlation was derived for maximum downwind concentration in terms of bund area and wind velocity.

15.37.8 Porton Down trials

In 1976–78, a series of trials were conducted at the Chemical Defence Establishment for the HSE involving the virtually instantaneous release of a dense gas at atmospheric pressure and temperature. These have been described by Picknett (1981), and also by McQuaid (1979b).

Five sites were used: two flat with short grass, one flat with rough grass and two sloping with short grass. The source was a 40 m^3 collapsible tent. The gas released was a mixture of Refrigerant 12 and air, with relative densities varying from 1.03 to 4.2. Coloured smoke was also added to make the cloud more visible.

Selected experiments in the Porton trials are given in Table 15.47. The experiments were designed to yield pairs of trials which differed in one feature only and thus

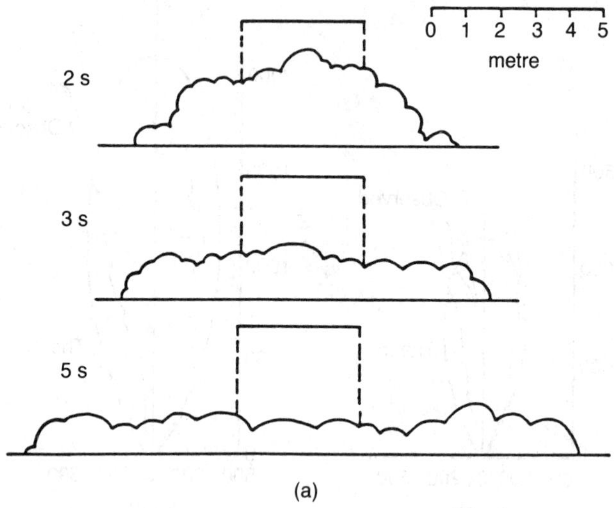

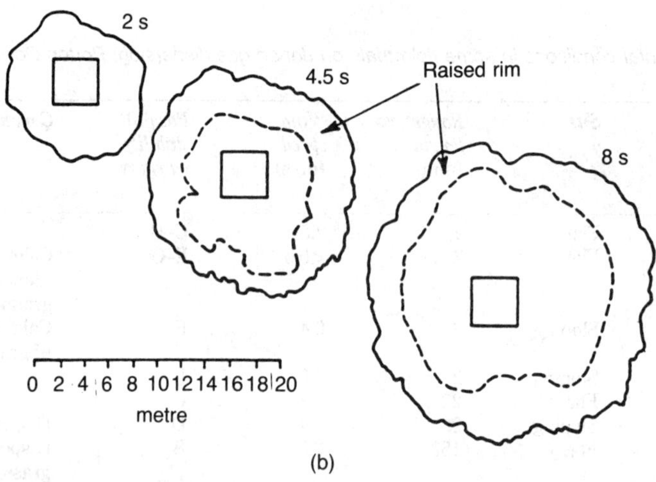

Figure 15.93 *Porton Down trials on dense gas dispersion with Refrigerant 12 – growth of cloud in trial 8 – (Picknett, 1981): (a) elevation view; (b) plan view (Courtesy of Pergamon Press)*

demonstrated the effect of that feature. Picknett gives tables of such trial pairs showing the effect of relative density, roughness length, ground slope and wind speed.

Picknett describes the behaviour of the dense gas in these trials. The effect of density is seen in trial 8 which was conducted in still air. The cloud kept its original shape momentarily after the collapse of the tent and then gravity caused an accelerating flow downwards to the ground, followed by a billowing, outwards motion involving considerable mixing with air. The cloud then expanded as a horizontal disc but with a raised annular rim. This rim was highly turbulent and persisted for some seconds before the turbulence and the rim itself collapsed. These features are illustrated in Figure 15.93. There followed a phase in which the cloud height remained virtually constant, but the area of the cloud grew almost linearly, as shown in Figure 15.94.

The effect of wind, as in trial 36, was to cause the cloud to spread initially upwind somewhat and to assume downwind a horseshoe shape, as shown in Figure 15.95. At first the width exceeded the length, but this then reversed. Initially the cloud velocity was much less than the wind speed, but the cloud underwent a gradual acceleration.

The initial violent motion of the cloud caused rapid dilution. In trial 8 the gas concentration fell from an

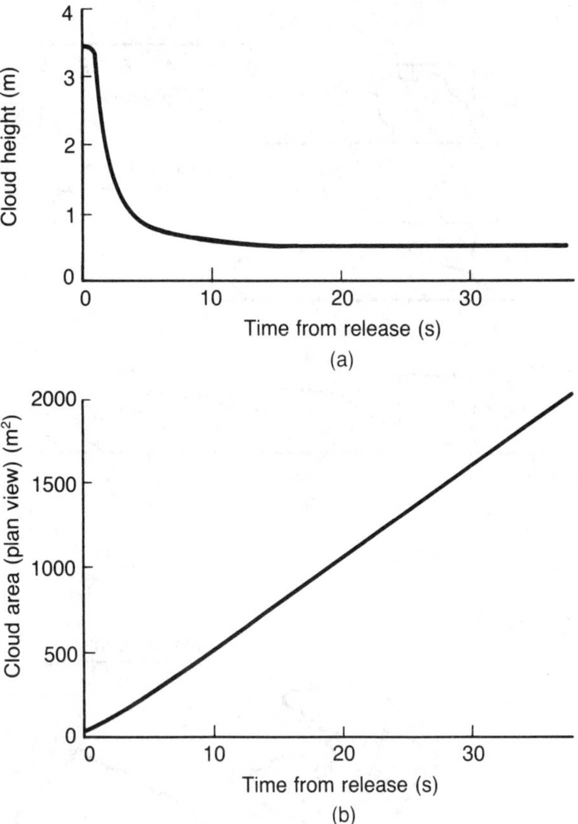

Figure 15.94 *Porton Down trials on dense gas dispersion with Refrigerant 12 – growth of cloud in trial 8 (Picknett, 1981) (Courtesy of Pergamon Press)*

initial value of 30% to 3% in 6 s. In wind the rate of dilution was even higher.

The concentration of gas in the cloud initially was very inhomogeneous. In trial 8 in still air the concentration measured near the ground fluctuated by a factor of 2 or more within 10 s. In trials 6 and 12 with wind the behaviour of the concentration varied with the measurement distance. In trial 6, at a point near the source the vertical concentration profile at 5 s was concave, but eventually a profile was established in which the concentration decreased monotonically with height, as shown in Figure 15.96(a). In trial 12, at a distance of 25 m the profile at 10 s was convex, but again in due course a monotonically decreasing profile was established, as shown in Figure 15.96(b).

The effect of slope was very large for a release in still air. In wind the effect was less marked. The main feature was that with wind blowing uphill and the cloud flowing downhill, the two opposing effects suppressed the formation of the raised rim and gave instead a tongue flowing along the ground and a reverse, uphill flow above the tongue. The cloud soon became dilute enough for the wind to carry it uphill.

15.37.9 China Lake trials: Avocet, Burro and Coyote series

The Naval Weapons Center (NWC) test facility at China Lake, California, has been the site of several series of dense gas dispersion trials. Accounts of the facility have been given by Ermak *et al.* (1983). Much of the work at this site has been sponsored by the Department of Energy (DoEn) and conducted by the Lawrence Livermore National Laboratory (LLNL).

The Avocet series was a set of four trials involving spillage onto water of some 4.5 m^3 of LNG over about 1 min. The main purpose of these trials, described by Koopman, Bowman and Ermak (1980), was the development of the test facility.

In 1980, the Burro series of trials was performed involving the continuous release onto water of up to 40 m^3 of LNG over periods up to 3.5 min. These tests are therefore classed as continuous releases. They have been described by Ermak, Chan *et al.* (1982) and Koopman *et al.* (1982). Selected trials are summarized in Table 15.48.

The experiments were well instrumented and obtained a fairly comprehensive set of measurements, including turbulence and concentration fluctuations in the plume.

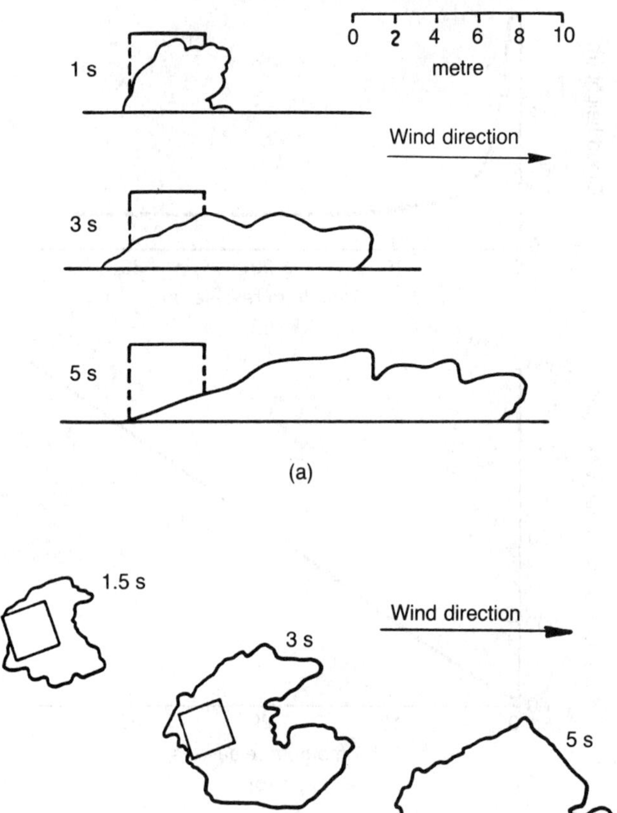

Figure 15.95 *Porton Down trials on dense gas dispersion with Refrigerant 12 – growth of cloud in trial 36 (Picknett, 1981): (a) elevation view; (b) plan view (Courtesy of Pergamon Press)*

Figure 15.97(a) shows the plume in Burro 8 after 30 s and Figure 15.97(b) the plume in Burro 6 after quasi-steady state had been reached. Analyses of the results of the Burro trials have been given by Koopman *et al.* (1982) and by Ermak, Chan *et al.* (1982). The latter have given comparisons of the results with predictions from the Germeles–Drake model and from SLAB and FEM3.

Further analyses have been reported by Koopman, Ermak and Chan (1989). Concentration contours in the Burro trial 8 are shown in Figure 15.98. Figure 15.98(a) gives the observed crosswind elevation view contours at a downwind distance of 140 m at 180 s. Figures 15.98(b) and (c) give the predictions of FEM3 for flat terrain and variable terrain, respectively. Figures 15.98(d) and (e) show the observed and predicted plan view contours.

The Burro tests were followed in 1981 by the Coyote series which were concerned primarily with rapid phase transition (RPT) and with ignition.

15.37.10 Maplin Sands trials

In 1980, a series of trials was carried out by Shell at Maplin Sands. Accounts of the work has been given by Puttock and co-workers (Puttock, Blackmore and Colenbrander, 1982; Puttock, Colenbrander and Blackmore, 1983, 1984; Colenbrander and Puttock, 1984).

Figure 15.99 shows the site used for the trials. The liquefied gases were released onto water from a pipeline terminating in a downwards-pointing pipe and measurements were taken by instruments located on pontoons placed in arcs around the release point.

Two series of trials were carried out, one with propane and one with LNG. The experiments were mainly continuous spills but, for both gases, some trials with instantaneous spills were conducted. A total of 34 trials were performed, of which 11 involved ignition. Table 15.49 gives details of some of the trials selected by the authors as providing results useful for dispersion work.

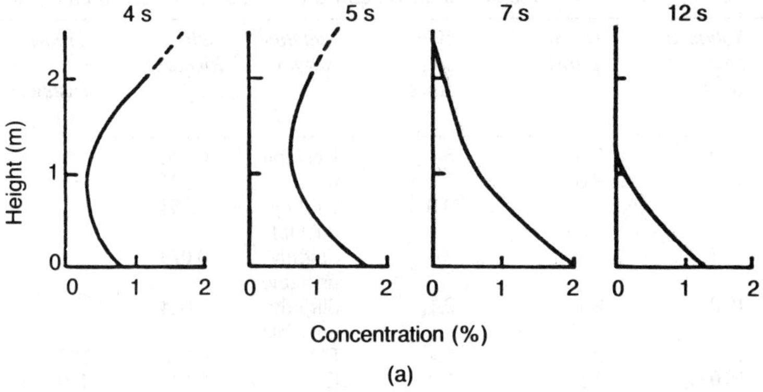

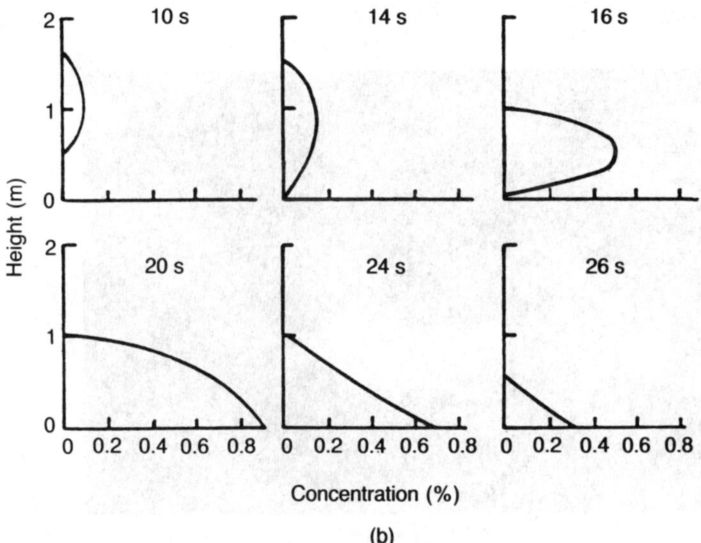

Figure 15.96 *Porton Down trials on dense gas dispersion with Refrigerant 12 – vertical concentration profiles in trials 6 and 12 (Picknett, 1981); (a) at point close to source in trial 6; (b) at 25 m downwind in trial 12 (Courtesy of Pergamon Press)*

The propane trials with continuous spills, of which there were 11, have been described by Puttock, Colenbrander and Blackmore (1983). The two trials not shown in Table 15.49 are trial 42, an underwater release, and trial 50, an ignited release. Two experiments in particular are described. In trial 46 the wind speed was 8.1 m/s and in trial 54 3.8 m/s. Plates 5(a) and (b) show the plumes from these trials. Figures 15.100(a) and (b) show the corresponding comparisons between the outline of the observed plume and the contour predicted by HEGADAS.

An account of the three propane trials with instantaneous spillage has been given by Puttock, Blackmore and Colenbrander (1982). Trial 60 was performed with a very low wind speed of 1.2 m/s and gave an almost circular cloud. In trial 63 the wind speed was 3.5 m/s, and thus also fairly low. The cloud started off roughly circular and very shallow, then moved off with the wind, with a raised head around the edge, and growing in height as it moved. Plate 6 shows one of these trials.

The LNG trials have been described by Puttock Blackmore and Colenbrander (1982), Puttock, Colenbrander and Blackmore (1984) and Puttock (1987c). There were 13 continuous spills, of which the authors describe three in particular: trials 15, 29 and 56. Plates 7(a) and (b) show the plumes in trials 29 and 56 with wind speeds of 5 m/s and 4.8 m/s, respectively, and Figure 15.101 shows a comparison for the latter between the outline of the observed plume and the contour predicted by HEGADAS.

Of the instantaneous LNG spills, trials 22 and 23 were considered of particular value. The wind speed in these trials was 5 m/s and thus higher than in the propane instantaneous spills. Compared with the latter the cloud

Table 15.48 Experimental conditions in some field trials on dense gas dispersion: China Lake, Burro series

Trial no.	*Volume released* (m^3)	*Volumetric flow* (m^3/min)	*Duration*[a] *(min)*	*Wind speed* (m/s)	*Stability category*	*Bulk Richardson* No.[b]	*Travel*	
							Distance (m)	*Concentration* (5)
2	34.3	11.9	2.9	5.4	Unstable	−0.178		
3	34.0	12.2	2.8	5.4	C	−0.221	255	LFL
4	35.3	12.1	2.9	9.0	Slightly unstable	−0.054		
5	35.8	11.3	3.2	7.4	Slightly unstable	−0.079		
6	27.5	12.8	2.1	9.1	Slightly unstable	−0.044		
7	39.4	13.6	2.9	8.4	D	−0.018	200	LFL
8	28.4	16.0	1.8	1.8	E	+0.121	420	LFL
9	24.2	18.4	1.3	5.7	D	−0.014	325	LFL

Sources: Ermak, Chan *et al.* (1982); Koopman *et al.* (1982); Havens (1992).
[a] Calculated from volume and volumetric flow.
[b] Richardson No. at 2 m height.

Figure 15.97 Nevada Test Site trials, Burro series, on dense gas dispersion with LNG (Ermak, Chan et al., 1982): (a) plume in Burro trial 8 after 30 s; (b) plume in Burro trial 6 at quasi-steady state (Courtesy of Elsevier Science Publishers)

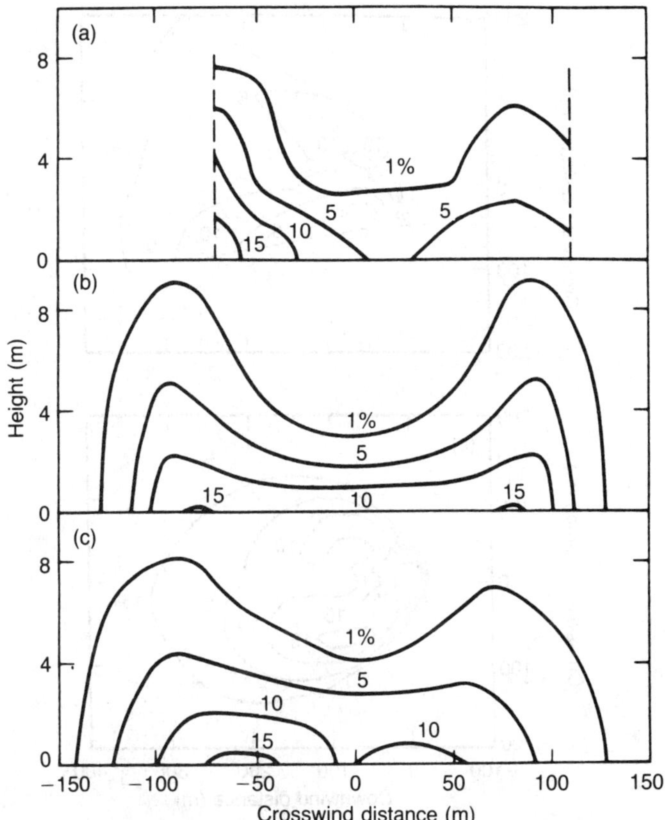

Figure 15.98 *Nevada Test Site trials, Burro series, on dense gas dispersion with LNG (Koopman, Ermak and Chan, 1989) – trial 8: (a) observed crosswind elevation view of concentration contours 140 m downwind after 180 s; (b) predictions of FEM3 of crosswind elevation view of concentration contours 140 m downwind after 180 s with flat terrain; (c) predictions of FEM3 of crosswind elevation view of concentration contours 140 m downwind after 180 s with variable terrain; (d) observed plan view of concentration contours at 1 m above ground level at 180 s; (e) predictions of FEM3 of plan view of concentration contours at 1 m above ground level at 180 s with variable terrain (Courtesy of Elsevier Science Publishers)*

produced was very elongated and more like that from a continuous spill. Figure 15.102 shows the development of the cloud in trial 22 together with a comparison of the contour predicted by HEGADAS.

15.37.11 Thorney Island trials

The Thorney Island trials were a progression from the Porton Down trials and again involved release of a dense gas at atmospheric pressure and temperature. Most of the releases were instantaneous. The second series involved obstructions and a building. A third series was done with continuous releases. The Thorney Island trials are described in Section 15.38.

15.37.12 NTS trials: Desert Tortoise and Eagle series

The experimental work described so far has been concerned with dispersion of dense gases in general and of hydrocarbons in particular. More recently, there has been growing interest in other hazardous materials, especially toxic liquefied gases.

Much of this work has been done at the Nevada Test Site (NTS). This site includes the area of Frenchman Flat which is the location of the Liquefied Gaseous Fuels Spill Test Facility (LGFSTF). This facility, and some of the test programmes conducted there, have been described by Leone (1990), and also by Koopman, Ermak and Chan (1989).

In 1983, the Desert Tortoise series of trials involving liquid ammonia were carried out at the NTS by the LLNL for the US Coast Guard and The Fertilizer Institute (TFI). These have been described by Goldwire (1986) and Koopman *et al.* (1986), and also by Koopman, Ermak and Chan (1989). A summary of the Desert Tortoise trials is given in Table 15.50. There were four trials, all approximating continuous spills.

The accounts concentrate mainly on the largest release, trial 4. In this experiment 60 m^3 of liquid

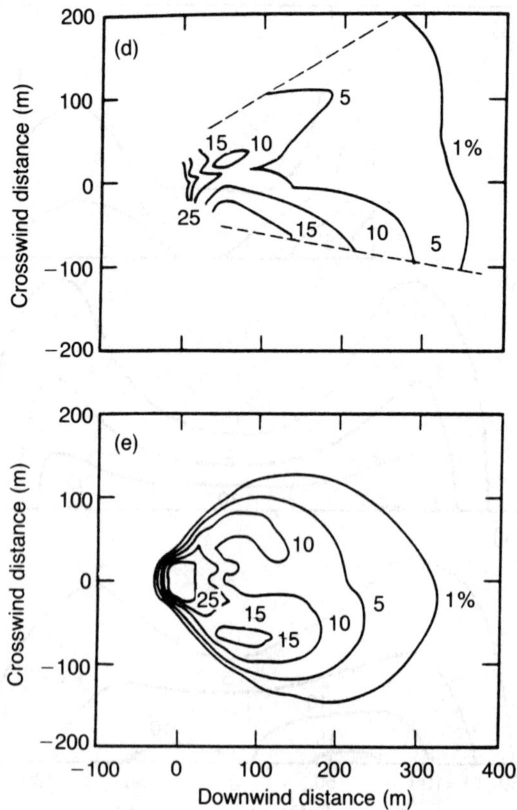

Figure 15.98 continued

ammonia at 171 psia and 24°C was released at a rate of 9.5 m^3/min into air at 30.8°C and relative humidity 21% with a wind speed of 4.5 m/s and stability category E. The release was from a horizontal pipe 0.8 m above the ground surface and pointing downwind. The adiabatic flash was estimated to give 17% vapour and 83% spray.

Most of the ammonia remained airborne as vapour and spray but, in contrast to the smaller releases, a large pool of liquid formed and persisted for about 8 min. The pool temperature was measured as −52°C at 3 m and −63°C at 9 m downwind. Mass balances on the plume showed that the proportion of material released which passed the 100 m and 800 m arcs was about 70%; 30% was unaccounted for. At the 100 m point, small puffs of boiled-off vapour were seen on and off at least 1100 s after the start.

The jet persisted beyond the 100 m arc. The movement of the plume was dominated by the jet and dense gas features. At the 100 m arc the plume was about 70 m wide and generally less than 6 m high. At the 800 m arc the width had increased to about 400 m, but the height was still about 6 m. The ammonia fog persisted beyond 400 m.

The values given for the maximum gas concentrations vary slightly. Koopman *et al.* state that this concentration was 10% at 100 m, 1.6% at 800 m and 0.53% at 2800 m. The crosswind elevation of the concentration contours at the 800 m arc given by Goldwire is shown in Figure 15.103.

The boiling point of ammonia is −33.4°C. The temperature of the plume rose to −7°C at 100 m and to 23°C at 800 m, compared with the ambient air temperature of 32.8°C. The surface heat flux was determined as 2.2 kW/m^2.

The Eagle trial series, performed by LLNL for the US Air Force, involved spillages of nitrogen tetroxide and followed straight on from the Desert Tortoise series. It has been described by McRae and co-workers (McRae *et al.*, 1984; McRae, 1986).

15.37.13 NTS trials: Goldfish series

In 1986 the Goldfish series of trials involving liquid hydrogen fluoride was carried out at the Frenchman Flat area of the NTS for Amoco by LLNL and Amoco. An account of these has been given by Blewitt *et al.* (1987a), and also by Koopman, Ermak and Chan (1989).

A summary of the Goldfish trials is given in Table 15.51. There were six trials, all approximating continuous spills, three concerned with dispersion and three with water spray barriers. Of the former group, the first was an exploratory test. The main dispersion trials, therefore, were trials 2 and 3.

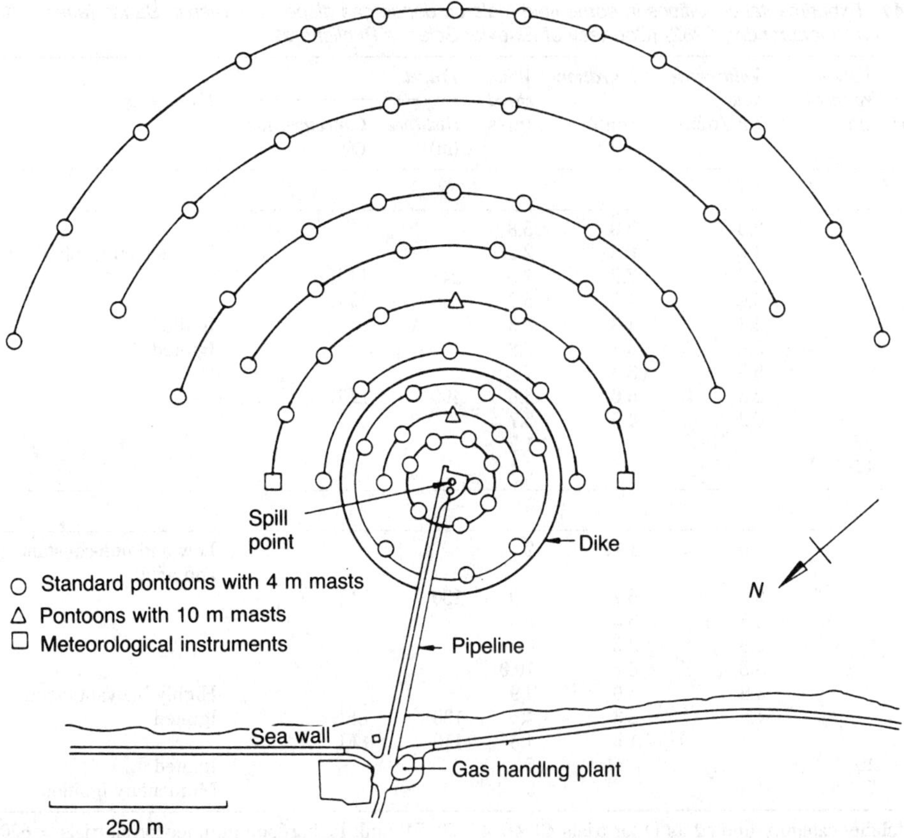

Figure 15.99 *Maplin Sands trials on dense gas dispersion – site layout (Puttock, Blackmore and Colenbrander, 1982) (Courtesy of Elsevier Science Publishers)*

The liquid hydrogen fluoride was released as a horizontal jet. The adiabatic flash was estimated to give some 20% vapour and 80% spray, although the thermodynamics of hydrogen fluoride are complex. In none of these tests was any liquid collected in the spill pond. Special facilities were provided to give a capability to increase the humidity of the ambient air. These were located upwind and consisted of a large pond and of an array of nozzles half operating on steam and half on water spray.

Trial 2 was performed at low relative humidity, the dewpoint temperature being 1.1°C. In this experiment liquid hydrogen fluoride at 115 psig and 38°C was released at a rate of 175 US gal/min (0.011 m^3/s) into the ambient air at 36°C with a wind speed of 4.2 m/s and stability category D. The crosswind elevation of the concentration contours at 300 m at 54 s and 121 s are shown in Figures 15.104(a) and (b).

Trial 3 differed in that the relative humidity was raised artificially to give a dewpoint temperature of 6.6°C. In this experiment the wind speed was 5.4 m/s and the air ambient was at 26.5°C, the other conditions being similar to those mentioned above for trial 2. The plume was visible at 400–500 m. The crosswind elevation concentration contours at 300 m at 113 and 225 s and at 1000 m at 292 and 358 s are shown in Figures 15.104(c)–(f).

The boiling point of hydrogen fluoride is 20°C. In trial 2 the temperature of the plume at 20 m downwind dropped within seconds to about −30°C below ambient temperature, then fell more gradually to about −45°C below over 350 s, whilst in trial 3 it fell sharply to about −48°C below and remained at about −45°C below again for about 350 s. In both trials the plume temperature then rose rapidly, stayed at about −10°C below until about 600 s and then rose to eliminate the temperature differential by about 700 s.

Trials 4–6 on water spray barriers are described in Section 15.53.

15.37.14 NTS trials: Hawk series

In 1988 another series of trials, the Hawk series, was carried out on hydrogen fluoride at the Nevada Test Site for the Industry Co-operative Hydrogen Fluoride Mitigation and Assessment Program (ICHMAP, or HMAP). These trials investigated not only dense gas dispersion but also the use of vapour barriers and water spray barriers. An account has been given by Diener (1991). The Goldfish trials caused some concern due to

Table 15.49 Experimental conditions in some field trials on dense gas dispersion: Maplin Sands (after Puttock, Blackmore and Colenbrander, 1982) (Courtesy of Elsevier Science Publishers)

Trial No.	Trial Type	Volume released (m^3)	Volumetric flow (m^3/min)	Duration (min)	Wind speed (m/s)	Travel Distance (m)	Travel Concentration (%)	Comments
A Propane[a]								
43	C		2.3	5.8	5.8			
45	C		4.6	6.8	2.2			Unusual atmospheric conditions
46	C		2.8	7.8	7.9	245	LFL	
47	C		3.9	4.5	5.2	340	LFL	
50	C		4.3	4.0	8.3			Ignited
51	C		5.6	3.7	7.8			Ignited
52	C		5.3	3.3	7			
54	C		2.3	5.0	3.6	400	LFL	
55	C		5.2	4.0	5.7			
60	I	27			1.2			
63	I	17			3.4			
B LNG[b]								
12	C		1.0	10.7	2			Low and non-constant spill rate and wind
15	C		2.7	6.7	3.9	150	LFL	
29	C		3.5	5.4	5			
34	C		2.9	3.5	8.5			
35	C		3.8	4.5	10.0			
37	C		3.9	5.0	4.9			Highly buoyant cloud
39	C		4.5	2.3	4.5	130	LFL	Ignited
56	C		2.5	1.5	4.8	110	LFL	
22	I	10			5			Ignited
23	I	7			5			Momentary ignition

[a] Pasquill stability category quoted as D for trials 43, 46, 47, 50, 54; bulk Richardson number for all trials = 600.
[b] Pasquill stability category quoted as D for trials 29, 34, 35, 39, 56; bulk Richardson number for all trials = 40.

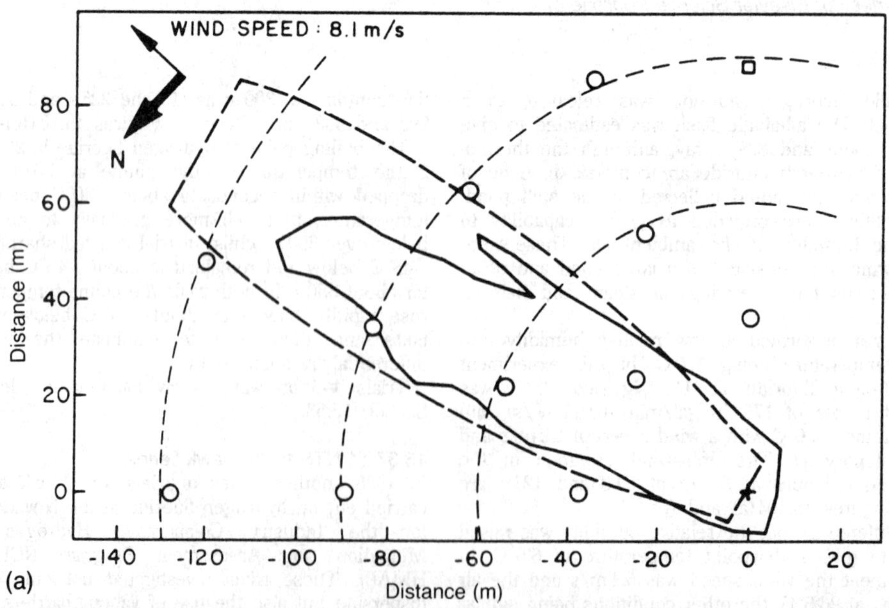

Figure 15.100 Maplin Sands trials on dense gas dispersion – continuous spills of propane: comparison of observed plumes with limits of visible plumes predicted by HEGADAS (Puttock, Colenbrander and Blackmore, 1983). (a) Trial 46 (wind speed = 8.1 m/s); (b) trial 54 (wind speed 3.8 m/s) (Courtesy of Reidel Publishing Company)

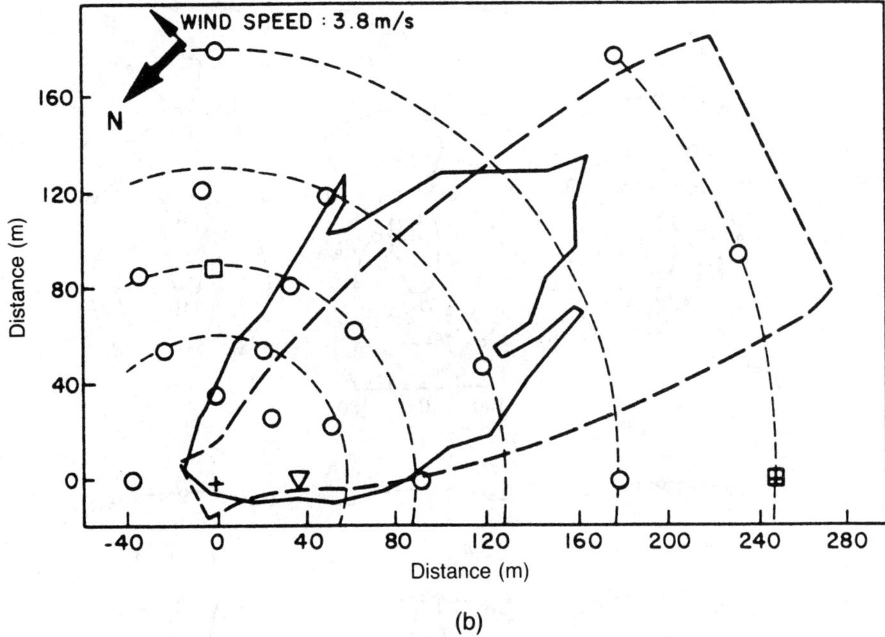

(b)

Figure 15.100 continued

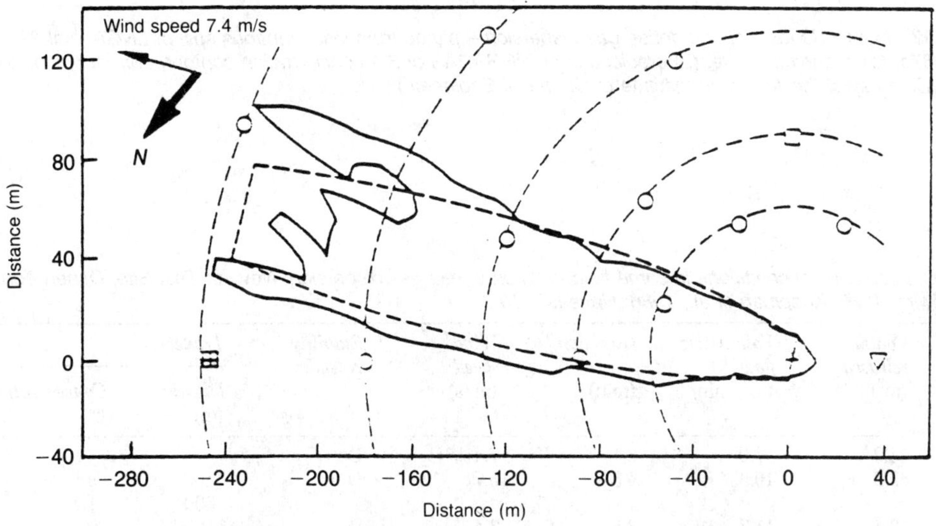

Figure 15.101 *Maplin Sands trials on dense gas dispersion – plume from continuous spill of LNG: Trial 29: comparison of observed plume with limits of visible plume predicted by HEGADAS (Puttock, Colenbrander and Blackmore, 1984) (Courtesy of the American Institute of Chemical Engineers)*

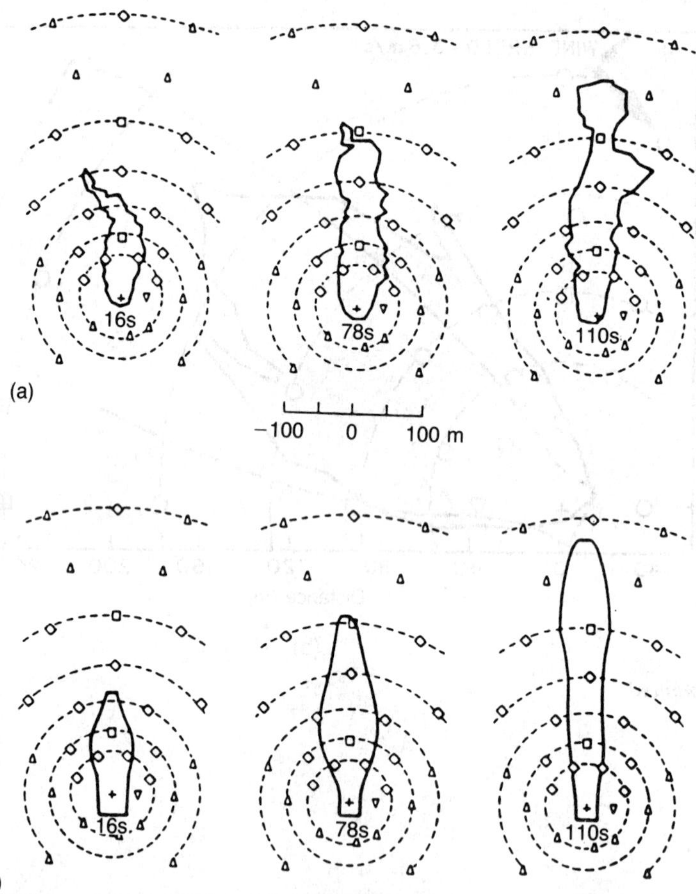

Figure 15.102 *Maplin Sands trials on dense gas dispersion – plume from instantaneous spill of LNG – trial 22 (Puttock, 1987c) (a) observed plume; (b) predictions by HEGADAS of 3% concentration contour, corresponding to visible limit (Courtesy of the American Institute of Chemical Engineers)*

Table 15.50 *Experimental conditions in some field trials on dense gas dispersion: Nevada Test Site, Desert Tortoise series (Goldwire, 1986; Koopman et al., 1986; Havens, 1992)*

Trial no.	*Volume released* (m^3)	*Volumetric flow* (m^3/min)	*Duration* (min)	*Wind speed* (m/s)	*Stability category*	*Travel*	
						Distance (m)	*Concentration* (%)
1	14.9	7.0	2.1	7.4	D		
2	43.8	10.3	4.3	5.7	D	100	9
						800	1.4
3	32.4	11.7	2.8	7.4	D	100	9
						800	1.6
						2400	0.22
4	60.3	9.5	6.4	4.5	E	100	10
						800	1.6
						2400	0.53

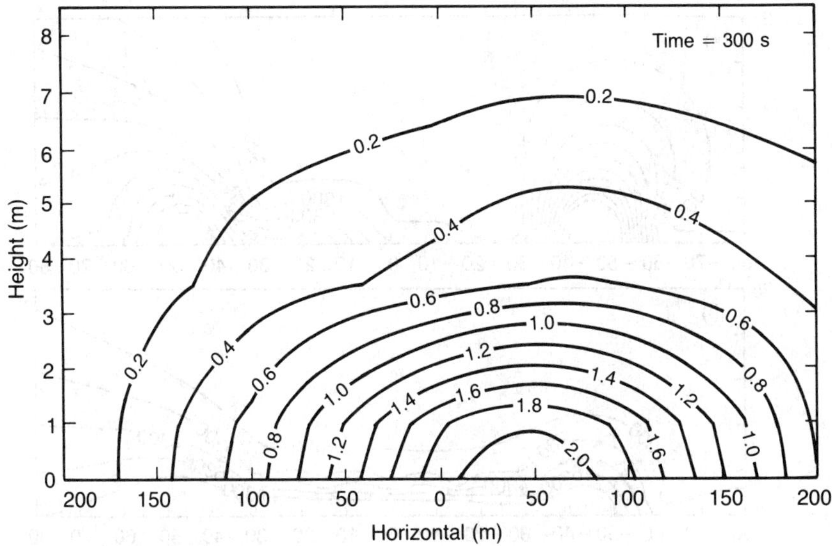

Figure 15.103 Nevada Test Site trials, Desert Tortoise series, on dense gas dispersion with ammonia – trial 4: crosswind elevation view of concentration contours at 800 m arc (Goldwire, 1986) (Courtesy of the American Institute of Chemical Engineers)

Table 15.51 Experimental conditions in some field trials on dense gas dispersion: Nevada Test Site, Goldfish series (after Blewitt et al., 1987a) (Courtesy of the American Institute of Chemical Engineers)

Trial No.	*Volume released* (m^3)	*Volumetric flow* (m^3/min)	*Duration* (min)	*Wind speed* (m/s)	*Stability category*	*Comment*
1	3.7	1.77	2.1	5.6	D	System check and dispersion test
2	4.0	0.66	6.0	4.2	D	Dispersion test
3	3.9	0.65	6.0	5.4	D	Dispersion test
4	3.6	0.255	14	6.8	D	Air/water mitigation test
5	2.0	0.123	16	3.8	C/D	Water spray (upflow) test
6	2.0	0.123	16	5.4	C/D	Water spray (downflow) test

the ineffectiveness of a dike for containing a pressurized release and to the hazard ranges obtained.

The work included the development of a complete model system for emission, vaporization and dispersion of hydrogen fluoride. The HEGADAS model was selected for development as the basis of the system, HFSYSTEM, described in Section 15.31. The trials on vapour barriers and water spray barriers are described in Section 15.53.

15.38 Dispersion of Dense Gas: Thorney Island Trials

In 1982–84 the HSE organized a series of heavy gas dispersion trials (HGDTs) at Thorney Island. The trials were a major undertaking involving over a number of years the collection, processing and reconciliation of data from the trials and the comparison of the results with those from theoretical models and from physical modelling. They are described in *Heavy Gas Dispersion Trials at Thorney Island* (two volumes) by McQuaid (1985a, 1987).

The main trials were with instantaneous releases and were conducted in two series, Phases I and II. Phase I involved unobstructed releases over a flat surface, Phase II releases over a flat surface with obstructions. A short series of trials with continuous releases was also performed. In addition, the Thorney Island site was subsequently used by the US Coast Guard and Department of Transportation to conduct their own series of experiments, termed the Phase III series. A summary of the Thorney Island trials is given in Table 15.52.

15.38.1 Phase I trials

An account of the Phase I trials has been given by McQuaid (1984b,c, 1985b,c). The organization of the trials has been described by A.G. Johnston (1985) and McQuaid (1985b). Other accounts cover the following

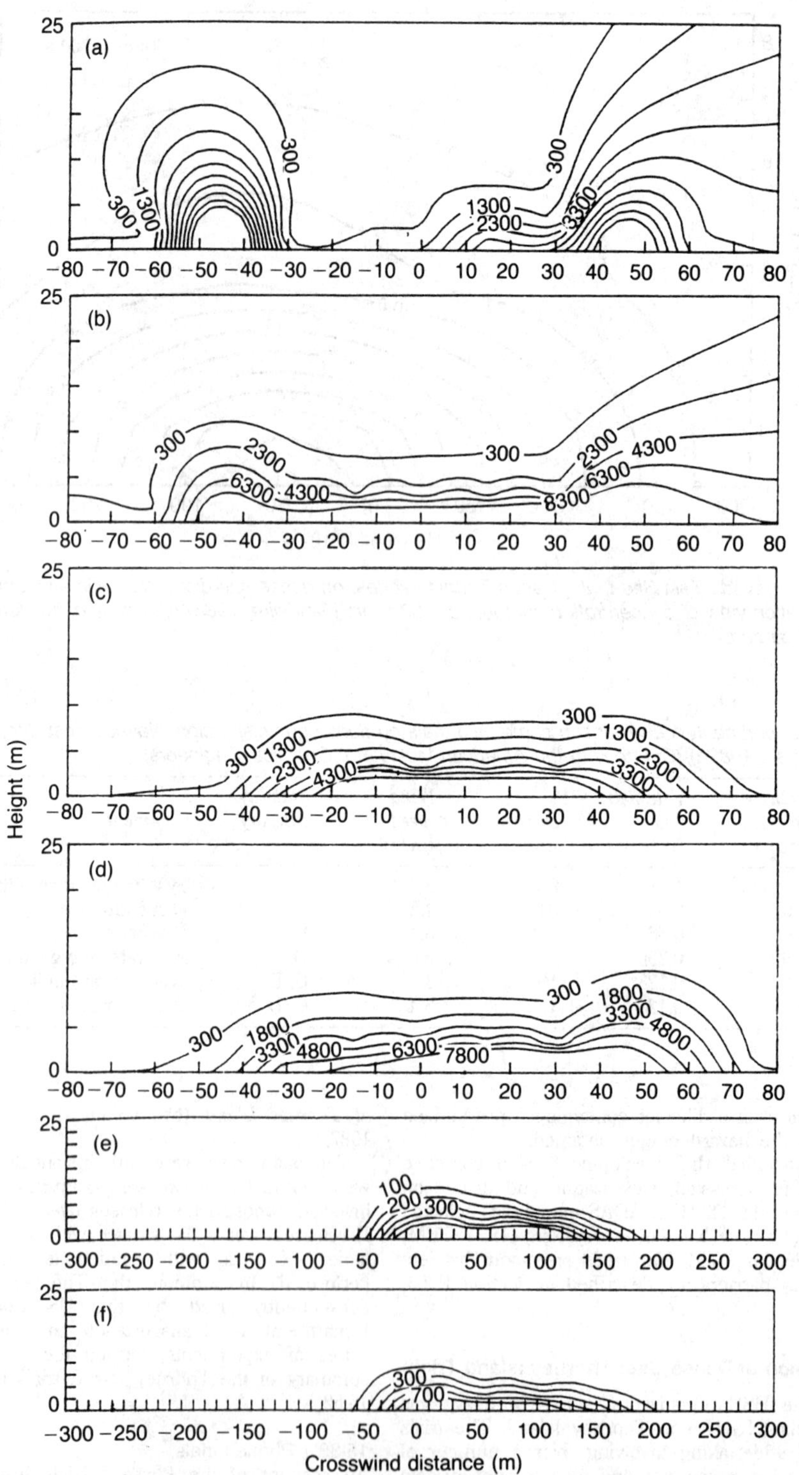

Figure 15.104 *Nevada test Site trials, Goldfish series, on dense gas dispersion with hydrogen fluoride – crosswind elevation views of concentration contours for trials 2 and 3 (Blewitt et al., 1987a): (a) at 300 m downwind at 54 s in trial 2 ; (b) at 300 m downwind at 121 s in trial 2 ; (c) at 300 m at 113 s in trial 3; (d) at 300 m downwind at 113 s (sic) in trial 3 ; (e) at 1000 m downwind at 225 s in trial 3 ; and (f) at 1000 m downwind at 358 s in trial 3 (Courtesy of the American Institute of Chemical Engineers)*

Table 15.52 *Experimental conditions in some field trials on dense gas dispersion: Thorney Island*

Trial No.	Type[a]	*Volume released* (m³)	*Volumetric flow* (m³/min)	*Duration* (min)	*Relative density*	*Wind speed* (m/s)	*Stability category*	*Bulk Richardson No.*	*Comments*
005	I/1	1320			1.65	4.6	B		
006	I/1	1580			1.60	2.6	D/E	8.9	
007	I/1	2000			1.75	3.2	E	9.3	
008	I/1	2000			1.63	2.4	D	15.5	
009	I/1	2000			1.60	1.7	F	26.5	
010	I/1	2000			1.80	2.4	C	17.7	
011	I/1	2100			1.96	5.1	D	4.9	
012	I/1	1950			2.37	2.6	E	24.7	Flat unobstructed terrain
013	I/1	1950			2.00	7.5	D	2.2	
014	I/1	2000			1.76	6.8	C/D	2.1	
015	I/1	2100			1.41	5.4	C/D	1.9	
016	I/1	1580			1.68	4.8	D	3.0	
017	I/1	1700			4.20	5.3	D/E	12.5	
018	I/1	1700			1.87	7.4	D	1.7	
019	I/1	2100			2.12	6.6	D/E	3.5	
034	I/1	2110			1.8	1.4	F	60.4	
020	II/2	1920			1.92	5.7		3.5	5 m high, solid fence at 50 m
021	II/2	2050			2.02	3.9		8.8	5 m high, solid fence at 50 m
022	II/2	1400			4.17	5.9		8.1	5 m high, solid fence at 50 m
023	II/2	1960			1.80	5.8		3.0	Two rows of permeable screens, 10 m high
024	II/2	1925			2.03	6.8		2.7	Four rows of permeable screens, 10 m high
025	II/2	2000			1.95	1.4		61.8	5 m high, solid fence at 50 m
026	II/2	1970			2.00	1.9		34.8	9 m cube building, 50 m downwind
027	II/2	1700			4.17	2.2		71.0	9 m cube building, 50 m downwind
028	II/2	1850			2.00	9.0		1.5	9 m cube building, 50 m downwind
029	II/2	1950			2.00	5.6		4.0	9 m cube building, 50 m upwind
045	C/3	1972	260	7.6	2.0	2.1	E/F	380	Flat unobstructed terrain
046	C/3	1490	260	5.7	2.0	3.2	D	160	
047	C/3	1938	250	7.8	2.05	1.5	F	740	Flat unobstructed terrain
030	III/4	1603	260	6.2	1.4	4.3	E	0.2	Release inside fenced enclosure
033	III/4	1870	340	5.5	1.6	2.8	D/E	1.2	
036	III/4	1963	310	6.3	1.6	2.4	F/G	1.6	
037	III/4	1891	255	7.4	1.6	3.2	E	0.7	
038	III/3	1867	280	6.7	1.6	3.8	F	0.5	Flat unobstructed terrain
039	III/4	1808	350	5.2	1.4	5.8	D	0.1	Release inside fenced enclosure
040	III/4	1860	310	6.0	1.2	3.1	D	0.2	
042	III/4	1973	185	10.7	1.6	3.4	D	0.5	
043	III/4	1899	265	7.2	1.3	1.5	F	2.5	
049	III/4	1907	260	7.3	1.6	2.5	F	1.3	
050	III/4	1800	270	6.7	1.4	1.74	F	2.1	

Sources: McQuaid (1985b, 1987); M.E. Davies and Singh (1985a); M.E. Davies and Inman (1987)
[a] Phase C, continuous trials. Type: 1, instantaneous release, flat unobstructed terrain; 2, instantaneous release, flat obstructed terrain; 3, continuous release, flat unobstructed terrain; 4, continuous release, flat obstructed terrain.
[b] Nominal value.

aspects: systems development and operations procedures (D.R. Johnson, 1985); the site geography and meteorology (M.E. Davies and Singh, 1985b); the gas sensor system (Leck and Lowe, 1985); the meteorological data (Puttock, 1987a); the determination of the cloud area and path from visual records (Brighton, Prince and Webber,

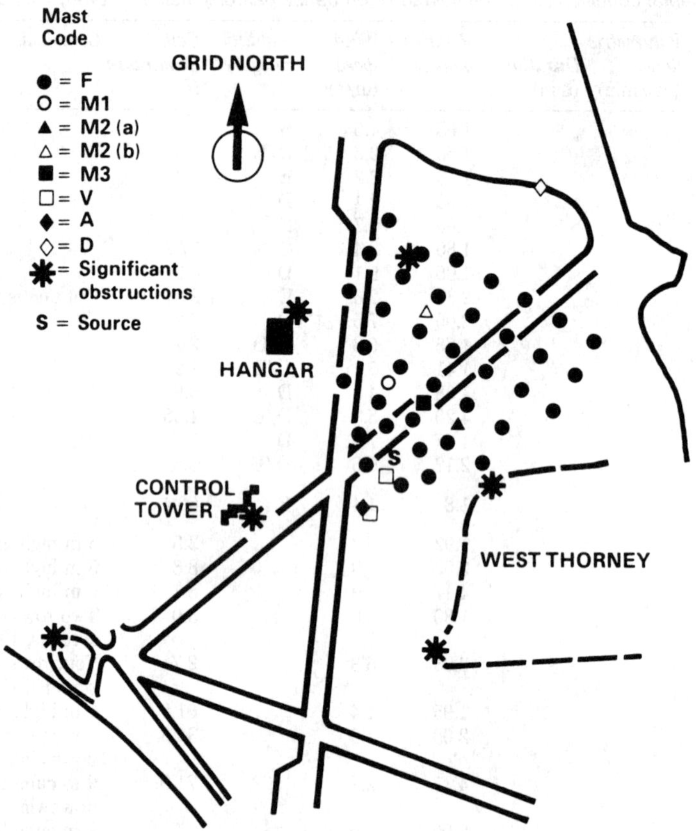

Figure 15.105 *Thorney Island trials on dense gas dispersion – site layout (McQuaid, 1985b) (Courtesy of Elsevier Science Publishers)*

1985; Prince *et al.*, 1985 SRD R318); the determination of the cloud path from concentration measurements (Brighton, 1985 SRD R319); the computer processing of the visual records (Riethmuller, 1985); the cloud advection velocities (Wheatley and Prince, 1987); the effect of averaging time on the statistical properties of the sensor records (Nussey, Davies and Mercer, 1985); the mean concentration and variance (Carn, 1987); the area-averaged concentration, height scales and mass balances (Brighton, 1985a); the computer processing of the data for comparison with models (Pfennning and Cornwell, 1985); the presentation and availability of the data (Roebuck, 1985); and users' views of the data (N.C. Harris, 1985; Brighton, 1987). Critiques which discuss the contribution to dense gas dispersion modelling have been given by Hartwig (1985) and Puttock and Colenbrander (1985).

The aim of the Phase I trials was to obtain data which might be used to increase understanding of the mechanisms of dense gas dispersion and to validate models. The trials were designed with the needs of modellers very much in mind. Consideration was given to the requirements for the meteorological parameters and the parameters used in box and *K*-theory models. In particular, attention was given to the variables required to define stability and to characterize both Richardson number and Monin–Obukhov similarity, and thus in particular to wind speed, roughness length, friction velocity and sensible heat flux. These aspects are discussed by McQuaid (1985b).

The site was an airfield flat to within about 1 in 100. The layout of the site is shown in Figure 15.105. Some 250 instruments were deployed. The gas sensors were fixed to masts, the lowest level being located 0.4 m above ground and the highest levels being 4, 10 and 14.5 m in the near, mid and far fields, respectively. Photographic records were also taken, the visibility of the cloud being enhanced by the use of cartridges of orange smoke.

The Phase I trials involved an instantaneous release of 2000 m^3 of a gas–nitrogen mixture. The gas used was Freon 12, the proportion being varied to give mixtures of different relative density up to 4.2. The container was a 12-sided container 14 m across by 13 m high. The container was held up by rigging until the operation of a release mechanism which allowed it to fall to the ground.

On release, the gas cloud slumped rapidly and spread out radially. There was some movement upwind until this

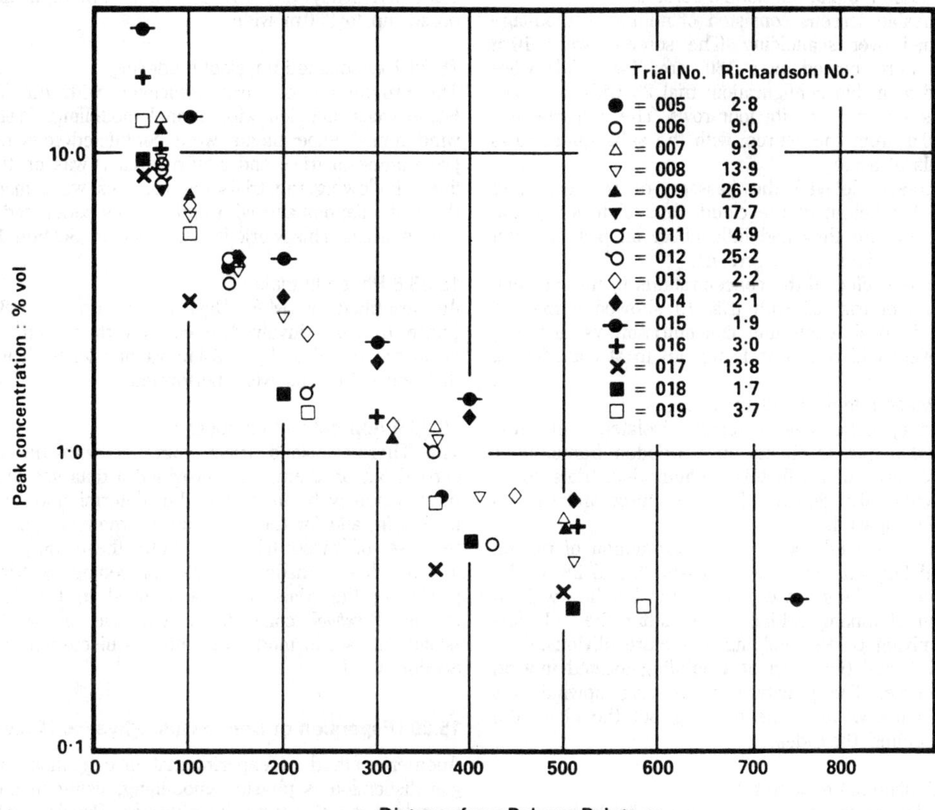

Figure 15.106 *Thorney Island trials on dense gas dispersion – peak concentrations on path of cloud centroid for instantaneous release trials (McQuaid, 1985b) (Courtesy of Elsevier Science Publishers)*

was stopped by the wind. The cloud had a raised front and assumed the characteristic doughnut shape. It then travelled in this form for several hundred metres. Eventually a low-lying cloud formed which covered a large part of the site.

A total of 15 trials was performed, as shown in Table 15.52. In each trial measurements were taken of the concentration vs time profiles at the gas sensors within the cloud. These raw data have subsequently been processed to yield a variety of results. Plate 8 shows the development of the gas cloud in one of these trials. Figure 15.106 shows the peak concentration on the path of the cloud centroid in the Phase I trials.

15.38.2 Phase II trials

The Phase II trials have been described by M.E. Davies and Singh (1985a). The following aspects of the Phase II have been covered in other accounts: the overall properties of the clouds (Brighton and Prince, 1987); and a user's view of the data (Brighton, 1987).

The Phase II series was concerned with dense gas dispersion at obstructions. Three types of obstruction were used:

(1) solid fence;
(2) permeable screen;
(3) building cube.

Details of the Phase II trials are given in Table 15.52.

15.38.3 Phase II trials: solid fence

The solid fence consisted of tarpaulin sheets hung over scaffolding. The fence was 5 m high and was located on a 180° arc at 50 m from the source. Four trials were conducted with this configuration: trials 20–22 and 25.

In trial 25 there was complete blocking of the cloud by the fence, in trial 21 partial blocking and in trial 20 virtually no blocking. There was also appreciable blocking in trial 22 in which the cloud density was much higher.

The results of these trials were compared with Phase I trials carried out under similar conditions but without a fence. The fences affected the concentrations in the far field. Comparison of trial 20 with trial 19 exhibited a marked reduction in peak centreline concentration vs distance, with a several-fold reduction factor out to at least 500 m.

15.38.4 Phase II trials: permeable screens

The permeable screens consisted of military camouflage nets draped over scaffolding. The screens were 10 m high and were located on a 180° arc. Two trials were performed with this configuration: trial 23 with two rows of screens and trial 24 with four rows. The first row was located 50 m from the source, with the subsequent rows at intervals of 3.3 m.

The screens delayed the passage of the gas, and increased the height of the cloud, thus distributing the gas vertically, but they had little effect on peak ground level concentrations at the screens.

The screens affected the concentrations in the far field. Comparison of trial 23 with trial 19 showed a marked reduction in peak centreline concentration vs distance, with a several-fold reduction factor out to at least 500 m.

15.38.5 Phase II trials: building cube

The building cube was a single, isolated 9 m cube constructed of plastic sheets on a wooden frame. Four trials were performed with this configuration: trials 26–28 with the cube 50 m downwind of the source and trial 29 with it 27 m upwind.

The effect of wind speed on the promotion of mixing by the building was examined in trials 26 and 28. In the former trial the cloud passed around the building without much sign of mixing, whilst in the latter the enhancement of mixing by the building was more obvious.

Trial 27 tested the effect of a building located upwind of the source. The gravity front moving upwind was drawn into the wake of the building but the cloud did not pass around the sides.

15.38.6 Continuous release trials

A description of the continuous release trials has been given by McQuaid (1987). Accounts have also been given of the following aspects: the mass and flux balances (Mercer and Nussey, 1987); the turbulence records (Mercer and Davies, 1987); and a user's view of the data (Brighton, 1987). The aim of the trials was essentially similar to that for the instantaneous releases, i.e. to obtain data which might be used to increase understanding of the dispersion mechanisms and to validate models.

The gas used was a mixture of air and Freon 12 with a relative density of 2.0 in all trials. The source was designed to give a release with zero vertical momentum, the gas being released through an aperture in the ground with a cap above it. The release was at a rate of 5 m^3/s maintained for 400 s. Details of the continuous trials are given in Table 15.52. Three trials were conducted: trials 45–47.

The conditions for trials 45 and 47 were similar, except that the wind speeds were 2.1 and 1.5 m/s, respectively. In trial 45 at the higher wind speed the lateral spread was less and the downwind extent was greater than in trial 47.

The concentration vs time profiles showed that at 36 m downwind a steady-state profile existed for some 360 s, whereas at 250 m no steady profile was obtained. The shape of the concentration profile at 36 m was a sharp rise followed by a slow fall, whilst at 90 m there was a nearly symmetrical concentration rise and fall. With regard to the vertical concentration profile, at 36 m the concentration exhibited intermittency. The plume was characteristically very shallow, about 2 m high, and very broad, up to 300 m wide.

15.38.7 Associated physical modelling

The programme of work associated with the Thorney Island trials included wind tunnel modelling. Preliminary wind tunnel experiments were conducted to screen the programme of trials and confirm that it was on the right lines. Following the trials comparisons were made with these simulations and with earlier ones conducted for the Porton trials. This work is described in Section 15.39.

15.38.8 Phase III trials

As described by M.E. Davies and Inman (1987), the Phase III trials involved releases with a 54 m × 26 m enclosure bounded by a 2.4 m vapour fence. They state that some 40 trials were performed.

15.38.9 Application to models

The Thorney Island trials, together with the analysis carried out on them, have provided a data set which has been extensively used for the determination of parameters in, and for the validation of, models. Reference to the use of these trials for both these purposes has already been made in the preceding sections. In particular, the trials have been used in the derivation of the *Workbook* correlations. The use of the Thorney Island data set in model validation is discussed further in Section 15.41.

15.39 Dispersion of Dense Gas: Physical Modelling

Another method of experimental investigation of dense gas dispersion is physical modelling, using mainly wind tunnels and water tunnels or flumes. Physical modelling makes it possible to perform dispersion experiments not only for flat unobstructed surfaces, but also with different ground slopes and other topographical features and with all kinds of obstructions, including fences and buildings, and even complete process plants, sites or terminals. It is also sufficiently economical to allow experiments to be repeated, either exactly or with minor variations.

The principal drawback is that it is not possible to scale so as to achieve complete similarity between the physical model and the scenario simulated, so that there is always a degree of distortion. However, the effects of such distortion are fairly well understood.

Accounts of physical modelling have been given by Snyder (1981), D.J. Hall, Hollis and Ishaq (1982), Meroney and Lohmeyer (1984), Meroney (1987a) and in the *Workbook* by Britter and McQuaid (1988).

15.39.1 Guidelines for physical modelling

Guidelines for physical modelling of dense gas dispersion have been given by Meroney (1987). They include guidance on formulation of the information to be obtained, the choice of modelling medium, the design of the experiments, the reporting of the experiments, the performance envelope within which experiments may be conducted and the scaling of the experiments.

The experiments should be matched to the information required. It may be required to determine the distance beyond which a hazardous concentration will not exist, or the distance at which the hazardous condition will occur more than a certain time with a certain probability.

Or again, it may be required to determine the maximum concentration at a location near the source, or the concentration vs time profile at that location.

The information sought should be limited to what is really needed and overelaboration should be avoided. The data set specified should accord with what is realistically obtainable. It is useful to carry out preliminary experiments with a simple box model.

The modelling medium may be air or water. Broadly, water is often convenient for qualitative studies and visualization, but air is more often used for quantitative work. Air allows the use of a wide range of density ratios. Using heat it is comparatively easy to stratify. In theory, water should allow a Reynolds number some 16 times that of air, but in practice this advantage is not easily realized.

15.39.2 Wind tunnel facilities

A wind tunnel not only represents a major investment but also requires expertise in its use. It is therefore normal to contract work to an organization operating the wind tunnel with its own specialist team. A list of some of the principal wind tunnel facilities is given by Meroney (1987a).

15.39.3 Scaling for wind tunnel modelling

Physical modelling involves simulation of a full-scale situation by scaling down the dimensions in such a way as to retain similarity in the essential features.

Accounts of scaling for physical simulation include those given by Meroney (1979b, 1982), D.J. Hall, Hollis and Ishaq (1982), Cheah, Cleaver and Milward (1983a,b), Krogstad and Pettersen (1986), and Schatzmann, Koenig and Lohmeyer (1986) and the one given in the *Workbook* by Britter and McQuaid (1988).

The usual procedure is to list the relevant variables and then to derive from these the appropriate dimensionless groups. The number of dimensionless groups should be equal to the number of variables minus the number of dimensions. The lists of the variables given by different workers tends to vary slightly, in part because some list a variable only to disregard it, whilst others do not list it in the first place.

The conditions for similarity may be defined in stages. Thus the *Workbook* defines first the requirements for the atmospheric boundary layer, then those for release of a gas with the same properties and temperature as air, and finally those for release of a dense gas. The variables commonly listed for similarity of the atmospheric boundary layer are

u characteristic velocity
L characteristic length scale
ρ density of atmosphere
μ viscosity of atmosphere
g acceleration due to gravity

The *Workbook* includes also

$\Delta\rho/\rho$ characteristic potential density difference due to atmospheric stability/instability
f Coriolis parameter

The dimensionless groups listed by the *Workbook* are:

$$\mathrm{Re} = \frac{uL\rho}{\mu} \qquad \text{Reynolds number}$$

$$\mathrm{Ri} = g\frac{\Delta\rho}{\rho}\frac{L}{u^2} \qquad \text{Richardson number}$$

$$\mathrm{Ro} = \frac{u}{fL} \qquad \text{Rossby number}$$

This gives only three dimensionless groups, not the four which the rule quoted above would indicate. The explanation given in the *Workbook* is that the Boussinescq approximation has been made, so that the effect of $\Delta\rho/\rho$ on inertia is neglected but that on buoyancy is not.

Proceeding straight to the dense gas case, the further variables introduced in the *Workbook* are:

q_o volumetric release rate, or
Q_o volume released
ρ_a density of air
ρ_o density of source gas
x, y, z downwind, crosswind, vertical distance

The further dimensionless groups given are then:

$$\frac{q_o}{uL^2} \qquad \text{source number (continuous source), or}$$

$$\frac{Q_o}{L^3} \qquad \text{source number (instantaneous source)}$$

$$\frac{\rho_o}{\rho_a} \qquad \text{density ratio}$$

$$\mathrm{Fr} = \frac{u^2}{gL} \qquad \text{Froude number}$$

$$\frac{x}{L}, \frac{y}{L} \text{ and } \frac{z}{L} \qquad \text{dimensionless downwind, crosswind and vertical distance}$$

If a degree of distortion is accepted, it is practice in modelling buoyant plumes to make the following approximations, which the *Workbook* tentatively proposes for dense plumes also. The Froude number and the density ratio may be combined using the Boussinesq approximation to give

$$\frac{u^2}{g(\Delta\rho_o/\rho_a)L} \qquad \text{densimetric Froude number}$$

Similarly, the source number and density ratio may be combined to give

$$\frac{\rho_o q_o^2}{\rho_a u^2 L^4} \qquad \text{source momentum ratio}$$

If the source momentum is negligible, this term may be replaced with the source number q_o/uL^2.

If the fine scale structure of the concentration field is to be preserved, the similarity of the Schmidt number Sc is relevant

$$\mathrm{Sc} = \frac{\nu}{D} \qquad \text{Schmidt number}$$

where D is the diffusion coefficient and ν is the kinematic viscosity.

This aspect may also be expressed in terms of the Peclet number Pe:

$$\mathrm{Pe} = \frac{uL}{D} \quad \text{Peclet number}$$

since

$$\mathrm{Pe} = \mathrm{Re} \cdot \mathrm{Sc}$$

If non-neutral temperature stratification is being modelled, the similarity of the Prandtl number Pr is relevant:

$$\mathrm{Pr} = \frac{\nu}{\mathrm{k}} \quad \text{Prandtl number}$$

where k is the thermal conductivity.

Additional quantities mentioned by some workers include:

α ground slope
z_o roughness length
u_* friction velocity
μ_o viscosity of source gas

and additional dimensionless groups are

$\frac{z_o}{L}$ dimensionless roughness length

$\frac{u_*}{u}$ dimensionless friction velocity

The conditions for flow to be fully aerodynamically rough depend on the group $u_* z_o/\nu$. In other words, for such flow the role of the kinematic viscosity ν is taken over by the group $u_* z_o/\nu$. The *Workbook* therefore defines a more relevant Reynolds number as

$\frac{uL}{u_* z_o}$ modified Reynolds number

It may be noted that this modified Reynolds number is the product of the reciprocals of the dimensionless roughness length and friction velocity just defined.

The source volumetric flow q_o is sometimes represented by the product $u_o A_o$, where u_o is the source velocity and A_o its area.

Turning to the application of these similarity criteria to the scaling problem, it is not possible to achieve similarity of all the relevant groups. Complete similarity is not attainable and some degree of distortion has to be accepted.

Some of the groups quoted are of secondary importance, and some are frequently not mentioned by modellers. Since the full scale and model fluids are generally both air, the Prandtl number is the same. Rossby number similarity is not attainable, but a limit can be set. The *Workbook* suggests that it would be reasonable to neglect rotational effects for Ro $\geq$ 10, but that a value of Ro $\approx$ 1 should be justified.

It is not possible to obtain Reynolds number similarity. However, since the properties of turbulent flow are independent of the Reynolds number provided it is large, the approach generally taken is to make the Reynolds number as large as practical.

The *Workbook* gives a detailed discussion of the setting of the Reynolds number. The essential requirement is similarity of the mean and turbulence velocity profiles and in practice the boundary layer is adjusted until this is achieved. It suggests that the condition $u^* k/\nu > 50$ gives Reynolds number independence, though for sharp-edged elements particularly $u^* k/\nu > 5$ may suffice. Since typically $z_o \approx 0.1$, these conditions are equivalent to $u^* z_o/\nu > 5$ and $u^* z_o/\nu > 2$, respectively.

As far as concerns Froude number similarity, this is often difficult to achieve because it tends to involve an air velocity that is too low to give stable flow. Generally the approach taken is to combine the density ratio and Froude number into a densimetric Froude number, as described above. The *Workbook* recommends that priority be given to achieving similarity of two groups, the densimetric Froude number and the source momentum ratio.

If the fine scale structure of the concentration field needs to be modelled, and molecular diffusion effects taken into account, the Schmidt number should be similar and the Reynolds number as large as possible. Failure to achieve adequate Reynolds number results in loss of fine scale structure and is not compensated by a correct Schmidt number.

Reports of physical modelling frequently discuss aspects of scaling, either in relation to the design of the experiments or in explanation of the results, and many of those quoted below include such discussions.

15.39.4 Wind tunnel performance envelope

There are a number of practical features which set a limit to the conditions under which a wind tunnel may be operated and to the applications for which it may be used. There are problems in achieving a steady low velocity. A velocity of 1 m/s is quoted as a practical lower limit, although velocities half this or less are sometimes used. This tends to be a major constraint.

Dense gas clouds are shallow but wide, and the width of the tunnel may be a constraint. The length of the tunnel can also constrain the concentrations which can be studied; there is usually little difficulty in obtaining concentrations down to typical flammability limit levels but it may not be possible to achieve concentrations at levels of interest for some toxic gases.

The use of a very small model can cause problems of spatial resolution. There is a practical limit to the fineness of size of measurement probes.

Attainment of a Reynolds number high enough to give correct wake flows can be a difficulty, particularly for smooth structures.

The performance envelope of a wind tunnel, with particular reference to LNG dispersion experiments, has been described by Meroney and Neff (1980) and Meroney (1987a).

15.39.5 Universal plume profile

An early wind tunnel study of dense gas dispersion was that by D.J. Hall, Barrett and Ralph (1976). These workers addressed the determination of the distance for the gas concentration to fall to 2% and expressed their results in the form of a universal plume, as shown in Figure 15.107. The plume is defined by the dimension D, which for a cloud boundary concentration of 2% is given by the empirical equation

$$D\left(\frac{u}{Q}\right)^{\frac{1}{2}} = 107 \qquad [15.39.1]$$

where Q is the volumetric release rate and u is the wind speed. The results apply only to a fully developed plume, which may take some hours to develop.

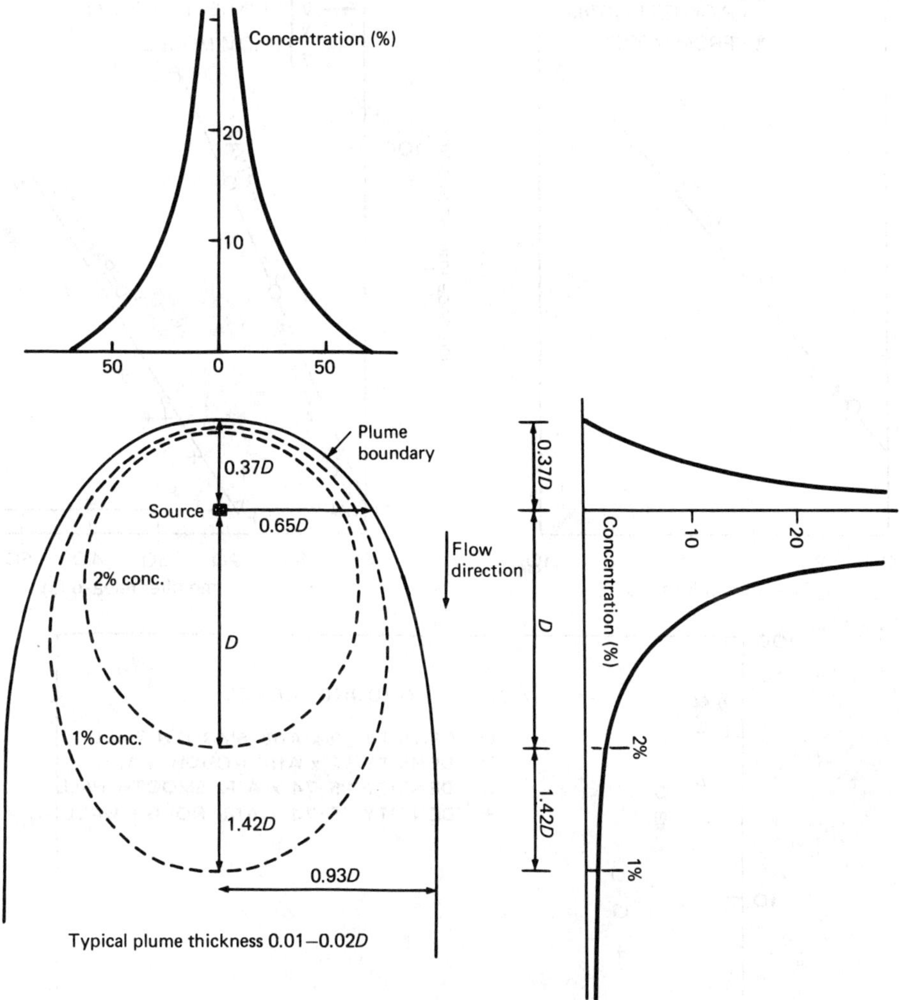

Figure 15.107 *Wind tunnel simulation of dense gas dispersion – universal plume profile (D.J. Hall, Barrett and Ralph, 1976) (Courtesy of Warren Spring Laboratory)*

The work illustrated the fact that a dense gas forms a wide, very shallow plume with a strong resistance to vertical dispersion and that its behaviour is quite different from that of neutral buoyancy gas. The authors present calculations which show that for stable atmospheric conditions the travel distances given by passive dispersion models considerably exceed those obtained from their model.

15.39.6 Porton Down trials

A wind tunnel investigation of some of the Porton Down trials has been carried out by D.J. Hall, Hollis and Ishaq (1982, 1984). The aim of the work was to assess the applicability of wind tunnel modelling to dense gas dispersion.

Figures 15.108 and 15.109 show some of the results obtained for the Porton trials. Figures 15.108(a)–(c) give, respectively, the cloud width, the cloud arrival and departure times and the profile of the peak ground level concentration vs downwind distance for trial 37. Figures 15.109(a)–(d) show the concentration–time profiles at four points in this trial. The figure shows the profiles of the physical model tending to lead, and to be less smoothed than, the full-scale profiles.

Figure 15.109(e) shows the variability of the concentration–time profile in repeat runs at a single point in trial 52. The figure brings out the high degree of variability between individual runs and the desirability of performing enough runs to obtain a stable characterization of the concentration. Since such repeat runs are not practical in field trials, but are so in a wind tunnel; this is an important contribution of the latter. The authors give comparative photographs for the field trials and the wind tunnel. This format for presenting results

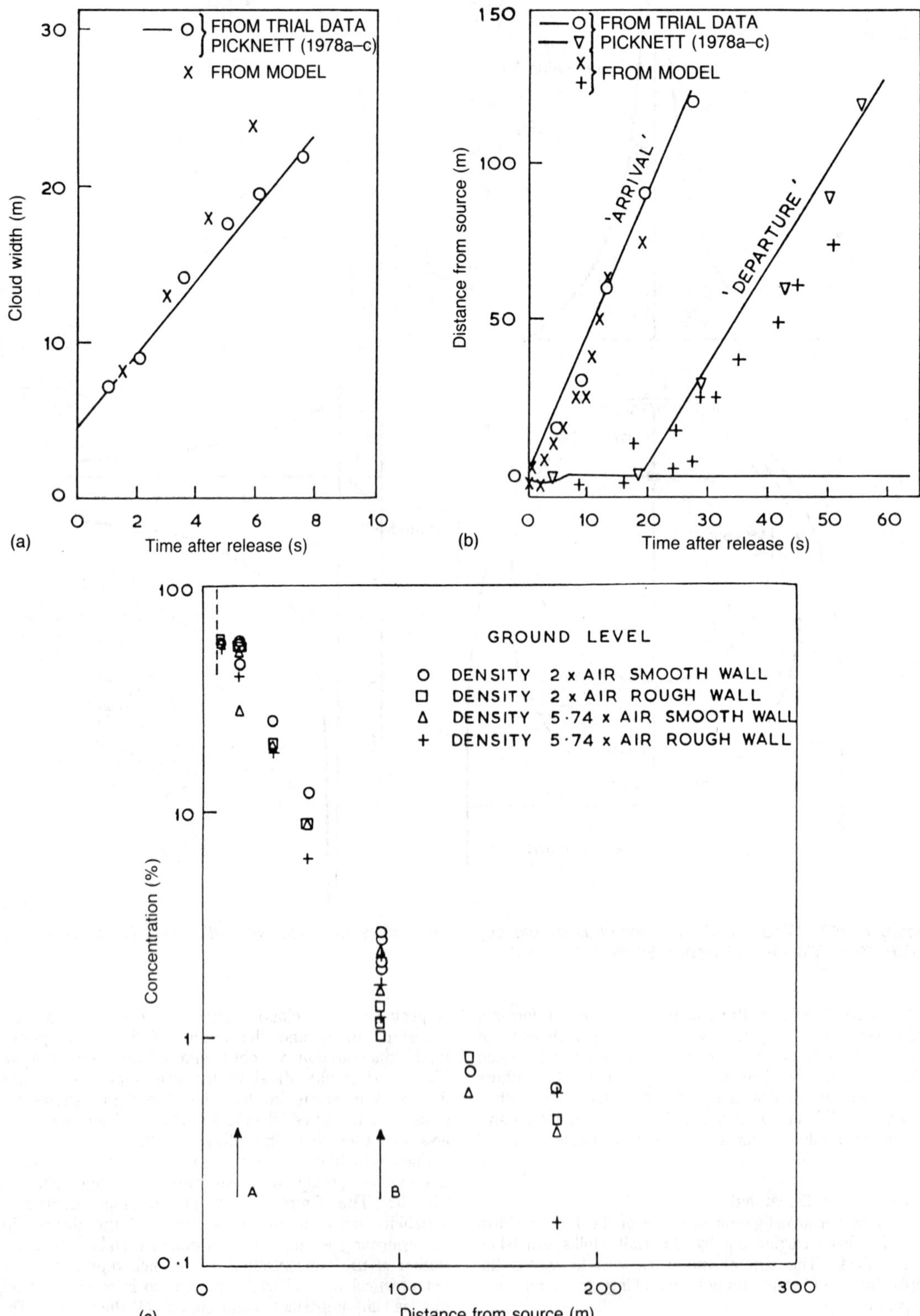

Figure 15.108 *Wind tunnel simulation of Porton Down trials on dense gas dispersion – comparison of observed and simulated features (D.J. Hall, Hollis and Ishaq, 1984): (a) cloud width in trial 37; (b) arrival and departure times in trial 37; (c) peak ground level concentration (Courtesy of Springer Verlag)*

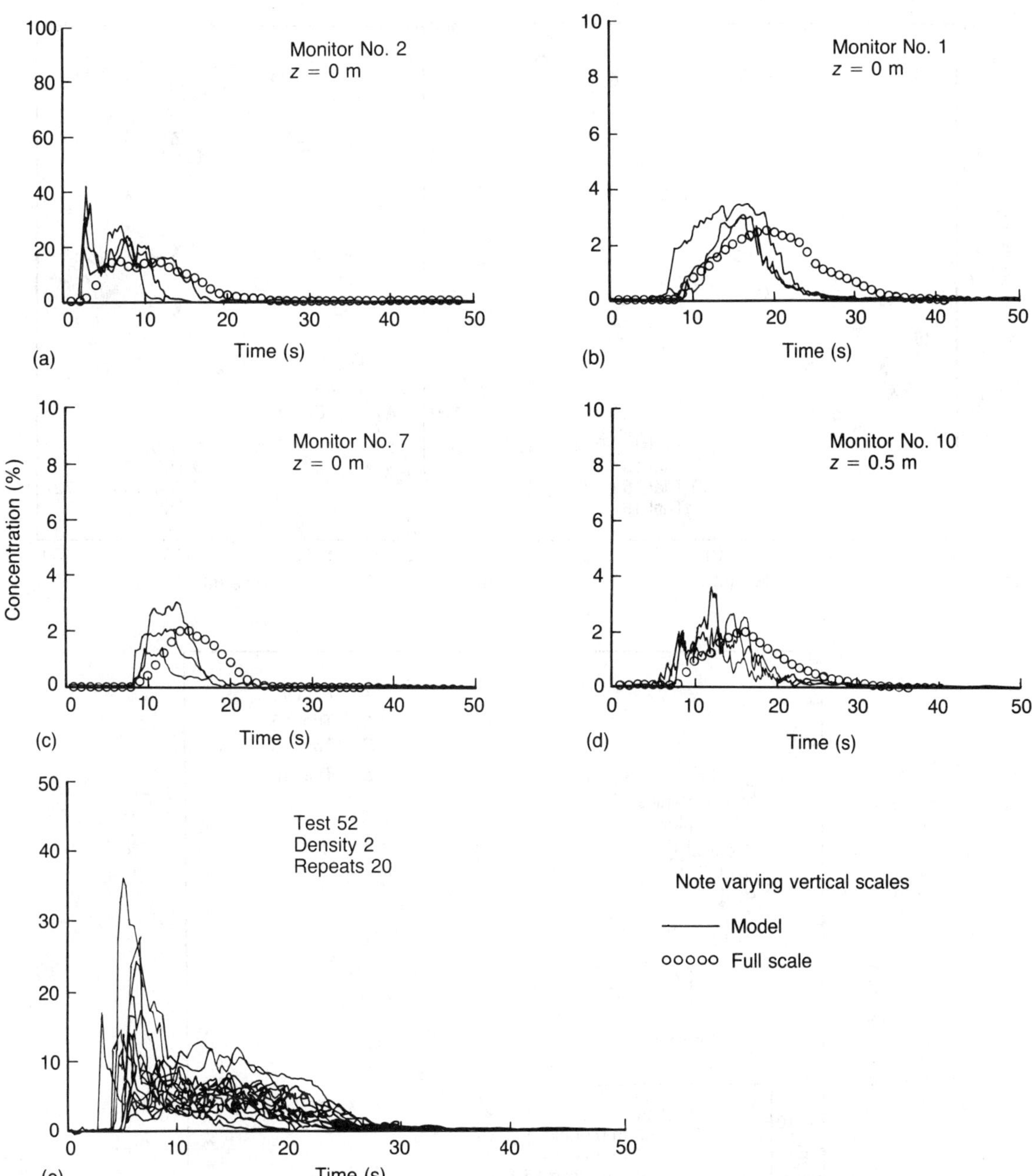

Figure 15.109 *Wind tunnel simulation of Porton Down trials on dense gas dispersion – comparison of observed and simulated features for trials 37 and 52 (D.J. Hall, Hollis and Ishaq, 1984): (a)–(d) concentration profiles at four points in trial 37; (e) variability of concentration profile in repeat simulation runs for trial 52 (Courtesy of Springer Verlag)*

has also been used in subsequent work by the authors and others.

The authors also proposed a relationship between the distance to the 2% concentration level and the Richardson number. They investigated this further in relation to the Thorney Island trials, as discussed below.

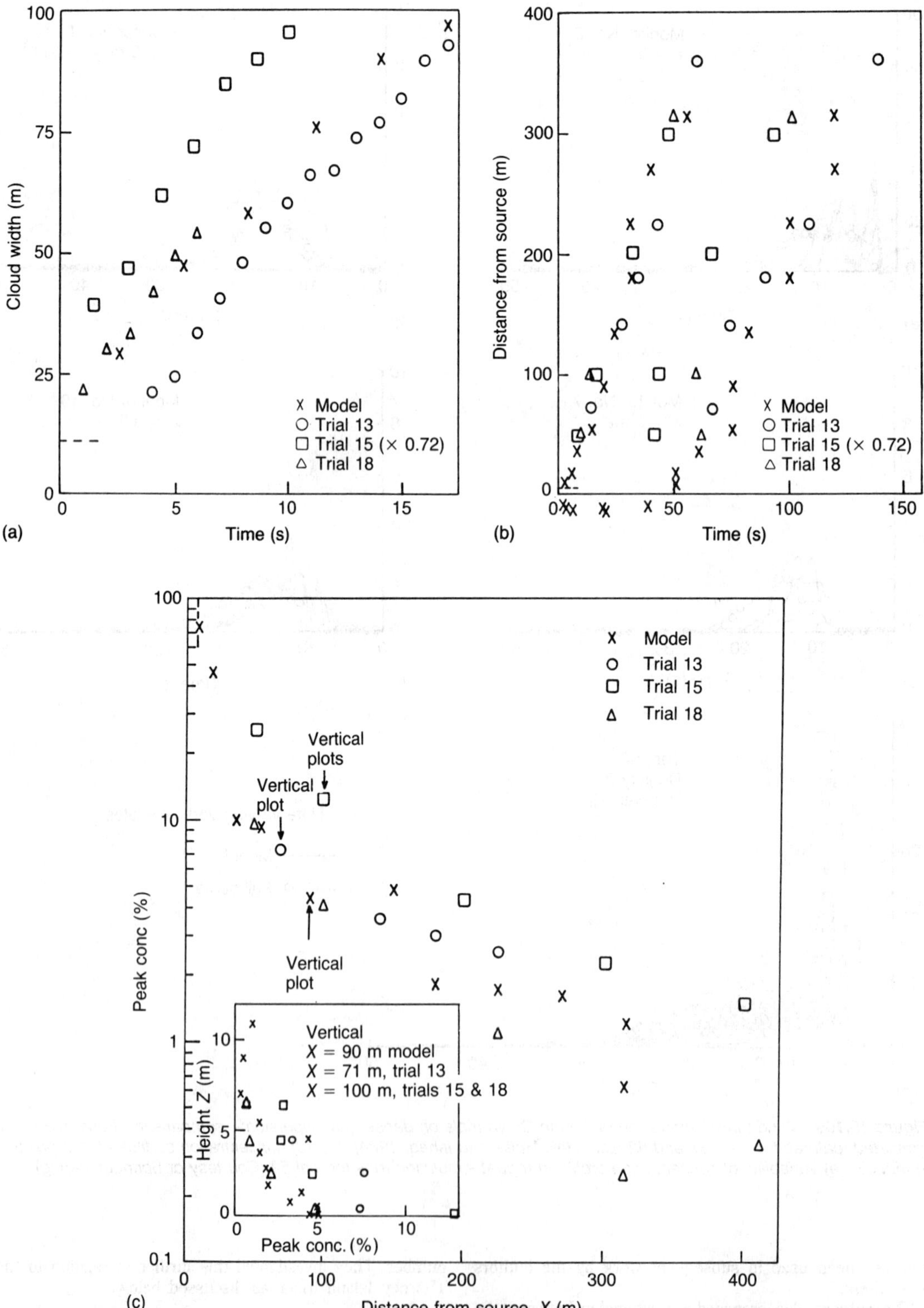

Figure 15.110 *Wind tunnel simulation of Thorney Island trials on dense gas dispersion – comparison of observed and simulated features for (D.J. Hall and Waters, 1985): (a) cloud width; (b) cloud travel time; (c) peak ground level concentration (Courtesy of Elsevier Science Publishers)*

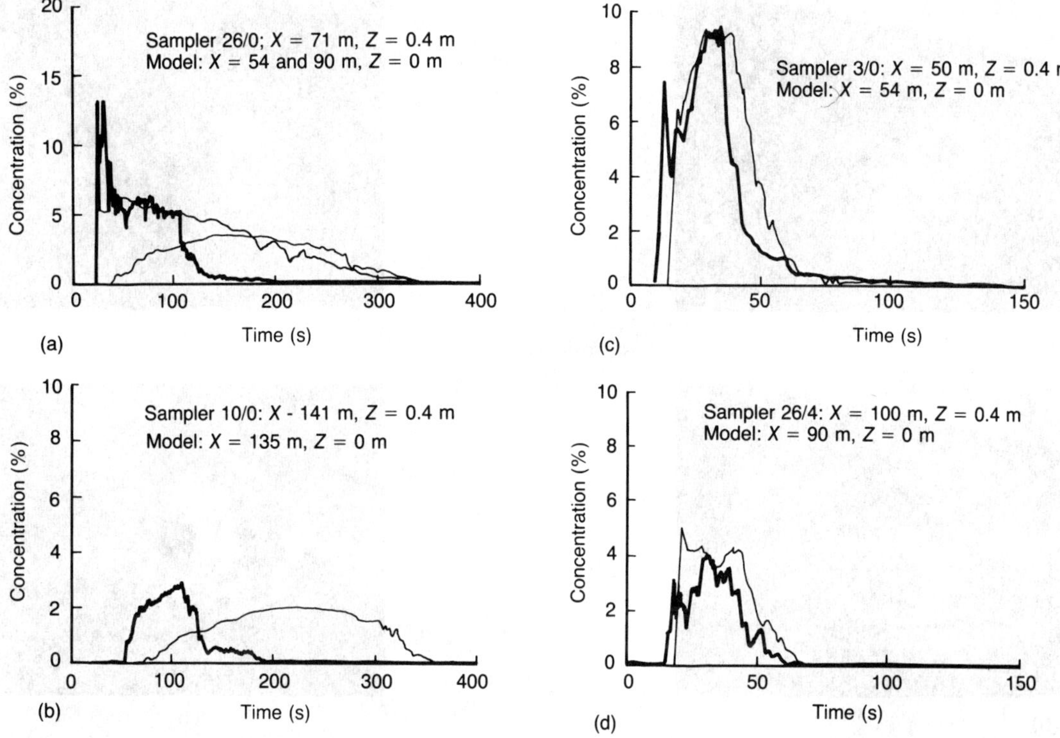

Figure 15.111 *Wind tunnel simulation of Thorney Island trials on dense gas dispersion – comparison of observed and simulated features: (a) and (b) concentration profiles at two points in trial 7; (c) and (d) concentration profiles at two points in trial 18 (D.J. Hall and Waters, 1985) ——— model; ——— trial samplers. Note the varying vertical scales (Courtesy of Elsevier Science Publishers)*

15.39.7 Thorney Island trials

At the time when the above work was being done, the Thorney Island experiments were being designed and some additional, exploratory wind tunnel work was carried out to assist in this. Following the trials, the simulations done on the Porton trials and this further work for the Thorney Island trials were revisited. Several of the simulations were sufficiently closely matched to particular Thorney Island trials to allow comparisons to be made. These have been described by D.J. Hall and Waters (1985).

Figure 15.110 shows some of the results obtained for the Thorney Island trials. Figures 15.110(a)–(c) again give the cloud width, the cloud arrival and departure times and the profile of the peak ground level concentration vs distance, respectively. Figure 15.111 shows the profile of concentration vs time at particular downwind locations. Figures 15.112 and 15.113 show, respectively, plan and elevation views of the full scale trial and of the wind tunnel model.

Figure 15.111 shows the profiles of the physical model tending in this case to lag, and to be more smooth than, the full-scale profiles. This contrasts with the results described above for Porton trial 37, where the reverse was the case. This 'persistence' of the concentration was the main discrepancy. The authors suggest that it may be explained in part by surface roughness present in the wind tunnel but not at full scale, and perhaps in part by the need to use a low wind speed of 0.4 m/s. Nevertheless, the concentrations obtained in the wind tunnel work were within a factor of 2 of those at full scale.

The authors also consider a correlation derived from their Porton simulations between the distance to a 2% concentration and the reciprocal of the Richardson number. This is shown in Figure 15.114. The distance is shown as having a maximum at $Ri \approx 3$ ($1/Ri \approx 0.3$). For high values of Ri the cloud is advected by the wind with relatively little dispersion. For lower values atmospheric turbulence increases dispersion above that obtained from gravity-driven flow. If valid, the relationship is of some interest, but it does not appear to be supported by the Thorney Island results.

M.E. Davies and Inman (1987) describe a programme of 86 simulations of the Thorney Island trials. They give numerous sample plots, particularly concentration–time profiles, concentration–distance correlations and plan view concentration contours.

Figure 15.115 shows photographs of full scale trials and wind tunnel simulations for instantaneous and continuous releases, in both cases without and with fences. The decay of peak concentration with distance for the continuous releases, without and with fences, is correlated in the form shown in Figure 15.116, after

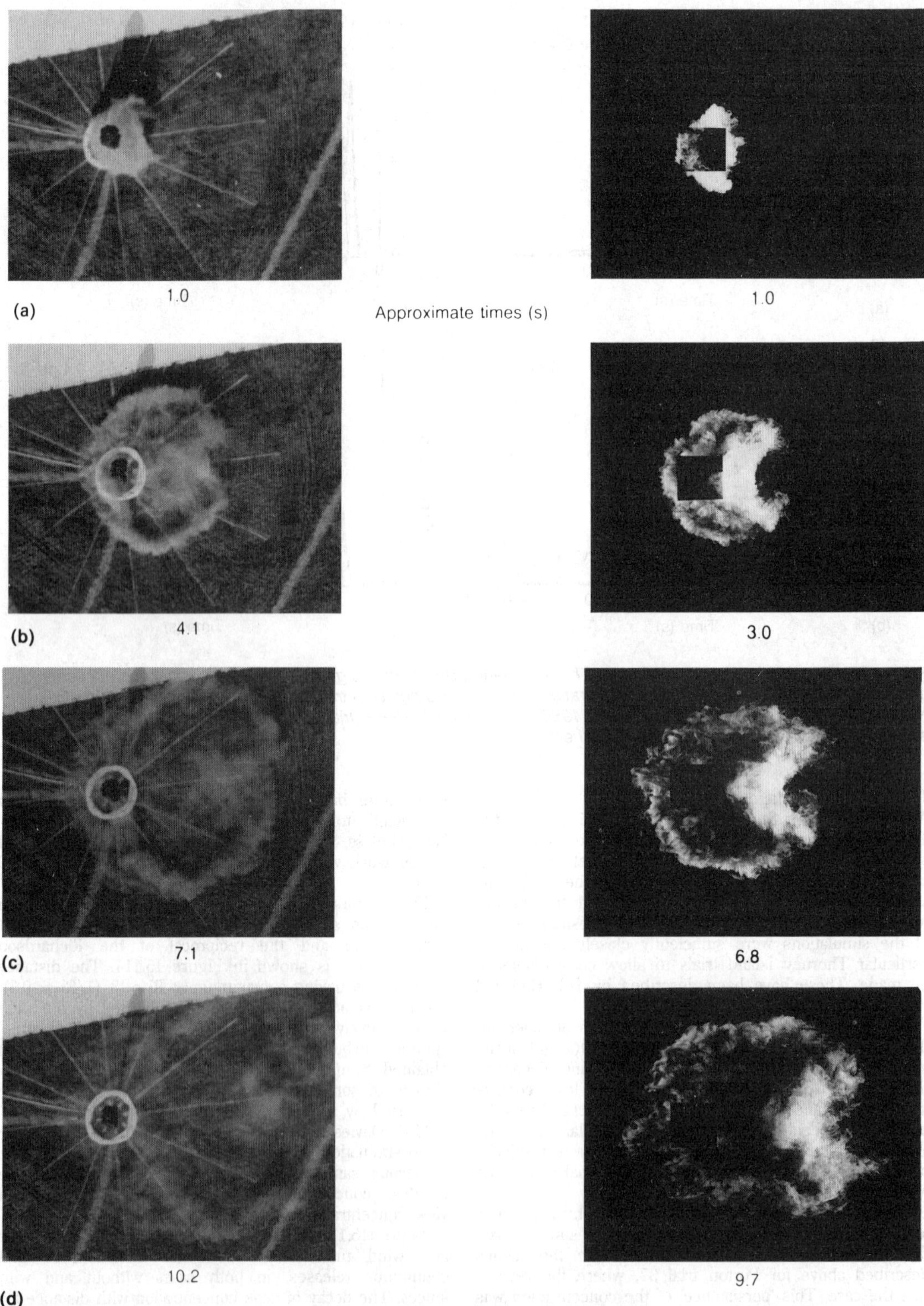

***Figure 15.112** Wind tunnel simulation of Thorney Island trials on dense gas dispersion – comparison of simulated and observed plan views of cloud in trial 13 (D.J. Hall and Waters, 1985): (a) both after 1.0 s; (b) observed after 4.1 s and simulated after 3.9 s; (c) observed after 7.1 and simulated after 6.8 s; (d) observed after 10.2 s and simulated after 9.7 s. All times are approximate (Courtesy of Elsevier Science Publishers)*

Figure 15.113 *Wind tunnel simulation of Thorney Island trials on dense gas dispersion – comparison of observed and simulated elevation views of cloud in trial 13 (D.J. Hall and Waters, 1985): (a) both at time zero; (b) observed after 3.1 s and simulated after 2.9 s; (c) observed after 6.1 s and simulated after 5.8 s; and (d) observed after 9.2 s and simulated after 8.7 s; (e) observed after 11.2 s and 11.6 s; all times approximate (Courtesy of Elsevier Science Publishers)*

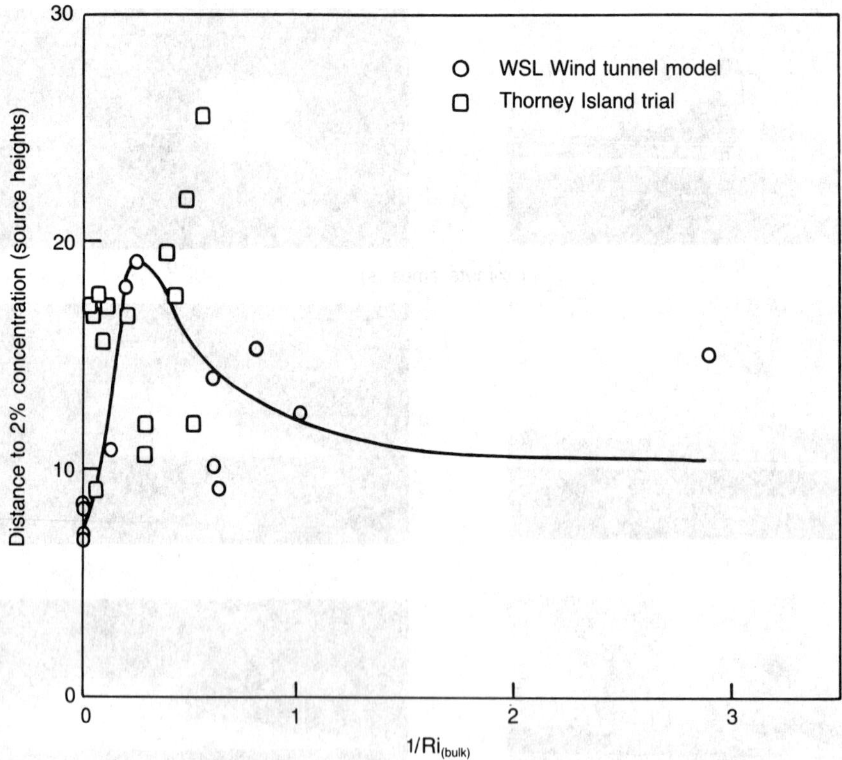

Figure 15.114 *Wind tunnel simulation of Thorney Island trials on dense gas dispersion – effect of bulk Richardson number on concentration downwind (D.J. Hall and Waters, 1985) (Courtesy of Elsevier Science Publishers)*

Meroney and Neff (1980). A buoyancy length scale l_b is defined as

$$l_b = \frac{g'Q}{u^3} \qquad [15.39.2]$$

and the further length scale l_q as

$$l_q = \left(\frac{Q}{u}\right)^{\frac{1}{2}} \qquad [15.39.3]$$

where Q is the volumetric release rate.

The plot gives

$$\frac{cul_b^2}{Q} \propto \left(\frac{x}{l_b}\right)^{-2} \qquad [15.39.4]$$

and hence

$$c \propto \left(\frac{x}{l_q}\right)^{-2} \qquad [15.39.5]$$

where c is the volumetric concentration.

The distance to the 2% concentration was investigated. It was found that the ratio of this distance in the model to that at full scale fell to about unity, though with some scatter. The authors suggest that the full scale distance may be estimated with 90% confidence as lying within the range delimited by a lower limit of half the model value and an upper limit of twice that value. The peak concentrations were found to be insensitive to the value of the Richardson number.

Several other workers have performed wind tunnel simulations of the Thorney Island trials. Figure 15.117 shows a computer graphics simulation of trial 8 by S.T. Chan, Ermak and Morris (1987); the computer simulation has the appearance of a wind tunnel simulation.

Van Heugten and Duijm (1985) investigated trial 8. They obtained reasonable agreement between the full-scale trials and wind tunnel model and between both of these and the TNO model described by van den Berg (1978).

Thorney Island trial 20, with a solid fence, has been simulated in a wind tunnel by Knudsen and Krogstad (1987). Again there was a tendency for the wind tunnel concentration profile to persist longer than that for the full-scale trial. The authors state that a possible explanation is the difference in Reynolds number and thus in the wake at the fence.

15.39.8 Ground slope

Physical modelling is particularly suited to investigation of the effects of terrain, obstructions and buildings. The

Figure 15.115 *Wind tunnel simulation of Thorney Island trials on dense gas dispersion – comparison of observed and simulated releases (M.E. Davies and Inman, 1987): (a) instantaneous release (Type 1); (b) instantaneous release with fence (Type 2); (c) continuous release (Type 3); (d) continuous release in fenced enclosure (Type 4) (Courtesy of Elsevier Science Publishers)*

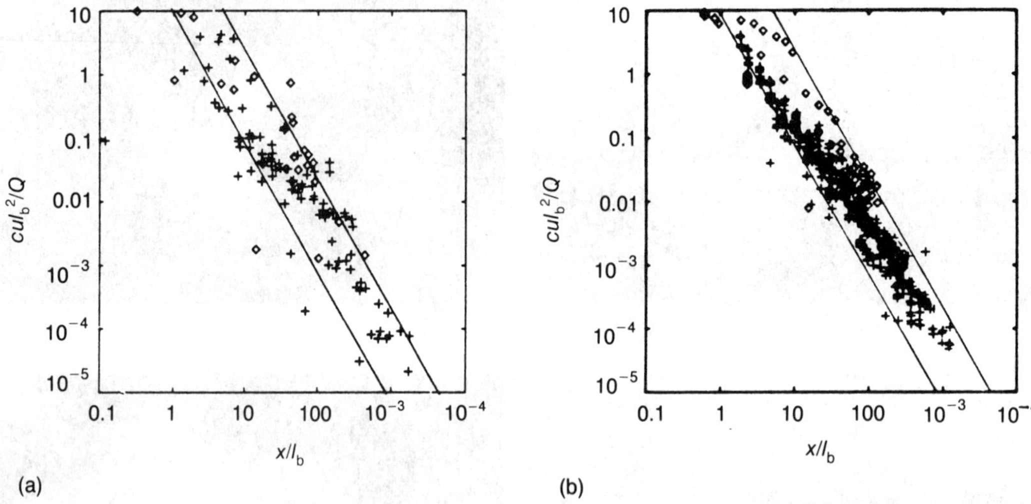

Figure 15.116 *Wind tunnel simulation of Thorney Island trials on dense gas dispersion – distance to 2% concentration (M.E. Davies and Inman, 1987): (a) observed; and (b) simulated. The solid lines represent the limits of data in the work of Meroney and Neff (1980) (Courtesy of Elsevier Science Publishers)*

dispersion of dense gas on a slope was another feature investigated in the study by D.J. Hall, Barrett and Ralph (1976) already described. Some of their results for the universal profile on a slope are shown in Figure 15.118.

15.39.9 Obstructions and buildings

There have been several investigations of dense gas dispersion in the presence of obstructions and buildings. Cheah, Cleaver and Milward (1983a) have used a water tunnel to make an *ad hoc* study of the effect of a simple barrier on the dispersion of a dense gas plume. Their results give a clear illustration of the effect of the gravity-driven phase. The authors comment that Figure 15.119, which is a plot of mean concentration vs downstream distance (the distance being expressed in terms of the ratio of the distance to the source diameter and with the relative density as a parameter) shows that for the plumes of higher relative density (and Richardson number) there is an initial phase with gravity slumping but little entrainment followed by a further gravity-driven phase where entrainment does occur and that the slope of the curve in this latter phase is not the same as for passive dispersion, whilst for the plume of low relative density (and Richardson number) the slope in the second region progressively approximates that of a passive plume.

The work also indicated the effectiveness of a barrier in reducing the downstream concentration. Even with the least effective barrier distance the concentration downstream was reduced by an order of magnitude.

The dispersion of a dense gas plume around a building has been investigated by Krogstad and Pettersen (1986). The presence of the building caused a strong modification of the plume.

C.I. Bradley and Carpenter (1983) have described an assessment of physical modelling which included the comparison of results from such modelling with results from models for the dispersion of 1000 te of LNG spilled onto land at a tanker terminal and onto water at the terminal. Experiments were done using both a wind tunnel and a water flume.

They discuss the problem of scaling up in respect of the temperature of the gas. Heat transfer from the surface to the cloud has a greater influence at low wind speeds, but at higher wind speeds is less significant. They followed Neff and Meroney (1981) in taking the gas density as the source value, but restricted their study to higher wind speeds (≥5 m/s).

Comparisons were made with results from the Germeles and Drake, Cox and Carpenter, Fryer and Kaiser, and ZEPHYR models. For the land spills a global roughness length of 0.2 m, corresponding to rough terrain, was used in the box models. The ZEPHYR model, being a *K*-theory model, modelled the effect of the structures directly.

Broadly, the distances to the extinction of the flammable plume in both the wind tunnel and water flume modelling were reduced by the presence of the terminal structures by a factor of about 2. Further, the dispersion of the land spill was dominated by the presence of the single tank from which the spill originated. The models tended to give higher distances to cloud extinction than the physical modelling, but the distances generally agreed to within a factor of 2 and again showed the reduction caused by the terminal structures.

Guldemond (1986) performed wind tunnel simulations of releases of 15 ton of liquefied ammonia at release rates in the range 6.5–52 kg/s. Air entrainment was assumed such that the air/ammonia mass ratio at the source was 10 : 1 and that there was complete evaporation of the liquid spray formed. Figure 15.120(a) shows the model of

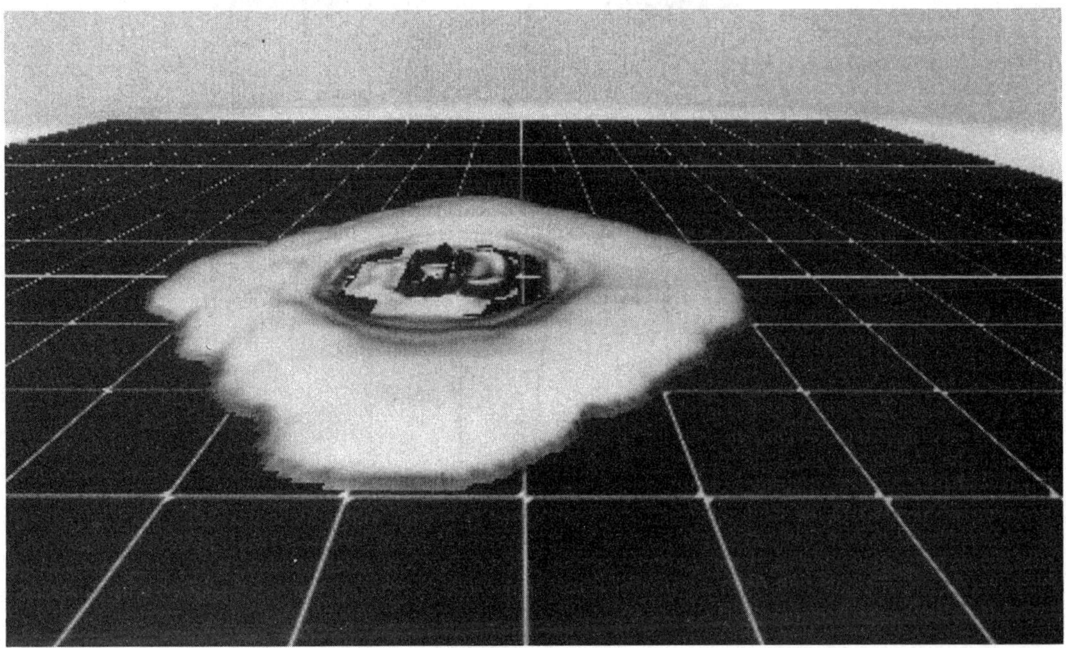

(a)

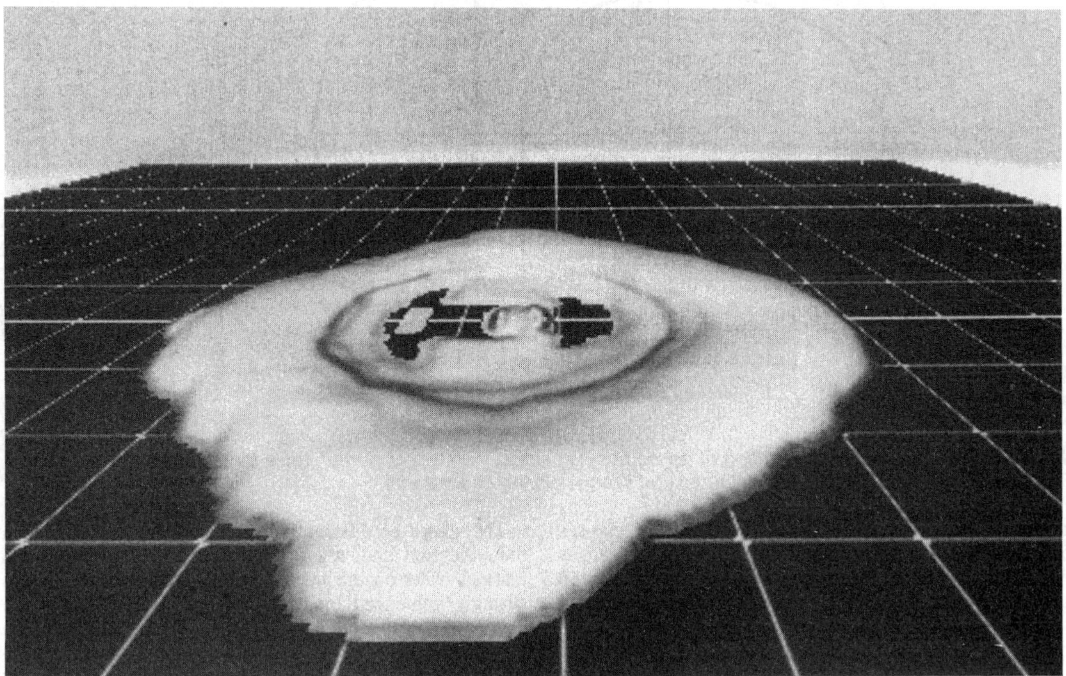

(b)

Figure 15.117 *Wind tunnel simulation of Thorney Island trials on dense gas dispersion – simulation of trial 9 (S.T. Chan, Ermak and Morris, 1987) (a) time 20 s; (b) time 30 s (Courtesy of Elsevier Science Publishers)*

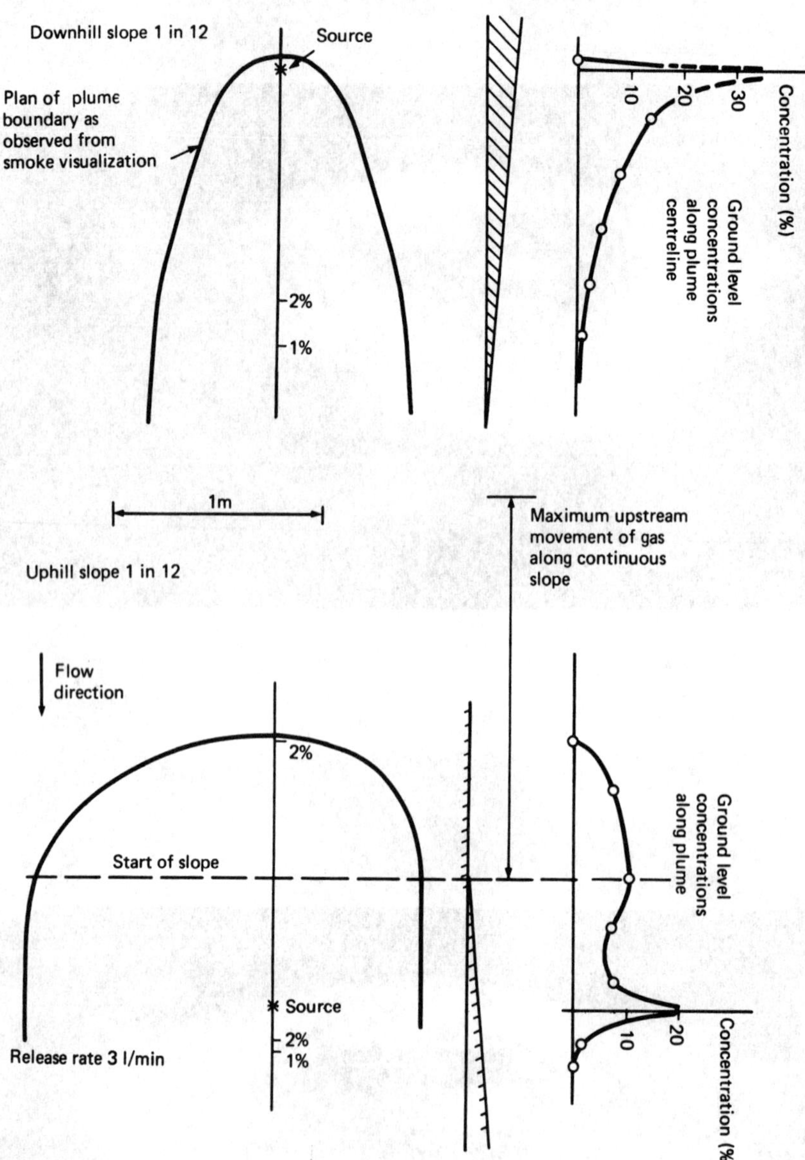

Figure 15.118 Wind tunnel simulation of dense gas dispersion – universal plume profile for dispersion on a slope (D.J. Hall, Barrett and Ralph, 1976) (Courtesy of Warren Spring Laboratory)

the industrial site in the wind tunnel and Figures 15.120(b)–(e) show the plumes formed by a dense gas on flat terrain and on the industrial site and by neutral buoyancy gas on these two types of terrain, respectively.

On flat terrain for the neutral buoyancy gas the upwind dispersion was 10 m and the initial cloud width was 120 m. For the dense gas the values were 90 and 540 m, respectively. For the industrial site the corresponding figures for the neutral buoyancy gas were 60 and 250 m and for the dense gas 210 and 540 m.

The effect of different locations of the source within the industrial site was investigated. The location had a marked effect on the neutral buoyancy plume, which was strongly influenced by particular buildings. It had much less effect on the dense gas plume. Mixing of the dense gas plume was caused not only by atmospheric turbulence but was also strongly affected by mechanically induced turbulence due to the obstructions. The cloud tended to mix vertically up to the average height of the buildings.

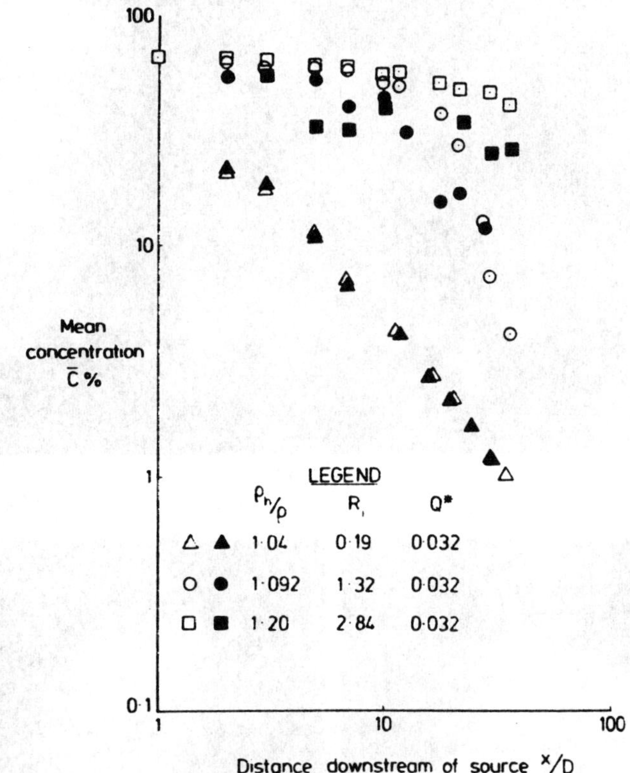

Figure 15.119 *Water tunnel simulation of dense gas dispersion at an obstacle – variation of mean concentration at ground level with distance for gas plumes of different density $\bar{C}$, mean concentration; D, diameter of plume source; Q*, non-dimensional plume spill rate; Ri, Richardson number; x, distance downstream of source; ρ, density of surrounding fluid; ρ_h, density of dense fluid (Cheah, Cleaver and Milward, 1983a) (Courtesy of the Institution of Chemical Engineers)*

The author gives concentration vs distance profiles and data on times to particular concentrations. Measurements were made at distances of 100, 400 and 450 m downwind. The concentrations at the first point were in the range 1000–3000 ppm, except for the very low release rate. Thereafter, the slope of concentration decay with distance on a log–log plot was about 1.7, which is characteristic of passive dispersion.

15.39.10 Concentration variability

As already mentioned, one of the strengths of wind tunnel modelling is that it furnishes an economical way of making a large number of repeat runs at the same nominal conditions. The variability in the concentration vs time profiles obtained in the simulation of the Porton trials by D.J. Hall, Hollis and Ishaq (1982, 1984) has already been described.

A systematic investigation of this variability in terms of the intensity of concentration fluctuations has been made by Meroney and Lohmeyer (1984). Their work is discussed in the context of concentration fluctuations in Section 15.45.

15.40 Dispersion of Dense Gas: Terrain, Obstructions and Buildings

The account of dense gas dispersion given so far has concentrated on dispersion over flat, unobstructed terrain. It is necessary now to consider more complex and obstructed terrain, including that containing buildings and obstructions. The corresponding treatment for passive dispersion is given in Section 15.17.

Some particular types of terrain which have been the subject of investigation include:

(1) slopes and ramps;
(2) fences and vapour barriers;
(3) water spray barrier and steam curtains;
(4) buildings – release upstream;
(5) buildings – release into wake;
(6) buildings – release inside;
(7) industrial sites;
(8) complex terrain.

The three principal approaches are the use of: (1) analytical models, (2) three-dimensional models and (3) physical modelling. The most versatile methods are

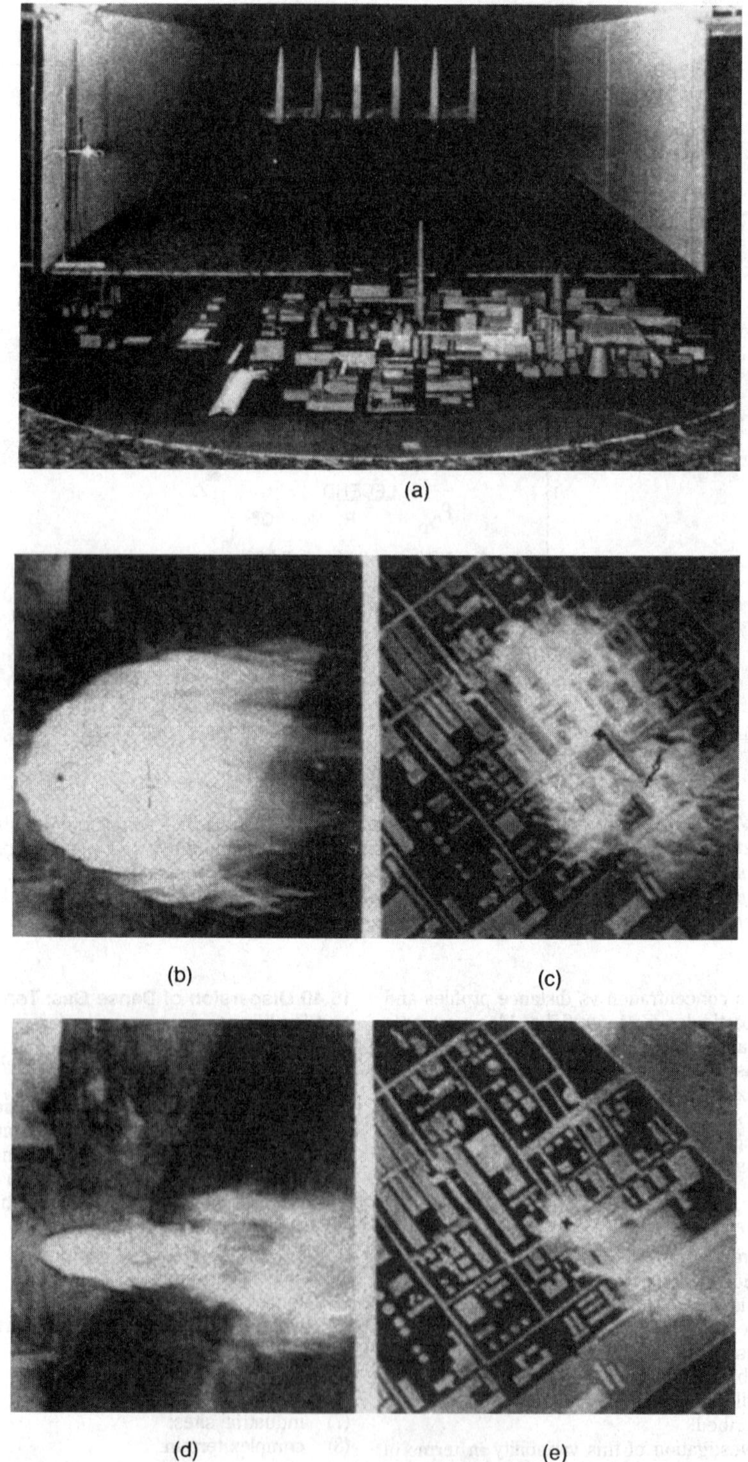

Figure 15.120 *Wind tunnel simulation of dense gas dispersion over an industrial site – continuous release of ammonia (Guldemond, 1986); (a) model in wind tunnel; (b) dense gas plume on flat terrain; (c) dense gas plume on industrial site; (d) neutral buoyancy gas plume on flat terrain; (e) neutral buoyancy plume on industrial site (Courtesy of the Institution of Chemical Engineers)*

three-dimensional models and physical modelling, but there is a growing body of analytical models for such features as slopes and obstructions which generally draw on the theory of gravity-driven flow and which may be used alone or in combination with the other methods.

General discussions of the effect of more complex and obstructed terrain have been given by Britter and Griffiths (1982), Gunn (1984), McQuaid (1986) and Britter and McQuaid (1988).

15.40.1 Slopes and ramps

An account of the behaviour of a dense gas on a slope in the field trials at Porton Down has been given by Picknett (1978a), as described in Section 15.37. Analytical models of the dispersion of a dense gas on a slope have been given by several workers.

A gravity current flowing down a slope develops a characteristic thickening at the downhill edge, or 'head'. Ellison and Turner (1959) conducted an experimental and theoretical investigation of the gravity current on a slope. For the experiments they used salt water flowing down an inclined plane in pure water. They found that for the continuous current well behind the head the mean velocity is independent of distance, but that the thickness of the fluid layer increases at a constant rate due to entrainment. The density excess decreases, maintaining a constant buoyancy flux down the slope.

Another experimental and theoretical study is described by Britter and Linden (1980). The experiments again involved the flow of salt water in pure water down an inclined plane. These workers investigated particularly the head of the gravity current on a slope. They found that the behaviour of the head depends on the angle θ of the incline to the horizontal. For $\theta \gtrsim 5°$ the head has a constant velocity, for $\theta \lesssim 0.5°$ its velocity is strongly affected by friction and decreases with distance, whilst in the range $0.5° \lesssim \theta \lesssim 5°$ there is a perceptible but weaker frictional effect.

They also found that the velocity u_f of the front is proportional to the volumetric flow per unit width:

$$u_f \propto (g'_o Q)^{\frac{1}{3}} \qquad [15.40.1]$$

with

$$g'_o = 2g\frac{\rho_2 - \rho_1}{\rho_2 + \rho_1} \qquad [15.40.2]$$

where g'_o is a reduced gravity, Q is the volumetric flow per unit width, u_f is the velocity of the front, or head, ρ is the density, and subscripts 1 and 2 denote less and more dense fluid, respectively.

They obtain the relation

$$\frac{u_f}{(g'_o Q)^{\frac{1}{3}}} = S_2^{\frac{1}{3}}\left[\frac{\cos\theta}{\alpha} + \frac{\alpha \sin\theta}{2(E + C_D)}\left(\frac{\sin\theta}{E + C_D}\right)^{-\frac{2}{3}}\right. \qquad [15.40.3]$$

with

$$\alpha = u_s/u \qquad [15.40.4]$$

$$E = \mathrm{d}h/\mathrm{d}x \qquad [15.40.5]$$

$$= S_2\frac{\mathrm{Ri_n}\tan\theta - C_D}{1 + 0.5S_1\mathrm{Ri_n}} \qquad [15.40.6]$$

$$\mathrm{Ri_n} = \left(\frac{g'h\cos\theta}{u^2}\right)_n \qquad [15.40.7]$$

$$g' = g\frac{\rho - \rho_1}{\rho_1} \qquad [15.40.8]$$

where C_D is a drag coefficient due to stress at the lower surface, E is a dimensionless rate of entraiment, or entrainment coefficient, g' is a reduced gravity, h is the height of the following flow, u is the mean velocity of the following flow, u_s is the velocity on a defined streamline, x is the downhill distance, S_1 and S_2 are parameters, θ is the angle of incline (°), and subscript n denotes 'normal'. The numerical values of the parameters are typically as follows: $C_D < 0.02$; $0.1 \le S_1 \le 0.15$; $0.6 \le S_2 \le 0.9$; $\alpha \approx 1.2$. Ellison and Turner obtained typical values of $E = 0.001\theta$ and $S_2 = 0.75$.

This value of the entrainment coefficient E should be valid for dense gases also, provided the density difference is not too large. For flow of a dense gas on slope there is a significant dilution which quite quickly produces a small density difference.

Britter and Linden found that for the velocity u_f of the head

$$\frac{u_f}{(g'_o Q)^{\frac{1}{3}}} = 1.5 \pm 0.2 \qquad \theta \ge 5° \qquad [15.40.9]$$

The authors give expressions for the mean velocity u of the following flow for limiting cases. For large slopes ($\theta \ge 30°$) they obtain with a small density excess $u_f \approx 0.6u$ and with a large density excess $u_f \approx u$ as $\theta \rightarrow 90°$. For small slopes they quote the results of Middleton (1966) to the effect that at very small angles $u_f = u$ and that u_f/u decreases rapidly as θ increases.

De Nevers (1984a) has given a model based on a force balance in which the force in the downhill direction due to gravity, less buoyancy, is balanced by three forces acting in the uphill direction: the shear forces at ground surface and at the top of the cloud, and the form force at the leading edge of the cloud. The force balance on a slope of angle θ is then

$$\pi R^2 H\rho_a g'\sin\theta = \pi R^2\rho_g\frac{u^2}{2}(f_b + f_t) + 2RH\rho_a\frac{u^2}{2}C_D \qquad [15.40.10]$$

and hence

$$u = \left[\frac{2\sin\theta}{\frac{\rho_g}{\rho_a}(f_b + f_t) + \frac{2HC_D}{\pi R}}\right]^{\frac{1}{2}}(g'H)^{\frac{1}{2}} \qquad [15.40.11]$$

where C_D is the form drag coefficient, f is the friction factor, g' is the reduced gravity, H is the height of the cloud, R is the radius of the cloud, u is the velocity downhill, ρ is the density, and subscripts a and g denote air and gas and b and t the bottom and top of the cloud, respectively.

De Nevers concludes that for even a modest slope the downhill velocity can be as large as the lateral velocity

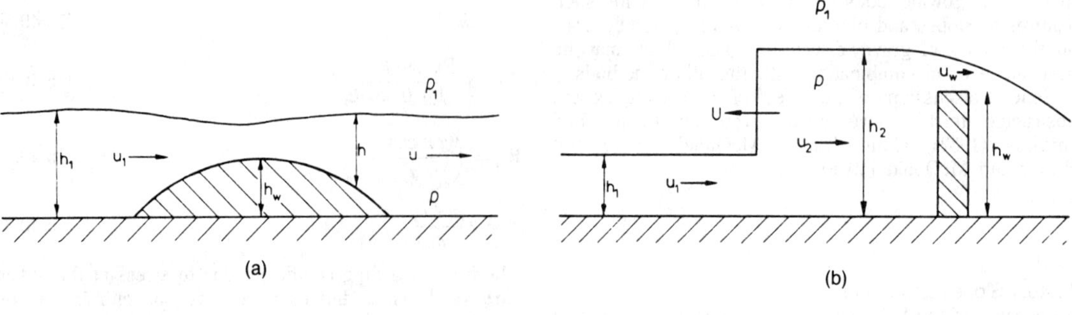

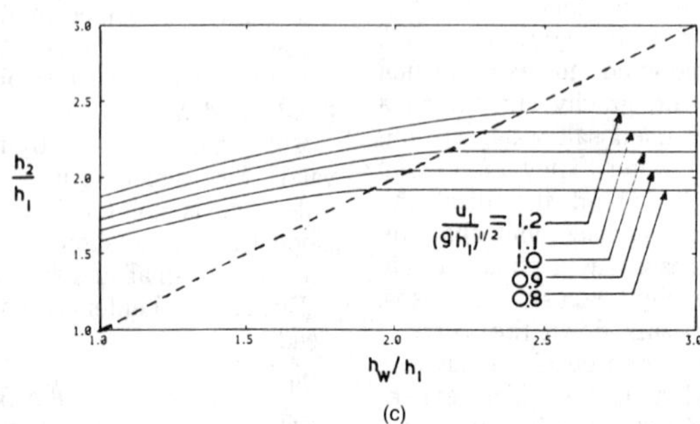

Figure 15.121 Dispersion of dense gas over an obstacle (Rottman et al., 1985): (a) shallow two-layer flow over an obstacle; (b) shallow two-layer flow over a wall; and (c) solutions of shallow water equations for depth of fluid behind the hydraulic jump as function of wall height (Courtesy of Elsevier Science Publishers)

due to gravity spreading. The effect of a degree of slope in the Burro field trials has been investigated by Koopman, Ermak and Chan (1989) using FEM3, as described in Section 15.37.

The universal plume profile of a dense gas on a slope has been studied by D.J. Hall, Barrett and Ralph (1976). This work was described in Section 15.39. A model of the dispersion of a dense gas up a ramp has been given by Britter and Snyder (1988).

The effect of a slope in mitigating the dispersion of a dense gas from a source to an uphill target has been studied by Heinhold, Walker and Paine (1987) using both an analytical model and wind tunnel work. They start by quoting the work of Meroney *et al.* (1977) who carried out wind tunnel tests on the flow of dense gas on a slope with an uphill wind. After the initial gravity slumping, the flow of the gas is governed by the wind speed. There is a critical uphill wind speed below which the dense gas flows downhill and above which it flows uphill.

The authors model this situation as an energy balance between the potential energy which must be supplied to raise a plume element up the slope and the kinetic energy of the wind, and derive the relation

$$\frac{u^2}{2} = g\frac{z}{\theta}\frac{d\theta}{dz} + g_o'z\frac{A_o}{A_p} \qquad [15.40.12]$$

with

$$g_o' = \frac{\rho_p - \rho_a}{\rho_p} \qquad [15.40.13]$$

where A_o is the initial cross-plume area of the plume element, A_p is the final cross-plume area of the element, g_o' is the initial reduced gravity, u is the wind speed at ground level, z is the height from the source to the target, θ is the potential temperature, ρ is the density, and the subscript p denotes the plume. The left-hand side term is the kinetic energy. The first term on the right-hand side is the potential energy of the atmosphere, assuming constant lapse rate; this is zero for neutral conditions and a maximum for stable conditions. The second term on the right-hand side is the change in potential energy of the element. For a given density, and hence concentration, Equation 15.40.12 may be used to determine the height uphill which the plume can reach. The initial and final cross-plume areas were determined using the authors' own AIRTOX dense gas dispersion model.

Heinhold, Walker and Paine describe the application of this model to releases of liquefied ammonia, assuming complete evaporation of all liquid spray formed. Two release rates were studied, 10 and 100 kg/s. At the lower flow rate the plume became passive at 100 m. Only atmospheric stability inhibited dispersion uphill. At the higher flow rate the plume remained dense beyond 1000 m. In the latter case, the concentration at uphill locations decreased as the slope increased. For slopes of 50:1, 10:1, 5:1 and 2:1 the concentration was the following proportion of its value for flat terrain: 98, 90, 83 and 45% at 100 m; 86, 65, 35 and 20% at 200 m and 50, 28, 14 and 8% at 400 m uphill. The authors conclude that a slope can afford an appreciable degree of protection to a target uphill of a source. This is so not only for dense gas but also for passive dispersion.

Deaves (1987b) has described the treatment of the case of dense gas flowing over a river and encountering a river bank on the far side. For this use was made of the model of Rottman *et al.* (1985) for an obstacle, given below.

15.40.2 Fences and vapour barriers

One of the simplest obstacles to the dispersion of a dense gas is a fence, wall or similar barrier. This may be a feature which is already there, but in others a vapour barrier may be expressly erected to effect a reduction in downstream concentrations.

An account of the behaviour of a dense gas encountering a fence in the field trials at Thorney Island has been given by M.E. Davies and Singh (1987a), as described in Section 15.38. Field trials on dispersion of a dense gas at a fence have also been described by M. Nielsen (1991). A simple treatment of the effect of a fence has been given by Jensen (1984).

A model for the behaviour of a dense gas encountering an obstruction such as a fence or vapour barrier has been given by Rottman *et al.* (1985). The situation considered is shown in Figure 15.121(a). The model is based on the shallow water equations. At steady state the governing equations are

$$\frac{\partial(uh)}{\partial x} = 0 \qquad [15.40.14]$$

$$u\frac{\partial u}{\partial x} + g'\frac{\partial(h_w + h)}{\partial x} = 0 \qquad [15.40.15]$$

with

$$g' = \frac{\rho - \rho_1}{\rho_1} \qquad [15.40.16]$$

where g' is the reduced gravity, h is the height of the dense fluid layer, h_w is the height of the obstacle, u is the velocity of the fluid, ρ is the density of the fluid and subscript 1 denotes upstream.

These equations are solved to give

$$uh = u_1 h_1 \qquad [15.40.17]$$

$$\frac{u^2}{2g'} + h + h_w = \frac{u_1^2}{2g'} + h_1 \qquad [15.40.18]$$

The regimes of interest are those of blocked flow and partially blocked flow. The equation of the boundary between these two regimes is

$$\frac{u_1^2}{g'h_1} = \frac{1}{2}(H_w - 1)^2\left(\frac{H_w + 1}{H_w}\right) \qquad [15.40.19]$$

with

$$H_w = \frac{h_w}{h_1} \qquad [15.40.20]$$

where H_w is a ratio of heights.

If the flow is completely blocked, the gas does not disperse downwind at all. The dispersion downwind for the case of partial blockage is obtained as follows. This situation is shown in Figure 15.121(b). Conservation of mass and momentum at the jump requires the conditions

$$(u_1 - U)h_1 = (u_2 - U)h_2 \qquad [15.40.21]$$

$$(u_1 - U)^2 = \frac{1}{2}(g'h_1)\frac{h_2}{h_1}\left(1 + \frac{h_2}{h_1}\right) \qquad [15.40.22]$$

and, assuming that flow over the obstacle is critical

$$u_w = (g'\Delta h)^{\frac{1}{2}} \qquad [15.40.23a]$$

with

$$\Delta h = h_2 - h_w \qquad [15.40.23b]$$

where Δh is the head difference, U is the velocity of the wave travelling upstream, u_w is the velocity over the obstacle, and subscript 2 is the fluid in the space between the upstream-moving wave and the obstacle.

By mass conservation Equation 15.40.23 can be expressed in terms of u_2 to give

$$u_2^2 = (g'h_2)\left(1 - \frac{h_w}{h_2}\right)^3 \qquad \frac{h_w}{h_2} < 1 \qquad [15.40.24a]$$

$$= 0 \qquad \frac{h_w}{h_2} \geq 1 \qquad [15.40.24b]$$

It can be shown that Equations 15.40.21, 15.40.22 and 15.40.24 yield for $h_w/h_2 < 1$

$$u_1 - (1 - H_b)U = (g'h_1)^{\frac{1}{2}}\left[H_b\left(1 - \frac{H_w}{H_b}\right)\right]^{\frac{3}{2}} \qquad [15.40.25]$$

$$(u_1 - U)^2 = \tfrac{1}{2}(g'h_1)H_b(1 + H_b) \qquad [15.40.26]$$

and for $h_w/h_2 \geq 1$

$$u_1 = (1 - H_b)U \qquad [15.40.27]$$

$$U^3 = u_1U^2 - (g'h_1)^{\frac{1}{2}}U + \tfrac{1}{2}(g'h_1)^{\frac{1}{2}}u_1 = 0 \qquad [15.40.28]$$

with

$$H_b = \frac{h_2}{h_1} \qquad [15.40.29]$$

where H_b is a ratio of heights.

Solutions of these equations are shown in Figure 15.121(c). The range $1.5 < h_2/h_1 < 2.5$ covers most cases of interest and as a rule-of-thumb $h_2/h_1 = 2$.

Release upstream of an obstacle such as a fence is one of the two main cases of more complex terrain treated in the *Workbook* by Britter and McQuaid (1988). The treatment is based on the work of Britter (1986 SRD R407). Using the *Workbook* notation given in Section 15.34, for a plume from continuous release encountering a fence height H normal to the wind direction the height h of the plume in the absence of the fence is obtained from

$$C_g uh = C_o q_o^* \tag{15.40.30}$$

where C_g is the maximum volumetric concentration at ground level, C_o is the volumetric concentration of the initial release, h is the height of the cloud, q_o^* is the volumetric release rate per unit width and u is the mean velocity at the fence.

The following regimes are distinguished. For $u/(g_o' q_o^*)^{\frac{1}{3}} > 5$ and $H/h \leq 30$, the plume is not blocked by the fence. In the range $4.5 \leq H/h \leq 15$ the ground level concentration is as if the plume were not dense and

$$\frac{C}{C_o}\frac{uH}{q_o^*} = 1.7 \pm 0.3 \tag{15.40.31}$$

whilst in the range $3.5 \leq u/(g_o' q_o^*)^{\frac{1}{3}} \leq 5$

$$\frac{C}{C_o}\frac{uH}{q_o^*} = 2.2 \pm 0.4 \tag{15.40.32}$$

where C is the volumetric concentration.

For $h_{0.5}/H \geq 2$ and $u/(g_o' q_o^*)^{\frac{1}{3}} \geq 3.5$, where $h_{0.5}$ is the height at which the concentration is half the ground level value, the ground level concentration is little affected by the fence.

As far as concerns plume width, the relations given for the plume widths W_f and W_{nf} with and without a fence, respectively, are as follows. The ratio W_f/W_{nf} increases with increase in fence height:

$$\frac{W_f}{W_{nf}} = 3 \qquad \frac{h}{H} \approx 0.05 \tag{15.40.33a}$$

$$\frac{W_f}{W_{nf}} = 1.5 \qquad \frac{h}{H} \approx 0.15 \tag{15.40.33b}$$

W_f/W_{nf} also increases weakly with $u/(g_o' q_o/W_{nf})^{\frac{1}{3}}$.

The following relation for concentration in terms of fence width is also given:

$$\frac{C}{C_o}\frac{uHW_f}{q_o} = 2.1 \pm 0.3 \qquad 0.04 < \frac{h}{H} < 0.2; \quad \frac{u}{(g_o' q_o/W_{nf})^{\frac{1}{3}}} \geq 4 \tag{15.40.34}$$

where q_o is the volumetric rate of release.

Meroney (1991) has described the use of models for the simulation of the effects of a fence on the dispersion of a dense gas. Carissimo *et al.* (1989) have described the use of MERCURE-GL to study the effect of a fence on dense gas dispersion. The water tunnel study by Cheah, Cleaver and Milward (1983a) of dense gas dispersion at a simple barrier has been described in Section 15.39.

15.40.3 Water spray barriers and steam curtains

Work on water spray barriers and steam curtains is described in Section 15.53. Accounts of the use of HEAVYGAS to simulate the effects of the water spray barrier used in the field trials described by Moodie (1981) have been given by Deaves (1983b, 1984). This work is described in Section 15.53. In the work just mentioned on dense gas dispersion at a fence, Meroney (1991) has also treated the case of a water spray barrier.

15.40.4 Buildings: release upstream

Another well defined obstacle to the dispersion of a dense gas is a single building or a defined sequence of buildings. The effect of a building on an upstream release is well treated for passive dispersion, but not so well for dense gas dispersion.

McQuaid (1987) has given an account of the behaviour of a dense gas encountering a building in the field trials at Thorney Island, as described in Section 15.38. A review of turbulent diffusion near buildings, including a treatment of dense gas dispersion, has been given by Meroney (1982). Krogstad and Pettersen (1986) used a wind tunnel to investigate the flow of a dense gas around a rectangular building. The effect of a sequence of rectangular buildings on the dispersion of a dense gas has been studied by Deaves (1989) using an adaptation of the model by Brighton for release into a building wake, described below.

15.40.5 Buildings: release into lee

Release into the immediate lee of a building is another topic where treatments are available for passive dispersion, but where there is less guidance for dense gas dispersion. This case is relevant, however, to a number of scenarios which may arise in practice. In particular, it applies to the case where a spillage occurs from a storage tank and then evaporates in the lee of the tank.

A model for the dispersion of a dense gas in the lee of a building has been given by Brighton (1986). He takes as his starting point the models for passive dispersion in a building lee given by Vincent (1977, 1978) and Fackrell (1984b), which were described in Section 15.17. A dense gas will tend to reduce the turbulent mixing and, if the flow is large, to modify the mean wake flow, and a separate treatment is necessary.

He considers a rectangular building of height h, width w, length l and frontal area A

$$A = wh \tag{15.40.35}$$

and a recirculation region, or wake, in the lee with height h, width w, length l_w, surface area S_w and volume V_w. The dimensionless wake length λ_w is defined as

$$\lambda_w = l_w/h \tag{15.40.36}$$

The basis of the model is the existence of two layers, a lower and an upper layer, in the lee. It is assumed that air flows directly into the upper layer but not into the lower one:

$$F_A = \alpha_o S_w u \tag{15.40.37}$$

where F_A is the flow of air into the wake, u is the wind speed and α_o is a constant. The concentration of contaminant C resulting from a source of strength Q in the lee is

$$C = Q/F_A \tag{15.40.38}$$

Then from Equation 15.40.37

$$C = Q/\alpha_o S_w u \tag{15.40.39}$$

The heights of the two layers are

$$\bar{h}_L = h_L/h \qquad [15.40.40]$$

$$\bar{h}_U = h_U/h \qquad [15.40.41]$$

$$h_L + h_U = h \qquad [15.40.42]$$

where h_L and h_U are the heights of the lower and upper layers, and $\bar{h}_L$ and $\bar{h}_U$ are the corresponding normalized values, respectively. Hence

$$\bar{h}_L + \bar{h}_U = 1 \qquad [15.40.43]$$

The treatment utilizes two main dimensionless parameters:

$$\bar{Q} = \frac{Q}{Au} \qquad [15.40.44]$$

$$B = \frac{g'_o Q}{u^3 w} \qquad [15.40.45]$$

with

$$g'_o = g\frac{\rho - \rho_a}{\rho_a} \qquad [15.40.46]$$

where B is the dimensionless buoyancy flux, g'_o is the reduced gravity at the source, $\bar{Q}$ is the dimensionless release rate, ρ is the density of the cloud and ρ_a is the density of air.

The flow from the lower to the upper layer F_{LU} is equal to that from the upper to the lower layer F_{UL}, given that there is no air flow directly into the lower layer. These flows are taken as

$$F_{LU} = F_{UL} = \frac{\alpha_I \lambda_w}{\text{Ri}} \qquad \text{Ri} > \text{Ri}_T \qquad [15.40.47a]$$

$$= \alpha_M \lambda_w \qquad \text{Ri} \leq \text{Ri}_T \qquad [15.40.47b]$$

where α_I and α_M are mixing coefficients.

The Richardson number Ri is defined as

$$\text{Ri} = \frac{(C_L - C_U)g'_o h}{u^2} \qquad [15.40.48]$$

which at the source becomes

$$\text{Ri} = \frac{g'_o h}{u^2} \qquad [15.40.49]$$

The transition Richardson number Ri_T is

$$\text{Ri}_T = \alpha_I/\alpha_M \qquad [15.40.50]$$

From analysis of the flows and concentrations in the wake, Brighton derives for one regime $\text{Ri} \leq \text{Ri}_T$:

$$C_L = \frac{(\alpha_M \lambda_w + 0.7\bar{h}_U)\bar{Q}}{0.7\bar{h}_U \alpha_M \lambda_w + (\alpha_M \lambda_w + 0.7\bar{h}_U)\bar{Q}} \qquad [15.40.51]$$

$$C_U = \frac{\alpha_M \lambda_w C_L}{\alpha_M \lambda_w + 0.7\bar{h}_U} \qquad [15.40.52]$$

where C_L and C_U are the volumetric concentrations in the lower and upper layers, respectively.

At the limit as $\bar{h}_L \to 0$ and $\bar{Q}_L \to 0$

$$\frac{C_U}{\bar{Q}} \to 1.4 \qquad [15.40.53]$$

whilst at the other limit as $\bar{h}_U \to 0$

$$C_L \to 1 \qquad [15.40.54]$$

This latter case corresponds to the situation where the source strength is so great that the gas released moves with a velocity comparable to the wind speed and occupies the whole of the wake region.

For the other regime, with $\text{Ri} > \text{Ri}_T$:

$$C_L = 1 - \frac{\alpha_I \lambda_w}{B} \qquad B \geq \alpha_M \lambda_w \qquad [15.40.55]$$

$$C_U = \frac{\alpha_I \lambda_w}{0.7\bar{h}_U \text{Ri}_o} \qquad [15.40.56]$$

where Ri_o is the initial Richardson number before mixing.

Brighton describes the matching of this model to unpublished experimental work by Britter. He takes for λ_w a value of 2.08, as given by Fackrell; $\alpha_I \lambda_w$ is taken as 0.036; and α_M as unity, although the influence of this parameter is weak.

This model gives as outputs the width, height and concentration at the end of the wake region. The inputs required for a dense gas dispersion model are these parameters plus the velocity. Brighton suggests matching may be achieved by retaining the width and concentration given by his wake model, but adjusting the height to give mass conservation. This will generally involve a reduction in the height; the plume contracts as it accelerates up to the wind speed.

If the buoyancy flux is large with $B \gg 1$, there is a tendency for the cloud at the source to spread upwind and sideways. For this condition Brighton quotes the relations obtained by Britter (1980):

$$S_U = \tfrac{1}{2}D + 2L_B \qquad [15.40.57]$$

$$S_L = \tfrac{1}{2}D + 4L_B \qquad [15.40.58]$$

with

$$L_B = \frac{g'_o Q}{u^3} \qquad [15.40.59]$$

where D is the diameter of the source orifice, L_B is the buoyancy length scale, S_L is the lateral spread of the cloud and S_U is the upwind spread of the cloud. This model has been used by Deaves (1987b, 1989), as described above, in modelling for safety cases.

Release into the lee of a building is the other main case treated by Britter and McQuaid (1988) in the *Workbook*. The treatment for a continuous release is based on the work of Britter (1982, 1986 SRD R407) on a square flat plate of side H normal to the direction of flow.

For the concentrations in the immediate lee, the following regions or limits are distinguished:

Region 1 $\quad \dfrac{g'_o q_o}{u^3 H} < 4 \times 10^{-3}$

Region 2 $\quad 4 \times 10^{-3} < \dfrac{g'_o q_o}{u^3 H} < 4 \times 10^{-2}$

Limit 3 $\qquad \dfrac{g'_o q_o}{u^3 H} > 2 \times 10^{-1}$

Limit 4 $\qquad \dfrac{g'_o q_o}{u^3 H} > 1$

where g'_o is the reduced gravity at the source, H is the height of the obstacle, q_o is the volumetric rate of release of material and u is the mean velocity at the obstacle height.

In region 1 the release rate is so low that the density of the release has virtually no effect on the concentration. In region 2 the lateral width of the plume is maintained within the recirculation zone in the immediate lee. At limit 3 there is some enhanced dilution in the immediate lee but the effect is weak. By limit 4 the effect of the obstacle has become negligible.

In both regions 1 and 2 by a distance of $x/H = 2$ the concentration is given by the relation

$$\frac{C}{C_o}\frac{uH^2}{q_o} \approx 1.5 \qquad [15.40.60]$$

where C is time mean ground level volumetric concentration and C_o is the volumetric concentration at the source.

These results apply for the condition (q_o/uH^2) or (q_o^*/uH) less than about 0.1, where q_o^* is the volumetric rate of release per unit length. Further downwind, the effect of the density difference is liable to reassert itself. For concentrations further downwind, one approach is to work in terms of a new source with values of q_o and g'_o determined from the value of C.

15.40.6 Buildings: release inside

A release of gas may occur inside a building and then escape into the atmosphere. A study of this scenario has been made by Brighton (1986, 1989 SRD R468). The particular case of interest was a release of liquefied chlorine.

For a continuous release of gas into a space ventilated by air, the molar flow of gas is

$$\frac{q_G}{M_G} = \frac{N_G Q}{V} \qquad [15.40.61]$$

and that of air is

$$\frac{Q_A \rho_A}{M_A} = \frac{N_A Q}{V} \qquad [15.40.62]$$

where M is the molecular weight, N is the number of moles, Q is the volumetric flow, V is the volume of space, ρ is the density, and subscripts A and G denote air and gas, respectively. The variables Q and ρ without a subscript refer to the gas–air mixture.

The natural ventilation flow Q_A is given by the ventilation equation

$$Q_A = bA(\Delta P/\rho_A)^{\frac{1}{2}} \qquad [15.40.63]$$

where A is the area of the opening, ΔP is the pressure difference between the outside and the inside, and b is a constant. Equation 15.40.63 is an application of Bernouilli's equation and is a standard relation given in various ventilation codes such as BS 5925: 1980. The value of constant b is taken by Brighton as 0.88.

The pressure drop is a function of the wind speed u:

$$\Delta P = \tfrac{1}{2} C \rho_A u^2 \qquad [15.40.64]$$

where C is a pressure coefficient. Values of C for cuboid buildings are tabulated in BS 5925: 1980. They are in the range 0.7–0.8 for an upwind face, −0.8 to −0.5 for the sides, and −0.4 to −0.1 for the lee.

The cases considered are (1) release under gravity alone and (2) release under the combined influence of gravity and wind. For the first case, Equation 15.40.63 is used for the air flow; Equation 15.40.64 is not used. For this case, the gas–air mixture flows out through an area at the bottom of the space and air flows in through an area at height h at the top of the space. The hydrostatic pressure is

$$= (\rho - \rho_A) gh \qquad [15.40.65]$$

where ρ is the density of the gas–air mixture and ρ_A is the density of air.

The pressure at the top openings is

$$= P_o - P' \qquad [15.40.66]$$

and that at the bottom opening is the sum of these

$$= P_o - P' + (\rho - \rho_A) gh \qquad [15.40.67]$$

where P' is the pressure difference between atmospheric pressure and the internal pressure and P_o is the atmospheric pressure. Applying Equation 15.40.63 to the two areas, the inflow of air is

$$Q_A = bA_i (P'/\rho_A)^{\frac{1}{2}} \qquad [15.40.68]$$

and the outflow of gas–air mixture is

$$Q = bA_o \left(-P'/\rho + \frac{\rho - \rho_o}{\rho} gh \right)^{\frac{1}{2}} \qquad [15.40.69]$$

where A_i is the area for inflow and A_o is the area for outflow.

Eliminating P' from Equations 15.40.68 and [15.40.69 and then utilizing Equations 15.40.61 and [15.40.62 yields

$$\tilde{q}_G = \tilde{N}_G \left(\frac{\rho/\rho_A - 1}{n^2 \tilde{N}_G^2 + a^2 \rho/\rho_A} \right)^{\frac{1}{2}} \qquad [15.40.70]$$

with

$$\tilde{q}_G = \frac{M_A q_G}{b M_G \rho_A A_i (gh)^{\frac{1}{2}}} \qquad [15.40.71]$$

$$\tilde{N}_G = \frac{N_G R T_A}{P_o V} \qquad [15.40.72]$$

$$a = A_i/A_o \qquad [15.40.73]$$

$$n = N_A/N_G \qquad [15.40.74]$$

where a is the ratio of areas, n is the ratio of molar flows, $\tilde{N}_G$ is a dimensionless quantity, q_G is the flow of gas, $\tilde{q}_G$ is the dimensionless flow of gas, R is the universal gas constant and T_A is the absolute temperature of the air.

Treating the contents of the space as a mixture of perfect gases or a two-phase mixture with vapour fraction f:

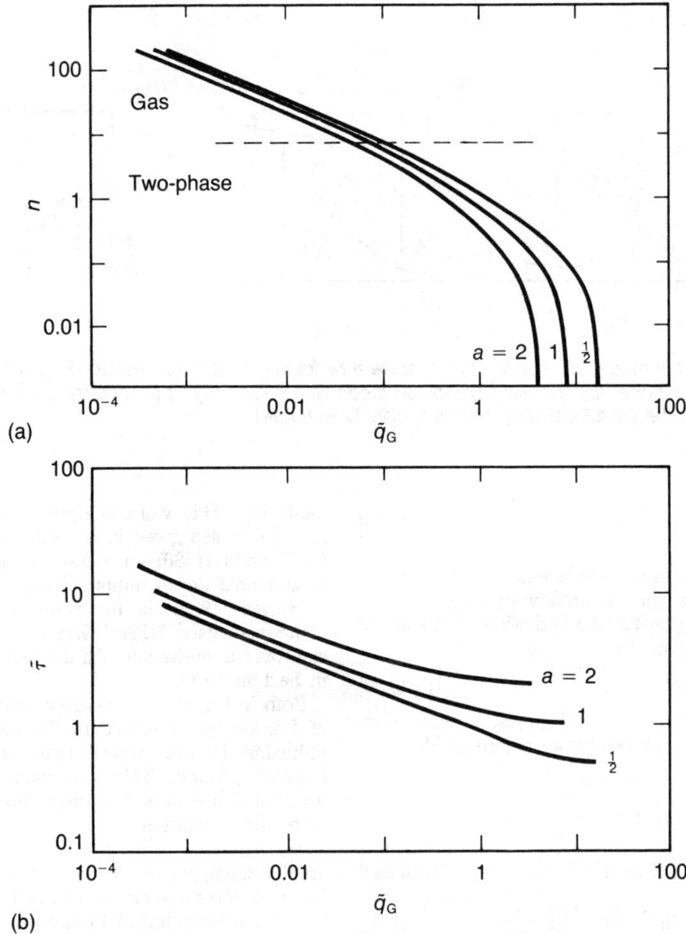

Figure 15.122 *Brighton model of release of gas inside a building – continuous release of chlorine under gravity alone (Brighton 1989 SRD R468): (a) air/chlorine molar flow ratio; (b) dimensionless time (Courtesy of the UKAEA Safety and Reliability Directorate)*

$$\tilde{N}_G = \frac{1}{1+f}\,\frac{T_A}{T} \quad [15.40.75]$$

where f is the vapour fraction and T is the absolute temperature of the gas–air mixture.

The time τ for the concentration of the gas to reach steady state if outflow is neglected is

$$\tau = \frac{M_G N_G}{q_G} \quad [15.40.76]$$

$$= \frac{\bar{\tau} V}{b A_i (gh)^{\frac{1}{2}}} \quad [15.40.77]$$

with

$$\bar{\tau} = \frac{\tilde{N}_G}{\tilde{q}_G} \quad [15.40.78]$$

For this case n and $\bar{\tau}$ are shown in Figures 15.122(a) and (b), respectively.

For the second case, that of release under the combined influence of gravity and wind, use is made of both Equations 15.40.63 and 15.40.64. For this case, it is necessary to consider several different situations since, depending on the configuration, the ventilation flow may assist or oppose the gravity flow. This is illustrated in Figure 15.123. For Case (a) in Figure 15.123, the inlet opening is high up on the upwind face and above the outlet opening and hence the ventilation effect always assists the gravity effect. For Cases (b) and (c) the inlet opening is low down on the upwind face and below the outlet opening and hence the ventilation effect opposes the gravity effect. In Case (b) the ventilation effect is strong enough to overcome the gravity effect, whilst in Case (c) it is not. Case (a) is termed the 'co-operative combined' mode, whilst Cases (b) and (c) are termed the 'opposing combined' mode.

A Richardson number is defined as

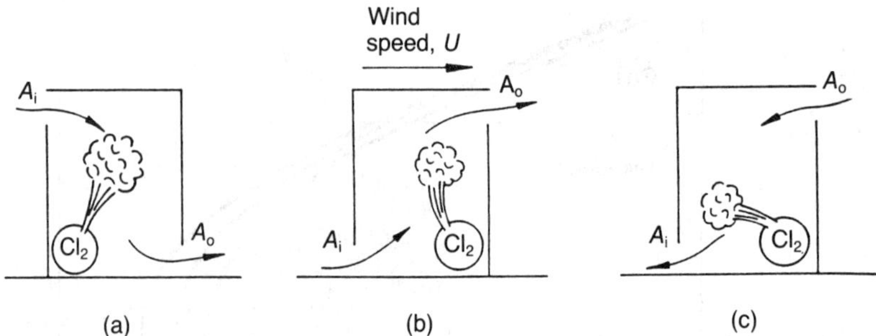

Figure 15.123 *Brighton model of release of gas inside a building – modes of release (Brighton 1989 SRD R468): (a) co-operative combined mode; (b) opposing combined mode Ri < $(\rho/\rho_o - 1)^{-1}$; (c) opposing combined mode Ri > $(\rho/\rho_o - 1)^{-1}$ (Courtesy of the UKAEA Safety and Reliability Directorate)*

$$\mathrm{Ri} = \frac{2gh}{(C_i - C_o)u^2} \qquad [15.40.79]$$

where C_i is the pressure coefficient for the inflow opening and C_o is that for the outflow opening.

The point at which gravitational and wind effects are of equal importance is given by

$$\mathrm{Ri}^{-1} = \rho/\rho_A - 1 \qquad [15.40.80]$$

Then, incorporating the additional pressure term from Equation 15.40.64 and proceeding as before yields

$$\tilde{q}_G = \tilde{N}_G \left(\frac{1 - \rho/\rho_A + \mathrm{Ri}^{-1}}{n^2 \tilde{N}_G^2 + a^2 \rho/\rho_A} \right)^{\frac{1}{2}}$$

$$\mathrm{Ri} < (\rho/\rho_A - 1)^{-1} \text{ (Case (b))} \qquad [15.40.81a]$$

$$\tilde{q}_G = \tilde{N}_G \left[\frac{\rho/\rho_A - 1 - \mathrm{Ri}^{-1}}{n^2 \tilde{N}_G^2 + (1/a^2)\rho/\rho_A} \right]^{\frac{1}{2}}$$

$$\mathrm{Ri} > (\rho/\rho_A - 1)^{-1} \text{ (Case (c))} \qquad [15.40.81b]$$

For this case, n and $\bar{\tau}$ are shown in Figures 15.124(a) and (b), respectively. The plots have a rather peculiar shape. At values of $\tilde{q}_G$ below a critical value, there are three steady-state solutions, of which one is unstable. For the case of a space containing originally zero concentration of the gas, it is the higher value of n in Figure 15.124(a) which represents the eventual steady state. At values of $\tilde{q}_G$ above the critical value there is only one steady state.

Accounts of the use of this work in the modelling of dense gas dispersion for safety cases have been given by Deaves (1987b, 1989).

15.40.7 Industrial sites

There have been several studies of the effect on the behaviour of a dense gas when it disperses over an industrial site containing features such as process plants, storages or terminals.

C.I. Bradley and Carpenter (1983) have conducted an investigation of dense gas dispersion at a storage and terminal site using both ZEPHYR and wind tunnel modelling. This work is described in Section 15.39. An account is also given in the same section of the work of Guldemond (1986) on a wind tunnel study of dispersion of ammonia on an industrial site.

Deaves (1989), in the work on safety cases already mentioned, used HEAVYGAS to study dense gas dispersion over a works site. An account of this study is given in Section 15.33.

Both in the work of Bradley and Carpenter and in that of Deaves the presence of the industrial site caused a reduction in the hazard range compared with unobstructed ground. This reduction was typically of the order of 2, although this value should be regarded as no more than a pointer.

15.40.8 Complex terrain

Complex terrain such as hills and valleys appears not to have been investigated to any great extent in relation to dense gas dispersion. Such terrain is of practical importance for passive dispersion of pollutant gases where the contamination distances are typically some kilometres, but of less concern for dense gas dispersion where the hazard ranges tend to be shorter. Likewise, coastal regions and urban areas have not received much attention as far as dense gas dispersion is concerned.

15.41 Dispersion of Dense Gas: Validation and Comparison

In the account given so far of dense gas dispersion, the various models have been presented, often with some mention of comparisons made with experimental work, and physical modelling has been described, much of it simulating field trials.

The degree of scale-up between laboratory experiments (or even field trials) and incident scenarios, the differences in the definitions of concentration used and the variability of concentration between different realizations of an ensemble of experiments, the large number of models and the appreciable differences in the predictions which they give, and the variety of outputs which may be of interest, all underline the need for methods of evaluating models.

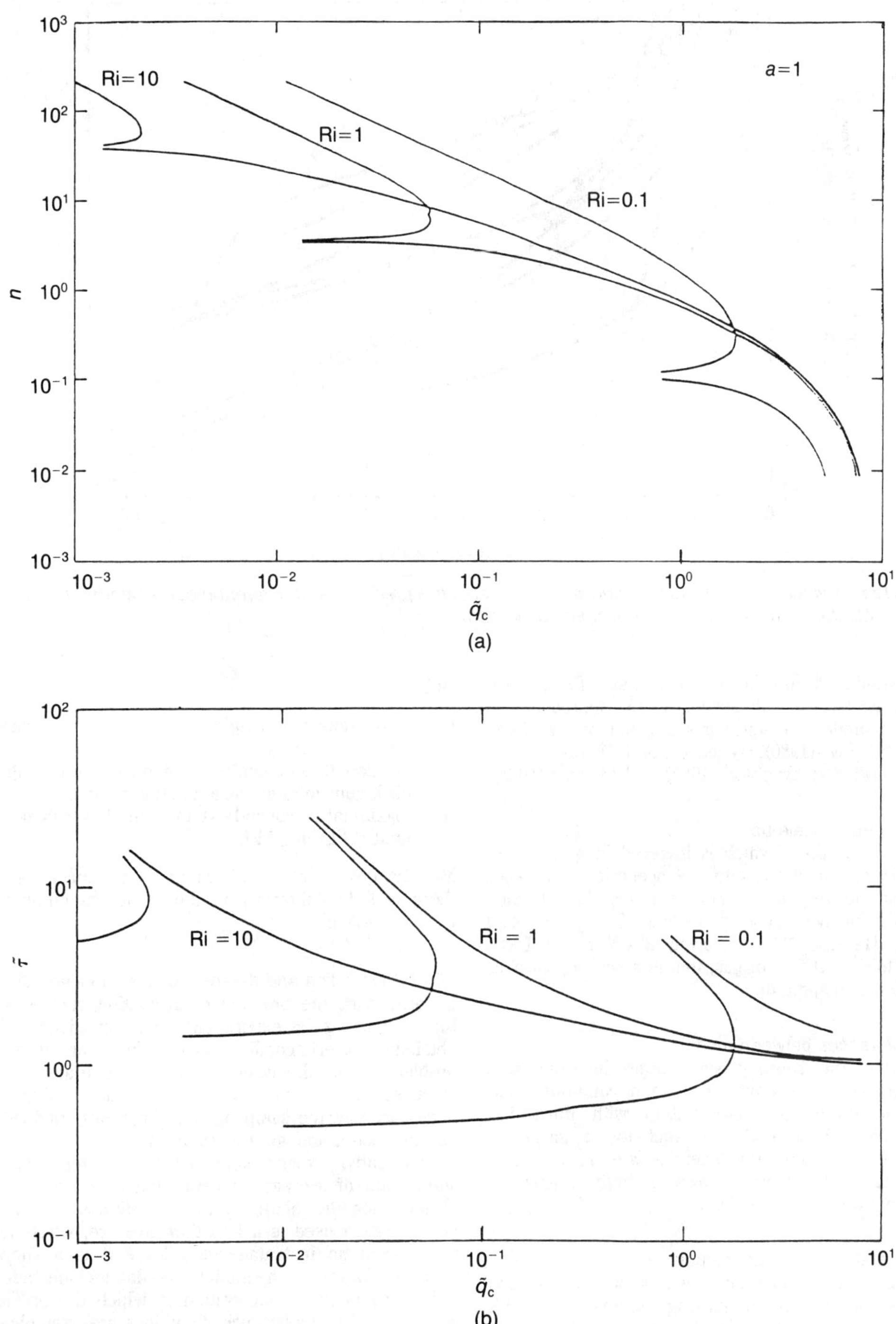

***Figure 15.124** Brighton model of release of gas inside a building – continuous release of chlorine in opposing combined mode with a = 1 (Brighton 1989 SRD R468): (a) air/chlorine molar flow ratio n; and (b) dimensionless time (Courtesy of the UKAEA Safety and Reliability Directorate)*

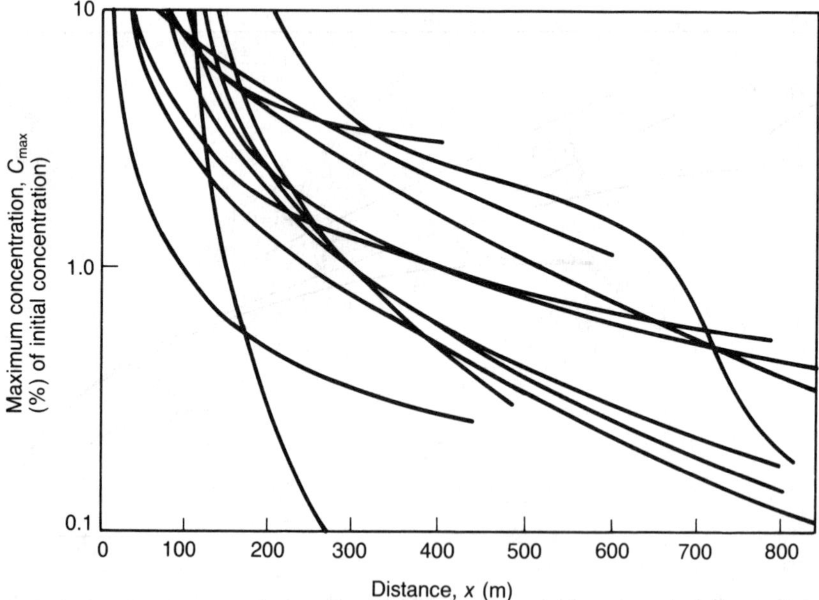

Figure 15.125 *Evaluation of models for dense gas dispersion – predictions of different models prior to the Thorney Island trials (McQuaid, 1984b) (Courtesy of Springer Verlag)*

Model evaluation has been a live issue for passive dispersion, as described in Section 15.15. Approaches to validation of models of dense gas dispersion have been described by Fay (1980), Wheatley and Webber (1984 CEC EUR 9542 EN), Brighton (1987) and Mercer (1988).

15.41.1 Degree of scale-up

The degree of scale-up which is involved in going from even the largest field trials to the scenario of a major incident can be very large. For example, the Thorney Island trials involved 2000 m^3 of gas. The LNG spill studied by Havens (1978) involved 25 000 m^3 of LNG, equivalent to 6×10^6 m^3 of gas; this is a scale-up of over three orders of magnitude.

15.41.2 Differences between models

At the start of the Thorney Island trials the organizers invited modellers to submit for defined conditions and releases the profile of concentration with downwind distance which their models predict for a particular scenario. The results of this exercise are illustrated in Figure 15.125. The figure shows a high degree of variability between the models.

15.41.3 Differences in model outputs

There is large variety of outputs which have been used by authors of models in making comparisons with experimental results. A review of these outputs has been given by Mercer. They include the concentrations

(1) bulk concentration as function of distance or time;
(2) peak concentration as function of distance or time;
(3) concentration at a given distance and time;

and

(4) cloud radius or height as a function of distance or time;
(5) location of cloud centre, leading edge or trailing edge, peak concentration as a function of time;
(6) maximum downwind extent of the lower flammability limit (LFL) or $\frac{1}{2}$ LFL.

Mercer gives details of comparisons made, showing there is little uniformity in the choice of outputs used for comparison.

15.41.4 Definition and determination of concentration

In large part, the problem of evaluation centres around that of defining the experimental concentrations and then obtaining experimentally stable values for these. One problem is that the values of the concentration obtained in a single trial vary with the sampling time. It is necessary that the sampling time be clearly defined both for the model and for the experiment.

Frequently, experimental results are reported using some kind of average concentration at a fixed location. Usually, for the plume from a continuous release the concentration used is a long-time average, whilst for the cloud from an instantaneous release it is a short-time average. Likewise, a model should include adequate definitions of the concentration to which the predictions are intended to apply. Such definitions are exemplified by those given in the *Workbook*. For a continuous release the correlations are for a concentration averaged over 10 min, whilst for a instantaneous release they are for the ensemble average of the maximum of 0.6 s mean concentrations.

A second problem is that there is a high degree of variability between the values of a concentration measured in experiments under the same nominal conditions. In principle, it is necessary to perform sufficient experiments to obtain a stable ensemble value. However, this is impractical with large field trials. Yet it is highly desirable that the predictions of models be compared with large-scale experiments.

A third problem is that the maximum concentration of a plume occurs on the centre line, but in relation to a fixed location this centre line may be continuously displaced due to plume meander. An alternative approach described by Puttock, Blackmore and Colenbrander (1982) is therefore to reconstruct from experimental measurements the variation of the location of the centre line with time and to utilize for comparison the concentrations on this reconstructed centre line.

A more detailed discussion of concentration is given in Section 15.45. The work described there, particularly that of Chatwin and co-workers, points to a fundamental approach in which the concentration is defined in terms of concentration probability contours.

15.41.5 Methods of evaluation

Some of the methods available for checking a model are described by Mercer. The predictions of the model may be compared with experimental work, either by the authors or by others. The data may also be compared with those from other models. The sensitivity of the predictions may be investigated. The variables and the dimensionless quantities used in the model may be examined. The constants in the submodels may be optimized by comparison with observation.

15.41.6 Comparison between models and experiments

Most authors of models have presented some comparison of predictions from their model with results of experimental work, ranging from laboratory-scale experiments to field trials. However, as just described, the problem of evaluation is a complex one, and such comparisons are frequently pointers rather than full validations.

A review of comparisons between models and experiments has been given by Mercer (1988). He defines a number of variables which are commonly compared (as described above), tabulates those which have been used in particular comparative studies and lists the data sets used in these studies. He comments that there is little commonality in the variables chosen for comparison and little justification given for the choice of experiments with which to make the comparison.

15.41.7 Comparison between models and experiments: some studies

Some mention of comparisons made between models and experiments has been made above in the accounts given of the individual models. In most cases these are comparisons made by the authors themselves. There have also been a number of comparative studies involving a number of models and trial series. Reviews of studies involving comparison of models with experiments have been given by Blackmore, Herman and Woodward (1982) and the CCPS (1987/2).

The CCPS (1987/2) describes evaluation studies by a number of workers. Quantities compared include: cloud dimensions; concentration contours, both in plan and crosswind elevation views; concentration vs distance; and mean relative error.

J.L. Woodward *et al.* (1982, 1983) have compared the Germeles and Drake model, the Eidsvik model, HEGADAS, ZEPHYR and MARIAH with the Dutch Freon, Matagordo Bay and Porton Down trials. Quantities compared include cloud height, cloud concentration contour (for LFL), and the concentration vs time profile at a fixed point.

A comparison has been given by Havens, Spicer and Schreurs (1987a) of SIGMET-N, ZEPHYR, MARIAH-II and FEM3 with the Burro trials, the quantity compared being concentration vs distance.

Koopman, Ermak and Chan (1989) have described a comparison of the Gaussian model, SLAB, FEM3 and HEGADAS and the proprietary model CHARM with the Burro, Coyote, Desert Tortoise, Eagle, Goldfish, Falcon, Maplin Sands and Thorney Island trials. Quantities compared include: concentration contours, both in plan and crosswind elevation views; concentration vs distance; and cloud temperature vs distance.

S.R. Hanna, Strimaitis and Chang (1991b) have compared the Gaussian model, INPUFF, HEGADAS, DEGADIS and the *Workbook* model and the propriety models AIRTOX and CHARM with the results of the Burro, Coyote, Desert Tortoise, Goldfish, Porton Down, Maplin Sands and Thorney Island trials and also with the passive dispersion Hanford and Prairie Grass trials. The quantities compared were the fractional bias and the normalized mean square error, evidently for the concentrations at different distances.

Touma *et al.* (1991) have compared SLAB, HEGADAS and DEGADIS and also AIRTOX and CHARM with the results of the Burro, Desert Tortoise and Goldfish trials, the quantity compared being concentration vs distance.

In some cases (e.g. CCPS, 1987/2) investigators have found the performance of the Gaussian model to be virtually as good as that of dense gas dispersion models, though this finding should be treated with caution.

15.41.8 Comparison between models and experiments: some problems

The use of data from field trials to validate models involves a number of problems. An account of these in the context of the Thorney Island trials has been given by Brighton (1987). He describes in detail the sources of error in the data obtained from such trials.

The data required depend on the type of model. The data yielded by a trial are a set of measurements of profiles of concentration vs time at each sensor within the cloud together with the parameters.

For a three-dimensional model the outputs are again concentration vs time profiles and thus comparison is relatively straightforward. A problem can arise, however, if the experimental measurements are taken very close to the ground. The grid size in the three-dimensional model may be too coarse to give an accurate prediction of these concentrations.

The comparison is more difficult for a box model. It is not a simple matter to extract from the experimental data the quantities required by modellers. The extent of the processing necessary to convert the Thorney Island results into a data set that conforms with the requirements of modellers was indicated in Section 15.38.

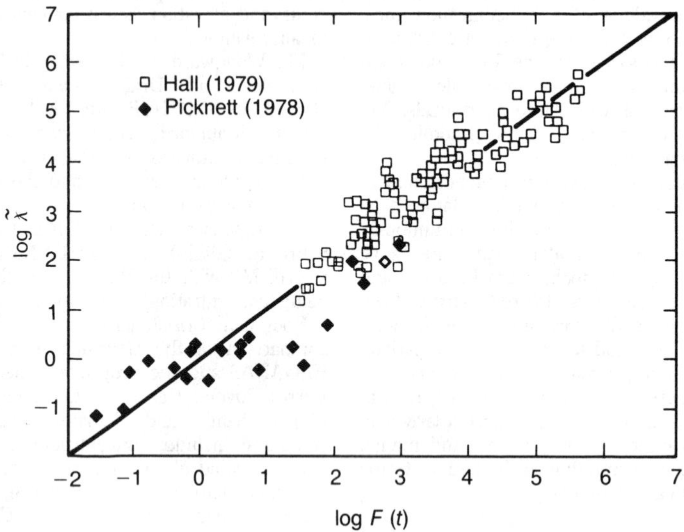

***Figure 15.126** Evaluation of models for dense gas dispersion – comparison of observed peak concentrations with those predicted by models τ, time of passage of peak concentration; $\tilde{\chi}$, peak cloud concentration. (Fay and Ranck, 1983) (Courtesy of Pergamon Press)*

Quantities required are those such as mean concentration, cloud dimensions and advection velocity. Brighton gives examples of the differences in the estimates of these quantities given by different workers.

In view of the work involved in establishing the data sets required to conduct model evaluations, it is valuable, where the data permit it, to have an archive of data in standard format. S.R. Hanna, Strimaitis and Chang (1991b) describe the creation of such a modeller's data archive (MDA).

15.41.9 Comparison between different models: some studies

In addition to comparisons between predictions of models and results of experiments made by the authors of the particular model, comparisons have also been made between a number of models. A series of studies of the performance of available models has been carried out by Havens, starting with the LNG spill studies described in Section 15.22. Mention of some of the other comparisons made between models has been made in the accounts of the various models.

The CCPS *Workbook of Test Cases for Vapor Cloud Source Dispersion Models* (CCPS, 1990/8) gives results from a number of models, principally the Gaussian dispersion model, SPILLS, SLAB and DEGADIS, and thus gives a comparison of models in a set of typical hazard assessment applications.

15.41.10 Measures for evaluation

There have been, therefore, quite a large number of model evaluation studies. Most of these studies have relied on essentially *ad hoc* criteria of evaluation. One problem in developing evaluation criteria is to identify those parameters which are the best discriminators. Thus, for example, one of the main features which a model should predict is the degree of air entrainment. The cloud radius tends not to be a good discriminator for this. It is the height which is much more strongly affected by entrainment.

Another problem is the formulation of criteria which allow comparisons to be made between models. Here criteria such as cloud dimensions or concentration profiles may be abandoned in favour of statistical measures. There is no generally accepted set of measures for validating models, but several authors have developed particular methods.

Fay (1980) has utilized a scatter diagram technique in which for the variable of interest the experimental value is plotted against the model value. By normalizing the variable, the method can be extended to allow comparisons to be made between the experimental value and the values predicted by different models. One of the plots using a normalized variable given by Fay is shown in Figure 15.126.

Work on goodness-of-fit measures (GFMs) between area-averaged concentrations obtained in experiments and concentrations given by box models has been described by Wheatley, Prince and Brighton (1985).

S.R. Hanna, Strimaitis and Chang (1991b) have utilized two statistical measures: the fractional bias (FB) and the normalized mean square error (NMSE). They give a series of plots of FB vs NMSE, each showing the performance of a number of models for a particular series of trials.

In wind tunnel work there is evidence of a degree of uniformity in comparing results of field experiments and of physical modelling, in that several workers have adopted the practice of D.J. Hall, Hollis and Ishaq (1982, 1984) in utilizing cloud width, arrival and departure times, and concentration vs time profiles.

15.41.11 Extent of validation attainable

For gas dispersion modelling generally, it has to be recognized that there may be limits to the degree of agreement which can be expected between model predictions and experimental results. A discussion of the problem is given by Mercer (1988). In large part, it centres on the variability of concentrations just described. Mercer quotes the following statement by Lamb (1984) made in the context of air pollution and its regulation:

> The predictions even of a perfect model cannot be expected to agree with observations at all locations. Consequently, the goal of 'model validation' should be one of determining whether observed concentrations fall within the interval indicated by the model with the frequency indicated, and if not, whether the failure is attributable to sampling fluctuations or is due to failure of the hypotheses on which the model is based. From the standpoint of regulatory needs the utility of a model is measured partly by the width of the interval in which a majority of observations can be expected to fall.

For heavy gas dispersion specifically, the high degree of inertia of the cloud results in some reduction in the variability which would otherwise occur. Mercer suggests that a realistic expectation for a model may be that its predictions be accurate within a factor of 2–3.

In order to reduce the degree of uncertainty, the experimental work should cover a wide range of conditions and scales and should include repeat experiments, which may most conveniently be obtained by physical modelling.

15.41.12 Validation of models against Thorney Island trials: Phase I

Mention has already been made of the exercise carried out before the Thorney Island trials in which modellers were invited to submit predictions from their models. A large number of comparisons have been presented between the results of the Thorney Island Phase I trials and those from model simulations. Some of these comparisons are listed in Table 15.53.

15.41.13 Validation of models against Thorney Island trials: Phase II

The number of models which are capable of simulating dense gas dispersion over other than flat, unobstructed terrain is limited, and the number of comparisons between the results of the Thorney Island Phase II trials and those from model simulations is correspondingly smaller. Such comparisons have been given for SLUMP and HEAVYGAS by Deaves (1985, 1987a,b).

15.41.14 Comparison of physical models with Thorney Island trials

Comparisons have also been presented between the results of the Thorney Island trials and those from physical modelling, predominantly in wind tunnels. Mention has already been made of the exploratory work done by D.J. Hall, Hollis and Ishaq (1982, 1984) using a wind tunnel to assist in the design of the trials. Some of the trials conducted were sufficiently close to these pre-trial experiments to allow comparison to be made. These comparisons have been reported for trials 7, 11, 13, 15 and 18 by D.J. Hall and Waters (1985). Further comparisons include those for trial 8 by van Heugten and Duijm (1985) and for trial 20 by Knudsen and Krogstad (1987). A programme of wind tunnel simulations for all the Thorney Island trials has been described by M.E. Davies and Inman (1987).

Table 15.53 *Some comparisons made between results of Thorney Island Phase I trials and results of model simulations*

Model	*Authors*
van Ulden	van Ulden (1987)
BG/C&W	Carpenter et al. (1987); Cornwell and Pfenning (1987)
FEM3	S.T. Chan, Ermak and Morris (1987)
HEGADAS	Cornwell and Pfenning (1987)
HEGABOX/HEGADAS	Puttock (1987b)
DEGADIS	Spicer and Havens (1985, 1987)
Fay and Ranck	Sherrell (1987)
Eidsvik	Gotaas (1985); Cornwell and Pfenning (1987)
Picknett	Sherrell (1987)
Webber and Wheatley	Webber and Wheatley (1987)
CIGALE2	Cabrol, Roux and Lhomme (1987)
MARIAH II	Cornwell and Pfenning (1987); Havens, Schreurs and Spicer (1987)
MERCURE-GL	Riou (1987)
Jacobsen and Magnussen	Jacobsen and Magnussen (1987)

15.42 Dispersion of Dense Gas: Particular Gases

So far the account given of dense gas dispersion has been a general one. It is now necessary to consider some gases of industrial interest for which the dense gas behaviour is governed by the specific characteristics of the gas. The gases considered are:

(1) propane;
(2) LNG;
(3) chlorine;
(4) ammonia;
(5) hydrogen fluoride.

The source terms for some of these gases are considered in Section 15.23. The validation of models against experimental work on, and the application of these models to simulations for, these gases are treated in the sections on the individual models, Sections 15.25 to 15.36. Mitigation of releases of the gases is dealt with in Sections 15.53 and 15.54.

The account in this section deals with the effect on the density of the gas cloud of the following factors:

(1) molecular weight;
(2) boiling point;

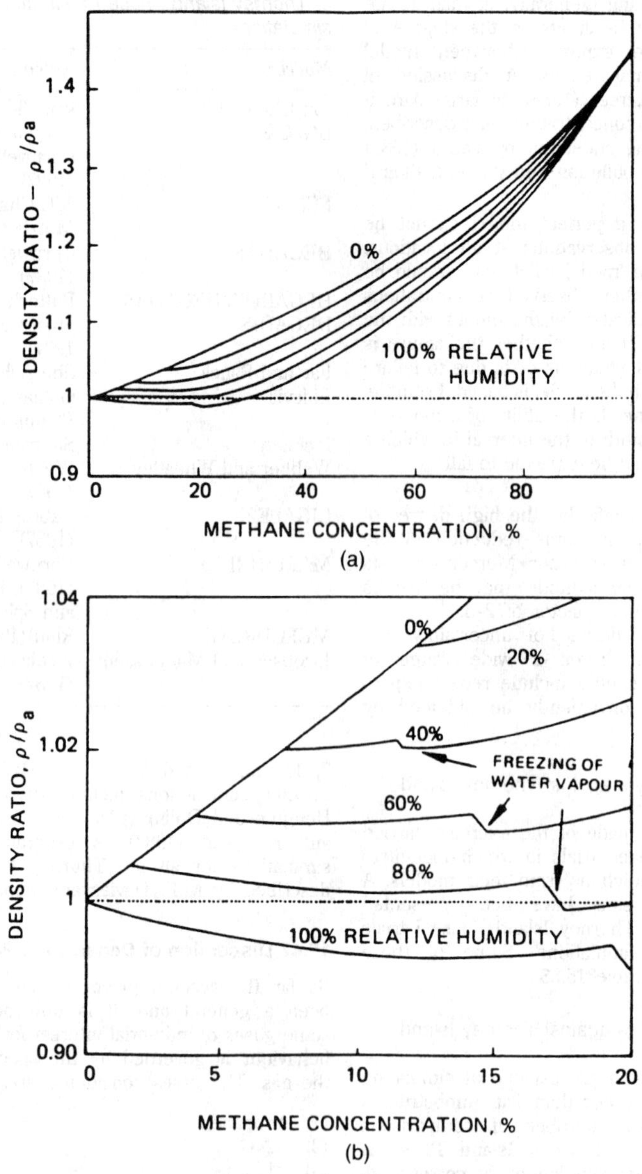

Figure 15.127 *Density of a methane–air mixture formed by release of LNG into the atmosphere (Puttock, 1987c): (a) main plot; and (b) enlargement of lower left-hand section of (a) (Courtesy of the American Institute of Chemical Engineers)*

(3) chemical behaviour;
(4) atmospheric humidity;
(5) surface heat transfer;
(6) liquid spray.

The modelling of the dispersion of these materials raises two questions: whether the gas cloud is dense immediately following the initial release; and, if so, how the cloud changes in density and whether it becomes buoyant. The main interest is in the behaviour of the gas cloud from release of a liquefied gas.

15.42.1 Propane

Propane has a molecular weight of 44. It is dense by virtue of its molecular weight. This is so whether the release is one of propane gas or liquefied propane.

The normal boiling point of propane is –42°C. If the release is one of liquefied propane, the low temperature

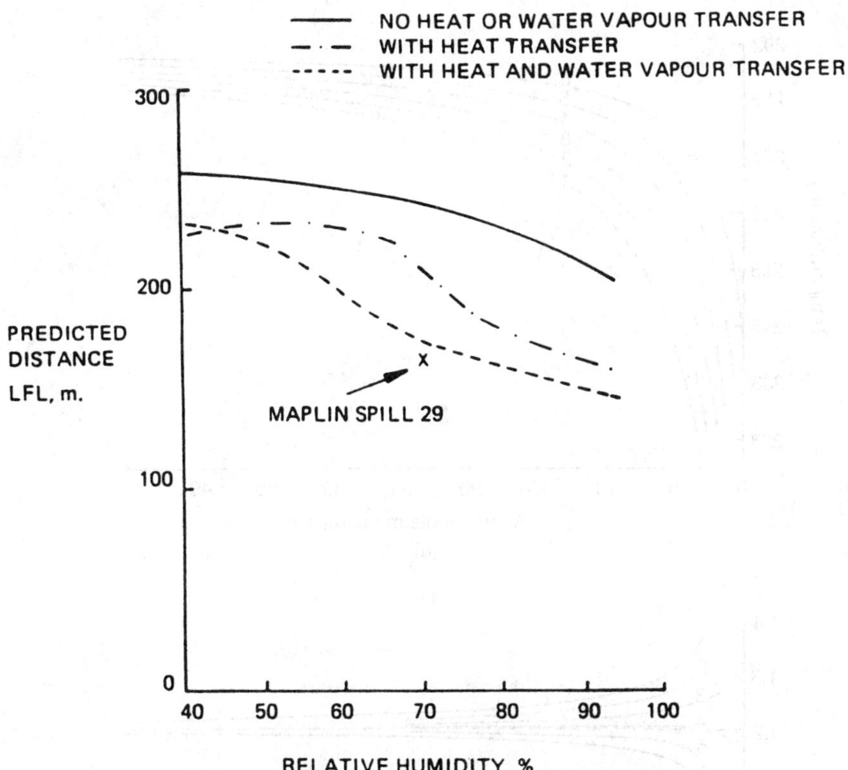

Figure 15.128 *Effect of heat transfer from surface on dense gas dispersion – predictions of HEGADAS for distance to the LFL for the Maplin Sands LNG spill trial 29 under three different assumptions (Puttock, 1987c) (Courtesy of the American Institute of Chemical Engineers)*

makes the gas cloud initially somewhat more dense. In principle, heating up of the gas will render it less dense. However, the initial temperature difference between the gas at its boiling point, and the ambient features, the ground or water surface and the atmosphere, is not great and heat transfer to the gas does not play a major role. Any density changes due to temperature changes tend to be small.

The density of the gas cloud will also be affected by atmospheric humidity and by any liquid present as spray. The effect of these factors on the density of propane gas clouds does not appear to have been much studied, but some indication may be obtained from their effect on clouds of the other gases, as described below. Propane is generally modelled as a 'simple' dense gas.

15.42.2 LNG

The characteristics of the gas cloud from a release of LNG, mainly methane, have been described by Puttock (1987c). Broadly, LNG releases normally give rise to dense gas clouds. However, under conditions of very high atmospheric humidity the initial cloud may be buoyant, whilst in other cases the cloud, initially dense, may become buoyant with time due to surface heat transfer.

LNG is mainly liquefied methane and, as such, has a molecular weight of 16 and a normal boiling point of −161°C. Insofar as it behaves as a dense gas, it does so by virtue of factors other than its molecular weight. Of these the principal factor is temperature. The gas is sufficiently cold that on release it is dense.

The temperature difference between the cold gas and the ambient features is significant. Heat transfer from the atmosphere and from the surface is appreciable and can have a significant effect on the density. The gas may be heated up sufficiently that it becomes buoyant.

Another factor which affects the density of the gas cloud is the humidity. The cold gas causes condensation, and also freezing, to occur and the heat release consequent on these phase changes causes heating of the cloud. If the humidity is high, this can just cause the cloud to become buoyant.

Figure 15.127, from Puttock, gives relative density vs concentration curves for mixing of methane at its boiling point with air with relative humidity (RH) as parameter. They show that with dry air (RH = 0%) the gas cloud remains dense at all concentrations, whereas with wet air (RH = 100%) the cloud is slightly buoyant at lower concentrations.

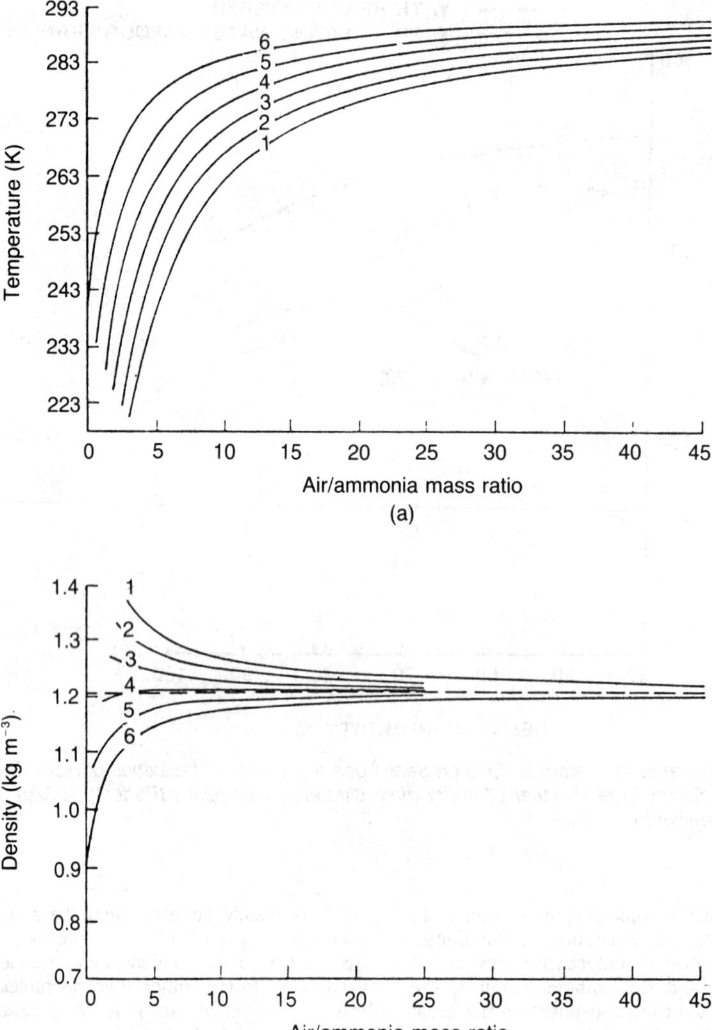

Figure 15.129 *Temperature and density of an ammonia–air mixture formed by release of ammonia into the atmosphere (after Haddock and Williams, 1978 SRD R103): (a) temperature for release into dry air; (b) density for release into dry air: and (c) density for release into wet air. (1) f = 20%; (2) f = 16%; (3) f = 12%; (4) f = 8%; (5) f = 4%; (6) f = 0, in all three plots (Courtesy of the UKAEA Safety and Reliability Directorate)*

The effect of the above variables is illustrated in the modelling work done on the Maplin Sands trials, described by Puttock. Using a version of the HEGADAS model in which surface heat transfer and atmospheric humidity were neglected gave downwind distances to the LFL which were large and exceeded those observed, whilst inclusion of these effects in the model reduced the distances appreciably and brought them closer to the observed values. This is illustrated in Figure 15.128 for trial 29, for which the wind speed was quite high (7.4 m/s). Comparative results for this and other trials are tabulated by Puttock.

Some accounts of work on mathematical modelling and physical modelling of LNG dispersion suggest that surface heat transfer can be neglected, except for low wind speed conditions. These results of Puttock indicate that failure to take it into account can lead to appreciable error.

The density of an LNG cloud will also be affected by any liquid spray which may be present. The source

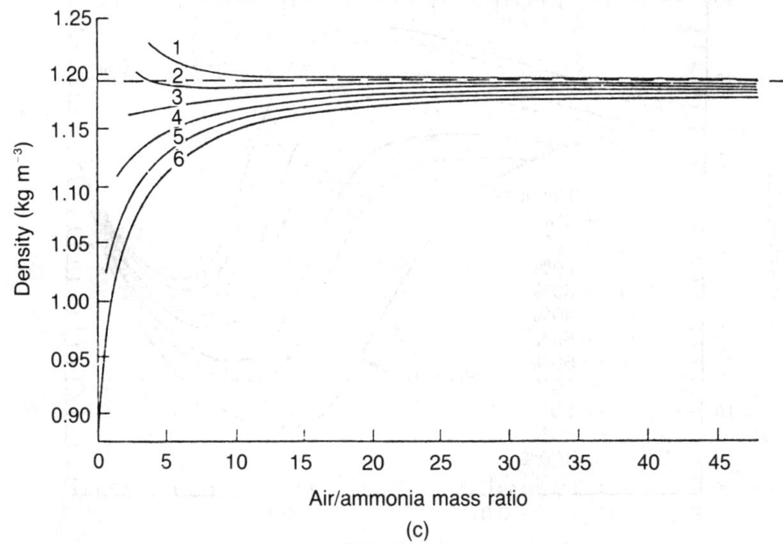

Figure 15.129 continued

invariably considered for an LNG cloud is a spillage of the refrigerated liquid followed by evaporation of the liquid pool. By comparison with the quantity of spray generated by a typical flashing liquid jet, the quantity entrained in the vapour evolved will be small; it may be expected to be rather greater from a spillage on sea than from one on land. In any event, accounts of the density of gas clouds from LNG spillages tend to disregard the effect of liquid spray.

15.42.3 Chlorine

As a dense gas, chlorine has some of the features of propane. The molecular weight of chlorine is 71 so that it is dense by virtue of its molecular weight. This is so whether the release is one of chlorine gas or liquefied chlorine.

The normal boiling point of chlorine is −34°C. If the release is one of liquefied chlorine, the low temperature makes the gas somewhat more dense so that, in principle, heating up of the gas will render it less dense, but in practice heat transfer to the gas does not play a major role and any density changes due to temperature changes tend to be small.

With regard to atmospheric humidity and liquid spray, the points made above in relation to propane apply also to chlorine. Chlorine is another gas which tends to be modelled as a 'simple' dense gas.

15.42.4 Ammonia

The characteristics of the gas cloud from a release of anhydrous liquefied ammonia have been described by Haddock and Williams (1978 SRD R103, 1979) and Blanken (1980). Broadly, a release of liquid anhydrous ammonia normally gives rise to a dense gas cloud. The cloud will tend to be dense if the atmospheric humidity is low and/or the fraction of liquid spray is high. A cloud with little or no liquid spray in air with high humidity may be buoyant.

The molecular weight of ammonia is 17 and its normal boiling point is −33°C. Thus the molecular weight is not responsible for any dense gas behaviour. A systematic investigation of the factors governing the density of clouds of ammonia in air has been made by Haddock and Williams (1978 SRD R103, 1979). They consider chemical interactions, atmospheric humidity and liquid spray.

Possible chemical interactions with atmospheric air would appear to be formation of ammonium hydroxide (NH_4OH), ammonium bicarbonate (NH_4HCO_3) and ammonium carbonate ($(NH_4)_2CO_3$). The authors conclude that formation of these compounds directly from ammonia vapour, water vapour and carbon dioxide would be negligible, but that there would be dissolution of ammonia into water droplets formed due to cooling of the air by the cold ammonia gas and dissolution of carbon dioxide in these droplets. The effect of these processes on the density and temperature of the cloud would be minimal, but the effect on toxicity could be appreciable. Thus the volume of air required to convert a cloud of pure ammonia to one with an ammonia concentration of 2000 ppm contains enough carbon dioxide to reduce the concentration of free ammonia to about 1400 ppm.

A model of the thermodynamics of the NH_3–H_2O–air system is used to derive plots of the density and temperature of a cloud formed from the mixing of ammonia and air, as shown in Figure 15.129, whilst Figure 15.129(a) shows the temperature for the dry air case. Figure 15.129(a) shows a plot of the density vs the air/ammonia mass ratio with the liquid fraction f as the parameter, for ammonia initially at its boiling point, for dry air (RH = 0%) and Figure 15.129(b) that for wet air (RH = 100%).

The work showed that if there is no liquid spray, the density of the cloud, for dry or wet air, rises asymptotically to that of air, and therefore never exceeds that of

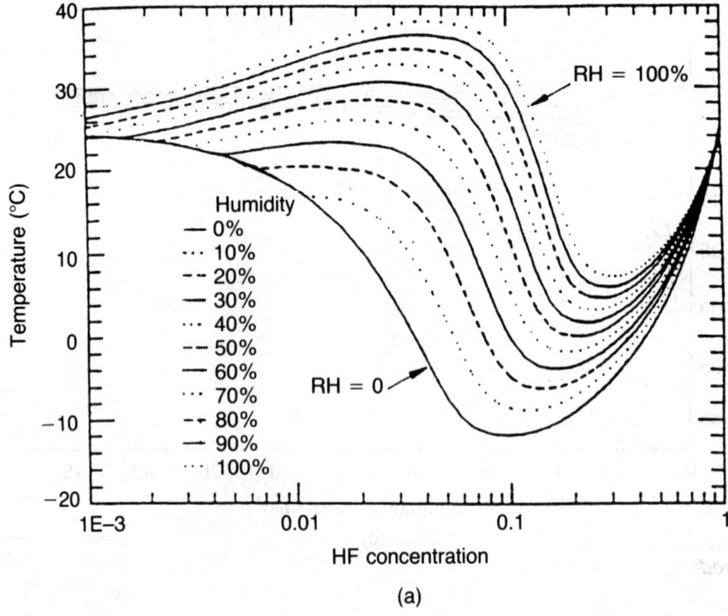

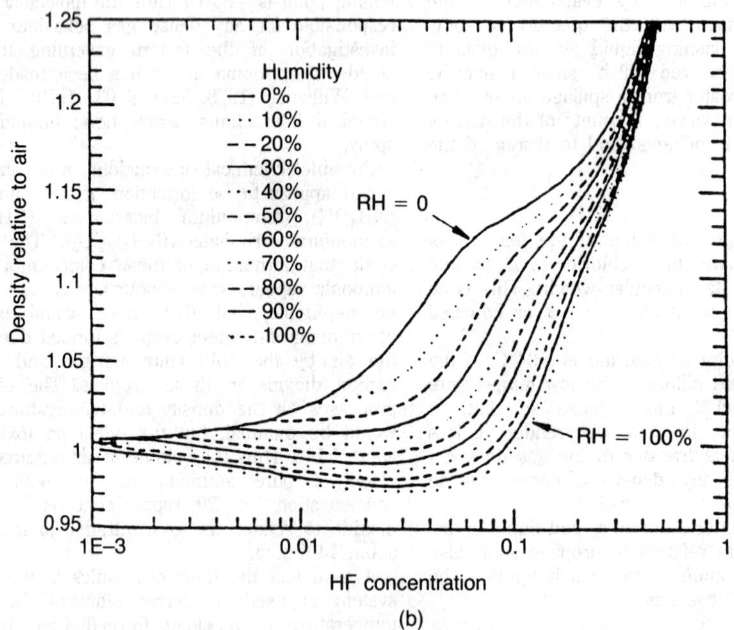

***Figure 15.130** Temperature and density of a hydrogen fluoride–air mixture formed by release of hydrogen fluoride into the atmosphere (Puttock, MacFarlane, Prothero, Roberts et al., 1991): (a) temperature for release into air of varying humidity; and (b) density for release into air of varying humidity (Courtesy of the American Institute of Chemical Engineers)*

air. If liquid spray is present, the cloud may be denser than air. There are limits to the liquid fraction below which the cloud density will not exceed that of air, the limits being different for dry and wet air. For dry air this limit on the liquid fraction is about 8% and for wet air it is about 12%.

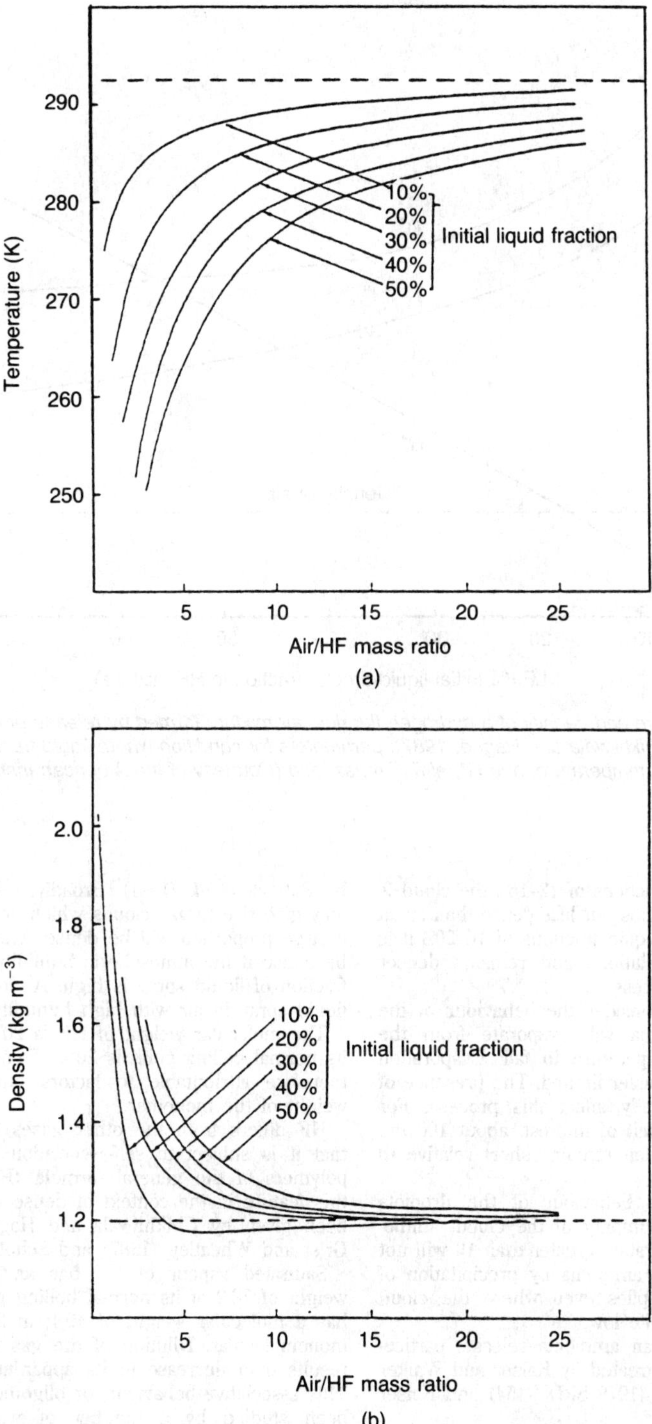

***Figure 15.131** Temperature and density of a hydrogen fluoride-air mixture formed by release of hydrogen fluoride into the atmosphere – 1 (Chikhliwala and Hague, 1987): (a) temperature for release with different liquid fractions; (b) density for release with different liquid fractions (Courtesy of the American Institute of Chemical Engineers)*

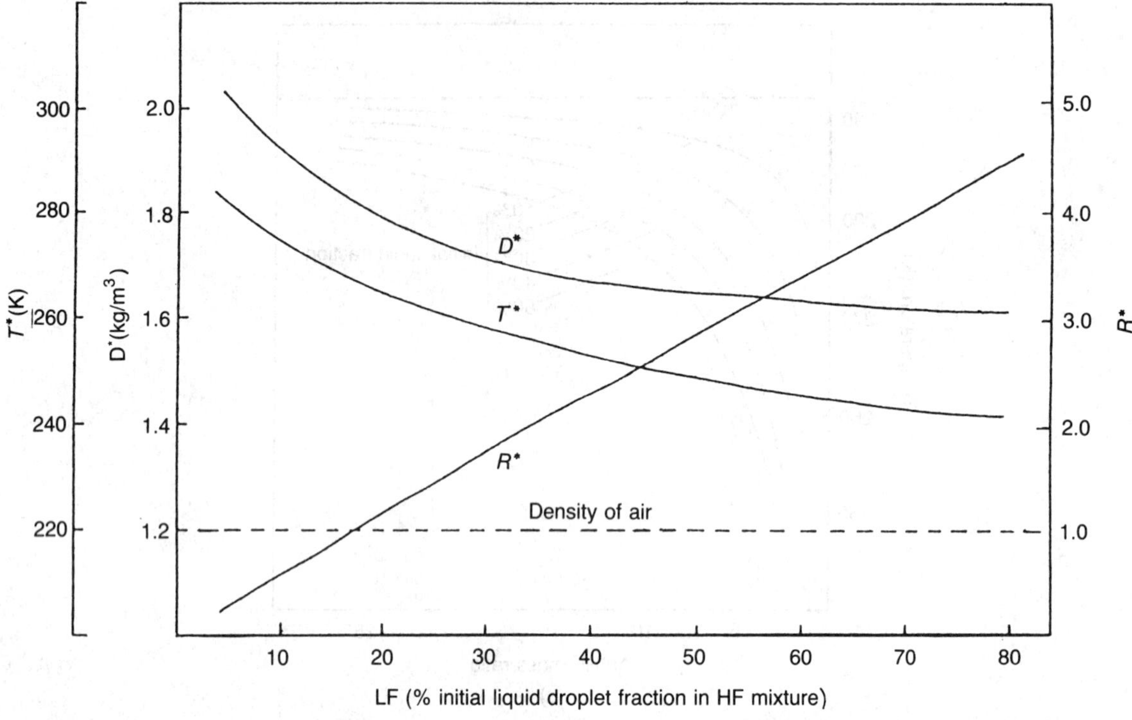

Figure 15.132 *Temperature and density of a hydrogen fluoride–air mixture formed by release of hydrogen fluoride into the atmosphere – 2 (Chikhliwala and Hague, 1987): parameters for condition where liquid fraction (LF) just evaporates. D*, density; T*, temperature; and R*, air/HF mass ratio (Courtesy of the American Institute of Chemical Engineers)*

For wet air with liquid fractions of 12–16% the cloud is denser than air at low dilutions but less dense than air at high dilutions, whilst with liquid fractions of 16–20% it is denser than air at low dilutions and remains denser throughout the dilution process.

Haddock and Williams consider the behaviour of the ammonia droplets. Ammonia will evaporate from the droplets, causing their temperature to fall. Evaporation becomes mass and heat transfer limited. The presence of water vapour will not greatly affect this process. For droplets up to the upper limit of interest, about 100 μm, the time scale for evaporation remains short relative to that for precipitation.

As far as concerns the behaviour of the droplets during the initial gravity slumping of the cloud, ammonia–air mixtures with mass ratios greater than 10 will not be significantly depleted of ammonia by precipitation of ammonia droplets. This applies even where the cloud height after slumping is quite low.

The broader aspects of an ammonia release, particularly dispersion, have been treated by Kaiser and Walker (1978), Griffiths and Kaiser (1979 SRD R154) and Kaiser and Griffiths (1982).

15.42.5 Hydrogen fluoride

The gas cloud characteristics for a release of liquefied anhydrous hydrogen fluoride (HF) have been described by Puttock *et al.* (1991). Broadly, releases of liquid HF may give rise to gas clouds which are dense or buoyant; a large proportion will be dense. The cloud will tend to be dense if the atmospheric humidity is low and/or the fraction of liquid spray is high. A cloud with little or no liquid spray in air with high humidity may be buoyant.

The molecular weight of HF is 20 (as monomer) and its normal boiling point is 20°C. Dense gas behaviour is therefore attributable to factors other than molecular weight of the monomer.

HF differs from the other gases considered here in that it is subject to self-association and tends to form polymers of the general formula $(HF)_n$. Treatments of this feature in the context of dense gas dispersion have been given by Chikhliwala and Hague (1987), Clough, Grist and Wheatley (1987) and Schotte (1987).

Saturated vapour of HF has an apparent molecular weight of 78.2 at its normal boiling point. At 100°C, HF has a molecular weight of 49.1; at 300°C it is virtually monomolecular. Dilution of the gas cloud with air also results in a decrease in its apparent molecular weight. This associative behaviour, or oligomerization, of HF has been studied by a number of workers and various schemes have been proposed. For example, MacLean *et al.* (1962) give a monomer–dimer–hexamer model, and Beckerdite, Powell and Adams (1968) give a monomer–trimer–hexamer model.

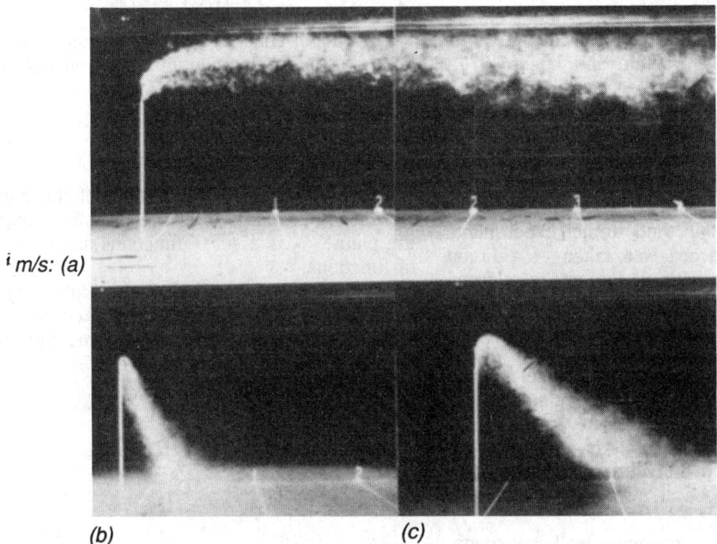

Figure 15.133 *Dispersion of dense gas from an elevated relief: wind tunnel tests (Bodurtha, 1961): Full scale conditions: stack diameter 610 mm, stack height 30.5 m, wind velocity 1.34 m/s, gas exit velocity 6.1 m/s: (a) gas specific gravity 1 (air); (b) gas specific gravity 5.15; and (c) gas specific gravity 1.52 (Courtesy of the Air Pollution Control Association)*

Thus, in order to have a complete description of the behaviour of an HF cloud it is necessary to take account of self-association, atmospheric humidity and liquid fraction. A model of HF fog formation is that proposed, and revised, by Schotte (1987, 1988). The model treats the self-association of HF and the formation of HF fog, or liquid HF–H_2O droplets.

Another model of HF which takes into account self-association and atmospheric humidity has been described by Clough, Grist and Wheatley (1987); the model is entitled WETAFH. It comprises for self-association the monomer–dimer–hexamer model of MacLean *et al.*, which includes the chemical equilibrium relations, and a model for the equilibrium and thermodynamic relations of the HF–air–water system. The model shows that in moist air an HF–H_2O liquid phase will generally form and that in some cases as much as 50% of the HF will be in this liquid phase. It is assumed that the liquid will be in the form of an aerosol. The density of the aerosol would be low, of the order of 0.05 kg/m^3. Provided the droplet size is not too large (<20 μm), the aerosol would be expected to persist for several hours. The authors give plots of cloud density and temperature vs concentration of HF, with relative humidity of the air as parameter.

A model which yields similar information and which is incorporated in HFSYSTEM has been described by Puttock, MacFarlane, Prothero, Roberts *et al.* (1991). The model is based on that of Schotte. Plots from this model are shown in Figure 15.130. This figure gives relative density and temperature curves with relative humidity as parameter for the mixing of HF and air, both initially at 25°C, thus close to the boiling point of HF. They show that with dry air (RH = 0%) the gas cloud remains dense at all concentrations, whereas with saturated air (RH = 100%) the cloud is slightly buoyant at lower concentrations.

The effect of liquid present as spray is treated in the model given by Chikhliwala and Hague (1987). The authors refer to the method of Haddock and Williams for ammonia, and adopt a somewhat similar approach. They use the self-association model of Beckerdite, Powell and Adams and a model for the equilibrium and thermodynamic relations of the HF vapour–HF liquid system in dry air. Figure 15.131 gives plots of density and temperature vs the air/HF mass ratio with the liquid fraction as the parameter for the mixing of HF and air, the initial temperatures of the HF and of the air being 292.7 and 293 K, respectively.

Figure 15.132 gives the parameters for the condition where the liquid fraction just evaporates: the density D^*, the temperature T^* and the air/HF mass ratio R^*.

15.43 Dispersion of Dense Gas: Plumes from Elevated Sources

The treatment of dense gas dispersion given so far has been largely confined to releases from sources at ground level. It is now necessary to consider elevated releases. A review of work on the dispersion of an elevated plume of dense gas has been given by Ooms and Duijm (1984).

In this and the following section an account is given of some of the principal models for the dispersion of an elevated release of a dense gas. A further discussion related to safe dispersion from reliefs and vents is given in Section 15.48.

15.43.1 Experimental work

Experimental work on dispersion of dense gas from an elevated source has been mainly by physical modelling. Bodurtha (1961, 1980, 1988), Hoehne and Luce (1970), Hoot, Meroney and Peterka (1973), Giesbrecht, Seifert and Leuckel (1983), Li Xiao Yun, Leidens and Ooms (1986) and Schatzmann, Snyder and Lawson (1993) used a wind tunnel; Holly and Grace (1972), J.L. Anderson, Parker and Benedict (1973), Chu (1975) and Badr (1984) used a water flume.

Bodurtha (1961) carried out wind tunnel experiments in which a photographic record was taken of a Freon 114–air mixture made visible with oil–fog smoke released from a stack. Measurements taken from the photographs include the maximum initial plume rise, distance to touchdown, velocity of plume descent and plume radius. Figure 15.133 shows the effect of the specific gravity of the gas on the behaviour of the plume.

The wind tunnel work of Hoot, Meroney and Peterka (1973) yielded data on, and correlations of, the maximum initial plume rise, distance to touchdown, and maximum concentrations, including those at maximum rise and at touchdown.

Work in the field has been reviewed by Schatzmann, Snyder and Lawson (1993). The main data sets are those of Bodurtha (1961) and Hoot, Meroney and Peterka (1973). The authors comment that the data of Bodurtha do not lend themselves to generalization and that those of Hoot, Meroney and Peterka are regarded as the most reliable and complete. However, the latter set has certain limitations. The experiments were performed in a laminar crossflow, whereas the real atmosphere is turbulent. The densimetric Froude numbers investigated did not cover the whole range of interest. The data of Bodurtha (1961) and Li Xiao Yun, Leidens and Ooms (1986) are also for laminar crossflow and low-to-medium Froude numbers. The work of Giesbrecht, Seifert and Leuckel (1983) does cover larger Froude numbers, but for quiescent atmospheres. The investigation which the authors report extends to a turbulent atmosphere and to higher Froude numbers.

Schatzmann, Snyder and Lawson found that a dense gas jet released into a turbulent atmosphere tended, by comparison with one released into a laminar atmosphere, to have a reduced rise height and an increased touchdown distance with lower concentrations both at the point of maximum rise and at touchdown.

The concentration of interest for a toxic release is generally much lower than that for a flammable release. The relative effect of atmospheric turbulence increases with distance from the release point. The authors suggest that, whereas it may not be overly conservative to neglect this factor for concentrations of interest in respect of their flammability, it appears essential to take it into account if a realistic estimate is to be obtained for toxic concentrations.

15.43.2 Ooms model

A model for dispersion of a dense gas plume from an elevated source which has been widely used is that described by Ooms (1972) and Ooms, Mahieu and Zelis (1974). Whereas previous models were based on the assumptions that the gas is of neutral density and that the plume velocity is equal to the wind speed, the Ooms model does not rely on these assumptions but gives a more realistic description of entrainment into and drag on the plume.

The model consists of a set of equations for conservation of mass, momentum and energy and for air entrainment of the general form

$$\frac{\mathrm{d}}{\mathrm{d}s}\int_0^{\sqrt{2}.b} \phi_i(r)2\pi r\mathrm{d}r = \psi_i \qquad [15.43.1]$$

where b is a characteristic width of the plume, r is the radius of the plume, s is the distance along the axis of the plume, ϕ_i is a set of functions and ψ_i is a further set of functions.

The functions ϕ_i is given by the following relations for conservation of total mass, mass of contaminant, momentum in the horizontal direction, momentum in the vertical direction and energy, respectively:

$$\phi_i(r) = \rho u \qquad [15.43.2a]$$

$$= cu \qquad [15.43.2b]$$

$$= \rho u^2 \cos\theta \qquad [15.43.2c]$$

$$= \rho u^2 \sin\theta \qquad [15.43.2d]$$

$$= \rho u\left(\frac{1}{\rho} - \frac{1}{\rho_{ao}}\right) \qquad [15.43.2e]$$

where c is the concentration of the contaminant, u is the velocity along the axis of the plume, θ is the angle between the plume axis and the horizontal, ρ is the density of the plume and ρ_{ao} is the density of the atmosphere at the stack outlet.

The functions ψ_i are functions primarily of velocity and density.

For example, the relation used for the conservation of mass draws on the work of Albertson *et al.* (1950) and Abraham (n.d.) for entrainment into a turbulent jet and of Abraham (1970) for entrainment into a plume and is

$$\frac{\mathrm{d}}{\mathrm{d}s}\int_0^{\sqrt{2}.b} \rho u 2\pi r\mathrm{d}r = 2\pi b\rho_a[\alpha_1 \mid u_*(s) \mid + \alpha_2 U_a \mid \sin\theta \mid \cos\theta + \alpha_3 u'] \qquad [15.43.3]$$

where u' is the entrainment velocity due to atmospheric turbulence, U_a is the mean wind velocity, ρ_a is the density of the air and α_1, α_2 and α_3 are the entrainment coefficients for a free jet, a line thermal and due to turbulence, respectively. The values taken for the constants α_1, α_2 and α_3 are 0.057, 0.5 and 1.0, respectively. The set of equations represented by Equation 15.43.1 is integrated with respect to s numerically.

The model has been compared with various sets of experimental results and gives good agreement for the position of the plume axis. The Ooms model has been encoded for the US Coast Guard in the computer code ONDEK. For determination of the distances within which vented hydrocarbon releases may be assumed for design purposes to have fallen below the lower flammability limit, Ooms, Mahieu and Zelis give the relations shown in Figure 15.134(a).

They also give the illustrative example shown in Figure 15.134(b) for the discharge of butane from two different sized vents, one giving an exit velocity of

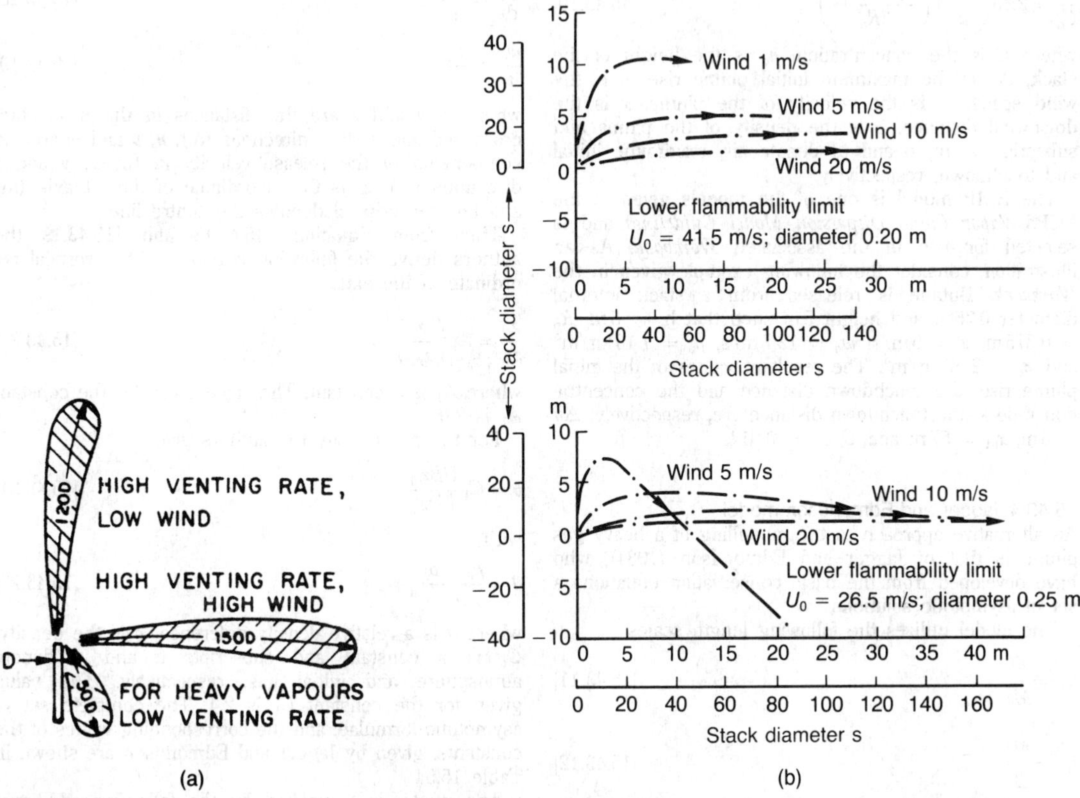

Figure 15.134 Dispersion of dense gas from an elevated relief: (a) approximate distances for dispersion of hydrocarbons; (b) dispersion of butane vapour (Ooms, Mahieu and Zelis, 1974) (Courtesy of Elsevier Science Publishers)

26.5 m/s and the other one of 41.5 m/s. For this heavy gas the lower outlet velocity leads to a much more rapid descent to ground level than does the higher velocity.

15.43.3 Hoot, Meroney and Peterka model

Another model is that given by Hoot, Meroney and Peterka (1973) (HMP model). This model too has found widespread use. The basic equations of the model are those of conservation of mass, of horizontal and vertical momentum and of entrainment, and are as follows:

$$\frac{d(R^2\rho_o u_s)}{ds} = \rho_a \frac{d(R^2 u_s)}{ds} \quad [15.43.4]$$

$$\frac{d(R^2\rho_o u_s^2 \cos\theta)}{ds} = \rho_a u \frac{d(R^2 u_s)}{ds} \quad [15.43.5]$$

$$\frac{d(R^2\rho_o u_s^2 \sin\theta)}{ds} = (\rho_a - \rho_o)R^2 g \quad [15.43.6]$$

$$\frac{d(R^2 u_s)}{ds} = \alpha_1 R \mid u_s - u\cos\theta \mid + \alpha_2 R u \mid \sin\theta \mid \quad [15.43.7]$$

where R is radius of the plume, s is the distance along the plume axis, u_s is the velocity along the plume axis, θ is the angle of the plume axis to the horizontal, ρ is the density of the plume, ρ_a is the density of the air, ρ_o is the density of the outlet gas and α_1 and α_2 are the entrainment coefficients. Values of α_1 and α_2 from wind tunnel work are 0.09 and 0.9, respectively.

The authors divide the plume path into three regions and solve the model Equations 15.43.4–15.43.7 analytically in each region. The HMP model is then as follows. For the initial plume rise

$$\frac{\Delta h}{2R_o} = 1.32\left(\frac{w_o}{u}\right)^{\frac{1}{3}}\left(\frac{\rho_o}{\rho_a}\right)^{\frac{1}{3}}\left[\frac{w_o^2}{2R_o g(\rho_o - \rho_a)/\rho_o}\right]^{\frac{1}{3}} \quad [15.43.8]$$

for the touchdown distance

$$\frac{x_{td}}{2R_o} = \left[\frac{w_o u}{2R_o g(\rho_o - \rho_a)/\rho_o} + 0.56\left\{\left(\frac{\Delta h}{2R_o}\right)^3\left[\left(\frac{2 + h_s}{\Delta h}\right)^3 - 1\right]\right\}^{\frac{1}{2}} \times \{u^3/[2R_o w_o(\rho_o - \rho_a)/\rho_a]\}^{\frac{1}{2}}\right. \quad [15.43.9]$$

and for the maximum concentration on the plume axis

$$\frac{C}{C_o} = 2.43\left(\frac{w_o}{u}\right)\left(\frac{h_s + 2\Delta h}{2R_o}\right)^{-1.95} \quad [15.43.10]$$

where C is the concentration, h_s is the height of the stack, Δh is the maximum initial plume rise, u is the wind speed, w is the velocity of the plume, x is the downwind distance, ρ is the density of the plume and subscripts a, m, o and td denote air, maximum, initial and touchdown, respectively.

The HMP model is one of the models given in the CCPS *Vapor Cloud Dispersion Model Guidelines* and is selected for use in the associated *Workbook*. As an illustration, consider the following example given in the *Workbook*. Butane is released from a stack internal diameter 0.75 m and height 5 m such that h_s = 5 m, R_o = 0.375 m, u = 5 m/s, w_o = 12.7 m/s, ρ_a = 1.16 kg/m^3 and ρ_o = 2.67 kg/m^3. The resultant values of the initial plume rise, the touchdown distance and the concentration ratio at the touchdown distance are, respectively, Δh = 6 m, x_{td} = 57 m and C/C_o = 0.014.

15.43.4 Jagger and Edmondson model

An alternative approach to the modelling of a heavy gas plume is that of Jagger and Edmondson (1984), who have developed from the basic conservation equations a set of asymptotic solutions.

The model utilizes the following length scales

$$l_Q = \frac{Q}{M^{\frac{1}{2}}} \quad [15.43.11]$$

$$l_m = \frac{M^{\frac{3}{4}}}{B^{\frac{1}{2}}} \quad [15.43.12]$$

$$z_m = \frac{M^{\frac{1}{2}}}{U} \quad [15.43.13]$$

$$z_B = \frac{B}{U^3} \quad [15.43.14$$

with

$$M = Qu_o \quad [15.43.15]$$

$$B = Qg' \quad [15.43.16]$$

$$g' = \frac{\rho - \rho_a}{\rho_a} g \quad [15.43.17]$$

where B is the initial buoyancy flux (m^4/s^3), g is the acceleration due to gravity (m/s^2), g' is the reduced gravity (m/s^2), M is the initial momentum flux (m^4/s^2), Q is the initial volumetric flow (m^3/s), u_o is the outlet velocity (m/s) and U is the wind speed (m/s).

The length l_Q is a function of the source and only influences flow near the source. l_m indicates the point of transition from momentum to buoyancy control. z_m and z_B indicate the influence of crossflow. For large values of z_m and z_B the jet is essentially vertical, while for small values it is bent over.

The method allows the calculation of the trajectory of the vertical co-ordinate z_s of the jet or plume axis and, in particular, of heights for the jet–plume transition and for the virtual source of the resultant effective plume.

For a vertical jet the relations given include:

$$\frac{dz_s}{dx} = \frac{w_{cl}}{u} \quad [15.43.18]$$

$$\frac{w_{cl}}{u} \approx \frac{z_m}{z_s} \quad [15.43.19]$$

where x, y and z are the distances in the downwind, crosswind and vertical directions (m), u, v and w are the components of the release velocity in the x, y and z directions (m), z_s is the co-ordinate of the jet axis (m) and the subscript cl denotes the centre line.

Then from Equations 15.43.18 and [15.43.19 the authors derive the following relation for the vertical co-ordinate of the axis:

$$\frac{z_s}{z_m} = A_1\left(\frac{x}{z_m}\right)^{\frac{1}{2}} \quad [15.43.20]$$

where A_1 is a constant. The value given for the constant A_1 is 2.0.

For the gas density the authors give

$$\theta = C_1 \frac{UBz_m}{Mgz_s} \quad [15.43.21]$$

with

$$\theta = \frac{\rho_a - \rho}{\rho_o} \quad [15.43.22]$$

where θ is a relative density difference, ρ is the density, C_1 is a constant and subscripts a and o denote atmosphere and initial gas, respectively. The value given for the constant C_1 is 2.4. The complete set of asymptotic formulae, and the corresponding values of the constants, given by Jagger and Edmondson are shown in Table 15.54.

The model is completed by the following additional relations. The maximum rise of the plume with weak crossflow is:

$$z_{max} = A_5 \frac{M^{\frac{3}{4}}}{B^{\frac{1}{2}}} \qquad \frac{M^{\frac{1}{2}}}{B} U \ll 1 \quad [15.43.23]$$

and in strong crossflow conditions

$$z_{max} = A_6 \left(\frac{M^2}{BU}\right)^{\frac{1}{3}} \qquad \frac{M^{\frac{1}{2}}}{B} U \gg 1 \quad [15.43.24]$$

where z_{max} is the maximum rise and A_5 and A_6 are constants. The value given for constants A_5 ands A_6 is about 2.0 for each.

It is convenient from this point for the purposes of exposition to introduce several quantities additional to those given in the original paper and to use a modified notation.

The distance of the virtual source of the jet below the stack outlet is given by

$$z_{vs} = \frac{d_o}{2\alpha} \quad [15.43.25]$$

where d_o is the diameter of the stack outlet (m), z_{vs} is the distance of the jet virtual source below the stack outlet (m) and α is an entrainment coefficient. The value given for α is 0.15.

The model also utilizes a second virtual source, that of the plume falling to earth. Use is made of two sets of coordinates. The first set is the horizontal and vertical coordinates (x_{je},z_{je}) of the jet end point. The second set

***Table 15.54** Formulae and constants for asymptotic plume model (after Jagger and Edmondson, 1984) (Courtesy of the Institution of Chemical Engineers)*

A Formulae

Flow type	*Range of validity*	(A) w_{cl}/u	(B) z_s/z_m	(C) z_s/z_B	(D) $Mg\theta/UB$
Vertical jet	$z_s \ll z_m$	$\approx z_m/z_s$	$A_1(x/z_m)^{1/2}$	–	$C_1(z_m/z_s)^2$
Bent-over jet	$z_s \gg z_m$	$\approx (z_m/z_s)^2$	$A_2(x/z_m)^1$		$C_2(z_m/z_s)^2$
Vertical plume	$z_s \ll z_B$	$\approx (z_B/z_s)^{1/3}$	–	$A_3(x/z_B)^{3/4}$	$C_3(z_B/z_s)^{5/3} . (z_m/z_B)^2$
Bent-over plume	$z_s \gg z_B$	$\approx (z_B/z_s)^{1/2}$	–	$A_4(x/z_B)^{2/3}$	$C_4(z_B/z_s)^2 . (z_m/z_B)^2$

B Constants

A_1	A_2	A_3	A_4	C_1	C_2	C_3	C_4
2.0	2.0	1.6	2.0	2.4	2.4	2.4	3.1

is the horizontal and vertical coordinates ($x_{\text{veff}}, z_{\text{veff}}$) of the plume virtual source. The second set of coordinates is obtained from the first by applying the distance corrections ($\Delta x_v, \Delta z_v$).

The following equations apply:

$$z_{je} = h_s - z_{vs} + z_{max} \quad [15.43.26a]$$

$$x_{\text{veff}} = x_{je} - \Delta x_v \quad [15.43.26b]$$

$$z_{\text{veff}} = z_{je} + \Delta z_v \quad [15.43.26c]$$

$$x_{td} = \phi x'_{td} + x_{\text{veff}} \quad [15.43.27]$$

$$c = f(\theta) \quad [15.43.28]$$

where c is the concentration of the plume at touchdown (v/v), h_s is the height of the stack (m), x_{je} is the horizontal coordinate of the jet end point (m), x_{td} is the horizontal distance from the base of the stack at which touchdown occurs, or the touchdown distance (m), x'_{td} is an uncorrected touchdown distance (m), Δx_v is the horizontal distance between the jet end point and the plume virtual source (m), x_{veff} is the horizontal coordinate of the plume virtual source (m), z_{je} is the vertical coordinate of the jet end point (m), Δz_v is the vertical distance between the jet end point and the plume virtual source (m), z_{veff} is the vertical coordinate of the plume virtual source (m) and ϕ is a correction factor. The coordinates $x_{je}, x_{\text{veff}}, z_{je}, z_{\text{veff}}$ are measured from the base of the stack. x_{veff} may assume a negative value.

The correction factor ϕ is introduced to allow for the fact that the prior calculation applies to the centre line of the plume, whereas the touchdown distance at which a significant concentration occurs is that of the edge of the plume nearest the base of the stack. An empirical value of 0.85 is assigned to ϕ.

The application of the model to the case where there is wind is to determine by repeated use of the asymptotic formulae given in Table 15.54 first the coordinates of the jet end point, then those of the plume virtual source and finally the touchdown distance and concentration. The procedure is illustrated in the example given below.

There is a separate treatment for calm conditions, or zero wind. For this case the method is to determine first the maximum rise of the plume. The plume is then envisioned as falling from this point to earth so that the plume fall distance is

$$z_{pf} = h_s - z_{vs} + z_{max} \quad [15.43.29]$$

where z_{pf} is the plume fall distance (m). The falling plume has a radius b and downward velocity w given by

$$b = \frac{6\alpha z}{5} \quad [15.43.30]$$

$$w = \frac{5}{6\alpha}\left(\frac{9\alpha B}{10\pi}\right)^{\frac{1}{3}} z^{-\frac{1}{3}} \quad [15.43.31]$$

and covers an area A_{pgr} and has a volumetric flow V_{pgr}, which are

$$A_{pgr} = \frac{\pi}{4} b^2 \quad [15.43.32a]$$

$$V_{pgr} = wA_{pgr} \quad [15.43.32b]$$

where A_{pgr} is the ground area centring on the base of the stack covered by the falling plume (m^2), b is the radius of the plume (m), V_{pgr} is the volumetric flow of the plume at ground level (m^3/s) and w is the vertical downwards velocity of the plume (m/s). The concentration is then obtained from

$$c = Q/V_{pgr} \quad [15.43.33]$$

The application of the model where there is a wind is as follows. The fluxes M and B, the wind speed U and the length scales l_m, l_Q, z_B and z_m are determined. The maximum height of the plume z_{max} is then obtained from Equation 15.43.24. The distance z_{vs} of the jet virtual source below the outlet is calculated from Equation 15.43.25.

The procedure involves first the determination of coordinates (x_{je}, z_{je}). The coordinate z_{je} is obtained from Equation 15.43.26a and the coordinate x_{je} from the relation given in Table 15.54 for a bent-over jet $z_s/z_m = A_2(x/z_m)^{\frac{1}{3}}$, setting $z_s = z_{max}$ and $x = x_{je}$. Next, an estimate is made of the distances ($\Delta x_v, \Delta z_v$). The first step is to obtain from the relation given in Table 15.54 for a bent-over jet the group $Mg\theta/UB = (z_m/z_s)^2$, setting $z_s = z_{max}$. Then the distance Δz_v is determined from the relation in Table 15.54 for a bent-over jet $Mg\theta/UB = C_4(z_B/z_s)^2(z_m/z_B)^2$, setting $z_s = \Delta z_v$. The distance Δx_v is calculated from the relation in Table 15.54 for a bent-over plume $z_s/z_B = A_4(x/z_B)^{\frac{2}{3}}$, setting $z_s = \Delta z_v$ and $x = \Delta x_v$. The coordinates (x_{veff}, z_{veff}) are then obtained from Equations 15.43.26b and 15.43.26c.

For the touchdown conditions, the uncorrected touchdown distance x'_{td} is obtained from the relation in Table 15.54 for a bent-over plume $z_s/z_B = A_4(x/z_B)^{\frac{2}{3}}$, setting $z_s = z_{veff}$ and $x = x'_{td}$. The corrected touchdown distance x_{td} is then calculated from Equation 15.43.27. The touchdown concentration is obtained from the relation in Table 15.54 for a bent-over plume $Mg\theta/UB = C_4(z_B/z_s)^2(z_m/z_B)^2$, setting $z_s = z_{veff}$, which gives the relative density difference θ. For low concentrations the relation between the relative density difference θ and the volumetric concentration c is $c \approx \theta[(\rho_o - \rho_a)/\rho_a]$.

For the case of calm conditions the model is applied as follows. The maximum rise of the plume z_{max} is obtained from Equation 15.43.23 and the plume fall distance from Equation 15.43.29. The touchdown radius b, velocity w, ground area A_{pgr}, volumetric flow V_{pgr} and concentration c are then determined from Equations 15.32.30–15.43.33.

Jagger and Edmondson give a set of illustrative examples of their method, in which they determine the touchdown distances and concentrations for a release of butane from an elevated pressure relief valve. The base conditions are

Height of vent $h_s = 30\,m$
Diameter of outlet $d_o = 0.2\,m$
Mass flow of butane = 7.1 kg/s
Density of butane = $2.54\,kg/m^3$

Hence

Volumetric flow of butane $Q = 2.8\,m^3/s$
Initial momentum flux $M = 255.6\,m^4/s^2$
Initial buoyancy flux $B = 27.8\,m^4/s^3$
Distance below outlet of jet virtual source $z_{vs} = 0.7\,m$

Cases considered are calm conditions and wind speeds of 2 and 4 m/s. For these cases:

	Wind speed (m/s)		
	0	2	4
l_Q	0.18	0.18	0.18
l_B	12.1	12.1	12.1
z_m		8.0	4.0
z_B		3.5	0.4

For the case of a wind speed of 2 m/s, the method described yields

$z_{max} = 20.6\,m$
$x_{je} = 17.1\,m$; $z_{je} = 49.9\,m$
$Mg\theta/UB = 0.36$
$\Delta x_v = 21.5\,m$; $\Delta z_v = 23.5\,m$
$x_{veff} = -4.4\,m$; $z_{veff} = 73.4\,m$
$x'_{td} = 118.8\,m$; $x_{td} = 96.6\,m$
$\theta = 8.2 \times 10^{-4}$
$c(\approx 2\theta) = 0.00164\,v/v = 0.164\%\,v/v$

For the case of calm conditions

$z_{max} = 24.4\,m$
$z_{pf} = 53.7\,m$
$b = 9.7\,m$; $A_{pgr} = 296\,m^2$
$w = 1.56\,m/s$; $V = 462\,m^3/s$
$c = 0.0061\,v/v = 0.61\%\,v/v$

The touchdown conditions obtained for all three cases are:

	Wind speed (m/s)		
	0	2	4
Touchdown distance (from stack base) (m)	0	96.6	239
Touchdown concentration (v/v)	0.0061	0.0016	0.0011

(These values differ from those given in the original paper but have been checked with one of the authors (S.F.J.)).

15.43.5 Emerson models

Another model is that given by Emerson (1986a). The model is a modification of the Ooms model already described. The Emerson model differs from the Ooms model in several ways. One is that, whereas in the Ooms model the vertical component of the axial velocity tends to a constant value, in the Emerson model it is multiplied by the Gaussian term and therefore tends to zero. This obviates the need for the use of $2^{\frac{1}{2}}b$ as the limit of integration. Another difference is in the selection of a more appropriate mass flow. This avoids problems with situations where the jet emission is in the upwind direction. The Emerson model is the basis of the computer code TECJET (Technica, 1987).

Emerson (1986b, 1989) has also given a further, separate model for a jet of dense gas undergoing transition to, and then dispersing as, a dense gas cloud. As already described, models for dense gas dispersion are sensitive to the source term used. It is the object of this model to provide an improved source term where the release occurs due to a jet, which is the usual situation.

The dispersion of the gas is assumed to occur in three phases, as shown in Figure 15.135. The first phase is a momentum jet, the second phase is a hybrid one in which slumping and transition occur and the third phase is a dense gas cloud. A box model approach is used in which the dimensions of the gas cloud are calculated and concentration within this envelope is assumed to be constant.

For the momentum jet phase Emerson quotes the work of Ricou and Spalding for the entrainment of air with distance along the jet axis and gives for a jet in still conditions the relation

$$Q_a = k_1(\rho_a I_o)^{\frac{1}{2}} \qquad [15.43.34]$$

where I_o is the momentum flux, Q_a is the entrainment flow and k_1 is a constant. The value of the constant k_1 is

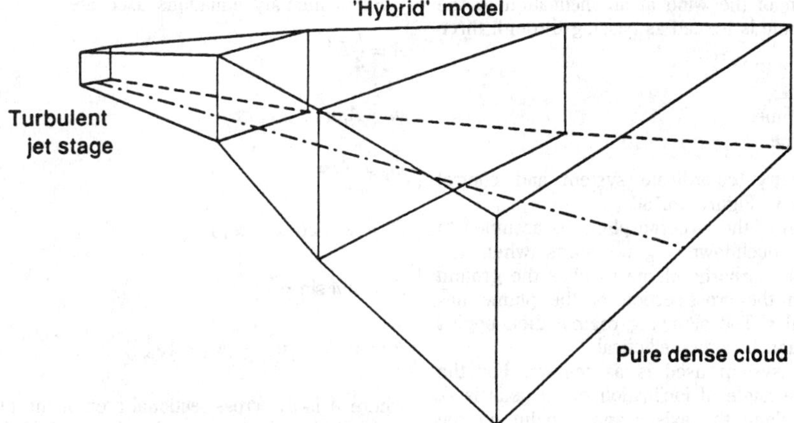

Figure 15.135 *Emerson model for dispersion of dense gas from elevated jets and plumes – idealized cloud shape (Emerson, 1986a) (Courtesy of the Clarendon Press)*

0.282. Equation 15.43.34 is equivalent to Equation 15.20.13a of Ricou and Spalding.

Then, for a jet in windy conditions, Emerson assumes that Equation 15.43.34 can be modified to read

$$Q_a = k_1(\rho_a I_o)^{\frac{1}{2}} \frac{v - u_w}{v} \quad [15.43.35]$$

where I_o is now the excess momentum flux, u_w is the wind speed and v is the velocity of the gas in the jet.

The jet is assumed to have a square cross-section. The rate of lateral growth is determined solely by the entrainment of air into the cloud. Transition to the second phase occurs when this rate of spreading falls below that which would occur due to gravitational slumping.

The second, hybrid phase combines features of both the preceding and the succeeding phases. It is assumed that the rate of air entrainment is as in the previous stage, but the rate of lateral spreading is based on dense gas behaviour. The relation used is

$$\frac{dR}{dt} = (k_2 g H \Delta\rho/\rho)^{\frac{1}{2}} \quad [15.43.36]$$

where H is the height of the cloud, R is the half-width of the cloud, ρ is the density of the cloud, $\Delta\rho$ is the density difference between the cloud and the air and k_2 is a constant. The value of the constant k_2 is approximately unity. The height of the cloud in this second phase is obtained from the mass of air entrained in the jet and the resulting necessary cross-section.

In the third phase the dispersion is that of a dense gas cloud. This is described using the model of R.A. Cox and Carpenter (1980).

15.43.6 Woodward model

J.L. Woodward (1989a) has given a model for a release from an elevated source of a two-phase flashing fluid of pressurized liquefied gas, forming a jet of vapour and liquid droplets. The model is derived from that of Ooms, Mahieu and Zelis. The model is the basis of the computer code EJET.

15.43.7 Other models

There are several other models for dispersion of an elevated release of a dense gas. The Environmental Protection Agency (EPA) (1989a) has developed a relief valve screening model. Other models include those by Giesbrecht, Seifert and Leuckel (1983) and Havens and Spicer (1985).

15.43.8 Criterion for dense gas dispersion

A criterion for deciding whether a dense gas treatment is necessary has been given by Fay and Zemba (1986) as follows:

$$\mathrm{Ri_o} \gg 1$$

with

$$\mathrm{Ri_o} = \frac{w_o D_o}{(u_*^3)} g \frac{\rho_o - \rho_a}{\rho_a} \quad [15.43.37]$$

where D is the diameter, w is the velocity of the plume, ρ is the density of the plume and the subscripts a and o denote air and initial, respectively. This criterion is adopted in the CCPS *Vapor Cloud Dispersion Model Guidelines*.

15.44 Dispersion of Dense Gas: Plumes from Elevated Sources – PLUME

A model for dispersion from a jet or plume has been developed by Shell as part of the HGSYSTEM package. This model, PLUME, is one of the front-end models for the main dispersion model, HEGADAS. It has been described by McFarlane (1991). PLUME is a generalization of a more specific model HFPLUME developed for the dispersion of hydrogen fluoride.

15.44.1 Plume development

The scenario modelled is that of a jet or plume issuing along the direction of the wind at an inclination to the horizontal. The plume is treated as passing through three stages:

(1) airborne plume;
(2) touchdown plume;
(3) slumping plume.

The plume geometry, co-ordinate system and control volume are shown in Figure 15.136.

The cross-section of the airborne plume is assumed to be circular. The touchdown regime starts when the circumference of the airborne plume touches the ground and it ends when the cross-section of the plume has become semi-circular. The slumping regime then begins and the cross-section becomes elliptical.

The co-ordinate system used is as follows. For the airborne plume the angle of inclination of the axis is ϕ, and the distance along the axis s and a point on the cross-section of that plume is given by the polar co-ordinates r and θ. For the plume in all regimes the centroid of the plume is located at distances x and z from the sources in the horizontal, vertical directions. For the touchdown and slumped plumes the height of the centre of the plume is z_c.

The core of the model is a complex set of integral equations which are then simplified to the sets of equations for the three regimes given below.

15.44.2 Airborne plume

The model for the airborne plume consists of the following set of six differential equations:

$$\frac{d\dot{m}}{ds} = \text{Entr} \qquad [15.44.1]$$

$$\frac{d\dot{P}_x}{ds} = \mathbf{Drag} \cdot \mathbf{e}_x + \text{Shear} \qquad [15.44.2]$$

$$\frac{d\dot{P}_z}{ds} = \mathbf{Drag} \cdot \mathbf{e}_z + \text{Buoy} \qquad [15.44.3]$$

$$\frac{d\dot{E}_x}{ds} = \text{Ener} \qquad [15.44.4]$$

$$\frac{dx}{ds} = \cos\phi \qquad [15.44.5]$$

$$\frac{dz}{ds} = \sin\phi \qquad [15.44.6]$$

where $\mathbf{e}_x$ is a unit vector in the horizontal, wind aligned direction, $\mathbf{e}_z$ is a unit vector in the vertical direction, $\dot{E}$ is the excess energy flux, $\dot{m}$ is the mass flux, $\dot{P}_x$ is the excess horizontal momentum flux, $\dot{P}_z$ is the excess vertical momentum flux and s is the distance along the axis of the plume.

The term 'Buoy' is the section averaged buoyancy force in the plume; **'Drag'** is the drag force acting on the plume as a result of vortex formation in the plume wake; 'Ener' is the change in energy in the plume resulting from vertical gradients of temperature and wind speed; 'Entr' is the entrainment rate for unit distance along the plume axis; and 'Shear' is the shear force associated with the vertical gradient of wind speed.

Supplementary equations used are

$$A = \frac{\pi}{4} D^2 \qquad [15.44.7]$$

$$\dot{m} = A\rho u \qquad [15.44.8]$$

$$\dot{m}_g = Acu \qquad [15.44.9]$$

$$\dot{P}_x = \dot{m}(u \cos \phi - u_\infty) \qquad [15.44.10]$$

$$\dot{P}_z = \dot{m}u \sin \phi \qquad [15.44.11]$$

$$\dot{E} = \dot{m}[(h + \tfrac{1}{2}u^2) - (h_\infty + \tfrac{1}{2}u_\infty^2)] \qquad [15.44.12]$$

where A is the cross-sectional area of the plume, c is the concentration (mass per unit volume), D is the diameter of the plume, h is the specific enthalpy, ρ is the density, u is the mean flow velocity and subscripts g and ∞ denote material released and ambient conditions at centroid height.

15.44.3 Touchdown plume

Touchdown occurs when the plume envelope first touches the ground. The plume cross-section is then modelled as a circular segment, the centroid of which falls until the cross-section becomes a semi-circle, at which point the touchdown regime ends.

The model for the touchdown plume also consists of a set of six differential equations. These are Equations 15.44.1, 15.44.2 and 15.44.4–15.44.6 with:

$$\frac{d\dot{P}_z}{ds} = \mathbf{Drag} \cdot \mathbf{e}_z + \text{Buoy} + \text{Foot} \qquad [15.44.13]$$

where 'Foot' is a reactive pressure force function.

Supplementary equations used are Equations 15.44.8–15.44.12 and

$$A = \frac{D^2}{4}[\cos^{-1}(-\eta_c) + (1 - \eta_c)^{\frac{1}{2}}] \qquad [15.44.14]$$

$$\eta_c = \frac{\eta}{2\left(1 + \left\{1 - \frac{2}{3[\eta^2(1-\eta^2)(1-\eta^2)^{\frac{1}{2}}]}\right\}^{\frac{1}{2}}\right)} \qquad [15.44.15]$$

$$\eta = \frac{2z}{D \mid \cos \phi \mid} \qquad [15.44.16]$$

$$\eta_c = \frac{2z_c}{D \mid \cos \phi \mid} \qquad [15.44.17]$$

where η is a parameter and subscript c denotes centre.

15.44.4 Slumped plume

The model for the slumped plume consists of the same set of six differential equations as for the touchdown plume. Supplementary equations used are Equations 15.44.8–15.44.12 and

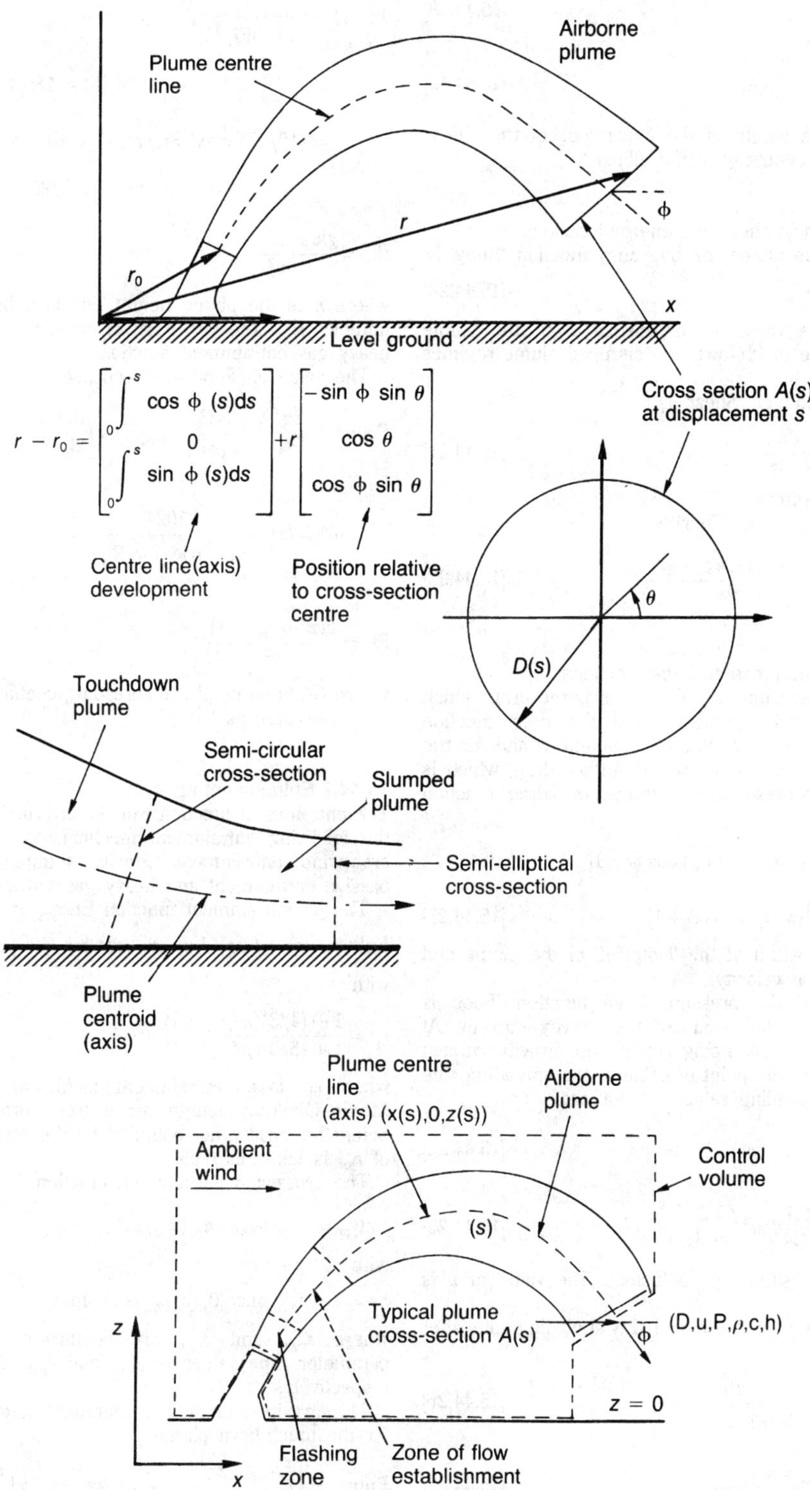

***Figure 15.136** PLUME model for dispersion of dense gas from elevated jets and plumes plume geometry, coordinate system and control volume (McFarlane, 1991) (Courtesy of the American Institute of Chemical Engineers)*

$$A = \frac{e\pi}{8} D^2 \qquad [15.44.18]$$

$$e = \frac{3\pi}{2} \frac{(z/D)}{|\cos\phi|} \qquad [15.44.19]$$

where D is the length of the major axis of the ellipse and e is the eccentricity of the ellipse.

15.44.5 Buoyancy, shear and energy functions

For the airborne plume the buoyancy function 'Buoy' is

$$\text{Buoy} = A(\rho - \rho_g)g \qquad [15.44.20]$$

This relation is valid, for expansion about the plume centroid, for the touchdown and slumped plume regimes also.

The shear function 'Shear' is

$$\text{Shear} = \dot{m} \sin\phi \frac{du_\infty}{dz} \qquad [15.44.21]$$

for all three regimes.

The energy function 'Ener' is

$$\text{Ener} = \dot{m} \sin\phi \frac{d(h_\infty + \frac{1}{2}u_\infty^2 + gz)}{dz} \qquad [15.44.22]$$

15.44.6 Drag and pressure force functions

For the airborne plume the drag is airborne drag, which is not considered significant, and the drag function **'Drag'** is taken as zero. For the touchdown and for the slumped plume there is a ground surface drag, which is significant. For these latter regimes the drag function **'Drag'** is

$$\mathbf{Drag} = l\rho_\infty u_*^2[(\rho/\rho_\infty)^{\frac{1}{2}}(u/u_\infty)\cos\phi - 1]$$

$$[(\rho/\rho_\infty)^{\frac{1}{2}}(u/u_\infty)\cos\phi + 1] \qquad [15.44.23]$$

where l is the width of the 'footprint' of the plume and u_* is the friction velocity.

The value of the pressure force function 'Foot' is governed by the behaviour of the gravity current. At some point in the slumping regime the gravity current collapses. Before the point of collapse the spreading rate and the corresponding value of 'Foot' are

$$\frac{1}{2}\frac{dD}{ds} = (k/u)[gh(1 - \rho_\infty/\rho)]^{\frac{1}{2}} \qquad [15.44.24]$$

$$\text{Foot} = \left(\frac{3\pi^2}{32}\right)^3 \frac{1}{k^2} l\rho u^2 \left(\frac{1}{2}\frac{dD}{ds}\right)^2 \qquad [15.44.25]$$

where k is the slumping coefficient. The value of k is taken as 1.15.

The criterion of collapse and the spreading rate after collapse are

$$(\rho_\infty/\rho)^{\frac{1}{2}}(D/h) > \frac{16}{3\kappa(\text{Ri}_*)^{\frac{1}{2}}\Phi(\text{Ri}_*)} \qquad [15.44.26]$$

$$\frac{1}{2}\frac{dD}{ds} = \frac{\rho_\infty u_* h}{3\kappa C_D u D \text{Ri}_* \Phi(\text{Ri}_*)} \qquad [15.44.27]$$

with

$$\Phi(\text{Ri}_*) = \frac{1}{(1 - 3\text{Ri}_*/5)^{\frac{1}{2}}} \qquad \text{Ri}_* < 0 \qquad [15.44.28a]$$

$$= 1 \qquad 0 \leq \text{Ri}_* < 189/80 \qquad [15.44.28b]$$

$$= (10/17)\max[(\text{Ri}_*/7), (1 + 4\text{Ri}_*/5)^{\frac{1}{2}}]$$

$$\text{Ri}_* \geq 189/80 \qquad [15.44.28c]$$

$$\text{Ri}_* = \frac{gh(\rho/\rho_\infty - 1)}{u_*^2} \qquad [15.44.29]$$

where h is the plume height, Ri* is a bulk Richardson number, κ is the von Karman constant and Φ(Ri*) is a heavy gas entrainment function.

The value of 'Foot' after collapse is

$$\text{Foot} = \left(\frac{3\pi^2}{64}\right)^3 \frac{3\kappa C_D}{\Phi(\text{Ri}^*)} l_{*\rho u u_*}(D/z)\frac{dD}{ds} \qquad [15.44.30]$$

for

$$(\rho_\infty/\rho)^{\frac{1}{2}}(D/z) > \frac{1024}{(9\pi^2)(\text{Ri}_*)^{\frac{1}{2}}\Phi(\text{Ri}_*)} \qquad [15.44.31]$$

with

$$\text{Ri}_* = \frac{2gz(\rho/\rho_\infty - 1)}{u_*^2} \qquad [15.44.32]$$

where C_D is an empirical spreading coefficient. The value of C_D is taken as 5.0.

15.44.7 Entrainment function

The entrainment function Entr is, in principle, the sum of the following entrainment mechanisms: jet entrainment, crosswind entrainment, gravity slumping entrainment, passive entrainment and heavy gas entrainment.

The jet entrainment function Entr_{jet} is

$$\text{Entr}_{\text{jet}} = e_{\text{jet}}\eta_\rho L_{\text{fs}}\rho_\infty |u - u_\infty \cos\phi| \qquad [15.44.33]$$

with

$$\eta_\rho = \frac{1 + (4/3)(\rho/\rho_\infty - 1)}{1 + (5/3)(\rho/\rho_\infty - 1)} \qquad [15.44.34]$$

where e_{jet} is an entrainment coefficient and L_{fs} is the Monin–Obukhov length for a free surface, this latter being the surface not bounded by the ground. The value of e_{jet} is taken as 0.08.

The crosswind entrainment function Entr_{cw} is

$$\text{Entr}_{\text{cr}} = e_{\text{u,cr}}\eta L_{\text{fs}}\rho_\infty u_\infty (u_\infty/u)^{\frac{1}{2}} |\sin\phi| \qquad [15.44.35a]$$

with

$$\eta_\rho = 1 + e_{\rho,\text{cr}} \max[0, (\rho/\rho_\infty - 1)\sin\phi] \qquad [15.44.35b]$$

where $e_{\text{u,cr}}$ and $e_{\rho,\text{cr}}$ are coefficients and η_ρ is a parameter. The values of $e_{\text{u,cr}}$ and $e_{\rho,\text{cr}}$ are 0.6 and 7.5, respectively.

The gravity slumping entrainment function Entr_{gs} is, for the touchdown plume

$$\text{Entr}_{\text{gs}} = \left[1 - \frac{2z_c}{D|\cos\phi|}\right] e_{\text{gs}}\rho_\infty z u |\cos\phi| \frac{dD}{ds} \qquad [15.44.36]$$

and for the slumped plume

$$\text{Entr}_{gs} = e_{gs}\rho_\infty zu \mid \cos\phi \mid \frac{dD}{ds} \qquad [15.44.37]$$

where e_{gs} is a coefficient. A value of 0.85 is given for e_{gs}, but with some qualification.

The passive entrainment Entr_{at} due to atmospheric turbulence is

$$\text{Entr}_{at} = (1 - 1/D)\pi e_{at}\rho_\infty \epsilon^{\frac{1}{3}}(l_y^{\frac{4}{3}} + l_z^{\frac{4}{3}}) \qquad [15.44.38]$$

with

$$l_y = \min[D/2,\ 0.88(z + z_o)(1 - 7.4\kappa\zeta/(1 - 5\kappa\zeta)]$$

Stability category A – C [15.44.39a]

$$= \min[D/2,\ 0.88(z + z_o)]$$

Stability category D [15.44.39b]

$$= \min[D/2,\ 0.88(z + z_o)/(1 + 0.1\zeta)]$$

Stability category E, F [15.44.39c]

$$l_z = \min[D/2,\ 0.88(z + z_o)(1 - 7.4\kappa\zeta/(1 - 5\kappa\zeta)]$$

Stability category A – C [15.44.40a]

$$= \min[D/2,\ 0.88(z + z_o)]$$

Stability category D [15.44.40b]

$$= \min[D/2,\ 0.88(z + z_o)/(1 + 4\zeta)]$$

Stability category E, F [15.44.40c]

$$\epsilon = (1 - 5\kappa\zeta)u_*^3/[\kappa(z + z_o)]$$

Stability category A – C [15.44.41a]

$$= u_*^3/[\kappa(z + z_o)]$$

Stability category D [15.44.41b]

$$= (1 + 4\kappa)u_*^3/[\kappa(z + z_o)]$$

Stability category E, F [15.44.41c]

$$\zeta = \frac{z + z_o}{L} \qquad [15.44.42]$$

$$L = \frac{u_*^3}{\kappa(g/T_{gr})/u_* T^*} \qquad [15.44.43]$$

where e_{at} is a coefficient, l_y and l_z are the turbulent length scales in the horizontal and vertical directions, L is the Monin–Obukhov length, T_{gr} is the temperature of the ground, $u_* T^*$ is the ground–air heat flux, ϵ is the dissipation rate of turbulent kinetic energy and ζ is a parameter. The value of e_{at} is 1.0.

The heavy gas entrainment Entr_{hg} is

$$\text{Entr}_{hg} = \frac{1}{\Phi(\text{Ri}_*)} L_{fs}\kappa\rho_\infty u_* \qquad [15.44.44]$$

with $\Phi(\text{Ri}_*)$ given by Equations 15.44.28 and 15.44.29, but with z substituted for h in the latter.

The jet entrainment and heavy gas entrainment each modify the level of turbulence and are not independent mechanisms. The combined contribution of these two effects is taken as the greater of the each of the two effects considered separately. Further, the contribution of each of the two effects is taken as proportional to the value of that effect divided by the sum of the values of the two effects. The passive entrainment and gravity slumping entrainment constitute another pair which interact and which are treated in a similar manner.

15.44.8 Atmospheric model

The plume behaviour is a function of the atmospheric conditions, and an atmospheric model is defined which includes the wind speed and temperature profiles and the Monin–Obukhov length.

The wind speed profile is

$$\frac{\kappa(z + z_o)}{u_*}\frac{du_\infty}{dz} = \psi_u(\zeta) \qquad [15.44.45]$$

with

$$\psi_u(\zeta) = (1 - 22\zeta)^{-\frac{1}{4}} \qquad \zeta < 0 \qquad [15.44.46a]$$

$$= 1 + 6.9\zeta \qquad \zeta \geq 0 \qquad [15.44.46b]$$

The temperature profile is

$$\frac{\kappa(z + z_o)}{T^*}\frac{dT_{\infty pot}}{dz} = \psi_T(\zeta) \qquad [15.44.47]$$

with

$$\psi_T(\zeta) = (1 - 13\zeta)^{-\frac{1}{2}} \qquad \zeta < 0 \qquad [15.44.48a]$$

$$= 1 + 9.2\zeta \qquad \zeta \geq 0 \qquad [15.44.48b]$$

where the subscript pot denotes potential.

The Monin–Obukhov length is

$L = -10.99 z_o^{0.0993}$ Stability category A [15.44.49a]

$= -13.44 z_o^{0.1631}$ Stability category B [15.44.49b]

$= -113.5 z_o^{0.2946}$ Stability category C [15.44.49c]

$= \infty$ Stability category D [15.44.49d]

$= 125 z_o^{0.3090}$ Stability category E [15.44.49e]

$= 26.12 z_o^{0.1667}$ Stability category F [15.44.49f]

15.45 Concentration and Concentration Fluctuations

The account of gas dispersion given so far has touched only lightly on the question of concentration and concentration fluctuations and has dealt mainly with average concentration. It is now necessary to go into this more deeply.

Concentration fluctuations are relevant both to flammable and toxic gas releases. For flammable gas clouds, fluctuations may cause the gas concentration equivalent to the lower flammability limit to reach further than the boundary given by the mean value. For a toxic gas cloud, fluctuations may result in toxic effects associated with concentrations higher than the mean value.

Accounts of concentration fluctuations have been given by a number of authors, including Gosline (1952), Gifford (1960b), Long (1963), Slade (1968), Barry (1971), Ramsdell and Hinds (1971), Csanady (1973), Chatwin (1982b), C.D. Jones (1983), S.R. Hanna (1984), Ride (1984a,b) and J.K.W. Davies (1987, 1989).

It is convenient to deal primarily with the concentration in a plume from a continuous release, but that in a cloud from an instantaneous release is also considered.

15.45.1 Experimental studies and empirical features

It has been shown in experiments on plumes from a continuous point source by a number of workers that the instantaneous concentration measured at a fixed point fluctuates about the mean value and that the ratio of the maximum instantaneous value to the time mean value, or peak-to-mean ratio, can be quite large.

An account of experimental work on the peak-to-mean (P/M) ratio has been given by Gifford (1960b). He describes the results obtained by six previous workers. The ratio of the peak concentration P to the mean concentration M, or P/M, is a function of the ratio of the corresponding sampling times t_p and t_m from which these concentrations are derived. As shown in Figure 15.137 the P/M ratio tends to increase with increase in t_m/t_p, or decrease in t_p/t_m.

As Figure 15.137 indicates, the P/M ratio measured at ground level is appreciably greater for an elevated source than for one at ground level. This is illustrated also in Figure 15.138, which gives P/M as a function of distance. Gifford states that for measurements at ground level P/M can be expected to be about 1–5 for a ground level source but as much as 50–100 for an elevated source. Work by Ramsdell and Hinds (1971) has shown that the P/M ratio tends to be greater off the centre line.

Other workers have measured the proportion of time for which the concentration at a point is zero, or the intermittency. Work on concentrations in a plume by C.D. Jones (1983) has shown that the levels of intermittency can be very high, some 80–90%, with P/M ratios in the range 30–150.

Physical modelling using wind tunnels makes it possible to perform repeated experiments at the same nominal conditions. Figure 15.109(e) shows an ensemble of 21 concentration vs time profiles obtained by D.J. Hall, Hollis and Ishaq (1982) in wind tunnel work at Warren Spring Laboratory (WSL). The variability from one run to another is pronounced.

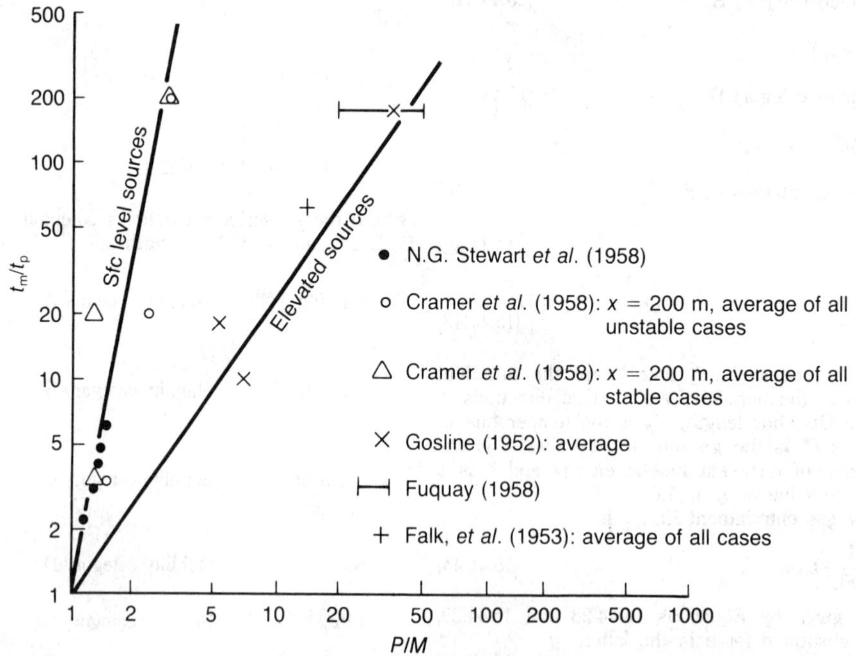

Figure 15.137 *Concentration fluctuations in a meandering plume: effect of mean sampling time t_m and peak sampling time t_p on the peak-to-mean concentration (Gifford, 1960b) (Courtesy of Pergamon Press)*

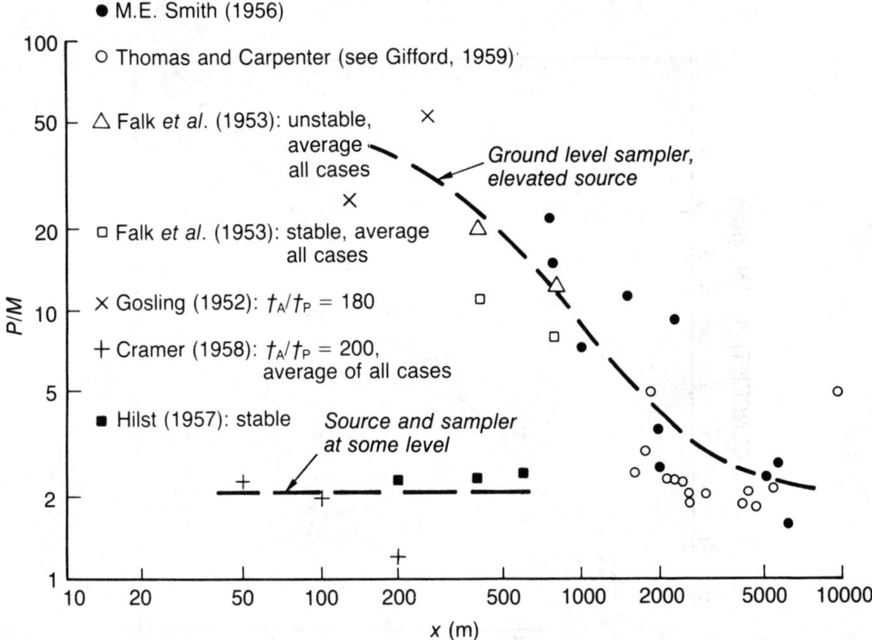

***Figure 15.138** Concentration fluctuations in a meandering plume: effect of distance (Gifford, 1960b) (Courtesy of Pergamon Press)*

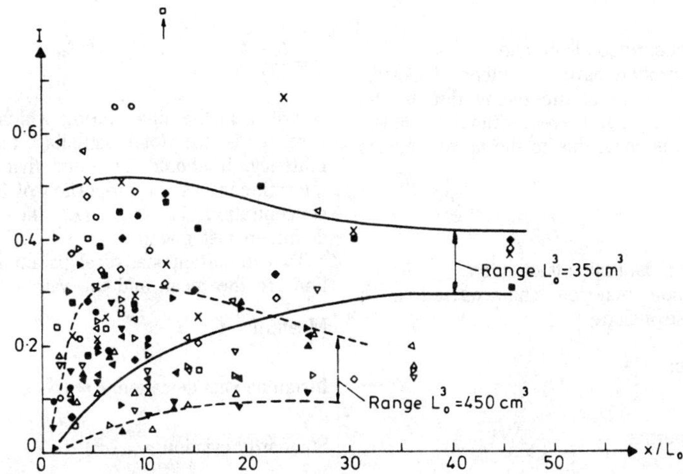

***Figure 15.139** Concentration fluctuations in gas dispersion – concentration fluctuation intensities measured in a wind tunnel (Meroney and Lohmeyer, 1984). Different symbols represent different release volumes, each being the result of at least five runs; I, intensity, L_o volume released; x, downstream distance (Courtesy of Kluwer Academic Publishers)*

The intensity of the concentration fluctuations, defined below, is another widely used measure. An intensity value of zero corresponds to zero fluctuation. Figure 15.139 shows intensities obtained in wind tunnel work by Meroney and Lohmeyer (1984).

Such concentration fluctuations are not limited to gas clouds. Birch, Brown and Dodson (1980, 1981) have found large fluctuations in jets.

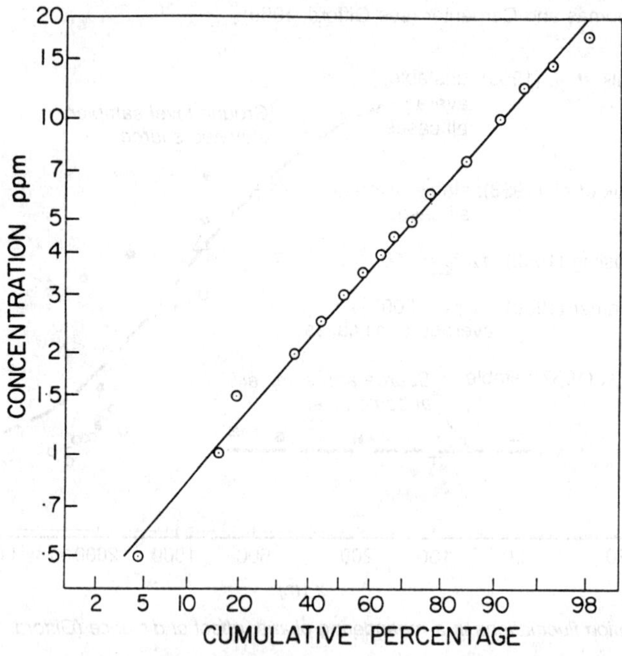

***Figure 15.140** Concentration fluctuations in a meandering plume: distribution function of concentrations (Csanady, 1973) (Courtesy of Reidel Publishing Company)*

15.45.2 Causes of concentration fluctuation

Fluctuations of concentration have a number of causes. One of these is meandering of the plume due to the variability of wind direction. But even without meander, concentration fluctuations occur due to the normal effects of turbulence.

15.45.3 Measures of concentration fluctuation

Concentration fluctuation may be characterized in a number of ways. These include

(1) peak-to-mean ratio;
(2) intermittency;
(3) intensity;
(4) distributional properties.

The peak-to-mean ratio is the ratio of the maximum instantaneous concentration to the time mean value:

$$\frac{P}{M} = \frac{C_{max}}{\bar{C}} \quad [15.45.1]$$

where $\bar{C}$ is the time mean concentration and C_{max} is the maximum instantaneous concentration.

Csanady (1973) introduced the concept of intermittency. The intermittency of concentration fluctuations, or simply intermittency, is the proportion of time during which the concentration is non-zero:

$$\gamma = \frac{t_s - t_p}{t_s} \quad [15.45.2]$$

where t_p is the time during which the concentration is zero, t_s is the total sampling time and γ the intermittency. It should be noted that some authors define intermittency as the proportion of time during which the concentration is *zero* (i.e. the complement of the definition just given).

Two important statistical parameters of the concentration are the mean and the variance:

Mean: $\mu = \bar{C}$

Instantaneous deviation: $c = C - \bar{C}$

Standard deviation $\sigma = (\overline{c^2})^{\frac{1}{2}}$

The relative intensity i of concentration fluctuations, or simply intensity, is the ratio of the standard deviation to the mean of the concentration:

$$i = \frac{(\overline{c^2})^{\frac{1}{2}}}{\bar{C}} \quad [15.45.3]$$

The distributional properties are the distribution, and its parameters, which best describe the concentration fluctuations. The distribution most favoured is the log-normal distribution, which is characterized by the mean and log standard deviation. Alternatively, the mode may be used instead of the mean.

15.45.4 Time mean concentration

As discussed in Section 15.16, the mean concentration obtained by averaging over a particular sampling interval is a function of that interval. The correction used by D.B. Turner (1970) for this effect is

$$\frac{C}{C_r} = \left(\frac{t_r}{t}\right)^p \qquad [15.45.4]$$

where C is the concentration, t is the sampling time interval, p is an index and subscript r denotes reference. The value of p is between 0.17 and 0.2 where the sampling reference time t_r is of the order of 10 min and the other sampling time t is up to 2 h.

Another approach is that described by the CCPS (1987/2). On certain assumptions it can be shown that

$$\frac{\sigma^2(T)}{\sigma^2(0)} = 2\left\{\frac{T_I}{T}\left(1 - \frac{T_I}{T}\right)\left[1 - \exp\left(-\frac{T}{T_I}\right)\right]\right\} \qquad [15.45.5]$$

where T_I is the integral time scale. They give for T_I a typical value of 10 s.

Other approaches have been used as described by D.J. Wilson (1991a).

15.45.5 Models for concentration fluctuation

Most of the approaches used to model dispersion may also be applied to the modelling of concentration fluctuations. A review by S.R. Hanna (1984) describes Gaussian, gradient transfer, similarity, meandering plume and distribution function models.

15.45.6 Distribution function for concentration

The form of the distribution function of the instantaneous concentrations at a fixed point in a passive plume has been studied by several workers, including Barry (1971) and Csanady (1973).

As described above, since concentration is zero for a proportion of the time, Csanady introduced the concept of intermittency. The intermittency is close to unity on the centre line of the plume and to zero at the cloud edges.

Csanady adduces theoretical arguments that the distribution function for the non-zero concentrations should be log–normal. He also obtains the log–normal distribution for experimental results given by several authors. Figure 15.140 shows his plot of Gosline's results on log–probability paper, confirming the log–normal form for the concentration distribution.

Barry (1971) has found that the following exponential relation gives a good fit to experimental results:

$$P(C) = P(0)\exp\left(-P(0)\frac{C}{M}\right) \qquad [15.45.6]$$

where $P(0)$ and $P(C)$ are the probabilities that the concentration exceeds 0 and C, respectively, and C and M are the concentrations averaged over the peak sampling time t_p and the total sampling time t_m, respectively.

The form of the distribution function for concentration is also an issue in the work of Wilson, of Chatwin and of Davies, as described below.

15.45.7 Peak-to-mean concentration models

The distribution functions just described may be used as the basis for models of the peak-to-mean ratio.

Barry (1971) points out that the peak-to-mean ratio is simply a special case of the ratio of any short-lived concentration. By definition, the peak concentration is that which occurs only once. Hence the probability $P(P)$ of the peak value is

$$P(P) = t_p/t_m \qquad [15.45.7]$$

But from Equation 15.45.6

$$P(P) = P(0)\exp\left[-P(0)\frac{P}{M}\right] \qquad [15.45.8]$$

Then from equations 15.45.7 and 15.45.8

$$\frac{P}{M} = \frac{1}{P(0)}[\ln P(0) - \ln(t_m/t_p)] \qquad [15.45.9]$$

Csanady gives the following approximate treatment of the case where the intermittency factor is unity and the concentration distribution is log–normal. This restriction on the intermittency factor implies that the plume does not meander. The probability density function of the concentration dF/dC is

$$\frac{dF}{dC} = \frac{1}{(2\pi)^{\frac{1}{2}}\sigma_1 C}\exp\left[-\frac{[\ln(C/C_{med})]}{2\sigma_1^2}\right] \qquad [15.45.10]$$

and the probability distribution function F is then

$$F = \frac{1}{2}\left\{1 + \text{erf}\left[\frac{\ln(C/C_{med})}{2^{\frac{1}{2}}\sigma_1}\right]\right\} \qquad [15.45.11]$$

where C_{med} is the median concentration and σ_1 is the logarithmic standard deviation. The mean concentration C_m is then

$$C_m = \int_0^1 C\frac{dF}{dC}dC \qquad [15.45.12a]$$

$$= C_{med}\exp(\sigma_1^2/2) \qquad [15.45.12b]$$

Then from Equation 15.45.11 the probability $P(= 1 - F)$ that a given peak concentration will not be exceeded is

$$P = 1 - \frac{1}{2}\left\{1 + \text{erf}\left[\frac{\ln(C_p/C_{med})}{2^{\frac{1}{2}}\sigma_1}\right]\right\} \qquad [15.45.13]$$

and hence the peak concentration C_p is

$$C_p = C_{med}\exp[2^{\frac{1}{2}}\sigma_1\text{erf}^{-1}(1 - 2P)] \qquad [15.45.14]$$

Then from Equations 15.45.12b and 15.45.14

$$\frac{C_p}{C_m} = \exp\left[2^{\frac{1}{2}}\sigma_1\text{erf}^{-1}(1 - 2P) - \frac{\sigma_1^2}{2}\right] \qquad [15.45.15a]$$

$$= \frac{P}{M} \qquad [15.45.15b]$$

15.45.8 Gifford meandering plume model

As stated above, concentration fluctuations in a passive plume are due partly to turbulent movements and partly to meandering. The effect of these two influences is shown in Figure 15.141.

A model for a meandering plume has been given by Gifford (1960b). The basis of the model is Equation 15.16.41 for a release Q from a continuous point source at ground level. He treats the dispersion variance σ_y^2 as

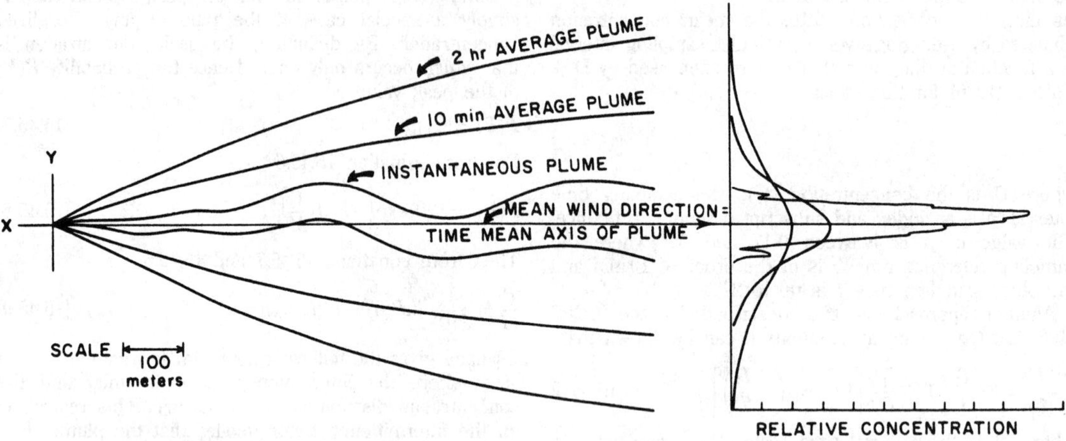

Figure 15.141 *Concentration fluctuations in a meandering plume: effect of combination of meandering and of turbulence (Slade, 1968)*

the sum of two parts: Y^2, the variance attributable to the fluctuations about the instantaneous centre line; and D_y^2, the variance attributable to the meandering. Similarly, the variance σ_z^2 is taken as the sum of the fluctuation variance Z^2 and the meandering variance D_z^2.

$$\sigma_y^2 = Y^2 + D_y^2 \quad [15.45.16a]$$

$$\sigma_z^2 = Z^2 + D_z^2 \quad [15.45.16b]$$

From Equation 15.16.41 the mean concentration χ_m is then taken to be

$$\chi_m = \frac{Q}{2\pi u(Y^2 + D_y^2)^{\frac{1}{2}}(Z^2 + D_z^2)^{\frac{1}{2}}} \times \exp\left\{-\left[\frac{y^2}{2(Y^2 + D_y^2)} + \frac{z^2}{2(Z^2 + D_z^2)}\right]\right\} \quad [15.45.17]$$

and the instantaneous concentration χ_i to be

$$\chi_i = \frac{Q}{2\pi u(Y^2 Z^2)^{\frac{1}{2}}} \exp\left\{-\left[\frac{(y - D_y)^2}{2Y^2} + \frac{(z - D_z)^2}{2Z^2}\right]\right\} \quad [15.45.18]$$

The peak value of the instantaneous concentration χ_p occurs at a point on the centre line of the instantaneous plume ($y = D_y$, $z = D_z$), and hence

$$\frac{\chi_p}{\chi_m} = \frac{(Y^2 + D_y^2)^{\frac{1}{2}}(Z^2 + D_z^2)^{\frac{1}{2}}}{(Y^2 Z^2)^{\frac{1}{2}}} \times \exp\left\{\left[\frac{y^2}{2(Y^2 + D_y^2)} + \frac{z^2}{2(Z^2 + D_z^2)}\right]\right\} \quad [15.45.19]$$

At ground level on the centre line ($y = z = 0$)

$$\frac{\chi_p}{\chi_m} = \frac{(Y^2 + D_y^2)^{\frac{1}{2}}(Z^2 + D_z^2)^{\frac{1}{2}}}{(Y^2 Z^2)^{\frac{1}{2}}} \quad [15.45.20a]$$

$$= \frac{P}{M} \quad [15.45.20b]$$

If conditions are such that

$$Y^2 = Z^2 \quad [15.45.21a]$$

and

$$D_y^2 = D_z^2 = D^2 \quad [15.45.21b]$$

then

$$\frac{P}{M} = \frac{Y^2 + D^2}{Y^2} \quad [15.45.22a]$$

$$= 1 + \frac{D^2}{Y^2} \quad [15.45.22b]$$

But, at large distances, D^2 approaches a constant value, whereas Y^2 continues to grow, and hence

$$\frac{P}{M} \to 1 \quad [15.45.23]$$

15.45.9 Ride model

A model for the intermittency effect in a passive plume has been given by Ride (1984b). The model envisages the cloud as a number of spheres of equal size and uniform concentration in a matrix with zero concentration between the spheres. It can be shown that the peak-to-mean concentration is

$$\frac{C^*}{\bar{C}} = \frac{1}{1 - \gamma} \quad [15.45.24]$$

and that

$$\frac{C^*}{\overline{C}} = \left(\frac{\sigma}{\overline{C}}\right)^2 + 1 \qquad [15.45.25]$$

where C^* is the concentration in the sphere, $\overline{C}$ is the true mean concentration, γ is the intermittency and σ is the standard deviation of the instantaneous concentration. This author defines intermittency as the proportion of time the concentration is below a specified value.

If the maximum concentration which can occur over time τ is C_τ, the peak-to-mean ratio for this time is $C_\tau/\overline{C}$. In the limit, C_τ tends to C^* as τ tends to zero and to $\overline{C}$ as τ tends to infinity. The intermittency is then γ_τ.

Equation 15.45.25] holds for spheres of uniform size. Rodean (1982) has derived the more general relation

$$\frac{C_\tau}{\overline{C}} = k\left(\frac{\sigma}{\overline{C}}\right)^\lambda + 1 \qquad [15.45.26]$$

where k is a constant and λ is an index.

From the foregoing it is possible to derive the relation

$$\frac{C_\tau}{\overline{C}} = k\left(\frac{\sigma}{\overline{C}}\right)^\lambda \psi^{\lambda/2} + 1 \qquad [15.45.27]$$

with

$$\frac{\sigma_\tau}{\overline{C}} = \frac{\sigma}{\overline{C}}\psi^{\frac{1}{2}} \qquad [15.45.28]$$

$$\psi = \left(\frac{\beta}{ut + \beta}\right) \qquad [15.45.29]$$

where β is a constant. From the data of C.D. Jones (1983), Ride (1984b) assigns the values: $k = 11$, $\lambda = 1.5$ and $\beta = 0.4x$, where x is the downwind distance. He gives for $C_\tau/\overline{C}$ a comparison of predicted values with observations made at short ranges by Jones.

In a further treatment, Ride (1984a) omits the factor ψ, equivalent to setting $\psi = 1$.

15.45.10 Wilson model

A model for the intensity of concentration fluctuations in a passive plume has been given by D.J. Wilson (1982).

The concentration fluctuation c' is defined as

$$c' = c - \bar{c} \qquad [15.45.30]$$

where c is the concentration, c' is the concentration fluctuation about the mean and $\bar{c}$ is the mean concentration. The concentration variance is $\overline{c'^2}$.

The model is based on the Gaussian description of a plume:

$$\bar{c} = \frac{Q}{2\pi u_c \sigma_y \sigma_z} G_m F_m \qquad [15.45.31]$$

with

$$G_m = \exp\left[-\frac{1}{2}\left(\frac{y}{\sigma_y}\right)^2\right] \qquad [15.45.32]$$

$$F_m = \exp\left[-\frac{1}{2}\left(\frac{z-h}{\sigma_z}\right)^2\right] + \exp\left[-\frac{1}{2}\left(\frac{z+h}{\sigma_z}\right)^2\right] \qquad [15.45.33]$$

where F_m is a height correction, G_m is a crosswind distance correction, h is the height of the source, Q is the mass rate of release and u_c is the advection velocity of the plume.

From the work of D.J. Wilson, Robins and Fackrell (1982), the concept of a variance source strength q is introduced

$$\overline{c'^2} = \left(\frac{q}{2\pi u_c \sigma_y \sigma_z}\right)^2 G_v F_v \qquad [15.45.34]$$

with

$$G_v = \frac{1}{2}\exp\left[-\frac{1}{2}\left(\frac{y}{\sigma_y} - \phi\right)^2\right] + \frac{1}{2}\exp\left[-\frac{1}{2}\left(\frac{y}{\sigma_y} + \phi\right)^2\right] \qquad [15.45.35]$$

$$F_v = \exp\left[-\frac{1}{2}\left(\frac{z-h_v}{\sigma_z}\right)^2\right] - \alpha\exp\left[-\frac{1}{2}\left(\frac{z+h_v}{\sigma_z}\right)^2\right] \qquad [15.45.36]$$

$$\frac{h_v}{\sigma_z} = \left[\left(\frac{h}{\sigma_z}\right)^2 + \phi^2\right]^{\frac{1}{2}} \qquad [15.45.37]$$

where F_v is a height correction, G_v is a crosswind distance correction, h_v is the effective height of the variance source, q is the variance source strength, α is the sink strength factor and ϕ is a location parameter for the variance source. The value of the parameter ϕ is 0.706.

The variance source strength is given by the relation

$$\frac{q}{Q} = \frac{0.38\lambda}{\lambda^2 + \lambda_o^2} \qquad [15.45.38]$$

with

$$\lambda = \frac{(\sigma_y \sigma_z)^{\frac{1}{2}}}{H} \qquad [15.45.39]$$

$$\lambda_o = \frac{0.119d}{0.033z_e + d} \qquad [15.45.40]$$

$$z_e = h\exp\left[-\left(\frac{\sigma_z}{h}\right)^2\right] \qquad [15.45.41]$$

where d is the diameter of the source, H is the height of the neutrally stable atmospheric boundary layer, z_e is a scale height, λ is the normalized downwind distance and λ_o is the virtual origin of the variance source. The value of H is taken as 650 m.

The intensity i is defined as

$$i = (\overline{c'^2})^{\frac{1}{2}}/\bar{c} \qquad [15.45.42]$$

Hence from Equations 15.45.31–15.45.36 and 15.45.42

$$i = \left(\frac{q}{Q}\right)\frac{(G_v F_v)^{\frac{1}{2}}}{G_m F_m} \qquad [15.45.43]$$

The intensity just derived is the total intensity. Use is also made of the conditional intensity i_p based on the time when the concentration is non-zero.

A corresponding set of conditional concentrations is defined

$$c'_p = c - \bar{c}_p \qquad [15.45.44]$$

where c'_p is the conditional concentration fluctuation about the mean, $\bar{c}_p$ is the conditional mean concentration and subscript p refers to the time when the concentration is non-zero. The conditional concentration variance is $\overline{c_p'^2}$. The conditional intensity i_p is

$$i_p = \frac{\overline{(c_p'^2)}^{\frac{1}{2}}}{\bar{c}_p}$$

The ratio of the mean concentration to the conditional mean concentration is the intermittency γ

$$\gamma = \frac{\bar{c}}{\bar{c}_p} \qquad [15.45.46]$$

This author defines intermittency as the proportion of time the concentration is non-zero.

Then from Equations 15.45.42 and 15.45.44–15.45.46

$$i^2 = \frac{i_p^2}{\gamma} + \frac{1-\gamma}{\gamma} \qquad i_p > 0 \qquad [15.45.47]$$

$$i = \left[\frac{1-\gamma}{\gamma}\right]^{\frac{1}{2}} \qquad i_p = 0$$

On certain assumptions it can be shown that the conditional intensity i_p is given by

$$i_p = i_{p\infty}\frac{\left[1 - \alpha\exp\left(-2\frac{h_v z}{\sigma_z^2}\right)\right]^{\frac{1}{2}}}{\left[1 + \exp\left(-2\frac{hz}{\sigma_z^2}\right)\right]} \qquad [15.45.49]$$

with

$$i_{p\infty} = \frac{q_p}{Q}\exp(-\phi^2/4) \qquad [15.45.50]$$

$$\frac{q_p}{Q} = \frac{0.38\lambda}{\lambda^2 + (0.119)^2} \qquad [15.45.51]$$

where $i_{p\infty}$ is the conditional intensity on the centre line with negligible surface interaction and q_p is the conditional variance source strength. The value of α is taken as 0.6. The results are sensitive to the value used for α.

From Equation 15.45.47 the intermittency γ is

$$\gamma = \frac{i_p^2 + 1}{i^2 + 1} \qquad [15.45.52]$$

The probability of exceeding a particular concentration c is then obtained using an appropriate distribution for the concentration c. The cumulative distribution function F_c for the concentration is

$$F_c = \int_0^c p(c)\mathrm{d}c \qquad [15.45.53]$$

where $p(c)$ is the probability density function for concentration. With intermittency this latter is

$$p(c) = (1-\gamma)\delta(c) + \gamma p_p(c) \qquad [15.45.54]$$

where $p_p(c)$ is the probability density function for conditional concentration. The delta function $\delta(c)$ is a spike at $c = 0$ and zero elsewhere with the further property that

$$\int_{-\infty}^{\infty}\delta(c)\mathrm{d}c = 1 \qquad [15.45.55]$$

Hence from Equations 15.45.3–15.45.55

$$F_c = (1-\gamma) + \gamma\int_0^c p_{p(c)}\mathrm{d}c \qquad [15.45.56]$$

For the log–normal distribution

$$p_p(c) = \frac{1}{(2\pi)^{\frac{1}{2}}\sigma_1 c}\exp\left[-\frac{\ln^2(c/c_{mp})}{2\sigma_1^2}\right] \qquad [15.45.57]$$

with

$$c_{mp} = \frac{\bar{c}_p}{(1+i_p^2)^{\frac{1}{2}}} \qquad [15.45.58]$$

$$\sigma_1 = [\ln(1+i_p^2)]^{\frac{1}{2}} \qquad [15.45.59]$$

Applying Equation 15.45.46, Equations 15.45.58 and 15.45.59 become

$$c_{mp} = \frac{\bar{c}}{\gamma[\gamma(1+i^2)]^{\frac{1}{2}}} \qquad [15.45.60]$$

$$\sigma_1 = \{\ln[\gamma(1+i^2)]\}^{\frac{1}{2}} \qquad [15.45.61]$$

Then from Equations 15.45.54, 15.45.56 and 15.45.57

$$F_c = (1-\gamma) + \frac{\gamma}{2}\left\{1 + \mathrm{erf}\left[\frac{\ln(c/c_{mp})}{2^{\frac{1}{2}}\sigma_1}\right]\right\} \qquad [15.45.62]$$

It is more convenient to work in terms not of the conditional median concentration c_{mp} but of the conditional mean concentration $\bar{c}_p$. The relation between them is

$$c_{mp} = \bar{c}_p\exp(-\sigma_1^2/2) \qquad [15.45.63]$$

Then from Equations 15.45.46 and 15.45.63 it can be shown that

$$\ln\left(\frac{c}{c_{mp}}\right) = \ln\left[\frac{\gamma c}{\bar{c}\exp(-\sigma_1^2/2)}\right] \qquad [15.45.64]$$

But from Equation 15.45.31 the relation between the mean concentration and the mean concentration $\bar{c}_o$ on the centre line is

$$\bar{c} = \bar{c}_o\exp(-y^2/2\sigma_y^2) \qquad [15.45.65]$$

Hence from Equations 15.45.62], [15.45.64] and [15.45.65]

$$F_c = 1 - \gamma\left(1 - \frac{1}{2}\left\{1 + \mathrm{erf}\left[\frac{\ln(\gamma c/\bar{c}_o) + (y^2/2\sigma_y^2) + (\sigma_1^2/2)}{2^{\frac{1}{2}}\sigma_1}\right]\right\}\right) \qquad [15.45.66]$$

Equation 15.45.66] is the basic equation of the model, giving the distribution function of c as a function of the intensity i, the conditional intensity i_p and the intermittency γ.

Wilson also discusses the use of alternative distributions. The two distributions which are most appropriate for physical reasons are the log–normal and the exponential. The exponential distribution has advantages for high intermittency, but overall he favours the log–normal distribution.

For the exponential distribution the probability density function in terms of the conditional mean concentration is

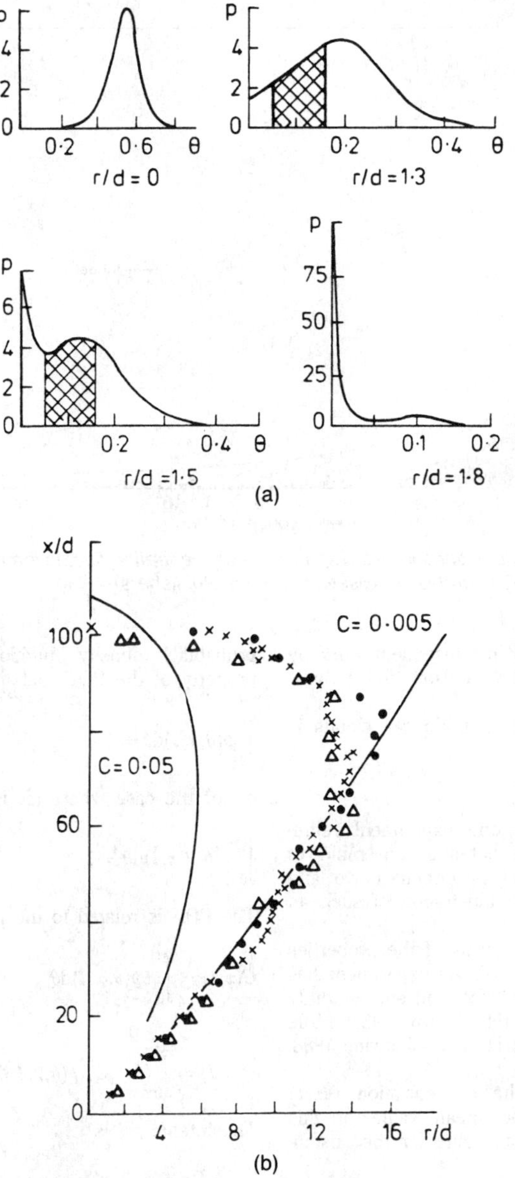

Figure 15.142 *Concentration fluctuations in gas dispersion – concentration and ignition probability in gas jets: (a) probability density functions of concentration in a methane jet (Birch et al., 1978) Courtesy of Cambridge University Press; (b) ignition probability in radial plane of a natural gas jet (Birch, Brown and Dodson, 1980; reproduced by permission) C, concentration; d, diameter; p, probability; r, radial distance; x, axial distance. In (b) measurements made at Reynolds number 12 500 (△), 16 700 (×) and 20 900 (•). Curve of measured ignition probabilities is different from that of the LFL (C = 0.05)*

$$p_p(c) = \frac{1}{\bar{c}_p}\exp(-c/\bar{c}_p) \quad [15.45.67]$$

and that in terms of the mean concentration is

$$p_p(c) = \frac{\gamma}{\bar{c}}\exp(-c\gamma/\bar{c}) \quad [15.45.68]$$

Then from Equation 15.45.56 and Equations 15.45.67 and 15.45.68

$$F_c = 1 - \gamma + \gamma[1 - \exp(-c/\bar{c}_p)] \quad [15.45.69]$$

$$F_c = 1 - \gamma + \gamma[1 - \exp(-c\gamma/\bar{c})] \quad [15.45.70]$$

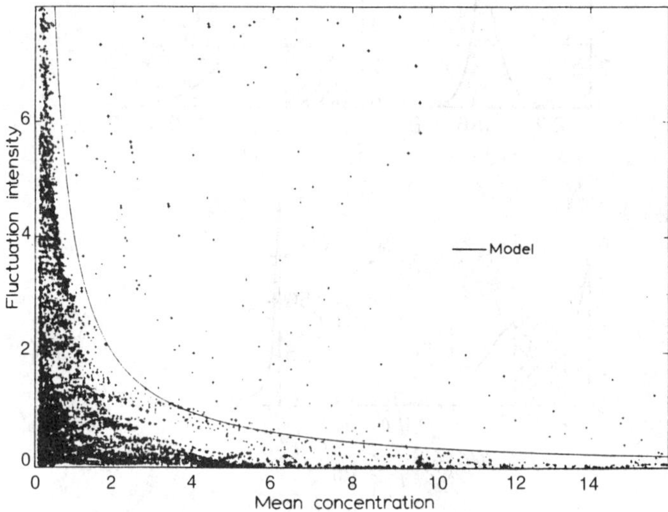

Figure 15.143 Concentration fluctuations in gas dispersion – concentration fluctuation intensity factor for Thorney Island Phase 1 trials (Carn, 1987) (Courtesy of Elsevier Science Publishers)

This model has been developed in subsequent work by the authors and co-workers (D.J. Wilson and Simms, 1985; D.J. Wilson, 1991a).

The application of this model to toxic gas clouds is described in Section 15.47.

15.45.11 Concentration ensemble

Concentration measurements in one experiment represent a single realization of a potential ensemble of experiments. The WSL data set described above and shown in Figure 15.109(e) is an illustration of such an ensemble.

The importance of working in terms of the properties of the ensemble rather than of a single experiment has been brought out by Chatwin [1982b), and subsequently developed (Carn and Chatwin, 1985; Carn, 1987; Mole and Chatwin, 1987; Carn, Sherrel and Chatwin, 1988; Chatwin, 1991).

Chatwin (1982b) discusses the concentration $\Gamma(x,t)$ from a single experiment. The mean value of this concentration is $\bar{\Gamma}(x,t)$. The mean value for the ensemble $C(x,t)$ is defined as

$$C(x,t) = \lim_{n\to\infty}\left[\frac{1}{n}\sum_{r=1}^{n}\Gamma^{(r)}(x,t)\right] \quad [15.45.71]$$

The concentration fluctuation $c(x,t)$ is then

$$c(x,t) = \Gamma(x,t) - C(x,t) \quad [15.45.72]$$

This fluctuation has a mean value $\bar{c}(x,t) = 0$. The ensemble mean square fluctuation $\overline{c^2}(x,t)$ satisfies

$$\overline{c^2}(x,t) = \bar{\Gamma}^2(x,t) - C^2(x,t) \quad [15.45.73]$$

The quantities $C(x,t)$ and $c^2(x,t)$ are the mean and variance of the ensemble, respectively.

In a given realization, there is a certain probability of a particular concentration $\Gamma(x,t)$. Let this concentration be $\theta = \Gamma(x,t)$. The ensemble may be characterized by a probability density function (PDF) $p(\theta,x,t)$. It is a property of the PDF that

$$\int_0^1 p(\theta,x,t)\mathrm{d}\theta = 1 \quad [15.45.74a]$$

or for the case where air is entrained so that $\theta_{max} < 1$

$$\int_0^{\theta_{max}} p(\theta,x,t)\mathrm{d}\theta = 1 \quad [15.45.74b]$$

The PDF is related to the mean and variance as follows:

$$C(x,t) = \int_0^1 \theta p(\theta,x,t)\mathrm{d}\theta \quad [15.45.75]$$

$$\overline{c^2}(x,t) = \int_0^1 (\theta - C)^2 p(\theta,x,t)\mathrm{d}\theta \quad [15.45.76]$$

The intensity i is

$$i = \frac{(\overline{c^2}(x,t))^{\frac{1}{2}}}{C(x,t)} \quad [15.45.77]$$

If the PDF can be characterized by one of the standard distributions, such as the normal or log–normal distributions, the probability may be obtained analytically. Otherwise it must be obtained by numerical integration of the PDF. Thus if the PDF fits the normal distribution, then, for example, there is 90% confidence that the concentration θ lies between the limits $C(x,t) \pm 1.65[\overline{c^2}(x,t)]^{\frac{1}{2}}$.

A typical application of the PDF approach is the determination of the probability P that a concentration in a steady state cloud or jet will lie between the lower flammability limit θ_L and the upper flammability limit θ_U. Then

$$P = \int_{\theta_L}^{\theta_u} p(\theta, x) d\theta \qquad [15.45.78]$$

Such an application is illustrated in the work of Birch, Brown and Dodson (1980) on the concentration fluctuations, and hence ignition probabilities, in methane jets. Figure 15.142(a) shows PDFs as a function of radial distance measured in such jets and Figure 15.142(b) shows the probability of ignition of the jets.

The experimental determination of a concentration ensemble from which statistical properties are to be derived requires considerable care. The experimental variables which can cause variability in concentration must be fully defined.

The properties of the ensemble are affected by the framework, absolute or relative, in which the measurements are made. If an absolute framework is chosen, the properties depend significantly on any feature which influences the motion of the cloud as a whole. This implies the need for a much larger number of experiments.

The relationships between the intermittency and the other quantities in this scheme have been given by Carn and Chatwin (1985). If the concentration θ is assumed to have only one of two values, θ_o or 0:

$$p(\theta, x, t) = \gamma(x, t)\delta(\theta - \theta_o) + [1 - \gamma(x, t)]\delta(\theta) \qquad [15.45.79]$$

where γ is the intermittency and δ is the delta function. These authors define intermittency as the proportion of time the concentration is non-zero. Then from Equation 15.45.79 and Equations 15.45.75 and 15.45.76 the mean and variance are

$$C(x,t) = \theta_o \gamma(x,t) \qquad [15.45.80]$$

$$\bar{c}^2(x,t) = \theta_o^2[\gamma(x,t) - \gamma^2(x,t)] \qquad [15.45.81]$$

Hence from Equations 15.45.77 and 15.45.81 the intensity is

$$i(x,t) = \left[\frac{\theta_o}{C(x,t)} - 1\right]^{\frac{1}{2}} \qquad [15.45.82a]$$

$$= \left[\frac{1}{\gamma(x,t)} - 1\right]^{\frac{1}{2}} \qquad [15.45.82b]$$

15.45.12 Concentrations for dense gas: intensity parameter

In an extension of this work, Carn (1987) has addressed the problem of concentration fluctuations in dense gas clouds. He defines an intensity parameter R which is the square of the intensity I so that

$$R(C) = \frac{\bar{c}^2(x,t)}{\bar{C}^2} \qquad [15.45.83]$$

Thorney Island trials 7–19 were analysed and empirical values of this intensity parameter were obtained as shown in Figure 15.143. It was found that $R(C)$ was constant for a substantial proportion of the total record of $C(t)$ and constant over much of the area of the cloud, but there were also some substantial variations of $R(C)$ even at constant $\bar{C}$.

Carn quotes the following empirical relation, due to Chatwin (1982b), for the bound on $R(C)$:

$$R(C) = \epsilon(1/\bar{C} - 1) \qquad [15.45.84]$$

where ϵ is a constant. The suggested value of ϵ is 0.04.

The behaviour of $R(C)$ was further investigated by obtaining an area-averaged value. This started at a peak of about 5, fell over some 30 s to about 1 and held this value for a period of 30–150 s and then rose to a plateau of about 4 some 160 s after the release.

15.45.13 Concentrations for dense gas: *Workbook*

The *Workbook* by Britter and McQuaid (1988) is primarily concerned with the provision of a method of estimating the mean concentrations in dense gas dispersion. It does, however, include a discussion of concentration fluctuations. It points out that, even for passive dispersion, there is uncertainty about the definition and quantification of concentration. The treatments available for dense gas dispersion are rudimentary.

It is possible, however, that for dense gas dispersion the problem may be simpler than for passive dispersion. The more deterministic nature of dense gas dispersion may result in smaller concentration fluctuations. Moreover, in dense gas dispersion the release is usually at ground level and the cloud remains near the ground. Work on passive dispersion has shown that fluctuations are less for ground level and near ground level releases.

15.45.14 Jets and plumes

The concentration fluctuations in jets have been measured by a number of workers. On the basis of this work Long (1963) has proposed that for the estimation of the maximum instantaneous concentration in a jet the value of the constants k_2 and k_3 to be used in Equation 15.20.8 should be 9 and 2, respectively, and that for the maximum instantaneous concentration in a plume the value of the constant k_4 to be used in Equation 15.20.96 should be 17; he does not propose a value for the constant k_5.

The work of Brown, Birch and Dodson (1980, 1981) on the probability density function approach to concentration fluctuations in jets has been described above.

15.46 Flammable Gas Clouds

It is often necessary to be able to estimate the mass of gas within the flammable range in a cloud from a flammable gas release. Workers who have studied this problem include Burgess *et al.* (1975), Eisenberg, Lynch and Breeding (1975), V.J. Clancey (1977c), van Buijtinen (1980) and J.G. Marshall (1977, 1980).

The treatments are mainly for jets and plumes at the point of release or for clouds with passive dispersion. The cases considered here are

(1) instantaneous source;
(2) continuous source;
(3) momentum jet;
(4) buoyant plume.

The flammable mass in a dense gas cloud is also discussed.

Approaches to dealing with the effect of concentration fluctuations on the determination of the flammable cloud are described.

15.46.1 Instantaneous source

A simple treatment for the mass of gas in the cloud from the instantaneous release of a neutral density gas has been given by V.J. Clancey (1977a), who uses the following equations derived from Equation 15.16.40 with Equation 15.14.5:

$$\chi = \frac{2Q^*}{(2\pi)^{\frac{3}{2}}\sigma^3}\exp\left(-\frac{1}{2}\frac{r^2}{\sigma^s}\right) \qquad [15.46.1]$$

$$\chi_{cc} = \frac{2Q^*}{(2\pi)^{\frac{3}{2}}\sigma^3} \qquad [15.46.2]$$

where Q^* is the mass released, r is the radial distance, σ is the dispersion coefficient, χ is the concentration and the subscript cc denotes the cloud centre.

The mass W of flammable gas between the radii r_1 and r_2 may then be calculated as

$$W = \int_{r_1}^{r_2} 2\pi r^2 \exp\left(-\frac{1}{2}\frac{r^2}{\sigma^2}\right)\mathrm{d}r \qquad [15.46.3]$$

where r_1 is the inner radius, r_2 is the outer radius and W is the mass of flammable gas. For $r_1 = 0$ and $r_2 = \infty$, Equation 15.46.3 gives $W = Q^*$.

The maximum quantity of flammable mixture, and hence the maximum hazard, occurs when the concentration χ_{cc} at the centre of the cloud equals the upper flammability limit χ_U:

$$\chi_U = \chi_{cc} \qquad [15.46.4]$$

$$= \frac{2Q^*}{(2\pi)^{\frac{3}{2}}\sigma_m^2} \qquad [15.46.5]$$

where σ_m is the value of σ at maximum hazard. The lower flammability limit χ_L and the radius r_L at which this occurs are given by the equations

$$\chi_L = \chi_U \exp\left(-\frac{1}{2}\frac{r_L^2}{\sigma_m^2}\right) \qquad [15.46.6]$$

$$r_L = [2\sigma_m^2 \ln(\chi_U/\chi_L)]^{\frac{1}{2}} \qquad [15.46.7]$$

The mass F of flammable gas between the flammability limits at maximum hazard is

$$F = \int_0^{r_L} 2\pi r^2 \chi_U \exp\left(-\frac{1}{2}\frac{r^2}{\sigma_m^2}\right)\mathrm{d}r \qquad [15.46.8]$$

The evaluation of Equation 15.46.8 is assisted by noting that it may be rewritten as

$$F = \int_0^{r_L} \frac{2Q^* r^2}{\sigma_m^2}\left[\frac{1}{\sigma_m(2\pi)^{\frac{1}{2}}}\exp\left(-\frac{1}{2}\frac{r^2}{\sigma_m^2}\right)\right]\mathrm{d}r \qquad [15.46.9]$$

where the expression inside the square brackets is the Gaussian distribution.

This model indicates that the ratio χ_U/χ_L of the upper and lower flammability limits is an important quantity in determining the fraction of the cloud within the flammable range and thus available for combustion.

For a typical hydrocarbon the ratio of the upper to the lower flammability limit is in the range 4–6, so that the ratio of gas within the flammable range to the total gas in the cloud at maximum hazard is

$$F/Q^* \approx 2/3 \qquad [15.46.10]$$

The growth of the cloud prior to this point may be determined from Equations 15.46.1–15.46.3. As a first approximation, however, the mass of gas within the flammable range may be taken as rising linearly from the initial value to the value at maximum hazard with a rapid fall-off thereafter.

A further treatment for an instantaneous release from a point source has been given by van Buijtenen (1980). Formulating his model in terms of mass and mass concentrations of gas, the starting point is the following relation for an instantaneous release from an elevated source, of height H, derived from Equations 15.16.40b and 15.16.43:

$$\chi(x,y,z,t) = \frac{Q^*}{(2\pi)^{\frac{3}{2}}\sigma_x\sigma_y\sigma_z}\exp\left[-\frac{(x-ut)^2}{2\sigma_x^2}\right] \times \exp\left(-\frac{y^2}{2\sigma_y^2}\right)\left\{\exp\left[-\frac{(z-H)^2}{2\sigma_z^2}\right] + \exp\left[\frac{(z+H)^2}{2\sigma_z^2}\right]\right\} \qquad [15.46.11]$$

Then from Equation 15.46.11 the fraction of gas between the flammability limits is

$$\frac{Q_e^*}{Q^*} = \mathrm{erf}[(\ln\chi_{cc}/\chi_L)^{\frac{1}{2}}] - \mathrm{erf}[(\ln\chi_{cc}/\chi_U)^{\frac{1}{2}}] - \frac{2\chi_L}{(\pi)^{\frac{1}{2}}\chi_{cc}}[(\ln\chi_{cc}/\chi_L)^{\frac{1}{2}}] + \frac{2\chi_U}{(\pi)^{\frac{1}{2}}\chi_{cc}}[(\ln\chi_{cc}/\chi_U)^{\frac{1}{2}}] \quad \chi_{cc} > \chi_U \qquad [15.46.12a]$$

$$= \mathrm{erf}[(\ln\chi_{cc}/\chi_L)^{\frac{1}{2}}] - \frac{2\chi_L}{\pi^{\frac{1}{2}}\chi_{cc}}[(\ln\chi_{cc}/\chi_L)^{\frac{1}{2}}] \quad \chi_{cc} > \chi_U \qquad [15.46.12b]$$

where Q^* is the mass of gas released and Q_e^* is the mass of gas within the flammable range. Equation 15.46.12 applies to a cloud completely free of the ground or, alternatively, to a cloud at ground level ($H = 0$).

The maximum value of the fraction Q_e^*/Q^* of gas within the flammable range is obtained by differentiating Equation (15.46.12a) with respect to χ_{cc}, which gives

$$\ln\chi_{cc} = \frac{\chi_U^2\ln\chi_U - \chi_L^2\ln\chi_L}{\chi_U^2 - \chi_L^2} \qquad [15.46.13]$$

Substituting for χ_{cc} from Equation 15.46.13 into Equation 15.46.12a gives

$$\left(\frac{Q_e^*}{Q^*}\right)_{max} = \mathrm{erf}(v_2^{\frac{1}{2}}) - \mathrm{erf}(v_1^{\frac{1}{2}}) - \frac{2\exp(-v_2)v_2^{\frac{1}{2}}}{\pi^{\frac{1}{2}}} + \frac{2\exp(-v_1)v_1^{\frac{1}{2}}}{\pi^{\frac{1}{2}}} \qquad [15.46.14]$$

with

$$v = \chi_U/\chi_L \qquad [15.46.15a]$$

$$v_1 = \ln(v)/(v^2-1) \qquad [15.46.15b]$$

$$v_2 = v^2\ln(v)/(v^2-1) \qquad [15.46.15c]$$

where the subscript max denotes the maximum value.

The influence of the ratio χ_U/χ_L of the flammability limits on the fraction Q_e^*/Q^* of gas between the

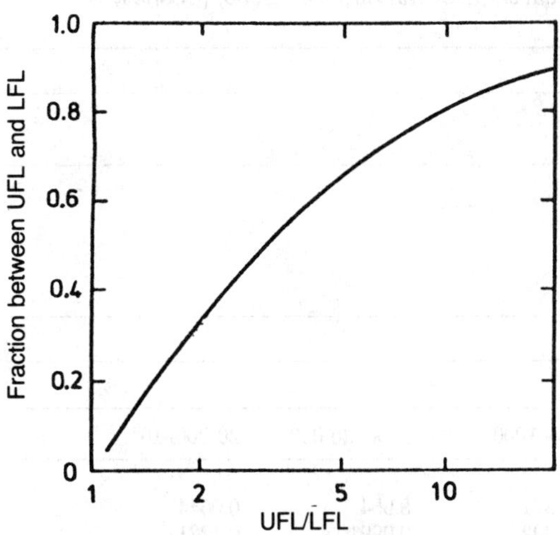

***Figure 15.144** Flammable gas clouds – effect of ratio of flammability limits on fraction of gas within flammable range for release from an instantaneous point source (van Buijtenen, 1980). LFL, lower flammability limit; UFL, upper flammability limit (Courtesy of Elsevier Science Publishers)*

flammability limits according to Equation 15.46.14 is shown in Figure 15.144.

The application of this model can be extended if necessary by the use of the virtual source method. Van Buijtenen also gives an equation for an instantaneous source of finite size.

The problem of the amount of gas available for combustion has also been considered by Eisenberg, Lynch and Breeding (1975). These authors list three assumptions which may be made about this quantity:

(1) all the fuel above the lower flammability limit;
(2) all the fuel between the upper and lower flammability limits;
(3) all the fuel between the upper and lower flammability limits for which oxygen is available.

The third assumption is the most accurate. For this latter case the mass of gas available for combustion is

$$W_F = \int_{V_{LS}} \chi \mathrm{d}V + \int_{V_{SU}} \chi F(c) \mathrm{d}V \qquad [15.46.16]$$

where $F(c)$ is the weighting function giving the fraction of fuel present that enters into the combustion reaction, V_{LS} is the volume of the cloud between the lower flammability limit and the stoichiometric concentration (m^3), V_{SU} is the volume of the cloud between the stoichiometric concentration and the upper flammability limit (m^3) and W_F is the mass of fuel available for combustion (kg).

The weighting function $F(c)$ is defined as

$$F(c) = \frac{n_c}{n_p} \qquad [15.46.17]$$

where n_c is the number of moles of fuel consumed and n_p is the number of moles of fuel present. Then if 1 mole of fuel reacts with L mole of oxygen

$$n_c = \frac{0.21 n_a}{L} \qquad [15.46.18]$$

where n_a is the number of moles of air in the rich mixture. Hence

$$F(c) = \frac{0.21}{L} \frac{n_a}{n_p} \qquad [15.46.19]$$

15.46.2 Continuous source

Van Buijtenen also gives a model for a continuous source, derived from Equation 15.16.43. In this case it is necessary to use a correlation for the crosswind and vertical dispersion coefficients. The relations used are

$$\sigma_y = ax^b \qquad [15.46.20a]$$

$$\sigma_z = cx^d \qquad [15.46.20b]$$

where a and c are constants and b and d are indices.

Then the ratio Q_e/Q of the mass of gas within the flammability limits to the mass of gas released is

$$\frac{Q_e}{Q} = \frac{b+d}{b+d+1} \frac{x_L - x_U}{u} \qquad [15.46.21]$$

where u is the wind speed, and x_L and x_U are the maximum distances to the lower and upper flammability limits, respectively. Equation 15.46.21 applies to a cloud completely free of the ground or, alternatively to a cloud at ground level.

An alternative formulation of Equation 15.46.21 may be obtained by substituting for x_L and x_U from Equation 15.16.43 in Equation 15.46.21 to obtain for a plume at ground level

$$\frac{Q_e}{Q} = \frac{b+d}{(b+d+1)u} \left(\frac{Q}{\pi uac}\right)^{1/(b+d)} \times \left[\left(\frac{1}{\chi_L}\right)^{1/(b+d)} - \left(\frac{1}{\chi_U}\right)^{1/(b+d)}\right] \qquad [15.46.22]$$

For a plume free of the ground the term πuac in Equation 15.46.22 is replaced by $2\pi uac$. If the concentration at the source is below the upper flammability limit χ_U, x_U in Equation 15.46.21 and the term in χ_U in Equation 15.46.22 are zero.

Values given by van Buijtenen of the parameters in Equation 15.46.20 for use in this model are shown in Table 15.55.

For typical values of the indices b and d, the rate of increase of the ratio Q_e/Q with Q exceeds a linear rate. The application of this model can be extended, if necessary, by the use of the virtual source method. This is discussed by van Buijtenen.

A rather similar model for a continuous source has been given by J.G. Marshall (1980), who writes Equation 15.16.41 in the form

***Table 15.55** Meteorological parameters for van Buijtenen's equation (after van Buijtenen, 1980) (Courtesy of Elsevier Science Publishers)*

A Values of a and b

Pasquill stability category	*Parameters* a	b
B	0.371	0.866
C	0.209	0.897
D	0.128	0.905
E	0.098	0.902
F	0.065	0.902

B Values of c, d and e

Pasquill stability category	*Distance* (m)				
	1–10	10–100	100–1000	1000–20 000	20 000–10^5
Values of c					
B	0.15	0.1202	0.0371	0.054	0.0644
C	0.15	0.0963	0.0992	0.0991	0.1221
D	0.15	0.0939	0.2066	0.9248	1.0935
E	0.15	0.0864	0.1975	2.3441	2.5144
F	0.15	0.0880	0.09842	6.5286	2.8745
Values of d					
B	0.8846	0.9807	1.1530	1.0997	1.0843
C	0.7533	0.9456	0.9289	0.9255	0.9031
D	0.6368	0.8403	0.7338	0.5474	0.5229
E	0.5643	0.8037	0.6865	0.4026	0.3756
F	0.4771	0.7085	0.7210	0.2593	0.3010
Values of e					
B	0	0	3.1914	2.5397	0
C	0	0	0.2444	1.7383	0
D	0	0	−1.3659	−9.0641	0
E	0	0	−1.1644	−16.3186	0
F	0	0	−0.3231	−25.1583	0

$$\chi(x,y,z) = \frac{\dot{m}_o}{\pi\sigma_y\sigma_z u}\exp\left[-\frac{1}{2}\left(\frac{y^2}{\sigma_y^2}+\frac{z^2}{\sigma_z^2}\right)\right] \quad [15.46.23]$$

where $\dot{m}_o$ is the mass rate of release. Thus for the concentration on the centre-line at ground level ($y = z = 0$)

$$\chi(x,0,0) = \frac{\dot{m}_o}{\pi\sigma_y\sigma_z u} \quad [15.46.24]$$

Use is made of the following correlation for the product of the crosswind and vertical dispersion coefficients

$$\sigma_y\sigma_z = Dx^f \quad [15.46.25]$$

where D is a constant and f an index.

On the assumption that the crosswind isopleths are semi-elliptical in shape, substitution for $\sigma_y\sigma_z$ from Equation 15.46.25 in Equation 15.46.23 yields

$$Q_{FL} = \left(\frac{1}{\pi D}\right)\frac{1}{f}\frac{f}{f+1}\left(\frac{\dot{m}_o}{u}\right)^{(f+1)/f}\left(\frac{1}{\chi_L^{1/f}}-\frac{1}{\chi_U^{1/f}}\right) \quad [15.46.26]$$

where Q_{FL} is the mass of gas within the flammability limits. Values given by Marshall of the parameters in Equation 15.46.25 for use in this model are shown in Table 15.56.

15.46.3 Momentum jet

J.G. Marshall (1977) has also treated the cases of a flammable momentum jet and a buoyant plume. The basis of the model for a jet is the work of Long (1963).

For a momentum jet the concentration in the jet is given by the equation

$$c(z,r) = \frac{k_1 d_o}{z}\frac{M_o T_a}{M_a T_o}\frac{\bar{M}}{M_o}\exp\left[-\left(\frac{k_2 r}{z}\right)^2\right] \quad [15.46.27]$$

where c is the concentration (volume fraction), d_o is the effective diameter of the discharge (m), M is the molecular weight, $\bar{M}$ is the mean molecular weight of

Table 15.56 *Meteorological parameters for Marshall's equation (after J.G. Marshall, 1980)*

Pasquill stability category	*Parameters*	
	D	f
A	3.06×10^{-3}	2.4
B	1.38×10^{-2}	1.9
C	8.9×10^{-3}	1.8
D	6.0×10^{-3}	1.7
E	3.88×10^{-3}	1.7
F	1.43×10^{-3}	1.7

the gas mixture at an axial distance z, r is the radial distance from the axis (m), T is the absolute temperature (K), z is the axial length or vertical height (m), k_1 and k_2 are constants, and subscripts a denote air and o the original discharge.

Marshall states that d_o is the effective diameter of the jet after it has expanded to atmospheric pressure but before any entrainment has occurred. For a vent this will be the diameter of the vent, whereas for a plant failure it will be related to the effective 'hole' size and internal pressure.

The distance at which transition from jet to plume occurs is governed by the internal Froude number Fr:

$$z_{tr} = k_3 Fr d_o \quad [15.46.28a]$$

$$= \frac{k_3 w_o d_o}{(g' d_o)^{\frac{1}{2}}} \quad [15.46.28b]$$

with

$$g' = g\frac{M_a T_o - M_o T_a}{M_a T_o} \quad [15.46.29]$$

where w is the velocity (m/s), k_3 is a constant and subscript tr denotes transition.

Equation 15.46.27 may be used to determine an axial length and may be integrated for volume and for quantity of flammable material. In order to simplify the integration Marshall assumes that $\bar{M}$ is equal to M_a. The effect of this simplification for a gas of lower molecular weight than air is to overestimate the axial distance at which the upper and lower limit concentrations are reached, the former more than the latter. For a gas of higher molecular weight than air the opposite is true. For most flammable gases the error in axial distance or quantity is much less than 20%.

The equations for the jet are then

$$z_L = \frac{k_1 d_o}{c_L}\left(\frac{M_a T_a}{M_o T_o}\right)^{\frac{1}{2}} \quad [15.46.30]$$

$$V_{FL} = \frac{\pi k_1^3}{9k_2^2}\left(\frac{M_a T_a}{M_o T_o}\right)^{\frac{3}{2}} d_o^3\left(\frac{1}{c_L^3} - \frac{1}{c_U^3}\right) \quad [15.46.31]$$

$$Q_{FL} = \frac{\pi k_1^3}{6k_2^2}\frac{273}{22.4}\left(\frac{M_a}{T_o}\right)^{\frac{3}{2}}\left(\frac{T_a}{M_o}\right)^{\frac{1}{2}} d_o^3\left(\frac{1}{c_L^2} - \frac{1}{c_U^2}\right) \quad [15.46.32]$$

where Q_{FL} is the mass of flammable material in the cloud between the flammability limits (kg), V_{FL} is the volume of the cloud between the flammability limits (m^3) and subscript L denotes the lower flammability limit and U the upper flammability limit.

The values of the constants used by Marshall are as follows. For k_1 and k_2 the values recommended by Long (1963) as giving the maximum instantaneous concentrations are 9 and 2, respectively. But these two constants are not independent. If the radial spread of velocity and concentration in the jet are the same

$$k_2 = (2)^{\frac{1}{2}} k_1 \quad [15.46.33]$$

Marshall uses values of 9 and 12.7 for k_1 and k_2, respectively. For k_3, values of 1.39 and 1.74 have been obtained by B.R. Morton (1959) and by J.S. Turner (1966) for the case where momentum and buoyancy forces are directly opposed. Marshall takes k_3 as 1.55, which is the mean of these two values.

For this model a comparison can be made between the length of the jet to the lower flammability limit and the length of the corresponding jet flame. The length of a turbulent jet flame has been given by Hawthorne, Weddell and Hottel (1949). Their equation is quoted in Chapter 16 as Equation 16.3.3. Rewriting this in Marshall's notation

$$z_F = \frac{5.3 d_o}{c_s}\left\{\frac{T_F}{\alpha T_o}\left[c_s + (1 - c_s)\left(\frac{M_a}{M_o}\right)\right]\right\}^{\frac{1}{2}} \quad [15.46.34]$$

where α is the ratio of the number of moles of unreacted and reacted gas and subscript F denotes the flame and s the stoichiometric mixture. Then, making the following assumptions,

$$T_o = T_a \quad [15.46.35a]$$

$$\frac{T_F}{\alpha T_o} \approx 8 \quad [15.46.35b]$$

$$c_s \approx 2c_L \quad [15.46.35c]$$

$$c_s(1 - c_s)\frac{M_a}{M_o} \approx \frac{M_a}{M_o} \quad [15.46.35d]$$

it can be shown from Equations 15.46.30] and [15.46.34] that

$$\frac{z_F}{z_L} = 0.83 \quad [15.46.36]$$

15.46.4 Buoyant plume

For the buoyant plume the basis of the model given by Marshall is the work of B.R. Morton, Taylor and Turner, (1956), Yih (1951) and Rouse, Yih and Humphreys (1952). The basic treatment is for neutral stability conditions, but other stability conditions are also taken into account, as described below.

For a buoyant plume the concentration in the plume is given by the equation

$$c(z,r) = \frac{k_4 \dot{v}_o^{\frac{2}{3}}}{z^{\frac{5}{3}} g^{\frac{1}{3}}}\left(\frac{M_a T_o}{M_a T_o - M_o T_a}\right)^{\frac{1}{3}} \exp\left[-\left(\frac{k_5 r}{z}\right)^2\right] \quad [15.46.37]$$

where g is the acceleration due to gravity (m/s^2), $\dot{v}$ is the volumetric flow (m^3/s) and k_4 and k_5 are constants.

The equations for the length and volume of the flammable zone in the plume and for the mass of flammable gas in it are

$$z_L = \frac{k_4^{\frac{3}{5}} \dot{v}_o^{\frac{2}{5}}}{g^{\frac{1}{5}} c_L^{\frac{3}{5}}} \left(\frac{M_a T_o}{M_a T_o - M_o T_a} \right)^{\frac{1}{5}} \quad [15.46.38]$$

$$V_{FL} = \frac{5\pi k_4^{\frac{9}{5}} \dot{v}_o^{\frac{6}{5}}}{27 k_5^2 g^{\frac{3}{5}}} \left(\frac{M_a T_o}{M_a T_o - M_o T_a} \right)^{\frac{3}{5}} \left(\frac{1}{c_L^{\frac{9}{5}}} - \frac{1}{c_U^{\frac{9}{5}}} \right) \quad [15.46.39]$$

$$Q_{FL} = \frac{5\pi M_o k_4^{\frac{9}{5}} \dot{v}_o^{\frac{6}{5}}}{12 T_a k_5^2 g^{\frac{3}{5}}} \frac{273}{22.4} \times \left(\frac{M_a T_o}{M_a T_o - M_o T_a} \right)^{\frac{3}{5}} \left(\frac{1}{c_L^{\frac{4}{5}}} - \frac{1}{c_U^{\frac{4}{5}}} \right) \quad [15.46.40]$$

The dispersion is affected by three separate but related aspects of the atmospheric conditions. These are (1) the degree of turbulence, (2) the wind speed and (3) the stability conditions.

These characteristics are not independent. Inversions in which the temperature increases with height occur more frequently in conditions of low turbulence and low wind speed. This is the combination of conditions least conducive to dispersion.

The effect of temperature gradient may be taken into account by using the parameter G:

$$G = \frac{g}{T_a}(\mathrm{d}T_a/\mathrm{d}z + \Gamma) \quad [15.46.41]$$

where Γ is the dry adiabatic lapse rate (= 0.0098°C/m). Equation 15.46.41] is related to the Richardson number.

It is not uncommon in the UK to have temperature gradients $\mathrm{d}T_a/\mathrm{d}z$ of the order of 0.01°C/m, while gradients of 0.1°C/m are not unknown. The corresponding values of the parameter Γ are:

$\mathrm{d}T_a/\mathrm{d}z$ (°C/m)	G (s^{-2})
0 (isothermal)	0.000 334
0.01	0.000 674
0.1	0.003 74

For an atmosphere in which the temperature gradient is greater than that of a neutral adiabatic atmosphere ($\Gamma > 0$) and in which these conditions continue sufficiently far upward, there will come a height at which the density of a plume of positive buoyancy equals that of the surrounding air. Marshall gives the following equation for the height of a plume based on the work of B.R. Morton, Taylor and Turner (1956):

$$z_{max} = \frac{k_6 \dot{v}_o^{\frac{1}{4}}}{G^{\frac{3}{8}}} \left(\frac{M_a T_o - M_o T_a}{M_a T_o} \right)^{\frac{1}{4}} \quad [15.46.42]$$

where k_6 is a constant.

Equation 15.46.42 sets a limit on the rise of the plume in stability conditions where the temperature increases with height. The plume height z_{max} calculated from Equation 15.46.42 may be compared with the height z_L at which the concentration reaches the lower flammability limit as given by Equation 15.46.38. If z_{max} is less than z_L, the plume does not rise sufficiently high to be diluted below the lower flammability limit.

The values of the constants used by Marshall are as follows. For k_4 the value recommended by Long as giving the maximum instantaneous concentration is 17, and this is used by Marshall. For k_5 Marshall assumes that the ratio k_5/k_4 is the same as for time mean values and thus obtains for k_5 a value of 13; for k_6 he uses the value of 6.65 given by B.R. Morton, Taylor and Turner (1956).

15.46.5 Comparative cloud sizes

J.G. Marshall (1977, 1980) has presented a number of illustrative calculations of the dimensions of and quantities of flammable material in clouds from a continuous source, a jet and a buoyant plume for hydrocarbon gases. J.G. Marshall (1980) gives the following summary of his equations.

Continuous source

$$Q_{FL} = \frac{0.321 \dot{m}_o^{1.59}}{D^{0.59} u^{1.59}} \left(\frac{1}{\chi_L^{0.59}} - \frac{1}{\chi_U^{0.59}} \right) \quad [15.46.43]$$

Equation 15.46.43 is obtained from Equation 15.46.26 with the value of the index f applicable to Pasquill stability categories D–F, which is 1.7.

Momentum jet

$$Q_{FL} = 3.4 \rho_a^{1.5} \left(\frac{\dot{m}_o}{w_o} \right)^{1.5} \left(\frac{1}{\chi_L^2} - \frac{1}{\chi_U^2} \right) \quad [15.46.44]$$

Buoyant plume

$$Q_{FL} = \frac{1.45 T_o^{1.8}}{T_a^{1.8}} \left(\frac{\rho_a^{0.6} \rho_o^{0.6} \dot{m}_o^{1.2}}{\Delta\rho_o^{0.6}} \right) \left(\frac{1}{\chi_L^{0.8}} - \frac{1}{\chi_U^{0.8}} \right) \quad [15.46.45]$$

with

$$\Delta\rho_o = \rho_a - \rho_o \quad [15.46.46]$$

where $\dot{m}_o$ is the mass rate of release (kg/s), u is the wind speed (m/s), $\Delta\rho_o$ is the density difference between air and the gas (kg/m³) and χ is the concentration (kg/m³).

The results shown in the table below were obtained by J.G. Marshall (1977) for jets of methane and ethylene at 288 K, from Equations 15.46.30] and [15.46.32]:

	M_o	c_L (v/v)	c_U (v/v)	z_L/d_o	$\dot{m}/d_o^2$ (kg/m²s)	Q_{FL}/d_o^3 (kg/m³)	z_F/z_L
CH_4	16	0.05	0.15	243	236	1388	0.83
C_2H_4	28	0.027	0.36	340	304	4020	0.73

For buoyant plumes of the same gases at the same temperature, from Equations 15.46.38 and 15.46.42:

	$\dot{m}$ (kg/s)	z_L (neutral atmosphere) (m)	z_{max} (stable atmosphere) $G = 0.000\,334$ (m)	$G = 0.001$ (m)
CH_4	100	181	374	248
	1000	455	665	441
C_2H_4	100	346	175	116
	1000	870	311	206

Marshall also gives results for jets of ethane, propane and n-butane and at temperatures of 423 and 573 K.

For buoyant plumes the results show the relation between the maximum height attained by the plume z_{max} and the height required for dilution to the lower flammability limit z_L. For methane, which has a large positive buoyancy, $z_{max} > z_L$ approximately in all the cases considered. For ethylene, which has only a small positive buoyancy, $z_L > z_{max}$ in all the cases considered.

Both for jets and buoyant plumes the residence time is of interest as giving the lower limit for the time required for the flammable cloud to form. For the cases of jets considered by Marshall the residence times within the flammable envelope were in the range 2.5–11 s for mass flows of 100 kg/s and 8–24 s for mass flows of 1000 kg/s. For the cases of plumes the minimum residence times were 3 s and 5 s for mass flows of 100 and 1000 kg/s, respectively. The results also show that emission rates of 100–1000 kg/s can give rise to vapour clouds containing at least 1 te of flammable material.

In a second study extending the comparison to a continuous source also, J.G. Marshall (1980) gives the emission rates necessary to produce clouds containing given amounts of gas as follows:

Type of release	*Emission rate* (kg/s) *to give cloud containing* 1000 kg	10 000 kg
Continuous source	6–100	25–440
Jet	13–110	50–600
Buoyant plume	16–110	100–700

He also gives the graphical comparison shown in Figure 15.145. In each of the three cases the critical parameter is different: for the continuous source in Figure 15.145(a) it is the stability category and wind speed u; for the jet in Figure 15.145(b) it is the pipe velocity w_o; and for the buoyant plume in Figure 15.145(c) it is the hydrocarbon species. In the case of the continuous source calculation assumes a hypothetical hydrocarbon with lower and upper flammability limits of 0.039 and 0.176 kg/m^3, respectively. In the case of the jet, which is for a typical lower paraffin, the lower pipe velocity is close to the limiting value below which jet dispersion would no longer be the controlling mechanism, while the upper velocity is close to the sonic velocity. In the case of the plume, methane has been taken as the most positively buoyant material likely to be of interest, other than hydrogen, while ethylene at 288 K represents the limiting gas density above which other modes of dispersion would predominate. This study indicates that in respect of the mass of gas within the flammable limits for a given size of release the three types of release appear remarkably similar.

15.46.6 Dense gas clouds

For dense gas clouds, the approach taken depends on the type of model used. A box model gives the cloud dimensions directly. The *Workbook* method gives correlations for the area within a given concentration.

15.46.7 Empirical relations

An empirical equation for the radius of a vapour cloud has been given in the *Second Report* of the Advisory Committee on Major Hazards (ACMH) (Harvey, 1979b):

$$R = 30M^{\frac{1}{3}} \qquad [15.46.47]$$

where M is the mass of vapour (e) and R is the radius of the cloud (m).

Equation 15.46.47] is based on the assumption that the ratio of the cloud radius to cloud height is 5:1. It appears to agree reasonably well with the estimated size of the clouds at Flixborough and at Beek. If it is assumed that the quantities released in these two incidents were 40 te and 5.5 te, respectively, the cloud sizes as calculated from Equation 15.46.47 are 103 and 53 m, which compare with the estimated cloud sizes of 102 m (kidney-shaped cloud) (Sadée, Samuels and O'Brien, 1976—77) and of 50 m (Ministry of Social Affairs, 1976), respectively.

15.46.8 Concentration fluctuations

It has long been recognized that the effect of concentration fluctuations is to extend beyond the nominal mean lower flammability limit (LFL) concentration values the range and the area of the cloud which is susceptible to ignition.

Feldbauer *et al.* (1972) allowed for this by the use of an effective limit, replacing the LFL value by an effective value, which they took as the 0.5 LFL value. This approach has been widely utilized.

It has been suggested by Pikaar (1985) that the Maplin Sands vapour cloud fire tests indicate that for this purpose use may be made of a practical peak/mean concentration ratio of 1.4, which translates to a 0.7 LFL value. The account given by this author is discussed in more detail in Chapter 17.

More sophisticated approaches may be used, based on the modelling of concentration fluctuations, as described in Section 15.45. In the work of Birch, Brown and Dodson (1980, 1981) on ignition in jets, the relationship between the gas concentration and the occurrence of ignition is expressed as a probability distribution.

A discussion of the problem in the context of dense gas dispersion has been given by McQuaid (1984a). The deficiencies of applying averaging to turbulence phenomena have been discussed by Spalding (1983). For instance, the average obtained depends on the averaging time selected. There is no simple solution.

15.46.9 Cloud size estimation procedures

A number of companies have developed procedures for estimating the size of the flammable vapour cloud to be considered in plant design. Essentially these define the scenarios to be treated, such as the size and duration of

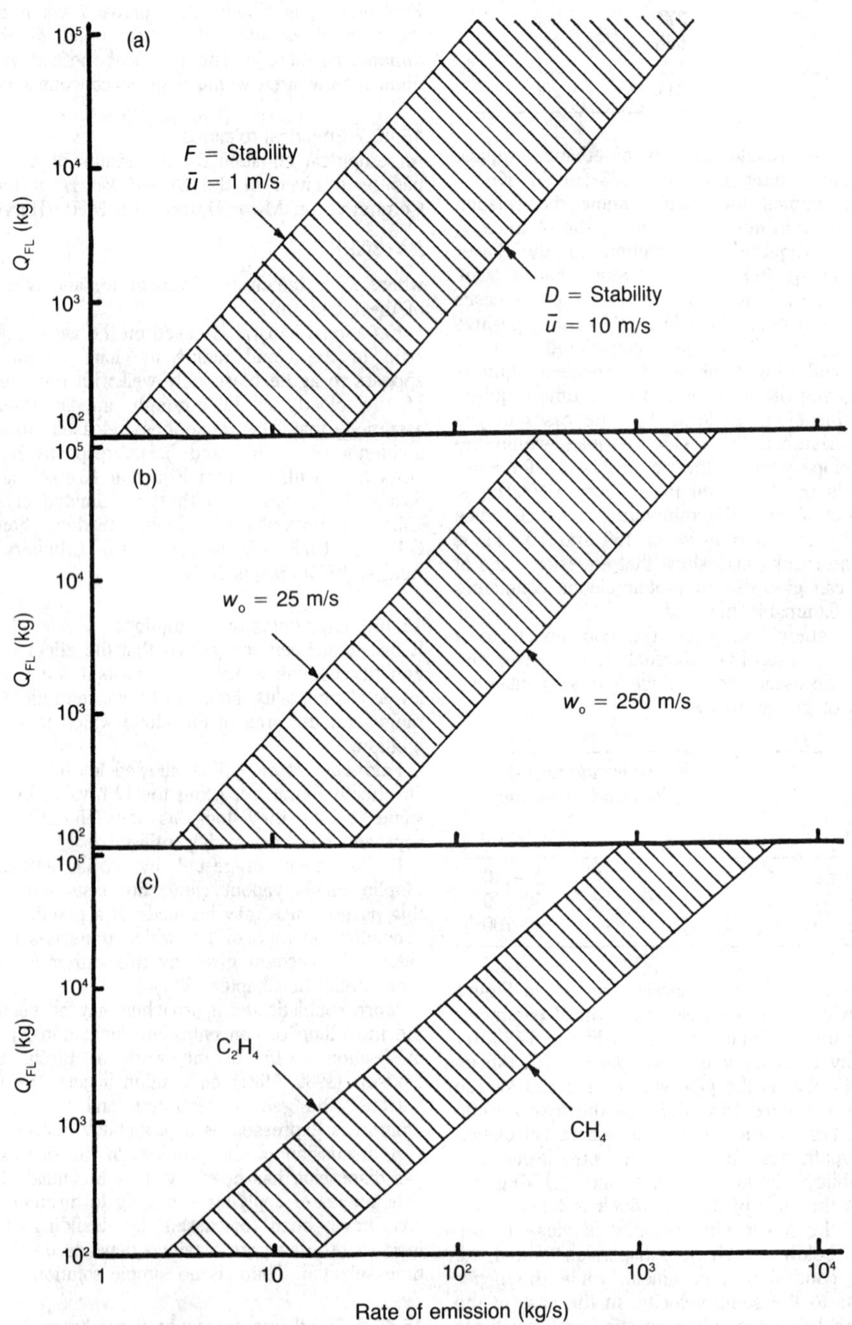

Figure 15.145 *Flammable gas clouds – mass of gas within flammable limits in gas cloud in some principal dispersion situations (J.G. Marshall, 1980): (a) passive gas dispersion; (b) jet dispersion; (c) plume dispersion. $\bar{u}$, mean wind speed; ω_0, jet discharge velocity (Courtesy of the Institution of Chemical Engineers)*

the leak. Accounts of such procedures include those by Exxon (n.d.), Brasie (1976a), FMRC (1990) and Industrial Risk Insurers (IRI) (1990), and a review is given by the CCPS (1994/15).

The procedure of Brasie (1976a) for a continuous leak utilizes for the leak rate from a short nozzle or pipe the relation

$$W_{\mathrm{lr}} = 2343P^{0.7} \quad [15.46.48]$$

where P is the operating pressure (bar) and W_{lr} is the mass flow per unit area (kg/m^2 s). The flow of vapour into the cloud is obtained from the flash fraction. The mass of vapour in the cloud above the LFL is a function of time. The maximum time is the time span for the cloud concentration to fall below the LFL. For this use is made of the relation

$$t = 516(W/Mf) \quad [15.46.49]$$

where f is the lower flammability limit (%), M is the molecular weight, W is the mass flow (kg/s) and t is the time span (s).

The CCPS (1994/15) describes an unpublished procedure of Exxon (n.d.). This distinguishes between releases of (1) gas, (2) liquid and (3) two-phase or flashing liquid. For a gas release, the mass of vapour entering the cloud is taken as the lesser of (1) the total inventory or (2) the product of the flow and the time to stop the leak. For a liquid release, the quantity spilled is taken as the lesser of (1) the total inventory or (2) the product of the flow and the time to stop the leak. The mass of vapour entering the cloud is then taken as the product of the liquid evaporation rate and the time to find a likely source of ignition. For a liquid which is two-phase or flashing, the mass of vapour entering the cloud is the lesser of (1) the total inventory times twice the flash fraction or (2) the product of the rate of release and the time to stop the leak and twice the flash fraction.

The procedure of the Factory Mutual Research Corporation (FMRC, 1990) is oriented towards the estimation of the loss potential in insurance. In this method the credible releases are taken as: (1) 10-minute releases from (a) the largest connection on the largest vessel or vessel train, (b) atmospheric or pressure storage, credit being given for the operation of excess flow valves, or (c) above-ground pipelines conveying material from a remote, large capacity source; and (2) the total inventory of mobile tanks, such as tank trucks or rail tank cars.

Another insurance-related procedure is that of the Industrial Risk Insurers (IRI, 1990). For the catastrophic loss potential, the mass released is based on the content of vessels or vessel trains, storage tanks and pipelines to storage. For the latter a guillotine fracture is assumed with a release duration of 30 minutes. No credit is allowed for shut-off valves between vessels. For the probable maximum loss potential, a credible release is determined, allowing for the operation of emergency shut-off and blowdown. The minimum release is the inventory of the largest process vessel and the maximum that of the largest train of vessels which are interconnected and not isolated.

15.47 Toxic Gas Clouds

The effect of exposure to a cloud of toxic gas is a function of the concentration–time profile and of the toxic characteristics of the gas. An account of the concentration behaviour of gas clouds was given in Section 15.45, whilst the toxic characteristics of gases are described in Chapter 18. This section describes the attempts made to marry these two features.

15.47.1 Toxic load and probit

As described in Chapter 18, the inhalation toxicity of a gas is characterized by the toxic load L which produces some defined toxic effect which is typically of the form

$$L = \int_0^T C^n \mathrm{d}t \quad [15.47.1a]$$

$$= C^n T \quad [15.47.1b]$$

$$= \sum_{i=1}^{p} C_i^n t_i \quad [15.47.1c]$$

where C is the concentration, L is the toxic load, t is the time, T is the time of exposure and n is an index.

The proportion of the exposed population which will suffer, or the probability that a single individual will suffer, this defined degree of injury is generally expressed in terms of a probit equation of the form

$$Y = k_1 + k_2 \ln L \quad [15.47.2]$$

where Y is the probit and k_1 and k_2 are constants. The probability P of injury is a function of the probit Y. The use of probit equations is discussed in Chapter 9.

Equation 15.47.1 is the form typically taken by correlation of experimental work. There is usually little information on its applicability to combinations of values of (C,t) where C is high and t is short. There must therefore be doubt as to the appropriateness of applying it in a mechanistic way.

15.47.2 Intermittency and its effects

Griffiths and co-workers (R.F. Griffiths and Megson, 1984; R.F. Griffiths and Harper, 1985) have drawn attention to the implications of concentration intermittency when combined with the toxic load relation 15.47.1. The intermittency γ is defined as the fraction of time for which the concentration is zero:

$$\gamma = \frac{t_{\mathrm{s}} - t_{\mathrm{pe}}}{t_{\mathrm{s}}} \quad [15.47.3]$$

where γ is intermittency and subscripts pe and s denote exposure and sampling, respectively. These authors define intermittency as the proportion of time the concentration is zero. The average concentration $\bar{C}$ is

$$\bar{C}t_{\mathrm{s}} = C_{\mathrm{p}}t_{\mathrm{pe}} \quad [15.47.4]$$

where $\bar{C}$ is the average concentration and C_{p} is the concentration over time t_{pe}. Then from Equations 15.47.2, 15.47.3 and 15.47.4

***Table 15.57** Effect of variations in concentration on toxic load (Ride, 1984a) (Courtesy of Elsevier Science Publishers)*

Intensity of fluctuations $\sigma_\tau/\bar{C}$	*Enhancement factor,* ω		
	HCN n = *1.8*	*Ammonia* n = *2.0*	*Chlorine* n = *2.75*
0	1.0	1.0	1.0
1	7.2	12.0	77.4
2	16.1	32.1	433
5	47.3	124	4607

$$Y = k_1 + k_2 \ln\left[\left(\frac{1}{1-\gamma}\right)^{n-1} \bar{C}^n t_s\right] \quad [15.47.5]$$

For the usual case of $n > 1$, the probit Y, and hence the probability P of injury, increase as the intermittency increases. The effect can be pronounced.

15.47.3 Ride model

The model give by Ride (1984b) for concentration fluctuations in a passive plume was described above. He has applied this model to the determination of toxic load.

From Equation 15.45.27, with $\psi = 1$

$$\frac{C_\tau}{\bar{C}} = k\left(\frac{\sigma_\tau}{\bar{C}}\right)^\lambda + 1 \quad [15.47.6]$$

From this equation, with Equation 15.45.24]

$$[k(\sigma_\tau/\bar{C})^\lambda + 1]^{-1} = 1 - \gamma_\tau \quad [15.47.7]$$

Then for a receptor with response time τ Ride obtains from Equation 15.47.1a for toxic load with Equations 15.47.6 and 15.47.7

$$L = C_\tau^n(1-\gamma_\tau)^T \quad [15.47.8]$$

$$= \bar{C}^n T\omega \quad [15.47.9]$$

with

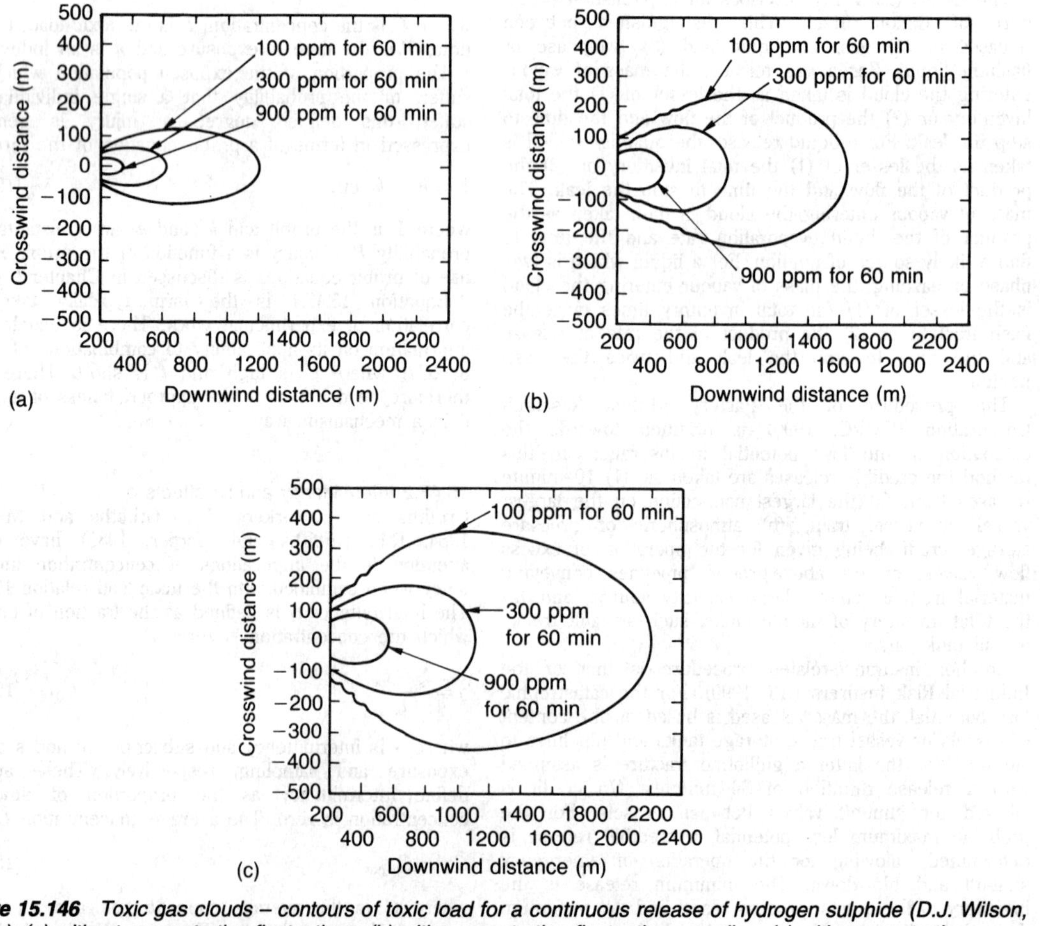

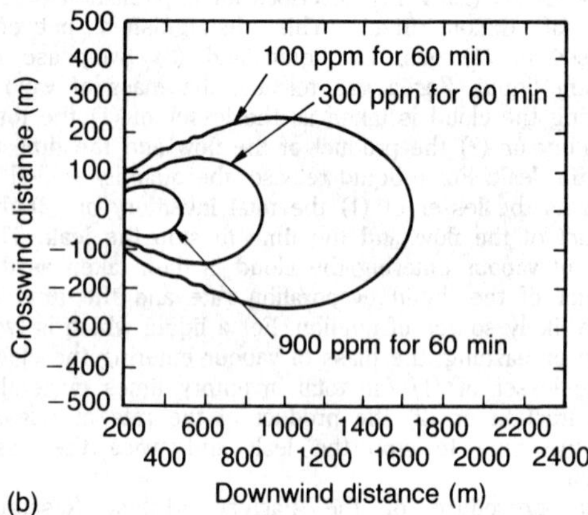

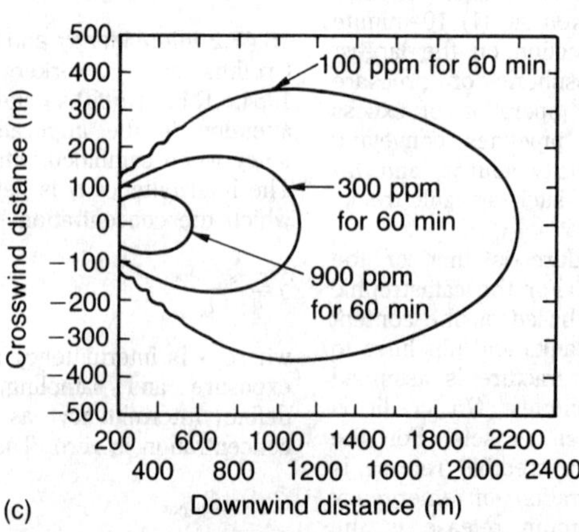

***Figure 15.146** Toxic gas clouds – contours of toxic load for a continuous release of hydrogen sulphide (D.J. Wilson, 1991a); (a) without concentration fluctuations; (b) with concentration fluctuations, median; (c) with concentration fluctuations, 99% percentile (Courtesy of the American Institute of Chemical Engineers)*

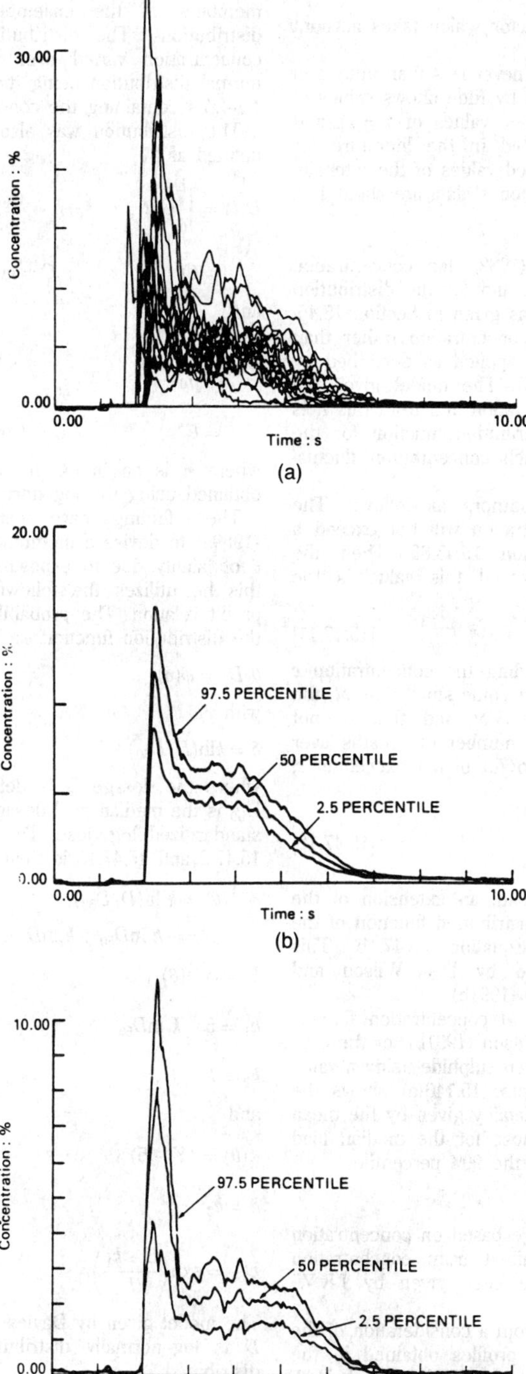

***Figure 15.147** Concentration fluctuations in gas dispersion – statistical parameters of a data set (J.K.W. Davies, 1987): (a) set of 21 concentration vs time traces obtained in repeat experiments in a wind tunnel (the WSL data set); (b) 95% confidence limits on mean concentration; (c) 95% confidence limits on ensemble standard deviation (Courtesy of Elsevier Science Publishers)*

$$\omega = [k(\sigma_\tau/\bar{C})^\lambda + 1]^{n-1} \quad [15.47.10]$$

where ω is an enhancement factor which takes account of concentration fluctuations.

The enhancement factor ω is never less than unity and can be large. Table 15.57 given by Ride shows values of this factor obtained taking Jones' values of $k = 11$ and $\lambda = 1.5$ with values of n quoted in the literature for several gases and with postulated values of the intensity $\sigma_\tau/\bar{C}$; the values of $\sigma_\tau/\bar{C}$ from Jones' data are about 1–2.

15.47.4 Wilson model

The model of D.J. Wilson (1982) for concentration fluctuations in a passive plume and for the distribution function of the concentration was given in Section 15.45. If the relevant effect is toxic concentration rather than toxic load, that model may be applied as described by D.J. Wilson and Simms (1985). The model gives the distribution function for concentration and from this it is possible to determine the distribution function for the probability of fatality given such concentration fluctuations.

This is obtained by the authors as follows. The probability F_c that the concentration will not exceed a value c is given by Equation 15.45.66. Then the probability V_c that it will exceed this value is the complement:

$$V_c = 1 - F_c \quad [15.47.11]$$

V_c is the probability of exceeding the concentration c over the exposure time T_e. Over some small time δt, the probability of exceeding c is $V\delta t$ and that of not exceeding it is $1 - V\delta t$. If the number of breaths over time T_e is N_b, the probability $S(T_e)$ of not exceeding c, and thus of surviving, is

$$S(T_e) = (1 - VT_e/N_b)^{N_b}$$
$$= \exp(-VT_e) \qquad N_b \to \infty \quad [15.47.12]$$

If the effect is one of toxic load, an extension of the model is required, giving the distribution function of the toxic load as defined by Equation 15.47.1b. This extension has been discussed by D.J. Wilson and Simms (1985) and D.J. Wilson (1991b).

An illustration of the effect of concentration fluctuations has been given by D.J. Wilson (1991b) for the case of continuous release of hydrogen sulphide using a value of n in Equation 15.47.1a. Figure 15.146(a) shows the contours for the probability of fatality given by the mean toxic load, Figure 15.146(b) those for the median load and Figure 15.146(c) those for the 99% percentile.

15.47.5 Davies model

A model for the effective dosage based on concentration and dosage distributions obtained from concentration profiles for an ensemble have been given by J.K.W. Davies (1987, 1989a).

J.K.W. Davies (1987) starts from a consideration of the ensemble of 21 concentration profiles obtained in the work at Warren Spring Laboratory (WSL) by D.J. Hall, Hollis and Ishaq (1982) shown in Figure 15.147(a). Analysis of the WSL data for concentration C showed that at the 95% confidence level the mean $\mu(C)$ is constant over the period 2.4–4.4 s and the standard deviation $\sigma(C)$ is constant over the range 2.4–5.4 s, as shown in Figures 15.147(b) and (c), respectively. The variations of concentration at a given time between the members of the ensemble were fitted to various distributions. The distributions which best fitted the concentration varied over the time period, the log–normal distribution being the best for the critical period 1.7–2.7 s containing the concentration peak.

The distribution was also determined for dosage D, defined as

$$D(t) = \int_0^t C\mathrm{d}t \quad [15.47.13a]$$

$$= Ct \qquad C = \text{constant} \quad [15.47.13b]$$

and

$$D(t) = \int_0^t C^n \mathrm{d}t \quad [15.47.14a]$$

$$= C^n t \qquad C = \text{Constant} \quad [15.47.14b]$$

where n is an index. In both cases the best fit was obtained using the log–normal distribution.

These findings have been applied by J.K.W. Davies (1989a) to devise a method of estimating the probability θ of fatality due to exposure to a toxic gas cloud. For this he utilizes the following alternative form of the probit relation. The probability θ of fatality is given by the distribution function ϕ:

$$\theta(D) = \phi(\delta) \quad [15.47.15]$$

with

$$\delta = k\ln(D/D_{50}) \quad [15.47.16]$$

where the dosage D is defined by Equation 15.47.14], D_{50} is the median fatal dosage, k is a constant and δ is a standardized log dose. The relation between Equations 15.47.2 and 15.47.15 is then

$$\phi^{-1}(\theta) = k\ \ln(D/D_{50}) \quad [15.47.17a]$$

$$= -k\ \ln D_{50} + k\ \ln D \quad [15.47.17b]$$

$$Y = \phi^{-1}(\theta) + 5 \quad [15.47.18]$$

$$k_1 = 5 - k\ \ln D_{50}$$

$$k_2 = k$$

and

$$\phi(\theta) = (Y - 5)^{-1} \quad [15.47.19]$$

$$k = k_2 \quad [15.47.20]$$

$$D_{50} = \exp\left(\frac{5 - k_1}{k}\right) \quad [15.47.21]$$

The model given by Davies is then as follows. The dose D is log–normally distributed. Thus $\ln D$ is normally distributed:

$$\ln D = N[\ln\bar{D},\ \sigma^2(\ln D)] \quad [15.47.22]$$

where N[...] signifies the normal distribution and D^- is the mean dose. Then the standardized log dose δ, defined by equation [15.47.16], is also taken as normally distributed

Table 15.58 Effect of variations in dose between members of an ensemble on predicted fatality: $\phi(\delta^*) - \phi(\bar{\delta})$ (J.K.W. Davies, 1989a) (Courtesy of Elsevier Science Publishers)

$\sigma(\delta)$	$\phi(\delta^*) - \phi(\delta)$ $\bar{\delta}$			
	0	1	2	3
0	0	0	0	0
0.2	0	0	0	0
0.5	0	–0.03	–0.02	0
1	0	–0.08	–0.06	–0.02
2	0	–0.17	–0.17	–0.09
5	0	–0.20	–0.28	–0.29

Table 15.59 Effect of variations in dose between members of an ensemble on predicted fatality: 20 te chlorine release[a] (J.K.W. Davies, 1989a) (Courtesy of Elsevier Science Publishers)

Crosswind distance (m)	$\ln C^n t$	Y	δ	σ	$\phi(\delta)$	$\phi(\delta^*) - \phi(\delta)$
52	28.80	8	3	0.5	1	0
215	26.84	6.04	1.04	1	0.85	–0.08
235	26.32	5.52	0.52	1.2	0.70	–0.04
287	24.96	4.16	–0.84	2	0.20	–0.17

[a] Table relates to conditions 1000 m downwind of an instantaneous release of 20 te chlorine in $D/3$ conditions, assuming a source term in which 400 te of air is entrained. The cloud radius, defined as 10% of the centre line concentration, at this distance is 417 m.

[b] Toxicity parameters: $n = 3.49$; $k = 1$; $\ln D_{50} = 25.8$

$$\delta = N[\bar{\delta}, \sigma(\delta)] \qquad [15.47.23]$$

with

$$\bar{\delta} = k \ln(\bar{D}/D_{50}) \qquad [15.47.24]$$

$$\sigma^2(\delta) = k^2 \ln(\ln D) \qquad [15.47.25]$$

where $\bar{\delta}$ is the mean standardized log dose.

An effective dose D^* is defined as the dose which gives rise to the same mean number of fatalities as would be obtained by averaging the number of fatalities observed in successive realizations of the ensemble in which dose D is allowed to vary log–normally.

An effective standardized log dose δ^* is also introduced, defined as the standardized log dose which gives rise to the same mean number of fatalities as would be obtained by averaging the number of fatalities observed in successive realizations of the ensemble in which the standardized log dose δ is allowed to vary normally. It is defined as

$$\delta^* = k \ln(D^*/D_{50}) \qquad [15.47.26]$$

An ideal gas dispersion model will produce for the standardized log dose a prediction $\bar{\delta}$ and the corresponding probability of fatality will be $\phi(\bar{\delta})$. The true probability of fatality is $\phi(\delta^*)$. The difference $\phi(\delta^*)$—$\phi(\bar{\delta})$ is a measure of the error in the prediction of fatality, even by an ideal model due to variations between members of the ensemble.

The equivalent standardized log dose δ^* may be determined by selecting values of δ from the normal distribution, calculating the corresponding values of the probability of fatality $\phi(\delta)$ and taking the mean value of this probability as $\phi(\delta^*)$. Table 15.58 (from Davies) shows the results of such a calculation.

The case of $\sigma(\delta) = 0$ corresponds to no concentration fluctuation, whilst $\sigma(\delta) = 1$ is typical of the value in the body of the cloud and $\sigma(\delta) \gg 1$ of values at the edges. For $\bar{D} > D_{50}$ the use of $\bar{\delta}$ instead of δ^* results in an overestimate of the probability of fatality, whilst for $\bar{D} < D_{50}$ it results in an underestimate. This misestimate is less than 10% in absolute terms for $\sigma(\delta) < 1$, but for $\sigma(\delta) > 1$ it becomes progressively greater, reaching a value of about 30% at $\sigma(\delta) = 5$.

An illustrative example is shown in Table 15.59. The case considered is the toxic effects 1000 m downwind of a 20 te release of chlorine in $D/3$ conditions. The index n is taken as 3.49, which is at the upper end of the range of published estimates. The effect of concentration fluctuations on toxic load is most marked towards the edges of the cloud.

15.48 Dispersion over Short Distances

It is necessary in some applications, particularly in relation to plant layout, to estimate dispersion over short distances.

15.48.1 Dispersion of passive plume

The relationships for passive dispersion are commonly used over relatively large distances, but the question of the range of his model was specifically considered by O.G. Sutton (1950). He describes experiments on the dispersion of a vertical jet of hot gas for which he obtained a value of the diffusion coefficient C. This value is similar to that obtained in the dispersion of smoke from a smoke generator over a distance of 100 m, provided that for comparability use is made in the latter case of the instantaneous rather than the time width of the plume. Sutton states:

> This leads to the striking conclusion that the same coefficient of diffusion is valid for the spread of smoke over hundreds of meters as for the mixing of a tiny column of hot air with the cold air of a laboratory.

The use of the Sutton model for estimation of dispersion in relation to hazard areas around vents has been treated by Long (1963), who devotes an appendix to the applicability of the model over short ranges. He discusses both the basic assumptions and the available experimental results. He reviews the assumptions systematically and finds no fundamental inapplicability for short ranges. In particular, the semi-angle of divergence of the roughly conical plume predicted using the values given by Sutton for the diffusion coefficient is about 10° and 16° in the vertical and horizontal directions, respectively, which compares well with semi-angles in the range 9–16° found for jets and plumes. The only experimental work found by Long was that of Katan (1951). Using values of the Sutton coefficient based on a 3-minute sampling time, Long found that the Sutton equations tend to overestimate the concentration somewhat.

15.48.2 Dispersion of dense gas plume

Some guidance on the dispersion of a dense gas plume over short ranges has been given by McQuaid (1980) in the context of the design of water spray barriers. He takes the plume as issuing at an angle of 35° and having a height of 1 m.

15.48.3 Dispersion from vents and reliefs

It is common practice in the process industries to discharge material to atmosphere. Intended discharges occur through chimneys, vents and relief devices. The treatment here is mainly confined to the problem of the safety of discharges occurring in an emergency, usually as a result of the operation of relief devices. Discharges to flare are discussed in Chapter 12 and disposal of material vented from chemical reactors in Chapters 12 and 17.

Recent years have seen a marked tightening up in the extent to which discharge to atmosphere is permissible and, although the situation varies as between countries, the trend is clear. Accounts of safe discharge include those given in the various editions of API RP 520 and RP 521 and by Long (1963), Loudon (1963), Bodurtha, Palmer and Walsh (1973) and Gerardu (1981). The topic is also frequently discussed by authors of models developed to assist with this problem.

An early review is that by Loudon (1963), who identifies the hazards arising from such a release as including, for flammable or toxic material

(1) hazardous concentration, particularly at ground level;

and, for flammables

(2) vapour cloud explosion;
(3) jet flame.

Another hazard is flashback of flame into a vent.

For discharges, a distinction is to be made between those which have high momentum and/or buoyancy and those which do not. If the release has a high velocity, and thus high momentum, this promotes entrainment of air and dilution of the cloud. Another aspect is the height to which the material discharged travels before it starts to descend to the ground, the effective 'stack' height. This is increased if the release has high buoyancy and, for a discharge directed upwards, high momentum.

Safe discharge is generally addressed by a combination of engineering measures and hazard assessment. The former centre mainly on the height of the vent and on the velocity of the discharge, whilst the latter involves the use of suitable models to assess the hazards listed above.

Widespread use has been made of the equation in API RP 520 for the estimation of concentration of contaminant as a function of distance for use in high velocity discharge calculations based on the recommendations on discharge velocity. This equation is given in Chapter 12.

There are a considerable number of models for the estimation of the ground level concentration resulting from an elevated release. They include: the early models of Bosanquet (1935) and Bosanquet and Pearson (1936); O.G. Sutton (1952); the Pasquill model (Pasquill, 1961) and the Pasquill–Gifford model (Pasquill, 1961, 1962; Gifford, 1961b); the model of Cude (1974b); and that of G.A. Briggs (1965, 1969). Other models have been developed to obtain the ground level concentration for elevated releases of a dense gas. Amongst these are the models of Hoot, Meroney and Peterka (1973), Bodurtha, Palmer and Walsh (1973), Ooms, Mahieu and Zelis (1974), Jagger and Edmunson (1984), Emerson (1986a), J.L. Woodward (1989a) and McFarlane (1991). Accounts of many of these models are given in Sections 15.20, 15.43 and 15.44.

The mass of gas from a discharge which is within the flammability limits and thus available to participate in a vapour cloud fire or explosion may be obtained from models of jets and plumes. There are also some models which specifically address this problem. These include the families of models given by Hess and co-workers (Hess and Stickel, 1970; Hess, Leuckel and Stoeckel, 1973), J.G. Marshall (1977, 1980), Palazzi *et al.* (1984) and Stock and co-workers (Stock and Geiger, 1984; Stock, 1987; Stock, Geiger and Giesbrecht, 1989). The effects of ignition may be estimated using models for vapour cloud fires or explosions, which are dealt with in Chapters 16 and 17, respectively.

Ignition of a sustained high momentum discharge gives a jet flame. Early models of jet flames developed for the assessment of such situations include those by Craven (1972, 1976) and Kovacs and Honti (1974). There are now a number of jet flame models, as described in Chapter 16.

Several authors have described sets of models used for the estimation of the hazards from discharges. They include Kovacs and Honti (1980) and Gerardu (1981). The models cover ground level concentration, jet flame radiation and vapour cloud overpressure.

Mention has already been made of the work of Katan (1951) on safe dispersion distances for aircraft fuelling. More recent work in this area has been described by

Thorne (1986), who gives models for the plumes from circular, slot and elliptical vents. The results from these models show distances for dilution of the jet fuels below the LFL in the range 1.5–4 m.

15.48.4 ASME equation

A formula for plume rise from a momentum source widely used in design for safe discharge is that given by the American society of Mechanical Engineers (ASME, (1969/1). This is

$$\Delta h = D\left(\frac{v_s}{u}\right)^{1.4} \qquad [15.48.1]$$

where D is the internal diameter of the stack (m), Δh is the plume rise (m), v_s is the velocity of the stack gas (m/s) and u the wind speed at stack height (m/s).

The equation quoted for the maximum concentration at ground level is Equation 15.16.55]. The applicability of the ASME equation is to a stack gas which is essentially of neutral density.

15.48.5 Bodurtha and Walsh method

An early method for safe discharge which applies to a dense gas mixture is that of Bodurtha and Walsh, described in Bodurtha, Palmer and Walsh (1973). Figure 15.133 shows photographs of the simulation of a dense gas plume in a wind tunnel obtained by these workers. The difference between the three tests lies in the density of the gas discharged relative to that of air, the relative densities in Figures 15.133 (a)–(c) being 1.0, 1.52 and 5.17, respectively.

On the basis of these admittedly limited tests, Bodurtha and Walsh obtained the following design equation:

$$h_s = 1.33ED^2 \qquad [15.48.2]$$

where D is the diameter of the vent tip (ft), E is the number of dilutions with air of the stack gas required to reduce its concentration to the lower flammability limit and h_s is the height of the vent above the exposure level (ft). As a design equation, the relation is conservative. The authors state their belief that it contains ample safety factors. They also say, however, that it is intended only for short period releases of dense gases.

15.48.6 Hanzevack method

Another experimentally based method is that by Hanzevack (1982), who conducted tests in which he measured the ground level concentration from elevated releases of both ambient and high density gas mixtures. He gives a model for the ground level concentration due to a continuous emission which is based on a modification of the ASME plume rise formula described in Section 15.48.4. The modification allows for differences in the density of the stack gas mixture and includes a constant to improve the fit to his experimental data.

The equations given by Hanzevack for the plume rise from a momentum source and for the maximum ground level concentration are

$$\Delta h = C_1 D\left(\frac{v_s}{u}\right)^{1.4}\left(\frac{\rho_a}{\rho_s}\right)^{1.4} \qquad [15.48.3]$$

and

$$\chi_{cr} = C_2 \frac{Q}{h^{1.29}D^{0.71}v_s}\left(\frac{\rho_s}{\rho_a}\right) \qquad [15.48.4]$$

where D is the internal diameter of the stack (m), h is the height of the stack (m), Δh is the plume rise (m), Q is the mass rate of release (g/s), u is the wind speed (m/s), v_s is the stack gas outlet velocity (m/s), ρ is the density of the gas mixture (kg/m^3), χ_{cr} is the critical ground level concentration (μg/m^3) and subscripts a and s denote air and stack gas, respectively. The relations apply to a 1-hour time-averaged concentration.

C_1 and C_2 are empirical constants obtained from the experimental data. In adjusting concentrations averaged over other time periods to the 1-hour average, use was made of Equation 15.16.45 with a value of the index p measured as 0.5.

For design, a more conservative approach is suggested with the two constants C_1 and C_2 replaced by constants C_3 and C_4, respectively. The values given for the four constants are:

	C_1	C_2	C_3	C_4
$\rho_s \leq \rho_a$	6	0.012	3	0.02
$\rho_s > \rho_a$	2	0.027	1	0.044

The more conservative values thus incorporate a safety factor of about 2.

15.48.7 Method of de Faveri, Hanzevack and Delaney

Recognizing that in many instances the release of interest is not a continuous one, but a relatively short one, de Faveri, Hanzevack and Delaney (1982) have described an extension to Hanzevack's method which allows for this. The authors conducted experiments in which 10–minute time-averaged concentrations were measured for (1) continuous releases and (2) 10–minute releases, and obtained the ratio of the two time-averaged concentrations. This ratio should theoretically lie between 1 and 3. Their results indicate a ratio of approximately 2.

15.48.8 Bodurtha method

In addition to the method of Bodurtha and Walsh, described in Section 15.48.5, Bodurtha (1988) has also given another method for safe discharge of dense flammable gases.

The method is based on an adaptation of the Hoot, Meroney and Peterka (HMP) model. As discussed by Bodurtha (1980), the touchdown concentration passes through a maximum with stack gas outlet velocity, which the author terms the critical stack gas outlet velocity $v_{s,cr}$. The expression given by the author for this velocity is

$$v_{s,cr} = \frac{3.73D^{0.684}(\beta C_s/L)^{1.05}}{u^{0.368}A^{2.05}} \times 10^{-3} \qquad [15.48.5]$$

with

Table 15.60 *Constants for correlation of downwind extent of flammability limits, and other cloud distances, for instantaneous releases of propane and butane (HSE, 1981a) (Courtesy of HM Stationery Office)*

Constant	*Propane*				*Butane*			
	LFL		$\frac{1}{2}$ *LFL*		*LFL*		$\frac{1}{2}$ *LFL*	
	D/5	*F/2*	*D/5*	*F/2*	*D/5*	*F/2*	*D/5*	*F/2*
K	0.12	0.17	0.14	0.21	0.10	0.14	0.12	0.16
k_1	0.8	1.2	0.8	1.2	1.0	1.6	1.0	1.5
k_2	0.5	0.4	0.4	0.4	0.4	1.2	0.4	0.2
k_3	0.1	0.25	0.1	0.3	0.2	0.6	0.2	0.6

$$A = s^{0.67}/(s-1)^{0.33} \quad [15.48.6]$$

$$s = M_w T_a / 29 T_s \quad [15.48.7]$$

$$C_m = L/\beta \quad [15.48.8]$$

where C_m is the maximum permissible value of the touchdown concentration of the contaminant (% v/v), C_s is the concentration of the contaminant in the stack gas (% v/v), D is the internal diameter of the stack (mm), L is the lower flammability limit (% v/v), M_w is the molecular weight of the stack gas mixture, s is the specific gravity of the stack gas mixture relative to air, T_a is the absolute temperature of the ambient air (K), T_s is the absolute temperature of the stack gas (K), u is the wind speed (m/s), v_s is the velocity of the stack gas (m/s), A is a parameter, β is the applicable peak/mean concentration ratio, and subscript cr denotes critical. The value of the touchdown concentration C_m passes through a maximum with specific gravity, this maximum occurring at a value of $s=2$, and thus $A=1.59$.

Substitution of the critical value $v_{s,cr}$ of the stack gas outlet velocity in the HMP model yields the following expression for the required height of the stack:

$$h_s = \frac{43.8 D^{1.35}(\beta C_s/LA)^{1.05}}{u^{0.702}} \times 10^{-6} \quad [15.48.9]$$

where h_s is the height of the stack (m).

The model is applicable to a release directed vertically upwards and only for stack gas specific gravity $s > 1.15$ and for stack gas outlet velocity v_s less than the choked, or sonic, value. The author suggests that for design purposes the value of the wind speed be taken as $u = 1$ m/s.

For the peak/mean concentration ratio β, Bodurtha quotes minimum and maximum values of 2.5 and 5, respectively, and states that the former is used in current practice. This method is conservative in that the calculation of the required stack height is based on the critical value $v_{s,cr}$ of the stack outlet velocity, whereas use of the actual stack outlet velocity v_s will generally give a much lower stack height.

Bodurtha gives several sample calculations. As an illustration, consider the first of his examples. The problem is stated in terms of the following values of the variables:

$$C_s = 100\,\text{v/v};\ L = 2.2\%\,\text{v/v};\ M_w = 44;\ T_a = 298\,\text{K};$$

$$T_s = 298\,\text{K};\ u = 1\,\text{m/s};\ v_s = 152.3\,\text{m/s};\ \beta = 2.5$$

Hence $s = 1.52$, $A = 1.64$, $v_{s,cr} = 8.6$ m/s and $h_s = 6.6$ m. The degree of conservatism is indicated by the fact that the actual velocity v_s is very much greater than the critical value $v_{s,cr}$.

15.49 Hazard Ranges for Dispersion

It is convenient for certain purposes, such as comparison of models or preliminary hazard assessment, to have simple expressions for the downwind extent of particular concentrations, such as lower flammability limits or toxic concentrations; in other words, for the hazard range.

The variables which enter into the simpler correlations are generally the concentration, the downwind distance, the mass rate of release and the wind speed, and the treatment here is confined to these.

15.49.1 Passive dispersion

For passive dispersion of an instantaneous release from a point source the relevant expression in the Pasquill–Gifford model is Equation 15.16.40b. Considering concentration at the centre of the cloud at ground level so that $x = ut$, $y = 0$ and $z = 0$, this equation reduces to

$$\chi \propto \frac{Q^*}{\sigma_x \sigma_y \sigma_z} \quad [15.49.1]$$

From the correlations of Slade given in Table 15.31 and noting that it is usual to take $\sigma_x = \sigma_y$:

$$\sigma_x = \sigma_y \propto x^{0.92} \quad [15.49.2a]$$

$$\sigma_z \propto x^{0.7} \quad [15.49.2b]$$

Then relations [15.49.1] and [15.49.2] give for the hazard range x

$$x \propto Q^{*0.39} \quad [15.49.3]$$

In this case the hazard range is a function of the mass released, but not of the wind speed.

For passive dispersion of a continuous release from a point source the relevant expression in the Pasquill–Gifford model is Equation 15.16.41. Considering concentration on the centre-line at ground level so that $y = 0$ and $z = 0$ this equation reduces to

Table 15.61 *Constants for downwind extent of flammability limits, and other cloud parameters, for instantaneous and continuous releases of propane and butane (Considine and Grint, 1985) (Courtesy of Gastech)*

A Dispersion of instantaneous release over land

Constants for	*Pressurized propane*		*Pressurized butane*		*Refrigerated propane*		*Refrigerated butane*	
	D/5	*F/2*	*D/5*	*F/2*	*D/5*	*F/2*	*D/2*	*F/2*
Value of k								
R_{LFL}	145.7	112.5	153.1	143.1	121.1	147.2	116.5	139.8
U	13.1	15.7	12.2	18.1	29.6	64.5	31	71.3
R_{UFL}	23.9	10.2	12.8	15	60.7	56.5	54.4	44.5
Y_{max}	29.7	32.1	35.5	41.7	49.6	92.4	50.2	96.9
h	5.6	4.2	4.8	2.1	3.0	0.94	3.12	1.33
Value of n								
R_{LFL}	0.294	0.248	0.319	0.273	0.383	0.408	0.386	0.415
U	0.356	0.374	0.375	0.374	0.364	0.351	0.363	0.346
R_{UFL}	0.241	0.226	0.382	0.239	0.392	0.423	0.393	0.442
Y_{max}	0.34	0.322	0.349	0.334	0.390	0.384	0.391	0.384
h	0.335	0.340	0.264	0.353	0.220	0.237	0.174	0.175

B Dispersion of instantaneous release over water (refrigerated gas)

	k				*n*			
Constants for	*Propane*		*Butane*		*Propane*		*Butane*	
	D/5	*F/2*	*D/5*	*F/2*	*D/5*	*F/2*	*D/5*	*F/2*
R_{LFL}	174.1	270.8	104.9	268.3	0.351	0.352	0.355	0.356
U	22.5	36	21.9	38.1	0.387	0.405	0.394	0.404
R_{UFL}	109.4	156	98.64	145.4	0.337	0.332	0.342	0.332
Y_{max}	49.8	91.9	50.2	96.2	0.389	0.385	0.390	0.385
h	1.4	0.843	1.94	0.683	0.313	0.277	0.269	0.276

C Dispersion of continuous release over land (pressurized and refrigerated gas)

	k				*n*			
Constants for	*Propane*		*Butane*		*Propane*		*Butane*	
	D/5	*F/2*	*D/5*	*F/2*	*D/5*	*F/2*	*D/5*	*F/2*
R_{LFL}	12.1	44.2	11.23	41.9	0.557	0.571	0.582	0.574
R_{UFL}	6.25	15.8	5.53	12.9	0.553	0.486	0.553	0.481
h	0.51	0.34	0.46	0.29	0.327	0.323	0.329	0.330
k^*	1.85	5.47	2.1	6.0	0.159	0.145	0.149	0.142

D Dispersion of continuous release over water

	k				*n*			
Constants for	*Propane*		*Butane*		*Propane*		*Butane*	
	D/5	*F/2*	*D/5*	*F/2*	*D/5*	*F/2*	*D/5*	*F/2*
R_{LFL}	19.4	76.4	18.71	73.6	0.524	0.503	0.526	0.513
R_{UFL}	10.31	31.6	9.47	27.65	0.499	0.408	0.496	0.404
h	0.403	0.243	0.390	0.196	0.343	0.361	0.349	0.371
k^*	0.805	2.16	0.845	2.25	0.253	0.262	0.252	0.248

$$\chi \propto \frac{Q}{\sigma_y \sigma_z u} \qquad [15.49.4]$$

From the correlations given in Table 15.31

$$\sigma_y \propto x^{0.9} \qquad [15.49.5a]$$

$$\sigma_z \propto x^{0.85} \qquad [15.49.5b]$$

Then relations 15.49.4 and 15.49.5 give for the hazard range x

$$x \propto \left(\frac{Q}{u}\right)^{0.57} \quad [15.49.6]$$

In this case the hazard range is a function both of the mass rate of release and of the wind speed.

15.49.2 Dense gas dispersion

It is less easy to derive analytical expressions for the hazard range of a dense gas. One approach is to correlate results from a number of runs of a dense gas dispersion model. The Second *Canvey Report* gives for an instantaneous release the following correlation derived from DENZ for propane and butane:

$$R = kM^{0.4} \quad [15.49.7]$$

where M is mass of release (te), R is the downwind range (km) and k is a constant. It also gives the following additional relations for distance (km): maximum width $= k_1R$; distance to maximum width, $X = k_2R$; and upwind range $= k_3R$. The values of the constants are shown in Table 15.60.

A more detailed treatment on the same lines has been given by Considine and Grint (1985), who used DENZ and CRUNCH to obtain for propane and butane correlations for instantaneous and continuous releases, respectively. For a quasi-instantaneous release they give for any given distance or dimension L the relation

$$L = kM^n \quad [15.49.8]$$

where L is a distance/dimension (m), M is the mass released (te), k is a constant and n is an index for the distance/dimension L. These parameters are given for the following dimensions: R_{LFL}, the downwind range to the LFL (m); R_{UFL}, the downwind range to the UFL (m); U_{LFL}, the upwind range to the LFL (m); Y_{max}, the maximum crosswind range of the LFL (m); and h the height of the cloud (m).

The crosswind distance Y at any distance R downwind is given by

$$Y = Y_{max}\left\{1 - 4\frac{[R - (R_{LFL} - U_{LFL})/2]^2}{(R_{LFL} + U_{LFL})^2}\right\}^{\frac{1}{2}} \quad [15.49.9]$$

For the cloud radius R_c at any downwind distance R

$$R_c = R_{LFL} - R \qquad R > R_{cr} \quad [15.49.10]$$

$$R_c = [Y_1^2 + (R - X_1)^2]^{\frac{1}{2}} \qquad R < R_{cr} \quad [15.49.11]$$

with

$$R_{cr} = R_{LFL} - \frac{2Y_{max}^2}{R_{LFL} + U_{LFL}} \quad [15.49.12]$$

$$X_1 = \left[R - \frac{4Y_{max}^2(R_{LFL} - U_{LFL})}{2(R_{LFL} + U_{LFL})^2}\right] \Big/ \left[1 - \frac{4Y_{max}^2}{(R_{LFL} + U_{LFL})^2}\right] \quad [15.49.13]$$

$$Y_1 = Y_{max}\left\{1 - \frac{4[X_1 - (R_{LFL} - U_{LFL})/2]^2}{(R_{LFL} + U_{LFL})^2}\right\}^{\frac{1}{2}} \quad [15.49.14]$$

where R is the downwind distance (m), R_c is the radius of the cloud (m), R_{cr} is a range transition criterion value (m), X_1 is a downwind dimension (m), Y is the crosswind distance (m) and Y_1 is a crosswind dimension (m).

For a continuously formed cloud the authors give for any given distance or dimension L the relation

$$L = k\dot{m}_v^n \quad [15.49.15]$$

where L is a distance/dimension (m), $\dot{m}_v$ is the mass rate of release of vapour (kg/s), k is a constant and n is an index for the distance/dimension L. These parameters are given for the following dimensions: R_{LFL}, the downwind range to the LFL (m); R_{UFL}, the downwind range to the UFL (m); h, the height of the cloud (m); and k^*, a constant (defined below). Then for a downwind distance R the crosswind distance Y is

$$Y = k^*R^{\frac{2}{3}} \quad [15.49.16]$$

The values of the constants are shown in Table 15.61.

The flammable inventory of the cloud is

$$I = \frac{\dot{m}_v R_{LFL}}{u} \quad [15.49.17]$$

where I is the inventory (kg) and u is the wind speed (m/s).

The authors also give guidance for the case of a flammable gas cloud where the release is neither truly instantaneous nor continuous, but transient. The procedure suggested is to evaluate R_{LFL} and the arrival time T_a from

$$T_a = R_{LFL}/u \quad [15.49.18]$$

where T_a is the arrival time (s). The release is treated as continuous if $T_d > T_a/4$, where T_d is the duration of the release (s); otherwise the release is treated as an instantaneous one, based on the total mass released over the duration time.

Correlations obtained in this way are only as good as the model from which they are derived. The reliability of the hazard ranges predicted by dense gas dispersion models has been discussed in Section 15.41.

Explicit expressions for the downwind extent, or hazard range, are given in the *Workbook* by Britter and McQuaid (1988). For an instantaneous release Equation 15.34.21 gives for the downwind distance x

$$x \propto Q_0^{0.25} u_{ref}^{0.5} \quad [15.46.19]$$

where Q_o is the volume released, u_{ref} is the wind speed at the reference height and x is the downwind distance. For a continuous release Equation 15.34.10 yields

$$x \propto q_o^{0.4} \quad [15.49.20]$$

where q_o is the volumetric rate of release. In this case there is no dependence on wind speed. Further details, including the range of applicability of these equations, are given in Section 15.34.

15.50 Transformation and Removal Processes

There are various transformation and removal processes which a gaseous or particulate pollutant may undergo in the atmosphere. These may need to be considered in relation to acute accidental releases, although in general they are more relevant to long-term pollution. Accounts of these processes include those given by Chamberlain (1953), van der Hoven (1968), Csanady (1973), Hosker

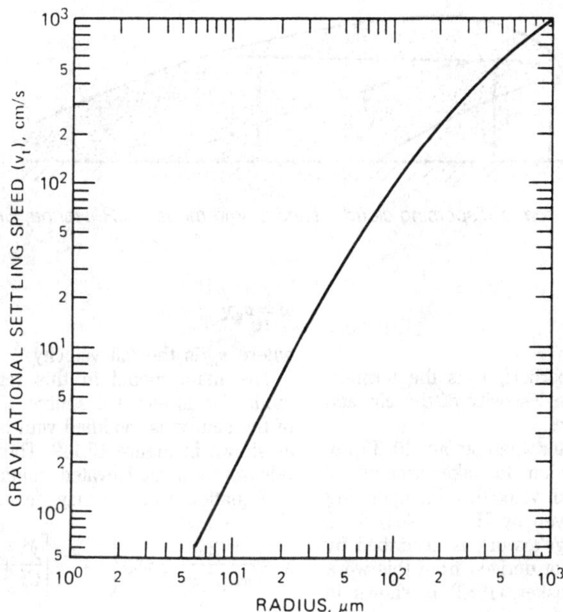

Figure 15.148 Dry deposition from a dispersing cloud – gravitational settling speed of particles near earth's surface (S.R. Hanna, Briggs and Hosker, 1982). Particle density 5 g/cm^3

(1974), Gifford (1976a), Hales (1976) and S.R. Hanna, Briggs and Hosker (1982).

Processes of transformation and removal include:

(1) chemical transformations;
(2) physical transformations;
(3) dry deposition –
  (a) gravitational settling,
  (b) small particle deposition;
(4) Wet deposition –
  (a) rainout,
  (b) washout.

An account is now given of the models and parameters for these various processes.

15.50.1 Chemical transformations

Atmospheric reactions may occur in the gaseous or aqueous phase and may be conventional reactions or photochemical reactions. There is a large literature on atmospheric chemistry which includes *Chemistry of the Atmosphere* (McEwan and Phillips, 1975), *Atmospheric Chemistry* (Heicklen, 1976) and *Atmospheric Chemistry* (Meszaros, 1981). Information on conventional reactions is given in standard texts and information on photochemical reactions is given by Leighton (1961), Calvert and Pitts (1966) and Demerjian, Kerr and Calvert (1974). A review of atmospheric photochemical reactions is given by Hales (1976).

The decay of concentration due to chemical reaction is frequently modelled by a first order decay model:

$$\chi(t) = \chi(0) \exp(-t/\tau_c) \qquad [15.50.1]$$

where χ is the concentration of the pollutant and τ_c is a chemical decay time constant.

From Equation 15.50.1], the half-life $t_{\frac{1}{2}}$ of the concentration is

$$t_{\frac{1}{2}} = 0.693/\tau_c \qquad [15.50.2]$$

15.50.2 Physical transformations

Physical transformations are relevant principally to liquid particles, or aerosols. They include: nucleation, in which a liquid particle is created by condensation of vapour; condensation, in which further vapour condenses onto an existing particle; and coagulation, in which particles come together to form a larger particle. Another relevant physical process is the absorption of soluble gas in water particles.

A case of particular relevance here is the behaviour of ammonia in humid air. A detailed discussion is given by Haddock and Williams (1978 SRD R103). An outline of their treatment of ammonia is presented in Section 15.42.

15.50.3 Dry deposition

In dry deposition of particles from a plume there are two distinct regimes. Deposition of larger particles is governed essentially by gravitational settling, while that of smaller particles is determined by turbulent motion.

Gravitational settling of the larger particles, say of radius > 5 μm, is given by Stokes law, or modifications of it. Stokes law, for spherical particles, is

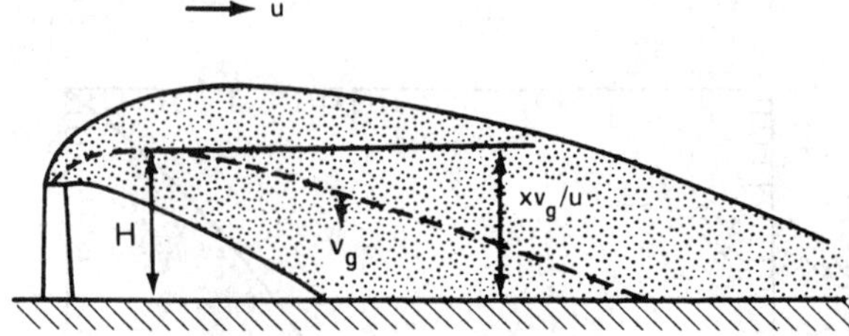

Figure 15.149 Dry deposition from a dispersing cloud – tilted plume model (S.R. Hanna, Briggs and Hosker, 1982)

$$v_t = \frac{2r^2 g \rho_p}{9\mu} \qquad [15.50.3]$$

where r is the radius of the particle, v_t is the terminal velocity of the particle, μ is the viscosity of the air, and ρ_p is the density of the particle.

For particles of radius greater than about 10–30 μm, Stokes' law requires modification to take account of inertia and slip flow. Terminal velocities incorporating these corrections have been given by Hage (1964) for a particle of density 5 g/cm^3. Hage's work is described by van der Hoven (1968), and a plot derived from this work by S.R. Hanna, Briggs and Hosker (1982) is shown in Figure 15.148. For a particle of another density ρ_{p2} the corresponding value of the terminal velocity v_{t2} may be obtained from the standard value of the density ρ_{p1} and from the velocity v_{t1} as given by the figure using the approximate relation, based on Equation 15.50.3:

$$v_{t2} = v_{t1}\frac{\rho_{p2}}{\rho_{p1}} \qquad [15.50.4]$$

For a non-spherical particle, an equivalent radius r_e may be defined as

$$r_e = \left(\frac{3V_p}{4\pi}\right)^{\frac{1}{3}} \qquad [15.50.5]$$

where V_p is the volume of the particle. The terminal velocity v_{tn} of the particle may be estimated from the terminal velocity v_{ts} of a spherical particle with radius r equal to its equivalent radius r_e using an empirical dynamical shape factor α defined by

$$v_{tn} = v_{ts}/\alpha \qquad [15.50.6]$$

Values of the dynamical shape factor are given by Chamberlain (1975). For a cylinder with a 1:1 ratio of axes the value of the factor is 1.06 and for a ratio of 4:1 it is 1.32.

As stated above, the gravitational settling regime starts at a particle radius of about 5 μm, which corresponds to a terminal velocity of about 1 cm/s. There are a number of models for dry deposition and the appropriate model depends on the regime, and hence on the particle radius. For deposition of particles with a terminal velocity > 100 cm/s, which corresponds to a radius of about 100 μm, it is suggested by van der Hoven that gravitational settling is so fast that it may be determined from the terminal velocity and the wind speed using a conventional ballistic approach.

For particles with terminal velocity in the range 1–100 cm/s the deposition rate ω is defined as

$$\omega = v_g \chi \qquad [15.50.7]$$

where v_g is the fall velocity.

The main model in this regime is the tilted plume model for an elevated source. In this model the height H of the source is modified with distance by the term xv_g/u as shown in Figure 15.149. Then the Sutton relation for a release from an elevated continuous point source given in Equation 15.16.29 may be modified to give

$$\chi = \frac{2Q}{\pi C_y C_z u x^{2-n}} \exp\left\{-x^{n-2}\left[\frac{y^2}{C_y^2} + \frac{(H - xv_g/u)^2}{C_z^2}\right]\right\} \qquad [15.50.8]$$

The effect of depletion of the cloud may be taken into account by modifying Equation 15.50.7 by a depletion factor F, so that the deposition rate becomes

$$\omega = F v_g \chi \qquad [15.50.9]$$

This depletion factor has been given by Csanady (1955) as

$$F = 1 - \frac{1}{(1 - n/2)(Hu/xv_g - 1) + 2} \qquad [15.50.10]$$

The tilted plume model is applicable only in the well mixed boundary layer such as occurs in daytime adiabatic conditions.

For smaller particles with a terminal velocity of < 1 cm/s, the deposition rate is defined in terms of the deposition velocity v_d:

$$\omega = v_d \chi \qquad [15.50.11]$$

Models available for the deposition of such particles are discussed by van der Hoven (1968). One is the model by Chamberlain (1953). For a ground level continuous point source this author modifies the Sutton relations to obtain for an initial source strength Q_o an effective source strength Q_x at distance x using the depletion term

$$\frac{Q_x}{Q_o} = \exp\left(-\frac{4v_d x^{n/2}}{nu\pi^{\frac{1}{2}}C_z}\right) \qquad [15.50.12]$$

This correction is then applied to Sutton's relation for a release from a ground level continuous point source given by Equation 15.16.27 to yield for the ground level concentration

Table 15.62 *Source depletion fraction Q_x/Q_o given by the van der Hoven model for dry deposition (after van der Hoven, 1968)*

Pasquill stability category	*Source height* (m)	*Distance* (m)			
		10	*100*	*1000*	*10 000*
A	0	0.90	0.82	0.75	
	10	1.0	0.96	0.90	
	100	1.0	1.0	1.0	
B	0	0.90	0.80	0.65	
	10	0.98	0.95	0.80	
	100	1.0	1.0	1.0	
C	0	0.90	0.70	0.60	0.45
	10	1.0	1.0	0.80	0.60
	100	1.0	1.0	1.0	0.80
E	0	0.90	0.60	0.15	0.012
	10	1.0	1.0	0.80	0.10
	100	1.0	1.0	1.0	0.80
F	0	0.90	0.65	0.40	0.07
	10	1.0	1.0	0.80	0.20
	100	1.0	1.0	1.0	0.80

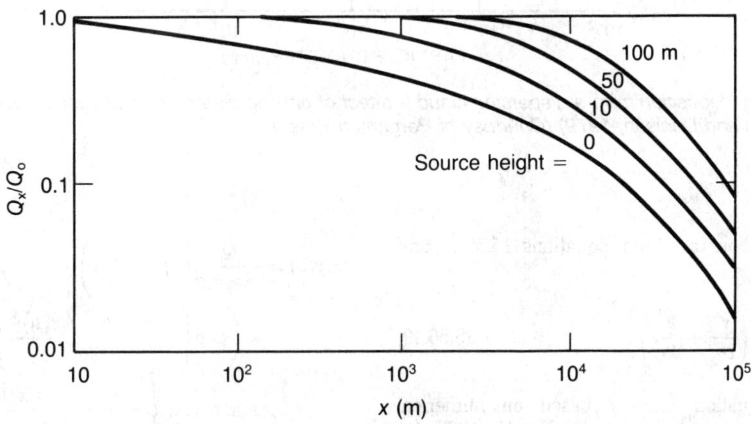

Figure 15.150 *Dry deposition from a dispersing cloud – effect of distance on source depletion fraction (van der Hoven, 1968). Weather conditions D/1; wind speed = 1 m/s; deposition velocity 10^{-2} m/s*

$$\chi(x, y, 0) = \frac{2Q_x}{\pi C_y C_z u x^{2-n}} \exp\left(-x^{n-2}\frac{y^2}{C_y^2}\right) \quad [15.50.13]$$

and for the deposition rate from Equations 15.50.11 and 15.50.14

$$\omega = \frac{2Q_x v_d}{\pi u C_y C_z^{2-n}} \exp\left(-x^{n-2}\frac{x^{n-2}y^2}{C_y^2}\right) \quad [15.50.14]$$

Chamberlain has also given a model for an elevated source.

An alternative model given by van der Hoven is derived from the Pasquill–Gifford equations. The depletion of the source is obtained as follows:

$$\frac{dQ_x}{dx} = \int_{-\infty}^{\infty} \omega dy \quad [15.50.15]$$

$$= -\left(\frac{2}{\pi}\right) \frac{v_d Q_x}{u\sigma_z} \exp\left(-\frac{h^2}{2\sigma_z^2}\right) \quad [15.50.16]$$

which on integration yields

$$\frac{Q_x}{Q_o} = \left[\exp \int_0^x \frac{dx}{\sigma_z \exp(h^2/\sigma_z^2)}\right]^{-(2/\pi)^{\frac{1}{2}}(v_d/u)} \quad [15.50.17]$$

This correction is then applied to the Pasquill–Gifford equation for a release from an elevated continuous point source (Equation 15.16.43) to yield for the ground level concentration

$$\chi(x, y, 0) = \frac{Q_x}{\pi\sigma_y\sigma_z u} \exp\left[-\left(\frac{y^2}{2\sigma_y^2} + \frac{h^2}{2\sigma_z^2}\right)\right] \quad [15.50.18]$$

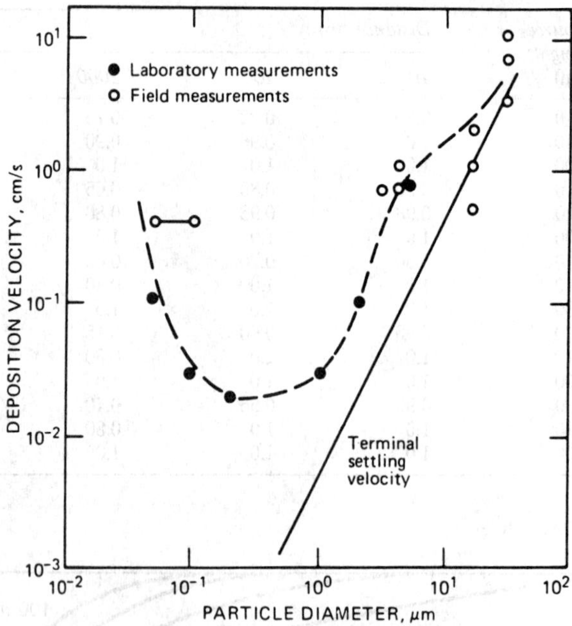

Figure 15.151 Dry deposition from a dispersing cloud – effect of particle diameter on deposition velocity of particles to grass (McMahon and Denison, 1979) (Courtesy of Pergamon Press)

and for the deposition rate from Equations 15.50.11 and 15.50.18

$$\omega = \frac{Q_x v_d}{\pi \sigma_y \sigma_z u} \exp\left[-\left(\frac{y^2}{2\sigma_y^2} + \frac{h^2}{2\sigma_z^2}\right)\right] \qquad [15.50.19]$$

Evaluations of Equation 15.50.17 based on numerical integration have been made by van der Hoven (1968) for Pasquill categories A to F for a wind speed $u = 1\,\mathrm{m/s}$ and a deposition velocity $v_d = 10^{-2}\,\mathrm{m/s}$. Figure 15.150 shows his plot for D conditions, while Table 15.62 summarizes the other plots in terms of the times at which Q_x falls to half its initial value of Q_o. For other wind speeds and deposition velocities use may be made of the relation

$$\left(\frac{Q_x}{Q_o}\right)_2 = \left(\frac{Q_x}{Q_o}\right)_1^{u_1 v_{d2}/u_2 v_{d1}} \qquad [15.50.20]$$

where subscripts 1 and 2 refer to the standard conditions used by van der Hoven, as given in Figure 15.150, and to the conditions of interest, respectively.

Another model used for small particles is the partial reflection model of Overcamp (1976). The model is based on the Pasquill–Gifford relations and incorporates both the tilted plume approach to deal with the gravitational settling of larger particles and a reflection coefficient α which qualifies the 'image' term, or term containing $(z + h)$. The basic equation is

$$\chi(x,y,z) = \frac{Q}{2\pi\sigma_y\sigma_z u}\exp\left(-\frac{y^2}{2\sigma_y^2}\right) \times \left(\exp\left\{-\frac{[z-(h-v_t x/u)]^2}{2\sigma_z^2}\right\} + \alpha(x_G)\exp\left\{-\frac{[z+(h-v_t x/u)]^2}{2\sigma_z^2}\right\}\right) \qquad [15.50.21]$$

where

$$\alpha(x) = 1 - \frac{2v_d}{v_t + v_d + (uh - v_t x)\sigma_z^{-1}(d\sigma_z/dx)} \qquad [15.50.22]$$

and where x_G is obtained from the implicit equation

$$\left(h - \frac{v_t x_G}{u}\right)\frac{\sigma_z(x)}{\sigma_z(x_G)} = z + \left(h - \frac{v_t x}{u}\right) \qquad [15.50.23]$$

Another model is the surface depletion model of Horst (1977).

A comparison of the models available has been made by Horst (1979), who found that in the near field the partial reflection model was reasonably accurate and easy to use, while in the far field the source depletion model performed better.

These models incorporate the deposition velocity which, as described below, is subject to considerable

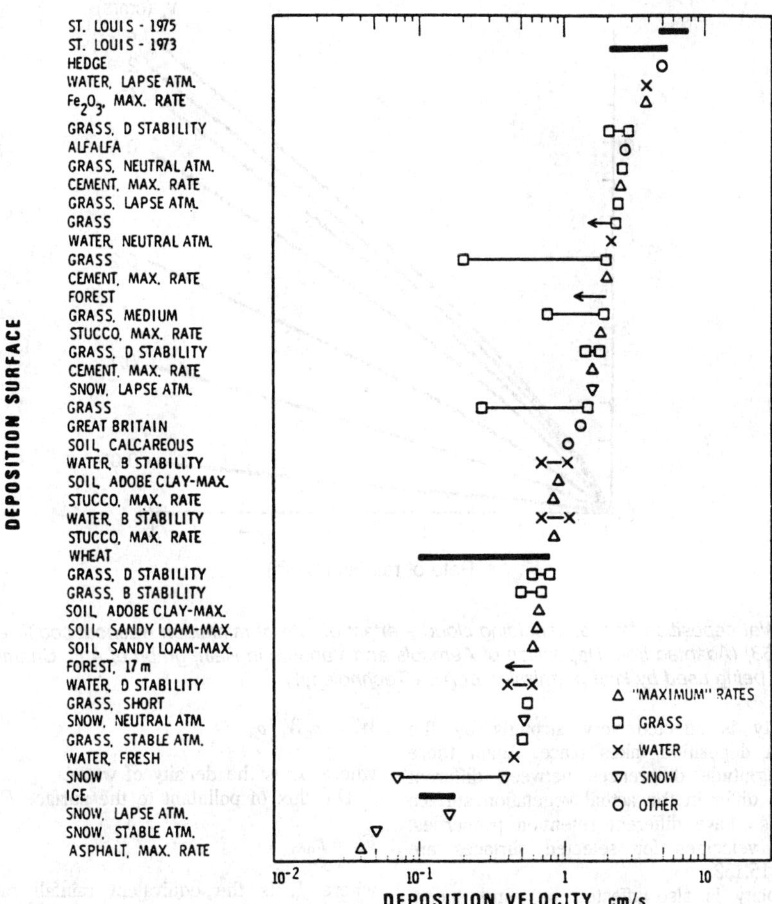

Figure 15.152 *Dry deposition from a dispersing cloud – effect of deposition surface on deposition velocity of sulphur dioxide (Sehmel, 1980) (Courtesy of Pergamon Press)*

uncertainty and may be at least as great a source of error as inadequacies in a particular model.

15.50.4 Deposition velocity

Reviews of the deposition velocity, which include extensive data, have been given by McMahon and Denison (1979) and by Sehmel (1980). The deposition velocity is defined by Equation 15.50.11. It should be noted that the surface area for deposition implied in this equation is the normal ground surface area, not the actual vegetation surface, which can be much greater.

Sehmel lists a large number of variables which affect deposition velocity, divided into four categories: meteorology, surface, particle and gas. For particles of diameter $>1\,\mu$m gravitational settling is dominant, while the movement of particles of diameter $<0.1\,\mu$m is governed by Brownian motion. The deposition velocity tends to pass through a minimum with particle diameter and it may vary by two orders of magnitude. These effects are illustrated in Figure 15.151, which shows the variation of deposition velocity with particle diameter for grass surfaces.

Deposition velocity is applicable to gases also. Inert gases have almost negligible deposition velocity, say 10^{-4}–10^{-3} cm/s, but chemically or biologically active gases tend to have a much larger deposition velocity, typically in the range 0.5–3 cm/s. Of interest in the present context are the deposition velocities of toxic gases. From data given by Sehmel, the ranges of deposition velocities of some typical toxic gases are as follows:

Gas	*Deposition velocity* (cm/s)
Cl_2	1.8–2.1
SO_2	0.04–7.5
HF	1.6–3.7
H_2S	0.015–0.38

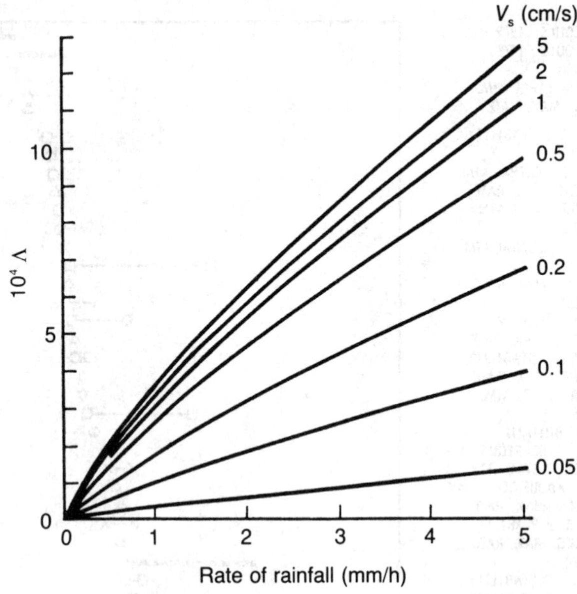

Figure 15.153 *Wet deposition from a dispersing cloud – effect of rate of rainfall on washout coefficient (Chamberlain, 1953) (Adapted from Deposition of Aerosols and Vapours in Rain, graph of A.C. Chamberlain (1955), the original graph being used by kind permission of AEA Technology)*

Deposition velocity is affected very strongly by the surface on which deposition takes place. Again there are order of magnitude differences between different surfaces. Surfaces differ in the actual vegetation surface exposed. They also have different retention properties. Some deposition velocities for selected surfaces are shown in Figure 15.152.

Deposition velocity is also affected by wind speed, roughness length and friction velocity, but these effects are less readily characterized and are a matter of some debate.

A method for prediction of deposition velocity has been given by Sehmel and Hodgson (1978), but prediction remains subject to considerable uncertainty.

15.50.5 Wet deposition

In wet deposition due to rain, a distinction may in principle be made between scavenging within clouds (or rainout) and below clouds (or washout), but in practice these two are generally treated as a single process.

Wet deposition of a pollutant is often treated empirically in terms of the washout ratio W_r:

$$W_r = k_o/\chi_o \qquad [15.50.24]$$

where χ is the concentration of pollutant in the air, k is the concentration of pollutant in the rainwater and subscript o denotes the reference height.

Use is also made of an alternative washout ratio W_r'

$$W_r' = \rho_a k_o'/\chi_o \qquad [15.50.25]$$

where k' is the concentration of pollutant in the rainwater and ρ_a is the density of air. The difference between k and k' is that the first has units of mass per unit volume and the second has units of mass per unit mass. The relation between the two washout ratios is:

$$W_r' = \rho_a W_r/\rho_w \qquad [15.50.26]$$

where ρ_w is the density of water.

The flux of pollutant to the surface F_w is

$$F_w = k_o J_o \qquad [15.50.27]$$

where J_o is the equivalent rainfall rate. Hence from Equation 15.50.24

$$F_w = W_r \chi_o J_o \qquad [15.50.28]$$

A wet deposition velocity v_w may be defined

$$v_w = F_w/\chi_o \qquad [15.50.29a]$$

$$= W_r J_o \qquad [15.50.29b]$$

An alternative approach is that based on the washout coefficient, or scavenging coefficient Λ. The decay of the concentration of pollutant due to scavenging is frequently modelled by a first-order decay model

$$\chi(t) = \chi(0)\exp(-\Lambda t) \qquad [15.50.30]$$

It may readily be shown, by deriving Equation 15.50.8 from the unsteady-state mass balance, that the same washout coefficient may also be defined in terms of the relation between the interphase transport rate w and the concentration χ:

$$w = \Lambda\chi \qquad [15.50.31]$$

The flux of pollutant F_w through a wetted plume of height z_w is

Table 15.63 *Washout coefficient for Lycopodium spores (after May, 1958) (Courtesy of The Royal Meteorological Society)*

Type of rain	*Rate of rainfall* (mm/h)	*Wind velocity* (cm/s)	*Washout coefficient,* Λ 10^{-4}/s	
			Observed	*Theoretical*
Frontal	3.91	320	10.2	9.7
Frontal	1.12	543	4.2	3.6
Heavy frontal	14.1	845	30.8	26.8
Frontal	1.01	334	3.2	3.2
Continuous rain of showery type	3.64	332	8.9	9.2

$$F_w = \int_0^{z_w} w \mathrm{d}z \qquad [15.50.32a]$$

$$= \int_0^{z_w} \Lambda \chi \mathrm{d}z \qquad [15.50.32b]$$

For a Gaussian plume from a continuous point source the concentration χ may be obtained from Equation 15.16.41. Then, integrating this equation with respect to z from 0 to ∞, the effective concentration χ_e over the height z_w is

$$\chi_e = \frac{Q}{(2\pi)^{\frac{1}{2}} \sigma_y u z_w} \exp\left(-\frac{y^2}{2\sigma_y^2}\right) \qquad [15.50.33]$$

Assuming constant Λ and substituting χ_e from Equation 15.50.33 for χ in Equation 15.50.32b gives

$$F_w = \frac{\Lambda Q}{(2\pi)^{\frac{1}{2}} \sigma_y u} \exp\left(-\frac{y^2}{2\sigma_y^2}\right) \qquad [15.50.34]$$

On the centre-line

$$F_w = \frac{\Lambda Q}{(2\pi)^{\frac{1}{2}} \sigma_y u} \qquad [15.50.35]$$

The use of a constant washout coefficient is a rather crude approximation. Strictly, it is applicable only to single size, or monodisperse, rain drops. Data on the washout coefficient for polydisperse systems have been given by Dana and Hales (1976).

It may be noted that for a polydisperse system the use of the geometric mean drop size tends appreciably to underestimate the washout coefficient. Another problem in the use of a constant washout coefficient concerns the absorption of gas in rain drops. If an absorbed gas does not undergo an irreversible reaction, it is liable to be desorbed as it passes through a zone of lower gas concentration.

15.50.6 Washout ratio and washout coefficient

The washout ratio W_r is an essentially empirical parameter. Data on the washout ratio have been given by McMahon and Denison (1979). The washout ratio tends to decrease with rainfall. As a rule of thumb, it halves for each order of magnitude increase in rainfall. The washout coefficient, or scavenging coefficient, Λ has been estimated theoretically for certain situations. The following relation has been given by Chamberlain (1953):

$$\Lambda = \int_0^{s_{max}} A_s N_s v_s E \, \mathrm{d}s \qquad [15.50.36]$$

where A_s is the cross-sectional area of a drop of radius s, N_s is the number of drops per unit volume per unit time, v_s is the velocity of the drops and E is the collection efficiency and subscript max denotes maximum.

Chamberlain used Equation 15.50.18] to determine theoretical values of the washout coefficient as a function of the velocity v_s and the rainfall, as shown in Figure 15.153. Experimental work by F.G. May (1958) on the washout of *Lycopodium* spores has shown good agreement with theory, as shown in Table 15.63.

The washout coefficient is a function of the physical and chemical characteristics and of the raindrop size distribution and the rainfall. In many cases where the system is not so well defined as that just described theoretical estimates may be less satisfactory. In such cases it is necessary to fall back on empirical values. Data on the washout coefficient have been given by McMahon and Denison (1979).

15.51 Infiltration into Buildings

In most toxic release scenarios the major part of the population in the path of the gas cloud is likely to be indoors. The mitigating effect of such shelter can be appreciable. It is therefore necessary to be able to estimate the indoor concentrations. Discussions of air infiltration relevant to hazard assessment have been given in *Meteorology and Atomic Energy* (Slade, 1968) and the *Rasmussen Report* (AEC, 1975) and by Haastrup (1984) and P.C. Davies and Purdy (1986).

The degree of protection given by a building depends on the frequency of the air changes, or ventilation rate. The latter is a function of the construction and topographical situation of the building, of the wind speed and direction, of the difference between the outside and inside temperatures and of the number of doors and windows left open.

15.51.1 Models of infiltration

The infiltration of air into a building is usually modelled by assuming that the air in the building space is perfectly mixed so that

$$\frac{\mathrm{d}c_i}{\mathrm{d}t} = \lambda(c_o - c_i) \qquad [15.51.1]$$

where c is the concentration (various units), t is the time (h) and λ is the ventilation constant (h^{-1}) and subscripts i

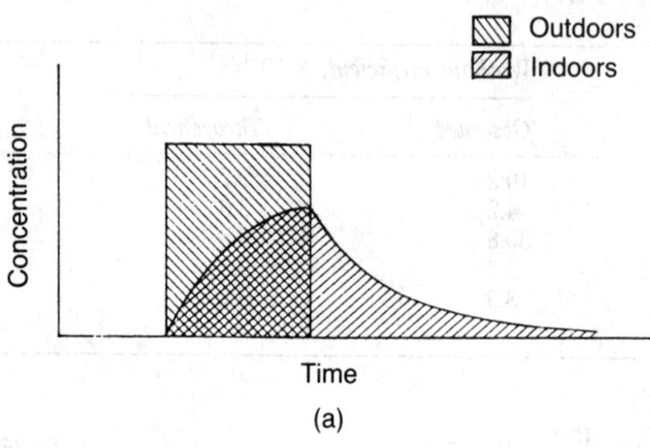

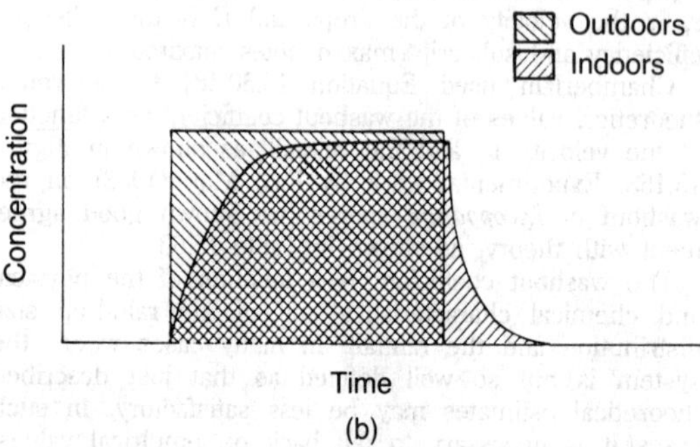

Figure 15.154 *Infiltration of gas into buildings – outdoor and indoor concentration profiles for a rectangular pulse forcing function: (a) short pulse, low ventilation rate; and (b) long pulse, high ventilation rate*

and o refer to indoor and outdoor, respectively. The ventilation constant is equal to the frequency of the air changes, is generally also referred to as the ventilation rate and is most often expressed as air changes per hour (ach). Imperfect mixing may be allowed for, as described below.

In the general case, where the outdoor concentration profile $c_o(t)$ is arbitrary, the indoor concentration is obtained from Equation 15.51.1 numerically.

For the case where the outdoor concentration profile, or forcing function, takes the form of a pulse, with

$$c_o(t < t_o) = c_o \quad [15.51.2a]$$

$$c_o(t > t_o) = 0 \quad [15.51.2b]$$

as shown in Figure 15.154(a), an analytical solution is available. If the outdoor concentration undergoes a step change from zero to c_o and then persists to time t_o, the rise of the indoor concentration is obtained by integrating Equation 15.51.1 to give

$$c_i = c_o[1 - \exp(-\lambda t_o)] \quad [15.51.3]$$

If the outdoor concentration then falls to zero, the decay of the indoor concentration from its maximum value $c_i(t_o)$ is again obtained by integrating Equation 15.51.1:

$$c_i = c_i(t_o)\exp[-\lambda(t - t_o)] \quad [15.51.4]$$

$$= c_o[1 - \exp(-\lambda t_o)]\exp[\lambda(t - t_o)] \quad [15.51.5]$$

This yields the concentration profile shown in Figure 15.154(b).

This rectangular pulse model may be used to obtain an approximate estimate of the indoor concentrations both for a puff from a nominally instantaneous release and for a plume from a short-duration continuous release. For a long-duration continuous release the appropriate model is a step forcing function and the indoor concentration is given by Equation 15.51.3.

In assessing the effect of a toxic gas, however, it is frequently necessary to work in terms of toxic dose or, more generally, toxic load rather than of concentration. The toxic dose is defined as

$$D = \int c \, dt \quad [15.51.6]$$

and the toxic load is generally defined as

$$L = \int c^n dt \quad [15.51.7]$$

where D is the toxic dose (units h) and L is the toxic load (units^n h).

It is shown below that for the pulse model the ratio of the indoor dose D_i to the outdoor dose D_o is unity. For the step model it is readily shown from Equation 15.51.3 that

$$\frac{D_i}{D_o} = 1 - \frac{1 - \exp(-\lambda t)}{\lambda t} \quad [15.51.8]$$

An extension of the pulse model in terms of toxic load has been given by Haastrup (1984), who has derived the following relations:

$$L_o = \int c_o^n dt \quad [15.51.9]$$

$$L_i(t < t_o) = \int_0^{t_o} c_i^n dt \quad [15.51.10a]$$

$$= c_o^n \int_0^{t_o} [1 - \exp(-\lambda t)]^n dt \quad [15.51.10b]$$

$$L_i(t > t_o) = \int_{t_o}^{\infty} c_i^n dt \quad [15.51.10c]$$

$$= c_o^n \frac{[1 - \exp(-\lambda t_o)]^n}{\lambda^n} \quad [15.51.10d]$$

The ratio R of the indoor to the outdoor toxic load is then

$$R = \frac{[1 - \exp(-\lambda t_o)]^n}{n\lambda t} + \frac{1}{t}\int_0^{t_o} [1 - \exp(-\lambda t)]^n dt \quad [15.51.11]$$

It may be noted that if $n = 1$, Equation 15.51.11 simplifies to

$$R = \frac{D_i}{D_o} = 1 \quad [15.51.12]$$

For low values of the group λt_o it can be shown that

$$L_i(t > t_o) \gg L_i(t < t_o) \quad [15.51.13]$$

In other words, the total indoor load is dominated by the load during the decay period, or tail. For this case

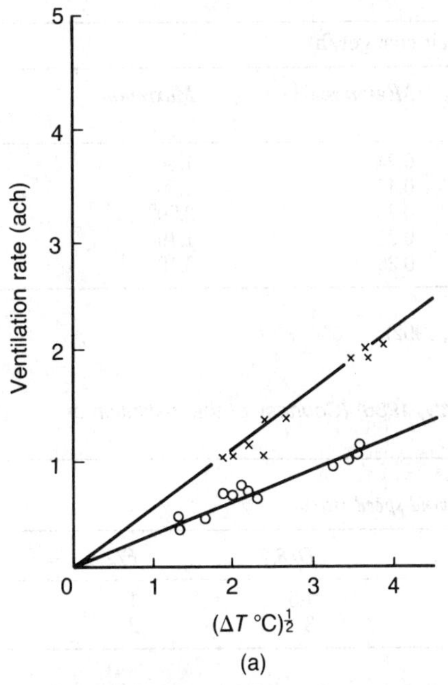

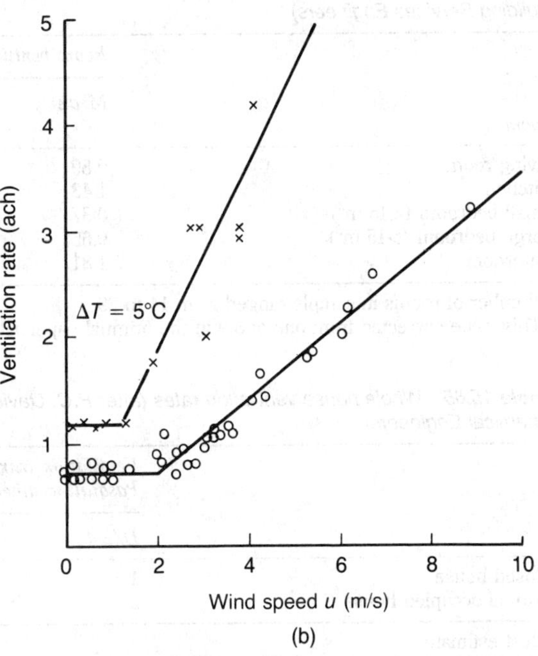

Figure 15.155 *Infiltration of gas into buildings – effect of buoyancy and wind on house ventilation rate (Dick and Thomas, 1951): (a) effect of buoyancy (temperature difference); (b) effect of wind speed. o, house with no vents open; x, house with three bedroom vents open. (Courtesy of the Chartered Institution of Building Services Engineers)*

$$\exp(-\lambda t) \approx 1 - \lambda t \qquad \lambda t < 0.1 \qquad [15.51.14]$$

Then

$$L_i \approx L_i(t > t_o) \qquad [15.51.15]$$

and from Equations 15.51.10b and 15.51.14

$$L_i = \frac{c_o^n t_o^n \lambda^{n-1}}{n} \qquad [15.51.16]$$

If the building space is not perfectly mixed, the effect of imperfect mixing may be taken into account in the model by substituting for the ventilation rate λ an effective ventilation rate λ'. The effective ventilation rate may be determined by applying to the ventilation rate a mixing efficiency factor k:

$$\lambda' = k\lambda \qquad [15.51.17]$$

15.51.2 Ventilation rate

The ventilation of a building may be natural or mechanical. For the assessment of toxic hazard it is normally natural ventilation which is relevant. Information on ventilation is available from a number of sources. These include *Principles of Modern Building* (Building Research Establishment, 1959), the CIBS *Guide* Part 1 A4 *Air Infiltration* (Chartered Institute of Building Services, 1976) and BS 5925: 1991 *Code of Practice: Ventilation Principles and Designing for Natural Ventilation.*

The equations for air infiltration given by these sources have been reviewed by Brighton (1986). They are all based on following relation:

$$Q = k_1 A \left(\frac{\Delta p}{\rho_o}\right)^{\frac{1}{2}} \qquad [15.51.18]$$

where A is the area of the aperture (m^2), Δp is the pressure drop (N/m^2), Q is the volumetric flow of air (m^3/s), ρ_o is the density of the ambient air (kg/m^3) and k_1 is a constant. Equation 15.51.18 is based on Bernouilli's theorem and the constant is related to the contraction ratio of the emerging jet. For this constant k_1 Brighton gives a compromise value of 0.88.

Work on air infiltration into houses is done for a number of purposes. These include the comfort of the occupants, the safety of gas appliances and the saving of energy. In the latter case the emphasis is on eliminating infiltration, and work in this area illustrates the high degree of leak-tightness which can be achieved.

Studies of air infiltration into houses in Britain have been done at the Building Research Establishment (BRE). Work has been published by Dick (1949, 1950a,b), Dick and Thomas (1951), Brundrett (1977) and Warren and Webb (1980a,b).

Air infiltration is due to the existence of pressure difference between the outside and inside of the building and this pressure difference may be caused by wind or by air temperature, and hence air density, or buoyancy,

Table 15.64 Room ventilation rates[a] (after P.R. Warren and Webb, 1980b) (Courtesy of the Chartered Institute of Building Services Engineers)

Room	Room ventilation rate (air changes/h)		
	Mean	*Minimum*	*Maximum*
Living room	0.89	0.24	1.64
Kitchen	1.43	0.43	3.50
Small bedroom (<15 m^3)	0.87	0.28	2.90[b]
Large bedroom (>15 m^3)	0.65	0.25	1.19
Bathroom	1.81	0.25	3.19

[a] Number of rooms in sample ranged from 14 to 29.
[b] This value corrected from one of 3.9 in the original paper (P.R. Warren, 1992).

Table 15.65 Whole house ventilation rates (after P.C. Davies and Purdy, 1986) (Courtesy of the Institution of Chemical Engineers)

	Ventilation (air changes/h)[a] Pasquill weather category/wind speed (m/s)			
	D/2.4	*D/4.3*	*D/6.7*	*F/2.4*
Closed house	1	1	1.5	1
Normal occupied house	2	2	3	2

[a] Best estimate.
[b] Values used by Pape and Nussey (1985) for these four stability category/wind speed combinations are 0.7, 1, 1.5 and 0.5 air changes/h, respectively.

difference. The ventilation rate due to wind effects may be expressed as

$$\lambda_w = k_2 u \qquad [15.51.19]$$

and that due to buoyancy effects as

$$\lambda_b = k_3(|\Delta T|)^{\frac{1}{2}} \qquad [15.51.20]$$

where ΔT is the temperature difference between outdoors and indoors (°C), u is the wind speed (m/s), λ_b is the ventilation rate due to buoyancy (m^3/h), λ_w is the ventilation rate due to wind (m^3/h), and k_2 and k_3 are constants. Figure 15.155 shows data obtained by Dick and Thomas (1951) in work at the BRE on the effect of temperature difference and wind speed on ventilation rate.

In considering ventilation rate it is necessary to distinguish between closed houses and occupied houses, the difference being that the latter typically have some windows open. The BRE work includes correlations for the ventilation rate of a closed house as a function of wind speed. P.C. Davies and Purdy (1986) have presented versions of these equations modified so that the wind speed is that at the standard 10 m height. These are:

Exposed site

$$\lambda = 0.87 + 0.13u \qquad [15.51.21]$$

Sheltered site

$$\lambda = 0.88 \qquad u < 4.2 \qquad [15.51.22a]$$

$$\lambda = 0.22u \qquad u > 4.2 \qquad [15.51.22b]$$

Davies and Purdy give the following relation as the most recent BRE infiltration model:

$$\lambda = \lambda_s(k_4 \Delta T^{1.2} + k_5 u^{2.4})^{0.5} \qquad [15.51.23]$$

where λ_s is the ventilation rate measured at standard conditions (h^{-1}) and k_4 and k_5 are constants.

The distribution of infiltration between the different apertures in the building has been investigated by Warren and Webb (1980b). For a typical house with a pressure difference of 50 Pa and an air infiltration rate of 3000 m^3/h they found the following distribution:

	Contribution to air infiltration	
	(m^3/h)	(%)
Windows and WC fan	940	31.3
Back door	280	9.3
Skirting boards	575	19.2
Lights, plugs, pipes, etc.	85	2.8
Other sources	1120	37.4

For occupied houses it is necessary to take into account the effect of open windows. The relation given by Dick and Thomas (1951) for houses with open windows is

$$\lambda = 0.87 + 0.13u + 0.23(n + 1.4m) + 0.05(n + 1.4m)u \quad [15.51.24]$$

where m is the number of open casement windows and n is the number of open top-vent windows. Equation 15.51.24 is based on a study on an estate at Abbotts Langley. The following data were obtained on the number of windows open:

	Abbotts Langley	*Bucknalls Close*
Casement windows open, m	0.25	0.4
Vent windows open, n	1.75	3

Abbotts Langley is an exposed site and Bucknalls Close a sheltered one.

The number of windows open depends on a number of factors, including the degree of exposure, the season, the time of day, the wind speed, the outside temperature and the number of occupants. These factors are discussed in detail by Davies and Purdy. Many of the factors tend to cancel out. For example, the number of windows open on an exposed site tends to be less than on a sheltered site. Likewise, the number open in summer is greater than in winter, but the buoyancy effect is less.

Information on room ventilation rates from the work of Warren and Webb is given Table 15.64. The best estimates of whole house ventilation rates given by Davies and Purdy are shown in Table 15.65.

Data for air infiltration into British houses are not necessarily applicable in other countries. The ventilation rates for American houses are less and those for Swedish houses much less. Data given by Kronvall (1978) indicate that ventilation rates for Swedish houses are about a quarter those in Britain. Data on ventilation rates for buildings in the USA have been given by Handley and Barton (1973). The work of Handley and Barton also shows, however, that there is a wide range of ventilation rates. They found values from 0.07 to 3.0 air changes/h.

15.51.3 Hazard assessments

The mitigating effect of shelter is considered in the First *Canvey Report* in Appendix 7 by Beattie, who quotes Equation 15.51.3 and suggests that a suitable value of the ventilation rate λ is 1 change/h for a modern building, but that this may rise to 2–3 changes/h if doors or windows are left ajar.

In the *Rijnmond Report* credit is taken for the effect of shelter using equation 15.51.1. A ventilation rate of 2 changes/h is used, but it is stated that this could be as low as 1 or as high as 3.

15.52 Source and Dispersion Modelling: *CCPS Guidelines*

The CCPS has issued two publications giving practical guidance on emission and dispersion of hazardous materials. These are now described.

15.52.1 *Guidelines for Use of Vapor Cloud Dispersion Models*

The CCPS *Guidelines for the Use of Vapor Cloud Dispersion Models* (the CCPS *Vapor Cloud Dispersion Model Guidelines*) (1987/2) covers both emission and dispersion. The *Guidelines* give an overview of release scenarios, particularly pipe ruptures, vessel rupture and reactor venting, review the dominant phenomena of initial acceleration and dilution, buoyancy and atmospheric turbulence, and give a decision tree for handling the scenarios. A selection of release scenarios is given, and amplified in an appendix. The characteristics of routine and accidental releases are compared.

Models are given for the determination of emission flows. These include: two-phase flow; flow from tanks under liquid head as well as pressure; and flow from pipelines. Models for various types of jet are covered. Models for vaporization from pools include both heat and mass transfer limited cases and spreading pools. A critical review is given of pool vaporization models.

For gas dispersion, separate treatments are given for ground level and elevated releases. An overview is given of passive gas dispersion, including the Pasquill–Gifford model, the Pasquill stability categories and correlations for dispersion coefficients. The set of six models contained in the EPA UNAMAP6 system is outlined.

The variety in type and origin of dense gas dispersion models is described. The authors state that there are some 100 models, with 10 further models being created annually. Many existing models continue to undergo development and to acquire new features.

Dense gas dispersion is considered first in the absence of heat effects. The box model of van Ulden is outlined. The authors then derive for dense gas dispersion a number of simplified relations for cloud dimensions and concentration, and compare them with the Thorney Island trials data.

Accounts are then given of the thermodynamic and heat transfer phenomena and of correlations used for air entrainment. The three-dimensional models considered are FEM3 and MERCURE-GL. The review given of the criteria utilized by various workers for effectively passive dispersion brings out the large disparities in the criteria adopted. Physical modelling is described, particularly in respect of plume lift and complex terrain. The implications of concentration fluctuations are treated and the effect of sampling time and volume are discussed.

A review is given of some principal dispersion models, covering some 40 models and giving a tabulation of the characteristics of each model. Four models are described in more detail: AIRTOX, DEGADIS, FEM3 and INPUFF.

For elevated releases of dense gas the *Guidelines* give an overview of models for momentum jets and buoyant plumes. The use is described of an effective height in the Pasquill–Gifford model for handling plume rise and gas density. Momentum jet scenarios are classified and the model of D.J. Wilson (1979b) for a momentum jet from a pipeline is described. For plumes reference is made to the models of G.A. Briggs (1984) which are utilized in UNAMAP. An account is given of the models of Ooms, Mahieu and Zelis (1974) and of Hoot, Meroney and Peterka (1973). A criterion is given for effectively dense gas dispersion in an elevated release.

The *Guidelines* consider the extent to which models have been validated against field trials and the uncertainty in model estimates, and list some of the comparisons made between models and trials. They refer to the Environmental Protection Agency set of model evaluation procedures for passive dispersion and give a number of measures of performance.

The *Guidelines* give one of the few quantitative treatments of uncertainty in modelling. The sources of uncertainty considered are model physics, random variability and input data errors. A treatment is given of the effect on uncertainty of the number of model parameters.

15.52.2 *Workbook of Test Cases for Vapor Cloud Source Dispersion Models*

The CCPS *Vapor Cloud Dispersion Model Guidelines* are supplemented by the CCPS *Workbook on Test Cases for Vapor Cloud Source Dispersion Models* (the CCPS *Vapor Cloud Source Dispersion Model Workbook*) (1989/8). The *Workbook* gives a set of emission, pool vaporization and jet models, and describes methods of matching the output of these models with the input required by selected dispersion models. The principal models treated are: for emission, the Fauske–Epstein models for liquids, subcooled liquids and saturated liquids; for vaporization, the Fleischer model for an evaporating pool; for jets, the Briggs momentum jet model and the Hoot, Meroney and Peterka dense gas jet model; and for gas dispersion, the Pasquill–Gifford model, CAMEO, the Ooms model, SPILLS, SLAB and DEGADIS.

The criterion used to determine whether it is necessary to use for a continuous release a dense gas dispersion model is $Ri \gg 10$ with

$$Ri = \frac{\pi}{4} g' \frac{u_o D}{u_*^2 u} \qquad [15.52.1]$$

$$u_* \approx 0.065u \qquad [15.52.2]$$

where D is the initial diameter of the plume, g' is the reduced gravity, u is the wind speed, u_o is the velocity of release and u_* is the ambient friction velocity.

Five release scenarios are treated:

(1) continuous jet release of gaseous butane from a vent stack;
(2) continuous release of liquefied ammonia from a storage vessel–
    (a) nozzle failure (liquid release),
    (b) line rupture (two-phase release);
(3) continuous jet release of high pressure carbon monoxide gas from a low level source;
(4) continuous two-phase release of liquefied chlorine from a storage vessel;
(5) continuous release of liquid acetone into a bund.

These are all continuous releases, the choice being deliberate, since the models for instantaneous or transient releases are subject to more uncertainty. Likewise, the first four cases were selected because the excess density is significant and it is necessary to use a dense gas dispersion model.

The *Workbook* gives detailed working for these five cases, with hand calculation of the sections on emission, vaporization, jet behaviour and interface to the dispersion model being followed by results of the gas dispersion computer codes. Several sample outputs are given from CAMEO, SPILLS, SLAB and DEGADIS.

Table 15.66 *Some values of the heat transfer parameter $(k_s\rho_s c_s)^{\frac{1}{2}}$ for different substrates (after Dilwali and Mudan, 1987) (Courtesy of the American Institute of Chemical Engineers)*

Material	*Heat transfer parameter* $(k_s\rho_s c_s)^{\frac{1}{2}}$ $(Ws^{\frac{1}{2}}/m^2\ K)$
Soil (dry)	2570
Sand (dry)	2660
Sand (wet: 3% moisture)	2335
Uninsulated concrete	3750
Insulated concrete	230–440
Polyurethane	140
Other insulating materials:	
Celofoam	99
Foamglas	74

15.53 Vapour Release Mitigation: Containment and Barriers

There are a number of methods of preventing or mitigating the dispersion of gases. They include the use of:

(1) bunds;
(2) foam;
(3) solid barriers;
(4) fluid barriers–
    (a) water spray barriers;
    (b) steam curtains.

An account of the effect of some of these devices on gas dispersion is given in Section 15.40. The results of field trials and of physical modelling are given in Sections 15.37–15.39. Guidelines for vapour release mitigation have been given by Prugh (1985–, 1987b,c). The various methods of prevention and mitigation are now considered.

15.53.1 Bunds

The rate of evaporation of a spill, particularly that of a refrigerated liquefied gas, can be greatly reduced if the spill is contained within a bund. In turn, the evaporation from the bund may be much reduced by appropriate design. Two principal features are the geometry and the substrate of the bund. A bund around a storage tank is generally designed to contain the contents of the tank. A bund with a smaller floor area but higher walls presents a smaller surface to heat up and vaporize the liquid. The higher wall also acts as a barrier to the flow of the vapour.

The use of a suitable material for the substrate can greatly reduce the rate of heat transfer to the liquid. The relations for this heat transfer process are given in Section 15.10. From Equation 15.10.30 the rate of heat transfer is proportional to the heat transfer parameter $(k_s\rho_s c_s)^{\frac{1}{2}}$. Table 15.66 lists some values of this parameter for different substrates, as given by Dilwali and Mudan

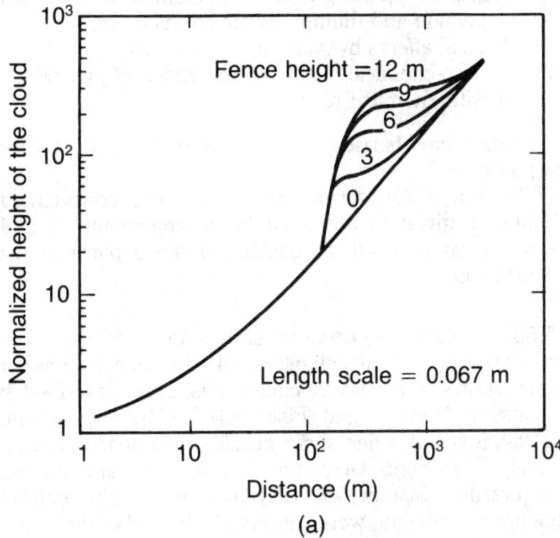

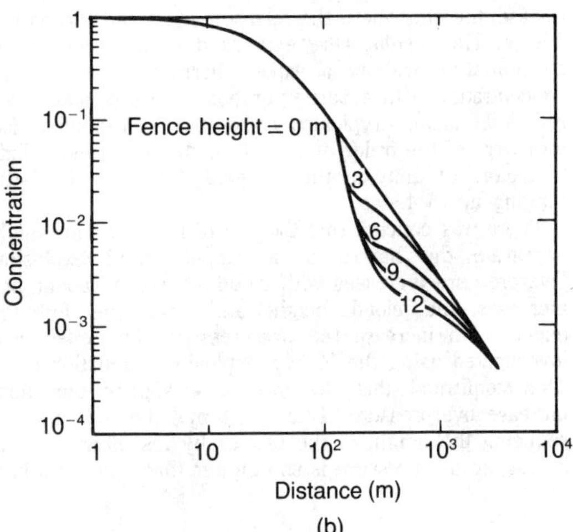

***Figure 15.156** Dispersion of dense gas at obstacles – predictions of FENCE62 for dispersion of hydrogen fluoride across barrier for Goldfish trial 1 (Meroney, 1991): (a) effect on plume height; (b) effect on centre line concentration (Courtesy of the American Institute of Chemical Engineers)*

(1987). As the table indicates, there are differences of up to 16 in the values of the parameter. These authors describe a hazard assessment of the dispersion of the vapour from a spill of liquid chlorine into a bund. The study considered various designs of bund and showed that the distance to the concentration immediately dangerous to life and health could be reduced from 2.6 km for the base case to 320 m for an alternative design. Further, the provision of a bund facilitates the use in suitable cases of foam, which can effect a further large reduction in evaporation.

15.53.2 Foam

The use of foam is an effective means of reducing the rate of evaporation, where a suitable foam is available. Accounts of the testing of foams at facilities in 1986 at Pueblo, Colorado, and in 1988 at the Nevada Test Site (NTS) have been given by Dimaio, Norman and co-workers (Dimaio and Norman, 1988, 1990; Norman and Dimaio, 1989; Norman and Swihart, 1990).

Foams tested included: Hazmat 2, an acid resistant foam; ARAFFF, an alcohol resistant AFFF foam; and a fluoroprotein foam. Chemicals tested included bromine, chlorine, hydrogen fluoride, monomethylamine, phosphorous oxychloride, phosphorous pentachloride and sulphur dioxide. Using Hazmat 2 the proportional vapour suppressions achieved were as follows: bromine >95%; chlorine 67–82%; hydrogen fluoride 70–85%; sulphur dioxide >61%.

Foam acts by insulating the surface of the spill and preventing vaporization. Other modes of action include absorption of the vapour and scrubbing out of aerosol and particulate matter. Any foam used must be suitable for the application; the wrong foam can do more harm than good. A guide to the use of foam on hazardous materials has been published by Norman (1987). Likewise, water should only be used advisedly; incorrect use can make things worse.

Foam should be applied gently, possibly with continuous application or frequent reapplication. Personnel should be trained and should have suitable protective equipment, including self-contained breathing apparatus.

15.53.3 Solid barriers

A solid barrier such as a fence or wall can serve either to contain a gas cloud entirely or to effect an appreciable dilution. An impermeable barrier designed for this purpose and erected within the works is generally know as a vapour barrier. A barrier does not need to be impermeable to mitigate a gas release. A plantation of trees may provide a worthwhile degree of dilution of a gas cloud.

15.53.4 Vapour barriers: Hawk programme

As described in Section 15.37, two series of field trials have been conducted at the NTS on the mitigation of hydrogen fluoride releases. These trials were largely concerned with water spray barriers and are considered in Section 15.53.13. The second investigation, the Hawk programme, under the auspices of ICHMAP, also included a study of vapour barriers. An account of this work is given here.

The programme on vapour barriers included a review of the effectiveness of vapour barriers and wind tunnel tests on such barriers. The review concluded that, whilst a vapour barrier reduces the near field concentrations, in the far field this effect decays.

In the wind tunnel work, two types of configuration were tested. One consisted of vapour barriers or fences, with only partial enclosure of the gas cloud, and the other of vapour boxes, with complete enclosure. In addition, the effect of other types of plant obstacle was investigated.

The work demonstrated that the effect of a vapour barrier is extremely dependent on the nature of the release, the site where the release occurs and the barrier design. The results were expressed as the ratio of the concentration without a vapour barrier or box to the concentration with a barrier or box – the concentration reduction factor. Typically, concentration reduction factors were of the order of 2 to 10 in the near field, falling to factors of unity in the far field, but with the latter ranging up to 4.

There was concern that the use of a vapour box might aggravate the hazard of a vapour cloud explosion. Overpressure increases with cloud height; a vapour box increases the cloud height and may therefore be expected to increase the overpressure. The effect was investigated using the FLACS explosion simulation code. This confirmed that the use of a vapour box does increase overpressure. This effect may be mitigated by reducing the height of the box or by installing vents in it, but the overpressure is still higher than without a box.

15.53.5 Vapour barriers: FENCE62

Meroney (1991) has described the use of a set of models with the generic title FENCE to simulate the behaviour of a dense gas cloud at a fence used as a vapour barrier, in particular FENCE62.

Models for the simulation of the dilution of a cloud of hydrogen fluoride are adaptations of the box model DENS62 and the slab model DENS23. For simulation of the effects of a vapour barrier the corresponding models are FENCE62 and FENCE23 and those for the effects of a water spray barrier are SPRAY62 and SPRAY23; SPRAY 65 is a modification of SPRAY62.

These models were used to simulate the Goldfish field trials on the dispersion by a fence of a continuous release of hydrogen fluoride. Figure 15.156(a) shows the effect of fence height on the height of the cloud and Figure 15.156(b) the effect on the downwind concentration. These effects decay so that by about 200 fence heights downwind they have largely disappeared.

The fence was found to be more effective in diluting the gas cloud if located near the source. For a fence placed within 400 m of the source, the dilution effects did not persist beyond about 1000 m. For a fence located 100 m from the source there was little effect of wind speed in the range 1–8 m/s on either downwind cloud height or concentration.

15.53.6 Water spray barriers

A good deal of work has been done on the development of water spray barriers, or water curtains, as a means of mitigating the hazard from a gas release. The systems used include fixed water spray installations and mobile water spray monitors. Fixed installations are typically a set of spray nozzles several metres off the ground with the spray directed downwards. Such systems are used both in the open air and in buildings. A typical mobile monitor system is a set of spray nozzles inclined at an angle of approximately 45° to the horizontal. Water sprays and curtains are used mainly against flammable gases, but may be applied against toxic gases also.

A water spray directed vertically through a gas cloud will have a number of effects. These include:

(1) mechanical effects of acting as a barrier to the passage of gas, of imparting upward momentum to a gas or of dispersion and dilution by air entrainment;
(2) thermal effects by warming of cold gas;
(3) physico-chemical effects of absorption of gas, without or with chemical reaction.

A water spray barrier may be used for any of these purposes.

Where a water spray barrier is used, consideration should be given to the need for arrangements to drain away the water used, particularly if the application may be prolonged.

15.53.7 Water spray barriers: early work

An experimental investigation of the effectiveness of water sprays against gas clouds has been described by Eggleston, Herrera and Pish (1976). The experimental rig was a set of water spray nozzles mounted on a frame initially 15 ft high, later modified to 20 ft, and directed downwards. Clouds of ethylene and vinyl chloride monomer vapour were directed towards the spray system at ground level.

The water spray was effective as a vapour barrier in some experiments but not in others. The spray entrains air and pumps it down and horizontally outwards. If the velocity of this air exceeds that of the vapour cloud, the spray acts as a barrier, otherwise it does not. Thus vapour clouds with velocities of 3 and 9 mile/h were stopped by sprays with 100 psig nozzle pressure, but only the former was stopped when the nozzle pressure was reduced to 40 psig. The vapour tended to spread out at ground level in front of the spray, but did not go over the top.

The air entrainment rates obtained were in the range 87–213 ft^3/min air per US gal/min water with nozzle pressures in the range 20–90 psig. Total entrained air rates were in the range 1242–7107 ft^3/min. The velocity of the air pumped by the spray was low and the flow was therefore sensitive to any obstruction. In cases where the vapour cloud passed through the spray barrier there was still an appreciable dilution effect. The gases used are highly insoluble and the absorption effects were negligible.

The effect of the water spray on the ignition of a flammable vapour cloud and on the flame speed in the cloud was also investigated. It was found that the water spray did not prevent ignition occurring and that the flame speed actually increased, probably due to the greater turbulence.

Further experimental work on water sprays has been described by J.W. Watts (1976). Again the experimental rig was a set of nozzles mounted on a 20 ft high frame and directed downwards. Clouds of propane vapour were directed towards the spray system at ground level.

Some experiments were done in which ignition sources in the form of cans of burning gasoline were located on the far side of the sprays. In cases where the spray was not an effective barrier the vapour cloud ignited and the flame passed back through the spray.

Further experiments were carried out in a 13 ft high open rig with a single nozzle with the introduction of ethylene vapour through a pipe just below the nozzle. The flow of ethylene was increased until the concentration measured by gas detectors near the ground rose to

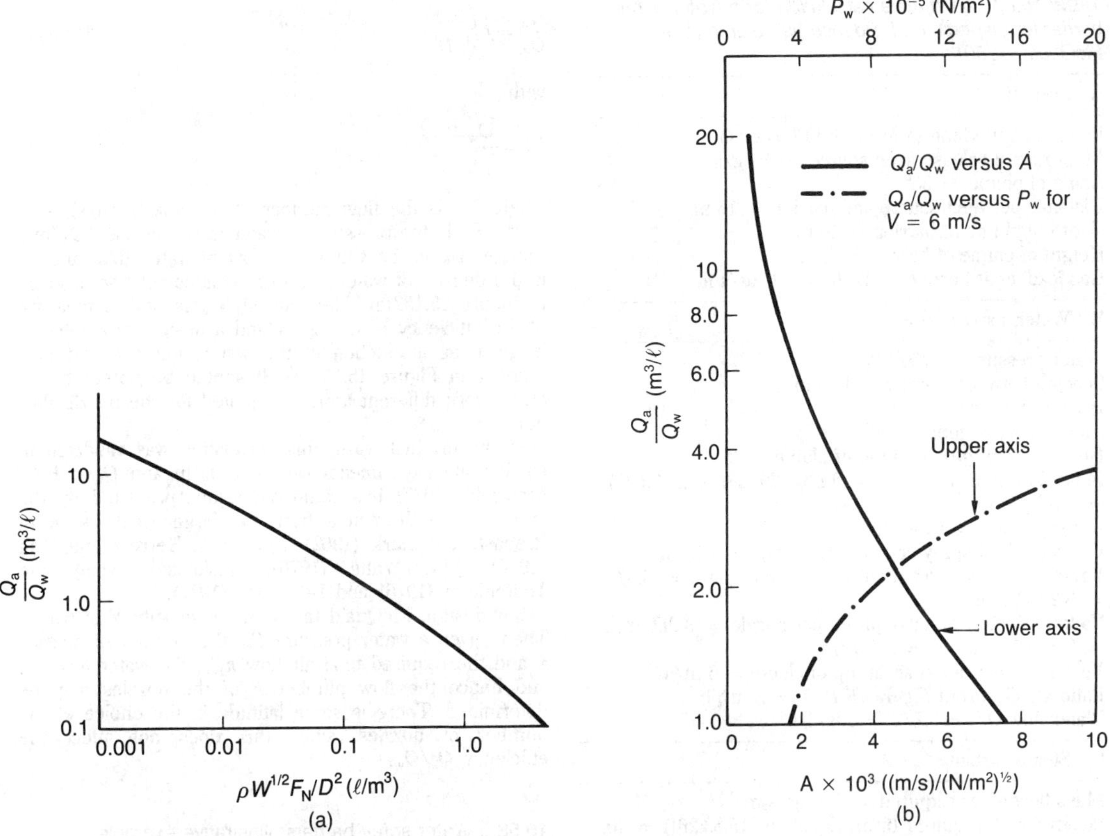

Figure 15.157 *Mitigation of gas dispersion by water spray barriers – water spray nozzle relationships (McQuaid, 1977): (a) and (b) air entrainment relationships.* $A = \pi V/4P_w^{\frac{1}{2}}$ *(Courtesy of HM Stationery Office)*

just below the lower flammability limit. The maximum flow of ethylene controllable by the spray was 2609 lb/h with a calculated air entrainment rate of 23 600 ft³/min.

Other experiments were done in which volumes of ethylene–air mixture held in polyethylene chambers were ignited by pentaerithrytol tetranitrate (PETN) and observations were made of the conditions under which deflagration and detonation were obtained. The volumes used were 500, 4000 and 10 000 ft³. The authors state that given a sufficient charge of PETN detonation could be obtained, that the explosion yield was generally much higher with detonation than with deflagration, that the detonation limits were 5–10.5% and were thus narrower than the flammability limits, and that with a water spray in the chamber the strength of the PETN charge necessary to obtain detonation was considerably increased.

The effectiveness of water sprays in suppressing combustion in mists of heat transfer media, specifically Dowtherm A, has been investigated by Vincent *et al.* (1976a) in a 6.5 ft diameter spherical vessel. It was shown that combustion could be suppressed and it was tentatively concluded that, for such mist droplets, the mechanism was one of scrubbing rather than quenching.

Although in Watts' work water sprays were not effective in arresting the passage of flames, Brasie (1976b) has described work at the Bureau of Mines in which flames in gas clouds were quenched by the use of very high nozzle pressures in the range 1000–2000 psig, which give a much finer spray. In this case the mechanism was considered to be one of quenching, the distance between the drops being less than the quenching distance of the gas.

Experiments on the use of a water curtain generated by mobile monitors against a chlorine gas cloud have been described by the Ministry of Social Affairs (1975) in the Netherlands. A water curtain was successfully used to protect a small area against the gas cloud.

15.53.8 Water spray barriers: air entrainment

The design and application of a water spray barrier to entrain air into a gas cloud has been described in a series of studies by McQuaid. These include: the entrainment characteristics of a water spray barrier (McQuaid, 1975); the use of a water spray barrier to effect ventilation in mines (McQuaid, 1976a); the design of a water spray barrier to dilute flammable gas clouds (McQuaid, 1977); a comparative review of water spray barriers and steam curtains for dilution of flammable gas

***Table 15.67** Illustrative calculation for a water spray barrier for dispersion of a dense gas plume (after McQuaid, 1980)*

A Scenario
Release of ethylene (MW = 28; LFL = 2.7%)
Mass rate of release from source = 1 kg/s
Angle of plume = 35°
Distance between source and barrier = 16 m
Width of plume at barrier = 10 m
Height of plume at barrier = 1 m
Gas loading at barrier = 1/10 = 0.1 kg/s m
B Water spray barrier
Water pressure = 700 kPa
Spacing between nozzles = 1.15 m
Height of nozzles = 2m
Angle of spray cone = 60°
Distance from nozzle to top of plume = 2 − 1 = 1 m
Mass flow of air required = 0.1 × (29/28) × (1/0.027) = 3.8 kg/s m
Density of air = 1.2 kg/m^3
Diameter D of spray cone at plume top = 1.15 m
Volumetric flow of air required per unit length = 3.8/1.2 = 3.2 m^3/s m
Volumetric flow of air required per nozzle = 3.2D m^3/s = 3.2 × 1.15 m^3/s = 3.7 m^3/s
Velocity of entrained air at top of cloud = 6 m/s
Ratio Q_a/Q_w (from *Figure 15.157*) = 2 m^3/l
Water flow Q_w = 3.7/2 = 1.85 l/s
C Steam curtain
Mass flow of air required = 3.8 kg/s m
Steam flow S required (from Equation [15.53.26]) = (0.1 × 2 × 3.8)/200 = 0.0385 kg/s m

clouds (McQuaid, 1980); and the application to dense gas clouds. These aspects are described below.

An account of work on the entrainment of air by a water spray barrier has been given by McQuaid (1975), who has also described the application of this work to the design of water spray barriers for dispersion of gas clouds (McQuaid, 1977).

In order to dilute a gas cloud it is necessary to deliver air not only in sufficient quantity but also at a sufficient velocity. The relation for the flow of air Q_a and the air velocity V at the base of the cone of spray cone with base diameter D is

$$D = \left(\frac{4Q_a}{\pi V}\right)^{\frac{1}{2}} \quad [15.53.1]$$

where D is diameter of the spray cone (m), Q_a is the volumetric flow of air (m^3/s) and V is the velocity of the entrained air (m/s).

McQuaid obtained the following correlation for entrainment of air into a spray nozzle:

$$\frac{Q_a}{Q_w} = f\left(\frac{\rho_w^{\frac{1}{2}} F_N}{D^2}\right) \quad [15.53.2]$$

with

$$F_N = \frac{Q_w}{P_w^{\frac{1}{2}}} \quad [15.53.3]$$

where F_N is the flow number of the nozzle ((l/s)·(kN/m^2)$^{\frac{1}{2}}$), P_w is the pressure of water at the nozzle (kN/m^2) (gauge) Q_w is the volumetric flow of water (l/s) and ρ_w is the density of water (kg/m^3). This correlation is given in Figure 15.157(a). The ratio Q_a/Q_w is both a measure of the efficiency of the spray and a design parameter. It is given as a function of the water pressure and flow number in Figure 15.157(b). It should be noted that in other work different units are quoted for the nozzle flow number.

In the original work, this correlation was validated in small scale experiments with $D = 0.3$ m and $Q_w < 1$ l/s. McQuaid (1977) has compared values predicted by the correlation with results from the larger scale work of Rasbash and Stark (1962), Eggleston, Herrera and Pish (1976), J.W. Watts (1976), Heskestad, Kung and Todtenkopf (1976) and Beresford (1977).

For design, McQuaid takes an air velocity V of 6 m/s. Then, given a water pressure P_w, the number of nozzles n and the required total air flow nQ_a, the water flow Q_w and hence the flow number F of the nozzles may be determined. There is some latitude in the choice of the number of nozzles, since this does not affect the efficiency Q_a/Q_w.

15.53.9 Water spray barriers: illustrative example

An illustrative example of the design of a water spray barrier is given by McQuaid (1980). This example is summarized in Table 15.67. The problem considered is the design of a barrier for the dilution of a continuous release of ethylene gas escaping at a rate of 1 kg/s and forming a plume with an angle of 35° and a height of 1 m. The table also treats the case of a steam curtain, as described below.

15.53.10 Water spray barriers: Moore and Rees model

P.A.C. Moore and Rees (1981) have given a model, or rather a set of models, for the forced dispersion of gases by water or steam curtains. The model is applied by the authors to dilute gas plumes and to dense gas plumes. The situation considered is that shown in Figure 15.158.

Following Bosanquet (1957), the authors consider an elemental slice of the gas cloud

$$\frac{d(\delta V)}{dt} = cSu \quad [15.53.4]$$

where c is an entrainment parameter for the wind, S is the surface area of an elemental slice, u is the velocity of the gas and δV is the volume of the slice.

It is assumed that the cloud velocity u is equal to the wind speed w:

$$u = w \quad [15.53.5]$$

Also

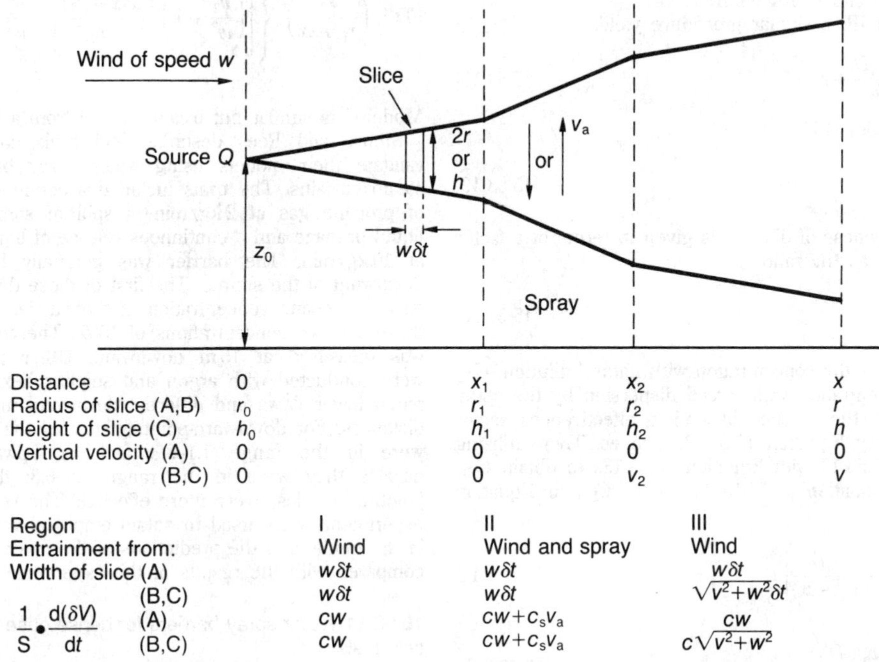

***Figure 15.158** Mitigation of gas dispersion by water spray barriers – elevation view of dense gas cloud passing through a water spray barrier (P.A.C. Moore and Rees, 1981) (Courtesy of the Institution of Chemical Engineers)*

$$\frac{dx}{dt} = w \qquad [15.53.6]$$

where x is the distance along the plume.

The first model (model A) is for a downward pointing spray. The three regions (I–III) are considered in turn. For region I, the source has a volumetric rate of release Q. In time δt a volume $Q\delta t$ is emitted which at distance x is contained within the volume δV with radius r:

$$\delta V = \pi r^2 w \delta t \qquad [15.53.7]$$

The area S is

$$S = 2\pi r w \delta t \qquad [15.53.8]$$

By conservation of mass

$$Q\delta t = \pi r_o^2 w \delta t \qquad [15.53.9]$$

$$= C\pi r^2 w \delta t \qquad [15.53.10]$$

where C is the volumetric concentration and subscript o denotes the source.

From Equations 15.53.4, 15.53.5, 15.53.7 and 15.53.8

$$\frac{d(r^2)}{dt} = 2crw \qquad [15.53.11]$$

Hence from Equations 15.53.6 and 15.53.11

$$\frac{dr}{dx} = c \qquad [15.53.12]$$

Integrating Equation 15.53.12 with respect to x over region I and at the boundary of the region

$$r = r_o + cx \qquad [15.53.13a]$$

$$r_1 = r_o + cx_1 \qquad [15.53.13b]$$

where subscript 1 denotes the end of region I.

In region II

$$\frac{d(\delta V)}{dt} = c \cdot 2\pi r w \delta t \cdot w + c_s \cdot 2\pi r w \delta t \cdot v_a \qquad [15.53.14]$$

where c_s is an entrainment parameter for the spray and v_a is the velocity of the air entrained in the spray. In this equation the first term on the right-hand side represents the entrainment of air by the wind and the second term represents the entrainment by the spray. Proceeding as before yields

$$\frac{dr}{dx} = c + \frac{c_s v_a}{w} \qquad [15.53.15]$$

and integrating gives

$$r = r_1 + \left(c + \frac{c_s v_a}{w}\right)x - x_1 \qquad [15.53.16a]$$

$$r_2 = r_1 + \left(c + \frac{c_s v_a}{w}\right)D \qquad [15.53.16b]$$

with

$$D = x_2 - x_1 \qquad [15.53.17]$$

where D is the width of the spray and subscript 2 denotes the end of region II.

In region III, a similar procedure yields

$$r = r_o + \frac{c_s v_a D}{w} + cx \qquad [15.53.18]$$

From Equation 15.53.10

$$C = \frac{Q}{\pi r^2 w} \qquad [15.53.19]$$

The effectiveness of dilution is given in terms of a factor FD defined as the ratio

$$\mathrm{FD} = \frac{C_{\mathrm{ND}}}{C_{\mathrm{FD}}} \qquad [15.53.20]$$

where C_{ND} is the concentration with natural dilution, C_{FD} is the concentration with forced dispersion by the spray and FD is the forced dispersion effectiveness factor without spray Equation 15.53.13a applies. Then utilizing Equation 15.53.19 with Equation 15.53.13a to obtain C_{ND} and with Equation 15.53.18 to obtain C_{FD} in Equation 15.53.20:

$$\mathrm{FD} = \left[1 + \frac{c_s v_a D}{w(r_o + cx)}\right]^2 \qquad [15.53.21a]$$

$$= \left(1 + \frac{c_s v_a D}{cwx}\right)^2 \qquad cx \gg r_o \qquad [15.53.21b]$$

Bosanquet gives a value of $c = 0.13$. The authors also derive from Katan's formula a value of $c = 0.14$.

The two other models (models B and C) are for an upward pointing spray. Taking model B, it is assumed that in region II the elemental slice gains a vertical velocity v in the spray and that in region III the horizontal velocity is given by

$$u = (w^2 + v^2)^{\frac{1}{2}} \qquad [15.53.22]$$

This model contains an additional equation for dv/dx. The result for FD is

$$\mathrm{FD} = \left(\frac{r_2}{r_1 + cx}\right)^2 \left\{\left[\left(\frac{v_2^2}{w^2} + 1\right)^{\frac{1}{2}} + \frac{c(x - x_2)}{r_2}\right]^4 - \frac{v_2^2}{w^2}\left(\frac{v_2^2}{w^2} + 1\right)\right\}^{\frac{1}{2}} \qquad [15.53.23]$$

Model C is similar but treats a cloud from a line source.

Moore and Rees describe field trials conducted to validate their models using water spray barriers and steam curtains. The trials included a continuous release of propane gas at 2 kg/min, a spill of some 60 kg of liquid propane and a continuous release of liquid propane at 10 kg/min. The barrier was generally located 1 m downwind of the source. The first of these three sources gave a propane concentration at the barrier of 10% and the other two concentrations of 100%. The concentration was measured at 15 m downwind. Other experiments were conducted with argon and sulphur hexafluoride at much lower flows and different downwind measurement distances. For downward-pointing nozzles the FD factors were in the range 1.1–3.3 and for upward-pointing nozzles they were in the range 2.5–6.3; the upward-pointing nozzles were more effective. The results of the experiments were used to obtain empirical values of c_s/c in the model and the predictions of the model were then compared with the results of the experiments.

15.53.11 Water spray barriers for dense gases: HSE field trials

Experimental work by the HSE on the dispersion of a dense gas plume from a continuous source has been described by Moodie (1981, 1985) and McQuaid and Moodie (1983). Three configurations of water spray barrier were used: sprays pointing vertically downward, sprays pointing downward but angled forward at 45° and sprays pointing vertically upward. The downward-pointing nozzles were at a height of 3 m and were spaced at 0.33 m intervals along a 17 m pipe; the upward-pointing nozzles were at the same interval on an 18 m long pipe. Solid cone, hollow cone and flat fan sprays were investigated. The distance between the barrier and the source was varied, one distance quoted being 5 m. The gas released was carbon dioxide at release rates of 2 and

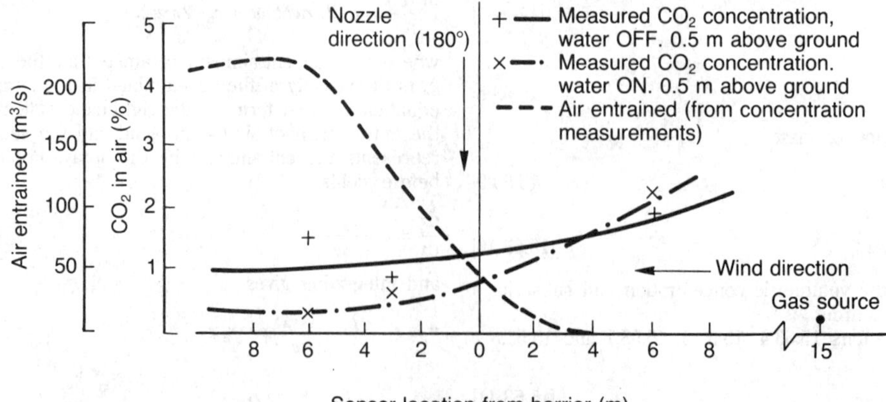

Figure 15.159 *Mitigation of gas dispersion by water spray barriers – variation of entrained air and concentration of carbon dioxide across barrier in HSE trials (Moodie, 1981) (Courtesy of the Institution of Chemical Engineers)*

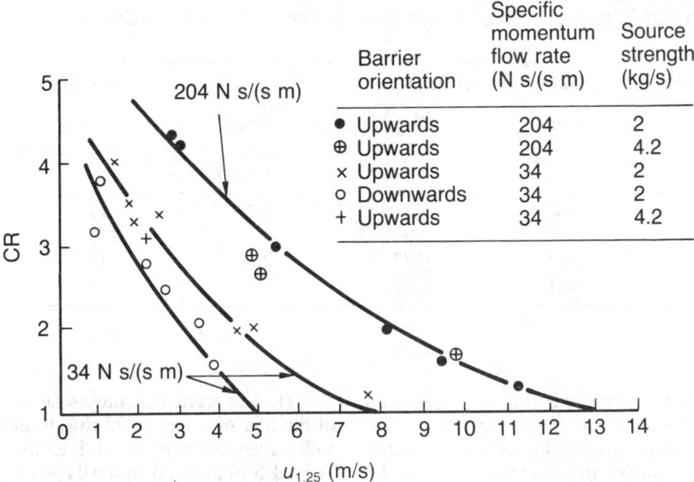

Figure 15.160 *Mitigation of gas dispersion by water spray barriers – effect of wind speed and specific momentum on concentration ratio (CR) for carbon dioxide in HSE trials (Moodie, 1985). Flat fan nozzles spaced 1 m apart each with momentum flow number $M_N = 4.9$ N/(s bar) (Courtesy of the Institution of Chemical Engineers)*

4.2 kg/s in wind speeds of 0.6–11 m/s and with water consumptions at the barrier of 0.4–7.6 l/(s m). Wind speeds were measured at, and quoted for, 1.25 m.

Figure 15.159 shows the variation of the entrained air and of the concentration of carbon dioxide across the barrier. The results are correlated in terms of the momentum of the spray. Moodie (1985) defines a momentum flow number M_N as

$$M_N = KF_N \quad [15.53.24]$$

where F_N is the flow number (l/(min bar$^{\frac{1}{2}}$)), M_N is the momentum flow number (N s/(s bar)) and K is a constant. The value of the constant K is 0.236. Use is also made of the momentum flow M of the barrier (N s/(s m)).

The results are presented as a concentration ratio CR, defined as the ratio of the concentration with the water sprays off to that with the sprays on. Figure 15.160 shows the variation of CR with wind speed with specific momentum as the parameter. Values of CR are mainly in the range 1–4. The ratio increases with the momentum of the spray and decreases with wind speed.

Compared with the sprays pointing vertically downwards, those angled at 45° showed no apparent improvement in performance, but the upwards-pointing sprays did, being some 17% more effective at a wind speed of 2 m/s.

For upwards-pointing barriers only, and for the range of parameters studied, the authors give the following correlation:

$$M = u_{1.25}^2[0.65 \exp(0.5CR) - 0.23]^2 \quad [15.53.25]$$

where M is the specific momentum flow (N s/(s m)) and $u_{1.25}$ is the wind speed at 1.25 m height (m/s).

McQuaid and Moodie (1983) have reviewed the implications of this work for the design of a water spray barrier. The barrier should be located near the source, though not so near as to promote increased evaporation of any liquid pool. A barrier near the source gives a considerable reduction in gas in the first few tens of metres downwind, but this effect decays quite rapidly. It is most effective at low wind speeds (<3–4 m/s).

Water consumption at a water pressure of 700 kPa is about 1.6 l/(s m). A standard fire tender giving 60 l/s can thus supply a 35 m length of barrier.

15.53.12 Water spray barriers for dense gases: HSE modelling studies

McQuaid and Fitzpatrick (1981, 1983) of the HSE have described the modelling of the dispersion of a dense gas plume by a water spray barrier. One question addressed was the extent to which the effect of the water spray barrier falls off with downwind distance. For a dense gas cloud the effect of the density difference is to increase the area of the cloud top by lateral spreading, but to reduce vertical mixing. As far as concerns entrainment of air, the former effect tends to outweigh the latter so that the net effect is to increase entrainment and to shorten the length of the plume to a given concentration. A water spray barrier gives an addition of air and hence an immediate dilution, but it also gives a reduction of density and hence in lateral spread. Another question considered was the effect of the distance between the source and the barrier. If a barrier is close to the source it can be made smaller, but it will reduce the density of the cloud just at the point where this density is most effective in promoting entrainment.

The dispersion of a dense gas from a continuous release was modelled using CRUNCH. The experimental work showed that the entrained air mixes with the gas cloud within a distance roughly equal to the width of the spray. This effect was therefore modelled by assuming that the entrainment of the air at the barrier effects a change both in the geometry of the cloud and in its

Table 15.68 Predictions of CRUNCH of effect of water spray barrier on plume from continuous release of carbon dioxide (after McQuaid and Fitzpatrick, 1983). Air entrainment rate = 3.6 kg/s m (Courtesy of the Institution of Chemical Engineers)

Barrier downwind distance (m)	*Pasquill stability category*	*Plume dimensions* (m)			*Air flow* (kg/s) *due to*	
		Width	*Height before barrier*	*Height after barrier*	*Plume entrainment*	*Spray barrier*
11	D	14.4	1.2	2.3	27	53
30	D	25.6	2.5	3.4	172	93
11	F	38.2	0.75	2.94	15	143
30	F	88.3	0.66	2.87	30	319

concentration. The plume modelled was a continuous release of 2 kg/s of carbon dioxide, representative of the sources used in the experiments of Moodie (1981). Some predictions given by the model are shown in Table 15.68 and Figures 15.161(a)–(d), which illustrate, respectively, the effect on plume height and concentration of the stability category, the wind speed, the downwind distance of the barrier and the specific air entrainment rate. The plume widths at the barrier vary, but it is assumed that they are not altered by passage through the barrier and that the only effect of this is to alter plume height. The results show that a water spray barrier gives a higher dilution if located close to the source, that a barrier with high specific air entrainment rates gives higher dilutions and that a barrier is particularly effective in stability category F conditions. The large air flows in this case are due to the large plume widths. The results also show that the effect of the barrier decays and that some distance downwind it disappears altogether.

As just described, the assumption in Figures 15.161(a)–(d) is that the effect of the barrier is to change the height of the plume but to leave its width unaltered. Figures 15.161(e)–(f) illustrate the results of the alternative assumption that it is the plume width which changes whilst the height remains unaltered. They show that it is much more effective to change the plume width. Quite how as a practical matter this may be effected is not obvious, but the general message is clear that any outward movement which can be imparted to the plume is highly beneficial.

These HSE field trials have also been modelled by Deaves (1983b, 1984) using HEAVYGAS. Figure 15.162 shows a comparison of the observed and predicted concentration profiles in one experiment. Other results from the model show that the gas, which had been hugging the ground, is lifted up and mixed by an upwind recirculation of the entrained air.

15.53.13 Water spray barriers for dense gases: Goldfish and Hawk trials

The Goldfish field trials at the NTS in 1986 on the dispersion of hydrogen fluoride included three trials involving the use of forced dispersion barriers. The main natural dispersion trials (trials 1–3) were described in Section 15.37 and the forced dispersion trials (trials 4–6) are described here. Accounts of these latter trials have been given by Blewitt, Yohn, Koopman and Brown (1987) and Blewitt, Yohn, Koopman, Brown and Hague (1987). The work constitutes an investigation of the case of dilution of a gas cloud due to absorption of the gas as well as entrainment of air into the cloud.

Trial 5 utilized an upwards-pointing water spray barrier and trial 6 a downwards-pointing one. A barrier with the water spray atomized by air was used in trial 4. The general conditions of the trials were described in Section 15.37. In trial 5 the flow of liquid hydrogen fluoride was 32.5 USgal/min with a wind speed of 3.8 m/s and stability category C/D, and trial 6 was essentially similar but with a wind speed of 5.4 m/s. The ambient temperature was 21°C in all three tests. In none of these three tests was any liquid collected on the spill pad.

The interpretation of trials 5 and 6 was complicated by shifts in wind direction during both trials, but the use of mass balances as well as concentration measurements allowed an estimate to be made of the proportion of hydrogen fluoride removed. The reduction in maximum hydrogen fluoride concentration 300 m downwind of the barrier was estimated as between 36% and 49%; the concentration at this point in trial 5 before the spray was turned on was 2028 ppm. The atomized water spray used in trial 4 was not effective in the form tested.

As already mentioned, a further set of field trials, the Hawk series, was performed at the NTS in 1988 under the auspices of ICHMAP. The gas dispersion aspect of these trials was considered in Section 15.37 and the vapour barrier work was described in Section 15.53.4. An account of the work on water spray barriers is given here.

A total of 87 trials were performed to investigate the effect of various parameters on the removal of hydrogen fluoride by a water spray barrier. The water/HF liquid volume ratio was found to be the dominant factor. Hydrogen fluoride removal efficiencies of 25% to $\geq$90% were achieved at water/HF liquid volume ratios of 6:1 to 40:1.

Of the other factors, the liquid droplet size also had some influence. Upward-pointing sprays were rather more effective than downward-pointing ones. The pressure and temperature of the hydrogen fluoride had little effect and the same applied to the atmospheric humidity and wind speed. Little difference was observed in removal efficiency of the spray as between hydrogen fluoride and alkylation unit acid.

Trials were done with the water augmented by various additives, including sodium bicarbonate, sodium hydroxide, calcium chloride and surfactant, but the only benefit

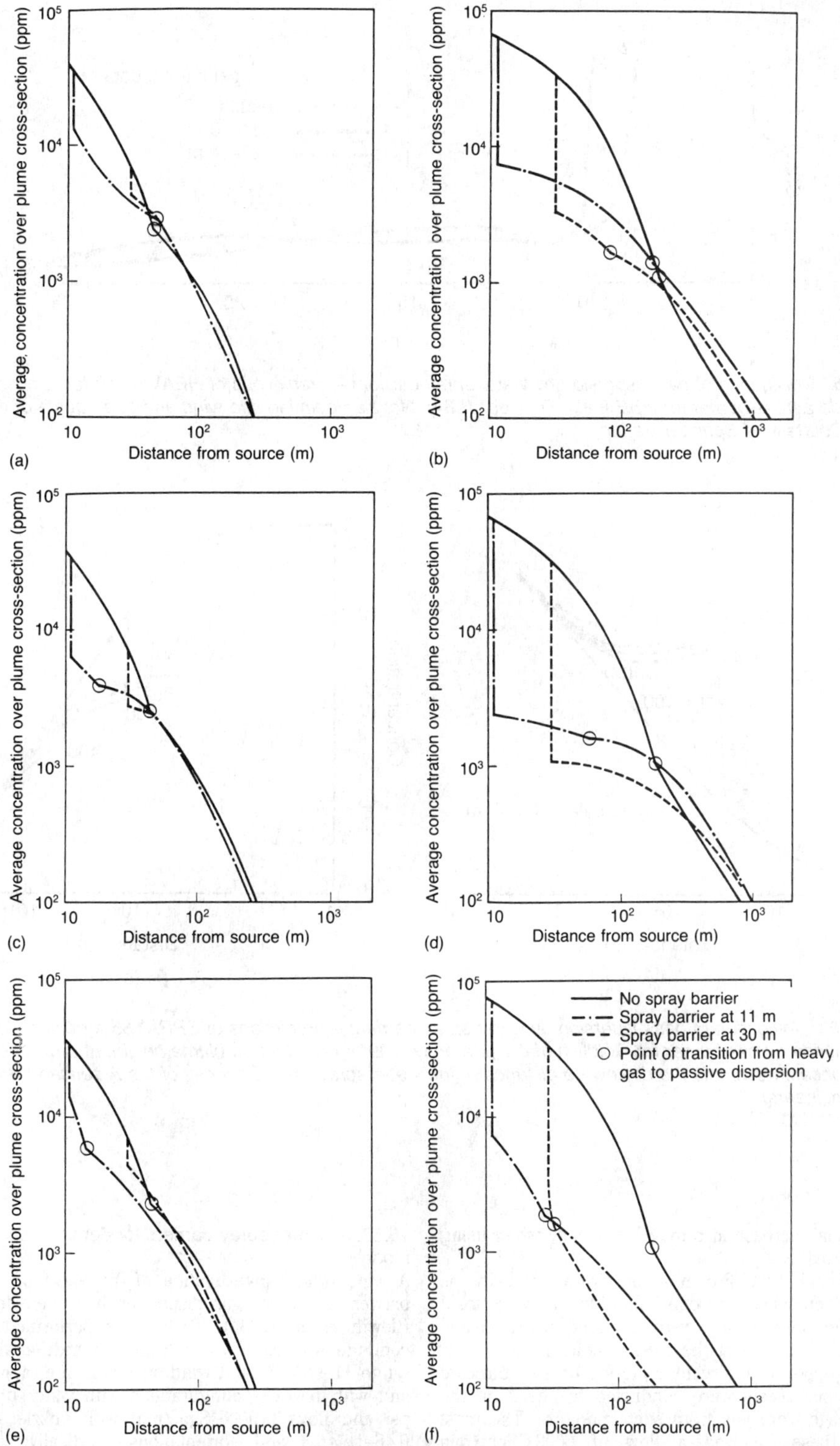

Figure 15.161 *Mitigation of gas dispersion by water spray barriers – predictions of CRUNCH for concentrations of carbon dioxide downwind of source for trials (McQuaid and Fitzpatrick, 1981): (a) air flow at barrier = 3.6 kg/s m, weather conditions D/4.3; (b) air flow at barrier = 3.6 kg/s m, weather conditions F/2; (c) air flow at barrier = 12 kg/s m, weather conditions D/4.3; (d) air flow at barrier = 12 kg/s m, weather conditions F/2; (e) air flow at barrier = 3.6 kg/ s m, weather conditions D/4.3; (f) air flow at barrier = 3.6 kg/s m, weather conditions F/2. Plume width maintained constant through barrier in (a)–(d) and plume height maintained constant in (e)–(f). The key to all six figures is given in graph (f). (Courtesy of the Institution of Chemical Engineers)*

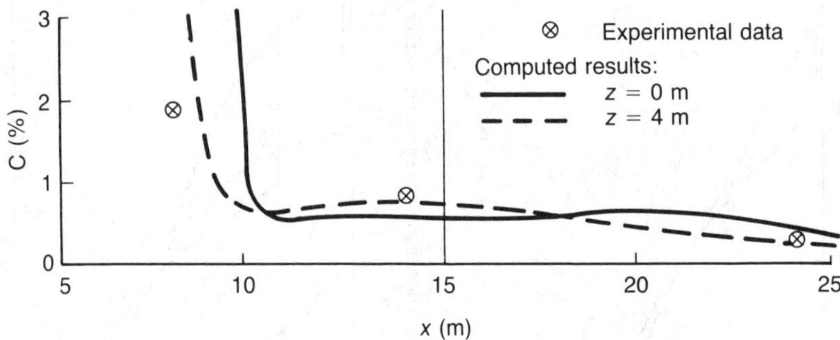

Figure 15.162 Mitigation of gas dispersion by water spray barriers – predictions of HEAVYGAS for concentrations of carbon dioxide across barrier for HSE trials (Deaves, 1984). Nozzles pointing into wind and at an angle of 45° to horizontal. (Courtesy of Springer Verlag)

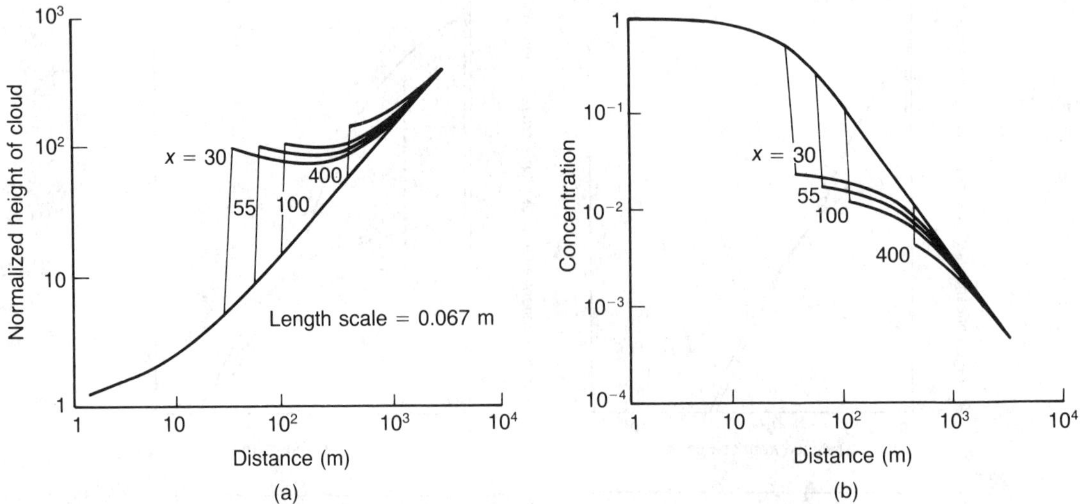

Figure 15.163 Mitigation of gas dispersion by water spray barriers – predictions of SPRAY65 for dispersion of hydrogen fluoride across barrier for Goldfish trial 1 (Meroney, 1991): (a) effect on plume height; and (b) effect on centreline concentration. x is the downward distance of the water spray (m). (Courtesy of the American Institute of Chemical Engineers)

was a marginal increase in removal efficiency when using sodium hydroxide.

It was found that fire monitors were virtually as effective in removing hydrogen fluoride as were fixed barriers. The best results were obtained using coarse droplets aimed directly at the release point from a short distance, as opposed to a wide angle fog or a jet directed from a distance. Tests were conducted to check on the risk of damage from jets from such monitors. The most severe of these involved a flow of 6000 USgal/min applied as a narrow jet at a distance of 7 ft from the target; damage was minimal.

15.53.14 Water spray barriers for dense gases: physical modelling

A wind tunnel investigation of the effect of water spray barriers on dense gas dispersion has been described by Blewitt *et al.* (1991). Tests were performed simulating continuous releases of hydrogen fluoride at flows in the range 11–96 kg/s and wind speeds in the range 3–10 m/s and with barrier configurations with water pressure 172 psi and flow 33 000 USgal/min with nozzle spacings of 0.46–1.83 m and orientations vertically downwards, angled, horizontal and vertically upwards. The most effective dilution was obtained with nozzles directed horizontally against the wind. It was possible to obtain good mixing with nozzles spaced as far apart as 1.83 m, the widest spacing studied. The authors discuss the principles of scaling water sprays in wind tunnel work.

15.53.15 Water spray barriers for dense gases: SPRAY65

The work of Meroney (1991) on the use of models of dense gas dispersion at vapour barriers was mentioned earlier. He has also reported work on models for dense gas dispersion at water spray barriers using the model SPRAY65.

This model was used to simulate the Goldfish field trials on the dispersion of a continuous release of hydrogen fluoride by a water spray barrier. Figure 15.163(a) shows for Goldfish trial 1 the effect of the water spray barrier location on the height of the cloud, and Figure 15.163(b) shows the effect on the downwind concentration. The water spray barrier is more effective in diluting the gas cloud if located near the source. The effects decay with distance and eventually disappear.

15.53.16 Water spray barriers: leak containment

The use of a water spray barrier by a works fire brigade to deal with escapes of gas has been described by Beresford (1981). The approach adopted is to hold the gas cloud inside a 'chimney' formed by a set of four mobile water spray barriers. These form a barrier to the passage of the gas, entrain air and dilute the gas, and impart to it upwards momentum so that it rises and disperses harmlessly. It is possible to see inside the chimney and for personnel to enter to effect valve isolation.

Two monitors are placed some 30 ft apart and arranged to give an inverted V some 40–50 ft above ground. At this point the water spray feathers and heavy drops fall to ground. Two fan-shaped sprays are set at right angles to the monitor curtains to give a vertical water curtain some 40 ft wide × 40 ft high. The overall effect is to surround the gas with an upward-moving wall of water so creating the chimney effect. The gas disperses some 50 ft or more above ground. The monitors are 250–1200 UKgal/min and the fan sprays 180–200 UKgal/min devices and the water pressure used is 150 psig.

This technique is suitable for quite serious leaks but not for catastrophic failures. It may be used for flammable or toxic releases. It requires that the equipment be set up rapidly. Ignition of a flammable vapour cloud, for example, is liable to occur within 3–5 minutes of the start of the leak. It is necessary to consider the drainage of the water if application is likely to be prolonged.

Beresford gives examples of a number of escapes handled by this method. They include a leak of propylene from a 2 in. diameter drain line at 400 psig discharging at a rate of about 100 te/h which could not be isolated for several hours; one of ethylene gas from a flanged joint on a 14 in. diameter pipe at 500 psig which lasted some 50 minutes; one of methane escaping at about 50 te/h for 30 minutes from a fracture on a 6 in. diameter pipe at 60 psig; and various other leaks of benzene, ethylene, ethylene oxide, naphtha and hydrocarbon solvents.

15.53.17 Water spray barriers: enclosed spaces

Work on the use of water sprays to promote entrainment into, or act as a barrier to, instantaneous and continuous leaks of gas in an enclosed space has been described by van Doorn and Smith (van Doorn and Smith, 1980; J.M. Smith and van Doorn, 1981).

15.53.18 Steam curtains

A steam curtain is used as a permanent installation to contain and disperse leaks of flammable gas heavier than air. A steam curtain system has been described by Cairney and Cude (1971) and Simpson (1974).

The arrangement is that the plant is surrounded by a solid but lightly constructed wall. A horizontal steam pipe with a row of small holes is mounted near the top of the wall. The pipe is divided into sections which are individually supplied with steam from distribution mains with remotely controlled, quick-acting valves. The associated monitoring of leaks is carried out by the flammable gas detector system, as described in Chapter 16. In the system described by the authors mentioned, Sieger diffusion-type detectors are used.

The wall constitutes an initial barrier to the cloud and thus serves both to provide a time delay during which the detection system can operate and to spread the gas along the length of the curtain. The relative spacing of the potential sources of escape, the detectors and the wall are chosen with these factors in mind. The wall is 1.5 m high, which is sufficient to check a heavy vapour cloud, but does not prevent dispersal of small leaks by natural air movement.

The steam pipe is designed so that the individual jets combine to form a planar jet which entrains sufficient air to dilute the vapour cloud below its lower flammability limit. The division of the pipe into sections allows steam to be supplied only to those sections near the leak. The activation of the system is controlled by the process operators.

The steam curtain was designed to dilute heavy flammable vapours. It is possible in principle to apply it to the dispersion of toxic gases, but in this case it is necessary to effect dilution to much lower concentrations.

The use of steam jets, particularly jets containing wet steam, can create a hazard of ignition by static electricity. Therefore particular care is taken that all pipework and other metal objects near the jets are properly earthed.

A method for the design of a steam curtain has been given by McQuaid (1980). The quantity of steam required is determined from the relation

$$S = \frac{G + 2Q_a}{200} \qquad [15.53.26]$$

where G is the mass flow of gas, Q_a is the mass flow of air and S is the mass flow of steam. The equation is based on conservation of momentum, a wind velocity of 2 m/s and a steam velocity equal to the velocity of sound in steam of 400 m/s. The illustrative example given in Table 15.68 also covers the case of a steam curtain.

P.A.C. Moore and Rees (1981), in the work already described, also conducted experiments on steam curtains and obtained empirical correlations for the FD value.

The generation of static electricity in steam curtains and in water curtains has been investigated by Napier (1974b). He concluded that a degree of static electricity hazard exists both with steam and with water curtains.

Table 15.69 *Some principal topics treated in the CCPS* Vapor Release Mitigation Guidelines *and their treatment in this text*

	Chapter
1. Overview	
2. Inherently Safer Design	11
3. Engineering design	
3.1 Plant Integrity	
3.1.1 Design practices	11, 12
3.1.2 Materials of construction	12
3.1.3 Inspection and testing (in construction)	12, 19
3.2 Process integrity	
3.2.1 Identification of process materials	20
3.2.2 Process operating limits	11, 20
3.2.3 Process control system	13
3.2.4 Pressure relief system	12
3.3 Emergency control	
3.3.1 Emergency relief disposal	11, 12
3.3.2 Emergency abort systems	12
3.3.3 Emergency isolation systems	12
3.3.4 Emergency material transfer	12
3.3.5 Spill control	15, 23
4. Process Safety Management	6
5. Early Vapour Detection and Warning	
5.1 Detectors and sensors	13, 16, 18
5.2 Detection by personnel	20
5.3 Alarm systems	13, 24
6. Countermeasures	
6.1 Vapour-liquid release	
6.2 Vapour release countermeasures	
6.2.1 Water sprays	15
6.2.2 Water curtains	15
6.2.3 Steam curtains	15
6.2.4 Air curtains	–
6.2.5 Deliberate ignition	17
6.2.6 Ignition source control	10, 16
6.3 Liquid release countermeasures	
6.3.1 Dilution	
6.3.2 Neutralization	
6.3.3 Covers	15, 16
6.3.4 Avoidance of spill aggravation	15, 16
7. On-site Emergency Response	24
8. Off-site Emergency Response	4, 24
9. Selection of Measures	9

15.53.19 Water spray barriers vs steam curtains

Several authors have compared the use of water spray barriers and steam curtains. A detailed comparison of the two devices has been made by McQuaid (1980), as illustrated in Table 15.68. For the water spray barrier the water flow required is 1.6 kg/(s m) and the steam flow is 0.0385 kg/(s m). The ratio of the required mass of water to that of steam is therefore some 40 : 1.

P.A.C. Moore and Rees (1981) state that in their experimental work the FD value for steam curtains was less than 4, whereas for upwards-pointing water spray barriers it was in the range 4–16.

15.54 Vapour Cloud Mitigation: *CCPS Guidelines*

The CCPS has published guidance on the mitigation of vapour clouds and this is now described.

15.54.1 *Guidelines for Vapor Release Mitigation*

The CCPS *Guidelines for Vapor Release Mitigation* (the CCPS *Vapor Release Mitigation Guidelines*) (1988/4) cover a wide range of approaches to mitigation. Prugh (1985–, 1987b,c) has given a number of accounts of vapour release mitigation which foreshadow the *Guidelines*.

Table 15.70 Selected references on fugitive emissions

R.K. Palmer (1957); Anon. (1958); Steigerwald (1958); API (1959 Bull. 2513, 1962 Bull. 2516, 2518, 1967 Bull. 2522, 2533, 1980 Publ. 2517, 4322, 1981 Publ. 2514A, 1983 Publ. 2519); Schwaneke (1965, 1968, 1970); Chieffo and McLean (1968a,b); G.F. Bright (1972); Lillis and Young (1975); Mack (1975); Radian Corporation (1975, 1979); Boland *et al.* (1976); Chicago Bridge and Iron Co. (1976b); Kremer (1976); Meteorological Research Inc. (1976); Burklin, Sugarek and Knapf (1977); Bierl *et al.* (1977); Jonker, Porter and Scott (1977a,b); Rosebrook (1977); Taback *et al.* (1977); Western Oil and Gas Association (1977); Zanker (1977); Bierl (1978, 1980); Bierl and Kremer (1978); Cavaseno (1978a); EPA (1978a,b, 1982b, 1984c, 1988a); Kittleman and Akell (1978); Monsanto Research Corporation (1978); Bochinski, Schoultz and Gideon (1979); Hesketh (1979); T.W. Hughes, Tierney and Kahn (1979); Hydroscience Inc. (1979); Morgester *et al.* (1979); M.J. Wallace (1979); Schroy (1979); Kenson and Hoffland (1980); Kunstmann (1980); Beckman and Gillmer (1981); Siegel (1981); Wetherold, Rosebrook and Tichenor (1981); Blackwood (1982); Freeburg and Aarni (1982); Gwyther (1982); Anon. (1983 LPB 49, p. 28); Cleaver (1983); Hanzevack, Anderson and Russell (1983); Harbert (1983, 1984); Hesketh and Cross (1983); Hess and Kittleman (1983); Kusnetz and Phillips (1983); Wetherold *et al.* (1983); Beckman (1984); BOHS (1984, 1987); Flanagan (1984); Goltz (1984); A. Jones (1984); R. Powell (1984); Rhoads, Cole and Norton (1984); Sivertsen (1984); Recio and Rhein (1985); Beckman and Holcomb (1986); CONCAWE (1986 2/86, 1987 87/52); Anon. (1987o); Berglund and Whipple (1987); Lipton and Lynch (1987, 1994); Lipton (1989, 1990, 1992); Strachan *et al.* (1989); Early and Eidson (1990); J.C. Edwards and Bujac (1990); Kaufman *et al.* (1990); M. King (1990); Samdani (1990b); Surprenant (1990); Anon. (1991h,i,z); Adams (1991, 1992); Brestal *et al.* (1991); Gardner (1991); Jacob (1991b); Myerson (1991); Schaich (1991); Shiel, Siegel and Jones (1991); Stucker (1991); Bruere and Coenen (1992); CIA (1992 RC58); Crowley and Hart (1992); Gardner and Spock (1992); IBC (1992/96, 1993/108); Key (1992); Crowley (1993); Fruci (1993); Moretti and Mukhopadhy (1993); P.A. Ross (1993); Ruddy and Carroll (1993); Spock (1993); J.B. Wright (1993)

The main headings of the *Guidelines* may be summarized as

(1) overview;
(2) inherently safer design;
(3) engineering design;
(4) process safety management;
(5) early vapour detection and warning;
(6) countermeasures;
(7) on-site emergency response;
(8) off-site emergency response;
(9) selection of measures.

The interpretation of the concept of mitigation in the *Guidelines* is a broad one and covers both prevention and protection. It constitutes, but is more than, a comprehensive checklist of measures. Many of the topics listed above are dealt with in other parts of the present text. Table 15.69 gives an indication of the chapters or sections in which these topics are treated here.

Table 15.71 Estimated sources of emission in a refinery (Freeberg and Aarni, 1982) (Courtesy of the American Institute of Chemical Engineers)

Source	*Proportion of total emissions* (%)
Combustion	2
Solvents, organics	2
Tanks	9
Bulk loading	2
Fugitive emissions	85
Total	100

Table 15.72 Estimated sources of emission in chemical plants (British Occupational Hygiene Society, 1984 TG3)

Source	*Proportion of total emissions* (%)
Waste disposal	2–5
Storage tanks, transport vents	8–10
Process vents, stacks	65–70
Fugitive emissions	15–20
Total	100

The *Guidelines* give as a base case a system consisting of an unloading rail tank car, pump transfer, storage vessel, pump transfer and reactor. This system is then used throughout the text to illustrate the various points under consideration.

It starts with a discussion of the broad features of a vapour release: the forms which a release may take; the features of a flashing release; the causes of release; and the consequences of release. The different types of release are exemplified and categorized and an appendix is given which contains a checklist of release scenarios.

The principles of inherently safer design treated are: limitation of inventory; substitution of less hazardous materials; substitution of a less hazardous process; and siting. Consideration is given to both on-site and off-site exposure. There is an appendix on the properties of hazardous materials.

The contribution of engineering design is considered under the headings of plant integrity, process integrity, and emergency control. There is a fairly extensive treatment of the latter, dealing with emergency relief disposal, abort systems, isolation systems, material transfer, and spill containment. Emergency relief disposal covers active and passive scrubbers, stacks and flares, catchtanks, incinerators, and absorbers, adsorbers and condensers. There is an appendix on catchtank design and another on vendors of emergency equipment.

The treatment of management in the *Guidelines* concentrates on: operating procedures; training for emergencies; audits and inspections; equipment testing; maintenance programmes; plant and process modifications; methods of stopping a leak; and security.

Early warning is treated in terms of detectors for flammable and toxic gases, including their response time and positioning, detection by personnel, and alarm systems.

Countermeasures against vapour releases are water sprays, water, steam and air curtains, deliberate ignition, and ignition source control, whilst those against liquid releases are dilution, neutralization, and liquid, foam and solid covers, together with the method of application and avoidance of actions which may aggravate vaporization.

On-site emergency response embraces: on-site communications; emergency shut-down equipment and procedures; site evacuation; havens; escape; personal protective equipment; medical treatment; and emergency plans, procedures, training and drills. There is an appendix on the capacity of a haven.

Off-site emergency response as such is not the direct responsibility of the company, and the account is confined to those aspects which are its responsibility: alerting systems; roles and lines of communication; and information to be communicated.

The approach described to the selection of mitigation measures is based on a formal system of hazard identification and assessment, by which the scenarios for loss of containment and potential countermeasures are identified and the nature of the hazards and the effectiveness of the countermeasures are assessed.

15.55 Fugitive Emissions

Process plants have numerous potential sources of leaks and, although the emission rate from any one source may be small, the cumulative effect of these leaks may be such as to require action. One effect of such emissions is to raise the concentration of toxic chemicals in the plant and thus make it difficult to meet the workplace limit values set. Another effect is to increase the level of pollutants outside the plant. There is also a cost due to the loss of the chemicals.

Some of the emissions from the plant are attributable to particular sources or operations such as combustion and solvents handling, storage and terminal operations, process vents and stacks. The others are classed as fugitive emissions and consist of leaks from flanges, valves, pump and compressor seals, pressure relief valves, and so on.

An account of fugitive emissions is given in *Fugitive Emissions of Vapours from Process Equipment* by the British Occupational Hygiene Society (BOHS) (the BOHS *Guide)* (1984 TG3), *Health Hazard Control in the Chemical Process Industry* (Lipton and Lynch, 1987), *Fugitive Emissions and Controls* (Hesketh and Cross, 1983) and *Handbook of Health Hazard Control in the Chemical Process Industry* (Lipton and Lynch, 1994). There have been a number of publications by the Environmental Protection Agency (EPA) (1973, 1978a,b, 1982b, 1984c). Other accounts include those by Rosebrook (1977), Bierl and co-workers (Bierl *et al.*, 1977; Bierl, 1978, 1980; Bierl and Kremer, 1978), T.W. Hughes, Tierney and Khan (1979), M.J. Wallace (1979) and Wetherold *et al.* (1983). Selected references on fugitive emissions are given in Table 15.70.

Estimates of sources of emission in a typical refinery and in chemical plants are given in Tables 15.71 and 15.72, respectively.

Table 15.73 *Fugitive emission sources in a hypothetical refinery (Environmental Protection Agency)*

Source	*Emission rate*	
	(lb/h)	(%)
Pump seals	60	48.7
Valves	305	4.0
Flanges	25	9.6
Compressor seals	29	4.6
Relief valves	21	7.7
Drains	48	3.4
API separators	138	22.0
Total	626	100

15.55.1 Regulation of fugitive emissions

In the USA, early work on the control of fugitive emissions was the Los Angeles County project in 1958. Los Angeles has a particular problem of smog to which hydrocarbons are a main contributor.

A study was carried out of fugitive emissions in 11 refineries in the Los Angeles area in which measurements were made of the emission factors for various types of equipment. These constituted the original data for the compilations of emission factors produced by the EPA and were published as AP-42, which then became the basis for regulatory controls.

The EPA subsequently commissioned a number of studies, including in 1979 studies by the Radian Corporation and by Monsanto. The Radian project covered total emissions on refineries, but fugitive emissions received particular attention. The Monsanto project covered similar ground for chemical plants. The EPA has issued a number of documents giving emission factors.

The EPA has promulgated as amendments to the Clean Air Act 1970 regulations limiting fugitive emissions from refineries and chemical plants. These controls are concerned mainly with pollution rather than occupational exposure.

In the Federal Republic of Germany the Federal authorities maintain a register of emission factors (Bierl, 1980), which are used by the inspectorate of the States. In Nord Rhein-Westfalen, for example, these emission factors are incorporated in the local regulations.

In the UK there is no specific requirement on emission factors as such. Fugitive emissions are regulated as an aspect of pollution and of the control of toxic concentrations in the workplace.

15.55.2 Measurement of fugitive emissions

The existence of a leak may be detected using the soap bubble method. However, this method is not generally regarded as suitable for quantitative measurement. The concentration at the point of leak may be used as a measure of the leak rate, but the values obtained are highly dependent on factors such as the size of leak, the wind conditions, the plant geometry, etc.

In the Radian study the maximum concentration obtained around an emission source was termed the 'screening value'. The EPA has attempted to relate

Table 15.74 *Fugitive emission sources in four chemical plants (after T.W. Hughes, Tierney and Khan, 1979) (Courtesy of the American Institute of Chemical Engineers)*

Source	*Plant type*							
	Monochloro-benzene		*Butadiene*		*Dimethyl terephthalate*		*Ethylene oxide/ glycol*	
	(ton/year)	(%)	(ton/year)	(%)	(ton/year)	(%)	(ton/year)	(%)
Pumps	1	3.1	96	8.6	8		2	
Valves	0.2	0.6	1000	90.0	?		?	
Flanges	29	91.0	0	0	?		?	
Other	1.7	5.3	16	1.4	7.9		8.6	
Total	31.9	100.0	1112	100.0				

Table 15.75 *Emission factors in Los Angeles County project (AP-42) (after Environmental Protection Agency, 1973)*

Source	*Items inspected*	*Items tested*	*Emission rate* (g/h)
Pumps	473	75	79
Valves	9521	100	3.8
Flanges	326	129	0
Compressors	326	90	161
Relief valves	165	165	55

screening values to mass emission rates, but the correlation shows appreciable scatter.

The two principal methods of measuring emission rates both involve enclosing the source in a container. In the bag method the bag is placed around the source, air is passed through the bag and the outlet steady-state concentration of the chemical in the air stream is measured. In the box method there is no air flow and the measurement taken is the rate of rise of concentration of the chemical in the air in the box.

15.55.3 Monitoring of fugitive emissions

The arrangements for monitoring emissions depend on the purpose of the monitoring. If the aim is to reduce pollution outside the works, periodic spot checks on emission sources may be sufficient. The control of toxic concentrations in the atmosphere of the workplace normally requires additional measures. In the latter case it is necessary to implement continuous sampling of the atmosphere. The data obtained by such sampling must then be interpreted statistically. A discussion of sampling and interpretation of sampled data is given in the BOHS *Guide*.

15.55.4 Emission factors for equipment

The contribution of the various sources of fugitive emissions to the total plant emission has been the subject of several studies. Table 15.73 shows the fugitive emission sources assigned in a hypothetical refinery in an EPA study. Table 15.74 gives the actual sources in the Monsanto study (T.W. Hughes, Tierney and Khan, 1979).

Information on emission factors is available from several sources. Mention has already been made of the Los Angeles County study and of the studies commissioned by the EPA. The California Resources Board has carried out a project involving six refineries to check the 1958 data. Rockwell International has carried out a study for the American Petroleum Institute on oil production operations.

Table 15.75 lists emission factors from the Los Angeles County project as given in the 1973 edition of the EPA publication AP-42. Tables 15.76 and 15.77 show emission factors obtained in the work by Radian (Wetherold, Rosebrook and Tichenor, 1981) and by Monsanto (T.W. Hughes, Tierney and Khan, 1979).

Work by Bierl (1978, 1980) at BASF is shown in Table 15.78. The literature values referred to in this table are those from earlier work by Schwaneke (1965, 1968, 1975). Bierl expresses his own measured values as percentages of these published ones.

A composite set of emission factors given by Schroy (1979), based mainly on those of AP-42 and Bierl, are shown in Table 15.79.

The spread of emission rates is illustrated by the data for valves in light liquid service given by A.L. Jones (1984) and shown in Table 15.80.

The effect of maintenance on emission rates is shown in the data in Table 15.81 given by Kunstmann. In this a set of valves was selected at random and the leak rates were measured, the average value being 2.37 g/h. Some 80% of the total emission was attributable to the valves with leak rates greater than 2 g/h. The leak rates of these valves were reduced by maintenance and their leak rates were remeasured with the result that the average leak rate for all the valves was reduced to 0.44 g/h.

15.55.5 Emission losses from storage tanks

Losses from storage tanks have been the subject of a series of API bulletins, including:

API Bull. 2513: 1959 *Evaporation Loss in the Petroleum Industry – Causes and Control* (reaffirmed 1973)

API Bull. 2516: 1962 *Evaporation Loss from Low Pressure Tanks*

Table 15.76 *Emission factors in Radian project (Wetherold, Rosebrook and Tichenor, 1981)*

Source	*Emission rate* (g/h)		
	Mean	*95% confidence limits*	
		Lower	*Upper*
Pump seals:			
Light liquid service	113	73	170
Heavy liquid service	20.9	8.6	50
Valves:			
Gas/vapour service	26.8	14	49
Light liquid service	10.9	7.7	16
Heavy liquid service	0.23	0.1	0.68
Hydrogen service	8.16	3.2	20
Flanges:	0.25	0.1	1.1
Compressor seals			
Hydrocarbon service	635	300	1300
Hydrogen service	49.9	20	100
Pressure relief valves	86.2	32	220
Drains	31.8	10	91
Open-ended lines	2.3	0.73	7.3

Table 15.77 *Emission factors in Monsanto project (after T.W. Hughes, Tierney and Khan, 1979) (Courtesy of the American Institute of Chemical Engineers)*

Source	*Emission rate by plant type (g/h)*[a]			
	Monochloro benzene	*Butadiene*	*Dimethyl terephthalate*	*Ethylene oxide/ glycol*
Pump seals	7.7 (23)	63 (160)	3.3 (20)	13 (82)
Valves	0.05 (1.5)	17 (120)	1.5 (32)	0.07 (1.6)
Flanges	2.2 (82)	0	3.4 (110)	0.03 (1.0)
Compressor seals	–	54 (59)	–	5.9 (11)
Relief valves	–	5 (14)	0	0
Agitators	200 (200)	–	145 (218)	–
Drains	–	–	–	40 (68)
Sample valves	–	–	40 (91)	–

[a] The figures given are average emission rates, those without parentheses being for all potential sources and those within parentheses being for significant sources.

Table 15.78 *Emission factors in BASF project (after Bierl, 1980)*

	Items tested	*Emission rates*		
		Literature value (g/h)	*Measured value*	
			(% of literature value)	(g/h)
Pumps (mechanical seals)	67	60	9	5.4
Valves:				
General	1800	2.8	20	0.56
High pressure valves	400	2.8	18	0.50
Ball valves and stop-cocks	125	2.8	0.5	0.014
Flanges:				
Gas service	5000	0.2 g/h m	2.5	0.005
Liquid service	600	2.0 g/h m	2.0	0.04
High pressure flanges	950	0.2 g/h m	4.8	0.01
Reciprocating compressors	13	104	12.5	13
Relief valves	50	55	1.1	0.61
Agitators (mechanical seals)	8	60	0.2	0.12

Table 15.79 Some composite emission factors (after Schroy, 1979)

Source	*Emission rate*[a] (g/h)
Flanges:	
Gas service (10–20 bar)	0.02 g/h m
Liquid service (10–20 bar)	0.2 g/h m
High pressure (40–80 bar)	0.01 g/h m
Valves:	
Gate and control valves	
10–20 bar	6
40–80 bar	0.1
Ball valves	0.015
Pressure relief valves	2.8
Pressure relief valve + bursting disc	0
Pumps and centrifugal compressors:	
Packed seals	80
Single mechanical seals	
Process fluid flush	6
Water flush	0.02
Double mechanical seals	0
Reciprocating compressors:	
Rod single packing	161
Rod double packing	13
Agitators	–[b]
Tanks:	
Floating roof	–[c]

[a] Emission rate is expressed as g/h per metre of outer flange circumference.
[b] Use packed or double mechanical seal data for pumps with the relation:
Loss for agitator = Loss for pump × u/1.9
where *u* is the face velocity (m/s).
[c] Loss 60% of value estimated from method given in API Bull. 2517.

Table 15.80 Distribution of emission rates for valves in light liquid service (A.L. Jones, 1984) (Courtesy of the Institution of Chemical Engineers)

No. of valves	*Emission rate* (g/h)
573	0
4	0.0076
10	0.021
14	0.056
28	0.152
40	0.414
58	1.125
50	3.06
52	8.32
46	22.6
20	61.4
12	167
4	454
2	1234

Table 15.81 Effect of maintenance of emission rates of valves (after Kunstmann, 1980) (Courtesy of Elsevier Sequoia SA)

A Valves before maintenance

Valves tested No.	(%)	*Emission rate* (g/h)
31	79	0–2
6	16	2–10
2	5	10–50
0	0	>50

B Valves after maintenance

Valves tested No.	(%)	*Emission rate* (g/h)
36	92	0–2
3	8	2–10
0	0	10–50

API Pub. 2517: 1980 Evaporation Loss from External Floating Roof Tanks

API Bull. 2518: 1962 *Evaporation Loss from Fixed Roof Tanks*

API Pub. 2519: 1983 *Evaporation Loss from Internal Floating-roof Tanks*

API Bull. 2521: 1966 *Use of Pressure–Vacuum Vent Valves for Atmospheric Pressure Tanks to Reduce Evaporation Loss*

API Bull. 2522: 1967 *Comparative Test Methods for Evaluating Conservation Mechanisms for Evaporation Loss*

API Bull. 2523: 1969 *Petrochemical Evaporation Loss from Storage Tanks*

These publications give equations for the estimation of evaporation losses.

Extensive work has been done on emissions from floating roof tanks. The background to this work is discussed by Zabaga (1980), by Hanzevack, Anderson and Russell (1983) and by Laverman (1984).

15.55.6 Control of fugitive emissions

The information on emission factors given above indicates that for a given type of emission source there is a wide variability in emission rates but also that for almost all types the technology exists to reduce the emission rate to a low level. A detailed review of the methods of obtaining leak tight plant is given in the BOHS *Guide*. A further account is given by Bierl *et al.* (1977).

The BOHS *Guide* describes the following types of seal:

(1) static seals –
  (a) gasketed joints
    (full faced flanges,
    narrow faced flanges),
  (b) automatic face seals;
(2) dynamic seals –
  (a) contact seals,
  (b) clearance seals (bushing seals, labyrinth seals);
(3) miscellaneous seals –
  (a) O-rings,

(b) flexible lip seals.

The *Guide* also gives a detailed account of packed glands (stuffing boxes) and mechanical seals.

For flanges, the *Guide* describes the elements of good practice necessary to reduce leakage. Its principal proposal for cases where high leak-tightness is required is the elimination of flanges by the use of all-welded pipe.

For valves, leakage of the rising stem may be reduced by the use of 100% graphite packing. For high leak-tightness use may be made of bellows-sealed or diaphragm-sealed valves. Bellows-sealed valves were developed for use in the nuclear industry. The *Guide* gives the following emission rates for rising stem valves:

	Emission rate ($\mu g/s$)
Regular packing	
< 20 bar	100
> 20 bar	2
100% graphite packing	
< 20 bar	12
> 20 bar	0.2

For pumps, the *Guide* gives the following emission rate index:

Seal type	*Shaft emission rate index*
Regular packing	100
Regular packing with lantern ring	10
100% Graphite packing	10
Single mechanical seal	1.2
Double mechanical seal	0.004

For bellows seal pumps, diaphragm pumps and canned pumps the index is zero. Details of the seal arrangements are given in the *Guide*. A further discussion of packed glands is given by Hoyle (1978).

For agitators the *Guide* states that the pump data may also be used as a guide to the relative effectiveness of the different seal types. A discussion of agitator seals is given by Ramsey and Zoller (1976).

The sealing of centrifugal or reciprocating compressors is appreciably more difficult than the sealing of pumps. Labyrinth seals are commonly used on both types, but leakage rates are high, being typically 1–2% of the flow through the compressor. Carbon seals give lower leak rates, but for a more positive seal the *Guide* suggests oil film seals or mechanical seals with oil flush. A further discussion of centrifugal compressor seals is given by W.E. Nelson (1977).

Flange protection arrangements for the control of emissions are shown in Plate 2.

15.55.7 Economics of emission control

Several studies have been published on the economics of emission losses and of control measures, including those by Kittleman and Akell (1978), Wetherold *et al.* (1983) and Rhoads, Cole and Norton (1984). The cost of control measures is also considered by Kunstmann (1980). These studies tend to be concerned with the costs of reducing leaks and with minimization of these costs rather than with savings obtained by leak elimination. The general conclusion is that effort should be concentrated on individual sources with a high leak rate, the outliers, and that thereafter the law of diminishing returns sets in strongly.

15.56 Leaks and Spillages

Leaks and spillages are main causes of fire and explosion in chemical plants. They may give rise to a local fire or explosion, a wider flash fire or a large vapour cloud explosion. Some of the situations which can give rise to leaks and spillages have been described by Kletz (1975b). They include:

(1) leaks from pump glands and seals;
(2) leaks from joints, valves and fittings;
(3) spillages from drain valves;
(4) spillages occurring during maintenance work;
(5) spillages from road and rail tankers.

Leaks have occurred not infrequently on pumps handling cold liquid ethylene. In some cases vapour clouds have been formed, while in others the leak has ignited, apparently due to static electricity.

On pumps handling C_3 and C_4 hydrocarbons at ambient temperature and moderate pressure there is little trouble with seal leakages, but failures have occurred on high pressure injectors.

Joints, valves and fittings are common sources of leakage due to features such as loose bolts, ruptured gaskets, failed instrument connections.

Another fertile source of spillage is drain valves. Typically a drain valve blocks due to ice or hydrate formation, the blockage is cleared in some way, but it then proves impossible to shut the valve off.

The disaster at the Rhone Alps refinery at Feyzin in 1966 (Case History A38) began with blockage in a drain line from a propane storage sphere which was almost certainly caused by ice or hydrate. The drain line was a vertical pipe below the sphere and had two valves on it. The standing instruction was to drain by opening the top valve and controlling the flow with the lower one so that, if necessary, the former could be used to shut the flow off. But the operator in fact turned the bottom valve wide open and controlled draining from the top one. When this latter valve blocked, he opened it further, the blockage gave way and a jet of propane came out full bore through the 2 in. line, struck the ground, rebounded into the operator's face and knocked the handle off the valve. Attempts to replace the handle on the top valve or to shut the bottom one failed. There was a massive escape of propane, the vessel was engulfed by fire and 18 people were killed.

Maintenance work sometimes gives rise to spillages. This is usually the result of failure to ensure isolation of equipment. Isolation procedures are described in Chapter 21. Failures of connecting hoses on road and rail tankers at filling terminals are not unusual. These are discussed in Chapter 22.

The volume of vapour generated by a liquid spillage can be very large. Moreover, only a small percentage of vapour is needed to give a flammable mixture. Thus one volume of liquid can become 10 000 or more volumes of a flammable vapour–air mixture. Leaks of liquefied flammable gas (LFG) of different sizes are discussed in the ICI *LFG Code* (ICI/RoSPA 1970 IS/74).

Small leaks such as those from pump and valve glands or pipe joints are quite probable, but usually involve little risk, provided that the leak is discovered quickly, the gas is dispersed rapidly and there is no nearby source of ignition. Dispersion is assisted by ventilation and may be promoted by steam from a steam lance.

Medium leaks are defined as those large enough to go beyond the plant boundaries. The leak should be isolated if possible. Isolation by the use of hand valves near the leak may be hazardous and the use of remotely operated isolation valves is preferable if the latter are available. It may be noted that excess flow valves do not operate at lower leakage rates.

Attempts should also be made to disperse the leakage and to prevent it finding a source of ignition. Use may be made of steam curtains, if installed, or of mobile water curtains. A leak of LFG which is water soluble can be dealt with by diluting it with large amounts of water.

Major leaks are defined as those of 1 ton/h or more. The methods just described are appropriate, but a major leak usually gives rise to a large pool of liquid. There is typically a rapid flash-off of vapour followed by slower vaporization from the pool. It is necessary, therefore, to decide how the liquid is to be handled. One method is to assist vaporization and dispersion, another is to suppress vaporization and to dispose of the liquid. Vaporization may be suppressed by laying foam on the surface. In some cases the formation of an ice layer effects complete suppression.

Leaks are also discussed by J.R. Hughes (1970) who states that a leak of more than 1 UKgal/h (4.5 l/h) would be regarded as unusual on a modern pump seal and would not be allowed to continue, and that experiments on this size of leakage with motor gasoline have shown that in the open air the concentration of vapour produced is so small that a flammable concentration exists only within a few inches of the leakage point.

Hughes also describes experiments involving the spillage of two pints (1 l) of gasoline. In a light wind of 2 mile/h (3.2 km/h) the vapour cloud floated downwind and gave a flammable concentration near the ground at 20 ft (6 m) downwind, but at a few inches above ground the concentration was undetectable. For spillage near a low wall the distance at which the cloud gave a flammable concentration increased to 30 ft (9.1 m).

Leaks in confined spaces, such as an enclosed and poorly ventilated compressor or pump house, are another matter and can easily build up to flammable concentrations, giving rise in due course to a fire or explosion.

15.57 Notation

Bo	Bowen ratio
Fr	Froude number
Gr	Grashof number
Ma	Mach number
Nu	Nusselt number
Pe	Peclet number
Pr	Prandtl number
Re	Reynolds number
Ri	Richardson number
Ro	Rossby number
Sc	Schmidt number
Sh	Sherwood number
St	Stanton number
We	Weber number
c_p	specific heat
g	acceleration due to gravity
g'	reduced gravity
M	molecular weight
P	absolute pressure
R	universal gas constant
t	time
T	absolute temperature
γ	ratio of gas specific heats
Δ	reduced density
μ	viscosity
ρ	density
$\Delta\rho$	density difference

Superscript:

.	time derivative

Subscripts:

max	maximum
r or ref	reference

Note:

(a) Some of these variables have different local definitions, for example; P (probability), R (radius of cloud), T (time interval), γ (edge entrainment coefficient).
(b) There are different definitions of the reduced gravity g'.

Section 15.1

a	cross-sectional area of pipe
a_c	cross-sectional area of vena contractor
a_o	cross-sectional area of orifice
A	cross-sectional area of pipe or orifice; constant defined by Equation 15.1.62
B	constant defined by Equation 15.1.63
c	velocity of sound
C	effective coefficient of discharge
C_c	coefficient of contraction
C_{cv}	coefficient of contraction at vena contracta (external mouthpiece)
C_D	coefficient of discharge
C_v	coefficient of velocity
d	diameter of pipe or orifice
e	pipe roughness
f	friction factor
f_D	Darcy friction factor
f_F	Fanning friction factor
F	total frictional loss
F_c	frictional loss due to sudden contraction
F_f	frictional loss due to pipe
F_{ft}	frictional loss due to fittings
G	mass velocity
h	head
h_c	head loss due to contraction
h_{ft}	head loss due to fittings
H	enthalpy

k expansion index
k_h number of velocity heads
K total number of velocity heads lost
K_e number of velocity heads lost at entrance
K_f number of velocity heads lost due to friction
K_{ft} number of velocity heads lost due to fittings
l distance along pipe, length of pipe
q heat absorbed from surroundings
Q total volumetric flow
R shear stress at pipe wall wall (Equation 15.1.15)
R_o shear stress at pipe wall (Equation 15.1.11)
S entropy; perimeter of pipe (Equation 15.1.11)
u velocity
u_c velocity at vena contractor
v specific volume
W total mass flow
W_s work done on surroundings, shaft work
x mass fraction of vapour in wet vapour, or quality
z vertical distance, height
Z compressibility factor
γ ratio of gas specific heats
η pressure ratio
ϕ friction factor

Subscripts:
a acceleration al
c critical
f frictional
g gravitational
l liquid
v vapour
1 intial, upstream, stagnation
2 final, downstream, outlet

Subsection 15.1.9
d diameter of pipe (m)
G mass velocity of gas (kg/m^2 s)
G^* critical mass velocity of gas through an adiabatic nozzle (kg/m^2 s)
k expansion index
L length of pipe (m)
N pipe resistance factor
p absolute pressure (N/m^2)

Section 15.1.10
A cross-section area of vessel at height h (m^2)
A_o cross-sectional area of orifice (m^2)
C constant defined by Equation 15.2.95
C_D coefficient of discharge
D diameter of cylinder (m)
g acceleration due to gravity (m/s^2)
h height of liquid above bottom of vessel (m)
h_o height of orifice, or hole, above bottom of vessel (m)
h_p equivalent liquid height
H height of vessel (m)
L length of horizontal cylindrical vessel (m)
ΔP pressure difference between liquid surface and atmosphere, imposed pressure (Pa)
Q outflow through orifice (m^3/s)
t time (s)
V volume of vessel (m^3)
x dimensional hole height
θ half-angle of cone of vertical conical vessel (°)
ρ density of liquid (kg/m^3)
τ dimensionless drainage time

Section 15.2
A cross-sectional area of pipe
C parameter defined by Equation 15.2.47
C_d coefficient of discharge
d diameter of pipe
f friction factor
F friction loss
G mass velocity
h specific enthalpy
k slip ratio
l distance along pipe, length of pipe
n polytropic index
dP_a/dl pressure drop due to acceleration
dP_f/dl pressure drop due to friction
$(dP_f/dl)_g$ pressure drop which would occur if gas phase were flowing alone in pipe
$(dP_f/dl)_{go}$ pressure drop which would occur if whole fluid were flowing as gas in pipe
$(dP_f/dl)_l$ pressure drop which would occur if gas phase were flowing alone in pipe
$(dP_f/dl)_{lo}$ pressure drop which would occur if whole fluid were flowing as liquid in pipe
dP_g/dl pressure drop due to gravitational forces
Q volumetric flow
R_o shear stress at wall
S perimeter of pipe
u velocity
v specific volume
W mass flow
x mass fraction of vapour, or quality
X parameter defined by Equation 15.2.41
z vertical distance, height
α fraction of volume occupied by vapour, void fraction, or voidage; parameter defined by Equation 15.2.3
β vapour phase volumetric flow fraction; parameter defined by Equation 15.2.4
Γ parameter defined by Equation 15.2.42
η critical pressure ratio
λ parameter defined by Equation 15.2.1
σ surface tension
ϕ parameter used in Equations 15.2.40–15.2.43
ψ parameter defined by Equation 15.2.2

Subscripts:
a air
c critical
E equilibrium
g gas
go gas only
h homogeneous
l liquid
lo liquid only
L all liquid condition

m	two-phase mixture
o	stagnation
s	two-phase
w	water

Section 15.3

A_a	cross-sectional area of aperture
A_p	cross-sectional area of pipe
c_l	specific heat of liquid
C	specific heat of liquid
C_a'	modified cavitation number
C_d	coefficient of discharge
C_{DM}	modified coefficient of discharge
d	diameter of pipe
f	friction factor
f^*	friction factor (two-phase flow)
f_m	average friction factor
F	flow correction factor
F'	friction loss
G	mass velocity
k	slip ratio
K	coefficient of discharge
l	distance along pipe, length of pipe
l_e	equilibrium, or relaxation, length
n	polytropic index
N	non-equilibrium parameter
P	absolute pressure
P^*	absolute effective downstream pressure
ΔP	pressure drop
P_a	absolute atmospheric pressure
P_c	absolute back pressure in choked flow
P_o	absolute stagnation pressure
P_v	absolute vapour pressure
R_g	fraction of cross-sectional area occupied by vapour
s	specific entropy
U	average velocity
v	specific volume
x	mass fraction of vapour, or quality
α	void fraction
β	parameter defined by Equation 15.3.12
ζ	parameter defined by Equation 15.3.34
η	parameter defined by Equation 15.3.16
ρ_l	density of liquid
ρ_m	average density

Subscripts:

b	downstream
c	critical
E	equilibrium
ERM	equilibrium rate model
f	liquid
fg	transition from liquid to vapour
g	vapour
h	constant enthalpy
l	liquid
m	two-phase mixture
n	two-phase annular
t	throat

Section 15.4

C_f	specific heat of liquid
D	diameter of pipe
f	friction factor
Fi	flow inclination number
h	specific enthalpy
G	mass velocity
G^*	dimensionless mass velocity
G_o	mass velocity through nozzle
L	length of pipe
N	pipe resistance factor
s	specific entropy
v	specific volume
x	mass fraction of vapour, or quality
α	void fraction
η	pressure ratio
θ	angle between pipe axis and vertical

Subscripts:

c	critical
f	liquid
fg	transition from liquid to vapour
lm	limiting
o	stagnation
r	reduced
s	saturation

Section 15.5

a	parameter defined by Equation 15.15.53
A	area of vent (m^2)
A_x	cross-sectional area of vessel (m^2)
C_o	distribution parameter
D	diameter of vessel (m)
F	geometric factor
h_{fg}	latent heat of vaporization (kJ/kg)
H	height of liquid (m)
H^*	dimensionless liquid height
H_{BL}	height of interface below which there is no vapour carryunder (m)
H_o	height of liquid without swell (m)
H_x	height of vessel (m)
j	total volume flux, or superficial velocity (m/s)
j_g	superficial velocity of gas (m/s)
j_{gl}	superficial velocity of gas phase relative to liquid phase, drift flux (m/s)
j_l	superficial velocity of liquid (m/s)
l_{lg}	superficial velocity of liquid phase relative to gas phase, drift flux (m/s)
$j_{g\infty}$	vapour superficial velocity (m/s)
J_o	dimensionless heat flux
k	constant
L	length of vessel (m)
m	index
n	index
q	heat input per unit mass (kW/kg)
Q_{ex}	external heat input flux (kW/m^2)
S	volumetric source strength (m^3/s)
U	velocity (m/s)
U_c	downward liquid recirculating velocity (m/s)
U_g	area average velocity of gas (m/s)
U_l	area average velocity of liquid (m/s)
U_s	upward liquid superficial velocity (m/s)
U_∞	bubble rise velocity (m/s)
V_l	volume of liquid (m^3)
x_e	mass fraction of vapour at vent
z	vertical distance from bottom of vessel (m)

α void fraction
α_o initial void fraction, or freeboard
δ thickness of boundary layer (m)
ϵ geometric factor
λ parameter
ρ density (kg/m^3)
σ surface tension (N/m)
ψ dimensionless source strength, dimensionless velocity
Ω ratio of vessel area to vent area

Superscript:
- average

Subscripts:
c continuous phase
g gas
l liquid
∞ rise

Section 15.6
See Section 15.4 plus:
A cross-sectional area of valve (m^2)
A_{in} cross-sectional area of inlet pipe (m^2)
A_2 cross-sectional area of outlet pipe (m^2)
C_f specific heat of liquid (J/kg K)
D_{in} diameter of inlet pipe (m)
D_2 diameter of outlet pipe (m)
f Fanning friction factor
g acceleration due to gravity (m/s^2)
G mass velocity (kg/m^2 s)
h specific enthalpy (J/kg)
H change in elevation (m)
K valve discharge coefficient
L_2 length of outlet pipe (m)
L_{in} length of inlet pipe (m)
ΔP_{in} total pressure difference across inlet pipe (Pa)
P absolute pressure
P_b absolute back pressure at orifice (Pa)
P_c absolute critical pressure (Pa)
P_o absolute stagnation pressure in vessel (Pa)
P_s absolute saturation pressure of liquid (Pa)
T absolute temperature (K)
v specific volume (m^3/kg)
$\bar{v}_{in}$ average specific volume in inlet pipe (m^3/kg)
W mass flow (kg/s)
x mass fraction of vapour, or quality
α void fraction
η pressure ratio defined by Equation 15.6.5
ρ density (kg/m^3)
ω parameter

Subscripts:
act actual
c critical
f liquid
fg liquid–vapour transition
g vapour
o stagnation
req required
s saturation
1 upstream end of outlet line
2 downstream end of outlet line

Section 15.7
a velocity of sound in gas at choke (m/s)
A cross-sectional area of choke (m^2)
C_D coefficient of discharge
H enthalpy (J/kg)
P absolute pressure (Pa)
P_{ms} absolute pressure at which metastability is no longer supported (Pa)
P_s absolute saturated vapour pressure of liquid (Pa)
P_x absolute orifice pressure (Pa)
R universal gas constant (kJ/mol K)
S entropy (kJ/kg K)
T absolute temperature (K)
T absolute temperature upstream of orifice (K) (Equation (15.7.7))
T_c absolute critical temperature of liquid (K)
V volume of vessel (m^3)
W mass flow from vessel (kg/s)
Π depressurization rate (N/m^2 s)
ρ density of gas (kg/m^3) (Equations 15.7.1–15.7.6); density (Equations 15.7.7 and 15.7.8)
ρ density of gas (kg/m^3) (Equations 15.7.1–15.7.6); density (Equations 15.7.7 and 15.7.8)
σ surface tension of liquid (N/m)

Subscripts:
c choke
d downstream
g gas
i vessel
l liquid
o initial
s isentropic
u upstream

Section 15.8

Subsection 15.8.1
h height of liquid in vessel
h_b height reached by liquid on depressurization
h_o initial height of liquid
h_v height of vessel
ΔH_v latent heat of vaporization
K ratio of vent area to vessel cross-sectional area
m mass
ΔT superheat
δ diameter of vapour nucleus
ρ density
σ surface tension

Subscripts:
l liquid
nb nucleate boiling
o initial
onb onset of nuclear boiling
r retained
sat saturation value

v vapour

Subsection 15.8.2

A_1 area of aperture
A_2 area of cloud perpendicular to direction of expansion at end of second stage (Hardee and Lee model)
c constant
h height of cloud; specific enthalpy (Equation 15.8.4)
r radius, radial distance
s specific entropy
u velocity
V volume of cloud
W initial mass of fluid
ρ density
ψ momentum of cloud

Subscripts:

a air
d dump
f liquid
v vapour
1, 2 initial and final states

Subsections 15.8.3 and 15.8.4

a radius of initial cloud formed on completion of bursting process
c concentration
l mixing length
r radial distance
u' fluctuating velocity
V_g effective volume of vapour
w propagation velocity of cloud boundary
X_c parameter characterizing mass fraction of gas in core
ϵ eddy diffusion coefficient
τ dimensionless time

Subscripts:

c core concentration
g core radius
ign within flammable limits
o initial cloud

Subsection 15.8.5

c concentration at transition
M initial mass of liquid
N transition criterion
r distance
t_g time to transition
V_g volume of vapour released
w radial velocity
α momentum per unit mass
Δ buoyancy term
ρ density

Subscripts:

a air
v vapour

Section 15.9

A cross-sectional area of pipeline
d diameter of pipeline
f friction factor
f_D Darcy friction factor
l length of pipeline
m mass flow from pipeline
N correction factor
u velocity
v specific volume
W mass hold-up in pipeline
Y void fraction
z distance along pipeline from rupture point
τ_1, τ_2 time constants
ϕ_g parameter defined by Equation 15.9.14
ψ parameter defined by Equation 15.9.7a

Subscripts:

e exit
f liquid
fs superficial value for liquid
g vapour
i two-phase interface
o initial
s sound

Section 15.10

Subsections 15.10.1–15.10.3

a albedo
A_p area of pool
C_{Bm} logarithmic mean concentration difference
C_T total concentration
d pool diameter or, more generally, characteristic length
D diffusion coefficient
h heat transfer coefficient
ΔH_v latent heat of vaporization
j_h j factor for heat transfer
j_m j factor for mass transfer
k thermal conductivity
k_m mass transfer coefficient
m_v mass vaporization rate per unit area
p partial pressure
p^o vapour pressure
q heat flow per unit area
q_{cd} heat flow per unit area by conduction from ground to pool
q_{cn} heat flow per unit area by convection from atmosphere to pool
q_r net heat flow per unit area by radiation from pool
q_{ra} heat flow per unit area by radiation from atmosphere to pool
q_{rl} heat flow per unit area by radiation from pool to atmosphere
q_{rs} heat flow per unit area by solar radiation to pool
q_v heat required per unit area for vaporization
Q total heat transferred per unit area
Re_x point Reynolds number
u velocity
w mass of liquid
z vertical distance down into soil
α thermal diffusivity
ϵ emissivity

θ temperature above datum, or soil, temperature
λ constant
σ surface tension (Equation 15.10.4); Stefan–Boltzmann constant (Equations 15.10.35 and 15.10.36)
ϕ fraction of liquid vaporized, flash fraction
ψ solar constant

Subscripts:
a air, atmosphere
b normal boiling point
d drop
i initial
l liquid
o initial
s soil
v vapour
∞ atmosphere beyond influence of pool

Equations 15.10.10 and 15.10.11:
k constant
M_i molecular weight of component i
M_{To} initial mass of liquid
n_T moles of liquid
p_{is} vapour pressure of component i
x_{io} initial mol fraction of component i in liquid

Section 15.10.4
a constant defined by Equation 15.10.41
D diffusion coefficient of vapour in air (cm^2/s)
E evaporation rate (g/s)
f_1, f_2 parameters defined by Equations 15.10.45 and 15.10.46
k von Karman constant
K constant defined by Equation 15.5.39
K' constant defined by Equation 15.5.40
M molecular weight
n diffusion index
p^o vapour pressure of liquid (dyn/cm^2)
r radius of pool (cm)
R universal gas constant (erg/gmol K)
T absolute temperature (K)
u wind speed (cm/s)
u_1 wind speed at height z_1 (cm/s)
x_o downwind length of pool (cm)
y_o crosswind width of pool (cm)
z height (cm)
z_1 height at which wind speed u_1 is measured (cm)
λ parameter (cm^2/s)
ν kinematic viscosity of air (cm^2/s)
χ_o concentration (g/cm^3)

Equations 15.10.50 and 15.10.51:
As for Subsection 15.10.4, but with SI units plus
a, a′ constants

Section 15.10.5
C_D drag coefficient
u wind speed
ρ_g density of gas at 10 m from pool
ρ_{sg} saturation density of gas at smbient conditions

Section 15.10.6
c specific heat
C constant
h heat transfer coefficient
j_m j factor for mass transfer
k constant
k_m mass transfer coefficient
N mass transferred per unit area
p partial pressure
q_{cn} heat flow per unit area by convection from atmosphere to pool
u wind speed
χ diameter of pool

Subscripts:
a air, atmosphere
l liquid
v vapour
∞ atmosphere beyond influence of pool

Section 15.10.7
c constant
C constant
C_L constant (for land)
C_W constant (for water)
f constant
F friction term
F_g gravity force
F_i inertia force
h height of pool
r radius of pool
s shape factor
u velocity of pool edge
V volume of pool
β shape factor
Δ buoyancy term
ν kinematic viscosity

Subscripts:
l liquid
w water

Section 15.10.8
As Subsections 15.10.1–15.10.3 plus:
A_v vaporization parameter

Equation 15.10.75:
t time (s)
v vaporization, or regression, rate of liquid surface (cm/s)

Section 15.10.9
As Subsections 15.10.1–15.10.3 plus:
A volume of liquid for instantaneous spill
k_3 constant
m total mass vaporized
r radial distance
v vaporization, or regression, rate of liquid surface

Equations 15.10.77 and 15.10.78:
E_i mass vaporized per unit area within 1 min (g/cm^2)
E_s steady-state mass vaporization rate per unit area (g/cm^2 s)

ΔH_v latent heat of vaporization (cal/g)
k_1, k_2 constants

Equation 15.10.79:

m_v mass vaporization rate per unit area
q_w heat flow per unit area from water to spill
ΔT temperature difference between substrate and spill
β coefficient of cubical expansion (Equation 15.10.79)
ν kinematic viscosity

Subscript:

w water

Equation (15.10.80):

m_v mass vaporization rate per unit area (kg/m^2 s)
t time (s)

Subsection 15.10.10

A volume of liquid for instantaneous spill
A_p area of pool
B volumetric flow of liquid for continuous spill
ΔH_v latent heat of vaporization
k thermal conductivity
m total mass vaporized
M_s mass of spill liquid
r radius of pool
v vaporization, or regression, rate of liquid surface
V volume of pool
X surface roughness factor
α thermal diffusivity
β_1 constant defined by Equation 15.10.102
β_2 constant defined by Equation 15.10.106
θ parameter defined in Equation 15.10.98

Subscripts:

b normal boiling point
e end of evaporation
o initial
s soil
1, 2 at time t_1, t_2

Subsections 15.10.11 and 15.10.12

A_o area of confined spill
c specific heat
ΔH_v latent heat of vaporization
k thermal conductivity
L characteristic length
m_v mass vaporization rate per unit area
q heat flux per unit area
r radial distance
S parameter defined by Equation 15.10.114
t_{ch} characteristic evaporation time
t_{cr} time for transition from instantaneous to continuous spill, crossover time
t_o time for spill to cover confined area
ΔT temperature difference between liquid at distance r_1 and soil
v vaporization, or regression, rate of liquid surface
V volume of pool
V_c volumetric flow of liquid into continuous spill
V_i volume of liquid in instantaneous spill
α thermal diffusivity
ξ dimensionless maximum spread
τ dimensionless time

Subscripts:

e end of evaporation
l liquid
s soil
sp spill duration
1 value at radial distance r_1

Subsection 15.10.13

p fraction which dissolves in water, partition coefficient
r_e final evaporation radius (ft)
t_e final evaporation time (s)
v regression rate (ft/s)
V volume spilled (USgal)
Δ buoyancy term defined by Equation 15.10.63

Sections 15.11–15.56

c_E slumping coefficient
x, y, z distances in downwind, crosswind, vertical directions
u wind speed
u, v, w wind velocity components in downwind, crosswind, vertical directions
u_* friction velocity
x_i mixed layer height
z_o roughness length
Γ dry adiabatic lapse rate
$\sigma_u, \sigma_v, \sigma_w$ standard deviations of wind velocity in downwind, crosswind, vertical directions
$\sigma_x, \sigma_y, \sigma_z$ standard deviations, or dispersion coefficients, in downwind, crosswind, vertical directions
χ concentration (mass per unit volume)

Subscripts:

a air
cl centreline
cr critical
mgl maximum ground level
tr transition
x, y, z in x, y, z directions

Section 15.11

h height of cloud
p index

Section 15.12

Subsection 15.12.4

f Coriolis force
p pressure
X, Y accelerations in downwind, crosswind directions representative of forces not due to pressure and gravity
ϕ latitude
ω angular velocity of earth

Subsections 15.12.5–15.12.28 and 15.12.32–15.12.34

a	constant
c	constant
c_g	drag coefficient
c_p	specific heat of air
d	zero plane displacement, displacement height
E	vertical flux of water vapour, evaporation rate
f	Coriolis force
F_s	vertical flux
h'	scale height of neutral atmospheric boundary layer
H	vertical heat flux, sensible heat flux
i_y, i_z	intensities of lateral, vertical turbulence
k	von Karman constant
K	coefficient of exchange, turbulent exchange coefficient
K_H	coefficient of exchange for heat, eddy thermal diffusivity
K_M	coefficient of exchange for momentum, eddy diffusivity for momentum, eddy viscosity
K_{Mo}	adiabatic value of K_M
K_W	coefficient of exchange for water vapour, eddy diffusivity
l	mixing length
L	Monin–Obukhov length
L_e	latent heat of vaporization
L_W	latent heat of water vapour
n	index
p	index (Subsection 15.12.8 and 15.12.25); absolute pressure (Subsection 15.12.13)
p_o	absolute pressure at surface
P_u	P value
P_θ	P value
q	concentration of water vapour, or specific humidity
Q_o	surface heat flux
Ri_B	bulk Richardson number
Ri_F	flux Richardson number
Ri_{mod}	modified Richardson number
s	environmental stability
S	non-dimensional windshear (Equation 15.12.25); mean quantity of property per unit mass of air (Equation 15.12.55)
S^*	scale parameter
T_o	absolute average temperature of mixed layer
u_g	geostrophic wind speed
U_R	wind speed ratio
u, v, w	wind speeds in downwind, crosswind, vertical directions
$\bar{u}, \bar{v}, \bar{w}$	mean wind speeds in downwind, crosswind, vertical directions
u', v', w'	wind speed fluctuations in downwind, crosswind, vertical directions
$\overline{u'^2}, \overline{v'^2}, \overline{w'^2}$	variances of wind speed in downwind, crosswind, vertical directions
w^*	convective velocity
X	parameter defined by Equation 15.12.108
z	height
α	Monin–Obukhov coefficient
α_1, α_2	constants
β	ratio of specific heats of air; constant (Equation 15.12.49b)
ϵ	mean height of roughness elements (Subsection 15.12.12); eddy dissipation rate (Subsection 15.12.28)
θ	potential temperature
ζ	Monin–Obukhov parameter
μ	Kazanski–Monin parameter
ν	kinematic viscosity of air
ρ	density of air
$\Delta\rho$	density difference between cloud and air
σ_A	standard deviation of azimuth
σ_E	standard deviation of elevation
σ_θ	standard deviation of horizontal direction of wind
σ_ϕ	standard deviation of inclination to horizontal of wind
τ	mean momentum flux per unit area, Reynolds stress
ϕ_H	similarity factor for heat flux
ϕ_M	similarity factor for momentum flux
ϕ_s	similarity factor for property s
ϕ_W	similarity factor for water vapour flux

Subscripts:

s	property specified
1, 2	at height 10 m and 50 m
Equation 15.12.40:	
H	sensible heat flux (W/m^2)
S	incoming solar radiation (W/m^2)
Equation 15.12.80:	
p	index
t	sampling period
Equations 15.12.112–15.12.114:	
A, B	constants
Q_m	maximum surface heat flux
z_{im}	maximum mixed layer height
τ	time after sunrise

Subsection 15.12.29

i	intensity of turbulence
r	radius of eddy
t_E	Eulerian time scale
t_L	Lagrangian time scale
β	ratio of Lagrangian to Eulerian time scales
ω	tangential velocity

Subsections 15.12.30 and 15.12.31

$R(x)$	correlation coefficient
$R(\tau)$	correlation coefficient
t	time
u	velocity
u'	component of velocity
x	distance
σ_x	standard deviation in downwind direction
τ	time

Subscripts:

1, 2	points 1, 2

Sections 15.14 and 15.15

Subsection 15.14.1

K	diffusion coefficient (m^2/s)

K_x, K_y, K_z diffusion coefficients in downwind, crosswind, vertical directions (m^2/s)
r radial distance (m)
t time (s)
u, v, w wind speeds in downwind, crosswind, vertical directions (m/s)
x, y, z distances in downwind, crosswind, vertical directions (m)
χ concentration (kg/m^3)

Subsection 15.14.5
D diffusion coefficient, diffusivity
k thermal conductivity
l characteristic dimension
u velocity
ν kinematic viscosity

Section 15.16

Subsections 15.16.1, 15.16.2 and 15.16.5
As Subsection 15.14.1 plus:
C diffusion parameter ($m^{(1/2)n}$)
C_x, C_y, C_z diffusion parameters in downwind, crosswind, vertical directions ($m^{(1/2)n}$)
h height of source (m)
H height of source (m)
K diffusion coefficient (m^2/s)
K_o modified Bessel function of second kind
n diffusion index
p index
Q continuous mass rate of release (kg/s)
Q' continuous mass rate of release per unit length (kg/m s)
Q^* mass released instantaneously (kg)
y_o, z_o cloud bondary in y, z directions (m)
σ_x, σ_y, σ_z standard deviations, or dispersion coefficients, in downwind, crosswind, vertical directions (m)

Subsection 15.16.3
C ground level concentration ($units/m^3$)
C_c ground level concentration axis of plume for elevated source ($units/cm^3$)
C_o ground level concentration axis of plume ($units/m^3$)
d distance in downwind direction (km)
d' half-distance for development of uniform vertical concentration profile (km)
F_1 stack correction factor
F_2 off-axis correction factor
h vertical spread (m)
h' limiting vertical spread (m)
H height of stack (m)
Q mass rate of release (units/min)
x downwind distance (m)
u mean wind speed (m/s)
α deviation from axis (°)
θ lateral spread (°)
σ_y standard deviation of concentration in crosswind direction (m)
σ_θ standard deviation of wind direction (°)
σ_ϕ standard deviation of wind inclination (rad)
ϕ wind inclination (rad)

Sections 15.16.6–15.16.8
As Sections 15.16.1–15.16.4 plus:
D dosage
D_{tid} total integrated dosage
e base of natural logarithms
L load
n index (Equation 15.16.67)

Subscripts:
cb cloud boundary
cc cloud centre
ex extinction

Subsection 15.16.9
C volumetric concentration (v/v)
v_1 distance between first virtual source and actual source (m)
v_2 distance between second virtual source and actual source (m)
V volume of gas released (m^3)
x downwind distance (m)
x_b distance between actual source and works boundary (m)
x_s distance between actual source and point of interest (m)
x_{v1} distance between first virtual source and works boundary (m)
x_{v2} distance defined by Equation 15.16.72

Subsections 15.16.7 and 15.16.8
K parameter
m, n indices
t_e time for total discharge or evaporation

Section 15.17

Subsection 15.17.1
Q_a mass rate of release per unit area
Δx incremental length
χ_a concentration above mixed layer height
χ_b concentration upwind of city, background concentration

Subsection 15.17.3
As Sections 15.16.1–15.16.4 plus:
a radius of hill
D_y, D_z parameter defined by Equations 15.17.5a and 15.17.5b
h actual stack height
Δh plume rise
h_e equivalent stack height
h_o effective stack height
h_1, h_2 height of hills 1 and 2
h_p local height of plume
ζ parameter defined by Equation 15.17.7
η parameter defined by Equation 15.17.6
ψ stream function

Subscripts:
f flat terrain
o effective source height

Subsection 15.17.4
D_1 larger of dimensions height h and width w

D_s smaller of dimensions height h and width w
h height of building
h_s height of source
l alongwind length of building
R scaling factor
w crosswind width of building

Equations 15.17.11–15.17.15:
h_c height of roof cavity
l_c length of roof cavity
x downwind distance
Z_{II} upper boundary of high turbulence layer
Z_{III} upper boundary of roof wake

Equations 15.17.6 and 15.17.17:
A characteristic area
A_v vent area
K dimensionless concentration
Q mass rate of release
w_o effluent velocity

Equations 15.17.18 and 15.17.19:
h actual stack height
h' effective stack height after downwash
h_d downwash
h_o effective plume height

Equations 15.17.20–15.17.22:
A, B parameters defined by Equations 15.17.21, 15.17.22
x_r, y_r, z_r length, width, height of wake cavity

Equations 15.17.23–15.17.26:
A cross-sectional area of building normal to wind
c constant
D_A atmospheric dilution factor
D_B building dilution factor
D_{tot} total dilution factor
Q mass rate of release
x downwind distance
σ_y, σ_z dispersion coefficients in crosswind, vertical directions

Equations 15.17.27–15.17.44:
$\bar{C}$ mean concentration
H parameter defined by Equation (15.17.28)
k constant
k_f characteristic energy of free stream turbulence
l_f characteristic length of free stream turbulence
Q source strength
S dimension of block
u mainstream air velocity
V volume of wake bubble
X length of wake bubble
Λ free stream turbulence parameter

Equations 15.17.33–15.17.44:
A frontal area of building
b crosswind breadth of building
C_w average concentration in near-wake region
F flow out of wake
h height of building
L_r length of recirculation region
Q source strength
S surface area of near-wake region
t_r residence time in near-wake region
u_h wind speed at building height in undisturbed flow
V volume of near-wake region
α flux constant
β wake shape parameter
λ_r dimensionless length of recirculation region
τ_r dimensionless residence time in recirculation region
χ_w dimensionless average concentration in near-wake region

Section 15.18
H height of 'lid'
Q mass rate of release

Section 15.19

Subsection 15.19.1

As Subsections 15.16.1–15.16.4 plus:
f_1, f_2 $f_2$universal functions
$R(\tau)$ autocorrelation coefficient
t_L Lagrangian time scale
T_i dispersion time
T_y, T_z time scales for lateral, vertical dispersion
δ delta function
σ_θ standard deviation of wind direction

Subsection 15.19.3

As Subsection 15.16.3 plus:
d distance
h vertical spread
θ lateral spread

Subsection 15.19.6

As Subsections 15.16.1–15.16.4 plus:
a, c, f, h constants
b, d, g, j indices
$F(z_o, x)$ roughness length correction factor
P P value
T duration of release (h)
u_{10} wind speed at 10 m height (m/s)
σ_y dispersion coefficient in crosswind direction (m)
σ_{yt} turbulent diffusion term (m)
σ_{yw} wind direction fluctuation term (m)
σ_θ standard deviation of horizontal wind direction (rad)

Subsection 15.19.7

As Subsections 15.16.1–15.16.4 plus:

Equation 15.19.25:
a, c constants
b, d indices

Equation 15.19.26:
a, b, c constants
σ dispersion coefficient

Subsection 15.19.8

As Subsections 15.16.1–15.16.4:

Subsections 15.19.9–15.19.11

As Sections 15.16.1–15.16.4 plus:
p index
R reference distance
α, β indices

Subsection 15.19.12

As Subsections 15.16.1–15.16.4 plus:
c_1, c_2, c_3 constants
ϵ eddy dissipation rate
σ dispersion coefficient
σ_o initial size of puff

Subscripts:
o initial
plume plume
puff puff

Section 15.20
d_o diameter of outlet
r_o radius of outlet

Subscripts:
e exit
o outlet

Subsections 15.20.2–15.20.3
u velocity of gas

Equations 15.20.1–15.20.23:
b_o half-width of slot
c volumetric concentration
d width of slot
k constant
k_1–k_7 constants
m mass flow
M momentum flux
r radial distance
r_p radius of pipe (for radial jet)
u velocity of gas
w wind speed
x distance along axis of jet
β half angle of jet
θ air-fuel mass ratio

Subscripts:
m centre line, and hence maximum
tr transition (from jet to plume)
x mean value at axial distance x
xo centre line
xr at axial distance x and radial distance r

Equations 15.20.24–15.20.37:
D diameter
M mass flow
u velocity of gas

Subscripts:
a atmospheric
E exit, or outlet
N nozzle, or outlet
s shock plane
v vessel

Equations 15.20.38–15.20.46:
b_1–b_3 constants
d_{eq} equivalent diameter
j volumetric concentration
x distance along axis of jet
x' distance between virtual source and outlet
x'' distance defined by Equation 15.15.41
y radial distance
$\rho_{ga'}$ normalized density of gas at atmospheric conditions
$\rho_{go'}$ normalized density of gas at outlet
$\rho_{oeq'}$ normalized density of gas at equivalent diameter

Subscripts:
m centre line, hence maximum
xy at point x, y

Equations 15.20.47–15.20.62:
A cross-sectional area of flow
C_D coefficient of discharge
d diameter
k compressible eddy coefficient
k' eddy viscosity coefficient
m mass fraction
N pressure ratio
r radius
$r_{0.5}$ radius at which velocity is half centre line value
u velocity of gas
x_c dimensionless core length
z distance along axis of jet
z_b barrel length
z^* modified axial distance

Subscripts:
eq equivalent
m centre line, hence maximum
v vessel

Subsection 15.20.4
a radius of outlet
c constant
d diameter of outlet

k constant representing air resistance
l height of outlet
x distance in horizontal direction
u horizontal velocity of liquid
z distance in vertical direction
Z breakup length
α angle between jet and horizontal (Equations 15.20.63–15.20.78); constant (Equations 15.20.79–15.20.86)
δ amplitude of disturbance
δ_o initial amplitude of disturbance
σ surface tension

Subsection 15.20.6
c volumetric concentration
F buoyancy flux
k_4–k_5 constants
l characteristic length
M momentum flux
Q volumetric flow
r radial distance
u velocity of gas in jet or plume
x distance along axis
β coefficient of cubical expansion
$\Delta\rho_o$ density difference

Subscripts:
lt transition (from laminar to turbulent)
xr at axial distance x and radial distance r

Subsection 15.20.7

As Subsections 15.16.1–15.16.4

Subsection 15.20.8
C constant
F buoyancy flux
M momentum flux
u velocity of gas
z_{max} maximum plume rise

Subscripts:
1 ambient fluid

Subsection 15.20.9
A time parameter (defined by Equation (15.20.105))
D_c drag coefficient
$F_1(x_1)$ function expressing change in density of plume due to dilution by air
h_{pl} total effective height of plume
H height of outlet (m)
i rise of fall of plume at distance wt_1
J volumetric flow from outlet referred to ambient temperature (m^3/s)
k_1, k_6–k_8 constants
l height of jet (above outlet)
m_a mass flow of air
m_l mass flow of gas–air mixture at distance l
m_o mass flow at outlet
M molecular weight of gas
M_a molecular weight of air
N volumetric flow from outlet
Q volumetric flow at outlet
p_a concentration along axis at distance x (mole fraction)
p_{mgl} maximum ground level concentration at distance x (mole fraction)
r radial distance from axis
t_1 arbitrary chosen time interval
u velocity of gas
u_b buoyancy velocity
u_σ velocity of plume along axis of falling plume from virtual source to ground
v velocity at outlet
w wind speed
x downwind distance, downwind distance from virtual source
x_{max} downwind distance from virtual source at which maximum rise of plume occurs
x_{mgl} downwind distance to maximum ground level concentration
x_o downwind distance of transition point from virtual source
x_1 downdwind distance from virtual source at which plume axis reaches ground
x_1 distance defined by Equation (15.20.106)
x_θ distance from virtual source to point where plume reaches ground measured along axis of plume
z vertical distance above transition point
z_{max} maximum plume rise
θ angle between axis of falling plume and horizontal
ρ_g density of gas at outlet
ρ_l mean density of gas–air mixture at distance l
ϕ angle between axis and edge of plume, half-angle of plume
ψ momentum flux in still air
ψ_w momentum flux in windy conditions

Subscripts:
g gas at outlet
trw transition in windy conditions
w windy conditions

Subsection 15.20.10
C_1, C_2 constants
e base of natural logarithms
F buoyancy flux
h_e effective stack height
h_s stack height
Δh plume rise (above stack outlet)
H surface buoyancy flux
k von Karman's constant
m molecular weight
M momentum flux
Q mass rate of release
r radius
s environmental stability
s' modified stability
S ratio of effective area influenced by plume momentum to cross-sectional area of so-called thermal plume
u wind speed
V volumetric flow
w velocity in vertical direction

w' parameter defined by Equation (15.20.180)
z vertical distance above stack outlet; height of jet or plume
z_{max} maximum plume rise
α constant
β constant
ϵ ambient eddy dissipation rate
η constant
σ dispersion coefficient

Subscripts:
c critical
eff effective
en environment
eno initial environment
eq equilibrium
fb fallback
p plume
po initial plume

Section 15.21

Subsection 15.21.1
ϵ constant

Subscripts:
sat saturation
sh shattering

Subsection 15.21.2
A cross-sectional area of jet
C concentration (mass per unit volume)
G mass flow in jet
R radius of jet
u velocity of jet
z distance downstream from source
α entrainment coefficient
β half-angle of jet

Subscripts:
d end of deposition zone
∞ asymptotic value

Subsection 15.21.3
A cross-sectional area of jet
B buoyancy flux
C concentration of contaminant
f momentum flux
L length scale
p dimensionless variable defined by Equation 15.21.32
q dimensionless variable defined by Equation 15.21.33
R radius of jet
u velocity of jet
u_E entrainment velocity
U velocity scale
x downstream distance
α entrainment coefficient
ϕ ratio of density of jet to that of air

Subsection 15.21.4
A cross-sectional area of jet
e entrainment rate
E_o entrainment coefficient
h specific enthalpy
P pressure in jet
R radius of jet
u velocity of jet
x vapour fraction, or quality
z distance along jet axis

Subscripts:
a at end of depressurization zone
b at orifice
bp boiling point
f liquid
fg liquid–vapour transition
g vapour
o stagnation conditions
tp two-phase
∞ atmospheric conditions

Section 15.24

Subsections 15.24.6–15.24.9
k van Karman constant
K eddy diffusivity
L Monin–Obukhov length
p index
z height
Δ reduced density difference defined by Equation (15.24.3)
Δ' reduced density difference (alternative formulation) defined by Equation (15.24.5)
$\Delta\rho$ absolute density difference
ϵ energy dissipation rate
κ turbulence energy
ψ profile function

Subsection 15.24.10
C_f friction factor
G mass velocity
h heat transfer coefficient
j_H j factor for heat transfer
k thermal conductivity
L characteristic dimension
St_H Stanton number for heat transfer
ΔT temperature difference for heat transfer
u velocity of cloud
β coefficient of thermal expansion

Subscripts:
f film
gr between ground and cloud

Subsection 15.24.11
c velocity of wave
h height of dense fluid
n symmetry index
u velocity of dense fluid
x distance

Subsection 15.24.12
B negative buoyancy
H height of cloud
R radius of cloud
u_e top entrainment velocity

V volume of cloud
w_e edge entrainment velocity

Section 15.25

As Subsections 15.16.1–15.16.4 plus:

c volumetric concentration
c_p specific heat of gas in source cloud
E heat source in cloud
H height of cloud
L width of source
m mass of gas in source cloud
R radius of cloud
u_e top entrainment velocity
$\dot{V}$ volumetric rate of release
W_v velocity of vapour source
x_v distance of virtual source upwind of actual source
ϵ_v heat released by condensation and freezing of water vapour
ϵ_w surface heat flux

Subscript:

i initial

Section 15.26

A, B constants
B rate of buoyancy destruction of turbulent energy
c_l constant
c_t constant
C volumetric concentration
D rate of dissipation of turbulent energy into internal heat
F_a force due to reaction of ambient fluid to outward radial acceleration of cloud edge
F_d drag force on cloud edge due to presence of quiescent ambient fluid
F_s static pressure force due to negative buoyancy of cloud
F_v force due to vertical acceleration in cloud and reaction of ambient fluid
G rate at which gravity transforms potential energy into mean kinetic energy
H height of cloud
H_f height of fluid front
I_E internal heat
K_E kinetic energy of mean radial and vertical motions
L half-width of plume
M_{rd} radial momentum integral of cloud
P_E potential energy
Q volumetric rate of release
R radius of cloud
Ri_t bulk turbulent Richardson number
s profile shape parameter
S cross-sectional area of plume from continuous release (Equations (15.26.16)–(15.26.18)); rate of shear production of turbulent energy (Equations (15.26.30) and (15.26.31))
T_E turbulent kinetic energy
u_e top, or vertical, entrainment velocity
u_f velocity of fluid front
u_y cloud velocity in crosswind direction
$\overline{u_t}$ bulk turbulent velocity
V volume of cloud
w_e edge, or horizontal, entrainment velocity
W_o volumetric flow from source
γ edge entrainment coefficient
Δ reduced density difference
Δ'_c reduced density difference for cloud
ρ_1, ρ_2 density of fluid behind, ahead of wavefront
$\bar{\rho}_c$ mean density of cloud
$\overline{\Delta\rho_c}$ mean density difference of cloud
$\overline{\Delta\rho_o}$ mean initial density difference for cloud

Superscript:

- average

Subscripts:

c cloud
o time zero, hence source
u transition

Subsection 15.26.4

Subscripts:

c continuous
i instantaneous

Section 15.27

C_d drag coefficient
C_f friction factor
c_{pc} specific heat of cloud
D drag force on cloud
f wind speed coefficient
F friction force
h specific enthalpy
h_{grn} heat transfer coefficient between ground and cloud by natural convection
H height of cloud
k_E spreading constant
l turbulence length scale
l_{vs} width of virtual source
L half-width of cloud
m total mass in cloud
m mass of component in cloud (with subscript)
m_e mass of air entrained at edge of cloud
m_t mass of air entrained at top of cloud
Q continuous mass rate of release
Q^* mass released instantaneously
Q_e volumetric flow of air entrained at edge of cloud (total value for instantaneous release; value per unit downwind distance for continuous release)
Q_{gr} heat flux from ground to cloud
Q_t volumetric flow of air entrained at top of cloud (total value for instantaneous release; value per unit downwind distance for continuous release)
$r(n)$ distance downwind from source n
R radius of cloud
S^* parameter defined by Equation 15.27.20
u' magnitude of vector sum of slumping and wind velocities
u_{av} average wind speed cloud height

u_c velocity of cloud
u_e top, or vertical, entrainment velocity
u_h wind speed at cloud height H
u_r wind speed at reference height (normally 10 m)
u_l longitudinal turbulence velocity
V volume of cloud per unit distance downwind (total value for instantaneous release; value per unit downwind distance for continuous release)
w_e edge, or horizontal, entrainment velocity
α top entrainment coefficient
β limiting value of group (u_e/u_l)
γ edge entrainment coefficient
σ dispersion coefficient
τ shear stress

Subscripts:
c cloud
gr ground
l liquid
o source
v vapour
wl water liquid
wv water vapour
y crosswind

Section 15.28
C^* constant
g_{gr} heat transfer coefficient between ground and cloud
H height of cloud
l turbulence length scale
k constant
k thermal conductivity (with subscript)
m total mass in cloud
m_a mass of air
m_g mass of gas
Q_{gr} heat flux between ground and cloud
R radius of cloud
ΔT_a temperature difference between air and cloud
ΔT_{gr} temperature difference between ground and cloud
u_c velocity of cloud
u_e top, or vertical, entrainment velocity
u_r wind speed at reference height (normally 10 m)
u_l longitudinal turbulence velocity
V volume of cloud
w_e edge, or horizontal, entrainment velocity
x_{vy} distance downwind of virtual source for horizontal dispersion
x_{vz} distance downwind of virtual source for vertical dispersion
X parameter defined by Equation 15.28.61
z_r reference height (normally 10 m)
Z parameter defined by Equation 15.28.11
β coefficient of volumetric expansion
γ edge entrainment coefficient
Δ reduced density difference
$\Delta\rho$ density difference

Subscripts:
c cloud
f fluid film
g gas, material release
gr ground
H Hosker scheme
o initial
vy virtual source for crosswind effects
vz virtual source for vertical effects

Subsection 15.28.1
A, B, C parameters
L characteristic dimension of cloud

Subsection 15.28.2
A, B, C parameters
A' parameter defined by Equation 15.28.62
L half-width of cloud

Section 15.29
c concentration (mass fraction)
C_d drag coefficient
h specific enthalpy
H_{min} height at breakup
K_h, K_v horizontal, vertical eddy diffusivities
l turbulence length scale
L_o latent heat of water
$\dot{m}$ mass rate of evaporation
$\dot{m}_{min}$ mass rate of evaporation at breakup
p pressure
p_s pressure at surface
p_t pressure at top
q heat flux
t_{max} time at breakup
U heat transfer coefficient
W mass of water vapour per unit mass of air
π parameter defined by Equation 15.29.7
σ dimensionless pressure co-ordinate
$\dot{\sigma}$ substantial derivative in x,y,σ,t coordinates
σ_ϵ standard derivation of wind direction
τ_o shear stress at surface
ϕ geopotential height
ω substantial pressure derivative

Subscripts:
c cloud
g gas
l liquid
s surface

Section 15.30
b cloud shape parameter
B cloud width parameter
C volumetric concentration
C_1, C_2 concentration functions
h cloud height parameter
k von Karman constant
K_a ambient vertical diffusion coefficient
K_h horizontal diffusion coefficient
K_v vertical diffusion coefficient
K_ρ dense-layer diffusion coefficient
$\mathbf{K}^m$ eddy diffusion tensor for momentum
$\mathbf{K}^\theta$ eddy diffusion tensor for energy
$\mathbf{K}^\omega$ eddy diffusion tensor for mass fraction

Q_{gr} heat flux between ground and cloud
p pressure deviation from an adiabatic atmosphere at rest
p' index
S temperature source term
u_c velocity of cloud
u_e vertical entrainment velocity
u_o velocity of material at source
u_y velocity of cloud in crosswind direction
u_{*c} cloud friction velocity
$\mathbf{u}$ velocity tensor
w_e horizontal entrainment velocity
w_{*c} cloud 'convection velocity'
β cloud shape parameter
β^* constant
θ potential temperature deviation from an adiabatic atmosphere
ρ_h density corresponding to pressure p
ϕ profile function
ω mass fraction of material released

Subscripts:
c cloud
g material released
o source

Section 15.31
A_i half-area of source for observer i
b half-width of middle part
B half-width of gas blanket
B_{eff} effective half-width of plume
B_p half-width of pool
B_s half-width of source
B_{si} half-width of source for observer i
B'_{si} local half-width of source for observer i
c concentration in mass per unit volume
c_A ground level concentration on centre line
c'_A corrected ground level concentration on centre line
c_{gr} concentration at ground level
C_f friction factor
H height of cloud
H_b height of 100% gas blanket
H_e surface heat flux
H_{eff} effective height of cloud
i counter for 'observers'
k von Karman constant
K eddy diffusion coefficient
K_o constant
K^*_o vertical eddy diffusion coefficient at reference height z_r acting on concentration profile of width $2B$
L length of gas blanket
L_p length of pool
L_s length of source
L_{si} length of source for observer i
Q_{gr} heat flux between ground and cloud
Q_p total mass rate of evaporation
Q'' mass rate of take-up into atmosphere per unit area
Q''_i mass rate of release per unit area for observer i
Q''_{max} mass rate of take-up into atmosphere per unit area
Q''_o mass rate of evaporation per unit area
Q''_p mass rate of evaporation per unit area
r shape factor for crosswind concentration profile
R radius of source
R_b radius of gas blanket
R_1 radius of spreading pool
Ri_* Richardson number defined by Equation 15.31.34
Ri_{*I} internal Richardson number defined by Equation 15.31.65
s shape factor for vertical concentration profile
S dispersion parameter
t_s specified time
t_{1i}, t_{2i} times when observer i passes upwind, downwind edges of source
u wind speed; gas velocity (Equation 15.31.58)
u_e top entrainment velocity
u_{eff} effective local cloud velocity
u_A effective air velocity
u_B bulk cloud velocity
u_i velocity of observer i
u_{*I} internal velocity scale
w_e edge entrainment velocity
W_b half-width of concentration profile
x_i downwind distance of observer i at time t
x_{tr} downwind distance of transition point
x_v downwind distance of virtual source
y_{Ls} limit of integration of y
y_{w3} mole fraction of surface water vapour
α_T thermal diffusivity
α' index
β_c coefficient of volumetric expansion of cloud
β' constant
γ edge entrainment coefficient
γ' constant
δ' constant
ν kinematic viscosity
ρ_E density at emission conditions
ρ_{gr} density downwind
τ_{gr} shear stress acting on downwind ground surface
τ_o shear stress acting on upwind ground surface
ϕ profile function

Subscripts:
c cloud
g material released
gr ground, ground level
i observer i

Section 15.32

As Subsection 15.24.10 and Section 15.31 plus:
H height
k von Karman constant
Ri_* modified Richardson number
u_e vertical entrainment velocity
α' index
ϕ profile function

Section 15.33

c_ϕ constant or function
p pressure
S_ϕ function
u velocity
x distance
Γ_ϕ function
ϵ energy dissipation
κ turbulence energy
ϕ scalar quantity

Subscripts:
i, j i, j direction

Section 15.34

A constant given in Figure 15.85(b)
B constant given in Figure 15.87(b)
C volumetric concentration
$\bar{C}_m$ maximum time mean volumetric concentration on the centre line
C_o volumetric concentration at initial release
D characteristic dimension of source
g'_m reduced gravity corresponding to maximum concentration
g'_o reduced gravity of initial release
G source geometry
h measure of plume height
h_c height of crest
H height of cloud
l_b buoyancy length scale
l_i length scale of turbulence in the atmospheric boundary layer
L_h half-width of plume
L_{ho} half-width of plume at source
L_u upwind extent of plume
L_v height of plume
n fraction of initial concentration ($=\bar{C}_m/C_o$); index (Equation 15.34.42)
q_o volumetric rate of release
Q volume of cloud
Q_o volume of material released
R radius of cloud
R_o initial radius of cloud
t_a arrival time of leading edge of cloud
t_c time between arrival time t_a and departure time t_d
t_d departure time of trailing edge of cloud
x_c distance of leading edge
x_n distance at which the fraction of initial concentration is n
y_c half-width of cloud
u_{ref} characteristic mean velocity
ρ_a density of air
ρ_o density characteristic of material released
ϕ stability parameter
ψ stability parameter

Subsections 15.34.4 and 15.34.5

T_o duration of release

Subsection 15.34.7

T_a absolute temperature of air
T_o absolute temperature of initial volume
V_a volume of air
V_o non-isothermal initial volume

Subscripts:
i isothermal
ni non-isothermal

Section 15.35

a_1–a_3 parameters
C concentration
C_m ground level centre concentration
F_h parameter defined by Equation 15.35.5
F_v parameter defined by Equation 15.35.6
L_s Monin–Obukhov length scale
s parameter
x' relative coordinate
X downwind coordinate of cloud centre
w parameter
Z vertical coordinate of cloud centroid
ζ measure of skewness of cloud
ξ downwind coordinate relative to skewed cloud

Section 15.36

C volumetric concentration
$\hat{h}$ normalized height of cloud
H height of cloud
$\hat{r}$ normalized radius of cloud
R radius of cloud
$\hat{t}$ normalized time
t_o initial time scale
u_E edge entrainment velocity
u_f cloud front velocity
u_T top entrainment velocity
$\hat{v}$ normalized volume of cloud
V volume of cloud
α parameter defined by Equation 15.36.13a
β parameter defined by Equation 15.36.14a

Superscript:
$\hat{}$ normalized value

Subscript:
o initial

Section 15.39

A_o area of source
D diffusivity
f Coriolis parameter
L characteristic length scale
q_o volumetric release rate
Q_o volume released
u characteristic velocity
u_o source velocity
α ground slope
κ thermal conductivity
μ viscosity of atmosphere
μ_o viscosity of source gas
ν kinematic viscosity
ρ density of atmosphere
ρ_a density of air
ρ_o density of source gas
$\Delta\rho/\rho$ potential density difference

Subscript:
o source gas

Subsection 15.39.5
D plume dimension
Q volumetric release rate

Subsection 15.39.7
c volumetric concentration
l_b buoyancy length scale
l_q length scale
Q volumetric release rate

Section 15.40

Equations 15.40.1–15.40.9:
C_D drag coefficient
E entrainment coefficient
g'_o reduced gravity
h height of following flow
Q volumetric flow per unit width
S_1, S_2 parameters
x downhill distance
u mean velocity of following flow
u_f velocity of front
u_s velocity on a defined streamline
α parameter defined by Equation (15.40.4)
θ angle of incline to horizontal

Subscripts:
n normal
1, 2 less, more dense fluid

Equations 15.40.10 and 15.40.11:
C_D form drag coefficient
f friction factor
H height of cloud
R radius of cloud
u velocity downhill
θ angle of slope

Subscripts:
b bottom of cloud
g gas
t top of cloud

Equations 15.40.12 and 15.40.13:
A_o initial cross plume area of plume element
A_p final cross plume area of plume element
g'_o initial reduced gravity
u wind speed at ground level
z height from source to target
θ potential temperature

Subscript:
p plume

Subsection 15.40.2

As Section 15.34.4 plus:
C_g maximum volumetric concentration at ground level
C_o volumetric concentration at initial release
h height of dense fluid layer
Δh head difference
h_w height of obstacle
$h_{0.5}$ height at which concentration is half ground level value
H height of fence
H_b ratio of heights defined by Equation 15.40.29
H_w ratio of heights defined by Equation 15.40.20
q_o volumetric rate of release
q_o^* volumetric release rate per unit width
u velocity of fluid
u_w velocity over obstacle
U velocity of wave travelling upstream (hydraulic jump)
W plume width

Subscripts:
f with fence
nf without fence
1 upstream
2 space between upstream moving wave and obstacle

Subsection 15.40.5
A frontal area of building
B dimensionless buoyancy flux
C time mean ground level concentration
C_L concentration in lower layer
C_o volumetric concentration at source
C_U concentration in upper layer
D diameter of source orifice
F_A flow of air into wake
F_{LU} flow from lower layer to upper layer
F_{UL} flow from upper layer to lower layer
g'_o reduced gravity at source
h height of building
h_L height of lower layer
h_U height of upper layer
$\bar{h}_L$ dimensionless height of lower layer
$\bar{h}_U$ dimensionless height of upper layer
H height of obstacle
l length of building
l_w length of recirculation region, or wake
L_B buoyancy length scale
q_o volumetric rate of release
q_o^* volumetric rate of release per unit length
Q release rate, source strength
$\bar{Q}$ dimensionless release rate
Ri_o initial Richardson number before mixing
Ri_T transition Richardson number
S_L lateral spread of cloud
S_U upwind spread of cloud
S_w surface area of wake
u wind speed; mean velocity at obstacle height (Equation (15.40.60))
V_w volume of wake
w width of building
α_I mixing coefficient
α_M mixing coefficient
α_o constant
λ_w dimensionless length of wake

ρ density of cloud
ρ_a density of air

Superscript:
- normalized

Subsection 15.40.6
a ratio of areas defined by Equation (15.40.73)
A area of opening
A_i area for inflow
A_o area for outflow
b constant
C pressure coefficient
C_i pressure coefficient for inflow opening
C_o pressure coefficient for outflow opening
f vapour fraction
h height of inlet opening
n ratio of molar flows
N number of moles
$\tilde{N}_G$ dimensionless quantity defined by Equation 15.40.72
P' pressure drop between atmospheric pressure and internal pressure
ΔP pressure difference between outside and inside
P_o atmospheric pressure
q_G flow of gas
$\tilde{q}_G$ dimensionless flow of gas
Q volumetric flow
Q_A volumetric air flow of air
T absolute temperature of air–gas mixture
T_A absolute temperature of air
u wind speed
V volume of space
τ time for concentration of gas to reach steady state if outflow is neglected
$\tilde{\tau}$ ratio defined by Equation 15.40.78

Subscripts:
A air
G gas

Section 15.42
D^* density of cloud
R^* air/HF ratio of cloud
T^* absolute temperature of cloud

Section 15.43

Subsection 15.43.2
b characteristic width of plume
c concentration
r radius of plume
s distance along axis of plume
u velocity along axis of plume
u' entrainment velocity
U_a mean wind velocity
θ angle between axis of plume and horizontal
ρ density of plume
ρ_a density of air
ρ_{ao} density of atmosphere at stack outlet
α_1–α_3 entrainment coefficients
ϕ_i set of functions defined by Equation 15.43.2
ψ_i set of functions used in Equation 15.43.1

Subsection 15.43.3
C concentration
Δh maximum initial plume rise
h_s height of stack
R radius of plume
s distance along plume axis
u_s velocity along plume axis
w velocity of plume
x downwind distance
α_1, α_2 entrainment coefficients
θ angle of plume axis to horizontal

Subscripts:
m maximum
o initial
td touchdown

Subsection 15.43.4
A_1–A_6 constants
b width of plume (m)
B initial buoyancy flux (m^4/s^3)
C constant
C_1–C_6 constants
d_o diameter of outlet (m)
g acceleration due to gravity (m/s^2)
g' reduced gravity (m/s^2)
h_s stack height (m)
l_m length scale, related to source (m)
l_Q length scale, related to transition from momentum to buoyancy control (m)
m index
M initial momentum flux (m^4/s^2)
Q initial volumetric flow (m^3/s)
u, v, w wind velocities in downwind, crosswind, vertical directions (m/s)
u_o outlet velocity (m/s)
U wind speed (m/s)
w^* entrainment velocity (m/s)
x, y, z distances in horizontal, crosswind and vertical directions (m)
x_{TD} horizontal distance from virtual source of final plume to touchdown point (m)
x^*_{td} horizontal distance from stack to touchdown point (m)
x_{tr} horizontal distance from stack to jet transition point (m)
x_v horizontal distance from stack to virtual source of final plume (m)
z_B length scale, related to effect of crossflow on plume (m)
z_{effB} effective stack height with strong crossflow (m)
z_{effv} effective stack height with weak crossflow (m)
z_m length scale, related to effect of crossflow on jet (m)
z_{max} maximum rise (m)
z_s vertical coordinate of jet or plume axis (m)
z_v distance of virtual source of jet below stack outlet (m)

α	entrainment coefficient
θ	relative density difference

Subscripts:

o	source
v	virtual source

Subsection 15.43.5

H	height of cloud
I_o	momentum flux (Equation 15.43.34); excess momentum flux (Equation 15.43.35)
k_1, k_2	constants
Q_a	flow of entrained air
R	half-width of cloud
u_w	wind speed
v	velocity of jet
$\Delta\rho$	density difference between cloud and air

Subsection 15.43.8

D	diameter
w	velocity of plume

Subscript:

o	initial

Section 15.44

A	cross-sectional area of plume
Buoy	buoyancy function
c	concentration (mass per unit volume)
C_D	spreading coefficient
D	diameter of circle, major axis of ellipse
Drag	drag function
e	eccentricity of ellipse
e_{at}	coefficient
e_{gr}	coefficient
e_{jet}	entrainment coefficient
$e_{u,cr}$	coefficient
$e_{\rho,cr}$	coefficient
$\mathbf{e}_x$, $\mathbf{e}_z$	unit vector in horizontal (wind aligned), vertical direction
$\dot{E}$	excess energy flux
Ener	energy function
Entr	entrainment function
Foot	reactive pressure force function
h	plume height; specific enthalpy (Equation 15.44.12)
k	slumping coefficient
l	width of 'footprint' of plume
l_y, l_z	turbulence length scales in crosswind, vertical directions
L	Monin–Obukhov length
$\dot{m}$	mass flux
$\dot{P}_x$	excess horizontal momentum flux
$\dot{P}_z$	excess vertical momentum flux
r	polar radius coordinate of cross-section of airborne plume
Ri_*	bulk Richardson number
s	distance along axis of plume
Shear	shear force function
T_{gr}	absolute temperature of ground
T_*	u_*T_* = ground to air heat flux
u	mean flow velocity
x, z	horizontal, vertical distance of centroid
z_c	vertical distance of centre
ϵ	dissipation rate of turbulent kinetic energy
η	parameter defined by Equation 15.44.16
η_c	parameter defined by Equation 15.44.17
η_ρ	parameter defined by Equation 15.44.34
θ	polar angle coordinate of cross-section of airborne plume
ζ	parameter
κ	von Karman constant
ϕ	angle of inclination of airborne plume
$\Phi(\mathrm{Ri}_*)$	heavy gas entrainment function
ψ_T	temperature function
ψ_u	velocity function

Subscripts:

at	atmospheric turbulence
c	centre
cw	crosswind
fs	free surface
g	material released
gs	gravity slumping
hg	heavy gas
jet	jet
pot	potential
∞	ambient conditions at centroid height

Section 15.45

M	mean value of concentration
P	peak value of concentration

Subsections 15.45.1–15.45.7

$\underline{c}$	instantaneous deviation of concentration
$\overline{c^2}$	variance of concentration (see text)
c	concentration
C	mean concentration
C_{max}	maximum concentration
C_{med}	median concentration
C_p	peak concentration
$\bar{C}$	time mean concentration
$F(C)$	probability distribution function of concentration C
i	intensity
M	mean value of concentration
p	index
P	peak value of concentration
$P(C)$	probability that concentration exceeds C
t	sampling time
t_m	sampling time for mean value M
t_p	time during which concentration is non-zero (Equation 15.45.2); sampling time for peak value P (Equations 15.45.6–15.45.9)
t_s	total sampling time
T_I	integral time scale
γ	intermittency
μ	mean of concentration
σ	variance of concentration
σ_l	standard deviation (logarithmic)

Subsection 15.45.8

As Sections 15.16.1–15.16.4 plus:

D_y, D_z	crosswind, vertical standard deviations attributable to meandering

Q continuous mass rate of release
Y crosswind standard deviations attributable to fluctuations about centre line
Z crosswind standard deviations attributable to fluctuations

Subscripts:
i instantaneous
m mean
p peak

Subsection 15.45.9
C^* concentration in sphere
C_τ maximum concentration which can occur over time period τ
$\bar{C}$ time mean concentration
k constant
x downwind distance
β constant
γ intermittency
λ index
σ standard deviation of instantaneous concentration
τ time period
ψ parameter defined by Equation 15.45.29

Subscript:
τ limiting value

Subsection 15.45.10

As Subsections 15.16.1–15.16.4 plus:
c concentration
c' concentration fluctuation about mean
$\bar{c}$ mean concentration
$\overline{c'2}$ concentration variance
c_{mp} condition median concentration
$\bar{c}_o$ mean concentration on centre line
c'_p conditional concentration fluctuation
$\bar{c}_p$ conditional mean concentration
$c_p'^2$ conditional concentration variance
d diameter of source
F_c cumulative distribution function for concentration
F_m height correction
F_v height correction
G_m crosswind distance correction
G_v crosswind distance correction
h height of source
h_v effective height of variance source
H height of neutrally stable atmospheric boundary layer
i intensity
i_p conditional intensity
$i_{p\infty}$ conditional intensity on centre line with negligible surface interaction
$p(c)$ probability density function for concentration
$p_p(c)$ probability density function for conditional concentration
q variance source strength
q_p conditional variance source strength
Q mass rate of release
u_c advection velocity of plume
z_e scale height
α sink strength
δ delta function
γ intermittency
λ normalized downwind distance
λ_o virtual origin of variance source
σ_1 standard deviation (log–normal)
ϕ location parameter for variance source

Subscript:
p conditional

Subsection 15.45.11
$c(x,t)$ concentration fluctuation about mean
$\bar{c}(x,t)$ mean value of concentration fluctuations $c(x,t)$
$\overline{c^2}(x,t)$ variance defined by Equation (15.45.76)
$C(x,t)$ mean value of ensemble of concentrations $\bar{\Gamma}(x,t)$
$\bar{C}$ mean concentration
i intensity
I intensity
$p(\theta,x,t)$ probability density function for θ
P probability that concentration θ lies between lower and upper flammability limits
$R(C)$ intensity parameter
$\gamma(x,t)$ intermittency
$\Gamma(x,t)$ concentration from a single experiment
$\bar{\Gamma}(x,t)$ mean value of $\Gamma(x,t)$
δ delta function
ϵ constant
$\theta(x,t)$ particular value of $\Gamma(x,t)$
θ_L lower limit of flammability
θ_o limiting value of θ
θ_U upper limit of flammability

Subsection 15.45.14
k_2–k_5 constants

Section 15.46

Subsection 15.46.1

As Sections 15.16.1–15.16.4 plus:
F mass of flammable gas at maximum hazard
H height of source
Q^* mass of gas released instantaneously
Q_e^* mass of gas within flammable range
r radial distance
v ratio of upper to lower flammability limit
v_1, v_2 parameters defined by Equations 15.46.15b and 15.46.15c
W mass of flammable gas
σ dispersion coefficient

Subscripts:
cc cloud centre
L lower flammability limit
m at maximum hazard
U upper flammability limit
1, 2 inner, outer

Equations 15.46.16–15.46.19:

F_c	weighting function
L	number of moles of oxygen reacting with one mole of fuel
n_a	number of moles of air in rich mixture
n_c	number of moles of fuel consumed
n_p	number of moles of fuel present
V_{LS}	volume of cloud between lower flammability limit and stoichiometric concentration (m^3)
V_{SU}	volume of cloud between stoichiometric concentration and upper flammability limit (m^3)
W_F	mass of fuel available for combustion (kg)

Subsection 15.46.2

a, c	constants
b, d	indices
D	constant
f	index
$\dot{m}_o$	mass rate of release
Q	mass of gas released
Q_e	mass of gas within flammability limits
Q_{FL}	mass of gas within flammability limits
x_L, x_U	distances to lower, upper flammability limits

Subsection 15.46.3–15.46.5

c	concentration (volume fraction)
d_o	effective diameter of discharge (m)
D	constant
f	index
g	acceleration due to gravity (m/s^2)
G	parameter defined by Equation (15.46.41)
k_1–k_6	constants
$\dot{m}_o$	mass rate of release (kg/s)
M	molecular weight
$\bar{M}$	mean molecular weight of gas mixture at distance z
Q_{FL}	mass of flammable material in cloud between flammability limits (kg)
r	radial distance from axis (m)
T	absolute temperature (K)
u	wind speed (m/s)
$\dot{v}$	volumetric flow (m^3/s)
V_{FL}	volume of cloud between flammability limits (m^3)
w	velocity (m/s)
z	axial length or vertical height (m)
α	ratio of number of moles of unreacted and reacted gas
Γ	dry adiabatic lapse rate (°C/m)
$\Delta\rho_o$	density difference between gas and air
χ	concentration (kg/m^3)

Subscripts:

F	flame
L	lower flammability limit
o	original discharge
s	stoichiometric mixture
U	upper flammability limit

Subsection 15.46.7

M	mass of vapour (t)
R	radius of cloud (m)

Subsection 15.46.9

f	lower flammability limit (%)
P	operating pressure (bar)
t	time span (s)
W	mass flow (kg/s)
W_{lr}	mass flow per unit area (kg/m^2)

Section 15.47

Sections 15.47.1–15.47.3

C	concentration
C_p	concentration over time t_p
k_1, k_2	constants
L	toxic load
n	index
P	probability of injury
T	exposure time
Y	probit
λ	intermittency

Subscripts:

pe	exposure
s	sampling

Equations (15.47.8)–(15.47.10):

As Subsection 15.45.9 plus:

ω	enhancement factor

Subsection 15.47.4

c	concentration
F_c	probability that concentration will not exceed c
n	index
N_b	number of breaths
$S(T_e)$	probability that concentration will not exceed c over exposure time T_e
T_e	exposure time
V_c	probability that concentration will exceed c

Subsection 15.47.5

C	concentration
D	dosage
$\bar{D}$	mean dose
D^*	effective dose
D_{50}	median fatal dose
k	constant
k_1, k_2	constants
n	index
$N[\,]$	normal distribution

Y probit
δ standardized log dose$\bar{\delta}$
mean standardized log dose
δ^* effective standardized log dose
θ probability of fatality
$\mu(C)$ mean of concentration
$\sigma(C)$ standard deviation of concentration
ϕ distribution function

Section 15.48

Section 15.48.4
D internal diameter of stack (m)
Δh plume rise (m)
u wind speed (m/s)
v_s velocity of stack gas (m/s)

Section 15.48.5
D diameter of vent tip (ft)
E number of dilutions to lower flammability limit
h_s height of vent above exposure level (ft)

Section 15.48.6
C_1–C_4 constants
D internal diameter of stack (m)
h height of stack (m)
Δh plume rise (m)
Q mass rate of release (g/s)
u wind speed (m/s)
v_s stack gas outlet velocity (m/s)
ρ density of gas mixture (kg/m^3)
χ_{cr} critical ground level concentration (μg/m^3)

Subscript:
s stack gas

Section 15.48.8
A parameter
C_m maximum permissible value of touchdown concentration of contaminant (%v/v)
C_s concentration of contaminant in stack gas (%v/v)
D internal diameter of stack (mm)
h_s height of stack (m)
L lower flammability limit (%v/v)
M_w molecular weight of stack gas mixture
s stack gas specific gravity relative to air
T_a absolute temperature of ambient air (K)
T_s absolute temperature of stack gas (K)
u wind speed (m/s)
v_s velocity of stack gas (m/s)
β peak/mean concentration ratio

Section 15.49

Section 15.49.1

As Sections 15.16.1–15.16.4

Section 15.49.2

Equation (15.49.7):
k constant
k_1–k_3 constants
M mass released (te)
R downwind range (km)
X distance to maximum width (km)

Equations (15.49.8)–(15.49.18):
h height of cloud (m)
k constant
L distance/dimension (m)
M mass released (te)
n index
R downwind distance (m)
R_c radius of cloud (m)
R_{cr} range transition criterion (m)
R_{LFL} downwind range to lower flammability limit (m)
R_{UFL} downwind range to upper flammability limit (m)
X_1 downwind dimension (m)
Y crosswind distance at downwind distance R (m)
Y_{max} maximum crosswind range of lower flammability limit (m)
Y_1 crosswind dimension (m)
U_{LFL} upwind range to lower flammability limit (m)

Equations (15.49.15)–(15.49.18):
I inventory (kg)
k constant
k^* constant
L distance/dimension (m)
$\dot{m}_v$ mass rate of release of vapour (kg/s)
n index
R downwind distance (m)
T_a arrival time (s)
T_d duration of release (s)
u wind speed (m/s)
Y crosswind distance at downwind distance R (m)

Equations (15.49.19) and (15.49.20):
q_o volumetric rate of release
Q_o volume released
x downwind distance

Section 15.50

Sections 15.50.1–15.50.3

As Sections 15.16.1–15.16.4 plus:

F depletion factor
H height of source
Q_o initial source strength
Q_x effective source strength
r radius of particle
r_e equivalent radius
$t_{1/2}$ half-life of concentration
x_G distance defined by implicit Equation 15.50.23
v_d deposition velocity
v_g fall velocity
v_t terminal velocity of particle
V_p volume of particle
α dynamical shape factor (Equation 15.50.6); reflection coefficient (Equations 15.50.21 and 15.50.22)
μ viscosity of air
ρ_p density of particle
τ_c chemical decay time constant
ω deposition rate

Subscripts:

n non-spherical
s spherical
1, 2 standard value, other value

Subsection 15.50.5 and 15.50.6

A_s cross-sectional area of drop
E collection efficiency
F_w flux of pollutant to surface
J_o equivalent rainfall rate
k concentration of pollutant in rainwater (mass per unit volume)
k' concentration of pollutant in rainwater (mass per unit mass)
N_s number of drops
s diameter of drop
v_w wet deposition velocity
v_s velocity of drop
w interphase transport rate
W_r washout ratio
W_r' washout ratio (alternative formulation)
z_w height of wetted plume
Λ washout coefficient, scavenging coefficient
ρ_a density of air
ρ_w density of water
χ concentration of pollutant in air
χ_e effective concentration over height of wetted plume

Subscripts:

o reference height
s drop

Section 15.51

c concentration (various units)
D toxic dose (units h)
k mixing efficiency factor
L toxic load (units^n h)
n index
R ratio of indoor to outdoor toxic loads
t time (h)
t_o duration of finite outdoor concentration (h)
λ ventilation rate, ventilation constant (h^{-1})
λ' effective ventilation rate (h^{-1})

Equation 15.51.18:

A area of aperture (m^2)
k_1 constant
Q volumetric flow of air (m^3/s)
Δp pressure drop (N/m^2)
ρ_o density of ambient air (kg/m^3)

Equations (15.51.19)–(15.51.24):

k_2–k_5 constants
m number of open casement windows
n number of open top-vent windows
ΔT temperature difference between outdoors and indoors (°C)
u wind speed (m/s)
λ ventilation rate (h^{-1})
λ_b ventilation rate due to buoyancy (m^3/h)
λ_s ventilation rate measured at standard conditions (h^{-1})
λ_w ventilation rate due to wind (m^3/h)

Subscripts:

i indoors
o outdoors

Section 15.52

D initial diameter of plume
u_o velocity of release

Section 15.53

Subsection 15.53.1

As Section 15.10.8

Section 15.53.8

D diameter of spray cone (m)
F_N flow number of nozzle ($(l/s)\cdot(kN/m^2)^{1/2}$)
P_W pressure of water at nozzle (kN/m^2) (gauge)
Q_a volumetric flow of air (m^3/s)
Q_W volumetric flow of water (l/s)
V velocity of entrained air (m/s)

Subsection 15.53.10

c entrainment parameter
c_s entrainment parameter for spray
C volumetric concentration
C_{FD} concentration with forced dispersion

C_{ND}	concentration with natural dilution
D	width of spray
FD	forced dispersion effectiveness factor
Q	volumetric rate of release
r	radius of plume
S	surface area of elemental slice of plume
u	velocity of plume
v	vertical velocity in spray
v_a	velocity of air entrained in spray
δV	volume of elemental slice of plume
w	wind speed
x	downwind distance

Subscripts:

o	source
1, 2	end of region I, II

Subsection 15.53.11

CR	concentration ratio
F_N	flow number (l/(min bar$^{1/2}$))
K	constant$M$
	specific momentum flow (m^2/s^3)
M_N	momentum flow number (Ns/(s bar))
$u_{1.25}$	wind speed at 1.25 m height (m/s)

Subsection 15.53.18

G	mass flow of gas
Q_a	mass flow of air
S	mass flow of steam

CW00689434

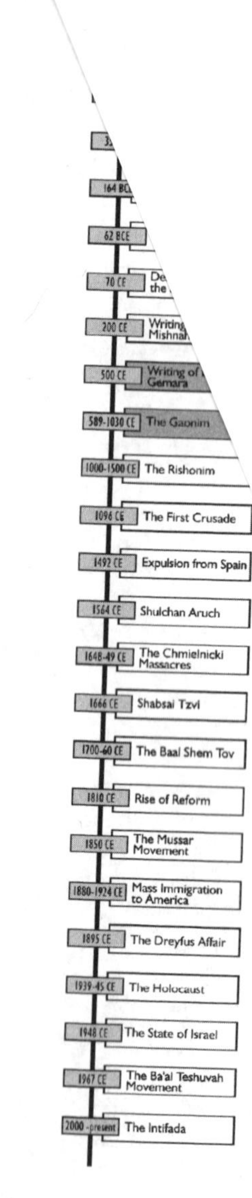

164 BCE
62 BCE
70 CE
200 CE
500 CE
Gemara
589-1030 CE
The Gaonim
1000-1500 CE
The Rishonim
1096 CE
The First Crusade
1492 CE
Expulsion from Spain
1564 CE
Shulchan Aruch
1648-49 CE
The Chmielnicki Massacres
1666 CE
Shabsai Tzvi
1700-60 CE
The Baal Shem Tov
1810 CE
Rise of Reform
1850 CE
The Mussar Movement
1880-1924 CE
Mass Immigration to America
1895 CE
The Dreyfus Affair
1939-45 CE
The Holocaust
1948 CE
The State of Israel
1967 CE
The Ba'al Teshuvah Movement
2000 -present
The Intifada